Temperature Conversions

Fahrenheit-Celsius

°F	°C		Some Conversions of Common Interest
212°F	100°C		Water boils at standard temperature and pressure
160°F	71°C		Flash pasteurization of milk
131°F	55°C		Many enzymes inactivated
98.6°F	37°C		Human body temperature
68°F	20°C		Standard room temperature
32°F	0°C		Water freezes at standard temperature and pressure
−40	−40	−40°, F° = C°	

Scale: °F 230, 220, 210, 200, 190, 180, 170, 160, 150, 140, 130, 120, 110, 100, 90, 80, 70, 60, 50, 40, 30, 20, 10, 0, -10, −20, −30, −40; °C 110, 100, 90, 80, 70, 60, 50, 40, 30, 20, 10, 0, -10, -20, -30, -40

To convert temperature scales:

Fahrenheit to Celsius $°C = \frac{5}{9}(°F - 32)$

Celsius to Fahrenheit $°F = \frac{9}{5}(°C) + 32$

Length Conversions

Centimeter-Inches

Centimeters: 15, 14, 13, 12, 11, 10, 9, 8, 7, 6, 5, 4, 3, 2, 1, 0

Inches: 6, 5, 4, 3, 2, 1, 0

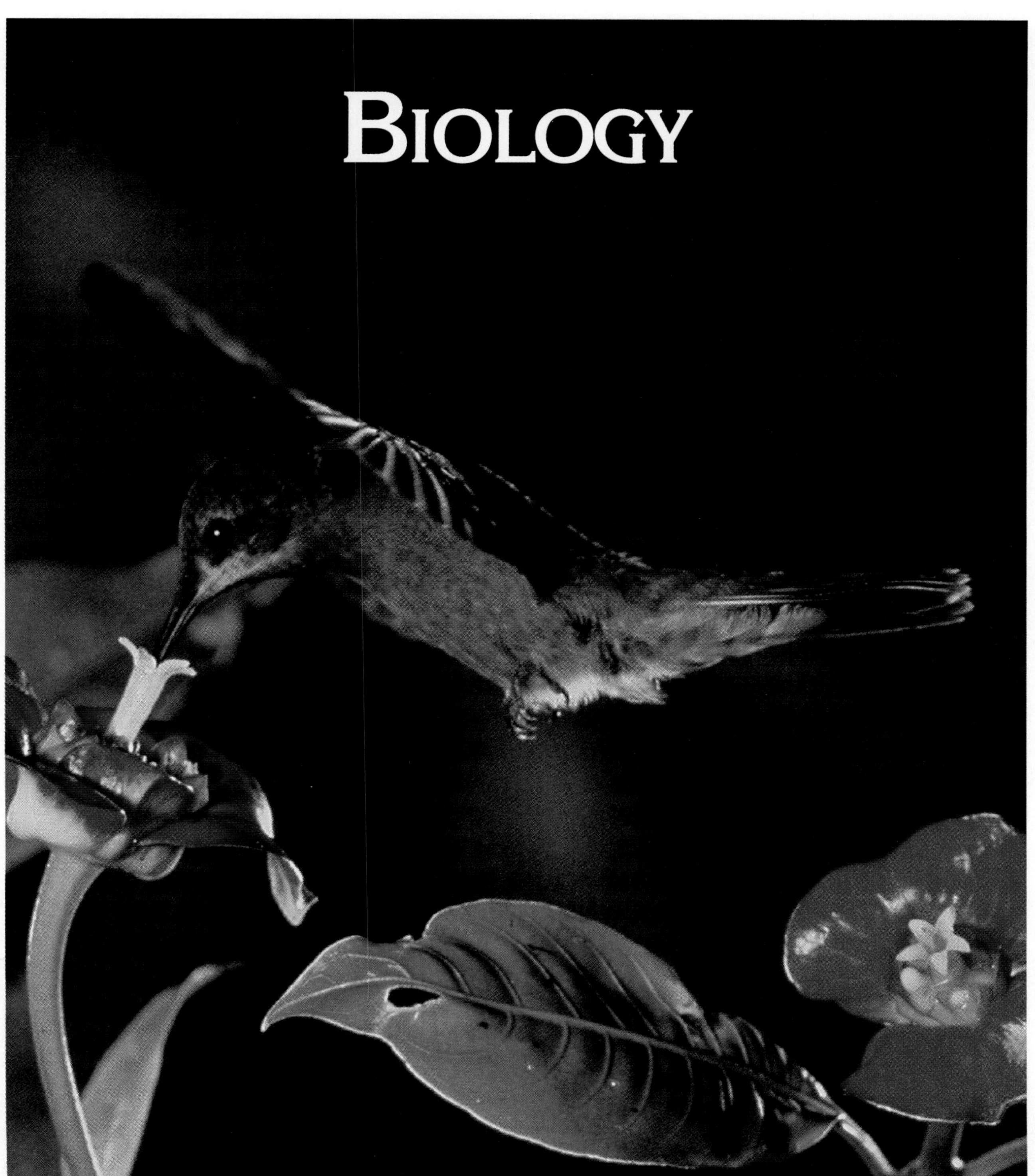
BIOLOGY

BIOLOGY
The Realm of Life

THIRD EDITION

ROBERT J. FERL
UNIVERSITY OF FLORIDA

ROBERT A. WALLACE
UNIVERSITY OF FLORIDA

GERALD P. SANDERS
GROSSMONT COLLEGE
Contributing Author

HarperCollins*CollegePublishers*

Editor in Chief: *Glyn Davies*
Sponsoring Editor: *Liz Covello*
Developmental Editor: *Richard Koreto, Thomas Henry Moore*
Project Editor: *Robert Ginsberg*
Design Manager: *Jill Little*
Cover Designer: *Lesiak/Crampton Design: Cindy Crampton*
Front and Back Cover Photographs: *Art Wolfe*
Art Studio: *J/B Woolsey Associates, Inc.*
Electronic Production Manager: *Michael Kemper*
Electronic Desktop Manager: *Heather A. Peres*
Project Coordination and Page Makeup: *Electronic Publishing Services Inc.*
Printer and Binder: *R. R. Donnelley & Sons Company*
Cover Printer: *Phoenix Color Corp.*

On the cover: Any of about 25 species of birds belonging to the Meropidea family, commonly found in southern Europe, Africa, Asia, and Australia. As their name suggests, a bee-eater's main source of food comes from insects, such as bees and wasps. These brightly colored birds range in length from 6 to 14 inches.

Biology: The Realm of Life, Third Edition

Library of Congress Cataloging-in-Publication Data
Wallace, Robert A.
Biology : the realm of life / Robert A. Wallace, Robert J. Ferl. — 3rd ed.
p. cm.
Originally published: Biosphere: the realm of life : Glenview, Ill. : Scott, Foresman, c1984.
Includes index.
ISBN 0-673-46624-8
1. Biology. I. Ferl, Robert J. II. Wallace, Robert A. Biosphere. III. Title.
QH308.2.W355 1996
574--dc20 94-32575
CIP

95 96 97 98 9 8 7 6 5 4 3 2 1

About the Authors

Rob Ferl was born and raised in the small town of Conneaut, Ohio. He received a B.A. in Biology from Hiram College, and a Ph.D. in Genetics from Indiana University, before moving to the University of Florida, where he continues his work on gene structure and function. At the University of Florida he is a Professor in the Institute for Food and Agricultural Sciences and Assistant Director of the Biotechnology Program.

At the University of Florida, he has taught introductory biology for both majors and non-majors; he received the Distinguished Teaching Award and was named Teacher of the Year for his work in these courses. He has also served as Director of the Biological Sciences Teaching Program.

Rob's research program is centered around the structure of genes and the mechanisms of gene regulation, and his research results are widely published. He is funded by grants from NSF, USDA, NIH, and NASA, supporting studies that include a diverse array of projects dedicated toward an understanding of the way genes work and applying that knowledge to real problems. His laboratory has hosted a number of dedicated students and postdoctoral associates. His studies on DNA are heavily oriented toward laboratory technologies, making him a "lab rat" in the happiest sense of the term. Application of those studies has, however, led to many interesting places, such as the beaches of Costa Rica, where he uses DNA sequencing to identify and tag nesting marine turtle populations, and to the mid-dock of the space shuttle, where he uses genetically engineered plants to study gene responses to microgravity environments.

He and his wife, Mary-B, and their son, Evan, live in Gainesville, Florida.

Bob Wallace was raised in Arkansas. He received a B.A. in Fine Art and Biology from Harding College, an M.A. dealing with muscle histochemistry from Vanderbilt University, and a Ph.D. focusing on the behavioral ecology of island birds from the University of Texas at Austin. He has taught at a number of colleges and universities in the United States and Europe.

He is the author of many textbooks in general biology and animal behavior, as well as nonfiction books on science.

Bob has worked as a nickel prospector in Alaska, a longshoreman, a martial arts instructor, a scuba diver, a janitor, and an art teacher. (He specializes in 17th-century painting techniques.) His interests include marathoning, fly-fishing, surf-casting, boating, shooting, skiing, and cooking.

He is a fellow of the Explorers Club based in New York, and a Contributing Editor of the *Explorers Journal*. He is a fellow of the Royal Geographical Society, and has explored in the Arctic, the Mediterranean, the Indian Ocean, the West Indies, and the Amazon. For the past several years, working with the shamans and healers of Ecuador's often-feared Waorani tribe, the nearby Sacha Runes, and the Shar headhunters, he has collected medicinal plants to be tested against AIDS, cancer, malaria, and leishmaniasis. In recent years he has guided small expeditions into these areas and has lectured on his work extensively in colleges and universities both here and abroad. He has been active in promoting the conservation of the world's rain forests, and was recently awarded a medal by the government of Ecuador for his work.

He and his wife, Jayne, live with Stormy the cat in Steamboat Springs, Colorado, and Amelia Island, Florida.

To my parents, and to Mike Kimmel, Ed Kicey, Bill Laughner, Drew Schwartz and Bill Stern—friends, wonderful people and dedicated teachers each, together constituting as fine a progression of mentors as any academic scientist could hope for.

R. J. Ferl

To John A. Murphree, good friend and true southern gentleman, who knows all about shotguns, bird dogs and pocket knives—
And to Kay, who explains things to him.

R. A. Wallace

PREFACE

It is gratifying to be writing this preface. It means that the long years of working on this edition are almost over. And what a time it's been—a time, it seems, marked mostly by change. There have been changes not only in biology, but also in the way biology is taught. We have rewritten—indeed, rethought—the new edition of this book to reflect these changes. Of course, one thing that has not changed is our goal to provide a student-friendly text: We continue to explain difficult concepts with interesting examples, and to show how seemingly unrelated topics are part of the same realm of ideas that characterizes biology.

The most obvious change, especially to those who used the last edition of *Realm of Life,* is in the table of contents. The new organization is more reflective of the way biology is taught today. Part I introduces the basic chemistry of science. Part II covers genetics and molecular biology. Part III combines evolution and diversity in a logical sequence. Parts IV and V explore the principles of botany and zoology, respectively. Part VI covers behavior and ecology, from the fundamentals to the late-breaking issues (see the section on Organization of the Text, below, for a detailed discussion of each of these parts).

The next big change is that the entire text has been rewritten. This is due partly to the changing nature of the field itself, and partly to the exhaustive reviews that constantly stimulated new ways of looking at things. To our minds, the explanations in biology could be made simpler, clearer, briefer, and friendlier. We felt our goal was not so much to impress students with how serious and difficult science could be, but to engage them and invite them along for the intellectual time of their lives. We think biology is fun, often astonishing, and usually proves that truth is stranger than fiction. We also believe that, in these critical times, a solid grounding in biological principles is increasingly important to the voters and decision-makers of tomorrow. We don't expect every student who takes this course to continue in the life sciences (though, from our experience, many will), but we do expect them to come away with an understanding of some fundamental principles concerning life, and a sense that it's all meaningful at a personal level.

And finally, the last big change will jump out at you by even casually thumbing through the pages. Look at what's happened to the art! Realizing that today's students are becoming more sophisticated *visual* learners, we decided to develop a completely new art program. Almost every piece of line and reflective art in this edition is brand new, and every piece was reviewed and developed in close collaboration between authors and artists to achieve pedagogical effectiveness, scientific accuracy, and aesthetic appeal. In addition, we combed photo libraries for hundreds of new, clearer, and more beautiful photographic images. The result is a fresh look with heightened explanatory powers, and a tight coordination of the art and photo program with the textual material.

Those familiar with previous editions of *Biology: The Realm of Life* may also notice changes on the cover and title page: This is the third edition of a book once called *Biosphere: The Realm of Life.* We think the new title more aptly describes the content. In addition, the authors have changed from Wallace, King and Sanders to Ferl and Wallace, our Jerry Sanders having taken leave of us to pursue other projects (although he was still instrumental in helping to develop what you see here).

So after years of hard work and a lot of fun and excitement, *Biology: The Realm of Life* is the result. Our reviewers are divided. Some believe the book is most appropriate for majors, others for non-majors, and that's just the way we would have it. Its use, of course, will depend on the instructors and their individual approaches to the course. We hope that students enjoy the course and that they find the book informative, interesting and friendly. We invite student comments, because it is the student, after all, whom we feel we work for.

THE ORGANIZATION OF THE TEXT

As noted above, *Biology: The Realm of Life* now has six parts. Here is a brief discussion of what each part provides.

Part I: The first part first shows the student how science is done and then moves on to a discussion of the universal foundations that determine the structure and function of all living things. We discuss the major principles of molecular organization, placing chemical details in context so that they do not end up as tedious, unconnected facts. A logical progression of increasing levels of organization leads the student to an understanding of the basic components of the cell and the processes that keep the cell alive.

Part II: One of the hallmarks of living things is that they reproduce. In these chapters we examine the mechanisms by which cells reproduce, especially the means by which all of the information that it takes to grow and live is imparted to new cells. An integral part of the discussion is the concept of biological information as stored and transferred using DNA and RNA. We end our discussion of DNA and genetics, and prepare for Part III, by considering how changes occur in biological information.

Part III: The first four chapters of this part describe the processes of evolution from the modern perspective. We begin with the processes of natural selection, move through evolution in populations and the current view of speciation, and then look at the history of our own species. Although we didn't want to overwhelm the student by discussing all the latest theories in this volatile field, we have introduced the most commonly accepted theories with their pros and cons. From there, we begin an exploration of the variety of life on Earth. It is here that we focus on the major evolutionary events that have shaped the great lines of living things, firmly establishing each of these events before describing the characteristic traits of each form.

Part IV: In these chapters we explore the realm of plants, from their reproduction to their growth and organization, mechanisms of transport and hormonal responses. Recognizing the subtle yet often critical changes taking place in plant biology, we invited a specialist, Dr. Lisa Baird of the University of San Diego, to review and contribute to this material. Her expertise and insight have been instrumental in developing simpler and more straightforward explanations of plant processes and in advancing the current view in those cases still in hot debate.

Part V: In this part we explore the world of animals by examining each major system while focusing on the evolutionary view. Reviewers and students alike requested that we shorten much of the material to make the evolutionary themes more pronounced and to eliminate extraneous material that, while interesting, may not be critical to a basic understanding of animal systems. The result is more concise descriptions that enhance and underline the major themes. In each case, we updated the material in keeping with new findings and searched the current literature for the best examples to illustrate our points.

Part VI: In the final part we first take a look at current concepts in animal behavior, focusing on both the principles of behavioral development as well as the influences of ecology and evolution. Turning to ecological concepts, we begin with an overview of the biosphere, then move on to discussions of ecosystems, communities, and the dynamics of populations, including humans. We have made an effort to present the latest data, theories, interpretations and projections, and to do so in an interesting and optimistic manner.

The organization of the six parts is based on the assumption that the principles of biology are most easily understood if one goes from the small picture (micro) to the big picture (macro). Simply put, we feel it is best to learn about bricks before one builds a cathedral. And there is no greater cathedral of ideas, we think, than bringing the sciences together into an ecological framework.

THEMES AND FEATURES

The two most significant threads holding the fabric of this book together are **ecology** and **evolution.** We have made an effort to show the adaptiveness of the characteristics of living things, and how those characteristics might have come about through natural selection.

Parts and chapters begin with an **introductory paragraph** that introduces key topics. Most chapters then continue with a story or vignette designed to catch the student's interest.

Each chapter closes with a brief section called **Perspectives** that places the material in the chapter into its proper context and shows how that material leads into the next discussion.

Within the chapters are **Essays** that focus on interesting sidelights. These essays are self-contained, and we believe they offer an opportunity for enrichment and interest. Also within the textual material are leading discussions and clearly stated opinions

followed by questions to promote **critical thinking,** the exercise of evaluation and the development of conclusions on the part of the student. We feel it is important to include such prompting with the scientific discussions themselves, and not isolated at the end of the chapters.

The **end of chapter material** furnishes the student with learning aids that can help in learning the material through review. This material is organized as:

1. **Perspectives,** a preview of the following chapter that emphasizes the continuity between one topic and another.
2. **Summary,** a review of the chapter's major topics.
3. **Key Terms,** a list of the boldfaced terms appearing in the chapter.
4. **Review Questions,** a self-quiz based on the information found in the chapter.

At the end of the book are further learning aids. These include:

1. **Classification of Organisms** (Appendix), a listing of current thinking regarding the grouping and relationship of living things.
2. **Suggested Readings,** a list of books and articles that have been critical to modern thinking, that have been important in the more recent development of ideas, and that explore the leading edge of current thought.
3. **Glossary,** definitions of the terms that are boldfaced in the text and that appear in the Key Terms.
4. **Index,** which lists all the Key Terms as well as all topics.

On the front endpapers of the text you will find a **Biological Lexicon,** a list of the roots of biological terms and the Greek, Latin, or other source from which they come. On the back endpapers are **Conversion Tables** for temperature, length, and volume.

Additional Aids Provided by the Publisher

HarperCollins has developed a number of resources to aid in the presentation of this material. These resources have been carefully correlated with the material included in the text, both as pedagogical aids—ways to enhance learning, and as enrichment aids—ways to stimulate interest and add a matrix into which major ideas can be introduced. These include:

1. The **Instructor's Manual** contains outlines, overviews, objectives, a detailed lecture outline, supplemental readings, and other information.
2. The **Test Bank** and the **Computerized Test Bank** have over 2000 questions at different levels in multiple-choice, matching, true/false, and essay formats.
3. The **Student Study Guide** includes additional summary material, review questions designed to test students' comprehension, biological crossword puzzles, and original art.
4. The **Transparency Acetates and Slides** reproduce 200 key pieces of line art for classroom discussion.
5. 100 of the above pieces are reproduced in black and white, without labels, in the **Transparency Masters.** These are useful for tests or student study aids, and they can be photocopied as necessary.
6. The **HarperCollins Biology Encyclopedia.**
7. The **HarperCollins Biodisc.**

Reviewers

We are particularly grateful to our reviewers, those specialists who labor in the trenches, working daily with their students, and who know what works and how it needs to be said. They have indeed played a critical role in the development of this book and we thank them for their contributions.

David Aldridge, *North Carolina A&T State*
Sylvester Allred, *Northern Arizona University*
Andy Anderson, *Utah State University*
Lisa Baird, *University of San Diego*
Robert Beckmann, *North Carolina State Unviversity*
Helen Benford, *Tuskegee University*
Carol Biermann, *Kingsborough Community College*
Dennis Bogyo, *Valdosta State College*
Charles Brueske, *Mount Union College*
Bruce Chase, *University of Nebraska-Omaha*
David Cochrane, *Tufts University*
Gale Davis, *University of Alabama*
D. G. Davis, *University of Alabama*
Peggy Rae Dorris, *Henderson State University*
Patrick Doyle, *Middle Tennessee State University*
James Fawcett, *Kingsborough Community College*
Carl Finstad, *University of Wisconsin-River Falls*
Richard Firenze, *Broome Community College*
Clyde Fisher, *Cisco Junior College*
Elizabeth Gardner, *Pine Manor College*
Elliot Goldstein, *Arizona State University*
Donald Green, *New Mexico Junior College*
Robert Griffin, *City College of San Francisco*
Thomas Griffiths, *Illinois Wesleyan University*
Kent Holsinger, *University of Connecticut*
Robert Hurst, *Purdue University*
Fred Janzow, *Southeast Missouri State University*
Gene Kalland, *California State-Dominquez Hills*
Judy Kaufman, *Monroe Community College*
Valerie Kish, *Hobart and William Smith*
Cliff Knight, *East Carolina University*
Joe Leverich, *Saint Louis University*
Jerri Lindsey, *Tarrant County Junior College*
Shaun McEllin, *New Mexico Highlands University*
James Moroney, *Louisiana State University*
Sallie Noel, *Austin Peay State University*
Catherine Oliver, *Manatee Community College*
Anne Packard, *Plymouth State College*
C. O. Patterson, *Texas A&M*
Robert Paul, *St. Mary's College of Maryland*
Ronald Pfohl, *Miami University*
Barbara Pleasants, *Iowa State University*
Michael Renfroe, *James Madison University*
Elaine Ross, *Cecil Community College*
David Sadava, *Claremont McKenna College*
Edward Saiff, *Ramapo College*
Gary Sarinsky, *University of Nebraska-Omaha*
Aubrey Scarbrough, *Towson State University*
Edwin Schreiber, *Mesa Community College*
R. W. Search, *Thomas College*
David Senseman, *University of Texas-San Antonio*
Joel Sheffield, *Temple University*
S. L. Sides, *University of Texas, Pan American*
John Smarrelli, *Loyola University-Chicago*
Jerry Smith, *St. Petersburgh Junior College*
Robert Smith, *Skyline College*
R. L. Smith, *West Virginia University*
Kathleen Steinert, *Bellevue Community College*
Marshall Sundberg, *Louisiana State University*
Robin Tyser, *University of Wisconsin—La Crosse*
James Whitfield, *University of Missouri-St. Louis*
William Wissinger, *St. Bonaventure University*
John Zimmerman, *Kansas State University—Manhattan*

ACKNOWLEDGMENTS

No textbook these days is done entirely by the authors. Biology textbooks, especially, are the result of a team effort, and we, the authors would like to acknowledge, here, the other members of our team. We are indeed grateful for their dedication, talent, intelligence, and insight.

First, we would like to thank Susan Katz, Publisher of HarperCollins Educational Publishing, and Susan Driscoll, Vice-President and Associate Publisher. We are grateful to them for their confidence in us and for their commitment to this project.

Glyn Davies, Vice-President and Editor in Chief, expertly made the day-to-day decisions, acting as organizer, ombudsman and friend. We are deeply thankful that he was so often there for us. In many ways, this is Glyn's book.

Richard Koreto, our Development Editor, very capably managed all the traffic, handled the reviews and kept us on track—no easy task.

Bob Ginsberg, our Project Manager, conscientiously saw the book through its production stages. He worried about all the right things and didn't have a single nervous breakdown. He attended to every detail and we owe him a great deal.

We can't say enough about John Woolsey, either personally or professionally. His knowledge of biology is impressive, his artistic talent unparalleled, and he always points his skis downhill.

Thom Moore, Senior Development Editor, saw the book through the final stages before production, managed the art, manuscripts, and text traffic with a skillful hand. He also helped us deal rationally with our computers. Thanks, Thom.

Kathi Kunz, Supplements Editor, very capably signed and developed the supplements program for this edition of *Realm*. Her energy and good cheer have been helpful and refreshing.

Finally, we must thank the people at the interface, the sales managers and representatives. If you, the student or teacher, are reading this book, it is probably because one of these capable people got it into your hands. Some of them we've worked with more closely than others, and some we haven't even met yet, but the names that pop into our minds include National Sales Manager Bob Carlton, Sales Managers Arnold Parker, Bill Cornett, Jim Marshall, Greg Odjakjian, Meg Holden, Dick Stratton, David Horwitz, Judith Allen, David Shea, Vonalaine Crowe, and Kathi Callahan. Some of the energetic, talented people they depend on (their troops, as it were) include many of our friends we've worked with personally such as Otis Taylor, Kelly Bell and Laura Stowe (regional consultants) and Frank Capek, Pat Kelley, Melissa Martin, Larry Sifford, Kent Merrill, Sheila Abruzzo, Bob Andrews, Cindy Yates, John Cross, Nan Williams, Chuck Hickman, Erin Kelly, Joan McKee, Lucinda Cutrer, Edison Diest, Dave Fleming, Pat Quinlan, Jim Northington, Glenn Russell, and Dan Cooper. Don't be surprised if one of them introduces us to each other sometime.

Robert J. Ferl
Robert A. Wallace

Contents in Brief

Contents

BIOLOGY

MOLECULES AND CELLS

SCIENCE AND ITS DESCENT

THE BASIC CHEMISTRY OF LIFE

THE MOLECULES OF LIFE

THE STRUCTURE AND FUNCTION OF CELLS

CELL TRANSPORT

ENERGY AND THE CELL

PHOTOSYNTHESIS

GLYCOLYSIS AND RESPIRATION

WE BEGIN OUR STUDY OF BIOLOGY, NATURALLY, WITH A DISCUSSION OF CHEMISTRY. What? Chemistry? You may well be thinking that this cannot be true, and that, in fact, the reason you took biology was to avoid chemistry. These days, however, chemistry and the other sciences often play a critical role in biology. After all, biology is the study of life, and life is nothing if not complex and interrelated. Part I examines the logical foundations of chemical principles, so that we can eventually arrive at a view of life that is partly derived from an understanding of the organization of simple chemical structures.

CHAPTER

1

Science and Its Descent

THE STUDY OF BIOLOGY IS A SCIENCE, AND THE WAY THAT WE UNDERSTAND *biology is based on the general principles of scientific inquiry. But what is it that is so special about science, anyway? Do you have to be a scientist to think like one? How does thinking like a scientist help one understand biology? These are legitimate questions that deserve some answers. Thus, we introduce our study of biology with a discussion of science.*

The old man was clearly wrong. Any educated person of his time could easily point out the error in his logic—and he was a heretic as well. The Church taught that all the heavenly bodies revolved around a central, unmoving earth. Galileo Galilei, however, believed in a bold new idea that had been put forth by the Polish astronomer, Copernicus—the idea that the earth and the planets revolved around the sun. The churchmen hadn't stopped Galileo from teaching Copernicus's heliocentric theory, as long as it was presented as merely an interesting idea; but Galileo had overstepped those bounds. His mistake was in trying to prove the theory with his own astronomical observations. The year was 1633. Galileo was 69 years old.

Now, as he stood before the court of the Papal Inquisition in Rome, his bald head was bowed in forced obedience, humiliation, and rage (Figure 1.1).

Recant! he was ordered. The inquisitors handed him a sheet of paper on which were written the words that he was required to say: that he "objured, cursed, and detested" his erroneous claim that the earth moved around the sun. Galileo reluctantly mouthed the words. He was given a prison sentence anyway, based on a forged document purporting to show that he had previously been expressly forbidden to publish his ideas. Later the sentence was commuted to lifelong house arrest. As the story goes, when he left the tribunal Galileo was heard to mutter, "Nonetheless, it moves."

GALILEO AND THE ROOTS OF SCIENCE

What had been Galileo's error? First, he had confronted the powerful Church. But he had also chosen an unaccepted method of attempting to establish truth.

The scholars of the Middle Ages had developed a method for determining truth that was based on the teachings of Aristotle, heavily interpreted by medieval theologians such as Aquinas and Abelard. This was the formal method of deductive reasoning known as *Aristotelian logic,* which starts with a few established truths and deduces from them other, less obvious truths. The chain of reasoning is presented in the form of a *syllogism,* of the "*if . . . then . . . therefore*" variety that we will encounter shortly in our discussion of deductive reasoning. The formal proofs of plane geometry are the finest products of this approach. The Church, which controlled the schools and the universities, still taught in Galileo's time that Aristotelian logic was the only acceptable method of determining truth. In particular, the philosophers taught that the senses were not to be trusted. Galileo nevertheless was trying to determine truth with his senses, through observation, by looking through a telescope.

Observation? Why? Observation was clearly unnecessary, he was told, because observation depends on our senses, and our senses can mislead us. Only rigorous, abstract *logic* could rise above the meaningless distractions of the imperfect, rough-and-tumble world to reveal underlying truths. So said the educated men of the time. As an example of where their "logic" led them, two accepted truths of the time were that *the heavens are perfect,* and *the moon is in the heavens.* It followed, then, that *the moon is perfect.* Another supposedly self-evident truth was that *a sphere is the only perfect shape.* From this it followed that the moon must be a sphere. The problem was that Galileo, using the telescope he helped develop, saw imperfections on the moon—jagged craters and lofty mountains. He had begged the members of the Papal court to look through his telescope and see for themselves, but they refused. The craters and mountains were not there, and that was that.

FIGURE 1.1 GALILEO AND THE CHURCH.
Galileo's approach to science brought him into conflict with the Church.

But Galileo persisted. The old man belonged to the coming modern age, not the fading medieval age that still held the minds of the Church philosophers. In the face of philosophical objections, he continued to observe and measure.

Galileo had spent his life devising ways of measuring things. He measured the speed of sound and had even attempted to measure the speed of light; he measured the speed and acceleration of falling weights. He applied mathematics to the real world, boldly proclaiming that "the book of Nature is written in mathematical characters."

Galileo had been a convinced Copernican since his youth, although he had kept his belief a secret to avoid ridicule. However, with his reliable telescope, he believed he saw a way to test the Copernican theory so that others would realize the truth of it. This notion alone put him on the road to modern science. He could make a *prediction* based on the heliocentric (sun-centered) theory. (A **prediction** is a statement concerning something that is going to happen, or that has already happened but has not yet been observed.) With a prediction the Copernican theory became *testable.* For example, Galileo could predict that if the earth and planets revolved around the sun, the planet Venus should have phases like those of the moon; that is, it should always be illuminated on the side facing the sun. Here was a prediction that could be tested through new observations.

Over several years Galileo pointed his long, thin telescope toward Venus and maintained careful records. He found that the planet was, in fact, illuminated on the side that, according to Copernican calculations, should be facing the sun. Galileo thought he had found clear evidence for the heliocentric theory.

However, the religious inquisitors and scholars were of another mind; their formal training stood squarely in the way. They could understand Galileo's argument only if they put it into the form of one of their own syllogisms:

1. *If* the planets and the earth circle the sun,
2. *then* Venus should have phases like those of the moon. Venus does have phases like those of the moon.
3. *Therefore,* the planets and the earth circle the sun.

Phrased this way, the inquisitors could see that the problem lay in the observation (No. 2). By the conventions of Aristotelian logic, this was clearly a classic example of erroneous, faulty, and invalid reasoning! Obviously, the old man, with his strange machines and cantankerous spirit, simply did not understand how real science was done. The inquisitors noted that other explanations could just as easily account for Galileo's assertions. Some said that one could just as well argue:

1. *If* the world is an illusion designed to deceive us,
2. *then* Venus could have phases like those of the moon, *and* Venus does have phases like those of the moon.
3. *Therefore,* the world is an illusion designed to deceive us.

This line of reasoning was actually more acceptable than that offered by Galileo. The idea of an illusory world had been taught by many of the early philosophers (who, though living in an illusory world, were usually on time for dinner just the same). By comparison, the idea that the earth revolves around the sun was much more difficult to accept. It violated both philosophy *and* common sense (even though the senses were not to be trusted, they did clearly seem to indicate that the sun traveled across the sky over the unmoving land).

Part of the Church's problem was that, years before, its leaders had unknowingly asked for trouble. They had told Galileo that he could publish a book on his ideas if he also stated that the Copernican theory was merely an unproved idea. The problem was this: When Galileo's book was completed, the required provision was there, but his arguments for the Copernican theory were so strong that they made the dictated conclusion of the Church look weak and foolish. Galileo's real crime was not in disagreeing with the Church over whether the sun or the earth was at the center of the universe but in daring to test the real world with a new and dangerous tool. His crime was in being a scientist.

Actually, the refusal of the Papal court to look through Galileo's telescope was a credit to their own intellectual integrity. However, they were fighting a doomed battle. They kept Galileo under house arrest until his death eight years later, but in the end the world became convinced of the validity of the heliocentric theory. And today the Church agrees that the earth revolves around the sun.

How did the world become convinced? By the same kind of evidence and reasoning that Galileo had offered. Measurement after measurement supported the theory that the earth and the planets revolve around the sun. Prediction after prediction had been tested and had supported the Copernican hypothesis. Today the guidance and navigational systems of space probes are based on incredibly precise predictions of where each planet will be at any given point in time (Figure 1.2). Perhaps the astronomers and engineers were deceived; the world may yet turn out to be an illusion, but for now we are convinced that it is not. Our beliefs are based, not on formalized proofs, but on the sheer weight of evidence gained through observation. Galileo's way of looking for truth—hopelessly flawed by 17th-century standards—has been called many things, among them the *scientific method.*

The Scientific Method

In its broadest sense the **scientific method** might be described as the way one gets at the truth or, as one prominent scientist put it, "doing one's damnedest with one's mind, no matter what." Nonetheless, even in our free-wheeling world, there are rules to getting at the truth, and some methods are held in greater esteem than others. The two most fruitful methods that have been used for the last hundred years or so are *inductive reasoning* and *deductive reasoning* (Table 1.1).

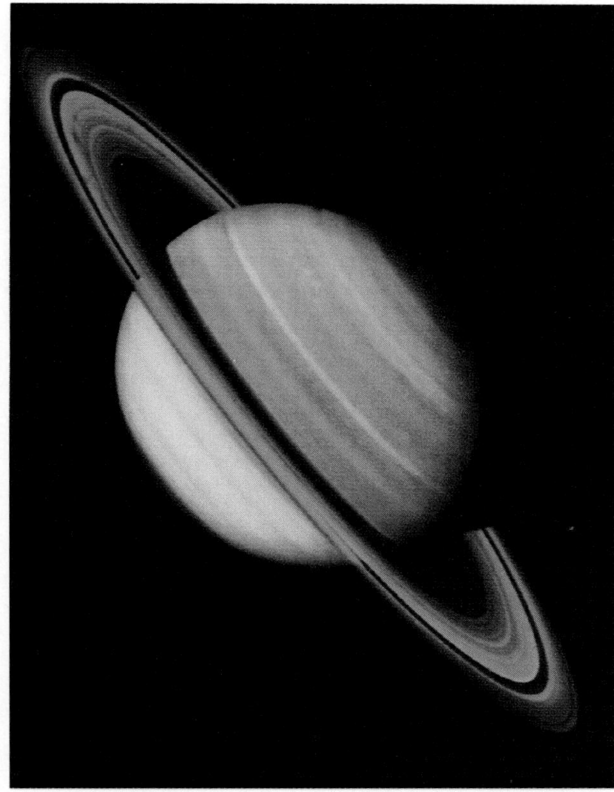

Figure 1.2 Galileo and Astronomy. Galileo's telescope allowed him to see not only the craters of the moon but the rings of Saturn, as he searched for truth.

Inductive Reasoning

Inductive reasoning involves reaching a conclusion based on observations as it moves from the specific to the general. For example, Galileo made a number of specific observations, and from these he drew a general truth regarding the relationship of the earth and the heavens. He *induced* this truth from the available evidence.

Deductive Reasoning

Unlike the inductive process, **deductive reasoning** involves drawing specific conclusions from some larger assumption. It leads from the general to the specific, that is, from a broad idea to one or more specific statements. Deductive reasoning leads logically to predictions, which are often described as a form of "if . . . then" reasoning. Recall Galileo's procedure: "*If* the planets revolve about the sun, *then* Venus should have phases as we see in the moon." He was using deductive reasoning, and the "if" part of his proposition stemmed from a *hypothesis.*

Hypothesis, Theory, and Law

A **hypothesis** is a provisional explanation that can be used to form predictions, which can then be tested. If what was predicted does, in fact, come about, we can say that the new observation is *consistent with the hypothesis* and thus supports it; the observation increases our confidence in the likelihood that the hypothesis is true. The hypothesis passes a test that *it might have failed.* (This is important—the hypothesis must have the possibility of failure.) On the other hand, if the predicted event is not observed, the new findings are *inconsistent with the hypothesis,* and thus weaken or even destroy it. At the very least, the original hypothesis may have to be reformulated to take account of the new observation.

Einstein's thinking on relativity led to the hypothesis that light is bent by gravity. One resulting prediction was that the apparent position of a distant star would change when a closer, massive star came between it and the earth; that is, the light would be bent by the gravity of the massive star. Soon after this prediction was made, there was an opportunity to test it. Special observatories were set up. The moment came; the astronomers' photographic plates were analyzed and, sure enough, light had been bent just as Einstein had predicted. Thus the hypothesis was supported, but still it was not *proved.* In fact, Einstein's theory of relativity remains unproved to this day, although it has passed a great number of such tests. Although it is now generally accepted by physicists, it is still subject to further testing and might be disproved or modified in the future.

It is important to understand that a scientific observation (or a thousand scientific observations) can support a hypothesis but cannot prove it to be true. On the other hand, a single observation can prove a hypothesis to be false. For example, if the light passing the star did not bend, Einstein's hypothesis regarding the relationship of gravity and light would be rejected.

Some hypotheses generate predictions that are more difficult to test, such as the prediction that

TABLE 1.1 Examples of Inductive and Deductive Reasoning

	Inductive Reasoning
Observation:	Coral atolls usually consist of a circle of islands.
Observation:	Coral atolls form from the deposits of living animals.
Observation:	Coral animals without direct access to fresh seawater tend to die.
Observation:	The interior of an atoll seems to consist of sunken coral.
Generality:	Coral atolls form as coral animals secrete deposits. Animals in the center, lacking nutrient-laden water, die and sink. This leaves a ring that, in turn, breaks apart to form a circle of islands.
	Deductive Reasoning
Generality:	Coral atolls form as coral animals secrete deposits. Animals in the center, lacking nutrient-laden water, die and sink. This leaves a ring that, in turn, breaks apart to form a circle of islands.
Deduction:	Coral atolls will have a sunken center.
Deduction:	Coral animals need contact with fresh seawater.
Deduction:	Seawater contains something that coral animals need.
Deduction:	Coral atolls comprised of more nearly complete rings of land are probably recently formed.

continued cigarette smoking causes lung cancer. As the tobacco industry is quick to point out, there are both healthy octogenarian smokers and young, cancer-stricken abstainers. Therefore the hypothesis must be modified to state that cigarette smoking has a *significant tendency* to cause lung cancer. Such a hypothesis can be tested only by using massive numbers of subjects and sophisticated statistical analysis to show that the connection does or does not exist.

As a hypothesis gains weight by the sheer accumulation of evidence, the statement may finally enter the realm of *theory.* A **theory** is an unproven explanation of nature that is supported by a considerable body of evidence. A theory has considerably more supporting evidence than does a hypothesis. Ideally, such evidence represents the results of testing a variety of predictions through many experiments or observations. (The difference between a hypothesis and a theory is more a matter of confidence than anything else.)

As more and more evidence comes in that is consistent with a given theory, it becomes virtually (but not quite) irrefutable and is called a **law.** Laws enjoy more confidence than theories and, in fact, seem to withstand any kind of testing we can invent. Examples are the law of gravity and the laws of thermodynamics, both very lofty, universal statements from physics. Biology is notoriously short on laws because life is by nature shifty, elusive, and hard to define. Although biology has its "laws" (such as Mendel's laws and the Hardy-Weinberg law), biological statements wear their titles provisionally and uneasily. By the way, most biological "laws" are based on mathematical descriptions rather than simple observation.

The Experiment

Scientists are curious souls, and one of their favorite tools is the experiment. Because their reputations as well as their conclusions depend on the nature of their experiments, they carry them out according to time-tested and accepted rules.

The "rules" say that an experiment should begin with a hypothesis. In fact, an experiment often can be considered a contrived situation that tests a hypothesis.

Let's consider an example based on events of recent history. A group of bird-watchers reported that many of the eggs in a certain area were failing to hatch. They notified an esteemed biologist in the area, and he found the problem to be even greater than was originally believed. One of his first questions was, "Have the conditions in the area recently changed in some way, thereby causing the eggs to fail?" He found that the eggs stopped hatching about the time that foresters began spraying the area with herbicides (plant-killing agents) causing trees to drop their leaves, creating fire-breaks—barren strips that impede the progress of forest fires. His hypothesis therefore became, "The herbicides are somehow killing the eggs." His first prediction was, "If the foresters stop using the herbicide, eggs will begin to hatch again."

However, the foresters had problems of their own, and they refused to cooperate; so the esteemed biologist had to devise another testable hypothesis. The new hypothesis became, "If I treat normal, healthy eggs with the herbicide, the eggs will not hatch." He now had the hypothesis and the experimental design. However, the weather had also changed about the time the foresters began spraying, and the days during this period had been unusually warm. The question then became, was the herbicide killing the eggs, or was the warm weather the cause? To separate any two possibly contributing factors, the scientist must hold one of them constant while varying the other.

In the laboratory, the biologist incubated three batches of eggs. The first was cared for under what would be normal conditions in the wild. The conditions for the second batch duplicated those of the first except that the air in the incubator was warmed to duplicate the unusually warm weather at the time. The third group was incubated normally, except that the air in the incubator contained about the same level of herbicide as the air surrounding failed nests in the wild. The scientist found that the first and second group hatched normally, but the third group failed completely; none of the eggs exposed to the herbicide hatched. He concluded that the herbicide was killing the eggs.

This is obviously a rather simple experiment with rather predictable results. However, there was a twist: The scientist had to rule out the effects of hot weather. This was done by heating one group of eggs without adding herbicide and by maintaining a second group of eggs at the normal incubation temperature while adding herbicide. In this experiment the added heat and herbicide are called *variables.* A **variable** is the nonconstant part of the experiment that may be responsible for the observed results. The constant part of the experiment (here, the eggs that are incubated under normal conditions) is called the **control.**

In this experiment, if all three groups of eggs had hatched normally, the next step would be to change the variables so that the eggs were subjected to heat and herbicide simultaneously. Experimental design can indeed become complicated, but our example illustrates how the process works.

What we described here is a **controlled experiment.** There may be many variations in such experiments, but the principle is the same. A number of subjects, in this case eggs, are randomly divided into two (or more) groups, and each group is treated exactly the same in all respects except one, which becomes the variable. The group subjected to the variable is called the **experimental group.** The others comprise the **control group,** the group to which the treated subject is compared. Since only one factor is varied at a time, we can find reliable answers to a question. For example, if we want to find out why a cake will not rise, we should not use less flour, turn down the oven, and add more eggs all at the same time. We have to alter each condition individually. Figure 1.3 illustrates a more complex experiment.

The Workings of Science

Reductionists and Synthesists

Scientists, like any other group of cheerful, well-adjusted people, tend to approach problems in different ways. One group, for example, may look at the small picture, while another may focus on the big picture. Those scientists who reduce any problem to its simplest level and study it at this level are called **reductionists.** They often make the best experimenters because they are interested in solving single aspects of one problem at a time. Sometimes this powerful principle is itself described as being "the" scientific method, but the reductionist approach, as valuable as it is, is only one way of doing science.

In every scientific field there are also **synthesists.** These people tend to take the available information

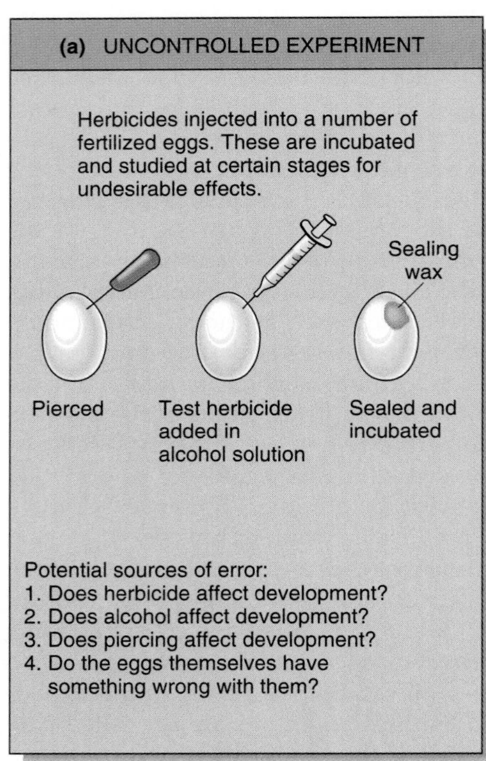

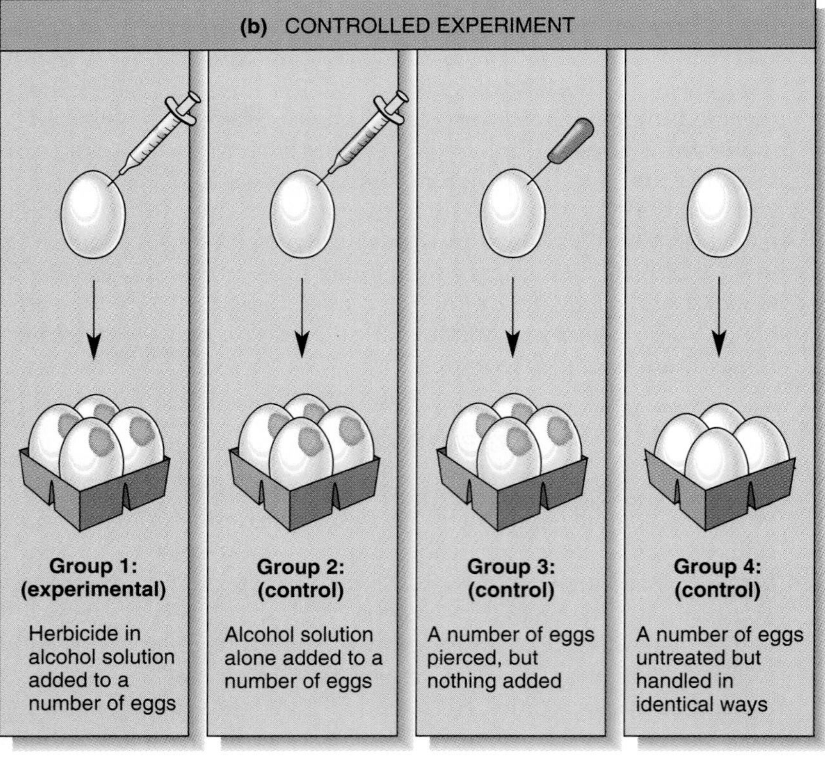

Figure 1.3 **Scientific Experimentation.**
An experiment intended to determine the effect of an herbicide (plant killer) on bird development prior to its use on plants. The uncontrolled experiment (**a**) contains several potential sources of errors. To correct for such errors, a controlled experiment must be devised that addresses these errors. In the controlled experiment (**b**) the results of the three groups are compared, and the results point to the effects of the herbicide.

(often from a variety of sources) and try to formulate some sort of grand, encompassing principle. These are the "big picture" people. Any active scientist can be either a reductionist or a synthesist at different times, of course. The synthesists tend to be theorists who operate by induction, developing theories that can then be tested through deductive reasoning. Theorists and synthesists generate new ways of looking at old observations (an "old" observation being anything that has already been published). A grand idea or theory usually is formulated from a large body of seemingly unrelated data, often with the stimulus of some new, at first inexplicable, finding. Copernicus was one such theorist; Einstein was another.

Revolutions and Paradigms

The sorts of questions scientists ask and the way they go about answering them depend on a set of encompassing views called *paradigms.* A **paradigm** is a general, accepted, and internally consistent view of the world. When an existing paradigm is replaced, the effect may be shattering to the scientific community, and the replacement may meet with stiff resistance. We saw this in the case of the Copernican revolution: Galileo's critics really could not understand the point he was trying to make. Others were all too aware of what the acceptance of Galileo's arguments would mean, and they were determined that Galileo fail in order to preserve the current paradigm, the one they had supported all those years.

Major scientific revolutions have played havoc with accepted paradigms, and there have been a number of such revolutions. Physics has seen two major revolutions in thought—relativity and quantum mechanics—in just this century. In the 20th century the science of geology was rocked to its foundations by the idea of continental drift. What is a major paradigm change in one field of science may create only a minor ripple of change in another field. Physicists were only mildly affected by the consequences of the new paradigm in geology, and neither group had to reorganize its thinking as a result of biology's revolutionary paradigm, the *theory of evolution,* presented by Charles Darwin.

All of the great "-ologies" of science are artificial constructs. They were devised, in part, so that scientists could deal with narrowly defined tasks. A geologist might think, "I like rocks, and at least I have some hope of being successful in understanding them as long as I don't have to deal with cats." If the biologists, on the other hand, who deal with living things, were to try to explain why the earth's cats exist where they do, they would have to include the geologist's paradigm of continental drift.

Since the role of science is to explain the natural world—all of it—scientists have had to talk increasingly to each other and to carefully monitor the revolutions and developing paradigms of their colleagues in other areas.

Vitalism and Mechanism

In the 17th and 18th centuries, two quite different notions developed regarding the nature of life: *vitalism* and *mechanism.* **Vitalism** stated that living organisms have special properties and are governed by special laws that are not applicable to inanimate objects or to the chemistry of nonliving substances. These special properties have been termed *vital forces.* What is the difference between a person and a corpse, it was argued, except for the presence or absence of a vital force? Unfortunately, because the "vital force" hypotheses were usually untestable, they haven't been subject to the sorts of investigations that comprise "good science."

The other school of thought, **mechanism,** stated that a living body is a machine, subject to physical and chemical laws and depending only on those laws. Understand the laws and you will understand life, it said, for life has no laws of its own. In fact, the mechanists cheerfully explained that if any vital force existed, it too could be explained in terms of physics and chemistry, which drove the vitalists right up the wall.

The notoriously unromantic mechanistic view was slow in taking hold. However, a new focus in the ongoing controversy arose in the 19th century over the question of **spontaneous generation,** the notion that nonliving material can give rise to life. Historically, vitalists believed that the unexplained appearance of molds, bacteria, and even maggots in rotting foods could be attributed to spontaneous generation by the material itself; in other words, life was being created by the "vital forces" of nature. However, the noted scientist Louis Pasteur showed that some food-spoiling organisms simply multiplied from preexisting life present in microbe-laden dust (Figure 1.4). Biologists are now convinced that, today, life arises only from preexisting life (see Chapter 4). The key term is "today," since life must have originated spontaneously at least once, but under far different circumstances (see Chapter 22).

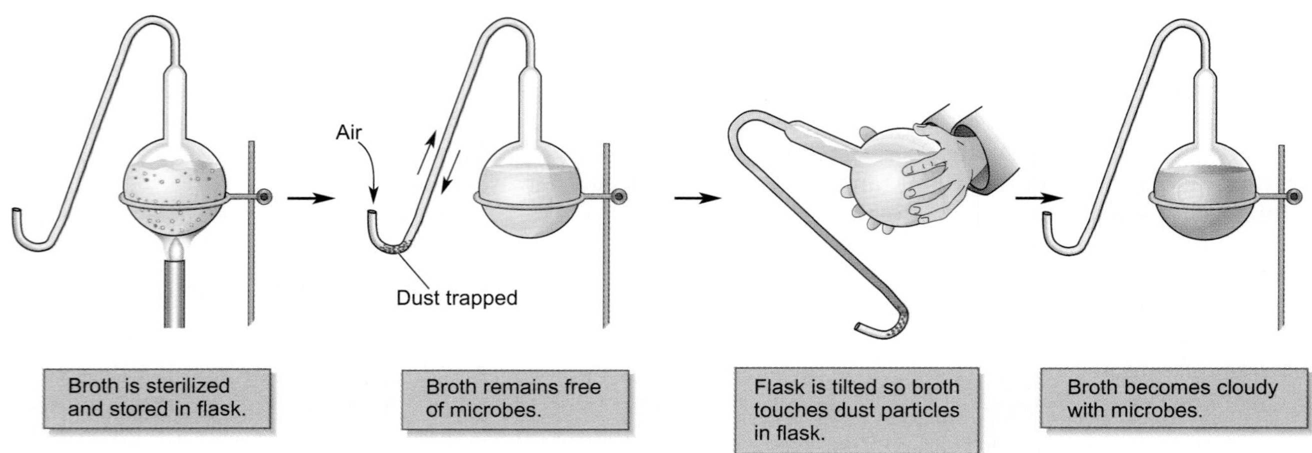
PASTEUR'S EXPERIMENT
Air
Dust trapped
Broth is sterilized and stored in flask.
Broth remains free of microbes.
Flask is tilted so broth touches dust particles in flask.
Broth becomes cloudy with microbes.

FIGURE 1.4 SPONTANEOUS GENERATION.
The question of the spontaneous generation of life has a long history in science. On June 22, 1864, the idea was dealt a fatal blow. Louis Pasteur, the brilliant French biologist, was convinced that decay bacteria were simply carried from place to place by air currents and did not arise spontaneously. He also knew from past experiments that by boiling a rich broth and sealing the container he could preserve the contents indefinitely. His chief antagonist, F. A. Pouchet, quite logically pointed out that without the presence of air—which is essential to life—spontaneous generation could not occur. Pasteur met the challenge. He created a flask in which broth could be sterilized by boiling and air safely admitted. The question of spontaneous generation was laid to rest when it was discovered that if no microbe-laden dust reached the broth, no bacteria appeared.

THE UNITY AND DIVERSITY OF LIFE

One very basic observation in biology is that there are fundamental similarities among very different kinds of living things. This "unity of life" is manifest in numerous ways. For example, the genetic mechanisms of all living things are remarkably similar. Inherited traits are passed along in much the same way in such diverse groups as bacteria, dandelions, and real lions. All living things tend to use similar sorts of molecules as food and to shuffle them along very similar sorts of routes as the foods are drained of their energy. That energy is then used in very similar ways. These sorts of similarities suggest a rather narrow range in the chemical makeup of quite diverse species.

We also find that the membranes covering the cells of living things are likely to be quite similar from one group to the next. A shark searching the ocean floor has cells bounded by membranes that are similar to those surrounding the cells of a desert mouse. If one were to examine the cells of a shark and a mouse, one would find similar tiny structures within the cells. Most of the life on this planet bears many of the common stamps that suggest relatedness. The notion of relatedness, of course, implies common ancestry. Even very different forms of life on the planet today may well be descended from the same distant ancestor. (The more related the life forms, the less distant the ancestor.) This notion is, of course, the central idea of evolution. We can define evolution simply as change in living things over time. Living things change as they adapt to an ever-changing environment—and through these adaptive changes, new species constantly arise. We will return to this central theme of biology again and again.

The Characteristics of Life

You may occasionally run across a discussion called the "Definition of Life." You won't find anything like that here, though, simply because we have no illusions of our being up to the job. In a nutshell, life is notoriously difficult to define. Instead we can tell you some of the "Characteristics of Life," and in so doing, various general principles may emerge that reflect the underlying Unity of Life. These are the common threads in the fabric of living things (Figure 1.5):

Life generally is organized into cells. Nearly everything alive is organized on the cellular principle; that is, organisms are comprised of one or more cells. Across the spectrum of life, cells have many complex molecules and highly regimented chemical processes in common. The same key molecules and reactions are present in nearly all life forms.

Life requires a constant input of energy and materials. Maintaining a highly ordered state is costly. Much of the material and energy initially entering the living realm is captured by *autotrophs,* plants and plantlike organisms that make food and energy from sunlight. Heterotrophs, on the other hand, feed on autotrophs and other heterotrophs, using their valuable, hard-won molecules as a source of energy and building materials.

Life is homeostatic. Homeostasis is the maintenance of a "steady state," a rather constant condition in a changeable world. The maintenance of such a constancy requires sensitivity and appropriate response mechanisms.

Life responds to stimuli in the surroundings. Common examples of stimuli to which various forms of life respond are light, heat, cold, sound, movement, touch, and, of course, other organisms. The ability to respond to stimuli is fundamental to maintaining homeostasis. The total pattern of responses is known as *behavior.*

Life reproduces. Perhaps the most obvious and unique characteristic of life is its overwhelming focus on reproduction. All of the chemical and physical qualities that characterize each life form are replicated and preserved in new generations.

Life evolves. Populations of living things change as time passes, the passage marked by successive generations. The change is often an adaptation, a beneficial response to a changing environment.

Biological Diversity and the Wider View of Life

Although common threads run through the realm of life, one is particularly struck with life's great diversity. The opportunism of life has allowed the environment to shuffle, massage, enhance, and eliminate various traits of living things in such a way that those life forms may take advantage of whatever opportunities the environment affords. As life has probed every nook and cranny of our planet, each habitat has placed certain demands on its denizens; and as life has changed to meet these demands, diversity has increased.

We should add that a knowledge of the unity and diversity of life has quite useful applications. For example, when we realize just how a four-chambered heart functions in birds and mammals, we can draw conclusions about how the human heart operates. In fact, medical research has taken advantage of the critical similarities between humans and other animals and has in a sense regarded other animals as cheap, expendable, and silent humans. Most modern techniques in heart surgery were first performed on dogs, pigs, or baboons—species similar to us in many ways (Figure 1.6).

Of all the sciences, biology perhaps is the least subject to rigid laws. It has a fascinating and frustrating element of unpredictability, perhaps even unruliness. Part of the reason lies in the constantly changing realm of life—always shifting, adapting, retreating, advancing, and modulating. Part of the problem (with apologies to our philosopher friends) also must lie in the phenomenology of biological explanations. Phenomenology encompasses the notion of the interpretation of observations, and biology is an interpretive science. Just as bread is flavored by the peculiar flora of the baker's own hands, so the essence of the biologist enters into his or her own explanations of the world of life. The biologist also has the problem of an eye trying to see itself because, after all, we are indeed a part of the realm of life.

(a)
(b)
(c)
(d)
(e)
(f)

(g)

(h)

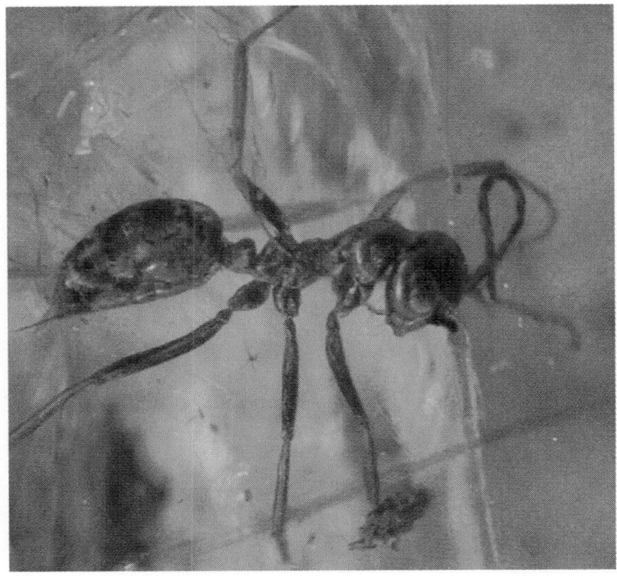
(i)

FIGURE 1.5 ORGANIZATION IS THE CATCHWORD OF LIFE.
(a) Even seemingly simple organisms such as the radiolaria are highly organized. The radiolarian skeleton is essentially made of glass, and its sculptured appearance is testimony to how beautifully complex life forms can become. **(b)** Life must exist in a steady state within rather narrow limits, and many forms have devised remarkable ways of regulating their internal environments. These bumblebees are vigorously fanning the nest, a common practice among bees when temperatures rise. **(c)** Life takes in and uses energy in order to retain its own highly organized state. Most energy enters the living realm through photosynthesis in green plants. **(d)** All organisms respond to their surroundings, but some living things are often extraordinarily responsive to external stimuli. The chameleon's lightning "tongue-flick" is a response to the presence of a moth. **(e)** Living things reproduce in an endless variety of ways. The aphid is giving birth to live offspring. **(f)** The embryos of some terrestrial vertebrates develop in a protective egg as did this emerging gavial. **(g)** Living things adapt to their environment in quite surprising ways. For instance, chimpanzees are known to make and use simple tools. The chimp carefully strips a slender stem off its leaves and uses it to gently probe a termite's nest. The stick is carefully withdrawn, and any clinging termites are eaten. **(h)** When not soaring through majestic mountain passes, these bald eagles pass the time freeloading in a garbage dump located on Adak Island in the Aleutians. Such visits occur on a regular basis when the usual food supplies become scarce. **(i)** Life evolves. Much of what we understand about evolution comes from the study of fossils, but the fossil record is notoriously incomplete, especially in providing clues to transitional organisms—extinct forms representing steps in the evolution of today's lines. This fossil insect is a rare and exciting find, a transition species representing a stage in the evolution of ants from waspy ancestors. Its head, eyes, abdomen, and sting are typical of wasps, but its thorax (middle body) and waist are decidedly antlike, while its antennas have characteristics of both insects.

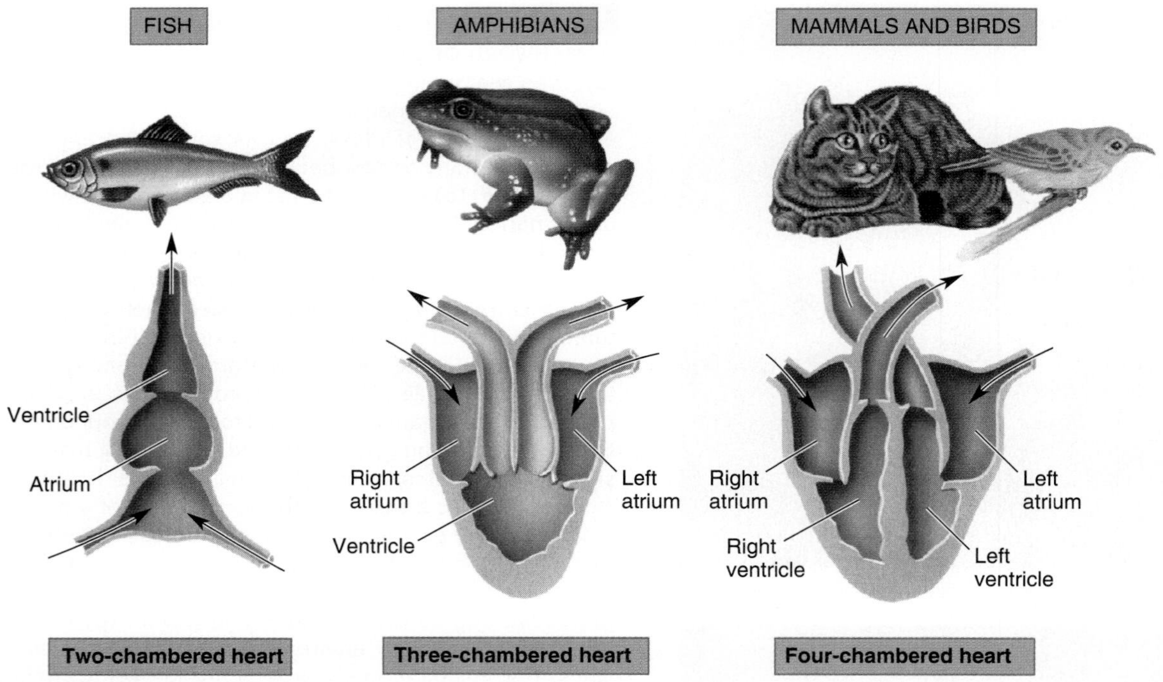

FIGURE 1.6 **THE HEART AS AN EXAMPLE OF THE UNITY AND DIVERSITY IN LIFE.** The hearts of vertebrates reveal certain structural and functional similarities, yet an inescapable trend over time—a progressive evolutionary change—is seen when the hearts are arranged in a series from the simplest to the most complex. The earliest vertebrates were the fishes, with their two-chambered hearts. Most amphibians and all reptiles are air breathers, using the lungs for gas exchange. The hearts of these cold-blooded amphibians and reptiles are mainly three-chambered, a condition that persisted over millions of years. Birds and mammals have evolved as warm-blooded creatures that are able to survive quite well in the coldest terrestrial environments. Part of their adaptive success can be attributed to the four-chambered heart, which meets the increased oxygen demands of an elevated metabolic rate.

PERSPECTIVES

We have seen that as a science, the study of biology follows a particular process of thinking, a process that involves observations, conclusions, and predictions. Through science, we can characterize life and living things, and move toward an understanding of what life is.

In the next few chapters we will turn our attention to the chemical basis of living things. We will see that, using the principles of science, it is possible to arrive at an understanding of how life can be characterized as an organization of chemical building blocks.

Summary

Galileo and the Roots of Science

Although others in his time searched for truth through Aristotelian logic, Galileo tested ideas through prediction and observation. Religious inquisitors disputed Galileo's findings because his methods threatened established procedures.

The Scientific Method

The scientific method can be described as the way one gets at truth. Truth can be sought through both inductive and deductive reasoning.

With inductive reasoning, specific observations lead to larger generalizations, whereas in deductive reasoning, generalizations give rise to specific statements.

Hypotheses are provisional explanations that can help form predictions. To be useful, they must be testable. While hypotheses cannot be proved (only supported), they can be disproved. Theories have more supportive evidence than do hypotheses. Where confidence is extremely strong, laws may emerge.

Controlled experiments involve a control group (the constant part of the experiment) and an experimental group in which some single factor (the variable) differs from conditions in the control group. Any changes in the experimental group from the control group are considered due to the variable.

The Workings of Science

Reductionist scientists work with single, often minute, aspects of a problem. Synthesists use information from many sources to formulate larger principles.

When new, major paradigms arise in science, they often meet strong resistance from more traditional forces.

Whereas vitalism is the notion that vague "vital forces" are responsible for biological functions, mechanism is the conviction that all biological phenomena can be explained through chemical and physical laws. A vitalistic belief known as spontaneous generation was replaced by the biogenetic law.

The Unity and Diversity of Life

Fundamental similarities or commonalties in organisms—the so-called "unifying themes"—are seen at all levels, from molecules onward, and include both structure and function.

Common, unifying characteristics of life include: cellular organization, the need for energy and materials, homeostasis, responsiveness, reproduction, and evolution.

The diversity of life is largely due to the exploitation of different opportunities. The unity, or common traits, of many life forms enables us to substitute one form for another experimentally. The phenomenology in biological explanations may influence results because of individualized effects that can bias results.

Key Terms

prediction • **6**
scientific method • **7**
inductive reasoning • **8**
deductive reasoning • **8**
hypothesis • **8**
theory • **9**
law • **9**
variable • **9**
control • **9**
controlled experiment • **10**
experimental group • **10**
control group • **10**
reductionist • **10**
synthesist • **10**
paradigm • **11**
vitalism • **11**
mechanism • **11**
spontaneous generation • **11**

Review Questions

1. Compare Galileo's approach to problem solving with the more traditional Aristotelian approach of his day.
2. Explain why the Church was so opposed to the manner in which Galileo reached conclusions. Why was the Copernican theory itself so vehemently opposed?
3. Using examples, distinguish between deductive and inductive reasoning. Discuss the status of each type in science today.
4. What is a hypothesis? What characteristic must a hypothesis have if it is to be a useful part of the scientific method?
5. In general, how do hypotheses, theories, and laws differ in science?
6. Why is it essential that a large number of subjects be used in controlled experiments? Why are controls critical? Distinguish between a control and a variable.
7. Using examples, compare the contributions of synthesists and reductionists.
8. What is a paradigm in science?
9. List six characteristics shared by all organisms.
10. Briefly suggest why diversity arises in spite of the unifying characteristics of life.

2

The Basic Chemistry of Life

WHAT ARE THE BUILDING BLOCKS OF LIFE? ESSENTIALLY, THEY ARE THE *same as the basic building blocks of the universe. The key, as we will see, is how these building blocks are joined together to form structures of ever-increasing complexity. We begin here by describing things that are clearly chemical in nature and would not by themselves be confused with living things. But these inanimate structures are the stuff of life, imbuing living things with every biological characteristic.*

Since biology is the study of life, it can occupy the attention of all sorts of scientists. Some, for example, wear muddy boots and ask where ducks go. In fact, biology was once almost the exclusive domain of people out tramping around in the woods and fields, telling us about the natural history of the world. But now they must share the arena. New faces, new ideas, and new techniques have brought other kinds of scientists into the world of biology. Some deal with numbers and are enchanted by odds. Others generate fascinating figures on computer screens to tell us how an organism grows, or how two species might interact. Among the most welcome of modern biologists, too, are those who can tell us about the most fundamental properties of living things—the chemistry of life. We will find that a great deal depends on the behavior of the chemicals that make up life.

SPONCH

Chemists would start by defining elements, atoms, and molecules. An **element** is a substance that cannot be divided into simpler substances by chemical means. There are 92 naturally occurring elements (and 18 more that have been made in the laboratory). Of the naturally occurring elements, just six—*S*ulfur, *P*hosphorus, *O*xygen, *N*itrogen, *C*arbon, and *H*ydrogen—make up about 99% of living matter. Notice that the initials of these elements spell SPONCH, the sequence indicating the increasing proportion of each molecule among the stuff of life.

The smallest unit of any element is the **atom** (*atom* is Greek for "indivisible"). Should the atoms of an element be broken down by some means, they would no longer be representative of the element.

ATOMS AND THEIR STRUCTURE

Let's begin our look at atomic structure by noting that atoms consist of three principal particles—**protons, neutrons,** and **electrons.** Protons and neutrons form a large, dense core known as the **nucleus.** Protons and neutrons together make up most of what is called the **atomic weight.** In contrast, electrons are much less massive. (If you weigh 150 pounds, your body is made up of about one ounce of electrons and about 149 pounds, 15 ounces of protons and neutrons.) Protons have positive electrical charges, while neutrons have no charge.

The electrons, comparatively minute particles, are found outside the nucleus, traveling in paths known as **orbitals.** Each electron has a negative charge equal to the proton's positive charge. Since *like* charges repel each other and *unlike* charges attract, the positively charged protons tend to hold the negatively charged electrons in their orbitals. If the number of electrons equals the number of protons, the atom bears no net charge and is said to be electrically balanced.

The smallest and lightest atom, hydrogen, consists of a single proton with one electron revolving around it. Since the number of protons in the nucleus determines its **atomic number,** we say that the atomic number of hydrogen is 1. One of the largest and heaviest naturally occurring atoms is uranium-238, the nucleus of which contains 92 protons and 146 neutrons (Figure 2.1). The atomic number of uranium-238 is 92; its mass is 238 (92 + 146).

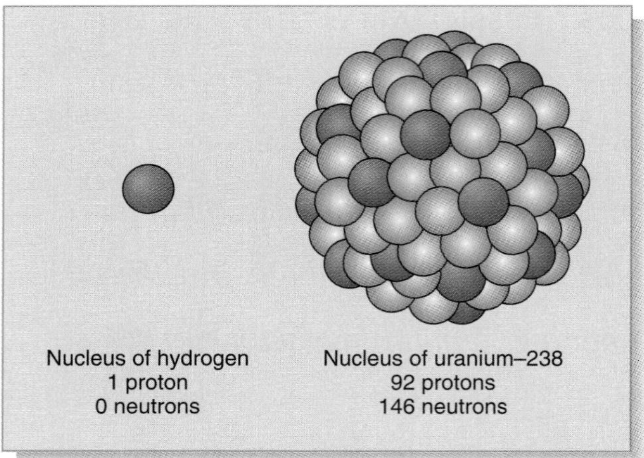

FIGURE 2.1 THE LIGHTEST AND ONE OF THE HEAVIEST ATOMIC NUCLEI.
The atomic weight of uranium is approximately 238 times that of hydrogen. Interestingly, the properties of both of these elements have come to the attention of those making nuclear weapons.

ISOTOPES

The number of protons in an atom (its atomic number) is constant, but the number of neutrons in an element often varies. Atoms with the same atomic number but different atomic masses are called **isotopes.** Carbon-14 is an isotope of carbon-12, having two extra neutrons.

Of the more than 320 known natural isotopes, about 60 are unstable, or **radioactive.** *Unstable* refers to an isotope's tendency to disintegrate spontaneously, or decay, releasing radiation in some form. In the process of decay, the radioactive isotope usually changes from one element to another. (Such a change is brought about by alteration of the number of protons in the nucleus.) The new element may or may not be radioactive.

The time required for half of the atoms of any radioactive material to decay is its **half-life.** Half-lives can vary considerably and depend on which isotope of which element is being considered. Many naturally occurring radioisotopes are extremely durable; some half-lives are billions of years long. Uranium-238 has a half-life of about 4.5 billion years, during which half of the atoms decay to form an isotope of lead called lead-206. Some artificial radioisotopes have a fleeting half-life of only seconds.

The longer-lived isotopes often are used in determining the age of rocks from the earth's crust. In medicine radioisotopes are used to destroy cancer-ridden tissues and in diagnosis (Figure 2.2). Especially since World War II and the advent of nuclear weapons, scientists have been vitally interested in the destructive effects of radiation.

Shorter-lived radioisotopes are used as tracers to determine the role of certain chemicals in living cells and organisms. This is possible because their chemical behavior is identical to nonradioactive atoms. The radioactive isotopes, however, can be traced by their telltale radiation as they move through living systems. Tracking radioisotopes of carbon, phosphorus, and hydrogen, for example, has helped unlock the secrets of photosynthesis, the process whereby plants use sunlight energy to convert carbon dioxide and water into sugars. The use of radioactive phosphorus, hydrogen, nitrogen, and

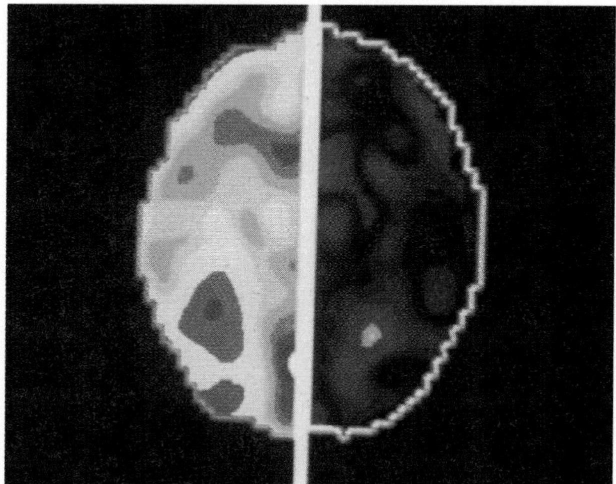

FIGURE 2.2 MEDICAL USE OF RADIOACTIVITY.
Radioactive tracers can pinpoint abnormalities in body structure or function. Here a PET scan of the brain of a patient who received an infusion of a radioactive substance illuminates an abnormality in blood flow to his brain. This patient, later diagnosed as suffering from panic disorder, had an asymmetry of blood flow (right greater than left) in a specific region of the brain.

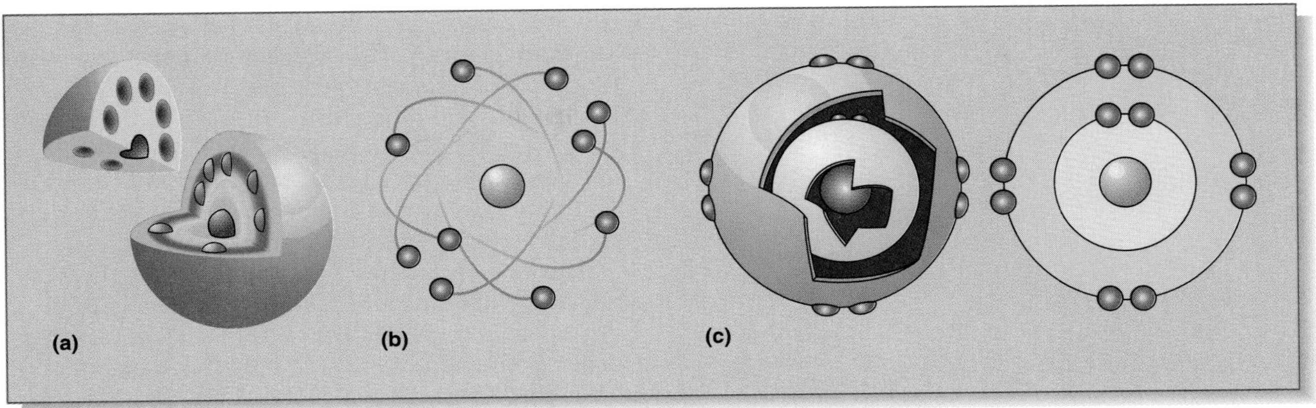

FIGURE 2.3 **THREE MODELS OF ATOMIC STRUCTURE.**
(a) The "watermelon" model, an attempt to show that atoms contain electrons (depicted as the seeds). **(b)** A later model showing electrons orbiting around a central nucleus, much as planets orbit around the sun. **(c)** The concentric circles of the Bohr atom on the right indicate that different energy levels or electron shells are occupied by different electrons as they move around a central nucleus.

sulfur has been vital in determining the structure and function of DNA, the gigantic molecule that determines inheritance.

DEPICTING THE ATOM

An atom is too small to be seen, even with the most powerful microscope; therefore, much of what we know is derived deductively from circumstantial evidence. In spite of our information, depicting an atom has proved to be a major problem because any diagram is bound to be simplistic. Some notions of how atoms must look have been with us a long time, and most have been proved to be quite wrong, although in their time they were useful (Figure 2.3).

FIGURE 2.4 **ELECTRON DENSITY CLOUD MODEL OF HYDROGEN WITH ITS ONE ELECTRON.**
The density of the reddish color indicates the expected frequency of the electron in that region near the central nucleus over a given period of time.

Atoms are still diagrammed in a number of ways. Each can be useful, depending on the point of the discussion. For example, the orbitals that we have shown as lines can also be depicted as hazy clouds indicating where there is a certain probability that an electron *might* be at any given point in time. These are called **electron cloud models** (Figure 2.4). In such depictions the orbitals may appear hazy and indistinct, but they have definite shapes, each a precise distance from the nucleus.

It is sometimes useful to describe atoms in terms of a model developed by the great physicist, Niels Bohr. In the **Bohr atom** (Figure 2.3), the nucleus is shown being circled by electrons (much as the planets circle the sun). Bohr's model suggests that electrons move in a single plane, but this is not the case. Electrons can and do move in any plane and at various distances from the nucleus.

Energy Levels and Shells. Scientists have known for some time that there are several distinct energy levels at which the electrons of any element may be found. Such energy levels are also known as **energy shells.** The first (nearest the nucleus) and lowest energy shell contains a maximum of two electrons. The second shell can hold as many as eight electrons, as can the third. The electrons in the second shell are at a higher energy level than those of the first, and those in the third are at a still higher level.

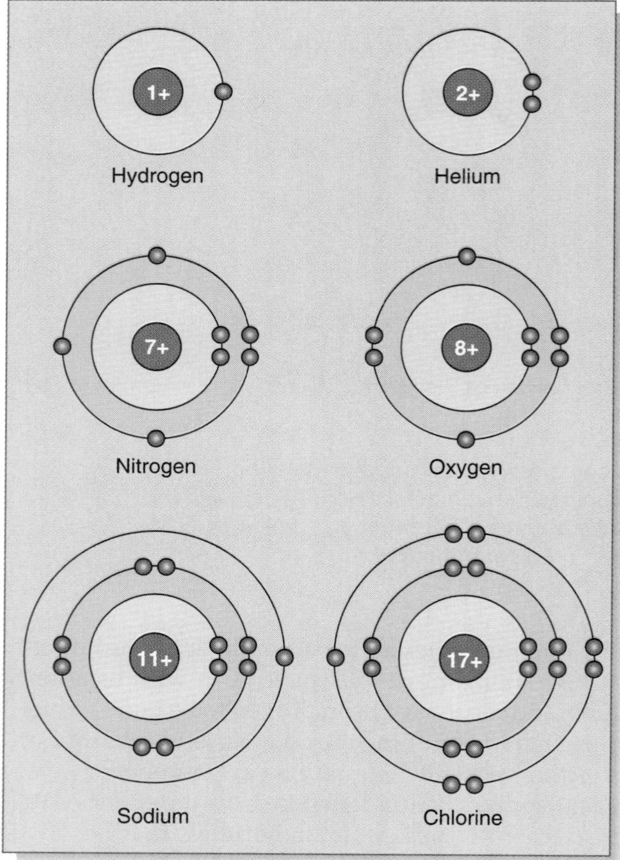

FIGURE 2.5 BOHR ATOM DIAGRAMS OF SEVERAL COMMON ELEMENTS.
Note the regular progression in which the first, second, and third electron shells are filled.

Although even higher electron energy levels exist (and these hold more than eight electrons), we will be primarily interested in the lighter elements in which electrons occupy only the three innermost energy levels. We will see that when all of the shells are filled with their maximum number of electrons they tend to be stable and nonreactive.

Atoms fill their shells from the innermost shell outward (Figure 2.5). For example, hydrogen's one electron occurs in the first shell, as do the two electrons of helium. Helium, because its single shell is full, is nonreactive, or inert. On the other hand, hydrogen is much more reactive because its shell is not full.

Nitrogen (atomic number 7) fills its inner shell with the usual two electrons, and its other five electrons partially fill the second shell. With its 17 electrons, chlorine nearly fills three shells, which contain two, eight, and seven electrons, respectively. The arrangement of electrons and energy levels in the SPONCH elements are given in Table 2.1.

Electrons can move from one shell to the next in either direction if their energy level is increased or decreased by a specific required amount. In addition, if an electron in the outer shell is sufficiently energized, it can escape the atom completely.

This is all fascinating stuff, but one might wonder, what does all this have to do with biology? A great deal, it turns out. For example, it is the shifting of electrons energized by sunlight that makes it possible for plants to make the food that supports nearly all forms of life on the earth (see Chapter 7).

MOLECULES AND COMPOUNDS

SPONCH atoms are almost never found singly; nor are the atoms of most other elements. In fact, atoms join in all sorts of combinations to form *molecules.* A **molecule** is two or more chemically joined atoms. Oxygen gas, for example, is a molecule that is composed of two oxygen atoms—hence the symbol O_2.

Molecules may also consist of different elements. A **compound** contains more than one element, occurring in fixed proportions. For instance, sodium chloride (NaCl)—common table salt—always has equal proportions of sodium and chloride atoms. Water (H_2O) contains twice as much hydrogen as oxygen. Carbon dioxide (CO_2), on the other hand, has two oxygen atoms for each carbon atom. We will see that even these simple molecules have enormous implications for life (Essay 2.1). Carbon dioxide is the primary building block by which all of life's organic molecules are built, beginning in the process of photosynthesis (Chapter 7), and it is a waste product in the breath from our own bodies (Chapter 39). Some molecules of interest to biologists may consist of thousands or even millions of atoms. We will encounter some of these in Chapter 3.

INTERACTION AND SHIFTING ENERGY

Atoms interact with other atoms, forming molecules in the process of **chemical reactions.** Through chemical reactions, atoms join with each other to form molecules, or molecules split into smaller molecules or atoms. These reactions form the basis of all life functions—including whatever is going on in your mind this instant. Chemical reactions give you the strength to scratch your head as you ponder a

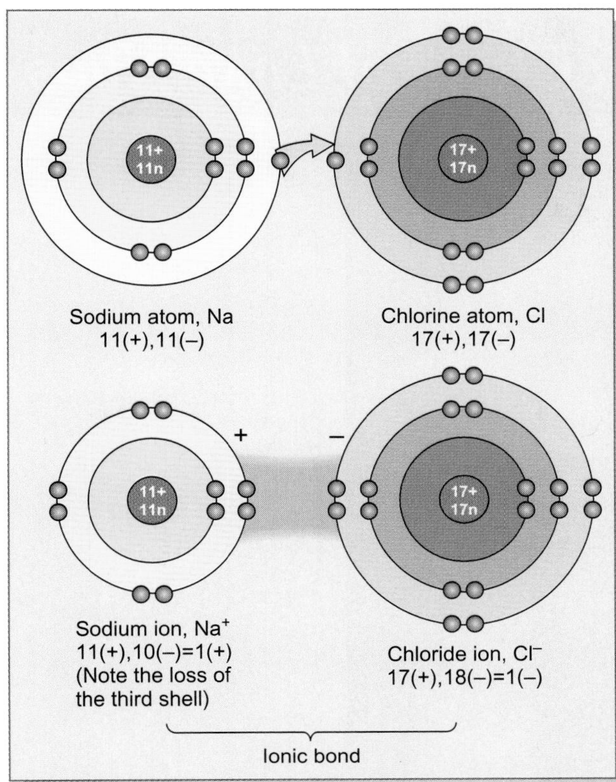

FIGURE 2.6 IONS AND THE TRANSFER OF ELECTRONS.
The loss of an electron renders sodium (shown here in yellow) positive while the gain of an electron by chlorine (shown here in green) renders it negative, and the two form ions. The resulting electrostatic attraction between the two ions is called an ionic bond.

sentence like this. There are three fundamental tendencies that contribute to chemical reactions.

1. Atoms tend to have a balance between positive and negative charges.
2. Electrons within atoms and molecules tend to pair.
3. Electron shells of atoms and molecules tend to become filled.

The different tendencies of an atom are sometimes in conflict. For example, the tendency of an atom to have balanced electrical charges usually leaves the outer shell unfilled, and when the outer shell is filled, the protons and electrons are usually not in balance. Oxygen, for example, has eight protons. But note that after the first shell is filled with two electrons, there are only six electrons left for the next shell, although eight are required to fill it. Therefore, the outer shell of an oxygen atom lacks two electrons. If the oxygen atom were to fill its outer shell with electrons, it would have 10 electrons and only eight protons, thereby losing the balance between protons and electrons. How are these competing tendencies accommodated? Through the formation of molecules by chemical reactions, a compromise is reached.

CHEMICAL BONDS

An atom can fill its outer shell in one of three ways: (1) it can gain electrons from another atom; (2) it can lose all of the electrons in its outer shell to another atom, exposing the underlying filled shell; or (3) it can share electrons with another atom. The result will be formation of a molecule through a **chemical bond.**

THE IONIC BOND

Let's first see how atoms can gain or lose electrons (Figure 2.6). Sodium (Na) has 11 protons. It also has 11 electrons: two in the first shell, eight in the second shell, and only one in the third: the outer shell is seven electrons short (see Table 2.1). It is not energetically possible for an atom (in this case, sodium) to gain seven electrons; but if it can lose one electron, then the full second shell will become the outer shell, and its shell requirements will be met. In contrast, chlorine (Cl) has 17 protons and 17 electrons. Thus its third shell has seven electrons—one short of being full. Chlorine, therefore, can complete its outer shell by accepting just one more electron.

TABLE 2.1 *ELECTRON ARRANGEMENTS OF SPONCH AND OTHER SELECTED ELEMENTS*

Element	Atomic Number	First Shell (2)	Second Shell (8)	Third Shell (8)
		Electrons in Each Shell		
Hydrogen	1	1	0	0
Helium	2	2	0	0
Carbon	6	2	4	0
Nitrogen	7	2	5	0
Oxygen	8	2	6	0
Sodium	11	2	8	1
Phosphorus	15	2	8	5
Sulfur	16	2	8	6
Chlorine	17	2	8	7

Essay 2.1 The Death Cloud of Lake Nyos

Who could have thought that the seemingly peaceful Lake Nyos would suddenly and tragically release a cloud of carbon dioxide? How many other such lakes are there?

The villagers saw the cloud hovering around the edges of Lake Monoun. It was a curious sight, but they had no reason to fear the cloud until those who walked into it fell dead. Before the day was over, the cloud had killed 37 villagers in the West African country of Cameroon. The year was 1984.

The outside world paid little attention to the curious event. But then one August evening in 1986, the villagers of Nyos, not far away, heard a crashing roar. The sound came from a lake about a mile away, high up in the hills. They could not see the cloud of gas that erupted from the lake, and they had no way of knowing that it was coming down the valley toward them. The people who were outside their huts were joined by those whom the sound had awakened. As they stood there trying to guess what might have happened, the cloud reached them and they began falling to the ground, tearing at their clothing, trying to breathe. Those asleep inside their huts succumbed more slowly as the gas quietly seeped into their sleeping quarters. But in minutes, everyone in the town of Nyos lay dead, even as the cloud moved on to Cha and Subum, taking more than 1,200 lives within an hour.

News of the disaster was first regarded as an unlikely rumor, but then people outside began to realize it was true. Within a few days, an international team of scientists began to reach the area, and it soon became clear that Lake Nyos, in an almost unbelievable event, had belched a cloud of carbon dioxide. Geologists soon determined that the gas had been produced deep in the earth, and had emerged through the lake. Even though an enormous amount of carbon dioxide has been released, scientists believe that the danger has not passed.

In 1994 there were indications that the gas was building up again. Scientists busily began searching for ways to avert another disaster, such as by inserting pipes into the lake bottom to relieve the gas little by little—a gas that living things normally produce in their own bodies, and one necessary for the earth's plants to grow. Yet, as with so many things in the biological realm, too much can be deadly. However, our bodies require oxygen. With so much CO_2 around, there was simply no oxygen for the people of Nyos.

FIGURE 2.7 **CRYSTALS.**
Sodium chloride forms crystals of indeterminate size rather than discrete individual molecules. The crystal can grow from microscopic proportions to gigantic size, such as those seen on the shores of the salt-laden Dead Sea, as the sodium and chloride ions form ionic bonds because of the attraction of their opposite charges.

Because of their particular structures, sodium and chlorine react with each other easily and swiftly. In this chemical reaction, sodium is called an **electron donor,** and chlorine is an **electron acceptor.** In its pure state, each element is a deadly chemical because of its tendency to react, but when bound together, sodium and chlorine form table salt (NaCl).

In forming salt, both chlorine and sodium become *ions.* An **ion** is any atom or molecule that has a net electrical charge, either negative or positive. In table salt the sodium has only 10 electrons and 11 protons, so it now has a net positive charge of +1. Chlorine has 18 electrons and only 17 protons; it takes on a net negative charge of –1.

Because of the opposite charges on the ionized sodium and chlorine, they are attracted to each other. They join through an **ionic bond** to form sodium chloride (Figure 2.7).

Ions and Life. Ions commonly form in the watery fluids of living things. Examples include the simple ions of hydrogen (H^+), calcium (Ca^{2+}), potassium (K^+), and magnesium (Mg^{2+}) as well as more complex ions, such as the phosphate ion ($H_2PO_4^-$), the ammonium ion (NH_4^+), and the sulfate ion (SO_4^{2-}). Carbon, one of the most important elements of life, as we will see, often forms complex molecules, parts of which are ionized. As you might expect, these charged particles are quite interactive in living systems and often help with vital functions.

THE COVALENT BOND: SHARING ELECTRONS

Another way an atom can fill its outer electron shell is by sharing electrons with other atoms. Two atoms sharing electrons form what is called a **covalent bond.** A covalent bond is generally written as a single dash or a pair of dots.

Consider hydrogen gas (H_2, H—H or H:H), formed from two hydrogen atoms. Each atom comprises one proton and one electron. Since electrons tend to be in pairs, the two hydrogen atoms can pool their single electrons and satisfy this requirement. The shared pair of electrons form a new molecular orbital that includes both atomic nuclei. In this way, not only are the two electrons paired, but also both atoms satisfy the requirement of filling their electron shell. Furthermore, they still maintain the balance between protons and electrons (Figure 2.8). For oxygen gas (O_2), two pairs of electrons are shared.

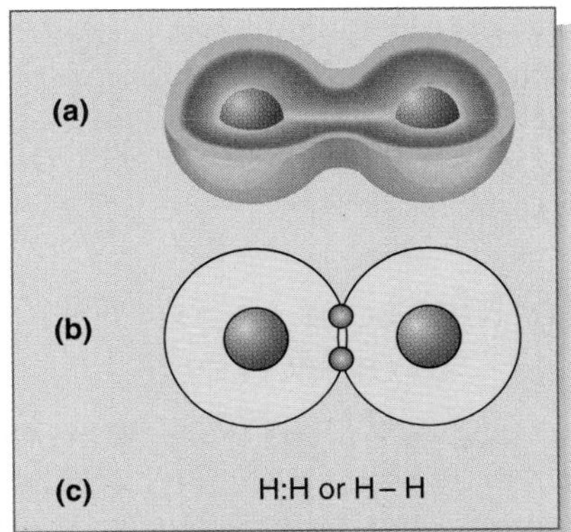

FIGURE 2.8 **COVALENT BONDS.**
(a) Two hydrogen atoms share their lone, unpaired electrons, forming a covalent bond as H_2 gas. **(b)** The covalent bond can be indicated by the sharing of electrons in Bohr atom models. **(c)** However, we often use these shorthand designations for the sharing of electrons in covalent bonds.

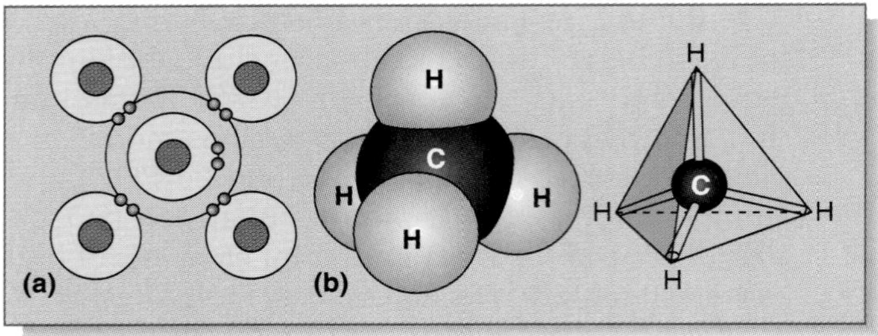

FIGURE 2.9 **ELECTRON ORBITALS AND SHAPES OF MOLECULES.**
(a) In methane (CH_4), the carbon's outer shell electrons pair with four hydrogen electrons. **(b)** In various types of space-filling models, the hydrogens occupy the points of a tetrahedron, while the carbon sits in the center.

When two pairs of electrons are shared, the result is a **double bond,** written O═O.

As atoms join to form molecules, their orbitals may change shape, and shape can be very important in terms of how molecules interact. Figure 2.9 shows electron orbitals of a simple molecule, methane gas (CH_4). The four hydrogen atoms are arranged around a single carbon atom. Note that their orbitals project outward in different directions, each the farthest possible distance from the others. The four tips of a methane molecule form the corners of a perfect tetrahedron, a fairly common molecular shape.

HYDROGEN BONDS AND THE PECULIAR QUALITIES OF WATER

We are generally aware that water is necessary for life, but we may not know that much of its magic lies in its molecular bonding. You may recall that hopes for discovering life on Mars faded when interplanetary probes failed to find significant amounts of water there. Let's see why this finding bred such discouragement.

THE MOLECULAR STRUCTURE OF WATER

A water molecule is made up of two hydrogen atoms covalently bonded to one oxygen atom. All three atoms fill their outer shells by sharing electrons in the formation of single covalent bonds. The four electron pairs in the outer shell of the oxygen atom move in orbitals with the same tetrahedral shape that we saw in methane. In the water molecule, however, only two of the four molecular orbitals surround the hydrogen nuclei; the other two do not (Figure 2.10). This makes the water molecule strangely and magnificently lopsided. With two of the four points of the tetrahedron absent, the water molecule is a planar triangle. Although the positive and negative charges of the molecule are balanced numerically, they are not evenly distributed—the water molecule is polar because the shared electrons are more attracted to the oxygen atom. In essence, a water molecule has a positive side and a negative side. This simple configuration had a great deal to do with the appearance of life on this planet.

The Hydrogen Bond. Because water molecules are polar, with positive and negative regions, they interact with each other. The positive region of one water molecule forms a weak, very temporary attraction for the negative region of another water molecule. This weak attraction between the slightly positive-charged hydrogen atom of one molecule and the slightly negative region of another water molecule is called a **hydrogen bond.** Hydrogen bonds break and re-form rapidly, giving water many of its peculiar characteristics. For instance, despite the weakness of individual hydrogen bonds, their sheer number causes water molecules to cling with surprising tenacity. This stickiness produces the cohesive properties of water.

Water Temperature and Life. One of the most important qualities of water is its temperature stability. Whoever was watching the proverbial pot that gave rise to the adage was probably amazed by the enormous amount of heat necessary to raise the temperature of water.

Specific heat is the amount of heat required to raise the temperature of a substance to some specified amount. If that substance is water, the specific heat is very high. Raising the temperature of water just 1°C requires 33 times as much energy as is

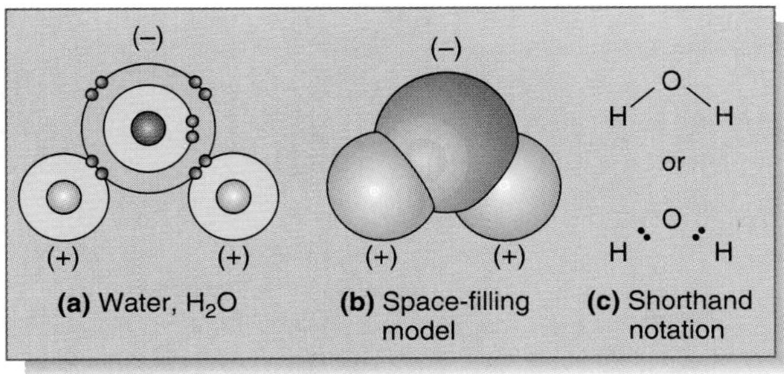

FIGURE 2.10 THE POLAR WATER MOLECULE. **(a)** In water, an uneven sharing of electrons from the two hydrogen atoms produces negative and positive sides to the water molecule **(b)**. The angle of the bonds contributes to the polarity of the molecule, and the angles of the bonds are often, but not always, reflected in the shorthand representations of water **(c)**. The resulting charges are the basis for the formation of hydrogen bonds and give water its life-supporting properties.

required to produce a similar temperature rise in lead. Why should water be so difficult to heat? Keep in mind that as a substance absorbs energy, its molecules move much faster, and its temperature rises. Water's resistance to increased molecular motion is a product of its countless hydrogen bonds that form and re-form. How is this important to life?

The resistance of water to changing temperatures means that creatures composed largely of water, like ourselves, have a certain built-in temperature stability. It also means that organisms that live in water are buffered to some degree by the resistance of water to changes in temperature.

However resistant water might be to changes in temperature, such changes do occur, and changes in temperature affect the physical properties of water. As water cools, its molecules slow down and move closer as hydrogen bonds become more stable. It condenses and becomes heavier. Thus, cool surface water in a lake sinks (causing currents and mixing). Water's density is greatest at 4°C. As it cools below 4°C, its density decreases, causing the very coldest water to rise to the surface again. The density of water reaches its minimum in the open molecular lattice called ice (Figure 2.11). Since ice represents water in a lighter configuration, it forms only on the lake surface, which makes skating much more pleasurable. In addition, the movement of water and the formation of ice have dramatic influences on the forms of life living in association with bodies of water (see Chapter 25).

Water the Solvent. Water is well-known as a cleaning agent, partly because it is not expensive and partly because it is one of the best solvents known. Water's properties as a solvent lie, once again, in its molecular properties. Because it is polar, it has an affinity for a number of substances, such as salt (NaCl), with its positive and negative parts, and sugars, with their own areas of positive and negative charges. Any substance that interacts readily with water is described as **hydrophilic** (water-loving).

Ionic compounds such as sodium chloride dissolve in water by **dissociation** (dissociation is a separation into ions), with water molecules clustering around the resulting ions (Figure 2.12). This happens because negative chloride ions attract the positive parts of polar water molecules; and positive sodium ions attract the negative parts of the water molecules. The same applies to the many other ions that form in our bodies. Other molecules such as sugar may also be polar, with slightly negative and slightly positive regions. Therefore, water will form hydrogen bonds with regions of polar molecules, which is why sugar dissolves almost as readily as salt.

Nonpolar molecules, those lacking charged regions—such as oils, petroleum products, and certain fats—do not interact with water and, therefore, do not dissolve. Instead, the affinity of water molecules for each other causes the excluded molecules to cluster into their own masses. That's why you have to keep shaking Italian dressing.

The behavior of nonpolar molecules in water is called *hydrophobic* interaction. **Hydrophobic** means "fear of water." Hydrophobic interactions account for the fairly strong apparent attraction between nonpolar molecules (or between the nonpolar *portions* of molecules) that occurs in the presence of water. For example, if melted chicken fat (or any liquid fat or oil) is mixed with water, the fat will join into globules and the globules will merge. The fat molecules are simply being *excluded by the water.* Water molecules, as we've seen, have a strong attraction for each other and for other polar molecules, but they have no attraction for the nonpolar molecules of chicken fat.

ACIDS, BASES, AND THE pH SCALE

Most of the chemical reactions of life take place in watery solutions, and their specific reactions are strongly influenced by whether the solution is acidic, basic, or neutral.

Acids, as you might know, have a sour taste (such as the citric acid of lemons or the acetic acid of vinegar). The acid present in your car battery is quite

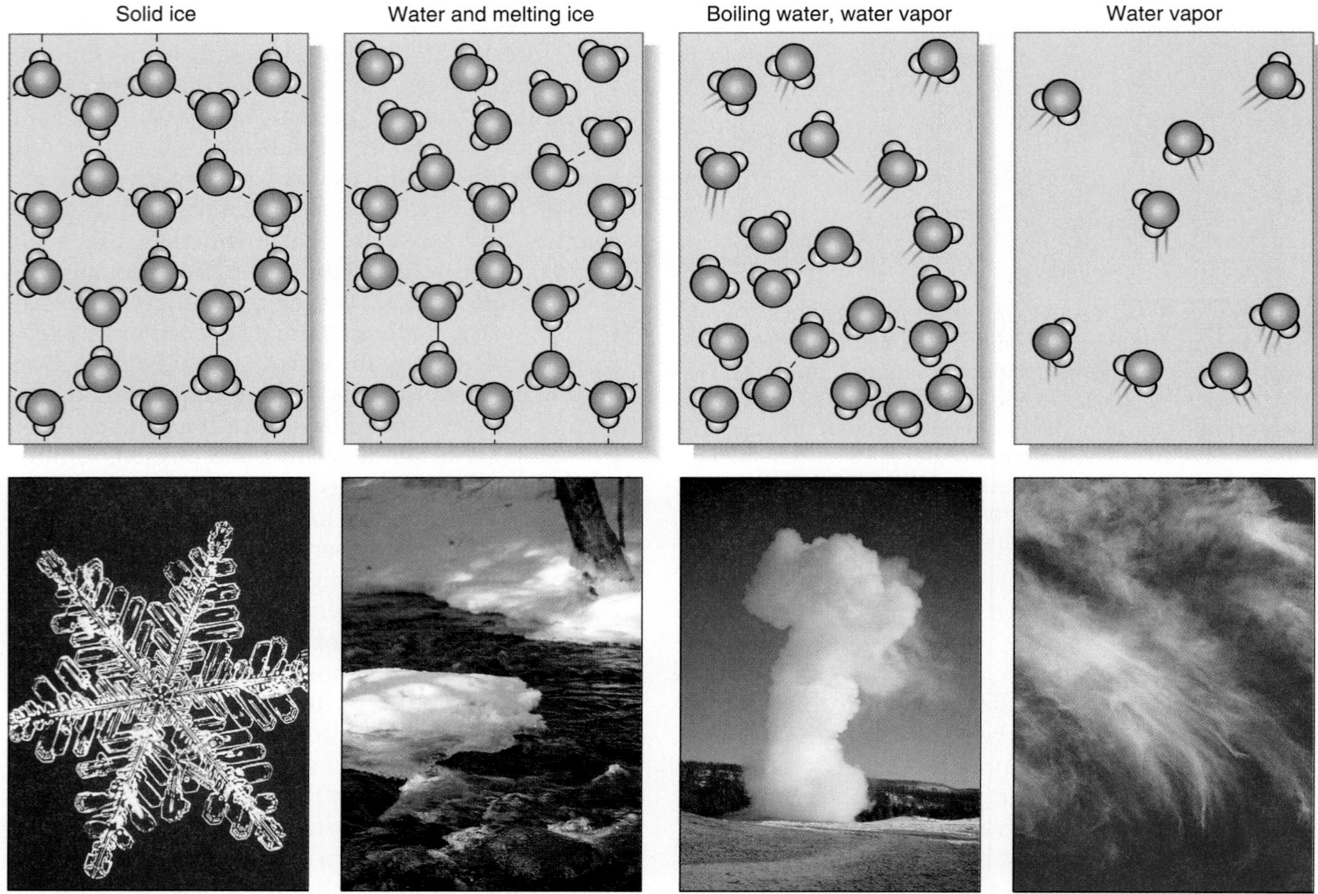

FIGURE 2.11 THE VARIOUS STATES OF WATER.
As water approaches 0°C (32°F), the dynamic sliding lattice of its liquid state changes to the rigid, expanded, crystalline array known as ice. The dashed lines are hydrogen bonds. As energy is added to ice, it melts—as the hydrogen bonds can no longer keep the water molecules in place. As water evaporates, the hydrogen bonds become transient, allowing the water molecules to move freely and form a gas.

potent and will even produce holes in your clothes if it gets on them. Bases, or *alkalis,* as basic substances are known, have a slippery consistency like soap. Some bases, such as oven cleaner, household ammonia, and drain cleaner, are pungent, and can be as dangerous as acids.

Pure water is **neutral,** neither acidic nor basic. Although nearly every molecule of pure water is in its molecular form (H_2O), a very minute portion (one molecule in about 550 million) dissociates spontaneously, forming positively charged hydrogen ions (H^+) and negatively charged hydroxide ions (OH^-). The ionized particles quickly rejoin to form H_2O, even as other molecules dissociate. A solution increases in acidity when the number of H^+ exceeds the number of OH^-. **Acids** are substances that *increase* the H^+ ion concentration. Basicity (or alkalinity) increases with a rise in OH^-, and a rise in OH^- concentration results in a decreased H^+ because H^+ and OH^- combine to form water. Correspondingly, any substance that *decreases* the hydrogen ion concentration is a **base.**

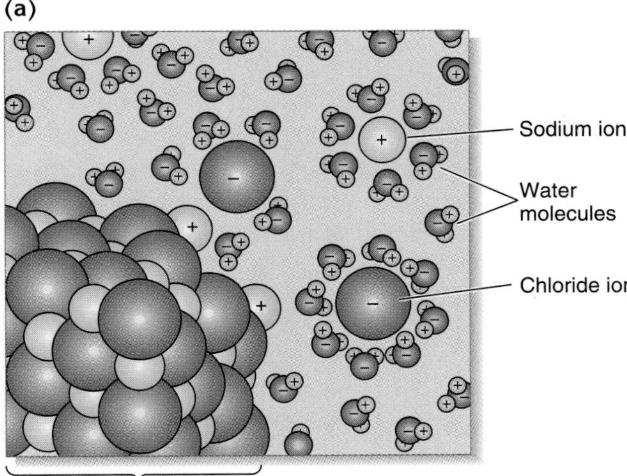

FIGURE 2.12 WATER AS A SOLVENT FOR POLAR BUT NOT OILY SUBSTANCES.
(a) Because of its polar structure, water readily interacts with sodium and chloride ions. Note the specific manner in which the positive and negative poles of water orient to the negative and positive ions of chlorine and sodium. **(b)** Because oils are not polar, they are hydrophobic and fail to interact with water. They will not dissolve in water, and that is a large part of what makes ocean oil spills so nasty.

An example of a reaction that increases acidity is:

$$HCl \longleftrightarrow H^+ + Cl^-$$
(hydrochloric acid)

A reaction that increases basicity is:

$$NH_4OH \longleftrightarrow NH_4^+ + OH^-$$
(ammonium hydroxide)

(The reason that acidity is decreased is that the OH^- ion can join with an H^+ ion to form water, thus decreasing the H^+ concentration.)

Acidity is measured on the **pH scale,** a measure of hydrogen-ion concentration. The scale extends from pH 0 to pH 14, where the pH number is the negative of the log of the H^+ ion concentration. Stronger, concentrated acids are assigned the lowest numbers (pH 1 or 2), and the stronger, concentrated bases the highest pH numbers (pH 13 or 14). The pH 7 refers to the neutral condition found in pure water (see Figure 2.13). Note that the pH scale is deceiving. It is not linear, as it might appear; it is *logarithmic.* That is, an increase or decrease of one pH number actually represents a tenfold increase or decrease in hydrogen ion concentration.

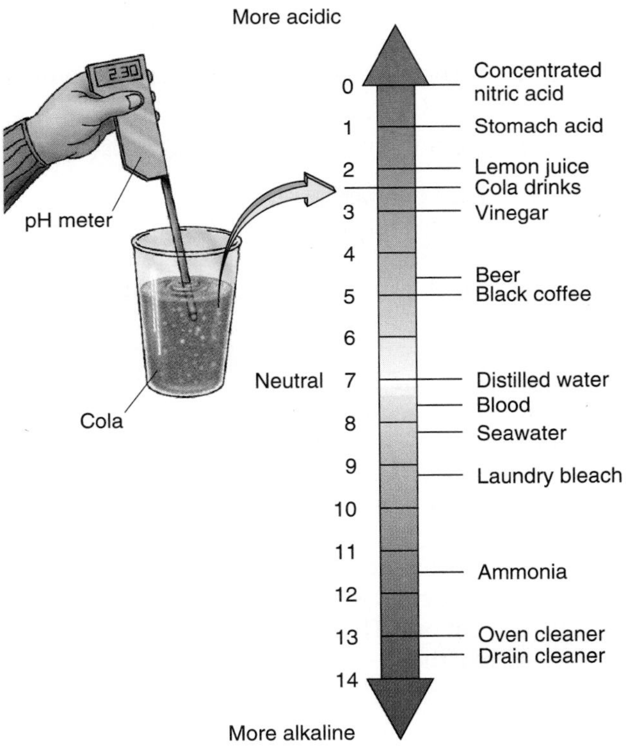

FIGURE 2.13 THE pH SCALE.
Some common chemicals, including several foodstuffs, demonstrate the wide range of the pH scale. The number value assigned to the pH refers to the negative log of the H^+ ion concentration, so each unit change in pH represents a tenfold change in H^+ ion concentration.

PERSPECTIVES

Atoms are the basic chemical unit. They are made up of smaller particles, and can in turn be joined together to form larger structures called molecules. This fundamental feature of chemicals, the ability to be joined together in various ways, is the reason why molecules can be so complex and have such seemingly complicated characteristics.

In the next chapter, we will see this idea continued. We will find that the chemical joining of molecules into even larger structures is how the even more complicated characteristics of biology emerge from inanimate chemicals.

SUMMARY

SPONCH

Elements are substances that cannot be chemically simplified. The most common to life are sulfur, phosphorus, oxygen, nitrogen, carbon, and hydrogen (SPONCH). The basic units of elements are atoms.

ATOMS AND THEIR STRUCTURE

Atomic nuclei contain positively charged protons and uncharged neutrons, which together comprise atomic weight. The much smaller, negatively charged electrons travel in orbits outside the nucleus. The atoms in each element have a specific number of protons, which is also the atomic number.

Atoms with the same atomic number but different numbers of neutrons are called isotopes. Radioisotopes emit energy as they break down into simpler elements. The rate of decay is measured in half-lives. Some are used as tracers in research and medicine.

Electrons occur at distinct energy levels or shells, in direct proportion to their energy. Shells fill from the innermost out. Those atoms with unfilled outer shells are reactive.

MOLECULES AND COMPOUNDS

Molecules are chemical combinations of two or more atoms. When two or more different elements are combined, compounds are produced.

INTERACTION AND SHIFTING ENERGY

The tendencies leading to bond formation include: balancing electrical charges, pairing of electrons, and the filling of outer shells. Competition among these tendencies leads to compromises that become chemical bonds.

CHEMICAL BONDS

In the formation of ionic bonds, electrons move from electron donors to electron acceptors. The loss or gain of electrons produces oppositely charged ions. The resulting attraction is an ionic bond. Ions readily separate again in water.

Covalent bonds are formed as elements share electrons and form new molecular orbitals.

HYDROGEN BONDS AND THE PECULIAR QUALITIES OF WATER

The hydrogen bond accounts for many of the characteristics of water.

In water, two hydrogens are covalently bonded to one oxygen. Because of unequal sharing of hydrogen electrons, water molecules have slightly positive and negative regions. The oppositely charged regions of adjacent water molecules attract each other, forming weak hydrogen bonds.

Because of its hydrogen bonds, water has high specific heat—that is, it has greater resistance to temperature changes than most substances. This makes it a favorable environment for life.

As liquid water cools, its density increases, reaching its maximum at 4°C. As ice forms, water becomes much lighter.

Water is a good solvent because its charged regions interact with ions and other polar substances. Water repels nonpolar (uncharged) substances, which form isolated clusters.

Pure water is neutral, neither acidic nor basic. The presence of hydrogen ions (H^+) in excess of hydroxide ions (OH^-) in water constitutes an acid. The presence of an excess of hydroxide ions characterizes a base. A pH scale of 0 to 14 is used to describe the number of hydrogen ions relative to hydroxide ions. Acids range from pH 0 to 7. Pure water is pH 7. Bases range from above 7 to 14.

Key Terms

element • **19**
atom • **19**
proton • **19**
neutron • **19**
electron • **19**
nucleus • **19**
atomic weight • **19**
orbital • **19**
atomic number • **19**
isotope • **20**
radioactive • **20**
half-life • **20**
electron cloud model • **21**
Bohr atom • **21**
energy shell • **21**
molecule • **22**
compound • **22**
chemical reaction • **22**
chemical bond • **23**
electron donor • **25**
electron acceptor • **25**
ion • **25**
ionic bond • **25**
covalent bond • **25**
double bond • **26**
hydrogen bond • **26**
specific heat • **27**
hydrophilic • **27**
dissociation • **27**
hydrophobic • **27**
neutral • **28**
acid • **28**
base • **28**
pH scale • **29**

Review Questions

1. Define an element and list the six that are most common to life.
2. Name the three principal particles making up atoms, state where they are found and the charges they have. Explain how they relate to atomic weight and number.
3. List the number of protons, neutrons, and electrons in the elements copper, iodine, and chlorine.
4. In what way do the isotopes of an element differ from each other? Mention two uses for radioisotopes.
5. Prepare drawings of atoms of the elements helium, oxygen, and chlorine.
6. Clearly distinguish between the terms *molecule* and *compound*, and cite examples of each.
7. Using sodium and chlorine, illustrate the formation of an ionic bond between the two. Carry out the same task with atoms of hydrogen and chlorine and with atoms of magnesium and chlorine.
8. Explain how the behavior of electrons differs in the formation of ionic and covalent bonds.
9. Draw a simple model of the water molecule, show its charged regions, and describe the manner in which several molecules of water interact through hydrogen bonding.
10. Explain how the presence of hydrogen bonds accounts for the great resistance of water to temperature change. Why is this quality important to life?
11. Describe the relationship between temperature and density in water. Include the two critical temperatures in a body of water.
12. What characteristic of water makes it such a good solvent? Describe the behavior of ionic compounds in water.
13. Describe, in terms of hydrogen and hydroxide ions, neutral, acidic, and basic solutions.

CHAPTER

3

The Molecules of Life

In this chapter we get our first real look at the organization of *chemicals into molecules, especially those molecules that characterize biological systems. One of the guiding principles, as we will soon see, is the building up of larger molecules by the systematic joining of smaller molecular units. Why is this so important? First, it is this logical approach to organization that essentially allows these large molecules to exhibit characteristics of obviously living things, such as the ability to grow and reproduce. Second, once the principles are understood, the structures of the larger molecules can be remembered simply by knowing the constituent subunits and the means by which they are joined together.*

From one point of view, our planet can appear to be an essentially hostile place, its very nature disruptive to the processes of life. In a sense, life exists not because of the earth's benevolence, but in spite of its constant dangers and frustrations. How has life managed to survive in such a place? Largely because it has developed ways to manipulate its molecules. The molecules of life are able to interact with each other in ways that evade, ignore, conquer, and even use the hostile features of the earth that otherwise would threaten the critical and precise organization of living systems.

We are here, in large part, because of such seemingly mundane factors as the number of electrons surrounding the nucleus of an element called carbon. We are here because chemical bonds store energy. We are here because some molecules have fatty tails that hate water. We are composed of molecules, and we owe our existence to their precise and predictable behavior. Clearly, then, if we are to understand this improbable thing called life, we need to learn about its tiny constituents, the molecules.

Carbon: Backbone of the Molecules of Life

Carbon is a fascinating element that has prompted chemists, not known for their poetry, to refer to its "magic." (We will assume that learning about its specific properties in no way diminishes its magic.) Carbon is found in just about every molecule important to living things. In fact, it is the backbone, or framework, upon which organisms build a range of essential molecules. Its importance is a result of the fact that it forms four covalent bonds at once.

Carbon is quite versatile in the way it forms bonds. It can, for example, form four single covalent bonds with hydrogen to become methane, sharing an electron with each. It also can form double bonds, in which two pairs of electrons are shared with another atom. Carbon dioxide, in fact, contains two sets of double bonds, with the carbon in the middle and double-bonded oxygens on either side.

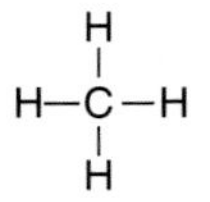

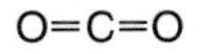

Carbon atoms can also be linked with each other by single or double bonds to form long chains; they can form rings, chains of rings, and a whole range of other complex structures. There seems to be no limit to how large an organic (carbon-bearing) molecule can be (some contain millions of atoms).

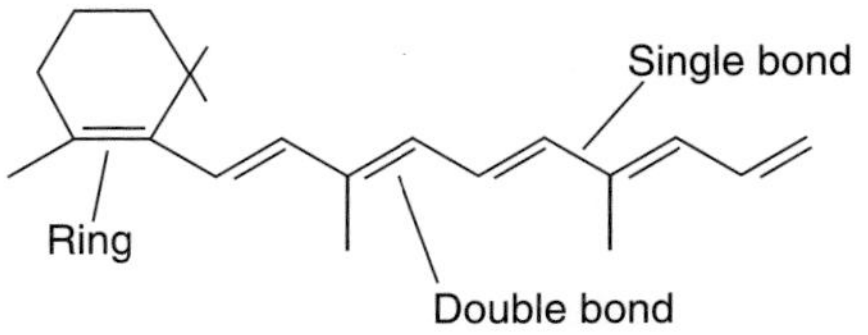

Many such large molecules, or **macromolecules,** are *polymers.* **Polymers** are made by the joining of large numbers of individual subunits called **monomers.** Interestingly, all biologically important types of monomers are chemically bonded to each other in a similar fashion. Each time two monomers join, a molecule of water is released by removal of two hydrogens and an oxygen from the two monomers. This is called **dehydration synthesis**. Conversely, whenever a polymer is degraded into its monomer components, water must be added to the molecules. This is called **hydrolysis**. As we describe each type of biological macromolecule, be sure to note where the water comes and goes.

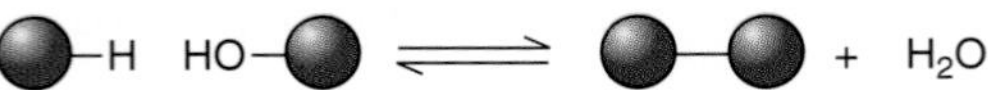

The important macromolecules fall into one of four classes, characterized in part by the identity of the monomer subunits they contain. The macromolecules are: carbohydrates (polymers of monosaccharides or simple sugars), lipids (some of which are polymers of fatty acids), proteins (polymers of amino acids), and nucleic acids (polymers of nucleotides).

Carbohydrates

Carbohydrates are a type of organic molecule containing carbon, hydrogen, and oxygen, normally with two hydrogens to one oxygen and one carbon. They are found in all living things, but are especially abundant in foods such as fruits, grains, and potatoes. In fact, over two thirds of the diet of the people on this planet is from the carbohydrate-rich cereal of wheat, rice, and corn. Carbohydrates are also often the prime ingredient of snack foods, because they are relatively easy to acquire and process. For the same reason, carbohydrates comprise a disproportionate share of the diet of the poor. There are several classes of carbohydrates, the most familiar being sugars and starches. Carbohydrates get their name from their chief elements, carbon, hydrogen, and oxygen (carbon and water). The formula of most carbohydrates is, as mentioned, $(CH_2O)_n$, meaning that if a molecule is $(CH_2O)_6$ it has 6 carbons, 12 hydrogens, and 6 oxygens and can be written $(C_6H_{12}O_6)$.

Monosaccharides: Simple Sugars

Monosaccharides are the smallest carbohydrates, and may have from three to eight carbon atoms. By far the most familiar is the six-carbon sugar, **glucose** (also called *dextrose*) (Figure 3.1). Glucose units are often linked to form large polysaccharides. In addition, as a building block of polysaccharides, glucose is itself an important energy source. Other important six-carbon simple sugars are **fructose,** a common sugar of grapes and other fruits, and **galactose,** found in milk. Incidentally, all three of the six-carbon sugars have the same chemical formula, $C_6H_{12}O_6$, but they differ in geometry and arrangements of atoms.

Other monosaccharides critical to life are the *pentose* sugars with five carbons. Two of these, **ribose** and **deoxyribose,** are constituents of the huge genetic molecules called DNA and RNA, which we'll get to later in the chapter.

Glucose can take on a straight chain or a closed ring configuration (see Figure 3.1). Extending from its carbon backbone are atoms that compose three types of side groups: hydrogen (—H), **hydroxyl** (—OH), and **aldehyde** (—CHO). Note also that glucose occurs in two forms—alpha and beta, which differ only in the orientation of one of the hydroxyl groups. All of these factors become important as we consider how the subunits are linked to form larger molecules.

Disaccharides: Linking the Monosaccharides

Logically enough, two monosaccharides covalently bonded together are called a **disaccharide** ("two sugars"). There are several common examples. Fructose and glucose, when covalently linked, form **sucrose,** common table sugar. Two glucose molecules join to form the disaccharide **maltose. Lactose,** or milk sugar, is formed from glucose plus galactose. (If you suffer from lactose intolerance, it is this sugar that causes your distress after eating milk products.)

As for all biological polymerizations, simple sugars are linked through *dehydration synthesis.* Remember, dehydration synthesis involves the removal of water. A look at the two reacting glucose subunits in Figure 3.2 shows us where the water comes from. In two adjacent hydroxyl groups, one gives up a hydrogen, the other gives up a hydroxyl group, and a molecule of water is released. A new covalent bond forms between the two glucose molecules. When we digest carbohydrates and other foods, this bond is broken in a process called *hydrolytic cleavage* or *hydrolysis* (see Chapter 6). As it is broken, a molecule of water is split to restore the lost hydrogen and hydroxyl groups.

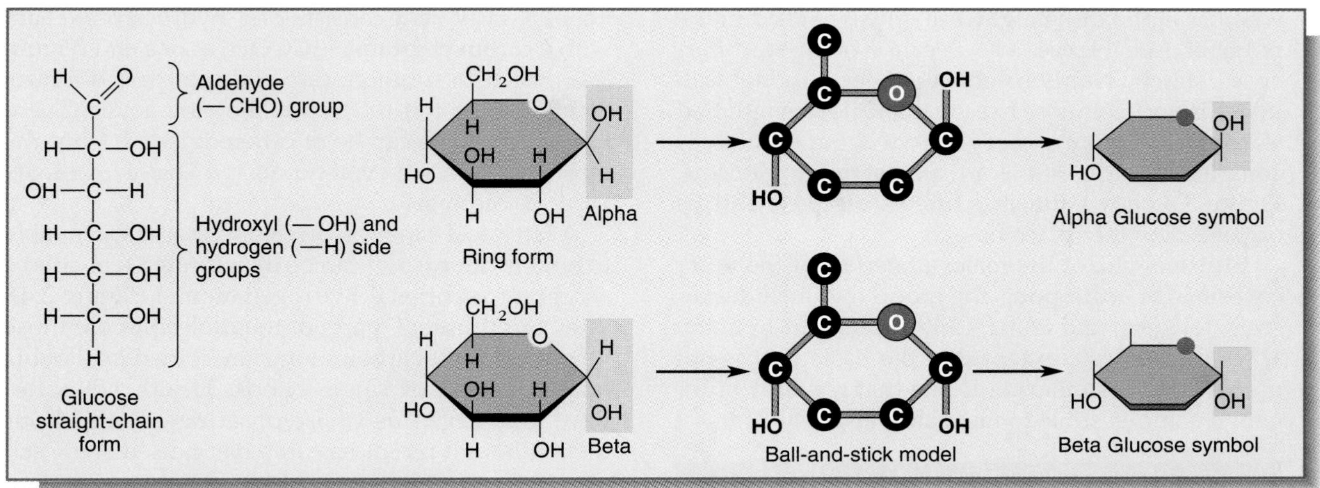

FIGURE 3.1 GLUCOSE OCCURS IN A STRAIGHT-CHAIN AND A RING FORM.
Note the opposite orientation of the —H and —OH groups in the alpha and beta rings. A space-filling model of glucose reveals some of the true geometry of this important molecule that is the chemical basis of much of life's energy. Because of its importance in metabolism, glucose is often a component of energy replenishment drinks.

POLYSACCHARIDES: THE LARGER CARBOHYDRATES

Polysaccharides ("many sugars") are long chains of monosaccharides. The simplest polysaccharides are polymers of glucose, joined by dehydration synthesis. Such polysaccharides include starches such as *amylose* and *amylopectin* (storage products in plants), and *cellulose,* a carbohydrate that lends strength to plants. Animals also assemble polysaccharides such as *glycogen* (animal starch) and *chitin,* a major structural material in some animal groups (and some fungi as well). Starches are important storage materials because their energy is so readily available (Figure 3.3). Starches traditionally have formed a large part of our diet.

Cellulose is a principal constituent in plant cell walls. Like starch, it is composed of long chains of glucose molecules. However, there are important differences between the two that are based on

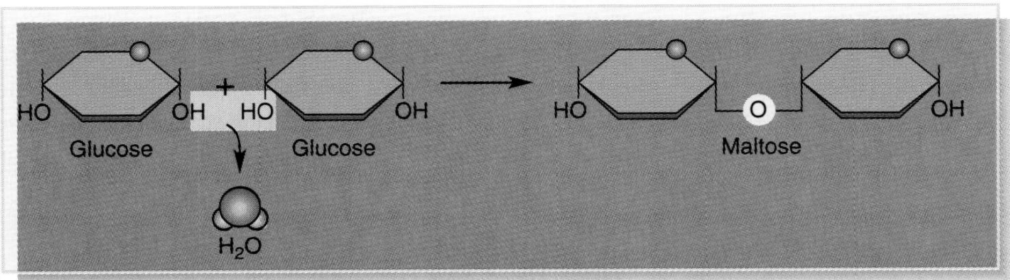

FIGURE 3.2 DISACCHARIDE FORMATION.
Two glucose subunits react in the presence of the enzyme to form the disaccharide maltose and release water.

whether alpha or beta glucose units are used in the polymer (see Figure 3.1). Because of these differences, starch is fairly soluble in water, although cellulose is not. Cellulose has great tensile strength that starch lacks; starch is easily digested, but cellulose is completely indigestible to almost all organisms. Figure 3.3 shows the structure of cellulose and its organization into plant fibers.

Chitin is one of the major materials of the outer coverings of arthropods, the group that includes insects, lobsters, and crabs. Chitin is similar in many ways to cellulose, except that the basic unit is not glucose, but a similar molecule that contains nitrogen. It is indigestible to most animals.

Lipids

The **lipids** are a diverse group of molecules, defined as compounds that cannot be dissolved in water. They are soluble in nonpolar solvents such as ether and chloroform. Lipids may be very small, or large and complex. They function as energy-storage reservoirs, insulators, lubricants, and even as hormones. They are also an important part of the membranes that surround cells. Lipids include animal fats, vegetable oils, waxes, steroids, and an interesting group called the *phospholipids.* Your own brain (as perhaps people have told you) consists largely of fat—more accurately, it contains large amounts of fat-soluble phospholipids.

Triglycerides and Their Subunits

Animal fats and vegetable oils are familiar lipids. Generally, fats are solid at room temperature, whereas oils are liquid. At an animal's own normal body temperature, however, fats are usually liquid.

Both fats and oils are **triglycerides**—compounds with three fatty acids covalently bonded to one molecule of *glycerol.* Triglycerides are important as storage lipids.

Glycerol is a small three-carbon molecule, a carbohydrate, in fact, with three hydroxyl (—OH) groups. Glycerol provides the backbone for all triglycerides. The differences among various triglycerides depend on what kinds of *fatty acids* are attached to that backbone. A **fatty acid** consists of a hydrocarbon chain with a **carboxyl group** (—COOH) at one end (Figure 3.4). When in solution, carboxyl groups can ionize, yielding a proton (H^+); thus they are acidic. These fatty acid chains can be of different lengths, but the most common are even-numbered chains of 14, 16, 18, or 20 carbons.

A fatty acid can be **saturated** with hydrogen (able to hold no more) or it can be **unsaturated** (capable of accepting additional hydrogen atoms) (Figure 3.4). The "unsaturated" parts of the molecules are those where adjacent carbon atoms are joined by double bonds instead of single bonds. Health advocates have long urged us to use unsaturated fats in our diet, although recent research has caused some scientists to question this advice. In general, plant triglycerides are likely to be more unsaturated than those found in animal fat and are therefore considered to be more healthy.

Triglycerides are produced through dehydration synthesis (Figure 3.5), which means that, as each fatty acid joins the glycerol, a molecule of water is released.

Triglycerides contain a great deal of stored energy. In fact, they yield about twice as much energy per gram as do carbohydrates. Plants store triglycerides in seeds, and animals may build up fat as reserves to be used as nourishment in lean seasons ahead and as insulation against temperature extremes.

Humans also tend to store fat under their skin and around their internal organs. The tendency to build fat reserves may have been advantageous to our hunting ancestors, but improvements in agriculture, food storage, and efficient transportation have largely exempted us from the rigors of seasonal food depletion. Still, many of us seem to be taking no chances.

Phospholipids

Phospholipids are structurally similar to the triglycerides; however, the two differ in one important respect. Whereas triglycerides contain three fatty acids covalently linked to one glycerol, phospholipids contain only two fatty acids. In place of the third is a negatively charged phosphate group. The end of the molecule that has the phosphate group is

Figure 3.3 Starches Can Be Straight-Chained or Branched.
Glycogen, the animal starch, is a branched-chain storage carbohydrate. Amylose is a straight-chain plant starch, while amylopectin is a branched-chain plant starch. Cellulose, a structural polysaccharide, is formed from numerous beta glucose subunits. Cellulose chains form lengthy microfibrils that can be readily resolved by the electron microscope. Their laminated arrangement explains the great strength of plant cell walls—for example, those of tall trees.

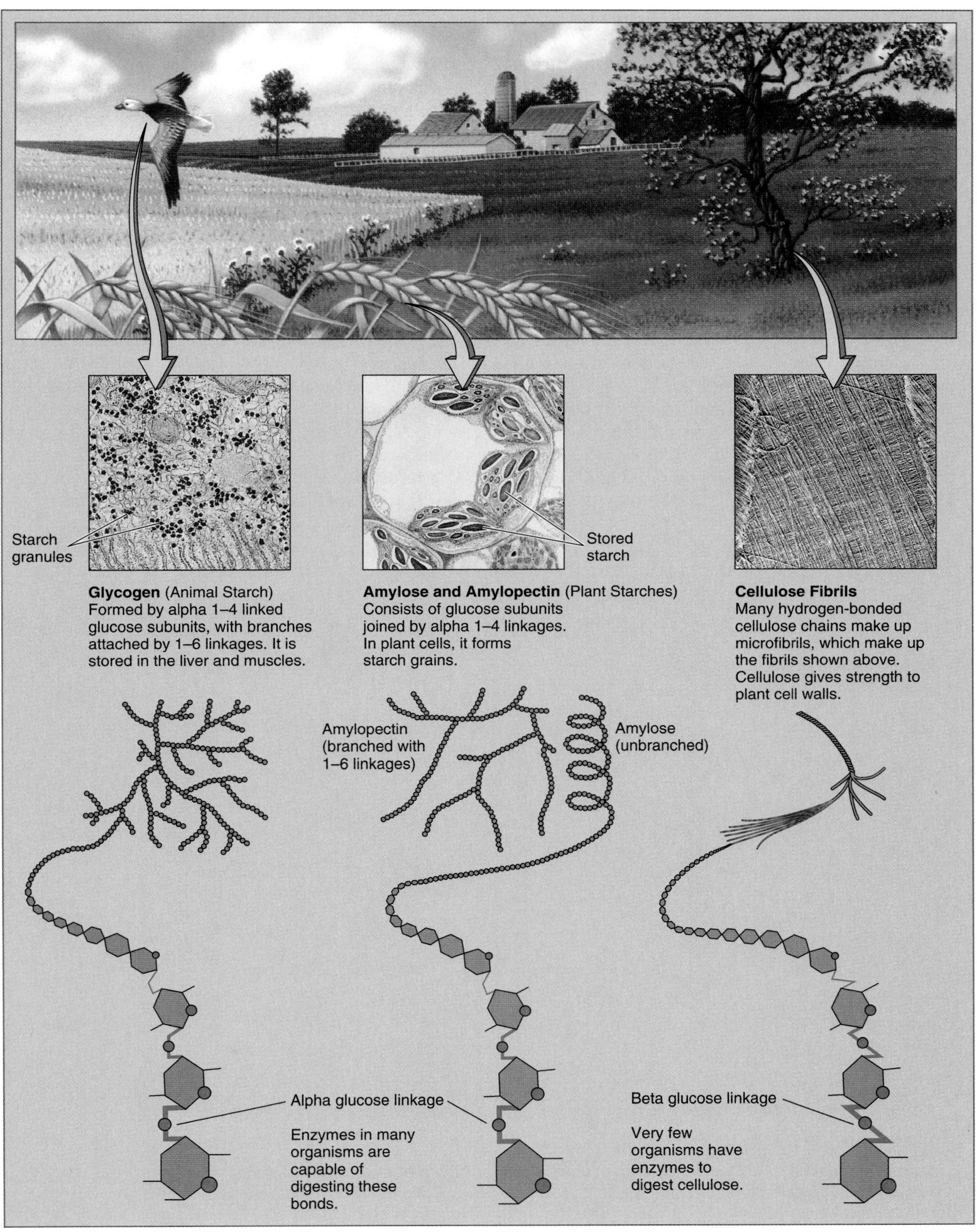
Starch granules
Stored starch
Glycogen (Animal Starch)
Formed by alpha 1–4 linked glucose subunits, with branches attached by 1–6 linkages. It is stored in the liver and muscles.
Amylose and Amylopectin (Plant Starches)
Consists of glucose subunits joined by alpha 1–4 linkages. In plant cells, it forms starch grains.
Cellulose Fibrils
Many hydrogen-bonded cellulose chains make up microfibrils, which make up the fibrils shown above. Cellulose gives strength to plant cell walls.
Amylopectin (branched with 1–6 linkages)
Amylose (unbranched)
Alpha glucose linkage
Enzymes in many organisms are capable of digesting these bonds.
Beta glucose linkage
Very few organisms have enzymes to digest cellulose.

hydrophilic, meaning that it can dissolve in water. The end of the phospholipid with the fatty acid chains is uncharged and hydrophobic. Because of this arrangement of charges, phospholipids placed in water form spherical clumps, the nonpolar "tails" pointing in and the polar phosphate "heads" pointing out (Figure 3.6).

Molecules that have polar parts and nonpolar parts can simultaneously form hydrogen bonds with water and hydrophobic interactions with nonpolar substances. For example, laundry detergents consist of molecules with long nonpolar hydrocarbon tails and ionized heads. In practical terms, the hydrocarbon tail forms hydrophobic interactions with gravy on shirts, while the ionized head forms bonds with water. Thus the detergent-gravy complex dissolves into the water.

Lecithin, a natural detergent, is a phospholipid of egg yolk. Like laundry detergent, it can dissolve fat in water to some degree. In fact, the lecithin in egg yolk forms a bridge between salad oil and vinegar. The result is mayonnaise.

The dual personality of phospholipids makes them excellent building materials for the membranes of living cells. As we will see in Chapter 4, these membranes are made up in large part of two layers of phospholipids. Their nonpolar, fatty acid tails point in toward one another, forming a water-resistant core. Their charged phosphate heads face out, forming the inner and outer surfaces (Figure 3.7).

Other Important Lipids

Other lipids include *waxes* and a peculiar group called the *steroids.* **Waxes** contain one fatty acid, but instead of being bound to glycerol, it is bound to a long-chain molecule with a single hydroxyl group. Because waxes have powerful water-repellent properties, they are common in organisms that must conserve water. Insect bodies generally are covered by a waxy layer, and many plants—especially those in drier areas—have waxy leaves.

Steroids are structurally quite different from fatty acids, but since they are generally hydrophobic, they

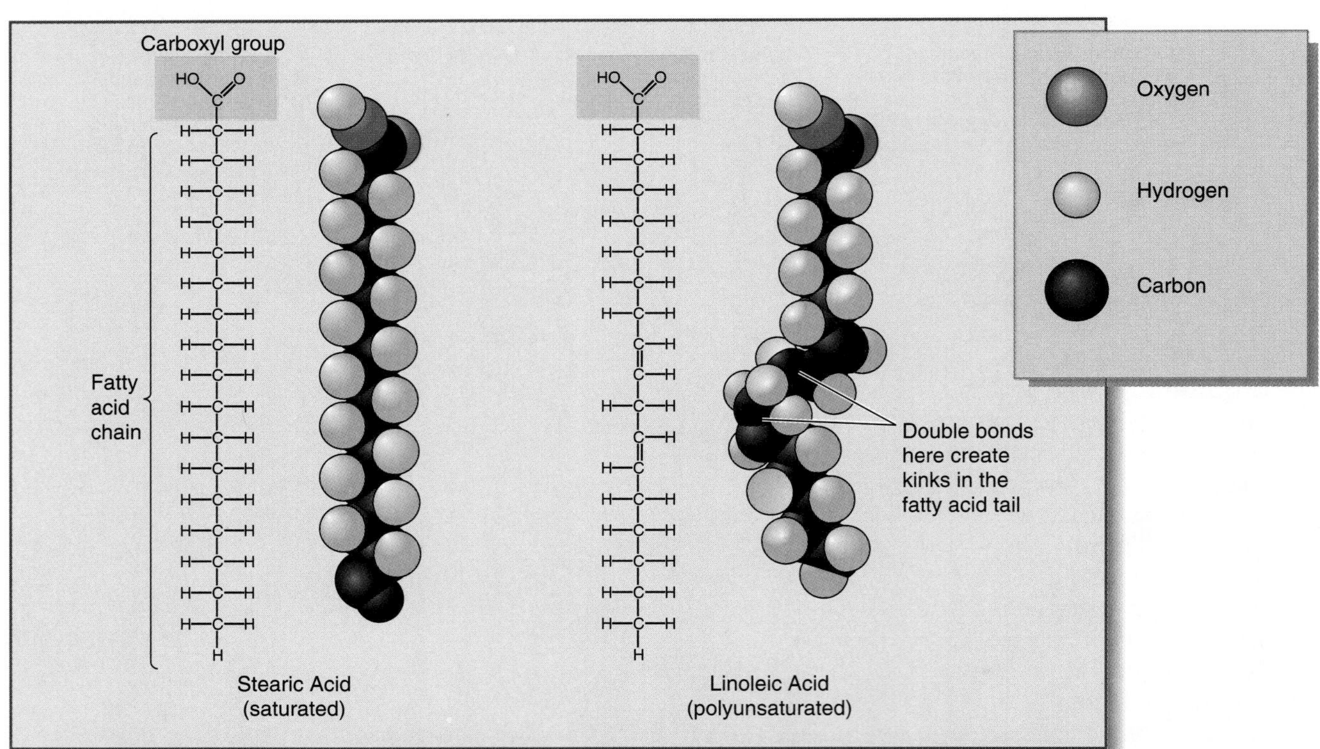

Figure 3.4 Fatty Acids.
Stearic acid, an animal fatty acid, contains the maximum number of hydrogen groups possible, so it is saturated. Linoleic acid, a plant fatty acid, is unsaturated. Since there is more than one double bond, the term *polyunsaturated* is used. Note that both stearic acid and linoleic acid have a carboxyl group "head."

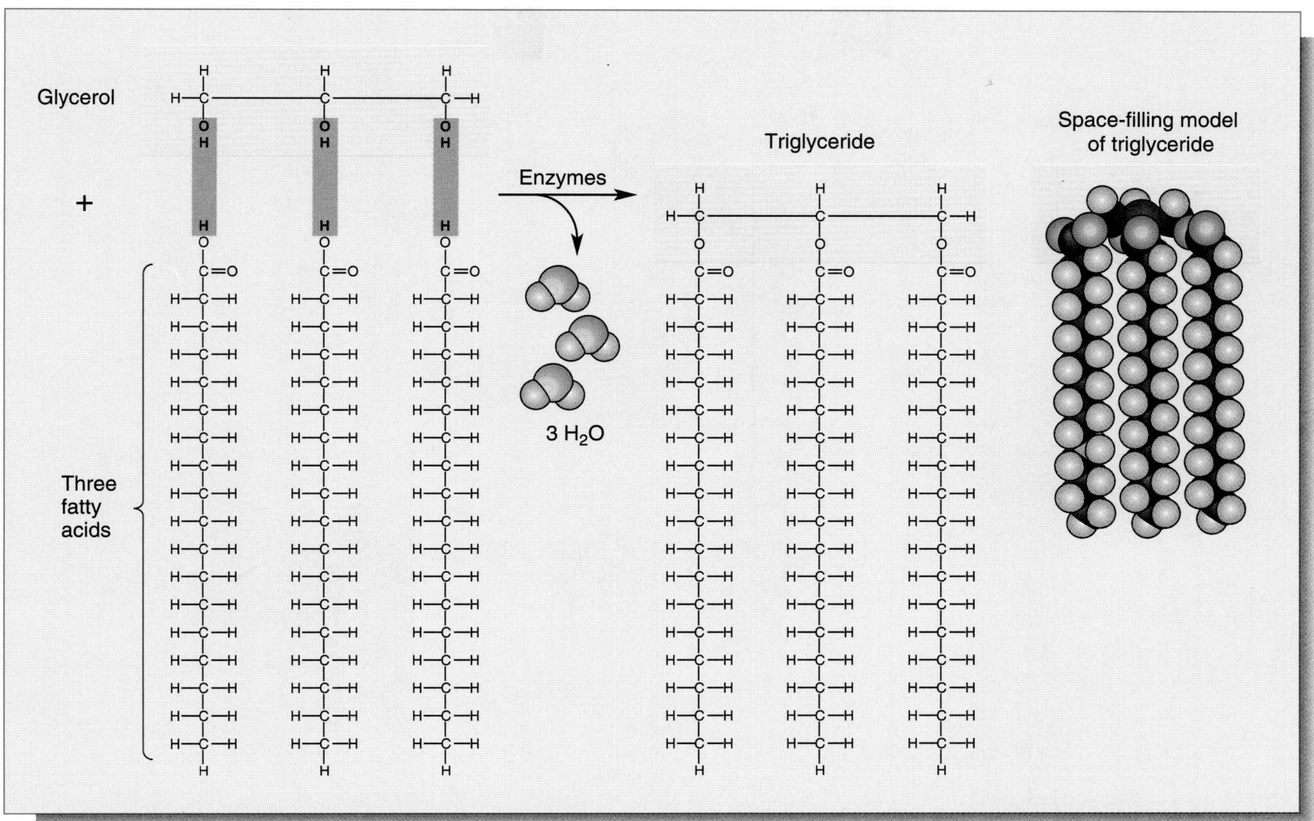

FIGURE 3.5 **TRIGLYCERIDE SYNTHESIS.**
Triglycerides such as animal fats consist of a backbone of glycerol joined to three fatty acids by dehydration linkages. Note the three water molecules that form in the process.

fall within the lipid category. All steroids have a basic structure of four interlocking rings. The differences in their biochemical activities are a function of the side groups that protrude from the rings. Some steroids are very hydrophobic, some less so. We will see later that many of the regulatory chemicals circulating in the blood, the ones called hormones, are steroids, including those influencing sex.

We are all aware of something called **cholesterol** because food manufacturers don't hesitate to boast that their products are cholesterol-free. But why should we care? Cholesterol, we are told, leads to arterial plaques, abnormal thickenings of the walls of arteries, which can raise blood pressure dangerously (Figure 3.8). It cannot be denied that people with high levels of cholesterol in the blood have a greater risk of some circulatory disease.

However, recent findings suggest that cholesterol is not all bad. Our bodies, in fact, manufacture it. We need it for a number of vital functions, such as the development of cell membranes. Blood cholesterol—in the form of *high-density lipoproteins*—is actually beneficial in warding off bacterial disease. Also, bile salts, which are necessary for fat digestion, are modified cholesterol. When irradiated with ultraviolet light, cholesterol becomes vitamin D, which is necessary for normal bone growth. The human sex hormones—as well as the material that gives lustre to your hair—start out as cholesterol.

PROTEINS

Proteins are polymers of subunits called amino acids. Proteins are twisted and fascinating molecules with a great variety of shapes and roles in the pageant of life. As you might suspect of any molecule with such important and varied responsibilities, there are many kinds of proteins.

Proteins can function as structural molecules, as food reserves, and even as hormones (just as can carbohydrates and lipids). Some proteins form

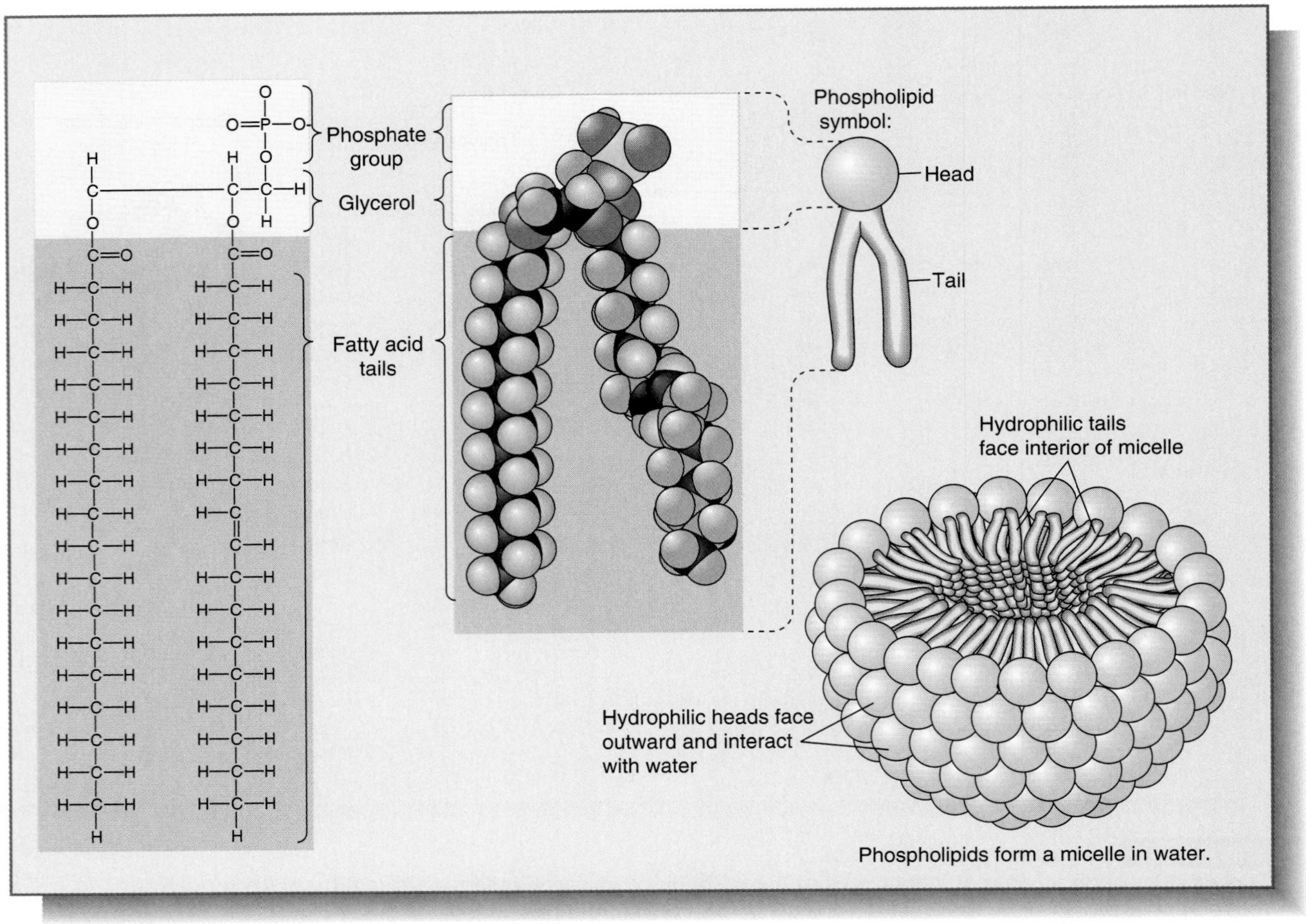

Phosphate group
Glycerol
Fatty acid tails
Phospholipid symbol:
Head
Tail
Hydrophilic tails face interior of micelle
Hydrophilic heads face outward and interact with water
Phospholipids form a micelle in water.

FIGURE 3.6 PHOSPHOLIPIDS IN WATER.
When dispersed in water, phospholipids tend to aggregate with their fatty-acid tails clumped together by hydrophobic interaction. Their charged polar heads interact with water, forming a droplet called a micelle.

FIGURE 3.7 PHOSPHOLIPIDS IN MEMBRANES.
Phospholipids are the principal components of most cell membranes. The phospholipids' hydrophobic tails dissolve into one another; the polar heads face the watery medium both inside and outside the cell.

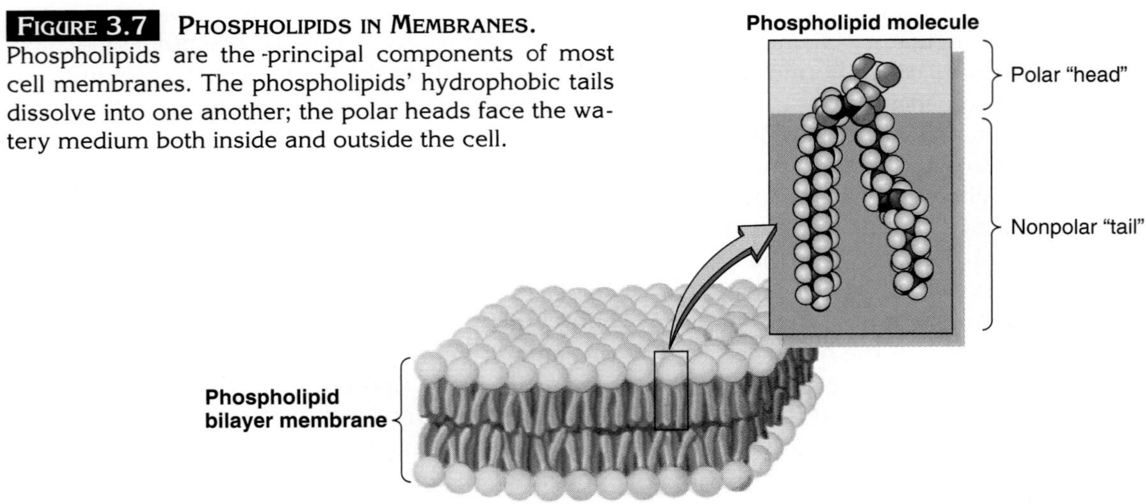

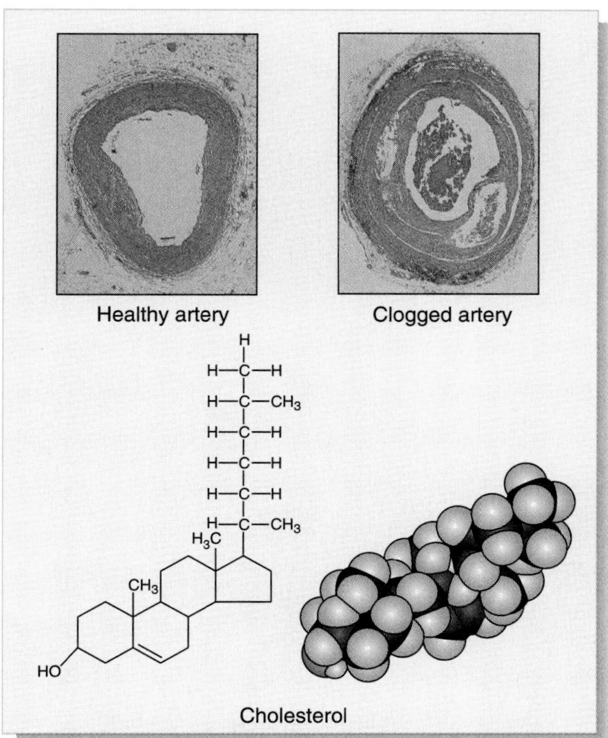

FIGURE 3.8 CHOLESTEROL AND ARTERIAL DISEASE. Accumulation of cholesterol in plaques on the interior of arteries greatly reduces the inside diameter of the artery and increases the pressure of the blood flowing through it.

enzymes—vital chemical catalysts that speed up certain chemical reactions in the cell; still others make movement possible. A vast army of protein antibodies helps the body to fight off infections. One unique protein found in an Antarctic fish acts as an antifreeze.

There are about 100,000 different proteins in the human body alone, all of which function in important ways. No matter how complex they are, proteins are all formed from the same 20 or so different amino acids. (But then, all of the elegantly arranged words in this book are formed from only 26 letters.)

AMINO ACIDS

All amino acids have a few critical traits in common (Table 3.1). For example, every **amino acid** has a carboxyl group (—COOH), which is an acid, and an amino group (—NH_2) (Figure 3.9). Hence the name *amino acid.* Each amino acid also has one carbon atom, the alpha carbon, that links the amino and carboxyl groups together. Also attached to the alpha carbon is a hydrogen atom and what is called an *R group.* The R is simply a shorthand reference to any one of the 20 different chemical groups that could be in that position. The 20 different amino acids differ only in which R group is present (Table 3.1).

The covalent bond between two amino acids is called a **peptide bond.** The bond is formed just as in polysaccharides and lipids, by a dehydration synthesis reaction. The formation of each peptide bond yields one molecule of water.

PROTEIN ORGANIZATION

Joining two amino acids together to form a dipeptide is a first step in the formation of proteins, which are composed of one or more polypeptides (chains of amino acids linked by peptide bonds). Actually, proteins can have as many as four levels of structure (Figure 3.10). The first, or **primary,** level is the sequence of amino acids. The **secondary** level of organization forms as a result of hydrogen bonds between amino acids. These bonds may twist the protein into a right-handed coil called an alpha helix. In other instances, the secondary level forms "beta-pleated sheets" when adjacent sections of the protein lie side by side.

In the **tertiary** (third) level of protein structure, polypeptides undergo an elaborate but highly specific folding that allows the R groups to interact chemically with each other. The fourth, or **quaternary,** level of structure occurs when two or more separate polypeptides become associated in a larger complex. Hemoglobin, for example, contains four polypeptides, a pair of alpha proteins, and a pair of beta proteins. (Each alpha and beta polypeptide has an iron atom incorporated in a complex structure called a *heme group.* Hence the name, *hemo*globin.)

The shape of a molecule, especially a protein molecule, is very important. The structure of the sheets and coils of certain proteins determines, for example, the texture of your hair and whether or not you can digest lactose.

NUCLEIC ACIDS

Our final category in the molecules of life includes the **nucleic acids, RNA** *(ribonucleic acid)* and **DNA** *(deoxyribonucleic acid).* RNA is composed of nucleotides containing the sugar ribose, while DNA contains the sugar deoxyribose. While some nucleic acids can be small, this category includes the largest of all biological macromolecules. In most organisms, the hereditary units called genes are

TABLE 3.1 AMINO ACIDS COMMONLY FOUND IN PROTEINS

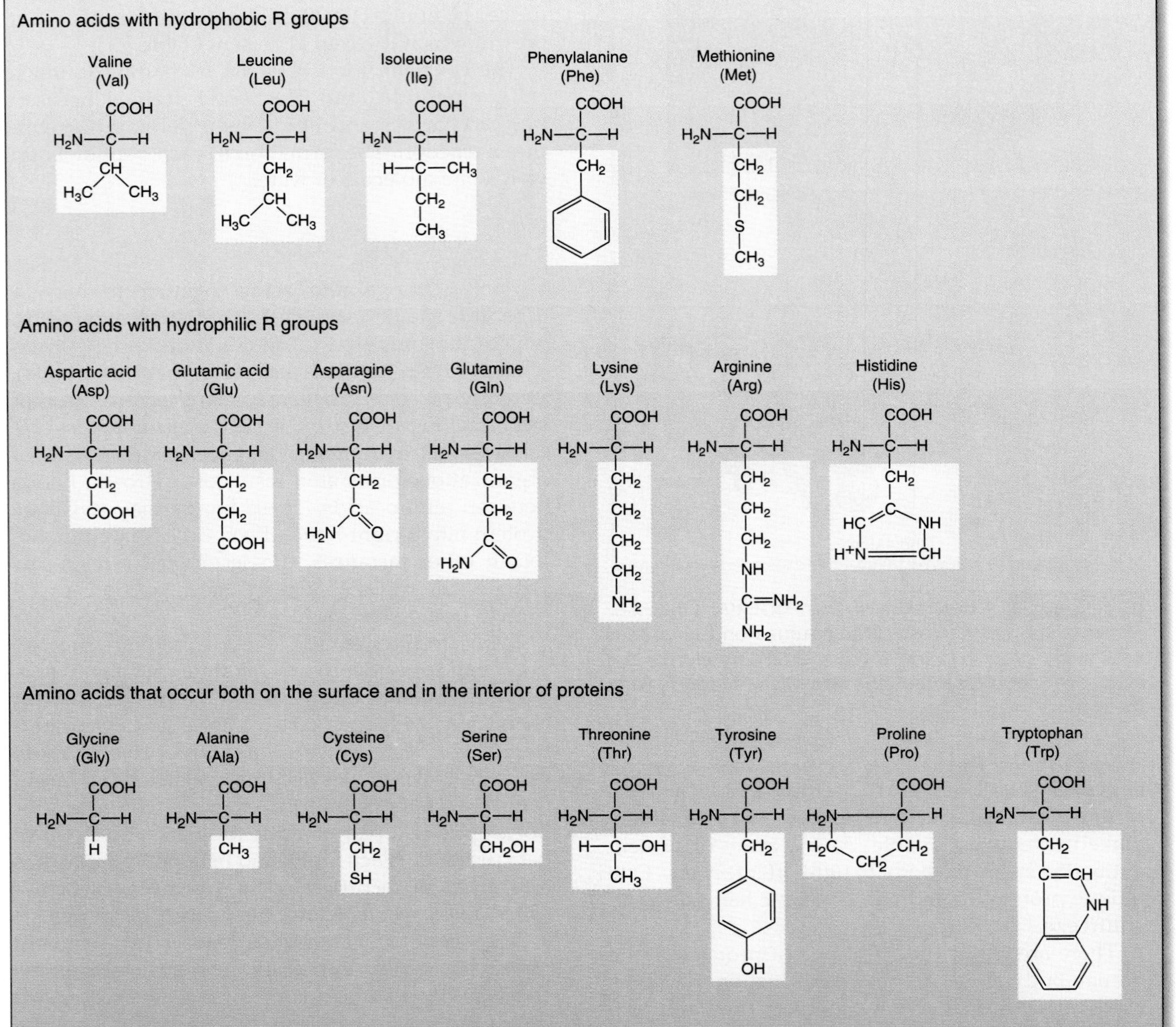

Amino acids with hydrophobic R groups

Valine (Val), Leucine (Leu), Isoleucine (Ile), Phenylalanine (Phe), Methionine (Met)

Amino acids with hydrophilic R groups

Aspartic acid (Asp), Glutamic acid (Glu), Asparagine (Asn), Glutamine (Gln), Lysine (Lys), Arginine (Arg), Histidine (His)

Amino acids that occur both on the surface and in the interior of proteins

Glycine (Gly), Alanine (Ala), Cysteine (Cys), Serine (Ser), Threonine (Thr), Tyrosine (Tyr), Proline (Pro), Tryptophan (Trp)

The portion of the amino acid that varies is highlighted in yellow. Note that some of the amino acids contain more than one amino or acid group, giving them greater basic or acid qualities then the others. Cysteine contains a sulfur-hydrogen group at its R-terminal. This has special importance in determining the shapes of proteins. The abbreviations given in parentheses are used for convenience in writing protein formulas.

composed of DNA. DNA molecules can be immensely long polymers.

As in all of the large molecules of life, DNA and RNA are composed of monomers; in this case the monomers are *nucleotides*. **Nucleotides** are chemical structures that contain a five-carbon sugar, a phosphate group, and ringlike **nitrogenous bases** (Figure 3.11). Despite the length of nucleic acids, the organization of their nucleotides is surprisingly simple. In fact, it was DNA's deceptive simplicity that, until the early 1950s, kept biologists in the dark as to the chemical nature of the gene. Although proteins can contain as many as 20 different amino acid subunits, there are only four different subunits (nucleotides)

ESSAY 3.1 Sweet, Sour, Salty, Bitter—and Umami

"Aged Beef" reads the menu. The restaurant is not referring to old cows, but to meat with a superior flavor because it has been allowed to age in a cool refrigerator, thereby improving its taste. No one quite understood why aged beef tastes better until recently. Now we know it's because beef contains an enzyme that breaks down certain protein into peptides that are only eight amino acids long. It is the accumulation of these peptides that improves the taste of beef, and some researchers believe that the peptides stimulate a somewhat mysterious fifth kind of taste receptor called umami. The peptides may also improve the richness of the taste by interacting in complex ways with the four known taste sensations.

Why, then, doesn't leftover steak usually taste very good? One reason is that the peptides, as they deteriorate, may form yet smaller molecules that stimulate the sour or bitter taste receptors on the tongue. Another reason is that the fat normally found in beef quickly turns rancid as iron molecules in the meat itself promote oxidation of fat. This process can be retarded, oddly enough, by adding chitin to the steak. The chitin binds the iron and renders it ineffective. However, since few people are interested in adding ground insect exoskeletons to their steaks, a search is on for ways to manufacture peptides that can be sprinkled on one's leftover steak to overcome these two problems.

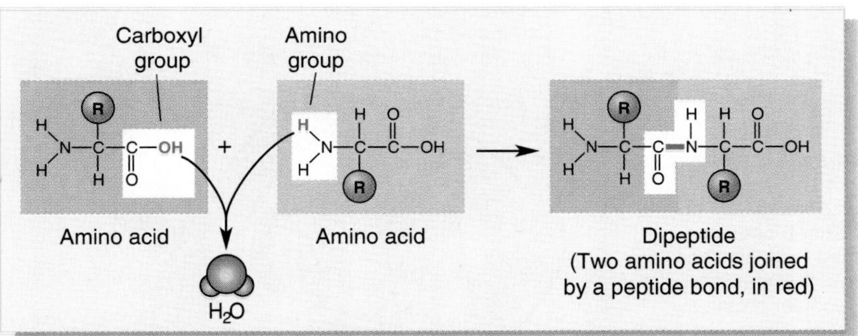

FIGURE 3.9 AMINO ACIDS AND PEPTIDE BONDS. All amino acids have an amino group, a carboxyl group, and an R group. It is the R group that is different among the many amino acids, as shown in Table 3.1. Amino acids are joined by peptide bonds (dehydration linkages) between adjacent amino and carboxyl groups. Shown here is a simple dipeptide, but the same process accounts for the addition of many amino acids together to form polypeptides.

in DNA and four in RNA. We will see later that DNA and RNA store and carry information in the *order* of these four nucleotides along the polymer chain.

In the assembly of a DNA polymer, the nucleotides are linked, one atop the other, into two opposing strands. Following the principles we have seen for the other large molecules, a water molecule is released when a nucleotide subunit is added to the polymer. The two strands are wound around each other in such a way as to form the well-known **double helix** (Figure 3.11). In the assembly of RNA, nucleotides are also linked in a similar fashion, but as we will see later, there are some important differences between DNA and RNA.

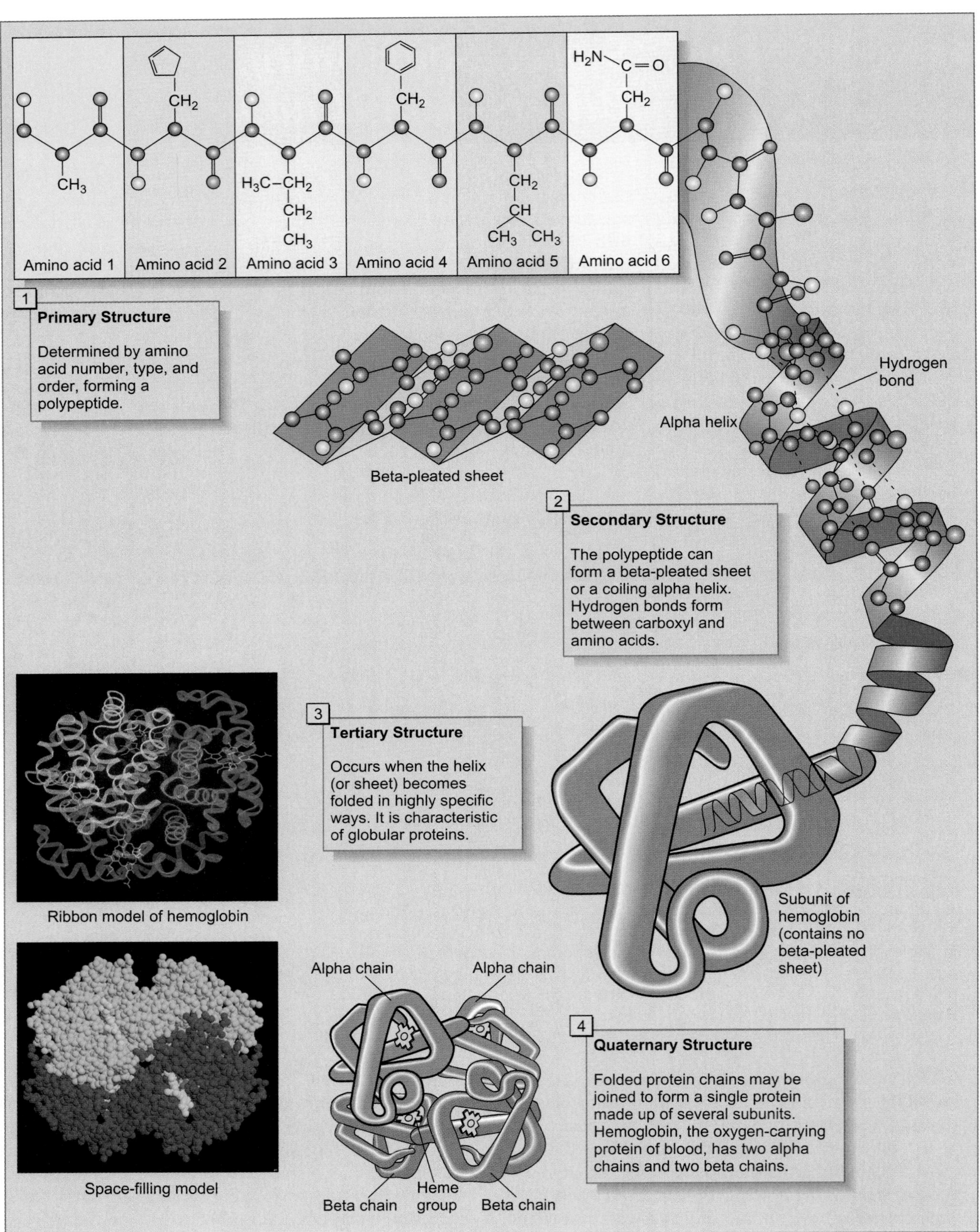

Amino acid 1
Amino acid 2
Amino acid 3
Amino acid 4
Amino acid 5
Amino acid 6
1
Primary Structure
Determined by amino acid number, type, and order, forming a polypeptide.
Beta-pleated sheet
Alpha helix
Hydrogen bond
2
Secondary Structure
The polypeptide can form a beta-pleated sheet or a coiling alpha helix. Hydrogen bonds form between carboxyl and amino acids.
Ribbon model of hemoglobin
3
Tertiary Structure
Occurs when the helix (or sheet) becomes folded in highly specific ways. It is characteristic of globular proteins.
Subunit of hemoglobin (contains no beta-pleated sheet)
Space-filling model
Alpha chain
Alpha chain
Beta chain
Heme group
Beta chain
4
Quaternary Structure
Folded protein chains may be joined to form a single protein made up of several subunits. Hemoglobin, the oxygen-carrying protein of blood, has two alpha chains and two beta chains.

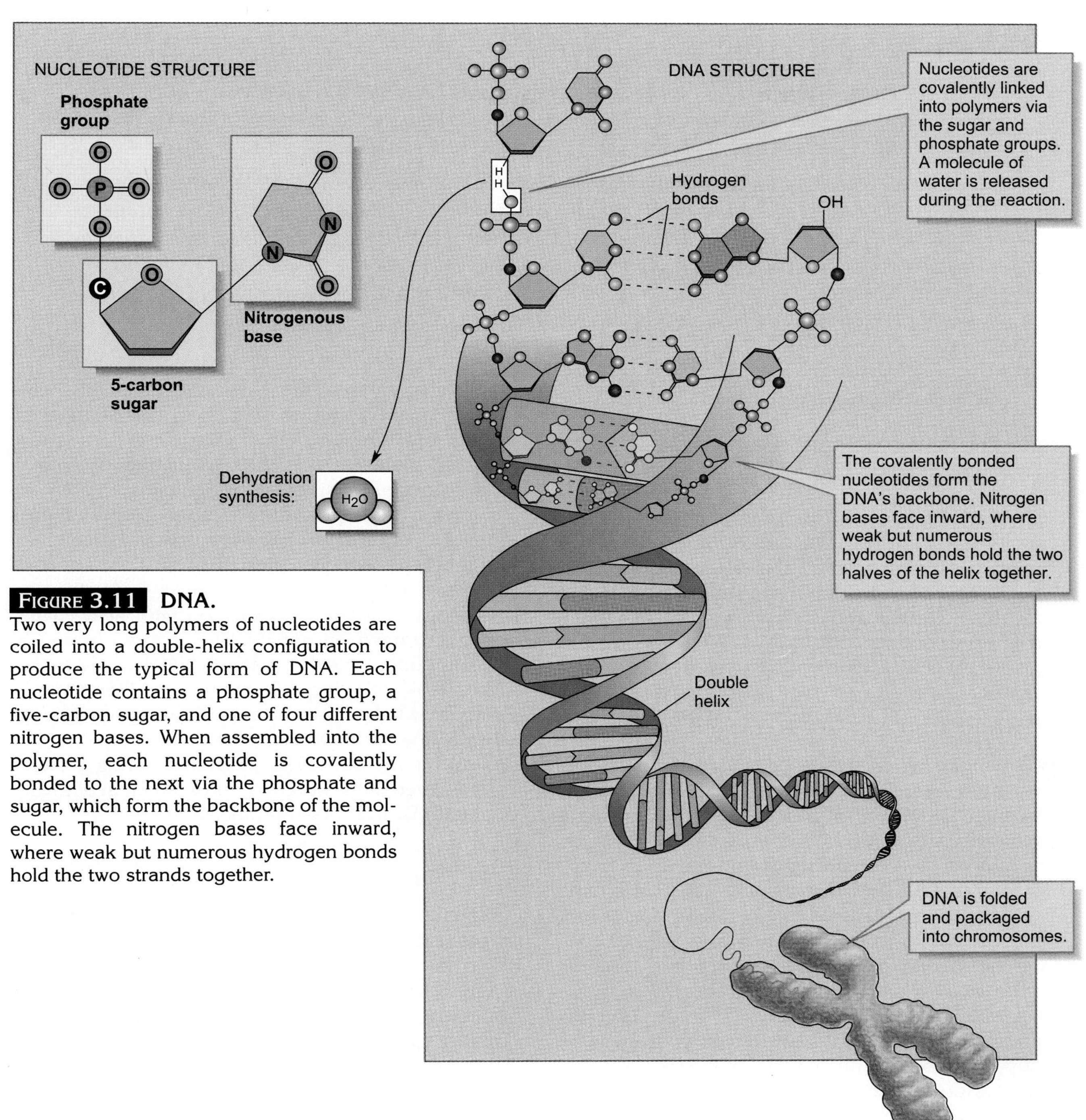
NUCLEOTIDE STRUCTURE
Phosphate group
Nitrogenous base
5-carbon sugar
Dehydration synthesis:
H_2O
DNA STRUCTURE
Nucleotides are covalently linked into polymers via the sugar and phosphate groups. A molecule of water is released during the reaction.
Hydrogen bonds
OH
The covalently bonded nucleotides form the DNA's backbone. Nitrogen bases face inward, where weak but numerous hydrogen bonds hold the two halves of the helix together.
Double helix
DNA is folded and packaged into chromosomes.

FIGURE 3.11 DNA.
Two very long polymers of nucleotides are coiled into a double-helix configuration to produce the typical form of DNA. Each nucleotide contains a phosphate group, a five-carbon sugar, and one of four different nitrogen bases. When assembled into the polymer, each nucleotide is covalently bonded to the next via the phosphate and sugar, which form the backbone of the molecule. The nitrogen bases face inward, where weak but numerous hydrogen bonds hold the two strands together.

FIGURE 3.10 PROTEINS MAY HAVE FOUR LEVELS OF STRUCTURE.
The primary level is determined by the amino acid number, content, and order, forming the polypeptide. In the secondary level the polypeptide forms a coiling alpha helix or a beta-pleated sheet. The tertiary level occurs when the helix becomes folded in highly specific ways. Such folding is seen in hemoglobin, an oxygen-carrying pigment occurring in muscles. The quaternary level of organization is characterized by two or more polypeptides joined by cross-bridges. Hemoglobin has two alpha and two beta polypeptide chains along with four iron-containing heme groups.

PERSPECTIVES

The large molecules of life are generated by the addition of subunit molecules. While the underlying rules governing their assembly are rather straightforward, we have seen that the end result is a diverse set of molecules that can be magnificently complex.

As we move into the next chapter, we shall see that the properties of these large molecules are responsible for the shape and function of the cell, the first truly living thing that we encounter in our discussions. Later, we will investigate the other roles that large molecules, especially proteins and nucleic acids, play in the regulation of biological energy, growth, and reproduction. Virtually everything that is discussed in the rest of this text will be governed by the characteristics of the molecules we have seen in this chapter.

SUMMARY

CARBON: BACKBONE OF THE MOLECULES OF LIFE

Carbon forms the structural backbone for important molecules of life including carbohydrates, lipids, proteins, and nucleic acids. Carbon atoms join other carbon atoms through single and double bonds. They form chains, rings, and chains of rings. All macromolecules are polymers, created by joining monomers together via dehydration synthesis reaction.

CARBOHYDRATES

Carbohydrates, including sugars and starches, are common constituents of foods. They contain carbon, hydrogen, and oxygen in a 1:2:1 ratio, occurring as monosaccharides, disaccharides, and polysaccharides.

Glucose, a common six-carbon monosaccharide, is the structural unit of many polysaccharides. Other important simple sugars are six-carbon fructose and galactose, and five-carbon ribose and deoxyribose.

The linking of monosaccharides into disaccharides such as maltose and sucrose occurs through dehydration synthesis, in which —OH and —H groups are removed and water is formed. Hydrolysis, the opposite process, utilizes water in the splitting of disaccharides, restoring the —OH and —H groups.

The most common polysaccharides—starches, glycogen, cellulose, and chitin—are lengthy polymers of glucose. While the first two contain alpha glucose and are readily digestible, cellulose contains beta glucose and is indigestible to most organisms. Chitin is similar to cellulose but contains nitrogen.

LIPIDS

Lipids include fats, oils, waxes, steroids, and phospholipids.

Triglycerides—animal fats and plant oils—contain glycerol with three fatty acid chains attached. They join through dehydration synthesis. A fatty acid is a hydrocarbon chain containing a carboxyl (acid) group. Fatty acids vary in length and may be saturated (the maximum number of hydrogens present) or unsaturated (less than the maximum number of hydrogens present).

Phospholipids, major components of cell membranes, consist of glycerol, two fatty acid chains, and a charged phosphate. The charged heads are hydrophilic, attracting water, while the tails are hydrophobic, rejecting it. In membranes the nonpolar, fatty acid tails form a water-resistant core while the heads line the inner and outer surfaces.

Waxes are used chiefly in waterproofing. Steroids are utilized in the lipoproteins of blood, in bile salts, vitamin D, and some hormones.

PROTEINS

Proteins are composed of one or more chains of amino acids that are arranged as polypeptides.

The various kinds of proteins differ in their amino acid content. There are 20 or so different amino acids, each with at least one carboxyl group and at least one amino group. One amino acid differs from the next in its R group. Amino acids join through dehydration synthesis in which covalent, peptide bonds are formed.

Proteins may have four levels of organization: The primary level includes the kinds of amino acids used and their sequence in the polypetide. The secondary level includes alpha helix coiling and beta sheets formed through hydrogen bonding between amino acids within the same peptide chain. The tertiary level is a folding stabilized by linkages between R groups of amino acids. The quaternary level is reached when two or more folded polypeptides become associated into a larger complex.

Nucleic Acids

Nucleic acids are polymers consisting of structural units called nucleotides. Nucleotides are made up of a phosphate group, a five-carbon sugar, and nitrogenous bases. The DNA molecule contains two chains of nucleotides that twist into a double helix.

Key Terms

macromolecules • **33**
polymer • **33**
monomer • **34**
dehydration synthesis • **34**
hydrolysis • **34**
carbohydrate • **34**
monosaccharide • **34**
glucose • **34**
fructose • **34**
galactose • **34**
ribose • **34**
deoxyribose • **34**
hydroxyl• **34**
aldehyde • **34**
disaccharide • **34**
sucrose • **34**
maltose • **34**
lactose • **34**
polysaccharide • **35**
cellulose • **35**
chitin • **36**
lipid • **36**
triglyceride • **36**
glycerol • **36**
fatty acid • **36**
carboxyl group • **36**
saturated • **36**
unsaturated • **36**
phospholipid • **36**
wax • **38**
steroid • **38**
cholesterol • **39**
protein • **39**
peptide bond • **41**
primary structure • **41**
secondary structure • **41**
tertiary structure • **41**
quaternary structure • **41**
nucleic acid • **41**
RNA • **41**
DNA • **41**
nucleotide • **42**
nitrogenous base • **42**
double helix • **43**

Review Questions

1. List three 6-carbon and two 5-carbon monosaccharides. Write the formulas for the following side groups: hydrogen, hydroxyl, and aldehyde.
2. Name and describe the reaction through which monosaccharides are joined to form disaccharides and polysaccharides.
3. List three polysaccharides found in plants and two found in animals. What purposes does each serve?
4. List the components of a triglyceride and name the reaction through which they become linked.
5. State two or three ways in which the many kinds of triglycerides differ from each other.
6. Draw a simple model of a phospholipid, showing the fatty acid tails and phosphate group.
7. Using a simple drawing, illustrate the arrangement of phospholipids in a membrane. Which regions are hydrophobic? Hydrophilic?
8. Describe the structure of a steroid and list three important uses.
9. Write the structural formula for a generalized amino acid, labeling the following: carboxyl group, amino group, alpha carbon, R group. How does one kind of amino acid differ from another?
10. Explain the four levels of protein structure. Name a protein with all four levels present.
11. Describe the general appearance of DNA. List the parts of its structural unit.
12. What is the role of DNA?

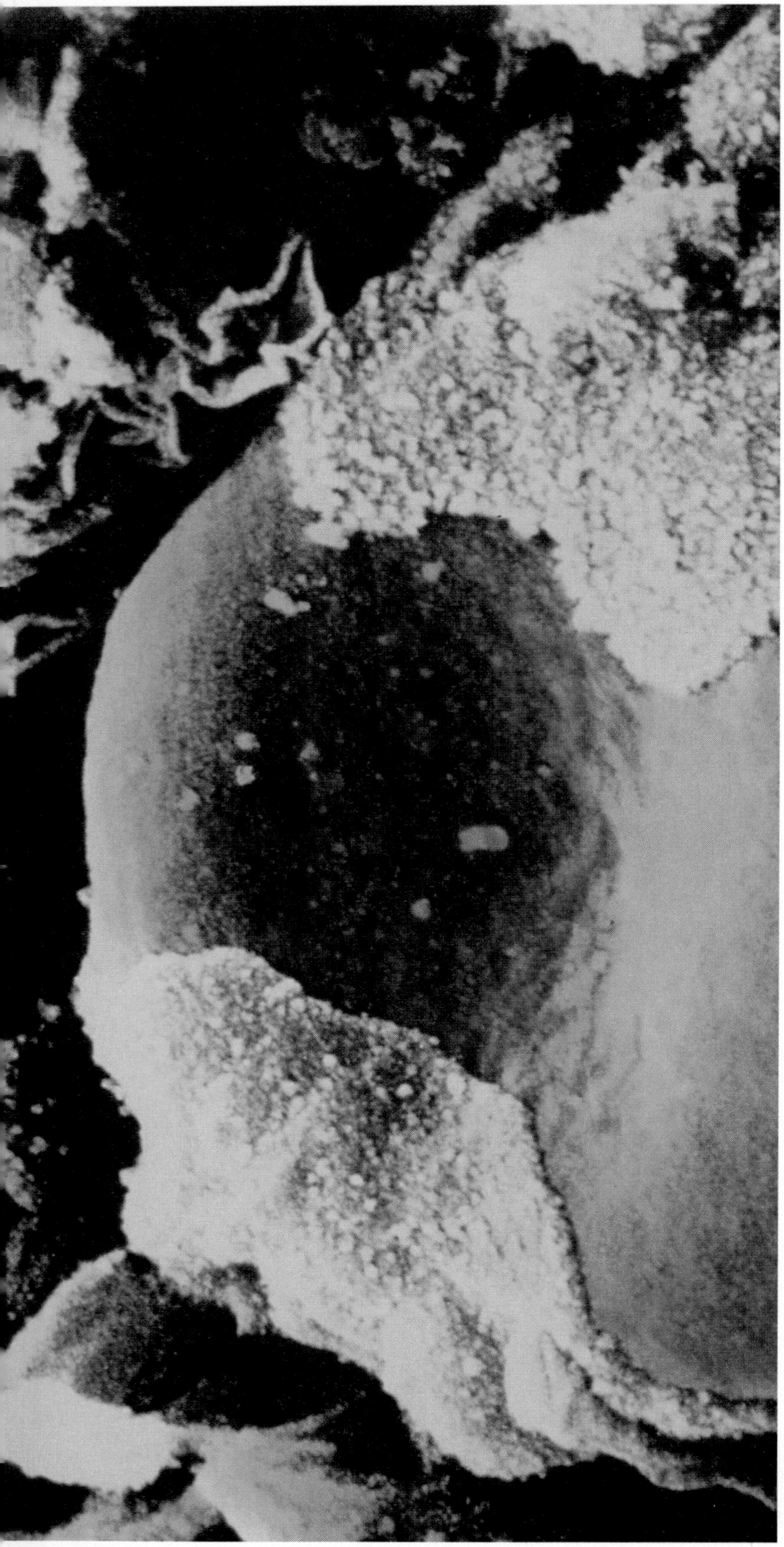

4

The Structure and Function of Cells

NOW WE BEGIN OUR DISCUSSION OF THINGS THAT ARE UNARGUABLY ALIVE. *In the last few chapters we have shown how simple elements and straightforward chemical processes combine to produce molecules of increasing size and complexity. Here we show how those complex molecules can be organized in the basic biological entities known as cells. What is the big deal about cells? Just that cells are the units of life, of which all organisms are composed. As we will see, all animals and plants are composed of cells, cells that are organized into the tissues and organs that characterize living organisms. Isn't it interesting that the term* organism *is similar to the word* organized?

Robert Hooke realized that he had a problem. He had just been appointed Curator of Experiments for the prestigious Royal Society of London, and one of his first tasks was to devise some sort of demonstration for the next weekly meeting. He wanted something that would enlighten, entertain, and impress. He also wanted to make the Society members aware of his own abilities. Neither would be easy. The problem was, these were some of the brightest, crustiest, most argumentative, skeptical, and jaded people in all of 17th-century England—the elite of British science.

Hooke considered a number of possibilities, working and fretting until he struck upon a solution. Obviously, he had to show them something new, and the most exciting new technology of his day was curved and polished glass. He would demonstrate the lens.

The scientific world was buzzing with talk of lenses. With their ability to magnify, they revealed an entirely new world. Things no one had suspected existed were suddenly visible. Through new eyes, people could again see things once forgotten. With a pair of lenses held in a frame, people who had been nearly blind were able to see again—a miracle come true. Old men who had been unable to read had their books and letters restored to them. Of course, magnification by lenses was not a new idea. Earlier in the century, Galileo had pointed a lens toward the sky and had drawn some conclusions, as well as the wrath of the Church. Hooke, however, had a different intellectual appetite. He wanted to see things that were too small to be seen without a lens. He was fascinated with the idea of using the microscope to explore the world of the minuscule (Figure 4.1).

Obviously, it is difficult to impress people if you're standing on *their* turf; but microscopy was *Hooke's* turf. So he decided to use his lens in some novel way. But what should he arrange for them? What would they like to see? Maybe cork. Cork was a mystery, appearing to be solid, yet able to float. Perhaps it was not so solid after all. Hooke aimed his microscope at the cork, and what do you think he saw? Wrong. He saw nothing, because microscopes do not work very well with reflected light. So with a penknife he cut a very thin sliver of cork from the bottom, and shined a bright light upward through it. This time, what he saw puzzled him and was sufficiently interesting to please the Society members. Hooke wrote that the cork sliver seemed to be composed of "little boxes." These, he surmised, were full of air, accounting for the ability of corks to float. Hooke called the little boxes *cells* because they reminded him of the rows of monks' cells in a monastery. And from this modest beginning, a new scientific field was born.

CELL THEORY

The birth, however, was slow. What was one to do with such knowledge? The group that week was pleased, but a full century would pass before the scientific world would understand the importance of Hooke's cells.

(a)

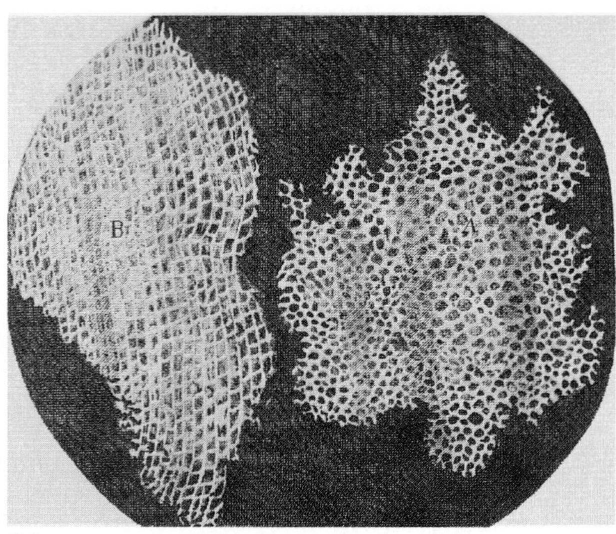

(b)

FIGURE 4.1 **THE FIRST MICROSCOPE.** Hooke's microscope was a primitive instrument, but it was elegant in the way of scientific instruments of the era. **(a)** It consisted simply of a tube that held lenses in a proper configuration that magnified the image of the sample held near the bottom of the tube. **(b)** Hooke looked at sections of cork, seeing what he called "cells."

One of the first to try to use the knowledge was the German naturalist Lorenza Oken, who focused his primitive microscope on just about everything he could think of. Finally, in 1805, he wrote, "All organic beings originate from and consist of vesicles or cells." This simple statement became known as the **cell theory.** Then, in 1838, two other Germans, the botanist Matthias Jakob Schleiden and the zoologist Theodor Schwann, independently published the conclusion that all living things are composed of cells. (Schleiden and Schwann usually get the credit for the idea, even today.) Almost 20 years later, another German, Rudolf Virchow, added the statement, "omnis cellula e cellula." Anything written in Latin is, of course, very important—even if it only means "all cells come from cells."

Once Hooke had described his little boxes, the art of microscopy blossomed. Everyone in science wanted a microscope. And nothing—literally nothing—remained sacred. Everything was a fair target. Curious souls were anxious to be the first, the very first to see . . . whatever. At first, most of these efforts were prompted by sheer curiosity. In time, however, the more serious among them used microscopes to answer questions. They noticed certain common themes and differences. They began to make generalizations. For example, they suggested that although organisms may be vastly different, certain cells and tissues could be nearly identical. They noted that muscle, nerve, and reproductive cells in most animal species are quite similar. Thus people began to think of cells in terms of function.

WHAT IS A CELL?

Cells are minute, highly organized, living units, surrounded by membranes and containing a variety of tiny structures, each with a specific role. Biologists distinguish two principal types of cells, those of **prokaryotes** (*before* the *kernel,* or *nucleus*) and **eukaryotes** (*true kernel,* or *nucleus*). The first group includes bacteria—single-celled forms of life that have little visible, internal structure, although they may be biochemically complex. We will come back to the prokaryotes later in the chapter.

The other group, the eukaryotes, have cells with elaborate internal structures and includes all other cellular life. In the larger, multicellular organisms, similar cells with the same function cluster into **tissues,** which, when grouped together, become **organs** that carry on specific tasks. Groups of organs form **systems,** or organ systems, which are responsible for major body functions.

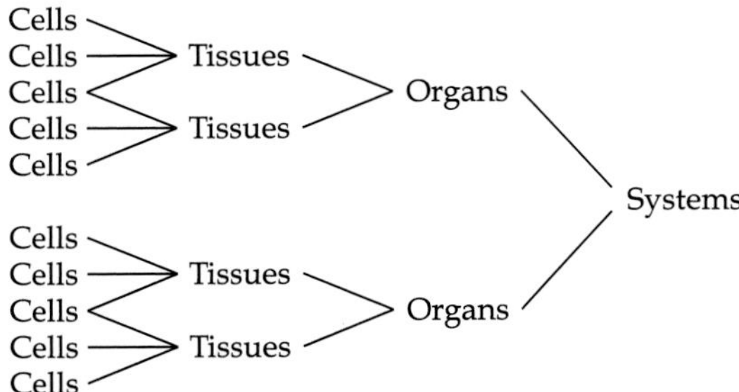

Cells carry out virtually all the processes that make life possible. Thus it should be no surprise to learn that they are complex. Eukaryotic cells contain numerous membrane-surrounded **organelles** ("little organs"), each with a separate role in carrying out life processes. In most cases we know what that role is, but in a few, we still aren't sure.

We do know that certain organelles specialize in synthesizing complex molecules such as proteins. Others convert food stored in the cell into energy, and still others store waste products either permanently or temporarily. Some organelles may transport materials to the cell's membrane, where they can be released to the outside. Later we will see that almost all cellular activities are directed by yet another membrane-bounded organelle, the cell nucleus.

Cell Size

Can cells be seen with the naked eye? They can if the naked eye can see a chicken egg yolk, which is a single cell; but most cells are far too small to be seen without some sort of magnification. It is very difficult to generalize about cell size and shape because there are so many kinds of cells, each specialized for a particular function. For example, there are cells virtually invisible to the human eye that are over a meter long, such as the nerve cells that extend the length of a giraffe's leg. Most plant and animal cells are about 90 micrometers (μm) long. Almost none are smaller than 10 μm in diameter, and only a few are larger than 100 μm. A micrometer is one millionth of a meter (see Table 4.1). To give you an idea of the size of such cells, about fifty 10-μm cells set end-to-end would reach across the period at the end

TABLE 4.1 A Comparison of Some Cell and Cell Structure Dimensions

The range of cell size is enormous, extending from the bacterium to the meter-long nerve cell, a difference of about a millionfold. Yet most plant and animal cells are about 80 micrometers (μm) across. The smallest known cells, those of* Mycoplasma, *sometimes are not visible through the light microscope. Viruses can be seen only with an electron microscope.

Abbreviations used in table: m = meter; mm = millimeter = 0.001 (10^{-3}) meter; μm = micrometer = 0.000001 (10^{-6}) meter; nm = nanometer = 0.000000001 (10^{-9}) meter.

Human eye	
1 meter	← Some nerve cells
	← Ostrich egg
100 mm	
	← Hen's egg
10 mm	
	← Frog's egg
1 mm	
	← Diameter of squid giant nerve cell
Light microscope	
100 μm	← Human egg
	← Ameba
	← Human hair
	← Human smooth muscle cell
	← Human bone cell
	← Human liver cell
10 μm	← Human milk-secreting cell
	← Human red blood cell
	← Chloroplast
	← Nucleus of human liver cell
	← Mitochondrion (length)
	← Bacterium *(E. coli)*
1 μm	← Diameter of human nerve cell process
	← Lysosomes
	← *Mycoplasma*
Electron microscope	
	← Cilium (diameter)
	← Centriole
100 nm	← Large virus
	← Nuclear pore
	← Microtubule
	← Ribosome
10 nm	
	← Cell membranes, including plasma membrane (thickness)
	← Microfilaments
	← Globular protein
	← Diameter of DNA double helix
1 nm	
	← Diameter of protein α-helix
	← Amino acid
0.1 nm	← Diameter of hydrogen atom

of this sentence. Plant cells tend to be somewhat larger than animal cells, perhaps because they contain water-filled cavities called *vacuoles.* In general, the amount of **cytoplasm** (living material) in plant and animal cells is about the same.

Confronted with such information, biologists are likely to wonder what determines cell size. Why are most of them so small? Why do they first go through a rapid growth phase, then slow down as some critical size is reached, divide in half, and start over? One way to phrase the question is to ask: What is the advantage of having small cells?

The Surface-Volume Hypothesis. The generally accepted explanation for the small size of cells is that size strongly affects the ratio of the surface area to the volume. Specifically, smaller cells have larger surface areas in proportion to their volume.

In biology there always seems to be a next question. This time the next question is: Why is a large surface area necessary? The answer centers on the fact that the surface area of cells is covered by a membrane.

A large area of membrane covering a small cell ensures that the material within is never far from the cell's exterior, and therefore the cell is more likely to be able to move nutrients in and wastes out in an efficient manner (Figure 4.2). However, a large cell has relatively less membrane, so the problem of getting things in and out gets worse as cells get bigger.

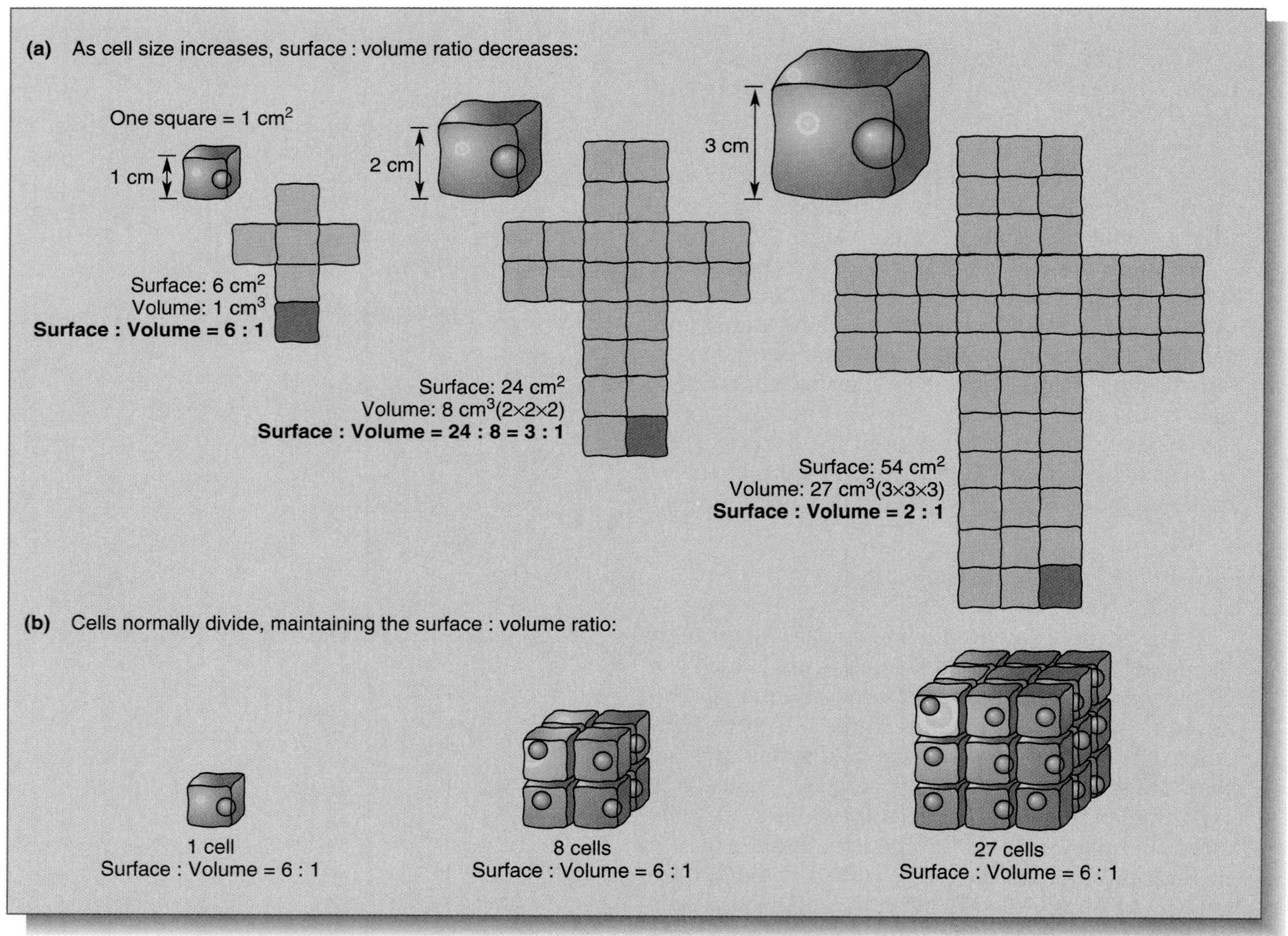

FIGURE 4.2 **SURFACE-TO-VOLUME RATIO.**
In two bodies of similar shape, the larger single cell **(a)** has more volume for its surface area than a body composed of an equal volume of single cells. If the bodies depend on materials moving through the covering surface, the body composed of many small cells will be at an advantage since it has a comparatively large surface area to service its interior **(b)**.

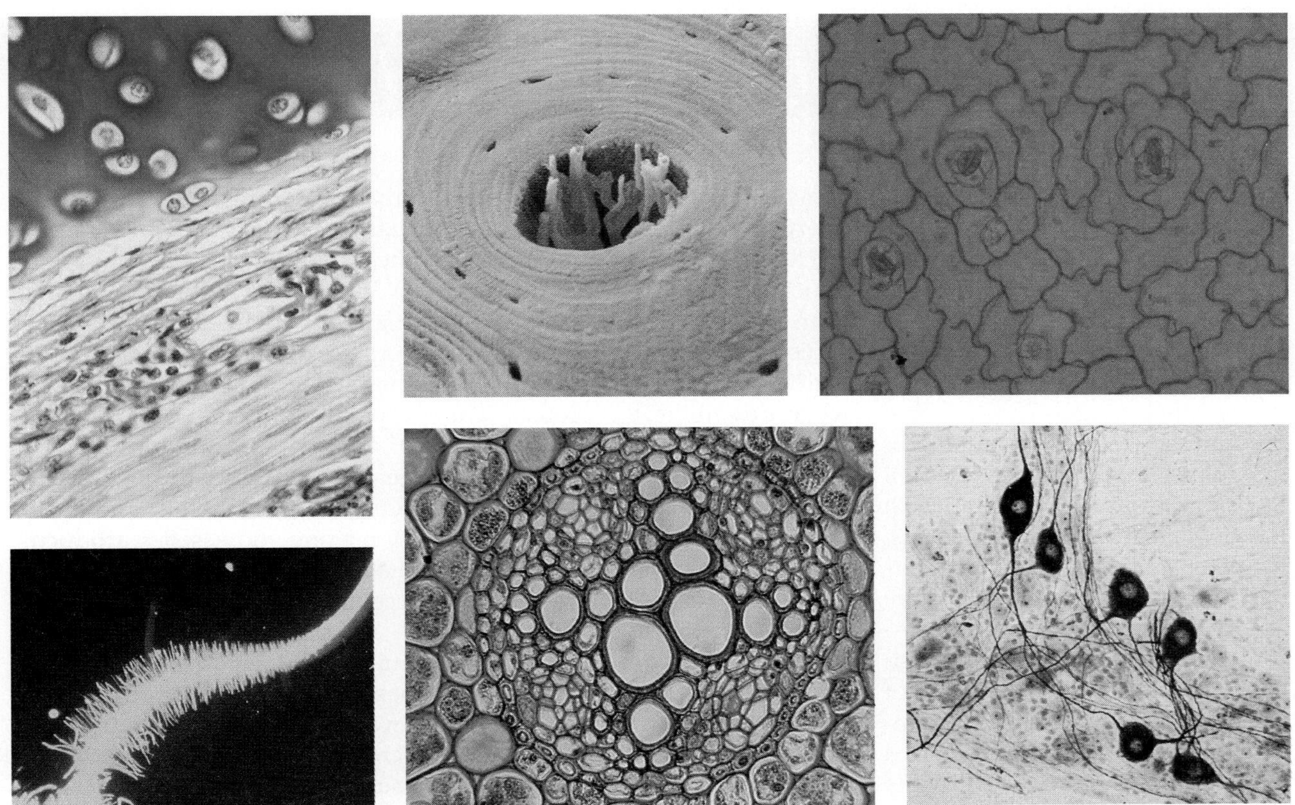

FIGURE 4.3 EXAMPLES OF THE WIDE VARIETY OF CELL TYPES, SHAPES, AND FORMS.

The question then arises: Why aren't cells even smaller? No one really knows, but biochemists suggest that the lower limits to cell size may be restricted simply by the size of critical macromolecules such as proteins, DNA, and RNA. Cells must have certain numbers of these giant molecules, and there must be room for them. The smallest known cells are those of bacteria known as *Mycoplasmas*, which range in size from 0.1 to 0.3 μm (micrometer). Laid out side by side, it could take as many as 1,000 of these bacteria to reach across the period at the end of this sentence. Contrary to the original cell theory, some life, as we see in Essay 4.1, is not composed of cells.

A LOOK AT CELL DIVERSITY

Cells come in many different sizes and shapes, and each has its own limits and abilities. This is not surprising, since we would not expect the cells of a carrot to be identical to those of a clam, or those of a muscle to those of a nerve (Figure 4.3).

Cells with different appearances may also behave differently. For example, some are highly irritable—that is, they respond quickly to environmental changes. Some can contract, others can secrete fluids, and still others have long tails and can swim. Such variety indicates the great advantage of multicellularity: *specialization*. A many-celled organism is equipped to meet many demands if it has different kinds of cells, each with its own special abilities. Imagine the advantage of having some cells that can react to light, others that can distinguish pain, and others that can contract and therefore move you toward the light or away from the pain. From the perspective of evolution, we could expect groups of cells to become highly specialized for very specific functions, increasingly adapting the organisms they comprise to better exist in a competitive world.

Essay 4.1 Life Without Cells

Life on our small planet presents few opportunities for those who like tidy categories. There are few pigeonholes in the real world of adaptive, changing, opportunistic life. In fact, life is so diverse that there are exceptions to virtually every rule. (This is probably among the most reputable of the rules.) Exceptions to the cell theory are found, for example, in the tiny creatures called protists. If, as some researchers believe, certain of these protists *are* cells, they are certainly very large ones—sometimes large enough to be visible to the naked eye. The length of one common single-celled organism, the *Paramecium,* may reach 0.3 mm, and even it is dwarfed by the giant ameba, whose length often measures 5 mm (about 3/16 of an inch). These giant single-celled organisms are 50 times longer than the typical animal cell.

However, protists' cells are constructed along lines very different from the cells of multicellular organisms. Because of this, some scientists prefer the term *acellular* when describing the protists. The implication is that they are not single-celled organisms, and that they simply are not organized along cellular principles. There are also organisms such as some flatworms and some fungi that undergo a strange sort of division in which the nucleus may divide, perhaps many times, but the total cell does not. The result is a very large "cell" with many nuclei.

At the other end of the size spectrum, far too small to be seen with the light microscope, are the viruses. Viruses are believed to be the smallest "organisms," although many scientists do not consider them to be alive.

Viruses lack most cell-associated traits, so perhaps they shouldn't be compared with cells. Structurally, they are very simple—consisting of a protein container with a supply of enzymes surrounding a core of genes. In essence, they are self-reproducing, in that they are capable of making more self-reproducing material. However, viruses require a cellular host in which to replicate.

Although the cell theory applies quite well to most organisms, life has a disconcerting way of not allowing easy generalizations. We can see that some forms of life have found it advantageous to take another route. The result is the vast array of life on this planet.

Cell Structure

Now that we have some idea of the size and variation of cells, let's take a closer look at their structure. For convenience, we have grouped the structures of the eukaryotic cell according to their functions. We'll begin at the cell surface and work our way in. Figure 4.4 shows representative views of plant and animal cells.

Organelles of Support and Transport

Plasma Membrane. The **plasma membrane** is simply the membrane that surrounds each cell. Now let's see if "simply" is a good choice of words. We learned earlier that a basic function of the membrane is to control the movement of materials into and out of the cell. Since the processes of life must be tightly regulated, the plasma membrane must be selective about what it allows to pass in and out. It must freely accept some substances and utterly reject others. It may let things through passively, like a sleeping gatekeeper, as it does with water and gases; or it may take an active part in transporting materials, alertly and with great discrimination using cellular energy to force other, more reluctant substances in or out. Because of such discrimination, plasma membranes are described as *selectively permeable* (see Chapter 5).

The two major components of the plasma membrane are proteins and phospholipids. The phospholipids exist in a double layer, with all the uncharged fatty acid tails directed inward. Under the electron microscope, the membranes look like two dark lines separated by a clear area about 5 nm (nanometers) wide.

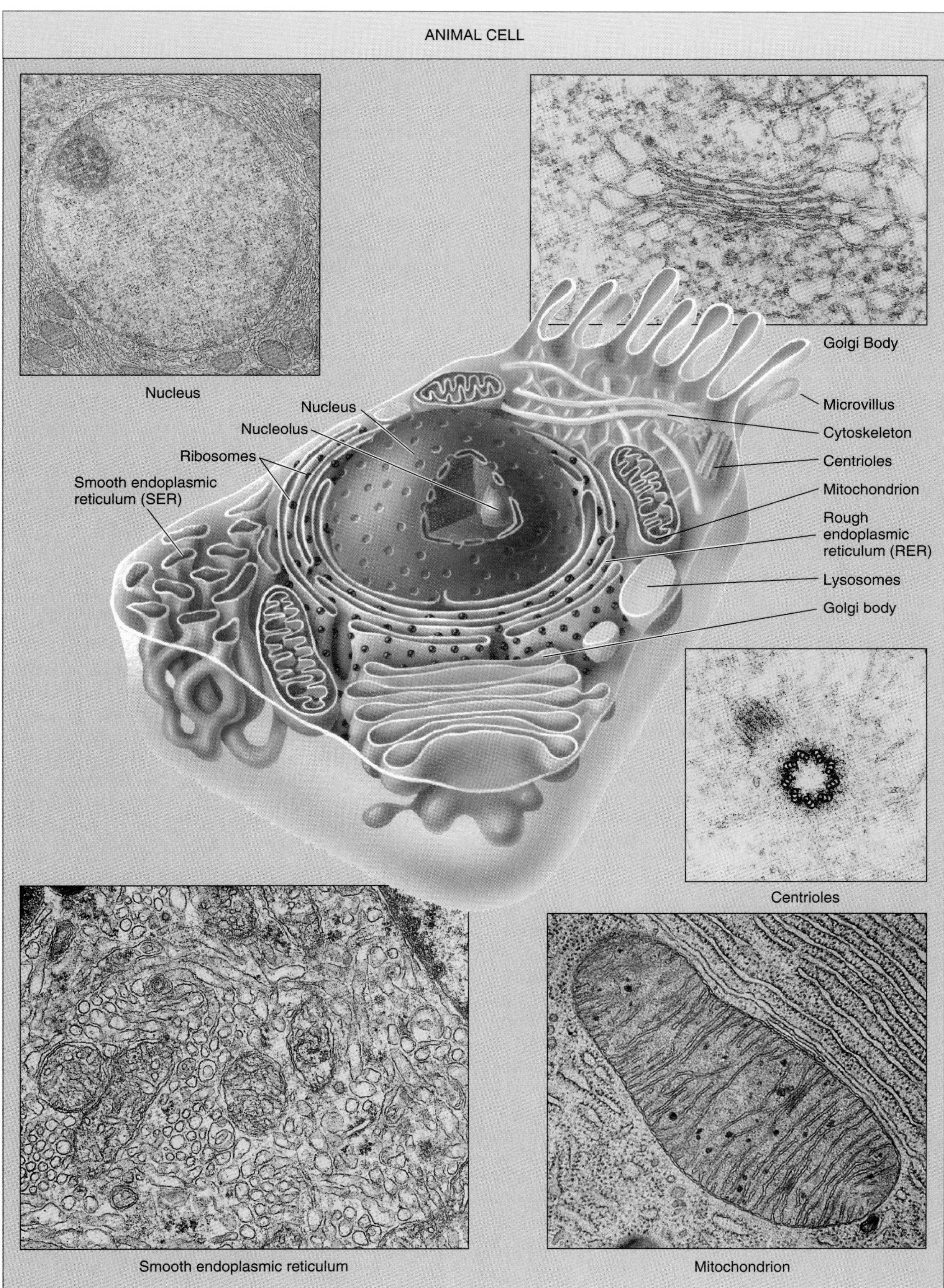

FIGURE 4.4 REPRESENTATIVE ANIMAL AND PLANT CELLS

Fluid Mosaic Model. It has been possible to deduce the structure of the cell membrane by pulling together everything known about it and developing a model that accounts for all of its properties. It is called the **fluid mosaic model** and was initially described by S. J. Singer. In this model (Figure 4.5) the phospholipids are visualized as small spheres and tails. The spherical heads are composed of glycerol, phosphate, and perhaps other organic groups. Each head has two tails that point inward.

This arrangement forms a hydrophilic ("water-loving") surface that interacts readily with water and other charged substances. The tails, however, are nonpolar or uncharged and cling, forming a more-or-less oily hydrophobic ("water-fearing") core (see Chapters 2 and 3). This arrangement is called the **phospholipid bilayer.**

Specific proteins are associated with the plasma membrane. Large proteins may completely traverse the phospholipid bilayer. Some of these proteins are involved in transport of materials through the membrane. Other proteins may be confined to only the inner or outer lipid layers. Those on the outer side may interact with hormones and other messengers reaching the cell, or they may help cells recognize others of their type. Proteins associated only with the inner side of the plasma membrane may interact with cytoplasmic structures important to cell shape and movement (see Figure 4.5), or they may serve other functions. Proteins within the plasma membrane are mobile, and can travel laterally.

The fluid mosaic model is supported by freeze-fracture preparations (Figure 4.5). When frozen cells split between the tails of phospholipids, the membrane proteins remain on one of the surfaces, while the other surface is pitted with indentations where the proteins had been.

Special Molecules of the Membrane Surface. The membrane is composed not only of phospholipid and protein, but also carbohydrates and lipids that, with certain proteins, form **glycoproteins** (sugar proteins) and **lipoproteins.** Other carbohydrates may form part of the outer plasma membrane in ways that are not well understood but that may influence how cells interact with each other.

Let's return to a point just mentioned: Cells apparently recognize one another on the basis of the carbohydrates and proteins of the membrane. That recognition is important in many processes. Cells must be sensitive to foreign cells to build immunity against them. Also, as an embryo develops, recognition of various cell types assists in the coordinated and orderly development of the different body parts. One startling experiment shows the remarkable ability of cell types to recognize each other. If cells from different animal tissues are mixed together and allowed to grow on a nutrient medium, the cells will slowly begin to move about until each has found others of its own type. The result is distinct masses of specific tissue types. Such specificity of cell membranes plays a part in blood groupings, cancer defense, and the rejection of transplanted organs.

Plant Cell Wall. One of the most obvious structures in all plant cells is the **cell wall,** a rigid, nonliving layer just outside the plasma membrane and composed primarily of cellulose and other polymers (see Figure 3.3). The primary cell wall has tiny, randomly oriented fibers of cellulose chains called *microfibrils.* Secondary cell walls are composed of fibers arranged in layers. Each layer lies at an angle with respect to the one below it, forming a laminated, strong, and porous covering for the cell. Its basic structure is strengthened as it becomes impregnated with hardening substances such as pectin. Pectin is also the substance largely responsible for holding adjacent plant cells together.

Other substances may be added to the wall matrix, depending on the function of the cell. For example, **lignin** is continually secreted into cells in the stems of woody plants, forming a hard, thick, decay-resistant wall. **Suberin,** a waxy substance, is secreted into the outer layer of some plant cells, forming a protective, waterproof layer. The cells of the upper surface of leaves, on the other hand, are further waterproofed by a thickened layer of another waxy secretion called **cutin.** Scientists are continually learning more about the structure of living things as advances are made in microscopy, as we see in Essay 4.2.

Organelle of Cell Control: The Nucleus

The nucleus is not easily described because of the complexity of its functions. To begin, we should note that the nucleus is often the most prominent organelle of the cell. It was also one of the first structures to be seen inside the cell. In fact, the word *nucleus* ("kernel") was first used in 1831, about the time of the development of the cell theory.

The nucleus has two vital functions: *reproduction* and *control.* The first involves preserving the hereditary information and copying it when it is to be passed along to new generations of cells. The

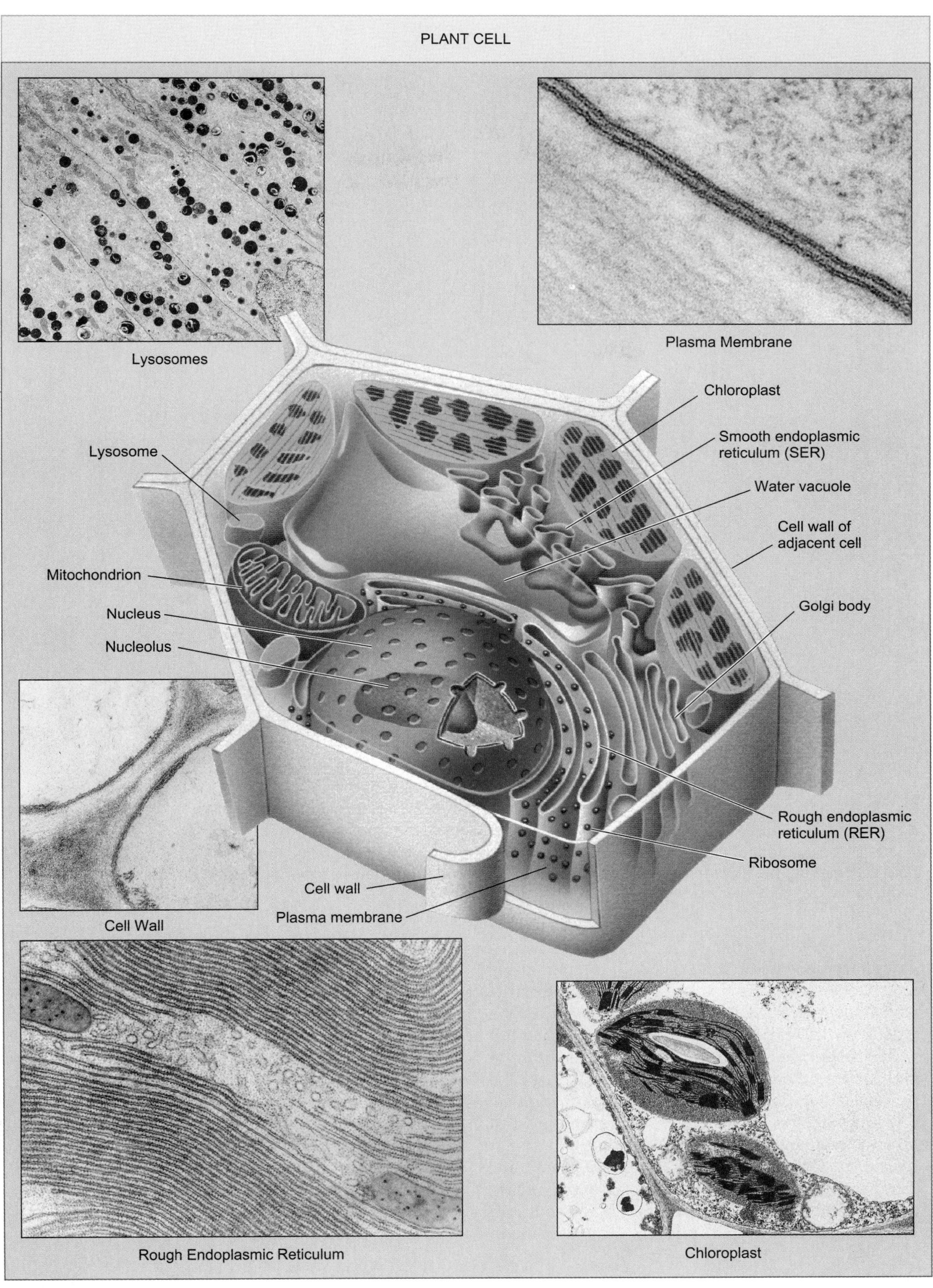
PLANT CELL
Lysosomes
Plasma Membrane
Chloroplast
Lysosome
Smooth endoplasmic
reticulum (SER)
Water vacuole
Cell wall of
adjacent cell
Mitochondrion
Golgi body
Nucleus
Nucleolus
Rough endoplasmic
reticulum (RER)
Ribosome
Cell wall
Plasma membrane
Cell Wall
Rough Endoplasmic Reticulum
Chloroplast

FIGURE 4.4 (CONTINUED)

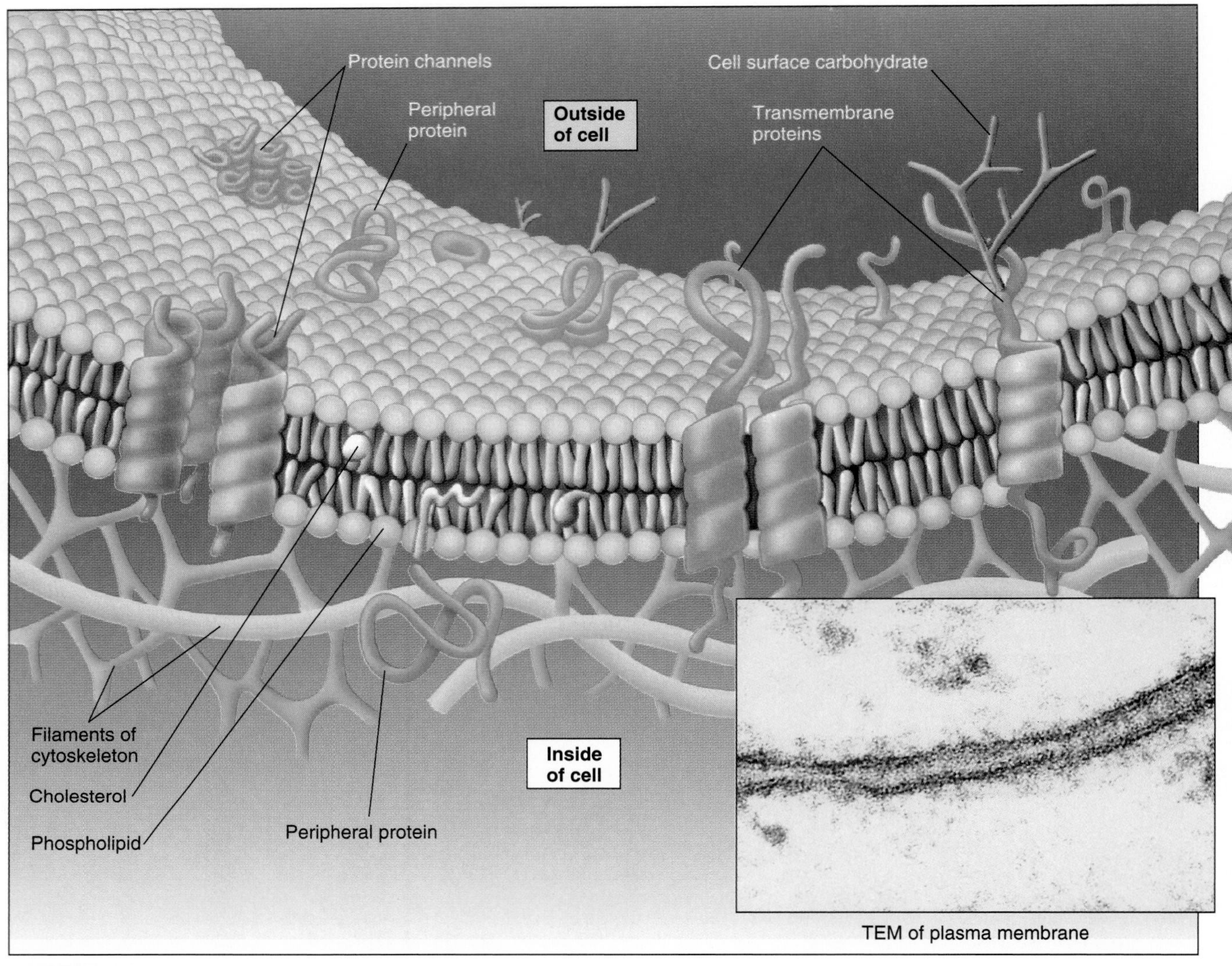

FIGURE 4.5 FLUID MOSAIC MODEL OF A PLASMA MEMBRANE
The basic structure includes a phospholipid bilayer, numerous globular proteins. Freeze-fracture studies help substantiate the globular protein component of the membrane. Holes are left by the transmembrane proteins as the lipid layer is raised.

hereditary information is encoded in immensely long DNA molecules that are organized into large protein-containing structures called *chromosomes*. We will deal with chromosomes in detail in Chapter 9.

The nucleus also uses its repository of information to direct virtually everything the cell does. The nucleus accomplishes this controlling function by directing the synthesis of enzymes, hormones, and other active molecules.

The nucleus is bound by a double membrane called the **nuclear envelope.** It consists of two tightly adjoined membranes, each of which is formed from the familiar lipid-protein bilayers. The two membranes pinch together in scattered places over the nuclear surface to form **nuclear pores** (Figure 4.6). The word "pore" implies hole, but these are not holes. Instead, they are indentations that are filled with special proteins that permit a controlled passage of materials. The pores apparently connect the interior of the nucleus with the rest of the cell.

Many nuclei include dark, prominent bodies known as *nucleoli* ("little nuclei"). Nucleoli are rich in the nucleic acid RNA and have the specific task of assembling a special kind of RNA found in small, round structures called *ribosomes,* which we will consider in more detail later.

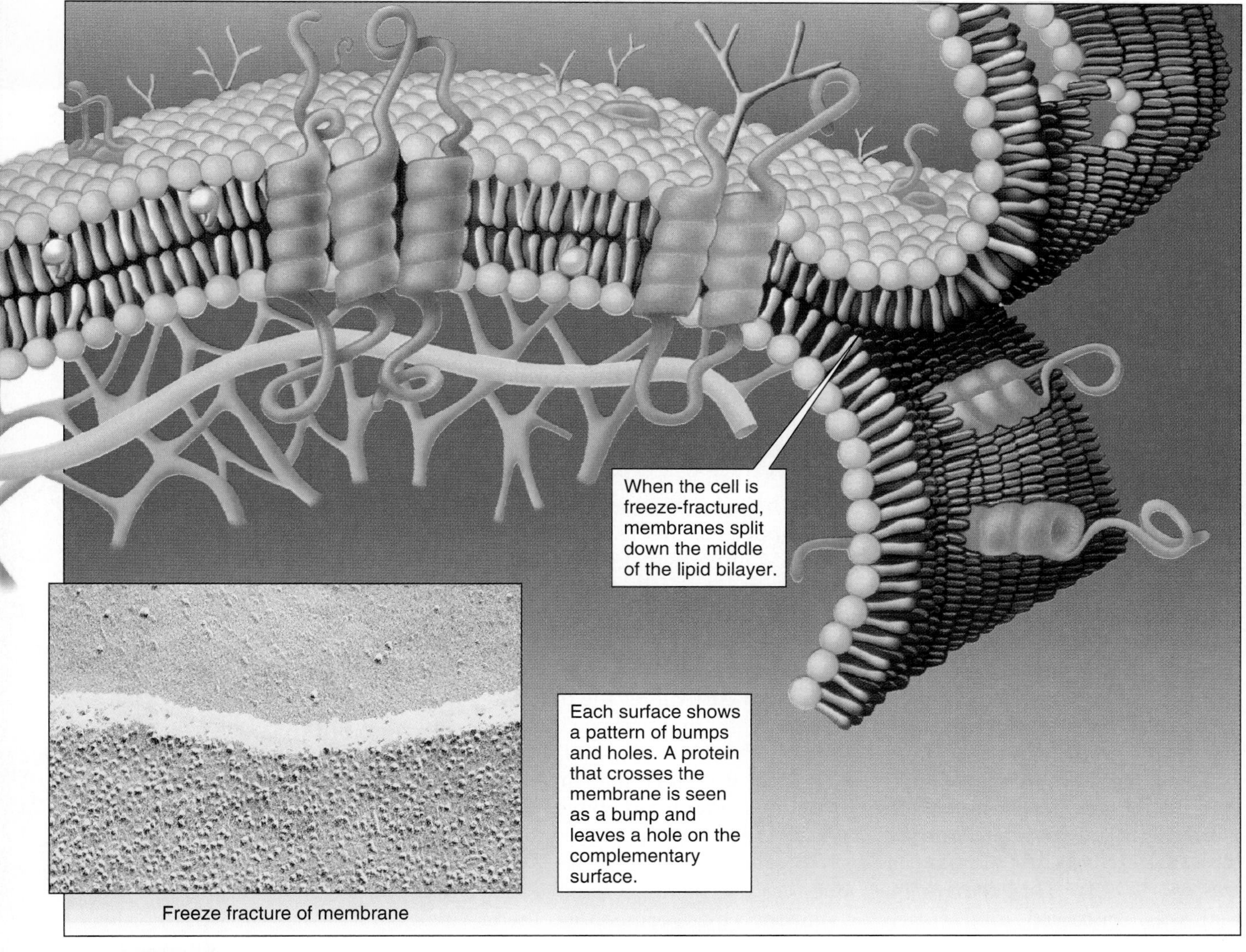

When the cell is freeze-fractured, membranes split down the middle of the lipid bilayer.
Each surface shows a pattern of bumps and holes. A protein that crosses the membrane is seen as a bump and leaves a hole on the complementary surface.
Freeze fracture of membrane

FIGURE 4.5 (CONTINUED)

ORGANELLES OF CELL SYNTHESIS, STORAGE, DIGESTION, AND SECRETION

Many of the membranes within a cell are actually part of a single membrane network that interconnects with the nuclear envelope. The various parts of this network are actively moving, fusing with one another, and pinching off different components. Each membrane component has its own name and function.

Endoplasmic Reticulum. The **endoplasmic reticulum,** or **ER,** is a complex membrane system that takes up a large part of the cytoplasm of eukaryotic cells, particularly those involved in the synthesis of proteins (Figure 4.7). Parts of highly folded ER are known to connect directly to the outer membrane of the nuclear envelope (see Figure 4.4). The ER is involved primarily in the synthesis, modification, and transport of substances produced by the cell.

There are two types of ER: rough and smooth. Rough endoplasmic reticulum is common in cells that manufacture proteins to be secreted outside the cell. Rough ER gets its name from the ribosomes that appear tightly adhered to one side of the membrane, making it look like coarse sandpaper (Figure 4.7).

Essay 4.2 How Cytologists See Cells

Cytologists, the biologists who study cells, use an arsenal of highly sophisticated devices and precise techniques in their efforts to develop exact descriptions of cells. Their most venerable tool is the direct descendant of Hooke's primitive apparatus. It is the *compound light microscope* (see the illustration). Developed in the 19th century, it is still very useful and marvelously precise; but it is limited by the nature of the energy it uses—light. Any two objects closer together than 250 nm (nanometers) merge as a single, blurred image, because of the properties of visible light. The limitations of any microscope are, therefore, defined as *resolving power*—the ability to distinguish close objects as being separate from one another.

Fortunately, we need no longer be limited by the resolving powers of the light microscope. We now have the *transmission electron microscope* (TEM). This remarkable tool was conceived in the early part of the 20th century, but was not perfected until the 1950s. With the TEM, electrons—not light—are the energy source. The electrons are emitted from a heated coil and are focused so that they pass through the object. The electrons' great advantage is that their oscillations (wavelengths) are substantially smaller than the wavelengths of light. They can pass between the most finely separated objects in the cell. As they pass through matter of different densities, they cast an image on a screen or photographic film. The magic of the TEM lies in the fact that its resolving power is practically unlimited as far as cells are concerned. The TEM can magnify an object up to hundreds of thousands of times. Whereas the finest light microscopes can produce only clear outline images of a single bacterium, the TEM can produce a detailed image of its inner structure.

Despite its great advantages, the TEM initially met strong resistance among traditional cytologists because the preparation of cells for the TEM involved extremely harsh, disruptive steps that could easily alter the material so drastically that no one could be sure of what he or she was actually seeing. But since that time, the structures identified have been verified in many ways.

To prepare a cell for the TEM, bits of tissue are "fixed" and "stained" by permeating them with chemicals, including heavy metal salts that solidify the cell's protein. If the material to be viewed is tiny, as are virus particles, the metals are sprayed on at an angle, in what is known as *shadow-casting* (see the illustration). This treatment increases the tendency of cell structures to absorb electrons, making their images darker and sharper.

One of the early techniques—freeze-fracturing—has continued to yield particularly valuable information. Quick-freezing the tissue produces natural fracture lines wherever two lipid regions meet. Because of this, membrane-bounded structures within cells become clearly visible (see photograph).

Today an entire family of electron microscopes has emerged, each demanding its own sophisticated techniques. The *scanning electron microscope* (SEM) has produced rather startling three-dimensional images of whole subjects. A shower of electrons sweeps back and forth across the subject, and electrons are scattered in different ways from the subject's surface. They land on image-producing plates, where they are detected and analyzed electronically.

An even newer way to probe the incredible and unseen sea of minutiae around us is the *high-voltage electron microscope.* Its penetrating power is so great—1 million volts—that it is not even necessary to slice the cell in order to see inside it. This three-story-high tool also produces three-dimensional images, revealing incredible details not visible even with the best standard electron microscopy techniques. Only a few of these gigantic microscopes exist at present, and there is a long line of cytologists—eager to probe more deeply into the world of the cell—awaiting their turn to use them.

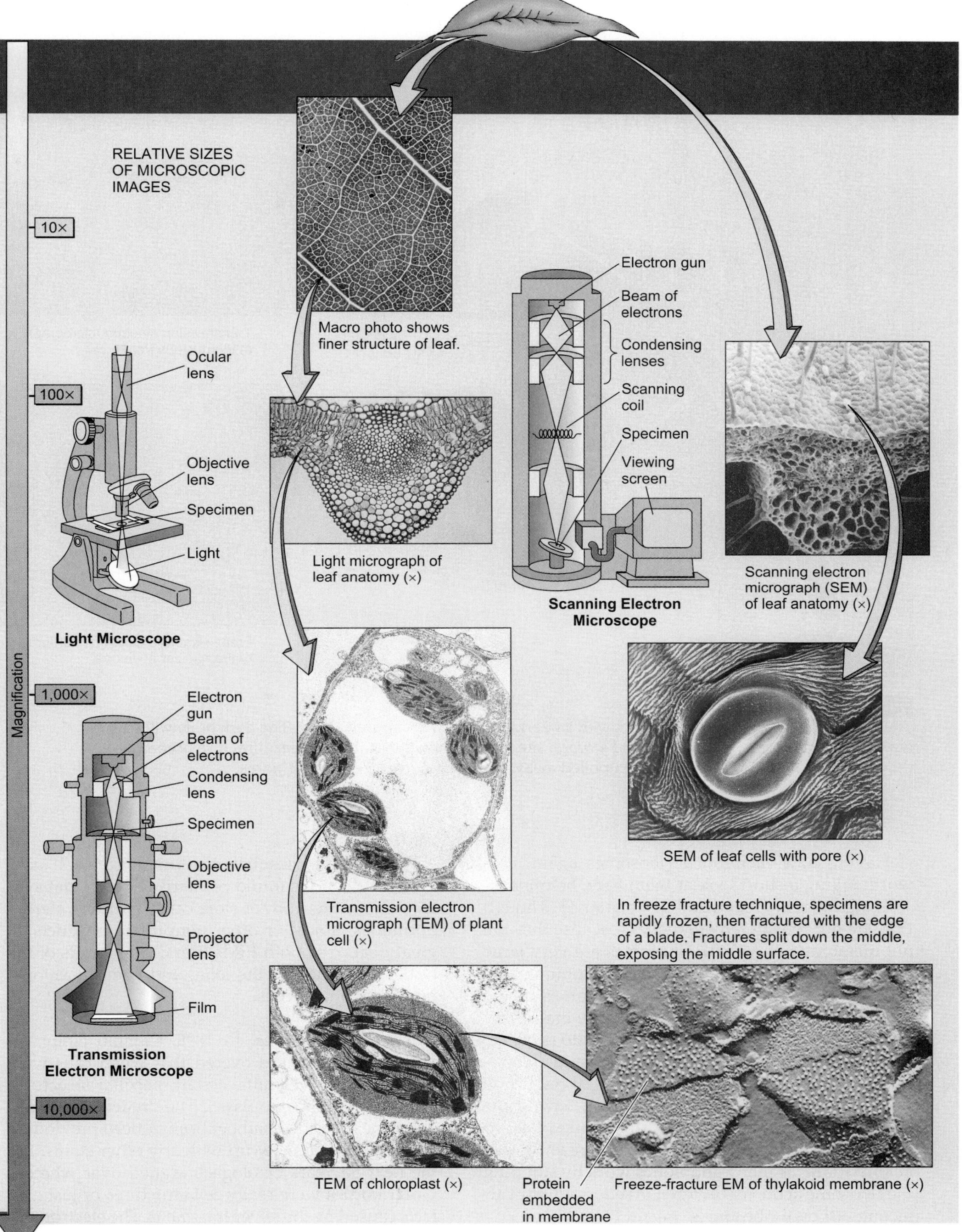

RELATIVE SIZES OF MICROSCOPIC IMAGES
10×
100×
1,000×
10,000×
Magnification
Ocular lens
Objective lens
Specimen
Light
Light Microscope
Electron gun
Beam of electrons
Condensing lens
Specimen
Objective lens
Projector lens
Film
Transmission Electron Microscope
Macro photo shows finer structure of leaf.
Light micrograph of leaf anatomy (×)
Electron gun
Beam of electrons
Condensing lenses
Scanning coil
Specimen
Viewing screen
Scanning Electron Microscope
Scanning electron micrograph (SEM) of leaf anatomy (×)
SEM of leaf cells with pore (×)
Transmission electron micrograph (TEM) of plant cell (×)
In freeze fracture technique, specimens are rapidly frozen, then fractured with the edge of a blade. Fractures split down the middle, exposing the middle surface.
TEM of chloroplast (×)
Protein embedded in membrane
Freeze-fracture EM of thylakoid membrane (×)

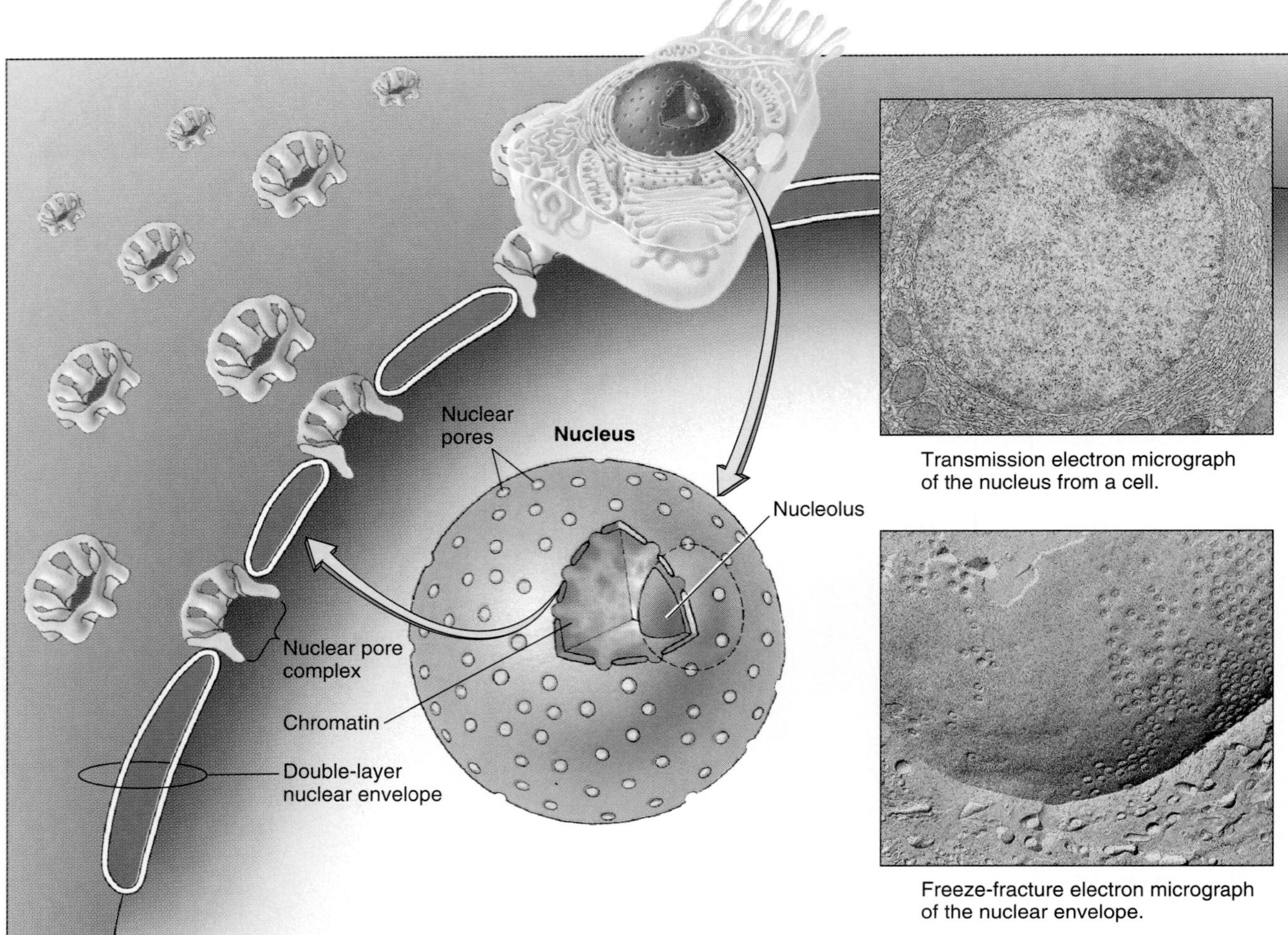

FIGURE 4.6 THE NUCLEUS.
The nucleus is the large central sphere with its dark-staining DNA spread about. The dark body within the nucleus is a nucleolus, a synthesis and storage site for ribosomal RNA. In the freeze-fracture preparation, the outer nuclear membrane has been pulled away, revealing numerous nuclear pores.

Having now mentioned ribosomes again, it is worth taking a short look at them here, before considering them in great detail in Chapter 15. Though they are not considered organelles (because they are not membrane-bounded), ribosomes are very large molecular structures composed of two subunits that each have both RNA and protein. Ribosomes are the sites of protein synthesis, containing the machinery necessary for assembling amino acids into proteins. The ribosomal subunits of prokaryotes and eukaryotes are quite similar in function, though the prokaryotic ribosomes are smaller and have some chemical properties different from the eukaryotic ribosomes. In fact, these differences in ribosomes can be important, as many antibiotics work by stopping the ribosome action of bacterial invaders without affecting our own ribosomes.

Smooth endoplasmic reticulum (Figure 4.7) lacks ribosomes, and is found primarily in cells that synthesize, secrete, and/or store carbohydrates, steroid hormones, lipids, or other nonprotein products. A great deal of smooth ER is found in the cells of the testis, oil glands of the skin, and some hormone-producing gland cells.

The Golgi Bodies. In 1898, Camillo Golgi, an Italian cytologist, discovered that when cells were treated with silver salts, certain peculiar structures appeared in the cytoplasm. The "reticular apparatuses" he described had not been noticed previously, and they didn't show up with any other stains. For the next 50 years cytologists argued over whether **Golgi bodies** were really cell structures or just artifacts caused by the silver treatment. The electron mi-

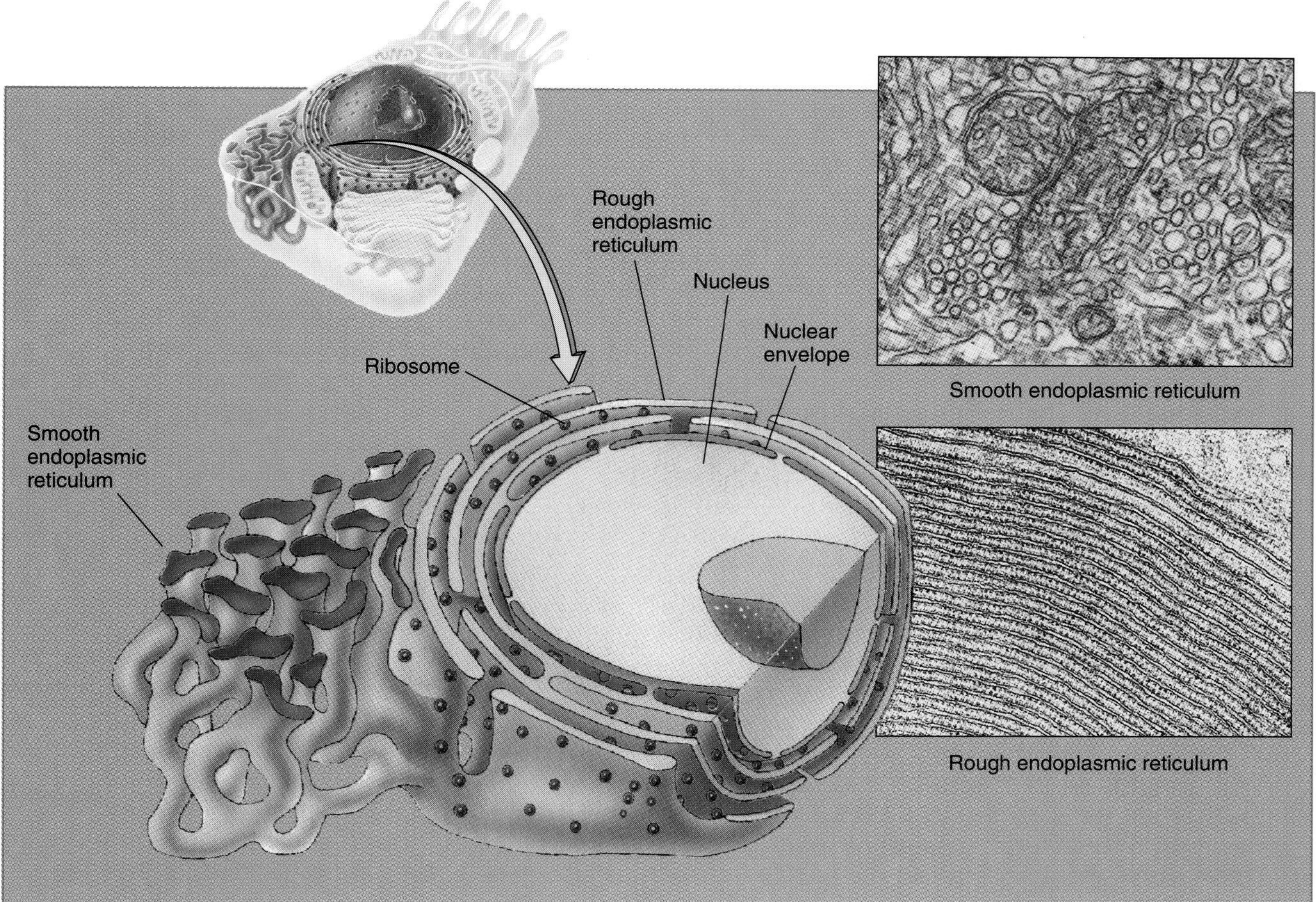

FIGURE 4.7 THE ENDOPLASMIC RETICULUM.
Endoplasmic reticulum (ER) consists of parallel rows of membranes surrounding deep channels. The small round bodies along the rough ER membranes are ribosomes. The smooth ER lacks ribosomes.

croscope came to the rescue by showing that Golgi bodies are indeed real (Figure 4.8). Furthermore, they have a characteristic structure, regardless of the kind of cell they are found in. They are stacks of flattened, membranous sacs, or **cisternae.** They are also highly dynamic organelles, with the membrane area continually forming and dismantling.

Golgi bodies continue some of the chemical functions of the endoplasmic reticulum. Material-laden transport vesicles bud off the ER and fuse with the forming face of nearby Golgi sacs. Within the Golgi bodies the materials are sorted, modified, and packaged in various ways. For instance, the addition of carbohydrates to proteins can occur in the Golgi, and proteolytic enzymes are safely isolated for future use in resulting **storage vesicles** called *lysosomes.* Other substances enter **secretion vesicles,** spherical containers that periodically break away from the maturing face of the Golgi and move to the plasma membrane, where the vesicles fuse with the membrane and the substances carried are released, or secreted, outside the cell.

Lysosomes. The **lysosomes** are roughly spherical, membrane-bounded sacs (Figure 4.9) that contain powerful digestive enzymes, synthesized and packaged by the Golgi apparatus. If these digestive enzymes were released into the cell's cytoplasm, they would quickly digest the cell. Christian de Duve, who first described lysosomes, called them "suicide bags." His poetic fancy was not entirely unwarranted, since lysosomes sometimes do destroy the cells that bear them. However, this destruction is not necessarily disruptive to the organism. Cell death is a normal part of embryonic development. For example, as the fingers form from paddlelike tissue, the cells between them must die and disappear. In addition, lysosomes may rupture and release their deadly enzymes into a superfluous cell or one that is not functioning well. Lysosomes may also aid in digestion within the cell by releasing enzymes into food-containing vacuoles (see below) where digestion can safely occur.

Microbodies. The **microbodies** are small organelles found in a great variety of organisms, including plants and animals. There are two types of microbodies, *peroxisomes* and *glyoxysomes.*

Peroxisomes appear as very dense bodies with a crystalline core, somewhat resembling a cross section through a honeycomb. (In animals, they are most common in liver and kidney cells.) The peroxisomes contain enzymes that, like those of the lysosomes, are important in certain chemical reactions. For example, the enzymes of peroxisomes are known to break down hydrogen peroxide into oxygen and water, protecting cells from its corrosive effect.

Glyoxysomes are microbodies that commonly occur in the lipid-storing regions of seeds. During germination (sprouting) of plant seeds, enzymes of the glyoxysomes use the stored lipids to provide energy to maintain the young seedling.

Vacuoles. The term **vacuole** means "empty cavity" in Latin. In cytology, however, the term refers to a membrane-bounded body with little or no inner structure. Of course, vacuoles are not empty; they hold water and other things, but those things can vary widely, depending on the cell and the organism.

Plant cells generally have more and larger vacuoles than animal cells. In fact, the vacuoles of many types of plant cells dominate the central part of the cell, crowding the other organelles against the cell wall (see Figure 4.4).

The fluids within a plant vacuole may be solutions that include inorganic salts, organic acids, atmospheric gases, sugars, pigments, or any of several other materials. Sometimes the vacuoles are filled with a colorful sap containing blue, red, or purple pigments, some of which will grace the petals of flowers. Some plants store poisonous compounds in vacuoles. It has been suggested that this stored material helps protect the plant against grazing animals.

Returning to the idea that these membrane organelles are part of a coordinated system, Figure 4.9 summarizes some of the interrelationships among the membrane-bounded organelles of a cell.

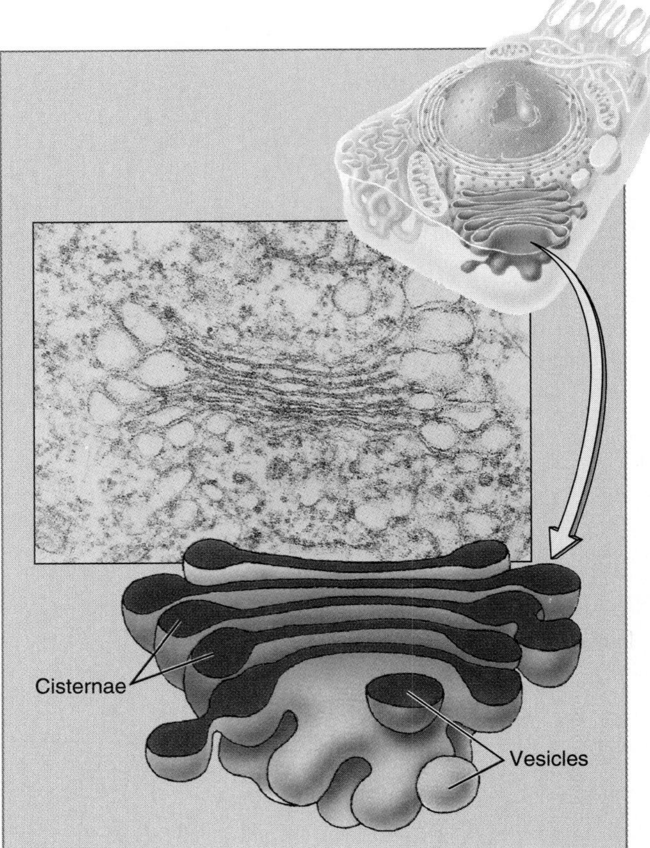

FIGURE 4.8 THE GOLGI AND VESICLES.
Golgi bodies appear as flattened stacks of membranes. The three-dimensional drawing suggests the function of the bodies. Note that the flattened membranous sacs seem to fill at either end and "bud off," forming vesicles.

Organelles Involving Energy

Chloroplasts. The **chloroplasts** of plants are large, green, round or oval organelles, easily seen through the light microscope. Most obvious in leaf cells, they are present in all photosynthetic eukaryotes. (*Photosynthesis* is the process by which plants make food by using the energy of light, and the key enzymes for carrying out photosynthesis are organized within the chloroplast.)

Like the nucleus, chloroplasts are surrounded by a double membrane. Inside the double membrane are layers of flattened, membranous disks known as **thylakoids,** which contain green, light-capturing pigments called chlorophylls and carotenoids. A stack of the thylakoids is known as a **granum** (plural, *grana*). There are many grana in a chloroplast, each of which is connected to its neighbors by membranous extensions called **lamellae.** The clear fluid

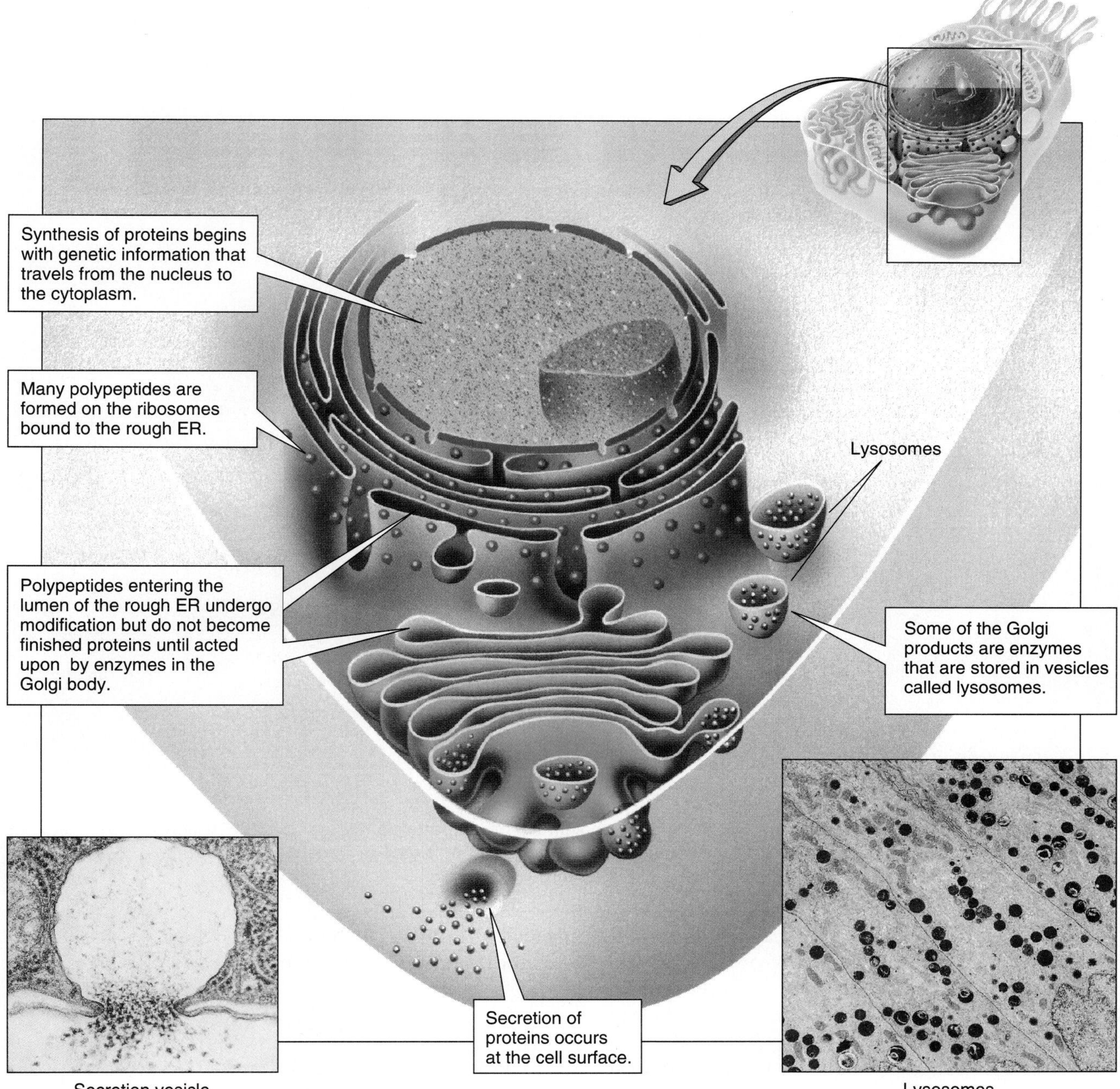

Synthesis of proteins begins with genetic information that travels from the nucleus to the cytoplasm.
Many polypeptides are formed on the ribosomes bound to the rough ER.
Polypeptides entering the lumen of the rough ER undergo modification but do not become finished proteins until acted upon by enzymes in the Golgi body.
Lysosomes
Some of the Golgi products are enzymes that are stored in vesicles called lysosomes.
Secretion of proteins occurs at the cell surface.
Secretion vesicle
Lysosomes

FIGURE 4.9 MEMBRANE RELATIONSHIPS WITHIN A CELL.
Activity in a specialized secretory cell suggests a functional relationship among several organelles. Note the transport vesicles budding from the rough ER and joining the Golgi. Note also the secretion vesicles budding from the face of the outermost cisterna. Lysosomes—the dark spheres in these cells—are storage bodies for powerful hydrolytic enzymes. Hence, their membranes are exceptionally strong.

area surrounding the grana and lamellae is the **stroma** (Figure 4.10).

Obviously, the chloroplast is a complex, highly organized structure. As a general rule, there is a close relationship between the structure and function of an organelle. In this case, the intricate process of photosynthesis depends so much on the chloroplast structure that if the structure is altered, the process cannot continue.

Chloroplasts are actually specialized **plastids,** a group of plant organelles that also includes **leukoplasts** (starch storage bodies) and **chromoplasts** (pigmented bodies that lend their colors to flowers, fruits, and autumn leaves).

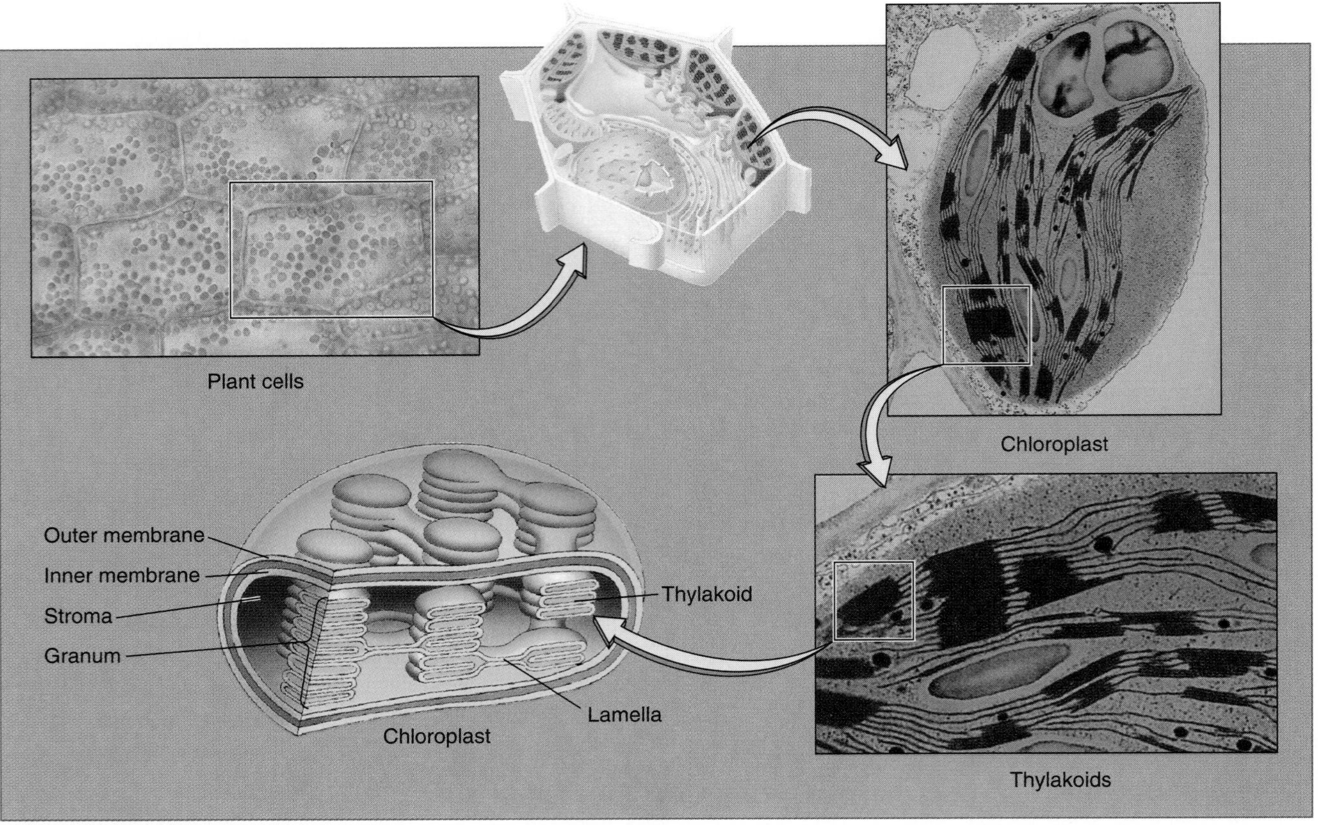

FIGURE 4.10 THE CHLOROPLAST.
Chloroplasts in an intact plant cell appear as numerous minute green spheres. At low electron microscope magnification, the inner structure becomes visible. The dark, neat stacks consist of thylakoids, each stack forming a granum.

Mitochondria. The **mitochondria** (singular, *mitochondrion*) are complex, energy-producing organelles that are found in every eukaryotic cell. Unlike chloroplasts that use light as an energy source, mitochondria release energy from the chemical bonds in foods. Like chloroplasts, mitochondria are enclosed in double membranes. Mitochondria are usually smaller than chloroplasts. In electron micrographs a mitochondrion usually appears as an oval structure, with the inner membrane folded, and with shelflike extensions reaching into the organelle's interior (Figure 4.11).

The folds of the inner mitochondrial membrane are known as **cristae,** and they greatly increase the inner surface area of the organelle. This is important since most of the mitochondrion's biochemical work is done on the cristae, as we will see in later chapters. As for the chloroplast, this internal structure is critically important to mitochondrial function.

Although mitochondria are found in all eukaryotic cells, there are more in some cells than in others. This should be expected since mitochondria are involved in metabolism and since some cells are more metabolically active than others. For example, mitochondria are quite abundant in muscle cells.

CELL SHAPE AND MOVEMENT

Cytoskeleton. Most cells have a particular shape that is characteristic of that cell type; muscle and nerve cells are long and thin, red blood cells are round and flat. The shape of a cell, as well as the distribution of organelles within the cell, is determined by the **cytoskeleton,** a weblike system of fibers of various kinds (Figure 4.12). The most prominent fibers are microtubules and microfilaments.

The **microtubules,** as the name implies, are very tiny tubes (Figure 4.12). They not only help maintain the shape of the cell, but also for some cells, they are important for movement. Microtubules are also important parts of *centrioles, cilia,* and *flagella*—all organelles involved in certain types of cell movement.

Microtubules are made up of a common protein called **tubulin.** Each tubulin molecule consists of two spheres of slightly different polypeptides linked together into a figure-eight shape. In the

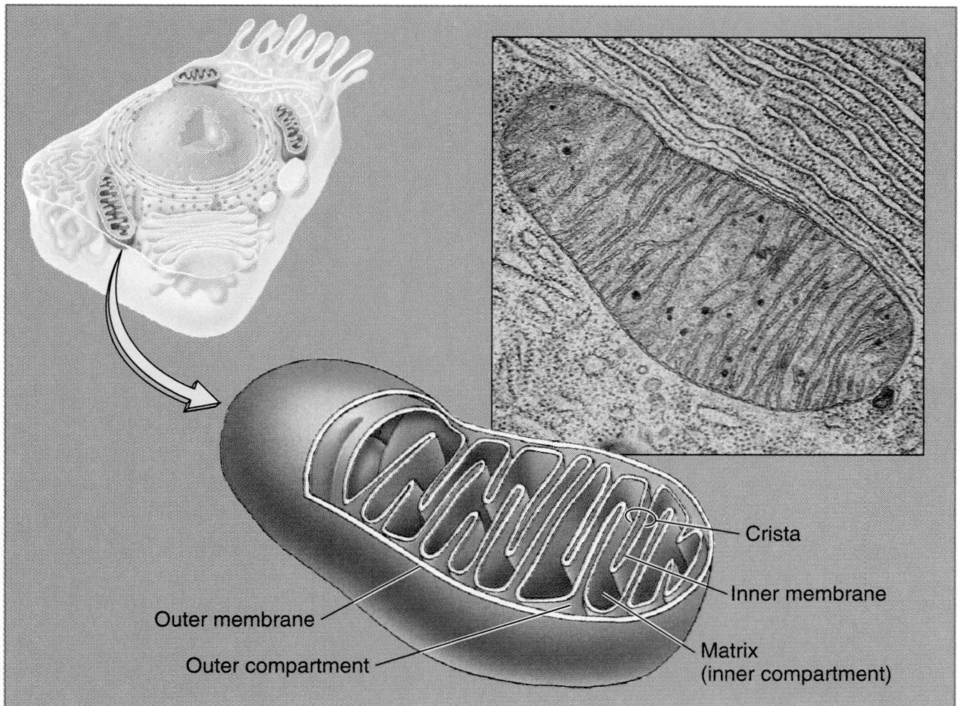

FIGURE 4.11 **THE MITOCHONDRION.**
In thin sections such as this, the long, somewhat tubular mitochondria appear as irregular ellipses and circles. Each has two membranes, with the inner one folded repeatedly to form cristae.

microtubule, the tiny figure eights join in a regular way to form a hollow tube. Microtubules have the remarkable ability to assemble quickly when needed and, just as quickly, to disintegrate when they are no longer required.

Microfilaments are also important in the cytoskeleton and in many structures of movement, such as animal muscle. The microfilament consists of a double strand of the globular protein **actin,** with each strand wound about the other in the helical conformation so common in protein. We will come to actin again when we discuss vertebrate muscle.

Cilia and Flagella. The **cilia** and **flagella** are fine, hairlike, movable projections extending from the surfaces of some cells. Both the cilia and flagella appear, superficially, to be outside of the cell—but the cell membrane actually protrudes, covering each of them.

Structurally, the cilia and flagella are almost identical to one another, differing only in length, numbers per cell, and patterns of motion. Cilia are short, numerous, and move in a characteristic rowing pattern (Figure 4.13). The cilia of all eukaryotes that have cilia—from protists to humans—are identical in size, structure, and movement. Flagella are more variable, but they are always long, fewer in number, and move by undulation (in waves). Both cilia and flagella serve to move the cell through its environment or to move the fluid of the environment past the surface of the cell.

In cross section, a cilium or a flagellum is seen to consist of a regular array of microtubules. Two microtubules run down the center of the shaft and are surrounded by nine additional pairs of microtubules. This universal arrangement is called the "nine-plus-two" pattern. The microtubules in each pair are connected by short arms, and the bending movements of the cilium or flagellum are the result of coordinated sliding movements between microtubule pairs.

Basal Bodies. Beneath each cilium or flagellum, in the cytoplasm of the cell, is a **basal body.** The two central microtubules do not extend into the basal body. The nine paired microtubules do, and each pair is joined by yet a third short microtubule; in

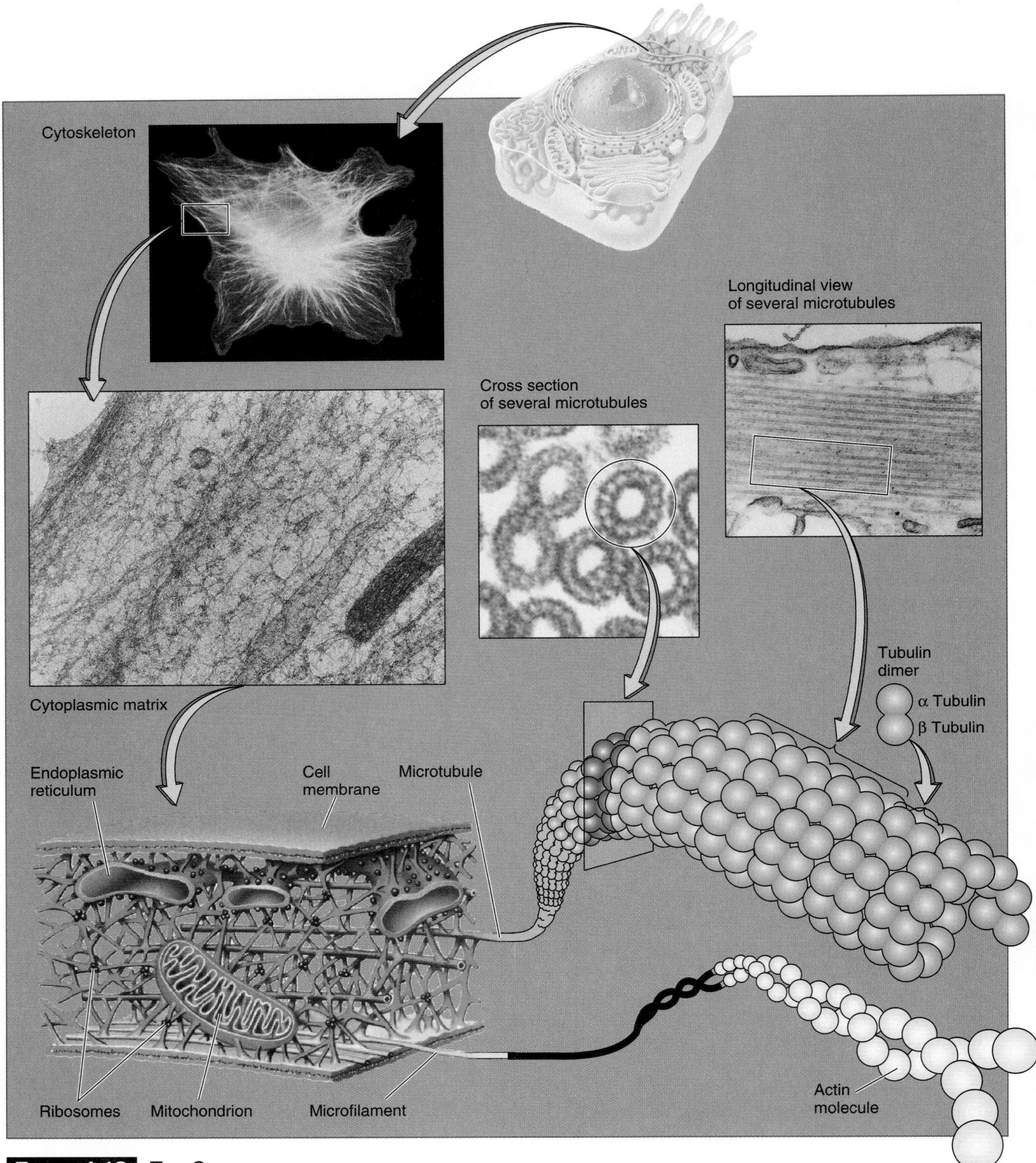

FIGURE 4.12 **THE CYTOSKELETON.**
The cytoskeleton is an extensive network of tubular elements that extends throughout the cell. Note that the major organelles of the cell are suspended in this lattice. Microtubules appear as long rods, which—in cross section—are found to be hollow. The tubes consist of spherical units of the globular protein tubulin. Microfilaments also contain globular proteins, but they form a twisted filament.

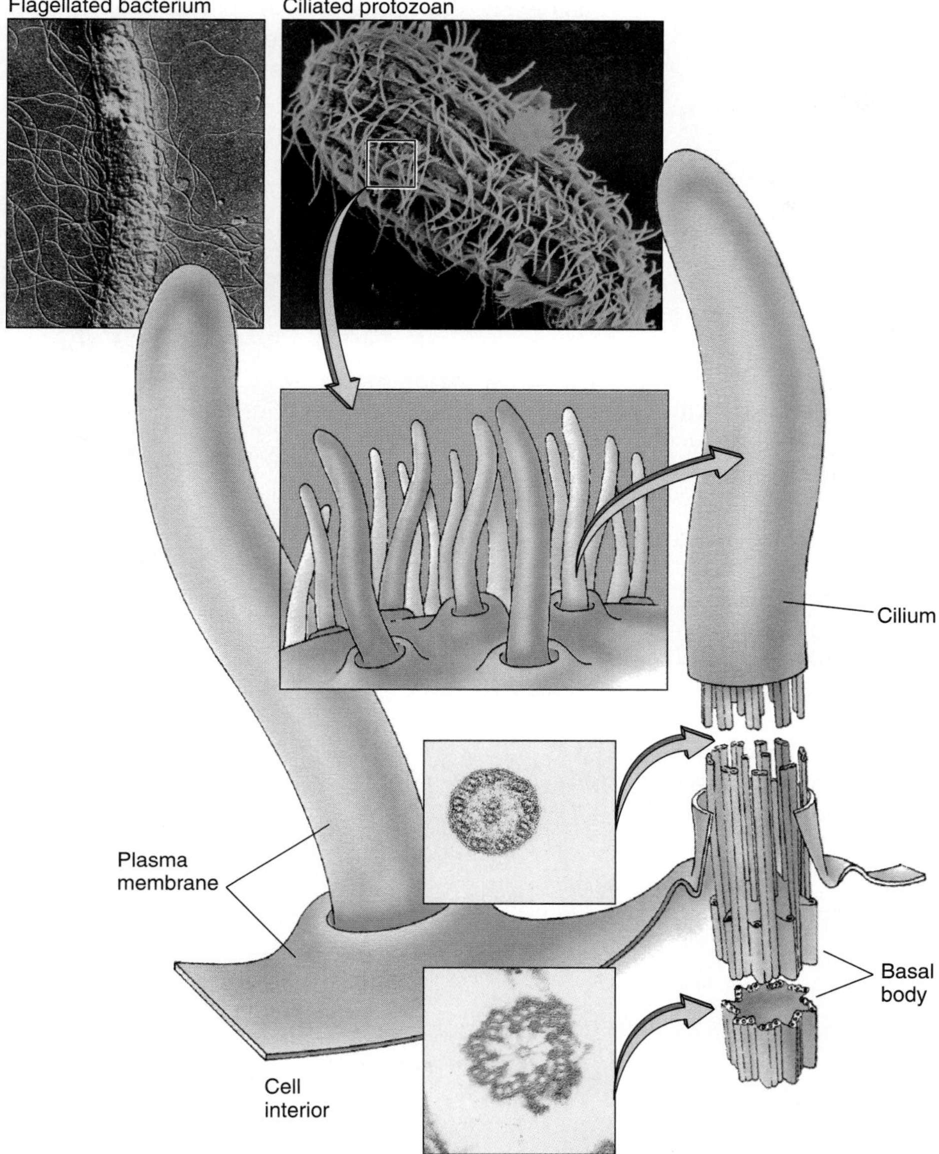

FIGURE 4.13 **THE CILIA AND FLAGELLA.**
Flagella can move in a variety of ways, creating an undulating motion that pushes or pulls the cell body through the medium. Cilia move some single-celled organisms by a highly coordinated rowing action. Various combinations can produce wildly spinning and gyrating movements. The cross section of the cilium reveals the nine-plus-two arrangement of microtubules and the structure of the basal body.

cross section, therefore, the basal body shows a ring of nine triplets of microtubules (Figure 4.13), called the "nine-plus-zero" arrangement. Cilia and flagella form from outgrowths of the microtubules in the basal body.

Centrioles and Spindle Fibers. The **centrioles** have the same appearance, in cross section, as basal bodies; but centrioles are found deep within the cytoplasm and have a very different function. They are thought to organize the specialized microtubules called **spindle fibers** that serve to separate chromosomes during cell division (Figure 4.14). Each centriole actually consists of two short "nine-plus-zero" cylinders, held somehow at right angles to one another.

The cells of higher plants have no cilia, flagella, basal bodies, or centrioles. They manage to organize their chromosome-separating fibers perfectly well all the same.

The Prokaryotic Cell: A Different Matter

As you may recall, the prokaryotes are **bacteria** (we'll often use the terms interchangeably). A detailed treatment of this group will be made in Chapter 22, so here we will be brief.

Prokaryotes are a diverse group of minute, usually single-celled organisms whose origins can be traced to the most ancient forms of life known. All other forms of life (including human) are eukaryotes. You are already quite aware of prokaryotes—for among them are the familiar "germs" that linger in kitchens and bathrooms and inspire fascinating TV commercials.

In a more objective light, let's note that in terms of their energy sources, prokaryotes (like other organisms) fall into two major categories: **heterotrophs** ("other-feeding") and **autotrophs** ("self-feeding"). The heterotrophs require complex organic food sources, generally those produced by other organisms. Heterotrophs include the aforementioned "germs," clinically known as **pathogenic** (disease-causing) bacteria. Pathogens are parasites, of course. But most heterotrophic bacteria are from a group known best as **decomposers** (decay bacteria) that use nonliving organic matter as a food source. They are often the cause of those pungent odors (or perhaps stenches) associated with body wastes and

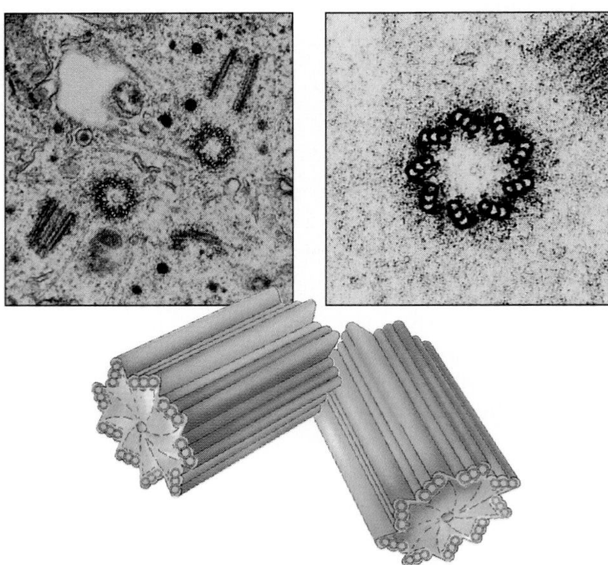

FIGURE 4.14 The Centrioles.
Centrioles are paired cylindrical bodies that generally appear at right angles to each other. Note the nine sets of triplet microtubules in a thin section by the electron microscope across the central axis of one of the centriole's paired bodies.

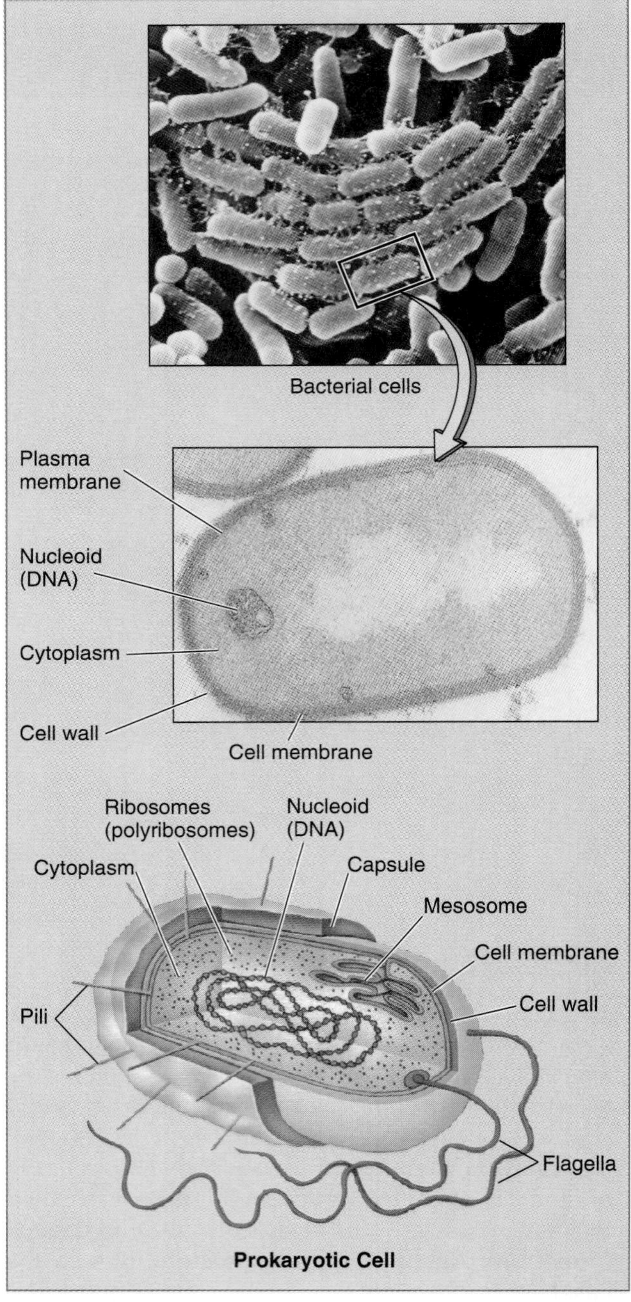

FIGURE 4.15 Prokaryotes.
In contrast to the eukaryotes, prokaryotes lack membrane-surrounded organelles. Nevertheless, all life functions, including self-replication, occur in these cells.

decay; and while they thrive in our foods and other goods, they generally do not invade our bodies. In fact, decomposers are vital to life because they recycle essential elements such as carbon and nitrogen.

Whatever their energy source, prokaryotes tend to be structurally simple (Figure 4.15). They generally range in size from one to ten μm, only a tenth the diameter of representative eukaryotic cells. One group, the cyanobacteria, is exceptional, often being larger than other bacteria. Cyanobacteria have an elaborate inward extension of their surrounding membrane that contains light-absorbing pigments necessary for photosynthesis. This is the prokaryote version of the eukaryote thylakoid.

In addition to their surrounding membrane, virtually all bacteria have a protective cell wall. (So do many eukaryotic cells, but their walls are composed of cellulose or chitin rather than the peptidoglycan and protein of prokaryotic cell walls.) Some bacteria surround the wall with a sheath, and others, with a slimy protective capsule. Within the plasma membrane is the cytoplasm, the living material of the cell—a veritable cauldron of chemical activity (actually, not unlike what goes on in eukaryotic cytoplasm).

The genetic material of bacteria, their DNA, lies free in the cell in an area called the **nucleoid.** (In bacteria there is no organized cell nucleus.) Bacterial DNA occurs as one continuous circular molecule, which—while referred to as a chromosome—is quite unlike the linear eukaryotic chromosomes. Some bacteria have small amounts of DNA incorporated into small circular chromosomes, or plasmids, separate from the main chromosome.

Finally, bacteria have structures that provide movement. Many aquatic and soil species move about through the action of a flagellum, which in prokaryotes is a novel, S-shaped structure. The unique prokaryotic flagellum spins on its axis like a propeller and is the only spinning structure seen in cells. The key differences among prokaryotes and eukaryotes are listed in Table 4.2

TABLE 4.2 *KEY DIFFERENCES AMONG THE CELLS OF PROKARYOTES, PLANTS, AND ANIMALS*

Feature	Prokaryotic cell	Higher plant cell	Animal cell
Plasma membrane	Present	Present	Present
Supporting structure	None seen	Protein cytoskeleton	Protein cytoskeleton
Nuclear envelope	Absent	Present	Present
Membrane-bounded organelles	Absent	Many	Many
Endoplasmic reticulum	Absent	Present	Present
Ribosomes	Smaller, free	Larger, some membrane-bounded	Larger, some membrane-bounded
Cell wall	Peptidoglycan	Cellulose	None
Flagella or cilia	Tubular, rotating	Absent*	Microtubular (9+2)
Ability to engulf solid matter	Absent	Absent*	Present
Centrioles	Absent	Absent*	Present

* *Sometimes present in primitive plants.*

PERSPECTIVES

The cell is a living unit made up of a number of discrete subunits called organelles. It is the organization and integration of these organellar parts that together describe the structure and function of an organism.

In the next chapters, we will describe some of the functions of the major organelles. We will find that biological structure and biochemical function are closely related and integrated, and that the integration between structure and function is a clear hallmark of all living things.

SUMMARY

The term *cell* was first applied to life by Robert Hooke when he described the microscopic structure of cork.

CELL THEORY

The cell theory proposes that living organisms are composed of cells. It was suggested by Oken but formally proposed by Schleiden and Schwann in 1838. Virchow stated that cells arise only from preexisting cells.

Cells are the organizational units of life, capable of carrying out the life processes. Two fundamental kinds occur: prokaryotic and eukaryotic. In multicellular organisms, similar cells join to form tissues, which in turn contribute to organs and organ systems.

While cells range greatly in size, most plant and animal cells are about 90 μm (micrometers) in diameter. Since the relative membrane area does not increase as fast as volume, a small size favors efficient membranal transport. Cells have a lower limit set by the size of their macromolecules.

Cells in multicellular organisms undergo specialization whereupon each is suited for a specific task.

CELL STRUCTURE

The plasma membrane surrounding the cell is selectively permeable, favoring the entry and exit of certain substances. In accordance with the fluid mosaic model, it contains a double layer of phospholipids and a number of dispersed proteins and other molecules. The outer surfaces are polar and hydrophilic, while the core is hydrophobic. Some proteins pass through the membrane; others penetrate partway or perch on the surface. These imbedded membrane proteins are mobile.

The plant cell wall consists of layered microfibrils of cellulose, impregnated with strengthening materials.

The nucleus is a prominent sphere surrounded by the nuclear envelope (two membranes). Nuclear pores permit substances to cross the nuclear envelope.

An extensive membranous channel, the endoplasmic reticulum (ER), is the site of synthesis and transport. Rough ER gets its name from the presence of numerous protein-synthesizing bodies called ribosomes. The synthesis of various other substances occurs in the smooth ER.

Golgi bodies consist of flattened, membranous sacs and rounded vesicles. They form continuously from sections of endoplasmic reticulum that contain newly synthesized proteins ready for further chemical modification. Some Golgi bodies secrete their contents outside the cell.

Lysosomes, former Golgi vesicles, are tough sacs containing hydrolytic enzymes to be used in digestion.

Peroxisomes are crystalline bodies containing enzymes that break down hydrogen peroxide.

Vacuoles are simple membranous containers that hold many substances, including water, salts, acids, sugars, foods, and pigments.

Chloroplasts are complex membranous organelles that contain chlorophyll pigments. They capture light energy and use it for making foods. Mitochondria are membranous bodies that contain enzyme systems and mechanisms for transferring chemical energy from foods to a form that is useful to the cell.

Microtubules and microfilaments are responsible for cell shape and movement. Microtubules occur in cilia and flagella where their coordinated movement causes the required rowing and undulating motion. Microfilaments are associated with cytoplasmic movement and muscle contraction. While cilia and flagella have a nine-plus-two arrangement of microtubules, they originate from basal bodies that have a nine-plus-zero microtubular arrangement. Centrioles, organizers of spindle fibers, are paired versions of basal bodies.

The Prokaryotic Cell: A Different Matter

Prokaryotes (bacteria) lack membrane-bounded organelles, have cell walls, one circular chromosome, and tubular flagella that rotate.

Key Terms

cell theory • **50**
cell • **50**
prokaryote • **50**
eukaryote • **50**
tissue • **51**
organ • **51**
system • **51**
organelle • **51**
cytoplasma • **52**
plasma membrane • **54**
fluid mosaic model • **55**
phospholipid bilayer • **55**
glycoprotein • **55**
lipoprotein • **55**
cell wall • **55**
lignin • **55**
suberin • **55**
cutin • **55**
nuclear envelope • **58**
nuclear pore • **58**
endoplasmic reticulum (ER) • **59**
Golgi body • **62**
cisterna • **63**
storage vesicle • **63**
secretion vesicle • **63**
llysosome • **63**
microbody • **64**
peroxisome • **64**
glyoxysome • **64**
vacuole • **64**
chloroplast • **64**
thylakoid • **64**
granum • **64**
lamella • **64**
stroma • **65**
plastid • **65**
leukoplast • **65**
chromoplast • **65**
mitochondrion • **66**
crista • **66**
cytoskeleton • **66**
microtubule • **66**
tubulin • **66**
microfilament • **67**
actin • **67**
cilia • **67**
flagellum • **67**
basal body • **67**
centriole • **69**
spindle fiber • **69**
bacteria • **70**
heterotroph • **70**
autotroph • **70**
pathogenic • **70**
decomposer • **70**
nucleoid • **71**

Review Questions

1. Summarize the propositions of the cell theory. Name the chief contributors.
2. Clearly distinguish between the terms *cell, tissue, organ, organ system,* and *organelle.* In what type of organism are tissues and organs seen?
3. Summarize the size range of cells—the largest and smallest. What dimensions are seen in most animal and plant cells?
4. Suggest two advantages to small cell size. Explain the peculiar relationship between cell surface area and volume.
5. Explain how selective permeability is an important trait for the plasma membrane.
6. Prepare a simplified drawing of the cell membrane showing the arrangement of the phospholipids and proteins. Add a glycoprotein.
7. Which groups of organisms produce cell walls? Describe how cellulose is used in the formation of such walls.
8. List and briefly explain two functions of the nucleus.
9. Describe the arrangement of the two types of endoplasmic reticulum. In what materials does each specialize?
10. Describe the Golgi body.
11. Explain why de Duve thought of lysosomes as "suicide bags."
12. List several substances carried or stored in plant vacuoles.
13. What is the relationship between the ribosome, rough ER, Golgi body, lysosome, and secretion vesicle?
14. Prepare a simple illustration of a chloroplast, labeling the outer membrane, stroma, grana, lamellae, and thylakoids.
15. What special function does the chloroplast have? In what organisms are these bodies seen?
16. What is the function of the mitochondrion?
17. Describe the general shape and organization of the mitochondria. What are two ways in which the inner membrane is very special?
18. Briefly describe the structure of the microtubule. List three organelles in which they are seen.
19. What special functions do cilia and flagella have? Compare their movement.
20. List five ways in which cells of prokaryotes differ from those of eukaryotes.

5

Cell Transport

Passive Transport
Diffusion
Water Potential and Bulk Flow
Osmosis: A Special Case
Facilitated Diffusion
Active Transport
Membrane Pumps
Endocytosis and Exocytosis
Direct Transport Between Cells

IN HOOKE'S TIME, THE WORD "CELL" WAS USED TO DESIGNATE THE TINY *chamber in monasteries where each monk slept. Hooke adopted the term for his purposes by analogy because it seemed to him that the "little boxes" he could see in slivers of cork were regular little rooms, with walls that clearly separated the inside of the cell from the outside. We now realize that this separation—of the living inside from the nonliving outside—imparts a physical definition to a living cell. Yet the inside of a cell must communicate with its environment. Substances must* cross *the cell's membrane—the very structure designed to keep the inside* separate *from the outside! It is time that we investigated the cell membrane and the properties of the membrane that allow it to achieve this delicate balance of function.*

The cell is a lively thing. Inside, in its roiling fluids, the business of life goes on with astonishing precision and swiftness. Molecules are bound and released, ions ebb and flow, chemicals appear and disappear in a constantly shifting scenario. All this is made possible, in large part, by the precise movement of materials in and out of cells and from one cell to another. Such materials must cross the *plasma membrane* which, as we will see, is exquisitely adapted to its function of accepting some substances while excluding others. The membrane performs its gatekeeper tasks in a variety of ways that generally fall within two categories: (1) *passive transport* and (2) *active transport.*

The two categories involve very different energy sources. One is energy-costly to the cell; the other is not. In **passive transport,** thermal (heat) energy of the cellular environment provides all of the energy, much of it quite random. **Active transport,** however, requires the cell to do work, and it costs something in terms of the cell's precious energy reserves.

PASSIVE TRANSPORT

Passive transport includes four distinct kinds of movement, none of which requires an energy expenditure on the part of the cell. The four processes are *diffusion, facilitated diffusion, bulk flow,* and *osmosis.*

DIFFUSION

Diffusion is the *net* movement of molecules down their concentration gradient from regions of higher concentration to regions of lower concentration. In any system with random movement, particles may move in any direction. If particles move in one direction more than any other, there is said to be a net movement in that direction. In biological systems, it is an especially important way for ions and small molecules to get around. Diffusion enables substances to cross cell membranes and to move within the cytoplasm.

Diffusion occurs because the molecules of any liquid or gas move constantly and randomly, bumping into each other and rebounding into new paths. This random movement is called *Brownian motion.* It is due to the inherent kinetic energy of the particles. The warmer the gas or liquid is, the faster its molecules move. The movement of individual molecules is still random, but heat accelerates the net movement of those molecules, always in accord with the same physical principles—from the region of higher concentration to regions of lower concentration. This net movement continues until the distribution of molecules is uniform. Once the molecules are uniformly dispersed, there will be no further net movement in any direction. Diffusion, then, is a random process: Molecules move away from an area of higher concentration by their own thermal energy until equilibrium is reached (Figure 5.1).

The rate at which ions and molecules diffuse depends on many factors, but principal among these are temperature, the steepness of the diffusion gradient (difference in concentration from one end of the system to the other), and the size of the particles in motion (smaller particles move faster than bigger ones). The rate of diffusion also depends on the physical state of the medium; diffusion is fastest in gases, slower in liquids, and even slower (but it does happen) in solids.

Diffusion and the Plasma Membrane. In living systems, other factors influence diffusion. Cells are surrounded by membranes that have specific chemical and physical properties that determine the membrane's **permeability** to various substances. Permeability refers to the tendency of a membrane to allow specific kinds of molecules and ions to cross. The permeability of the plasma membrane, then, depends on both its own characteristics and those of the substances in transit. Such membranes are often described as **selectively permeable** because of the limited kinds of molecules that can pass through.

Recall that in the fluid mosaic model (see Figure 4.5) the membrane is basically a bilayer of phospholipids interrupted in places by surface and transmembranal proteins and other molecules. The charged heads and uncharged tails of the phospholipids result in the two sides of the molecule bearing polar charges. Sandwiched between these charged layers is an oily, nonpolar core layer. Because of the polar nature of the membrane, the electrical charges of molecules and ions are critical to their passage across the membrane. Nonpolar molecules such as oil-soluble steroids (sex hormones, for example) readily cross the phospholipid bilayer, as do molecules of oxygen, carbon dioxide, and nitrogen, which are nonpolar and relatively small.

Ionized particles cross at different rates. For example, the membranes of red blood cells more readily accept positive ions than negative ions. Further, singly charged ions cross the membrane more easily than do doubly charged ones. Although water is decidedly polar, with asymmetric charges, it readily crosses plasma membranes. Since the middle of the membrane is hydrophobic, this permeability has long represented an enigma to cell biologists. They theorized that plasma membranes must have protein-lined pores. Such pores would indeed account for the membrane's permeability. But electron microscopic studies haven't revealed any such pores, and experiments with both protein-free artificial membranes and living cells with little protein in the membrane reveal that water crosses both with ease. Scientists are still left wondering exactly how water crosses plasma membranes so easily.

Membranes from various tissues differ considerably in their permeability. Even within the same tissues, permeability to various solutes can change. For example, kidney cells can alter their permeability to water according to the body's need to conserve water (see Chapter 37).

FIGURE 5.1 DIFFUSION.
The high concentration of dye in a solution eventually becomes distributed throughout the vessel, as the dye molecules move randomly about by the process of diffusion.

Water Potential and Bulk Flow

The development of life on this planet is intimately associated with the abundance of a simple molecule, H_20. Water is, indeed, important to both the development and sustenance of life. So you can imagine that evolution has provided adaptations for moving it around, shifting it to places where it will be of greatest advantage in various life forms.

Water potential is simply the tendency of water to move from one place to another. The rule is that water moves from regions of higher to regions of lower potential. This "potential for movement" is influenced by such commonplace forces as gravity and pressure as well as more subtle effects, such as solute concentration. (A solute, you may recall, is any substance that is in solution in a solvent such as water or other liquid. Salt and sugar are common solutes.)

The simplest analogy of water potential is seen in a dam. A great deal of water can accumulate behind a dam where the water level is higher than

1 Fill a graduated cylinder with water. Add a concentrated sugar solution to the reservoir of a long thistle tube and cover the end with a semipermeable membrane which permits water, but not sugar, to cross.

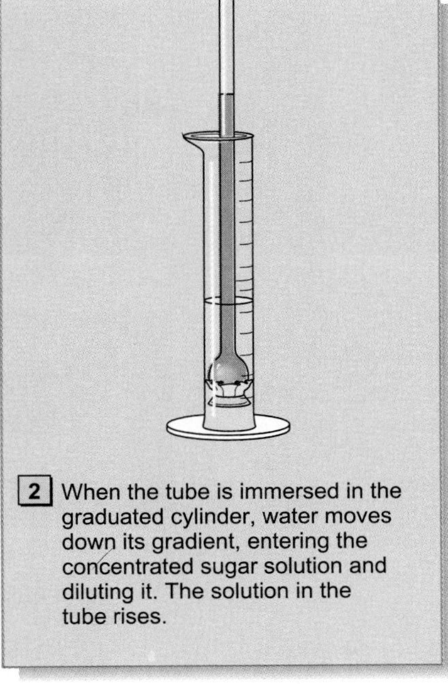

2 When the tube is immersed in the graduated cylinder, water moves down its gradient, entering the concentrated sugar solution and diluting it. The solution in the tube rises.

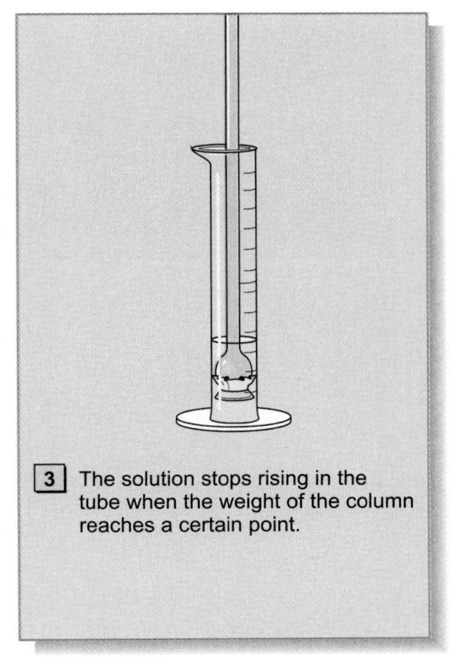

3 The solution stops rising in the tube when the weight of the column reaches a certain point.

FIGURE 5.2 OSMOSIS.
In a demonstration of osmosis, a tube containing a 3% sugar solution and covered with a selectively permeable membrane is immersed in distilled water. The membrane will permit water, but not sugar, to cross. As expected, water moves *down its concentration gradient,* crossing the membrane and entering the sugar solution to raise the level of liquid in the tube.

that in the riverbed below. Thus, the dammed river has a great water potential. (The potential of that water will become quite clear if the dam should give way.) Simple pressure can also increase water potential. A mechanical pump forcing water up into a storage tank increases the tendency for that water to move downward.

The movement of fluids such as water by such forces is often referred to as *bulk flow.* **Bulk flow** is the movement of fluids in which all molecules move in the same direction. (Diffusion, by contrast, involves random movement, but with the *net* movement of molecules along a gradient.)

Bulk flow is important biologically for a number of reasons. For instance, bulk flow is behind blood movement in our own circulatory systems. Our pumping heart and elastic arteries increase the bulk flow. In plants, bulk flow moves sap through the tubular elements that comprise the food transport system.

OSMOSIS: A SPECIAL CASE

Osmosis is a word that has been borrowed from science and then twisted and misused beyond recognition. The next time you hear the word in casual conversation, ask the speaker to define it. You will probably make a lifelong friend if you smirk, look around the room, and say, "Wrong! Osmosis is the diffusion of water across semipermeable membranes, from an area of greater water potential to one of lesser potential." Go on by saying, "You see, when two solutions are separated by a selectively permeable membrane—that is, a membrane that allows only water to pass—the net movement of water will be from the solution with the greater concentration of water molecules, through the membrane, to the solution with the lesser concentration of water molecules. Keep in mind also that the concentration of water molecules is lower on the side of the membrane that contains the *higher* solute concentrations. Further, it doesn't really make any difference what kind of ion or molecule is in solution; it could be sugar, amino acids, or any other soluble substance. The water moves only according to the relative number of water molecules on either side of the membrane" (Figure 5.2). By now, of course, everyone will probably have left.

Osmosis is a simple principle that, at first, may seem hard to understand. For example, someone

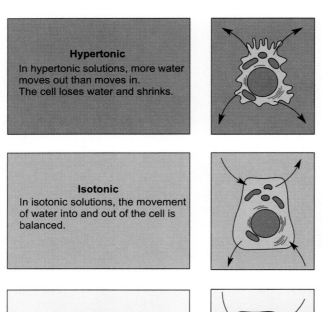

Hypertonic
In hypertonic solutions, more water moves out than moves in.
The cell loses water and shrinks.

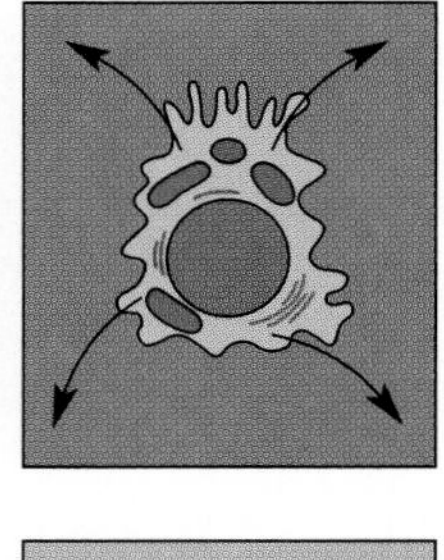

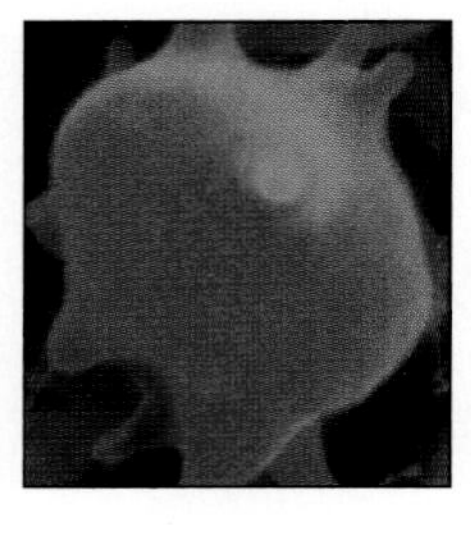

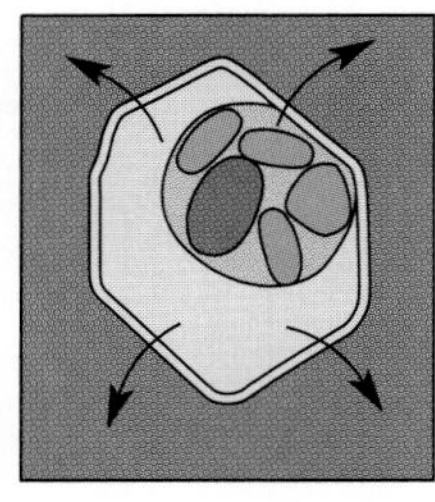

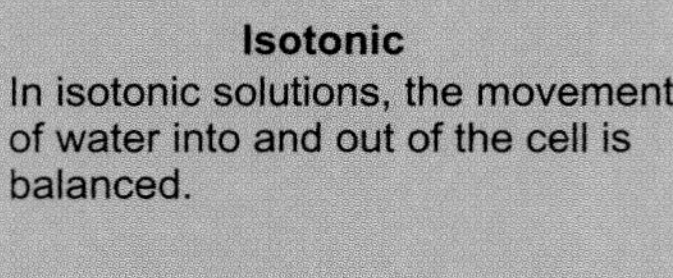

Isotonic
In isotonic solutions, the movement of water into and out of the cell is balanced.

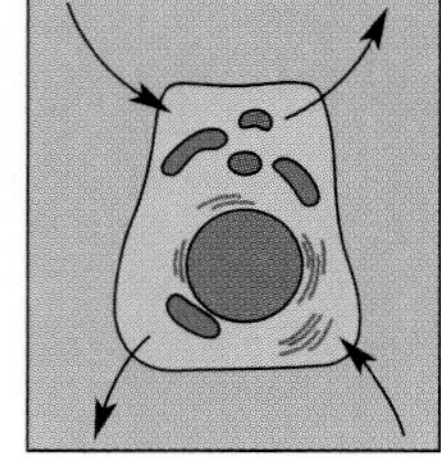

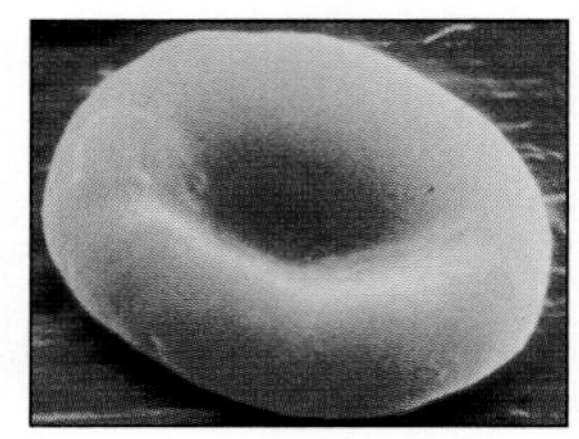

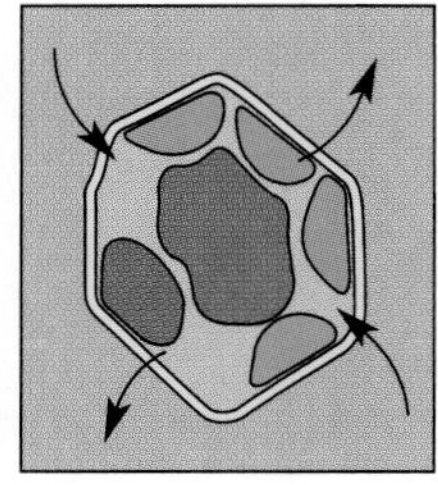

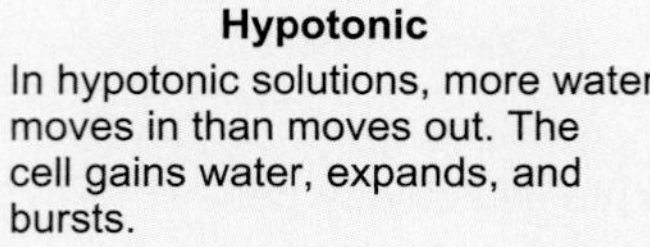

Hypotonic
In hypotonic solutions, more water moves in than moves out. The cell gains water, expands, and bursts.

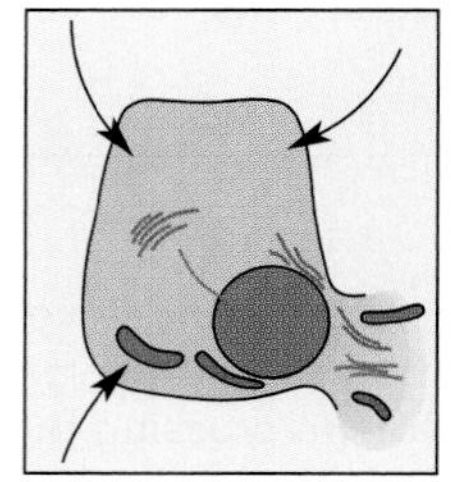

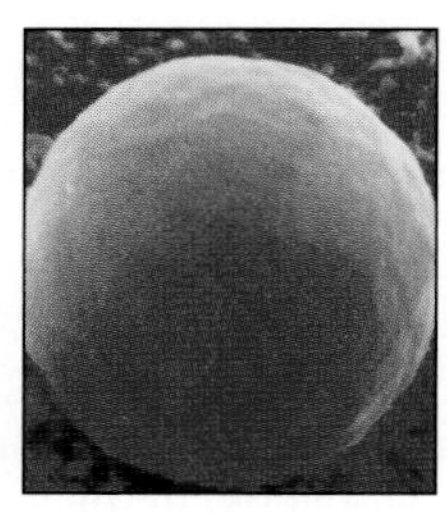

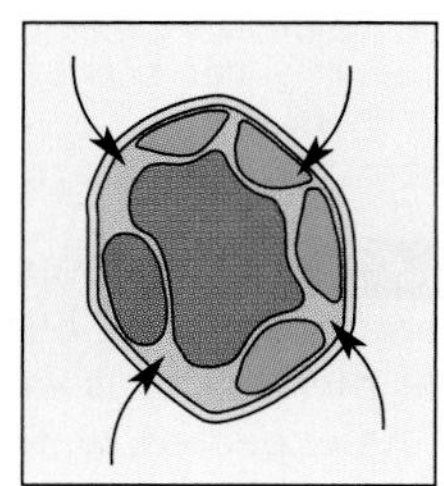

FIGURE 5.3 MOVEMENT OF WATER IN AND OUT OF CELLS.
Changes in their watery environment affect plant and animal cells differently. Although animal cells cannot tolerate a severe water gain or loss, plants make use of water gain to support softer structures such as leaves and younger green shoots.

might ask how long does osmosis go on? Theoretically, it continues until the water potential on both sides of the membrane is equal. This equilibrium rarely occurs between cells and their surrounding fluids, however. Metabolism demands constant change, and there is a constant ebb and flow of molecules across any living membrane. In the system shown in Figure 5.2, water potential will drive the movement of water across the membrane and into the tube until the weight of the water in the tube counteracts the water potential. The net movement of water will then stop. At this point we are able to measure **osmotic pressure,** the amount of force necessary to equal the water potential. One of the clearest examples of the effects of osmotic pressure in nature occurs in plants, as we will see next.

Turgor and Wilting. The large central vacuoles of certain plant cells contain water and various solutes. Each vacuole is surrounded by a selectively permeable membrane. When the concentration of water outside the cell is greater than that in the vacuole, water enters by osmosis, which causes the vacuole to swell, pressing the rest of the cell contents against the cell wall. Animal cells lack walls and will simply swell and burst. But the plant cell wall is extremely strong, so it holds. As the size of the vacuole continues to increase, it meets increased resistance until, finally, the pressures are equalized and no more water can enter. This special kind of osmotic pressure from within the cell is called *turgor pressure.* **Turgor pressure** is the force that holds leaves and soft stems of plants erect. Should turgor pressure decrease because of a reduction of water within the plant's vacuoles, the result is **wilting.**

Cells, Solute Conditions, and Tonicity. The osmotic environment of cells is described in terms of tonicity. For example, environmental conditions are **isotonic** (*iso,* same; *tonic,* tension) when the relative concentrations of water and solutes on either side of the plasma membrane are equal—which, of course, means the water potential is equal. In isotonic systems there is no *net* movement of water molecules across a membrane, though water molecules will be moving in both directions equally.

On the other hand, when the water outside a cell contains less solute than does the water inside (meaning the water potential outside is greater), the environment is called **hypotonic** (*hypo,* low). Cells immersed in hypotonic solutions tend to swell. For example, blood cells in tap water will swell and rupture.

When the water outside the cell contains more solutes than that on the inside (resulting in the water potential outside being lower), the environment is **hypertonic** (*hyper,* over). Cells in a hypertonic solution tend to lose water. Note that as we use the terms "isotonic," "hypotonic," and "hypertonic" we refer to the relative solute concentration in water surrounding the cell or organism (Figure 5.3).

Cells in organisms make constant use of the osmotic mechanism, taking in or releasing water and shifting it from cell to cell as needed. While there is no known mechanism for actively transporting water, cells can and do actively transport solutes, and where solutes go, water follows. For example, plants can increase the uptake of water into the root by first actively transporting mineral ions into the root cells. This sets up the desired water potential gradient, and water flows in through osmosis. Similarly, in our own kidneys and those of other water-conserving mammals, concentrations of sodium and chloride ions across the plasma membranes of certain cells set up the osmotic conditions needed for reclaiming water that would otherwise escape in the urine.

Facilitated Diffusion

Facilitated diffusion (also called facilitated transport) is similar to simple diffusion in that it is passive, it involves no cellular energy, and molecules follow the usual diffusion gradient (from high concentration to low). However, it differs from simple diffusion in that the movement of selected molecules is significantly accelerated by **carriers,** proteins that are embedded in the plasma membrane and designed to carry a molecule from one side of the membrane to the other (Figure 5.4).

Active Transport

Since life is such a delicate and constantly adjusting process, it is not surprising to find cells moving, shifting, adding, and expelling molecules. Accumulations of some molecules often occur inside or outside a cell. For example, mammalian red blood cells work against the concentration gradient to move sodium ions out and accumulate potassium ions. Similarly, many marine fish expel sodium ions from salt-secreting glands in their gills (chloride ions follow passively). They must work against a powerful concentration gradient, because the sea is saltier than their blood. So simple an act as moving molecules against a concentration gradient might crudely be compared to rolling boulders uphill. Work must be done, and work requires the expenditure of cellular energy. This is why this type of movement is called *active transport*—defined as a form of transport that requires the active input of energy to move substances against a concentration gradient. Let's see how molecules are shifted around at an energy cost.

Membrane Pumps

Some substances are carried across membranes against a gradient by proteins embedded in the membranes. For instance, many cells have **sodium-potassium ion exchange pumps,** membranal carriers that alternately bring in potassium ions and usher out sodium ions. Such pumps are found in a number of kinds of cells, such as those of the nerves and kidneys. They are powered by a special energy-storing molecule called adenosine triphosphate (ATP) (see Chapter 6). For each molecule of ATP expended, the cell captures two potassium ions and

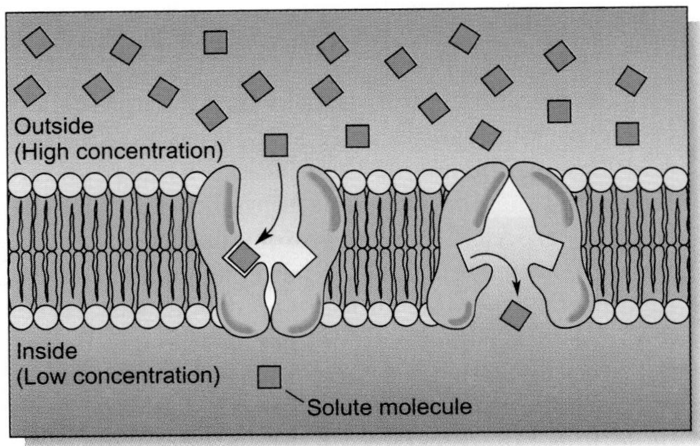

Figure 5.4 Facilitated Diffusion.
In facilitated diffusion, membrane proteins utilize conformational changes to accelerate selected molecules or ions down their gradient. This model suggests that the entry of a molecule triggers one event (the shape change), and its release triggers another (the change back).

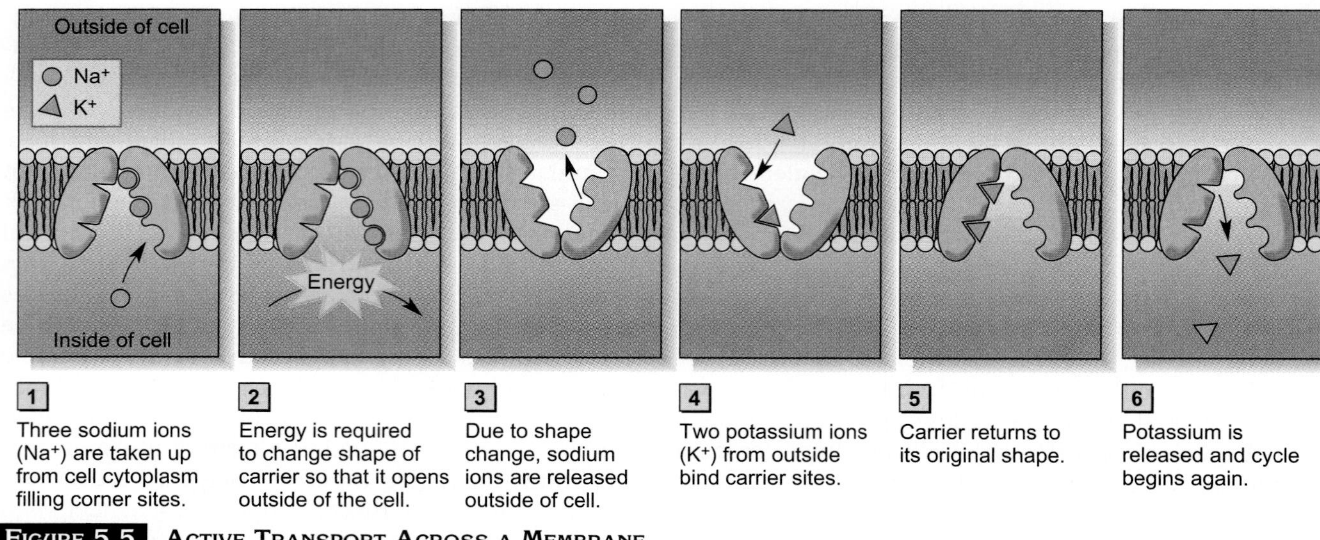

Outside of cell
Na+
K+
Energy
Inside of cell
1
Three sodium ions (Na+) are taken up from cell cytoplasm filling corner sites.
2
Energy is required to change shape of carrier so that it opens outside of the cell.
3
Due to shape change, sodium ions are released outside of cell.
4
Two potassium ions (K+) from outside bind carrier sites.
5
Carrier returns to its original shape.
6
Potassium is released and cycle begins again.

FIGURE 5.5 ACTIVE TRANSPORT ACROSS A MEMBRANE.
A carrier molecule picks up ions such as sodium on one side of the membrane, then rotates and releases them on the other side, where it picks up other ions, such as potassium, for the reverse journey. The pump in this case is called a sodium-potassium pump.

ejects three sodium ions, and creates steep gradients of the two ions. Although much remains to be learned about such pumps, the general principle is described in Figure 5.5.

Active transport mechanisms, such as ion exchange pumps, were once perceived as being rather unusual; but research has revealed that active transport is probably common in most membranes. Researchers have also discovered calcium pumps and special pumps for hydrogen ions.

ENDOCYTOSIS AND EXOCYTOSIS

In some animal cells and in many single-celled creatures, one can actually see materials entering and leaving the cell. If the material is surrounded by the cell membrane as it is being brought into the cell, the process is called **endocytosis** ("inside the vessel"); if the material is being expelled from a membrane sac, the process is called **exocytosis** ("outside the vessel"). In both cases, the membrane behaves similarly (see the discussion of the Golgi body in Chapter 4). These processes can move large molecules into or out of the cell without damaging the membrane.

Endocytosis includes two slightly different processes, **phagocytosis** ("cell eating"), in which solid materials are involved, and **pinocytosis** ("cell drinking"), in which the substances taken in are in solution (Figure 5.6). Pinocytosis is commonly observed in the walls of animal capillaries where bulk materials are moved in and out of the bloodstream. Materials entering capillary cells through pinocytosis are taken in via channels called **pinocytic vesicles,** which, upon filling, pinch off into vacuoles.

Endocytosis and exocytosis were first observed in the feeding of single-celled animals called amebas. When an ameba (also spelled *amoeba*) contacts a food particle, phagocytosis begins as its membrane buckles inward, taking on a cuplike appearance and trapping the particle or prey. The cup then pinches off completely, forming a membranous sac called a **food vacuole.** Digestive enzymes from lysosomes enter the food vacuole and break down its contents. The digested products then diffuse into the surrounding cytoplasm, leaving behind undigested residues.

The ameba's digestive wastes are eliminated through exocytosis. In a reversal of the endocytic events, the waste-laden vacuole fuses with the plasma membrane, and the wastes are expelled. In addition to helping the ameba get rid of what it no longer needs, exocytosis is also a mechanism commonly used for cell secretions in other animals (see Chapter 4).

Phagocytosis is common not only in single-celled creatures but also in simple animals, including sponges, jellyfish, and flatworms. In complex animals it is the mechanism by which white blood cells engulf invading microorganisms and clean up the debris from dead cells (Figure 5.7).

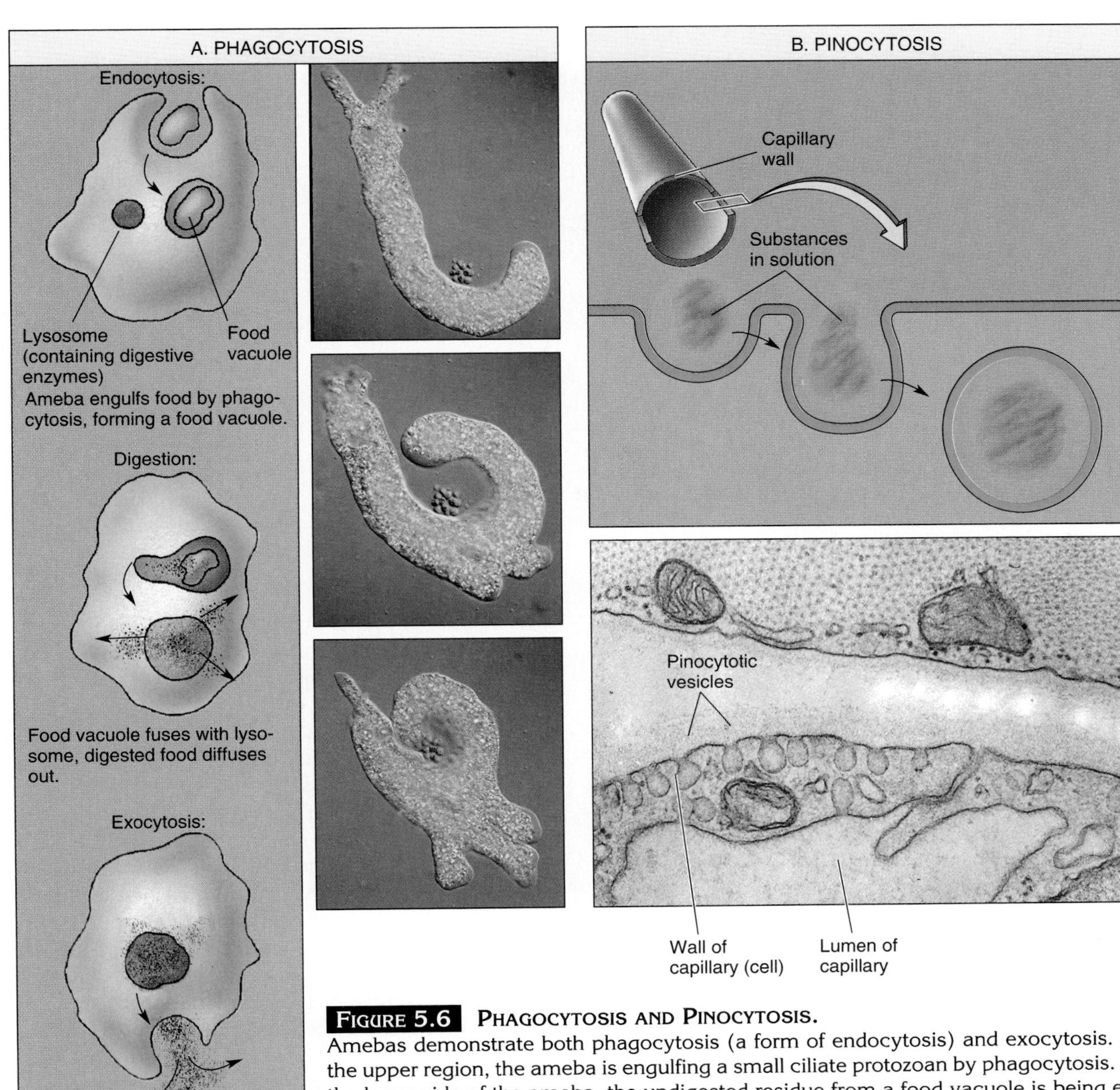

FIGURE 5.6 **PHAGOCYTOSIS AND PINOCYTOSIS.**
Amebas demonstrate both phagocytosis (a form of endocytosis) and exocytosis. At the upper region, the ameba is engulfing a small ciliate protozoan by phagocytosis. At the lower side of the ameba, the undigested residue from a food vacuole is being expelled by exocytosis. Pinocytosis involves engulfing smaller molecules, and can occur at specialized areas in membranes such as those of vascular capillaries, where many substances are absorbed by cells from the blood.

DIRECT TRANSPORT BETWEEN CELLS

One problem in describing cellular processes piecemeal, as we are doing, is that it gives an artificial impression of simplicity. For example, it may seem that cells are isolated entities that simply shift for themselves and maintain independence from surrounding cells. This may be true for one-celled creatures, but for other organisms it is not. In multicellular organisms the cytoplasm of adjacent cells is often in direct contact through special passages, such as the **gap junctions** of animal cells or the **plasmodesmata** of plants (Figure 5.8). Their presence, of course, facilitates intercellular transport (between cells) and increases the efficiency of cellular coordination.

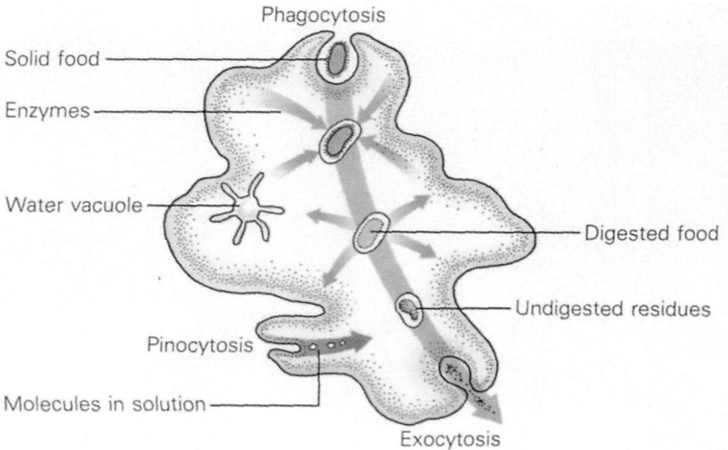

FIGURE 5.7 PHAGOCYTOSIS.
Phagocytosis is seen in a variety of organisms, such as the ameba. Here food enters through phagocytosis, is acted upon by enzymes, and wastes leave the body through exocytosis.

Such cellular passages were once thought to exist in only a few kinds of cells, but they have been found in so many tissues that they are now believed to be the rule rather than the exception. For example, gap junctions are found in human tissues from the heart, epithelial linings, liver, urinary bladder, pancreas, and kidneys. Their presence in heart tissue is believed to help speed impulses that bring about contraction of that tireless organ. Plasmodesmata are also widespread, occurring commonly in leaf cells, sugar-transporting tissues, and root tip cells.

The openings between cells are formed quite differently in animals and plants (see Figure 5.8). The gap junctions of animals are formed by clusters of proteins arranged in circles that surround minute cytoplasmic channels, and, like elaborate rivets, firmly hold adjacent membranes together. The plasmodesmata of plants are long, slender, cytoplasmic extensions that pass from cell to cell through membrane-lined pores in adjacent cell walls.

The constant influx of essential substances and the exodus of manufactured products and wastes are critical aspects of cell function. We have seen a number of such processes at work, but they are actually only a fraction of the processes involving the cell membrane. In later discussions we will find, for instance, that animal cell membranes have highly specialized receptors that bind with chemical messengers, prompting a cascade of responses (Chapter 36). In addition, we will learn of one of life's most fundamental membrane responses in our discussion of the surprisingly active participation of the egg membrane when nudged and prodded by a lashing sperm (Chapter 42). The stage for such discussions, however, has been set by what we've seen here.

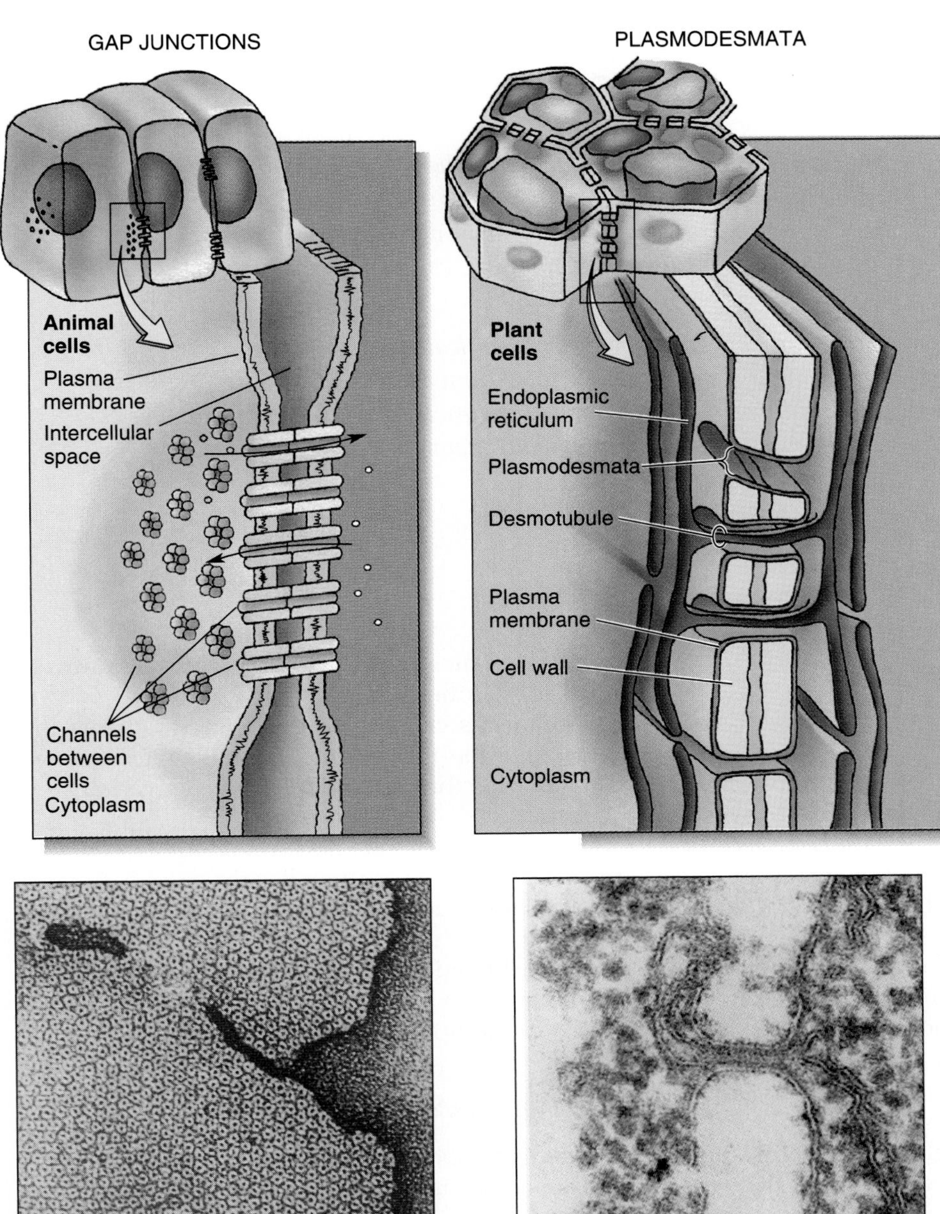

FIGURE 5.8 JUNCTIONS BETWEEN CELLS.
Materials move between certain animal cells through pores called gap junctions. Each pore is formed through the special assembly of proteins. The assemblies in each membrane are arranged back-to-back, creating a minute pipeline connecting the cytoplasm of the two cells and squeezing the two membranes together. The presence of porelike plasmodesmata between plant cells permits the ready passage of materials from one cell to the next. The cytoplasmic connection often includes desmotubules—tubelike segments of endoplasmic reticulum that pass between the cells.

PERSPECTIVES

We have seen that the cell membrane—the boundary layer between the living cell and the nonliving environment—is a surprisingly active entity. We have noted some of the physical rules that govern the movement of molecules into and out of cells, across that membrane. In later chapters we will consider additional roles for the cell membrane, especially when we discuss cell division and growth.

In the next chapters we turn our attention to other chemical reactions that occur within cells, chemical reactions that provide the energy that allows living things to remain organized. Some of these reactions occur within membrane-bounded organelles, and involve proteins attached to and embedded in membranes. So don't forget about membranes as we proceed.

SUMMARY

PASSIVE TRANSPORT

Passive transport is the movement of substances across the plasma membrane without an investment of the cell's own energy. It is based on random movement caused by heat. Passive transport includes diffusion, facilitated diffusion, bulk flow, and osmosis.

Diffusion is the net movement of molecules down their concentration gradient, from regions of higher to regions of lower concentration, until uniformity is reached. Diffusion rates depend upon temperature, concentration gradient, particle size, and charge.

Plasma membranes are selectively permeable, admitting some substances freely, slowing the passage of some, and excluding others. Small uncharged particles such as gases cross readily, while charged particles may be slowed. With some important exceptions, membranes readily admit water.

Water potential—the tendency of water to move—is affected by gravity, pressure, and solute concentration. Bulk flow is the movement of fluids in which all the molecules move in one direction.

Osmosis is the diffusion of water across a semipermeable membrane from areas of greater to areas of lesser water potential. In osmosis, water potential is determined by the concentration of solute inside or outside a membrane. Osmosis continues until it is blocked by counterforces, such as gravity.

Organisms create useful osmotic gradients that encourage the movement of water where needed. They do this through the selective, active transport of solutes.

Isotonic, hypotonic, and hypertonic are relative terms. An isotonic solution has the same concentration of solutes as a cell; hypotonic solution has less, hypertonic has more.

In facilitated diffusion, carrier proteins in the membrane move selected molecules and ions down the concentration gradient.

ACTIVE TRANSPORT

In active transport, the cell's energy is used to move substances against a concentration gradient.

Membrane pumps are proteins embedded in plasma membranes that move substances by active transport. The best-known membrane pump, the sodium-potassium ion exchange pump, uses ATP energy to eject sodium ions and take in potassium ions. Other pumps include those that move calcium, protons, and chloride ions.

In endocytosis, segments of membrane actively trap materials and move them into the cell. The intake of solids is phagocytosis, while that of material in solution is pinocytosis. Digestive residues are carried to the membrane and expelled through exocytosis.

Direct Transport Between Cells

Gap junctions and plasmodesmata form passages between cells in animals and plants, respectively. The passages greatly facilitate transport and communication in many kinds of tissues.

Key Terms

passive transport • **75**
active transport • **75**
diffusion • **75**
permeability • **76**
selectively permeable • **76**
water potential • **76**
bulk flow • **77**
osmosis • **77**
osmotic pressure • **78**
turgor pressure • **78**
wilting • **78**
isotonic • **78**
hypotonic • **79**
hypertonic • **79**
facilitated diffusion • **79**
carrier • **79**
sodium-potassium ion exchange pump • **79**
endocytosis • **80**
exocytosis • **80**
phagocytosis • **80**
pinocytosis • **80**
pinocytic vesicle • **80**
food vacuole • **80**
gap junction • **81**
plasmodesma • **81**

Review Questions

1. Explain how the source of energy differs in passive and active transport.
2. Explain the physical basis for molecular movement. How can such movement have a net direction? At what point does diffusion cease?
3. List three factors that affect the rate of diffusion.
4. List several characteristics of the plasma membrane that might affect the passage of molecules or ions.
5. Why is it that small, nonpolar molecules pass so readily through the phospholipid layers of the membrane?
6. List three factors that affect water potential. How does water potential affect the direction in which water moves?
7. How does bulk flow differ from diffusion?
8. Using the following terms, write a concise definition of osmosis: water potential, solute concentration, semipermeable membrane, diffusion.
9. Contrast the effects of hypertonic and hypotonic solutions on plant and animal cells. In what way do plants rely on turgor pressure?
10. Why is it important for biologists to maintain cultures of cells or tissues in an isotonic solution?
11. List two ways in which one could determine whether active transport was going on.
12. Summarize the steps followed in a complete cycle of the sodium-potassium ion exchange pump. Where, in our own bodies, are such pumps located?
13. Explain the role played by the plasma membrane in endocytosis. Contrast the way solids and fluids are taken in through endocytosis, and name the processes.
14. What happens to the plasma membrane when exocytosis occurs? Of what value is exocytosis to an ameba?
15. Briefly describe the structure of gap junctions and plasmodesmata, and explain how they are important to animals and plants.

6

Energy and the Cell

WE HAVE SEEN THAT THE CELL IS AN ORGANIZED ENTITY. IT HAS SHAPE, *structure, and functions—and all of these things take energy to maintain. Where does that energy come from? And how is that energy used to maintain biological form and function? These are the questions we will consider in the next three chapters. We begin our discussion in this chapter by considering just what energy is, and how it is formed and stored.*

We humans often seem to enjoy maligning ourselves, being particularly fond of stressing our unusual savagery. But if humans did not walk the earth, would the planet be a gentler place? Probably not. After all, *life demands energy,* and many living things must derive energy from other living things. The problem is, if an organism wishes to harvest the energy stored in another organism's body, that body must be disrupted, damaged, and, very likely, killed. The unending search for energy can indeed be brutal. Those organisms seeking energy and those avoiding being exploited have developed many ways to carry out these tasks. Dainty plants growing silently on a flower-strewn hillside may relentlessly engage in a battle for survival as they compete for the sun's rays and the earth's minerals. Yet after they have been blessed by the sun and are able to manufacture food, they must often yield that sequestered energy to some casual grazer. In turn, the grazers are likely to fall prey to some sharp-toothed carnivore seeking the energy held in the grazers' bodies (Figure 6.1), energy previously derived from those hillside plants. Eventually, though, even muscular creatures with sharp teeth answer the ultimate call, and the energy they once stored becomes the salvation of small microbes as they break down a ponderous corpse.

We will return to flow of energy through food chains in later chapters. Here our interest is in the energy itself.

ENERGY

Energy is needed because, in this world, matter tends to become disorganized. We see this in a rotting corpse of a rabbit beside the road. That body once had been highly organized in the fashion of rabbits. Now, though, its molecules are being broken down and scattered in a final disorganization.

Life can exist only when molecules resist that tendency and remain organized. The organization, though, is only possible with the expenditure of energy. In a nutshell, this is why any consideration of life must involve the nature of energy.

What are the characteristics of energy? How does it behave? The concept of energy is extremely elusive, but we can begin with a simple definition: **Energy** is the ability to do work. Work, as physicists tell us, is the movement of mass against an opposing force. This work, this energy expenditure, keeps the processes of life organized.

POTENTIAL AND KINETIC ENERGY

Energy essentially exists in two states—*potential energy* and *kinetic energy* (Figure 6.2). **Potential energy** is stored energy: It is not doing anything. For example, a huge boulder may rest on a hill above a house. This boulder represents a considerable store of potential energy (although the homeowner may describe the situation differently, especially if the house is paid for). The boulder may have gained its potential energy long ago as it was raised by some great, geological upheaval or glacier.

Should the soil around the boulder give way, we would quickly learn that potential energy can give rise to *kinetic energy.* **Kinetic energy** is energy in motion. As the boulder rolled down the hill, its effect on the house would attest to its potential energy being transformed into kinetic energy.

The behavior of energy has been described in highly reliable, time-tested observations known as the *laws of thermodynamics.* These physical principles apply equally to the living and nonliving worlds.

FIGURE 6.1 **PART OF THE FIGHT FOR LIFE IS OVER ENERGY.**
A grizzly and a young cougar fight over an elk kill. Energy is passed from one organism to the next, oftentimes through dramatic predator-prey interactions.

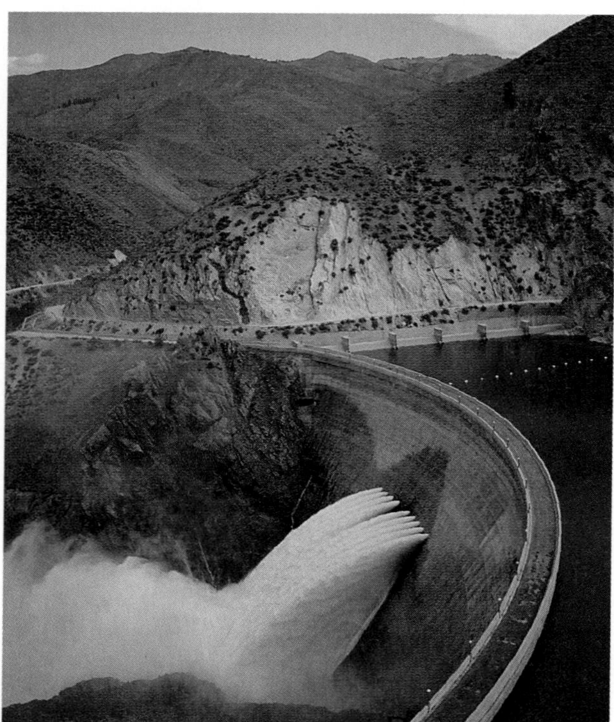

FIGURE 6.2 **ENERGY TRANSFORMATIONS.**
Energy can be stored and can change from one form to another. As water falls to a lower level, its potential energy becomes kinetic energy. In this example, the kinetic energy of the falling water turns electrical generators, transforming mechanical energy into electrical energy, which can then be transformed into light and other forms of energy.

THE LAWS OF THERMODYNAMICS

Living systems involve matter and chemical processes, and the laws of physics and chemistry are as valid in living systems as anywhere else. Two of the laws of thermodynamics are essential to our understanding the incredibly complex, delicate, and sensitive processes of life.

THE FIRST LAW

The **first law of thermodynamics** states that *energy can neither be created nor destroyed.* This means, simply, that the total amount of energy in the universe remains constant. Energy cannot be created or destroyed, but it can change form and be channeled in different directions. The earth receives energy from the sun and radiates some energy back into space. The total energy on the planet may remain relatively constant, so that the earth is in a *steady state.* But it is a dynamic steady state. All activity occurs because energy can change from one kind to another, it can be stored and released, and it ebbs and flows in countless directions, changing the face of the planet.

Energy can change in a number of ways. To illustrate, energy is locked in the chemical bonds of gasoline. When the fuel combines with oxygen and a spark inside an engine, the gasolene's chemical bond energy is transformed into the energy of heat, noise, and motion. That is, the potential energy of gasoline becomes kinetic energy expressed as heat. The sudden release of heat expands gases in the cylinder, forcing a piston to move a shaft that causes the machine to roll over your foot. When this happens, certain energy shifts may occur in your own body, and the sound you generate (and later regret) represents chemical energy that was changed to the mechanical energy of vibrating air we know as noise. Once released, much of the energy in such an engine (or in you) is dissipated as useless heat, which brings us to the next law.

The Second Law

The **second law of thermodynamics** states that *energy transfer leads to less organization.* Expressed another way, systems tend to become more random and disorganized as time goes on. This second law is sometimes called the "law of entropy." **Entropy** is a measure of the tendency toward disorganization.

The two laws make it clear that organized systems—those with a complex molecular nature and great amounts of potential energy—change, in time, to a simpler, more disordered or randomized molecular state (Figure 6.3).

The concept of entropy can also be illustrated by fuel in an engine. Gasoline molecules are long hydrocarbon chains. They possess both great molecular organization and abundant potential energy in that many atoms are precisely bonded to one another. As the engine alters those complex molecules, combining them with oxygen and releasing stored energy, the lengthy molecular chains end up as simple carbon dioxide and water, their organization and potential energy virtually depleted. Essentially, the potential energy is dissipated as heat. One can say, then, that both matter and energy have reached a state of maximum entropy. There is no way to convert all of that heat to useful potential energy again; therefore, the increase in entropy is irreversible. Nevertheless, as we will see, useful work can be accomplished as entropy is increased—that is, as systems become more disorganized.

Figure 6.3 Death and Loss of Organization. Organisms maintain their complex molecular state by taking in energy. When energy input ceases and the organism dies, the molecular organization yields to the inexorable second law of thermodynamics. Entropy increases as the remains yield to the unpleasantness of decay. Decay is accomplished by physical, chemical, and biological processes. Here we see the remains of a large mammal—and a dead ant that is being degraded by a growing fungus.

Thermodynamics and the Delicate Processes of Life

Now we will see how work can be accomplished as energy changes form and entropy increases. Every movement we make, every thought we generate, and every chemical reaction within our cells involves a quiet shift in energy states and a measurable loss of energy.

A biologically vital application derives from the second law, and it may be evident to you by now. Transfers of energy in life—shifts from one kind of energy to another or from one organism to another—are never completely efficient; with every transfer some energy is lost, having been converted to an unusable form. For this reason, the myriad chemical activities within our own bodies require that considerably more energy be taken in than is necessary to simply carry out the reactions. This is because much of the energy ends up as heat (one of the reasons we have warm bodies). The transfer of energy from one group of organisms to the next (from the consumed to the consumers, in what ecologists call food chains) is always less than perfect, with less than 10% of the energy available to work after such a transfer.

Chemical Reactions and Energy States

We sometimes find the laws of thermodynamics at work in startling ways. Consider the simple case of combining hydrogen gas and oxygen gas (Figure 6.4), the results of which can be dramatic. A virtual explosion takes place, and water is produced.

Why is the water formed so explosively? Because of energy states. Water (H_2O) is at a lower energy state than an equivalent amount of H_2 and O_2. Thus, a mixture of molecular oxygen and hydrogen contains more potential chemical energy than does water. When, for example, the hydrogen in the zeppelin *Hindenburg* burned, the chemical mix of the hydrogen in the zeppelin and the oxygen in the air moved to a lower energy state, producing water and releasing the excess energy as heat and light.

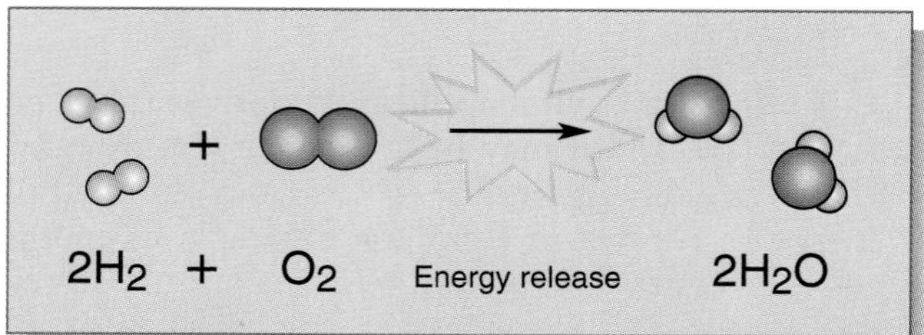

Figure 6.4 The Hydrogen-Filled Zeppelin *Hindenburg* Exploded in 1937.
Hydrogen is so reactive that the smallest spark or flame can provide the impetus for a rapid chain reaction in which hydrogen combines with oxygen to form water.

So, oxygen and hydrogen will combine explosively to produce energy and water, but if you tried to demonstrate the principle by simply mixing oxygen with hydrogen, nothing would happen. Molecular hydrogen and molecular oxygen tend to remain as they are, with their outer orbitals nicely filled. What now? Perhaps a spark of intuition.

With a tiny spark, the mixture will blow your hat off. But why doesn't the mixture blow up without the spark? Molecular oxygen and molecular hydrogen do not combine with each other at room temperature because they must first be energized. The atoms making up each molecule must separate and then rejoin violently to be able to react. The spark provides the energy that forces the molecules apart, allowing the reaction. The atoms separated from each other now have free electrons; that is, the electrons are no longer tied up in covalent bonds. Once the first molecules are disrupted, they are able to enter into chemical reactions and release energy. This energy then provides the energy to separate the atoms of other molecules. The spark sets off a chain reaction, and all of the molecules in the system quickly separate—and then rejoin to form water.

Any reaction in which energy is released and the products end up with *less* energy than the reactants is called an **exergonic reaction.** The violent reaction between hydrogen and oxygen is a good example, as is the combustion of gasoline in the engine mentioned earlier. Exergonic reactions also occur as foods are oxidized in mitochondria of cells, thereby releasing the energy stored in their chemical bonds. In contrast, **endergonic reactions** create products that have *more* energy than the reactants. Plants carry on endergonic reactions when they use carbon dioxide and water to form sugars. The first law of thermodynamics should alert us that some outside energy source spurs endergonic reactions. For plants, that outside energy source is virtually unlimited: It is sunlight (see Chapter 7).

Activation Energy and Enzymes

The energy required to start a reaction is called **activation energy** (Figure 6.5). Hydrogen and oxygen can combine to form water at room temperature if a spark provides the energy of activation. Such reactions can also proceed by the addition of **catalysts**—substances that speed up chemical reactions

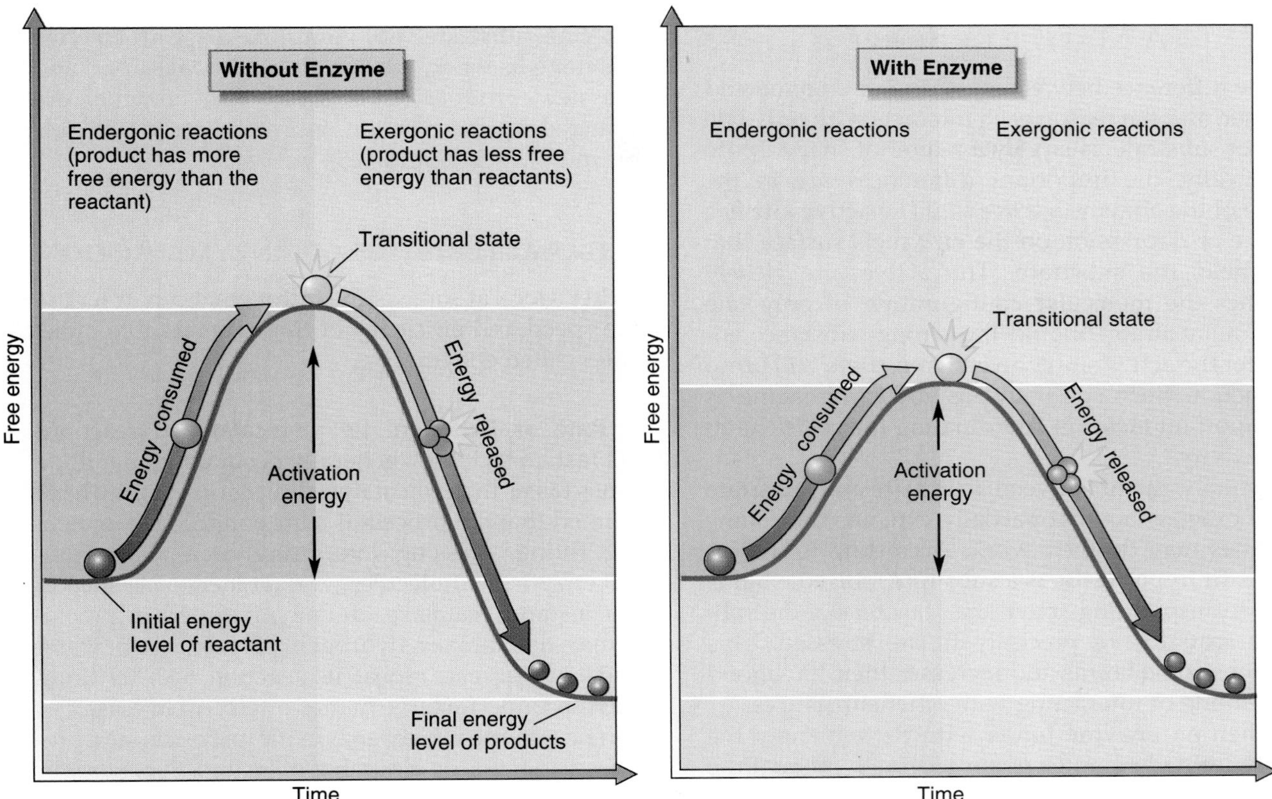

FIGURE 6.5 ACTIVATION ENERGY AND ENZYMES.
The energy required to initiate a chemical reaction is called the activation energy. The reactant will remain inactive until its energy level is raised, much like the balls here that need to be pushed up the hill in order to fall to a lower energy level on the other side. The important role of enzymes is to lower the activation energy necessary to begin the reaction. Note that both with and without enzymes, the products end up at a lower free-energy level than the initial reactant.

but are unchanged by the reactions. Catalysts work by lowering the required activation energy. For example, hydrogen and oxygen will combine explosively if powdered platinum is present as a catalyst. (In the case of this catalyst, the hydrogen first combines with the platinum and then with the oxygen, leaving the platinum in its original state, unchanged by the reactions.)

ENZYMES: BIOLOGICAL CATALYSTS

In the highly orchestrated interior of a cell, the use of intense heat would be a disruptive way to initiate biochemical processes. Instead, intricately coordinated and somewhat vulnerable living systems depend on the gentler activity of *enzymes.* Actually, **enzymes,** the catalysts of cells, are a special class of biologically active, globular proteins that accelerate biochemical reactions. As soon as they interact, they emerge unchanged, ready to interact again—cycling with dazzling speed. Many different enzymes comprise any cell's biochemical arsenal, but each enzyme functions in only one kind of reaction. That is, it recognizes and interacts with its own specific *substrate.* In the language of biochemistry, a **substrate** is simply a substance that can be chemically altered by an enzyme.

We have now mentioned three important characteristics of enzymes: (1) They are proteins, (2) they are highly specific in their substrates, and (3) they decrease the level of activation energy necessary to bring about chemical reactions. Figure 6.5 compares the required activation energy with and without the intervention of an enzyme. Let's look further into the factors behind an enzyme's specificity.

A Matter of Shape

The difference between one kind of enzyme and another, and the reason each interacts with only one kind of substrate, is largely a matter of shape. More specifically, the important differences are in the shape of the enzyme's *active site.* The **active site** is a groove or depression on the enzyme's surface that will hold the substrate. The active site closely matches the molecular configuration of only one kind of substrate. Should the wrong substance encounter the active site of an enzyme, there will be no interaction. Such specificity, as you can imagine, is an important factor in coordinating the cell's chemical activity.

Actually, the fit between active site and substrate is not exact, a fact that partially explains how some enzymes may do their work. According to the **induced-fit hypothesis,** as a substrate molecule binds to its corresponding active site, the areas of the substrate that do not precisely fit are stressed. This weakens some bonds and increases their likelihood of breaking or interacting with other substances.

When an enzyme binds with its substrate, the newly formed complex is appropriately named the **enzyme-substrate (ES) complex.** After an enzyme has done its work, the ES complex breaks apart and the new product drifts away. The enzyme, unchanged by the reaction, is immediately available for more work (Figure 6.6).

Characteristics of Enzyme Action

Let's look at some of the factors that can influence the speed and efficiency of these remarkable molecules called enzymes.

Rate of Reaction. Enzyme-catalyzed reactions are fast; in fact, they generally occur about a million times faster than uncatalyzed reactions. It may have seemed that the process is a bit ponderous—matching, fitting, stressing, interacting, breaking away—but enzymatic molecules work at incredible speeds. For instance, catalase, an enzyme found in peroxisomes that breaks hydrogen peroxide into water and oxygen, can repeat its reaction 600,000 times each second. (Enzyme names often end in "-ase.")

The rate at which enzymatic products are produced depends on a number of factors. For example,

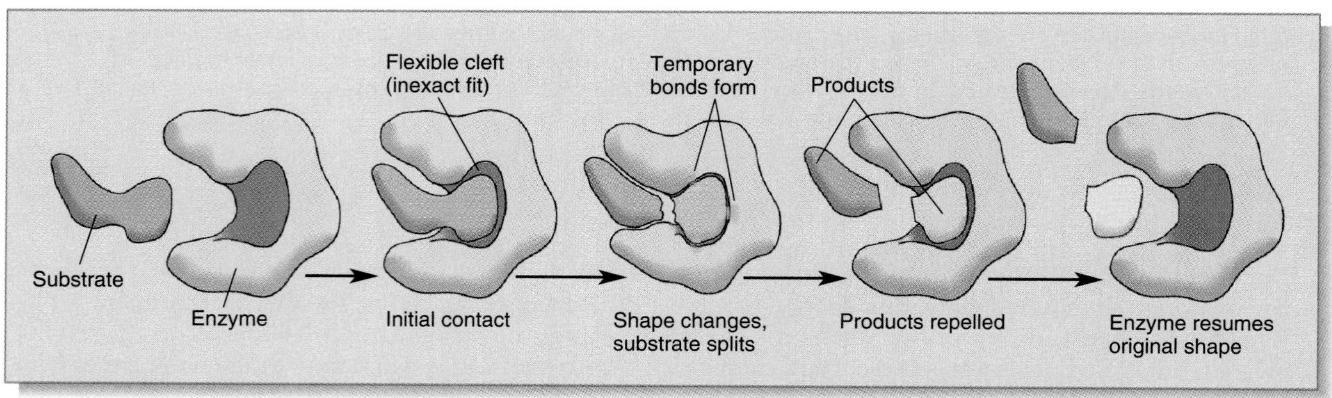

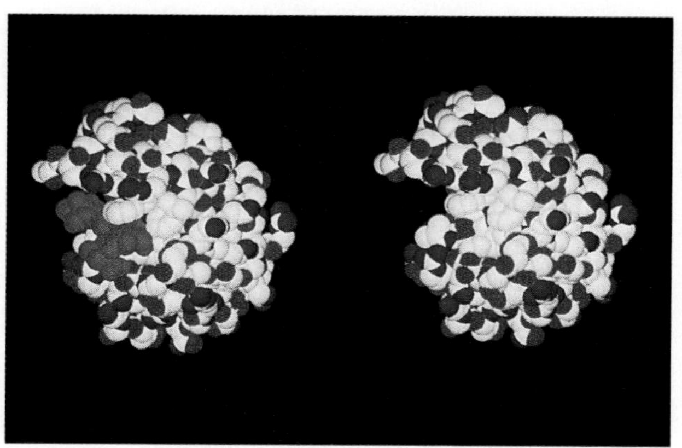

Figure 6.6 The Action of an Enzyme. A substrate entering the active site of an enzyme forms an imperfect fit, held in place chiefly by hydrogen bonds and other subtle forces. The enzyme changes shape slightly, placing stress on the substrate at key points and cleaving it into products. After cleavage, the enzyme returns to its original form. These computer-generated space-filling models show that the fit of enzyme and substrate really is based on the forms of the molecules, as the purple substrate *(left)* fits nicely into the cleft of the enzyme itself *(right).*

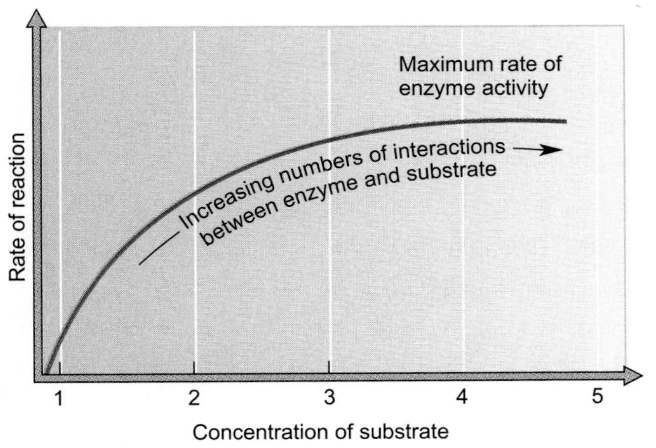

FIGURE 6.7 **REACTION RATE VS. SUBSTRATE CONCENTRATION.**
This shows the rate of reaction when the quantity of enzyme remains constant while the concentration of substrate is increased. The reaction rate increases until such time as the amount of enzyme becomes limiting; then no matter how much more substrate is added, the rate of reaction remains the same.

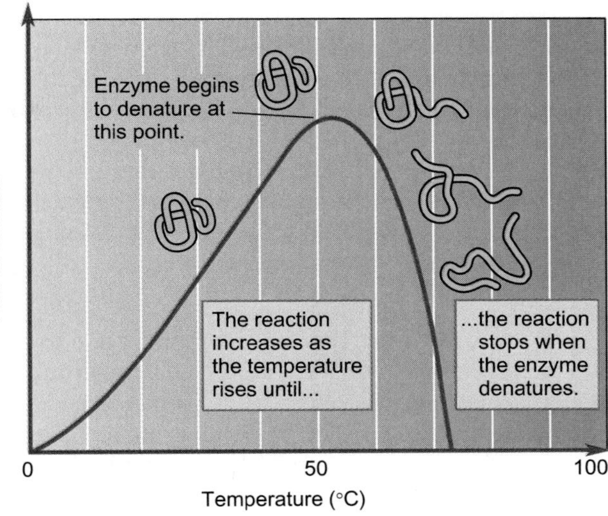

FIGURE 6.8 **REACTION RATE VS. TEMPERATURE.**
As thermal energy (heat) is added to an enzyme-substrate mixture, the rate of the reaction increases. However, too much heat can denature an enzyme, altering its properties and rendering it inactive.

products may be formed faster when more substrate is present. There is no mystery about this, since an enzyme and its substrate join purely by chance collision. Naturally, when more substrate molecules are present, the chances of collision increase. But eventually, enzymes become saturated with substrate. Increasing substrate even further without increasing enzyme concentration will not increase the rate of formation of the product (Figure 6.7).

The Effect of Temperature. As a rule of thumb, an increase in temperature of 10°C doubles the rate of chemical reactions. Furthermore, the rule generally applies to enzyme activity (Figure 6.8). The effect of increased temperature in such cases is easy to understand. *Heat* is molecular motion. An increase in the movement of substrate molecules and enzyme increases the chances of their collision. Of course, there is a limit to how high the temperature can get without damaging the enzyme. Enzymes, which are proteins, are said to be "*denatured*" when altered or destroyed by high temperatures.

The Effect of Acids and Bases. The pH (see Chapter 2) within a cell can also influence enzyme activity. This is because each enzyme operates at an optimum pH. A few enzymes perform best in strongly acidic surroundings, but most require a more neutral condition (Figure 6.9). Apparently, improper levels of acidity can interfere with the very specific folding of enzymes. In some instances, an enzyme's actions can even be reversed by changes in the acidity of a cell. For example, the same enzyme that helps synthesize glycogen from glucose at a high pH will reverse the process at a low pH, breaking glycogen down to glucose.

Teams of Enzymes: Metabolic Pathways. Any enzymatic reaction is likely to be only one simple link in a long sequence of reactions that keep the cell

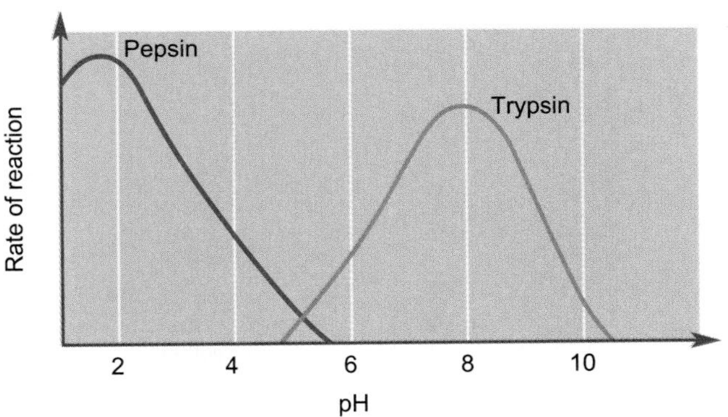

FIGURE 6.9 **ENZYMES ARE SENSITIVE TO PH.**
Pepsin, the protein-digesting enzyme of the stomach, is inactive in all but fairly strong acidic conditions (low pH). Trypsin, a protein-digesting enzyme of the small intestine, requires an alkaline environment. Most enzymes work optimally at near-neutral (pH 7) conditions.

alive and functioning. This sequence is called a **metabolic pathway** (Figure 6.10). In such chains, each product becomes the starting material for the next reaction. Clearly, such an interdependent system must be very organized.

Some metabolic pathways involve the breakdown of molecules. Such processes are called **catabolic.** Pathways that build molecules by joining smaller ones together are called **anabolic.**

Enzymes and Cofactors. Many enzymes are assisted in their work by agents known as **cofactors,** which are ions or molecules that must be associated with an enzyme in order for the enzyme to function properly. Cofactors include inorganic ions such as iron (Fe^{2+}), manganese (Mn^{2+}), and zinc (Zn^{2+}). Others are complex organic molecules known as **coenzymes,** many of which are actually modified vitamins.

Allosteric Sites and Enzyme Control. In addition to its active site, an enzyme may have an **allosteric site,** a second binding site that, when filled, changes the shape of the enzyme's active site. The result is that the enzyme will no longer bind to its usual substrate. The allosteric site is sometimes specific for one of the enzyme's products (Figure 6.11). In this way, as a product accumulates, the allosteric sites will fill, inhibiting more enzymes from producing even more of the product. Often, enzymes with allosteric sites will be part of a team of enzymes that act sequentially on the same substrate. It is important to know that a molecule doesn't permanently bind to the allosteric site. As soon as the unbound product molecules become scarce—used up or enzymatically altered to a new product—the allosteric sites remain unfilled, and the enzyme can again act.

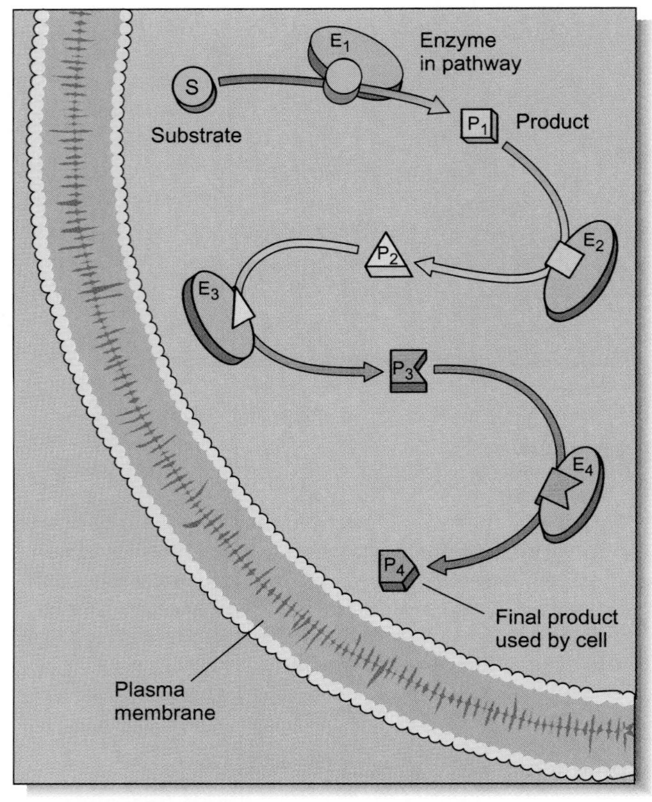

FIGURE 6.10 A METABOLIC PATHWAY.
This is a pathway in which several enzymes work, one after another, to convert a substrate into its final product.

The operation of the allosteric site is an example of an important type of biological regulation called

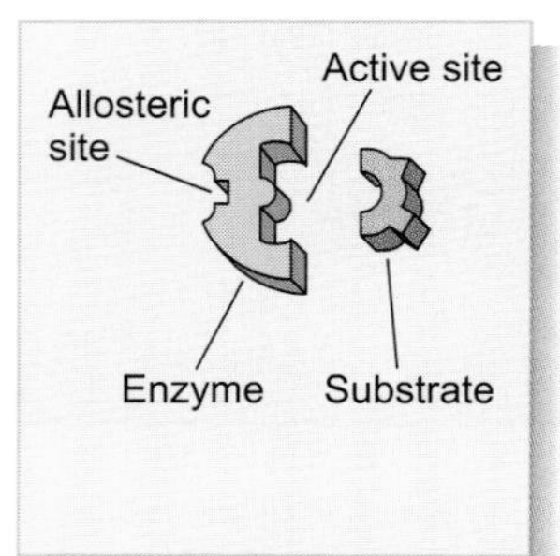

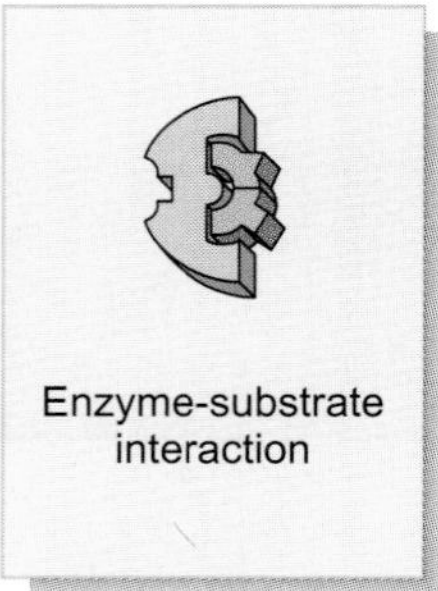

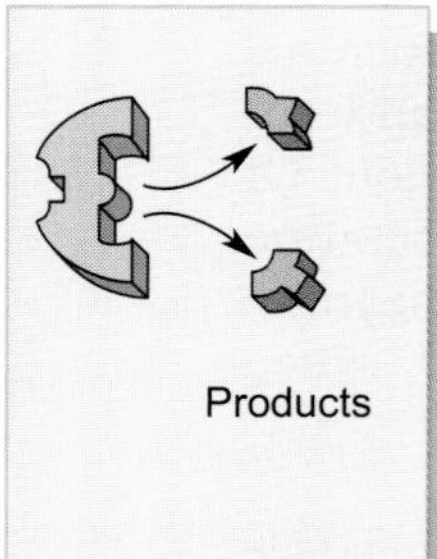

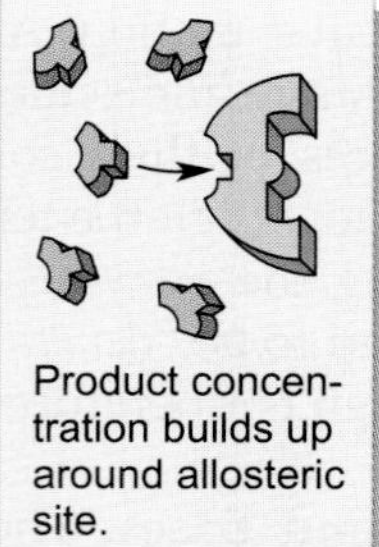

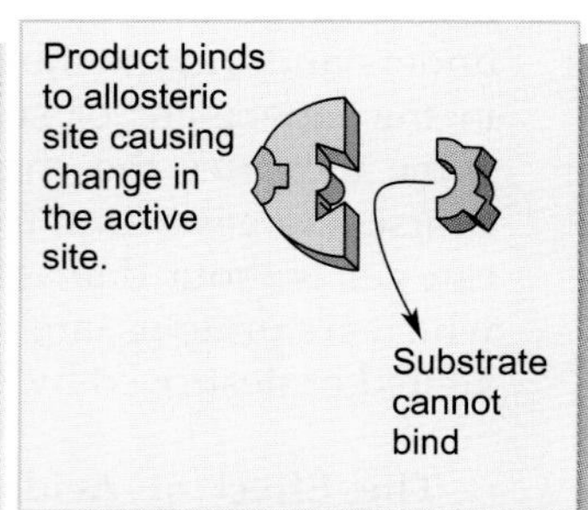

FIGURE 6.11 SOME ENZYMES HAVE TWO DIFFERENT BINDING SITES.
The active site binds with the substrate molecule as usual, but the allosteric site can bind with a different molecule, here one of the products of the reaction. When the allosteric site becomes occupied, the shape of the enzyme changes enough to render it incapable of forming an enzyme-substrate complex. This provides a built-in means to slow the enzyme down when the products reach a high concentration.

negative feedback. We will consider it in detail later, but the concept bears mentioning here. In negative feedback, the product of an action inhibits the action that produced it in the first place. As the amount of product increases, more of the enzyme is inhibited from producing more product. Such regulatory mechanisms are elegant in that they are both simple and automatic. They are also vital to the overall process of maintaining constant internal conditions (Chapter 37). Here we see that molecular mechanisms exist that keep enzymatic reactions operating within a defined set of limits. Never too much, never too little.

We are all aware that cells must have energy to do their work. So now we will be introduced to a fascinating molecule called adenosine triphosphate (ATP). This simple molecule is intimately involved with the cell's ability to carry out its tasks.

ATP: The Energy Currency of the Cell

Adenosine triphosphate (ATP) has been called the cell's "energy currency," because it can be "spent" to get things done. The more energy that is required, the more ATP that must be spent.

ATP (Figure 6.12) is particularly interesting because it is used by all forms of life; it is a nearly universal molecule of energy transfer. Energy produced in photosynthesis or respiration is eventually stored in ATP. When that energy is needed, it is released by ATP. You may recall that cells can store energy in molecules such as carbohydrates, lipids, and proteins. But before energy can be retrieved from such molecules, it must first be used to synthesize ATP.

To illustrate the activity of ATP in our daily lives, consider this: The average adult male uses about 190 kilograms (419 lb) of ATP each day. Since few of us even weigh 190 kilos, we obviously don't have that much ATP on hand. In fact, our bodies contain only about 50 grams (1.8 oz) of ATP at any time. This means that this small amount must recycle furiously, as spent ATP is recharged.

Structure of ATP

The secret of ATP's abilities lies in its structure (Figure 6.12). **Adenosine triphosphate** consists of three parts: a double ring of carbon and nitrogen called **adenine;** a simple, five-carbon sugar called ribose; and three phosphate units (triphosphate),

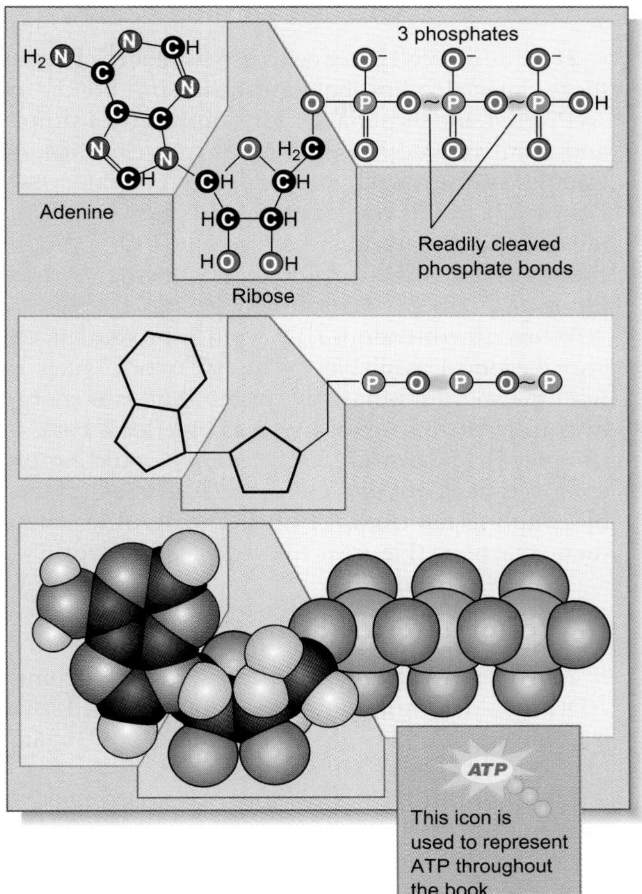

FIGURE 6.12 REPRESENTATIONS OF ADENOSINE TRIPHOSPHATE (ATP).
ATP consists of adenine covalently bonded to ribose, which is in turn covalently bonded to three phosphates. The last two contain energy-rich phosphate-to-phosphate bonds. We will often use a somewhat fanciful icon to show the presence of ATP in a reaction, without having to show all the details of its structure.

which form a kind of tail. The phosphates are linked together by oxygen atoms.

The Phosphate-to-Phosphate Bond. Notice in Figure 6.12 that the phosphates forming ATP's tail are linked by bonds shown as wavy lines. These wavy lines indicate bonds that are readily cleaved to produce energy. Their energy is released when a phosphate bond is broken and the inorganic phosphate ion, P_i, is released. Such bonds can be formed only by enormous energy input, which means that a great deal of energy is released when those bonds are broken.

ATP and Cellular Chemistry

How does a cell manage to use the energy held in those special phosphate-to-phosphate bonds of ATP? First, the terminal bond is broken by a simple and familiar process called hydrolytic cleavage, or *hydrolysis*—the enzymatic addition of water (see Chapter 3). In this way, the terminal phosphate is removed and replaced by a hydroxyl (—OH) group, changing the ATP to **ADP, or adenosine diphosphate** (*di*, "two").

When a high-energy ATP bond is broken, its energy becomes available for useful work, such as with the sodium-potassium pump that uses energy to transport ions against a gradient (look back to Chapter 5). Unharnessed, that energy would simply escape as heat, but the cell is prepared to avoid this, by coupling the cleavage of ATP with other enzymatic reactions (Figure 6.13) to do cellular work.

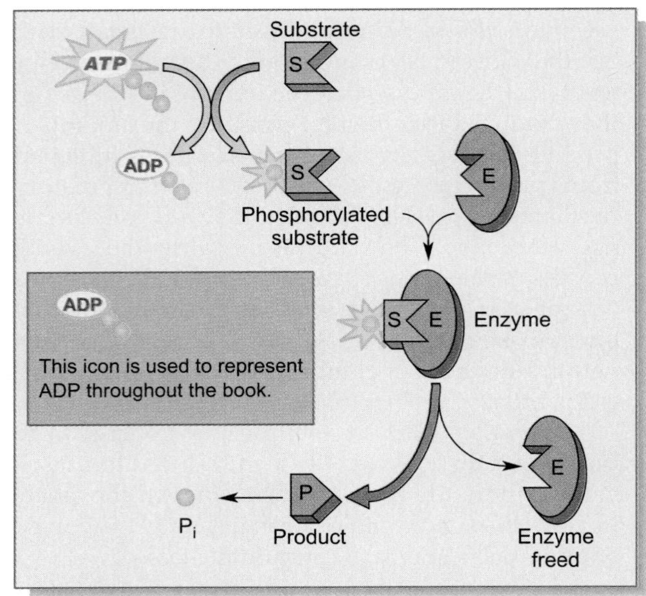

Figure 6.13 **ATP Cleavage.**
Interaction between ATP and a substrate molecule results in cleavage of ATP's terminal phosphate. The energy of the phosphate molecule increases the energy in the substrate, sometimes by being attached to the substrate. ATP can also enter into reactions and be cleaved to ADP without transfer of the phosphate to the substrate.

The ATP Cycle

As ATP loses one phosphate ion (P_i) and becomes ADP, the ADP molecules are simply recycled; that is, they regain phosphate, becoming ATP again, ready to supply the cell's energy.

The problem here is obvious: Because breaking those phosphate bonds *releases* a great deal of energy, making the bonds *requires* a great deal of energy. That energy is made available through two vital processes: cellular respiration and photosynthesis—processes that in mitochondria and chloroplasts. ATP molecules constantly lose phosphates (releasing energy) and regain them (requiring energy) in one of the most vital cycles of life (Figure 6.14).

We have taken a look at two kinds of molecules that are important to cellular energetics—enzymes and ATP. Their roles are closely intertwined, since enzymes are catalysts for chemical reactions and ATP can provide the energy for those reactions. Now let's consider those operating partners of many enzymes, the coenzymes.

Coenzymes, Oxidation, and Reduction

In order to understand how high-energy ATP is generated from ADP, we need to know something about the movement of electrons by a special group of cofactors called coenzymes.

Coenzymes work in close association with enzymes and substrate. While there are many kinds of coenzymes, our interest here is in three that are directly involved in making ATP. Two of these are **nicotinamide adenine dinucleotide (NAD)** and **nicotinamide adenine dinucleotide phosphate (NADP).** The third is **flavin adenine dinucleotide (FAD).**

Structure of NAD$^+$, NADP$^+$, and FAD

Note in Figure 6.15 that NAD and NADP are chemically similar to ATP: There is a nitrogen base (adenine), two ribose sugars, and a number of phosphates (NADP has one more phosphate than NAD). These molecules also have a nitrogen-containing ring called *nicotinic acid*, which is the chemically active part of these two coenzymes. In FAD, the nitrogen-containing ring is called *riboflavin*. Both nicotinic acid (also called *niacin*) and riboflavin are derived from B vitamins. In fact, as we go through the role of these coenzymes, you may gain a new respect for vegetables.

Although each of these coenzymes works in its own way, they all do essentially the same thing.

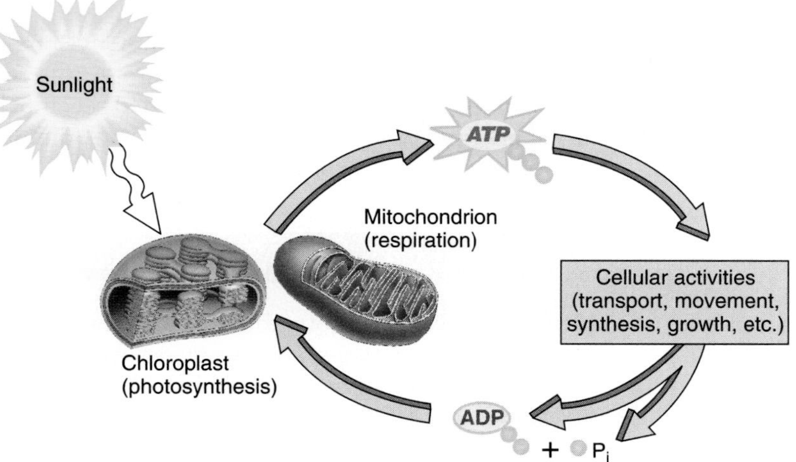

FIGURE 6.14 **ATP AND ENERGY RELEASE.**
As ATP provides the energy for cellular activities, it is broken down during cellular activities to ADP and P_i. These, in turn, are in effect recycled by the actions of the chloroplasts and the mitochondria, whereby they again become ATP.

Working closely with enzymes, they accept electrons and transfer them to other molecules. Chemically, the removal of electrons from a substrate is called **oxidation.** The addition of electrons to a substrate is called **reduction.**

As oxidizing enzymes strip electrons (sometimes along with protons) from a substrate, they pass them to coenzymes. Thus, when a substrate loses electrons, a coenzyme gains them: As the substrate is oxidized, the coenzyme is reduced. Coenzymes don't remain reduced very long; they quickly pass their newly acquired electrons to another substrate, thus reducing it. In their oxidized state, the three coenzymes mentioned above are written as NAD^+, $NADP^+$, and FAD. In their reduced state they become $NADH + H^+$, $NADPH + H^+$, and $FADH_2$.

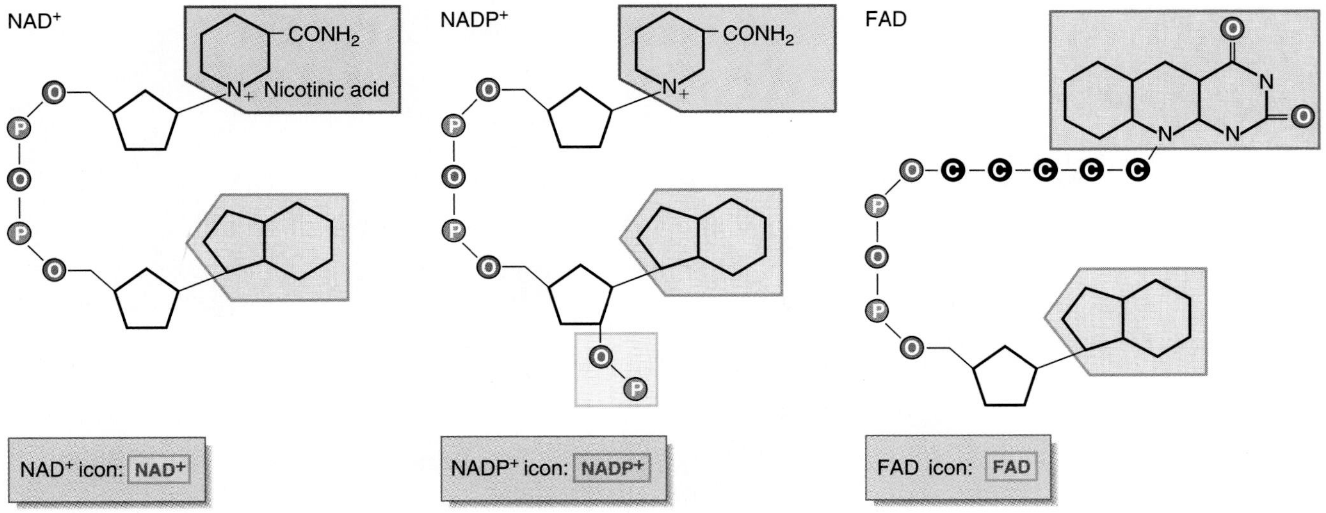

FIGURE 6.15 **COENZYMES.**
The coenzymes often participate in reactions where energy is involved. Note the similarities among the coenzyme molecules, and their overall similarity to ATP. NAD^+ and $NADP^+$ are very similar, but note the third phosphate of $NADP^+$. The active group is nicotinic acid. As nicotinic acid is reduced, it accepts a hydrogen ion and two electrons.

Reduced NAD and NADP are often written simply as NADH and NADPH.

Electron Transport Systems

NADH, NADPH, and $FADH_2$ can also deliver their electrons and protons to special sites on the membranes of chloroplasts and mitochondria. Here they reduce membrane-bounded proteins known as **electron carriers.** Many of the carriers are complex iron-containing proteins called **cytochromes.** Such carriers, arranged in an orderly series in certain membranes, form the **electron transport systems (ETS),** systems of proteins that will extract energy from those electrons and produce ATP.

Members of electron transport systems pass energy-rich electrons from carrier to carrier in a sequence of reductions and oxidations. As electrons move along, their energy is progressively drained and used to move protons across a membrane. They finally emerge from the electron transport system in a substantially depleted state (Figure 6.16). So, where does their energy end up? As we'll see, much of the energy ends up in ATP. Let's look at the details of one of life's most vital processes.

Formation of ATP in Cells

Now that we know something about energy, enzymes, ATP, coenzymes, and electron transport systems, we can look at how cells make ATP. ATP production is cyclic; that is, the phosphorylation of ADP forms ATP, which is then broken back down to ADP. ADP can be "recharged" to ATP in two ways: *substrate-level phosphorylation* and *chemiosmotic phosphorylation.*

Substrate-level phosphorylation is the generation of ATP directly from the chemical bonds of cellular fuels such as glucose. Although this ATP-gen-

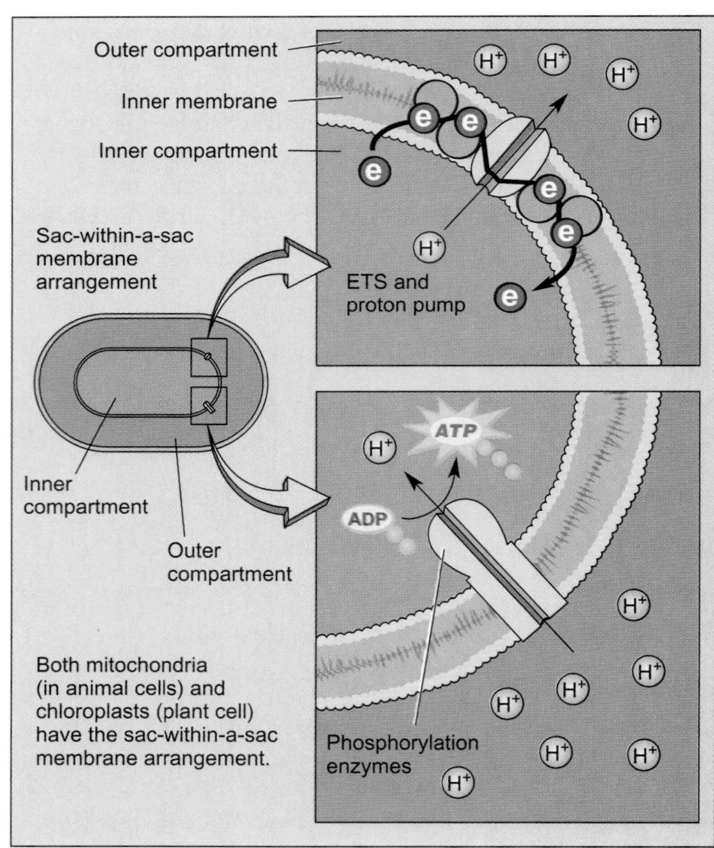

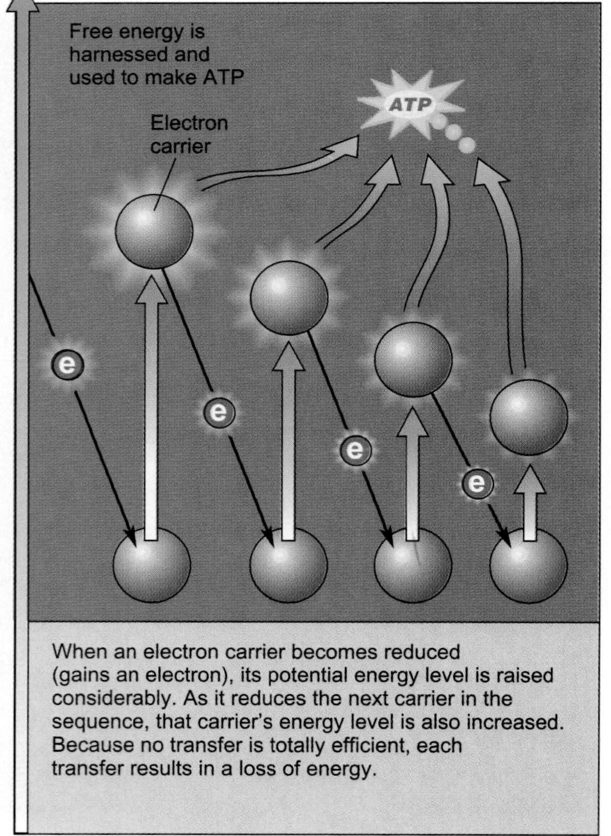

Free energy is harnessed and used to make ATP
Electron carrier
ATP
When an electron carrier becomes reduced (gains an electron), its potential energy level is raised considerably. As it reduces the next carrier in the sequence, that carrier's energy level is also increased. Because no transfer is totally efficient, each transfer results in a loss of energy.

Figure 6.16 The Electron Transport System (ETS) and Chemiosmosis.
Electron transport is the process of passing high-energy electrons from one carrier molecule to another, extracting the electrons' energy, and using that energy to form a high concentration of H^+ ions. The H^+ gradient powers the formation of ATP as the H^+ ions are released through the F1 particles.

erating mechanism is commonplace, it delivers far less ATP than chemiosmotic phosphorylation. Substrate-level phosphorylation is thought to be an ancient process, probably used by the earth's first heterotrophic organisms. We will return to the process in Chapter 8; here we will concentrate on chemiosmotic phosphorylation.

CHEMIOSMOSIS AND CHEMIOSMOTIC PHOSPHORYLATION

Chemiosmosis is the creation of very steep proton gradients (also called *chemiosmotic gradients*) between membrane-bounded compartments in certain organelles. To form these gradients, immense numbers of protons (hydrogen ions: H^+) are actively transported, or "pumped," into membrane-surrounded compartments, leaving behind a low H^+ concentration and relative large numbers of hydroxide ions (OH^-). This produces great potential energy, since there is a very strong tendency for these oppositely charged ions to join. The ions, in fact, are allowed to rejoin, but in a very special way. The great potential energy of the ions is gradually released to power enzyme-directed reactions that generate ATP. This release of energy and the formation of ATP takes place in complexes called F1 particles in mitochondria (CF1 particles in chloroplasts). The use of energy from the chemiosmotic gradient to form ATP from ADP and P_i is called **chemiosmotic phosphorylation** (Figure 6.16).

We marvel at the evolution of such complex systems, but in one form or another, they may have been around a long time, and have been molded by the processes of natural selection. Both mitochondria and chloroplasts are believed to be highly evolved descendants of what were once free-living prokaryotes (bacteria). These ancient, comparatively simple invaders likely employed some form of chemiosmosis as their means of generating ATP. Today's prokaryotes have advanced chemiosmotic systems, so we can't look to them for answers. Interestingly, aerobic (oxygen-using) bacteria simply use their surroundings for their outer compartment, pumping protons out and then letting them back in through phosphorylating enzymes in their plasma membranes. Such bacteria are functionally similar to mitochondria.

PERSPECTIVES

The energy required for maintaining biological form and function is created and stored in specific chemical bonds according to the rules of thermodynamics. There are specific molecules involved in energy production, storage, and use—chief among them ATP and enzymes. However, one of the most interesting things that we have seen in this chapter is the involvement of membranes and proteins in the process of segregating ions in the production of energy.

In the next two chapters, we will see these principles continued and expanded. We will see how the membranes of the chloroplast are involved in capturing light energy, and how the membranes of the mitochondria are involved in the conversion of that captured energy for use by the cell. We will look into some of nature's best-kept secrets—how simple sunlight is used by our green partners to form new chemical bonds in foods, and how cells use the energy in those bonds to form their own ATP.

SUMMARY

ENERGY

Energy is the ability to do work. Potential energy is stored energy. Kinetic energy is energy in motion. Energy can change from one state to another.

THE LAWS OF THERMODYNAMICS

The first law of thermodynamics states that energy can neither be created nor destroyed. The second law of thermodynamics, or the law of entropy, states that energy transfer leads to less organization. The state of disorder in a system is its entropy. The second law also indicates that each energy transfer is accompanied by a loss of useful energy, ordinarily in the form of heat.

Chemical Reactions and Energy States

Hydrogen and oxygen gases react because the two have greater energy than the product, water. Whereas energy must be applied to start the reaction, the subsequent release of energy forms a self-sustaining chain reaction.

Exergonic reactions are those in which the products end up with less energy than the reactants. Endergonic reactions are those in which the products end up with more energy than the reactants.

The energy required to start a reaction is called activation energy. Reactions may require substantial activation energy, but in the presence of a catalyst, this energy requirement is greatly reduced. Such catalysts emerge unchanged, and are able to recycle rapidly.

Enzymes: Biological Catalysts

Enzymes are biologically active protein catalysts that reduce the activation energy requirement, thus accelerating cell reactions. Each kind of enzyme reacts only with its own specific substrate.

The shape of an enzyme's active site accounts for enzyme-substrate specificity. The induced-fit hypothesis maintains that an imperfect fit in the active site aids the reaction by placing physical stress on the substrate. Following the reaction, the product leaves the site, and the enzyme is ready to act again.

Enzyme reaction rates can be increased by the addition of substrate until saturation occurs. Higher temperature increases the reaction rate, but too high a temperature causes denaturation. Most enzymes work best at a neutral pH.

Enzymes typically work in teams forming metabolic pathways. Reactions may be anabolic (building molecules) or catabolic (tearing down molecules). Enzymes often require cofactors—ions or molecules that enzymes need to function properly—or coenzymes, complex organic molecules. One kind of control system makes use of allosteric sites on the enzyme. Excess product fills the allosteric site, rendering the enzyme inoperable. As product is depleted, the sites clear and reactions resume. A product repressing its own production is an example of negative feedback control.

ATP: The Energy Currency of the Cell

ATP (adenosine triphosphate), the major source of energy in the cell, is produced through energy transfers from foods in cellular respiration and through the use of sunlight energy in photosynthesis.

ATP contains adenine, ribose sugar, and three phosphates. Two of the connecting phosphate linkages are energy-rich bonds.

ATP becomes ADP (adenosine diphosphate) and P_i when its terminal high-energy bond is hydrolyzed. In a phosphorylation reaction, the phosphate plus some energy is transferred to substrate, preparing it for further reactions.

ADP can be recycled through energy-yielding processes wherein new high-energy bonds are formed, restoring ATP.

Coenzymes, Oxidation, and Reduction

In the making of ATP, the coenzymes NAD, NADP, and FAD work in close association with enzymes. NAD, NADP, and FAD have oxidation/reduction roles. They become reduced when they take in electrons and/or protons, and they become oxidized when they pass them along. The oxidized forms of the coenzymes are NAD^+, $NADP^+$, and FAD. The reduced forms are $NADH + H^+$, $NADPH + H^+$, and $FADH_2$. The latter are in a higher energy state.

Coenzymes commonly pass their electrons and/or protons to membranal carrier proteins that make up electron transport systems. As electrons pass through such systems, some of their energy is used to form ATP.

Formation of ATP in Cells

ATP is produced through substrate-level phosphorylation involving the direct transfer of high-energy phosphate from substrate to ADP, and chemiosmotic phosphorylation that occurs at specially structured membrane sites.

Chemiosmosis is the formation of steep proton gradients between membrane-bounded organelles in some cells. The concentration gradient formed has great potential energy, which is used in generating new ATP through chemiosmotic phosphorylation. Chemiosmotic systems have membrane-bounded compartments, proton pumps powered by electron transport systems, proton and energy sources, and enzyme complexes (F1 and CF1 particles). Energy from these systems is used to phosphorylate ADP.

Key Terms

potential energy • **87**
kinetic energy • **87**
first law of thermodynamics • **88**
second law of thermodynamics • **89**
entropy • **89**
exergonic reaction • **90**
endergonic reaction • **90**
activation energy • **90**
catalysts • **90**
enzyme • **91**
substrate • **91**
active site • **92**
induced-fit hypothesis • **92**
enzyme-substrate (ES) complex • **92**
metabolic pathway • **94**
catabolic • **94**
anabolic • **94**
cofactor • **94**
coenzyme • **94**
allosteric site • **94**
adenosine triphosphate (ATP) • **95**
adenine • **95**
adenosine diphosphate (ADP) • **96**
nicotinamide adenine dinucleotide (NAD) • **96**
nicotinamide adenine dinucleotide phosphate (NADP) • **96**
flavin adenine dinucleotide (FAD) • **96**
oxidation • **97**
reduction • **97**
electron carrier • **98**
cytochrome • **98**
electron transport system (ETS) • **98**
substrate-level phosphorylation • **98**
chemiosmosis • **99**
chemiosmotic phosphorylation • **99**

Review Questions

1. State the formal definitions for *energy* and *work*. How do these definitions apply to familiar situations such as moving the limbs, washing down the driveway, and the pumping of blood by the heart?
2. Carefully distinguish between energy that is potential and energy that is kinetic.
3. According to the first law of thermodynamics, what happens to energy in a system such as the earth?
4. Starting with the potential energy of water behind the dam in a hydroelectric facility, suggest a number of energy conversions as that energy is utilized by consumers. In what form does most end up?
5. According to the second law of thermodynamics, what is the fate of the earth? The solar system? The universe?
6. Cite a familiar example of entropy increasing in your surroundings. How might this be reversed?
7. Carefully explain how the second law of thermodynamics applies to energy passing through food chains (reread the chapter introduction).
8. Using the energy states of hydrogen gas, oxygen gas, and water as a background, explain the chemistry of the *Hindenburg* disaster.
9. Distinguish between exergonic and endergonic reactions. Explain, in light of the first law of thermodynamics, how product energy can exceed reactant energy.
10. Describe how the presence of a catalyst affects the usual way chemical reactions occur.
11. Briefly discuss the chemical structure, specificity, and value of enzymes.
12. Using the induced-fit hypothesis as a background, explain how an enzyme does its work.
13. Explain why adding more substrate in an enzymatic reaction increases the amount of product, but only up to a point?
14. Why does increasing the temperature around an enzyme increase reaction rate? What limits the effect of temperature?
15. Explain the role of enzymes in metabolic pathways. How do catabolic and anabolic pathways differ?
16. Explain how negative feedback works and how it pertains to allosteric enzyme control.
17. Draw a simplified ATP molecule, labeling its three components. Add the high-energy bonds.
18. List the two products formed when ATP's terminal phosphate bond is broken.
19. Describe a phosphorylation reaction, and explain how this might be used in molecule building.
20. Draw a simple ATP cycle, indicating how ATP is used and how it is regenerated.
21. What is the role of the coenzyme oxidation and reduction reactions?
22. Write the oxidized and reduced forms of the coenzymes NAD, NADP, and FAD.
23. What are electron transport systems? What is the electron energy generally used for?
24. Chemiosmosis involves the concentrating of great numbers of protons in membrane-bounded compartments. What does this mean in terms of potential energy, and how can such a system be put to use?
25. Summarize the requirements of a chemiosmotic system.

7

Photosynthesis

VIRTUALLY ALL BIOLOGICAL ENERGY COMES FROM THE SUN. A SIMPLE *statement, but one with profound implications. How does energy from the sun, arriving as light, become entwined in biological systems? The goal of this chapter, quite simply, is to describe that process. You will notice principles and components that have been discussed in previous chapters, so keep your eyes open for things that you have seen before—such as ATP, proteins, membranes, and organelles.*

Interest in solar power as an energy source to replace fossil fuels has an almost touching naiveté about it. We seem to view solar power as a new concept, an untouched source of energy for human endeavors. In fact, sunlight has long provided us with our most fundamental source of energy. Such fossil fuels as oil, gas, and coal are simply releasing solar energy stored away in the bodies of long-dead plants and algae. So while we continue to wander in the maze of engineering problems associated with harnessing the sun's energy to produce electricity, perhaps we should turn to the real experts—and the real experts are likely to be green (Figure 7.1).

A great variety of living things on the planet, including green plants and algae, make food from simple molecules, such as carbon dioxide and water, through a process called *photosynthesis.* This process is powered by the powerful energy of sunlight. In this chapter we will consider how *phototrophs* ("light-feeders") are able to use something as ethereal as sunlight to manufacture their own nutrients. This happens as the energy of sunlight is captured by certain molecules, such as the chlorophyll of green plants.

OVERVIEW OF PHOTOSYNTHESIS

Photosynthesis is the conversion of light energy into chemical energy. The energy of sunlight is first used to sharply increase the energy level of certain chlorophyll electrons. Such energy-rich electrons then flow through highly ordered carriers in electron transport systems, powering proton pumps as they go, and finally reducing $NADP^+$ to NADPH. Proton pumps, as we saw in Chapter 6, generate the great energy of the chemiosmotic systems responsible for forming the energy-rich bonds of ATP. Both ATP and NADPH provide energy that is used to form food (such as glucose) from carbon dioxide and water. (An imaginative biochemist, pondering this delicate electron emission, once philosophized, "Life seems to begin with a little flow of electricity.")

But let's ignore the electrons and energetics for the moment and look at photosynthesis through a time-honored equation. Photosynthesizers use carbon dioxide and water to form glucose, yielding oxygen as a by-product:

$$\underset{\text{Carbon dioxide}}{6\,CO_2} + 12\,H_2O \xrightarrow{\text{sunlight}} \underset{\text{Glucose}}{C_6H_{12}O_6} + 6\,H_2O + 6\,O_2$$

The equation is a nice overview, but hides a great amount of detail. For instance, it is not apparent from the equation that the 12 oxygen atoms ($6O_2$) to the right of the arrow come from the water molecules and not from carbon dioxide to the left of the arrow.

The equation also does not show the actual roles of **photons,** discrete packets of sunlight energy that move in a wavelike path. The energy of the photons, when captured by a pigment such as *chlorophyll,* powers the photosynthetic process, generating energy-rich ATP and hydrogen-rich NADPH. Both of these work with teams of enzymes in the glucose-synthesizing pathway.

FIGURE 7.1 **CAPTURING LIGHT ENERGY.** Photosynthetic organisms are undoubtedly the dominant forms of life on earth. Nearly all of the earth's creatures depend on their ability to capture the sun's energy.

Also not apparent is the importance of structure to the process. It is no coincidence that the chloroplast is an intricately structured organelle. In fact, its organization is so essential, that we should begin with a closer look at it.

THE CHLOROPLAST

Chloroplasts occur in the cells of leaves and green stems of plants and in the cells of algae (Figure 7.2). Recall from Chapters 4 and 6 that chloroplasts are relatively large organelles encasing a watery, protein-rich fluid, the stroma. The *stroma* of plant chloroplasts contain many tiny membrane-bounded structures resembling stacks of coins. These are the *grana,* each composed of disklike thylakoids. Each *thylakoid* contains a minute inner compartment known as the lumen. The lumen acts as the proton reservoir in the chloroplast's chemiosmotic system (see Chapter 6). The membranes that form the thylakoids often extend at intervals as lamellae, forming interconnections between neighboring grana.

The clear, watery stroma of the chloroplast contains the enzyme teams responsible for the synthesis of glucose. Thylakoids are responsible for the capture of the light energy needed to drive the process. Embedded in the thylakoid membranes are the light-capturing pigments—organized into the two kinds of *photosystems* described below. Also present are *electron transport systems* and *proton pumps.* They drive chemiosmotic phosphorylation (the charging of ATP) and the reduction of $NADP^+$. The system works well because of the precise organization of the chloroplast: The various sites of chemical activity have a spatial relationship that contributes to their function.

PHOTOSYSTEMS

Photosystems are clusters of light-absorbing pigments and their associated molecules (Figure 7.2). For example, researchers studying the thylakoids from spinach chloroplasts have found that each cluster, or photosystem, contains about 200 molecules of chlorophyll *a* and *b* (Figure 7.3) and about 50 molecules of **carotenoids,** another family of light-absorbing pigments. Other accessory pigments may also be present. The presence of several kinds of pigments enables the photosystem to absorb light over much more of the visible light spectrum (see Essay 7.1).

The energy of the absorbed light is converted to chemical energy in the **reaction center** (Figure 7.2). This special region consists of one molecule of chlorophyll *a* and closely associated proteins. Since the rest of the pigment cluster only "gathers" light energy, shunting it to the reaction center, it has been aptly named the **light-harvesting antenna.** When an incoming photon is captured, the energy of the photon is absorbed and increases the energy level of the reaction center. The rest of photosynthesis is dedicated to channeling and changing that energy.

Photosystems I and II. The two photosystems within the thylakoids are designated **photosystem I** and **photosystem II.** Their reaction centers are called P700 and P680, respectively; the numbers refer to the wavelength of light best absorbed by the reaction center.

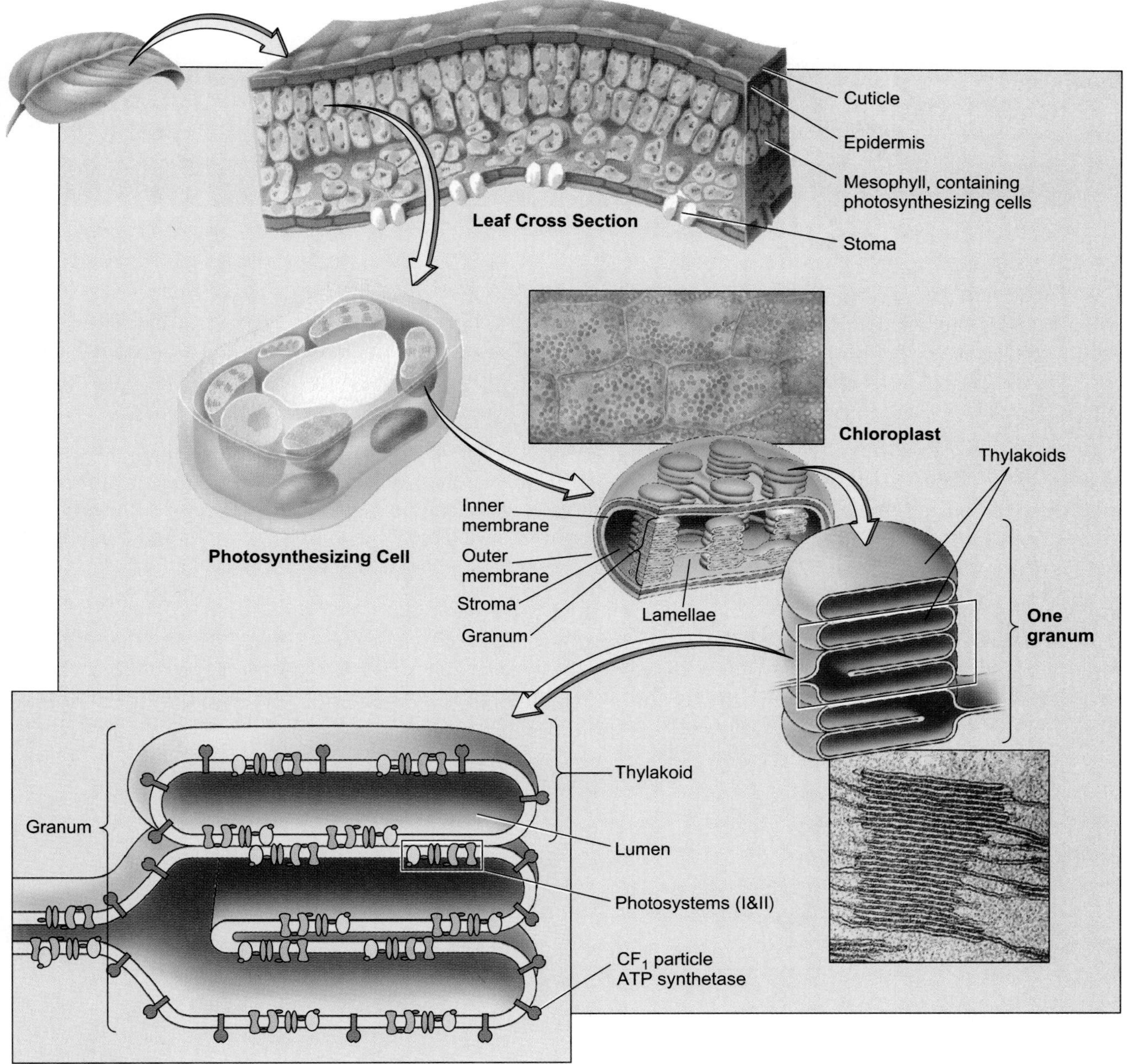

FIGURE 7.2 **LEAF TISSUE.**
Tissues within the leaf contain vast numbers of photosynthetic cells, each with numerous chloroplasts. Within the chloroplasts are membranous grana. Each thylakoid is bound by two complex membranes that are alternately pressed together to form the *lamellae* and bulged outward to form the inner compartments, or *lumina* (singular, *lumen*). Each membrane contains many light-harvesting antennas associated with photosystems, as well as associated electron carriers involved in chemiosmosis.

ELECTRON TRANSPORT SYSTEMS AND PROTON PUMPS

In Chapter 6, we learned that electron transport systems are sequences of carriers within membranes that form pathways for the movement of energy-rich electrons. This energy powers proton pumps. Such pumps actively transport protons into the lumen, thus generating the all-important chemiosmotic gradient. The energy established by the isolated protons is then put to work, making ATP. This is accomplished by using the proton gradient to drive the CF1 particle to combine ADP and P_i to form ATP.

Essay 7.1

The Visible Light Spectrum and Photosynthesis

The earth is constantly bathed in radiation emanating not only from the sun, but also from a host of other celestial bodies. Part of the radiation that reaches us is visible light. Visible light, however, is only part of an electromagnetic spectrum that includes (in order of increasing energy) radio waves, microwaves, infrared radiation, visible light, ultraviolet radiation, X rays, and gamma rays. The only real difference in these types of radiation is in their wavelength. The energy of any electromagnetic radiation is the inverse of its wavelength, such that the shortest wavelengths have the most power.

Visible light is visible because it interacts with special pigments (light-absorbing molecules) in our eyes. It also interacts with pigments such as chlorophyll, which are the molecules that absorb the energy of light and provide power for photosynthesis. Visible light, of course, interacts with the retinas of our eyes, but it also provides the energy that enables green plants to grow. More energetic wavelengths are usually too powerful for most of life to utilize, since they tend to disrupt molecules, especially proteins and DNA. Ultraviolet light (UV), for example, burns our skin and damages the retinas of our eyes. More penetrating radiation, such as the extremely short-waved gamma rays and man-made X rays, is called ionizing radiation, because it can break up the water molecules within cells.

The specific light-absorbing qualities of the chlorophylls and carotenoids can be determined by using a device known as a spectrophotometer. First, the pigments are extracted and dissolved in a solution. Next, light of a known wavelength is passed through the solution and measured on the other side. The wavelength of the entering light can be varied to see which wavelength is most absorbed by the solution. Finally, the data are plotted on a graph to form what is called an absorption spectrum. Note the absorption spectra for chlorophylls *a* and *b* on the graph shown here. The peaks represent

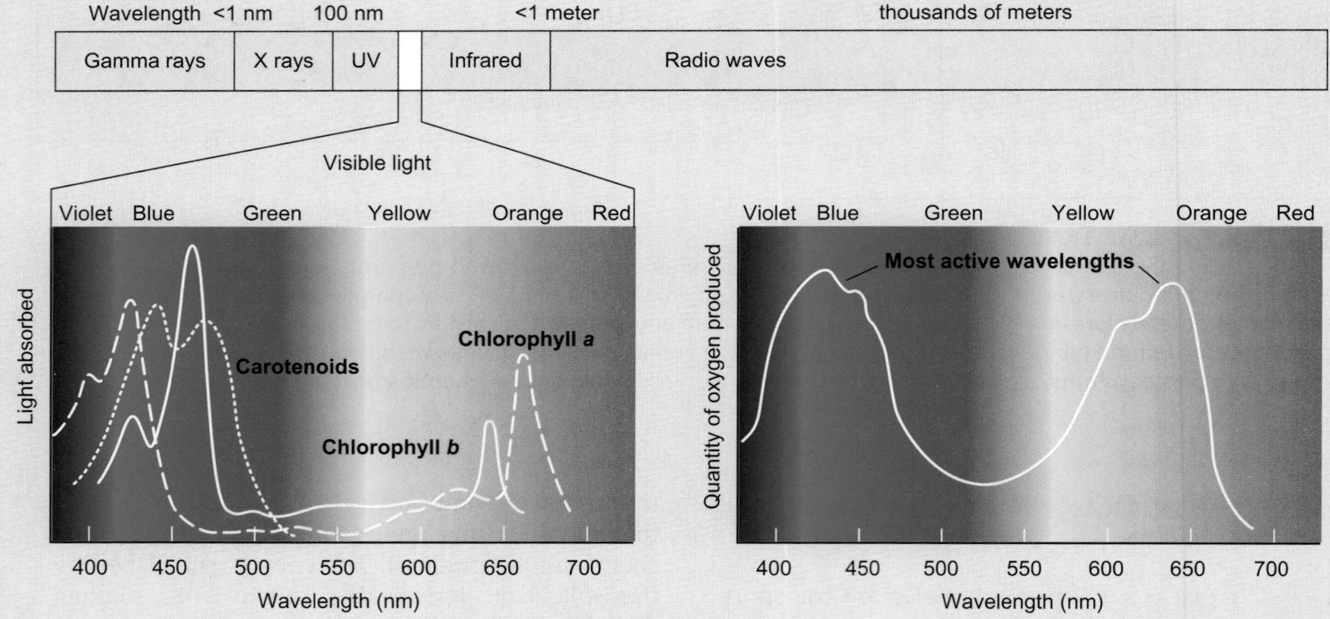

light that is maximally absorbed by the pigment; the valleys represent light that passes through. The green and yellow hues are least absorbed. These are the colors that are reflected—the ones that we see when we look at a chlorophyll solution or a leaf.

Certain wavelengths, such as violet-blue and orange-red, are strongly absorbed by chlorophyll. This suggests that these wavelengths participate in the light reactions, but such evidence is circumstantial. It is possible, however, to get more direct evidence. One way is to discover the rate at which some product of photosynthesis is produced when a plant is subjected to monochromatic light (light of one color only). Since photosynthesis produces oxygen, we can measure the rate of photosynthesis by measuring the rate of production of this product under various wavelengths of light. With these data we can plot what is known as an action spectrum, which turns out to be very similar to the absorption spectrum. As a result, we can be much more confident in the proposition that the light absorbed by the pigments is the light that drives photosynthesis.

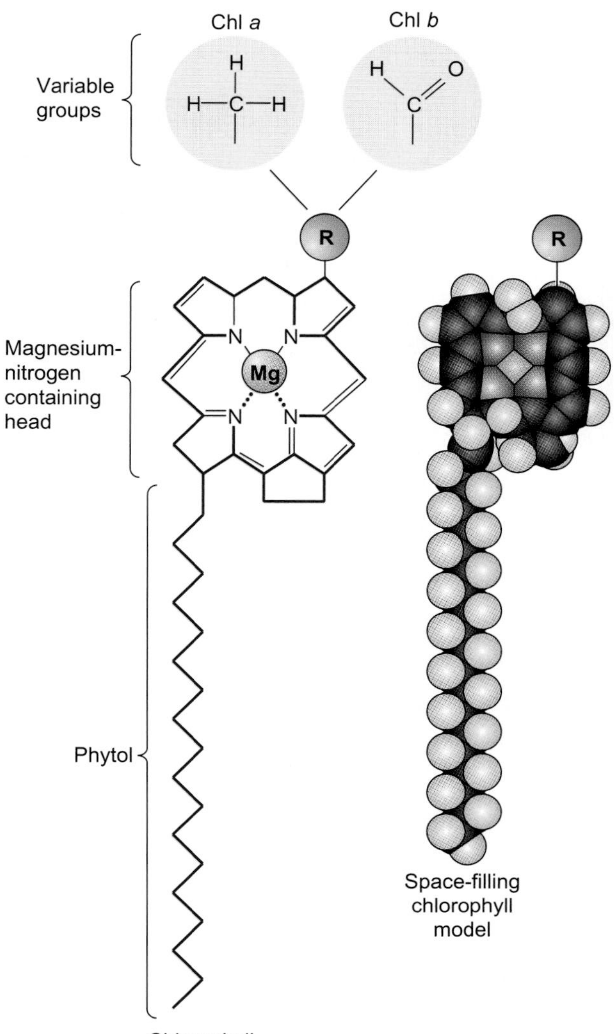

FIGURE 7.3 CHLOROPHYLL.
The chlorophylls consist of a complex head, containing a multiple carbon-nitrogen ring structure with a single magnesium atom at its center. The lengthy tail is a lipid-soluble hydrocarbon chain. The difference between chlorophyll *a* and *b* is seen at the top.

THE PHOTOSYNTHETIC PROCESS

The notion of gentle sunlight falling peacefully on a green leaf belies the complex activities of photosynthesis that go on inside that leaf. Many of the mysteries of this process remain unsolved, but we do have solid information about much of it. For instance, we know that some reactions require light and that others do not. Thus the process of photosynthesis can be conveniently divided into two parts: the **light-dependent reactions** (also called the

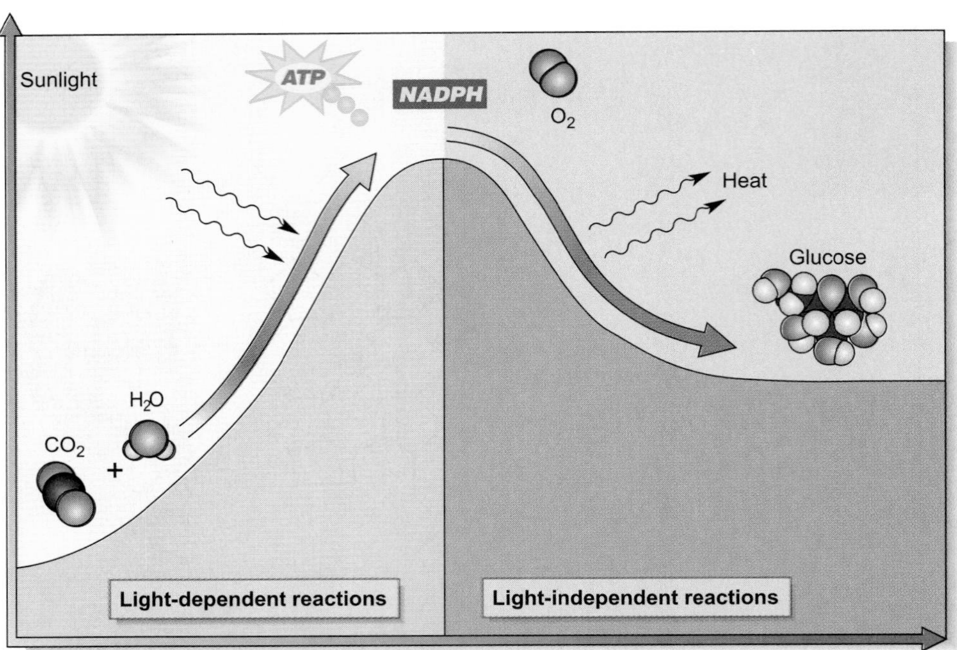

FIGURE 7.4 ENERGY AND PHOTOSYNTHESIS
In a general sense, the light-dependent reactions charge the system, increasing its free-energy state. The light-independent reactions are essentially downhill, with the decreasing free energy used to accomplish work. That work can be viewed as the production of sugars such as glucose, which have a higher energy level than the original chemical constituents of CO_2 and H_2O.

light reactions) and the **light-independent reactions** (also called the *dark reactions*). We can consider the light reactions as a way to increase energy of the system. Light-independent reactions use that energy to form glucose (Figure 7.4).

THE LIGHT-DEPENDENT REACTIONS

The increase in energy made available in the light reactions is used to: (1) build the chemiosmotic or proton gradient, (2) generate ATP, and (3) reduce $NADP^+$ to NADPH.

ATP can be generated in two ways, through *noncyclic photophosphorylation* and *cyclic photophosphorylation* (*photo* here refers to "light-driven"). We will refer to the two simply as the "noncyclic" and "cyclic" events. The two differ in the route taken by light-activated electrons and in some of the products formed.

NONCYCLIC EVENTS

The **noncyclic events** produce ATP using a linear noncyclic pathway. They begin as light is absorbed in photosystem II, its energy shunted into the P680 reaction center. This starts a flow of light-activated electrons from chlorophyll *a* of the reaction center, which passes to the neighboring electron transport system (Figure 7.5).

Once in the electron transport system, energy-rich electrons are passed from carrier to carrier, eventually entering photosystem I. The electron's energy progressively dissipates as it moves along the chain of acceptors. Some of the energy will be used to "pump" protons across the membrane into the thylakoid lumen. In fact, each light-activated electron enables the proton pump to move one proton across the membrane.

Partially spent electrons then make their way into P700, which has also been actively absorbing light (see Figure 7.5). Thus the electrons receive a second boost, now reaching their highest energy level yet as they progress through a second electron transport system. This time, though, as they are passed from carrier to carrier, there will be no proton pumping. Their energy is reserved for one purpose only—the reduction of $NADP^+$. At the last stage, two electrons will join $NADP^+$, giving it a net negative charge that immediately attracts two protons from the stroma.

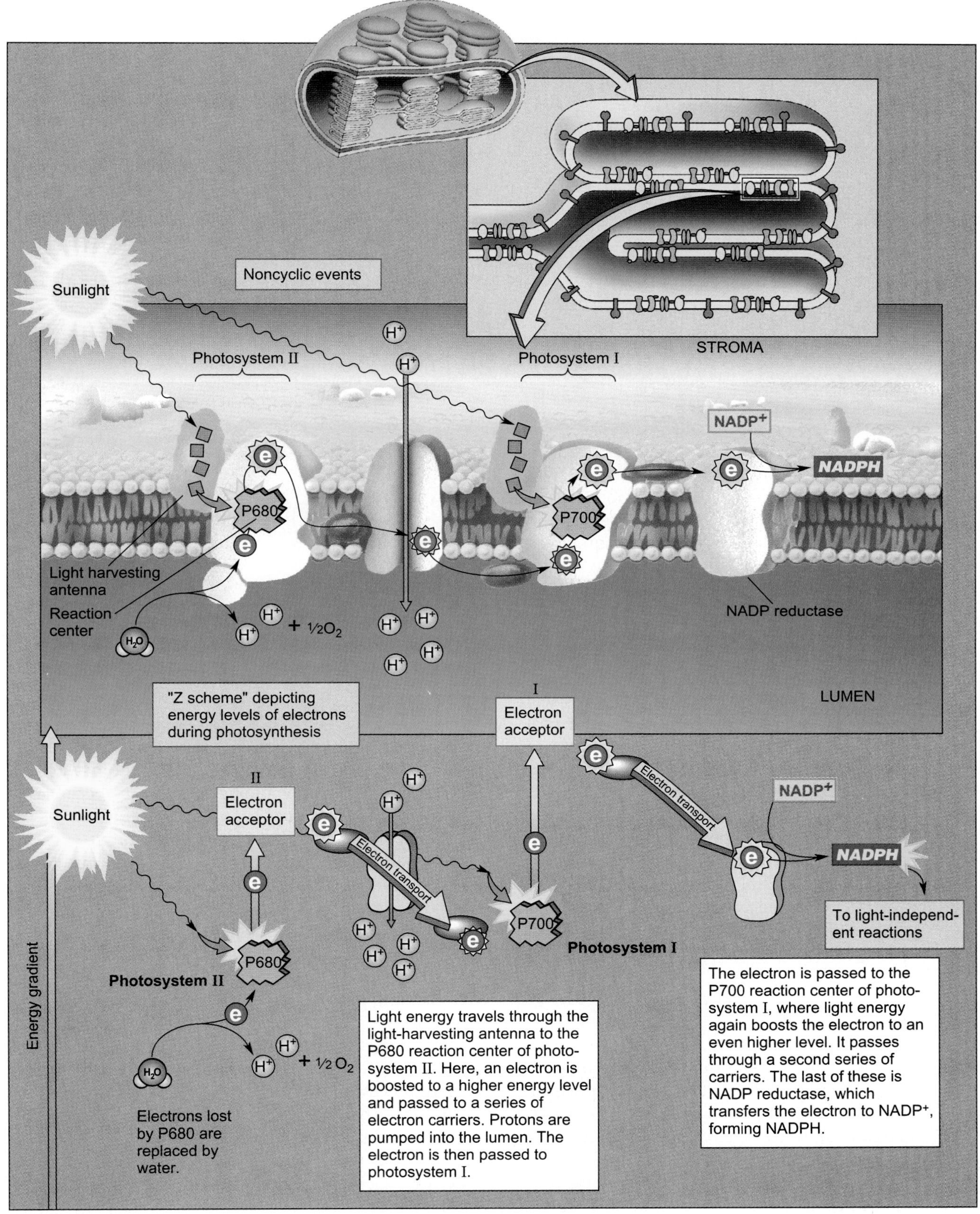

FIGURE 7.5 NONCYCLIC REACTIONS AND PHOTOSYNTHESIS.
In the light-dependent reactions, sunlight boosts electrons from water along photosystem II to an electron acceptor that passes them along a series of molecules at increasingly lower energy levels to photosystem I, which boosts them to another electron acceptor. They are then passed to $NADP^+$, forming NADPH, a high-energy molecule that, with ATP, powers the light-independent reactions, where CO_2 is used to make glucose.

Thus $NADP^+$ is reduced to the highly energetic NADPH.

The electron flow that began in the P680 reaction center leaves the chlorophyll *a* short on electrons, but water provides a ready source of replacement electrons.

The Splitting of Water. Electrons needed to restore chlorophyll *a* (Chl *a*) are made available by the splitting of water molecules. This can be represented as:

$$H_2O \longrightarrow 2\,e^- + 2\,H^+ + \tfrac{1}{2}O_2$$

$$2\text{ Chl }a^+ + 2\,e^- \longrightarrow 2\text{ Chl }a$$

(2 oxidized Chl *a*) (2 reduced Chl *a*)

Because two electrons are released in each event, two chlorophyll *a* molecules can be reduced. Splitting the ordinarily stable and resistant water molecule is no easy matter. It is possible only because oxidized chlorophyll *a* in P680 creates a significant "pull" that sets the water-disrupting events in motion. We can look at the electron flow through the photosystems as a "pulling" force as well as a light-driven "push."

In the splitting of water, two protons are released. This is quite important because they are "dumped" directly into the thylakoid lumen, which serves to increase the chemiosmotic gradient.

The other product of the disrupted water molecule is oxygen. Two oxygen atoms join to form molecular oxygen (O_2), a critical gas of the atmosphere. If we balance the noncyclic events in terms of one disrupted water molecule, we can see that for each *two electrons* yielded by water, *four protons* enrich the chemiosmotic gradient (two pumped by electron transport and two dumped directly into the lumen), and one NADPH molecule is formed. Two of these events provide the earth's atmosphere with a molecule of oxygen.

Cyclic Events

The **cyclic events** produce ATP using a cyclic biochemical pathway. These brief events involve only photosystem I (the P700 center)—water, photosystem II, and $NADP^+$ are not involved. The excited, light-activated electrons leave the reaction center, pass through part of an electron transport system, and then return to P700 again. What does the cyclic process accomplish? As you see in Figure 7.6, each cycling electron reaches the proton pump, powering the transport of one proton into the lumen. The abrupt cyclic process helps to enrich the proton gradient, leading ultimately to the generation of ATP.

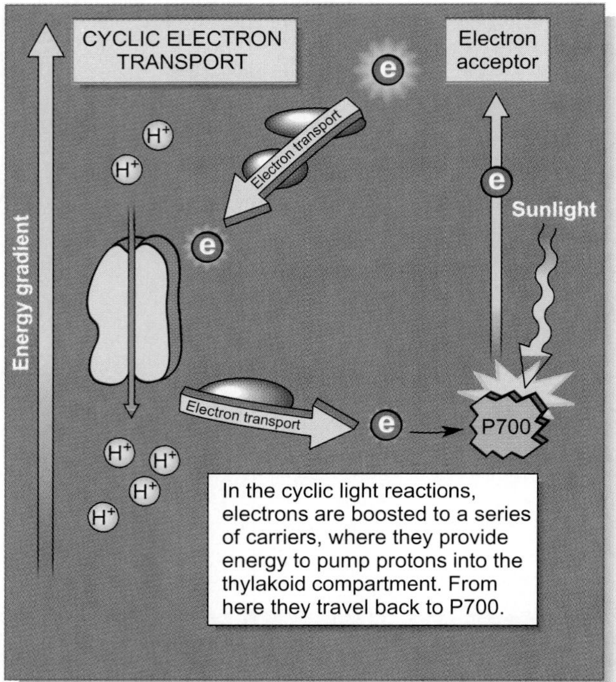

FIGURE 7.6 **CYCLIC REACTIONS.** In the cyclic light reactions, energetic, light-activated electrons from photosystem I flow outward from P700, provide the energy for pumping protons into the thylakoid lumen, and cycle back to P700 once again.

Cyclic photophosphorylation by photosystem I was an enigma to biologists for years because it seemed only to duplicate part of the noncyclic process. Most agreed that it was a primitive system, and some theorized that perhaps the cyclic reactions were a form of "photosynthetic engine-idling" that occurred when $NADP^+$ was in short supply or when NADPH simply wasn't needed. But with the development of the chemiosmosis theory, researchers began to see that such a pathway could be useful just for adding protons to the chemiosmotic differential. The idea gained strong support when cytologists learned that the components of photosystem I often occur alone; they are not necessarily physically linked to those of photosystem II as had once been thought. Thus the cyclic process may be far more significant than was once believed.

Keep in mind that we have considered only the activity in one pair of photosystems. There are many photosystems in the membrane of each thylakoid (see Figure 7.2), and there are many, many thylakoids in each chloroplast. Consider, further, the numerous chloroplasts in each photosynthetically active leaf cell and the thousands of such cells in a single leaf. (And how many leaves are there on an oak?) Such considerations could seriously alter any idea about plants leading quiet, inactive lives. One might even gain a new, abiding respect for spinach.

We have considered various ways that protons can accumulate on one side of a membrane, and we are aware that this gradient serves as an energy source for making ATP. Now let's see just how this happens in the chloroplast.

Chemiosmotic Phosphorylation

The proton concentration in the thylakoid lumen may reach 10,000 times that of the stroma. The protons make the lumen quite acidic. Outside in the stroma the presence of hydroxide ions (OH^-) renders this environment basic. This system has great potential energy: An acid is on one side, a base is on the other, and they have a strong tendency to join. If the protons are permitted to escape and join the hydroxide ions in a controlled manner—the energy of their passage can be harnessed to do work.

There is only one escape route for the sequestered protons: the channels leading into the phosphorylating sites that stud the separating membrane. In the chloroplast these are the CF1 particles, rich in phosphorylating enzymes (Figure 7.7). Energetic protons filing through CF1 particles join hydroxide ions, forming low-energy water and releasing their pent-up energy to power the formation of new high-energy ATP bonds. It has been calculated that each pair of escaping protons powers the formation of one energy-rich ATP bond.

The ATP thus formed in the light-dependent reactions, along with the NADPH formed in the noncyclic events, will provide the energy and reducing power necessary for the synthesis of glucose in the light-independent reactions out in the stroma (look back at Figure 7.5). But before getting into that, let's summarize the main events of the light-dependent reactions:

1. Light-activated electrons from P680 pass through the electron transport system and power the transfer of protons into the thylakoid lumen.
2. The missing electrons of chlorophyll *a* of P680 are restored by electrons from the hydrogen of water. The protons of water are released in the lumen, directly increasing the chemiosmotic differential.
3. Electrons passing through photosystem I, along with protons from the stroma, reduce $NADP^+$ to NADPH.
4. Protons passing down the chemiosmotic gradient through the CF1 particles power the phosphorylation of ADP, yielding ATP.

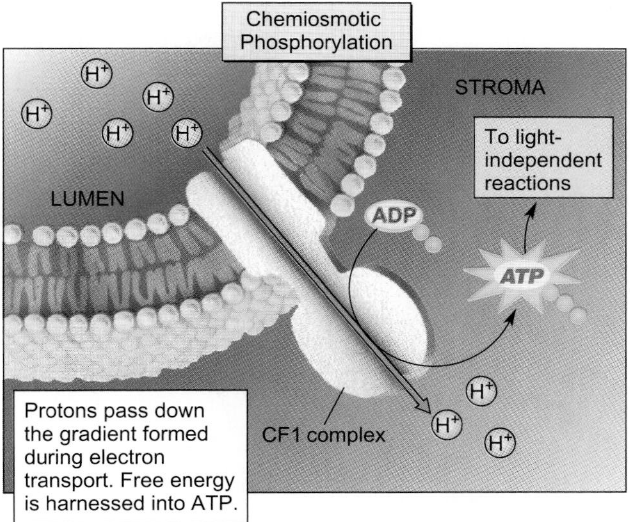

Figure 7.7 Chemiosmosis in Photosynthesis. Chemiosmotic phosphorylation uses the free energy of the osmotic gradient to produce the energy-rich bond of ATP. The process begins as protons escape from the thylakoid lumen and enter CF1 particles. The ATP is formed as protons, which have been pumped into the thylakoid lumen by the energy released as electrons move down the electron-transport system, move out of the lumen into the stroma. They must pass through particles contain ATP synthase. an enzyme able to join ATP and inorganic phosphate (P_i) to make ATP.

The Light-Independent Reactions

The first half of photosynthesis, the light-dependent reactions, produces energy-rich ATP and the powerful reducing agent NADPH. These molecules will provide the energy and hydrogens needed in the light-independent reactions, in which glucose and other carbohydrates are made. This phase of photosynthesis occurs in the watery stroma.

Considerable amounts of ATP and NADPH are used each time a single glucose molecule is formed. The process of incorporating carbon dioxide into glucose is aptly called **carbon dioxide fixation.** The fixing is done by the giant enzyme **ribulose bisphosphate carboxylase (RuBP carboxylase),** which is the most abundant enzyme on earth. Its ability to

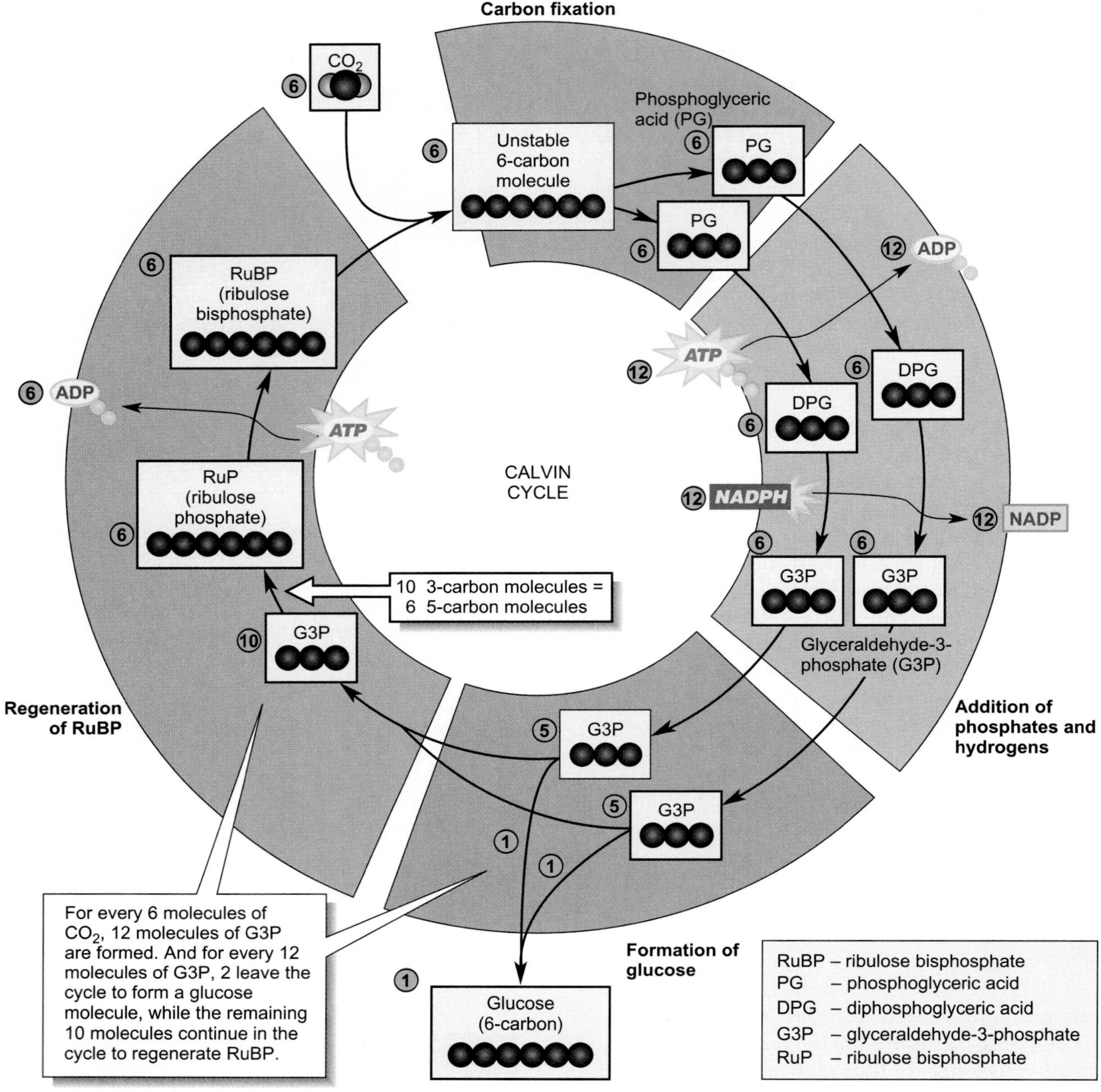

FIGURE 7.8 **THE CALVIN CYCLE.**
In this simplified view of the Calvin cycle, the products of the light reactions and carbon dioxide—powered by energy from ATP—are used to form glucose. Highlights include the fixing of carbon dioxide, the phosphorylation of PG (phosphoglycerate) by ATP and the reduction of DPG (1,3-diphosphoglycerate) by NADPH, and the formation of glucose by 2 out of each 12 molecules formed. The cycle keeps going as the remainder of the G3P is recycled to regenerate RuBP.

fix carbon dioxide represents a vital, unique link between the physical and biological worlds.

Actually, an entire battery of enzymes is involved in glucose production, each catalyzing a specific step. Much of what we know about the process was unveiled by Nobel Prize winner Melvin Calvin. Today, the pathway is called the **Calvin cycle** (Figure 7.8).

THE CALVIN CYCLE

The first thing we see in Figure 7.8 is that as each carbon dioxide enters the cycle it is joined to a five-carbon compound called **ribulose bisphosphate (RuBP).** RuBP might be called a "resident molecule," because it cycles continuously.

As a carbon dioxide is added to a RuBP, the newly forming product cleaves simultaneously into two 3-carbon molecules of phosphoglyceric acid (PG). The energy of ATP comes into play as two ATP molecules then react with the two PG molecules, phosphorylating them. Next, two NADPH molecules reduce them to form two molecules of glyceraldehyde-3-phosphate (G3P).

Two Fates of G3P. With the emergence of G3P, we come to an important branch in the Calvin cycle, one in which the reactions can go in either of two directions. In one case, two molecules of G3P are used to generate one molecule of glucose. In the other, the G3P molecules are returned to the cycle so that more carbon dioxide can be fixed.

The Calvin cycle involves pools of molecules, not just one at a time as shown in Figure 7.8. Perhaps the easiest way to view this cycle is to consider 6RuBP and $6CO_2$ at the start (six 5-carbon molecules plus six 1-carbon molecules). These form 12G3P molecules. Two of the G3Ps can be used to make a glucose, while the remaining ten are rearranged into six 5-carbon RuBPs ready for the next round of the cycle (six 5-carbon molecules plus one 6-carbon molecule).

This is an interesting cyclic process whereby a renewal of the starting reactant (RuBP) and a yield of glucose are produced. The spent ADP and P_i, and the $NADP^+$ (depleted of its reducing power), recycle into the thylakoid where they will again be available for the light-dependent reactions. After their energetic capabilities are restored, both will once again enter the stroma to power the Calvin cycle. (Essay 7.2 describes an intriguing metabolic alternative to the Calvin cycle, the C4 pathway.)

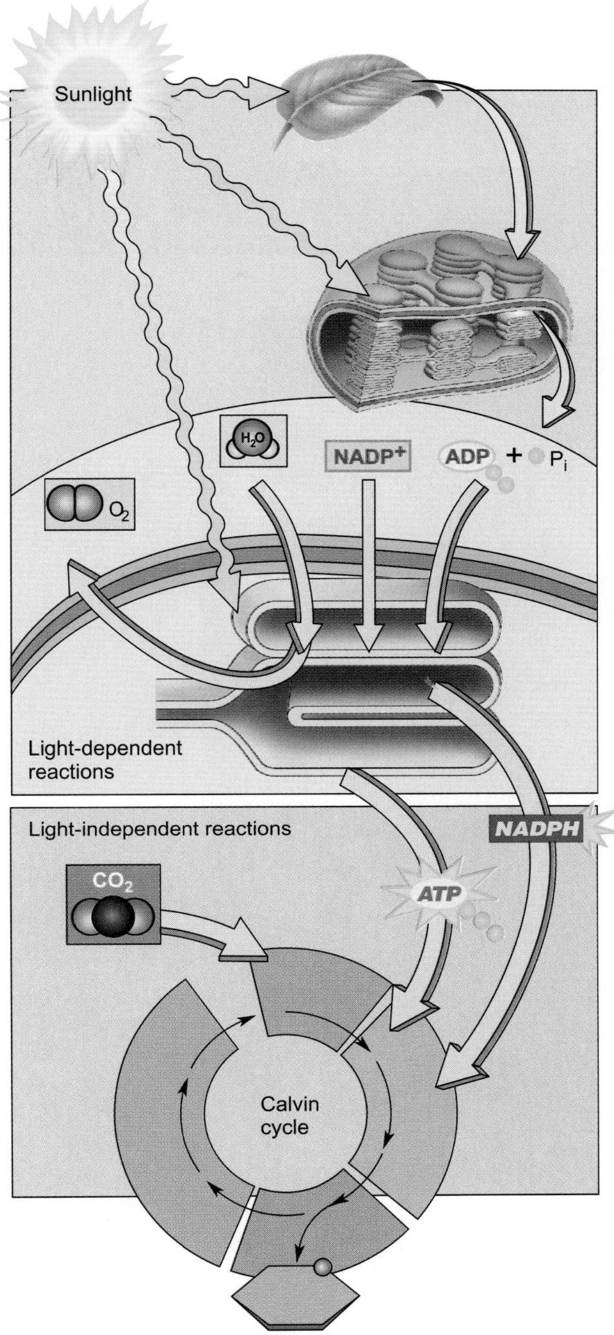

FIGURE 7.9 SUMMARY OF PHOTOSYNTHESIS. The main aspects of the photosynthetic process are summarized here in graphic form. Note the inputs—CO_2, H_2O, and light—as well as the products, O_2 and glucose.

What Does a Plant Do with Its Glucose?

Photosynthesis is a precisely orchestrated sequence of remarkable complexity (Figure 7.9). Yet it occurs with dazzling speed in the chloroplasts of countless cells. This efficiency results in a great deal of glucose production. What happens to it?

Plants, like animals and most other organisms, use glucose as an energy source; glucose is broken down through elaborate metabolic pathways, and its chemical bond energy is then used to form more ATP (see Chapter 8). But glucose is also important structurally.

In the growing plant, a great deal of glucose is incorporated into cellulose in newly emerging cell walls. Much glucose is stored, as plant cells link the glucose units and form starches. The potato plant, for instance, stores excess starch in swollen underground tubers. Most fruits are sweet because they are laden with sugars. Sugar beets and sugar cane store

Essay 7.2

Problems with the Calvin Cycle—and the C4 Alternative

In the Calvin cycle, CO_2 is fixed by the giant enzyme ribulose bisphosphate carboxylase (RuBP carboxylase). The six-carbon product yields two molecules of 3-phosphoglycerate (PG), a three-carbon product. All plants use the Calvin cycle. But it was discovered in the mid-1960s that some plants first fix CO_2 in an alternative pathway.

Australian researchers M. D. Hatch and C. R. Slack determined that in some plants CO_2 enters the leaf mesophyll cells and first joins a substance called phosphoenol pyruvate (PEP) to form a four-carbon intermediate. The four-carbon intermediate passes into other cells, where it is acted on by enzymes, and CO_2 is released. The CO_2 then enters the Calvin cycle in the usual manner.

Plants with this capability became known as C4 plants, because of the four-carbon intermediate (the others were dubbed C3 plants). This alternative pathway, the Hatch-Slack pathway, occurs in over 100 genera, including Bermuda and crab grasses, corn, and sugar cane (see photograph). These plants are well adapted to hot, dry climates, where they thrive in the intensely bright sunlight that tends to stunt the growth

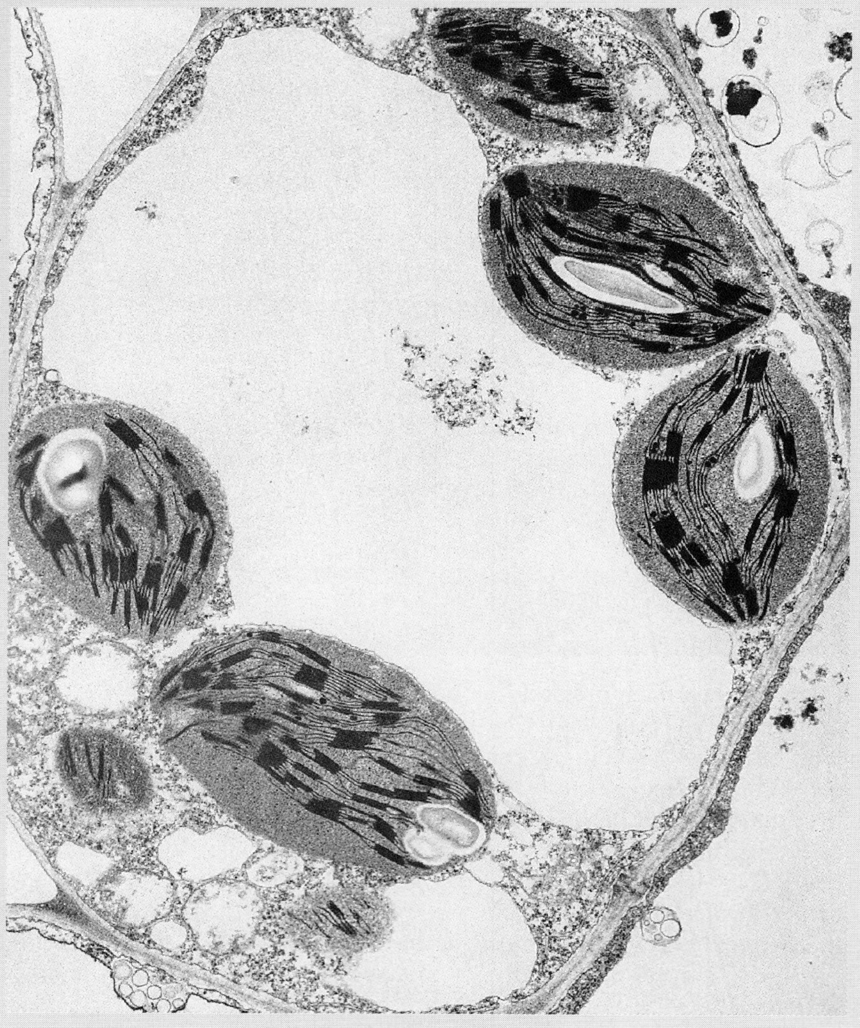

of C3 plants. Why should this be? Our first clue is climate, but what does it have to do with the alternative pathway?

The answer lies in the peculiar conditions that arise when photosynthesis proceeds at a maximum rate, perhaps too fast. Intensely lighted conditions trigger a form of biochemical stress known as photorespiration. During this process, the CO_2-incorporating enzyme, ribulose bisphosphate carboxylase, reduces the CO_2 concentration to the point that it starts incorporating O_2 instead. Photorespiration can produce energy, but at lower levels than photosynthesis.

C3 plants can survive this trauma; but if it continues, their metabolism and growth substantially slow. By contrast, photorespiration is virtually absent in C4 plants. C4 plants can incorporate CO_2 through the action of an enzyme called PEP carboxylase, which converts phosphoenol pyruvate (PEP, 3 carbons) to four-carbon oxaloacetate. PEP carboxylase has a much higher affinity for CO_2 than does RuBP carboxylase, and will bind to it even at very low concentrations. Thus, in spite of its limited quantity, carbon dioxide is rapidly taken up as it enters the leaf, and a steep, inward diffusion gradient is thus established with the surrounding air. C4 plants, therefore, are much better than C3 plants at capturing carbon dioxide.

The by-product—pyruvate, a three-carbon acid—then recycles to the mesophyll cells. Through this cycle, the plant manages to concentrate CO_2 in the vicinity of the sluggish giant, RuBP carboxylase, and the Calvin cycle goes on full speed in spite of the climatic conditions.

Whereas C3 plants struggle to grow under the intense illumination of the hot deserts and tropics, C4 plants take advantage of the endless supply of sunlight to grow at a maximum rate. In hot climates, Bermuda grass has an advantage as it encroaches on your lawn, threatening to strangle your expensive C3 greenery.

As we see, PEP does its magic in the leaf mesophyll cells where its successor, oxaloacetate, is converted to four-carbon malate. Malate enters the bundle sheath cells, a tissue surrounding the leaf veins. There, at the expense of NADPH, the reactions are reversed, and carbon dioxide is released for uptake into the regular Calvin cycle.

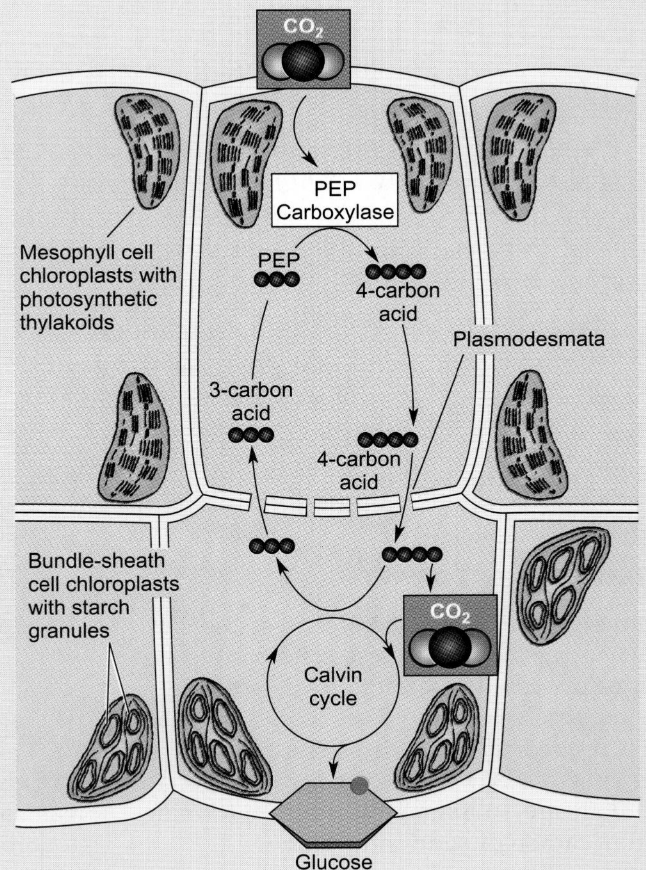

sucrose (which we use as table sugar) in roots and stems, respectively. Humans are highly dependent on the ability of plants to make and store such carbohydrates, one of our primary food sources.

Plants also make other sorts of molecules from the products of the Calvin cycle. For example, G3P is used by most seed plants to fashion a number of lipids and amino acids. Phospholipids are vital to the construction of cell membranes, and triglycerides are stored in great quantities as oils in seeds and grains. Amino acids are joined to make proteins or are further modified in the formation of nitrogenous bases in DNA, RNA, ATP, and coenzymes.

Have You Thanked a Plant Today?

Nearly all life on this planet depends on capturing the energy radiating from its nearest star. Energy is essential to living things because there is a strong tendency for increases in entropy. Life is highly organized and needs energy to maintain that level of organization.

Deep in the ocean floor there are organisms that derive all of their energy from sulfurous compounds emitted from volcanic vents. Perhaps on other worlds the energy to sustain life comes from different sources, such as heat emanating from the planet's core. But for the majority of life on earth, energy comes from our sun and can be captured only by organisms that are often rudely or cavalierly treated by our species. If plants did not turn solar energy into food, this would be a far, far different place.

Let's add one more parting thought. Photosynthesis is an energetically costly process. Its molecular reactants, water and carbon dioxide, are in a low potential energy state—that is, they are stable—and are quite reluctant to participate in any chemical events. Fortunately for life as we know it, sunlight is plentiful, and evolution has provided some organisms with the ability to use it. With this enormous and seemingly inexhaustible energy source, photosynthesizers can power their elaborate biochemical machinery without much concern for fuel economy. In our own energy-hungry world, we have much to learn from plants.

Perspectives

The ability of plants to convert the energy of the sun into the chemical energy currency of the cell is certainly one of the most fundamental biological processes. This process makes use of fundamental chemical principles and cellular components that we have already seen in previous chapters, and some familiarity with these components simplifies our understanding of what could be a mystifying process.

In the next chapter, we will complete our direct discussions of cellular energy by considering the "other side of the coin"—the process of energy utilization. You shouldn't be surprised when you see—again—ATP, membranes, and organelles!

Summary

Overview of Photosynthesis

Photosynthesis is the conversion of light energy into chemical energy. In photosynthesis, photons of sunlight energy absorbed by chlorophyll generate a flow of energy-rich electrons. They in turn provide the energy needed to establish chemiosmotic systems that generate ATP and to reduce $NADP^+$ to NADPH. Both are subsequently used in forming glucose from carbon dioxide and water.

A simplified formula for the overall process is:

$$\text{Water} + \text{Carbon dioxide} \xrightarrow[\text{chlorophyll}]{\text{sunlight}} \text{Glucose} + \text{Water} + \text{Oxygen}$$

The Chloroplast

The chloroplast provides the required structural organization for photosynthesis. Within is the stroma (outer compartment) containing the enzymes for glucose synthesis. Included are numerous grana consisting of light-capturing thylakoids, each enclosing a minute lumen (inner compartment). Thylakoid membranes contain photosystems, electron transport systems, and proton pumps, all used in establishing the chemiosmotic gradient and reducing $NADP^+$.

Photosystems I and II contain chlorophylls and carotenoids that act as light-harvesting antennas. They shunt energy into reaction centers designated P700 and P680. Reaction centers release energy-rich electrons.

Light-activated electrons from reaction centers enter electron transport systems where their energy is used to pump protons into the thylakoid lumen.

The Photosynthetic Process

Light is captured in the light-dependent reactions, while glucose is synthesized in the light-independent reactions.

The Light-Dependent Reactions

Light energy provides for building the chemiosmotic gradient, the synthesis of ATP and the reduction of $NADP^+$ to NADPH. The light reactions occur through noncyclic and cyclic events. The use of light energy in forming ATP is called photophosphorylation.

Light absorbed in photosystem II produces a flow of electrons into the electron transport system. As they pass to photosystem I, their energy is used for pumping protons into the lumen. Light absorbed in photosystem I provides a new energy boost to electrons, which then pass through a second electron transport system, to $NADP^+$, reducing it to NADPH.

The electrons lost from the P680 reaction center are replaced by those from water, which is split, yielding electrons, protons, and oxygen. The oxygen escapes as O_2, while the protons add directly to the chemiosmotic gradient.

Light absorbed in photosystem I produces a flow of electrons from P700 that reach the proton pump where their energy is used to transport protons into the lumen. The depleted electrons return to photosystem I. Water and $NADP^+$ are not involved.

As the chemiosmotic gradient builds, the lumen of each thylakoid becomes rich in protons. Protons escape from the lumen through the CF1 particles where their energy is used to form new high-energy bonds between ADP and P_i.

The Light-Independent Reactions

With the aid of ATP and NADPH, carbon dioxide entering the stroma is fixed into glucose.

In the major steps of the Calvin cycle, (a) carbon dioxide combines with ribulose bisphosphate, which (b) breaks down to two 3-carbon phosphoglycerates (PG). (c) The two PG molecules are phosphorylated by ATP and reduced by NADPH to form G3P.

For every 12 molecules of G3P formed in the Calvin cycle, two are available to form glucose. The remainder recycle, forming ribulose phosphate, which is joined by ATP to form ribulose bisphosphate, the starting molecule.

What Does a Plant Do with Its Glucose?

Uses of glucose include fuel for ATP production and incorporation into cellulose, sugars, and starches. G3P can be converted into amino acids, fatty acids, and nitrogenous bases.

Have You Thanked a Plant Today?

Nearly all life on earth depends on chemical bond energy initially captured by photosynthesizers. Although photosynthesis is energy-costly, organisms make common use of the process because sunlight energy is seemingly unlimited.

Key Terms

photosynthesis • **103**
photon • **103**
carotenoid • **104**
reaction center • **104**
light-harvesting antenna • **104**
photosystem I • **104**
photosystem II • **104**
light-dependent reaction • **107**
light-independent reaction • **108**
noncyclic events • **108**
cyclic events • **110**
carbon dioxide fixation • **112**
ribulose bisphosphate carboxylase (RuBP carboxylase) • **111**
Calvin cycle • **112**
ribulose bisphosphate (RuBP) • **112**

Review Questions

1. Electricity is often described as a flow of electrons. How is photosynthesis similar?
2. Write a balanced general formula for photosynthesis. Comment on the fate of carbon, hydrogen, and oxygen atoms in each of the reactants.
3. Prepare a simplified illustration of a chloroplast, labeling the following structures and areas: surrounding membrane, stroma, granum, thylakoid, lamellae, lumen.
4. Prepare an illustration of one thylakoid, showing the organization of photosystems I and II, the two electron transport systems, and the proton pump.
5. Explain how the stroma and lumen relate to the chemiosmotic system (see Chapter 6).
6. What are the functions of each of the following: chlorophylls and carotenoids, reaction center, electron transport system, proton pump?
7. Summarize the main purposes of the light-dependent reactions.
8. Using your illustration of a thylakoid, add the noncyclic flow of electrons beginning with P680. What two things does this flow accomplish?
9. Explain the role of water in restoring chlorophyll *a* of P680. In what way does the breakdown of each water molecule help increase the energy of the chemiosmotic gradient?
10. Trace the flow of electrons in the cyclic events of the light reactions. What, precisely, is accomplished by a cyclic flow?
11. The chemiosmotic gradient in a chloroplast is often referred to as a proton gradient, a pH gradient, and an electrical gradient. How can it be all three?
12. Explain how the energy of the chemiosmotic gradient is put to work forming ATP. A diagram might help.
13. List the three reactants that gather in the stroma for the light-independent reactions.
14. Why might the enzyme RuBP carboxylase be thought of as the most abundant enzyme on earth?
15. Explain how carbon dioxide enters the Calvin cycle. What does it join, and what happens immediately afterward?
16. What are the precise roles of ATP and NADPH in the Calvin cycle? What happens to their spent forms, ADP and $NADP^+$?
17. What are the two fates of G3P? What fraction of their number goes to each?
18. From a close look at Figure 7.8, account for the 12 molecules of NADPH and the 18 molecules of ATP required for the net yield of just one molecule of glucose.
19. List five ways in which glucose is used by the plant.

8

Glycolysis and Respiration

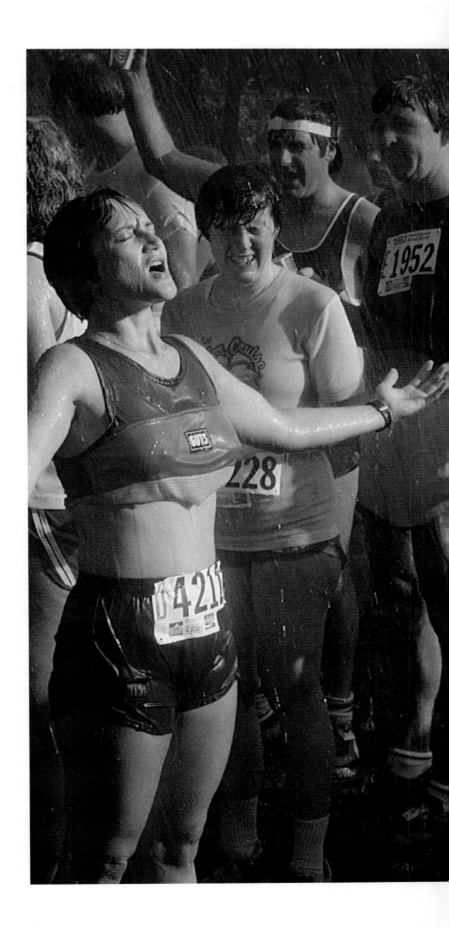

GLYCOLYSIS

CELLULAR RESPIRATION

Citric Acid Cycle

The Electron Transport Systems and Proton Pumping

Chemiosmotic Phosphorylation

THE FERMENTATION ALTERNATIVE

Alcohol Fermentation and Yeast

Lactate Fermentation and Muscles

AS WE COME TO THE END OF PART I OF THIS TEXT, WE FINISH OUR *discussion of the biology and chemistry of cells by considering the formation of usable cellular energy. In the previous chapter, we saw how light energy is captured by photosynthesis and stored in the form of carbohydrate molecules. But how is the energy stored in those carbohydrates recovered by cells for their use in organizing their life? In this chapter, we will see that the process of getting energy out of carbohydrates is interestingly complex, but involves—once again—some of the molecules, organelles, and processes that we have already come to know.*

As this small planet makes its rounds through space, it is constantly bathed in light from the sun. This light literally brings its life-giving energy to earth. This energy maintains the precise and delicately balanced organization of life despite universal tendencies toward disarray. Living things capture this light energy and store it in the chemical bonds of newly formed molecules such as glucose. This, however, is only half of the problem that confronts each living organism. Once the energy is stored, there must be ways of getting it back, releasing it, making it available for life's business. Living things have developed complex and precisely controlled means of releasing this stored energy. Power-laden molecules are handled in elaborate ways, as their energy is gradually drained to support the many processes of life.

Organisms such as plants, which can harness the elusive power of sunlight and store it in new chemical bonds, are called **photosynthetic autotrophs** (light-using self-feeders). Other kinds of organisms, such as humans, which must rely on the stored energy of autotrophs, are called **heterotrophs** (other-feeders). All animals, many single-celled creatures, all fungi, and many bacteria are heterotrophic.

This chapter describes the way cells break down glucose and other cellular fuels, using the potential energy of their chemical bonds to form ATP. Keep in mind that the fuel for energy comes from food, which is basically converted to glucose in order to enter the energy-extracting pathways we will explore in this chapter. We heterotrophic animals eat to obtain food, and we need to remember that the ultimate source of food is photosynthesis. In fact, all life can be viewed (in a biological sense) as the ultimate conversion of sunlight into work (Figure 8.1). In photosynthesis, carbon is fixed (reduced) into carbohydrates to store the energy that was captured from sunlight. We will now see how that energy is released to form ATP while glucose is oxidized to form CO_2 and H_2O. Glucose metabolism in most organisms involves two major processes, *glycolysis* and *cellular respiration* (Figure 8.2).

Glycolysis is the breakdown of glucose into three-carbon molecules. It occurs in the fluids of the cytoplasm and is an **anaerobic** process; *it does not require oxygen.* Glycolysis is a prerequisite for all cellular respiration. **Cellular respiration** is the energy-releasing pathway that occurs within the mitochondrion. Since *it requires oxygen,* cellular respiration is **aerobic.** Cellular respiration consists of two parts. Part I, the citric acid cycle, essentially takes the products of glycolysis and finishes breaking them down into CO_2 and energy stored as NADH and $FADH_2$. Part II is an electron transport system (ETS) that converts the energy in NADH and $FADH_2$ into ATP. (At this point, be sure not to confuse cellular respiration with breathing or gas exchange. Cellular respiration occurs *within* cells.)

You may have heard the old cliché that food is the fuel "burned" by the body. It's a catchy idea, but any such one-step process would be disastrous. Cells must use very gradual, multistepped, enzymatically controlled energy transfers for two closely re-

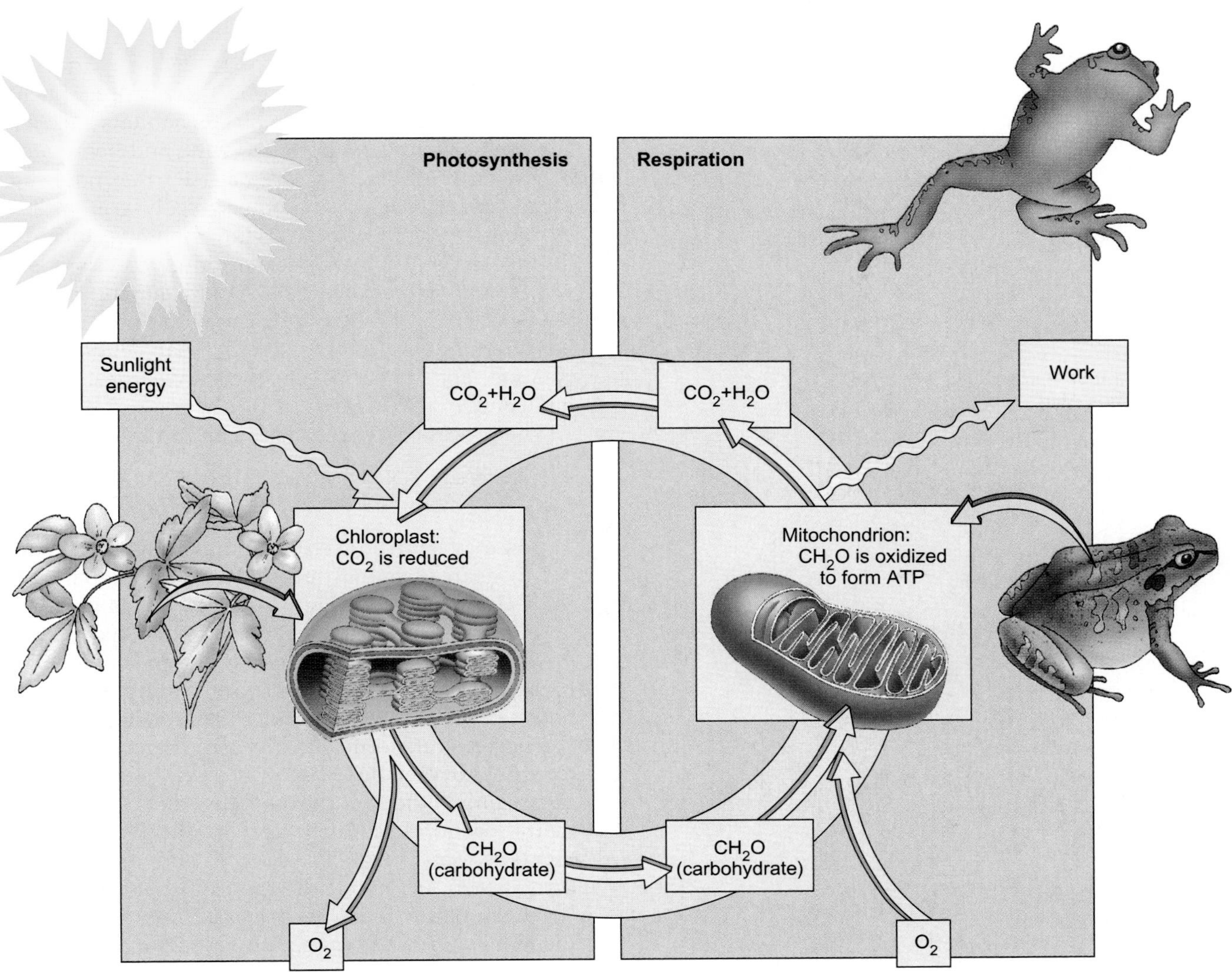

FIGURE 8.1 THE BIOLOGICAL CONVERSION OF SUNLIGHT INTO WORK.
In this basic biochemistry of life, sunlight is captured and converted into energy-rich carbohydrates by the process of photosynthesis, and the production of oxygen is a by-product. The carbohydrate is then used by all forms of life to power the production of work, in many cases using oxygen in the process and converting the carbohydrate to CO_2 and water.

lated reasons. First, such transfers must be tightly controlled to avoid unwarranted energy losses. Second, living things hold a great deal of energy within their bodies, and that energy must be released in a stepwise fashion to avoid production of excessive heat. (Were the energy in your body to be released all at once, you would be incinerated.)

GLYCOLYSIS

Glycolysis, as we've seen, refers to the breakdown (lysis) of glucose. Let's see just how much destruction really goes on. Upon entering the glycolytic pathway, glucose receives an energy boost—two ATP molecules are invested just to get the stable molecule ready to react. Glucose is then broken apart and altered by at least six additional enzymes. Through them it is *phosphorylated, cleaved, oxidized,* and *dephosphorylated* (Figure 8.3).

In the oxidation step, hydrogen is transferred to NAD^+, reducing it to NADH (actually NADH + H^+, but we will use the shorter version; see Chapter 6). Reduced coenzymes, you'll recall, have greatly increased energy levels and can perform useful work. Even more significantly, two reactions lead to **substrate-level phosphorylations,** reactions in which ADP is converted to ATP directly.

Finally, two molecules of a three-carbon molecule called *pyruvate* emerge from the glycolytic pathway.

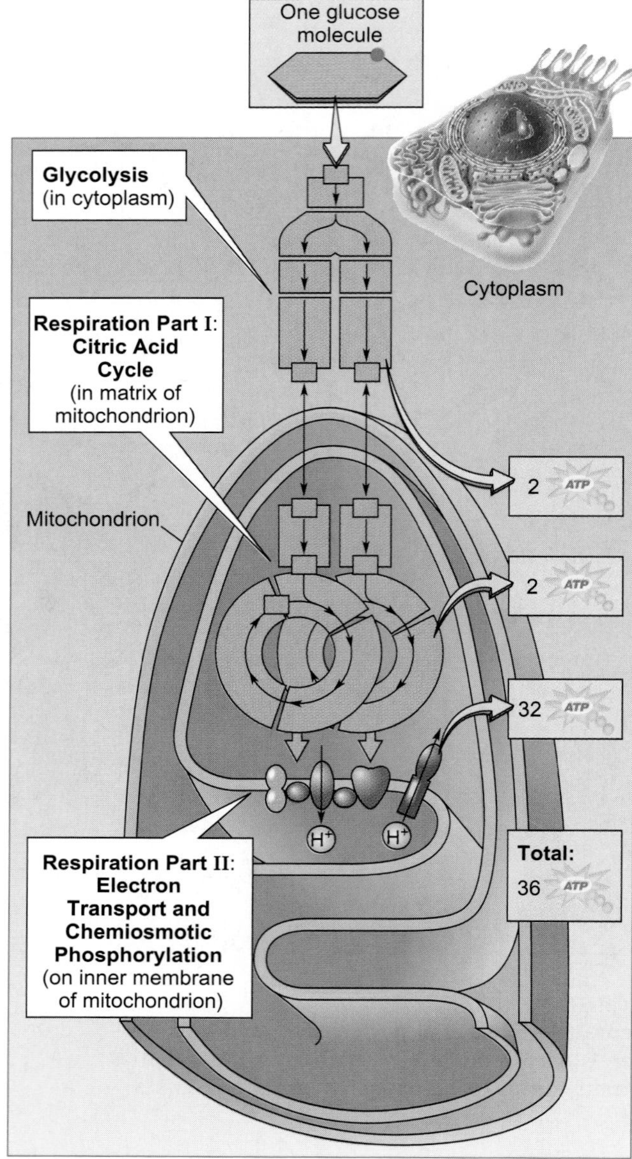

FIGURE 8.2 THE MAJOR PARTS OF GLUCOSE METABOLISM.
Most ATP production in aerobic cells occurs through chemiosmosis within the mitochondrion; however, some ATP is produced in the cytoplasm by glycolysis and by the citric acid cycle.

These are high-energy molecules, containing most of the free energy originally in the glucose. In fact, they will be used to make as many as 34 additional ATP molecules in the next phase.

Note that the glycolytic pathway does not yield much energy (in the form of ATP molecules). Nonetheless, glycolysis is the sole source of energy for many of the earth's anaerobic heterotrophs. Such anaerobes commonly live in airless habitats such as swamps, mucky lake bottoms, wine and beer vats, and even the bowels of animals. Furthermore, in certain circumstances, anaerobic glycolysis is a principal energy-yielding process in vertebrates, including ourselves. Perhaps one reason glycolysis is so widespread is because it is so old, having established itself in the early forms of life. Indeed, it is considered to be far more ancient than cellular respiration, and perhaps even the very earliest metabolic pathway to evolve.

The process of glycolysis can be summarized as:

$$\text{Glucose} + 2\ \text{NAD}^+ + 2\ \text{ATP} + 2\ \text{P}_i + 2\ \text{ADP} \longrightarrow 2\ \text{Pyruvate} + 2\ \text{NADH} + 2\ \text{H}^+ + 4\ \text{ATP}$$

You may have noticed that parts of the glycolytic pathway are essentially a reversal of certain parts of the Calvin cycle (see Figure 7.8), in which glucose was the final product, ATP was consumed, and NADPH was oxidized. This "reversed" metabolic pathway is just one of several interesting contrasts between photosynthesis and glucose metabolism.

The pyruvate formed in glycolysis can take any of a number of metabolic pathways, depending on the organism. Although pyruvate is the product of many reactions and alterations, it still has most of the potential energy of glucose, and energy-rich compounds are not to be ignored. Let's first see what happens to pyruvate in aerobic organisms, and then look at two common pathways in anaerobes.

CELLULAR RESPIRATION

Following glycolysis in aerobic organisms, the scene shifts from the open cytoplasm to the mazelike interior of the mitochondrion. There, pyruvate and NADH enter the processes of cellular respiration wherein most of the cell's ATP is produced. The mitochondrion is a sausage-shaped organelle with an inner and outer membrane (see Figure 4.11). The inner membrane is highly folded into cristae, which greatly increases its surface area. Surface area is quite important since the inner membrane contains numerous electron transport systems (ETS) and proton pumps. In addition, the inner membrane is studded with numerous F1 particles containing the vital ATP-forming enzymes. It is apparent that structure and function are quite interdependent in the mitochondrion, just as they are in the chloroplast. (For a preview of mitochondrial organization, see Figure 8.5.)

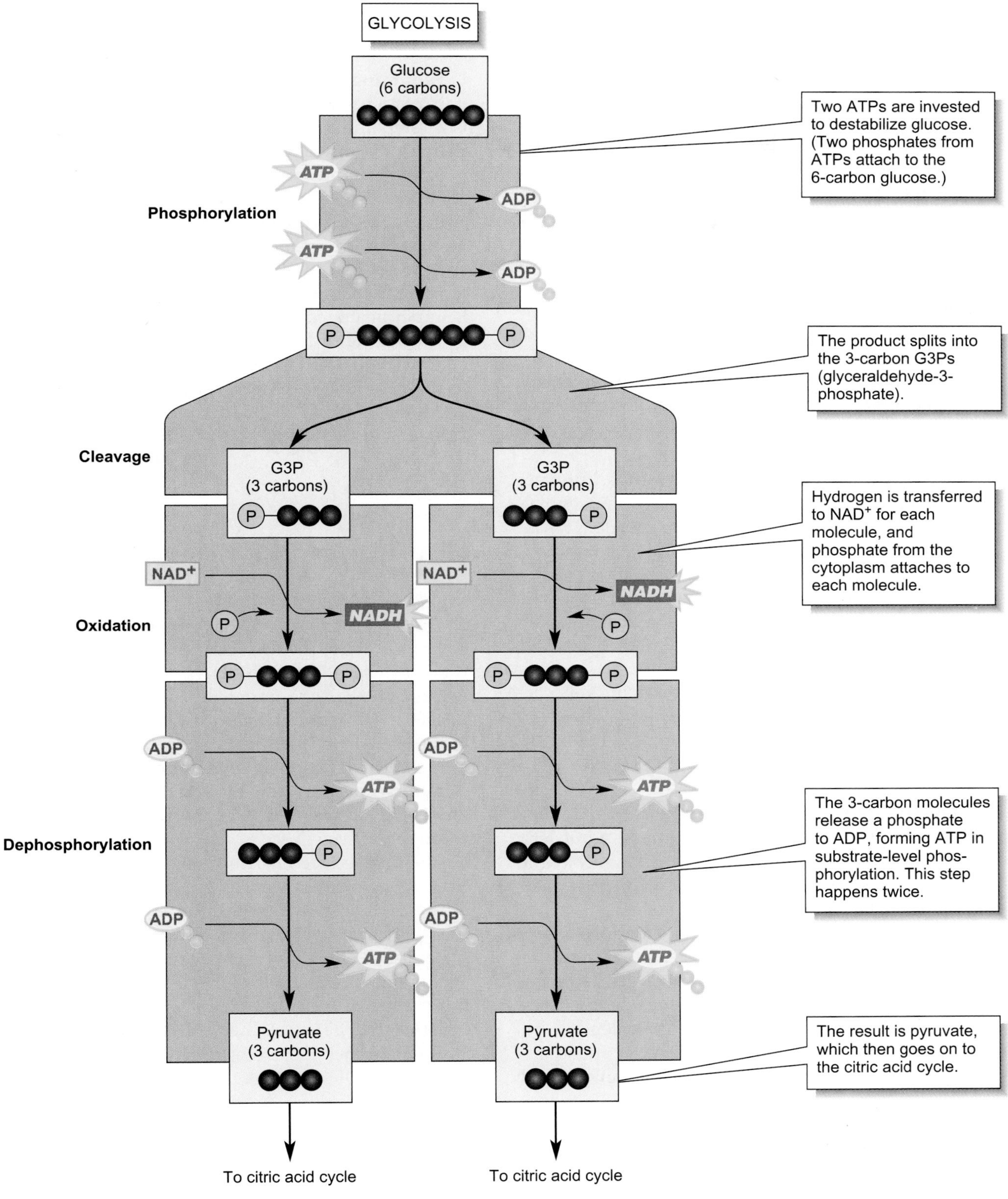
GLYCOLYSIS
Glucose
(6 carbons)
Two ATPs are invested to destabilize glucose. (Two phosphates from ATPs attach to the 6-carbon glucose.)
ATP
ADP
Phosphorylation
ATP
ADP
P
P
The product splits into the 3-carbon G3Ps (glyceraldehyde-3-phosphate).
Cleavage
G3P
(3 carbons)
G3P
(3 carbons)
Hydrogen is transferred to NAD+ for each molecule, and phosphate from the cytoplasm attaches to each molecule.
NAD+
NADH
NAD+
NADH
Oxidation
P
ADP
ATP
ADP
ATP
Dephosphorylation
The 3-carbon molecules release a phosphate to ADP, forming ATP in substrate-level phos-phorylation. This step happens twice.
ADP
ATP
ADP
ATP
Pyruvate
(3 carbons)
Pyruvate
(3 carbons)
The result is pyruvate, which then goes on to the citric acid cycle.
To citric acid cycle
To citric acid cycle
Summary: Net yield is 2 ATP, 2 NADH, 2 pyruvate per glucose molecule.

FIGURE 8.3 **GLYCOLYSIS, THE PATHWAY FROM GLUCOSE TO PYRUVATE.**
Shown here are the major steps involved in processing the 6-carbon glucose molecule to two 3-carbon molecules of pyruvate. Note the input of two ATPs and the recovery of four ATPs for each glucose molecule. Thus we say that glycolysis produces a net gain of two ATPs.

Citric Acid Cycle

The citric acid cycle was first described in 1937 by the late Oxford biochemist, Sir Hans Krebs. (Hence, it is often called the Krebs cycle.) The pathway is a cycle in that it both begins and ends with a four-carbon molecule called *oxaloacetate* (Figure 8.4). So why isn't it called the "oxaloacetic acid cycle"? Aside from the verbal trauma this might generate, Krebs noted that the first new substance formed by the cycle happens to be citrate (the familiar *citric acid* of oranges and lemons). Let's follow one molecule of pyruvate through the cycle, bearing in mind for bookkeeping purposes that two pyruvate molecules were generated from each molecule of glucose that originally entered glycolysis.

Acetyl-CoA Gets the Cycle Going. Pyruvate readily enters the mitochondrion, but it must be altered before it can enter the citric acid cycle. This alteration results in the formation of a compound called acetyl-CoA, and the release of $NADH^+ + H^+ + CO_2$ (see Figure 8.4). The three-carbon pyruvate loses CO_2 to become the two-carbon acetyl group, and coenzyme A is added (to become the CoA part of acetyl-CoA).

Now, fully prepared, acetyl-CoA joins oxaloacetate to form six-carbon citrate, whereupon the coenzyme A is released to recycle. (For a look at other fuels entering the citric acid cycle, see Essay 8.1.)

Energy Transfers and the Release of CO_2. The citric acid cycle is outlined in Figure 8.4, but let's make a few important observations. First, each of the major reactions is facilitated by its own specific enzyme. More importantly, energy transfers occur during certain crucial steps. Most are oxidations where NAD^+ is reduced to NADH or FAD is reduced to $FADH_2$. Armed with their newly acquired reducing power, both coenzymes will deliver high-energy electrons to nearby electron transport systems.

Second, with each turn of the cycle, substrate-level phosphorylation generates one molecule of ATP. (Actually, another triphosphate, **GTP** (guanosine triphosphate), is generated in the cycle, but it transfers its terminal phosphate to ADP in a second reaction.)

Finally, note where carbon dioxide leaves the citric acid cycle. This is the carbon dioxide all aerobic cells release during respiration and, incidentally, the same carbon dioxide we are exhaling right now. After the release of two molecules of CO_2, only four-carbon molecules remain. In the final step, four-carbon oxaloacetate is regenerated.

Summing Up the Citric Acid Cycle. For each pair of pyruvate molecules completing the acetyl-CoA step and passing through the citric acid cycle, there is a total yield of two molecules of ATP, eight of NADH, and two of $FADH_2$. To obtain a total yield from glucose so far, we add the two molecules of ATP and two of NADH from glycolysis. The total yield so far is four molecules of ATP, ten of NADH, and two of $FADH_2$. Cytoplasmic NADH cannot enter the mitochondrion because the outer membrane is impermeable to NADH. Instead, its electrons and protons are shuttled in by carriers that reduce mitochondrial NAD^+. The cytoplasmic NAD^+ is left free to recycle, thus keeping glycolysis going. But what about the yield of ATP? Four molecules of ATP seem meager, but that is not yet the whole story.

The Electron Transport Systems and Proton Pumping

Once NAD^+ and FAD have been reduced in the citric acid cycle, the final events of cellular respiration can begin. The NADH now reduces the first carriers of the electron transport system (Figure 8.5). $FADH_2$ has less reducing power than NADH, but it reduces a carrier further along in the system. Each reduced carrier can use its new reducing power to pump two protons directly into the outer compartment and to shunt two energetic electrons along through the system.

The first reduced carriers pass high-energy electrons along from carrier to carrier. Then, at key proton-pumping sites, the electron energy is used to transport more protons across the membrane from the matrix to the outer compartment. Note that this arrangement of compartments is opposite to that of the chloroplast's thylakoid (see Chapter 7). Also notice that the ETS here has three proton-pumping sites where each ETS in the thylakoid has only one.

Each NADH molecule accounts for the transfer of six protons into the outer compartment; each $FADH_2$ molecule accounts for four. We will be summing up respiration soon, whereupon we will equate the number of NADHs and $FADH_2$s and protons pumped with the final generation of ATP.

We have seen the chemiosmotic system at work in our model (Chapter 6) and in the thylakoids of the chloroplast (Chapter 7), and here the results are similar. But before we move in for a closer look at chemiosmosis in the mitochondrion, let's answer a question that may have occurred to you.

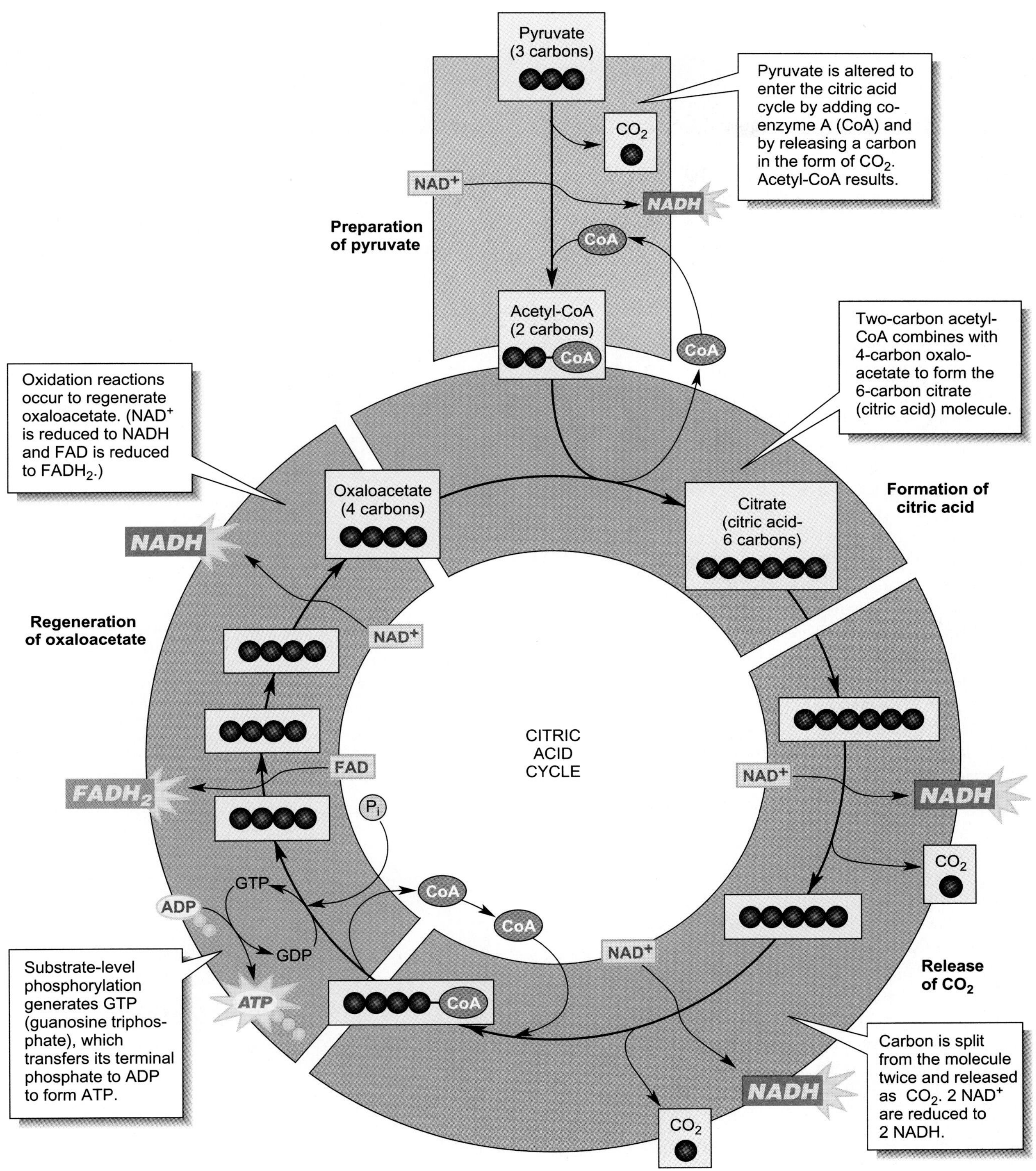

FIGURE 8.4 **DETAILS OF THE CITRIC ACID CYCLE.** The major steps include the preparation of pyruvate, the formation of citric acid (which can also be referred to here as citrate), the release of CO_2, and the regeneration of oxaloacetate. Note also the production of one ATP per molecule of pyruvate. This means that there are two ATPs produced by the citric acid cycle for each glucose molecule (see Figure 8.3).

Summary: Each glucose molecule produces two pyruvates, so the cycle turns two times. Hence:

2 pyruvate $\longrightarrow$ 2 ATP + 8 NADH + 2 $FADH_2$ + 6 CO_2
(includes 2 NADH from acetyl-CoA step)

Essay 8.1 Alternative Fuels of the Cell

Glucose furnishes much of the energy that living things need to generate ATP, but it is not the only usable fuel. Fats and proteins are also important sources of energy. As a matter of record, fats yield twice the energy of carbohydrates, on a gram-for-gram basis.

However, before fats and proteins can eventually enter the respiratory pathway, they must be simplified through the digestive process. During digestion, fats are broken into fatty acids; proteins, into amino acids. These are then distributed to the cells by the circulatory system.

The fatty acids destined for oxidation are transported into the mitochondrion, where they are fragmented into a number of two-carbon acetyl-CoA molecules. The latter process also yields a bonus—a number of NADH and $FADH_2$ molecules. Acetyl-CoA enters the citric acid cycle; NADH and $FADH_2$ go directly to the inner membrane, where they contribute electrons to the electron transport systems and yield protons to the chemiosmotic gradient. The ATP yield from fatty acids is considerable. For instance, the complete oxidation of palmitic acid, a 16-carbon fatty acid, yields over 130 molecules of ATP.

If the amino acid products of protein digestion are to be used as fuels, they must first be deaminated—have their amino groups removed. Since the amino groups form ammonia (which can be harmful), they must be rapidly eliminated or rendered harmless through additional reactions. The useful modification product from the 20 amino acids may differ, with some occurring as pyruvate, others as acetyl-CoA, and still others as the four-carbon acids of the citric acid cycle.

How much ATP can be derived from amino acid fuels depends on where, after modification, they enter the cycle. The greatest yields of NADH (and $FADH_2$) and ATP come from those amino acids that are converted to pyruvate.

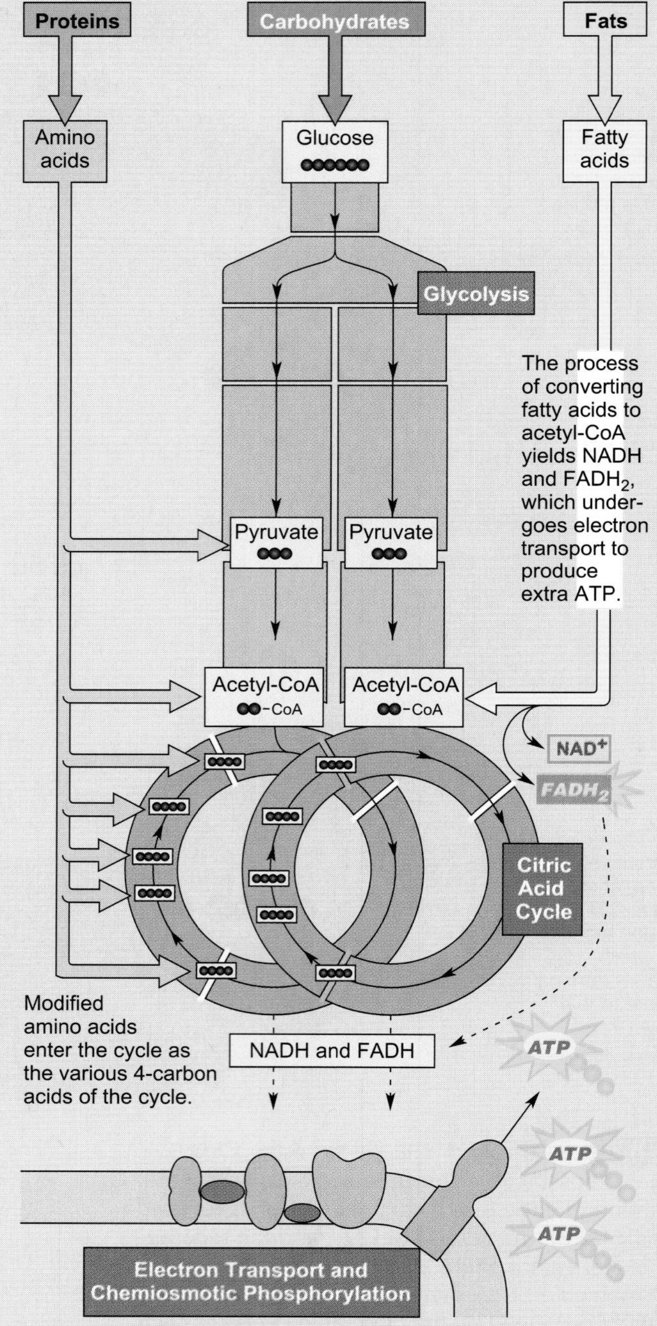

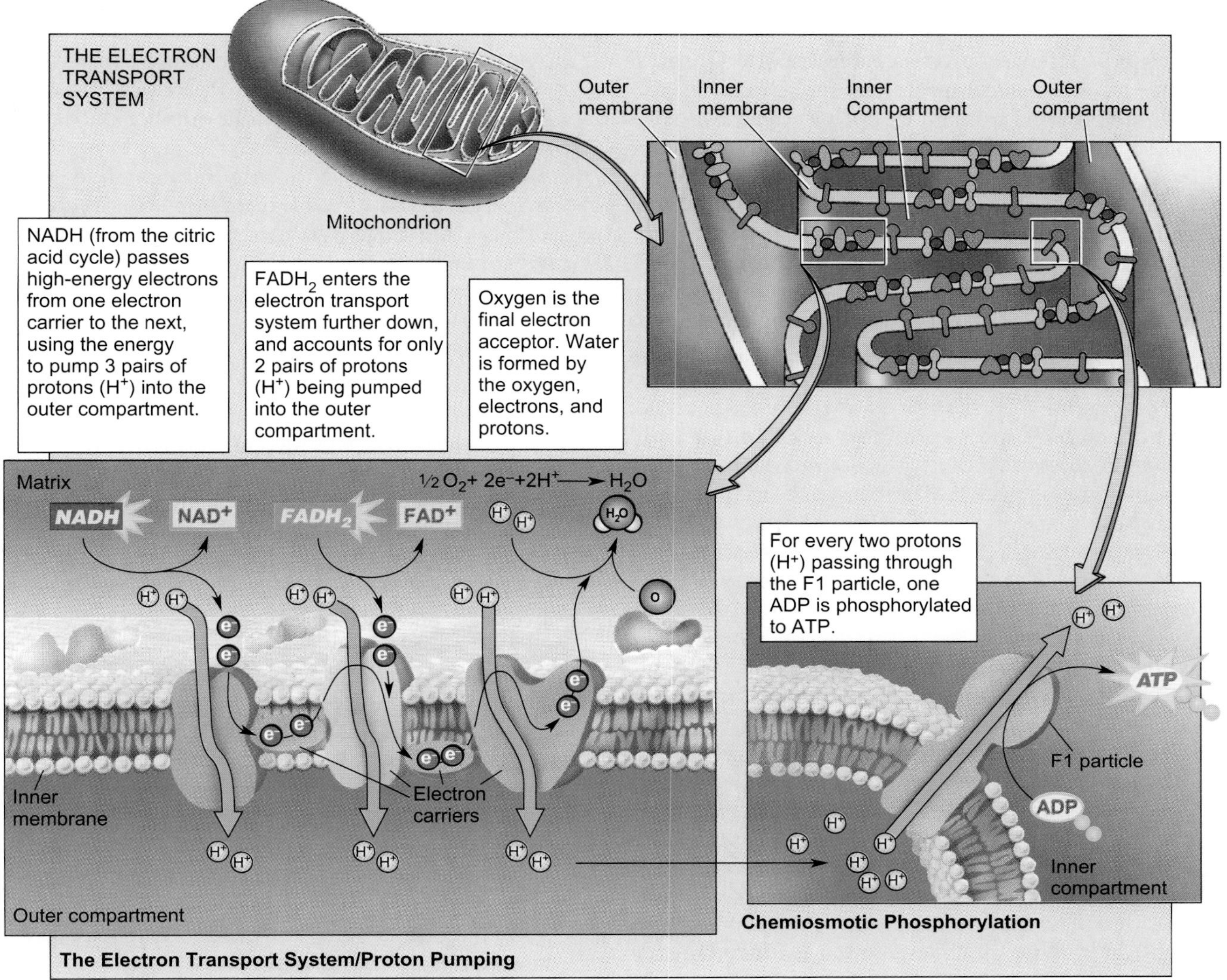
THE ELECTRON TRANSPORT SYSTEM
Outer membrane
Inner membrane
Inner Compartment
Outer compartment
Mitochondrion
NADH (from the citric acid cycle) passes high-energy electrons from one electron carrier to the next, using the energy to pump 3 pairs of protons (H+) into the outer compartment.
FADH2 enters the electron transport system further down, and accounts for only 2 pairs of protons (H+) being pumped into the outer compartment.
Oxygen is the final electron acceptor. Water is formed by the oxygen, electrons, and protons.
Matrix
1/2 O2+ 2e−+2H+ → H2O
NADH
NAD+
FADH2
FAD+
Inner membrane
Electron carriers
Outer compartment
The Electron Transport System/Proton Pumping
For every two protons (H+) passing through the F1 particle, one ADP is phosphorylated to ATP.
ATP
F1 particle
ADP
Inner compartment
Chemiosmotic Phosphorylation

FIGURE 8.5 CHEMIOSMOSIS IN THE MITOCHONDRION.
Here is where most ATP comes from. NADH and $FADH_2$ that were produced in the citric acid cycle add their hydrogens to the electron transport system at two key sites. In both instances, protons pass across the membrane, and energetic electrons pass through the electron transport system. In all, there are three proton-pumping sites available, each of which passes two protons across simultaneously to eventually power ATP formation at the F1 particle. Each NADH molecule powers three sites, while each $FADH_2$ powers two. Finally, note the fate of energy-depleted electrons. Each pair of electrons is joined by an oxygen atom and two protons from the matrix to form water.

What About the Oxygen? Electrons passing through the carriers of the electron transport system reach the last carrier in an energy-depleted condition. This is where oxygen comes in. Specifically, oxygen acts as the final electron acceptor. Oxygen combines with the electrons, along with free protons from the matrix, to form water (H_2O).

Oxygen, then, in its role as a final electron acceptor, serves as an electron dump. And that is why we breathe: to provide a receptacle for spent electrons that were once parts of high-energy food molecules. Without oxygen, the flow of electrons stops, and energy production by the mitochondria quickly ceases.

CHEMIOSMOTIC PHOSPHORYLATION

The phosphorylation of ADP in the mitochondrion (Figure 8.5) bears a general similarity to the process that occurs in the thylakoid (Figure 7.6). Interestingly, the spatial orientation of the compartments is reversed. Thylakoid protons are transported from an outer to an inner compartment; mitochondrial protons are pumped from an inner to an outer compartment. But the important aspect of chemiosmosis is only that a proton gradient be established. Once the mitochondrial chemiosmotic gradient is established, and ADP and P_i are present, all that remains is for protons to pass down their gradient through the F1 particles. In these particles, the energy of the chemiosmotic system is used by phosphorylating enzymes to generate new high-energy ATP bonds.

Cellular respiration in the mitochondrion generates far more ATP than does anaerobic glycolysis: The anaerobic process produces a net gain of only two molecules of ATP per glucose molecule, and the aerobic process nets 36. There is some disagreement over how many escaping protons it takes to make one molecule of ATP, but the consensus is two. That is, for every two protons passing though an F1 particle, one ADP is phosphorylated to ATP. Each NADH contributes the formation of three ATPs, while each $FADH_2$ contributes two ATPs. The overall yields of coenzyme, proton, and ATP from glucose are shown in Figure 8.6.

If glucose is completely oxidized experimentally by burning it and measuring the heat, the energy yield is higher than that recovered by glycolysis and oxidative respiration. The 36 ATP molecules from glycolysis and oxidative respiration actually represent an ATP efficiency of about 41%. All things considered, a fuel efficiency of 41% isn't bad. A well-tuned automobile engine has a fuel efficiency of less than 25%.

What happens to the other 59% of glucose energy? In this instance, 376 kcal of heat is released for each mole of glucose passing through the respiratory mill. The heat eventually escapes from the organism, but in warm-blooded creatures, the heat is not entirely wasted. In fact, heat from respiration is vital in maintaining the constant body temperature that is part of our adaptation to a wide variety of environmental temperatures. (We'll have more to say about this subject in Chapter 37.)

The Fermentation Alternative

We've seen how the glucose molecule is ravaged by glycolysis, and how the resulting pyruvate molecules are taken through the equally active citric acid cycle. But this route is not always the destiny of pyruvate. Some pyruvate goes through **fermentation pathways,** additional metabolic reactions that do not utilize oxygen. There are two major fermentation pathways, and their end products are quite distinct. In vertebrates and in a number of bacteria the end product is *lactate (lactic acid).* In yeast, it is *ethanol (ethyl alcohol).* In both fermentative pathways, the NADH formed during glycolysis reduces (adds hydrogen to) pyruvate (Figure 8.7). This is a vital step since it frees NAD^+, permitting it to recycle and glycolysis to continue.

Alcohol Fermentation and Yeast

The common brewing and baking yeasts and some bacteria use the alcohol fermentation pathway. The yeasts, however, are metabolically quite versatile. They are metabolic "switch-hitters" and can metabolize glucose aerobically as well as anaerobically. When glucose is limited but oxygen is available, yeasts become aerobic. They break down glucose all the way to carbon dioxide and water, just as we do, thus gaining a maximum yield of ATP.

When glucose is abundant, however, yeasts take the anaerobic alternative, even in the presence of oxygen. This seemingly wasteful process enables the yeast to avoid the high metabolic cost of producing the more complex enzyme system needed for oxidative respiration.

The fermentative activity of yeast is obviously important to the brewer who needs the ethanol, but it is also critical to the baker who uses the gaseous carbon dioxide to make bread rise. The chemical reaction that they both depend on is:

$$\text{Pyruvate} + \text{NADH} + \text{H}^+ \longrightarrow$$
$$\text{CO}_2 + \text{Ethyl alcohol} + \text{NAD}^+$$

Lactate Fermentation and Muscles

Most people are familiar with the taste of ethyl alcohol, but lactate also has a familiar taste. It is lactate that gives sauerkraut its tangy flavor. (The two products have quite different physiological effects, as witnessed by the fact that drivers are not pulled over to be tested for "cabbage-breath.") Lactate also forms in other foods. Lactose (milk sugar), converted to lactate, lends its flavor to buttermilk and yogurt. Essentially, the food and beverage industries make good use of what are really anaerobic waste products. We should keep in mind, though, that from the organism's point of view, the important thing is the recycling of NAD^+.

Perhaps of greater interest is the formation of lactic acid by cells in your own body. The formula for

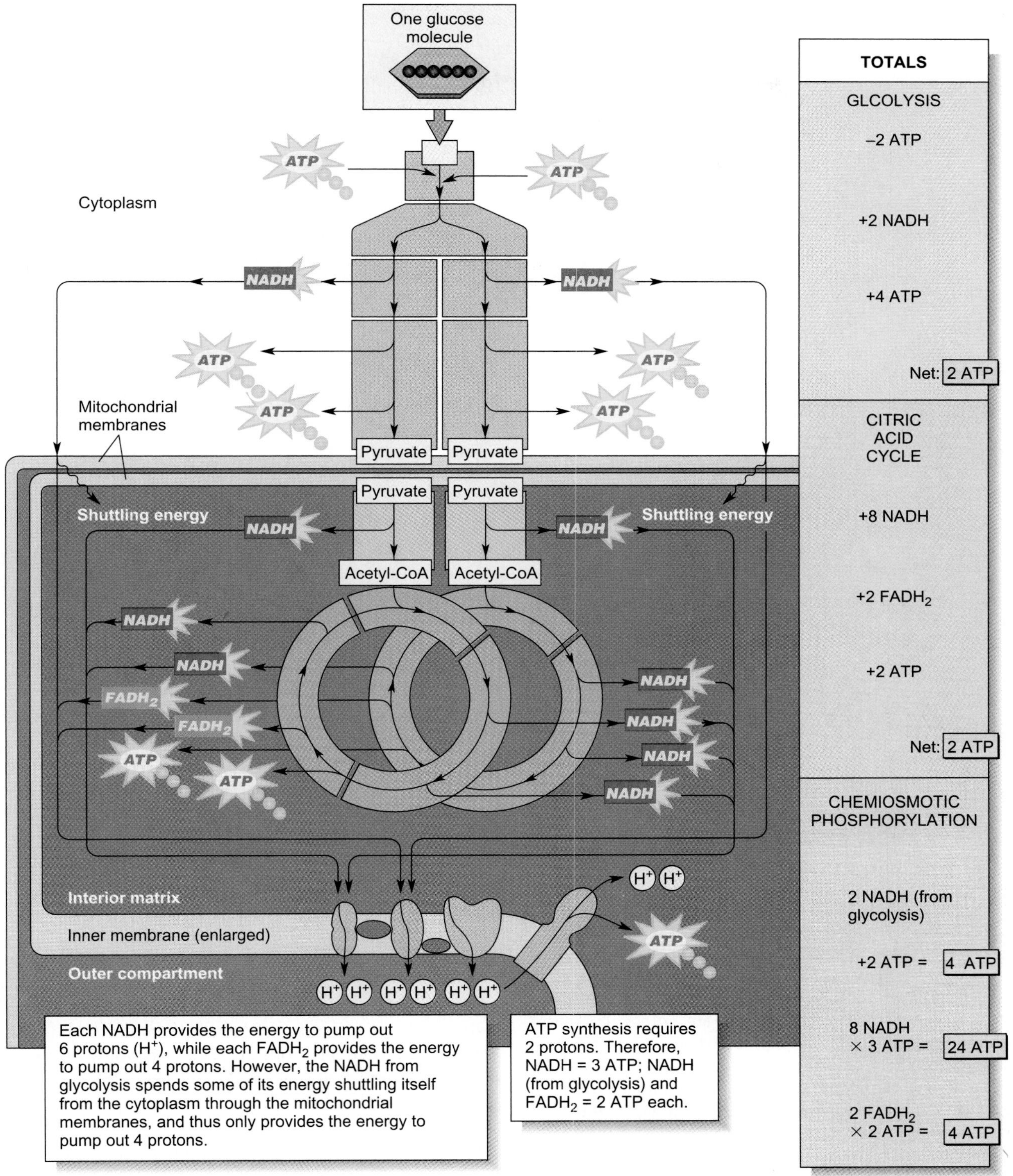

FIGURE 8.6 THE ENERGY PRODUCTION BALANCE SHEET.
The general outline of glucose metabolism that was outlined in Figure 8.2 is expanded here to show the detail of where the ATPs are derived. Two ATPs are derived from glycolysis, two from the citric acid cycle, and the rest from chemiosmotic phosphorylation.

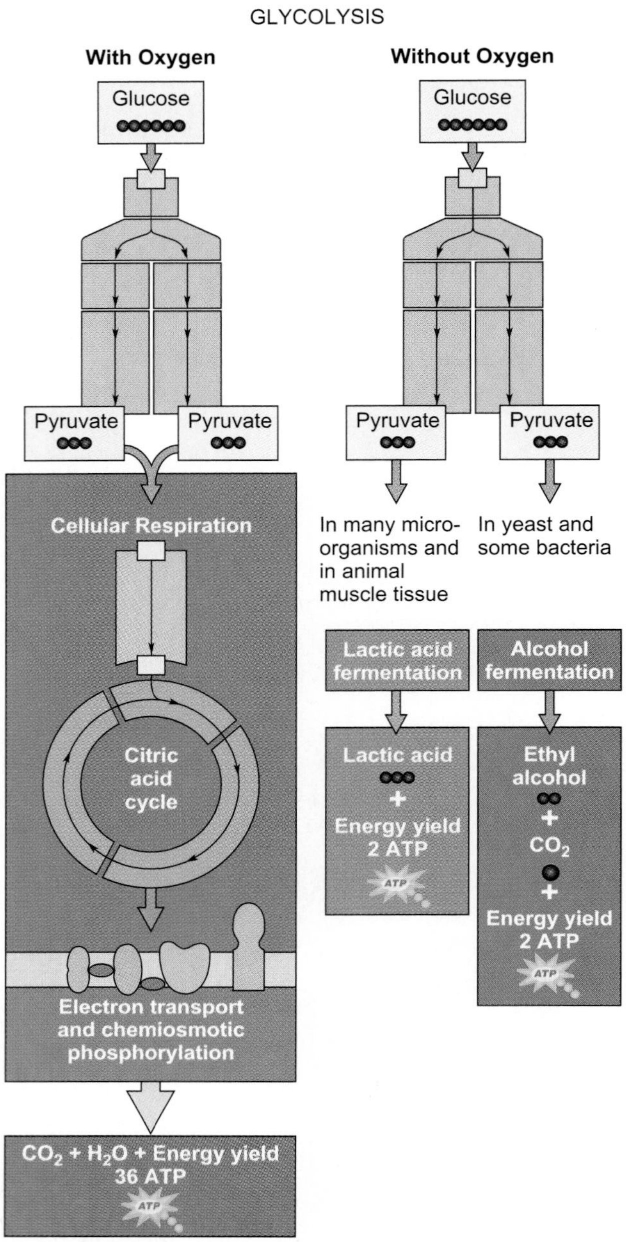

the lactate fermentation pathway, starting with pyruvate, is:

$$\text{Pyruvate} + \text{NADH} + \text{H}^+ \longrightarrow \text{Lactate} + \text{NAD}^+$$

As in the alcohol fermentation pathway, when NADH is oxidized, the NAD^+ formed can recycle to be used again.

In your muscles, aerobic respiration provides adequate ATP during periods of light activity, such as casual strolling. But should you spot the neighbor's big, mean dog clearing its fence, you may wish to consider a faster gait, say sprinting! Under such conditions the human circulatory system simply cannot provide sufficient oxygen to sustain the sudden new requirement for ATP. So you "go anaerobic," rapidly converting muscle glycogen (animal starch) to glucose and sending the glucose through glycolysis and the simpler and much faster lactate fermentation pathway. In that way you can meet your elevated ATP requirements, at least for a time.

How long a person can rely on glycolysis depends on physical conditioning. Conditioning is partially a matter of how much glycogen is stored and how efficiently the cardiovascular system can deliver oxygen to the muscles. Switching to the anaerobic lactate fermentation pathway can be vital at times, but it has its price—something called an **oxygen debt.** This can be defined as the amount of oxygen (processed through cellular respiration) needed to clear lactate from the muscles, replenish muscle glycogen, and restore the muscles to their former state of readiness.

Oxygen debt builds because, with vigorous muscular activity, ATP reserves quickly dwindle, lactate rapidly accumulates in the muscle tissue, and this leads to fatigue. Lactate must be removed from the muscles by transporting it via the bloodstream to the liver where, through a process called *gluconeogenesis,* it is converted back to glucose. The glucose is then returned to the muscles where it is restored to glycogen. Gluconeogenesis is an energy-costly process, requiring six ATP molecules for the conversion of two molecules of lactate to one molecule of glucose. However, from an adaptive point of view, it's worth the price, especially if survival depends on sudden, vigorous movement (as in staying ahead of the dog).

Since recovery is essentially an aerobic process, extra oxygen is required. This explains the familiar gasping and panting that follow the sprint. Even a well-conditioned athlete may breathe a bit more heavily after a 100-meter dash. The heavy breathing eventually repays the oxygen debt as muscle glycogen reserves are replenished.

FIGURE 8.7 THREE COMMON FATES OF PYRUVATE. In the presence of oxygen, pyruvate enters the mitochondrion where it is first converted to acetyl-CoA and then completely oxidized for a maximum ATP yield. In the absence of oxygen, there are two possibilities for recovering some additional energy (but not nearly as much as with oxygen). Pyruvate entering the lactate fermentation pathway is reduced by NADH, yielding lactate and NAD^+. In the alcohol fermentation pathway, pyruvate is decarboxylated, yielding CO_2, and reduced by NADH, yielding ethyl alcohol and NAD^+.

FIGURE 8.8 **ENERGY COSTS.**
Crocodiles can move extremely rapidly, but only for short bursts, and the cost to them is long periods of rest while their energy stores are revitalized.

Interestingly, many smaller animals generate all or most of their muscle ATP aerobically and have no problem with lactate accumulation and oxygen debt. Examples include seemingly tireless migratory birds and small, fast-running mammals. But the great body mass of larger animals does not allow their circulatory systems to keep up with the increased oxygen demands. Nile crocodiles, for instance, are generally sluggish creatures. But when threatened or when stalking prey on land, they can make astonishingly fast charges and can lash their tails about with results that are legendary. Following such outbursts, the giant reptiles must remain still, requiring many hours to repay the oxygen debt (Figure 8.8).

PERSPECTIVES

Now our look at energetics is complete. We have seen how glucose is produced by autotrophs that use light energy to form this precious commodity. Its great potential energy, stored in stable chemical bonds, is then tapped in complex respiratory processes by both autotrophic and heterotrophic cells as they meet the energy requirements of life.

So not only is our discussion of cellular energy complete, but so is our discussion of basic cellular structures and processes. Don't worry that we will never encounter some of these "cell" things again, as we certainly will! In later chapters, we will see how different cell types are gathered together to form tissues and organs—in both plants and animals. We will also see how energy is derived from food by the actions of some organs and systems, how oxygen is delivered by other systems, and how that energy is used for motion by yet other systems.

However, in the very next unit of the text, we will turn our attention to how cells reproduce. Again we will see that cellular organelles play major roles, as do the molecules and chemicals that we have come to know.

SUMMARY

Photosynthetic autotrophs (plants, algae, and certain bacteria) use light energy and simple inorganic molecules for food-making, whereas heterotrophs (animals, fungi, and many protists and bacteria) rely on organic compounds from plants or other organisms.

The extraction of energy from glucose and other fuels occurs in two ways: an anaerobic, cytoplasmic process called glycolysis—and an aerobic, mitochondrial process called cell respiration. Both are multistep metabolic pathways.

GLYCOLYSIS

Glycolysis is the breakdown of glucose. It requires the investment of two molecules of ATP. The final products of glycolysis are four molecules of ATP, two of NADH, and two of pyruvate. The net yield is two molecules of ATP.

CELLULAR RESPIRATION

High-energy pyruvate next enters the mitochondrion, where most of an aerobic cell's ATP is produced through cell respiration.

Each pyruvate joins coenzyme A in preliminary reactions that yield carbon dioxide, NADH, and acetyl-CoA. The latter enters the citric acid cycle, joining oxaloacetate to form citric acid. Further reactions in the cycle yield two more carbon dioxide molecules, one of ATP, and—through four oxidation reactions—three molecules of NADH and one of FADH.

Electron transport systems (ETS) in the inner membrane contain three proton-pumping sites. Each NADH adds two energy-rich electrons to the first carrier of the ETS, thereby powering the transport of six protons into the outer compartment. Each $FADH_2$ reduces a carrier further along, powering the transport of only four protons into the outer compartment. NAD^+ and FAD recycle.

Proton pumping creates a steep proton gradient (representing a high potential-energy state). The energy is used to generate ATP as protons pass down their gradient through the F1 particles. There, phosphorylating enzymes join P_i to ADP. Each ATP produced requires the transit of two protons from the outer to the inner compartment. The total yield for glucose (including glycolysis) is 36 molecules of ATP per molecule of glucose, an efficiency of 41%. The remaining energy is given off as heat.

THE FERMENTATION ALTERNATIVE

Ethyl alcohol fermentation and lactate fermentation are two major anaerobic pathways pyruvate may follow. An important aspect of each is the reduction of pyruvate by NADH, which frees NAD^+ for recycling to glycolysis.

Yeasts can metabolize glucose aerobically as well as anaerobically. In the anaerobic process pyruvate is decarboxylated and reduced, yielding carbon dioxide and ethyl alcohol and freeing NAD^+ for recycling. During baking, the carbon dioxide is essential to the "rise" of bread dough.

The food industry uses lactate from bacterial fermentation to produce a tangy flavor in many foods. The lactate fermentation pathway also occurs in vertebrate muscle. During heavy muscular exertion, vertebrates often rely on anaerobic glycolysis for rapid ATP generation. The end product, pyruvate, is reduced by NADH to lactate, freeing NAD^+ for recycling. The accumulation of lactate produces oxygen debt, whereupon cell respiration must be used in the clearing of lactate and restoration of the muscles. Lactate is transported to the liver where, during gluconeogenesis, ATP is expended to convert lactate back to glucose. Small vertebrates can rely almost entirely on aerobic respiration. Larger vertebrates cannot do this, and fatigue is a common problem.

Key Words

photosynthetic autotroph • **120**
heterotroph • **120**
glycolysis • **120**
anaerobic • **120**
cellular respiration • **120**
aerobic • **120**
substrate-level phosphorylation • **121**
GTP • **124**
fermentation pathway • **128**
oxygen debt • **130**

Review Questions

1. Compare glycolysis and cellular respiration as follows: the site where they occur, the need for oxygen, and the yield of ATP.
2. In general, why are such multistepped, biochemical pathways used by the cell for extracting energy from its fuels?
3. Briefly summarize the steps of glycolysis. What happens to the molecule, and how does the reaction lead to a useful energy transfer?
4. List the final products of glycolysis. What is the net amount of ATP yielded?
5. Develop an illustration of a mitochondrion, labeling the inner and outer membranes and compartments and several F1 particles.
6. Relate the diagram from Question 5 to the chemiosmotic system (review Chapter 6 as necessary).
7. Where in the mitochondrion do the following occur: conversion of pyruvate to acetyl-CoA, citric acid cycle, electron transport and proton pumping, chemiosmotic phosphorylation?
8. Summarize the acetyl-CoA step, listing the reactants and products. Which products are of no further value to the cell?
9. State the numbers (per glucose) of ATPs (net) and reduced coenzymes yielded in the following: (a) glycolysis, (b) acetyl-CoA step, (c) citric acid cycle, (d) grand total.
10. Prepare an illustration of a mitochondrial electron transport system. Label the electron carriers and proton pumps and indicate the carriers that are reduced by NADH and $FADH_2$.
11. How many protons does the chemiosmotic gradient gain by the action of NADH? $FADH_2$?
12. Oxygen is often described as the final electron acceptor in the mitochondrial electron transport system. Explain.
13. Explain how the free energy of the proton gradient is put to work making ATP. A diagram might be helpful. What is the ratio between protons pumped and ATP generated?
14. Carry out the calculations that reveal the percent efficiency of glucose metabolism in the cell.
15. Summarize the reactions leading to ethyl alcohol in the alcohol fermentation pathway. Under what conditions do yeasts use this pathway?
16. What organisms besides vertebrates make use of the lactate pathway? How does the food industry make use of these organisms?
17. Under what conditions do our bodies make use of the anaerobic production of ATP? Of what advantage is this?
18. What is oxygen debt, and how is it repaid? What makes it possible for small animals to rely heavily on lactate fermentation without fatigue?

CONTINUITY OF LIFE

Of the several obvious characteristics of life, the ability to reproduce is one of the most universal and fundamental. Virtually every individual organism has arisen from a preexisting organism. Here we begin our investigation of cellular reproduction, and in the process we will examine biological information. What is biological information? It is the data from which cells and organisms are constructed, and it lies resident within our DNA. We will see that it is this information that directs everything from the color of our eyes to the taste of sweet corn.

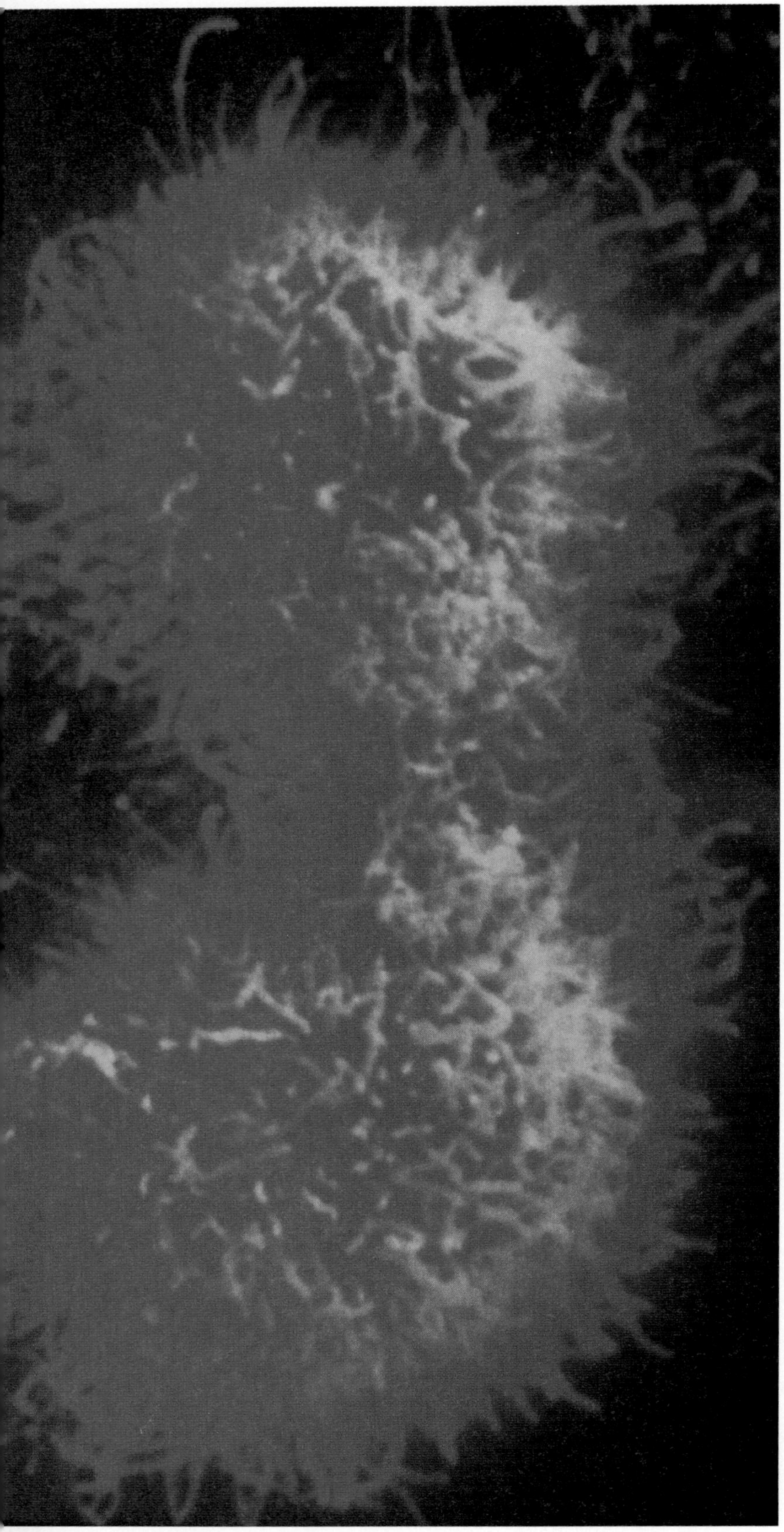

9

Cell Division: Life Continues

DNA Replication: A Brief Look

Bacterial Division

Eukaryotic Division

Chromosome Structure

Chromosomes, Division, and the Cell Cycle

Mitosis

AT THE HEART OF BIOLOGY IS THE IDEA THAT LIFE COMES FROM PREEX-*isting life. So any given cell came from some cell that existed before, and that cell came from one before it, and so on. Later we will look at how it all began, but here we see how one cell forms two. We will see that the process is remarkably similar for all cell types in virtually all living things—and is designed to ensure that the two resulting cells have all the necessary genetic and cellular components to carry on an independent existence.*

Even as you sit reading this fascinating account, your body is changing at an amazing rate. Certain cells are being created, others are dying. As a result of cellular reproduction, you are producing millions of red blood cells, as well as several million other cells in your body. Since you have probably ceased physical growth, an equal number of cells must have died over the same period. The process by which new cells arise from preexisting cells is called cell division.

For single-celled organisms such as bacteria, cell division is equivalent to reproduction of the organism. For multicellular organisms, cell division allows for growth and maintenance of tissues. In addition, many reproductive cells such as sperm, eggs, spores, and pollen are also produced by a special kind of cell division (see Chapter 10).

In this chapter, we will consider just how this very important process is carried out. We will concentrate on the main theme of division—that is, the (nearly equal) separation of the necessary material into the newly forming daughter cells. It turns out that most of the cellular contents are simply parceled out randomly by the division process, so that each daughter cell receives approximately equal allotments of cell organelles. But the distribution of DNA is something different altogether. In fact, the precise delivery of a full complement of the cell's DNA to each of its daughter cells is the single most important result of cell division.

The most straightforward way to view this whole process is to consider cell division as simply a part of the life cycle of the cell. A cell grows, metabolizes, duplicates its DNA, and divides—producing two daughter cells that themselves continue the process.

In order for each of the daughter cells to receive a full complement of DNA, DNA replication must precede cell division. While we will consider the details of DNA replication in Chapter 14, a brief look at the process will highlight the important concepts as they apply to cell division.

DNA REPLICATION: A BRIEF LOOK

Recall from Chapter 3 that DNA is an extremely long, double-stranded polymer consisting of subunits called nucleotides. The two strands are held together by numerous hydrogen bonds between nucleotides on the opposite strands. The hydrogen bonds spontaneously wind the two strands into a double helix (Figure 9.1).

During *DNA replication*, a complex of enzymes moves along the double helix, unwinding it at certain places and breaking the hydrogen bonds between the two strands. Where the bonds are broken, the two strands untwist, exposing the individual nucleotides (Figure 9.1b). The exposed nucleotides then join with new nucleotides floating free in the nucleoplasm. The additional nucleotides are not added at random, however; nucleotides join in a highly specific manner known as **base pairing.** Because of this specificity, base pairing replaces the precise nucleotides that were present before the molecule was unwound and the two DNA strands separated.

In a sense, the DNA molecule is split in half along its length, and each part replaces the half it lost. Newly forming hydrogen bonds then draw the new partners together, and the double helix is restored. In the meantime, the enzyme complex moves

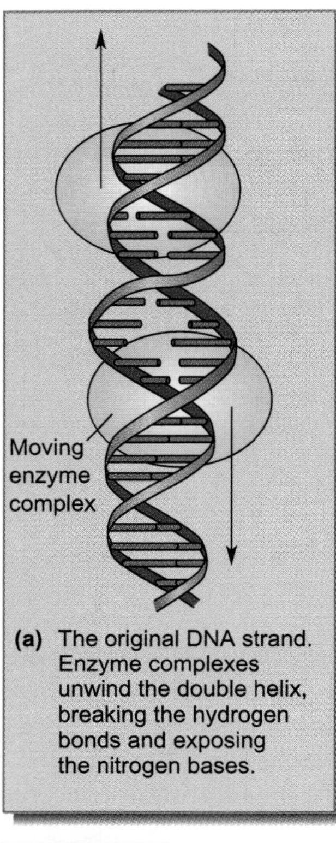

Moving
enzyme
complex
(a) The original DNA strand. Enzyme complexes unwind the double helix, breaking the hydrogen bonds and exposing the nitrogen bases.

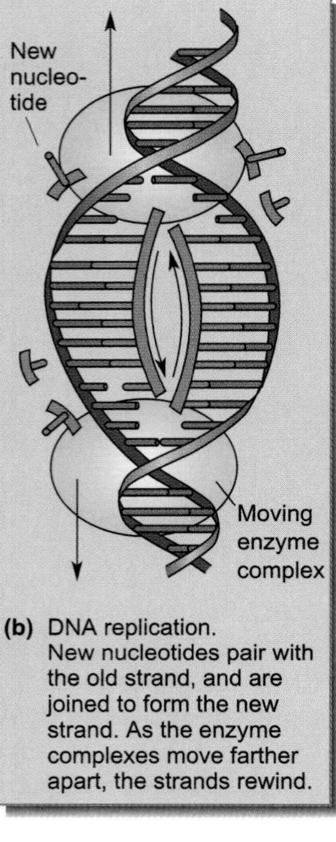

New
nucleo-
tide
Moving
enzyme
complex
(b) DNA replication. New nucleotides pair with the old strand, and are joined to form the new strand. As the enzyme complexes move farther apart, the strands rewind.

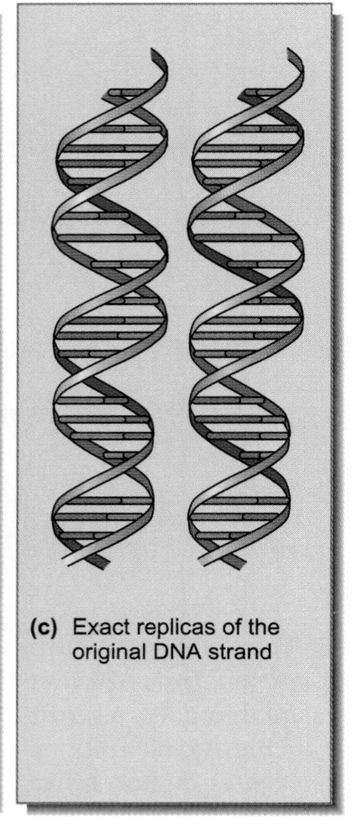

FIGURE 9.1 DNA REPLICATION.
(a) Enzyme complexes unwind the double helix, breaking its numerous hydrogen bonds and exposing the nitrogen bases. **(b)** New nucleotides pair with one of the old strands, and are joined to form a new strand that replaces the old strand. As the replication enzyme complexes move along, the strands rewind. **(c)** Eventually two exact replicas of the original DNA emerge.

ahead, and replication continues along the lengthy molecule. At the completion of the replication process there will be two DNA molecules identical to the original.

The cell has now completed the first requirement for cell division: It has replicated its DNA, making one complete copy available for each of the daughter cells. As the cell divides, each daughter cell must receive one of the copies. Let us now see how this is done. We can begin by looking at bacteria, where the cell must manage the distribution of only one, relatively small piece of circular DNA. Then we will move on to eukaryotes, which must cope with much larger amounts of DNA.

BACTERIAL DIVISION

Bacterial division, you will be glad to see, is rather straightforward. There is only a single, circular piece of DNA that must be replicated and distributed. The DNA is connected to the plasma membrane at a specific attachment site (Figure 9.2). Two copies must be segregated from each other, so that each of the daughter cells receives a copy. Thus, as replication of the DNA is completed, the attachment site splits so that each copy of the DNA is connected to a different point on the plasma membrane.

After replication, the membrane grows and extends between the two attachment sites. As the attachment sites physically separate, the DNA copies attached to the sites are segregated to opposite ends of the bacterium. Membrane growth continues between the two copies of DNA, completely separating the two copies of DNA into what are now two distinct daughter cells.

EUKARYOTIC DIVISION

Cell division in eukaryotes is, let us say, more challenging. Basically this is because eukaryotes must manage hundreds of times more DNA in each cell than do bacteria. Eukaryotes partially solve the problem by packaging the DNA into compact, manageable units called chromosomes, which are complex associations of DNA with proteins packaged into the nucleus. The rest of the solution to the problem is called *mitosis,* the process by which eukaryotic cells distribute their chromosomes.

Chromosome Structure

If you looked at a cell under the microscope, you probably would not be able to see chromosomes at any time other than when a cell is in mitosis. At times other than mitosis, the genetic material is dispersed throughout the nucleus in a loose form known as **chromatin.** Chromatin consists of **nucleosomes,** bead-like structures of DNA wrapped around proteins called **histones.** However, in preparation for mitosis, the chromatin condenses into visible chromosomes. The structure of the fully condensed chromosome is essentially a system of coils within coils (Figure 9.3).

Mitotic condensation is the ultimate in packaging. The length of DNA double helix in a human X chromosome, one of the larger of our set of chromosomes, would presumably be about 13 cm (5 in.) long if fully stretched out. In its fully coiled and condensed state, however, it is 30,000 times shorter. This tight package can be readily moved about the cell. Without this condensation, mitosis would demand far too much space to be feasible.

The fully condensed mitotic chromosome is diagrammed in Figure 9.4. It consists of the two copies of DNA together with associated proteins, coiled and compressed into two short rodlike structures, called **chromatids,** that are joined together at the **centromere.** They are called chromatids only while they are held together by the centromere. During mitosis, the centromere will divide, whereupon each chromatid will become a single, full-fledged chromosome. As we will see, the centromere, much like the membrane attachment sites in bacteria, serves as the hook by which the two copies of the DNA (chromatids) will be separated from one another.

How many chromosomes are in a cell? It depends. Members of each species have a precise number of chromosomes (Table 9.1). For example, human beings have 46, but the lowly ameba has 50, and Rover sleeping at your feet has 78. Essay 9.1 explains

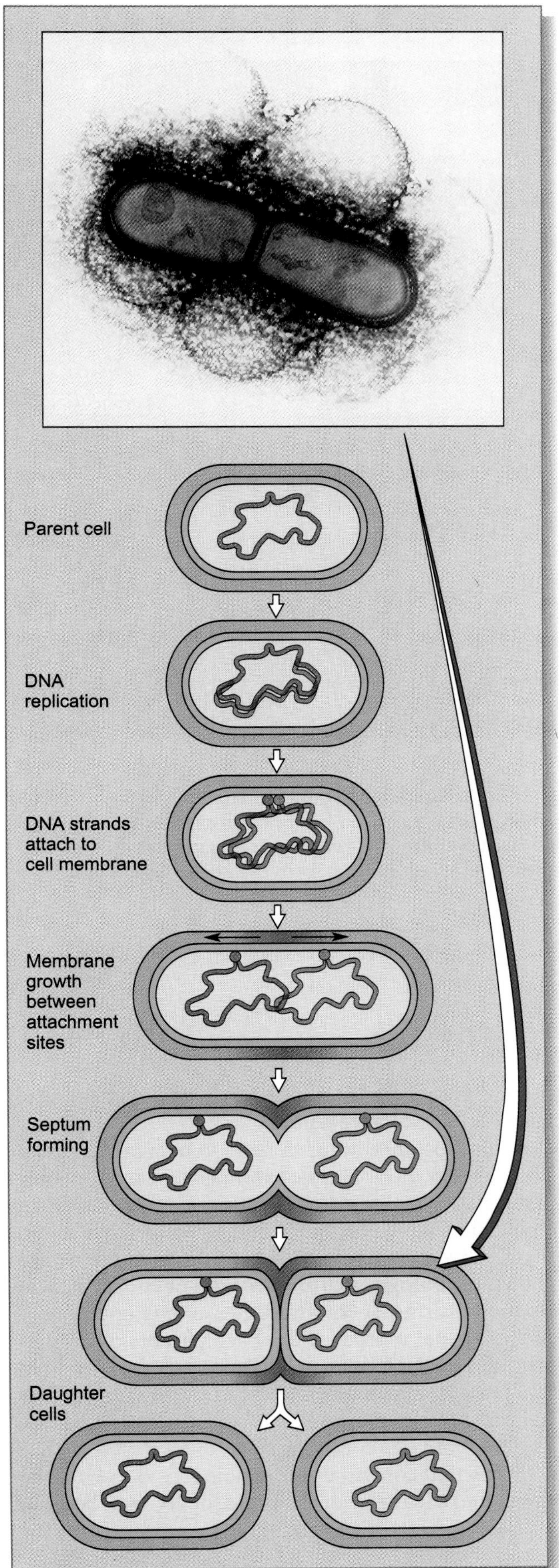

FIGURE 9.2 BACTERIAL FISSION.
In bacterial fission, DNA replication occurs first, and each replica attaches to the cell membrane. The cell then divides, and identical daughters emerge.

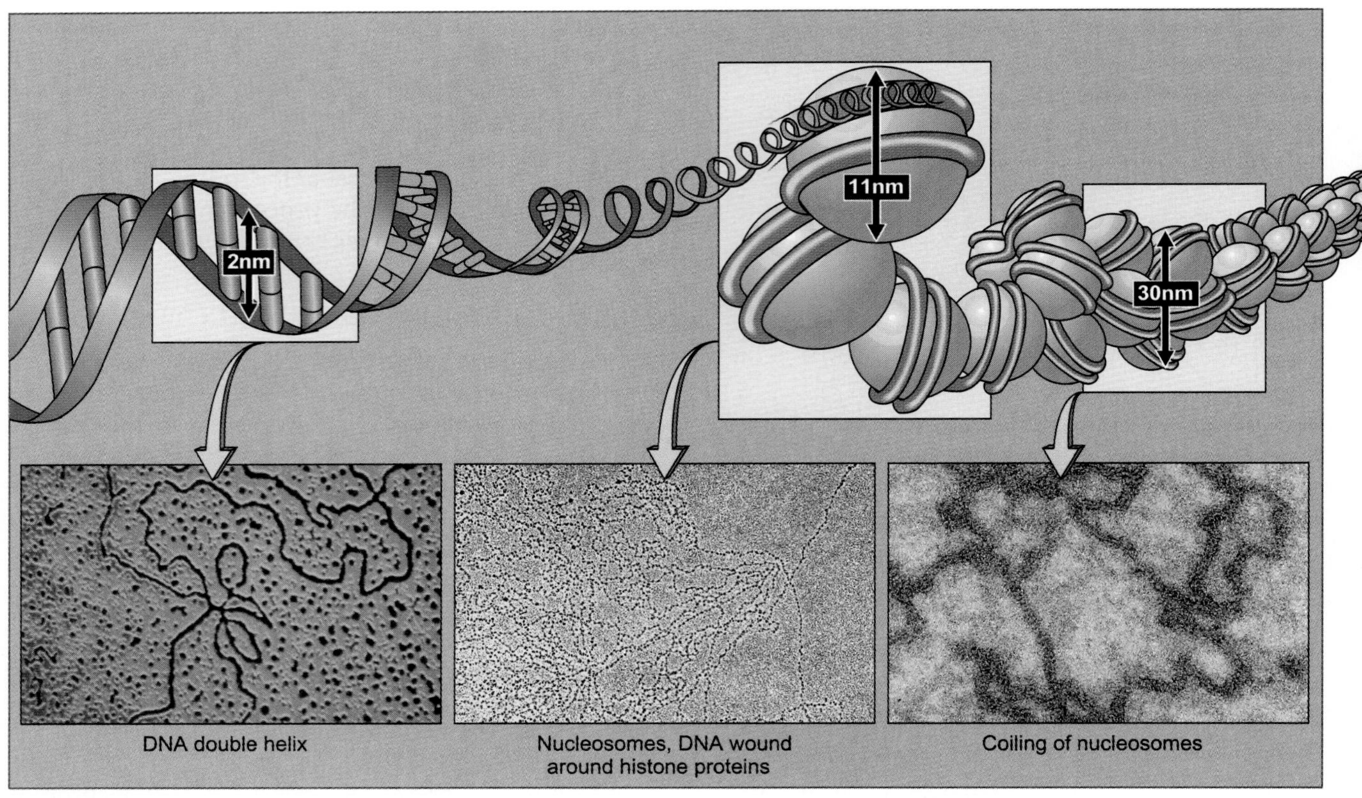

FIGURE 9.3 DNA AND THE CHROMOSOME.
The familiar chromosome is the ultimate result of several levels of organization and condensation. Naked DNA associated with histone proteins forms nucleosomes that give a "bead-on-a-string" impression. Coiled packing of nucleosomes forms the 30-nm fiber. Further twisting of the 30-nm fiber forms increasingly denser coils, which eventually intertwine to form the chromosomal strands that form the arms of the chromosome.

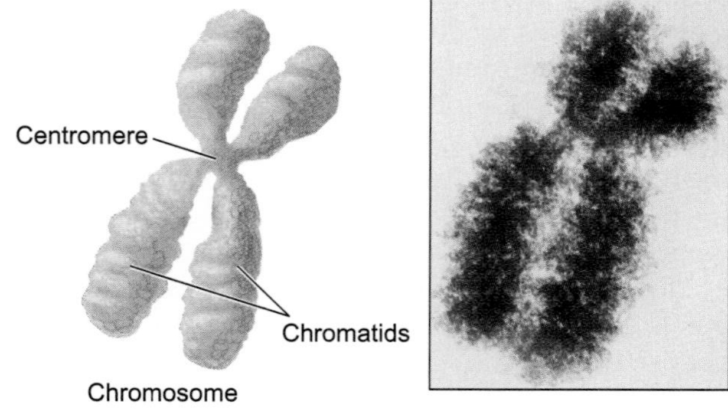

FIGURE 9.4 CHROMOSOME STRUCTURE.
Following replication, a chromosome consists of two chromatids joined at the centromere. The centromere will divide during mitosis, and each chromatid will then be a fully qualified chromosome.

how chromosomes are identified and counted in a process called **karyotyping.**

Chromosomes occur in pairs in nearly all human cells and in those of other animals and most plants. That is, for every chromosome (or complete DNA molecule) there is another one virtually identical to it. The two members of the chromosome pair are called **homologous chromosomes,** or simply **homologues.** Pairing of chromosomes is the logical product of sexual reproduction: Each parent contributes one member of each pair. Thus, your own paired chromosomes can be traced to your parents, each of whom contributed one member of each pair at the moment of your conception.

Cells that contain both homologues of each chromosome pair are termed **diploid,** or **2N,** cells because

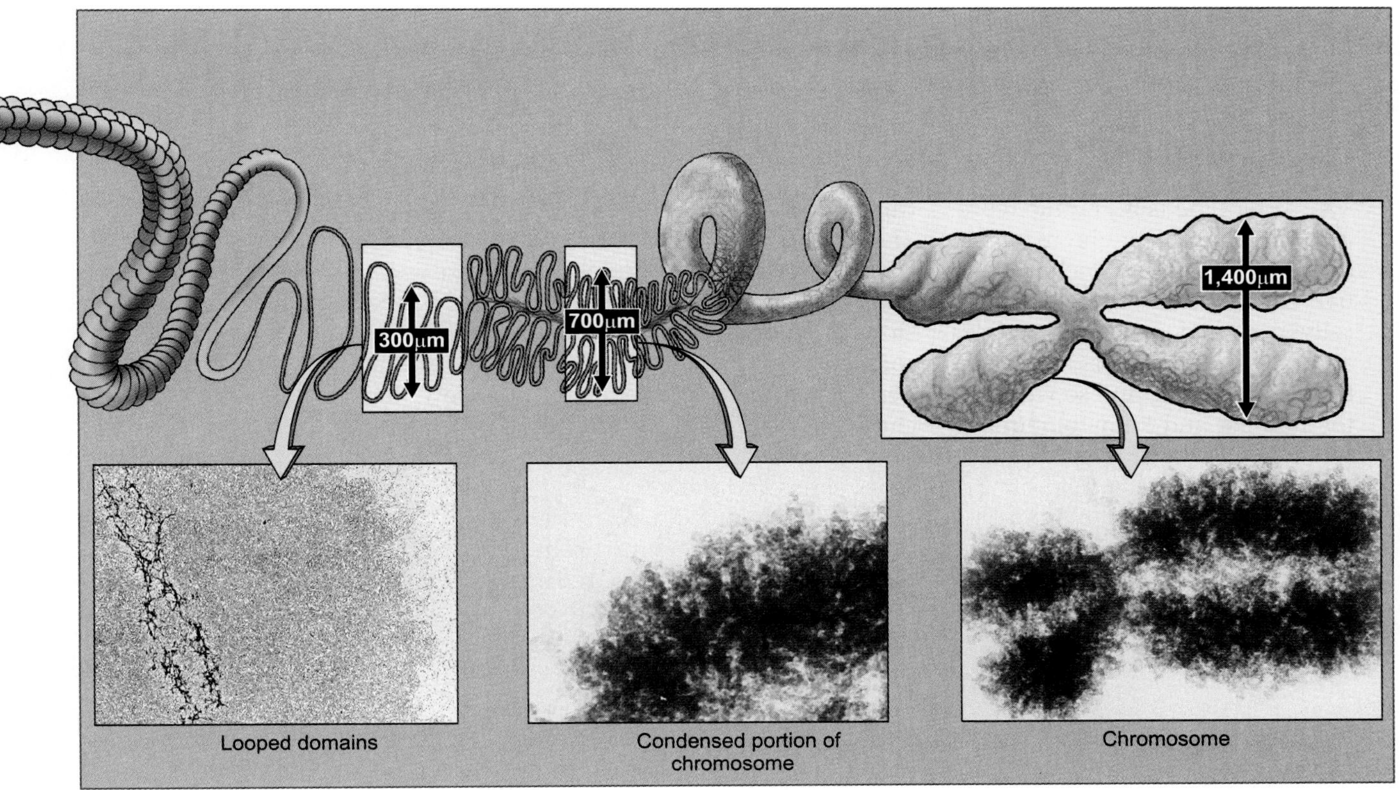

they contain two copies of each chromosome. Not all cells are diploid. Some have only one homologue from each pair, and these cells are **haploid,** or **N.** In humans and other animals, the only haploid cells are **gametes** (sperm and egg cells). Of course, their mission in life is to join during fertilization, producing new individuals whose chromosome state will then be diploid. However, many simpler organisms, particularly protists and fungi, spend much of their life cycle in a haploid state. Some familiar plants, while

TABLE 9.1 CHROMOSOME NUMBERS

There is no apparent significance to chromosome number as far as biologists can determine. Note the variation in some plant and animal species. Plants undergo spontaneous doubling and tripling of chromosome number, so their numbers may vary.

Alligator	32	English holly	40	Opossum	22
Ameba	50	Fruit fly	8	Penicillia	2
Brown bat	44	Garden pea	14	Pheasant	82
Bullfrog	26	Goldfish	94	Pigeon	80, 79
Carrot	18	Grasshopper	24	Planaria	16
Cat	32	Guinea pig	64	Redwood	22
Cattle	60	Horse	64	Rhesus monkey	42
Chicken	78	House fly	12	Rose	14, 21, 28
Chimpanzee	48	Human	46	Sand dollar	52
Corn	20	Hydrae	32	Sea urchin	40
Dog	78	Lettuce	18	Tobacco	48
Earthworm	36	Marijuana	20	Turkey	82
Eel	36	Onion	16, 32	Whitefish	80

Essay 9.1 Karyotyping

A karyotype is a graphic representation of the chromosomes of any organism in which individual chromosomes are systematically arranged according to size and shape. Each species has its particular karyotype, and so we know the number and kinds of chromosomes found in carrots, fruit flies, and people. Karyotyping in humans is done by a simple and straightforward method.

A blood sample is drawn, and the white cells are separated and transferred to a culture medium. The medium contains not only nutrients, but also chemical agents that first induce mitosis and then stop the process when the chromosomes are at their maximal condensation (the stage at which the chromatids are seen most easily). The cells are then put on microscope slides and stained.

In the (not so) old days, cells showing all of the chromosomes were photographed through a microscope. Then a large print was made and the chromosomes were cut out with scissors, sorted by size and shape, and mounted with rubber cement.

Computer-aided image analysis has modernized the process. Now technicians can view the chromosomes, identify them, and sort them all electronically to produce the karyotype.

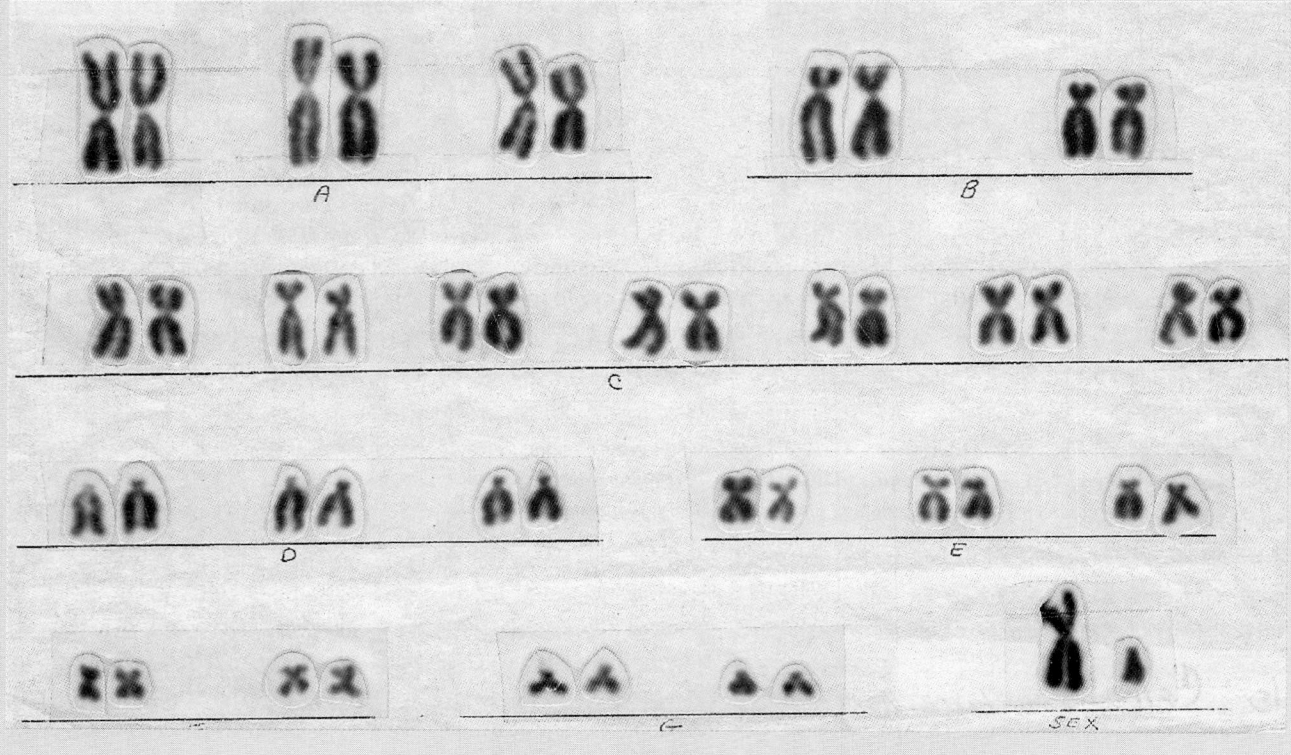

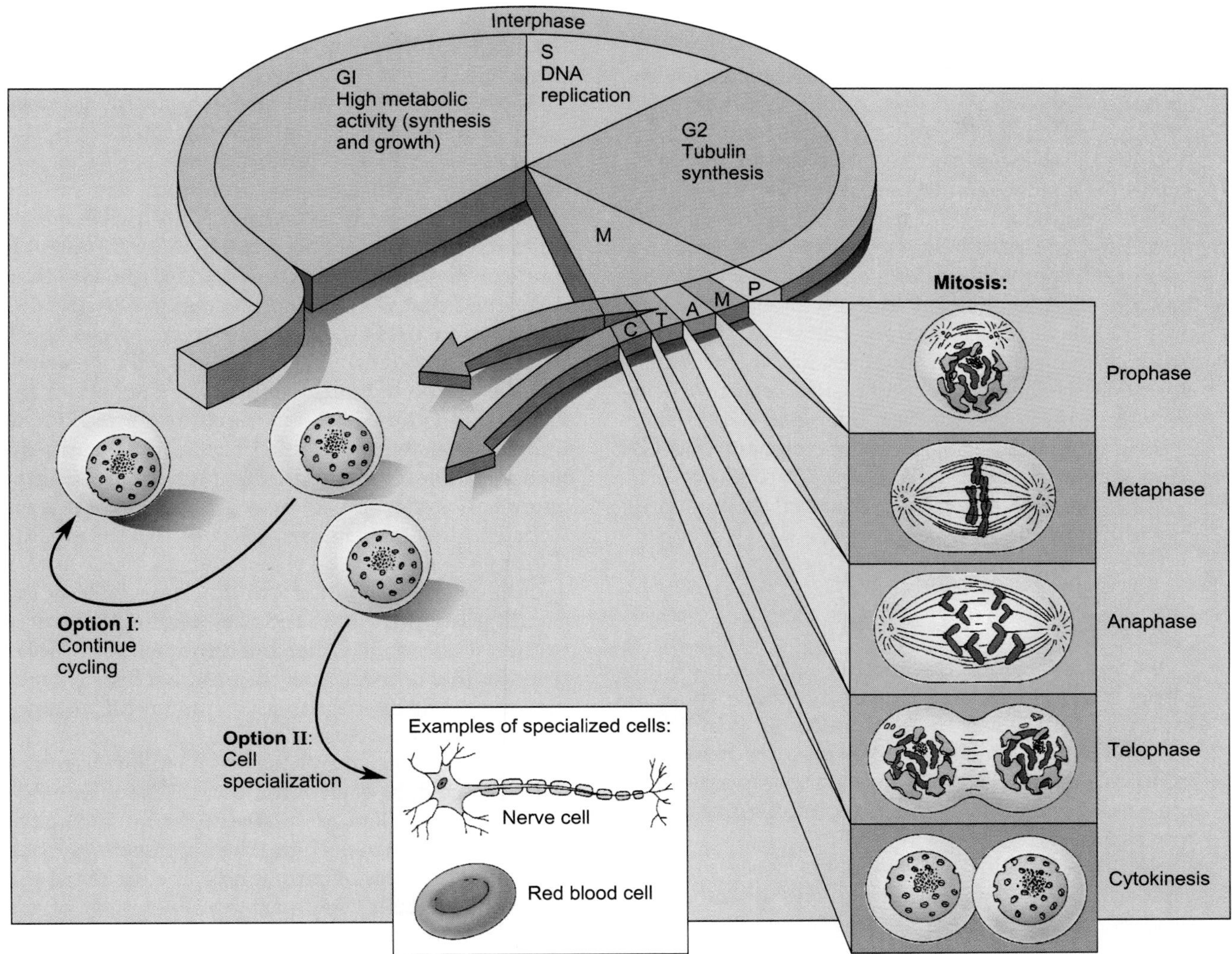

FIGURE 9.5 **TYPICAL CELL CYCLE.**
In this generalized representation of a cell cycle, the major phases of G1, S, G2, and M are shown as pie-piece parts of a circle. The cycle continues if the cells resulting from mitosis take option I. However, many cells that take option II leave the cell cycle to become specialized cells that will die (after serving their function) rather than reenter the cell cycle.

primarily diploid, have many haploid cells, some of which eventually give rise to gametes.

CHROMOSOMES, DIVISION, AND THE CELL CYCLE

The life of a cell is often described as a cycle, and the **cell cycle** can be divided functionally into two major parts and four subparts. The major parts are *interphase* and *mitosis* (Figure 9.5). **Interphase**—the period between cell divisions—has three subparts, *G1, S,* and *G2.*

In human cells the first part of interphase, **G1** (for gap-one, meaning the gap in time between divisions), lasts about 8 hours in most cells that are actively dividing. During this time the cell grows to about twice its original size. Growth includes protein synthesis, organelle construction, and storage of materials.

DNA replication occurs in the *S phase* (for DNA synthesis). The **S phase** is a vital prerequisite for cell division in which each DNA molecule forms a replica of itself, as we saw earlier in this chapter. Following DNA replication, **G2** (gap-two) begins—a period of renewed protein synthesis in which much of the effort is placed on making *tubulin,* which will later be assembled into numerous microtubules, structures important to chromosome separation.

The duration of the cell cycle can vary tremendously. Newly fertilized sea urchin and frog embryos, for instance, may complete cell cycles in less than 30 minutes. Certain mouse connective tissues in culture require as long as 22 hours, while root tip cells of the common garden pea spend only 3 hours between divisions. Our own red blood cells live only 120 days. The enormous battery of cells that produce them must, all together, undergo 2.5 million cell divisions per second just to keep up. Cells such as those in the muscles, liver, and nerves stop cycling altogether and remain permanently in G1 during their life span.

G2 completes interphase. The cell is now prepared for mitosis.

Mitosis

During interphase the nucleus is, indeed, a busy place. It was once called the resting phase, though, for a very simple reason: All this activity is invisible when viewed with the light microscope. However, in mitosis the activity becomes starkly apparent, as the nuclear material condenses, reorganizes, and moves around.

We simply define **mitosis** as the process of nuclear and cell division. Although it is a continuous process with one event leading gradually to the next, biologists have felt compelled to organize things, so the process is described in five parts: *prophase, metaphase, anaphase, telophase,* and *cytokinesis.* (Some biologists view cytokinesis, or division of the cytoplasm into two cells, as a separate event, not actually a part of mitosis.) We will look briefly into the visible events of each phase and take a closer look at a few of these processes. A simplified view of mitosis is seen in Figure 9.6.

Prophase. The onset of mitosis in nearly all cells is called **prophase,** the stage marked by the condensation of chromosomes. Remember that there are already two copies of DNA for each chromosome, called chromatids, that are joined together at the centromere. Generally, the individual chromatids, connected by their centromeres, are not visible until late in prophase. As condensation of the chromosomes proceeds, the nucleoli disappear; and in the final stages of prophase the nuclear envelope disappears and microtubules form the **spindle**—the scaffolding upon which chromosome movement will occur.

The centrioles (where present, as in animals, many protists, and a few plants) migrate to opposite sides of the cell. The centrioles then become surrounded by a cluster of microtubules that radiate outward, taking on a "starburst" form called an **aster.** Keep in mind that centrioles and asters are not found in the cells of fungi and most plants, a key distinction that we will explore later.

As prophase continues, the chromosomes begin to move, apparently being tugged about by the spindle. Following the breakdown of the nuclear envelope, each pair of sister chromatids is attached via their centromeres to spindle fibers from opposite directions. A special structure known as the **kinetochore** acts as the "hook" by which the spindle is attached to the centromere region of the chromatids (Figure 9.7). One of the chromatids is attached to a spindle fiber emanating from the "east" pole of the dividing cell; the other becomes attached similarly to the "west" pole. Other microtubules form polar spindle fibers, fibers that are not attached to chromosomes, but overlap to span the entire distance between the two poles.

Lastly, in late prophase, movement becomes more directed, as each chromosome heads towards the cell's center (Figure 9.6), finally coming to rest at the equatorial plane. There, the chomosomes, each consisting of a pair of chromatids, line up across the center of the cell, thus marking the arrival of the next phase, metaphase.

In summary, prophase is an incredibly busy time in which the chromosomes condense, the spindle apparatus forms, the nucleolus disappears, the nuclear envelope breaks down, and the chromosomes become arranged across the cell. Perhaps the key feature is the precise attachment of centromeric spindle fibers to replicated chromatids. This sets the stage for equitable distribution of the replicated DNA into the daughter cells.

Metaphase. In the stage of mitosis called **metaphase,** the chromosomes arrive at an equatorial plane in a region of the spindle known as the **metaphase plate** (Figure 9.6). Here the chromosomes are momentarily held in place. However, while the chromosomes appear to be at rest, their static position is maintained by strong opposing forces. Observers note that stray cytoplasmic particles tumbling into the metaphase plate are immediately swept poleward. If you've a mind for fanciful analogies, think of each chromosome and its opposing

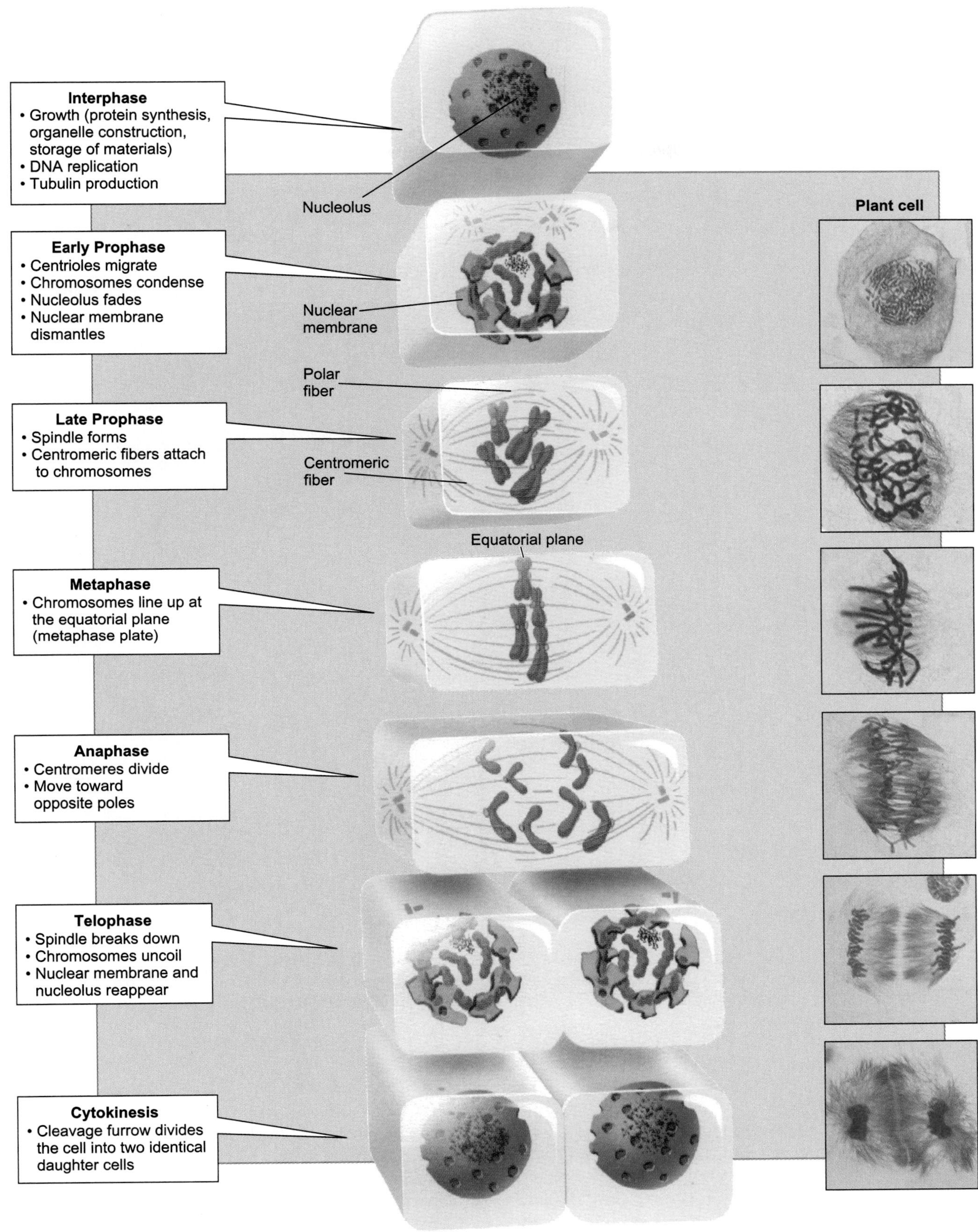
Interphase
• Growth (protein synthesis, organelle construction, storage of materials)
• DNA replication
• Tubulin production
Nucleolus
Early Prophase
• Centrioles migrate
• Chromosomes condense
• Nucleolus fades
• Nuclear membrane dismantles
Nuclear membrane
Polar fiber
Late Prophase
• Spindle forms
• Centromeric fibers attach to chromosomes
Centromeric fiber
Equatorial plane
Metaphase
• Chromosomes line up at the equatorial plane (metaphase plate)
Anaphase
• Centromeres divide
• Move toward opposite poles
Telophase
• Spindle breaks down
• Chromosomes uncoil
• Nuclear membrane and nucleolus reappear
Cytokinesis
• Cleavage furrow divides the cell into two identical daughter cells
Plant cell

FIGURE 9.6 MITOSIS.
The stages of mitosis are diagrammed and aligned with photomicrographs of examples from plants.

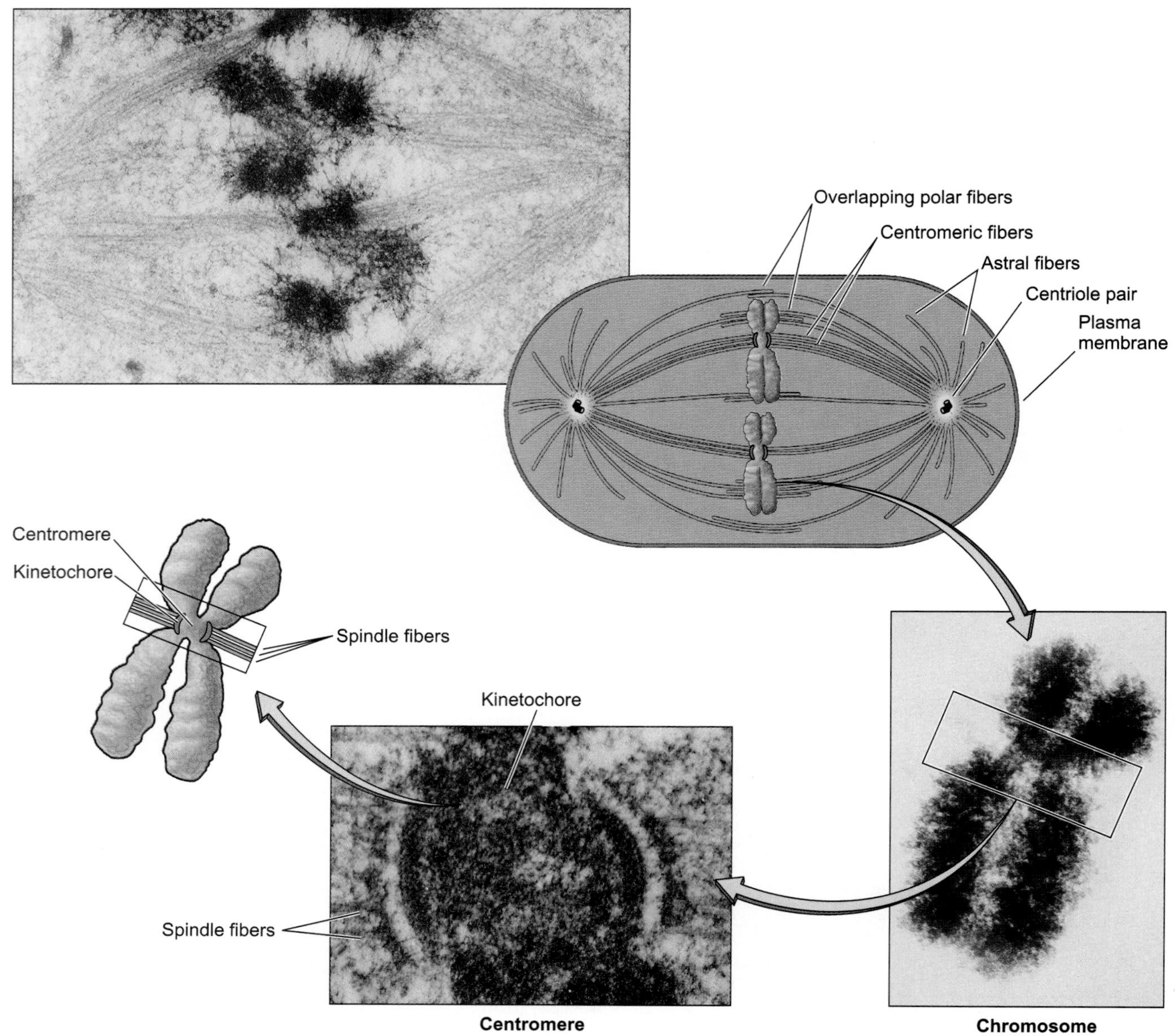

FIGURE 9.7 THE MITOTIC SPINDLE APPARATUS AND SPINDLE FIBER ATTACHMENT.
Note the relationship of the spindle fibers, asters, centrioles, and chromosomes. The microtubules of the spindle fibers are seen attached to the curved kinetochore, which is physically associated with the centromere of the chromosome.

microtubules as a horse roped and held tenuously in place by two struggling but determined handlers.

Anaphase. Metaphase ends abruptly, as **anaphase** begins when the centromeres holding the chromatids together split apart, separating the two chromatids, each containing a copy of the original DNA. The chromatids can now be referred to as daughter chromosomes, and take on rough V-shapes as they are drawn to their respective poles, centromere first, by the spindle fibers, which are attached to the centromeres by the kinetochores. Chromosome movement requires several minutes. Both polar and centromeric fibers are thought to be responsible for this movement, the centromeric fibers literally pulling the chromosomes along and the polar fibers elongating, thereby pushing the poles apart (see Essay 9.2).

It is in anaphase that the precise and equal separation and distribution of chromosomes takes place, even though the cell has yet to divide. One sister chromatid, now a daughter chromosome, has been brought to each end of the cell. This must happen with utterly faithful precision if the two emerging cells are to maintain their genetic integrity.

Telophase. Many of the events of prophase are essentially reversed in the stage called **telophase.** The chromosomes begin to decondense, their sausagelike form fading into the diffuse chromatin we saw at interphase. The spindle is dismantled, its proteins to be recycled in forming the new cytoskeleton. A new nuclear envelope is formed from remnants of the old, and soon the nucleoli are reestablished.

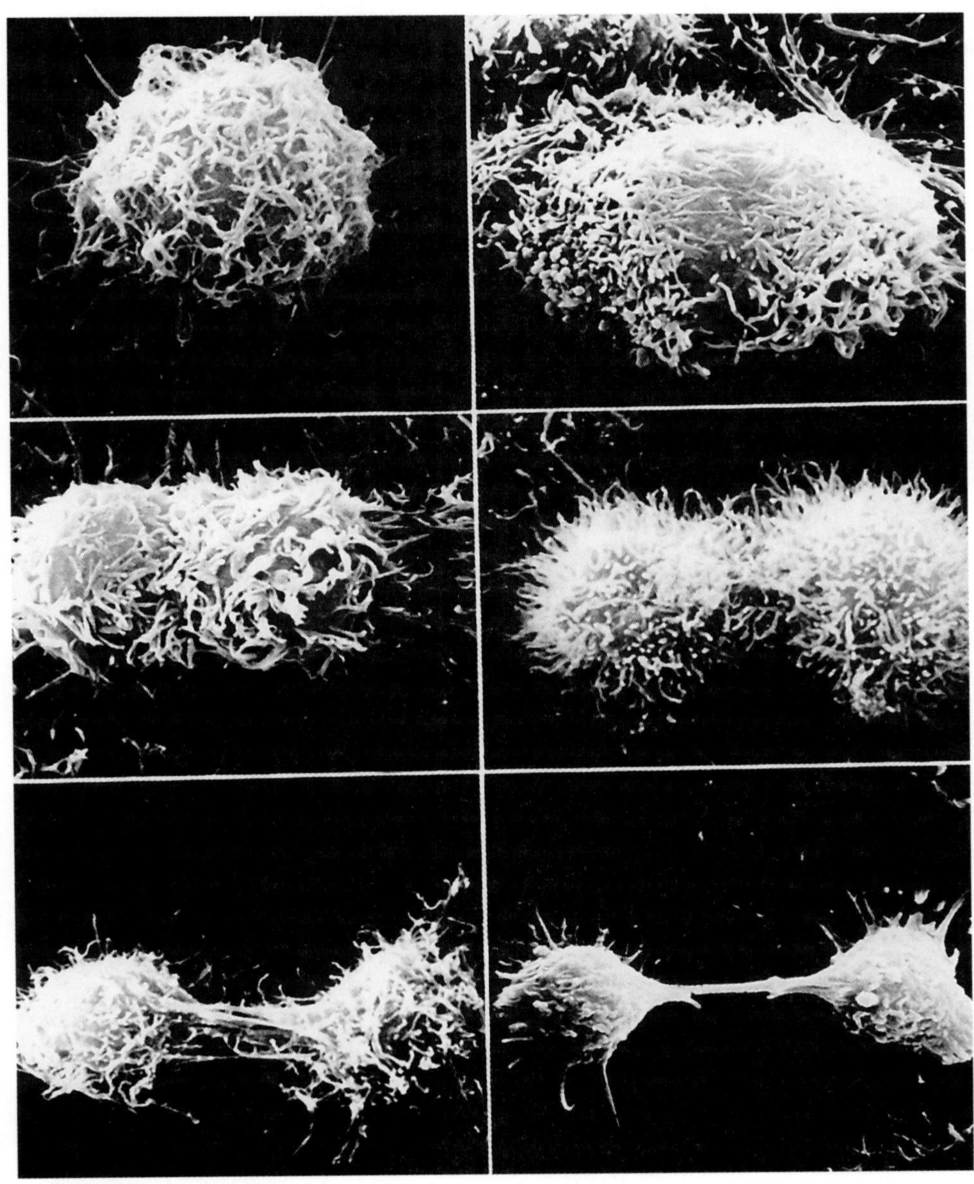

FIGURE 9.8 **CYTOKINESIS IN ANIMAL CELLS.**
Cell division in animal cells finishes with the formation of a cleavage furrow.

Essay 9.2

The Roles of the Centriole and Spindle

We have noted that in early prophase, the two centrioles migrate to opposite spindle poles from a position near the nucleus. But what is their role in mitosis and cell division? Until recently, they were ascribed important mitotic roles, but this notion has been discarded. After all, plants and fungi lack centrioles, yet they form perfectly good spindles and go faultlessly through mitosis. It is far more likely that their separation in early prophase is just the cell's way of making sure one pair ends up in each daughter cell. Centrioles, as we saw in Chapter 4, are identical in structure to basal bodies, which in turn organize the formation of cilia and flagella.

Whereas they have no apparent role in chromosome movement, centrioles may influence cytokinesis. Evidence suggests that they play some role in establishing the plane of cell division, at least in animal cells. In experimental situations, anything that affects the position of the centrioles affects the division plane accordingly. Researchers are still trying to learn why.

We've seen that the spindle apparatus is architecturally at the very core of mitotic movement. It consists primarily of microtubules composed of the protein tubulin, which is also a component of cilia and flagella. In fact, in the late G2 phase of the cell cycle, tubulin accounts for as much as 10% of the cell's protein. The assembly of tubulin into the microtubules of the spindle occurs in early prophase, but the details of this assembly are not known. The locale of microtubule assembly is a region at each spindle pole. Curiously, these regions, aptly called mcrotubule organizing centers (MTOCs), lie directly around the centrioles in animal cells. Yet a functional relationship between the MTOC and centriole has yet to be established.

Centromeric spindle fibers actively separate the chromosomes by pulling on them during anaphase, and then the polar spindle fibers somehow elongate the spindle, apparently aiding in this separation. While we don't know exactly how they work, there are some hypotheses. Let's begin with the microtubules of the centromeric spindle fibers.

Careful observations of the centromeric spindle fibers clearly indicate that as the chromosomes are pulled to the poles, the connecting

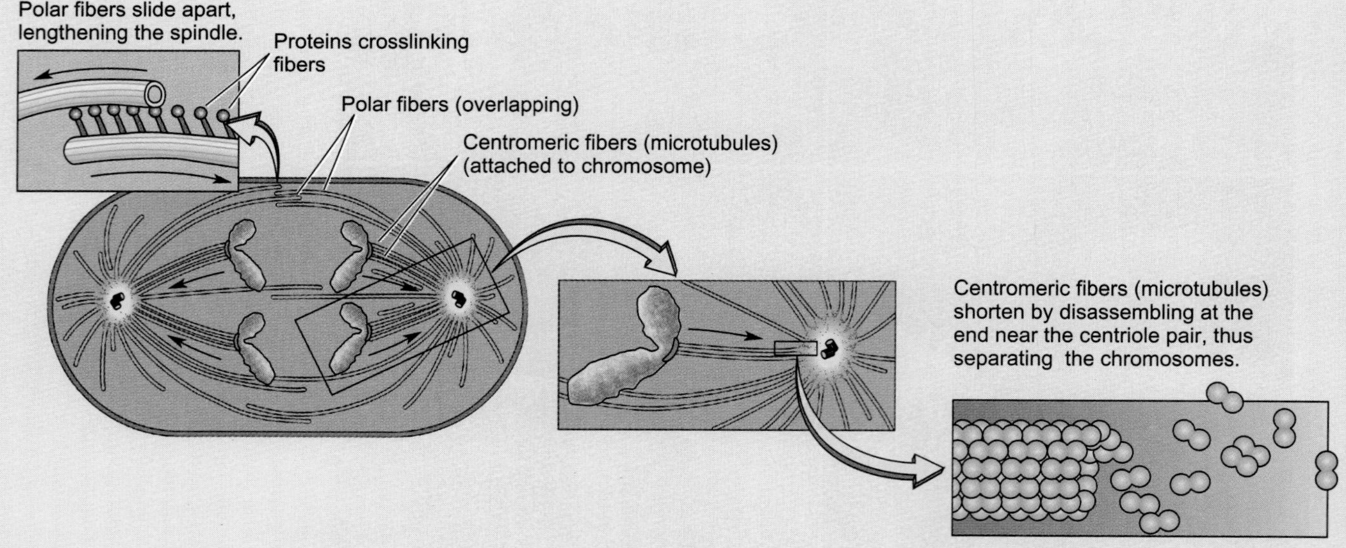
Polar fibers slide apart, lengthening the spindle.
Proteins crosslinking fibers
Polar fibers (overlapping)
Centromeric fibers (microtubules) (attached to chromosome)
Centromeric fibers (microtubules) shorten by disassembling at the end near the centriole pair, thus separating the chromosomes.

microtubules get shorter. This observation is explained by the microtubular disassembly hypothesis. Apparently, these spindle fibers are continually disassembled at the microtubule organizing center, disappearing entirely at the end of anaphase. Researchers have found that when substances known to inhibit microtubule disassembly are applied, chromosome movement slows. Conversely, when substances known to accelerate microtubule disassembly are applied, chromosome movement speeds up.

Apparently, the polar spindle fibers operate quite differently. First, let's note that the microtubules do not actually extend from pole to pole, but rather, they extend out from each pole, meeting and overlapping each other. The discovery of this overlapping relationship immediately suggested a mechanism such as that seen in the cilium and flagellum. There the protein dynein hydrolyzes ATP, bringing on a sliding and bending action (see Chapter 4). The application of dynein ATPase inhibitors effectively stops the sliding in polar microtubules. Thus, movement of polar microtubules appears to operate on a very similar principle to those of cilia and flagella.

Cytokinesis. The division of the cytoplasm is called **cytokinesis.** In other words, this is the point at which two new, separate cells are formed. The process is fundamentally different in plants and animals.

In animals (and many protists), cytokinesis begins with a furrow along the cell's equator. The furrow eventually becomes a deep groove. Finally, the cell is pinched in two, forming two daughter cells. The entire sequence is often referred to as **cleavage** (Figure 9.8). The **cleavage furrow** is produced by contracting rings of actin microfilaments that are just under the plasma membrane.

Cytokinesis in plants must occur within the confines of a cell wall. The plant cell produces a new wall between the telophase nuclei. The new wall begins with the formation of a number of tiny vesicles originating from Golgi bodies (see Chapter 4), which form in the plane of the metaphase plate. The vesicles fuse to form a **cell plate,** which extends to the original cell wall. The cell plate fills with pectin, a carbohydrate, to form the **middle lamella.** This is a gummy layer that will come to lie between the mature cell walls of adjacent plant cells. New plasma membranes assemble over the middle lamella, and cellulose is deposited through the membrane of each daughter cell, forming a new **primary cell wall** (Figure 9.9).

A summary of the major features of mitosis is presented in Table 9.2.

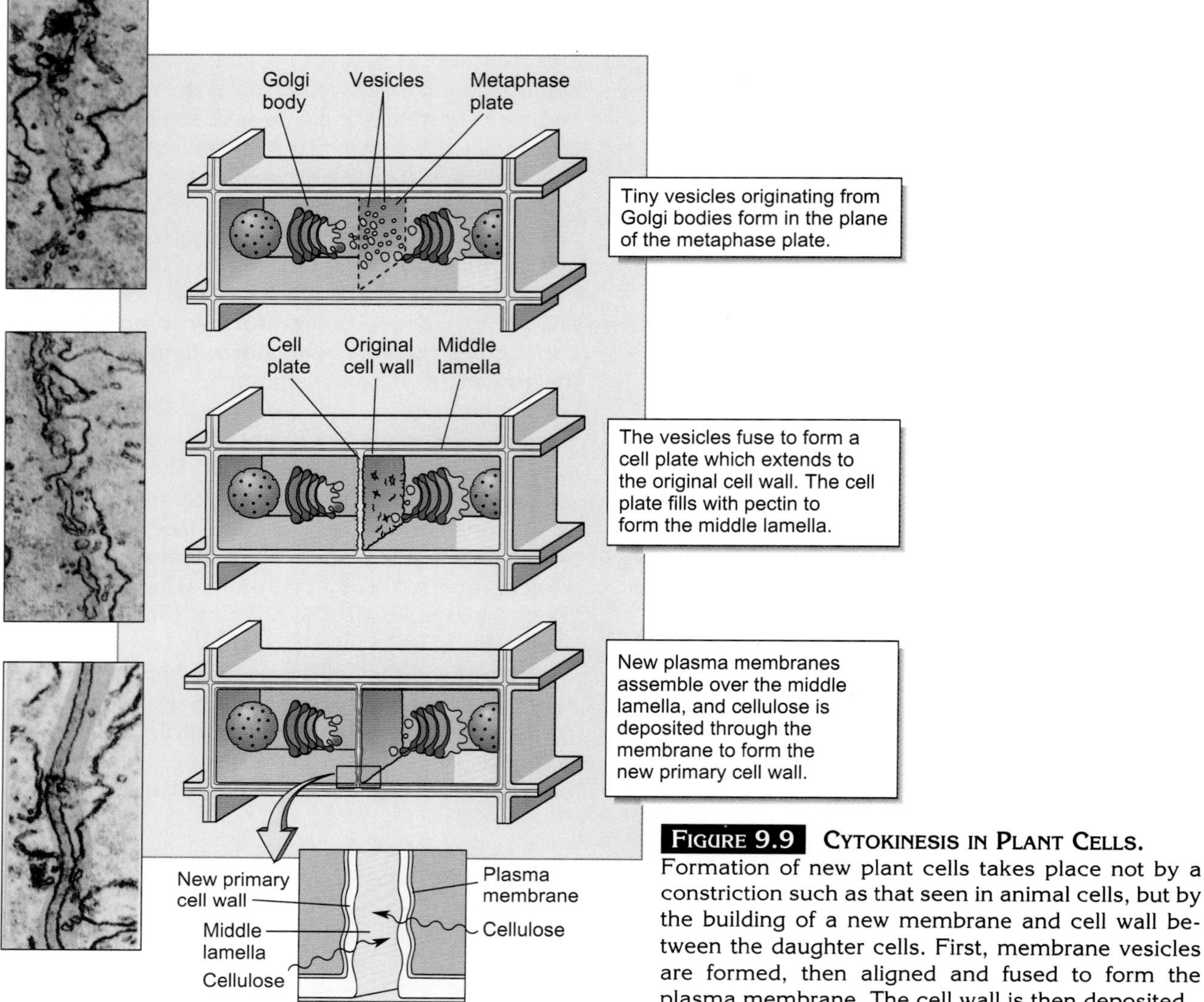

Golgi body
Vesicles
Metaphase plate
Tiny vesicles originating from Golgi bodies form in the plane of the metaphase plate.
Cell plate
Original cell wall
Middle lamella
The vesicles fuse to form a cell plate which extends to the original cell wall. The cell plate fills with pectin to form the middle lamella.
New plasma membranes assemble over the middle lamella, and cellulose is deposited through the membrane to form the new primary cell wall.
New primary cell wall
Middle lamella
Cellulose
Plasma membrane
Cellulose

FIGURE 9.9 CYTOKINESIS IN PLANT CELLS. Formation of new plant cells takes place not by a constriction such as that seen in animal cells, but by the building of a new membrane and cell wall between the daughter cells. First, membrane vesicles are formed, then aligned and fused to form the plasma membrane. The cell wall is then deposited.

TABLE 9.2 MITOSIS

Interphase	Prophase	Metaphase	Anaphase
Chromosomes are decondensed S phase: DNA and chromosomal proteins synthesized Spindle proteins and other mitotic proteins formed in G2 phase Cell increases in volume	Chromosomes condense Separate chromatids may become visible Asters and mitotic spindle begin to form Nuclear membrane breaks down Nucleolus disperses Centromeres become attached to the centromeric spindle fibers Chromosomes migrate toward the metaphase plate	Spindle is fully formed Chromosomes are aligned with their centromeres on the metaphase plate Centromeres begin to divide	Centromeres divide Sister chromatids separate to become chromosomes Daughter chromosomes go to opposite poles Centromeric spindle fibers shorten; the spindle as a whole elongates
Telophase	**Cytokinesis (Cytoplasmic Cell Division) in Animals**	**Cytokinesis (Cytoplasmic Cell Division) in Plants**	**Interphase**
Chromosomes begin to decondense Nuclear membranes re-form Nucleoli reappear Spindle disappears Cytokinesis (cytoplasmic cell division) usually occurs	Microfilaments associated with the cell membrane form a circular band around the cell Microfilaments contract, pinching apart daughter cells	Small membrane vesicles fuse in the plane of the previous metaphase plate to form the cell plate Cell plate grows to separate daughter cells Cell walls are laid down	Daughter cells are in the G1 stage of interphase (one DNA molecule per chromosome) after mitosis Chromosomes become diffuse

PERSPECTIVES

We have looked at cell reproduction in three widely different types of organisms—bacteria, animals, and plants. In each case, we noted the underlying requirement of reproduction: the accurate delivery of duplicate copies of DNA to each of the daughter cells. Different kinds of organisms may exhibit variations in mechanisms, but the end result is the same.

In the next chapters, we will investigate the details of just why the information in the DNA is so crucial to the survival of the cell and the organism, and how that information is maintained, managed, and utilized by the cell.

SUMMARY

Cell division permits organized growth and development to occur in multicellular organisms, replaces aging or damaged cells, and is a means of reproduction in single-celled organisms. Genetic continuity is provided by DNA replication—chromosome copying—and mitosis, the organized division of replicas.

DNA REPLICATION: A BRIEF LOOK

In replication, the DNA molecule is unwound from its helix, and sections of its two strands are separated. Through base-pairing, new matching nucleotide strands are assembled. Two identical molecules emerge.

BACTERIAL DIVISION

In bacteria, the single, circular DNA molecule is attached to the plasma membrane. After DNA replication, the attachment sites of the two DNA molecules get physically separated by membrane growth, which also eventually forms a division between the daughter cells.

EUKARYOTIC DIVISION

Chromosomes are made up of DNA molecules and proteins, forming chromatin. In early mitosis, chromosomes condense, forming coils upon coils and becoming readily visible. The condensed form aids in separation during mitosis.

Chromosome numbers are consistent within the same species, but may vary among species. Chromosomes in most cells occur as homologues (pair members), a contribution from each parent at conception. While most cells in animals are diploid (all homologues present), gametes are haploid (one homologue from each pair present). The diploid condition is restored at fertilization.

Interphase is divided into Gl (growth), S (DNA replication), and G2 phases. Interphase is followed by mitosis (with cytokinesis).

In early prophase, the centrioles migrate, chromosomes condense, nucleoli fade, the nuclear membrane is dismantled, and the spindle appears. In late prophase, the chromosomes are moved to a central region forming the metaphase plate.

At metaphase, the chromosomes pause at the plate, each centromere attached to opposing centromeric spindle fibers. At anaphase, the centromeres divide and the chromatids separate, moving to opposite poles. Movement is believed to occur through a sliding mechanism between overlapping polar fibers and through the shortening of centromeric spindle fibers.

In telophase, the chromosomes uncoil, the spindle breaks down, and the nuclear membrane and nucleoli reappear.

Cytokinesis is the process by which cytoplasmic material is divided between daughter cells. In animal cells, cytokinesis occurs by the contraction of microfilaments, which form a deepening cleavage furrow and divide the cell.

Although centrioles are absent in most plants, spindle formation and chromosome movement are similar to that of animals. Cytokinesis occurs through the establishment of new cell walls. Vesicles form a cell plate, which takes in pectin, establishing a middle lamella. With the deposition of cellulose, a primary cell wall emerges.

Key Terms

base pairing • **137**
chromatin • **139**
nucleosome • **139**
histone • **139**
chromatid • **139**
centromere • **139**
karyotyping • **140**
homologous chromosome (homologue) • **140**
diploid (2N) • **140**
haploid (N) • **141**
gamete • **141**
cell cycle • **143**
interphase • **143**
G1 • **143**
S phase • **143**
G2 • **143**
mitosis • **144**
prophase • **144**
spindle • **144**
aster • **144**
kinetochore • **144**
metaphase • **144**
metaphase plate • **144**
anaphase • **146**
telophase • **147**
cytokinesis • **149**
cleavage • **149**
cleavage furrow • **149**
cell plate • **149**
middle lamella • **149**
primary cell wall • **149**

Review Questions

1. List five important purposes for cell division.
2. List the three subparts of interphase, and describe the activity in each.
3. Using simple diagrams of DNA, explain how replication occurs. What assures that the two emerging molecules are replicas?
4. What is the composition of chromatin?
5. Describe the organization of a mitotic chromosome. Why is such a compact organization necessary?
6. Compare the chromosome number in every fifth species listed in Table 9.1. What is the significance of the chromosome number?
7. What are homologous chromosomes? What event in the life of an organism brings homologous chromosomes together?
8. Where in humans would one find cells with the diploid number of chromosomes? The haploid number?
9. Describe the spindle.
10. List five changes that occur in the cell at the onset of prophase.
11. Describe the arrangement in the cell of the asters, polar spindle fibers, and centromeric spindle fibers.
12. Briefly describe the attachment of the centromeric spindle on each chromosome along the metaphase plate. What is the effect of such an attachment when anaphase begins?
13. Summarize the two mechanisms that account for the migration of chromosomes from the metaphase plate.
14. List five events associated with telophase.
15. What is cytokinesis? Explain how it occurs in animals.
16. What effect, if any, might the absence of centrioles in higher plants have on mitosis? What problem does this seem to introduce?
17. List the events surrounding cytokinesis in plants. What factors make plant cytokinesis different from animal cytokinesis?

10

Meiosis and the Life Cycle

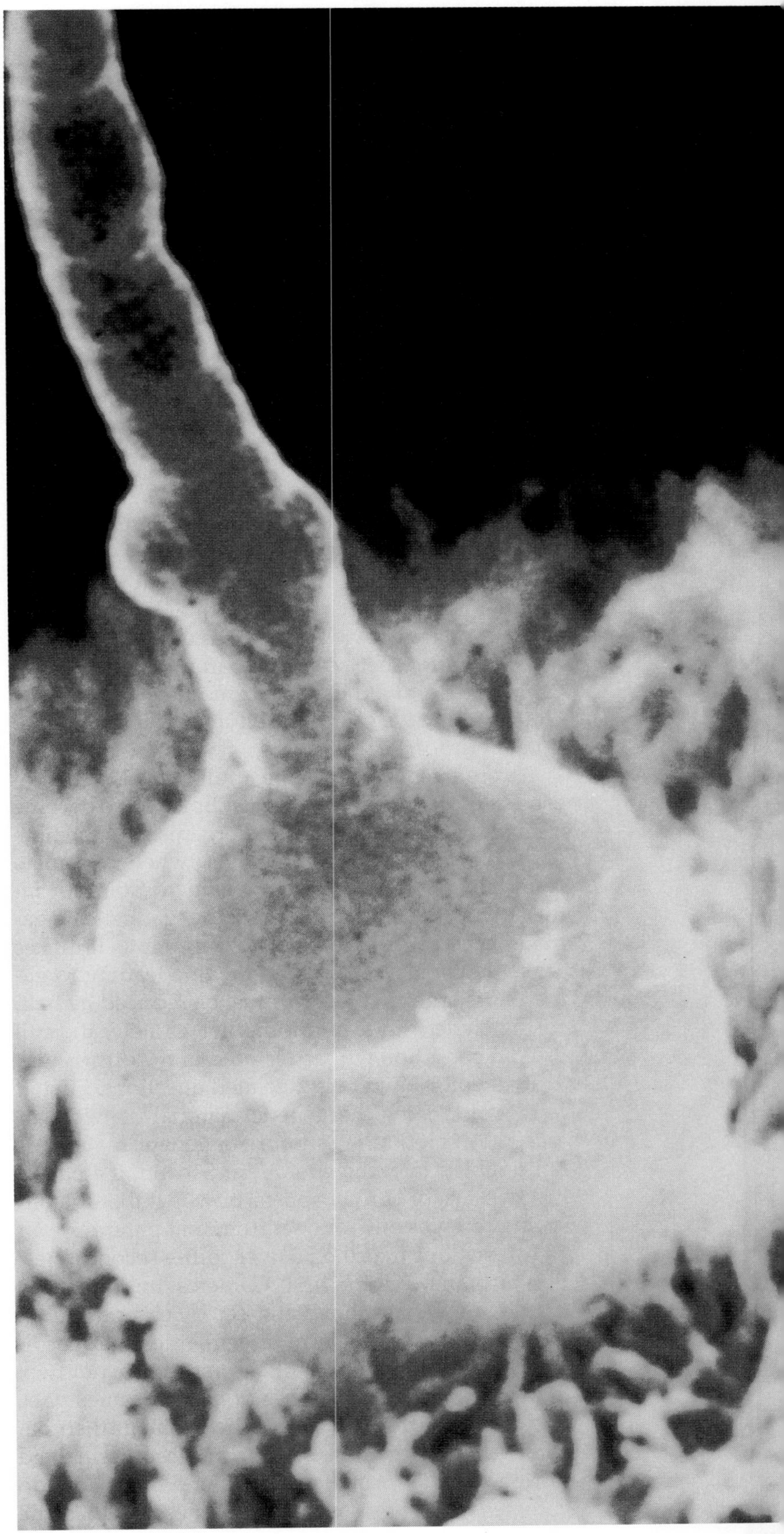

WE HAVE SEEN THAT MITOSIS IS THE ORDERLY DIVISION OF THE CELLULAR *contents, especially the DNA, into the two cells that result from the division process. In this chapter, we look at a slightly more complicated process that methodically reduces the amount of DNA in the resulting cells. Why do cells do this? As we will soon see, it is the cellular prelude to sexual reproduction, the mechanism that results in unique combinations of biological information (and people).*

We are all individuals. Each of us has our own identity, and we each take a certain amount of satisfaction in knowing that we are unique. However, it is also reasonably obvious that each of us is the product of a union between our parents, and that we carry some of the traits of our mothers and fathers. How is it that we can be individuals, yet a combination of our forebears?

As organisms, we receive a full set of DNA from each of our parents, making us diploid. We become a separate organism, a diploid organism, at the time of fertilization (see Figure 10.1), when a gamete from our mother fuses with a gamete from our father to form a **zygote,** a fertilized egg. Since the gametes each contribute DNA to the zygote, it makes sense that somewhere along the line leading to gametes the amount of DNA must be reduced by half, so that when fused with another gamete, the full diploid condition will be restored in the offspring.

It all boils down to two related questions. First, if fertilization joins the genetic material of two parents, why is it that the amount of genetic material is not doubled in each generation? Second, if a parent has two sets of chromosomes, how is it that the offspring receives only one set from that parent?

It turns out that the answer to the second question also answers the first. Gametes are **haploid**—that is, they have only one set of chromosomes. Thus, each gamete (sperm or egg) enters into fertilization with only one set of chromosomes, not two. Because of this, the number of chromosomes does not double each generation. The problem then becomes, how is the number of chromosomes halved during gamete formation? In other words, how do gametes with one set of chromosomes per cell come from cells with two sets of chromosomes? The answer involves an orderly set of steps called *meiosis.*

OVERVIEW OF MEIOSIS

Meiosis is the process whereby a diploid set of chromosomes is reduced to a haploid set of chromosomes in a cell, resulting in the production of a gamete. A highly simplified scheme is outlined in Figure 10.2. Take a look at the basic process of meiosis, and see how it fits into the basic life cycle presented in Figure 10.1.

In diploid organisms, meiosis guarantees (1) that the chromosome number will remain stable from generation to generation and (2) that each sexually reproduced offspring will receive two complete sets of genetic instructions, one from each parent. We will soon see that the meiotic process also virtually guarantees that each gamete will be different from every other gamete. In humans, there are 46 chromosomes in nearly every cell of the body—23 originating from the father's sperm (the paternal chromosomes) and 23 from the mother's egg (the maternal chromosomes). It is in meiosis that the 46 chromosomes per cell are reduced to 23 chromosomes per gamete, so that when fertilization occurs a new individual of 46 chromosomes will result.

The meiotic process actually involves two divisions. The first division, imaginatively named **meiosis I,** is the critical phase of the process. Unlike mitosis, the centromeres do not divide in meiosis I. Instead, the homologous chromosomes line up together then separate from each other, and each whole chromosome (a pair of chromatids) moves to

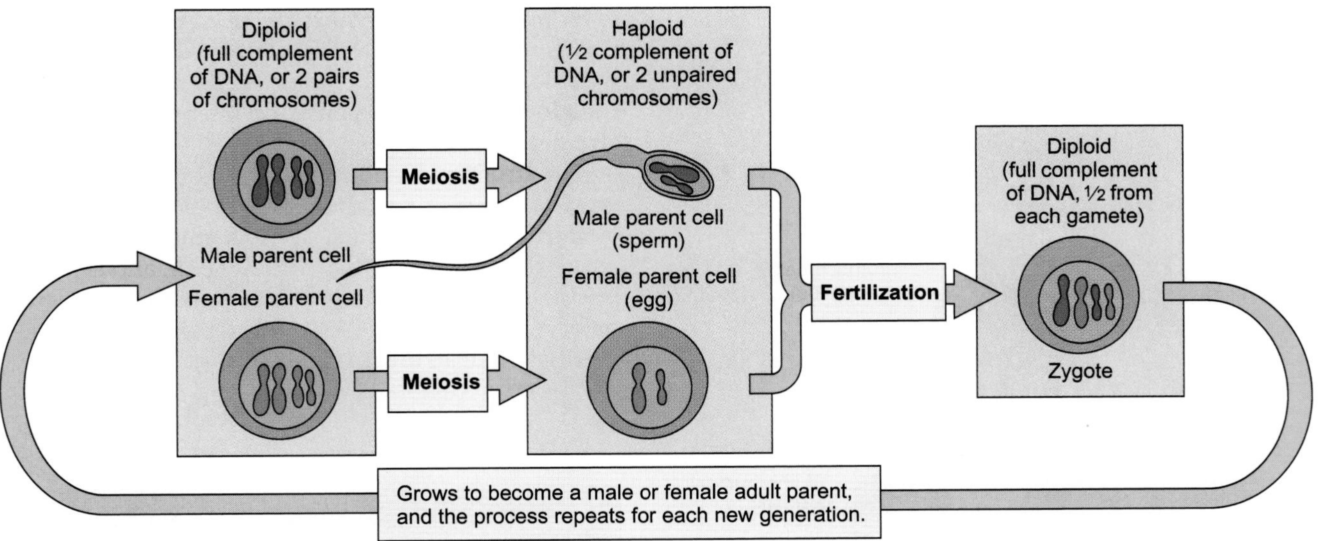

FIGURE 10.1 A SIMPLIFIED VIEW OF A LIFE CYCLE.
An individual is the result of fertilization, the process whereby the haploid gametes of the two parents combine to form the diploid zygote. It is the process of meiosis that reduces the complement of DNA to haploid in the gamete, so that each new generation doesn't have double the number of chromosomes.

a pole (Figure 10.2). Do you see why this is the critical phase of meiosis? If you are following this so far, you can see that it is at this point that the chromosome number is halved.

The second division of meiosis, **meiosis II,** on the other hand, is functionally identical to mitosis. Here, just as in mitosis, each chromosome lines up by itself on the metaphase plate, the centromere divides, and the chromatids (now chromosomes) are separated to opposite poles.

In meiosis I, homologues of paternal and maternal origin not only recognize each other but become intimately associated, intertwining and actually fusing together at various places along their length. (Remember that the two homologues contain the same information, but the homologue from one parent may contain a slightly different *version* of the same information.) During **synapsis,** as the pairing and fusing process is known, **crossing over** occurs. During crossing over, homologous chromatids exchange equivalent segments of DNA. Simply put, they swap parts. The homologues then separate, but by then each chromatid contains a new mix of DNA from each parent.

We will take a closer look at this exchange of DNA in the next section, but for now, let's summarize the differences between meiosis and mitosis.

1. Mitosis requires one division; meiosis requires two.
2. In mitosis the chromosome number is retained, but in cells emerging from meiosis the number is halved.
3. Although centromeres divide in anaphase of mitosis, they do not divide in anaphase of meiosis I. Instead, paired homologues separate from one another, one of each moving to opposite sides of the cell.
4. In mitosis homologues remain separate from each other, but in meiosis they synapse and crossing over occurs.
5. Most significantly, mitotic daughter cells are genetically identical, but meiotic daughter cells are genetically diverse.

MEIOSIS I: THE FIRST DIVISION

During the mitotic cell cycle, a cell preparing for mitosis grows, replicates its DNA, and forms spindle proteins. The same thing happens to a cell preparing for meiosis. Because of DNA replication, a human cell entering the first prophase of meiosis has 92 DNA molecules in its nucleus. With two DNA molecules per centromere, that makes 46 chromosomes. After meiosis I, each cell will have only 23 chromosomes, each still consisting of two DNA molecules and a single centromere.

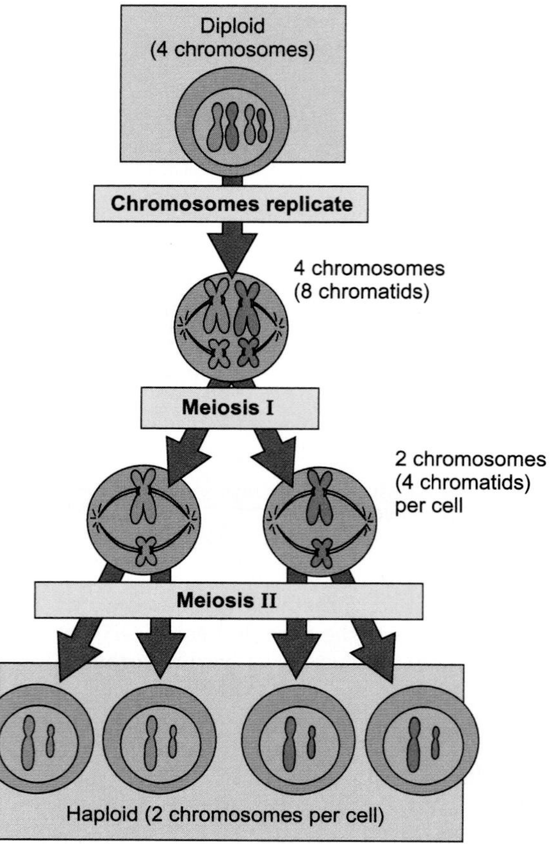

FIGURE 10.2 OVERVIEW OF MEIOSIS.
In preparation for meiosis, the chromosomes replicate. The first of two successive meiotic divisions then occurs. Note that the centromeres do not divide in meiosis I, and homologues have separated. Each daughter cell receives a member from each pair. Note also that we have drawn the process so that, of the two chromosome pairs, the paternal members go to the same cell. This need not be the case, as which goes where is totally random. With the division of meiosis II, centromeres divide and daughter cells receive two chromosomes per cell. This means they are now haploid (compare these to the diploid cell at the beginning).

PROPHASE I

As the cell enters **prophase I,** the chromosomes begin to condense, just as we would expect. But in meiosis, the process occurs slowly (see Figure 10.3). We are eventually able to detect the long, spindly chromosomes as they first begin to shorten, but we may also see that they are moving. Each chromosome must find its counterpart, its homologue. The meiotic cell nucleus slowly begins to roll or tumble in such a way that the chromosomes within are moved randomly about. When two homologues finally bump into each other in the right way, homologous *regions* of the chromosomes will adhere, side-by-side. The regions of side-by-side fusion grow as the two homologues come together much like the two halves of a zipper.

The structure responsible for this zipperlike pairing is a complex proteinaceous organelle called the **synaptonemal complex.** It is synthesized on the chromosomes before they actually pair, and acts as a bridge between homologues. In electron micrographs, the synaptonemal complex even looks surprisingly like a zipper (Figure 10.4a).

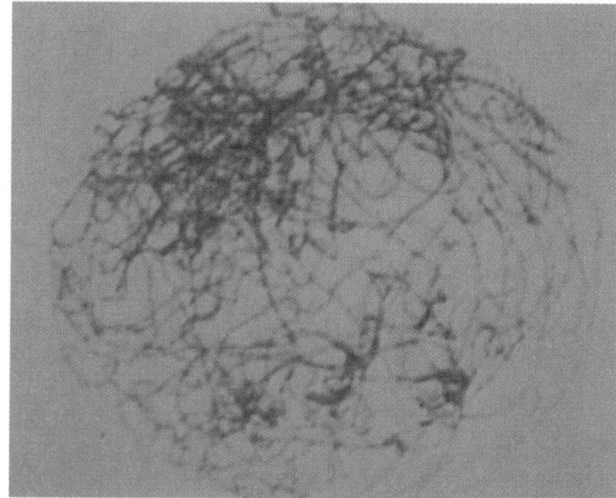

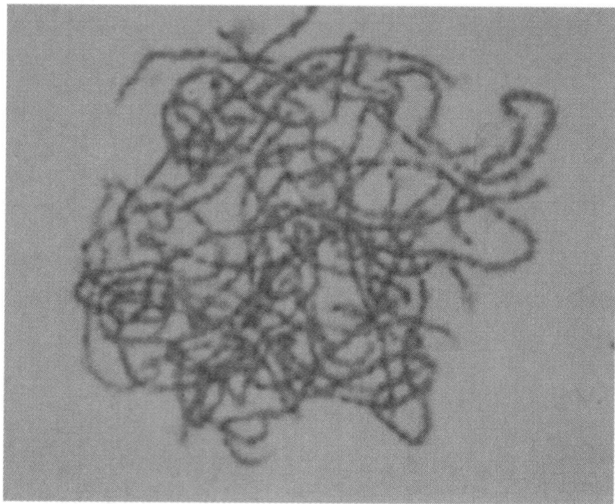

FIGURE 10.3 PROPHASE.
In earliest prophase I of meiosis, chromosomes only partially condense, appearing as long, spindly strands. Chromosomes begin to move, and homologues find each other and fuse, permitting crossing over to occur.

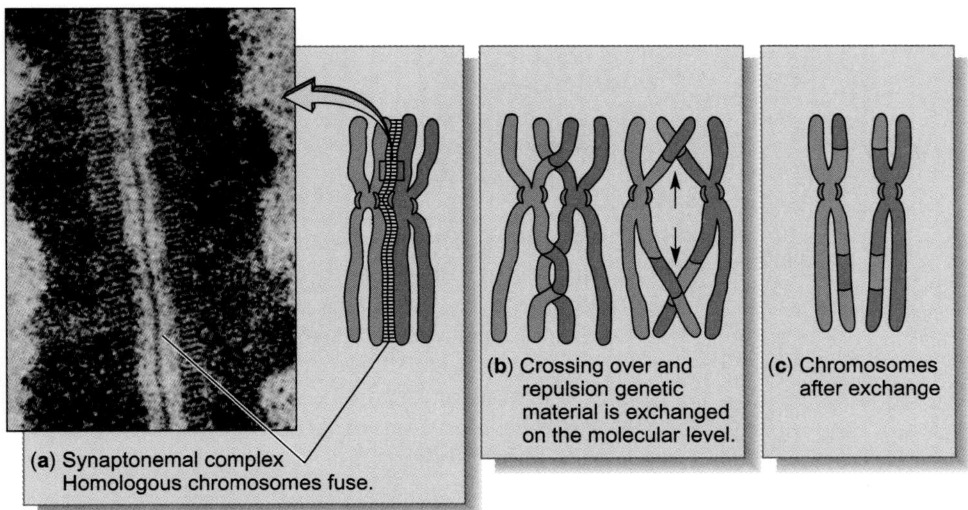

(a) Synaptonemal complex
Homologous chromosomes fuse.
(b) Crossing over and
repulsion genetic
material is exchanged
on the molecular level.
(c) Chromosomes
after exchange

FIGURE 10.4 CHROMOSOME PAIRING AND CHIASMATA. Early in prophase I, synapsis occurs as homologous chromosomes fuse through the formation of a zipperlike synaptonemal complex **(a)**. In this state, crossing over occurs **(b)**. Following synapsis, the four chromatids are clearly visible. Clinging regions form X-shaped chiasmata that will mark crossover regions. The four emerging chromatids have undergone genetic recombination **(c)**.

It is within the synaptonemal complex that crossing over—the actual exchange of DNA—takes place (Figure 10.4b). In human meiosis, there is an average of about ten exchanges among the four chromatids of each chromosome pair. How does it happen? The details are still a subject of intense scientific investigation. The most important thing to realize about crossing over is that it is in no way an accident or a by-product of other processes; yet where it occurs seems to be totally random. It is such an important process that at least 50 enzymes are responsible for seeing that it happens.

After crossing over, things begin to happen a bit more rapidly. The chromosomes continue to shorten and condense, and then at some point the synaptonemal complex breaks down. The homologues lose their affinity for each other, and they move apart.

The separation of the chromosomes doesn't occur as completely as it might because the homologues persist in clinging together at scattered points called **chiasmata** (singular, *chiasma*). They mark the places where crossing over had taken place earlier (Figure 10.4b).

As the first metaphase of meiosis approaches, the chiasmata slide to the ends of the chromosomes, much like a ring being pulled off two napkins. Finally, the chromosomes are joined only at their tips. They continue to touch for this last moment as they move together to the metaphase plate where they prepare to be separated forever.

METAPHASE I AND ANAPHASE I

At **metaphase I,** when the chromosomes form the metaphase plate, there is a brief pause in activity. Then the last of the clinging homologues completely disengage and **anaphase I** begins. But as emphasized previously, unlike what happens during mitotic anaphase, *the centromeres do not divide.* Instead, sister chromatids remain joined and move together, with homologues drawn toward opposite poles of the cell (Figure 10.5). Let's be sure this is clear. Members of the chromosome pair become segregated, ending up at opposite sides of the cell, and following cytokinesis, ending up in different cells. This process must be precise; it *must be homologues that separate in anaphase I.* Any other kind of separation would mean that the daughter cells would end up with some chromosomes missing and extra copies of others. Both situations can be lethal.

How is the precision of anaphase I guaranteed? What assures that all homologues move in opposite directions? The answer brings up yet another difference between meiosis and mitosis. Recall that in mitosis (see Figure 9.6), each individual chromosome received centromeric spindle fibers from both poles. Not so in meiosis I. At metaphase I of meiosis, each pair member receives only one set of centromeric spindle fibers, and these are from the pole opposite those received by the other pair member (Figure 10.5). This arrangement guarantees the segregation

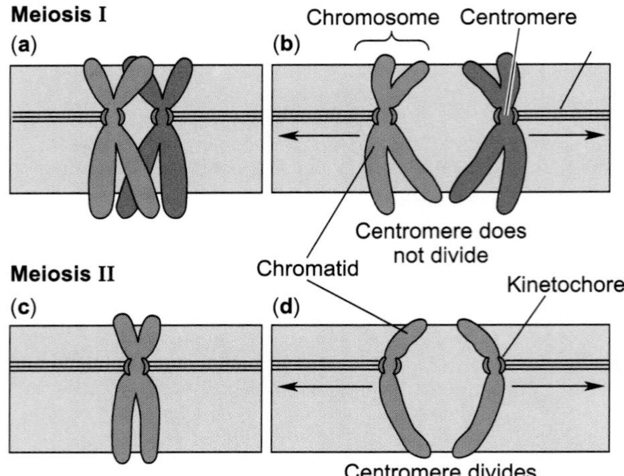

FIGURE 10.5 CENTROMERES AND SPINDLE FIBERS.
The arrangement of centromeric spindle fibers explains how homologues become segregated during meiosis I and how chromatids finally separate during meiosis II. In anaphase I, the spindle fibers that attach to each homologue *originate at opposite poles* (**a**). When the pulling occurs, homologues become separated and end up segregated in the two daughter cells (**b**). In anaphase II, as in mitosis, centromeric spindle fibers attach to *each side of each chromosome* (**c**). When the pull occurs, the centromeres divide and the chromatids separate (**d**).

of homologous chromosomes in anaphase I. Once the chromosomes have gathered at their respective poles, telophase I begins.

TELOPHASE I

Telophase I is similar to mitotic telophase. The chromosomes uncoil (sometimes only partially), the nuclear membrane forms around them, centrioles replicate, and cytokinesis occurs. Of course, the daughter cells are unlike any produced by mitosis, since chromosome pairs no longer exist. Remember, the maternal and paternal homologues were separated forever as they moved away from the metaphase plate.

After telophase I, the daughter cells enter their **meiotic interphase,** the period before the second stage of meiosis begins. In this interphase there is *no DNA replication,* so the chromosomes entering meiosis II will appear exactly as they did at the end of meiosis I. Each chromosome will be composed of two recombined chromatids still attached by a centromere.

FIGURE 10.6 MEIOSIS.
The stages of meiosis are diagrammed. Representative stages of meiosis in the lily are shown in the photomicrographs.

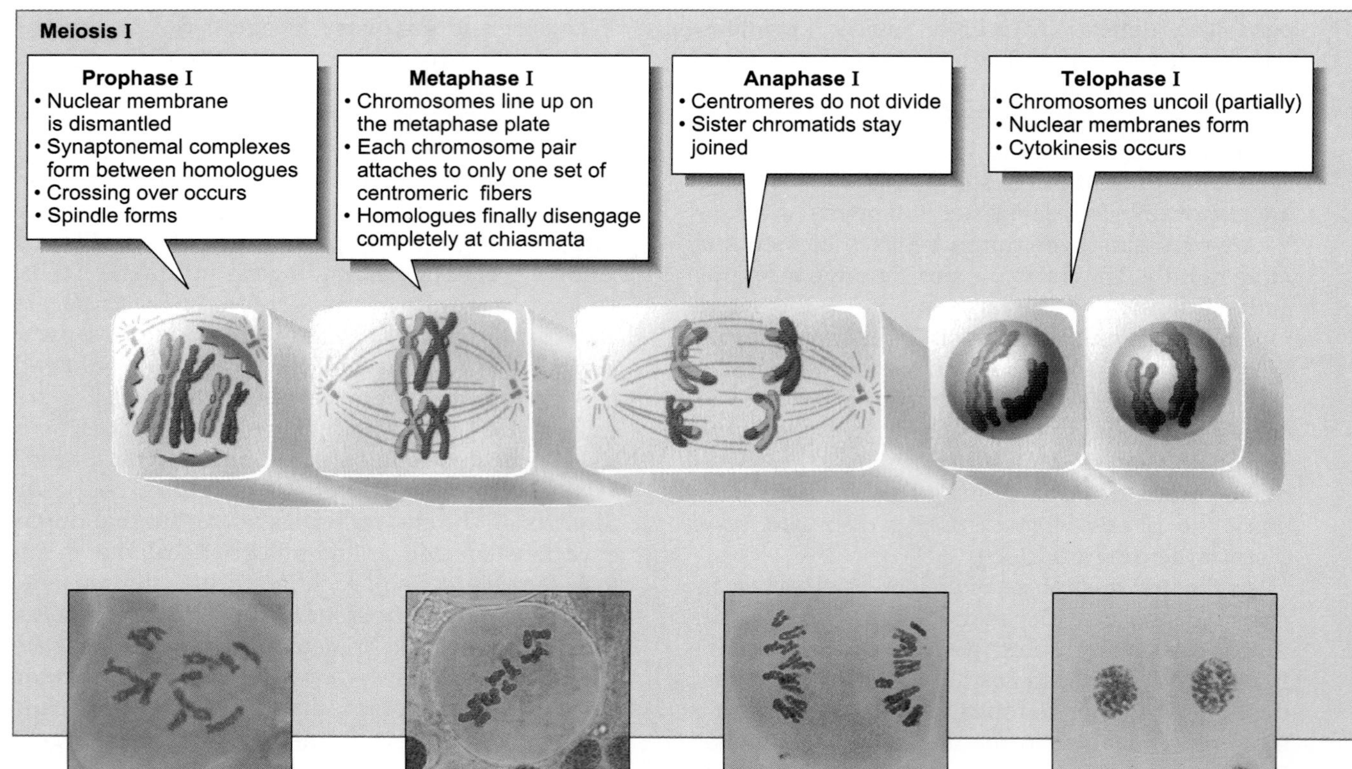

Meiosis I
Prophase I
• Nuclear membrane is dismantled
• Synaptonemal complexes form between homologues
• Crossing over occurs
• Spindle forms
Metaphase I
• Chromosomes line up on the metaphase plate
• Each chromosome pair attaches to only one set of centromeric fibers
• Homologues finally disengage completely at chiasmata
Anaphase I
• Centromeres do not divide
• Sister chromatids stay joined
Telophase I
• Chromosomes uncoil (partially)
• Nuclear membranes form
• Cytokinesis occurs

Meiosis II: The Second Division

Meiosis II is a bit easier to follow since it proceeds much as does mitosis (see Figure 10.2). Both daughter cells from meiosis I enter **prophase II.** At this stage, the centromeric spindle fibers from each pole attach to either side of the chromosomal centromeres, just as they do in mitosis. The chromosomes then line up on the metaphase II plate in preparation for being separated. The centromeres finally divide as **anaphase II** begins, and the chromatids, now daughter chromosomes, are drawn to opposite poles. Following **telophase II,** the two daughters of meiosis I divide into four cells.

As shown in Figure 10.6, the final four daughter cells of meiosis each contain half the chromosome number of the original cell. In addition, each is unique, its chromosomes bearing a different mix of genes from those of the original cell. Figure 10.7 makes a final comparison between mitosis and meiosis. Notice how similar mitosis and meiosis II are, while meiosis I is quite special.

Where Meiosis Takes Place

In multicellular organisms, meiosis usually takes place in the **germinal tissues.** In animals, the organs in which germinal tissues are found are the *ovaries* (in females) and the *testes* (in males). In flowering plants, the equivalent structures are the flower's *ovaries* and *anthers.* Curiously, the germinal tissues of animals begin to form early, while the individual is still an embryo. Flowering plants, however, maintain regions of uncommitted tissues, which begin to change following an environmental cue. In response to these cues, some cells break out of the usual cycling and develop into the floral parts.

Meiosis in Humans

In humans, as well as most other animals, the gonads (ovaries and testes) are formed during embryonic development. In males, the germinal tissue forms the lining of the long, highly coiled tubules that make up most of the mass of the testes. However, once these tissues are formed, activity

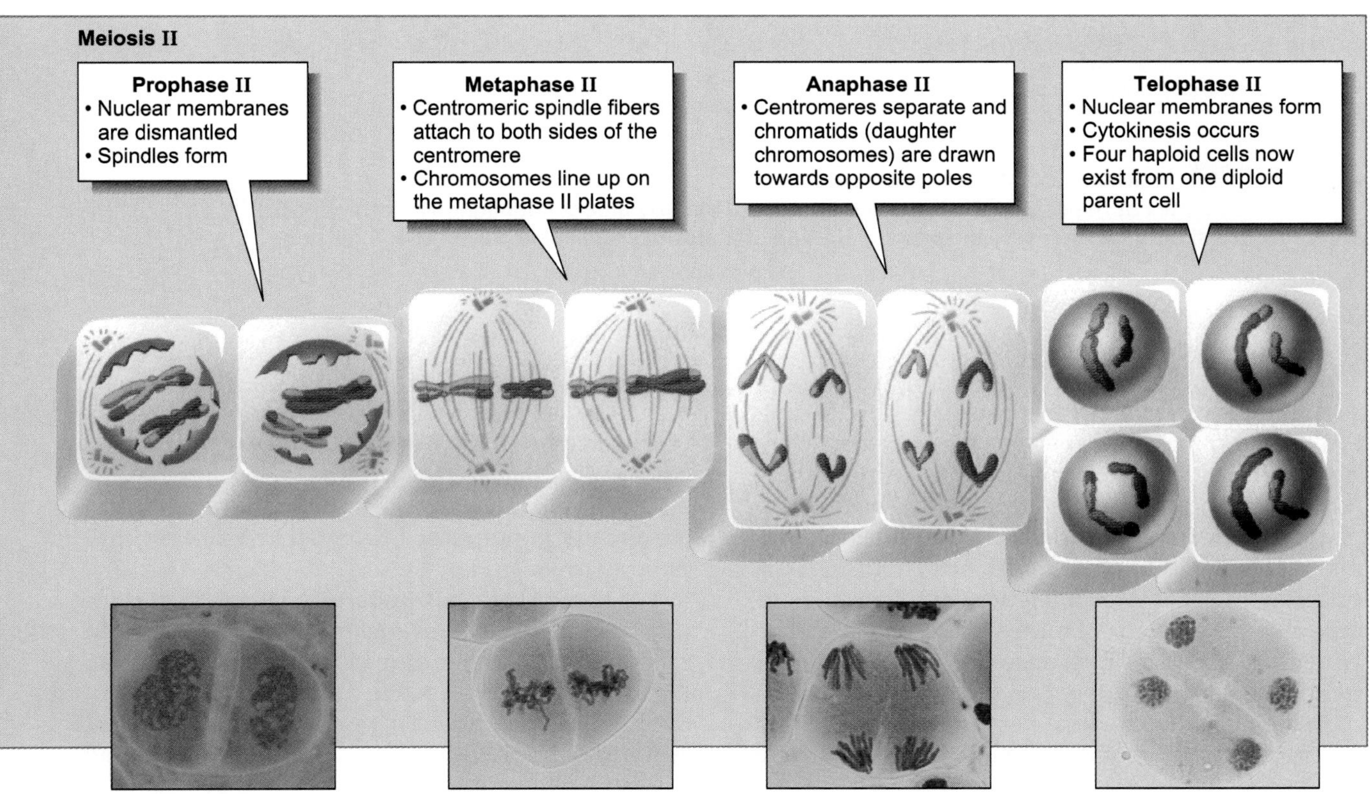
Meiosis II
Prophase II
• Nuclear membranes are dismantled
• Spindles form
Metaphase II
• Centromeric spindle fibers attach to both sides of the centromere
• Chromosomes line up on the metaphase II plates
Anaphase II
• Centromeres separate and chromatids (daughter chromosomes) are drawn towards opposite poles
Telophase II
• Nuclear membranes form
• Cytokinesis occurs
• Four haploid cells now exist from one diploid parent cell

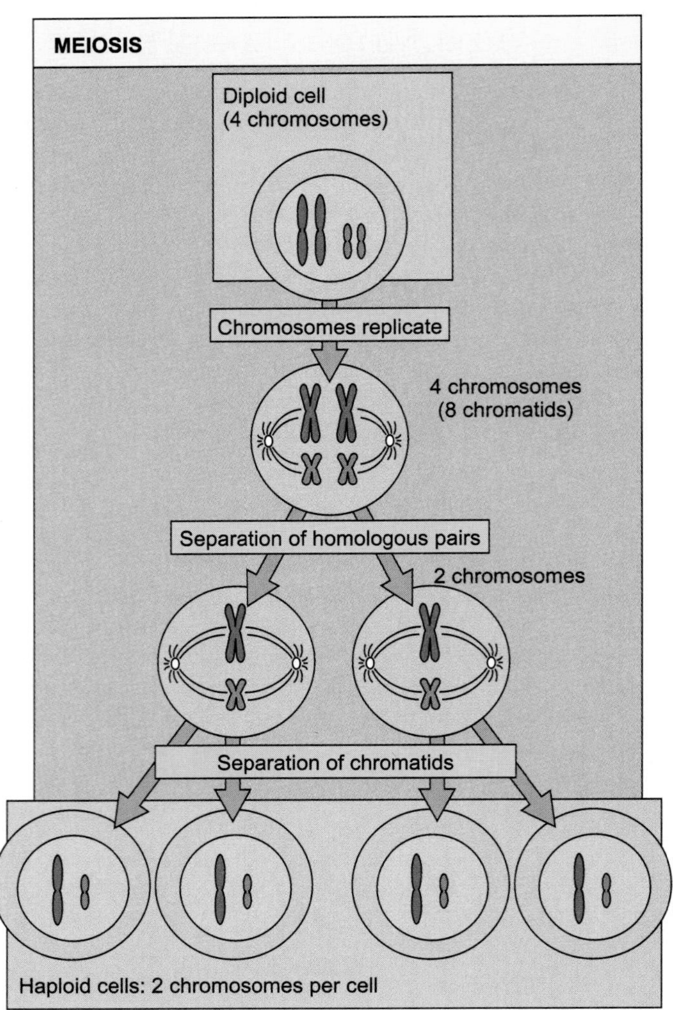

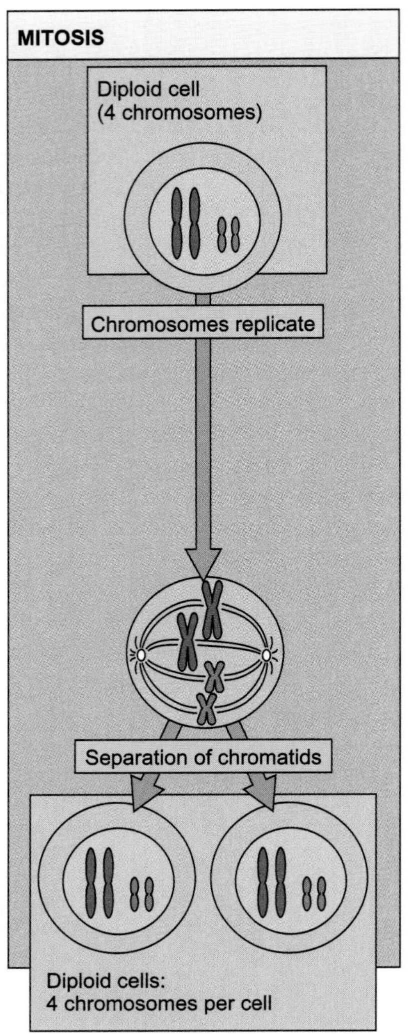

FIGURE 10.7 MEIOSIS COMPARED TO MITOSIS.
Simplified schemes of meiosis and mitosis are compared side by side to highlight similarities and differences. It should be clear why the second division of meiosis is often called a "mitotic" division.

ceases until puberty—when the cells begin dividing again. Some of the new cells produced in these divisions will begin to undergo meiosis, while others continue with mitosis, forming a ready reserve supply (for future sperm). Male meiosis during **spermatogenesis,** the formation of sperm, holds no surprises. In each complete meiotic event, four haploid cells are formed. Each of these will become a sperm. However, meiosis in females is quite different (Figure 10.8).

In women (and in females of other vertebrate species as well), the cells in the ovaries that will give rise to eggs take a somewhat unexpected developmental route. Meiosis in females during **oogenesis** (the formation of eggs) is well under way during the embryonic stage. In fact, a newborn girl already has all the developing eggs she will ever have; and most are already in prophase I, where they will stay until puberty or long after.

When a girl reaches puberty, one or two oocytes (future eggs, or ova) resume meiosis each month, in preparation for **ovulation** (release of the egg from the ovary). This process will be repeated throughout the reproductive life of a woman. Although an infant girl is born with several thousand developing eggs in her ovaries, only about 400 to 500 actually

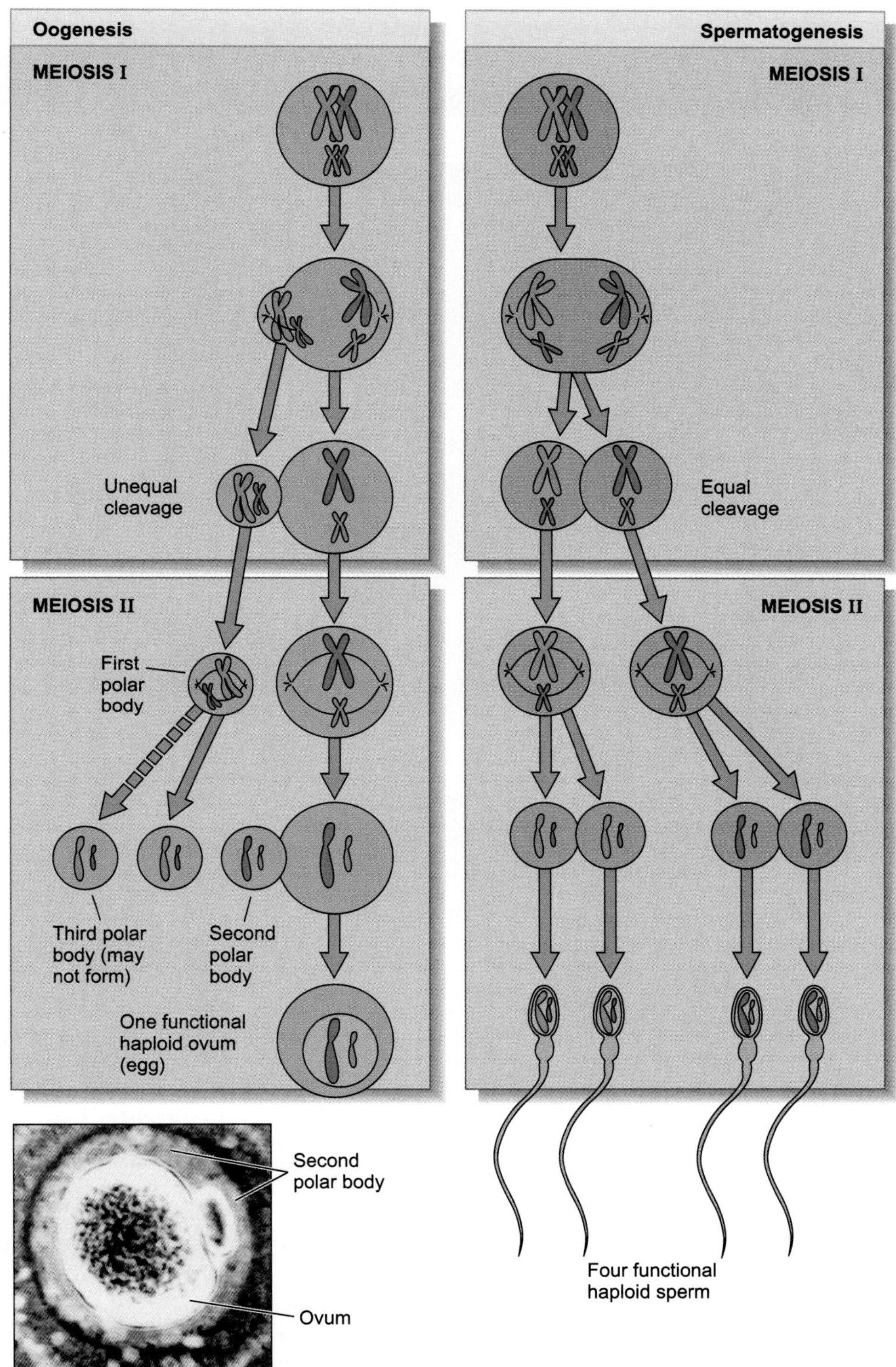

FIGURE 10.8 **FORMATION OF EGGS AND SPERM.**
Spermatogenesis occurs, as might be expected, from meiosis—with one cell yielding four haploid products that will each become sperm. However, oogenesis is different. In each of the two divisions, the metaphase plate forms to one side of the oocyte, and unequal cleavage occurs. Thus, only one large ovum results from each meiotic event. The ovum in the photo shows a polar body on its right side.

Essay 10.1 When Meiosis Goes Wrong

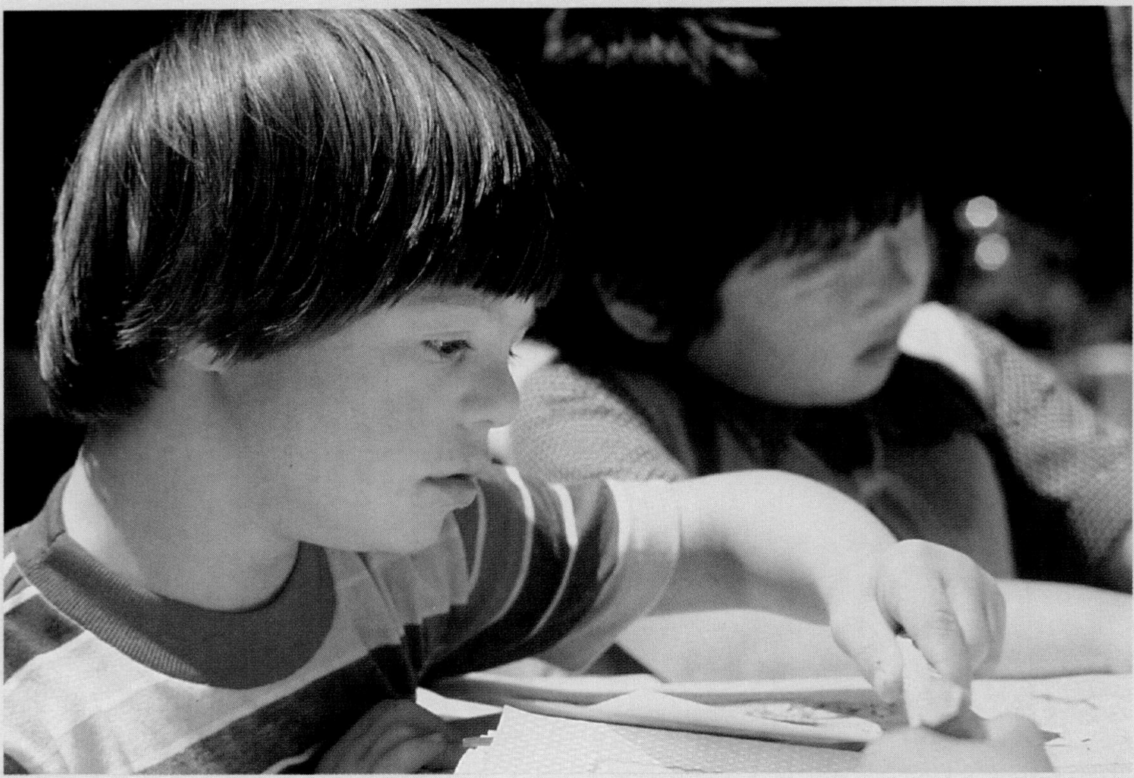

Meiosis is a much more complicated process than mitosis. When you consider the lengthy prophase with all the subphases of chromosome pairing and crossing over, in addition to the presence of two cell divisions, you shouldn't be surprised that frequently something goes wrong. In humans, for instance, about one-third of all pregnancies spontaneously abort in the first two or three months. When the expelled embryos can be examined, it turns out that most of them have the wrong number of chromosomes. Failure of the chromosomes to separate correctly at meiosis is termed nondisjunction.

Not all failures of meiosis result in early miscarriage. There are late miscarriages and stillbirths of severely malformed fetuses. Even worse, about one live-born human baby in 200 has the wrong number of chromosomes, accompanied by severe physical and/or mental abnormalities.

mature. So, although sperm are produced continuously after puberty in men, women are born with all the oocytes already formed.

There are other significant differences between meiosis in males and females. The end product of meiosis in males is four gametes for each cell going through meiosis, but the result for the meiosis in females is only one potential egg cell. What happens to the other three cells we have come to expect?

In females, a cell entering oogenesis is called a **primary oocyte.** Meiosis I results in two cells as expected, but they are far different in size. It seems that the cleavage plane in the female oocyte is off to one side of the cell, and even though chromosome separation occurs as expected, one daughter cell gets nearly all of the cytoplasm and is carried forward to the next steps. The other, much smaller second cell becomes known as a **polar body.** The polar

About one baby in 600 has three copies of tiny chromosome 21. Such persons may grow into adulthood, but they have all kinds of abnormalities. The syndrome is known both as *trisomy-21* and *Down syndrome,* after the 19th-century physician who first described it. Characteristics of the syndrome are general pudginess, rounded features, and a rounded mouth in particular, an enlarged tongue which often protrudes, and various internal disorders. Often, a peculiar fold in the eyelids is seen. Trisomy-21 individuals also have a characteristic barklike voice and unusually happy, friendly dispositions. The "happiness" is a true effect of the extra chromosome and not a result of their usually low IQs; those with serious mental impairment are usually, by most indications, miserable.

Trisomy-21 occurs most frequently among babies born to women over 35 years old. At that age the incidence is about 2 per 1,000 births. By age 40, this climbs to about 6, and by age 45, 16 children with Down syndrome are born for each 1,000 births. The age of the father apparently has little, if any, effect. We can guess that the much-prolonged prophase I of the human oocyte might have something to do with this difference.

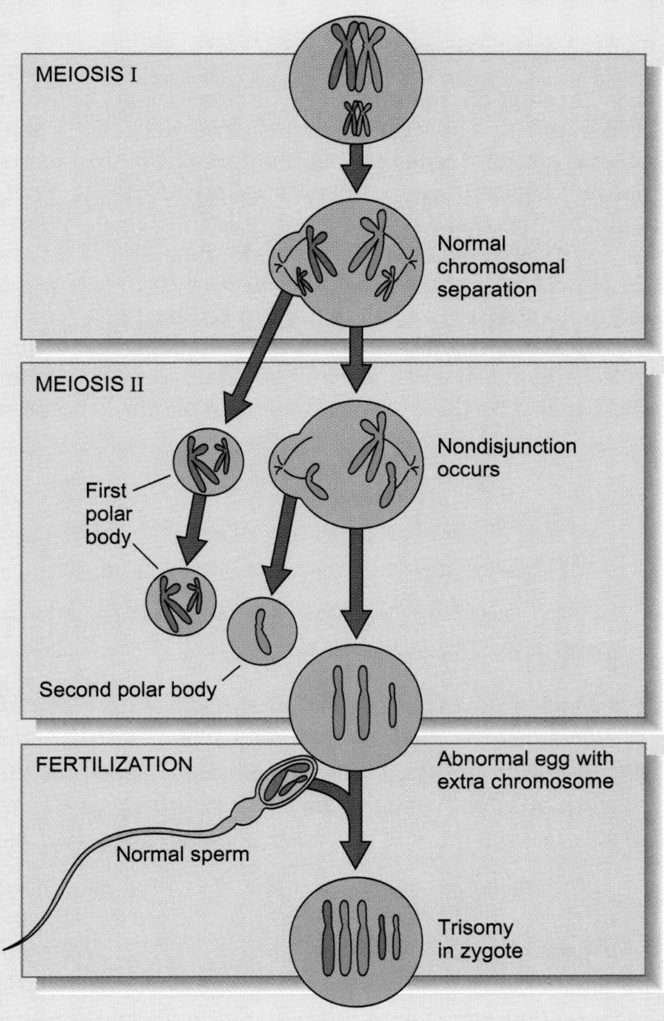

body acts only as a receptacle for half the homologues, an important task, although the cell has no real future. Once produced, polar bodies may or may not divide again.

The larger cell, now called a **secondary oocyte,** enters meiosis II and divides as expected, but again, a highly unequal cytoplasmic division produces a large cell and another polar body. The large cell is now haploid and is the mature egg or **ovum.** Meiosis in females produces only one functional haploid gamete. The two or three polar bodies simply disintegrate.

Although we have emphasized the orderly progression of meiosis, you shouldn't be surprised to learn that such a complex mechanism sometimes fails. Errors in chromosome and chromatid separation do occur. For example, the centromeres may not release the chromosomes, and one cell will end up short one chromosome while the other has an extra one. Should such an abnormal gamete become fertilized and the embryo survive, the results are usually tragic. Such an error, known as **nondisjunction,** occurs all too frequently in our own meiotic mechanisms (Essay 10.1).

Meiosis in Plants

Gamete production in animals (humans being rather typical examples) is the direct result of meiosis. However, this isn't true in most plants. In flowering plants, meiosis in the anthers is followed by mitosis, so that there are two haploid cells in each pollen grain (in such plants, pollen houses the male gamete). However, only one of the two cells in pollen is a potential sperm cell. Another peculiarity of flowering plants is that just before fertilization, the potential sperm cell will enter mitosis again, producing two sperm. In the ovary of flowering plants, meiosis proceeds as usual, but only one of the haploid products will survive. Rather than becoming an egg, it enters three rounds of mitosis, producing eight nuclei. One nucleus divides and produces two, which produce four, which produce eight, only one of which will become an egg cell (see Chapter 28).

Ferns, like a number of nonflowering plants, live double lives. The familiar, graceful plant we admire does not itself engage in sex, although it does carry on meiosis. Its meiotic products, millions of tiny, dust-sized spores containing the haploid number of chromosomes, are carried away by wind and water. Should they find themselves in a favorable situation, the spores will begin to grow and divide, producing tiny (but independent) plants called gametophytes. The primary role of each of these haploid plants is to produce sperm and egg cells by mitosis. Once fertilization occurs, a new, graceful, but celibate individual will emerge—to produce more spores (see Chapter 24).

Perspectives

Meiosis is the process that allows sexual reproduction to occur in diploid eukaryotes. Without meiosis, fusion of gametes would result in a doubling of the number of chromosomes with each generation. Through meiosis, however, the number of chromosomes carried by gametes is reduced in half. When gametes fuse during fertilization, the normal diploid state is restored. Once one understands this main concept, the actual mechanism of meiosis makes sense.

So why bother with sex? Your personal observations notwithstanding, over the next few chapters we will develop the means to answer that question. (The question is, as you will see, much more involved than it first seems.) Along the way we will discover and examine the means by which the DNA in chromosomes actually carries the information that results in the myriad of traits that marks each individual as a unique organism. Because, as we will eventually find in Part III of this book, variation in the spectrum of traits carried by individuals is the mechanism that helps ensure the continuance of life.

Summary

Overview of Meiosis

Through meiosis, cells become haploid gametes, and—due to crossing over—gametes receive a mix of parental DNA. Fertilization combines the chromosomes from the two gametes to restore the diploid condition. Meiosis consists of two parts. In meiosis I, the homologous chromosomes separate from each other and the centromeres remain intact. This is the division that reduces the chromosome number. Meiosis II is very similar to mitosis in that the centromeres split and what was a single chromatid (now a chromosome) is delivered to each pole.

Meiosis I: The First Division

Early in prophase I, homologues synapse, synaptonemal complexes form between homologues, and crossing over occurs. The separation of homologues follows, except at crossover points called chiasmata, where they still cling.

During prophase I, each homologue has received centromeric spindle fibers from an opposite pole. The homologues line up on the metaphase plate at metaphase I. The result is that homologues move to opposite poles during anaphase I. Centromeres do not divide.

In telophase I, the chromosomes uncoil, the nuclear membrane and nucleolus reappear, centrioles replicate, and cytokinesis occurs. But there is no DNA replication in the ensuing interphase.

Meiosis II: The Second Division

Meiosis II is similar to mitosis, including the attachment of centromeric spindle fibers from each pole to opposite sides of each chromosome. Centromeres divide, and chromatids finally separate. Following telophase II, four haploid cells emerge. Each will carry one homologue from each chromosome pair. Each chromosome carries a mix of parental DNA.

Nondisjunction is the failure of chromosomes to separate properly during meiosis, resulting in gametes that carry abnormal numbers of chromosomes.

Meiosis occurs in germinal tissue in the ovaries and testes of animals and in the ovaries and anthers of plants.

Meiosis in males, during spermatogenesis, occurs in the walls of tubules within the testes. It begins at puberty, and each complete meiotic event produces four haploid sperm.

Meiosis in females, during oogenesis, occurs in the germinal tissue of the ovaries. It begins in the embryo, but the oocytes become arrested in prophase I. Meiosis resumes at puberty, occurring in one or two oocytes each month. Cytokinesis in the oocytes is highly uneven—with large, functional cells and minute polar bodies forming in both divisions. Because of this, the product of each complete meiotic event is one egg instead of four.

Meiosis in flowering plants yields haploid cells that later produce gametes through mitosis. Following meiosis in the anther, each haploid product in the anther enters mitosis, forming two cells, one of which will divide again to form two sperm cells.

Following meiosis in the ovary, three of the haploid cells are lost, but the remaining cell undergoes mitosis three times. One of the resulting cells formed becomes the egg.

In ferns, the products of meiosis are haploid aerial spores, some of which, upon germination and development, produce minute, independent plants in which sperm and egg cells form through mitosis.

Key Terms

zygote • **154**
haploid • **154**
meiosis • **154**
meiosis I • **154**
meiosis II • **155**
synapsis • **155**
crossing over • **155**
prophase I • **156**
synaptonemal complex • **156**
chiasmata • **157**
metaphase I • **157**
anaphase I • **157**
telophase I • **158**
meiotic interphase • **158**
prophase II • **159**
anaphase II • **159**
telophase II • **159**
germinal tissue • **159**
spermatogenesis • **160**
oogenesis • **160**
ovulation • **160**
primary oocyte • **162**
polar body • **162**
secondary oocyte • **163**
ovum • **163**
nondisjunction • **164**

Review Questions

1. What is the logic in the "arithmetic" of meiosis? What would happen at fertilization if chromosome reduction did not occur?
2. Is there any special way that pairs of chromosomes are reduced and assigned to gametes during meiosis? What would happen if the assignment were random?
3. Discuss the results of having two divisions for meiosis to be completed.
4. How does the behavior of centromeres differ between mitosis and meiosis I?
5. Briefly discuss synapsis and crossing over in meiosis. What effect does crossing over have on the genetic makeup of the daughter cells?
6. Describe the specifics of crossing over. What evidence of crossing over remains after chromosomes begin to separate?
7. With the aid of a drawing, show the specific manner in which centromeric spindle fibers are attached to homologues prior to anaphase I. What effect does this have on their movement during anaphase I?
8. List a fundamental difference between mitotic interphase and interphase between meiosis I and meiosis II. Why would this make sense in meiosis?
9. Describe the attachment of centromeric spindle fibers to each chromosome in metaphase II. How does this affect the behavior of the chromatids in anaphase II? Compare this to the similar events in mitosis.
10. Compare the following aspects of spermatogenesis and oogenesis: site of germinal tissue, time of life in which meiosis begins, size of cells, numbers of gametes.
11. Develop a diagram illustrating human oogenesis, with emphasis on polar body formation. What is the only purpose of the polar body?
12. In animals, meiosis gives rise directly to gametes. Contrast this to the formation of gametes in flowering plants.

CHAPTER

11

Mendel's Factors

HERE WE GET OUR FIRST REAL INSIGHTS AS TO THE REASONS WHY DNA *and chromosomes are important. After spending the last few chapters discussing these structures, we now begin to demonstrate just how they function. However, rather than simply presenting the conclusions, we will look at how the conclusions were reached. The reason? Because the story of genetics is so logical and so clear in its development that it serves as a terrific framework for illustrating scientific reasoning.*

The silent dance of chromosomes at meiosis is precise and intricate, but the implications of this delicate shuffling are profound. Since chromosomes carry the genetic material of an individual, their segregation and combinations directly influence the traits of the offspring. Indeed, the relationship between chromosomes and inherited traits was deduced long before the discovery of DNA itself. As you wend your way through these next few chapters, you have the benefit of knowing beforehand the relationship of DNA and chromosomes. Keep in mind though, that scientists of the 19th century could only go by what they could see.

As the preeminent biologist of his time, Charles Darwin was led to wonder about just how traits are inherited. Like other biologists of those days, he was particularly struck by the observation that offspring tended to look like a combination of the two parents.

This observation led Darwin to accept the only intellectually respectable theory of his day, which was that of *blending inheritance.* According to this theory, the "blood" or hereditary traits of both parents blended in the offspring, just as two colors of ink blend when they are mixed. The blending theory appeared superficially to work reasonably well for some traits—such as height or weight, where there is a continuous gradation of possible values—but it couldn't account for others.

It's always surprising, in hindsight, to recognize the degree to which a strongly held belief will blind its proponents to obvious contradictions. The blending theory would predict that the offspring of a white horse and a black horse should always be gray, and that the original white or black should never reappear if gray horses continued to be bred. In reality, of course, the offspring of a white horse and a black horse are not always gray, and black or white descendants do appear. Something was obviously wrong with the theory, but no one seemed to notice—or if they did, they tried to ignore it.

Darwin had become embroiled in the aftermath of his new book, *On the Origin of Species by Natural Selection* (1859). His most trying problem was to counter the arguments of his sharper critics, who logically pointed out that natural selection and blending inheritance are opposing concepts; any new traits that might arise would be swamped by existing traits and simply be "blended away." Darwin was never able to answer this criticism adequately. But across the Channel, deep in the European continent, a bright and dedicated monk was starting the groundwork for a new revolution by crossing strains of garden peas. He was also answering Darwin's most vexing questions, but Darwin would die ignorant of the monk's work.

The monk, Gregor Johann Mendel (Figure 11.1), was a member of an Augustinian order in Brunn, Austria. Early in his life Mendel began training himself, and he became a rather competent naturalist. To support himself during those early years, he worked as a substitute high-school science teacher. The professors at the school, noting his unusual abilities, suggested that he take the rigorous qualifying examination to become a regular member of the high-school faculty. Mendel took the test and did reasonably well; but he failed to qualify, so he joined a monastic order.

In 1851, his superiors, confident of his abilities, sent him to the University of Vienna for two years of concentrated study in science and mathematics. There he learned about the infant science of statistics. He was to use this information when he returned to his old

FIGURE 11.1 **MENDEL.**
Gregor Johann Mendel (1822–1884) developed the basic laws of heredity.

hobby of plant breeding. This time, though, he had specific questions in mind, and he thought he knew how to go about finding the answers.

MENDEL'S CROSSES

Mendel began by trying to find the effects of crossing different strains of the common garden pea, but he carried out his research with more precision than mere casual curiosity would call for. To begin with, he based it on a very carefully planned series of experiments, and more importantly, he attempted to analyze the results statistically. The use of mathematics to describe biological phenomena was a new concept. Clearly, Mendel's two years at the University of Vienna had not been wasted.

Mendel was able to purchase 34 *true-breeding* strains of the common garden pea for his experiments. (**True-breeding** strains are those that consistently, generation after generation, yield offspring with the same traits.) These strains differed from each other in very pronounced ways, so that there could be no problem in identifying the results of a given experiment. Mendel decided to work with seven different pairs of traits:

1. Seed form—round or wrinkled
2. Color of seed contents—yellow or green
3. Color of seed coat—white or gray
4. Color of unripe seed pods—green or yellow
5. Shape of ripe seed pods—inflated or constricted between seeds
6. Length of stem—short (9–18 inches) or long (6–7 feet)
7. Position of flowers—axial (along the stem) or terminal (at the end of the stem)

MENDEL'S FIRST LAW: PAIRED FACTORS AND THEIR SEGREGATION

Remember that in the first meiotic division, homologous chromosomes bearing gene pairs are segregated from each other, one going to each of the two daughter cells (see Chapter 10). Mendel didn't know anything about meiosis, chromosomes, or DNA—but somehow he concluded that factors come in pairs, and that each member of the pair goes to different daughter cells. He called this separation of pairs of factors into different gametes **segregation.** The concepts of paired factors controlling heredity and their segregation in gametes are included in Mendel's first law. The first law states that *each genetic character is produced by a pair of factors, and factors segregate into different gametes.* We now know that the segregation (separation) occurs during meiosis. Mendel was truly operating at the frontier of science, with little to go on except his own ample intuition and creativity.

But to see how Mendel started, we must first know something about peas. Each pea in a pod is essentially a unique plant, with its own genes and traits, or, in the language of genetics, its own *genotype* and *phenotype.* (The total combination of an organism's genes is called its **genotype,** and the combination of its visible traits is called its **phenotype.** Other terms used in this chapter are defined in Table 11.1.) Therefore, the first three traits in Mendel's list of pea traits can be categorized by simply examining the peas in their pods.

TABLE 11.1 GLOSSARY OF TERMS IN GENETICS

Term	Definition	Term	Definition
allele	one of two or more alternative forms of a gene (**A** and **a**, **R** and **r**)	**monohybrid cross**	a cross in which one pair of alternative alleles is under consideration
dominance	where the expression of one allele masks the expression of its alternative	**phenotype**	an individual's visible traits (e.g., **tall, round, green**)
dihybrid	a cross in which two different pairs of alleles are under consideration	**progeny testing**	determining the genotype of offspring by self-pollinating or inbreeding
genotype	the combination of genes producing a trait (e.g., **Tt, Rr, Yy**)	**Punnett square**	a grid used to record the possibilities in a cross
heterozygous	an individual with two alternative forms of a pair of alleles (e.g., **Tt, Rr, Gg**)	**recessiveness**	where an allele's expression is masked by its alternative allele
homozygous	an individual with identical alleles for a gene (e.g., **TT, rr, gg, GG**)	**segregation**	the physical separation of alleles when meiosis occurs
independent assortment	where the segregation of one pair of alleles has no effect on the segregation of another	**test cross**	determining the genotype of a dominant individual by crossing it to a recessive (**T_** × **tt**)
linked genes	genes situated on the same chromosome	**true-breeding**	individuals whose offspring are genetically identical to themselves (homozygous)

The Experimental Procedure

Mendel's approach, a novel one at that time, was to cross two true-breeding strains that differed in only one characteristic. Mendel began by asking simple questions, such as: What will happen if I cross a true-breeding yellow-seed pea plant with a true-breeding green-seed pea plant?

Pea breeding, by the way, is extremely tedious work. To carry out a cross, Mendel first had to select and plant his seeds, and then wait for them to grow and flower. That waiting gives one plenty of time to read, file one's nails, and practice accents. But later, things become a bit more hectic. A garden pea plant, if left alone, will generally self-pollinate, each flower fertilizing itself (Figure 11.2). In this way, garden peas go on happily producing their own true-breeding kind. But Mendel was interested in crosses. To cross two strains, he had to open the flowers early in their growth and cut off the pollen-producing anthers of particular plants. Then, using a fine brush, he had to transfer pollen from other flowers—a laborious task. The plants selected for the cross would be called the $\mathbf{P_1}$ (first parental) generation and their offspring the $\mathbf{F_1}$ (first filial) generation. The offspring of the F_1 would be called the $\mathbf{F_2}$, and so on.

F_1 Generation and the Principle of Dominance. When Mendel crossed his original P_1 plants, he found that the characteristics of the two plants didn't blend, as prevailing theory said they should. When plants grown from yellow seeds were crossed with those grown from green seeds, their F_1 offspring did not have yellowish-green seeds. Instead, all of them had yellow seeds, indistinguishable from the yellow seeds of the true-breeding parental strain. Mendel termed the trait that appeared in the F_1 generation the **dominant** trait, and he described the one that had failed to appear as the **recessive** trait. But Mendel was left with a vexing question. What happened to the trait for green seeds?

Mendel had quite a puzzle on his hands, but he was apparently quite good at puzzles. His next step, a stroke of intuition, was to allow his F_1 plants to self-pollinate. In this second filial generation (the F_2 generation), Mendel found that roughly ¼ of the peas were green and that ¾ were yellow. The recessive trait had reappeared (Figure 11.3)! He repeated the experiment with other pea strains, with comparable results. When he crossed a round pea strain with a wrinkled pea strain, all of the F_1 peas were round; but in the F_2 generation, about ¼ of the peas were wrinkled. The constancy of the ratios did not escape the tenacious Mendel, who was determined to keep tackling the problem until he could make some sense of his results.

Two Kinds of Yellow Peas: Homozygous and Heterozygous. From his experiments thus far, Mendel realized that there were two kinds of yellow peas: the true-breeding kind, which—like the original parent stock—would grow into plants that would bear only yellow peas; and another type,

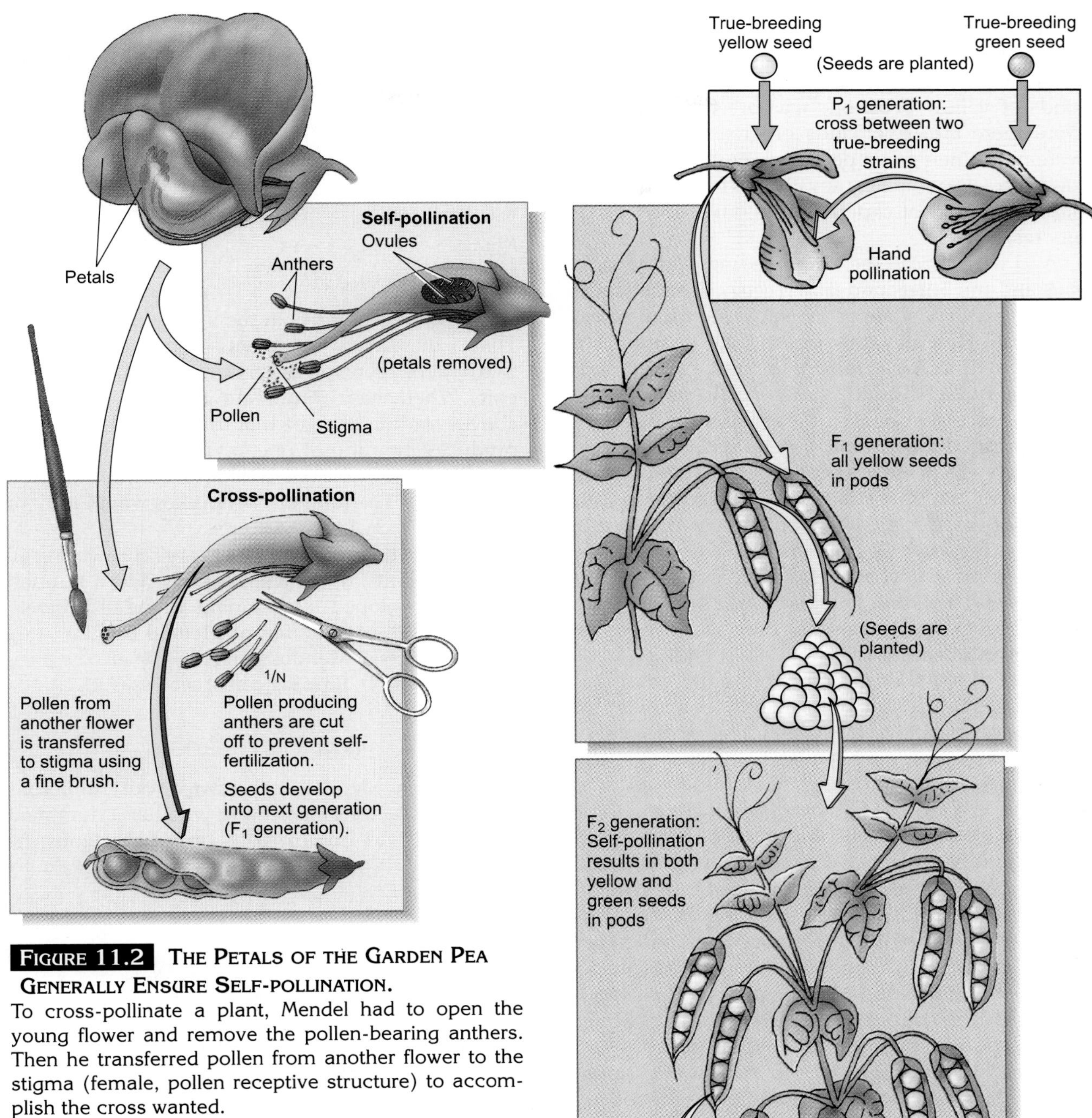

Petals
Self-pollination
Ovules
Anthers
(petals removed)
Pollen
Stigma
Cross-pollination
1/N
Pollen from another flower is transferred to stigma using a fine brush.
Pollen producing anthers are cut off to prevent self-fertilization.
Seeds develop into next generation (F_1 generation).
True-breeding yellow seed
True-breeding green seed
(Seeds are planted)
P_1 generation: cross between two true-breeding strains
Hand pollination
F_1 generation: all yellow seeds in pods
(Seeds are planted)
F_2 generation: Self-pollination results in both yellow and green seeds in pods
(39)
(13)
¾ yellow seeds
¼ green seeds

FIGURE 11.2 **THE PETALS OF THE GARDEN PEA GENERALLY ENSURE SELF-POLLINATION.**
To cross-pollinate a plant, Mendel had to open the young flower and remove the pollen-bearing anthers. Then he transferred pollen from another flower to the stigma (female, pollen receptive structure) to accomplish the cross wanted.

FIGURE 11.3 **YELLOW AND GREEN PEA SEEDS.**
In this P_1 generation, Mendel crossed true-breeding yellow peas with true-breeding green peas. All the seeds in the F_1 generation were yellow. These seeds were planted and, when grown, were allowed to self-pollinate, producing an F_2 generation. F_2 seeds included both yellow and green peas in a ratio of ¾ to ¼, or 3 : 1.

those that appeared in his F_1 generation, which—when grown and self-pollinated—would produce pods containing both yellow and green peas. Two kinds of yellow peas: one true-breeding, one not. Were there also two kinds of green peas? There were not. When green peas were cultivated and allowed to self-pollinate, they always bore only green peas. This kind of experiment is now called **progeny testing.**

And we now call the true-breeding peas *homozygous* and the other kind *heterozygous.* **Homozygous** means that the organism bears two identical factors for a trait. **Heterozygous** means that the organism bears two different factors for the trait, regardless of its appearance. Thus, a yellow F_1 or F_2 pea *could* be carrying recessive genes for green color.

It appeared that the factor determining the recessive form was passed down from the true-breeding recessive P_1 parental strain through the hybrid (offspring of crossbreeding) F_1 generation to the true-breeding recessive F_2 generation. Whatever it was, however, it was not being expressed in the F_1 generation. Mendel, at that time the world's only mathematical biologist, thought he could use symbols to express his dilemma.

He let a capital letter, **Y,** represent the factor that determines the dominant form, and let a lowercase letter, **y,** represent the factor that determines the recessive form. (The letter symbol for a gene is usually taken from the first letter of the dominant form; in this case, Yellow.) The F_1 hybrid, he concluded, must have both factors present, and could be represented as **Yy.** Since there are two parents, Mendel figured that in the F_1 hybrid, **Y** comes from one parent and **y** from the other. (If this sounds simplistic, considering what you know about chromosomes, remember that Mendel didn't know about chromosomes.) Using various crosses, Mendel determined experimentally that it didn't matter which parental strain bore the peas and which provided the pollen.

If heterozygous plants get a **Y** from one parent and a **y** from the other parent, and are symbolized **Yy,** it makes sense that the true-breeding dominant forms get two **Y** factors—one from each parent—and can be symbolized **YY.** In the same way, the true-breeding recessive forms get **y** factors from both parents and can be symbolized **yy.** We can use **YY** to symbolize the *dominant homozygote* (when the dominant factor from each parent is identical), **yy** to symbolize the *recessive homozygote* (when the recessive factor from each parent is identical), and **Yy** to represent the *heterozygote* (when the factor from each parent differs from the other).

Let's use these symbols to take a closer look at Mendel's first crosses. Mendel had by now deduced:

True-breeding P_1:	**YY** (yellow)	crossed by	**yy** (green)
F_1 hybrid progeny:		All **Yy** (yellow)	
F_1 self-pollinated:	**Yy**	crossed by	**Yy**
F_2 progeny:	**YY** (yellow)	**Yy Yy** (yellow)	**yy** (green)

The ratio of genotypes in the F_2 progeny is 1 : 2 : 1. That is, there is one dominant homozygote for every two heterozygotes and every one recessive homozygote. When these results are grouped as they are above, you can also see that this ratio of genotypes produces the ratio of phenotypes that Mendel observed. In the F_2, he noted that ¾ were yellow and ¼ were green. The ratio of phenotypes was ¾ to ¼, or more simply 3 : 1 (yellow : green).

Actually, there is a simple way of representing all of these results, using what is known as a **Punnett square,** developed by Reginald Crandall Punnett, an early-20th-century fan of Mendel (Figure 11.4). (Some aspects of Mendel's first law can also be illustrated by simply flipping coins; see Essay 11.1.)

The Test Cross

Although Mendel had carried out numerous progeny tests for determining whether a dominant individual was homozygous or heterozygous, he

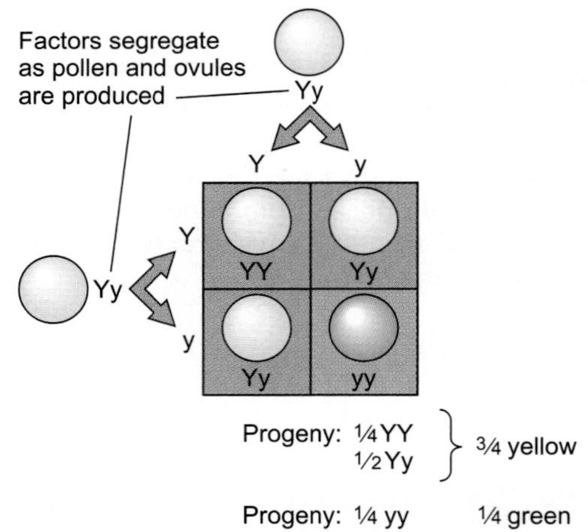

Figure 11.4 **Punnett Squares.**
Punnett squares are useful for keeping track of the outcomes of a cross. The gametes are placed outside the box as shown, and new combinations are written inside.

soon devised a much simpler procedure to find out if a **Y_** plant was **YY** or **Yy**. (We use an underline symbol to indicate that we don't know what sort of allele it is. Here, **Y_** could be **YY** or **Yy** and still be yellow.) In what was called the **test cross,** such an individual is simply crossed with a homozygous recessive individual. The predictions are straightforward, and shown in Figure 11.5:

1. If the dominant yellow individual in question is homozygous, then the test cross becomes **YY** × **yy**, and all of the progeny will be yellow **(Yy).**
2. If the dominant yellow individual is heterozygous, then the test cross becomes **Yy** × **yy**, and half the offspring will be heterozygous yellow (**Yy**) and half will be green (**yy**).

Yellow

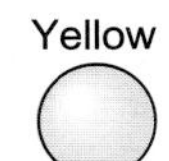

Q: Is the genotype YY or Yy?

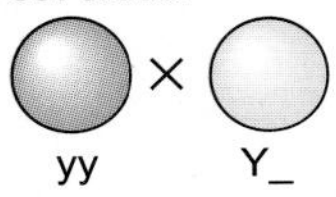

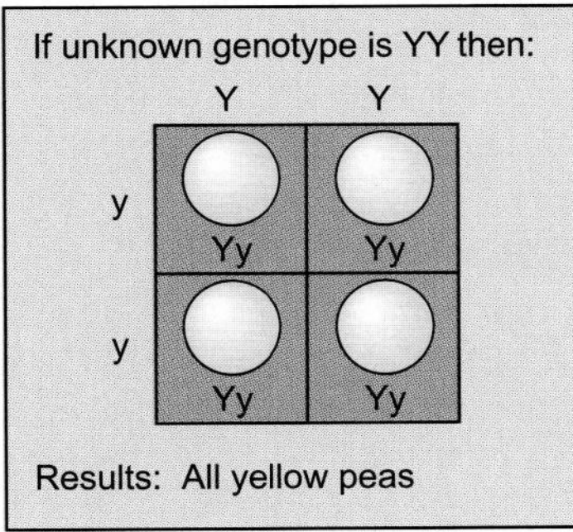

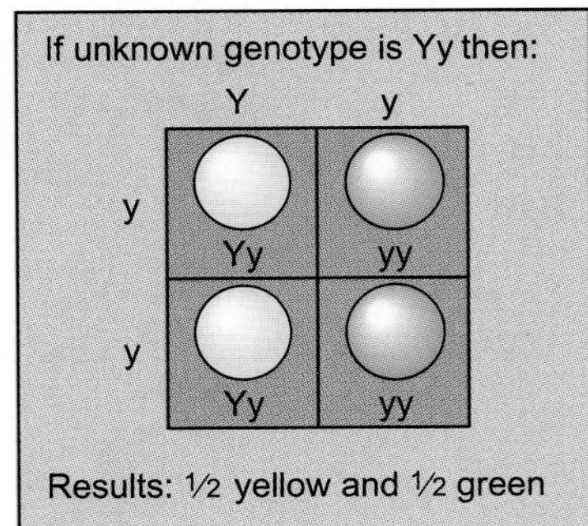

FIGURE 11.5 TEST CROSS.
The two types of yellow seeds (heterozygous, homozygous) can be distinguished by the results of a cross by homozygous recessive green. The homozygous **YY** produces all yellow offspring. The heterozygous **Yy** produces 1 : 1 (yellow : green).

SUMMING UP MENDEL'S FIRST LAW

Mendel's discoveries in the crosses described so far have been brought together into a number of principles that form what we call Mendel's first law (or the law of paired factors and segregation). An expanded form of this law states:

1. A trait is produced by at least a pair of factors. The pair may be homozygous or heterozygous. These factors segregate during pollen and ovule formation.
2. A gamete receives one of a pair of factors.
3. For each trait, offspring receive only one factor from each parent. If one parent's factors are heterozygous, the offspring has an equal chance of receiving either factor from that parent.
4. Where dominance is found, the dominant factor will be expressed over the recessive; the recessive will be expressed only when two recessive factors come together in offspring.

These statements are not in Mendel's words, except perhaps for our use of the word *factor.* In modern terminology, the term **allele** is used to designate the alternative forms of a gene. As you may have noticed, most of what the first law contains actually is due to meiosis and the fact that each of the alleles (or factors) is carried on one member of a homologous pair of chromosomes. But remember that Mendel had no knowledge of meiosis. He arrived at his conclusions about inheritance through sheer logic.

SCIENCE AND MODELS

We have called Mendel a mathematical biologist not just because he was trained in both mathematics and biology or because he was the first biologist to use statistical analysis in his work, but because of the way he arrived at his conclusions. What does a mathematical biologist do? Usually, he or she starts with a set of observations. In Mendel's case, it was the dominance of one trait in the first generation and the reappearance of the recessive trait in a subsequent generation. Through a mental process involving both intuition and logic, the scientist then constructs a *model.* A **model** is an imaginary biological system based on the smallest possible number of assumptions, and it is expected to yield data consistent with past observations. New experiments are then performed to test further predictions of the model. If the new data don't fit the predictions, the model is discarded or adjusted to fit the new observations so that further experiments can be done.

ESSAY 11.1 Genes, Coins, and Probability

Suppose someone flips a penny and says, "call it." If the penny is a real one, you can assume that there is a statistical probability of ½ that it will land "heads." This can be written in algebra as follows: $P(\text{H}) = ½$; that is, the probability of heads = ½. If you didn't know anything about coins and tried to generalize from the single toss, you might assume that coins *always* come up heads. But the more times you toss the coin the closer the number of tails will be to ½. That's why a large sample size (large numbers) is better in the statistical game.

Predicting the results of simple Mendelian crosses has a lot in common with flipping coins. For example, we begin Mendelian crosses by determining what types of gametes are possible. Obviously, there's no chance involved when a homozygote produces gametes; it's like flipping a two-headed coin. Heterozygous individuals, however, have two different alleles for a trait, say **A** and **a;** an **Aa** individual forms two kinds of gametes, **A** and **a,** in equal numbers. The probability that any given gamete will carry the **A** allele is ½. The probability that a gamete will carry **a** is also ½. All of Mendelian genetics is based

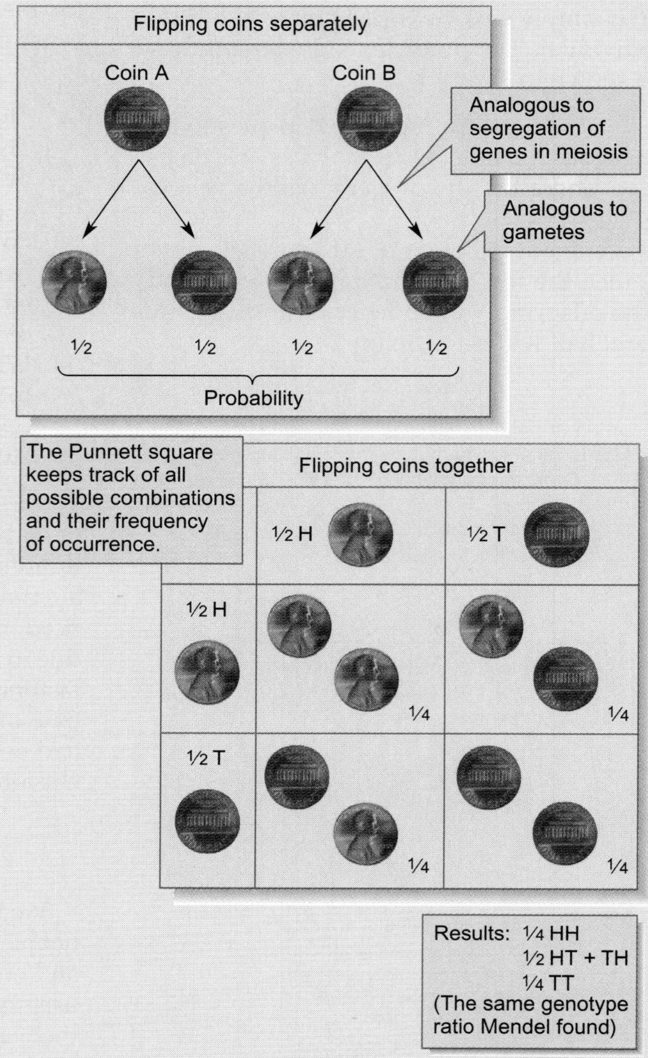

Models cannot be proved with experimental data. We can only say that the data are *consistent* with the model. Mendel did not prove his first law, but its simplicity and consistent usefulness in predicting the results of experiments enabled him to make and test new predictions. Even then, others were unconvinced until after the discovery of chromosomes and meiosis. Have you noticed how well Mendel's findings fit with what you already know about meiosis? Imagine how elated Mendel would have been if meiosis had been discovered in his own lifetime!

We've mentioned that Mendel extended his crosses to all seven selected traits. How well did his results fit the mathematical model? Table 11.2 shows that in each cross, his results were amazingly close to the 3 : 1 phenotypic ratio predicted by his models (see Figure 11.4). The closeness between Mendel's

on this 50-50 segregation of alternate alleles.

So far we have only seen how flipping a penny can represent segregation in the gametes. Let's see how it applies to a cross between two gamete-forming individuals. For this you need a partner and two coins. Since each penny has two sides, a head **(H)** and a tail **(T),** the result of:

HT × HT

will be any of four *possible outcomes:*

1. both coins come up heads **(HH);**
2. both coins come up tails **(TT);**
3. your coin is *heads,* your partner's is *tails* **(HT);** or
4. your coin is *tails,* your partner's is *heads* **(TH).**

All four of these possible outcomes are equally likely; each has a probability of ¼. Why is this? It's an example of the multiplicative law of probability, which states that *the probability of two independent outcomes both occurring is equal to the product of their individual probabilities.* (Independent outcomes are outcomes that don't influence one another; we assume that the way your coin lands has no influence on the way your partner's coin lands, and vice versa.) The law states that the probability of **HH** equals the probability of your coin landing heads *times* the probability of your partner's penny landing heads, which is $\frac{1}{2} \times \frac{1}{2} = \frac{1}{4}$.

The third and fourth possible outcomes are more interesting and can be used to illustrate another law. First, let's see how their probabilities are determined. The probability of:

- your penny landing heads and your partner's landing tails is $\frac{1}{2} \times \frac{1}{2} = \frac{1}{4}$;
- your penny landing tails and your partner's landing heads is $\frac{1}{2} \times \frac{1}{2} = \frac{1}{4}$.

What is the probability of heterozygous offspring **(Aa)** in our cross? To determine the answer, the probabilities of the last two possible outcomes are simply added together ($\frac{1}{4} + \frac{1}{4} = \frac{1}{2}$). There is a ½ chance that the two pennies will come up one head and one tail.

This basic law, called the additive law of probability, states that *the probability that any one of two (or more) mutually exclusive outcomes will occur, is equal to the sum of their individual probabilities.* Mutually exclusive outcomes means that if one happens, the other can't; for instance, if the outcome is: your coin, *heads;* your partner's coin, *tails;* this excludes the possibility of the outcome being the other way around.

Now, what is the probability that at least one head will be showing? Here, we combine the three mutually exclusive outcomes (1), (3), and (4). The probability that at least one head will come up is $\frac{1}{4} + \frac{1}{4} + \frac{1}{4} = \frac{3}{4}$.

The multiplicative and additive laws can be applied to independent assortment, where two or more traits are considered simultaneously (the dihybrid cross). It's like tossing two pennies and two quarters at the same time. If you toss two pennies and two quarters simultaneously, what is the probability of seeing both Lincoln and Washington (at least once each)? The probability of seeing Lincoln is ¾ (the probability of heads, see the paragraph above); the probability of seeing Washington is also ¾; the chance of seeing both is $\frac{3}{4} \times \frac{3}{4} = \frac{9}{16}$. The chance of seeing Abe and *not* seeing George is $\frac{3}{4} \times \frac{1}{4} = \frac{3}{16}$, the same as for seeing George and not Abe; and the probability of seeing neither president is $\frac{1}{4} \times \frac{1}{4} = \frac{1}{16}$. Of such simple stuff are Mendelian ratios made.

expectations and observations did not escape the attention of certain skeptical statisticians. In 1936, R. A. Fisher, a noted statistician and geneticist, concluded that Mendel's data were literally too good to be true. Did the good monk fudge his data? Or did he see what he had expected to see, an all-too-human trait? The question has generated great controversy, but we will leave all that to the historians. Whatever the case, Mendel's first law was found to apply to animals as well as plants, and it has held up under the most rigorous scrutiny.

Mendel's Second Law: Independent Assortment

We have noted Mendel's success in breaking his problem down to its smallest parts—that is, studying

TABLE 11.2 MENDEL'S F_2 GENERATIONS

The dominant and recessive traits analyzed by Mendel are shown, along with results of F_1 and F_2 generations. Note the large numbers he worked with. How does a large sample size (large numbers) improve the validity of the conclusions? How well do his numbers in the last two columns agree with what we would expect in the crosses (see Essay 11.1)? The proportion of the F_2 generation showing recessive forms is in the far-right column.

Dominant form of trait in one parent plant	×	Recessive form of trait in one parent plant	Number of dominant plants in F_2	Number of recessive plants in F_2	Total examined	Ratio of dominant to recessive (avg. is 3:1)	Proportion of F_2 which is recessive (avg. is 25%)
Round seeds		Wrinkled seeds	5,474	1,850	7,324	2.96:1	25.3
Yellow seeds		Green seeds	6,022	2,001	8,023	3.01:1	24.9
Gray seed coats		White seed coats	705	224	929	3.15:1	24.1
Green pods		Yellow pods	428	152	580	2.82:1	26.2
Inflated pods		Constricted pods	882	299	1,181	2.95:1	25.3
Long stems (tall)		Short stems (dwarf)	787	277	1,064	2.84:1	26.0
Axial flowers and fruit		Terminal flowers and fruit	651	207	858	3.14:1	24.1

TABLE 11.3 MENDEL'S PREDICTIONS AND RESULTS FOR F_2 PHENOTYPES (556 SEEDS OBSERVED)

Phenotype of F_2	Fraction Predicted	Number Predicted Out of 556	Number Actually Observed
Round and yellow	9/16	312.75	315
Wrinkled and yellow	3/16	104.25	101
Round and green	3/16	104.25	108
Wrinkled and green	1/16	34.75	32

only one characteristic at a time. His next step was to consider two characteristics at a time. He therefore crossed a true-breeding strain that bore round, yellow peas with another true-breeding strain that bore wrinkled, green peas. He wanted to see if there was any relationship between the inheritance of one gene and the inheritance of another.

The F_1 offspring (which, remember, could be categorized and counted while still in the pod) were all round and yellow. We can symbolize this as follows:

RRYY × **rryy** ⟶ **RrYy**

We are now considering two factors at the same time, and from now on we will refer to contrasting factors as alleles. We will also refer to the *locus* of an allele. The term **locus** (plural, *loci)* derives from our present knowledge that each allele occupies a specific place, or locus, on the chromosome. In the first crosses we will look at, **R** and **r** will be symbols for the two alleles of the round-or-wrinkled locus, and **Y** and **y** will be the symbols for the two alleles of the yellow-or-green locus.

Now let's see what happened in the F_2 generation when Mendel crossed plants that were different in two ways (called a **dihybrid cross**) (see Table 11.3). Remember that the F_1 peas were uniformly round and yellow. In the F_2 generation—the offspring of $F_1 \times F_1$ (**RrYy** × **RrYy**)—Mendel found and classified 556 peas. He was able to divide them into four groups, or phenotypic classes:

315 round and yellow	**R_Y_**
101 wrinkled and yellow	**rrY_**
108 round and green	**R_yy**
32 wrinkled and green	**rryy**

(Again, use an underline symbol to indicate that we don't know, or can't tell, if these are heterozygous or homozygous genotypes. For example, **R_Y_** could be **RRYY**, **RrYY**, **RRYy**, or **RrYy**; either genotype would be the same phenotype, round and yellow.)

Note that a total of 133 peas were wrinkled (101 + 32), and 140 peas were green (108 + 32). In either case, this comes close to 139, which is ¼ of 556. So about ¼ of the F_2 peas were wrinkled and ¼ were green, while ¾ were round and ¾ were yellow, which demonstrates Mendel's first law in both cases. But the data indicated more than that. Mendel also found that the two pairs of contrasting characters were inherited *independently*. This means that a gamete's receipt of an **R** or an **r** has nothing to do with its receipt of a **Y** or a **y**. How did Mendel arrive at that? He looked at the numbers. As Figure 11.6 reveals, two pairs of alleles assort independently, producing four kinds of gametes in both the pollen and the ovules. The chances of any particular combination of alleles occurring in any of the progeny is equal to that of any other combination. In this way he developed what would be called the **law of independent assortment**—alleles for different traits are inherited independently of each other. Again, the Punnett square comes to the rescue, helping us to better see what Mendel was getting at.

THE 9 : 3 : 3 : 1 RATIO

When Punnett squares are used, the gametes possible from both individuals must be determined first. If two traits are being considered in plants and both individuals being crossed are heterozygous for both traits, there will be four kinds of pollen and four kinds of ovules. We can use this information to sum up a cross between plants differing in two traits. The phenotypic results are 9/16, 3/16, 3/16, and 1/16.

These readily convert to the ratio 9 : 3 : 3 : 1, which is as familiar in genetics as $E = mc^2$ is in physics.

As we see in Table 11.3, when Mendel classified his 556 peas according to the four possible phenotypes and counted the peas in each group, the ratio was remarkably close to $^9/_{16}$: $^3/_{16}$: $^3/_{16}$: $^1/_{16}$, as would be expected if two gene loci were segregating independently, each with a $^3/_4$: $^1/_4$ phenotypic ratio.

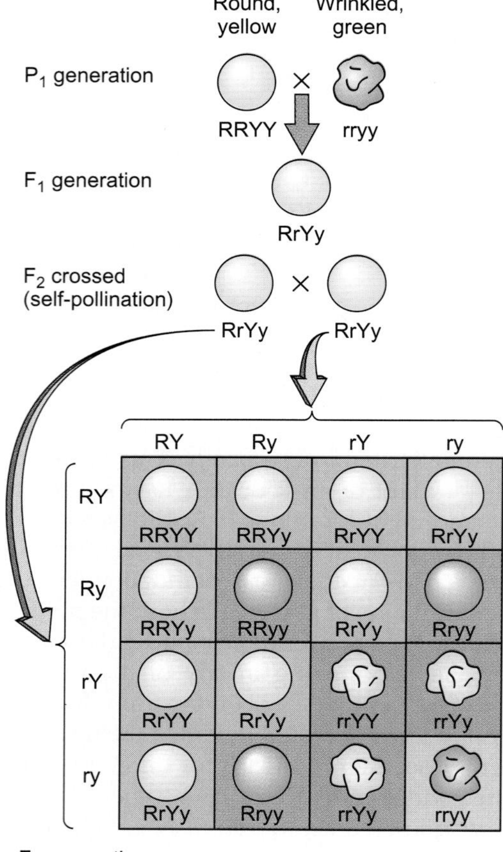

FIGURE 11.6 INDEPENDENT ASSORTMENT OF TWO PAIRS OF ALLELES.
The scheme illustrates the crosses from P_1 through F_2, using the Punnett square to show the results of the F_1 self-pollinated cross. The traits being considered are round and wrinkled shape and yellow and green color. The F_2 generation comprises four distinct phenotypes, which include every possible color and shape combination. These occur in a 9 : 3 : 3 : 1 ratio.

But the monk wasn't finished yet. A good scientist is never satisfied with explanations that merely account for observations that have already been made. To be useful, an explanation must lead to new predictions that can then be tested. Mendel's theory predicted that when there were four phenotypes in a 9 : 3 : 3 : 1 ratio, there should be a total of nine genotypes in a 1 : 2 : 1 : 2 : 4 : 2 : 1 : 2 : 1 ratio. Mendel knew that he could determine the genotypes of his peas by letting them grow to adult plants and examining their progeny; so the 556 peas went into the soil, and Mendel waited another year. A total of 529 of the resulting plants fertilized themselves and produced a new crop of peas, the F_3 generation. Now the genotypes could be determined by observing the F_3 produced by each F_2 plant. (The **rryy** F_2 plants should produce only wrinkled green peas, while **RRYy** plants would produce round peas, $^3/_4$ of which would be yellow. Can you see that each potential F_2 genotype can be identified by the looks of its F_3?) Mendel's data are shown in Table 11.4. Mendel must have been elated—his prediction had held!

SUMMING UP MENDEL'S SECOND LAW

Mendel's own statement of the principle encompassed in his second law, in translation, is as follows:

- The inheritance of each pair of differing traits is independent of any other pair of traits.

Or, in terms that we can now correlate with meiosis:

- The way in which a pair of alleles on one set of homologous chromosomes segregates during meiosis has no effect on the way a pair of alleles on another set of homologous chromosomes segregates.

Another modern version of the principle of independent assortment is:

- If an organism is heterozygous at loci on different chromosomes, each locus will assort independently of the others.

Mendel didn't know everything. And while it is true that most pairs of gene loci follow his second law, some don't. The ones that do are called *unlinked,* and the ones that don't are called *linked.* As you may have guessed, **linked genes** are genes that are located on the same chromosome so that they are usually passed along together. Linkage will be discussed further in the next chapter.

TABLE 11.4 MENDEL'S PREDICTIONS AND RESULTS FOR F_2 GENOTYPES (529 PEAS CLASSIFIED BY PROGENY TEST)

Genotype of F_2	Fraction Expected	Number According to the Hypothesis	Number Actually Observed by Progeny Testing
RRYY	1/16	33	38
RRYy	2/16	66	65
RRyy	1/16	33	35
RrYY	2/16	66	60
RrYy	4/16	132	138
Rryy	2/16	66	67
rrYY	1/16	33	28
rrYy	2/16	66	68
rryy	1/16	33	30

Note: Keep in mind that Mendel had to perform a progeny test for each of the 529 F_2 peas in order to learn their genotype. There is no other way to prove the genotype of a heterozygote.

THE DECLINE AND REBIRTH OF MENDELIAN GENETICS

In 1866, after seven years of experimentation (at the very time Darwin was pondering the enigma of heredity), Mendel presented his results to a meeting of the Brunn Natural Science Society. His audience of local science buffs sat there politely, probably not understanding a word of what they were hearing. The minutes of the meeting, which still exist, record that not a single question was asked. Instead, the restless audience launched into a discussion of the "hot" topic of the time—Darwin's *Origin of Species*. Mendel's single paper was published in the society's proceedings the following year and was distributed widely. The learned scientists of the day were just as baffled and uninterested as was Mendel's original audience.

Historians have come up with a small, sad, but remarkable piece of information. In Darwin's huge library, which is still intact, a one-page account of Mendel's work with peas appears in a German encyclopedia of plant breeding. Some relatively obscure work is described on the facing page, and it is covered with extensive notes in Darwin's handwriting. The page describing Mendel's work is clean. Darwin must have seen the paper that could have clarified his theory of natural selection, saving him years of agony and uncertainty. But even Darwin was not ready, and he too failed to grasp Mendel's simple but profound ideas.

Mendel's work continued to be ignored until 1900, the year his work was suddenly revived with great fanfare. In that year, three biologists in three different countries, each trying to work out the laws of inheritance, searched through the old literature and came up with Mendel's paper. They immediately recognized its importance. Science had changed in 35 years. The 20th century had arrived, and the obscure monk became one of the most famous scientists of all time. But he had been dead for 16 years.

PERSPECTIVES

Gregor Mendel's insights were a beautiful combination of observation and logic. He distilled the rather vague idea of a physical trait down to the entity we now consider a gene. His notions of how genes are transmitted form the cornerstone for our understanding of the patterns of inherited features.

In the next chapter, we will see that the relationship between the physical behavior of chromosomes in meiosis and Mendel's mathematically based laws is pleasingly straightforward.

SUMMARY

Because he subscribed to the notion of blending inheritance, Darwin could not successfully explain how new genetic variation arose. This was important in arguing his case for natural selection. Mendel's background in mathematics, particularly elementary statistics, was a significant factor in his success at analyzing his crosses.

MENDEL'S CROSSES

In planning his crosses, Mendel made use of true-breeding strains of peas, each with easily recognized traits. True-breeding pea strains are those whose traits always appear without variation in their offspring.

MENDEL'S FIRST LAW: PAIRED FACTORS AND THEIR SEGREGATION

Part of Mendel's first law, the segregation of alternative factors into different daughter cells can be anticipated through a consideration of genes, homologous chromosomes, and meiosis. Genes occur on chromosomes, and each has its counterpart on the homologous chromosome. When homologues separate during meiosis I, gene pairs (Mendel called them factors) must also separate (Mendel used the term *segregate*), going to opposite daughter cells.

An individual's genotype is the combination of genes it carries, while its phenotype is the outward, commonly visible expression of those genes.

Mendel began by crossing true-breeding peas differing in one trait. This was his P_1 generation. The dominant trait appeared in the F_1 generation, while the recessive trait was not seen.

When the F_1 peas self-pollinated, the recessive trait reappeared in the F_2 generation, but in only ¼ of the offspring; the remaining ¾ displayed the dominant trait. Through progeny testing (self-pollinating) of the F_2 generation, Mendel found that some of the dominant individuals were true-breeding, or homozygous, while others were not. The latter were heterozygous (carried both a dominant and a recessive factor). The recessive F_1 individuals were all homozygous. In such crosses, the phenotypic ratio is 3 : 1, and the genotypic ratio is 1 : 2 : 1.

Individuals with the dominant phenotype may be homozygous or heterozygous. A test cross with homozygous recessives will determine whether they are homozygous or heterozygous.

Traits are produced by two factors—now called alleles—that segregate when gametes are produced. Chance determines which two gametes unite in fertilization, and therefore chance determines the combination of alleles in offspring. Only one dominant allele is required for the dominant trait to appear, but a recessive trait requires two recessive alleles.

MENDEL'S SECOND LAW: INDEPENDENT ASSORTMENT

By crossing P_1 individuals differing in two traits, Mendel discovered the law of independent assortment—that the segregation of factors governing one trait occurred independently of the segregation of factors governing another.

In the cross of round/yellow × wrinkled/green, the F_1 were all round and yellow, the dominant traits. However, four phenotypes appeared in the F_2. The phenotypic ratio was 9 : 3 : 3 : 1.

9/16 round/yellow
3/16 round/green
3/16 wrinkled/yellow
1/16 wrinkled/green

When the alleles **RrYy** segregate (in meiosis), each pollen or ovule will receive one of each kind of allele, and all of the possible combinations will be found in equal numbers. Thus there are:

Four kinds of pollen: **RY, Ry, rY,** and **ry**
Four kinds of ovules: **RY, Ry, rY,** and **ry**

If all possible combinations of the four kinds of pollen and ovules are taken into account (a Punnett square helps), the result will always be a 9 : 3 : 3 : 1 phenotypic ratio.

When Mendel tested his new model with various combinations of true-breeding peas, it was clearly supported by the results.

When gametes form, the alleles governing one trait assort independently of the alleles governing any other trait. The genes responsible for traits reside at specific loci on chromosomes. Where two gene loci occur on different chromosomes, the separation of one pair of alleles in meiosis I has no effect on the separation of any other pair of alleles. However, if two loci are on the same chromosome, the genes are said to be linked and independent assortment does not apply.

The Decline and Rebirth of Mendelian Genetics

Although they were published in his lifetime, Mendel's findings were largely ignored until after his death. In 1900, his achievements were rediscovered and their importance proclaimed by three independent researchers.

Key Terms

true-breeding • **169**	heterozygous • **172**
segregation • **169**	Punnett square • **172**
genotype • **169**	test cross • **173**
phenotype • **169**	allele • **173**
P_1 • **170**	model • **173**
F_1 • **170**	locus • **177**
F_2 • **170**	dihybrid cross • **177**
dominant • **170**	law of independent assortment • **177**
recessive • **170**	linked genes • **178**
progeny testing • **172**	
homozygous • **172**	

Review Questions

1. Briefly explain why the theory of blending inheritance opposed rather than supported natural selection.
2. What does the term *true-breeding* mean? What might have happened had Mendel used heterozygous individuals to start his work?
3. Aside from being a careful and perhaps lucky researcher, what other special qualities or training did Mendel have that might help account for his success?
4. Part of Mendel's first law actually describes the behavior of one pair of chromosomes in meiosis. To illustrate this, draw a pair of chromosomes as they would appear in prophase I, but add a pair of factors or alleles (say an uppercase *A* to one and a lowercase *a* to the other), and complete meiosis with successive drawings (review Chapter 10 as necessary).
5. Using Punnett squares as needed, carry out a cross between pure-breeding tall and short pea plants. Then, cross two of the F_1 individuals and determine the F_2. State the phenotypic and genotypic ratios.
6. Carry out a cross between a heterozygous yellow-seeded pea plant and a green-seeded pea plant. State the phenotypic ratio in the offspring. What are the chances (expressed as a fraction) of getting heterozygous offspring in such crosses?
7. An experimenter cross-pollinates two plants bearing red flowers, plants the seeds produced, and in the next generation counts 289 plants with red flowers and 112 with white flowers. What genotypes would you suspect in the parent plants? Test your answer.
8. Assuming that your calculation of the parental genotypes was correct in the cross above (Question 7), what numbers of red and white would you have *expected* in the offspring? Give a logical explanation for the difference between the *expected* and the *observed.*
9. Which method is more efficient for determining the genotype of dominant F_2 offspring, progeny testing or test crossing? Explain your answer.
10. Briefly explain the term *model.* Explain how models are used by scientists.
11. Two sets of traits with which Mendel worked were pod shape (inflated versus constricted) and flower location (axial versus terminal). Inflated and axial are dominant. Consider the following cross involving both sets of traits:

    IiAa × IiAa

    a. Show the cross using a Punnett square.
    b. List the phenotypes and their fractions, and then convert this to a phenotypic ratio.
12. Carry out a cross between a pea plant that is homozygous for round seeds and heterozygous for yellow with one that produces wrinkled seeds and is also heterozygous for yellow. State the expected phenotypes of the offspring and their phenotypic ratio.
13. An experimenter carries out a dihybrid cross of pea plants and finds the following in the offspring:

    32 have round, yellow seeds
    29 have round, green seeds
    9 have wrinkled, yellow seeds
    11 have wrinkled, green seeds

    Working back to the parents, determine their most probable genotypes. (Begin by considering the total numbers for each trait; e.g., there are 61 round and 20 wrinkled, etc.)

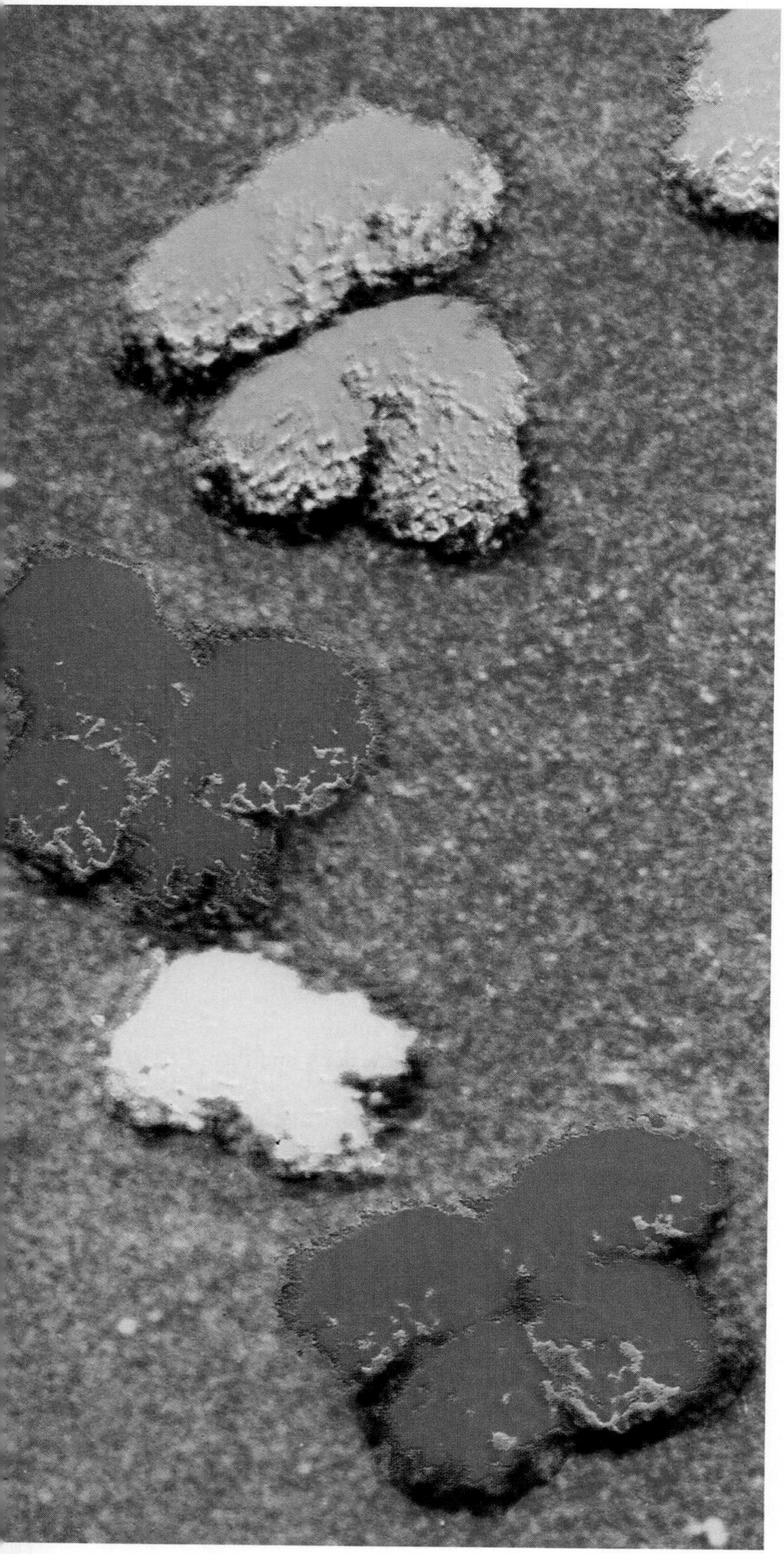

12

Genes and Chromosomes

MENDEL'S FACTORS INDEED BEHAVE IN MATHEMATICALLY PRECISE WAYS. *But what are these factors, and more importantly, why do they behave with such numerical precision? In this chapter, we will see that the inheritance patterns of Mendel's factors are due to the factors being located on chromosomes, those long bundles of DNA that are so carefully duplicated and partitioned as part of the processes of mitosis and meiosis. We will see that an understanding of the movement of chromosomes leads directly to an understanding of inheritance.*

Long, idyllic days of thoughtful puttering in a Moravian monastery once marked the leading edge of the field that would be called genetics. The coming of warm spring days would, year after year, signal new growth, new experiments, and new ideas by the talented monk, Gregor Mendel. Unfortunately, he was almost ignored in those days. Science itself was not ready for him.

Nevertheless, even in Mendel's final years there were rapid improvements in the microscope and in various techniques for studying cells. Biologists were able to watch the puzzling pageant of mitosis and meiosis, wherein strange, twisted bodies go about their slow dances. They decided that chromosomes must be important, but they had no idea why. The closing years of the 19th century marked a very busy, exciting, and often baffling time for biologists.

The stuff of genes, DNA, was isolated and characterized in Mendel's own lifetime, although a more precise understanding of the chemical nature of the gene was more than 50 years away. By the first year of the 20th century, the world was at last ready for Mendel. After years of obscurity, Mendel's work was thrust upon 20th-century science. Biologists of the time were keenly interested in heredity, and some were experimenting with plants. Then the inevitable happened. Several researchers independently rediscovered the monk's findings, grasping at once the meaning of Mendel's ratios. The Mendelian revival was followed by an era of intense activity, in which 3 : 1 and 9 : 3 : 3 : 1 ratios dominated the conversations of turn-of-the-century geneticists.

MENDEL AND MEIOSIS

The phenomenon of pairing and separation of homologous chromosomes in meiosis wasn't worked out until the turn of the century and the rediscovery of Mendel's work. Some microscopists already suspected that chromosomes were the carriers of inheritance; and soon after Mendel's work was republished, Theodore Boveri in Europe and Walter Sutton in America published influential papers pointing out the relationship between Mendelism and meiosis.

It isn't hard to follow their thinking. Suppose a pair of alternate alleles (for instance, **R** and **r**) are carried on a pair of homologous chromosomes. When the homologous chromosomes separate during the first division of meiosis, exactly half of the resulting haploid cells will receive one of the alleles **(R)**, and exactly half will get the other **(r)**. And that, quite simply, is the physical basis of Mendel's first law.

The interpretation of Mendel's second law—the law of independent assortment—is almost as clear. (Recall from Chapter 11 that the second law involves alelles that are located on different chromosome pairs.) Suppose one pair of homologous chromosomes carries one pair of alternate alleles—**R** and **r**—and a second pair of chromosomal homologues carries another pair of alleles—**Y** and **y** (Figure 12.1). When the chromosomes line up on the metaphase plate, either of two things can happen. In one possible way of lining up, the **R** and **Y** carrying chromosomes will go to one pole of the dividing

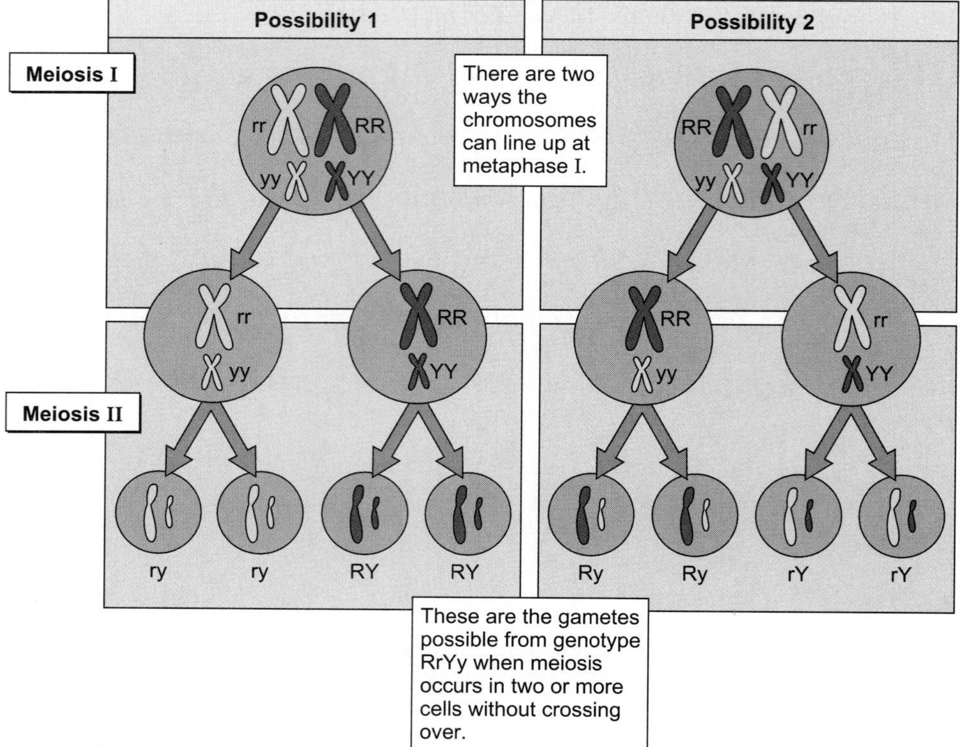

FIGURE 12.1 **MENDEL'S LAWS AND MEIOSIS.**
Mendel's first and second laws of heredity also describe the behavior of chromosomes in meiosis. A cell entering meiosis with two pairs of genes on two different chromosome pairs has two alternatives for lining up at metaphase of meiosis I. There are four equally likely combinations of chromosomes and genes in the gametes of each alternative configuration.

cell, and the chromosomes with **r** and **y** will go to the other pole. In the second way of lining up, the chromosomes carrying **R** and **y** move to one pole and the ones with **r** and **Y** move to the other. There is exactly the same chance of one possibility as of the other. The overall result, when many such cells undergo meiosis, is that the genotypes of the haploid cells formed will be **RY, Ry, rY,** and **ry** in equal numbers—just as Mendel claimed.

Before continuing, we should add that as the story unwinds, we will include the concept of the gene. Genes can be variously defined, as you will see, but for our purposes here we will say that a **gene** is a hereditary unit, located at a specific locus along a chromosome, that can have alternate forms. (These alternate forms are the alleles.)

Walter Sutton, one of the cytologists who recognized a relationship between the inheritance of traits and meiosis, went on to predict linkage. He reasoned that since there are only relatively few chromosomes in any one cell, and there are many hereditary factors, each chromosome must carry many genes.

GENE LINKAGE AND RECOMBINATION

The problem that Sutton (and others) encountered was that Mendel's principle of independent assortment shouldn't, and didn't, work with all combinations of genes. Sutton decided that genes on the same chromosome would stay together during assortment, moving with the chromosomes as a group. Thus, a **linkage group** came to be defined as any group of genes found on the same chromosome that tend to be inherited together.

LINKED GENES

Linked genes, of course, are exceptions to independent assortment. And soon enough, some exceptions to the second law began to surface. William

TABLE 12.1 THE BATESON-PUNNETT TEST CROSS

Offspring		Expected Frequency		
Phenotype	Genotype	Hypothesis I*	Hypothesis II†	Observed Frequency
Blue, Long	BbLl	25%	50%	44%
Blue, round	Bbll	25%	0	6%
red, Long	bbLl	25%	0	6%
red, round	bbll	25%	50%	44%

**Mendel's independent assortment*
†Sutton's complete linkage

Bateson and R. C. Punnett, in trying to extend Mendel's findings, for example, got some puzzling results. They started with two sweet pea strains: one with blue flowers **(BB)** and long pollen grains **(LL),** the other with red flowers **(bb)** and round pollen grains **(ll).** The offspring of this cross had blue flowers and long pollen grains, so "blue" and "long" were Mendelian dominants. This is their original cross:

BBLL × **bbll** ⟶ **BbLl**

Following Mendel, they tried to reconfirm the law of independent assortment by crossing the double heterozygote **BbLl** back to the double recessive **bbll** parental strain (a test cross; see Table 12.1):

BbLl × **bbll**

According to Mendel's second law, independent assortment in the offspring should have produced four different phenotypes, all occurring in equal numbers:

25% **BbLl** : 25% **Bbll** : 25% **bbll** : 25% **bblL**
(blue-long) (blue-round) (red-long) (red-round)

If, however, the gene for blue flowers was on the same chromosome as that for long pollen grains, the offspring should have been of only two phenotypes:

50% **BbLl** : 50% **bbll**
(blue-long) (red-round)

However, they didn't get either ratio. Although all four phenotypes showed up in the offspring (blue-long, blue-round, red-long, and red-round), suggesting independent assortment, *the numbers were all wrong!* When the phenotypes were counted, almost all were either blue-long (44%) or red-round (44%). Blue-round and red-long accounted for only 6% each. Bateson and Punnett were puzzled by the strange ratio (see Table 12.1).

The two Mendel fans were stumped. They couldn't come up with anything that made sense. In Bateson and Punnett's time, people knew very little about meiosis, and without that information they were hard-pressed to explain such aberrant test results. But let's take what we know about meiosis and see what really happened to their sweet peas. The real explanation involves that remarkable process called crossing over, in which homologous chromosomes exchange parts during prophase of meiosis I (see Chapter 10).

The genes for flower color and pollen shape are indeed linked on the same chromosome, so only the two P_1 phenotypes, blue-long and red-round, should have shown up in the test cross. But the blue-round and red-long offspring did show up, albeit only 12% of the time. So 12% of the progeny were the products of exchanges between homologous chromosomes. These exchanges permit the new combination of alleles to appear in the offspring. That is, genetic recombination occurred. But why the odd percentages? The answer can be found in a closer look at genetic recombination.

Genetic Recombination

The formation of new associations of linked genes requires the mutual exchange of DNA segments. Thus, alleles that were once located on different homologous chromosomes end up on the same chromosome. This is the process of crossing over (see Chapter 10). The molecular details of the process are still not entirely clear, so it's no wonder the early geneticists had a hard time of it. Amazingly, they worked out the concepts of gene linkage and crossing over using only crosses, test crosses, and progeny counts.

It was confusing at first. Some genes showed much stronger tendencies to remain linked than others, remaining together through generation after generation of test crosses (Figure 12.2). Others, such as Bateson's and Punnett's **B** and **L** and **b** and **l** alleles, regularly broke out of their linked state, with their

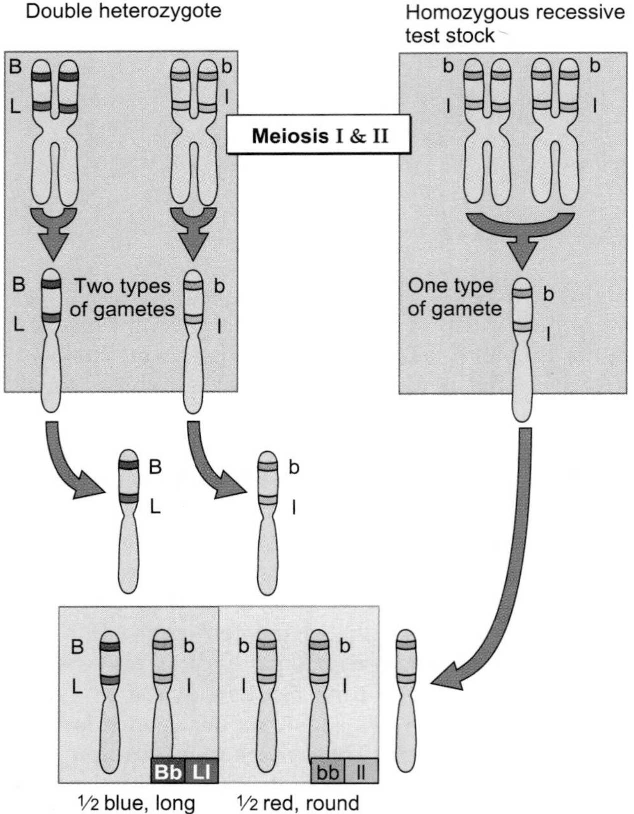

FIGURE 12.2 **LINKAGE.**
In this example situation, genes B and L and genes b and l are permanently linked as though crossing over could not occur. In this test cross, a heterozygous individual is crossed with a homozygous recessive. Since both genes are passed as a unit, only two phenotypes are possible in the offspring, and these occur in a 1 : 1 ratio.

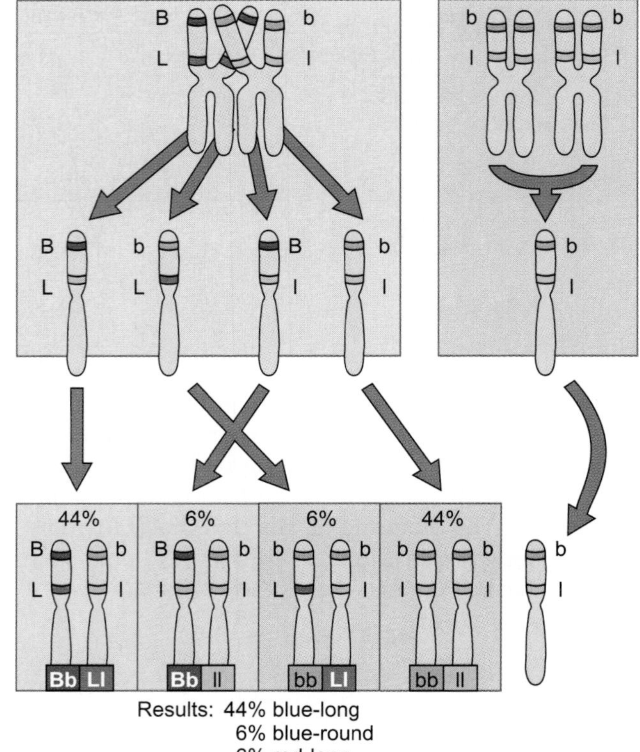

FIGURE 12.3 **LINKAGE WITH RECOMBINATION.**
Let's reconsider the test cross seen in Figure 12.2, this time in a more realistic situation allowing for a limited amount of crossing over between the two loci. Here, 12% of the chromatids undergo reciprocal exchange between the two genes for flower color and pollen shape. The remaining 88% of the chromatids do not undergo exchange. There are two kinds of *recombinant* chromosomes produced and two kinds of *nonrecombinant* chromosomes produced. Note the approximate 7 : 1 : 1 : 7 ratio, but be aware that any such ratio would depend upon the amount of recombination between genes.

recombination progeny dependably showing up in the test cross results (Figure 12.3). Gene pairs that had very low percentages of recombination were described as *tightly linked,* and those with higher percentages of recombination were described as *loosely linked.* These test cross numbers had none of the appeal of Mendel's wonderfully precise ratios. If the terms "loosely" and "tightly" linked genes sound vague, consider the dilemma of the geneticists of that period. A considerable time would pass before they would be able to equate linkage groups with the linear structure of DNA.

Working with fruit flies, A. H. Sturtevant eventually reasoned that the more tightly linked gene pairs (those more likely to stay together in crosses) were very close together on a chromosome, while the loosely linked gene pairs (those more likely to separate) were spaced farther apart on the same chromo-

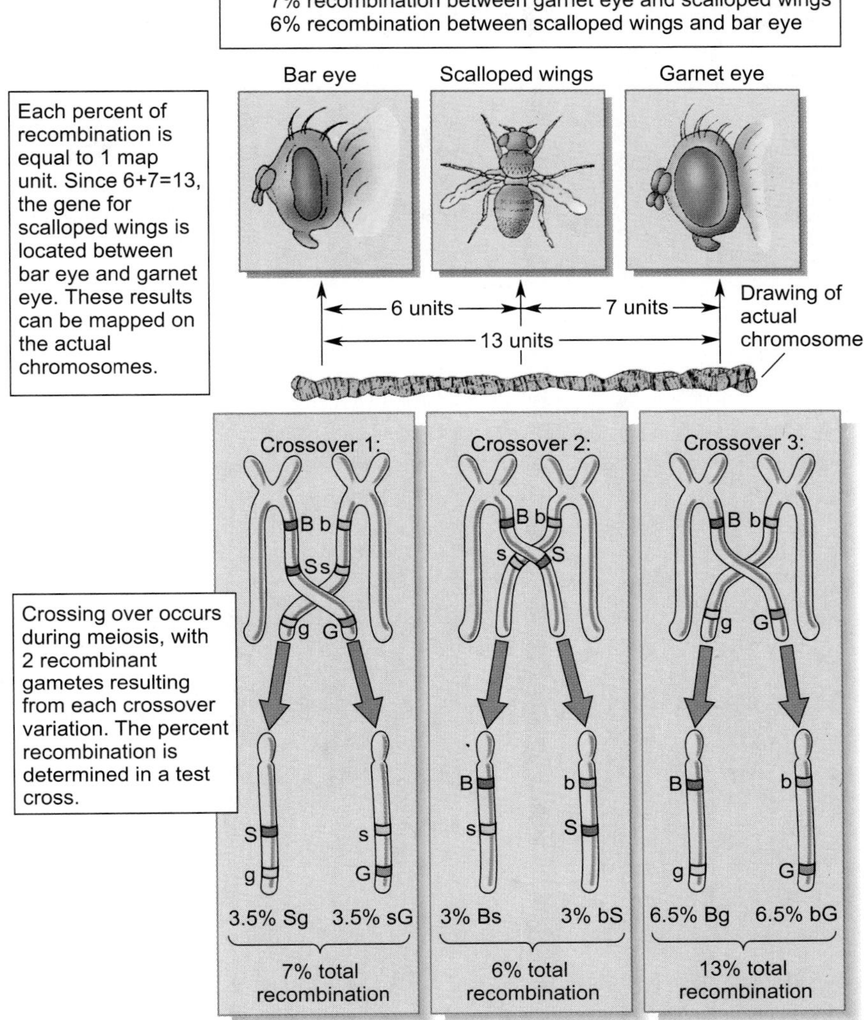

FIGURE 12.4 **GENETIC MAPS.**
In recombination mapping, each percent of recombination is equal to one map unit. In this example, the locus of scalloped wings was determined to be between Bar eye and garnet eye, all mutant characters of *Drosophila*. The percent recombination was determined by test crosses between heterozygous individuals and homozygous recessives.

some. The farther apart two genes were, Sturtevant reasoned, the greater was the chance that the length of chromosome separating them would break and rejoin in a crossover.

Eventually, some patterns began to emerge. Two mutant genes, *Bar eye* and *scalloped wings* in the fruit fly *(Drosophila melanogaster)* were linked. In addition, *Bar eye* and *garnet eye* also were found to be linked. The inference, as mentioned previously, was that all these genes were on the same chromosome. The numbers began to make some sense, too. For instance, in the example given, garnet eye and scalloped wings showed 7% recombination in a test cross; scalloped wings and Bar eye showed 6% recombination; and in another test cross, Bar eye and garnet eye showed just slightly less than 13% recombination progeny. (According to our calculations, 6% plus 7% equals 13%.)

Sturtevant believed that these data meant not only that all three genes were on the same chromosome, but also that they were in a line, with scalloped wings in the middle (Figure 12.4). He made maps of all of the linkage groups, with each **map unit** being equal to 1% recombination. Each site on the gene map was henceforth called a **gene locus,** and a given gene locus could be occupied by one allele. Genetic maps eventually were made of all of the *Drosophila* chromosomes and included hundreds of known genes (Figure 12.5). The technique of gene mapping was soon used for other organisms.

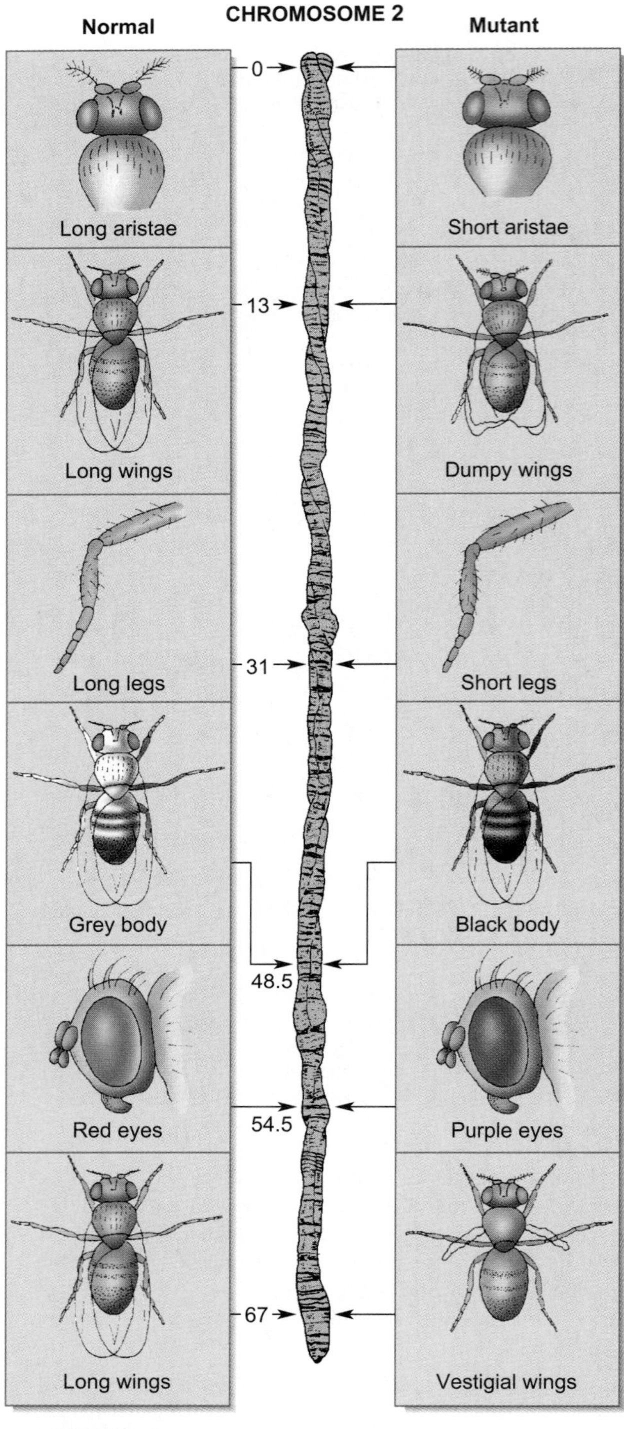

FIGURE 12.5 GENETIC MAP AND A PHYSICAL REPRESENTATION OF A *DROSOPHILA* CHROMOSOME. Recombination map distances are seen along with the names and phenotypes of some mutant traits.

SEX AND CHROMOSOMES

The theory that chromosomes were the carriers of genetic information was strengthened, in the early years of this century, when it was shown that chromosomes could determine sex. In mammals and many insects (including fruit flies), there are two different-shaped chromosomes designated X and Y. The X and Y determine the sex of the individual and so are called **sex chromosomes.** They can be readily distinguished from **autosomal chromosomes (autosomes)** by their appearance (Figure 12.6). (Autosomes include all chromosomes other than sex chromosomes.)

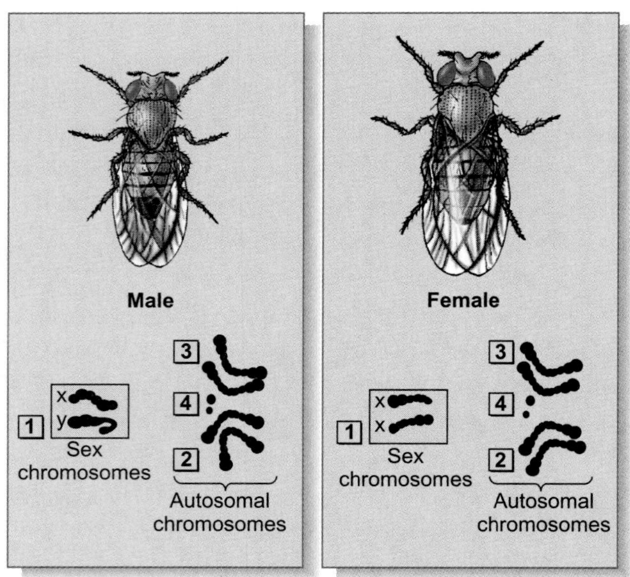

FIGURE 12.6 THE FOUR PAIRS OF CHROMOSOMES IN *DROSOPHILA MELANOGASTER*.
In females, each pair contains identical pair members. In males, however, only three pairs are truly homologous. The fourth pair consists of an X chromosome and a J-shaped Y chromosome.

ESSAY 12.1 Sex Chromosome Abnormalities

As with trisomy-21, or Down's syndrome, abnormal numbers of sex chromosomes are brought about by nondisjunction—the failure of chromosomes to assort properly during meiosis. A leading factor causing this problem is believed to be aging in the oocytes. While most nondisjunctions in autosomal chromosomes are fatal to the embryo, wrong numbers of sex chromosomes in humans result in live babies and abnormal adults. There are many varieties of sex chromosome conditions. The shorthand designations of the normal individuals and the most common abnormalities are:

XX	Normal female (two X chromosomes)
XY	Normal male (one X, one Y)
XO	Turner's syndrome female (one X, no homologue)
XXY	Klinefelter's syndrome male (two Xs, one Y)
XYY	Extra Y, or XYY syndrome male
XXX	Trisomy-X, or XXX female

In addition to the above, there are many more extreme situations, such as XXYY, XXXY, XXXYY, XXXX, XXXXX, XXXXYY, and so on, each syndrome having its own distinguishing characteristics. However, we can make four generalizations: first, one must have at least one X chromosome to live. Second, the presence of a Y causes the individual to develop as a male, and the absence of a Y causes the individual to develop as a female. Third (and this is probably why these syndromes are not fatal), all but one of the X chromosomes will condense into heterochromatin and be visible as a Barr body when stained, so that XO females lack a Barr body, XXY males have one, XXX females have two, XXXXXYY males have four, and so on. Fourth, the more sex chromosomes a person has, the taller he or she will be, so that XO females are tiny, while on the average XXX, XXY, and XYY individuals are usually much taller than chromosomally normal men and women, and XXYY men are huge.

XO (Turner's) individuals are phenotypically female but do not develop ovaries. They remain sexually immature as adults unless given hormones. XXY (Klinefelter's) males are tall and have small, imperfect testes and low levels of male hormones. They may have femalelike breast development and somewhat feminine body contours. XXX females are tall and frequently sterile but otherwise appear normal. XYY males appear normal except for their extreme height and for a tendency toward severe acne. They are also generally sterile. On the average, they have somewhat reduced IQs and, in common with other low-IQ groups, they average significantly increased criminal arrest records. At one time there was speculation that XYY males had "genetic criminal tendencies," but other analysis suggested that an XYY male is no more likely to be arrested than an XY or XXY male of the same IQ.

Turner's syndrome karyotype

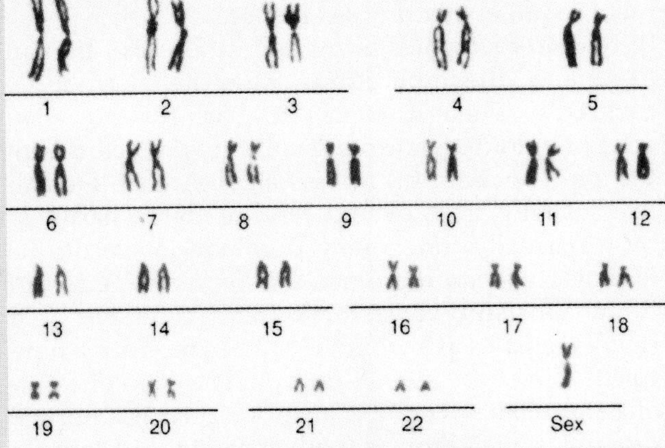

Normal female sex chromosome complement (XX) missing one X chromosome

Klinefelter's syndrome karyotype

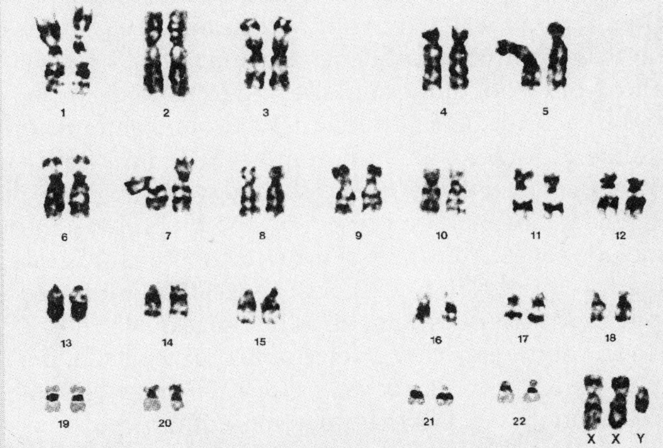

Normal male sex chromosome complement (XY) has an extra X chromosome

But, even though X and Y chromosomes appear different, they pair with one another during meiosis. (See also Essay 12.1.)

Sex Linkage

In humans and many other animals, males are designated as XY. At conception each male receives an X chromosome from his mother and a Y chromosome from his father. Females are designated XX. At conception every female gets an X chromosome from her mother and another X chromosome from her father.

Since a female has only X chromosomes, every egg she produces must carry an X chromosome. In males, half of the sperm produced will carry an X chromosome and half will have a Y. The sex of the offspring will depend on whether the fertilizing sperm is X-bearing or Y-bearing. (So now you know who to blame!) An XX combination in the zygote produces a female child and an XY combination, a male.

Of the two human sex chromosomes X and Y, the X is the more interesting, because it carries thousands of genes that affect all aspects of the phenotype. Also, its pattern of inheritance is a little more complex, since it is present in both sexes. But before we try to understand the genetics of the X chromosome, we should take a brief look at its smaller partner, the Y chromosome.

The Y Chromosome

The Y chromosome's pattern of inheritance is simplicity itself: it is passed from fathers to sons, period. If you are a man, your Y chromosome was inherited from your father, from your father's father, and from your father's father's father and so on, right back to the first human Y chromosome. And unlike all other human chromosomes, the Y chromosome never undergoes crossing over.

The genes that are present on the Y chromosome include several genes for sperm function. In humans, the Y chromosome also has genes that influence height. In humans and all other mammals, the Y chromosome carries the determinant of sex itself, the **TDF gene.** At about six weeks the TDF gene ("testis-determining factor") in human male embryos stimulates development of the testes, which respond by secreting male hormones. Under their influence the male reproductive organs develop. In the absence of male hormones, female structures develop.

The X Chromosome and Sex-Linked Genes

In earlier chapters we emphasized that diploid organisms have two copies of every gene—one from the father and one from the mother. This is perfectly true for women, who get two X chromosomes, but it is not true for men. About 10% of all human genes are carried on the X chromosome, and for each of these genes, a man (XY) gets only one copy, and he gets that copy from his mother. From his father he gets a Y chromosome. Those genes found on the X chromosome are said to be *sex-linked* or *X-linked.*

This means that human males are effectively haploid for approximately 10% of their genes. (The rest of the genes, carried on the autosomes, are present in two copies.) The lone copies of a mother's X-linked genes in her son can therefore be referred to as neither homozygous nor heterozygous; geneticists have coined a special term for them: **hemizygous** (*hemi,* half); they have no allelic counterparts.

In 1910, T. H. Morgan showed that a gene, the determinant of white eyes in *Drosophila,* was not inherited independently of sex, but showed a pattern of inheritance known as **sex linkage,** because the gene is located on the X chromosome (see Figure 12.7). Since the trait is on the X chromosome, males will always have white eyes if they inherit the gene from their mother. Since the trait for white eyes is recessive, females must inherit two copies (one on the X from their mother, one on the X from their father) in order to be homozygous and have white eyes. This is why sex-linked genes give the appearance of affecting males more often than females. Morgan eventually received the Nobel Prize for his achievement in understanding sex linkage.

Both in theoretical terms and in terms of human experience, the implications of this are enormous—especially when it comes to harmful, recessive genes. Diploid organisms usually are protected from the effects of harmful recessive genes since such alleles usually are "covered" and rendered harmless by a normally functioning dominant allele on the homologous chromosome. This protective effect of dominance still works for X-linked recessives, but only in females. If a woman (XX) has received a nonfunctioning, recessive, X-linked allele from one parent, she still has a good chance of receiving a normal dominant allele from the other parent. But when a man (XY) receives a detrimental, recessive, X-linked gene from his mother, it will always be expressed. His Y chromosome will be of no help.

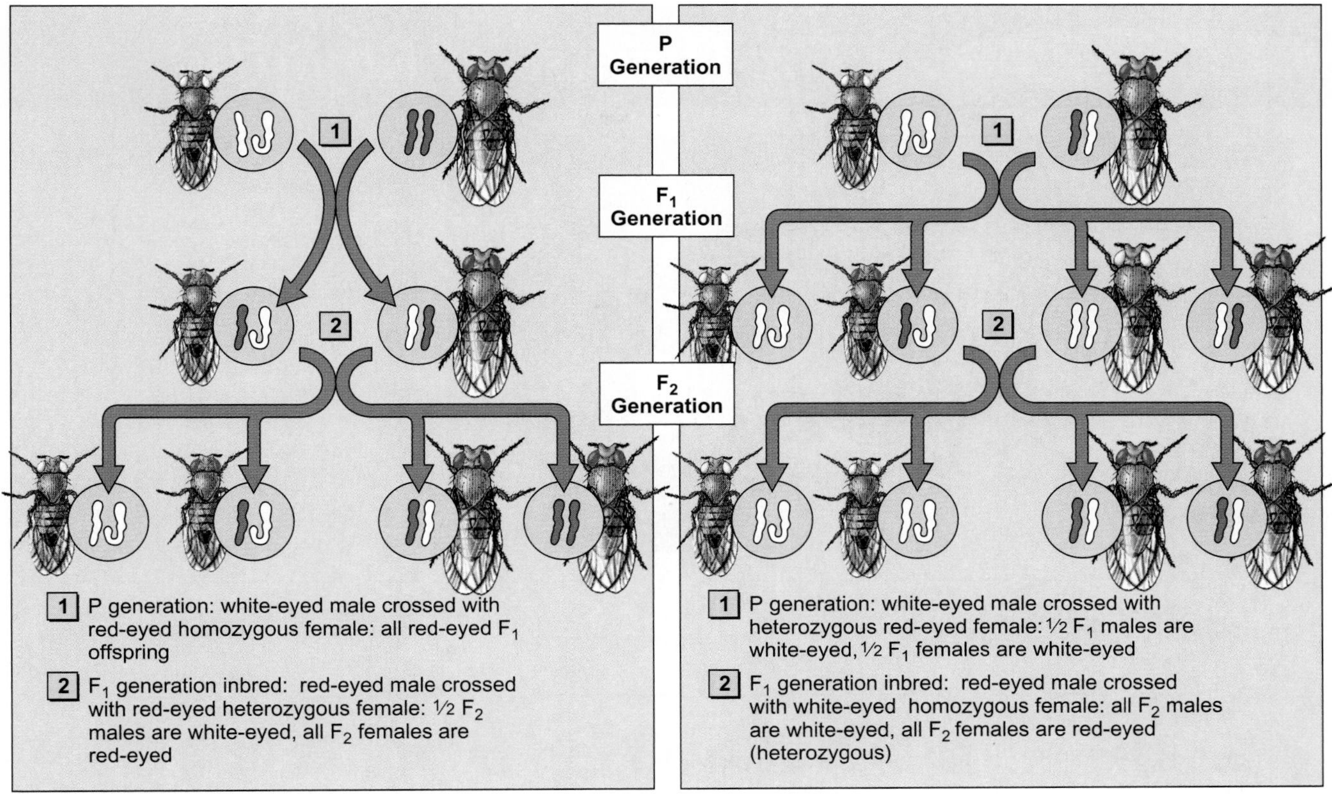

P Generation
F1 Generation
F2 Generation
1
2
1 P generation: white-eyed male crossed with red-eyed homozygous female: all red-eyed F1 offspring
2 F1 generation inbred: red-eyed male crossed with red-eyed heterozygous female: 1/2 F2 males are white-eyed, all F2 females are red-eyed
1 P generation: white-eyed male crossed with heterozygous red-eyed female: 1/2 F1 males are white-eyed, 1/2 F1 females are white-eyed
2 F1 generation inbred: red-eyed male crossed with white-eyed homozygous female: all F2 males are white-eyed, all F2 females are red-eyed (heterozygous)

FIGURE 12.7 **WHITE EYES AND SEX LINKAGE.** In his first cross, Morgan mated his newly discovered white-eyed male with a normal, red-eyed female. All the F_1 offspring were red-eyed. Inbreeding the F_1 produced an F_2 that was three-fourths red-eyed and one-fourth white. But all the white-eyed flies were males. A test cross between the original white-eyed male and a red-eyed female from the F_1 showed that the female was heterozygous. Another test cross was done with a red-eyed male and a white-eyed female. All the male offspring were white-eyed.

This is not to say that women cannot express sex-linked traits. They can and do, but the probability is quite reduced. Men get the dubious distinction of being the ones who are most often affected by X-linked genetic pathologic conditions. The list of these is long and depressing and includes three kinds of muscular dystrophy, two kinds of hemophilia (bleeder's disease), Lesch-Nyhan syndrome, three types of hereditary deafness, pituitary dwarfism, testicular feminization, and several kinds of congenital blindness, in addition to less life-threatening conditions such as color blindness (Figure 12.8). For a look at a sex-linked disease that has had a substantial effect on history, see Essay 12.2.

Barr Bodies, Drumsticks, and the Lyon Effect

A simple chromosome study has been used in women's athletic competitions to determine whether a competitor is, in fact, a woman (see Essay 12.3). The lining of the athlete's mouth is gently scraped, and the cells are stained. A quick study of cell nuclei will establish genetic sex (XX or XY). The nucleus in female mouth cells reveals a dark-staining body, the **Barr body,** which is actually a condensed X chromosome (Figure 12.9a). Because we all need a functioning X chromosome, Barr bodies are absent in normal males.

Essay 12.2 The Disease of Royalty

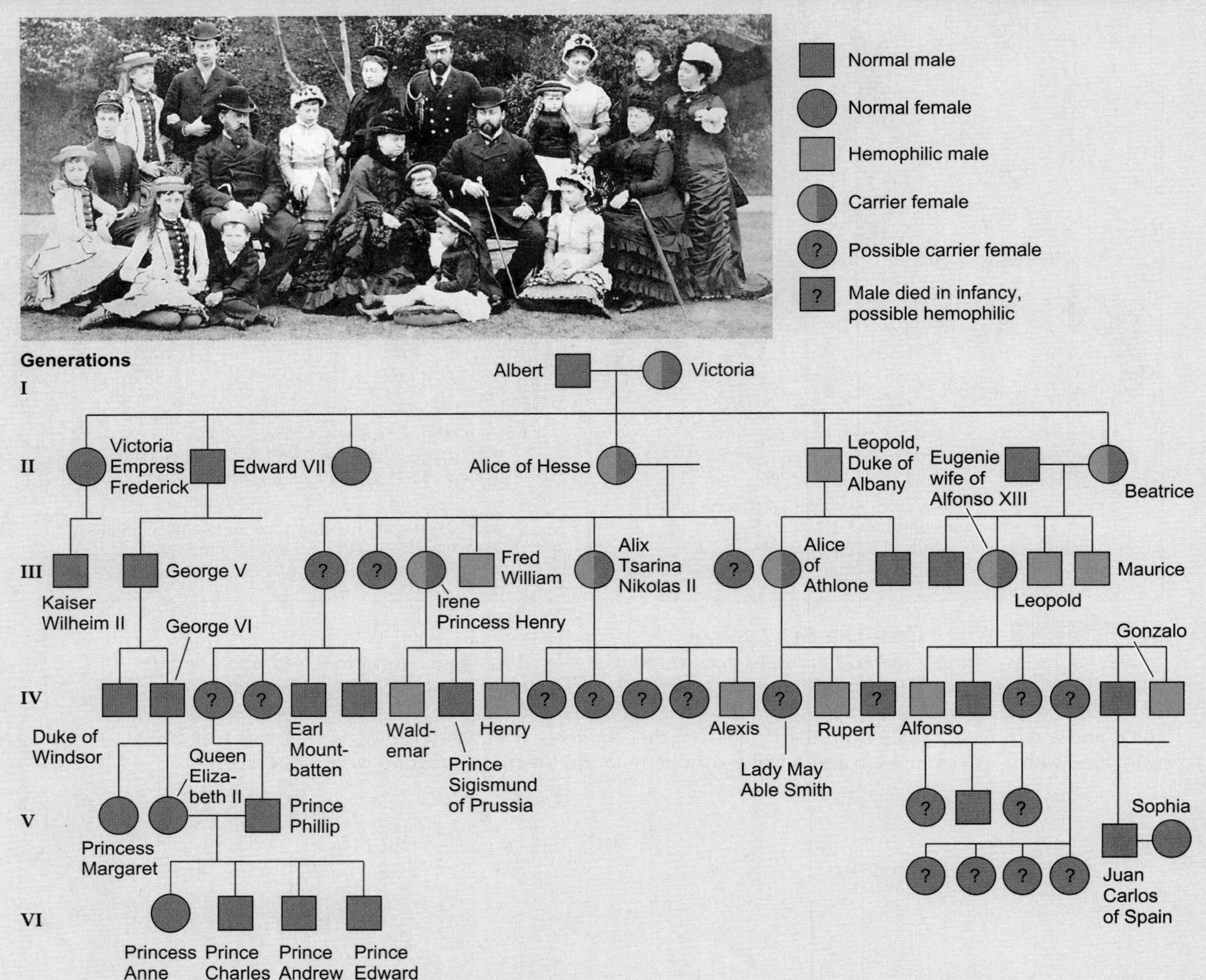

Present generation and their children are free from hemophilia.

Because ruling monarchs consolidated their empires through marriage alliances, hemophilia was transmitted throughout the royal families of Europe. Hemophilia is a sex-linked recessive condition in which the blood does not clot properly; any small injury can result in severe bleeding—and, if the bleeding cannot be stopped, in death. Hence, it has sometimes been called the bleeder's disease.

The hemophilia allele has been traced back as far as Queen Victoria, who was born in 1819. One of her sons, Leopold, Duke of Albany, died of the disease at the age of 31. Apparently, at least two of Victoria's daughters were carriers, since several of their descendants were hemophilic. Hemophilia played an important historical role in Russia during the reign of Nikolas II, the last Czar. The Czarevich, Alexis, was hemophilic, and his mother, the Czarina, was convinced that the only one who could save her son's life was the monk Rasputin—known as the "mad monk." Through this hold over the reigning family, Rasputin became the real power behind the disintegrating throne.

Why is an X chromosome condensed and visible when the other chromosomes are spread out and impossible to detect? Every female cell has twice as many X-linked genes as are present in a male cell. Presumably, there would be a physiological imbalance if both X chromosomes were active. In any case, evolution has solved the problem through what is called the **Lyon effect** (first discovered by a British geneticist, Mary Lyon). One of the two X chromosomes in each XX cell is deactivated during embryonic development. Which of the two X chromosomes is inactivated seems to be a matter of chance.

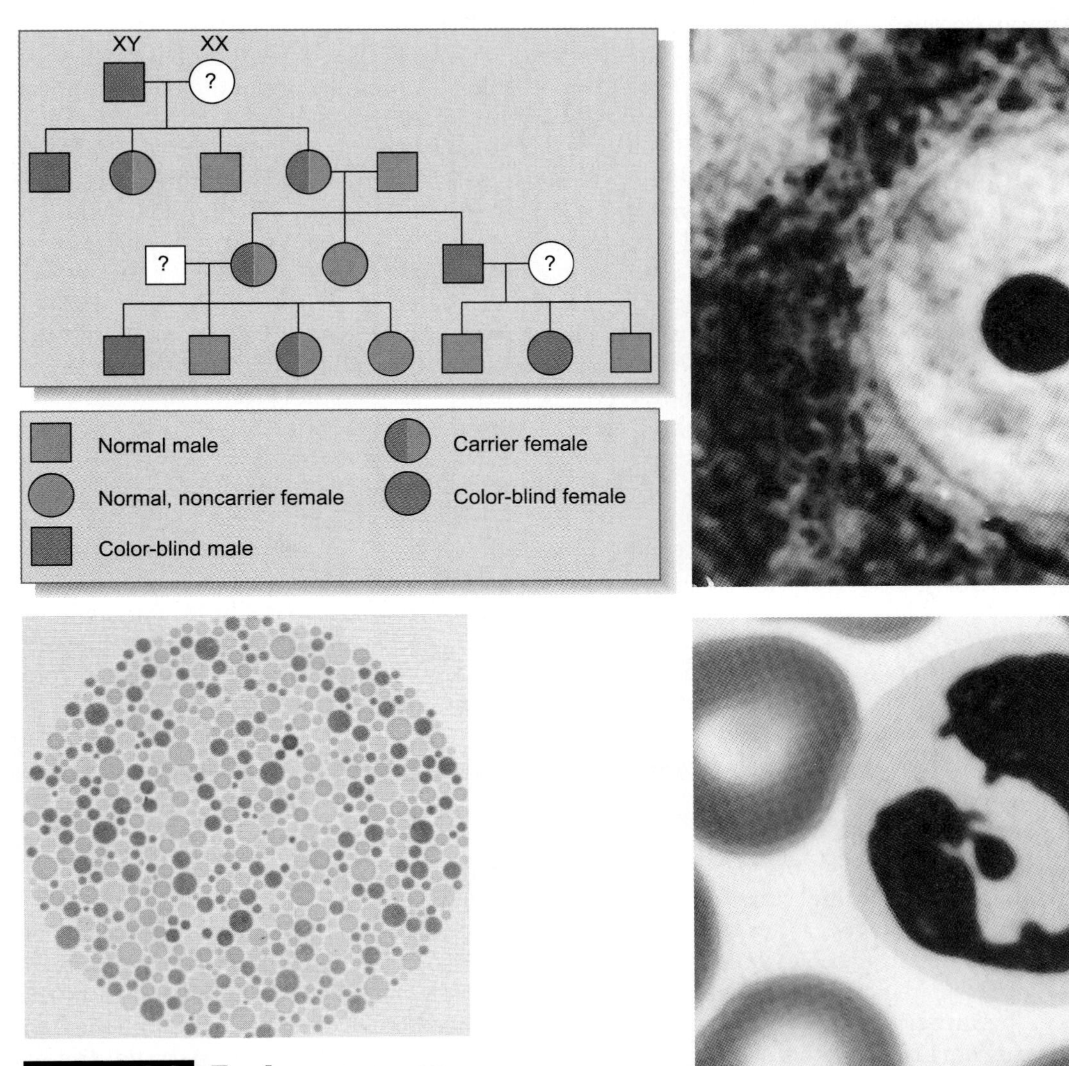

FIGURE 12.8 **THE INHERITANCE OF COLOR BLINDNESS.**
The color blindness allele *(solid color)* in this family tree has been traced through four generations. Using the patterns of inheritance, it is possible to determine the most probable genotypes of the three phenotypically *normal* individuals (indicated with question marks). What are the clues that sex linkage is involved? Color vision is tested using colored plates such as the one shown here. Actually, several plates are required for the complete test. If you are having trouble seeing the number *9* (in red and orange), you may want to take the complete test.

FIGURE 12.9 **CONDENSED AND INACTIVE CHROMATIN.**
The inactive X chromosome is seen in the Barr body **(a)** and the drumstick of cells from women **(b)**. Barr bodies are visible in specially stained cells; the drumstick is seen in stained white blood cells.

ESSAY 12.3 Testing for Sex

In the 1936 Olympics, a man competed as a woman for Nazi Germany. The Nazis clearly expected him to win simply because men are generally bigger, stronger, and faster than women. They apparently neglected the fact that there are big, strong, fast women, and the man came in fourth.

In those same Olympics, a Polish woman won the silver medal in the 100-meter dash (she had won the gold four years earlier). She went on to collect 11 world records and retired to the United States as a naturalized citizen. In 1980, she went grocery shopping in Cleveland and was gunned down in the crossfire during an attempted robbery. An autopsy revealed that the woman had both male and female chromosomes and ambiguous genitalia. When the findings were made public, there was a demand, unmet, that her records be stricken from the books.

Both cases reveal an underlying assumption that men have the edge in sports—and, in fact, examinations to verify sex ("gender verification") have been in effect in sporting events for decades. Beginning in 1968, the test involved examining a sample of cells scraped from inside the cheek and then looking for two X chromosomes in women and an X and Y chromosome in men. The testers found that a very small percentage of female athletes were feminine in virtually every respect, but carried a Y chromosome. Some had a condition called androgen insensitivity—which meant that, although they had an XY chromosomal makeup, their cells were unable to react to testosterone and so the female condition prevailed although testes were usually found hidden within the body and the vagina ended blindly.

Actually, there are perhaps a dozen so-called "intersex states" in which there can be a mixing of male and female traits. Do we test for them all? What should be the standards for competition?

Some sports organizations examine the genitalia of women "to determine that a penis or scrotum is not present," and in 1992 the International Olympic Committee used newer techniques of molecular biology to amplify DNA segments for easy, convenient testing. Still, some see inequities and even bad science. Some genetically normal women, for example, produce large amounts of testosterone and develop heavier musculature than other women. Is this unfair? Is it unfair to test only women? And then there are those who argue that tall basketball players have an unfair advantage and could damage the self-confidence of shorter players. Where does it end? What is your opinion?

So every tissue in the adult female is a mosaic of cells in which one or the other X chromosome remains condensed and inactive. However, the condensed X is replicated and passed from cell to daughter cell. It is inactive only with regard to metabolic function. For instance, the muscle tissue of women who are heterozygous carriers of X-linked muscular dystrophy is made up of patches of normal muscle and degenerate muscle. The normal cells expand in size and strength, fully compensating for the defective ones, so that such women appear to be normal.

The condensed X chromosome also can be viewed microscopically in certain white blood cells, where it forms a characteristic projection from the cell nucleus called the **drumstick.** Although the Barr body was named after Murray L. Barr, its discoverer, the drumstick got its name because it reminded someone of a chicken leg (Figure 12.9b).

PERSPECTIVES

Now we have seen that the somewhat abstract, mathematical concepts of Mendelian genetics is a rather simple, direct result of the movements of chromosomes during meiosis. In fact, once the connection between the "genes" and the "chromosomes" is established, it is obvious that the patterns of inheritance observed by Mendel are a necessary outcome of meiosis.

We have also seen that the simple ratios don't quite hold for genes that are on the same chromosome. Linkage, of course, does not invalidate or contradict the basic principles of Mendel; it merely extends those principles. In the next chapter, we will look at several other cases that would seem to violate Mendel's simple rules, but in fact do not.

SUMMARY

MENDEL AND MEIOSIS

Although Mendel knew nothing about meiosis, upon the rediscovery of his work, the relationship between his laws and the meiotic process was soon established. In the dihybrid cross **RrYy** × **RrYy** (see Chapter 11), the way alleles combine in the gametes (**RY, Ry, rY, ry**) is explained in the two ways two pairs of chromosomes bearing such alleles align and separate during meiosis I.

GENE LINKAGE AND RECOMBINATION

Variations from Mendel's genetic ratios are sometimes due to the presence of linkage groups—genes linked together on a single chromosome.

A dihybrid P_1 cross carried out by Bateson and Punnett produced the usual F_1 phenotypes. Progeny testing of the F_1 yielded four phenotypes, which might have meant that independent assortment was occurring, but a surprising ratio emerged instead of the expected 1 : 1 : 1 : 1. Had the alleles been fully linked, the test cross would have produced a 1 : 1 phenotypic ratio. The peculiar results can be explained through crossing over.

Crossing over, which results in genetic recombination, occurs through the exchange of DNA segments between homologous chromosomes. Genes with high percentages of recombination are called loosely linked, and those with low percentages of recombination are called tightly linked. Geneticists determined that the frequency of crossover was directly proportional to the distance between gene loci. Using recombination frequencies from fruit fly crosses, Sturtevant developed chromosome maps. They indicated the chromosome on which a given gene locus occurred, the relative order of loci, and their distances from each other.

Sex is, as a rule, chromosomally determined. In many species the sex-determining pair are heteromorphic, designated as X and Y. The XX combination produces females, and the XY combination, males. Observers tracking two pairs of heteromorphic chromosomes through meiosis were able to relate meiosis to Mendel's laws.

SEX LINKAGE

Genes may be linked on a sex chromosome and generally be passed along together. In humans, far more genes are carried on the X chromosome than on the Y chromosome.

The human Y chromosome carries genes related to sperm function, height, and male sex. The TDF gene stimulates testis development in the early embryo.

The human X chromosome carries 4,000 to 5,000 genes that have no counterpart on the Y. Thus all X-linked genes act as dominants in males, a condition called hemizygous. Predictably, the frequency of sex-linked abnormalities in males greatly exceeds that in females.

Female cells can be identified by the presence of a condensed X chromosome (the Barr body in mouth cells and the "drumstick" in certain white blood cells). One of the two X chromosomes in each cell is randomly but permanently deactivated (the Lyon effect).

Key Terms

linkage group • **184**
map unit • **188**
gene locus • **188**
sex chromosome • **188**
autosomal chromosome (autosome) • **188**
TDF gene • **190**
hemizygous • **190**
sex linkage • **190**
Barr body • **191**
Lyon effect • **193**
drumstick • **194**

Review Questions

1. Using two pairs of chromosomes with the markers **A** and **a** on one pair and **B** and **b** on the other, carefully illustrate the principle of independent assortment through the events of meiosis. Bear in mind that demonstrating this relationship requires drawing two cells going through meiosis. (Why?)
2. Explain why two pairs of differently shaped chromosomes must be present in meiosis for independent assortment to be directly observed.
3. Review the Bateson-Punnett F_1 test cross, **BbLl** × **bbll.** Carefully work out the cross so that you can determine (a) what the results would have been if independent assortment had been at work, (b) what the results would have been had the genes been completely linked, and (c) the precise way the crossover occurred that produced the new combinations seen.
4. Distinguish between the terms "loosely linked" and "tightly linked."
5. What is a genetic map? Explain the rationale behind using recombination frequencies to determine gene locus distances in the construction of such maps.
6. What genes does the human Y chromosome carry? Compare this to the human X chromosome.
7. Explain why all alleles on the X chromosomes act as dominants in males. Can females express sex-linked traits at all?
8. Using the symbols XX and XY to represent female and male sex, suggest why the ratio of sexes in offspring (at birth) is about 1 : 1. (Use a Punnett square.)
9. Who can males blame their sex-linked abnormalities on? Is this always true? Explain both answers.
10. Carry out a cross between a color-blind man and a color-blind carrier (heterozygous) woman. In what fraction of the children should color blindness occur? Are there color-blind females? Carrier females? (You might symbolize the cross as **Cc** × **co,** or **Cc** × **cY,** just to remind yourself of hemizygosity in the male.)
11. In a family with two male and two female children, one of the boys is color-blind, but the rest of the children are normal. List the *most probable* genotypes of the other children and the parents.
12. In a family of six children, there is a hemophilic boy with normal vision, two color-blind boys with normal clotting times, two completely normal girls, and one color-blind girl with normal clotting time. Determine, as accurately as you can, the genotypes of the parents.

CHAPTER

13

Going Beyond Mendel

IF GENETICS IS SO SIMPLE, AND INHERITANCE FOLLOWS SUCH STRAIGHT-*forward rules, why is it that most of the variations we see among, say, the children in a single human family are not so easily explained? Why are there more than three sizes of humans? Why do eyes come in so many colors? In this chapter, we will see that such things can indeed be explained with the knowledge we have of genetics. It's just that a lot of genes may be involved—genes that interact with each other and the environment to produce an individual.*

Mendel indeed had a string of remarkable successes. How could he be so fortunate? Was it just plain luck? Or did his luck reside in his experimental design? Mendel frankly attributed his success to his deliberate decision to work only with "factors" that always produced large, dramatic effects with clear and distinct phenotypes. He examined such traits, one or two at a time, in highly inbred, genetically unvarying, true-breeding strains. Only in such simple systems could he have worked out his famous ratios. However, not all genetic variation is so simple. Mendel's discoveries were valuable because nearly all genes have proved to be Mendelian in their behavior and their transmission from generation to generation. But most genetic variation is not Mendelian because of the complexities of development and gene expression that Mendel had so carefully avoided.

Most of the genetic differences we see in our friends—differences in height, weight, body build, skin color, temperament, facial features, athletic ability, intelligence, and hairiness—are due to normal allelic variation (along with modifying effects by environment). These normal phenotypic differences, which add so much to human interest, seldom show up in the usual Mendelian ratios, although the genes responsible segregate and assort with faithful Mendelian precision. Even blue and brown eye color, a popular example of Mendelian inheritance in humans, turns out to be quite complex and unpredictable. People do not merely have blue or brown eyes; they may have gray, light blue, deep blue, hazel, flecked, or green eyes. In this chapter, we will look beyond Mendel to learn more about how Mendelian genes behave in what seem to be non-Mendelian ways.

EXTENDING MENDEL'S PRINCIPLES

So the actual expression of genes into phenotypes in natural populations is influenced or determined by many kinds of complications that Mendel deliberately avoided. Whole courses in genetics are devoted to these complications, but we can list the major categories and then give some examples of each. The question might be: Why don't we find Mendel's simple 3 : 1 ratios everywhere? And the answers are:

1. *Dominance relationships.* Mendel considered only cases in which the effect of one allele completely masks the effect of the other. Actually, there is a whole range of ways in which two homologous alleles can interact, producing intermediate effects.
2. *Multiple alleles.* Up to this point, we have considered only two alleles at each gene locus. There may actually be many different alleles that can occupy a specific gene locus.
3. *Epistasis.* The genes at one locus may affect the phenotypic expression of genes at another locus.
4. *Polygenic inheritance.* Many different gene loci may have an additive effect on the same phenotypic trait, especially if the trait is one that varies continuously, such as height or weight. The phenomenon is called **polygenic inheritance** ("many origins"), where many genes interact to determine a single trait.
5. *Other sources of variability.* Also important are environmental interactions (the same genotype may be expressed differently in different environments), sex influences (the same trait may be expressed to different degrees in males and females), and effects of development and aging

(some genes begin to exert their effects only in adulthood or old age).

It is important to remember that none of these examples violate any of Mendel's principles. All genes segregate and assort just as expected. The apparent problems we may perceive in the ratios are due to these five issues that extend, but do not violate, Mendel's principles.

Dominance Relationships

The various ways in which two alleles at one gene locus can affect the phenotype are called **dominance relationships.** Remember that the two alleles come from the two parents. Each parent contributes to the effects controlled by each locus.

Complete Dominance

What happens when the gene from the father and the corresponding gene from the mother give conflicting instructions? Say, an allele for red flowers comes from one parent and an allele for white flowers comes from the other. What happens then? Actually, many things can happen—but one of the simplest results is that the information from one allele (the recessive) will appear to be ignored through **complete dominance,** where one allele completely determines the phenotype. In that case, the other allele is dominant. You will recall that this is what Mendel saw in his peas. Dominance, as you will also recall, results in the phenotype of the heterozygote (the individual with the conflicting genetic instructions) being exactly like that of one of the homozygotes. The allele that is expressed, no matter what its partner is, is called the dominant allele, and the allele that is not visibly expressed in the heterozygote is called the recessive allele.

So far we have given only observations and definitions, not explanations. So now we can ask, what really happens when two different alleles occur together? How is one expressed and one ignored?

Actually, dominance can occur in several ways. In most cases the recessive allele isn't expressed because it simply isn't doing anything; its code is nonfunctional. In a relatively benign example shown in Figure 13.1, albinos lack an enzyme that is necessary to make melanin pigments. They didn't get a functioning gene from either their father or their mother, so their cells simply cannot make pigment. Such individuals, then, are homozygous for recessive alleles, which in this case are alleles that aren't functioning. Heterozygotes for albinism, on the other hand, have one working allele and one that doesn't work, and produce only half the usual amount of

FIGURE 13.1 ALBINISM IS A STRAIGHTFORWARD RECESSIVE TRAIT. The absence of pigmentation in hair and a light skin color in albino individuals is a striking, simple recessive genetic trait in humans. Albinism also results in visual sensitivity to bright light.

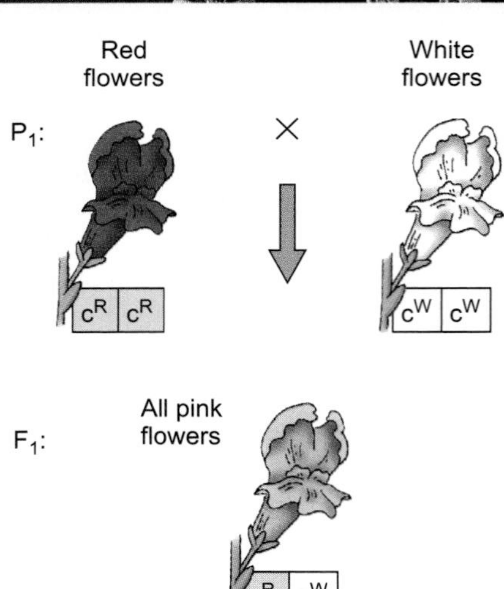

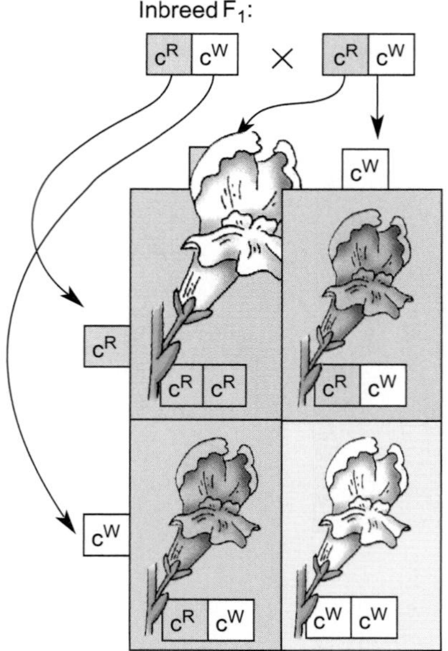

enzyme; but in most cases half the normal amount of enzyme is still enough to produce needed pigment, and the phenotype of the heterozygote will be perfectly normal.

Sometimes the relationship between alleles is not so simple. There are two cases where two alleles interact such that the heterozygote is distinctly different from either homozygous combination.

Incomplete Dominance

Incomplete dominance (sometimes called *partial dominance*) is where the heterozygote appears to be a blend of the two homozygotes. A classic example of partial dominance is found in crosses between snapdragons with red and white flowers (Figure 13.2). Here, the two alleles can be symbolized by $\mathbf{c^R}$ and $\mathbf{c^W}$. (Historically, incompletely dominant alleles are usually designated by small letters. Hence, we use $\mathbf{c^R}$ and $\mathbf{c^W}$ here, instead of **R** and **r,** which would imply complete dominance.) When the homozygous $\mathbf{c^Rc^R}$ red-flowered snapdragons are crossed with homozygous $\mathbf{c^Wc^W}$ white-flowered snapdragons, the plants in the F_1 generation, the $\mathbf{c^Rc^W}$ heterozygotes, all have pink flowers. Since all three genotypes are easily distinguished (by the phenotypes of Red, Pink, and White), we can see that the F_2 phenotypic ratio is exactly the same as the F_2 genotypic ratio (Figure 13.2). In this example it is fairly easy to see why the term *partial* is used. The heterozygote pink color gives the impression that the single dominant allele is not quite strong enough to produce enough pigment to make the color red, so the flower ends up not quite red, but pink.

Codominance

On the other hand, **codominance** is the equal expression of both alleles that results in a mixed phenotype. This kind of coexpression is most easily seen in some biochemical phenotypes like the blood types that we will discuss in the next section. But, for now, consider roan cattle. The roan phenotype

Figure 13.2 Partial Dominance.
Flower color in snapdragons presents a good example of partial or incomplete dominance. The hallmark of partial dominance is that the heterozygotes appear to have a phenotype that is somewhere in between the homozygous dominant and the homozygous recessive. Another distinguishing feature is that the phenotypic ratio of the progeny is always the same as the genotypic ratio. Can you see why?

results from a cross of red cattle and white cattle. The phenotype of the heterozygote is, yes, a kind of pink color called roan (Figure 13.3). But if you look closely at a roan-colored cow, you will find not pink hairs, but a mix of red and white hairs. So here we have an equal expression of the two alleles.

Perhaps now you see that dominance is a question of degree. At one end of the spectrum we have completely dominant and completely recessive alleles. In between we have alleles that interact as partial or as codominant alleles.

FIGURE 13.3 CODOMINANCE. The roan-colored cow is an example of codominance, the equal expression of two alleles.

MULTIPLE ALLELES

So far we have been looking at traits that are governed by two alternative alleles. However, many traits are the product of **multiple alleles,** three or more allelic forms of one gene. (Keep in mind that in this situation, each *individual* still receives *just one pair* of alleles.)

Multiple alleles are of great interest today because of the growth of organ and tissue transplant surgery. A primary requirement for a successful transplant—one that will not be rejected—is that the tissues of the donor and recipient be as genetically and immunologically compatible as possible. This gets to be quite a problem because some of the genes that create incompatibilities have a great many allelic forms. (One gene has 12 different allele possibilities, and another has more than 20!) Even close matches, however, cannot prevent rejection.

The example of multiple alleles we will look at in detail is from the ABO blood group system, where there are three alleles possible at a single locus. There are several alleles responsible for blood types, with A, B, and O being the most common types. Blood type is due to the presence of certain molecules on the surface of red blood cells. These molecules are called antigens, and they serve to mark or identify the cell as a particular type.

Persons who are homozygous for the $\mathbf{I^A}$ allele produce one kind of cell-surface antigen and are said to be type **A.** Persons who are homozygous for the $\mathbf{I^B}$ allele produce a different kind of cell-surface antigen and are said to be type **B.** Heterozygotes, $\mathbf{I^AI^B}$, carry both alleles, producing both kinds of cell-surface antigens, and are said to be type **AB.** In this situation, because both alleles are expressed equally, $\mathbf{I^A}$ and $\mathbf{I^B}$ are codominant.

Persons who have neither the A antigen nor the B antigen belong to a fourth type, **ii**—type **O.** Type **O** is the most common blood type. It seems that i is a third allele at the same gene locus, an allele that produces neither antigen. It is recessive when paired with either $\mathbf{I^A}$ or $\mathbf{I^B}$. With three different possible alleles, there are six different possible genotypes, as shown in Figure 13.4.

Although $\mathbf{I^A}$ and $\mathbf{I^B}$ are codominant with respect to each other, they are simply dominant with respect to the i allele. Note that although there are three alleles of this gene locus in the human species, no one individual can have all three alleles at once. Diploid organisms like ourselves are limited to two at a time.

Essay 13.1 discusses another aspect of blood types.

EPISTASIS

When Mendel crossed true-breeding round, yellow peas with true-breeding wrinkled, green peas, the F_2 generation yielded $\frac{9}{16}$ round and yellow, $\frac{3}{16}$ round and green, $\frac{3}{16}$ wrinkled and yellow, and $\frac{1}{16}$ wrinkled and green. Such ratios are possible only in straightforward situations in which the genes do not influence each other. When two different gene loci do interact, the tidy 9 : 3 : 3 : 1 ratio doesn't show up. One such example is due to **epistasis,** wherein the expression of one gene influences or masks the expression of another.

Mouse coat color, for instance, would have driven Mendel up the wall. Consider two gene loci that affect the coat color of mice. In one case, black (**B**) is dominant to brown (**b**). If a true-breeding black strain (**BB**) is crossed with a true-breeding brown strain (**bb**), the F_1 generation will all be heterozygous and black (**Bb**), and the F_2 will show a 3 : 1 ratio of black to brown. No surprises here.

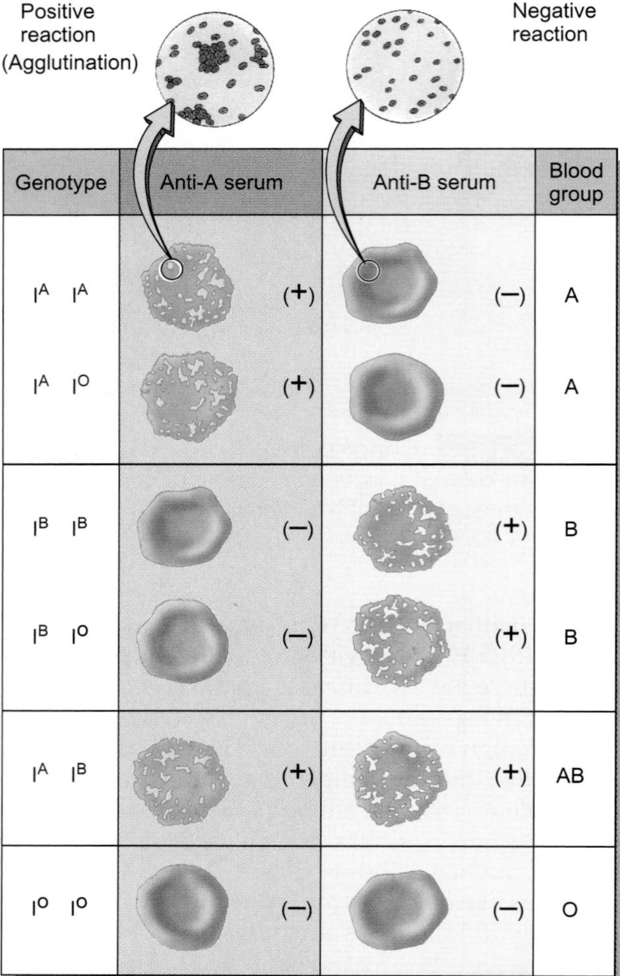

Positive reaction (Agglutination)
Negative reaction
Genotype
Anti-A serum
Anti-B serum
Blood group
I^A I^A (+) (−) A
I^A I^O (+) (−) A
I^B I^B (−) (+) B
I^B I^O (−) (+) B
I^A I^B (+) (+) AB
I^O I^O (−) (−) O

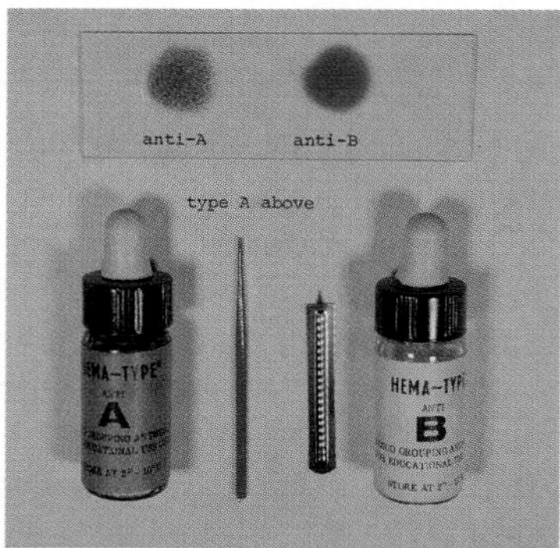

FIGURE 13.4 MULTIPLE ALLELES IN HUMAN BLOOD TYPES.
The three different alleles of the I gene ($\mathbf{I^A}$, $\mathbf{I^B}$, and **i**) can be found in six possible combinations, and this results in four different blood groups. The phenotype of a blood sample can be determined by testing the blood with antibodies. A positive reaction with the anti-A serum indicates the presence of the A antigen, whereas a positive reaction with the anti-B serum reveals the presence of the B antigen.

But at yet another gene locus there is another pair of alleles that can also affect coat color. These are the alleles at the albino locus. The recessive homozygote (**cc**) is a snow-white albino mouse. (The recessive albino gene (**c**) is defective and does not supply one of the enzymes necessary for the production of pigment.) The dominant homozygote (**CC**) and the heterozygote (**Cc**) have normal pigment, although it can be black or brown, depending on the alleles (**B** or **b**) at the first locus.

Now consider a mating between an albino mouse from a true-breeding white strain (**cc**) and a mouse from a true-breeding brown strain (**bb**). What would you expect? You might *not* expect the litter to be entirely black. But here is one way it could happen.

CCbb (brown) × **ccBB** (white) ⟶ **CcBb** (black)

The dominant B alleles were carried by the white strain, completely hidden. In this example of epistasis, the recessive albino genotype is epistatic to the brown/black genotype. Epistasis is, then, a relationship among genes. A gene that blocks the apparent action of another gene is said to be epistatic to that gene.

Considering the genotypes of the participants in the cross above, the only possible outcome in the offspring would be **CcBb.** Since there is both an active allele for pigment formation and a dominant color allele present (**C** and **B**), the entire litter is black.

Now consider random crosses among the doubly heterozygous offspring, **CcBb × CcBb.** Will the usual 9 : 3 : 3 : 1 ratio appear in the offspring (providing you raise enough of them)? No. The F_2 generation will approximate a 9 : 3 : 4 ratio: $\frac{9}{16}$ black, $\frac{3}{16}$ brown, and $\frac{4}{16}$ albino. Figure 13.5 confirms that the two gene loci assort independently, just as Mendel would have predicted. But because of epistasis, the last two terms of the 9 : 3 : 3 : 1 ratio are combined to produce the 9 : 3 : 4 ratio. After all, once a mouse is albino, it doesn't much matter whether the coat *could* have been brown or black.

POLYGENIC INHERITANCE

Many of the phenotypic traits that are most important to biologists, and especially to plant and animal breeders, do not fit into "either-or" categories. Instead, these traits occur in a gradient of phenotypes, a situation known as **continuous variation.** Differences among individuals are still caused by different alleles at specific gene loci on chromosomes, and these alleles segregate and assort according to the usual Mendelian laws. But in these

Essay 13.1 Incompatible Rh Blood Types

At one time some women who knew their blood type was Rh negative actually sought out Rh negative men as potential husbands. They did so in an effort to keep from destroying their own children in the womb.

The basis of their concern was that the Rh positive allele (Rh^+) is dominant over the Rh negative allele (Rh^-). Therefore, when an Rh^- woman is impregnated by an Rh^+ man, the baby she carries may be Rh^+. If the man is homozygous, this will certainly be the case, but if he is heterozygous, the child has a 50 : 50 chance of being Rh^+. (In either case, the Rh^+ baby will definitely be heterozygous.)

There is a certain element of risk in such matings since a serious incompatibility may arise if some of the baby's Rh^+ red cells should cross the placenta and enter the mother's bloodstream (or should she receive Rh^+ blood in a transfusion). When this happens, the mother's immune system responds by producing Rh antibodies against the cell-surface antigens carried by the invading red cells. There is little danger to the mother, but should her antibodies cross the placenta, they can begin agglutinating the fetus's red cells, causing a severe, even fatal, condition called *erythroblastosis fetalis.* It involves erythrocyte (red blood cell) destruction, which can lead to severe anemia and jaundice.

The problem generally doesn't arise with the first pregnancy, since most blood mixing occurs during delivery, too late for the mother's immune system to react and affect that baby. But in subsequent pregnancies, the sensitized mother's newly formed Rh antibodies can readily cross the placenta, as many useful antibodies do, late in pregnancy. Once in the baby's circulatory system, they begin their deadly role. If the baby survives delivery, a massive blood transfusion may be necessary for its survival.

It is now standard medical practice to treat Rh^- mothers of newborn Rh^+ babies with a single injection of Rh antibodies shortly after delivery. The antibodies destroy any of the baby's invading red blood cells before they can trigger the immune response. In the treated mother, the injected antibodies soon dissipate, and subsequent pregnancies are as safe from the Rh incompatibility problem as was the first.

cases, many different gene loci will contribute to a single phenotypic trait, with each individual allele having a relatively small effect. This is called *polygenic inheritance.* Common examples of such continuous variation in humans are height, skin color, foot size, nose size, weight at birth, and aspects of intelligence. At first glance, continuous variation may seem to lend belated support to the old notion of blending inheritance, but it can be explained quite satisfactorily in Mendelian terms.

Many gene loci determine height, but for simplicity we'll assume that height is determined by only three loci. Also, in reality there may be multiple alleles possible at each gene locus, but in our example we'll assume only two alternatives are available: "short" alleles and "tall" alleles. Now let's assume that the presence of a "tall" allele rather than a "short" allele increases adult height by about 5 cm (2 ½"). People with only the "short" alleles (six in all) will then grow to the basal human height of about 160 cm (5' 3"). Those with only the "tall" alleles (again, six) grow to 190 cm (6' 3"), which is the basal height of 160 cm plus the added influence of 5 cm from each of the six tall alleles. In the middle, with three short alleles and three tall alleles, are individuals about 175 cm tall (5' 9"). Figure 13.6 summarizes the seven height categories possible in this model. Note that the distribution approximates a normal distribution that is, in fact, observed in real populations for adult height and for nearly every other kind of continuous variation. Most people are of intermediate height; few are very tall or very short.

Other Sources of Variability

It may have occurred to you that the polygenic model of the genetic control of adult height can't possibly explain everything. For one thing, men (as a group) tend to be taller than women (as a group).

And what about diet, general health, and other environmental factors? Modern-day humans are taller than their medieval and 19th-century ancestors (have you noticed how small the suits of armor in museums are?), and young Japanese adults are noticeably taller than their own parents because of rapid improvements in the Japanese standard of living. Furthermore, don't we all get shorter in our old age? We do, indeed. The point is that nongenetic effects have very marked influences on many genetic traits.

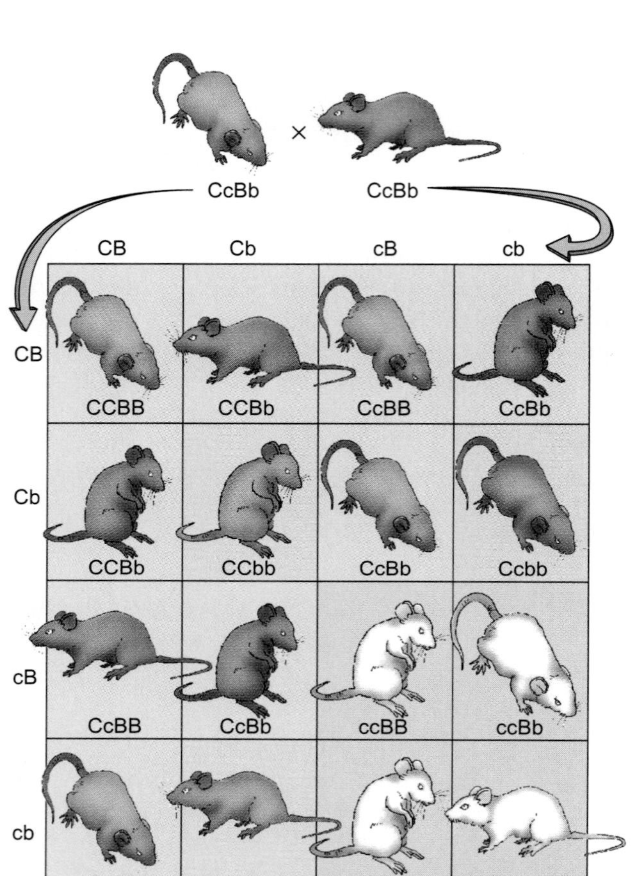

cc = white, regardless of other gene
CC or Cc = black or brown, depending on other gene as follows:
BB or Bb = black
bb = brown
Resulting ratio = 9 : 3 : 4

FIGURE 13.5 EPISTASIS.
This diagram shows the inheritance of two pairs of genes that influence mouse coat color. The **B** (black) and **b** (brown) alleles occur at one locus, and the **C** (color) and **c** (albino) alleles occur at another locus. The two pairs assort independently in typical Mendelian fashion, but produce seemingly strange ratios when heterozygous F_1 mice are inbred (**CcBb** × **CcBb**). The resulting ratio is 9 : 3 : 4 (9 : 3 : 3 + 1) because of an epistatic interaction that makes two different genotypes phenotypically identical.

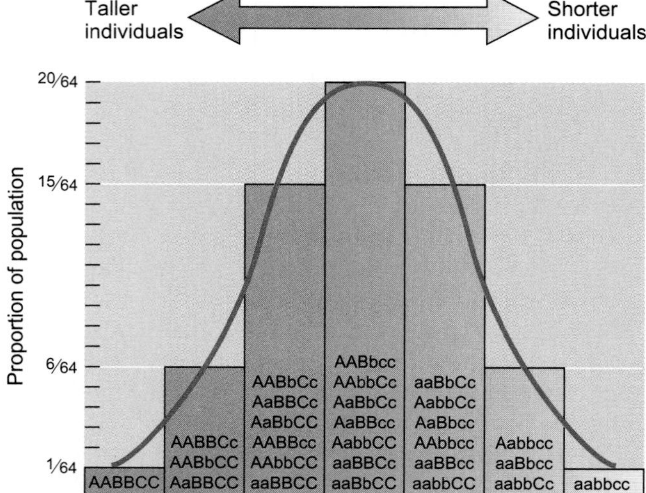

FIGURE 13.6 POLYGENIC INHERITANCE.
Polygenic genetic traits, when plotted by phenotypes gathered from large samples, tend to form curves approximating a normal distribution. The photograph shows a good example. The people are arranged in rows according to height, with the tallest person at the left. This distribution is the result of the fact that many different genotypes can result in average height classes, while extreme heights require more specialized, and therefore more rare, combinations.

ENVIRONMENTAL INTERACTIONS

Environmental interactions may be very direct or very subtle. Let's consider a straightforward example: the Siamese cat. One of the enzymes in its pigmentation pathway has mutated so that it is now sensitive to temperature; the enzyme functions only below a certain temperature. As a result, Siamese cats are darker in their cooler extremities: the nose, ears, tail, and feet (Figure 13.7).

FIGURE 13.7 ENVIRONMENTAL INFLUENCE ON PHENOTYPE.
Coat color in the Siamese cat is an example of the influence of environment on gene expression. Where the skin is coolest, in the extremities, dark pigment is produced in the hair.

INCOMPLETE PENETRANCE AND VARIABLE EXPRESSIVITY

In some cases an individual may have a dominant genotype without showing it, a condition known as **incomplete penetrance.** For instance, the gene that gives one the ability to roll his or her tongue is dominant, yet not all persons with the dominant allele will be able to roll their tongues (Figure 13.8). But for any one person it is all or nothing: There are no people who can roll their tongue only partway. In fact, in a room full of people with the dominant allele, only about $^{8}/_{10}$ will actually be able to roll their tongue. Therefore we say that the tongue-rolling allele has 80% penetrance. Penetrance is, then, the percentage of individuals (out of those with the proper genotype) that actually show the trait.

There is a related situation called **variable expressivity,** where the mutant phenotype will be expressed to varying degrees. For instance, a rare dominant trait in human genetics is polydactyly, the tendency to have extra fingers or toes (Figure 13.8). Persons carrying this dominant allele show variable expressivity in that all four extremities may be

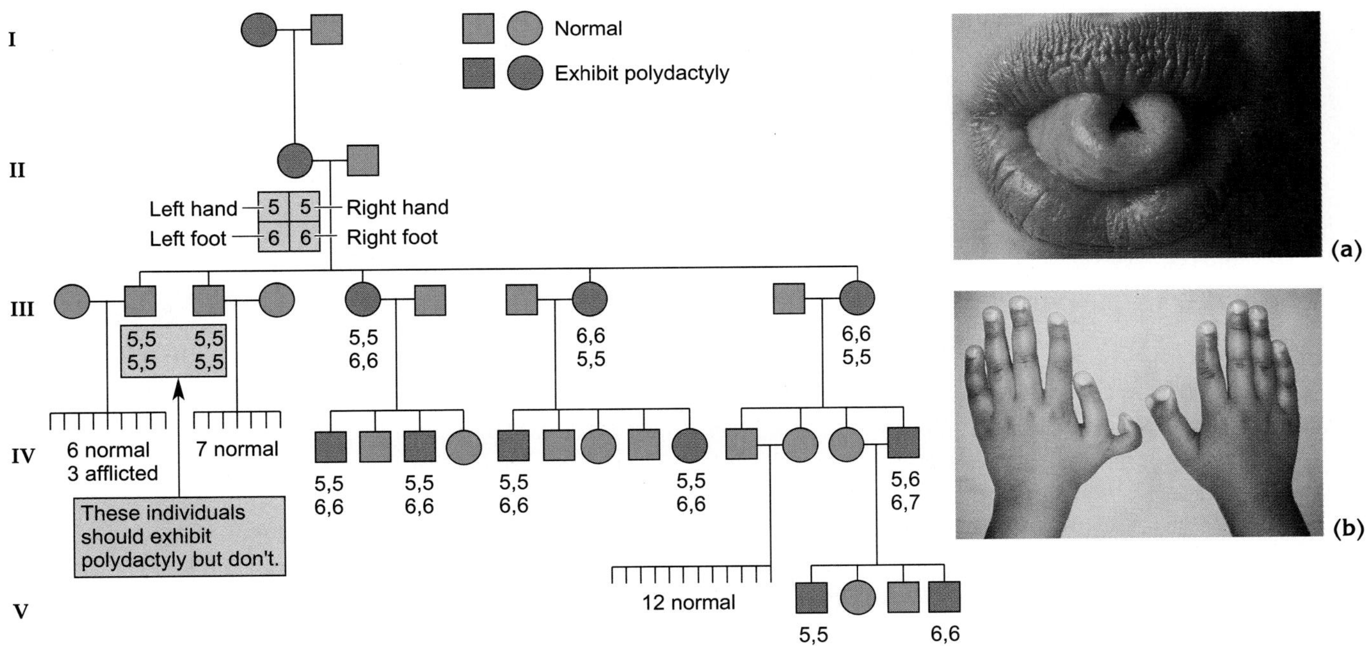

FIGURE 13.8 INCOMPLETE PENETRANCE AND VARIABLE EXPRESSIVITY.
(a) The ability to roll your tongue is caused by a dominant allele. However, you may not be able to roll your tongue even if you have the dominant allele. **(b)** Polydactyly, the inheritance of extra fingers or toes, is a dominant trait. However, it shows variable expressivity in that a person with the allele may show polydactyly on all four extremities, hands only, feet only, or any combination—or they may not show the trait at all. The numbers below the individuals of this pedigree indicate the number of digits on the hands and feet.

affected, or only the one hand or one foot. Both hands and both feet of a given carrier have the same genes and the same environment but may have either normal or abnormal numbers of digits, indicating that some form of developmental chance is at work. Of course, this means that a person carrying the allele may just be lucky enough to have only five toes on each foot and only five fingers on each hand. In such a case, fortune has smiled four times, and the person would be unaware of carrying this allele if it weren't for the fact that unless fortune continued smiling, about half of his or her children would have extra fingers and toes.

Sex-Limited and Sex-Influenced Effects

A dominant gene is known to be responsible for a rare type of cancer of the uterus. However, this gene can be inherited by a female from her father. This is a **sex-limited trait** (affects only one sex) since, needless to say, it has no influence on men. Still other genetic conditions affect one sex more often, or more severely, and are said to be **sex-influenced traits.** The most common kind of middle-age baldness is an example. It behaves like a dominant in men and a recessive in women, perhaps due to the different set of sex hormones present in men and women. Pyloric stenosis is also sex-influenced. It is a serious but correctable problem in newborn babies, in which abnormally thick muscle development restricts the movement of food into the small intestine. It runs in families, but affects five times as many boys as girls.

Variable Age of Onset

Baldness due to a dominant gene, in addition to being sex-influenced, also has a **variable age of onset.** Even among brothers carrying the same dominant gene, one man may lose his top hair in his 20s, while the other loses it in his 40s or even later. A more serious example of a dominant gene with a late, variable age of onset is that for Huntington's disease, or Huntington's chorea. This disease is characterized by a severe neuromotor condition that begins with subtle personality changes and progresses through tremors and paralysis to death. The dominant gene does not usually begin to show its effects until sometime in adulthood. The average age of onset of symptoms is about 40, but it can begin affecting heterozygous men or women as early as age 15 or as late as age 60. Unfortunately, many victims learn of the defective gene's presence only after they have children.

Many of the more common diseases of old age are believed to have a genetic basis, at least in part. The genes for some "old-age disorders" may be considered to have a delayed and variable age of onset. At some point, it is impossible to tell what is genetic and what is environmental, but we do know this: Some people seem old at 40, while others are surprisingly young at 70.

Perspectives

Mendel's simple phenotypic ratios of 3 : 1 and 9 : 3 : 3 : 1 don't seem to hold for many kinds of inherited traits. In this chapter, we've seen examples of the various influences that result in nonsimple ratios, such as dominance relationships, multiple alleles, epistasis, and polygenic inheritance. The message here is that even these complex situations are governed by Mendel's basic laws of segregation and assortment. Keep in mind that virtually every gene in the nucleus is inherited strictly by Mendelian rules. Complications arise because genes can interact with each other and with the environment. Nonetheless, Mendelian principles hold.

We've spent the last few chapters dealing with genes as inherited factors that produce a phenotype. But so far, we have skirted the issue of just what it is about DNA that constitutes a gene, and how a gene produces its trait. As you can guess, then, these are among the things we will consider in the next few chapters.

Summary

Extending Mendel's Principles

Not all patterns of inheritance produce Mendel's simple ratios because of such influences as dominance relationships, multiple alleles, epistasis, polygenic inheritance, and other sources of variability.

Dominance Relationships

Dominance relationships refer to the various ways alleles interact. Simple Mendelian dominance and recessiveness refer to the masking of an inoperative (recessive) allele by a normally functioning (dominant) allele. The enzyme deficiency responsible for the absence of pigment in the albino is an example.

Incomplete dominance occurs when alternative alleles produce a compromising effect. Pink snapdragons are an example.

In codominance both alternative alleles fully express themselves. People with the A and B blood group allele express both, producing both the A and B red blood cell antigens.

Multiple Alleles

The term *multiple alleles* applies to genetic traits produced by more than the usual two alternative alleles. In the human ABO blood system, each person carries two alleles from the three alternatives $\mathbf{I^A}$, $\mathbf{I^B}$, and **i**. Because of codominance between the I^A and I^B alleles and recessiveness of the i allele, four phenotypes (A, B, AB, and O) and six genotypes ($\mathbf{I^AI^A}$, $\mathbf{I^Ai}$, $\mathbf{I^BI^B}$, $\mathbf{I^Bi}$, $\mathbf{I^AI^B}$, and **ii**) are possible.

Epistasis

In mouse coat color one allele pair determines black and brown color in the usual Mendelian manner. The dominant form of a second pair of alleles does not affect color, but its recessive form is epistatic; that is, it inhibits the alleles for color formation. Both pairs follow Mendel's law of independent assortment, but epistasis results in non-Mendelian phenotypic ratios.

Polygenic Inheritance

In polygenic inheritance two or more pairs of alleles contribute to a single phenotypic trait such as human body height. The effect is a gradient of phenotypes called continuous variation.

Other Sources of Variability

Diet, climate, chance, sex, and age are among the variables that often influence the expression of a trait. For example, better diet increases body height; in cold temperatures Siamese cats get darker; and sex affects the expression of some traits. Sex-limited traits involve one sex only, whereas sex-influenced traits occur more frequently in one sex. Certain genetic abnormalities only arise later in life.

Key Terms

polygenic inheritance • **198**
dominance relationships • **199**
complete dominance • **199**
incomplete dominance • **200**
codominance • **200**
multiple alleles • **201**
A • **201**
B • **201**
AB • **201**
O • **201**
epistasis • **201**
continuous variation • **202**
incomplete penetrance • **205**
variable expressivity • **205**
sex-limited trait • **206**
sex-influenced trait • **206**
variable age of onset • **206**

Review Questions

1. Review the terms *gene, locus, homologue, allele, dominant, recessive, phenotype,* and *genotype* before answering the questions below.
2. The absence of pigment in albinism, a recessive condition, provides us with a fairly clear-cut example of Mendelian dominance and recessiveness. Are people who are heterozygous for albinism visibly affected? Why?
3. Carefully distinguish between the three forms of dominance: complete dominance, partial dominance, and codominance.
4. In a flower called the four o'clock, a cross between white and red produces pink. What form of dominance is this? Carry out a cross involving a pink- and a white-flowered plant, stating the phenotypic and genotypic ratios.
5. A grower wants to produce a line of pure-breeding pink snapdragons. What advice would you give the person? Explain why.
6. List the three genetic concepts represented in the genetics of the human ABO blood group.
7. Three alleles—$\mathbf{I^A}$, $\mathbf{I^B}$, and **i**—are responsible for the human ABO blood groups. How many alleles does each person carry? List all of the genotypes and phenotypes possible.
8. Jones, Smith, White, and Doe each claim to be heirs to a fortune left by a wealthy couple who had only one child. The couple's blood types are known to have

been AB and A; the father of the type A individual was known to have been type O. Which of the following blood types, if any, are ruled out. Jones—Type AB; Smith—Type O; White—Type B; Doe—Type A.

9. Carry out this version of the epistatic cross involving mouse coat color: **CcBb** × **ccBb.** State the phenotypic ratio in the offspring. (Remember that **BB** and **Bb** are black, and **bb** is brown. Also, where **cc** appears, no pigment is produced.)

10. Careful study of a variety of tomatoes reveals that there are five distinct colors (yellow, orange, light red, dark red, and deep red) present in the ripened fruit. How might the concept of polygenic inheritance be applied here? Assuming that random mating was involved, which color would you expect to be most common?

11. Considering the tomato color problem in Question 10, carry out the following cross representing random mating in the population: **AaBb** × **AaBb**. Determine the phenotypic ratio of the offspring. Graph the results, and describe the shape of the resulting curve.

12. Why is it difficult to predict accurately the appearance of extra fingers and toes in a family carrying the dominant allele for polydactyly?

13. Carefully distinguish between sex-limited and sex-influenced traits. Provide an example of each.

Chapter 14

But What *Is* DNA?

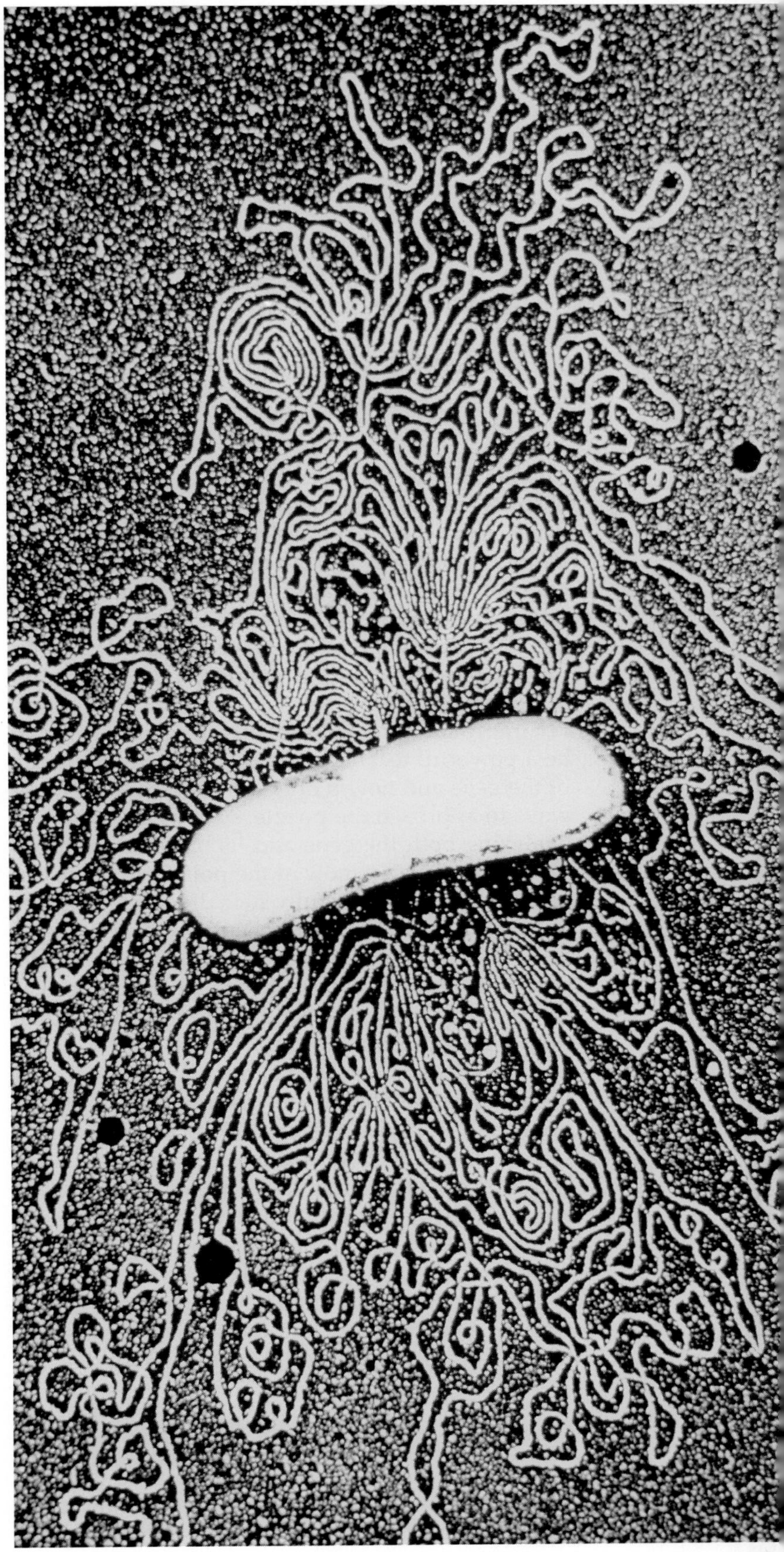

WE HAVE DISCUSSED MANY THINGS THAT HAVE TO DO WITH DNA, SUCH *as alleles, genes, and chromosomes. We have also considered ideas relating to biological information and inheritance. While we have described the role of DNA in such matters, we have not yet considered just what DNA is and how it plays its many roles in the pageant of life. So let's take a closer look at this remarkable molecule. We will quickly find that the DNA story is—like the Mendel story—very logical, indeed.*

Mendel and the early geneticists dealt with genes as abstract entities, but they knew that, at some level, genes must have a physical reality. As William Bateson put it in 1906: "But ever in our thoughts the question rings, what are these units [genes]? How the pack is shuffled and dealt, we are beginning to perceive; but what are they—the cards? Wild and inscrutable the question sounds, but genetic research may answer it yet."

Simply counting kinds of offspring has proved to be a powerful technique for discovering the nature of the gene and how it operates. But there are other ways to approach the puzzle, such as isolating and analyzing both the gene and its biochemical products. In fact, we are now at the point where we can begin to merge successfully such diverse techniques as progeny counts and biochemistry. Since the inception of molecular biology (which includes the study of genes and gene action at the molecular level), geneticists have actually learned more than they expected about the gene.

In this chapter and the next, we'll review some of the experiments that led to our modern concept of the gene. We'll take a close look at its biochemical structure, and at how it directs protein synthesis. Finally, we will consider how the genes themselves are controlled, and how changes in DNA (mutations) can affect gene function.

Historically, genetics consists of two stories that at first take their own tracks and finally merge: the story of gene function, and the story of gene structure (that is, the chemical structure of DNA).

DISCOVERING DNA: THE STUFF OF GENES

It seems incredible, but DNA was isolated and characterized in Mendel's lifetime. In 1869, Johann Friedrich Miescher, a Swiss chemist, first isolated the nuclei of concentrated, dead white blood cells—acquired from pus taken from bandages gathered while visiting hospital wards (biologists will look at anything). He then chemically dissolved the nuclear membrane and most of the remaining protein. This left a phosphorus-rich material he called *nuclein,* which we now call *chromatin* (DNA and its associated proteins were discussed in Chapter 9). That same year Miescher also found nuclein in a number of other cell types. He speculated that the material might simply serve as a way for the cell to store phosphate, and he also suggested that it might have something to do with heredity.

By 1889, other European chemists, who had further purified nuclein by removing the last traces of protein, found that they were left with a gummy, acidic substance, which they named *nucleic acid.* Researchers eventually determined that there are two kinds of nucleic acid, which we now call *DNA* (deoxyribonucleic acid) and *RNA* (ribonucleic acid). The researchers of that day, however, had no way of putting their findings into a context that would clarify the basic mechanisms of genetics. Then along came some people named Griffith, Avery, Hershey, and Chase, who set the stage for others named Wilkins, Franklin, Watson, and Crick. Their names have become synonymous with the great advances in modern genetics.

GRIFFITH: TRANSFORMATION

In 1928, Fred Griffith, a British bacteriologist, conducted what seemed at first to be an oddball experiment, but it proved to be a classic. He studied the virulence (disease-producing capacity) of a bacterium that includes a strain that causes pneumonia and another strain that is quite harmless. The virulent, disease-producing strain formed a smooth, gelatinous polysaccharide coating around each cell that apparently protected it from the host's defenses. The harmless strain lacked a gummy coat. When grown in the laboratory, the virulent strain produced smooth, glistening colonies, whereas the harmless strain (which lacked the proper enzymes to make a gummy coat) produced rough colonies.

When Griffith injected the dangerous smooth-strain bacteria into mice, the mice died, as expected. When he injected the harmless rough-strain bacteria into other mice, the mice did not die. Furthermore, when he injected smooth-strain bacteria that had been heated (sterilized), the mice did not die. But then Griffith mixed dead smooth-strain bacteria with live rough-strain bacteria—both of which had proved to be harmless—and injected the mixture into another group of mice. These mice died. What had happened?

Autopsies showed that the dead mice were full of virulent, living smooth bacteria! Where did they come from? Apparently the genetic material of the harmless, living rough-strain bacteria had somehow been *transformed* by something in the dead smooth-strain bacteria, something that made them deadly also. Moreover, the rough-strain's progeny were all smooth (Figure 14.1). Griffith had discovered the phenomenon of **transformation,** the process in which cells incorporate foreign DNA into their own genetic material.

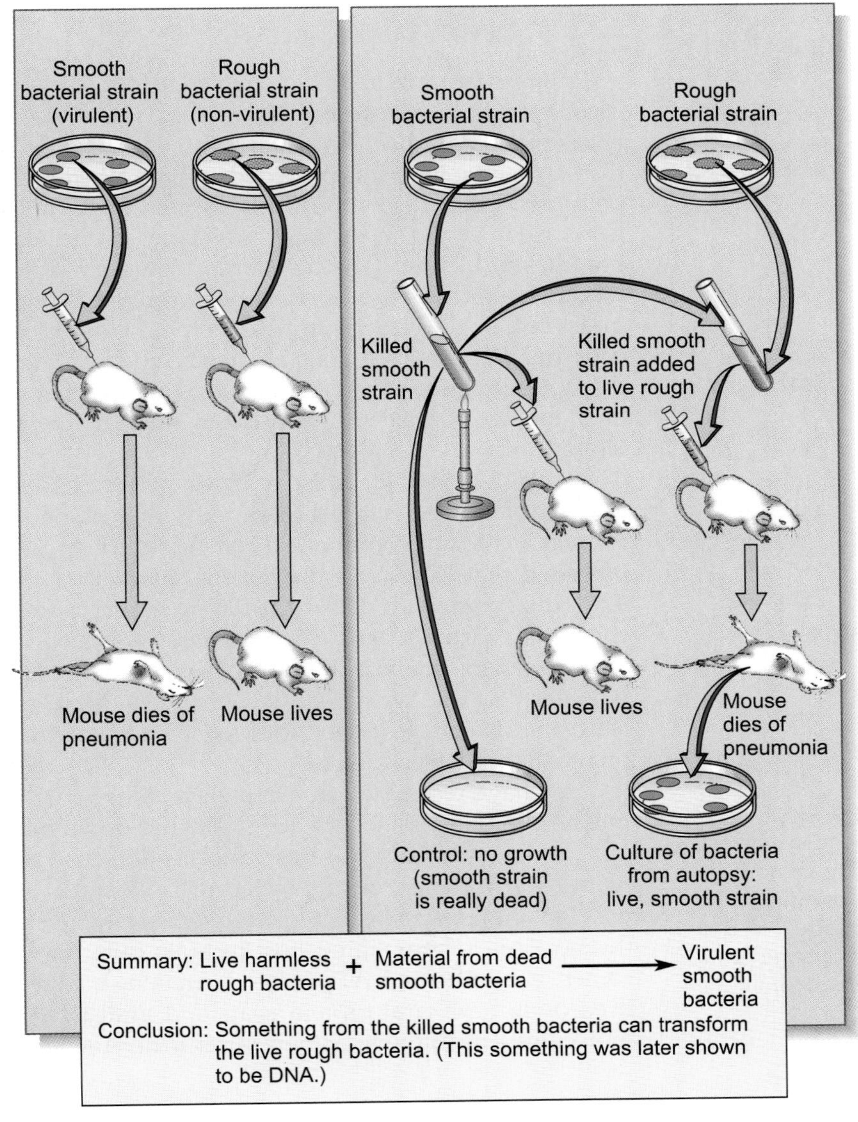

FIGURE 14.1 TRANSFORMATION. Twenty-five years before the function of DNA was finally resolved, Griffith's experiments clearly laid the groundwork for the idea that DNA was the genetic material.

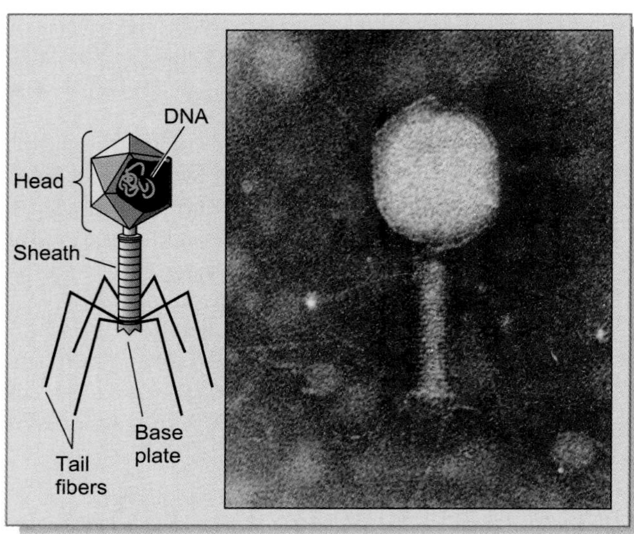
DNA
Head
Sheath
Base plate
Tail fibers

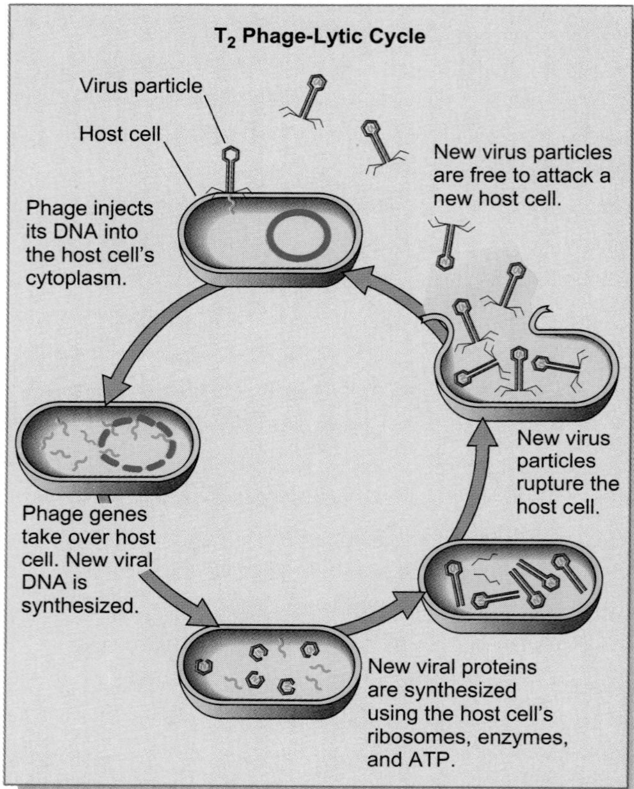
T₂ Phage-Lytic Cycle
Virus particle
Host cell
Phage injects its DNA into the host cell's cytoplasm.
New virus particles are free to attack a new host cell.
New virus particles rupture the host cell.
Phage genes take over host cell. New viral DNA is synthesized.
New viral proteins are synthesized using the host cell's ribosomes, enzymes, and ATP.

FIGURE 14.2 **THE BACTERIOPHAGE.**
The bacteriophage consists of a head and a narrow, hollow tail ending in several tail fibers. The head contains viral DNA protected by its protein covering. The phage attaches via its tail fibers and injects its DNA into the host, which dutifully replicates the phage in a process that is called the lytic cycle of infection.

AVERY: DNA IS THE TRANSFORMING SUBSTANCE

Sixteen years later, O. T. Avery and his colleagues set out to learn more about transformation. After finding that the phenomenon could take place in test tubes as well as in living mice, they decided to try to discover just what substance was causing it. Various substances derived from the dead smooth bacteria were isolated to see whether they might be the mysterious "transforming substance." In 1944, they finally found it. Avery and his colleagues discovered that it was DNA from the deadly smooth-strain bacteria that could transform a rough-strain bacterium, giving it the ability to synthesize the necessary enzymes for making the protective smooth coat.

HERSHEY AND CHASE: DNA MAKES THE NEXT GENERATION

Progress in molecular biology often follows the development of new technology. Given a new tool, researchers can expect new information. In this case, the new technology was provided by Fred Waring, a popular bandleader of the 1950s. When he was not leading his group in song, he was busy inventing the kitchen blender. The blender, of course, whips food into a mush. The gourmet cook probably is not interested in the fact that, in so doing, it disrupts cells. But biologists happened to need a good way to disrupt cells, and so they crept away with their families' kitchen blenders and installed them in their laboratories.

In 1952, Alfred Hershey and Margaret Chase, who had followed Avery's work with interest, performed a crucial experiment. Their work firmly established that DNA was the genetic material of at least one organism. The *bacteriophage—phage,* for short—is a virus that attacks bacteria. It was a new tool for genetic analysis. If you have ever wondered whether a germ can get sick, you will be glad to learn that it can. It can be infected by a bacteriophage that can even kill it.

The phage resembles, in some ways, a hypodermic needle (Figure 14.2). When it touches down on the surface of its bacterial host, tail (needle) first, it pierces the bacterial cell and injects its DNA. Then a peculiar thing happens. The phage genes take over the host cell. The obliging host puts its own ribosomes, enzymes, and ATP to work, making a hundred or so new viral protein coats and viral DNA molecules. The burgeoning new viruses then reward the bacterial host by rupturing it, and in the process free the new viruses to infect other cells.

Only the viral DNA is injected into the host cell. The original empty protein body stays outside, like a discarded overcoat. Hershey and Chase discovered this by growing bacteriophages on bacteria that had been fed radioactive sulfur (^{35}S) or radioactive phosphorus (^{32}P). They knew that proteins contained sulfur and no phosphorus, whereas nucleic acid contained phosphorus and no sulfur; so they had uniquely "tagged" each kind of molecule (Figure 14.3).

Hershey and Chase infected bacteria with the radioactive bacteriophages and allowed enough time for them to attach themselves and inject their DNA, but not enough time for the production of new bacteriophages. Then they put the mixture into their Waring blender.

The empty viral coats, which were pure protein, were broken loose from the bacterial surfaces and could be separated from them by centrifugation. That is, the whole mixture, consisting of infected bacteria and loose empty viral coats in liquid, was put into a high-speed centrifuge. The whirling force settled the heavy bacterial cells at the bottom of the centrifuge tube, leaving the viral coats suspended in the remaining liquid. All of the radioactive sulfur was found in the liquid with the empty protein

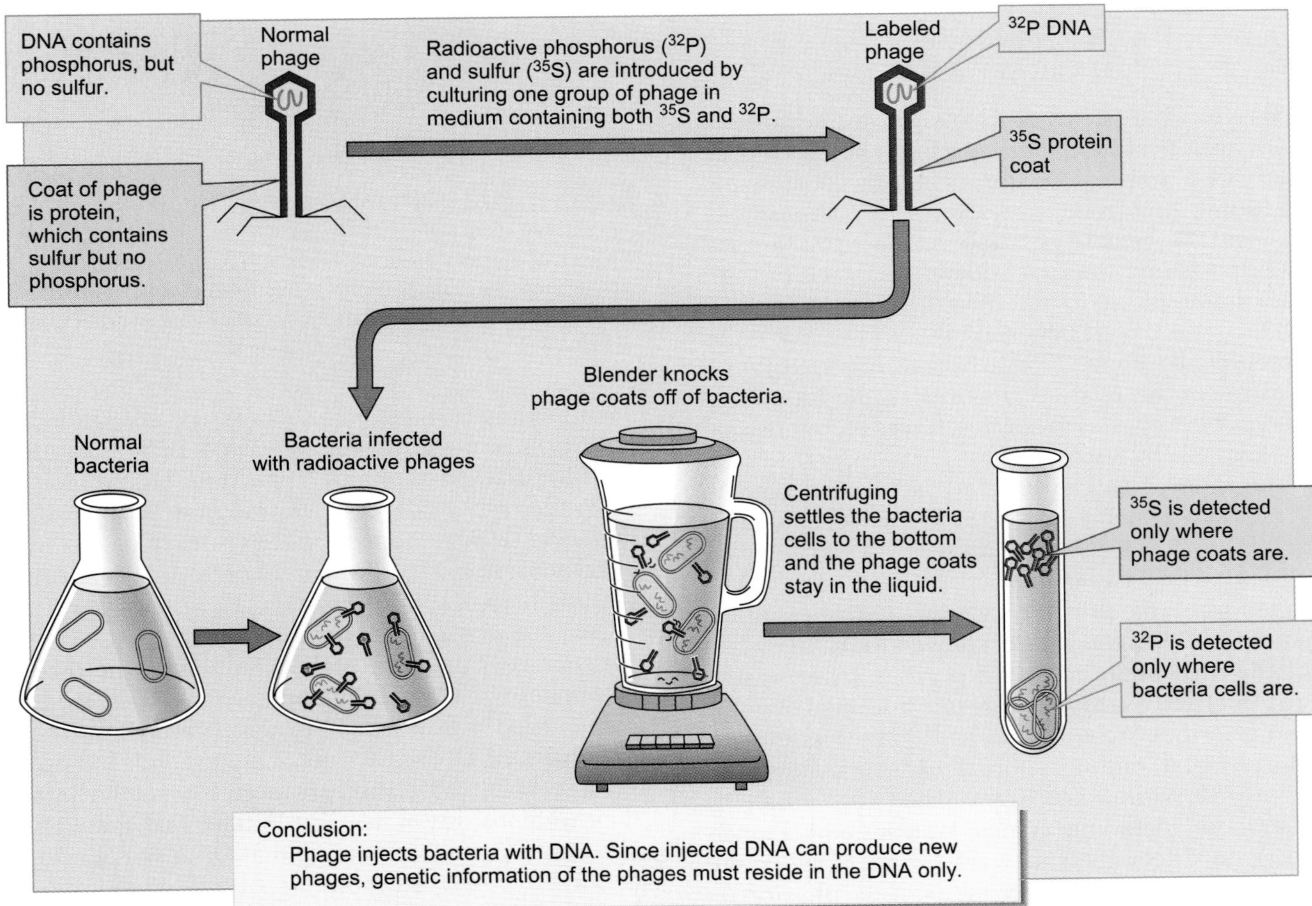

FIGURE 14.3 HERSHEY-CHASE EXPERIMENTAL PROCEDURE.
Hershey and Chase were able to grow phage particles that had their DNA labeled with radioactive phosphorus (^{32}P) and their protein coats labeled with radioactive sulfur (^{35}S). They were then able to show that only the phage DNA enters the cell to carry out its genetic activity, since only the ^{32}P was found in infected cells.

coats, but *radioactive phosphorus was found inside the infected bacteria.* It seemed logical that what was transferred to the bacteria was the source of information used to create the next viral generation. What was transferred to the bacteria was DNA. Since the infected bacteria could still produce complete, virulent virus particles, Hershey and Chase had shown that all of the genetic information of the tiny organisms resided in its DNA—and only in its DNA.

Gene Structure: Closing In on DNA

DNA was now known to be the stuff of genes. What was its structure? How was it formed, and how did it carry out the work of the gene? How was it able to make precise copies of itself?

Chargaff's Rule

Erwin Chargaff tried to answer such questions with analytical biochemistry. He broke down purified DNA from many sources into its constituents, including the four nucleotides of *adenine (A), thymine (T), guanine (G),* and *cytosine (C).* He carefully measured the exact quantity of each of these four subunits. Up to that time, the four nucleotides A, T, G, and C were believed to exist always in equal amounts. By about 1953, Chargaff reported that it wasn't so; the relative amounts of the four bases varied from species to species. It was therefore clear that not all DNA was alike.

In spite of this, a regular pattern emerged in Chargaff's data that came to be known as **Chargaff's rule:** *The amount of adenine was always equal to the amount of thymine, and the amount of guanine was always equal to the amount of cytosine.* Regardless of the source of DNA, the pattern showed that exactly half of the nucleotide bases are **purines** (adenine and guanine) and exactly half are **pyrimidines** (thymine and cytosine). We will come back to the significance of this vital finding shortly. Let's just note that Chargaff's work was being watched very closely by everyone, including James Watson and Francis Crick in England, who were trying to work out DNA structure. Eventually, they saw the physical relationship behind Chargaff's rule.

Wilkins and Franklin: Sizing Up the Molecule

X-ray crystallography is a technique that helps to determine the structure of crystals. It involves shooting a narrow beam of X rays at a crystal and noting how the rays are diffracted (bent) before reaching a target of photographic film (Figure 14.4).

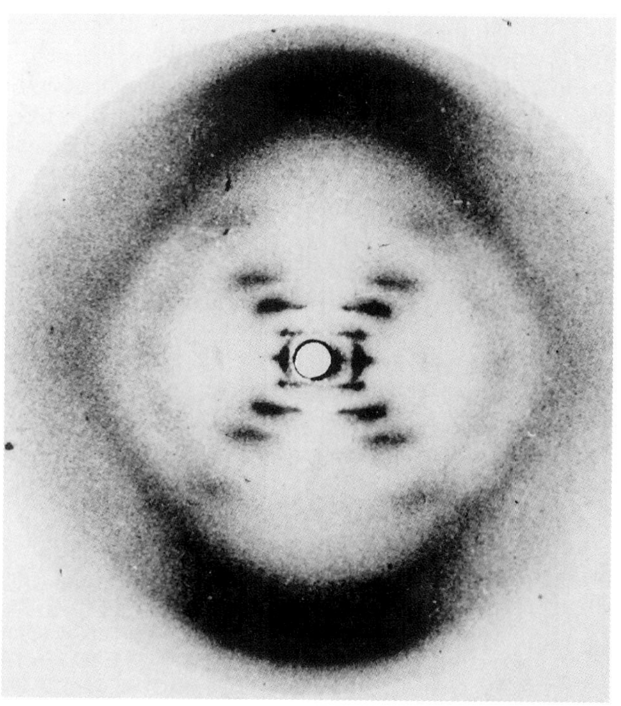

Figure 14.4 Crystallographic Analysis of DNA.
From the pattern produced through X-ray crystallography analysis of DNA, Wilkins and Franklin were able to determine its helical form and critical intramolecular distances.

By the early 1950s, only a few organic substances had been studied this way. DNA was much on everyone's mind, and a few laboratories were beginning to look at DNA crystals.

Among those engaged in such work were Maurice Wilkins and Rosalind Franklin (who died before her work was fully appreciated). Their studies revealed a few repeating distances within the molecules of DNA: 0.34 nm, 2.0 nm, and 3.4 nm. We'll see what these numbers mean later; at the time they were just mysteries. But Wilkins and Franklin also saw a pattern indicating that DNA was a *helix* (a corkscrew-shaped molecule). The idea that biological molecules could have helices was a popular notion at the time.

Franklin and Wilkins had almost figured out the structure of DNA. Franklin had even argued that there were probably two strands in each molecule, not one or three. They were very close to solving the great problem. Yet the solution was to fall to the researchers James Watson and Francis Crick. Interestingly, they did no biochemical experiments of their own, but they integrated what was known of the chemistry of DNA with the structural data of Wilkins and Franklin to deduce the structure of DNA. Thus, this story of DNA is an example of an interesting and common phenomenon in science, where data from one laboratory is gathered, incubated, and extended by a different and perhaps competing research group.

DNA Chemistry

In order to appreciate what happened next, let's jump ahead and say that DNA is now known to be a polymer, that is, a chain of repeating subunits. The repeating subunits are called *deoxynucleotides* or, simply, *nucleotides.* A **nucleotide** consists of a sugar, a phosphate, and a ringed structure called a nitrogen base. The first diagram below represents one of the four nucleotides of DNA, the adenine nucleotide, as it appears in DNA. The pentagon-shaped portion in the lower center is *deoxyribose,* a sugar. Attached to the sugar are both a phosphate ion and adenine, a nitrogen base. The five carbons of the sugar molecule are numbered. The five deoxyribose carbon atoms are numbered 1' (one prime) through 5' (five prime). These *"prime"* designations are worth remembering, as they will later help us orient the DNA molecule.

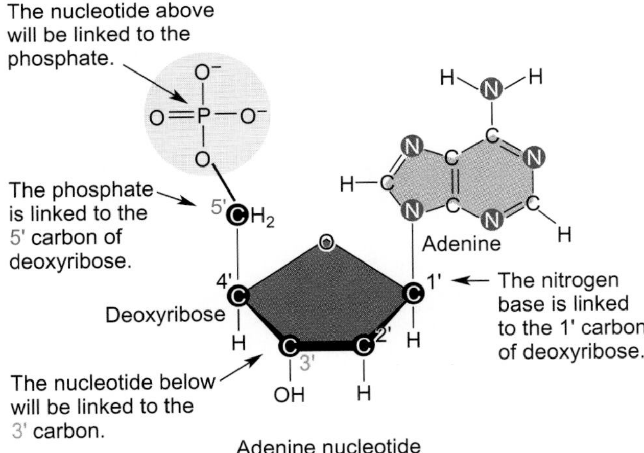

The nucleotide above will be linked to the phosphate.
The phosphate is linked to the 5' carbon of deoxyribose.
Deoxyribose
The nucleotide below will be linked to the 3' carbon.
Adenine
The nitrogen base is linked to the 1' carbon of deoxyribose.
OH

Adenine nucleotide

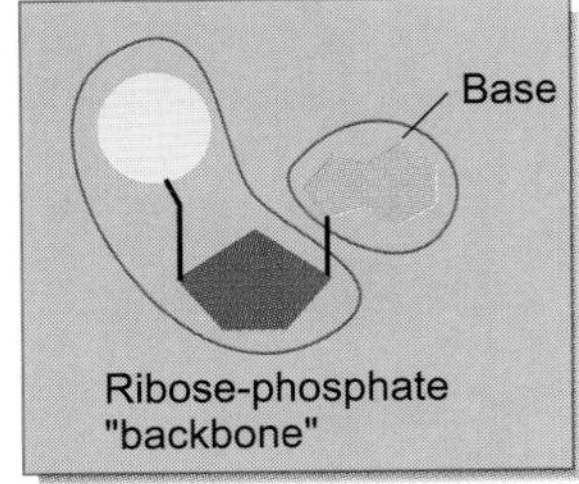

Base
Ribose-phosphate "backbone"

The deoxyribose molecule serves as a link holding the other parts of the DNA molecule. Its important bonding points are the 1', 3', and 5' carbons. The nitrogen bases are always covalently bonded to the 1' carbon of deoxyribose, as shown in the diagram. In free-floating nucleotides, the phosphate is linked to the 5' carbon only; but when the nucleotides are joined to make a strand of DNA, the 5' phosphate of one nucleotide is joined to the 3' carbon of the previous nucleotide. The alternating phosphates and sugars form what is called the backbone of the DNA strand; the nucleotide bases hang off the sides of the molecular backbone.

All four of the DNA nucleotides follow this organization, with only the identity of the nitrogen base portion being different. The four DNA bases, as we have seen, are adenine, thymine, guanine, and cytosine (A, T, G, and C, respectively). The nitrogen bases of the purines (guanine and adenine) consist of two attached rings: one ring with five sides and one with six sides, the two rings sharing a common side. The pyrimidines (thymine and cytosine) each consist of a single six-sided ring.

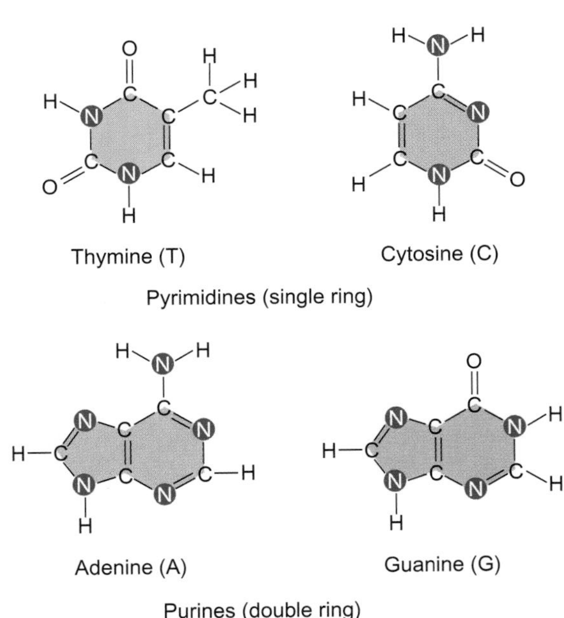

Thymine (T)
Cytosine (C)
Pyrimidines (single ring)
Adenine (A)
Guanine (G)
Purines (double ring)

The two halves of DNA—the two strands of the double helix—fit together in an interesting way. But that's Watson and Crick's story; let's return to them and see how they finally figured out the structure of DNA.

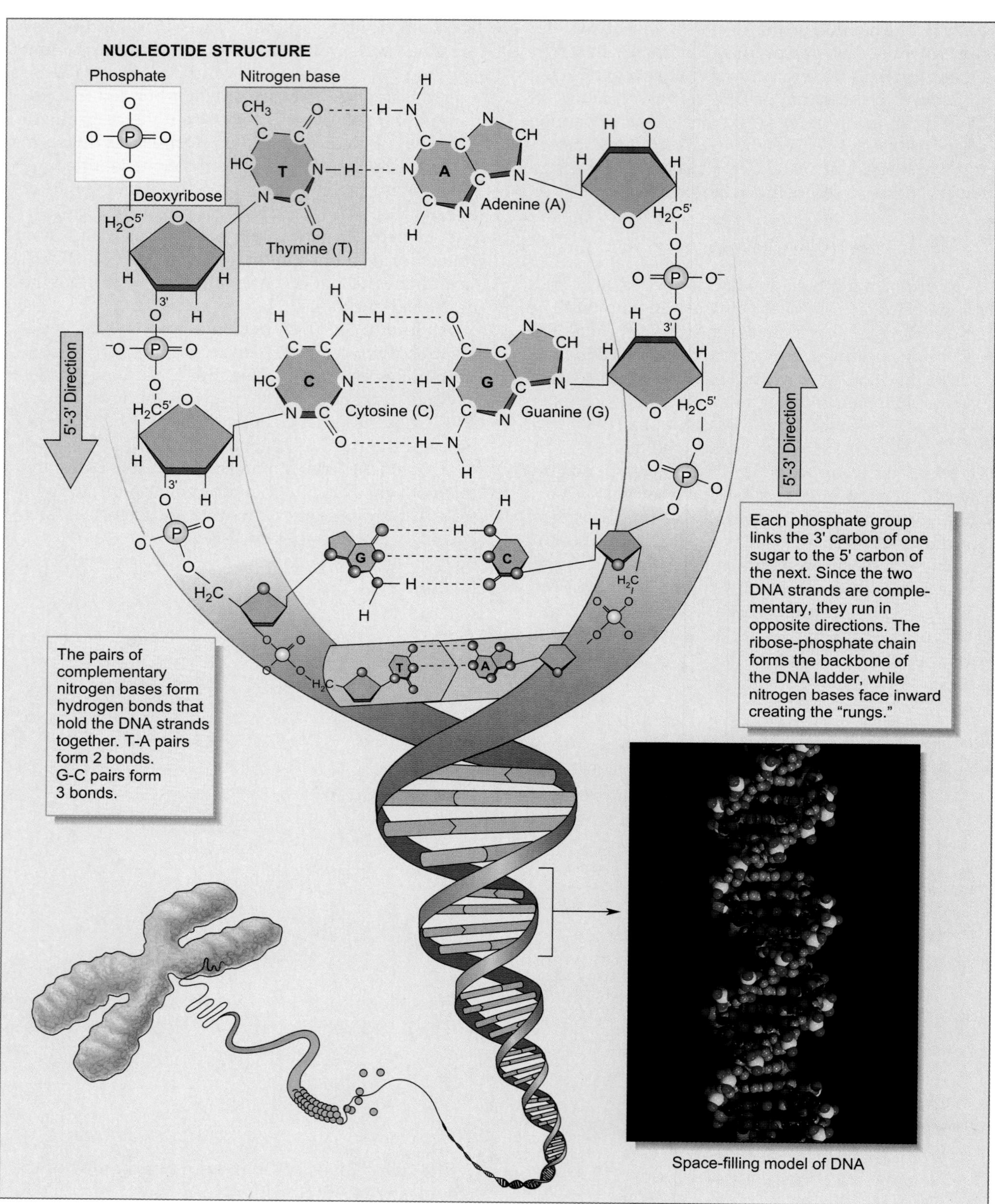
NUCLEOTIDE STRUCTURE
Phosphate
Nitrogen base
Deoxyribose
Thymine (T)
Adenine (A)
Cytosine (C)
Guanine (G)
5'-3' Direction
5'-3' Direction
Each phosphate group links the 3' carbon of one sugar to the 5' carbon of the next. Since the two DNA strands are complementary, they run in opposite directions. The ribose-phosphate chain forms the backbone of the DNA ladder, while nitrogen bases face inward creating the "rungs."
The pairs of complementary nitrogen bases form hydrogen bonds that hold the DNA strands together. T-A pairs form 2 bonds. G-C pairs form 3 bonds.
Space-filling model of DNA

Watson and Crick: The Great Synthesis

Let's consider first what Watson and Crick knew when they tackled the DNA problem in 1953. They knew: (1) that DNA was a polymer containing great numbers of nucleotides; (2) the chemical structures of the nucleotides; and (3) that the deoxyribose and the phosphate alternated in long chains, with the nucleotide bases hanging off the sides of each chain like signal flags on a single rope. (4) From Wilkins and Franklin's work they knew that DNA formed some kind of a helix, with three intramolecular distances: 0.34 nm, 2.0 nm, and 3.4 nm. (5) They knew Chargaff's rule: The number of adenines equalled the number of thymines, and the number of guanines equalled the number of cytosines. That was where everything came together.

Watson and Crick's understanding was based on real models made of wire, sheet metal, and (literally) nuts and bolts. These models were designed to provide a graphic representation of the DNA molecule and its constituent parts. They began to fiddle with these models, to see how the parts might fit together. (Sometimes the fingers can grasp what the mind cannot!)

Biological intuition and fitting the pieces this way and that seemed to indicate that there might be two strands wrapped around one another, with the phosphate-sugar backbone on the outside and the nucleotide bases inside, facing one another. But how were the bases arranged inside? Wilkins and Franklin's numbers began to make sense:

- 2.0 nm: their molecular model was consistent with the entire double helix being just 2.0 nm wide.
- 0.34 nm: the nucleotide bases were just about 0.34 nm thick, and perfectly flat. If the bases were stacked one on top of the other, like pennies in a roll, the layers would be 0.34 nm apart.
- 3.4 nm: with a gentle curve to the twisting helix, the double backbone would make one complete turn every 3.4 nm along the axis of the molecule; in other words, there would be exactly ten ($0.34 \times 10 = 3.4$) nucleotide pairs in each helical turn (Figure 14.5).

It was Watson who insightfully grasped the true meaning of Chargaff's mysterious rule. In the sheet metal and wire model, two purines would not fit opposite one another within the 2.0 nm confines of the double helix; and two pyrimidines would leave a gap. But one purine and one pyrimidine *could* fit opposite one another. Watson saw that adenine (a purine) and thymine (a pyrimidine) in a flat plane would form two hydrogen bonds with one another; guanine (a purine) and cytosine (a pyrimidine) would similarly form three hydrogen bonds in a plane (Figure 14.5). And this was the *only* way the DNA molecule would hold together. Opposite every adenine in one strand there had to be thymine in the other strand; every guanine in the one strand could be firmly hydrogen-bounded to a cytosine in the other strand, and so on. This was the key to DNA structure.

Watson and Crick's elaboration of the matchup between purines and pyrimidines, called **base pairing,** was one of their major findings. The linear order of nucleotides in any one strand might seem perfectly arbitrary, but whatever nucleotides were present in one strand rigidly fixed the sequence of nucleotides in the other strand. We say, then, that the two strands are *complementary*—they are not identical, but if you know the nucleotides on one strand you can easily deduce the nucleotides on the other strand. Thus Chargaff's empirical rules suddenly made logical sense: every A with a T, every T with an A, every G with a C, every C with a G. From this belatedly obvious conclusion sprang a whole new era in science, the era of molecular biology.

As we have said, the paired bases lie perfectly flat and are stacked one on another much like a stack of coins. The thickness of the layers and the molecular distances along the sugar-phosphate

Figure 14.5 DNA Structure.
In the space-filling model of the DNA double helix we can clearly distinguish the sugar-phosphate backbone of the two intertwining strands. The pairs of nitrogen bases are represented by lines of spheres lying stacked within the helix. Each measurement was determined through X-ray crystallography studies. Base pairing in DNA always occurs between adenine and thymine and between cytosine and guanine. The pairs are joined by hydrogen bonds *(dashed lines)*, and as you see, the bonding is very specific.

backbone produce the characteristic coiling of the molecule. The flat base pairs are not quite aligned with the one below, but are slightly offset, like the steps in a winding staircase.

The two chains of the DNA molecule run in opposite directions. That is, if you pictured a DNA molecule vertically on a page, one of the chains would appear upside-down relative to the other. (We can tell the "direction" of a DNA strand by noting the orientation of the sugars and phosphates, and recalling the numbering scheme for sugars. In short, the 5' end of a DNA strand has a phosphate attached to the 5' carbon of the sugar. The 3' end has simply the free—OH group.) Thus if one strand has its 3' end at the top of the page and its 5' end at the bottom of the page, the other strand will have its 5' end at the top and its 3' end at the bottom.

DNA molecules can be quite long. The average chromosomal DNA molecule in a human cell nucleus consists of about 140 million nucleotide pairs (two strands of 1.4×10^8 nucleotides each). However, because atoms are so small, a stretched-out DNA double helix of that size would be only about 5 cm long. But the 46 DNA molecules in each human cell nucleus total some 6.4 billion nucleotide pairs (6.4×10^9 pairs) which, stretched out end-to-end, would measure a little more than 2 meters!

DNA Replication

Watson and Crick very much appreciated a simple fact that had been known about genetic material since the beginning of studies on the nature of DNA. It must be able to direct its own replication. One of the things that pleased them most about their structure for DNA was that it set up a mechanism for making copies. In fact, they ended their classic 1953 paper by stating, "It has not escaped our notice that the specific pairing we have postulated immediately suggests a possible copying mechanism for the genetic material." Their suggestion was soon shown to be correct.

The complementary pairing of nucleotide bases in DNA suggests the analogy of a positive and a negative photographic film. The same information is present in both; one can make a positive print from a negative and a negative film from a positive. Similarly, a DNA "Watson" strand can be produced from a "Crick" strand (an "inside joke"—a Watson strand or a Crick strand can be either one of the two halves of a DNA molecule). In DNA replication the Watson and Crick strands unwind, a new Crick strand is produced from the information in the Watson strand, and a new Watson strand is made from the information in the Crick strand. Thus there are two complete DNA molecules where before there was only one. Since each new polymer contains one intact strand and one newly assembled strand, the process is called **semiconservative replication** (see Essay 14.1).

Replication Enzymes at Work

Note in Figure 14.6 that the first step in DNA replication is the unwinding of the double helix. This involves breaking the weak hydrogen bonds holding the two strands together. The process requires energy and a special "unwinding enzyme" called **helicase.** Helicase is joined by a second enzyme, **DNA polymerase,** to form a large **replication complex.** After helicase unwinds and opens the DNA polymer, DNA polymerase base-pairs each exposed nucleotide with a new complementary nucleotide, thus generating two new opposing strands. The energy comes from ATP and from the nucleotides themselves. Raw nucleotides occur in a nuclear pool as triphosphate nucleosides. Like ATP, each has two high-energy bonds available for use in replication.

Replication complexes work in pairs, proceeding in both directions along DNA, and forming two Y-shaped **replication forks** (see Figures 14.6 and 14.7). One of the new strands being assembled is known as the **lagging** (or discontinuous) **strand.** The names come from the fact that new DNA strands grow *only* from their 5' end toward the 3' direction. In the **leading strand**, nucleotides are added one after another in a smooth, *continuous* manner. But since the other strand runs in the opposite direction, assembly is discontinuous. Nucleotides are first assembled in short sections (in the 5' to 3' direction). With the help of another enzyme called **ligase,** the pieces, called **Okazaki fragments** (for their discoverer), are joined together into the lagging strand.

Replication in Prokaryotes and Eukaryotes

In eukaryotes each of the chromosomes consists of one very long DNA molecule, with a large number of histone spheres and other proteins. In prokaryotes most or all of the cell's DNA is in a circular molecule of DNA. It is not nearly as long as a typical eukaryote DNA molecule. In fact, you have about 1,400 times more DNA per cell than does the

Essay 14.1 Meselson and Stahl: Semiconservative Replication

Semiconservative replication predicts that a new double helix will be composed of one old strand and one new one. Hence the term "semiconservative"; half of each DNA molecule is conserved from the parent molecule. To test this prediction, Meselson and Stahl needed a way to tell the difference between new DNA and old DNA. What they did was grow bacteria on a medium that was composed of ^{15}N, an atom of nitrogen that makes the resulting DNA molecules more dense than normal. After several generations, all of the DNA within the bacteria was composed of DNA containing the heavier ^{15}N. Since DNA isolated from bacteria grown on normal ^{14}N is lighter, it is possible to separate the two kinds of DNA in a cesium chloride solution that sets up a density gradient when subjected to centrifugation. The ^{14}N DNA simply "floats" more than the heavier ^{15}N DNA.

The real test came when they took bacteria that had been growing in ^{15}N media, washed them, and let them grow in ^{14}N media for one generation. Semiconservative replication would predict that the DNA isolated from these bacteria would be of an intermediate density, composed of half ^{14}N DNA and half ^{15}N DNA. That is exactly what they found.

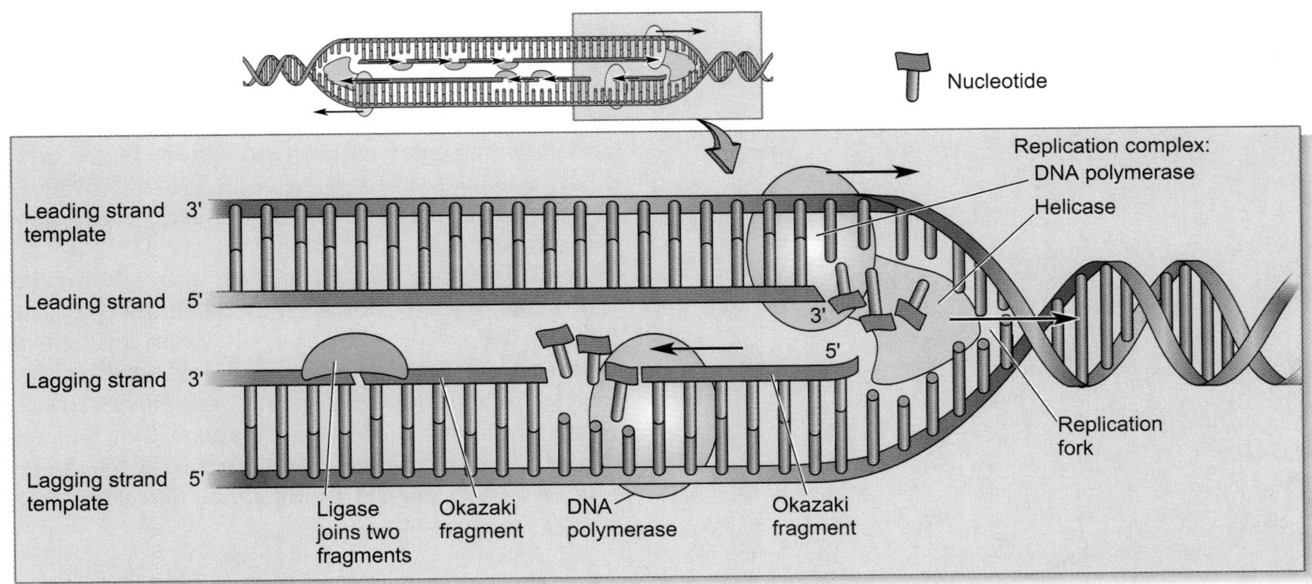

Figure 14.6 DNA Replication.
During replication the original DNA chain is opened by traveling replication complexes that unwind the helix, separate the opposing strands, and build two complete strands by base pairing. Note the use of ligase to seal Okazaki fragments in the lagging strand. At the completion of replication, each replica will consist of an old and a new strand. Replication is thus semiconservative.

average bacterium. As you might expect, these differences affect the way replication is done.

In prokaryotes replication begins at some specified spot in the circular DNA molecule, and just two replication forks form. The two travel in opposite directions, eventually backing into each other and completing replication (Figure 14.7). Replication in eukaryotes is not a single event. Numerous replication forks simultaneously produce hundreds of spreading, bubblelike replication events all along each chromosome. Eventually all of the bubbles run into each other, and the new replicas emerge (Figure 14.8).

How does a bacterium get along with just two replication forks? It's a matter of speed. Prokaryotes can add nucleotides at the rate of 1 million per minute. Eukaryotes are infinitely slower, adding only 500 to 5,000 nucleotides per minute. If they used just one replication site, it would require weeks instead of hours for most eukaryotes to complete DNA replication. *Drosophila,* the fruit fly, may be a eukaryote record holder. Using some 50,000 replication sites, its newly fertilized egg can complete replication of the four pairs of chromosomes in about three minutes.

DNA and Information Flow

We have seen that DNA can contain information in the sequence of nucleotides, and that it has the capacity to replicate itself precisely. These properties of DNA are a direct result of its structure. Yet as the story of DNA began to unfold, it was realized that somehow the information held within DNA *must* be expressed as proteins. After all, it is the proteins of a cell that give it structure and enable it to perform its functions.

As it turns out, the world was not ready for the first major contribution to the biochemistry of the gene. In 1908, A. E. Garrod, a physician influenced by Mendel's work, published a book called *Inborn Errors of Metabolism.* His subject was *Homo sapiens,* which, at the time, was an organism that had received little attention from genetic researchers. Garrod was interested in metabolic defects, break-

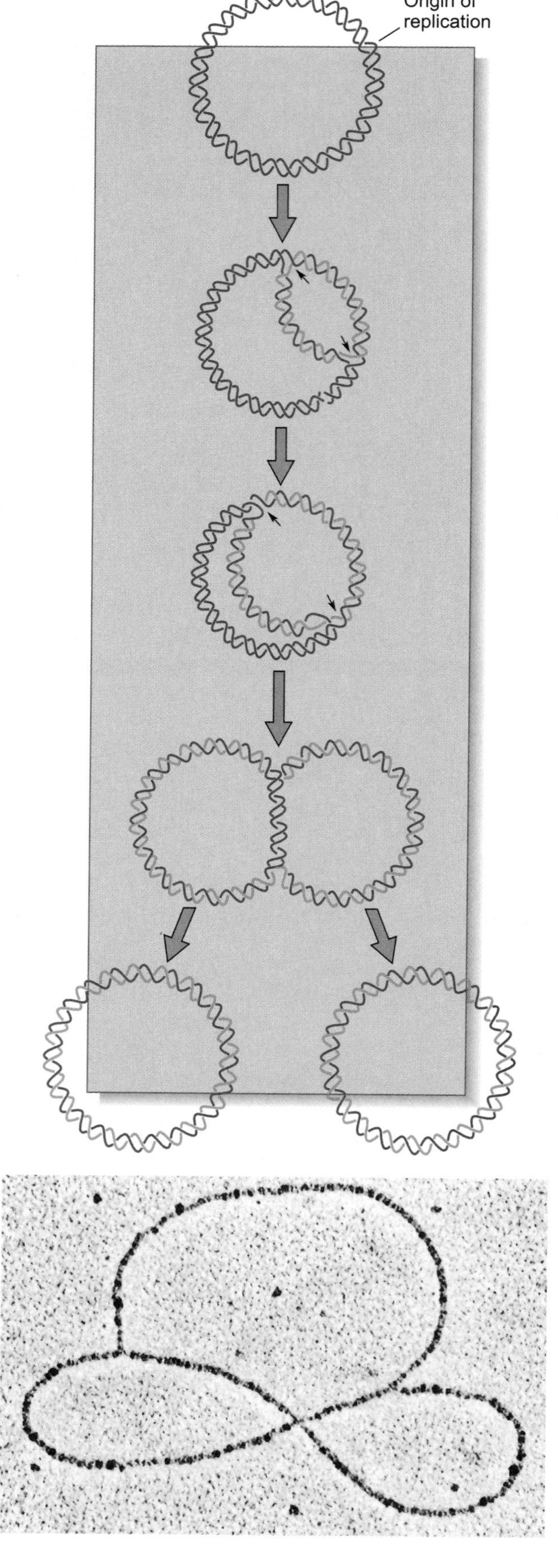

Figure 14.7 Replication of Circular DNA. Many prokaryotic chromosomes are circular. As replication proceeds, the two replication forks travel in opposite directions, eventually meeting at the opposite side of the circle to produce the two replicas.

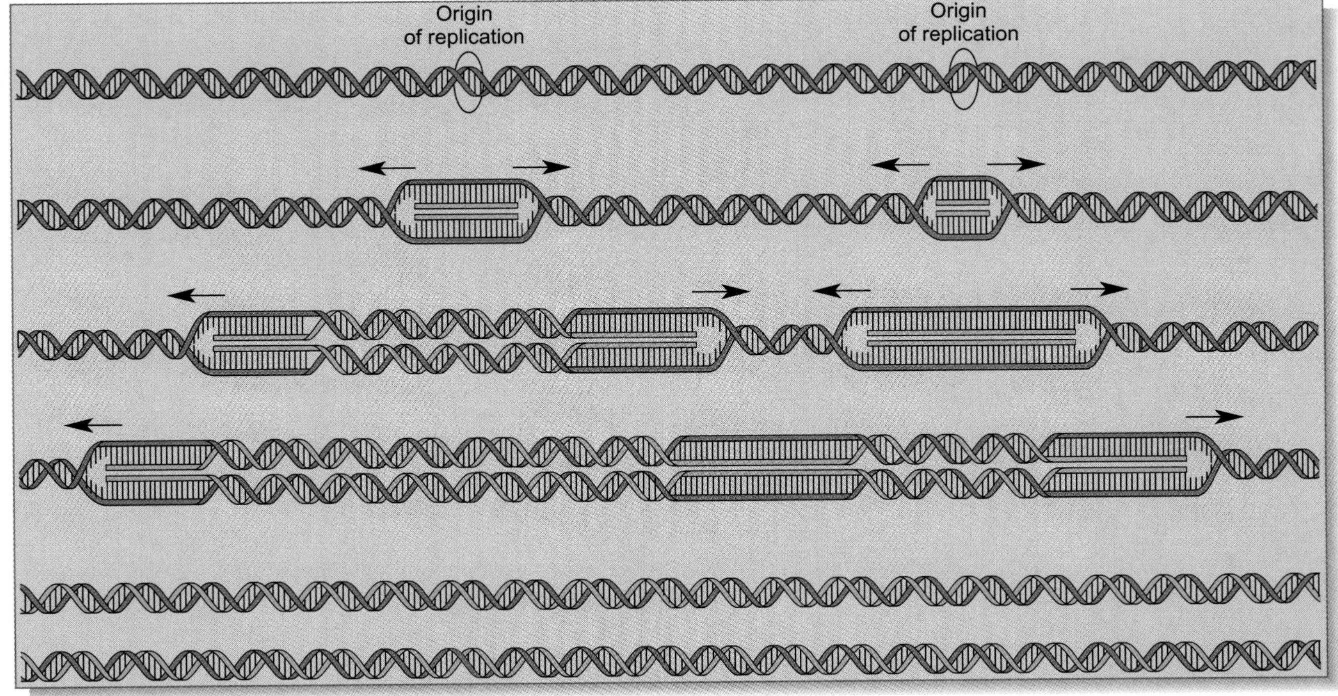

FIGURE 14.8 SIMULTANEOUS REPLICATION.
The task of replicating all of the very long DNA molecules in a eukaryotic nucleus is immense. Part of the problem is solved by having replication begin in many places, as these several replication bubbles on one chromosome demonstrate.

downs in the complicated biochemical processes of life. He searched for the abnormal products of such defects in the urine, where many metabolic products end up. Of special interest to Garrod was alkaptonuria, a disease in which the urine contains products of metabolism that turn black, and so are easily revealed. Infants with alkaptonuria are usually detected as soon as their diapers start turning black. As the child grows older, black pigments begin to settle in cartilage and other tissues, blackening the ears and even the whites of the eyes. Another more serious effect is a form of arthritis, caused by the accumulation of the metabolite in the cartilage of the joints.

Garrod observed that the disease tended to be found in brothers or sisters in a single family. By studying family histories, he correctly inferred that alkaptonuria and certain other inborn errors of metabolism were genetic in origin. The problem, he deduced, is caused by the absence of specific enzymes that are necessary for the long chains of biochemical reactions to occur. If an enzyme for a particular reaction is absent, no reactions can take place past the point where the enzyme is normally a part of the chain, and the substance that the enzyme acts on builds up.

A →	B →	C →	D
enzyme 1	enzyme 2	enzyme 3	

Normal

A →	B → ✕ →	(No C or D)
enzyme 1	no enzyme 2	Buildup of B

Abnormal

Other inborn errors of metabolism create albinism (a complete or partial lack of melanin pigment) and phenylketonuria (which also affects hair and skin pigmentation but has a much more severe effect on mental development because of the accumulation of toxic metabolites in the nervous system). As it turns out, albinism, phenylketonuria, and alkaptonuria are all caused by defects in enzymes that act in the metabolism of the amino acids phenylalanine and tyrosine (Figure 14.9).

Significantly, Garrod discovered that heredity played a role in enzyme activity and that there was a definite connection between heredity and the

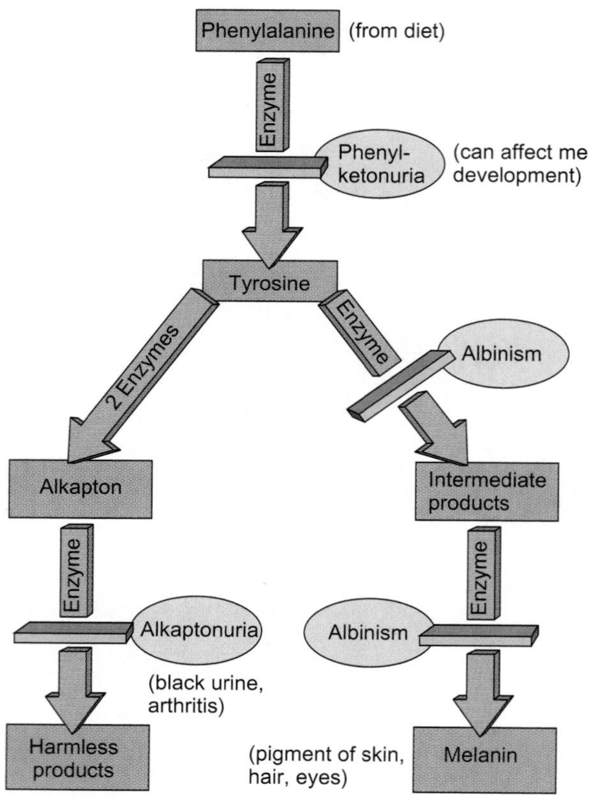

FIGURE 14.9 THE GENETICS OF A METABOLIC PATHWAY.
Garrod's work established the relationship between genes and enzymes. Three conditions caused by the blocking of enzymes are albinism, alkaptonuria, and phenylketonuria. All the blocked enzymes operate on intermediate breakdown products of the amino acid phenylalanine. Look again at Figure 13.1. The relationship between loss of an enzyme and the recessive nature of albinism should now be quite clear.

presence of normal or abnormal enzymes—quite an accomplishment for his day. But he was essentially ignored. In fact, it wasn't until 30 years later that geneticists George Beadle and Edward Tatum extended his ideas to the concept of "one gene, one enzyme," firmly establishing the link between genes and protein (enzymes, recall from Chapter 2, are proteins). In Garrod's time biochemical science was still emerging, and geneticists were more interested in how the genes influenced morphology.

What, then, are genes? There is no easy definition. But consider that their expression results in genetic traits, they are inherited as factors with specific locations on chromosomes, they are encoded by segments of DNA, and they act through proteins.

So, even after the discovery of the genetic secrets of DNA, the remaining issue was, How does the information stored in the DNA molecules get expressed as proteins? That issue would lead researchers finally to the core of gene structure and function. The conceptual framework for the expression of genes was provided by Watson and Crick soon after their discovery of the structure of DNA, and has become known as the Central Dogma of molecular biology. (In science, there are no dogmas, so this was a little joke by biologists at that time.) The primary aspect of the Central Dogma is that DNA (because of the complementary strands) directs its own replication. The remainder of the Central Dogma concerns how the genetic information stored in DNA is finally expressed as proteins, and is the subject of the next chapter.

PERSPECTIVES

How is it that DNA is capable of being the genetic material? It became fairly obvious that DNA could carry information, simply in the sequence of nucleotides that were contained in the strands of the double helix that we have just described. The question then arose, How does that information end up as proteins that make up a phenotype? We will soon see that it is the sequence of nucleotide bases that contain the genetic information that is expressed as proteins, and therefore genes can be simply viewed as segments of DNA.

It was Crick who made the most contributions to the conceptual leap between DNA structure and DNA function. Before there were data to support the notions, Crick proposed the Central Dogma, establishing the flow of genetic information from DNA to proteins. In the next chapter, we will see just how accurate Crick's ideas were.

Summary

Discovering DNA: The Stuff of Genes

Miescher isolated DNA in 1869, naming it nuclein and describing its phosphate-rich quality. Its acidic nature and the two forms of nucleic acid, DNA and RNA, were described in the late 1800s.

Griffith found that if live cells from a harmless strain of bacteria were mixed with dead cells of a virulent, pneumonia-causing strain, transformation would occur and the harmless strain would become virulent.

Avery essentially repeated the experiments of Griffith, but used purified DNA instead of dead cells from the virulent source. He determined that the transforming substance was DNA and that DNA had genetic capabilities.

Hershey and Chase knew that bacteriophages injected their genes into host bacterial cells. They introduced into bacterial cultures bacteriophages whose proteins had incorporated radioactive sulfur and whose DNA had incorporated radioactive phosphorus. Allowing time for gene injection, they dislodged the viruses and then determined that it was the DNA that entered the cells whereas the protein (viral coats) remained outside. DNA was again shown to be the genetic material.

Gene Structure: Closing In on DNA

Through a study of nitrogen bases from the DNA of many kinds of cells, Chargaff determined that adenine, thymine, guanine, and cytosine occur in specific proportions: A = T and G = C (Chargaff's rule). This relates to the fact that purines and pyrimidines always pair in DNA.

Using a technique called X-ray crystallography, Wilkins and Franklin studied the molecular structure of DNA, making specific measurements and determining that it was helical and double-stranded.

Each nucleotide is made up of deoxyribose (a sugar), a phosphate ion, and one of four nitrogen bases. The pyrimidines (thymine and cytosine) are single rings of nitrogen and carbon, whereas the purines (adenine and guanine) are double rings.

Watson and Crick: The Great Synthesis

From the available information on DNA structure, Watson and Crick deducted its molecular structure.

In developing their nuts-and-bolts model of DNA, Watson and Crick determined the meaning of X-ray studies: the molecule was 2.0 nm wide, its base pairs 0.34 nm thick, and a complete turn of the helix 3.4 nm (10 base pairs). From Chargaff's rule, they derived base-pairing, the idea that each pyrimidine in the DNA chain is opposed by a purine; specifically adenine opposes thymine and guanine opposes cytosine. The bases lie in a flat plane, each hanging off of the sugar-phosphate backbone.

Because of base pairing, the order of bases in one strand determines the order of bases in the other. Each strand has a 5' (phosphate) end and a 3' (deoxyribose) end. Further, one strand is upside-down relative to the other. The average human DNA molecule has 140 million base pairs.

DNA Replication

In replication the double helix unwinds, and the strands separate. The order of bases in each strand is followed in assembling a new partner strand. The two DNA molecules formed are replicas of the former DNA molecule. Because the new DNA molecule is made up of one "old" strand and one "new" strand, replication is called semiconservative.

Replication requires a replication complex consisting of the unwinding enzyme (helicase) and the nucleotide-assembling enzyme (DNA polymerase). Replication complexes proceed in opposite directions, forming two replication forks. The assembly of nucleotides occurs in the 5' to 3' direction, a requirement that results in a different means of assembly in the leading and lagging strands.

Eukaryotes form numerous replication forks, carrying on simultaneous replication. Prokaryotes form just two replication forks that travel in opposite directions around the circular chromosome.

DNA and Information Flow

Garrod provided pioneering insight into the fact that genes must act through proteins (enzymes). His conclusions were based on observations of inherited diseases, which he determined to be caused by enzyme deficiencies. Later, Beadle and Tatum provided further evidence that each gene acted through a protein ("one gene, one enzyme").

Key Terms

transformation • **211**
Chargaff's rule • **214**
purine • **214**
pyrimidine • **214**
nucleotide • **215**
base pairing • **217**
semiconservative replication • **218**
helicase • **218**
DNA polymerase • **218**
replication complex • **218**
replication fork • **218**
lagging strand • **218**
leading strand • **218**
ligase • **218**
Okazaki fragment • **218**

Review Questions

1. List two DNA discoveries that were made before the 1900s.
2. Summarize the observations on transformation made by Griffith. How was the transforming substance finally identified?
3. Describe a bacteriophage, and explain how it enters a bacterial cell and how it reproduces.
4. Explain the experimental procedure that led Hershey and Chase to conclude that the genetic material was DNA. Did they prove this to be true of all life? Explain.
5. What did Chargaff find about the amounts of the four nitrogen bases in cells? Did he immediately understand the meaning of these quantities?
6. Summarize the findings of Wilkins and Franklin. What technique did they apply to study DNA?
7. List the molecular subunits of a nucleotide. What is important about the 1', 3', and 5' carbons of the sugar? How does one DNA nucleotide differ from the next?
8. Name the four DNA bases and state whether they are purines or pyrimidines.
9. How did Watson and Crick interpret the molecular distances 2.0 nm, 0.34 nm, and 3.4 nm discovered earlier by Wilkins and Franklin?
10. Explain the specific pairing of bases in DNA. Summarize the intuitive thinking that led Watson to the discovery of base pairing.
11. Describe the arrangement of nucleotide bases along the DNA molecule and the orientation of the two strands.
12. What did the double-stranded structure of DNA and the newly discovered base-pairing principle suggest to Crick about DNA replication?
13. Explain how the structure of DNA might make it an informational molecule.
14. Using the following nucleotide sequence, show how DNA replication occurs (ignore the replication complexes):

    ATGCCGTTATAATCG
    TACGGCAATATTAGC

15. Name two enzymes of the replication complex and summarize their roles.
16. Briefly explain why Okazaki fragments occur in replication of the lagging strand. How does this differ from replication in the leading strand?
17. Summarize a major difference between prokaryote and eukaryote replication. Suggest a reason for the difference.
18. Explain how Garrod made the connection between genes and metabolic disorders such as phenylketonuria.

15

How Genes Are Expressed

WE HAVE SEEN THAT MENDEL'S FACTORS ARE ACTUALLY SEGMENTS OF *the DNA molecule within a chromosome. So DNA is the genetic material, the stuff entrusted with carrying the genetic information from one generation to the next. But we have also seen that it is proteins that are actually responsible for phenotypes. The question arises, then, How does the genetic information that is carried in DNA end up as proteins?*

We have followed the history of science's long search for the stuff of heredity. We saw it unveiled as DNA and found that the key to its signals to the cell might be the sequence of the nucleotides along its length. We have also found that DNA replicates, and that the precision of the replicating mechanisms explains the constancy of the DNA from one generation to the next as well as patterns of inheritance. So let's now see how this fascinating molecule dictates the business of the cell by encoding enzymes and other proteins.

We will find that genes encode proteins in an indirect manner (Figure 15.1). Although the protein-specifying code is preserved in DNA, the information in the code works through intermediaries. These intermediaries are molecules of ribonucleic acid, or RNA. We will see that the process occurs in two steps. The first step is *transcription* of the genetic information from DNA into RNA. The second step is the *translation* of the RNA information into protein. We will then look at how a gene is controlled—told when and where it is to be expressed.

THE RNA MOLECULE

The discovery that genes are composed of DNA and that their information lies in the linear arrangement of DNA nucleotides was an enormous breakthrough, but it also left many questions unanswered. For example, exactly how are the genetic instructions stated? And how do they determine the order of amino acids in proteins? It didn't take researchers long to determine that DNA is not *directly* involved in making proteins. It couldn't be, because in eukaryotes DNA remains in the nucleus, while protein assembly occurs in the cytoplasm. So DNA directs protein synthesis by long distance. But how?

The answer, quite simply, is: by sending its instructions on molecules of RNA.

DNA and RNA are very similar. Both DNA and RNA are lengthy polymers of nucleotide subunits that are held together by covalent bonds between phosphates and sugars. The nucleotide bases of both extend from the familiar sugar-phosphate backbone like signal flags on a line. But in spite of their fundamental similarity, there are marked differences between them.

1. In RNA, the five-carbon sugar is *ribose* instead of *deoxyribose.* As you see in the molecule shown below, the 2' carbon of RNA has a single hydrogen side group, but opposing it is a hydroxyl (OH) side group. So deoxyribose has one less oxygen atom than ribose (*deoxy* refers to "less oxygen").

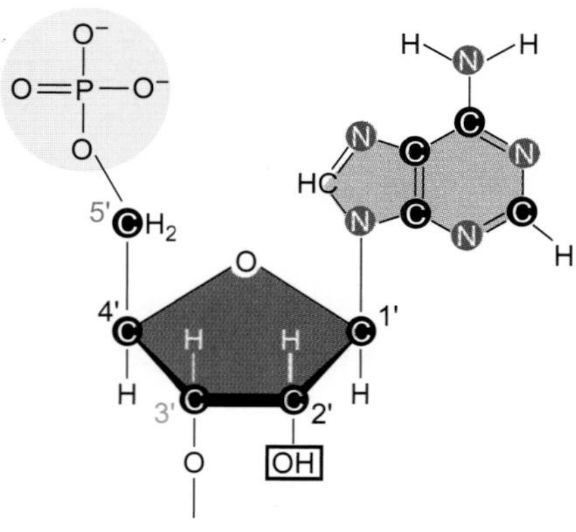

Ribose (RNA)

2. Both RNA and DNA contain adenine (A), guanine (G), and cytosine (C), but the thymine (T) of DNA is replaced in RNA by a base called **uracil** (U). Uracil is structurally very similar to thymine and base-pairs with adenine in the same way.

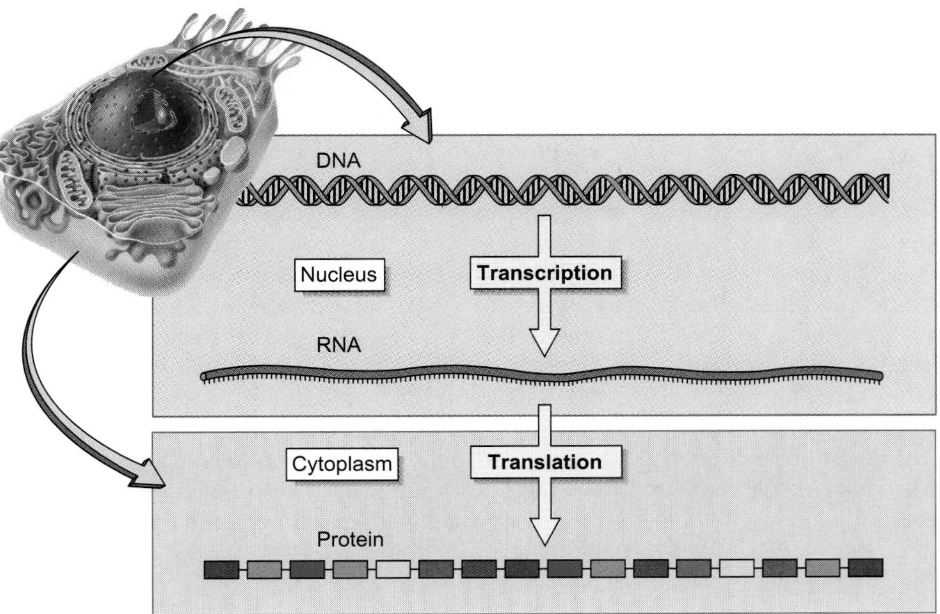

FIGURE 15.1 **FLOW OF GENETIC INFORMATION.**
This scheme of information flow is the cornerstone of our understanding of the way in which the hereditary data is stored and replicated as DNA, yet enacted and expressed as protein. The three processes involved are replication, transcription, and translation. We have already discussed replication in Chapter 9. Transcription is the process by which the information stored in DNA is converted to RNA, while translation (which quite logically involves changing languages—from nucleotides to amino acids) is the process that converts the information from RNA into protein.

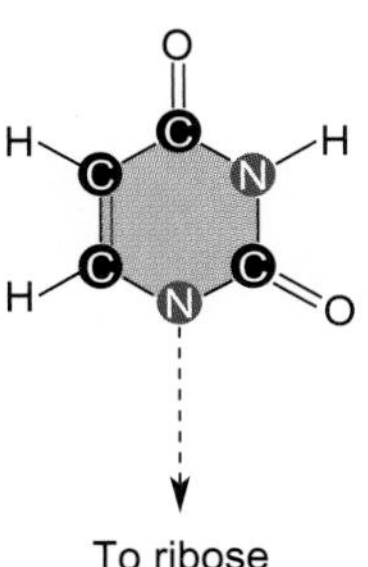

Pyrimidine base: Uracil

3. DNA almost always occurs as a double-stranded helix with a very regular structure. RNA almost always occurs as a single-stranded molecule, and some forms have complex secondary (coiled) and tertiary (folded) structure, twisting and folding on themselves, much as proteins do, forming a precise three-dimensional shape.
4. DNA molecules are typically much longer than RNA molecules. DNA molecules are millions of nucleotides long, whereas the RNA nucleotides number only in the thousands or less.
5. DNA is more stable than RNA—that is, it is more resistant to chemical breakdown.
6. There are several different classes of RNA molecules, each with different functions, whereas DNA functions in storing and transmitting information.

TRANSCRIPTION: THE SYNTHESIS OF RNA

Transcription is the process of making an RNA copy of one strand of the DNA molecule, and is at first reminiscent of DNA replication. DNA is unwound and its bases are exposed, just as if it were about to begin replication. Instead of each strand of DNA making a complementary strand, however, one strand lies dormant while short sections of the other act as a template for the formation of RNA. And, as we mentioned, uracil is used instead of thymine in RNA; but all base-pairing rules remain the same. During RNA synthesis, uracil pairs with DNA's adenine all along the strand being transcribed. After the copying is over, the RNA is called a **transcript** and drifts free while the DNA rewinds (Figure 15.2).

Just as the replication complex of DNA synthesis consists of the enzymes helicase and DNA polymerase, transcription is accomplished by a **transcription**

complex consisting mainly of the enzyme **RNA polymerase.** In eukaryotes the length of DNA along which a single RNA molecule is transcribed is roughly equivalent to a gene.

Transcription begins at special DNA sequences called **promoters** that bind RNA polymerase, and ends at termination sequences that release RNA polymerase from the DNA. As soon as the first few bases of an RNA sequence have been formed, the DNA and RNA strands begin to separate, and the growing RNA chain, or *transcript,* begins to dangle off to the side. The DNA to which it was attached is then free for further transcription. Many RNA molecules can be transcribed in a very brief time from the same DNA transcription unit. This is known as simultaneous transcription (Figure 15.3).

Three Kinds of RNA

We have stated, in general terms, that RNA plays a vital role in the process that results in genes being expressed as proteins. Let's now examine the different classes of RNA to see what specific role each of them plays. Once we have described these players, we will put them all on stage in an unfolding pageant that describes the process itself.

The three major kinds of RNA in most cells are *ribosomal RNA (rRNA), transfer RNA (tRNA),* and *messenger RNA (mRNA).* Each is transcribed from DNA, as described above. Ribosomal RNA and transfer RNA may be thought of as bit players, but in their supporting roles they function in the expression of

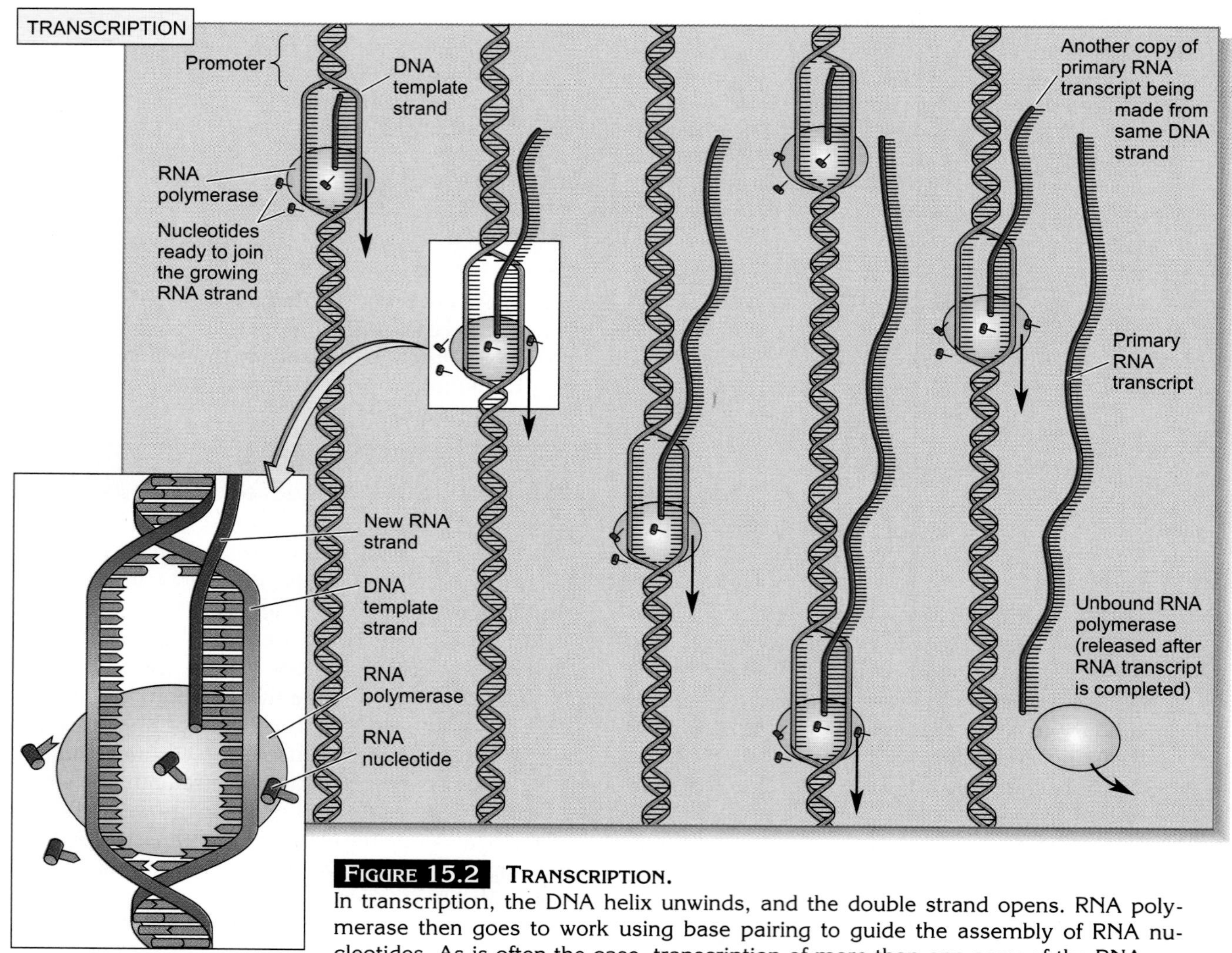

Figure 15.2 Transcription.
In transcription, the DNA helix unwinds, and the double strand opens. RNA polymerase then goes to work using base pairing to guide the assembly of RNA nucleotides. As is often the case, transcription of more than one copy of the RNA occurs at the same time.

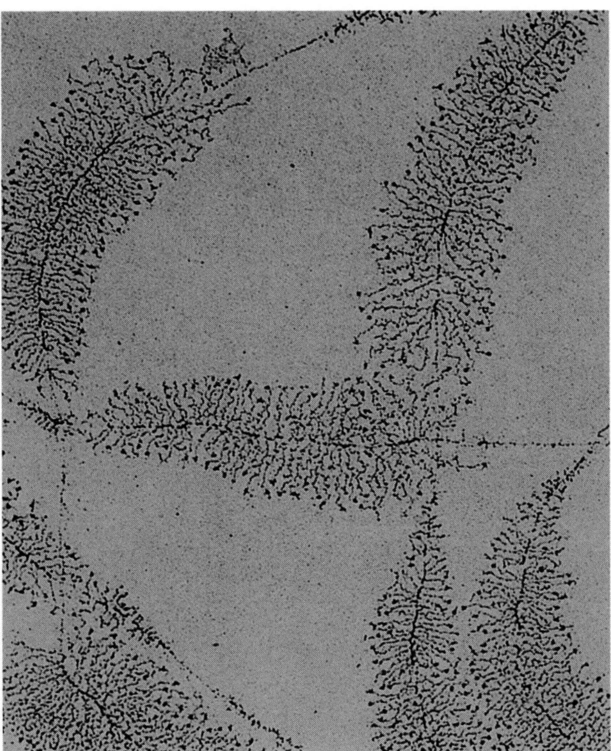

FIGURE 15.3 SIMULTANEOUS TRANSCRIPTION. Evidence of simultaneous transcription is seen in chromosomes from an amphibian egg as the cell builds its protein reserves. The shorter, feathery strands of RNA are just starting, while the longer ones are well along in the transcription process.

virtually every gene. **Ribosomal RNA** is the main constituent of ribosomes (the site of protein synthesis, as we will see). **Transfer RNA** is responsible for bringing the proper amino acid to be put into the protein. Messenger RNA, however, is the star. It is in the direct line of information flow between DNA and protein. As its name implies, **messenger RNA** carries the coded message from the gene that will determine the polypeptide to be produced in the cytoplasm. The messenger RNA for each gene is unique to that gene, so there are many thousands of different mRNAs.

MESSENGER RNA AND THE GENETIC CODE

Messenger RNA carries the specific information of a gene that has been stored in the sequence of nucleotides in DNA. In DNA that information is coded in the form of nucleotide base sequencing—the linear order in which A, T, C, and G are arranged in DNA strands. The equivalent sequence is copied into RNA, with U replacing T.

RNA Processing: Introns and Exons in Eukaryotes. Raw mRNA starts out as a transcript—a complete copy of a DNA transcription unit (or gene). Before this RNA transcript can be considered mRNA and before it can move out into the cytoplasm, it must undergo **posttranscriptional processing.** One of the first steps is adding a string of A's to the 3' end (Figure 15.4). This is called the *poly A tail.* The processed mRNA contains the coded message that will then undergo *translation.* Recall that **translation** is the synthesis of protein from the message contained in mRNA. The portion of mRNA to be translated into protein is preceded by a *5′ leader* and a *3′ trailer.* Eukaryote genes typically include **introns**—*intervening sequences* that must be removed as RNA is processed. After their removal, the introns are simply disassembled into nucleotides and recycled for further use. What remains are called the **exons,** or *expressed sequences.* They comprise the actual genetic message.

Introns, then, are DNA sequences within genes that are removed from the message. Curiously, introns are not trifling amounts of DNA. For instance, they make up 85% of the DNA in a gene known to code for ovalbumin, a protein of bird eggs. A new mRNA transcript may contain as many as 50,000 nucleotides, but after processing it is reduced to only about 2,000 on average. Some of the clipped-out introns contain as many as 10,000 nucleotides—quite a stretch of RNA. The removal of introns is a precise and complex process accomplished by an RNA-Protein complex called a *spliceosome* (Figure 15.4).

There is no proven advantage to the presence of introns, although biologists are willing to guess. Some think that introns may be adaptive in that they absorb spontaneous chemical change. Thus, some potentially damaging mutations would be actually removed before protein synthesis begins. Other biologists maintain that introns greatly facilitate successful crossing over during meiosis. With the gene effectively divided into many "domains," they say there are simply many more chances for recombination to occur. This increased recombination helps the species to maintain a certain level of genetic variability.

Codons. The message from the gene that produces proteins is encrypted in what is called the **genetic code.** The messenger RNA molecule carries its message in the form of **codons,** which are groups of three nucleotides. Each codon corresponds to one of the 20 different amino acids commonly found in

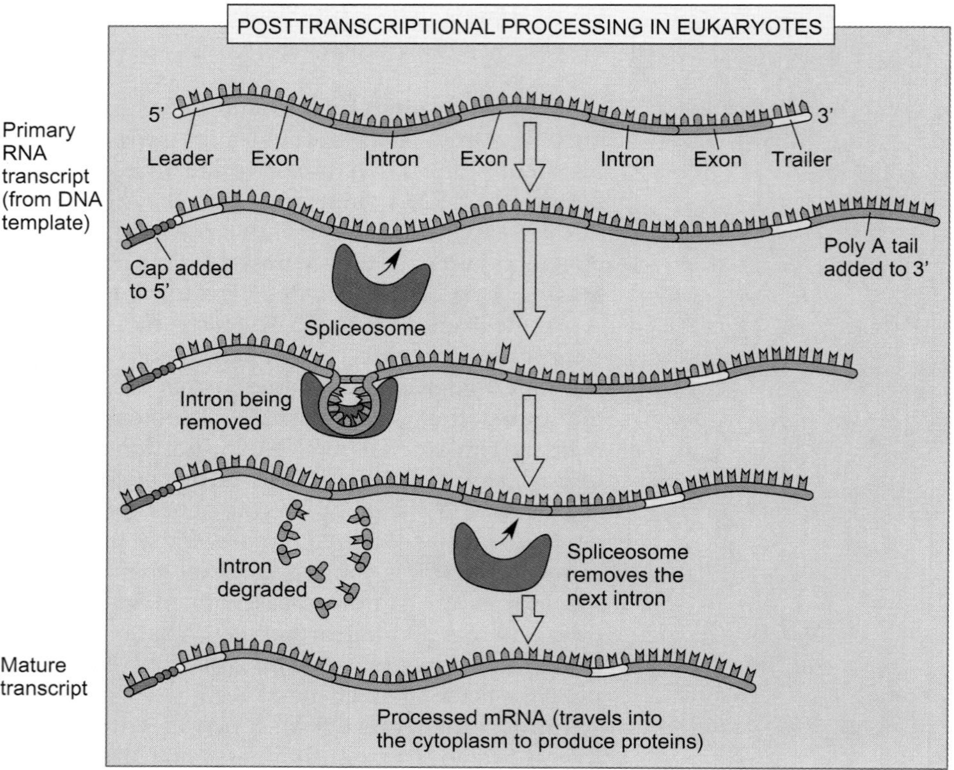

FIGURE 15.4 PROCESSING MESSENGER RNA.
Eukaryotic mRNA differs from prokaryotic mRNA in that eukaryotic mRNAs are processed after transcription, through steps involving adding a cap to the 5' end, adding a poly A tail to the 3' end, and removing introns. Prokaryotic mRNAs do not undergo this processing.

proteins. For example, the three-nucleotide sequence GAG (guanine-adenine-guanine) is a codon specifying *glutamic acid.* The 64 codons (representing all of the ways in which four kinds of nucleotides can be arranged in a sequence of three) are shown in Table 15.1, together with the amino acids they specify. The unveiling of the *genetic code* is a major discovery of 20th-century science.

To understand better how the code works, let's consider the *principle of colinearity,* a concept that Francis Crick so elegantly proposed long before much was known about RNA. DNA, RNA, and protein are each linear molecules consisting of subunits, and there is a clear relationship among the units. According to the **principle of colinearity,** the linear ordering of nucleotides in DNA specifies the order of codons in mRNA, and the linear ordering of codons in mRNA specifies the linear ordering of amino acids in the protein (Figure 15.5).

It is now well established which amino acids are specified by which codons, but working out the genetic code required the efforts of some of the best minds in molecular biology and a great deal of painstaking experimental verification. Once the genetic code table was finally worked out, the code proved to be exactly the same for such diverse groups as humans, bacteria, and yeast—underscoring the basic relationships and unity of life—and so it is often known as the "universal" genetic code. (However, it has been shown that our very own mitochondria use a slightly different genetic code—and yeast mitochondria have yet other slightly different codon assignments; but these are more like dialects of the genetic language, not new languages.)

A Closer Look at the Genetic Code. You have probably wondered why there are 64 codons to code for only 20 amino acids. Obviously, some of the amino acids must be coded for by more than one codon. That is indeed the case, and different codons that specify the same amino acid are called **synonymous codons.** As you can see in Table 15.1,

TABLE 15.1 SYNONOMOUS CODONS

First base	Second base: U	Second base: C	Second base: A	Second base: G	Third base
U	UUU Phe	UCU Ser	UAU Tyr	UGU Cys	U
	UUC	UCC	UAC	UGC	C
	UUA Leu	UCA	UAA Stop	UGA Stop	A
	UUG	UCG	UAG Stop	UGG Trp	G
C	CUU Leu	CCU Pro	CAU His	CGU Arg	U
	CUC	CCC	CAC	CGC	C
	CUA	CCA	CAA Gln	CGA	A
	CUG	CCG	CAG	CGG	G
A	AUU Ile	ACU Thr	AAU Asn	AGU Ser	U
	AUC	ACC	AAC	AGC	C
	AUA	ACA	AAA Lys	AGA Arg	A
	AUG Met Start	ACG	AAG	AGG	G
G	GUU Val	GCU Ala	GAU Asp	GGU Gly	U
	GUC	GCC	GAC	GGC	C
	GUA	GCA	GAA Glu	GGA	A
	GUG	GCG	GAG	GGG	G

Amino acid abbreviations: alanine, Ala; arginine, Arg; asparagine, Asn; cysteine, Cys; glutamic acid, Glu, glutamine, Gln; glycine, Gly; histidine, His; isoleucine, Ile; leucine, Leu; lysine, Lys; methionine, Met; phenylalanine, Phe; proline, Pro; serine, Ser; threonine, Thr; tryptophan, Trp; tyrosine, Tyr; valine, Val.

most (but not all) synonymous codons come in blocks and differ only in the third position.

Three of the 64 codons do not specify amino acids at all but indicate STOP. These are UAA, UAG, and UGA; they are, in effect, punctuation marks, and translate into "this is where to end the polypeptide." There is also a START codon, AUG; it also specifies the amino acid methionine. This means that all newly synthesized polypeptides have to start with methionine. If a methionine in the first position of a protein doesn't suit the needs of the organism—and apparently it often doesn't—then it will have to be removed enzymatically later on. Since AUG is the only codon for methionine, when it occurs in the middle of a message, it is ignored as a START codon and is simply read as a methionine-specifying codon.

We've seen that the genetic message is transcribed from DNA into mRNA, which, in turn, carries the encoded message to the cytoplasm for translation. Before we can understand how the sequence of codons in mRNA is translated into a sequence of amino acids in a polypeptide, we must look at the other two types of RNA involved in protein synthesis: ribosomal RNA and transfer RNA.

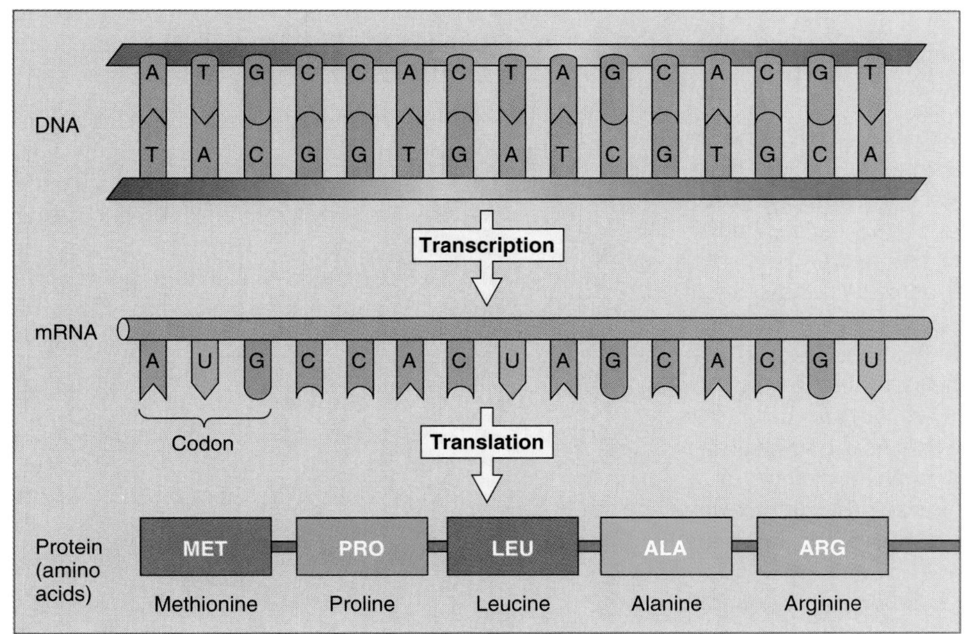

FIGURE 15.5 THE CONCEPT OF COLINEARITY AMONG DNA, MRNA, AND PROTEIN. The triplet nucleotides of DNA are transcribed into the codons of mRNA, which in turn specify the order and kind of amino acids to be inserted into protein. This specification of order occurs through the action of tRNA, which has an anticodon sequence and also holds an amino acid.

RIBOSOMAL RNA AND THE RIBOSOME

Ribosomal RNA (rRNA) occurs in ribosomes. Unlike other RNA, most of it is formed within the nucleolus. This region of the nucleus contains multiple copies of rRNA genes carrying the ribosomal RNA coding sequences.

Large and small rRNAs are joined by 75 to 80 special protein molecules forming the ribosome. Ribosomes are the vital organizing centers that assist in assembling polypeptides. Ribosomes have been likened to gigantic enzyme complexes (or to mobile factory workbenches), but they are like no other molecular complex or cellular organelle in all of life. Intense electron microscopic and biochemical studies of ribosomes have begun to clarify their unique structure.

Ribosomes have two distinct parts—a *large subunit* and a *small subunit,* each with its own characteristic shape (Figure 15.6). Each functioning (assembled) ribosome contains several important attachment sites. One site holds the lengthy mRNA in a groove formed between the two subunits. Another site binds an enzyme active in linking pairs of amino acids. There is an mRNA binding site, important to the initial activation of the ribosome. Finally, there are two sites that are pockets in which transfer RNAs with amino acids fit as they are assembled.

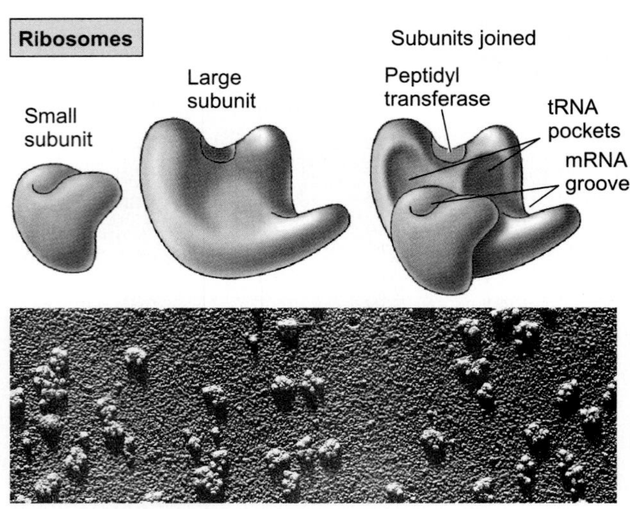

FIGURE 15.6 RIBOSOMES.
The two ribosome subunits join during the process of protein synthesis. Each ribosome has three attachment sites, identified here as two tRNA binding pockets and a groove to which the leader of an mRNA will attach when translation begins. In addition, each ribosome has a site where peptidyl transferase operates. This enzyme is instrumental in joining new amino acids to the growing strand. Ribosomes, as seen in the electron micrograph, often appear in clusters.

TRANSFER RNA

We've seen that messenger RNA carries the instructions for polypeptide assembly to the cytoplasm and that ribosomal RNA is used in constructing ribosomes, the "traveling workbench" upon which the actual assembly occurs.

Transfer RNA, in a sense, gets the amino acids over to the workbench. There is a type of tRNA for each of the codons (except the STOP codons). If you think of messenger RNA as the assembly line and ribosomes as the workbench where workers assemble the product, then perhaps the tRNAs are specialized forklifts, scurrying back and forth between stockroom and assembly line delivering the parts to be assembled. Let's see how all of this really happens in the cell.

Transfer RNA's role in the assembly of the polypeptide can be broken down into three parts:

1. Each tRNA becomes *charged;* that is, it is enzymatically attached to its own specific amino acid.
2. Charged tRNAs join binding sites on the ribosome where a polypeptide is being assembled.
3. Charged and bound tRNAs recognize a specific codon in mRNA and present their amino acids to be assembled accordingly.

The proper connections between tRNA and specific amino acids are made by **charging enzymes.** There is a highly specific charging enzyme for each variety of tRNA. Each enzyme has three binding sites, one for the tRNA, a second for the proper amino acid, and a third for ATP. The enzyme uses the energy of ATP to form the bond connecting the amino acid to the tRNA.

Transfer RNA is formed from a linear strand of about 100 nucleotides whose sequence is transcribed from tRNA genes in DNA just as mRNA sequences are transcribed from other genes. After tailoring, the strand becomes twisted and folded into a three-dimensional shape, forming three "loops" and a "stem." In its finished form the whole molecule takes on sort of an upside-down L shape (Figure 15.7).

The stem and loops in the final tRNA configuration play individual roles. The 3' end, or **CCA stem,** common to all tRNAs is the site of amino acid attachment. The D and T loops form base pairing attachments to ribosomal RNA in the ribosome. Most interesting is the **anticodon loop,** which contains a three-base sequence known as an **anticodon.**

As the name suggests, anticodons "oppose" codons. That is, the three bases of a tRNA anticodon will base-pair with the three bases of an mRNA codon. For example, the mRNA codon GCA (according to the genetic code table) specifies alanine. One of the alanine-specific tRNA molecules has the anticodon CGU, which makes the proper base pairing with GCA (Figure 15.8).

TRANSLATION

We now have all of the elements needed for a step-by-step description of protein synthesis. The process will be divided into three parts: *initiation, elongation,* and *termination.*

INITIATION

Initiation is the beginning of the process of translation. It starts when the elements of the **initiation complex** are brought together: mRNA (with its AUG START codon), the ribosome, and a charged tRNA molecule (Figure 15.9). Since methionine is always the first amino acid in any protein being synthesized, this will be a methionine-charged tRNA. Its anticodon—$^{3'}$UAC$^{5'}$—will match the codon—$^{5'}$AUG$^{3'}$—on messenger RNA.

All elements of the initiation complex must be present before the two ribosome subunits can join to form an intact ribosome. When the initiation complex is assembled, the methionine-charged tRNA will occupy the left-hand pocket of the larger ribosome subunit, and the AUG START codon of the mRNA will fit into the ribosomal groove. All is now ready for the next step, elongation.

ELONGATION

With the initiation complex in place, **elongation,** the addition of amino acids, can proceed. Recall that the left-hand pocket of the ribosome was occupied by methionine-charged tRNA; the right-hand pocket was empty. Below the right pocket, however (as seen in Figure 15.9), was the next mRNA codon waiting for a charged tRNA with a matching anticodon. Many charged tRNA molecules, in random motion, may bounce in and out of the empty pocket, but only one with the correct anticodon will bind. When the match occurs, the right-hand pocket will be filled.

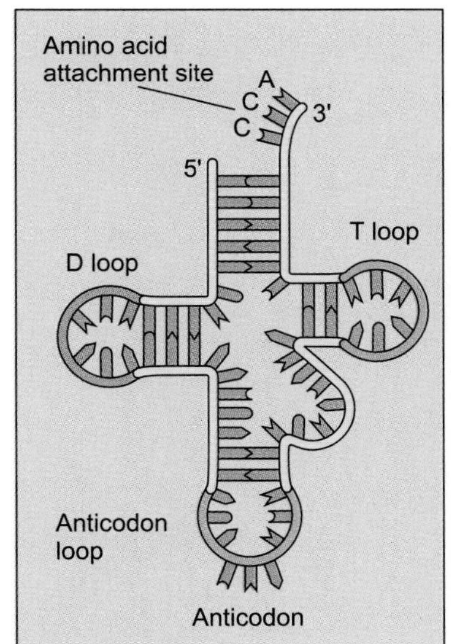

Cloverleaf model showing basic structure

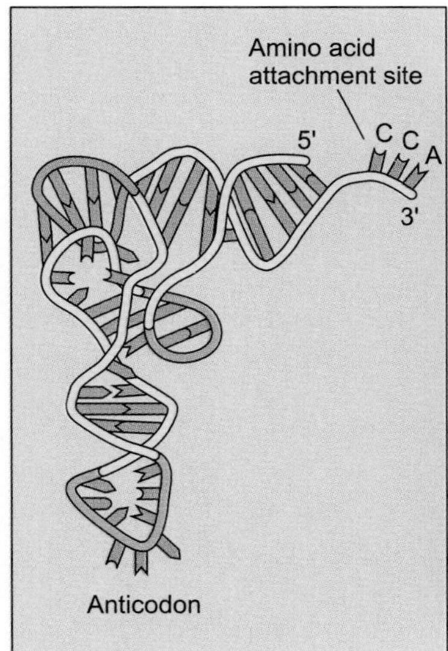

Model showing tertiary structure and folding

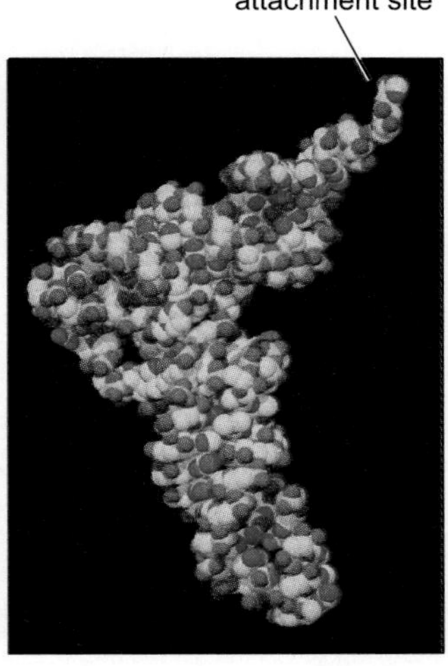

Space-filling model

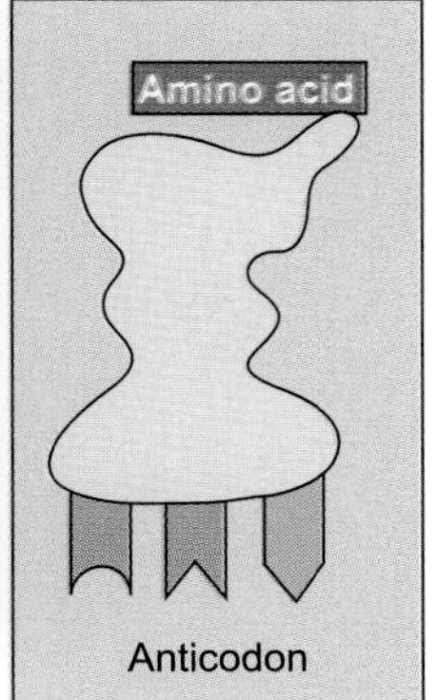

Generalized icon used in this chapter

FIGURE 15.7 TRANSFER RNA.
In its secondary form, the nucleotides of tRNA base-pair to produce a multilooped figure. The 3' CCA end is the attachment site for an amino acid, whereas one of the side loops interacts with rRNA on the ribosome. The lower, or anticodon, loop holds three nitrogen bases that pair or match up with codons in mRNA. In its final or tertiary form, tRNA twists and folds into the shape of an inverted L.

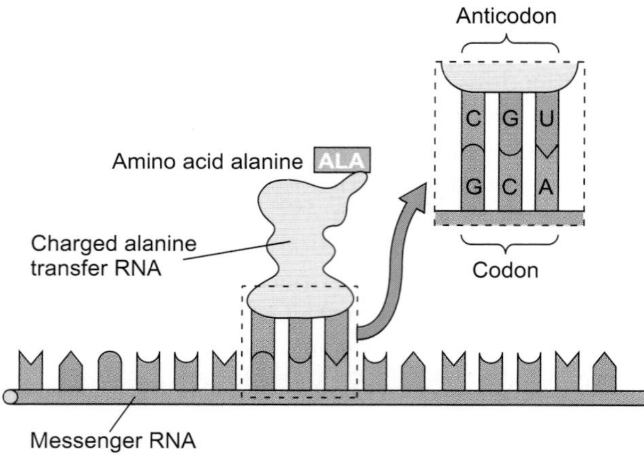
Anticodon
C G U
G C A
Codon
Amino acid alanine
ALA
Charged alanine transfer RNA
Messenger RNA

FIGURE 15.8 **ALANINE AND tRNA.**
In this example of how tRNAs function, alanine-tRNA contains the anticodon CGU, which will pair up with the codon GCA. This is but one step in the translation of genetic code into protein, but it demonstrates the role of tRNAs in translation of mRNA into protein.

Once the second charged tRNA is in place, the two amino acids will be joined, producing the first of many covalent peptide bonds. The enzyme responsible for this event, peptidyl transferase, is integrated into the larger ribosomal subunit. When the peptide linkage has formed, a crucial step called *translocation* can occur.

In **translocation** the ribosome moves one codon to the right, and in so doing the tRNA in the left-hand pocket, now uncharged, leaves the scene to be recycled for future use. The right-hand pocket is now occupied by the next codon on the mRNA, waiting for the next proper tRNA to bind. Figure 15.10 shows the action of elongation. As the process repeats, and the ribosome works its way down the mRNA, a number of amino acids will have been joined to produce the growing protein (or polypeptide).

Thus the action continues, with the ribosome clumping along its mRNA in short jumps of one codon at a time. Charged tRNA molecules arrive constantly as the right-hand pocket keeps emptying. With every translocation step, the protein grows one amino acid longer. You can see how the ribosome plays its essential role in keeping things organized. By moving along the mRNA strand exactly one codon at a time, it ensures the continued translation of the triplet code.

All of this may seem a bit complex and tedious, so how long does it take? The fastest time known for the completion of an entire protein is six seconds (in *E. coli*).

TERMINATION

Termination is the end of the translation process. As the ribosome moves along its mRNA to the end of

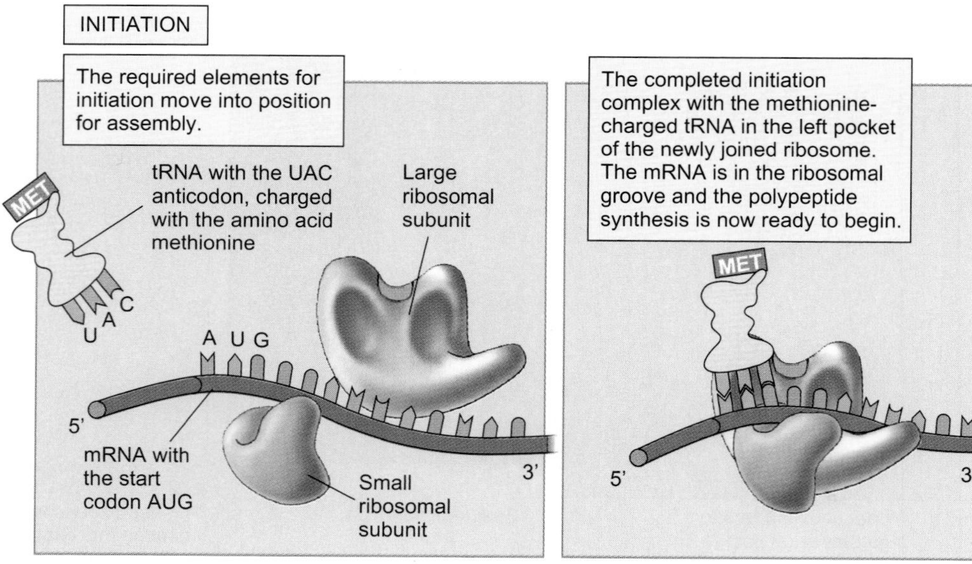
INITIATION
The required elements for initiation move into position for assembly.
tRNA with the UAC anticodon, charged with the amino acid methionine
Large ribosomal subunit
MET
U A C
A U G
5'
mRNA with the start codon AUG
Small ribosomal subunit
3'
The completed initiation complex with the methionine-charged tRNA in the left pocket of the newly joined ribosome. The mRNA is in the ribosomal groove and the polypeptide synthesis is now ready to begin.
MET
5'
3'

FIGURE 15.9 **INITIATION.**
The required elements for initiation are mRNA, the two ribosomal subunits, and methionine-charged tRNA. The initial event is base pairing between the mRNA initiator codon, AUG, and the anticodon UAC. As the base pairing occurs, the smaller ribosomal subunit joins the RNAs.

the message, it runs into one or another of the three STOP codons: UAA, UAG, or UGA. Sometimes there are double stops (for example, UAAUAG), just to be sure that the ribosome gets the message to terminate elongation (Figure 15.11).

There are no tRNAs with anticodons that correspond to any of the STOP codons. Instead there are specific proteins that move into position when the STOP codon is on the ribosome. There they seem to clog the works and bring about the release of the last tRNA, which is then removed enzymatically from the end of the now-completed protein (Figure 15.11).

Finally, with no charged tRNAs in place, the ribosome separates into its large and small subunits,

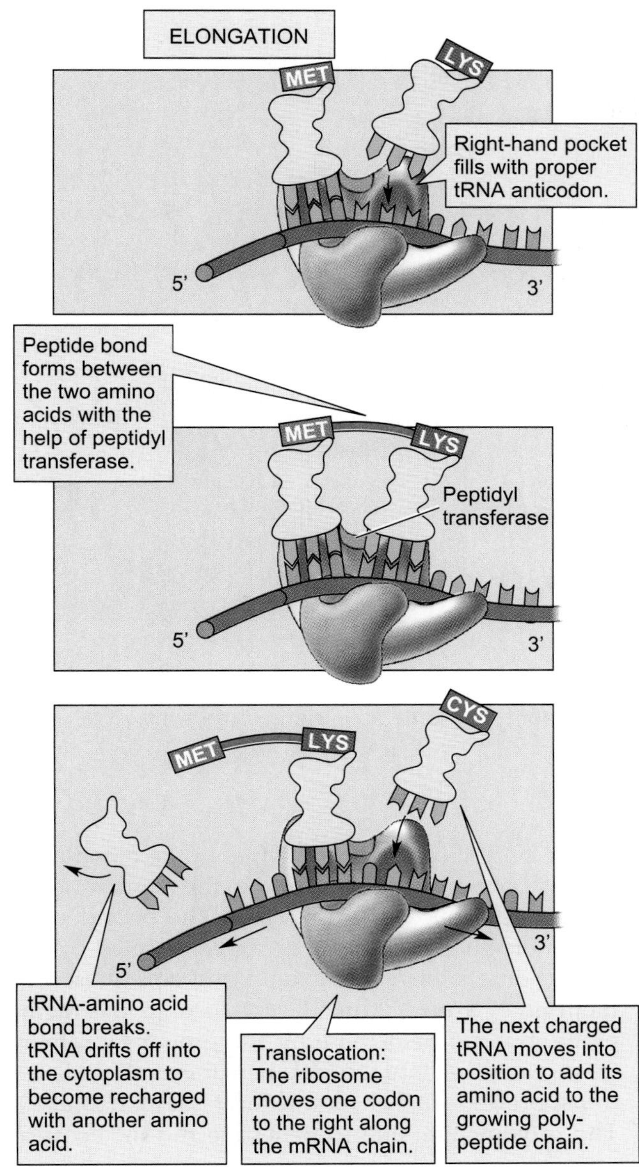

tRNA-amino acid bond breaks. tRNA drifts off into the cytoplasm to become recharged with another amino acid.

FIGURE 15.10 TRANSLOCATION AS TRANSLATION CONTINUES.
The activity represents the translation process partway through the synthesis of a protein (or polypeptide). The left-hand pocket is occupied by methionine tRNA. Lysine tRNA has landed in the right-hand pocket. Next, with the aid of the enzyme peptidyl transferase, a covalent bond forms between lysine and glycine. During translocation, the ribosome moves along one codon to the right, bumping methionine-tRNA out as lysine-tRNA now occupies the left pocket. The right pocket is again empty, ready for the next tRNA to drop into the empty pocket.

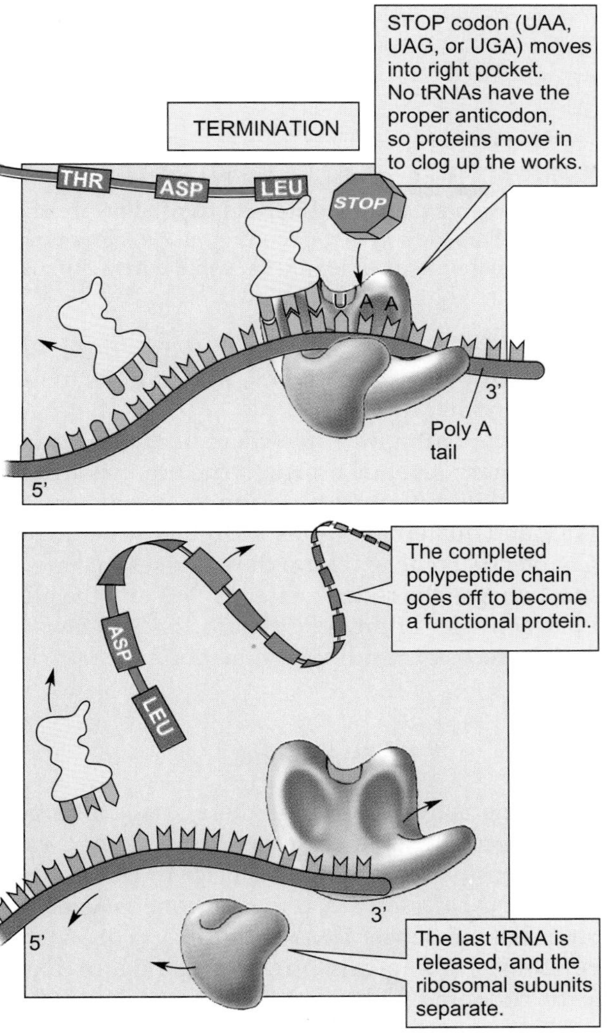

TERMINATION
STOP codon (UAA, UAG, or UGA) moves into right pocket. No tRNAs have the proper anticodon, so proteins move in to clog up the works.
THR
ASP
LEU
STOP
U A A
3'
Poly A tail
5'
The completed polypeptide chain goes off to become a functional protein.
ASP
LEU
3'
5'
The last tRNA is released, and the ribosomal subunits separate.

FIGURE 15.11 TERMINATION AT STOP CODONS.
When no charged tRNA appears for the STOP codon, termination factors (represented by the stop sign) occupy the pocket and derail the translation process and release the translated protein.

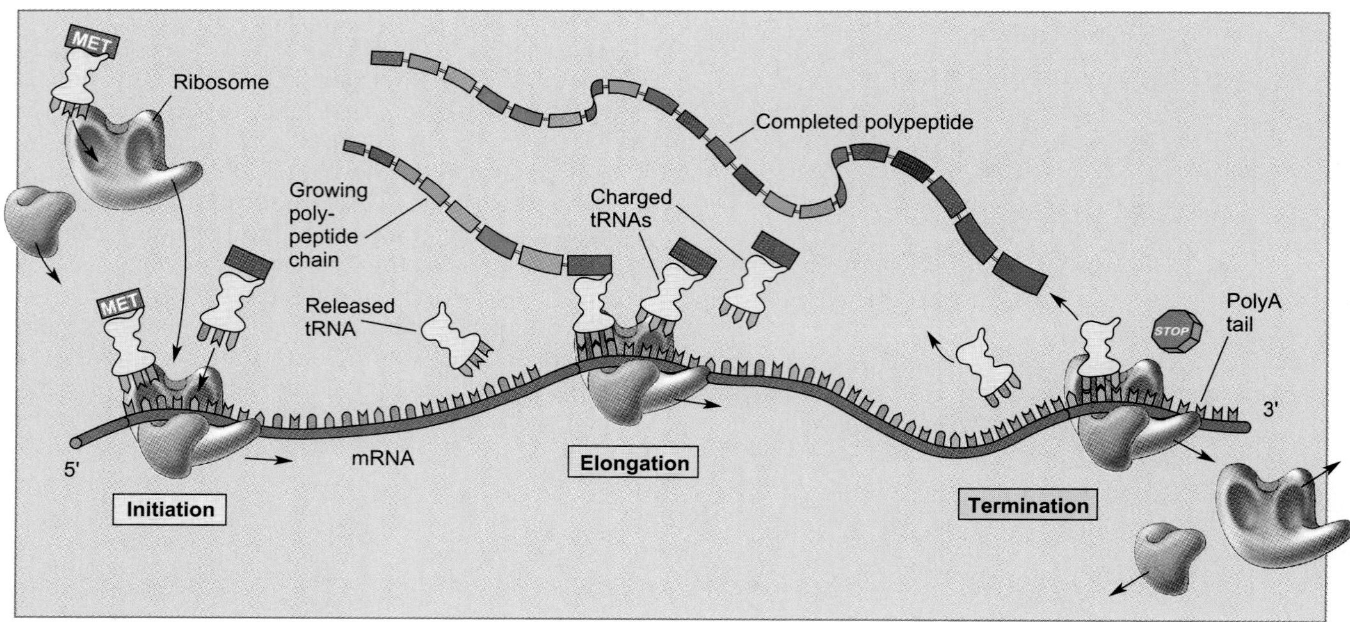

FIGURE 15.12 **SUMMARY OF TRANSLATION.**
In this summary, all of the steps of translation are shown together, and this is appropriate since usually all of these steps are in fact occurring at the same time in the cell. As one ribosome is finishing, others are still translating the same mRNA, while others are just initiating.

which drift away from the messenger RNA to recycle for another run.

The protein, now a free chain of amino acids at the primary level of protein structure, has its own fate. It will automatically form its secondary, tertiary, and quaternary levels of protein structure (see Chapter 3). Some will form the enzymes that are the link between the coded genes of DNA and the metabolic activities of the cell. (Figure 15.12 reviews the entire process, including termination.)

POLYRIBOSOMES

During active protein synthesis, ribosomes generally occur in small clusters, with perhaps 2 to 10 ribosomes per cluster. Each cluster of ribosomes is called a *polyribosome.* A **polyribosome** is actually a strand of messenger RNA with a group of attached ribosomes, like pearls on a string (Figure 15.13). Each ribosome will travel the whole length of the mRNA, from the START codon to the STOP codon, then each will fall apart and drop off, as others assemble at START. Thus several ribosomes can be producing copies of the same protein at the same time, each working on a different portion of the mRNA strand.

The polyribosomes can be free in the cytoplasm, or they can appear to be bound to the membranes of the endoplasmic reticulum (ER). The bound ribosomes produce proteins that can be secreted by the cell at some later time. Actually the polyribosomes themselves are not directly attached to the membrane, but are held there by the growing proteins that are moving into the ER where they will undergo transport and further modification (see Chapter 4). The proteins are able to penetrate the ER because

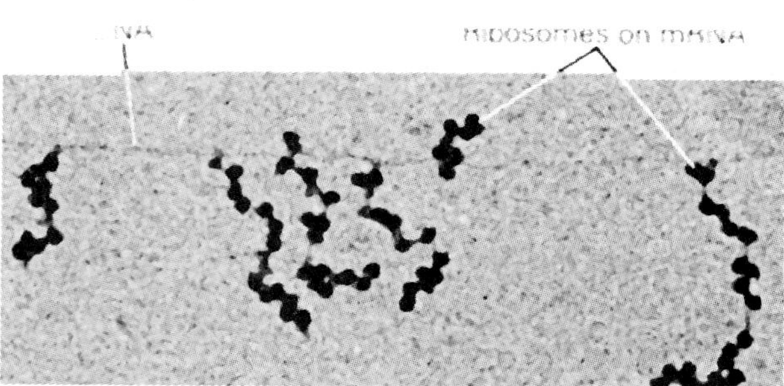

FIGURE 15.13 **POLYRIBOSOMES.**
Multiple ribosomes are seen lying along a strand of mRNA (just visible) in this highly magnified view through the electron microscope.

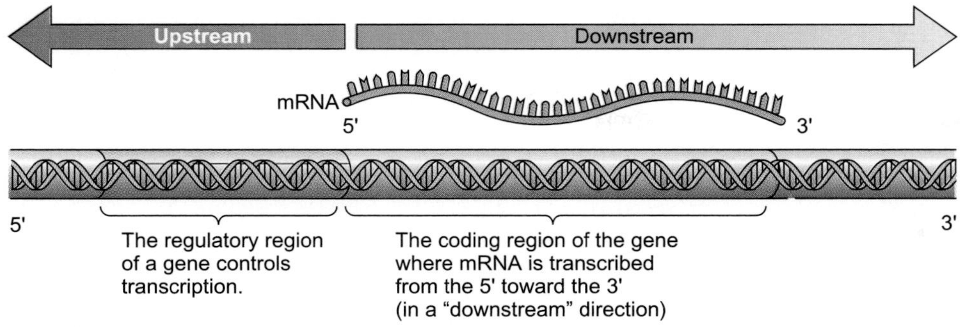
Upstream
Downstream
mRNA
5'
3'
5'
3'
The regulatory region of a gene controls transcription.
The coding region of the gene where mRNA is transcribed from the 5' toward the 3' (in a "downstream" direction)

FIGURE 15.14 **GENERALIZED GENE STRUCTURE.**
Most genes consist of two separate parts—a coding region that actually is transcribed into mRNA, and a regulatory region that controls the process of transcription. The 5' end of the mRNA is the boundary between the regions. Movement toward the 3' end of the mRNA is in the "downstream" direction, while movement in the opposite direction is "upstream."

their leading ends contain a sequence of amino acids that recognize and bind to ER receptor sites.

MECHANISMS OF GENE CONTROL

The concepts of transcription and translation answer many difficult questions about the nature of genes. Yet even after determining the sequence of events leading from DNA to protein and the chemical structure of the hereditary material, several important questions remained, and continue to remain today. One question involves gene control: What controls gene transcription? How is it that an organism expresses only those genes it needs at the present time, saving its resources by not expressing all genes at all times? If all genes were expressed in all cells, wasteful or even dangerous combinations of proteins would result. It certainly would not do to have hair protein produced on your teeth or digestive enzymes produced on your scalp!

Most of what we have learned about gene control comes from studies of bacteria. From those studies we have learned that genes are composed of component parts. The two main components are the *coding region* and the *regulatory region.* The **coding region** contains the information to encode the mRNA that in turn encodes the protein. The **regulatory region** directs the transcription of the coding region. One component of the regulatory region would be the promoter that we discussed earlier in the chapter. As shown in Figure 15.14, this division of a gene actually imparts a directionality to gene structure, allowing us to refer to "upstream" and "downstream" relative to the direction of transcription. While there are certainly variations on this theme, most control apparently occurs at upstream regulatory regions.

GENE CONTROL IN PROKARYOTES

While geneticists were still trying to work out the genetic code and the mechanisms of protein synthesis, two French microbiologists were conducting gene regulation experiments. In 1961, François Jacob and Jacques Monod unveiled the results of their work, a model called the **operon** (Figure 15.15), which is a regulated cluster of genes.

The first major feature of an operon is that a number of coding regions that code for proteins needed at the same time can be grouped together under the control of a single regulatory region. This simple organizational feature guarantees that proteins needed at the same time will be produced at the same time, especially since all the coding regions will be transcribed onto a single mRNA. Such an mRNA (encoding more than one protein) is said to be **polycistronic.** The experiments of Jacob and Monod involved the *lac* operon, which controls expression of the proteins of a simple biochemical pathway—the enzymes that allow a bacterium to live off lactose, or milk sugar. It turns out that three enzymes are needed simultaneously to digest lactose. These are beta galactosidase, galactose permease, and thiogalactoside transacetylase—encoded by the *z*, *y*, and *a* areas of the coding region (Figure 15.16).

A second major feature of the operon, and the feature by which it is named, is a part of the regulatory region known as the *operator.* The **operator** is a small segment of DNA that lies between the promoter and

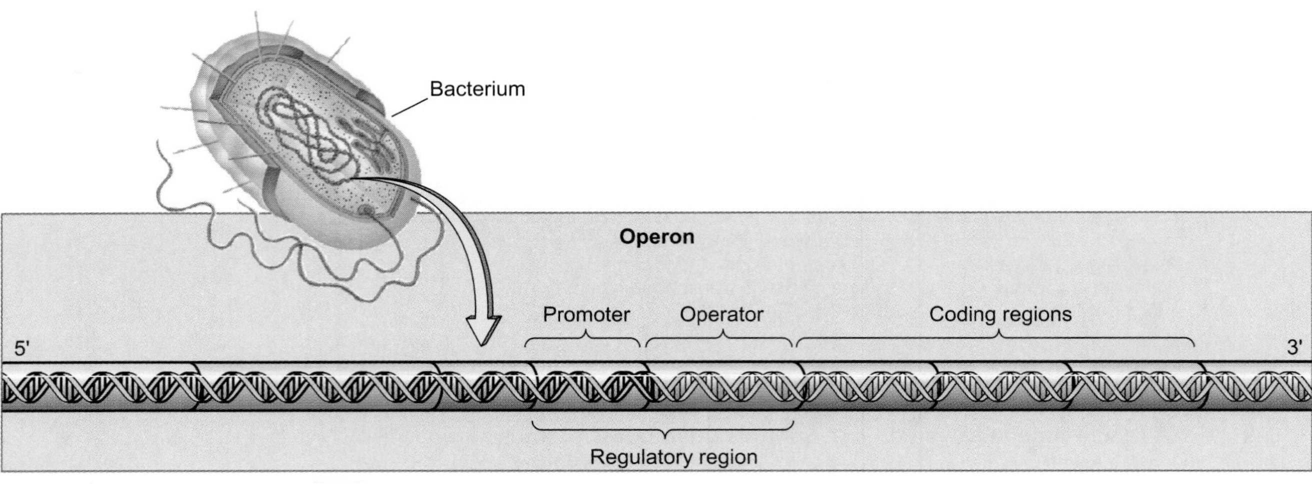

FIGURE 15.15 **OPERONS.**
The key features of an operon are the grouping of several protein coding regions under the control of a single regulatory region, and a regulatory region that consists of a promoter and an operator. The operator is the site where a repressor molecule may bind, preventing RNA polymerase from beginning transcription from the promoter.

the coding region. It is the operator that basically decides whether or not transcription will occur. How does it accomplish this regulatory feat? The answer is really quite simple and involves a regulatory protein called a *repressor.* The **repressor** is a DNA-binding protein (produced by a gene named *i*) that is capable of attaching to the operator and blocking transcription. Simply enough, if the repressor is bound to the operator, no transcription occurs. On the other hand, if there is no repressor bound to the operator, RNA polymerase moves right through the operator and transcription occurs.

But how does the repressor know whether or not to bind to the operator? Let's look more carefully at the *lac* operon to find out.

Normally, bacteria such as *E. coli* grow on sugars like glucose. *E. coli* does not usually have the enzymes for digesting lactose available in case the bacterium encounters some lactose. Instead, the gene encoding the enzymes for lactose metabolism are *induced* (turned on) when lactose appears in the medium. So under most conditions, the *lac* operon is repressed, with the *lac* repressor occupying the operator and preventing transcription. Repressor, by the way, is kept at a constant concentration of about ten molecules per cell by low but consistent expression of the regulatory gene *i*.

It is the repressor molecule itself that is the key to both sensing when lactose is present and allowing transcription. You see, the repressor actually has two binding sites, one for the operator DNA and one for lactose. It can bind either one, but not both at the same time. When lactose enters the cell in reasonably high concentrations, all of the repressor molecules become bound to lactose instead of the operator. Transcription of the *z, y,* and *a* coding regions on a single polycistronic mRNA then proceeds, precisely when the enzymes are useful (Figure 15.16).

Suppose the concentration of lactose diminishes to the point where the enzymes are no longer needed. What happens then? As the concentration of lactose is reduced, more and more repressor molecules are free to bind to the operator, thus shutting down expression of the operon now that it is no longer needed.

Many other operons have been discovered over the years since the presentation of the original operon model. There are subtle variations in the structure of promoters, operators, and repressors, each of which imparts unique and varied aspects of regulation to a particular gene. But the main idea of having DNA-binding proteins that interact with regulatory regions of genes remains a cornerstone for understanding the basic mechanism for all gene control mechanisms.

GENE CONTROL IN EUKARYOTES

Gene regulation is relatively simple in prokaryotes. But multicellular eukaryotes consist of a variety of cell types with capacities to transcribe different sets of genes and so the problem of regulation in eukaryotes is much more complex.

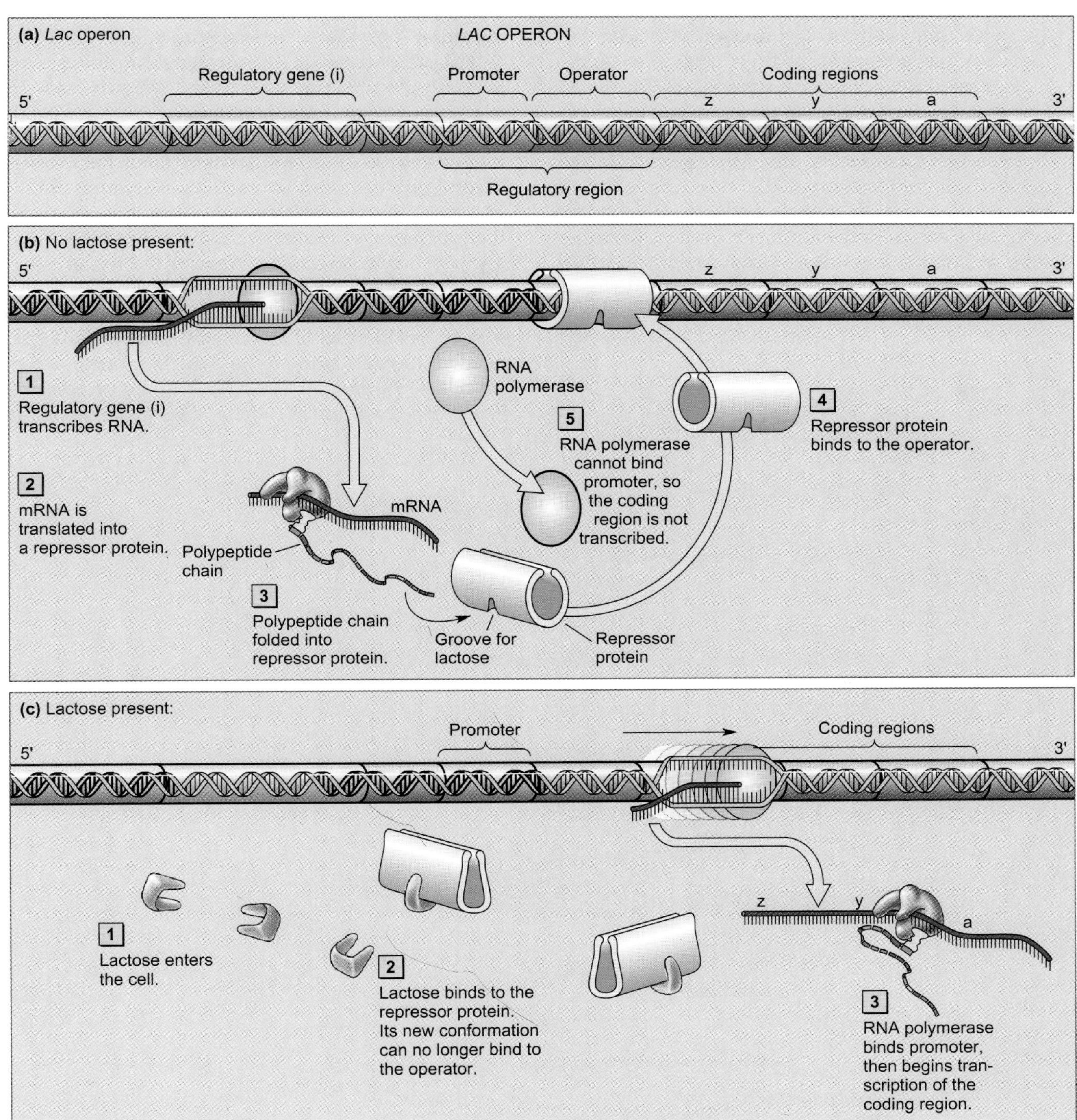
(a) Lac operon
LAC OPERON
Regulatory gene (i)
Promoter
Operator
Coding regions
5'
z
y
a
3'
Regulatory region
(b) No lactose present:
5'
z
y
a
3'
1
Regulatory gene (i) transcribes RNA.
2
mRNA is translated into a repressor protein.
Polypeptide chain
mRNA
3
Polypeptide chain folded into repressor protein.
Groove for lactose
Repressor protein
RNA polymerase
5
RNA polymerase cannot bind promoter, so the coding region is not transcribed.
4
Repressor protein binds to the operator.
(c) Lactose present:
Promoter
Coding regions
5'
3'
1
Lactose enters the cell.
2
Lactose binds to the repressor protein. Its new conformation can no longer bind to the operator.
z
y
a
3
RNA polymerase binds promoter, then begins transcription of the coding region.

FIGURE 15.16 THE *LAC* OPERON.

The production of the three inducible enzymes responsible for the metabolism of lactose is under control of a system known as the *lac* operon. **(a)** The operon consists of three principal parts, as shown in the diagram: the regulatory gene *i*, the promoter/operator region, and the structural genes *z, y,* and *a*. **(b)** The repression of the structural genes is accomplished by an interaction between the product of regulatory gene *i* and the operator. *(1)* The regulator gene transcribes messenger RNA that *(2,3)* is translated into a repressor protein, which *(4)* binds the operator, blocking the action of RNA polymerase along the promoter. Thus the transcription of mRNA in the coding region *(5)* is inhibited. **(c)** When lactose enters the cell *(1)*, it acts as an *inducer, (2)* tying up the repressor protein. Transcription is now allowed *(3)*.

We have seen that the bacteria have elegant mechanisms for gene control, and we have also seen that these mechanisms are fairly direct. That is, when the messenger RNA is made, the protein is translated immediately. In fact, the protein translation can begin even before the mRNA is completely assembled.

Eukaryotes cannot do this. The presence of the nuclear membrane necessitates the completion of transcription and the transport of the mRNA to the cytoplasm before translation can occur. But rather than an impediment, these additional processes offer new opportunities to control the eventual expression of a gene. After all, we don't see the effects of the gene just because it is transcribed, we see it only after it produces an active protein.

The presence of the nuclear membrane offers eukaryotes the ability to separate the two processes of transcription and translation in both time and space. The various steps that accompany this separation of translation and transcription, then, offer several opportunities for regulation (Figure 15.17). For example, one of these opportunities involves the processing of introns from the initial RNA transcript. Since mRNA is not transported to the cytoplasm until the introns are removed, some transcripts could be held in the unprocessed state as a sort of ready reserve. When the gene product is needed, the only thing to be done is to remove the introns and transport the mRNA to the cytoplasm. There are even cases where alternative splicing

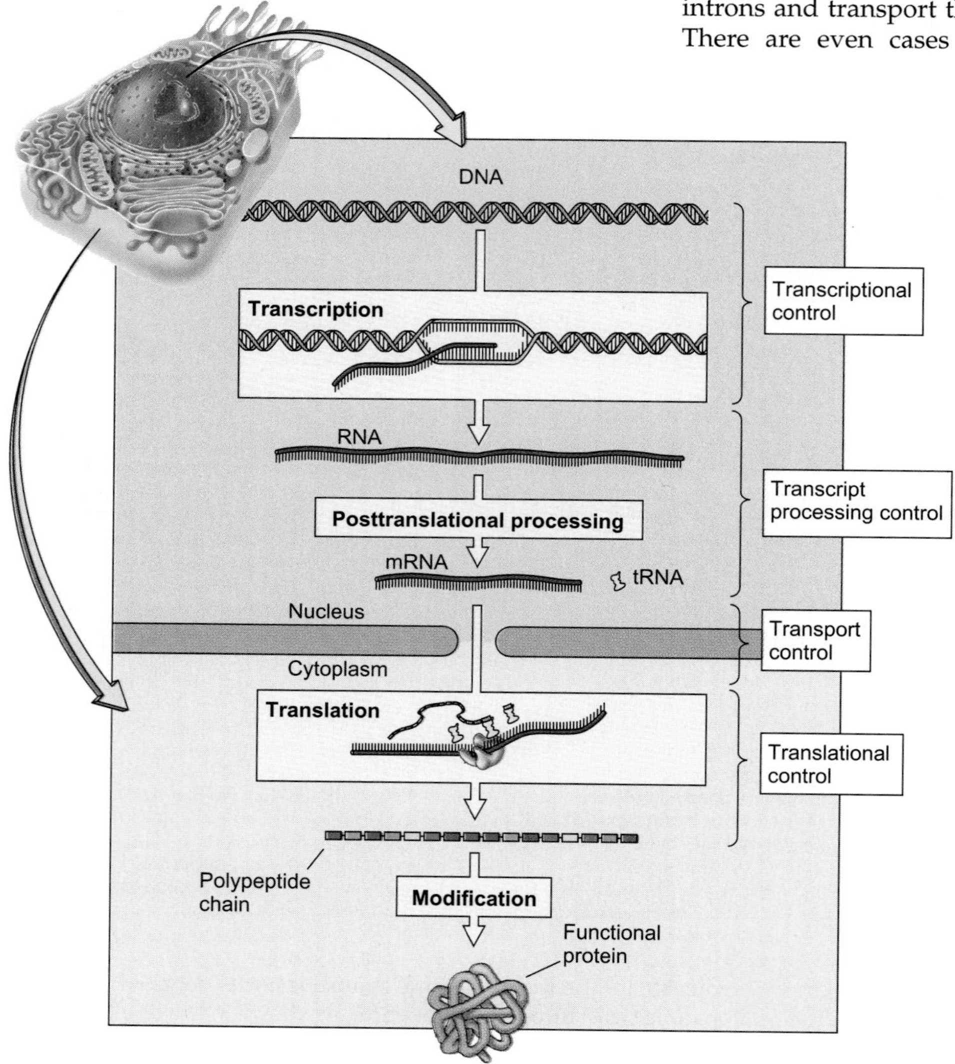

FIGURE 15.17 OPPORTUNITIES FOR GENE CONTROL IN EUKARYOTES. This chart parallels our description of the flow of genetic information and points out the several steps in the production of a eukaryotic protein where regulation may occur.

Essay 15.1

Homeotic Genes and the Homeobox

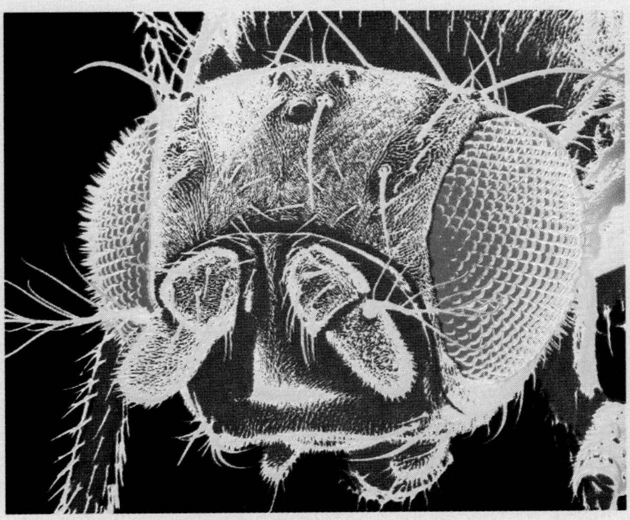

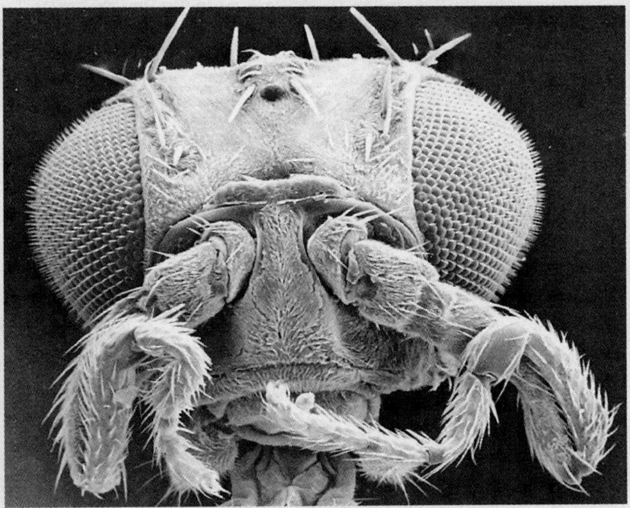

As we'll see in later chapters, there are many questions that can be asked as an organism develops from the fertilized egg. After just a few cell divisions, we could ask, Which end will be the head and which the tail? How many segments will be made, and which will carry wings?

It has been known since the early work in *Drosophila* genetics that there are several genes that can drastically alter the answers to these developmental questions. In fact, we now know that there are some eight different genes that control critical developmental steps in fruit fly maturation. These genes are generally referred to as *homeotic* genes (*homeo,* same), since they can result in the replacement of organs by structures similar to those found elsewhere on the fly. For example, the antennapedia gene can result in the production of legs on the head where the antennae should be. Other homeotic genes can control the number of segments in the body, the number of wings, and the head-to-tail orientation of body segments. In short, the eight different homeotic genes are related in that they control key developmental processes, processes that must require the coordinated action of many different genes. The homeotic genes are considered, then, to be a class of regulatory genes, genes that control the expression of other genes.

Researchers are now beginning to understand the way in which these regulatory genes exert their controlling effects. While the protein products of the various homeotic genes are distinctly different, they all share one feature. They are all regulatory proteins that contain a section of approximately 60 amino acids, called the *homeobox.* The homeobox, it turns out, is a protein structure that can bind to DNA.

So part of the mystery of how homeotic genes act is solved. The homeotic genes apparently exert their regulatory functions through protein products that directly bind to DNA, probably the promoter DNA of the many genes required to accomplish the elaborate tasks of building organs and body parts. It turns out that not only all animals (including humans) but also plants have homeotic genes, and that the use of homeotic genes to control development is an evolutionarily concerned theme.

schemes can be used to generate two different proteins from the same transcript!

There are many instances where a gene may be expressed in only certain cell types or under certain environmental conditions. Consider one example, where estrogen, a steroid hormone released into the bloodstream, can activate a gene. As steps in the process were discovered, it became clear that the system was rather complicated (Figure 15.18), but boiled down to a situation reminiscent of bacterial genes.

Estrogen target cells have a special estrogen receptor protein. These are the only cell types that have this protein, and therefore only these cells respond to the hormone. When an estrogen molecule contacts such a receptor, it joins with it. Then the estrogen-receptor complex apparently migrates to the regulatory regions of target genes. Once there, it binds to specific control regions that stimulate transcription.

Regarding gene regulation in eukaryotes, scientists are currently unraveling complex questions such as how regulatory proteins recognize and bind to specific sequences of DNA, and how the intricate changes in development and differentiation occur. (See Essay 15.1.)

We saw in Chapter 9 that eukaryotic DNA is organized within the nucleus as chromatin, bound to histone proteins. Now let's briefly take a closer look at the chromatin structure and see how its packaging has a direct impact on gene regulation.

Recall that visible chromosomes are chromatin that is condensed into large, highly visible bodies during cell division. During interphase, when the chromosomes "unwind" (or, more properly, become diffused), they form very long, thin threads. In fact, only these unwound threads are able to transcribe RNA; condensed portions are inactive. This should not be surprising. Imagine RNA polymerase molecules trying to reach specific DNA sequences that are buried within the highly condensed chromatin network. They just can't get in.

Two striking examples illustrate how condensation is important to gene activity. The first is the condensation of one of the X chromosomes in mammalian females to form the Barr body, which is transcriptionally inert (see Chapter 12). The other conspicuous example of the importance of chromatin decondensation in gene regulation comes from studies of *Drosophila* chromosomes during their larval stages. In fruit flies, and many other insects, the larval stage is a period of intense gene activity. In Figure 15.19, we see that gene activity in larval salivary glands is signaled by visible changes in chromatin structure. The figure shows the regions of transcribing genes as unwound "puffs" along a *Drosophila* chromosome. Puffs actually consist of loops of DNA that are decondensed such that the DNA is diffuse and accessible to RNA polymerase and regulatory proteins.

Thus we see that gene expression is not a simple matter of the gene being turned "on" or "off." Indeed, there is a whole host of regulatory possibilities that do not directly involve transcription at all, and it is very likely that the ultimate expression of all eukaryotic genes is subjected to some degree of management at each of these levels.

PERSPECTIVES

We have come a long way in our understanding of Mendel's "factors" that we came to call genes. At first, genes were mysterious factors that were located on chromosomes and somehow caused a phenotype. Then we learned that genes are just segments of DNA. Now, of course, we know that these segments of DNA contain the information of the genetic code, the information necessary to specify the proteins that are ultimately responsible for a phenotype. Not only that, these DNA segments also contain the information to specify how and when they will be expressed. All this happens through a complex set of protein-DNA interactions, followed by the complex interplay of the ribosomes, mRNA, and tRNA.

The story of what genes are and how they work is now essentially complete for our purposes. What remains for us to look at is, first, what all of this new information has meant to us, and second, how this information changes over evolutionary time. So let's go on now to consider the intriguing topics of genetic engineering and mutation.

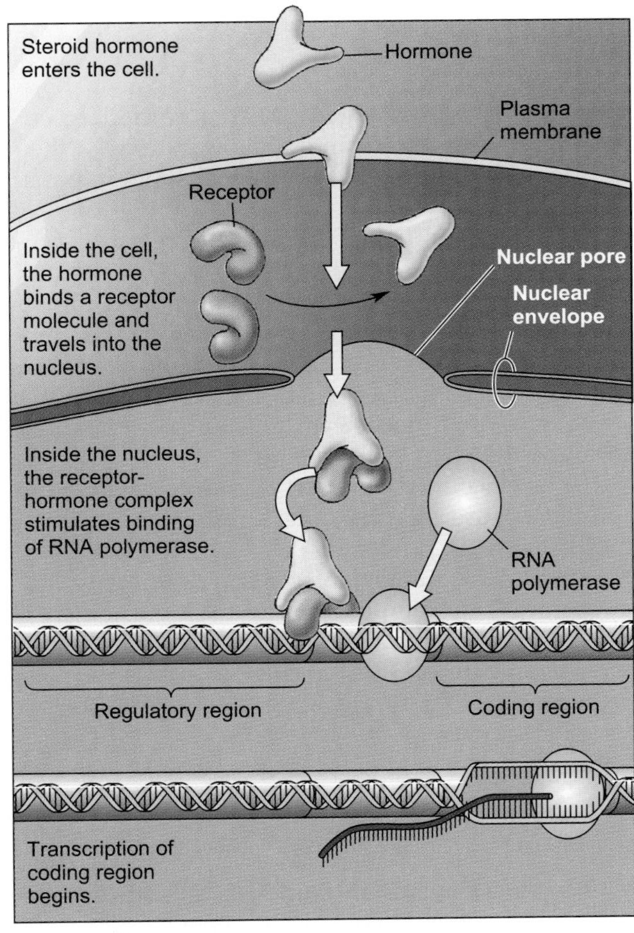

Steroid hormone enters the cell.
Hormone
Plasma membrane
Receptor
Inside the cell, the hormone binds a receptor molecule and travels into the nucleus.
Nuclear pore
Nuclear envelope
Inside the nucleus, the receptor-hormone complex stimulates binding of RNA polymerase.
RNA polymerase
Regulatory region
Coding region
Transcription of coding region begins.

FIGURE 15.18 GENE CONTROL BY STEROID HORMONES.
In this model, steroid hormones, such as estrogen, bind to cellular receptor proteins. The protein-hormone complex can stimulate transcription of genes containing the proper recognition sequence.

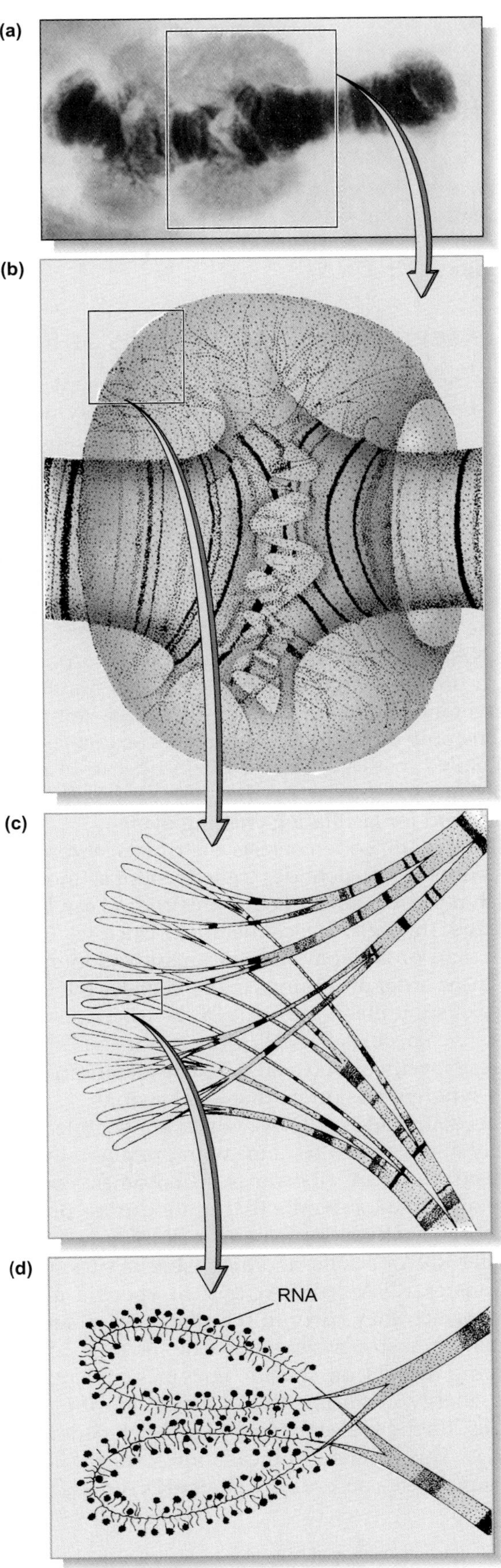

(a)
(b)
(c)
(d)
RNA

FIGURE 15.19 CHROMOSOME CONDENSATION AND GENE CONTROL.
Puffs are regions of the polytene salivary chromosomes where transcription is known to occur. **(a)** An actual photomicrograph of a puff. **(b–d)** Interpretations of the structure of a puff, showing that several stages of decondensation accompany transcription of this area of the chromosome.

Summary

The RNA Molecule

RNA is a single-stranded polymer composed of nucleotides that utilize ribose sugar and the nitrogen base uracil (instead of deoxyribose and thymine, as in DNA).

Transcription: The Synthesis of RNA

Transcription is similar to replication, except it occurs along one DNA strand only. RNA polymerase may carry on simultaneous transcription along one DNA segment.

Three Kinds of RNA

RNA occurs as messenger RNA, ribosomal RNA, and transfer RNA.

The nucleotide sequence of messenger RNA (mRNA) specifies the amino acids to be incorporated into a protein. Raw transcripts contain a number of intervening sequences (introns) that are removed, leaving only the expressed sequences (exons), the actual coded message. Intervening sequences may be important for nullifying the effects of mutagenic agents and for facilitating crossing over.

The genetic code consists of triplets, three-base sequences of nucleotides called codons, most of which represent amino acids. This coding can be visualized through the colinearity principle: DNA, RNA, and protein have similar organization in that the linear order of subunits in one dictates the linear order of subunits in the next. Of the 64 codons possible, one specifies START, three specify STOP, and the rest specify amino acids. Nearly all amino acids have synonymous (more than one) codons.

Ribosomal RNA (rRNA), which is assembled primarily in the nucleolus, joins with protein in the cytoplasm to form ribosomes. Ribosomes contain large and small subunits that unite during protein synthesis. Each has a messenger RNA groove, two transfer RNA pockets, and other special sites.

Transfer RNAs (tRNA) link to specific amino acids, which they carry to the ribosome where the protein is being assembled. The linkage, called charging, occurs on specific enzymes and utilizes ATP energy. Transfer RNAs have specific anticodons, three-base sequences that match codons on mRNA. This provides a means for correctly inserting amino acids according to the mRNA message.

Translation

Messenger RNA (with its START codon in position), methionine-charged tRNA, and the large ribosomal subunit form an initiation complex. When it is joined by the smaller ribosomal subunit, translation begins.

In elongation, both ribosomal pockets are occupied with charged tRNAs. Next, a peptide bond forms, joining the two amino acids and permitting translocation to occur. In translocation, the ribosome moves one codon to the right, and the tRNA from the right pocket enters the left pocket. This frees the right pocket for the entry of another charged tRNA.

Upon encountering a STOP codon, termination proteins cause the ribosomal subunits to separate, and the protein is released.

Typically, a number of ribosomes, or polyribosomes, carry on simultaneous translation along one mRNA. Bound ribosomes, those of the rough endoplasmic reticulum (ER), are held in place by their polypeptide chains, which are themselves directed into ER lumen.

Mechanisms of Gene Control

Control mechanisms are responsible for the activation and deactivation of specific genes.

Jacob and Monod proposed a mechanism of prokaryote gene control they called the operon. In the inducible *lac* (lactose) operon, three genes transcribe mRNA that directs the synthesis of three lactose-metabolizing enzymes. However, transcription only occurs when lactose is present. Otherwise, the genes are inhibited by repressor protein, which blocks the promoter by binding with the operator. When lactose is present, the repressor protein binds with lactose instead of the operator, and transcription begins. As the enzymes break the lactose down, inhibition begins once more.

Eukaryotic transcriptional control is in some ways similar to prokaryotes, in that regulatory regions of DNA interact with regulating proteins. However, the situation is complicated by the packaging involved in chromatin, and decondensation is a necessary prerequisite of transcription. Also, the presence of the nuclear envelope and the many processing steps in the maturation of a eukaryotic mRNA offer additional points of potential control of gene expression.

Key Terms

uracil • **226**
transcript • **227**
transcription complex • **227**
RNA polymerase • **228**
promoter • **228**
ribosomal RNA (rRNA) • **229**
transfer RNA (tRNA) • **229**
messenger RNA (mRNA) • **229**
posttranscriptional processing • **229**
translation • **229**
intron • **229**
exon • **229**
genetic code • **229**
codon • **229**
principle of colinearity • **230**
synonymous codons • **230**
charging enzyme • **232**
CCA stem • **232**
anticodon loop • **233**
anticodon • **233**
initiation • **233**
initiation complex • **233**
elongation • **233**
translocation • **234**
termination • **234**
polyribosome • **236**
coding region • **237**
regulatory region • **237**
operon • **237**
polycistronic mRNA • **237**
operator • **237**
repressor • **238**

Review Questions

1. List three differences between RNA and DNA, involving sugars, nitrogen bases, and the polymer itself.
2. Using simple diagrams of DNA and RNA, describe the transcription process. What is simultaneous transcription?
3. What is the role of messenger RNA? Explain why posttranscriptional modification is necessary. In what ways might introns be important to the gene?
4. Describe the basic organization of the genetic code. What makes up a codon? What does a codon represent in protein?
5. In looking at the genetic code table, explain the roles of the codons UAA, UAG, UGA, and AUG. In how many instances are amino acids represented by four codons? What term describes these codons?
6. Where is most of the ribosomal RNA formed? What assures an adequate supply of these vital molecules? What general role do ribosomes play in protein synthesis?
7. Explain the organization of the ribosome. Include its molecular composition, its subunits, and its principal external features.
8. The analogy of a factory forklift was used to describe the role of tRNA. Explain this.
9. Describe the events in charging of tRNA.
10. List the three important regions of any tRNA molecule. Explain how a charged tRNA manages to deliver its amino acid to the right place in the protein.
11. List the participants that form the initiation complex. What is the last event prior to the start of elongation?
12. Describe the events leading up to elongation. How does translocation occur? What happens to uncharged tRNA?
13. Describe the events that end translation. What happens to the mRNA and ribosomal subunits?
14. Describe polyribosomes.
15. List the major elements of the *lac* operon and very generally explain what each does.
16. How does the substrate lactose overcome the repressed state in the *lac* operon? Once activated, how is the *lac* operon shut down?
17. Describe the general similarities and differences in gene structure and regulation between prokaryotes and eukaryotes.
18. List the possible points where the regulation of gene expression could take place, other than at transcription.

CHAPTER

16

Genetic Engineering

OFTEN IN SCIENCE, THE SEARCH FOR BASIC UNDERSTANDING OF FUNDA-*mental processes eventually leads to our using that understanding in some beneficial way. Certainly that has been true for genetics. The basic principles of heredity, often discovered by those who simply wanted to know, have been routinely utilized for improving our lot and now molecular biologists have entered the scene and begun to manipulate the gene itself. Genetic engineering, as we will see, now offers our best hope for solving some of humankind's most vexing problems.*

The field of genetics is presently undergoing sweeping changes largely because of a simple introduction: Big business has met the microbe. Already we find that the most conservative of financiers are interested in the mating habits of a lowly bacterium that is found in the bowels of everyone (including their employees). How did this all come to be? How did a generally harmless germ become the focus of such unexpected attention?

Perhaps because the bacterium has become one of the potentially most powerful tools known to science. Here, we will learn—in conjunction with our new understanding of how DNA works—just how innocuous microbes have presented scientists with the ability to quite literally alter the course of humankind. In a nutshell, we will see that we now have the ability to find specific genes, to cut them away from their chromosomes, to insert them into the chromosomes of other species, and then—after they have been duplicated countless times along with the normal genes of those species—to harvest them or their protein product in large quantities.

It has been done. For instance, we can now excise from human chromosomes the genes that make human growth hormone. We can also open up certain bacterial chromosomes and insert the human gene into the bacterial chromosome. There, as the bacterial genes go about their business of making bacterial products, the human product (growth hormone) is made right along with them. Each time the bacteria reproduce, they reproduce the human gene along with their own. At some point, the human product can be extracted in large quantities from millions of altered bacteria.

DNA TECHNOLOGY

Since the late 1970s, scientists engaged in molecular biology research have developed a wide variety of tools to manipulate DNA. Some of these tools are mechanical or electrical, but clearly the most powerful are biological. These biological tools are natural enzymes and DNA segments that are so well understood that their properties can be used to create new genes and gene arrangements. The process of manipulating DNA to form new genes, or putting altered genes in different organisms, is called **genetic engineering.**

THE TOOLS OF GENETIC ENGINEERING

Before we get to a description of the processes involved in genetic engineering, let's define a few of the basic components. These will be the players on our new stage.

Recombinant DNA is a hybrid of DNA from two (or more) sources. Often, it will be composed of source DNA (sometimes called foreign DNA) and host DNA.

Source DNA contains the gene of interest, or the part of the DNA molecule that is of interest in the project at hand. It is this DNA that is to be inserted into the host DNA.

Host DNA is the DNA that will be opened up to allow the insertion of the source DNA.

Vector DNA is the name given to the chromosome that will serve to carry the source DNA into its new host.

Restriction enzymes are the enzymes that cut DNA (such as the vector and source DNA) at specific places so that DNA segments can be rejoined in specific combinations.

Clones are groups of genetically identical organisms that are descended from a single ancestor. For example, an individual bacterium on a plate of nutrient medium grows to become a clone colony of cells. Genes can also be cloned. Once a particular gene is inserted into a bacterium, that bacterium is cloned and the gene and its product can be recovered from the bacterial descendants.

Genetic engineering techniques are simply an extension of natural biological processes. Over time the techniques have become vastly simplified. In fact, at one time, researchers would have to prepare each of the enzymes they needed from a biological source. Now all they need to do is order everything they need from a catalog (Figure 16.1). As we describe the processes involved in genetic engineering, you might want to look them up in Figure 16.1, and think back to where you might have seen them described in previous chapters. (Also see Essay 16.1.)

Finding and Isolating a Gene

The first thing to do in cloning a particular gene is to find that gene. This first step is rather simple. If an organism produces a certain substance—a protein, for instance—one can be sure that it has a gene for producing that protein. The trick is in rummaging through its chromosomes to find just where that gene is, so that it can be excised and reproduced.

One way to approach the problem is to subject the cells containing the gene to restriction enzymes. These, as we will see next, will chop the chromosomes apart, cutting them in specific places and leaving fragments that can then be inserted into the DNA of host chromosomes (the host DNA).

Only some cells will carry the gene of interest in their chromosomes. From among this pool, then, the researcher must isolate those cells containing that gene. This can be done by identifying those cells that manufacture the protein produced by the gene.

How DNA Is Cut

Every species of bacteria produces at least one specific restriction enzyme. They obviously don't go to all this trouble just to help recombinant researchers, so of what use are the enzymes to the bacteria? The enzymes are a defense against invasion by viruses that insert their own chromosomes into bacteria, causing them to make new viruses. The restriction enzymes can chop up and render harmless invading viral DNA.

The fact that restriction enzymes cleave DNA at very specific places along its length makes the enzymes very valuable to researchers. For example, the restriction enzyme *ECO* RI produced by the intestinal bacterium *E. coli* recognizes the following sequence.

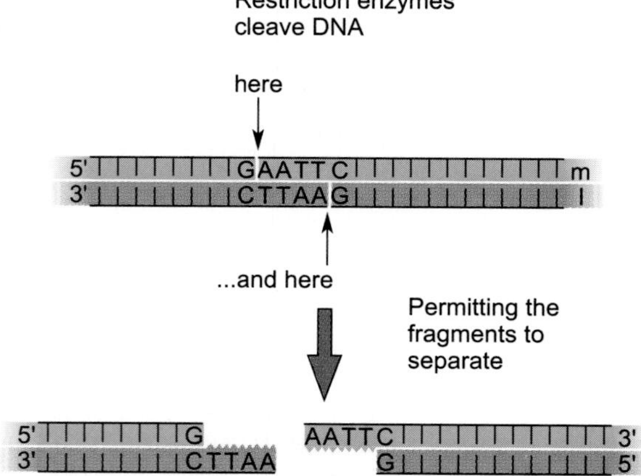

Restriction enzymes cleave DNA
here
5' GAATTC m
3' CTTAAG l
...and here
Permitting the fragments to separate
5' G
3' CTTAA
AATTC 3'
G 5'
Resulting DNAs have complementary or "sticky" ends.

The result is two DNA molecules with **sticky ends**—ends that are uneven, or staggered. Furthermore, these ends tend to join with other molecules with a complementary sequence of nucleotides in the ends—that is, with any other fragments with matching ends cut with the same enzyme.

In the laboratory, fragments of DNA from different organisms, even from different species, can be joined together at their sticky ends, thereby producing recombinant DNA. The enzyme that is used to join two DNA fragments together is called **ligase.** Remember, the ligase and restriction enzymes don't care about the source of the DNA. Wherever this specific target sequence of nucleotides exists, restriction enzymes will cut it. And whenever complementary sticky ends encounter each other, they can be joined by ligase.

So, after ascertaining the existence of a particular gene within the cells of an organism, the DNA from those cells can be cleaved by a restriction enzyme. This results in a pool of DNA fragments with very specific sticky ends. That pool of fragments can then be ligated (joined) to vector DNA molecules.

Splicing and Cloning Genes

The next step is to get those fragments, some carrying the gene of interest, into a bacterial population

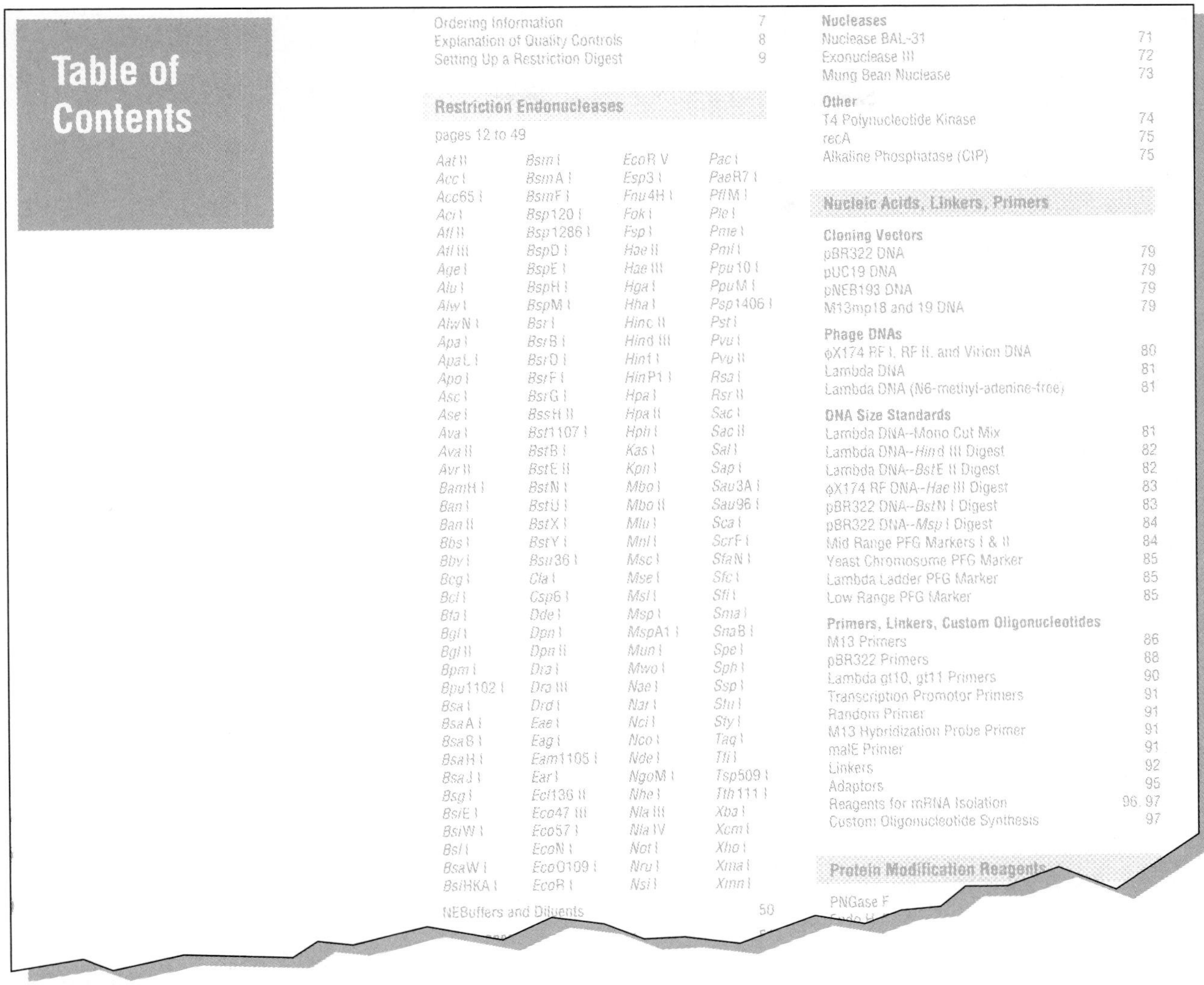
Table of Contents

Ordering Information 7
Explanation of Quality Controls 8
Setting Up a Restriction Digest 9

Restriction Endonucleases

pages 12 to 49

Aat II	*Bsm* I	*Eco*R V	*Pac* I
Acc I	*Bsm*A I	*Esp*3 I	*Pae*R7 I
*Acc*65 I	*Bsm*F I	*Fnu*4H I	*Pfl*M I
Aci I	*Bsp*120 I	*Fok* I	*Ple* I
Afl II	*Bsp*1286 I	*Fsp* I	*Pme* I
Afl III	*Bsp*D I	*Hae* II	*Pml* I
Age I	*Bsp*E I	*Hae* III	*Ppu*10 I
Alu I	*Bsp*H I	*Hga* I	*Ppu*M I
Alw I	*Bsp*M I	*Hha* I	*Psp*1406 I
*Alw*N I	*Bsr* I	*Hinc* II	*Pst* I
Apa I	*Bsr*B I	*Hind* III	*Pvu* I
*Apa*L I	*Bsr*D I	*Hinf* I	*Pvu* II
Apo I	*Bsr*F I	*Hin*P1 I	*Rsa* I
Asc I	*Bsr*G I	*Hpa* I	*Rsr* II
Ase I	*Bss*H II	*Hpa* II	*Sac* I
Ava I	*Bst*1107 I	*Hph* I	*Sac* II
Ava II	*Bst*B I	*Kas* I	*Sal* I
Avr II	*Bst*E II	*Kpn* I	*Sap* I
*Bam*H I	*Bst*N I	*Mbo* I	*Sau*3A I
Ban I	*Bst*U I	*Mbo* II	*Sau*96 I
Ban II	*Bst*X I	*Mlu* I	*Sca* I
Bbs I	*Bst*Y I	*Mnl* I	*Scr*F I
Bbv I	*Bsu*36 I	*Msc* I	*Sfa*N I
Bcg I	*Cla* I	*Mse* I	*Sfc* I
Bcl I	*Csp*6 I	*Msl* I	*Sfi* I
Bfa I	*Dde* I	*Msp* I	*Sma* I
Bgl I	*Dpn* I	*Msp*A1 I	*Sna*B I
Bgl II	*Dpn* II	*Mun* I	*Spe* I
Bpm I	*Dra* I	*Mwo* I	*Sph* I
*Bpu*1102 I	*Dra* III	*Nae* I	*Ssp* I
Bsa I	*Drd* I	*Nar* I	*Stu* I
*Bsa*A I	*Eae* I	*Nci* I	*Sty* I
*Bsa*B I	*Eag* I	*Nco* I	*Taq* I
*Bsa*H I	*Eam*1105 I	*Nde* I	*Tfi* I
*Bsa*J I	*Ear* I	*Ngo*M I	*Tsp*509 I
Bsg I	*Ecl*136 II	*Nhe* I	*Tth*111 I
BsiE I	*Eco*47 III	*Nla* III	*Xba* I
*Bsi*W I	*Eco*57 I	*Nla* IV	*Xcm* I
Bsl I	*Eco*N I	*Not* I	*Xho* I
*Bsa*W I	*Eco*O109 I	*Nru* I	*Xma* I
*Bsi*HKA I	*Eco*R I	*Nsi* I	*Xmn* I

NEBuffers and Diluents 50

Nucleases
Nuclease BAL-31 71
Exonuclease III 72
Mung Bean Nuclease 73

Other
T4 Polynucleotide Kinase 74
recA 75
Alkaline Phosphatase (CIP) 75

Nucleic Acids, Linkers, Primers

Cloning Vectors
pBR322 DNA 79
pUC19 DNA 79
pNEB193 DNA 79
M13mp18 and 19 DNA 79

Phage DNAs
φX174 RF I, RF II, and Virion DNA 80
Lambda DNA 81
Lambda DNA (N6-methyl-adenine-free) 81

DNA Size Standards
Lambda DNA–Mono Cut Mix 81
Lambda DNA–*Hind* III Digest 82
Lambda DNA–*BstE* II Digest 82
φX174 RF DNA–*Hae* III Digest 83
pBR322 DNA–*BstN* I Digest 83
pBR322 DNA–*Msp* I Digest 84
Mid Range PFG Markers I & II 84
Yeast Chromosome PFG Marker 85
Lambda Ladder PFG Marker 85
Low Range PFG Marker 85

Primers, Linkers, Custom Oligonucleotides
M13 Primers 86
pBR322 Primers 88
Lambda gt10, gt11 Primers 90
Transcription Promotor Primers 91
Random Primer 91
M13 Hybridization Probe Primer 91
malE Primer 91
Linkers 92
Adaptors 95
Reagents for mRNA Isolation 96, 97
Custom Oligonucleotide Synthesis 97

Protein Modification Reagents
PNGase F

FIGURE 16.1 **GENETIC ENGINEERING CATALOGS.**
This is a reproduction of the table of contents of a catalog from one of many companies in the business of selling reagents for molecular biology research. Note the large number of different restriction enzymes available. The catalog also lists many of the other enzymes mentioned in this and prior chapters, such as DNA and RNA polymerases and DNA ligase.

where they can be cloned. This is done by using a vector molecule. *E. coli,* and many other bacteria, contain a large, circular chromosome and may also contain small circles of DNA called **plasmids** (Figure 16.2). Many plasmids have been modified by scientists to be used as vectors for molecules of DNA.

By using some of the same restriction enzyme that broke apart the source DNA, the plasmid rings can be cleaved at those places where the appropriate nucleotide sequence occurs. If the source DNA fragments and the opened plasmids are then mixed together, the source fragments can be spliced into the plasmid molecule, as their complementary sticky ends fuse (in the presence of ligase).

Now the plasmids, with their new DNA insertions carrying the gene of interest, can be inserted into host bacterial cells. This can be done by subjecting the new host cells to calcium, which renders their cell wall permeable to the altered plasmids. Once the altered plasmids are inside their new host cells, they will reproduce themselves, and their new DNA sequence, each time the host cell reproduces.

ESSAY 16.1 How Bacteria Have Sex

Now let's consider a topic you may not have thought much about: how bacteria have sex. Many kinds of bacteria, including *E. coli,* normally reproduce through a procedure quite similar to mitosis. That is, they reproduce their genetic component and then pinch in two, each half receiving identical genes. But *E. coli* also has ways of recombining its genes in a process somewhat akin to sex.

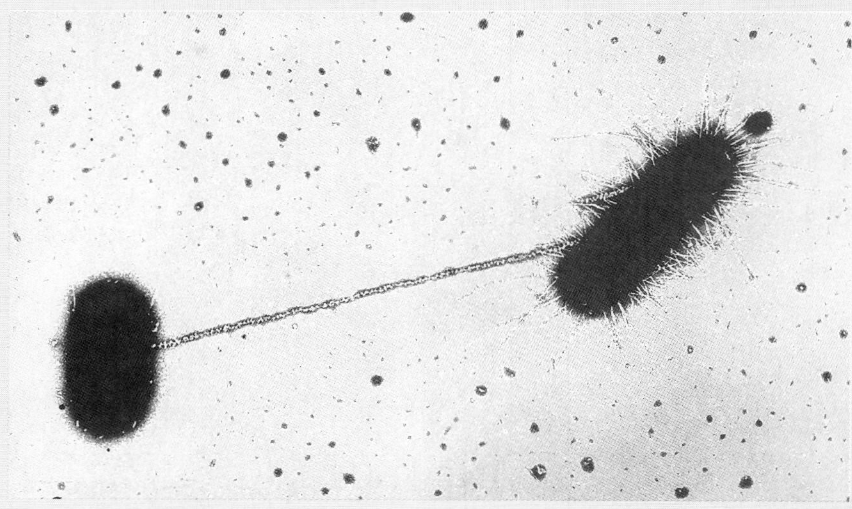

Two bacteria involved in the transfer of genetic material from one organism to another through a conjugation tube.

According to those who have seen bacterial sex, *E. coli* doesn't really have the hang of it (but then, who does?). However, in about one case in a million, two bacteria will undergo something called *conjugation.* In conjugation, one bacterium forms a long, slender tube that penetrates the body of an adjoining bacterium. Then, a type of genetic material moves from the bacterium that produced the bridge through the bridge into the body of the second bacterium. In this way, genes from one organism can join the genes of another organism, and as the genes are shuffled about, recombination can take place.

Now the story gets even stranger. In a nutshell, maleness has turned out to be a kind of disease (long suspected anyway). It is now known that populations of *E. coli* are composed of two "mating types" that were (a bit hastily) labeled "male" and "female." Each type will conjugate only with the other type. The label "male" was given to the donors of genetic material during conjugation, and the label "female," to the recipients. In conjugation, once the tubelike bridge between the two mating types has formed, certain of the "male's" chromosomal rings break, and the linear results slowly begin to move across the tube from the male into the female.

But what's this about maleness being a kind of disease? Experiments have shown that when male and female strains are mixed, the males remain males, but about a third of the females also become male. On the surface it seems that maleness is, indeed, somewhat contagious.

The reason for this contagion was a great puzzle until it was learned that maleness is due to the presence of plasmids (see illustration), those small circles of DNA found inside bacteria. Some kinds of plasmids contain only a few genes, and their activity is intensely directed toward their own repro-

In this way the recombinant DNA (source DNA plus vector DNA) is cloned, making many copies that can later be retrieved. The process is straightforward, as we see in Figure 16.3. However, it is not particularly efficient, and not all the host bacteria absorb the altered plasmids, so the researcher has to separate out the recombinants.

SORTING THROUGH THE LIBRARY

Because not all potential host cells will absorb the altered plasmids, at this point the researcher has a population of bacteria, some of which contain plasmids and some of which don't. This population of cells is often called a *library,* because it is a collection

duction. In *E. coli,* the plasmid not only contains the genes that confer "maleness," but it also directs the bacterium to form a conjugation tube. As the tube forms, the plasmid reproduces, forming linear copies of itself. These copies pass through the bridge into the female. The female is, in this way, changed to a male and rendered impermeable to penetration by any other male. (This accounts for the growing number of males in the mating group.) Scientists by now know enough to discard the old "male" and "female" labels. The males simply had an F plasmid and the females didn't. So the males are now considered F^+ and the females F^-.

Conjugation begins with an F^+ cell developing a conjugation tube and penetrating a receptive F^- cell. The DNA of the plasmid then begins to replicate, remaining circular as the copied strand, produced in linear form, breaks away and passes through the tube and into the F^- cell. Once the entire sequence of DNA is inside the recipient and replication is completed, the strand becomes circular, and a new male (F^+) is formed. Now you know about the private life of a common bacterium. And remember, you learned it here.

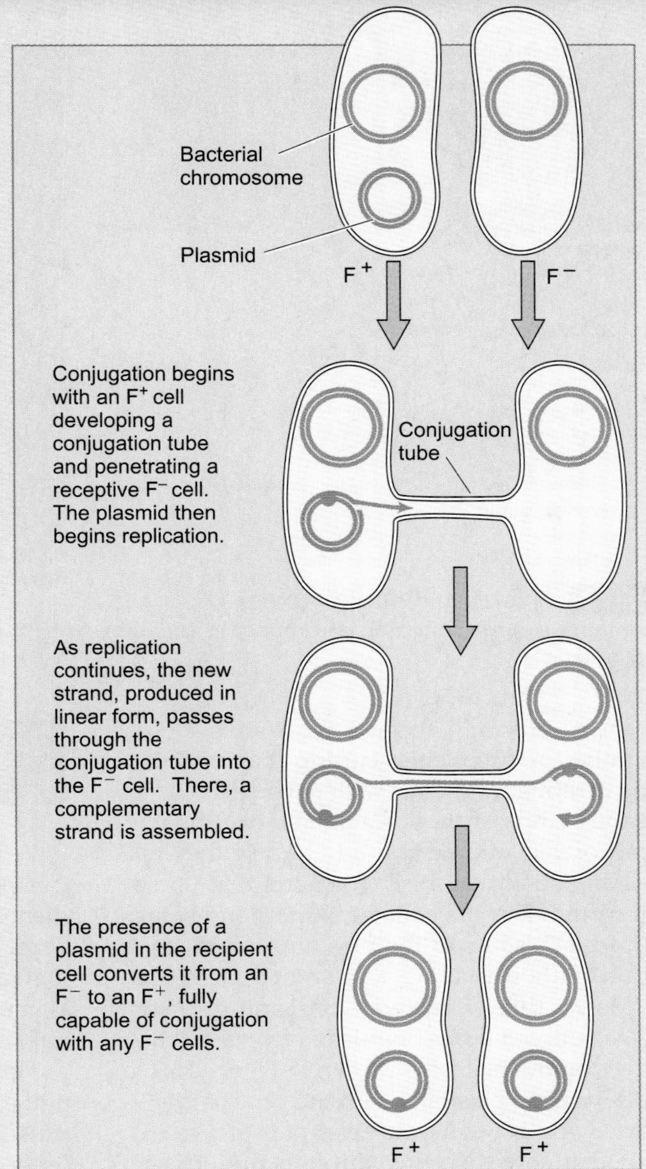

Bacterial chromosome
Plasmid
F+
F−
Conjugation begins with an F+ cell developing a conjugation tube and penetrating a receptive F− cell. The plasmid then begins replication.
Conjugation tube
As replication continues, the new strand, produced in linear form, passes through the conjugation tube into the F− cell. There, a complementary strand is assembled.
The presence of a plasmid in the recipient cell converts it from an F− to an F+, fully capable of conjugation with any F− cells.
F+
F+

of different information. But, also like a library containing books, finding the volume of interest can be a difficult job. The first problem is to find those cells that now contain plasmids and to separate these from those that do not. This is not as difficult as it sounds because most plasmids used in cloning experiments have genes that confer antibiotic resistance to their hosts. So the host population is subjected to antibiotics, such that only those bacteria that now harbor the plasmid can grow (Figure 16.4). Those that don't have the plasmid die.

The next problem is more difficult. Now, the researcher must somehow find those cells carrying plasmids that have the gene of interest. Remember,

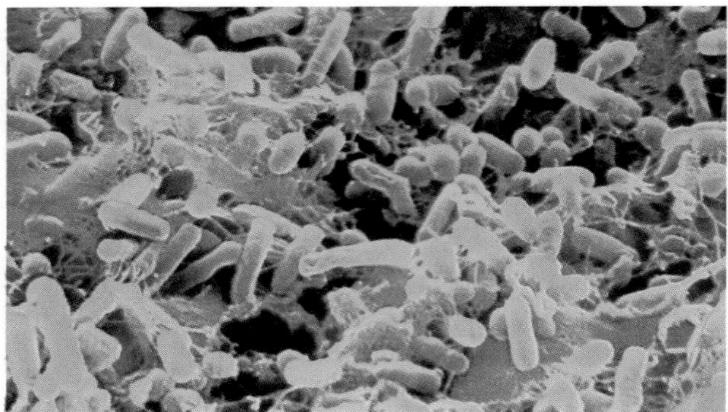

Photomicrograph of *E. coli*

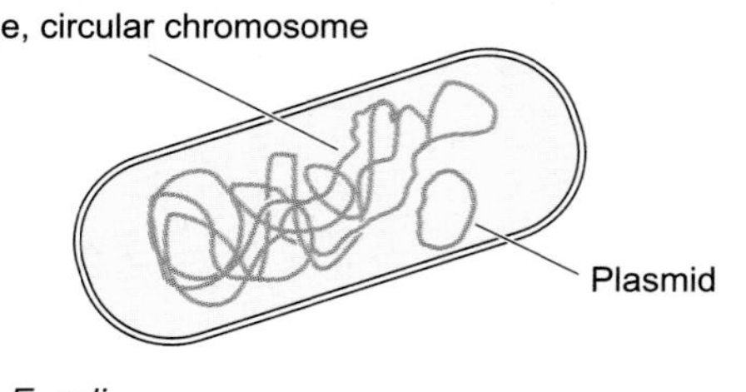

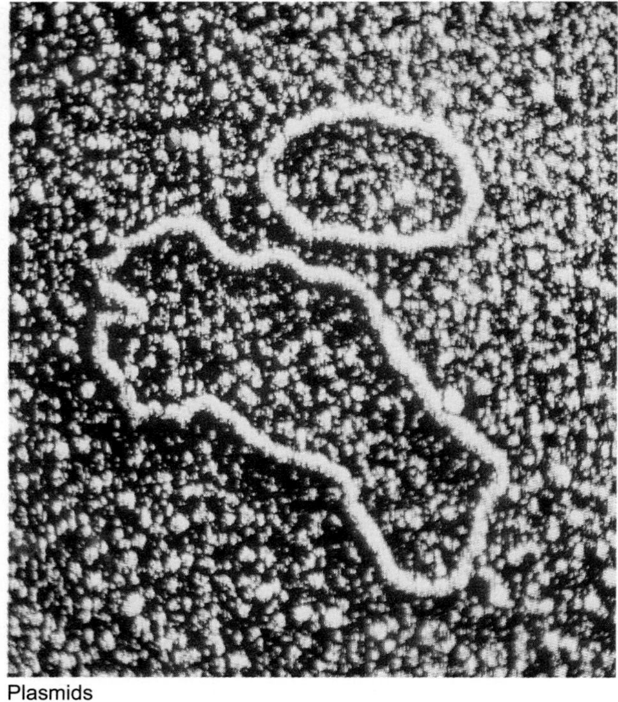

Plasmids

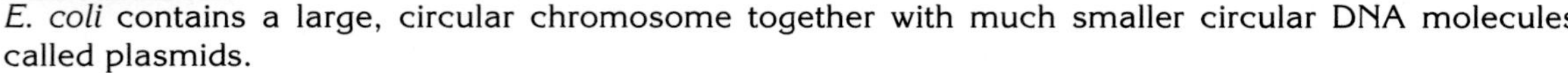

FIGURE 16.2 BACTERIAL CHROMOSOMES.
E. coli contains a large, circular chromosome together with much smaller circular DNA molecules called plasmids.

the restriction enzyme chopped the source DNA into a number of fragments, cutting it in specific places and creating fragments with particular sticky ends. Various kinds of these fragments could have been incorporated into a plasmid because they all had the same sticky ends. This means that some plasmids would bear the gene of interest and some wouldn't. In fact, some of the plasmids could simply reclose, without having any source DNA inserted. (Many times, bacteria containing these "empty" vectors can be distinguished by special indicator dyes on the plates, as shown in Figure 16.4.)

The cells bearing plasmids with the gene of interest are sorted out through rather complex processes. One method involves sorting through the library, looking for the protein that the gene is known to produce. Another method is to look for the DNA sequence itself. If the nucleotide sequence of the gene is known (or can be deduced by analyzing the amino acid sequence of its protein), then complementary nucleic acid segments can be chemically synthesized and used to identify the gene by bonding to its matching sequences. Some laboratories carry such nucleic acid sequences in stock, and the molecules can simply be ordered. (This is called "cloning by phone.")

TAMPERING WITH GENES, OR GENETIC ENGINEERING?

Genetic engineering, as the name implies, involves manipulating genes to achieve some particular goal. Some people object to the entire idea of tailoring molecules with such profound implications for life. Where could it lead? Would we have the wisdom not to unleash something terrible on the earth?

Perhaps the greatest threat of recombinant techniques, some would say, lies in its very promise. The possibilities of such genetic manipulation seem limitless. For example, we can mix the genes of anything—say, an ostrich and a German shepherd. This may bring to mind only images of tall dogs, but what would happen if we inserted cancer-causing genes into the familiar *E. coli* that is so well adapted to living in our intestines? What if the gene that makes botulism toxin, one of the deadliest poisons known, were inserted into the DNA of friendly *E. coli* and then released into some human population? One might ask, "But who would do such a terrible thing?" Perhaps the same folks who brought us napalm and nerve gas.

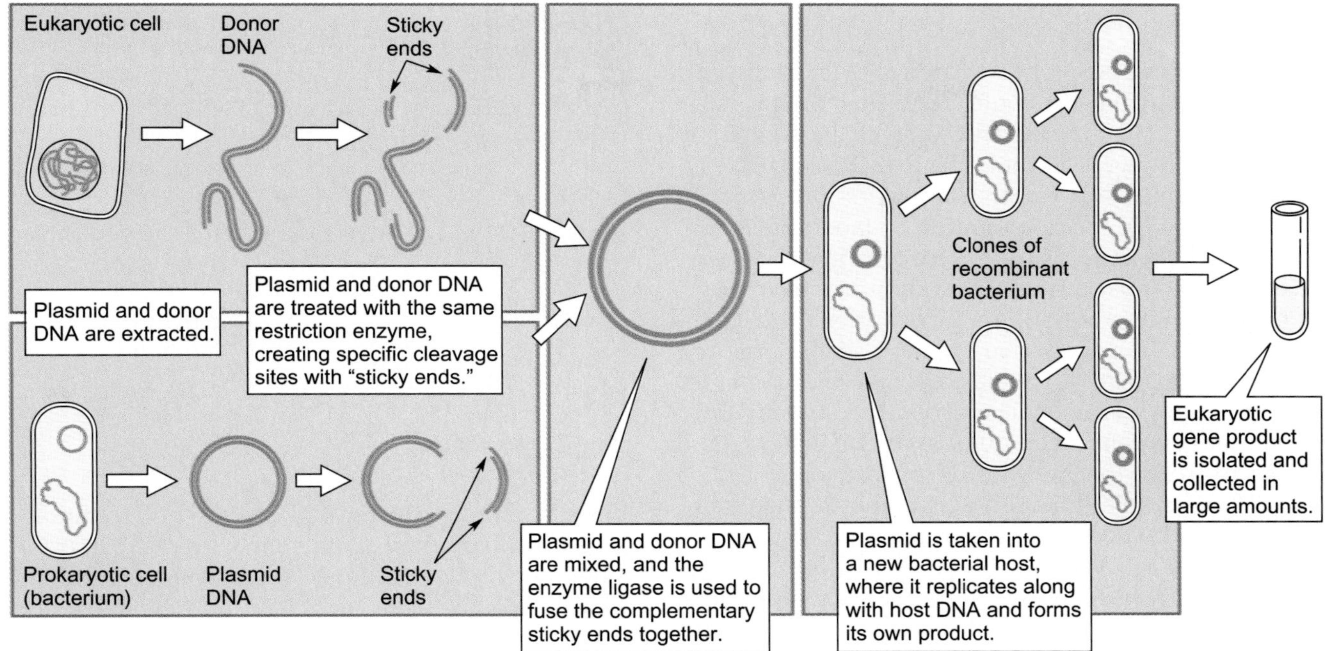

FIGURE 16.3 RECOMBINANT DNA.
Donor DNA from nearly any source can be inserted into a plasmid, where it will be replicated and transcribed as part of the bacterial genome. Bacterial plasmids and segments of donor DNA are obtained by using the same restriction enzyme. The plasmid and donor DNAs are mixed, and the enzyme ligase is then used to fuse the matching sticky ends of the donor DNA to the sticky ends of the plasmid. Next, the plasmid is transformed into a bacterial host, where it will be replicated along with the bacterial DNA.

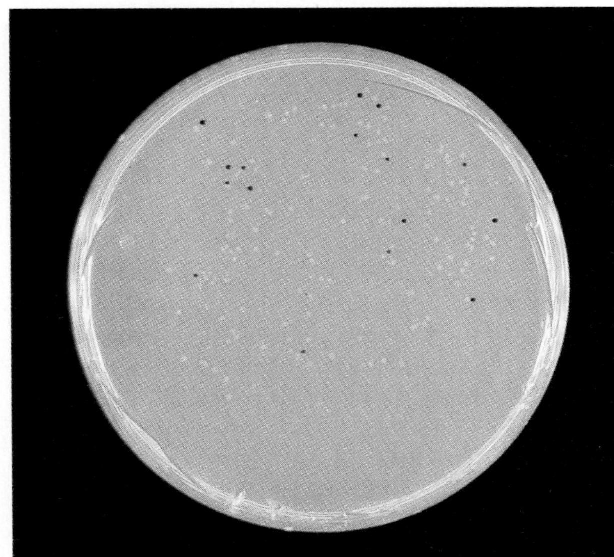

FIGURE 16.4 ENGINEERED BACTERIA.
Because this petri plate contains antibiotic, all of the bacterial colonies contain the plasmid since the plasmid is necessary to confer antibiotic resistance to the cells. In this case, certain of the cells contain foreign DNA in their plasmid such that the colonies are white instead of dark blue.

Another, less cynical concern is that well-intended scientists could mishandle some deadly variant and allow it to escape from the laboratory. Some variants have been weakened to prevent such an occurrence; but we should remember that even though smallpox was "eradicated" from the earth, there were two minor epidemics in Europe caused by cultured experimental viruses that had escaped from a lab. One person died of a disease that technically didn't exist.

Such things, of course, are the stuff of nightmares, and when gene splicing became a real possibility, there was immediate concern. Many of the scientists involved met together in Asilomar, California, in February, 1975, and in some rather heated sessions forged a set of operating guidelines to keep such potential catastrophes from ever becoming a reality. These guidelines were self-imposed and have evolved into quite firm rules at this point. The rules have worked fairly well, though there have been occasional problems. The ethical implications of this technology are at times contentious, and remain an ongoing concern for scientists and nonscientists alike.

How Great a Promise?

There were people who feared the results of gene splicing, but no one could deny that the promise was great and that the successes of the technique were beginning to mount. The beneficial aspects of recombinant DNA research were first proposed in the late 1970s, when a number of laboratories announced preliminary results suggesting that gene splicing could be a practical solution to a great many human problems. As these various small laboratories announced their findings, Wall Street reacted with great verve and fervor, and millionaires were made overnight. The hooks and ladders of financial gain tempted a number of dedicated scientists to escape the ivory towers of academia.

These people—imaginative, highly trained, and highly motivated—saw all kinds of possibilities in recombinant DNA research. They realized that the major ability of genetic engineering is simply to produce great amounts of rare DNA by, for example, inserting the DNA into bacteria (Figure 16.5). It is possible to take a human gene, insert it into a plasmid, grow the bacteria that harbor the plasmid, and then recover the products of the human DNA from vatloads of bacteria.

Figure 16.5 Large-Scale Growth of Bacteria. Industrial fermenters allow the production of essentially unlimited amounts of bacteria containing recombinant DNA.

Table 16.1 Genetically Engineered Products

Product	Function
Human growth hormone	Promotes growth in children with hypopituitarism
Interferon	Helps cells resist viruses
Interleukin	Stimulates the proliferation of white blood cells that take part in immune responses
Insulin	Treats diabetes by enabling cells to take up glucose
Renin inhibitor	Decreases blood pressure

Take, for example, insulin. Insulin had traditionally been harvested from human tissue, or from slaughtered pigs and cows (which yielded a similar kind of insulin that worked in most humans, but caused allergic reactions in others). Unfortunately, very little insulin could be extracted by such methods. Now, though, the gene that codes for human insulin has been isolated, inserted into a plasmid, and cloned—so authentic human insulin is now essentially available for anyone in need.

A good many such success stories now exist. A sample of therapeutic drugs presently manufactured through recombinant DNA is found in Table 16.1. In many cases, treatment with recombinant DNA products has been every bit as good as expected (Figure 16.6).

Advances in Agriculture

Some of the greatest promise of recombinant research lies in agriculture. For centuries farmers have been waging war on plant pests and diseases, infertile soil, and an increasing need to produce more food for the world's burgeoning human population. Genetic engineering offers new hope as it raises the possibility of making plants resistant to insects, bacteria, fungi, and unsuitable growing conditions. Researchers are also working to increase the size of plants' edible parts, such as roots, seeds, and fruits. Further, plant foods can now be made more nutritious through alterations in their amino acid content.

There are both advantages and disadvantages in doing recombinant research on plants. The good news is that several important species (carrots, cabbage, citrus, and potatoes) can be grown from single cells. So once a gene is introduced to a cell, a clone of that cell can produce countless altered progeny. The

FIGURE 16.6 SUCCESSFUL USE OF ENGINEERED PROTEINS.
Human growth hormone, one of many proven therapeutic agents produced through recombinant DNA techniques, has allowed the girl on the left to attain a near-normal height.

bad news is that most plant characteristics that need improvement (such as growth rate, size of edible parts, and amino acid balance) are polygenic—controlled by many genes. The problem is, we haven't identified most of the genes responsible for such traits; and even if we had, the replacement of even five or ten genes would be terribly difficult.

Still, there have been some impressive feats of genetic engineering in plants. For example, not long ago it was discovered that herbicides were ineffective against certain plants. When an area was sprayed, everything died except a few plants that were found to have special protective enzymes. The genes that produced the enzymes were cloned and are now being transferred into crop plants (Figure 16.7). The idea is that, soon, cropland can be sprayed with herbicides, killing noxious weeds but not killing crops. And because specific resistance genes can be developed essentially as needed, researchers can tailor crops to be resistant to herbicides chosen on the basis of reduced or benign effects on the environment.

DIAGNOSIS AND DECISION IN HUMANS

At least 2,000 genetic errors in human metabolism are known. Some disorders appear in the young and kill them or mark them for life. Yet others appear later in life, perhaps after the individual has reproduced and passed on the defective gene. In the past, we could look only at family trees to predict the odds of someone turning up with one of these diseases.

Now, we are able to look at the genes themselves, at least in certain cases (Essay 16.2). If a mutation results in a specific disease, and that mutation occurs in a restriction site (and since there are many restriction enzymes, the odds are pretty good), then the corresponding restriction enzyme will no longer work there. An example involving the detection of sickle cell alleles is shown in Figure 16.8. Here, the restriction enzyme Dde I has three recognition sites within the β-globin gene, such that two fragments are normally formed, one of 170 base pairs and one of 200 base pairs. The size of the DNA fragments is determined by a process called *electrophoresis* (Figure 16.8). However, the mutation that results in sickle cell also destroys one of the Dde I sites. Digestion of the β-globin gene from sickle cell homozygous persons results in only one fragment, which is 370 base pairs long. It is thus possible to tell people if they are heterozygous (carriers) for a

FIGURE 16.7 GENETICALLY ENGINEERED PLANTS.
One successful application of genetic engineering in plants is the production of plants that are resistant to certain herbicides. The plants on the left were engineered for resistance while those on the right (marked by a stake) were not.

ESSAY 16.2 The Gene Machine

New techniques based on plasmid transfer now make it possible to determine the exact nucleotide sequence of any isolated section of DNA or RNA. Already, the entire base sequences of DNA in bacteriophages, plasmids, polio viruses, and human mitochondria have been determined. They produce pages and pages filled with A, T, C, and G. (There are exactly 5,315 in the case of the first bacteriophage sequence to be discovered.) Your own DNA sequence of some 6.4 billion nucleotide pairs would fill a thousand volumes the size of this one. And, incredible as it sounds, the task to sequence all the DNA in a human, the so-called Human Genome Project, is well underway.

And now we have the gene machine. This is a device with Madison Avenue overtones, but it is not expensive, and it does just what it says. An operator can create any DNA sequence simply by typing it out. The computerized machine does the rest. For example, one might know a protein amino acid sequence but be unable to isolate the corresponding mRNA. If the investigator knows the sequence of even a short segment of the gene, it is typed into the gene machine, and a probe locates the rest of the gene from a DNA library consisting of cloned chromosome segments of the species in question. If all one has is a tiny amount of protein, a highly sensitive protein-sequencing accessory is also available. Protein samples placed in the automatic sequenator are dismantled chemically, one amino acid at a time from one end of the polypeptide. These fragments are identified automatically, and the sequence is printed out.

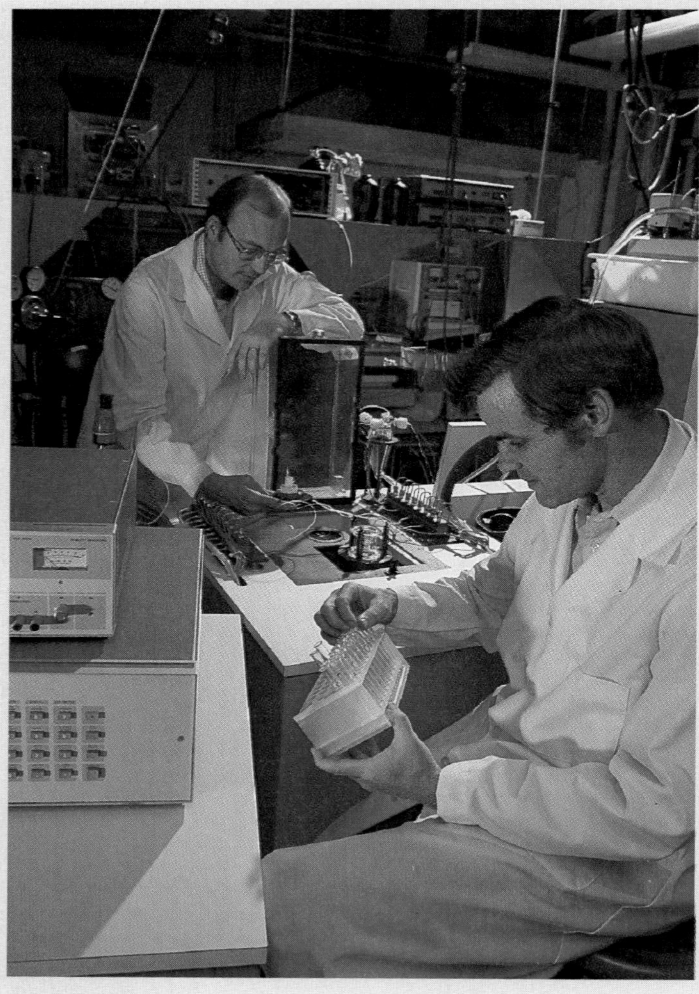

In DNA sequencing, scientists can, for example, analyze a human gene to look for congenital disease. With a gene machine, scientists can create any DNA sequence simply by typing it out—the machine does the rest.

So, once a probable nucleotide sequence is determined, it is typed into the gene machine, which then goes through the steps of that DNA's synthesis. In about a day's time, a small quantity of the desired DNA is produced. Sections are cloned in a bacterial plasmid, which is then taken up by a receptive bacterium. Standard biological gene cloning methods then take over. Potentially, then, any enzyme the experimenter can imagine—or any other kind of protein or gene—can be made to order.

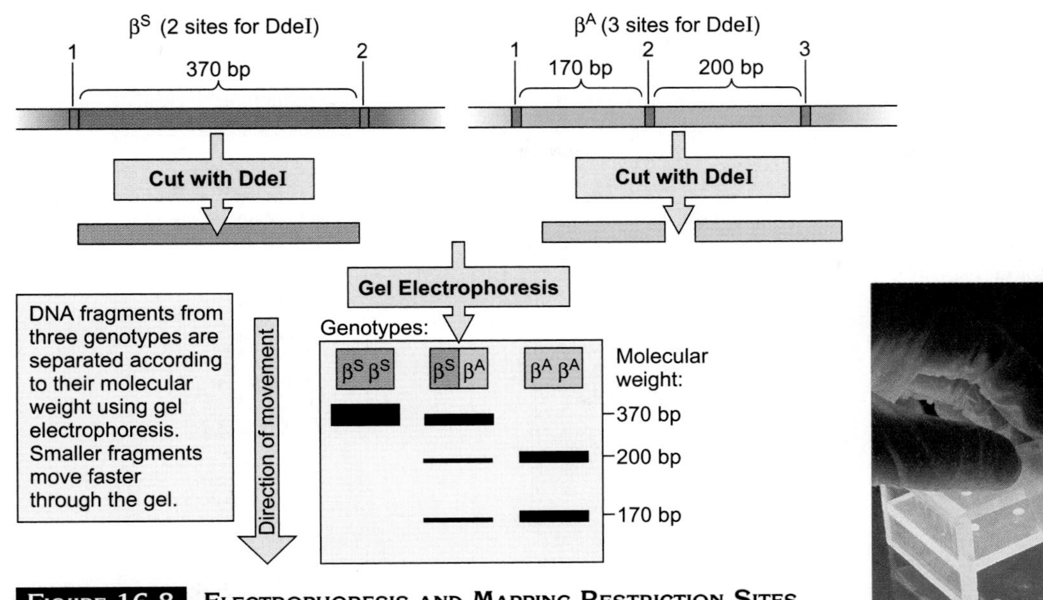

FIGURE 16.8 ELECTROPHORESIS AND MAPPING RESTRICTION SITES. Electrophoresis of DNA fragments separates them by size, with the smaller fragments moving further through a gel. A map of the β-globin gene from humans shows that the β^A globin allele region contains three sites for the restriction enzyme Dde I. In the β^S allele that is responsible for the sickle cell trait, the base sequence does not contain the central Dde I site.

condition (if only half of the restriction sites are broken), because they will show a combination of all three fragments. It is also possible to tell pregnant women whether their fetuses bear certain inborn errors of metabolism.

This ability to determine, in essence, the molecular genotype of an individual has other applications that extend the utilization of DNA technology. For example, forensic scientists have applied the DNA technologies to identify criminals—rapists are the prime example. Using DNA that can be isolated from minute quantities of blood, semen, or even hair, molecular biologists can use restriction enzymes to create a molecular "fingerprint" of the criminal (Figure 16.9). These **DNA fingerprints,** precise patterns of DNA fragments, are extraordinarily accurate—it is essentially impossible for two people (except identical twins) to have the identical DNA fingerprint. DNA fingerprints have already played a major role in convicting criminals. It is equally important that DNA fingerprint analysis has also already resulted in the acquittal of inmates mistakenly imprisoned for rape.

In some cases, such information can result in our having to make very hard decisions. As examples, Alzheimer's disease and Huntington's disease are among the "delayed onset" disorders. If the gene for the disease is present, a terrible fate awaits the bearer of the gene in the middle or late years. There may be two reasons for someone wanting to know if he or she bears the gene for these terrible conditions. One reason is to know whether or not to have children. Huntington's disease, for example, is due to a dominant gene, so if a person bears the gene, there is an even chance that the offspring will be afflicted. Another reason is that, with a dismal and dependent future ahead, one might like to prepare for the worst, even while living each day with such terrible knowledge.

Whereas most people would want to protect any children they might have, many do not want to know about their own fate. Think about it. Would you want to be told that sometime after your thirty-fifth birthday you will begin a period of rapid neural deterioration leading to an early death (Huntington's disease), or that during old age you may begin to lose your memory and soon after most of your mental abilities (Alzheimer's)? A recent study at Johns Hopkins revealed that 60% of us don't want to know.

But what if your condition could be treated by providing you with the normal gene product? Would you change your mind? Such treatment is becoming increasingly likely. Or how about a much more direct treatment—simply replacing the defective gene? We are, in fact, on the threshold of being able to replace defective genes with normally functioning genes.

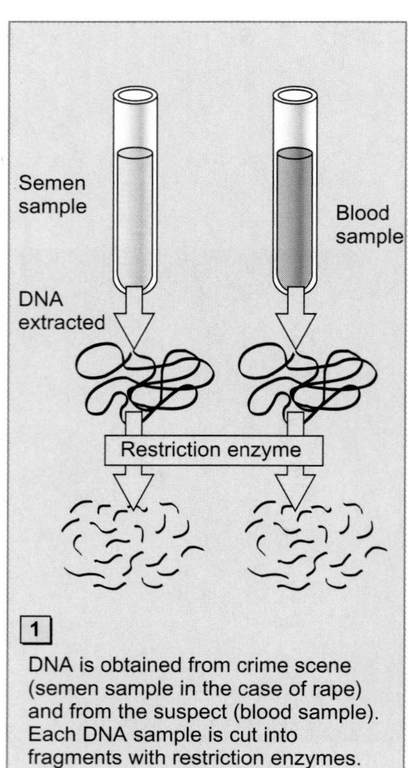

1 DNA is obtained from crime scene (semen sample in the case of rape) and from the suspect (blood sample). Each DNA sample is cut into fragments with restriction enzymes.

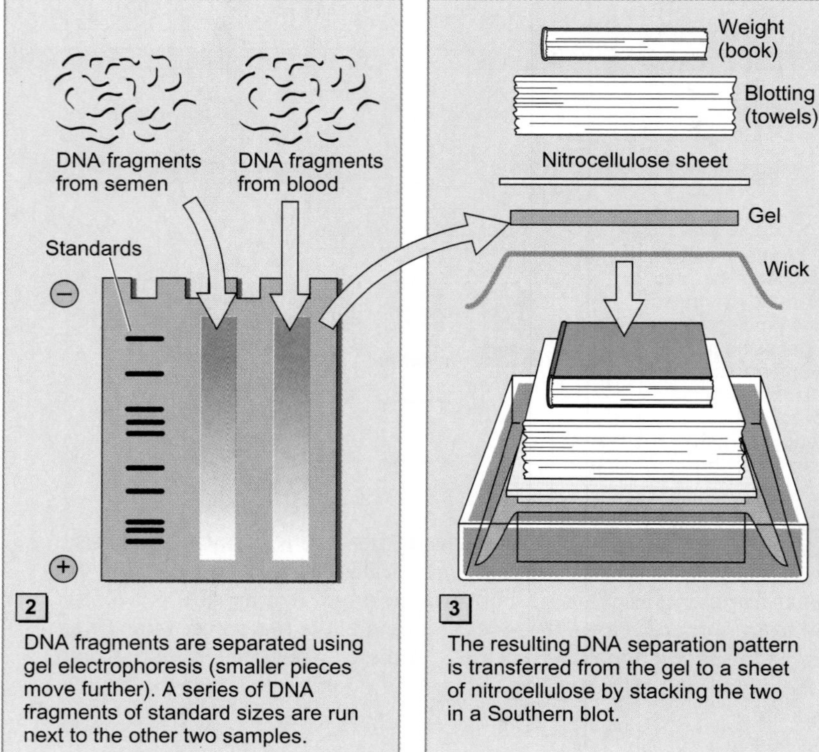

2 DNA fragments are separated using gel electrophoresis (smaller pieces move further). A series of DNA fragments of standard sizes are run next to the other two samples.

3 The resulting DNA separation pattern is transferred from the gel to a sheet of nitrocellulose by stacking the two in a Southern blot.

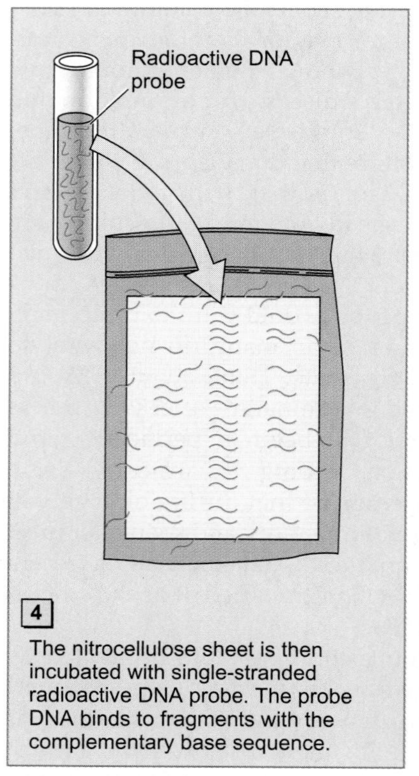

4 The nitrocellulose sheet is then incubated with single-stranded radioactive DNA probe. The probe DNA binds to fragments with the complementary base sequence.

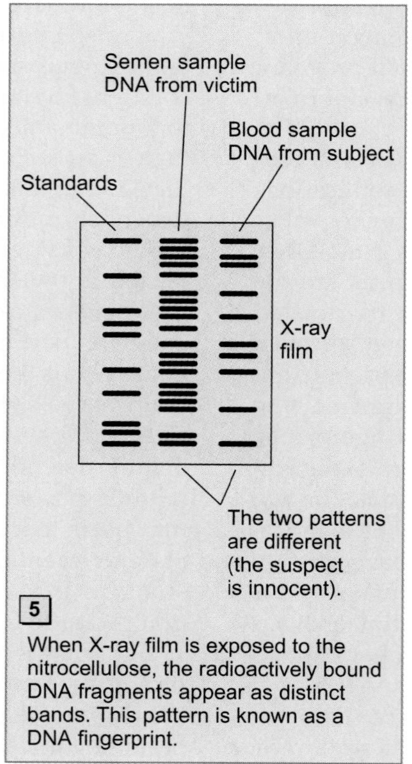

5 When X-ray film is exposed to the nitrocellulose, the radioactively bound DNA fragments appear as distinct bands. This pattern is known as a DNA fingerprint.

FIGURE 16.9 **DNA FINGERPRINTING.**
DNA sequences from a suspect are to be compared with DNA recovered from the scene of a crime. In this example, DNA recovered from a rape victim is compared to DNA recovered from the blood of a potential suspect to test for a match of certain restriction fragments.

Gene Replacement Therapy

Hypogonadism is a relatively common recessive condition in mice and is also found in humans. Homozygotes have underdeveloped gonads, are sterile, and (in mice anyway) seem to have no idea of how to go about mating. The problem is, they are unable to produce a particular hormone (called gonadotropin-releasing hormone, the messenger that stimulates the pituitary to prompt the formation of gametes and sex hormones by the gonads). The condition has been cured in mice by injecting eggs with the normal gene (Figure 16.10). In about 20% of the cases, the egg's chromosomes incorporated the normal genes, and later, when fertilized and reimplanted into a surrogate mother, the altered embryos grew into normal fertile mice.

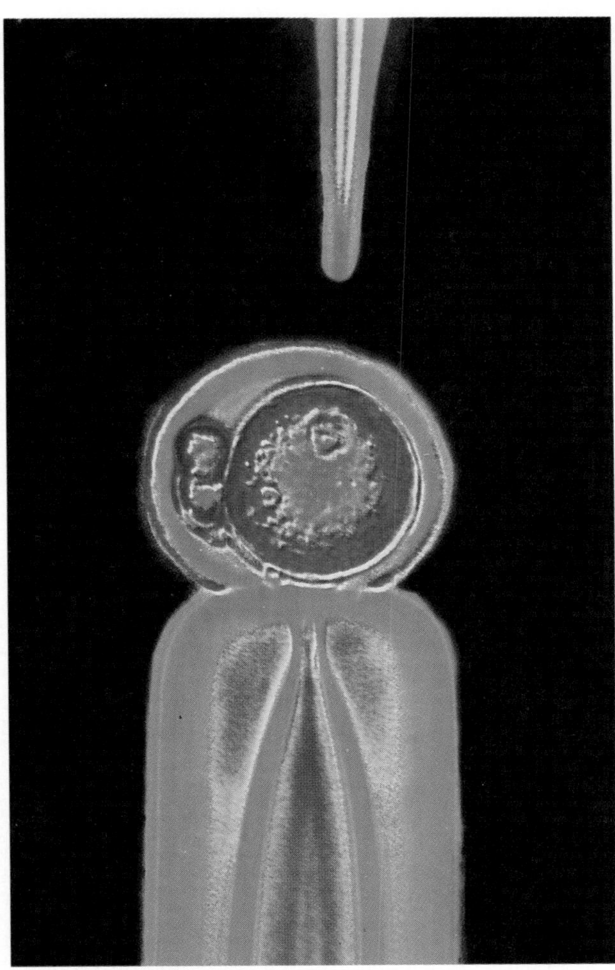

Figure 16.10 Putting DNA into an Egg.
The injecting needle is at the top. The egg itself is held in place by a suction pipette from the bottom.

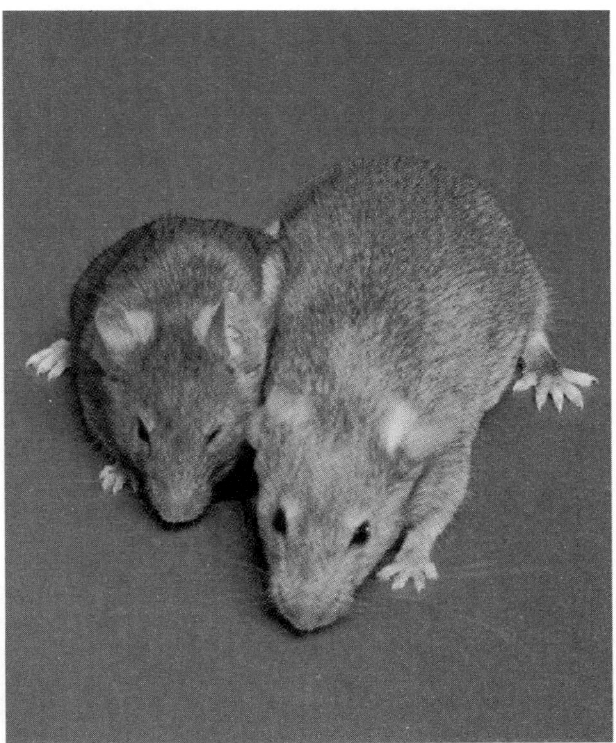

Figure 16.11 Gene Replacement Therapy in Mice.
Here a dwarf mouse has been "cured" by being transformed with a gene for a growth hormone.

Replacing defective genes with normal genes is called **gene replacement therapy.** There have been some apparent successes in experimental animals, such as the widely publicized genetically engineered "cure" of dwarf mice (Figure 16.11). This treatment is, however, the newest application of genetic engineering, so there are many unknowns. Yet the possibilities are very intriguing. No one is seriously considering gene replacement therapy on gametes (eggs and sperm), but replacing defective genes in somatic (body) cells is a real possibility. For example, researchers are working on a treatment for immune deficiencies. Cells are obtained from a healthy person's bone marrow and grown in the laboratory (bone marrow is the source of cells of the immune system). Genes from these normal bone marrow cells can then be introduced by a vector into the cells of an immune-deficient person's bone marrow. Preliminary work with mice indicates that such introduced genes will function normally, correcting the deficiency. (But remember, since the reproductive cells aren't treated, the defective gene would still be passed on to any offspring.)

PERSPECTIVES

We see, then, that the potential of genetic engineering is immense. There probably isn't a genetic defect that can't be reversed in some way, a protein that cannot be made in bacterial cultures, or a gene that cannot be replaced—someday. Of course, most genes will not be altered, for the obvious reasons; but it is important to know that we have the technology to do so.

We have also seen that genetic engineering is nothing more than the directed application of enzymatic and molecular processes that already occur in nature. But given the implications, the question remains as to the degree of our abilities to use such powerful technological tools wisely and well.

SUMMARY

DNA TECHNOLOGY

Gene splicing involves producing chosen DNA sequences (genes), inserting them into plasmids, reintroducing the plasmids into bacterial cells, cloning the cells, and harvesting large amounts of the gene products (such as insulin).

Recombinant DNA is a hybrid of DNA molecules. It is created by cleaving DNA with restriction enzymes, then joining the resulting fragments together in new combinations. Clones are groups of identical organisms derived from a single ancestor.

The problem in cloning a gene is to cleave the DNA into fragments, obtain clones for those fragments, then identify the clone that has the gene of interest.

Restriction enzymes are capable of cleaving DNA at specific recognition sequences. Many restriction enzymes leave sticky ends, which allows fragments to be joined to other DNA with the same sticky ends.

DNA fragments produced by restriction enzymes can be joined together by the enzyme ligase. When cloning a gene, the DNA fragment of interest is usually spliced into a vector molecule, such as a bacterial plasmid.

After a collection of clones exists, it is called a gene library. Often that library must be searched through in order to find bacteria that harbor the gene of interest. This can be accomplished by several means, and it is made easier by some artificial changes in the plasmid vector molecules.

TAMPERING WITH GENES, OR GENETIC ENGINEERING?

The manipulation of genes is a potentially troublesome issue, with ethical, moral, and safety questions. Scientists have conducted recombinant DNA research under strict, self-imposed guidelines.

There are many predicted benefits to genetic engineering. Several examples of genetically engineered products exist, including human insulin. Three areas with particular promise are agriculture, the diagnosis of human genetic diseases, and gene replacement therapy.

KEY TERMS

genetic engineering • **247**
recombinant DNA • **247**
source DNA • **247**
host DNA • **247**
vector DNA • **247**
restriction enzyme • **248**
clone • **248**
sticky end • **248**
ligase • **248**
plasmid • **249**
DNA fingerprint • **257**
gene replacement therapy • **259**

REVIEW QUESTIONS

1. What is meant by the term *clone?* How does it apply to organisms and molecules?
2. Outline the general steps involved in cloning any particular piece of DNA.
3. Define the enzymatic activity and application in genetic engineering of restriction enzymes and ligase. Be specific.
4. What is ligase? How is it used in recombinant DNA techniques? Where and when is it used in normal, living cells?
5. What is a gene library? How does one search through a gene library?
6. What is a gene machine? What are its uses in genetic engineering?
7. Why might insulin produced by genetic engineering be better than that produced by earlier techniques?
8. Describe the process of cloning the gene of a particular protein, and then the engineering of that gene for industrial-level expression of the protein.
9. What is gene replacement therapy, and how might it be accomplished?
10. Why is it unlikely that traits corrected by gene replacement therapy will be passed to the next generation in corrected form?

CHAPTER

17

Changes in the Genetic Information

IN THESE LAST FEW CHAPTERS, WE HAVE EXAMINED THE CELLULAR AND *molecular machineries for the storage, transfer, and expression of genetic information. DNA replication and mitosis virtually insure that each new cell receives the same genetic blueprint as the cell before it. Yet the story of life on this planet is a story of change and diversity. So, one wonders, how can diversity arise when these biological processes so precisely produce exact duplicates?*

It is hard to imagine the genetic diversity of life on this planet. Not only are all the species different, but within the species—even within populations and families—each individual varies genetically from the next. It is just this variation that is so important to the process of natural selection in evolution. In terms of reproductive success, some genetic combinations prove more fortuitous than others, and so come to predominate in future generations. Where does all this genetic variation come from? What causes such diversity? Fundamentally, it begins with a genetic change called a *mutation.* A **mutation** is, quite simply, an alteration in the genetic material.

We should first note that mutations can have quite different effects, depending on where and when they appear. Mutations in any cells that will develop into sperm or eggs can result in seriously ill, malformed (or dead) offspring. If the mutations do not prevent reproduction, the changes will be passed on to the next generation and are called **heritable mutations.** Mutations in other cells of the body are called **somatic mutations** (Figure 17.1). These occur in, say, skin, liver, or muscle cells and will not be passed on to the next generation, although in some cases they can cause cell death or even cancer. Fortunately, many mutations are benign, that is, they neither help nor harm the individual. In general, though, it is important for DNA to be resistant to change.

THE STABILITY OF DNA

The double helix itself provides a measure of protection from the damaging effects of chemicals, since the encoded nucleotide bases are tucked inside and tightly hydrogen-bonded to one another. Further, in the nucleus of eukaryotes, DNA is tightly bound to protective histone molecules and packaged into beadlike nucleosomes. The most important protection that DNA has against random chemical change, however, is the base-pairing organization itself. The specific ordering of bases in one strand uniquely determines the specific ordering of bases in the other, so that there are actually two copies of the coded information in each DNA molecule. If one strand is altered, the cell is usually able to correct the damaged strand and to make a new, perfectly good replacement by using the other strand as a template. Still, damage occurs, sometimes due to specific agents called *mutagens.*

A **mutagen** is any agent that causes DNA damage and produces a mutation. The best-known mutagens are radiation and certain chemicals. Forms of radiation that can be mutagenic are gamma and X rays, alpha and beta particles, as well as the more familiar ultraviolet (UV) light radiation, such as that from the sun or tanning parlors. Thoughts of chemical mutagens may conjure up foul-smelling ooze or dumping sites, but the potential problem is not limited to such things. For example, sodium nitrate, a common food additive that prevents spoilage of hot dogs and lunch meat, has the potential for conversion into a powerful mutagen.

Damaged areas of DNA are known as **primary lesions.** Primary lesions are actually quite common, but fortunately most are quickly corrected by enzyme complexes known as **DNA repair systems** (Figure 17.2). Such repair systems may contain the replication enzymes, DNA polymerase and ligase. Like railroad repair crews, DNA repair systems rove along DNA strands, uncovering irregularities such as broken phosphate bonds and altered or unmatched nucleotide bases. When such irregularities are encountered, segments of DNA are routinely removed and a new, correct sequence is assembled. Although

FIGURE 17.1 SOMATIC MUTATIONS.
Mutations that occur in nongerminal cell lineages are called somatic mutations and will not be passed on to the next generation. Here we see an example of a somatic mutation in a human being. A mutation here produced a patch of cells that have grown a shock of white hair for this person's entire life, but his trait will not be passed on to his children.

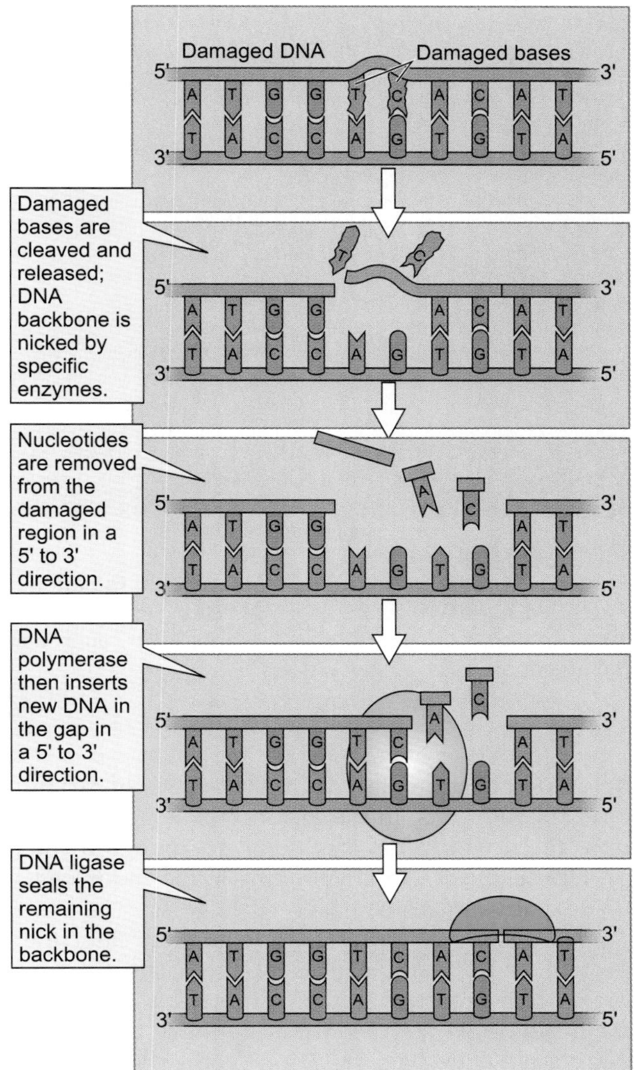

Damaged DNA
Damaged bases
5'
3'
Damaged bases are cleaved and released; DNA backbone is nicked by specific enzymes.
Nucleotides are removed from the damaged region in a 5' to 3' direction.
DNA polymerase then inserts new DNA in the gap in a 5' to 3' direction.
DNA ligase seals the remaining nick in the backbone.

FIGURE 17.2 DNA REPAIR SYSTEMS.
When a damaged portion of DNA is encountered, the faulty section is removed, and a new strand is assembled by matching the nucleotide bases in the remaining strand.

repair systems are quite efficient, some lesions escape detection and sometimes repair systems fail entirely. When this happens the primary lesions become mutations. It turns out that many skin cancers (Figure 17.3) result from failure to repair DNA damaged by UV light from the sun.

MUTATIONS AT THE MOLECULAR LEVEL

Mutations involving single base changes in the DNA sequence are called **point mutations.** These are subtle but potentially quite destructive. They stand a good chance of being passed on from generation to generation. The effects may be neutral or even beneficial, but the effects may also be disastrous. Point mutations fall into two general categories: *base substitutions* (changing one base to another), and *insertions and deletions* (adding or removing bases).

BASE SUBSTITUTIONS

Base substitutions involve a single DNA nucleotide being replaced by another nucleotide. Such changes in a protein-coding region of DNA can replace one of the 64 codons with another. Such changes may be *silent,* and have no effect at all if the new codon happens to be synonymous with the old (as occurs when the third base of many codons is changed; see Table 15.1 on p. 231). Figure 17.4 shows part of the coding region for the β-globin gene in or-

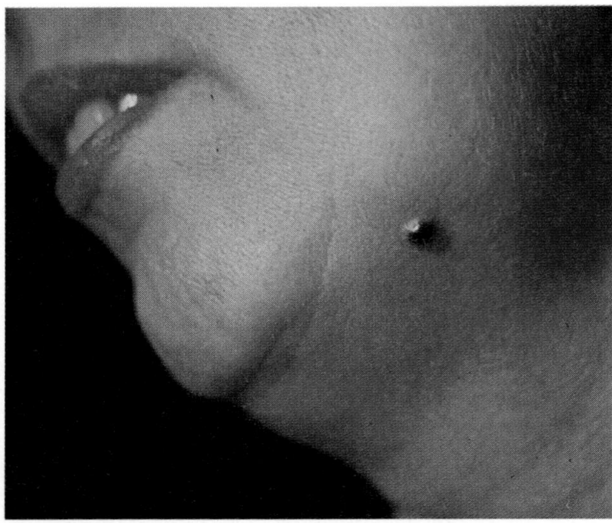

FIGURE 17.3 **SUN AND CANCER.**
The sun is a potent source of ultraviolet radiation. Extended exposure to its DNA-damaging effects can result in sun-damaged skin and skin cancer, such as this melanoma.

der to demonstrate the nature and effects of certain base substitutions. In the instance of a silent mutation, there would be absolutely no change in the protein (Figure 17.4b). Synonymy in the codons has distinct advantages here.

Other base substitutions may actually alter the protein, causing single amino acid substitutions in the protein product. A change from GAA to GUA, for instance, would mean that glutamic acid is replaced by valine (Figure 17.4c). Such a change may produce a severely damaging effect.

In the genetic disease sickle-cell anemia, for example, this single base substitution changes glutamic acid to valine in the sixth amino acid of one of the polypeptides of human globin; the small change causes the red blood cells to distort into a sickle shape and also reduces the ability of the hemoglobin to carry oxygen. These damaged cells can have a drastic effect on the health of the individual, and can eventually lead to fatal blood clots that form in small blood vessels (see Essay 17.1).

If a base substitution happens to create a STOP codon, the mutation is called (not surprisingly) a **chain-terminating mutation.** When the mRNA carrying this new stop codon is translated, the protein will terminate at this new stop codon rather than at the normal stop. A shorter, likely nonfunctional protein is the result. It turns out that some kinds of hemophilia are due to chain-terminating, base-substitution mutations in the genes specifying certain critical blood-clotting proteins.

INSERTIONS AND DELETIONS

An **insertion** is the addition of a base pair into the DNA sequence, and a **deletion** is the removal of a base pair from the DNA sequence. What would you expect to happen when extra nucleotides are inserted into or deleted from the coding region of a gene? The usual effect of inserting or deleting a single nucleotide is to cause a **frame-shift mutation,** a mutation that causes an incorrect sequence of codons to be read by the ribosomes. Ribosomes can read mRNA molecules only three nucleotides at a time. As shown in Figure 17.5, a simple insertion or deletion of a single base can have a complex impact on the protein. The translated protein in this instance would be completely different after the point of the insertion or deletion and would be nonfunctional. Not surprisingly, frame-shift mutations, when homozygous, are usually lethal.

There is a different class of insertion and deletion that deserves special attention, and will lead us into discussion of larger-scale mutations. Biologists have discovered that the DNA can be infected by **retroviruses.** Like all viruses, they are complete parasites and must infect living cells in order to carry on the ordinary life functions. The retrovirus is special in that it inserts its DNA into the host chromosome, where it may remain for some time. Incidentally, the name "retrovirus" ("reverse virus") refers to the peculiar nature of this group. Their chromosome consists entirely of RNA rather than DNA. Upon penetration of the host cell, the virus uses a unique enzyme called *reverse transcriptase* to transcribe its RNA genes into DNA, which can then be inserted into the host chromosome.

What has this to do with mutation? We have seen that the insertion of a single base can throw off the process of translation. Imagine the havoc produced when a retrovirus inserts its DNA into the middle of a gene! This king-sized insertion will completely disrupt the gene, making the production of a functional protein very unlikely.

It has recently been shown that many naturally occurring mutations—studied so intently over the past decades—have, in fact, been changes produced by gene-wrecking retroviruses and other movable pieces of DNA (see below). Included are T. H. Morgan's famous white-eyed fruit fly mutants of 1910 (see Chapter 42). Much more recently, immunologists have determined that the agent of the

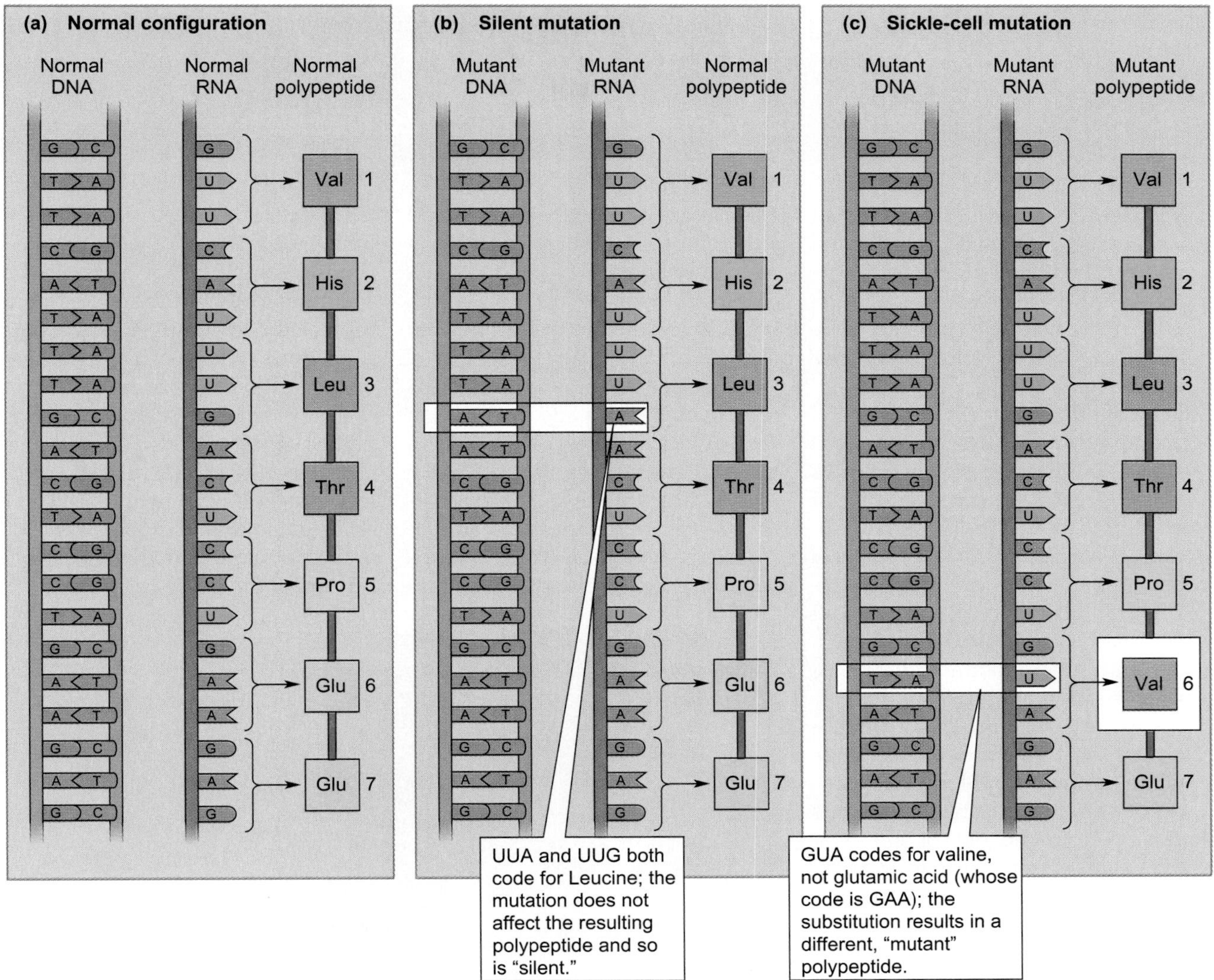

FIGURE 17.4 **THE β-GLOBIN GENE AND THE RESULTS OF BASE SUBSTITUTIONS.**
These examples are taken from real life—they are actual sequences of alleles found in human beings. **(a)** The first 21 bases of the coding region of the normal β-globin gene, and the resulting first 7 amino acids. **(b)** A silent mutation, where the change of a base pair has no effect whatsoever on the resulting amino acids. **(c)** Unfortunately, the base change that results in the substitution of a valine for the glutamic acid at amino acid position 6 has a drastic effect on the ability of hemoglobin to carry oxygen.

frightening disease AIDS is a retrovirus. We'll have much more to say about AIDS in Chapter 40.

MUTATIONS AT THE CHROMOSOME LEVEL

Like DNA itself, chromosomes are relatively stable as cellular structures. Nonetheless, they are subject to wear, tear, and disruptive influences such as heat, radiation, viruses, and mutagenic chemicals. We have seen that very small changes in DNA sequence can cause serious effects in the resulting proteins. Here we are interested in large-scale changes in entire portions of chromosomes, called **chromosomal mutations.** Again, such mutations become especially important when they occur in cells that are destined to enter meiosis and form gametes. You will recall that the early meiotic process—with its highly organized synapsis and crossing over—is quite precise, as is the unique chromosome segregation that follows. It isn't surprising that altered chromosomal structures can play havoc with such highly orchestrated events.

Essay 17.1

Linus Pauling and Sickle-Cell Anemia

In 1949, Linus Pauling, the great theoretical chemist, investigated and purified hemoglobin from persons with sickle-cell anemia, which is characterized by strangely misshapen red blood cells that are inefficient at carrying oxygen. Pauling found that the net electrical charge of purified hemoglobin from sickle-cell-affected homozygotes was slightly different from that of healthy people. The difference in charge was due to a difference of one amino acid out of 287! Heterozygous carriers of the sickle-cell gene had both kinds of hemoglobin.

Pauling demonstrated that very small molecular shifts could be responsible for large-scale phenotypic effects. This one small change in an amino acid sequence produces extremely destructive effects and has initiated one of the most intense research efforts in medical history. The condition is generally lethal in homozygotes, since the sickled cells cannot efficiently bind oxygen and they tend to form clots within the blood vessels. The heterozygotes bear both normal and sickled genes, and their blood may have only about 70% of the hemoglobin content of normal blood. For this reason they tend to suffer from anemia.

Sickle-cell anemia is most common in east Africa and in Ghana in west Africa. It would seem that in killing off its victims sickle-cell anemia would decline in the population; however, it persists because of a peculiar biological twist. Persons heterozygous for the condition are highly resistant to malaria, a disease that kills many people who produce only normal hemoglobin. The people with affected cells are protected because the malaria parasite cannot live within these abnormal cells. With natural selection favoring the heterozygotes, it is apparent how the sickled condition is able to persist in the population. In fact, in some of the more severely stricken malarial regions, over 40% of the population are heterozygous carriers of the sickle-cell gene.

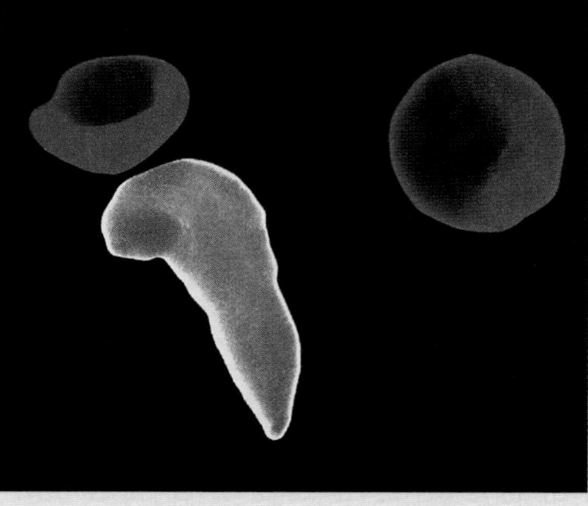

Sickle-cell victims go through periodic crises during which the normally circular red blood cells (top left and right) take on the sickled form (bottom).

Massive chromosomal changes involve the breaking of one or more chromosomes and their rejoining in different combinations. Chromosome breakage is not unusual, and most breaks are quickly repaired. Studies of *Drosophila,* the fruit fly, reveal that more than 99% of chromosome breaks are successfully repaired. Unfortunately, some breaks are not repaired at all, and some are repaired incorrectly.

If a chromosome breaks two or more times, a whole segment may be left out in the repair process. The repaired but shortened chromosome is said to have undergone a *deletion.* Needless to say, cells or offspring with deletion chromosomes, should they survive, will not be normal. A very serious abnormality in humans, called cri-du-chat (cat's cry) syndrome, has been traced to a deletion in part of the fifth chromosome. The syndrome includes, in addition to the plaintive mewing voice the name suggests, serious mental retardation, gross facial abnormalities, and motor difficulties.

Another change in chromosome structure resulting from a double break is an **inversion.** In an inversion, the middle fragment is turned end-for-end before the repair occurs. Although all of the genes may be intact, an inversion chromosome presents complications when it attempts to pair up with its normal homologue in the next meiotic prophase (Figure 17.6).

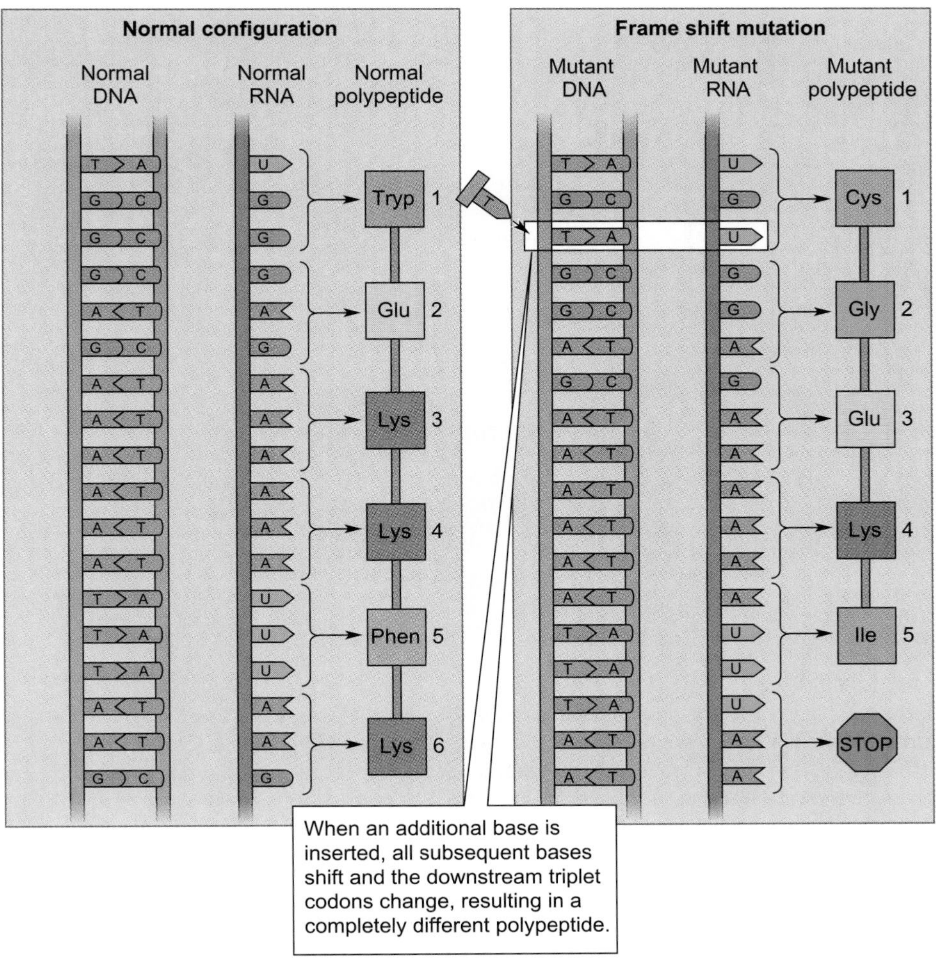

FIGURE 17.5 FRAME-SHIFT MUTATIONS.
Addition or deletion of bases usually has a very drastic effect on the resulting protein. For example, the addition of a base in this sequence creates a totally different protein, since each codon is "shifted" in the reading frame. Almost all of the sequence of the mRNA is unchanged, but the resulting protein bears no resemblance to the original protein.

Even more exotic is the repair error that results in a **ring chromosome.** A large chromosome segment, complete with its centromere, has its two broken ends fused together to form a circle. The genetic effects are variable, depending in part on how vital the lost fragments were. There are reasonably healthy people walking around today with ring chromosomes in every one of their cells. However, ring chromosomes are in some cases associated with hereditary mental retardation. Ring chromosomes, needless to say, also complicate the processes of mitosis and meiosis.

When two chromosomes break, fragments from the nonhomologous chromosomes may be joined, producing what is called a **translocation.** There are several medical problems associated with particular translocations. For example, there is a type of illness called chronic myelogenic leukemia that has been shown to be the result of a translocation involving chromosomes 9 and 22 in humans. Over 90% of patients with this leukemia have what is called the Philadelphia Chromosome. The Philadelphia Chromosome is an altered chromosome 22, which carries a small portion of chromosome 9 attached to its long arm (Figure 17.7). It turns out that the breakpoints of the translocation involve an oncogene, a gene whose action can lead to cancer (see Essay 17.2).

TRANSPOSONS

In somewhat belated recognition of a startling idea and the exhaustive work that explored that idea,

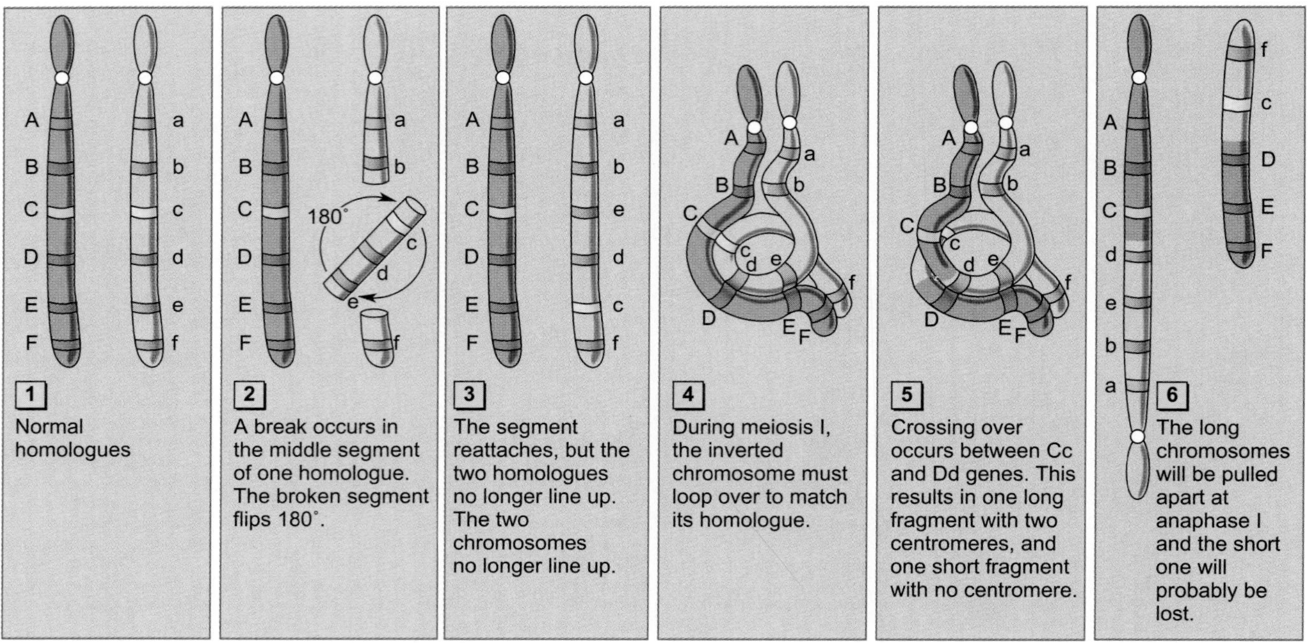

FIGURE 17.6 **CHROMOSOME INVERSIONS.**
The reversal of chromosome segments may have no immediate effects, but at meiosis a different story emerges. The chromosomes become highly distorted as pairing up occurs in prophase of meiosis I. Crossing over complicates matters even further, and when the homologues separate, some segments end up with extra centromeres and others without centromeres. As a result, anaphase of meiosis II becomes hopelessly tangled.

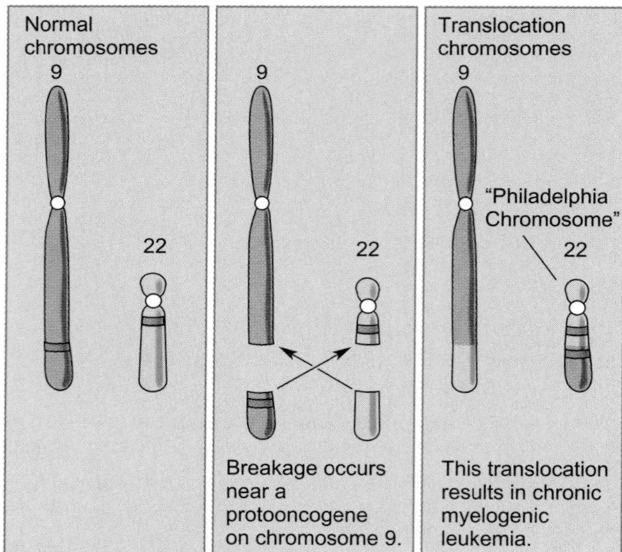

FIGURE 17.7 **CHROMOSOME TRANSLOCATIONS.**
In a documented medical genetic example, a translocation between chromosomes 9 and 22 has resulted in what is called the Philadelphia Chromosome. This chromosome is mainly chromosome 22, but it carries a small part of the long arm of chromosome 9. Unfortunately, the break point of the translocation on the Philadelphia Chromosome often causes leukemia, because the area of translocation involves an oncogene.

geneticist Barbara McClintock (Figure 17.9) was awarded a Nobel Prize for her discovery of the so-called "jumping genes" of corn. The idea of movable or transposable genes called **transposons** (or *transposable elements*) had come to her some 30 years earlier as she observed the appearance of "impossible" combinations of traits in her crosses of maize plants. Their occurrence could be explained only by the movement of genes from one locale to another. Since that time, other researchers have learned that such gene mobility is widespread in both prokaryotes and eukaryotes.

But why have these mobile control elements become so important to our understanding of genetics and mutations? What are they? What do they do? Transposons (there are hundreds of kinds) consist, basically, of two parts—as shown in Figure 17.8. The central part is the gene for *transposase,* an enzyme that enables a transposon to insert a copy of itself elsewhere. On either side of the transposase gene are sequence elements called *inverted repeats* (because the nucleotide sequence of these regions is identical, but in opposite orientations). Simple transposons may consist of these inverted repeats and the transposase gene only. Others, though, are more complex, containing several complete, functional genes in addition to the transposase.

Essay 17.2 Oncogenes and Heritable Cancers

For most of this century, cancer research has focused on viruses as possible cancer-causing agents. Although they have now been ruled out as a major cause of human cancer, virus research has helped identify certain human genes that are implicated in at least some cancers.

Researchers discovered that some cancerous cells contain genes that can be incorporated into viruses and then transmitted to normal cells. Once there, they can help initiate the development of cancer. It has also been discovered that virtually those same genes are actually a normal part of the cell's genome. That is, they occur in normal cells. In normal cells, however, the *protooncogenes,* as the cancer-related genes are known, are under tight control. So far, about 30 protooncogenes have been identified.

Since the hallmark of cancer is uncontrolled cell proliferation, it is not surprising that most protooncogenes function in normal cell division. Apparently, these genes can be mutated or activated inappropriately, converting them to functional, cancer-causing *oncogenes* and causing the rapid and unrestrained growth of cancer cells, which subsequently invade and displace normal tissue.

How does the virus-transported oncogene affect the normal cells? One possibility is the viruses may also pick up and transmit the DNA sequence that can activate uncontrolled cell proliferation. Or it may be that the viruses transmit the protooncogenes into healthy cells without also passing along the DNA control segments, those that restrain the gene and keep it from acting inappropriately (as an oncogene). Thus, while viruses themselves are not actually cancer-causing agents, their control makes them important accessories.

The mechanism of triggering cancer suggested by viral research could also help explain how some chemical carcinogens and ultraviolet rays operate to cause cancer. In fact, we now suspect that several different cancers (particularly leukemias) begin when X rays, UV light, or certain chemical carcinogens cause chromosome breaks or mutations near or within certain protooncogenes.

Chromosome-level mutations such as inversions, translocations, or deletions may convert protooncogenes to oncogenes by severely affecting their expression. Point mutations can also convert protooncogenes to oncogenes. For example, researchers think that genes called *ras* code for regulatory proteins that normally switch back and forth between active and inactive states, depending on conditions. But one of two different substitution mutations (in either the twelfth or the sixty-first codon) is all that is needed to create the oncogene whose product is locked in the active state. Mutation of *ras* is likely to lead to common forms of lung, pancreas, and bladder cancer. *Ras* is particularly susceptible to mutation by chemicals found in tobacco smoke.

Such findings are helpful, but final answers are not around the corner. Before we can say that we truly understand cancer, some hard questions have to be answered. For instance, the notion that cancer begins when a single oncogene is moved from one place to another or mutated is incompatible with the long lapse between exposure to a carcinogen and development of the disease. (Many researchers feel that cancer begins only after two or more oncogenes have been created. This helps explain things, but more evidence is needed.) It might also be possible that the stage is set for carcinogenesis when a single oncogene is created, but that cancer initiation occurs only when normal changes in our bodies—such as those associated with aging—provide the right conditions for oncogene expression.

How do transposons act as mutagens? Obviously, because they move around, and in doing so may insert into a critical part of a gene. It is easy to see how insertion of DNA elements within a chromosome can disrupt gene function. First, insertion into promoter sequences could completely inactivate transcription, or even turn a gene on when it should be off. Second, insertion of large elements could physically separate the promoter from its coding region by thousands of bases. And finally, insertion into coding regions could cause frameshifts or insertions of completely new amino acids.

In addition, the process of moving DNA can itself be mutagenic. Often the excision of a transposable

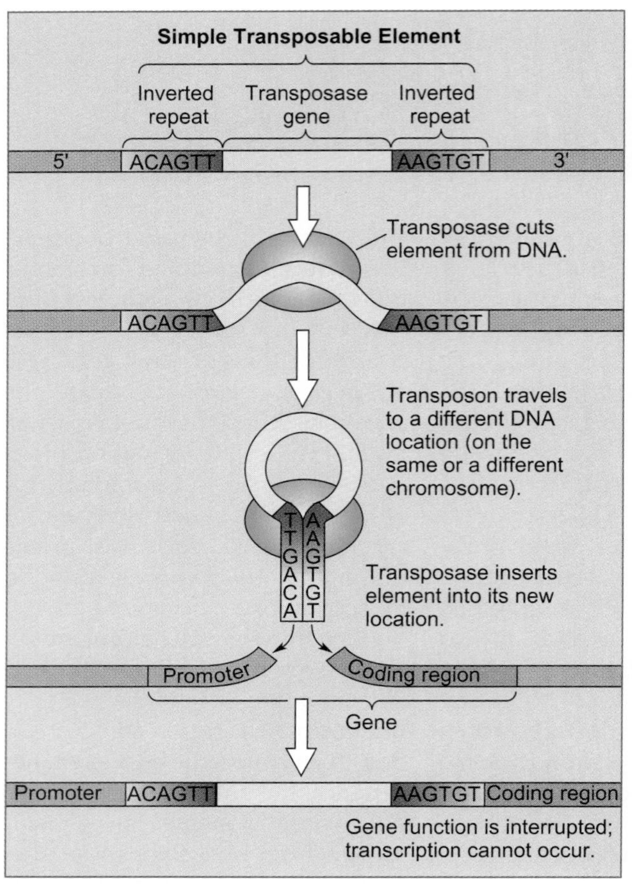

FIGURE 17.8 TRANSPOSABLE ELEMENTS. The simplest form of a transposable element is composed of a pair of inverted repeat sequences (same DNA sequence, but reversed in one of the pairs) that flank a transposase gene. This transposase gene encodes a protein that is capable of recognizing the inverted repeat sequences of the element, cutting the element out of the surrounding DNA, and reinserting the element at a new location.

FIGURE 17.9 BARBARA MCCLINTOCK. It was the careful observation of the genetics and development of color patterns in corn that led Barbara McClintock to the discovery of transposable elements. As transposable elements leave certain color-determining genes during the growth of the maize kernels, colored spots and stripes can occur.

element is not clean, and small insertions of several base pairs may be left behind. Also, some transposable elements leave behind broken chromosomes, resulting in large-scale chromosomal mutations.

Much of the time, transposon movement appears to be random. However, Barbara McClintock (Figure 17.9) suggested in her Nobel lecture that, since movement of transposable elements increases manyfold when cells are placed under life-threatening stress, the process may be an organism's last ditch attempt to generate new variability. Thus, under some circumstances, jumping genes may be vital to the continuity of life, and the movement of genes around the genome is an important aspect of evolutionary change (see Essay 17.3).

ESSAY 17.3

Mutation and the Evolution of Gene Structure

We can estimate the number of genes in the genome of various kinds of organisms, such as *E. coli* (about 2,000), *Drosophila* (about 5,000), and humans (about 120,000). Obviously, gene numbers today vary tremendously from species to species.

But at some time in the distant past there were fewer forms of life with fewer genes each. What is responsible for the changes from this primordial condition, and by what means have they occurred? Intuitively, we might deduce that increases in organism complexity were associated with more enzymes, structural proteins, and regulatory proteins, and therefore more genes. So the question arises, How did the number of genes in evolving populations increase? Most of the answers are based on what we know of transposition and gene duplication. Consider the human hemoglobin gene complex as an example.

The hemoglobin molecules of adult humans are made up of the products of separate genes that code for the α– and the β-globin polypeptide chains. You may recall that in its quaternary level of organization, each hemoglobin molecule consists of two α and two β subunits (see Chapter 3). But at some time in the past, an ancient version of animal hemoglobin is thought to have consisted of only a pair of α-like subunits from a single globin gene. So the initial step in the evolution of hemoglobin, as we know it, must have been gene duplication of a sort that resulted in an extra copy of the α gene. This could occur by the action of translocations or transposons.

Following the creation of two separate α genes (by whatever means), point mutations must have accumulated in the separate genes to change the resulting structures of the two polypeptides slightly, finally creating a pair of molecules that could combine and work together efficiently to carry oxygen in some primordial creature's body fluids.

But the story of the development of hemoglobin doesn't end there. Mammals, you see, have at least four different kinds of hemoglobin; each is used during a different period of development and life. Embryonic hemoglobin appears soon after the embryo is formed. This molecule is well suited to pick up oxygen from the surrounding fluids. Then, somewhat later, following the development of the placenta, embryonic hemoglobin is replaced by fetal hemoglobin, which is well suited to exchange gases with the mother's circulation. Finally, a little before birth, red blood cells carrying one of two forms of adult hemoglobin begin to appear. Adult hemoglobin is especially well suited for picking up oxygen from the lungs and transporting it to the tissues. Just how these important changes evolved was a good mystery until it was discovered that each of the human globin genes is actually a tightly linked gene complex. There are four β-globin genes on one chromosome and three α-globin genes on another.

Thus, gene duplication, transposition, and evolution have worked to allow an elegant solution to the human body's changing oxygen-carrying demands during development.

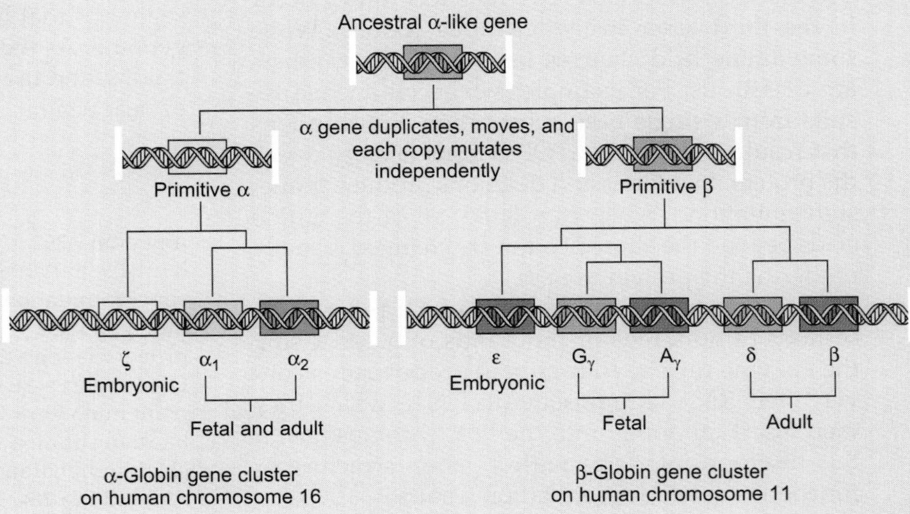

PERSPECTIVES

We come now to the end of Part II. We have covered a lot of ground since our discussion of mitosis and meiosis in Chapter 9. Yet each new discussion—be it gene structure, gene function, or mutation—has enriched our ideas about how the molecular information that guides life is stored, expressed, and passed on to new generations.

In this chapter, we have discussed those mechanisms that allow changes in genetic information to creep into the collective books of life. If a mutation is passed on to the next generation, a new phenotype appears among the individuals that will contribute to future generations. Thus mutations are the agents of change, the source material for the evolutionary process.

As we enter Part III, where we will concern ourselves directly with the evolutionary process, try to remember the basic principles of the role of DNA. Keep in mind the mechanisms that turn the abstract concept of genetic information into the biological reality of a living being.

SUMMARY

THE STABILITY OF DNA

Because the order of nucleotides in one strand of DNA dictates that in the other, the molecule is easily repaired. DNA repair systems detect lesions that are readily removed and replaced by correct nucleotide sequences.

MUTATIONS AT THE MOLECULAR LEVEL

Point mutations—changes in nucleotide sequences in DNA—include base substitutions, insertions, and deletions. Substitutions have no effect if they occur in the third letter in synonymous codons. While some amino acid changes have little effect, others are disastrous. For example, sickle-cell anemia results from a single base substitution. Substitutions that result in the early STOP codons usually wreck the protein. Insertions and deletions produce frame-shift mutations. When translated, all of the amino acids beyond the altered codon are changed, usually rendering the protein useless.

Retroviruses, such as the agent of AIDS, create insertions of hundreds or thousands of bases. Using the enzyme reverse transcriptase, retroviruses convert their RNA chromosome to DNA, which they then insert randomly into the host chromosome. If this insertion interferes with a gene's structure or function, a deleterious mutation results.

MUTATIONS AT THE CHROMOSOME LEVEL

Although most chromosome breakages are correctly repaired, translocations, deletions, inversions, and rings can form. Translocations involve the union of nonhomologous chromosomes, while deletions are lost segments. In an inversion, a segment is turned end-for-end before being rejoined. The union of two broken ends of the same chromosome form a ring chromosome. All chromosome rearrangements create problems when normal and affected chromosomes pair in meiosis.

TRANSPOSONS

Transposons, or transposable elements, are DNA segments that have the ability to move around the genome. As they move, they can leave behind mutations, and they can cause mutation by inserting in or near a gene.

KEY TERMS

mutation • **262**
heritable mutation • **262**
somatic mutation • **262**
mutagen • **262**
primary lesion • **262**
DNA repair system • **262**
point mutation • **263**
base substitution • **263**
chain-terminating mutation • **264**
insertion • **264**
frame-shift mutation • **264**
deletion • **264**
retrovirus • **264**
chromosomal mutation • **265**
inversion • **266**
ring chromosome • **267**
translocation • **267**
transposon • **268**

Review Questions

1. Define mutation. In what ways are mutations significant to the individual? What must happen before they are significant to a population or species?

2. Summarize the work of DNA repair systems.

3. How does a point mutation differ from a chromosomal mutation? List three types of point mutations.

4. How do synonymous codons provide some protection against base-substitution mutations?

5. Explain the mutational basis for the disease sickle-cell anemia.

6. What effects can the insertion or deletion of a single nucleotide base have on the protein produced?

7. Review the manner in which retroviruses insert themselves into host chromosomes.

8. Using simple diagrams, explain how each of the following chromosomal abnormalities might occur: translocation, deletion, inversion, ring.

9. Using one or two examples, explain how chromosome mutations might affect meiosis. Why is this effect so important?

EVOLUTION AND DIVERSITY

The next four chapters DESCRIBE ONE OF THE FUNDAMENTAL CONCEPTS IN MODERN SCIENCE, EVOLUTION. *We begin with the mechanism of natural selection, then examine evolution in populations and the origin of species, and end with a look at our own history on the planet. Then we take a step back and consider some of the ideas about how life began, with natural selection as the mechanism for its perpetuation. From there we begin our tour of life, starting with the tiniest forms and continuing to our own species. Along the way we address each major evolutionary event in the formation of life as it exists today.*

18

The Development of Darwinian Evolution

THIS IS THE STORY OF HOW CHARLES DARWIN SET SAIL IN THE LAST *century and changed the course of science. We will see how Darwin and those who worked around him wrestled mightily with the concept of* natural selection. *Remember, Darwin didn't understand Mendel's work and had no idea what a gene was. We will see how the ideas of Darwin and Mendel were joined with the New Synthesis. Finally, we will consider several lines of evidence for evolution, some of which Darwin knew about, some of which he certainly did not.*

During the early 1800s, while the United States was rationalizing its legalized slavery, the British—under the reign of Queen Victoria—were enjoying a period of relative tranquillity. The British upper classes occasionally took time to peer down their noses at the unseemly behavior of the colonies, but mostly they were busy with their horses, hounds, and gardens. As a nation, they were also setting the stage for world trade. Their ships were everywhere—and their military was everywhere, often accompanied by the more adventurous of British scientists, who were probing, watching, and describing, as well as winning over new friends and colleagues. Among those young scientists of the day was one Charles Darwin (Figure 18.1).

Young Darwin was a well-bred, well-connected fellow of great energy and uncommon curiosity whose social position had helped him land a position on an extended (he thought, overextended) voyage around the world. During the voyage, Darwin saw things that no other British scientists of his time had seen. After his return to England, he put his ideas together and developed a new paradigm that was to shake the intellectual world to its foundations. His grand, new scheme sought to explain the diversity of life through a process called *evolution.* Although the concept of evolution did not originate with Darwin, he was the first to provide an overwhelming amount of observational evidence to support it. He was also the first to suggest a mechanism through which evolution might occur. That mechanism was called "natural selection."

EVOLUTION THROUGH NATURAL SELECTION

Evolution has been described in many ways, but we will define it as *genetic change in a population over time.* Obviously, this is a modification of Darwin's definition, since he had no information about genes. We will, however, adhere to Darwin's notion of how natural selection can result in evolution.

Darwin was the first to propose that evolution can proceed through a process of *natural selection.* Simply stated, **natural selection** is the process by which individuals with certain traits tend to survive longer and leave more offspring than individuals with other traits. The result is that, in time, succeeding populations will come to be predominantly comprised of individuals with those certain traits. Natural selection operates when four conditions are met:

1. *Overproduction.* When organisms produce greater numbers of offspring than can survive.
2. *Limited resources.* When populations of organisms tend to overutilize the resources available to sustain them, resulting in increased levels of competition.
3. *Variation.* When the individuals that make up a population vary one from another.
4. *Survival of certain types.* When individuals with certain traits are more likely to survive and reproduce than are individuals with other traits.

FIGURE 18.1 **CHARLES DARWIN (1809–1882).** Here Darwin is shown at age 31, four years after returning from the voyage of the *Beagle.*

Regarding this last point, the more successful individuals are better suited to a given environment and will therefore tend to leave more offspring than others. Each generation is, then, made up of the offspring of the most successful reproducers of previous generations. In a sense, we see, *nature selects* which lines of living things will survive and reproduce, and which lines will fade and die. (Hence, *natural selection.*)

An earlier biologist, Jean-Baptiste de Lamarck (1744–1829), had suggested that species change through some "force of life" that causes the organism to generate new structures to be passed along to the offspring. Lamarck also developed a "law of use and disuse," which stated that those traits an animal used more would be more strongly passed along to its offspring than those it didn't. ("Use it or lose it," in today's vernacular.) He also believed in the inheritance of acquired characteristics whereby physical changes in an organism could be inherited. For example, if a working man developed heavy calluses, his son would have rough hands. Because Darwin's description of the mechanism of natural selection had greater explanatory power, it replaced Lamarck's notions. The two ideas are contrasted in Figure 18.2.

Scientists today generally accept the idea of evolution; but as biology has grown more sophisticated, various esoteric arguments arise over just how it operates or what mechanisms are involved. For example, some researchers argue strongly over whether evolution occurs *primarily* through natural selection. Because of such arguments (which can quickly become rather technical), it may appear that there is disagreement over *whether* evolution occurs, when actually the argument is usually over *how* it occurs.

Replacing Some Old Ideas

The theory of evolution challenged the idea that each species was a permanent, fixed entity. Evolution implied change, after all, not fixity. The idea of *perfection* in nature also fell with the acceptance of the theory of evolution. Naturalists had hitherto assumed that biological organisms were perfectly adapted to their environments; Darwin showed that living things, no matter how admirable or wonderful, were not perfect but always open to change and improvement. The acceptance of Darwin's theory meant that the assumption that nature was full—that there was a place for every creature, and that every creature was in its place—was discarded. Gone was the idea of the "balance of nature," the deep-seated, almost mystical belief that all creatures interacted in a harmonious way that ensured that all would prosper. Gone was the idea that some organisms existed for the benefit of others—the lamb for the wolf or the flower for the bee. Swept away, in fact, was the more recent, hard-won conviction that biology, like physics, was governed by fixed laws, and that all of nature was predictable if only the laws were known. In the place of these comforting ideas, Darwin offered only the cold arithmetic of inequality and differences in reproductive success.

The Sources of Darwin's Ideas

The young Charles Darwin was something of a problem to his wealthy family. He had already failed at medical school and would not be following in the footsteps of his father and grandfather, both successful and famous physicians. He had trained for the clergy, but he showed little inclination for ecclesiastical pursuits. It seems that he was most interested in collecting beetles and in gathering rocks. Natural history and amateur geology were acceptable hobbies for a young gentleman, and the family had plenty of money—but Charles was 22, and his father felt that he should have a job. (Does that sound familiar?) Darwin in those days was given to carousing with his friends, enjoying fine wine and

FIGURE 18.2 **A COMPARISON OF EXPLANATIONS.**
Lamarck believed that giraffes have long necks as a result of reaching for leaves. The necks grew long because they were stretched upward during feeding and those traits were then passed along to the offspring. Darwin believed that in each generation, some individuals would have longer necks than others and so be able to reach more food, which would give them an advantage in survival and reproduction.

playing blackjack until the wee hours, and his father was not amused.

Meanwhile, another young, wealthy, and rather more successful fellow, Captain James Fitzroy of the Royal Navy, needed a companion of his own social class to accompany him aboard the H.M.S. *Beagle* (Figure 18.3) on a five-year expedition around the world (1831–1836). The captain's companion would have the official position of Naturalist, and was expected to collect and study plants and animals

FIGURE 18.3 THE BEAGLE.
A small vessel, the *Beagle* was just under 100 feet in length, but her strong-willed captain, James Fitzroy, an expert navigator, guided her unerringly through a five-year voyage around the globe.

along the way. The position was to be without pay, and of course only gentlemen need apply. In those days the "real" naturalist on such voyages was usually the ship's doctor, and so the title was normally only a formality.

Charles, at this point, was trying to decide whether to enter the clergy. But one of Darwin's professors knew of the position, talked it over with Charles (who agreed it would be a wonderful opportunity and great fun to boot), and recommended him to Captain Fitzroy. Fitzroy almost rejected Darwin because he believed a man's character is reflected in the shape of his nose, and Darwin's nose just didn't show much character.

But when Darwin presented the idea to his father, it went nowhere. Dr. Darwin, whose girth matched his great prestige, was dead set against such a waste of time. Finally, though, he told Charles that if he could find one person of good judgment to agree that the trip was a good idea, he could go. Charles found such a man in the person of his uncle, Josiah Wedgwood, of the Wedgwood pottery firm. So Charles's father reluctantly agreed to let him go. Dr. Darwin had no way of knowing just what events he had set in motion.

Darwin himself had some trepidation about going, as witnessed by a letter he sent to his sister Susan:

> Fitzroy says the stormy sea is exaggerated; that if I do not choose to remain with them, I can at any time get home to England; and that if I like, I shall be left in some healthy, safe, and nice country; that I shall always have assistance; that he has many books, all instruments, guns, at my service. . . . There is indeed a tide in the affairs of men, and I have experienced it. Dearest Susan, Goodbye.

The H.M.S. *Beagle* set out from Davenport in 1831, with Darwin's bunk crammed into the captain's map room. Darwin quickly made an important discovery: He tended to get seasick. With Darwin hanging over the rail, the boat continued relentlessly on toward South America, and at every landfall a grateful Darwin would leap ashore and head off on some terrestrial investigation. Among his numerous observations, Darwin was struck by the fact that many distinctly local mammals and birds were doing the same things in the same ways as other mammals and birds in other parts of the world. For instance, the pampas grasslands were the home of the mara (Figure 18.4), a mammal that looked like a rabbit and behaved like a rabbit; it was not a rabbit, however, but a rodent related to other South American rodents such as the guinea pig. Why would it be behaving like a rabbit? Where were the rabbits? It is from just such basic questions that great science can arise.

Some naturalists, if pressed, might have said that the pampas was the right place for this creature just as Europe was the right place for rabbits. Darwin, however, thought that there must once have been

FIGURE 18.4 THE MARA.
Otherwise known as the Patagonian hare *(Dolichotis patagonum)*, the mara is a South American rodent that lives much as a rabbit does.

an empty place in nature, an opportunity for a rabbit, or rabbitlike creature, to survive and flourish. But since European or North American rabbits had no way to cross the ocean, their place had been taken by a South American rodent whose descendants became increasingly suited, over time, to the grassland habitat—that is, they became increasingly rabbitlike.

Darwin reasoned that the mara was able to take this adaptive route because rabbits simply could not cross the ocean and establish themselves there. It began to dawn on him that chance plays a large role in the history of life. He would conclude that evolution depends not only on the chance occurrence of heritable variations on which natural selection could act, but also on such chance factors as what kinds of plants and animals were available to exploit opportunities in nature, or what kind of geographic features or barriers would prevent (or facilitate) the spread of living things from one place to another.

The picture became clearer to Darwin as the ship pressed up the west coast of South America and out to a group of remarkable little islands off the coast of Ecuador, the Galápagos Islands (Essay 18.1). They rose as volcanos from the Pacific Ocean floor in fairly recent geological times (on average about 100,000 years ago). The animals and plants that live there now are apparently the descendants of a chance assemblage of random migrants that floated or were blown there—where they encountered a variety of new opportunities. Darwin, as usual, began making observations, taking notes, and with his shotgun collecting examples of the life he found there, particularly birds. When Darwin collected the birds, he assumed that they were an unrelated motley collection of blackbirds, finches, and wrens. But later analysis by experts in England revealed that these ecologically very different birds were all finches that were rather closely related to one another. It is clear from Darwin's journal that this belated revelation—two years after the *Beagle*'s return—was the shock that suddenly opened his mind to the idea of evolution. He realized that all of the Galápagos land birds had probably descended from mainland finches that had been storm-blown to the islands tens of thousands of years previously. The descendants of these pioneers eventually diverged into a number of distinct species (Figure 18.5).

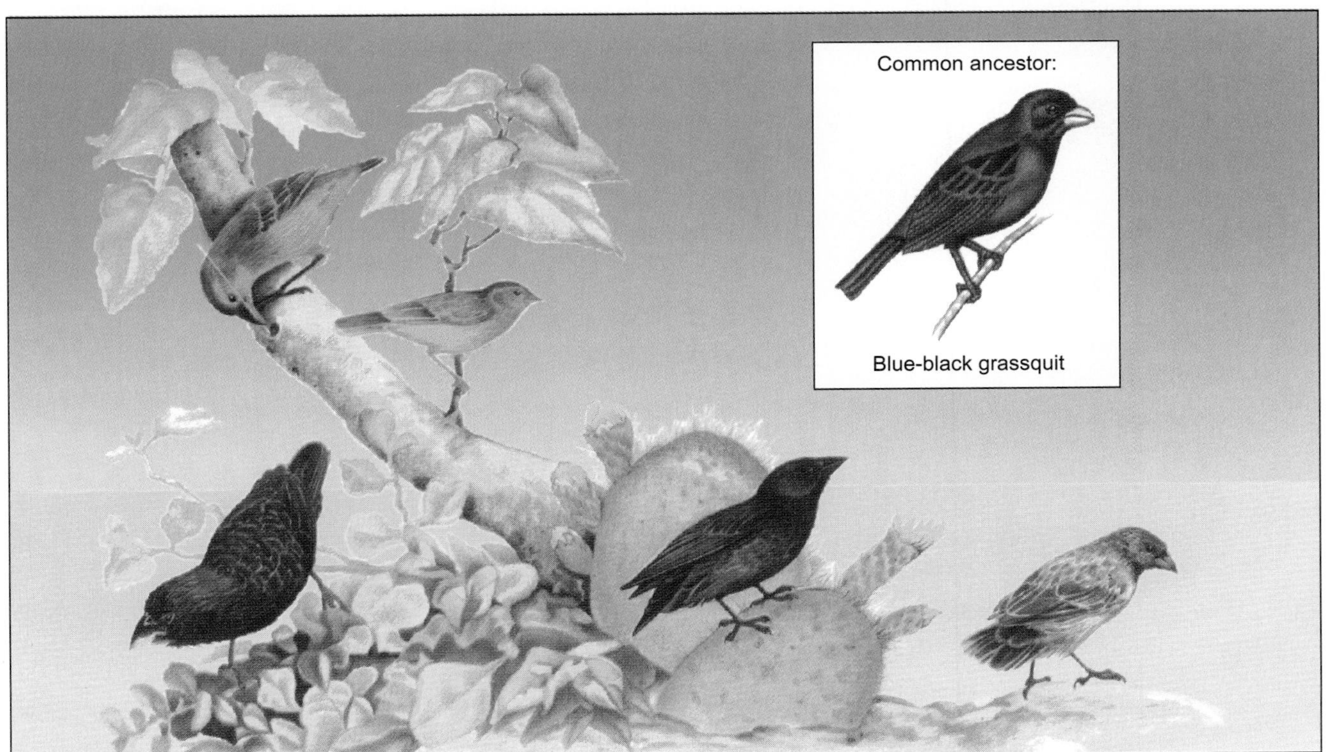

FIGURE 18.5 DARWIN'S FINCHES.
Although they are now considered separate species, evidence from many studies clearly indicates that the different kinds of Darwin's finches evolved from a common finch ancestor.

ESSAY 18.1 The Galápagos Islands

The Galápagos Islands include habitats ranging from dry, lowland deserts to wet, species-rich highlands. The islands are home to a bizarre collection of life, from grazing lizards and giant tortoises to a variety of bird life and shore dwellers. Darwin, who despised the place, was to make it famous because he saw it as an experiment of nature in progress.

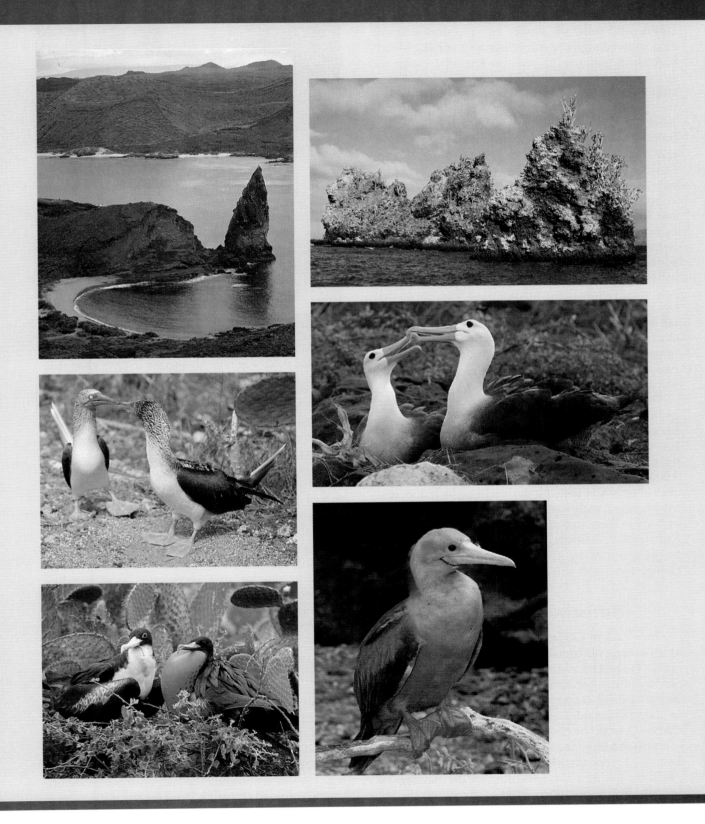

The Impact of Lyell

Darwin was not the only scientist asking new questions about life. Less hampered than other disciplines by religious tenets, the physical sciences also showed signs of rumblings and stirrings—disturbances that would one day make it easier for some of their most notable authorities to fall into step with Darwin as he slowly marched away from the parade. Among the physical scientists also marching to a different drummer and developing bold new ideas about the earth was Charles Lyell (1797–1875), himself only a few years older than Darwin (Figure 18.6). Lyell had set forth many of his new ideas in his book *Principles of Geology*, the first volume of which was published before the *Beagle* set sail. Darwin had acquired a copy and had asked to have the second volume sent to him en route. Lyell (who was to become a good friend of Darwin) had some rather startling things to say about the physical evolution of the earth. He said that the world was much older than anyone had imagined; that over long periods of time, continents and mountains rose slowly out of the sea; and that they just as slowly subsided again or were washed away. Most importantly, Lyell claimed that the very forces that had so changed the earth in the past were still at work, and that the world was still changing.

Darwin's own observations of South American geology seemed to confirm Lyell's position at every hand. In his adventurous climbs in the Andes, he found fossil clam shells at 10,000 feet. Below them, near what appeared to be an ancient seashore at 8,000 feet, he found a petrified pine forest that had clearly once lain beneath the sea because it, too, was interspersed with seashells. In fact, the *Beagle* had arrived in Peru just after a strong local earthquake had destroyed several cities, in some places *raising the ground level by two feet.* Clearly, the earth had changed and was still changing.

FIGURE 18.6 CHARLES LYELL.
A close friend of Darwin and perhaps the greatest geologist of their time, Lyell said that the earth was ancient, and thus there was time for species to have changed.

The Impact of Malthus

Once back in England, Darwin was grateful to be again among his friends, his new colleagues, and his books. He had been enthusiastically accepted into the world of British science, and after a flurry of activity, he married his cousin, Emma Wedgwood, retreated to the countryside, and began to enjoy the quiet mornings when he could find time to work and read.

In his reading, he came across an old essay by the Reverend Thomas Malthus (1766–1834) that was probably the first clear warning of the dangers of human overpopulation. In the essay, which appeared in 1798, Malthus pointed out that populations tended to increase in a geometric (exponential) progression, and that if humans continued to reproduce at the same rate, they would inevitably outstrip their food supply and create a teeming world full of "misery and vice" (Figure 18.7). Malthus's message may have been theological, but Darwin applied the idea to his own work and concluded that environmental factors could keep populations from growing out of control—factors such as the starvation and disease that accompany overcrowding. Further, some individuals would succumb sooner than others, and those who survived would propagate the generations to come.

The Emergence of a Theory

Beginning in 1837, Darwin started to keep a journal entitled *Transmutation of Species* and was soon making entries referring to "my theory." Reading that journal today, and the journal of the voyage of the *Beagle*, is like reading a detective story after you already know whodunit: The suspense is in watching the detective sift through clues for the right answer.

FIGURE 18.7 **A CROWDED ENGLISH CITY.** The hunger and misery of the English poor at the end of the 18th century helped Malthus to form his conclusions.

A penciled manuscript of 1842 laid out the entire theory of the origin of species through natural selection. Darwin did not publish it, however. He knew that the idea would arouse fierce resistance, and that he would have to back up every part of his idea with evidence. He set about preparing to defend his argument.

But Darwin's ordinarily robust health mysteriously began to fail. He began to vomit frequently, and he complained of headache, nausea, fatigue, and heart palpitations. In time, because he tired quickly, he virtually stopped seeing anyone, even his valued scientific colleagues. He resigned as secretary of the Geological Society. His father bought him a country home, and his wife became his nursemaid. Strangely enough, he continued to look completely healthy. His mind was as vigorous as ever, but his strength was gone. He continued to work, but only for a few hours a day.

GATHERING EVIDENCE

When he was able to work, Darwin set about compiling information to back up his theory. He even suggested further tests. Along with all his other scientific virtues, Darwin was also an experimental biologist (a fact sometimes forgotten), and his experiments showed that life can spread from mainlands to islands, such as the Galápagos, quite easily. He found that a variety of species can survive extended periods of time in seawater (and so some species could have floated or "rafted" there), and also that airborne organisms can easily move from one place to another (and so some could have been transported by wind or storm). Such evidence was important because much of the foundation of Darwin's developing theory depended on the ability of species to disperse.

Working only a few hours a day, Darwin also published lengthy works on the classification of barnacles (still a major reference work), on a theory of the origin of coral islands, and on geology. But all the while he continued to develop his theory of natural selection. He showed his early manuscript on natural selection to only one close friend, the botanist Joseph Hooker, who remained doubtful. Twenty years passed, and finally in 1856 Darwin began writing his major work, to be called *Natural Selection.* It was planned as an enormous six-volume monograph. But in 1858 his work on the manuscript was suddenly interrupted. Unexpectedly, another manuscript, a very brief one, arrived in the mail from Alfred R. Wallace, a young naturalist working in Malaya. Wallace asked politely whether Darwin would care to make any comments on the manuscript. The article was a short but well-written statement about natural selection and its role in evolution—Darwin's ideas had been quite independently deduced by someone else! Darwin was mortified. In a letter to his friend Hooker he lamented,

> So all my originality, whatever it may amount to, will be smashed. . . . Do you not think that this sketch ties my hands? . . . I would far rather burn my whole book, than that he or any other man should think that I have behaved in a paltry spirit.

Darwin was about to be scooped—not an uncommon experience in science. Would his own work now be thought to have been based on the ideas of another?

Darwin's friends persuaded him to allow his previously unpublished 1842 summary and the text of a long letter describing his ideas to be presented, together with Wallace's paper, before the Linnean Society of London. Because of Darwin's much more substantial evidence, his paper was given first. The two were presented in July and published in August 1858. Now Darwin went furiously to work, and—putting aside the idea of a huge, definitive monograph (which never was written)—he quickly finished an "abstract" (1500 pages!) of it and published it in 1859 (Figure 18.8). The first edition sold out on the first day, in spite of its full title: *On the Origin of Species by Means of Natural Selection, or the Preservation of Favoured Races in the Struggle for Life.*

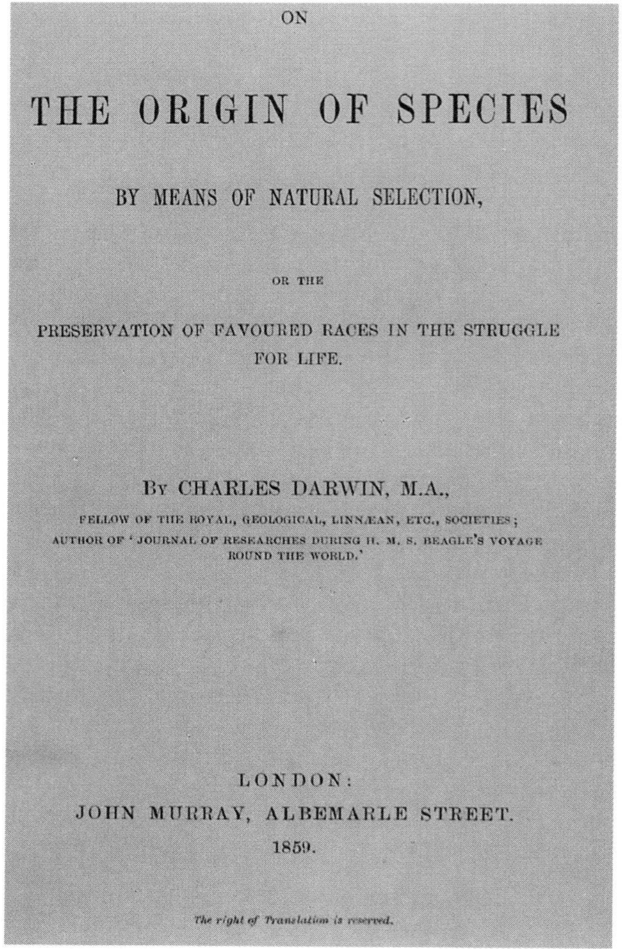
ON

THE ORIGIN OF SPECIES

BY MEANS OF NATURAL SELECTION,

OR THE

PRESERVATION OF FAVOURED RACES IN THE STRUGGLE FOR LIFE.

By CHARLES DARWIN, M.A.,

FELLOW OF THE ROYAL, GEOLOGICAL, LINNÆAN, ETC., SOCIETIES;
AUTHOR OF 'JOURNAL OF RESEARCHES DURING H. M. S. BEAGLE'S VOYAGE ROUND THE WORLD.'

LONDON:
JOHN MURRAY, ALBEMARLE STREET.
1859.

The right of Translation is reserved.

FIGURE 18.8 THE ORIGIN.
The title page of what came to be called *The Origin of Species.*

ON THE ORIGIN OF SPECIES

In the first chapter of *On the Origin of Species,* Darwin sought to establish that animals and plants under domestication are extremely variable, that the variation is heritable, and that the many domestic varieties have arisen from wild ancestors through breeders determining which animals should reproduce. In the second chapter, he drew together evidence that plants and animals in nature were variable too. In the third chapter, he discussed Malthus's idea of the struggle for existence—the idea that the natural reproductive capacities of living things greatly outreach the ability of the environment to support them, so that many organisms perish by starvation or disease without reproducing. In the fourth chapter, he introduced natural selection:

> How will the struggle for existence . . . act in regard to variation? Can the principle of selection, which we have seen is so potent in the hands of man, apply in nature? I think we shall see that it can act most effectively. . . . If such (variations) do occur, can we doubt (remembering that many more individuals are born than can possibly survive) that individuals having any advantage, however slight, over others, would have the best chance of surviving and of procreating their own kind? On the other hand, we may feel sure that any variation in the least degree injurious would be rigidly destroyed. This preservation of favorable variations and the rejection of injurious variations, I call Natural Selection.

THE NEW SYNTHESIS

Charles Darwin developed a cornerstone of modern biology in his description of evolution through natural selection. But he had a great deal of trouble convincing his contemporaries of his ideas. Most scientists of his time were quick to accept the fact that evolution occurs, but many of them failed to see just how natural selection was involved. Darwin was unable to convince them, primarily because he had no *mechanism* whereby natural selection could proceed. The problem stemmed from the fact that Darwin didn't even know genes *existed.* So as he set about trying to explain how natural selection might work, he began grasping at straws. He continued to believe that inherited factors were somehow blended in the offspring. Furthermore, with each revision of *On the Origin of Species,* he lapsed deeper into Lamarckian explanations that even his contemporaries knew didn't hold water.

We now know that not far away—in what was then a part of the Austro-Hungarian empire and came to be Czechoslovakia—the monk Gregor Mendel was generating precisely the information that would clear up Darwin's great dilemma. But since Mendel was the only modern geneticist in the world, he had trouble making people—including Darwin—see the importance of his work. In fact, a copy of Mendel's paper was found in Darwin's library, completely unmarked, with an adjoining article from the *Source* journal covered in Darwin's handwriting.

As we saw in Chapter 11, Mendel's work was rediscovered in the early part of this century, long after the hardworking monk had gone to his reward. The scientific world, newly fascinated with genes and how they are carried along from generation to generation, began to wonder just what genes have to do with natural selection. By the 1930s, the broad principles of the relationships between genes and natural selection had been worked out. Then in the 1940s scientists, including the great evolutionary biologist Ernst Mayr (Figure 18.9), finally produced what they called the *New Synthesis,* the integration of the principles of genetics and evolution.

The New Synthesis explained a great deal and laid the framework for further investigation. That investigation, not unexpectedly, has resulted in the alteration of some of the original statements, and has in fact produced arguments not only over rates of evolution but also just how evolution proceeds (see Chapter 19). Many scientists see such disagreements as not weakening the general premise at all, but simply as fine-tuning an established principle that will remain, perhaps altered, but alive and well, and a linchpin of modern scientific thinking. With this critical integration of ideas in mind, then, let's see just what kinds of evidence exist today for the theory of evolution.

FIGURE 18.9 ERNST MAYR.
As one of the most important evolutionary biologists of this century, Mayr was important in developing the New Synthesis.

EVIDENCE OF EVOLUTION

So how good is the evidence supporting the theory of evolution? How well has Darwin's idea held up? Pretty well, it turns out, with some fine-tuning along the way. Let's consider some of the most powerful supporting evidence for the theory, in particular the fossil record, biogeography, comparative anatomy, and artificial selection.

THE FOSSIL RECORD

Darwin is best known as a biologist, but he was also one of the foremost geologists of his time. In fact, his interest in rocks and strata and his enthusiasm for discovering fossils contributed powerfully to his developing evolutionary theory. For example, in his "geologizing" in South America, he uncovered the fossil remains of an extinct giant armadillo (Figure 18.10) deep in ancient rocks, deposited long ago. Above, on the surface, lived smaller armadillos of a different sort, scurrying over the graves of the giants. To Darwin, this suggested change over time. The extinct species of giants, he thought, must have been ancestral to the living species.

Many fossils have been discovered since the time of Charles Darwin, and the record is now immense. Fortunately, for an aging Darwin, one of the most amazing records was uncovered during his lifetime. In 1879, three years before Darwin's death, Yale University paleontologist Othniel C. Marsh published *Equus,* a comprehensive study on the evolution of the modern horse. Marsh's findings lent considerable credibility to a troubled Darwin, whose book *On the Origin of Species,* published 20 years earlier, had created a storm of controversy. Thomas Huxley, scientist, master debater, and Darwin's greatest defender, put these findings to good use in arguing the case of evolution through natural selection.

Saga of the Horse. Tracing the origin of the horse has taken paleontologists back some 65 million years to the Eocene epoch of the Cenozoic era (see the geological table, Appendix A). In that distant time lived *Hyracotherium,* a timid, dog-sized, woodland creature that browsed on the soft parts of plants and literally tiptoed around the forest (like the dog, it had multiple toes and footpads). From this decidedly "unhorselike" creature arose a succession of forms that came to take on the appearance of the familiar horse (Figure 18.11). The changes represented a continuing adaptation to a new source of energy, grass. Grasses first appeared in the Eocene epoch and began their gradual spread

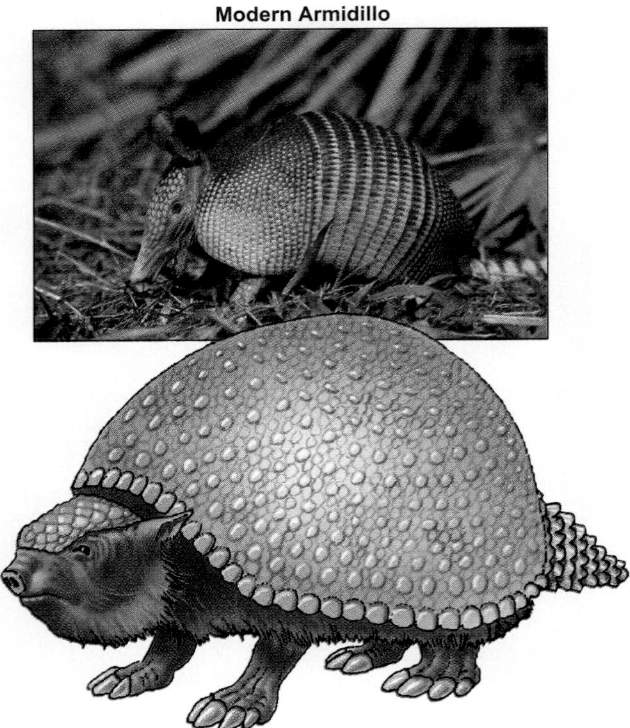

Modern Armidillo
The glyptodon is thought to be the extinct ancestor to the modern armadillo

FIGURE 18.10 GIANT AND MODERN ARMADILLOS. Fossil of the extinct giant armadillo from South America and a modern armadillo.

over great expanses of the earth. However, the changes in the horse was not a linear progression. Instead, many forms arose, most dying out as the horse continued to be represented by lines that were, for whatever reason, successful.

The horse's anatomy exhibits many adaptations to grassland—not the least of which is in the dentition (teeth). After all, grasses are highly abrasive, and so eating them requires large, tough incisors for nipping and molars with broad, hardened surfaces for grinding. Of course, as the primitive horses left the forests in search of grass, predators followed. Because concealment was more difficult in the grassy, open spaces, natural selection favored "early warning systems" in the form of long necks and improved vision and hearing. It also favored the ability to respond with great speed—thus we see major changes in the lower legs and feet, particularly the consolidation of the toes into the hooved foot.

Small Steps or Great Leaps? The excellent fossil record of the horse, where one form leads to the next, with many of the transitional forms known, is probably an exception to the rule. Yet Darwin was convinced that this is the way evolution occurs—slowly and step-by-step. Darwin's assumption that species change slowly through time because of the accumulation of small changes is called **gradualism.** Yet even Darwin was fully aware that there were certain problems with this concept, *particularly* as evidenced in the fossil record. He was aware that there were notable absences of intermediate forms. He was convinced that the problem lay in the "incompleteness" of the fossil record. After all, he reasoned, we are likely to find only pitifully small remnants of the vast array of life that once existed. Further, the remnants themselves are often incomplete and so severely treated by the ravages of time that it can be hard to judge what the living thing might have looked like, or just how rapidly its kind had changed.

Recently, Niles Eldridge of the American Museum of Natural History and Stephen Jay Gould of Harvard University (Figure 18.12) have proposed that these abrupt changes in the fossil record are not due to its incompleteness, but because of the fact that evolution is not a slow, gradual process after all. They claim, quite simply, that evolution generally proceeds in fits and starts; that species, once formed, go unchanged for great periods; and that when they do change, they are likely to change markedly. This concept is called **punctuated equilibrium**—long periods of inactivity, or equilibrium, punctuated by rapid evolutionary change. The "suddenness" of change, they stress, must be viewed in geological time, perhaps tens of thousands of years. This is a markedly rapid period of time, considering that species last, on the average, several million years. Gradualism and punctuated equilibrium are compared in Figure 18.13.

BIOGEOGRAPHY

Biogeography is the study of the distribution of living things over the earth. It was the biogeography of animals that first suggested common descent to Darwin. Why does each geographically isolated region, he wondered, have its own peculiar assemblage of plants and animals? Why would the Galápagos be populated with plants and animals similar to those of the South American mainland instead of those of other islands elsewhere? Obviously, the ancestors of those animals must have come from South America. But why were they, then, different in some ways from the mainland forms? Had they changed after reaching the islands? Didn't that change imply evolution?

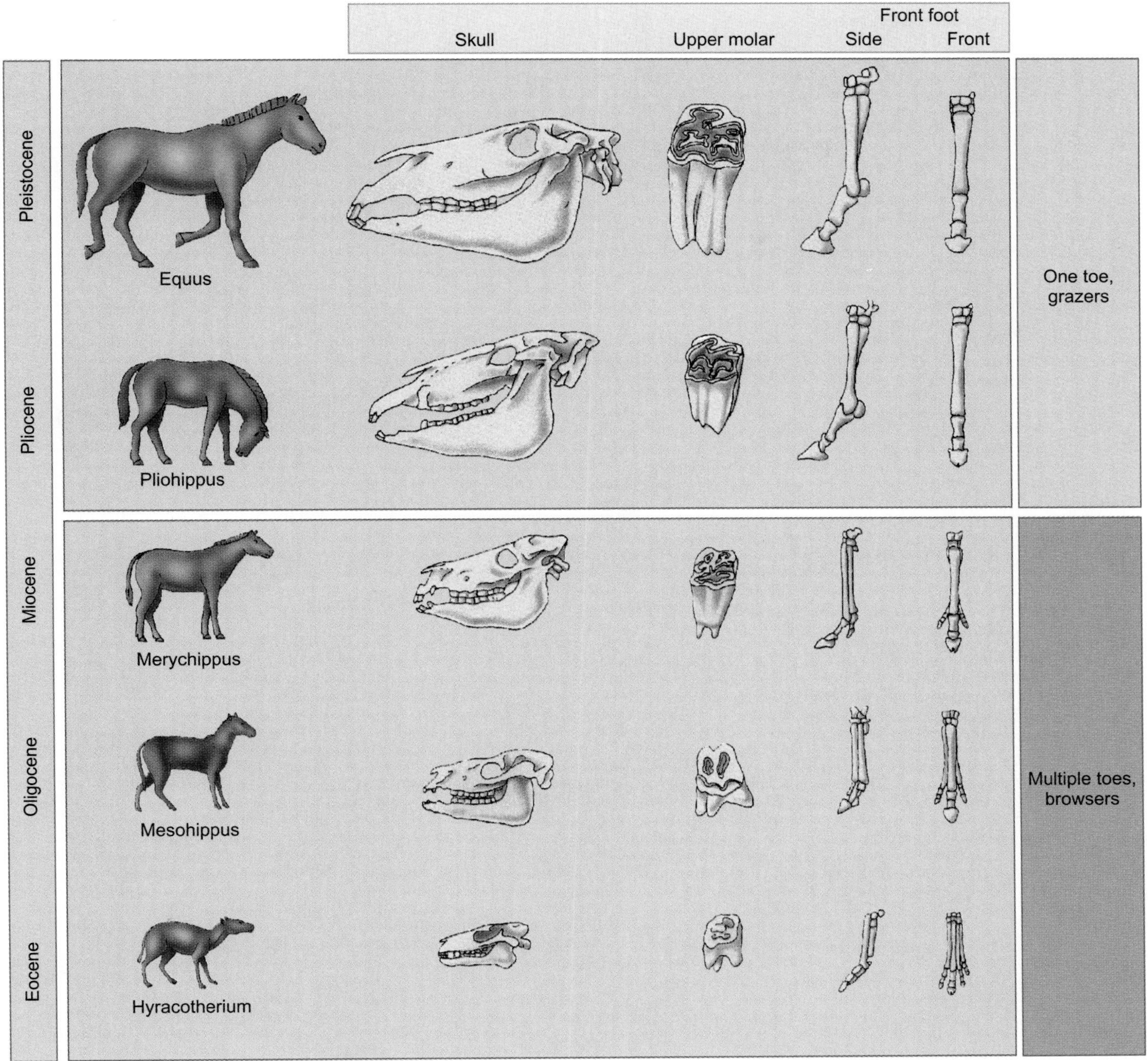

FIGURE 18.11 **HISTORY OF THE HORSE.**
The reconstruction of the ancestral history of the modern horse reveals changes in size, skull, feet, and teeth. The modern horse may stand 152 cm (60 in.) tall at the shoulder, but the first known horse, *Hyracotherium*, was only one-sixth that size (about 11 inches). Adaptations to grassland living began sometime during the Miocene epoch. Note the loss of some toe bones and enlargement of the remainder as the hoof became consolidated. Along with the enlargement of the skull, the teeth enlarged and the hardened enamel surfaces increased (seen here are the upper molars).

Australia provides another example of how geography can influence life forms. With the exception of bats and introduced animals, Australia, an isolated island continent, is populated only by marsupial (pouched) mammals (those lacking a true placenta) (Figure 18.14). Marsupials are rare over the rest of the earth (the opossum and a few pouched rats are exceptions). Since studies of today's marsupials show that they are related to each other, their distribution lends support to the idea that animals are where they are because of their evolutionary history. As a final line of evidence, related animals are

GRADUALISM
Early division produces two lines and undergoes gradual transitions; accumulation of minute changes leads to modern representatives.
Okapi line
Okapi
Common ancestor to okapi and giraffe
Giraffe line
Giraffe

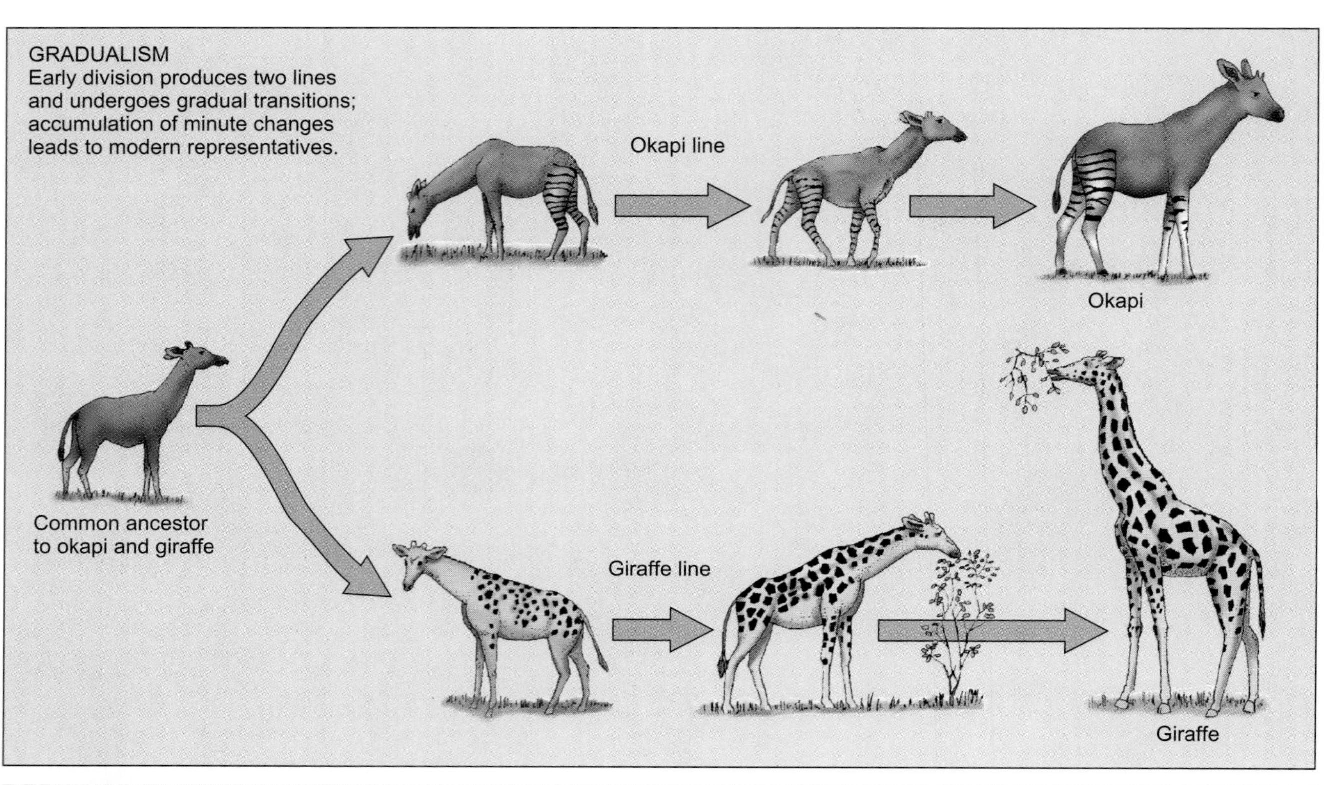

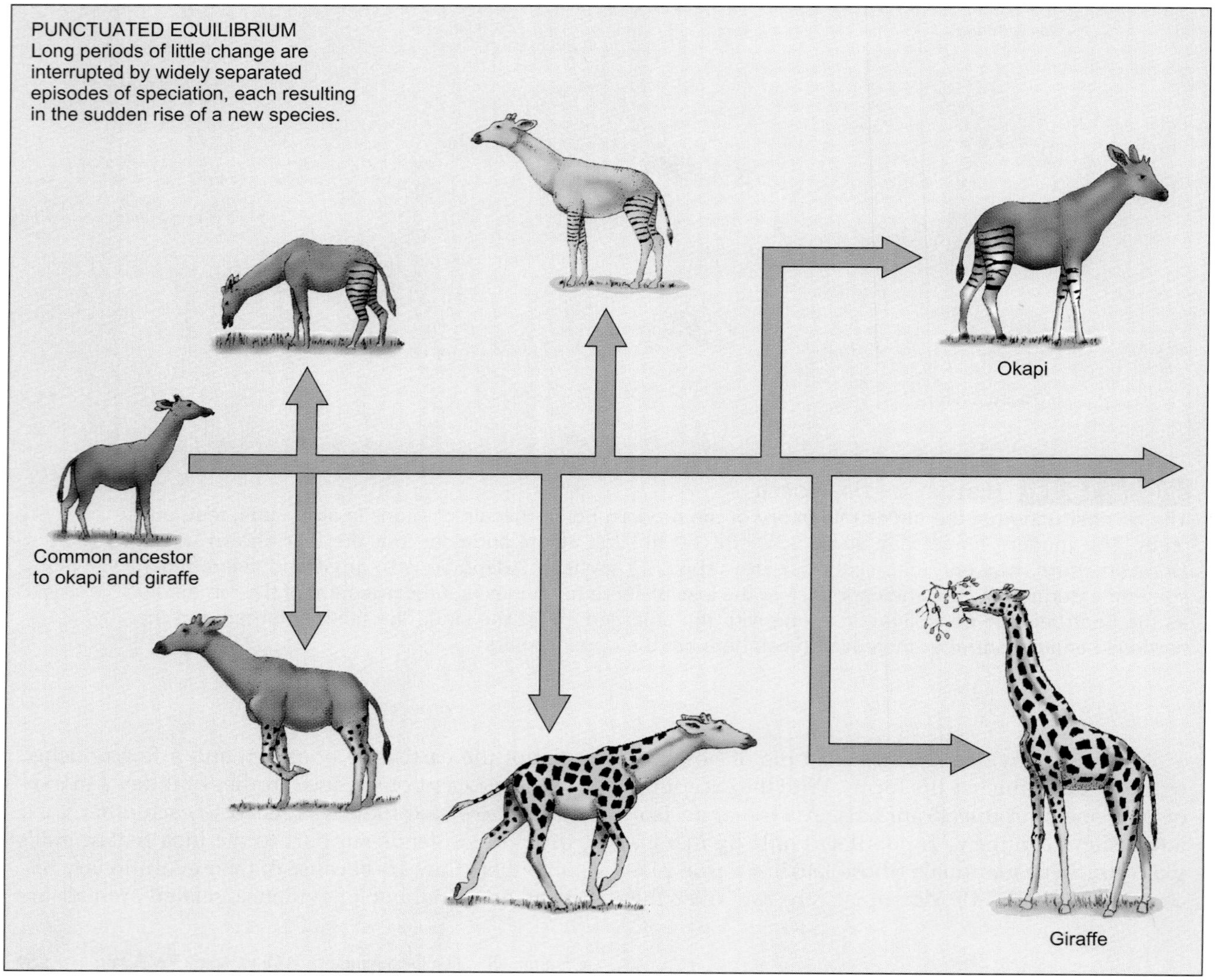

PUNCTUATED EQUILIBRIUM
Long periods of little change are interrupted by widely separated episodes of speciation, each resulting in the sudden rise of a new species.
Okapi
Common ancestor to okapi and giraffe
Giraffe

FIGURE 18.12 NILES ELDRIDGE AND STEPHEN JAY GOULD.

likely to be found in the same area, suggesting that they are descended from a common ancestor. Essay 18.2 discusses how movements of the earth's crust have resulted in the location of combinations of life we see today.

COMPARATIVE ANATOMY

An important source of information about evolutionary relationships comes from the study of **comparative anatomy,** in which the structures of modern species are compared. Such studies are particularly useful where the fossil record is poor, but often they essentially reinforce our interpretation of adequate fossil records. For instance, studies of the skulls of various vertebrates (animals with backbones) have provided many clues to the evolutionary trends that led to today's species. The animal skeleton abounds with examples. The relatedness of mammals is supported by the fact that virtually all of them, from bats to whales to giraffes, have exactly seven cervical (neck) vertebrae. Also consider the gill arches (see Chapter 27): All vertebrate embryos (including humans) have gill arches. Studies of development reveal that, although gill arches give rise to gills in fishes, they became highly modified into wholly unrelated structures in other vertebrates (ours become part of the lower jaw, middle ear, larynx, and tongue) (see Chapter 42). Thus, if we know something about how fish gills and the human jaw develop embryologically, we are led to the existence of a common, if distant, ancestor.

In the comparative study of anatomy it is necessary to distinguish structures that are *homologous* from those that are *analogous.* **Homologous structures** are those with common origins in the embryo (such as fish gills and vertebrate jaws), indicating common evolutionary origin. Today, though, such structures may look quite different and may even have taken on different functions. **Analogous structures,** on the other hand, often appear similar, and even have similar functions; but they are unrelated in evolution because they form differently in the embryo. The wings of insects and birds are often cited as examples of analogous structures. Homologous structures are much more useful, we can see, in determining evolutionary histories.

Studies of vertebrate anatomy provide many examples of homologous organs and, along with the fossil record, have given biologists a clear idea

FIGURE 18.13 GRADUALISM AND PUNCTUATED EQUILIBRIUM.
Here the two ideas are contrasted in the evolution of the okapi and the giraffe.

ESSAY 18.2 Continents Adrift

In 1912, Alfred Wegener published a paper that was triggered by the common observation of the good fit between South America's east coast and Africa's west. Could these great continents ever have been joined? Wegener coordinated this jigsaw-puzzle analysis with other geological and climatological data, and proposed the theory of *continental drift.* He suggested that about 200 million years ago, all of the earth's continents were joined together into one enormous land mass, which he called *Pangaea.* In the ensuing millennia, according to Wegener's idea, Pangaea broke apart and the fragments began to drift northward to their present location.

Wegener's idea received rough treatment in his lifetime. His geologist contemporaries attacked his naiveté as well as his supporting data, and his theory was neglected until about 1960. At about that time, a new generation of geologists revived the idea and subjected it to a new scrutiny based on recent findings. These findings buttressed Wegener's old notion and breathed new life into it.

The most useful data have been based on magnetism in ancient lava flows. When a lava flow cools, metallic elements in the lava are oriented in a way that provides a permanent fossil of the earth's magnetic field at the time, recording for future geologists both its north-south orientation and its latitude. From such maps it is possible to determine the ancient positions of today's continents. We now believe that not only has continental drift occurred, as Wegener hypothesized, but that it continues to occur today.

Geologists have long maintained that the earth's surface is a restless crust, constantly changing, sinking, and rising because of incredibly powerful, unrelenting forces beneath it. These constant changes are now known to involve large, distinct segments of the crust known as *plates.* At certain edges of these masses, immense ridges are being thrust up, while other edges sink lower. Where plates are heaved together, the buckling at the edges has produced vast mountain ranges. When such ridges appear in the ocean floor, water is displaced and the oceans expand. (Astoundingly precise satellite studies reveal that the Atlantic Ocean is 5 cm wider each year.)

In addition to its fascinating geological implications, an understanding of continental drift (or *plate tectonics*) is vital to the study of the distribution of life on the planet today. It helps to explain the presence of fossil tropical species in Antarctica, for example, and the unusual animal life in Australia and South America.

As the composite maps indicate, the disruption of Pangaea began some 230 million years ago, in the Paleozoic era. By the Mesozoic era, the Eurasian land-mass, called *Laurasia,* had moved away to form the northernmost continent. *Gondwanaland,* the mass that included India and the southern continents, had just begun to divide. Finally, during the late Mesozoic era, after South America and Africa were well divided, what was to be the most recent continental separation began, with Australia and Antarctica drifting apart. Both the North and South Atlantic Oceans continued to widen considerably through the Cenozoic era, a trend that is continuing today. So we see that although the bumper sticker "Reunite Gondwanaland" has a trendy ring to it, it's an unlikely proposition.

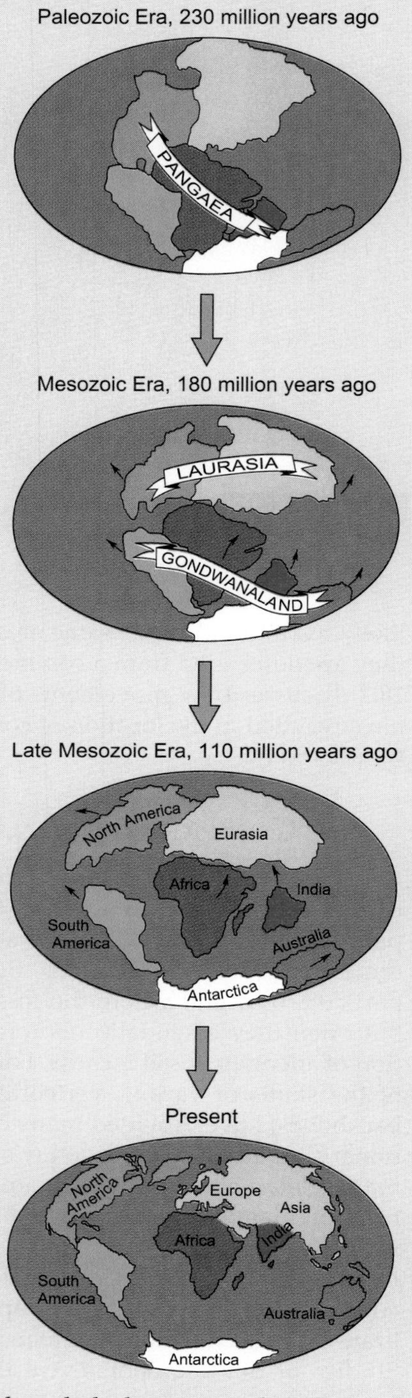

Paleozoic Era, 230 million years ago

Mesozoic Era, 180 million years ago

Late Mesozoic Era, 110 million years ago

Present

about vertebrate evolution. For instance, what inferences can we make from a consideration of the forelimbs of various animals and the upper arm of a human (Figure 18.15a)? A study of the embryos of these vertebrates reveals that in each case, the limbs emerge from similarly formed limb buds.

A look at the whale's anatomy and that of the python (Figure 18.15b) presents another slant to the idea of homology. Both lack hindlimbs, but tucked away in the body are useless vestigial bones that are clearly remnants of what was once a functional pelvic (hip) girdle. There are even reduced, nonfunctional hindleg bones. These findings suggest that both animals are descended from four-legged land dwellers. Such structures are looked upon as rare but welcome records of past evolutionary associations.

Artificial Selection

Finally, the principles of artificial selection have been as valuable to us in understanding natural selection as they were to Darwin when he developed the idea. We have relentlessly molded the genes of almost every species we have had an interest in, sometimes for practical reasons (as when we create stronger and faster horses), sometimes whimsically (as when we developed hairless, mouselike dogs or wrinkled rats). The result of some of our selection has occasionally been unfortunate, as we see in extremely aggressive and unpredictable pit bulls. Nonetheless, we have proven time and again that by selecting individuals with certain traits for breeding, those traits can become magnified in the population. The primary difference between our choices and nature's is that nature selects only those traits that lead toward successful reproduction, while we are often more capricious (Figure 18.16).

Comparative Molecular Biology

Molecular biology came of age in the early 1950s, with great fanfare and marching to the bugles of Watson and Crick. Since that time there has been an unimagined increase in the technical capability of the molecular biologist, culminating in the dazzling new technologies of genetic engineering. The powerful new investigative tools have now been focused, with considerable success, on questions of evolution. As a result, biologists routinely describe evolutionary relationships based on the deep investigation and comparison of computer-generated sequences of amino acids in proteins and sequences of DNA nucleotides. The underlying logic is that the amino acid sequences and nucleotide sequences of

Figure 18.14 Marsupials.
These marsupials evolved on an ancient isolated island and are believed to have a common ancestor.

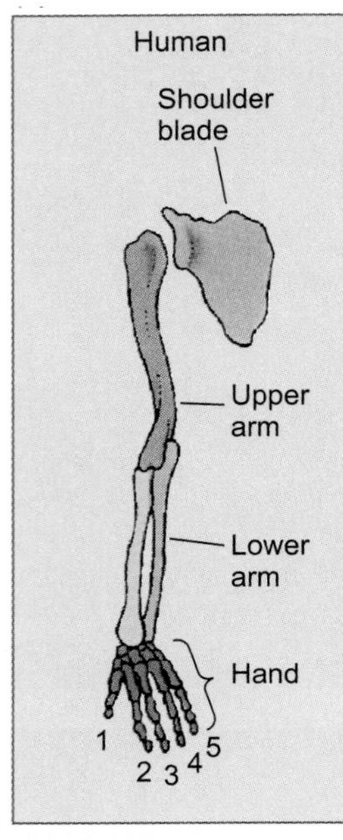

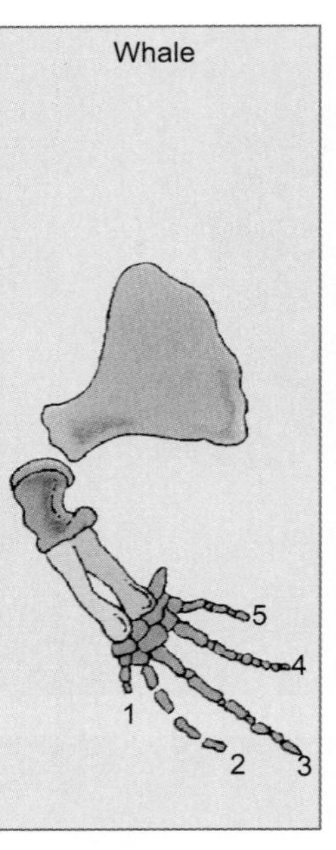

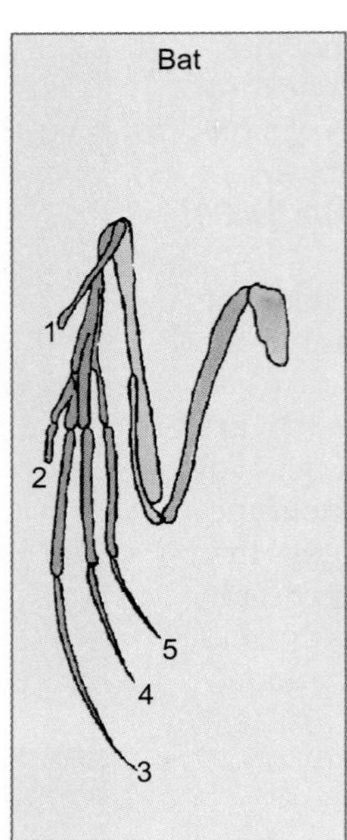

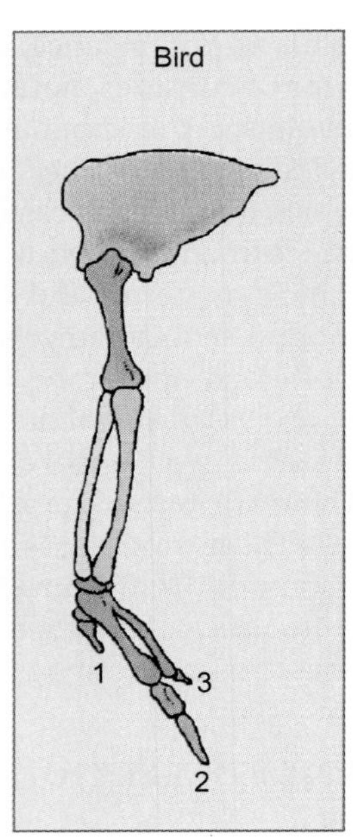

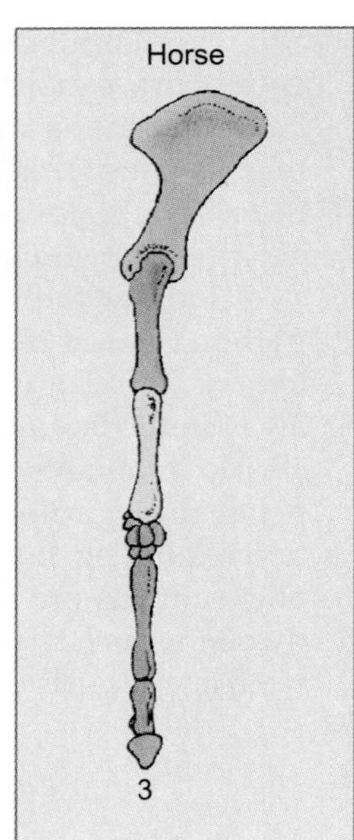

(a)

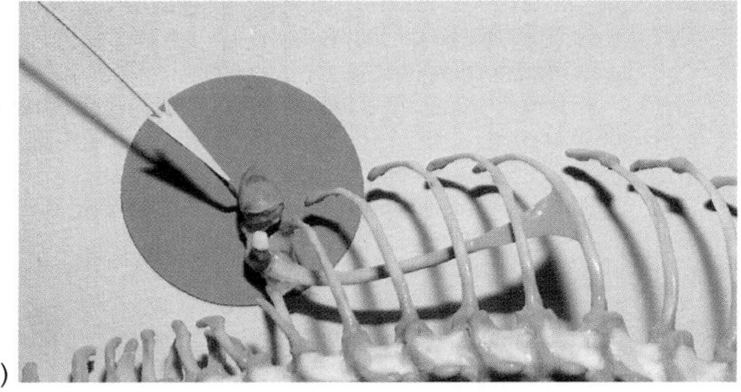
(b)

FIGURE 18.15 HOMOLOGOUS STRUCTURES. (a)The forelimbs of several representative vertebrates are compared here with those of the suggested primitive ancestral type. Since they all have the same embryological origin, they are said to be homologous. The most dramatic changes can be seen in the horse, bird, and whale. In these animals, many individual bones have become smaller and some have even fused. Another interesting modification is seen in the greatly extended finger bones of the bat. (b) Remnant pelvic girdles and hindlimbs in the python.

closely related organisms are more similar than are those of more distantly related organisms. Of course, scientists still use the older and still powerful methods of systematics and taxonomy to try to determine relatedness; and as the molecular work goes on, they still ponder over such things as leg length and bill size, and they carefully trace the organism's embryological development. The similar results yielded by these diverse investigative techniques is powerful evidence of evolution.

In fact, it is often revealing (and just plain fun—as well as comforting) to compare the conclusions of the more traditional approaches with newer molecular techniques, and see how often the two agree. As the molecular data come in, crusty old biologists who know their bones have time and again been shown to have been on the right track all those years.

Let's see, then, how comparative analysis of the amino acid sequences in proteins and the nucleotide sequences of thousands of species has revealed fascinating evidence for the existence of a *molecular clock.* The molecular clock is based on the idea that point mutations (single nucleotide changes in DNA) and the resulting amino acid changes in proteins (see Chapter 17) occur with clocklike regularity. If such regularity exists, not only can we use protein differences to determine relatedness of species, but we can also tell when the divergences occurred. In other words, the data generated from comparative molecular studies can not only be used to organize phylogenetic ("family") trees, they can also tell us when and where each branch in the tree occurs. This means that we can draw such trees with a once-unimagined accuracy. The length of each branch fol-

FIGURE 18.16 BREEDING AND SELECTION. We have managed to breed very specific traits into a variety of animals, sometimes for a reason, sometimes capriciously. One of these was bred to attack; one was not.

lowing a division can now reflect the actual time elapsed since the two species diverged.

Sometimes the findings from such studies are startling. For instance, in his comparisons of serum albumin, a primate blood plasma protein, Vincent M. Sarich came up with some surprising data on human divergence. Most researchers had concluded that humans diverged from other primates from 20 to 30 million years ago, but Sarich's data suggested that the divergence occurred only 5 million years ago. (This was later supported by nucleotide sequencing studies of globin genes in which the nucleotide arrangement in those genes was compared.)

From comparative amino acid analyses, molecular biologists have developed what they call a **minimum mutation tree** (Figure 18.17), a phylogenetic scheme based on the smallest number of nucleotide changes that could have produced the observed amino acid differences. Such trees are designed to reveal the order in which various organisms appeared and how they are related to each other.

One of the earliest minimum mutation trees involved cytochrome *c*, a respiratory pigment. Cytochrome *c*, which consists of 108 amino acids in most vertebrates and slightly more in other organisms, is believed to have changed very slowly throughout evolutionary history. For example, humans and chimpanzees have identical sequences—and humans and rhesus monkeys have only one amino difference—in spite of the 20 million years that have elapsed since the divergence of our ancestral lines.

Other amino acid sequences have also been analyzed in this manner, including those from the respiratory enzyme carbonic anhydrase (see Chapter 39), which appears to have had a much faster rate of change than cytochrome *c*. Minimum mutation trees have also been developed using data from DNA sequencing studies (described in Chapter 16).

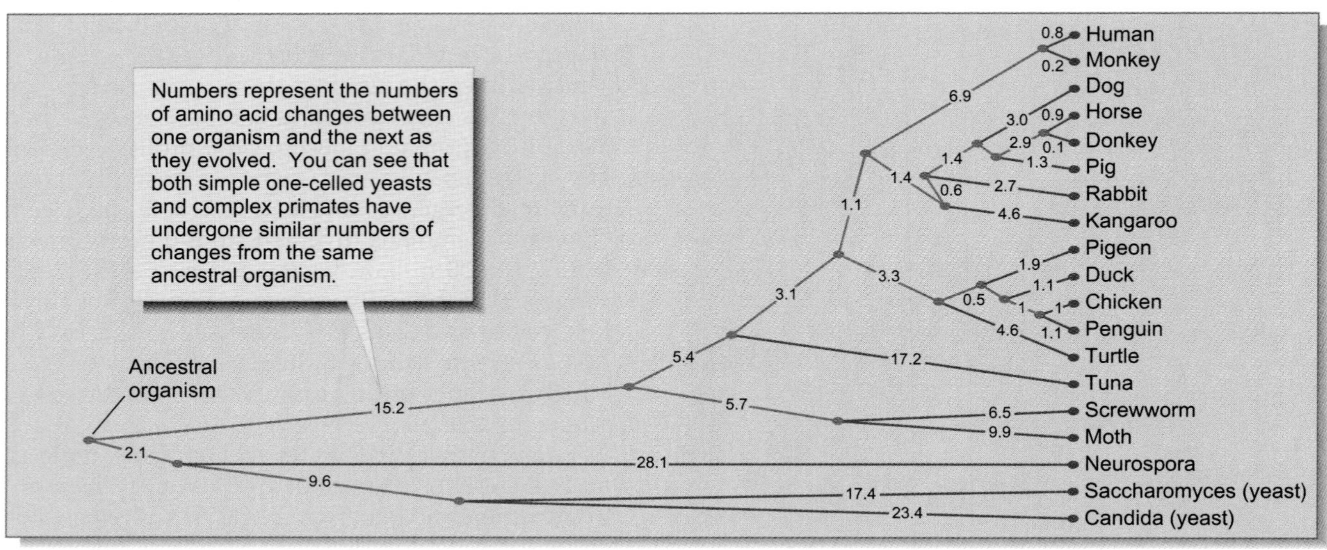

FIGURE 18.17 EVOLUTIONARY TREE BASED ON MUTATIONS IN DNA.
Minimum mutation trees include those developed by amino acid sequencing and through DNA sequencing. This computer-generated phylogenetic tree represents amino acid differences in the polypeptide sequence of respiratory pigment cytochrome c, which is found in all aerobic organisms. Each amino acid difference represents at least one DNA mutation that has been "accepted" in evolution. The numbers along the branches of the tree represent the numbers of differences in amino acid sequences. In general, the tree derived from one short protein is consistent with other trees based on morphology or the fossil record, although the computer seems to think that the turtle is a bird.

PERSPECTIVES

Although the notion of evolution is one of the most pervasive themes in biology, we are still far from solving some of the most fundamental questions. A virtual barrage of problems in phylogenetic organization exists from the species to the kingdom level. But with the small army of researchers now attacking these problems, new and exciting ideas are constantly emerging to be tested by science's own selective processes.

In this chapter, we have followed the concept of evolution from the time the idea was developed to the theory's acceptance and integration into biological thought. We focused particularly on the fascinating history of Darwin himself as he left England as a young man to return as a respected scientist—until, eventually in ill health and full of self-doubt, he began to modify his theory to meet his critics' demands. Finally, we considered the modern evidence that supports the theory. Next, we will consider the development of the earth's far-flung species in terms of the accumulation of small changes at the molecular level, and see just how genes behave in populations. But to get at the basic mechanisms of evolution we must return to the allele, the hereditary unit itself. From the allele we will learn something of how genetic variation arises and, once in place, how it is tested by the pervasive forces that characterize natural selection.

Summary

Charles Darwin developed a new paradigm in which the diversity of life was explained through evolution. Evolution, which involves change over time, implies that all organisms have a common line of descent.

Evolution Through Natural Selection

As a mechanism of evolution, Darwin proposed natural selection. Its four main aspects are overproduction of offspring, limited resources, variation, and survival of certain types.

In accepting evolution we discard such notions as perfection in nature, fixed or unchanging species, and balance and harmony in nature.

Darwin's observations during his voyage on the *Beagle* and his studies afterward convinced him that life underwent change or evolution. He recognized the importance of past geological events and the time that would have been necessary to achieve evolutionary change. Lyell wrote that life on earth had indeed existed for great periods of time. The puzzle of the Galápagos finches played a central role in Darwin's development of the concept of natural selection.

The Emergence of a Theory

Beginning in 1837, Darwin began to refer in his writings to "my theory." By 1842, he had laid it out—even as he began to fall ill—showing the manuscript to only a few friends. Prodded by competition from Alfred R. Wallace, he presented his theory in 1859. Later, it was fused with Mendelian genetics to form what is called the New Synthesis.

Evidence of Evolution

The most compelling evidence for evolution is the fossil record. The fossil history of the horse can be traced throughout the Cenozoic era. Changes in its feet, limbs, and dentition reveal ongoing adaptations to the newly emerging grasslands.

The frequent absence of transitional forms in the fossil record has led some paleontologists away from the idea of gradualism and toward punctuated equilibrium. The latter proposes that long periods of evolutionary inactivity (equilibrium) lead to brief but far-reaching episodes of extinction and change (punctuation). In these periods new species arise and rapidly replace existing species.

Biogeography is the study of living things on earth. The uniqueness of animal groups in different geographic regions supports the idea that each group had its own line of descent from a common ancestor. Each group's members are more closely related to each other than to other such groups.

Comparative anatomy involves the comparison of the structures of modern species. Evidence of common ancestry is seen in a comparison of homologous structures. For instance, the limbs in various vertebrates appear quite different, yet common features are usually still visible, and when the development of each is compared in the embryo, they are seen to have similar origin. Analogous structures, such as bird and insect wings, differ in developmental origin and thus are not a sign of common evolutionary origin. Vestigial organs are signs of past evolutionary associations.

Artificial selection—that is, using human decisions to choose the breeders from a population—can mimic the effects of natural selection. Darwin believed the same process acts in natural selection except that nature does the "choosing."

The regularity with which mutations occur enables molecular biologists to equate nucleotide and amino acid differences among species to past evolutionary time. Thus evolutionary schemes represent both relatedness and time of divergence.

Minimum mutation trees are phylogenetic schemes based on the smallest number of nucleotide changes that could have produced the observed amino acid differences. Molecules such as cytochrome *c*, carbonic anhydrase, and DNA have been used to construct such trees.

Key Terms

evolution • **277**
natural selection • **277**
gradualism • **288**
punctuated equilibrium • **288**
biogeography • **288**
comparative anatomy • **291**
homologous structure • **291**
analogous structure • **291**
minimum mutation tree • **295**

Review Questions

1. Although the Darwinian concept of evolution represents a new paradigm, others before Darwin proposed the idea of evolution. Why is he given so much credit?
2. Why was the concept of natural selection so important to Darwin's theory of evolution?
3. List the four main aspects of natural selection.
4. What is the status of the evolution concept among scientists today? Which parts of Darwin's theory remain controversial?
5. Which prevailing idea did the notion of chance challenge most directly?
6. Of what importance were the Galápagos finches to Darwin's emerging theory?
7. In general, how does the fossil record support the concept of evolution?
8. Briefly discuss the idea of punctuated equilibrium. What prompted its development?
9. What is there about the geographical distribution of animals that supports the idea of common ancestry?
10. Carefully distinguish between analogous and homologous structures. Which indicates common ancestry?
11. Considering forelimb structure and function in the horse, bat, bird, and whale, why would anyone conclude that these animals have common ancestry? In what way do embryos provide clues?
12. Considering, again, the limb structure of various vertebrates, what might the limbs of the common ancestor of today's vertebrates have looked like?
13. What is a vestigial organ? How do such organs support the idea of evolution?
14. How might DNA sequencing provide clues to past evolutionary events?
15. Describe the importance of each of the following in the development of the theory of evolution: Jean-Baptiste de Lamarck, Charles Lyell, Thomas Malthus, Alfred R. Wallace.
16. What is the logical basis for using nucleotide sequences in DNA and amino acid sequences in proteins for establishing phylogenetic schemes?
17. What is the "molecular clock," and what does it add to phylogenetic schemes derived from nucleotide and amino acid sequencing?

CHAPTER

19

Evolution in Populations

WE HAVE JUST EXAMINED THE BASIC PRINCIPLES OF EVOLUTION, AND *now we will consider the specific aspects of genetic change that occur during the evolutionary process. In what ways must we think differently about genes and alleles when considering populations instead of individuals? How can we determine the genetic changes that occur within populations? How do the evolutionary changes arise?*

Among any group of offspring, there will be some variation. Look your little brother directly in the eye and you might develop a new appreciation for this principle. Biologically, the differences between you and him may be a way of ensuring that at least one of you will be able to cope with the unpredictable world in which you find yourselves. One of you is simply likely to be more suited than the other to whatever environment might arise. But keep in mind that there are biological implications of the differences between you and him. For instance, your little brother's close-set eyes, receding chin, prominent brow, sloping forehead, and the tendency to snort may interfere with his success in the world of fraternities. But his unusually long arms may prove beneficial if things should change so that climbing and throwing become more important. The prevailing environment, then, can "select," or favor, the traits of one of you over the other. The process, as described in the last chapter, is called *natural selection.* It is distinguished from *artificial selection,* the process by which a breeder chooses which traits to perpetuate in his stock. A cattle rancher, for example, might wish to breed only the animals that yield more beef (Figure 19.1).

In *On the Origin of Species,* Charles Darwin used the example of artificial selection as an analogy to natural selection. Even though artificial selection involves human intervention, and often human whim, it certainly shows how radically the phenotype can be altered in a short time.

THE POPULATION AND HOW IT EVOLVES

Evolution is defined as descent with modification. It involves the principle that *genes* evolve. That is, they mutate and change form. But not all the genes in a population mutate and of those that do, not all change in the same way, so there may be many alleles for the same gene. The question then arises, does the proportion of various alleles for the same gene in a population change over time? That is, do populations evolve? They do and we will see how here.

First we should establish a few terms in our evolutionary lexicon. A **population** is a group of interbreeding individuals of a single species. All of the alleles found in such a population make up the **gene pool** of that population. As certain alleles in a population increase or decrease, the character of the gene pool changes (Figure 19.2). We'll begin with a few fundamental ideas in **population genetics,** the study of changes in the representation of alleles within populations.

THE HARDY-WEINBERG LAW AND POPULATIONS IN EQUILIBRIUM

G. H. Hardy, an eminent mathematician, had few professional interests in common with R. C. Punnett, the young Mendelian geneticist, but they frequently met for lunch or tea at the faculty club of Cambridge University. One day in 1908, Punnett was telling his colleague about a small problem in genetics. Someone had noted that there was a rare dominant gene for abnormally short fingers, while the allele for normal fingers was recessive. In view of the dominance of short fingers and the famous 3:1 Mendelian ratio, shouldn't short fingers become more and more common with each generation, until no one in Britain had normal fingers at all? Punnett didn't think this argument was correct, but he couldn't explain why.

Hardy thought the problem was simple enough, and wrote a few equations on his napkin. He showed that the relative numbers of people with normal fingers and people with short fingers ought to stay the same for generation after generation, as long as there were no outside forces—such as natural selection—to change them.

Punnett was excited. He was amazed that his friend had solved so complex a puzzle so casually and wanted to have the idea published (on something besides a napkin) as soon as possible. But Hardy was reluctant. He felt that the idea was so simple and obvious that he didn't want to have his name associated with it and risk his reputation as one of the great mathematical minds of the day. But Punnett prevailed, and the relationship between genotypes and phenotypes in populations quickly became known as *Hardy's law.* As fate would have it, Hardy the mathematician is now best known for his reluctant contribution to biology.

FIGURE 19.1 **ARTIFICIAL SELECTION.**
Today's cattle are the result of generations of intense artificial selection. The longhorn (*top*) is a rangy beast, rough and well suited to the sparse western grassland. It is very similar to its wild ancestors. The Hereford *(bottom)* is the result of breeding for maximum beef yield.

FIGURE 19.2 **A HUMAN POPULATION.**
The degree of variability in a human population depends, to a large extent, on its size and whether new members are constantly incorporated, adding their varied alleles to the gene pool.

In Germany, Hardy's law was known as *Weinberg's law,* since it had been discovered independently by a German physician by that name. Weinberg, in fact, published his version within weeks of the publication of Hardy's short paper. Eventually the concept became known as the **Hardy-Weinberg law,** and it is now the basis of the population genetics of sexual organisms. Later, it was discovered that both had been scooped by an American, W. E. Castle, who had published a neglected paper on the relationship in 1903. Some people now refer to the Castle-Hardy-Weinberg law, but we will follow convention and also neglect our compatriot for now.

Genetic Equilibrium: An Implication of the Hardy-Weinberg Principle. To rephrase the problem of the Mendelians, let's consider human eye color. Although the genetics of eye color isn't completely understood (several gene loci are thought to be involved), we will ignore the complications and note that the allele for brown eyes is dominant over the allele for blue. So why are blue eyes still around?

From basic Mendelian genetics we know what happens in a cross between a homozygous blue-eyed person and a homozygous brown-eyed person. But in population genetics, we must consider not just one cross, but large numbers of matings and all their outcomes taken together. Imagine thousands of blue-eyed homozygotes (**bb**) mating with thousands of brown-eyed homozygotes (**BB**). The several thousand children resulting from such festivity will all be brown-eyed, since all the offspring will be **Bb** heterozygotes. Blue eyes have disappeared, as feared by the early Mendelians.

Now imagine that all these genetically identical F_1 heterozygotes grow up, choose mates, and manage to reproduce. It becomes apparent that the blue-eyed alleles didn't disappear after all. In fact, about one-fourth of their children will have blue eyes. Now if we let the F_2 generation grow up and choose mates, things start to get complicated.

But why? They weren't so complicated in Mendel's pea plants, which were entirely self-fertilizing. But humans are not self-fertilizing. And, despite songs and poems to the contrary, we can assume that people don't care all that much about blue or brown eyes, but will essentially mate at random insofar as eye color goes. So let's take a look at our F_2 generation. We find blue-eyed males, blue-eyed females, brown-eyed males, and brown-eyed females. And, in fact, there are two kinds of brown-eyed individuals of each sex—homozygotes and heterozygotes. We end up with six genetically distinct kinds of humans in our population: three genotypes of males and three genotypes of females. This means that there can now be *nine* kinds of matings.

We will let these types wander off into the sunset and begin pairing off. In due time the F_3 generation will appear. When we look them over we see that once again one-fourth of the children will be blue-eyed and three-fourths will be brown-eyed (Figure 19.3). And so it will continue for generation after generation.

We see, then, that the gene for blue eyes didn't disappear at all. What did happen? Since the heterozygotes harbored hidden blue-eyed genes, the actual *frequency* of the blue-eyed allele remained at 50%, just as at the start. **Frequency** is defined simply as the proportion of the whole, either in fraction or percent. Here, all the members of the F_1 have one **B** and one **b** allele, so the frequency of **b** in the F_1 is ½ or 0.5 or 50%.

The idea is, whether genes are expressed in homozygotes or hidden in heterozygotes, they are distinct physical entities that don't just go away without some reason. In the language of population genetics, eye color and the alleles for finger length are in **genetic equilibrium**—that is, their allele frequencies do not change over time. The crux of the Hardy-Weinberg law, then, is that in randomly mating populations allele frequencies remain in equilibrium and do not change in the absence of selection or other biasing factors.

F_2 Genotypes: ¼ BB, ½ Bb, ¼ bb	Ratio:	½ B	½ b

When one parental genotype is:	And the other parent is:	The offspring may be:	The resulting allele count is:	
BB	BB	BB	6B	2b
	Bb	BB or Bb		
	bb	Bb		
Bb	BB	BB or Bb	8B	8b
	Bb	BB,Bb,bB, or bb		
	bb	Bb or bb		
bb	BB	Bb	2B	6b
	Bb	Bb or bb		
	bb	bb		
		Total F_3 allele count:	16B	16b
		Ratio:	½ B	½ b

FIGURE 19.3 ALLELE DISTRIBUTIONS AND RANDOM MATINGS.
Random mating in F_2 individuals where the genotypes are **BB**, **Bb**, and **bb** and the phenotypic ratio is ¼, ½, and ¼ occurs in nine possible ways. The frequency of the two alleles was the same in the P_1, F_1, and F_2 generations. Is the frequency the same in the F_3 generation?

HARDY-WEINBERG: AN ALGEBRAIC APPROACH

In the example above, the two alleles start out at the same frequency—half **B** and half **b**. For those burning with a fierce love of mathematics, let's put it into algebra. First, let p be the frequency of allele **B**, while q is the frequency of allele **b**. If there are only two alleles, the following *must* be true:

$$p + q = 1$$

In the expression above, 1 is the same as 100%, and in our example 50% (frequency of **B**) plus 50% (frequency of **b**) equals 100%. The Hardy-Weinberg law also works in populations in which the alleles have different frequencies, such as 75% for the frequency of **B** and 25% for the frequency of **b**.

Let's consider all possible ways in which the **B** and **b** alleles can combine in a population: Males have two kinds of alleles, **B** and **b**, and females have the same two kinds, **B** and **b**, so all of the possible matings can be represented as a kind of Punnett square and an algebraic equivalent:

For Alleles:

	B	b
B	**BB**	**Bb**
b	**Bb**	**bb**

For Frequencies:

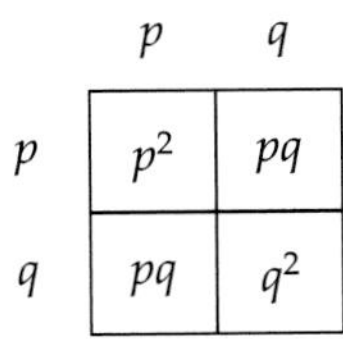

All matings:

$$1 = (p + q)^2 = p^2 + 2pq + q^2$$

The latter expression is often called the Hardy-Weinberg formula, and it is actually quite useful.

It tells us that we can find the frequency of the homozygous dominant (**BB**) individuals in a population by squaring the frequency of the dominant allele ($p \times p = p^2$).

Similarly, we can find the frequency of the recessive (**bb**) individuals by squaring the frequency of the recessive allele ($q \times q = q^2$).

Finding the frequency or proportion of heterozygotes is just a little different. To do this, we multiply the frequency of the dominant and recessive alleles and then multiply that answer by 2. (If you have forgotten where the 2 comes from, see the Punnett squares above.)

All of this begins to make considerably more sense when the Hardy-Weinberg formula is applied to real-life examples, so let's have a look at albinism in human populations.

Albinism: Real People and the Algebraic Equivalent. The Hardy-Weinberg law is not just a mathematical exercise. It has implications that could affect such diverse areas as politics, law, and sociology. For example, if we know the prevalence in the population of persons affected with a recessive condition such as albinism—the absence of normal melanin pigment—we can predict, within limits, the probability that any apparently normal couple will have an albino baby. Here's how this would work. Normal skin and eye pigment in humans is dominant over the albino condition; thus albinism is associated with a recessive allele, **a.** Normal pigmentation is associated with the corresponding dominant allele **A.** True albinos with the homozygous recessive genotype (**aa**) occur in about one of every 20,000 people. According to the Hardy-Weinberg law, the frequency of the **aa** genotype would be q^2, so

$$q^2 = 1/20{,}000$$

The frequency of the **a** allele (q) has to be the square root of this, or:

$$q = \sqrt{1/20{,}000} = 1/141$$

The frequency of the dominant **A** allele (p) would then be easy to figure since $p + q = 1$, such that $p = 1 - q$, or

$$p = 1 - 1/141 = 140/141$$

The heterozygous condition **Aa** would occur in the population with a frequency of $2pq$:

$$2pq = 2 \times 140/141 \times 1/141 = 1/70$$

or about 1.4%. Therefore, about one person in 70 is a *carrier* for this fairly rare condition.

Earlier we asked about the chances of an apparently normal couple having an albino baby. In this population the chances are 1 out of 19,600. We determine this by first recognizing that the man and woman must both be heterozygous, and secondly, by applying the multiplicative law (see Chapter 11). The chances of getting heterozygotes together in our randomly mating population is $1/70$ times $1/70$, or $1/4{,}900$. Since the probability of such a mating (**Aa** × **Aa**) producing an albino child (**aa**) is $1/4$, we must multiply again: $1/4{,}900 \times 1/4 = 1/19{,}600$.

We might note that there are actually several different kinds of albinism, due to different mutant alleles at different gene loci. That makes this kind of analysis very difficult unless the initial diagnosis is completely accurate; for the Hardy-Weinberg law to be of use, only one gene locus at a time can be considered.

A Closer Look at the Model. If it has occurred to you that mathematical models are often a little short on realism, you're right. In fact, the Hardy-Weinberg law has a number of stringent restrictions:

1. Mating must be completely random.
2. There can be no migration either into or out of the population.
3. The alleles in question must segregate according to Mendel's first law. This eliminates, for example, sex-linked alleles.
4. There can be no mutation.
5. There can be no selection operating on the population.
6. For exact results the population and the sample must be infinitely large.

Of these restrictions, the last is the least realistic, since no populations and no samples are infinitely large. But statisticians are willing to assume that smaller samples are representative of an infinite population.

Obviously, it's easy to violate the assumptions of the Hardy-Weinberg law. So is it of any real use? What happens when the assumptions of the Hardy-Weinberg law are violated? For many purposes, the Hardy-Weinberg predictions are the most useful when they *don't* come true. There are many conditions in natural populations that will give rise to departures from Hardy-Weinberg expectations. These departures themselves are interesting and can sometimes tell us such things as how much mutation, natural selection, or plain random change may be going on. These, of course, are the primary forces of evolution. The only way evolution can proceed is through such events; thus a population *out of equilibrium* is immediately interesting.

Looking at the population from a much broader perspective, let's note that the gene pool is the reservoir of genetic information of any population. This pool, though, should not be likened to a still pond. Its apparent stillness is deceptive; these waters change. As some waters flow over the dam to be lost, new genetic waters join the pool through myriad tributaries. The point is that although some gene pools change slowly and others change rapidly, they do change. Let's see now how the gene frequencies in such a pool can change and what this means in the evolution of living things.

Three major forces change gene frequencies: *natural selection, mutation,* and *genetic drift.* None of these forces acts alone, but together they direct the course of evolution.

NATURAL SELECTION AT WORK IN POPULATIONS

Natural selection can act in rather extreme ways, as in the case of certain *lethal* alleles. Lethal genes can kill you outright. Thus, the death removes the individual (and the allele) from the gene pool. Let's consider an unusual implication of this seemingly straightforward effect—the case of the Manx cat.

NATURAL SELECTION AND MANX CATS

Manx cats are peculiar animals, having rather large hind legs and practically no tails. No one has ever been able to develop a strain of true-breeding Manx cats for the simple reason that the tailless animals are all heterozygotes. Normal cats, with tails, are **TT** homozygotes; Manx cats, without tails, are **Tt** heterozygotes. The problem is, the homozygous **tt** genotype is an embryonic lethal; that is, these cats die while still embryos. Since death removes **tt** individuals from the population, this is an example of extreme selection. So when two Manx **(Tt)** cats mate, ¼ of the offspring will be normal cats, ½ will be Manx cats, and ¼ will be dead cats. The dead ones are reabsorbed by the mother's body as embryos, so the only litter you would see from such a cross would be ⅔ **Tt** Manx and ⅓ **TT** normal alley cats:

MATING OF MANX

Tt × **Tt**:	¼ **TT**	Normal
	½ **Tt**	Manx
	¼ **tt**	Die as embryos
Actual ratio in a litter:	⅓ **TT**	Normal
	⅔ **Tt**	Manx

Now suppose that someone should populate a remote island with a shipload of Manx cats, turning them loose to run wild, yowling and scratching and mating randomly, in the fashion of cats. What would happen? We know that the frequency of the Manx allele, **t,** starts out at $q = ½$, since all of the cats are Manx and therefore heterozygote. But then we find that, in one generation, the frequency of **t** is reduced to $q = ⅓$, because of selection removing the **t** alleles in the **tt** homozygotes. What happens thereafter? With random mating, the third generation will be $p^2 = 4/9$ **TT** (normal), $2pq = 4/9$ **Tt** (Manx), and $q^2 = 1/9$ **tt** (homozygous lethal).

But when the recessive homozygous lethal individuals are removed by natural selection, ½ of the remaining cats are now Manx and ½ are normal alley cats, making the frequency of the **t** allele now ¼. The frequency of the recessive lethal **t** allele has gone from ½ to ⅓ to ¼ in three generations, and in succeeding generations it will fall further to ⅕, ⅙, and so on. Meanwhile, the proportion of homozygous lethal **tt** zygotes will also follow the Hardy-Weinberg distribution: 1/4, 1/9, 1/16, 1/25, 1/36, and so on. As you can imagine, Manx cats will become increasingly rare on our fair island.

Figure 19.4 plots the course of the genotypes over 70 generations. Note that the allele frequency of **t** has fallen to 1.37% by the end of the 70 generations, and that about 2.8% of the cats will then be Manx. It would take another 70 generations to bring the recessive allele frequency down to 0.7%, but it could never reach zero.

Selection doesn't have to be so severe, of course. As a general rule, the rate of change in allele frequency in a population is proportional to the amount of selection against some genotype. For instance, in the case of Manx cats, suppose that the recessive genotype **tt** weren't lethal, but merely reduced the individual cat's reproductive ability by ⅒. Then it would take 700 generations, rather than 70, to go from $q = 50\%$ to $q = 1.37\%$.

You might wonder why there are any Manx cats at all. It's because people tend to be impressed by anything bizarre in cats, and they are likely to keep Manx kittens while disposing of the alley kittens. So artificial selection keeps the Manx gene going quite well in real life. Let's continue in the real world now to consider a case of natural selection operating in nature—the classic case of the peppered moth.

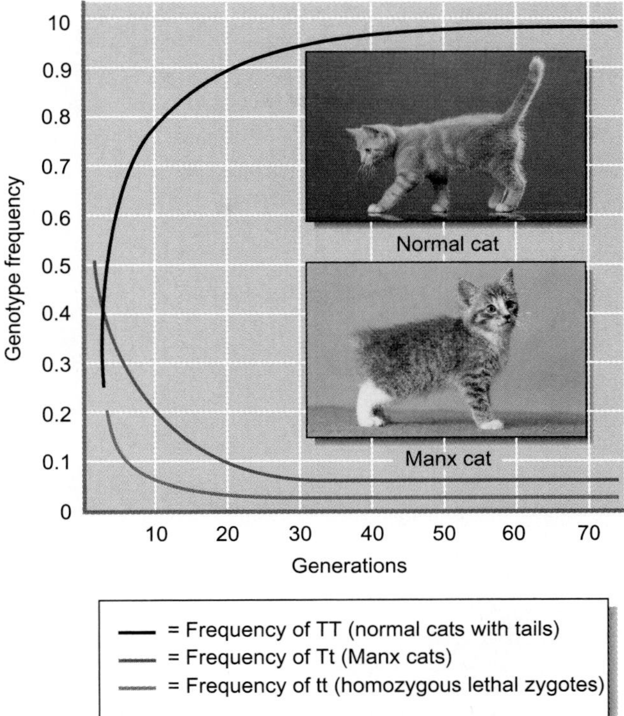

FIGURE 19.4 SELECTION IN A MANX CAT POPULATION.
Where lethal alleles are involved, selection is extremely strong at first, but as the frequency of the lethal allele decreases, it slows considerably. Why doesn't the Manx tailless allele simply disappear from a wild population?

NATURAL SELECTION AND PEPPERED MOTHS

The British peppered moth, *Biston betularia*, occurs in two forms, or *morphs* (Figure 19.5). One morph is light and mottled (or peppered), and the other is black. The British, who have a long tradition of collecting butterflies and moths, have kept excellent records on the peppered moth for two centuries. The black morph, whose color is controlled primarily by a single dominant gene, originally showed up in 18th-century collections as a rare, highly prized variant, or mutant.

In the early days of the industrial revolution (in the 1840s), the black morph began to appear more and more often, especially near cities. In fact, the black morph became so common in industrialized areas that it eventually outnumbered the light peppered morph. (In Manchester, England's industrial center, the black morph came to comprise 98% of the population.) Meanwhile, the light peppered morph remained the predominant form in rural areas.

What had happened? It seemed clear that the species was adapting to some environmental change. That change was due to the burning of coal in the factories of heavy industry. Industrial England in the 19th century was quietly submitting to a dark cloak of carbon. As the countryside darkened, the frequency of the black morph increased in the population. It was hypothesized that this was because the black morphs were harder for predatory birds to see against the soot, while the lighter, mottled individuals stood out in sharp contrast and were quickly taken.

From the historical data, J.B.S. Haldane calculated that the black morph survived and reproduced twice as well as the mottled in the industrial environment. A British naturalist, H.B.D. Kettlewell, performed the crucial experiment that validated the hypothesis. Kettlewell released known numbers of marked black and light peppered moths in unpolluted woodlands and two similar groups in polluted, soot-blackened woodlands. Later, he recaptured a portion of the released moths. The following are some of Kettlewell's mark-and-recapture data:

Dorset, England (unpolluted woodland)	**Light morph**	**Black morph**
Marked and released	496	473
Recaptured	62	30
Percentage recaptured	12.5%	6.3%
Relative survival	1.00	0.507

Birmingham, England (soot-blackened woodland)	**Light morph**	**Black morph**
Marked and released	137	447
Recaptured later	18	123
Percentage recaptured	13.1%	27.5%
Relative survival	0.477	1.00

In the first set of moths, released in the unpolluted woodland, twice as many light forms survived as black forms. In the second set, selection favored the black morph and acted against the light morph.

FIGURE 19.5 SELECTION OF THE PEPPERED MOTH. In the absence of soot from industry, the light speckled variety of peppered moth is well concealed from predators, while its black counterpart is easily seen. In the sooty environment the opposite is true.

About twice as great a percentage of the favored black type survived long enough to be recaptured.

Incidentally, England has been doing rather well lately in its battle against pollution. The woodlands near the cities are once again becoming covered with lichens and the soot is disappearing. And, as one might predict, the black morphs of *Biston betularia* are becoming scarce again.

POLYGENIC INHERITANCE AND NATURAL SELECTION

The stories of the Manx cats on imaginary islands and peppered moths in polluted woodlands are instructive, but their simplicity can be a little misleading. To be sure, evolution does sometimes proceed by the rapid sweep of a dramatically advantageous allele through a population. But most of the genetic differences between individuals in a population are not caused by a few genes with great effects, but by numerous genes, each with a small effect. A trait controlled by more than one gene is said to be under the influence of *polygenic inheritance* (see Chapter 13).

Many traits—such as height, weight, blood pressure, length of limbs, skin color, and swiftness of foot—are largely controlled by the cumulative action of many different gene loci, each gene with a vanishingly small effect. The cumulative effect of polygenic inheritance on the phenotypic variation in a population is enormous, and it is on this sort of variation that natural selection usually works.

PATTERNS OF NATURAL SELECTION

We see, then, that the individual gene obediently follows Mendel's rules, but there are so many effects of gene interaction and other complications that the response of a population to natural selection is seldom one of simple allele increase and replacement. Let's take a look at how natural selection works with polygenic traits. To approach this problem, we should first be aware that most individuals will be about average for most traits. Thus, for any given measurement, there will be relatively few individuals with extremely high or low values. If we group all of the individuals in a population according to a single trait, the resulting graph will almost always form a bell-shaped curve—the statistician's *normal distribution* (see Figure 13.6). In considering the effects of natural selection, the question is, How do the individuals in the middle of the distribution thrive compared with those on either extreme? Depending on how they fare, we can find three trends: *directional selection, stabilizing selection,* and *disruptive selection* (Figure 19.6).

Directional selection is a selection that favors one extreme of the phenotypic range—one end of the curve. This is the kind of selection practiced by dairy breeders who want only the offspring of the cows that give the most milk. In nature, directional selection may be a response to a change in the environment that begins to favor individuals at one extreme. For example, a population may find itself in an unfamiliar territory, a new place offering new challenges. Or the species may suddenly lose a competitor and have new food sources open to it. In such cases, the formerly aberrant individuals at one end of a curve may be better adapted to the new conditions than are those at the center of the curve. Their previously unfavored traits may then become the new optimum. Subsequent evolution can be rapid, as the population quickly adapts to its new opportunities. The peppered moth story, discussed earlier, is an excellent example of directional selection at work. Also, the early evolution of giraffes represents a classic case of directional selection, as shown in Figure 19.7.

Stabilizing selection is a selection usually associated with a population that has become well adapted to its particular surroundings. These surroundings usually are rather stable. Any genetic change in the population, therefore, is likely to be harmful. Although genetic variability still exists, selection tends to favor the mean, or average, individ-

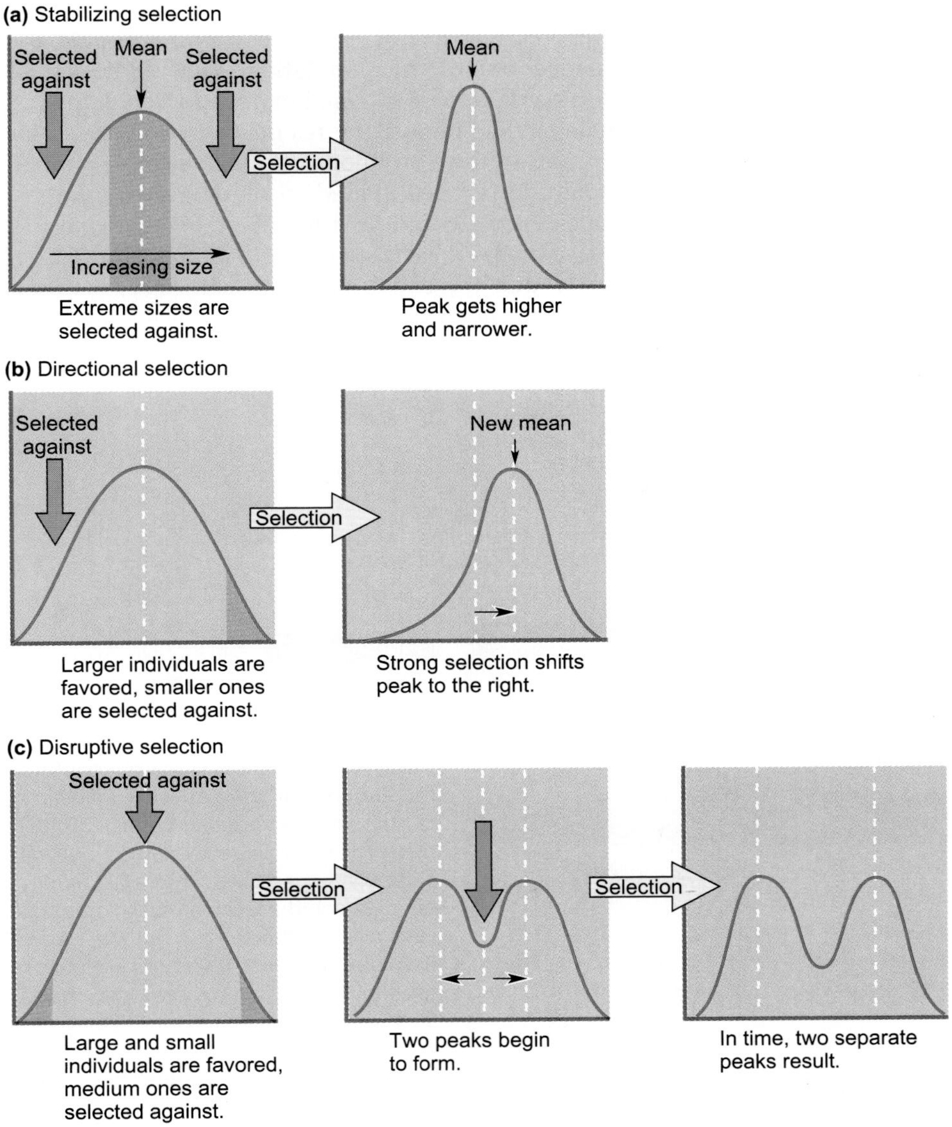

(a) Stabilizing selection

Selected against
Mean
Selected against
Selection
Increasing size
Mean

Extreme sizes are selected against.

Peak gets higher and narrower.

(b) Directional selection

Selected against
Selection
New mean

Larger individuals are favored, smaller ones are selected against.

Strong selection shifts peak to the right.

(c) Disruptive selection

Selected against
Selection
Selection

Large and small individuals are favored, medium ones are selected against.

Two peaks begin to form.

In time, two separate peaks result.

FIGURE 19.6 **TYPES OF NATURAL SELECTION.**
(a) Stabilizing selection for a trait maintains a cluster about the mean. The shape of the resulting bell-shaped curve depends on the strength of selection for the mean. **(b)** In directional selection, the mean shifts to the right or left, and some alleles become scarce and others more common. **(c)** Disruptive selection can result in the two extremes in phenotypic expression being emphasized in a population.

ual. Because most populations are well adapted to their environments most of the time, stabilizing selection is the most common kind of natural selection.

Consider again the giraffe. As far as necks and legs go, giraffes are now well adapted to their environment and are no longer subject to directional selection. In fact, they haven't changed much at all in 20 million years. The only selection that occurs now keeps giraffes as they currently are—unusually tall or short giraffes would be selected out. (Very tall giraffes might gain little feeding advantage while placing much of their metabolic energy into growth.)

Perhaps the best-studied example of selection for an intermediate condition is that of birth weight in human babies. The data are readily available from hospital obstetric wards. If we plot survival rate against birth weight, we find that abnormally small babies have relatively low rates of survival, a fact that is not too surprising. But abnormally large babies also have lower survival rates (Figure 19.8). The highest survival rate is for babies around 3.4 kg (7.4 lb). In this case, the *optimal* birth weight (as determined by survival rate) is almost exactly the *average* birth weight. Selection works against genes for both high and low birth weight. In essence, the average tends to be the best, and "survival of the fittest" becomes "survival of the most medium."

In **disruptive selection** the intermediate types are selected against, and those at *both* the extremes are favored. In this way, disruptive selection produces a *bimodal* (or two-humped) distribution curve. For example, it may be advantageous for males to be large and females to be small, as is the case with elephant seals (Figure 19.9). In such a situation, individuals of intermediate sizes would be weeded out of the population. That is, selection would not favor very small males and very large females. When bimodality results from differences in the sexes, the condition is called **sexual dimorphism.** If females tended to mate with the larger males, there would be a tendency for males to increase in size. The disparity of size, then, would be partly due to sexual selection, a form of natural selection operating through mate choice. Now let's

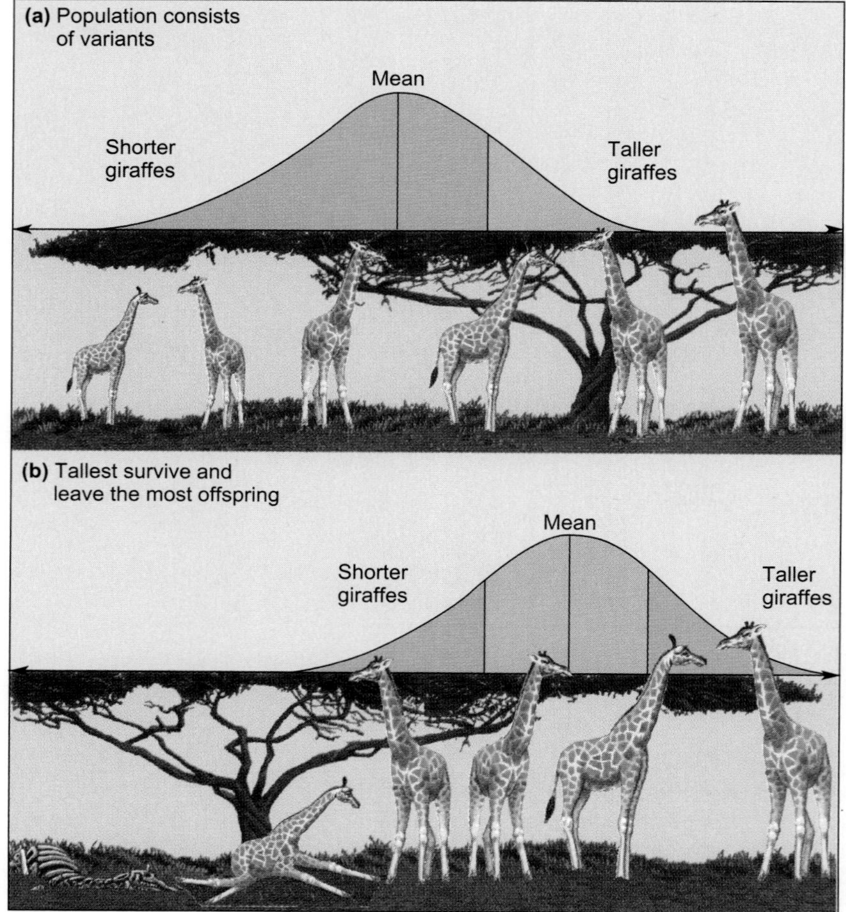

FIGURE 19.7 SELECTION AND EVOLUTION IN GIRAFFES. In the evolutionary history of the giraffe, an animal that browses on tall trees, height has been the critical factor. Among the antelopelike ancestors **(a)** height was variable, but as competition for the foliage increased natural selection began to favor taller individuals that could reach the still-untapped resources of the trees. (Natural selection was probably favoring taller trees as well.) **(b)** As we see, the mean shifted. The taller individual survived, leaving the most offspring. The offspring of the survivors tend to resemble their successful parents, although there is some regression to the mean. The average height increases over the course of one generation (exaggerated here). Over many generations, giraffes become taller and taller. (And so, incidentally, do the trees, as only the tallest trees escape defoliation by giraffes.)

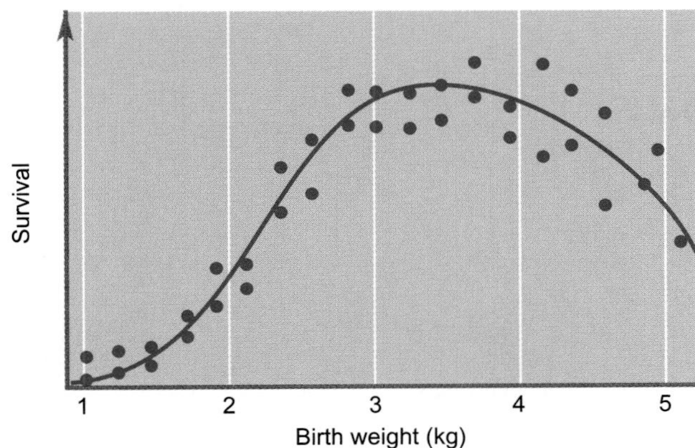

FIGURE 19.8 SELECTION FOR BIRTH WEIGHT IN HUMAN BABIES.
Plotting the survival rate of babies of different birth weight clearly indicates that there is an optimal weight of about 2.7 to 3.8 kg (6 to 8 lb).

briefly consider a type of selective process that is not related to some advantage that depends on one's position along a population curve.

Frequency-dependent Selection. Frequency-dependent selection occurs when the reproductive success of a genotype depends on its frequency in the population. This happens if a genotype has a net advantage when it is rare and a net disadvantage when it is more common. The problem is, if a rare organism has a net advantage, its frequency will increase in the population until it is no longer rare, and its advantage is diminished.

For example, some tropical freshwater fish populations have a common gray morph and a relatively rare red morph. The red fish are rare because they are easier for predatory birds to see. But their red coloration is intimidating to the duller gray fish, so they will almost always win in any competitive encounter and thus tend to mate more frequently. They are not any larger or stronger; they win simply because they are red. However, as the red form becomes more common, they are selected out of the population more often by the predatory birds. The red forms are kept at an equilibrium frequency at which the competitive benefits of being red are just balanced by the disadvantages of being conspicuous (Figure 19.10).

FIGURE 19.9 SELECTION FOR BODY SIZE IN ELEPHANT SEALS.
Plotting the range of body weights for male and female elephant seals produces a decidedly bimodal curve.

MUTATION: THE RAW MATERIAL OF EVOLUTION

If the environment is to select the "best" individuals from a population, individuals within the group must have unequal traits. Variation implies inequality, and some individuals will have traits that give them an advantage over other individuals. The question is, How does this variation arise? Actually, variation can arise in a number of ways, such as through meiosis, crossing over, and the chance combination of gametes. These sorts of variation, however, involve the shuffling of genes already present in the gene pool. What about the appearance of new kinds of genes? How do *they* arise? Even the best gamblers must draw new cards. In living things, gambling through evolution, new kinds of genes must arise from mutation, the random alteration of DNA. Earlier we saw how mutations arise through rare, unrepaired changes in base sequences and through chromosome breakage and rejoining. Let's find what happens to those mutations after they appear in the gene pool of populations.

MUTATION RATES: THE CONSTANT INPUT OF NEW INFORMATION

Any given gene in a population is bound to mutate at one time or another. In fact, each genome undergoes mutation in a statistically regular and predictable manner, producing a constant and measurable input of new genetic information into the gene pool. The emphasis here is on the term *constant*. Each specific gene mutation occurs at its own

By virtue of their color, the red variety of fish are more reproductively successful.

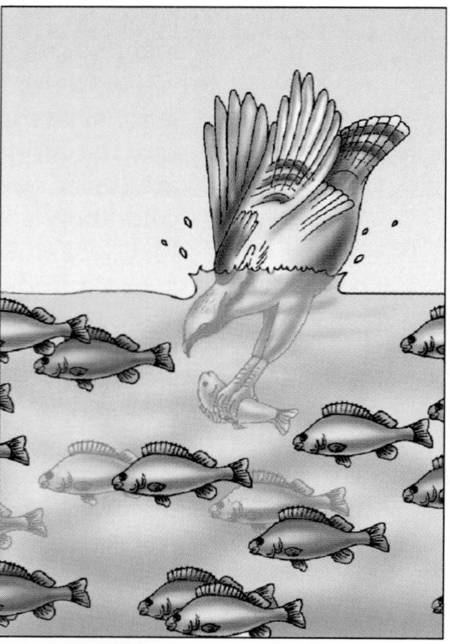

However, as their numbers increase, their bright colors make them easier for predators to find.

Thus the numbers of red fish never overwhelm the gray.

FIGURE 19.10 FREQUENCY-DEPENDENT SELECTION.
Frequency-dependent selection is a type of selection that occurs when a certain form has a reproductive advantage when it is rare, but is at a disadvantage when it is more common.

detectable rate. *Typically*, mutations arise at the rate of one per gene locus per 100,000 gametes. The range in humans, according to some authorities, is from 0.1 to 10 per 100,000 gametes. With this constant input of new genetic material, why do gene pools remain so stable? The reason is that each new mutation is tested by natural selection. A few mutations are retained; most are discarded.

BALANCE BETWEEN MUTATION AND SELECTION

Any gene bearing a mutation that has an immediate beneficial effect can expect to rapidly become more common in a population. However, as we have emphasized before, such changes are extremely rare. After all, our genes work remarkably well and most mutations can't be expected to make a gene work better than it did before. Most changes will have the opposite effect. It's as though you raised the hood on your new Mercedes and let your neighbor's kid randomly bang the internal workings with a hammer. He may make precisely the adjustment needed to make the car run better; all in all, though, you may not want to take the chance.

The point is, most mutations are harmful (decreasing the productive success of the individual) or neutral (neither harmful nor helpful). In some cases the harmful effect is subtle, with the mutation causing a slight decrease in reproductive output; in other cases the effect may be more dramatic with total reproductive failure. Interestingly, the most harmful mutations may go undetected. It has been estimated that one-third to one-half of all human zygotes fail to develop because they bear such deleterious mutations. The potential parents are usually quite unaware that anything has happened. The lethal gene carries the newly formed embryo into oblivion, its delicate tissues simply absorbed by the mother's body. Such lethal mutations are virtually impossible to detect in human populations, but they can be counted accurately in such experimental organisms as yeasts and fruit flies.

It is important to realize that, genetically, mutations that prove lethal at the embryonic stage will have about the same effect on bearers as mutations that allow them to be born but kills them before they reproduce, or one that renders them sterile. Those persons' genes are not passed along; they die with them.

So although lethal alleles are constantly fed into the gene pool by mutation, they remain rare because they are not passed along. However, if the mutant allele is only *partially* limiting in its effect, and the afflicted individuals reproduce but at a reduced level, then the mutant form may become more common. From the known frequencies of recessive genetic diseases, it can be calculated that the average person carries about seven recessive alleles for serious genetic diseases. However, these alleles don't hurt the carriers at all, and only cause problems in the rare event of matings with another carrier.

Mutation and Natural Selection in Humans

The easiest mutations to study, of course, are those dominant mutations that cause *visible* changes. An invisible change might be an alteration in some enzyme that had little or no effect. One example of a visible change, a type of human dwarfism known as achondroplasia, occurs in one out of about 12,000 births to couples of typical stature (Figure 19.11). In achondroplasia, the person's head and trunk are of normal dimensions, but the arms and legs are stunted or fail to grow at all. Because the condition is dominant, each new incident represents a newly mutated gene; the mutation for achondroplasia is one per 24,000 gametes. When achondroplastic dwarfs (who are, necessarily, heterozygous) mate with normal individuals, about half of their children are afflicted by the condition.

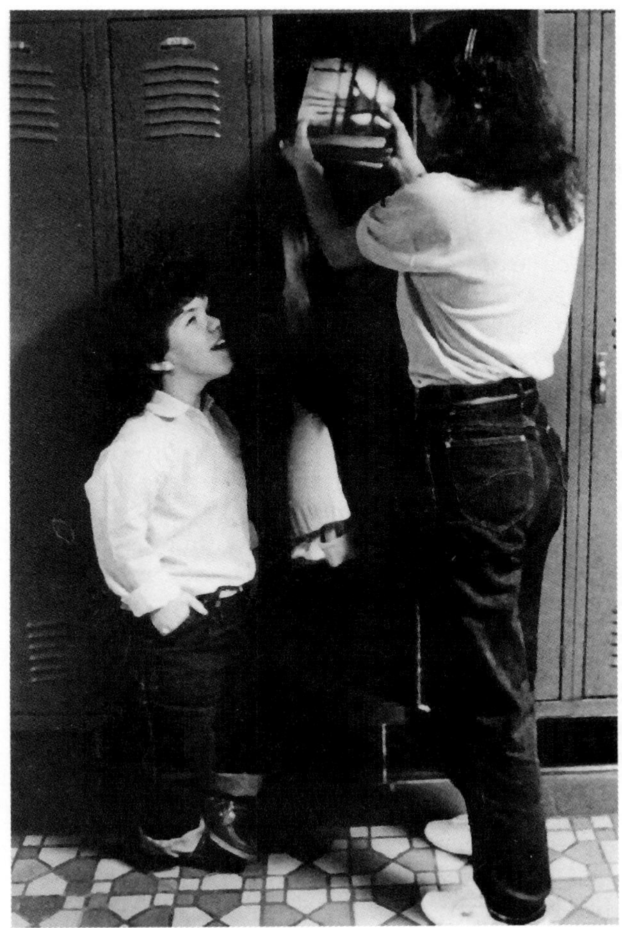

Figure 19.11 Achondroplasia in Humans. Since achondroplasia (a form of dwarfism) is dominant in humans, its occurrence in a family with normal parents represents a new mutation in the father or mother. (*Milwaukee Journal* photo.)

Consider this: With an average mutation rate of about one mutation per gene locus per 100,000 gametes, and with an estimated 40,000 to 50,000 gene pairs per human zygote, *everyone* is likely to be carrying a newly arising mutation. The beneficial or neutral mutations, of course, will readily be passed on. If the mutation is harmful, it is more likely to be transmitted if its effects are minor. Some of the more common of less-than-ideal genetic conditions are nuisances such as missing teeth, dental malocclusion, nearsightedness, and deviated nasal septum.

But, as we've just seen, some genetic conditions are severe. To achondroplasia we can add hemophilia, albinism, sickle-cell anemia, and phenylketonuria—all discussed in earlier chapters. And to these we can add schizophrenia, manic-depressive syndrome, early-onset diabetes, hereditary deafness, cystic fibrosis, Tay-Sachs disease, and thousands of other known genetic disorders. You should remember, however, that in spite of such devastating potential, mutation is a prime source of the variation that has permitted the array of life seen on the planet today, and is one reason the planet is not almost entirely inhabited by creatures that look like amebas.

Human Intervention and Our Genetic Future. This brings us to a philosophical and moral question about genetic variation in future human populations. The question inevitably arises: If better medical and social care saves the lives of persons with adverse genetic conditions, thus allowing them to reproduce, what is to become of us?

It's not hard to find examples that illustrate the problem. One genetic condition, pyloric stenosis, an abnormal overgrowth of a stomach valve muscle, was once invariably fatal in infancy. Since the 1920s a simple surgical procedure has saved the lives of nearly all affected infants in developed countries. But about half of the offspring of the saved people are also affected, and they also need surgery. Since there are also just as many new mutations as ever, the condition constantly appears anew, adding to the descendants of those people who have had the surgery. Therefore, the genes that cause the condition have actually increased in frequency in the human population. One might ask, then, What will be the frequency of pyloric stenosis among humans in 10,000 years?

What has happened, in effect, is that a *severe* genetic condition—which 60 years ago meant certain death in infancy by intestinal obstruction—has been transformed to a relatively *mild* genetic condition that can be readily corrected surgically. Suppose that the remaining risks of the condition, including possible delayed diagnosis and surgical mishap, are such that now only about one affected infant in 20 fails to survive. The mutant genes for the condition will continue to increase. In the meantime, many lives will have been saved, and many new people—with slightly aberrant genotypes—will have joined the human population. Is that good or bad?

The increase in frequency of once-detrimental conditions has already happened repeatedly in a variety of cultures. For example, in primitive hunting and gathering tribes, myopia (nearsightedness) must be a nearly lethal condition. Presumably, nearsighted aborigines can't even find roots and berries efficiently, let alone a zebra. But long before the invention of corrective lenses, the stable social conditions that came with villages and agriculture allowed myopics to survive. In some cases they may have even benefited. For example, male myopics would have been of no value in hunting or in war, and might have stayed safely home to make tools, weave baskets, tell stories, and help the women around the hut. In any case, myopia is much more common among people with a long history of agriculture, writing, and urban civilization than among groups that have more recently given up nomadism or the hunting-gathering life.

Genetic Drift

Genetic drift is evolution produced by random changes in gene frequency. Here, certain genes accumulate by chance—not because of the effects of selection. The phenomenon is seen most easily in small populations. For example, should some catastrophe, such as a flood or a plague, suddenly wipe out most of a population, the survivors would then begin to reproduce. Just by chance, however, the presence of certain alleles and the frequency of others in this small population might be different from that of the former, larger group. In the jargon of genetics, the population would have gone through a **bottleneck**—a brief period in which a population becomes drastically reduced, causing a random change in gene frequencies (Figure 19.12). The harmful alleles, which were once rare, may now be not so rare. Similarly, if a few individuals stray out of their normal habitat and establish a successful colony in a new place, any rare genes they happen to carry may become common among their descendants. This special kind of bottleneck is referred to as the **founder effect** (the genetic effect of the chance assortment of a few individuals from a larger population).

There are many examples of genetic bottlenecks and founder effects in human history. The Afrikaaners of South Africa are all descended from some thirty 17th-century families. (Remember, all humans carry their share of harmful, recessive genes, but they are seldom expressed, except in homozygotes.) The Afrikaaners' genes, having gone through a bottleneck of only 30 families, cause present-day Afrikaaners to suffer from a unique set of recurrent, recessive genetic diseases that are seldom seen in other populations. Among these is a version of porphyria, a metabolic disease that results in liver damage and bright red urine. Conversely, Afrikaaners are almost completely free of other recessive genetic diseases; the 30 families obviously did not carry those particular recessive genes to Africa. Similarly, Jews of Eastern European ancestry harbor a different but equally distinctive array of recessive genetic diseases, one of which is the tragic Tay-Sachs disease—an enzyme deficiency that leads to blindness, mental deterioration, and death in infants. These high gene frequencies can be taken as circumstantial evidence for one or more population bottlenecks in the history of this group.

Neutralists: Holding a Different View of Evolution

We've built a strong case for natural selection as *the* molding force of evolution, and most biologists would agree with this position. We have seen, however, that evolutionary change can occur in the absence of selection, as in the Afrikaaners and in the Eastern European Jews. In these cases, genetic

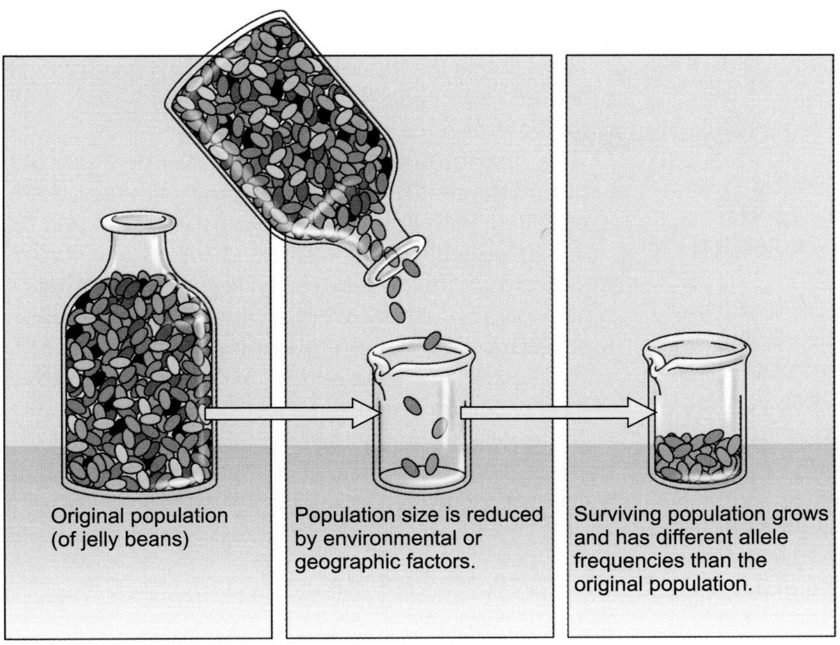
Original population
(of jelly beans)
Population size is reduced
by environmental or
geographic factors.
Surviving population grows
and has different allele
frequencies than the
original population.

FIGURE 19.12 GENETIC BOTTLENECK.
A bottleneck occurs anytime that conditions reduce the numbers of a population such that the remaining individuals no longer represent the distribution of the original population.

change exists that is clearly nonadaptive. Other genetic changes may be neither harmful nor beneficial, but completely neutral with regard to natural selection. The question of just how significant nonselective evolution is has led to raging arguments. In recent years, there has been some controversy over to what degree natural variation is due to *neutral mutations* and *random drift.* **Neutralists** say that much variation on the molecular level is simply incidental and nonadaptive. For example, there are frequently several slightly different forms of any enzyme in a population. These enormous molecules may differ from each other by only one or two amino acids. Neutralists hold that most of this kind of variation has no effect on the function of the enzyme. They believe that the responsible alleles are equivalent and appear by chance mutation, and that the allele frequencies "drift" meaninglessly. **Selectionists,** on the other hand, contend that virtually all variation is due to natural selection. They add that it must have some adaptive basis, even if we don't happen to know what it is.

PERSPECTIVES

In this chapter, we have discussed ways to look at genes within populations and have seen how changes in the representation of those genes within populations can result in material for evolution. We have also noted how certain events can lead to the formation of new populations with different allele frequencies, and how mutation delivers the original changes that contribute to variation.

We now turn to broader aspects of the evolutionary process, and consider how the types of changes that we have just encountered can result in the formation of new species.

Summary

The Population and How It Evolves

Evolution can be thought of as occurring at three levels: the gene, the population, and the species. A population is an interbreeding group whose alleles make up the total gene pool.

The Hardy-Weinberg law, developed independently by Hardy, Weinberg, and Castle, proposes that if mating in a population is random and there is an absence of selective forces, the frequency of dominant and recessive alleles will remain constant. This explains why recessive traits do not simply disappear, swamped as it were, by dominants.

Allele frequencies can be represented as p and q, where p is the frequency of one allele and q is the frequency of the other, and $p + q = 1$. In the expanded form, this becomes $p^2 + 2pq + q^2$, the Hardy-Weinberg formula that gives the frequencies of the homozygote (p^2 and q^2) as well as the heterozygote ($2pq$) genotypes.

In order for the formula to hold, the following conditions must be met:

1. Mating must be completely random.
2. There can be no migration either into or out of the population.
3. The alleles in question must segregate according to Mendel's first law. This eliminates, for example, sex-linked alleles.
4. There can be no mutation.
5. There can be no selection operating on the population.
6. The expectations are exact only if the population and the sample are infinitely large.

Natural Selection at Work in Populations

In the Manx cat example, cats with normal tails, **TT,** and Manx, **Tt,** survive, while the homozygous Manx, **tt,** dies as an embryo. Thus, in an interbreeding population of Manx cats, the **t** allele is subject to strong negative selection. Except for lethal alleles, selection is much less severe.

In past years, soot from coal burning in England's industrial centers blackened tree trunks in nearby forests. The dark coloring hid the dark morph of *Biston betularia* from bird predators, but it exposed the light morph, whose numbers dwindled. Studies involving the release and recapture of the two morphs in both sooty and clean forests showed that gene frequencies were changing due to natural selection operating through predatory birds.

Since most traits are polygenic (the product of many genes), when the genotypes are plotted, they form a continuous distribution.

The distribution of phenotypes can be analyzed for trends in natural selection. Three such trends are directional, stabilizing, and disruptive selection.

In directional selection, individuals at one end of the spectrum are favored; thus a less common phenotype becomes the new average phenotype. In stabilizing selection, the average phenotype is most favored, and selection is against the extremes. Disruptive selection occurs when the most common frequency becomes less advantageous than the less common extremes. The extremes then become favored, and their traits increase in the population.

Mutation: The Raw Material of Evolution

New variation arises through mutation—genetic change brought about by changes in base sequences in DNA and through chromosome rearrangement.

Typical mutation rates are one per gene locus per 100,000 gametes. Most are quickly selected out by natural selection.

Most mutations are harmful or neutral; few are helpful. While helpful alleles tend to increase in frequency, the frequency of harmful alleles decreases. The rate of decrease depends on how strongly they affect reproduction. Each human carries about seven recessive alleles for serious genetic diseases.

In modern human societies the effects of natural selection are greatly moderated through medical intervention. Under primitive conditions the reproductive fitness of individuals with new harmful mutations would be decreased. Medical treatment not only assures the survival of such persons but also permits them to reproduce. Thus certain harmful alleles are likely to increase.

Genetic Drift

Genetic drift refers to random changes in allele frequencies, changes not produced through natural selection. In a bottleneck a population may suffer a drastic reduction wherein, just by chance, certain alleles may be reduced or lost and others increased. The same may happen through founder effect, where small groups establish colonies.

Neutralists attribute a significant amount of evolutionary change to mutations with neutral effects and to random drift. Although they do not discount natural selection as an important evolutionary force, they argue that most minor changes in alleles are simply meaningless drift. Selectionists say that virtually all variation is due to natural selection.

Key Terms

population • **300**
gene pool • **300**
population genetics • **300**
Hardy-Weinberg law • **301**
frequency • **302**
genetic equilibrium • **302**
directional selection • **306**
stabilizing selection • **307**
disruptive selection • **308**
sexual dimorphism • **308**
bottleneck • **312**
founder effect • **312**
neutralist • **313**
selectionist • **313**

Review Questions

1. Carefully define the terms *population, frequency,* and *gene pool.*
2. Summarize the problem that vexed Punnett. What conclusion did Hardy reach when he applied his math skills to the problem?
3. From your knowledge of cells, explain why recessive alleles cannot simply disappear. (In reality, what are alleles?)
4. Basically, how does the approach of the population geneticist differ from that of the Mendelian geneticist?
5. Briefly state a way to prove that in the absence of selection, the frequency of recessive alleles remains constant. What term describes such stability?
6. Briefly explain each of these algebraic formulae:

   $$p + q = 1$$
   and
   $$p^2 + 2pq + q^2$$

   What do the letters p and q represent? What does each expression actually represent in a population?
7. How does one actually go about determining the value of q^2 in a real population? How would the number be expressed?
8. Consider a population in which 400 people have the recessive trait (a hypothetical one) hairless nostril (**hh**). The remainder, 9,600 people, have normal hairy nostrils (**HH** and **Hh**). What is the frequency of the hairless genotype? Find the frequency of each allele and the other two genotypes.
9. In a certain population, 36% cannot roll their tongues. People without this "talent" cannot produce tongue-rolling children, and children who can roll their tongues have at least one tongue-rolling parent. Is the trait dominant or recessive? How many heterozygotes might we expect in this population?
10. Find the genotype frequency of both kinds of dominant individuals (**RR** and **Rr**) in a population where the frequency of recessive Rh-negative individuals **(rr)** is 16%.
11. The term *random mating* has been used several times in this chapter. List as many of the attributes of random mating as you can. What effects might nonrandom mating have on alleles that are in equilibrium?
12. Since the Hardy-Weinberg concept has such stringent requirements, how can it be of any use to geneticists?
13. What do changes in allele frequencies indicate about a population?
14. Using the representative letters **T** and **t,** carry out the cross between two Manx cats. State the expected phenotypic ratio. What ratio actually appears? How might this represent natural selection at work?
15. In the peppered moth story (pp. 307–308), what was the actual selective force? How did industrialization affect the fate of the dark morph? What happened after the curbs on air pollution became effective?
16. In the absence of air pollution, would the black morph be considered a lethal, neutral, or helpful mutation? Explain.
17. Using a simple graph, characterize directional selection. Relate this kind of selection to the peppered moth episode.
18. Giraffe neck-length is now believed to be under the influence of stabilizing selection. What does this mean? What condition might change this?
19. Under what geological conditions would one expect disruptive selection to occur? How might the loss of a major predator (say, through bounty hunting) bring on disruptive selection?
20. What happens to most newly formed mutations? What obvious factor determines the strength of selection against an allele?
21. How would you measure the effects of natural selection against a slightly harmful allele? What would you actually observe? In other words, what finally happens to the frequency of slightly harmful alleles?
22. Briefly describe how humans overcome the usual effects of natural selection. How might this affect the human gene pool in generations to come? How might genetic engineering head off this effect?
23. Using bottleneck and founder effect as extreme examples, explain what genetic drift is about.

20

Evolution and the Origin of Species

NOW THAT WE HAVE SEEN HOW ALLELES BEHAVE IN POPULATIONS, WE *can consider how the accumulation of changes in the alleles within a population can lead to the formation of new species. Interestingly, although the origin of species must be a direct result of evolution, it is rarely observed in nature. Why do you suppose this is?*

One of the most interesting things about biology, and probably other sciences as well, is that although a great deal of attention is given to working out details, many of the larger, more basic questions remain unanswered. For example, some biologists may peer endlessly at the pattern of bristles on a fruit fly's back and ask why they grow that way, while other biologists stroke their chins and ask where new species come from. And as if that question weren't basic enough, others point out that we still don't even know what a species *is.*

First, let's see how living things are named, then ask what a species is, after all, and finally, see how species arise. (*Species,* by the way, is both singular and plural.) Before we go any further, though, it is important to realize that **speciation** (the formation of new species) is not the same as evolution but that it is an important *result* of evolution.

NAMING NAMES

It would seem that naming new species would be easy, even fun. But it's not a simple thing. As with so much else in life, there are rules. The rules were first set down by a Swede by the name of Karl von Linné (1707–1778), or Carolus Linnaeus, as he preferred to call himself, latinizing even his own name. Linnaeus took upon himself the incredible task of naming all the plant and animal species known. His discipline, naming new species, is called **taxonomy,** and Linnaeus himself was the first modern taxonomist.

Linnaeus was also the first to fully apply the system of **binomial nomenclature**—the practice of giving species a *scientific name.* A scientific name is composed of two Latin names: a *generic name* and a *specific name.* The generic name is the name of the **genus,** or group of closely related and ecologically similar species; the specific name identifies the *species* within the genus. Long-standing tradition and international regulations state that:

1. Both names must be in Latin or at least latinized.
2. Both are to be written in italics.
3. The first (generic) name is always capitalized, while the second (specific) name is never capitalized.

Once assigned, this combination of names may never be used for anything else. Also, the generic name is generally a noun, and the specific name should be an adjective. These rules are bent a little to allow a latinized adjectival version of the name of a friend, colleague, or authority figure (the rules prohibit anyone naming a species after himself or herself). Thus do taxonomists seek to impose order on a disorderly world.

As an example, we humans are called *Homo sapiens. Homo* is our generic name, and *sapiens* is our specific name. *Homo* means "man," while *sapiens* means "wise" or "discerning." (Make of that what you will.) Other species in the genus are *Homo habilis* (*habilis* means "able to do or make") and *Homo erectus* ("upright"), both extinct (Figure 20.1). Also, once the genus has been stated, the first letter may thereafter be used, as in *H. sapiens.*

When necessary, the binomial nomenclature is extended to a third Latin term, the **subspecies,** or variety of the species. Thus contemporary humans are sometimes referred to as *Homo sapiens sapiens,* to distinguish us from our recently departed cousins, *Homo sapiens neanderthalensis.* And one race of apes is called *Gorilla gorilla gorilla.* (Using the same name for genus and species is acceptable in zoology, but not in botany. A botanist friend was once heard to ask disparagingly, "Really, now, what kind of information is conveyed by a name like *Gorilla gorilla gorilla?*" To which a zoologist present muttered "All you need to know.") For a look at other taxonomic categories, see Essay 20.1.

ESSAY 20.1 The Kingdoms

Kingdoms are the largest taxonomic category. The number of kingdoms varies in different schemes according to what is considered to be important, but at the moment, most biologists would go along with five.

If we were describing political organizations, kingdoms would be the equivalent of nations. They are generally subdivided into what taxonomists call *phyla* (singular, *phylum*)—equivalent, roughly, to our own states, if we continue the political analogy. Each phylum is itself subdivided into *classes* (counties), and so the subdivisions continue. It is easier at this point to look at the total organization as we make our way down to the species level. The increasingly finer divisions (and a way to remember them) are as follows:

Kingdom	King
Phylum*	Philip
Class	Came
Order	Over
Family	From
Genus	Greece
Species	Singing
Subspecies	Songs

We will find that there is no single "correct" way to place an organism into a scheme such as this. In fact, there is often a great deal of argument over such placement. Obviously, however, the smaller the group into which two kinds of organisms are placed, the more similar they must be. Those placed in the same species are likely to be so similar that they can interbreed. Those in the larger categories may be quite different. For example, even though humans and sponges are not alike at all (your opinion of your next-door neighbor notwithstanding), both are animals.

The organization of life into the five kingdoms is as follows:

1. *Kingdom Monera.* The prokaryotes or bacteria are chiefly single-celled or colonial, and nearly all lack membranous organelles. Included are chemotrophs, phototrophs, and heterotrophs. Heterotrophs include ecologically vital decomposers and numerous disease-causing parasites.
2. *Kingdom Protista.* The protists include the **protozoa** (single-celled, animal-like heterotrophs) and the *algae* (plantlike, photosynthetic autotrophs that may be single-celled, colonial, or multicellular).
3. *Kingdom Fungi.* The fungi are multicellular heterotrophs with lengthy, tubular cells, cell walls composed of chitin or cellulose, and elaborate sexual structures. Included are many ecologically important decomposers and a large number of parasites.

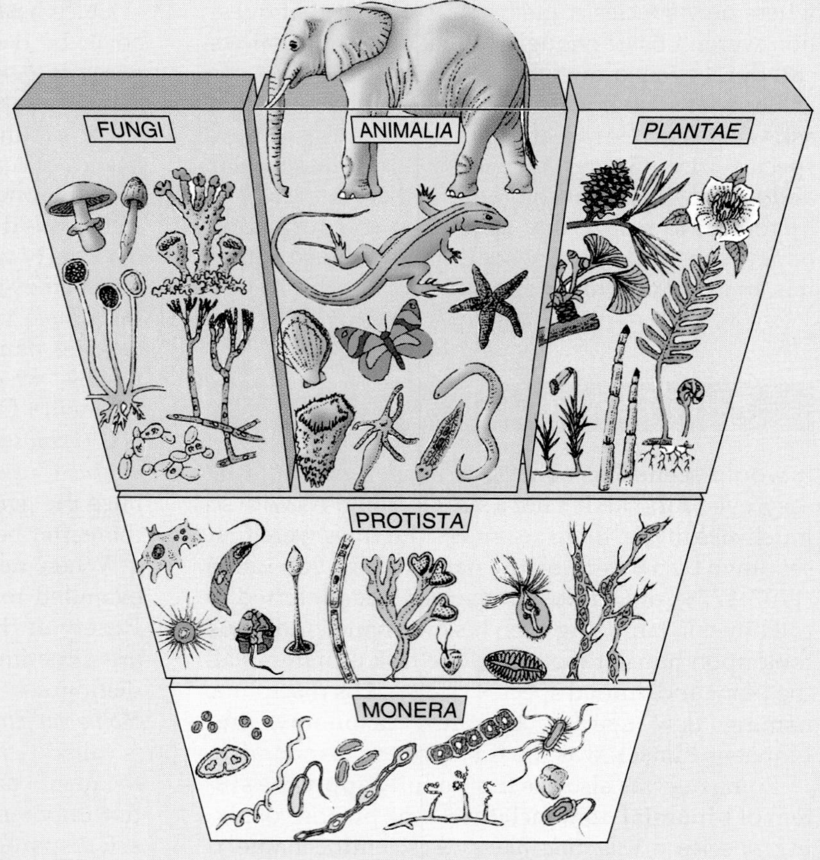

*In botanical terms, the phylum is replaced by the division.

4. *Kingdom Plantae.* Plants are nonmotile (stationary), multicellular, photosynthetic autotrophs with cell walls of cellulose. Most have highly specialized tissues and organs.
5. *Kingdom Animalia.* Animals are motile multicellular heterotrophs that lack cell walls and have highly specialized tissues and organs.

Within the five-kingdom scheme there are some problems, specifically with regard to the protists. Kingdoms are supposed to be cohesive units with all of their members traceable to a common evolutionary ancestor. Such descent is called *monophyletic.* Yet there is no evidence that the various protozoa and algal protists are descended from a single line. In fact, they are definitely *polyphyletic,* having evolved from different ancestors. However, they will be lumped together provisionally until their evolutionary origins can be sorted out.

FIGURE 20.1 HUMAN SPECIES. Today's humans all fall within a single genus and species, *Homo sapiens.* We share our genus with two extinct species—*Homo habilis* and *Homo erectus.* We share many traits with these two, but the differences are sufficient to warrant individual species designations.

What Is a Species?

You can rattle a graduate student preparing for a doctoral examination in biology by asking, "Quick, what is a species?" This may be surprising because the question seems simple enough. But the simplicity is deceiving.

There usually isn't much difficulty in telling one species from another when you are out in the field—on safari, for instance. A giraffe is quite distinguishable from a lion. A tree fern clearly belongs to one species, and an African violet clearly belongs to a different species, and no one would try to saddle up an oyster (Figure 20.2).

However, definition by simple observation can quickly run into trouble. How about those species that appear similar but are actually quite distinct? Taxonomists stress that in order to be of the same species, two groups must be so similar that they can successfully interbreed, producing healthy and viable offspring. But what about groups that are *capable* of interbreeding but are separated by geography (Figure 20.3)? Now the issue becomes sticky. Animals that interbreed successfully when confined together in zoos (Figure 20.4) may ignore or never encounter

FIGURE 20.2 EASILY DISTINGUISHED SPECIES.
For most purposes, it is not difficult to distinguish among species. Here, for example, the zebras are a single species, as are the wildebeest and the rushes and grasses along the bank. But perhaps not far away are zebras of another stripe. Are the two groups of the same species? How do we know?

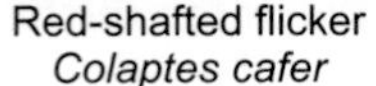

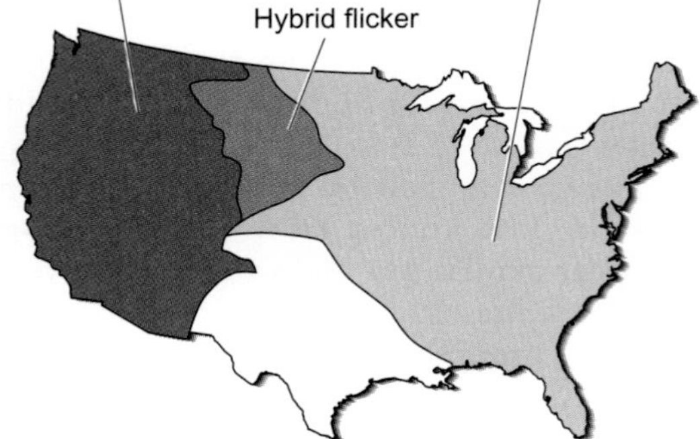

FIGURE 20.3 HYBRID SPECIES.
The red-shafted flicker and the yellow-shafted flicker normally occupy different habitats and so do not interbreed. But where their ranges overlap, hybrids may appear.

one another in nature. Plants that set seed when pollinated by hand wouldn't necessarily cross-fertilize without our help. So the interbreeding criterion is generally useful but far from absolute. Furthermore, it doesn't help at all in identifying species in asexual organisms, or in plants that commonly hybridize in nature. Here, then, we will offer a definition by the famed zoologist Ernst Mayr:

> A **species** is a group of actually or potentially interbreeding natural populations that is reproductively isolated from other such groups.

With this definition in mind, let's turn to our main subject—how species arise. We will begin with a brief overview of the processes, and then see how geographic influences exert their effect.

MECHANISMS OF SPECIATION

The formation of new species can arise through two processes called *phyletic speciation* and *divergent speciation* (Figure 20.5). **Phyletic speciation** is the formation of a new species as an existing species gradually changes. The result of phyletic speciation is that a species may change so much that an individual might not even recognize its own forebears. Until recently many anthropologists believed that much of hominid (human line) evolution occurred in this "straight-line" fashion, at least from the advent of *Australopithecus* on to modern humans (see Chapter 21). Currently, though, anthropologists favor a branching tree account of human evolution, based on **divergent speciation**—that is, the formation of species as an existing species gives rise to two. In fact, evolutionists agree that *most* species arise through divergent speciation.

Divergent speciation theoretically occurs in two principal ways. The first is **allopatric speciation** (*allo,* apart), the formation of two new species as an existing population becomes geographically divided. Allopatric speciation is believed to be the far more common means by which new species arise. Less frequently, **sympatric speciation** (*sym,* together) may occur. In this case a new species springs from a single, interbreeding group, while staying with the original group. Sympatric speciation is probably most common among plants.

ALLOPATRIC SPECIATION

Since geographic separation is generally so important to the origin of species, let's take a closer look at allopatric speciation and explore some of its causes and effects.

FIGURE 20.4 **TIGLON.**
Captive animals, subjected to conditions that do not exist in the wild, may behave differently from the ways they behave in nature. For example, African lions and Siberian tigers would not interbreed in the wild partly because they would never encounter each other. But such mating can occur in captive animals, as witnessed by this hybrid *tiglon* from a male tiger and a female lion.

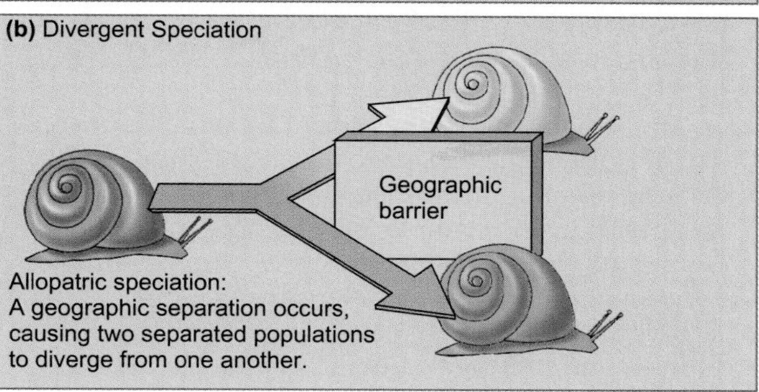

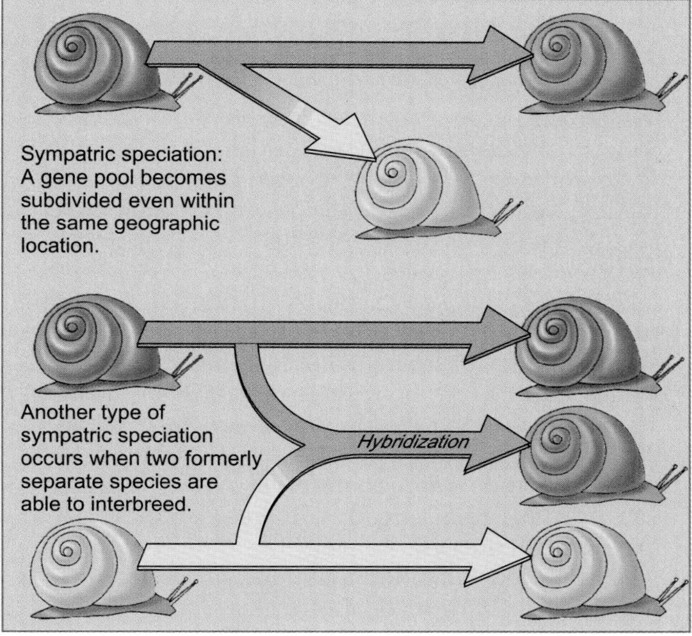

Sympatric speciation:
A gene pool becomes subdivided even within the same geographic location.
Another type of sympatric speciation occurs when two formerly separate species are able to interbreed.
Hybridization

FIGURE 20.5 **TYPES OF SPECIATION.**
(a) Phyletic speciation involves the gradual accumulation of changes in a line of descent, wherein a species simply replaces itself. **(b)** Divergent speciation includes allopatric and sympatric speciation.

First, it is obvious that a population can be separated into two subpopulations that are then geographically isolated by the formation of a new mountain range or a new river or some other geographic barrier. Such isolated groups would be unable to interbreed and would be subjected to different selective factors. These conditions, then, would effectively set the stage for speciation as the isolated groups became more and more different. Ultimately, should the opportunity arise for the separated populations to interbreed, it would no longer be possible. Thus there would be two species in the place of one (Figure 20.6).

Geographic barriers can also be set up as land masses themselves move, dividing populations of organisms. The most fundamental changes of the planet's surfaces began some 230 million years ago when the most recent worldwide process of **continental drift,** the movement of continental land masses, started (see Essay 18.2), leading, for example, to the separation of South America and Africa some 65 million years ago. The uniqueness of species on the present-day continents, particularly those of Australia and South America, has been the result of millions of years of allopatric speciation made possible by the separation of continents.

There are other ways that populations of a species might become geographically isolated. One possibility is illustrated by a seed, an inseminated female, or a group of individuals finding itself—by some happenstance—in a new, yet hospitable, place. Ocean islands, for instance, are occasionally populated by the descendants of unwilling and thoroughly drenched passengers on driftwood logs. Similarly, birds and flying insects may be blown to some island by particularly violent storms, or perhaps through an error in navigation.

The Galápagos Islands and Darwin's Finches. The finches of the Galápagos Islands (Figure 20.7), called Darwin's finches (see Chapter 18), provide the best-known example of allopatric speciation. Compared to the giant tortoises, strange flightless cormorants, and impish sea iguanas living on these islands, the finches may not seem particularly interesting—until the saga of their evolution is revealed.

All the finches are 10 to 20 cm long, and both sexes are drab brown and gray. Six are ground-dwelling species, feeding on different-sized seeds or cactus, and eight are tree finches. In all the species, the bill has become modified, adapted to a specific diet. (In one case, the behavior, rather than the bill, has become modified. The woodpecker finch lacks the long, piercing tongue of the woodpecker, so it uses cactus spines to pry insect grubs out of cracks and crevices in the trees.)

On the *Beagle*'s historic visit to the Galápagos Islands, Darwin collected everything he could find, shoot, or catch, including these little brown finches. At the time, he took no special interest in them. But after his return, a London bird taxonomist examined the specimens and noted that the inhabitants of different islands—though they were very similar to one another—were clearly different species; yet they were all finches.

Darwin concluded (and his conclusions are supported by ensuing years of careful research by others) that the different species of birds he observed were all descended from the same stock (Figure 20.7). It is now believed that about 10,000 years ago these volcanic islands were colonized by South American finches that were probably blown out to sea by a storm. Apparently conditions on the islands were favorable, and the "castaways" flourished. Their descendants began island-hopping and eventually populated all the islands. However, the island-hopping was rare enough to ensure the virtual isolation of each group.

The little birds had the islands to themselves, as far as they were concerned. They found food of all sorts everywhere. They were already well adapted for foraging for small seeds on the ground, but now there were other plentiful food resources—such as large seeds, insects, and a variety of other life among the branches, food not ordinarily eaten by finches.

Soon enough, the expanding populations were depleting the supply of available small seeds. Thus natural selection began to favor birds that could also cope with larger seeds, as well as other types of food. In time, the birds' bill sizes began to change as each population began to adapt more closely to the food found on its particular island. Not only were bill sizes changing, but so were allele frequencies, as each relatively isolated group began to accumulate its own kinds of alleles (see Chapter 19). Eventually, the birds of the different islands differed genetically to the extent that any island hoppers would find themselves reproductively incompatible with the residents of other islands. In this way, then, new

FIGURE 20.6 SPECIES VARIATION.
Species that cover vast ranges encounter a variety of environmental influences. Thus they may vary from place to place. Where the environment varies gradually, the population may change gradually from one place to the next, so that individuals at the extremes of the range may be quite dissimilar, as we see here in these rat snakes.

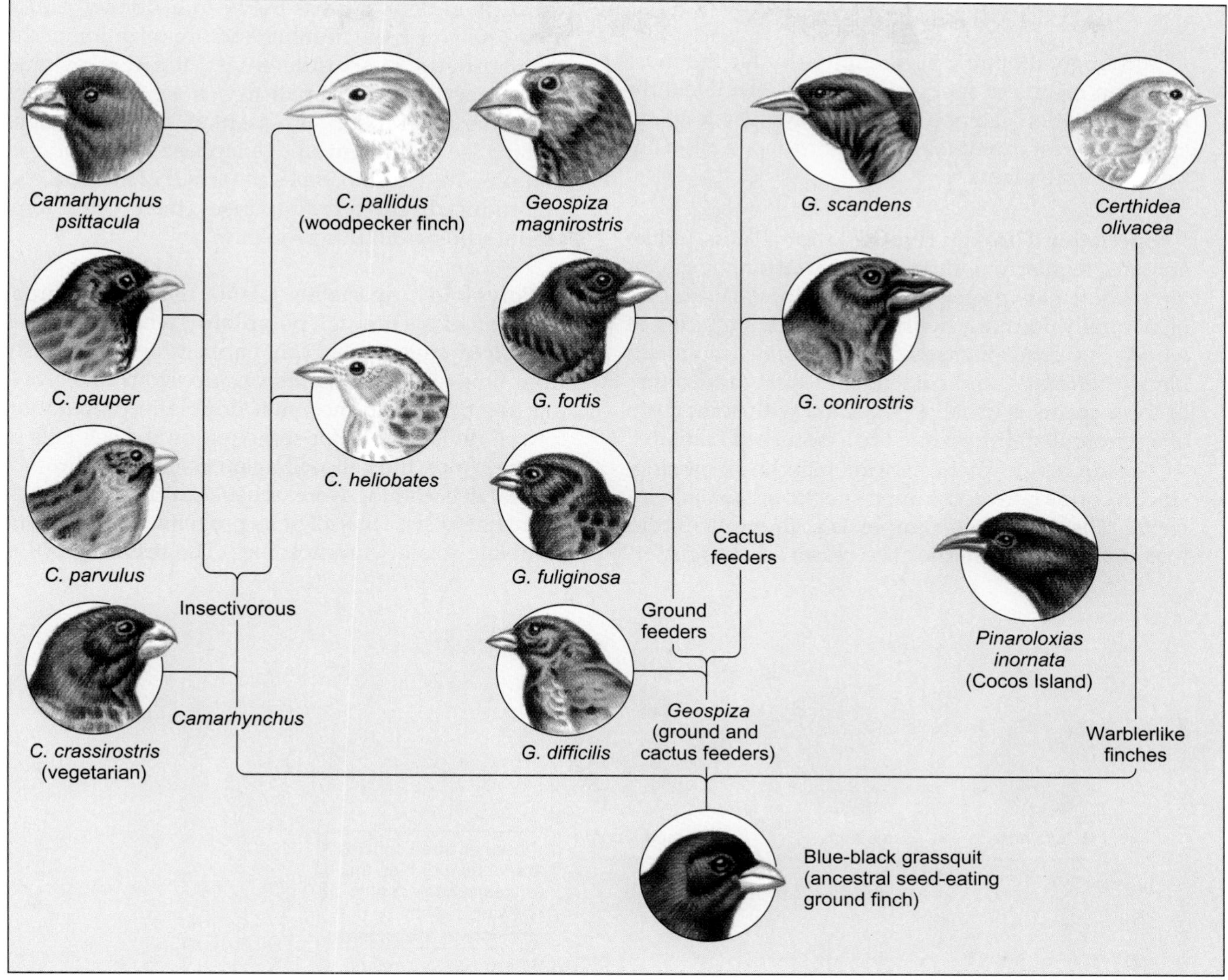

FIGURE 20.7 DARWIN'S FINCHES.
Darwin's finches include six species of ground foragers and eight species of tree foragers. What does a comparison of the size and shape of beaks suggest about individual diets? Intensive studies of Galápagos finches have led to a scheme suggesting a number of divergences in their evolutionary past.

species were formed, through **adaptive radiation**—the spread of a population into new environments accompanied by adaptive evolutionary changes.

What if some species invaded an island and found a genetically incompatible species already there? They might be able to coexist with the residents if the two groups tended to use different resources. With two species of finches trying to survive on one small island, natural selection would favor the individuals in *each* population that were as different as possible from those in the other population (since such individuals would be less affected by the competition from the other population). And as each group became a specialist in utilizing the environment in a particular way, there would be a tendency toward further separation and divergence of the two groups. This tendency for differences in competing species to become exaggerated is called **character displacement** (Figure 20.8).

After thousands of years of occupying these remarkable islands and separating, changing, specializing, and rejoining, a number of species now exist together that are totally unable to interbreed. In fact, several of the species now coexist on every island, each species utilizing the resources of the island in its own unique manner, in some cases occupying places that are filled by other kinds of birds on the mainland.

Sympatric Speciation

Although allopatric speciation is by far the most common means of speciation, under certain conditions speciation can occur through sympatric speciation. The best examples of sympatric speciation are found among plants.

Speciation Through Hybridization. Plants, unlike animals, regularly undergo **hybridization,** or the interbreeding of two species. The best-known examples of naturally forming hybrids come from species of woody flowering plants such as *Eucalyptus,* mountain lilacs (*Caenothus*), and oaks (*Quercus*). Hybridization in these plants is usually associated with some form of environmental stress such as a disturbed habitat.

The success of plant hybrids may be surprising, since hybrids between animal species are usually infertile. The mule, for example, is a vigorous hybrid that is completely sterile. The reason is straightforward: A mule is a cross between a donkey and a horse; however, the number and organization of the chromosomes is so different in the donkey and horse that they cannot pair in meiosis (see Chapter 10). This pairing up is essential for the proper metaphase alignment and anaphase separation of homologous chromosomes if normal gametes are to be produced. The question arises, then, Why don't plants suffer from this problem?

Polyploidy. Speciating plants may avoid problems of meiosis through **polyploidy,** where multiples of diploid genes exist. This happens spontaneously from time to time in the mitotic divisions of a growing plant. The chromosomes double in preparation for cell division, but for some reason the cell fails to divide. Later, the cell will again double its chromosomes, as if nothing were unusual, and proceed with normal mitosis. Thus all of its progeny will have four complete sets of chromosomes. The resulting cell is

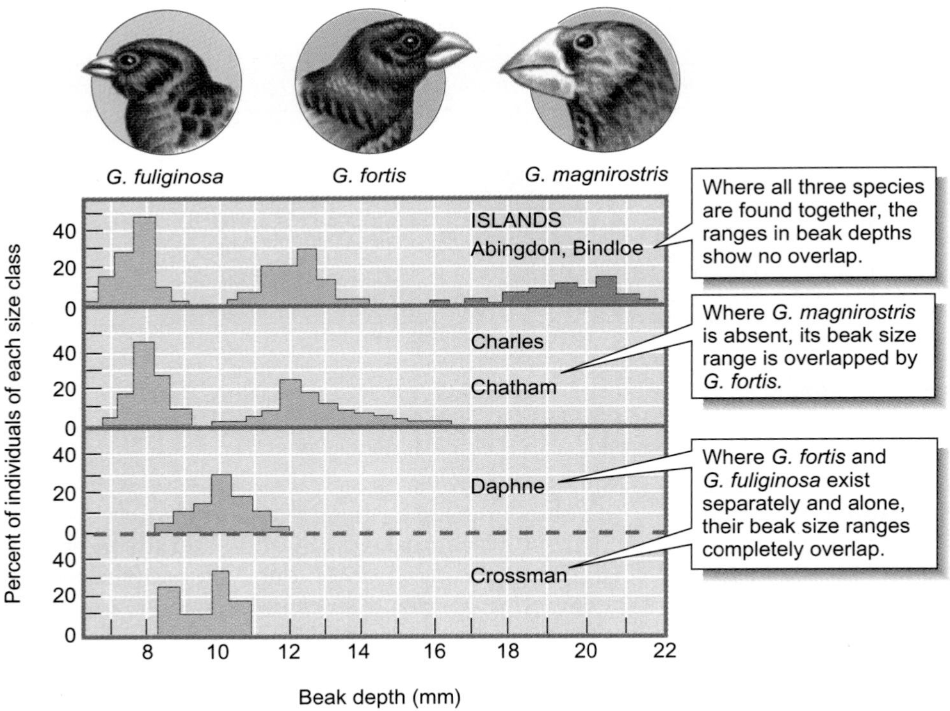
G. fuliginosa
G. fortis
G. magnirostris
ISLANDS
Abingdon, Bindloe
Charles
Chatham
Daphne
Crossman
Percent of individuals of each size class
Beak depth (mm)
Where all three species are found together, the ranges in beak depths show no overlap.
Where G. magnirostris is absent, its beak size range is overlapped by G. fortis.
Where G. fortis and G. fuliginosa exist separately and alone, their beak size ranges completely overlap.

Figure 20.8 Character Displacement.
Character displacement is revealed in beak size changes in three species of Galápagos finches, *Geospiza fuliginosa, G. fortis,* and *G. magnirostris.* Where all three species are found together, the ranges in beak depths are clearly separated with no overlap. Where *G. magnirostris* is absent, its beak size range is overlapped by *G. fortis.* However, notice that it and *G. fuliginosa* remain distinct. Where *G. fortis* and *G. fuliginosa* exist separately and alone, their beak size ranges completely overlap.

called a *tetraploid* cell, and its progeny may form tetraploid tissue and even tetraploid flowers.

Sexual reproduction in hybrids may fail early in meiosis because the hybrid cell has two different haploid chromosome sets. But when spontaneous tetraploidization occurs, there will suddenly be two different complete diploid chromosome sets. Such a tetraploid is called an *allotetraploid* ("other-tetraploid"). When flowers form in the allotetraploid hybrid, there is no longer a compatibility problem in meiosis, since every chromosome now has a homologue, and meiosis can proceed normally. Without such homologues, pairing is not possible (Figure 20.9).

The allotetraploid plants may immediately be reproductively isolated from the parental stock. While they can successfully reproduce through self-fertilization, if they are crossed with either parental species, they produce only *triploid* seeds (since one parent has only a single set of chromosomes), and these seeds are infertile. This reproductive isolation means that the new allotetraploid plant constitutes an "instant" species. The new species will soon fail, of course, unless it happens to establish a new niche or unless it can outcompete another plant species—perhaps one of its parents. But this has been known to happen.

Plant breeders can repeatedly bring about polyploidy through the application of colchicine, a chemical agent that prevents chromatid separation during mitosis (see Chapter 9). Thus it is possible to produce *hexaploids* (six genomes, or three doubled copies of three different parental genomes), *octaploids,* and so on. Strange as it seems, polyploidy is rather common among many wild and domestic flowering plants. Wheat, for example, is actually an allohexaploid, a genetic combination of the chromosome complements of three different species of wild grasses. *Triticale* (Figure 20.10) is a human-engineered cross between wheat and rye grasses.

For reasons not fully understood, polyploid species often have a tolerance for harsh climatic conditions. In one study only 26% of the plants sampled in lush tropical regions were polyploids, while 86% of all the flowering plant species in the raw environs of northern Greenland were polyploids.

We see, then, that hybridization followed by polyploidization creates instant sympatric species of flowering plants, ready to be tested by the forces of natural selection. This versatility helps explain how flowering plants arose so abruptly in the Cretaceous period (see Chapter 24), and how they spread so quickly over the landscape to create the incredible diversity of plant species that we see today.

Sympatric Speciation in Animals. Although there are polyploid species of animals, hybridization is not as likely in species that have sex chromosomes. This is because in polyploid meiotic cells, the Xs tend to pair with Xs and the Ys with Ys, instead of the normal XY pairing. The resulting gametes are therefore unbalanced for sex chromosomes. Polyploidy is frequently found among fish, amphibians, and reptiles because sex in many of these groups is influenced by factors such as temperature rather than by chromosome ratios.

Reproductive Isolating Mechanisms

Once new species arise, they generally develop **reproductive isolating mechanisms,** means of impeding successful matings outside the species. Some are **prezygotic** (before the zygote) reproductive isolating mechanisms—that is, they prevent fertilization. For example, the breeding seasons of the two new species may change enough so that they enter their reproductive cycles at different times. Also, the two populations might not attempt to interbreed if their courtship behavior changes so that they do not appropriately signal each other. Or perhaps even if mating is successful, the sperm might encounter a less-than-ideal environment in the female's reproductive tract. Prezygotic isolation, then, can include behavioral, anatomical, and physiological conditions that prevent fertilization.

Postzygotic (after the zygote) reproductive isolating mechanisms occur after fertilization. In such cases, offspring may be produced, but may be physically weak or may die as embryos. Perhaps the offspring survive but grow into sterile adults, commonly because the chromosomes fail to synapse (pair up) properly during prophase I of meiosis (see Chapter 10). In any case, even closely related species (or, perhaps, *especially* closely related species) normally develop mechanisms that foil interbreeding and thereby preserve their accumulated genetic adaptations for a particular environment.

Major Evolutionary Trends

How can we put the concept of speciation into the larger context of evolutionary history? Let's begin by taking a look at some of the major trends in evolution as revealed by the fossil record, and by comparing living animals.

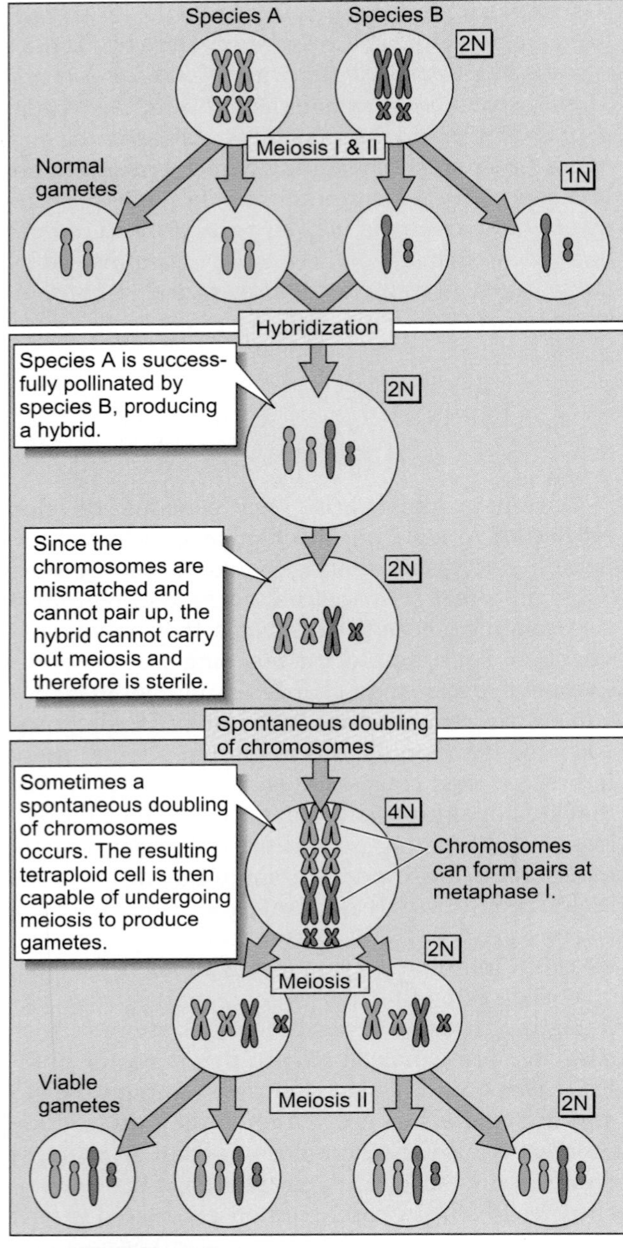
Species A
Species B
2N
Meiosis I & II
Normal gametes
1N
Hybridization
Species A is successfully pollinated by species B, producing a hybrid.
2N
Since the chromosomes are mismatched and cannot pair up, the hybrid cannot carry out meiosis and therefore is sterile.
2N
Spontaneous doubling of chromosomes
Sometimes a spontaneous doubling of chromosomes occurs. The resulting tetraploid cell is then capable of undergoing meiosis to produce gametes.
4N
Chromosomes can form pairs at metaphase I.
2N
Meiosis I
Viable gametes
Meiosis II
2N

FIGURE 20.9 HYBRIDIZATION.
Species A is successfully pollinated by species B, producing a hybrid. Since there is no match in the chromosome complement, the hybrid cannot carry out meiosis and therefore is sterile. However, should a spontaneous doubling of chromosomes occur, as it sometimes does, normal pairing-up and meiosis occur and normal gametes are produced. Following self-fertilization, a new viable species is produced.

DIVERGENT EVOLUTION

We used the term *adaptive radiation* to describe the manner in which speciation occurred in the Galápagos finches. We found that the finches developed differences in how they interacted with their environment, and in the process, they became less similar to one another. Thus adaptive radiation depends on **divergent evolution**—the process in which species become increasingly different from each other. (Don't confuse divergent evolution with divergent speciation. Evolution is a process; speciation is a result of that process.)

Divergent evolution is strongly dependent on ecological forces, because differences in populations are often associated with differences in utilizing resources. If the resources available to two diverging populations are markedly different, those populations will diverge more quickly as each changes to take advantage of what its environment offers.

Such divergence is often represented in branching schemes called phylogenetic trees. The tree is a good analogy, and we will look at examples later in this chapter and in chapters to come.

CONVERGENT EVOLUTION

We may be struck by the tendency toward divergence in evolution, but it is not the only possible trend. **Convergent evolution** occurs when different species grow more *alike* as they adapt to similar environmental conditions.

Darwin was impressed by the evidence of convergent evolution in his travels. You will recall, for example (Chapter 18), his observations that the South American mara was quite similar, in both appearance and ecological niche, to the European rabbit. But then close examination of the mara revealed that it was not a rabbit at all, but a rodent (closely related to the guinea pig) that had taken on rabbitlike characteristics because its environment was similar to that of rabbits in other places.

Convergent evolution is dramatically evident in comparisons between placental mammals and the distantly related marsupial (pouched) mammals of Australia. These two groups have taken on a striking resemblance to one another. For example, in Australia we find the rabbit bandicoot, the marsupial mouse, the marsupial mole, the flying phalanger, and the banded anteater. All of these have a placental counterpart on other continents. There are other examples (Figure 20.11).

Coevolution

Coevolution occurs when two species influence the evolution of each other. It is particularly likely to occur if either species is dependent on the other. The most obvious examples of coevolution are seen in predator-prey relationships. As natural selection improves the predator's skill, it also favors the development of traits that improve the prey's ability to escape. As the predator population gets better, so does the prey population—which in turn causes the predator population to change to meet the new challenge.

Another example of coevolution is seen in flowering plants and their insect pollinators. They may, indeed, be interdependent because many flowering plant species are selectively pollinated by only one kind of insect, and that insect visits only one kind of plant. For example, yucca is entirely dependent on the yucca moth for pollination; the yucca moth, in turn, is entirely dependent on the yucca flower, where it lays its eggs and where its larvae grow on a diet of yucca seeds. The yucca moth must ensure pollination, because only properly pollinated flowers will produce the seeds the moth larvae need. Thus the two species have developed a variety of signals and physiological responses that help ensure harmonious cooperation.

As a remarkable example of herbivore-plant coevolution, consider the relationship between passionflower vines and a *helioconius* butterfly whose caterpillars specialize in eating them (Figure 20.12). This plant has succeeded in manufacturing poisons that prevent most other insects from devouring its leaves and young shoots, but these butterflies have evolved the ability to detoxify the poisons. The butterfly lays bright yellow eggs on the young leaves, which act as a warning to other female *helioconius* butterflies and causes them to move away in search of their own leaves, thereby reducing any competition between the larvae of the females. Through natural selection, the plant has taken advantage of this behavior and has now developed rather good mimic eggs—little round lumps of bright yellow tissue scattered randomly on the leaves and shoots of the vine. To top it off, the vines with the fake eggs have also mimicked the shriveling of the leaves that marks larval infestations. The butterflies, in response to these "eggs" and shriveled leaves, fly away to lay their eggs elsewhere. One might guess that the butterflies are currently developing ways to spot fake eggs.

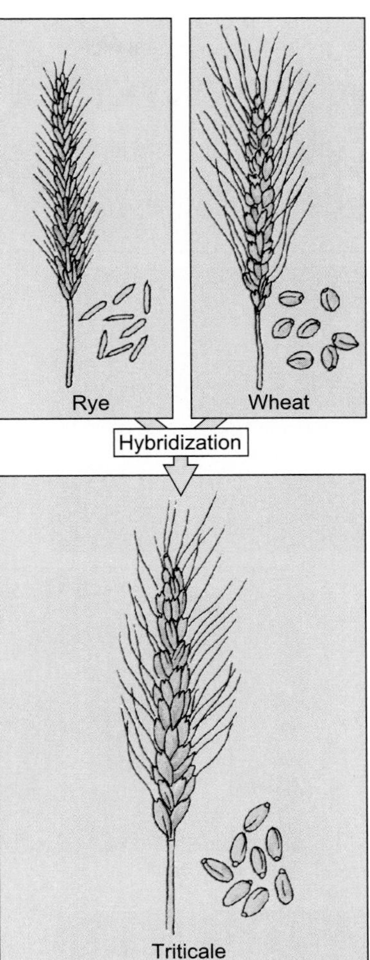

FIGURE 20.10 HYBRIDIZATION AND THE PRODUCTION OF *TRITICALE*.
Triticale is a hybrid grain produced by crossing wheat and rye (the name is a combination of *Triticum* and *Secale*, the genus names of the two). Agricultural scientists solve the problem of infertility by treating the hybrid with colchicine, a mitosis inhibitor that causes synthetic tetraploidy. *Triticale* combines the vigor of rye with the high-grain yield of wheat.

Extinction and Evolution

Just as speciation marks the birth of species, extinction heralds the death of species. Extinctions have always been a part of the evolving drama of life on the planet, from those first failed droplets in that primordial soup to the much-chronicled modern species that have fallen into oblivion as we watched. In ad-

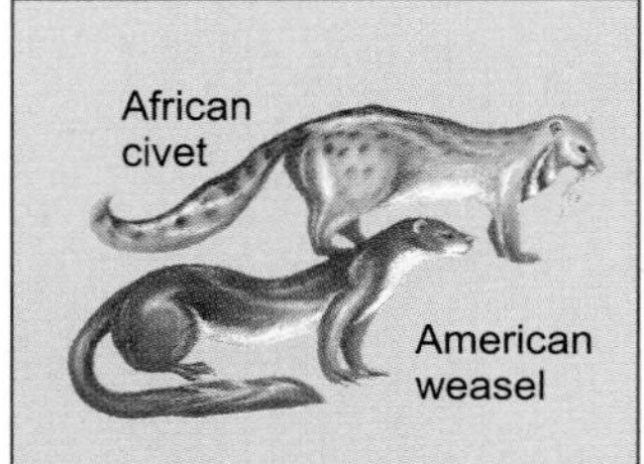

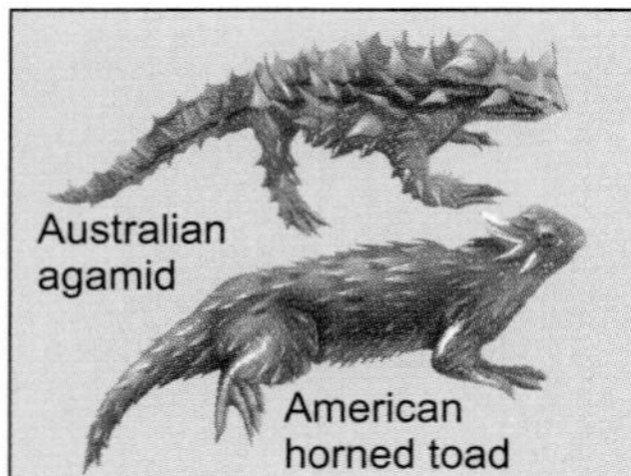

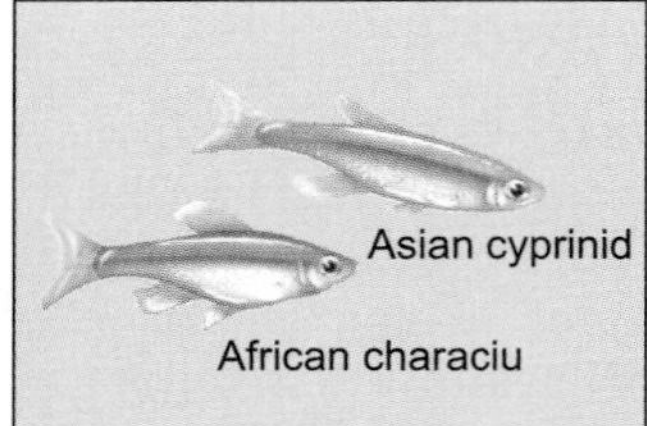

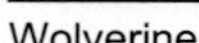
Wolverine

Marsupial cuscus

Tasmanian devil

Placental sloth

Numbat

Short nosed bandicoot

Giant anteater

Norway rat

FIGURE 20.11 **CONVERGENT EVOLUTION.**
The concept of convergent evolution is well supported by observations of similar animals that live on separate continents. Although phylogenetically unrelated to their counterparts on other continents, they bear a striking resemblance.

FIGURE 20.12 COEVOLUTION.
The passionflower vine *(Passiflora)* continually evolves new strategies to resist the parasitic activity of the *Helioconius* larvae. Displaced nectar glands (yellow spots) resembling the butterfly's eggs discourage "additional" egg laying. (In this case the mimicry has not worked. The brighter spot is the egg of *Helioconius* amid the less colorful spots produced by the plant.) Variation in leaf shape may also mislead egg-laying females searching for a specific shape and size.

dition to such isolated extinctions, there are rather constant, ongoing, and low-level "background" extinctions that have always marked life on earth.

Here, though, we're interested in the great extinctions—dramatic die-offs involving a large number and variety of species. Some scientists estimate that there have been five great extinctions since life appeared. These were enormously important not least because they were critical to the evolution of the remaining life. The first extinction occurred at the end of the Cambrian period, during the Paleozoic era (about 500 million years ago). The second great extinction occurred at the end of the Ordovician period (about 425 million years ago), and the next at the end of the Devonian period (some 345 million years ago).

Then came the fourth great extinction. This one was perhaps the most sweeping, marking the demise of a great many species at the close of the Permian period (some 225 million years ago). It is estimated that more than 90% of marine life went extinct at that time. The fifth, and so far the last, major extinction occurred at the end of the Mesozoic era (some 65 million years ago), in which the dinosaurs and many other groups disappeared.

In 1983, D. Raup, J. Sepkowski, and D. Daveys reported evidence of great die-offs every 26 to 28 million years throughout life's tenure on earth. Their conclusions have been heavily debated, and a number of other theories have been proposed to explain the occurrences. Some believe the fatal events were celestial, due to the earth being hit by a great meteor, while others believe that natural changes on earth were sufficient to cause the die-offs. There are also reasons to believe that human behaviors are currently

Essay 20.2 Extinction and Us

There is disconcerting evidence that we are apparently in the midst of another truly devastating mass extinction, and this time there is little question of the cause. As Pogo once said, "We have met the enemy, and he is us." Indeed, the extinction rate is presently estimated by a number of experts to be between a thousand and several thousand species per year. If we continue on this course of environmental destruction, by the year 2000 the extinction rate could be about 100 species per day. Within the next several decades, we could lose a quarter to a third of all species now alive. This rate of loss is unprecedented on the planet. Furthermore, the problem is qualitative, as well as quantitative. Not only are we losing *more* species, but we're also losing different *kinds* of species. The earlier mass extinctions involved only certain groups of species, such as the cycads and the dinosaurs. Other species were left more or less intact. This time, however, species are dying out across the board. That is, the current extinction affects all the major categories of species. Of particular note, this time the terrestrial plants are involved. In the past, such plants provided resources that the surviving animals could use to launch their comeback. With the plants also devastated, any comeback (usually marked by a period of rapid expansion and speciation) by animals will be greatly slowed.

The present great extinction will also prove to deter a resurgence of species for another reason. This time we are killing the systems that are particularly rich in life, including tropical forests, coral reefs, saltwater marshes, river systems, and estuaries. In the past, these systems have provided genetic reservoirs from which new species could spring and replenish the diversity of life on the planet. But now, in effect, we are drying up the wellspring of future speciation.

In the past, extinctions have been due to two major processes: environmental forces (such as climatic change) and competition. For the first time, a single species (our own) has had the opportunity to cause mass extinctions at both levels. Much of the habitat destruction has been due to humans needing the land where other species lived, forcing them into extinction as they failed in their competition with us. And we also now find ourselves in the remarkable position of being able to alter environmental forces. As the Amazon basin is destroyed, the trees are no longer available to cycle water back to the atmosphere, causing many experts to predict sweeping changes in the weather. And we are continuing to interact in new ways with the environment. For instance, as we continue to release chlorofluorocarbons into the atmosphere, the ozone holes grow larger, and the delicate veil of life on earth becomes increasingly bathed in destructive radiation. And because we continue to use the great waters as dumps for chemical waste, we are learning that the oceans have become perilous places for many forms of life.

There are those who say that our course, by now, is irreversible, that the damage is done and that we must now sit back and prepare to reap what we have sown. Others, though, argue that there is still time to alter our course, to salvage much of what is left, and to protect it from further damage. The danger is in accepting the first alternative if the second is really the case.

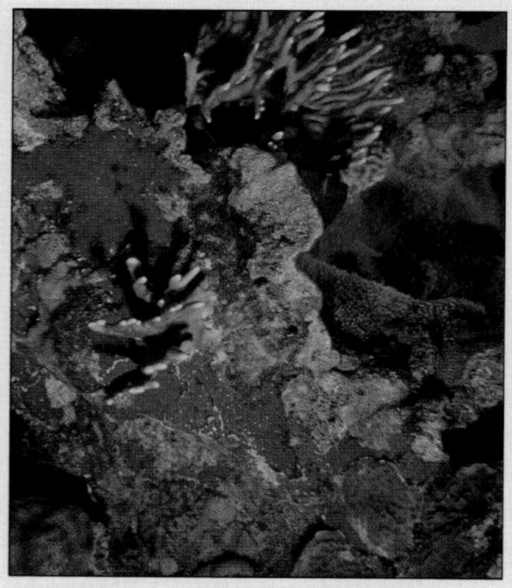

hastening the next great extinction onto the stage, and that this one offers perils not seen before (Essay 20.2).

However frequently and for whatever causes the great extinctions occurred, they certainly took a great toll on existing life and opened new evolutionary directions for the survivors.

In this chapter, we have looked at some of the great issues in evolutionary theory, such as the birth of species, how they change, and why they die. But to better understand how life changes we must focus yet more tightly, descending to the molecular level where many of the greatest questions lie.

PERSPECTIVES

The concept of evolution, we see, encompasses a range of complex and interactive ideas. Since its inception, it has also provided us with a host of questions, enduring puzzles that continue to resist our intellectual attacks. Someday we may know the answers to all our vexing questions about how and why life changes and how the various forms are related. And yet, perhaps not. It is possible that some of the events of our biological past will never be unraveled, and that these same questions will remain forever as unweathered monuments to human limitation.

Indeed, as we next turn the light on ourselves, we will see that it is a mottled light, with some areas brighter than others, even as scientists continue to bolster its intensity and fill in the darkened areas with the glow of hard-won information.

SUMMARY

NAMING NAMES

Linnaeus, an early naturalist and taxonomist, applied binomial nomenclature to the naming of species. Each scientific name includes genus and species, generally a noun and an adjective written in Latin with the first always capitalized. Subspecies designations may follow.

WHAT IS A SPECIES?

A species is a group of actually or potentially interbreeding natural populations that is reproductively isolated from other such groups.

Compounding the problem of delimiting a species is the fact that some species form a continuously changing cline across their range. The degree of change that occurs in semi-isolated pockets depends on the amount of gene flow between the pocket and the main group.

MECHANISMS OF SPECIATION

In phyletic speciation, one species changes enough to become a new species. In divergent speciation, main lines diverge forming two new species. Included in divergent speciation are allopatric speciation, where geographically isolated groups diverge into new species, and sympatric speciation, which occurs without geographic isolation. Reproductive isolation is a necessary prerequisite to both.

Geographic barriers may include newly formed rivers, islands, and mountain ranges, but the grandest example is the result of continental drift.

The Galápagos finches represent one of the best-known examples of allopatric speciation. A few finches blown from the South American mainland eventually populated all the Galápagos Islands, undergoing adaptive radiation. The earliest groups, isolated on separate islands, changed enough to become reproductively isolated from other such groups. Upon reestablishing themselves on the same island, character displacement occurred, further promoting differences.

Plants can undergo sympatric speciation through hybridization and polyploidy. Hybrids are commonly sterile because of incompatible chromosomes that fail to synapse at meiosis. Spontaneous doubling of chromosomes forms polyploids, which solves the problem of pairing. The progeny, immediately reproductively isolated, may form a new species if conditions are favorable. Producing successful hybrids through induced polyploidy is a common practice among plant breeders.

Where reproductive isolation occurs, two groups can no longer interbreed. Prezygotic isolation includes physical, physiological, or behavioral changes that preclude successful mating. Postzygotic isolation includes incompatibilities that interfere with the survival of the embryo or offspring or that render them infertile.

Major Evolutionary Trends

In divergent evolution, species become increasingly different from their ancestral type. Such changes help relieve competition.

In convergent evolution, unrelated or distantly related species adapt to similar kinds of environments and come to resemble each other. Many examples are seen in the similarities between marsupial and placental mammals.

In coevolution, the adaptive changes of one species adjust to (track) those of another. Coevolution commonly occurs between predator and prey species. For example, the defensive measures taken by the passionflower vine against insects have led to countermeasures by a butterfly whose larvae feed on the leaves.

Extinction and Evolution

Extinctions have been part of the evolutionary processes of life on earth. There have been low-level, constant "background" extinctions and five great extinctions that have marked the passage of many species in a relatively brief period. Human behavior may now be contributing to the arrival of the next great extinction.

Key Terms

speciation • **317**
taxonomy • **317**
binomial nomenclature • **317**
genus • **317**
subspecies • **317**
species • **320**
phyletic speciation • **320**
divergent speciation • **320**
allopatric speciation • **320**
sympatric speciation • **320**
continential drift • **321**
adaptive radiation • **323**
character displacement • **323**
hybridization • **324**
polyploidy • **324**
reproductive isolating mechanism • **325**
prezygotic • **325**
postzygotic • **325**
divergent evolution • **326**
convergent evolution • **326**
coevolution • **327**

Review Questions

1. Describe binomial nomenclature, and provide an example.
2. Summarize the problems in delimiting *species*. In particular, explain why the criterion of interbreeding isn't always dependable.
3. List some of the complications in identifying species along an extensive range.
4. State Ernst Mayr's definition of species. For what contingency does the term *potentially interbreeding* provide?
5. Using simple diagrams, compare phyletic and divergent speciation. Which is representative of gradualism as discussed in previous chapters?
6. Why is reproductive isolation so important to speciation? What happens to populations that rejoin before reproductive isolation is complete?
7. Make a technical distinction between prezygotic and postzygotic reproductive isolation.
8. List several ways in which geographic isolation of populations might occur. What must happen before the formation of separate species would be assured?
9. List the hypothetical events that led to allopatric speciation in the Galápagos finches. What is character displacement, and what role did it play in finch speciation?
10. Essentially how does sympatric speciation differ from allopatric speciation?
11. What generally happens reproductively when meiosis occurs in a hybrid (such as the mule)? How have some plants managed to overcome the problem?
12. Prepare a diagram to show how polyploidy might create instant reproductive isolation in a hybrid.
13. Explain how the polyploid state can be artificially induced. Suggest a practical application for this practice.
14. Is polyploidy simply an oddity, without value, in plants? In what regions is it most commonly seen?
15. Define coevolution and provide three examples.

CHAPTER

21

Human Evolution

NOW THAT WE HAVE LEARNED SOMETHING ABOUT THE PRINCIPLES OF *evolution and how evolutionary changes proceed, we can turn our attention to ourselves. We will begin where the primates appeared, in the Paleocene; then we will trace the primates, ourselves among them, to modern times. Pay particular attention to how the various human traits, of necessity, developed together, each accenting and supplementing the other. As we reach the marvelous evolutionary crescendo that resulted in the modern human, we will have learned something about our forebears, and we may have raised questions about where we go from here.*

Now let's put together some of the principles we have just discussed and bring another species—ourselves—under the explanatory umbrella of natural selection. As you are undoubtedly aware, this simple shift in subject can bring on a barrage of questions. Some of the more common questions are:

When did the first humans appear?
Did we really evolve from apes?
Are we still evolving?

The answers are, respectively:

It depends on what you mean by human.
Probably not in the way you're thinking.
Yes, but things have gotten very complicated.

If these answers seem vague, it's because the questions are a lot more complex than they seem. Moreover, we now find ourselves in an area not only rife with preconceptions and high emotions, but also marked by a long history of conflicting ideas. So let's take it one step at a time and consider some of the prevailing ideas regarding human evolution and some of the evidence behind those ideas. We can begin by considering primates in general.

PRIMATE ORIGINS

Humans are in the order *Primates* (*prima,* first), so we might best begin to trace our origins by reviewing what we know about our taxonomic order. Primate fossils, like those of most other placental mammals, first appear in the Paleocene deposits, formed about 60 to 65 million years ago, soon after the dinosaur extinction and the onset of the great mammalian expansion. The earliest primates, with their long snouts and claws, somewhat resembled modern tree shrews or even rodents. In fact, they may have occupied ecological niches similar to those of modern rodents, with some primate species living in trees and others scurrying about on the ground or living in burrows. People in technologically advanced societies may be a bit embarrassed to think of how many grubs our ancestors ate.

By the beginning of the Eocene epoch (about 60 million years ago), primates began to resemble those existing today. They had become primarily fruit-eating and tree-dwelling, and there was a general shortening of the snout, a more forward location of the eyes, and a more definite primate tooth structure. Toward the middle of the Miocene epoch, about 20 million years ago, lines that lead to all modern primate families (including **Hominoidea**—the family of apes and humans) were well established (Figure 21.1).

MODERN PRIMATES

Primates differ from other mammals in a number of ways—for example, in their limbs. Most primates are well adapted for **arboreal** (tree-dwelling) life, as evidenced by their **prehensile** (grasping) hands and feet, long arms, and in the case of the New World monkeys, prehensile tails.

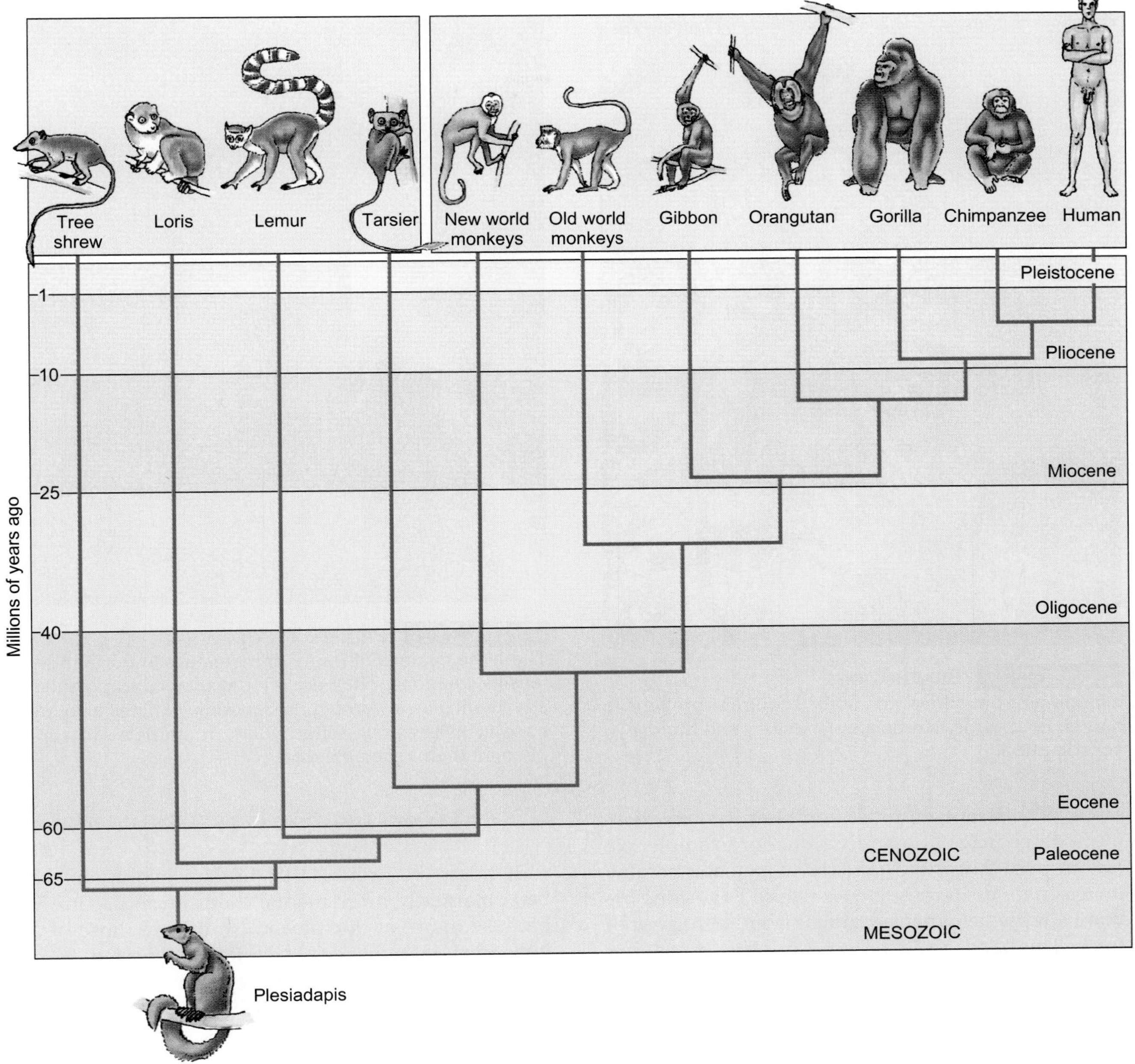

FIGURE 21.1 **TENTATIVE PHYLOGENY OF THE PRIMATES.**
This phylogeny places *Plesiadapis* at the base. The earliest branches produced the tree shrew and prosimian lines, followed next by the New World monkeys. These two groups represent the greatest divergence in the living primates. Old World monkeys diverged some 10 million years later, subsequently followed by the great apes. The most recent divergence occurred between humans and chimpanzees.

But life in the trees gave rise to other changes as well. Selection favored a more frontal location of the eyes, which improved both binocular and stereoscopic vision. With both eyes focused on an object, a fine, three-dimensional image could be produced that permitted more accurate estimates of distance. Coupled with this visual precision was an acute eye-hand coordination, which undoubtedly continued to develop as visual information became more and more "informative." These traits are obviously important if one is to swing through the trees (Figure 21.2).

Such anatomical specializations would require a substantial degree of simultaneous brain develop-

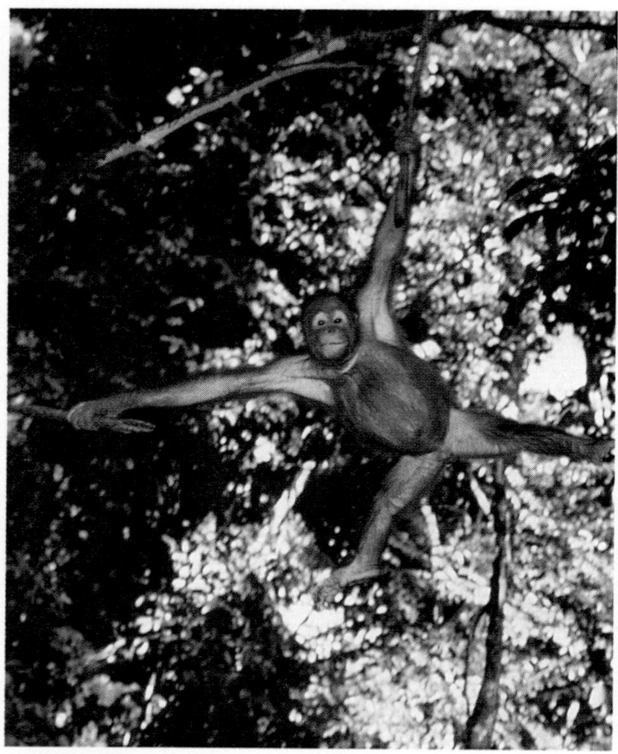

FIGURE 21.2 **BRACHIATION.**
Swinging from limb to limb requires prehensile hands, long arms, stereoscopic vision, and hand–eye coordination.

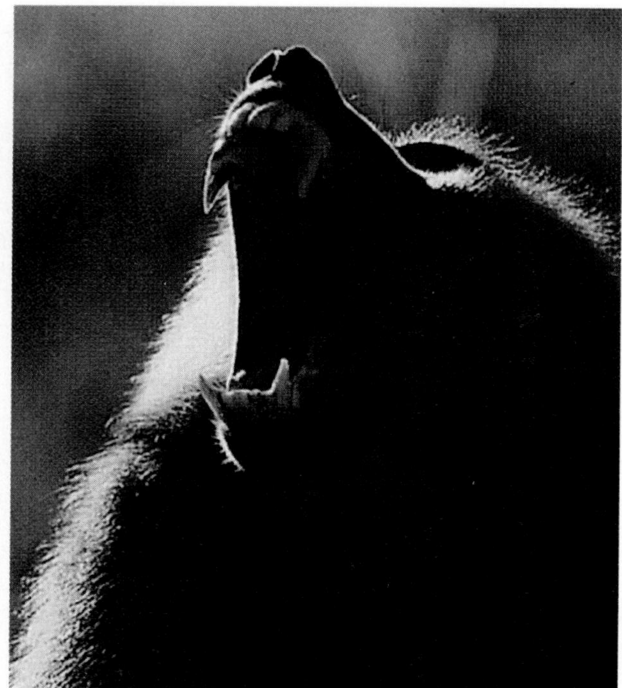

FIGURE 21.3 **PRIMATE TEETH.**
The large canines of many primates are important defensive weapons. They are often displayed to potential adversaries as a warning. In baboons, canines also are used in killing prey—often young mammals—to supplement their vegetable diet.

ment, with strong emphasis on brain centers dealing with coordination and vision. For example, we can propose that as the hand developed and became increasingly dexterous, its use would have gone beyond simply keeping the animal from falling out of trees. The hand could be used in other ways, but learning these would have demanded a correspondingly more complex (larger) brain. Such a brain might then discover new uses for the hand—but then an even more advanced brain would have been required. Thus hand and brain development may have occurred in a kind of positive feedback system in early primate evolution.

Primates tend to be omnivores, eating all sorts of food, and their mouths are relatively unspecialized. However, with the exception of humans, primates—particularly the males—have rather large canine teeth. These are used primarily in aggressive encounters with other male primates and in defense against predators (Figure 21.3). Human canines have become reduced in size, probably an adaptation to a changing diet. Their loss has since been more than compensated for by the development of such tools as rocks, spears, knives, and missiles.

HUMAN SPECIALIZATIONS

It might be argued that human evolution has been markedly more mental than physical. This is because many of our physical traits are those of a generalist (a nonspecialist)—traits expected in a species with a high intelligence that tends to live an opportunistic existence. So, as our intelligence presents us with many opportunities, our generalized bodies are able to help us take advantage of them.

Our principal evolutionary specializations, outside of the nervous system, are our **bipedal** (two-legged) **gait** and our **opposable thumb,** which can touch any finger of the same hand. The bipedal gait, a result of our upright posture, is quite unlike that of any other living primate. Savannah chimpanzees sometimes stand to see over tall grass, and they may walk bipedally a short way, but normally the apes are **quadrupedal,** walking on all fours (Figure 21.4). It is amusing to see an ape running on its rear legs, partly because it is so humanlike and yet so ungainly. Humans, however, are beautifully adapted to bipedal walking and running. We owe this ability to our enlarged *gluteus maximus* (buttock) and, in

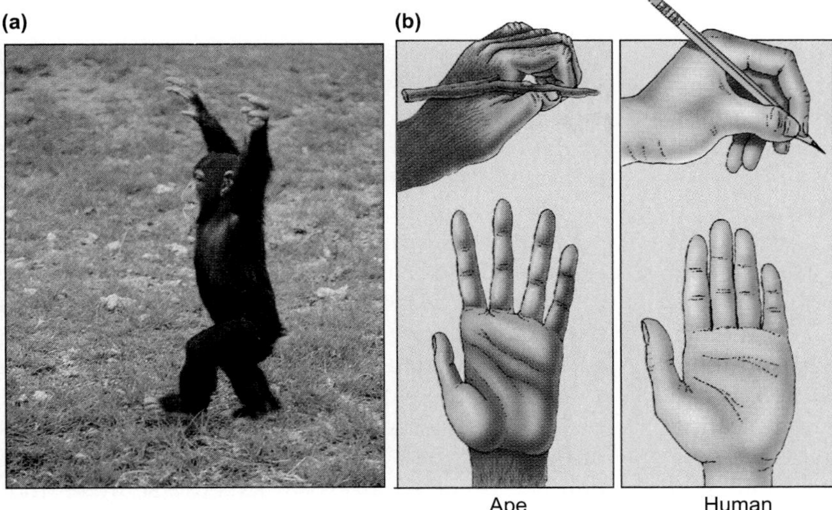

FIGURE 21.4 **MAJOR EVOLUTIONARY SPECIALIZATIONS OF HUMANS.** **(a)** Apes can assume a bipedal posture, but neither their hip nor leg structure supports this posture very well. In bipedal walking, the weight is borne on the outer edge of the archless foot. **(b)** The hands of humans and chimpanzees appear generally similar, but there are important differences. Chief among these are the length and musculature of the thumb. In the chimpanzee, the thumb doesn't quite reach the base of the forefinger, while in humans, the thumb extends nearly to the middle joint.

part, to our uniquely specialized foot with its springlike arched construction and broad first toe. Interestingly, human sprinters have been known to outrun a horse over 100 m (horses are slow getting started), and a marathon runner can outrun a horse in a 50-km race. (Horses tire first.) But on a 2-km (1.25-mi) run, bet on the horse.

As for the human thumb, it is a refinement of a specialization that is possessed, to some degree, by other primates. Originally, the opposable thumb evolved as an adaptation that allowed for grasping branches. But in most primates it is relatively short, poorly muscled, and much less capable of precise movement. The human thumb has evolved special musculature and can be rotated readily. Apes can be trained to use simple tools and to open beer cans, but they are comically clumsy all the same. With our much greater manual dexterity, we can make precise tools, even from rocks and sticks if we must, and many of us have been known to open beer cans with great dexterity. Other animals have been reported to use or even make tools, although the definition of "tool" must be stretched to include such things as the stone upon which a sea otter smashes a clam, or the twig that a chimpanzee uses to probe termite nests and to extract the excited insects. But no other species can build a watch that works or make tools that are used just to make other tools. To worry about how much of this has to do with thumbs and how much has to do with brains is to miss the point—which is that our intelligence and our manual dexterity evolved together.

It is important to understand that our intelligence has not evolved without costs. Like other mammals, we are born in a helpless state, but human infants seem to be particularly helpless, with fewer built-in adaptive responses than infants of many other species. A baby hare, for example, will lunge and hiss at an intruder. Newborn antelopes follow their mothers within minutes, and infant baboons quickly learn to ride on their mother's back. Virtually all baby primates will hold onto their mother's hair, so that she can move with ease. In comparison, our newborns seem witless and almost completely helpless. (And in what other species do 20-year-old offspring demand so much from their parents?) This early dependency of humans allows us time to learn all the things that will be important for living under the specific conditions in which our own complex culture exists.

Even the extensive development (some say *over*development) of the human brain merely extends a long-standing trend in mammalian and especially primate evolution. Over the millions of years, the consistent trend in most mammalian orders has been toward larger and larger brains. This is accompanied by a greater capacity for learning and for versatile behavior at the expense of stereotyped, genetically based, instinctive behavior.

The importance of intelligence to humans, coupled with our complex and highly interactive social system, has led to another unusual trait in our species: the development of language. Of course, wolves howl, birds sing, and bees "tell" each other the location of food by complex dances. And there are controversial reports that apes or dolphins can learn various nonverbal means of communicating with humans. However, humans, so far as we now know, can claim exclusive rights to the ability to communicate abstract ideas.

Humans also have an enormous capability for mathematics, some of us more than others. Why or how this ability evolved is not easy to fathom. Careful neurological measurements reveal that the human brain has a highly localized site for the process of multiplication—and an entirely different, equally localized site for long division. What these specialized parts of the brain were doing just a few thousand years ago, before mathematics had been invented, is anybody's guess.

The Human Line

In the early 1960s, comparisons of proteins clearly indicated that humans are related more closely to African chimpanzees and gorillas than to Asian orangutans. Researchers were aware that such molecules as these change at a constant and predictable rate, and thus were able to provide us with kind of a molecular "clock." (See Chapter 18.) This clock suggested that humans and African apes diverged from a common ancestor not more than 5 million years ago. The suggestion was startling at the time; anthropologists had always assumed that the separation had occurred much earlier. And so these findings were met with skepticism and in some cases, outright hostility. However, other data—both biochemical and paleontological—began to appear that supported the newer findings, and so we have adjusted our notions of human lineage.

The current theory is that about 3.5 to 4 million years ago, humanlike forms began to appear in the open grasslands of eastern and southern Africa. Fossils suggest that the line eventually consisted of four species, all of which have now been assigned to the genus *Australopithecus*. The name, which means

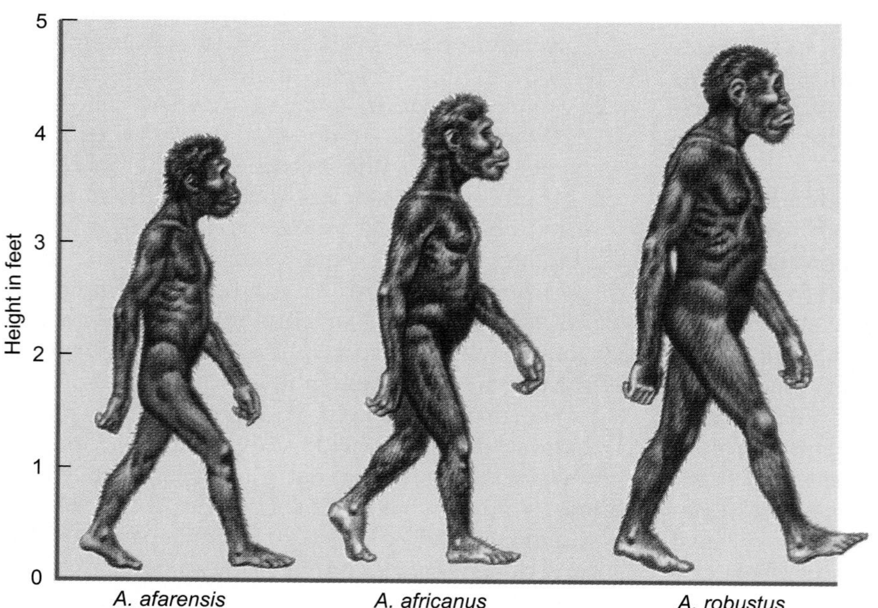

FIGURE 21.5 THE AUSTRALOPITHECINES.
The australopithecines were rather small, from 1 to 1.5 m (3.5 to 5 ft) tall. They were heavy boned, suggesting strong muscularity and a weight of up to 68 kg (150 lb). Their heads appeared more apelike than humanlike, with a low cranial profile and little or no chin. However, the jaws were large and forward-thrusting.

"southern ape," is unfortunate in light of what we now know—but let's see what these species were like and what became of them.

The Australopithecines

The **australopithecines** (members of the genus *Australopithecus*) were rather small-boned, light-bodied creatures, about 1 to 1.5 m (3.5 to 5 ft) tall. They had humanlike teeth and jaws with small incisors and canines, and they walked upright (Figure 21.5). They apparently hunted baboons, gazelles, hares, birds, and giraffes. The australopithecine *cranial capacity* (a measure of brain size) ranged from 450 to 650 cc, as compared with 1200 to 1500 cc for modern adult humans.

To date, four species of australopithecines have been named. While their phylogeny is still tentative, a number of physical anthropologists go along with the scheme seen in Figure 21.6. There we see *Australopithecus afarensis* placed in a position of extreme importance, that of the common *hominid* (in the human line) ancestor. We see that the *A. afarensis* line produced two or three branches, one of which led to the genus *Homo,* which includes modern humans. The other line or lines produced *A. africanus, A. robustus,* and *A. boisei* (Essay 21.1).

A. boisei and *A. robustus,* as the latter's name implies, were larger boned and heavier jawed than *A. africanus,* the oldest of the three. Most strikingly, the two later arrivals had much larger jaws and teeth with greatly expanded cheekbones to accommodate the massive jaw muscles. These features reach their extreme in *A. boisei.* From a traditional view this progression may seem to be backward—toward an apelike condition—but other physical evidence and dating are to the contrary. The shape of the teeth and the curve in the jaws are decidedly human, not apelike. The tooth wear, incidentally, suggests that much of the diet of these australopithecines consisted of plant food.

The first specimen of *A. afarensis,* the oldest known hominid (dubbed "Lucy" after a Beatles song from the 1960s), was found in the northern Ethiopian desert region in 1974 by Donald Johanson (Figure 21.7). The fossil remains, estimated to be over 3 million years old, consisted of little more than half of the skeleton of an upright-walking female. The following year, Johanson's luck blossomed. He unearthed the remains of 13 more "Lucy" types, all in one area. After several years of puzzling over these slightly built creatures, Johanson reached a decidedly unorthodox yet insightful conclusion. He proposed that *A. afarensis* was indeed more primi-

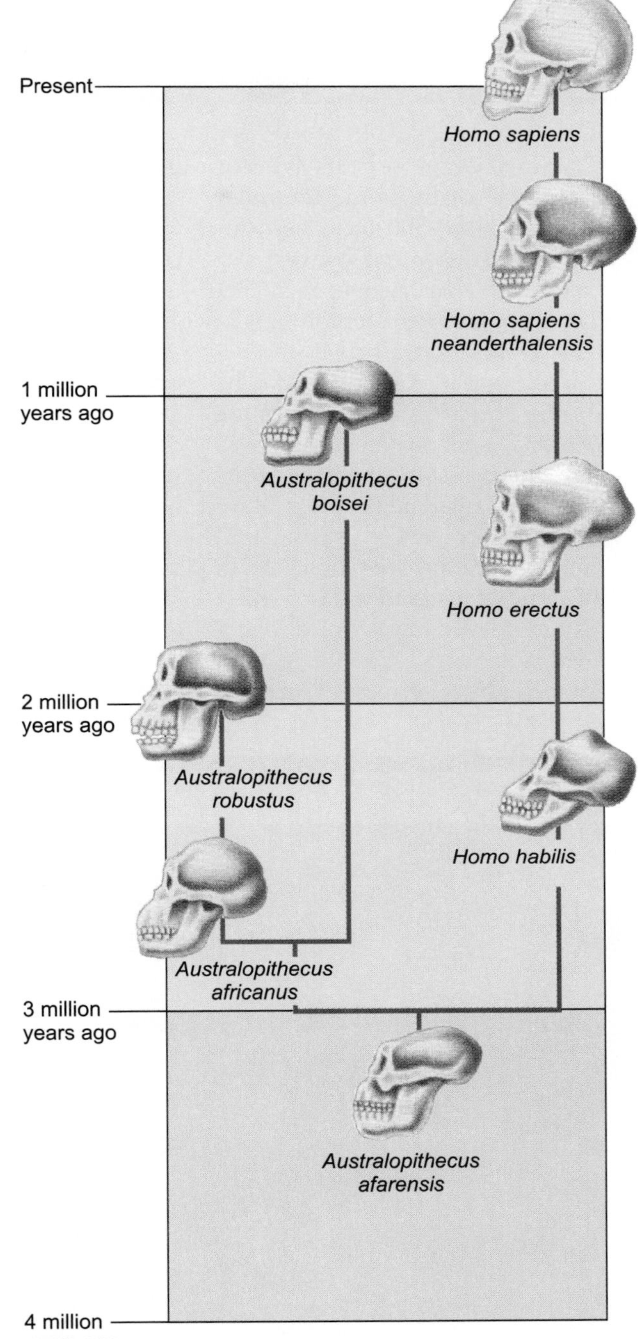

Figure 21.6 The Known Hominid History. The history of hominids spans a period of nearly 4 million years. In a scheme based partly on the conclusions of Donald Johanson, a line represented by *Australopithecus afarensis* produced a branch containing other australopithecines and one leading to *Homo.* There are other schemes besides this one. For example, *A. africanus* is sometimes placed in its own line.

ESSAY 21.1 Discovery at Olduvai

Olduvai Gorge appears as a tear in the earth, some 25 miles long and about 300 feet deep (for your own amusement, convert that to metric). It was discovered in 1911 when a German entomologist pushed back some brush and almost fell into it. Along its walls he saw the strata deposited over the millennia, and an expedition was soon launched to investigate its rich animal fossil beds. The preliminary findings were so unusual that a new expedition was launched in 1913, and it yielded a fascinating array of fossils until World War I ended the work. Unable to find funding, the German expedition leader asked a young museum curator in Kenya, Louis Leakey, to take over the work, but Leakey couldn't find the money either, until 1931.

Leakey was overwhelmed by his animal findings and by his discovery of some primitive tools, but for 28 years he couldn't find fossil evidence of the toolmaker. Then one morning, on July 17, 1959, when Louis awoke feverish and with a splitting headache, his wife Mary insisted he stay in camp—but the season was growing short, so Mary drove out to the work site alone. As she walked along, her educated eye spotted a rounded shape she recognized as part of a skull. Searching further, she found two big, brown, shiny teeth. She marked the spot and sped back to camp.

Louis heard the car racing back and, alarmed, jumped up, certain Mary had been bitten by a snake. But Mary jumped out of the car yelling "I've got him! I've got him!" Louis forgot his illness and raced back with her under the hot African sun. The skull turned out to be that of a young male, dense-boned and heavy-featured, a species now called *Australopithecus boisei* in honor of Charles Boise, who had helped to finance the Leakeys' early work.

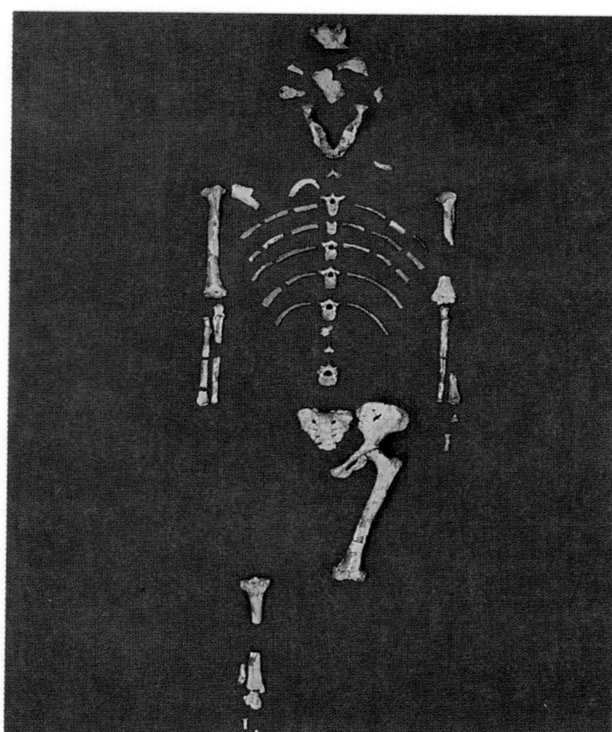

FIGURE 21.7 LUCY.
Australopithecus afarensis, the oldest known hominid and the area in which the fossil was found.

tive than other australopithecines, placing it at the base of the hominid tree. The idea, hotly contested at first, is now well accepted by many researchers.

Homo habilis

The earliest group to be placed squarely in the *Homo* line was *Homo habilis*. The group was represented by certain fossils found in Olduvai Gorge in Tanzania by Louis Leakey (Figure 21.8). The fossil remains show *H. habilis* to be clearly associated with toolmaking; crude tools were unearthed with the fossil skeletons. Critics have argued that the new find is simply a variant of the genus *Australopithecus*, and that Leakey was unjustified in trying to include his fossils within the genus *Homo*. For the moment, though, *H. habilis* is placed in the early *Homo* line (see Figure 21.6).

Homo erectus

Even more recently, fossil beds of the eastern shore of Lake Turkana yielded fossils that are more similar to the modern human. The new species, called *Homo erectus*, is regarded as an extinct member of our own genus. The earliest African *H. erectus* skulls are known to be more than 1.5 million years old. What is most interesting about them is that they

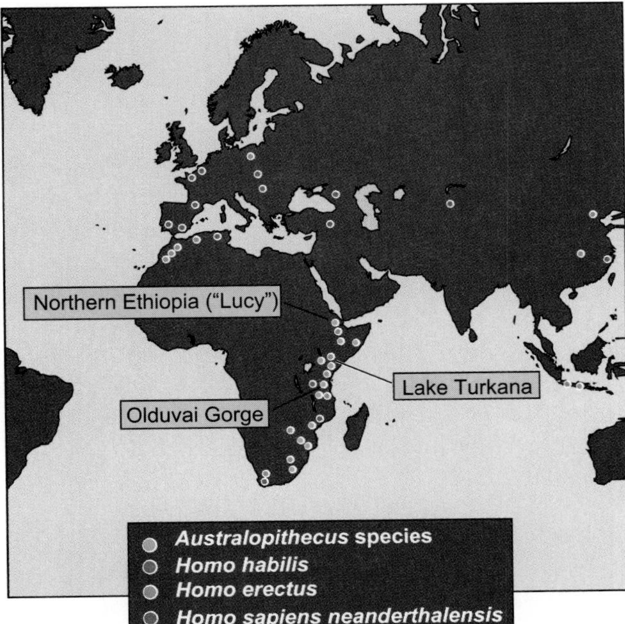

FIGURE 21.8 FOSSIL HOMINID DISTRIBUTION. The areas show the locations of some of the major finds of fossil hominids.

appear to be of the same species as some of the first hominid fossils ever found—those that were once called "Java Ape Man" and "Peking Man." The problem is that these fossils were only about half a million years old. Other *H. erectus* fossils have been found that may be even less than 200,000 years old. In other words, *H. erectus* flourished, relatively unchanged, for well over a million years. Furthermore, during its first 300,000 years it coexisted with other more primitive hominid species, including the australopithecines and, possibly, *Homo habilis.* And during at least the final 100,000 to 200,000 years of its existence, persistent members of *H. erectus* shared the planet with yet another species, *Homo sapiens.* Fossils of *H. erectus* have been found in China, Europe, and southern Africa, and, of course, East Africa. The evidence left by *H. erectus* provides fascinating grist for the mills of our imaginations.

H. erectus was obviously successful, but in time its tenure on the planet came to an end. Most populations of *H. erectus* passed from the earth just after the first glaciers of the Pleistocene epoch receded. But at least one line of descendants is believed to have persisted, quietly changing and adapting and eventually leading to us. The distribution of fossil hominids is seen in Figure 21.8. But there is still one more story to tell before we come to modern humans.

NEANDERTHALS AND US

In 1856, while Darwin was puzzling over natural selection at his country estate, workmen in a steep gorge in the Valley of Neander (in German, *Neanderthal*) were pounding at a rock they didn't recognize. They finally saw that it was a skeleton, but by then it had been smashed to bits. Fortunately, they had left enough for researchers to study, and soon it was announced that there was clear evidence of a new and different kind of human. (Interestingly, a similar skull had been unearthed at Gibraltar a few years earlier but had not created much of a stir.)

Scientists are rarely at a loss for words, so an explanation of this peculiar find was immediately forthcoming. Professor E. Meyer of Bonn examined the heavy-browed skull, cleared his throat, and proclaimed that the skull and bone fragments belonged to a Mongolian cossack chasing Napoleon's retreating troops through Prussia in 1812. An advanced case of rickets had deformed his legs and had caused him great pain. The pain had furrowed his brow and had produced the skull's great ridges. Because he was so distraught, he had crawled into a cave to rest, but alas, he died.

This scenario has now been rejected in favor of the idea that the bones were those of an early form of human that became extinct—the so-called **Neanderthal** (*Homo sapiens neanderthalensis*)—and that the skeleton's unusual features were actually common to the whole group. (Interestingly, the individual who originally owned that skeleton did indeed have rickets. Perhaps it was hard to get enough sunshine while living in a cave in Northern Europe during the Ice Age.)

At first, Neanderthal fossils seemed to show exactly what anyone of the 19th century would expect of an "ape-man." Neanderthals were depicted as being heavy boned and heavy browed, with low foreheads, wide faces, and little or no chin. "Reconstructions" drawn in the 19th and early 20th centuries always showed ugly, stooped, hairy, apelike bodies.

Later, it was found that, although the Neanderthals did indeed have large brow ridges, sloping foreheads, and small chins, their necks and bodies were like ours (Figure 21.9). There is no way to know whether they had a spoken language, but we do know, from fossil pollens found with their remains, that Neanderthals sometimes covered their dead with flowers before burial. Whether or not that's sophisticated, it certainly is human—touchingly so.

Why did the Neanderthals disappear? Perhaps we had something to do with it. Keep in mind that the classic (or more primitive) Neanderthal line and the line of modern humans diverged from each

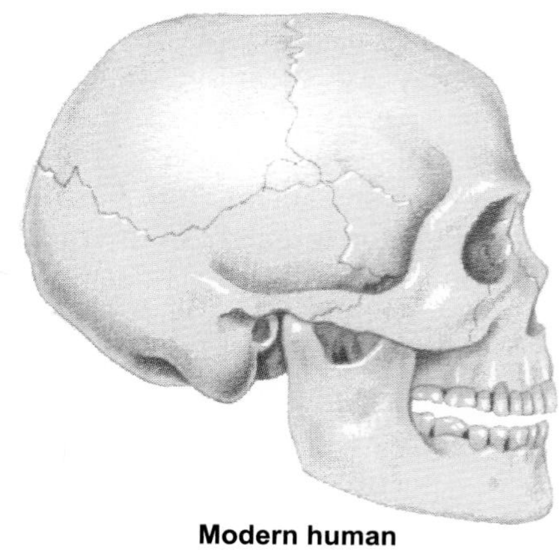

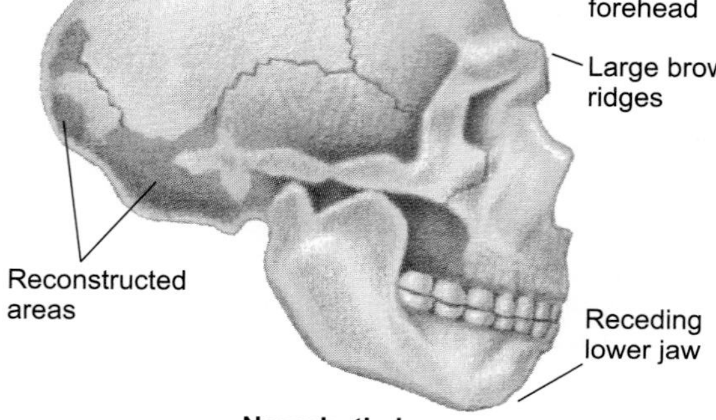

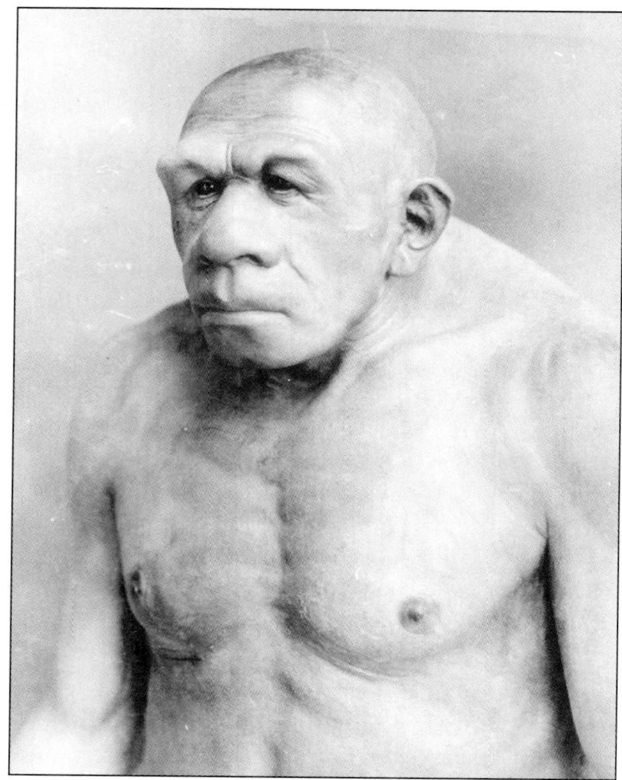

FIGURE 21.9 MODERN HUMAN VERSUS THE NEANDERTHAL.
A modern human and the Neanderthal reveal many general similarities and some striking differences. In the Neanderthal, note the sloping of the forehead, the large brow ridges, and the receding lower jaw. The reconstruction of the head and upper body produces an image quite different from the traditional view of the Neanderthal, long considered an apelike creature.

other as long as 250,000 years ago. For some reason, the Neanderthals flourished at the time the developing modern human types were few in number. Then, about 40,000 years ago, the modern human line (*Homo sapiens sapiens*) suddenly burgeoned and expanded its range, just as the Neanderthal line became extinct. One theory holds that they may have died off from attacks by *H. sapiens.*

A second theory suggests that the differences between the two groups were insignificant, because *H. sapiens* in the Neanderthal age was a highly variable species that included both the classic Neanderthal types as well as the modern human line, individuals much like ourselves. This theory implies that the Neanderthals were squarely in the mainstream of human evolution and not a dead-end side branch. Therefore, the theory states, the Neanderthals didn't die out, but merely changed—into us.

Whatever the case, *Homo sapiens* became rather well established on earth, and, at the point record-keeping began, our account is called *history.* We will have more to say about our history, our present, and our future later. We deserve the attention, not only from an egotistical view, but because of the impact we had—and are continuing to have—on our small planet.

PERSPECTIVES

We have, in these last few chapters, examined the principles of evolution. The underlying mechanism, we have learned, is natural selection. In the next section, we will take a step backward, in a sense, while maintaining our theme. We will see how natural selection operated much earlier in the development of life by noting its effect on mindless molecules and organized droplets—molecules and droplets that, over time, would lead to life itself.

SUMMARY

PRIMATE ORIGINS

Primate divergence from other mammals began in early Cenozoic, the earliest members resembling rodents. The divergence of modern primate families began just 20 million years ago.

Primate characteristics include arboreal adaptations; prehensile hands, feet, and tails; improved frontal, stereoscopic vision; and good hand–eye coordination. Hand and brain evolution may have positively reinforced each other. The canine teeth in humans are smaller than in other primates.

Human specializations include a higher level of intelligence; an upright, bipedal posture and gait; and greater dexterity because of the opposable thumb. The upright posture and gait is made possible by enlarged buttock muscles and the foot arch. Humans are not unique as tool users but are the only animal that uses tools to fashion other tools.

Human development is slower than that of other primates, but the learning capacity is greater. Unique to humans is the use of complex language and mathematics.

THE HUMAN LINE

Newer studies based on mutation rates and the molecular clock place the origin of the human line at only 5 million years. The australopithecines are known to have existed in the plains of Africa some 3.5 to 4 million years ago.

Australopithecines were smaller than modern humans, with cranial capacities generally about half that of modern humans. *A. afarensis,* the oldest, gave rise to two lines, one including *A. africanus, A. robustus,* and *A. boisei.* The other began with *Homo habilis* and led to *H. erectus* and finally to *H. sapiens. A. afarensis* was the smallest. The other australopithecines were heavier boned and had larger jaws.

One of Louis Leakey's discoveries, *H. habilis,* made crude tools and was a contemporary of the australopithecines.

The oldest fossils of *H. erectus* date at 1.5 million years, and they are known to have persisted until some 200,000 years ago. In their last years they coexisted with *H. sapiens.* One line gave rise to *H. sapiens.*

Neanderthals closely resemble modern humans in stature and build, but the skull has larger brow ridges and a sloping forehead, and the chin is smaller. *H. sapiens neanderthalensis* ranged widely over Europe and Asia, persisting up until about 40,000 years ago. All newer fossils are of *H. sapiens sapiens.* Some anthropologists theorize that modern humans are one offshoot of *H. erectus* and that Neanderthal is another. Others state that modern humans are simply one of many variants of Neanderthal.

KEY TERMS

Hominoidea • **334**
arboreal • **334**
prehensile • **334**
bipedal gait • **336**
opposable thumb • **336**
quadrupedal • **336**
australopithecine • **339**
Neanderthal • **341**

REVIEW QUESTIONS

1. From a close study of Figure 21.1, list the main divergences in the primate line and indicate when they occurred. What was the ancestor of all primates like?
2. List four important traits of most primates.
3. Explain how evolutions of the eyes and hands of primates reinforced each other over time.
4. What is the theoretical significance of the large canines in most primates?
5. Discuss human physical characteristics that go along with true upright posture and the bipedal gait. Do any other primates have these abilities?

6. In what specific ways is the human thumb different from that of a chimpanzee? How might this have affected toolmaking and tool usage?
7. How does human tool use differ from that of other animals?
8. How has the human mental capacity affected the ability of newborn babies? Compare these abilities with those of other primate newborns.
9. List three or four specific capabilities of humans that are not developed in other animals.
10. How long, according to recent molecular studies, have the hominids been around? What part of the world did hominids first inhabit?
11. Of the four species of australopithecines, which is probably the oldest? What happened to the australopithecine skull and jaw as evolution continued?
12. With which hominid does the human line actually begin? How might this placement represent a departure from traditional thinking about the physical traits of the earliest humans?
13. Through what time period did *Homo erectus* extend?
14. What is the traditional or historical notion of Neanderthal's appearance? Compare this to the modern view (Figure 21.9).
15. How long ago did Neanderthal cease to exist? Summarize the two theories about its demise.

CHAPTER

22

Bacteria, Viruses, and the Origin of Life

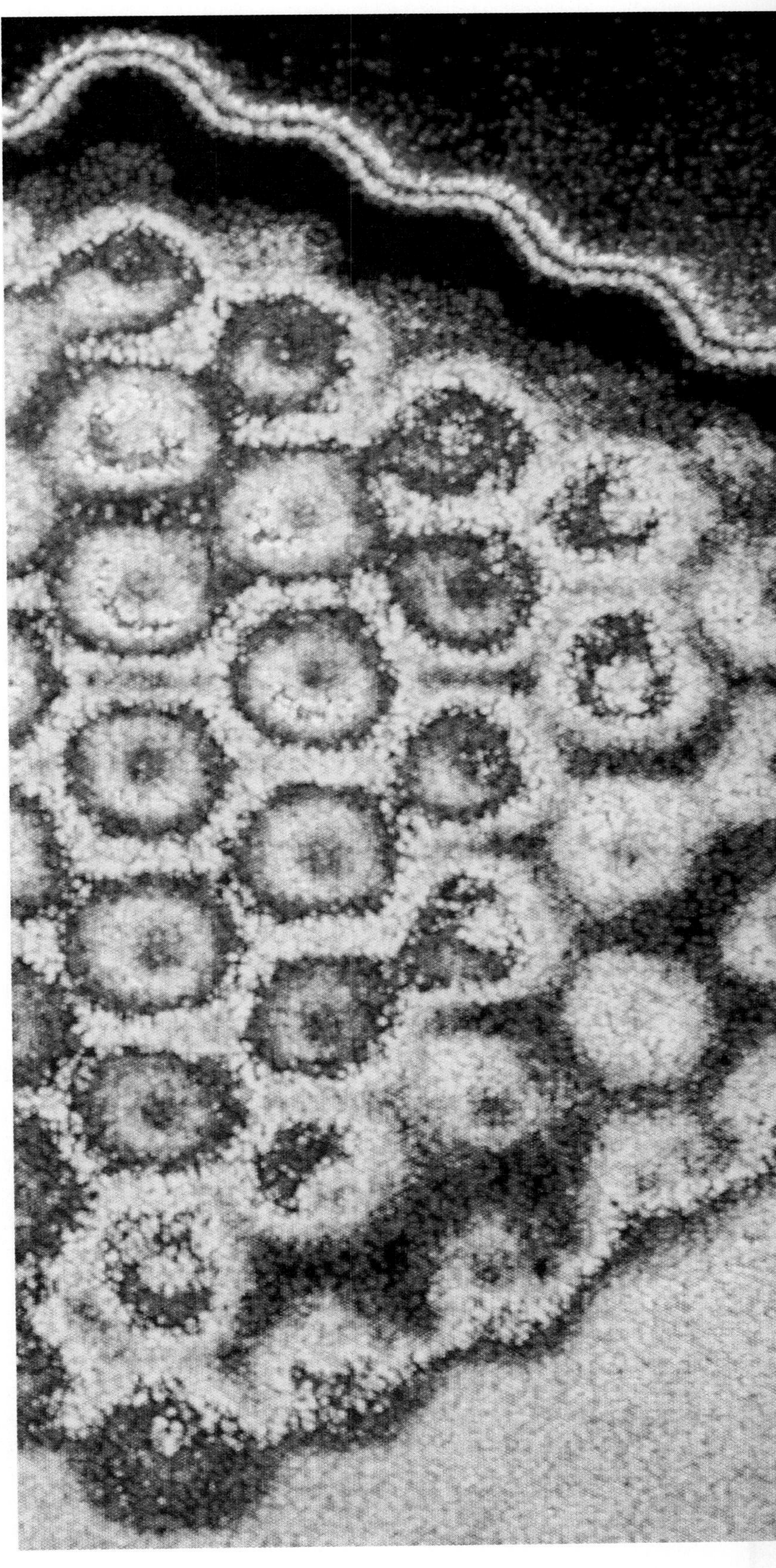

IT IS HERE THAT WE APPROACH THAT MYSTERY OF MYSTERIES—HOW LIFE *began. You will see that an impressive body of evidence has been marshalled to support the most current notions, but that some very fundamental questions remain—such as did the precursors of life really fall to earth in dirty snowballs? As we consider the transition from lifelike droplets and bits of clay to living things, we are led to the oldest forms of life on earth and their modern descendants, the bacteria. Finally, we consider that puzzling and dangerous realm whose members can turn to crystals—the viruses. The question to ask yourself is, Are they alive?*

It's difficult to ponder the complexities of life for long without arriving at the ultimate question: How did life begin? The question is as old as reason. In fact, the ancients throughout the world were absorbed by the mystery, and their conclusions have formed the bases for practically all religions. Those revered philosophers, though, didn't have some of the tools we possess today, and they didn't have the instant communication we enjoy, the means to get our hands on each other's data and ideas (if not throats) almost immediately.

We sometimes like to think that modern science is always coolly objective in its straightforward search for truth. Alas, science is done by scientists and scientists are people (they are *too*), and so science is fettered with all manner of restraints, prejudices, hunches, and hopes. It may be said that in the sciences, faith is as strong as it is in religion or philosophy. Furthermore, even in science, faith must grow increasingly strong as solid evidence grows weaker. Thus, in searching for an explanation of how life began, each researcher must have a measure of faith that he or she is on the right track. But faith in an idea is only a starting point; it must eventually be bolstered by data. So how does one collect data about an improbable event that may have happened only once, several billion years ago? Perhaps with a touch of humility, because it's clear that we can never prove how life really first came to be on this planet. We can, however, examine a number of seemingly plausible notions of how life *might* have arisen. Most of these speculations about the possible origin of life involve hypotheses that lead to predictions, and these, in turn, can lead to testing. Many such hypotheses have been trotted out, and many have been found wanting. But one proposed hypothesis offers enough promise to occupy the energies of a growing list of researchers.

THE ORIGIN OF LIFE

Although Darwin speculated that life might have arisen in a warm, phosphate-rich pond, the first serious proposals concerning the spontaneous origin of life (those that were based on sound biochemical and geological information) began to appear only 50 years ago. Similar schemes were presented by J. B. S. Haldane, a Scottish biochemist, and by A. P. Oparin, his Soviet counterpart. They proposed that shortly after the earth's formation, under conditions quite different from those of today, a period of chemical synthesis occurred in the warm primeval seas. During this era, the precursors of life's molecules—amino acids, sugars, and nucleotide bases—formed spontaneously from the hydrogen-rich molecules of ammonia, methane, and water. Such synthesis was possible because there was no destructive oxygen in the atmosphere, and there was an abundance of energy in the form of electrical discharges, ultraviolet light, heat, and radiation (Figure 22.1). Since there were no organisms or oxygen to degrade the spontaneously formed organic molecules, they accumulated until the sea became a "hot, thin soup."

FIGURE 22.1 THE EARLY EARTH.
The birthplace of life (as far as we know).

Haldane and Oparin suggested that continued synthesis and increasing concentrations led to the formation of polypeptides from amino acids, and, eventually, to a diversity of molecules, including the first enzymes. These enzymes would have catalyzed more synthesis, producing more interactive proteins. Collections of these new catalysts then became enclosed by simple, water-resistant lipid shells that allowed certain molecules to pass through them.

These self-forming collections, termed **coacervates** (Figure 22.2) by Oparin, supported and perpetuated themselves by making use of the energy-rich nutrients of the sea.

To continue with the Haldane-Oparin scenario, the coacervates initially were more or less random collections of molecules. They grew to a critical size and simply divided into smaller parts, only to increase again. But eventually the molecular structure of some became specific, and new, self-perpetuating forms arose. Included were molecules that encouraged energy-yielding reactions and others that used such energy to assemble replicas of the unique molecules. Such systems became more efficient at capturing and using the precursor molecules. Further, several distinct lines began competing for the energy sources. The transition from coacervate to **protocell,** a primitive early cell (*proto,* early), was under way.

After presenting his coacervate hypothesis in the 1930s, Oparin spent many years experimenting with versions of coacervate droplets. He and his coworkers tested great numbers of different combinations of molecules and succeeded in simulating simple versions of cell metabolism in their droplets. They added starch-synthesizing enzymes and glucose, and starch appeared. They switched to starch-digesting enzymes and got their glucose back. Of course, these experiments must be kept in proper perspective. Coacervates are not "alive," and the impressive feats demonstrated by Oparin and his followers ultimately depended on the droplet being supplied with enzymes extracted from living cells. The experiments, however, do provide some information about what might have been a critical stage in the origin of life. After all, with these developments, the fundamental essentials of life are in place. This scenario provides both energy-capturing

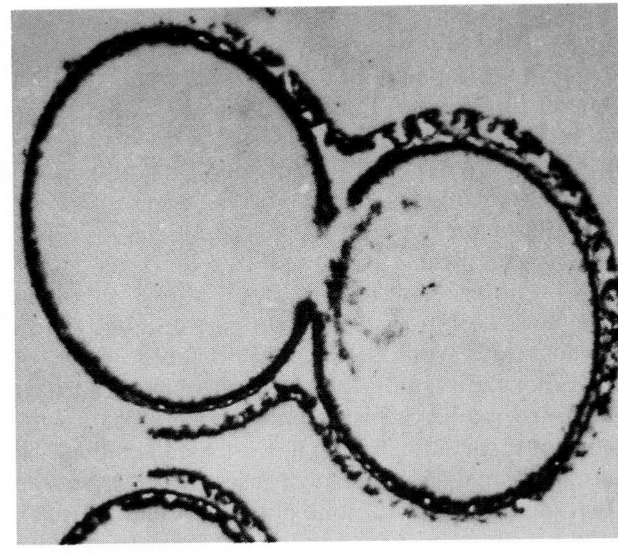

FIGURE 22.2 COACERVATE DROPLETS.
These nonliving droplets show many of the characteristics of living cells.

and replicating mechanisms. Further, there is variation and competition for limited resources, setting the stage for evolution through natural selection, evolution that would produce the first protocells—true, if crude, living forms.

Haldane and Oparin's hypothetical scheme remained a neglected intellectual curiosity until 1952, when Nobel laureate Harold Urey and a younger colleague, Stanley Miller, began testing some of the assumptions in earnest. (Urey gave Miller most of the credit for the experiments.)

FIGURE 22.3 STANLEY MILLER.
In the classic Miller-Urey experiment, the heated gases thought to comprise the primitive atmosphere were subjected to electrical discharges in a sealed, sterile environment. Residues were collected in the lower chamber and analyzed. Results indicated that some of the simple monomers of life could be produced spontaneously under test conditions.

THE MILLER-UREY EXPERIMENT

The crucial supposition of the Haldane-Oparin hypothesis was that *organic* (carbon-containing) molecules would form spontaneously under the primitive conditions. This was a key assumption that remained to be tested. Then Urey and Miller created a laboratory apparatus at the University of Chicago that attempted to simulate what were then believed to be some of the primitive conditions of the earth (Figure 22.3). They introduced a small amount of water and a mixture of gases including methane, ammonia, water vapor, and hydrogen sulfide—but no free oxygen—into the apparatus. Energy was provided in the form of repeated electrical discharges (lightning?) through the atmosphere of the upper flask. After a week-long run, analyses of the sediments that collected in the lower flask revealed the presence of aldehydes, carboxylic acids, and most interestingly, amino acids—all common constituents of living cells.

Although these small molecules were a far cry from anything alive, their production under simulated primitive conditions provoked a lively revival of interest in the Haldane-Oparin hypothesis. Until recently, the special atmosphere developed by Miller and Urey was believed to be representative of conditions on the early earth, conditions produced chiefly by volcanic activity.

Computer simulations now show that any methane, ammonia, and hydrogen sulfide in those ancient days would be rapidly broken down by ultraviolet radiation, and that most of the hydrogen liberated would be lost to outer space. According to current theory, the principal constituents of the early atmosphere were a variety of gases: water vapor, carbon dioxide, carbon monoxide, molecular nitrogen, and possibly some free hydrogen. Recent repeats of the Miller-Urey experiment, this time using the revised version of the probable primitive atmosphere, happily give even greater yields of appropriate small organic molecules. Included among these are the nucleotide bases of RNA and DNA.

In the 1980s, Thomas Cech at the University of Colorado suggested that RNA may have been critical in the development of early life. He and his coworkers found that RNA can behave as an enzyme, just as do protein enzymes—that is, they can accelerate certain biological processes by acting as catalysts. Modern cells, they found, use RNA catalysts (called *ribozymes*) to process new RNA. They further suggested that before molecules of proteins or DNA existed on the planet, there was RNA that could direct the synthesis of new molecules much like itself. Furthermore, unlike DNA which exists in the form of a double helix, RNA can exist in a variety of complex forms, providing the variety so crucial to the effective forces of natural selection. Even this RNA replication, though, would rely on outside sources of energy to help drive its processes.

We should keep in mind that there was no shortage of energy in the primitive atmosphere. Ultraviolet (UV) light was plentiful since the ozone layer—today's protective screen in the upper atmosphere, shielding the life below from damaging ultraviolet light—had not yet formed. (That thin layer of ozone is currently threatened by certain artificial chemicals being released into the atmosphere, as we see in Chapter 46.) Other forms of readily available energy included lightning, heat, and radioactivity. Further, volcanic activity was intense. So the consensus of scientists today is that the major requirements of chemical evolution—reactive gases and concentrated energy—were abundant.

The Polymerization Problem

The spontaneous generation hypothesis is vigorously upheld today, even as scientists focus on its troublesome aspects. For example, whereas the spontaneous formation of monomers offers few conceptual problems, there are difficulties at the higher levels of the scenario, those involving polymerization—the joining together of monomers, the subunits.

The polymerization of monomers into the familiar macromolecules of life—proteins, carbohydrates, lipids, and nucleic acids—without the assistance of enzymes (themselves polymers) presents many difficult problems. It is here that the original "hot, thin soup" hypothesis is weakest. Biological polymers in water slowly dissociate back into monomers, and heat accelerates the process. The reaction moves in the direction of spontaneous polymerization *only* when the concentration of monomers is very high and the concentration of water is very low.

How can such conditions be achieved in nature? There have been many suggestions. Carl Woese of the University of Illinois, for example, has proposed that life began not in the sea, but in the hot, extremely dense atmosphere of the *very* early earth. His proposal has been supported by Sidney Fox of the University of Miami, who has demonstrated that amino acid polymerization (the joining together of amino acids, forming peptides, polypeptides, and perhaps proteins) occurs readily under hot, drying conditions, such as might be found in small pools alongside volcanos and on the hot rocky beaches of ancient seas. As the pools dried, the material suspended in them would have become more concentrated, the molecules having more opportunities to interact. Using drying heat and a solution of amino acids, Fox has succeeded in producing polypeptides composed of 200 or more amino acids. Fox calls these spontaneously generated polymers *thermal proteinoids.* Importantly, when placed in water, the thermal proteinoids cluster into rounded *proteinoid microspheres* (Figure 22.4). Such spheres automatically form two-layered membranes, isolating their internal structure from their watery surroundings much in the manner of coacervates.

Using a variation of Fox's methods, Israeli scientists have polymerized amino acids on the surfaces of fine clay particles. The clay particles provided a substrate on which the molecules could be concentrated. (They suggest that the first living, successfully reproducing organisms might have been made mostly of clay!)

Perhaps the most vexing aspect of the spontaneous-generation idea is how nucleotides, once formed, would have spontaneously joined to produce nucleic acids even remotely resembling DNA. Investigators have boiled and dried concentrated solutions of energy-rich nucleoside triphosphates in the presence of single strands of DNA, loading all the dice toward the successful production of a second DNA strand. But without the appropriate enzymes, no recognizable polymers are formed. Spontaneous linkages can be forced, but they occur in the wrong places. Supporting evidence for the spontaneous formation of nucleic acids remains to be found.

A Nurturing Rain of Comets

There are those who argue that those first organic compounds that nurtured the development of life on earth came not from materials here but from outer space—from the comets that struck the planet in its infancy. Their arguments are persuasive enough to have generated an international meeting

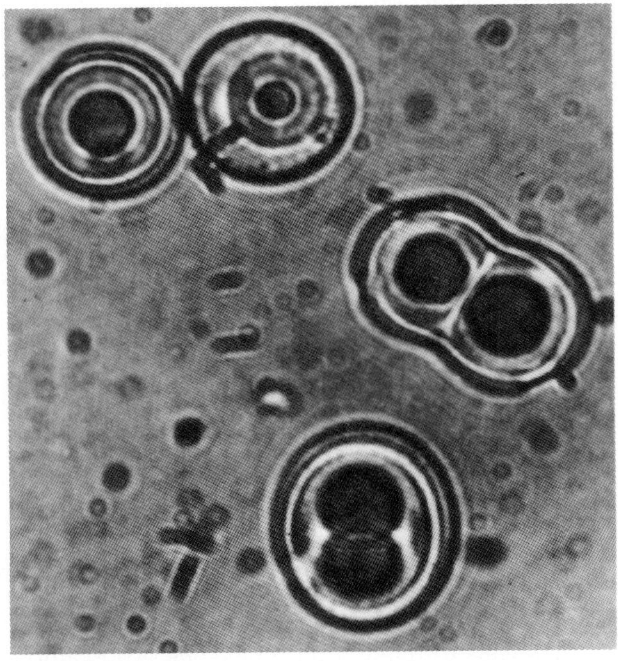

Figure 22.4 Proteinoid Microspheres. Photomicrograph of individual proteinoid microspheres.

at the University of Wisconsin in 1991 called "Comets and the Origins and Evolution of Life." At the meeting, astrophysicist J. M. Greenberg acknowledged that the organic constituents of those "dirty snowballs" called comets—constituents such as formaldehyde, hydrogen cyanide, and more complex molecules—are indeed fragile. But he also argued that cometary ice has great insulating properties, and that some of the material could have survived the great heat that would have been produced as a comet entered the earth's atmosphere. Stanley Miller disagreed, holding to the traditional view of the development of organic matter on earth. Carl Sagan of Cornell University and his colleagues showed in computer simulations that only small parts of comets could survive the inferno of atmospheric entry, probably too little to have seeded the earth with the stuff of life. Still, the question remains open. Many of the answers may come in the year 2002 when the European Space Agency plans to land a device on a passing comet, to drill a core sample from the comet, and to return the sample to earth for direct laboratory analysis. That should be fun.

The Earliest Cells

Having presumptuously taken the giant step between chemically active aggregates to self-reproducing protocells—leaving huge gaps for future theorists to deal with—we can apply some informed speculation about early cellular life. The first question is: What were the earliest cells like?

Those ancestral living beings probably relied heavily on the simple, anaerobic process of fermentation (see Chapter 8), as they broke down certain kinds of monomers still being spontaneously formed. Such cells also may have evolved ways to prey on each other, but with increasing competition for prey, natural selection would have strongly favored organisms that could exploit other energy sources. And so those cells that could make their own energy-laden molecules, however crudely, would have gained an advantage. Such life-forms are called *autotrophs* ("self-feeders"—see Chapter 7) and they are common today as we see, for example, in the green plants.

The Early Autotrophic Cells. Autotrophs are of two main types: phototrophs, which make food molecules using the energy of light, such as do algae and plants; and chemotrophs, which use the energy in chemicals to make food molecules. An example of the latter is the communities surrounding vents on the ocean floor. Both require carbon dioxide as a source of carbon, and simple, inorganic molecules such as water or hydrogen sulfide as a source of hydrogen to reduce the carbon dioxide. They also require a source of energy.

The most primitive living phototrophs obtain their hydrogen from dissolved hydrogen sulfide, and it's a good bet that the earliest successful phototrophs also employed this source, using light energy in the simplest of photosystems to pry the hydrogen away (see Chapter 7). But the number of places in the world that provide both hydrogen sulfide and abundant sunlight are severely limited. At some point, a variant line of ancestral phototrophic bacteria began to obtain their hydrogen from a far more available source: water. This giant step was energetically difficult and probably required complex chemical pathways. However, the accomplishment was a smashing success and it was to change the earth forever.

Oxygen: A New Atmosphere and the End of an Era. A waste product of photosynthesis is molecular oxygen. As the early phototrophs flourished, they began to slowly bathe the earth in oxygen, a highly corrosive gas (Figure 22.5). This meant that as new monomers were formed, they were quickly degraded by the oxygen. Further, many of the early life forms could not carry out their metabolic activities in the presence of oxygen, and so they were eventually restricted to anaerobic pockets in the sea bed and the earth's crust. Still, the autotrophs continued to pour out the deadly gas as they went about their increasingly efficient way of sustaining themselves in a newly competitive world.

New Kinds of Heterotrophs. Although many oxygen-sensitive organisms undoubtedly became extinct—or were literally relegated to the anaerobic mud—new forms emerged through mutation and natural selection. These new forms would have had to develop means of surviving in the presence of oxygen. The next step would be to actually harness the corrosive power of oxygen, to use the oxygen in metabolic pathways, marking the advent of oxidative respiration.

The burgeoning autotrophs themselves would have represented an abundant new source of food for any heterotroph that could capture them and assimilate their bodies; thus the world saw the emergence of the first herbivores. Later, organisms would appear that preyed on herbivores, as the chain of life became dizzyingly complex. All the while, the remaining anaerobes were relegated to a backwater in the progression of life, hidden away

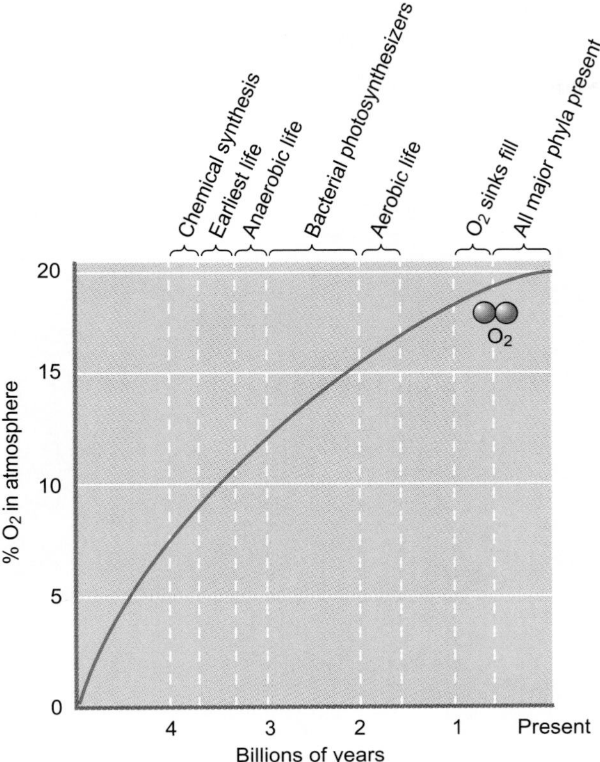

FIGURE 22.5 OXYGEN ACCUMULATION. The accumulation of oxygen in the atmosphere of the early earth was a slow process at first, requiring about 2 billion years to approach the concentration that now exists. Many *oxygen sinks* absorbed photosynthetic oxygen as fast as it could be produced (elemental iron, elemental sulfur, and iron sulfide). Only when these oxygen sinks were finally saturated could free oxygen increase in the air.

from poisonous oxygen in places such as pockets of the earth's crust, nutrient-rich muds, or deep recesses of stagnant waters.

In this scenario, we have seen two important forms of life emerging: the oxygen-producing autotrophs, and the oxygen-using heterotrophs. With these forms established, variations would be produced that would lead to their own specialized lines, growing ever more complex until finally, their bodies were durable enough for their corpses to leave fossils. It is at this point that our hypotheses about early life grow stronger, buttressed by concrete evidence.

As we see, much of the origin of life remains unexplained, and what explanations we do have are based on conjecture and on our imperfect knowledge of what the primitive earth was like. But this is the way of science, and whether current theory thrives and grows or dwindles into oblivion will ultimately depend on imaginative experiments and observations still to come.

We now move to one of the smallest, simplest, and most successful of life forms—the prokaryotes. These, of all the life today, probably resemble most those cells that graced the planet in the earliest days of life on earth.

KINGDOM MONERA: THE PROKARYOTES

The prokaryotes are found only in the kingdom Monera and are marked by cells lacking membranes around the nucleus and other cell organelles. The prokaryotes include bacteria and the cyanobacteria (photosynthesizing prokaryotes). Prokaryotes are distinguished from the eukaryotes (protists, plants, fungi, and animals), the groups which generally have membrane-enclosed cell nuclei and organelles. There is no longer any doubt that the prokaryotes were among the earliest forms of life. In fact, there are clear indications of their presence in deposits 3.5 billion years old, a period marking the beginning of what can rightly be called the Age of Prokaryotes, the longest period to be dominated by a single kingdom. (By contrast, the first eukaryotes probably arose only 1 to 1.2 billion years ago as we see in Figure 22.6.) The most widespread evidence of prokaryote antiquity is seen in strange columnlike deposits known as **stromatolites** (see Essay 22.1). These highly laminated deposits of sedimentary rock feature layers produced by dense populations of bacteria—probably cyanobacteria—that were infiltrated over the eons by sand-bearing high tides.

Some paleontologists doubted that these strange geological formations were evidence of early life until the discovery of still-active stromatolite-forming cyanobacteria at work. They can be found in Yellowstone National Park in the United States, on the shores of Shark Bay on Australia's west coast, and in Baja California.

BACTERIAL CHARACTERISTICS

Most people, it seems, tend to define the word *bacteria* as "germs." In a sense, they're not far off, because "germ" means beginning, and there are those who believe that bacteria, indeed, represent the beginning of real life. Technically, however, bacteria are defined as single-celled prokaryotes.

Bacteria are extremely diverse. Whereas most forms exist singly, a few reveal signs of *colonial* organization, that is, clusters of cells living in some

ESSAY 22.1 The Oldest Life

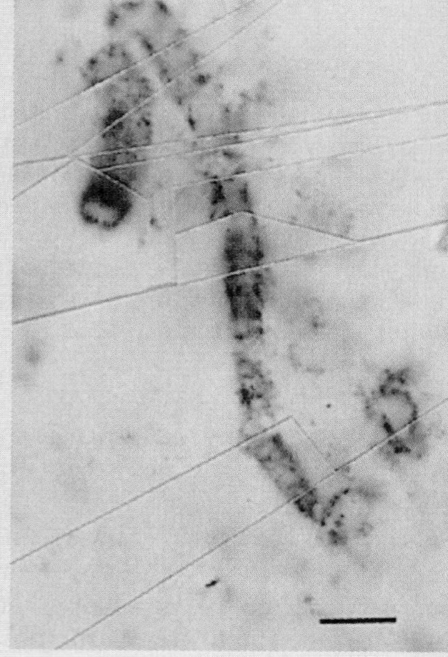

There is an area of western Australia so forbidding that only the most hardened of miners and scientists go there. The miners are there for the usual reasons; the scientists go to dig for some of the oldest known rocks on earth. The rocks, called stromatolites, are interesting precisely because of their age. In 1980, a surprising announcement was made: The rocks contained fossilized remains of life that existed about 3.5 billion years ago—that is, a billion years before any other known life.

Moreover, these rocks contained at least five different life forms. The fossils were tiny, to be sure. Most of the forms were elongated, or strand-like, and the cells lacked any sign of a nucleus, as do bacteria today. Furthermore, the organisms probably lived under a thin layer of warm water, and, since they used carbon dioxide, they were probably photosynthetic. It appeared that scientists had found one of the earth's earliest pond scums.

The oldest known rocks on earth are 3.8 billion years old, and some scientists think that they may contain evidence of life also. If this is so, since the earth is estimated to be only some 4.5 billion years old, life may have appeared very suddenly after the earth's surface cooled. According to some scientists, this means that there is an increased likelihood of life appearing wherever physical conditions permit. This, in turn, suggests to some that there is greater likelihood that life has appeared on other planets in the vast, far-flung universe.

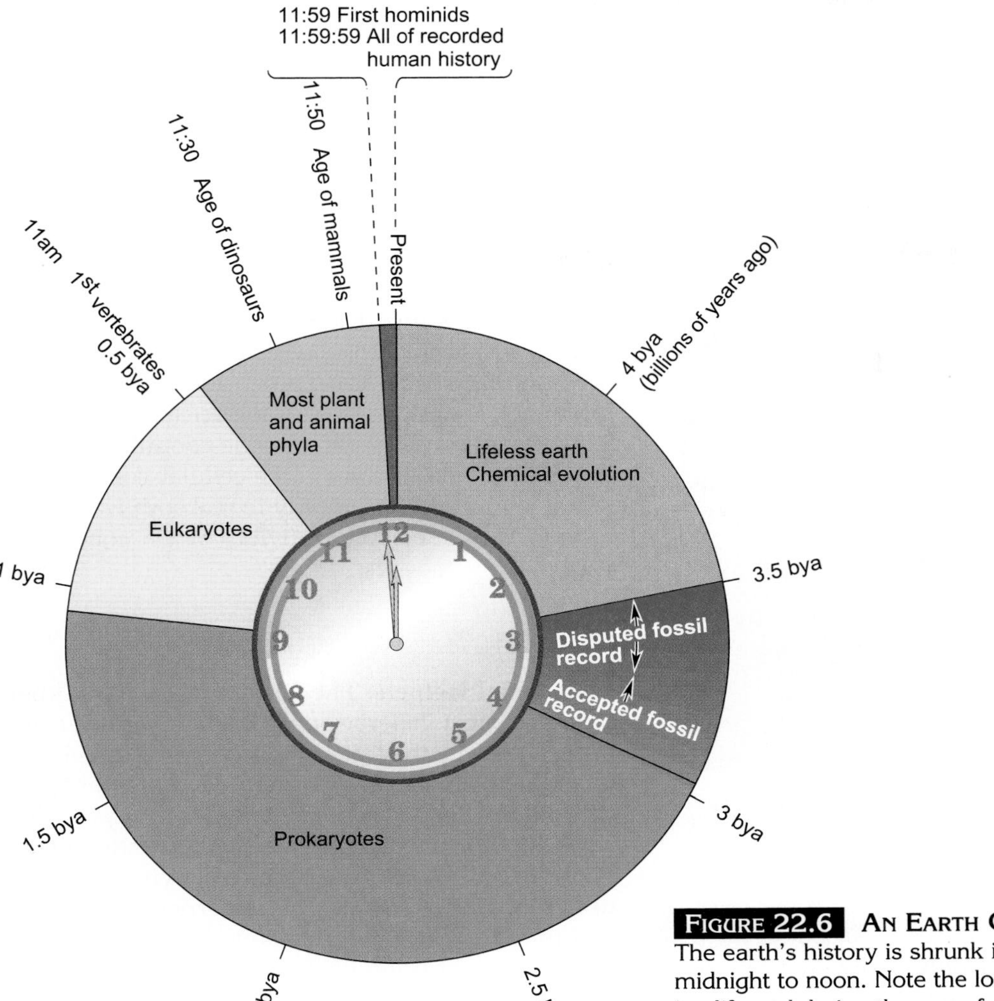

FIGURE 22.6 AN EARTH CALENDAR. The earth's history is shrunk into a period of 12 hours—midnight to noon. Note the long periods of time preceding life and during the age of prokaryotes.

state of interdependence. Most bacteria are heterotrophs, relying on other organisms for their materials and energy. Of these, an immense number are *decomposers* (also called *reducers*)—soil and water bacteria whose decay activities are vital in recycling essential mineral ions. Others are *pathogens* (disease-causing parasites) of plants and animals (including humans). Several groups are autotrophic, either phototrophs or chemotrophs. The phototrophs utilize light-gathering pigments in photosynthesis. Chemotrophs, as we have seen, metabolize inorganic substances such as iron and sulfur compounds from the earth's crust.

BACTERIAL CELL

The bacterial cell is unique in a number of ways. Most are small, about one-tenth the size of typical eukaryotic cells (Figure 22.7) and their internal structure is much simpler (Figure 22.8). As we saw in Chapter 4, membrane-surrounded organelles are rare among the bacteria. Exceptions include the well-formed thylakoid of cyanobacteria and the *mesosome*, a common structure in bacteria. The mesosome is an extension of the plasma membrane that may be involved in cell wall formation. There is no organized nucleus, and the single chromosome is essentially a naked, circular molecule of DNA, lacking the protein of eukaryotic chromosomes.

In most bacteria, the molecular building block of the cell wall is *peptidoglycan*, a complex polymer containing short amino acid chains interwoven with modified glucose units. The antibiotic penicillin kills bacteria by interfering with peptidoglycan synthesis.

Many bacteria move about through the use of one or more flagella, as do some eukaryotes. However, unlike the microtubular, undulating eukaryotic flagellum, the bacterial version is an S-shaped,

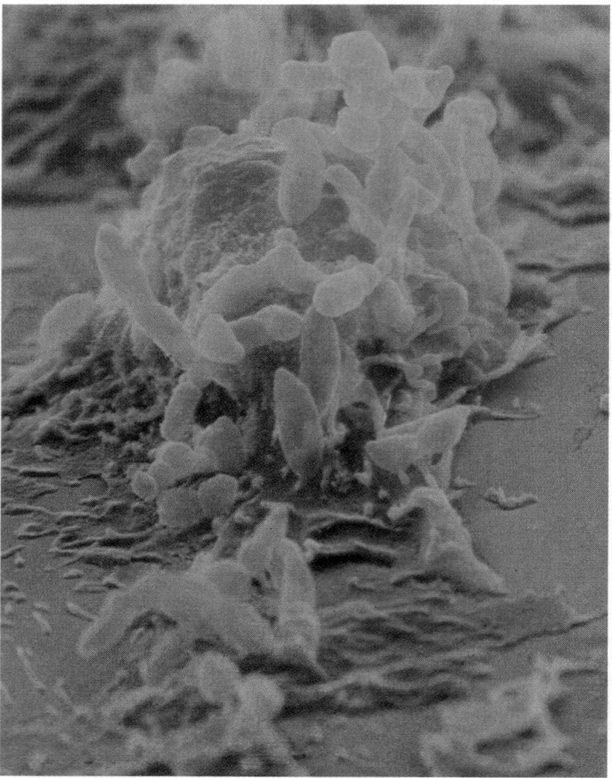

FIGURE 22.7 RELATIVE SIZES OF A PROKARYOTE AND A EUKARYOTE.
Here, a neutrophil (a eukaryotic cell) is seen alone with several bacteria (prokaryotic cells).

rotating, hollow tube—the only known rotating organelle in any species (Figure 22.9).

Bacterial Reproduction. You may recall from Chapter 16 that some bacteria make use of a connecting sex pilus in a primitive and rare form of sexual reproduction called conjugation. However, asexual reproduction is far more common. It occurs through *fission,* a primitive form of cell division that does not employ a microtubular spindle apparatus (see Chapter 9). Following chromosome (DNA) replication, the two replicas attach to separate sites on the plasma membrane, and the cell wall is laid down between them.

Using fission, bacteria can multiply at a phenomenal rate. For instance, under ideal growth conditions, the colon bacterium *Escherichia coli* can divide every 20 minutes (Figure 22.10). Potentially, 72 generations could form in just one day. (That's 2^{72}, or 40 with 21 zeroes behind it.)

Diversity in Form and Arrangement. The first criterion for classifying bacteria has been *shape.* The three primary shapes—rod, sphere, and spiral—are known, respectively, as **bacillus, coccus,** and **spirillum** (Figure 22.11).

The bacillus forms occur as single cells and in chains enclosed in sheaths. The spherical forms may occur singly (simply called coccus), in pairs (diplococcus), in beadlike chains (streptococcus), or in grapelike clusters (staphylococcus) (see Figure 22.11). The spirillum forms occur singly. Many of these have lengthy flagella. (These bacteria are responsible for Lyme disease, a condition carried by ticks—see Essay 22.2.)

Many bacilli are also known for their ability to form highly resistant, thick-walled *endospores* (Figure 22.12) in response to unfavorable conditions. These dehydrated bodies contain the cellular components in a state of dormancy, ready to reabsorb water and resume their metabolic activities when conditions improve.

HETEROTROPHIC BACTERIA

Bacterial Helpers: The Decomposers and Other Friends. Most heterotrophic bacteria are **decomposers,** that is, they gain their energy through the breakdown of organic matter in dead organisms or their wastes. This activity is ecologically crucial since it releases such key ions as nitrates, phosphates, and sulfates back into the environment for use by other organisms. This is particularly important to the growth and metabolism of phototrophs, organisms that bring energy into the living realm.

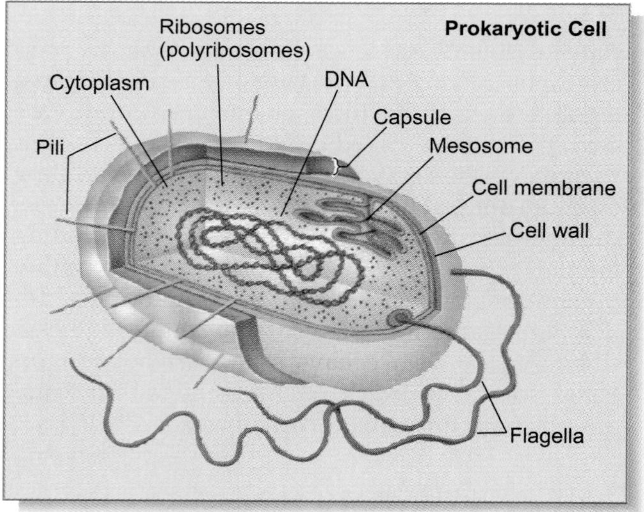

FIGURE 22.8 THE PROKARYOTIC CELL.
Here, a bacterium demonstrates many of the special characteristics of the prokaryotes.

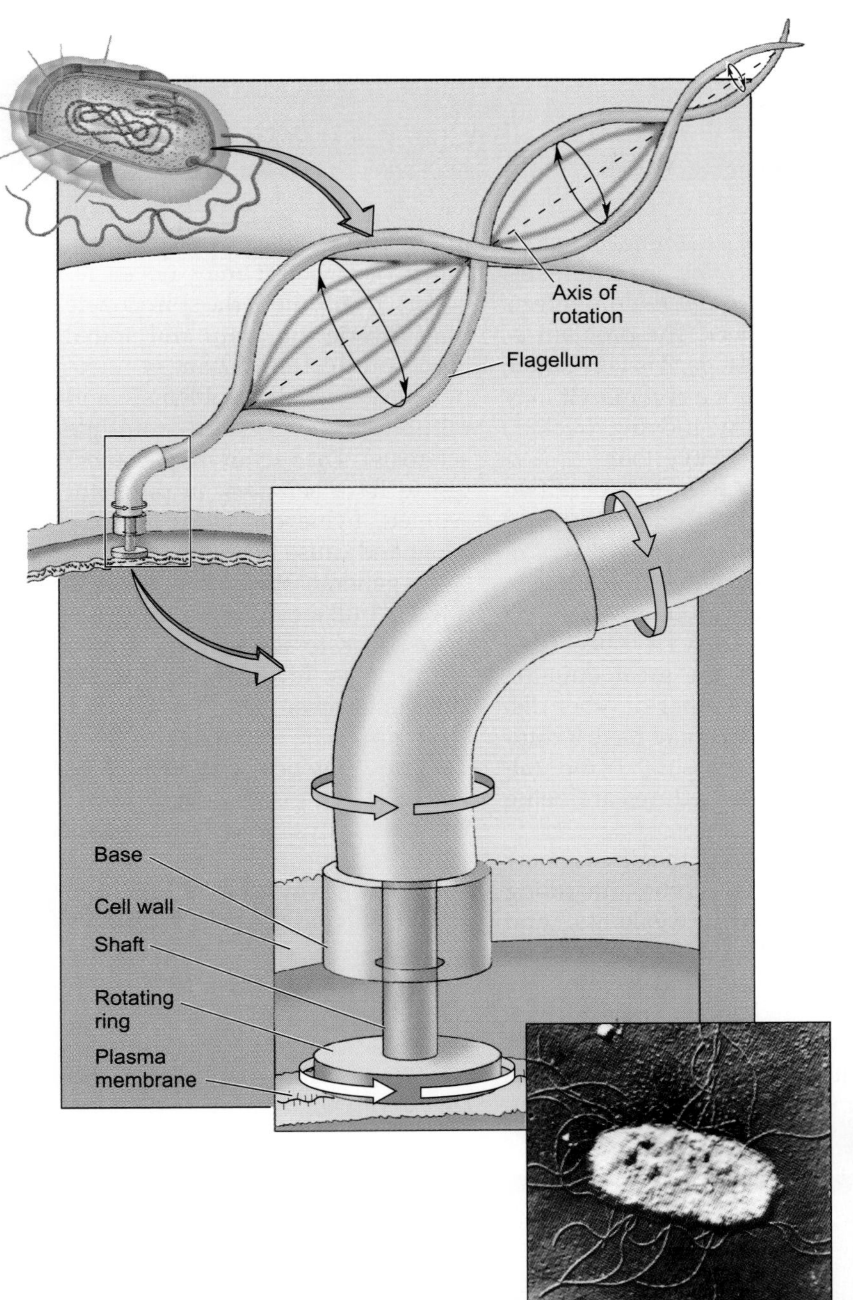

FIGURE 22.9 PROKARYOTIC FLAGELLA. Flagella are tubular structures, permanently bent into a gentle S-shape. The base inserts through the wall and into the plasma membrane, where it joins a ring of spherical proteins. The ring rotates in place, spinning the flagellum.

(Of course, if the organic material being broken down is tonight's prime rib, we may lose this objective view of decomposer activity.)

Bacterial species become quite specialized in their role as decomposers. Each has its own nutritional requirements, and quite often the energy-rich products of one group are used by the next. Thus a succession of different bacterial decomposers may be found in a food source. (And thus today's neglected prime rib will take on a sequence of appearances and fragrances on succeeding days.) Nowhere is this better illustrated than in the nitrogen cycle, where the conversion of dead organisms to useful nitrates requires the sequential activities of several specialists (see Chapter 46).

We often go to a lot of trouble to keep decomposers from getting at the prime rib. The problem is not an easy one, however, because in general, ideal growth conditions for such heterotrophs include only food, warmth, moisture, and (for many) oxygen. Thus our food is as attractive to decomposers as it is to us. Accordingly, there are many familiar ways of preventing or slowing their growth. We can interfere with their success in a number of ways,

ESSAY 22.2 Lyme Disease

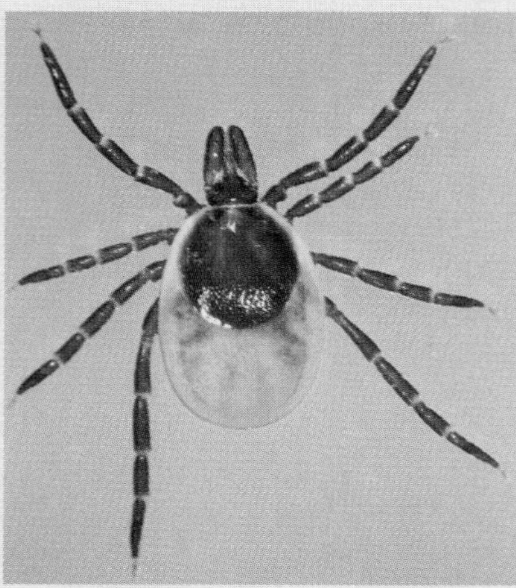

As if you didn't have enough to worry about, now there's Lyme disease. (If you haven't heard of it, raise your hand.) What it is, is a disease caused by a spirochaete, a corkscrew-shaped bacterium, *Borrelia burgdorferi*, and carried by a tick, *Imodes dammini* (at least in the Northeast and Midwest). The tick has three stages in its life cycle. In its larval stage, it mainly infects the white-footed mouse. In its nymph stage, it infects a range of mammals, including dogs, raccoons, and humans. In its adult stage, it mainly infects white-tailed deer. The problem is, in the nymph stage, it is very small, about the size of a comma. (It may look like a tiny moving freckle.) Thus it is easy to overlook.

The disease was first identified in 1975 in Lyme, Connecticut, and has since spread swiftly throughout the United States. In fact, there may have been 50,000 cases in 1989 alone. We say "may have been" because Lyme is the great imitator, mimicking a number of other diseases. Physicians may have a difficult time diagnosing Lyme (although new procedures are being implemented).

The first sign of the disease may be flulike symptoms, including headache, fever, weakness, and stiff joints. If antibiotics, such as tetracycline or doxycycline, are administered immediately, the victim is likely to recover. However, as time passes without diagnosis and proper treatment, a debilitating arthritis may appear (from which the victim may never recover). In addition, the spirochaete may invade the brain and spinal cord, producing dizziness, incoherency, visual problems, and numbness (suggesting multiple sclerosis). The victim may also begin to have seizures. In pregnant women, Lyme can cross the placenta and cause fetal damage.

In general, the warning signs are a "bull's-eye" rash that appears days to weeks after a bite, followed by fever, chills, fatigue, and headache. The joints may swell and ache and in some cases the heart may beat arrhythmically, with the legs growing very weak. Facial paralysis and numbness may also appear.

The best way to avoid the disease is to avoid the tick. Keep your pets tick-free, stay out of brushy and grassy areas, wear long clothing (tuck your pants into your socks). Use a repellent that contains DEET and permanone, and check yourself for moving freckles.

such as by sterilization (canning or irradiation), cooling and freezing, and drying, salting, and sugaring. The last three effectively remove water from the food or set up unfavorable osmotic conditions for the bacteria. Heat can kill the decomposers as well. We are aware, for example, that pasteurization, the brief application of heat, kills certain disease-causing bacteria in milk. Chemical preservation, a favorite in the food industry (some food labels can be almost frightening), includes the use of small amounts of sorbic acid, sodium nitrate, calcium propionate, and other chemicals to retard the growth of microbes.

Although a lot of effort ordinarily goes into the prevention of bacterial growth in foods, there are some instances in which bacteria are encouraged. Certain harmless bacteria provide many appetizing flavors in foods. For instance, we owe the tartness of pickles and sauerkraut to lactic acid bacteria. Other foods whose taste comes from bacteria include yogurt, buttermilk, sour cream, and many popular cheeses (Parmesan, Cheddar, Swiss, and Camembert, to name a few). There are, indeed, many commercial and industrial uses for bacteria, but let's turn now to some bacterial activities we definitely want to discourage.

Pathogenic Heterotrophs: The Villains. Many of the best-known bacteria are **pathogens.** That is, they cause disease. Pathogens are found among all three

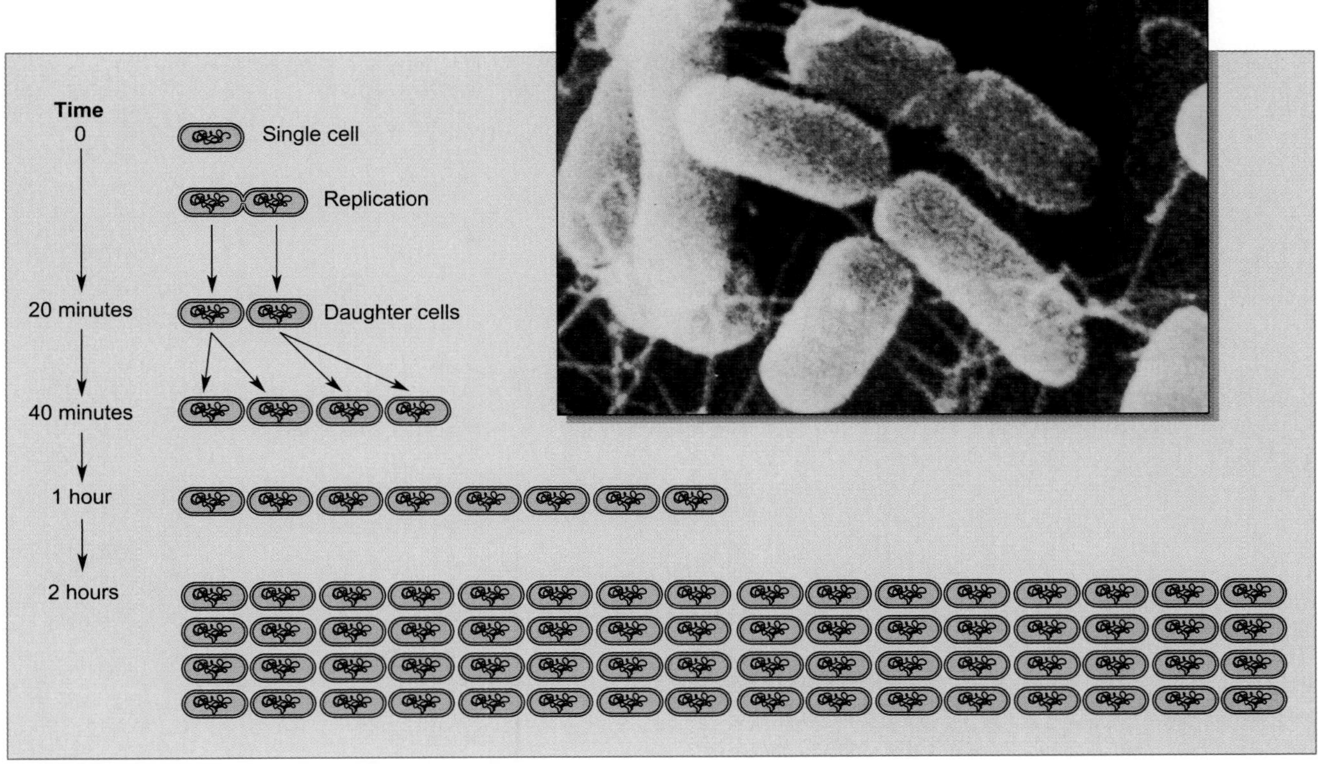

FIGURE 22.10 **BACTERIAL FISSION.**
DNA replication occurs first. Then the cell divides and identical daughters result. The process takes only 20 minutes, so a large number of cells can be produced in a short time.

bacterial forms. Pathogenic bacilli, for example, include the agents of such dread diseases as leprosy, typhus, black plague, diphtheria, and tuberculosis. These dangerous diseases have been the focus of intensive research over the years, and effective means of combatting them have been developed. However, a new and very resistant strain of tuberculosis is currently making an appearance, stimulating a burst of research into just how to control it.

In spite of such notable progress with some pathogenic bacteria, we are still threatened by others. For instance, when the highly resistant spores of the soil anaerobe *Clostridium botulinum* get into foods during the canning process, if the foods are not sterilized by heating, they can germinate and thrive in the nutrient-rich, airless environment. When the contaminated food is eaten, the deadly bacterial secretions can cause a potentially fatal food poisoning called botulism. With the decline of home canning, botulism is rare today, but the threat is always present (Figure 22.13).

Most coccoid pathogens are not as life-threatening as *Clostridium,* but they can also cause serious problems. Staphylococcus, the rounded form that occurs in grapelike clusters, is commonly involved in minor skin infections, boils, and pimples. One virulent strain of staphylococci, called "hospital staph," may sweep through hospital nurseries and cause numerous deaths before it can be contained.

One highly persistent coccoid bacterium that is increasingly reported today is *Neisseria gonorrhoeae,* the diplococcus of gonorrhea, a sexually transmitted disease (Figure 22.14). At one time, gonorrhea was readily cured with antibiotics; but even as we produce new drugs, the bacteria themselves evolve and adapt to the challenges we present them. Now new and highly resistant strains of the gonorrhea organism exist. And recently a harmful strain of the normally benign *E. coli* (*E. coli* 0157) has been turning up everywhere (see Essay 22.3).

Strep throat is caused by the streptococcus bacteria that occur in chains. Because of the widespread use of antibiotics, strep throat isn't nearly as serious a health threat in the United States today as it was early in this century, or as it still is in less developed parts of the world. The body's own reaction to the streptococcus is the real threat: Allergic reactions to the bacterium can produce scarlet fever, rheumatic heart disease, or kidney inflammation. All can be fatal. In the last few years, a particularly gruesome

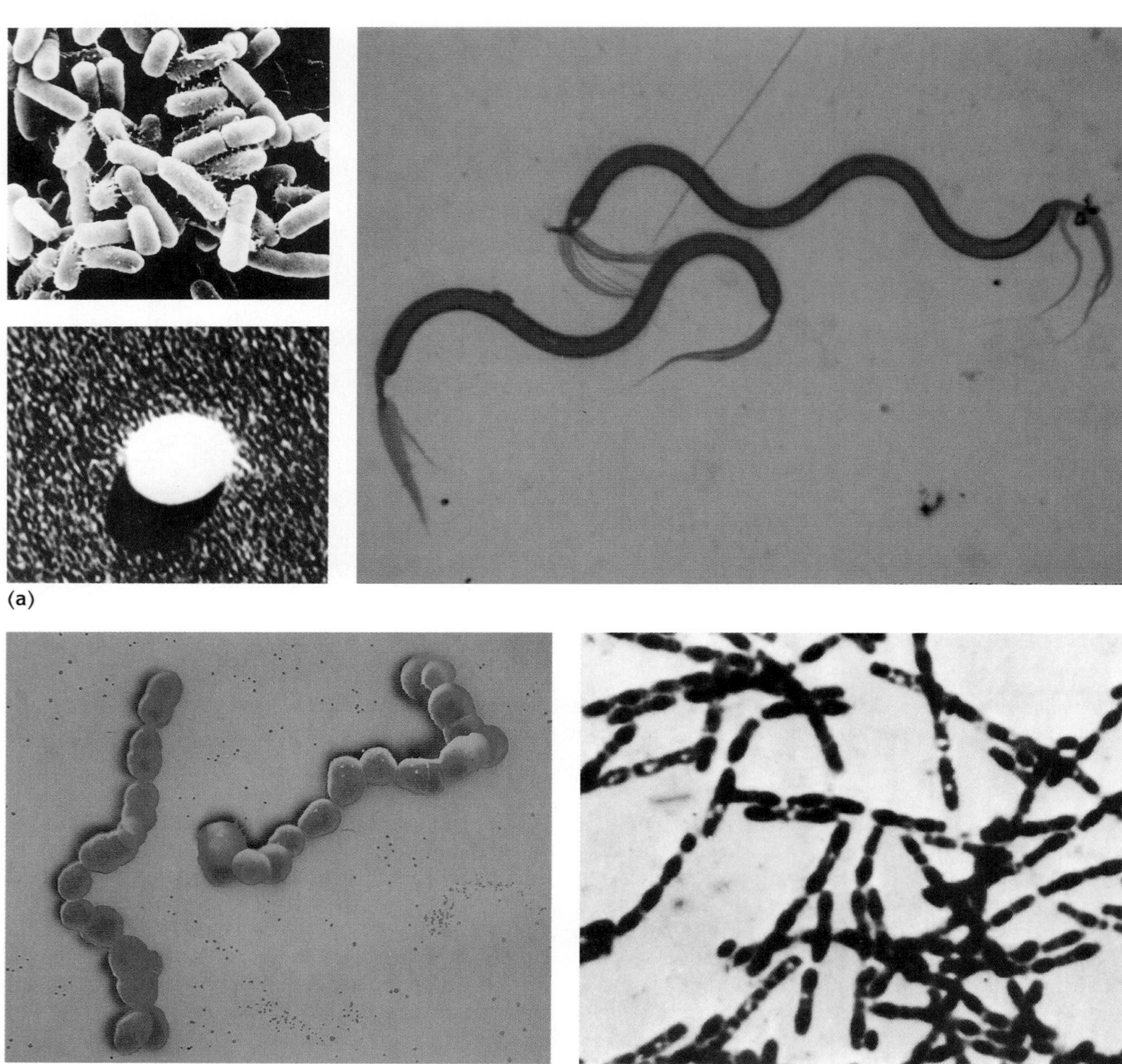

FIGURE 22.11 **THE THREE BASIC SHAPES OF BACTERIA.**
(a) The three main shapes of bacteria are the rodlike bacillus, the rounded coccus, and the corkscrew-shaped spirillum. (b) Chains of bacilli and pairs of cocci.

strep disease has been making a comeback since its last surge in the 1950s.

The most notorious of the spirillum form of bacteria is *Treponema pallidum*, the corkscrew-shaped organism of syphilis, another common sexually transmitted disease. In its later stages, syphilis has been called the "great pretender" because its effects on the body are widespread and it mimics the symptoms of a wide range of diseases. If the disease is untreated, the spirillum eventually enters the central nervous system, permanently damaging brain tissue and causing blindness, insanity, and death. Sadly, the spirillum can cross the placenta, enter a growing fetus, and cause tragic birth defects. Before the development of antibiotics, syphilis was a serious threat, and many people died of the disease.

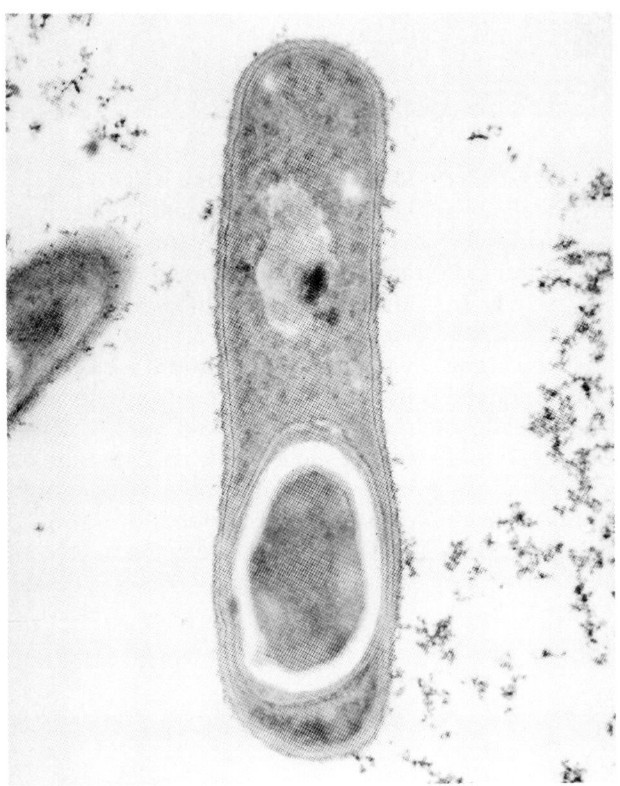

FIGURE 22.12 ENDOSPORE.
Many bacteria, particularly those in the soil and water, survive unfavorable conditions by forming tough-shelled, resistant endospores.

FIGURE 22.13 A WARNING SIGN.
This swollen can warns of possible botulism poison in its contents. Tasting the contents is a bad idea.

AUTOTROPHIC BACTERIA

These are the species that make their own food, either from the energy of light (the photosynthetic bacteria) or from chemicals (the chemosynthetic bacteria).

Photosynthetic Bacteria. The best known of the photosynthetic bacteria are the cyanobacteria, which are mostly aquatic organisms that live singly or in colonies (Figure 22.15). Although the most ancient fossils, the 3.5-billion-year-old stromatolites, may have been produced by cyanobacteria, the cyanobacteria are nevertheless considered to be highly specialized organisms. This is because they capture light with bacteriochlorophyll *a* and carotenoids, pigments similar to those in plants, that are embedded in a complex thylakoid. Further, they use water as a source of hydrogen to attach to carbon during carbohydrate synthesis (see Chapter 7).

Another photosynthetic group, the **halobacteria** (*halo*, salt), get their name from the salty environment in which they thrive. In the simplest of photosystems, the halobacteria capture light energy with the purple pigment bacteriorhodopsin, using the energy to pump protons out of the cell. This creates a steep chemiosmotic gradient that the bacterium uses to generate ATP (see Chapter 8). The ATP is then used in the synthesis of other molecules.

Chemosynthetic Bacteria. Chemotrophic bacteria, we have seen, obtain their energy from the oxidation of inorganic compounds in the earth's crust. (Recall from Chapter 6 that oxidation is the removal of electrons from substrate.) The bacteria may use a variety of inorganic compounds. One group, for example, breaks down sulfur compounds, producing sulfuric acid, which ionizes in soil water, releasing hydrogen ions and sulfate ions. Another group gets its energy by oxidizing hydrogen sulfide (H_2S) to sulfur, while yet another may oxidize reduced forms of iron and manganese.

Cyanobacteria, along with some soil bacteria and certain bacteria that live in association with flower-

Essay 22.3 Good Bugs, Bad Bugs

At this very moment, your intestine is laden with *E. coli* bacteria. This is the natural condition of the human body. The bacteria are not only harmless but may help make available to us certain vitamins from our diet. Imagine, then, the concern when in 1982, people in two states became ill from eating hamburgers contaminated with a strain of *E. coli* called *0157*. Today, the bacterium is turning up everywhere and has been found in pork, lamb, chicken, and even potatoes.

E. coli 0157 doesn't just pass through our intestines like its more common cousin does. Instead, it clings to the intestinal wall and produces a toxin that permeates the body, causing widespread internal bleeding. Most victims recover in about a week, but up to 7% suffer from hemolytic uremic syndrome (HUS), a complication in which red blood cells are destroyed, leading to kidney failure, seizures, and eventually death.

In 1993, 477 people fell ill from eating contaminated hamburgers from a fast-food outlet. At least one child died. That same year 23 New Englanders fell ill from drinking local apple cider that had somehow become contaminated. Because *0157* is increasingly found in a variety of foods, it is being regarded as an "emerging pathogen" and preventive strategies against it are not yet in place—but at the very least it is a good idea to eat only ground meat that is cooked well. Cuts of meat, such as steaks, are not as likely to carry the pathogen because all the surfaces are seared. That's good news because, as everyone knows, steaks cooked rare taste best!

ing plants, are **nitrogen fixers.** That is, they take in nitrogen gas (N_2) from the atmosphere and reduce it to ammonia (NH_3) and ammonium ions ($NH^+{}_4$) that they can then use. This nitrogen, of course, is necessary for the synthesis of amino acids, nucleotides, and other essential molecules. But more significantly, the fixed nitrogen is also released into the environment where it becomes available to plants and algae. Thus *nitrogen fixation* is an important source of new usable nitrogen for other organisms. (The nitrogen cycle is discussed in Chapter 46.)

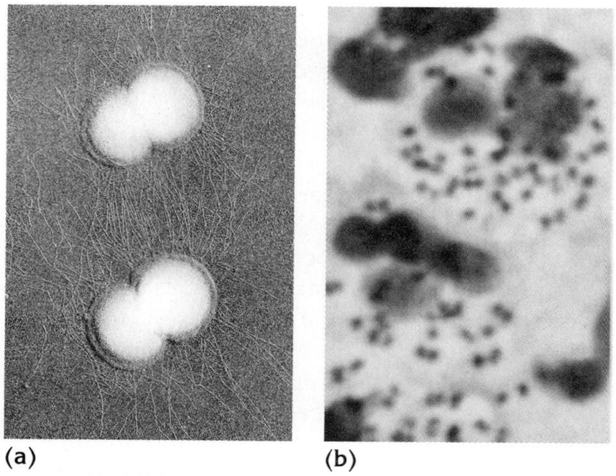

(a)

(b)

FIGURE 22.14 THE DIPLOCOCCUS THAT CAUSES GONORRHEA.
(a) EM showing many pili, extensions that aid in attachment to host cells. (b) Diplococci (small paired dots) in host white blood cells.

Prokaryote Origins and Phylogeny

Until about 1980, most biologists divided the prokaryotes into two major groups: the *bacteria* and the *cyanobacteria* (the cyanobacteria, in fact, were once not considered to be bacteria at all, and were called blue-green algae). Together, they composed the kingdom Monera. As you are well aware, however, notions about how things should be grouped can change rapidly in biology. New molecular techniques for establishing phylogenetic relationships have changed the way we classify microorganisms (see Chapter 18). Carl Woese, a leader in the new efforts, first suggested that the prokaryotes be divided into two kingdoms: Eubacteria ("true bacteria") and Archaebacteria ("first, or ancient, bacteria"), a designation that would give us six kingdoms. Both qualify as prokaryotes, but they differ in enough ways to suggest, if not prove, that they evolved separately. Although the cells of the two bacterial groups look very much alike, they are as biochemically different from each other as they are from eukaryotes (Table 22.1). We have reviewed the group referred to as eubacteria, so let's briefly look now at the archaebacteria.

ARCHAEBACTERIA

Although the name archaebacteria suggests a greater antiquity than the eubacteria, it is clear that these tough little creatures are about equally ancient. The two groups probably diverged from each other over 3.5 billion years ago in an anaerobic world, most of them evolving as strict anaerobes but with aerobes arising among them as well.

One reason archaebacteria have only recently been recognized as distinct is their peculiar living conditions. They live in hard-to-get-at places such as acidic hot springs, ammonia-rich muck, salty ponds, mud flats, and dark, sediment-rich, anaerobic lake bottoms.

Their habitat is not the only aspect of archaebacterial life that holds the microbiologists' attention. Archaebacterial cell walls contain little-known proteins, and their cell membranes include unusual *branched* fatty acids. Archaebacterial ribosomes look more like those of eukaryotes than do those of eubacteria. And, as we've seen, phototrophic forms use a peculiar light-capturing pigment called bacteriorhodopsin instead of bacteriochlorophyll (Table 22.1). All in all, archaebacteria seem to be a strange lot indeed, distinctly different from the eubacteria.

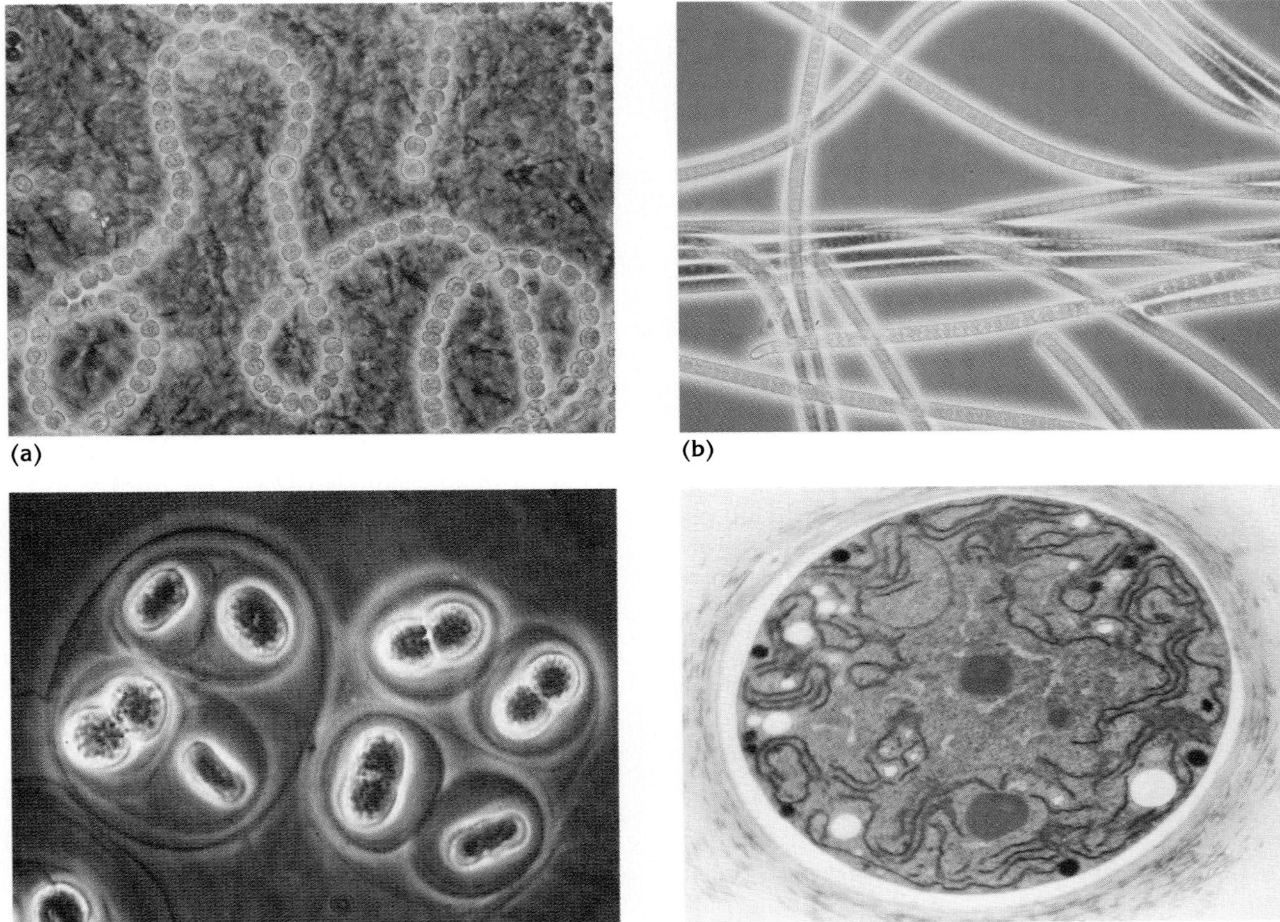

FIGURE 22.15 **THE DIVERSE CYANOBACTERIA.**
Cyanobacteria include the beadlike *Nostoc* **(a)**, the quietly undulating, filamentous *Oscillatoria* **(b)**, and the spherical *Gleocapsa* **(c)**, enclosed in its gelatinous wall. All are inhabitants of stagnant fresh water. The larger cells of *Nostoc* are known as *heterocysts*, which specialize in nitrogen fixation. Electron microscopic studies of cyanobacteria **(d)** reveal that their cellular organization can be quite complex. Note the membranous thylakoid (wavy structure).

The largest group of these archaebacteria, the **methanogens** (methane-generators), are found in habitats where carbon dioxide and hydrogen are readily available, but where there is little or no oxygen. Among these habitats are mucky, anaerobic marshes, sewage treatment plants, anaerobic sea and lake bottoms (such as in the Black Sea), and the airless bowels of animals, including humans. There they ferment carbohydrates and reduce carbon dioxide, producing a mixture of methane gas (CH_4), or "marsh gas" as it was first called, and hydrogen gas (H_2). Well-designed sewage treatment plants can supply their own energy by utilizing the methane gas produced by methanogenic archaebacteria.

THE VIRUSES

As we muster our intellectual forces and try to come to grips with the nature of life, we are inexorably led to the realm of the *virus.* A **virus** is a minute, biologically active particle made up of a nucleic acid core, a covering of protein, and sometimes one or more enzymes. While viruses can indeed be biologically active in cells, as anyone with a simple cold can tell you, some can also become crystallized, like common table salt, and remain inactive for a seemingly indefinite period. Viruses range in size from 20 to 300 nm, a range that extends from the size of the smallest bacterial cells to that of large molecules (Figure 22.16), and they come in a variety of shapes.

Many scientists are willing to treat them as odd crystals and say they are not living at all. And even those scientists who place them among living organisms are uneasy about where they fit in the organizational scheme of things. Adding to the biologist's dilemma, there are many apparently unrelated viruses that share only a few features, such as their extremely small size and highly limited mode of life.

All things considered, it is patently clear that viruses aren't cells. They lack just about everything cells have, except the genetic instructions—the coded nucleic acids (DNA or RNA)—needed to ensure their own perpetuation. Because of this, all viruses are obligate parasites. They can neither carry on metabolic activity nor can they reproduce unless they are in a host cell.

Virtually all forms of life are susceptible to attack by viruses (see Essay 22.4). Human viral miseries include rabies, polio, smallpox (now existing only in laboratory cultures, due to a worldwide vaccination program), encephalitis, and yellow fever—each of which can be fatal. Less severe, but still dangerous, are the many forms of influenza, as well as the common cold, measles, chicken pox, and herpes. Viruses are also implicated in some animal cancers and are the agent of AIDS (acquired immune deficiency syndrome, discussed in Chapter 41).

VIRAL GENETIC VARIATION

No one knows precisely how viruses originated, but it seems probable that they arose as aberrant genetic material of normal cells. Different kinds of viruses must have arisen independently of one another, since their basic mechanisms are so different. For example, in some viruses the genetic material is standard double-stranded DNA, just as it is in all metabolizing organisms. But in other viruses we find (1) single-stranded DNA, (2) double-stranded RNA, (3) a single large molecule of single-stranded RNA,

TABLE 22.1 *COMPARISONS BETWEEN ARCHAEBACTERIA AND EUBACTERIA*

	Archaebacteria	Eubacteria
Cell wall	Variety of substances, often proteinaceous	Peptidoglycans
Plasma membrane lipids	Modified branched fatty acids	Straight-chain fatty acids
DNA	Naked (without proteins attached), circular	Naked, circular
Membrane-surrounded organelles	Absent	Absent (except for mesosome and thylakoid in some)
Ribosomes	30S, 50S subunits; structural similarity to eukaryotic	30S, 50S subunits, unlike archaebacteria and eukaryotic
Flagella	Unknown	Tubular, rotating
Photosynthetic pigments	Bacteriorhodopsin	Primarily bacteriochlorophyll (chlorophyll a)
Cell division	Fission	Fission

or (4) several small strands of RNA. The nucleic acid can be circular or linear, and the single-stranded nucleic acid can be either the transcribed strand or the nontranscribed strand. Furthermore, once the virus is inside its host cell, it may even change the kind of nucleic acid it uses: The RNA viruses often make DNA copies of their own genomes.

How Viruses Behave—or Misbehave

Some viruses enter the host cell by being ingested by the host, but most viruses contain one or more specialized enzymes that facilitate attachment to and penetration of a host cell. In some viruses, notably the bacteriophages ("bacteria eaters"), only the nucleic acid enters the host cytoplasm; but in other viruses, some enzymes are carried in as well. Once inside the host cytoplasm, the virus can take over the cell's metabolic apparatus, using it for its own ends, in what is called the *lytic cycle*. Or it can simply make a home for itself within the host cell, at least for a while, in what is called the *lysogenic cycle* (Figure 22.17).

The Lytic Cycle. In the **lytic cycle,** the virus replicates and ruptures the host cell. In a typical host-killing lytic cycle, a virus penetrates a cell and within minutes stops the host's protein synthesis by destroying the host's DNA. Following this, the virus transcribes its own genes, forming viral messenger RNA. Using the host's protein-assembling machinery, the virus makes new penetrating enzymes and protein coats. Next, using either the host's replicating enzymes or its own, the invader assembles new viral nucleic acid molecules.

Finally, the new infectious viral particles are assembled, and viral enzymes lyse, or dissolve, the remains of the host cell. The new viruses are then liberated, each fully capable of infecting and killing another cell. Some of the mammalian viruses

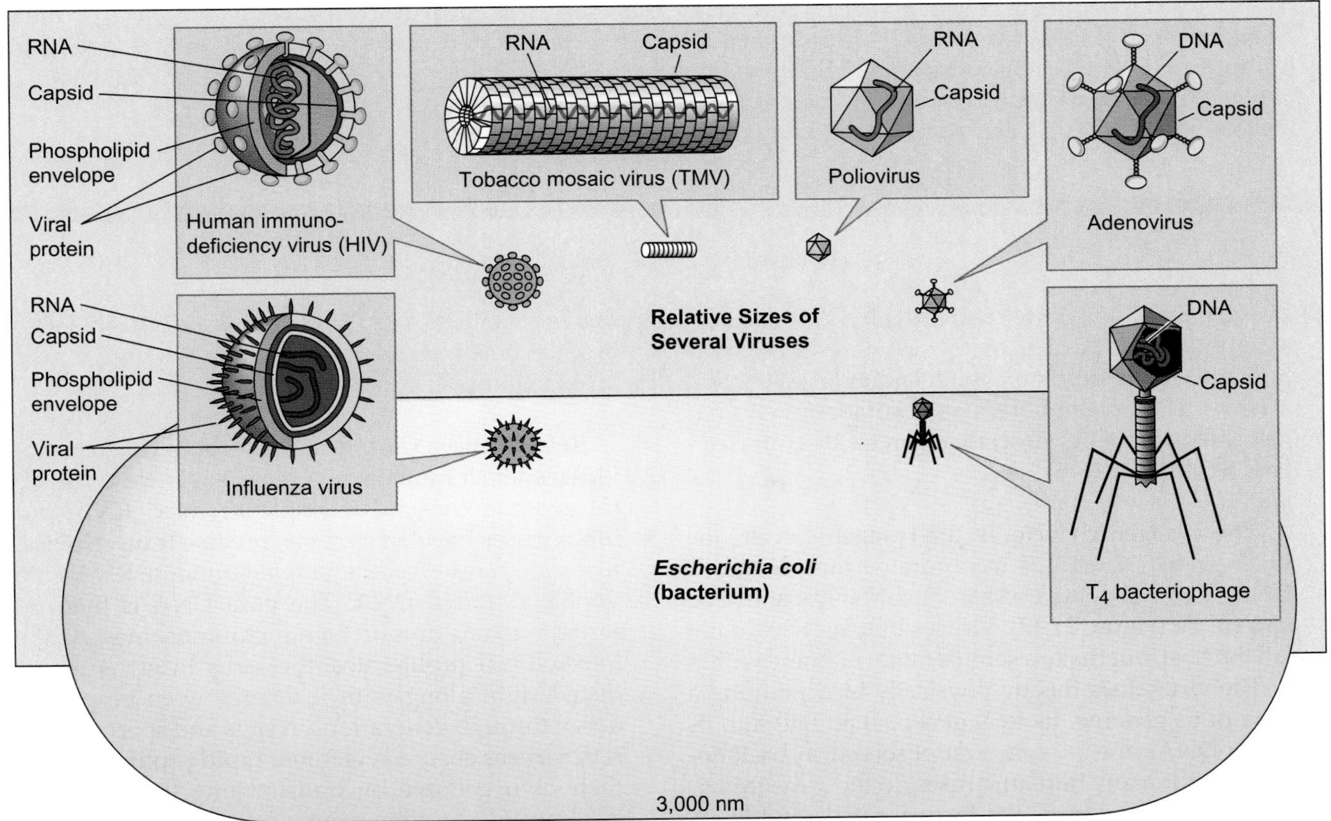

Figure 22.16 Viruses.
A comparison of several representative viruses with the familiar bacterium *E. coli* shows how small viruses are, as well as the range of sizes and shapes of viruses. *E. coli* measures about 3,000 nm in length (about half the diameter of a human red blood cell). The smallest virus, a bacteriophage, measures a minute 20 nm, while one of the largest, the tobacco mosaic virus, is just visible with the light microscope. The smallpox virus is even a little larger.

ESSAY 22.4 A New Virus Comes of Age

On the morning of May 14, 1993, a young Navajo climbed into the back seat of the car, and the family headed to town. The 19-year-old hadn't been feeling well for the last few days, and suddenly he began to gasp for air. The family called a hospital from a convenience store, and an ambulance soon arrived. In spite of all the crew could do on the way to the emergency room, the boy died. The autopsy would show that his lungs were full of straw-colored plasma, the fluid of blood. Only a month earlier a 30-year-old Navajo woman had died the same way. Tests were run for all the known organisms that could cause such symptoms. They all came back negative. Within a short time, five more deaths occurred. All were marked by flulike symptoms and sudden respiratory distress, quickly followed by death.

The medical establishment was put on alert, with descriptions of "clusters" of the mysterious illness in the Southwest as the death toll continued to rise. Researchers quickly found that the disease was not transmitted from one person to another, but the disease organism, identified as a relative of an Asian hantavirus (first isolated from an area near the Hantaan River of Korea), was found in mice. Viruses generally infect only specific species, and this virus's host was the deer mouse *Peromyscus maniculatus.* The virus didn't harm the mice, but when contaminants from the urine, bedding, or blood of the mice was inhaled by humans as dust, the result was often death.

Navajo legend warns of contact with mice, and ritual demands no contact with mice. If a mouse even touches someone's clothes, those clothes must be burned. Perhaps, it is thought, the virus has been around a long time. So why has it only recently caused trouble? No one knows; but researchers are uneasily aware that the deer mouse is found over much of the United States, and without more information, they fear that other areas may also be vulnerable to infection. And always, in the back of their minds, lurks a single question: "What other bugs are out there?"

provide a final insult to the host cell. Each partially assembled viral particle merges with a section of the host plasma membrane, which is then used to form a new viral envelope. (It's as if someone who has just eaten all of your groceries then steals your overcoat on the way out.)

The Lysogenic Cycle. In the **lysogenic cycle,** the viral genetic material is incorporated into the host's DNA. The lysogenic cycle is remarkably controlled and subtle (Figure 22.17). Viruses in this cycle do not kill the host, but make a semipermanent home within it. The virus does this by physically incorporating a copy of its genome (its total genetic material) into its host's DNA, in a process first observed in bacteriophages, but many human viruses do the same thing.

At some later time, the incorporated viral DNA cuts loose from the host chromosome and reverts to the lytic cycle, taking over the cell and making more infectious virus particles, thereby killing the host. The lytic phase cannot be predicted but some researchers believe that viruses may switch to this behavior whenever something goes wrong with normal host cell DNA replication. It is as if this is a sign that the host is in trouble and that it is time for the virus to jump ship.

Retroviruses. Certain RNA viruses called **retroviruses** infect humans in a remarkable way. The viral genome consists of single-stranded RNA, but retroviruses have an enzyme, **reverse transcriptase,** that can transcribe the single-stranded RNA into double-stranded DNA. The viral DNA is then inserted into one or more of our chromosomes. As the infected cell proliferates, the virus insert is reproduced right along with it. It may even be passed down through generations in eggs and sperm. Some retroviruses ensure even more rapid proliferation of their own genome by transforming the host cells into rapidly dividing cancer cells. A retrovirus has been identified as the agent of AIDS.

Herpes Simplex. Before AIDS became seared into the public consciousness, the sexual disease most people worried about was herpes. (An earlier generation worried about syphilis, now making a big comeback.) But since AIDS is a fatal disease, herpes—an often somewhat painful nuisance—has almost been pushed aside.

There are two types of the herpes virus. *Herpes simplex 1* causes troublesome fever blisters (cold sores). These are often activated by some trauma, such as tension or direct sunlight on the lips. *Herpes simplex 2* causes the more serious genital herpes. (There can be some transmission of type 1 to the genitals and type 2 to the lips, so sexual activity should be limited during either outbreak.)

So what is herpes all about? Surprisingly, the herpes virus resides in nerve cells, but it doesn't usually kill them. When the host's immune defenses are down, often because of stress, the virus buds off from its sanctuary, perhaps deep in the spinal cord, and travels along certain nerves to the skin. There it infects surface cells of mucous membranes, killing them and liberating numerous infectious virus particles. Time and again, the host builds up antibodies, causing the virus to become quiescent and retreat into the nerve cells, where it is safe from the antibodies. Then the antibody level falls, and sooner or later herpes strikes again. (An old joke goes, What is the difference between love and herpes? Herpes is forever.) Herpes infections, indeed, last a lifetime and are presently incurable, although intensive research is under way that may soon lead to at least a partial remedy. For example, the drug acyclovir, a DNA replication suppressant, shows some promise in controlling active herpes, but it has no apparent effect on herpes's dormant state. A promising new vaccine is also being tested, a combination of alum and a genetically engineered protein called glycoprotein D. A lot of people have their fingers crossed.

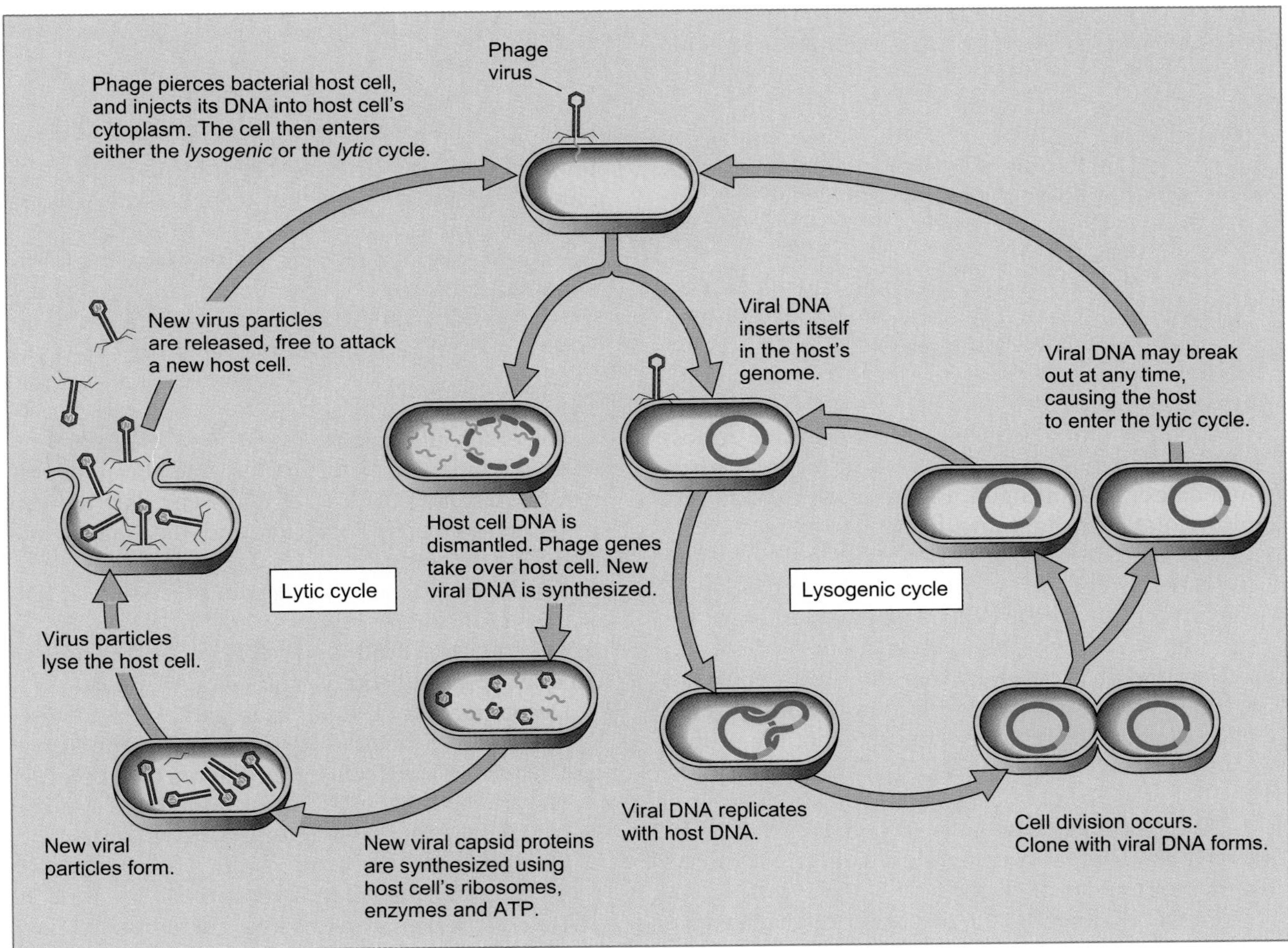

FIGURE 22.17 **SOME VIRUSES ARE CAPABLE OF BOTH A LYTIC AND A LYSOGENIC CYCLE.**
(left) The virus invades, reproduces, and lyses the cell. *(right)* The virus invades and inserts its nucleic acid into the host genome. It may later break out of the lysogenic cycle and enter the lytic cycle.

PERSPECTIVES

We have begun our survey of life in the biosphere with an introduction to what may seem to be the simpler forms. It should be apparent, however, that the "simpler forms" are not so simple after all. Life at any level is complex, and our understanding of it is riddled with unknowns that, for some, translate into exciting challenges. We move now from the realm of crystals and the smallest and oldest living things to discussions of other single-celled organisms and the trend toward multicellularity with the increased specialization it brings. As we will see, this specialization involves greater complexity than we have seen before.

SUMMARY

THE ORIGIN OF LIFE

Ideas on the spontaneous generation of life were first formalized by J. B. S. Haldane and A. P. Oparin. They proposed that the conditions present on the primitive earth permitted the chemical precursors of life (such as sugars, amino acids, and nucleotide bases) to form spontaneously. From there, continued polymerization would form enzymes and other proteins, and collections of these surrounded by molecular shells might have formed coacervates. The growth of coacervates and inclusion of self-perpetuating systems might eventually have led to the first cells or protocells.

S. Miller and H. Urey developed an apparatus in which they simulated the conditions of the primitive earth. Methane, ammonia, water vapor, and hydrogen were exposed to electrical discharge. Among the chemical products were amino acids.

Theorists today believe that the primitive atmosphere contained water, carbon dioxide, carbon monoxide, and nitrogen. Energy sources included ultraviolet light, lightning, heat, and radioactivity.

The weakest part of the spontaneous-generation hypothesis involves polymerization—formation of the macromolecules. While some polymers form spontaneously, they generally fall apart in water. To get around this, some suggest that such chemical synthesis may have taken place in the atmosphere or on clay particles. S. Fox suggests that the proper concentrations of reactants could have occurred in small, hot, isolated pools. He has succeeded in producing polymers he calls thermal proteinoids, which in water form spherical collections called proteinoid microspheres.

Theorists suggest that the first organisms were simple anaerobes, heterotrophs that utilized spontaneously forming monomers for energy and raw materials. Early phototrophs utilized sunlight energy and hydrogen from hydrogen sulfide in forming carbohydrates. Later, the means for extracting hydrogen from the water molecule developed. Oxygen was released as a by-product and began to accumulate, causing the spontaneous breakdown of monomers.

Heterotrophs adapted by evolving the means for detoxifying oxygen, eventually putting it to use in aerobic respiration. Their source of food became the autotrophs, along with other heterotrophs.

KINGDOM MONERA: THE PROKARYOTES

Prokaryotes were the first known forms of life. The oldest fossils, probably cyanobacteria, formed columnar stromatolites.

Bacteria can be heterotrophs or autotrophs. The heterotrophs include decomposers and pathogens (disease producers). Autotrophs include phototrophs and chemotrophs. Phototrophs have the uniquely bacterial pigments bacteriochlorophyll and bacteriorhodopsin.

Bacterial cells are small, and with the exception of thylakoids and mesosomes, membranous organelles are absent. The DNA is circular; the cell walls contain peptidoglycan; and the flagella are tubular and rotating.

Bacteria reproduce through simple fission, which follows replication.

The three most common cell forms are the rodlike bacillus, spherical coccus, and spiral spirillum.

Decomposers are ecologically vital since they recycle mineral nutrients. The food industry cultures certain bacteria in foods for the flavors they provide.

The bacterial agents of most human disease can be controlled, but new pathogens are appearing.

Many cyanobacteria are quite advanced; they produce chlorophyll pigments similar to those in plants and have an extensive, membranous thylakoid. Halobacteria utilize bacteriorhodopsin in a simple chemiosmotic system.

Chemotrophs oxidize inorganic materials in the earth's crust and take part in the nitrogen cycle.

Some are nitrogen fixers, incorporating atmospheric nitrogen into useful compounds and ions.

In newer schemes based on biochemical and molecular data, prokaryotes have been assigned to two kingdoms, Archaebacteria and Eubacteria. They differ in cell wall makeup, plasma membrane lipids, ribosomal RNA, and other factors.

Archaebacteria inhabit acidic hot springs, anaerobic lake-bottom mud, salty ponds, and the gut of many animals.

The Viruses

Viruses consist of a protein coat and a nucleic acid core (the genes). All are unable to reproduce outside the host cell. Many are important disease agents of humans.

The nucleic acid core of viruses may be single- or double-stranded DNA, double- or single-stranded RNA, or small strands of RNA.

Within the host, viruses may enter a lytic cycle in which they reproduce and lyse (destroy) the cell. The host's protein-synthesizing and replicating mechanisms are used by the virus. In the lysogenic cycle, the viral genome is inserted into a host chromosome, and it may pass from generation to generation of cells and individuals in a dormant state. Later, it may enter a lytic cycle.

Retroviruses are RNA viruses that use the enzyme reverse transcriptase to convert their single strand of RNA to a double strand of DNA. The DNA is then inserted into the host's chromosomes. This disrupts many gene functions. A retrovirus causes AIDS.

The *Herpes simplex 2* virus causes genital herpes. The virus infects genital and mouth-lining cells, forming blisters as it reproduces and lyses the cells.

Key Terms

coacervate • **347**
protocell • **347**
stromatolite • **351**
bacillus • **354**
coccus • **354**
spirillum • **354**
decomposer • **354**
pathogen • **356**
halobacteria • **359**
nitrogen fixer • **360**
methanogen • **362**
virus • **362**
lytic cycle • **363**
lysogenic cycle • **364**
retrovirus • **364**
reverse transcriptase • **364**

Review Questions

1. Summarize Haldane's and Oparin's ideas on the spontaneous synthesis of molecular precursors. Why was an absence of oxygen essential?
2. Describe the experiments on coacervates. What is the main problem their with leading to life?
3. Discuss the experiment of Miller and Urey. Include their apparatus, the substances, the energy source, and the results.
4. Summarize what today's atmospheric scientists believe were the gases and energy sources in the primitive earth.
5. Discuss the problem of spontaneous polymer formation. What tends to happen when polymerization occurs in water?
6. Summarize the findings of Sidney Fox. Did he succeed in synthesizing cells? Explain.
7. Briefly summarize the four hypothetical steps that lead from the protocell to more advanced autotrophs and heterotrophs.
8. Why was the evolution of a photosystem that used water for its hydrogen source so significant to life? Was this a sudden development? Explain.
9. What are stromatolites? How do we know they once lived?
10. Prepare a simple illustration of a prokaryote cell. Label the flagellum, cell wall, cell membrane, and any internal structures.
11. How do the following differ between prokaryotes and eukaryotes: flagellum, cell wall, chromosomes, membrane-surrounded organelles?
12. Describe the fission process in bacteria. How does it differ from cell division in most eukaryotes?
13. Prepare simple drawings of the three forms of bacterial cells.
14. What is the single most ecologically important activity of heterotrophic bacteria?
15. List ways of preventing food spoilage by bacteria, and explain how each works.
16. List three activities of bacterial chemotrophs.
17. What changes are taxonomists making in the classification of prokaryotes? List four differences between members of the two new kingdoms.
18. Describe the structure of a virus. Why is it not considered to be a cell?
19. List five important human viral diseases. What is the latest important disease to be added to the list?
20. Into what taxonomic grouping do the viruses fall? Why?
21. Briefly summarize the events of the lytic cycle.
22. Explain how the lysogenic cycle operates.
23. Explain how retroviruses enter a lysogenic cycle. Why is there a greatly renewed interest in the retrovirus?

23

Protists and Fungi

WHAT IS A PROTIST?

PROTOZOANS: ANIMAL-LIKE PROTISTS

Flagellates and Amebas (Phylum Sarcomastigophora)

Sporozoans (Phylum Apicomplexa)

Ciliates (Phylum Ciliophora)

ALGAE: PLANTLIKE PROTISTS

Dinoflagellates (Phylum Pyrrophyta)

Golden-Brown Algae and Diatoms (Phylum Chrysophyta)

Euglenoids (Phylum Euglenophyta)

Red Algae (Phylum Rhodophyta)

Brown Algae (Phylum Phaeophyta)

Green Algae (Phylum Chlorophyta)

SLIME MOLDS AND WATER MOLDS: FUNGUSLIKE PROTISTS

Slime Molds (Phyla Acrasiomycota and Myxomycota)

Water Molds (Phylum Oomycota)

FUNGI

Bread Molds (Phylum Zygomycota)

Sac Fungi (Phylum Ascomycota)

Club Fungi (Phylum Basidiomycota)

Fungi Imperfecti (Phylum Deuteromycota)

LICHENS

As we leave the realm of organized molecules, viruses, and the *prokaryotes, we move to the other living things, the eukaryotes. We find a great deal of diversity here also, with perhaps the greatest diversity in that grab bag called protists—all the living things that aren't plants or animals or fungi. Later on, as you might guess from the chapter title, we consider fungi—and it is in these groups that we first encounter the importance of life cycles in evolutionary history. In our walk through the various phyla we will consider the special traits of each group. (You are probably anxious to learn about slime molds.)*

The eukaryotes are a diverse lot. They range from fungal spores to Washington lobbyists. Perhaps the oldest living forms among the eukaryotes are the protists and fungi, the groups we will consider here. We will find that these venerable species are fascinating, indeed. For example, some emit tiny bursts of light, setting aglow warm tropical seas. Others poison those waters and kill marine creatures by the millions. Yet others are tiny filamentous predators that can ensnare worms and feed on their bodies. And then we will see species that form great undersea forests.

Some people go beyond studying the protists for their own sake, and try to use the protists to gain a clearer picture of how life may have evolved. For example, they look at how the simplest eukaryotes arose from distant prokaryote ancestors. (A provoking theory on the emergence of eukaryotes is presented in Essay 23.1.) Further, protists may tell us how multicellularity arose, since this group harbors simple single-celled eukaryotes, others showing a hint of multicellularity, and still others that are *clearly* multicellular. The advent of multicellularity is important, for it marks the setting of the evolutionary stage for the specializations necessary for species to explore new resources and survive in an increasingly complex and competitive world.

What Is a Protist?

The kingdom Protista contains the eukaryotic organisms that are not plants, animals, or fungi. So we know what they *aren't*, but describing what they *are* takes a bit of doing, because there is no simple definition of a protist. In essence, protists are a grab bag of disparate creatures. Whereas many protists exist as single organisms, others form loose associations called *aggregates* in which each maintains its independence from the others. In some species of protists, groups of individuals show some degree of interdependence, and these are called *colonies*. By contrast, the cells in multicellular species have reached a state of total interdependence and, for the most part, cannot function separately.

The protists can be divided into three groups. The *protozoans,* the *algae,* and the *funguslike protists* (the water molds and slime molds). The most diverse of these are the algae, since they range from the green, single-celled organisms to giant ocean kelp over 50 meters in length (included by some people in the plant kingdom).

Protozoans: Animal-like Protists

Proto means "first" and *zoa* refers to "animals"—and indeed many protozoans are clearly similar to animals. For example, they are heterotrophic, which means they derive their energy and carbon by feeding on other organisms or their products. Protozoans require amino acids, carbohydrates, lipids, and vitamins in their diet, much as we do. Further, like animals, they show a wide range of feeding specializations. Many are parasites, some are simple decomposers of dead organic material, and others actively capture and digest living prey.

The 65,000 known species of protozoans have been assigned into seven phyla, but we will focus on the three that contain the most numerous and best-known species. Phylum **Sarcomastigophora** includes the flagellates and amebas; Phylum **Apicomplexa**

ESSAY 23.1

The Serial Endosymbiosis Hypothesis

The evolutionary transition from the prokaryotic cell to the much more complex and elaborate eukaryotic cell has perplexed biologists for some time. One hypothesis proposes that eukaryotic organelles evolved when infoldings of the cell membrane pinched off and enclosed various cellular functions. Hypotheses such as this often seem reasonable enough, but they simply can't be easily tested.

In 1967, a totally different hypothesis was proposed by Lynn Margulis of the University of Massachusetts at Amherst. Using information from a wide variety of fields (a trait said to be characteristic of genius), Margulis concluded that the primitive eukaryotic cell developed in at least three separate stages that involved the union of four different prokaryotic lines. Her conclusion has become known as *serial endosymbiosis*. Margulis proposed that mitochondria, which are found in all eukaryotic cells, are the descendants of once free-living bacteria that long ago formed a mutualistic (mutually beneficial) relationship with a host cell. She also proposed that chloroplasts and cilia are the descendants of other bacteria that also entered the eukaryotic cell in what began as a mutualistic association.

Margulis's symbiosis hypothesis was testable: Specifically, it allowed her to predict that mitochondria, chloroplasts, and cilia should be biochemically and genetically more similar to certain free-living prokaryotes than they are to other parts of the eukaryotic cell. These predictions have held up remarkably well for mitochondria and for chloroplasts, but not for eukaryotic flagella. So far there has been no way to equate the ro-

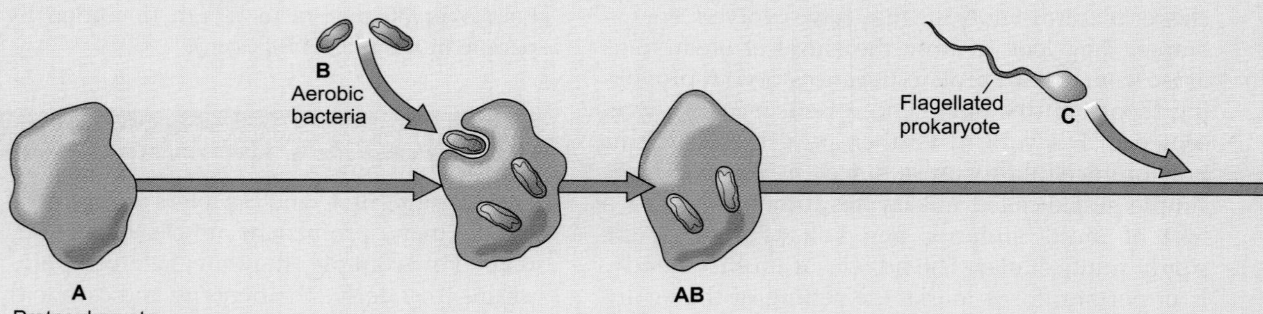

Two hypothetical prokaryotes, now extinct, included a protoeukaryote (line A) which was a simple anaerobe with the ability to phagocytize its food. One organism it fed upon was an aerobic bacterium from line B.

Eventually, line B cells became incorporated, and a symbiotic relationship developed. They became interdependent, with the mitochondria-like bacteria carrying an aerobic respiration within the line A cells. The line B cells lost their ability to live outside their hosts. As the host reproduced, the line B cells did likewise, just as mitochondria do in eukaryotic cells today.

The new AB symbiont now took a step closer to a eukaryotic status by incorporating into its cytoplasm a cell from line C, a flagellated prokaryote.

tating prokaryote flagellum with the undulating, microtubular eukaryote counterpart.

Margulis's case seems strongest for the origin of chloroplasts. Chloroplasts arise only by division of other chloroplasts. Both chloroplast and mitochondrial DNA are susceptible to certain antibiotics that do not affect the eukaryote nucleus —further evidence of similarity between the inclusions and prokaryotes. When the chloroplasts of *Euglena* have been destroyed by such antibiotics, the parent cell line can be kept growing indefinitely in a nutrient broth, but the chloroplasts never reappear.

The case for the bacterial origin of mitochondria is almost as good. True, mitochondria have internal membranes (cristae) unlike anything seen in bacteria, and they seem to have very few functional genes. But at least one mitochondrial protein, cytochrome *c*, is recognizably similar in shape and sequence—presumably by descent —to the cytochromes of certain bacteria, especially the photosynthetic nonsulfur purple bacteria. It now seems likely that the original mitochondrial symbiont was a photosynthetic symbiont, and that it later became restricted to respiratory roles.

A lightweight ribosomal RNA is found in mitochondria, chloroplasts, eukaryote cytoplasm, bacteria, and cyanobacteria (but not in cilia): A computer analysis of such RNA sequences produces exactly the relationships that are uniquely predicted by Margulis's hypothesis. The symbiotic origins of mitochondria and chloroplasts are now as thoroughly established as any "fact" in biology today.

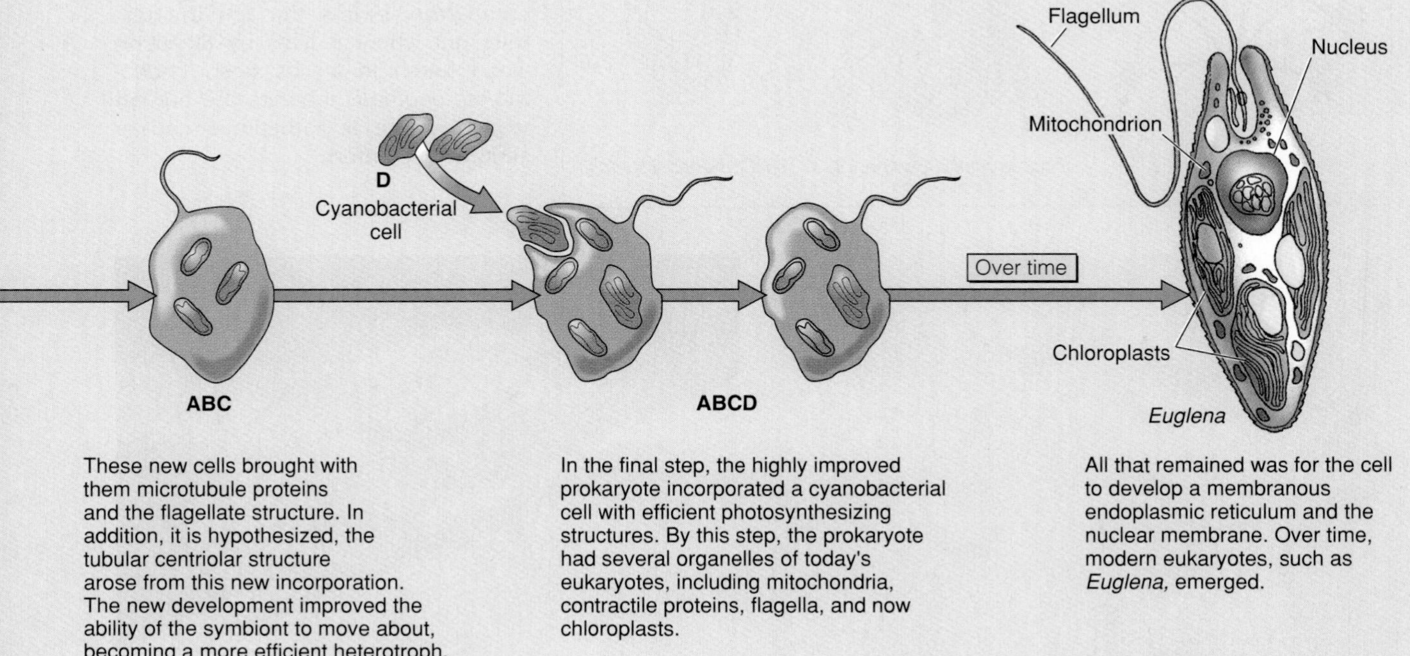

These new cells brought with them microtubule proteins and the flagellate structure. In addition, it is hypothesized, the tubular centriolar structure arose from this new incorporation. The new development improved the ability of the symbiont to move about, becoming a more efficient heterotroph.

In the final step, the highly improved prokaryote incorporated a cyanobacterial cell with efficient photosynthesizing structures. By this step, the prokaryote had several organelles of today's eukaryotes, including mitochondria, contractile proteins, flagella, and now chloroplasts.

All that remained was for the cell to develop a membranous endoplasmic reticulum and the nuclear membrane. Over time, modern eukaryotes, such as *Euglena,* emerged.

contains the nonmotile sporozoans, most of them parasites; and Phylum **Ciliophora** contains the active, fast-moving ciliates.

Flagellates and Amebas (Phylum Sarcomastigophora)

Flagellates (Subphylum Mastigophora). *Flagellates* have one or more whiplike, undulating flagella, and may be free-living or parasitic. Flagella occur singly, in pairs, or in greater numbers, and they can either push or pull the cell about or help the cell acquire food (Figure 23.1). As a rule, feeding is through either simple absorption or phagocytosis (see Chapter 5).

Like so many protists, flagellates reproduce asexually, employing mitosis and cell division through animal-like cleavage. Sexual reproduction occurs, but it is apparently infrequent. Curiously, sex in flagellates does not involve gametes; two flagellates simply fuse to form a zygote. Further, such fusions are not preceded by meiosis; there is no need for chromosome reduction beforehand, since flagellates, like so many protists, are haploid throughout most of their lives. As you would expect, the fusion of two haploid cells produces a diploid chromosome number as the genetic material from the two parents is allowed to commingle. The newly formed zygote soon undergoes meiosis, forming four haploid cells. The life cycle of flagellates is referred to as a *zygotic life cycle.* It contrasts with the *gametic life cycles* of animals, in which, except for the haploid sperm and eggs, cells are diploid throughout life (see Essay 23.2).

Some parasitic flagellate protozoans cause severe diseases in humans. Among them are the *trypanosomes,* members of the genus *Trypanosoma.* A trypanosome can be identified by an undulating flap—actually a modified flagellum—that runs the length of its spindly body (Figure 23.2). One particularly dangerous form is *Trypanosoma gambiense,* the cause of African sleeping sickness. The bite of the infamous tsetse fly injects the parasite into the bloodstream of a mammalian host. Multiplying trypanosomes eventually enter the host's central nervous system, whereupon they may bring on coma and death.

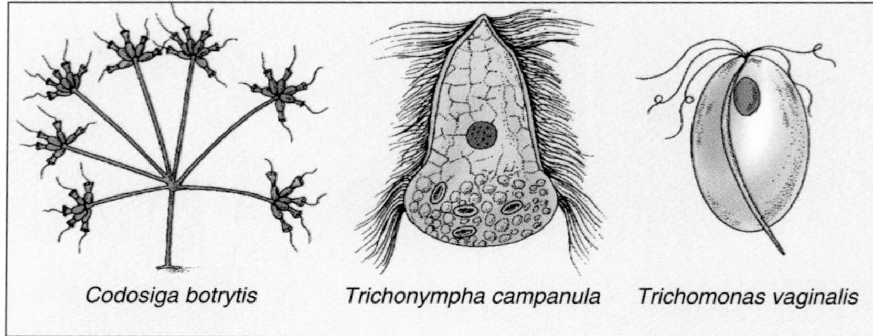

FIGURE 23.1 **FLAGELLATES.**
Codosiga botrytis forms a colony of stationary, stalked cells that use their flagella for feeding. *Trichonympha campanula* swims through the termite gut where it lives by digesting wood taken in by its host. *Trichomonas vaginalis* inhabits the human vagina where it sometimes causes serious discomfort.

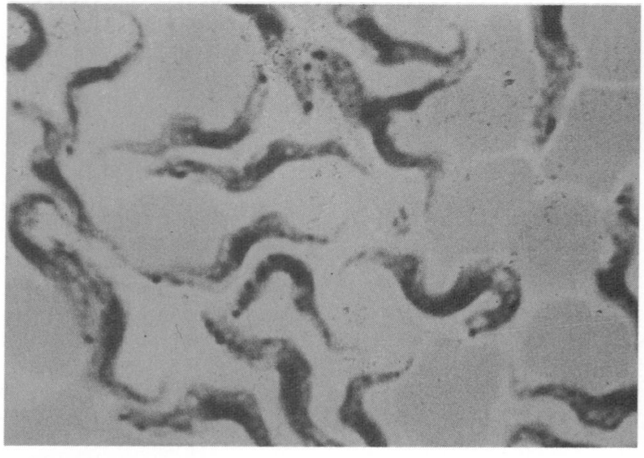

FIGURE 23.2 **A TRYPANOSOME AND ITS CARRIER.**
A complex flagellum attached to an undulating membrane characterizes the flagellate *Trypanosoma gambiense*, the agent of African sleeping sickness. It is spread by the bite of a tiny tsetse fly, a second host.

Amebas, Heliozoans, Radiolarians, and Foraminiferans (Subphylum Sarcodina). Amebas are fascinating creatures. Some of them are quite large as protists go. A few giants, such as *Pelomyxa,* may reach 5 mm (3/16 in.) in length, making them easily visible to the unaided eye (Figure 23.3a). (Their size may explain why they require numerous nuclei for protein synthesis.) Most amebas are free-living predators that feed by surrounding and engulfing prey through phagocytosis and digesting it in food vacuoles. Other amebas are dangerous parasites. *Entamoeba histolytica,* for example, infects cells lining the human colon, causing amebic dysentery. Throughout history, many an invading army has had to hastily retreat from areas harboring *E. histolytica.*

Typically, amebas move about and capture prey through the use of *pseudopods* ("false feet"). Pseudopods are temporary, lobelike extensions of the cytoplasm that involve the reorienting of microfilaments in the cytoskeleton, thus softening the cytoplasm. When this happens, a pseudopod can extend in any direction, and the rest of the cytoplasm then flows into it. Obviously, there isn't much point in describing the "shape" of an ameba.

Other Sarcodina are more structured. The dramatically beautiful *heliozoans* or "sun animals" (Figure 23.3b) live in fresh water and are partially encased in glassy capsules of silicon dioxide. The radiating rays consist of lengthy *axopods,* pseudopods of motile cytoplasm, stiffened with long, straight bundles of microtubules. The microtubule bundles can quickly disassemble and re-form later as they are needed. *Radiolarians* of the open sea (Figure 23.3c) have a similar body plan, and their glassy corpses litter the ageless ocean floor, forming deposits known as radiolarian ooze.

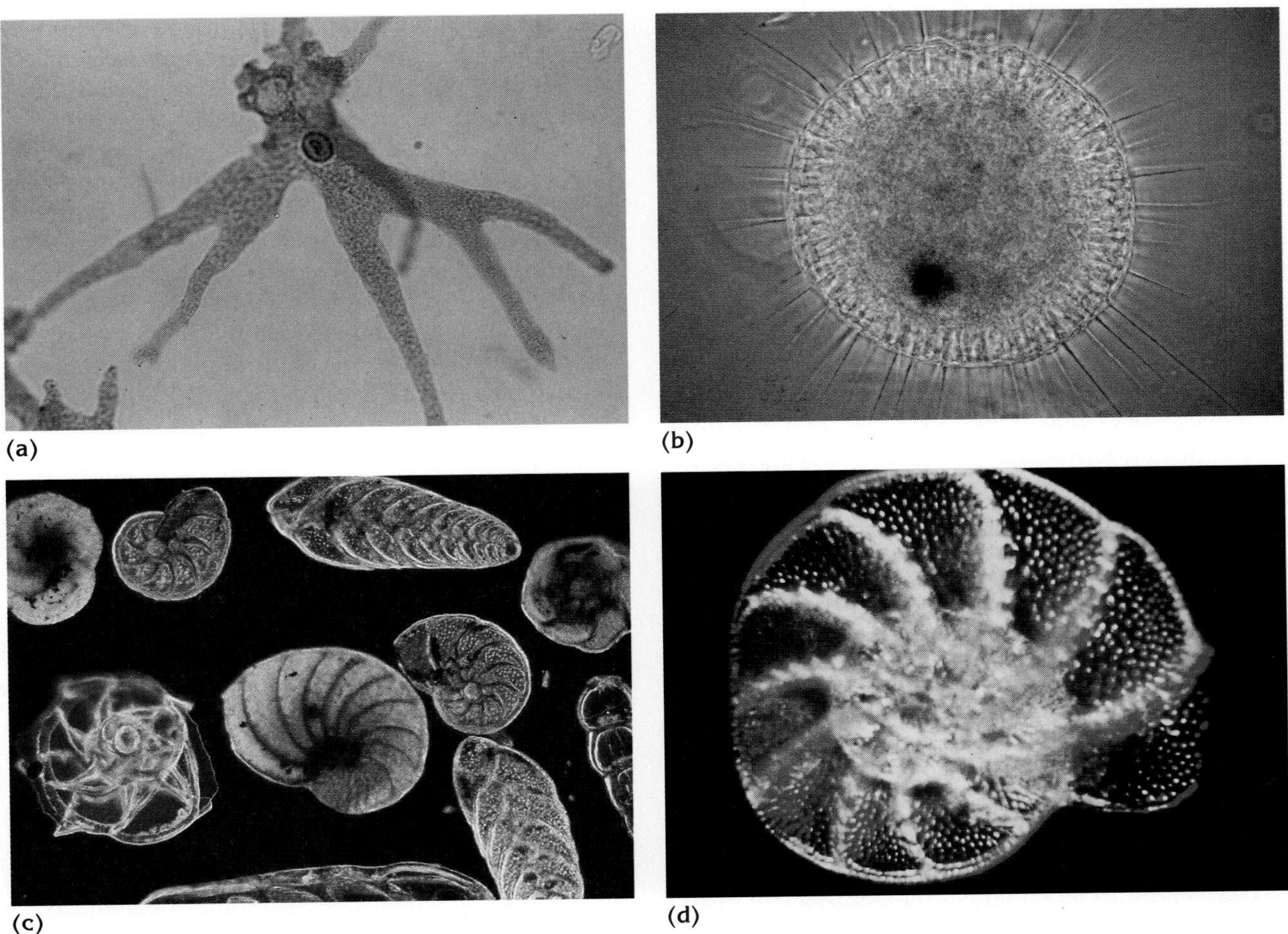

FIGURE 23.3 AMEBAS, HELIOZOANS, RADIOLARIANS, AND FORAMINIFERANS.
(a) *Ameba* can reach several millimeters in diameter. **(b)** The heliozoan ameba *Actinosphaerium* has numerous spiny axopods, each with an overlayer of moving cytoplasm that actively engages in phagocytosis. **(c)** Radiolarians are often spherical with highly sculpted glassy skeletons. Radiolarians, along with the coiled foraminiferans **(d)**, contribute significantly to ocean bottom sediments.

Foraminiferans belong to another group of planktonic ameboid protozoans. They produce chalky, not glassy, skeletons, sometimes in shapes remarkably reminiscent of the shells of garden snails (Figure 23.3d). England's famed White Cliffs of Dover are formed largely from ancient foraminiferan skeletons.

Sporozoans (Phylum Apicomplexa)

Sporozoans have three major characteristics: (1) they form spores, (2) they are parasites, and (3) except for gametes, they have no means of movement. Many sporozoans have complex life cycles, reproducing in both sexual and asexual phases, often completed in different hosts.

One group of sporozoans is of great importance to humans since it is the agent of malaria, a disease that has been called humanity's greatest curse, and currently a growing threat as new forms appear that are resistant to traditional chloroquine treatments. Malaria is spread from person to person by the female *Anopheles* mosquito, a tiny insect harboring the even tinier sporozoan parasite *Plasmodium vivax.* (The female *Anopheles* is recognized by her habit of "standing on her head," hind legs in the air, as she draws blood.) Actually, there are some 50 species of *Plasmodium,* all carried by mosquitoes and capable of infecting vertebrates, primarily birds.

The life cycle of *P. vivax* and the course of malaria are summarized in Figure 23.4. Note particularly the organism's asexual stages in the malaria victim and the events that bring on cycles of fever and chills. Sexual reproduction, we see, takes place in the mosquito. (For a strange twist involving malaria and the sickle-cell allele, see Essay 15.1.)

Ciliates (Phylum Ciliophora)

Among the most complex cells on earth are those of the *ciliates* (Figure 23.5). Ciliates range in size from about 10 µm to 3 mm, which is about the same relative difference that exists between a mouse and a blue whale. The ciliates' most obvious characteristic is that they are covered with cilia, in longitudinal or spiral rows. Like flagella, cilia can move; but they are so numerous that it wouldn't do for each cilium to beat wildly, independent of the rest. Instead, their movement is highly coordinated by a network of connecting fibers that run beneath the surface of the cell. Other ciliates are sessile (fixed in place); their cilia are used only for feeding.

Many ciliates feed by using cilia to sweep bacteria and organic debris into the *cytostome,* a mouthlike opening where phagocytosis occurs. The food then enters a *food vacuole* into which enzymes are secreted, and, as the vacuole breaks away and moves through the cytoplasm, the digested food is distributed (Figure 23.6). Undigested residues are discharged from the cell through exocytosis, which occurs at another specific location called the *cytopyge.* (The specifics of phagocytosis and exocytosis are discussed in Chapter 5.)

Water regulation in freshwater ciliates and many other protozoans is carried out by *contractile vacuoles.* These specialized, permanently located organelles take in water from the cytoplasm and squeeze it out through the surrounding membrane. Contractile vacuoles fill and empty periodically, using ATP to power the contractions of surrounding microfilaments that do the squeezing.

Ciliates are unusual protists in that they maintain their DNA in two kinds of nuclei. A germlike DNA, set aside for replication, is kept in one or more tiny *micronuclei.* Micronuclei produce larger *macronuclei,* metabolically active bodies that specialize in RNA transcription and protein synthesis.

During asexual reproduction, the micronuclei undergo replication and mitosis, the macronucleus pinches in two, and cytokinesis occurs. Typically, sexual reproduction in ciliates occurs through **conjugation,** a fusion of two individuals of compatible *mating types* (there are no males and females, as such). Following cell fusion, the micronuclei undergo meiosis, and some of the haploid products are exchanged. As the haploid nuclei join, the diploid state in each partner is restored.

Algae: Plantlike Protists

There are about 18,000 named species of *algae.* The oceans are laced with vast beds of seaweeds and kelps, and astronomical numbers of simpler algae congregate in drifting populations referred to as **phytoplankton.** Other algal protists live as symbionts within the cells of corals, giant clams, sea slugs, and protozoans.

In biological relationships, symbiosis simply refers to living in an intimate association, usually one organism within another. Symbiosis includes *mutualism,* in which both symbionts benefit; *commensalism,* in which one partner benefits while the other is unaffected; and *parasitism,* in which one partner benefits while the other is harmed.

Altogether, marine algae fix more carbon by photosynthesis than do all of the land plants and other algae combined. But not all algal protists are marine;

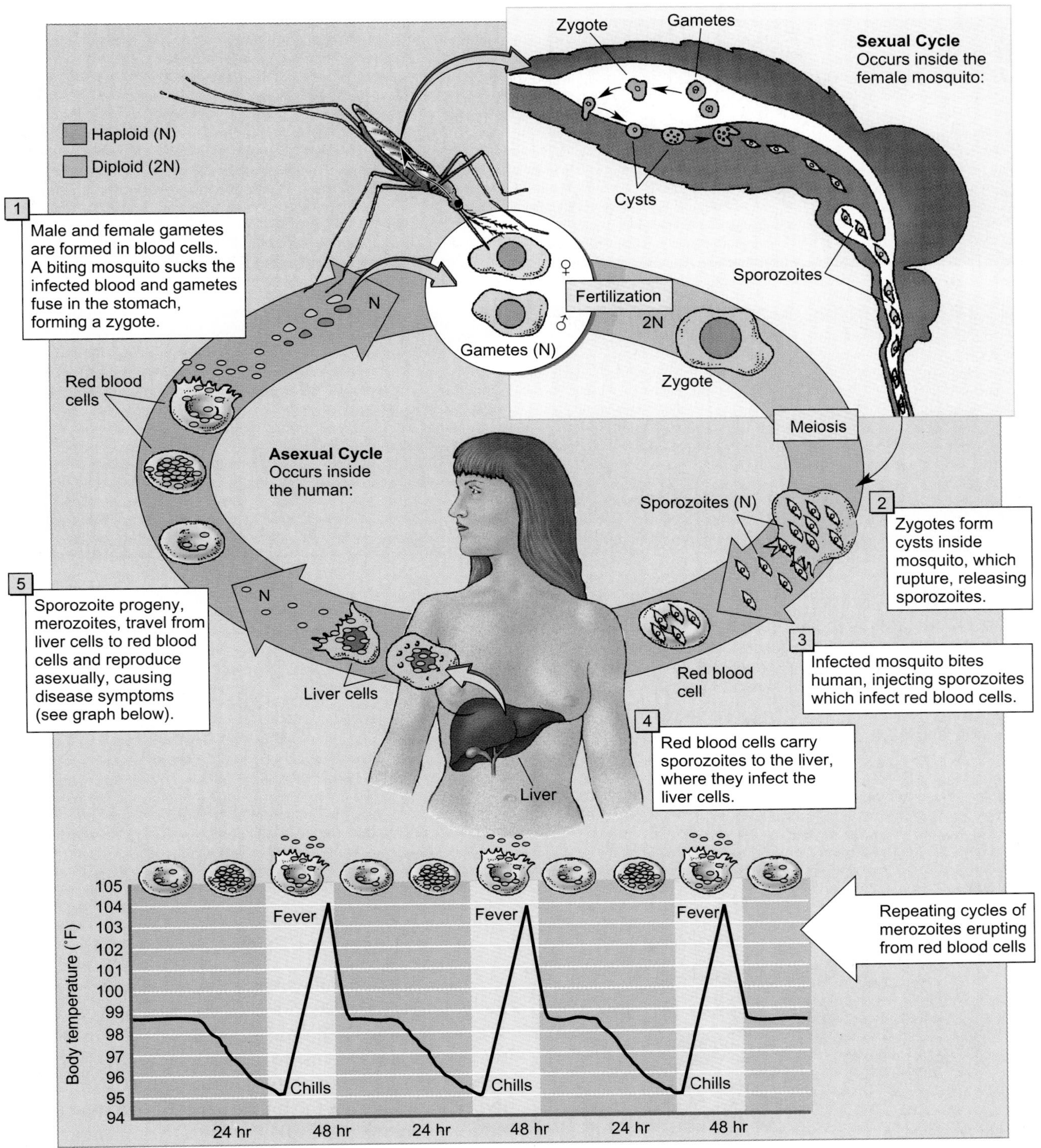

Zygote
Gametes
Sexual Cycle
Occurs inside the female mosquito:
Cysts
Haploid (N)
Diploid (2N)
1
Male and female gametes are formed in blood cells. A biting mosquito sucks the infected blood and gametes fuse in the stomach, forming a zygote.
Sporozoites
N
Fertilization
2N
Gametes (N)
Zygote
Red blood cells
Meiosis
Asexual Cycle
Occurs inside the human:
Sporozoites (N)
2
Zygotes form cysts inside mosquito, which rupture, releasing sporozoites.
5
Sporozoite progeny, merozoites, travel from liver cells to red blood cells and reproduce asexually, causing disease symptoms (see graph below).
N
3
Infected mosquito bites human, injecting sporozoites which infect red blood cells.
Red blood cell
Liver cells
4
Red blood cells carry sporozoites to the liver, where they infect the liver cells.
Liver
Body temperature (°F)
105
104
103
102
101
100
99
98
97
96
95
94
Fever
Fever
Fever
Chills
Chills
Chills
24 hr
48 hr
24 hr
48 hr
24 hr
48 hr
Repeating cycles of merozoites erupting from red blood cells

FIGURE 23.4 THE MALARIAL CYCLE.
Malaria is caused by a protistan called a *Plasmodium* that develops a form called a sporozoite within the mosquito. When Plasmodium is injected into the human body, it enters the liver and multiplies repeatedly there, releasing the rapidly increasing agents into the bloodstream, where they enter red blood cells. As agents multiply within the cells, they are sporadically released, thereby infecting new red blood cells. These releases mark times of high fever, followed by chills. Some of these released agents occasionally produce male and female forms. When a mosquito draws blood from an infected person, it may consume these male and female gametes, which join in the mosquito's body to form new sporozoites.

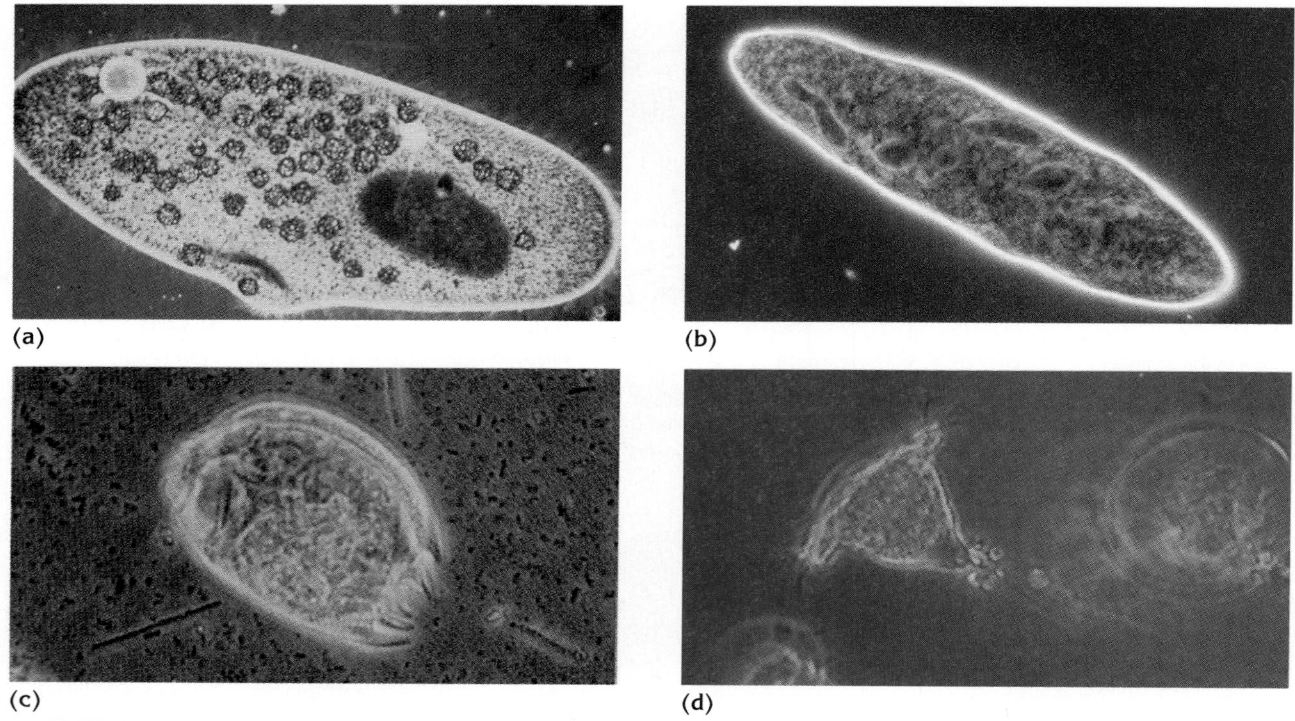

FIGURE 23.5 **CILIATES.**
Ciliates range greatly in size from that of **(a)** *Diplodinium dentatum,* a commensal in the cow's stomach (20 to 40 μm), to **(b)** *Spirostomum ambiguum,* a protozoan giant whose length is measured in millimeters (3 mm or 3,000 μm). (This size difference is equivalent to the relative size of mice and blue whales.) Of interest is **(c)** the funnel-shaped *Vorticella campanula,* which pops up and down on its contractile stalk, using its cilia for sweeping in food particles. **(d)** *Paramecium multimicronucleatum,* as its polysyllabic name implies, has numerous micronuclei.

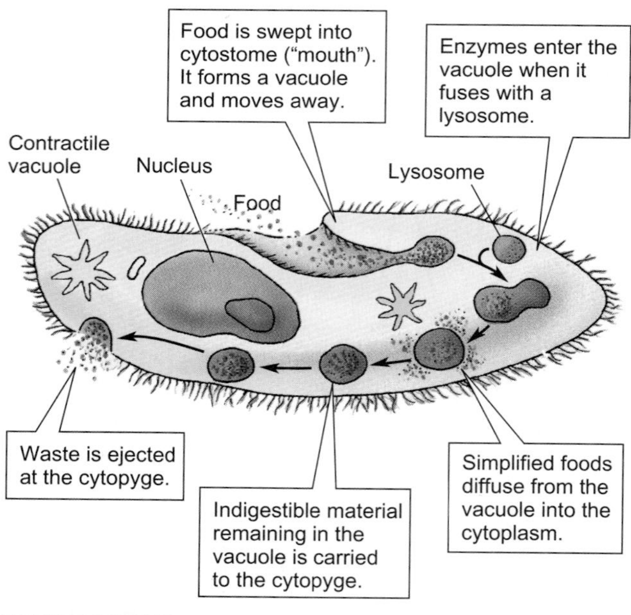

Food is swept into cytostome ("mouth"). It forms a vacuole and moves away.
Enzymes enter the vacuole when it fuses with a lysosome.
Contractile vacuole
Nucleus
Food
Lysosome
Waste is ejected at the cytopyge.
Simplified foods diffuse from the vacuole into the cytoplasm.
Indigestible material remaining in the vacuole is carried to the cytopyge.

FIGURE 23.6 **DIGESTION IN CILIATES.**
In *Paramecium,* as in many protozoans, digestion occurs in food vacuoles.

many live in fresh water, and others occur on land, some of the latter in partnerships with terrestrial organisms, where both partners benefit.

DINOFLAGELLATES (PHYLUM PYRROPHYTA)

Visitors to tropical waters are often surprised and delighted as they are rowed in the evening between ship and shore. Each time the paddle slides into the water, the sea seems to explode with iridescent light. Even the wake of the canoe is aglow. The tiny, twinkling lights are the magic of minute living creatures, the microscopic *dinoflagellates,* or fire algae.

However, when conditions are favorable, their numbers can increase explosively, and these same little creatures can be responsible for another, equally dramatic, and far more dangerous phenomenon: the *red tide.* Teeming trillions of dinoflagellates turn the sea a bloody color, and fish, subjected to deadly toxins, die by the thousands. Clams, oysters, and mussels survive, but accumulate the poisons as they filter-feed on the dinoflagellates. At these times—generally, from May through August—the poisoned mollusks can prove fatal to humans who eat them. (An old axiom states that you should eat shellfish only in months with an "r" in them.)

Dinoflagellates are flagellated, photosynthetic protists, covered with plates of cellulose and quite unlike anything else alive. Each typically has two flagella, one trailing and one confined to a groove that encircles its waist (Figure 23.7). Some dinoflagellates live most of their lives as photosynthetic symbionts in the cells of marine protozoa or animals, often imparting a peculiar green color to their larger symbiotic partners.

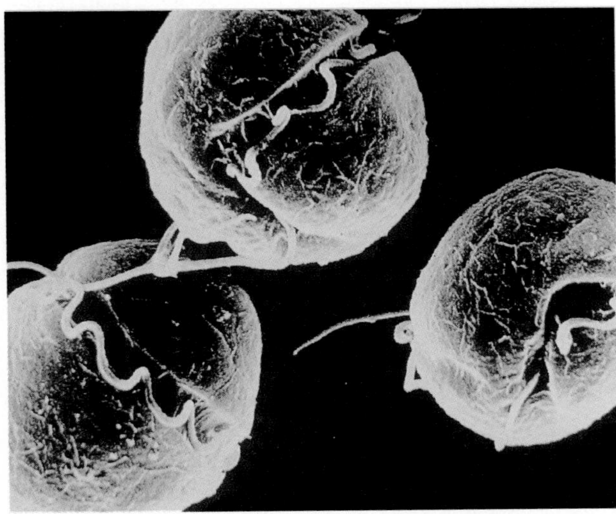

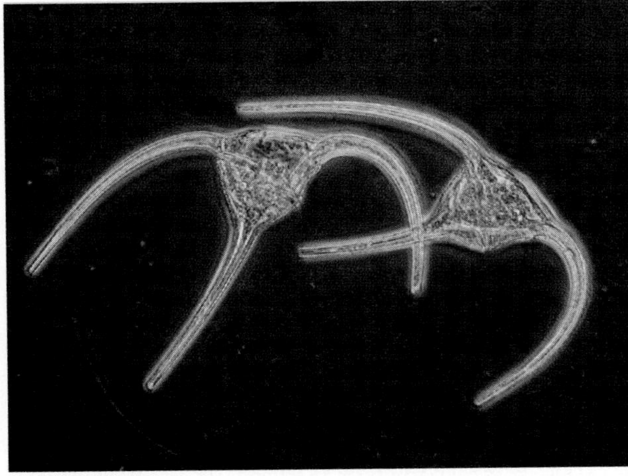

FIGURE 23.7 DINOFLAGELLATES.
Among the marine phytoplankton are the dinoflagellates, which are primitive, single-celled photosynthesizers.

Golden-Brown Algae and Diatoms (Phylum Chrysophyta)

The **golden-brown algae** are single-celled and colonial algae of both marine and fresh waters. You can imagine what color they are, color derived chiefly from carotenoid pigments (accessory photosynthetic pigments; see Chapter 7). As photosynthesizers, the golden-brown algae join the diatoms in helping to provide the energy base of marine food chains.

The **diatoms** produce cell walls of glass (silicon dioxide). Over time, their numbers have been so immense that today we routinely mine rich diatomaceous earth deposits for use in filters and cleaning agents. The glassy walls of diatoms come in a variety of patterns, the result of numerous tiny perforations needed for permitting gases to enter and leave the cell (even glass houses need windows; Figure 23.8). The glassy walls are like tiny pillboxes, each with a lower lid and an overlapping upper lid.

The glassy walls form excellent fossils, so we know that diatoms appeared in the late Mesozoic era but didn't really become numerous until the beginning of the Cenozoic era (65 million years ago).

Euglenoids (Phylum Euglenophyta)

Euglenoids are single-celled, flagellated algae that live in fresh water, often in nutrient-enriched or polluted waters where their numbers during algal "blooms" can turn the surface bright green. *Euglena gracilis* (Figure 23.9) is probably the best-known euglenoid. However, *Euglena,* as beautiful as it is, defies and confounds the taxonomist. With its flexible body, active flagella, and photoreceptor (the "eyespot"), it may well seem to be a protozoan flagellate. However, also included among *Euglena*'s organelles is a well-formed chloroplast and numerous starch storage bodies—both traits characteristic of algae. Studying its life cycle isn't much help because it reproduces asexually through mitosis and cell division, and sexual reproduction hasn't been observed. Thus we find *Euglena* in both the zoologist's and the botanist's taxonomic schemes.

Red Algae (Phylum Rhodophyta)

The 4,000 or so species of *red algae* are primarily seaweeds, although a few live on the land and in fresh water. Seaweeds attach to the rocky sea floor by fingerlike structures known as *holdfasts.* The algae's name can be deceptive, because whereas some

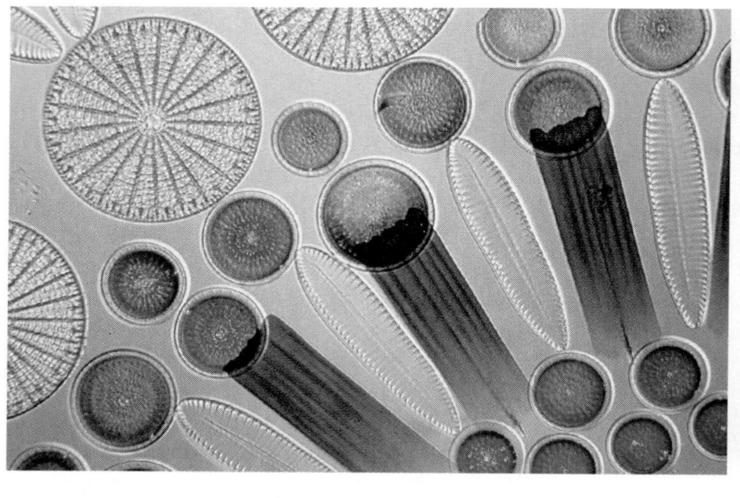

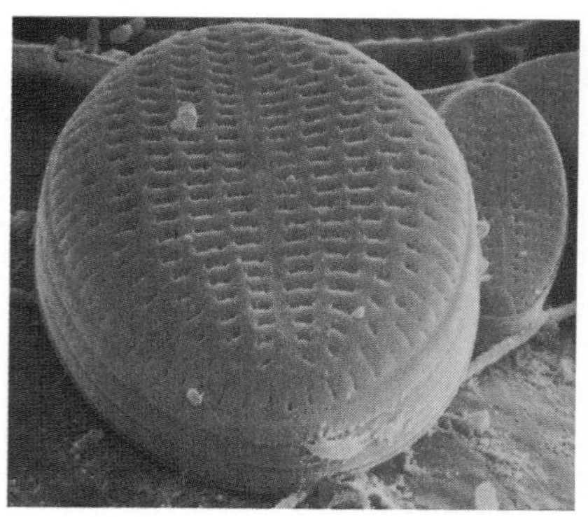
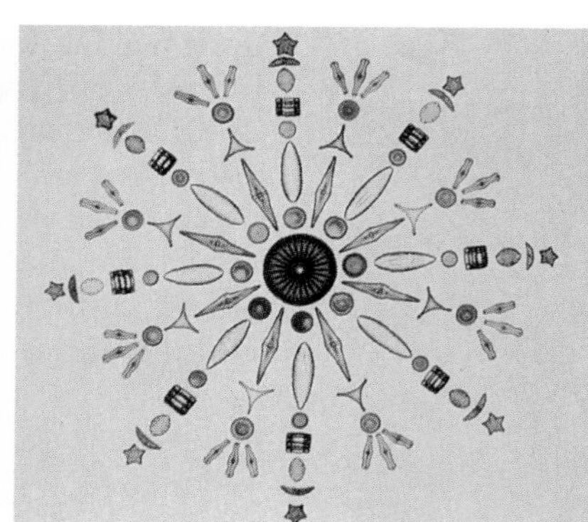
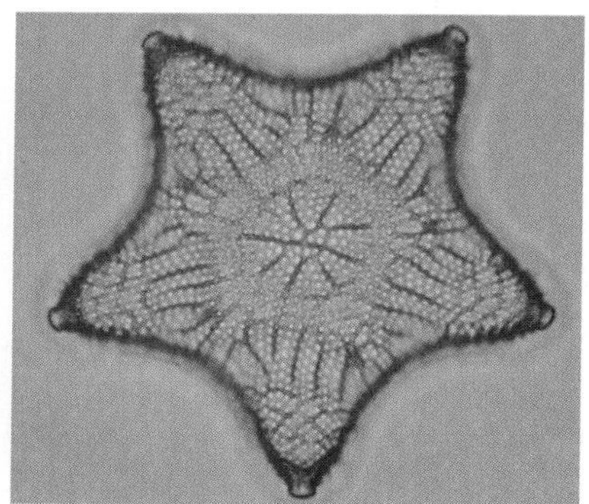

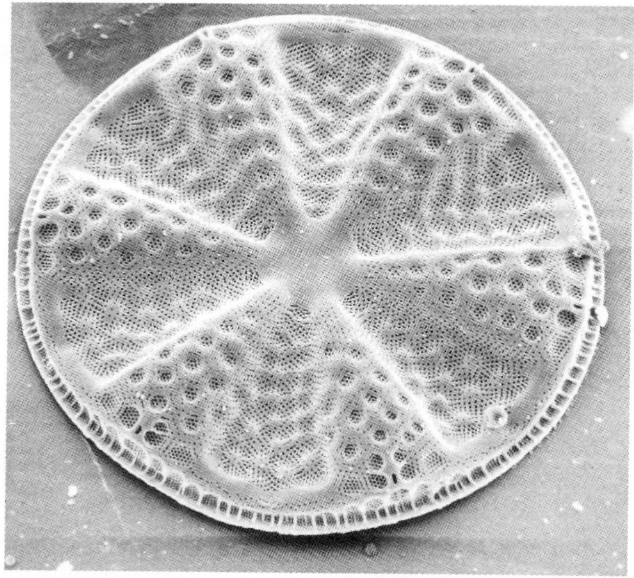

FIGURE 23.8 DIATOMS.
The most spectacular members of the Chrysophyta are the 5,000 or so species of diatoms. Their glassy walls contain numerous types of finely etched patterns so regular and clear that mounted specimens are used to test the quality of microscope lenses.

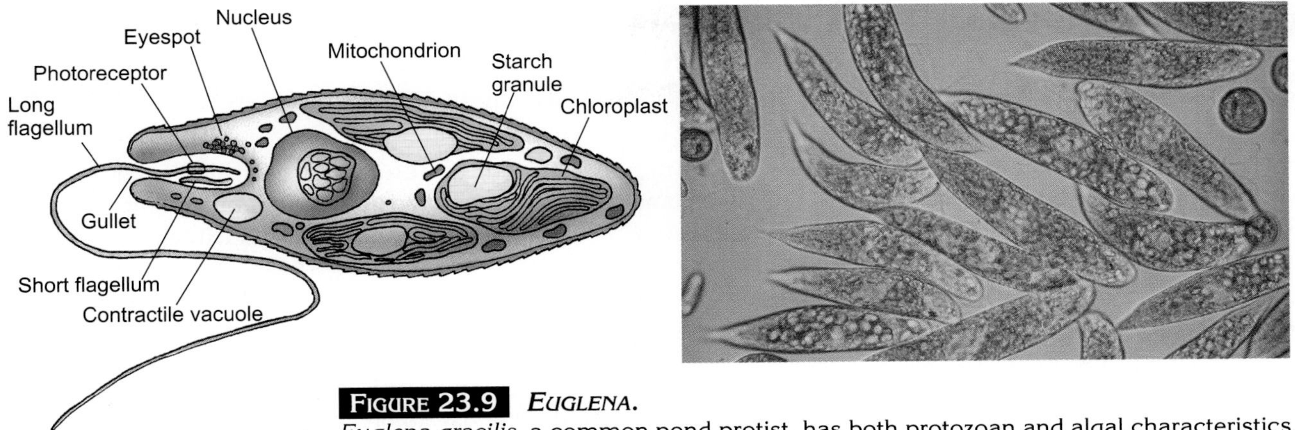

FIGURE 23.9 *EUGLENA.*
Euglena gracilis, a common pond protist, has both protozoan and algal characteristics.

red algae are actually red, others are green, black, blue, or violet. Most are small and frilly, with lengths up to perhaps a meter (Figure 23.10). *Sushi* lovers are familiar with the thin, greenish-black edible red alga that is used to wrap the raw fish delicacy.

As in plants, the storage polysaccharide of many red algae is starch. Red algae are commercially important as a vegetable in the Asian diet and as a source of the polysaccharides *agar* and *carrageenan.* Agar is used in cosmetics, medicine capsules, and in agar-agar, the jellylike material upon which microbiologists grow bacteria and fungi. Carrageenan is used as a smoothing agent in paints and cosmetics, and it gives chocolate-flavored dairy drinks their fake creaminess.

When red algae reproduce, they form nonmotile, drifting sperm rather than the usual flagellated form of other algae. The life cycle is plantlike, or sporic. That is, a diploid, spore-producing **sporophyte** phase alternates with a haploid, gamete-producing **gametophyte** phase. The two-part life cycle is described as an **alternation of generations,** a term we will be saying more about shortly (see also Essay 23.2).

BROWN ALGAE (PHYLUM PHAEOPHYTA)

The *brown algae* include the large kelps, some of the seaweeds, and a few microscopic forms. They owe their color to the brown pigment *fucoxanthin.*

Most of the 1,000 or so named species of brown algae live in cold coastal waters, but one genus, *Sargassum,* grows in shallow tropical and subtropical waters. Storms regularly break *Sargassum* bodies from their holdfasts, and the broken plants are washed to sea, where they continue to photosynthesize and grow, although they can no longer successfully reproduce. In the Middle Atlantic, vast tangled masses of *Sargassum* form the Sargasso Sea, which has spawned many salty legends about trapped ships and sea monsters.

The giant kelps can grow over 50 m (164 ft) in length (Figure 23.11). Their large flattened, leaflike *blades* can be close to the sunlit surface, while their tangled, elaborate holdfasts anchor them far below. Connecting the two ends of the plant are stemlike *stipes,* and the blades are kept near the surface by the buoyancy of hollow *floats.* Although they are listed among the protists, kelps actually have evolved food-conducting tissues very similar to those of higher

FIGURE 23.10 **RED ALGAE.**
Red algae are most common in warm marine waters, where they attach to rocks on the seabed. The branched body of *Polysiphonia* is frilly and delicate.

ESSAY 23.2 Diversity in Life Cycles

There are three basic themes in the life cycles of eukaryotes. Of course, we are most familiar with our own, so we will review it first. The other two may seem a bit strange by comparison.

(a) THE GAMETIC LIFE CYCLE. The gametic life cycle occurs in some protists and all animals, including humans. It is quite direct and uncomplicated. In the gametic cycle, the cells are diploid except for the gametes—the sperm and egg cells or their counterparts (isogametes)—which are haploid. In animals, gametes are produced as certain diploid cells undergo meiosis. The diploid state is restored as gametes unite in fertilization. While this is a conceptually comfortable cycle, many of the world's creatures have evolved quite different systems. The next theme, in fact, is pretty much the opposite of the animal life cycle.

(b) THE ZYGOTIC LIFE CYCLE. The zygotic life cycle evolved much earlier than the gametic cycle. In fact, it is the most primitive of all the cycles. We find it to be the reproductive theme in some of the green algae, some protozoa, and the fungi. In these species nearly all of the individual's life is spent in the haploid state. At some point, certain haploid cells enter conjugation, their nuclei fusing through this primitive form of fertilization, thus forming diploid zygotes. The resulting zygote may enter an extended period of dormancy, but it is more likely that meiosis will immediately follow fertilization, restoring the usual haploid state.

(c) THE SPORIC LIFE CYCLE. The sporic life cycle can be quite varied, so we'll concentrate on its characteristics in plants. Meiosis in plants does not directly result in the formation of gametes. Instead haploid *spores* are formed. If conditions are suitable, the spores grow and divide, giving rise to a specialized multicellular phase in which gametes are produced. Because this phase is already haploid, the gametes are produced by simple mitosis. This phase of the life cycle is sensibly called the *gametophyte* ("gamete plant").

The fusion of gametes at fertilization produces a diploid zygote. Through growth and cell division, the zygote gives rise to a diploid, multicellular generation, which will again enter meiosis and give rise to more spores. The diploid, spore-producing phase is aptly called the *sporophyte* ("spore plant"). For most plants this is the prominent phase, the one most easily found. Thus the sporophyte includes the plant body, the familiar roots, stems, and leaves. The gametophyte is tucked away in the flower or cone. Biologists call this alternation between the sporophyte and gametophyte phases an **alternation of generations.**

In its more primitive state (some algae and plants such as ferns), the alternating gametophyte and sporophyte phases occur in physically separated individuals. In a few instances, we find a highly dominating gametophyte and a very reduced sporophyte.

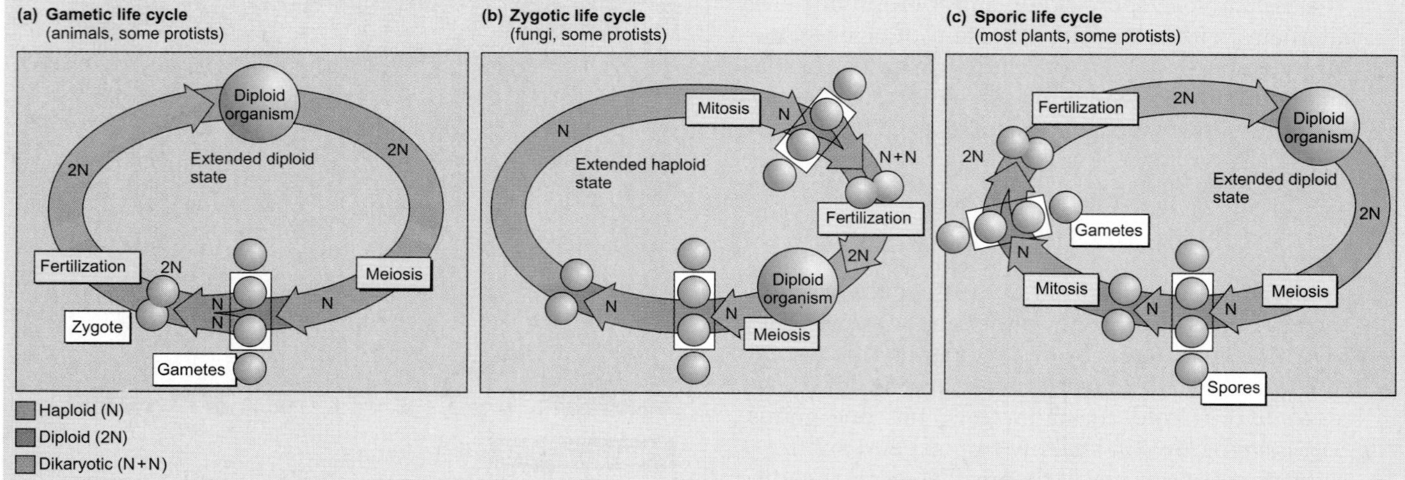

(a) Gametic life cycle (animals, some protists)

Diploid organism; Extended diploid state; 2N; Fertilization; Zygote; Gametes; Meiosis; N

(b) Zygotic life cycle (fungi, some protists)

Mitosis; Extended haploid state; N; N+N; Fertilization; 2N; Diploid organism; Meiosis

(c) Sporic life cycle (most plants, some protists)

Fertilization; 2N; Diploid organism; Extended diploid state; Gametes; Mitosis; Meiosis; N; Spores

Haploid (N)
Diploid (2N)
Dikaryotic (N+N)

plants. These conducting tissues allow photosynthetic products to be transported throughout the alga.

Like the red algae, the brown algae are commercially important. Kelps are harvested along the southern California coast on a regular basis for their rich sodium, potassium, and iodine content. Their structural polysaccharide, *algin,* is used as a thickening and stabilizing agent for foods and paints and as a paper-coating agent. In Asia some brown algae are eaten as a vegetable.

Also like the red algae, many browns have a sporic life cycle, exhibiting a clear alternation of generations. In some species, the two alternating phases are amazingly plantlike. They have a highly dominating sporophyte that retains and protects its spores while they subsequently develop into tiny, protected gametophytes (see Essay 23.2). In addition, some species produce **heterogametes** (*hetero,* different), which means they have large, stationary eggs and tiny, flagellated sperm. The presence of dominant sporophytes, vascular tissue, and heterogametes in brown algae and plants is attributed to convergent evolution. That is, although the two groups are not closely related, they have made similar adaptive changes.

FIGURE 23.11 GIANT KELPS.
Giant kelps thrive in cold oceans where they form undersea "forests." After storms, their remains are a common sight along California beaches.

Green Algae (Phylum Chlorophyta)

Most species of *green algae* live in fresh water, although a sizable number are marine and a few are terrestrial. Some are no longer independent and must live within the cells of certain ciliates and invertebrate animals. Other green algae live in a mutual association with fungi, forming some of the many varieties of rock-encrusting lichens. The 7,000 or so species of green algae range from single-celled to colonial and multicellular.

Green algae are biochemically similar to plants. They make use of the light-gathering pigments chlorophyll *a* and *b* and carotenoids; they form typical plant starches, and build their cell walls of cellulose.

Single-celled Green Algae. Single-celled green algae may be nonflagellated or flagellated. One of the best-known flagellated forms is *Chlamydomonas reinhardi,* which has been studied extensively by geneticists. Each tiny, oval cell has two anterior (pulling) flagella, a red, light-sensitive eyespot, and a single cup-shaped chloroplast that takes up much of the cell's volume. For most of its life cycle *Chlamydomonas* is haploid and reproduces asexually by mitosis and cell division, but it has a sexual phase also, in which its **isogametes** fuse. (Isogametes—*iso,* same—are so called because they are identical in appearance; see Figure 23.12.)

Colonial Green Algae. *Volvox,* a rather remarkable green alga, is comprised of an organized spherical colony of flagellated cells (Figure 23.13). As a colony, *Volvox* is a community of cells behaving as a single organism. Although the cells are not wholly dependent on each other, their association appears to offer a degree of increased efficiency, as each cell type fulfills a different role in the life of the colony, through a kind of "division of labor." Whereas *Volvox* is not directly related to any truly multicellular higher plant, nor even to truly multicellular green algae, its organization does suggest something about how multicellularity itself *might* have begun.

Multicellular Green Algae. Many green algae are multicellular. Some form thin filaments of cells, and others exist as leaflike blades. *Ulothrix,* a common freshwater filamentous alga (Figure 23.14a), is characterized by a zygotic life cycle. Following a long haploid phase, *Ulothrix* produces motile, flagellated isogametes that fuse to form a diploid zygospore. The zygospore enters meiosis to liberate

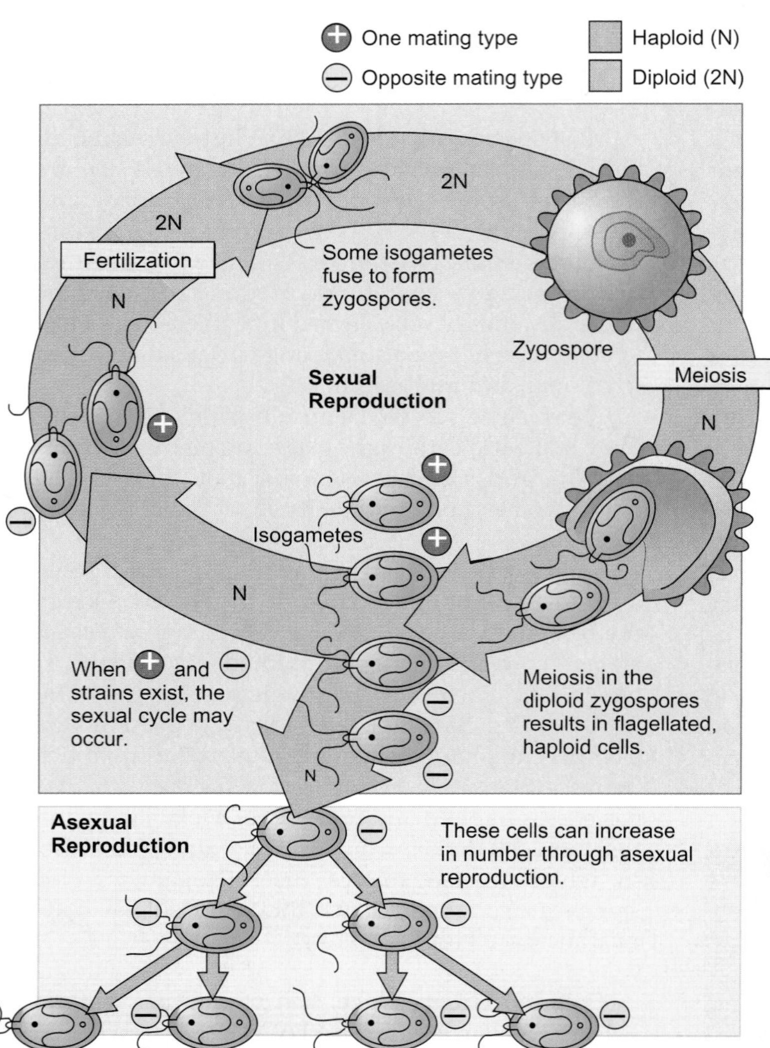

One mating type
Opposite mating type
Haploid (N)
Diploid (2N)
2N
2N
Fertilization
N
Some isogametes
fuse to form
zygospores.
Zygospore
Meiosis
N
Sexual
Reproduction
Isogametes
N
When + and −
strains exist, the
sexual cycle may
occur.
Meiosis in the
diploid zygospores
results in flagellated,
haploid cells.
N
Asexual
Reproduction
These cells can increase
in number through asexual
reproduction.

FIGURE 23.12 ALGAL REPRODUCTION. Meiosis in the diploid zygospore of *Chlamydomonas reinhardi* results in flagellated, haploid cells that can increase their number through asexual reproduction or enter a sexual cycle where they form isogametes. Different mating types fuse, forming new zygospores.

haploid cells that will produce a new filamentous body (Figure 23. 15).

Another multicellular green alga, *Ulva* ("sea lettuce"), is quite different (Figure 23.14b). It has a sporic life cycle with a clear alternation of generations. The sporophytes and gametophytes take the form of large, leaflike blades that are two cells thick. Although their roles in the life cycle and their chromosome numbers differ, sporophytes and gametophytes are identical in size and shape.

Green algae are interesting in their own ways, but a primary reason for focusing on the green algae is that they have relatives in "high places." About 400 million years ago, during the Paleozoic era, one or more kinds of delicate but adventuresome multicellular green algae began what was to be the first successful transition to terrestrial life. It was from this humble beginning that today's great array of terrestrial plant life is believed to have emerged.

SLIME MOLDS AND WATER MOLDS: FUNGUSLIKE PROTISTS

Now let's consider those protists that resemble fungi in certain ways, while retaining their protistan membership (although some people, we should add, still consider them to be fungi). These are the poetically named slime molds and water molds.

SLIME MOLDS (PHYLA ACRASIOMYCOTA AND MYXOMYCOTA)

Although they are unrelated and dramatically different, the two groups of slime molds, *cellular slime molds* (Phylum Acrasiomycota) and *acellular slime molds* (Phylum Myxomycota), have something in common: They both lead double lives.

In one portion of their lives, cellular slime molds multiply and grow over their food supply (often a rotting log), each cell moving along like an ameba. But when their food source is depleted, they change drastically. The individual cells then crawl together and coalesce, forming a single body called a *slug*. The slug wanders blindly about like a mindless worm until, eventually, it stops and vertical stalks begin to arise from its mass. The stalks bear fruiting bodies that give rise to airborne spores, just as any respectable mold might do (Figure 23.16).

Acellular (*a*, without) slime molds produce huge multinucleate masses, forming what looks like a giant ameba—large enough to cover an entire log. The mass, however, may be only a millimeter thick. As it creeps along, it engulfs bits of organic matter and shows some sensitivity in that it avoids obstacles. When its habitat dries out, it produces slender stalks that will release haploid spores. These are released into the environment to later fuse and begin dividing, producing the diploid, amebalike mass again.

Water Molds (Phylum Oomycota)

Water molds include those unsightly furry growths that plague the fish in our aquariums. Others are parasites of insects and plants, and one is blamed for the exodus of many Irish after the infamous potato famine of the 1840s. (If you are an American of Irish descent, you may owe your citizenship to that very parasite.) The threadlike body of the water mold is reminiscent of some fungi, but there are important differences. For instance, water molds are essentially diploid, which means their life cycles are gametic, a characteristic of animals (see Essay 23.2). Now to the fungi themselves.

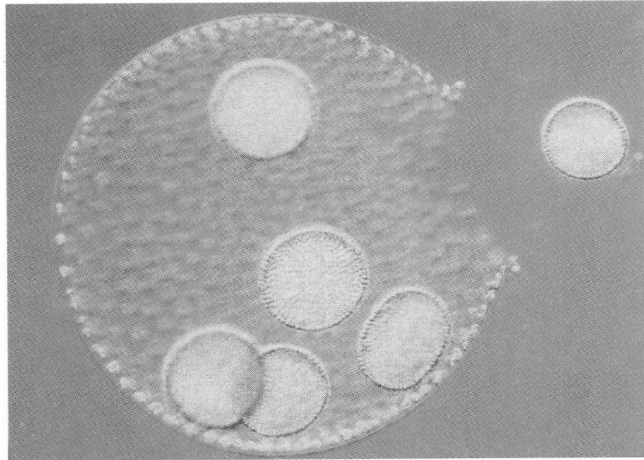

Figure 23.13 **Colonial Green Algae.** *Volvox* forms a spherical colony made up of an outer layer of interconnected swimming cells and smaller spheres representing new colonies within. The smaller spheres begin their development as zygotes after sexual reproduction has occurred. The spheres then grow through cell division and eventually emerge from the parent colony to begin life on their own.

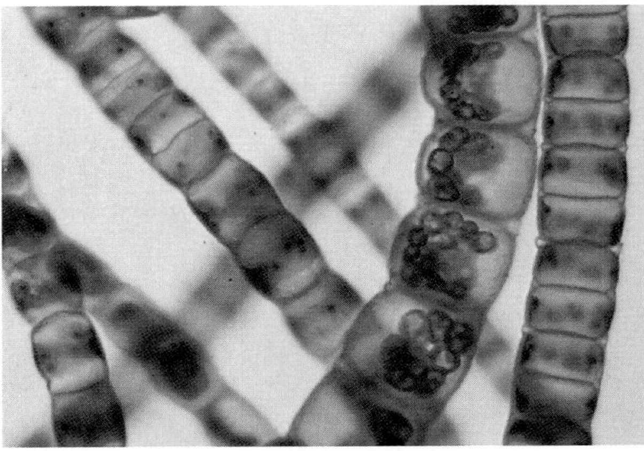

(a)

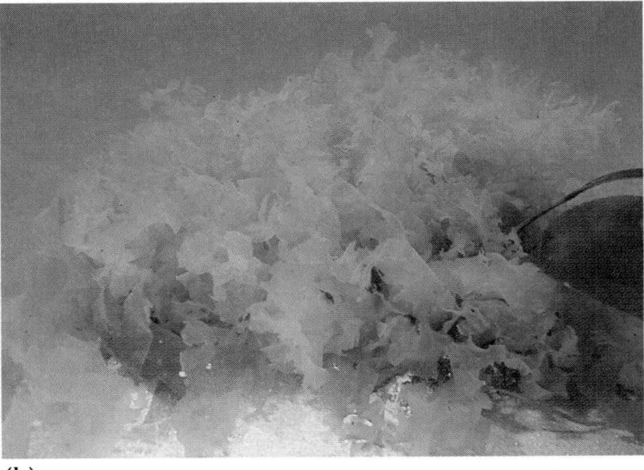

(b)

Figure 23.14 **Multicellular Green Algae.** (**a**) The freshwater filamentous alga *Ulothrix*. (**b**) The leafy marine alga *Ulva*, known as "sea lettuce."

Fungi

Fungi are a fascinating lot, in many ways quite unlike other organisms. They are stationary, multicellular, and heterotrophic; most live as parasites, decomposers, and mutualists. To some of us the most familiar fungi are the delicate mushrooms and enchanting toadstools that grace our forests, while others might think at once of furry growths on leftovers. Actually, however, most species live as parasites on other organisms. Thousands of species are

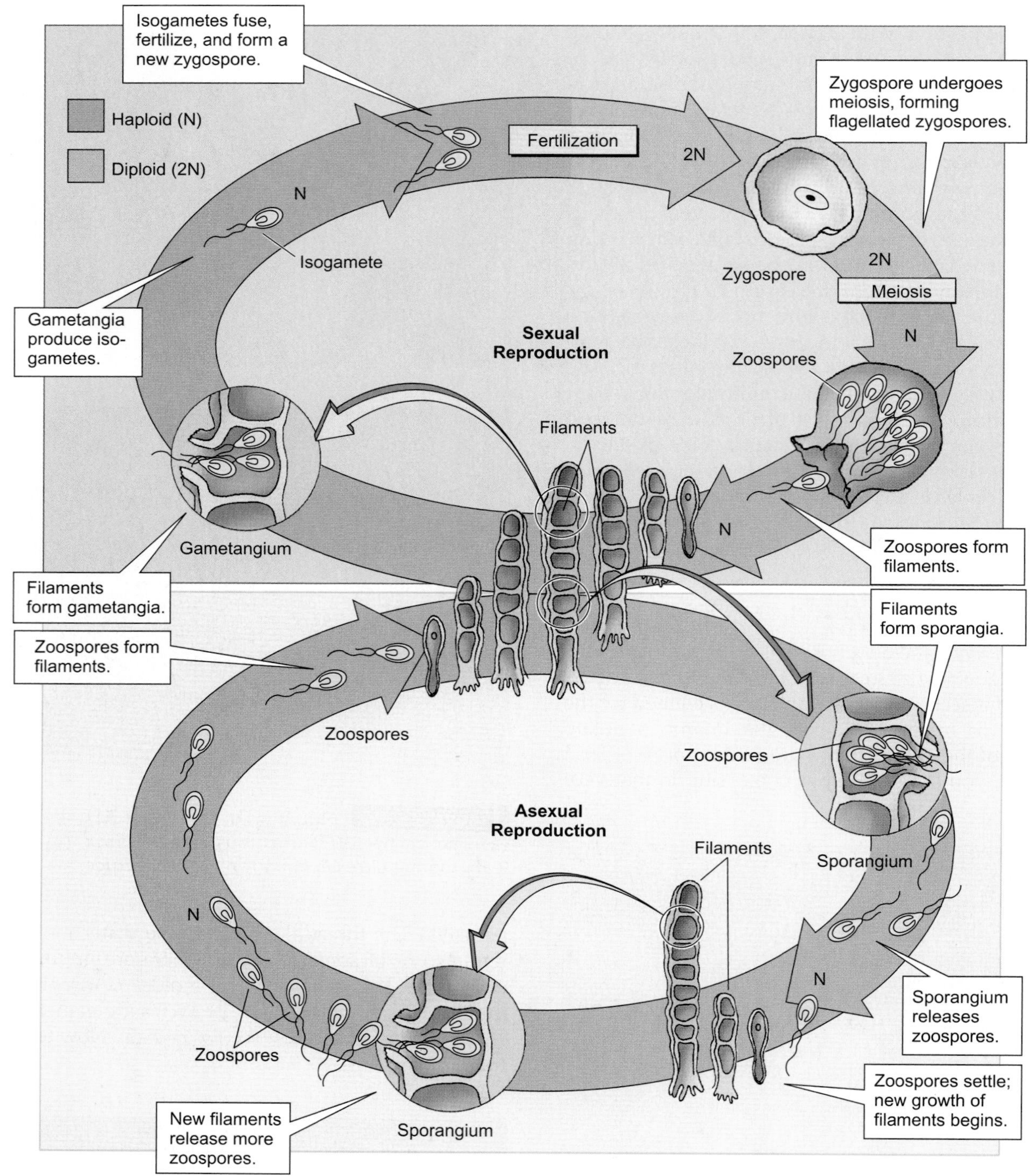

FIGURE 23.15 LIFE CYCLE OF *ULOTHRIX*, A GREEN ALGA.
During the extremely brief sporophyte generation, the zygospore (the result of gametes joining in fertilization) undergoes meiosis and becomes filled with flagellated spores called zoospores. The zoospores produce filaments that form either sporangia (spore-forming structures) or gametangia (gamete-forming structures). The gametangia produce flagellated isogametes (gametes of the same appearance; *iso,* same) that can join in fertilization with gametes from another filament. This union results in a new zygospore. The sporangium, on the other hand, releases spores that simply settle to begin the growth of new filaments in a form of asexual reproduction. Under certain (usually adverse) conditions, these filaments may stop reproducing asexually and begin to form gametangia, leading to sexual reproduction.

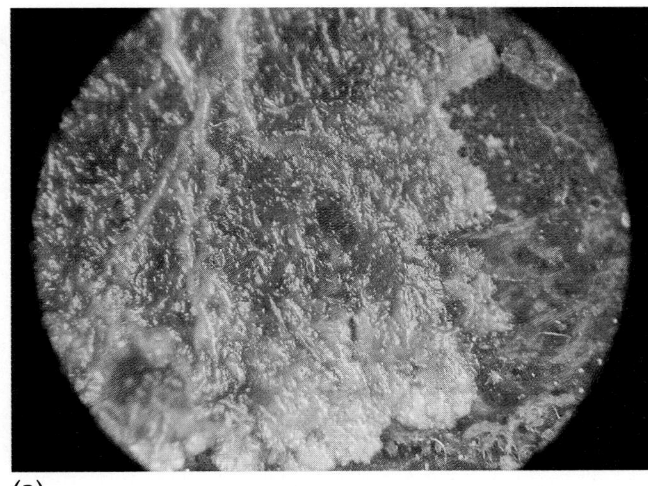
(a)

FIGURE 23.16 **SLIME MOLDS.**
Cellular slime molds **(a)** live as independent, unicellular amebas, phagocytizing soil bacteria and reproducing asexually during the vegetative part of their life cycles. In the sexual phase **(b)**, individual cells coalesce into a heap, which differentiates into a motile, nonfeeding "slug." **(c)** Eventually the slug stops crawling; a stalk appears first as a bump which then lengthens, lifting a ball of cells to its tip **(d)**. The ball (sporangium) develops spores as it rises higher with the further elongation of the stalk. Eventually the spores are released into the wind.

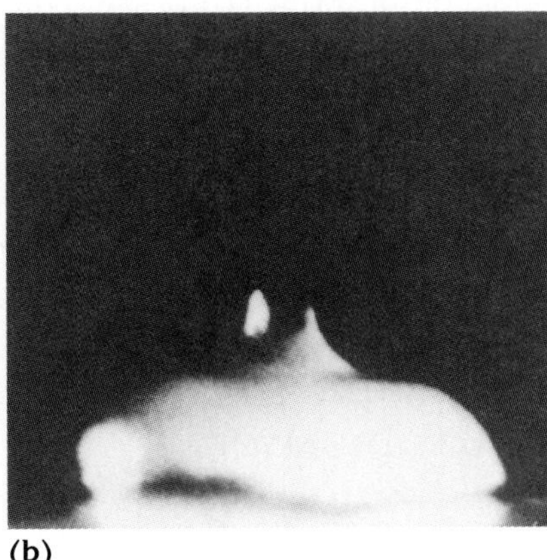
(b)

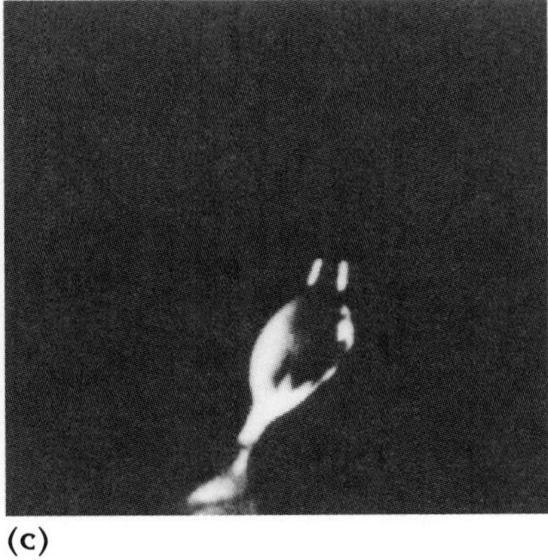
(c)

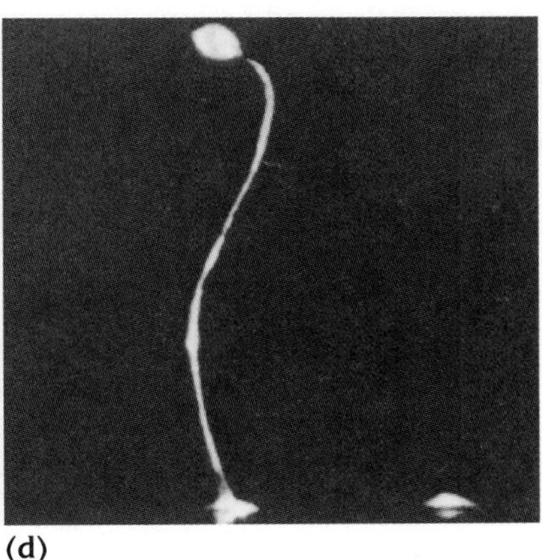
(d)

mutualists; they help form lichens and *mycorrhizae*, the latter of which are fungus-plant associations in the roots of many plants. The fungi help absorb minerals and water, and the roots provide sugars (see Chapter 28).

Fungi spread by forming tiny spores that can be carried by air currents. Under favorable conditions—when there is adequate moisture, food, and warmth, the spores germinate, and a *vegetative* (feeding) *stage* emerges. In most fungal species, the **vegetative stage** is marked by an extensive, spreading *mycelium*, which consists of numerous branching, threadlike growths called *hyphae* that secrete digestive enzymes into food and absorb the products. Some hyphae grow upward, later forming and releasing spores, as mentioned earlier. Individually, the hyphae are minute tubes of cytoplasm, surrounded by a cell wall that is usually composed of the complex carbohydrate *chitin* (the same tough, flexible material that is found in the skeletons of insects and their relatives). In some fungi, the cytoplasm is not organized into cells, but is continuous and contains many nuclei. Other fungi exhibit a typical cellular state, with one nucleus per cell.

Fungi spend most of the time in a haploid state typical of the zygotic life cycle (see Essay 23.2). As with many protists, sexual reproduction involves the union of different mating types, referred to here as "plus (+) strains" and "minus (–) strains." As usual, the union produces the diploid state, but in a most unusual way in some species. Each mating partner contributes a haploid nucleus, but the two nuclei may not fuse in fertilization for a long time. When fertilization is delayed, the zygote isn't really diploid. Such a cell is said to be in a **dikaryotic** state (two nuclei—*di,* two; *kary,* kernel). We'll see shortly how such dikaryotic cells can develop into elaborate fruiting bodies that give rise to the next generation of spores. (Mushrooms, toadstools, and shelf fungi are familiar fruiting bodies.)

The fungal kingdom contains four phyla: **bread molds** (Phylum Zygomycota), **sac fungi** (Phylum Ascomycota), **club fungi** (Phylum Basidiomycota), and the **Fungi Imperfecti** or imperfect fungi

(Phylum Deuteromycota). (The first three are also known more commonly as *zygomycetes, ascomycetes,* and *basidiomycetes.*)

The origin of fungi from protist ancestors is obscure, but disputed funguslike fossils occur in Precambrian strata some 900 million years old. This places them among the oldest eukaryotes.

Bread Molds (Phylum Zygomycota)

One well-known bread mold, *Rhizopus stolonifer,* is a fuzzy, black invader that is bound to turn up in everyone's refrigerator at one time or another. *Rhizopus* grows vigorously on many foods but is partial to the starchy types (Figure 23.17). The haploid mycelium develops long, horizontal hyphae that lack cell walls, so the continuous cytoplasm is multinucleate. The tiny black dots on moldy bread are the **sporangia,** asexual bodies that generate many tiny haploid spores through mitosis. Some of the spores may be dispersed over great distances by air currents.

Sexual reproduction in the bread molds occurs when a plus strain meets a minus strain (Figure 23.17). The hyphae of the two types connect, permitting the haploid nuclei from each strain to join. Their union brings about the development of a tough, black, diploid **zygospore.** The brief diploid state ends as soon as meiosis occurs in the zygospore. When the zygospore germinates, the hyphae that emerge are once again haploid.

Sac Fungi (Phylum Ascomycota)

The ascomycetes (and the basidiomycetes) have an essentially cellular structure, with *septa* (crosswalls) isolating each nucleus in the tubelike hyphae. The septa have large pores through which the nuclei are free to migrate. The names *ascomycetes* and *sac fungi* both refer to a characteristic, microscopic, saclike reproductive structure, the **ascus** (plural, *asci*), in which sexual spores are produced.

The sac fungi include the powdery mildews sometimes seen on plant leaves. Two of these, the chestnut blight and the agent of Dutch elm disease, have raised havoc in American forests. Edible morels and truffles and the red, blue-green, and brown food spoilers are also sac fungi. One sac fungus that holds rather ghastly implications for humans is *Claviceps purpurea,* a fungal parasite that causes a disease called ergot in rye and other grasses. When ergotized rye flour is eaten, it has the strange property of causing hallucinations and burning sensations in the hands and feet, where blood vessels become constricted. (Some historians believe the Salem "witches" were suffering from ergot poisoning. The good citizens there hanged them anyway.)

Not all sac fungi are harmful, however; the pink mold *Neurospora* helped geneticists discover the biochemistry of the gene in past years (see Chapter 14). Commercially important yeasts used in winemaking, brewing, and breadbaking are also sac fungi. (The troublesome "yeast" of skin and vaginal infections is actually a yeastlike member of another phylum, the Fungi Imperfecti.)

Sac fungi reproduce asexually through the production of simple spores called **conidia.** Sexual reproduction includes the fusion of plus and minus strains and the formation of a dikaryotic state, the odd form of delayed fertilization mentioned earlier. Such binucleate hyphae are joined by haploid parental hyphae in forming an elaborate sexual structure, the **ascocarp** (Figure 23.18). (The edible truffle and morel are ascocarps—see Essay 23.2). Union of the plus and minus nuclei is followed immediately by meiosis and the formation of haploid **ascospores.**

Club Fungi (Phylum Basidiomycota)

The basidiomycetes, or club fungi, include the common forest mushroom, the shelf fungus, the puffball, and other fleshy species, in addition to the parasitic wheat rusts and corn smuts (Figure 23.19). The names *basidiomycetes* and *club fungi* both refer to a characteristic microscopic reproductive structure from which the sexual spores are budded off: the **basidium** ("little pedestal"). To some, the structure resembles a club.

The familiar mushroom growing on the forest floor is only a part of the organism. Below the soil surface lies an extensive, unseen mycelium, silently secreting enzymes, digesting organic matter, and absorbing nutrients. The mycelium may also form the mycorrhizal relationship with plant roots, as was described earlier. The mushroom's aboveground portion—the part sold in supermarkets—is known, technically, as the **basidiocarp** ("pedestal fruit"). It is produced after hyphae of plus and minus strains have conjugated below ground (Figure 23.20). Each cell produced after the union is dikaryotic, containing a haploid nucleus from each parental strain. The dikaryotic state persists in the mushroom, just as it did in the ascocarp of the sac fungus, its hyphae contributing to the growing

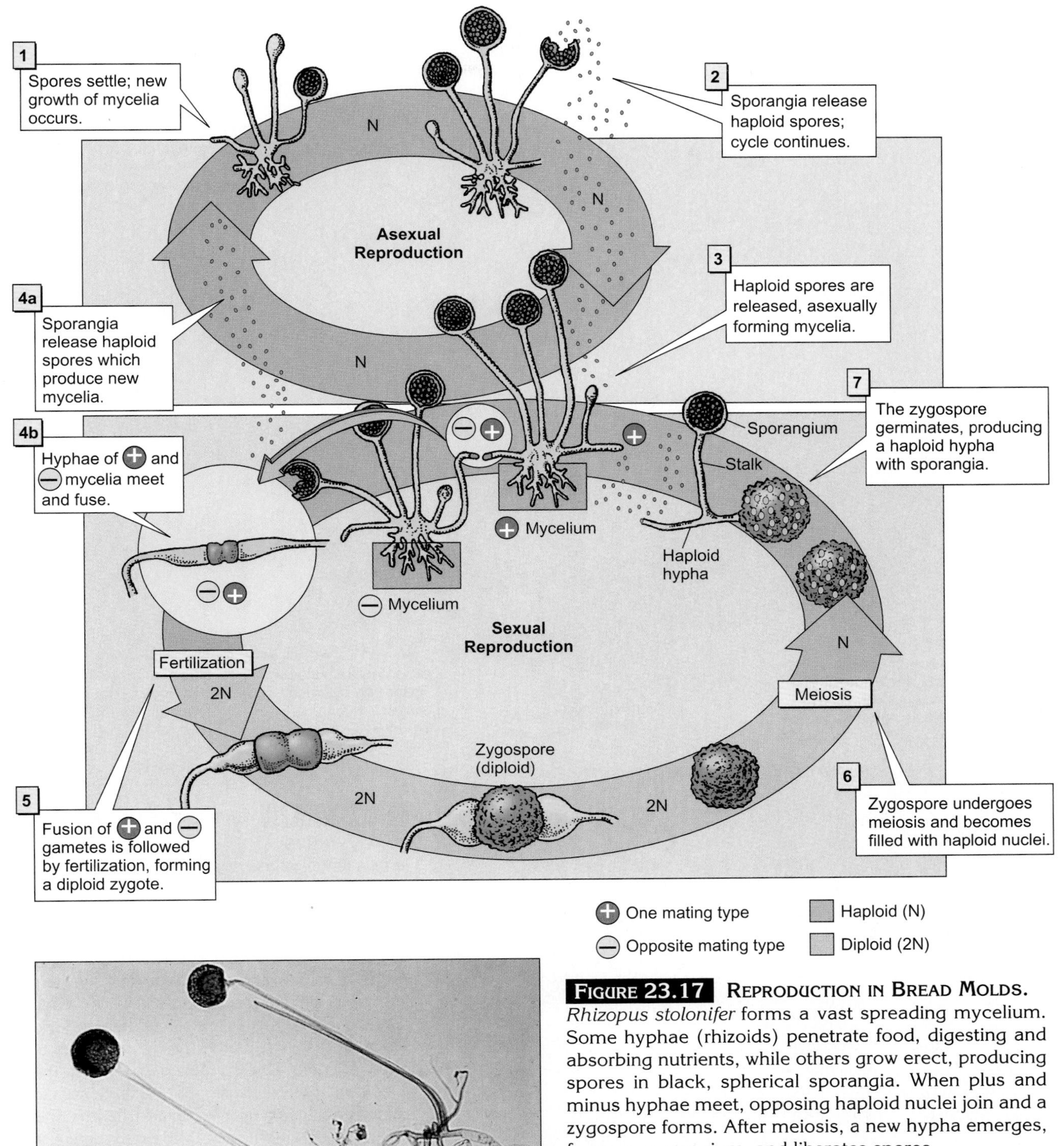
1
Spores settle; new growth of mycelia occurs.
N
2
Sporangia release haploid spores; cycle continues.
N
Asexual Reproduction
3
Haploid spores are released, asexually forming mycelia.
4a
Sporangia release haploid spores which produce new mycelia.
N
7
The zygospore germinates, producing a haploid hypha with sporangia.
Sporangium
Stalk
4b
Hyphae of + and − mycelia meet and fuse.
+ Mycelium
Haploid hypha
− Mycelium
Sexual Reproduction
N
Fertilization
2N
Meiosis
Zygospore (diploid)
5
2N
2N
6
Zygospore undergoes meiosis and becomes filled with haploid nuclei.
Fusion of + and − gametes is followed by fertilization, forming a diploid zygote.
+ One mating type
− Opposite mating type
Haploid (N)
Diploid (2N)

FIGURE 23.17 **REPRODUCTION IN BREAD MOLDS.** *Rhizopus stolonifer* forms a vast spreading mycelium. Some hyphae (rhizoids) penetrate food, digesting and absorbing nutrients, while others grow erect, producing spores in black, spherical sporangia. When plus and minus hyphae meet, opposing haploid nuclei join and a zygospore forms. After meiosis, a new hypha emerges, forms a sporangium, and liberates spores.

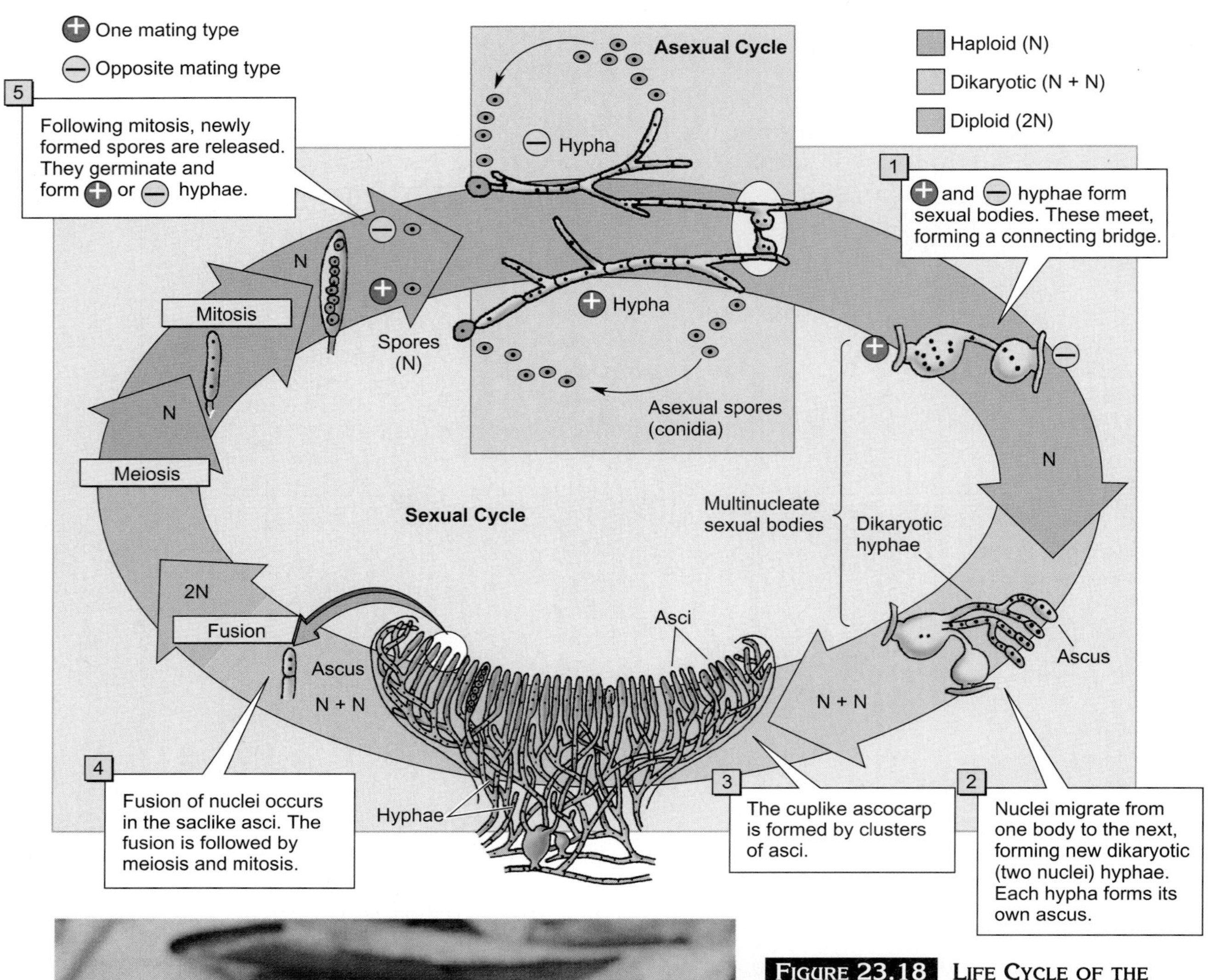
One mating type
Opposite mating type
Asexual Cycle
Haploid (N)
Dikaryotic (N + N)
Diploid (2N)
5
Following mitosis, newly formed spores are released. They germinate and form or hyphae.
Hypha
1
and hyphae form sexual bodies. These meet, forming a connecting bridge.
N
Mitosis
Hypha
Spores (N)
N
Asexual spores (conidia)
Meiosis
N
Sexual Cycle
Multinucleate sexual bodies
Dikaryotic hyphae
2N
Fusion
Asci
Ascus
Ascus
N + N
N + N
4
Fusion of nuclei occurs in the saclike asci. The fusion is followed by meiosis and mitosis.
Hyphae
3
The cuplike ascocarp is formed by clusters of asci.
2
Nuclei migrate from one body to the next, forming new dikaryotic (two nuclei) hyphae. Each hypha forms its own ascus.

FIGURE 23.18 LIFE CYCLE OF THE SAC FUNGUS.
In its asexual phase, this sac fungus produces spores that form new hyphae as they germinate. Sexual activity begins when plus and minus strains form bulbous, multinucleate sexual bodies. Nuclei migrate from one body to the next, forming new dikaryotic hyphae. In a special cell (ascus) at the tip of each dikaryotic hypha, the two nuclei fuse to yield a diploid nucleus, as the hypha continues to grow, forming a fleshy ascocarp. Within the ascocarp *(left),* the fusion of nuclei occurs in saclike asci. Fusion is followed by meiosis and mitosis, which leads to the emergence of eight ascospores. These will be released to begin the new asexual generation.

FIGURE 23.19 CLUB FUNGI.
(a) Bracket fungus. (b) The edible mushroom *Lepiota* looks too much like the poisonous *Amanita* ("death angel") (c) even for experts to take chances. Corn smut (d) can be devastating to crops.

basidiocarp. Nuclear fusion finally takes place in the numerous basidia lining the thin **gills** on the underside of the mushroom cap.

The diploid state is brief, followed at once by meiosis and the formation of haploid spores. People who count such things tell us that a single mushroom can produce billions of spores in this manner—a sizable investment in its future.

FUNGI IMPERFECTI (PHYLUM DEUTEROMYCOTA)

Imperfect is not a derogatory term in botany. Instead, it refers to the presumed absence of sexual reproduction. There are thousands of poorly understood species of Fungi Imperfecti, which apparently will occupy the energies of taxonomists for decades to come. Undoubtedly, many of these species will eventually be placed among either the sac or club fungi. The Fungi Imperfecti include a number of human parasites such as *Trichophyton mentagrophytes,* which causes athlete's foot. Another causes ringworm, and yet another is the agent of vaginal yeast infections. A fascinating but presumably unrelated species, *Dactylaria,* is a predator that catches certain roundworms that live in the soil (Figure 23.21).

LICHENS

We include *lichens* among the fungi, but they are actually composed of a symbiotic association of a fungus and either a green alga or a cyanobacterium. A number of fungi may be involved in such associations, but the most common are the sac fungi. Whatever the species, however, the association is mutual; both partners benefit. The fungus provides moisture, mineral nutrients, and strong attachment to the substrate. The alga or cyanobacterium provides food produced through photosynthesis. This unique partnership permits lichens to grow in unusual places where there is little competition from other organisms. We commonly find them on rocky outcrops, where, because of their erosive effects on the rocks, they may begin soil formation. Lichens also grow on trees and as luxurious ground cover in the frigid tundra (Figure 23.22). They rarely live in cities, since they are highly susceptible to the toxic effects of air pollution.

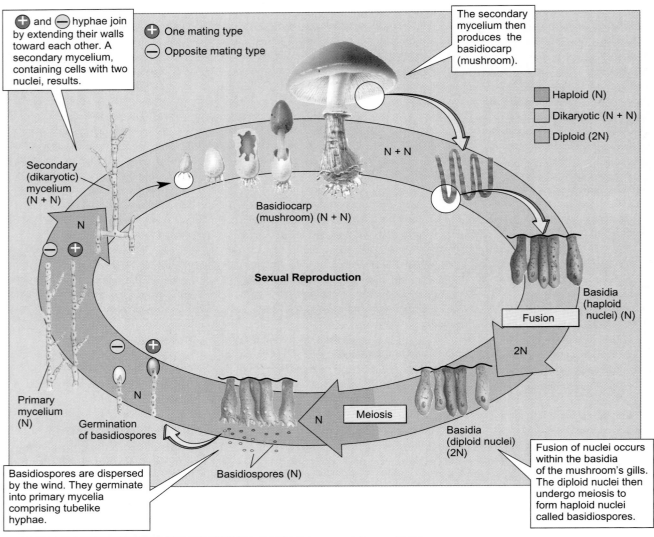

⊕ and ⊖ hyphae join by extending their walls toward each other. A secondary mycelium, containing cells with two nuclei, results.

⊕ One mating type
⊖ Opposite mating type

The secondary mycelium then produces the basidiocarp (mushroom).

Haploid (N)
Dikaryotic (N + N)
Diploid (2N)

N + N

Secondary (dikaryotic) mycelium (N + N)

Basidiocarp (mushroom) (N + N)

N

Sexual Reproduction

Basidia (haploid nuclei) (N)

Fusion

2N

Primary mycelium (N)

N

Germination of basidiospores

N

Meiosis

Basidia (diploid nuclei) (2N)

Basidiospores (N)

Basidiospores are dispersed by the wind. They germinate into primary mycelia comprising tubelike hyphae.

Fusion of nuclei occurs within the basidia of the mushroom's gills. The diploid nuclei then undergo meiosis to form haploid nuclei called basidiospores.

FIGURE 23.20 LIFE CYCLE OF THE COMMON EDIBLE MUSHROOM, A BASIDIOMYCETE.
Spores are dispersed by wind. If they land in a suitable place, they will germinate into primary mycelia, which are comprised of tubelike hyphae. If two hyphae of appropriate types come together, they will join by extending their walls toward each other. These extensions contain nuclei, which end up in the same cell at fertilization. A secondary dikaryotic mycelium (containing cells with two unfused nuclei) now spreads through the soil, and some may eventually form the visible mushrooms. In sexual fusion, the diploid nucleus undergoes meiosis to form four haploid nuclei. Each nucleus is expelled (with some cytoplasm) through an extension at the tip of the basidium, a clublike structure (for which the club fungi are named). A tough wall forms around it and it becomes a spore. *(left)* The underside of a mushroom cap showing the thin, bladelike gills on which are the basidia where nuclear fusion takes place.

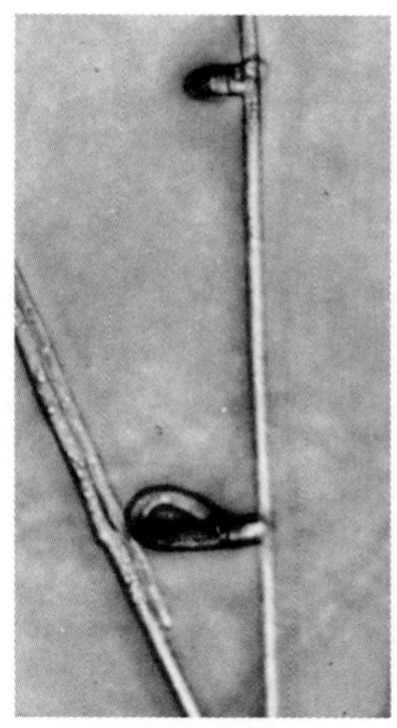
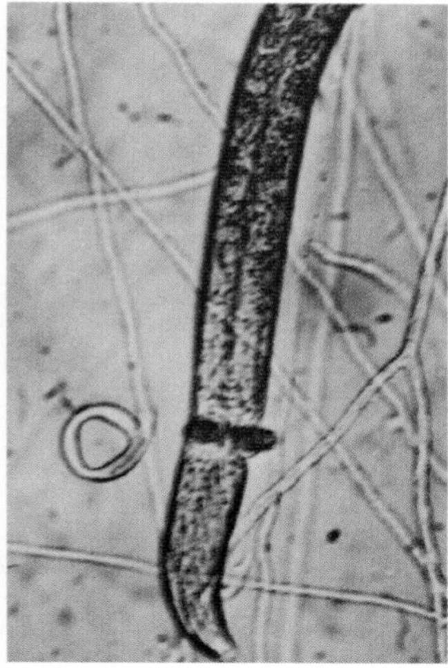

FIGURE 23.21 FUNGI IMPERFECTI.
Among the Fungi Imperfecti is the predatory species *Dactylaria*. Its bizarre mycelium contains many looplike snares, each capable of swelling rapidly on touch, to close the loop on anything passing through. Soil nematodes, moving through the soil water, are its usual victims. Once captured, they are penetrated by fast-growing fungal filaments that secrete digestive enzymes.

FIGURE 23.22 LICHENS.
Lichens are a mutualistic association of fungi and algae or cyanobacteria. They sometimes occur as a colorful though crusty growth on bare rock surfaces.

PERSPECTIVES

We have made our way through some of the simplest cellular life on the planet. We have seen how some of these simple organisms have played important evolutionary roles in the development of more complex life forms, and we have noted the role of interaction among life forms, from symbiotic cellular inclusions to the interdependence found in lichens. We saw the importance of ciliated protists to the development of animals, and of green algae to the development of plants. Next, then, let's look at plants and their dramatic evolutionary history among life on earth.

SUMMARY

WHAT IS A PROTIST?

The protist kingdom is wide-ranging and includes single-celled, colonial, and multicellular forms. The size range is enormous, and most protists are aquatic.

PROTOZOANS: ANIMAL-LIKE PROTISTS

Protozoans are heterotrophic with animal-like nutritional requirements. Three major phyla contain the flagellates and amebas, the sporozoans, and the ciliates.

The flagellates bear flagella, are free-living or parasitic, and feed through absorption or phagocytosis. Reproduction is most commonly asexual. Most of the zygotic life cycle is spent in the haploid state. Sexual reproduction is through cell fusion, followed by meiosis.

Sporozoans are nonmotile, parasitic spore-formers. The best known, *Plasmodium vivax,* is the mosquito-borne agent of malaria. Fever and chills follow regular cycles of the parasite's asexual activity, which is marked by red blood cell lysis. Sexual stages form in the host but must be completed in the mosquito.

Ciliates, the most complex protozoans, move and feed through the coordinated movement of cilia. In some, food is swept into a cytostome, whereupon it enters a food vacuole, where it is joined by digestive enzymes. The products diffuse into the cytoplasm, and the residues leave the cell through exocytosis at the cytopyge.

Water regulation is carried out by the contractile vacuole, which fills and empties through the action of microfilaments.

Micronuclei contain germlike DNA reserved for reproduction, whereas macronuclei contain replicated DNA for use in transcription and protein synthesis. The cell is gametic.

Algae: Plantlike Protists

Included among the abundant marine algae are seaweeds, kelps, phytoplankton, and intracellular symbionts. Many algae live in fresh water, and a few are terrestrial.

Dinoflagellates (fire algae, Phylum Pyrrophyta) are mainly flagellated and photosynthetic. Many create an iridescent glow at night, and some, during their periodic blooms, cause red tide. The accumulation of their toxins in clams, oysters, and other bivalves can render these shellfish poisonous to humans.

The diatoms and golden-brown algae (Phylum Chrysophyta) make up an important part of the marine food chain. Diatoms have been so numerous in the past that their glassy walls have formed dense deposits called diatomaceous earth.

Euglena (Phylum Euglenophyta) is single-celled, flagellated, and photosynthetic, with a flexible body, well-formed chloroplast, starch storage bodies, and a photoreceptor. Taxonomists cannot agree whether it is a protozoan or an alga.

Red algae (Phylum Rhodophyta) are frilly, multicellular, marine seaweeds. Holdfasts anchor them to the seabed. Agar and carrageenan are extracted from them for commercial use. They have an alternating, two-part, sporic life cycle made up of diploid, spore-forming sporophytes and haploid, gamete-forming gametophytes.

Brown algae (Phylum Phaeophyta), multicellular marine kelps and seaweeds, are often large and structurally complex. Some have elaborate holdfasts, large leaflike blades, stipes with conducting tissue, and bulbous floats. Algin and minerals are extracted from them for commercial use. Many have a plantlike, sporic life cycle that includes a dominating sporophyte and highly reduced gametophyte. The latter gives rise to heterogametes (minute sperm and large eggs).

Most green algae (Phylum Chlorophyta) live in fresh water and are biochemically similar to plants. They may be single-celled, colonial, or multicellular.

Single-celled green algae include *Chlamydomonas reinhardi,* whose zygotic life cycle is principally haploid. A brief diploid state, called a zygospore, enters meiosis, restoring the haploid state.

Volvox, a spherical colony of interacting, flagellated cells, suggests one way in which multicellularity may have begun.

Multicellular green algae include the filamentous *Ulothrix,* with a primitive zygotic life cycle, and the more recently evolved marine seaweed *Ulva,* which has a sporic life cycle. Plants are believed to have evolved from one of the multicellular green algal lines.

Slime Molds and Water Molds: Funguslike Protists

Both acellular and cellular slime molds (Phyla Acrasiomycota and Myxomycota) are protistlike in their ameboid vegetative or feeding stages and funguslike in their asexual reproductive stages. In the latter, they give rise to stalks tipped by spore-forming fruiting bodies.

Water molds (Phylum Oomycota) are economically important because some (e.g., potato blight) damage crops, and they are biologically important because of their advanced sexual structures. They resemble fungi in growth and asexual reproduction, but unlike fungi, their life cycles are gametic.

Fungi

Fungi are multicellular heterotrophs. Many are parasites, and some are free-living decomposers. Some associate with plant roots, forming mineral-absorbing mycorrhizae. Typically the vegetative or feeding fungal mycelium consists of threadlike, spreading, and branching hyphae whose walls are of chitin. The life cycle is zygotic. Some retain separate plus and minus nuclei—a dikaryotic state—for a time, before actual fusion occurs. Large, elaborate, and fleshy fruiting sexual bodies are often formed. Fungi are among the oldest eukaryotes.

Among bread molds (Phylum Zygomycota), *Rhizopus stolonifer* has typical spreading hyphae with rootlike rhizoids that secrete enzymes and absorb nutrients. Vertical hyphae produce sporangia in which asexual spores are formed. Union of plus and minus hyphae form diploid zygospores that undergo meiosis and produce new, sexually recombined, haploid hyphae.

Hyphae in sac fungi (Phylum Ascomycota) have septa (crosswalls). The group includes important parasites, edible forms, and common food spoilers. Sexual reproduction includes fusion of plus and minus strains, a lengthy dikaryotic state, and the formation of a cuplike ascocarp. Following nuclear fusion, meiosis and ascospore formation occur in saclike asci.

Club fungi (Phylum Basidiomycota) include mushrooms, toadstools, wheat rusts, corn smuts, and others. Fusion of plus and minus hyphae in the extensive underground fungal mycelium results in a dikaryotic growth that emerges to form the basidiocarp, a fruiting body. Flattened gills within the cap contain numerous clublike basidia, in which nuclear fusion occurs. Meiosis ensues, and numerous basidiospores form.

The Fungi Imperfecti (Phylum Deuteromycota) are species with no known sexual stages. Members include the agents of athlete's foot, ringworm, vaginal yeast infections, as well as a nematode-trapping predator.

Lichens

Lichens are a mutualistic fungus-alga or fungus-cyanobacterium association. They are ecologically important because of their soil-building activities.

Key Terms

Sarcomastigophora • **369**
Apicomplexa • **369**
Ciliophora • **372**
conjugation • **374**
phytoplankton • **374**
golden-brown alga • **377**
diatom • **377**
sporophyte • **379**
gametophyte • **379**
alternation of generations • **379**
heterogamete • **381**
isogamete • **381**
vegetative stage • **385**
dikaryotic • **385**
bread mold • **385**
sac fungus • **385**
club fungus • **385**
Fungi Imperfecti • **385**
sporangium • **386**
zygospore • **386**
ascus • **386**
conidium • **386**
ascocarp • **386**
ascospore • **386**
basidium • **386**
basidiocarp • **386**
gill • **389**

Review Questions

1. Generally characterize the range of organisms in the kingdom Protista. What do they have in common?
2. Distinguish among the following: the single-celled state, the aggregate, the colony, and the multicellular state.
3. List two characteristics that protozoans have in common with animals.
4. List several characteristics of flagellates. What form of life cycle do they exhibit? Explain what this means.
5. Name an important parasitic flagellate, and describe the disease it causes.
6. Explain the manner in which free-moving amebas move about and capture food. What is the physical basis for this movement?
7. Compare the habitat, body form, and method of feeding in the heliozoans and foraminiferans.
8. Relate the cyclic symptoms of malaria to the activities of the sporozoan agent. What must happen for the parasite to reproduce sexually?
9. Discuss movement, feeding, digestion, and water regulation in the representative ciliate *Paramecium.*
10. Describe the activity in the micronuclei during sexual reproduction in a ciliate. Would you expect the macronuclei to participate? Explain.
11. What are the phytoplankton? What is their major importance to marine life?
12. List the plantlike and animal-like characteristics of *Euglena gracilis.*
13. Describe the general body form, habitat, and economic importance of the red algae.
14. List the three types of green algae, and provide an example of each.
15. Describe the life cycle of *Ulva.* In what way or ways is *Ulva* advanced over the green algae mentioned above?
16. Describe ways in which the slime molds are similar to both protists and fungi.
17. What kind of life cycle do the fungi follow? Explain what a dikaryotic state is.
18. With the aid of a simple diagram, describe the vegetative stages and asexual reproduction in the bread mold *Rhizopus stolonifer.*
19. List three groups of club fungi that are on the list of dangerous parasites.
20. Describe the sexual cycle of a mushroom, starting with the union of plus and minus hyphae below ground. When does the fusion of plus and minus nuclei finally occur?
21. What is the basis for assignments to the phylum Fungi Imperfecti?

24

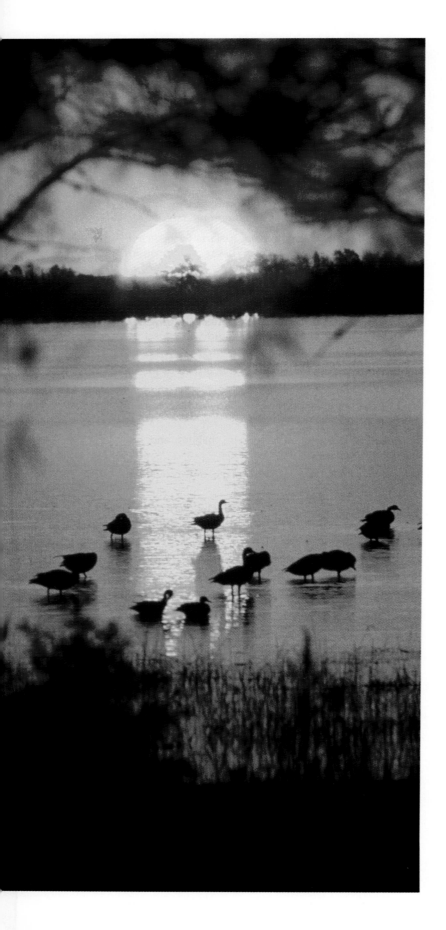

Evolution and Diversity in Plants

WHAT IS A PLANT?

ALTERNATION OF GENERATIONS IN PLANTS

NONVASCULAR PLANTS: DIVISION BRYOPHYTA

VASCULAR PLANTS

Vascular Plant Divisions

FERNS: DIVISION PTEROPHYTA

THE MESOZOIC ERA: TIME OF GYMNOSPERMS (DIVISIONS CYCADOPHYTA, GINGKOPHYTA, GNETOPHYTA, AND CONIFEROPHYTA)

Gymnosperm Specializations

The Gymnosperms Today

THE CENOZOIC ERA: THE TIME OF THE FLOWERING PLANTS

The Angiosperms Today

WE BEGIN BY TACKLING THE THORNY OLD QUESTION, "WHAT IS A PLANT?" *As is so often the case, we will settle for a working definition. After reviewing the alternation of generations in this group, we move to those plants virtually without supporting and conducting tissue—the bryophytes. Then we consider the rest of the plants, the tracheophytes, which boast both of these tissues. As we encounter the various groups—each highly specialized according to its lifestyle—we will focus on the adaptiveness of their traits, and how and when those traits arose in the evolutionary history of plants.*

We humans sometimes take delight in wondering aloud just who has inherited the earth: cockroaches, rats, houseflies, or us. Of course, it's a nonsensical question, but it is interesting that we never consider our silent partners, the plants. After all, just about everywhere we look we see green. Not only are plants the most prominent kind of land life on our planet, but also they form the base of nearly every terrestrial food chain. In other words, they manufacture the organic molecules that are eventually passed along through both plant-eaters and predators, the creatures that roam the earth in search of the food that is stored in the bodies of others.

As we will see, the plant kingdom includes two groups of plants, the *bryophytes* with their reduced vascular (conducting) tissue, and the *tracheophytes* with their well-developed vascular tissue—a critical evolutionary event, as we shall see.

What Is a Plant?

If you ask an old man watering his garden what a plant is, he is not likely to say, "Plants are multicellular photosynthesizers with roots and that have ways of protecting their embryos from drying out." Although he might not come up with this technical definition, you would be hard pressed to convince him he doesn't know what a plant is. Furthermore, as he stoops to straighten a stem, he is likely to be unimpressed if you continue the definition by stating that the photosynthetic pigments of plants include chlorophyll *a* and *b* and the carotenoids, that plants store food in the form of starch, that their most common structural material is cellulose, and that the plant embryos are protected within a special gametophyte structure, an important adaptation for living on land. By that time he would be likely to turn the hose on you.

Alternation of Generations in Plants

As we saw in Chapter 23, the plant life cycle has an alternation of generations and is sporic (see Essay 23.2 and Figure 24.1). Briefly, in the sporic life cycle the haploid products of meiosis are spores and not gametes (unlike in the gametic animal life cycle). In plants, haploid spores form multicellular gametophytes, which in turn give rise to tissues in which gametes are formed. When gametes fuse in fertilization, the diploid sporophyte phase reappears and a new cycle begins. Certain cells in the multicellular sporophyte enter meiosis once more, and another cycle of spores and gametophytes begins.

In its most primitive state, as seen in some filamentous green algae, the gametophyte phase is quite dominating and the sporophyte phase brief and simple. Further, the two phases tend to occur in separate individuals. But recall that although the green alga *Ulva* and the red alga *Polysiphonia* maintain separate sporophytes and gametophytes, the phases are about equal and are even similar in appearance. In the brown algae and in almost all plants, the sporophyte dominates the life cycle, and the gametophyte is incorporated into the sporophyte body. The trend is most striking in the seed plants. In fact, in the most recently evolved plants—

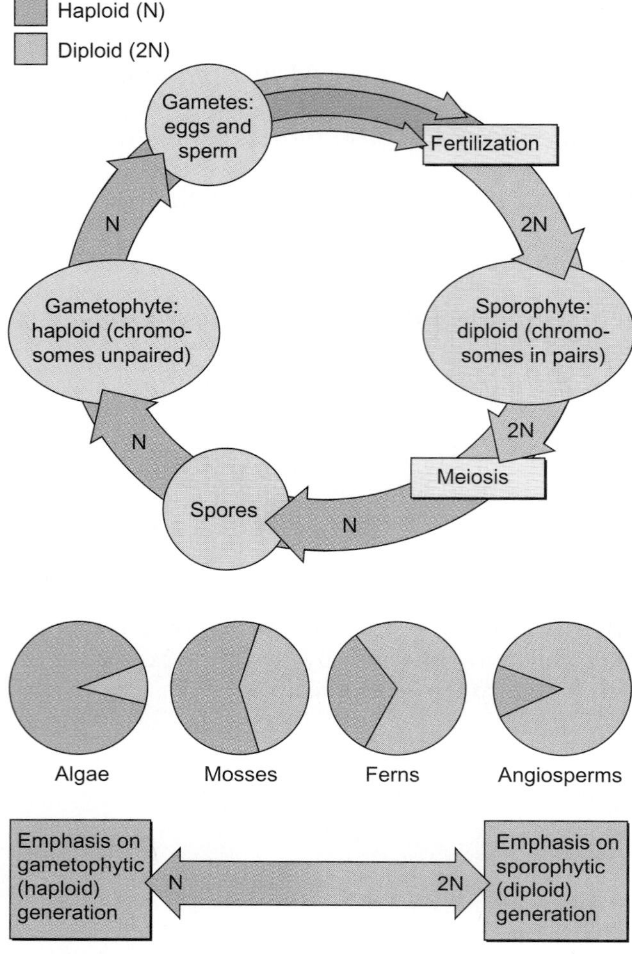

FIGURE 24.1 **ALTERNATION OF GENERATIONS.** Some plants show a pronounced alternation of generations. The sporophyte generation begins when the eggs and sperm are joined in fertilization. The mature sporophyte then produces spores (male or female) whose subsequent growth comprises the gametophyte generation. The gametophytes then produce gametes ("eggs" or "sperm"), which join in fertilization to form a new sporophyte. The relative importance of diploidy differs in the generations of various groups of plants. We see, then, that the most conspicuous generation in algae is the gametophytic (haploid), while in angiosperms it is sporophytic (diploid). What about mosses and ferns? (There are exceptions to this trend within each group.)

the angiosperms, or flowering plants—the gametophyte is restricted to a relatively few haploid cells within the flower.

The earliest plants were **homosporous** (*homo,* same)—that is, they produced only one kind of spore, as do some plants today. Most more recently evolved plants, however, produce two kinds of spores and are called **heterosporous** (*hetero,* different). The two different kinds of spores are, in most plants, called **microspores** ("small spores") and **megaspores** ("large spores"). These descriptive terms are not really precise, because microspores and megaspores are often similar in size. They are better distinguished by what they produce. Microspores give rise to male gametophytes (called **microgametophytes**), which in seed plants are minute pollen grains. Megaspores give rise to female gametophytes (**megagametophytes**), the large ovules of seed plants. Eventually, selected cells in the two will develop into many sperm and only one egg.

With these two evolutionary trends in mind, let's look into an important exception, the *bryophytes,* which are essentially nonvascular evolutionary holdouts with highly dominating gametophytes and with spores that all look alike.

NONVASCULAR PLANTS: DIVISION BRYOPHYTA

The bryophytes (Division Bryophyta)—the mosses, liverworts, and hornworts—are multicellular terrestrial plants with specialized tissues that divide the tasks of life. There are about 23,000 named species, most of which are mosses.

Bryophytes generally lack the specialized supporting and conducting tissues of vascular plants, although many mosses have primitive versions of such tissues. For the most part, water and nutrients move through the bryophyte body via the tedious process of cell-to-cell transport in these small ground-hugging plants (Figure 24.2).

The absence of vascular tissues in bryophytes also means that they lack *true* roots, stems, and leaves, each of which, by definition, requires such tissues. They do have leaflike scales that contain chloroplasts in which photosynthesis occurs, as well as stemlike structures and **rhizoids,** which are rootlike hairs.

Bryophytes anchor themselves in the soil by these rhizoids. Unlike true roots, the rhizoids are not involved in the absorption and transport of water; instead, water is absorbed through the entire organism. Many mosses can get by in fluctuating environments because they can survive a temporary drying out, even becoming dry and brittle; but they revive and begin metabolizing again when moist conditions return. In fact, some bryophytes do quite well in deserts and high mountains.

Despite their successful transition to terrestrial life, the mosses and their relatives have some of the

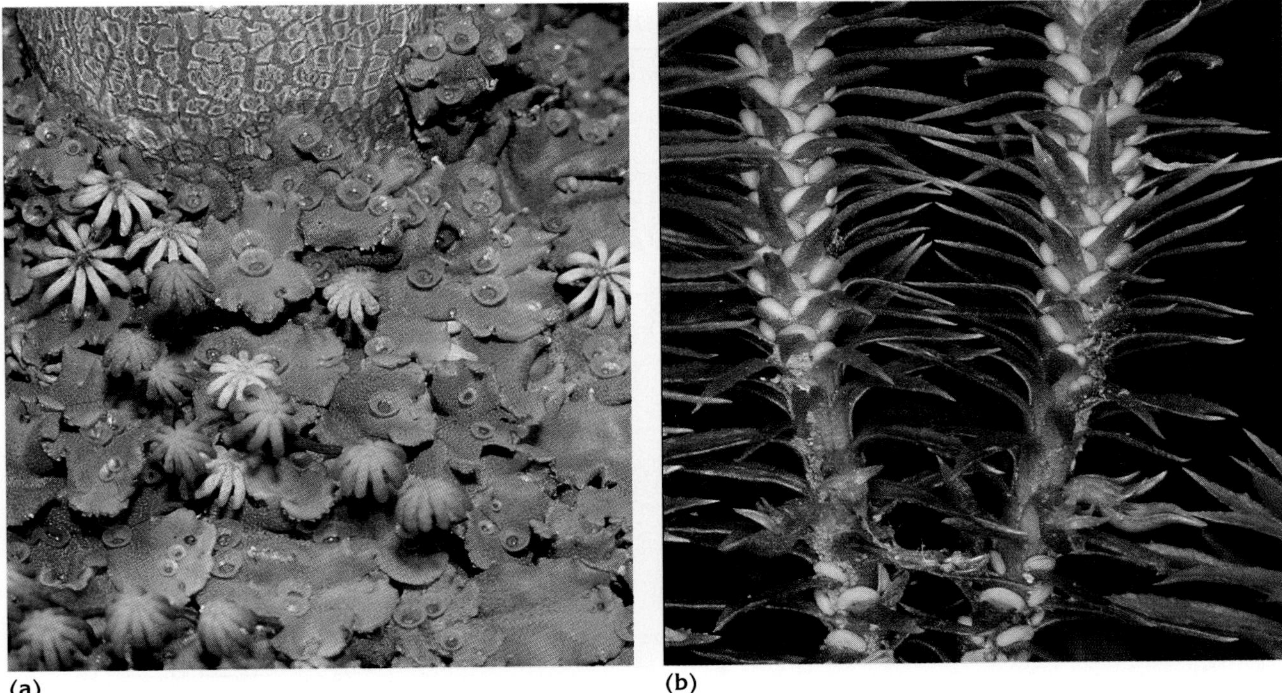

(a) (b)

FIGURE 24.2 BRYOPHYTES.
The gametophytes of *Polytrichum* (**a**) form the green leafy carpet we generally think of as moss. Emerging from gametophytes are the sporophytes, consisting of a brown stalk and capsulelike sporangia. The capsules will open to scatter the haploid spores. The liverwort gametophyte (**b**) is a flattened, ground-hugging growth. The erect bodies that look like miniature palms are egg-producing structures.

same requirements for reproduction as did the ancient green algae from which they ultimately descended. Their sperm have flagella and must swim in order to reach the egg, which lies protected within the parent gametophyte in a vaselike structure called the **archegonium.** In some bryophytes, swimming to the egg is not much of a trick, because the egg and sperm are produced on the same gametophyte. In other cases, the sperm must travel from the **antheridium** (where it is produced) of one gametophyte to the archegonium of another gametophyte. It accomplishes this more frequently by being splashed by falling raindrops than by its own flagellar activity, but it is a perilous trip all the same. The embryo then develops within the archegonium, where it is shielded from the drying air, an important evolutionary advancement in the development of land plants. The life cycle of a moss is seen in Figure 24.3.

While some botanists consider the bryophytes to have primitive traits, closely resembling the earliest plant life, others disagree. They maintain that bryophytes branched from the vascular plant line some 350 million years ago, well after vascular plants arose (Table 24.1). (The oldest known plant fossils are clearly those of vascular plants.) Bryophytes, they assert, are not primitive after all, but owe their unique traits to evolutionary simplification, the loss of certain ancestral characteristics.

VASCULAR PLANTS

The vascular plants (tracheophytes), comprising the vast majority of today's plant species, have two types of specialized conducting tissues: **xylem,** which transports water and minerals, and **phloem,** which distributes organic food from one part of the plant to another. The xylem often has an important secondary function: It provides support for the plant body. Wood consists primarily of xylem. Xylem and phloem are discussed in detail in Chapter 29.

The earliest tracheophytes (appearing in the fossil record some time before the earliest bryophytes) belong to an extinct group, called the Rhyniophytes, a division that flourished in the marshes of the early Paleozoic era (see Table 24.1). They were simple plants that lacked leaves and roots. A **rhizome** (underground stem) produced vertical branches and

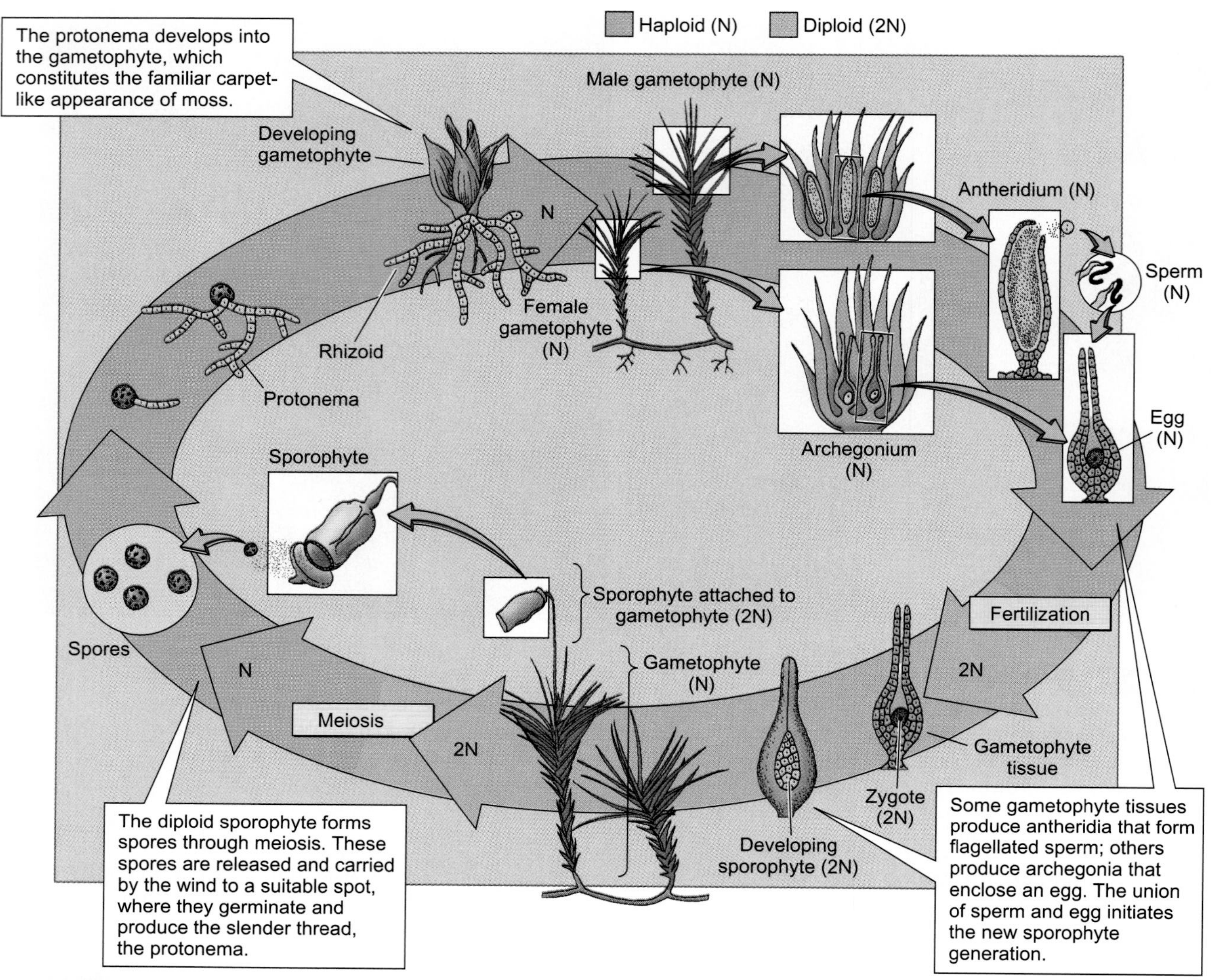
Haploid (N)
Diploid (2N)
The protonema develops into the gametophyte, which constitutes the familiar carpet-like appearance of moss.
Male gametophyte (N)
Developing gametophyte
N
Antheridium (N)
Sperm (N)
Female gametophyte (N)
Rhizoid
Protonema
Egg (N)
Archegonium (N)
Sporophyte
Sporophyte attached to gametophyte (2N)
Fertilization
Spores
N
Gametophyte (N)
2N
Meiosis
2N
Gametophyte tissue
Zygote (2N)
The diploid sporophyte forms spores through meiosis. These spores are released and carried by the wind to a suitable spot, where they germinate and produce the slender thread, the protonema.
Developing sporophyte (2N)
Some gametophyte tissues produce antheridia that form flagellated sperm; others produce archegonia that enclose an egg. The union of sperm and egg initiates the new sporophyte generation.

FIGURE 24.3 THE MOSS LIFE CYCLE.
The haploid generation dominates. In fact, the diploid sporophyte grows out of the gametophytic tissue and depends on it for nourishment. The diploid sporophyte is essentially a stalked capsule that forms spores. These are released and borne by the wind to some suitable, moist place where they germinate and produce the slender thread, the protonema. This develops into the familiar carpet we think of as moss. Some tissues produce antheridia that form flagellated sperm, others produce archegonia that enclose a stationary egg. The union of the sperm and egg initiates the new sporophyte generation.

simple, anchoring rhizoids (Figure 24.4). Within their stems lay a system of crude tubules, true xylem elements called **tracheids,** which are specialized for water conduction. With this innovation, the evolutionary future of the tracheophytes was assured.

Vascular Plant Divisions

According to most taxonomists, there are nine divisions of vascular plants, four of which release spores that develop into inconspicuous—but independent—gametophytes. The great 18th-century taxonomist Linnaeus was so confounded by the apparent lack of reproductive structures in these plants that he named them *cryptogams,* which quaintly translates as "hidden marriage." The rest of the plants, with their highly visible cones and flowers, came to be known as *phanerogams,* or "apparent marriage." The nine vascular plant divisions are shown in Table 24.2.

The first three divisions of seedless plants in Table 24.2—the whisk ferns, club mosses, and horse-

TABLE 24.1 *Geological History of Plants*

Era	Period(s) (Millions of years ago)	Conditions	Plant History	
Cenozoic	Quaternary Tertiary	Glaciation, mountain building, cooling	Extensive grasslands Herbaceous plants Deciduous forests	AGE OF ANGIOSPERMS
	65			
Mesozoic	Cretaceous	Rocky mountains Extensive lowlands	Flowering plants	AGE OF GYMNOSPERMS (seed plants)
	135			
	Jurassic	Lowlands, inland seas	Conifers Cycads	
	197			
	Triassic	Mountains, drying		
	225		Most recent continental drift begins	
Paleozoic	Permian	Glaciers Inland seas dry up	Gingkos Earliest conifer fossils Earliest fern fossils	GREAT PALEOZOIC FORESTS (seedless plants)
	280			
	Carboniferous	Mountain building	Sphenophytes Earliest bryophyte fossils	
	345			
	Devonian		Lycophytes	PLANT LIFE BEGINS
	405			
	Silurian	Extensive shallow seas, mild climate	Earliest vascular plant fossils (Rhyniophytes)	
	425	Plants invade land		
	Ordovician		First plant fossils	
	500			
	Cambrian		Algae, fungi	
	570			
Precambrian			Algae, bacteria	

tails—were once an important part of the earth's flora. They were the giants of the Paleozoic forests. Vast coal deposits in the earth today are convincing evidence of the dominating size and vast numbers reached by these plants. But the heyday of such plants is over, and only a few stragglers remain, such as the psilophytes, lycophytes, and sphenophytes (Figure 24.5). The fourth seedless plant division, the pterophytes, or ferns, is a different story. A close look at this holdout will help you to visualize the unusual life cycle of the cryptogams.

Ferns: Division Pterophyta

The ancient Paleozoic forests were also graced with a much more familiar plant, the fern. Ferns (Division Pterophyta) have survived in great numbers and have adapted to both a vastly changing environment and competition from seed plants (Figure 24.6). They are widespread, thriving in both temperate and tropical climates. Some even live in the desert, and a few manage to survive in the dim recesses of smoky lounges.

TABLE 24.2 THE NINE VASCULAR PLANT DIVISIONS

I. **SEEDLESS PLANTS** (tiny independent gametophytes)
  Division Psilophyta ("naked plants"): whisk ferns
  Division Lycophyta ("spiderlike plants"): ground pines or club mosses
  Division Sphenophyta ("wedge plants"): horsetails
  Division Pterophyta ("winged plants"): ferns

II. **SEED PLANTS** (tiny gametophyte incorporated into sporophyte)
  A. **GYMNOSPERMS** (naked seeds)
  Division Cycadophyta: cycads
  Division Ginkgophyta (maidenhair tree): ginkgo
  Division Gnetophyta: gnetophytes
  Division Coniferophyta ("cone-bearers"): pines and other conifers
  B. **ANGIOSPERMS** (seeds with fruit)
  Division Anthophyta ("blossom or flower plant"): the flowering plants

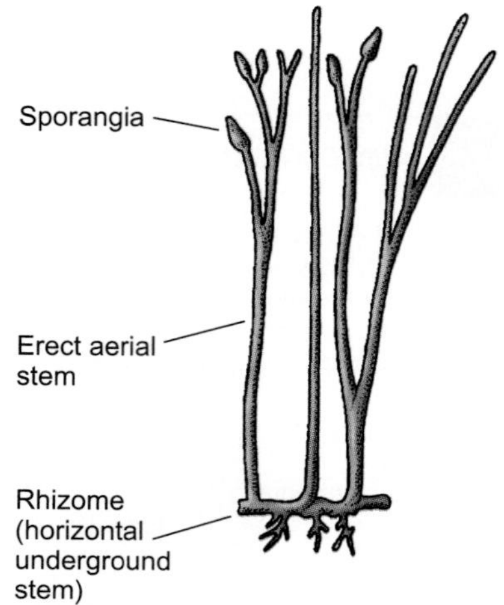

FIGURE 24.4 RHYNIA, THE OLDEST KNOWN FOSSIL VASCULAR PLANT.
Rhynia flourished some 350 to 400 million years ago. It lacked leaves and true roots but had photosynthetic stems that grew up from an underground stem, or rhizome. Spore-producing sporangia developed on the tips of the erect stems. True conductive tissue is evident in the fossil traces. This is the sporophyte; the gametophyte is unknown.

(a)

(b)

(c)

FIGURE 24.5 REMAINING PLANTS OF THE TYPE THAT ONCE DOMINATED THE EARTH.
(a) *Psilotum,* a psilophyte, produces an erect, sometimes branched stem with scattered sporangia. **(b)** Lycophytes, also called ground pines and club mosses, are neither pines nor mosses. Some produce complex, conelike sporangia at their tips. **(c)** *Equisetum,* a sphenophyte, also called a horsetail, may have different photosynthetic and reproductive shoots rising from a single rhizome. Note the multiple sporangia clustered at the tip of the reproductive shoot.

FIGURE 24.6 FERNS.
Although many ferns are restricted in size and are confined to the forest floor, tree ferns *(left)* may grow to a height of 15 m (about 50 ft).

The fern sporophyte often consists of a number of lush, finely divided leaves that arise from a thick rhizome, or underground stem. The leaves function in both photosynthesis and spore production. As the plant grows, the leaves first appear as the familiar coiled *fiddleheads,* each of which then unrolls into a large, complex leaf. The stem also gives rise to a profusion of fine, hairy roots that absorb water and minerals (Figure 24.7).

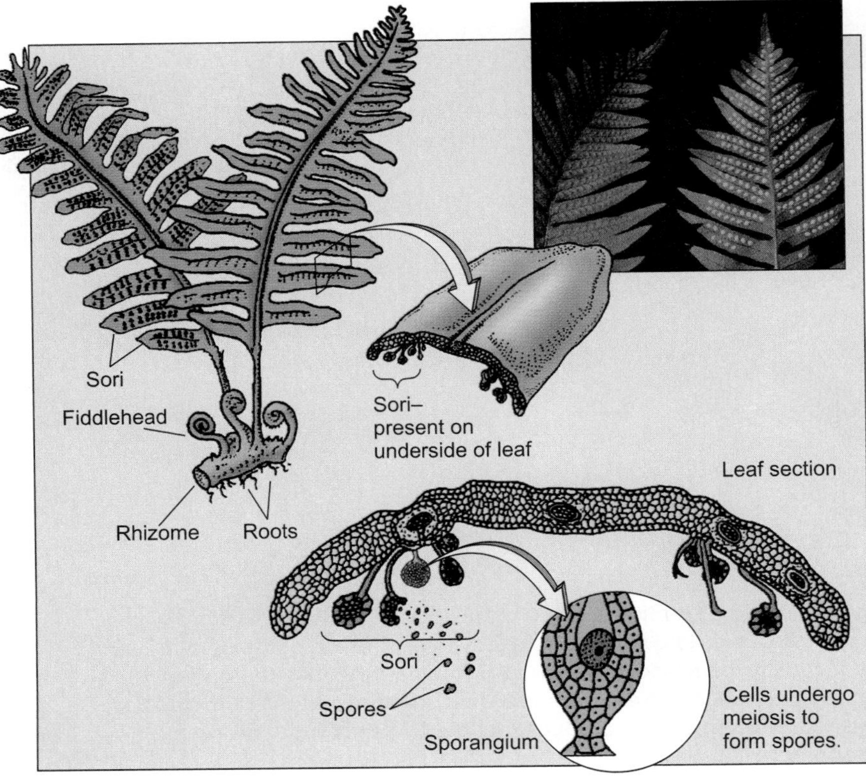

FIGURE 24.7 ANATOMY OF A FERN.
Below the soil, an extensive underground rhizome produces root growth and sends leaves above ground to form the cluster shown here. The leaves emerge from the rhizome as tightly curled fiddleheads, which gradually open. Fern sporangia occur in a variety of forms and patterns (see photo), but the sorus shown here is common. Spores are produced meiotically in the sporangia, and remain there until the sporangium splits and the spores are ejected.

Most ferns are homosporous. The simple haploid spores are produced by meiosis in structures called *sori,* often seen on the underside of leaves. The sori may be scattered, or they may occur in neat rows, appearing as brown or red dots (that have often sent alarmed gardeners to the store for pesticides). Each sorus is a group of *sporangia* (spore-forming organs), sometimes hidden under a scalelike cover, or *indusium* (Figure 24.7). As the sporangia rupture, the spores are released, some becoming windborne and traveling great distances.

When a fern spore germinates, a tiny, heart-shaped, photosynthetic gametophyte, the *prothallus,* develops and produces gametes. Sperm and egg form in antheridia and archegonia, respectively—structures that develop on the gametophyte's lower

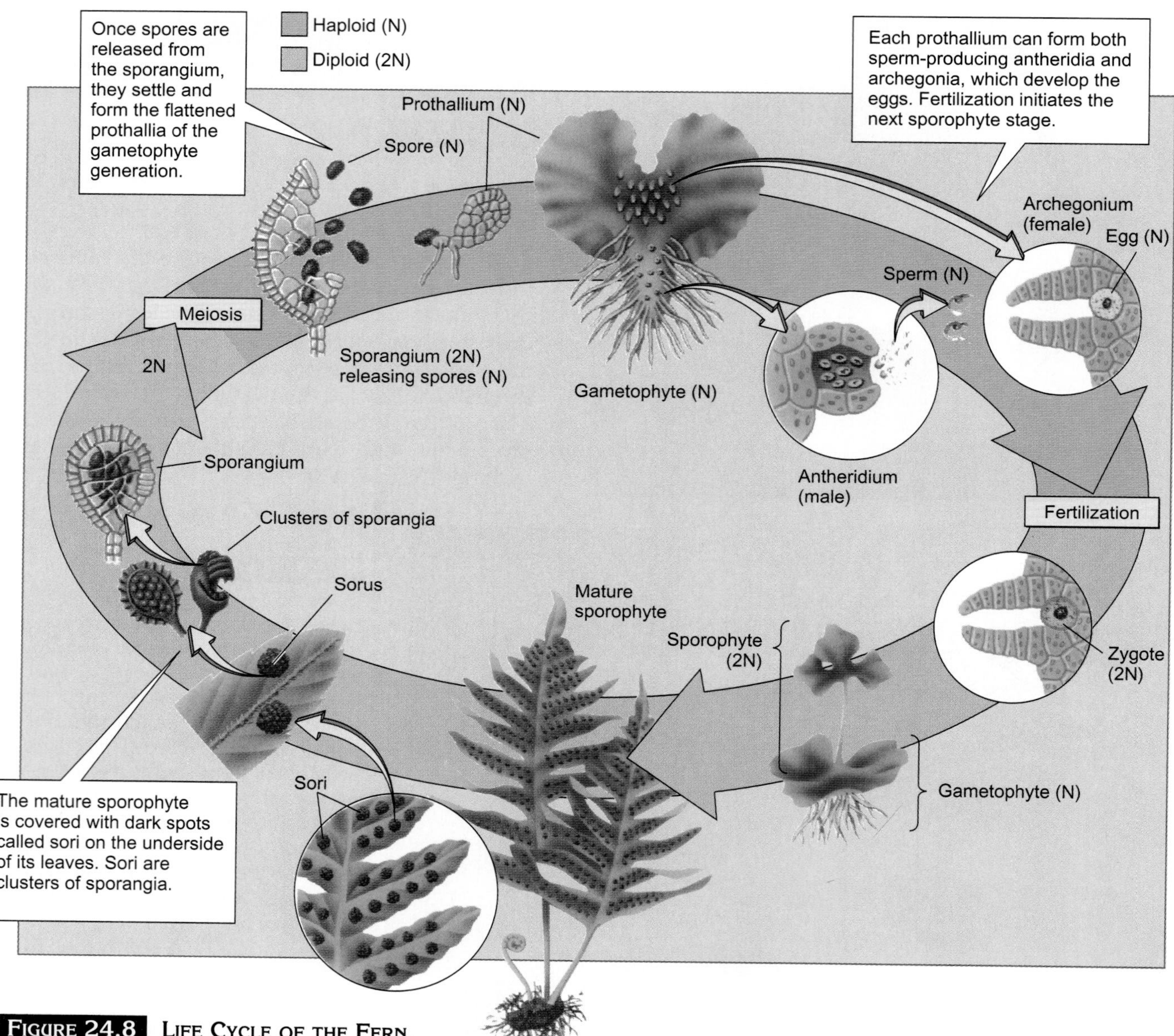

FIGURE 24.8 LIFE CYCLE OF THE FERN.
In ferns, the sporophyte generation is highly dominant. After fertilization, the sporophyte begins to grow from the gametophytic archegonium. It soon produces its own roots and leaves, as the old gametophyte withers and dies. Underneath the fronds form the dark spots called *sori,* within which are the spore-producing sporangia. The spores are released and, once settled, form flattened structures of the gametophyte generation, each called a *prothallus.* These form sperm-producing antheridia and egg-developing archegonia. Fertilization initiates the next sporophyte stage.

surface. After fertilization, the young, diploid sporophyte grows from the nurturing tissue of the gametophyte but soon becomes independent (Figure 24.8).

The Mesozoic Era: Time of Gymnosperms (Divisions Cycadophyta, Gingkophyta, Gnetophyta, and Coniferophyta)

The primitive vascular plants persisted through the Paleozoic era. Then, at its close some 225 million years ago, much of the earth experienced vast geological upheaval, and plant life also underwent dramatic change.

Throughout the first half of the Paleozoic era, the earth was a rather smooth globe, and the low-lying continents were largely covered with warm, shallow seas. Later in the Paleozoic era, and on into the next era—the Mesozoic—the surface of the earth itself began to shift as the soggy lowlands rose up and vast mountain ranges emerged. Great deserts formed in the shadows of these newborn mountain ranges. The monotonously warm global climate changed, with gradual cooling at first and then fluctuation between extremes as ice ages came and went. Even the great landmasses were restless, marked by the most recent episode of continental drift.

Toward the close of the Paleozoic era, for reasons not entirely understood, an estimated 90% of the animal species rapidly died out. These were the greatest mass extinctions of life the earth has ever witnessed. As the animal species perished, most of the great plants of the Paleozoic era died out with them, leaving behind only scattered remnants.

With the demise of the ancient forests, the competitive edge passed to the emerging Mesozoic seed plants. Up to that time, they had existed inconspicuously among the Paleozoic giants; but with the extinction of the older giants, the newly changed earth presented new opportunities. Soon the seed plants appeared everywhere over the Mesozoic landscape. These were primitive **gymnosperms**—plants in which the seed develops without a covering—the forerunners of today's conifers.

Gymnosperm Specializations

Why were the gymnosperms so successful? First, they had evolved extensive root systems that could draw water from the earth and transport it through the increasingly well-developed xylem tissue, upward to the leaves—just as the nutrients manufactured in the leaves were transported via the phloem throughout the plant. Furthermore, both the root and the stem were capable of extensive *secondary growth*—growth in diameter by the addition of rings of vascular tissue.

In addition, the gametophyte had become incorporated into the sporophyte. So this vital stage of the life cycle was protected from the rigors of a drying habitat, a factor that may have contributed to the demise of the Paleozoic giants. Further, the male gametophyte (gymnosperms are heterosporous, with male and female gametophytes) had evolved into the **pollen grain,** which could travel for miles and withstand severe drying, thereby remaining viable for years. Fertilization did not require an outside source of water, since the windblown male gametophyte could form a deeply penetrating **pollen tube** that could deliver sperm directly to the hidden female gametophyte.

Perhaps most importantly, there was the evolution of the **seed** itself (Figure 24.9). The female gametophyte came to remain within the parental sporophyte tissue, where it could be supported, protected, and nourished. Then, after fertilization, the seed itself developed and could be released for dispersal and propagation. The gymnosperm seed consists of tough, nearly waterproof outer seed coats surrounding a mass of stored nutrients and a dormant, partly formed plant embryo—the next sporophyte generation. Seeds, like pollen, can survive harsh conditions and may successfully germinate after years of dormancy.

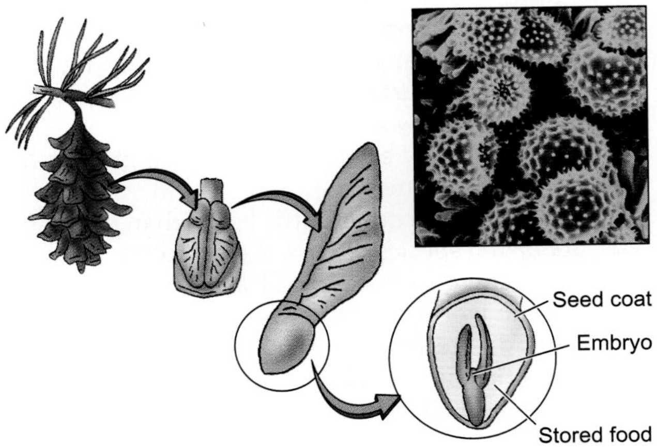

Figure 24.9 Pine Pollen and Seed. The pollen in the scanning electron micrograph is that of the northern white cedar. Pine seeds contain a tiny plant embryo surrounded by stored food and a tough, resistant seed coat. The seeds are sometimes winged.

(a)

(b)

(c)

FIGURE 24.10 **GYMNOSPERMS.**
(a) The ginkgo *(Ginkgo biloba)*. Male sporophytes (sexes are separate) produce bright yellow pollen cones. **(b)** Palmlike cycads have separate sexes also. Note the bright yellow pollen cone of this male. **(c)** *Welwitschia mirabilis*, a gnetophyte, is seen in the remote Kalahari desert. It produces two ragged, splitting leaves with seed cones on the edges.

THE GYMNOSPERMS TODAY

The word *gymnosperm* means "naked seeds," a fitting term since the seed is not surrounded by fruit as is the case with the angiosperm. Further, while the reproductive structure of the angiosperm is the flower, in most gymnosperms it is the cone. Four divisions of gymnosperms survive today, but they are not closely related to one another. These are the *ginkgos* (Division Gingkophyta), the *cycads* (Division Cycadophyta), the strange *gnetophytes* (Division Gnetophyta), and the familiar conifers (Division Coniferophyta).

The ginkgo, or maidenhair tree, qualifies as rare because there is only one surviving species, *Ginkgo biloba* (Figure 24.10a). Earlier, individuals of this single species were rare even in their native China. Now, however, the ginkgo has become a common decorative tree throughout the world—partly because it is quite pretty, partly because of its unusual appearance, and partly because it is relatively resistant to insects, disease, and even air pollution. The large, rather smelly seed is considered a great delicacy by some gourmets.

The cycads have done a little better than the ginkgos; about 100 species of cycads exist today. Most of these are native to the tropical and subtropical regions of the world. In hothouses and botanical gardens, cycads are often mistaken for palms, which they superficially resemble, just as ginkgos superficially resemble ashes or sycamores. But ashes, sycamores, and palms are all flowering plants, unrelated to either cycads or ginkgos. Furthermore, palms produce flowers, and cycads produce massive cones (see Figure 24.10b).

The 70 or so named species of gnetophytes are a bizarre, diverse lot. Some botanists believe that the flowering plants evolved from this group, or at least that the two groups are related. The evidence is sparse but interesting. For example, plants of the genus *Gnetum* have leaves with netlike veins, rather resembling the leaves of a cherry tree. All gnetophytes, including the rare and remarkable *Welwitschia mirabilis* (see Figure 24.10c), have xylem vessels. *Vessels* are specialized xylem cells that are common in flowering plants (see Chapter 29), but do not exist in the xylem of any other gymnosperm. Finally, gnetophytes, like conifers and angiosperms, have nonmotile sperm.

The Conifers. The conifers are the cone-bearers and include the pines, firs, spruces, redwoods, hemlocks, junipers, and larches. There are about 500 named species—not very many, considering the vast coniferous forests of the world. Still, the conifers comprise the largest group of gymnosperms. Many conifers thrive in the cold climates of the earth (Figure 24.11), but some also do quite well in warm regions. Typically, they produce needles: narrow, tough leaves with thick, water-resistant cuticles (waxy coverings). Needles are an adaptation to arid conditions, whether the habitat is simply dry or the water is periodically bound up as ice. Needles also resist high winds better than do flattened leaves. Nearly all conifers are "evergreen"—that is, the leaves do not die and fall off all at one time, leaving the tree leafless. However, some trees continually shed their leaves, keeping each of them for 2–4 years. The bristlecone pine, however, is known to keep the same leaves for up to 50 years.

The most common and best-known conifer group is the pine family, Pinaceae. Pines have separate male (pollen-bearing) and female (ovule-bearing) cones, and each tree is bisexual (that is, each sporophyte

FIGURE 24.11 CONIFERS.
Coniferous forests are found chiefly in the cold regions of the earth. They form a continuous belt across northern Asia, North America, and Europe.

bears both male and female cones). The shapes and sizes of the cones are usually the most distinctive features of the different pine species. The seeds—pine nuts—remain in the female cones for about two years as they slowly mature (Figure 24.12). During this period, the seeds of many pines develop a winglike structure that will catch the wind, so that when the cone opens and they are released, they will float far from the parent tree. In some pines, the cones will not open to release their seeds until they are first scorched by fire and then cool off; each new generation of such pines must await its own forest fire.

In ancient Mesozoic times, the dominance of the gymnosperms and that of the dinosaurs were strangely intertwined, and both groups fell from prominence toward the end of the Mesozoic era. The giant reptiles—except for the lines that led to monitor lizards, alligators, and crocodiles—were to pass into oblivion, while the conifers, although managing to survive, decreased drastically in number.

The Cenozoic era, which followed the Mesozoic era and began 65 million years ago, was the time of the flowering plants. The flowering plants came to dominate the landscape from the tropics to the temperate regions, leaving mainly the colder northern regions to most of the surviving gymnosperms. The conifers were well on their way to being rare, archaic relics, like the gnetophytes and the ginkgos, but in the Pleistocene period, which began only 2 million years ago, they began an amazing comeback. In these relatively recent times both their geographic range and the number of individuals, if not the number of species, have greatly increased.

THE CENOZOIC ERA: THE TIME OF THE FLOWERING PLANTS

Flowering plants (Division Anthophyta) are **angiosperms**—that is, their seeds are surrounded by *fruit.* The development of fruit was a remarkable evolutionary event, as we shall see. But first, we might ask, when did the flowering plants evolve? Were they inconspicuously sprinkled among the dominant, if primitive, conifers, cycads, and ginkgos of the Mesozoic era, or were they absent altogether in those days? What kinds of conditions promoted their rather sudden explosion to dominance?

One theory is rather simple: The flowering plants had evolved two advantages, *animal pollination* and *seed dispersal.* Insects had been around for a very long time, but the gymnosperms had depended on wind pollination alone. The earliest flowering plants, on the other hand, took advantage of the animals of the time, particularly the insects. Even today, most flowering plants depend on insects for pollination, although many have reverted to wind pollination. Animal pollination, being efficient and rather specific, may have given the early flowering plants a decided advantage in the Darwinian struggle. As for seed dispersal, the key event was probably the emergence of birds. Birds were attracted to the fleshy, tasty, sugar-laden tissue encasing resistant seeds, and seeds passed through their digestive systems in viable condition, often after being carried to distant but suitable locations. Again, this attractiveness to birds was a major advantage to early flowering plants, as it is to many flowering plants today.

Other theories abound as well. Some investigators believe that the relatively rapid emergence of the flowering plants can be explained on the basis of geological and climatological changes; temperatures of the Cenozoic era were colder than those of the Mesozoic era, for instance. Any such change brings new adaptive opportunities, and flowering plants did well under these more rigorous conditions. Also, at the end of the Mesozoic era, 65 million years ago, the drifting continents had reached approximately their present positions, and thus life could begin adapting to the relatively constant conditions that prevailed, based on their stable positions on the globe (see Essay 20.1).

THE ANGIOSPERMS TODAY

The fruit of angiosperms is often of the familiar soft and tasty kind, but also of the dry, hardened form such as those surrounding the seeds of grasses, hickory trees, and oaks. **Fruit** simply refers to the mature ovary in which fertilization and development have occurred. Angiosperms, like other seed plants, are heterosporous, producing megaspores and microspores. The female gametophyte is highly reduced, generally just seven cells in all, tucked away in the ovules of the flower. (Reproduction in angiosperms is discussed in detail in Chapter 28.)

All you have to do is look around to see that the angiosperms—the flowering plants—have "inher-

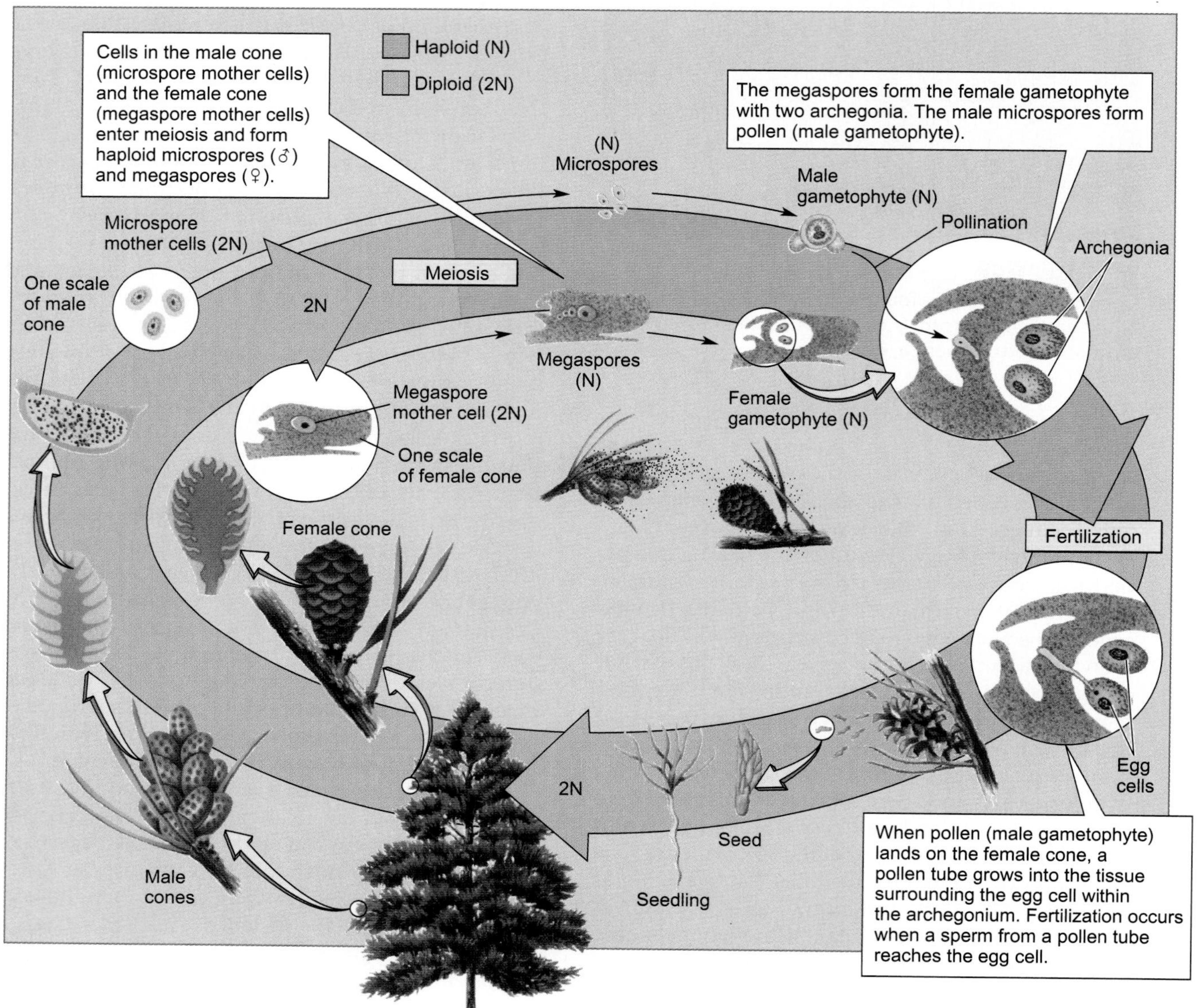

FIGURE 24.12 ALTERNATION OF GENERATIONS IN A CONIFER.
With the exception of cells hidden in their reproductive structures and an occasional yellow cloud of wind-borne pollen, all of the conifer we see is the sporophyte. The conifers have no truly separate gametophyte generation. Cells in the male cones, known as microspore mother cells, enter meiosis and form haploid microspores. These microspores form a winged pollen grain and, later, the male gametes. Cells in the female cone, the megaspore mother cells, enter meiosis and produce haploid megaspores. These produce the female gametophyte with two archegonia. When pollen (the male gametophyte generation) lands on the female cone, a pollen tube grows into the tissue surrounding the egg cell within an archegonium. Fertilization occurs when a sperm from a pollen tube reaches the egg cell.

FIGURE 24.13 **ANGIOSPERMS.**
The flowering plants, or angiosperms, are among the most striking of plant species.

ited the earth," at least for the time being (Figure 24.13). No one knows how many kinds of flowering plants there are, but about 250,000 different species have been named.

In spite of such diversity, the flowering plants can all be divided into two great classes. These are the **Monocotyledonae** ("monocots," for short)—those with one seed leaf—and the **Dicotyledonae** ("dicots")—those with two seed leaves. The monocots are in the minority with only about 50,000 named species. Among them are all of the grasses—including corn, wheat, and all the other cereal grains we depend on—and palm trees, orchids, tulips, lilies, yuccas, and many other familiar plants. Most other flowering plants are dicots. Monocots and dicots are rather distantly related; and although they share many key features, they also differ from each other in several important aspects. The most obvious differences are in their leaf patterns, seeds, floral arrangement, and the pattern of their vascular systems, as we see in Figure 24.14.

As we might expect of such a successful, varied, and numerous group, flowering plants have developed a dazzling variety of life cycles. Nonetheless, because of their common heritage and their essentially similar challenges, their cycles are based on underlying unifying themes, as we will discover in Part IV.

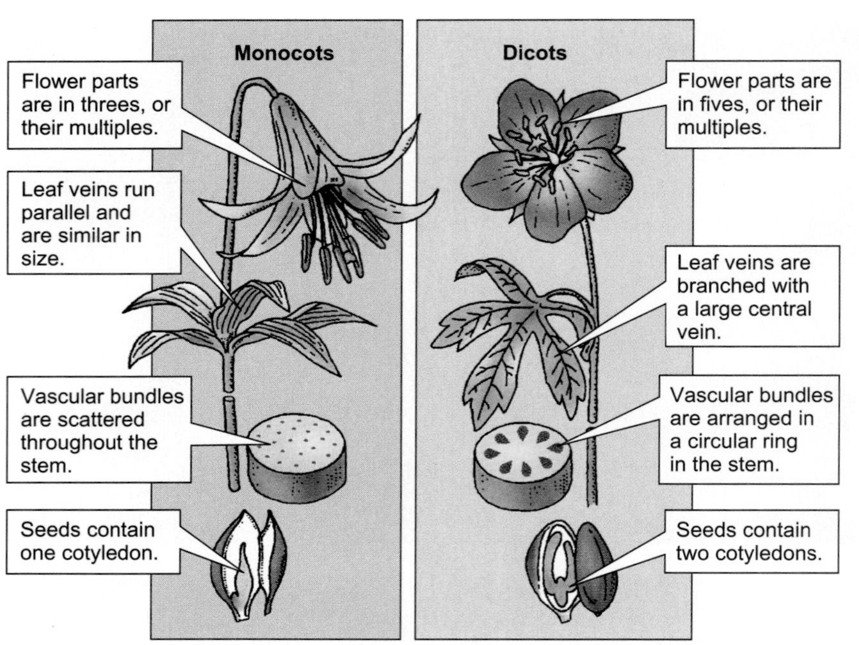

Monocots
Dicots
Flower parts are in threes, or their multiples.
Leaf veins run parallel and are similar in size.
Vascular bundles are scattered throughout the stem.
Seeds contain one cotyledon.
Flower parts are in fives, or their multiples.
Leaf veins are branched with a large central vein.
Vascular bundles are arranged in a circular ring in the stem.
Seeds contain two cotyledons.

FIGURE 24.14 MONOCOTS AND DICOTS. Angiosperms are divided into monocots and dicots according to four major structural differences. The most obvious difference is in the pattern of veins in the leaf, which is netlike in the dicot (such as a sugar maple) and parallel in the monocot (such as corn). The seeds also differ in that two *cotyledons* (seed leaves) are seen in the dicots, while only one is present in the monocots. The flowers of dicots have four or five floral parts, or multiples of these numbers. Monocots have three floral parts, or multiples of three. The vascular system of dicots is generally a neat ring of vascular bundles arranged in a circle around the stem. In the monocots, the vascular system commonly occurs in scattered bundles.

PERSPECTIVES

After learning what a plant is, after all, we learned something about how its gametophyte and sporophyte generations can alternate. Then we moved from the ground-hugging bryophytes to the tall-standing trees, focusing first on the great cycad and gymnosperm forests, and then on the familiar angiosperms. It is important to relate the changes in plants to geological events and to other species, especially the animals, since the histories of plants and animals are so inexorably linked. Next, then, as we begin our survey of the animals, keep in mind what you have just learned, and try to see the two kingdoms as part of one grand evolutionary pageant.

SUMMARY

WHAT IS A PLANT?

Some plant characteristics are multicellularity; autotrophy; the presence of roots, stems, and leaves; the presence of chlorophyll *a* and *b* and carotenoids; starch as stored food; cellulose as a structural material; and embryonic development within a protective archegonium.

ALTERNATION OF GENERATIONS IN PLANTS

Plants follow a sporic life cycle in which the evolutionary trend in the alternation of generations has been from separate sporophytes and gametophytes toward a highly dominating sporophyte that incorporates the gametophyte within reproductive structures. A second evolutionary trend has been away from homospory and toward heterospory. Thus most plants produce megaspores and microspores, which give rise to megagametophytes (female) and microgametophytes (male).

NONVASCULAR PLANTS: DIVISION BRYOPHYTA

The bryophytes include mosses, liverworts, and hornworts.

Although bryophytes have some specialized tissues, most lack extensive conducting and supporting vascular tissue; accordingly, have no true roots, stems, and leaves; and remain small. Rhizoids anchor the body, but water must be absorbed by all surface cells. Water is required for the transit of sperm from antheridium to archegonium, where fertilization occurs. In the moss alternation of generations, the haploid gametophyte (the green leafy plant) is dominant.

There is strong disagreement over whether bryophytes today resemble the earliest plants. Certain fossil evidence suggests that bryophytes appeared after the tracheophytes were established, and that their apparently primitive traits are due to evolutionary simplification.

Vascular Plants

Vascular plants—tracheophytes—have water-conducting xylem and food-conducting phloem. Vascular tissue also provides structural support. The more primitive tracheophytes have dominating sporophytes but retain minute, separate gametophytes. The earliest were the simple Rhyniophytes of the Paleozoic era. They had an underground rhizome from which simple, forking branches arose. Their xylem included tracheids but no vessels.

The primitive seedless plants, or cryptogams, contain four divisions (encompassing whisk ferns, ground pines, horsetails, and ferns). Most are homosporous, releasing simple, haploid, aerial spores that form separate, minute gametophytes. The seedless plants dominated the Paleozoic forests but, except for ferns, are relatively rare today.

The phanerogams, or seed plants, make up five divisions. They have minute gametophytes incorporated into the sporophyte and are heterosporous. The male gametophyte is released as pollen. Seed plants have dominated the forests since the Mesozoic era, although ferns are widespread.

Ferns: Division Pterophyta

The fern sporophyte contains complex leaves that develop from fiddleheads. True roots absorb water. Most ferns are homosporous, forming simple, haploid spores in leaf structures called sporangia. The spores grow into minute gametophytes, which form antheridia and archegonia.

The Mesozoic Era: Time of Gymnosperms

Mountain building, a drying landscape, and a cooling climate led to the demise of most seedless plant species. Their extinction cleared the way for the first seed plants, the gymnosperms.

Success in the drier Mesozoic environment was aided by an extensive vascular tissue, made possible by secondary growth. Other adaptations included incorporation of the gametophyte in the large sporophyte and the advent of highly resistant pollen grains and tough, resistant seeds containing the delicate embryo.

Today's gymnosperms form pollen and seed cones, the latter of which bear naked seeds (without fruit). One species of the ginkgo (the maidenhair tree) remains. Cycad species are more numerous but are restricted to warmer climates. Some gnetophytes are desert inhabitants.

Conifers are widespread in colder regions and have adapted to drier conditions. Needles (or scaly leaves) are always present, but shedding of older leaves is continuous in some species. Pines produce separate male and female cones, in which pollen tube growth and seed maturation occur.

By the start of the Cenozoic era, conifers were replaced in many regions by newly expanding flowering plants, the angiosperms.

The Cenozoic Era: The Time of the Flowering Plants

The success of flowering plants can be attributed in part to animal-assisted pollination and seed dispersal. Rapid geological change may have also played a role.

Angiosperms or anthophytes, the flowering plants, surround their seeds with fruit (the mature ovary). They are heterosporous and maintain their gametophytes within the flower. Important to their success has been diversification; there are 250,000 species. They are grouped into two classes, Monocotyledonae (monocots) and Dicotyledonae (dicots).

Key Terms

homosporous • **396**
heterosporous • **396**
microspore • **396**
megaspore • **396**
microgametophyte • **396**
megagametophyte • **396**
rhizoid • **396**
archegonium • **397**
antheridium • **397**
xylem • **397**
phloem • **397**
rhizome • **397**
tracheid • **398**
gymnosperm • **403**
pollen grain • **403**
pollen tube • **403**
seed • **403**
angiosperm • **405**
fruit • **406**
Monocotyledonae • **407**
Dicotyledonae • **407**

Review Questions

1. List five characteristics of plants. Which, if any, are exclusive to plants?
2. Review the sporic life cycle. How does it differ from your own gametic life cycle?
3. What evolutionary trend do we see in sporophyte and gametophyte dominance during the evolution of plants? Where is the gametophyte in a flowering plant?
4. Define the term *heterosporous.* Name the two kinds of spores, and suggest where they are to be found in the flowering plant.
5. List three kinds of bryophytes. What is lacking in bryophytes that is present in most other plants?
6. Why should the absence of conducting tissues affect the size to which a plant can grow?
7. In what way has the bryophyte failed to fully adapt to the dry, terrestrial environment? How does it make up for this?
8. Starting with the zygote, summarize the alternation of generations in a moss. Be sure to indicate where meiosis and fertilization occur, and specifically what parts of the life cycle are haploid and what parts are diploid (see Figure 24.3).
9. If bryophytes did not descend from the most primitive plants, how can we explain the scarcity of vascular tissue, their dominant gametophyte, and their simple organization?
10. When do vascular plants first appear in the fossil record? Describe the earliest known.
11. What are cryptogams? In what three primary ways do they differ from the more advanced vascular plants?
12. Review the fern life cycle, beginning with the zygote. Be sure to explain where meiosis occurs, what the gametophyte is like, and when the diploid and haploid states are found.
13. List the physical changes that occurred at the end of the Paleozoic era. What effect did this have on animals? On the cryptogams?
14. What plant group became widespread during the Mesozoic era? List three or four important adaptations that helped these plants adapt to the new conditions.
15. List four groups of gymnosperms that made it into modern times.
16. Where do most conifers live today? Describe the leaf form, and explain how it is adaptive to conditions in their habitat.
17. Where do the microgametophyte and megagametophyte form in the pine? What other events go on in these structures?
18. What common fate fell to both gymnosperms and reptiles at the end of the Mesozoic era? What geological event in more recent times led the way for the comeback of the conifers?
19. In what two ways were animals important to the early success of flowering plants?
20. Compare the number of species of angiosperms and their distribution on the earth with that of conifers.

CHAPTER

25

Animal Origins and the Lower Invertebrates

In this chapter we will begin by learning what an animal is, *something you already thought you knew. Then we will take a look at the body plans of the various animal species before we begin to focus on the phyla. The invertebrates are a widely diverse and complex group, though, so we will cover them in two chapters, based on their body cavities. As we travel through the ranks of these species, beginning with the sponges, we will encounter several major evolutionary events. Even if they seem insignificant at first, pay attention, because they are not.*

What is an animal? This is one of those deceptively easy questions. Almost anyone can come up with an answer—but probably no one can satisfy all the experts. Animals are so varied that exceptions can be made to almost any definition. For example, most animals move, have mouths, and eat things. If the trait of multicellularity is added, that just about covers the animal kingdom. But, invariably, some creature stops us and shouts, "Not me!" One such exception is a peculiar giant red tube worm discovered near the Galápagos Islands. It lives near recently discovered vents in the ocean floor that spew mineral-laden water that is heated by the molten earth beneath. This creature has neither a mouth nor a gut, but derives its nourishment from bacterial symbionts living in its body. It doesn't move much, but just lies there, absorbing oxygen from seawater and hydrogen sulfide that spews from the vents. In almost every way, this creature is exempt from the animal kingdom. It's an animal, nevertheless.

Despite such exceptions, let's see if we can come up with a workable definition for animals. It turns out that we can cover *most* animals, at least for our purposes, with rules organized into just five categories.

1. Animals are multicellular and eukaryotic, with cells organized into tissues and organs.
2. Animals reproduce sexually.
3. Animals are essentially diploid throughout life, with a gametic life cycle.
4. Animals typically have well-organized nervous and muscular systems.
5. Animals are heterotrophic and store carbohydrate as glycogen.

Here's another deceptively easy question. Where did animals come from? Of course, there are few hard data to support any argument, but most scientists believe that animals as well as plants and fungi sprang from the ancient protists. Actually, it is generally believed that the animal kingdom evolved from two different protist lines. One, a flagellated protozoan, produced the *Parazoa,* a subkingdom with only one phylum—*Porifera,* the sponges. The rest of the earth's animals are placed in the subkingdom *Eumetazoa,* and are probably descended from a ciliated protozoan.

Body Organization

Animals have two basic types of bodies: those with *radial* symmetry, and those with *bilateral* symmetry. **Radial symmetry,** which is believed to have evolved earlier, is characterized by a cylindrical, spherical, or disk-shaped form. The body parts are arranged so that any radius extending from the center outward will pass through similar parts (Figure 25.1).

Only a few phyla today have radial symmetry. These include a few porifera (sponges), the cnidarians (the jellyfish, corals, hydroids, and sea anemones), and the ctenophorans (such as sea walnuts and comb jellies). The **echinoderms** (starfish, or sea stars; sea urchins; sand dollars) as adults have a pentaradial (five-part) symmetry.

All other animal phyla consist of bilaterally symmetrical animals. **Bilateral symmetry,** as its name suggests (*bi,* two; *lateral,* side), creates right and left sides. A section through the midline will divide the organism, roughly, into mirror images. The overwhelming number of bilaterally organized animal species may suggest to you that there are disadvantages in radial symmetry, but beware of such conclusions. Radial animals are well adapted to their niches and perform quite efficiently within them.

Within most animals there are four levels of organization: *cellular, tissue, organ,* and *organ system.* To begin with, we are aware, if all is going well, that animals are composed of cells. A tissue is a group of cells that have a common structure and function, and an organ is a group of tissues with the same function. Most animal species have achieved what is called the *organ system* level, in which a number of organs carry out some major function or life process. In animals, for example, digestion occurs in the digestive system (an organ system), a collection of various organs, including the mouth, gut (intestinal tract), pancreas, and liver. Such organs are, in turn, made up of various specialized tissues—groups of cells—performing specific tasks within the organs. For instance, the tissues lining the gut specialize in secreting digestive enzymes and absorbing digested food.

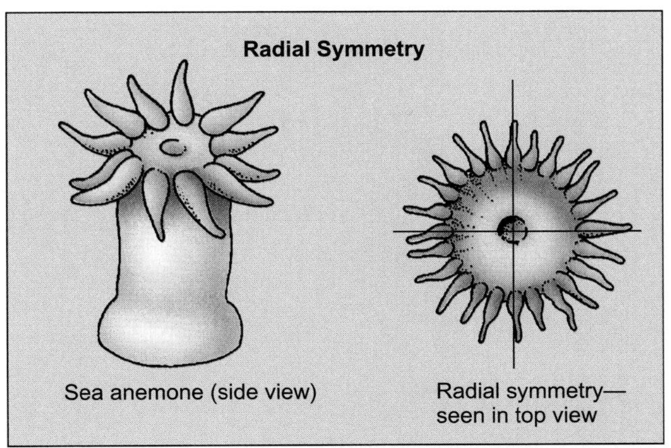

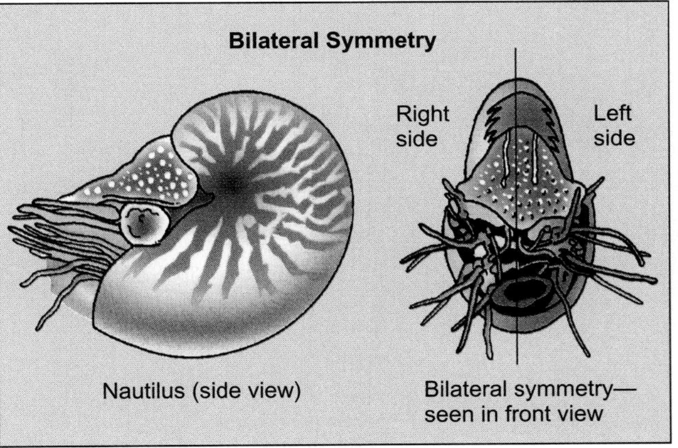

FIGURE 25.1 BODY SYMMETRY.
In radial symmetry, the body is essentially spherical, disk-shaped, or cylindrical. Any plane drawn through the center produces roughly equal left and right halves. Bilateral symmetry, on the other hand, means that the body can be equally divided by one plane only. This division produces mirror-image left and right halves that contain similar structures. Perfect examples of radial or bilateral symmetry are unusual. Can you think of structures in your own body that are asymmetric—that is, neither paired nor composed of mirror-image right and left halves?

PHYLOGENY OF THE ANIMAL KINGDOM

The history and relationship of animal phyla can be represented by phylogenetic trees that show the major milestones of animal evolution (Figure 25.2). As we saw in Chapter 19, such trees are constructed from several kinds of information, including the fossil record (paleontology) and the comparative anatomy, physiology, and embryology of living animals. More recently, data have been used from molecular and genetic sources. You will want to check Figure 25.2 as we discuss the development of each major evolutionary event.

The major branches of the tree represent the evolutionary events that produced the great variety of animals on earth today. The names shown in gray reflect the two major division of animals: the *protostomes* ("first mouth") and the *deuterostomes* ("second mouth"). (These terms refer to the embryological origin of the mouth. We will cover these terms later in this chapter, but for now, just keep in mind that such seemingly trivial embryological events as the development of the mouth can tell us a great deal about evolutionary history.) As a final point, note that this sort of tree implies little about evolutionary *time.*

THE EARLY FOSSIL RECORD

The earliest known animal fossil record, which appears to be fossil worm burrows, is estimated to be 700 million years old, but the burrows' authenticity is still being debated. The oldest undisputed animal fossils are found in rocks from 580 to 680 million years old. These rich beds, the *Ediacara fauna* of southern Australia, include abundant fossils of jellyfish, cnidarians, and worms (Figure 25.3). Interestingly, fossils in the extremely rich *Burgess Shale Formation* of western Canada represent all of the major living animal phyla and a number of extinct phyla as well (Figure 25.4), yet the Burgess Shale Formation is almost as old (570 million years) as the Ediacara fauna. This means that either the Ediacara fauna is not representative of animals of that period, or most animal phyla appeared over an incredibly brief period of evolutionary time, perhaps 10 to 100 million years.

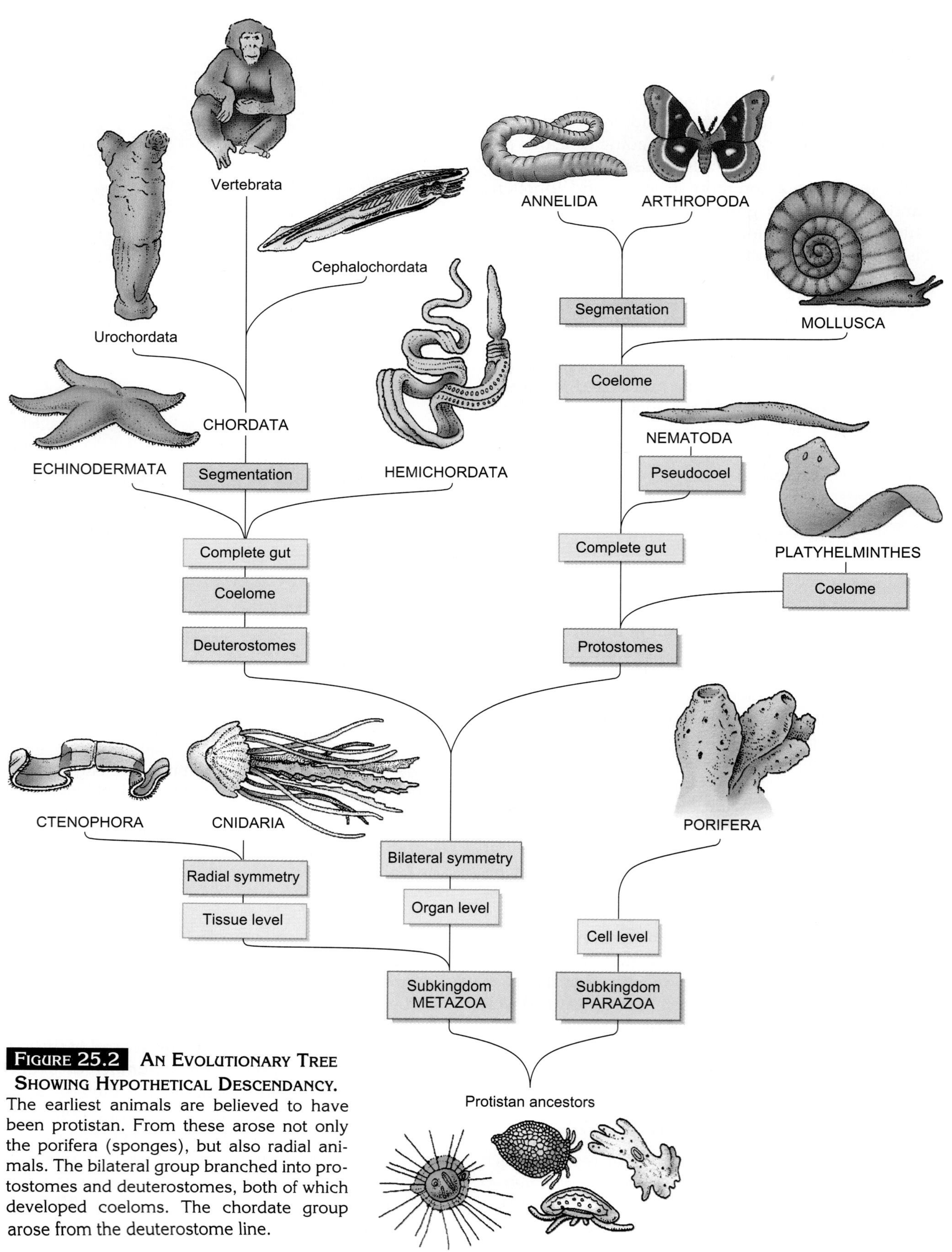

FIGURE 25.2 AN EVOLUTIONARY TREE SHOWING HYPOTHETICAL DESCENDANCY. The earliest animals are believed to have been protistan. From these arose not only the porifera (sponges), but also radial animals. The bilateral group branched into protostomes and deuterostomes, both of which developed coeloms. The chordate group arose from the deuterostome line.

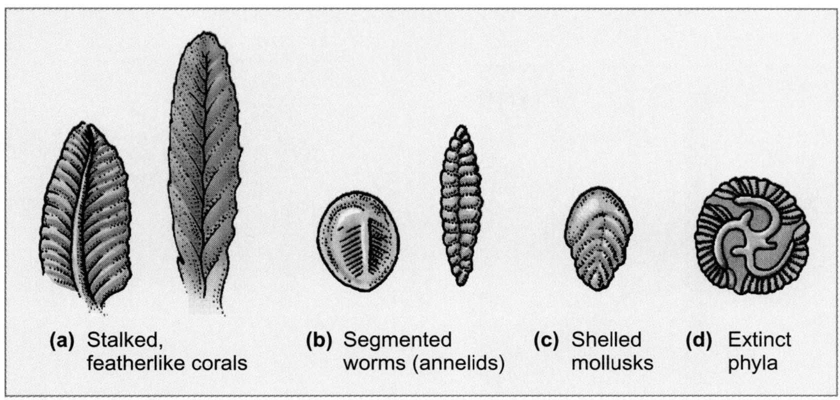

FIGURE 25.3 **THE EDIACARA FAUNA.**
The fossil record indicates that Ediacara fauna (animals) of the late Precambrian oceans probably was dominated by the thin-bodied cnidarians. Among these are **(a)** stalked, featherlike corals, nearly identical to today's "sea pens." The bottom dwellers include several species of annelids **(b)**, the segmented worms, identified by lines crossing the body. Representatives from the jointed-legged animals, the arthropods, are also present, as are a few shelled mollusks **(c)**. In addition, there are fossil animals of phyla that are now entirely extinct **(d)**.

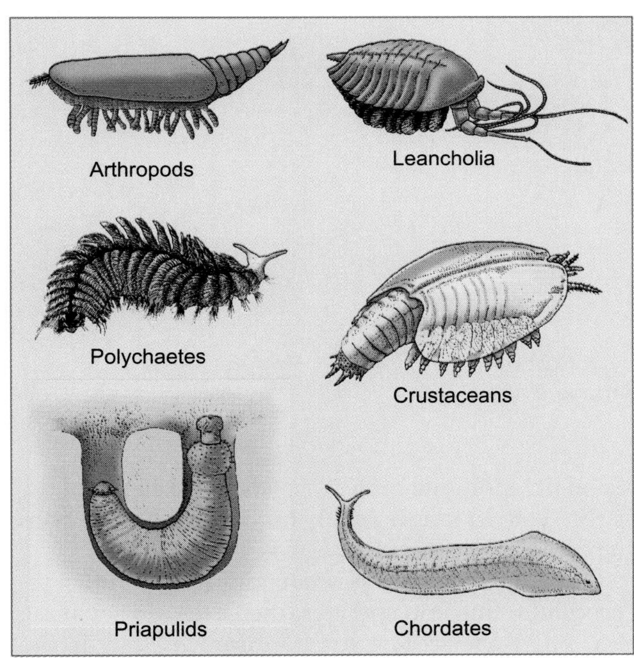

FIGURE 25.4 **THE BURGESS SHALE FAUNA.**
The fauna of the Burgess Shale of Canada includes many familiar phyla and some that are unknown. The complexity of many suggest a long evolutionary history. Interestingly, there has been some argument over the evolutionary interpretation of certain of these forms. Certain researchers believe that in some cases, the fossils were being viewed upside down and that, when righted, their similarity becomes evident and the fossil beds turn out not to be so diverse as was once believed.

SPONGES: PHYLUM PORIFERA

We stated that most animals have four levels of organization. One of the exceptions is the phylum Porifera—the sponges. Essentially, they consist of just a few types of cells, organized only at the cellular level, or at most, in loosely arranged tissues. They are called **parazoa,** meaning "beside the animals," indicating that they are markedly different from the other animal species, called **eumetazoa** which means "true multicellular animals."

Sponges are primarily ocean dwellers, although about 150 species live in fresh water. They live unobtrusively, fastened securely to the ocean or lake bed, where they busily filter tiny food particles, organic detritus, and microorganisms from the surrounding water. Sponges lack organized muscles and nerves, so they are incapable of rapid, whole-body responses.

The simplest sponges are the vaselike ascon sponges (Figure 25.5). Their hollow bodies are just a few cell layers thick and consist of only four or five types of cells. Like all sponges, they feed by filtering particles from the surrounding water. The food is carried past them by a current of water produced by the beating flagella of the *collar cells* (also called choanocytes) that line the inner body. (The water enters the *spongocoel,* or body cavity, through pore cells called *porocytes.*) The *collar,* which gives the cell its

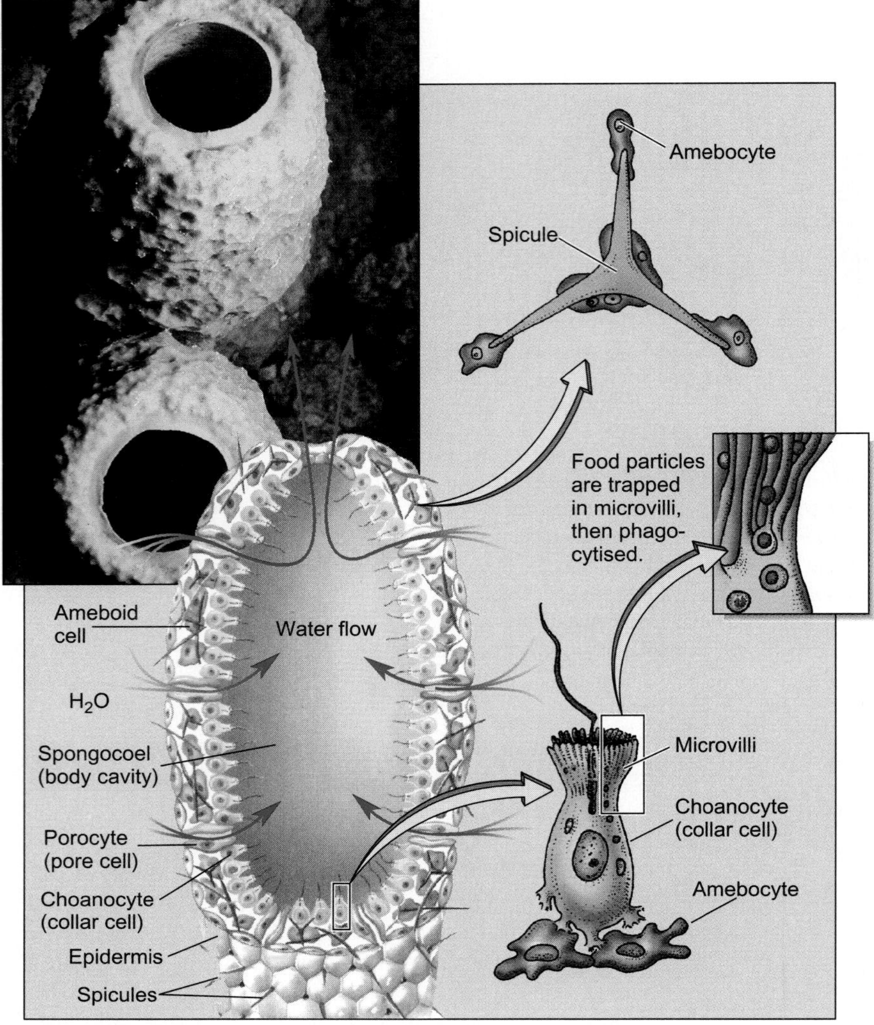

FIGURE 25.5 SPONGES.
The sponges are delicate animals that are different in so many critical respects from other animals that their placement among the kingdoms has been heavily debated.

name, consists of a ring of stiff *microvilli* (fingerlike membrane extensions) circling the base of the long flagellum. Food particles are trapped in mucus secretions on the collar cells and are carried down to the base of the collar, where they are engulfed through phagocytosis and later digested in food vacuoles. The partly digested food is then transferred to wandering *amebocytes* that creep about the tissues, digesting the food further and distributing it throughout the body.

Sponges maintain their shape because of a kind of skeleton formed of *spicules,* or protein fibers. Spicules are secreted by ameboid *mesenchyme* cells and, depending on the species, may consist of chalk (calcium carbonate), glass (silicon dioxide—the glassy sponge can be quite elaborate and beautiful), or *spongin.* Spongin, a fibrous protein, forms the skeleton of the large "bath" sponges (rarely used today, having been replaced by synthetic sponges).

Now let's consider how an immovable animal reproduces. In some sponges, the sexes are separate. But other species are **hermaphroditic;** that is, the same animal produces both sperm and eggs. The eggs are stationary, lying just below the collar cells. They are fertilized by clouds of sperm from a neighboring sponge that are drawn into the body cavity. The sperm must penetrate the recipient's body wall to fertilize the eggs. The zygote will develop into a flagellated, swimming *larva* (an immature stage quite different from the adult form). Sponge larvae finally settle to the bottom, become attached, and mature into adults.

POLYPS AND MEDUSAS: PHYLUM CNIDARIA

The phylum Cnidaria has radial symmetry. We have no firm ideas as to how the radial body form arose, but many zoologists theorize that the direct ancestor to all metazoans, including the radial ones, was probably a bilateral inhabitant of the muddy ocean bottom (where bilateral symmetry works well). It may have resembled the *planula larva,* an early developmental stage of cnidaria that we will look at shortly. The theorists go on to suggest that a radial offshoot from these simple animals did well as a swimming form and that this line gave rise to the radiates.

Cnidarians (formerly called coelenterates) barely rise above the tissue level of organization, although some organ development is present. The cnidarian body is essentially a thin-walled sac, with an inner and outer tissue layer, each only one cell thick. The outer layer, the *epidermis,* is the protective body covering. The inner layer, or *gastrodermis,* lines the *gastrovascular cavity,* which extends throughout the organism, even into the tentacles. Sandwiched between the two layers is a structureless, jellylike *mesoglea.* Wandering ameboid cells, nerve processes, and contractile fibers are found in the mesoglea. An example is *Hydra,* a freshwater cnidarian that stands a few millimeters high.

Hydra, like most cnidarians, has tentacles that are armed with stinging cells called *cnidocytes* (*cnid* means "nettle"), which are used in immobilizing prey and in defense. Cnidocytes are odd structures. Many contain stinging *nematocysts,* coiled hollow threads tipped with poisonous harpoons (Figure 25.6), while others release sticky, snaring threads that trap prey.

As with other radial animals, there is little centralization of the nervous tissue. (For example, there is no brain.) The nervous system, instead, is a *nerve net,* a diffuse network of nerve cells and their extensions covering the entire animal. In addition, impulses over these nerves can travel in either direction, unlike the one-way transmission of most nerve cells in bilateral animals with nervous systems.

Hydra are the terrors of their tiny world. Any small creature moving past triggers the *Hydra*'s stinging cells, and once the prey is immobilized, tentacles draw the stunned victim into the *Hydra*'s gastrovascular cavity. Once the victim is inside the cavity, it will undoubtedly be impressed with the very specialized lining it finds there. It will see that some gastrovascular cells produce digestive enzymes, others absorb digested food, some engulf whole food particles to be digested later in vacuoles, and still other cells have flagella with which they move particles around for greater ease of handling.

Two basic body patterns are found among the cnidarians, the **polyp,** or sedentary state (as in *Hydra*), and the **medusa,** a swimming jellyfish state. Both have radial symmetry. Interestingly, some groups alternate between the two body patterns, although one is usually the dominant or longer-lived state. In other groups, either the swimming medusa or the polyp state may be entirely absent. The organization of the cnidarians into their three classes is based principally on these factors.

The class Hydrozoa includes *Hydra,* along with many marine hydroids. The polyp is the dominant state. Typically, the tiny marine hydroids form highly branched colonies of polyps that feed on drifting plankton. Like *Hydra,* they produce new polyps on the sides of their bodies through an asexual process called *budding.* Unlike the buds of *Hydra,* which break away, the hydroid's buds remain attached, forming dense, branched colonies. When hydroids reproduce sexually, they bud off tiny, swimming medusas that form gametes and release them into the water. Following fertilization, the zygote develops into a ciliated planula larva. It soon settles to the ocean floor where it goes through a period of development, eventually forming a polyp (Figure 25.7).

In the class Scyphozoa, the dominant stage is the swimming jellyfish (medusa). Like the hydroid medusa, it releases its gametes into the water and produces a swimming planula larva. But this larva grows into a small, inconspicuous polyp that fastens to the underside of a rocky ledge and buds off young jellyfish that grow into adults.

The class Anthozoa includes the sea anemones and corals. In this class the medusa state is entirely absent; thus the life cycle is simple, essentially going from polyp to polyp with only a brief stage between.

Many corals form massive colonies, constantly secreting limestone and forming awesome and beautiful coral reefs. (Representatives of classes Scyphozoa and Anthozoa are seen in Figure 25.8.)

THE COMB JELLIES: PHYLUM CTENOPHORA

The phylum Ctenophora ("comb carriers," also called comb jellies) appears to be closely related to the cnidaria (Figure 25.9). Both show radial symmetry,

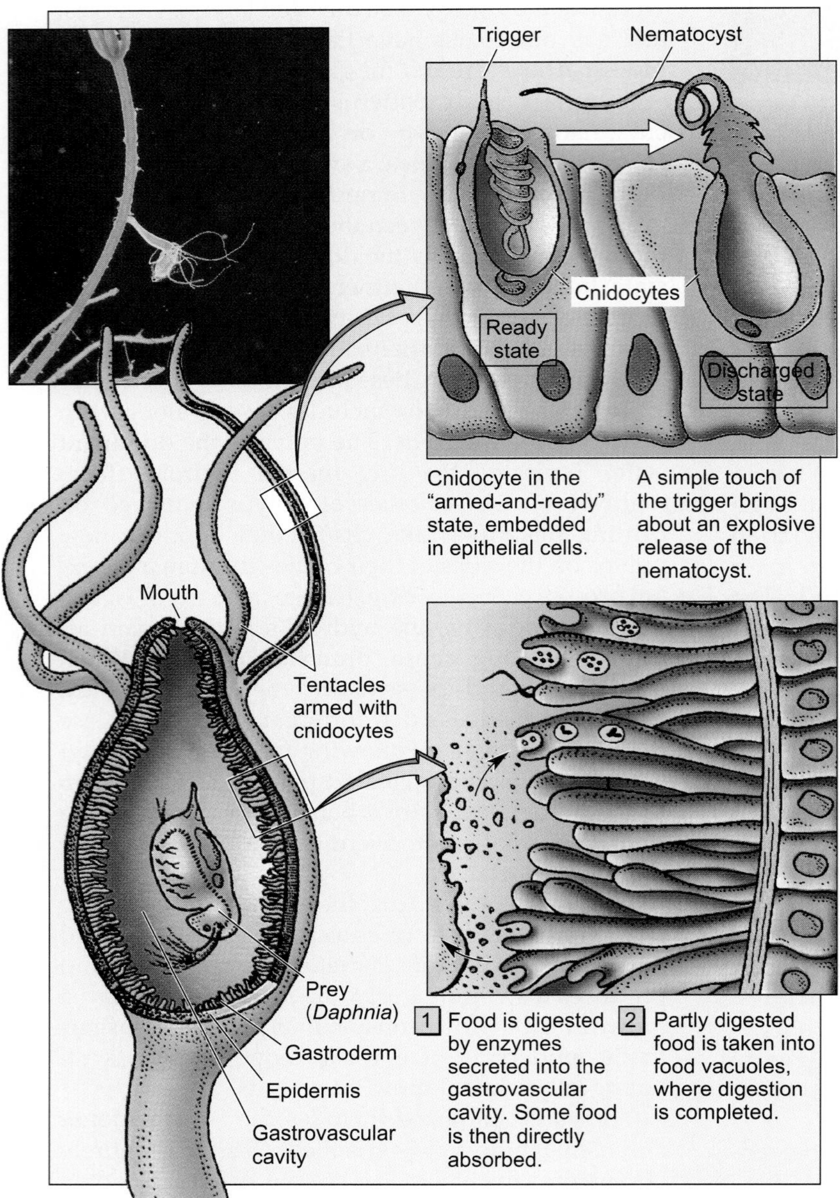
Trigger
Nematocyst
Cnidocytes
Ready state
Discharged state
Cnidocyte in the "armed-and-ready" state, embedded in epithelial cells.
A simple touch of the trigger brings about an explosive release of the nematocyst.
Mouth
Tentacles armed with cnidocytes
Prey (*Daphnia*)
Gastroderm
Epidermis
Gastrovascular cavity
1 Food is digested by enzymes secreted into the gastrovascular cavity. Some food is then directly absorbed.
2 Partly digested food is taken into food vacuoles, where digestion is completed.

FIGURE 25.6 A FRESHWATER HYDROZOAN.
The hydra stuns its prey with stinging cnidocytes. Using its tentacles, it draws the subdued prey into its gastrovascular cavity, where digestion occurs. A closeup of the cnidocytes shows both the "armed-and-ready" state and the discharged state. A simple touch of the trigger brings about an explosive release of the thread-like nematocyst.

but the two differ in that all comb jellies are free-swimming, and their tentacles lack stinging cells. Instead, their tentacles are coated with sticky secretions from what are known as *glue cells,* which are used to trap plankton. Some ctenophorans produce an intriguing—and sometimes eerie—luminescence that makes their watery home actually glow. Some are quite large. In fact, one species, *Cestum* ("Venus's girdle"), looks like a glowing ribbon nearly a meter long.

CLEAVAGE PATTERNS AND THE VERSATILE MESODERM

The radial, hollow-bodied cnidaria with their two layers of cells may strike you as being quite different from what you expected of animals. However, in the groups to come we will encounter not only heads and tails and bilateral symmetry but also vast

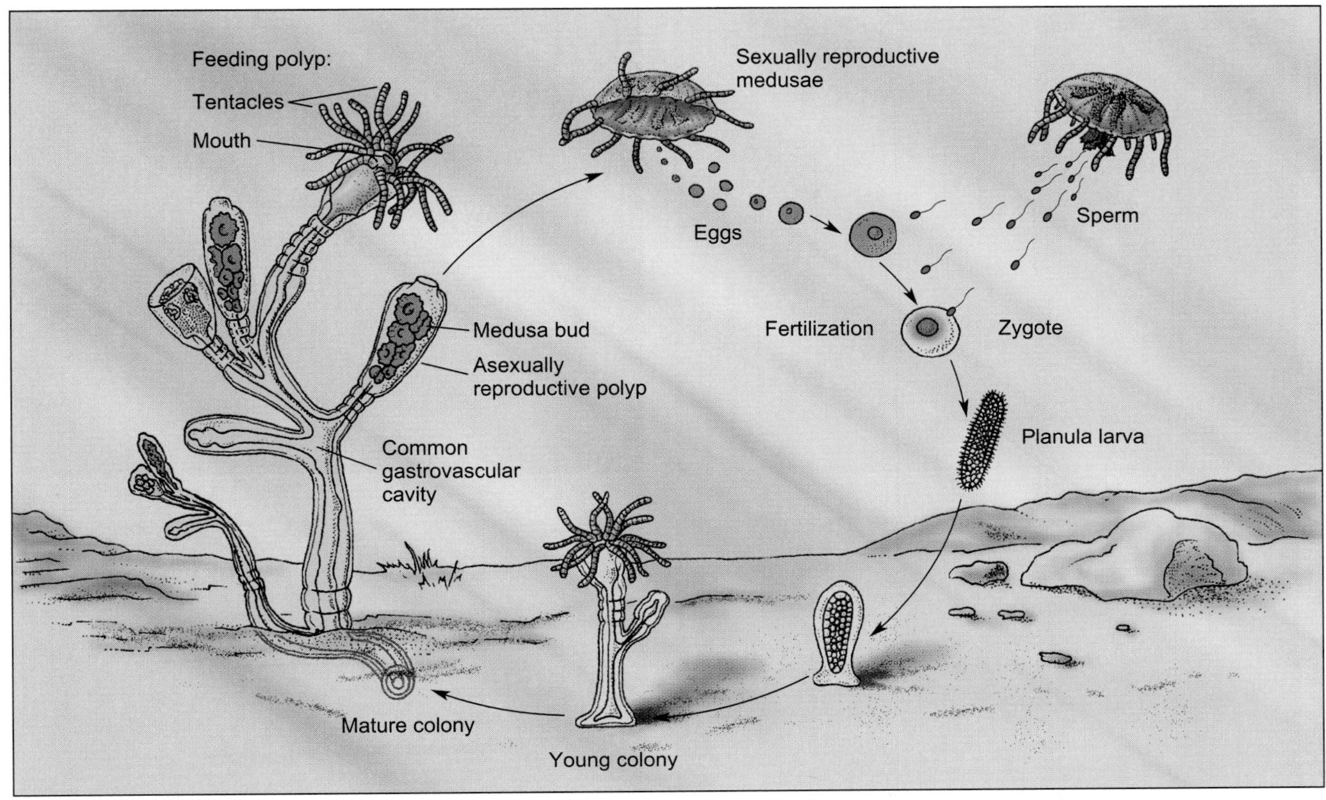

FIGURE 25.7 A MARINE HYDROZOAN.
The highly branched body of *Obelia*, a marine hydroid, contains feeding and reproductive polyps. Reproductive polyps bud off swimming medusas that release sperm and eggs into the water, where fertilization occurs. The zygotes develop into swimming planula larvae that eventually fasten to the seabed and develop into a polyp.

changes in internal structure. Bodies become dense, true muscle tissue appears, and soon we will come to those eumetazoans that can boast of skeletons that hold them erect, muscular hearts that pump blood, and centralized brains that integrate sensations and coordinate responses.

We can trace the appearance of these more characteristic animals to important changes in the embryo. In the development of cnidarian embryos, only two types of embryonic tissue ever form. These tissues, or germ layers, as they are known, are the endoderm and the ectoderm. During development, the endoderm ("inner skin") forms many internal linings. Ectoderm ("outer skin") is primarily involved in forming outer linings. It is the advent of mesoderm ("middle skin"), however, that sets the stage for momentous further changes. We see some evidence of the importance of mesoderm in the next phylum of animals, the group that includes the flatworms.

(a) (b) (c) (d)

FIGURE 25.8 SCYPHOZOA AND ANTHOZOA.
(a) The jellyfish (Class Scyphozoa) are free-swimming, trailing their stinging tentacles. A fish brushing against these will be stung until it is senseless and then digested. **(b)** Jellyfish larvae enter a brief polyp state, which buds off numerous young jellyfish. **(c)** Coral polyps (Class Anthozoa) are also colonial, with the many individual polyps secreting hardened calcareous walls around themselves. **(d)** The anemone (also Class Anthozoa) is a polyp frequently found in shallower waters.

FLATWORMS: PHYLUM PLATYHELMINTHES

With the phylum Platyhelminthes, the flatworms, we finally find bilateral animals with definite organ systems, and we can see how the mesoderm has increased their complexity (Figure 25.10). There are three major classes of flatworms, and we can't say that you will be thrilled to learn about all of them. You may feel okay about the free-living planarians of the class Turbellaria, but the other two groups are parasites, including some that attack humans. They are the tapeworms of the class Cestoda and the flukes (including leeches) of the class Trematoda.

But let's start on a cheerful note. Most streams and ponds are inhabited by free-living *planarians.* Most species of these fascinating little worms are just a few millimeters long. (One terrestrial relative grows to 60 cm (23 in.) and feeds on earthworms and leeches.) Planarians, like other flatworms, have flattened, bilateral bodies. The planarian moves slowly by means of cilia on its belly, and it steers with well-coordinated and well-defined muscles. The nervous system of this creature is rather simple and ladder-shaped, but here we see a major new evolutionary development. Most of the **neural ganglia** (masses of nerve cell bodies) are concentrated at the anterior (head) end. Two eyespots simply detect the presence and direction of light. While they do not form images, they enable the vulnerable creature to remain in the darker and safer regions of its domain.

The planarian's digestive system is a highly branched gastrovascular cavity. But it is still a saclike structure, with one opening. Nonetheless, it is rather specialized, with a protrusible pharynx (which means that the worm can extend its tubelike mouth out of its body). When feeding, it uses its muscular pharynx like a vacuum-cleaner hose, sucking up the juices and soft body parts of its prey.

Although some systems in the flatworm are very simple, the reproductive system is surprisingly complex. Planarians are hermaphroditic. They all have well-defined testes and ovaries, as well as long ducts through which gametes travel. The penis can extend from the chamber in which it rests, and a *genital chamber* acts as a vagina. Planarians usually don't self-fertilize; instead, they exchange sperm.

Parasitic Flatworms

The parasitic flatworms—the tapeworms and flukes—are quite unlike the planarians. As is true of most successful internal parasites, their evolutionary specialization has led to a sometimes startling emphasis on some systems, while other systems have become reduced or have even disappeared. For example, the tapeworm, commonly found in its host's small intestine, has no gut whatsoever, but it has a complex and specialized head with suckers and hooks that ensure a firm hold on the host's intestinal wall. It absorbs predigested food over its entire body surface, which is covered with tiny projections that increase the absorptive area. In a sense, the worm is inside out, with what amounts to a gut covering the outside.

FIGURE 25.9 CTENOPHORES.
Also called comb jellies, they have a gelatinous (jelly-like) body and eight rows of "combs" (fused cilia).

(a) Stained preparation

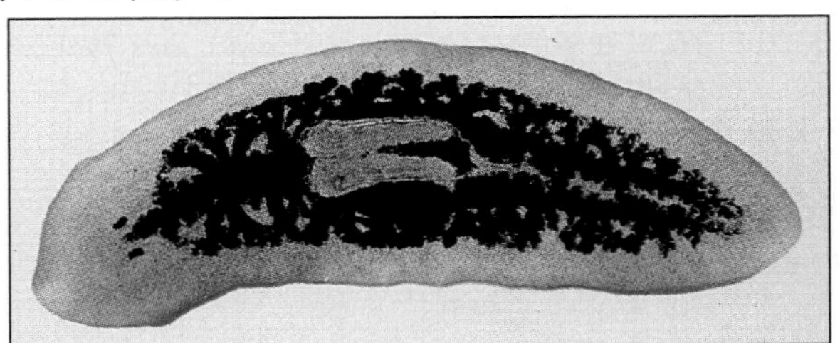

(b) Digestive system

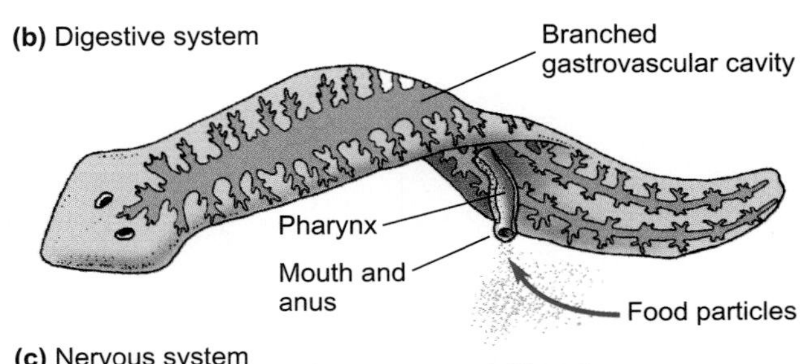

(c) Nervous system

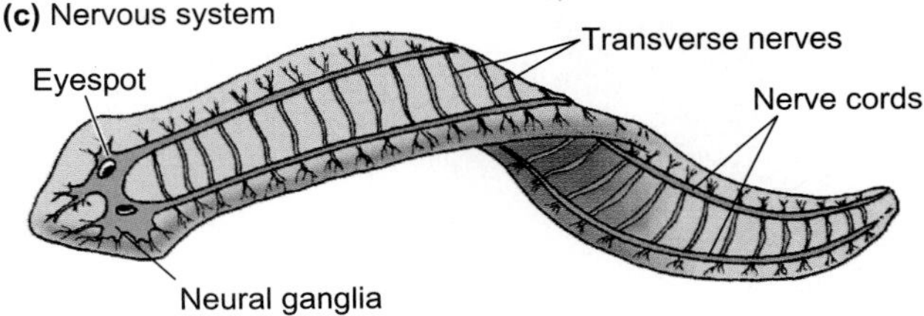

(d) Reproductive system

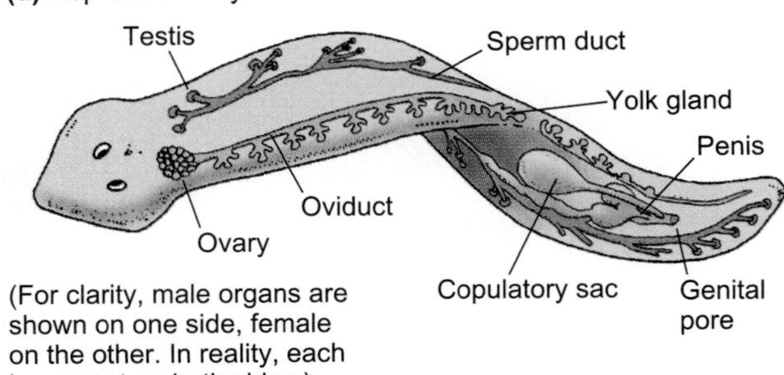

(For clarity, male organs are shown on one side, female on the other. In reality, each is present on both sides.)

FIGURE 25.10 ORGAN SYSTEMS AND THE PLANARIA.
The Planaria **(a)** demonstrated the specialization of organ systems. Digestion in planaria **(b)** occurs in a gastrovascular cavity, which is highly branched and includes a complex, muscular pharynx. Light-sensitive eyespots detect the direction of a light source, and a ladderlike nerve network **(c)** coordinates its movement. The reproductive system **(d)** includes well-defined testes and ovaries, along with related ducts and yolk glands. Since it is hermaphroditic, the organism has both penis and vagina.

With so many of life's daily problems solved, the tapeworm can devote much of its energy to reproduction. Its body, which may be several meters long, is composed of segments called **proglottids** that contain both ovaries and testes. The proglottids are continually produced just behind the creature's tiny head, or **scolex.** Each mature proglottid may contain thousands of eggs, and following fertilization (tapeworms, by the way, can self-fertilize), the proglottids break off and are passed out with the feces. When they rupture, the eggs are released, ready to renew the life cycle (Figure 25.11).

The eggs of beef and pork tapeworms must be swallowed by the cow or pig for the parasite to continue its life cycle. When an egg hatches, the minute offspring moves from the host's digestive tract to the bloodstream, thereby finding its way to the muscles. There it curls up and forms a protective, capsulelike cyst, remaining inactive until someone decides to have steak or pork chops a little on the rare side.

The life cycles of many parasites are extremely complicated. They may go through a number of physical changes and may inhabit a number of hosts. The timing and sequence of the changes must be precise, and a great deal is left to sheer luck. For example, consider the life cycle of the human liver fluke, as portrayed in Figure 25.12. The odds against making it to that final host are so great that natural selection has dictated that each worm maximize the number of eggs it produces, leaving the rest to chance.

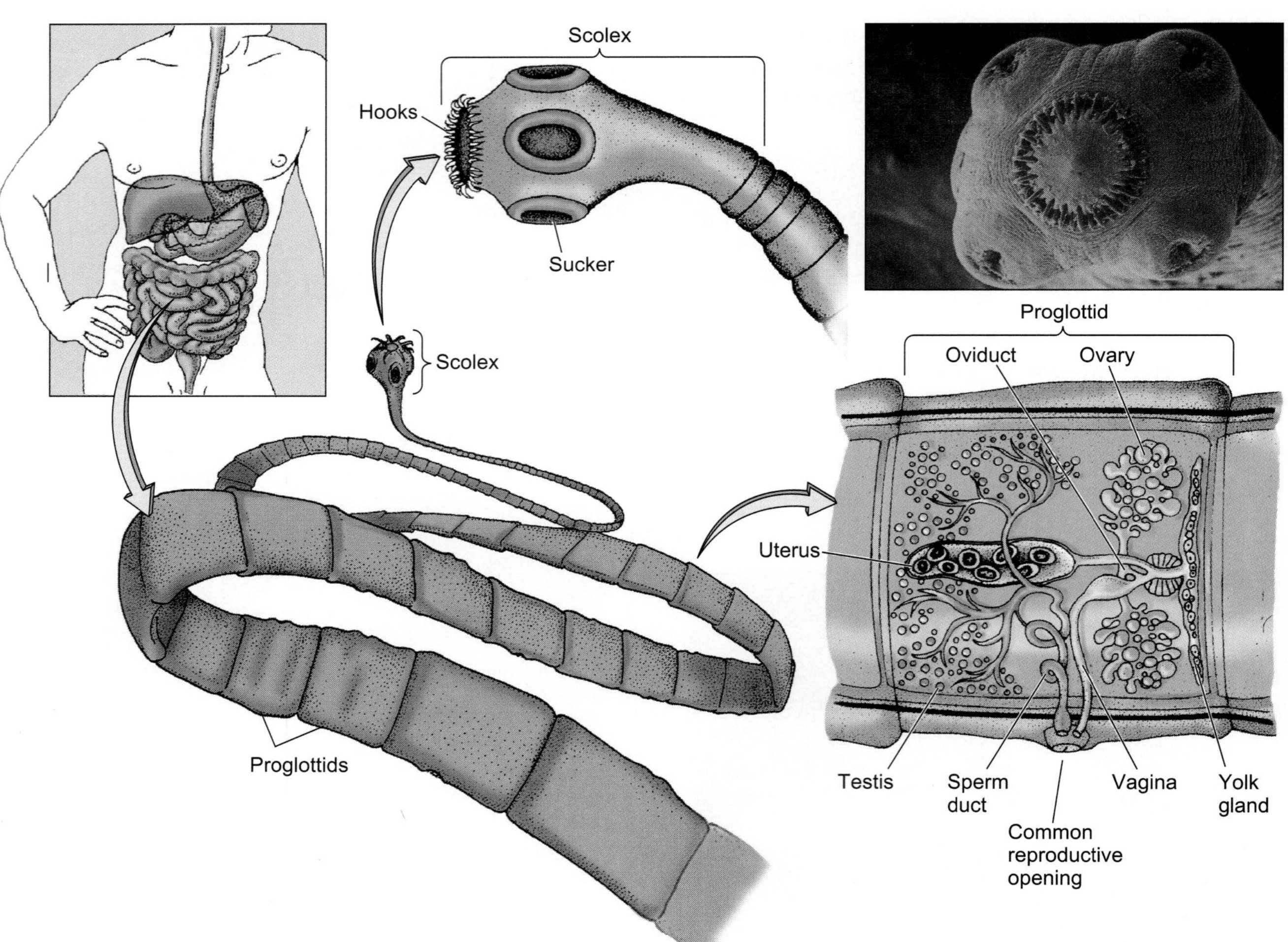

FIGURE 25.11 THE TAPEWORM, A COMMON INTESTINAL PARASITE.
The beef tapeworm of humans can grow to about 7 m (23 ft) in length, but the record is probably held by the broad (or fish) tapeworm of humans, which can exceed 18 m (60 ft). The photograph shows the scolex.

Key Evolutionary Events—Body Cavities, a One-way Gut, and a New Body Plan

Now let's review a few of the important evolutionary events that have proved to be key developments—changes that made possible new trends and directions, and altered life on this planet forever. (This is a good place to check back to Figure 25.2.)

With the development of bilateral symmetry, first seen in the flatworms, a significant and probably simultaneous trend in body organization arose. Animals developed a leading, or head, end and a trailing, or tail, end. At first, the head may well have been simply a concentration of muscles, an adaptation for burrowing in the soft sea beds of ancient oceans, which is where the oldest fossils of bilateral animals are found. These leading ends would have soon become equipped with sensory structures that could detect food, light, vibrations, and other stimuli, so that the animal could ascertain the nature of the environment before moving into it. Such structures require neural support and integration, so, as we might expect, clusters of nerve cells were located close by, in what would become the brain. The animals were now equipped to better detect the characteristics of the environments they were burrowing into and to more efficiently deal with that information. The flattened body produces a dorsal (back) surface and a ventral (belly) surface, each of which could become distinct from the other, allowing for greater specialization. The resulting two sides also offer opportunities for greater specialization, to the degree that they become different from each other, or redundancy (providing a kind of backup, each for the other) to the degree that they remain similar. With the more structured arrangement of the nervous system and the development of complex musculature, the animal is also now able to respond to environmental stimuli in increasingly complex ways.

With the bilateral plan and a head end well established, far-reaching evolutionary developments were in store. Specifically, these were the complete gut—tubelike, with a mouth and anus—and a cavity in which the gut would lie. The gut would, then, form a tube, surrounded by the body wall, and between the two was a cavity into which various organs arising from the gut would protrude. (Can you name a few in humans?) With a gut inside a body wall, the body plan of these animals took on a tube-within-a-tube organization. Apparently this organization was a huge success, paving the way for numerous modifications. Today, most animals have retained the tube-within-a-tube plan. Interestingly, animals can be divided according to the kinds of tissue that line this body cavity, as we see next.

Pseudocoelomates: Phyla Nematoda and Rotifera

Animals can be divided according to whether the body cavity is completely or incompletely lined with mesoderm. A body cavity completely lined with mesodermally derived tissue is called a **coelom.** We will look at invertebrates with such cavities in the next chapter. First, though, we will consider two groups with a **pseudocoelom** ("false coelom"), a body cavity lined with tissue not completely derived from mesoderm. These are the nematodes and rotifers.

Nematodes. Some people might find it hard to believe that the phylum Nematoda, the roundworms, could be either important or interesting, but they are both, as we shall see. The word *nematode* means "threadlike," a description that applies to most of the species, which are monotonously cylinder-shaped and tapered at both ends. Furthermore, although the gut in roundworms is complete (that is, it has both mouth and anus), it lies free and unattached in the fluid-filled pseudocoelom; also, its wall is not muscular and is usually only one cell layer thick; none of these conditions is typical of true coelomates. Taxonomists have already described about 12,000 of the estimated half million species of nematodes, many of which are efficient predators, winnowing their way among moist soil particles, paralyzing prey with their saliva, or piercing them with mouth parts to suck their body juices (Figure 25.13).

Virtually all plants and animals are parasitized by one or more kinds of nematodes. Some of these parasites, in fact, have been devastating to agriculture, and at least 10 species are dangerous to humans. On the other hand, about 50 species live in or on our bodies without doing us apparent harm. But let's take a brief look at a particularly dangerous species, the giant *Ascaris lumbricoides* (Figure 25.13).

Ascaris is an unusually large roundworm, often becoming longer than 20 to 35 cm (8 to 14 in.). It is one of those repulsive parasites whose habits could be considered the stuff of nightmares. Its powerful sucking mouth, with which it grasps the host's intestinal wall, leads to a flattened, ribbonlike, but very simple gut, a tube that extends from the mouth to the anus. Thousands of these creatures can live in the intestine of one person, distending the belly and draining the host of nutrients.

Like other parasitic worms, *Ascaris* is a prodigious reproducer, an adaptation to a life cycle filled with perils. Some mature females produce up to 200,000 eggs per day. The fertilized eggs pass out of the body with the feces, and new infestations occur when the eggs are accidentally swallowed. If you can't imagine this happening, it's probably because you haven't lived in places without toilets and haven't eaten vegetables from a garden fertilized with untreated "night soil" (human waste). However, many

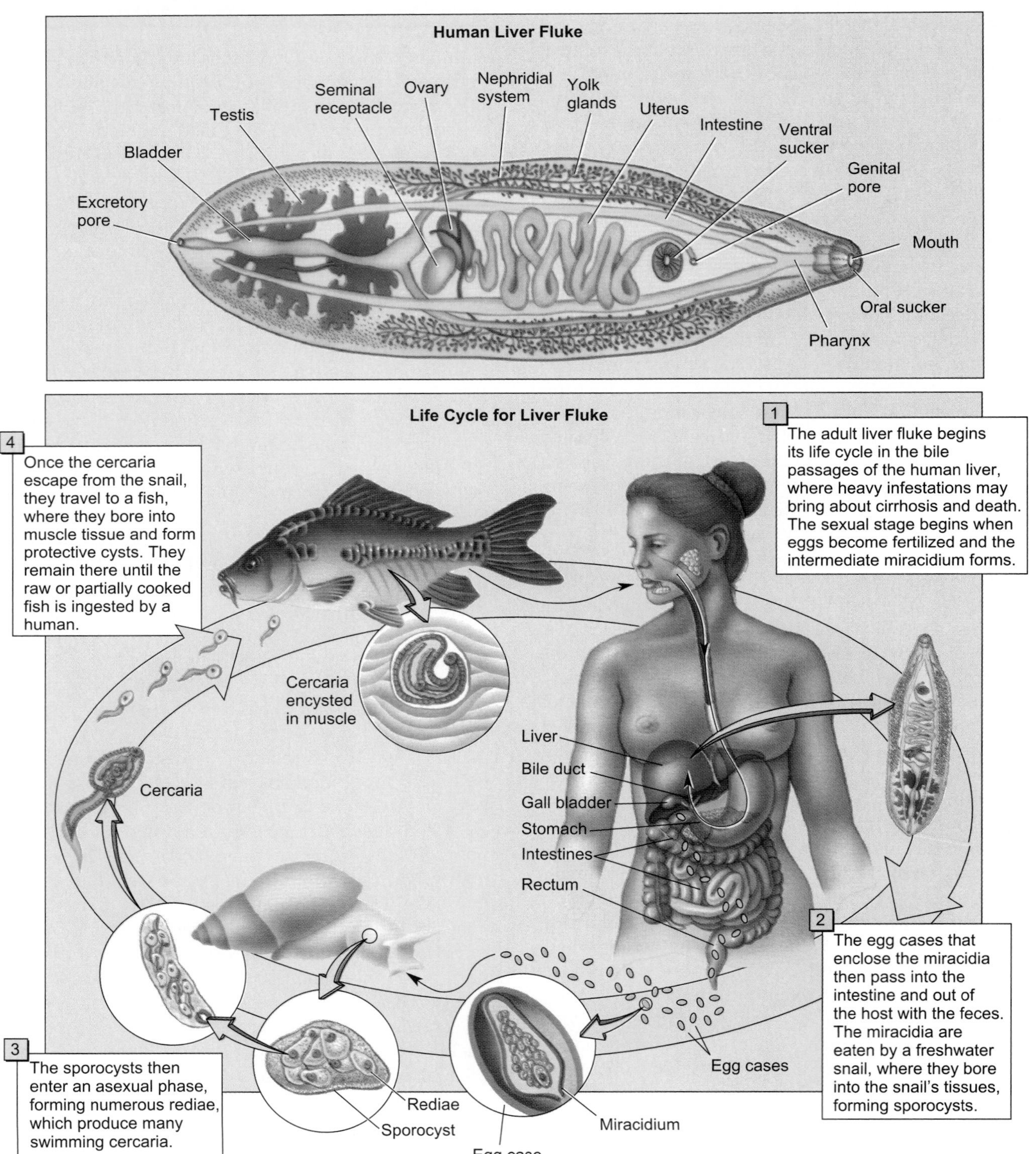

Human Liver Fluke
Testis
Seminal receptacle
Ovary
Nephridial system
Yolk glands
Uterus
Intestine
Ventral sucker
Genital pore
Bladder
Excretory pore
Mouth
Oral sucker
Pharynx
Life Cycle for Liver Fluke
1
The adult liver fluke begins its life cycle in the bile passages of the human liver, where heavy infestations may bring about cirrhosis and death. The sexual stage begins when eggs become fertilized and the intermediate miracidium forms.
4
Once the cercaria escape from the snail, they travel to a fish, where they bore into muscle tissue and form protective cysts. They remain there until the raw or partially cooked fish is ingested by a human.
Cercaria encysted in muscle
Cercaria
Liver
Bile duct
Gall bladder
Stomach
Intestines
Rectum
2
The egg cases that enclose the miracidia then pass into the intestine and out of the host with the feces. The miracidia are eaten by a freshwater snail, where they bore into the snail's tissues, forming sporocysts.
Egg cases
3
The sporocysts then enter an asexual phase, forming numerous rediae, which produce many swimming cercaria.
Rediae
Sporocyst
Egg case
Miracidium

of the earth's billions live under such conditions, and they are at greatest risk from *Ascaris.* When these eggs are swallowed, they hatch in the intestine, and the larvae eventually burrow through the intestinal wall and crawl to the liver, heart, and lungs, where they undergo developmental changes. Those in the lungs crawl up the windpipe and are swallowed. These worms then mate in the intestine, and their eggs pass out with the feces to contaminate the soil.

Rotifers. You will be glad to learn that the phylum Rotifera (Figure 25.14) is not parasitic. These species are microscopic and, in fact, are often confused with the protists whose watery environment they share. They are interesting because in spite of their tiny size, they are distinctly more complex than other pseudocoelomates.

Rotifers feed in a most unusual way. The double rings of cilia that circle their heads in a wheel-like manner (*rotifer* means "wheel bearer") sweep food into a grinding gullet, or gizzard. The rotifer's digestive system is unusually complex for such a small animal. Who would suspect that this tiny creature, almost invisible, has a mouth, a muscular, grinding pharynx, a stomach, two digestive glands, an intestine, and an anus? (Obviously, its cells are minute, indeed.) It also has a body cavity, and that cavity is incompletely lined with mesodermally derived tissue.

FIGURE 25.12 THE HUMAN LIVER FLUKE.
Clonorchis sinensis is common in many Asian regions. It lives in the bile passages of the liver, where heavy infestations can bring on cirrhosis and death. Its life cycle is one of the most complex known, requiring three separate hosts: human, snail, and fish. Its primary, or sexual, stage occurs in humans, where the eggs are fertilized and the first of several intermediate stages, the *miracidium,* forms. The egg cases then pass into the intestine and out of the host with the feces. When the miracidia are eaten by a freshwater snail (in rice paddies or other polluted waters), they hatch from the egg case and bore into the snail's tissues, forming *sporocysts.* The sporocyst then enters an asexual phase, producing numerous *rediae,* each of which, in turn, produces many swimming *cercaria.* The cercaria escape from the snail to seek out the next host, the fish, whereupon they bore into its muscles and secrete protective capsules (cysts) around themselves. The encysted cercaria (or *metacercaria*) remain there until some hapless human eats the fish—raw or partially cooked. The digestive enzymes of the human host weaken the capsules and the young flukes emerge and make their way up into the bile duct to the bile passages of the liver. The cycle then repeats.

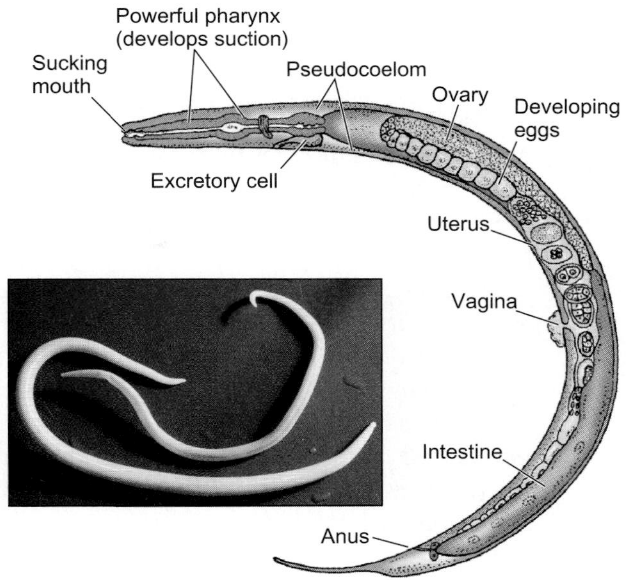

FIGURE 25.13 THE NEMATODE *ASCARIS LUMBRICOIDES.*
Nematodes are common in most soil, but abound in moist, fertile places. Many are scavengers or predators of soil organisms and are considered free-living, but a great number are agricultural parasites.

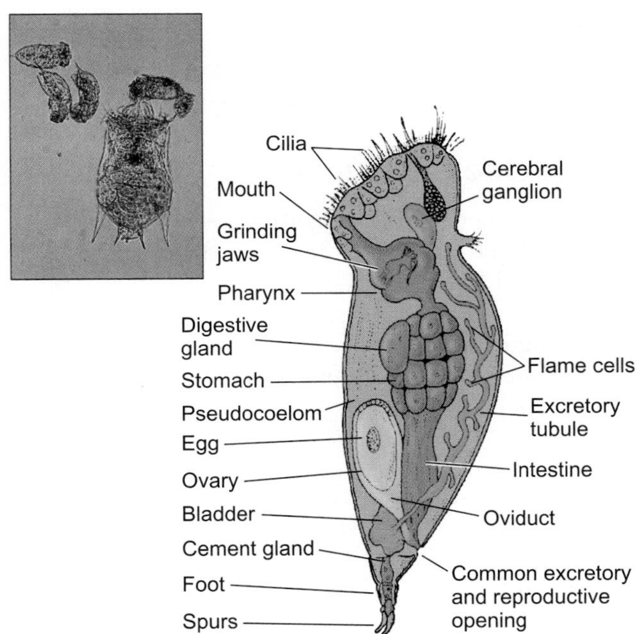

FIGURE 25.14 THE ROTIFERS.
Rotifers, extremely common freshwater animals, are about the same size as many protists (most less than 1 mm in length) but are much more complex. Note the well-defined "brain" and the complex digestive and reproductive systems. Cement glands at the base of the foot permit the animal to anchor itself to the bottom.

PERSPECTIVES

We've learned something about the lives of a few of the invertebrates. We may have also gained a new respect for their complexity, their life histories, and their place in the scheme of life. Such creatures are sometimes referred to as "primitive," but organisms cannot be classified in this way; only a particular trait can be described as primitive (retaining the ancestral condition of the group) or advanced (changed from the ancestral condition). We will soon see that there is no orderly progression toward the "advanced" condition. Each type of animal is uniquely adapted to its role; it is likely to be very good at doing those things necessary to its own existence. It may survive well with traits we find trivial or unnecessary, and it may have no use whatsoever for ours. As you review our discussion, keep in mind the major evolutionary events we have just seen, such as radial and bilateral body plans, the advent of mesoderm, the development of muscle, the appearance of a complete digestive tract, and when the mouth and anus form. Also recall the traits of the pseudocoelomates. Now we will move to those animals that still lack vertebrae, but at least can boast a true coelom.

SUMMARY

Animals are generally heterotrophic, multicellular, eukaryotic, with tissues and organs. They are sexual reproducers, essentially diploid with a gametic life cycle and well-developed neural and muscular systems. They store carbohydrate as glycogen.

Most theorists think that the parazoan ancestors were protozoan flagellates, and the eumetazoan ancestors, protozoan ciliates. Protozoans have animal-like characteristics and dietary requirements.

BODY ORGANIZATION

A few phyla have radial symmetry—any cuts in the form of "pie" segments made in their spherical or cylindrical bodies contain similar structures. Most phyla have bilateral symmetry—right and left, mirror-image sides. Most species have reached the organ-system level of organization.

PHYLOGENY OF THE ANIMAL KINGDOM

Phylogenetic trees, based on the fossil record, comparative anatomy, physiology, embryology, and biochemistry, portray major divergences that gave rise to the protostome and deuterostome lines.

Many animal phyla appear in the Ediacara fauna, from strata dated 580 to 680 million years old. All phyla appear in the Burgess Shale Formation, 570 years old, suggesting that the evolution of the major phyla was remarkably rapid.

SPONGES: PHYLUM PORIFERA

Sponges are roughly radial, containing a cellular level of organization (perhaps a rudimentary tissue level in some cases). They are filter-feeding aquatic animals with only a few cell types: collar cells (feeding), porocytes (admitting water), and amebocytes (ameboid feeding and reproduction), mesenchyme (producing spicules and other cells), and gametes (reproduction). Zygotes pass through a temporary ciliated larval stage. Sponges are parazoans; all other animals are eumetazoans.

POLYPS AND MEDUSAS: PHYLUM CNIDARIA

Radiates may be an offshoot of early bottom-dwelling bilateral animals. The earliest may have resembled the planula larva common to marine invertebrates.

Cnidarians have tissues and may have reached the organ level of organization. The hollow, thin-walled body contains two cell layers and surrounds an extensive gastrovascular cavity lined with digestive cells. Tentacles are armed with cnidocytes—stinging cells that discharge nematocysts (coiled, hollow threadlike structures) that subdue prey. Digestion begins in the gastrovascular cavity and continues in food vacuoles. Movement is through contractile fibers, and the neural structures have a netlike arrangement.

Cnidarian body forms include stationary polyps and swimming medusas. In hydrozoans, the polyp is prominent. It uses budding to produce swimming medusas that release sperm and eggs. The zygote forms a planula larva, which then develops into the polyp. In scyphozoans (jellyfish), the medusa dominates. Zygotes produce a brief polyp state that buds more jellyfish. Anthozoans (anemones and corals) have the polyp form only. Corals are ecologically important as reef builders.

The Comb Jellies: Phylum Ctenophora

Comb jellies are similar to cnidarians but use glue cells to trap prey. Included is the lengthy, ribbonlike "Venus's girdle."

Cleavage Patterns and the Versatile Mesoderm

Although the embryos of cnidarians have only two germ layers, ectoderm and endoderm, other eumetazoans have a third, the mesoderm.

Flatworms: Phylum Platyhelminthes

Flatworms have organ systems with dense, mesodermally formed muscle layers. Planarian turbellarians are small and free-living aquatic worms. They have a ladderlike nervous system, eyespots, a well-developed pharynx for feeding, and a highly branched gastrovascular cavity. Ciliated flame cells eliminate excess water. Planarians are hermaphroditic, each with well-developed male and female sexual structures.

The adult tapeworm lives in the host's intestine. Most of its systems are highly reduced, and specializations include a gripping head, or scolex, that continuously forms proglottids. Each proglottid has testes and ovaries, and is capable of producing great numbers of offspring. In many species, the fertilized eggs must be ingested by a specific animal, where the offspring upon hatching bore into muscle tissue and encyst. This tissue and the encysted worms must then be eaten if the sexually mature adult stage is to reoccur. The human liver fluke has three hosts—human, snail, and fish—each of which harbors a different stage of the parasite's life cycle.

Key Evolutionary Events—Body Cavities, a One-way Gut, and a New Body Plan

Bilateral animals evolved a head and tail end, the head an area of concentrated nerves. They also developed a tubelike (tube-within-a-tube), complete gut with separate mouth and anus. Bilaterality also provided both specialization and redundancy.

Pseudocoelomates: Phyla Nematoda and Rotifera

Nematodes (roundworms) and rotifers form a pseudocoelom, a body cavity incompletely lined by mesodermally formed tissue. All nematodes have threadlike cylindrical bodies with a pseudocoelom. Whereas most are minute, soil-dwelling, free-living predators, many are plant and animal parasites. The parasite *Ascaris lumbricoides* infects humans and hogs, where it can produce thousands of eggs daily.

Rotifers are tiny, but complex, free-living pseudocoelomates. They use a ciliated funnel in feeding and have a complex, complete gut.

Key Terms

radial symmetry • **412**
echinoderm • **412**
bilateral symmetry • **412**
parazoa • **415**
eumetazoa • **415**
hermaphroditic • **416**
polyp • **417**
medusa • **417**
neural ganglia • **421**
proglottid • **422**
scolex • **422**
coelom • **423**
pseudocoelom • **423**
nematode • **423**
rotifer • **425**

Review Questions

1. List six or seven important animal characteristics. Indicate which, if any, are unique to the animal kingdom.
2. Briefly, what are parazoans and metazoans? What are their theoretical origins?
3. List and define the four levels of organization in animals.
4. What major phyla are represented in the Ediacara fauna? How old is this part of the fossil record?
5. What additional phyla do we find in the Burgess Shale Formation? How old is this deposit?

6. Describe reproduction in the sponge. Suggest how a swimming larval stage might be adaptive to a sedentary animal.
7. Describe the general body form of the cnidarian, and name several specialized cells. What, if anything, qualifies as an organ?
8. Which of the two body forms predominates in the hydrozoan? The scyphozoan? The anthozoan?
9. Briefly contrast the life cycle of the hydrozoan with that of the scyphozoan.
10. Name the three germ layers in the three-layered embryo. What new kinds of tissue were made possible by the new middle layer?
11. Compare cleavage and gut development in protostomes and deuterostomes.
12. What systems are least apparent in the tapeworm? Most apparent? How does this make sense in terms of the worm's life history?
13. Why might successful reproduction in the beef or pork tapeworm be considered difficult? How does this affect its reproductive efforts?
14. Briefly summarize the stages and hosts in the life cycle of the human liver fluke. Where are the weak links in the cycle—vulnerable places where the chain of infestation might be broken?
15. Why was the evolution of a head and tail end in bilateral animals important?
16. Distinguish between a coelom and a pseudocoelom. Which appears in the nematodes?
17. Describe the life cycle of the ascarid worm. What difficulties does its enormous reproductive effort help overcome?

CHAPTER

26

The Coelomate Invertebrates

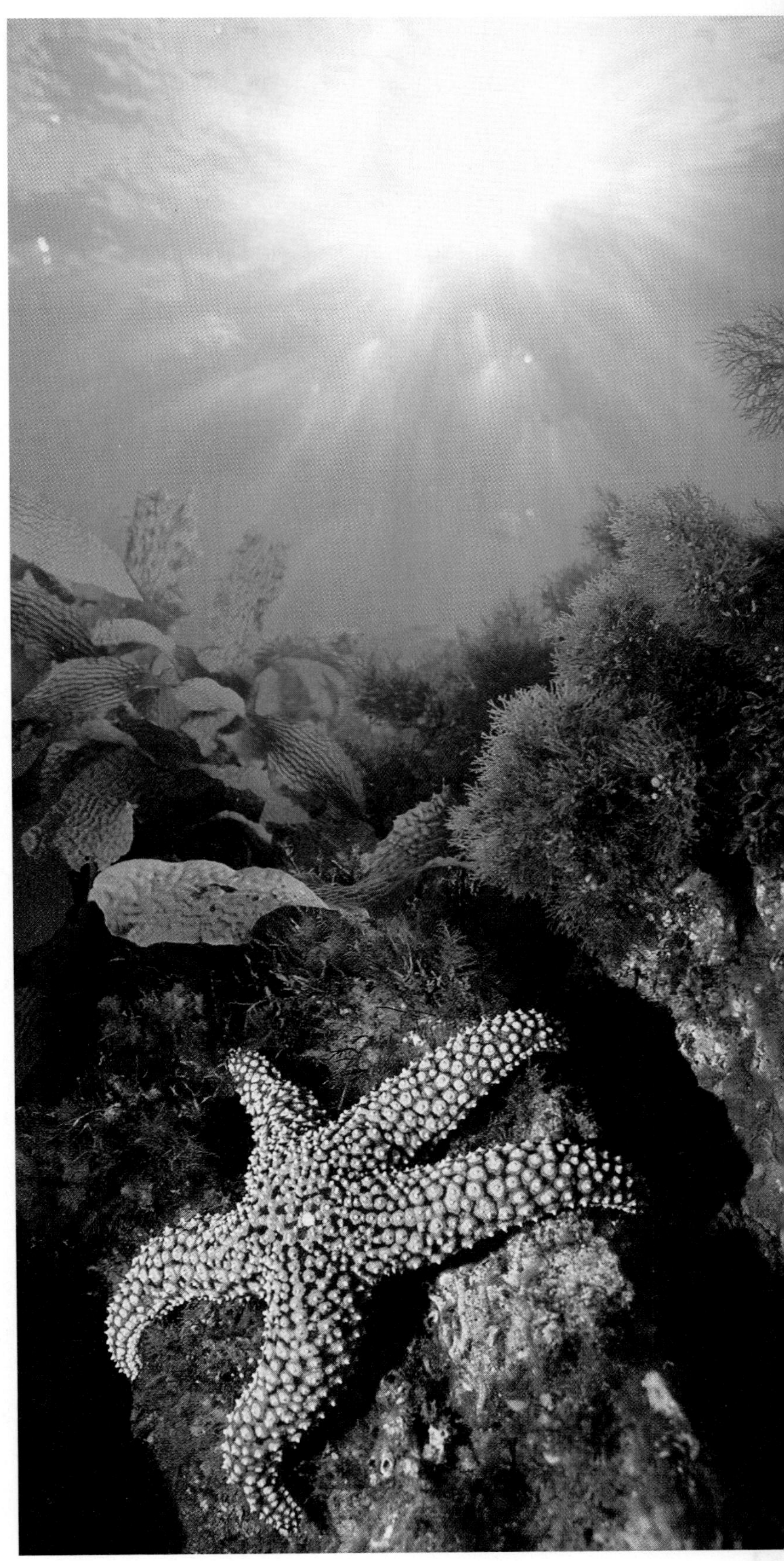

WE CONTINUE WITH THE HISTORY OF MAJOR EVOLUTIONARY EVENTS IN *animals by considering species that, while lacking a backbone, can at least boast a mesoderm-lined body cavity. As we move from one group to the next, watch for major evolutionary events such as segmentation, specialization of segments, and especially, when the mouth and anus form. Then we will review the chordate characteristics by considering some creatures you have perhaps never heard of. Remember, though, that we share their phylum and that we are on the evolutionary road, now, to animals with backbones.*

"Human beings are the most complex, the most advanced, and the most nearly perfect organisms on earth, just one cut above the majestic Indian elephant. Other animals are, by degrees, less advanced, less complex, and less perfect—which is to say, less and less human. Higher forms, such as birds and mammals, are more advanced—closer to perfection—than such lower forms as reptiles, fish, crabs, and worms."

We hope nothing so far has been touched by your yellow marking pen. These were the basic notions of an ancient idea called the *scala naturae,* or the natural scale, penned by Aristotle. Of course, they haven't a grain of truth in them—but somehow the idea has been subtly preserved, as is evident in the popular notion of "higher" and "lower" species (the higher ones, of course, being those with more traits in common with humans). In actuality, each species of life on this earth is continually adapting to better adjust to its place in nature, and that place may be quite different from our own.

Keep this in mind as we consider a second group of invertebrates. These are the ones that have been traditionally referred to as the "higher" invertebrates. On the other hand, it is generally the case that these species have a greater number of advanced traits than those discussed in the preceding chapter. You undoubtedly recall that an "advanced" trait is one that is considered to be more unlike those of ancestral organisms than a "primitive" trait. (So a primitive trait more closely resembles those of the animal's evolutionary forebear.) Actually, every species is really a mix of primitive *and* advanced traits. We'll see later how this is true of humans as well as other species.

THE COELOMATE BODY PLAN

The species we will consider in this chapter are the **coelomates** (see Figure 26.1). These are the animals with a mesoderm-lined body cavity. This cavity is called a **coelom.** The coelomates have a complex, complete gut lined with digestive and absorptive tissue, surrounded by muscle and covered by a smooth epithelium that permits freer movement of the gut within the coelom. In most coelomates the gut has specialized regions along its length. For instance, there are regions whose primary functions are grinding, swallowing, digesting, and absorbing food; there are others for the temporary storage of food; and yet others for the concentration and elimination of wastes. Such linear specialization permits food to be continuously processed, since, during a given time, food in different regions will be undergoing different stages of digestion (Figure 26.1).

The coelomate animals can be divided into two major groups—the protostomes and deuterostomes. Soon you will be able to amaze your friends and neighbors by telling them when the mouth and anus of various animals formed.

Protostomes and Deuterostomes

Notice that the great tree in Figure 25.2 is split into two major branches with *deuterostomes* ("second mouth") at the left and *protostomes* ("first mouth") at the right. What is this all about?

The answer is deceptively mundane. Early in its evolution, animal life diverged in the way the digestive tract formed in the embryo (Figure 26.2). Briefly, in **protostomes** the mouth forms early, in the region of the blastopore, and the anus later, at the other end. In **deuterostomes** the anus forms early, in the region of the blastopore, and the mouth later, at the other end. This simple divergence led to two vastly different types of animals. Today, the protostomes include the mollusks, annelids, and arthropods. The major deuterostome phyla are the echinoderms and chordates, although there are a few minor phyla as well.

Any such fundamental organizational difference as we see in protostomes and deuterostomes is likely to signal other basic changes as well. For example, the protostomes reveal their particular developmental route long before the mouth forms. We see it first in their **spiral cleavage** (Figure 26.2), in which daughter cells are offset from the underlying parent cell. This developmental route is characteristic of protostomes. In addition, in protostomes, the fate of cells is sealed early—that is, there is very little developmental flexibility. Once a cell is formed, it can take only one developmental route (called *determinate development*).

Deuterostomes, on the other hand, undergo **radial cleavage.** The cleavage planes are symmetrical, and each division produces daughter cells that lie directly above the parent cells (Figure 26.2). Further, the deuterostome embryo undergoes *indeterminate development.* That is, the fate of cells is not set in early cleavages. Each cell remains quite versatile for a considerable time. In fact, if the first four cells to form are carefully separated, four miniature but normal embryos will result.

Finally, in protostomes the mesoderm arises near the embryonic blastopore from tissue that then splits to form the coelom, while in deuterostomes the coelom arises from paired mesodermal pouches protruding from the wall of the primitive gut. It is believed that the protostome condition evolved earlier among the animals, and that the deuterostome condition evolved later and only once. If it did evolve only once, then the chordates and echinoderms are distantly but surely related. The radial cleavage and indeterminate growth they hold in common are further evidence of their relatedness.

The Segmented Body

Along with the development of the coelom and the linearly specialized digestive system, some coelomates achieved a third evolutionary milestone: the segmented body. In its simplest form, this means that the body was organized into serially repeated units or segments, as seen in the centipede and earthworm. Each segment is divided from the next by transverse *septa,* or crosswalls, and the structures within each are more or less repeated. In some cases, the basic segmentation has been modified to such a degree that it may no longer be apparent. For example, the segmentation in humans is relegated to repeated vertebrae, ribs, spinal nerves, and trunk muscles. Many zoologists believe that the coelom

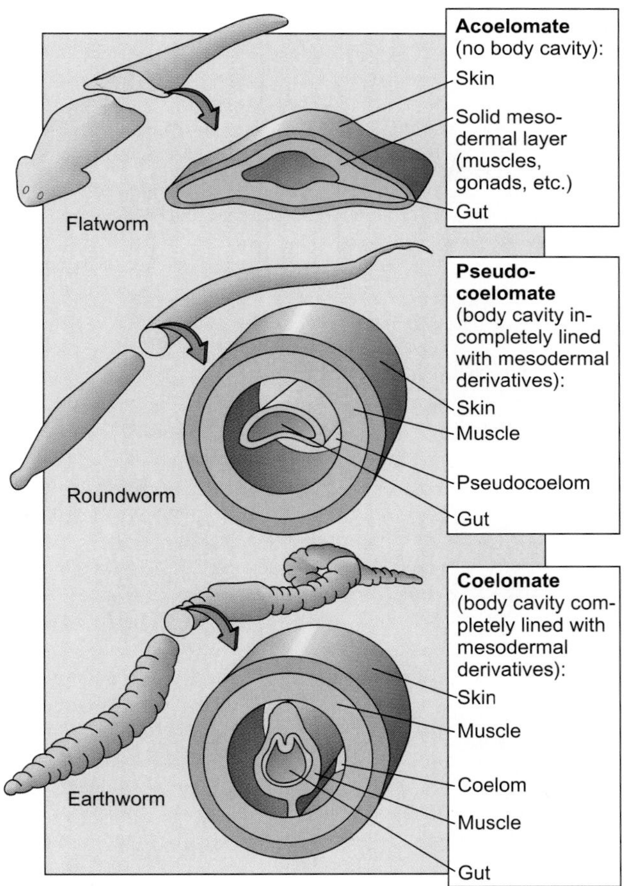

Figure 26.1 Acoelomates, Pseudocoelomates, and Coelomates.
Acoelomates lack a body cavity. Pseudocoelomates have a body cavity, but it is only partially lined with mesodermally derived tissue. Coelomates have a true coelom, a cavity completely lined with mesodermally derived tissue.

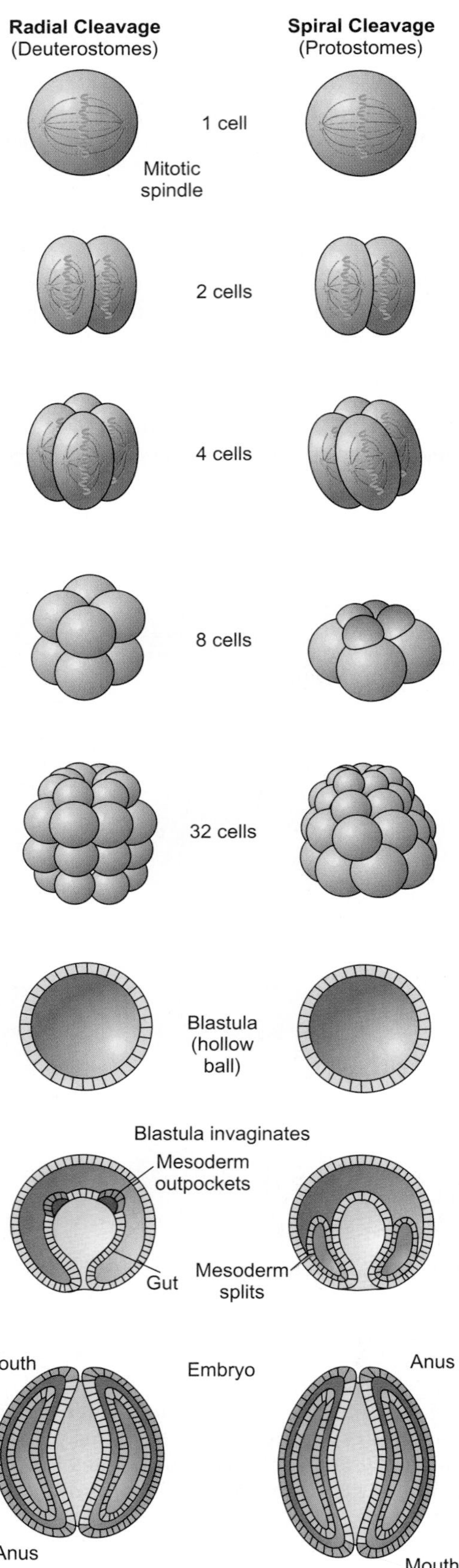

and segmentation arose independently in the protostomes and deuterostomes. Let's look now at the Mollusca, a phylum in which neither the coelom nor segmentation is obvious.

Shells and a Muscular Foot: Phylum Mollusca

The earth is burgeoning with mollusks, of the phylum Mollusca. In fact, biologists already know of about 100,000 species, which makes them third in numbers of species described—after the arthropods (such as insects) and nematodes. And, like the arthropods, the mollusks are a highly diverse phylum (Figure 26.3). Some mollusks are minute—tiny, inconspicuous creatures huddled in fragile shells—but others are truly gigantic. The North Atlantic squid, for example, may reach a length of 18 m (60 ft).

The fossil history of mollusks goes back to the Precambrian period and is found in the Ediacara beds (see Figure 25.3). Interestingly, the earliest fossils include an ancestral form very similar to *Nautilus,* a living shelled cephalopod. In fact, it seems that the Cambrian fossils include at least twenty times the number of mollusk species that now exist.

In many ways, the mollusks are a phylogenetic puzzle. For example, they show little, if any, evidence of segmentation, and the size of the coelom is so reduced that some scholars deny that mollusks have ever been coelomate animals. They hold that the mollusks share very little of their evolutionary history with the other protostomes.

Other biologists, however, believe that the mollusks are simply highly specialized relatives of the annelids and the arthropods. They note strong similarities in the earliest embryonic stages of mollusks and annelids, each with spiral, determinate cleavage—and their larvae, if they have them, are able to swim. Further, such theorists assert that mollusks and annelids both diverged early in the history of cnidarian protostomates.

FIGURE 26.2 GASTRULATION.
During embryonic development, an ingrowth of cells occurs through a process called gastrulation, producing the first outlines of the gut. In the protostomes, the initial site of gastrulation becomes the mouth; in deuterostomes, this early ingrowth marks the future site of an anus. Cleavage patterns in the early embryo are usually spiral in the protostome and radial in the deuterostome.

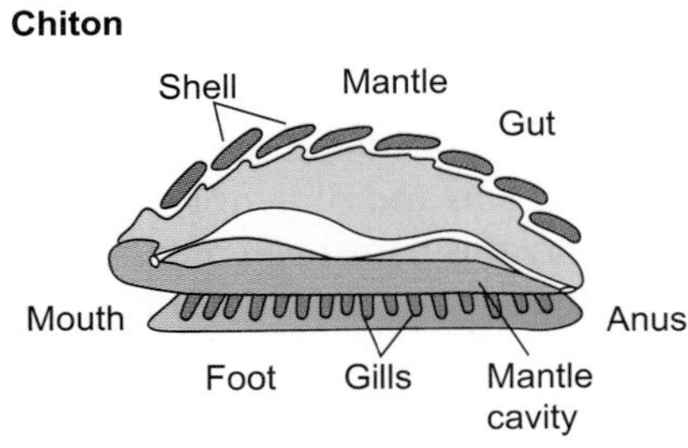

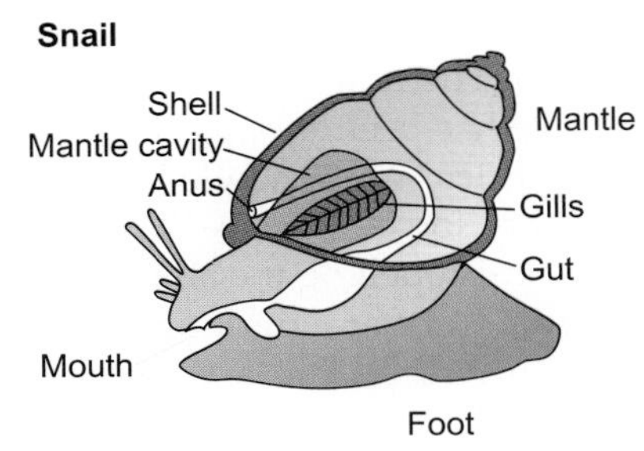

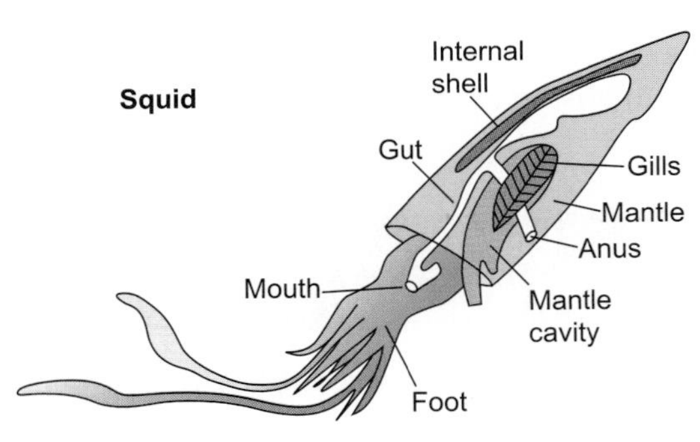

FIGURE 26.3 **SOME REPRESENTATIVE MOLLUSKS.** Membership in any of the major mollusk classes depends on specialization in the shell, mantle, foot, gut, and respiratory structures (gills).

Modern Mollusks

There are four major classes of modern mollusks and several minor ones. The major classes are the *chitons*, the *gastropods* (snails), the *bivalves* (clams and their relatives), and the *cephalopods* (octopuses and squids). As would be expected, no class of living animals conforms entirely to any simplified, generalized mollusk body plan. But all mollusk classes include at least some species with the basic molluscan characteristics: a muscular *foot*, a *shell*, a *mantle* and *mantle cavity*, and *gills* (Figure 26.3). Except for the bivalves, most mollusks have a highly specialized feeding device, the oscillating, rasplike *radula*, composed of hardened chitin. It is found in no other phylum.

All of these basic parts, however, have been subjected to intense evolutionary modification, and they may differ greatly from one group to the next. In some cases, a part may even be entirely absent.

Chitons. The chitons (Class Polyplacophora; Figure 26.3) are perhaps the least modified of the major groups of mollusks. This class has changed relatively little from the ancestral condition. The chiton's eight-part shell and repetitious rows of gill structures may reflect an ancient heritage.

Gastropods. The gastropods (Class Gastropoda) include the snails. They are more specialized than the chitons, and therefore have departed more from the ancestral line. For example, snails (Figure 26.3) have retained the simple gliding foot but, in most species, are able to retract it. They feed with a radula, but their shells, when present, are asymmetrical and quite different from the bilateral segmented shell of the chiton. Early in its development the snail embryo undergoes extreme *torsion*, or twisting (180 degrees), displacing the internal organs. The result is the anus ends up over the head, if you can imagine that. In addition, the genitals end up on one side of the head, in about the position of what would be the right cheek, if gastropods had cheeks. This torsion is then followed by an equally severe spiraling in the shell.

Many gastropods are hermaphroditic and have large penises with which they inseminate each other. A huge penis extending from the side of the head is a remarkable sight. Even more remarkable is the habit of garden snails of thrusting sharp "love darts" into the flesh of potential mates, seemingly as a means of arousal. These chalky arrows cause the potential mate to cringe reflexively, and although they do stimulate sexual behavior, they can also kill the object of affection.

Bivalves. The bivalves (Class Bivalvia—"two shells") include oysters, mussels, scallops, and clams (Figure 26.3). These species have followed yet another adaptive path: they have become *filter feeders*, filtering food from water, drawing the water through an *incurrent siphon* and directing it over and through their gills. The gills have been modified into sievelike feeding structures. Cilia covering the gill surfaces create the water currents that carry in food particles (minute algae), which become ensnared in a dense film of mucus that moves the food into the simple mouth. The water then passes on through the gills and exits through an *excurrent siphon.*

Cephalopods. The cephalopods (Class Cephalopoda—"head foot") have diverged the most from the basic molluscan body plan. This class includes the octopuses, squids (see Figure 26.3), and the chambered nautilus. The first two are swift predators, in which the foot has been modified into a number of grasping tentacles. The mantle, or covering, can rapidly contract, forcefully expelling water from beneath it, and jet-propelling the animal along its way.

The cephalopods are undoubtedly among the most fascinating of all the invertebrates. They all have excellent image-producing eyes, remarkably like those of vertebrates. Since cephalopods are not at all closely related to vertebrates, this is coincidence—an amazing example of convergent evolution. The nervous systems of these creatures have taken remarkable evolutionary directions. For example, the cephalopod brain is quite large in comparison to the brains of other mollusks, and we know that cephalopods can learn certain things with surprising ease.

Segmented Worms: Phylum Annelida

Earthworms are commonly used to represent annelids (Phylum Annelida), the segmented worms, because they are so familiar to most of us; but most annelid species are polychaete ("many-bristled ones"—Class Polychaeta) worms that live in the sea (Figure 26.4). Typical polychaetes are the sedentary, filter-feeding *fan worm* and the active, predatory *clam worm.* Both marine and terrestrial annelids have highly segmented bodies.

(a) (b) (c)

FIGURE 26.4 THE POLYCHAETE WORMS.
Many, such as the fan worms **(a)**, are tube dwellers that use feathery devices in feeding and respiration, popping them in and out of their tube houses. The oligochaetes—terrestrial annelids—lack complex outer structures that would complicate their burrowing activities. Segmentation, a prominent annelid characteristic, is readily evident in the earthworm **(b)**. Hirudineans **(c)** are external parasites, using their sucker devices to attach to the host animal.

THE EARTHWORM

The earthworms are land-dwelling oligochaetes ("few-bristled ones"—Class Oligochaeta). The "bristles," or *setae,* are chitinous spines that aid the worm in burrowing and also act as anchors, helping ensure that the early bird comes up short. Another group of annelids, the hirudineans (the leeches—Class Hirudinea) are blood-sucking, external parasites that inhabit moist or swampy environments (where they help liven up old Bogart movies). We will focus briefly on the earthworm.

Most of the earthworm's organ systems are well developed (Figure 26.5). The segmented body contains complex muscle layers. The coelom is fluid-filled, and the fluid is under pressure, forming what is called a **hydrostatic skeleton.** This means of support is common to many soft-bodied animals and provides a firmness that helps with movement and the maintenance of body form. As in other coelomate protostomes, the nervous system includes a ventral nerve cord that extends along the entire body, below the intestine. Two large ganglia (nerve cell clusters) make up the brain.

The earthworm's muscular, tubelike digestive system includes several unusual specializations. For example, the *typhlosole,* a large fold in the gut wall, aids digestion and absorption by greatly increasing the surface area. The excretory system, which is responsible for removing nitrogenous wastes and for water regulation, consists of paired, funnel-shaped *nephridia* that clear the wastes from the coelomic fluid in each segment and conserve body water by recycling it to the blood. Cell-by-cell diffusion of food and gases, which was quite sufficient in the smaller and simpler invertebrates, would not work in the comparatively large body of the earthworm. Here, then, we encounter, for the first time, an efficient circulatory system. Furthermore, it is a **closed circulatory system,** which means that blood remains within vessels rather than leaving vessels to percolate through open cavities and sinuses, as it does in the **open circulatory systems** of many other invertebrates. Since the blood contains hemoglobin, it is red, like ours. The blood is pumped through five pairs of "hearts," or *aortic arches,* which send it coursing through large blood vessels that branch into dense beds of finer, thin-walled capillary vessels.

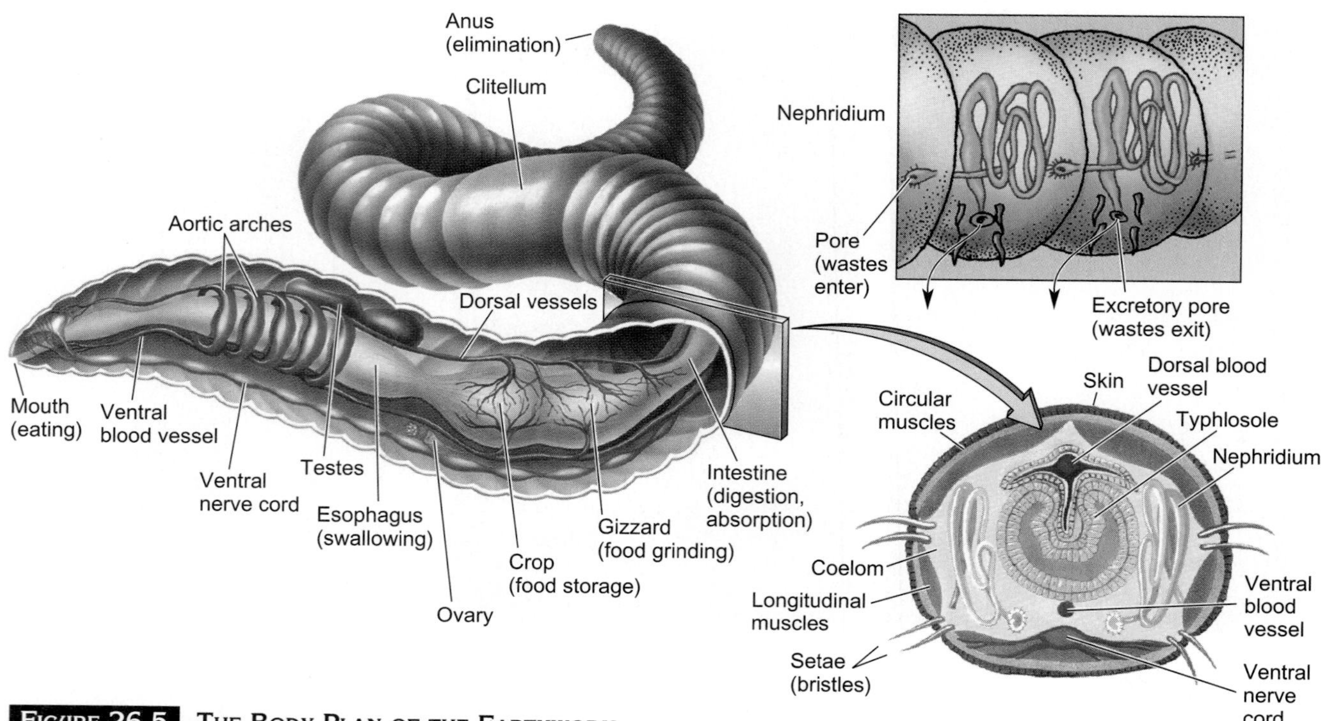

FIGURE 26.5 THE BODY PLAN OF THE EARTHWORM.
Note the prominent segmental body plan, the smooth, glandular clitellum, and the circular and longitudinal muscle groups in the body wall of the earthworm. Transport is carried out by a closed circulatory system that includes five paired "hearts" (aortic arches) and an extensive system of blood vessels. The gut, suspended in the coelom, contains several specialized regions. The nervous system consists of a pair of enlarged ganglia above the pharynx and a lengthy ventral nerve that gives rise to ganglia in each segment. Nearly all segments contain paired nephridia, complex tubular structures that eliminate nitrogen wastes and regulate body water.

The capillaries, passing through the thin, moist skin, permit oxygen to diffuse into the blood and carbon dioxide to diffuse out. The capillaries provide all tissues with food and oxygen.

Although the sexes are separate in many of the marine annelids, earthworms are hermaphroditic, with each individual possessing both testes and ovaries. Earthworms mate by lying head to tail and exchanging sperm, which are then stored in *seminal receptacles.* Later, a sleevelike mucus cocoon is secreted. It passes over ducts that release eggs and then over ducts that release sperm. The earthworm embryos develop in the cocoon.

A REMARKABLE SUCCESS STORY: PHYLUM ARTHROPODA

No one has any idea how many species of arthropods (Phylum Arthropoda) there are, but the most respected guesses range between 800,000 and 1,000,000. In any case, Arthropoda is the most diverse phylum on earth.

Arthropod means "jointed foot," so you will not be surprised to learn that arthropods have jointed legs and feet. They are also segmented, but their segments are not the simple repeating units characterizing the earthworms. Instead, the segments are specialized and modified for different tasks. Arthropods have exterior skeletons (exoskeletons) covering their bodies. The exoskeleton, or *cuticle,* is made up of chitin, a dense, flexible carbohydrate. In the aquatic crustaceans (Class Crustacea), it is often hardened with calcium salts; in insects (Class Insecta), it is hardened with various organic substances.

Chitin is also common to other protostomes, where it helps form many hardened structures. As they grow, all arthropods must periodically discard their hardened exoskeleton through *molting,* replacing it with a larger one. Molting (also called *ecdysis*) is under hormonal control.

Arthropods Today

Arthropods have been incredibly successful in their expansion over the earth, sometimes establishing very narrow ecological niches. (Consider, for example, the mayfly, a delicate creature that emerges as an adult without mouthparts and must mate and leave offspring in the few precious hours of life allowed it; or strange green insects that live only on year-round alpine glaciers that never melt). In fact, arthropod niches are often so specialized that many species can live in very close association without seriously competing with each other.

Arthropods include omnivores, herbivores, filter feeders, carnivores, scavengers, ectoparasites, endoparasites, and even a few opportunistic cannibals. How did they become so successful? For one thing, the use of jointed limbs was an immediate evolutionary success. Limbs were subject to all sorts of modifications that could eventually permit crawling, burrowing, jumping, grasping, feeling, and even hearing. The mouthparts themselves are derived from jointed appendages, and are also extremely specialized—modified for such actions as biting, chewing, sucking, stinging, or lapping. The development of wings in the insects also greatly increased their ability to disperse and to exploit previously unavailable habitats. In addition, a great many species produce larvae whose diet is entirely different from that of the adults, resulting in a subdivision of resources that helped to decrease parent-offspring competition.

Arthropod Diversity

Since there is no concise way to deal with the enormous diversity of arthropods, we'll have to pick and choose a few examples. The living arthropods are traditionally divided into two huge subphyla, the Chelicerata and the Mandibulata (Figure 26.6). The Chelicerata (arthropods that lack jaws)

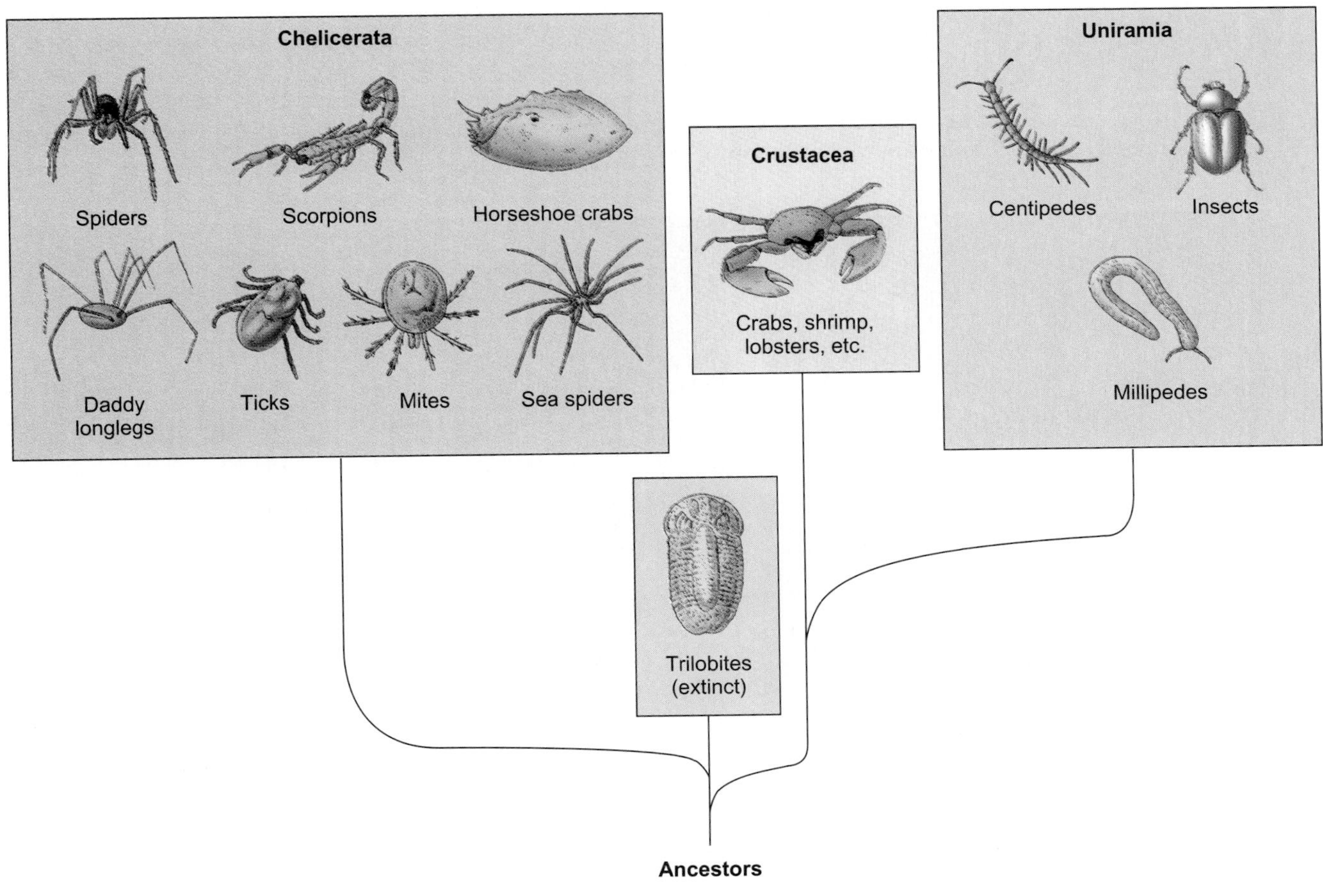

FIGURE 26.6 ARTHROPOD DIVERSITY.
The immense arthropod phylum contains two subphyla: the jawless Chelicerata and the jawed Mandibulata.

FIGURE 26.7 REPRESENTATIVE CHELICERATA.
Some arthopods have reputations as vicious hunters.

include the arachnids (Figure 26.7), the sea spiders, and the horseshoe crabs. The Mandibulata (the jawed arthropods) include the crustaceans, centipedes, millipedes, insects, and a number of minor groups (Figure 26.8). Some scholars claim that the mandibulates are actually three unrelated phyla that are lumped together only because of convergent evolution. In any case, we'll concentrate on just three arthropod groups: the arachnids, the crustaceans, and the insects.

Arachnids: Spiders and Their Relatives. Arachnids (Class Arachnida) include spiders, scorpions, ticks, mites, and daddy longlegs. Typically, arachnids have a two-part body: a thorax, which bears the appendages, and an abdomen, which contains digestive, reproductive, and silk-producing organs. They also have six pairs of appendages, including the specialized *chelicerae* (or fangs), which are followed by sensory *pedipalps,* and then by four pairs of legs.

Spiders are carnivores, but they lack jaws and must suck their foods. Typically, a spider first injects venom into its prey's body to kill or paralyze it, then pumps in digestive enzymes that liquefy the prey's tissues, enabling the spider to suck out the nutrients. Spiders breathe by *book lungs,* essentially sacs with slits that open to the outside. The sacs are lined

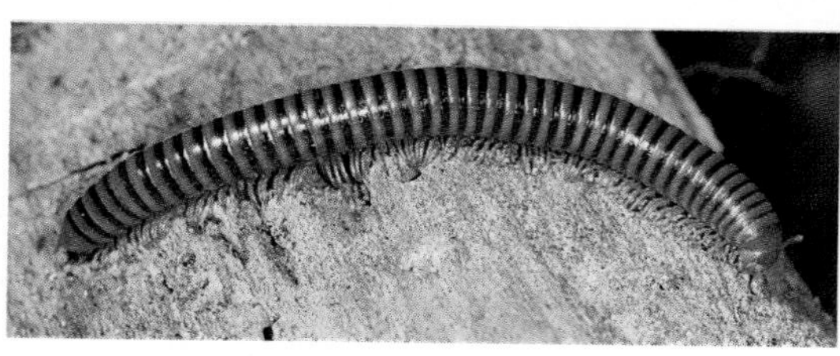

FIGURE 26.8 REPRESENTATIVE MANDIBULATA.
While insects make up the majority of mandibulates, crustaceans are also numerous. Minor groups include the millipedes and centipedes.

with flattened folds that resemble the pages of a book. Blood passing through the thin sheets exchanges gases with air in the sac.

Nearly all spiders have silk-producing glands connected to external appendages called *spinnerets.* The system is used to form both the web and the cocoon where some spiders store their fertilized eggs (Figure 26.9).

Crustaceans. Most species of crustaceans (Class Crustacea) are aquatic and are common in both marine and fresh waters. This class includes crabs, shrimp, crayfish, and lobsters, as well as terrestrial wood lice (also called sow bugs). Crustaceans range in size from the freshwater ostracods (1.1 mm, or about the size of one of these letters) to the gigantic Japanese spider crab of cold Pacific waters (4 m, or probably wider than your bedroom—or your whole house, if you live as students usually do).

Typically, crustaceans breathe through gills that are covered by the tough exoskeleton. Blood continuously flows through delicate, feathery gills, releasing carbon dioxide and taking in oxygen. Crustaceans often have highly developed sensory structures. Interestingly, the general plan of the arthropod nervous system is similar to that of the earthworm and the mollusk: a "brain" surrounding the esophagus, and paired, solid, ventral nerve cords with paired ganglia at intervals along their length.

Marine crustaceans and other saltwater inhabitants share the problem of life in a hypertonic environment—roughly a 3% salt solution. Some cope as *osmoconformers*—maintaining a salt content that roughly matches the surroundings. The *osmoregulators,* on the other hand, maintain a lower and relatively constant content in their bodies by secreting salt through special glands.

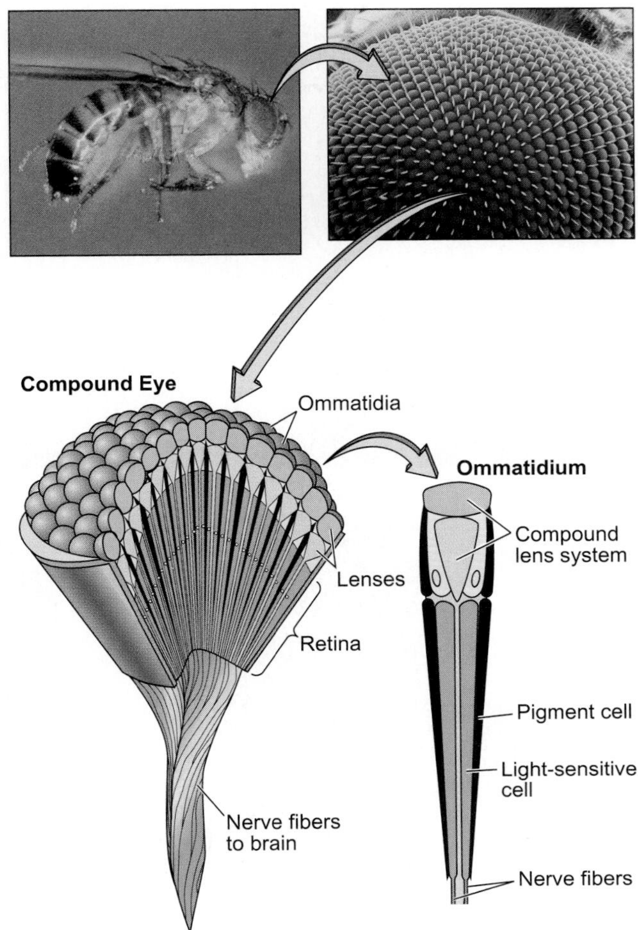

FIGURE 26.10 COMPOUND EYES.
The compound eyes of crustaceans and insects consist of many individual light-receiving units called *ommatidia.*

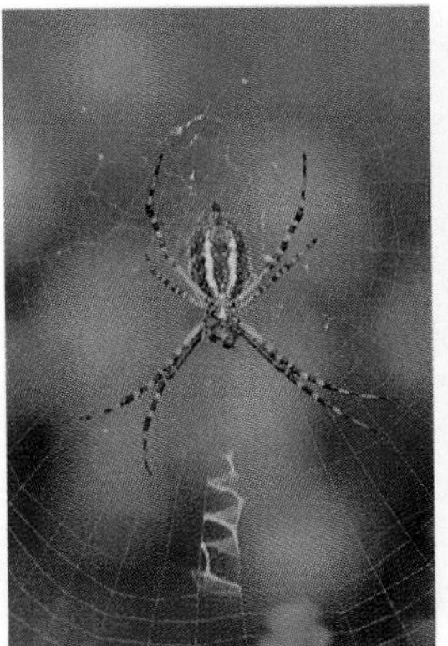

FIGURE 26.9 SILK GLANDS OF THE SPIDER.
The web in the photograph belongs to an orb spider. The spider can move about quickly on the web, but it must know where the sticky parts are or it will become ensnared in its own trap.

Among the most specialized of the crustacean sensory structures are the *compound eyes,* which are often borne on flexible stalks. The crustacean eye is similar to that of an insect, composed of a large number of visual units called *ommatidia.* Each ommatidium (singular) has a tough, transparent lens and pigmented, light-sensitive receptors (retinal cells) (Figure 26.10). With this arrangement the animal has a nearly 360-degree view of its surroundings, though it may see a shimmering mosaic of light and dark colors—quite different from the images we see.

Insects. Whereas the crustaceans are the most prevalent arthropods in the sea, insects (Class Insecta) dominate the land, both in number and in

(a)

(b)

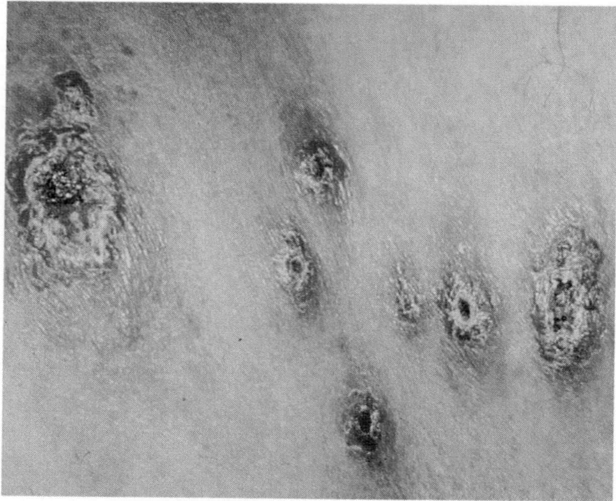
(c)

FIGURE 26.11 INSECT PLAGUES.
(a) Locust swarm in Africa. (b) The tsetse fly that carries sleeping sickness. (c) Lesions caused by leishmaniasis.

kind. Insects are of such ecological and economical importance that we literally could not have reached our own place in nature without understanding something about them. One reason is that we must continuously compete with them, and we don't always win. There are a great many humans who go hungry because their food is eaten or ruined by insects, and many live debilitated, shortened lives—victims of insect-borne diseases such as malaria, typhus, African sleeping sickness, and leishmaniasis (Figure 26.11), a disfiguring disease affecting the spleen and borne by sand flies. (Many of our Desert Storm troops were infected with this disease they had never heard of. One of your authors has collected at least twelve plants from the Amazon rain forest which are effective against leishmaniasis in the laboratory.) Such diseases are indeed a high price to pay to be able to share the planet with the butterflies.

The bodies of insects as we know them today (Figure 26.12) have undoubtedly changed considerably from the ancestral form. For one thing, many of the segments have become fused. Modern insects, for example, have a *head, thorax,* and *abdomen.* The segments of the head and thorax have become fused, but segments are still apparent in the abdomen. The sensory structures—the eyes, antennas, and sensory palps—are concentrated in the head region. The thorax gives rise to three pairs of legs, and commonly to two pairs of wings.

The wings of insects are thin, chitinous, and moved by powerful muscles in the thorax on elaborate, levered hinges. Wing movement in some insects is incredibly rapid. Fruit flies, for example, can maintain a continuous beat of 300 strokes per second for as long as two hours. (In case you're wondering how a muscle can contract and relax so quickly, actually it doesn't. A single muscle contraction can set up reverberations in the insect skeleton that, in turn, moves the wings.)

The open circulatory system of insects moves nutrients and waste, but is not the primary mechanism in respiration. Insects exchange gases with the environment by admitting air into a complex *tracheal system.* It consists of tubes (the *tracheae*) that pass throughout the body, branching to produce ever finer *tracheoles,* passages that end blindly in thin-walled sacs across which the gases can pass. The air enters through valvelike body openings called *spiracles.*

There is enormous variety among the specialized insect mouthparts. For example, the mouthparts of grasshoppers are suited for tasting, shearing, and chewing. Moths have long sucking mouthparts that they coil neatly when not in use. Mosquitos have piercing and sucking mouthparts, as we are all aware. The crass housefly salivates on your chocolate cake, stirs it with its bristled tongue, and then sucks up the resulting mess. It can't eat much. The rest is yours.

The insect digestive system includes three parts: the *foregut, midgut,* and *hindgut* (Figure 26.12). The foregut moves food along to the midgut. Food is digested in the midgut and then moves to the hindgut, where water is removed and fecal wastes are

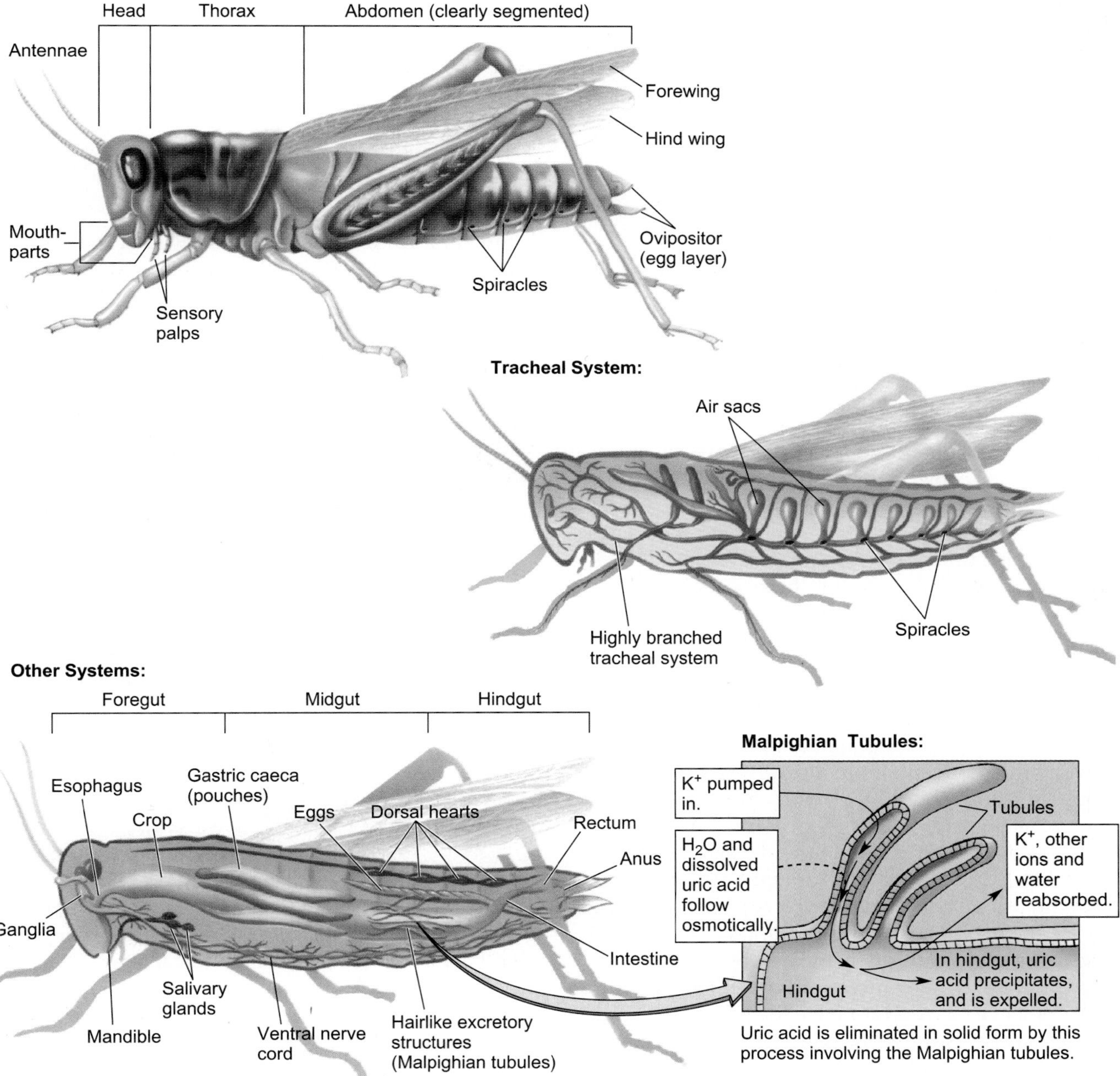

FIGURE 26.12 THE INSECT BODY PLAN.
Note the breathing system of the grasshopper. Air enters the body through the spiracles and diffuses through finely branching tubules, dead-ending in the tissues. In the insect digestive and excretory systems, the chewing mouthparts are well adapted for feeding on vegetation and the three gut regions have specialized functions. The Malpighian tubules absorb nitrogen wastes from surrounding body fluids.

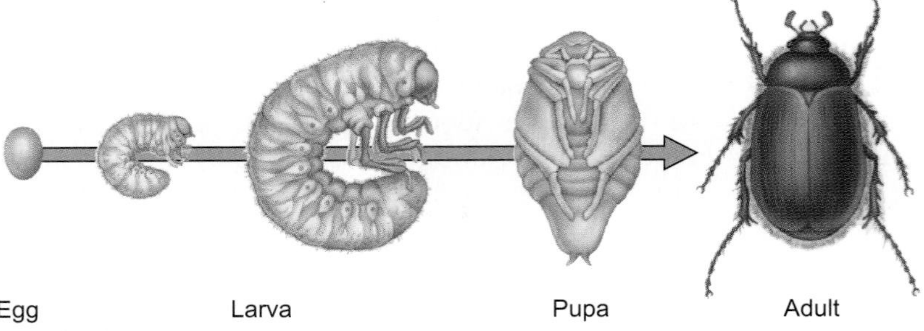

FIGURE 26.13 COMPLETE METAMORPHOSIS. Here the insect goes through both a larval and pupal stage in the transition from egg to adult. Because of their voracious appetite, many larvae are damaging agricultural pests, the target of relentless crop spraying and dusting. Pupae, while seemingly dormant, undergo great changes.

concentrated. The area where the midgut and hindgut join receives metabolic wastes from the excretory system, which consists of numerous thin tubes known as *Malpighian tubules.* The waste is eliminated as nearly dry crystals of uric acid, as described in Figure 26.12.

Insects, like other arthropods, have an open circulatory system. The main dorsal blood vessel has several contractile areas called "hearts." Openings in the vessel allow blood to be drawn in from the surrounding spaces and then pumped along sinuses and cavities. From there it sluggishly percolates through the tissue and back to the dorsal vessel, again to be moved along by the peristaltic (wavelike) contractions of the tubular heart.

The insects' reproductive adaptations contribute significantly to their success. The life span is typically short, but many species compensate by quickly producing very large numbers of offspring. (A female housefly has the potential to leave more than 5 trillion descendants in just seven generations, or one year's time!) In addition, each of the stages in the life cycles of many insects plays a different role. The *eggs* are often well concealed and resistant. When they hatch, the *larvae* go through *metamorphosis,* changing to *pupae* before entering adulthood (Figure 26.13). (Other species may not enter the pupal stage.) The larva and adult often utilize entirely different food sources, so they are not in direct competition. The pupa lives on food stored in the larval stage. Egg and pupa often survive the winter in dormant states.

Insects, indeed, abound over the earth. They have exploited virtually every available nook and cranny on the planet. Their success, however, shouldn't be surprising considering their varied diet, small size, rapid generation time, great reproductive output, and highly varied life cycles.

To sum up, we have pointed out that the protostome line produced three great coelomate phyla—Mollusca, Annelida, and Arthropoda—whose members make up the majority of animal species today with a notable range of diversity.

THE ECHINODERM–CHORDATE LINE—AN UNEXPECTED RELATIONSHIP

The earliest deuterostomes were small and simple, with extremely limited nervous systems. And yet there was an unsuspected potential in these animals. No one could be too surprised that the deuterostomes were to produce the witless starfish, but who would have imagined that they would also give rise to the chordates, and that from the chordates would come the vertebrates—the largest and brainiest animals on earth?

SPINY SKINS AND RADIAL BODIES: PHYLUM ECHINODERMATA

Echinoderms (Phylum Echinodermata) are exclusively marine animals, with many species living in the shallow waters of the continental shelf. Others, though, thrive in the deepest oceanic trenches. There are five major classes of echinoderms (Figure 26.14), each with its own variation on the basic five-part body plan.

Echinoderms are unusual in many respects. Their radial symmetry, for instance, is not like that of the simple coelenterates, but is usually a modified form called *pentaradial symmetry* (*penta,* five), in which the body could only be cut at certain places, beginning at the tip of an arm, to produce nearly identical sections. However, the embryonic stages are not radially symmetrical at all. The larvae (and, indeed, the ancient fossilized adult echinoderms) are bilaterally symmetrical.

Even the spiny echinoderm skeleton is a true *endoskeleton*—that is, it is of mesodermal origin and is located on the inside. Don't confuse the spines found in many echinoderms with exoskeletons. They form inside the animal and emerge from bony (mesodermal) plates located below its skin.

One of the most unusual features of the sea star or starfish (Class Asteroidea) is the **water vascular system** in which water is pumped through vessels, such as the ring canal and the radial canals, causing various parts to move. It is the principal mechanism of locomotion for some species (some sea urchins clamber about on their stiltlike, movable spines). The *madreporite* is the system's sievelike water inlet. The major parts of the internal plumbing, its vessels, are the hardened *stone canal, ring canal,* and *radial canals.* The movable parts are the *ampullas,* muscular "squeeze bulbs" that force water into the *tube feet,* thereby extending them. When water is withdrawn, each tube foot becomes a kind of suction cup, allowing the animal to grasp its substrate as it moves along (Figure 26.15).

Other systems in the sea star are much simpler. For example, respiration is carried out by simple ciliated extensions of the coelomic lining that stick out through the body wall and act in gas exchange. Dissolved gases can also readily cross the thin membranes of the tube feet, and coelomic fluid carries materials throughout the body. The nervous system of the sea star is a simple ring, with extensions reaching into each of its five arms. There is no sign of centralization, no clumps of ganglia, and no concentration of nerve cells. This may explain why the animal can move in any direction with equal ease and full coordination. Such decentralization may also contribute to its remarkable powers of regeneration. An entire animal can develop from an arm that has been removed, provided just a bit of the central disk is included, a fact that for a time escaped the oystermen of the Chesapeake who would take their vengeance on these oyster predators by chopping them up and throwing the pieces overboard. Many of the pieces formed new sea stars. The digestive tract is *complete*—that is, it includes a mouth, a gut, and an anus. The reproductive system is simple, consisting of testes or ovaries (sexes are separate) that lead to ducts. In most echinoderms the gametes are simply released into the water, where fertilization occurs.

(a)

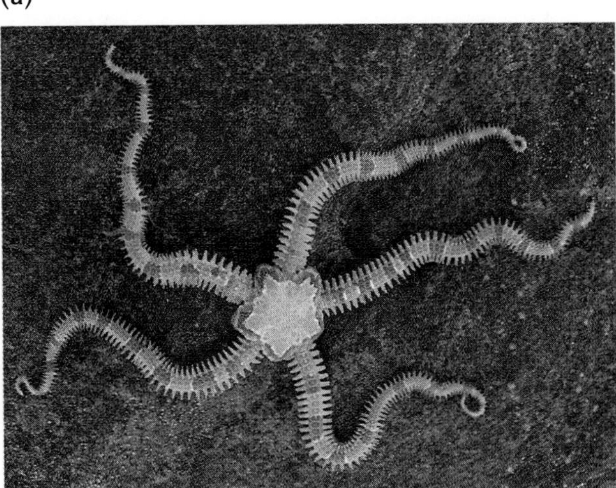

(b)

(c)

FIGURE 26.14 FIVE-PART RADIAL SYMMETRY. This is a basic feature of echinoderm anatomy, but many variations of this theme are seen in the phylum. **(a)** Sea star. **(b)** Brittle star. **(c)** Sea urchin.

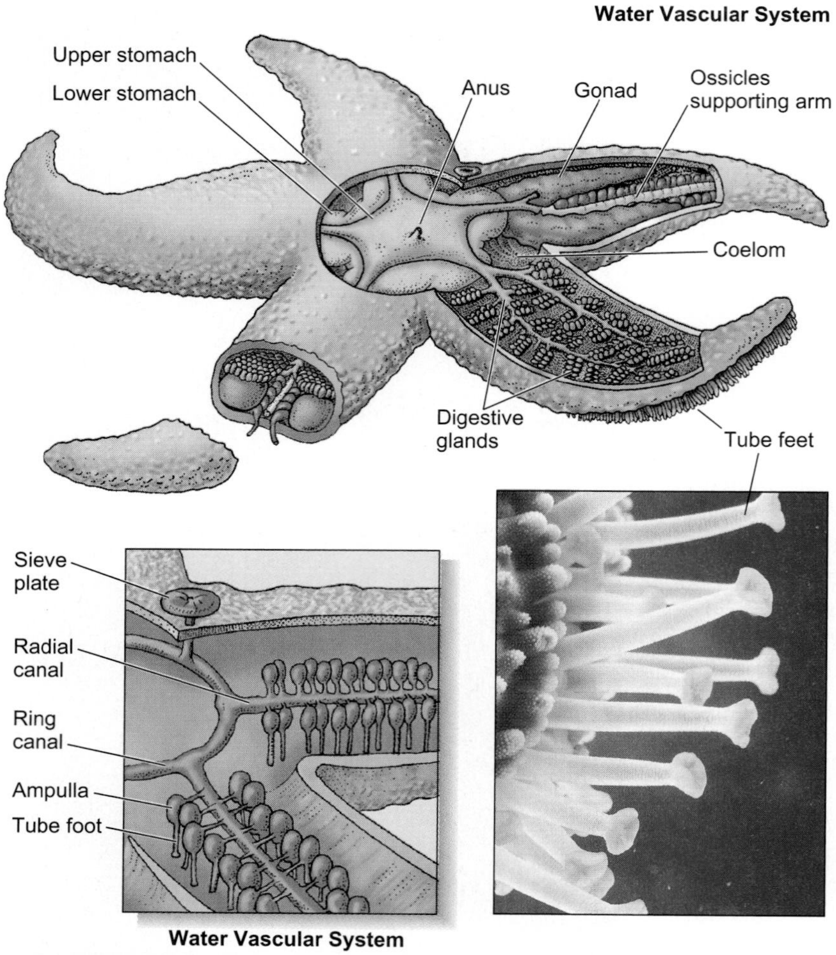

FIGURE 26.15 **WATER VASCULAR SYSTEM OF THE SEA STAR.** Water enters and leaves through the sieve plate. By shifting the water, the tube feet can be extended by pressure from the inside. When the tube feet are pressed against something solid, such as an oyster shell, strong suction can be developed as water is moved from the foot.

ACORN WORMS: PHYLUM HEMICHORDATA

The hemichordates (Phylum Hemichordata) comprise a group of rather uninspiring aquatic creatures including the *acorn worms.* The acorn worms of this minor phylum have two primitive chordate characteristics, *gill slits* and a *dorsal, tubular nerve cord* (which we will discuss in Chapter 27). Equally significant, the acorn worm's larvae are remarkably similar to those of some echinoderms (Figure 26.16).

SEA SQUIRTS TO HUMANS: PHYLUM CHORDATA

And now we come to the chordates (Phylum Chordata), the group that includes us. Some of the chordate traits may seem alien to you, but keep in mind that they may appear only briefly, perhaps at some embryonic stage.

Phylum Chordata includes three subphyla: Urochordata, Cephalochordata, and Vertebrata. In each

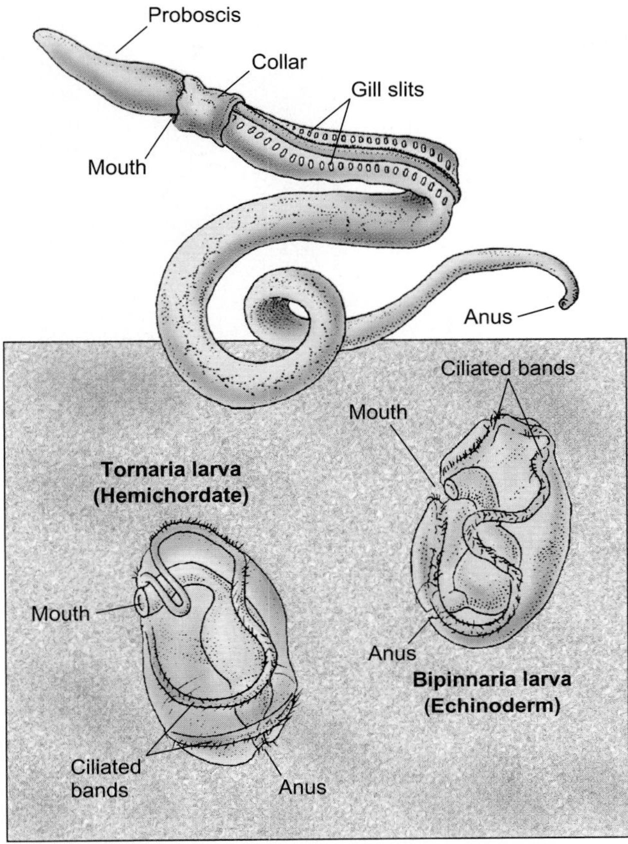

FIGURE 26.16 LARVAE OF HEMICHORDATES AND ECHINODERMS.
Many acorn worms spend their lives burrowing in saltwater mud flats, where they feed through the use of a ciliated proboscis. The relationship of hemichordates and echinoderms is well established on the basis of larval forms, which are remarkably similar, right down to the arrangement of the ciliated bands.

chordate subphylum, the following characteristics are shared:

1. All chordates possess, at least at some time in their lives, a **notochord**—an internal, flexible, turgid rod that runs along the dorsal (back) side. The phylum derives its name from this structure. (See Chapter 25.)
2. All chordates possess, at some time in their lives, a number of **gill arches** (between them are the **gill slits**).
3. All chordates have, at some time in their lives, a **dorsal tubular nerve cord.**
4. At some time (in an early embryonic stage, at least), all chordates possess **myotomes**—serially repeated blocks of muscle along either side of the notochord.
5. At some time in their lives, all chordates have a **postanal tail.**

Now let's briefly consider the urochordates and cephalochordates before moving on to the animals with backbones. The *urochordates* include the stationary *tunicates* (sea squirts); the transparent, free-swimming *salps;* and a minor, but highly significant group, the *larvacea,* planktonic forms with small, soft bodies.

Adult tunicates (Figure 26.17) have a hollow saclike body composed of a tough *tunic,* made up of a celluloselike material (rare in the animal world). A casual observer could confuse them with cnidaria—especially since, like sea anemones, they squirt seawater out of their bodies when they contract. But the larva of the tunicate unveils it as a chordate, with its bilateral form, a notochord, gill arches, myotomes, a dorsal tubular nerve cord, and a postanal tail that permits vigorous swimming. (The adult larvacea retain this body form.) Larval tunicates are called tadpoles, and indeed they look and act a lot like the larvae of frogs and toads.

Cephalochordates have been found fossilized in the Burgess Shale deposits of the early Cambrian period. They are very similar to the living cephalochordates, the lancelets. They belong to the genus *Branchiostoma,* but are perhaps known by their former scientific name, *Amphioxus.* They, at last, are clearly our relatives. In fact, the basic cephalochordate body plan seems almost like a simplified cartoon of the general vertebrate body plan. At the same time, it is similar in many ways to that of the urochordate larva (Figure 26.18). Like urochordates, cephalochordates are filter feeders. These similarities have strongly suggested how chordates may have evolved. (Such a scenario is seen in Figure 26.19.) Much later in the Cambrian period, ancestral cephalochordates gave rise to a new kind of animal in the oceans. At first, it was an awkward, slow-swimming, jawless creature, sucking up its nutrients from the mucky bottom sediments. But it *was* a fishy ancestor to a group that would someday change the very face of the earth.

FIGURE 26.17 **THE TUNICATE.**
As an adult, the tunicate hardly represents what we expect in a chordate, though the larva does have chordate characteristics **(a)**. The simple saclike body lacks nearly all of the typical chordate structures, except for the telltale chordate gill slits. As the larva matures **(b)(c)** the body plan is reorganized. The ciliated gill structure of the mature sea squirt (called a *gill basket*) is used in gas exchange and for straining food particles from the seawater drawn into the incurrent opening of the tunic **(d)**.

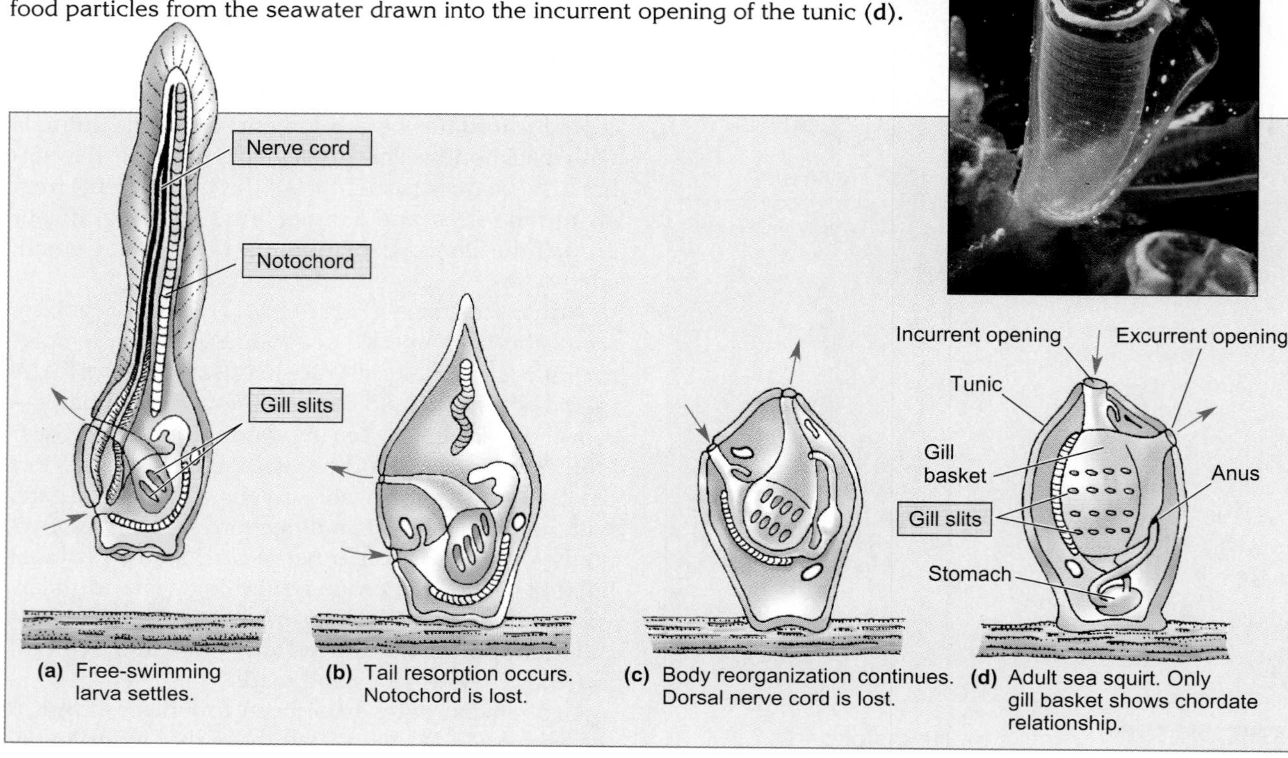

FIGURE 26.18 **THE LANCELET.**
Some zoologists believe that the lancelet, or *Branchiostoma,* is descended from larval tunicates that failed to metamorphose while becoming sexually mature. Adult lancelets are fishlike and they retain their notochord. They burrow into sandy bottoms, leaving only their mouth exposed. Food-laden water is drawn in and nutrient particles are swept along by cilia into the gut.

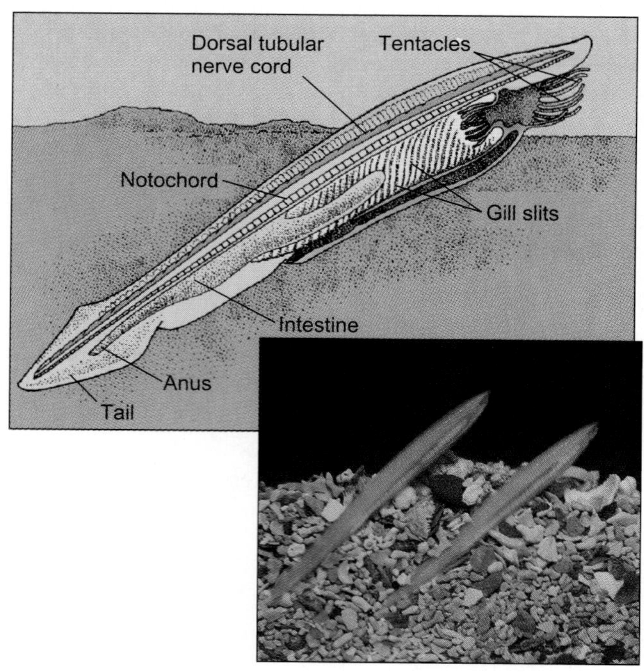

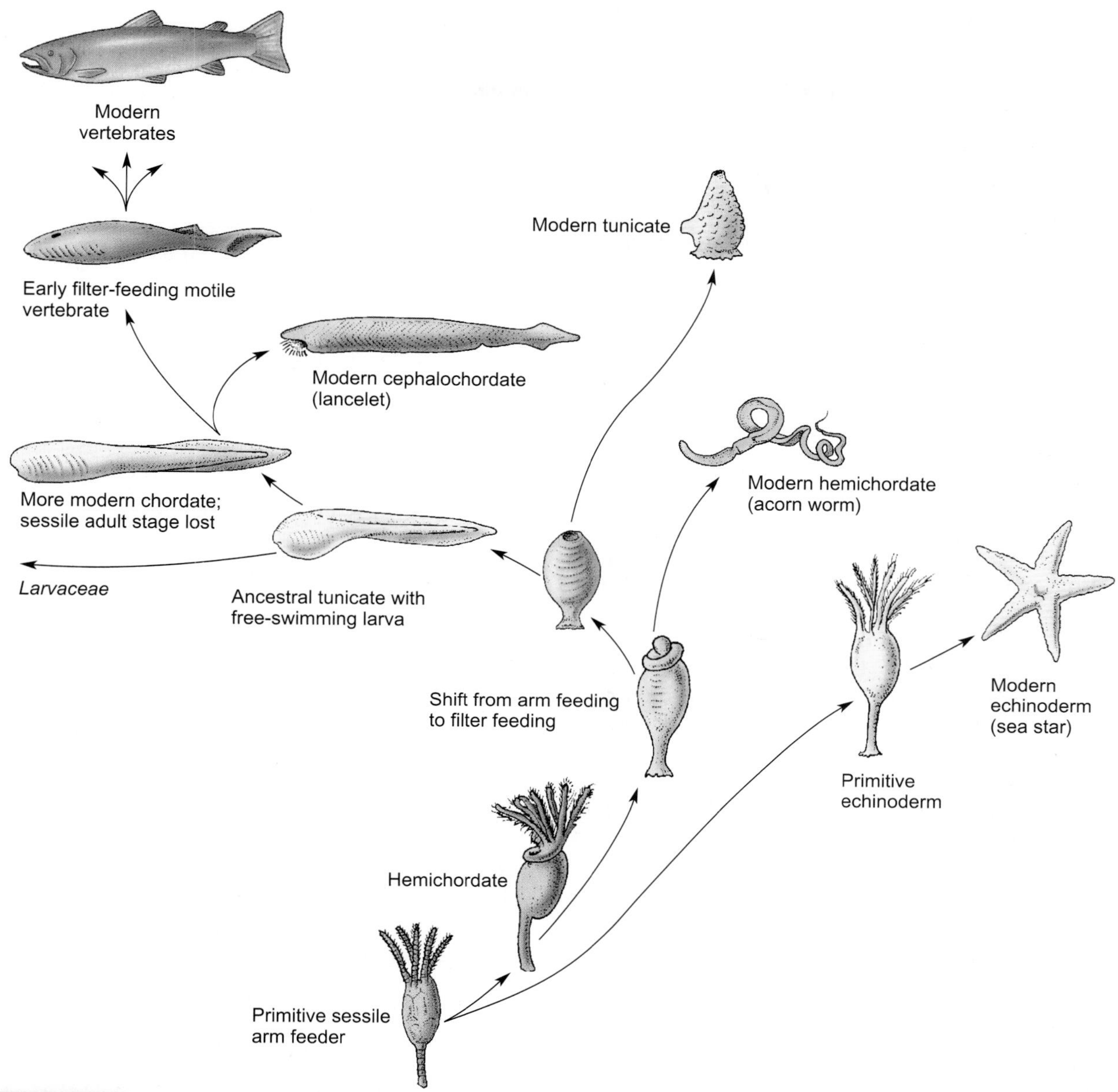

FIGURE 26.19 **THEORETICAL CHORDATE EVOLUTION.**
According to one theory, echinoderms, hemichordates, primitive chordates, and vertebrates arose from an ancestral arm-feeding deuterostome. From this ancestor arose the echinoderms and hemichordates, both of which still retain the free-swimming larva. Following these divergences, the deuterostomic line divided once more, but in a most unusual manner. One branch produced the urochordate line, today's tunicates and salps. The other arose from certain larvae that had somehow retained their juvenile body form while reaching sexual maturity. The best evidence of this strange capability is seen in *Larvaceae,* a group of living urochordates that remains permanently in a larval state. From this seemingly odd venture arose the cephalochordates and the vertebrates.

PERSPECTIVES

The coelomate invertebrates have given rise to a number of major evolutionary events, from the segmented body to segmentation with specialization. That specialization sometimes involved elaboration (as in the insect antennas) and sometimes simplification (as in the fused thorax). We learned here—by reviewing the significance of the protostomate and deuterostomate lines—the importance of embryology in determining relatedness. Finally, we moved to the chordate phylum, which sets the stage for a subphylum of this group, called the vertebrates—the animals with backbones.

SUMMARY

THE COELOMATE BODY PLAN

The two major animal types differ in gut formation. In protostomes (e.g., most vertebrates) the mouth forms earlier than the anus, in the region of the blastopore, while in deuterostomes (e.g., echinoderms and chordates) it forms later, and at the other end. In protostomes the earliest cleavages occur in a spiral pattern, and their development fate is set early (determinate development). In the latter, cleavages follow a radial pattern, and early cells remain uncommitted longer (indeterminate development).

Coelomates have a complete gut, most with regions specialized for swallowing, storing, grinding, digestion, and waste elimination. The body cavity is a mesodermally lined true coelom.

Segmentation, the division of the body into repeated segments, is highly pronounced in annelids and arthropods and is common throughout the coelomates.

SHELLS AND A MUSCULAR FOOT: PHYLUM MOLLUSCA

Molluscan fossils are numerous in strata of the Cambrian period. Today's species reveal little evidence of segmentation or a coelom, but other characteristics reveal their relationship to other coelomate protostomes.

The classes (chitons, gastropods, bivalves, and cephalopods) differ in specializations of the foot, shell, mantle, and mantle cavity. The presence of a radula for feeding is common.

Primitive characteristics in chitons include simple shell plates, a radula, and a simple mantle and foot. Advanced gastropod traits include body torsion, a twisted shell, and a retractable foot. Bivalves have two shells, powerful retracting muscles, and specializations for filter feeding. The fast-moving cephalopod predators have many advanced features, including tentacles (subdivided foot), jetlike use of the mantle, larger brains with marked learning capability, and image-producing eyes.

SEGMENTED WORMS: PHYLUM ANNELIDA

Annelids include marine polychaete worms (clamworms, fanworms, tubeworms) and terrestrial types, the oligochaetes (earthworms) and hirudineans (leeches).

Earthworms are highly segmented, with complex muscles, a hydrostatic skeleton (fluid-filled coelom), and well-developed systems. The nervous system includes large ganglia that form a brain and a ventral nerve cord. The digestive system has several specialized regions with an increase in the absorbing areas provided by a typhlosole. Paired nephridia in most segments remove nitrogen wastes. Hemoglobin-rich blood in the closed circulatory system is pumped by five pairs of aortic arches, with all exchanges occurring through capillary walls. Earthworms are hermaphroditic with well-developed reproductive organs.

A REMARKABLE SUCCESS STORY: PHYLUM ARTHROPODA

Members of the largest phylum, Arthropoda, have jointed appendages, modified segmentation, and chitinous exoskeletons. Many molt as they grow.

Through enormous structural diversification (especially limbs and mouthparts), arthropods have an enormous number of niches. Most reproduce prodigiously, and many have complex life cycles.

Subphyla include Chelicerata (spiders and others) and Mandibulata (crustaceans, insects, millipedes, and centipedes).

The spider body has two parts, a head and abdomen. Its six pairs of appendages include two chelicerae (fangs), two pedipalps, and eight walking legs. Spiders kill their prey with venom, later sucking up its body fluids. Gas exchange occurs through book lungs. Silk is formed in glands, and spinnerets spin webs and cocoons.

Crustaceans inhabit marine, freshwater, and terrestrial habitats. Gills are common in aquatic species. A brain and ventral nerve cord are present, as are compound eyes. The latter consist of ommatidia, visual units that detect movement. Species may be osmoregulators or osmoconformers.

Insects have had profound effects on human life, representing the most threatening animal competitors. The three-part body includes head, thorax, and abdomen. Many have two pairs of wings, and most have three pairs of walking legs. Insects have open circulatory systems and complex tracheal respiratory systems, with gas exchange occurring in thin-walled tubes called tracheae. Highly varied mouthparts permit great specialization. The digestive system includes a foregut, midgut, and hindgut. Malpighian tubules transfer nitrogenous wastes from body fluids to the digestive wastes.

Most insects have a short life span, but they produce great numbers of offspring. In species with complete metamorphosis, egg, larva, pupa, and adult stages form.

The Echinoderm–Chordate Line—An Unexpected Relationship

Deuterostomes arose from very simple animals with little neural development, yet this line produced the vertebrates, animals with the largest brains.

Spiny Skins and Radial Bodies: Phylum Echinodermata

Echinoderms include sea stars, brittle stars, sea urchins, crinoids, and sea cucumbers. Adults have a pentaradial (five-part) symmetry, but larvae are bilateral. The spiny endoskeleton emerges from mesodermal plates. A unique water vascular system, including numerous canals, "squeeze-bulb" ampullas, and tube feet, provides for movement and feeding. Gas exchange occurs through dermal branchiae and tube feet, and digestion occurs in a complete tract. The nervous system is a simple ring of nerves with little centralization. Fertilization is external.

Acorn Worms: Phylum Hemichordata

Hemichordates include acorn worms whose relationship to chordates is revealed by gill slits and dorsal hollow nerve cords.

Sea Squirts to Humans: Phylum Chordata

Chordates include urochordates (sea squirts and salps), cephalochordates (simple fishlike *Branchiostoma*), and vertebrates. All form notochords, gill arches and slits, and have a dorsal hollow nerve cord, myotomes, and a postanal tail. Sea squirts have hollow, thin-walled bodies and are stationary, but their larvae have typical chordate traits. Cephalochordates resemble vertebrates but in a highly simplified way, probably representative of the first vertebrates.

Key Terms

coelomate • **430**
coelom • **430**
protostome • **431**
deuterostome • **431**
spiral cleavage • **431**
radial cleavage • **431**
hydrostatic skeleton • **435**
closed circulatory system • **435**
open circulatory system • **435**
water vascular system • **443**
notochord • **445**
gill arch • **445**
gill slit • **445**
dorsal tubular nerve cord • **445**
myotome • **445**
postanal tail • **445**

Review Questions

1. List three important characteristics of coelomate animals.
2. In what way do theorists explain the presence of segmentation and a true coelom in both protostomes and deuterostomes?
3. What is segmentation? In what group of animals is it most obvious? List two or three examples of segmentation in our own bodies.
4. What factors suggest that mollusks are closely related to annelids and arthropods? Why do some authorities question this relationship?
5. List several examples of members in each of the four major mollusk classes. Describe the general appearance of the foot, shell, and mantle in each.
6. What are the gastropods? Describe a peculiarity in the formation of body and shell.
7. In which of the mollusk classes have the foot and mantle become the most specialized or advanced? Explain. List two or three other advanced characteristics of this group.
8. Name the three annelid classes, and cite a representative from each.
9. What is a hydrostatic skeleton? Suggest how it may function in the earthworm's movement.

10. List the regions of the earthworm digestive system. Why is specialization important to such an animal? Do humans share in this distinction?

11. The earthworm is described as "hermaphroditic." What does this mean?

12. List three unique characteristics of the arthropods.

13. Characterize the habitats, feeding, and reproductive capacity of arthropods as a group.

14. List five traits that spiders have in common.

15. Characterize the aquatic crustaceans in the following areas: exoskeleton, size range, respiratory system, and nervous system (include eyes).

16. Develop an outline that characterizes each of the following aspects of the insect body: major regions, appendages for movement, respiratory system, and feeding specializations.

17. What is the function of each of the following in insects: midgut, Malpighian tubules, open circulatory system?

18. Name the stages insects pass through in which metamorphosis is complete. What is the specialization of each stage?

19. Explain how a deuterostome skeleton differs from that of a protostome.

20. Name four or five examples of echinoderms. In which group does the body seem most different? Would this be a primitive or an advanced trait?

21. Describe the peculiar form of radial symmetry in echinoderms. Why are they considered actually to be bilateral?

22. Name the main parts of a water vascular system, and explain how it is used for movement.

23. What features of the hemichordate indicate its relatedness to echinoderms? To chordates?

24. Summarize the five important chordate characteristics. Do humans express all of these?

25. What is there about the urochordate larvae and cephalochordates that evolutionary theorists find so interesting?

CHAPTER

27

The Vertebrates

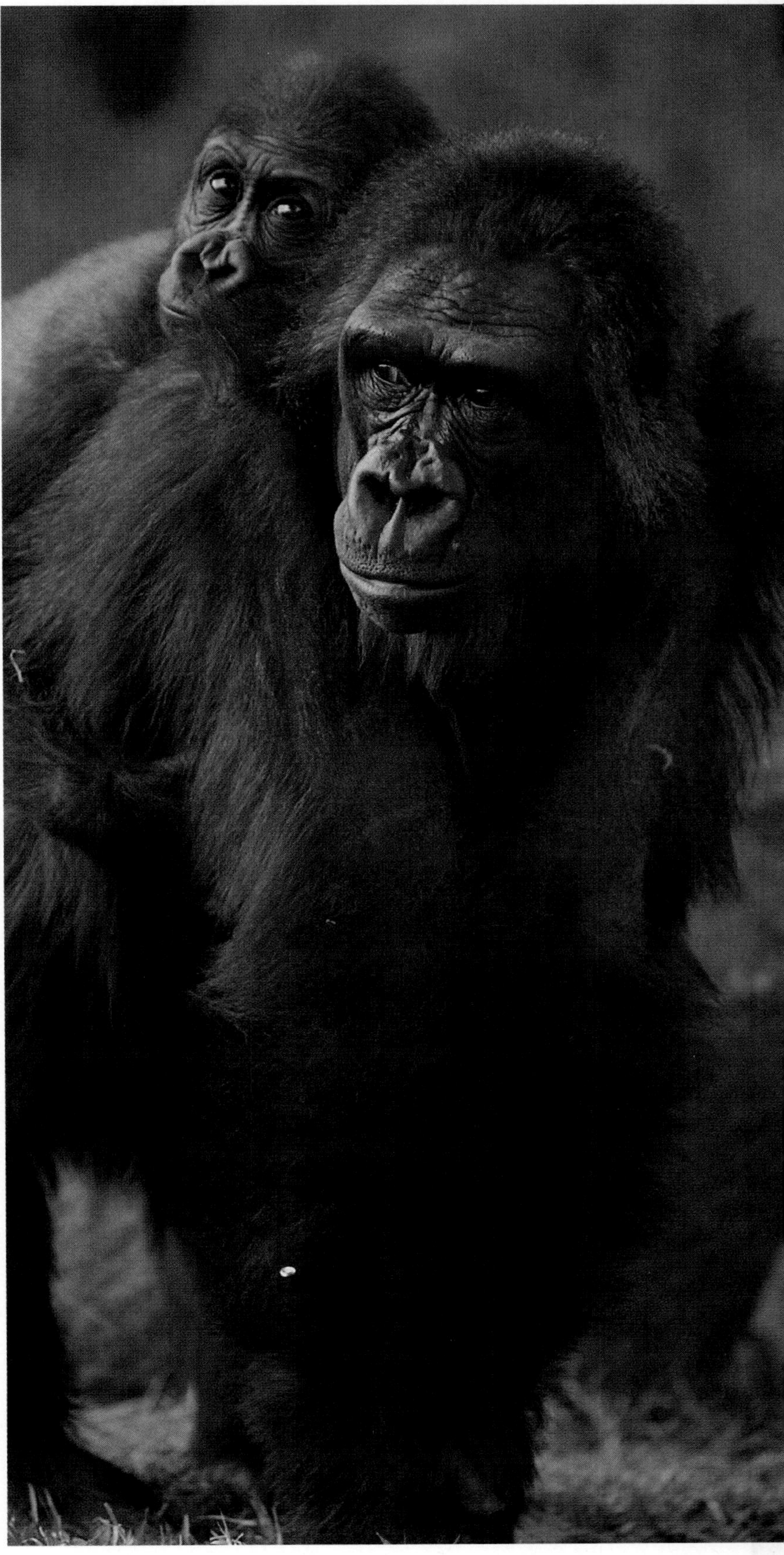

YOU ARE UNDOUBTEDLY MORE FAMILIAR WITH ANIMALS WITH BACK-*bones than those without, since you are one. But you may not know much about some of the beasts that share your subphylum. Not only are some of the modern classes, such as hagfishes and lamprey, likely to be a bit mysterious, but some of our extinct forebears were bizarre, indeed. Here, then, we will follow the trail of those distant fishy creatures as they invaded the terrestrial realm, and how they changed once there. We will also ask why so many of them died out all at once, and what sorts of changes are necessary for the survivors to continue their line.*

Now we come to the animals that usually cross our minds when we hear the word "animal." No ventral nerve cords here; no open circulatory systems; no bodies growing from arms; no anuses above the eyes. These are the chordates called vertebrates (Subphylum Vertebrata).

Most vertebrates share certain traits in addition to the standard chordate characteristics. Perhaps most striking is the **vertebral column** of bone or cartilage, the "backbone" that forms the body axis. In addition, vertebrates boast a highly centralized nervous system (brain), a closed circulatory system, a ventral heart, gills or lungs, two pairs of appendages, and a compact excretory system with paired kidneys. There are two distinct sexes in each species. Although exceptions to the rules are inevitable, they are minor.

In this chapter, we will focus on the evolutionary relationships of vertebrates—and we've got a lot to work with, because this group includes a staggering array of animals and lifestyles. We will begin where the vertebrates began—with the jawless fishes.

HUMBLE BEGINNINGS AND THE JAWLESS FISHES: CLASS AGNATHA

The earliest vertebrate fossils ever found are of the *ostracoderms.* Their remains have been found in strata from much of the Paleozoic era. The earliest findings are about a half billion years old, and they continued on for about 95 million years (Table 27.1). The ostracoderms whose fossils have survived were slow, heavy, armor-plated fishes living on the ocean bottom. They lacked jaws, and so they are placed in the class Agnatha (*a,* without; *gnath,* jaw). They obtained their food by sucking up the bottom sediments and sorting out the nutrients. They were fearsome-looking creatures with large heads and huge gill chambers. Their bluff occasionally might have been called, however, because, their appearance notwithstanding, most species were quite small.

The development of jaws may not seem at first like a momentous evolutionary event, but it is. In fact, almost all the jawless fish died out about 345 million years ago, possibly because they were displaced by the rapidly evolving jawed fishes. The only survivors are the eel-like *lamprey* and the *hagfishes* (Figure 27.1).

In many ways the lamprey has changed little since the Devonian period, and even today it is wide-ranging and quite successful. Its success belies its ungainly features and unseemly habits—habits that include clamping its round, jawless mouth onto the side of a larger fish; scraping away the skin and flesh with a tough, rasping tongue; and sucking out the blood and juices of its living victim, dropping off only when gorged. The larval lamprey almost seems to pretend that it is unrelated to the adult. A shy and retiring filter feeder, the larva retains both the ecological niche and most of the physical characteristics of its cephalochordate ancestors (see Figure 26.18).

TABLE 27.1 GEOLOGIC TIMETABLE

Era	Period (and epoch)	Years Before Present	Life Forms
Cenozoic	(Recent)	10,000	Increase in human population; rise of human civilization Angiosperms dominant; herbaceous (soft-bodied) plants increase
	Quaternary (Pleistocene: ice age)	3 Million	Human societies; extinction of many great mammals such as woolly mammoth Many plant species become extinct
	Tertiary	63 Million	First humans; mammalian divergence; land dominated by insects, birds, mammals Dominance of angiosperms pronounced
Mesozoic	Cretaceous	135 Million	Primates appear Dinosaurs reign, then become extinct; spread of insects; Angiosperms dominate; gymnosperms decline
	Jurassic	181 Million	First birds and mammals; dinosaurs common; cartilaginous fish decrease temporarily Gymnosperms dominate
	Triassic	230 Million	Mammal-like reptiles; first dinosaurs Burgeoning of gymnosperms and ferns; decline of club mosses and horsetails
Paleozoic	Permian	280 Million	Amphibians decline; reptiles expand Gymnosperms and possibly angiosperms evolve
	Carboniferous	345 Million	Amphibians reign; first reptiles appear; insects diverge rapidly Forests that will produce coal; ferns, mosses, and horsetails continue
	Devonian	405 Million	Fishes reign (jawed); amphibians appear Radiation of land plants; forests of ferns, club mosses, and horsetails
	Silurian	425 Million	First air-breathing terrestrial animals; fishes increase Vascular plants, fungi appear
	Ordovician	500 Million	Many marine invertebrates; first vertebrates (jawless fishes) Possible invasion of land by water plants
	Cambrian	600 Million	Marine invertebrates reign; first skeletal creatures; trilobites common Algae appear
Proterozoic	Precambrian	1.5 Billion	Multicellular animals arise Eukaryotes and protists appear
		2.5 Billion	Prokaryotes diversify Anaerobic and photosynthetic bacteria evolve
Archeo-zoic		4.5 Billion	Formation of solar system

(a)

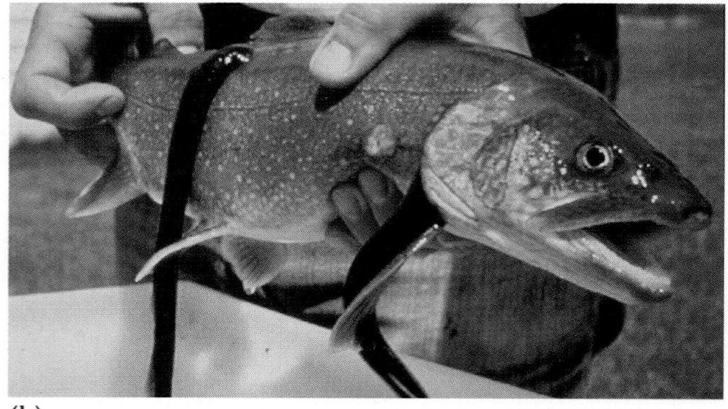

(b)

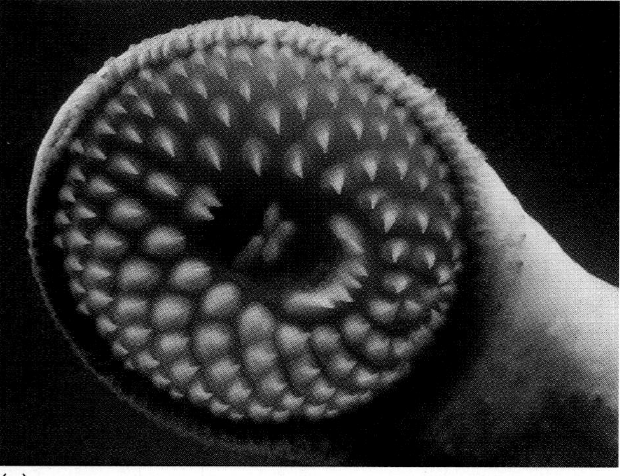

(c)

FIGURE 27.1 TWO BEAUTIES.
The hagfish **(a)** and the lamprey **(b)**. The notochord as a supportive structure is replaced or joined by vertebrae. These fish have no jaws and therefore must feed by sucking the juices out of other animals **(c)**. Jaws are believed to have developed from the support bars of the most anterior gill slit in some distant ancestor of these creatures. Think of how different the world would be if the simple ability to bite had never evolved.

THE FIRST JAWS: CLASS PLACODERMI

There is little doubt that the evolution of jaws was one of the most significant events in vertebrate history. It almost immediately changed the behavior of many species, greatly increasing the feeding niches and encouraging new variations, some of which have succeeded to the present. All sorts of toothy creatures can bite the daylights out of you these days.

Figure 27.2 depicts one hypothesis regarding the evolutionary development of the vertebrate jaw. Note that it is derived from the gill arches (the bony structures that support the gills) in embryonic development. For gills to give rise to jaws would, of course, require a great many changes both in the position and in the strength of the arches, as well as in the surrounding muscles.

Placoderms (Class Placodermi), now extinct, apparently appeared in the Silurian period, about 405

Jawless Ostracoderm

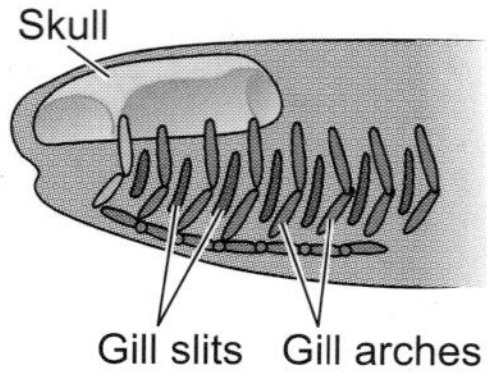

Placoderm

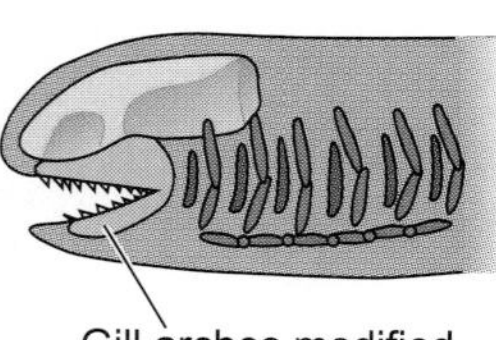

Modern Jawed Fish (Shark)

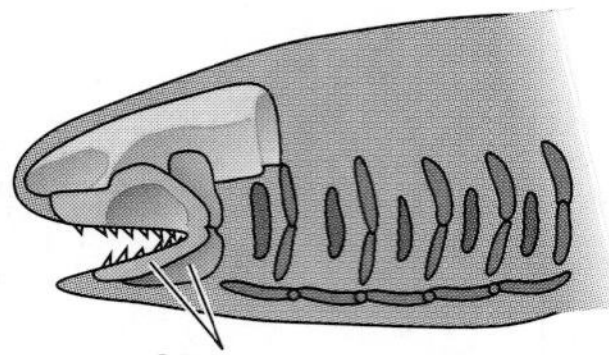

Gill arches modified into strong jaws and their supporting elements.

FIGURE 27.2 EVOLUTION OF THE JAW.
The forerunner of the jaw in the vertebrates was the primitive gill arch. In the jawless ostracoderms, the gill arches were all similar and unspecialized. As jaws evolved in the placoderms, the front gill arches became modified into very primitive upper and lower jaws. In the modern shark, the modifications involve more of the gill arches devoted to other specialized structures.

million years ago. Once established, their numbers increased rapidly, and the group persisted for some 150 million years—an obvious success story. With the placoderms, we find the first evidence of paired pelvic and pectoral fins, which later gave rise to the four limbs of terrestrial vertebrates. One of the more fascinating placoderms was the gigantic *Dunkleosteus* (Figure 27.3), which was about the size of a modern gray whale, and one of the most fearsome predators that ever lived.

Cartilaginous Fishes: Class Chondrichthyes

The cartilaginous fishes (Class Chondrichthyes), those with skeletons of cartilage, include the sharks, rays, and chimeras (Figure 27.4)—a large and successful group of predators and scavengers. The protective armor and heavy skeleton of the ancient species have been replaced in the modern species by a tough skin, slight frame, powerful muscles, and, in some cases, great speed. According to the fossil record, they first appeared in the early Devonian period, dwindled during the Jurassic period, and then began to increase again up to the present time.

The Shark

In addition to having cartilaginous skeletons, sharks are unusual in other ways. Their body and tail shapes are unlike those of most bony fishes (Figure 27.5), and their skin is rough—covered with minute, toothlike growths known as *placoid scales.* The shark's true teeth are not anchored into the jaw.

Figure 27.3 Dunkleosteus.
Giant whale-sized placoderms such as *Dunkleosteus* retained some of the armor of their predecessors, but most of the creature's huge body lacked such protection.

Figure 27.4 Representative Chondrichthyes.
Here the shark, ray, and chimera demonstrate diverse body plans, all with cartilaginous skeletons.

They originate in soft tissue, forming many rows deep in the mouth, and gradually moving forward to the jaw. They eventually fall out, to be replaced by new teeth from behind. Some parts of the ocean floor are covered with discarded shark teeth.

The shark's short, cylindrical intestine has a curious structure called the *spiral valve* (also seen in agnathans and primitive bony fishes). It is essentially a twisted flap resembling a spiral staircase and consisting of absorbing tissue. The valve greatly increases the surface area of the intestine and causes food to move through it slowly. These are both adaptations to a diet that includes sizable chunks of unchewed flesh (sharks cannot chew) that require more time for digestion. The shark's digestive system ends in a

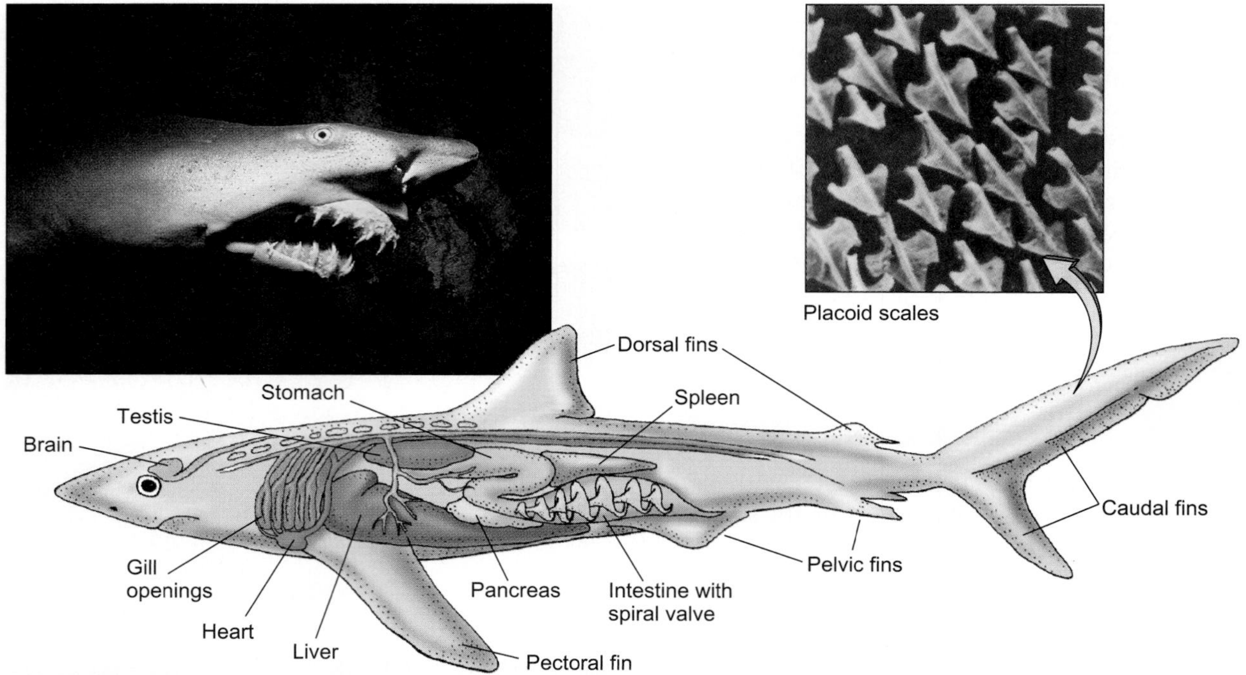

FIGURE 27.5 SHARK ANATOMY.
Sharks are often described as eating machines because of their ravenous appetites. There is no doubt about their success as ocean predators, and they show a number of adaptations to this mode of life. The body shape is, in fact, an excellent example of streamlining. Their sensory structures include an extensive lateral line organ running along the sides of the head and body, along with keen olfactory and visual senses. Within the intestine, a flaplike spiral valve increases the surface area and slows the movement of any chunks of food the shark swallows whole. The powerful jaws, with row after row of razor-sharp, replaceable teeth, tell their own story. The skin is tough and flexible, consisting of scales that resemble miniature teeth.

structure common to most vertebrates, the *cloaca.* This is a chamber that serves as a common passageway for solid and metabolic wastes. It also functions as the reproductive opening of the body, and is used by male and female sharks in copulation and by females in egg-laying.

The male shark lacks a true penis, but it has paired *claspers* on its pelvic fins that are grooved on the inner sides. When the claspers are brought together, they form a channel through which semen can flow into the female cloaca.

Sharks have keen senses and can locate prey by smell, by sight, and by vibration and water movement. Movements are detected in the *lateral line organ,* a sensory device found in bony fishes as well. The lateral line organ is composed of numerous sensory cells located in tiny canals that run along the head and body.

THE BONY FISHES: CLASS OSTEICHTHYES

The bony fishes (Class Osteichthyes) have hard skeletons made of bone. They are divided into two subclasses represented by the *ray-finned fishes* and the *lobe-finned fishes.* The ray-finned fishes possess the delicate fan-shaped fins seen so commonly. They are the ones that probably come to mind when you hear the word "fish." The ray-fins include such familiar fishes as perch, bass, tuna, swordfish, catfish, and sea horses. They are a very diverse group (Figure 27.6); presently, more than 20,000 species have been identified. The lobe-fin subclass contains only a few living species, all of which have fins with thick, heavy bases that have a profound evolutionary significance, as we shall see.

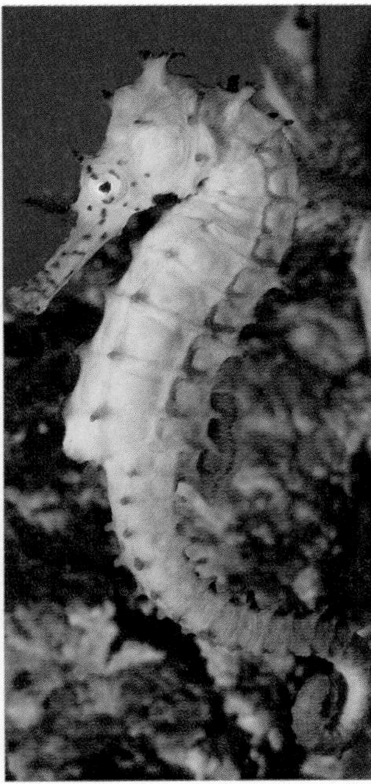

FIGURE 27.6 **RAY-FINNED FISHES.**
There is great diversity among the ray-finned fishes, as we see in the moray eel, the sea horse, the sturgeon, and the frogfish.

Some bony fishes have a "torpedo" appearance, like the shark, but the two are not closely related (Figure 27.7). For example, the fins of bony fishes, especially the ray-finned fishes, are generally much more delicate and movable, with fan-shaped supporting elements. Such fins permit great maneuverability. The quick, darting movements of the bony fish are not possible for the shark.

Many bony fishes have a *swim bladder.* By controlling the gas volume in the bladder, the fish adjusts its buoyancy and is able to remain stationary at any depth. The swim bladder, surprisingly enough, is an evolutionary remnant of paired lungs, dating back to the time when the distant ancestors of modern bony fishes evolved in shallow, stagnant waters.

The skin of most bony fishes is covered with scales and numerous mucous glands. The slimy mucus covering the body not only helps deter parasitic growths but is quite important to swimming. Studies reveal that its presence reduces water friction by up to 65%.

Sharks and bony fishes differ in how they respond to their salty environment. Sharks are *osmoconformers.* They retain enough of the nitrogen waste urea in their blood to balance the 3% salt of the surrounding sea. Bony fishes, like most vertebrates, are *osmoregulators.* They expend energy to actively secrete sodium chloride from special glands. In this way, they maintain a much lower salt content in their body fluids (about 1%).

Gas exchange in nearly all fishes occurs in the gills, where oxygen dissolved in the water crosses the thin gill membranes to enter the moving bloodstream, and carbon dioxide diffuses out into the water (see Chapter 39). In bony fishes there are usually five pairs of gills. These are located in *gill chambers,* each of which is covered by a protective bony flap called the *operculum.*

Reproduction in bony fishes occurs in a variety of ways, from simple, brief acts of spawning—where males and females come together just long enough to release their gametes—to intricate premating behavior, nest-building, internal fertilization (in some groups), and vigorous care of the eggs and young.

Development in bony fishes is also highly varied. In **oviparous** (egg-laying) fishes, such as the cod, the eggs are simply released to be fertilized, and the nutrients needed by the embryo are provided by the egg. In toothcarps (the family of the common aquarium guppy) and other **ovoviviparous** fishes, fertilization

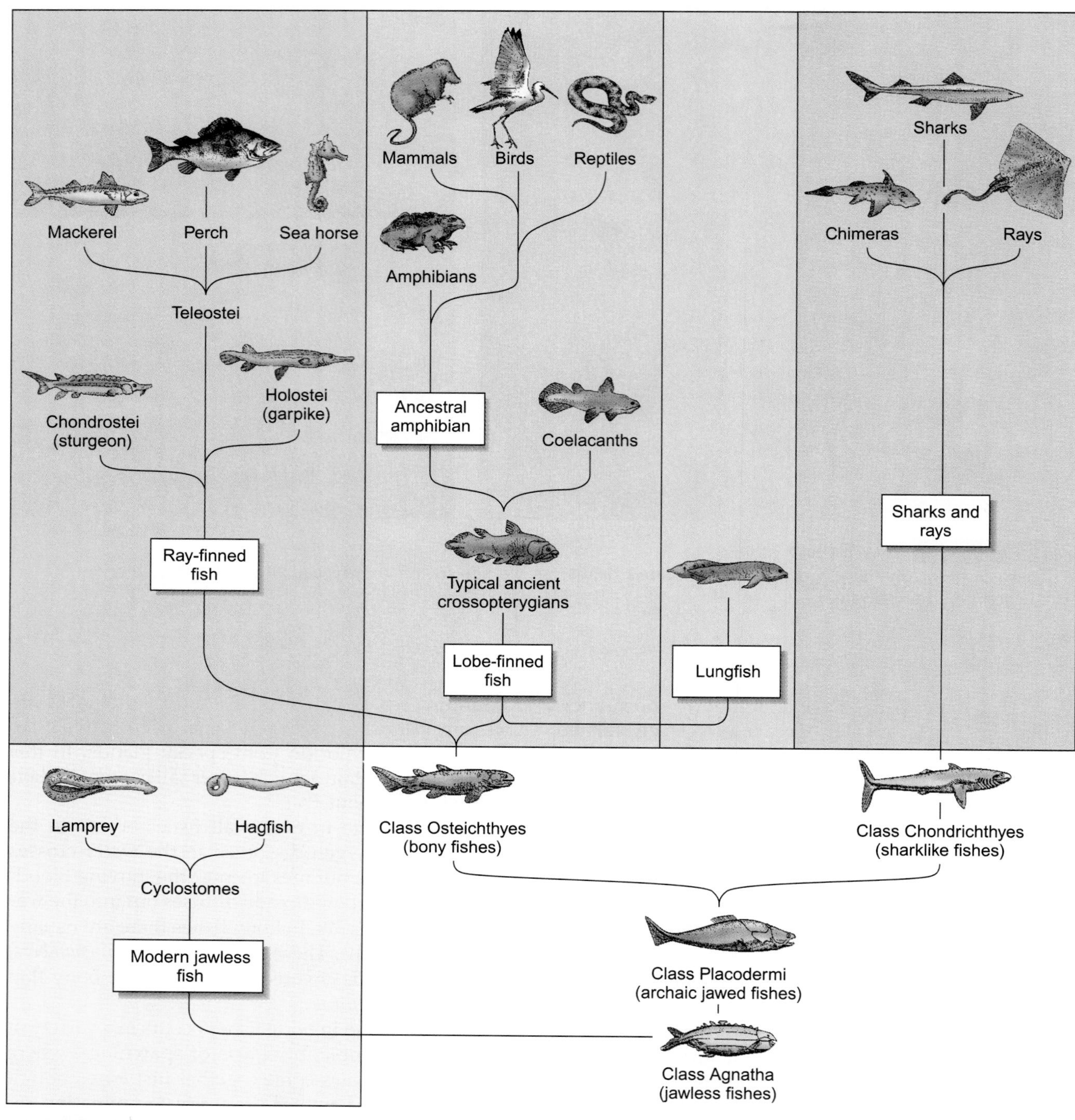

FIGURE 27.7 THE PHYLOGENETIC TREE OF FISHES.
Four branches gave rise to modern jawless fishes, sharks and rays, modern bony fishes, and lobe-finned fishes. One line of lobe-fins is ancestral to the amphibians, from which reptiles, birds, and mammals arose.

is internal—and the egg is retained within the female for development. The young emerge fully developed. However, in ovoviviparу, nourishment is still provided by food stores in the egg itself. Other species of toothcarps provide rare examples of fish that are truly live-bearing, or **viviparous,** a trait usually associated with mammals. In viviparity, some or all nourishment is provided by tissues in the mother's uterus.

Vertebrates Invade the Land

It was during the Devonian period that the bony fishes diverged into two separate groups: the ray-finned fishes and the lobe-finned fishes (see Figure 27.7). The ray-finned fishes quickly became the predominant group in waters around the world. The lobe-finned fishes didn't do as well and are rare today, but their descendants took some fascinating evolutionary turns.

Lobe-Finned Fishes: A Dead End and a New Opportunity

The Devonian lobe-finned fishes were unusual in that they had a nasal opening that extended into the mouth (instead of dead-ending in a nasal sac, as in most fishes), and they had lungs as well as gills. (It may seem strange that lungs first evolved in fishes.) The lobe-fins were also unusual in that their fins were heavy, fleshy appendages, probably best suited for resting on the muddy bottoms of shallow ponds or flapping along mud flats in search of deeper waters. In time, the early lobe-fins diverged into two lines. One line of lobe-finned fishes became the remarkable **lungfishes.** These sluggish, air-gulping oddities today live in highly restricted shallow ponds in Australia, Africa, and South America. They are unusual in that they have retained the air-breathing lung, which in most fishes has become the swim bladder. Thus these fish can hibernate in mud and remain active in water so stagnant that its oxygen concentration is too low for most other fishes, coming to the surface to gulp air.

The second line of lobe-finned fishes, the **crossopterygians** ("fringed-wing ones"), and their close relatives, the **coelacanths,** died out in the Cretaceous period—at least that's what scientists thought until 1939, when a group of puzzled fishermen caught a live coelacanth—a species known as *Latimeria chalumnae*—in deep waters off the east coast of South Africa. It was a startling find. These lobe-fins had survived for 80 million years. An intensive search began, and a second specimen was brought up in 1952 in the Comoros Islands. It turned out that the fishermen in the Comoros Islands, about 3,000 kilometers northeast of where the first specimen had been caught, were very familiar with the coelacanths and had been eating them regularly (Figure 27.8).

With their heavy fins and their primitive lungs, the early crossopterygians were in an ideal position

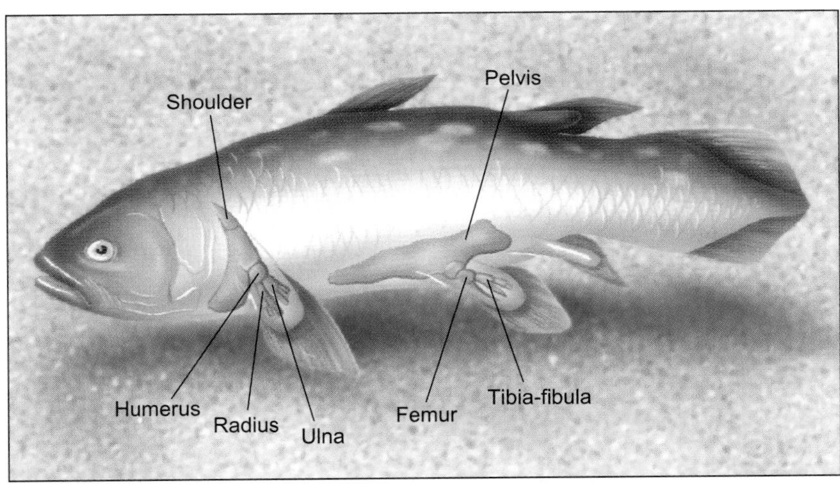

FIGURE 27.8 **A Living Fossil.**
Coelacanths still exist off the coast of South Africa. *Latimeria,* a direct descendant of the crossopterygians, was long thought to be extinct. Fishermen were alerted to be on the lookout for these fish and not to eat them. They ate some of them anyway.

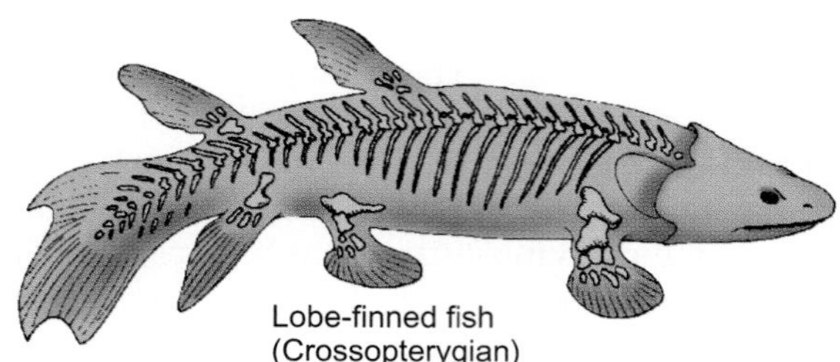

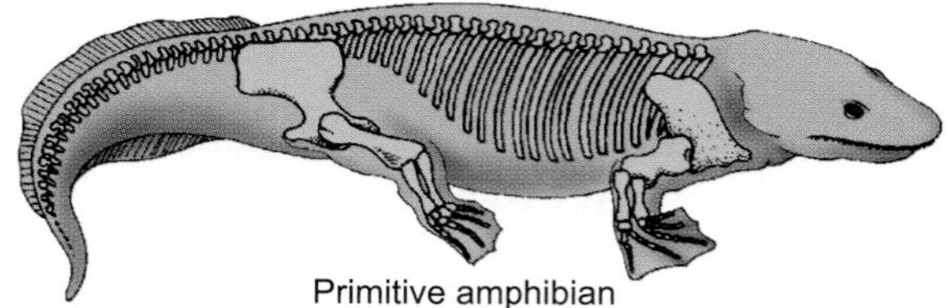

FIGURE 27.9 CROSSOPTERYGIANS.
By comparing ancient amphibian fossils and fossils of lobe-finned crossopterygians, we can see how the transition to land life might have occurred. For example, the first use of lobe fins as walking appendages may have simply involved getting from pond to pond, or even as a brief respite from relentless aquatic predators.

to press on to new opportunities—those on the land. There were, of course, great problems associated with living in the drying air, so the first land invaders were *amphibians,* animals that spend part of the time on land and part in the water. We see evidence of the transition in fossils of the early lobe-finned fishes and the first amphibians (Figure 27.9).

THE TRANSITION TO LAND: CLASS AMPHIBIA

Those first fishy attempts to crawl out of the water were undoubtedly rather awkward—but the unfamiliar land must have offered some rewarding new opportunities, and some species were soon getting around with a certain ease (Figure 27.10). Some of the earliest efforts at developing locomotor structures are preserved in the modern salamander. Its frail upper limbs protrude nearly straight out to the side before angling down, and its body drags on the ground. (Try doing push-ups with your arms in this position.) It moves with an undulating, fishlike motion (as if it were swimming on land), its body first curving to the left and then to the right—an impressive show of vigor, but not very efficient.

Amphibians (Class Amphibia) reveal ancestral habits in other ways as well. For example, the thin, moist skin of amphibians contains dense capillary beds. This permits gases to be exchanged, augmenting the work of the limited, hollow lung. For this to work efficiently, the skin must be kept moist, so most amphibians must avoid dry places (as we will see, there are exceptions). In addition, many amphibians require water for reproduction. Such species fertilize the eggs externally, and the young develop in the water.

The amphibian heart has three chambers, one muscular *ventricle* (lower pumping chamber) and

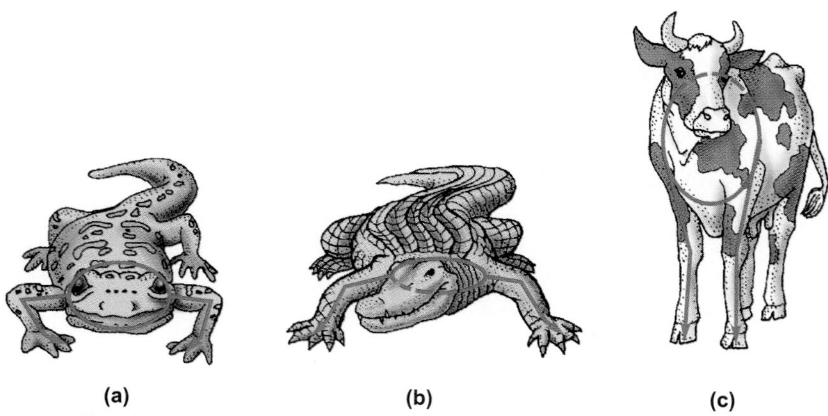

FIGURE 27.10 THE EVOLUTION OF LEG POSITION.
The adaptation to terrestrial life involves the positioning of limbs in four-legged creatures. Amphibians (**a**) such as the salamander and the newt have thin, lightly muscled legs splayed out to the side, unable to support the body for long. Compare this to reptiles (**b**) where we see the upper legs moving a little more underneath and powerful muscles that can support the body. The bodies of mammals (**c**), the fastest land creatures, are raised above the ground, with the limbs essentially below them.

two smaller *atria* (upper receiving and pumping chambers). This organization differs from what is essentially a two-chambered heart in fishes. In the fish, blood is pumped from the single ventricle to capillary beds in the gills, where it is oxygenated. It then passes throughout the body, entering many capillary beds once more before returning to the single atrium for another trip. In amphibians, a second circuit is added for oxygenation (Figure 27.11).

Modern Amphibians

There are three existing orders of amphibians. The first two have been mentioned: the *Urodeles* ("with tails"—the salamanders); and the *Anurans* ("without tails"—the frogs and toads). The third order is the *Caecilians,* an obscure group of legless, wormlike tropical amphibians (Figure 27.12). Some species of urodeles and anurans have evolved toward an increasingly terrestrial life, while others have returned completely to the water, where they spend their entire lives.

There are some remarkable examples of specialization among the amphibians. For example, they have developed very efficient ways of keeping their eggs from drying out. Most simply mate and lay their eggs in the water, whether a pond or pools collected in plants. The desert spadefoot toad, on the other hand, survives in its dry habitat by burrowing into the soil and waiting for one of the infrequent rains, whereupon it crawls out, quickly locates a temporary pond, and immediately mates. Eggs are laid directly into the water, where the offspring must develop quickly before the usual drought returns. As another example, the Surinam toad of South America carries its fertilized eggs in moist pouches on its back, where the offspring completely develop before leaving the parent.

Caecilians avoid the water problem altogether by actually copulating. The male has a copulatory organ and is able to deposit sperm directly into the female's cloaca. The eggs develop within the moist interior of the female, and the young caecilians are born fully formed.

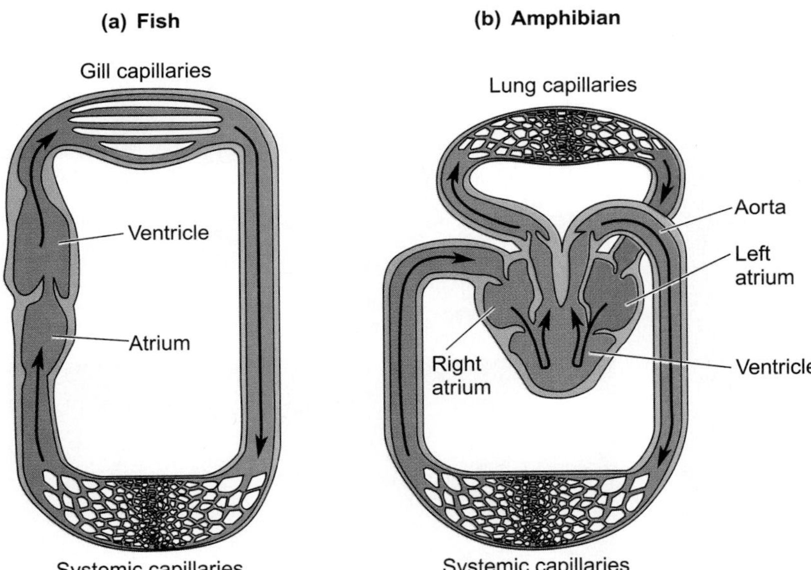

Figure 27.11 **Two- and Three-Chambered Hearts.** **(a)** The fish has a single circuit. Deoxygenated blood is received by the atrium, which delivers it to the ventricle. It is then pumped to the gill capillaries, where oxygenation occurs. From the gills the blood must pass through body capillaries before returning to the heart. **(b)** Two circuits are seen in the amphibian. Deoxygenated blood enters the right atrium of the three-chambered heart, from which it is pumped to the ventricle. Simultaneously, oxygenated blood enters the left atrium, and it too moves to the ventricle. The ventricle then pumps blood from the two sources to the lungs and body. Although it seems as though oxygenated and deoxygenated blood should totally mix in the single ventricle, a good separation is maintained by flaps and valves in the system.

(a)

(b)

(c)

FIGURE 27.12 THE THREE MODERN AMPHIBIAN ORDERS.
Salamanders (**a**) are probably the most primitive representatives because of their body shape and limbs; but their method of reproduction is quite advanced. Frogs and toads (**b**) are reproductively primitive, but their limb structure is advanced. Caecilians (**c**) are in the minority, comprising only 160 species. They are nearly blind and lack limbs, using their wormlike bodies to burrow into the soil, but their reproductive mode is better adapted to land life, an advanced trait.

TERRESTRIAL LIFE: CLASS REPTILIA

Amphibians and reptiles (Class Reptilia) evolved from the same common ancestor, and the first reptiles probably appeared soon after the amphibians. The reptiles, however, took up an entirely different lifestyle. While the amphibians continued to exploit the planet's aquatic niches, the reptiles moved to the drier environment, eventually adapting to a completely terrestrial life.

In so doing, the reptiles developed a tough, scaly, dry, water-repellent skin that controlled water loss and resisted wear. But this meant that the skin could no longer be used in gas exchange. Thus another adaptation included changes in the lung. Essentially, the lung's exchange surface was increased by the development of a spongy construction with many vascularized spaces, or *alveoli* (see Chapter 39).

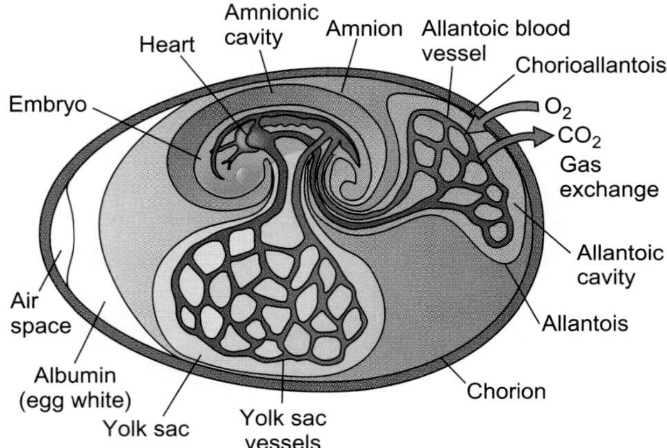

FIGURE 27.13 THE LAND EGG.
These eggs are actually complex enclosures that provide the delicate embryo with its developmental necessities: food and water supply, gas exchange, waste disposal, and protection from drying and injury. The extraembryonic membranes function in gas exchange, food and waste exchange, and protection.

The greatest adaptive changes demanded of terrestrial reptiles were in their methods of reproduction and development. External fertilization was not possible on the dry and hostile land, so, like the caecilians, the reptiles developed the capacity for internal fertilization through copulation.

A new, drastically modified reproductive device—the *land egg*—first introduced by the early reptiles, was a vital adaptation to complete terrestrial life (Figure 27.13). The land egg was surrounded by a tough leathery shell (later replaced by a calcium shell in birds), which, while admitting air, protected the embryo against desiccation and mechanical injury. Included within the egg was a supply of water and food, everything necessary for the embryo's development.

A second and equally important developmental adaptation involved the embryo itself. The developing reptile (and bird) produces a number of supporting *extraembryonic membranes,* with vast networks of blood vessels. The *yolk sac,* an extension of the embryo's gut, brings in food; the *chorion,* a thin but extensive membrane, exchanges respiratory gases; the *allantois,* an extension of the urinary bladder, receives and stores solid nitrogenous wastes (later in development, it fuses with the chorion forming the *chorioallantois,* which aids in gas exchange); and the *amnion* encloses the embryo, providing a protective, water-filled environment. All of these extraembryonic membranes are left behind when the animal hatches and leaves the egg.

The Age of Reptiles

Reptiles have faded somewhat from the stage, but they certainly had their day. In fact, the entire Mesozoic era, which lasted 160 million years (see Table 27.1), is referred to as the "Age of Reptiles." There were, indeed, many successful lines, each exploiting the planet in its own way.

By the start of the Mesozoic era (230 million years ago), the trend toward large size in reptiles was well under way, and as time went on, natural selection produced some huge creatures indeed. Even the flying reptiles—the *pterosaurs*—produced a giant called *Pteranodon,* which was about the size of a small airplane. Some of the great beasts were carnivores, and they were well equipped with huge claws and teeth like steak knives. The most fearsome of great reptiles was *Tyrannosaurus rex* ("tyrant lizard"), probably the largest terrestrial carnivore that ever lived, although larger, less violent herbivores shared the planet with them (Figure 27.14).

Exit the Great Reptiles

About 65 million years ago, at the end of the Mesozoic era, the great reptiles suddenly disappeared. The sudden passing of these reptiles has long presented a great puzzle. What could have caused such a massive extinction? Many hypotheses have been considered. For a time, the favored idea was that the dinosaurs were simply outcompeted by emerging numbers of smaller, more intelligent mammals of the time. Or perhaps, some thought, the furry little rascals ate so many eggs that the dinosaurs died out. Other ideas involved drastic climatic changes at the close of the Mesozoic era. The problem is that none of the explanations could account for the dramatic *suddenness* of the mass extinction or the fact that so many other species died out with them. (Essay 27.1 considers a rather startling answer to the riddle.)

Modern Reptiles

The four orders of modern reptiles are *Chelonia* (turtles), *Squamata* (snakes and lizards), *Crocodilia* (crocodiles and alligators), and *Sphenodonta* (tuatara) (Figure 27.15). *Sphenodon punctatus,* the tuatara of New Zealand, has a third eye, a crude light receptor (the *pineal eye*), on the top of its head, making it one of the world's most intriguing creatures.

Taking to the Air: Class Aves

The ancestors of modern birds (Class Aves) can be traced back about 180 million years to the early Jurassic period (see Table 27.1). *Archaeopteryx,* the oldest known fossil bird, had feathers and presumably could fly (Figure 27.16).

It takes a great deal of evolutionary change for a reptile to fly, and birds have certainly changed from their reptilian stock. However, beneath their obvious flight modifications lie many ancient reptilian traits. Their legs, for example, are still covered with reptilelike scales. And feathers, complex as they are in today's birds, can be traced back to reptilian scales. Today, the feathers provide not only an aerodynamic flight surface but also insulation for a body that must carefully regulate its temperature.

There are a number of other modifications that permitted flight in birds (Figure 27.17). The wrists and fingers have undergone extensive fusion and elongation, supporting the important primary flight feathers. The skeleton is light and strong. Many bones are hollow, containing extensive air cavities,

FIGURE 27.14 **RULING REPTILES.**
The so-called "ruling reptiles" of the Mesozoic include some of the largest animals ever to rove the land.

1. Araucarites
2. Ramphorhyncus
3. Allosaurus
4. Schizoneura

FIGURE 27.15 **FOUR ORDERS OF REPTILES TODAY.**
(a)(b) Lizards and snakes are members of the same order, squamata. **(c)** Turtles represent a second order, chelonia. **(d)** The crocodile is a third, crocdilia, and **(e)** the tuatara represents the fourth order of living reptiles, sphenodonta.

5. Matonidium
6. Archaeopteryx
7. Brontosaurus
8. Ginkgo
9. Pteranodon
10. Quercus
11. Cornus
12. Pandanus
13. Anatosaurus
14. Sassafras
15. Ankylosaurus
16. Palmetto
17. Sabalites
18. Salix
19. Struthiomimus
20. Magnolia
21. Triceratops
22. Stegosaurus
23. Tyrannosaurus
24. Neocalamites

and are crisscrossed with netlike, triangular bracings that add strength. Further weight reduction occurs in the gonads. Specifically, the weight of the testes is drastically reduced between breeding seasons (some 1,500 times in starlings), and the females have but one ovary. The largest flight muscles in active flyers such as the pigeon make up about half of the body weight. In soaring birds, these muscles may be greatly reduced, but the tendons and ligaments that hold the wings in position are considerably strengthened.

The ancestral reptilian jaw has been drastically lightened in birds, and teeth have been replaced with a light, horny bill. Bills vary enormously according to the feeding habits of the bird. The typical bird neck is long and flexible, and the bones of the trunk (pelvis, backbone, and rib cage) are fused into a semirigid unit. The sternum (breastbone) is greatly enlarged and has a *keel* (a flattened, vertical bone) from which the large flight (breast) muscles originate. The tail is reduced, consisting of only four vertebrae.

There are less obvious internal modifications for flight. Like mammals, birds are *homeothermic*—that is, they can maintain a relatively constant (and rather high) internal body temperature. Other vertebrates are typically *poikilothermic*—the body temperature can vary so that it is roughly the same as that of the surroundings (see Chapter 37). The constancy of temperature in birds and mammals is maintained at a metabolic cost, but it has permitted them to adapt to a wide range of climates. Helping to meet birds' high metabolic requirements is an efficient four-chambered heart that ensures that oxygenated and deoxygenated blood follow fully separated pathways in the circulatory system (Figure 27.18). The bird's respiratory system is unique in that the air moves in a one-way flow through the lung (Figure 27.19), as opposed to the in-and-out movement of air in other vertebrates.

The flow of air in the bird lung opposes the flow of blood, and so a **crosscurrent exchange** is established (Chapter 39). This results in a greater efficiency in the exchange of oxygen and carbon dioxide. Such efficiency is essential to flight at high altitudes where oxygen is less plentiful.

Birds continue the reptilian tradition of producing large, self-contained eggs. But as a rule, they produce fewer eggs than do reptiles, and tend to care for them more after laying them.

Essay 27.1

Asteroids and the Great Extinctions

No one knows what triggered the mass extinctions that heralded the end of the Mesozoic era. One explanation, though, springs from peculiar findings in the earth's crust. There are places in the crust where a stratum—a geological boundary—was formed 65 million years ago, marking the end of the Mesozoic era and the beginning of the Cenozoic era. In the waters off Gubbio, Italy, the marine deposits are particularly well defined. Below this geological boundary, in the Mesozoic strata, are carbonate rocks containing many types of plankton skeletons, particularly very large foraminifera—early protists.

But just above the boundary, in the rocks deposited in the early Cenozoic era, the life forms change drastically. Here, there are fewer and much smaller species. And there is something else in these rocks. At the boundary itself is a single layer of clay about a centimeter thick.

In 1980, paleontologist Walter Alvarez, along with his father, Luis Alvarez (a physicist and Nobel Prize winner), noted that this thin layer of clay has intriguing concentrations of some unusual elements. In particular, there is about 50 times more iridium and platinum in the clay than one would expect. Iridium is extremely rare on earth, but it is a common metal in meteors and asteroids. Why would iridium suddenly appear in the rocks being formed at the very time the dinosaurs were dying? The *Alvarez hypothesis,* as it has come to be known, suggests that an asteroid collided with the earth some 65 million years ago. The force was enough to vaporize the object, sending iridium into the atmosphere, later to fall, forming a sediment. Other research has supported the Alvarez hypothesis. Crystals that were at one time subjected to incredible stress have been discovered in several places around the world. These crystals contain fractures that could have been produced only by pressures of over 1.3 million pounds per square inch. Such pressures exist naturally on the earth only at the sites of meteor impacts. Apparently these small crystals were thrown out of the earth's atmosphere to reenter like ballistic missiles.

Judging from the amount of iridium found in that thin layer, the Alvarezes calculated that the Mesozoic asteroid was about 10 km (6 mi) in diameter. A rock of that size would certainly bring a quick end to any creature it happened to hit, but it may have also brought an end to the Mesozoic era.

The dust thrown up by such an impact would block out much of the sun's rays and darken the earth for months. With the darkness, photosynthesis could not occur, and the life that depends on it would begin to die. Some creatures would succumb to the freezing, wintry days on a darkened earth. Others have noted evidence of intense volcanic activity at about this time.

Only certain kinds of life would survive such events, and the clearing skies would have revealed a new kind of earth. Some species would have been untouched, such as marine plankton, which could form resistant spores, and the tough, protected seeds of plants. Many kinds of birds also somehow got through the event. Even some reptiles survived, although, for some reason, all species that weighed over 26 kg (about 60 lb) were eradicated.

The accident would have caused great shifts in the current of life as new niches opened over the entire planet. The remaining living things would have immediately spread out, taking advantage of new opportunities and adapting to their strange new world.

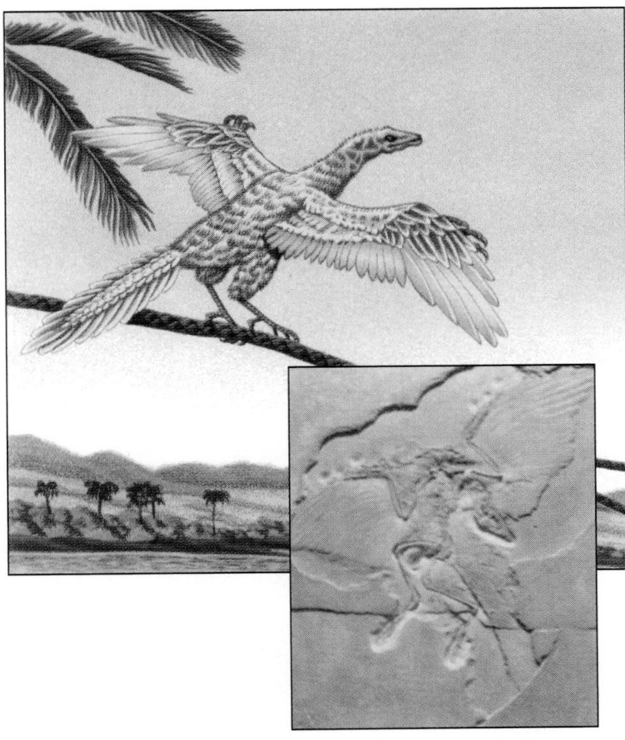

FIGURE 27.16 ARCHAEOPTERYX.
This is the earliest known bird—or is it? Note the numerous teeth, long tail, and clawed fingers. Whatever it was, this pigeon-sized animal was abundant in the Mesozoic era.

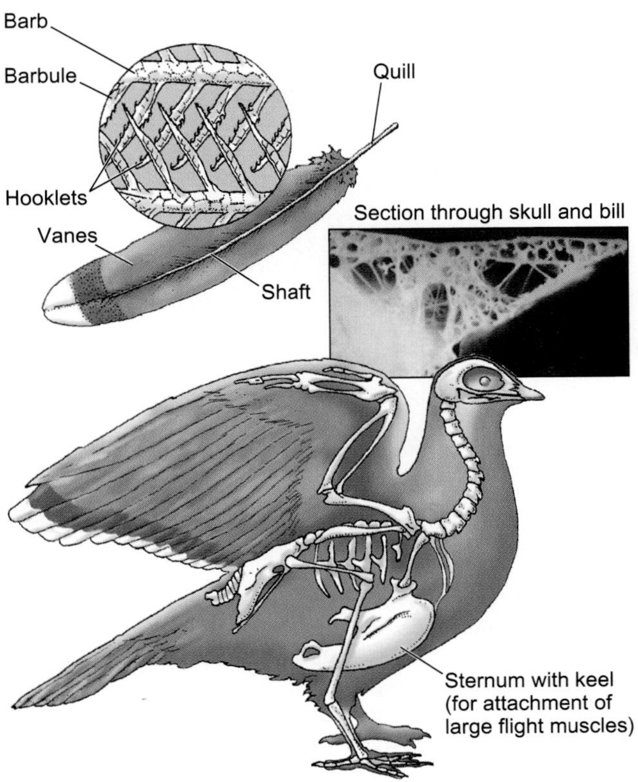

FIGURE 27.17 BIRD SPECIALIZATIONS.
Modifications that allow flight are seen in nearly every aspect of the bird's anatomy and physiology, from the streamlined form to the elevated metabolic rate. In spite of the demands placed on it, the skeleton is extremely light. In general, the slender, hollow bones of birds have a deceivingly delicate appearance; in fact, however, they are strong and flexible, containing numerous triangular bracings within (see photograph). Part of the skeletal strength is due to fusion, as is seen in the hip girdle, in the tail vertebrae, and, most spectacularly, in the long finger bones. Flight feathers, which can weigh more than the skeleton, owe their extreme strength and flexibility to numerous vanes. These have an interlocking arrangement of hooklike barbules.

ANOTHER SUCCESS STORY: CLASS MAMMALIA

The reptiles owned the Mesozoic era, but mammals (Class Mammalia) inherited the Cenozoic era, perhaps with a little help from an asteroid (see Essay 27.1). Whatever the cause, the survivors of that great extinction faced a new kind of world. With the extinction of the dinosaurs, countless new opportunities opened up for the mammals. They quickly took advantage of what the new earth had to offer. Rapidly diverging, and capitalizing on newly available resources, they established a variety of new niches for themselves (Figure 27.20).

Today, the surviving mammals are of three lines (Figure 27.21): the *monotremes,* mammals that lay eggs; the *marsupials,* pouched mammals without true placentas; and the *placentals,* which constitute the great majority of mammals on all continents other than Australia. The placentals nourish their embryos through a well-developed *placenta,* and give birth to relatively advanced young.

MODERN MAMMALS

How did mammals become so well established in the Cenozoic era? Aside from a possible sigh of relief from the mammals at the demise of the ruling reptiles and the new opportunities it brought, there is no explanation that doesn't arouse argument. But it does seem that mammals have evolved a particularly well-matched group of critical traits that have en-

Mammal and Bird

Lung capillaries

Right atrium

Left atrium

Right ventricle

Left ventricle

Systemic capillaries

FIGURE 27.18 **THE FOUR-CHAMBERED HEART.** Birds and mammals have an efficient four-chambered heart in which the oxygenated and deoxygenated blood is kept separate. In reptiles, the septum separating the left and right sides of the heart is incomplete and oxygenated and deoxygenated blood mixes, except in crocodilians, in which the septum is complete.

abled them to survive and flourish. If we concentrate on the placental mammals only, these traits include:

1. An efficient four-chambered heart supporting a constant body temperature and high metabolic rate
2. Internal development, with the embryo nourished through a placenta
3. Mammary glands (in the female) for milk production
4. Specialized teeth and efficient jaws
5. A muscular diaphragm separating the chest and abdominal cavities
6. Hair
7. A large and versatile brain

The Mammalian Brain. Above all, the hallmark of the modern mammal is its brain. Modern mammals are "smarter" than other vertebrates; they rely less on genetically programmed instincts, basing much of their behavior on parental guidance, individual experience, and other kinds of learning (Figure 27.22). Not only is the brain much larger in mammals than in reptiles, but the mammalian brain has parts not found in other vertebrates.

Keep in mind that increased learning capacity is only one evolutionary approach to survival; there are other adaptive routes. For example, the insects are our chief competitors for the earth's resources, and although they have a comparatively limited ability to learn, their alternative path of evolution works well for them.

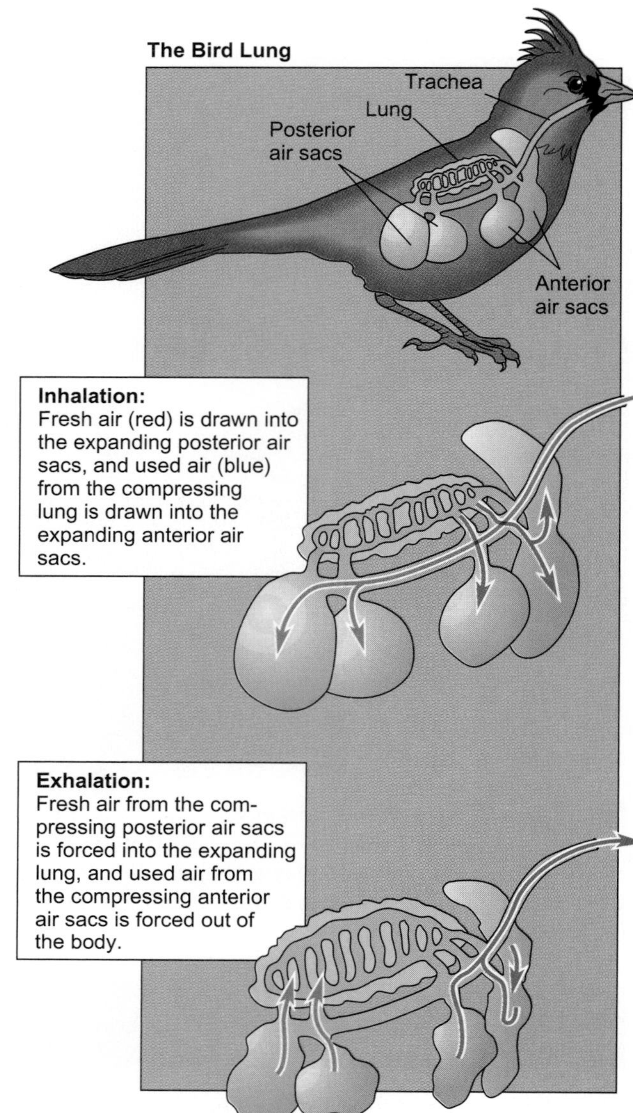

The Bird Lung
Trachea
Lung
Posterior air sacs
Anterior air sacs
Inhalation:
Fresh air (red) is drawn into the expanding posterior air sacs, and used air (blue) from the compressing lung is drawn into the expanding anterior air sacs.
Exhalation:
Fresh air from the compressing posterior air sacs is forced into the expanding lung, and used air from the compressing anterior air sacs is forced out of the body.

FIGURE 27.19 **THE UNIQUE BIRD LUNG.** The bird lung lies between anterior and posterior air sacs. We can follow the respiratory sequences in a highly simplified diagram. On inspiration, fresh air fills the posterior sacs, and air from the lungs is drawn into the anterior sacs. On expiration, air from the posterior sacs is forced into the lungs, and air from the anterior sacs leaves the body. Thus there is a one-way flow through the lungs. The enlarged view of a portion of the lung shows the cylinderlike parabronchi. Gas exchange occurs in a crosscurrent flow with capillaries that pass around the cylinders.

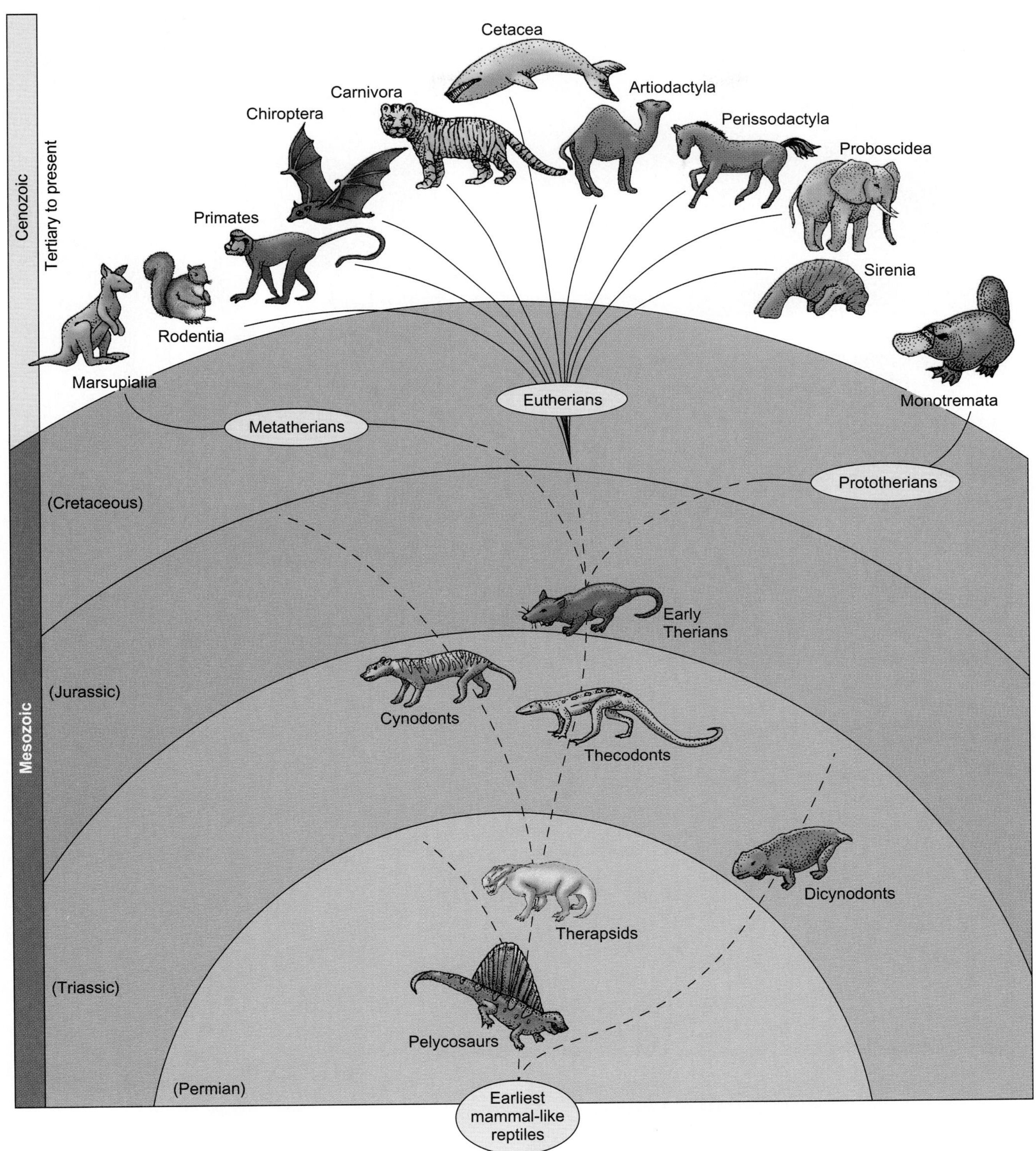

FIGURE 27.20 THE DIVERGENCE OF MAMMALS.
Today's mammals occupy 18 or 19 orders and are dispersed into innumerable niches over the earth. Mammalian roots reach back to the early Mesozoic era but do not appear to branch out significantly until the Cenozoic, an era of great climatic changes that saw the end of the ruling reptiles. Numbers of species are indicated by line thickness. Dashed lines represent hypothetical relationships.

(a)

(c)

(b)

FIGURE 27.21 THE MAMMALS.
The three lines of present-day mammals are the (**a**) monotremes, (**b**) marsupials, and (**c**) placentals.

FIGURE 27.22 LEARNING IN MAMMALS.
The young of mammals acquire many of their habits by being taught by older animals.

PERSPECTIVES

The story of the evolution of animals is indeed a fascinating one. Accounts such as this, though, must necessarily compress time, so that we may be left with visions of some finny creature crawling ashore and growing fur before its tail left the water. Instead, we should remind ourselves that evolution is primarily a story of failure—slow failure, as one "hopeful experiment" after another proved dead in the water (often at the water's edge). Those that succeeded, though, formed a line of backboned animals that showed more similarity within the group—from the agnatha to the mammals—than any group we have yet seen.

The successes of vertebrates, like those of other groups, were not marked by suddenness or fanfare. The backboned animals just continued on, changing with time as their sensitive protoplasm responded to a wheedling environment. Such processes indeed led to successes—that is, to their continued existence and to the panoply of adaptive strategies we see today.

SUMMARY

Vertebrate characteristics include the vertebral column, highly centralized nervous system, closed circulatory system, two pairs of appendages, image-forming eyes, compact excretory system, and distinct sexes. The seven living classes include jawless fishes, cartilaginous fishes, bony fishes, amphibians, reptiles, birds, and mammals.

HUMBLE BEGINNINGS AND THE JAWLESS FISHES: CLASS AGNATHA

Ostracoderms, the first jawless fishes and first vertebrates, lived in the Paleozoic era. They were small, bottom-dwelling, and armored. Survivors today include the lamprey and hagfish. The lamprey larva resembles the invertebrate chordates.

THE FIRST JAWS: CLASS PLACODERMI

The evolution of jaws in placoderms greatly expanded vertebrate niches. Jaws were derived from highly modified gill arches. Similar transitions can be seen in today's vertebrate embryos. Placoderms flourished for some 150 million years, some becoming whale-sized.

CARTILAGINOUS FISHES: CLASS CHONDRICHTHYES

Cartilaginous fishes lack bony skeletons but have tough, thin skins and powerful swimming muscles.

Sharks have toothlike, placoid scales; replaceable, unsocketed teeth; simple, hingelike jaws; and a short digestive system with a spiral valve. The cloaca is a common cavity into which the intestinal, urinary, and reproductive tracts open. Fertilization is internal. In addition to smell and vision, sharks can detect electrical fields and water movement. Water movement is detected by the lateral line organ.

THE BONY FISHES: CLASS OSTEICHTHYES

Bony fishes include ray-finned and lobe-finned types. A branch of the lobe-finned fishes (rare today) produced the first terrestrial vertebrates.

Bony fishes have small ray-fins (greater agility), a swim bladder (buoyancy control), osmoregulation, a bony operculum (gill flap), and lateral line organs. Reproduction in bony fishes is through simple ovipary (egg-laying), ovovivipary (retaining eggs until hatching), and (rarely) vivipary (retention and providing food and gas exchange for embryos).

VERTEBRATES INVADE THE LAND

Lobe-finned fishes include lungfishes—freshwater species that gulp surface air into lungs—and one marine species, the coelacanth. A related line produced the amphibians. The heavy, fleshy fins were forerunners of terrestrial limbs.

THE TRANSITION TO LAND: CLASS AMPHIBIA

Some amphibians have primitive limb structure; most exchange gases across the skin and in simple, hollow lungs; many require water for development—the young going through a larval or tadpole state. Amphibians have a three-chambered heart, with a separate respiratory circuit.

Amphibians include salamanders, frogs, toads, and caecilians (legless). Advanced terrestrial adaptations include desert survival by hibernation, estivation, rapid reproductive cycles, development in body pouches, and internal rather than external fertilization.

Terrestrial Life: Class Reptilia

Reptilian terrestrial adaptations include a dry, waterproof skin and a spongy lung (great surface area). Reptilian legs are strong and well adapted to walking and running. All fertilize internally. Typically, embryos develop in special land eggs, supported by extraembryonic membranes (yolk sac, chorion, allantois, and amnion).

Reptile diversity peaked during the Mesozoic era, the Age of Reptiles. Some flew, and others were the largest known terrestrial vertebrates.

The great reptiles and many other species died out at the end of the Mesozoic era, about 65 million years ago, just as mammals began to meet with greater success. Many different hypotheses attempt to explain the sudden extinction of ruling reptiles. Among the most controversial, but best supported, is one proposing that a huge asteroid, having struck the earth, produced enough dust to drastically alter the climate for many months.

Four orders of living reptiles include *Chelonia* (turtles), *Crocodilia* (crocodiles and alligators), *Squamata* (lizards and snakes), and *Sphenodonta* (the tuatara).

Taking to the Air: Class Aves

The oldest bird fossils are dated early Mesozoic. They arose from a line that also produced crocodilians. Flight modifications include feathers (flight surfaces and insulation), forelimb elongation and fusion, hollowing and crossbracing in long bones, loss of teeth and lengthening of neck, fusion and reduction in trunk bones, sternum enlargement and keel formation (attachment of flight muscles), reduction in weight of testes, loss of an ovary, and specialization in bills and feet. Physiological changes include homeothermy (warm-bloodedness) and a unique one-way, crosscurrent flow of air in the lungs. The bird embryo is supported by a reptilelike land egg and extraembryonic membranes.

Another Success Story: Class Mammalia

Mammals began their greatest divergence in early Cenozoic, forming three lines: egg-laying monotremes, pouched marsupials, and placental mammals.

Mammalian traits include a four-chambered heart; homeothermy; placental support of the embryo; mammary (milk) glands; specialized, socketed teeth; muscular diaphragm; hair. A large brain makes learning and complex behavior possible.

Key Terms

vertebral column • **452**
oviparous • **457**
ovoviviparous • **457**
viviparous • **458**
lungfish • **459**
crossopterygian • **459**
coelacanth • **459**
crosscurrent exchange • **465**

Review Questions

1. List six important vertebrate characteristics.
2. Describe the first vertebrates. Name the two surviving members.
3. In general, what effect did the evolution of jaws have on vertebrate opportunities? Did this innovation persist?
4. Explain the theoretic derivation of jaws. What supporting evidence do we find in today's vertebrate embryos?
5. Briefly describe the following aspects of shark anatomy: skin, teeth, digestive tract, reproductive structures.
6. Explain the differences between the oviparous, ovoviviparous, and viviparous modes of reproduction.
7. To which major grouping of bony fishes do lungfishes belong? In what ways are they different from other bony fishes?
8. List several characteristics of amphibians that seem to link them to aquatic ancestors.
9. What respiratory provision does the three-chambered heart make possible? Is a complete separation of oxygenated and deoxygenated blood possible? Explain.
10. List the modern amphibian orders and representatives. Which group appears most specialized for terrestrial life?

11. Briefly discuss two examples of terrestrial reproductive adaptations in amphibians.
12. Describe terrestrial adaptations in the reptilian skin and respiratory system. How are the two tied together?
13. Of what importance is copulation to terrestrial adaptation?
14. What provisions do the reptiles make for development of the embryo outside the mother?
15. In what era did the reptile peak? What was the apparent physical trend at this time? Provide examples.
16. List three possible reasons for the demise of the dinosaurs.
17. List five specific modifications for flight seen in the bird skeleton.
18. Describe the heart and respiratory system in the bird and explain how their advanced state relates to the bird's metabolic requirements.
19. When did mammals undergo their most dramatic divergence? Name the three lines they formed.
20. List seven important characteristics of mammals. Which of these is unique to the class?

PLANT BIOLOGY

IN THE NEXT FOUR CHAPTERS WE WILL LEARN EVERYTHING THERE IS TO KNOW ABOUT PLANTS. *Or maybe not, because the more one understands about plants, the more unexplored avenues open up, inviting further investigation. Focusing on flowering plants, we will, however, learn something about their reproduction, growth and organization, mechanisms of internal transport, and how they respond to stimuli. If your interest is piqued and you want to learn more, we can assure you that more information exists, alongside a host of unanswered questions—as we explore the fascinating world of our silent partners.*

28

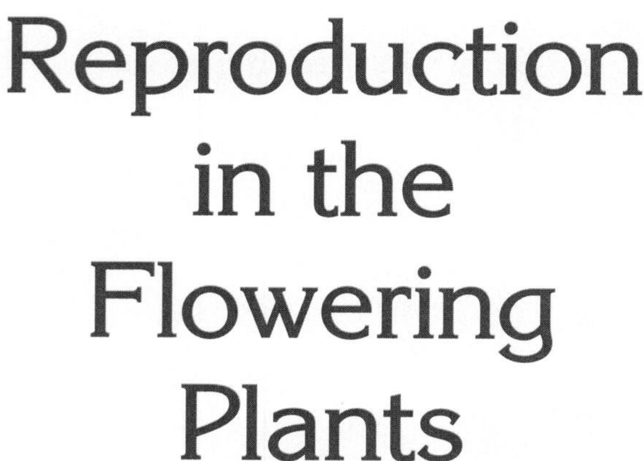

Reproduction in the Flowering Plants

ANATOMY OF THE FLOWER

THE SEX LIFE OF THE FLOWERING PLANT

Inside the Ovary: Ovule and Embryo Sac Formation

Inside the Anthers: Pollen Formation

Pollination and Fertilization

EMBRYO AND SEED DEVELOPMENT

The Mature Embryo

Seed Coats and Dormancy

Fruit

Seed Dispersal

Germination and Early Seedling Development

The Seedling

THE FLOWERING PLANTS— THE ANGIOSPERMS—HAVE ENJOYED A LONG *tenure on earth, and in that time they have exploited every possible opportunity and established themselves in every available habitat. The result is a staggering array of diversity, with an incredible variation in their reproductive structures, including their flowers. So we will begin with the flower and see what goes on in each of two kinds of plants, then trace the development of the seeds as they begin to germinate, forming the next generation of some of the most remarkable organisms on earth.*

The flowering plants, the angiosperms, began their tenure on earth with a rich heritage. By the time their great invasion of the earth began, they were already well prepared for the rigors of terrestrial life. Their vascular systems were well developed, permitting them to grow tall as they pushed their leaves toward the sun. Their massive but delicate roots were able to go deep below the earth's surface to draw precious water and minerals from the soil. They now produced pollen, and so fertilization itself no longer required water. And finally, the embryo was protected by a hardened seed coat that could resist most environmental rigors. But the solution to one problem—how to get the sperm to the egg—represents one of the flowering plants' most fascinating adaptations.

Early in their evolutionary history, seed plants had to rely primarily on the winds to get their male and female gametes together. Where great numbers of the same species exist side by side (as in grasslands or the great pine and spruce forests), wind works well in this role; but in mixed stands, or where individuals are widely dispersed, wind begins to lose its efficiency. Thus the emerging angiosperms began employing other forms of assistance, and the earth's animals, particularly its insects, were pressed into service. But insects were not in the business of raising plants. What was in it for them? How did plants enlist their little six-legged partners?

The earliest angiosperm insect lure was probably a device that oozed a sugary plant fluid, at least enough to interest some Mesozoic beetle that would crawl through the male parts, and then, loaded with male pollen, make its way to the waiting female parts. Once the trend was established, the race was on as new insect-tempting devices arose to be tested by the forces of natural selection. As a result, plants developed ever more attractive and efficient ways of making use of wandering insects, even as other insect species began to show up at the sweet offerings. At the same time, the insects were also undergoing changes that would enable them to exploit more efficiently whatever it was the plant was offering them. In time, many insects had evolved remarkably specific and efficient sensory devices for detecting flowers and long tubular mouthparts for sucking up nectar. Plants, of course, competed among themselves for the attention of these creatures. Some even began a garish form of advertising that attracted insects and other creatures that could carry pollen from one plant to another—those remarkable structures that Darwin referred to as "contraptions" and we will call *flowers* (see Table 28.1).

ANATOMY OF THE FLOWER

Although flowers seem to vary endlessly, most have structures in common (Figure 28.1). We can generalize by noting that flowers essentially have four parts: *sepals, petals, stamens,* and *carpels.* All of the parts are actually modified leaves, but the most leaflike are the sepals and petals. **Sepals** function in the bud stage, where they cover and protect the developing flower. Generally, **petals** function in attracting in-

TABLE 28.1 **ANIMAL POLLINATORS AND FLORAL ADAPTATIONS**

Animal Vector	**Beetles**	**Bees**	**Flies**
Visual Cues	Not significant, flowers dull colored, or white	Bright colors: yellow or blue (ultraviolet perception) Highly divided floral parts with uneven outline	Large flowers, dull, flesh-colored
Chemical Cues	Strong odors: fruity, spicy, or foul	Odors very significant, fragrant	Musky to rotting odors

sects or other pollinators (Table 28.1). For instance, typical bee-pollinated flowers are brightly colored and fragrant and form sugary nectar in *nectaries.* Plants that have evolved means of being pollinated by the wind, on the other hand, may have small, inconspicuous flowers. However, sepals and petals are accessory parts; the real business of reproduction goes on in the *stamens* and *carpels.*

Each **stamen** consists of a slender, stalklike **filament** ending in an enlarged **anther,** where, following meiosis, the male gametophyte is produced. The stamen, then, is a male floral structure.

The female parts taken together form the **pistil,** which is made up of one or more carpels (the exact number depends on the species). Each **carpel** contains three principal parts: **stigma, style,** and **ovary.** The stigma and style play a role in bringing the sperm and egg together, as we'll learn; but it is within the ovary, in small rounded bodies called **ovules,** that the female gametophyte forms. (Fruit, as we will also see, forms from an enlargement of the ovary.)

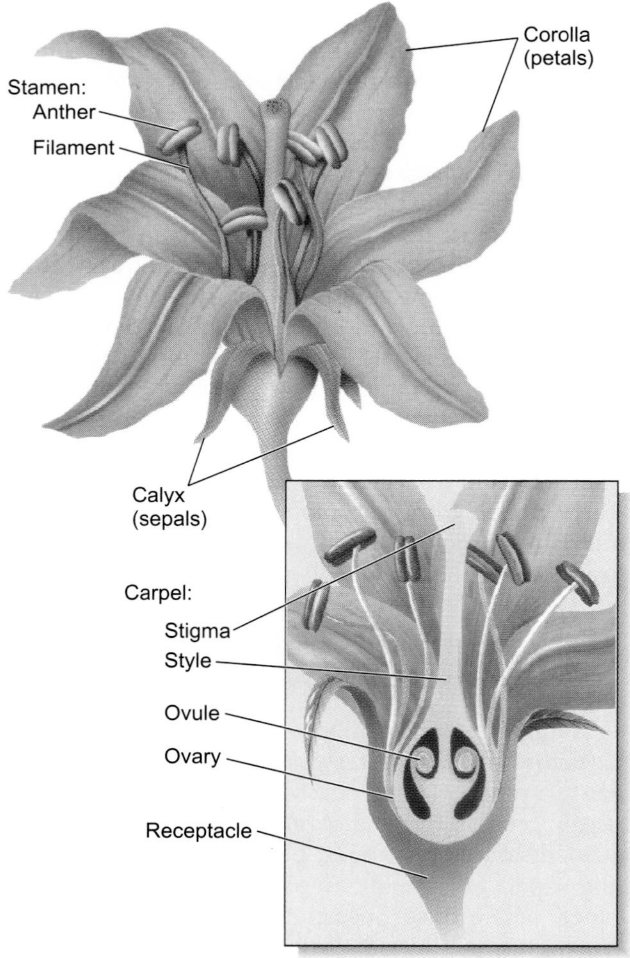

FIGURE 28.1 **A GENERALIZED FLOWER.**
Flowers usually consist of four major parts, occurring in spiraling whorls and supported on a receptacle. They are the calyx (formed of sepals), corolla (petals), stamens (filaments and anthers), and pistil (formed of one or more carpels, each composed of stigma, style, and ovary).

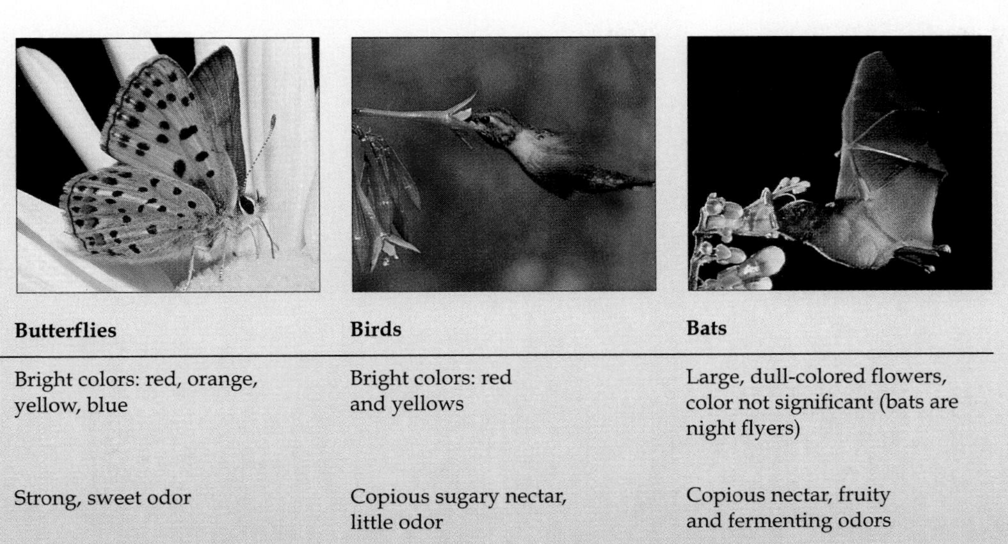

Butterflies	Birds	Bats
Bright colors: red, orange, yellow, blue	Bright colors: red and yellows	Large, dull-colored flowers, color not significant (bats are night flyers)
Strong, sweet odor	Copious sugary nectar, little odor	Copious nectar, fruity and fermenting odors

The Sex Life of the Flowering Plant

As a flowering plant prepares for sexual reproduction, changes begin in both the ovary and the anther. Certain of these diploid tissues contain special cells that will undergo meiosis as they usher in the inconspicuous gametophyte generation. As you might expect, flowering plants produce two types of spores, male and female.

Inside the Ovary: Ovule and Embryo Sac Formation

Within the soft tissues of the flower's young ovary, tiny ovules develop. Although the term *ovule* means "little egg," ovules are technically not eggs. Instead, in their early stages, this is where sporophyte cells reside that will produce the female gametophyte. Anatomically, each ovule consists of a **megaspore mother cell** surrounded by nutritive and protective tissues.

The megaspore mother cell is a large diploid cell that will produce the female gametophyte. This cell will undergo meiosis in the usual fashion and produce four haploid spores. Three of these will disintegrate, leaving one large cell, the **megaspore,** which then begins three rounds of mitosis and produces eight haploid nuclei (one round produces two nuclei, which produce four, and the four produce eight).

In most species, cell walls form and isolate six of the eight nuclei. The other two remain together in a large central cell. At one end, one of the isolated cells becomes the **egg,** the female gamete. The final seven-celled, eight-nucleate structure is the mature female gametophyte, or **megagametophyte,** which becomes the **embryo sac** (Figure 28.2). As we will see, both the egg cell and the binucleate central cell will enter into fertilization.

Inside the Anthers: Pollen Formation

While the embryo sac has been developing in the ovary, changes have been occurring in the anthers that will produce the male gametophyte (Figure 28.2). The anthers typically contain four chambers known as **pollen sacs.** Within the pollen sacs are numerous diploid cells, the **microspore mother cells.** It is these cells that undergo meiosis. Each meiotic event forms four haploid cells called **microspores.** Each of these cells then doubles by mitosis—just once—to produce the two-celled male gametophyte. Each of these two cells has its own role and name. One is called the **generative cell** and the other is called the **pollen tube cell.** The generative cell becomes completely enclosed within the cytoplasm of the pollen tube cell. Each gametophyte produces a tough, resistant coat and matures to become a small, light **pollen grain.** Most pollen is lost

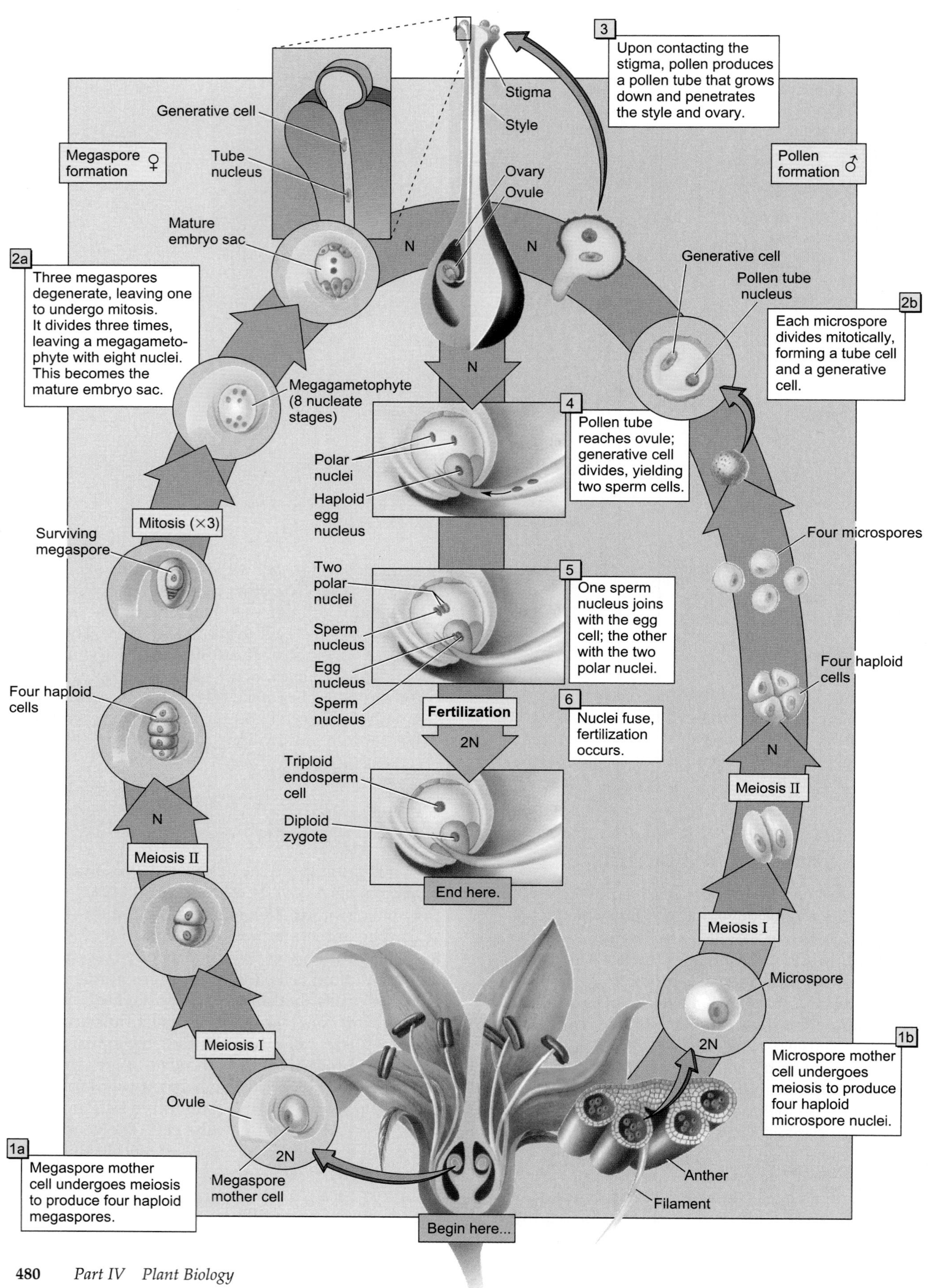

3
Upon contacting the stigma, pollen produces a pollen tube that grows down and penetrates the style and ovary.
Stigma
Style
Generative cell
Tube nucleus
Megaspore formation ♀
Pollen formation ♂
Ovary
Ovule
Mature embryo sac
N
N
2a
Three megaspores degenerate, leaving one to undergo mitosis. It divides three times, leaving a megagametophyte with eight nuclei. This becomes the mature embryo sac.
Generative cell
Pollen tube nucleus
2b
Each microspore divides mitotically, forming a tube cell and a generative cell.
N
Megagametophyte (8 nucleate stages)
4
Pollen tube reaches ovule; generative cell divides, yielding two sperm cells.
Polar nuclei
Haploid egg nucleus
Mitosis (×3)
Surviving megaspore
Four microspores
Two polar nuclei
Sperm nucleus
Egg nucleus
Sperm nucleus
5
One sperm nucleus joins with the egg cell; the other with the two polar nuclei.
Four haploid cells
Four haploid cells
Fertilization
6
Nuclei fuse, fertilization occurs.
2N
N
Triploid endosperm cell
Diploid zygote
Meiosis II
N
Meiosis II
End here.
Meiosis I
Microspore
Meiosis I
2N
1b
Microspore mother cell undergoes meiosis to produce four haploid microspore nuclei.
Ovule
2N
1a
Megaspore mother cell undergoes meiosis to produce four haploid megaspores.
Megaspore mother cell
Anther
Filament
Begin here...

in this uncertain world, but some grains will land on the stigma of a receptive flower.

In the angiosperms, the gametophyte generation (which was so prominent in the more primitive plants) is represented only by the pollen grain and the embryo sac.

Pollination and Fertilization

Technically, **pollination** occurs when pollen is deposited on a receptive stigma, but actual fertilization occurs somewhat later (Figure 28.2). The events leading to fertilization begin when pollen on the stigma germinates (emerges from dormancy). After the hard coat breaks open, a **pollen tube** emerges and grows through the stigma and down the style. The growing tube, under the direction of the pollen tube nucleus, produces enzymes that actually digest the soft tissues ahead of the tube, in some species following an existing channel in the style. The tube nucleus remains near the tip of the tube as it grows downward. During this growth the single generative cell nucleus undergoes one round of mitosis, producing two genetically identical, haploid **sperm.**

Finally the pollen tube penetrates the ovule at a tiny opening, the **micropyle.** The two sperm then enter the embryo sac, and a double fertilization will occur. One sperm fertilizes the egg cell, that is, male and female haploid nuclei fuse to form the new diploid zygote. The zygote will develop into the plant embryo and eventually form the plant body of the sporophyte generation. (The sporophyte generation, then, begins with fertilization.) The other sperm penetrates the large binucleate central cell and fuses with the two nuclei there to form a *triploid cell.* (This "double fertilization" is a unique feature of angiosperms.) This cell will undergo numerous mitoses to form a special nutritive tissue called the **endosperm.** The triploid endosperm usually contains the food reserves of the seed in the form of starch or oil.

Plant endosperm plays a critical role in human affairs, because it is the source of flour and meal produced from wheat, corn, rice, rye, millet, and oats.

Figure 28.2 Angiosperm Reproduction. The female gametophyte generation begins after meiosis in the megaspore mother cell. One of the haploid megaspores goes through three rounds of mitosis, eventually forming the seven-celled embryo sac. The embryo sac includes an egg cell and a central binucleate cell, both of which will participate in fertilization and embryo formation. The male gametophytes develop in pollen sacs within the anthers. Meiosis in a microspore mother cell produces four haploid microspores, each of which goes through one round of mitosis, producing a two-celled immature gametophyte. The tiny gametophyte secretes a thick, resistant wall and matures to become a pollen grain. On contacting the stigma, pollen produces a pollen tube that penetrates the style and ovary. The generative nucleus divides forming two sperm cells, which enter the ovule, fertilizing the egg and binucleate cell.

Embryo and Seed Development

As the endosperm grows, the embryo undergoes rapid cell division. Later, the embryonic cells will begin to differentiate (undergo changes leading to specialization) in preparation for different roles in the life of the plant. Among its first changes, the young embryo will give rise to one or two wings of tissue called **cotyledons.** An embryo will thus reveal at an early stage whether it is a *monocot* (that is, *monocotyledonous*) or a *dicot (dicotyledonous)* (see Chapter 24). Whereas monocots have one cotyledon, dicots have two. The cotyledon has a similar function in the two groups—feeding the growing and developing embryo—but there are differences between them. For instance, in corn and other grains, the single cotyledon lies alongside the endosperm, from which it absorbs food. While the cotyledon operates similarly in some dicots, in many plant families the endosperm is absorbed into the cotyledons during development; the cotyledons are essentially food storage organs.

The Mature Embryo

The embryonic development of the dicot called *shepherd's purse* is shown in Figure 28.3. As the embryo completes its development, we can see two large cotyledons and the massive **hypocotyl,** the region between the cotyledons and the root tip, which will have an important role in germination and early growth.

Plants also have several tissues known as **meristems,** areas of actively dividing cells that remain somewhat unspecialized, ready to take any of a number of developmental paths. The meristems are the source of cells for new growth throughout the life of the plant. One of these, the **shoot apical meristem,** is located in the **shoot tip,** a small mound between the cotyledon bases. As you would expect, it provides cells for continued shoot (stem) growth. In some embryos (the peanut and bean, for example), the shoot tip is quite developed, sporting a

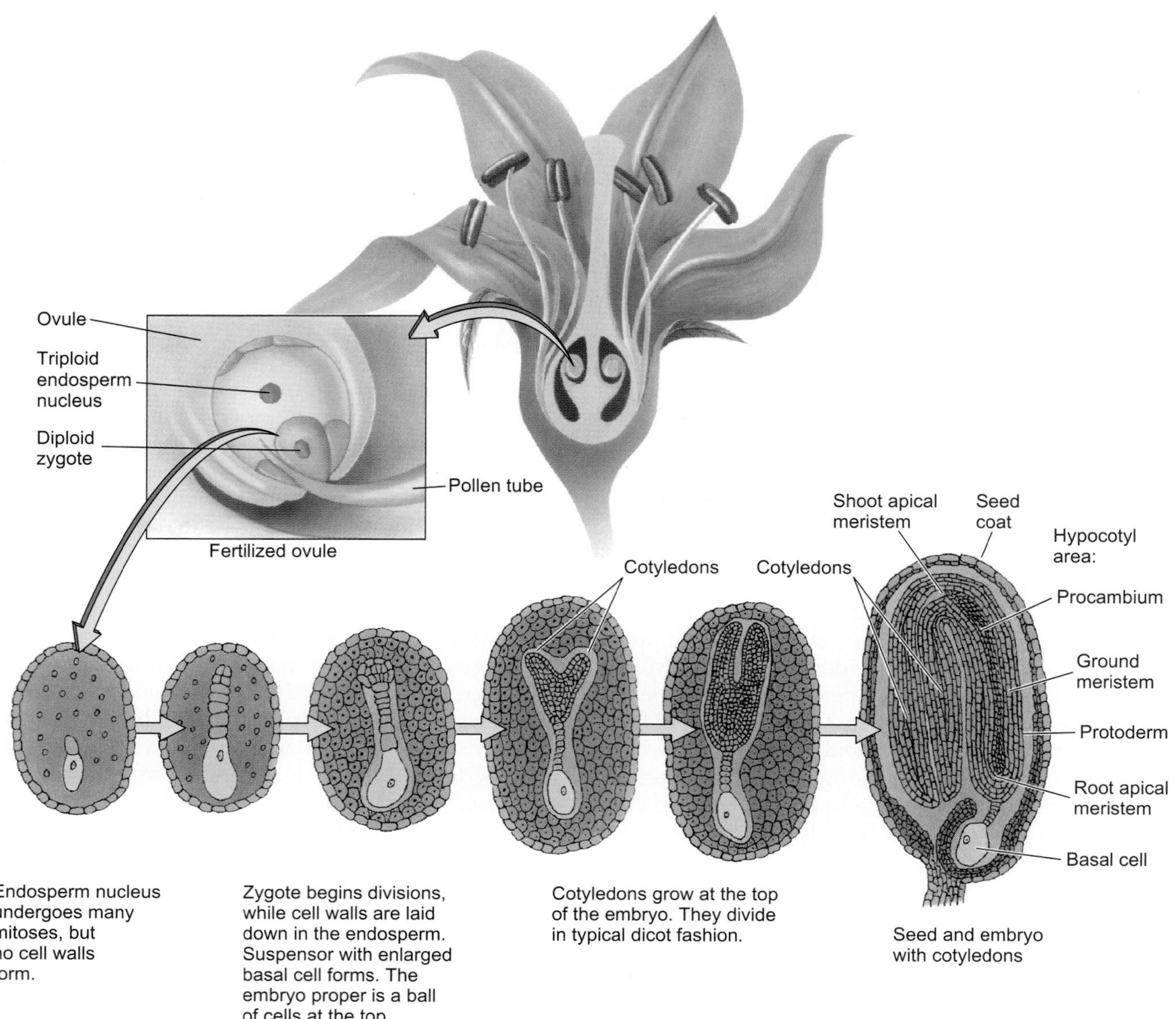
Ovule
Triploid endosperm nucleus
Diploid zygote
Pollen tube
Fertilized ovule
Cotyledons
Cotyledons
Shoot apical meristem
Seed coat
Hypocotyl area:
Procambium
Ground meristem
Protoderm
Root apical meristem
Basal cell
Endosperm nucleus undergoes many mitoses, but no cell walls form.
Zygote begins divisions, while cell walls are laid down in the endosperm. Suspensor with enlarged basal cell forms. The embryo proper is a ball of cells at the top.
Cotyledons grow at the top of the embryo. They divide in typical dicot fashion.
Seed and embryo with cotyledons

FIGURE 28.3 SHEPHERD'S PURSE.
Early development of the shepherd's purse *(Capsella)*, a dicot.

plumule, a tiny shoot complete with embryonic leaves. At the other end of the hypocotyl, a **root apical meristem** occurs in the developing **root tip.** Logically, this meristem provides for continuous root growth. In some embryos, root development begins early, and the well-defined embryonic root is referred to as a **radicle.** As we will see in the next chapter, three basic types of tissue—the ground tissue, the dermal tissue, and the vascular tissue—will form the complex bodies of the flowering plants.

SEED COATS AND DORMANCY

After its initial differentiation and development, the embryo becomes dormant. It will undergo no further changes until germination, the metabolic awakening process in the seed. At that time, cell division and differentiation will resume. Until then, the embryo lies protected by a hardened seed coat. In its dormant state, the seed will lose water, becoming dry and hard.

Interestingly, seeds can lie dormant but viable for a long time. Most seeds can sustain at least several years of dormancy; certain lotus seeds, found in a peat deposit near Tokyo, germinated after lying dormant for 2,000 years. The record, though, is held by a delicate flower of the Yukon, *Lupinus arcticus,* which grew into a fine plant after having lain in the frozen soil for over 10,000 years!

Fruit

In all angiosperms, the fruit develops along with the seed. **Fruit** is actually the mature ovary that surrounds the seeds. You may be surprised to learn that some foods you know as vegetables are, technically, fruits. Among these are squash, eggplant, cucumbers, and tomatoes, as well as corn, wheat, rye, and beans (if we include the pods). Essay 28.1 discusses the varieties of fruit.

Seed Dispersal

It is advantageous for any plant to be able to disperse its seeds. In this way, it does not compete with its own seedlings, and through them, it is able to invade new habitats. However, it is often the fruit, rather than the seed itself, that brings about seed dispersal. For example, such dry fruits as peas and beans may pop open and expel seeds forcefully, throwing them meters away. Other, less dramatic fruits may use wind and water, or birds, foxes, bats, and other animals to scatter the seeds even more widely (Figure 28.4).

Wind-dispersed fruit or seeds have structures such as plumes or wings that help them catch the breezes that carry them from the parent plant to some distant place. As examples, the winged maple fruit with its seed tends to drop straight down at first; but as its speed increases, it begins to spin horizontally, helicoptering away from the parent, thereby avoiding the shade and competition its parent would provide. The buoyant, water-borne coconut fruit (the husk) contains one huge, hollow seed. The coconut is tough and can float, so even the loneliest Pacific atoll is likely to boast a coconut tree. The cranesbill, foxtail, and burr clover are dry fruits adapted for clinging to the fur of animals, and some have done remarkably well in attaching to socks. Burrs may be carried for miles before they split open and release their seeds. The cranesbill fruit is remarkable in that it lodges itself into the ground once it has been dropped; as the humidity rises and falls, the spiral fruit opens and closes, turning and ratcheting itself into the ground. Other seeds are distributed in a

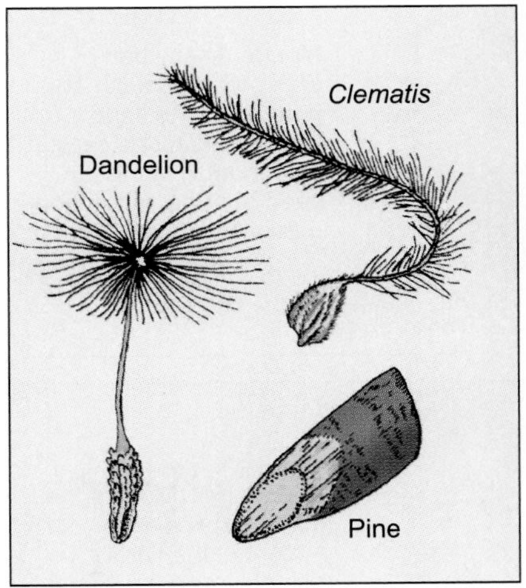

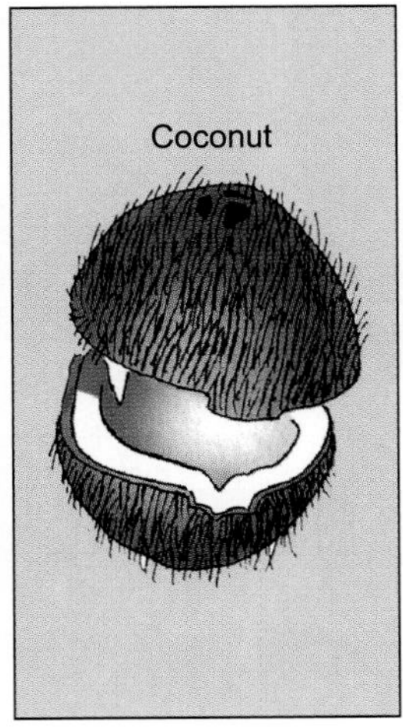

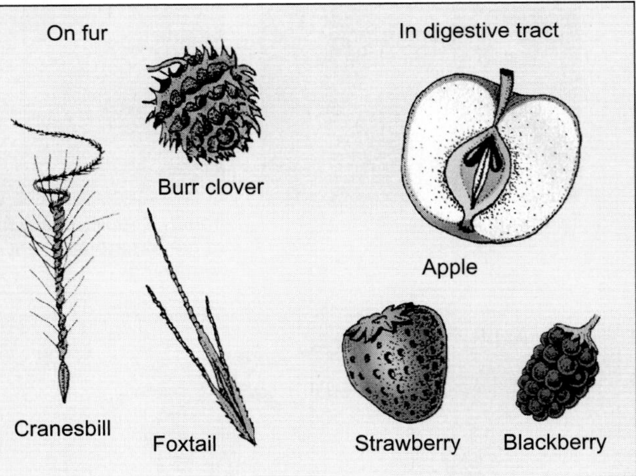

FIGURE 28.4 DIVERSITY IN SEED DISPERSAL.
Seeds and seed-bearing fruits are dispersed by wind, water, or animals. Plumes and wings help in lofting the seed-bearing structures of *Clematis* and dandelion into breezes that will carry the offspring well away from their parents. The coconut seed, well protected by its tough fruit (husk), is uniquely adapted for drifting in the sea. The cranesbill, foxtail, and burr clover have adaptations for clinging to the fur of animals. The seeds within blackberry and strawberry fruits are carried in an animal's digestive tract for a time, but are eventually deposited in the feces.

Essay 28.1 Flowers to Fruits, or a Quince Is a Pome

Is a tomato a fruit or a vegetable? Keep in mind that a fruit is a ripened ovary. Does that help? Okay, let's go further. There are three basic types of fruits: *simple, aggregate,* and *multiple,* depending on the number of ovaries in the flower or the number of flowers in the fruiting structure.

Simple fruits may be derived from a single ovary or, more commonly, from the compound ovary of a single flower. They can be divided into two groups according to their consistency at maturity: *simple fleshy fruits* and *simple dry fruits.*

Simple fleshy fruits include the *berry, pome,* and *drupe.* The berry has one or several united fleshy carpels, each with many seeds. Thus the tomato (a) is a berry, and each of the seed-filled cavities is derived from a carpel (containing the stigma, style, and ovary). Watermelons, cucumbers, and grapefruit are also berries (but, oddly enough, blackberries, raspberries, and strawberries are technically *not* berries, and in fact, what are usually thought to be seeds are actually fruits). *Pome* (b) means "apple," and the group includes apples, pears, and quinces. In the pome, only the inner chambers (roughly, the "core") are derived from the ovary, and most of the flesh comes from the receptacle. A *drupe*—what a wonderful word—is also derived from a compound ovary, but only a single seed develops to maturity. The ripened ovary consists of an outer fleshy part and a hard, inner *stone* containing the single seed. Peaches, olives, and cherries are drupes.

There are many kinds of simple dry fruits, but they are neatly categorized as follows: (1) those with many seeds, which split open and release their seeds, and (2) those with few seeds, which do not split open or release seeds. The first group is called *dehiscent* (c), from the verb *dehisce,* to split or to open, and includes poppies, peas, beans, milkweed, snapdragons, and mustard. The second group is called *indehiscent* (d) (nonsplitting). Its members include sunflowers, dandelions, maples, ash, and corn.

Aggregate fruits (e) are derived from numerous separate carpels of a single flower. Blackberries, raspberries, and strawberries are aggre-

Flower of tomato

Young, simple, fleshy fruit (berry) of the tomato with only sepals remaining

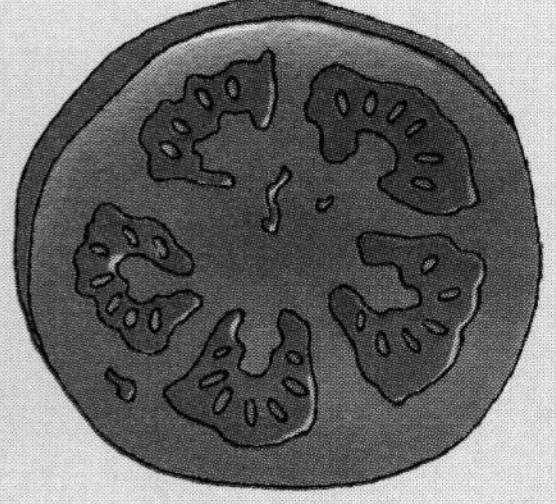

Mature, fleshy berry of the tomato. A cut at right angles to its axis reveals five fused carpels, each containing the seed-bearing, fan-shaped parts of the ovary.

(b) Apple

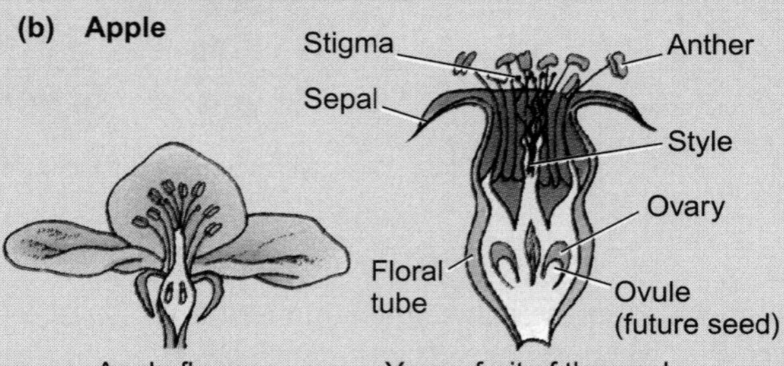

Apple flower

Young fruit of the apple

The organization of a pome becomes apparent in the young fruit, as the bases of corolla and calyx form the floral tube surrounding the ovary.

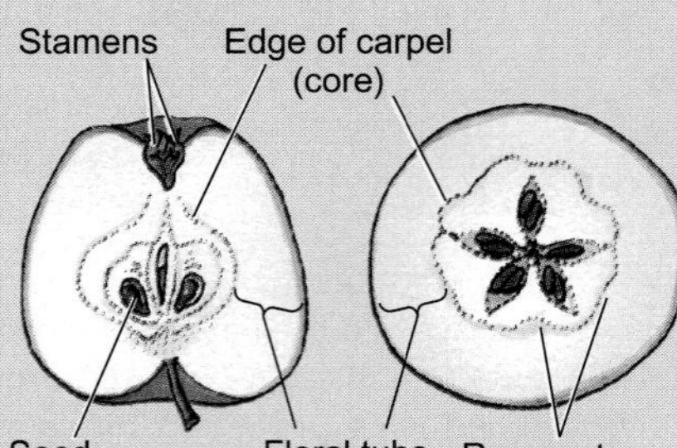

Mature fruit. Most of the floral parts have withered away, but the floral tube has enlarged, producing the edible portion of the fruit containing the ovary (core).The cross section on the right shows the remnants of the flower.

gate fruits. Aggregate fruits consist of many simple fruits clumped together on a common base.

Multiple fruits (f) are formed from the single ovaries of many flowers joined together, as seen in the fig, mulberry, and pineapple. The pineapple starts out as a cluster of separate flowers on a single stalk, but as the ovaries enlarge, they coalesce to form the giant multiple fruit. (The commercial variety, the kind we most commonly see, is a seedless hybrid.)

So is the tomato a fruit or a vegetable? A fruit is a mature ovary, while a vegetable is considered to be other edible parts of plants, including leaves, stems, roots, and flower buds. Because the taxes on imported fruits and vegetables may be quite different, in 1893 the United States Supreme Court was asked to rule on whether a tomato is a fruit or vegetable. Although informed that the tomato is a ripened ovary, they decided it wasn't sweet enough and ruled it a vegetable. So there you are.

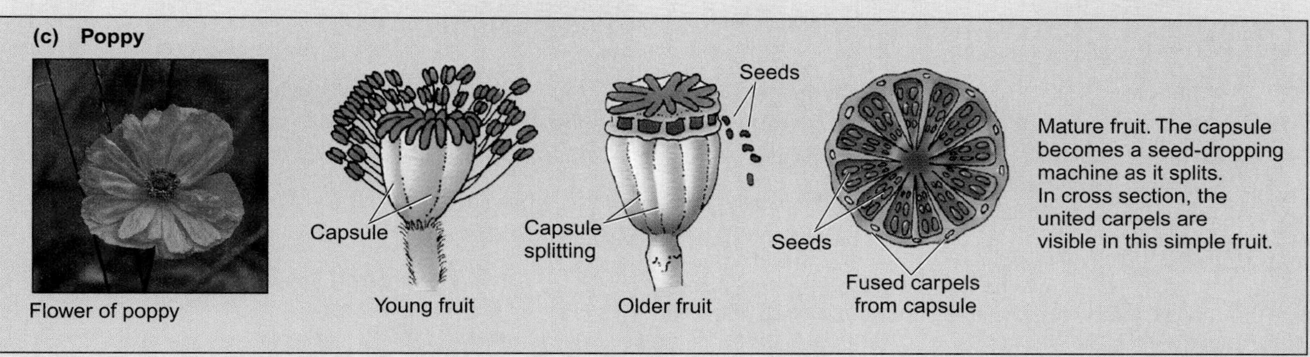

Mature fruit. The capsule becomes a seed-dropping machine as it splits. In cross section, the united carpels are visible in this simple fruit.

(d) Sunflower

Individual flower of sunflower (note the yellow ray flowers)

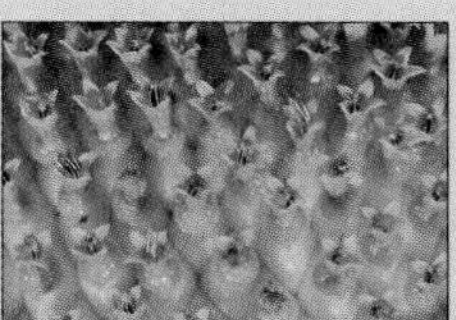
Individual sunflower fruits at different stages

The sunflower, a composite form, consists of many individual disk flowers, making up the head. Each flower is simple, consisting of one carpel that will hold a single seed.

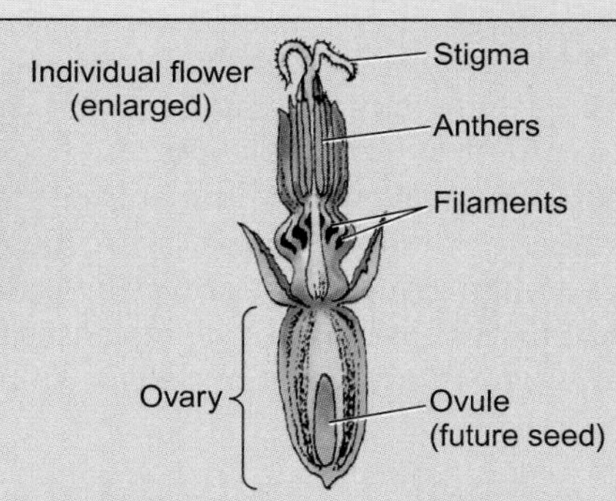

(e) Blackberry

Flower of the blackberry

Maturing aggregate fruit of the blackberry

(f) Fig

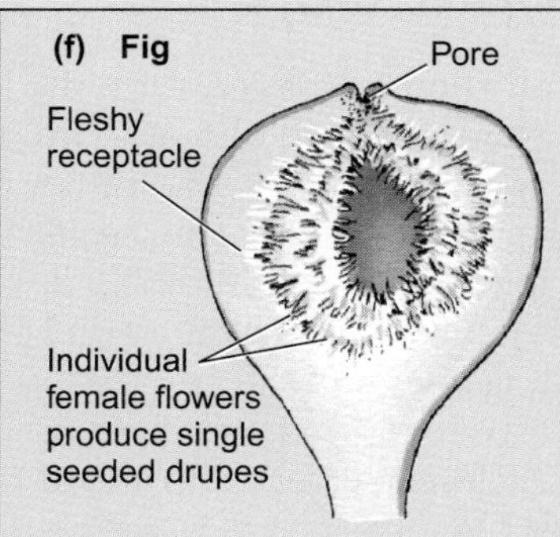

In the common fig, functional male and female flowers develop on different trees. The female flowers grow inside an odd, inverted receptacle. The "inside-out" receptacle becomes the fleshy, edible fig fruit.

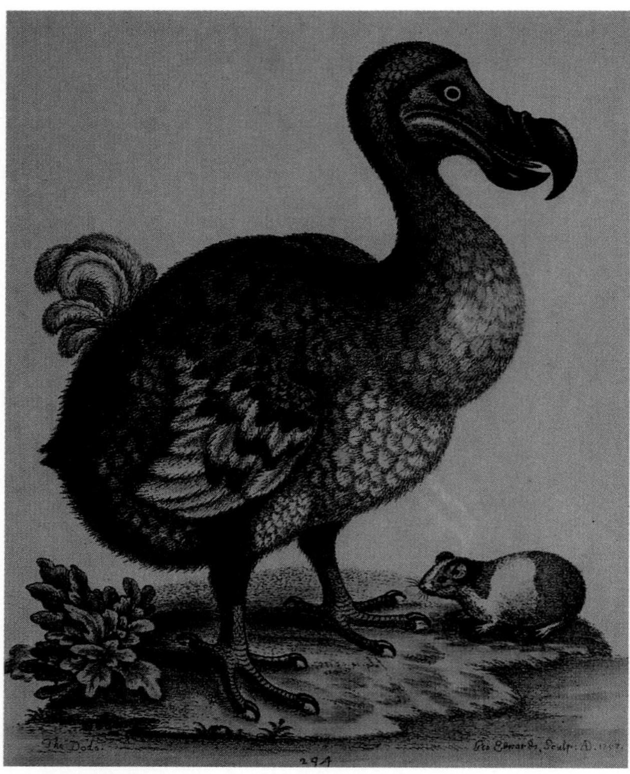

FIGURE 28.5 **A PROBLEM WITH COEVOLUTION.**
The Tombalocoque tree, having developed a dependency on the dodo for seed distribution, found itself without a partner when the bird was driven to extinction.

more familiar way. Their fleshy, sweet, and tasty fruit lures some animal into eating them. The seeds are either discarded or simply pass through the digestive tract, to be deposited with the animal's feces. (This explains why mulberry trees, with their succulent seeds, so often grow along power lines, the favorite roosts of many birds.) In some instances, in fact, the animal's digestive juices stimulate germination.

GERMINATION AND EARLY SEEDLING DEVELOPMENT

In most plants, germination is triggered by the presence of adequate water, proper temperature, and oxygen. But some species require some rather surprising conditions. For example, in some seeds, germination must be triggered by fire, in others by freezing temperature, and in yet others by the grinding action of water running over rocks.

The oddest cases involve seeds that won't germinate at all unless they have been subjected to an animal's digestive processes. On the island of Mauritius in the Indian Ocean, there are only 11 huge Tombalacoque trees, a species found nowhere else. All of the trees are about 300 years old. Every year they produce a crop of fruit containing huge seeds with thick seed coats. But the seeds never germinate. In this species, germination cannot occur unless the seed is passed through the crop of a bird native to the island—*Raphus cucullatus,* better known as the dodo (Figure 28.5). The problem is, the trusting and helpless dodos were killed in great numbers by 17th-century Europeans, and the last one died about 300 years ago. So year after year the seeds just lie around "waiting for the bird that will never come." (This tragic tale may yet have a happy ending, as the naturalist who discovered this curious phenomenon recently managed to get several of the seeds to germinate by passing them through a turkey.)

THE SEEDLING

Let's now consider what happens after germination by looking at the early growth of two kinds of seedlings: the green bean (a dicot) and corn (a monocot). These are arbitrarily chosen, and the two cannot be presumed to represent even most of the members of their respective groups because there is such diversity in the early growth of plants.

The Bean Seed and Seedling. The bean seed (a dicot—Figure 28.6) contains an embryo with two large cotyledons. When the bean absorbs enough water, the seed coats soften and split. The cotyledons open slightly as the embryo expands.

Growth in the bean embryo begins as the young root emerges from the seed and penetrates the soil, absorbing water and acting as an anchor. The rapidly expanding hypocotyl emerges next, and immediately forms a loop or hook that acts as a bumper as the lengthening hypocotyl "elbows" its way up through the soil, drawing the cotyledons after it (Figure 28.6). After the hook emerges, it straightens, exposing the cotyledons and plumule to the sunlight. Just below the plumule and above the cotyledon, lies the **epicotyl,** the embryonic stem region where cells rapidly elongate, raising the shoot tip further upward.

Drawing nourishment from the food reserves in the cotyledons, the plumule enlarges, producing leaves that unfold to the sun. Soon the little plant becomes independent, producing its own foods through photosynthesis. The cotyledons turn green and photosynthesize for a while, but eventually they wither and fall, as food reserves diminish.

The Corn Grain and Seedling. The corn grain (Figure 28.7) germinates in a different way. The single cotyledon takes the form of a food-absorbing structure called the *scutellum.* Surrounding the scutellum is the starchy endosperm. During germination the corn kernel produces enzymes that digest starch, and the products are absorbed by the scutellum.

The corn embryo contains roughly the same tissue organization as the bean embryo but with some

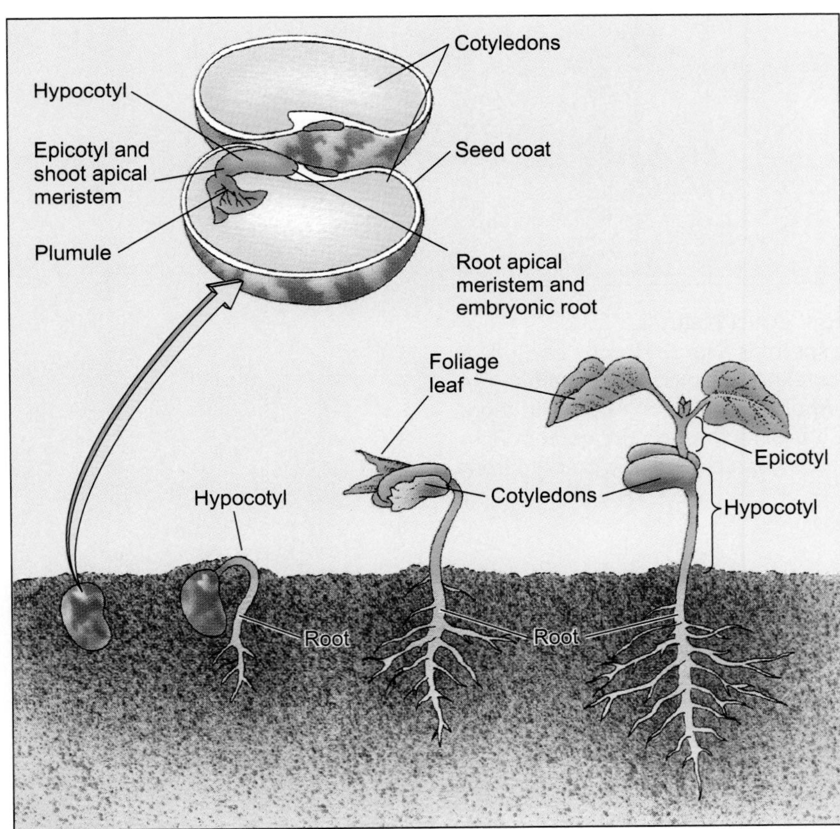

FIGURE 28.6 THE DICOTYLEDONOUS BEAN SEED.
The embryo, a miniature plant, contains an epicotyl and hypocotyl, both of which become very active at the time of germination and seedling growth. The rapidly growing root anchors the young plant. The epicotyl extends the shoot and raises the foliage leaves to sunlight.

significant differences. For instance, both the embryonic leaves and roots are surrounded by protective sheaths: the *coleoptile* and *coleorhiza,* respectively (Figure 28.7).

When germination and growth occur in corn, the kernel remains behind in the soil as the shoot emerges (Figure 28.7). At first the young shoot remains surrounded by the protective, green coleoptile. Soon the young leaf within breaks through, enlarged by rapid cell division in the shoot apical meristem below. Thus the first of several leaves is exposed to sunlight.

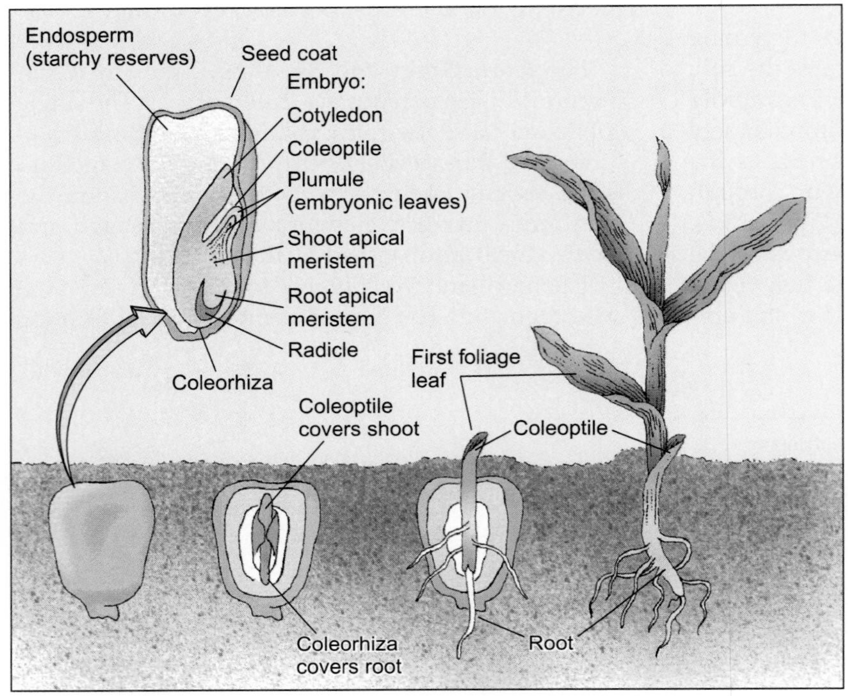

FIGURE 28.7 THE MONOCOTYLEDONOUS CORN KERNEL.
In the corn kernel, actually a fruit, the endosperm is a large starchy body. The embryo has a single cotyledon, the scutellum. Protective sheaths, the coleoptile and coleorhiza, surround the plumule and radicle. At germination, only the shoot reaches the surface, where it breaks through the coleoptile to unfurl as a foliage leaf. The fast-growing root must also quickly emerge from its sheath.

PERSPECTIVES

Here, we have learned something about how plants reproduce. Their means of reproduction, of course, is intimately tied to the evolutionary changes that brought them from the water to the land, and in some cases back again. In the next chapter, we will learn more about just how drastically cells can change, how they can specialize, act in concert, and even die to meet the challenges faced by plants on earth.

Summary

The earlier seed plants were commonly wind-pollinated, but as flowering plants emerged, part of their success was due to an increasing coevolution with animal pollinators. Included today are insects, birds, and bats.

Anatomy of the Flower

Accessory floral parts include leaflike sepals and petals. The latter are often brightly colored and contain nectaries. Reproductive parts include stamens (male) and pistils (female). The stamen holds the anther, in which the male gametophyte forms. The pistil is made up of one or more carpels, each of which contains a pollen-trapping stigma, a supporting style, and an ovary with one or more ovules. Female gametophytes form in the ovules.

The Sex Life of the Flowering Plant

Diploid megaspore mother cells within ovules enter meiosis. One of the four haploid megaspores continues into mitosis, giving rise to a seven-celled megagametophyte, the embryo sac, the female gametophyte.

Within the anther's many pollen sacs, diploid microspore mother cells enter meiosis, yielding haploid microspores. Each enters mitosis, forming a two-celled microgametophyte—made up of a generative and a pollen tube cell. Each microgametophyte will become a pollen grain, the male gametophyte.

Pollination is the transfer of pollen to the stigma. Following pollination, a pollen tube forms, penetrating the style and ovary, and entering the ovule through the micropyle. The generative cell divides, forming two sperm. One fertilizes the egg cell, forming a zygote, while the other fertilizes the binucleate central cell, forming a triploid cell that will give rise to the endosperm.

Embryo and Seed Development

The monocotyledonous embryo forms one cotyledon, whereas the dicotyledonous forms two. Cotyledons provide nourishment to the germinating embryo.

The shepherd's purse embryo, as an example, forms two large cotyledons and a large hypocotyl. Early tissue differentiation produces a shoot apical meristem and a root apical meristem, a source of unspecialized cells that later provide for stem and root growth. Some embryos have developing shoots and leaves called plumules and well-formed roots called radicles.

Following dehydration and hardening of seed coats, a dormant period ensues. Some seeds may be preserved almost indefinitely. Germination is marked by cell division and differentiation.

Fruit, the mature ovary surrounding the seeds, may take many forms, from fleshy to dry and hard. Seeds are adapted to a variety of dispersal agents, including self-expulsion, wind, water, and animals. Seeds or fruits may simply cling to animal fur, or they may be eaten and shed in the feces. Germination agents include water (with the proper levels of temperature and oxygen), fire, abrasion, freezing, and the action of animal digestive fluids.

Germination in the common bean includes the uptake of water, splitting of the seed coats, and rapid emergence of the radicle and hypocotyl. The latter breaks through the soil, drawing the cotyledons behind. Upon exposure to light, the epicotyl lengthens and the first leaves form. The cotyledons wither and fall after their food supply diminishes.

In germinating corn, the scutellum absorbs digested food from the endosperm. The young shoot emerges alone from the soil, embryonic leaves protected by the coleoptile, embryonic roots by the coleorhiza.

Key Terms

sepal • **477**
petal • **477**
stamen • **478**
filament • **478**
anther • **478**
pistil • **478**
carpel • **478**
stigma • **478**
style • **478**
ovary • **478**
ovule • **478**
megaspore mother cell • **479**
megaspore • **479**
egg • **479**
megagametophyte • **479**
embryo sac • **479**
pollen sac • **479**
microspore mother cell • **479**
microspore • **479**
generative cell • **479**
pollen tube cell • **479**
pollen grain • **479**
pollination • **481**
pollen tube • **481**
sperm • **481**
micropyle • **481**
endosperm • **481**
cotyledon • **481**
hypocotyl • **481**
meristem • **481**
shoot apical meristem • **481**
shoot tip • **481**
plumule • **482**
root apical meristem • **482**
root tip • **482**
radicle • **482**
fruit • **483**
simple fruit • **484**
aggregate fruit • **484**
multiple fruit • **485**
epicotyl • **487**

Review Questions

1. Explain how the sudden spread and success of angiosperms during late Mesozoic and early Cenozoic was tied in with animal evolution.
2. Using Table 28.1 as a guide, list several of the adaptations of flowers to insect pollinators.
3. Making a simple line drawing, label the following floral parts: sepals, petals, pistil (stigma, style, and ovary), stamens (filaments and anthers).
4. Give the function of each of the floral parts identified in Question 3.
5. Beginning with the megaspore mother cell, summarize the events leading up to and including the formation of an embryo sac.
6. Beginning with the microspore mother cell, summarize the events leading up to and including the formation of a pollen grain.
7. What constitutes the gametophyte in the flowering plant?
8. List the steps involved in pollination and fertilization. What must happen to the generative cell before fertilization is possible? Why is the term "double fertilization" appropriate?
9. What will the triploid cell become? Why is the triploid tissue of plants of such vast economic importance to us?
10. What structures in early seed formation indicate whether the plant is a dicot or monocot? What is the function of the structure in question?
11. List the developmental events in the shepherd's purse embryo from fertilization to seed formation.
12. Give the location of the following tissues in the embryo, and list their future roles: shoot and root apical meristems, ground meristem, protoderm, procambium.
13. Technically, what is fruit? Is fruit always sweet when ripe?
14. List three agents of seed dispersal and five ways seeds are adapted for dispersal.
15. What two factors do all seeds require for germination? List four germination conditions that probably represent special cases.
16. What are the roles of the hypocotyl and epicotyl in bean germination? What function do the cotyledons serve during germination, and what happens to them afterwards?
17. Contrast the germination and emergence of the corn seedling with that of the bean. What special tissues protect the young corn embryo?

CHAPTER

29

Growth and Organization in Flowering Plants

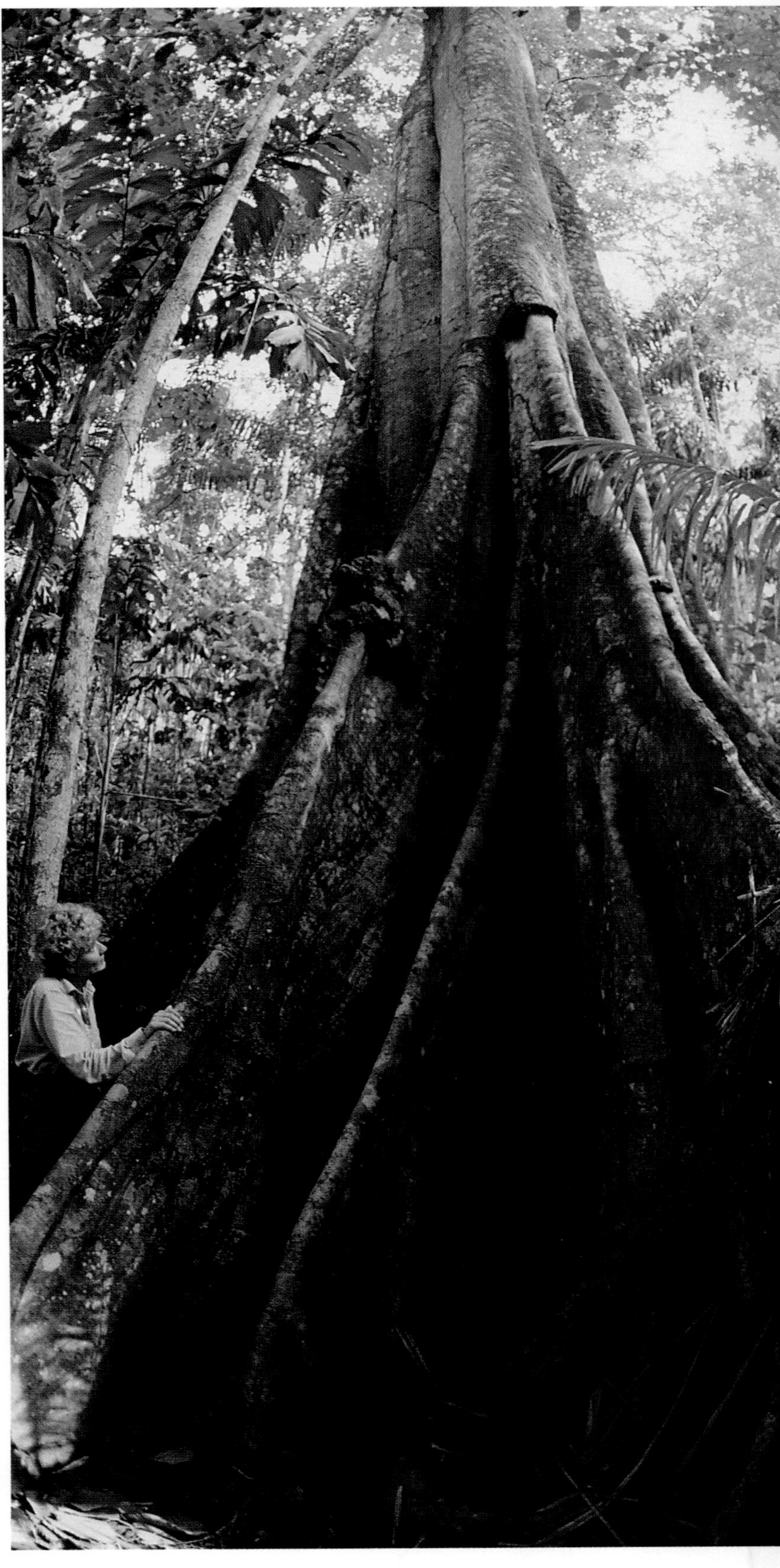

WHEN ONE CONSIDERS THE WILDLY DIVERSE ARRAY OF FLOWERING PLANTS *on earth, it is encouraging to learn that they are organized along some rather basic themes, and they develop according to certain common principles. Here, then, we will review those themes and principles, keeping in mind that barely visible plants growing in some pond are, after all, descended from the same ancestors as the towering tree that stands nearby.*

Eons of plant evolution have molded seeds into tough, resilient, and effective genetic repositories. We have seen how seeds form and what they contain, and we have learned something about how they are adapted to their roles. We have also launched the tiny seedlings on their way to becoming mature plants. The story does not end there, though, because many plants continue to grow throughout their lives. So let's see just how they grow, and how their tissues are organized.

INDETERMINATE GROWTH

Many of the familiar angiosperms are **annuals,** short-lived species that germinate, mature rapidly, reproduce, and die, all within a single season. Probably most of the delicate flowers that grace our meadows are destined to live but a few months. Others, the **biennials,** complete their life cycle in two seasons, with the second reserved for flowering and seed production. However, the largest and most dramatic plants, the great trees, are among the plants called **perennials.** Perennials have **indeterminate growth;** barring an untimely death from injury, infection, or predation, these plants grow as long as they live (Figure 29.1).

There are arguments over which tree holds the record for longest life. Among the oldest is a gymnosperm, a gnarled, 4,900-year-old bristlecone pine from the White Mountains in California (Figure 29.2). Of even greater antiquity are certain cottonwoods, some reportedly 8,000 years old. And a desert creosote bush is calculated to be 12,000 years old. In any case, we can be sure that some plant, somewhere, is the oldest living thing on earth.

The plant's continued growth and development is assured by its meristem. Meristematic tissue, as we've seen, is composed of undifferentiated and immature cells. Some of these cells are always held in reserve—in a sense, they are the plant's investment in its own future. When new growth occurs, some of the meristematic cells simply divide. Some of the daughter cells remain as meristematic tissue, while others enlarge and differentiate, producing a variety of new tissue (see Essay 29.1).

PRIMARY AND SECONDARY GROWTH

It is somehow sad to watch a developer clear a woodland lot of every single tree, build a house on the naked lot, plant some grass, and then wire a spindly "shade tree" in place on the manicured lawn. The

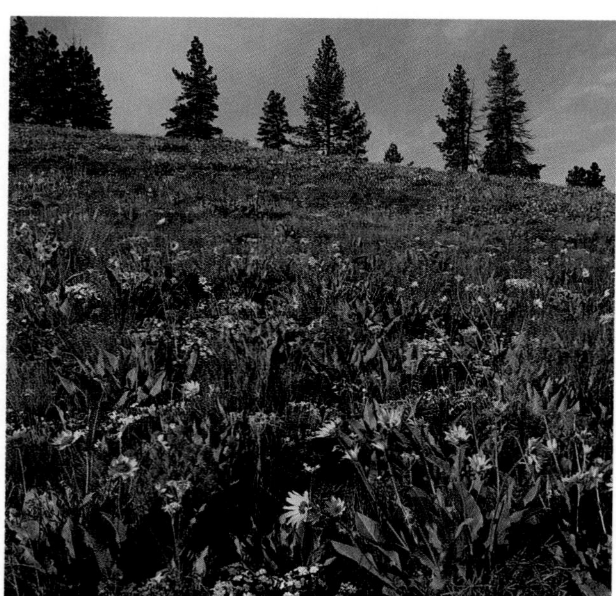

FIGURE 29.1 LIFE SPANS OF ANGIOSPERMS. Annuals, biennials, and perennials, all shown here, make up the familiar angiosperms.

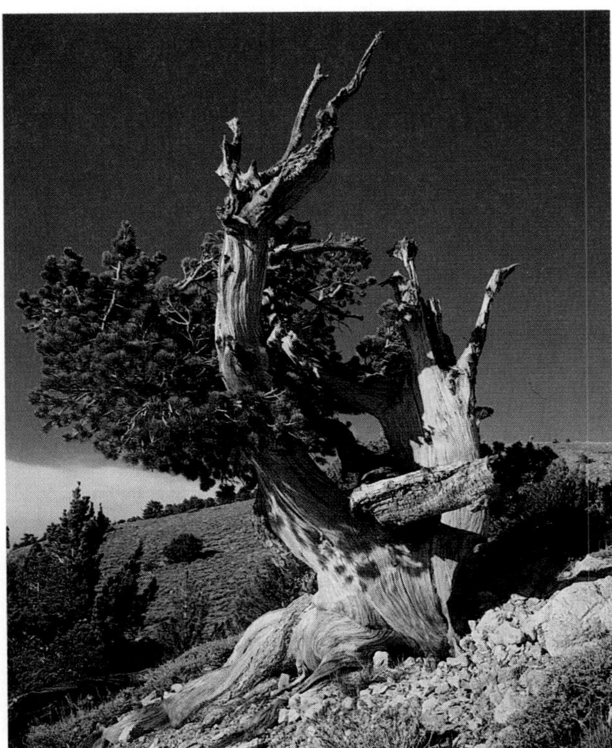

FIGURE 29.2 **THE OLDEST LIVING THING.**
This bristlecone pine tree holds the record for longevity, being over 4,900 years old.

house is likely to be paid off long before that tree casts its first shadow on the roof, and the original occupants will have probably died or moved on anyway. Trees seem to grow a lot slower than developers say or owners hope.

On the other hand, some tribal families in the Amazon Basin have difficulty keeping their mooring places for their dugout canoes cleared of grasses because some grasses there grow six inches a day! Growth is, indeed, variable among plant species. Yet, there are underlying principles involved in all plant growth, whether it involves the length or the girth of the plant. Let's take a look at some of these general principles.

First, we should establish that, essentially, **primary growth** increases the length of a plant, while **secondary growth** increases its girth or thickness. More specifically, primary growth occurs when cells developed through cell division in the shoot and root apical meristems elongate. Some of the specific products of primary growth are young roots, shoots (including new branches), leaves, and flowers. (Figure 29.3 shows the body and principal growth systems of a "generalized" plant.) Primary growth is, of course, responsible for the emergence of the young plant from the seed. Some of the cells produced through primary growth remain simple and unspecialized, capable of resuming activity later. Such cells, in some plants, will contribute to secondary growth.

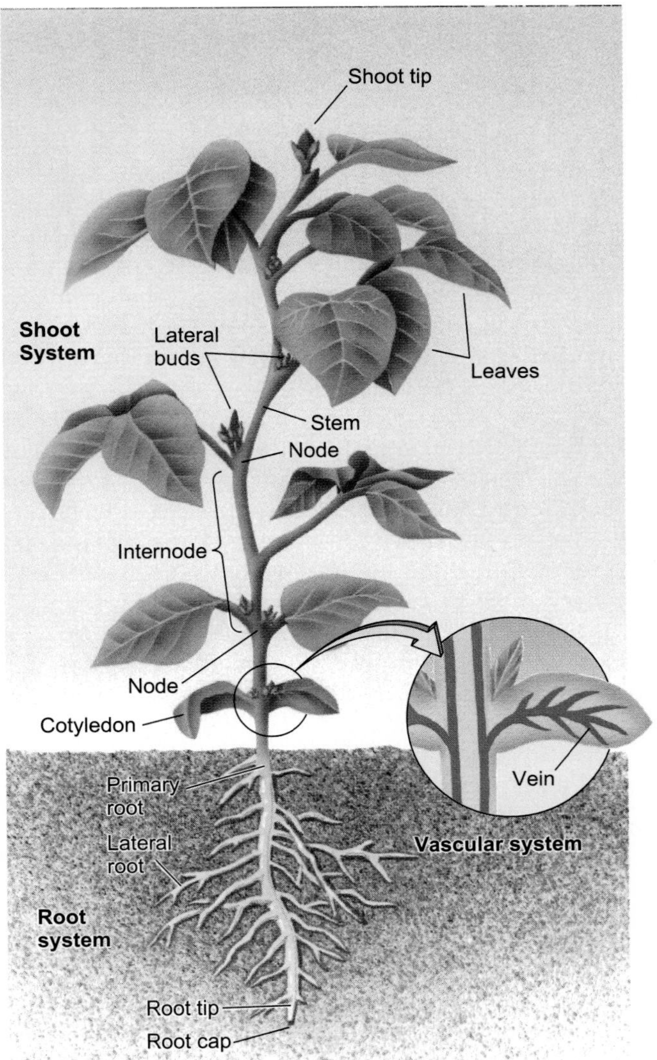

FIGURE 29.3 **A BASIC PLANT STRUCTURE.**
In this diagram, the root and shoot are shown to be composed of a complex array of cells and structures.

Secondary growth generally originates in two sources, *vascular cambium* and *cork cambium,* both of which emerge from the unspecialized reserves of meristematic tissues mentioned earlier, and produce a variety of tissue types (these will be discussed later). Although all plants carry on primary growth, secondary growth is not universal. The most obvious examples of secondary growth are

ESSAY 29.1 Totipotency in Plant Cells

Imagine taking a single cell of a St. Bernard and from it growing a whole new dog. You will have to imagine it, because it can't be done. But you *can* take a single *plant* cell and grow a whole new plant—and this fact has stimulated some of the most interesting research in plant biology.

From such regeneration studies, we have learned that many, perhaps most, plant cells are *totipotent*—that is, they have the ability to produce the entire organism from which they come. This is because, as their cells specialize for the role they will assume in the life of the plant, this specialization, or differentiation, is often reversible. Unlike virtually all animal cells, the plant cell can revert to an earlier condition and then take a different developmental pathway.

In the 1930s, it was discovered that carrot tissue could be grown from individual cells of the carrot embryo. Once separated, they were cultured in a medium made from coconut milk (which contains critical nutrients and hormones). In later experiments, mature phloem (food-conducting) cells from carrots, cultured in a similar medium, grew into rootlike structures that, when planted, produced entire carrots. More recently, there has been progress in culturing redwood trees and orchids in a similar manner. In addition, botanist James Shepard of Kansas State University dissolved the walls of mature potato cells, leaving behind the naked cells. He then grew individual cells in a nutrient medium and produced plants, proving again that mature plant cells have lost none of their genetic potency.

Although such work is in its early stages, the potential benefits are encouraging. An obvious outcome would be the use of *cloning* (the production of genetically identical individuals from a single cell) to produce selected crop plants. But even more in step with the new era of genetics would be the use of these techniques in gene splicing and recombinant DNA programs. Scientists have already succeeded in producing a potato-tomato hybrid by joining the nuclei from these plants, as well as developing lines specialized for beauty, disease-resistance, cold and drought resistance, and greater yield. Certain of the increased demands and special needs caused by a swelling human population may be met by creative scientists who understand the special nature of plant differentiation.

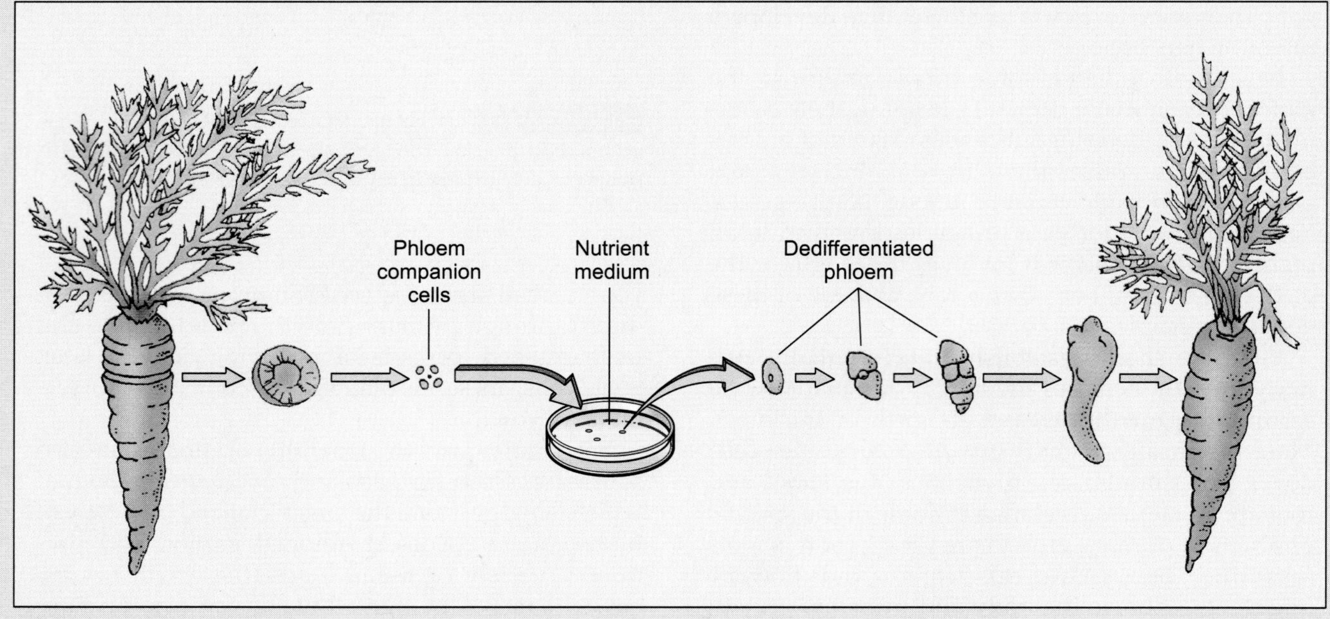

seen in the familiar trees, the woody dicots. Most monocots and many short-lived dicots show only primary growth.

Tissue Organization

Plants, our silent partners on the planet, may seem to be simple and rather inactive, but the appearance is deceptive. The truth is, they are highly organized and dynamic, interacting with their environment in very complex ways. This complexity is suggested in the variety and roles of the tissue comprising the bodies of flowering plants. Basically, however, these complex organisms arise from three primary meristematic (undifferentiated) tissues: the vascular tissues, the dermal tissues, and the ground tissues, which enable the plant parts to grow in length. Later, lateral meristematic tissue will enable the plant to grow in diameter. Table 29.1 lists the various meristematic tissues and the tissues they give rise to.

Vascular Tissue

Since terrestrial plants are not bathed in life-sustaining waters as were their ancestors, they have developed the means to transport the necessities of life within their bodies, primarily in the **vascular tissues** called the *xylem* and *phloem*. Let's take a look at these fascinating, fluid-filled channels.

Xylem is a vascular tissue that chiefly conducts water and minerals throughout the plant. Whereas xylem is composed of several types of fibers and cells, the cells most directly involved in transport are the *tracheids* and *vessel elements* (sometimes called vessel members). Tracheids, like vessel elements, are dead and hollow at maturity, when the cell constituents disintegrate. Tracheids are tapered at each end and pitted (indented) along the sides, allowing water to pass through the cell walls from one cell to an adjacent one. They also have pits in their tapered end walls through which water must pass. The cells of vessel elements are open at their ends, producing long tubes that allow the easy movement of water along their interiors (Figure 29.4). Their walls may thicken due to the deposit of a tough material called *lignin,* as cytoplasm disappears and the end walls are digested, forming the conducting tubes. These cell walls may contain pits, as do those of tracheids. Vast numbers of both tracheid and vessel cells, laid end-to-end, form minute tubes through which water can pass from root to stem and leaf. Their tough cell walls also help support the plant.

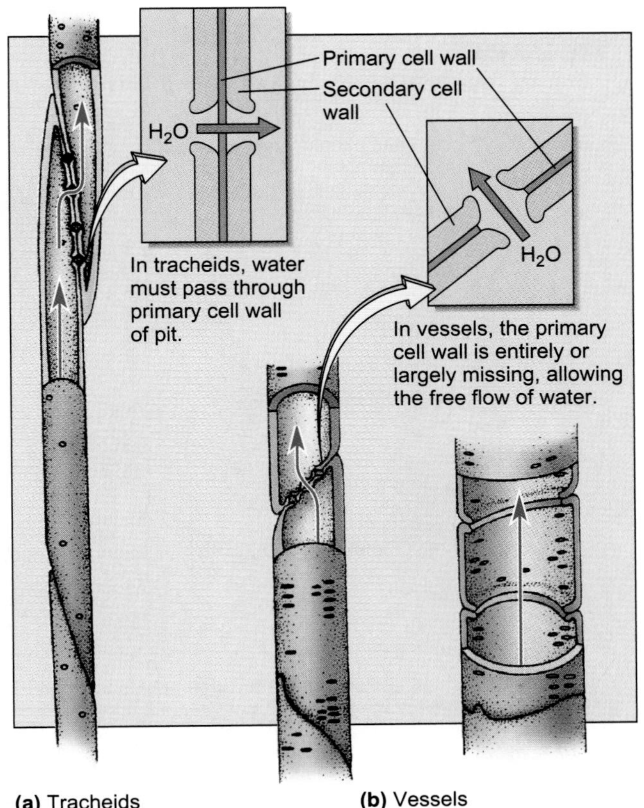

Figure 29.4 **Xylem.**
The conducting elements of xylem include tracheids **(a)** and vessels **(b)**.

Phloem conducts a solution containing organic compounds (foods) made by photosynthesis. This solution is called **sap.** Although the principal organic compounds are sugars, amino acids and even hormones are carried in the phloem stream. Unlike the xylem, phloem must remain alive to be functional.

Phloem is a complex tissue that includes *sieve elements, companion cells,* and *phloem parenchyma*—thin-walled cells that comprise most of the tissues of the plant (Figure 29.5). The sieve elements are the cells that actually carry out transport. Sieve elements include primitive sieve cells (found in most vascular plants) and the more advanced sieve tube members (restricted to flowering plants). We will consider primarily the latter.

The term *sieve* comes from the prominent pores that pock the walls of sieve tubes. They occur in both the side walls and the end walls. Through them run the *plasmodesmata* (see Chapter 5), where the cytoplasm of one sieve tube member is continuous with that of the next. Materials in the cytoplasm, then, can move easily from one member to the next. The largest pores occur in the *sieve plates,* located in the end walls.

TABLE 29.1 MERISTEMATIC TISSUES

Primary Growth During primary growth, primary meristem gives rise to:

Procambium
Procambium differentiates into xylem and phloem, the conducting vascular tissue of roots, shoots, and leaves. Xylem specializes in water and mineral transport, phloem, in food transport.

Protoderm
Dermal tissue differentiates into covering tissues, including the epidermis of roots, stems, and leaves. In addition, it forms guard cells, leaf hairs, and root hairs.

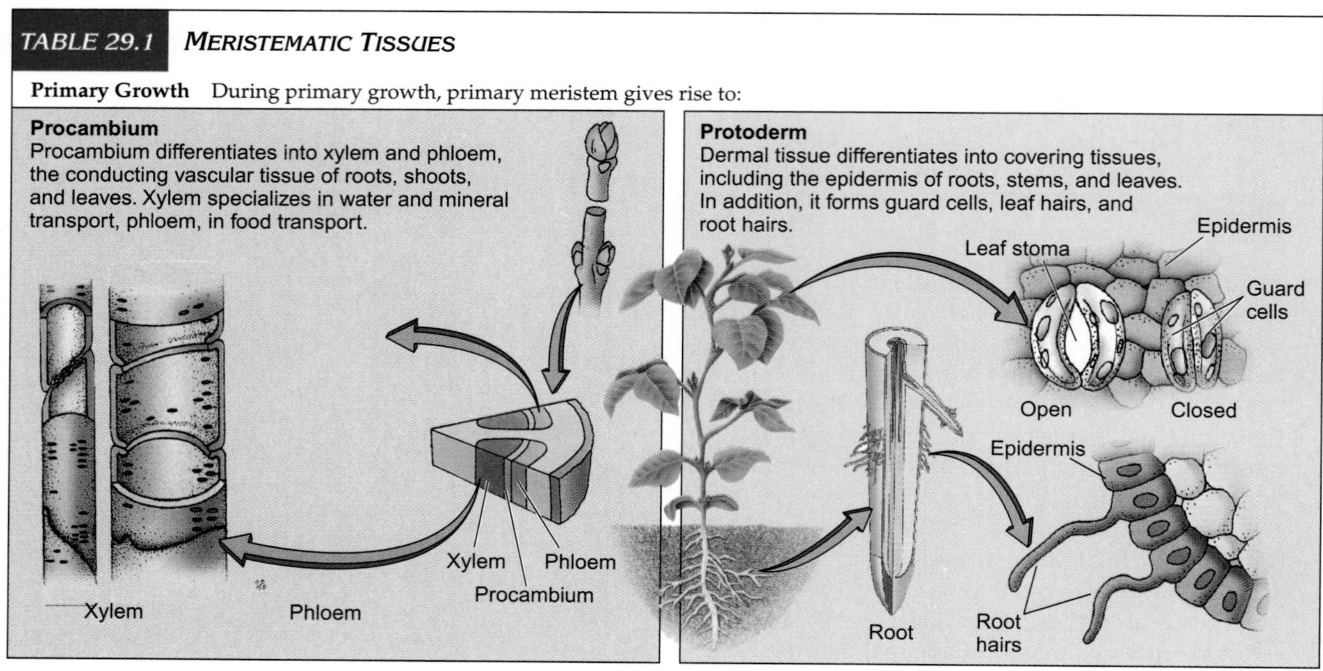

Ground meristem differentiates into three basic tissue types:

Parenchyma
Parenchyma is widely distributed in the stem and root and makes up the photosynthetic tissues of the leaf. The cells are large and thin-walled, often involved in storage.

Collenchyma
Collenchyma is primarily involved in support. Its thick-walled cells form a tough but flexible cylinder below the epidermis, within vascular tissue, and in supporting portions of the leaf.

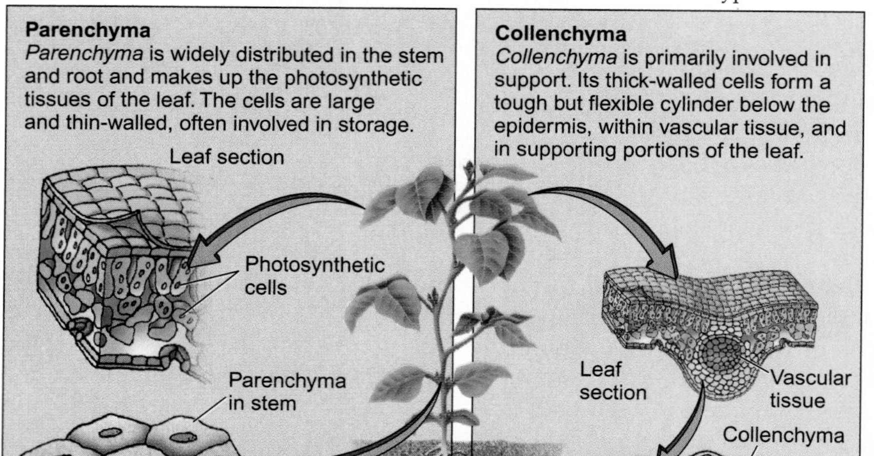

Sclerenchyma
Sclerenchyma, in its *fiber* form, strengthens shoots. In its *sclereid* form, it provides hardness for seed coverings and shells.

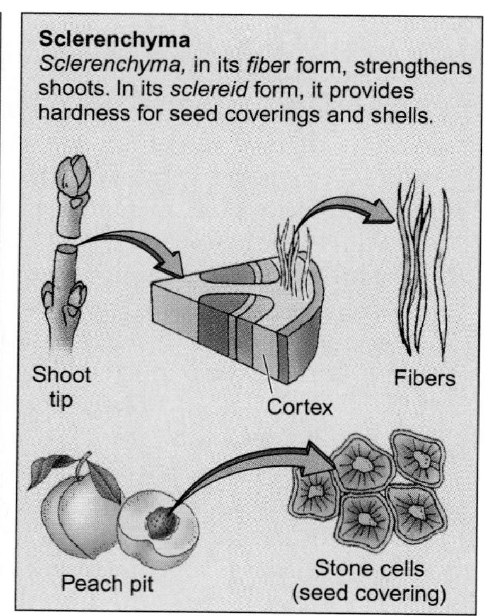

Secondary Growth In secondary growth, the lateral meristem tissue produces:

Vascular Cambium
Vascular cambium gives rise to secondary xylem and phloem as the diameter of the plant body increases.

Cork Cambium
Cork cambium gives rise to the protective covering that replaces epidermis in plants that are growing in diameter.

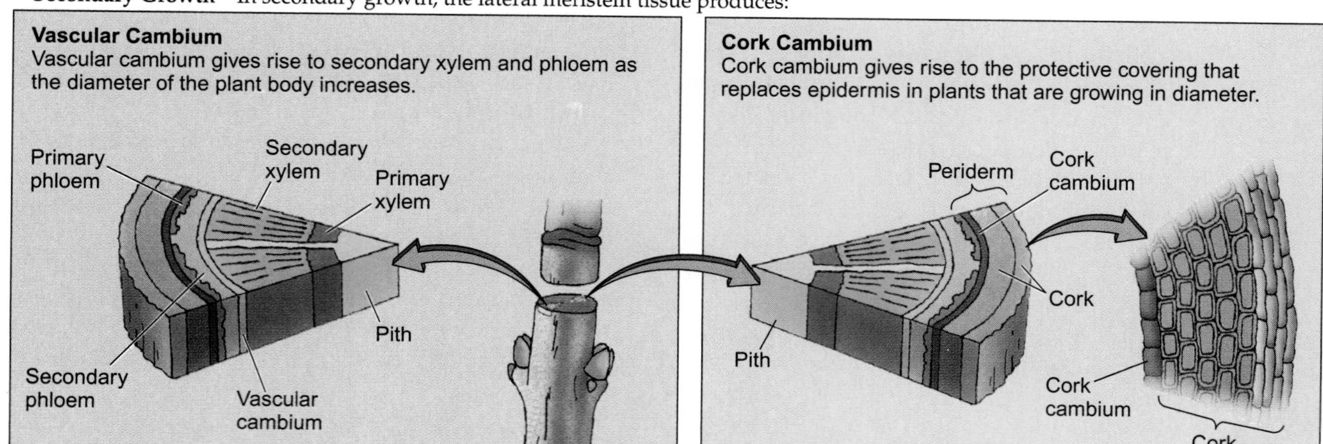

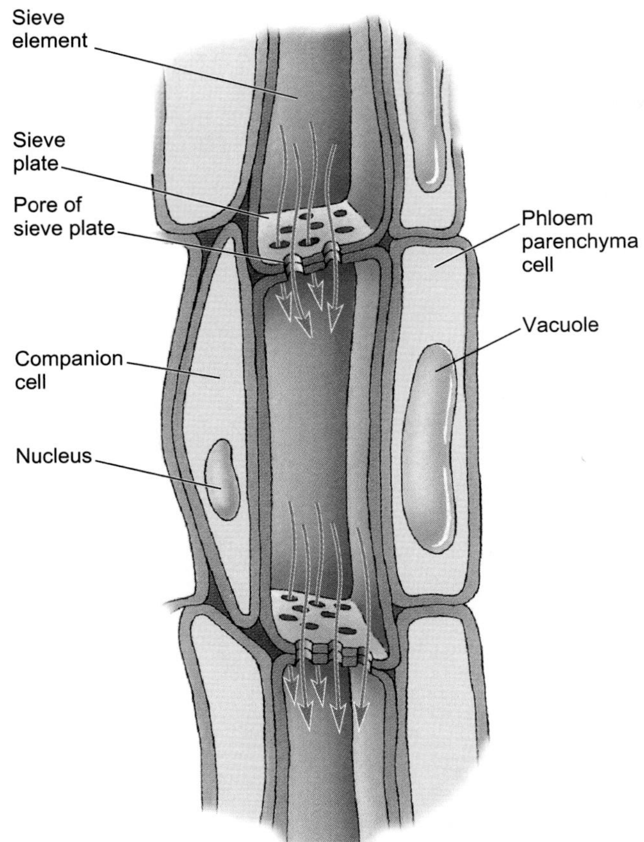

FIGURE 29.5 PHLOEM.
The major components of phloem are sieve elements, companion cells, and phloem parenchyma.

We said earlier that phloem is living. This may be stretching the definition a bit, however, since the cytoplasm of sieve tubes lacks a nucleus. How does a cell—that is, a sieve tube member—survive without a nucleus? In the case of phloem, each sieve tube member lies against a nucleated *companion cell,* and the cytoplasm from one can move into the other. Since sieve tube members lack nuclei, it is believed that most metabolic activities are carried out in the companion cells. We do know that the death of a companion cell signals the immediate death of its sieve tube member. Phloem parenchyma tissue primarily stores food that it actively transports into and out of sieve tube members.

DERMAL TISSUE

Dermal tissue covers the surface of the plant. It is formed from the embryonic **protoderm.** The surfaces of all parts of certain plants are covered by **epidermis**—plants that show primary growth only, such as peas, corn, and beans. Primary growth originates from the apical meristems (the regions of actively dividing cells at the tips of stems or roots). On the other hand, in stems and roots that show secondary growth—that is, growth in the lateral meristems that increase the girth of stems and roots—the epidermis may be sloughed off. In such cases, the plant is protected by the formation of **periderm,** comprised mostly of dead cork cells and cork cambium (with underlying parenchyma cells).

GROUND TISSUE

The embryonic tissue called **ground meristem** differentiates into the *ground tissue.* The **ground tissue,** as mentioned, includes all those tissues that are not procambium or protoderm. In particular, the ground tissue produces three basic tissues: parenchyma, collenchyma, and sclerenchyma. The last two function in support and protection (see Table 29.1). The parenchyma tissue at the center of the stem produces the *pith,* a region of large cells that may break down and form a hollow area. Nearer the surface, the parenchyma produces the *cortex,* which may be more extensive than the pith (although in woody plants the cortex may be crushed and replaced by new cells dividing from within). Both the pith and cortex cells may store food while they are alive—or, if they contain chloroplasts, they may produce food.

THE ROOT SYSTEM

A walk through a forest easily convinces us of the great diversity of plant shoots, stems, and leaves. But far from obvious is the forest beneath our feet, a hidden growth of vast root systems that are just as diverse as what we see aboveground. Different kinds of roots boast their own special properties, and different parts of roots have their own important functions.

The principal roles of the root are providing anchorage and support to the stems and foliage, taking in water and minerals, and to a varied extent, storing materials for later use.

PRIMARY GROWTH IN THE ROOT

The growing root includes three regions—the *root tip,* the *elongation region,* and the *maturing region* above (Figure 29.6). While the root tip and elongation regions are the sites of ongoing primary growth, events in the maturing region above set the stage for secondary growth.

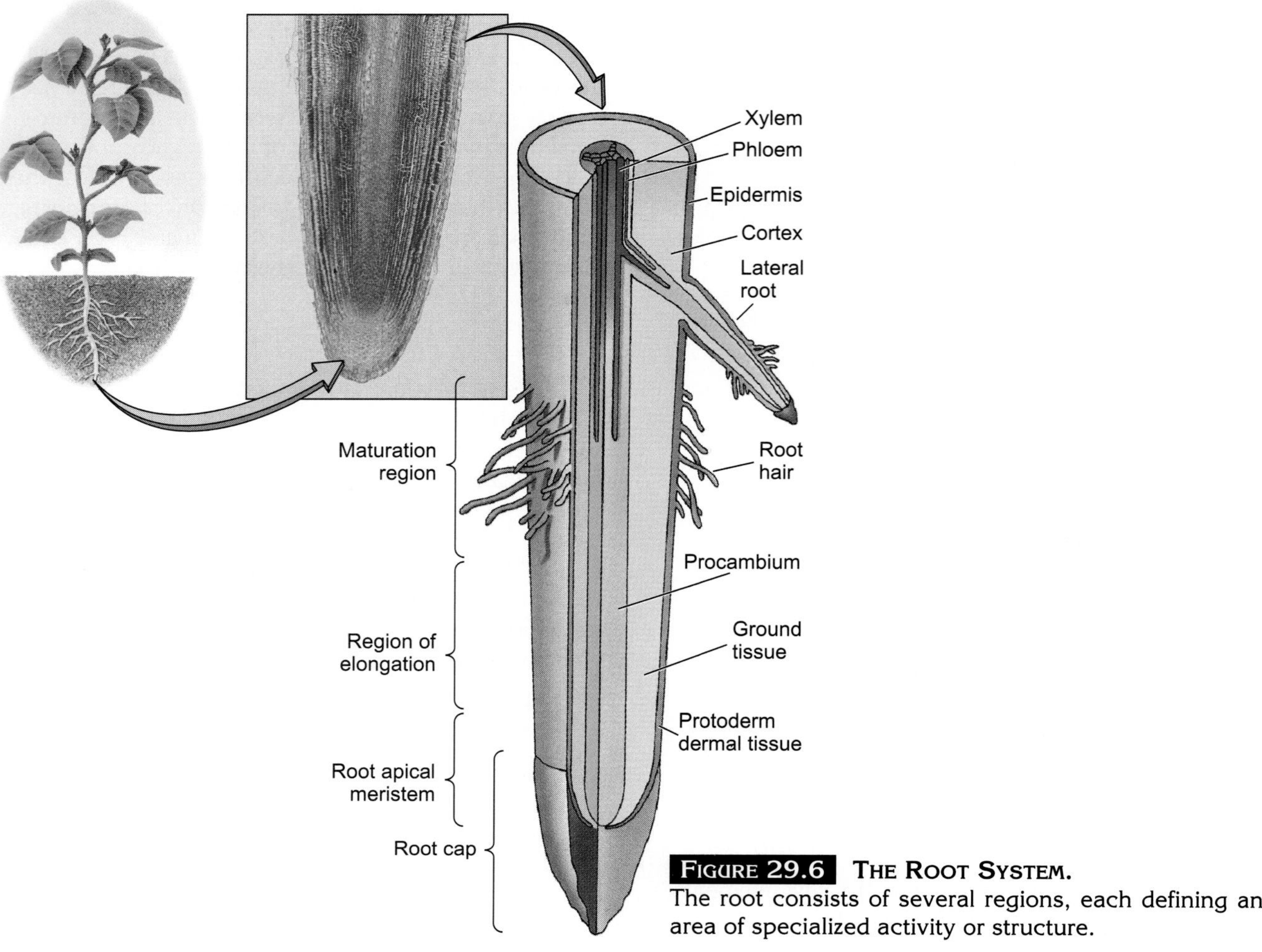

FIGURE 29.6 THE ROOT SYSTEM. The root consists of several regions, each defining an area of specialized activity or structure.

Figure 29.6 also illustrates intense activity in the **root apical meristem,** which is located just behind the very end of the root. Root apical meristem includes numerous tiny, undifferentiated cells that continually divide and redivide, contributing cells to the vascular tissue, the dermal tissue, and the ground tissue. As expected, the dermal tissue will produce the root covering, while the vascular tissue will form phloem and xylem. The ground tissue will later contribute to the cortex. Some of the cells from the root apical meristem replace those lost by the **root cap,** a protective cluster of cells at the very end of the root tip. Root caps are worn away as the root pushes through the abrasive soil, and the cells must be replaced.

Push is the right word, since root tips are literally forced through the soil. As cell division in the root apical meristem continues, the new cells left behind grow rapidly in length. Their elongation is what pushes the root tip along. Elongation is brought about by the absorption of water into the cells—which stretches the young, flexible primary walls and lengthens the cells in the direction of the root axis—and the growth of those cells by metabolic processes. Soon, firmer secondary walls will be laid down, and elongation will stop in those cells. In the meantime, a new generation of cells will develop and elongate, and the pushing continues.

While the elongation process takes place, tiny **root hairs** form as extensions of the epidermal cells. Root hairs grow in great profusion and provide an enormous surface area through which water and dissolved minerals can move into the plant (Figure 29.7). The mature root itself is largely impermeable to water because of a waterproof covering of waxy *suberin.*

At the uppermost region of the root tip, vascular tissue forms primary xylem and primary phloem, which forms the **stele** (Figure 29.8). In cross sections of dicot roots, the primary xylem may form a kind of cross or *X*. The smaller, thinner-walled primary phloem lies between the arms of the cross. Surrounding the primary xylem and phloem is a cylinder of cells called the **pericycle,** rather undifferentiated meristematic tissue that can give rise to *lateral roots,* branches of the primary root.

Just outside the pericycle lies the **endodermis**—a kind of "inner skin" that helps control the movement of water and minerals into the vascular tissue, as is described in the next chapter.

Outside the stele is an extensive cylinder of large, thin-walled parenchyma cells, often swollen with stored starches. Together with the endodermis, they make up the **cortex.** Around the cortex lies the root's epidermis, or outer skin.

FIGURE 29.7 ROOT HAIRS.
When germinated in a moist chamber, the seedling produces an enormous number of root hairs. Each is an extension of an epidermal cell. Their combined area provides the root with a great absorbing surface.

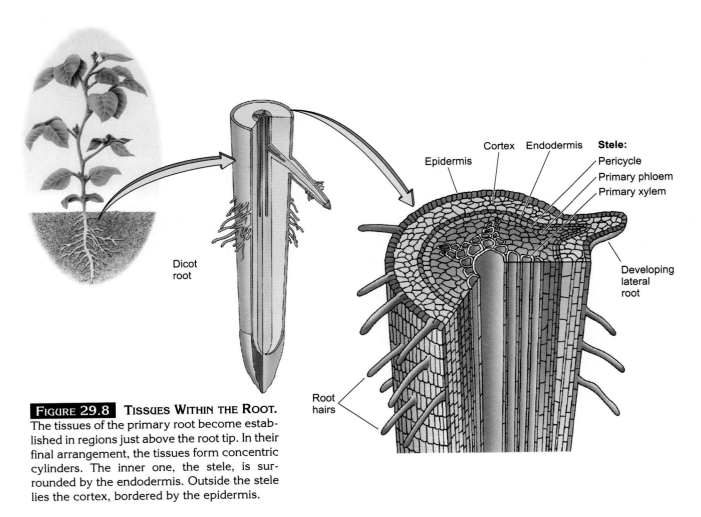

LATERAL ROOTS AND ROOT SYSTEMS

Most plants start out with a single primary root, which then may form side branches called **lateral roots.** Lateral roots originate in cells of the pericycle

FIGURE 29.8 TISSUES WITHIN THE ROOT.
The tissues of the primary root become established in regions just above the root tip. In their final arrangement, the tissues form concentric cylinders. The inner one, the stele, is surrounded by the endodermis. Outside the stele lies the cortex, bordered by the epidermis.

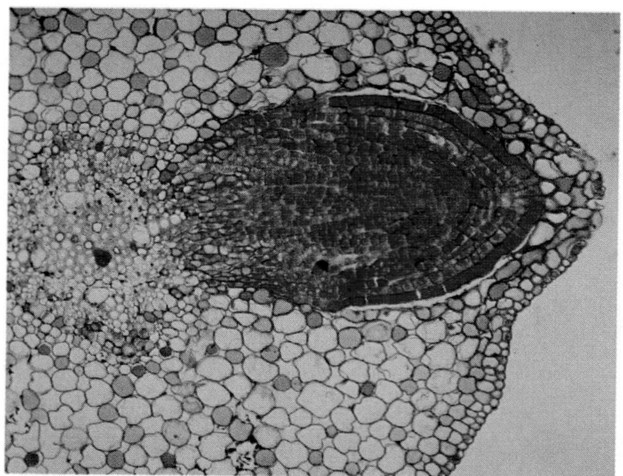

FIGURE 29.9 LATERAL ROOTS.
Lateral roots are initially produced in tap roots by the pericycle, where cells undergo repeated divisions. The mass of cells produced breaks through the endodermis and begins to push its way through the cortex.

well above the root tip. As they grow, they push across the cortex, eventually breaking through the epidermis (Figures 29.8 and 29.9). The vascular system of the young lateral root will join that of the primary root. As lateral roots mature, they produce side branches, which in time produce their own side branches, a process of subdivision that continues as long as the root grows. A large central root with many lateral roots is characteristic of the **taproot system.** Examples include the carrot, sugar beet, and dandelion (Figure 29.10a). Most dicots form taproots.

By contrast, in other kinds of plants, the primary root is very temporary. It quickly forms lateral roots that are roughly equivalent, producing a **fibrous root system.** These are characteristic of most monocots, grasses being a good example (Figure 29.10b). Examples of dicots with fibrous root systems are found in some rain forest trees, such as those of the Amazon Basin.

As an aside, it is unsafe to camp under large trees in the rain forest (although hard to escape). Because the soil is so poor, the trees must send out lateral roots, extracting minerals from newly decomposing material near the surface before the rains wash them away or they are taken by competing plants. The huge trees are thus anchored only superficially, and in the heavy rains and high winds so common in the jungle, they can be heard falling, seemingly randomly, throughout the forest as some storm passes.

Some plants have roots that grow outward from the base of the stem. These are called **adventitious roots.** Adventitious roots are those that arise in unusual places, such as along stems or leaves or buds. Adventitious roots that arise aboveground are referred to as aerial roots. An example is seen in the Algerian ivy that climbs along the sides of aging college dormitories (some claim the ivy is what holds the dorms up). Many plants can be propagated by cuttings that sprout adventitious roots when suspended in water.

(a)

(b)

FIGURE 29.10 THE SHAPES OF ROOT SYSTEMS.
(a) Taproot system of the dandelion. **(b)** Diffuse root system of grasses.

The Shoot System

The shoot system consists of the stem and the tissues that arise from it, such as leaves, buds, and branches. Growth at the shoot tip continues throughout the life of a plant, just as it does in the root tip. New tissues arise from the apical meristem and then elongate, resulting in stem growth. But aside from this, there are few similarities in the development of the root and the shoot.

Primary Growth in the Shoot

The shoot apical meristem (Figure 29.11) is somewhat dome-shaped and is covered by a protective layer called the *tunica.* The shoot meristem does not simply lengthen, leaving behind differentiating tissue. Instead, as it grows, it leaves behind both differentiating tissue and patches of various kinds of meristem. One of these kinds of patches, the *leaf primordia,* gives rise to leaves; another kind, *lateral bud primordia,* produces branches.

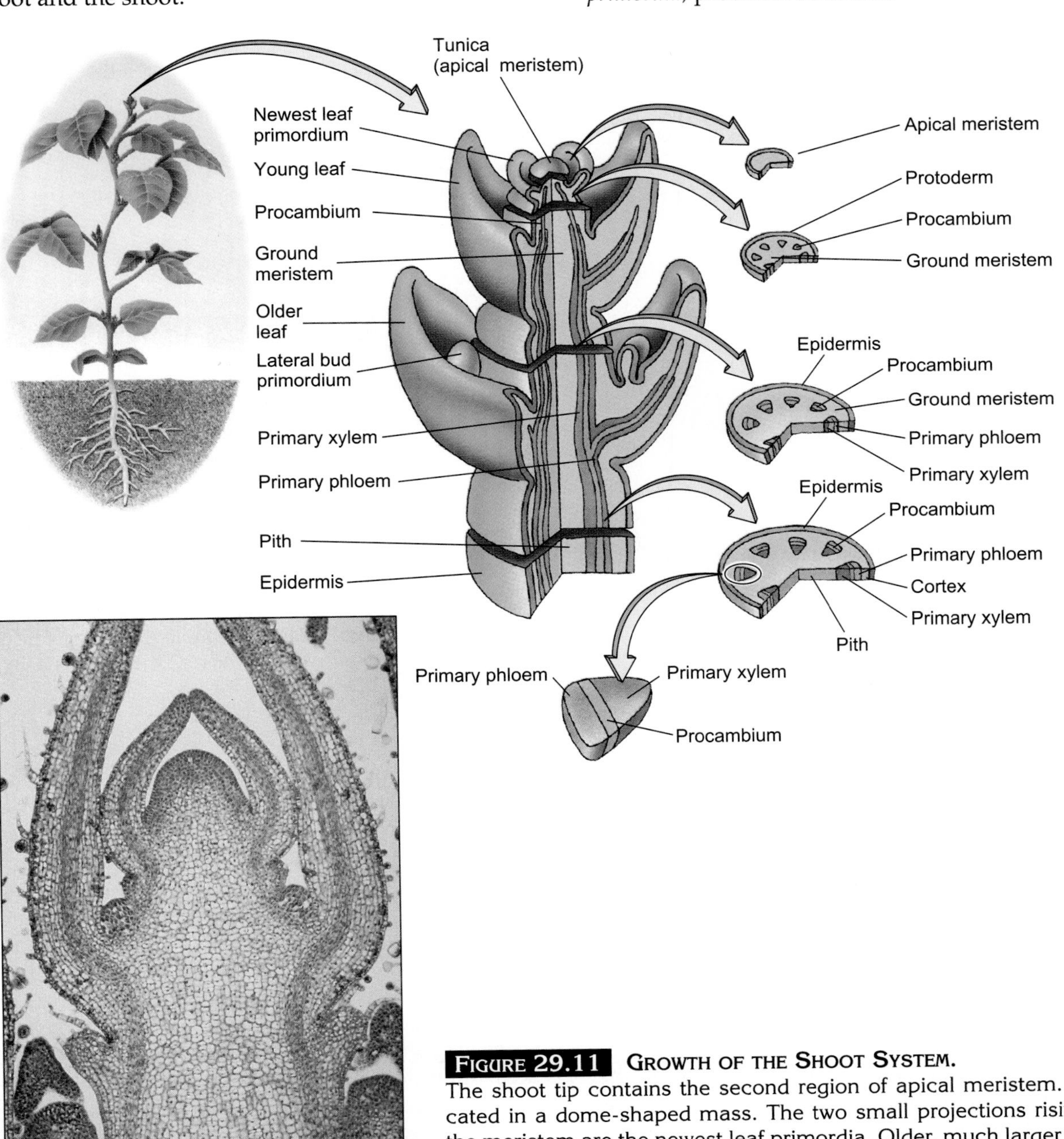

Figure 29.11 Growth of the Shoot System. The shoot tip contains the second region of apical meristem. It is located in a dome-shaped mass. The two small projections rising from the meristem are the newest leaf primordia. Older, much larger leaf primordia now rise up to cover the entire structure. In the axis, a patch of tissue marks the location of a lateral bud primordium.

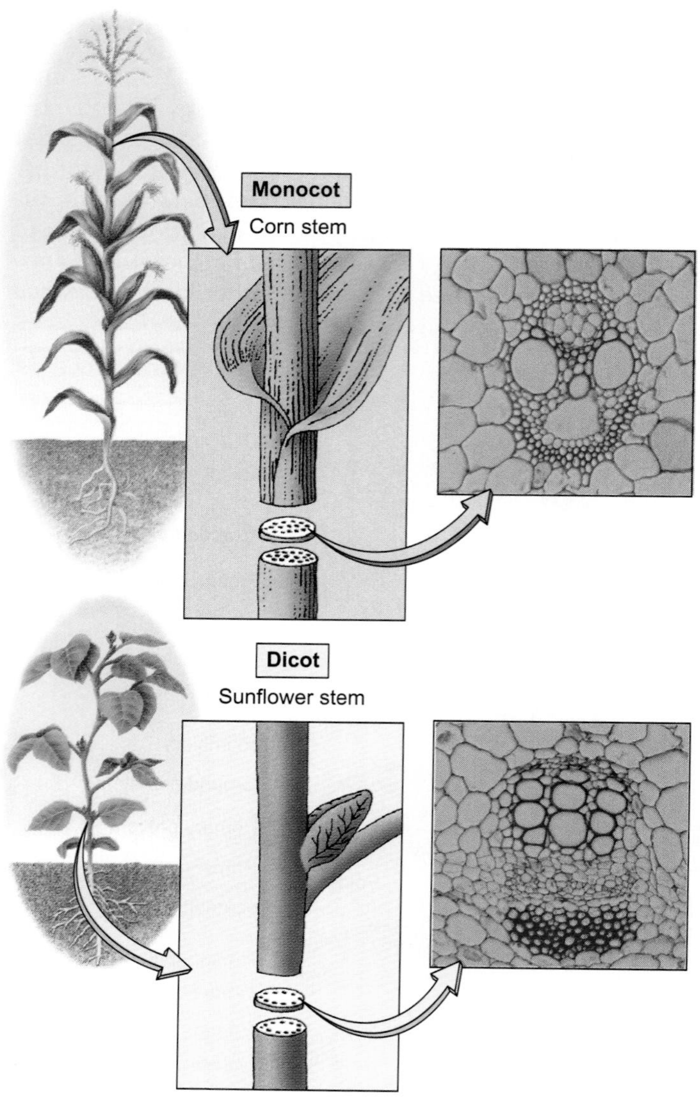

FIGURE 29.12 ORGANIZATION OF THE SHOOT VASCULAR SYSTEM. After primary growth, the vascular system in the young shoot is organized into bundles or cylinders. In the monocot corn, the vascular bundles are scattered in the stem. Each bundle contains xylem and phloem, surrounded by a sheath of thick-walled sclerenchyma. In the dicot bean, the bundles are arranged in an orderly circle, just below the epidermis. Each bundle contains xylem and phloem, separated by a patch of procambium. Most of the stem consists of soft pith.

Behind the Shoot Meristem. Tissues behind the shoot meristem remain undifferentiated for a time, but strands of procambium mark the sites of future xylem and phloem, and cells of the central region form ground meristem. Later, these will give rise to the cortex and *pith.* **Pith** is formed of large, thin-walled parenchyma cells at the shoot center. The outer cells of the young shoot, still green with chloroplasts and carrying on photosynthesis, have begun differentiating into the young epidermis.

In dicots, as the young tissues left behind by the apical meristem mature, they organize into specific patterns quite unlike those of the monocots. Bundles of vascular tissue are generally scattered in the monocots, whereas they tend to form ringlike patterns in the dicots (Figure 29.12). Also, in many dicots (but not in the monocots), a small region of procambium (a tissue that is critical to secondary growth) is retained between the xylem and phloem.

Secondary Growth in the Shoot

Since secondary growth in the root and shoot are quite similar, we will concentrate on the shoot. Secondary growth accounts for increases in thickness, but it does not occur in all plants. Instead, it is restricted to perennial dicots and gymnosperms. Thus the great size of trees is possible. However, a few plants such as the coconut palm, a monocot, grow quite large even without secondary growth.

Instead, they form an enormous apical meristem at the tip. (It looks like a greatly enlarged version of Figure 29.11.)

Transition from Primary to Secondary Growth. Secondary growth in the shoot begins just below the growing and differentiating shoot tip where primary growth ends. At this time, the young primary vascular tissues have collected into a ring of bundles containing xylem and phloem. Small patches of procambium remain between the two tissues, and it is this lingering tissue that plays a major role in secondary growth.

Procambium gives rise to **vascular cambium,** which produces **secondary xylem** and **secondary phloem.** In the mature stem, the vascular cambium forms a thin strip between the secondary xylem and phloem. Through continuing mitosis, the vascular cambium provides cells for xylem on its inner side and cells for phloem on its outer. Yet it perpetuates itself by holding some cells in reserve.

As the newly produced xylem grows, it pushes the vascular cambium and newly formed phloem outward. Likewise, the new phloem can only grow outward, crushing older tissue in its path. Secondary growth, the continuous outward expansion of the woody stem, is caused by an enlarging ring of dividing and growing cells within. The sequence of events in secondary growth is shown in Figure 29.13.

Older Woody Stem. Older regions of the stem, those that have gone through several seasons of secondary growth, are mostly composed of woody xylem tissue, referred to as "wood" (85% or more of the mature tree). The vascular cambium, the phloem, and the periderm (the outer bark where cork is found—described shortly) all lie outside the wood, and together they form a thin ring of living material.

The growth rate of the xylem tissue is highly dependent on environmental conditions. For example, it may stop or slow down drastically during winter seasons. A cross section of an older stem or trunk reveals a definite pattern in trees that undergo seasonal growth. During periods of rapid growth, the xylem tends to consist of large, relatively soft cells. As growth slows, the differentiating xylem cells do not expand as much and remain smaller and denser. The differences in seasonal rates of growth produce the **growth rings** of trees, most obvious in the trees of the earth's temperate zones where spurts of growth occur seasonally (Figure 29.14).

A second growth pattern seen in the older stem takes the form of radiating spokes that extend through the wood and secondary phloem. These are known as **vascular rays.** They consist of sheets of thin-walled parenchyma cells and their thicker-walled descendants called *collenchyma* (a supportive tissue; see Table 29.1). Both are produced by the vascular cambium. Vascular rays provide a means of transporting nutrients laterally, a trait required in younger stems, in which much of the tissue is still alive. In older stems, the growth of rays is essential in relieving the forces created by the expanding cylinder of xylem inside the trunk.

Secondary Growth at the Perimeter. Secondary growth in the vascular tissues is accompanied by changes in the shoot epidermis. Cells in the cortex near the epidermis now take on a new role. They begin to divide rapidly, producing what is known as the *cork cambium* (see Figure 29.14). The cork cambium produces layer upon layer of cells that are continually pushed outward, rupturing and replacing the old epidermis. These new tissue layers constitute the periderm. The outer layer of the periderm becomes impregnated with suberin, producing a waterproof covering. In this way, the plant has begun to produce **cork.** Cork is often mistakenly called bark, but technically, *bark* is everything outside the vascular cambium. The cork of the cork oak is used to make the corks for wine bottles.

The corks in wine bottles are true cork. But such cork does not occur naturally. Cork growers remove the periderm of the cork oak, causing the tree to respond to the injury by obligingly forming a new, smoother cork cambium. The new tissue is removed and cut into cylinders. (These cork cylinders are then cleverly placed into wine bottles in such a way that they crumble at any attempt to remove them.)

The Leaf

Leaves are so much a part of our lives that we tend to take them for granted, except perhaps when they signal a change of season (Figure 29.15), a time when we stand in awe of the spectacle. When leaves are alive and functioning, they are primarily devoted to the conversion of sunlight energy into chemical bond energy in organic compounds (food material). (As part of the intimate association between humans and plants, autumn leaves fall on lawns in order to get humans in shape for shoveling snow.)

The development of even the simplest of leaves necessitates contributions from several basic tissues. Protoderm contributes to the highly specialized epidermis, which slows water loss and admits air. The layers of light-trapping, photosynthetic parenchyma

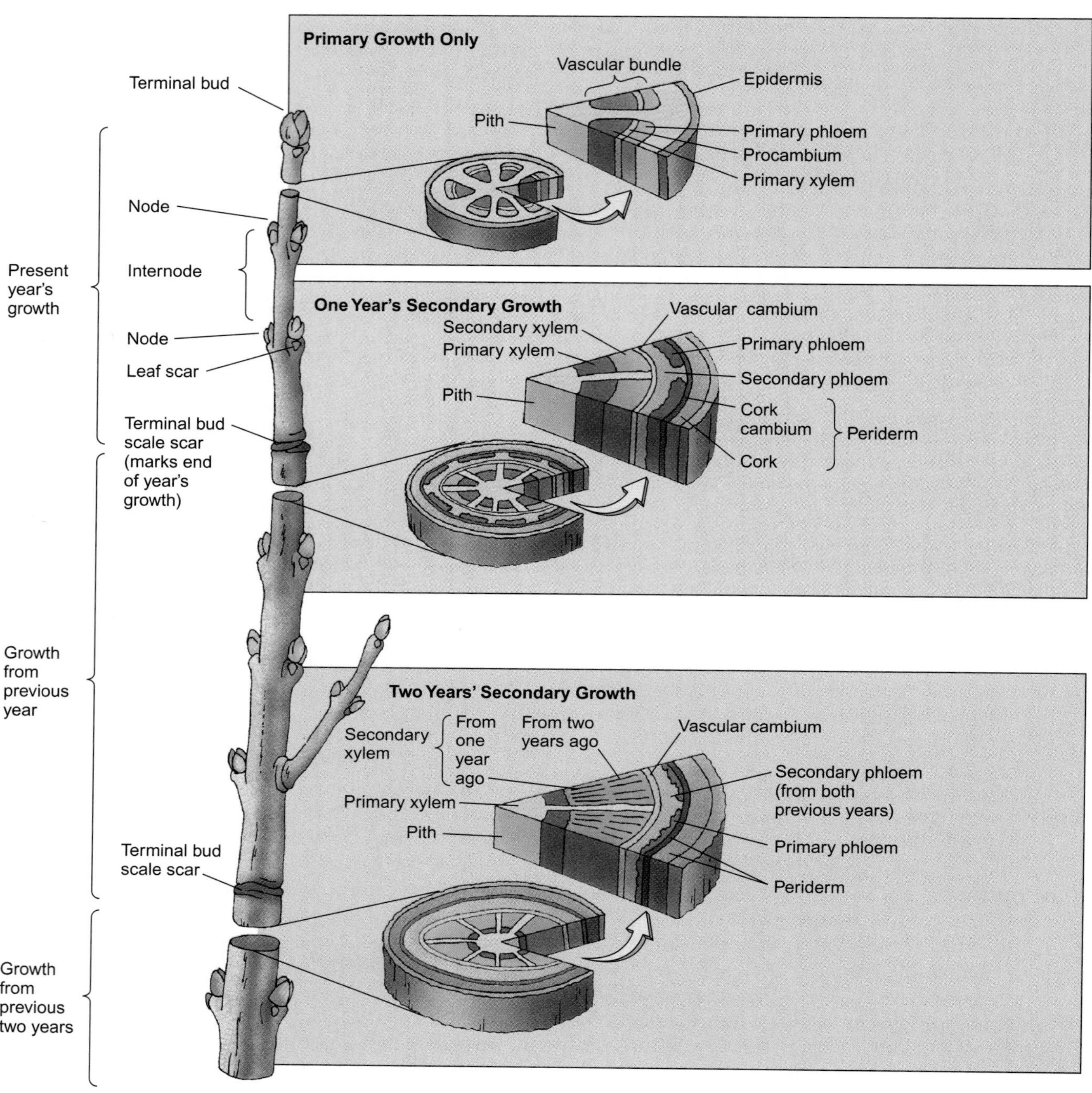

FIGURE 29.13 PRIMARY AND SECONDARY GROWTH.
In primary growth the vascular system of the dicot shoot matures into distinct vascular bundles. The vascular cambium produces secondary xylem on its inner side and secondary phloem on its outer side. This growth begins to push the primary xylem and phloem away, with the vascular tissue becoming a continuous ring. In an older region the results of continued activity in the vascular cambium are seen. Secondary growth has also replaced the epidermis with a new tissue, the periderm.

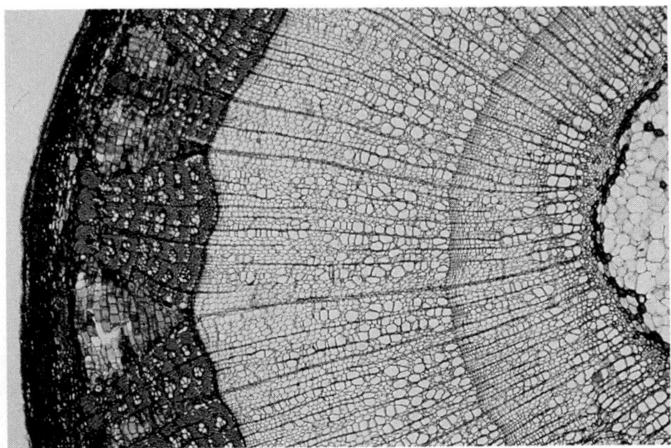

FIGURE 29.14 **A WOODY STEM.**
A photomicrograph of a cross section through a two-year-old woody stem clearly reveals regions of different tissues. The outermost tissue is the protective cork, a suberized layer of cells produced by the cork cambium lying just inside. The conducting tissue consists of a circle of phloem divided by rays. Just inside the phloem is a thin cylinder of vascular cambium, which seasonally produces phloem and xylem. Inside the vascular cambium and making up most of the stem tissue is the xylem, or wood. At the very center, a small region of pithy parenchyma remains.

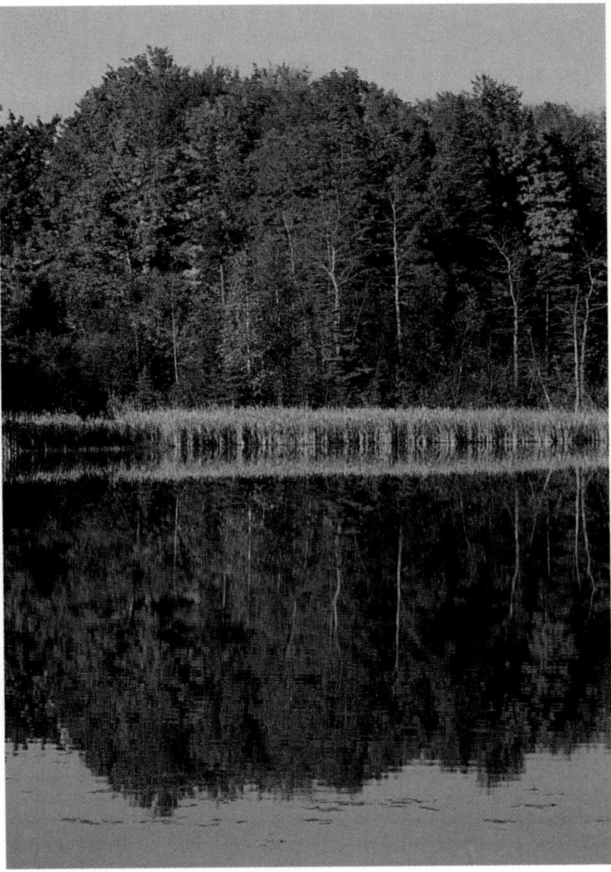

FIGURE 29.15 **AUTUMN LEAVES.**
Leaves become very noticeable in the fall, when chlorophylls are removed from the tissues, leaving behind the colorful array of accessory pigments normally obscured by the green.

cells within owe their presence to ground meristem, while the vascular tissues, so vital in bringing water and minerals in and carrying sugars out of the leaf, arise from the versatile procambium.

LEAF ANATOMY

The typical dicot leaf is a flattened *blade* attached to a stem by a stalklike *petiole.* The vascular system of the stem passes into the leaf through the petiole, and into the blade along a large central vein called the *midrib.* In typical dicots, the midrib supplies a network of smaller branches, or *veins,* that carry fluids through the blade. Within the blade, each smaller vein is surrounded by specialized cells that form the *bundle sheath.* Anything entering or leaving the veins must pass through these cells. We found in Chapter 7 that bundle-sheath cells have special significance to C_4 plants.

The organization of the monocot leaf is quite different. Such monocots as the grasses have no petioles. The leaves emerge from a tough sheath around the stem (which is why it is so frustrating to try to tear off a corn leaf). In most monocots, the veins do not branch from a central midrib; they lie in parallel rows, interconnected by short, smaller veins (Figure 29.16).

Organization of the Leaf. A cross section of a dicot leaf (Figure 29.17) reveals outer layers of cells of the *upper epidermis* and *lower epidermis.* The upper epidermis, the part most often exposed to light, consists of fairly large, flattened cells whose outer walls are coated with a layer of a wax called **cutin.** Most epidermal cells lack chloroplasts and are transparent.

Within the leaf there are several layers of parenchymal cells that contain chloroplasts. Those cells nearest the upper epidermis are arranged in one or more tightly packed layers and, because of their elongated shape, are called *palisade parenchyma.* They are the first to receive incoming light. Below this lie less orderly layers of cells that are more rounded. They are scattered in such a way that spaces form between them into which air can move. This layer is the *spongy parenchyma,* aptly named for its appearance. The two tissues are together called the *meso-*

phyll. The moist air spaces are important avenues of carbon dioxide diffusion, which is essential to photosynthesis. Air enters the leaf through minute openings called **stomata** (singular, *stoma*), which are found primarily in the lower epidermis—the underside of the leaf. Forming the stoma is a pair of kidney-shaped guard cells (see Figure 29.17).

Leaves are often covered with fine **leaf hairs,** each of which arises from a single epidermal cell. Some leaf hairs are soft and downy, but some are sharp and hooked and can be lethal to insect larvae that attempt to eat the leaf. Leaf hairs also have another function: They impede the movement of air over the leaf's surface, slowing the evaporation of water.

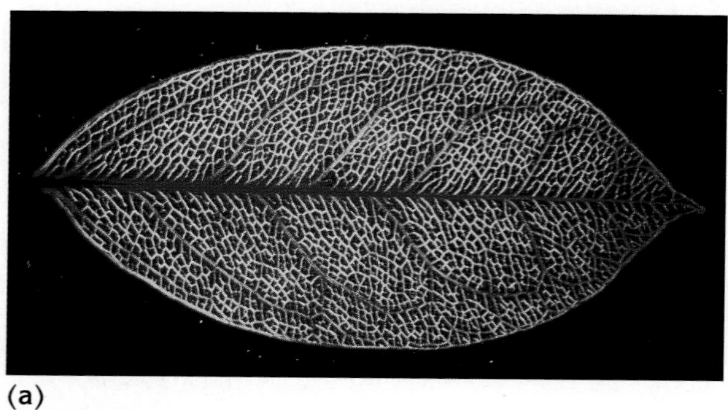

(a)

(b)

FIGURE 29.16 LEAF VEINS.
(a) Dicot leaves are typically net-veined, with a central midrib giving rise to numerous smaller, branching veins. A leaf petiole forms the attachment to the stem. (b) Monocot leaves, on the other hand, contain parallel veins.

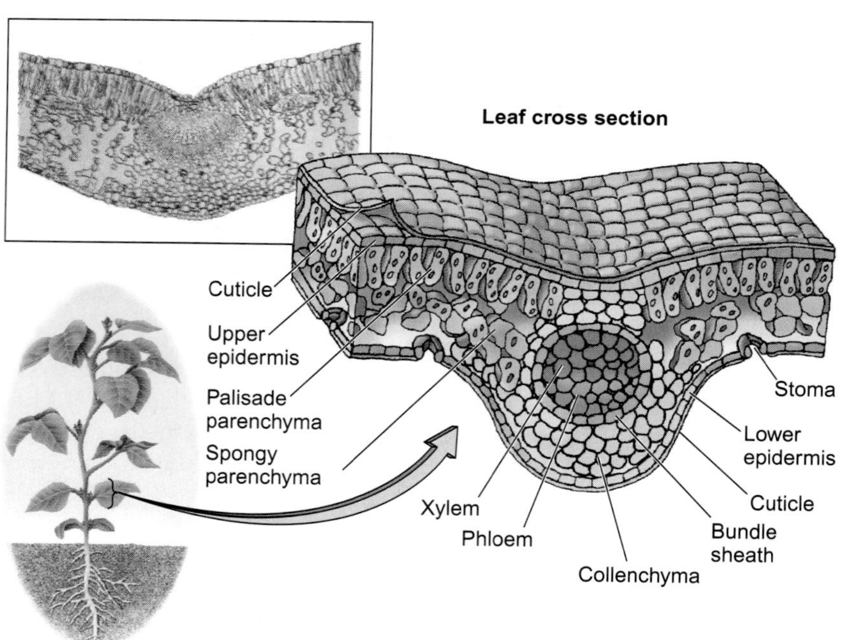

FIGURE 29.17 LEAF STRUCTURE.
The cross-sectional view of a leaf reveals its complex tissue organization. Many leaves have epidermal leaf hairs as well, as seen in the scanning electron micrograph.

PERSPECTIVES

We have seen that plants are highly organized and complex organisms. This complexity is revealed in the many specialized tissues of a single plant, as well as in their diversity when species are compared. Still, there are common themes: the general similarity of tissue types, primary growth mechanisms, and systems of transport. In the next chapter, we will take a closer look at these transport systems, and we will discover, in particular, just how the xylem and phloem are uniquely adapted to their vital tasks.

SUMMARY

INDETERMINATE GROWTH

In their life spans, plants are generally annuals, biennials, or perennials. The latter have open or indeterminate growth and an indefinite life span. This is made possible by a reserve of undifferentiated meristematic tissue.

PRIMARY AND SECONDARY GROWTH

Primary and secondary growth are growth in length and girth, respectively. Primary growth results in new shoots, leaves, flowers, and roots. Secondary growth originates from vascular cambium and cork cambium.

TISSUE ORGANIZATION

Mature plant tissues are derived from three primary meristematic (undifferentiated) tissues: vascular tissues, dermal tissues, and ground tissues.

The water- and mineral-conducting portion of xylem—vessels and tracheids—form continuous tubes throughout the vascular system. Mature xylem consists of nonliving elements. Water moves between adjacent elements via thin-walled pits. Tracheids have end-wall pits, whereas vessels have open ends through which fluid can pass.

Phloem, a living tissue, conducts sap (sugars and other substances). Included in phloem are sieve elements, companion cells, and phloem parenchyma. Sieve elements, arranged end to end, form the principal transport system. The presence of plasmodesmata in their sievelike end walls results in a continuous cytoplasm. The companion cells perform metabolic tasks, and along with phloem parenchyma, actively transport materials in and out of sieve elements.

Epidermis covers the surface of all plants. It continues in plants that show primary growth (from apical meristems), but is replaced in plants with secondary growth (from lateral meristems).

Ground tissue produces two kinds of parenchyma, pith and cortex. Both may store food while alive or make food if they have chloroplasts.

THE ROOT SYSTEM

Roots provide anchorage for stems, carry on water and mineral transports, and store foods.

The growing root includes three regions: the root tip, the elongation region, and the maturing region. Continuous cell division in the root apical meristem provides cells for the root cap ahead and the maturing region behind. Cell division and lengthening in the latter pushes the root through the soil. Epidermal cells and their root hairs absorb water and minerals. Maturing procambium forms the vascular tissues of the stele.

The stele includes the pericycle, xylem, phloem, and residual procambium. The cortex surrounding the stele begins with the endodermis, a cylinder of cells. The root epidermis surrounds the cortex.

Taproot systems have a main root with many branches. Cells of the pericycle in the primary root give rise to lateral roots, whose pericycles, in turn, produce additional roots. In diffuse or fibrous root systems, adventitious roots arise from stem regions above.

THE SHOOT SYSTEM

The shoot apical meristem provides cells for primary growth. Patches of meristem are left behind, forming leaf primordia and lateral bud primordia. Within the young shoot emerge future xylem, phloem, and simple pith parenchyma. Vascular tissue forms scattered bundles in monocots and distinct rings in dicots. In some dicots, reserves of procambium provide for secondary growth.

Secondary growth is restricted to perennial dicots (and gymnosperms). It begins in procambium which gives rise to vascular cambium, which then produces secondary xylem on its inner side and secondary phloem on its outer. The expanding cylinder of xylem crushes the pith and pushes residual vascular cambium outward. Expanding phloem crushes older, outer tissues.

Most of the older stem is nonliving xylem, or wood; living tissue, the bark, includes the tissues outside the vascular cambium. Differences in growth produce xylem of changing size, which is seen as an annual ring. Spokelike vascular rays, composed of parenchyma and collenchyma, provide lateral transport and relieve the stress of expansion.

Cork cambium divides continually, contributing cells that form the shoot periderm. Suberin-impregnated outermost cells form waterproof cork.

The Leaf

Leaves emerging from leaf primordia utilize protoderm, ground meristem, and procambium to form epidermis, photosynthetic parenchyma, and vascular tissue.

The dicot leaves consist of a stalklike petiole, a flattened blade, and a midrib. The latter contains a central vein. Veins are surrounded by bundle-sheath cells. Monocot leaves emerge from a broad sheath and have parallel veins.

A cross section through a dicot leaf reveals an upper epidermis (cutin-impregnated), photosynthetic tissues including a palisade parenchyma and spongy parenchyma, extensive air spaces, and a lower epidermis containing numerous stomata and leaf hairs. The stomata and leaf hairs inhibit water loss. The pore of each stoma is formed by paired guard cells.

Key Terms

annual • **492**
biennial • **492**
perennial • **492**
indeterminate growth • **492**
primary growth • **493**
secondary growth • **493**
vascular tissue • **495**
xylem • **495**
phloem • **495**
sap • **495**
dermal tissue • **497**
protoderm • **497**
epidermis • **497**
periderm • **497**
ground meristem • **497**
ground tissue • **497**
root apical meristem • **498**
root cap • **498**
root hair • **498**
stele • **498**
pericycle • **499**
endodermis • **499**
cortex • **499**
lateral root • **499**
taproot system • **500**
fibrous root system • **500**
adventitious root • **500**
pith • **502**
vascular cambium • **503**
secondary xylem • **503**
secondary phloem • **503**
growth ring • **503**
vascular ray • **503**
cork • **503**
cutin • **505**
stomata • **506**
leaf hair • **506**

Review Questions

1. List three plant categories according to life span. Which has indeterminate growth, and what provides for such "immortality"?
2. Distinguish between primary and secondary growth in terms of their effects on plant size. Are all plants capable of primary growth? Of secondary growth?
3. List the three kinds of primary tissue, and briefly state what they produce.
4. What is the general role of xylem? List the two principal types of xylem elements.
5. Name and describe the structures that permit water to move between tracheids and between vessel elements.
6. What is the general role of phloem? State a major difference between phloem sieve elements and vessels or tracheids.
7. List the two kinds of sieve elements. How do materials pass from one sieve element to the next?
8. What are the principal roles of the phloem parenchyma and companion cells?
9. List three primary functions of the plant root.
10. Prepare a simple, longitudinal outline drawing of a primary root and label the following: root tip, root cap, root apical meristem, epidermal cells, root hairs, elongation region, stele, and cortex.
11. Explain how a primary root continually moves through the soil. Essentially, what provides the push?
12. Describe the special feature of the endodermal cell walls. To what important function does this relate?
13. Describe the manner in which lateral branches form in the taproot system. How does repeated branching occur?
14. What are leaf primordia? Lateral bud primordia? From what tissue do they arise?
15. Compare the distribution of primary vascular tissues in dicots and monocots. In which does procambium persist?
16. What role does residual procambium play during secondary growth in the stem? To what specific tissues does it give rise?
17. Explain what happens when secondary xylem and secondary phloem increase in size. What must the vascular cambium provide if secondary growth is to continue over the years?
18. What can you determine by counting tree rings? Would this be true in tropical rain forests where rainfall occurred year-round?

19. List the tissues of a woody stem from the vascular cambium outward. Distinguish the term *bark* from the term *cork.*

20. Distinguish between the organization of dicot and monocot leaves. To which do the terms *midrib* and *petiole* apply?

21. Prepare a simple line drawing showing a cross section through a dicot leaf and label the following: the upper and lower epidermis, spongy parenchyma, palisade parenchyma, stomata, guard cells, and air spaces.

22. Explain how the arrangement of the spongy parenchyma provides for carbon dioxide uptake.

30

Mechanisms of Transport in Plants

The Movement of Water and Minerals
- Water Potential Revisited
- The Transpiration-Adhesion-Cohesion-Tension Hypothesis
- Water Movement in the Root
- Turgor and the Behavior of Stomata
- Mineral Uptake by Roots
- Mutualism and Mineral Uptake

Food Transport in the Phloem
- Flow from Source to Sink
- The Pressure Flow Hypothesis

A TINY PLANT LYING PRESSED AGAINST THE EARTH MAY NOT HAVE GREAT *problems in moving materials around within its body—nothing has far to go. But great trees and other large plants have had to develop ways to circulate water, food, and minerals throughout their mass. Some of the most fascinating questions in biology, in fact, have involved just how forest giants move water from their tiniest root hairs deep beneath the earth's surface to their loftiest leaves. As you will quickly see, all the answers are not in.*

The great trees of a forest are busier than you may think. Not only are they producing organic materials and drawing in water and minerals from the soil, but they are discharging that water into the air around them, and at a prodigious rate. Furthermore, in certain seasons, they do this day after day, dampening the forest air and lending authority to that glorious smell of the woodlands. The water is discharged mainly from the leaves, and those leaves may be hundreds of feet from the ground. Because the water they lose comes from the roots, deep in the earth (Figure 30.1), we may ask, How does the water, seeping into millions of tiny root hairs, reach those lofty leaves? The question is a good one.

And what about the other fluid that moves through such forest giants? In North America one often finds trees that have been girdled with a series of small holes, the work of a migratory little woodpecker called the sapsucker. The holes are shallow, but they penetrate the phloem and soon fill with sap that is intercepted on its way downward from the leaves. The sapsucker laps at the sugary fluid that seeps from the wounds. So, another question arises: What causes the sap to move along? The search for the answer involves another story from the saga of life.

THE MOVEMENT OF WATER AND MINERALS

In a very real sense, vascular plants have evolved around the behavior of water. That behavior is predictable to a degree because—even in living tissues—water obeys certain physical principles. Vascular plants, therefore, have evolved ways of taking advantage of those principles in developing ways to move water from the soil into the roots, and up through the plant body to the foliage. Keep in mind, then, that to understand how water moves in plants you have to keep in mind the physical principles involved in the behavior of water.

WATER POTENTIAL REVISITED

Earlier (Chapter 5), we said that *water potential* describes the tendency of water to move as a result of gravity, pressure, or solute concentration. These forces can account for such behavior as the swelling of cells due to osmotic pressure, the tendency of water to leave the swelling cells due to increasing turgor pressure, and the downward flow of water due to gravity.

Keep in mind that the net movement of water is from areas of higher water potential to areas of lower water potential. For example, water will leave tissues with a relatively low solute concentration (where the water potential is higher) and enter tissues where the solute concentration is high (and water potential is low). The water moves as a result of an osmotic gradient. This movement will continue until the swelling of the recipient tissue causes the internal pressure of those cells to force as much water out as is moving in. Remember that the transport of water within the plant body always occurs from regions of higher water potential to lower water potential. Water thus moves from cell to cell along a *water potential gradient.* Since we know that water moves from the roots to the leaves, we can infer that the roots have a higher water potential than the leaves. Why should a leaf cell have the lower water potential? Not much is used in photosynthesis, so where does the water go? It simply evaporates from the leaf. Let's see why.

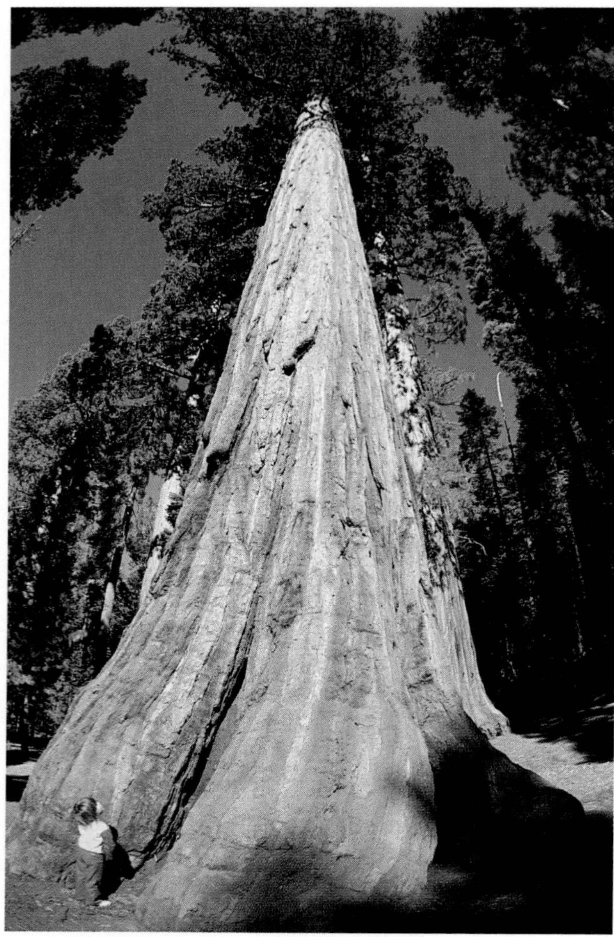

FIGURE 30.1 THE GIANT SEQUOIA.
This towering giant sequoia, *Sequoiadendron giganteum*, is one of the world's tallest trees at 83 m (272 ft). It is truly awe-inspiring to stand in one of the remaining groves of such giants. If you are aware of the problems of water transport, the marvel of it all increases.

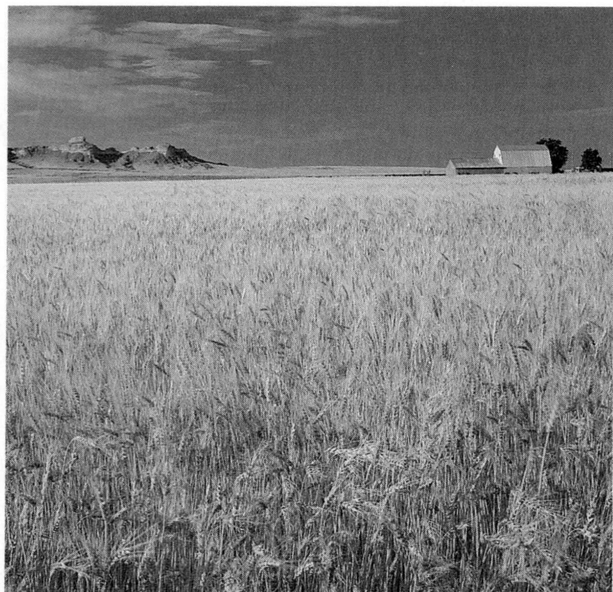

FIGURE 30.2 TRANSPIRATION.
Plants transpire far more water than they use in photosynthesis. The amounts can be staggering. During the growing season, each wheat plant in this field transpires about 100 liters (26.4 gal) of water.

TABLE 30.1 *TRANSPIRATION RATES*

Plant	Liters per Day
Cactus	0.02
Tomato	1
Sunflower	5
Ragweed	6
Apple tree	19
Coconut palm	75
Date palm	450

THE TRANSPIRATION-ADHESION-COHESION-TENSION HYPOTHESIS

The Problem of Water Loss. Plants lose water because the leaves leak—they have to. As you know, plants use carbon dioxide from the air in photosynthesis. It diffuses through the stomata into the leaf's moist air spaces as a gas, but it must be dissolved before it can move into cells. This means that plants must continually bathe their internal leaf cells in water. However, plants have not evolved a way to let carbon dioxide in without also letting water escape. Water escapes because of evaporation, and it is constantly replaced by water moving upward from the roots. The loss of water from the leaf openings (stomata) is known as **transpiration.**

Energy for Transpiration. The movement of water through a plant begins with transpiration in the leaf. The amount transpired daily is enormous (Figure 30.2 and Table 30.1), and all of it must be replaced if the plant is to remain active and healthy. This much movement will require the expenditure of large amounts of energy. Fortunately for plants, they do not have to provide this energy; the energy of evaporation is provided by the sun and the wind, and it is absolutely free. The sun's energy powers transpiration and subsequent water transport. But other mechanisms are also at work.

Transpiration Pull. As water evaporates from the moist air spaces in the leaf parenchyma, it is replaced by water from surrounding cells (Figure 30.3). The

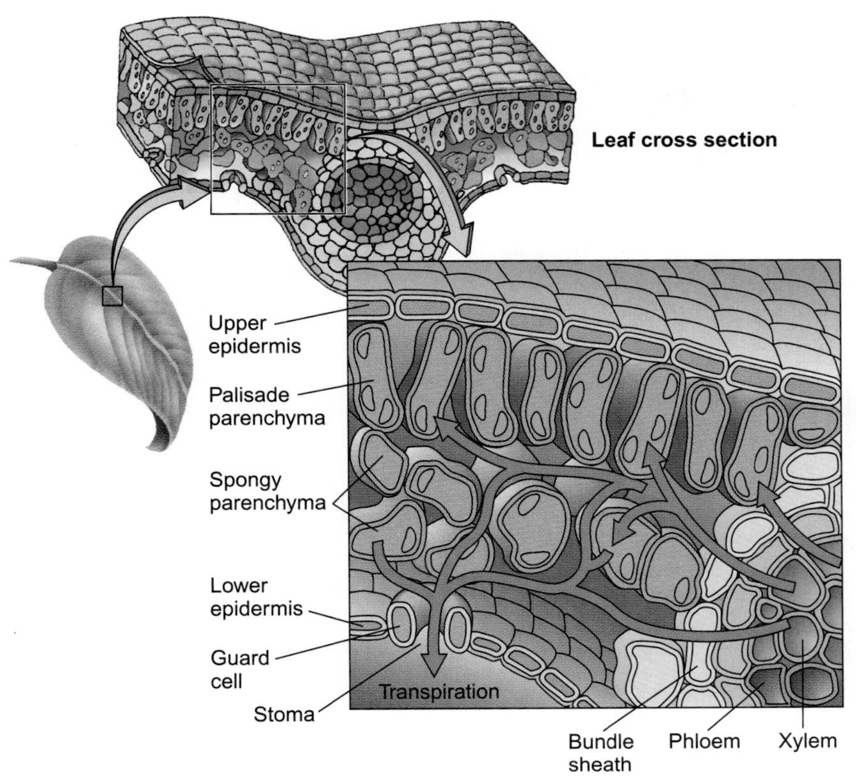

FIGURE 30.3 WATER MOVEMENT THROUGH THE LEAF. Movement of water (indicated by the blue arrows) occurs as a result of a gradient between the xylem and the air surrounding the leaf. The gradient is established by water evaporating from the leaf, which lowers the water potential in the air spaces within. This decrease in water potential starts a chain of events resulting in a constant flow from the xylem to the outside.

mechanism of water replacement isn't entirely clear, but the loss of water in cells adjacent to the air spaces would presumably decrease water potential in those cells. This would begin a water potential gradient that, as water begins to move, would spread across the adjacent cells. The leaf parenchyma cells react to changes in water potential more or less as a unit. After all, their cell walls are very porous, and their cytoplasm is connected by numerous plasmodesmata. The cell walls and plasmodesmata permit water to move rapidly from cell to cell along a water potential gradient to the site of transpiration.

Some authorities maintain that adhesion is a principal force in the movement of water through the leaf. **Adhesion** is the tendency for water molecules to adhere or cling to certain other substances, in this case cellulose, a major component of cell walls. These researchers suggest that water evaporating into the extensive air spaces of a leaf is replaced as water molecules literally creep by adhesion along cell wall boundaries. Since the combined cell surface area is so large, adhesion becomes a major pulling force. The pulling force (whether osmotic or adhesive or both) is called **transpiration pull.**

Of course, for water movement to continue in the leaf, there must be a constant supply, and this supply comes from the earth through the xylem (tracheids and vessels) of the leaf vascular system.

Cohesion and Tension in the Xylem. To understand what happens next, we must visualize the xylem as containing an unbroken *column* of water, extending from leaf to root. As water escapes through the stomata, its responding movement through the leaf cells exerts a pull, or tension, on the column. Any pull at the top of the column raises the entire column, permitting water to enter the leaf. The pull is substantial enough to raise a column of water to the foliage of the tallest trees. It is interesting that water in an ordinary hose or pipe cannot be lifted more than 10 m by any amount of suction, since suction is dependent on the force of atmospheric pressure *pushing* a heavy column of water upward. Yet transpiration can *pull* water 10 times higher than that. But let's look more closely at the notion of water being pulled through the vascular system, since it raises an important question: Why doesn't the column of water simply break up? There are at least two reasons.

Transpiration pull depends on both the extremely small diameter of the xylem elements and the cohesive quality of water. Water molecules, you may recall (see Chapter 2), cling together because of hydrogen bonding—the attraction between positively and negatively charged ends of adjacent molecules. This tendency for molecules to cling together is called **cohesion.** Since the forces of cohesion are inversely proportional to the diameter of the vessel,

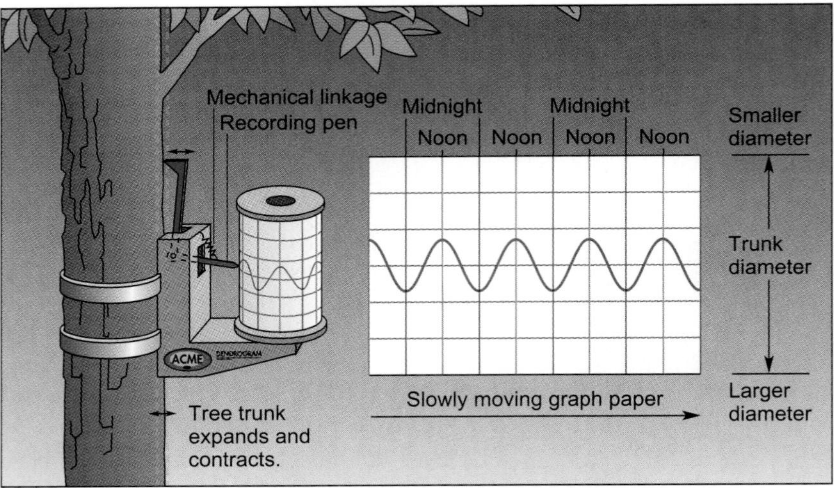

FIGURE 30.4 **TRUNK SIZE AND TRANSPIRATION RATE.** Physiologist D. T. MacDougal used a drum recording apparatus called a dendrograph to measure transpiration-related changes in tree trunk diameter. The tension on xylem during intense transpiration actually shrinks the trunk, as indicated by noon-time peaks in the graph.

the cohesive forces in water would be ineffective in a large tube, but because xylem tubes are microscopic, the water molecules within cling together tenaciously. In fact, the tensile strength of water in such a thin column approaches that of surgical steel wire of the same diameter!

The tension exerted on water in the xylem through transpiration pull is not hypothetical. It can be demonstrated through the use of a sensitive instrument known as a *dendrograph* (Figure 30.4). Precise measurements of tree trunks over 24-hour periods have revealed that trunk diameters decrease significantly at midday, when transpiration is most intense. This clearly indicates that the tension within the trunk is both real and enormous.

WATER MOVEMENT IN THE ROOT

With water transpiring from the leaf and being pulled out of the xylem for replacement, the root must be able to meet the demands by bringing in new water from the soil. As long as the water potential in the surrounding soil water is greater than that in the root cells, there will be an inward gradient.

Water moves through the root along two pathways. Much of it occurs along the vast network of porous cell walls while some occurs through the cytoplasm of the root cells. Cytoplasmic passage is aided by numerous plasmodesmata. However, the easy passage ends at the endodermis. As we see in Figure 30.5, *all* water entering the stele is directed by the Casparian strip across the endodermal membranes and through the cytoplasm. Plant physiologists believe that the endodermal cells can influence water uptake by varying their own concentration of mineral ions, thereby altering the osmotic pressure of their cell fluids. So the osmotic potential of the endodermal cells can help move water through roots. However, the osmotic mechanism can be overridden by even more powerful forces. During intense transpiration, the enormous pulling forces transferred to the root probably nullify the effects of osmosis. In fact, some plant physiologists believe that during intense transpiration most if not all water taken up by the root is through simple *bulk flow,* movement in response to a pressure gradient.

Plants can literally run out of water at such times. When soil water supplies dwindle, the leaves and softer shoots simply wilt, actually shrinking, as the central vacuoles empty of water. Unless the water shortage continues, though, they will recover. One way plants assure a continuous water supply is through rapid root growth, allowing new sources to be tapped. Some roots, in fact, grow with remarkable speed. For instance, prairie grass roots grow at a rate of 12 to 14 mm ($\frac{1}{2}$ in.) per day; those of the corn plant tear along at 50 to 60 mm (2 in.) per day.

In smaller plants, osmotic forces, with perhaps other influences, can produce *root pressure*—an upward push of water from the roots that is responsible

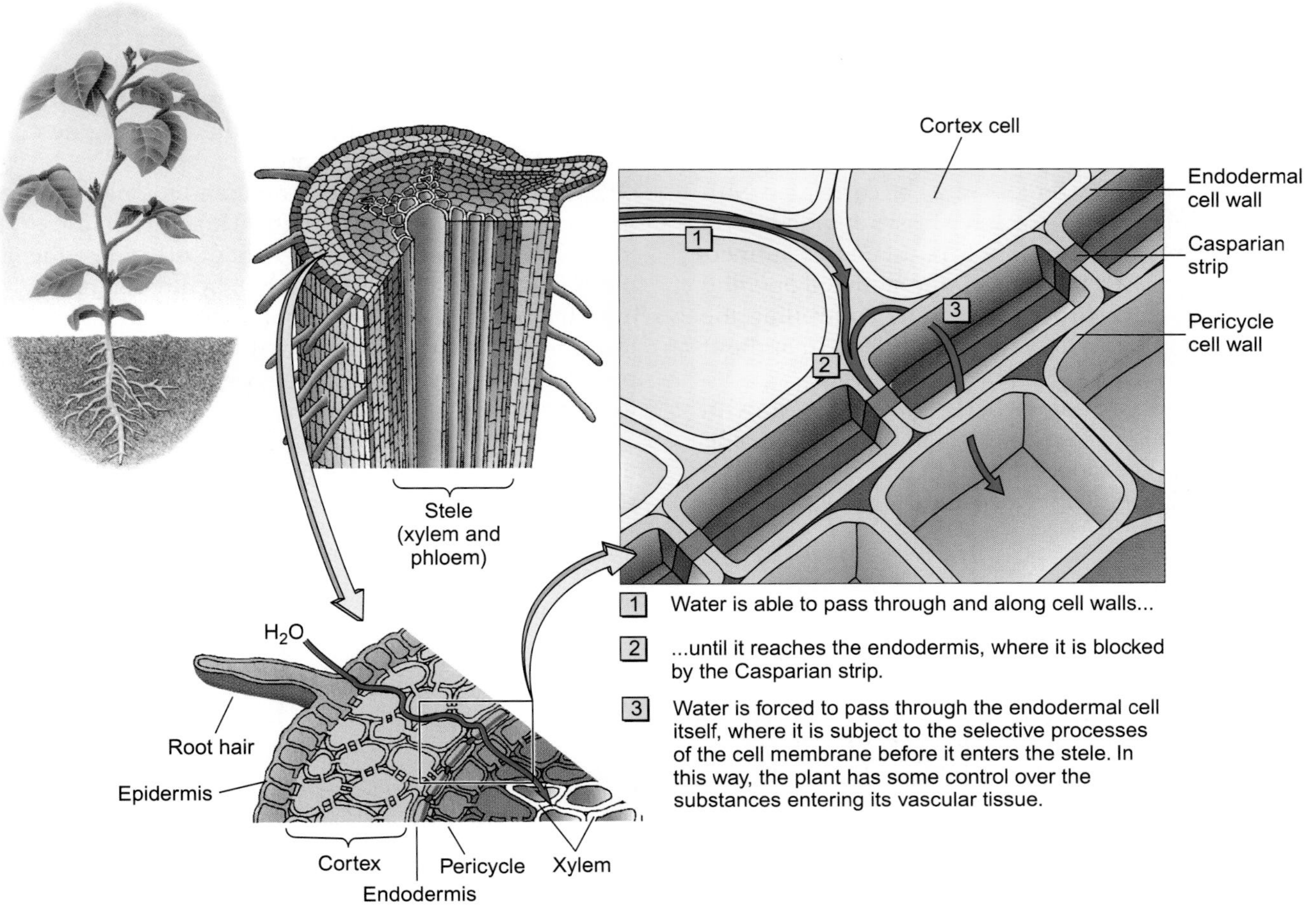

Cortex cell
Endodermal cell wall
Casparian strip
Pericycle cell wall
Stele (xylem and phloem)
H_2O
Root hair
Epidermis
Cortex
Endodermis
Pericycle
Xylem

1 Water is able to pass through and along cell walls...

2 ...until it reaches the endodermis, where it is blocked by the Casparian strip.

3 Water is forced to pass through the endodermal cell itself, where it is subject to the selective processes of the cell membrane before it enters the stele. In this way, the plant has some control over the substances entering its vascular tissue.

FIGURE 30.5 WATER MOVEMENT THROUGH THE ROOT.
Most water moving through the root cortex follows the cell wall route (**1**), but some moves through the cytoplasm. On reaching the endodermis (**2**), the Casparian strip directs all of the water across the endodermal membranes and into the cytoplasm (**3**) before it reaches the stele.

for **guttation,** the exuding of water from small openings at the tips of veins. For example, plants may be covered with water exuded in this way early in the morning before the sun dries them. These droplets are often mistaken for dew (Figure 30.6).

Using TACT. The combined focus of *transpiration, adhesion, cohesion,* and *tension* can be thought of as the TACT forces (Figure 30.7). To sum up the theory surrounding these forces, the energy needed to power water transport is provided by the sun and expressed in transpiration. As transpiration occurs at the leaf, the loss is translated into a water potential gradient there. Water moving along the gradient exerts a substantial pull or tension on water in the xylem elements. The adhesive and cohesive qualities of water keep the moving column intact, but this water must be continually replaced by uptake in the root.

FIGURE 30.6 GUTTATION.
The pressure developed by the root and stem water transport system can actually force water out the tips of veins in the leaf.

Turgor and the Behavior of Stomata

We've seen that water loss is actually an essential factor in the pulling of water upward through the xylem. However, water loss must be correlated with the water supply and excessive losses avoided if soil water is limited. Water loss is regulated by numerous paired *guard cells* that alter the stomatal opening by swelling and shrinking, thereby controlling the movement of air, with its water content, in and out of the plant.

Guard cells are somewhat sausage-shaped, with thin walls along their outer margins becoming thicker along the inner margins that line the stomatal opening (Figure 30.8). The guard cells are able to open and close the stomata because of the arrangement of cellulose fibers in the cell wall. The fibers are arranged like barrel hoops. When water enters the cells, they can't expand much, but they can lengthen, thereby forcing the stomata open.

Stomata are light-sensitive and tend to open during the day and to close at night. Thus carbon dioxide is allowed to pass into the plant mostly when the plant is actively involved in photosynthesis. However, the cells of a plant wilting for lack of water will lose turgor, and the stomata will close, conserving precious water even at the expense of being unable to carry on photosynthesis.

The precise mechanism of guard cell control remains unknown, but several hypotheses have been proposed. The most favored explanation at present

Transpiration in the leaf lowers the water potential gradient, causing water columns to move upward, out of xylem into leaf spaces.
Pull from water molecules leaving leaf surface and adhesion and cohesion of water in xylem create tension. This tension lifts water all the way from roots.
Water potential in soil is higher than in root, so water enters root.
Soil water
Soil water
Direction of water movement
Increasing water potential

Figure 30.7 Using TACT. Water potential gradient and water-moving forces in the plant can be summarized as Transpiration, Adhesion, Cohesion, and Tension.

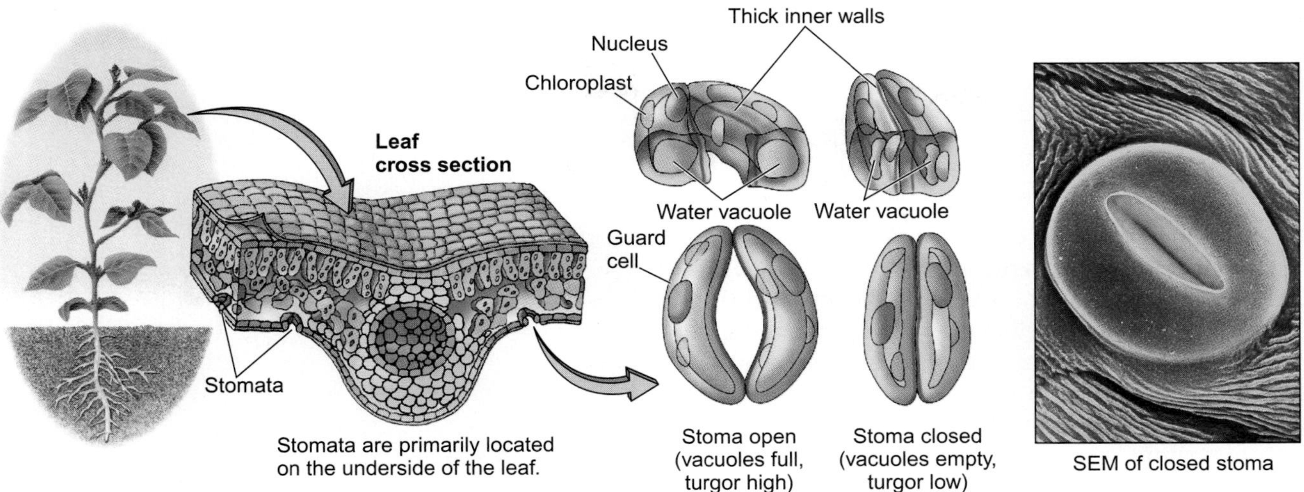

FIGURE 30.8 **STOMATA AND GUARD CELLS.**
The scanning electron microscope shows the slitlike stoma (closed). Note the paired guard cells with thick inner cell walls, and the larger, surrounding epidermal cells. When turgor is low in the guard cells, their thickened inner walls straighten, closing the stoma. Increased turgor, however, causes the guard cells to swell, which bends the thickened inner walls, opening the stoma.

involves a light receptor and the rapid active transport of selected ions. Under carefully controlled experimental conditions, the guard cells of some plants are shown to respond to light in the shorter, or blue, wavelength by pumping potassium ions into the cytoplasm. The ions are in solution, and their increased concentration brings about an inward movement of water by osmosis. Thus the guard cells elongate and the stomata open.

The potassium mechanism can be overridden under special circumstances, however. For instance, in times of severe water stress—as when soil water is depleted or evaporation from the leaf is too rapid—a plant hormone known as *abscisic acid* accumulates in the leaf. Since abscisic acid causes guard cells to lose their potassium ions, the solute concentration diminishes and water moves out. With the loss of turgor the stomata close.

In some plants the stomata close in the early afternoon, preventing excessive water loss in the heat of the day. In the desert, certain plants may store water in thickened leaves and some of these, notably the so-called CAM (crassulacean acid metabolism) plants, have evolved biochemical answers to the problem (Figure 30.9). Their stomata open only at night, admitting carbon dioxide that is stored in organic compounds until daylight. Then the stomata close, the biochemical reactions are reversed, and the carbon dioxide is released for photosynthesis.

MINERAL UPTAKE BY ROOTS

For most plants, the only sources of mineral nutrients are the ions dissolved in soil water. Since the plant has to take a great deal of water from the soil to make up for transpiration losses, you might expect that it would simply passively extract its needed minerals from the water. However, plants must actively pump some mineral nutrients from the soil, so ATP energy is expended. Once inside the

FIGURE 30.9 **DESERT ADAPTATIONS.**
Plants that live in the desert have developed certain structural and biochemical adaptations that allow them to conserve water.

root, minerals move from cell to cell by way of plasmodesmata. A second expenditure of ATP moves the ions across the endodermis where they readily enter the water flow in the xylem.

Mutualism and Mineral Uptake

The roots of most vascular plants form fungal associations called **mycorrhizae** (Figure 30.10). The fungal *mycelium* is vast, extending far out into the soil. Some of these tubelike filaments penetrate the root itself. Once there, the fungus can take advantage of the plant's food resources. But the association is mutualistic: Both parties benefit. The extensive mycelium concentrates mineral ions (Figure 30.11), particularly phosphates, in its mass. Thus the plant has ready access to certain scarce minerals.

Bacteria provide another example of mutualism. In addition to the more commonplace decomposing activities that help in the recycling of mineral nutrients, some bacteria are nitrogen fixers. They convert atmospheric nitrogen (N_2) into compounds such as nitrates and ammonium ions (see Chapter 46). In some instances, nitrogen-fixing bacteria actually invade the root, which responds by forming a confining nodule (cystlike growth) around the bacterial colony. As in the mycorrhizae, the bacteria obtain food and water and the plant gets the usable nitrogen. Nodules are common in the roots of such legumes as alfalfa, beans, peas, and clover.

Figure 30.10 The Importance of Fungal Associations.
The Calypso orchid, *Calypso bulbosa*, grows in mutualistic partnership with mycorrhizae fungi associated with its roots.

(a)

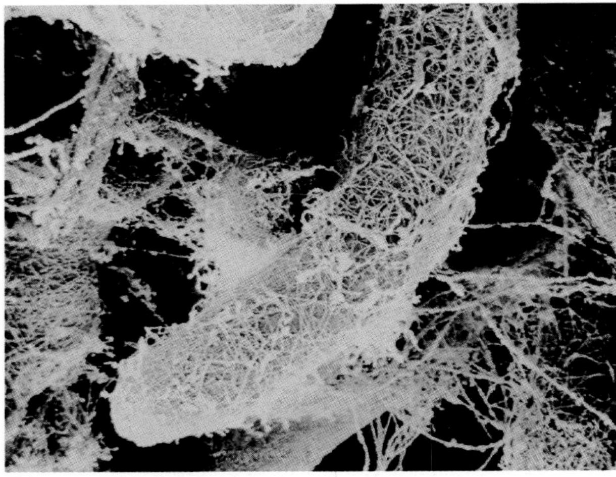

(b)

(c)

Figure 30.11 The Mycorrhizal Mycelium.
(a) Ectomycorrhizae of pine in three varieties of growth forms; **(b)** ectomycorrhizae of poplar (×400); **(c)** ectomycorrhiza of *Betula pendula* (×230).

Food Transport in the Phloem

Now that we have considered the movement of water and minerals through the plant, let's see how sugars and other manufactured products move through the phloem.

We learned earlier that transpiration creates a substantial *pull* on water columns in the xylem. But things are different in the phloem. The *phloem sap* is subjected to a *pushing* force in the form of high **hydrostatic pressure.** Hydrostatic pressure is the pressure that is exerted in all directions by a fluid at rest. (It might help to consider it as being similar to blood pressure in our vessels.) Aphids—insects that feed on plants—take advantage of the hydrostatic pressure of the sap in phloem by using their long, hollow mouthparts to drill tiny holes in individual sieve tube members. They just let the nutrient-laden sap flow into their bodies as if from an artesian well. In fact, researchers use aphids to obtain pure samples of phloem sap for analysis (Figure 30.12).

Sap is actually a rather thick fluid, especially when the plant is actively metabolizing and photosynthesizing. Nonetheless, it moves rapidly through the phloem, up to a meter per hour. Sucrose makes up about 90% of the solutes in sap, but it also carries other sugars, mineral nutrients, hormones, and amino acids.

Flow from Source to Sink

Nutrients originally produced in the leaf are widely distributed in the plant, moving to the actively growing regions, into the root, into maturing fruit, and to storage regions. This transfer, or *translocation,* as it is known, is multidirectional; it can occur in different directions in the same tissue at different times. We will refer to areas where nutrients are to be used or stored as *sinks,* and to areas where they originate as *sources.*

The flow of sap is *from source to sink:* Nutrients are actively transported into the sieve tubes at the source, and are actively transported out of the sieve tubes at the sink. The active transport at both sites requires an expenditure of ATP.

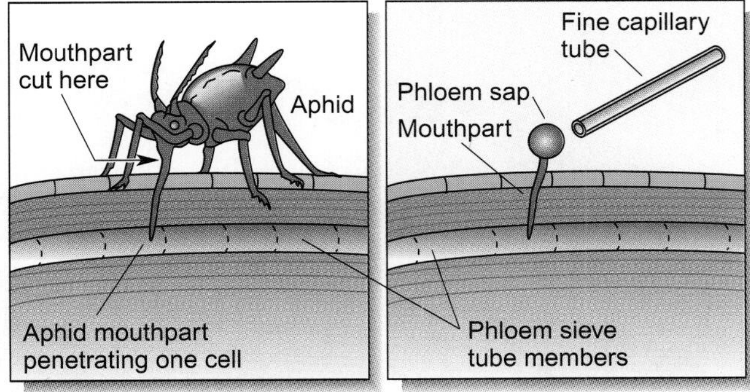

Figure 30.12 Aphids and Pure Sap. An ingenious means of obtaining pure sap involves the use of aphids. After the aphid has pierced the phloem, it is anesthetized and its mouthparts are severed so that they remain in place.

The Pressure Flow Hypothesis

Active transport loads nutrients into the sieve tubes at the source and unloads them at the sink, but what accounts for the flow between the two? There have been many hypotheses to account for this mechanism, but the one currently favored, the **pressure flow hypothesis,** is an older one, first proposed in 1927. It is based on the differences in water potential and osmotic pressure between the xylem and the phloem.

The idea is this: The active transport of sugars from the source into the phloem greatly decreases its water potential in comparison to the high water potential in the nearby xylem elements. The result is that water leaves the xylem and moves into the phloem sieve tubes, raising the hydrostatic pressure there. The increased hydrostatic pressure then forces the sap to move as a stream through the phloem tubes.

Meanwhile, at the various sinks, solutes are being unloaded. This results in an increase in the water potential within the sieve tube compared to the surrounding cells. Because of this, water leaves the sieve tube, entering the sink tissues or moving back into the xylem. Within the phloem, the sap follows the gradient in water potential that is created by active transport at both ends of the pipeline—that is, at the sink and at the source (Figure 30.13).

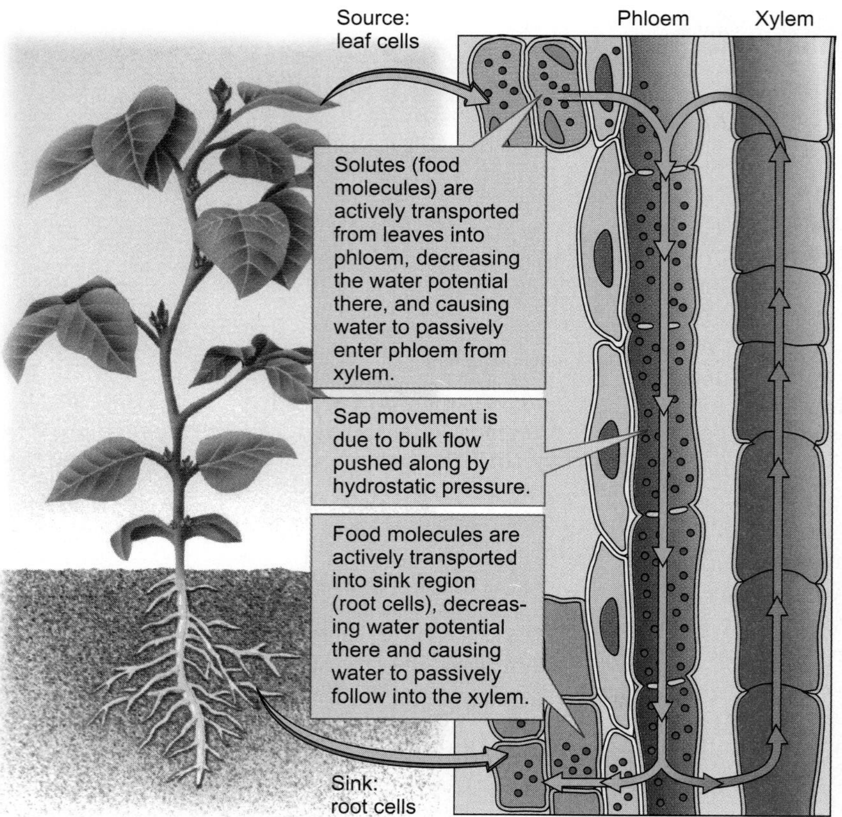

Source:
leaf cells
Phloem
Xylem
Solutes (food molecules) are actively transported from leaves into phloem, decreasing the water potential there, and causing water to passively enter phloem from xylem.
Sap movement is due to bulk flow pushed along by hydrostatic pressure.
Food molecules are actively transported into sink region (root cells), decreasing water potential there and causing water to passively follow into the xylem.
Sink:
root cells

FIGURE 30.13 THE PRESSURE FLOW HYPOTHESIS. According to the pressure flow hypothesis, phloem transport begins at the source with the active transport of sugars into the phloem elements. Water from nearby xylem follows passively, creating hydrostatic pressure, which produces the sap flow along the elements. At the sink, active transport removes the sugars, and water leaves passively, reentering the xylem.

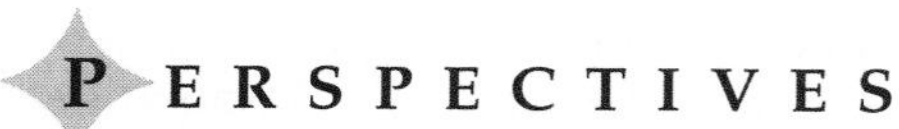

PERSPECTIVES

As you now see, the simple tree, standing among other simple trees, is not so simple after all. It is quietly and efficiently engaged in processes that dedicated scientists are only now beginning to understand. One of the most puzzling and intriguing of these processes is the movement of fluids through plants. These fluids move not only water, food, and minerals, but also the fascinating molecules that we will see next, the hormones.

Summary

The Movement of Water and Minerals

Plant evolution has been greatly influenced by the characteristics of water and the physical principles that determine its behavior.

Water potential refers to the potential movement of water due to gravity, pressure, or osmotic conditions. Differing solute concentrations in cells create water potential gradients, with water moving from regions of high water potential to regions of low.

In the process of transpiration, water evaporates through the stomata as carbon dioxide is admitted for photosynthesis. This loss starts the movement of water up through a plant.

Evaporation from the stomata sets up a pulling force—transpiration pull. Some of the force may be due to the water potential gradient, but much of it may be a product of adhesion, the attraction of water to cellulose cell walls. The pull is reflected on the xylem columns.

Transpiration creates a tension in the xylem water columns as they are raised up by the pulling forces from above. The water columns resist breaking because of cohesion—the attraction between water molecules. In the minute xylem columns the cohesive force or tension is enormous. During intense transpiration, tension actually causes a tree trunk to decrease its diameter, an observation made through dendrograph measurements.

Water moves from the soil into the root because of a water potential gradient there. Most movement occurs along cell walls in the cortex, but some is through the cytoplasm. At the endodermis, water must pass through the cell membrane and cytoplasm, where the uptake of both water and solutes may be controlled. During intense transpiration, bulk flow probably accounts for most movement. If the internal water pressure is lost, wilting occurs.

Stomata open when turgor pressure is great in the paired guard cells, and close when it is low. This movement usually corresponds to photosynthetic activity. Theoretically, the turgor-changing mechanism is a photoreceptor that reacts to light by causing potassium ions to be pumped into the cell. Water follows, and the guard cells swell, opening the stomata. In water stress, abscisic acid, a hormone, is believed to override the photoreceptor, turgor is lost, and stomata close. Desert CAM plants open their stomata at night and close them in daylight. At night they incorporate incoming carbon dioxide into organic compounds, releasing it during daylight for photosynthesis.

Minerals in soil water must be taken into the root cells by active transport. The principal route of transport is the cytoplasm, facilitated by plasmodesmata.

In mutualistic fungal-root associations called mycorrhizae, the fungus provides minerals and uses some of the plant's food supply. Mutualistic nitrogen-fixing bacteria live in root nodules, exchanging nitrogen compounds for plant food in the form of organic compounds.

Food Transport in the Phloem

Sap (primarily sucrose and water) is moved by hydrostatic pressure. It builds as water enters the phloem after sugars and other solutes have been pumped there through active transport.

In translocation, materials are moved from sources—tissues where foods are produced or stored—to sinks—tissues where they are used or stored. Food moves into and out of both regions through active transport.

In the pressure flow hypothesis, sugars actively transported into the phloem from the source cause the inward movement of water and a rise in hydrostatic pressure. The sap flow then occurs. When the sink is reached, active transport removes the sugars, and water subsequently follows.

Key Terms

transpiration • **512**
adhesion • **513**
transpiration pull • **513**
cohesion • **513**
guttation • **515**
mycorrhiza • **518**
hydrostatic pressure • **519**
pressure flow hypothesis • **519**

Review Questions

1. List three factors that can influence water potential.
2. Explain what a water potential gradient is, and how cells can create such gradients.
3. Why is it necessary for plants to permit a water loss to occur? What is this process called?
4. If water is pulled through a plant, what is the source of the energy required to exert the pull? Is ATP involved?
5. Describe the stepwise changes in water potential in the leaf, as water is lost through the stomata.

6. What is adhesion? How might adhesive forces add to transpiration pull? In what part of the leaf are the pulling forces finally reflected?
7. Compare the pulling force of transpiration pull with sucking forces that might be created by a vacuum. Can the latter lift water to the tops of tall trees?
8. Account for the great tension possible in xylem water columns. What do cohesion and xylem diameter have to do with this? Is there any experimental evidence of tension? Explain.
9. Describe the water potential gradient in the soil and root. Describe the two routes followed by water in the root cortex. Which is more important?
10. How does the Casparian strip affect water passage into the stele? How might this be important?
11. At what times does bulk flow overwhelm osmosis?
12. Describe the structure of guard cells in the stoma.
13. What force brings about opening and closing of the stomatal pore? What is the significance of the thick inner walls of guard cells?
14. Explain the relationship between light and guard cell shape. In what way does this correspond to photosynthesis?
15. Explain the peculiar adaptations of the CAM plants.
16. Describe the principal differences between the uptake and movement of mineral ions across the root cortex and the uptake and movement of water.
17. Describe the mycorrhizal association. What makes it mutualistic?
18. Generally compare the forces involved in water and sap transport. Which, if any, requires ATP energy?
19. Distinguish between the source of materials and their sink. Name the process of nutrient movement.

31

Response Mechanisms in Plants

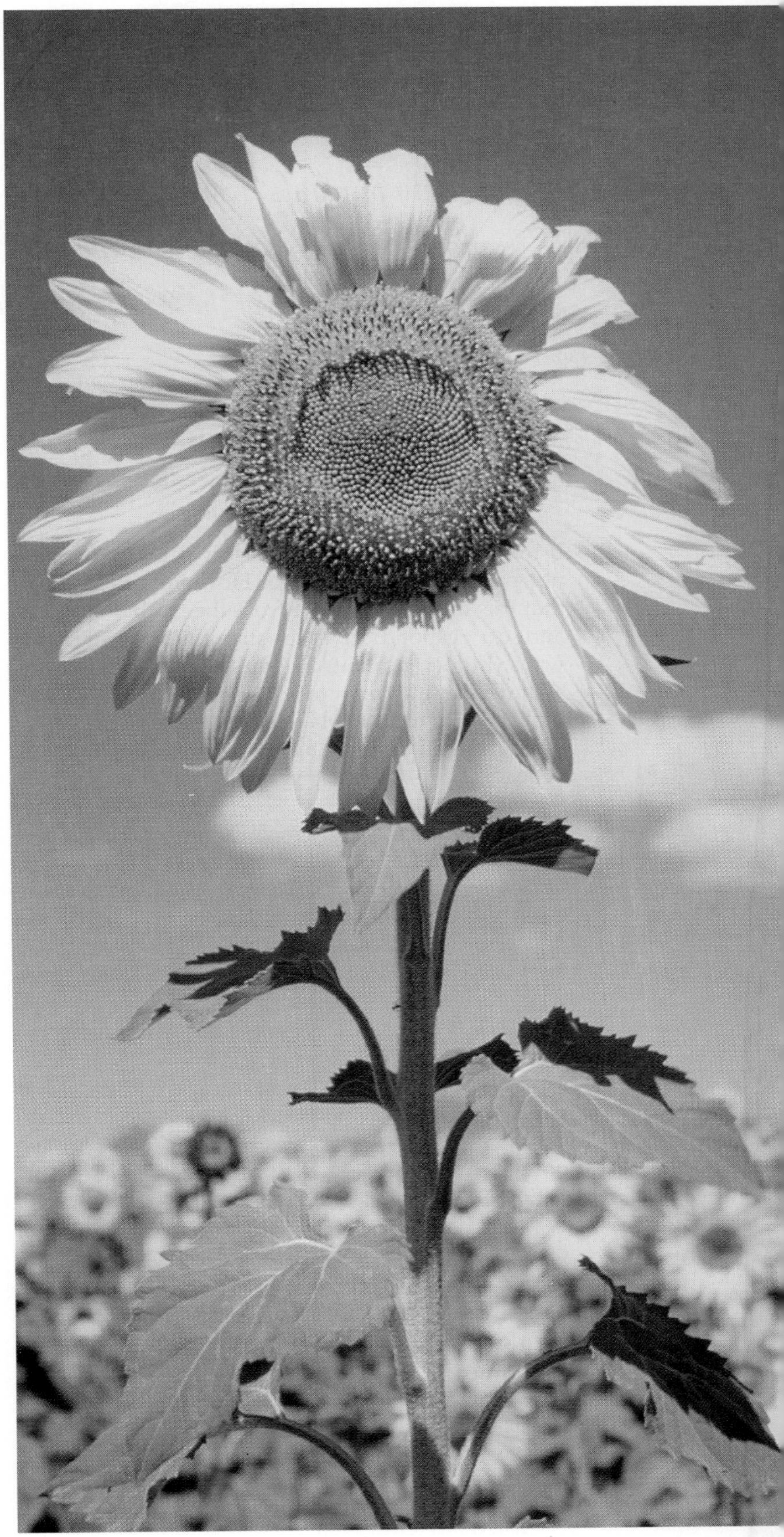

SOME PEOPLE SUGGEST THAT PLANTS ARE SHORT ON PERSONALITY. AFTER *all, they apparently respond to very few stimuli, and when they do respond, they do it excruciatingly slowly. Other people, though, are fascinated by the variety of plants' responses, and the search for the mechanisms involved has proven, in fact, to be one of the most intriguing searches in science. Here, we will look at some of those responses and the search for their mechanisms.*

It may seem to the casual observer that plants are not particularly responsive. However, we will see that this impression is far from the truth. Plants can and do respond to their surroundings in many ways. In fact, a plant, like any other organism, responds continuously to a variety of cues—signals both from within itself and from its environment. The impression may arise because a plant's responses are usually gradual and based on time-consuming chemical changes.

Regulation through chemical mechanisms frequently involves certain molecules known as *hormones.* Essentially, hormones are chemical messengers that are manufactured in one part of an organism and are transported to another part, where they can cause some sort of change. In some cases, the messengers are not transported very far and may even act in the same general area where they are manufactured.

Hormones are effective in minute amounts, and they are almost always short-lived. They do not accumulate in cells and tissues but are rapidly broken down. This turns out to be important in the coordination of hormonal activity.

Plant hormones are important in a number of ways. They are involved in growth, cell division, seed germination, flowering, tissue differentiation, dormancy, and other vital activities. You may recall from our earlier discussions, for example, that primary growth itself occurs through cell division and cell elongation, both of which require the presence of certain hormones. Let's now consider a few specific plant hormones and see how they function to regulate and coordinate the activities of the plant.

PLANT HORMONES AND THEIR ACTIONS

AUXINS

Auxins are a class of hormones that largely function in growth regulation in plants. They are very small molecules with far-reaching effects. The principal naturally occurring auxin is indoleacetic acid (IAA), and this is the one we will consider here, simply referring to it as "auxin." Auxin was chemically identified about 50 years after its presence was suspected for some time (Figure 31.1).

Auxin is best known as a growth hormone. It affects nearly every aspect of growth, from root tip to foliage. It doesn't cause plants to grow by increasing the rate of cell division, but by promoting cell enlargement in stems, particularly by elongation of cells behind the apical meristem (see Chapter 29). Although we still don't completely understand just how auxin works, it seems that, among other things, it promotes cell elongation by going to work on the soft primary cell walls of newly divided cells before they have reached their final, hardened state (that is, before they are impregnated with pectin and other wall hardeners). Auxin promotes the loosening of closely bound filaments of cellulose near the ends of the young cell walls, thus permitting turgor pressure to expand the cells in those areas. The increased turgor is due to enlarged central vacuoles, rather than to added cytoplasm.

Other Roles of Auxin. Working alone or in concert with other hormones, auxin also influences cell

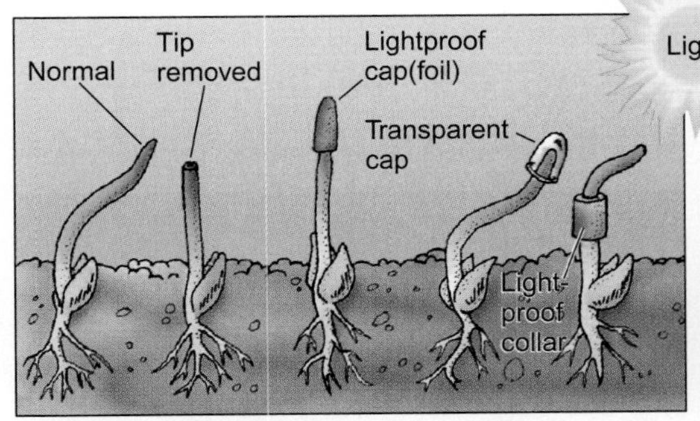

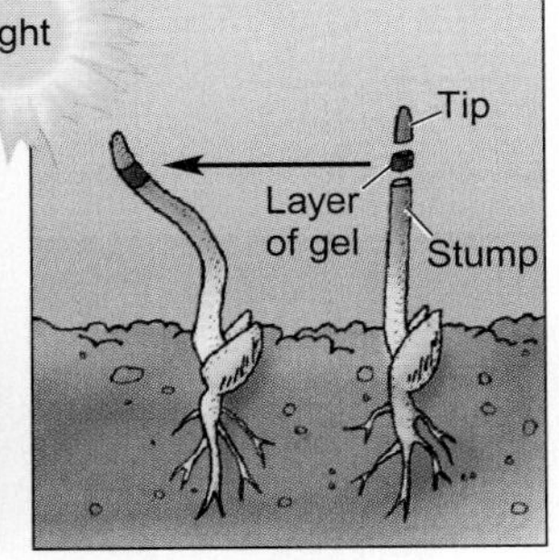

(a) Darwin's Experiment

Charles Darwin and his son Francis determined that a light-dependent agent in the tip of canary grass seedlings caused them to bend toward a light source.

(b) Boysen-Jensen's Experiment

P. Boysen-Jensen determined that the light-activated agent could diffuse through gelatin. This indicated that the agent was molecular.

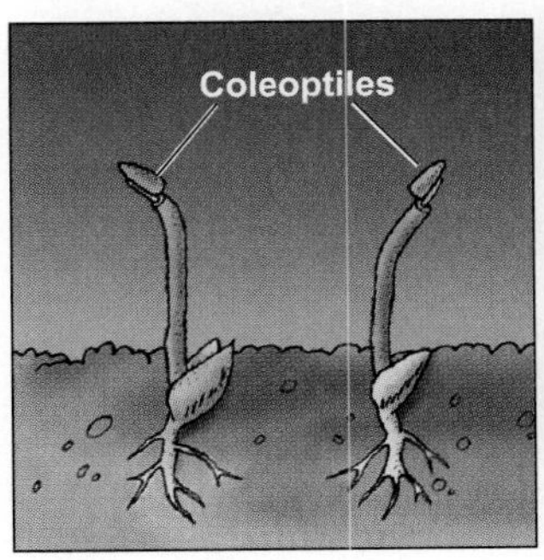

(c) Paal's Experiment

A. Paal carried out his key experiment in the dark. He used severed coleoptile tips to direct the bending of oat seedlings, showing that the coleoptile was the source of the active agent.

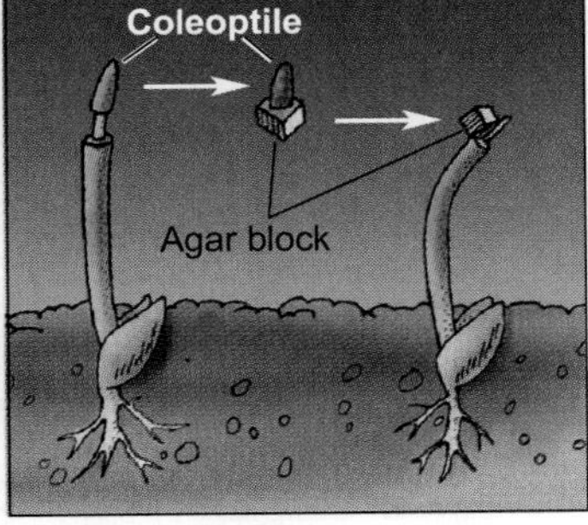

(d) Went's Experiment

Fritz Went, also working in the dark, removed coleoptiles and placed them on agar blocks for varying times. He then used the blocks to produce curvature, and related the amount of curvature to the amount of auxin present.

FIGURE 31.1 **DISCOVERING THE ROLE OF AUXIN.** **(a)** Charles Darwin and his son Francis determined that a light-dependent agent in the tip of canary grass seedlings caused them to bend toward a light source. **(b)** P. Boysen-Jensen determined that the light-activated agent could diffuse through gelatin. This indicated that the agent was molecular. **(c)** A. Paal carried out his key experiment in the dark. He used severed coleoptile tips to direct the bending of oat seedlings, showing that the coleoptile was the source of the active agent. **(d)** Fritz Went removed coleoptiles and placed them on agar blocks for varying times. He then used the blocks to produce curvature. By experimenting with various concentrations of auxin, he related the amount of curvature to the amount of auxin present.

differentiation, growth in the vascular cambium, fruit and leaf fall, and even the shape plants take as they grow. Two of these, leaf fall and tree shape, have been of great interest to naturalists.

In deciduous trees of the temperate zone, leaf fall (abscission) is a seasonal event accompanied by hormonal changes in the plant. Young leaves apparently produce hormones that inhibit the formation of specialized layers of cells at the base of the leaf stem. As the leaf ages, these hormones diminish, and two tissue layers form in the **abscission zone** (Figure 31.2), where the leaf separation will occur. The layer closer to the stem becomes heavily impregnated with wax that will protect the stem when the leaf falls away. In the other layer, the cells swell and become gelatinous. As temperatures drop and the days shorten, enzymes break down the middle lamella of cells in the area of separation. Now, only some strands of xylem hold the leaf to the stem, and the rustling wind or pattering raindrops are enough to dislodge the leaf, causing it to fall quietly to the earth below.

Some researchers believe that tree shape is often due to auxin's ability to both encourage and inhibit growth, working through a principle called **apical dominance.** You may have noticed that the uppermost growing tips of plants usually grow faster than those below. The reason may be because auxin, produced in the growing tip, stimulates growth there, but as it moves downward, it suppresses growth in the stems below. The result is the familiar triangular shape of conifers such as the pine and spruce

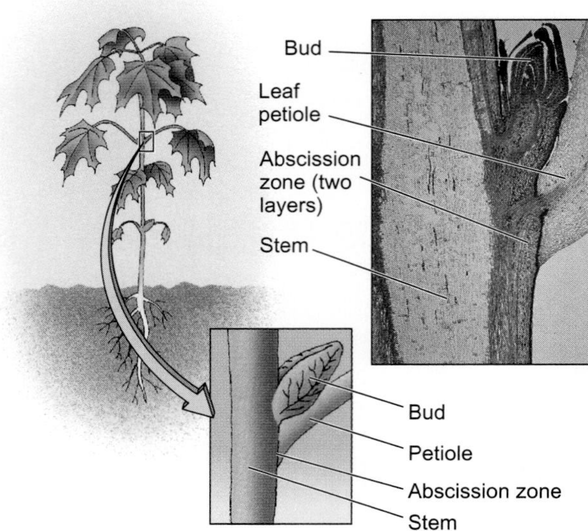

FIGURE 31.2 **THE ABSCISSION ZONE.** Before leaf fall, cells in the abscission layer die and harden. The formation of a scar prevents excessive water and sap loss and prevents infection.

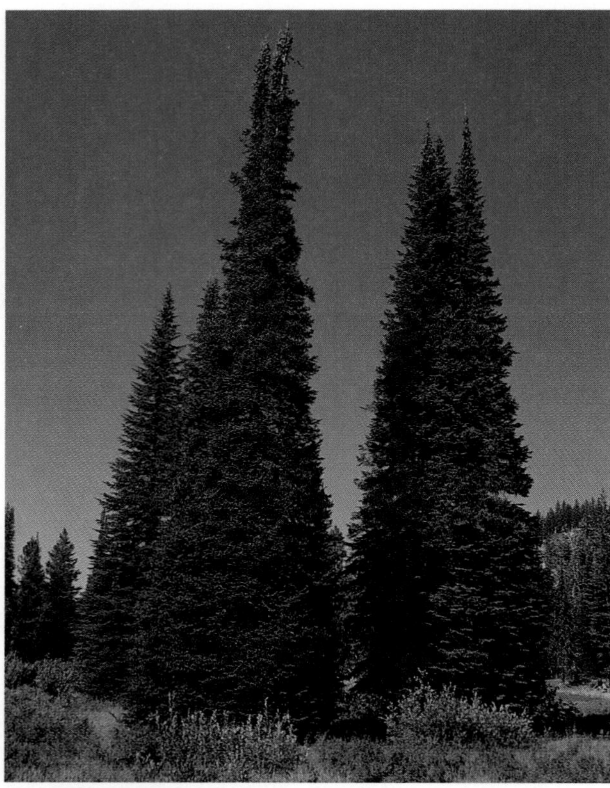

FIGURE 31.3 **APICAL DOMINANCE.** The triangular shape of many conifers is due to the growth-suppression effects of auxin as it makes its way down the trunk after being produced in the growing tip.

(Figure 31.3). Should the growing tip be lost, one of the lateral branches will take over, soon restoring the graceful triangular form.

GIBBERELLINS

The **gibberellins** are a class of plant hormones that promote stem elongation and function in plant reproduction. They received their name from the fungus in which they were first found, *Gibberella fujikuroi.* This fungus, which once threatened rice harvests in Japan, causes the rice plant stem to elongate strangely and does not permit the plant to produce normal flowers. There, it is called "foolish seedling disease." Gibberellins are now believed to be present in all plants and concentrated in the seeds.

The growth-promoting power of gibberellins has been dramatically illustrated in a number of experiments (Figure 31.4), particularly those involving genetic dwarfs. Dwarf corn, for instance, can be induced to grow to normal height after the application of gibberellins. Such experiments suggest that the dwarf corn is short only because of the lack of a hormone, the presence of which is genetically determined.

Gibberellins also have other roles. For example, they stimulate pollen germination and pollen tube growth. Later, during the germination of grain, the hormones stimulate the synthesis of the starch-digesting enzyme *alpha amylase.* Alpha amylase makes glucose available to the growing embryo. In other circumstances, gibberellins have been used to inhibit seed formation and to produce seedless fruits.

CYTOKININS

Cytokinins are a class of plant hormones that stimulate cell division and retard leaf aging. Plant researchers first learned about cytokinins by using coconut milk in cell cultures. They knew that some hormone in the coconut milk was encouraging cell division in their cultures. In 1964, the first of the cytokinins was isolated and named *zeatin* (first discovered in *Zea*—maize). Since then, three others have been identified.

Plant hormones often work cooperatively to produce a specific growth effect. Figure 31.5 shows how varying mixtures of auxin and cytokinins influence plant growth and differentiation.

FIGURE 31.4 THE EFFECTS OF GIBBERELLINS. Gibberellins have a dramatic effect on stem growth, as seen in these plants. The plant at the right was grown normally, while the one at the left was treated with gibberellins.

If freshly picked leaves are treated with the hormones, they wither and die far more slowly. In treated leaves the chlorophyll remains active, protein synthesis continues, and carbohydrates remain intact. Synthetic cytokinins have been applied to harvested vegetables to extend their storage life. (Unfortunately, cytokinins have no such effect on humans.)

ETHYLENE

Ethylene stimulates the ripening of fruit, triggers the separation of fruits, and influences the sex of flowers. It is literally a lightweight among the plant hormones—light enough, in fact, to escape from the plant as a gas. Ethylene's production is promoted by auxin, and it often operates in concert with auxin. Picked fruit will ripen much more rapidly if it is kept in an enclosed space, such as a paper bag, where the ethylene produced by the fruit itself can be concentrated. This finding has obvious commercial application. Fruits can be shipped in an unripe condition and, just before marketing time, exposed to synthetic ethylene gas, which hastens the ripening process. Ethylene is also known to promote fruit abscission (fruit fall), and growers who use mechanical harvesting devices sometimes apply the gas to loosen blackberries, cherries, and grapes.

Ethylene is also known to influence sex in species that produce separate male and female flowers. When young cucumber flower buds are treated with ethylene, most of the emerging flowers will be female (pistillate). In contrast, high levels of gibberellins encourage formation of male (staminate) flowers.

Ethylene may be important in maintaining the curved hypocotyl hook that acts as a bumper when some seedlings make their way out of the soil (Figure 31.6). The hormone prevents the hypocotyl

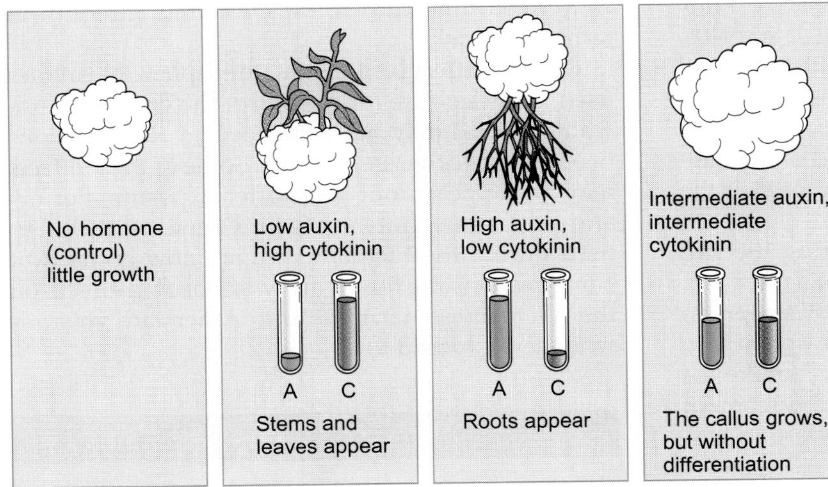

FIGURE 31.5 THE EFFECTS OF AUXIN AND CYTOKININ. The formless mass of tissue is called a callus. It has the ability to form on any of the plant structures. The effects of auxin and cytokinin on differentiation in the tobacco callus depend on the relative concentrations of the two hormones, here represented by test tubes labeled A (auxin) and C (cytokinin).

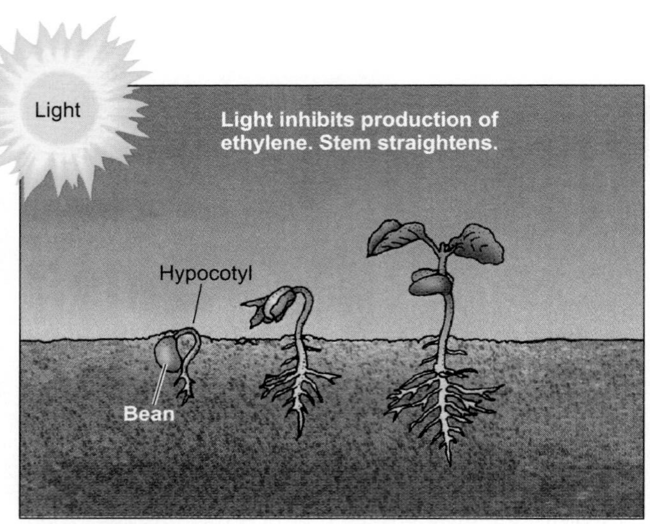
Light
Light inhibits production of ethylene. Stem straightens.
Hypocotyl
Bean

FIGURE 31.6 THE EFFECTS OF ETHYLENE. Ethylene is associated with shoot curvature in the seedling. Note the typical curvature of seedlings as they emerge from the soil. The curved hypocotyl in the bean is thought to act as a protective "bumper." As the concentration of ethylene is increased, the curvature also increases.

from straightening until it breaks through the soil surface. Once this happens, light apparently inhibits the production of ethylene, so that the stem straightens and the leaves unfold to the sun's rays.

ABSCISIC ACID

Abscisic acid causes stomata to close and buds and seeds to become dormant. In the late 1940s, plant researchers, encouraged by the discovery of growth-promoting hormones, found new growth-inhibiting substances. In the mid-1960s, one of the inhibitors was chemically identified and named abscisic acid (ABA). A chief source of ABA today is the cotton plant.

ABA is involved in wilting, by closing the stomata after excessive water loss. (See Chapter 30). ABA is also believed to inhibit growth by suppressing protein synthesis. Thus its effects are opposite to those of auxin and the gibberellins. Also, unlike the gibberellins, ABA is known to inhibit production of the starch-digesting enzyme alpha amylase in barley seeds, thus inhibiting germination. ABA shows some potential in agriculture, for example, in closing the stomata and causing dormancy in the buds and seeds so that they can be more easily stored. However, ABA is expensive and not available in large enough quantities for agricultural use.

ARTIFICIAL HORMONES AND PLANT CONTROL

As we humans have learned more about plant hormones, we've increasingly utilized these chemicals to cause plants to behave in the ways we want—such as to grow, to die, and to bear fruit. We've also looked for cheaper and better ways to control plants. One way has been through the production of synthetic or artificial hormones. For example, industry can now produce a number of inexpensive artificial auxins or other growth promoters. Some of these have even greater effects than the natural auxin because they are not rapidly broken down by the plant's enzymes. One growth promoter is the herbicide (plant killer) 2,4,5,-T (2,4,5-trichlorophenoxyacetic acid). Whereas small amounts of 2,4,5-T promote growth, high concentrations kill plants. This compound is very commonly used in weed control because it is highly selective. It is usually highly toxic to the dicotyledonous plants (such as clover, dandelions, and certain other "weeds") but spares the monocotyledonous plants (such as grasses, wheat, and oats). Fortunately, such herbicides are also *biodegradable;* they break down rapidly when exposed to the elements.

Use of 2,4,5-T is not exactly uncontroversial. There is some ominous evidence that it may interfere with the reproductive success of mammals, particularly by causing birth defects. Thus its value as a weed killer has to be measured carefully in terms of its risk.

Other artificially manufactured plant hormones used in agriculture (or to control the growth of unwanted flora) may have unexpected repercussions in organisms other than plants. Some of these effects may not appear until years after exposure. For example, contaminants in Agent Orange, a defoliant used during the Vietnam War, are now accused of being responsible for an array of horrible effects on the Vietnamese farmers and American veterans who were exposed to it.

PLANT TROPISMS

Since plants must generally stay put, their responses to different environmental stimuli are largely restricted to specific growth patterns. Such growth responses in plants—called **tropisms**—take many forms. For example, plants respond to light through

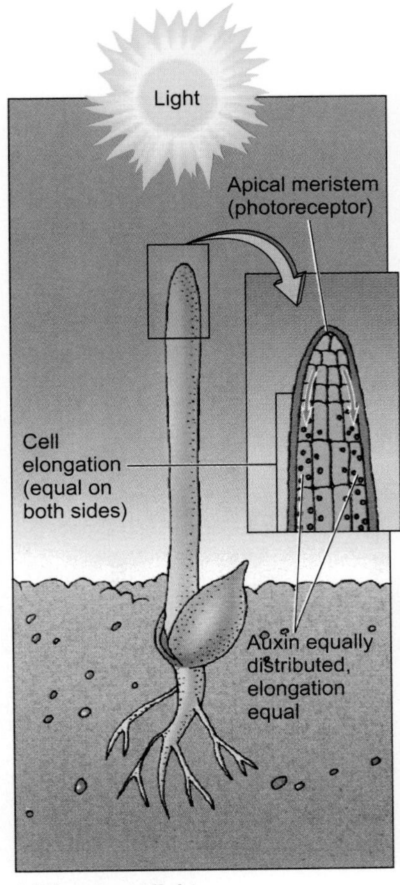

(a) Overhead light

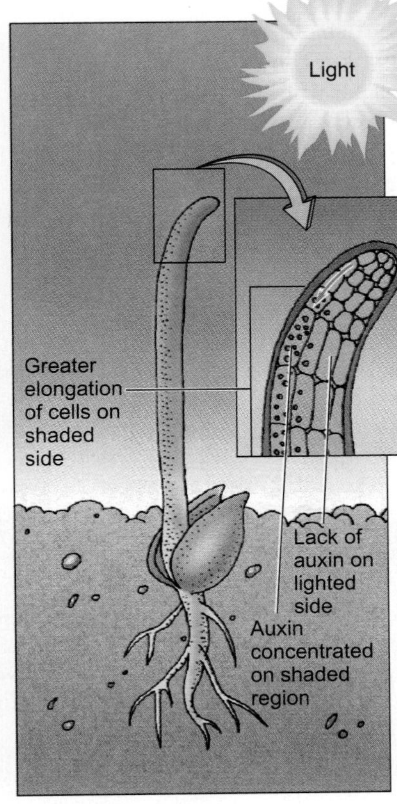

(b) Light from one side

FIGURE 31.7 PHOTOTROPISM. Cells in the apical meristem continually divide, producing cells which then elongate. **(a)** When light comes from overhead (all sides), elongation occurs equally around the entire stem. **(b)** When light comes only from one side, only the cells on the unlighted side elongate. This is because auxin diffuses or is transported to the unlighted side.

phototropism and to gravity through *gravitropism* (formerly called geotropism). Tropisms can be either positive or negative—a plant bending toward the light shows *positive* phototropism. A shoot growing away from gravity (up into the air) illustrates *negative* gravitropism. As we will see, hormones are usually involved in tropisms. In addition to having hormonally induced growth responses, some plants are capable of short-term responses involving rapid movement. One of the most fascinating of these is the touch response. It is apparent in the sensitive plant *Mimosa pudica* and in certain carnivorous plants that trap insects, as we will see.

PHOTOTROPISM

Phototropism is a plant's growth response to light. We see this, for example, when a plant bends toward light (see Figure 31.1). Obviously, something in the shoot tip must respond to light. But what is it? So far, no one has positively identified the specific mechanism, but plant physiologists are on the trail of a yellow-colored pigment, possibly one of the flavoproteins, that specializes in the absorption of light in the blue range. The receptor is believed to have an effect on the distribution of auxin as it is produced in the shoot tip. In one series of experiments, investigators learned that the total auxin production in the tip is the same under lighted conditions as it is in darkened conditions. But in the light, the distribution can be quite different. In plants exposed to light on the right side, for example, much of the auxin diffused or was transported to the left side. The cells on the left responded by elongating more rapidly, producing a curvature toward light (Figure 31.7). (This is the reason you have to keep rotating your window plants unless you favor the "windblown" look.)

This migration of auxin away from the light has been fully substantiated through the use of auxin that has experimentally incorporated carbon-14, a radioactive tracer. Its use permits the experimenter to follow the telltale radiation wherever the molecule migrates. The migration of auxin suggests that the light receptor somehow produces selective changes in the permeability of cell membranes in the shoot, favoring the diffusion, or perhaps the active transport, of auxin away from the lighted side.

If light brings about the changes in the distribution of auxin, you might expect that plants kept in the dark would grow straight and tall, at least until their energy reserves were depleted. And you would be right. For example, bean seedlings grown in a darkroom go through a surge of growth, becoming tall and spindly (as well as ghostly white, since the production of chlorophyll requires light). However, the photoreceptor is quite sensitive, and a mere pinpoint of light from any direction will initiate the phototropic response.

Gravitropism

Gravitropism is a plant's response to gravity. When a seedling is placed on its side, the shoot tip bends upward, and the root tip bends downward. A simple enough observation, but there are two distinct mechanisms at work here. In most cases, the shoot turns upward in response to processes we have already discussed, but the root turns downward in response to gravity. So the root's response is positively gravitropic.

Any explanation of gravitropism, until recently, was awash in a sea of contradictory hypotheses based on equivocal lines of evidence. For example, we know that auxin promotes elongation in shoot tips but for some reason seems to *inhibit* elongation in root tips. Physiologists were faced with the question, How can a hormone stimulate and inhibit the same response?

The key to such opposing responses, once again, is a matter of concentration—that is, low concentrations of auxin stimulate elongation, whereas high concentrations inhibit it. Before this peculiarity became known, however, scientists cast far and wide for other explanations. In the long run, the search paid off, since it brought to light several aspects of gravitropism that helped put pieces in place in a complicated puzzle.

In a seedling lying horizontally on its side, auxin accumulates in greater abundance in the lowermost cells and inhibits growth there. The uppermost cells, though, grow normally, and so the root tip bends downward (Figure 31.8). (When auxin is experimentally applied to cells on one side of the root apical meristem, those cells fail to elongate, and bending occurs.) In addition, calcium, an elongation inhibitor, may be involved. It would obviously accumulate differently than would auxin in the plant tissues.

When a seedling's root is in a vertical position, auxin moves down through the core of the root tip, eventually reaching the root cap. There, some of it simply diffuses out into the soil, but the rest moves back up the root tip, through cells of the root's perimeter. It is important to note that the return flow in the vertical root is evenly distributed around the root, and that the quantity is insufficient to inhibit elongation.

In addition, cellular organelles known as *statoliths* move under the influence of gravity. These are starchy grains that apparently act to mobilize auxin. If a potted plant is placed on its side, the statoliths will come to rest on the lower side (and within an hour or so in the root), and the cells on the upper side will begin to elongate, turning the root downward. The opposite occurs in stem cells.

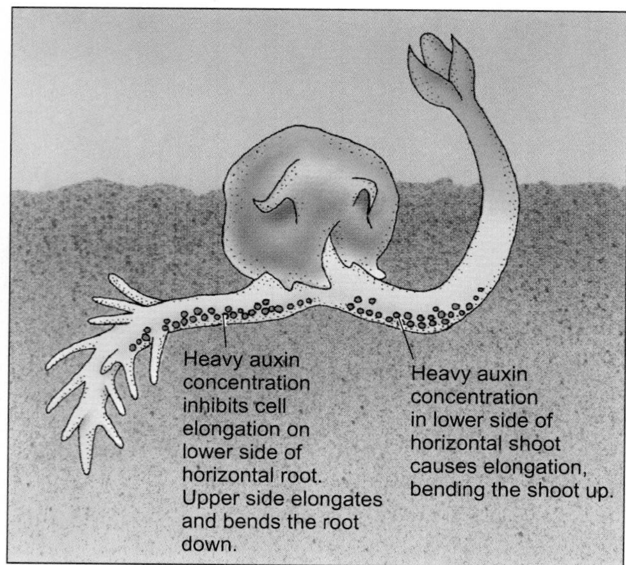

Heavy auxin concentration inhibits cell elongation on lower side of horizontal root. Upper side elongates and bends the root down.
Heavy auxin concentration in lower side of horizontal shoot causes elongation, bending the shoot up.

FIGURE 31.8 GRAVITROPISM. Seedlings placed in the dark and in the position shown respond by positive gravitropism in the root and negative gravitropism in the shoot.

Thigmotropism

Thigmotropism is a plant's response to touch. Touching the leaflets of the "sensitive plant," mimosa, *Mimosa pudica* (Figure 31.9a), causes the plant to "cringe." That is, the leaflets fold almost spasmodically, and the petioles droop. Some theorists suggest this drooping, a "thigmotropic" movement, tends to discourage browsing animals. Others claim it helps the plant avoid excessive water loss when the leaves are moved by hot, drying winds, which are common in its usual habitat.

(a)

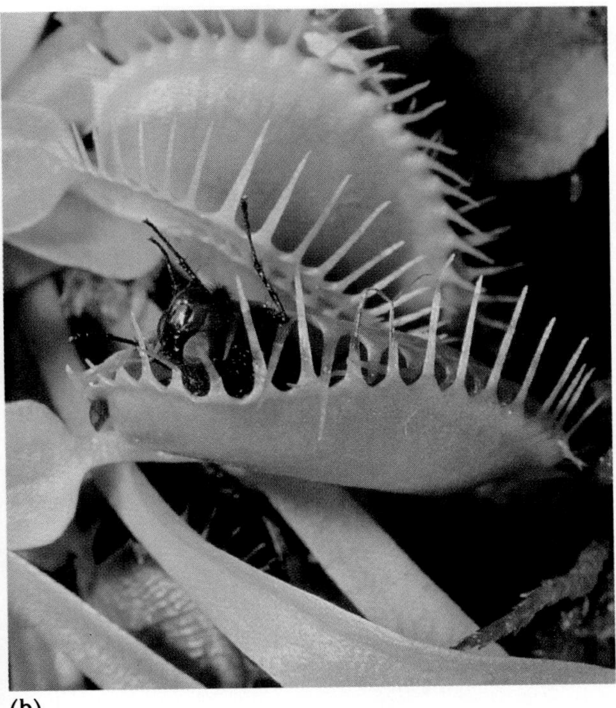

(b)

FIGURE 31.9 TOUCH RESPONSE. Examples of touch response are seen in the sensitive plant *Mimosa pudica* (**a**) and the Venus flytrap, *Dionaea muscipula* (**b**). The response may be due to active transport of ions that cause rapid changes in turgor.

Another example of the touch response is the Venus flytrap, *Dionaea muscipula* (Figure 31.9b). The trap, a highly modified leaf, lies open when at rest. Each half of the trap has three tiny, hairlike triggers that, when brushed by a wandering insect, spring the trap, quickly closing the leaf. The toothed leaf presses the insect against the digestive glands on its inner surface, and the plant gains nutrients, particularly nitrogen, from the insect's body. Like other insectivorous plants, the Venus flytrap is common to nitrogen-deficient bogs and marshes.

The ability of such plants to respond quickly to touch has presented an enduring problem to plant physiologists. The mechanism of sudden movement in plants is now thought to be based on the very rapid active transport of ions out of cells, lowering the water potential in the intercellular spaces and causing cellular water, moving from places of higher water potential to lower, to flood these spaces. Since leaves are often held erect by turgor pressure, its sudden loss could explain the rapid movement.

PHOTOPERIODISM AND FLOWERING PLANTS

The response of an organism to changes in relative lengths of light and dark is known as **photoperiodism.** A clear example of photoperiodism in plants is the production of flowers during their reproductive period, a period that is triggered by a specific day length. Flowering plants may respond in essentially three ways to changing day length. Depending on their response, the plants are called *short-day plants, long-day plants,* and *day-neutral plants.* The

(a)

(b)

Figure 31.10 **Short-Day and Long-Day Plants.** (a) The poinsettia is a short-day plant that produces flowers in the autumn. (b) The henbane requires long days (short nights) for flowering.

day-neutral plants can flower at any time of the year, and we won't bother with them here.

Short-day plants begin to flower as the light period becomes shorter than a certain critical length (depending on the species). For example, chrysanthemums and poinsettias (Figure 31.10a) typically begin flowering in autumn. **Long-day plants,** on the other hand, begin to flower when the light period lasts longer than a certain critical time period. The critical time period varies for different species, and none will react to any day length shorter than that specifically required. Thus long-day plants—such as potatoes, spinach, and henbane (Figure 31.10b)—flower at different times in summer. Spinach, incidentally, will not flower in tropical regions, because days are never long enough to exceed its 14-hour critical photoperiod.

Photoperiodism and the Length of Night

The long- and short-day terminology is unfortunately misleading. (And so, perhaps, is the notion we left you with in the preceding section.) This is because the terms *long-day* and *short-day* were already fixed in the scientific vocabulary before it was discovered that plant photoperiodism isn't determined by changes in day length at all. Instead, it's determined by the length of *night.*

By raising plants in darkness and controlling the time they are exposed to light, investigators have found that the length of darkness controls flowering. When the *light* periods are changed drastically, the plants show no response at all. If the *dark* periods are changed in the same way, however, the plants respond.

Another interesting experiment has provided us with yet more fascinating information. A long-night (short-day) plant can be fooled into behaving as if any night were short enough for flowering by being subjected to a flash of light in the middle of the night. The flash of light resets what has come to be called the plant's "dark clock." The dark clock behaves like a countdown timer that tells you when the soufflé is done. In long-night plants, flowering is triggered when the required length of darkness is reached. So it is the extended darkness (associated with winter) that represses the plant's flowering.

However, a flash of light (or, more precisely, the red wavelengths in the light) breaks up the dark period, setting the clock back to zero. The plant responds to the flash as if daylight had arrived. Then it is submerged in darkness again, running its countdown clock back to zero. The cool light of the real dawn soon comes and makes the night seem extremely short from the plant's viewpoint. Short nights mean long days, and so the flowering process begins. In this way, greenhouse gardeners can get their long-night plants to flower year-round.

PHYTOCHROME: THE LIGHT RECEPTOR

Plant physiologists have never been able to pin down the biochemical mechanisms of photoperiodism, although research has been intense and many major hypotheses have been proposed and discarded. Scientists have, however, established the nature of the light receptor involved in this phenomenon, and have named it *phytochrome.* **Phytochrome** is a protein associated with a light-absorbing accessory pigment that, like chlorophyll, absorbs light in the 660 nm range. Through several ingenious experiments, it has been conclusively shown that phytochrome is located in the leaves (Figure 31.11). There it responds to the light stimulus by setting the flowering process in motion, probably through the release of a hormone. Although no such hormone has been isolated, some botanists are confident enough that it exists to have tentatively named it (or them) *florigen.* Others, however, believe that no such hormone exists, and that flowering is controlled by certain combinations of auxins, gibberellins, and other known plant growth hormones. In addition to its role in flowering, phytochrome has also been shown to be the receptor often involved in leaf growth, seed germination, and the lengthening and branching of stems.

FIGURE 31.11 **PHOTOPERIOD AND THE COCKLEBUR.** This experiment with the cocklebur, a short-day plant, establishes that the phytochrome photoreceptor is in the leaf. Only one leaf in this series of six grafted cocklebur plants is exposed to the proper photoperiod, yet all six plants produce flowers.

PERSPECTIVES

It is apparent that plants are not only structurally complex, with a variety of very specialized tissues, but also that they are highly regulated, coordinated, and responsive organisms. In this chapter, we have explored some of the ways they are regulated and a few of the means by which they are able to respond to the environment. Of course, this has been only a brief glimpse at such responses, but it suggests the fascinating array of ways that plants have come to adapt to a complex environment. Next, we move to the realm of animals. This may seem like an abrupt change, but keep in mind that life on earth is an interdependent constellation of forms with underlying themes and abiding, if sometimes complex, relationships.

Summary

Plant regulation often involves hormones—chemical messengers that travel from their place of origin to target cells in which they initiate responses. Hormones have specific actions, are effective in minute amounts, and are short-lived.

Plant Hormones and Their Actions

Auxins (represented by indoleacetic acid or IAA) largely function in growth regulation apparently by loosening cellulose wall fibers, thus permitting turgor pressure to bring about cell elongations.

Working with other hormones, auxin influences differentiation, fruit and leaf fall, and tree shape. Specific combinations determine whether root or shoot tissue will form. In fruit and leaf fall, auxin withdrawal permits the breakdown of abscission zone cells and leaf separation. Apical dominance in conifers occurs as auxin promotes elongation in the uppermost shoot and partially inhibits those below.

Gibberellins promote stem elongation and function in plant reproduction. They also promote plant growth, pollen germination, and pollen tube growth, and, in germinating grains, stimulate the release of a starch-digesting enzyme.

Cytokinins, along with other hormones, stimulate cell division and differentiation and prevent leaf aging.

Ethylene, a gas, promotes fruit-ripening and fruit abscission and influences sex in flowers. It may also maintain the curved hypocotyl hook of emerging seedlings.

Abscisic acid (ABA) produces stomatal closure and slows growth by inhibiting protein synthesis. It also causes buds and seeds to become dormant.

Artificial Hormones and Plant Control

Artificial growth promoters such as the herbicide 2,4,5-T are used to kill weeds, but their use is controversial since it may adversely affect mammal reproduction. Humans exposed to Agent Orange herbicide contaminants in Vietnam reportedly developed serious health problems.

Plant Tropisms

Phototropism, the growth of shoots toward light, occurs when a highly sensitive flavoprotein photoreceptor somehow causes auxin to concentrate on the unlighted side, thereby promoting cell elongation and curvature on that side. In total darkness, the shoot grows straight.

In gravitropism—growth responses to gravity—roots respond positively (toward gravity), and shoots negatively (away from gravity). One hypothesis suggests that the different responses may be caused by high concentrations of auxin on the lower side of a horizontal root that inhibits cell elongation there. The auxin may be triggered by statoliths that sink to the downward side of cells.

Thigmotropic movement is the rapid response to touch, seen as a sudden drooping of leaflets and petioles in mimosa and in the closing of the Venus flytrap. The mechanism may be rapid, cell-to-cell, active transport of ions and subsequent losses of leaf-supporting water (turgor).

Photoperiodism and Flowering Plants

Flowering in short-day and long-day plants is photoperiodic—dependent on seasonal changes in the length of daylight periods. For this reason, different species produce flowers in specific parts of the year.

Studies support the idea that the length of darkness is actually the key factor in flowering photoperiodism. Interruptions in light periods have no effect on flowering, but even flashes of light (mimicking daybreak) during dark periods prevent flowering in long-night (short-day) plants, running the so-called dark countdown clock back to zero.

The light receptor of flowering is phytochrome, a protein and accessory pigment located in the leaves. Its operating mechanism is unknown, but a flowering hormone is suspected.

Key Terms

auxin • **524**
abscission zone • **525**
apical dominance • **525**
gibberellin • **526**
cytokinin • **526**
ethylene • **527**
abscisic acid (ABA) • **528**
tropism • **528**
phototropism • **529**
gravitropism • **530**
thigmotropism • **530**
photoperiodism • **531**
day-neutral plant • **532**
short-day plant • **532**
long-day plant • **532**
phytochrome • **533**

Review Questions

1. Define the term *hormone,* and list several characteristics of hormones.
2. What specific effect does auxin have on cells? What is the result of this effect?
3. List the steps involved in leaf fall. Explain how auxin can have a "negative" role.
4. Explain auxin's role in forming the triangular shape of some trees. How might such a shape be adaptive, say, in snowy regions?
5. How were the gibberellins first noticed? List three of their natural roles.
6. Describe the effects on plant differentiation of varying combinations of cytokinins and auxin.
7. List three of ethylene's effects on plants.
8. Compare the effects of gibberellins and abscisic acid on germinating seeds.
9. List two instances in which the application of artificial herbicides has had undesirable effects.
10. Summarize the phototropic mechanism in the shoot, starting with light and the pigmented photoreceptor.
11. In what ways do shoots and roots react to gravity? Why have the usual hormones been discounted as the agent responsible?
12. Describe the hypothetical role of auxin in gravitropism.
13. Describe two examples of a thigmotropic movement. What is the importance of each to the plant involved?
14. Suggest a mechanism that explains the sudden folding of leaflets in *Mimosa pudica.*
15. Define photoperiodism. Define short-day, long-day, and day-neutral.
16. Discuss the experimental evidence suggesting that night length is important in flowering.
17. Explain the mechanism believed to be involved in the flowering response. What evidence supports the idea that the response begins in the leaves?

Animal Biology

AS WE BEGIN OUR DISCUSSIONS OF THE VAST ARRAY OF SYSTEMS IN THE ANIMAL KINDGOM, WE MAY FIRST BE STRUCK BY THE AMAZING DIVERSITY IN THIS GROUP, AND OUR AMAZEMENT IS INDEED JUSTIFIED. *The animals have, in fact, developed a dazzling variety of ways to solve their problems and take advantage of their opportunities. However, confronted with many of the same problems and opportunities, they have also followed certain common themes. So as you learn about their diversity, keep in mind their similarities, many of which you may share.*

32

Animal Support and Movement

YOU MAY HAVE NOTICED THAT NOT ALL ANIMALS LIE FLATTENED AGAINST *the earth's surface and, furthermore, that some of them can move. Here, then, we will see the various ways that animals support themselves and furnish attachments for those remarkable tissues that are able to contract and thereby move the animal or its parts. As we go through these pages, keep an eye out for the great differences, such as those between animals that wear their support on the outside and those that have internal support. At the same time, look for themes that indicate a common ancestry, with variations according to that animal's particular needs.*

Now we will focus on the animal body. But we must begin with the niggling awareness that there is no such thing as "the animal body." A swan would have little in common with the mites that nestle among its feathers or the swimming planaria beneath its paddling feet. There is even great diversity within each distinctly related group, such as the vertebrates. The most skilled physician might have difficulty performing an appendectomy on a possum. Where would she begin to look? And she would look in vain for an appendix in a turtle. (It doesn't have one.)

On the other hand, there are common themes—fundamental similarities—that link the diverse representatives of some groups, such as the backboned animals. For example, the bodies of all vertebrates are composed of only a few types of tissues (about 20). The similarities continue to the molecular level where we will find that the cell constituents of all vertebrates are remarkably similar, and even the chemical reactions that power the muscles reveal a common heritage. In this chapter, we will focus on how animal bodies are supported and how they move—focusing on the vertebrates and, in particular, ourselves. First, though, let's review the basic kinds of tissues that compose virtually every animal body.

ANIMAL TISSUES

In spite of the great diversity and the complex organization of all the far-flung animal species, the tissues of most animals are composed of four fundamental types: *epithelial tissue, connective tissue, muscle tissue, and nervous tissue.* In our discussion, we will refer to their roles in the human body (Figure 32.1).

EPITHELIAL TISSUE

Epithelial tissue (also called epithelium) forms the interior and exterior linings of most surfaces of the vertebrate organism, and it is commonly separated from adjacent tissues by a fibrous, noncellular *basement membrane.* Not only is the skin, then, epithelial, but so are the linings of the mouth, the nasal cavities, the respiratory system, the coelom, the tubes of the reproductive system, the gut, and the interior of blood vessels. In addition, most glands of the body are composed of thickened tissue called *glandular epithelium* that produces secretions that are released through epithelial-lined ducts.

As you might expect, epithelial tissue differs greatly from place to place, both in the shape of the cell and in the numbers of cell layers present. There

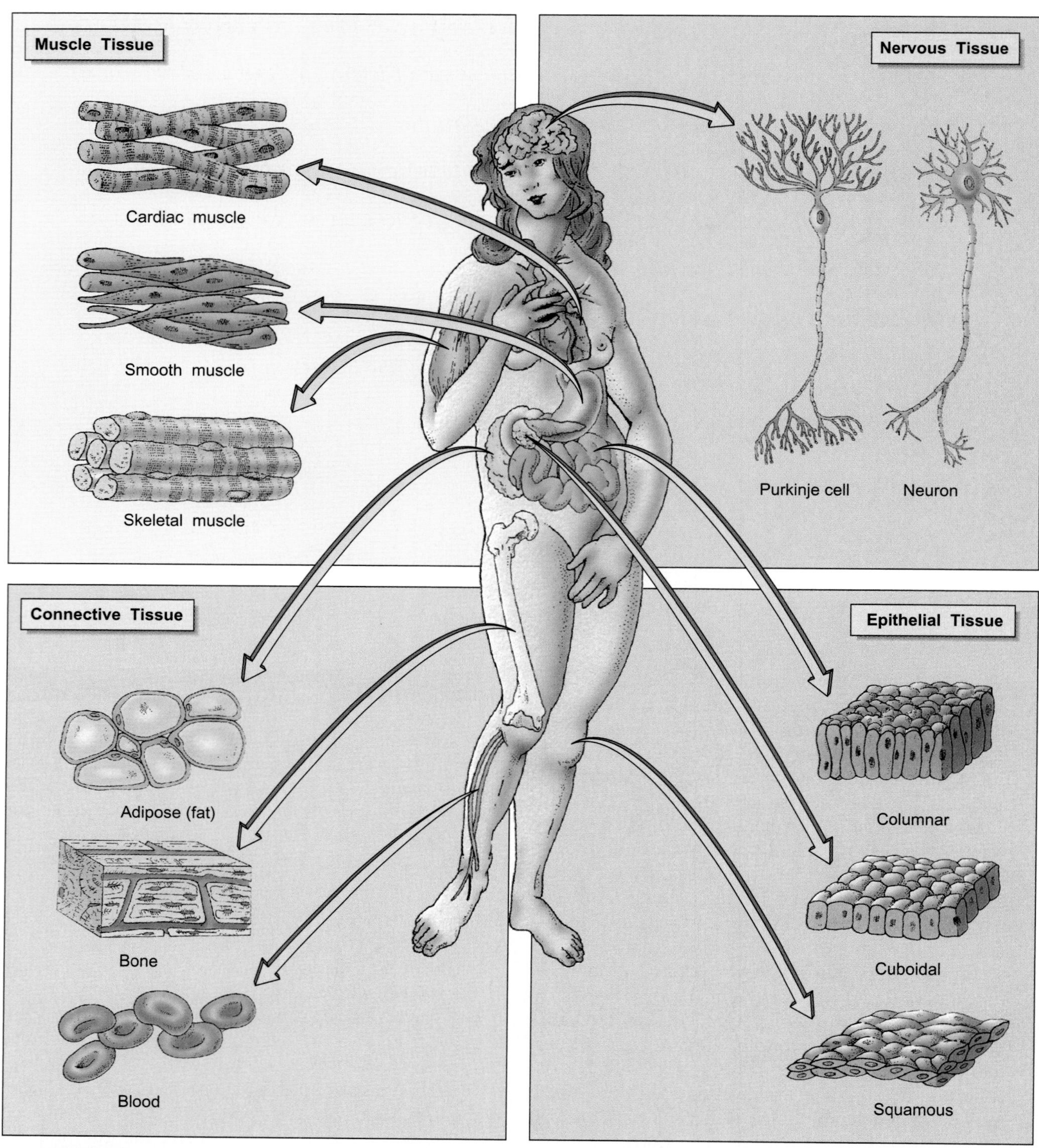

FIGURE 32.1 TISSUES OF THE HUMAN BODY.
These are some of the types of human tissues, examples of the cells that comprise them, and their locations in the body.

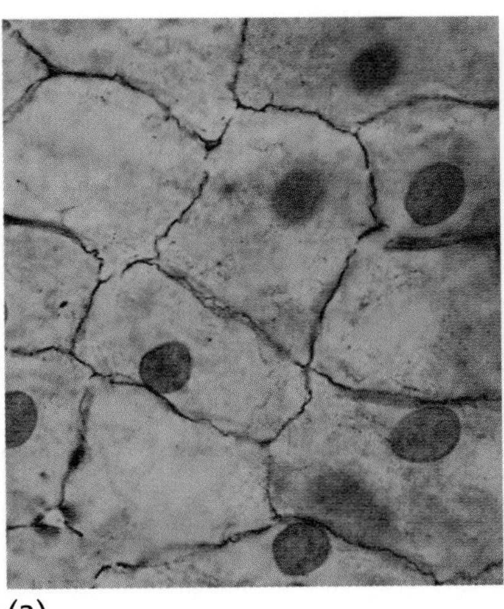

(a)

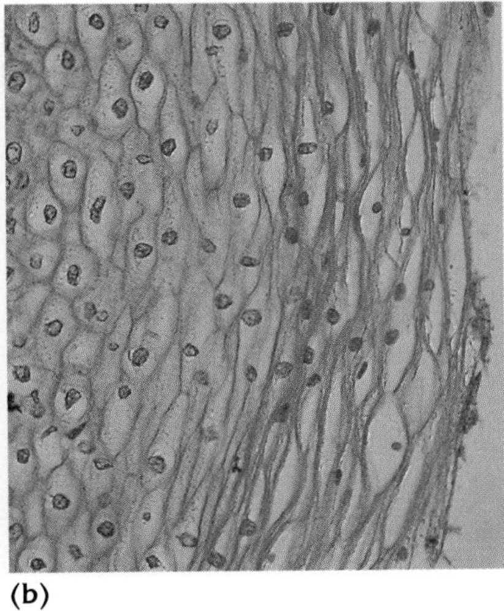

(b)

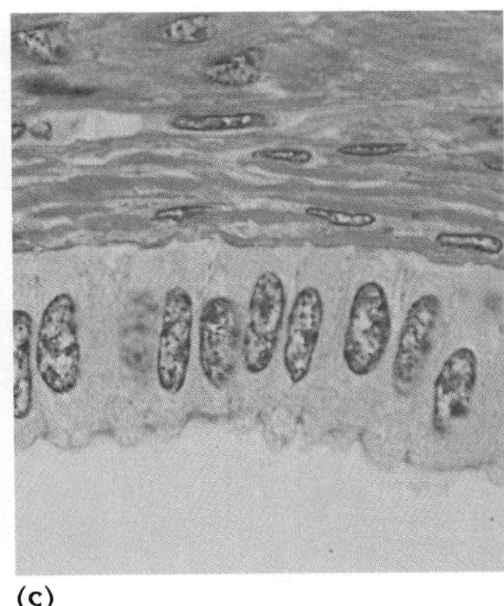

(c)

FIGURE 32.2 **EPITHELIAL CELLS.**
Epithelial cells form linings that may be arranged in a simple, flattened layer **(a)** or one that is stratified (layered) **(b)**, as seen in places of rapid wear. Cells lining the respiratory passages often contain columnar cells, some of which may be ciliated **(c)** or mucus secreting.

are three main shapes: *squamous* (flattened), *cuboidal* (cube-shaped), and *columnar* (elongated, column-shaped—these can be ciliated, such as those that form the secretory lining of the respiratory system) (Figure 32.2). Epithelial cells of any shape may form a single layer (simple epithelium) or multiple layers (stratified epithelium). Simple epithelium is very thin and is generally found in areas that are specialized for absorption or filtration, as in the kidney tubules. Stratified epithelium often forms a protective layer in areas that are subjected to high abrasion. For example, the surface of the skin, called the *epidermis,* has an outer stratified squamous epithelium containing many layers of flattened cells. The outermost of these cells are filled with the protein keratin and are hardened and dead. When these are shed as scales, they form that charming debris called dandruff as well as a startling proportion of household dust. The innermost cells of the epidermis, a living layer of relatively cuboidal cells, repeatedly divide to provide new cells that will become keratinized and flattened as they move to the surface.

CONNECTIVE TISSUE

Connective tissue basically holds the body together, but some types of connective tissue have other roles as well. There are four classes of connective tissue: (1) *connective tissue proper,* which includes the "packing material" between other tissues; (2) *bone;* (3) *cartilage,* and (4) *blood* (Figure 32.3). Although the connective tissues composing these

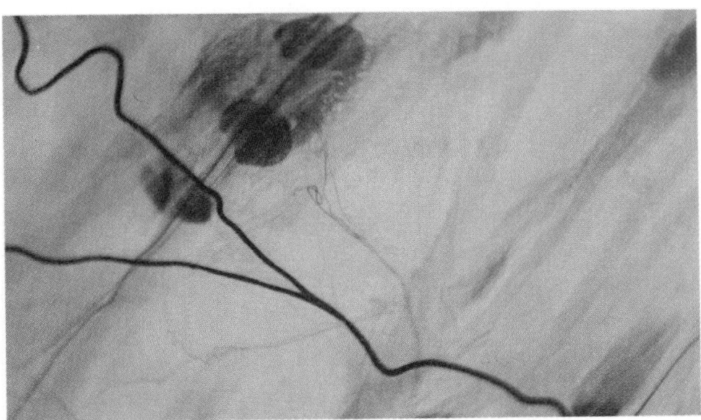

(a)

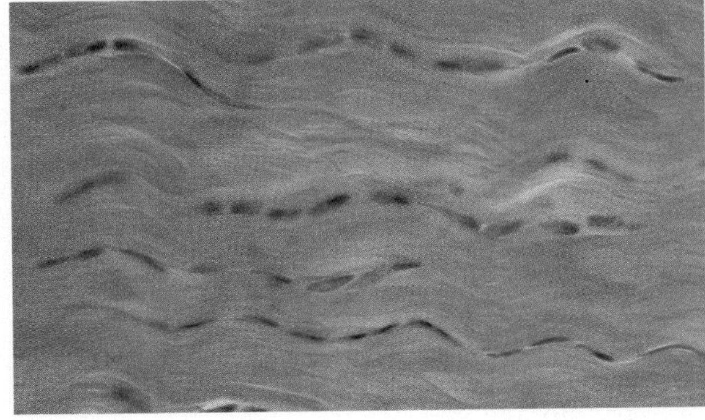

(b)

FIGURE 32.3 **CONNECTIVE TISSUE.**
Loose connective tissue **(a)** contains a loosely arranged matrix of collagen fibers and fine elastic fibers. It forms beneath the skin, in the walls of blood vessels, and around nerves. Dense connective tissue **(b)** has many collagenous fibers and forms tough ligaments and tendons.

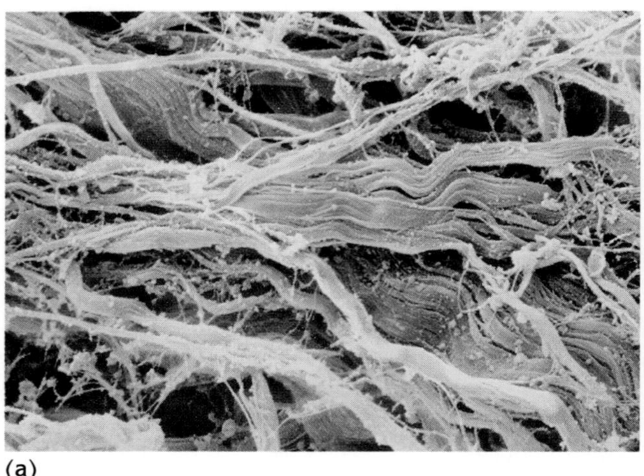

(a)

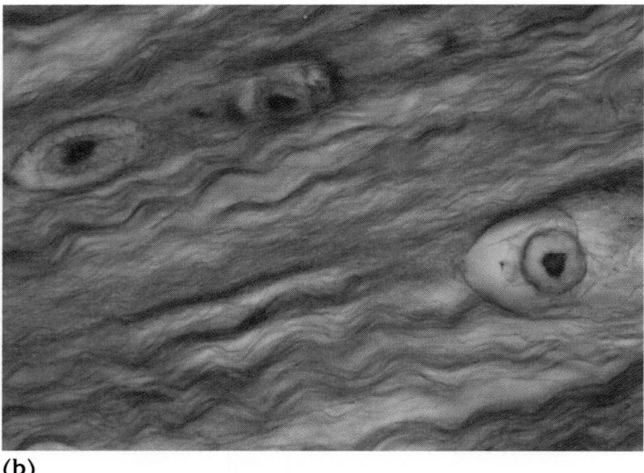

(b)

FIGURE 32.4 COLLAGENOUS FIBERS.
(a) Collagenous fibers form a tough binding material that literally holds the vertebrate body together. Logically enough, collagen is the primary material of tendons—strong, cordlike structures that connect muscles to bones. **(b)** Individual collagen fibers in tendons form a wavy pattern, with distinct cross-banding of dense protein.

four classes have various functions and appearances, they are all composed of cells and a **matrix** (intercellular material), which is secreted by the cells. The characteristics of each type of connective tissue are largely determined by the composition of the matrix, for example whether it is dense or loose.

The principal substance in most connective tissue matrices such as tendons, ligaments, basement membranes, and muscle coverings is **collagen** (Figure 32.4), a fibrous protein found in all animals that in larger vertebrates accounts for at least a third of the total body protein. Collagen has many roles; for example, it serves as the "glue" of the animal body, in that it binds various tissues together. It also forms such extremely tough structures as the cornea of the eye and the intervertebral disks (the cushions between the bones of the spine). In addition, collagen is a principal component of bone.

Connective tissue proper serves primarily to bind organs together. This type of connective tissue includes the tendons and ligaments. *Tendons* are flexible and cordlike, and bind muscle to muscle or muscle to bone. *Ligaments* are flexible and elastic. They bind one bone to another. Also considered a type of connective tissue proper is **adipose tissue,** a primary storage site of fats. These "fat cells" are found under the skin, between muscle fibers, in the breasts and buttocks, in intestinal membranes, and in other places in the body. Pinch tests provide one way to ascertain the amount and distribution of fat on your body. As a very simple test, if you can pinch more than one inch of fat in any area (usually the waist, the shoulder blade, and the back of the arm), you may be eating too much. Some parents consider a fat infant a healthy infant; but according to some nutritionists, overfeeding in infancy can build a surplus of adipose cells that may sentence the baby to a lifetime of obesity or constant dieting, since these cells, once formed, apparently do not go away. Instead, they simply store more or less fat according to one's nutrition.

Bone is a connective tissue that largely composes vertebrate support systems. Mature bone cells are located in spaces within a matrix that they secrete themselves. The matrix is hardened by calcium phosphate and calcium carbonate, with any flexibility largely due to collagen. We will discuss bone in more detail shortly.

Cartilage is important in some support systems. Cartilage cells, like those of bone, are located in small spaces within the matrix they secrete themselves. Cartilage is firm, flexible, resilient, and smooth, which makes it well suited for certain roles within the vertebrate skeletal system. For example, it is found in parts of the rib cage, where its flexibility allows the movement needed to inflate the lungs. Cartilage also forms smooth coverings on the ends of adjoining limb bones that allow the bones to slide easily past one another, and (with collagen) is an important part of intervertebral disks, the cushioning pads between the vertebrae. Cartilage has a remarkable ability to withstand pressure and still return to its original shape. You can demonstrate this by reaching over and twisting your neighbor's nose, a structure largely composed of cartilage. You may notice that the nose resumes its shape (although yours may not).

Blood is a connective tissue whose cells are suspended in a viscous matrix called the *plasma.* (This fascinating fluid is discussed in Chapter 40.)

Muscle Tissue

Muscle tissue is unique in that it has the ability to contract. This contraction, of course, is responsible for skeletal movement, as well as the movement in breathing, heartbeat, and digestion. As with bone and blood, we'll come back to muscle tissue shortly.

Nervous Tissue

Nervous tissue is excitable and irritable (Figure 32.5). These traits may at first seem to describe your roommate and to be associated with a certain unpleasantness. Here, however, they simply mean that nerve cells are responsive to external and internal stimuli. Nervous tissue responds by conducting *impulses* along the nerve cells, as is discussed in Chapter 33.

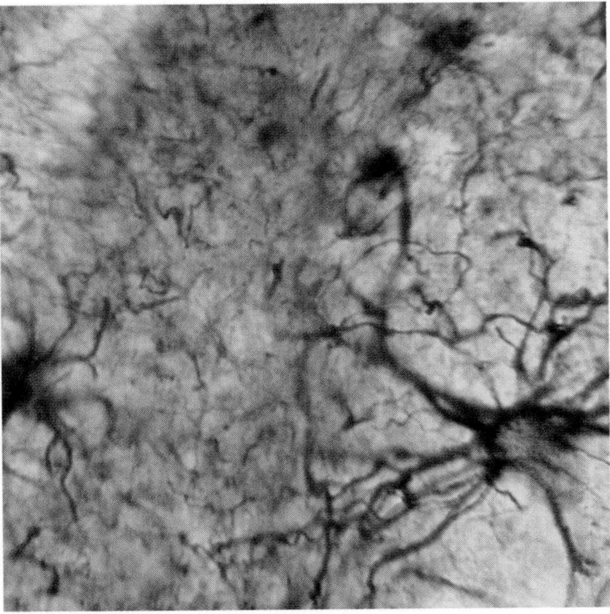

Figure 32.5 Nervous Tissue.
Seen here are groups of neurons. Although they have similar appearances, they may play very different roles in the life of an animal.

Skeletal Systems

Trees are able to stand erect in spite of their great weights because of the combined strength of their hardened cell walls. Animals, such as ourselves, though, lack cell walls. Still, we do not flounder about, collapsing here and there in quivering heaps, except perhaps on very special occasions. Generally, animals maintain their form and structure because of various sorts of support systems. The most familiar of these are *skeletal systems,* which provide support and assist movement by providing something firm for contractile fibers, like muscles, to pull against.

Types of Skeletal Systems

There are essentially three kinds of skeletons: *hydrostatic skeletons, exoskeletons,* and *endoskeletons.* One of the simplest skeletal systems is the **hydrostatic skeleton,** which keeps the animal firm as muscular contraction compresses body fluids. Hydrostatic skeletons are found in certain soft-bodied animals such as cnidarians (such as sea anemones), flatworms (such as planaria), and annelids (such as earthworms).

Species with an **exoskeleton,** a hardened body covering, include lobsters and snails as well as insects and clams. Opposite ends of muscles are attached to different parts of the exoskeleton, and when the muscles contract, the parts of the exoskeleton are drawn together and the animal moves. Exoskeletons also protect internal parts (just as do our skull and rib cage). Hard exoskeletons, however, present special problems as the animal continues to grow within the exoskeleton. These problems are solved in different ways by different species.

Snails and clams solve the problems by secreting extensions to their shells as they grow—so their shells simply grow with them. Lobsters, crabs, and crayfish, however, discard their old shells in a process called *molting* and grow new, larger ones. A molting animal usually retreats to a protected place because it is particularly vulnerable at this time. Before discarding its exoskeleton, the animal withdraws the valuable minerals from its shell, causing the shell to soften. The weakened shell, stressed by the force of the growing animal, then splits down the back. The animal crawls out, soft and vulnerable, its new shell incomplete. (This is where the "soft-shell crab" on the menu comes from.) The animal then swells by absorbing quantities of water as it grows, and proceeds to secrete a new shell into which it redeposits its hoarded minerals (Figure 32.6a).

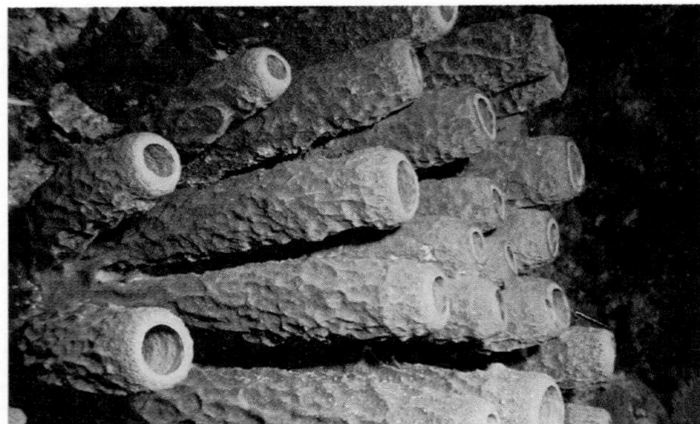

FIGURE 32.6 EXOSKELETONS AND ENDOSKELETONS. The molting arthropod illustrates the exoskeleton, while the sponge, a simpler creature, boasts an endoskeleton.

All vertebrates and some invertebrates possess an **endoskeleton,** a skeleton inside their bodies. Perhaps the simplest endoskeleton is that of the sponges, where we find bony spicules or rubbery spongin scattered through the body wall (Figure 32.6). In vertebrates, the endoskeleton serves not only as support but it moves when muscles attached to different parts of the skeleton contract, drawing two parts of the skeleton together. In the group of vertebrates that includes the sharks and rays, the skeleton consists entirely of cartilage, similar to that composing the bulb of your neighbor's nose, the one that you squeezed earlier. In most other vertebrates, the skeleton is made of bone—but even then, parts of the skeleton are formed by cartilage during embryonic development. These are largely replaced by bone as the body develops. Other parts of the skeleton, however, form directly as bone.

The bony skeleton in nearly all classes of vertebrates has five major functions: (1) it supports the body; (2) it provides a framework for muscle attachment and a resistant base for contraction; (3) it protects parts of the body; (4) it is a site where blood cells are formed; and (5) it is the site of most of the body's calcium and phosphate reserves.

Structure of Bone

Bones may have a variety of shapes and may form in different ways. Some bones, for example, form from the embryonic tissues directly as bone, while other bones, such as those of the skull and the arms and legs (the "long bones") pass through a cartilage stage and are later replaced by bone. Here, we will concentrate on the long bones.

In their general appearance, long bones (Figure 32.7) have expanded *heads* and longer, narrower *shafts*. Except at the joints, the long bones are surrounded by a living layer of connective tissue called the *periosteum* ("around the bone"). Its cells are important in both bone formation and in the repair of broken bones. Most of the shaft is composed of *compact bone.* As the name suggests, it is thick and dense. Within the shaft is a tubular central cavity containing a fatty material known as *yellow marrow.* The heads of long bones contain *spongy bone,* so called because the hardened matrix is weblike rather than solid. Spongy bone is remarkably strong for its weight because the bone is laid down along the bone's lines of stress. The spaces within spongy bone are filled with *red marrow,* a soft tissue that in some bones is the site of red blood cell production (Figure 32.7).

Microscopic Structure of Bone. Despite its nonliving, stonelike appearance, compact bone contains vast numbers of metabolically active bone cells called **osteocytes** and numerous nerves and blood vessels. Most of the bony mass (about 65%), however, consists of hardened mineral salts, such as calcium phosphate, through which run collagen fibers, much as fiberglass fibers form a mat within the epoxy matrix in the hull of a boat.

Microscopic sections of compact bone (Figure 32.7b) reveal intricate, repeated, structural units called **Haversian systems.** Each consists of a *central canal* that contains blood vessels, lymph vessels, and nerves surrounded by concentric *lamellae,* cylinders of calcified bone. The laminated arrangement, like layers of wood in plywood, imparts great strength and resilience to the bone.

Within the lamellae are the osteocytes. Each resides in a tiny cavity—a *lacuna* (plural, *lacunae*). Tiny canals, the *canaliculi,* pass through the hardened bone from one lacuna to the next. The osteocytes touch each other with long extensions projecting through the canaliculi. Some of these projections reach the central canal, permitting the living bone

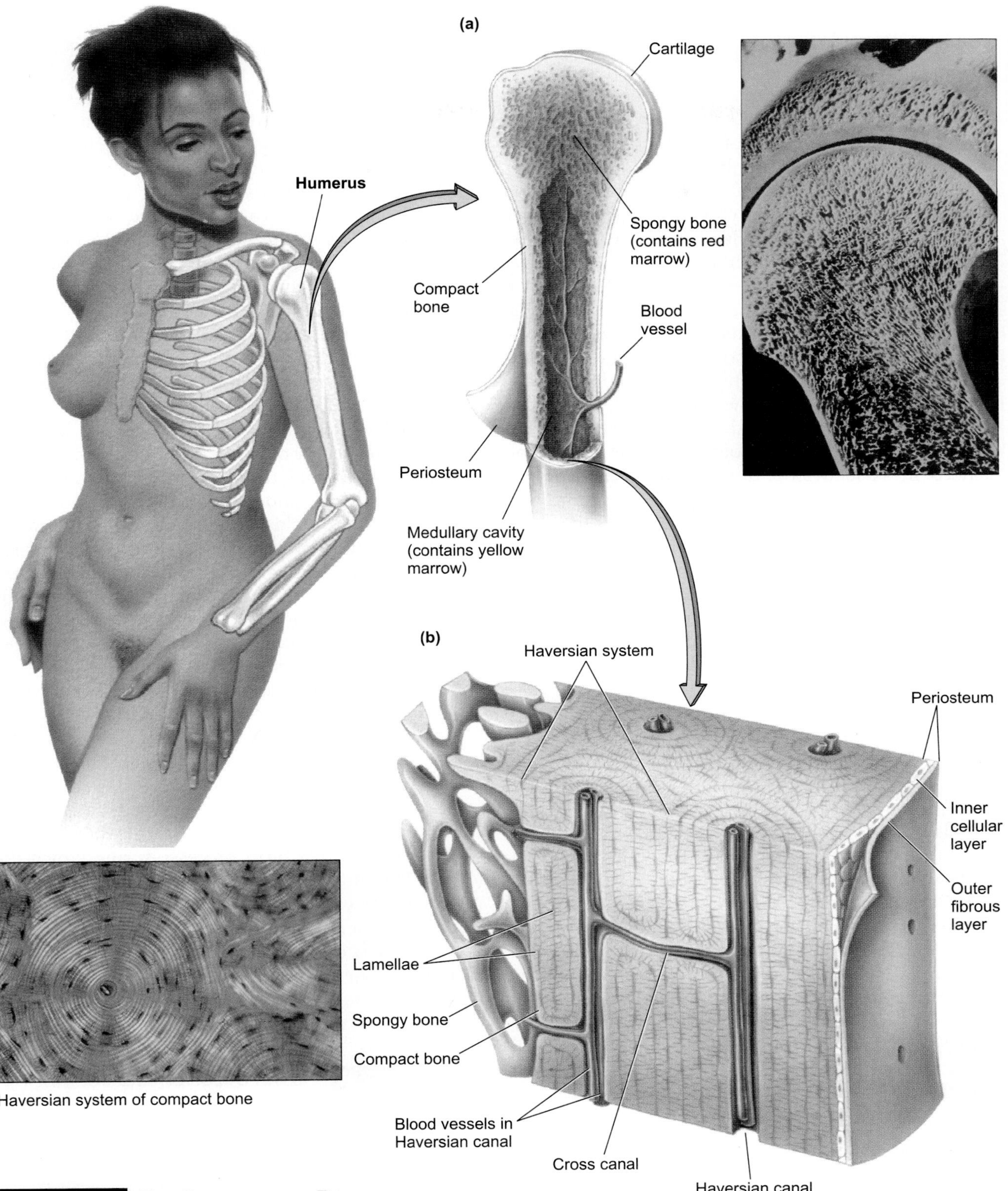

FIGURE 32.7 THE STRUCTURE OF BONE.
(a) The sectioned long bone consists of a hard shaft and a central cavity of yellow marrow. Spongy bone is seen near the joints. **(b)** Numerous central canals, through which pass blood vessels and nerves, are surrounded by hardened concentric rings, or lamellae containing bone cells (osteocytes). Such cells are entombed in lacunae, tiny cavities that communicate with each other via minute crevices called canaliculi. Such structures are grouped into Haversian systems.

cells to exchange materials with the circulatory system (Figure 32.7b).

Bone is a living tissue that is broken down and rebuilt all through life. Certain bone cells (called *osteoclasts*) withdraw calcium and phosphate salts from the matrix, and the bone would simply dissolve if it weren't for the fact that new bone is constantly being deposited by other bone cells (called *osteoblasts*) that are specialized to secrete these salts, forming new matrix. Without this turnover, for some reason, bones grow weak and brittle. In older people, the process of deposition slows, causing their bones to become increasingly fragile. The constant breakdown and deposition of bone also permits new growth when bone is most needed. For example, bone tends to grow most rapidly in places that are severely stressed. Certain cultures have used this principle to mold particular configurations in skeletal structures, as in certain African tribes in which the heads of women are bound in a tight metal ring that drastically changes the head shape. It is also interesting that experts can determine the muscular development of a person by skeletal examination only. Larger muscles place greater stress on the bones where they are attached, and these bones enlarge to accommodate the tension. So, as a general rule, the larger the area of attachment, the larger the muscle.

Organization of the Human Skeleton

The human skeleton (Figure 32.8) is something we generally prefer not to see, since the sight of it usually spells bad news for somebody. Nonetheless, we've seen it enough to know that it can be divided into two parts: the *axial skeleton* and the *appendicular skeleton*. The **axial skeleton** includes the skull, the vertebral column, and the bones of the thoracic (chest) region; the **appendicular skeleton** includes the pectoral girdle (shoulder) and pelvic girdle (hip), along with the limbs (the arms and legs). We'll take a closer look at all these shortly, but let's begin with a look at places where various bones meet.

Joints. Where bones of the skeleton meet, they form **joints**. There are a number of kinds of joints—for example, the free-ranging, rotating, *ball-and-socket joints* of the shoulder and hip, and the less versatile *hinge joints* of the fingers, toes, and knee. Far less mobile *gliding joints* are found in the wrist and ankle; so-called *slightly movable joints* occur in the pelvic bones. Even the skull has joints, but these immovable *sutures*, as they are known, are fused and virtually immovable in the adult.

The articulating (contracting) surfaces of the bones of movable joints are covered with a thin layer of smooth, glistening cartilage. The joint itself is bound together by **ligaments** made up of tough, flexible connective tissue. Some joints are surrounded by a sac or capsule formed from flattened ligaments. These are called *synovial joints*, and the lubricating fluid inside is called *synovial fluid*. Synovial fluid contributes to the silent efficiency of the active hip, elbow, shoulder, and knee joints. We can immediately see the evolutionary advantage of such a fluid. It wouldn't have done for primitive human hunters to go creaking around in the underbrush.

Axial Skeleton. The axial skeleton, as its name implies, forms the central axis of the body. It is particularly interesting to anthropologists because it includes the skull and jaw, structures that fossilize well and have yielded a great deal of information about human evolution (see Chapter 21). The role of protection falls largely to the axial skeleton, since the skull shields the brain, the rib cage surrounds the heart and lungs, and the vertebrae house the vulnerable spinal cord. You can see how the skeleton serves its protective function in Figure 32.8a.

The human *skull* consists of 28 bones. Most of its volume makes up the **cranium** ("brain case"), which surrounds and protects the delicate and baffling organ within. Since the cranial bones of the skull do not fully meet until a child is about 18 months of age, there is a vulnerable soft spot on top of a baby's head until then. This temporary flexibility of the skull helps in the birth process, during which the head is slightly compressed, permitting the baby to move more easily through the birth canal.

The **vertebral column,** or backbone, forms the flexible axis of the skeleton. It supports the weight of the upper body, delivering its load to the pelvic girdle below.

Figure 32.8 Human Skeleton and Joints. The skull, vertebral column, sternum, and ribs make up the central axial skeleton. The pectoral and pelvic girdles and the limbs form the appendicular skeleton **(a)**. Vertebrae **(b)** from different areas of the vertebral column have many features in common, but each has certain unique features (note the facets for rib attachments in the thoracic vertebra). Lumbar vertebrae support the most weight and are generally more massive than the others. Several kinds of joints **(c)** are formed by the skull, pelvis, arm bones, and hip. The knee, a hinge joint **(d)**, is quite complex, bound with several ligaments.

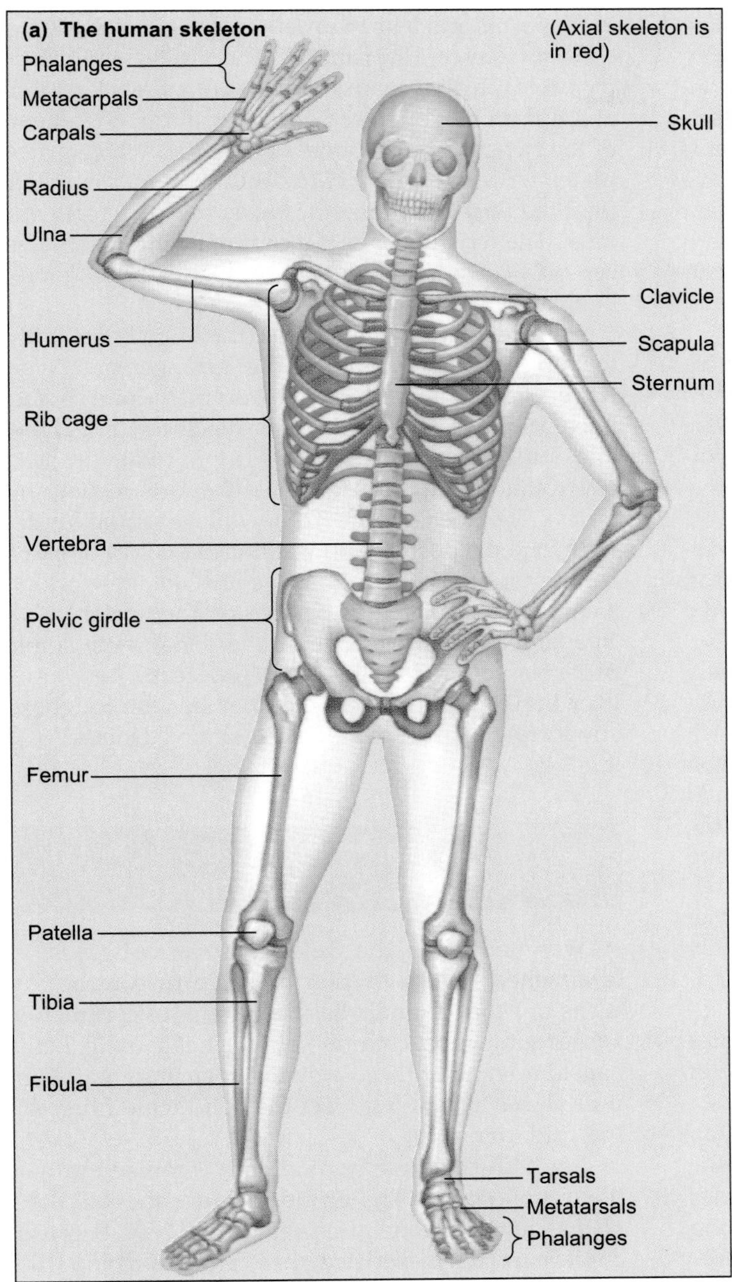
(a) The human skeleton
(Axial skeleton is in red)
Phalanges
Metacarpals
Carpals
Radius
Ulna
Humerus
Rib cage
Vertebra
Pelvic girdle
Femur
Patella
Tibia
Fibula
Skull
Clavicle
Scapula
Sternum
Tarsals
Metatarsals
Phalanges

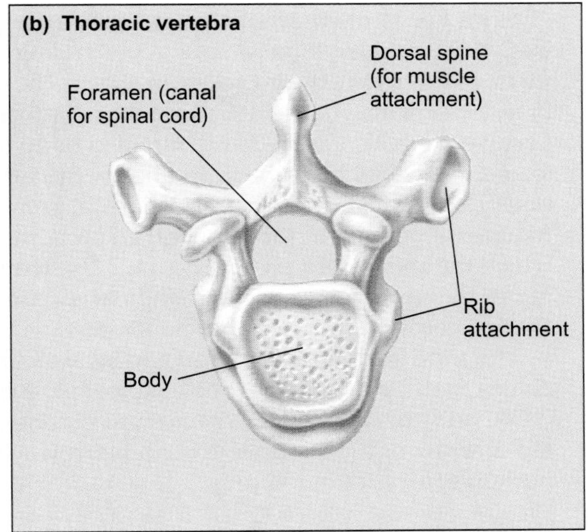
(b) Thoracic vertebra
Foramen (canal for spinal cord)
Dorsal spine (for muscle attachment)
Rib attachment
Body

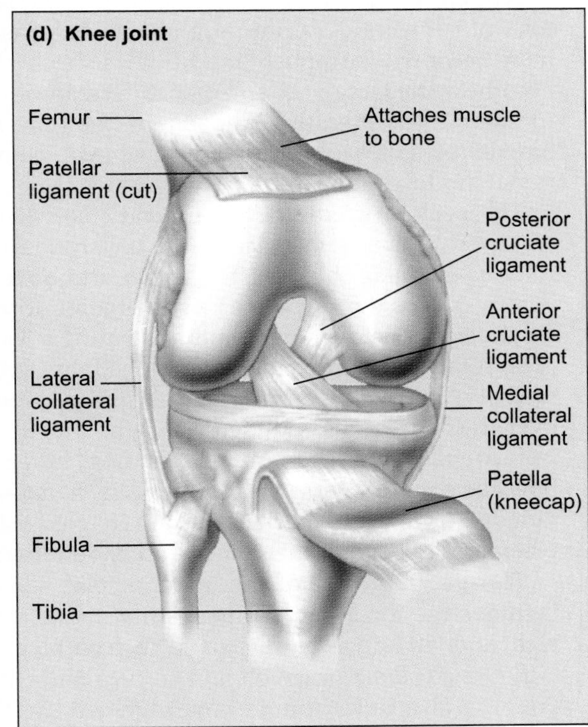
(d) Knee joint
Femur
Attaches muscle to bone
Patellar ligament (cut)
Posterior cruciate ligament
Anterior cruciate ligament
Lateral collateral ligament
Medial collateral ligament
Patella (kneecap)
Fibula
Tibia

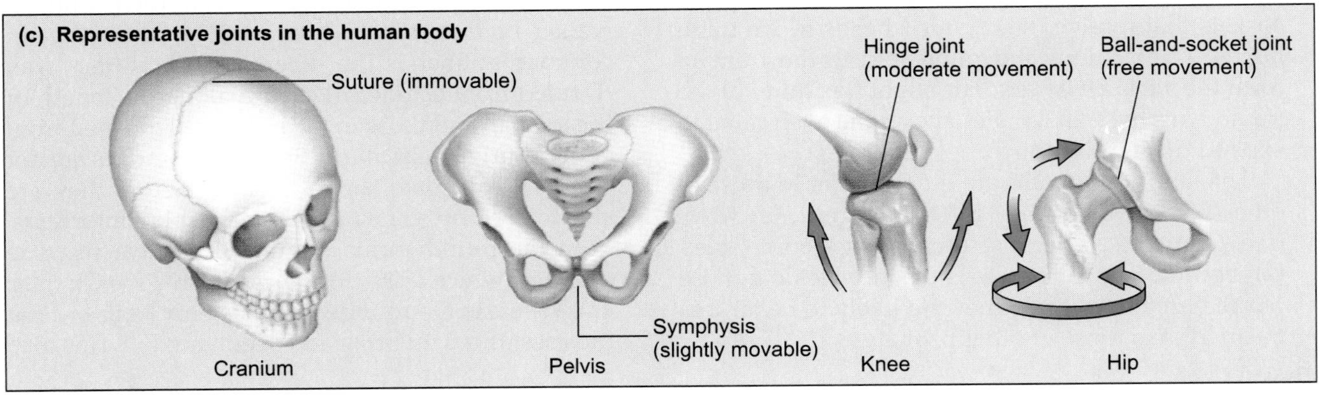
(c) Representative joints in the human body
Suture (immovable)
Hinge joint (moderate movement)
Ball-and-socket joint (free movement)
Symphysis (slightly movable)
Cranium
Pelvis
Knee
Hip

Since bones often serve as attachments for muscles, they usually have *processes* protruding from them, places where the muscles can attach. The vertebrae, in particular, have rather pronounced processes. Each vertebra also has a dense, rounded body that is joined to its neighbors above and below by cartilaginous disks. Most vertebrae also have a prominent foramen (opening) in the vertebral column, through which the spinal cord passes (Figure 32.8), terminating in the fused bones of the triangular sacrum and the tiny coccyx. The coccyx has no foramen.

The third region of the human axial skeleton includes the 12 pairs of ribs (one in 20 people have 13 pairs) and the shieldlike *sternum* (breastbone). The red marrow of the sternum and rib marrow are important sources of blood cells.

Appendicular Skeleton. The appendicular skeletons of vertebrates, from amphibians to primates, have become enormously modified as species have become increasingly adapted to a terrestrial environment. Also, because humans are the only truly bipedal primates, our skeletons have evolved additional modifications (see Chapter 21).

The **pectoral girdle** of the appendicular skeleton consists of the paired *clavicles* (the collarbones) and *scapulae* (shoulder blades). The upper arm bone (the *humerus*) joins the scapula at a socket joint, forming the most mobile of all our joints, providing a 360-degree rotation (a movement employed by softball pitchers). The gently curved scapula glides smoothly over the back of the rounded rib cage. This flexibility—coupled with the partially rotating, hinged elbow, the complex wrist, and the versatile hand—provides for a variety of motions that are characteristic of primates but beyond the range of other vertebrates.

The **pelvic girdle** of the appendicular skeleton includes the three pairs of bones that make up the right and left halves of the pelvis. Each half includes a flattened *ilium* (hip bone), an *ischium* (sitting bone), and a *pubis* (pubic bone). Where the three meet on each side, they form the two prominent sockets that receive the rounded heads of the thigh bones. The ischium and pubis join in the front to form the *pubic symphysis.* The slight flexibility of the pubic symphysis in females allows the birth canal to expand during childbirth.

The limbs of humans are quite similar to those of other primates, yet they differ in significant ways (see Chapter 21). Because our hands are free, they can become very dexterous; but since we do not use our feet much to climb, they are likely to be far less talented than those of other primates.

The arms and legs themselves also show the basic construction of all primates. The arm is extremely movable, partly because of the ball-and-socket joint at the shoulder, but also because of the arrangement of the two lower arm bones. The *radius,* which articulates with the wrist on the thumb side, can cross over the *ulna* (see Figure 32.8a) as the lower arm rotates. The wrist *seems* as if it can rotate but it actually can only do side-to-side and back-and-forth "good-bye" motions.

The articulation of the *femur* (the thigh bone) with the hip is another ball-and-socket arrangement. A far more peculiar and seemingly unlikely joint is the knee. While less versatile than the elbow hinge, the knee hinge does have some ability to rotate (in fact, overrotation is the cause of many sports-related injuries). The knee (Figure 32.8), while essential to upright, bipedal posture and walking, is such a complex contraption that its evolution seems like the work of a committee. The femur perches on the upright *tibia* (the larger lower leg bone or shinbone), with some protection offered by the *patella* (kneecap). Yet, however flimsy it may seem, the knee has the toughest supporting capsule of any of the synovial joints.

The Variety of Contractile Systems

As was mentioned, the skeleton can move because it is attached to tissues that are able to contract. It turns out that animals have a variety of contractile systems, the most common of which is muscle. Let's consider some of these systems, focusing finally on muscle but beginning with the contractile fibers of the cnidarians.

Cnidarians, you may recall, are organized rather like a hollow sac. The group includes the jellyfish, *Hydra,* and sea anemones (Figure 32.9). Because cnidarians lack mesoderm, they cannot develop true vertebrate-type muscles. However, their double-walled bodies contain two sets of rather efficient contractile fibers. The fibers in the outer wall (formed from ectoderm) extend along the length of the animal, and those in the inner wall (formed from endoderm) run circularly. Because anemones are not restricted by some sort of rigid skeleton, they are able to perform some rather remarkable contortions.

In the roundworms (Figure 32.10), we see true muscles (which arise from mesoderm—see Chapter 43). Whereas the roundworm has only longitudinal muscles, the flatworms and segmented worms also

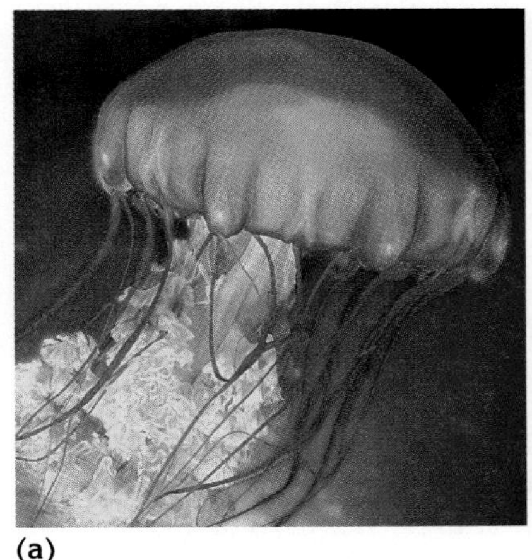

(a)

(b)

(c)

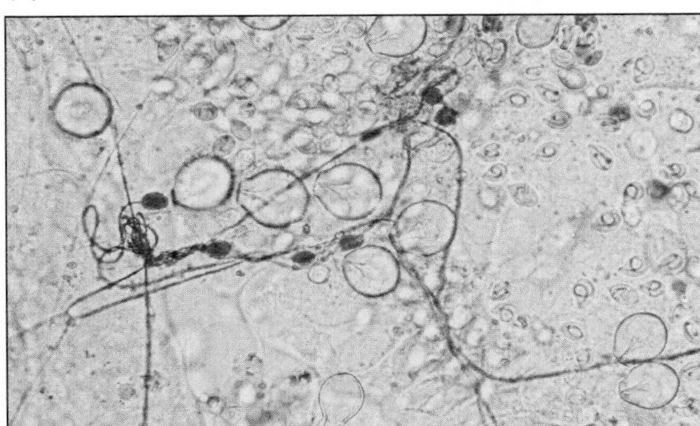

(d)

FIGURE 32.9 CNIDARIANS.
(a) Jellyfish, **(b)** *Hydra,* and **(c)** sea anemone are all cnidarians. Cnidarians lack mesoderm; hence they do not form true muscle. They can move slowly by contracting their outer ring of contractile tissue formed from ectoderm and an inner, radially arranged set formed from endoderm. **(d)** Cnidarian contractile fibers.

have circular muscles, and some species even have diagonal muscles (those that run at an angle between the dorsal and ventral body walls) and dorsoventral muscles (running directly across between the dorsal and ventral walls).

Arthropods, the joint-legged group that includes the insects, are supported by an exoskeleton. Here, we find a distinct arrangement of contractile fibers where the muscles lie *within* the skeleton. Figure 32.11 compares the arthropod and vertebrate systems.

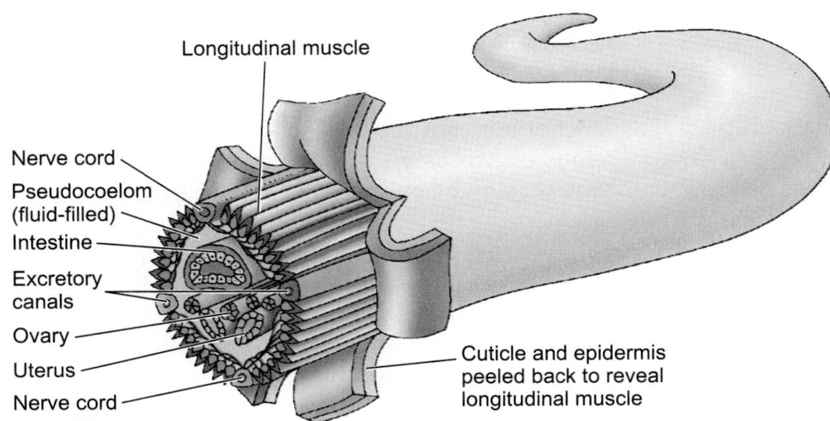

FIGURE 32.10 ROUNDWORM MUSCULATURE.
This cross section of a roundworm, *Ascaris,* a parasite of humans, shows the true muscles that arise from mesoderm.

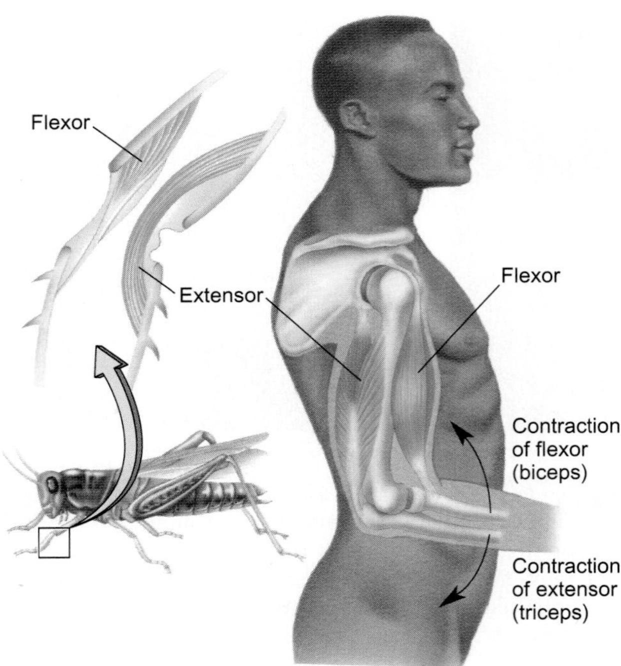

FIGURE 32.11 ARTHROPOD AND HUMAN MUSCULATURE

A comparison of the musculature of an insect and a human. Note, particularly, the relationship of muscles and joints. Also see the opposite arrangements of the flexors and extensors because of the differences in skeletal attachments.

VERTEBRATE MUSCLE AND ITS MOVEMENT

Any movement of cells, whether the waving of cilia, the contraction of the mitotic spindle fibers, or the flexing of an arm, relies on universal principles. Certain long, filamentous proteins, using energy derived from ATP, are able to slide past each other. The result is an effective shortening of the filament pair. When a great number of such filaments are organized into a muscle, and that muscle is attached to a structure such as bone, it can move that structure, sometimes with considerable force.

MUSCLE TISSUE TYPES

Vertebrate muscle can be classified into three types: *skeletal muscle* (also called "striated" or "voluntary" muscle), *cardiac muscle* (heart muscle), and *smooth muscle* (also called "visceral" or "involuntary" muscle). The three types of muscle tissue differ in appearance, location, function, and means of control.

Skeletal Muscle. Skeletal muscle is often called "striated" (striped) because of the prominent cross-banding visible under the microscope (Figure 32.12a). Its multinucleate cells are referred to as **muscle fibers.** Groups of fibers with the same action are called **muscle bundles.** Most skeletal muscles, as you may have guessed, move the skeleton. Generally, movement is voluntary, which means that there is conscious control over the action. Contraction in skeletal muscle can be rapid, powerful, and often sustained.

Cardiac Muscle. Cardiac muscle has traits in common with both smooth and skeletal muscle. For example, like smooth muscle, it is involuntary, and like skeletal muscle, it is striated. However, cardiac muscle differs from both in that its fibers are branched, forming a woven appearance. Furthermore, in cardiac muscle individual cells are separated by prominent *intercalated disks* (Figure 32.12b), which are not disks at all but interlocking, fingerlike foldings of adjacent plasma membranes. Because of the enormous interface created by this folding between adjacent cells, impulses that result in contraction can pass rapidly from one cell to the other.

Compared to skeletal contraction, contraction of cardiac muscle is highly rhythmic and moderately slow. Further, its rhythmic contraction is intrinsic (from within); it needs no external stimulus. Even cells removed and grown in the laboratory lie pulsing in their dishes for as long as they remain alive.

Smooth Muscle. Smooth muscle cells are long, spindly, and tapered. Each cell has one nucleus (Figure 32.12c). The cells commonly occur in flattened sheets and are found throughout the body—for example, in the walls of the gut, blood vessels, at the bases of hairs, in the iris of the eye, and in the uterus and other parts of the reproductive system. Smooth muscle is regarded as involuntary, meaning that there is little if any conscious control over its functioning; but there is evidence that some conscious control is possible. Smooth muscle tends to contract slowly and in a wavelike manner.

ORGANIZATION OF SKELETAL MUSCLE

Muscles are held together by connective tissue composed largely of extracellular material. Skeletal muscles are richly supplied with blood vessels that run between and within the bundles of fibers, supplying oxygen and nutrients and carrying away wastes. An efficient circulatory system is critical to the efficient functioning of muscle. (What happens when a working muscle does not receive enough

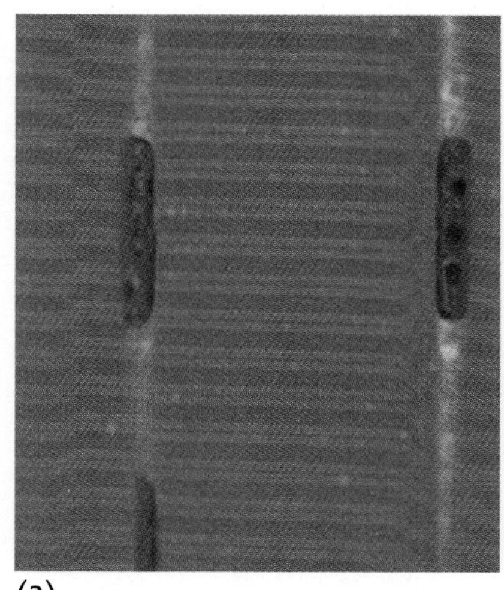
(a)

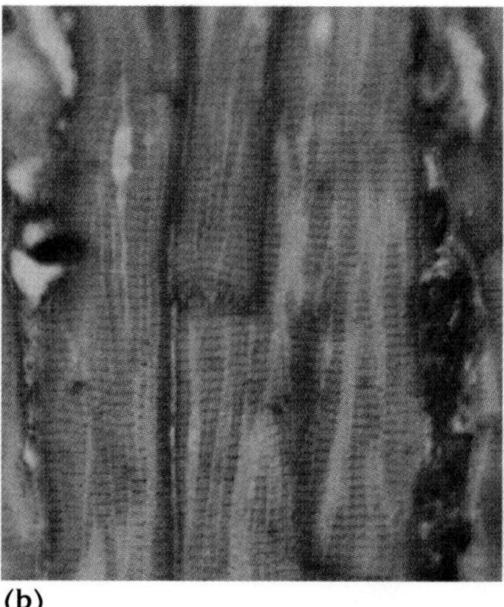
(b)

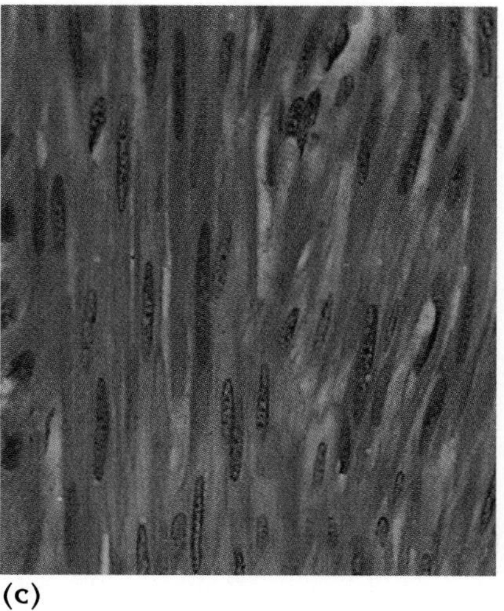
(c)

FIGURE 32.12 COMPARISON OF MUSCLE TISSUE.
Skeletal muscle **(a)** is heavily striated and multinucleate. Cardiac muscle **(b)** tissue branches and rebranches and is interrupted by intercalated disks. Like skeletal muscle it is heavily striated (striped). Smooth muscle **(c)** consists of long, spindly cells, each containing a single nucleus.

oxygen? See Chapter 8.) Nerves, too, run through the muscles, their branches dividing ever more finely until individual neurons finally innervate muscle fibers at what are called **neuromuscular junctions.** These nerves carry the messages that stimulate the muscles to contract.

Muscle bundles are enclosed by **fascia,** a tough casing of connective tissue. At both ends of the muscle, the fascia merges into increasingly dense collagenous tissue, forming cordlike **tendons** that attach muscle to bone. One end of a muscle is usually attached to a more-or-less stationary base, the **origin,** while the other end is attached to a more movable part, the **insertion.** (Can you find the origin and insertion in Figure 32.11?)

Not all skeletal muscles move bones. The muscles of your face, for instance, move your face. They enable you to produce all of your wonderfully attractive expressions. Tongue muscles also have complex origins and insertions, and so you can move your tongue in an interesting variety of ways. (Show your dexterity to the person sitting next to you on the bus.) Both smooth muscles and skeletal muscles may form rings around various passages or openings. These muscles, called **sphincters,** are found around the anus and in the gut, in the mouth, and even in many blood vessels. Other muscles form flattened sheets and have broad, thin tendons, such as the sheet of abdominal muscles that helps you pull in your stomach at the beach. Figure 32.13 shows some of the major muscles of the human body.

Muscles that move an appendage one way usually have opposing muscles that move it the other way. For example, the biceps flexes the arm and its opposing muscle, the triceps, straightens it. The muscle that contracts is called the **agonist.** The muscle with the opposite action is called the **antagonist.** The term suggests that they oppose or "fight" each other. In fact, however, they must cooperate in a highly coordinated fashion for most normal movements. Because of such interplay, we are able to rise gracefully from our chair and to walk over to meet our fiancée's parents rather than leaping from the chair and sprawling at their feet, ruining the afternoon for everyone.

ULTRASTRUCTURE OF SKELETAL MUSCLE

Now let's take a look at a more detailed level of muscle structure and see how such "ultrastructure" is important in muscle contraction. To get there, we must descend through several levels of organization (Figure 32.14), beginning with the whole muscle. Muscles are composed of muscle bundles, which subdivide into **fasciculi.** Within fasciculi lie the cellular units—the muscle fibers. As we have noted, the muscle bundle is made up of units called fibers, roughly the equivalent of cells. Thanks to the vast resolving power of the electron microscope we now know a great deal about the makeup of these fibers.

Skeletal muscle fibers range from a few millimeters in length to several centimeters—enormously

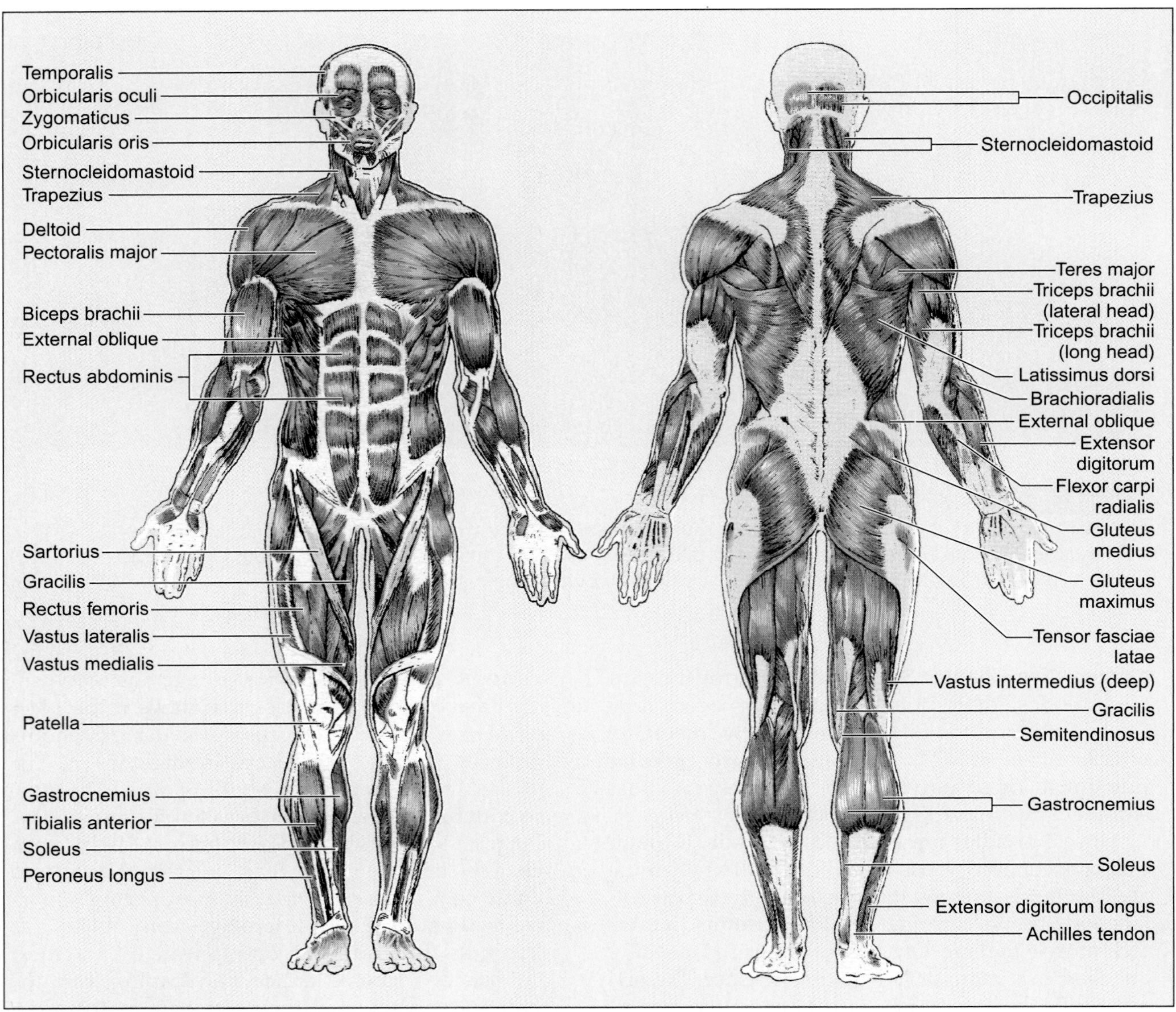

FIGURE 32.13 THE MUSCLES OF THE HUMAN BODY.
The human muscles illustrate the various types and arrangements and give us some idea about origins and insertions. Here, only the major muscles have been named.

long as cells go. Each muscle fiber is surrounded by a *sarcolemma,* the equivalent of a plasma membrane. Motor neurons (nerve cells carrying impulses to the muscle) join the muscle at neuromuscular junctions. The neuromuscular junctions are the nerve-muscle interfaces across which impulses pass, triggering contraction.

Within the muscle fiber, just below the sarcolemma, are found a number of nuclei and mitochondria—just what one would expect for such an active tissue. Lying just beneath the sarcolemma is the *sarcoplasmic reticulum,* which, like the endoplasmic reticulum of other cells, is a membranous, hollow

FIGURE 32.14 LEVELS OF MUSCLE ORGANIZATION.
A trip through the levels of muscle organization begins with the whole muscle, which is subdivided into muscle bundles, composed of fasciculi. Each muscle fiber contains, at its core, several clusters of contractile proteins called myofibrils. A dense sarcoplasmic reticulum and intermittent T-tubules are prominent, as are a large number of mitochondria. A single myofibril is subdivided by Z lines into sarcomeres or contractile units, each of which contains a great many myofilaments. The arrangement of myofilaments forms the striated pattern so prominent in skeletal muscle.

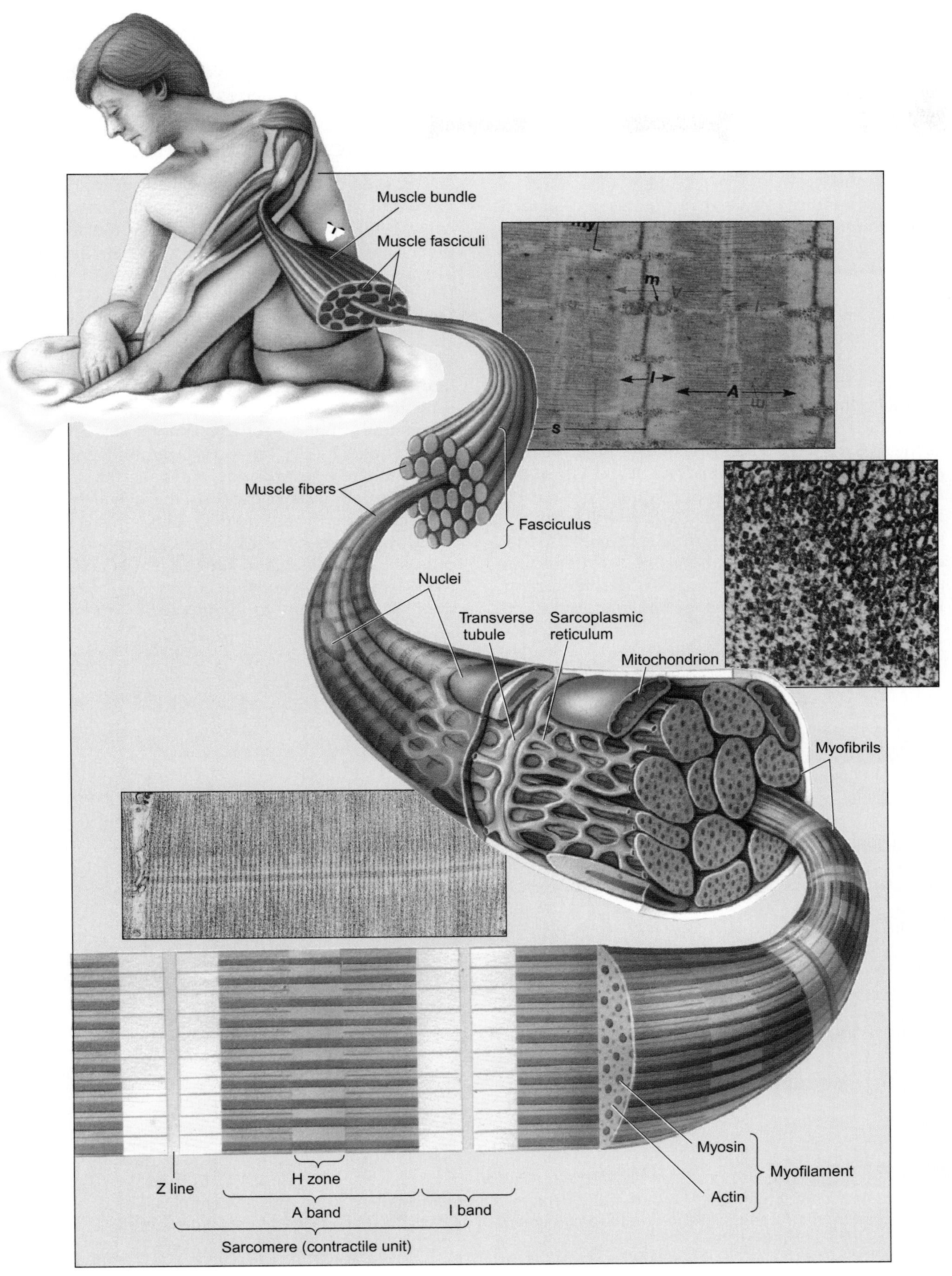
Muscle bundle
Muscle fasciculi
Muscle fibers
Fasciculus
Nuclei
Transverse tubule
Sarcoplasmic reticulum
Mitochondrion
Myofibrils
Myosin
Actin
Myofilament
Z line
H zone
A band
I band
Sarcomere (contractile unit)

structure. Somewhat larger than the sarcoplasmic reticulum are the *transverse tubules* (T-tubules).

The actual contractile parts of the muscle fiber are the rod-shaped **myofibrils** (Figure 32.14). Myofibrils form the visible pattern of striations characteristic of skeletal muscle.

Each myofibril contains numerous rod-shaped **myofilaments,** composed of the protein **myosin,** and an even greater number of thinner filaments of **actin.** When a myofibril is viewed in cross section, we usually find each thick myosin filament surrounded by six thinner actin filaments. The two kinds of filaments are arranged in a highly specific longitudinal manner.

Look at this arrangement from the surface, beginning with the very prominent *Z lines.* The region between the Z lines is the contractile unit, also known as the *sarcomere.* On either side of the Z line, we find a very light zone called the *I band.* It consists of the actin filament only, with no overlapping of myosin. Further inward is a broad, dark *A band,* with a smaller and lighter central strip, the *H zone.* The two darker parts of the A band consist of overlapping actin and myosin myofilaments. The lighter central H zone represents myosin alone unless the muscle is contracted. When muscle contracts, the H zone becomes markedly smaller, as we shall see.

Contraction of Skeletal Muscle

Painstaking studies of electron micrographs of relaxed and contracted sarcomeres reveal that, during contraction, the actin filaments move toward each other, sliding along the stationary myosin filaments. This is now called the **sliding filament model** (Figure 32.15). As seen through the electron microscope, the inward movement of actin from both sides brings the Z lines closer together and shortens the entire sarcomere. This movement continues until the actin filaments touch or even overlap. At that time the H zone becomes smaller, as the A band becomes uniformly dense. This is not a local event in contracting muscle, but occurs simultaneously in each sarcomere, producing a rapid shortening of the entire muscle (which explains the deepening furrows in your brow as you read this).

All this leaves us with a host of questions. What causes actin filaments to slide? What is actin, anyway? What is the source of energy for the contraction? The answers to many such fundamental

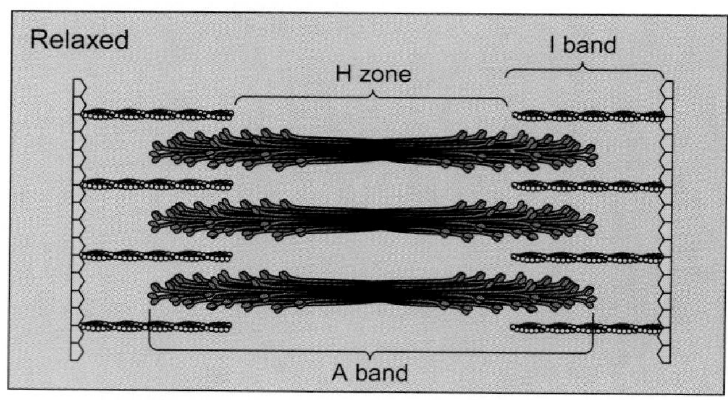

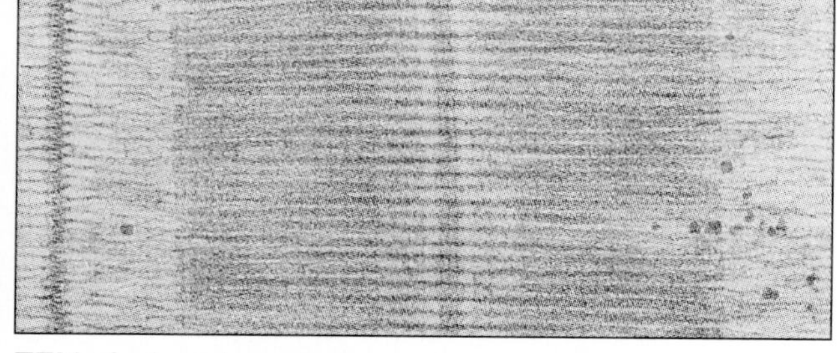

TEM of relaxed sarcomere

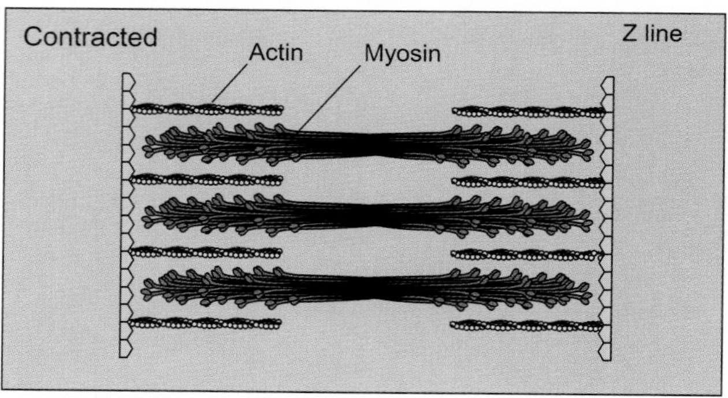

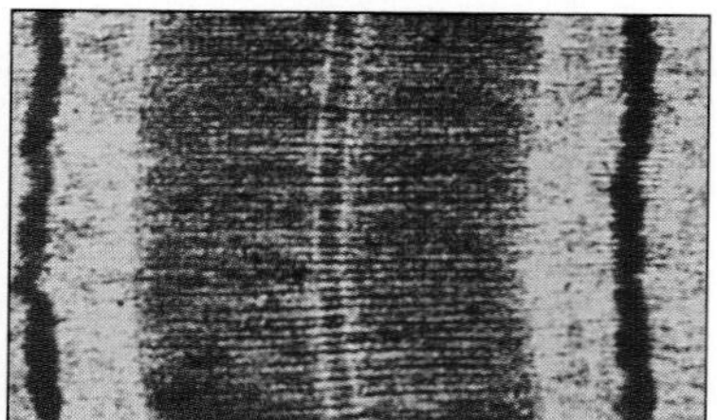

TEM of contracted sarcomere

FIGURE 32.15 **THE SLIDING FILAMENT MODEL.**
The two electron micrographs show what happens to the contractile unit when muscle contracts. The sliding filament model suggests the inward movement of the Z lines, the change in density of the A band, and the reduction of the H zone.

questions have been provided by the use of the electron microscope and the development of increasingly sophisticated biochemical analysis. For example, by use of the electron microscope, at very high magnification, researchers found numerous tiny cross-bridges extending from the myosin filaments to the surrounding actin filaments (Figure 32.15). The cross-bridges, we now know, are globular extensions of myosin, called *myosin heads.* During muscle contraction the myosin heads apparently extend outward toward the Z line and attach to the nearest actin filament, forming bridges; and then they bend inward toward the center of the sarcomere—and again, the myosin heads lean out along the length of the actin, attaching, bending inward, releasing, and then leaning out to get a new grip further along. The pulling action is not unlike the hand-over-hand movements a line of sailors might use to raise the mainsail.

As you can see in Figure 32.15, the pulling action occurs on each side of the H zone, which explains why the Z lines draw together as each sarcomere shortens. When the action is repeated throughout a muscle, the entire muscle shortens.

Control of Muscle Contraction

So, you may ask, what controls muscle contraction? What causes myosin bridges to form, and what prevents them from forming when they are not needed? The answer is on the actin filament, which actually has two additional proteins associated with the actin that composes the filament. In addition to actin, there are *tropomyosin* and *troponin.*

As we've seen, actin does the actual moving (sliding) during contraction. It provides binding sites for the myosin bridges, and it is this binding that triggers ATPase, the enzyme that breaks down ATP to ADP, activity in the myosin heads. Tropomyosin and troponin are able to control muscle contraction because they block the actin-myosin binding sites—thus inhibiting contraction (Figure 32.16). The sites will stay

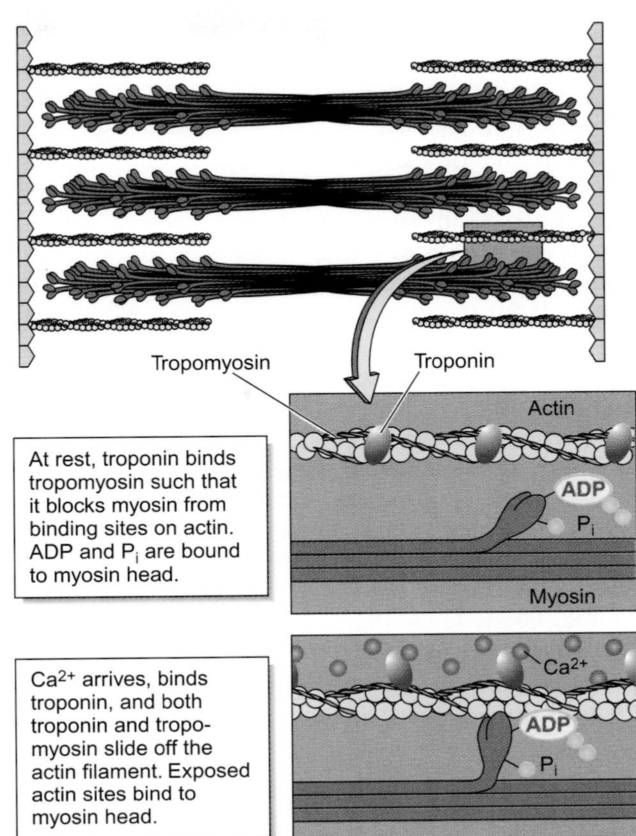

Tropomyosin
Troponin
Actin
At rest, troponin binds tropomyosin such that it blocks myosin from binding sites on actin. ADP and P_i are bound to myosin head.
ADP
P_i
Myosin
Ca^{2+} arrives, binds troponin, and both troponin and tropomyosin slide off the actin filament. Exposed actin sites bind to myosin head.
Ca^{2+}
ADP
P_i

ADP and P_i are released, causing myosin heads to bend, pulling actin past the myosin.

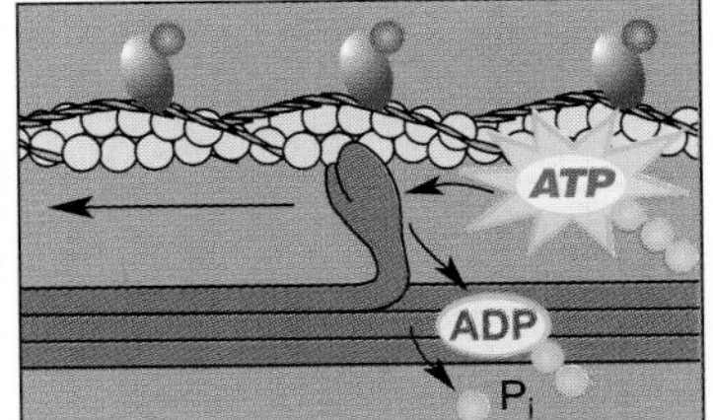

A new molecule of ATP joins myosin head, causing it to release actin.

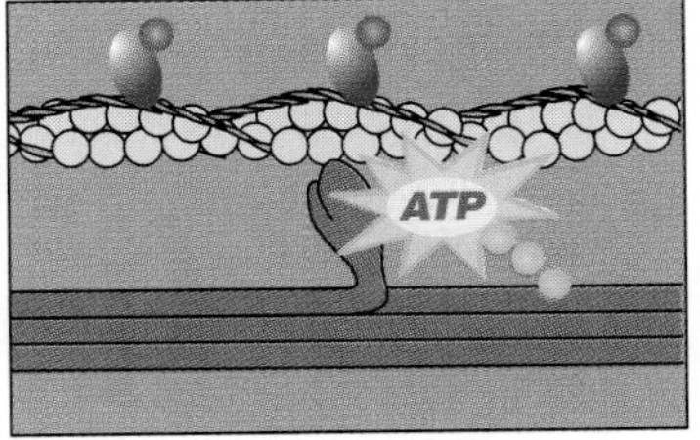

Hydrolysis of ATP returns myosin head to its original position. Removal of Ca^{+2} restores resting state.

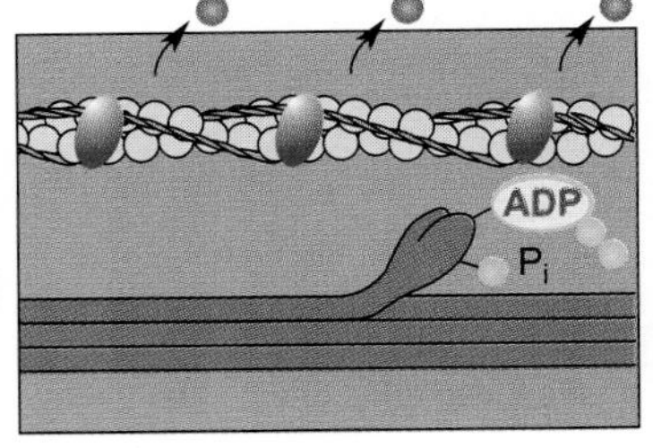

Figure 32.16 Mechanism of Muscle Contraction.
Calcium ions attach to troponin, causing the tropomyosin-troponin complex to shift, exposing the attachment site for the myosin head. The energy of ATP will then allow the myosin head to pull the actin and myosin molecules past each other. The calcium ions will be released and the actin filament will return to the resting stage.

blocked, in fact, until the appearance of the calcium ion (Ca^{2+}). When Ca^{2+} is present in the contractile unit, it binds to troponin, causing it to release its hold on the binding site of actin. This allows the tropomyosin to slide the troponin off the binding sites. Actin's myosin-binding sites are then exposed, and the myosin heads attach and become active. Upon the release of energy from ATP, the bridges draw the actin filaments inward, and the muscle contracts. It turns out that each complete action by a myosin head requires two ATPs, the first to straighten the myosin head, the second to break its attachment so that it may reattach to a new binding site.

The next question is, What controls the calcium? In the resting muscle, calcium ions are stored in the sarcoplasmic reticulum, continually transported there by ATP-powered membranal pumps. The pumps thus keep the contractile units clear of calcium until muscle contraction is needed.

We already know that neural impulses elicit muscle contraction, so all that remains is to fill in one more blank. When a neural impulse from a motor neuron reaches the neuromuscular junction on the sarcolemma, it travels along the sarcolemma, and then deep into the muscles via the T-tubules to the vicinity of the membranous sarcoplasmic reticulum—within which the calcium ions are stored. The disturbance causes calcium ion channels in the sarcoplasmic reticulum to open; and as the ions suddenly leak out, they diffuse rapidly into the surrounding contractile units. The calcium ions combine with the troponin, causing the tropomyosin-troponin complex to move, exposing the myosin binding sites. The muscle then contracts. All of this occurs so rapidly that all of the contractile units in each muscle fiber can contract almost simultaneously.

The muscle will contract as long as calcium ions are present, and calcium remains as long as the neural impulses continue. When the impulses cease, the events of contraction reverse themselves. The calcium ion pumps quickly to clear the ions from the contractile units (Figure 32.16), the tropomyosin-troponin complexes block the myosin-binding sites of actin, and the muscle relaxes.

We have seen how individual skeletal muscle fibers react to an impulse from a motor neuron, but we should point out that individual muscle fibers rarely act alone. In fact, usually at least 10 or 12 fibers are activated at once. One motor neuron usually branches into at least that many muscle fibers. When an impulse arrives, all of the fibers served by that motor neuron will react simultaneously; thus the reacting group is aptly called a **motor unit.** Motor units of just a few muscle fibers are usually found where precision is important, as in the muscles of the hands or eyes. On the other hand, motor units in the calf muscle, which tends to react as a whole, may include hundreds of muscle fibers.

PERSPECTIVES

We now have some ideas on the skeletal organization in a range of animals with special emphasis on the vertebrates, and we know some of the intricacies of muscle contraction. As with much of biology, however, it is difficult to separate one topic from another, because of the complexities, cooperation, and interdependencies of life. Thus we find that by discussing muscle, we have introduced the idea of nerve action. So in the next chapter, we will take a closer look at this fascinating phenomenon as we consider some of the intriguing new finds of neurobiology, one of the fastest growing fields of biology.

Summary

Animal Tissues

Epithelial tissue forms linings, separated from tissues below by a basement membrane. Examples include epidermis and the linings of the digestive, respiratory, and reproductive tracts. Glands are composed of secretory or glandular epithelium. Among the epithelial cells are the stratified squamous cells of the outermost epidermis and the continuously active cuboidal epithelium below. Columnar epithelium is found on many absorptive surfaces and forms much of the secretory lining of the respiratory system.

Connective tissue forms bone, cartilage, tendons, ligaments, and blood. It includes cells and the surrounding matrix they secrete. Muscle, a contractile tissue, has important functions in a variety of species. Nerve tissue responds to stimuli by conducting neural impulses. The functional units are called neurons or nerve cells.

Exoskeletons provide support from outside the body, and endoskeletons provide support from within. Muscle tissue of soft-bodied animals contracts against body fluids, creating pressure that forms a hydrostatic skeleton. Sponge skeletons consist of spicules or spongin scattered throughout tissue. Snails and clams grow by secreting extensions to their shells. Lobsters, crabs, and crayfish discard their old skeletons and grow new ones. The functions of the vertebrate skeleton include support, movement, protection, blood cell formation, and the storage of calcium and phosphate.

Microscopically, compact bone is organized into cylindrical Haversian systems, each with a central canal and concentric hardened layers or lamellae. Osteocytes (bone cells) reside within numerous tiny chambers called lacunae and communicate via minute canaliculi.

The axial skeleton includes bones of the skull, the vertebral column, and the thoracic or chest region. The limbs and the pectoral and pelvic girdles make up the appendicular skeleton.

Bones articulate at rotating ball-and-socket joints, hinge joints, gliding joints, slightly movable joints, and immovable sutures. Bone surfaces at joints contain smooth cartilage, and the joints are bound by tough ligaments. Some ligaments appear as capsules that form synovial joints.

The pectoral girdle includes the two clavicles and two scapulas, each pair forming ball-and-socket joints with a humerus (upper arm bone). The pelvic girdle has three pairs of bones (ilium, ischium, and pubis) joined at the rear by the sacrum. The three resulting joints are the pubic symphysis, sacroiliac, and hip joint.

Arm bones include the humerus, radius, and ulna. The latter two form a rotating hinge at both ends. In the leg, the upper end of the femur and the hip bone form a ball-and-socket joint, and the lower end of the femur—with the tibia and patella—form the hinged knee joint.

The Variety of Contractile Systems

Cnidarians lack mesoderm and so have no true muscle, but two sets of contractile fibers in their body walls, running longitudinally and circularly. Roundworms have true muscles that run longitudinally. Flatworms and segmented worms have longitudinal and circular muscles. Some have diagonal and dorsoventral muscles. Arthropods have an exoskeleton with muscles attached to the inside.

Vertebrate Muscle and Its Movement

Skeletal muscle (striated, voluntary) moves the skeleton and supports the trunk. Muscle fibers are the structural units. Contractions are rapid, forceful, and voluntary.

Cardiac (heart) muscle resembles skeletal muscle, but its fibers branch and are connected by intercalated disks. Its contractions are slow, rhythmic, and involuntary.

Smooth muscle (visceral, involuntary) occurs in the gut, blood vessels, iris, and uterus. Its contractions are slow and wavelike.

Skeletal muscle is bound by connective tissue and contains many blood vessels and nerves. Neurons form neuromuscular junctions on individual fibers. Fascia, enclosing muscle bundles, forms tough tendons that join muscle to bone. Such muscles have origins (stationary bases) and insertions (the part moved). Muscles with opposing functions—agonists and antagonists—work cooperatively in movement and posture.

The muscle fiber is enclosed by a sarcolemma, which receives motor neurons. Rod-shaped myofibrils are made up of thick myofilaments of myosin and thin, surrounding myofilaments of actin.

In the surface patterns, Z lines form the boundaries of sarcomeres. Within are lighter I bands of actin, followed by darker A bands of both actin and myosin, and central H zones of myosin alone.

The sliding filament model maintains that during contraction actin myofilaments slide past myosin myofilaments, which remain stationary. Accordingly, the Z lines move together, the I band decreases, and the H zone disappears. The sliding movement occurs as myosin heads join actin, forming cross-bridges and pulling the actin inward. In the resting state, bridge formation is inhibited.

Control of Muscle Contraction

Contraction is triggered by calcium ions whose presence alters the troponin-tropomyosin complex, the bridge-inhibiting factor. Bridge attachment triggers ATP action, and the myosin cross-bridges begin to contract.

Key Terms

epithelial tissue • **539**
connective tissue • **541**
matrix • **542**
collagen • **542**
connective tissue proper • **542**
adipose tissue • **542**
bone • **542**
cartilage • **542**
blood • **543**
muscle tissue • **543**
nervous tissue • **543**
hydrostatic skeleton • **543**
exoskeleton • **543**
endoskeleton • **544**
osteocyte • **544**
Haversian system • **544**
axial skeleton • **546**
appendicular skeleton • **546**
joint • **546**
ligament • **546**
cranium • **546**
vertebral column • **546**
pectoral girdle • **548**
pelvic girdle • **548**
muscle fiber • **550**
muscle bundle • **550**
neuromuscular junction • **551**
fascia • **551**
tendon • **551**
origin • **551**
insertion • **551**
sphincter • **551**
agonist • **551**
antagonist • **551**
fasciculi • **551**
myofibril • **554**
myofilament • **554**
myosin • **554**
actin • **554**
sliding filament model • **554**
motor unit • **556**

Review Questions

1. List four places where you would expect epithelial tissue to be located.
2. Describe the kinds of cells and their arrangement in the human epidermis. What lies below the epidermis?
3. Describe the appearance of collagen. List four places in the body where it is located.
4. List the four basic types of connective tissue.
5. What is the function of adipose tissue?
6. What is a hydrostatic skeleton?
7. Where are the two kinds of marrow located? What is the function of the red?
8. Using an outline drawing, illustrate the Haversian system. Label and give the role of the central canal, osteocytes, lacunae, lamellae, and canaliculi.
9. List the general parts that make up the axial and the appendicular skeleton.
10. List four types of movable joints and describe the range of movement in each.
11. Cite an example of a synovial joint and describe its makeup. (Figure 32.6)
12. What do the skull, vertebral column, and rib cage protect?
13. Why is the knee joint sometimes described as "improbable"? Upon what does this joint rely for its strength?
14. List three unique characteristics of cardiac muscle.
15. List three structures in the body where smooth muscle is located. Characterize its movement.
16. What roles do blood vessels, connective tissues, and motor neurons perform in skeletal muscle?
17. Prepare a simple drawing of a skeletal muscle and its accompanying long bone. Label *bundle, tendon of origin,* and *tendon of insertion.* Now add an antagonistic muscle.
18. List three places where sphincters are located. What is their function?
19. Briefly state what each of the following provides to a muscle fiber: sarcolemma, mitochondrion, neuromuscular junction, sarcoplasmic reticulum.
20. Describe the makeup of a myofibril and provide a simple drawing of an end view.
21. Prepare an illustration showing the surface pattern of a contractile unit. Label the parts and explain what they represent.
22. What happens to the pattern of bands and lines in the contractile unit when contraction occurs?
23. Explain contraction in terms of the formation and movement of myosin heads.
24. Summarize the role of calcium ions in contraction.
25. How does the muscle fiber restore its resting state?

CHAPTER

33

Neurons

ALL LIFE, AT WHATEVER LEVEL, FROM CELLULAR TO ORGANISMAL, MUST *be able to respond to its environment and to coordinate its internal activities. Thus we find that—within the body—various means of signaling have developed. As we will see, these can be chemical or electrical, or both. If you've ever left a cozy tent in the middle of the night and stepped on a burning ember from the campfire, you have some idea of just how rapidly these internal signals can move, and just how adaptive they can be.*

In the chill of morning, a browsing hare nips at tender buds on a hillside. Just inches away are other plants—fresh, dewy, and succulent. But the hare avoids these because its senses reveal that these plants contain toxins. The hare's movement attracts the attention of a bobcat that quickly crouches and begins a stealthy approach, ears forward, totally focused on the browsing hare. Each step brings the cat closer, and with each step it risks giving itself away. Although its attention is on the hare, it feels the ground before placing its weight on each tentatively advancing paw. Seconds later, both animals will be rushing at full speed, their bodies a wonderful coordinated symphony of interactions. If the bobcat is successful, the hare's symphony will end in the dissonance of death (Figure 33.1).

FIGURE 33.1 ANIMAL RESPONSES.
Animals generally respond rapidly to incoming environmental cues. This ability is important in a number of ways, such as for finding food and in recognizing danger, such as a bobcat.

Such sensitivity, coordination, and interaction are common in the animal world. After all, the earth is a variable and changing place, and from time to time and place to place it offers innumerable threats and gifts. Animals have evolved the ability, then, to recognize the nature of their surroundings and to respond in an adaptive manner. This is possible because of specialized nerve cells that are sensitive to specific aspects of the environment and that can stimulate the body to respond appropriately.

Nerve cells, called **neurons,** respond to stimuli and effect changes in a similar fashion in all animals. As we will see, the mechanisms (electrical and chemical) operate on the same principles, whether in a sea star, a sea lion, or a real lion.

THE NEURON

Neurons, or nerve cells, exist in many sizes and shapes, but every neuron has a cell body from which extend a number of processes. Some processes are extremely long, reaching from one part of the body to another some distance away. For example, a single neuron can reach from your foot all the way to your pelvic region.

The **cell body** (Figure 33.2) contains the nucleus and most of the cell's cytoplasm. The cytoplasm includes such typical cell organelles as ribosomes, an endoplasmic reticulum, and numerous secretory bodies. In addition, the cell body is a source of **neurotransmitters,** which are chemicals that move to special sites in the neuron where they can be released. When such chemicals reach adjacent neurons, the neurons cells may be stimulated to transmit an impulse or may be inhibited from doing so. Neurotransmitters may also stimulate an **effector**

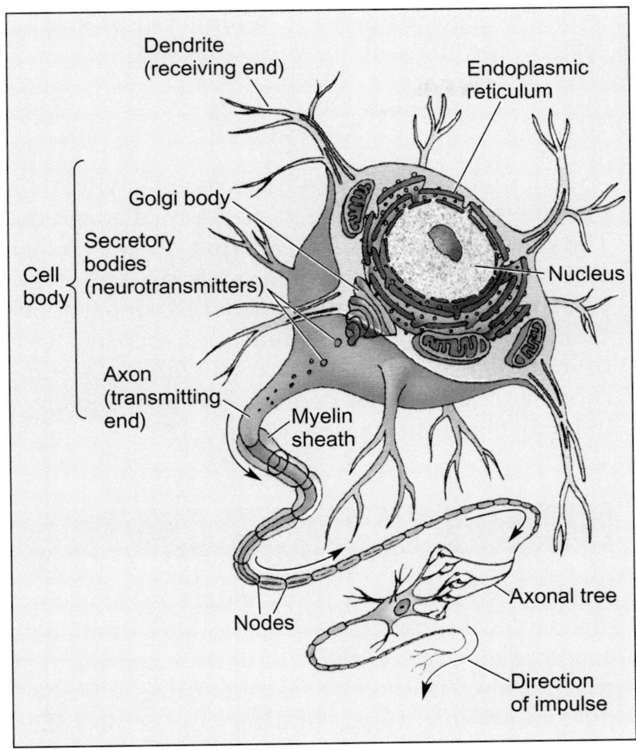

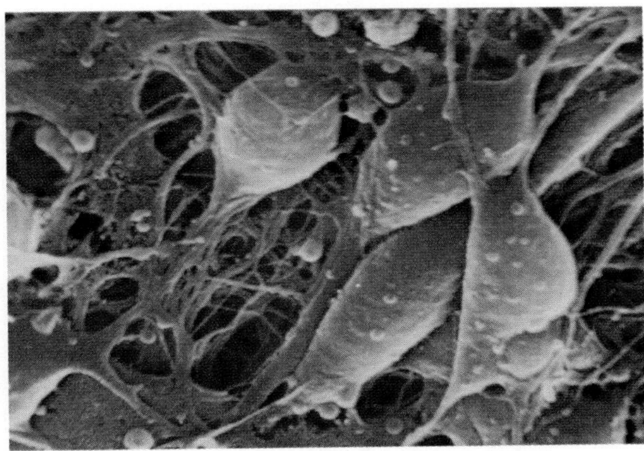

FIGURE 33.2 THE NEURON.
Neurons have cell bodies, dendrites, and axons; the latter are often myelinated. The direction of impulse movement is generally from dendrite to axon. An important function of the cell body is the synthesis of neurotransmitter molecules, which are transported into the axon, destined for the synaptic knobs.

(a structure capable of a response), such as a gland or a muscle.

Two major types of processes extend from the cell body (see Figure 33.2): the *dendrites* and the *axons.* The **dendrite** ("little tree") receives signals and, generally, transmits them *toward* the cell body.

The signal from the dendrite may reach the **axon,** which is a nerve fiber that transmits the neural impulse *away from* the cell body. It may communicate with other neurons or directly with an effector, such as a muscle. A neuron often has many dendrites, but it has only one axon. The single axon, however, may branch at any point along its length. An axon commonly divides and redivides at its tip, forming a terminal *axonal tree.* The axonal tree releases the neurotransmitters, thus chemically relaying the message to the next neuron or to an effector. The place at which an axon innervates a muscle fiber is called a *neuromuscular junction.*

The axons of many neurons outside the brain and spinal cord are covered by "flattened and rolled" *Schwann cells* that contain a fatty material called **myelin,** forming a **myelin sheath** (Figure 33.3). Like any lipid, myelin has great electrical resistance, so it acts as an insulator. Such wrappings are not complete. Between the cells wrapped around the axon are gaps called *nodes* where the axon remains exposed to its ion-rich, watery surroundings. As we will see, myelinated nerve fibers conduct impulses much more rapidly than nonmyelinated fibers because the impulse jumps from node to node.

TYPES OF NEURONS

There are three basic types of neurons—*sensory neurons, interneurons,* and *motor neurons*—that can transmit signals in a reflex arc (Figure 33.4).

Sensory neurons conduct impulses from sensory receptors to the central nervous system (the brain and spinal cord). We will say more about sensory receptors in Chapter 35, but let's note here that each type is specialized to detect specific environmental stimuli. Sensory neurons carry messages about conditions in the surroundings and in the body itself.

Interneurons communicate only with other neurons. They are largely responsible for integrating information, thereby coordinating responses. Interneurons make up much of the spinal cord and brain in humans.

Motor neurons transmit impulses to effectors such as muscles and glands. Thus activated motor neurons enable us to yell, jump, blink, secrete, sweat, blush, squint, and perform any number of other charming activities.

You should be aware that the brain and spinal cord contain other types of cells besides those that transmit impulses. In fact, one type, the **glial cells,** outnumber neurons ten to one. Glial cell functions

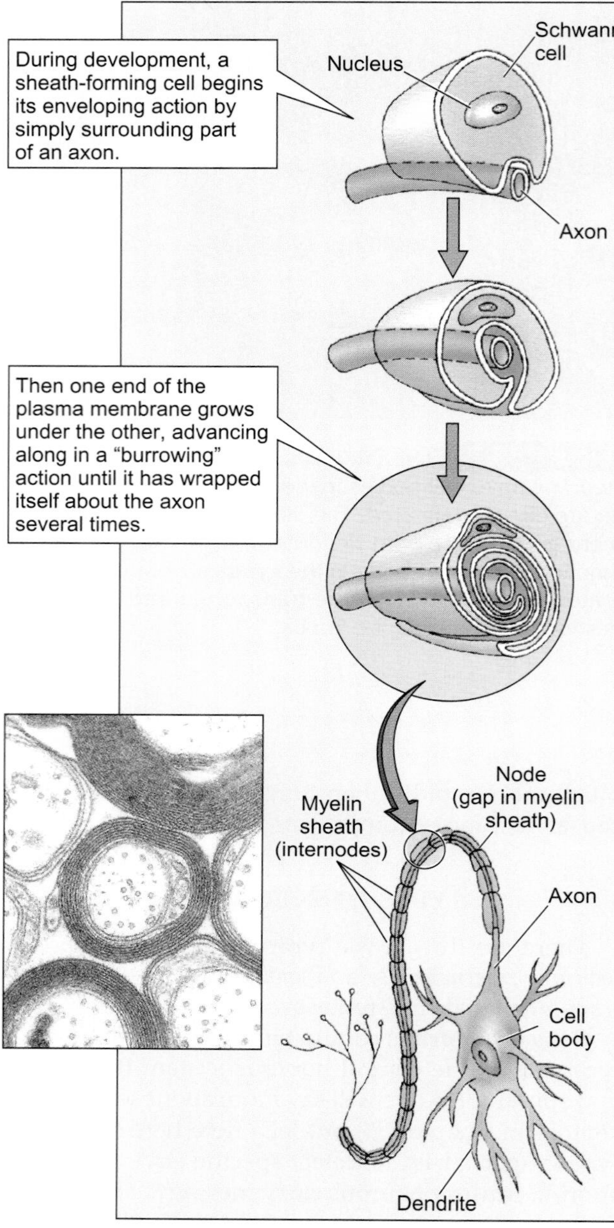
During development, a sheath-forming cell begins its enveloping action by simply surrounding part of an axon.
Nucleus
Schwann cell
Axon
Then one end of the plasma membrane grows under the other, advancing along in a "burrowing" action until it has wrapped itself about the axon several times.
Myelin sheath (internodes)
Node (gap in myelin sheath)
Axon
Cell body
Dendrite

FIGURE 33.3 **A MYELINATED NEURON.**
The electron micrograph is of a cross section through a myelinated axon. The encircling membranes are formed by Schwann cells. During development, a sheath-forming cell begins its enveloping action simply by surrounding part of an axon. Then one end of the plasma membrane grows under the other, advancing along in a "burrowing" action until it has wrapped itself about the axon several times.

are not well understood, but they may surround neurons, much as Schwann cells do, and they seem to serve some structural roles and provide metabolic support for the neurons. Neurobiologists have recently suggested possible information-processing roles for the glial cells as well.

NERVES

Bands of axons and dendrites, together with their protective and supportive tissues, are called **nerves.** Thus a spinal nerve, for example, might contain both dendrites of sensory neurons and axons of motor neurons. A nerve is somewhat like a telephone cable carrying many individual lines, each insulated from the other. Nerves appear as white, glistening cords, and are surrounded by their own coverings of tough connective tissue (Figure 33.5).

NEURAL ACTIVITY

While neurons can be said to conduct impulses from one part of the body to another, it's important not to confuse the term "conduct" with the usual electrical connotation. Neurons are far more than simple conductors, and a moving neural impulse is quite different from an electrical current passing through a wire conductor. A fundamental difference is that an electrical current passing through a conductor diminishes over time and distance, but a neural impulse, once started, does not diminish. As with a row of falling dominoes, in which the last falls with the same energy as the first, the impulse is transmitted with equal energy along the length of the neuron.

MEMBRANE PROPERTIES AND NEURAL ACTIVITY

We should begin our consideration of neural activity by looking more closely at the plasma membrane of the neuron because it holds the key to both the resting state and the impulse. As we saw in Chapter 5, a cell's plasma membrane regulates the passage of materials into and out of the cell. The neuron's plasma membrane is no exception. The substances of importance in neural activity are ions, most importantly sodium (Na^+) and potassium (K^+). One way that ions can cross the plasma membrane is through special protein-lined pores called **ion channels.** Ion channels tend to be highly specific, admitting only one or a few kinds of ions. For example, there are channels that admit only ions of sodium, potassium, calcium, or chloride (Figure 33.6).

Some of the channels are always open (*open channels*), but others have movable regions, called *ion gates,* that can open and close the channels. Those

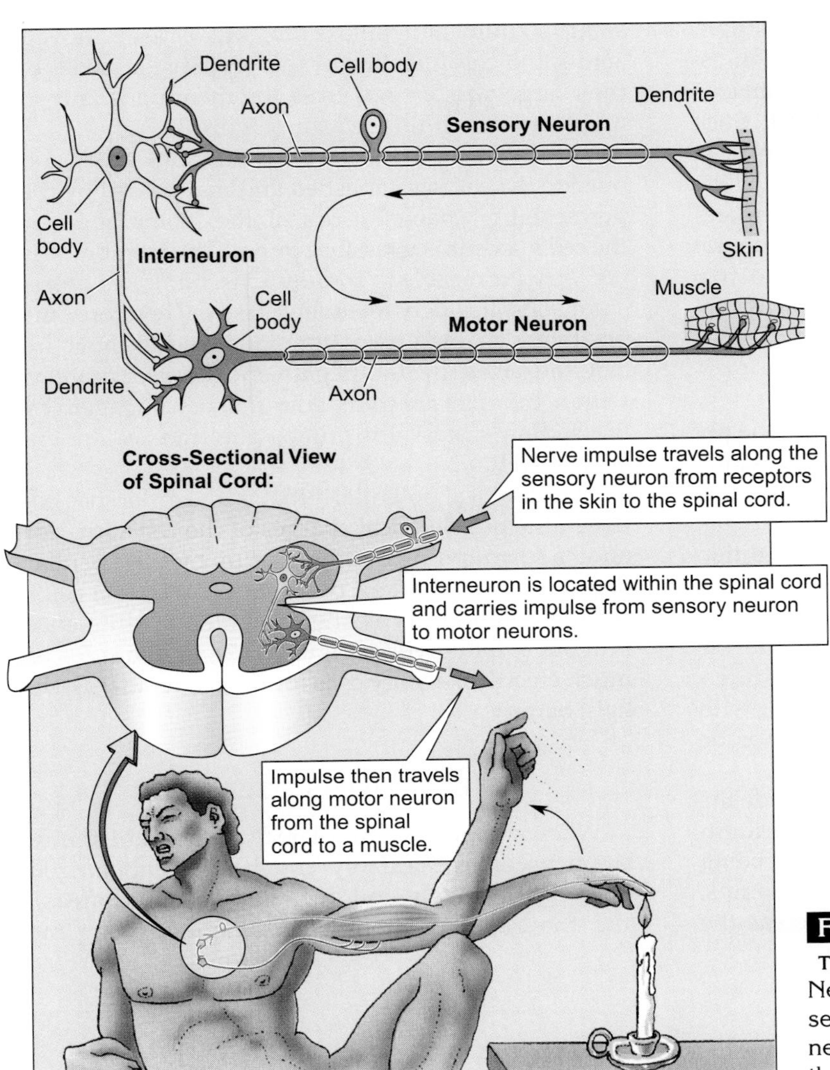
Dendrite
Cell body
Axon
Sensory Neuron
Dendrite
Cell
body
Interneuron
Skin
Axon
Cell
body
Muscle
Motor Neuron
Dendrite
Axon
Cross-Sectional View
of Spinal Cord:
Nerve impulse travels along the sensory neuron from receptors in the skin to the spinal cord.
Interneuron is located within the spinal cord and carries impulse from sensory neuron to motor neurons.
Impulse then travels along motor neuron from the spinal cord to a muscle.

FIGURE 33.4 TYPES OF NEURONS AND THE REFLEX ARC.
Nerve cells generally fall into three categories: sensory neurons, interneurons, and motor neurons. Interneurons are the most variable of the three types.

with such ion gates, logically enough, are called *gated channels.*

The forces that cause ions to move through the open channels are typically either a concentration gradient or an electrical gradient. Quite simply, ions tend to move away from the area of their greater concentration. So if a membrane separates a high concentration of certain molecules from a lower concentration, those molecules will show a net movement through the open channels in the direction of the lower concentration. And, because like charges repel while opposite charges attract, if there is a higher concentration of, say, positive ions on one side of a membrane than the other, positively charged molecules will tend to move away, through the open channels, to the other side of the membrane. The interaction of these two forces largely determines the passive distribution of ions through the open channels.

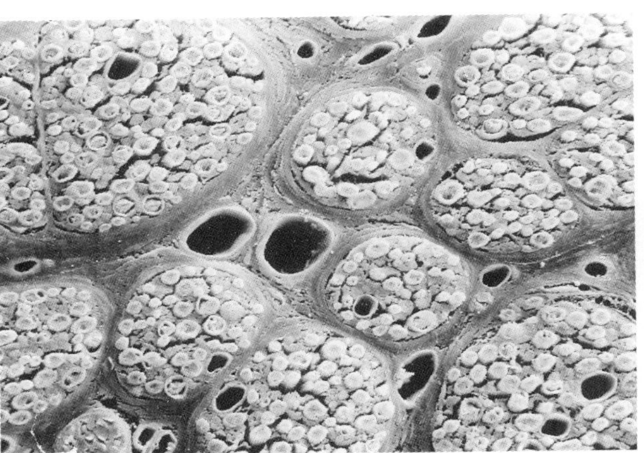

FIGURE 33.5 NERVES.
Nerve, as seen in cross section through the scanning electron microscope. The nerves are shown in bundles of neurons interspersed with blood vessels.

Another mechanism that determines the distribution of ions is the **sodium-potassium pump,** an assemblage of proteins that actively moves sodium ions out of the cell and potassium ions into the cell. Such pumps move three Na^+ out of the cell and two K^+ in for each pump cycle and require the energy of ATP.

Now that we know about channels, gates, and pumps, we can look at the events involved in neural transmission. We can divide the activity into two parts: *the resting potential* and *the action potential.*

The Resting Potential

The resting potential in a neuron refers to the period when no impulses are being transmitted. However, the term "resting" is misleading, because one thing that living neurons never do is rest. Instead, the "resting" neuron is *polarized*—that is, in a condition in which the electrical charges outside the axon's plasma membrane are different from those inside. Specifically, resting axons are more negative inside than outside. The maintenance of this charge difference involves the positioning of certain ions, with their electrical charges, on one side of the membrane or the other (Figure 33.7).

Sodium ions (Na^+) are more abundant outside the neurons, while potassium ions (K^+) are more abundant inside the neuron—their distribution being due to the actions of the sodium-potassium pumps, using ATP to power the mechanism. Whereas the small sodium and potassium ions can be moved across the cell membrane, the proteins, because of their large size, cannot cross the membrane and so must remain within the axon (Figure 33.7).

The difference in the ion concentrations inside and outside the cell is established and maintained by the permeability characteristics of the membrane and the cell's sodium-potassium pump. The membrane is not very permeable to sodium ions, but it lets potassium ions through relatively easily. Therefore, the positively charged potassium ions tend to move inside, attracted by the negatively charged proteins within, but they are pumped right back out again by the sodium-potassium pump, powered by the energy of ATP.

The electrical potential—that is, the difference between the net electrical charges of the cell's interior and its surroundings—that is set up by this peculiar distribution of charged particles is called the cell's **resting potential.** Typically, this potential difference is about –70 millivolts (Figure 33.7), or about 5% as much electrical energy as is found in a regular flashlight battery.

The Action Potential

When a neuron is stimulated, the point of stimulation becomes suddenly and briefly depolarized (without positively and negatively charged areas), and the depolarization sweeps rapidly along the

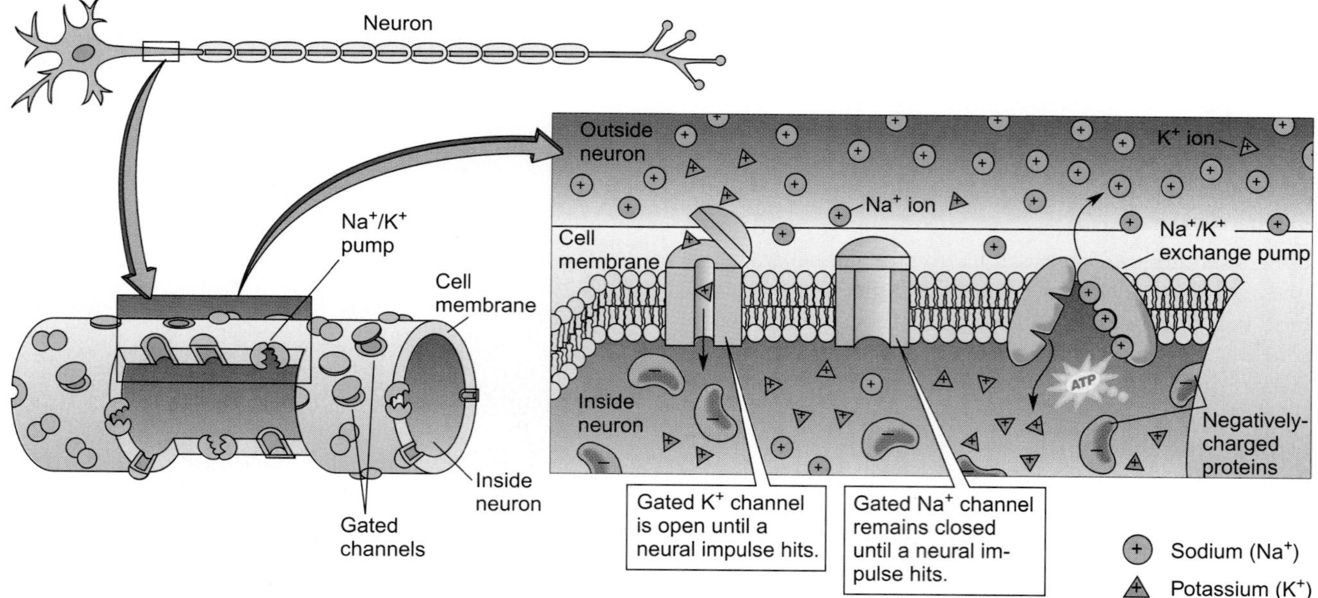

FIGURE 33.6 ION MOVEMENT ACROSS NERVE MEMBRANES.
Ions can cross neural membranes through open channels and gated channels, or they can be actively transported across by sodium-potassium pumps.

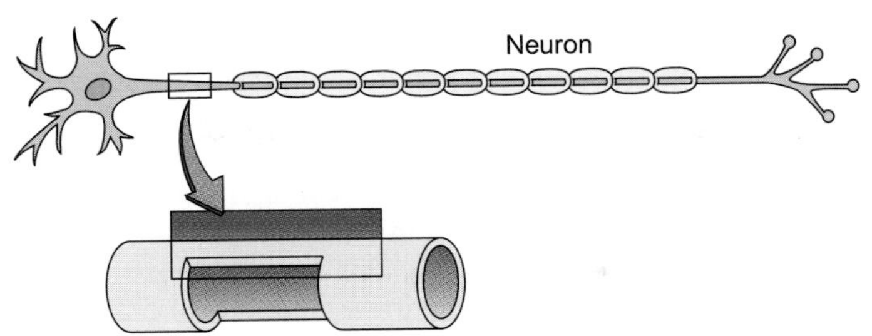

RESTING POTENTIAL: Membrane is polarized (positively charged outside, negatively charged inside); sodium ions (Na^+) are abundant outside and potassium ions (K^+) are abundant inside.

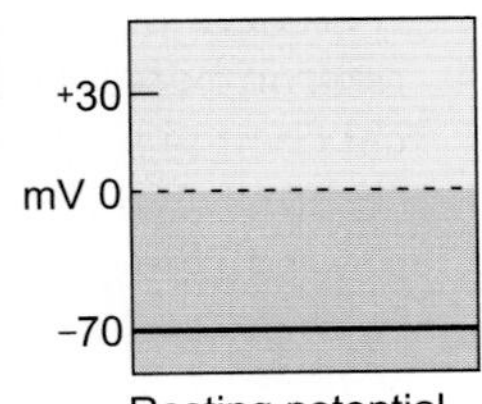

Resting potential

ACTION POTENTIAL: Sodium channel gates open, Na^+ ions rush in, depolarizing the membrane. Farther down the axon, the membrane is still in its resting potential.

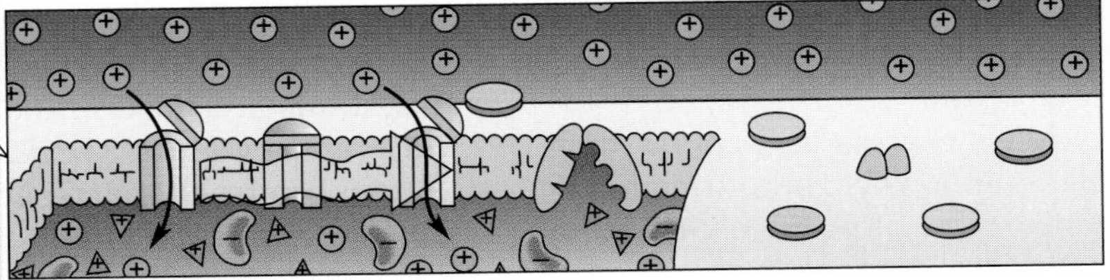

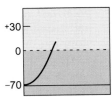

+30

0

–70

Action potential begins, depolarizing the membrane

REPOLARIZATION: The sodium channel gates close and the potassium channel gates open. K^+ ions rush out of the cell, reestablishing the resting potential. Further down the axon, the action potential causes sodium channel gates to open.

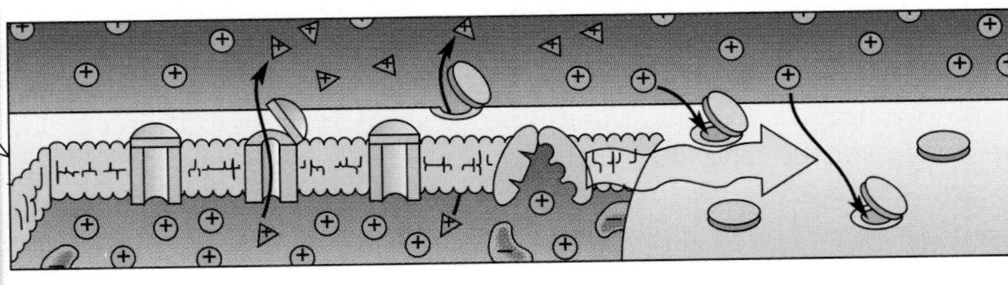

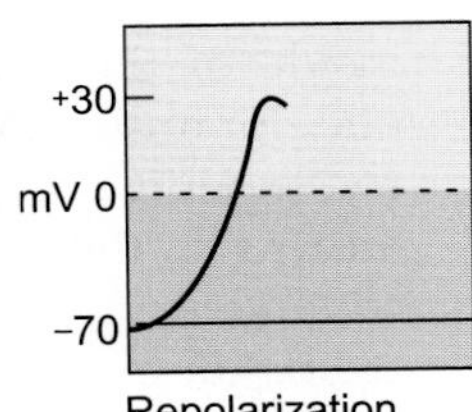

Repolarization begins

RESTORING Na^+/K^+ DISTRIBUTION: The sodium-potassium pump returns Na^+ ions outside and K^+ ions inside. Farther down the axon, repolarization continues.

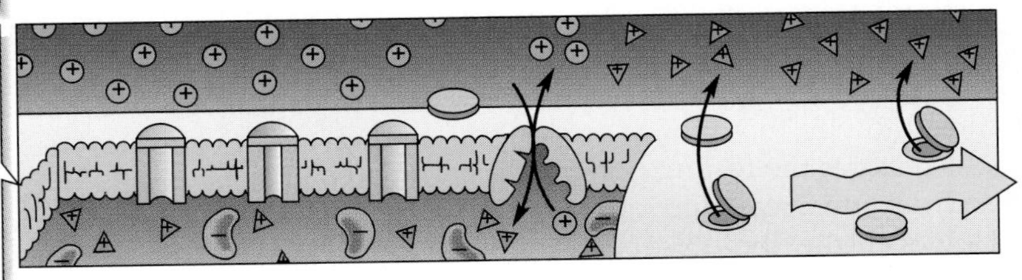

+30

mV 0

–70

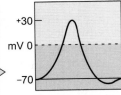

Repolarization is completed; return to resting potential

FIGURE 33.7 **RESTING AND ACTION POTENTIAL.**
In the resting, or polarized, neuron the sodium gates and the potassium gates are closed. Opposite sodium and potassium ion gradients are maintained, and a large number of immobile negative ions remain inside. An action potential begins as sodium activation gates open, and an inrush of sodium ions begins. Repolarization and recovery are almost immediate as the sodium inactivation gates close and the potassium gates open. Potassium ions rush from the interior, and the peak voltage of +30 mV drops rapidly toward –70 mV. As the resting, or polarized, state returns, the sodium inactivation gates open while the sodium activation gates and the potassium gates close.

length of the neuron, followed within about 1 millisecond (1/1,000 second) by repolarization. The depolarization is created by a rapid change in membrane permeability and a corresponding shift in the precarious balance of ions, which was maintained during the resting state. This shift of ions and electrical charges produces the **action potential,** also called the nerve impulse (Figure 33.7). Let's see what's behind these changes.

When a neuron is stimulated, the interior of the neuron becomes slightly less negative. If the electrical change is great enough, it causes the gates to the sodium channels to open, allowing sodium ions to move through. The positively charged sodium ions rush inward, driven by their concentration gradient and attracted by the negative charges within. The membrane potential shifts rapidly from –70 mV to +30 mV, as that area of the membrane becomes depolarized. The sodium channels close again about half a millisecond (1/2,000 second) after they open, stopping the sodium ion traffic.

When depolarization is near its peak, potassium gates open, and potassium ions begin to stream out of the cell. The outward flow of positively charged potassium ions repolarizes the membrane to near the resting potential (Figure 33.7).

Action potentials are very brief. The region just inside of the neuron becomes positively charged only for milliseconds. But as one area along the neuron experiences this shift in ions, it triggers the next area to do the same. Once started, the shift in ions continues in a cascading manner (like a row of falling dominoes)—in a *wave of depolarization followed by repolarization*—down the length of the axon (Figure 33.7).

As soon as the resting potential of –70 mV is restored, the neuron may fire again. The time during which it cannot fire is called the **refractory period.**

It may have occurred to you that while the exodus of potassium ions restores the resting potential, the distribution of sodium and potassium ions is not as it was before. However, the sodium-potassium ion-exchange pumps restore the earlier distribution of ions, and the neuron returns quickly to its former ion distribution, its resting potential.

Myelin and the Speed of the Impulse

There are two factors that increase the speed of impulse conduction in a neuron: *increased diameter of the axon* and *myelination.*

The first case can be stated simply: the larger the diameter of the axon, the faster the rate of conduction. For example, the giant axons of the squid are several millimeters thick and can conduct impulses at 30 m per second—a rate that is fast for an invertebrate, but still far from vertebrate capabilities.

Vertebrates have very complex nervous systems with a high density of neurons and a need for rapid impulse transmission. That need cannot be met by an increased neural diameter, though. There are too many neurons for that. So vertebrates rely on the fatty, insulating layers of myelin. The rapid transmission in the myelinated neuron is possible because of the nodes between the myelin wrappings (see Figure 33.3). It is here in these nodes that action potentials occur. An action potential in a node creates a minute flow of electrical current that passes, instantaneously, to the next node, activating the voltage-sensitive sodium gates and creating a new action potential there. The neural impulse literally jumps from one node to the next, like a rock skipping along the surface of a pond. This method of impulse transmission is called **saltatory propagation** (*salt-,* jump) (Figure 33.8). Impulses in myelinated neurons can reach 100 m per second.

As you can imagine, saltatory propagation is highly energy-efficient, since there are so few ions to be pumped in and out later. In one comparison between myelinated and nonmyelinated axons, it was estimated that the nonmyelinated axons required 5,000 times as much ATP as myelinated axons to restore the distribution of sodium and potassium ions.

Communication Among Neurons

Neurons communicate with each other at special sites called *synapses.* A **synapse** is the junction between one neuron and the next, across which impulses pass. (It can also be the junction between a neuron and an effector, such as a muscle.) Typically, the axon of one neuron will form a synapse with a dendrite or cell body of a second neuron. Although a neuron usually has but one axon, that axon's highly branched ending may provide hundreds to thousands of synaptic connections with other neurons. There are two distinct kinds of synapses: *electrical* and *chemical.* Chemical synapses are far more common, but both have unique advantages.

Electrical Synapses

Electrical synapses occur at gap junctions between neurons (Figure 33.9). You may recall from Chapter 4 that gap junctions are places where adjacent plasma membranes press together and minute

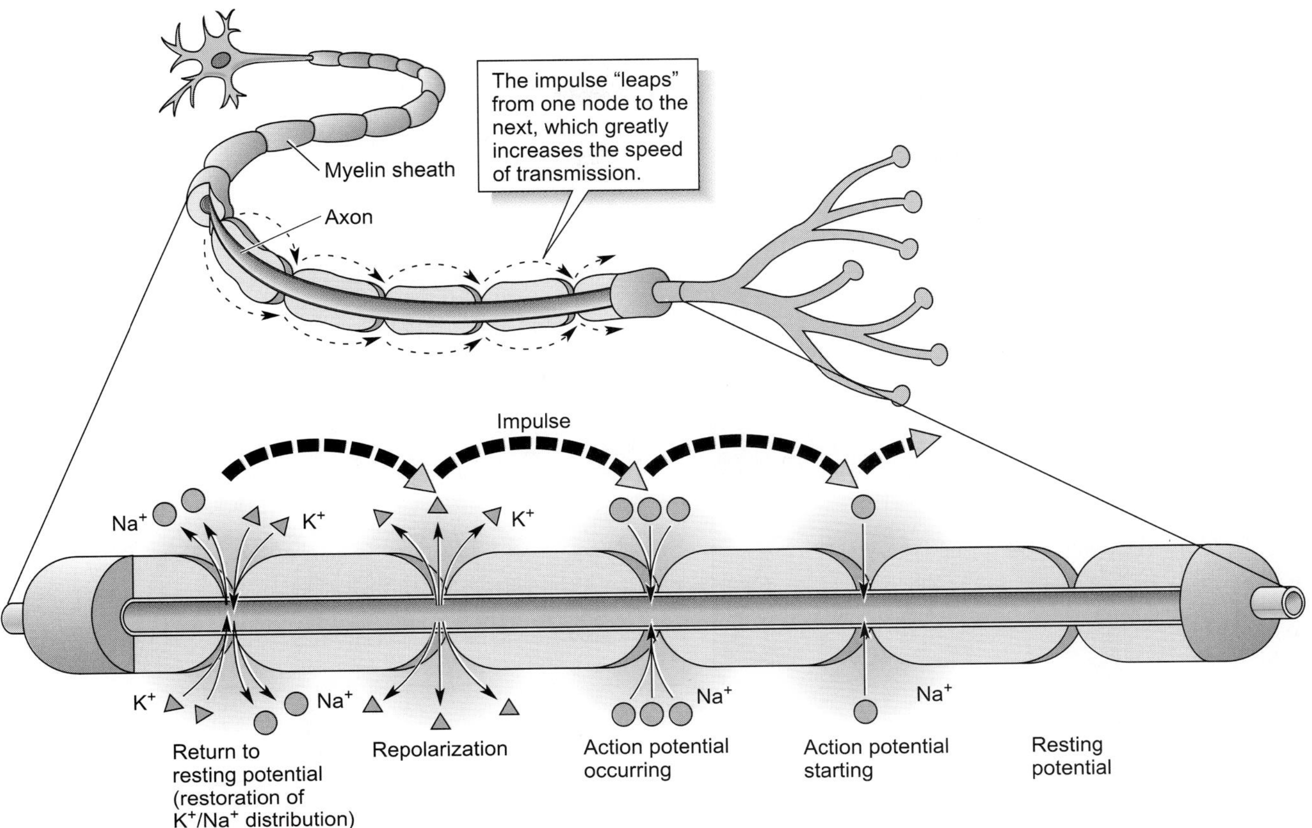

FIGURE 33.8 SALTATORY PROPAGATION.
In myelinated neurons, action potentials occur only at the nodes. The neural impulse leaps from node to node down the axon. This type of transmission, known as saltatory propagation, is considerably faster and requires less energy in terms of ATP than transmission in nonmyelinated neurons. Because of saltatory propagation, myelinated neurons transmit impulses up to 20 times faster than the fastest nonmyelinated neurons.

pores form, resulting in the cytoplasm of one cell being continuous with that of another. Gap junctions also occur in neurons, and the intimate relationship between neurons permits a wave of depolarization to pass readily from one neuron to another, prompting an action potential in the second.

A major advantage of the electrical synapse is speed. Chemical synapses require several steps for action potentials to be generated in the second neuron, and this requires more time than the current flow associated with electrical synapses. Electrical synapses have been found in several invertebrates and are associated with giant axons and escape movements. In the crayfish, for instance, electrical synapses occur between a giant nerve fiber from the brain and a large motor axon that activates the abdominal muscles. In response to a nerve stimulus, the muscles produce a powerful thrust in the paddlelike tail that propels the crayfish backwards, away from danger. Electrical synapses also occur in vertebrates. In fishes, for example, electrical synapses activate the sudden flip of the tail and make possible the quick starts and turns used in the feeding chase by both predator and prey.

Chemical Synapses

In chemical synapses, the neurons do not actually make physical contact, and action potentials do not simply pass from one neuron to the next, as they do in electrical synapses. Instead, they must be generated anew in receiving neurons. In chemical synapses, neurons are separated by a minute, 20-nm space known as the **synaptic cleft** (Figure 33.9). Communication occurs through the action of chemicals called neurotransmitters, which are molecules released by one neuron that stimulates the next neuron. Neurotransmitters released by the first or **presynaptic neuron** diffuse across the synaptic cleft, where they activate specialized receptors on the second or **postsynaptic neuron.** The transmission is, in most cases, one-way.

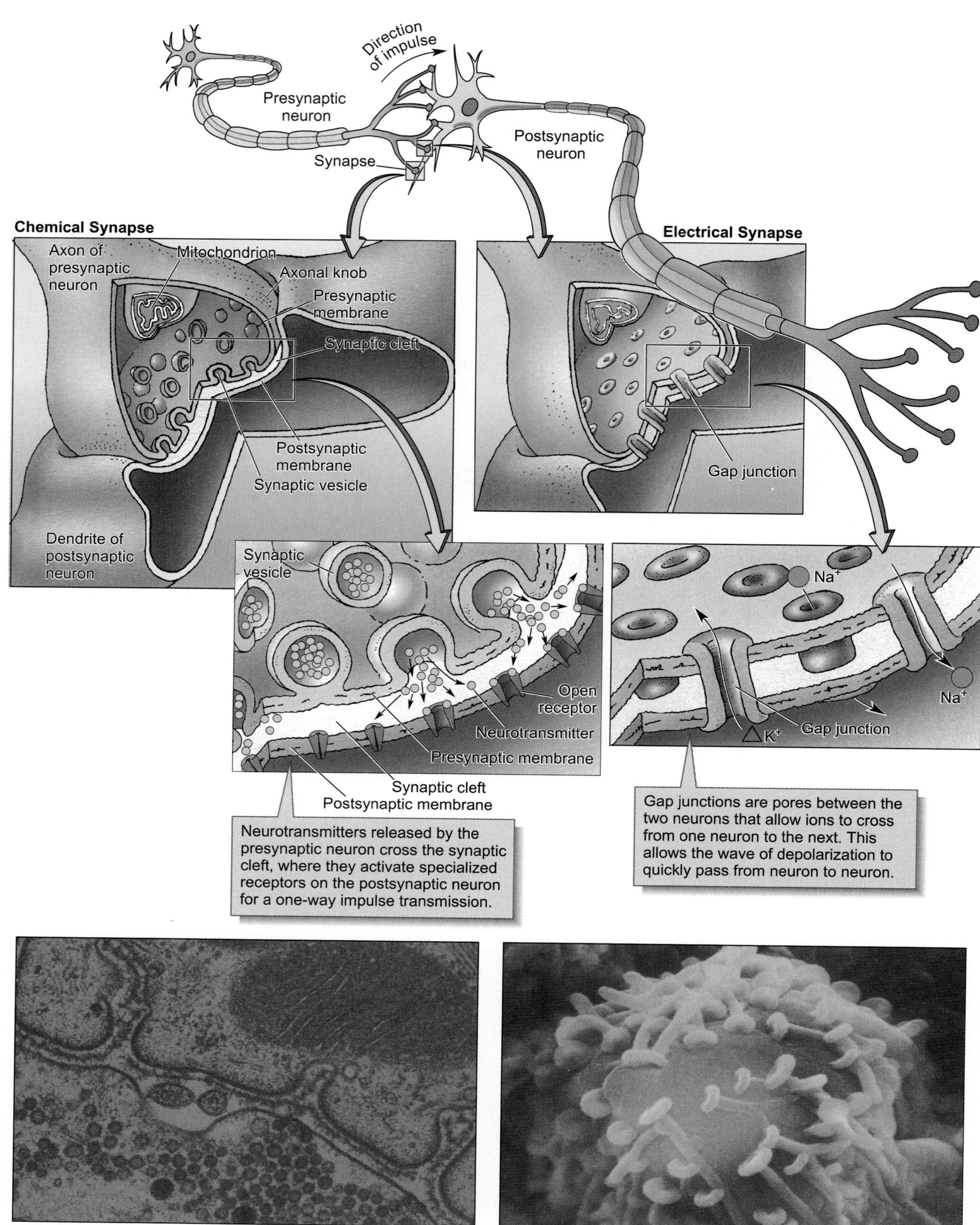

FIGURE 33.9 SYNAPSES.
The two kinds of synapses (the chemical synapse and the electrical synapse) make use of different kinds of connections between cells.

By its very organization, then, the synapse can act as a one-way valve between such neurons because, generally, the receiving membrane of the postsynaptic neuron has no neurotransmitters to release, and the secreting membrane of the presynaptic neuron has no receptor sites, places where the neurotransmitter might attach—initiating an impulse. One-way transmission is an important means of coordinating the work of the nervous system. Furthermore, although they are slower than electrical synapses, chemical synapses offer much more versatility in their responses, due to their variety of neurotransmitters and range of potential receptors.

Action at a Chemical Synapse

The action at a chemical synapse begins at the end of the axon. Axons branch near their ends, forming many minute swellings called **axonal knobs.** Many axonal knobs may be involved in the communication between just two neurons. In fact, some neurons have hundreds to thousands of axonal knobs, all synapsing with a single postsynaptic neuron.

Within each axonal knob are a large number of vesicles that store the chemical neurotransmitter. When an action potential reaches the axonal knob, voltage-sensitive gates of calcium-ion channels in its membrane open, and calcium ions diffuse inward. In their presence, the transmitter-laden vesicles fuse with the presynaptic membrane, where they rupture, releasing numerous neurotransmitter molecules into the synaptic cleft. The molecules then diffuse across the cleft and bind to specialized receptor sites on the postsynaptic membrane (Figure 33.9).

The neurotransmitter receptor sites of the postsynaptic membrane are *chemically* gated ion channels. That is, their gates respond to chemicals rather than to voltage changes. The activation of receptor sites opens certain ion channels, and the resulting flow of ions slightly changes the membrane potential of the second neuron. If it brings the membrane closer to the threshold value for generating an action potential, it is an **excitatory synapse.** When a sufficient number of such channels has opened in the postsynaptic membrane, the current generated by the shifting ions will spread toward the axon. If the resulting voltage surpasses the required threshold, an action potential will begin. Whereas one such event would probably have little effect, the presence of many excitatory synapses on a single receiving neuron helps ensure this reaction (Figure 33.10).

Another kind of synapse does not trigger action potentials, but *inhibits* them, making them less likely to occur. This is the **inhibitory synapse** (Figure 33.10). One way such inhibition occurs is through hyperpolarization, in which the net negative charge inside the neuron becomes considerably increased. This may happen when specialized receptors respond to a neurotransmitter by opening chemically gated chloride-ion channels. Once opened, they admit the negative chloride ion (Cl^-) into the neuron's interior, which increases its polarized state. Consequently, the hyperpolarized neuron requires considerably more than the usual threshold voltage for action potentials to start.

Inhibition can also occur if the neurotransmitter causes chemically gated potassium channels to open —channels that permit the escape of potassium ions. Their escape counteracts the uptake of sodium ions, brought about by nearby excitatory synapses, effectively canceling out depolarization.

Typically, a neuron or an effector will possess both inhibitory and excitatory synapses coming from other neurons. Whether or not the second neuron fires or the effector reacts depends ultimately on the net effect of both types of synapses. At times, inhibition can be vital. It is inhibition, for example, that permits sleep. By suppressing action potentials, certain inhibitory neurons can screen out routine incoming stimuli—background noises or random thought patterns—whatever might disturb your sleep. But a rustling sound under the bed might stimulate enough excitatory neurons to overwhelm the inhibitory neurons. In the next chapter, we will see that the cerebellum, the brain structure responsible for coordinating voluntary movement, does much of its work through inhibition.

Recovery at the Synapse

After a few impulses have traveled down a neuron, causing release of the neurotransmitter into the synaptic space, it might seem that the chemical would build up in the synaptic area and cause the next neuron to continue firing in the absence of a real stimulus. But do not despair. Neurotransmitters are quickly removed from the synapse, some by special enzymes. (For example, the enzyme that breaks down the common neurotransmitter acetylcholine is called acetylcholinesterase.) Other neurotransmitters are simply taken back into the tip of the axon and reused.

Humans sometimes accidentally come into contact with substances that inactivate these important neural enzymes—with horrible consequences. For example, certain insecticides, such as Malathion or Parathion, which are sprayed on crops, have the advantage of being short-lived, unlike DDT, which remains

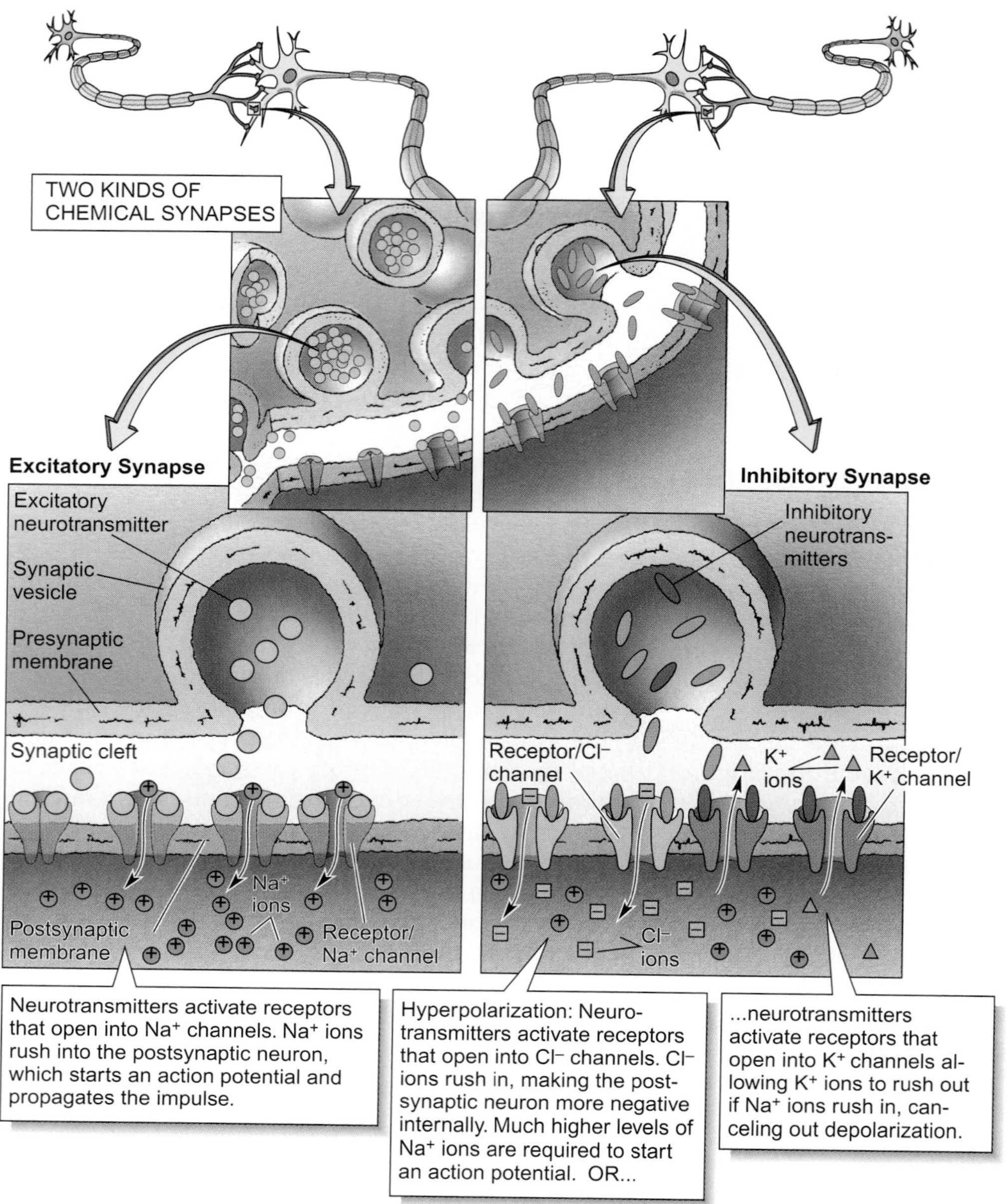
TWO KINDS OF CHEMICAL SYNAPSES
Excitatory Synapse
Excitatory neurotransmitter
Synaptic vesicle
Presynaptic membrane
Synaptic cleft
Na+ ions
Postsynaptic membrane
Receptor/ Na+ channel
Inhibitory Synapse
Inhibitory neurotrans-mitters
Receptor/Cl− channel
K+ ions
Receptor/ K+ channel
Cl− ions
Neurotransmitters activate receptors that open into Na+ channels. Na+ ions rush into the postsynaptic neuron, which starts an action potential and propagates the impulse.
Hyperpolarization: Neuro-transmitters activate receptors that open into Cl− channels. Cl− ions rush in, making the post-synaptic neuron more negative internally. Much higher levels of Na+ ions are required to start an action potential. OR...
...neurotransmitters activate receptors that open into K+ channels al-lowing K+ ions to rush out if Na+ ions rush in, can-celing out depolarization.

FIGURE 33.10 **EXCITEMENT AND INHIBITION AMONG NEURONS.** Shown here are the two kinds of chemical synapses: the excitatory synapse and the inhibitory synapse.

intact in the environment long after it is applied. Within a few days after spraying, the short-lived chemicals are believed to be relatively harmless (Figure 33.11). However, while they are active, they can block the effects of acetylcholinesterase. The enzyme functions not only to keep chains of neurons from firing blindly in response to accumulating transmitter substance, but also to keep acetylcholine from diffusing over to nearby neurons and causing them to fire as well. With this in mind, what do you think might be the effect on a farm worker exposed to the insecticides? They can, indeed, trigger spasms and convulsions. Several deaths have already been attributed to these chemicals.

FIGURE 33.11 INTERFERENCE WITH SYNAPSE FUNCTION.
Short-lived insecticides such as Malathion have largely replaced the use of DDT in the United States. Whereas DDT was assimilated into fatty tissue where it could cause long-term problems, Malathion destroys the enzymes that break down neurotransmitters, causing impulses to fire wildly and indiscriminately and bringing on convulsions and possibly death.

PERSPECTIVES

We've seen how muscles can move skeletons, close blood vessels, and pump blood. We've also seen that these activities must be controlled by nerves. So far, we've concentrated on the nervous system at the cellular level, and now we are prepared to see how nerve cells are organized into the nervous system itself.

First, we will look at the brain, the spinal cord, and the peripheral nervous system. Then, after laying the groundwork through examination of increasingly complex systems in other species, we will focus on the ways nerves can tell humans about their environment.

SUMMARY

Nerve cells work in a similar fashion in all animals.

THE NEURON

Neurons or nerve cells have three primary parts: a cell body (with cytoplasm and many organelles), dendrites (receiving structures), and axons (transmitting structures). Neurotransmitters are released by the axonal tree, permitting impulses to move from neuron to neuron and from neuron to effector.

Myelinated neurons are surrounded by glial cells or Schwann cells. Nodes are interruptions in the sheath.

Sensory neurons conduct impulses from specialized sensory receptors to the central nervous system. There, impulses are relayed to interneurons in the spinal cord and brain whose task is to integrate information. Responses are made through motor neurons that activate muscles or glands. In addition to neurons, the nervous system contains glial cells, which have structural and chemical supporting roles.

Nerves are cablelike collections of axons and dendrites surrounded by insulating connective tissue sheaths.

NEURAL ACTIVITY

The substances of importance in neural activity are ions, most importantly those of sodium (Na^+) and potassium (K^+). One way that ions can cross the plasma membrane is through special protein-lined openings called ion channels, which usually admit only one kind of ion. Some of the channels are always open, but many others have gates that open or close, thereby regulating the flow of ions through the channels and across the membrane. The forces that cause ions to move through their open channels are typically either a concentration gradient or an electrical gradient. The interaction of these two forces largely determines the passive distribution of ions.

Another important factor in determining the distribution of ions is sodium-potassium ion-exchange pumps, which move three Na^+ out of the cell and two K^+ in for each pump cycle. The pump is powered by ATP.

In the resting potential, the region immediately outside a neuron is positive to the region inside; thus the resting neuron is polarized. Sodium ions predominate outside, and potassium ions predominate inside. Large, negatively charged proteins help draw positively charged ions inward. Ion distribution is maintained by the characteristics of membrane permeability and by sodium-potassium ion-exchange pumps. At rest, the measured voltage potential or resting potential across the neural membrane is –70 mV.

An action potential, or neural impulse, is a wave of depolarization caused by the entry of sodium ions, followed by repolarization caused by the outflow of potassium ions. It begins when sodium gates open, admitting sodium ions. The voltage shifts from –70 mV to +30 mV. Near the peak of depolarization, the sodium gates close and potassium gates open, permitting potassium ions to cross. The outflow of potassium ions repolarizes the membrane, bringing the membrane potential close to the resting potential. An action potential moves along an axon because the current it produces opens sodium gates farther along the neuron.

The speed of conduction of a neural impulse can be increased by an increase in the diameter of the axon or by myelination. Myelinated neurons conduct impulses much faster than nonmyelinated neurons because action potentials occur only at nodes instead of all the way along the fiber. The current flow created by an action potential in one node instantaneously activates sodium gates in the next node; thus the impulse jumps from node to node (called saltatory propagation). This speeds up transmission and greatly reduces the amount of ATP used by sodium-potassium ion-exchange pumps.

Communication Among Neurons

At the synapse, one neuron can activate or inhibit another, or a neuron can activate an effector, across either electrical or chemical synapses.

Electrical synapses occur where gap junctions between neurons permit a wave of depolarization to pass from one neuron to another. The main advantage of such transmission is speed.

In chemical synapses, when an action potential reaches the axonal knob, calcium gates open, and calcium ions enter the cell. This causes the release of chemicals called neurotransmitters from synaptic vesicles into the cleft. The neurotransmitter diffuses across the cleft and joins specific receptors on the postsynaptic membrane.

Receptor activation opens certain ion channels, and the resulting flow of ions slightly changes the membrane potential of the second neuron. If it brings the membrane closer to the threshold value for generating an action potential, it is an excitatory synapse. If the depolarization reaches threshold, an action potential begins. If the flow of ions hyperpolarizes the membrane, it is an inhibitory synapse.

Neurotransmitters are quickly removed from the synapse by being broken down by enzymes or by reuptake into the axon.

Key Terms

neuron • **560**
cell body • **560**
neurotransmitter • **560**
effector • **560**
dendrite • **561**
axon • **561**
myelin • **561**
myelin sheath • **561**
sensory neuron • **561**
interneuron • **561**
motor neuron • **561**
glial cell • **561**
nerve • **562**
ion channel • **562**
sodium-potassium pump • **564**
resting potential • **564**
action potential • **566**
refractory period • **566**
saltatory propagation • **566**
synapse • **566**
synaptic cleft • **567**
presynaptic neuron • **567**
postsynaptic neuron • **567**
axonal knob • **569**
excitatory synapse • **569**
inhibitory synapse • **569**

Review Questions

1. Prepare an outline drawing of a motor neuron, labeling cell body, nucleus, dendrites, axon, axonal tree, and axonal knobs.
2. Add to your drawing a myelin sheath, labeling nodes and internodes. How do such sheaths form?
3. Summarize the general roles of sensory neurons, interneurons, and motor neurons. In what two structures are interneurons located?
4. Describe the structure of a nerve. How are its neurons protected from stimulating each other?
5. In what major way do neural impulses differ from electrical current? How are they similar?

6. Describe the concentration of positive and negative ions in the resting neuron.
7. What is an ion channel? What is a gated ion channel?
8. Explain the distribution of charges on each side of the membrane in and around the resting neuron in terms of sodium and potassium ions and protein. How do membrane permeability characteristics and the exchange pump maintain this?
9. What happens to the following when an action potential begins: the polarized state, the positive charges, the negative charges?
10. Explain in terms of membrane permeability and sodium ions how an action potential starts.
11. Repolarization immediately follows depolarization. Explain this in terms of behavior of specific ions. At what voltage potential does repolarization begin?
12. Explain, in terms of sodium gates, how an action potential can move along a neuron.
13. How does the presence of a myelin sheath affect the speed of an action potential? What other way is there to speed up an action potential?
14. Where do action potentials actually take place in myelinated neurons? How does the depolarizing wave move along a myelinated region?
15. Saltatory propagation saves a great deal of ATP. Explain why.
16. What is the role of the following at the synapse: calcium ions, neurotransmitter molecules, postsynaptic receptor sites, enzymes, neurotransmitter vesicles?
17. Why is it so essential that the neurotransmitter be removed from the synapse soon after it acts?

34

The Nervous System

HERE WE CONSIDER HOW THOSE MANY COMPLEX, SENSITIVE CELLS CALLED *neurons and their associated cells are organized into nervous systems. We focus first on the brain and spinal cord—the central nervous system. After considering the chemical and electrical activity in the brain, we move on to why you can't remember anything about our discussion. Finally we look at the remarkable peripheral nervous system. All the while we will be setting the stage for our next chapter on how you are able to perceive these events.*

We now come to the part of biology where the brain discusses itself. Here, then, we can expect great accolades and effusive praise. But here, also, we will find some of the greatest questions in the drama of life. Many of these questions are biological: How are thoughts generated? How do those thickened plaques associated with Alzheimer's arise? What are the effects of specific injury, or disease, on the brain? Can gray matter heal itself? How do the circuitries of learning form? But other questions lie outside the realm of biology: How is the brain associated with moral behavior? What is moral behavior, anyway? How early is personality established? And by what? What is intelligence? What is the brain's role in love? Such questions clearly lie beyond the reach of biology . . . or do they?

In any case, we will begin by considering how the brain has evolved. In the same light we will also look at the great thick cord that extends from beneath it, the spinal cord. In vertebrates, the brain and spinal cord, together, form the **central nervous system (CNS).** We will later consider the system of nerves that branch off the central nervous system—the peripheral nervous system (PNS).

EVOLUTION OF THE CENTRAL NERVOUS SYSTEM

Clues to the evolutionary development of central nervous systems can be gleaned from a cross-species survey from simple to complex animals (Figure 34.1). One of the simplest neural arrangements is that of the fresh-water *Hydra* (Figure 34.1a). It consists of a two-dimensional net of interconnecting neurons, or nerve cells, spread throughout the outer body layer. The entire surface of the animal is about equally covered. There is no part that controls the rest—no nerve center that functions in the regulation or coordination of the nerve net. Still, if one part of the animal is stimulated, the entire body responds and shows awareness of the stimulus.

The flatworm has a somewhat more centralized nervous system (Figure 34.1b). The neurons are arranged in two *longitudinal nerves* connected by *transverse nerves,* producing a "ladder" as opposed to the *Hydra*'s "net." Your keen eye will undoubtedly also have noted the aggregation of nerves in the head region. These form the "brain" and are composed of clumps of neural cell bodies.

A somewhat more complex nervous system is found in the earthworm (Figure 34.1c). The earthworm has a single longitudinal nerve, but it shows vestiges of a paired arrangement in that it is two-lobed, much like two cords pressed together. The nerve is ventral, with the heart and digestive tract lying dorsal to it. The distinct brain is formed from clumps of neural cells. Note the segmented nodes along the nerve cord, each with paired nerves reaching into a segment of the body.

The frog nervous system (Figure 34.1d) is relatively simple for that of a vertebrate, but it can be used to illustrate the basic neural plan in this group. Here, we find a distinct brain and spinal cord that together form a true central nervous system. In frogs, as in virtually all vertebrates, the longitudinal nerve (now called the spinal cord) is *dorsal, hollow, filled with fluid,* and *protected by bone.* (It is protected by cartilage in some fishes, such as sharks.) The brain is an elaboration of the ganglionic clumps of ancient forebears. The vertebrate brain, as exemplified by humans (Figure 34.1e), shows marked specialization—

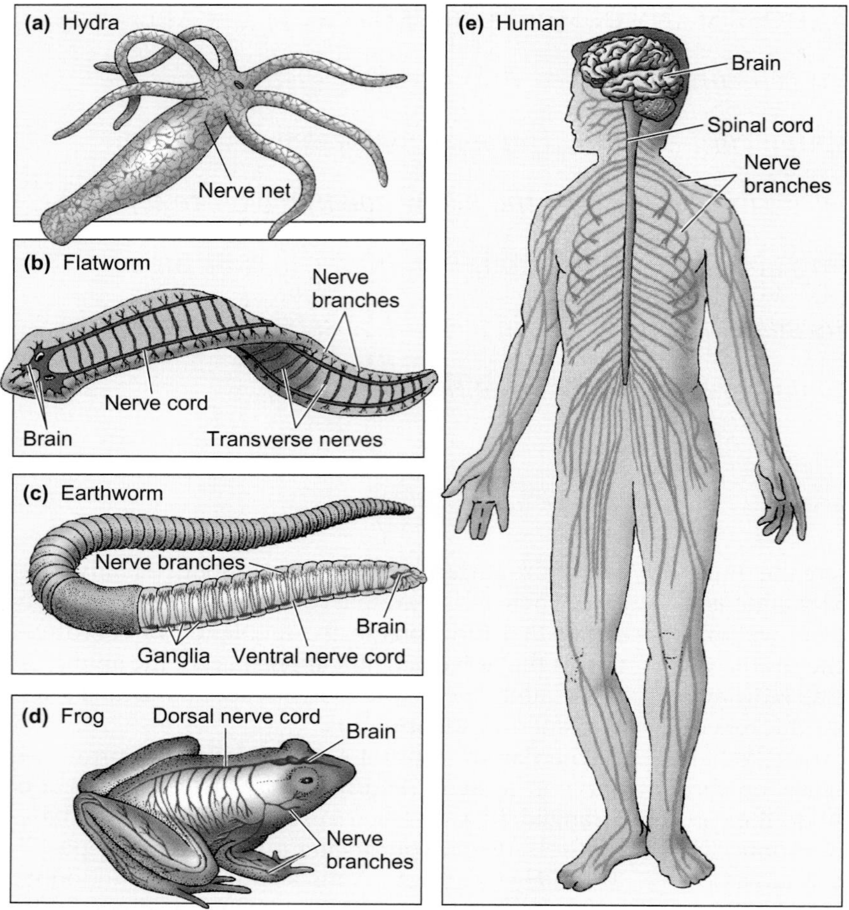

FIGURE 34.1 NERVOUS SYSTEMS.
Invertebrate nervous systems include **(a)** the simple nerve net of the cnidaria (*Hydra*), a radial animal. The ladderlike system in the planarian **(b)**, a bilateral animal, reveals some anterior consolidation and marked segmentation. The earthworm, an annelid **(c)**, has a sizable brain and a highly segmented arrangement of ganglia in its ventral nerve cord. The frog, a vertebrate **(d)**, has a highly consolidated brain with segmentation along the cord somewhat masked by specialization. The human **(e)** is less segmented, but shows a marked increase in the brain.

that is, different parts of it are associated with very specific functions. The central nervous system of vertebrates shows traces of the paired and segmental neural arrangements of their distant ancestors. For example, the brain is two-lobed, and paired nerves extend from it and from the spinal cord. However, the pair branching is no longer so regular and apparent because of specialization along the spinal cord as the vertebrate body plan became more complex. Even among vertebrates, there can be great differences in the organization of the brain. Figure 34.2 illustrates relative differences in parts of the brain from fish to reptile to mammal.

THE HUMAN BRAIN

The human brain (Figure 34.3) is a fascinating structure. To this day, we know far too little about it, and there are suggestions that some of the findings to come will be startling. For example, there is some evidence that every word you have ever said or heard is filed away in your brain, even though you will go to your grave having retrieved hardly any of that information.

Let's first consider some vital statistics. The human brain weighs about 1.4 kg (3 lb), has a volume

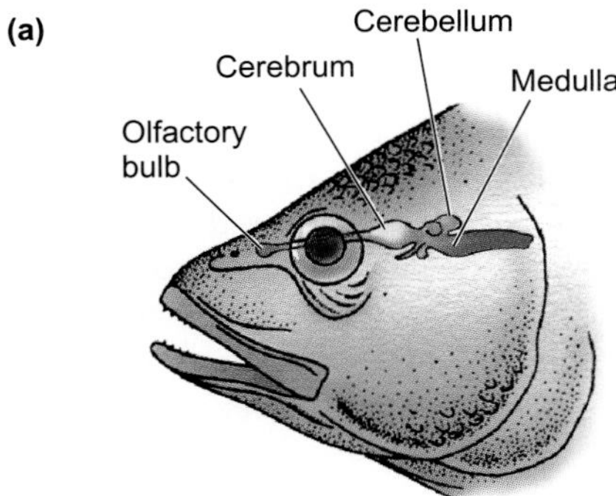

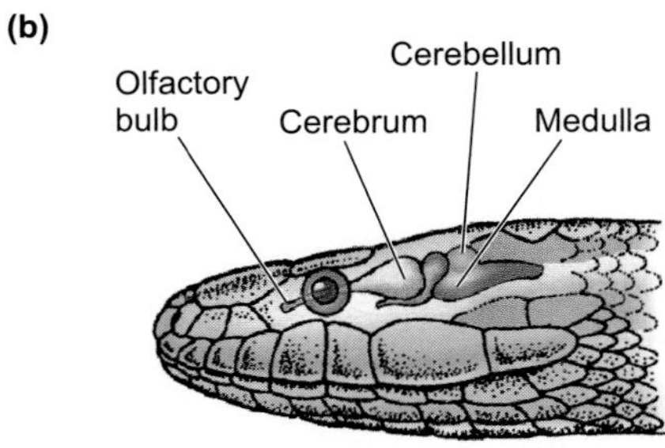

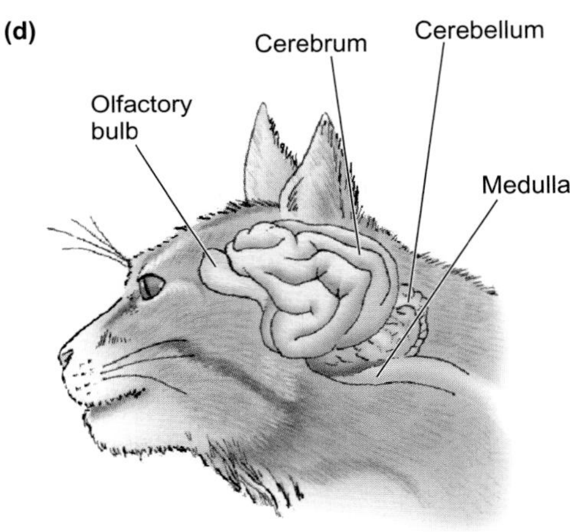

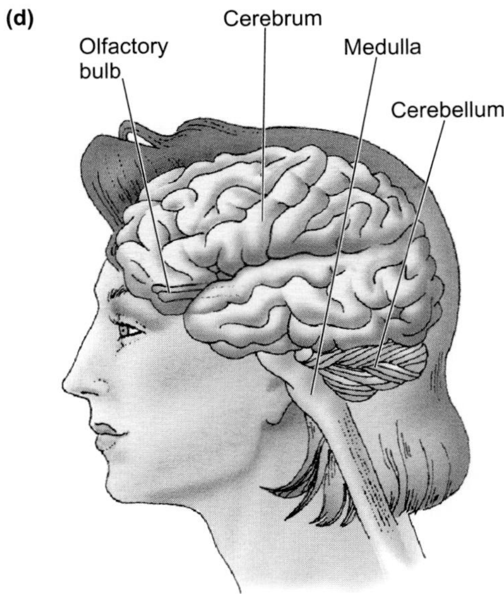

of 1,200 to 1,500 cc, contains over 100 billion neurons, and has about 10 times that number of supporting glial cells. Since each neuron may synapse with many others, there are a number of alternative pathways for impulses, and the coordination necessary to produce even a simple response must be due to a veritable neural symphony.

The brain, which has a consistency somewhat like gelatin, is obviously fragile, but it is well protected. In addition to the surrounding bony armor of the skull and vertebrae column, the brain and the spinal cord are directly enclosed by the tough, elastic coverings called the **meninges.** The spaces within various of these membranes, and the cavities within the brain itself, are filled with the pressurized, shock-absorbing **cerebrospinal fluid.** The billions of delicate neurons themselves may receive some protection by being embedded in the glial cells that make up much of the brain's mass.

The vertebrate brain consists of three regions: *hindbrain, midbrain,* and *forebrain.* In humans, the midbrain is not easily seen in the whole brain, since most of it is enclosed by the prominent forebrain (Figure 34.4). The structure and function of the parts of the human brain are summarized in Table 34.1.

The Hindbrain

The **hindbrain** consists of the *medulla oblongata,* the *pons,* and the *cerebellum.* As a rough generality, the more unconscious, involuntary, and mechanical processes are directed by these more posterior parts of the brain. The **medulla oblongata** (or, more simply, the medulla), the part of the brain that is directly connected to the spinal cord, has centers, or *nuclei,* that help regulate such functions as breathing rate, blood pressure, and heart rate. All communication between the brain and spinal column must pass through the medulla.

Figure 34.2 Comparing the Brains of Fish, Reptiles, and Mammals.
Notice the diminutive size of the cerebrum relative to the olfactory bulb in the fish **(a)**. Also note the relative mass of the lower brain (here, the cerebellum and medulla) compared to that of the cerebrum. Reptiles **(b)** have a somewhat larger, but still smooth cerebrum and a reduced olfactory area. The brain of the cat **(c)** is dominated by the convoluted cerebrum. The cerebellum, involved in coordination, is well developed in the cat, as is the olfactory bulb. The human brain **(d)** shows a reduced olfactory bulb with the enlarged cerebrum.

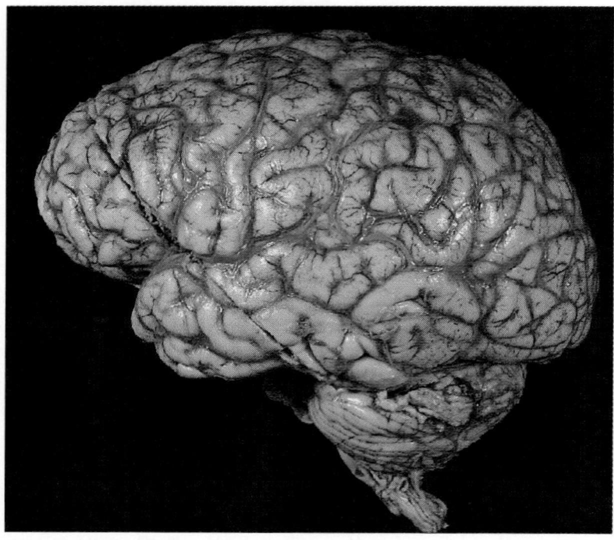

FIGURE 34.3 THE HUMAN BRAIN.
The cerebrum, with its two hemispheres and highly convoluted surface, dominates the human brain. Under the wrinkled exterior resides an incredibly complex array of neurons.

The **pons,** which lies just above (anterior to) the medulla, contains the ascending and descending nerve tracts that run between the brain and spinal cord. It also receives the fifth *cranial nerve,* one of the large nerves that extend from the brain itself to regions of the head and face. The pons also links the functions of the forebrain with those of the cerebellum.

The **cerebellum** is a paired, bulbous structure, with the general appearance of the two halves of an enlarged walnut. It lies above the medulla and somewhat toward the back of the head. The cerebellum coordinates all voluntary body movement in the hands and body and aids in the maintenance of posture and balance. (Because of its role, do you suppose that the cerebellum of a ballet dancer or tumbler might differ from that of others? It appears that it does, but the differences are subtle.) The cerebellum accomplishes its jobs chiefly through inhibition, limiting the force with which a muscle contracts and the distance a limb travels, and generally dealing with several muscle groups at the same time. To coordinate this activity, the cerebellum requires constant sensory input informing it about the degree of muscle contraction and the position of the limbs and body. Such input comes from the eyes, balance organs, and from the muscles themselves. Voluntary movement itself is directed by conscious centers in the cortex, which send impulses along motor pathways through the brain and spinal cord and out to the muscles. Such impulses are also shunted directly into the cerebellum by branches from the same motor pathways. So the cerebellum literally "bugs" the cortex, "listening in" on its activities. However, we are not consciously aware of the cerebellum's vital coordinating functions.

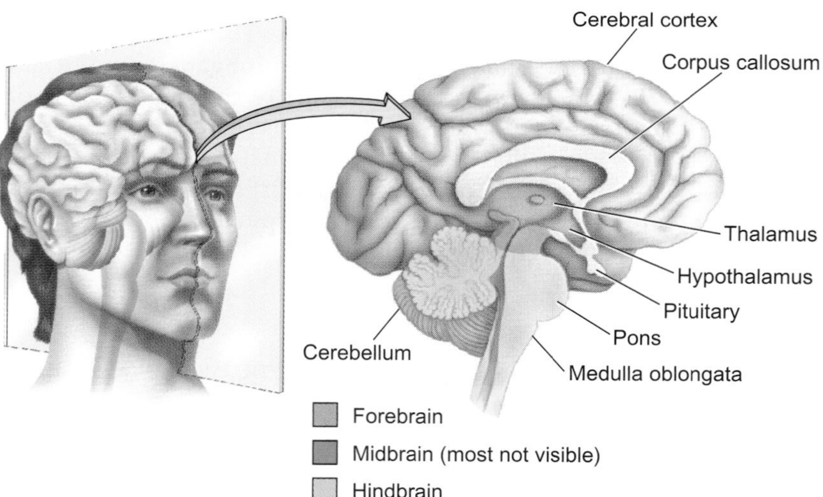

FIGURE 34.4 HUMAN BRAIN DIVIDED ALONG THE MIDLINE FROM FRONT TO BACK.
The dominant forebrain (thalamus, hypothalamus, and cerebrum) is obvious. The areas of the hindbrain (medulla oblongata, pons, and cerebellum) are also quite distinctive, but the midbrain that connects them is not well-defined. The corpus callosum connects the two halves of the brain.

TABLE 34.1 *STRUCTURE AND FUNCTION OF THE HUMAN BRAIN*

Structure	Function
Hindbrain	
Medulla	Control of subconscious activities such as breathing, heartbeat, swallowing, vomiting, and sneezing; connects the spinal cord with the brain
Cerebellum	Controls balance, equilibrium, and coordination
Pons	Connects the cerebellum and the cerebral cortex and other brain regions
Midbrain	Connects the hindbrain and forebrain; receives sensory input from the eyes
Forebrain	
Thalamus	The "great relay station"; connects various parts of the brain
Reticular system	Arousal of certain brain areas; filters impulses from sensory neurons
Hypothalamus	Regulates heart rate, blood pressure, body temperature, and the pituitary; controls drives such as hunger, thirst, and sex
Cerebrum	
Cerebral cortex (outer Cerebrum)	Gray matter; conscious thought; memory, intelligence; speech; association among senses
occipital lobe	Processes visual information
temporal lobe	Auditory reception and some visual information
frontal lobe	Regulates precise voluntary movement and the use of language and speech (includes the prefrontal lobe, which sorts information and orders stimuli)
parietal lobe	Receives information from the skin and processes information about body position
White matter (inner Cerebrum)	Nerve tracts allowing communication between parts of cortex; e.g., corpus callosum connects the two hemispheres and helps them communicate

THE MIDBRAIN

Essentially, the **midbrain** connects the hindbrain and forebrain and processes information from the ears and eyes. All auditory (sound) input of vertebrates is processed here before being sent to the forebrain. In most vertebrates, visual input is first processed here also, except in mammals. In mammals, visual information apparently is sent directly to the forebrain. Although the midbrain is involved in complex behavior in fishes and amphibians, many of these functions are assumed by the forebrain in reptiles, birds, and mammals.

THE FOREBRAIN

The **forebrain,** the largest and most dominant part of the human brain, is responsible for conscious thought, reasoning, memory, language, sensory decoding, and certain kinds of movement. The embryonic forebrain gives rise to such important structures as the *thalamus,* the *reticular system,* the *hypothalamus,* and the *cerebrum.*

THE THALAMUS

The **thalamus** is located at the base of the forebrain (see Figure 34.4). It consists of densely packed clusters of neurons, which provide connections between the various parts of the brain—between the forebrain and the hindbrain, between different parts of the forebrain, and between parts of the sensory system and the cerebrum. It has been rather unpoetically called the "great relay station of the brain"

THE RETICULAR SYSTEM

The thalamus also contains most of an extensive area called the **reticular system,** composed of interconnected neurons that are almost feltlike in appearance. These neurons run throughout the thalamus and into the midbrain and hindbrain. The reticular system is still somewhat of a mystery, but several interesting facts are known about it. For example, we know that it "monitors" the brain. Every pathway to and from the various portions of the brain sends side branches to the reticular system as it passes through the thalamus, so it virtually taps all incoming and outgoing communications. Also, reticular neurons appear to be rather unspecific. The same reticular neuron may be stimulated by impulses from the hand, foot, ear, or eye. As for its function, the leading hypothesis is that the reticular apparatus is something of an alarm system that serves to activate the appropriate parts of the brain upon receiving a stimulus. The portion of the reticular system involved in such arousal is logically called the *reticular activating system (RAS)*. The more messages it intercepts, the more the brain is aroused. If the RAS is severely damaged, the individual will go into an irreversible coma.

Because the reticular activating system serves to arouse the brain, its activity must be reduced to permit sleep; so sleep centers in other regions of the brain inhibit the RAS. Furthermore, you may have noticed that it is much easier to fall asleep when you are lying on a comfortable bed in a quiet room with the lights off than on a pool table in a noisy bar. (Besides, you would interrupt the game.) With fewer stimuli, the reticular system receives fewer messages, and the brain is allowed to relax.

The reticular system may also regulate which impulses are allowed to register in your brain. When you are engrossed in a television program, you may not notice that someone has entered the room. But when you are engaged in even more absorbing activities, it might take a *general* stimulus on the order of an earthquake to distract you, whereas the *specific* stimulus of a turning doorknob would immediately attract your attention. Such filtering and selective inhibition of stimuli apparently takes place in the reticular system.

The Hypothalamus

As the name implies, the **hypothalamus** lies below the thalamus (see Figure 34.4). It is densely packed with cells that help regulate the body's internal environment as well as certain aspects of behavior. The hypothalamus helps control heart rate, blood pressure, and body temperature by regulating the activity of other nervous centers, such as those in the medulla oblongata. It is also involved in such basic drives as hunger, thirst, sex, and rage. Electrical stimulation of various centers in the hypothalamus can cause a cat to act hungry, sexy, cold, hot, benignly, or angry. In humans, it is believed that a tumor pressing against the hypothalamus can cause a person to behave violently, even murderously.

A major function of the hypothalamus is its coordination of the nervous system with the endocrine (hormonal) system (Chapter 36). In fact, the hypothalamus has a certain control over the so-called "master gland," the pituitary (see Chapter 36), and produces some of the hormones that are released from the pituitary gland.

The Limbic System. The hypothalamus and the thalamus, along with certain pathways in the cortex, are functionally part of what is called the **limbic system** (Figure 34.5). The limbic system links the forebrain and midbrain and is composed of a number of nuclei that are also centers of emotion. The limbic system, then, appears to shape our emotional responses in ways that enhance our survival and

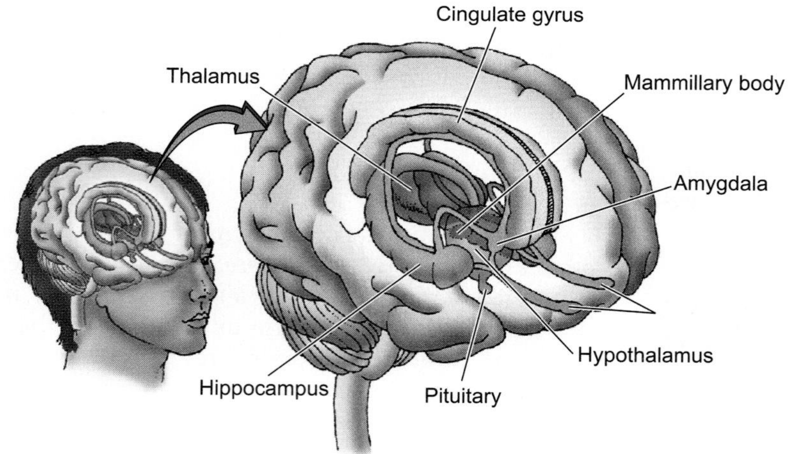

FIGURE 34.5 THE LIMBIC SYSTEM.
The major components of the limbic system include the amygdala, hippocampus, thalamus, hypothalamus, and certain parts of the frontal and temporal lobes of the cortex.

reproductive success. For example, the **amygdala** lying within the limbic system can produce rage if stimulated and docility if removed. The **hippocampus,** another limbic structure, may figure importantly in the memory of recent events. Without the hippocampus, people may be unable to complete a sentence because they forgot how it began.

The Cerebrum

To most people, the word *brain* conjures up an image of two large, convoluted gray hemispheres. Those hemispheres actually make up only part of the brain, the cerebrum—the largest and most prominent part of the human forebrain.

The left and right halves of the cerebrum are the **cerebral hemispheres,** and the outer layer of gray, unmyelinated cells is the **cerebral cortex.** (*Cortex* means "rind," and is a general biological term for the outer layer of any organ.) The cerebral cortex consists of a thin but extremely dense layer of about 15 billion nerve cell bodies and their dendrites. It overlies the whitish, more solid region of myelinated nerve fibers below (Figure 34.6).

The Lobes of the Cerebrum. In humans, each cerebral hemisphere is divided into four lobes: *occipital, temporal, frontal,* and *parietal* (Figure 34.7). At the posterior end is the **occipital lobe.** It contains a region that receives raw, visual sensory input from the optic

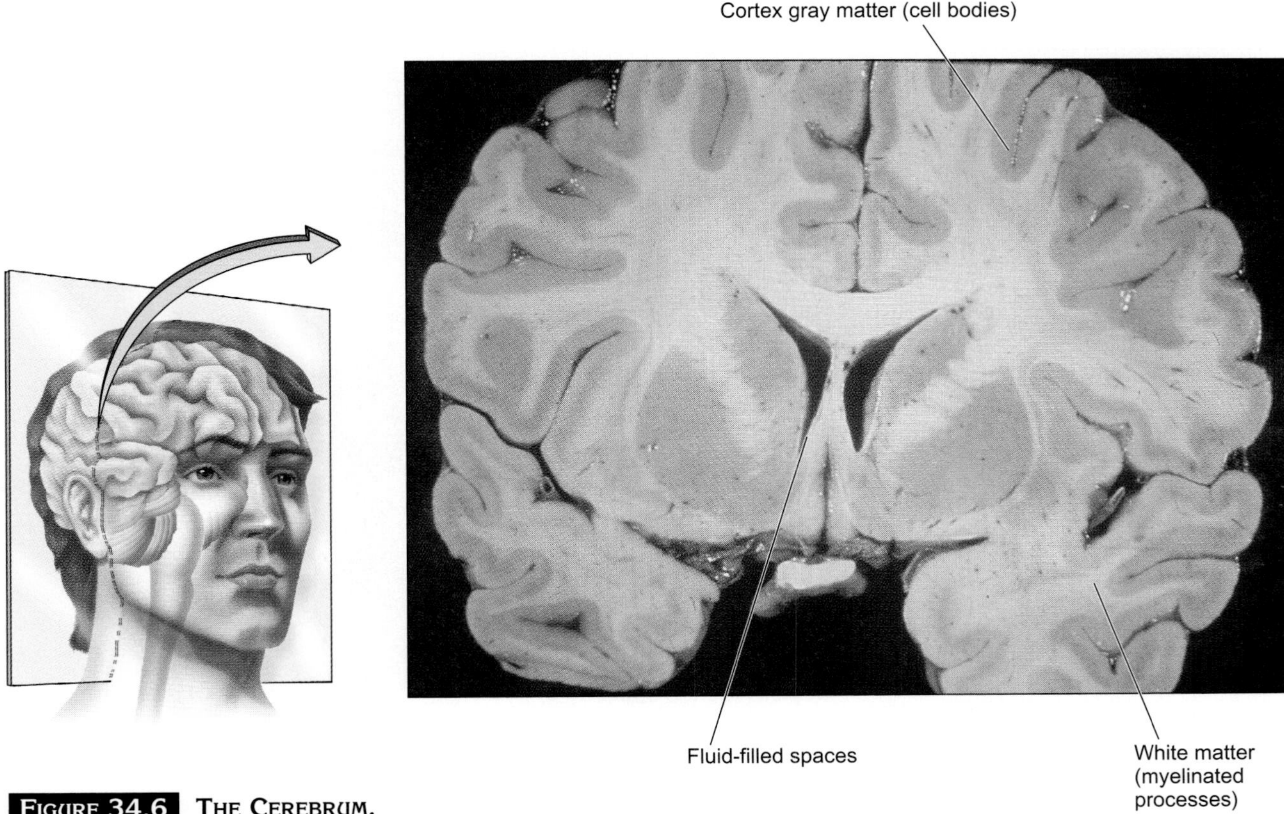

FIGURE 34.6 THE CEREBRUM.
A cross-section of the cerebrum reveals an outer layer of gray matter, the cortex, that contains several very dense layers of neurons.

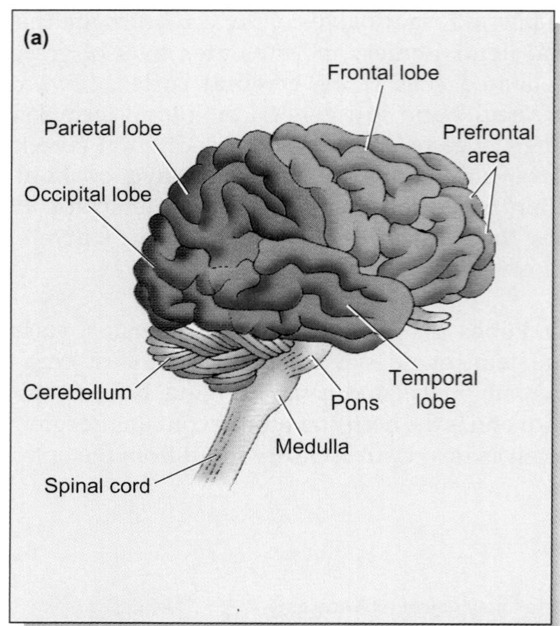
(a)
Frontal lobe
Parietal lobe
Prefrontal area
Occipital lobe
Cerebellum
Pons
Temporal lobe
Medulla
Spinal cord

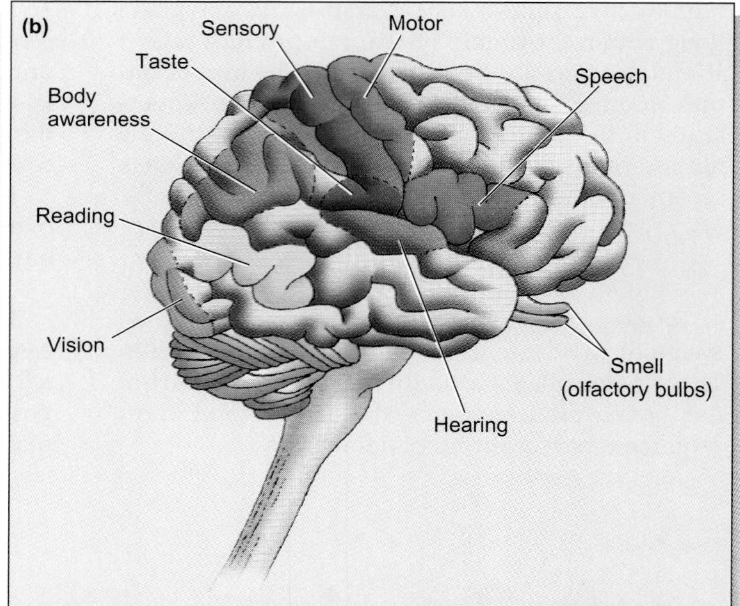
(b)
Sensory
Motor
Taste
Speech
Body awareness
Reading
Vision
Smell (olfactory bulbs)
Hearing

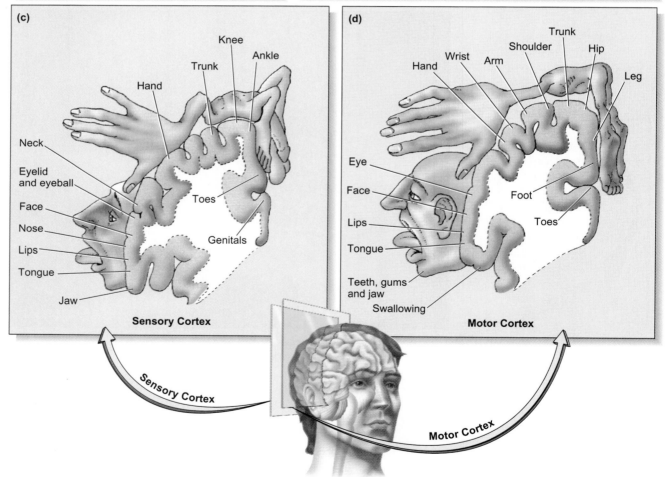

(c)
Knee
Ankle
Trunk
Hand
Neck
Eyelid and eyeball
Face
Nose
Lips
Tongue
Jaw
Toes
Genitals
Sensory Cortex

(d)
Trunk
Shoulder
Hip
Wrist
Arm
Hand
Leg
Eye
Face
Lips
Tongue
Teeth, gums and jaw
Swallowing
Foot
Toes
Motor Cortex

Sensory Cortex
Motor Cortex

nerve and begins the analysis of that input. If the occipital lobe is injured, black "holes" appear in the part of the visual field that is registered in that area.

The **temporal lobes** are at either side of the brain, under the temples. Each lobe roughly resembles the thumb of a boxing glove and is bordered anteriorly by a deep groove, the *lateral fissure.* The temporal lobe helps to process input from senses relating to hearing and smell. This lobe also helps with the processing of visual information, possibly constructing more comprehensive images from cruder information received from the occipital lobe.

The **frontal lobe** is right where you would expect to find it—at the front of the cerebrum. It underlies that part of the skull that people hit with the palm of their hand when they suddenly remember what they forgot. One part of the frontal lobe regulates precise voluntary movement. Another part (on one side) controls the bodily movements that produce speech and is considered to be part of the speech center.

The area at the very front of the frontal lobe is called the *prefrontal area.* This is the part behind the place you slap with the heel of your hand when you remember what you forgot. Whereas it was once believed that this area was the seat of the intellect, it is now apparent that its principal function is sorting out sensory information. In other words, it places information and stimuli into their proper context. The gentle touch of a mate and the sight of a hand protruding from the bathtub drain might both serve as stimuli, but they would be sorted differently by the prefrontal area.

The **parietal lobe** lies directly behind the frontal lobe, and the two are separated by a deep cleft called the *central sulcus.* The parietal lobe contains the sensory areas for the skin receptors and the cortical areas that detect body position. Even if you can't see your hands right now, you probably have some idea of where they are (it's a *real* good idea), thanks to receptors in your muscles and tendons that innervate centers in the parietal lobe (and the cerebellum as well). Damage to the parietal lobe can cause numbness and a sense that one's body is wildly distorted. In addition, the victim may be unable to perceive the spatial relationships of surrounding objects.

By probing the brain with electrodes, investigators have determined exactly which areas of the cerebrum are involved in the body's various sensory and motor activities and have mapped these functions on the cortex. Figure 34.7 shows the results of such mapping. The figures are distorted to demonstrate the relative area of the cerebrum devoted to each body part. The sensory areas of the cortex are largely devoted to integrating sensations from the face, tongue, hands, and genitals, while the motor areas are devoted primarily to the muscles of the tongue, face, and thumbs. (The genitalia, then, are extremely sensitive but without much control, a fact that may not surprise you.)

FIGURE 34.7 THE COMPONENT PARTS OF THE CEREBRUM.
The human cerebrum is divided into four prominent lobes: occipital, temporal, frontal, and parietal **(a)**. Among the distinct regions of the cortex **(b)** are the sensory and motor areas. The two are located on opposite sides of the central sulcus, a prominent dividing line between the frontal and parietal lobes. In the caricatures, the parts of the human body that are governed by each region are indicated directly on sections of the sensory cortex **(c)** and motor cortex **(d)** and are emphasized according to how much brain tissue is devoted to that part.

Right and Left Halves of the Brain. Although the two cerebral hemispheres are roughly equal in size and in potential, each is better than its partner at some functions. In most individuals (about 95 to 99% of right-handers and 60 to 70% of left-handers), for example, the language centers are located in the left hemisphere. Because of the importance we humans place on language, the hemisphere containing these centers is termed the dominant one. Left-hemisphere damage can result in aphasia—difficulty in speech. In addition to language, the left hemisphere is better at mathematics and logical reasoning than is the right hemisphere. It tends to process information analytically, by breaking the problem into its component parts.

In contrast, the right hemisphere tends to be more creative and artistic. Spatial perception is also a right-hemisphere specialization. Therefore, damage to the right hemisphere of the cerebrum may result in the inability to draw the simplest picture or diagram. It processes information more holistically—that is, by putting the pieces together to form an integrated whole. Although it usually does not control speech, the right hemisphere does understand simple speech.

In spite of these specializations, the right and left hemispheres operate as an integrated functional unit, connected primarily by the **corpus callosum** (Figure 34.8). It is because of this structure that the right side of the brain controls the left side of the body and vice versa. Furthermore, if one side of the brain learns something—for instance, when an object is touched with just one hand—the information will be transferred to the other hemisphere. However, if the corpus callosum has been severed, the left

Essay 34.1 The Great Split-Brain Experiments

Some years ago, Roger W. Sperry and Robert E. Meyers experimentally separated the two hemispheres of the brains of cats by severing both the corpus callosum and the optic chiasma. This meant that the right eye was now connected only to the right half of the cerebral cortex and the left eye to the left cerebral cortex. Intensive testing then showed that the animals behaved as if they had two separate brains. A cat could be trained to perform a task by using one eye (the other covered) and then, when presented with the same task viewed by the other eye, the animal would respond as if no learning had occurred.

Later, Sperry and Michael Gazzaniga severed only the corpus callosum in human epilepsy patients as treatment for their condition. With their optic chiasmas intact, they were tested in an apparatus that allowed a particular image to fall entirely on only one half of the retina (see illustrations). This meant an image could be effectively transmitted to only the right or the left hemisphere of the brain. The results were intriguing: The patients reported "seeing" only what was transmitted to the left hemisphere. If a word such as "heart" were shown but divided by a partition so that each eye could see only part of the word ("he" with the left eye and "art" with the right), then "he" fell on the right half of the brain, "art" on the left, and the subjects reported seeing "art.

It was quickly discovered that the results were not due to visual phenomena alone. The patients could only tell researchers what fell on the left halves of their brains because that is where the centers of speech are located. On the other hand, if they were allowed to express themselves nonverbally, they could immediately describe what the right half of the brain "saw." Because the right side of the cortex controls the left hand, and vice versa, a person can first feel an object, say with the left hand, and

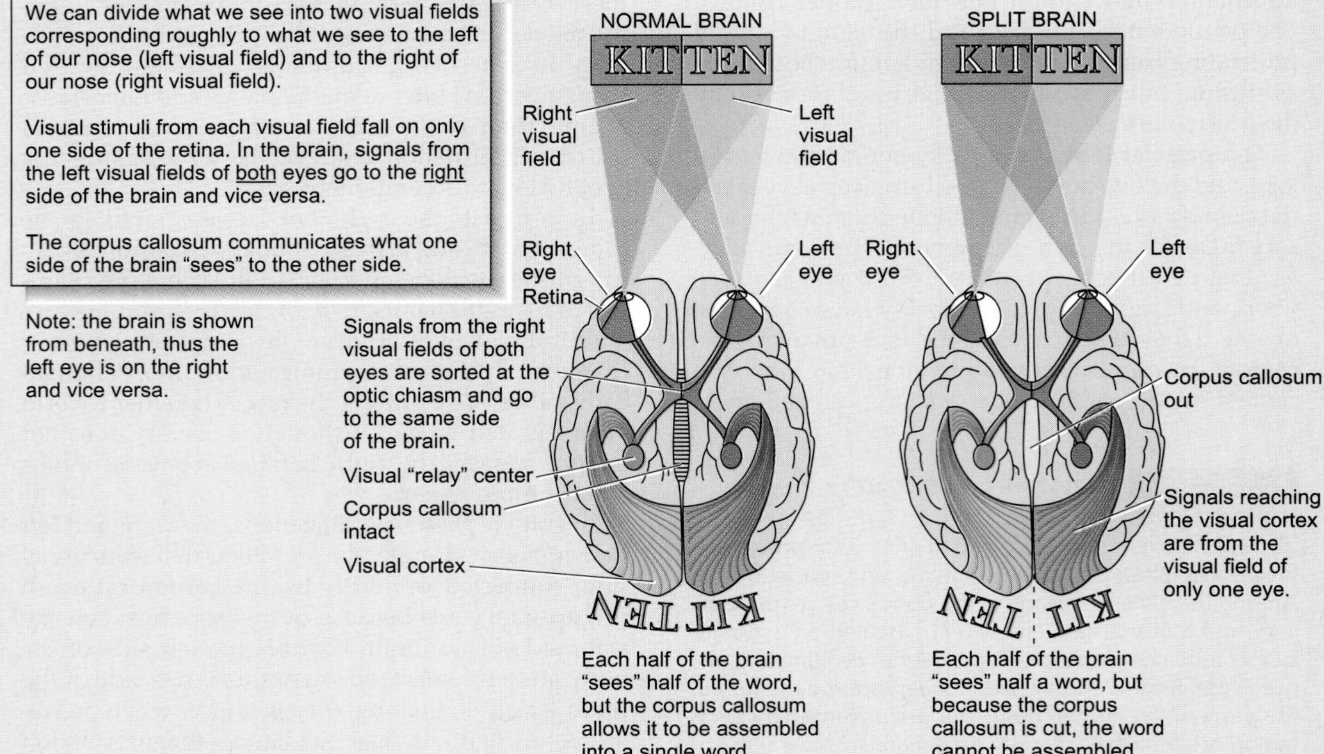

We can divide what we see into two visual fields corresponding roughly to what we see to the left of our nose (left visual field) and to the right of our nose (right visual field).
Visual stimuli from each visual field fall on only one side of the retina. In the brain, signals from the left visual fields of both eyes go to the right side of the brain and vice versa.
The corpus callosum communicates what one side of the brain "sees" to the other side.
Note: the brain is shown from beneath, thus the left eye is on the right and vice versa.
NORMAL BRAIN
KITTEN
Right visual field
Left visual field
Right eye
Retina
Left eye
Signals from the right visual fields of both eyes are sorted at the optic chiasm and go to the same side of the brain.
Visual "relay" center
Corpus callosum intact
Visual cortex
KITTEN
Each half of the brain "sees" half of the word, but the corpus callosum allows it to be assembled into a single word.
SPLIT BRAIN
KITTEN
Right eye
Left eye
Corpus callosum out
Signals reaching the visual cortex are from the visual field of only one eye.
KIT TEN
Each half of the brain "sees" half a word, but because the corpus callosum is cut, the word cannot be assembled.

then pick out that object with the same hand from a collection hidden behind a screen, simply by feel. However, because the information is traveling to the right half of the brain and that half cannot communicate with the left half, the person cannot name the object.

Interestingly, the person will be able to pick out the object faster by using the left hand than the right because the right cortex is far superior at dealing with spatial relationships. Further experimentation has revealed that the human brain is, in a sense, effectively divided into two halves—two brains, as it were. The left half, indeed, tends to deal more with rational, verbal, and logical information, while the right is more intuitive. Unfortunately, the information has fallen into the hands of pop psychologists, whom we now find breathlessly explaining that the problems of the world are due to too much dependence on the cold and logical left hemisphere and that we should be training ourselves to depend on the soulful and holistic right side. The secret, some say, is encompassed in the teachings of Eastern religions. Those who make such claims, however, seem to neglect the fact that the talents of each half of the brain are mutually dependent. The two halves work together—and have historically—to produce that peculiar combination of behaviors that are so distinctively human.

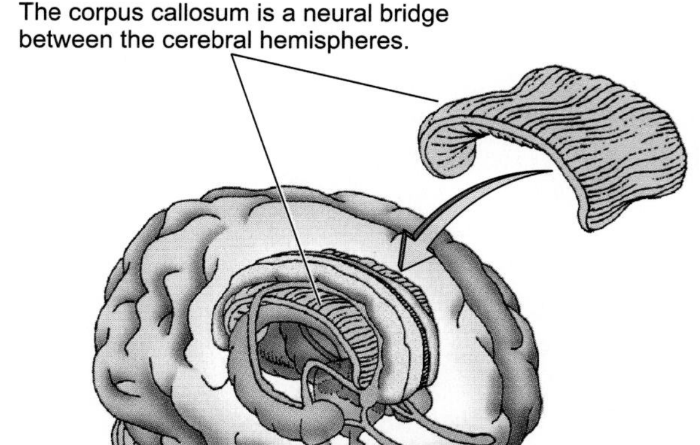
The corpus callosum is a neural bridge between the cerebral hemispheres.

FIGURE 34.8 THE CORPUS CALLOSUM.
The corpus callosum is a neural bridge between the cerebral hemispheres, relaying information from one half to the other.

side of the brain literally doesn't know what the left hand is doing. (Recall that neural tracts cross from one side of the brain to the other, so the right side of the body is controlled by the left side of the brain and vice versa.) Essay 34.1 describes some "split-brain" experiments that led to much of what we know about the hemispheres and their relationship.

THE SPINAL CORD

The **spinal cord** (Figure 34.9) is a cylinder of nervous tissue that extends from the medulla oblongata of the brain, passing through the *foramen magnum* ("big opening") at the base of the skull. It lies sheltered within the **vertebral canal,** a continuous channel lying within the vertebral column. Paired spinal nerves, which are part of the peripheral nervous system, emerge from the cord through the spaces between adjacent arches (Figure 34.9b).

One function of the spinal cord is to serve as the primary link between the brain and other parts of the nervous system. This is accomplished with the **spinal tracts,** bundles of myelinated nerve fibers grouped according to function and running parallel to one another in the outer regions of the spinal cord. Myelin sheaths covering the neurons give these tracts a white, glistening appearance.

The second main function of the spinal cord is to serve as a reflex center. Essential to this function is the butterfly-shaped gray region (the **gray matter**)

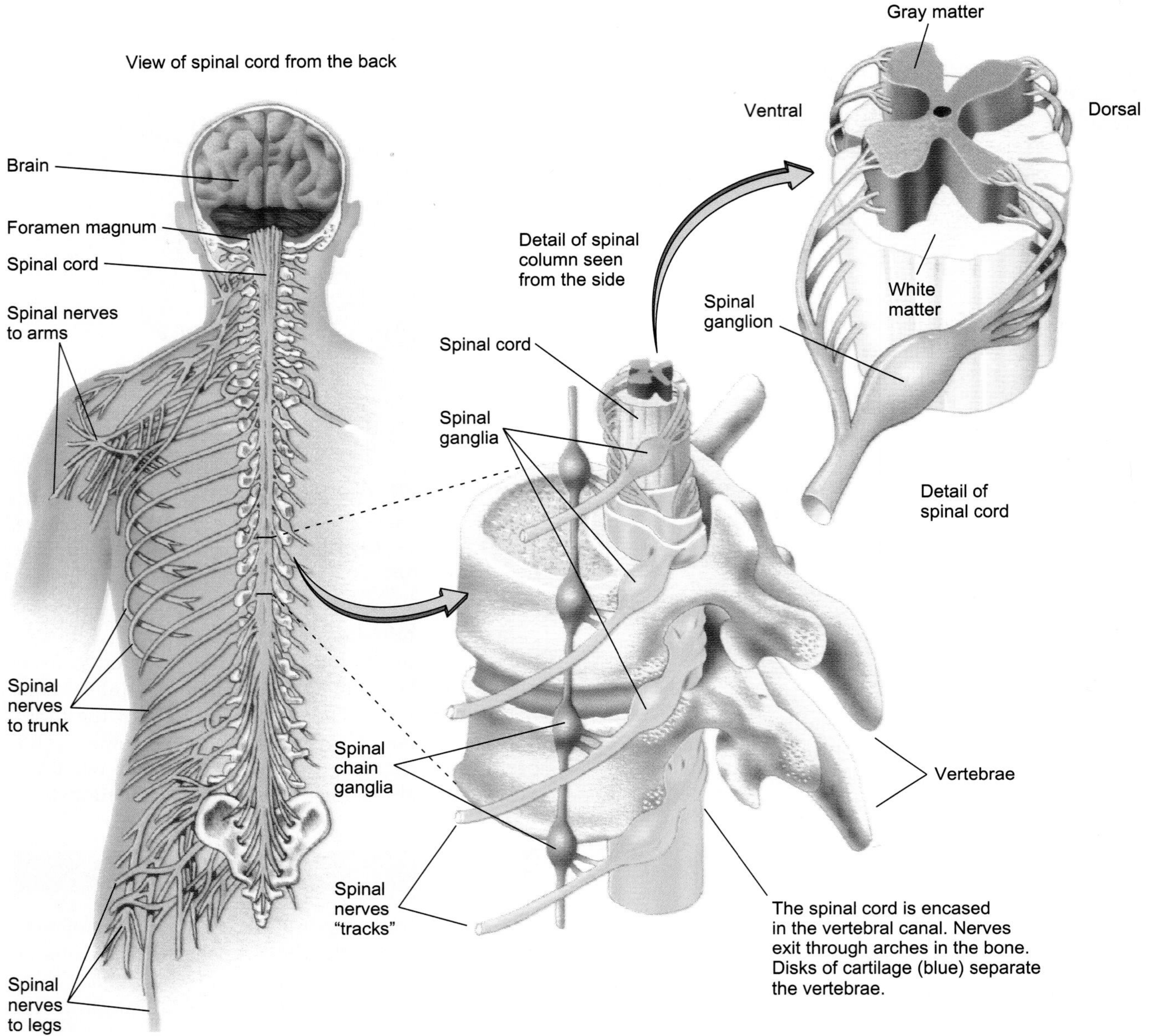

FIGURE 34.9 THE SPINAL CORD.
A three-dimensional view of the spinal cord reveals gray matter (cell bodies), surrounded by white matter (myelinated fibers). Incoming neural impulses from sensory neurons synapse in the dorsal root ganglia, passing then into the cord for transfer to the brain. Motor neural impulses emerge from the cord via the ventral roots, on their way to muscles, glands, blood vessels, and other effector organs.

in the central region of the spinal cord. This contains nonmyelinated neurons as well as the cell bodies of many spinal neurons.

If you pride yourself as being a "thinking animal," you may be a little disappointed to realize that your spinal cord can often receive information from the body's receptors, process it, and initiate the proper response before your brain even "knows" what has happened. The neural connections that permit this are called the **reflex arc.**

Physicians often like to tap the tendon below the knee, making the leg jump. This was once thought to be for the physician's amusement, but now we know that it is to test the patient's reflex arc. In the reflex arc shown in Figure 34.10, the message is generated in a special receptor called a stretch receptor that responds when it is elongated. The message is then transmitted to a sensory neuron, which enters the spinal column through the dorsal root of the spinal nerve. The sensory neuron then excites the proper motor neuron. The motor neuron leaves the spinal cord through the ventral root of the spinal nerve and travels outward to the effector, a muscle group. In most other reflex arcs the impulse is transmitted from sensory to motor neurons by an interneuron, a neuron that links two others.

Note that impulses from one side of the body can cross the spinal cord so that effectors in the other side of the body are stimulated. Direct neural routing from the spinal cord in a reflex arc saves time because the distance the impulse has to travel from receptor to effector is shorter. Furthermore, no time is spent in deliberating the decision. Of course, the brain is informed about what happened because the neurons involved also synapse with neurons leading to the brain. You are aware that your knee jerked, but after the fact.

Chemicals in the Brain

We know that the brain contains billions of neurons and that these neurons interact in a delicate and coordinated manner, integrating and shunting information from one place to another. This interaction includes both the excitation and inhibition of adjacent neurons through very specific actions in the trillions of synapses. The brain has at least 50 neurotransmitters, with more being discovered all the time. This number probably shouldn't be too surprising, considering the specificity necessary to orchestrate the brain's vast network of cells.

The brain's neurotransmitters may be simple *monoamines,* such as norepinephrine, dopamine, histamine, and serotonin. Other neurotransmitters may be modified or unmodified amino acids. One more recently discovered class of transmitters, the *neuropeptides,* are short chains of amino acids varying in number from 2 to about 40.

Among the more interesting brain neuropeptides are the *enkephalins* and *endorphins.* These modify our

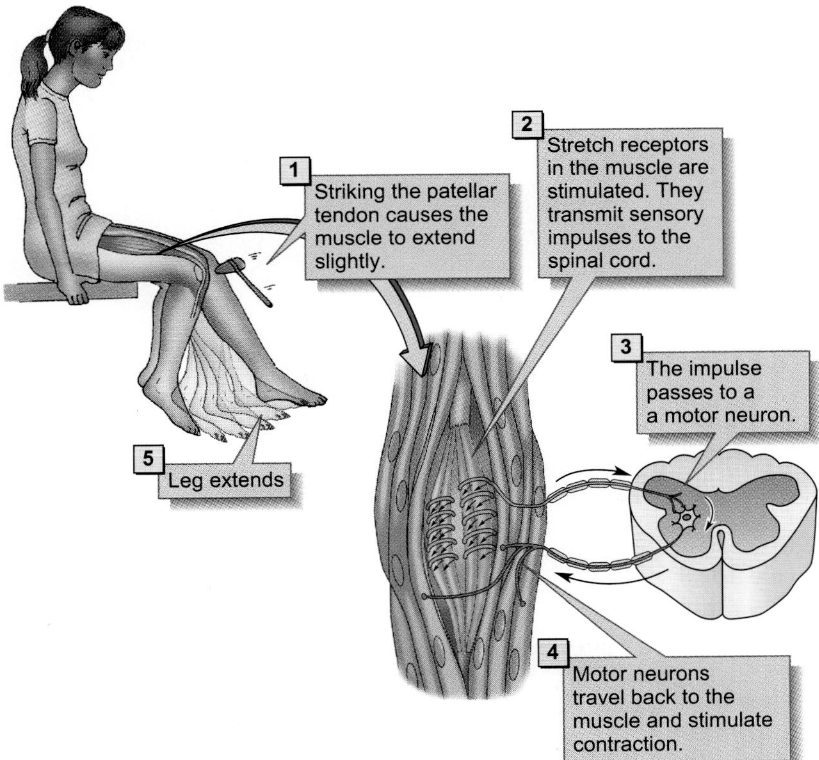

1
Striking the patellar tendon causes the muscle to extend slightly.
2
Stretch receptors in the muscle are stimulated. They transmit sensory impulses to the spinal cord.
3
The impulse passes to a motor neuron.
4
Motor neurons travel back to the muscle and stimulate contraction.
5
Leg extends

Figure 34.10 A Reflex Arc.
In the knee-jerk reflex, striking the tendon causes the muscle above to extend slightly. Stretch receptors, specialized sensory neurons in the muscles, are stimulated. They then transmit impulses to the spinal cord, where the appropriate motor neurons are activated, causing muscle contraction.

perception of pain and also have an elevating effect on mood. Anything acutely painful, such as running a marathon race or being shot in the foot, will stimulate the release of enkephalins. The "runner's high," the slight euphoria that may result after about a 10-mile run, is also attributed to the release of enkephalins (although some runners may simply be ecstatic at having covered that distance without dying in the process). These peptides are called *opioid neurotransmitters* because morphine and other opiates will also bind to the enkephalin neuron receptors and mimic their action.

Our rapidly expanding knowledge of the neurotransmitters is shedding some light on the action of certain drugs that affect the central nervous system. Amphetamines, for example, are believed to imitate norepinephrine, a neurotransmitter that stimulates the brain and accelerates blood pressure and heart rate. Cocaine has a different, more drastic effect. It amplifies the action of neurotransmitters such as serotonin, norepinephrine, and dopamine by blocking the enzymes and reuptake mechanisms that normally clear them from the synaptic cleft after they have acted. Drugs such as LSD, mescaline, and psilocybin mimic the role of other natural neurotransmitters. They produce their hallucinogenic effects by artificially stimulating action at the synapses. It is also being found that neurotransmitters are implicated in certain disease conditions (Essay 34.2).

Electrical Activity in the Brain

Each firing neuron in the brain creates impulses that generate a form of electrical activity. If, perhaps, a million neurons fire simultaneously, this electrical energy can be detected experimentally from outside the body. The instrument used in such detection is called an *electroencephalograph,* and the record obtained is called an *electroencephalogram (EEG).* Electrodes leading to the instrument are fastened at various places on the scalp, and the very faint electrical currents they pick up (primarily from the cortex) are amplified and recorded.

The EEG is not very useful in determining specifically what is going on in the brain, but it is useful in detecting changes in brain activity—the differences between wakefulness and sleep, for example, or certain abnormalities, such as certain kinds of brain damage. Figure 34.11 compares the EEG of a normal person with that of a person with epilepsy. As the recording shows, the brains of epileptics are subject to sudden, random bursts of electrical activity. In a few instances, this abnormality can be traced to physical defects, such as scars or lesions in the brain tissue, but most forms of epilepsy are simply not understood.

Sleep

Electrical activity in the brain does not cease with sleep—quite the contrary is true. EEGs reveal that sleep is accompanied by a considerable amount of brain activity. Sleep has four distinct phases (Figure 34.12). By far the most intriguing of these is called **rapid eye movement (REM) sleep.** During this period, the skeletal muscles are very relaxed—except for the eyes, which dart about beneath the eyelids. The EEG recording at this time is similar to that produced by wakefulness. A person awakened at this time will report vivid dreams (that will be forgotten if the person is allowed to waken naturally).

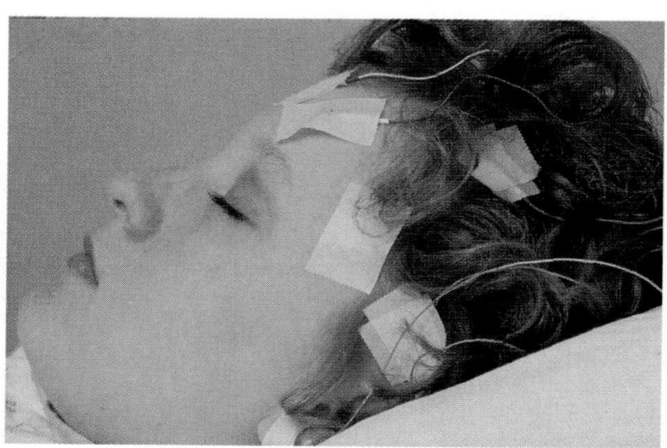

(a)

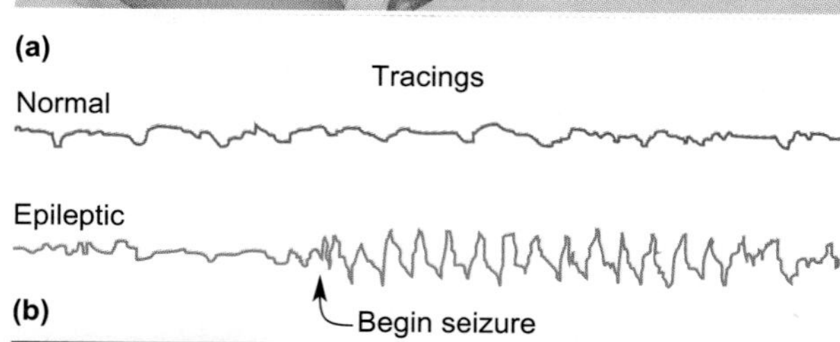

(b)

Figure 34.11 Brain Electrical Activity.
(a) An electroencephalogram (EEG) is produced by fastening a number of electrodes on the patient's scalp and then recording differences in the electrical potential between or among the leads. **(b)** The normal electrical activity of the brain is seen in EEG tracings from the cranium. In an epileptic person, the brain may show normal electrical activity between seizures, but when these episodes do occur, the electrical disturbances are obvious.

Essay 34.2 Alzheimer's Disease

We are all unhappily aware that some people may become forgetful, foolish, and incompetent as they age. They may regress to reliving the distant past, no longer functioning in the present. Such people suffer from *senile dementia,* the most prevalent form of which is called *Alzheimer's disease,* a relentlessly progressive condition that afflicts 5 to 10% of all people over 65. (It can also be found in younger people.) Many patients in nursing homes have been placed there because their cognitive abilities, and especially their memories, have deteriorated too far for family members to be able to care for them at home.

Alzheimer's disease first manifests itself as an inability to recall recent events. At this stage, afflicted people cannot remember what happened an hour or a week ago, although they can often remember childhood experiences in vivid detail. As short-term memory lapse continues, other cognitive functions begin to fail. In the second stage of the illness, victims gradually forget how to read or write or perform simple calculations, and their speech may become garbled and irrational. The impairment of these abilities is often accompanied by irritability, paranoia, and hallucinations. Yet, afflicted people remain alert (and aware of their degeneration) until the final stages of the disease.

Whereas any form of senile dementia was once considered to be simply a natural part of aging—the mind deteriorating as the body does—researchers now recognize certain variables associated with Alzheimer's disease. For example, the disease seems to be strongly influenced by genetic factors. However, this is difficult to fully substantiate, since the disease appears so late in life that there is a strong possibility that the individual may die before the traits ever have the opportunity to be expressed. One genetic indicator is expressed to a nondisjunction at meiosis resulting in a condition similar to Downs syndrome. The connection between the two conditions is being extensively investigated. Another indicator of Alzheimer's disease is a reduction in the level of choline acetyltransferase, an enzyme that helps synthesize the neurotransmitter acetylcholine in the brain cells of victims. In 1986, researchers reported a blood indicator of Alzheimer's. Some healthy people were reluctant to be tested for the indicator, however, since there is no cure. They preferred to live out their "normal" years not knowing what lay ahead. (Would you be willing to be tested?)

People suffering from Alzheimer's show a marked decrease in the number of neurons in the cortex of the brain (the gray matter) and especially in the *nucleus basalia,* a structure in the base of the forebrain that has many neural extensions to the cerebral cortex. (If you point your finger directly at your temple, you will be pointing at this somewhat mysterious structure.) Alzheimer's is also marked by dense tangles ("plaques") of neurons and their remnants in other parts of the brain.

Scientists continue to investigate the physiological causes of what was once thought to be a psychological condition. Their discoveries regarding Alzheimer's disease's effects on the brain raise possibilities of counteracting these effects with medication or surgery and offer new hope for treating this affliction of the elderly.

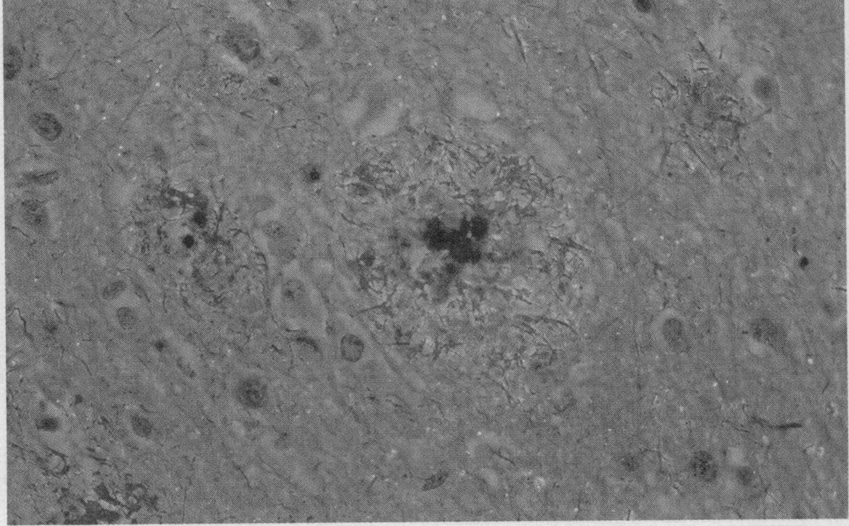

Psychologists and biologists generally agree that, for whatever reason, sleep appears to be essential, especially to mental concentration and balance, and is somehow restorative. Theorists suspect that sleep (and REM sleep, in particular) is a period of information sorting and storage, and may be essential to long-term memory storage.

MEMORY

Mammals have two distinct kinds of memory. **Short-term memory** lasts for just a few hours, whereas **long-term memory** is relatively permanent, and is believed to be associated with the formation of *engrams* once called "memory scars," physical changes in the brain, associated with learning. The two kinds of memory apparently function together, because the neural events that form memories must be encoded in the short-term memory first. These short-term memories may or may not be consolidated into long-term memory.

Short-term memory does not form engrams. In other words, there is no physical evidence of short-term memory formation in the brain. It is quickly extinguished, leaving no trace as far as we know. It has been hypothesized that short-term memory is due to temporarily reverberating neural circuits. In this model, a series of neurons would repeatedly fire along a specific pathway. Such a circuit would produce a feedback loop, so that when an impulse had passed over a circuit, it would be fed back to the beginning of the circuit (possibly over a "collateral neuron" running in the opposite direction), stimulating the circuit to fire again. Finally, the stimuli would weaken so that no more impulses could be initiated, and the neural pathway would come to rest, the matter that stimulated it forgotten. (Forgetting may have certain advantages, as we see in Essay 34.3.)

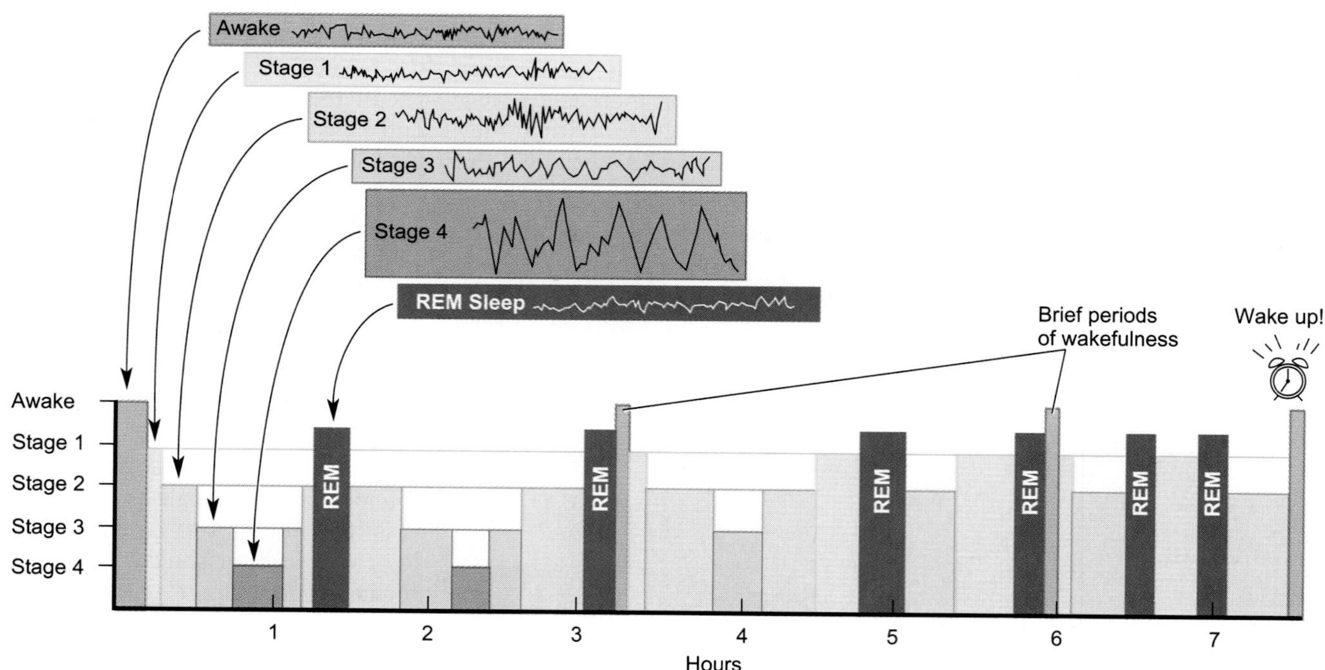

FIGURE 34.12 BRAIN ACTIVITY AND SLEEP.
By studying the electrical activity of the brain and eye movement during sleep, researchers have identified four sleep stages: Stage 1, drowsiness; stage 2, light sleep; stage 3, intermediate sleep; and stage 4, deep sleep. Each has its own pattern of electrical activity, as seen in the EEG. Typically, sleep is interrupted by bursts of electrical activity whose patterns resemble wakefulness. Such sleep is associated with rapid eye movement and is called REM sleep.

Long-term memory, on the other hand, *does* leave physical evidence—the engram. This has been clearly demonstrated by the formation of new receptors and new growth patterns in neurons as learning (involving memory) takes place. At least some kinds of learning apparently involve an increased likelihood that neural impulses will travel over specific routes. This happens when certain neurons initiate a new direction in growth or when they show an increased sensitivity to signals from the preceding (presynaptic) neuron.

According to one scenario, long-term memory takes place with repetition, causing a certain neural tract (pathway) to be repeatedly stimulated. Finally, the continual stimulation over that tract causes the postsynaptic neurons to be flooded with calcium. The calcium then activates a dormant enzyme called *calpain,* which now has the ability to break down certain proteins. One such protein, called *fodrin,* lends structural form to the dendrites of neurons. As the calpain breaks down the fodrin (sounds like an episode from *Star Wars*), the reaction exposes new receptors on the postsynaptic neuron, which can then be stimulated by neurotransmitters from the presynaptic cell. With these new receptors exposed, the postsynaptic neuron is much more sensitive to the transmitters (Figure 34.13).

Repeated stimulation of this tract causes the continued breakdown of fodrin by calpain until the protein structure of the postsynaptic neuron actually breaks down, leaving the neuron free to change its shape or to make new connections, thus facilitating the passage of new impulses along the pathway. This new "wiring" of the brain, then, is suggested as the physical basis of memory.

The Peripheral Nervous System

As was mentioned, the vertebrate nervous system is divided into two major parts: the central nervous

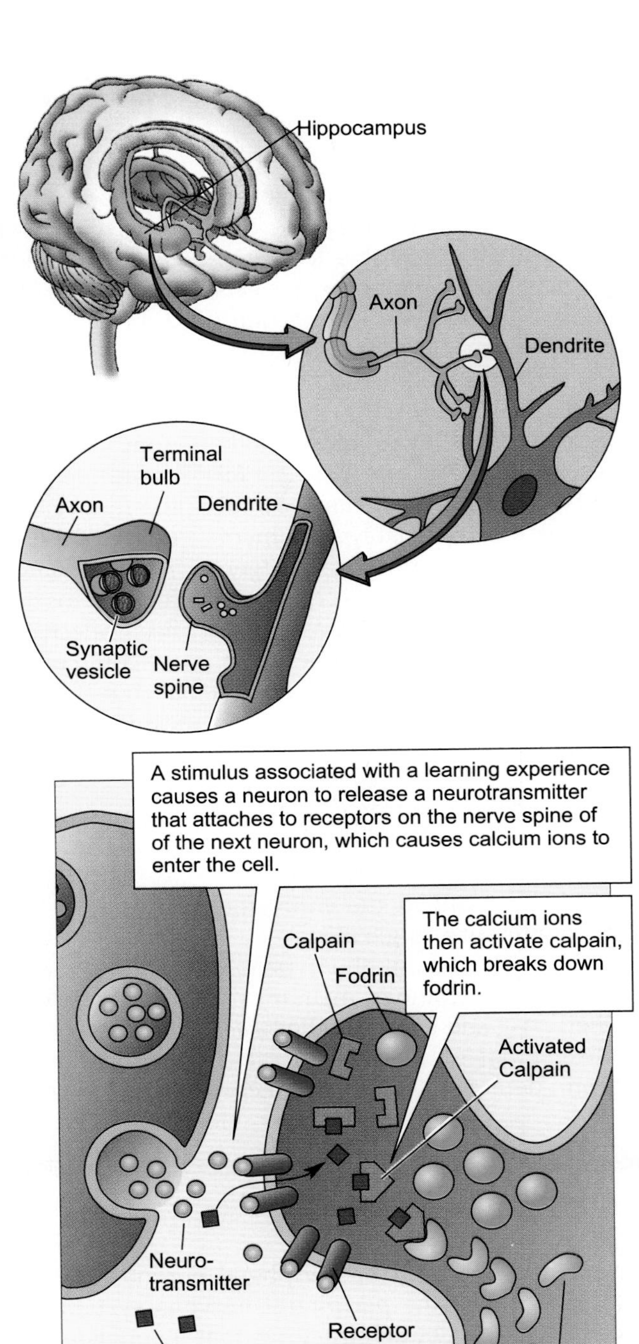

Figure 34.13 How Memories Are Formed. A learning experience causes changes in nerve spines of the hippocampus. A stimulus associated with the experience causes a neuron to release a neurotransmitter that attaches to receptors of the next neuron. This, in turn, causes calcium to enter the cell. The calcium then activates calpain, which begins to break down fodrin. As the fodrin degrades, more receptors begin to appear, allowing more calcium into the cell. The cell membrane weakens, and the cell begins to change shape as new spines develop. These new spines allow new neural connections that are believed to be the basis for memory.

ESSAY 34.3 The Advantage of Forgetting

There are all sorts of strange abilities associated with memory in humans. *Savants* have very low IQs, but some are able to accomplish incredible mathematical feats, such as multiplying two 5-figure numbers in their heads. (Dustin Hoffman played such a person in the movie *Rain Man.*) Others can immediately tell you the day of the week on which Christmas Day fell in 1492—or any day in any year (although this is probably not a memory feat). A normally intelligent Russian man made his living giving stage performances as a *mnemonist.* He would sometimes memorize lines of 50 words. Once, in just a few moments he memorized the nonsense formula:

$$N \cdot \sqrt{d^2 \cdot \frac{85^3}{vx}} \cdot 3\sqrt{\frac{276^2 \cdot 86x}{n^2 v \cdot 264}} = sv \cdot \frac{1624}{32^2} \cdot r^2 s$$

Fifteen years later, upon request, he repeated the entire formula without a single mistake.

And what about those people with "photographic memories"? They exist, and they are called *eidetikers.* Proof of their abilities has been demonstrated with stereograms. These are apparently randomized dot patterns on two different cards—but when they are superimposed by use of a stereoscope, a three-dimensional image appears. One woman was asked to look at a 10,000-dot pattern with her right eye for 1 minute. After 10 seconds, she viewed another "random" dot pattern. She then recalled the positions of the dots on the first card and conjured up the image of a *T.*

So, if such abilities are possible for our species, why haven't they been selected for—so that by now we could all, more or less, perform such feats? On the surface, the advantages seem enormous. The reason we can't is, in part, because there are serious drawbacks to remembering everything. Both the Russian mnemonist and the eidetiker could look at a barren tree, "recall" its leaves, and when they looked away, be confused over whether the tree was leafy or not. The mnemonist could watch the hands on a clock and not notice they had moved, remembering (or "seeing") only where they had been. Also, what about all those insignificant events it is of no advantage to remember? The energetically expensive neural apparatus would be wasted retaining such information for recall (assuming such retention takes more energy than the storage of material we normally can't recall). And the mnemonist had trouble with discerning time lapse. He recalled everything so well that it seemed to him as if every event of his life had just occurred.

Obviously, the reason we can't remember as well as the people in these examples is that we, as a species, have not found it necessary or useful. We don't need to recall every stone on the path to the place where we found food yesterday; we only need to remember the location of the path. In fact, remembering too much about our physical environment might mean that changes in our environment would not be adjusted to as quickly as when we were never quite sure what to expect. Perhaps it is best that we can't rely too strongly on our memories.

system (CNS) and the peripheral nervous system (PNS). In fact, the **peripheral nervous system** includes all the neural structures that lie outside the central nervous system. The peripheral nervous system can be further divided into two functionally different systems. One is the **somatic system,** which includes sensory pathways (which carry messages from sensory structures, to be discussed in the next chapter) and motor pathways (which carry impulses to effectors such as muscles). The other system is the **autonomic nervous system** (Figure 34.14). (Table 34.2 summarizes the parts of the CNS and PNS, and their functions.)

THE AUTONOMIC NERVOUS SYSTEM

The autonomic nervous system (ANS) is essentially a *motor system.* This means that it carries impulses from the brain and spinal cord to the organs it serves. In doing this, it works in concert with the central nervous system in regulating and sensing the activity of the internal organs. In addition to the prominent contents of the chest and abdominal cavities (the heart, lungs, digestive tract, kidneys, and bladder), the ANS also coordinates activities in such body parts as the arteries, the veins, the irises of the eyes, the nasal lining, the sweat glands, the salivary

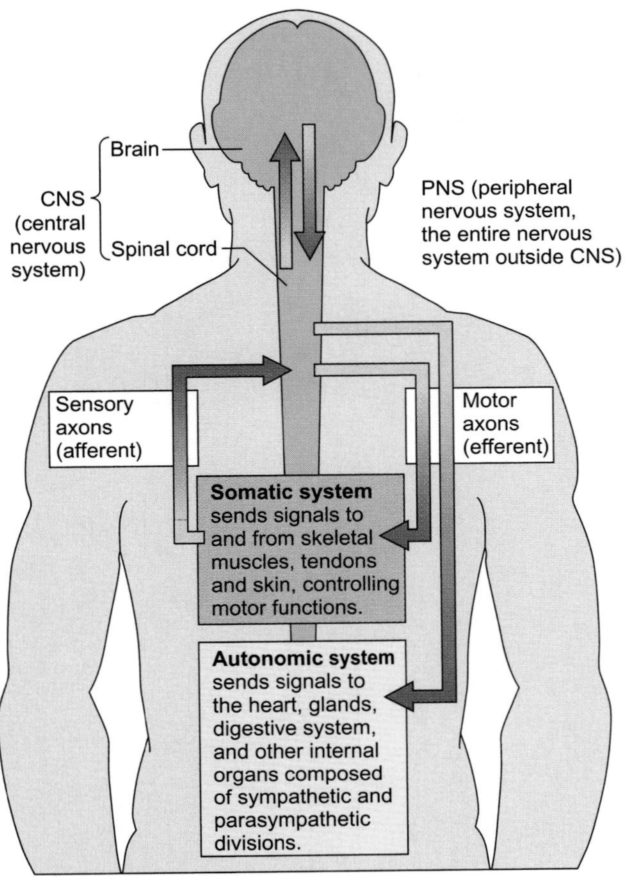

Brain
CNS (central nervous system)
Spinal cord
PNS (peripheral nervous system, the entire nervous system outside CNS)
Sensory axons (afferent)
Motor axons (efferent)
Somatic system sends signals to and from skeletal muscles, tendons and skin, controlling motor functions.
Autonomic system sends signals to the heart, glands, digestive system, and other internal organs composed of sympathetic and parasympathetic divisions.

FIGURE 34.14 THE CENTRAL AND PERIPHERAL NERVOUS SYSTEMS.
The central nervous system, together with the sensory and motor neurons of the peripheral system, work together to control the somatic and autonomic systems.

glands, and the tiny muscles that cause our hair to stand on end when we learn who's been elected.

The general function of the autonomic nervous system is to promote homeostasis. Essentially, homeostasis (Chapter 37) is the maintenance of stability or constancy in the face of changing conditions. The ANS constantly adjusts and coordinates the internal organs to meet changing demands. For such a system to function efficiently requires a considerable amount of sensory feedback from the organs served. This is carried out by sensory nerves of the somatic nervous system, which, by keeping the brain informed, also play an important role in homeostasis.

The autonomic nervous system itself is divided into two parts: the *sympathetic division* and *parasympathetic division* (Figure 34.15). The **sympathetic division** is generally active during emergency or threatening conditions. It is often referred to as the fight-or-flight system. In contrast, the **parasympathetic system** is active when the body is calm and at rest. These divisions often innervate the same organs, and when they do, they have opposite effects—with each bringing body systems to the optimum level of functioning for the situation at hand (Figure 34.15).

For example, if a bear rushes into the room where you are quietly reading, your body will react "sympathetically." The pupils of your eyes will dilate, the better to see the bear—and blood vessels to the skeletal muscles will increase in diameter, bringing oxygen-laden blood to these structures, the better to make your escape. Also, your peripheral blood vessels will decrease in diameter, so blood loss will be minimized in case the bear swats you on your way

TABLE 34.2 THE CENTRAL AND PERIPHERAL NERVOUS SYSTEMS

Central Nervous System		Peripheral Nervous System	
Brain	Integration; association; thought; directs most behavior	Somatic Nervous System	Primarily involved in sensations and actions of which we are conscious
		Cranial nerve	Services the head region (the vagus services the body region)
Spinal Cord	Carries messages to and from brain; center for spinal reflexes	Spinal nerves	Service the neck and body region
		dorsal root	Houses sensory neurons
		ventral root	Houses motor neurons
		Autonomic Nervous System	Controls involuntary activities of internal organs
		Sympathetic nervous system	"Fight or flight" reactions
		Parasympathetic nervous system	Returns the body to "normal" state after emergency; controls activity of organs under nonstressful conditions

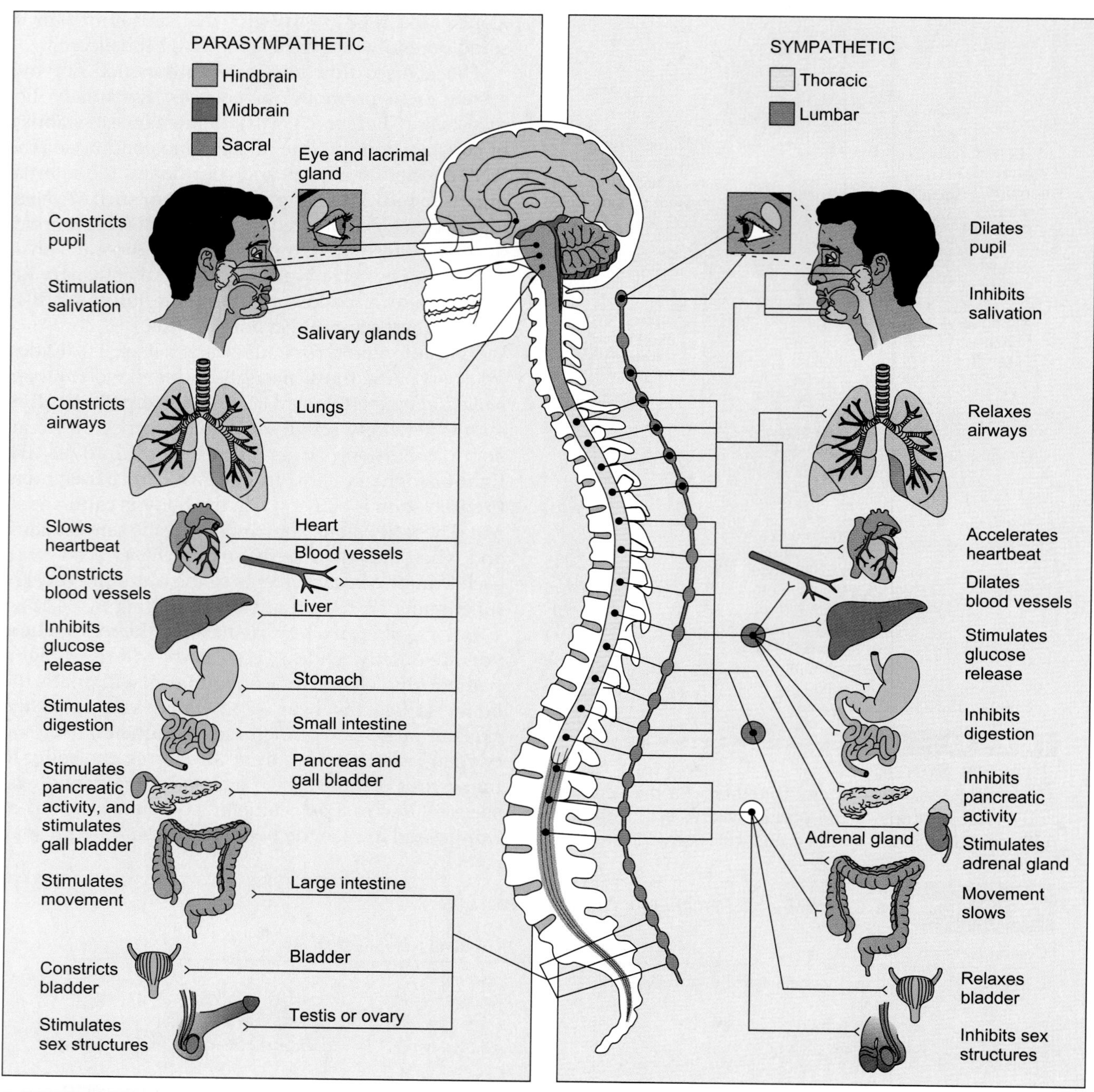

FIGURE 34.15 **THE AUTONOMIC NERVOUS SYSTEM.**
The sypathetic and parasympathetic componenets of the ANS are shown, with the central nervous system in the middle. The principal role of the ANS is to keep the internal organs operating in a carefully coordinated manner in response to shifting conditions and needs.

out; your heart rate will increase, bringing oxygen to your running muscles; your blood pressure and blood sugar will rise; your bronchial tubes will open, getting more oxygen to the muscles; your hair may stand on end; and your digestion will almost stop, since your blood is needed elsewhere. The blood supply to your brain may decrease, causing you to faint, in which case you will hope the bear is just there as part of some effort to help curb forest fires. If, upon awakening, you should discover the bear was only your roommate in a bear suit, your parasympathetic system will take over and reverse all these responses. You may then wish to activate the sympathetic system of your roommate.

Although our scenario implies that these systems are active in an all-or-none fashion, this is not usually the case. The shifting antagonism of the systems allows fine adjustments to be made by both.

How can neural impulses from neurons of the two branches of the autonomic nervous system have opposing effects on the same organ? The answer has to do with the specific neurotransmitters released by those neurons on the target organ. Most (but not all) of the sympathetic neurons secrete the chemical **norepinephrine** at their target organs, while neurons of the parasympathetic system secrete **acetylcholine** (as do motor neurons that move skeletal muscle.) We find, then, that norepinephrine accelerates heart rate, while acetylcholine slows it down, and that acetylcholine stimulates digestive activity, while norepinephrine inhibits it.

As you might expect, the divisions of the autonomic nervous system differ anatomically as well as physiologically. As shown in Figure 34.15, most of the parasympathetic nerves originate in the brain, with a few coming from the lower region of the spinal cord. Some go directly to their target organs; others end in external ganglia, which then relay their messages to nerves continuing toward the target organ. The sympathetic system originates only in the gray matter of the spinal cord. Furthermore, all of the sympathetic nerves leaving the cord pass through or synapse within rows of sympathetic ganglia just outside the cord. Some sympathetic nerves synapse once again in ganglia at various locations in the body.

PERSPECTIVES

We can see that the vertebrate body is never at rest. It is virtually abuzz with neural signals flashing in all directions, but in a highly coordinated fashion. In many cases, those signals are responses to stimuli from the environment. However, in order for the nervous system to respond appropriately to environmental opportunities or threats, it must have information from that environment. And so sensory systems have developed. It is important to know, however, that the environment can be both external and internal, and that special receptors have developed to generate information about both.

SUMMARY

EVOLUTION OF THE CENTRAL NERVOUS SYSTEM

The central nervous system consists of the brain and the spinal cord. The neurons in hydra lack consolidation and form a simple net. Planaria flatworms have a ladderlike system with transverse nerves and an anterior neural mass called the cephalic ganglia. Annelids and arthropods have a distinct ganglionic mass, and a paired but fused ventral nerve cord with segmental branches. The frog has a definite central nervous system; the brain has specialized regions. It still reveals a general paired construction. Also present is a complex, dorsal hollow nerve cord with segmental branching.

THE HUMAN BRAIN

The many billions of neurons and glial cells making up the brain occupy a volume of 1,200 to 1,500 cc. The fragile tissue of the brain and spinal cord is cushioned by cerebrospinal fluid and protected by both bone and the meninges, three tough membranes. In humans the usual vertebrate's hindbrain, midbrain, and forebrain are present, but the midbrain is obscured by the large forebrain.

THE HINDBRAIN

The hindbrain includes the medulla oblongata, pons, and cerebellum. The first controls many involuntary acts such as breathing and heart rate. The pons contains neural tracts from the brain, cord, and cranial nerves and also links the forebrain and cerebellum.

The cerebellum functions in balance, equilibrium, and muscle coordination.

The Midbrain

The midbrain in humans is small, acting primarily in processing audio input and forming connecting tracts between brain regions.

The Forebrain

The forebrain, the largest brain region in humans, is involved in both voluntary and involuntary activities.

The thalamus forms numerous relay connections between brain regions. It contains most of the reticular system, which monitors incoming and outgoing impulses. It contains the reticular activating system that sorts incoming stimuli. Activity of the RAS arouses the conscious centers. The RAS is inhibited by other brain regions during sleep.

The hypothalamus regulates internal functions such as heart rate, blood pressure, temperature, and basic drives such as sex, anger, and thirst. It also interacts with the endocrine system, regulating the pituitary gland and producing its own hormones.

The limbic system, consisting primarily of nuclei, links the forebrain and midbrain and is involved in emotions and recent memory.

The cerebrum is made up of two cerebral hemispheres. Its outermost gray layer, the cerebral cortex, contains dense layers of cell bodies. The cerebral lobes and their functions are as follows: (1) occipital—processes visual input from the eyes; (2) temporal—processes auditory and some visual input; (3) frontal—invokes voluntary movement and speech; the prefrontal area sorts sensory information; (4) parietal—receives sensory input from skin and body position receptors.

Right and left hemispheres have somewhat different functions. One side (usually the left) controls speech, mathematical ability, and logical reasoning. The other is more artistic and better at spatial perception. Physical control of each side of the body is centered in the opposite hemisphere. Information is transferred between hemispheres through the corpus callosum.

The Spinal Cord

The spinal cord exits the skull through the foramen magnum and passes through the vertebral canal. White matter (outside) represents myelinated fibers, while gray matter (inside) represents the cell bodies and nonmyelinated fibers.

One function of the spinal cord is to serve as the primary link between the brain and other parts of the nervous system. This is accomplished within the spinal tracts—myelinated nerve fibers in the outer regions of the spinal cord.

The second main function of the spinal cord is to serve as a reflex center. A spinal reflex arc can often receive information, process it, and initiate a response. In simple reflex arcs, sensory information from a receptor is carried to the spinal cord over a sensory neuron and enters the spinal cord over the dorsal nerve root. It synapses directly with a motor neuron which carries the impulse through the ventral nerve root to the effector, usually a muscle. In more complex reflex arcs, the impulse is transmitted from a sensory neuron to an interneuron and from an interneuron to a motor neuron.

Chemicals in the Brain

Relays of neural impulses in the brain are controlled by at least 50 neurotransmitters. Included are monoamines, modified and unmodified amino acids, and neuropeptides. The latter include enkephalins and endorphins, associated with the relief of pain and emotional well-being. They work similarly to opioids.

Electrical Activity in the Brain

Gross electrical activity in the brain is detected through electroencephalography. General patterns can be studied for evidence of abnormalities such as epilepsy.

EEG studies of sleep reveal several phases, one of which is REM (rapid eye movement) sleep, a time of vivid dreams.

Memory

Memory storage begins with short-term memory, where information is encoded. Most of this is temporary, but some events enter permanent long-term memory. Short-term memories are thought to involve reverberating circuits. Long-term memories leave physical evidence called engrams.

The Peripheral Nervous System

The peripheral nervous system includes all neurons outside the brain and cord. Its somatic branch includes primarily sensory pathways to the brain.

It has two functional parts. The somatic system deals mostly with sensory information received from outside the body, and with the control of voluntary movements.

The autonomic nervous system (ANS) is principally motor, its neurons affecting the functioning of many internal functions such as circulation, respiration, and digestion. The autonomic nervous system promotes homeostasis, a state of stability and constancy.

The sympathetic division of the ANS is active during emergencies, while the parasympathetic division is active during restful conditions. The two branches of the ANS generally have opposite effects on target organs. They do this through the secretion of different neurotransmitters.

Parasympathetic nerves emerge from the brain and lowermost spinal cord. Some go directly to target organs, while others relay their impulses at synapses. Sympathetic nerves emerge from the spinal cord only. They also follow direct and relay routes.

Key Terms

central nervous system (CNS) • **575**
meninges • **577**
cerebrospinal fluid • **577**
hindbrain • **577**
medulla oblongata • **577**
pons • **578**
cerebellum • **578**
midbrain • **579**
forebrain • **579**
thalamus • **579**
reticular system • **579**
hypothalamus • **580**
limbic system • **580**
amygdala • **581**
hippocampus • **581**
cerebral hemisphere • **581**
cerebral cortex • **581**
occipital lobe • **581**
temporal lobe • **583**
frontal lobe • **583**
parietal lobe • **583**
corpus callosum • **583**
spinal cord • **585**
vertebral canal • **585**
spinal tract • **585**
gray matter • **585**
reflex arc • **586**
rapid eye movement (REM) sleep • **588**
short-term memory • **590**
long-term memory • **590**
peripheral nervous system (PNS) • **592**
somatic system • **592**
autonomic nervous system • **592**
sympathetic division • **593**
parasympathetic division • **593**
norepinephrine • **595**
acetylcholine • **595**

Review Questions

1. Compare the nervous systems of *Hydra,* a planarian, and an earthworm. Why would you expect to find similarities between planarian and earthworm?
2. Describe the frog's central nervous system. In what ways is it similar to that of the earthworm? What are some striking differences?
3. In what ways are the meninges and the cerebrospinal fluid important to the brain? What group of cells also plays a supporting role?
4. List the principal parts of the hindbrain. In general, what kinds of acts does this region regulate?
5. What general function is ascribed to the thalamus?
6. Where is the reticular system located? List its two important functions.
7. What are two endocrine (hormonal) functions of the hypothalamus?
8. What specific parts of the limbic system are involved in emotions and memory?
9. Name a function that would most likely be impaired if severe damage occurred in the following: occipital lobe, temporal lobe, parietal lobe, prefrontal region.
10. Locate the cerebral regions responsible for voluntary movement and sensory input from touch receptors. What factor reveals the importance of movement and touch in a specific part?
11. In general, how does the control of body movements relate to specific cerebral hemispheres? What information about stroke victims supports this?
12. Which side of the brain usually contains the language and speech centers?
13. Specifically, how is information passed between hemispheres? How do we know this?
14. What are the general functions of the spinal cord? What are its white and gray regions made of?
15. List the events associated with the knee-jerk reflex.
16. Why is it not surprising to find a great number of neurotransmitters in the brain? List three main types.
17. When are enkephalins and endorphins released? What is their general effect?
18. What do electroencephalographs actually detect? How is such information used?
19. What is REM sleep?
20. Summarize the general functions of the autonomic nervous system. Why is it called a motor system?
21. Describe some effects of activity in the sympathetic division.
22. If both the sympathetic and parasympathetic divisions have identical action potentials, how can different effects on target organs be produced?
23. Summarize the flight-or-fight response. Which division is most involved?
24. Compare the sources and arrangement of the parasympathetic and sympathetic nerves.

35

The Special Senses

THERMORECEPTORS

CHEMORECEPTORS

MECHANORECEPTORS

Touch and Pressure Receptors

Auditory Receptors

Gravity and Movement Receptors

VISUAL RECEPTORS

LIFE EXISTS IN A CHALLENGING WORLD OF THREATS AND OPPORTUNITIES, *and living things must have information about such matters. If a lion is in the neighborhood, you need to know it. If a hamburger is in the neighborhood, you may need to know that, too. Lions and hamburgers, though, are more important in your life than in that of a mayfly, so your receptors are geared for detecting such things, while a mayfly is geared toward finding a mate. After all, a mayfly has only hours to live and in which to find that mate. The mayfly, then, has an entirely different set of receptors than you do. To a mayfly, hamburgers don't exist, and you will never know the ecstasy of smelling a mayfly mate. Here, then, we will look at some special receptors, keeping in mind that we are able to detect what we need to detect and that other stimuli may go completely unnoticed.*

If we were asked by some space traveler about the nature of our planet, what the earth is like, we would have a ready answer. We would talk about colors and feelings and textures and distances. We would describe sounds and music and pain. In essence, we would be trying to convey to the visitor what our senses convey to us. If the visitor was able to communicate with another species and ask the same question, he might become confused, however, because other species are likely to know a different kind of world. They have different impressions of the world because they have different kinds of receptors. Furthermore, they tend to perceive the things that are important to them. Because scent is more important to dogs than color is, for example, a dog might describe a gray world of stunning odors. A fly, on the other hand, might tell about a world of shimmering mosaics with swirling eddies and delicious surfaces. A tapeworm might know nothing of colors or sounds (except perhaps occasional rumbling vibrations), but might describe a watery world filled with food. The point is, each species is able to detect those things that are important to its survival.

We must keep in mind, then, that we are aware of only a small part of our environment—that the world is far richer than we imagine. Simply put, we, too, know what we need to know: After all, life is generally a conservative, waste-abhorring process; the irrelevant is soon discarded, with precious energy and materials invested mainly in paying propositions. Let's see, then, how living things become aware of those things that are important to them—through the functioning of their sensory receptors.

First, we should be aware that there is a great deal of variation in the complexity of sensory receptors. They range from the rather simple—as in the free neural endings in the skin—to the incredibly complex, as in the light-activated retina of the vertebrate eye.

Simple or complex, though, sensory receptors have several characteristics in common. For example, all receptors are highly specialized for certain stimuli, so that a taste receptor cannot respond to light wavelengths and vice versa. Also, all sensory receptors are transducers—that is, they convert (transduce) the varied stimuli they receive into electrical signals. In addition, sensory receptors, upon stimulation, like neurons, undergo changes in membrane potential; but instead of action potentials, they generate what are called *receptor potentials,* which in turn stimulate action potentials in sensory neurons. Unlike the "all-or-none" ("go" or "don't go") action potential, receptor potentials can vary in intensity. Increases in receptor potential intensity become translated into a higher frequency of action potentials in the sensory neuron, where the rate of firing has information value. Such "grading" of stimuli enables us to make very fine distinctions between bright and dim light, soft and loud sounds, and faint odors and damaging toxic fumes.

Thermoreceptors

Thermoreceptors are sensitive to temperature. They occur in a wide variety of animals. For example, thermoreception may be important to many invertebrates. Heat detection is especially important to the ectoparasites of birds and mammals—leeches, fleas, mosquitoes, ticks, and lice—parasites that must find a warm-blooded host. (Their thermoreceptors are generally located on the antennas, legs, or mouthparts.) Thermoreception among invertebrates is not well understood.

Most human thermoreceptors are located in the skin. These receptors respond not only to temperature, but particularly to *changes* in temperature (so that a change from cold to cool might be registered as warmth). Their precise mechanism of reception remains a mystery, although some physiologists ascribe thermoreception to the free nerve endings and to specialized skin receptors called *Ruffini corpuscles* and *Krause end bulbs* (see Figure 35.1). The latter are far more numerous and are stimulated by a greater temperature range than are the former.

Chemoreceptors

Chemoreceptors detect chemicals. Chemoreceptors are responsible for detecting chemicals not only in the external environment, but the internal as well, as they monitor the chemical condition of the body.

Chemoreception is common in invertebrates, particularly arthropods, where it is important to feeding, defense, and reproduction. In fact, the most powerful chemical sensitivity known is found in the male silk moth. It can detect a single molecule of the sex attractant *bombykol.* (Human chemoreceptors normally begin to respond only after stimulation by several million molecules.)

Among vertebrates, the most sensitive chemoreceptors are generally found in the mammals, especially carnivores and rodents, where scents are important in feeding and in reproductive activities (although no one wants to discount the ability of sharks to locate blood).

In humans, *gustation,* the sense of taste, is closely related to our sense of smell, and both forms of chemoreception are often active at the same time. In

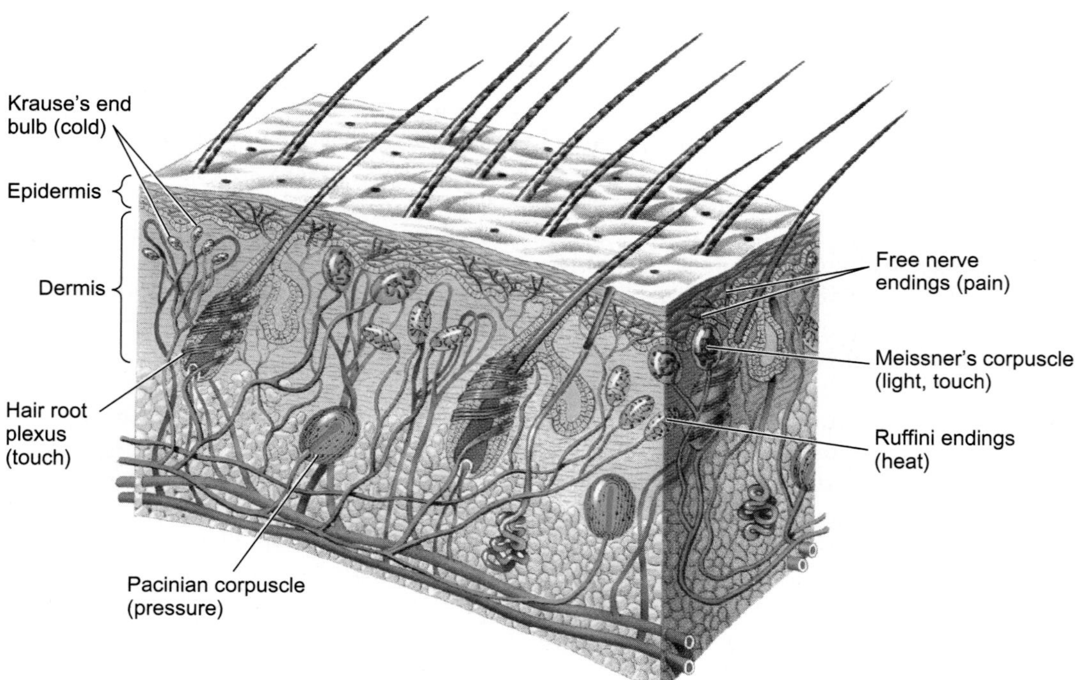

Figure 35.1 Specialized Touch Receptors in the Skin of Humans. Meissner's corpuscles are located close to the surface and register light touch. These are most numerous in the fingertips and around the lips. Pacinian corpuscles are in deeper skin locations, and their complex end bulbs register pressure. Generalized sensory neurons (no distinct structures) surround the hair follicles of mammals and are stimulated by hair movement. (Try to move a single hair on your forearm without feeling it.)

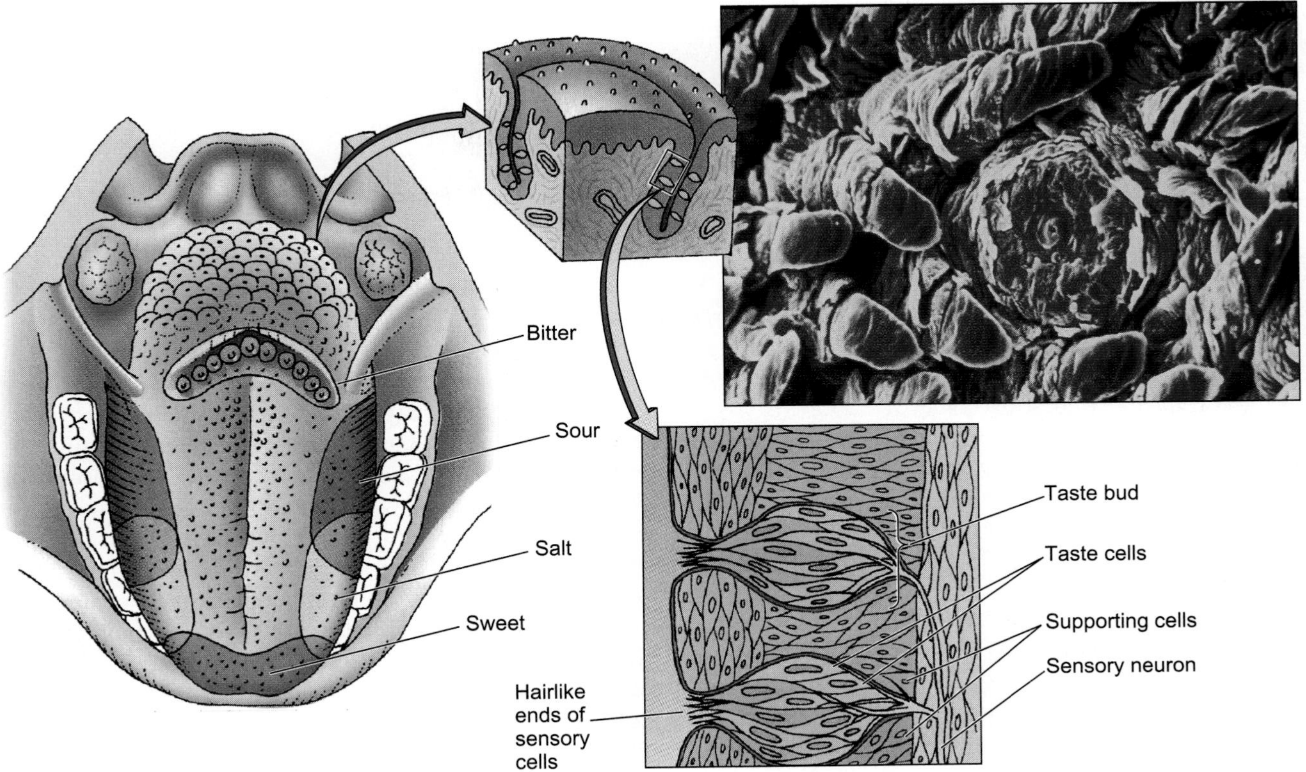

FIGURE 35.2 THE CHEMORECEPTORS OF THE TONGUE.
The arrangement of taste buds on the tongue is such that certain regions specialize in detecting each of the four primary tastes: bitter, sour, sweet, and salty. Within the taste buds, clusters of specialized receptor cells converge at tiny pits in which solutions form. Individual taste buds, as seen in the micrograph, are constantly being worn out and replaced.

fact, much of what we taste is actually being smelled. (You've probably noticed that food tastes a little flat when you have a cold.)

"True" taste receptors in humans are located in the taste buds of the tongue. There are generally believed to be four categories of taste: sweet, sour, salty, and bitter. (See Essay 35.1 for a discussion of a possible fifth taste.) Most complex tastes are really a mixture of two or more basic types. Sensitivity to each taste is somewhat localized on different parts of the tongue, as we see in Figure 35.2. (Keeping in mind that many poisons are bitter, can you suggest a reason for the location of the bitter receptors?)

Olfaction, the sense of smell, is not nearly as well developed in humans as it is in some other mammals. (There seem to be great differences among humans in the ability to taste and smell.) Human olfactory neurons (like those of other mammals) are located in the *olfactory epithelium,* where molecules that trigger olfaction are dissolved in the moist surface (Figure 35.3). The olfactory neurons send their signals to the *olfactory bulb* of the brain, where the information provided by the stimulating molecules is processed and interpreted.

MECHANORECEPTORS

Mechanoreceptors respond to distortions in the receptor itself or in nearby cells. Among the sensations they are responsible for are touch, hearing, equilibrium, and knowledge of body position.

TOUCH AND PRESSURE RECEPTORS

Tactile receptors responds to touch. They involve extremely sensitive, fast-firing neurons that respond when anything alters the shape of the neural membrane. In some invertebrates, the body surface overlying the touch receptor has bristles or hairs, and so objects in the environment can be detected before the rest of the body makes contact (Figure 35.4). The minute sensory hairs on the cockroach abdomen, for example, can detect the slight movement of air that might signal a descending foot.

Vertebrates have two kinds of tactile receptors. Distance receptors, such as the **lateral line organ** of fish, detect stimuli in the form of water disturbances from distant sources. Sharks, as we saw in Chapter

Essay 35.1 Fooling Mother Nature's Sweet Tooth

In West Africa, people can eat lemons just as people here eat oranges. The very idea makes our jaws pucker. How can they stand it? Do they like the sour taste? It turns out, there is no sour taste. This is because, before they eat the lemon, they chew a few "miracle fruits"—small red berries about the size of olives—from a plant, *Synsepalum dulcificum*, that grows there. Because of these berries, these people, in effect, are not eating sour lemons—they're eating *sweet* lemons. In fact, they can eat all sorts of local sour foods because chewing the miracle fruit first makes anything sour seem sweet for up to three hours after chewing the fruit.

It turns out that the active ingredient is a glycoprotein, a protein with sugar molecules bound to it. It is believed to work because the protein component can attach to a site close to where sweet receptors are located. The acidity of the sour food then alters the glycoprotein so that its sugar components are released. These then bind to the sweet receptor sites, stimulating them, and masking the sour taste.

Another substance, gymnemic acid, found in the leaves of an Indian plant, *Gymnema sylvestre*, can *remove* the sweet taste from any food. "It can make sugar taste like sand and sugar solution taste like tap water." The mechanism for generating the insensitivity is not entirely understood, but those who saw its potential usefulness as a dietary aid may have been disappointed to learn that it also masks a variety of nonnutritive sweeteners, such as cyclamates. And then there are those who would undoubtedly argue that the greatest taste deceiver of all is ketchup.

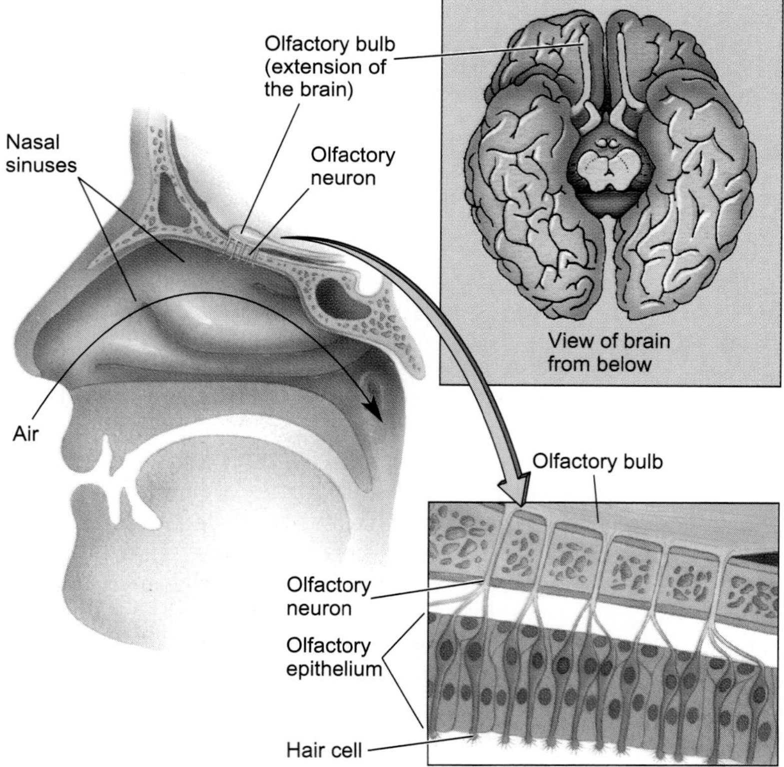

Figure 35.3 The Olfactory Receptors.
The olfactory receptors in the human nose connect to the olfactory bulb of the brain. The receptors are able to distinguish a wider variety of stimuli than the taste buds. The olfactory neurons are part of the nasal epithelium dispersed with other cells. Each neuron has numerous olfactory hairs that protrude from the epithelium. Each "hair" is a modified cilium. Interestingly, the olfactory neurons are easily damaged by a blow to the head, rendering the individual unable to smell.

FIGURE 35.4 TOUCH RECEPTORS.
Invertebrates commonly sense objects they contact through sensory hairs whose movement sends action potentials to the brain. Among the most acute are those of cockroaches.

27, have highly developed lateral line organs and are attracted to any unusual disturbances, such as the distress or panic movements of an injured fish. The second kind of tactile receptors detects touch, both direct touch and pressure.

Several types of tactile receptors are located in the human skin (Figure 35.1) Touch receptors include *Merkel's disks, Meissner's corpuscles,* and *free nerve endings,* all of which are located near the skin surface. (Free nerve endings also register pain.) Pressure is detected by bulbous *Pacinian corpuscles* that are located deeper in the skin and in some of the deep organs of the body, such as the pancreas. Human touch receptors are more concentrated in the fingertips, lips, nipples, face, and genitals. Body hairs can also transmit signals since at their base they are generally wrapped by free nerve endings. (Try moving one hair without feeling it.)

Until very recently, it was generally believed that human hairs, in their sensory function, were simply dead, mechanical levers that when touched would jostle the sensory nerves surrounding their roots. Then another theory was developed based on information that keratin, the hair protein, is so highly structured as to be essentially crystalline. In fact, it was found that each hair comprises a single *piezoelectric crystal.* A piezoelectric crystal discharges electricity when it is deformed. Researchers found that a perfectly dead hair generates a small electrical discharge when it is bent, and that nerve endings respond to this electricity.

Incidentally, the outer layer of skin is also made primarily of keratin, and it too has a piezoelectric effect. Bending or depressing the epidermis generates detectable electric discharges that are picked up by the touch-sensitive free nerve endings. We are very "touchy" creatures, indeed.

AUDITORY RECEPTORS

Auditory receptors register sound. *Audition,* or the sense of hearing, is similar in some respects to lateral line reception in fishes since, in both, a distant stimulus is transmitted to the receptors via a medium (air or water). In fact, evolutionary theory holds that the structure of balance and hearing in the vertebrates evolved from lateral line organs.

Most invertebrates don't have specialized sound receptors, but many are sensitive to vibrations in the air, water, or soil in which they live. One notable exception among the invertebrates is the insects, a group that boasts several kinds of sound receptors. For example, most species have sensory hairs that respond to the vibrations in air set up by low-frequency

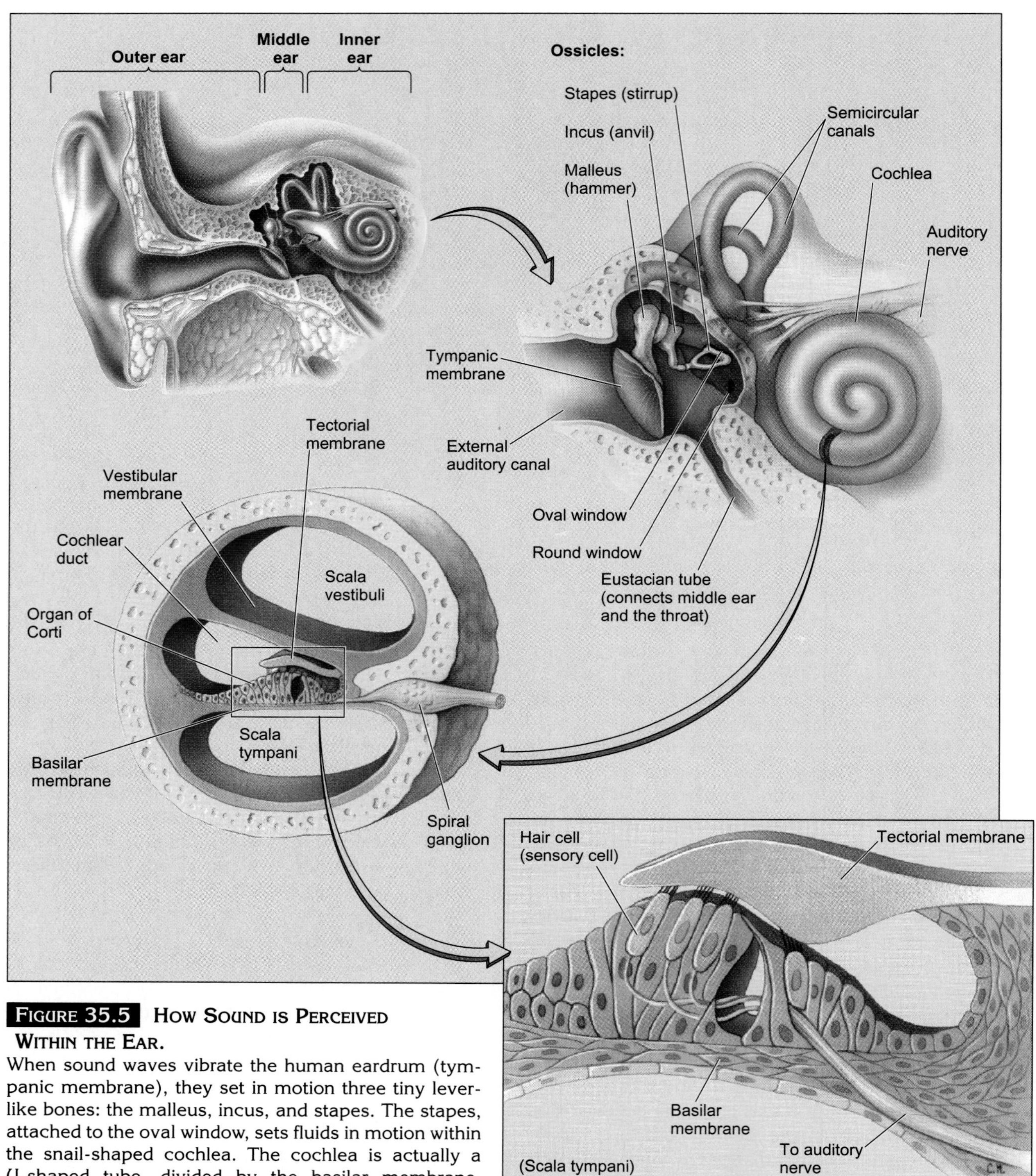

FIGURE 35.5 HOW SOUND IS PERCEIVED WITHIN THE EAR.
When sound waves vibrate the human eardrum (tympanic membrane), they set in motion three tiny lever-like bones: the malleus, incus, and stapes. The stapes, attached to the oval window, sets fluids in motion within the snail-shaped cochlea. The cochlea is actually a U-shaped tube, divided by the basilar membrane. Sensory hair cells of the membrane are embedded in the gelatinous tectorial membrane. The two membranes and the sensory hair cells are called the organ of Corti. Sound impulses are transmitted by the middle-ear bones from the tympanic membrane to the oval window of the fluid-filled cochlea and from there over the auditory nerve to the brain.

sound. Yet others, including grasshoppers, moths, and crickets, have *tympanal organs* that respond to certain high-frequency vibrations, much like the human eardrum. For a special case of insect hearing and its adaptiveness, see Essay 35.2.

Vertebrates show a wide range of auditory ability. Some reptiles show little, if any, ability to hear, while mammals generally hear remarkably well. We can take the human ear as an example.

The human ear, like that of other mammals, consists of three basic regions: the *outer, middle,* and *inner ear* (Figure 35.5). The structures of the ear and their functions are listed in Table 35.1. The *outer ear* includes the *pinna* and the auditory canal. The *pinna* directs sound waves inward, through the auditory canal to the middle ear.

Sound waves reach the *middle ear,* an air-filled cavity, and vibrates the *tympanic membrane,* or eardrum, which then moves three tiny bones: the *malleus, incus, and stapes.* (Malleus, incus, and stapes translate, respectively, as "hammer," "anvil," and "stirrup.") Acting as a jointed lever, these three bones transfer vibrations of the eardrum to the inner ear. The middle ear, therefore, transfers the energy of sound through an intricate lever system (via the bones of the middle ear) to the inner ear.

The *inner ear* consists of the *cochlea* (literally, "snail shell") and the *vestibular apparatus* (discussed below). The cochlea is a lengthy, fluid-filled tube that doubles back on itself and then coils like a snail in its shell. The coiling makes it hard to imagine how it works, but if we were to assume that it is straight (as it is, in fact, in birds), we would see a tube, divided into two chambers along its length by a complex membrane, containing the sensory neurons that are stimulated by the energy of sound (see Figure 35.5). One end of the U-shaped tube contains the *oval window,* to which the stapes is attached. The other end holds the flexible *round window.*

As sound waves strike the eardrum, it vibrates and moves the three middle ear bones that transfer the energy of sound to the oval window, causing it to vibrate rapidly. This vibration moves the fluid of the outer tube, and at its far end, it causes the round window to move back and forth. (The round window serves to dissipate the sound energy.) As the fluid pulsates within the tube, it activates what is called the *organ of Corti.* As shown in Figure 35.5, the organ of Corti consists of a *basilar membrane,* from which arise sensory *hair cells.* The tips of these hair cells are embedded in the overlying gelatinous *tectorial membrane.* As the basilar membrane moves, the sensory hairs are bent, creating receptor potentials that initiate impulses in the neurons. The impulses travel along the *auditory nerve,* eventually reaching hearing centers in the cortex.

Sounds, we know, vary in intensity and pitch. The perception of intensity, or loudness, seems to depend on the *number* of auditory neurons that fire, as well as the *frequency* of their firing. Difference in pitch (highness and lowness) depend on which auditory neurons are stimulated. The basilar membrane is narrow at the broad base of the cochlea and wide at its apex end. Hair cells at its thin, more rigid beginning respond better to higher frequencies, so input from hair cells in this region is interpreted by

TABLE 35.1 Structures of the Ear

Structure	Description	Function
Outer Ear		
Pinna	Part of ear outside the head	Catches and directs sound
Auditory canal	Tube between pinna and eardrum	Channels sound to middle ear
Middle Ear		
Tympanic membrane (eardrum)	Vibratory membrane	Vibrates in response to sound
Hammer (malleus) **Anvil (incus)** **Stirrup (stapes)**	Three small bones in middle ear	Transmits vibrations of eardrum to inner ear, increases force of vibration
Inner Ear		
Utricle, saccule, and semicircular canals	Two fluid-filled chambers and three bony canals at right angles to one another	Equilibrium
Cochlea	Fluid-filled, snail-shaped bone containing hair cells; auditory nerve starts here	Houses organ of hearing that generates nerve impulses

Essay 35.2 Moth-Bat Coevolution

Noctuid moths have two tympanums, one under each wing on either side of the thorax (the insect midsection), and each tympanum has only two receptor cells. Kenneth Roeder found that one, called the A1 cell, is sensitive to low-intensity (weak) sounds. The other, the A2 cell, responds only to loud sounds. Surprisingly, neither kind of receptor is very good at distinguishing frequencies (high versus low notes)—a sound of 20,000 Hertz (cycles per second) elicits the same neural action potential as one of 40,000 Hertz (a much higher sound). As any sound becomes louder, however, the A1 cell fires more frequently and with shorter lag time between receiving the stimulus and firing. Also, the A1 cell shows a greater firing frequency in response to pulses of sound than to continuous sounds; it fires increasingly slower if subjected to a continuous sound. And it just so happens that the bats that prey on noctuid moths emit pulses of sound.

In a sense, the moth has beaten the bat at its own game. Its very sensitive A1 cell is able to detect bat sounds long before the bat is aware of the moth. Not only can the moth detect the distance of the bat, but it can tell whether the bat is coming nearer, since the sound of an approaching bat would grow louder. In addition, the moth is able to determine the direction the bat is coming from. The mechanism is simple. If the bat is on the left side, the receptors in the left tympanum of the moth will be exposed to the sounds, while the receptors on the right will be shielded. Therefore, the left receptors fire sooner and more frequently than the right if the bat is on the left. If the bat is directly behind, both neurons will fire simultaneously. Thus, the moth can determine the distance and direction of the bat. But what about its altitude?

If the bat is above the moth, the loudness of the bat's sound will depend on the position of the moth's wings. When the moth's wings are up, the ears are exposed, and the bat's cries are loud. However, when the moth's wings are down, they cover its ears, and the bat's cries are muffled. In contrast, if the bat is below the moth, the loudness of the bat's cries will not be influenced by the position of the moth's wings. The moth, then, decodes the incoming data (probably in its thoracic ganglion—from which the auditory neurons emerge), so that it pretty well has the bat pinpointed.

What does it do with this information? If the bat is some distance away, the moth simply turns and flies in the opposite direction, thus decreasing the likelihood of ever being detected. The moth probably turns until the A1 cell firing from each ear is equalized. When the bat changes direction, so does the moth.

Bats fly faster than moths, though, and if a bat should draw to within 2.5 m (8 feet) of the moth, the moth's number is up—at least if it tries to outrun the bat. So it doesn't. If the bat and moth are on a collision course (that is, if the moth is about to be caught), the sounds of the onrushing bat will become very loud. At this point, the A2 fiber begins to fire—the signal of imminent danger. These messages are relayed to the moth's brain, which then apparently shuts off the thoracic ganglion that had been coordinating the antidetection behavior. Now the jig is up, and the moth changes tactics. Its wings begin to beat in peculiar, irregular patterns or not at all. The insect probably has no way of knowing where it is going as it begins a series of unpredictable loops, rolls, and dives. But it is also very difficult for the bat to plot a course to intercept the moth. The erratic course may take the moth to the ground where it will be safe since the echoes of the earth will mask its own echoes.

The noctuid moth's evolutionary response to the hunting behavior of the bat serves as a beautiful example of the adaptive response of one organism to another. Also, it shows clearly that the sensory apparatus of any animal is not likely to respond to elements that are irrelevant to its well-being. It is not important for moths to be able to distinguish frequencies of sound, but it is important that they be sensitive to differences in sound volume. Anyone who tried to train a moth to respond to different sound frequencies could only conclude that moths are untrainable.

the brain as higher-pitched sounds. The wider, more flexible apex of the snail-shaped chamber responds better to lower frequencies, so input from hair cells in this region is interpreted by the brain as those that translate into lower-pitched sounds (Figure 35.5).

Sound waves travel up one side of the U-shaped tube and down the other. Each particular sound frequency traveling up one side of the basilar membrane will be exactly in phase with the same frequency traveling down the other side of the membrane, only in one region. The in-phase resonance (or reinforced vibration) at that region vibrates the basilar membrane, producing the sensation of pitch.

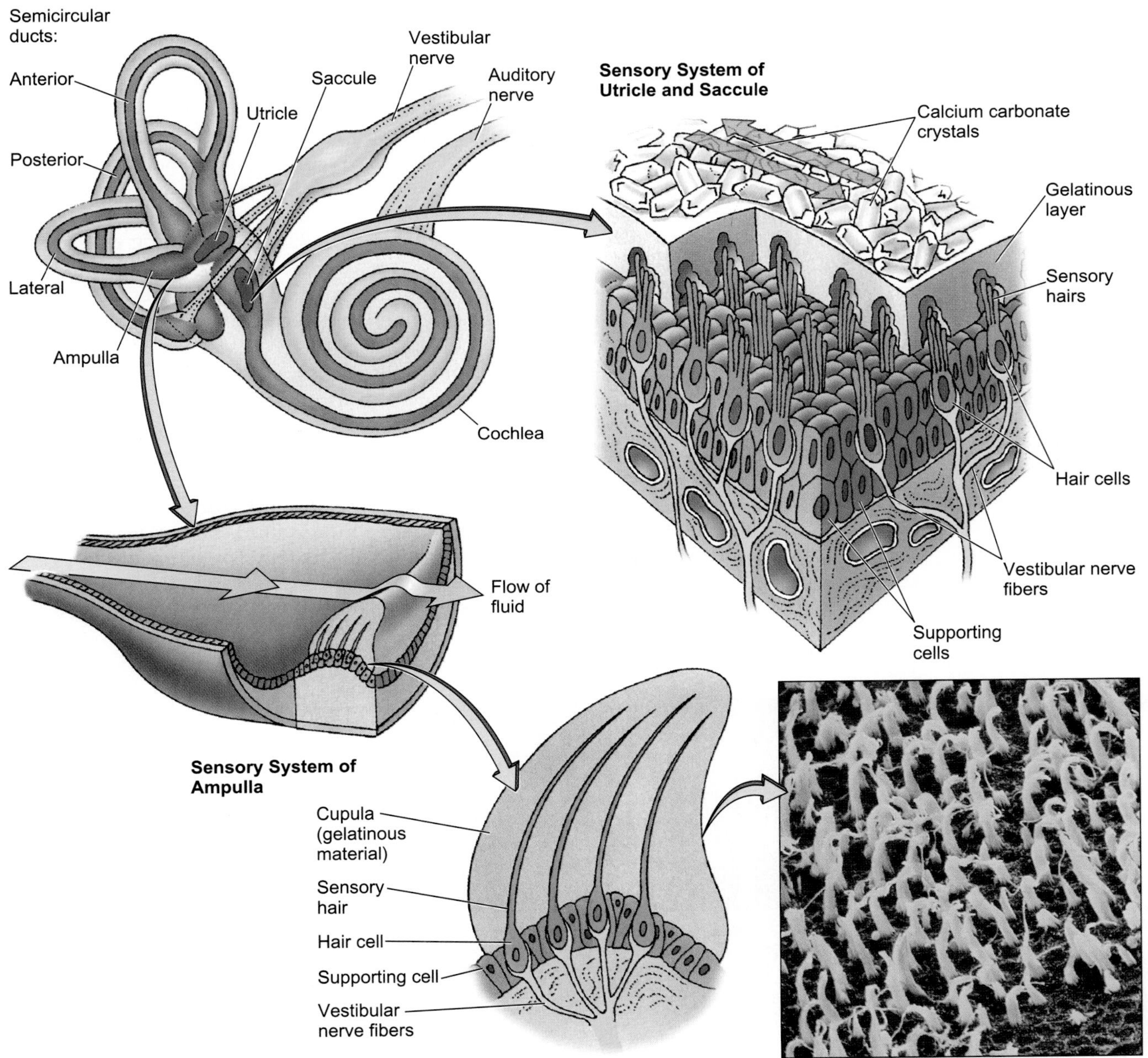

FIGURE 35.6 SEMICIRCULAR CANALS.
The inner ear is very sensitive to the body's position and movement. The semicircular canals lie at right angles to each other so that bodily movement in any direction shifts the fluid in at least one of them. The saccule and utricle are fluid-filled cavities in which grains are embedded in a jellylike matrix. Movement of these grains sends information to the brain regarding a change in the position of the head with respect to gravity.

Gravity and Movement Receptors

In humans and other mammals, body movement, position, and balance are detected by the *vestibular apparatus* of the inner ear (Figure 35.6). It is composed of the *semicircular canals,* the *saccule,* and the *utricle.* These three structures are closely associated with those of hearing. Each **semicircular canal** lies in a different plane, at right angles to each of the other two. This arrangement permits the sensing of movement—acceleration or deceleration—in any direction. Each canal is filled with fluid, and its movement jostles sensory hairs that extend into the canals. As the fluids move, they bend the hairs, creating receptor potentials, which in turn activate sensory neurons that send impulses to the brain.

The saccules and utricles contain sensory hairs coated with fine granules of calcium carbonate. Shifts in these granules pressing on the sensory hairs change the rate of neural impulses, providing information about the position of the head with respect to gravity. Some impulses travel to the spinal cord, where body position can be adjusted by reflex action; others are sent to the cerebellum, where other reflexive muscular coordination is orchestrated; and yet others move on to higher centers involved with the control of eye movement. Input from the eyes is important in maintaining balance. (Try to close your eyes and stand on one leg. This is said to be one of the tests in some states to test for drunk drivers. Thirty seconds, hands at sides. Reports are that almost no one, drunk or sober, can pass it.)

Visual Receptors

Visual receptors respond to light. In particular, most respond to "visible light" wavelengths—from about 430 to 750 nm, although some animals can detect ultraviolet or infrared light.

The invertebrates show a wide range of visual abilities. Among the simplest invertebrate visual systems are those of the planarian flatworms, which have two eyespots shaded by dense tissue on opposite sides. The animals can therefore tell which direction light is coming from according to which eye is being stimulated. Such ability is, of course, important to these bottom-dwelling and dark-seeking creatures.

Among invertebrates, the cephalopod mollusks —the octopus (Figure 35.7), squid, and others—are unique in that they have image-forming eyes, quite like those of vertebrates. Arthropods, such as spiders, crayfish, and insects, have exceptionally good vision, especially in detecting movement in their visual fields. Spiders have eight eyes (two rows of four each), but in most spiders the eyes lack enough photoreceptors to form clear images. The jumping spiders (*Salticidae*), though, have a relatively large number of photoreceptors. These are undoubtedly beneficial; the spiders leap from a distance onto their prey, and it wouldn't do for a weak-eyed spider to leap onto a hungry bird (Figure 35.7).

Crustaceans, such as crayfish, have two kinds of eyes: *simple eyes* and *compound eyes.* The simple eyes lack lenses and are found in the larval stages. In

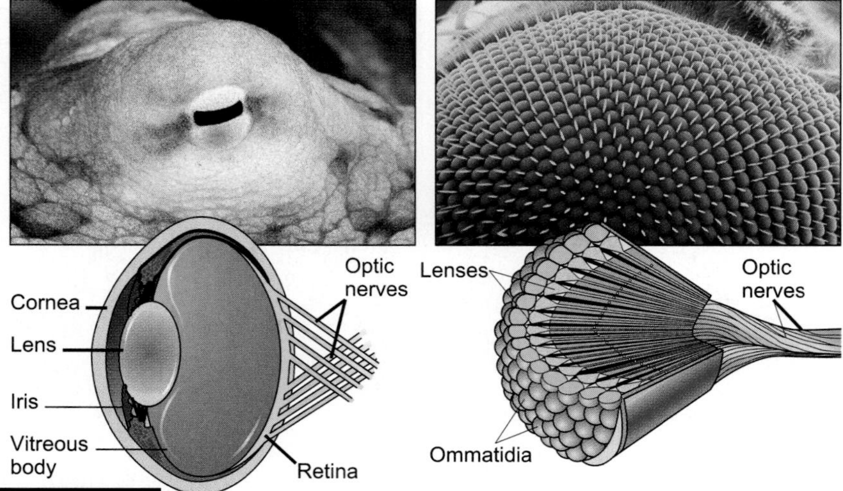

Figure 35.7 The Compound Eye and the Octopus Eye.
Compound eyes (*right*) are common among insects and crustaceans, and are particularly adept at detecting movement. The octopus eye (*left*) is remarkably similar to the human eye and is able to discern images clearly.

many species, the adult's sensitive compound eye rests on movable stalks. The term "compound" refers to the numerous visual units known as *ommatidia* (Figure 35.7). The **ommatidia** can't move to follow an image, so each stares blankly in its own direction until something moves into its visual field. Then it fires a signal to the central nervous system. Any movement across the animal's field of vision stimulates a series of ommatidia in turn, so that even very slight movements are detected. The convex surface of such eyes gives a visual field of about 180°.

Among the vertebrates, humans are particularly visual. We rely on sight in nearly every aspect of life, and we have quite a remarkable visual apparatus. The **eyeball** (Figure 35.8), a spherical, fluid-filled structure, contains three tissue layers: the outermost, tough, white **sclera,** to which muscles of eye movement attach; the middle **choroid coat,** rich in blood vessels; and the inner, light-sensitive **retina.** Light entering the eye first passes through the **cornea,** a transparent portion of the sclera that forms a slight bulge at the front of the eyeball, where the path of the light is first altered. It then passes through the clear fluid of the anterior chamber and on through the **iris,** an adjustable circle of pigmented tissue that controls the amount of light entering the eye through the opening in the iris, the **pupil.** Next, light passes through the **lens** where it is focused on the retina—but to reach this dense region of neurons it must cross the large, fluid-filled interior of the eye. The structures of the vertebrate eye and their functions are listed in Table 35.2.

A camera lens, as you may know, adjusts to distance by moving back and forth. Interestingly, sharks focus on objects in the same way. However, human eye lenses adjust to varying distances by changing their shape and by bending the light rays that pass through them.

The retina consists of four layers of cells (Figure 35.8). The deepest layer, attached to the inner surface of the choroid coat, is pigmented. It absorbs light that might otherwise be reflected and bounced around inside the eyeball. Overlying the pigmented layer are the **rods** and **cones,** the light receptors (Figure 35.8). You might expect them to be in the direct path of light, but they are covered by two more layers of rather transparent neurons. When the rods and cones are stimulated by light, they send their impulses to the overlying *bipolar cells;* these, in turn, synapse with the *ganglion cells* just above. Axons from ganglion cells gather from all parts of the retina to form the **optic nerve,** which carries visual information to the brain.

In many vertebrates, including cats and dogs and even some humans, the absorptive pigment layer can be altered at night by movement of the pigment granules within the cells—which unveils a layer of reflecting crystals behind the rods and cones. The reflective layer bounces the excess light back through the layer of rods and cones, which increases the sensitivity of the eye. This reflective layer in animals' eyes can be quite startling at night to travelers on lonely highways.

The slender rod cells far outnumber the cones by about 18 to 1 (125 million rods to 7 million cones in each eye). Impulses from the rods are decoded by the brain into black, white, and gray images; different colors (wavelengths of light) are not distinguished. However, the rods make up for this limitation by

TABLE 35.2 STRUCTURES OF THE EYE

Structure	Description	Function
Cornea	Transparent dome of tissue at front of eye	Bends light rays to help focus them on retina
Aqueous humor	Clear fluid between lens and cornea (in anterior cavity)	Transmits and bends light; pressure of fluid helps maintain shape of eye
Iris	Colored part of eye	Regulates the amount of light that enters eye
Pupil	Open center of iris	Entrance for incoming light
Lens	Semispherical transparent body of tissue	Adjustable focusing of light rays onto retina
Vitreous humor	Jellylike substance within cavity behind lens	Transmits and bends light; pressure of substance helps maintain shape of eye
Retina	Tissue containing rods and cones	Sensory area; receives light and generates nerve impulses
Fovea	Tiny pit on retina with a high density of cones	Most sensitive part of retina
Optic nerve	Bundle of nerve fibers leaving eye	Carries signals from retina to brain

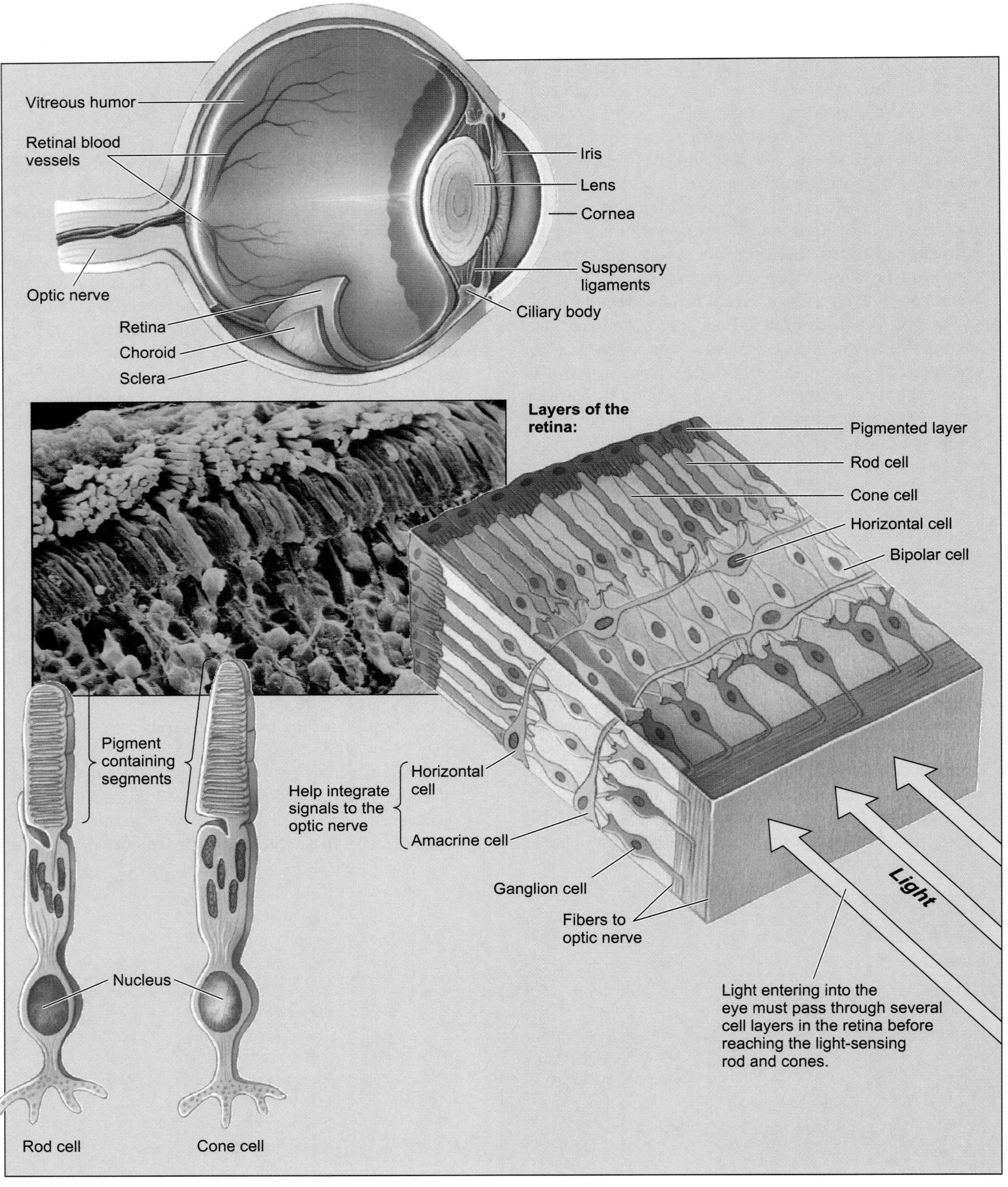

FIGURE 35.8 THE HUMAN EYE.
The eye is actually a rather tough structure. The anterior cavity in front of the lens contains a fluid called aqueous humor; the large chamber behind, the posterior cavity, is filled with vitreous humor. The white of the eye, the sclera, is modified in the front to form a transparent window, the cornea; the colored part of the eye is the iris. The lens is focused by the muscles that support it. The sensory area, the retina, is composed of rods and cones that send impulses to the brain over the optic nerve. Where the nerve enters the retina there are no receptors (the "blind spot"). The most sensitive part of the retina is a tiny pit with a very high density of receptors, the fovea. When threading a needle, we usually turn our heads in such a way that the image falls on the fovea. The rods can detect light at low levels; the cones can detect different colors.

being highly responsive to dim light, far more so than cones. In fact, the rods provide most of our night vision, which is why we don't see colors well at night.

The actual light-detecting regions in rods are the stacked disks that form the upper part of the cells. (Each rod may have up to a thousand such disks.) The disks contain rhodopsin, a photoactive pigment that breaks down in the presence of light into two colorless products: the protein *opsin* and a derivative of vitamin A called *retinal.* The chemical reaction itself sets events in motion that create neural signals in the bipolar and ganglionic cells above (see Figure 35.8).

The rods work in a peculiar manner—they are active only in the absence of light. At that time, their gated sodium channels are open, and they are in a continuing state of depolarization. They secrete an inhibitory neurotransmitter where they synapse with the bipolar cells above, thus inhibiting the bipolar cells from firing. When light is absorbed and rhodopsin is broken down, the sodium gates close, and the rods become hyperpolarized. The secretion of neurotransmitter slows considerably, and graded receptor potentials begin in the bipolar cells. The bipolar cells synapse with ganglion cells, and any resulting action potentials pass along the optic nerve to the brain.

Neurobiologists have yet to determine how such a negative system of operation might be adaptive, but they marvel over the acute sensitivity possible in rod cells. Apparently rod cells are excellent amplifiers. A single photon (unit of light) can produce a detectable electrical signal in the retina, and the human brain can actually "see" a cluster of five photons—a small point of light, indeed.

Opsin and retinal eventually enter into a chemical pathway in which they are recombined into rhodopsin. Apparently, dietary vitamin A must be supplied continually to keep the pathway going in the right direction. Severe night blindness can result from a lack of vitamin A. If one night a knock on the door brings you from a brightly lit room into near total darkness, you may experience a temporary form of night blindness. Under intense light, most of the rhodopsin in the rods bleaches to the opsin/retinal form, and a certain period of time is required for the reconversion to rhodopsin to catch up. While you are waiting for the rods to function, the cones—which aren't very useful at night—at least help you peer dimly into the darkness to see whatever brought you to the door in the first place.

While humans have reasonably good night vision, many day-flying birds lack rods altogether and are almost totally night-blind, which explains why so many birds come to roost at twilight. Owls and bats, on the other hand, avoid daylight and are quite at home in the dark. It is not surprising, therefore, to learn that their eyes have only rods.

You will not be startled to learn that cones are cone-shaped. Unlike the rods, each cone responds optimally to only its specific portion of the visible spectrum. Vision is most acute in the *fovea,* a slight depression directly in the path of light focusing on the retina. The fovea in mammals contains a rich concentration of cone cells with few other retinal cells. For this reason, cones are capable of much finer discrimination of detail in bright light than are rods, although cones hardly respond at all to dim light. Since cones are not very functional at night, it is easier to see a distant object, such as a faint star, if we look to the side of that object, avoiding the cone-rich fovea.

According to the most prevalent theory of color vision, there are three types of cones: red-sensitive, green-sensitive, and blue-sensitive. We can see so many variations of color because the sensitivity ranges of these receptors overlap. The color sensitivities of the cones depend on what sorts of visual pigments (light absorbers) they contain—or, more specifically, on variations in the proteins that join the visual pigments.

In addition to stimulating certain brain cells, each type of cone may prohibit those brain cells from receiving input from another type of cone. The interpretation of color is based on the relative output from each type of cone received by the brain.

PERSPECTIVES

The earth is a complex place, with an incredible interplay of energetic forces dancing across its surface. The various species of the planet have developed ways to detect some of these forces, particularly those that might be important to their survival, while remaining completely oblivious to others. Our world, then, is simplified by our ability to detect only limited aspects of it. Nonetheless, it remains a complex, fascinating, and often puzzling place, based on what we do know.

We move now to another kind of signal, the chemical signal. That enables some cells to influence others and thereby promote the delicate coordination so necessary for life to exist on a complex planet.

SUMMARY

Sensory receptors vary from simple to complex, but most operate along a common theme. They convert stimuli into receptor potentials, membrane potentials that can vary in intensity. Increases in receptor potential can be translated into a higher rate of action potentials in sensory neurons.

THERMORECEPTORS

Invertebrate heat detection is important to ectoparasites of warm-blooded hosts, but the receptors are not well understood. Human thermoreceptors include free nerve endings, Ruffini corpuscles, and Krause end bulbs.

CHEMORECEPTORS

Chemoreception in arthropods is important to feeding, defense, and reproduction. Male silk moth antennas are activated by trace amounts of sex attractants.

Vertebrate carnivores and rodents have keen chemoreception. Gustation (taste) in humans occurs in taste buds that specialize in sweet, sour, salty, and bitter. Olfaction in humans is the role of the olfactory epithelium in the nasal passages.

MECHANORECEPTORS

Mechanoreceptors respond to distortions in the receptor itself or in nearby cells. Among the sensations they are responsible for are touch, hearing, equilibrium, and knowledge of body position.

Tactile receptors include the lateral line organ of fish and various touch receptors in vertebrate skin. Keratin, in hairs, is essentially crystalline and discharges electricity when deformed, thereby triggering a neural impulse.

Auditory receptors are responsible for the sense of hearing. Insects perceive low-frequency sound through sensory hairs in the legs and higher frequencies through tympanal organs.

The human outer ear includes the pinna, auditory canal, and tympanic membrane, the latter of which is vibrated by sound. In the middle ear, three minute bones transfer tympanic vibrations to the oval window of the cochlea. The cochlea, part of the inner ear, is a coiled tube divided by a membrane that bears the sensory cells. The vibrating oval window moves fluids within the cochlea that activate hair cells in the basilar membrane of the organ of Corti. The bending hair cells generate receptor potentials that initiate impulses that travel to the auditory nerve.

Sound intensity seems to depend on the number of auditory neurons that fire and on the frequency of their firing. Ditch depends on which auditory neurons are stimulated.

Body movement, position, and balance are sensed by the semicircular canals, the saccule, and the utricle of the inner ear. Moving fluid in the canals activates sensory hairs whose impulses are perceived by the brain as acceleration or deceleration. Sensory hairs in the saccules and utricles are activated by head movement. Impulses inform the brain of head position.

VISUAL RECEPTORS

Visual receptors in most animals respond to the visible light spectrum. In invertebrates, receptors range from simple eyespots in planaria to complex image-forming eyes in cephalopod mollusks.

The human eye contains several tissue layers, including the light-sensitive retina. The iris adjusts light entering the eye, while the lens focuses images upon the retina. Focusing is done by changing the lens shape and by bending the light rays that pass through them.

The retina's light-activated cells are the rods and cones. They synapse with a layer of bipolar cells above, which activate overlying ganglion cells that form the optic nerve. A reflective layer below the rods and cones occurs in many nocturnal animals.

Rods detect all wavelengths and are sensitive to dim light. Rods contain rhodopsin that breaks down in light initiating action potentials, forming opsin and retinal. Rhodopsin is restored in a chemical pathway requiring vitamin A. Cones specialize in reds, greens, and blues, and work best in bright light. Overlaps in reception provide the intermediate colors.

Key Terms

thermoreceptors • **600**
chemoreceptors • **600**
olfaction • **601**
tactile receptor • **601**
auditory receptor • **603**
semicircular canal • **608**
visual receptors • **608**
ommatidia • **609**
eyeball • **609**
sclera • **609**
choroid coat • **609**
retina • **609**
cornea • **609**
iris • **609**
pupil • **609**
lens • **609**
rod • **609**
cone • **609**
optic nerve • **609**

Review Questions

1. How does the activation of a sensory receptor differ from that of the sensory neuron? How is this adaptive?
2. Transducers are electronic devices that convert mechanical movement into electrical current. Would the term transducer apply to sensory receptors? Explain.
3. To what specific group of invertebrate might thermoreception be highly essential?
4. Do human skin thermoreceptors actually detect specific temperatures? Explain.
5. Describe the sensitivity of the male silk moth to its sex attractant. What other animals are aroused by such scents?
6. Using a simple illustration, indicate the location of specific taste receptors on the human tongue.
7. Describe the arrangement of sensory cells in the human olfactory epithelium. Where are their signals sent?
8. What insect structures act as touch receptors? How do such receptors actually work?
9. How would a shark detect a swimmer that was beyond visual and olfactory range?
10. List three kinds of touch receptors in the human skin. How would the simplest of these be activated?
11. Explain how any touch on the skin surface of an animal might start an action potential in nearby free nerve endings.
12. List two kinds of auditory receptors in insects.
13. Trace the pathway of air disturbances from the environment to the oval window of the cochlea.
14. Suggest reasons why the use of the three inner ear bones to transfer vibrations might be adaptive. Why not a direct stimulation of the oval window by air?
15. Describe the organization of the human cochlea. What is the function of the fluid within?
16. Describe the organization of the organ of Corti. In what structures do receptor potentials originate?
17. What function do the semicircular canals serve? What is the significance of the arrangement of the three vessels?
18. Explain how changing the position of the head affects the structures within the saccule and utricle.
19. Trace the pathway followed by light in the human eyeball. Name each structure along the way and state its function.
20. Name the four layers of the retina and state their functions.
21. When do we rely most on the rod cells? What might be the effect of continuous bright light on these receptors? How does this explain the temporary night blindness that occurs when we go from a brightly lighted room into darkness?
22. Describe the chemical pathway generated by light striking the rod cells.
23. Under what conditions are the cone cells most effective? How are colors other than red, green, and blue produced?

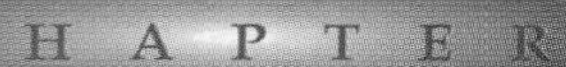

36

Hormones

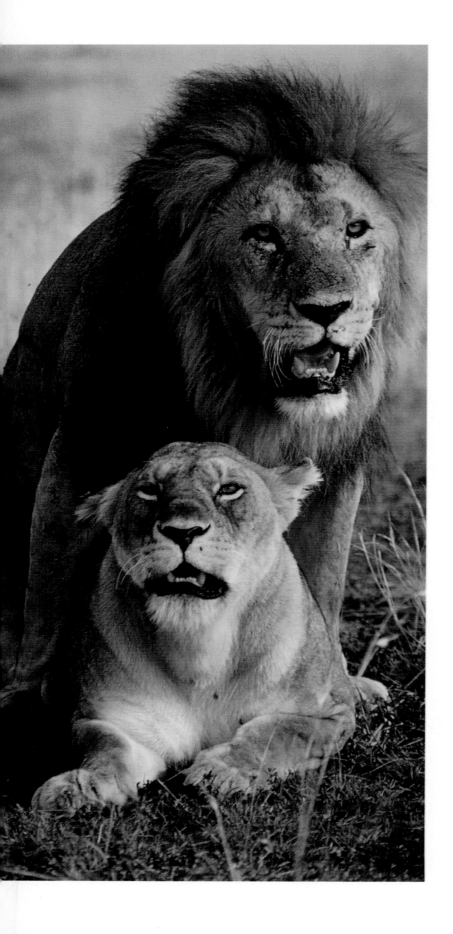

AS YOU SIT READING THIS, YOUR BODY IS AWASH IN CHEMICALS, EBBING *and flowing, surging and diminishing in a coordinated concert that in effect not only keeps you alive but also behaving optimally according to the world that is yours. These chemicals are signals, produced in one part of the body and acting in another part. The results of some of these signals will be apparent to you. Your height, for example, is related to these chemicals, as well as your voice quality and sexuality. Other of these molecules may play less obvious but no less important roles in your life. We will look at several of them, including some that you probably didn't know you had. Did you know, for example, that hormones may be involved in how a mother feels about her offspring?*

In a real sense, life is an unlikely condition. A smart celestial gambler would bet against its appearance. But the gambler might change the odds if he knew how much time was available. After all, life didn't appear all at once, full blown and operational. Instead, it apparently began with small, mindless molecules bouncing around, finally becoming associated in precise ways that could be called the earliest life. From these life forms, more complex types arose. The process was a slow one, and as it unfolded, life became ever more organized.

Maintaining the appropriate conditions for the continuance of complex life-forms presents problems that stagger the imagination. Yet life continues, its processes delicately balanced and coordinated in ways we are only beginning to understand. We do know, though, that each small action inevitably contributes to some larger process. Each time a cell goes about handling this ion or that, allowing this molecule through and barring that one, its actions must ultimately affect the body of which it is a part. Its behavior will eventually influence the activities of other kinds of cells in some distant part of the body. In fact, large, ponderous living things, such as ourselves, have come to help regulate the delicate processes of the body by actively producing chemical messengers called **hormones**—molecules produced in one part of the body that travel through the bloodstream and influence processes in other parts of the body.

THE CHEMICAL MESSENGERS

Until recently all chemical messengers within the body were called hormones. Today we know that not all such signals meet the traditional definition. For example, some chemical messengers do not enter the bloodstream at all. Some affect only adjacent or nearby cells, while others act within the very cell where they were produced. In this chapter, we will focus on the traditional examples of hormones and then briefly mention a few examples of other chemical messengers.

THE RELATIONSHIP OF HORMONES AND NERVES

Before we launch into our discussion, we should say that the great coordination of processes within living things is due, in large part, to hormones *and* nerves. However, these cannot be considered separate, independent, processes. On the contrary, the two systems are inextricably related to each other in three major ways.

First, hormones and nerves are *structurally* related. In fact, a number of glands that produce hormones are formed from nervous tissue. For example, the posterior pituitary forms from an extension of the brain, whereas the anterior pituitary is formed from the embryonic mouth.

Second, the two systems are *functionally* related. For example, when the nervous system detects an emergency situation, the adrenal medulla, part of the hormonal system, may release *epinephrine* (adrenalin) into the bloodstream, enabling the individual to react with an astonishing vigor. (The story of a little old lady lifting a car off her trapped husband appears periodically.)

Third, the two systems are *chemically* related. A number of hormones may also act as neural transmitters. Again, we can refer to epinephrine. The same chemical released by the adrenal medulla is used as a signal between certain kinds of nerve cells.

The Nature of Hormones

Hormones have a number of very special characteristics that underline their influence in the animal world. The most obvious characteristic, perhaps, is their inordinate power. They are effective in minute quantities—sometimes trace amounts. This enormous efficiency is due to their ability to trigger an amplification, so that the initial action sets up continuing and expanding reactions. For example, one of the effects of the hormone epinephrine in a liver cell is to cause glycogen to be broken down into glucose molecules. It has been estimated that a single molecule of epinephrine can trigger the formation of one million glucose molecules.

Another special trait of hormones is that they cause changes only in specific **target cells,** cells with the ability to respond to the hormone. Hormones encounter a variety of cells in their travels through the body, and it is up to the cell to identify the messenger. The identification is accomplished by the presence of very specific receptor sites on, or in, the target cells, sites that are able to bind only to certain hormones. Another important trait of hormones is that they are transient; once formed, they are quickly inactivated by enzymes. Thus, they act quickly and are gone; there is no lingering effect of the powerful messenger, an effect that could be potentially very disruptive. Indeed, hormones may bind to their targets only for a fraction of their short lives, and then float free.

Control of Hormone Secretion

The tiny hormonal molecules that ebb and flow through our bodies, lightly touching the very processes of life, causing them to shift this way or that, may seem to be a rather delicate and fragile way to control such critical processes. However, these tiny molecules have a sledgehammer impact. Because some of the changes they bring are swift, dramatic, and irreversible, they must be very precisely regulated.

Commonly, hormone secretion is regulated through negative feedback in which the producer of an action suppresses that action. When a certain level of hormone concentration is reached, or when the effects it causes in the target cells reach a given peak, the gland cells releasing the hormone are inhibited, and hormone release slows. As the hormone level falls, then, gland inhibition lessens and hormone secretion again increases. As an example, consider the interaction of the **trophic hormones** (those that regulate other endocrine glands) and their target glands. The anterior pituitary gland secretes a thyroid-stimulating hormone (TSH) that causes the thyroid gland to secrete the hormone thyroxine. As the thyroid responds and the thyroxine level rises in the blood, the thyroxine itself suppresses the secretion of TSH, which lowers the level

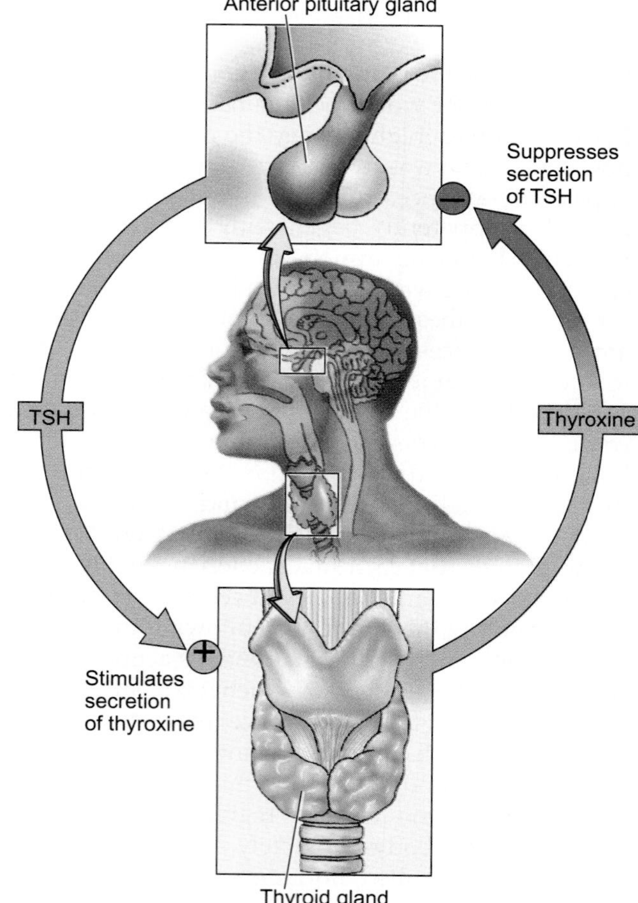

Figure 36.1 Feedback Between Hormones. The feedback mechanisms for thyroxine and TSH.

of thyroxine. As a result, more TSH is released. The levels of both hormones are thus kept within tight limits because of the influence they exert on each other (Figure 36.1). Such negative feedback is common throughout the endocrine system and furnishes some of the clearest and most fascinating examples of homeostasis in living things.

How Hormones Work

There are two general mechanisms by which hormones exert their effects. One pathway involves the formation of *second messengers,* chemicals that work inside the cell to bring about the hormone's effects. The other involves turning on genes that code for specific proteins.

Second Messenger Systems

Most of the hormones that work by second messengers are peptides—small, nitrogen-containing molecules. In this system the hormone itself is called the **first messenger,** because it travels from the endocrine gland to the cells where it will work; but it doesn't enter those cells. Instead, the hormone attaches to a receptor on the target cell and initiates a series of changes from there. It has been suggested that the hormone-receptor complex binds to an enzyme, adenylate cyclase, embedded in the membrane. This binding activates the adenylate cyclase, which then causes ATP to give up two phosphates and to form a molecule called cyclic AMP (adenosine monophosphate). Cyclic AMP (cAMP) actually operates within the cells, and so it is called a **second messenger** (Figure 36.2). A second messenger causes dramatic changes in the cell by triggering a series of events (called a biochemical cascade) (Figure 36.3). Since the basic scheme was discovered, a number of variations have been found, including other second messengers, such as inositol triphosphate or IP_3, which causes the release of calcium in muscle cells. (Since this calcium activates other proteins, it is called a third messenger.)

For an example of a hormone working through the second messenger cAMP, let's take a look at the mechanism of epinephrine in liver cells. It was in this system that the general mechanism of second messenger hormone action was first discovered by E. W. Sutherland, work for which he was awarded the Nobel Prize in 1971. As we have seen, epinephrine, released from the adrenal medulla, is carried by the blood to a target, such as the liver (Figure 36.3). It turns out that the liver cells have very specific receptor sites along their cell membranes. Once bound to the membrane, epinephrine activates the enzyme adenylate cyclase, which immediately converts ATP in the cytoplasm to cAMP. The cAMP then triggers the activity of certain enzymes and decreases the activity of others. In the case of epinephrine, the result is a decrease in the synthesis of glycogen (stored chains of glucose molecules) and

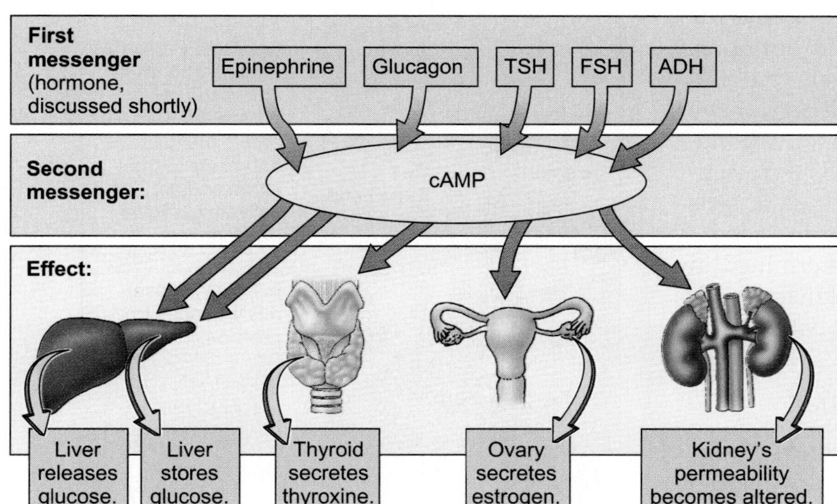

Figure 36.2 Action of Cyclic AMP.
When epinephrine binds to its plasma membrane receptor on a liver cell, it sets in motion a sequence of reactions that leads to the breakdown of glycogen and the transport of glucose out of the cell. Other hormones use cAMP to effect different changes in their own specific target cells.

Secondary Messenger Systems (peptide hormones)

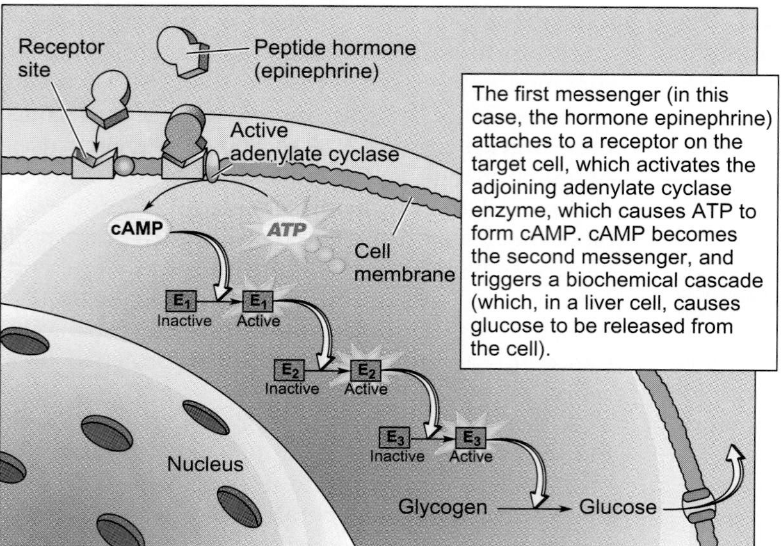
Receptor site
Peptide hormone (epinephrine)
Active adenylate cyclase
The first messenger (in this case, the hormone epinephrine) attaches to a receptor on the target cell, which activates the adjoining adenylate cyclase enzyme, which causes ATP to form cAMP. cAMP becomes the second messenger, and triggers a biochemical cascade (which, in a liver cell, causes glucose to be released from the cell).
cAMP
ATP
Cell membrane
E1 Inactive
E1 Active
E2 Inactive
E2 Active
E3 Inactive
E3 Active
Nucleus
Glycogen
Glucose

FIGURE 36.3 HORMONE ACTION: PEPTIDE HORMONES. The mechanisms of second messenger systems involve the binding of a hormone to a receptor on the plasma membrane, which in turn activates a cascade of events that lead to a biological action.

an increase in its rate of breakdown, with a net release of glucose into the blood.

At least 12 other hormones are known to stimulate the production of cAMP, utilizing it as a second messenger. But the question arises: How is the activity of so many hormones regulated if they all activate the same second messenger? The answer incorporates two important points. First, hormones have target cells, and a target cell has a specific binding site, or receptor, for a specific hormone. Thus, no other hormone (first messenger) can bind to that particular site (although a cell may have binding sites for more than one hormone). Second, while cAMP has a similar role in all cells—the activation of enzyme systems—different kinds of cells have different enzyme systems waiting for activation. All of this must, of course, be determined in development, when cells become specialized, each with its own receptors and biochemical battery.

GENE ACTIVATOR SYSTEMS

Basically, steroid hormones are lipid-soluble molecules with four interlocking carbon rings. Steroid hormones do not attach to the cell membrane. (Figure 36.4) Instead, they enter the cell. Once a steroid hormone enters a cell, it joins with a receptor molecule located in either the cytoplasm or the nucleus. Then, in the nucleus, the hormone-receptor complex binds to specific receptor sites on a chromosome and directs the synthesis of a certain kind of messenger RNA. The mRNA then leaves the nucleus and moves into the cytoplasm where it makes new proteins, many of which are enzymes, which bring about the

Gene Activators (steroid hormones)

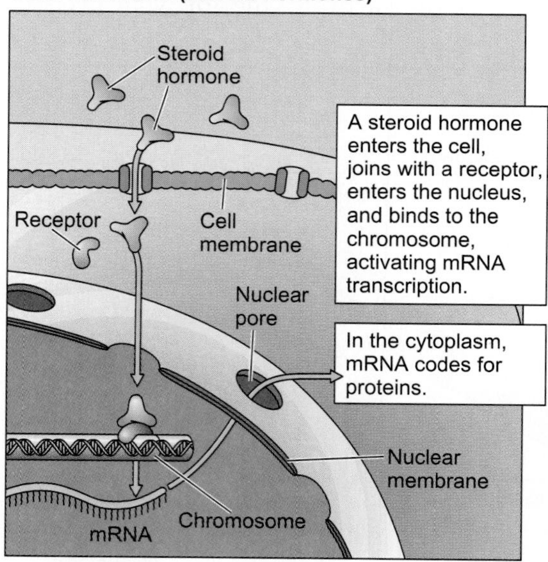
Steroid hormone
A steroid hormone enters the cell, joins with a receptor, enters the nucleus, and binds to the chromosome, activating mRNA transcription.
Receptor
Cell membrane
Nuclear pore
In the cytoplasm, mRNA codes for proteins.
Nuclear membrane
Chromosome
mRNA

FIGURE 36.4 HORMONE ACTION: STEROID HORMONES. Steroid hormones stimulate biological activity by binding to cellular receptor proteins, thus forming a complex that selectively activates gene transcription.

hormone's effects. Although it is not a steroid, thyroid hormone is also thought to work this way.

The steroid system was first worked out by Bert W. O'Malley and his coworkers. It was known that estrogen and progesterone treatments could cause the oviducts of female baby chicks to produce egg-white albumin and other proteins. These researchers labeled steroid hormones with radioactive tags and followed them through the sequence described above. At each step, the various intermediate compounds were isolated and identified, and at last they unraveled the story of how steroid hormones activate genes.

Now that we've seen the primary ways that hormones function at the cellular level, let's consider some specific human hormones, their sources, and their effects.

The Human Endocrine System

In humans, most hormones are produced in distinct glands and a number of scattered tissues that make up the **endocrine system** (Figure 36.5). The term *endocrine* means "ductless"; the products are not secreted into ducts but directly enter the blood for distribution. By contrast, there are many **exocrine glands.** These are the ducted glands, such as the salivary glands, sweat glands, and mucous glands, whose products enter ducts to be carried away. Table 36.1 summarizes the activities of the endocrine system, and the main functions are described below.

Figure 36.5 The Major Human Endocrine Glands, As Traditionally Identified.
A number of other tissues and organs produce chemical messengers that can be considered hormones (a few of these are indicated in parentheses).

The Pituitary

The *pituitary* is a tiny, bilobed structure about the size of a kidney bean, but its size and appearance belies its importance. The pituitary releases at least eight hormones important to the body's metabolism, growth, reproduction, and other activities. The pituitary lies at the end of a stalk called the *infundibulum* that emerges from the hypothalamus of the brain (Figure 36.6). (If you point a finger straight between your eyes and stick another finger straight into your ear, you will not only gain the full attention of bystanders, but you will also be pointing directly at your pituitary.) The stalk, as we'll see, represents one of the main functional connections that permit the nervous and endocrine systems to interact. The relationship

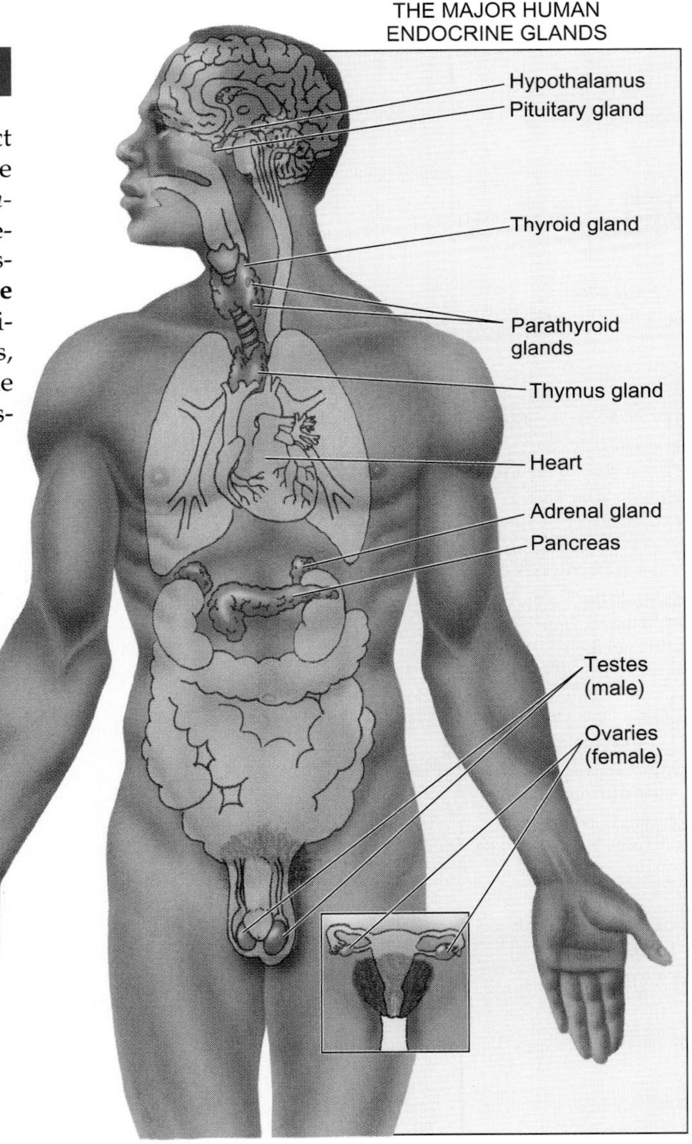

TABLE 36.1 THE ENDOCRINE SYSTEM

Structure and Secretion	Target	Action
Anterior Pituitary		
Adrenocorticotropic hormone (ACTH)	Adrenal cortex	Secretes steroid hormones; regulates fat metabolism
	Fat storage regions	Fatty acids released into blood
Growth hormone (GH)	All body cells, particularly bone and muscle	Stimulates growth; amino acid transport; growth
Thyroid-stimulating hormone (TSH)	Thyroid	Secretes T_3 and T_4
Prolactin	Breasts	Promotes milk production
Follicle-stimulating hormone (FSH)	Ovary and testis	Stimulates growth of follicle and estrogen production in females, spermatogenesis in males
Luteinizing hormone (LH)	Mature ovarian follicle	Stimulates ovulation, conversion of follicle to corpus luteum, and production of progesterone
	Interstitial cells of testis	Stimulates sperm and testosterone production
Melanocyte-stimulating hormone (MSH)	Melanocytes	Pigment dispersal (skin darkening)
Posterior Pituitary		
Oxytocin	Breasts	Stimulates release of milk
	Uterus	Contraction of smooth muscle in childbirth and orgasm
	Seminal vesicles	Ejaculation
Antidiuretic hormone (ADH)	Kidney	Increases water uptake in kidney (decreasing urine volume)
Thyroid		
Thyroxin (T_4)	General (no specific organs)	Increases oxidation of carbohydrates; stimulates (with GH) growth and brain development
Triiodothyronine (T_3)	same as above	same as above
Calcitonin	Kidney, bone	Decreases blood calcium level; increases excretion of calcium by kidney, inhibits release from bones
Parathyroid		
Parathyroid hormone (PTH)	Intestine, kidney, bone	Increases blood calcium level; decreases excretion of calcium by kidney and speeds absorption by intestine and release from bones
Islets of Langerhans		
Alpha cells: glucagon	Liver	Stimulates liver to convert glycogen to glucose; elevates glucose level in blood
Beta cells: insulin	Plasma membranes	Facilitates transport of glucose into cells; lowers glucose level in blood; causes fats and proteins to be stored
Delta cells: somatostatin	Unknown	Unknown
Adrenal Cortex		
Mineralocorticoids—aldosterone	Kidneys	Increased recovery of sodium and excretion of potassium and hydrogen ions
Glucocorticoids—cortisol, corticosterone	General (no specific organs)	Increases glucose synthesis through protein and fat metabolism; reduces inflammation
Cortical sex hormones	General (many regions and organs)	Promotes secondary sex characteristics
Adrenal Medulla		
Epinephrine Norepinephrine (both)	General (many regions and organs)	Increases heart rate and blood pressure; directs blood to muscles and brain; "fight or flight mechanism"
Ovaries		
Estrogen	General (many regions and organs)	Development of secondary sex characteristics; bone growth; sex drive (with androgens); regulates cyclic development of endometrium in menstruation;
Progesterone (ovarian source replaced by placenta during pregnancy)	Uterus (lining)	maintenance of uterus during pregnancy
Testes		
Testosterone	General (many regions and organs)	Differentiation of male sex organs; development of secondary sex characteristics; bone growth; sex drive
Heart		
Atrial natriuretic factor	Kidney	Increases sodium and water excretion
Pineal Body		
Melatonin	Melanocytes, other targets uncertain in humans—perhaps hypothalamus and pituitary	Pigment aggregation (blanching of skin); may have some influence over hypothalamus or pituitary in cyclic activity
Thymus		
Thymosin	Lymphocytes	Stimulates development of B-lymphocytes
Nonspecific Origin		
Prostaglandins	General (many regions and organs)	Presence in semen stimulates contraction in female genital tract, aiding sperm movement; stimulates ovulation in chickens, perhaps humans; aids birth through uterine contraction and cervical relaxation; affects blood clotting

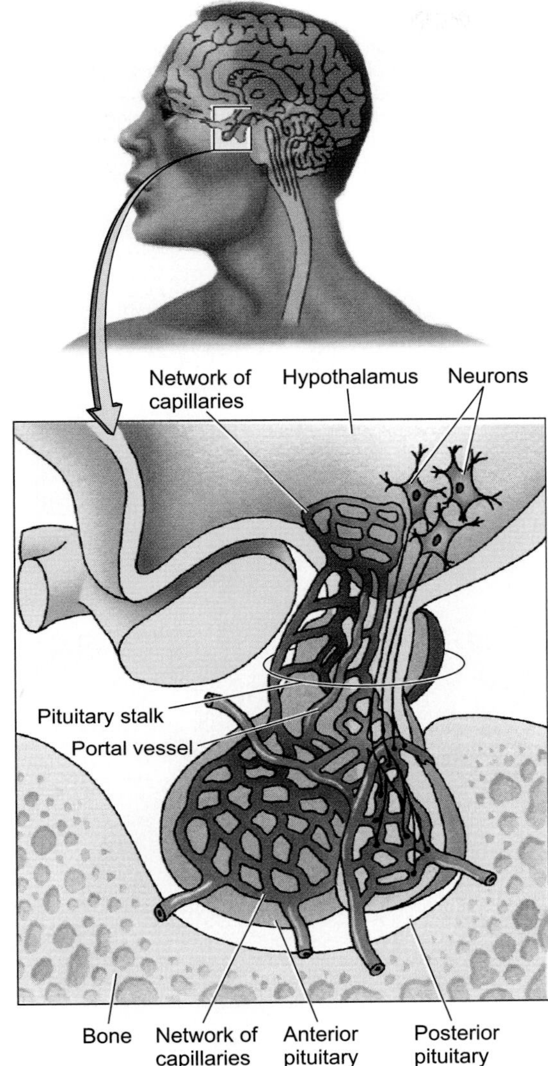

FIGURE 36.6 **THE HYPOTHALAMUS, ANTERIOR PITUITARY, AND POSTERIOR PITUITARY.**
Note that the hypothalamus communicates with the anterior pituitary by sending releasing factors through blood vessels. However, hormones of the posterior pituitary are actually produced in neurons of the hypothalamus and carried there via the axons of those neurons.

between the hypothalamus and the pituitary clearly illustrates this interaction.

The relationship of the Pituitary and the Hypothalamus. Both lobes (posterior and anterior) of the pituitary, under the influence of the brain's hypothalamus, release important but very different hormones (after all, the two lobes are actually different glands). The *posterior pituitary* is actually part of the nervous system; embryologically, it arises from tissue that will become the central nervous system. The posterior pituitary doesn't manufacture its own hormones; it simply stores and releases two hormones that are synthesized in nerve cell bodies in the hypothalamus and are transmitted to the posterior lobe by the nerve cell axons.

The *anterior pituitary* differs greatly from the posterior lobe in structure, function, and origin. Embryologically, it originates as a pocket of ectodermal epithelium on the roof of the mouth. It is a true hormone-synthesizing gland, producing and releasing six hormones, but it is also intimately influenced by the hypothalamus. The hypothalamus is drained by a network of capillaries that merge into a short blood vessel, which then travels down the pituitary stalk. There, the blood vessel divides again into a second capillary bed that spreads through the anterior pituitary. (Such an arrangement, with capillary beds at both ends of a vessel, is called a *portal* system.) It is through this circulatory connection that minute quantities, actually fleeting traces, of hormones travel from the hypothalamus to the anterior pituitary. When the hypothalamic molecules reach the anterior pituitary, they *stimulate* or they *inhibit* the release of major pituitary hormones. Describing the quantity of these messengers as "fleeting traces" is no overstatement. In the early search for them, 4 tons of hypothalamic tissue was needed to yield 1 *milligram* of the first of these hormones to be identified.

More importantly, it is clear that the pituitary, called the "master gland" by earlier endocrinologists, is largely under the influence of the hypothalamus. The hypothalamus is strongly influenced by other parts of the brain, and as we will see, it is modulated by a number of negative feedback loops. (So, who's running things?)

Hormones of the Anterior Pituitary. Of the six hormones produced by the anterior pituitary the first stimulates growth in a general way, the second stimulates milk production, and the other four influence other endocrine glands.

Let's begin with **growth hormone,** or GH. Proper levels of growth hormone in the body are essential to normal growth. Although bone and muscle are the major targets of growth hormone, it affects almost all body cells. Too much GH during the early years of development can produce **pituitary giants** (gigantism). People severely affected by this condition may grow to be 7 to 9 feet tall. Conversely, lower-than-normal GH levels in children result in **pituitary dwarfs** (dwarfism) (Figure 36.7). Pituitary dwarfism is now treatable with GH produced through recombinant DNA techniques (see Chapter 16).

Figure 36.7 Effects of Altered Growth Hormone Levels.
Problems in the pituitary can lead to altered levels of growth hormone, which can in turn lead to tremendous differences in height.

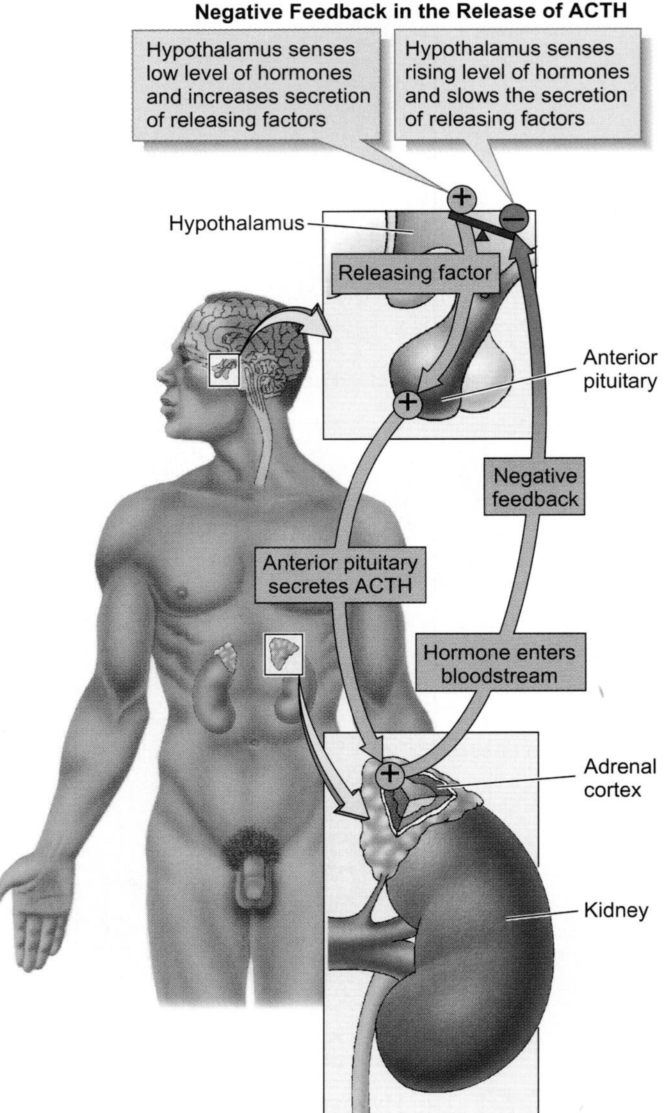

Negative Feedback in the Release of ACTH
Hypothalamus senses low level of hormones and increases secretion of releasing factors
Hypothalamus senses rising level of hormones and slows the secretion of releasing factors
Hypothalamus
Releasing factor
Anterior pituitary
Negative feedback
Anterior pituitary secretes ACTH
Hormone enters bloodstream
Adrenal cortex
Kidney

Figure 36.8 Negative Feedback in the Release of ACTH.
The release of ACTH by the anterior pituitary is controlled by a negative feedback loop. Feedback in this instance involves the rising level of a specific hormone of the adrenal cortex. The adrenal cortex synthesizes a variety of steroid hormones, while the adrenal medulla produces only epinephrine and norepinephrine.

Prolactin promotes milk production in mammals. Toward the end of pregnancy, the blood level of prolactin increases dramatically as the mother prepares to nurse her offspring.

The rest of the hormones from the anterior pituitary act on other endocrine glands. As shown in Figure 36.8, the production of **adrenocorticotropic hormone** (ACTH) is first stimulated by a hypothalamic releasing factor and is then regulated by a negative feedback control system. (Releasing factors stimulate the secretion of certain hormones from the pituitary.) The target of ACTH is the cortex (outer layer) of the adrenal gland. When stimulated by ACTH, the adrenal cortex secretes an entire battery of steroid hormones. As the levels of these hormones rise in the blood, they inhibit the hypothalamus, which then slows ACTH production in the pituitary. Besides activating the adrenal cortex, ACTH regulates the metabolism of fats. Under its influence, the body releases fatty acids into the bloodstream for redistribution through the body.

Thyroid-stimulating hormone (TSH, also known as *thyrotropin*) is another anterior pituitary hormone. It is responsible for stimulating the thyroid to release two thyroid hormones, *thyroxine* and *triiodothyronine,* discussed next.

Follicle-stimulating hormone (FSH) and **luteinizing hormone** (LH) are both involved in stimulating the production of gametes and sex hormones in the gonads, and are called *gonadotropins*. These hormones will be considered in Chapter 42. Finally, **melanocyte-stimulating hormone** (MSH) acts in darkening the skin.

Hormones of the Posterior Pituitary. Two hormones produced by hypothalamic neurons and stored in the posterior lobe are **oxytocin** and **antidiuretic hormone** (ADH). Oxytocin is important to reproduction: It stimulates the contraction of uterine muscle during labor, and after delivery it helps slow bleeding as the uterus continues to contract. Suckling the baby also stimulates the mother's posterior lobe to release oxytocin (the sensory nerves of the nipples communicate with the brain, in particular, the hypothalamus). Oxytocin then stimulates the smooth muscle around the milk ducts, and milk flow begins. Interestingly, oxytocin in women is also released during sexual stimulation and orgasm. Again, the uterus responds by contracting, and some researchers studying human sexual behavior have suggested that the contractions may cause the uterus to draw in the semen, assisting in fertilization. Oxytocin is also responsible for the pair-bonding between mother and offspring (at least in some mammals). Of course, some researchers are already asking, can a chemical be responsible for mother's love? Others cringe at the thought. (Do *you* think the question should be asked?)

Moving along to something less emotional, let's talk about urine formation. A diuretic is an agent that increases the water content, and thus the volume, of urine; an *antidiuretic* does the opposite. ADH decreases urine volume by bringing about an increase in the recovery of water by the urine-collecting ducts of the kidney, a part of the body's homeostatic maintenance. ADH is released when the hypothalamus detects a decrease in the blood's water content. The recovery of water that ADH causes adjusts blood pressure and osmotic conditions.

The Thyroid Gland

The *thyroid gland* is shaped somewhat like a bow tie and is located in an appropriate place—in front of and slightly below the larynx (Figure 36.9). Two of its three hormones, **thyroxine** (T_4) and **triiodothyronine,** (T_3), are very similar in structure; they are synthesized from the amino acid tyrosine. Thyroxine has four iodine atoms, and triiodothyronine has three (you can see why iodine is an essential mineral in the diet). T_3 and T_4 have the same targets and action but whereas T_3 is more abundant, T_4 is much more powerful.

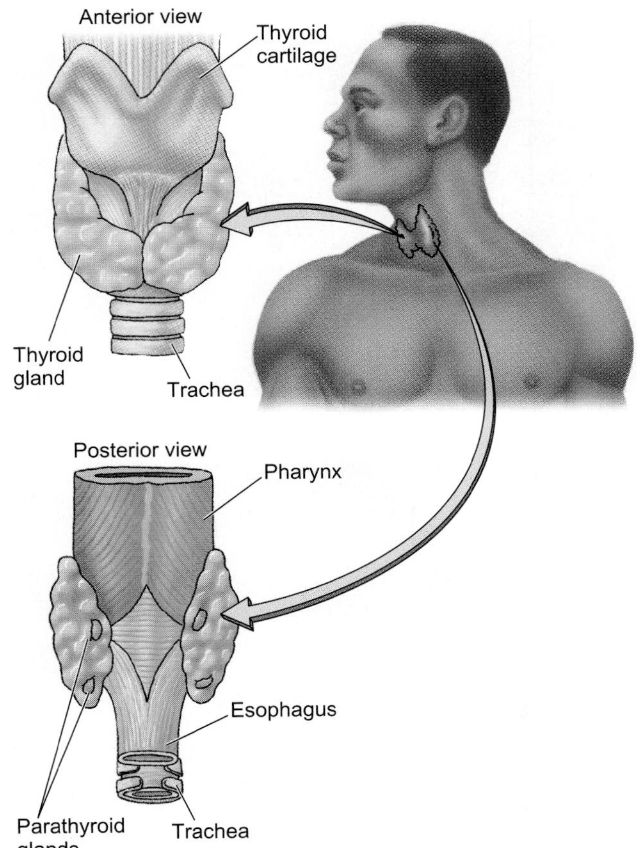

Figure 36.9 Human Thyroid and Parathyroids.
The thyroid gland is a bilobed structure nearly surrounding the trachea at a point just below the larynx. Its hormones include thyroxine, triiodothyronine, and calcitonin. Note the position of the parathyroid glands on the posterior side.

Thyroxine influences the rate at which carbohydrates are oxidized by cells throughout the body and, accordingly, the amount of body heat produced. The rate of oxidation, called the *basal metabolic rate* (BMR), is determined by measuring a resting subject's rate of oxygen consumption. Thyroxine is believed to work by activating genes that direct the synthesis of certain enzymes needed for glucose oxidation. It also increases the production of the sodium-potassium pumps in cell membranes. Interestingly, these pumps consume most of a cell's ATP. With more pumps operating, then, the cell must produce ATP even faster through its metabolic processes.

Thyroid abnormalities include **hyperthyroidism** (overactive thyroid), and **hypothyroidism** (underactive thyroid). A form of hyperthyroidism called Grave's disease is characterized by rapid metabolism,

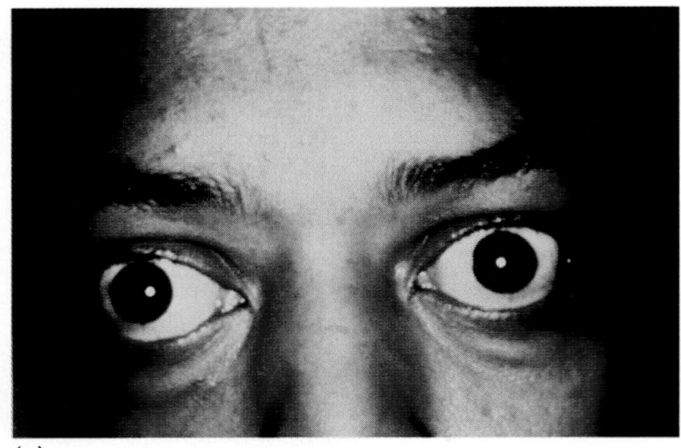

(a)

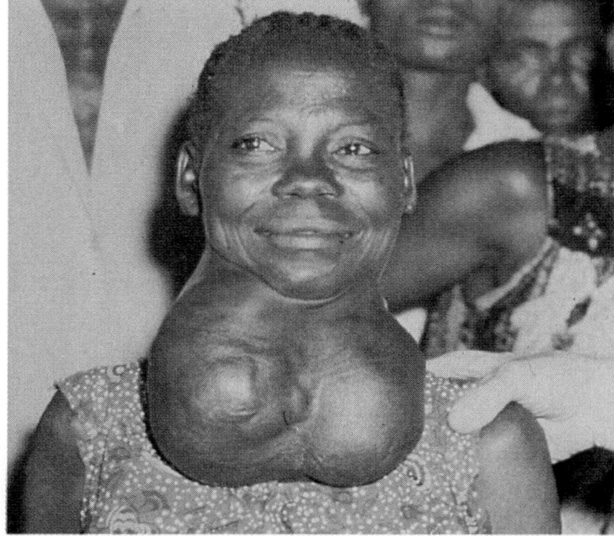

(b)

FIGURE 36.10 **GOITER.**
(a) Hyperthyroidism can produce a thin body, heavy perspiration, weakness, and exophthalmic goiter, protruding eyes due to the accumulation of fluid in the tissue behind them. **(b)** Iodine deficiency can lead to simple goiter, an enlargement of the thyroid.

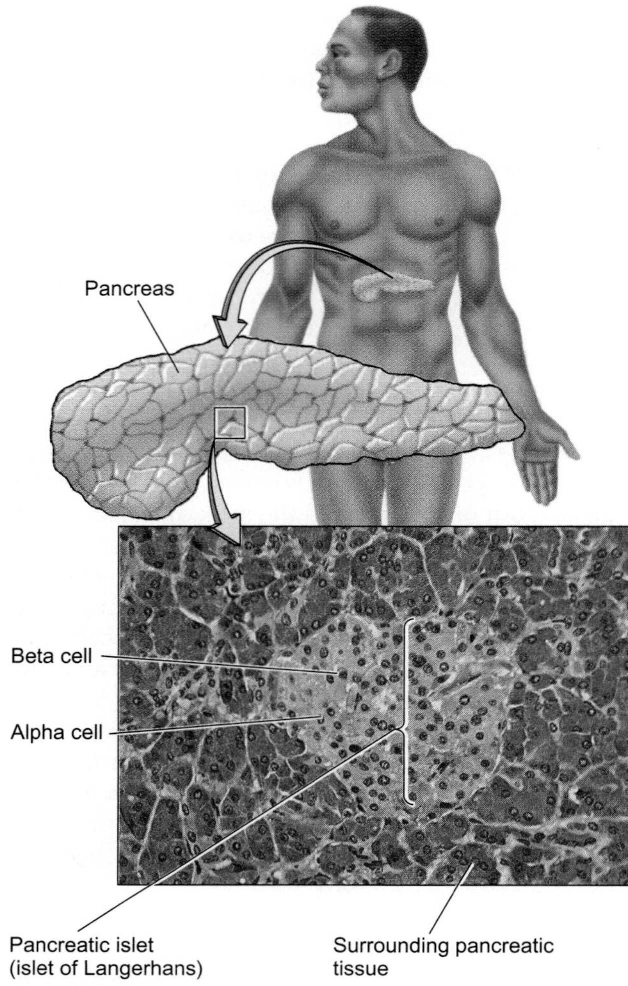

FIGURE 36.11 **THE PANCREAS AND THE ISLETS OF LANGERHANS.**
The islets of Langerhans are clearly distinguishable from the surrounding exocrine tissue. Within the clusters of endocrine tissue are patches of three types of secretory cells that produce the hormones.

weight loss, nervousness, and insomnia. (President George Bush was diagnosed as having Grave's disease.) In some instances, the eyes may bulge out noticeably, a condition known as exophthalmic goiter (Figure 36.10a). Mrs. Bush, in fact, suffered from this problem.

Hypothyroidism—thyroxine insufficiency—produces a general slowing of the metabolic rate. Hypothyroid adults are usually sluggish and overweight, have low blood pressure, and may have an enlargement of the thyroid called **simple goiter,** which is due to insufficient iodine in the diet; but since the availability of iodized salt, few cases have been reported in the United States (Figure 36.10b).

Calcitonin, the third thyroid hormone, works in concert with a hormone from the **parathyroid glands** in regulating calcium ion levels in the blood. Calcitonin inhibits bone breakdown and accelerates the uptake of calcium ions by bone, thereby lowering blood levels of calcium ions.

THE PARATHYROID GLANDS

The *parathyroid glands* are pea-sized bodies embedded in the tissue of the thyroid (Figure 36.8). The parathyroids secrete **parathyroid hormone** (PTH), a polypeptide that helps raise the calcium levels of the blood, acting as an antagonist of calcitonin. In their regulation of calcium, parathyroid hormone and calcitonin affect such things as bone deposition, release and reabsorption in the kidneys, and absorption by the intestine. Overall, the net effect of parathyroid hormone is to increase the calcium ion (Ca^{2+}) concentration in the blood, to decrease calcium excretion by the kidneys, speed calcium uptake by the intestine, and increase calcium release from bones into the bloodstream.

Abnormally low levels of parathyroid hormone cause muscle convulsions and, eventually, death. This can occur when the parathyroid glands are destroyed, as sometimes happens in an autoimmune reaction—that is, when the body's own immune system mistakenly begins to treat the parathyroid tissue as "foreign" and attacks it. The opposite condition, abnormally high levels of parathyroid hormone, results in a severe decalcification of bone (*osteoporosis*), with the subsequent formation of fibrous cysts in the skeleton. If the condition is severe, death follows.

The Pancreas and the Islets of Langerhans

The *pancreas* is, for the most part, a ducted exocrine gland, since its major secretions are digestive enzymes and sodium bicarbonate, which are released into the small intestine. Scattered through the pancreas, however, are groups of true endocrine cells that secrete their products directly into the bloodstream. These clumps of cells are the **islets of Langerhans,** or simply, the pancreatic islets (Figure 36.11). Each clump consists of at least three types of secretory cells, named *alpha, beta,* and *delta.* Alpha cells produce the hormone **glucagon,** while the beta cells secrete the more publicized hormone **insulin.** Each of these polypeptide hormones has a role in the regulation of blood sugar (Figure 36.12).

Delta cells produce **somatostatin,** a puzzling hormone whose specific action remains unknown. However, physiologists now suspect that pancreatic somatostatin inhibits the release of digestive hormones at times when they are not required. Let's briefly look at the roles of glucagon and insulin, the more familiar hormones.

Glucagon is a polypeptide consisting of a single chain of 29 amino acids. Its principal role is to stimulate the liver to break down glycogen into glucose. Glucagon is released into the bloodstream when blood glucose levels fall below a certain level, such as between meals. The glucose that is released from the liver restores the proper level of blood sugar, which in turn slows down the production of glucagon in the pancreatic islets—a classic negative feedback loop.

Insulin consists of a protein formed from 51 amino acids, arranged in two polypeptide chains joined by disulfide linkages (covalent bonding between sulfur side groups). Insulin acts on plasma membranes and helps move glucose into cells, where it can be metabolized, and it stimulates muscle and liver cells to store glucose as glycogen. Insulin also causes fats and proteins to be stored instead of being used as an energy source. In this way, they are available when

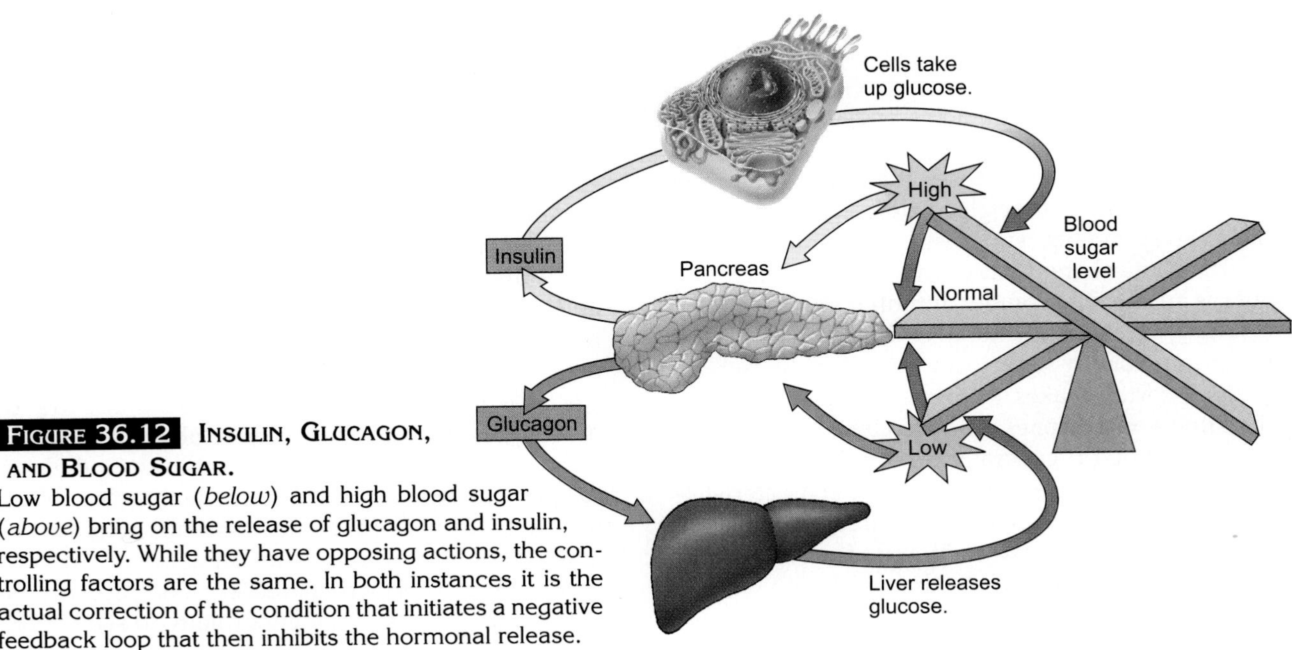

Figure 36.12 Insulin, Glucagon, and Blood Sugar.
Low blood sugar (*below*) and high blood sugar (*above*) bring on the release of glucagon and insulin, respectively. While they have opposing actions, the controlling factors are the same. In both instances it is the actual correction of the condition that initiates a negative feedback loop that then inhibits the hormonal release.

needed. Insulin also seems to be directly necessary for capillary function, and it stimulates cells (in tissue cultures in the laboratory) to divide.

Deficiency in the insulin system produces **diabetes mellitus.** In *juvenile-onset diabetes,* its most severe form, the deficiency is caused by the destruction of cells called beta cells, either by an infection or by an autoimmune response. In the more common *adult-onset diabetes,* affected persons have normal levels of insulin, but they may lack enough receptor sites on the membranes of the target cells. Diabetes mellitus, as well as some other blood sugar disorders, are considered further in Essay 36.1.

The Adrenal Glands

Ad means "upon," and *renal* refers to the kidney, so you will not be surprised to learn that the *adrenal glands* are located atop the kidneys. Each adrenal gland is essentially two glands. The outer layer, or *cortex,* produces a number of steroid hormones and has a different origin, structure, and function than the inner layer, the *medulla* (Figure 36.13).

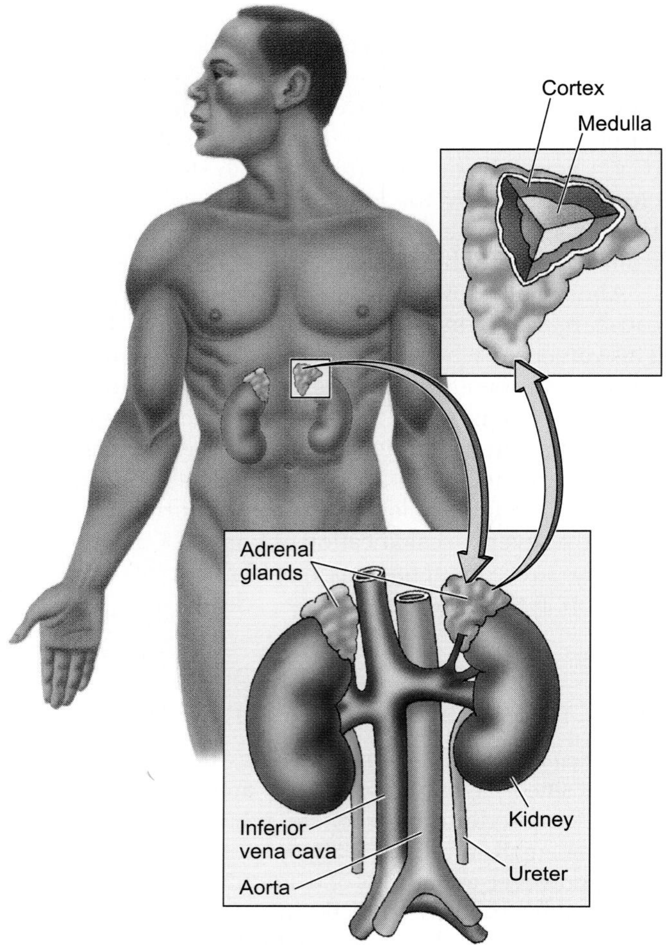

Figure 36.13 **The Adrenal Glands.** The adrenal cortex and medulla have different origins and release different hormones.

Adrenal Cortex. The hormones of the *adrenal cortex* fall into three categories according to their functions: **mineralocorticoids, glucocorticoids,** and **cortical sex hormones.** The mineralocorticoids regulate minerals. The best known, *aldosterone,* promotes the retention of sodium ions and the release of potassium ions by the kidney. The mineralocorticoids also promote inflammation as part of the body's immune defense reactions. In this role, they oppose—in a cooperative manner—the action of glucocorticoids.

Glucocorticoids are a group of hormones that influence metabolism and resistance to stress. One of the most common glucocorticoids, *cortisol,* promotes the conversion of amino acids and fatty acids into glucose. (The rapid appearance of this glucose may be stress-induced and causes the body to be more alert.) In addition, some glucocorticoids act as antiinflammatory agents, suppressing inflammation. (A pharmaceutical preparation, hydrocortisone, is available as an over-the-counter antiinflammatory.)

The cortical sex hormones of the adrenals play an important role in sexual development. In fact, the adrenal cortex makes estrogen and testosterone—the same sex hormones produced by the ovaries and testes—although generally in smaller quantities. Thus, they promote secondary sex characteristics, such as beards in men and enlarged breasts in women, and help to maintain reproductive function and libido. The principal adrenal steroid in both men and women is testosterone, with estrogens making up only a minor part of the total output. The effects of sex hormones are described further in Chapters 42 and 43).

Adrenal Medulla. The *adrenal medulla* produces two fascinating *amine* hormones: **epinephrine** and **norepinephrine** (commonly known as adrenaline and noradrenaline; *epinephros* in Greek and *adrenal* in Latin both mean "on top of the kidney"). Epinephrine and norepinephrine can cause very specific changes in the body, usually in response to sudden fright or anger. They reduce blood flow in the capillaries near the body surfaces, causing the skin to pale. The blood is diverted to the muscles, which may need to perform some strenuous activity, like getting you out of a car after an accident. At such times, digestive functions cease, blood pressure increases, the pupils of the eyes dilate (to see better),

ESSAY 36.1 Diabetes Mellitus and Other Blood Sugar Problems

Deficiencies in insulin activity produce hyperglycemia (high blood sugar), a condition that characterizes sugar diabetes (diabetes mellitus). In some diabetics, the ultimate cause of the disease is an actual deficiency of insulin, which can be caused by insufficient insulin production or by increased levels of insulinase, an enzyme that destroys insulin. However, most adult diabetics have normal levels of insulin but a deficiency of receptor sites on the membranes of the target cells. In some cases, the target cells have enough receptor sites but simply fail to respond properly.

The symptoms of diabetes include glucose in the urine, greatly increased urine volume, dehydration, constant thirst, excessive weight loss, exhaustion, fatty liver, ulcerated skin, local infections, blurred vision, and many other problems. Some symptoms are the result of glucose starvation of the cells. Though glucose may be present, an insulin deficiency prevents it from crossing the plasma membranes. In addition, the untreated diabetic has a remarkable halitosis. The exhaled breath contains ketones, especially acetone (one of the strong-smelling chemicals in some fingernail polish removers).

Mild cases of some forms of diabetes in adults can be controlled by a well-regimented diet in which carbohydrate and fats are carefully regulated. More advanced cases can often be controlled by insulin injections, as long as the problem does not involve insufficient cell surface receptors.

Diabetes in young people (those under 20), called juvenile-onset or insulin-dependent diabetes, accounts for about 10% of all cases of the disease. It is quite unlike adult-onset, or noninsulin-dependent diabetes, in that its cause is singular, specifically a degeneration of the insulin-producing beta cells of the pancreas. This leads to a chronically high blood sugar, since not enough insulin is produced to bring about glucose uptake by cells. With glucose unavailable, the body mobilizes its fat reserves, which brings on the accumulation of ketones and other organic metabolites in the blood. This accumulation lowers the blood pH, sometimes to a dangerous or even fatal level.

Juvenile-onset diabetes may not have a genetic basis, or at least not the strong genetic basis we find in adult-onset diabetes. Clues come from the study of diabetes in identical twins, where any genetic factors should be clear-cut. If the dominating factor is the gene, then when one twin becomes diabetic we can expect diabetes to appear soon in the other. This is exactly what happened in a study of twins over 50 years of age; when one became diabetic, the disease soon afflicted the other. Further, most cases of adult-onset diabetes involved causes other than insulin deficiency. But in younger diabetics, the second twin became diabetic in only about half of the cases. Further, the younger group of diabetics were clearly insulin-deficient. Juvenile-onset diabetes, then, is different from the adult-onset disease and seems to be strongly triggered by nongenetic factors. Ongoing studies suggest several possibilities, including autoimmune disease (where the immune system attacks the beta cells) and viral disease, where certain viruses preferentially attack beta cells.

Cells may also be glucose-starved because of hypoglycemia, or low blood glucose. A pancreatic tumor is sometimes at the root of this problem. As the pancreatic tissues grow out of control, too much insulin enters the blood. The insulin rapidly clears away any blood glucose and sequesters it as intracellular glycogen. The brain cells, which depend on glucose as a fuel, are the first to be affected. Low blood sugar in the brain results in dizziness, tremors, violent temper, and sometimes blurred vision and fainting.

In the absence of pancreatic tumors, some people can have milder (but sometimes serious) types of chronic hypoglycemia. In such cases, one of our much-praised negative feedback mechanisms goes haywire. An increase of blood glucose brings about an increase in insulin secretion, as you might suspect. The insulin in turn facilitates the conversion of blood glucose into tissue glycogen, bringing the glucose level back down. Unfortunately, there is a time lag, and the pancreatic beta cells may overproduce insulin, causing the blood glucose level to fall sharply.

and glucose is released from the liver into the bloodstream. Other changes also occur to prepare the body for emergencies through "flight-or-fight" responses—a capability of the autonomic nervous system as well (see Chapter 34).

Ovaries and Testes

The endocrine functions of the ovary and testes are discussed in detail in Chapter 42, so we'll note only some of the major points here. First, the two hormones of the ovary—*estrogen* and *progesterone*—and the principal hormone of the testis—*testosterone*—are all steroids. Estrogen is important in females in the development of secondary sex characteristics, in bone growth, sex drive, and the regulation of menstruation. Progesterone prepares the uterine lining to receive the fertilized egg and prepares the breast to produce milk. Testosterone influences male secondary sex characteristics and sex drive.

In addition to their reproductive and behavioral functions, the sex hormones have an important role in skeletal development. The sudden increase of their levels in the blood at the onset of puberty stimulates the lengthening of the long bones and causes a marked spurt of growth. Interestingly, it also causes a simultaneous spurt in mental growth, as measured by raw scores on IQ tests. (This undoubtedly explains the waves of intellectualism that sweep over teenagers.)

The Heart

The heart has not always been considered an endocrine gland. However, it is now known that the heart in fact produces a hormone called **atrial natriuretic factor** (ANF) when its upper chambers, the atria, are stretched. ANF promotes a reduction in the blood volume passing through the heart and in blood pressure. It accomplishes this, at least partially, by acting in the kidney to inhibit sodium ion reabsorption and in the adrenal cortex to inhibit the secretion of aldosterone. We will consider ANF again in the next chapter.

Other Chemical Messengers

Prostaglandins. The **prostaglandins** are the most recently discovered class of chemical messengers: The Nobel Prize was awarded to their discoverers in 1982. Although some prostaglandins are among the most potent biological materials known, perhaps one reason for their late discovery is that they are not produced by specialized organs. Various prostaglandins are produced by most kinds of tissues. For this reason they are not universally regarded as hormones.

Prostaglandins may be released by the activity of other hormones or by almost any irritation of the tissues. Some prostaglandins are involved in inflammatory responses and in the sensation of pain. Aspirin inhibits the synthesis of prostaglandins, which is why aspirin is effective against pain, inflammation, and fever.

Prostaglandins also have other actions, including effects on the excretory, circulatory, endocrine, reproductive, nervous, and immune systems. They are indeed a complex, critical, and wide-ranging group.

Growth Factors. Many—in fact, perhaps all—cells also produce peptide **growth factors** that regulate the rate of cell division. For instance, *nerve growth factor* is important in the development of the nervous system. It not only promotes survival and growth of neurons, but it may provide a chemical pathway that leads elongating neural processes to their target cells, thereby ensuring that the proper neural connections are made. *Epidermal growth factor* stimulates the growth of many types of cells. The brain is known to produce a *fibroblast growth factor* that stimulates cell division in some types of cells. Since cancer involves an unrestrained growth of cells, it is currently thought that growth factors may play some role in the development of certain cancers.

Melatonin. The pineal body produces **melatonin,** which influences melanocytes, the pigment cells of the skin, and may affect hypothalamus and pituitary. Recent findings suggest a role in the timing of biological clocks (internal rhythms) and sleep.

Thymosin. The thymus produces **thymosin,** which stimulates the development of B-lymphocytes, important in immunity (see Chapter 41).

We have certainly not exhausted the list of hormones here. For example, another class of chemical messengers called pheromones (discussed in Chapter 44) is produced by one animal and acts on another. Still other hormones are produced in the kidneys, liver, and skin. Yet our survey here has indicated something of the wide range of chemical regulation going on within our bodies.

PERSPECTIVES

A better understanding of chemical messengers can provide us with continuing insight into the precise coordination of life's processes. Because of such coordinated interaction, organisms are not solely dependent on the whims of the environment; they can, through precise chemical responses, adjust their internal environment as the need arises.

Now that we have some idea of the roles that nerves and hormones play in our bodies, we can examine the concept of how they—and other mechanisms—operate to keep our internal conditions relatively constant, thereby avoiding the fluctuations that could disrupt the harmony that marks our lives.

SUMMARY

Communication in the body occurs through neural impulses and through chemical messengers called hormones. Hormones occur throughout the animal kingdom.

THE CHEMICAL MESSENGERS

Until recently all chemical messengers were called hormones and were defined as chemical substances produced in one part of the body and transported by the blood, where they produced a specific effect. Today we know that many chemical messengers affect nearby cells and tissues, and others affect only the cell in which they are produced.

Hormones and nerves are related structurally (some hormone-producing glands are composed of nervous tissue), functionally (hormones and nerves may work together to accomplish the same job), and chemically (certain hormones are chemically identical to certain neurotransmitters).

THE NATURE OF HORMONES

Hormones work in minute amounts, have specific actions and target cells with specific receptor sites, and are short-lived. Chemically, they include steroids, proteins, polypeptides, and modified amino acids.

Hormone secretion is commonly regulated by a negative feedback mechanism in which the product of an action suppresses that action so that when the product builds up the action slows, and when it diminishes, the action increases. The ebb and flow of the tropic hormones illustrate the mechanism.

HOW HORMONES WORK

Hormones generally operate through second messenger systems (peptide hormones) or through gene activators (steroid hormones).

Many hormones prompt the formation of cyclic AMP, which initiates the response. When epinephrine attaches to liver cell receptor sites, it activates the membranal enzyme adenylate cyclase, which converts ATP to cAMP. The latter activates an enzyme system that breaks glycogen down into glucose. Many hormones can utilize the cAMP mechanisms because different target cells have different enzyme systems poised to act.

Steroid hormones may penetrate the cell, join cytoplasmic binding proteins, enter the nucleus, and activate certain genes.

THE HUMAN ENDOCRINE SYSTEM

The human endocrine system includes distinct, ductless glandular organs and tissues. Their products directly enter the blood for transport.

The pituitary is two glands in one, both of which are under the immediate influence of the hypothalamus. The posterior pituitary secretes two hormones that it receives from neurons originating in the hypothalamus.

The anterior pituitary is a true gland, producing and secreting six hormones. Their release is controlled by specific releasing and inhibiting hormones secreted by the hypothalamus. The hypothalamus is, in turn, controlled by the brain and by its own negative feedback loops.

The anterior pituitary hormones and their functions are as follows. Growth hormone (GH) promotes body growth. Oversecretion can produce pituitary giantism while undersecretion produces pituitary dwarfism. Adrenocorticotropic hormone (ACTH) stimulates the adrenal cortex to secrete steroid hormones. ACTH also regulates fat metabolism. Thyroid-stimulating hormone (TSH) prompts the thyroid to release its hormone. Prolactin promotes milk production. Follicle-stimulating hormone (FSH) and luteinizing hormone (LH) are involved in the production of gametes and sex hormones in the gonads.

The posterior pituitary receives oxytocin and antidiuretic hormone (ADH) from the hypothalamus. Oxytocin stimulates uterine contraction during delivery and sexual excitement and causes milk release. ADH increases the recovery of water during urine formation.

Thyroxine increases the basal metabolic rate, thereby increasing ATP energy and body heat. Oversecretion produces the hyperactivity of Grave's disease, and undersecretion produces hypoactivity and often goiter—thyroid gland enlargement.

Parathyroid hormone (PTH), along with calcitonin from the thyroid, regulates calcium uptake, excretion, and deposition in bone. Undersecretion of PTH causes convulsions and death. Oversecretion causes skeletal deterioration.

The pancreas is a ducted exocrine gland that secretes digestive enzymes, but it contains endocrine glands, the islets of Langerhans. When the blood glucose is low, islet cells secrete glucagon, which prompts the breakdown of glycogen and release of glucose. When the blood glucose level is high, other islet cells secrete insulin, which prompts glucose uptake by cells throughout the body. Diabetes mellitus occurs when beta cells fail to produce enough insulin or when receptor sites for insulin are too few. Somatostatin may inhibit the release of digestive enzymes.

The adrenal cortex secretes mineralocorticoids, which regulate mineral ions and promote inflammation, and glucocorticoids, which prompt the conversion of fatty and amino acids to glucose and are antiinflammatory. Cortical sex hormones prompt secondary sexual development and influence sex drive.

The adrenal medulla secretes epinephrine and norepinephrine, both of which cause responses that prepare the body for an emergency.

Ovaries and testes produce the steroid sex hormones estrogen, progesterone, and testosterone—all of which influence reproduction and sexual behavior. They also promote bone growth and mental development during puberty.

The heart secretes atrial natriuretic factor (ANF) when the atria are stretched. ANF reduces blood volume passing through the heart and blood pressure.

Prostaglandins act as antiinflammatory agents, cause uterine contraction, promote blood clotting in wounds (a function inhibited by aspirin), and prevent spontaneous clotting in vessels (a function promoted by aspirin). Growth factors are peptide hormones that regulate the rate of cell division. This class includes nerve growth factor, epidermal growth factor, and fibroblast growth factor.

Key Terms

hormone • **615**
target cell • **616**
trophic hormones • **616**
first messenger • **617**
second messenger • **617**
endocrine system • **619**
exocrine gland • **619**
growth hormone • **621**
pituitary giant • **621**
pituitary dwarf • **621**
prolactin • **622**
adrenocorticotropic hormone • **622**
thyroid-stimulating hormone • **622**
follicle-stimulating hormone • **622**
luteinizing hormone • **622**
oxytocin • **623**
antidiuretic hormone • **623**
thyroxine • **623**
hyperthyroidism • **623**
hypothyroidism • **623**
simple goiter • **624**
calcitonin • **624**
parathyroid gland • **624**
parathyroid hormone • **624**
islets of Langerhans • **625**
glucagon • **625**
insulin • **625**
somatostatin • **625**
diabetes mellitus • **626**
mineralocorticoid • **626**
glucocorticoid • **626**
cortical sex hormone • **626**
adrenal medulla • **626**
epinephrine • **626**
norepinephrine • **626**
atrial natriuretic factor • **628**
prostaglandin • **628**
growth factors • **628**

Review Questions

1. Explain how nerves and hormones are related structurally, functionally, and chemically.
2. List three factors that account for the highly specific action of hormones, and one factor that prevents an unending, uncontrolled action.
3. Explain how hormone levels are controlled by negative feedback. Give an example.

4. Draw a general scheme showing the work of a negative feedback loop.
5. What is a second messenger?
6. List the events surrounding the activity of epinephrine in liver cells.
7. How can cyclic AMP create so many different effects?
8. Explain how some steroids take their action directly to the gene.
9. Name and locate the major endocrine glands.
10. Explain how the anterior and posterior pituitary differ in origin and secretory activity.
11. Explain how the hypothalamus exercises chemical control over the anterior pituitary.
12. List the six hormones of the anterior pituitary and give an example of the action of each.
13. Name the two hormones of the posterior pituitary and summarize their actions.
14. How might thyroxine relate to body temperature regulation in humans?
15. Summarize the dual control mechanism for calcium regulation in the body.
16. List the hormones secreted by the islets of Langerhans cells and state the action of each.
17. Describe the glucose-regulating interplay between glucagon and insulin, including negative feedback aspects.
18. What is diabetes mellitus? List two abnormal situations that can bring on the condition.
19. List three families of hormones secreted by the adrenal cortex and state one action for each.
20. Describe an important growth effect of the cortical sex hormones. When is this most pronounced?
21. Explain, specifically, how the hormones of the adrenal medulla complement the work of the sympathetic par t of the autonomic nervous system.
22. What are growth factors? Give an example.

37

Homeostasis

AS YOU WAIT FOR THE BUS ONE WINTER MORNING, YOU REALIZE IT'S A LOT *colder out than you thought. The wind blows right through your light jacket, and you shiver. But you make it to school. There, the heat is up, and you begin to sweat. Then, to make matters worse, there's a quiz on homeostasis, and you've never heard of it. When the papers are returned, you learn that homeostasis is what kept you from freezing out on the sidewalks, and what caused you to sweat once inside (that and the quiz). You also learn that the body has a lot of ways to keep things stable internally so that the delicate processes of metabolism are not disrupted by extremes. Here, then, are some of the mechanisms involved in homeostasis and the reasons why the phenomenon is so important to us.*

Morning comes, and your eyes slowly open. You sit up and put your feet on the floor—the sure sign that the day has started. As you yawn and begin to rustle about, life seems simple and routine. You hardly notice what you are doing. How can this be? How can you handle mornings when you are barely awake? Perhaps it is because evolution has not burdened you with the awareness of what has been going on inside your body as you seek out your last pair of clean socks. While you slept, poisonous nitrogenous substances were washed from your bloodstream. After awakening, you felt certain urges—and while one part of your brain was deciding how to deal with them, another part was coordinating your muscular movements so that you didn't fall flat on your face. And you probably paid no attention to the remarkable fact that you were alive at all. Although it was a cool night, your body temperature did not fall dangerously. In fact, while you slept, various mechanisms were operating that kept your bodily processes within very precise limits. Your heart, silently and tirelessly pumping along, neither lagged nor surged, and when you stood up, scratching lazily, it did not allow blood to suddenly leave your brain, causing you to lose the consciousness you had just gained. When things are going well, your body's activities are, indeed, kept within certain critical limits. And, of course, we have a word for it. The maintenance of a relatively stable internal environment is called **homeostasis.**

MECHANISMS OF HOMEOSTASIS

The body has basically two types of homeostatic mechanisms—ways to keep the internal environment stable. The mechanisms are *physiological* and *behavioral.* Physiological responses generally are made without conscious intervention, and as we saw earlier, they often involve the autonomic nervous system and the endocrine system. An animal that has these capabilities does not really have to think much about physiological mechanisms such as increasing its metabolic rate to produce more body heat, or shunting blood into its extremities to cool its body. However, behavioral functions are generally conscious and may involve decisions. We see behavioral mechanisms at work when an animal curls up in a ball to retain more heat or moves into the shade to cool itself.

There are three elements operating in homeostatic mechanisms. One, the *receptor:* The receptor detects conditions in the environment, internal and external. Two, the *integrating center:* This center measures and evaluates the information coming from the receptors. (We have seen that the brain often handles this job.) Three, the *effectors:* These are the muscles and glands that respond to the environmental stimulus. Keep in mind that the stimulus sends signals over nerves, the coordinating center handles the signals over nerves, and then signals the effector over nerves. Obviously, homeostatic mechanisms strongly depend on the nervous system.

Homeostatic mechanisms are often based on a **negative feedback loop** where the output of the system diminishes the input (Figure 37.1). As we saw in Chapter 36, when a stimulus produces an action, that action reduces the stimulus. A cruise control on an automobile, the device that automatically keeps the car moving at a certain speed, works because as the motor runs faster, valves are closed, causing the car to slow down. The slowing of the car then causes the valves to reopen, and the car accelerates, causing the valves to begin to close again. Thus, the car's speed stays within certain limits.

Positive feedback works on the opposite principle: The product of a system increases the activity of that system. Using the example of an automobile again, with positive feedback, accelerating the motor would tend to open the carburetor and cause the engine to run even faster. Such an engine might be revved to such limits that it would explode. Obviously, then, homeostasis operates primarily through the more delicate mechanisms of negative feedback; but this is not to say that living things are not sometimes subjected to positive feedback.

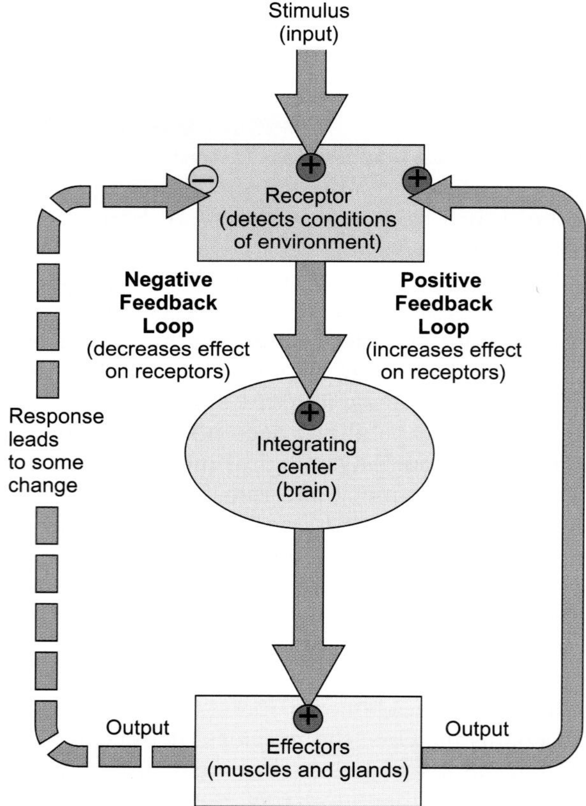

FIGURE 37.1 FEEDBACK CONTROL.
In the negative feedback loop, the product of an action reduces the level of that action. Here a stimulus is detected by a receptor that sends a signal to the brain which causes an effect to reduce the stimuli's effect on the receptor. In the positive feedback loop, the effect of a stimulus is to increase the receptor's response to it.

Positive feedback mechanisms in humans are sometimes associated with severe health problems. For example, if a person's temperature begins to rise above the normal 37°C, the body will activate corrective devices such as sweating and the opening of peripheral blood vessels (producing a heat-dissipating "flush"). However, at some point (usually at about 42°C), the negative feedback system breaks down, and a positive feedback begins. The high temperature begins to cause an increase in metabolic activity, which raises the heat, which increases metabolic activity, which can kill the unfortunate soul. (Positive feedback is the basis of the famed "vicious circle.")

To illustrate the adaptiveness and complexity of homeostatic mechanisms, we will consider two processes that involve multiple, interacting controls. One is *thermoregulation,* through which the animal controls its internal temperature. The other is *osmoregulation,* through which the animal controls its body-fluid and mineral-ion balances. The latter is closely associated with the excretory system, which rids the body of nitrogenous wastes.

THERMOREGULATION

Thermoregulation is the ability of an organism to maintain its body temperature within a certain range. There is great variation in this ability among the various animals. Some have the ability to thermoregulate precisely, while others have almost no ability to regulate their body temperatures at all. Most animals do have some such ability, though—so the question arises, why? Thermoregulation can be very metabolically expensive, so what advantages make it worth it?

WHY THERMOREGULATE?

The assumption is, there is no life in the sun's interior. The reason is, it is too hot there. And life probably doesn't exist in those celestial realms where the temperature approaches absolute zero. Intuitively, we seem to understand that life is likely only to exist within certain temperature limits. But what do extremes of temperature actually do to life? Under the most extreme conditions, low temperatures can freeze the water in living cells, resulting in ice crystals disrupting delicate membranes—and with the water tied up as ice, the solutes in the remaining fluids can become dangerously concentrated. High temperatures, on the other hand, can accelerate rates of biochemical reactions to unacceptable levels and can denature enzymes, rendering them biochemically inactive. Yet higher temperatures, of course, simply destroy the "stuff of life" with a crackling hiss. Since just such an episode is likely to do them in, most animals don't have to deal with such problems more than once. Instead, they generally have to solve more moderate, long-term problems associated with temperature.

Animals in an environment even only a little cooler than their bodies cannot completely avoid heat loss; and since the primary source of heat for animals is their cellular respiratory process (see Chapter 8), we find a paradoxical situation. As the body cools, greater metabolic heat is required. Yet lower temperatures slow chemical activity, including the cellular activity required to produce heat. The animal, then, must somehow raise its temperature to avoid going into a positive feedback loop in which lower temperature slows the metabolic processes that can produce heat, causing even greater heat loss.

Sources of Body Heat

There are two major ways that animals can warm their bodies: *ectothermy* and *endothermy.* In **ectothermy,** animals warm themselves by absorbing heat from the environment. Typically, an ectotherm warms itself by moving into warmer areas. For example, an ectotherm can gain heat by exposing itself deliberately to the sun (Figure 37.2a), or by moving to warmer parts of its environment—as when a fish living in a hot spring moves closer to the vent, or when a flea migrates to a dog's warmer parts. Conversely, whether fish or flea, when some animals become too warm, they can simply move to a cooler area.

Endothermy is the generation of heat through metabolic processes. You may recall that metabolic energy transfers are not 100% efficient, and that at each step some energy is lost as heat. That heat, when generated in sufficient amounts, can be used in thermoregulation. (In a sense, then, in endothermy, heat comes from the inside, and in ectothermy it comes from the outside.) Some animals have ways of retaining that heat, as we see in small birds that fluff their feathers to trap an insulating layer of air around their bodies (Figure 37.2b) or in humans who hold their body heat by adding clothing (Figure 37.2c). Note also in Figure 37.2c that animals can also generate more heat by increasing their activity.

If an animal can generate its own heat, then its internal environment has a means of remaining relatively stable in an unstable world, a place in which temperatures can vary widely. This constancy can be critical to survival. After all, if an animal can maintain a more constant temperature, its metabolism can proceed at a more constant rate. Thus, the animal can remain active at a wider range of environmental temperatures. Endothermy has been critical in the evolutionary history of animals quite simply because it provided a way for many kinds of animals to cope with colder or warmer climates, opening a wider range of potential habitats.

Homeotherms vs. Poikilotherms

Those animals that maintain a somewhat constant internal temperature are called **homeotherms.** On the other hand, those whose body temperatures tend to track that of their surroundings are called **poikilotherms.** The lay terms for these two conditions are "warm-blooded" and "cold-blooded," respectively. However, the lay terms are somewhat misleading, because they imply that the body temperatures of the two groups are different, while this is not necessarily so. For example, the blood of a poikilotherm basking in the sun may be just as warm as that of a homeotherm elevating its temperature by metabolic heat.

"Either-ors" are somewhat unusual in many realms of biology, and so we find that there are also animals called *heterotherms.* **Heterotherms** maintain a constant body temperature through metabolic mechanisms as homeotherms do, yet they can allow their bodies to cool to environmental temperatures. In this way, they can remain warm when they need to and then save energy by allowing themselves to cool. For instance, during the day, a hummingbird is homeothermic. The rapid fluttering of its wings, while it flies in search of food, generates body heat. But this means of maintaining a high temperature is a tremendous drain on energy resources. At night, then, the hummingbird enters a sleeplike torpor as its body falls to temperatures approaching those of the environment. Some species of bats may spend 80% of their lives hanging upside down in a heterothermic torpor. Other kinds of heterotherms conserve energy by warming only certain body parts. For instance, the temperature deep within the body of a bluefin tuna is remarkably warm as it swims with ice-cold extremities through frigid waters (see Essay 37.1).

(a)

(b)

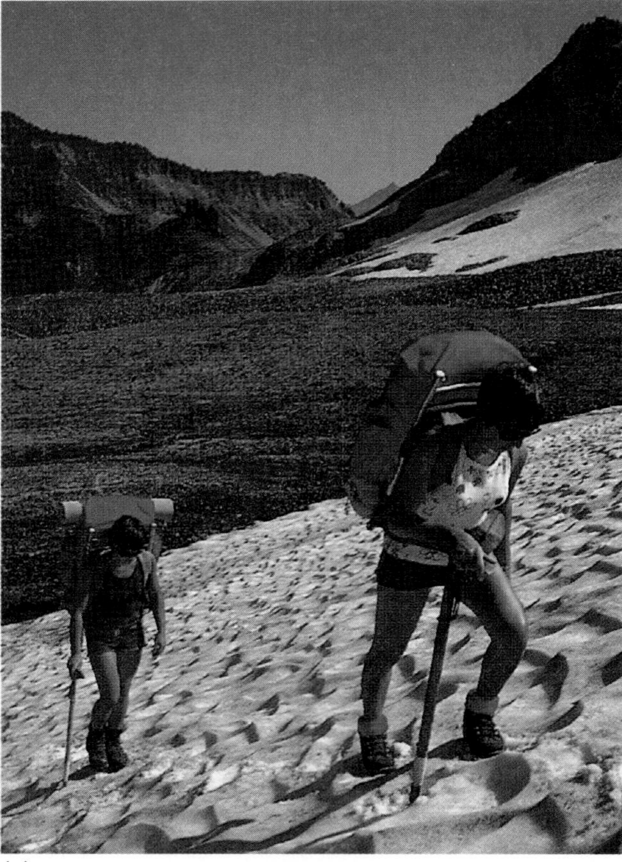

(c)

FIGURE 37.2 VERTEBRATE ADAPTATIONS TO COOLING EXTERNAL TEMPERATURES. **(a)** Reptiles may be baskers. Lacking an efficient physiological means of maintaining warmth, they often use the sun to raise their body temperatures. **(b)** Birds tend to crouch into a ball shape, drawing the extremities in. At this time their inner down feathers may be fluffed up, trapping air and improving their insulating quality. **(c)** Humans may pull clothing on or off and change their level of activity to regulate their temperatures.

THERMOREGULATION IN HUMANS

Humans, like other mammals, maintain internal body temperatures within rather narrow limits. Any consistent variation of more than a few degrees from the optimum usually means trouble. We adjust our body temperature physiologically by varying the rate of metabolic heat production and controlling heat loss or gain from our body surfaces. The two mechanisms work in close harmony through the efforts of the hypothalamus, the body's internal thermostat (Figure 37.3).

The Hypothalamus as a Thermostat. The temperature-regulating portion of the hypothalamus works a bit like an ordinary home thermostat. The hypothalamus, however, is far more sensitive and precise. In fact, physiologists have detected responses in the hypothalamus when blood temperature changed as little as 0.01°C. Information regarding the body's temperature comes from thermoreceptors in the skin and from the hypothalamus's own thermoreceptors, which monitor blood temperature (see Figure 37.3).

When the hypothalamus detects a drop in body temperature, it can respond in a number of ways. It can cause the autonomic nervous system to decrease blood flow to the skin, thereby shunting blood to the warmer internal areas and conserving heat. It can also cause the autonomic nervous system to increase heat production by stimulating the adrenal medulla, which releases epinephrine into the bloodstream and increases the conversion of glycogen to glucose in the liver and muscles. The glucose is the fuel that feeds the cellular metabolic processes and releases heat. Although the effects of epinephrine are short-lived, the hypothalamus can produce a longer lasting

Essay 37.1 A Warm-bodied Fish

The bluefin tuna was once believed to be a poikilotherm, like many other fishes. Yet it is among the fastest of all bony fishes and often lives in extremely cold water. This ability has always presented a puzzle; speed and cold bodies just don't go together. How could a poikilotherm generate the metabolic energy needed to swim so fast in its chilling habitat? For one thing, it turns out that the bluefin tuna is not the "cold fish" we once thought it was. In fact, even in the coldest water its internal temperatures can reach almost 32°C (almost 90°F). This is a far more efficient physiological regulation than we (as efficient homeotherms) are capable of. So how does the bluefin tuna keep its body warm in those frigid surroundings?

Much of the answer seems to be in the bluefin's circulatory system, which is quite different in some ways from that of most other bony fishes. In the bluefin tuna, the major arteries leaving the head and warm internal regions, and the veins returning blood from cool external regions, run paired with and parallel to each other; those of other fishes tend to branch individually. So here, then, we find the basis for the bluefin's unusual ability. The parallel vessels set up the mechanism for what is called a *countercurrent heat exchange,* in which blood moves through adjacent vessels in different directions and at different temperatures. It works this way: Warmer blood leaving the deeper tissues passes cooler blood coming back from the surface. As the two opposing streams pass each other, heat moves to the cooler returning blood, and thus much of this heat never reaches the extremities to be lost.

The flesh of the bluefin tuna has two distinctly different regions—essentially light and dark. The dark muscle is warmer (as much as 10°C warmer than the skin). In fact, its dark color is due to an immense network of blood vessels consisting of parallel arteries and veins. This vascularized region, known as the *rete mirabile* ("wonderful net"; see illustration), is a very dense countercurrent heat exchanger and conserver.[1] Because of its arrangement, the *rete mirabile* produces a comparatively warm, lively group of swimming muscles. And this is how the bluefin tuna (and a number of other fast-swimming predatory fish) can get the speed to chase their prey in very cold water.

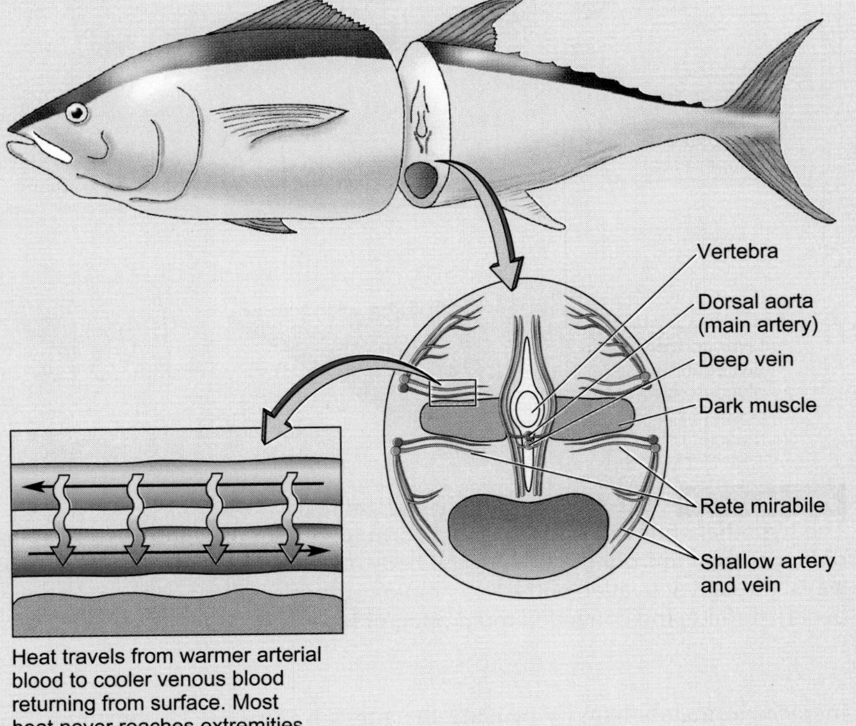

Heat travels from warmer arterial blood to cooler venous blood returning from surface. Most heat never reaches extremities.

[1] *In the canned tuna we enjoy, the very dark muscle at the center of the fish is scraped away and discarded. It has a "strong" taste and would tend to darken the rest of the canned flesh. The difference between better, more costly tuna and the less expensive brands may depend in part on how much of the dark, blood vessel laden muscle is used.*

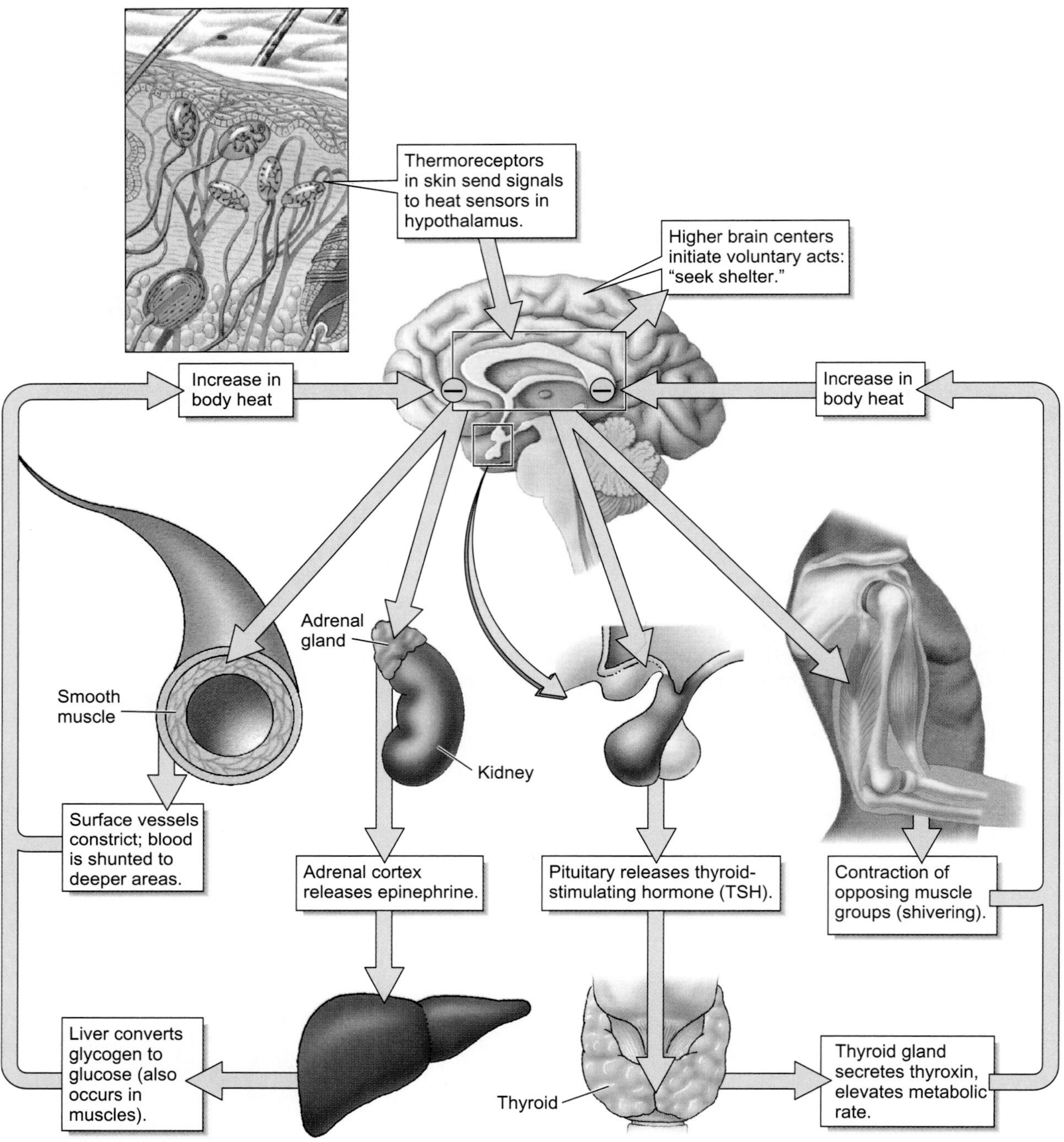
Thermoreceptors in skin send signals to heat sensors in hypothalamus.
Higher brain centers initiate voluntary acts: "seek shelter."
Increase in body heat
Increase in body heat
Adrenal gland
Smooth muscle
Kidney
Surface vessels constrict; blood is shunted to deeper areas.
Adrenal cortex releases epinephrine.
Pituitary releases thyroid-stimulating hormone (TSH).
Contraction of opposing muscle groups (shivering).
Liver converts glycogen to glucose (also occurs in muscles).
Thyroid
Thyroid gland secretes thyroxin, elevates metabolic rate.

FIGURE 37.3 THE CENTER FOR THERMOREGULATION.
The hypothalamus receives thermal information from thermoreceptors in the skin and by direct sensing of blood arriving from the core of the body. Its response to changing temperatures is carried out in two ways: through activation of the autonomic nervous system and the endocrine system. Negative feedback occurs through the continued monitoring of blood temperature.

increase in metabolism by causing the anterior pituitary to release thyroid-stimulating hormone (TSH), which stimulates the thyroid to release its metabolism-elevating hormones.

Shunting the blood away from the skin—and the release of epinephrine and TSH—act to warm the body, but sometimes these processes are not enough. If the body temperature continues to fall, the hypothalamus initiates "shivering"—rhythmic, rapid contractions of opposing skeletal muscle groups. The increased physical activity triggers a surge in metabolic activity and yet more fuel is burned, creating more heat.

When exposed to prolonged cold, some species of animals can generate heat through increased metabolic activity in brown fat. Brown fat is brown because it has numerous mitochondria that contain a brown enzyme, cytochrome oxidase. The energy produced in these mitochondria is not used to generate ATP, but is simply converted to heat. Unlike most other fat storage tissue, brown fat has a rich blood supply, so the heat generated in the fat is rapidly dispersed throughout the body. Interestingly, human babies are able to generate heat from brown fat, but adults cannot.

At the other end of the temperature spectrum, there is the problem of too much heat. How does the body cool itself? To begin with, as body heat increases, the hypothalamus senses the change and eases off on its heat-generating activity. Then it may initiate changes that will actually help to cool the body. For example, the hypothalamus can cause the vessels near the skin to dilate, bringing more blood to the skin, thereby permitting the heat to radiate into the cooler air. Heat escape is even faster if aided by the evaporation of perspiration. Humans, by the way, are unusual in their ability to sweat over virtually their entire bodies. In fact, sweat glands are found on only a few other mammals. Horses do have them over most of the body—but in dogs the glands are found on the footpads, and are associated with hair follicles. They are entirely lacking in reptiles and birds. (It wouldn't do for birds to try to fly around with soggy feathers.)

Skin and Lungs As Thermoregulators. We have seen that animals can gain or lose heat over their skin. Let's now see how the skin is able to perform this role so elegantly. We can see (Figure 37.4) that the blood flow to surface capillaries can be increased or decreased as determined by smooth muscle sphincters (circular muscles) in the tiny arteries leading to the capillary bed. Contraction of the sphincters is brought about by the hypothalamus,

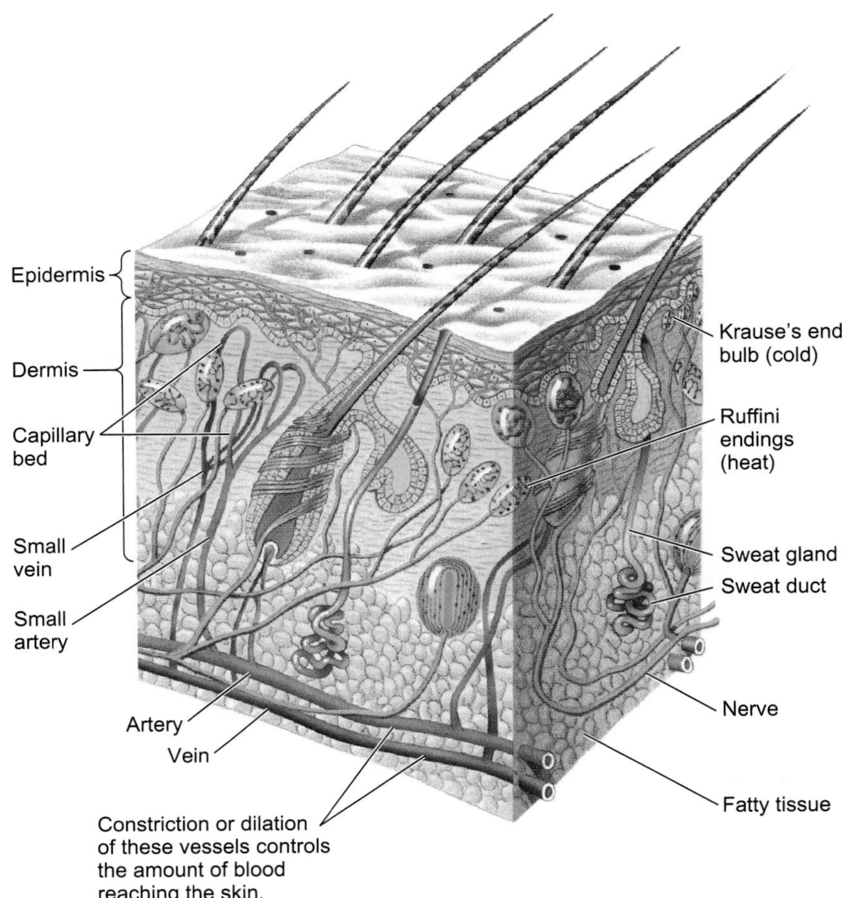

FIGURE 37.4 SKIN AS AN IMPORTANT THERMOREGULATORY ORGAN. Sweat glands secrete water through ducts. The evaporation of the sweat cools the skin. Sebaceous glands secrete oils that keep the hair and skin pliable and waterproof. Capillary beds below the surface can be opened, bringing blood near the skin for cooling, or alternately can be closed, keeping blood in deeper regions and thereby conserving heat. The direction of blood flow is controlled by sphincters (circular valves in the small arteries). Heat and cold receptors are abundant in the skin. Layers of fatty tissue provide heat-retaining insulation.

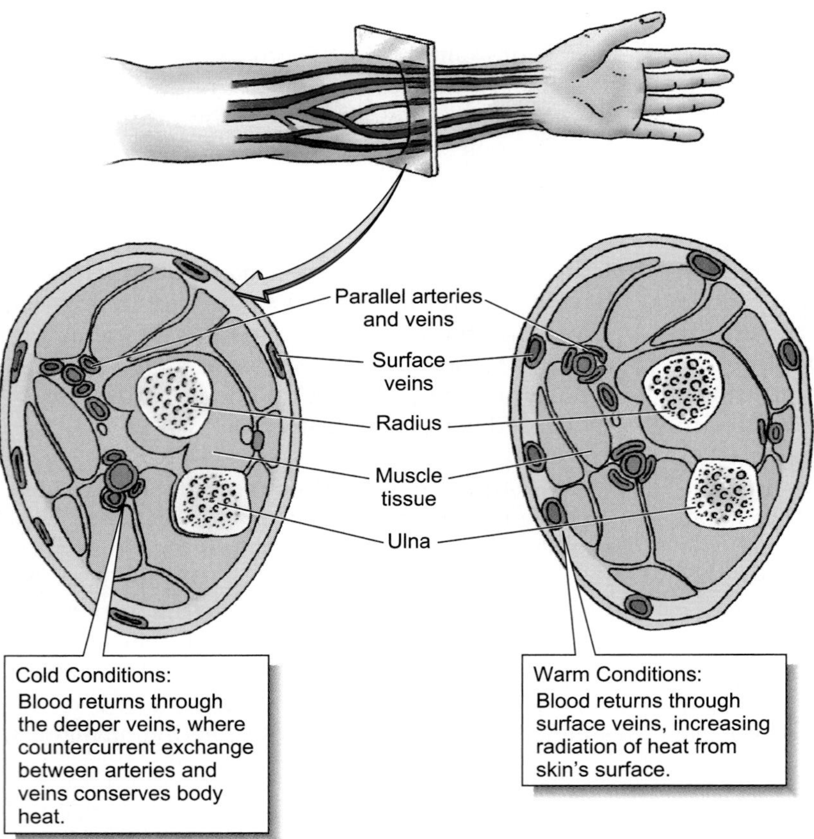

Parallel arteries and veins
Surface veins
Radius
Muscle tissue
Ulna
Cold Conditions:
Blood returns through the deeper veins, where countercurrent exchange between arteries and veins conserves body heat.
Warm Conditions:
Blood returns through surface veins, increasing radiation of heat from skin's surface.

FIGURE 37.5 COUNTERCURRENT HEAT EXCHANGERS IN HUMAN EXTREMITIES.
In the arm, for instance, the larger arteries and veins travel close together deep in the muscle. When the arm is cold, blood returns through the deeper veins, where the countercurrent exchange of heat between the arteries and veins conserves internal body heat. An alternate route for returning blood is through the veins that lie near the surface, just below the skin. When the arm is warm, returning blood follows the surface route, increasing the radiation of heat from the skin.

operating through the autonomic nervous system. When these sphincters close, blood moves deeper into the warmer parts of the body.

The lungs are also important in temperature regulation. Not only do animals cool themselves by panting (the rapid movement of air over the lung surfaces), but there is always some heat loss due to normal breathing.

Because breathing involves moving air into and out of the lungs, the lungs must be protected from extremely cold or hot air, and this is largely accomplished by passing the air over respiratory membranes, such as those in the nasal passages, before it reaches the lungs. For example, on a very cold day, the air reaching the lungs has usually been warmed to almost body temperature. Then the nasal passages themselves are rewarmed by the exhaled air.

Humans, like some other animals, are able to regulate body temperature by countercurrent heat exchangers, which make use of heated and cooled fluids passing in opposite directions in adjacent vessels so that heat is transferred along the entire route. As we see in Figure 37.5, humans have very effective countercurrent heat exchangers in the arms. When the body is warm, blood returns from the hands through veins near the skin (the ones you can see in your arm), thus permitting heat to escape. As the body becomes chilled, blood is shunted into deeper veins that run parallel to the arteries, thus enabling the body to conserve heat through a countercurrent heat exchange between the opposing vessels. Shunting blood deeper into the body to conserve heat is a costly defense. Cooler limb muscles do not work as efficiently, and the extremities, deprived of

TABLE 37.1 *MAMMALIAN RESPONSES TO TEMPERATURE STRESS*

Stimulus	Response	Effect
Hot temperature	Sweating or panting	Evaporative heat loss
	Dilation of peripheral blood vessels	Heat lost from warm blood at body surface
	Changes in behavior	Increase heat loss or reduce heat generated
Cold temperature	Constriction of peripheral blood vessels	Reduce heat loss at body surface
	Increased muscle activity (shivering)	Increase heat production
	Secretion of thyroxine from thyroid gland and epinephrine from adrenal medulla increases metabolic rate	Increase heat production
	Changes in behavior	Alter heat loss or heat production

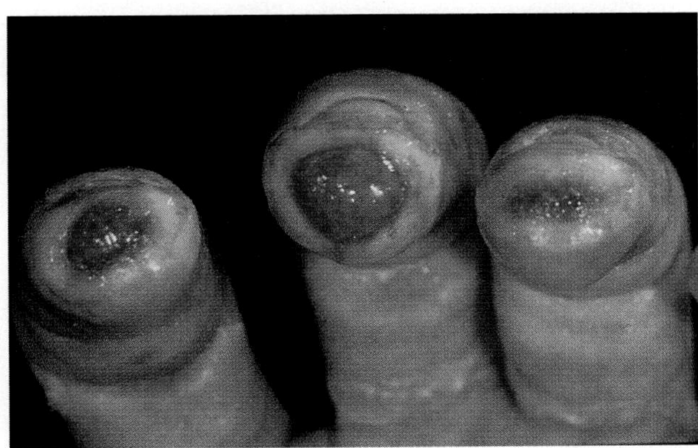

FIGURE 37.6 FROSTBITE.
Under cold conditions the body may shunt blood away from the extremities, leaving them vulnerable to tissue destruction.

warming blood, can become numb and eventually frostbitten. Thus, a person can actually lose fingers, toes, and ears before the core temperature reaches a danger level. The internal organs receive first priority in the battle for survival; the extremities are sacrificed (Figure 37.6). Table 37.1 summarizes responses to thermal stress.

OSMOREGULATION AND EXCRETION

Osmoregulation is the regulation of water and ion concentrations in the body. The mechanisms behind this regulation present a range of fascinating problems for physiologists. It is sometimes said that most cells are about two-thirds water, but to the animal, "about" won't do. In fact, for all living things the amount of intracellular water is critical, as is the relative abundance of various ions in the cell fluids—since the two problems are inseparable.

In particular, the question of how the balance of water and ions is maintained centers on excretion. **Excretion** is the removal of metabolic wastes from the body. Here, we are particularly interested in the removal of nitrogenous wastes produced from the metabolism of amino acids. As we will see, to a great degree, the animal's habitat dictates just how those nitrogenous wastes will be removed.

PRODUCING NITROGEN WASTES

In the next chapter, we will find that proteins can be broken down into individual amino acids during digestion so that they can be absorbed from the gut into the bloodstream. Those same amino acids can then be used to build new proteins in the cells; they may also be converted to fatty acids or carbohydrates for storage, or used as fuel in respiration. The processes by which the amino acids are prepared for their various roles routinely produce leftover fragments of nitrogen; and because accumulations of these nitrogen fragments can be extremely poisonous, they must be excreted (Figure 37.7).

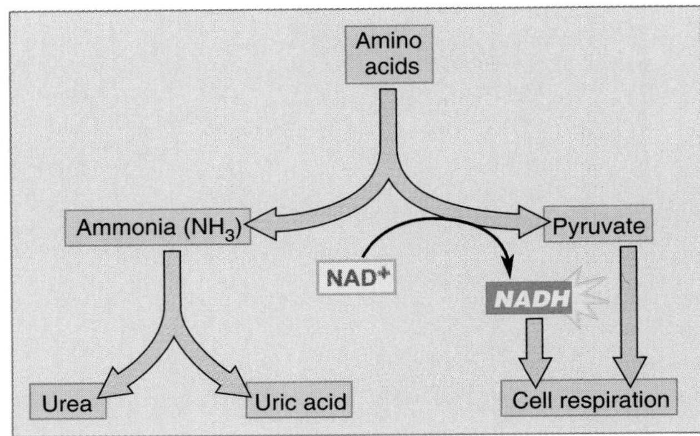

FIGURE 37.7 FORMATION OF NITROGEN WASTES.
In the metabolic breakdown of the amino acid alanine, pyruvate is formed, along with the ammonia. Ammonia may be excreted directly, or it may be converted to urea.

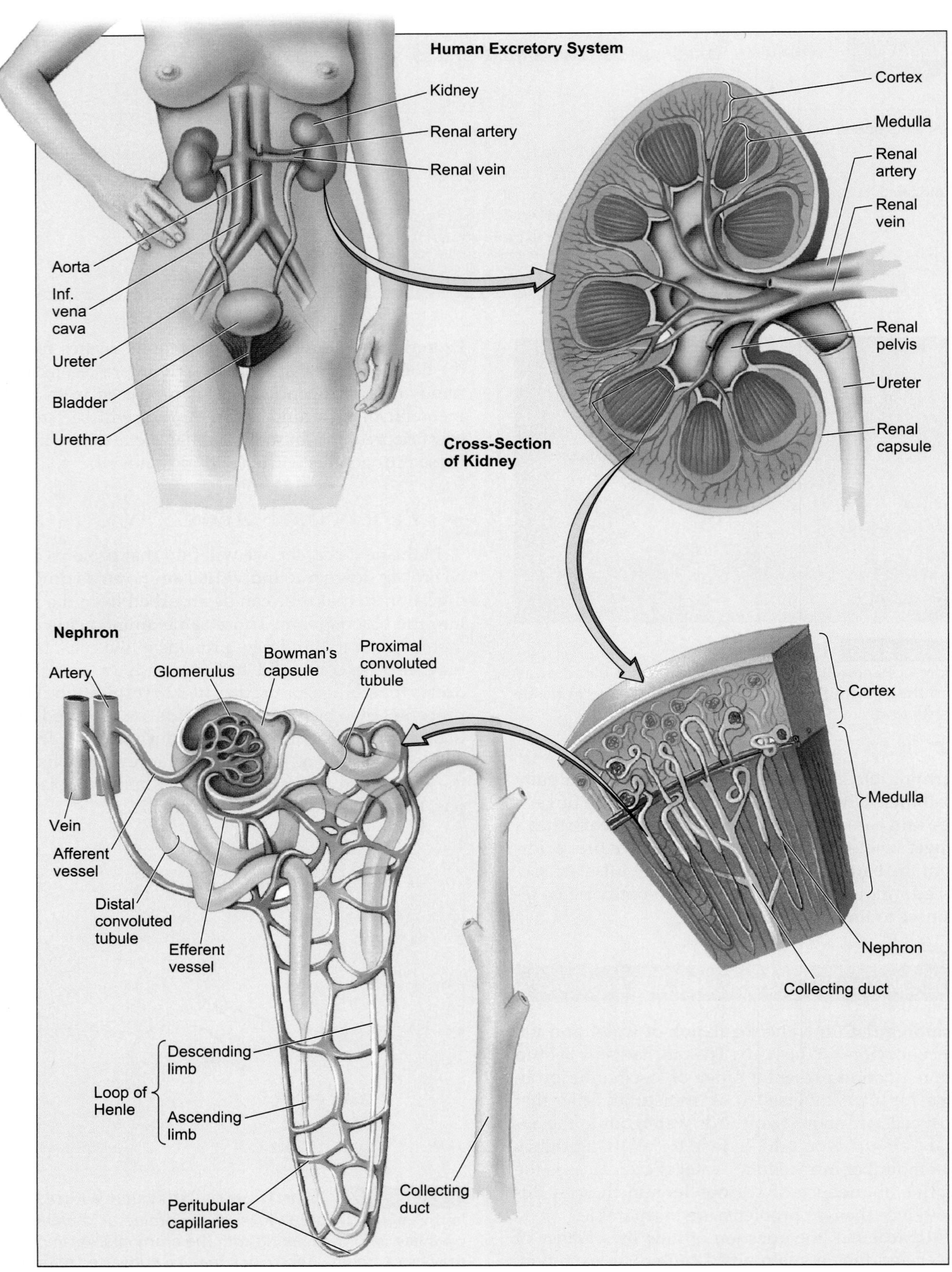
Human Excretory System
Kidney
Renal artery
Renal vein
Aorta
Inf. vena cava
Ureter
Bladder
Urethra
Cortex
Medulla
Renal artery
Renal vein
Renal pelvis
Ureter
Renal capsule
Cross-Section of Kidney
Nephron
Artery
Glomerulus
Bowman's capsule
Proximal convoluted tubule
Vein
Afferent vessel
Distal convoluted tubule
Efferent vessel
Loop of Henle
Descending limb
Ascending limb
Peritubular capillaries
Collecting duct
Cortex
Medulla
Nephron
Collecting duct

TABLE 37.2 *URINARY SYSTEM*

Structure	Function
Kidney	Filters blood and produces urine in nephron
Glomerulus	Filters blood
Bowman's capsule	Collects the filtrate
Proximal convoluted tubule	Actively pumps out sodium; chloride, other ions and water passively follow
Descending loop of Henle	Loses water and gains salt; passively concentrates urine
Ascending loop of Henle	Actively secretes chloride; sodium and potassium passively follow
Distal convoluted tubule	Sodium ions actively pumped out, followed by chloride ions and water
Collecting duct	Reabsorbs water under influence of ADH, releases urea into medullary tissues
Ureter	Carries urine to bladder
Urinary bladder	Stores urine
Urethra	Eliminates urine

These nitrogenous byproducts are produced by *deamination,* where an amine group, $-NH_2$, is removed from an amino acid. Usually the amine group is removed as NH_3, ammonia, a highly toxic compound in high concentrations. However, many animals can safely handle ammonia if they can dilute it enough. For example, freshwater fish have enough water available to excrete their wastes as ammonia. However, in many terrestrial animals (groups that generally must conserve water), the ammonia must be converted to something less toxic. In these species, ammonia is generally converted to uric acid or to urea. *Uric acid* is an insoluble product, primarily excreted by insects, reptiles, and birds, usually in the form of a dense paste that requires little water to be handled effectively. *Urea* is highly soluble, and because it is less toxic than ammonia, it can be handled more easily. Urea is the primary nitrogen waste of earthworms, many fishes and amphibians, and all mammals, including ourselves. Urine, then, contains urea. We see, then, that the ways nitrogen wastes are handled have evolved in response to the relative importance of conserving water.

FIGURE 37.8 THE EXCRETORY SYSTEM.
The human excretory system consists of the paired kidneys, their associated blood vessels, the renal arteries and veins, the uterus, urinary bladder, and urethra. The kidney contains an outer cortex, an inner medulla, and a final collecting region known as the renal pelvis. Note the relationship between the cortex and medulla of the kidney and the loop of Henle. The functional units of the kidney are the nephrons. Each nephron consists of four anatomical regions: Bowman's capsule, proximal convoluted tubule, loop of Henle, and distal convoluted tubule. Each nephron is joined to a nearby collecting duct. Note the blood supply to the nephron. Blood enters the nephron at the Bowman's capsule, where an afferent arteriole has branched into a mass of smaller vessels that comprise the glomerulus.

THE HUMAN EXCRETORY SYSTEM

As terrestrial creatures, we humans must conserve water. At the same time, we must wash potentially poisonous nitrogenous wastes from our bodies and regulate such ions as sodium, potassium, chloride, and hydrogen. As noted, human nitrogenous waste primarily occurs in the form of urea. So how do we manage this dilemma? How do we wash away urea while saving water? It happens in the kidneys. Let's take a closer look by noting that both the osmoregulatory and the excretory functions are handled by the kidneys (with assistance from the lungs and sweat glands).

Excretory Structures. The major human excretory structures are the *kidneys, ureters, urinary bladder,* and *urethra* (Figure 37.8). (The structures of the human excretory system and their functions are listed in Table 37.2).

The two kidneys lie in the dorsal area (at the back) and extend slightly below the protective rib cage. Actually, they lie behind the membrane that lines the abdominal cavity. (The venerable army jeep in World War II was built with metal seats that reached almost to the rib cage of most GIs, and a great number of servicemen returned home with kidney damage. Also, because the kidneys are relatively unprotected, blows to this area are generally illegal in boxing and the martial arts.)

Each human kidney contains about one million nephrons. A **nephron** is the excretory unit of the kidney. It includes a number of structures (Figure 37.8). We can describe the nephron in the sequence in which fluids pass through it after leaving the blood. The nephron begins as a hollow bulb or cup known as *Bowman's capsule.* From the Bowman's capsule, the nephron forms a slender tubule. The first part, the *proximal convoluted tubule* (*proximal,*

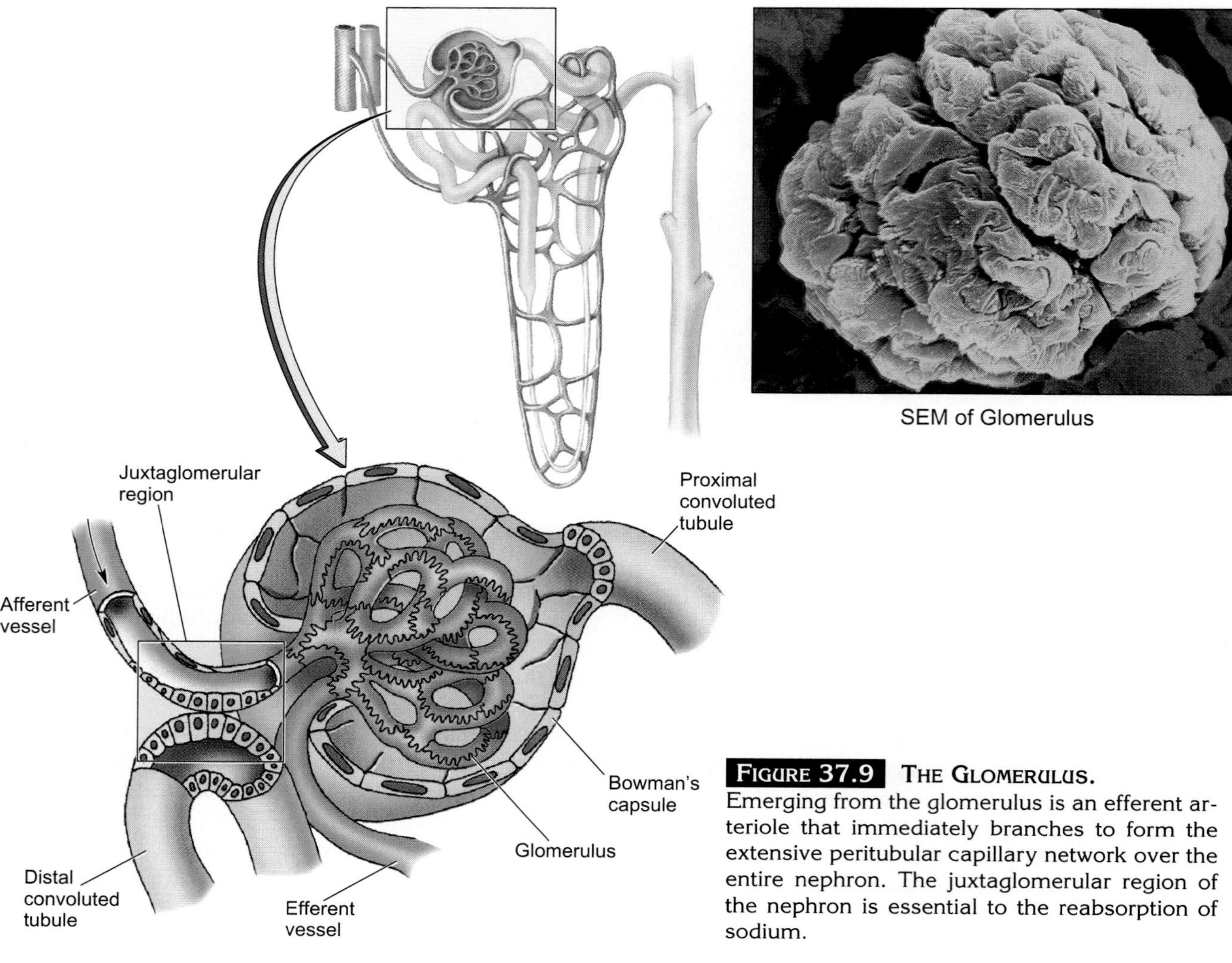

FIGURE 37.9 **THE GLOMERULUS.** Emerging from the glomerulus is an efferent arteriole that immediately branches to form the extensive peritubular capillary network over the entire nephron. The juxtaglomerular region of the nephron is essential to the reabsorption of sodium.

near), follows a twisted, contorted route and then descends into the second part, a long, hairpin structure known as the *loop of Henle.* The descending limb of the loop is called, not surprisingly, the *descending limb.* It dips down toward (or into) the renal medulla and then turns, forming the *ascending limb.* The ascending limb of Henle's loop then gives rise to the *distal convoluted tubule* (*distal,* far). Like its proximal counterpart, the distal convoluted tubule follows a twisting path. The tubule finally joins a *collecting duct,* along with the distal convoluted tubules of neighboring nephrons.

It is the Bowman's capsule that collects the fluids that pass from the blood into the nephron. These fluids are secreted from a ball of capillaries nestled within the Bowman's capsule known as the *glomerulus.* The ball is formed by the capillaries twisting around and looping back on themselves, giving the blood passing through plenty of time to secrete its load of watery waste and other materials into the Bowman's capsule. The glomerulus arises from a blood vessel called the *afferent vessel,* a tiny branch of the renal artery (see Figure 37.9). The capillaries of the glomerulus, having followed their twisted route within the Bowman's capsule, rejoin to form a small emerging *efferent vessel* that immediately divides again, this time into the *peritubular capillaries,* a network that surrounds the convoluted tubules and the collecting duct. Eventually these capillaries will merge into venules (little veins) that join to form the renal vein, through which filtered blood is returned to the general circulation of the body.

THE FORMATION OF URINE

The basic mechanism of **urine** formation involves three processes: filtration, reabsorption, and secretion (Table 37.3). Let's see how this happens. First,

TABLE 37.3	PROCESSES IN URINE FORMATION
Process	**Description**
Filtration	All but blood cells and large molecules (e.g., proteins) are filtered through glomerular capillaries into Bowman's capsule.
Reabsorption	"Valuable" materials (e.g., water, salts, nutrients) are selectively reabsorbed from the filtrate back into the blood.
Secretion	Specific chemicals are secreted from the blood into the filtrate.

the blood is filtered through the glomerulus into Bowman's capsule, pushed through pores in the glomerulus by the high blood pressure in the glomerulus (high because the efferent vessel leaving the glomerulus has a smaller diameter than the afferent vessel entering the glomerulus). The glomerulus simply works as a sieve, so this filtration is not a selective process; any constituents of the blood that can fit through the pores in the glomerulus enter Bowman's capsule in the same proportions as they are found in the blood. Some materials, however—such as blood cells and most proteins—are too large to pass through. Not everything that passes through into the Bowman's capsule is waste. The filtrate entering the capsule contains valuable materials, including water, salts, and nutrients that must be returned to the blood.

The retrieval largely takes place in the peritubular capillaries. It primarily happens as the filtrate moves through the convoluted tubules. The blood vessel leaving the glomerulus forms a network of capillaries, the peritubular capillaries, that surround the convoluted tubules and the loop of Henle through which the filtrate passes. Valuable materials, such as water and certain salts and ions are reabsorbed from the filtrate back into the capillaries, both through diffusion and by active transport (see Essay 37.2). Furthermore, certain waste materials are actively secreted from the capillaries into the convoluted tubules. The material now remaining in the tubules is urine. The process of urine formation is shown in Figure 37.10.

The Concentration of Urine

In a sense, the reclamation of water is accomplished by concentrating urine. As we will see, water conservation takes place largely in the collecting ducts under the influence of the antidiuretic hormone (ADH). Basically what happens is, a high salt concentration is set up in the tissue outside the collecting ducts and the loop of Henle. Water then leaves these passages because of the osmotic gradient between the fluid inside the passages and that outside, coupled with the collecting duct being highly permeable to water as a result of ADH. That water, then, enters the circulatory system and is restored to the body.

The high solute concentration in the tissues outside the nephron is increased as the proximal convoluted tubule actively pumps sodium out, which is followed by water and by chloride and other ions. The filtrate then enters the loop of Henle.

The principal role of the loop of Henle is to create that high salt concentration in its surroundings. You recall that the loops of Henle extend into the kidney medulla. So it is in the medulla that the highest salt concentration occurs. With a million nephrons participating, the salt concentration in the medulla becomes considerable. We learned earlier (Chapter 5) that water moves to places of its lower concentration—that is, to places where solute (e.g., salt) concentration is higher. Simply stated, where salt goes, water will follow. But how does the loop create this high salt concentration? The answer lies in its peculiar hairpin shape and some very capable cells in the ascending limb of the loop. As the filtrate (the fluid that entered Bowman's capsule from the glomerulus) makes its way through the proximal convoluted tubule and on through the loop, cells in the proximal convoluted tubule and the upper region of the ascending limb actively transport salt out of the filtrate into the fluid surrounding the loop. These salts are joined by urea that has diffused out of the collecting duct into the medullary tissues.

Water, then, leaves the filtrate through the descending loop of Henle, the distal convoluted tubule (under the influence of ADH), and through the collecting duct. In the distal convoluted tubule, sodium ions are actively pumped out, followed passively by chloride ions and water. In the collecting duct, water is reabsorbed (under the influence of ADH, which renders the duct permeable), and urea is released. What's left then, is urine, a waste fluid that contains urea, other solutes, and water. The result is that the tissues of the medulla are now bathed in water. This water, though, quickly reenters the body by moving into the peritubular capillaries.

At this point, then, let's review the path of the filtrate through the nephron, and add a few brief details to round out our account of how urine is formed. We can now say that water first enters the medulla through the wall of the descending limb of the loop of Henle, which is permeable to water. Here,

Essay 37.2 Pressure Gradients in the Kidney Tubule

You will be happy to learn that the mechanism of the kidney is a lot more complicated than it seems. You will also, of course, demand to know the details. You recall that in the human nephron, body fluids filter through the capillary wall into Bowman's capsule and from there through the proximal convoluted tubule (here straightened), the long loop of Henle (see Figure 37.10), the distal convoluted tubule, and then into the collecting duct. As the urine passes through the ascending loop of Henle (on the right), cells lining the loop pump salt out of the urine into the tissue surrounding the loop. In addition, urea moves out of the collecting ducts into the medulla, adding a second solute to the salty fluid there. It then enters the loop of Henle to return to the collecting ducts. This constant circulation means that the loop of Henle and the adjacent collecting duct are always bathed in a salt and urea solution, and it is this solution that sets up the osmotic gradient that withdraws water and solutes from the urine. The ascending loop is apparently impermeable to water since no water passes out with the salt. The descending limb of the loop of Henle and the collecting duct, however, is permeable to water, and water in the urine reaching it freely flows out in response to the high solute concentrations in the surrounding tissue. This water is picked up by the blood vessels that come from the glomerular region, and the hypertonic urine passes to the bladder. So there you are.

water is lost and salt is gained. As water diffuses out of the filtrate into the medulla in response to the osmotic gradient, the remaining filtrate becomes more concentrated. The filtrate then rounds the bend and up through the ascending limb of the loop of Henle. Interestingly, the lower region of the ascending limb is permeable to salt but not to water, and here the salt concentration of the filtrate within the loop is greater than that of the surrounding fluid—so salt diffuses out of the tubule into the medulla, where it enters the peritubular vessels and is carried back to the body. In the upper part of the ascending loop, salt is actively pumped out of the filtrate, thus increasing the solute concentration in the medullary tissue surrounding the loop. (The water cannot follow the salt because the walls of the ascending limb are impermeable to water.) The urine then reaches the collecting ducts.

Urea, now fairly concentrated in the collecting duct, diffuses out, thereby adding another solute to the inner medulla. Urea is, in fact, the solute responsible for most of the osmotic gradient in the inner region of the medulla. Some of the urea reenters the nephron at Henle's loop, returning to the collecting duct and thereby forming a cycle. As it continually moves out into the tissues of the medulla, however, it increases the osmotic gradient between the medullary tissues and the filtrate, causing yet more water to leave the filtrate and return to the body. So urea, the primary nitrogenous waste, also plays a role in water retention.

Just how much water is removed by the kidneys depends, of course, on the person's physiological state. If water intake has been excessive, the urine volume will be considerable and may be quite diluted. If a person is dehydrated, the urine volume may be small but highly concentrated. The amount of salt present in the urine also depends on the intake of sodium chloride in the diet. Controlling the reabsorption of water and salt, as we saw in the last chapter, involves the intervention of hormones.

Hormonal Control in the Nephron

Antidiuretic Hormone. One role of the hypothalamus is to monitor the osmotic state (the relative water and solute content) of the blood passing through its capillaries. If the blood is becoming hypertonic (thick), the hypothalamus secretes **antidiuretic hormone** (ADH) into the posterior pituitary, which then releases it into the bloodstream (Figure 37.11). ADH renders the walls of the collecting ducts more permeable to water, which flows freely from the ducts to reenter the bloodstream via the peritubular capillaries. As more water enters the blood-

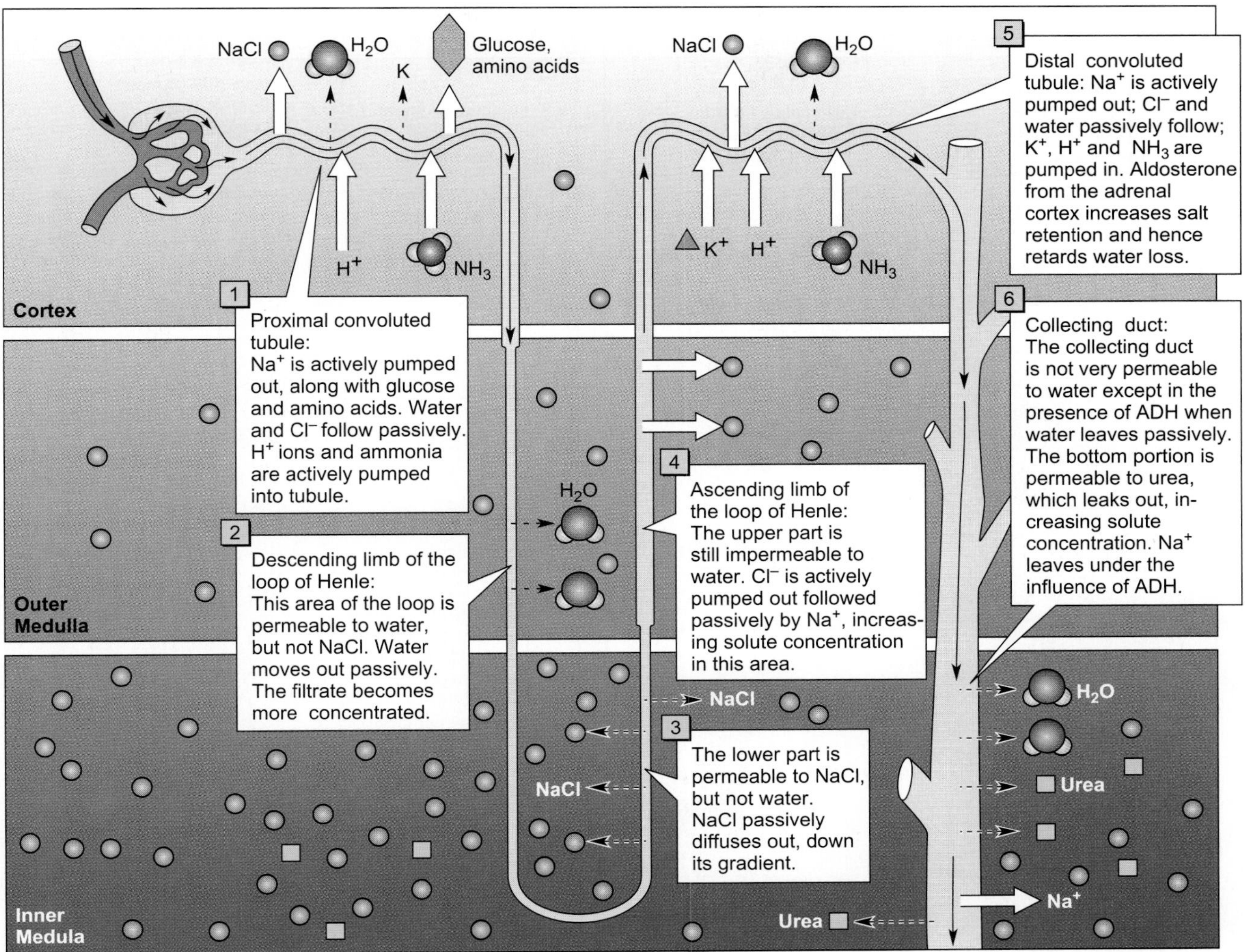

FIGURE 37.10 A DIAGRAMMATIC VIEW OF THE NEPHRON.
The biology of the nephron includes the various active and passive processes involved in filtration and reabsorption. Note the steep salt and urea gradient around the loop and collecting duct, important to the recovery of water.

stream, diluting and thinning the blood, less ADH is released, the tubules become less permeable, and the urine becomes more dilute.

Aldosterone. Another hormonal mechanism controls the retention of sodium chloride (Figure 37.11). In this case, the hormone is **aldosterone,** a mineralocorticoid (see Chapter 36). Aldosterone is released by the adrenal cortex. The link between sodium chloride retention and aldosterone release is indirect and not entirely understood.

We do know, however, that cells of the juxtaglomerular region of the nephron's distal tubule (see Figure 37.9) monitor minute changes in blood pressure in the nearby afferent vessel (the one carrying blood into the glomerulus). If blood pressure is below optimum, the juxtaglomerular region stimulates the release of the enzyme *renin* into blood passing through the afferent vessel. Renin has no effect on the nephron, but it does activate a blood protein that, on reaching the adrenal cortex, stimulates aldosterone secretion. Aldosterone's target cells are in the nephron's distal tubule and the collecting ducts. The target cells respond by speeding up their transport of sodium ions out of the tubule and back to the blood. As expected, water follows, and the additional water increases blood pressure enough to cause the juxtaglomerular region to reduce the release of renin, thereby reducing the release of aldosterone and the transport of sodium ions back to the blood.

This mechanism has two important aspects. First, it assures that the blood pressure in the kidney itself

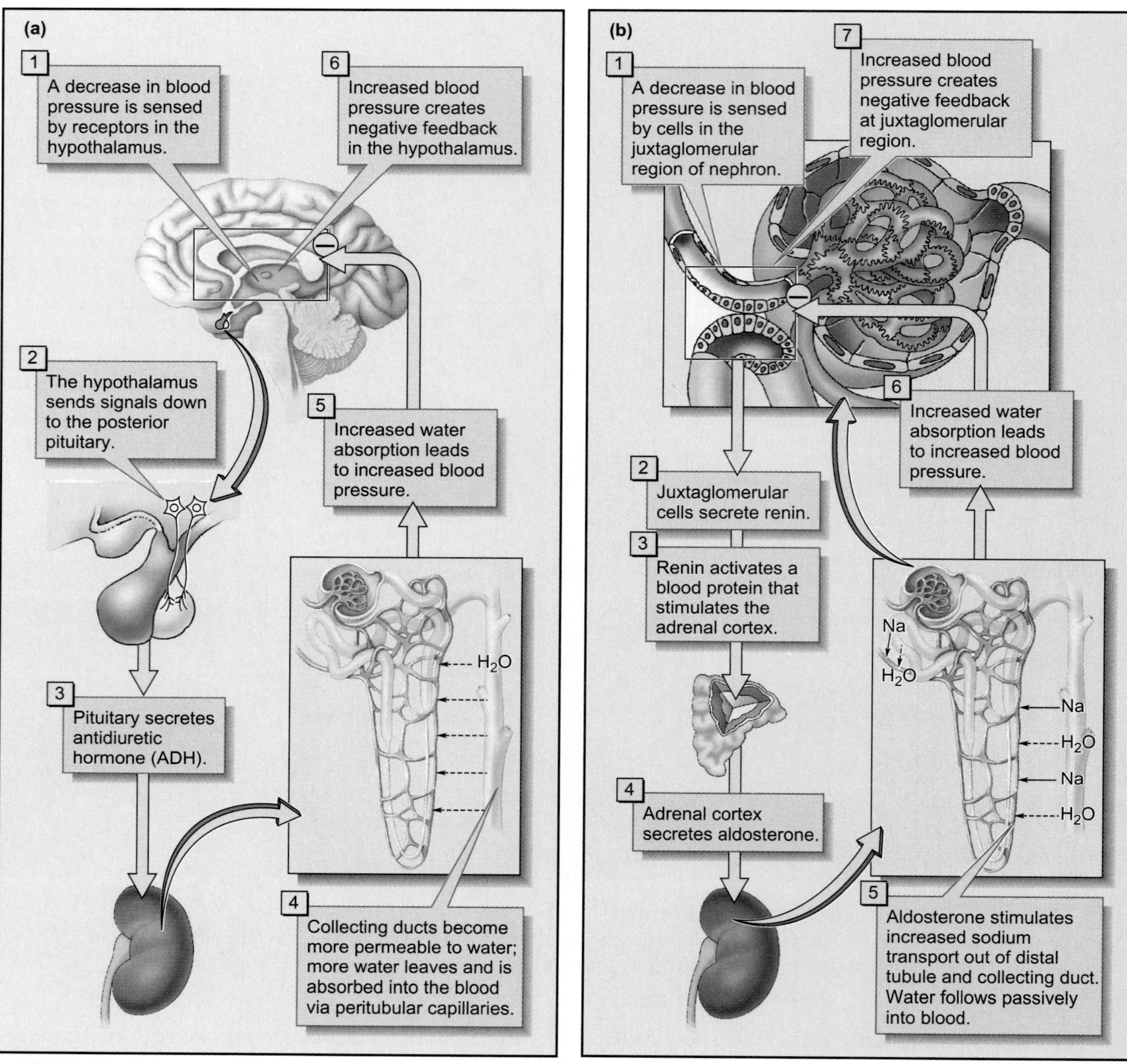

FIGURE 37.11 HORMONAL CONTROL OF WATER AND SALT.
(a) Water reabsorption is controlled in part by the hypothalamus, which secretes ADH via the posterior pituitary. Negative feedback is direct, since the osmoreceptors in the hypothalamus respond to the increased water content of the blood. **(b)** Salt reabsorption is influenced by the adrenal cortex hormone aldosterone, which is released when renin is present. Aldosterone increases the active transport of sodium in the distal tubule, which forms a negative feedback loop that slows renin secretion.

is great enough to maintain an efficient filtration phase. Second, the aldosterone-related increase in sodium transport also means an increase in the tubular secretion of potassium ions. The active transport mechanism involved is similar to the familiar sodium-potassium ion-exchange pump of blood and nerve cells.

Atrial Natriuretic Factor. When blood pressure rises, the heart secretes a hormone, **atrial natriuretic factor** (ANF), that promotes salt and water excretion by the kidneys. It is thought to act by blocking ADH, renin, and aldosterone secretion, and by directly inhibiting the ability of the distal tubule to reabsorb salt and water.

PERSPECTIVES

In this chapter, we have seen only a few of the many delicate, interacting, and highly coordinated mechanisms that keep the body's internal environment within the extremely precise limits critical to life. Remember, we live in what is essentially a disruptive environment. To remain organized and functional in the face of such potential disruption requires an uphill battle against a variety of forces. That battle is best fought under optimal conditions, which is what homeostasis is all about.

In the next several chapters, we will take a look at some of the processes that relatively constant internal conditions permit, beginning with digestion.

SUMMARY

MECHANISMS OF HOMEOSTASIS

Homeostasis involves maintaining stability through ongoing adjustments to change. Such responses can be physiological or behavioral.

THERMOREGULATION

Thermoregulation is the ability to keep body temperature within a certain range.

Excessively low or high temperatures can have damaging effects on cells. As heat is lost, chemical activity slows, setting up a positive feedback situation, which cools the body further. Some animals simply cool down as surrounding temperatures fall, eventually becoming immobile. Others have physiological or behavioral means of maintaining temperature.

Endotherms utilize metabolism to maintain body heat, whereas ectotherms utilize the heat of the surroundings.

Homeotherms maintain relatively constant, generally warm body temperatures. Poikilotherms permit their temperatures to change with surroundings. Homeotherms remain in a constant state of activeness, but at a cost of energy. Poikilotherms have no such energy expenditure, but are often inactive. Many animals utilize both means. Heterotherms may be homeothermic for a time and then switch to poikilothermy.

Humans utilize physiological mechanisms to generate and retain heat, to slow heat production, and to speed its escape. Body heat is monitored by the hypothalamus. As body temperature falls, the hypothalamus stimulates activity in the autonomic nervous system and the endocrine system. Circulatory changes shunt blood away from the skin and extremities. Epinephrine is released from the adrenal medulla and TSH from the anterior pituitary. As a result, metabolic heat output is increased through glucose release, and cell respiration speeds up. Shivering also increases heat production. Some animals, including human infants, use metabolic activity in brown fat to generate heat in response to prolonged cold. Highly vascularized air passages help avoid cooling the lungs by warming incoming cold air. Cooling is done by shunting more blood to the skin and generating sweat. Alternative surface arrangements and deeper countercurrent arrangements in blood vessels help in regulating body heat.

OSMOREGULATION AND EXCRETION

In humans, osmoregulation—the precise regulation of body water and ion content—is closely associated with excretion, the removal of metabolic wastes.

All animals must rid themselves of metabolic wastes, such as those containing nitrogen from the breakdown of protein. Different nitrogen-containing wastes are produced in species with different water conservation needs. Ammonia, a common waste product among fish, is poisonous and requires a great deal of water to flush it from the body. Urea, a soluble common waste product among mammals, is less toxic and can be excreted with less water loss. Insoluble uric acid is a common waste product among insects and birds, species with a particular need to conserve water.

Humans must conserve water and certain ions as they excrete wastes. Urea is the primary nitrogenous waste of humans, passing from the body in urine.

The excretory system includes the kidneys, ureters, urinary bladder, and urethra. The kidneys remove excess water and salts and also urea. Each kidney contains about a million filtering units, called nephrons.

Three processes are involved in urine formation: filtration, reabsorption, and secretion.

The parts of nephrons and their functions are as follows:

Bowman's capsule and the glomerulus: force filtration from the glomerulus to the capsule; create the crude filtrate.
Peritubular capillaries: surround nephron and receive reabsorption products.
Proximal convoluted tubule: begins reabsorption; through active transport and diffusion, returns water and other critical substances to capillaries; through tubular secretion, actively transports a variety of ions into the urine.
Loop of Henle: creates a dense salt (solute) concentration in the kidney outer medulla, prompting the exodus of more water from the filtrate. (Urea from the collecting ducts increases the solute concentration.)
Distal convoluted tubule: through further active and passive transport, returns more critical substances to capillaries; some waste materials are actively secreted from capillaries into the tubules.

A final adjustment of water and ions in the filtrate is made in the collecting duct as the urine forms.

Urine is concentrated as water is conserved. Water diffuses out of the filtrate in response to an osmotic gradient, moves through the medulla, and enters the surrounding capillaries. The osmotic gradient is created primarily by salt in the outer medulla and by urea in the inner medulla. The salt diffuses out of the lower ascending loop of Henle and is actively pumped out of the upper region of the ascending loop. The urea diffuses out of the collecting duct, enters the loop of Henle, and cycles back to the collecting duct. Water diffuses out of the descending loop of Henle and, if hormones permit, out of the collecting duct.

ADH is released through the posterior pituitary when the blood's water content is low. It alters permeability in the collecting duct, permitting water to reenter the bloodstream.

Aldosterone is released by the adrenal cortex when cells of the juxtaglomerular complex sense solute-related blood pressure changes. Aldosterone increases salt retention, which leads to increased water reabsorption and a favorable change in blood pressure. Low blood pressure triggers the release of renin, which stimulates aldosterone secretion.

Both hormones are regulated by negative feedback loops that influence the hypothalamus. Atrial natriuretic factor (ANF), secreted by the heart, promotes salt and water excretion by the kidneys. It acts by blocking ADH, renin, and aldosterone secretion and by inhibiting the reabsorption of salt in the distal tubule.

Key Terms

homeostasis • **633**
negative feedback loop • **634**
positive feedback • **634**
thermoregulation • **634**
ectothermy • **635**
endothermy • **635**
homeotherm • **635**
poikilotherm • **635**
heterotherm • **635**
osmoregulation • **641**
excretion • **641**
nephron • **643**
urine • **644**
antidiuretic hormone (ADH) • **646**
aldosterone • **647**
atrial atriuretic factor (ANF) • **648**

Review Questions

1. What is homeostasis?
2. Briefly discuss two kinds of homeostatic responses available to animals to conserve heat.
3. Explain the "paradoxical situation" arising when the animal body cools.
4. Distinguish between maintaining body heat through endothermy and ectothermy.
5. Make a general distinction between homeothermy and poikilothermy. In general, what animal groups would you expect to find in each category? List an apparent drawback to each.
6. Which of the four descriptive terms (homeothermy, poikilothermy, endothermy, ectothermy) is suitable when describing thermoregulation in humans? Support your answer.
7. What structure in humans acts as a central thermostat? What does it measure?
8. List several responses the human body makes to a lower than optimal body temperature.
9. Explain how certain arrangements of arteries and veins result in a countercurrent exchange. How does this affect body temperature?
10. What is osmoregulation, and why is it important to the individual?
11. What is the source of nitrogen waste in humans?
12. Prepare a large outline drawing of the nephron, labeling Bowman's capsule, proximal convoluted tubule, loop of Henle, and distal convoluted tubule. Finally, add a collecting duct.
13. Referring to the diagram you just drew, add in the following blood vessels: branch of the afferent vessel, glomerulus, efferent vessel, peritubular capillaries. Turning to the tubule, indicate with an *F* where the filtrate is formed, with an *Re* where reabsorption occurs, and with a *TS* where urine is formed.
14. Make a list of the nephron's parts, and after each, indicate what forces are at work, what substances (if

any) move into the nephron, and what substances (if any) move out.

15. What is the general purpose of the long hairpin-shaped loop of Henle? What familiar mechanism is suggested by the ongoing movement of substances from one part of the loop to the other?

16. Under what conditions does the hypothalamus cause ADH to be secreted? Where are its target cells? What is their response to ADH?

17. Describe the events leading up to aldosterone secretion. What does the juxtaglomerular complex actually measure?

18. If aldosterone causes salt retention to increase, how might this bring the blood pressure up to the optimum? What effect does this have on aldosterone secretion?

19. What causes ANF secretion? What are the effects of ANF?

38

Digestion

AS YOU SIT THERE, A LONG, WINDING TUBE RUNS THROUGH YOUR BODY. *That tube is technically, but unpoetically, referred to as the gut—and it is here that hamburgers (and soyburgers) are dealt with. A major role of that tube is to break down the molecules of whatever you have eaten, and to pass those smaller molecules across the membrane of the gut and into the bloodstream where they can be carried to where they are needed. The process can be quite complex—but fortunately you don't have to think about it, because the process was developed over long eons of life on the planet, and today it is normally handled without the conscious intervention that would probably ruin your meal.*

Within the thin veil of precious gases and water in which life exists on this planet, a deadly pageant constantly unfolds. The players are living things, and their roles involve the search for energy—the energy held in the molecules of food. Some species in the pageant play out their lives quietly making their food from the energy of sunlight, and a few even use the searing, mineral-laden water that boils from the ocean floor. And then some species seek out others—the predator or parasite availing itself of the carefully tended molecules in the bodies of other living things.

We humans are in this last group, and the fact that any of us are alive means that something else has died. As heterotrophs, we require the molecules held in other organisms' bodies. Some members of our species try to feed from low on the food chain and choose plants as victims. Others who eat flesh may try to distance themselves from the carnage, and sell real estate or become teachers, allowing others to do the killing, and trying not to think about it. And then there are those who think a great deal about it, and—whether using a curare-tipped arrow or a Beretta A303 with a Moneymaker barrel—they enjoy the hunt and plan much of their lives around it. The point is, no matter how we acquire those molecules, acquire them we must. When the dove, the calf, and the lettuce lived, they too needed energy-laden molecules to continue their existence. But their molecules now grace our bodies; they are gone, and our time will come, but life goes on. Here, then, we will look at how animals handle the molecules they acquire from other living things—how they break down and rearrange those molecules to drain them of their energy—energy that can then be used to promote the life of the animals on the planet. The problem for animals is twofold. First, they must find the proper kind of food. This addresses the problem of nutrition. Second, they must break it down efficiently so its components can be reorganized to fit their own needs. Essentially, this is what digestion is all about.

BASIC DIGESTIVE PROCESSES

Digestion is the process by which food molecules are broken down into their component parts, so that they can be absorbed by the body. In virtually all cases the chemical breakdown of bulk foods during digestion requires specific hydrolytic digestive enzymes (Table 38.1). As we have seen in Chapter 6, hydrolytic enzymes use molecules of water to break the chemical bonds linking the molecular building blocks of carbohydrates, fats, and proteins.

Often, but not always, this chemical digestion of food is assisted by mechanical action, in which the food is physically broken into smaller pieces. Some animals, including humans, accomplish this through chewing and by the churning that occurs in many digestive structures. This mechanical action simply creates more surface area by breaking the food apart, making it easier for enzymes to do their job.

TABLE 38.1 Digestive Enzymes and Their Functions

Sources and Enzyme	Substrate	Product
Salivary Glands		
Salivary amylase	Starch	Glucose
Stomach Lining		
Pepsin	Protein	Polypeptides
Renin	Casein (milk protein)	Insoluble curd
Gastric lipase	Triglyceride	Fatty acids + glycerol
Pancreas		
Proteolytic enzymes	Protein	Peptides
Ribonuclease	RNA	Nucleotides
Deoxyribonuclease	DNA	Deoxynucleotides
Pancreatic amylase	Starch	Glucose
Pancreatic lipase	Triglyceride	Fatty acids + glycerol
Carboxypeptidase	C-terminal bond	Shorter peptide and one free amino acid
Intestinal Lining		
Peptidase	Peptides	Amino acids
Nuclease	Nucleotide	Pentose + nitrogen
Maltase	Maltose	Glucose
Sucrase	Sucrose	Glucose + fructose
Lactase	Lactose	Glucose + galactose

Digestion in Invertebrates

There are several basic digestive schemes found among invertebrates (Figure 38.1). When the sponge, a simple filter feeder, moves water over its surface by the movement of the flapella of collar cells, and captures a morsel of food, the food is phagocytized (*phago,* eat; *cyte,* cell; therefore—cell-eating), entering the cell cytoplasm contained in a food vacuole. Lysozymes containing powerful digestive enzymes fuse with the food vacuole, and the food is broken down into simpler molecules. Wandering ameboid cells then distribute the digested food to other cells throughout the sponge's simple body (Figure 31.8a). The digestion within food vacuoles, called **intracellular digestion** (*intra,* within) is also common among protists, such as the ameba.

Other animals have evolved a primitive digestive surface, as we see in today's cnidarians (such as jellyfish and hydra) and flatworms (Figure 38.1b). Here food is taken into a saclike gastrovascular cavity (which is highly branched in the flatworm). While some food is captured by phagocytic cells and digested intracellularly, the rest is digested through **extracellular digestion** (*extra,* outside), outside the cell by enzymes released into the digestive cavity. The breakdown products of the food are then absorbed into the animal's body. This type of digestive system, essentially saclike, has a single opening through which food enters and undigested material leaves, and is called an **incomplete digestive system.** There is a certain inefficiency in the saclike digestive cavity because the two-way traffic allows for little specialization within the sac.

The development of the longitudinal body plan made a tubular gut possible. With a **complete digestive system**—a tubular gut that has an entrance (mouth) at one end and an exit (anus) at the other—digestion becomes not only decidedly more civilized but also more efficient. With this system, food can be handled serially, so that while some food is being digested, other food can enter the tract and begin its sequence. Here, enzymes are secreted into the digestive tract, and the food is broken down through extracellular digestion.

We find a highly specialized tubelike gut in many invertebrates, including the nematodes, mollusks, annelids, and arthropods. In earthworms (annelids), for example, various regions of the gut have different roles: food enters through the mouth, passes through the esophagus, is stored in the crop, and ground in the gizzard before chemical digestion begins. The grinding is an important advancement because it increases the surface area available to digestive enzymes, much as chewing does (Figure 38.1c).

In most invertebrates, digestion is extracellular, with ingested food exposed to powerful enzymes as it moves along an absorptive gut. As we will see, things are much the same in vertebrates.

Digestion in Vertebrates

The vertebrate digestive system follows a tubular plan that begins at the mouth cavity, in which teeth may be present to tear or grind food as it is manipulated by the tongue (although some species, such as wolves and sharks, simply gulp chunks of food). The food passes into a pharynx at the rear of the mouth and then through a muscular esophagus that squeezes it into a temporary storage and digestive organ—the stomach. The stomach empties into the small intestine, which is the major organ of digestion. From here, most products of digestion pass through the gut and into the bloodstream. Next, in the large intestine, excess water is absorbed into the blood, along with some vitamins. The residue of waste is stored in the rectum until it is emptied through the anus.

Any variation from the usual vertebrate gut, as you might expect, would be an evolutionary specialization for some particular diet. In no other group has the stomach become so extensive and specialized as in the *ruminants,* the cud-chewers—such as cattle, deer, giraffes, antelope, and buffalo. They are so specialized because they eat cellulose-containing plants, but they cannot digest cellulose. In fact, few animals can. (The beta linkages between the glucose units found in cellulose, you may recall, are difficult to break enzymatically). The ruminants, then, solve the problem by harboring in their gut immense numbers of protozoa and bacteria that *can* digest cellulose (Figure 38.2). (When these tiny partners die, their bodies are digested by the host.)

Some other mammals, including primates, horses, rabbits, ground squirrels, elephants, coneys (what's a coney?), and marsupials handle the problem of digesting plants a bit differently. Horses have an extraordinarily well-developed colon that can absorb nutrients across its walls. Rabbits, on the other hand, have cellulose-digesting bacteria in their colons, but their colons cannot absorb nutrients. Instead, they eat their green pellets and pass them through a second time. Many animals have a blind sac where the small intestine joins the large intestine. This is the *cecum,* a pouch that not only serves as a temporary storage area for digested material before it wends its way into the large intestine, but also as a kind of fermenting vat. The cecum is seething with bacteria that are able to break down many kinds of plant molecules that have managed to pass through the small

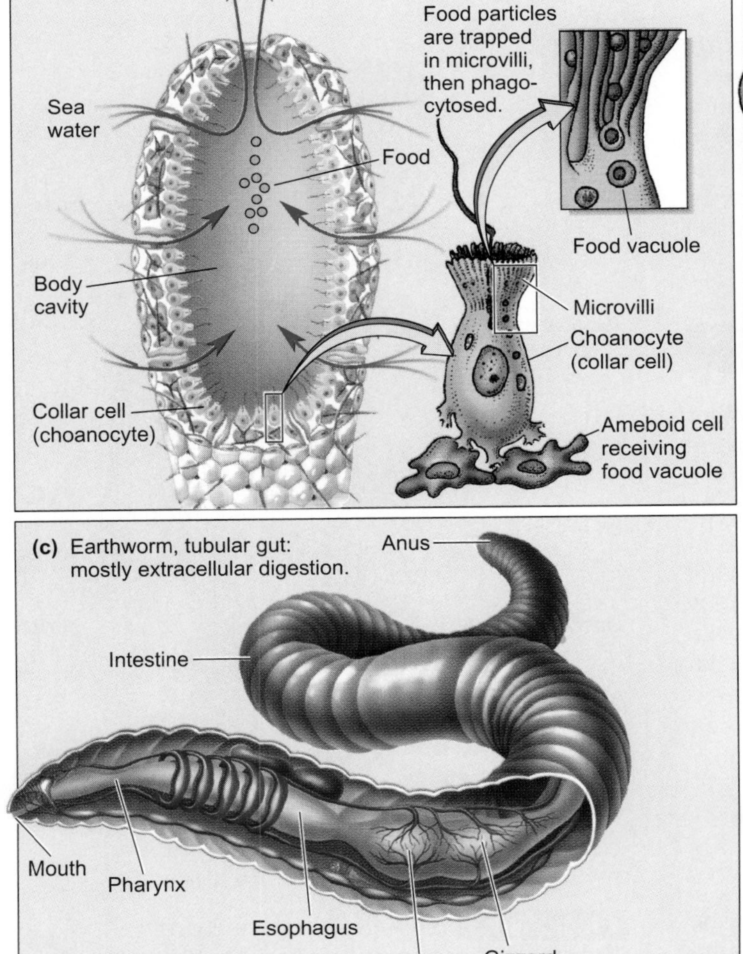

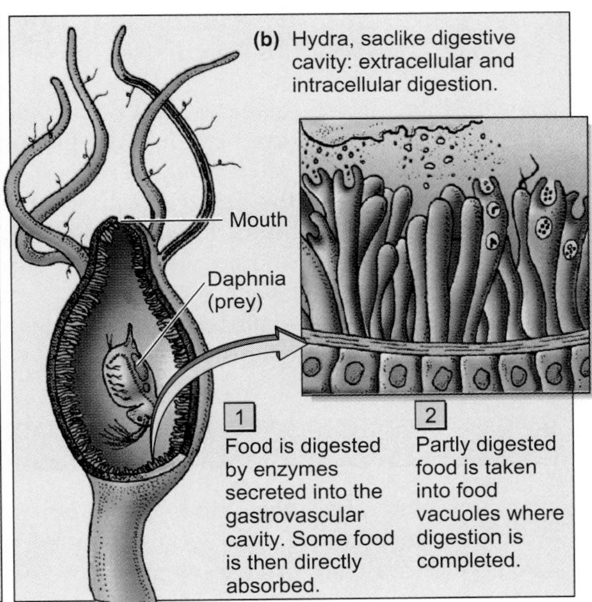

(b) Hydra, saclike digestive cavity: extracellular and intracellular digestion.
Mouth
Daphnia (prey)
1
Food is digested by enzymes secreted into the gastrovascular cavity. Some food is then directly absorbed.
2
Partly digested food is taken into food vacuoles where digestion is completed.

FIGURE 38.1 INVERTEBRATE DIGESTION PLANS. Invertebrates reveal several kinds of digestive arrangements: In the sponge (**a**), most digestion is intracellular. In the coelenterate (**b**), digestion is largely extracellular, but still simple. In the earthworm (**c**), the food passes through a gut with specialized regions.

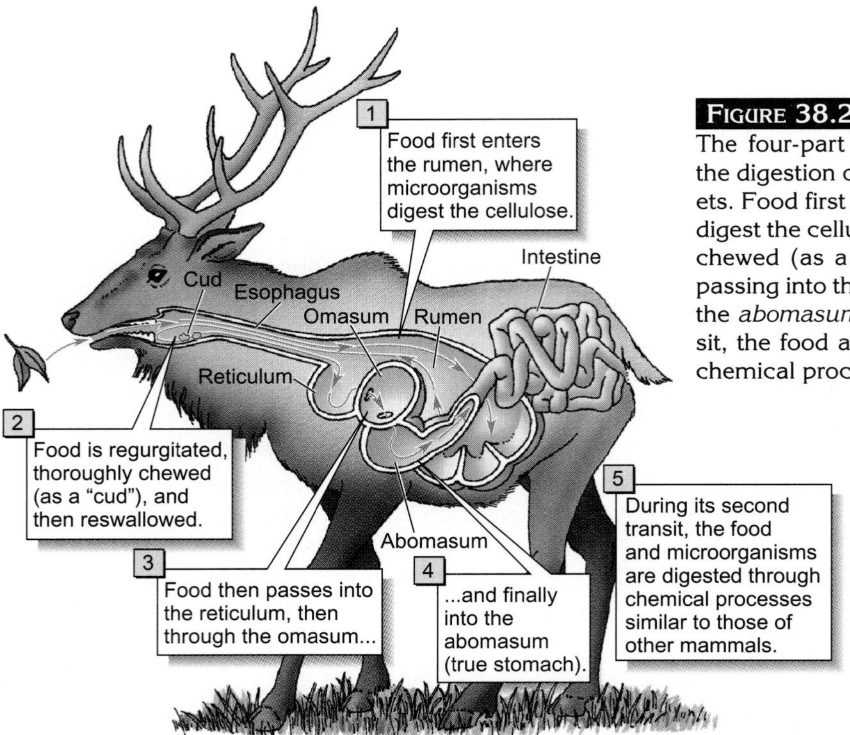

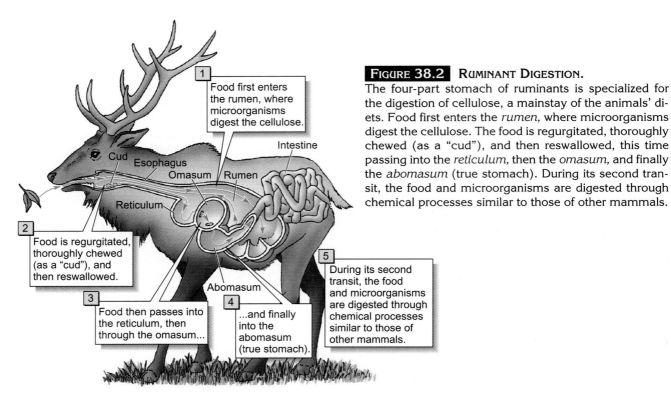

FIGURE 38.2 RUMINANT DIGESTION. The four-part stomach of ruminants is specialized for the digestion of cellulose, a mainstay of the animals' diets. Food first enters the *rumen,* where microorganisms digest the cellulose. The food is regurgitated, thoroughly chewed (as a "cud"), and then reswallowed, this time passing into the *reticulum,* then the *omasum,* and finally the *abomasum* (true stomach). During its second transit, the food and microorganisms are digested through chemical processes similar to those of other mammals.

intestine intact. The human gut has a rather small cecum, from which extends a small, fingerlike pouch, the *vermiform appendix* (*vermiform,* wormlike), which (partly because of all the bacterial activity there) is occasionally the site of infection (see Figure 38.3). If an infected appendix should weaken and rupture, those bacteria are allowed into the body cavity, which is normally free of bacteria. Without treatment the ensuing infection will result in death.

THE HUMAN DIGESTIVE SYSTEM

As mammalian digestive systems go, ours is rather ordinary, which allows us to serve as a good representative of our class. As we see in Figure 38.3, the human digestive system includes the mouth, esophagus, stomach, small intestine, large intestine, and accessory organs (Figure 38.3; Table 38.2). Lining the body cavity and the organs below the diaphragm is a membranous *peritoneum.* Loose folds of the peritoneum, called the *mesentery,* hold the organs in place and permit the movements associated with digestion. The mesentery carries blood vessels, nerves, and lymphatic vessels—all important to digestion.

THE MOUTH, PHARYNX, AND ESOPHAGUS

The digestive structures of the mouth include the lips, teeth, tongue, and salivary glands. In case you're wondering what the lips have to do with eating, try

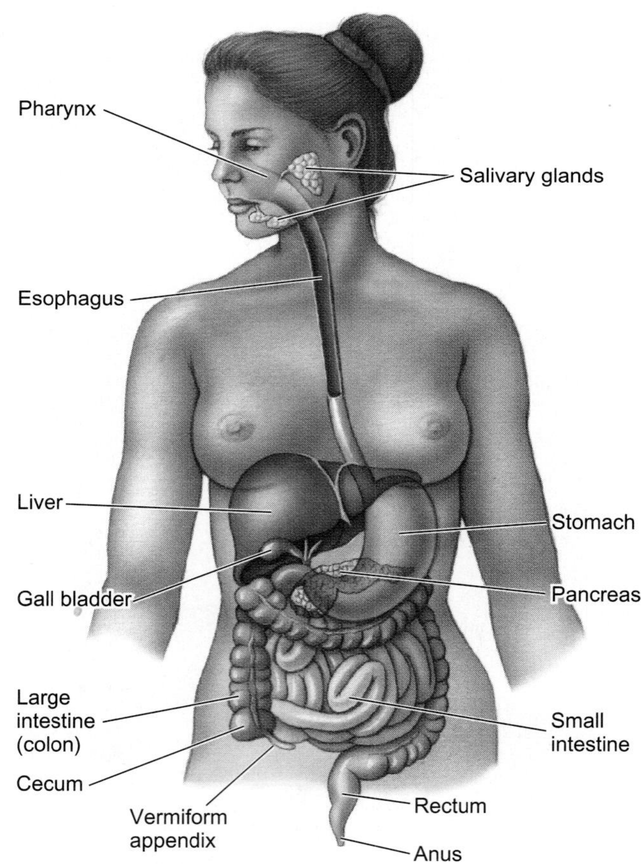

FIGURE 38.3 HUMAN DIGESTIVE SYSTEM. The human digestive system is typical of that of many mammals and is not particularly specialized.

TABLE 38.2	THE DIGESTIVE SYSTEM
Structure	**Function**
Mouth	Mechanically breaks down food during chewing; saliva from salivary glands moistens and binds food and begins the breakdown of starch
Pharynx	Moves food from mouth to esophagus
Esophagus	Chute between mouth and stomach
Stomach	Short-term storage; churns food mixing it with enzymes and HCl and liquefying it; begins digestion of proteins and some fat
Small intestine	Completes digestion of all nutrients using enzymes produced in small intestine and in the pancreas; absorbs digestive products
Large intestine	Absorbs water; forms feces; houses vitamin-producing bacteria
Accessory Organs	
Liver	Secretes bile for keeping fat in small droplets
Gall bladder	Stores bile before releasing it to the small intestine
Pancreas	Produces digestive enzymes that are released into the small intestine; secretes sodium bicarbonate to neutralize the acidic secretions of the stomach

eating and swallowing with your lips open —but do it somewhere else.

While it may not at first be evident, the tongue is also important in eating. In addition to its role in moving food into position for chewing, swallowing, and sorting out fish bones, it constantly monitors the texture and chemistry of foods. This is a valuable role, since it tells us about what's in our mouth in time to prevent us from swallowing something we shouldn't. In addition, the tongue can distinguish certain chemicals by specialized chemoreceptors located in the taste buds. These receptors can distinguish four basic tastes: salty, sour, sweet, and bitter—and perhaps umami (see Chapter 35).

As was mentioned, teeth are important in mechanical digestion. Adult humans usually have 32 teeth. Some have more, and others may have fewer due to decay or pugilistic endeavors. Teeth may be specialized for chewing, as we see in the pavement teeth of impala, or for tearing meat, as we see in lions (Figure 38.4). Human teeth are adapted for a variety of tasks. The front teeth, incisors, are chisel-shaped for cutting food. The second group, the canines, are useful for tearing food (betraying our carnivorous ancestry). Other carnivores use canines for capturing and killing prey. The last two groups of teeth, the premolars and molars, are specialized for grinding. They are large and fairly flat and are useful for breaking down the fibers and cellulose walls of plants, and for opening things.

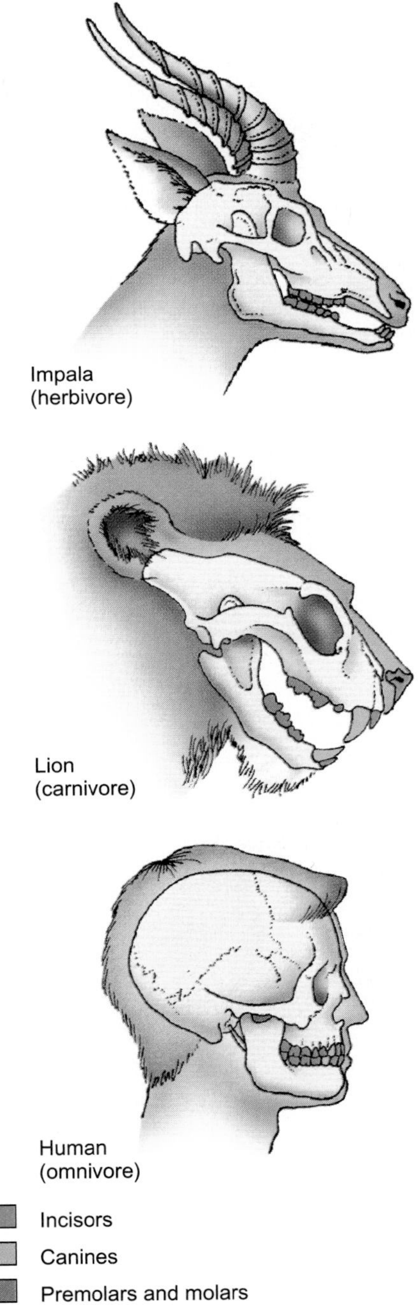

FIGURE 38.4 COMPARING TEETH. Impala teeth are wide and flattened ("pavement teeth") and are adapted to breaking down such tough plant material as grass, which the front incisors nip off. Lions, on the other hand, must be able to tear apart their food, and because they do not chew, even their jaw teeth are sharp and pointed. Human teeth include versions of these types of teeth.

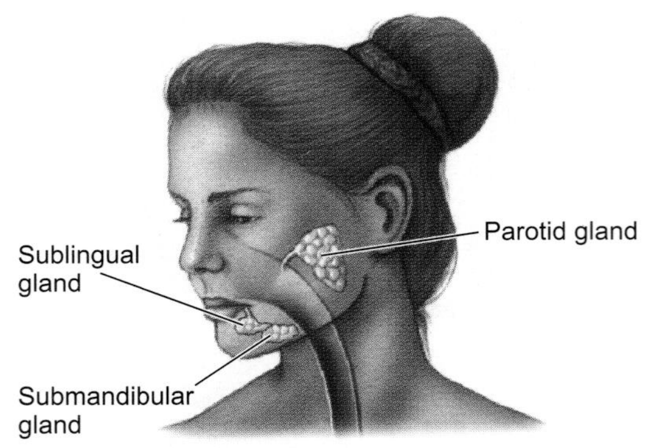

FIGURE 38.5 **SALIVARY GLANDS.**
Three pairs of salivary glands—the parotids, sublinguals, and submandibulars—secrete mucus and salivary amylase.

Saliva is the fluid in the mouth produced by three pairs of salivary glands: the *parotid, sublingual,* and *submandibular glands* (Figure 38.5). These glands secrete saliva through ducts that empty into the mouth. Saliva is about 95% water; the remaining 5% includes various ions, a slippery substance called *mucin,* (a lubricating glycoprotein), and an enzyme (salivary amylase) that begins starch digestion. Salivary amylase is probably less important in digestion than in oral hygiene, since it helps break down starchy food particles remaining in the mouth, depriving bacteria of a food source.

The *pharynx,* lying posterior to the mouth cavity, joins with the nasal cavity to form a common passageway. The pressure-equalizing *eustachian tubes* of the middle ear also open into the pharynx. (This is why a sore throat can lead to an ear infection.) Just below the root of the tongue the pharynx divides, marking the anterior boundary of the *larynx* and *esophagus.* The larynx is a complex, muscular structure containing the flattened cords of the voice box and several ringlike cartilages, along with the flaplike *epiglottis.* During swallowing, the larynx rises, and the epiglottis presses against it, directing food into the esophagus (Figure 38.6). If this fails, food will enter the larynx, producing violent choking or coughing spasms. In humans, the passageways for food and air are disconcertingly close together, and many deaths each year are caused by food entering the airway (see Essay 38.1).

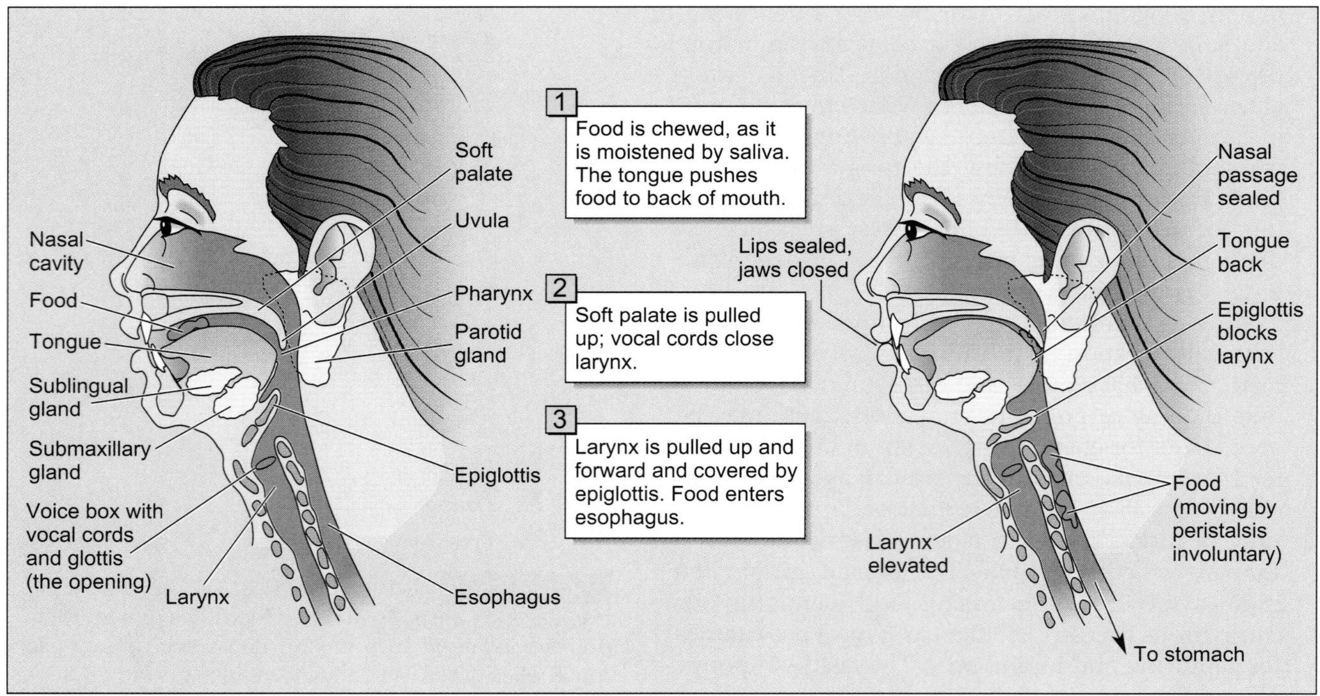

FIGURE 38.6
The positions of the important structures of swallowing. The act is initially voluntary and then becomes involuntary or reflexive.

ESSAY 38.1 The Heimlich Maneuver

Human evolution, unfortunately, has resulted in the openings of the trachea and esophagus being closer together than is the case in many other species. The result is a marked propensity for food to "go down the wrong way"—that is, for an occasional food particle to move into the air passages. This happens when the epiglottis is not completely closed during swallowing. In this case, the glottis spasmodically contracts and causes choking. In some cases, the food is simply coughed up, but in extreme cases the victim is completely unable to breathe. More than eight Americans die this way each day.

However, a rather simple action can save many of these lives. It is called the Heimlich maneuver, and it works as follows: (1) Stand behind the victim. (2) Wrap your arms around the waist. (3) Make a fist with one hand, knuckles directed upward and inward against the victim. (4) Place the knuckles between the rib cage and the navel. (5) Cup the other hand over the fist. (6) Quickly and sharply press inward and upward against the victim's abdomen. (7) Repeat if necessary.

If the victim is lying on his back, kneel, place your knees by his hips, and with the heels of your hands (one on top of the other) press upward on the abdomen with a quick thrust. (Repeat if necessary.) There are risks to the maneuver, such as damage to the diaphragm, but most people who are choking seem willing to take the risks.

The *esophagus* lubricates food and directs it to the stomach. Its upper region contains voluntary striated muscle; the rest, containing circular and longitudinal layers of smooth muscle, is involuntary. So, swallowing begins as a voluntary act but continues as an involuntary one.

Food is moved through the esophagus and the remainder of the digestive tract by **peristalsis**—coordinated, wavelike contractions of the smooth muscle in the wall of the digestive tract (Figure 38.7). The mass of food stretches the wall of the digestive tube slightly and triggers the contraction of circular muscles immediately behind it. The contraction pushes the food forward slightly, and the resulting expansion of the digestive tube wall triggers muscular contraction behind the food mass. Thus, a wave of contraction is created that pushes the food along the digestive tract.

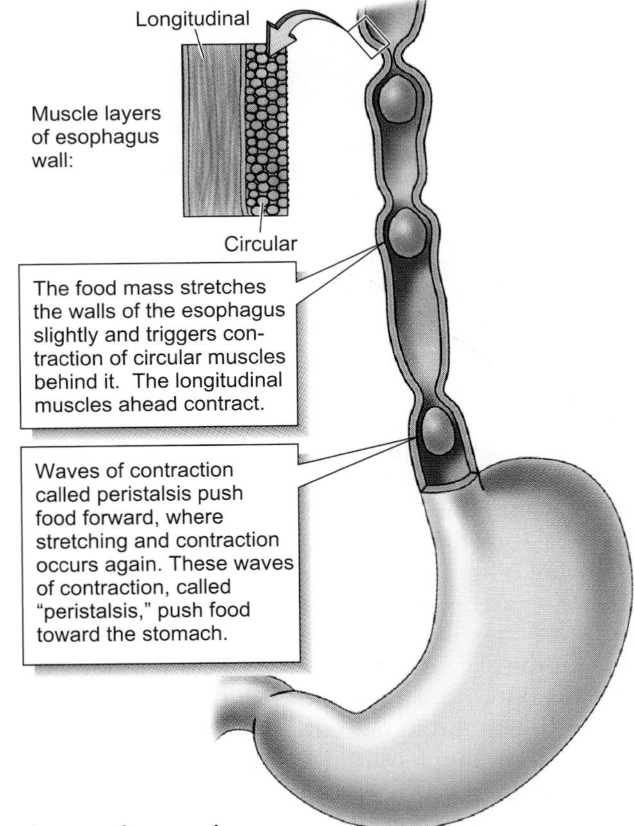

Longitudinal
Muscle layers of esophagus wall:
Circular
The food mass stretches the walls of the esophagus slightly and triggers contraction of circular muscles behind it. The longitudinal muscles ahead contract.
Waves of contraction called peristalsis push food forward, where stretching and contraction occurs again. These waves of contraction, called "peristalsis," push food toward the stomach.

FIGURE 38.7 PERISTALSIS.
As the food makes its way down the esophagus, the tube relaxes in front of it and is squeezed along by rings of peristaltic constrictions.

THE STOMACH

The human **stomach**—a muscular, J-shaped sac—has a complex glandular inner lining and is surrounded by three crisscross layers of smooth muscle

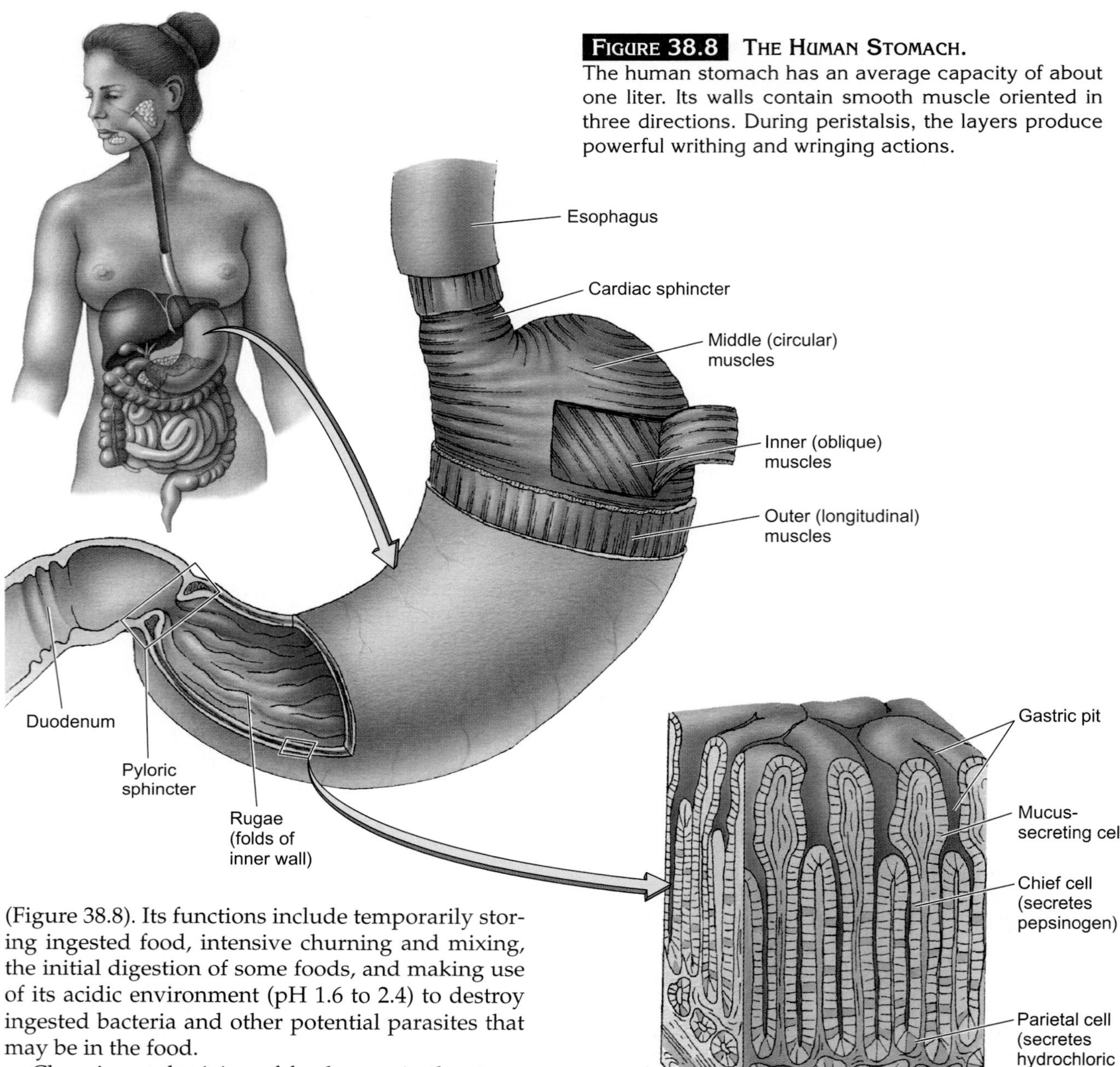

FIGURE 38.8 THE HUMAN STOMACH. The human stomach has an average capacity of about one liter. Its walls contain smooth muscle oriented in three directions. During peristalsis, the layers produce powerful writhing and wringing actions.

(Figure 38.8). Its functions include temporarily storing ingested food, intensive churning and mixing, the initial digestion of some foods, and making use of its acidic environment (pH 1.6 to 2.4) to destroy ingested bacteria and other potential parasites that may be in the food.

Churning and mixing of food occur in the stomach through a wringing, squeezing action made possible by the arrangement of its smooth muscle layers. This is sometimes accompanied by an intriguing growling sound, clearly audible when meeting your fiancé's parents. Ringlike *cardiac* and *pyloric sphincters* seal the stomach during churning, preventing the escape of food until it is ready to be released into the small intestine. (When the acidic contents of the stomach slip through the cardiac sphincter into the esophagus, the familiar condition called "heartburn" is produced.)

Gastric juices are secreted by specialized glands and cells of the stomach lining. *Chief cells* secrete the protein *pepsinogen,* and *parietal cells* secrete *hydrochloric acid* (HCl). The acid is required to activate the pepsinogen, which then forms the digestive enzyme *pepsin* that begins the chemical breakdown of proteins. Other glands secrete *rennin,* an enzyme that helps young mammals to digest milk, and *gastric lipase,* a fat-splitting enzyme that has a limited role in an adult's stomach.

The question arises, why don't we digest our own stomachs? One reason is that a layer of insoluble mucin, also found in saliva, coats the stomach lining.

THE SMALL INTESTINE

The **small intestine** is a convoluted tube about 6 m (20 ft) long, in which digestion is completed and through which most nutrient products enter the bloodstream (Figure 38.9). The longitudinal and circular muscle layers give the intestine a wide range of movement, including the peristalsis that moves the food along. Its inner surface is covered with tiny, fingerlike projections called **villi,** which increase the surface area of the intestinal lining. Furthermore, the surface area of each villus is increased by about 3,000 tiny projections called **microvilli** (Figure 38.9). The absorptive surface of the small intestine has been estimated at about 700 m^2, or about the floor area of four or five 3-bedroom houses. This great surface area is due to the coiling of the small intestine, the internal folding, and the presence of the villi and microvilli.

Within each villus is a minute lymph vessel, called a *lacteal,* which is surrounded by a network of blood capillaries. The digested products of certain fats move directly into the rather permeable lymph vessel, and on into the bloodstream, while the products of protein and starch digestion move into the blood capillaries. In fact, after a meal high in fat, the blood may take on a startling milky appearance.

The first ten inches or so of the small intestine comprise the *duodenum,* followed by the *jejunum* and *ileum.* Most digestion takes place in the duodenum, and it is here that digestive fluids from the pancreas empty into the digestive tract. Also, as we will see, bile from the liver enters the digestive tract here as well.

THE PANCREAS

The human **pancreas** is a long, glandular organ nestled in the mesentery between the stomach and the duodenum (Figure 38.10). It secretes sodium bicarbonate, which neutralizes the acid that enters the duodenum with the stomach contents. In addition, the pancreas secretes a battery of digestive enzymes that act in the digestion and metabolism of fats, carbohydrates, proteins, and nucleic acids.

One might wonder, then, why doesn't the pancreas digest itself? Largely because the protein-splitting enzymes are stored in an inactive form in the pancreas and are activated by the duodenal environment. Well then, the sharp-witted reader will ask, why isn't the duodenum digested? As a matter of fact, it would be if it were not "alive" with a membrane system that actively keeps harmful substances, such as digestive enzymes, from entering the cells. This is also why, in those unfortunate enough to be "wormy" (and many people are), when a worm dies, it loses the ability to repel the digestive enzymes and is immediately digested—the body's ultimate vengeance. In addition, the duodenum protects itself by secreting a protective shield of mucin (you recall the stomach does the same), while the pancreas secretes a high level of sodium bicarbonate that helps to neutralize the disruptive acids entering the small intestine from the stomach.

THE LIVER

Once the products of digestion move into the tiny capillaries within the villi, the nutrient-laden blood is carried to the *hepatic portal vessel,* which leads to the liver (a portal vessel, you may recall, is one that lies between two capillary beds). The blood then filters through tiny vessels and sinuses in the liver. In the **liver**, if the blood contains excess carbohydrates, some are removed by the liver and stored as glycogen. If the blood is low in carbohydrates, stored glycogen is broken down into its glucose subunits, which are then released into the bloodstream. Thus, proper glucose levels are maintained in the blood despite variations in food intake. Such shifts in blood glucose levels can result in various changes, including behavioral ones. For example, as the body's reserves of glucose drop below the amount needed to maintain a constant blood level, one has the sensation of hunger and is motivated to go out and find some glucose (in a variety of forms from rabbits to pizza).

In its digestive role, the liver secretes bile from the gall bladder into the duodenum. Bile contains sodium salts that act as detergents, breaking up large fat droplets into smaller ones, providing a greater surface area on which digestive enzymes can act.

The bile is stored in the gall bladder and is released by rather weak muscular contractions brought about by hormones whose release is triggered by fats in the small intestine. Bile reaches the duodenum through the common bile duct, which is joined by the pancreatic duct just before it reaches the intestine (Figure 38.10). An obstructed bile duct results in an inability to properly digest fats. Surgeons sometimes move the bile duct of a dangerously obese patient toward the lower end of the small intestine. The patient, unable to absorb dietary fat, loses weight dramatically.

As food passes from the duodenum through the rest of the small intestine, further digestion and absorption occurs (as we see in Table 38.1). The remaining matter, mostly composed of undigested foods, bacteria, and water, then move on into the colon.

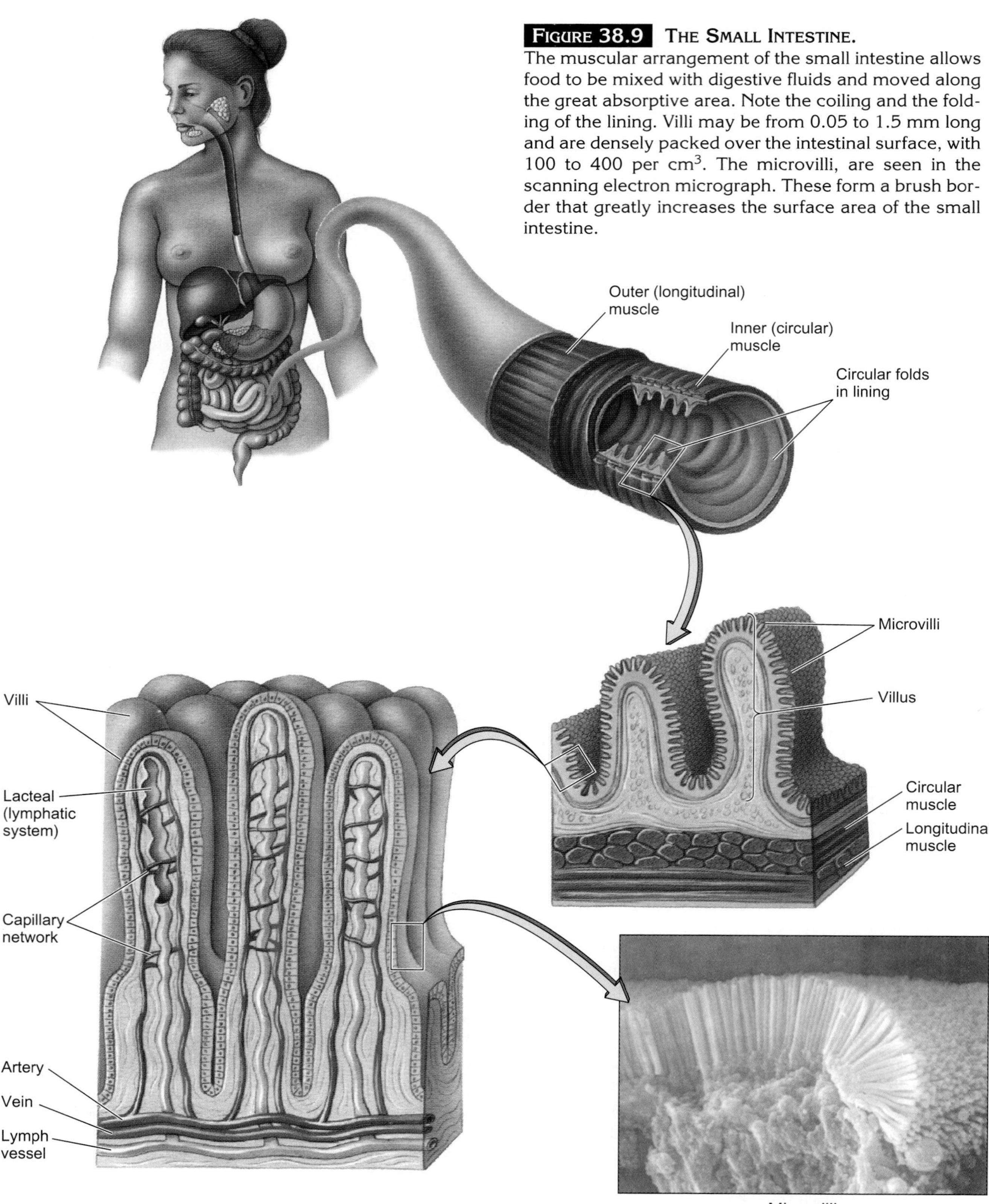

FIGURE 38.9 **THE SMALL INTESTINE.** The muscular arrangement of the small intestine allows food to be mixed with digestive fluids and moved along the great absorptive area. Note the coiling and the folding of the lining. Villi may be from 0.05 to 1.5 mm long and are densely packed over the intestinal surface, with 100 to 400 per cm^3. The microvilli, are seen in the scanning electron micrograph. These form a brush border that greatly increases the surface area of the small intestine.

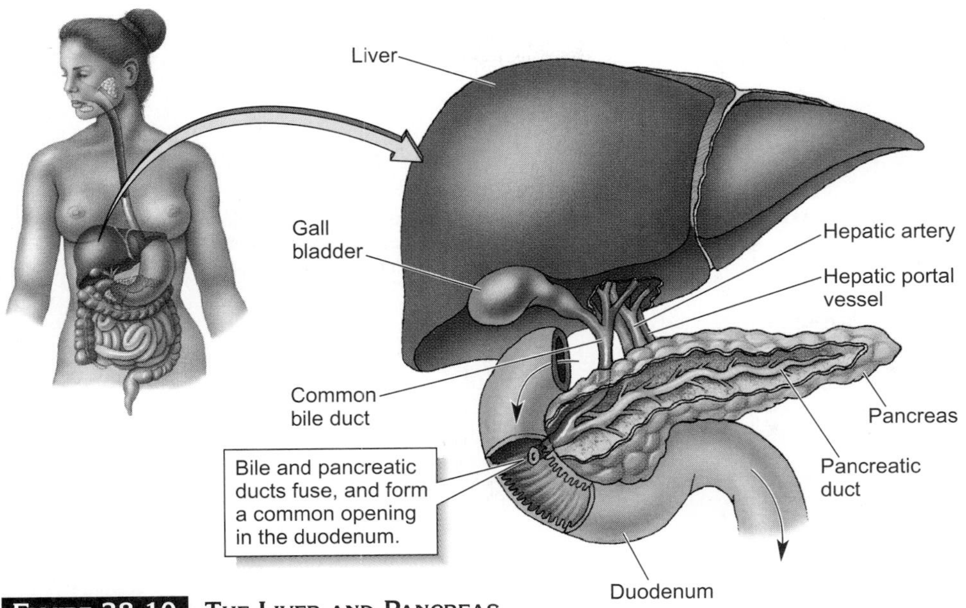

FIGURE 38.10 THE LIVER AND PANCREAS.
The liver and pancreas are accessory digestive organs. The ducts from both the gall-bladder and the pancreas meet to form a common duct that empties into the duodenum of the small intestine.

The liver is also responsible for metabolizing poisonous substances, such as alcohol, and removing them from the blood. In addition, the liver makes blood proteins, destroys old red blood cells, and produces urea from nitrogenous wastes that is then removed by the kidneys.

THE LARGE INTESTINE

The **large intestine** is divided into four principal parts: cecum, colon, rectum, and anal canal (Figure 38.11). The large intestine absorbs water and minerals, transferring them to the blood, and forms and stores feces before they leave the digestive tract via the anus.

Nutritionists argue over the role of the large intestine in digestion. The lining of the colon is glandular, but it secretes only mucus—no enzymes. The mucus protects the colon lining and lubricates it for the passage of feces. Recent evidence indicates that some digestion and absorption of food does occur in the large intestine, but it may be due solely to the activities of the vast numbers of bacteria that dwell there. You may be surprised—even shocked—to learn that about one-third of the fecal material actually consists of bacteria, living and dead. The remainder includes undigested residues, some nutrients, and inorganic material.

Actually, intestinal bacteria are quite useful. The bacteria provide humans with certain essential vitamins, including vitamin K, biotin, and folic acid, some of which are absorbed in the large intestine. Our relationship with our fecal bacteria is clearly mutually beneficial.

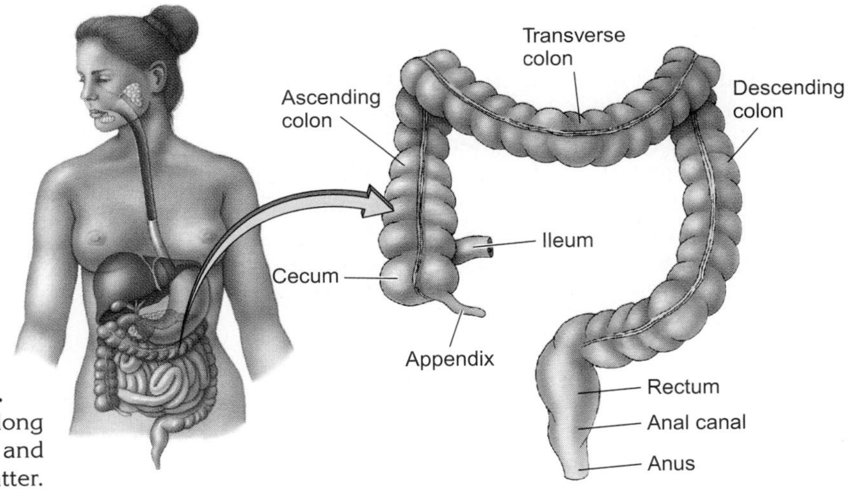

FIGURE 38.11 THE LARGE INTESTINE.
The large intestine is highly specialized along its lengths for the absorption of water and minerals and the preparation of waste matter.

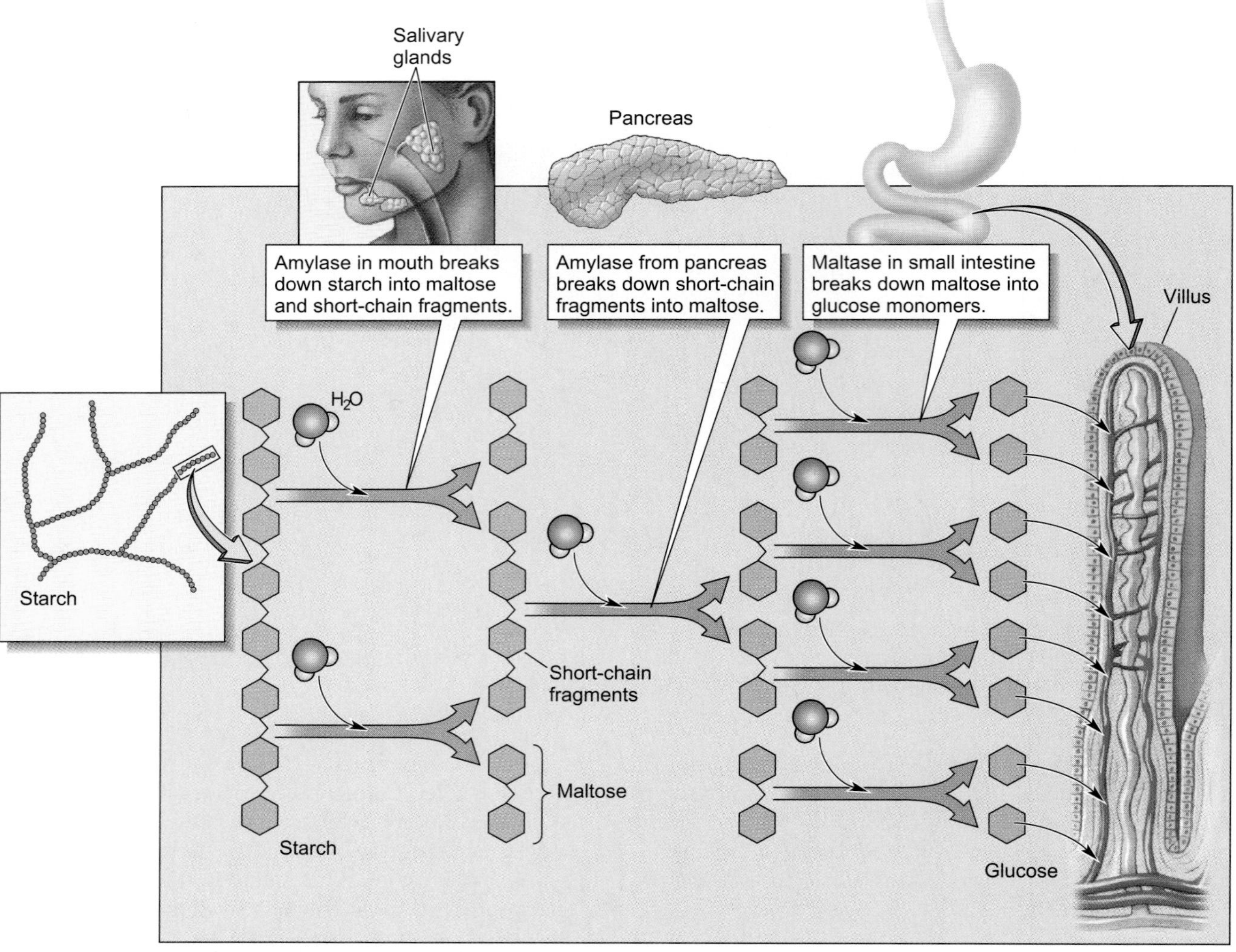

Salivary glands
Pancreas
Amylase in mouth breaks down starch into maltose and short-chain fragments.
Amylase from pancreas breaks down short-chain fragments into maltose.
Maltase in small intestine breaks down maltose into glucose monomers.
Villus
H_2O
Starch
Starch
Short-chain fragments
Maltose
Glucose

FIGURE 38.12 CARBOHYDRATE DIGESTION.
Amylose, our representative carbohydrate, is broken down into increasingly smaller fragments as shown here, finally to form monomers of glucose.

CHEMICAL DIGESTION AND ABSORPTION

We considered the molecular structure of proteins, fats, and carbohydrates in Chapter 3. Digestion can be viewed as the process of breaking these molecules down into smaller particles that can cross membranes and move into the bloodstream.

The bonds of the large molecules are broken through *hydrolysis* (see Chapter 3). Essentially, this means that enzymes add water to the linkages, disrupting them and yielding molecular subunits. Thus, in digestion, carbohydrates are dismantled into simple sugars, fats into fatty acids and glycerol, proteins into amino acids, and nucleic acids into nucleotides.

CARBOHYDRATE DIGESTION

Simple sugars, such as glucose, can be absorbed "as is" by the intestinal lining. But more complex carbohydrates, such as starch, must first be broken down into simple sugars. Even disaccharides, such as sucrose and lactose, must first be enzymatically split into their component parts. In fact, if sucrose is injected into the blood, it will be secreted by the kidneys unchanged, because *sucrase,* its enzyme, exists only as a membrane-bound enzyme of the gut epithelium.

Starch digestion begins in the mouth, where salivary amylase breaks some linkages, producing maltose and some larger fragments (Figure 38.12). In the small intestine, pancreatic amylase converts all starch into maltose, which is finally cleaved into two molecules of glucose by the enzyme *maltase.*

FAT DIGESTION

Fats, usually ingested in the form of triglycerides, are first dispersed into tiny droplets by bile salts from the gall bladder, and most are then hydrolyzed into fatty acids and glycerol (as well as some mono-

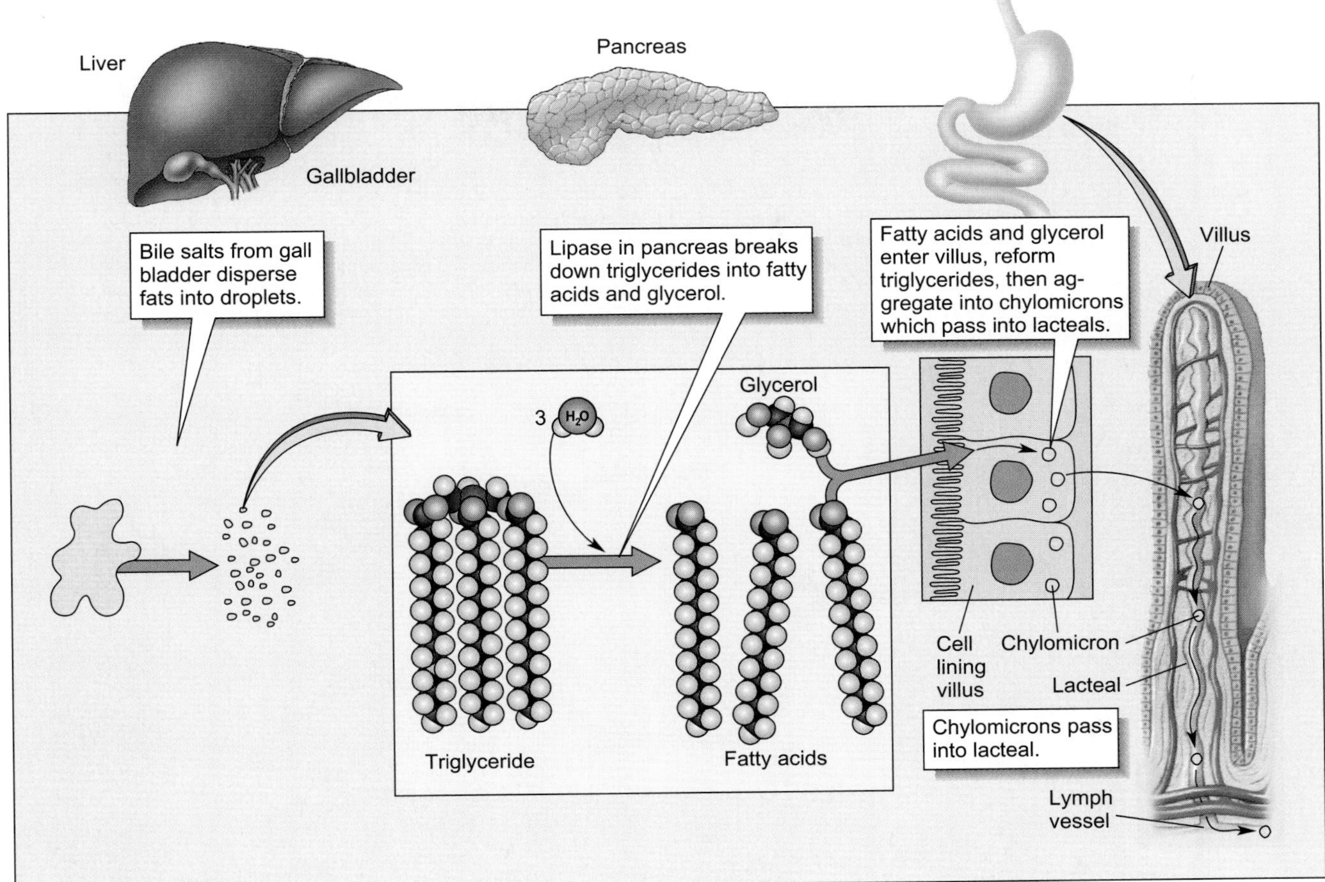
Liver
Pancreas
Gallbladder
Bile salts from gall bladder disperse fats into droplets.
Lipase in pancreas breaks down triglycerides into fatty acids and glycerol.
Fatty acids and glycerol enter villus, reform triglycerides, then aggregate into chylomicrons which pass into lacteals.
Villus
Glycerol
3 H_2O
Triglyceride
Fatty acids
Cell lining villus
Chylomicron
Lacteal
Chylomicrons pass into lacteal.
Lymph vessel

FIGURE 38.13 **FAT DIGESTION.**
Fats are broken down into tryglycerides by bile salts and then are broken down into fatty acids and glycerol by pancreatic lipase, only to be reassembled in the small intestine. These clump into larger droplets that pass into lymph vessels.

glycerides and diglycerides) by pancreatic lipase (Figure 38.13). The products—along with cholesterol, another lipid—then enter the cells lining the villi. Once inside, triglycerides are reformed and, along with cholesterol, are gathered into minute protein-surrounded bodies called *chylomicrons*—and in this form they pass into a nearby lacteal. The lacteals from each villus empty into lymph vessels that carry the lipids directly into the circulatory system for distribution throughout the body.

PROTEIN DIGESTION

Proteins are the largest and most complex of the food molecules, so it is not surprising that their digestion is complex. Essentially, protein digestion occurs in three steps, beginning in the stomach (Figure 38.14). Here, pepsin attacks the molecule, splitting it into peptide fragments of various lengths. The fragments are then subjected to a more specific disruption in the small intestine, where the pancreatic enzymes *trypsin* and *chymotrypsin* break only specific amino acid linkages. This leaves the protein in the form of peptide fragments only two to ten amino acids long. These are finally cleaved by enzymes of the small intestine into single amino acids. Following protein digestion, the amino acids enter capillaries in the villi and are carried directly to the liver, where they are used according to the body's metabolic needs.

SOME ESSENTIALS OF NUTRITION

Nutrition has become not only a trendy topic heard in cafeteria chatter and on television talk shows, but a critical one as well, as we are made aware of increasing numbers of contradictory options. Unfortunately, interest in food doesn't necessarily equal knowledge of nutrition. Many food faddists are vir-

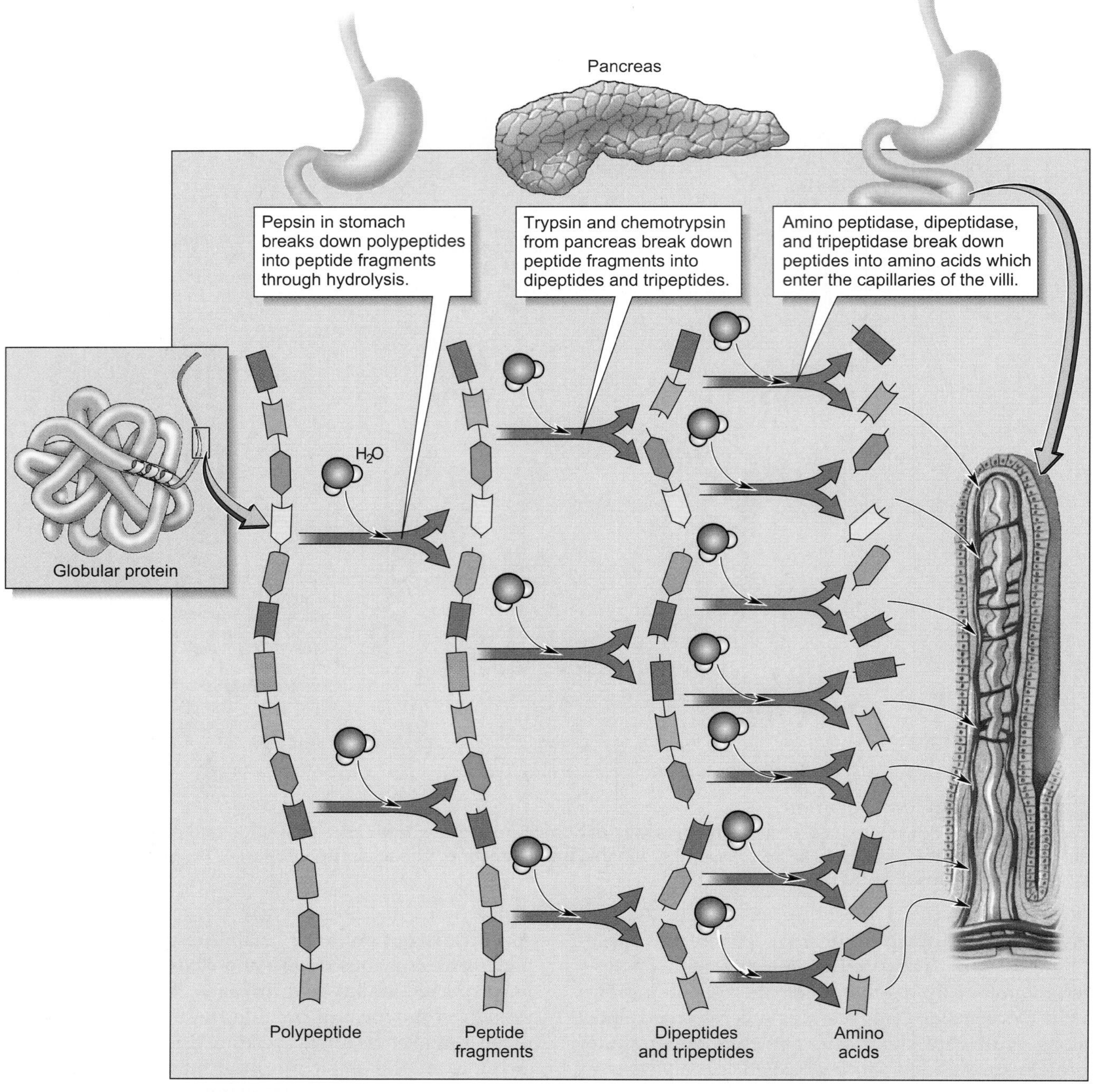
Pancreas
Pepsin in stomach breaks down polypeptides into peptide fragments through hydrolysis.
Trypsin and chemotrypsin from pancreas break down peptide fragments into dipeptides and tripeptides.
Amino peptidase, dipeptidase, and tripeptidase break down peptides into amino acids which enter the capillaries of the villi.
Globular protein
H_2O
Polypeptide
Peptide fragments
Dipeptides and tripeptides
Amino acids

FIGURE 38.14 PROTEIN DIGESTION.
Proteins contain hundreds—often thousands—of amino acids. The complete digestion of protein to free amino acids requires at least three chemical steps.

tually ignorant of the *science* of nutrition. Some do not seem to realize that organically grown food is no better for you than that grown with chemical fertilizers; that rose hip vitamin C is no different from synthetic vitamin C; that large doses of some vitamins can be harmful; that dietary protein that exceeds body needs is simply deaminated and converted to glucose or fat in the liver; or that high-protein diets have not been shown to be beneficial, and, in fact, can be both fattening and harmful. We can best make sense of such findings by reviewing certain large molecules in the context of their roles as food.

Carbohydrates

Carbohydrates are common sources of energy, as we saw in our discussion of their use in cell respiration (see Chapter 8). This usefulness, however, is quite varied and complex.

Once glucose makes its way from the gut into the circulatory system, it is immediately removed and stored as glycogen, primarily in the liver. From there it is meted out as glucose according to the body's needs. This keeps the blood glucose level rather constant, although there may be a temporary 10% to 20% increase immediately after a high-carbohydrate meal.

The skeletal muscles store a considerable amount of glucose as glycogen. During anaerobic activity, glycogen is again broken down into glucose, which is then used as an energy source. Long-distance runners build up muscle glycogen by continually depleting those reserves in long training runs and then building them back to even higher levels. Excess glucose is simply converted into body fat.

Fats

You are probably aware of the persistent controversy over the connection between saturated fats and cholesterol. You may have even switched to foods with high levels of unsaturated fats—the ones so heavily touted in advertisements for margarine and cooking oils (see Essay 38.2). Indeed people generally assume that it is good to avoid fats in the diet. Although this is generally a good practice for Americans, who consume too much fat, fats are not all bad. In fact, some amount of fat is essential in the diets of many animals, including humans.

Fats have a number of roles. First, they hold a great deal of energy. In fact, they have twice the energy value of carbohydrates. In addition, they increase intestinal absorption of the essential fat-soluble vitamins A, D, E, and K. Fats are also part of all cell membranes.

Some fats contain what are known as *essential fatty acids.* These must be provided in the diet. Other fats can be synthesized in the body.

Indeed, if you are trying to lose body fat, keep in mind that the body can synthesize certain fats not only from excess carbohydrates, but from excess proteins as well. For this reason, high-carbohydrate or high-protein diets may defeat their own purpose if intake is excessive.

Protein

Protein, an essential nutrient, must be constantly present in the diet since we don't store it in protein form for later release. This means that while some protein is maintained in the liver and muscles, there is a steady turnover of *total* body protein. We need even more protein when we are growing. (Low supplies of protein in the first two years of life can produce permanent brain damage.)

When protein is digested, its amino acids are utilized in several ways. For example, they are the building blocks of our own protein; they are used to form nitrogenous bases such as those of DNA; they can be oxidized for energy; and they can be stripped of their amino groups and converted to fats and carbohydrates.

Animals generally are able to convert certain amino acids into others. In fact, if necessary, we can convert 12 of the 20 amino acids we require into other amino acids. However, the remaining eight must be supplied in the diet. These are referred to as the *essential amino acids*—a phrase that delights advertising agencies. Actually, all 20 are needed; it's just that some are essential in the diet itself.

The quality of some proteins is lower than that of others, meaning that some kinds are deficient in one or more of the essential amino acids. The synthesis of any protein depends on the presence of all its required constituent amino acids. Thus, the usefulness of any protein in food, for anything other than calories, depends on the relative concentration of its most scarce essential amino acid. In general, animal proteins contain all the essential amino acids, but plant proteins lack some of them.

Plants are not in the business of raising animals, and so they use the proteins in their seeds for their own ends, such as for the development of new seedlings. The proteins stored in seeds such as oats, bran, wheat, and barley, then, are usually of comparatively low quality for human consumption. And the proteins in the plants themselves are typically deficient in tryptophan, methionine, or lysine. Nevertheless, a carefully managed vegetable diet can be quite sufficient for human needs, if the right *combinations* of foods are consumed (Figure 38.15). A combination of peas and beans, for example, provides the essential amino acids, but because of protein turnover, it must be constantly replenished. So these foods, or other combinations with a similar balance of amino acids, must become a regular part of a

Essay 38.2 Cholesterol and Controversy

Cholesterol! Just the word triggers fears of an impending lecture. Someone is about to tell us to stop doing something we like to do—like eating fatty foods. Cholesterol is, in fact, a lipid (the chemical group that contains fats and oils), and furthermore, it is a major constituent of the cell membranes throughout our bodies. So if it's useful to us, why all the fuss? The reason is, cholesterol can be transported in two ways—by low-density lipoprotein (LDL) and by high-density protein (HDL). LDL cholesterol is considered the "bad" cholesterol because it moves from the liver to the cells (including the cells lining arteries) and tends to attach to these cells, apparently contributing to the buildup of plaque, which can clog the arteries. HDL cholesterol, on the other hand, moves from the cells to the liver, helping to clear the arteries.

In the effort to reduce "bad" LDLs in the body, some researchers have suggested that people simply eat less food that is high in any cholesterol. Generally, animal food is high in both cholesterol and saturated fats (those fats saturated with hydrogens, so no double bonds exist between the carbon atoms in the molecule). Plants, on the other hand, contain a lot of healthful, unsaturated molecules and no cholesterol.

In a recent study, a large group of middle-aged men was divided into an experimental population and a control population. The experimental population was given a low-cholesterol, high-unsaturated fat diet. Sure enough, their blood cholesterol levels went down, and over the years, their rates of arterial disease and heart disease were lower than those of the control population. Case proven? Not quite. While the death rate due to heart attacks was lower in the experimental group, the overall death rate of the experimental group was significantly *higher* than that of the control group. It turns out that the experimental group had a higher cancer death rate.

What happened? One possibility is that the higher levels of cholesterol in the control group gave them healthier plasma membranes that protected them from cancer and other disease, or perhaps gave them a better-balanced steroid metabolism.

Another, more likely possibility is that the higher levels of unsaturated fats in the experimental group were actively toxic. Unsaturated fats spontaneously form free radicals—highly reactive molecules that contain unpaired electrons. Free radicals tend to react with and degrade other molecules in a random manner. Interestingly, ionizing radiation damages cells in a similar manner. It creates free radicals that attack DNA and other cellular constituents. In any case, the study indicates what can happen to people who radically change their diets in response to even a seemingly well-founded scientific opinion. To confuse the issue further, recent evidence indicates that moderate intake of alcohol (1 to 3 drinks a day) may lower LDL cholesterol and decrease the risk of heart attacks. And yet other researchers have indicated that diet is not nearly as effective in controlling cholesterol as are drugs designed for that purpose. A generation or two of Americans have avoided cholesterol like the plague and, overall, may have gained nothing from their efforts.

Cholesterol may have even far-more-reaching effects in women than in men. After puberty, women show higher levels of "good" HDL and lower levels of "bad" LDL; but after menopause, the reverse is true, and researchers believe the difference is due to the presence during the reproductive years of estrogen, a female hormone that causes changes in the way the body produces, handles, breaks down and eliminates cholesterol. In studies in which postmenopausal women were given synthetic estrogen supplements, the levels again reversed themselves, and LDL dropped as HDL surged in the bloodstream. Many doctors, these days, recommend estrogen replacement therapy for postmenopausal women.

FIGURE 38.15 HEALTH FOODS
Many Americans have the time, energy, and money to devote a great deal of effort to their diets. Unfortunately, they are often woefully misinformed.

healthy vegetarian diet. An intelligently planned vegetarian diet that is stretched to include fish as an honorary vegetable may be more healthful than the typical American diet, which, as we mentioned, includes an excess of animal fats.

VITAMINS

In addition to essential fatty acids and essential amino acids, a variety of *vitamins* are also essential in the diets of animals. **Vitamins** are organic compounds necessary for normal metabolism. Most of the vitamins we need have been clearly identified by biochemists so that we know both their molecular structure and their precise function (Table 38.3). A few are only vaguely understood, however; all we can say about them is that if you don't have them, you will develop something-or-other, and it will probably be bad. Shortages of niacin (vitamin B_6) and riboflavin, for example, can lead to serious illness because the coenzymes NAD and FAD, necessary for glycolysis and cell respiration, are derived from these vitamins. Vitamins C and E have a somewhat more general function: Both are antioxidants that remove free radicals from the body that would otherwise cause damage through random oxidation of the body's molecules.

Vitamins can be categorized as fat-soluble or water-soluble. The two behave quite differently in the body. For example, water-soluble vitamins function as coenzymes; fat-soluble vitamins do not. Whereas water-soluble vitamins function in most animals, the fat-soluble function only in vertebrates. Finally,

TABLE 38.3 *VITAMINS AND THEIR ROLES*

Vitamin	Source	Function	Result of Deficiency
A, retinol	Fruits, vegetables, liver, dairy products	Synthesis of visual pigments	Night blindness, crustiness about eyes
B_1, thiamine	Liver, peanuts, grains, yeast	Respiratory coenzyme	Loss of appetite, beriberi, inflammation of nerves
B_2, riboflavin	Dairy products, liver, eggs, spinach	Oxidative chains in cell respiration	Lesions in corners of mouth, skin disorders
Niacin, nicotinic acid	Meat, fowl, yeast, liver	Part of NAD and FAD cell respiration	Skin problems, diarrhea, gum disease, mental disorders
Folic acid	Vegetables, eggs, liver, grains	Synthesis of blood cells	Anemia, low white blood count, slow growth
B_6	Liver, grains, dairy products	Active transport	Slow growth, skin problems, anemia
Pantothenic acid	Liver, eggs, yeast	Part of coenzyme A of cell respiration	Reproductive problems, adrenal insufficiency
B_{12}	Liver, meat, dairy products, eggs	Red blood cell production	Pernicious anemia
Biotin	Liver, yeast, intestinal bacteria	In coenzymes	Skin problems, loss of hair and coordination
Choline	Most foods	Fat, carbohydrate, protein metabolism	Fatty liver, kidney failure, metabolic disorders
C, ascorbic acid	Citrus fruits, tomatoes, potatoes	Connective tissues and matrix antioxidant	Scurvy, poor bone growth, slows healing (colds?)
D	Fortified milk, seafoods, fish oils, sunshine	Absorption of calcium	Rickets
E	Meat, dairy products, whole wheat	Antioxidant	Infertility, kidney problems
K	Intestinal bacteria	Blood-clotting factors	Blood-clotting problems

excess water-soluble vitamins are excreted, but the fat-soluble ones are stored in the body fat. This is why it is dangerous to overdose on fat-soluble vitamins—they remain in the body's fat and can be released as that fat is metabolized.

Mineral Requirements

Animals also require a number of inorganic ions that are generally referred to as *minerals.* Those required in the largest amounts include calcium (a major constituent of bone and of many cellular processes), magnesium (necessary as a cofactor—see Chapter 6—for many enzymatic activities), and iron (a component of hemoglobin as well as the cytochromes of the electron transport chain). Sodium, potassium, and chloride are also needed in fairly substantial quantities in order to replace daily losses. Sodium and potassium are involved in nerve cell conduction and osmotic regulation.

Trace Elements. In addition to the various elements that are necessary in large amounts for life and good health, others are necessary only in very small amounts. Included are iodine (thyroid hormones), copper (hemoglobin formation and enzyme action), fluorine (protection of teeth), zinc (enzyme action and insulin synthesis), selenium and manganese (enzyme action), and cobalt (red blood cell formation). Recently added to the list of trace elements is silicon, a mineral constituent of rock and one of the commonest elements on the earth. But to be useful to the body, silicon must be in a soluble form, as it is in plant fibers. So, getting your "roughage" may be important for reasons other than the regularity so routinely discussed in TV ads during the evening news. Silicon is also now believed to be an important constituent in the elastic walls of arteries, and its absence in diets of highly processed foods may be a significant cause of arterial disease—another strong argument for eating fresh fruits and vegetables.

PERSPECTIVES

We've seen that low- and high-cholesterol diets are both potentially harmful; that the balance between saturated and unsaturated fats mustn't be tipped too far in either direction; and that too much protein is harmful, but too little is worse. Careless vegetarianism can be dangerous and so can a diet laden with red meat.

There is no simple answer to diet questions, and we find ourselves resorting to generalities and cliches, such as "moderation in all things," or "variety is the spice of life," or "shut up and eat your vegetables." At any rate, it is probably best to avoid diet fads, and to avoid saturated fats—but not fanatically.

Now we move on to an area that requires fewer decisions. We have no choice about whether to utilize oxygen, or how. We will, and we will do it according to certain rather strict physiological principles.

Summary

Basic Digestive Processes

As heterotrophs, animals have a two-fold problem: obtaining food that fulfills their nutritional requirements, and breaking it down through digestion for absorption into the body.

Digestion in Invertebrates

Chemical digestion is the breakdown of foods by hydrolytic enzymes into their molecular components. Mechanical digestion breaks food into smaller bits, facilitating chemical digestion.

Sponges carry on intracellular digestion, taking food in by phagocytosis and breaking it down within food vacuoles. Cnidarians continue the use of some intracellular digestion, but extracellular digestion occurs in the gastrovascular cavity. A gastrovascular cavity is a saclike, incomplete digestive system, one with a single opening through which food enters and undigested residues leave. A complete digestive system—a tubular gut with two openings—occurs in most invertebrates, which have specialized regions that process food in different ways.

Digestion in Vertebrates

The tubelike system of vertebrates often includes special adaptations along its length. Plant eaters may have an exceptionally long intestine. Ruminants have four-part stomachs housing microorganisms that digest cellulose. Other vertebrates have a large pouchlike cecum in which microorganisms also break down plant material.

The Human Digestive System

Humans have a generalized digestive system with typical vertebrate digestive organs.

The tongue is important to moving and mixing food, detecting foreign objects, and tasting. Teeth are important in mechanical digestion. The salivary glands secrete saliva that moistens and lubricates food and begins starch digestion.

Swallowing begins with voluntary movement in the pharynx. The larynx elevates, thereby causing the epiglottis to close off the glottis and helping direct food into the esophagus. Then, peristalsis, an involuntary action, propels food toward the stomach. Peristalsis is a wave of contraction of smooth muscle that moves food along the entire digestive tract.

The stomach stores and actively churns food and secretes gastric juices. Hydrochloric acid activates pepsinogen, forming pepsin, which begins protein digestion.

The small intestine completes digestion of all foods. Its great coiling length, circular folds, villi, and microvilli produce a large surface area for absorption. The villi contain capillaries and lacteals that receive digested foods carried in by active transport.

The pancreas secretes sodium bicarbonate (which neutralizes acid from the stomach) and digestive enzymes into the small intestine.

The liver produces bile, which breaks fat into small droplets, aiding chemical digestion. Bile is stored in the gallbladder and released when fats are present. The liver also metabolizes poisonous substances, makes blood protein, destroys old red blood cells and produces urea from nitrogenous wastes.

The large intestine includes the cecum, colon, rectum, and anal canal. The large intestine absorbs water and minerals, and forms and stores feces. Immense bacterial populations break down digestive residues and make useful vitamins available.

Chemical Digestion and Absorption

All foods are broken down into molecular subunits through hydrolytic cleavage. The primary products are glucose, fatty acids, glycerol, and amino acids.

Carbohydrate-digesting enzymes include salivary and pancreatic amylase, which yield maltose, and maltase, which yields glucose. Lactase and sucrase break down lactose and sucrose, respectively.

Pancreatic lipase breaks down fats into fatty acids and glycerol. Along with cholesterol, these products enter the villus where they are gathered into chylomicrons that are then carried into the lacteal for lymphatic transport.

Protein requires several enzymatic steps, proceeding from proteins to polypeptides, then to peptide fragments and dipeptides, and finally amino acids.

Some Essentials of Nutrition

Care must be taken in accepting nutritional information since much of it is misleading or wrong.

Carbohydrates are important as energy sources. Glucose is stored as glycogen and meted out by the liver. Skeletal muscles rely on glycogen, breaking it down into glucose for glycolysis. Excess glucose is converted to fat.

A certain level of fat intake is essential, but most Americans consume too much fat. Fats are used for energy, incorporated into structure, converted to glucose, and stored in adipose tissue. Some fats can be synthesized by the body. Those that must be included in the diet contain what are called essential fatty acids.

Protein must be part of the daily diet; it cannot be stored. Amino acids are used for protein and nucleotide synthesis, oxidized for energy, and the excesses converted to fats and carbohydrates. Eight of the twenty (essential amino acids) must be in the diet. They are most readily available from animal protein. Plant foods are usually deficient in certain amino acids and should be consumed in complementary combinations.

Vitamins are generally important as coenzymes and antioxidants.

Many mineral ions are needed as bone elements, enzyme cofactors, for heme groups in hemoglobin and cytochromes, in nerve action, and for osmotic regulation. Trace elements have varied uses. For instance, roughage is important to bowel regularity, and it is now known that the silicon it contains is important to arterial elasticity.

Key Terms

digestion • **653**
intracellular digestion • **654**
extracellular digestion • **654**
incomplete digestive system • **654**
complete digestive system • **654**
peristalsis • **659**
stomach • **659**
small intestine • **661**
pancreas • **661**
villi • **661**
microvilli • **661**
liver • **661**
vitamin • **669**

Review Questions

1. Write an accurate definition of digestion, contrasting it with dehydration synthesis (Chapter 3).
2. Using examples, distinguish between intracellular and extracellular digestion and complete and incomplete digestive system.
3. Why can many kinds of animal bodies be referred to as a tube-within-a-tube.
4. Briefly describe three vertebrate innovations for cellulose digestion. Why is such cellulose digestion a problem for most organisms?
5. What are the digestive functions of the following: salivary glands, tongue, pharynx?
6. Explain the voluntary and involuntary aspects of swallowing.
7. Describe the musculature of the stomach and its specific action. What keeps the contents inside?
8. List the components of gastric juice. What are two ways in which a very low pH is important?
9. What prevents the stomach lining from being eroded by harsh gastric juices?
10. What are the specific functions of the small intestine?
11. List four factors that provide great surface area to the small intestine. Why is a large surface important?
12. Prepare an outline drawing of one villus, labeling the smooth muscle, capillary bed, lacteal, and microvilli.
13. What digestive agent does the liver secrete? Explain how it is released, and what it does.
14. How is the pancreas important to digestion?
15. List three important functions of the colon. Why are its bacterial populations classified as mutualistic?
16. List the four enzymes responsible for carbohydrate digestion, and name the substrates and products of each.
17. List the general events in protein digestion, and state where they occur. Why are so many steps needed?
18. Where is most glycogen stored? To what use is most glycogen put?
19. What happens to excess glucose, fatty acids, and amino acids taken in? Why does a high-protein, weight-reducing diet defeat its own purpose?
20. List two common uses the body makes of vitamins.
21. List five specific ways the body makes use of minerals.

CHAPTER

39

Gas Exchange

Gas Exchange: An Evolutionary Perspective
The Respiratory Interface
Simple Skin Breathers
Complex Skin Breathers
Gill Exchangers
Tracheal Exchangers
Lung Exchangers
The Human Respiratory System
From the Pharynx to the Bronchial Tree
The Alveolar Interface
The Lungs and Breathing
The Exchange of Gases
Respiratory Control

"IN WITH THE GOOD AIR, OUT WITH THE BAD AIR," IS A PHRASE MOST OF *us probably have heard but have no idea where it came from. In any case, "good air" undoubtedly refers to oxygen-laden air entering the lungs, and "bad air" to carbon dioxide-laden air leaving the lungs. In these pages, we will see why oxygen is so highly regarded while carbon dioxide apparently is held in less esteem. We will also see why conditions were once different—why oxygen itself once would have been regarded as bad. We will focus on how it enters the bodies of a variety of living things, and what it does once there. As asides, we will consider the wonders of the bird lung and the horrors of smoking—all part of the fascinating topic of gas exchange.*

When the earth was young, the atmosphere was far different from what it is now. Most of the life that exists now couldn't have survived then, and by the same token, our present atmosphere would be deadly to those earliest forms of life. The difference between the atmosphere then and now can be summed up in a word: *oxygen.*

When life first appeared, there was little oxygen over the earth's surface, and that life had no use for it anyway. Furthermore, it had no defense against oxygen's corrosive forces. In time, though, as oxygen began to accumulate in the earth's atmosphere, all this would change. The accumulation was due to the production of oxygen by the cyanobacteria, the first plants, and by molecular interactions in the upper reaches of the atmosphere.

As the gas accumulated, it placed severe demands on the life that it touched. Many life forms fell before this strange, corrosive force. But, as natural selection worked its way, life forms emerged that had ways to deal with oxygen—ways to avoid being corroded by it and even ways to use it. Today, oxygen is an integral part of life on earth, and the species on the planet have developed a variety of means of handling it.

Here, then, we will consider just how various forms of life have come to deal with oxygen and another gas, carbon dioxide, formed as a by-product of oxygen use. We will first consider some general ideas about the exchange of gases in living things, and then review how the various kinds of animals actually go about it, focusing finally on our own species.

GAS EXCHANGE: AN EVOLUTIONARY PERSPECTIVE

As various forms of life began to find ways to use oxygen in their metabolism, a general problem arose: how to get it. The flip side of this problem was how to get rid of carbon dioxide, built up as a metabolic waste. Those earliest life forms did not have to develop complex mechanisms to get oxygen into their bodies and carbon dioxide out. All they had to do was to provide a surface across which the gases could pass by simple diffusion. Diffusion would work well simply because metabolic activities create the concentration gradients needed to encourage the inward movement of oxygen and the outward movement of carbon dioxide, and because the bodies of those early forms were small enough to use such a simple mechanism. As we will see, many kinds of animals even today rely on unassisted diffusion across their body surfaces to meet all their metabolic needs. Within these bodies, the gases diffuse, cell-to-cell, along whatever gradient exists, a slow process indeed. Now, however, other kinds of animals exist that are larger, with greater metabolic demands—and they have had to develop ways of more efficiently handling these gases, ways that usually involve the development of larger and more specialized surfaces, or interfaces, across which the gases can pass (Figure 39.1).

We will have to look at some of these interfaces next, but first we should make a distinction. *Cell respiration*—discussed in Chapter 8—is the metabolic process by which cells utilize oxygen in obtaining energy from fuels. *Organismic respiration*—which, we will consider here—is the term for the physical processes by which gases pass into and out of the body.

The Respiratory Interface

If gas is to move into and out of a body, it must cross a body covering, or *respiratory interface*—a living membrane that must be thin, moist, and of sufficient area to accommodate the animal's physiological needs. A thin membrane, of course, is easier for gases to cross than a thick one, and moistness is essential because gases normally must be dissolved in liquid in order to cross solid barriers easily.

Animals meet the requirements for a respiratory interface in a variety of ways, but three factors appear to be involved in determining its nature: the size and complexity of the organism, its metabolic rate, and the nature of the surrounding environment.

Simple Skin Breathers

Very different kinds of animals use the body surface, or skin, as a respiratory interface. Gases simply pass into and out of the body through the skin. It's a simple solution to the problem of gas exchange, but one that has severe restrictions as well. You may recall the general structure of small sponges. Their thin-walled, vaselike bodies permit a simple exchange of gases with both the surrounding seawater and a current of seawater carried through the hollow body by flagellated collar cells. However, not all sponges are small and thin. So, how do gases penetrate the mass of larger sponges? While the body surface is still the exchange interface, mass is no problem because the body, rather than being vaselike, is riddled with small canals. These canals are lined with collar cells that create currents of oxygen-laden water that reaches all parts of even the largest sponge body.

Similarly, although some species of cnidarians (such as jellyfish) attain considerable size, they also require no specialized respiratory structures. Their saclike bodies consist essentially of two layers that are virtually always in contact with the wet, oxygen-laden external environment. Their outer body wall and extensive gastrovascular cavity provide an adequate exchange interface.

Flatworms are considerably more dense than cnidarians. But their flat shape yields a large surface area for its mass. Flatworms also have an extensive, highly branched gastrovascular cavity into which water moves. Because of these characteristics, a flat body and a branching internal sac, much of the flatworm's body mass is exposed to its watery environment, and no cells are far from it. The flatworm can, therefore, use cell-by-cell diffusion as a means of exchanging gases with the environment. Of course, as long as it relies on diffusion alone to transport gases, it can never grow very large—its volume would proportionately outgrow its surface area (see Chapter 4).

Complex Skin Breathers

A few terrestrial animals, such as the earthworm and small lungless salamanders, manage to use their moist skin as a respiratory surface (the salamander also uses its highly vascular pharynx). But these animals, rather large and complex, can do so only by living under very special terrestrial conditions. Both must live in damp places—places that keep their respiratory surface moist. Thus, the earthworm lives underground, away from the drying air, and lungless salamanders spend their days underground or under logs, venturing out mainly at night when there is less risk of drying out. The skin of earthworms and lungless salamanders, by the way, is kept moist by the secretion of a slimy layer of mucus.

Although such skin breathers have no special structures with which to exchange gas with their environments, their relatively large mass prevents them from relying entirely on cell-by-cell diffusion of gases. Skin breathing in these creatures is enhanced by efficient circulatory systems that readily transport gases throughout their bodies.

Thus, skin breathing works for some larger terrestrial organisms—if they restrict themselves to a moist habitat, and if they have an efficient circulatory system. The earthworm and the lungless salamander, then, share an interesting combination of behavior and anatomy that permits their simple skin respiratory systems to work.

Gill Exchangers

Most large, complex aquatic animals also require a greater exchange of gas than can be accommodated by the skin. The most common development in these species is the **gill,** a respiratory structure with a great interface surface area in aquatic species. Gills are thin-walled extensions, or outpocketings, of the body wall. Their surface area is vastly increased by the presence of *lamellae,* flattened gill extensions over which water can pass. Gills are rich in

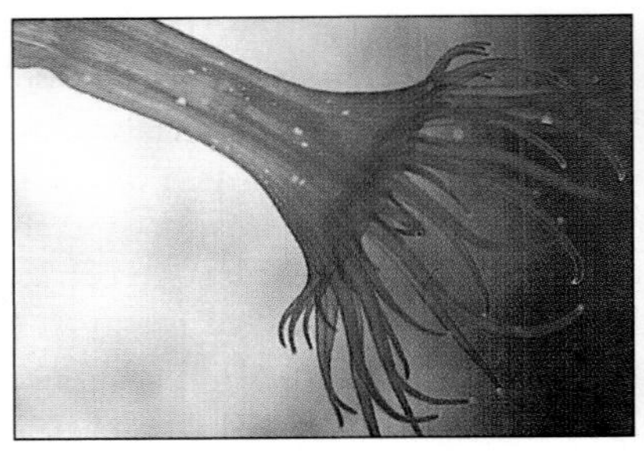

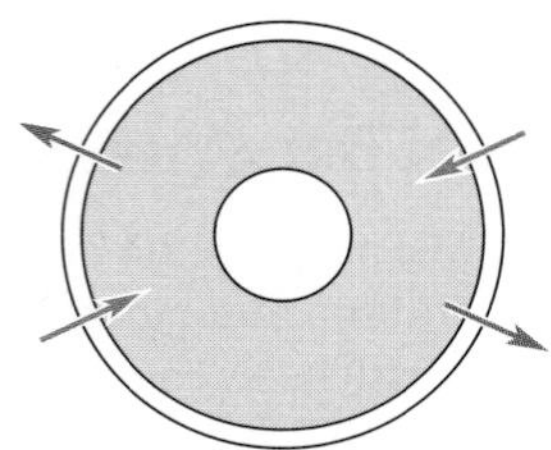

Because of its extremely thin body wall, each cell in the sea anemone can exchange gases directly with surrounding sea water.

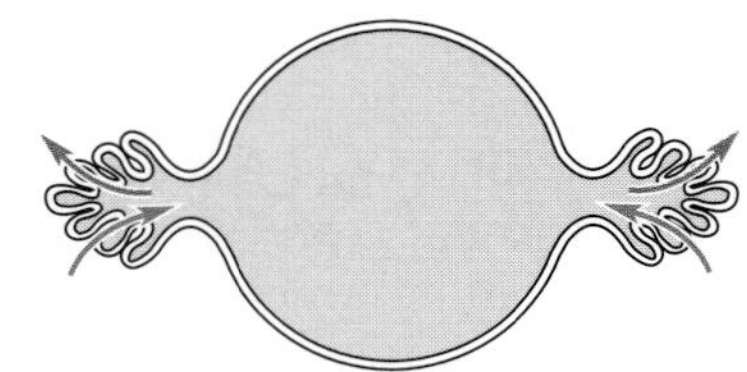

Some amphibians have permanent, thin-walled external gills in which gas exchange with the surrounding water occurs. Gases are transported to and from cells by the circulatory system.

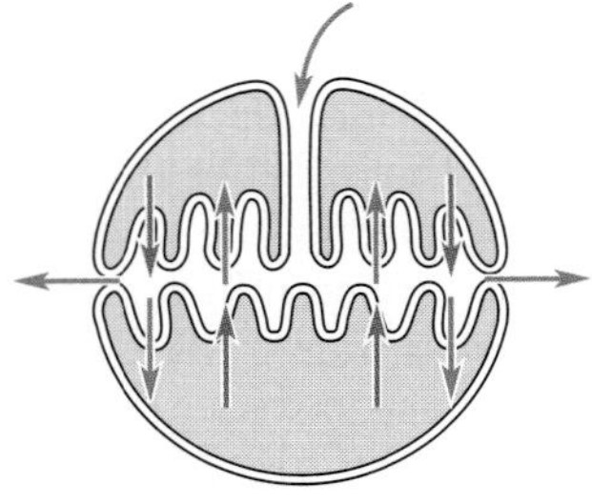

The fish gill carries on the initial exchange of gases with the blood (external respiration). The gases are transported to and from the body cells by the circulatory system. A second exchange occurs in the cells (internal respiration).

FIGURE 39.1 GAS EXCHANGE.
The simplest gas exchange interface occurs in skin breathers such as the sea anemone, where simple diffusion across a thin body wall is sufficient. Many aquatic animals utilize the gill, an outpocketing of the body surface. The external gill protrudes outward from the surface, as seen in *Necturus*, the mud puppy. The internal gill as seen in fish is enclosed in a covered gill chamber. Water and blood, moving oppositely, provide an efficient countercurrent exchange. Terrestrial insects make use of an extensive internalized tracheal system where minute thin-walled passages exchange gases throughout the body. The internalized lung of terrestrial vertebrates provides a protected, efficient gas exchange surface. In birds, the lung reaches its most specialized form, with a one-way flow and cross-current exchange.

tiny capillaries (thin-walled blood vessels). Because the capillaries passing through the gills have walls only one cell thick, the blood in these areas is very near the surrounding water, and gases are easily exchanged. *External gills* protrude from the body and are found in invertebrates, such as marine worms, and some salamanders. *Internal gills* lie within the body, usually with some mechanism to move oxygen-laden water over the interface. Such gills are common in fish.

Gills reach their greatest complexity in the fishes, where an efficient exchange is provided by a countercurrent flow of water and blood (Figure 39.2). In the gill, blood passing over the exchange network is constantly confronted by a fresh supply of water; thus, as long as blood is passing through the lamellae, there is an inward diffusion gradient for oxygen and an outward diffusion gradient for carbon dioxide. Such efficiency is especially important to active aquatic animals.

Water's low oxygen content means that active aquatic animals must devote considerable time and energy to moving large volumes of water over their respiratory surfaces. The faster-swimming bony fishes accomplish this by staying on the move. Other fishes pump water over their gills by contracting their pharyngeal muscles. To "breathe," the fish closes its opercula (gill covers), opens its mouth, and expands its mouth and gill chambers, drawing water in. It then closes its mouth, contracts the oral cavity, and opens its opercula to let the water flow out across the gills. That's why motionless fish look as though they are gulping.

TRACHEAL EXCHANGERS

Most terrestrial animals have developed specialized breathing surfaces, usually by enclosing the respiratory interface within the body. By internalizing their respiratory surfaces in this way, these land

The vast tracheal system in insects reaches all cells. It thus carries on a more-or-less direct exchange without the involvement of the circulatory system.

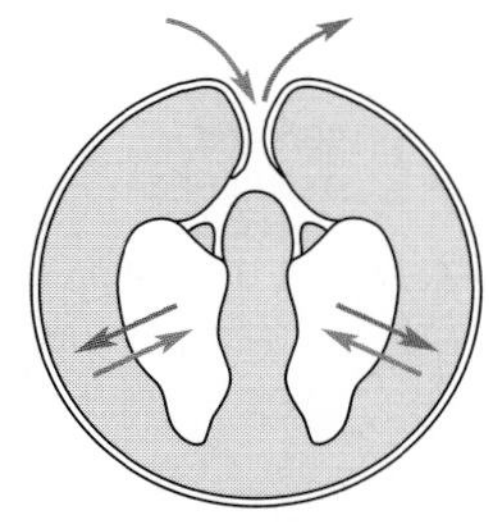

In most air-breathing vertebrates, the lungs carry on an exchange of gases with blood. Note the in-and-out movement of air. The gases are transported to and from the body cells by the circulatory system.

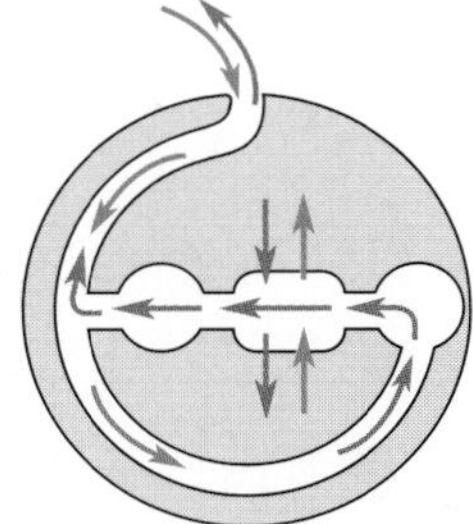

The bird lung is unique in that air flows through rather than in-and-out. As in other types of lungs, gases are exchanged first in the lungs and second in the body tissues. The circulatory system transports the gases between the two.

dwellers reduce the problem of excessive water loss. The internal pouches are usually highly folded, pocketed and convoluted, providing an increased area for gas exchange.

Among the variations of this internal arrangement are those of the arthropods, a group that does most things differently. Insects, for example, have a *tracheal system.* The insect body is riddled with a series of tiny, highly branched tubules called tracheae that branch further into tracheoles (Figure 39.3). These tubes open to the outside via tiny valvelike openings known as *spiracles,* commonly situated along the insect's sides.

In most species of insects, air simply diffuses through the tracheal system. Oxygen and carbon dioxide are exchanged through the walls of very fine tracheoles, which permeate the body and carry oxygen to the immediate vicinity of every active tissue. The finer branches of the tracheae are filled with fluid through which the respiratory gases can diffuse. In grasshoppers and some other larger insects, air is pumped in and out of the spiracles by a bellows action of the abdomen. In grasshoppers, bees, and some other insects, the branching tracheae terminate in elastic air sacs that expand and contract, and thereby move more air through the system.

Lung Exchangers

Most terrestrial vertebrates have solved the problem of keeping their respiratory membranes moist by bringing the respiratory interface inside the body where it is sheltered from the drying air. In terrestrial vertebrates, the *lungs* provide the moist internal interface for the exchange of gases. The lungs of most land vertebrates are paired structures consisting of inflatable, highly vascularized, and somewhat spongy tissue. (In amphibians, the lungs are simply hollow sacs, because much of their gas exchange occurs across the skin.) A vast capillary network spreads throughout the lung tissue, across which the oxygen must pass. The oxygen is transported from the capillaries of the lung throughout the body by a well developed circulatory system.

Air enters and exits the lung through a single, ventral tube, the *trachea,* and its branches, the *bronchi.* Various kinds of animals bring air into the lungs in different ways: in amphibians, by movements in the throat; in reptiles and birds, by contraction and expansion of the surrounding body wall (see Essay 39.1 regarding the special adaptations of the bird lung); and in mammals, by body-wall movement and contractions of a muscular *diaphragm* that

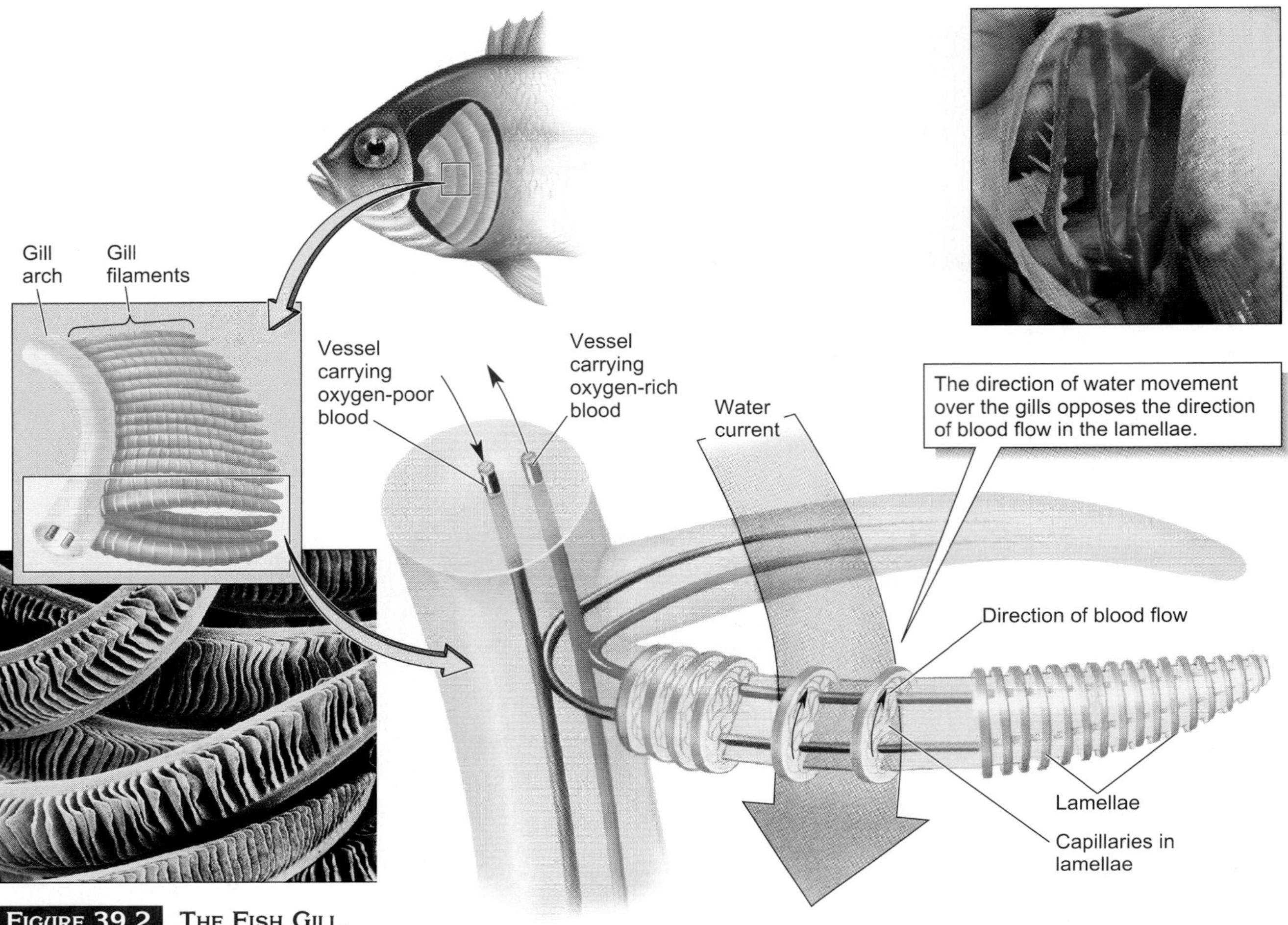

FIGURE 39.2 THE FISH GILL.
The supporting structures of the fish gills are the gill arches. At the outer curve, two rows of gill filaments protrude from each arch. The surface area of each filament is greatly increased by the presence of lamellae, as shown. Each has a rich supply of capillaries that branch from afferent vessels, carrying deoxygenated blood. Crossing a lamella, the blood loses its carbon dioxide and picks up oxygen before entering the efferent vessel leaving the filament. The opposing movement of blood and water in each lamella sets up a countercurrent exchange, greatly enhancing the exchange of gases.

lies stretched beneath the lungs. We will have a closer look at these structures in the human respiratory system.

THE HUMAN RESPIRATORY SYSTEM

The human respiratory system (Figure 39.4) is typical of that of most mammals. Its major parts include the *nasal cavity,* the *buccal* (mouth) *cavity,* the *pharynx,* the *larynx, trachea, bronchi,* and *lungs* (Table 39.1). Gases are exchanged in the lungs.

Air first enters the system through the nasal cavity or the mouth. Within the nasal cavity, it passes over a special nasal epithelium that filters, warms, and moistens the air before it enters the lungs. The nasal epithelium has mucus-secreting glands scattered among numerous ciliated cells. The mucus traps dust and other fine particles, and the ciliated cells sweep the dust-laden mucus toward the throat, where it is swallowed. Bacteria trapped in this mucus are usually killed by harsh stomach acids.

FROM THE PHARYNX TO THE BRONCHIAL TREE

Inhaled air passes through the nasal passages into the *pharynx,* a common passageway for food and air. From there it moves through the *larynx,* or voice box. The vocal cords are stretched over this

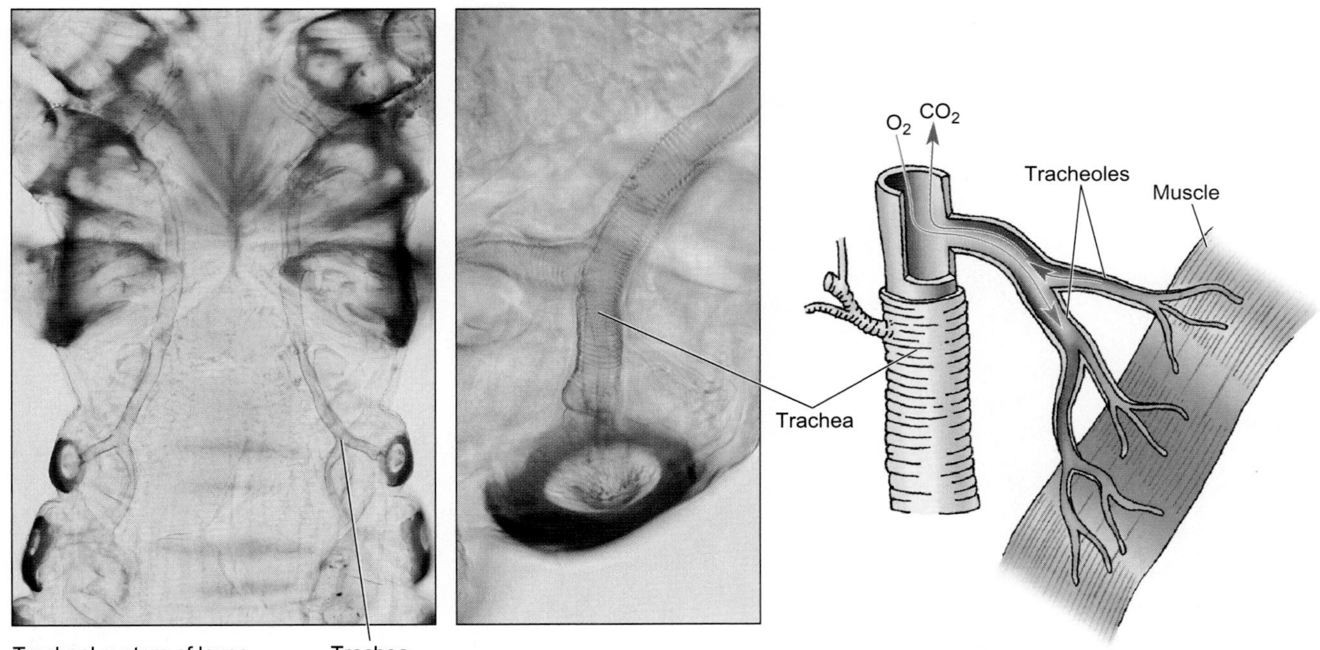

FIGURE 39.3 THE TRACHEAL SYSTEM.
The tracheal system of insects would be inefficient for larger animals as the gases must wend their way through increasingly smaller tracheal tubes to and from the tissue.

TABLE 39.1 *HUMAN RESPIRATORY STRUCTURES*

Structure	Function
Nasal passages	Filters, warms, and moistens incoming air
Pharynx	Adjustable passageway for air and food
Larynx	Passage of air, and sound production in mammals
Trachea	Passageway
Bronchi	Passage of air to each lung
Bronchioles	Passage of air to alveoli
Alveoli	Gas exchange

opening to form a triangle that narrows to a slit as the cords are tightened (Figure 39.4c) to allow us to say those things we can't believe we said.

Air moves past the larynx and on into the **trachea,** essentially a ribbed tube that is reinforced by C-shaped rings of cartilage. At the point where the trachea enters the chest cavity, it divides into right and left **primary bronchi.** The bronchi enter the lungs, where they branch and rebranch until they form the smallest branches, the **bronchioles.** These airways collectively make up what is called the **bronchial tree** (because it resembles an upside-down tree) (Figure 39.4). The trachea and larger passages of the respiratory tree are lined with an epithelium similar to that of the nasal passages, complete with mucus-secreting and ciliated cells (Figure 39.5d). The moving cilia sweep the mucous film upward, carrying trapped dust particles or other such substances out of the trachea where they, too, are swallowed. In persistent smokers, the cilia may have become paralyzed or destroyed by the smoke and the chemical toxins it contains, and the epithelium may have permanently degenerated, leaving the lungs vulnerable to a host of intruders (see Essays 39.2 and 39.3).

THE ALVEOLAR INTERFACE

The tiniest branches of the respiratory tree end in grapelike clusters of air sacs called **alveoli** (Figure 39.4e). Alveoli are extremely thin-walled and are surrounded by tiny capillaries, so here, the atmospheric air is brought close to the bloodstream. Oxygen can readily diffuse from the air in the lungs into the blood, as carbon dioxide diffuses in the opposite direction. The numerous alveolar clusters provide an enormous total surface area. In fact, our lungs hold some 300 million alveoli, with a combined surface area of nearly 100 m^2 (over 1,000 sq ft)—about the area of a tennis court.

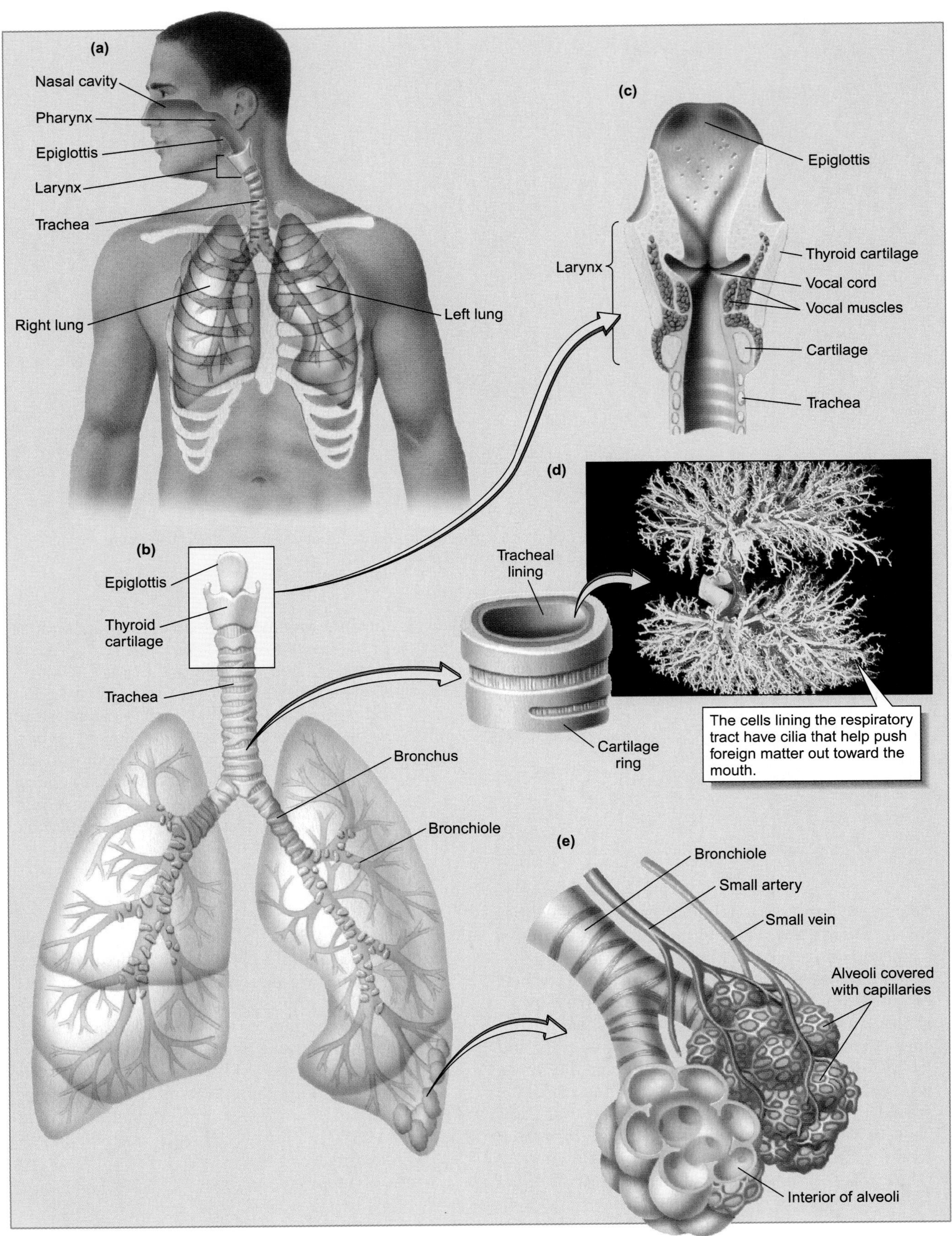
(a)
Nasal cavity
Pharynx
Epiglottis
Larynx
Trachea
Right lung
Left lung
(c)
Epiglottis
Larynx
Thyroid cartilage
Vocal cord
Vocal muscles
Cartilage
Trachea
(d)
(b)
Epiglottis
Thyroid cartilage
Trachea
Tracheal lining
Cartilage ring
The cells lining the respiratory tract have cilia that help push foreign matter out toward the mouth.
Bronchus
Bronchiole
(e)
Bronchiole
Small artery
Small vein
Alveoli covered with capillaries
Interior of alveoli

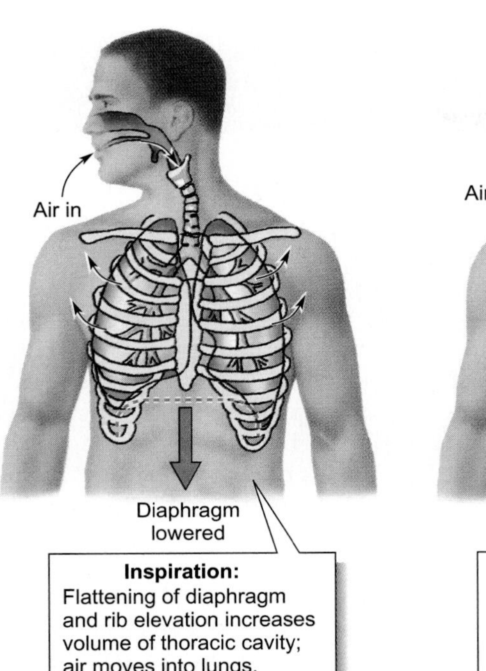

Air in
Diaphragm lowered
Inspiration:
Flattening of diaphragm and rib elevation increases volume of thoracic cavity; air moves into lungs.

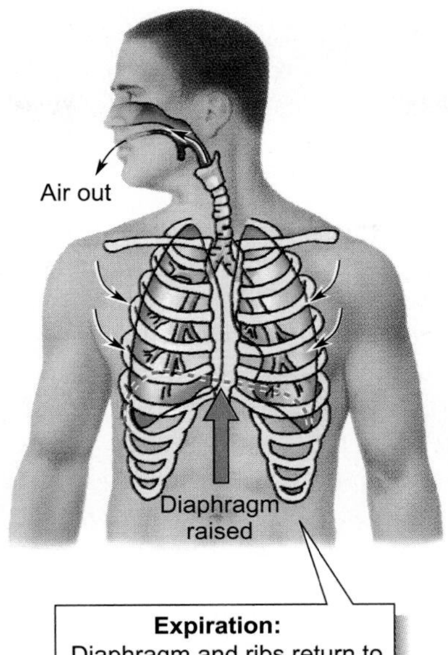

Air out
Diaphragm raised
Expiration:
Diaphragm and ribs return to normal positions decreasing volume of thoracic cavity; air moves out of lungs.

FIGURE 39.5 BREATHING.
At inspiration the flattening action of the diaphragm and the rib-elevating action of the rib muscles increase the volume of the thoracic (chest) cavity and the lungs fill. The opposite movements bring on expiration, which is generally a passive process in both the muscles and the lungs. The shelflife muscular diaphragm is strictly a mammalian characteristic.

THE LUNGS AND BREATHING

Human lungs are roughly triangular with a broad base. Each lung is enclosed by two layers of saclike membranes, the **pleurae.** The inner pleura is tightly attached to the spongy lung surface; the outer pleura is attached to the thoracic cavity. Inflammation of these membranes produces an extremely painful condition known as *pleurisy.*

At the base of the lungs is a muscular shelf, the **diaphragm,** which separates the abdominal and thoracic cavities. In its relaxed condition the diaphragm is dome-shaped and protrudes into the thoracic cavity. During inspiration the diaphragm is contracted, resulting in a lowering and flattening of its dome shape. In addition, the muscles between the ribs contract, causing the rib cage to rise and enlarge. These changes increase the volume of the thoracic cavity and cause the lungs to passively inflate. The inflation occurs because the expansion of the lungs causes the pressure within them to drop below atmospheric pressure, and air rushes in (Figure 39.5). The lungs deflate when the muscles relax and the air pressure within them becomes greater than atmospheric pressure. (No matter how hard you try, you can't force all the air out of your lungs. There is always about 1.5 liters left.)

FIGURE 39.4 THE HUMAN RESPIRATORY SYSTEM.
(a) The human respiratory system begins with the bronchial tree **(b)** where the trachea divides into the bronchi and then into the bronchioles. **(c)** A principal function of the epiglottis and larynx is to guard the entrance of the trachea, preventing foods from entering as they are swallowed. In swallowing, the larynx is elevated (you can check this with your fingers), and the epiglottis is folded over the respiratory openings, directing food into the esophagus. The larynx consists of cartilage and muscle. Toward the front is a bulging mass of cartilage, the so-called Adam's apple. In addition, the larynx houses the vocal cords. **(d)** The respiratory lining is also protected by the action of beating cilia that sweep debri upward. **(e)** The bronchioles terminate in grapelike clusters of alveoli—blind, thin-walled sacs whose surfaces contain extensive capillary beds where gas exchange occurs.

THE EXCHANGE OF GASES

As we've seen, the underlying mechanism of gas exchange in the body is diffusion, and in diffusion, as you recall, molecules move from regions of greater concentration to those of lesser concentration. There are two general areas in the body where major gradients in the concentrations of oxygen and carbon dioxide exist. One is in the body's active tissues, where concentrations of oxygen are low because it is used in cell respiration, and where concentrations of carbon dioxide are high because it is produced as a waste product. The other is in the lungs, where the blood, depleted of its oxygen and loaded with carbon dioxide, encounters air in the lungs with the reverse concentrations. Thus, in the

Essay 39.1 The Remarkable Bird Lung

The respiratory system of the bird is unlike that of any other vertebrate. The lung, like that of the reptile and mammal, is penetrated by many air passages and is thus quite spongy. But most of the similarity ends there. The passages in the bird lung do not end in alveoli (clusters of blind microscopic sacs), but occur as numerous open-ended tubes called parabronchi. There is no in-out movement of air as is the case in a human lung. Instead, air enters from the lung's posterior end—*a one-way flow.* Further, because of the arrangement of capillaries around the parabronchi, the blood passes at right angles to the air flow, forming what is called a *crosscurrent exchange.* A look at the entire bird respiratory system will help explain how all of this provides for an efficient exchange.

Air enters the respiratory system much the same as in other vertebrates, through the mouth and nostrils, and then moves into a tubular structure called a bronchus (in other vertebrates, the trachea). However, the air does not enter the lung directly at this point, but passes into a number of large air sacs. These air sacs are another unique feature of the bird respiratory system. There are three pairs of anterior air sacs and two pairs of posterior sacs (anterior and posterior to the lung). In addition, the air sacs form extensive branches, some passing into the major long bones. But these extensive air sacs have little to do with actual gas exchange. They function as air reservoirs and as a bellows that forces air in and out of the lung. As you can see in the illustration, the sacs inflate on inspiration, as the chest cavity of the bird expands (providing the force), and deflate on expiration, when the chest cavity contracts. Inflation fills the posterior sacs with fresh air and the anterior sacs with stale air (from the lung). Deflation forces fresh air into the lung and stale air into the bronchus and out of the body. Unlike the mammalian lung, which inflates and deflates during breathing, the bird lung changes very little in volume during the process.

The important points are that air flows in a one-way path through the lung—from the posterior air sacs, through the lung, to the anterior air sacs—and that this one-way flow provides for a very efficient exchange since it crosses the flow of blood through the lung capillaries. Such an arrangement required dramatic evolutionary changes from the simpler reptilian lung. How were such vast changes adaptive?

The primary advantage of the bird lung is that it adapts the bird to flight at high altitudes where oxygen levels are low. Maintaining rigorous activity at higher altitudes requires the most efficient gas exchange possible, and this, apparently, is where the unique bird respiratory system is most efficient. High-altitude flying is particularly important to birds that migrate, and while many birds migrate at an altitude of only 1,200 to 1,500 m, many fly higher. Pilots have reported birds flying at 6,000 m (19,685 ft), and radar tracking has found them at 7,000 m (about 23,000 ft). Mammals (including the flying bats) subjected to activity at this altitude would have great difficulty functioning at all and would quickly fall into a metabolic stupor. As you might expect, bats are low-altitude migrators (and humans lower still).

body's active tissues, the blood readily loses oxygen and picks up carbon dioxide—whereas in the lungs, the blood loses carbon dioxide and picks up oxygen (Figure 39.6).

Hemoglobin and Oxygen Transport. The key to the efficiency of oxygen transport in vertebrates (and some invertebrates) is a complex protein called *hemoglobin,* which makes up most of the content of red blood cells. The total hemoglobin content is so great that, if it were not contained in red cells, our blood would be too thick to circulate. Actually, oxygen could be carried in the fluid part of blood plasma without hemoglobin, but the maximum would be only about 0.3 ml of oxygen for each 100 ml of blood. Because of hemoglobin, the blood can carry about 20 ml of oxygen per 100 ml of blood—about 67 times as much.

What precisely is this magical molecule, and how does it work? **Hemoglobin** is a rather large

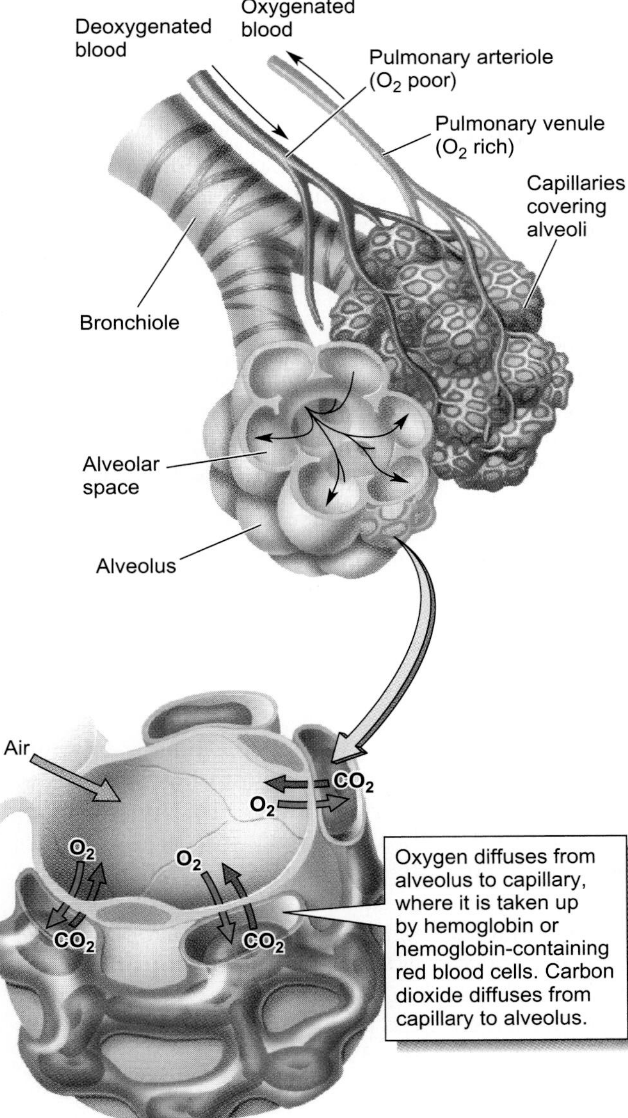

FIGURE 39.6 GAS EXCHANGE IN THE ALVEOLI. Diffusion gradients for oxygen and carbon dioxide in the alveoli and surrounding capillaries permit the required exchanges of gases. In a closer view, oxygen enters a capillary to be taken up by hemoglobin-containing red blood cells. The moving stream of red cells ensures the continuous uptake of oxygen until saturation occurs. Carbon dioxide, in solution in the plasma, escapes across the capillary wall into the alveolar space.

protein, made of four polypeptide chains and four *heme* groups, each containing an iron atom to which an oxygen molecule can attach (see Chapter 3). Importantly, the association is reversible. The four oxygen molecules are quickly released under the right conditions. The association and dissociation of O_2 and hemoglobin (Hb) can be written:

$$\underset{\text{Hemoglobin}}{\text{Hb}} + \underset{\text{Oxygen}}{4O_2} \longleftrightarrow \underset{\text{Oxyhemoglobin}}{\text{Hb}(O_2)_4}$$

This simply means than one hemoglobin molecule plus four oxygen molecules yields one oxyhemoglobin molecule, and the two-headed arrow shows that the reaction can go either way. The hemoglobin molecule is able to pick up a nearly full load of oxygen molecules in the lung and to release much of the load in the tissues, where they are needed, before it returns to the lungs for another load. (Oxygenated blood, such as that in the arteries, is usually bright red. Blood in the veins, lacking oxygen, is usually dark red.)

Carbon Dioxide Transport. Carbon dioxide transport is quite complex. To begin, carbon dioxide is readily soluble in water—in fact, it is 30 times more soluble than oxygen. Therefore, some of the carbon dioxide picked up by the blood (about 7%) goes into solution in the blood plasma. The rest of the carbon dioxide enters red blood cells, where it is transported in two ways. Some (about 23%) forms a loose, reversible association with hemoglobin the way oxygen does. However, carbon dioxide does not join the heme groups like oxygen, but reacts with amino acids of the huge protein, forming what is called *carbaminohemoglobin.*

The remaining carbon dioxide (about 70%) entering the red cells reacts with water, forming carbonic acid (H_2CO_3). Carbonic acid, in turn, dissociates into hydrogen ions (H^+) and bicarbonate ions (HCO_3^-). This reaction can occur in plasma, but there it is very slow. Red cells, however, contain the enzyme *carbonic anhydrase,* which not only speeds up the formation of carbonic acid but operates reversibly, and can rapidly convert carbonic acid back into carbon dioxide and water. This is quite important, since carbon dioxide must be reformed quickly if it is to leave the body when the blood reaches the lung.

When carbonic acid dissociates into hydrogen ions and bicarbonate ions in the red cells, the hydrogen ions are buffered (resisting pH changes) by the protein hemoglobin itself. The bicarbonate ions diffuse out into the plasma, where they are joined by sodium ions, forming sodium bicarbonate ($NaHCO_3$). In addition to providing a means of transporting carbon dioxide, sodium bicarbonate in the blood forms an important part of the body's acid-base buffering system—that is, it helps neutralize any acids or bases that might form, keeping the blood pH near neutral. (The sodium bicarbonate of the blood is identical to

Essay 39.2 The Joy of Smoking

It is interesting to watch an old film in which the lead characters smoke. We're not used to seeing it because lead characters today generally don't smoke. Smoking, Hollywood tells us, is not considered smart. Let's see why.

Tobacco is the dried leaf of the plant *Nicotiana tabacum.* It is usually rolled, shredded, or flaked, and then burned. The smoke is inhaled, allowing its products to cross the thin-walled alveoli of the lungs and to enter the bloodstream. Over 6,800 different chemicals are found in tobacco smoke, many of them carcinogens (cancer-causers). There is even evidence that the major psychoactive (mind-altering) product *nicotine* is carcinogenic. In larger doses, nicotine may also cause cramps, vomiting, diarrhea, dizziness, confusion, and tremors. In some cases respiratory failure precedes death, the ultimate lesson.

Smoke can permanently paralyze the tiny cilia that sweep the breathing passages clean, and can cause the lining of the respiratory tract to thicken irregularly. The body's attempt to rid itself of the toxins in smoke may produce a deep, hacking cough in the person next to you at the lunch counter. Console yourself with the knowledge that these hackers are only trying to rid their bodies of nicotines, "tars," formaldehyde, hydrogen sulfide, resins, and who knows what. Just enjoy your meal.

Smoking is hard enough to explain on the basis of smell and cost, but on the basis of health it is incomprehensible. The American Cancer Society tells us that a person aged 25 who smokes two packs of cigarettes a day will live about 8½ years less than a nonsmoker. Furthermore, the end of the smoker's life may be marked by extreme pain due to lung cancer, as well as cancer of the mouth, urinary bladder, pancreas, and esophagus. Even in the absence of cancer, the smoker may be severely debilitated by emphysema when thickened bronchioles and arteriole walls cause trapped air to permanently inflate the lungs. The inefficiency of the damaged lungs may affect both the heart (increasing the risk of coronary heart disease and heart attacks) and brain (bringing on behavioral changes, including sluggishness and irritability). Pregnant women who smoke increase the risk of stillbirths, and those who deliver tend to have smaller, sicker children.

Recent research has shown that a person need not be a smoker to suffer from the effects of smoking. Secondary smoke—that produced by people who do smoke, such as someone in the same room—can cause severe respiratory problems in a nonsmoker.

Fortunately, if you smoke, and if you quit in time, the damage is largely reversible. If you haven't done irreversible damage, within a year after quitting you are at markedly less risk for coronary heart disease, and 10 to 15 years later you are at no greater risk for premature death and coronary heart disease than a nonsmoker. The coughing may stop after a few weeks, and unless the damage is too great, lung function is likely to improve over time.

both commercial baking soda and the main buffering ingredient in many familiar stomach acid neutralizers.) The reactions, so far, are:

$$1)\ CO_2 + Hb \longleftrightarrow \text{carbaminohemoglobin}$$

$$2)\ CO_2 + H_2O \xleftrightarrow{\text{Carbonic anhydrase}} H_2CO_3 \leftrightarrow H^+ + HCO_3^-$$

$$3)\ Na^+ + HCO_3^- \longleftrightarrow NaHCO_3$$

As we have seen, the reactions are all reversible, and it is the quantity of carbon dioxide present that dictates the direction. This is typical of the way enzymes work. So in the active tissues, where carbon dioxide levels are high, the direction of the reactions is toward carbaminohemoglobin and toward the formation of hydrogen and bicarbonate ions and sodium bicarbonate. But in the capillaries of the alveoli, any free carbon dioxide escapes from the blood, so its concentration decreases. In the lungs, the reactions that occurred in the tissues are quite reversed: (1) The carbaminohemoglobin releases its carbon dioxide; (2) the bicarbonate of sodium bicarbonate in the plasma reenters the red cells; (3) it joins hydrogen ions to form carbonic acid; and (4) with a boost from carbonic anhydrase it is converted back to carbon dioxide and water.

ESSAY 39.3 A Brief History of Lung Cancer

Lung cancer claims about 145,000 lives in the United States each year, 85% of these cases due to smoking. There is no way we can show you the pain on these pages, or the hardship brought to loved ones—but we can show you what it looks like.

The normal ciliated epithelium **(a)** of the respiratory passages includes columnlike, ciliated, and mucus-secreting cells. In the smoker's respiratory lining **(b),** the cilia become partially paralyzed, and the mucus accumulates on the irritated lining. Where an early cancerous state exists, the lowermost basal cells divide more rapidly and begin to displace normal columnar cells. As the cancer progresses **(c),** most of the normal columnar cells are replaced by simple cancer cells that form a spreading tumor. In advanced cases, clusters of cancer cells may be carried away in the lymphatic system, spreading to other parts of the body. Wherever they stop, they begin to grow and multiply, producing new cancers and hastening death.

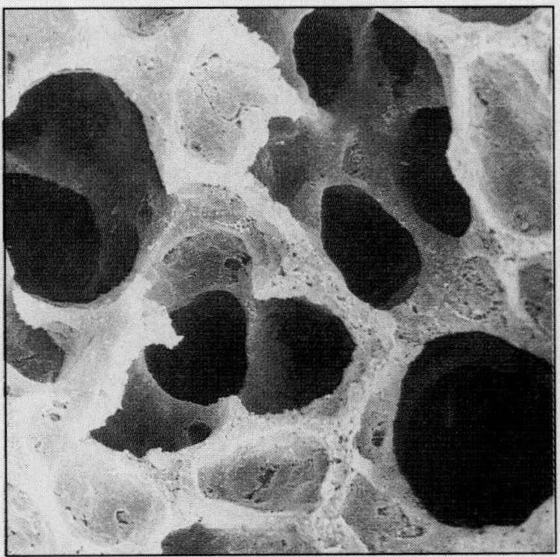

Passages in a normal lung.

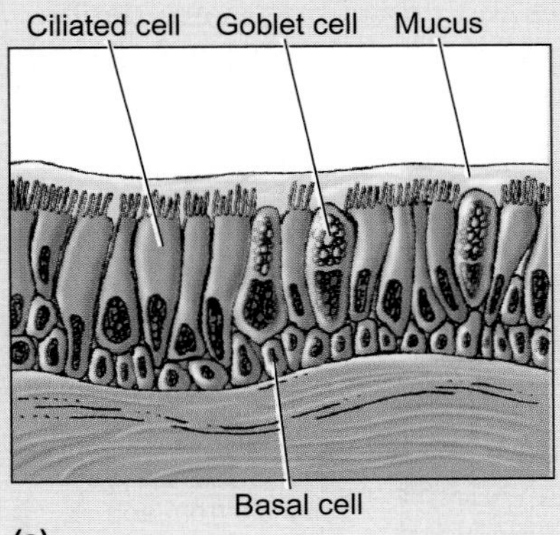

(a)

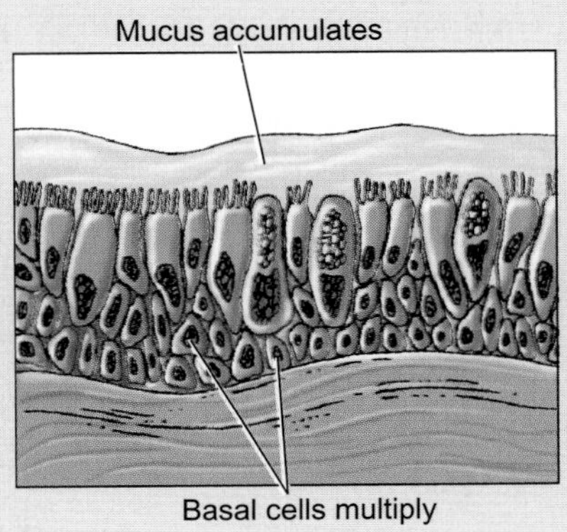

(b)

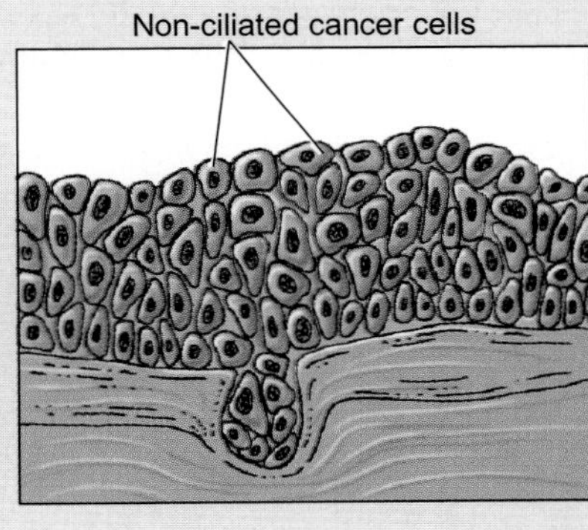

(c)

The Bohr Effect. The **Bohr effect** is named for its discoverer, Christian Bohr. It describes the effect of pH on the affinity of hemoglobin for oxygen. At low pH, such as occurs in tissues where CO_2 levels and acidic metabolites are high, hemoglobin has a lower affinity for oxygen (O_2) and hence gives it up more easily. Conversely, when at high pH (at low CO_2 levels), the hemoglobin binds more tightly to the oxygen (Figure 39.7). This means that oxygenated blood passing metabolically inactive cells does not tend to give up much oxygen. But in active cells, where CO_2 levels are high, oxygen is more readily released by hemoglobin. Actually, the Bohr effect is due more to the increase in acidity that accompanies rising CO_2 levels than to the CO_2 itself.

RESPIRATORY CONTROL

We can vary the rate and depth of our breathing, but only up to a point. If your little brother holds his breath to get his way, don't worry. He may begin to lose his rosy complexion, but the ruse won't work. No matter how hard he tries, as the CO_2 level in his blood rises, he will be forced to breathe. Thus, respiratory control has a chemical as well as a neural component. Both are only partially understood, but let's look at some of what we know.

As you see from your little brother's antics, the neural control of breathing is under both voluntary and involuntary neural control. Breathing movements may be influenced by voluntary centers in the cerebral cortex (those that permit him to make the threats about holding his breath) and by involuntary centers in the pons and medulla (those that defeat his strategy). While the anatomy of the involuntary centers is far from clear, it is apparent that the basic rhythm of breathing is set by the medulla, which contains *inspiratory centers* and *expiratory centers.* Activity by the expiratory center, as we'll see, is restricted to periods of strenuous breathing.

During quiet breathing, the inspiratory center is self-excited, creating impulses on its own. Its impulses pass through the spinal cord to the diaphragm and rib muscles, which contract and bring on inspiration. After about two seconds of activity, the inspiratory center spontaneously rests for three seconds when expiration (a passive process) occurs.

During periods of more strenuous activity, the breathing rate increases sharply. Whereas the inspiratory center still brings on inspiration, now at a faster rate, other factors come into play. For instance, the expiratory center now sends its messages to different muscles, some in the chest and some in the abdomen. Their contraction adds force to passive expiration, thus increasing the expulsion of air from the lungs. In addition, stretch receptors in the lungs—activated by prolonged inspiration—fire inhibiting signals back to the inspiratory center, thus permitting expiration and preventing overinflation.

Chemical control is based on input from several chemoreceptors, two in major arteries and one in

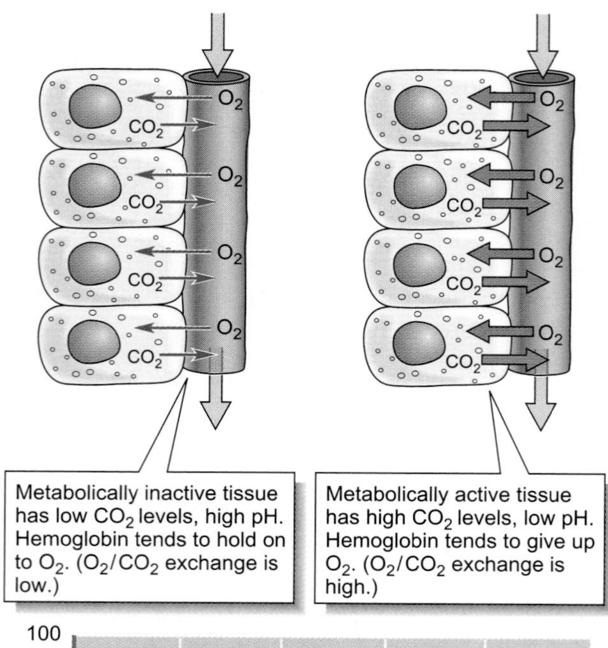

O_2
CO_2
Metabolically inactive tissue has low CO_2 levels, high pH. Hemoglobin tends to hold on to O_2. (O_2/CO_2 exchange is low.)
Metabolically active tissue has high CO_2 levels, low pH. Hemoglobin tends to give up O_2. (O_2/CO_2 exchange is high.)

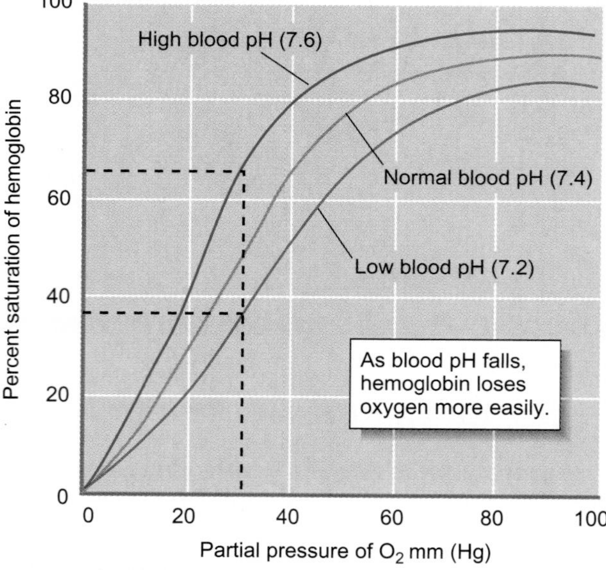

FIGURE 39.7 THE BOHR EFFECT.
In the Bohr effect, hemoglobin surrenders its oxygen load more readily in the presence of increasing amounts of carbon dioxide. In the organism, this chemical behavior means that metabolically inactive cells will receive less oxygen than metabolically active ones, regardless of the oxygen gradient. The Bohr effect is readily revealed when the rate of oxygen dissociation from hemoglobin versus carbon dioxide partial pressure is plotted on a graph.

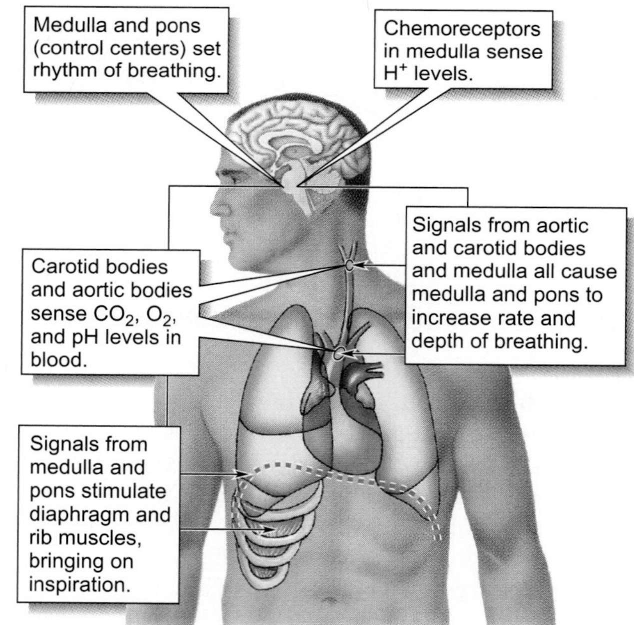

FIGURE 39.8 RESPIRATORY CONTROL.
Chemoreceptors in the aorta and carotid arteries monitor carbon dioxide, hydrogen ion (pH), and oxygen levels, communicating their information to the respiratory control centers in the medulla and pons. Additional receptors monitor the hydrogen ion content (pH) of cerebrospinal fluid in the medulla's spaces.

the brain (Figure 39.8). The arterial *carotid bodies* and *aortic bodies* are activated when blood CO_2 levels and the resulting acidity increase. They also sense decreases in oxygen, but, surprisingly, to a much lesser extent. Chemoreceptors in the fluid-filled spaces of the medulla oblongata in the brain monitor hydrogen ion levels in the cerebrospinal fluids. When neurons in any of these regions become active, their impulses are relayed to the respiratory centers in the pons and medulla, and the rate and depth of breathing increases. Such breathing, of course, decreases the level of CO_2 and acidity in the blood and increases the level of oxygen. The chemoreceptors, sensing these changes, slow their output of neural impulses (another example of negative feedback at work). Interestingly, during vigorous exercise this straightforward chemical regulatory mechanism is overridden by new input from the cerebrum and input from other kinds of sensors located in the joints and muscles.

PERSPECTIVES

Here, we set out to discuss the exchange of certain gases in the body. Few topics relating to life, though, can be discussed singly, apart from the rest. Life's processes are far too coordinated and interactive to focus on any process narrowly without losing sight of the overall picture. So we've found ourselves considering anatomy and physiology, gases and liquids, chemistry and neurology. The interplay, though, stands as one more testimony to the wondrous properties and harmonious balance of this thing called life.

Next we ask, once nutrients and oxygen enter the bloodstream, how do they get around? This leads us to the topic of circulation, that vital flow of fluids throughout the body sometimes referred to as "Rivers of Life."

SUMMARY

GAS EXCHANGE: AN EVOLUTIONARY PERSPECTIVE

The earliest aerobic life forms, like today's simpler animals, relied on simple cell-to-cell diffusion for exchanging oxygen and carbon dioxide with the environment. Diffusion remains a primary process of gas exchange, but complex animals have evolved more efficient transport mechanisms and exchange interfaces.

THE RESPIRATORY INTERFACE

The respiratory interface is generally an extensive, thin, moist membrane. Its complexity varies with the animal's size, metabolic activity, and the nature of the environment.

Many simpler invertebrates utilize simple environment-to-cell and cell-to-cell exchange. More complex species move oxygen and carbon dioxide between the respiratory surface and body cells via circulatory systems.

Most complex aquatic animals have gills—thin-walled, capillary-rich extensions of the body surface. The bony fish gill consists of rows of filaments whose blood flow opposes water flow, thus forming a countercurrent exchange. Water is moved past the gills by swimming motion or is pumped by the mouth chamber.

In tracheal systems gases pass through tiny, highly branched tubules that are filled with fluid. The gases diffuse into this fluid in passing to and from the cells.

Vertebrate lungs are saclike, spongy, internal structures with vast capillary beds that help form enormous, thin-walled exchange surfaces. Oxygen transport is greatly improved by the presence of hemoglobin. Air is moved in and out by movements of the body wall with the added assistance of the diaphragm in mammals.

THE HUMAN RESPIRATORY SYSTEM

The nasal epithelium includes cells that secrete dust-trapping mucus, and ciliated cells that sweep the mucus toward the pharynx, a common passageway for food and air.

The respiratory tree begins with the pharynx, the common opening for food and air. It continues on to the larynx. The trachea, a reinforced tube, branches into the two bronchi, which branch and rebranch many times into bronchioles forming the respiratory tree. The epithelial lining includes cleansing goblet and ciliated cells.

The lung interface is composed of grapelike clusters of thin-walled alveoli surrounded by capillaries through which oxygen and carbon dioxide pass.

Pleural membranes line and seal the chest cavity and form the surfaces of the paired lungs. The diaphragm closes the cavity below. At inspiration, the flattening diaphragm and elevating rib cage expands the chest cavity, decreasing pressure within the lungs below atmospheric pressure. Air rushes in through the bronchial tree, expanding the lungs. Relaxation at expiration produces the opposite effect, forcing air back out. The volume of air that cannot be exhaled is about 1.5 liters.

The necessary diffusion gradients for oxygen and carbon dioxide exchange in the active tissues are produced naturally by metabolic activity, with oxygen leaving the blood and carbon dioxide entering. The gradient reverses in the alveoli, with oxygen entering the blood and carbon dioxide leaving.

Because of hemoglobin the blood can carry 67 times the oxygen otherwise possible. Each hemoglobin molecule carries four molecules of oxygen (one per iron-containing heme), forming oxyhemoglobin, in an association that reverses in the tissues.

While some molecules of carbon dioxide enter solution in the plasma, most enter red blood cells. Some CO_2 joins hemoglobin forming carbaminohemoglobin, and some, with the aid of carbonic anhydrase, enters solution, forming carbonic acid. The acid ionizes, whereupon the hydrogen ions are buffered by hemoglobin and the carbonate ions join sodium, which becomes a pH buffer. The reactions simply reverse in the lungs, reforming water and carbon dioxide. The latter diffuses out.

In the Bohr effect, the rate at which oxygen disassociates from oxyhemoglobin is directly proportional to the amount of carbon dioxide present. As a result, acidity of the blood is determined primarily by the blood's CO_2 levels.

Breathing involves neural and chemical factors. In quiet breathing the neural inspiratory center in the medulla instigates rhythmic inhaling. Exhaling is passive. With strenuous activity, this action speeds up. The neural expiratory center also begins to act, producing more forceful expiration. Stretch receptors in the lung also act, preventing overinflation by inhibiting the inspiratory center. Breathing rate is influenced chemically by input from several chemical sensors that become active when carbon dioxide levels rise and pH falls (acidity increases).

Key Terms

gill • **675**
trachea • **679**
primary bronchus • **679**
bronchiole • **679**
bronchial tree • **679**
alveolus • **679**
pleura • **681**
diaphragm • **681**
hemoglobin • **682**
Bohr effect • **685**

Review Questions

1. Describe the manner in which gas exchange occurs in the simplest animals. How are favorable diffusion gradients produced?
2. Summarize the basic gas exchange problem faced by animals as their complexity increased.
3. List three characteristics of an efficient gas exchange interface.
4. List three groups of multicellular animals that rely on simple cell-to-cell diffusion for gas exchange.
5. Describe the tracheal system of an insect.
6. What structural features do most gills have in common?
7. Describe the structure of the bony fish gill, and explain how it provides for a countercurrent exchange with the water. (A simple diagram will help.)
8. Larger aquatic animals must devote a considerable amount of energy to gas exchange. What are the underlying reasons for this? How do bony fishes assure that their gills will be sufficiently ventilated?
9. What is the general makeup of the vertebrate lung? How does the breathing mechanism in the amphibians, reptiles, and birds differ from that of the mammal?
10. Describe the epithelium of the respiratory air passages and explain how it functions.
11. How is the respiratory tree protected from the accidental introduction of food or fluids during swallowing?
12. Describe the role of the diaphragm and rib muscles in inspiration. What actually causes the lungs to fill? In general, how does expiration occur?

13. Describe the diffusion gradients of the two respiratory gases in the metabolically active tissues and in the lungs.

14. Describe the hemoglobin molecule. In what part of the molecule is oxygen bound? How many molecules of oxygen can each molecule carry?

15. List the changes carbon dioxide goes through when it enters a red blood cell. Name the enzyme that facilitates both the formation and the breakdown of carbonic acid.

16. What factors determine the direction followed in the carbon dioxide reactions? List the events that occur as deoxygenated blood reaches the alveoli.

17. Describe the Bohr effect, and explain how it is adaptive.

18. Describe the neural control of breathing under restful conditions. List two additional factors that enter in when rapid forceful breathing is required.

19. Locate the three groups of chemoreceptors that monitor blood conditions. What activates these sensors? Explain how negative feedback might work here.

40

Circulation

WE HAVE BEEN AWARE THAT OUR BODIES ARE FILLED WITH FLUIDS SINCE *the first person on earth got cut. But it wasn't until 1628—thanks to experiments by the English biologist William Harvey—that we learned that these fluids circulate. Most people, though, still don't know much about how blood circulates, and where the vessels lie. Why is it dangerous to be shot in the shoulder (as we've seen in so many bad Western movies)? Where is a "better" place to be wounded? Why is it that most people don't know about that "other" circulatory system, although their very lives depend on it? Here we will try to gain a clearer picture of how fluids circulate within the bodies of living things, beginning with those species whose hearts lie on the opposite side of the body from ours.*

Even the crustiest and most jaded biologists are brought to a moment of reflection by something they have seen many times before: the formation of new life. An egg is taken from the incubator, the shell is cracked open, and the living chick embryo is exposed. It's a standard laboratory exercise, usually designed to demonstrate certain principles of development. Yet the sight of that tiny body in its watery home inevitably gives pause to each of us.

After two or three days of development, one can already distinguish familiar landmarks in this fragile form. There are the great, dark orbs that will form the eyes, and the bulbous lobes that will be the brain. And even now one can see vague and crudely defined channels where the first blood flows jerkily in halts and starts, moved along by the spasms of a simple and still tubular heart. The cells that will form other organs may still be migrating toward their posts, but the incipient heart is already in place and operating, even if feebly at first, pushing blood cells along through rough-hewn channels. Here, then, is the developing circulatory system—the array of pumps, tubes, and valves that moves fluids throughout the body—its early development emphasizing its critical role.

Why is the circulatory system so critical? Simply because it does just what its name suggests. It circulates substances from one part of the body to another—substances such as oxygen, carbon dioxide, nutrients, water, ions, hormones, antibodies, and wastes. Further, in homeothermic animals, the circulatory system transports heat, shunting it to surface parts to be radiated away or to core regions for retention (see Chapter 37). And then, as we will see, the circulatory system plays a vital role in the immune system, helping to combat bacteria, viruses, and other invaders (Chapter 41).

The circulatory systems of various kinds of animals have many traits in common, but as you might expect, there are also differences among the various groups. Let's next see how some of them have solved the problem of moving substances around throughout their bodies.

THE ARRAY OF CIRCULATORY SYSTEMS

THE INVERTEBRATE HEART

In the simpler-bodied invertebrate phyla—the sponges, cnidarians, flatworms, and a few others—there is little problem in moving substances from one part of the body to another. A relatively thin body with a large surface area and large internal spaces, such as gastrovascular cavities, preempts the need for a complex circulatory system. Cell-by-cell transport through diffusion and active transport are generally sufficient in these animals.

However, in other, more complex invertebrates such mechanisms will not suffice. The bodies are larger, the distances greater, and the tissues and organs can be quite dense. As in the simpler invertebrates,

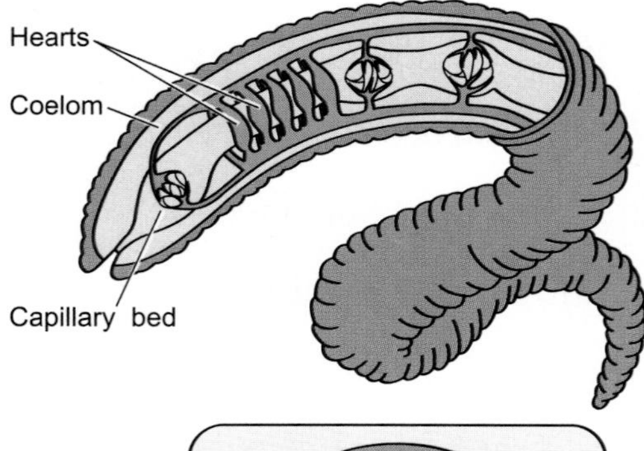

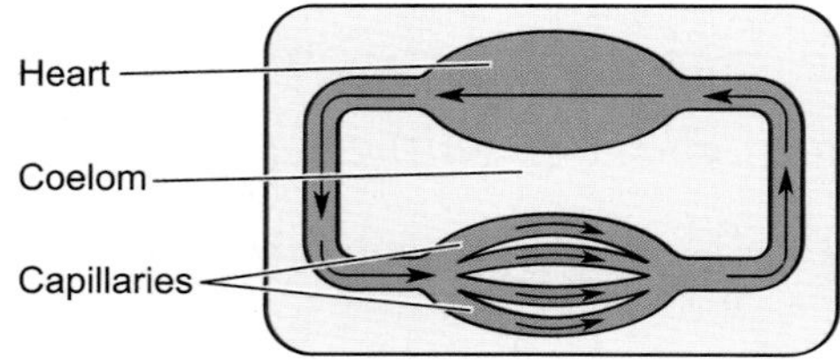

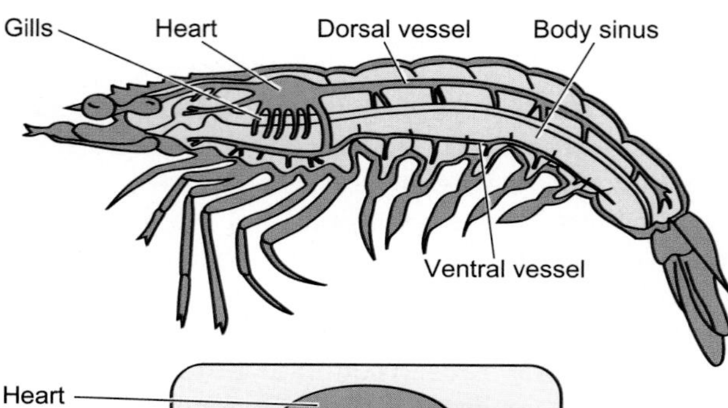

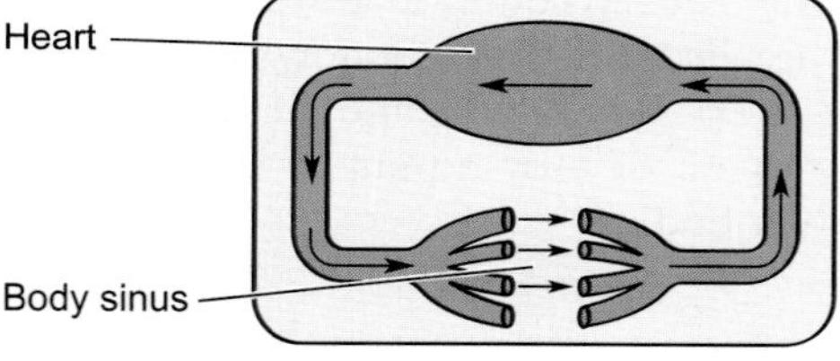

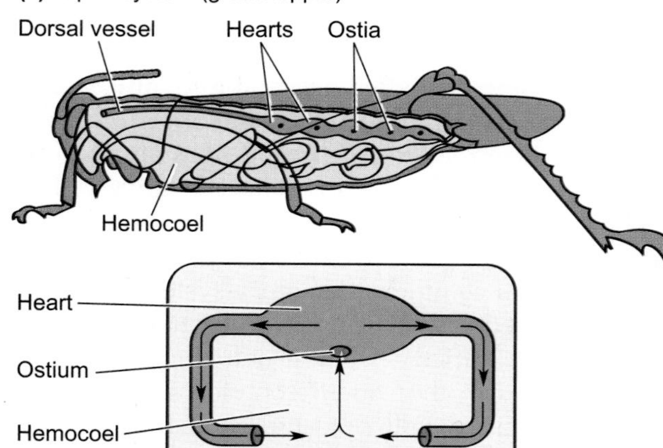

cell-by-cell transport is important, but it is simply too slow to do the job unassisted. For these reasons, the body works to shift molecules around, carrying them in channels to where they will be needed—as we see in annelids and arthropods.

Annelids—earthworms and their relatives—have a well-formed circulatory system (Figure 40.1a). The blood always remains within the vessels, so it is called a **closed circulatory system.** Interestingly, annelid blood contains hemoglobin, although it is chemically different from our own and floats free, rather than enclosed within red blood cells (see Chapter 39). Other respiratory pigments also occur in invertebrates, including a blue, copper-containing one called *hemocyanin (cyano,* copper).

The circulatory system of earthworms includes two great longitudinal vessels, one dorsal, one ventral. They are connected anteriorly by five pairs of *aortic arches,* whose tireless contractions move the blood along. Smaller vessels branch from the larger vessels, entering dense tissues where they rebranch, eventually forming extensive capillary beds. It is through the thin-walled capillaries that substances enter and leave the blood.

Earthworms and many other coelomate invertebrates (see Chapter 26) have essentially two circulatory systems since the coelomic fluid (in the space between the gut and the body wall) carries many of the same substances as does the blood and helps distribute them throughout the body.

Arthropods have an **open circulatory system**—that is, the blood is not always enclosed in vessels. It is carried along in places by vessels; but then the vessel ends, and the blood is free to move between cells (or "percolate") until it eventually collects into vessels again. Insects usually have only one large dorsal vessel (Figure 40.1b). As the dorsal vessel contracts, it forces blood into the few vessels that arise from it. From there the blood simply flows freely into tissue spaces, bathing the cells and exchanging materials as it percolates through the body. Eventually (largely pushed along to the insect's movements) it ends up back at the dorsal region, where it enters the heart through tiny openings called **ostia.** One reason a

FIGURE 40.1 VARIOUS CIRCULATORY SYSTEMS. A closed circulatory system, as in the earthworm (**a**), retains blood inside vessels, with exchanges occurring in capillary beds only. In the insect's and crustacean's open circulatory systems (**b** *and* **c**), blood is pumped through vessels to open sinuses through which it gradually makes a return to the heart. Before its return trip in the crustacean, it is shunted through the gills for oxygenation.

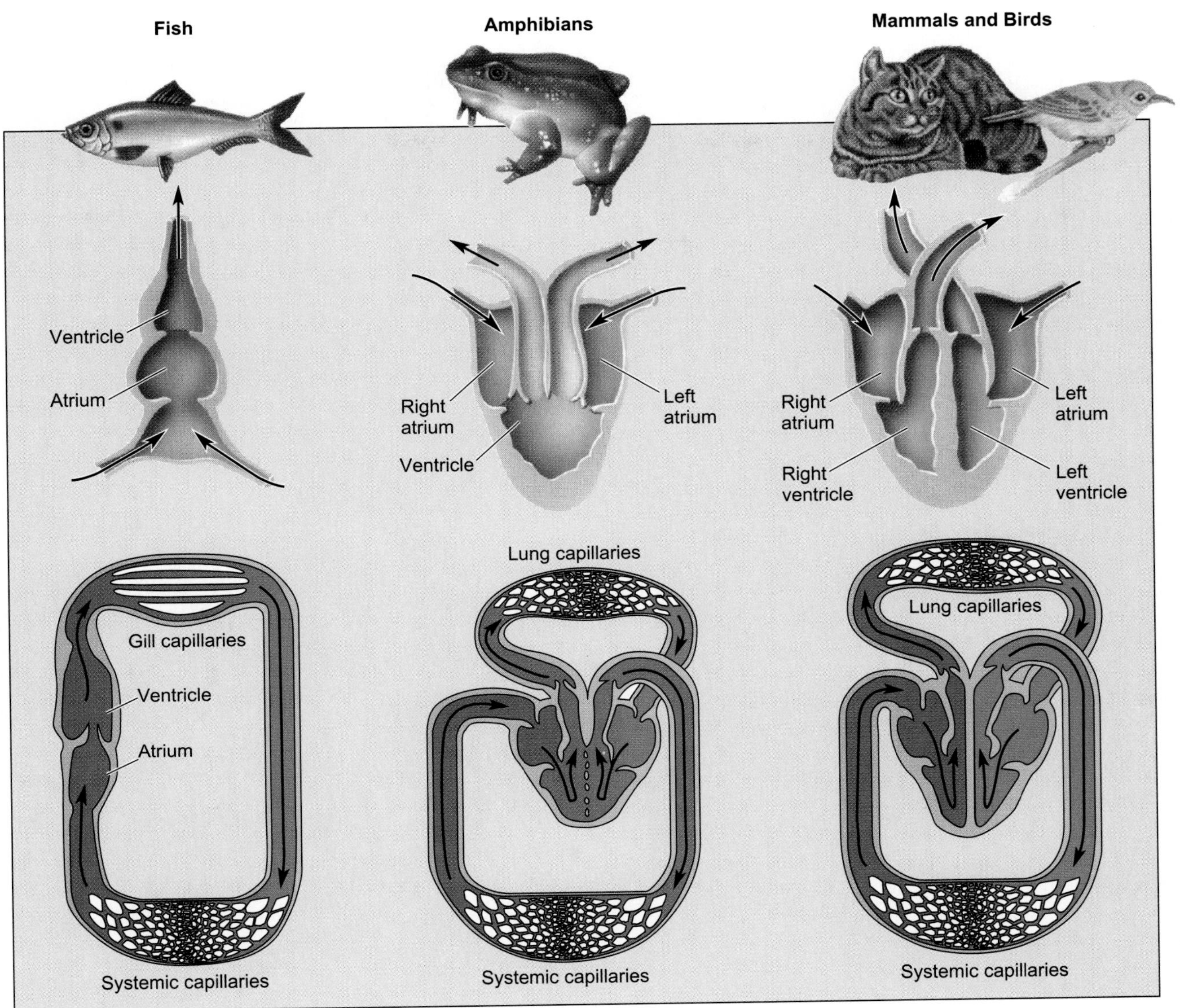

FIGURE 40.2 **TWO-, THREE-, AND FOUR-CHAMBERED HEARTS.**
The fish has a single circuit. Deoxygenated blood is received by the atrium, which delivers it to the ventricle. It is then pumped to the gill capillaries, where oxygenation occurs. From the gills the blood must pass through body capillaries before returning to the heart. Two circuits are seen in the amphibian. Deoxygenated blood enters the right atrium of the three-chambered heart, from which it is pumped to the ventricle. Simultaneously, oxygenated blood enters the left atrium, and it too moves to the ventricle. The ventricle then pumps blood from the two sources to the lungs and body. Birds and mammals have an efficient four-chambered heart in which the oxygenated and deoxygenated blood are kept in separate circuits.

sluggish circulatory system is sufficient in insects is that the blood functions primarily in carrying nutrients to the cells and carrying away waste. The circulatory system is not important in gas exchange—which is accomplished by a tracheal system in which tiny tubes communicate directly between the atmosphere and the insect's tissues (Chapter 39). Crustaceans on the other hand, actually have a fairly extensive system of vessels (Figure 40.1c) that keeps blood pressure high enough to assure a steady flow of blood through the gills.

THE VERTEBRATE HEART

The vertebrate heart is a muscular pump that moves the blood along. It varies widely in structure among invertebrates, its form being related to the life history of the organism. There is less variation in vertebrates, but there is an important progression regarding the number of chambers.

The fish heart has only two regions, roughly homologous to the atria and ventricles of mammals (Figure 40.2). One, the thin-walled region called the

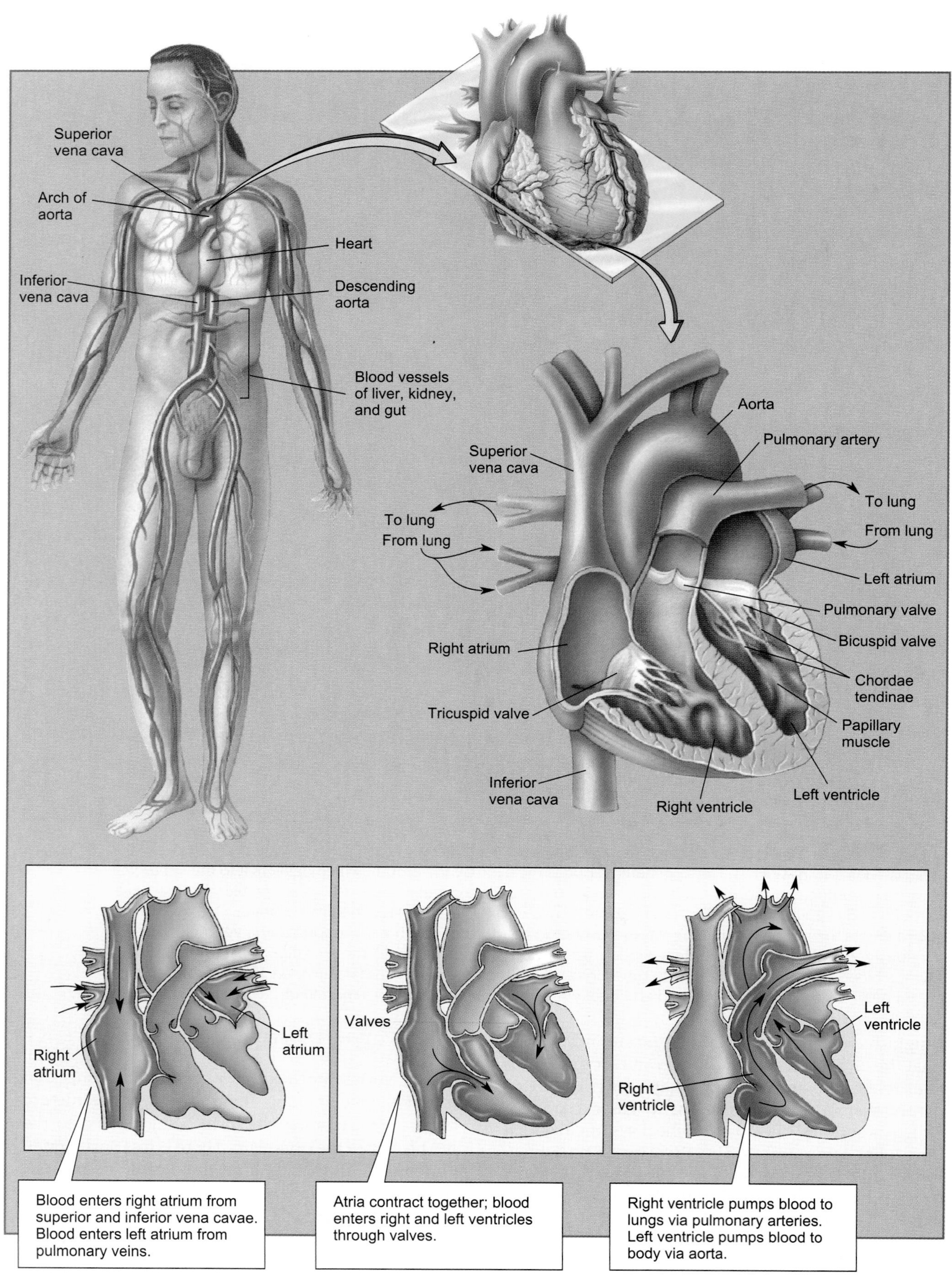

Superior vena cava
Arch of aorta
Heart
Inferior vena cava
Descending aorta
Blood vessels of liver, kidney, and gut
Aorta
Pulmonary artery
Superior vena cava
To lung
From lung
To lung
From lung
Left atrium
Pulmonary valve
Bicuspid valve
Right atrium
Chordae tendinae
Tricuspid valve
Papillary muscle
Inferior vena cava
Right ventricle
Left ventricle
Right atrium
Left atrium
Valves
Left ventricle
Right ventricle
Blood enters right atrium from superior and inferior vena cavae. Blood enters left atrium from pulmonary veins.
Atria contract together; blood enters right and left ventricles through valves.
Right ventricle pumps blood to lungs via pulmonary arteries. Left ventricle pumps blood to body via aorta.

atrium, receives blood after it has been circulated through the body and passed through the sinus venosus. Contraction of the atrium forces blood into the other chamber, called the **ventricle,** that moves the blood through the body. The atrium doesn't have to be very strong because it doesn't take much work to pump blood only to the next chamber. The ventricle, however, must generate tremendous pressure, since it must force the blood through the conus arteriosus and into the large artery called the **aorta,** and thence into the tiny mesh of the gill capillaries where gases are exchanged, and thence through the rest of the body. From the major tissues of the body, blood reenters the veins. (**Arteries** carry blood away from the heart; **veins** carry blood toward the heart.) The fish ventricle, then, must be able to pump blood through two capillary beds—one in the gill and one in the body. Friction is high in these beds due to the enormous surface area of the interior of the tiny vessels of such an extensive network. Not surprisingly, the ventricle is extremely thick and muscular.

The amphibian heart (Figure 40.2) is a bit more complex in that it has two atria. One receives oxygenated blood (blood high in oxygen) from the lungs and skin, and the other receives deoxygenated blood (high in carbon dioxide) from the rest of the body. The two may mix in the single ventricle, so the system is not particularly efficient; but it is efficient enough for the rather lethargic frog (lethargic till you try to catch one). Also, much of the gas exchange in frogs takes place across the skin, and so they have not needed particularly efficient lungs.

Reptiles go one step better in that the ventricle is partially divided by an incomplete wall. (The wall may be complete in crocodilians.) Because of the incomplete ventricular *septum* (wall), there is some mixing of blood from the two sides of the heart. Still, this system is more efficient than the frog's, since one side of the heart pumps deoxygenated blood through the lungs and the other side pumps oxygenated blood through the rest of the body.

In birds and mammals (Figure 40.2), the right and left sides of the heart are completely separated. With an efficient circulatory system, it is easier for these homeothermic animals to maintain their body temperature by moving the blood to the heat-generating cells and distributing the heat to the rest of the body when environmental temperatures are low.

FIGURE 40.3 THE FLOW OF BLOOD THROUGH THE HUMAN HEART.
Blood from the right atrium (deoxygenated and deep red) is pumped to the right ventricle and from there through pulmonary arteries to the lungs. It returns oxygenated and bright red to the left atrium and from there to the left ventricle to be pumped to the body via the aorta. Notice that the two atria contract simultaneously, as do the ventricles. Also, see how the coronary arteries service the heart itself.

STRUCTURES OF THE HUMAN HEART

In keeping with its critical role, the human heart is an enormously powerful and virtually tireless muscle. Since the left ventricle must pump blood throughout the body, it is particularly large and strong. This, coupled with the fact that the heart is slightly tilted, causes many people to assume the heart is located on the left side, whereas it is actually located in the center of the chest. Thus, people swear to all sorts of things with the right hand held over the left lung. Tracing the flow of blood through this magnificent muscle (Figure 40.3), we begin with the deoxygenated blood returning to the heart from the body. This blood is delivered through two great veins, the *superior vena cava* (from the upper part of the body) and the *inferior vena cava* (from the lower part of the body). They empty into the right atrium, a thin-walled receiving chamber. When the right atrium contracts, it forces blood into the right ventricle, the thicker-walled chamber below. Contraction of the right ventricle forces blood into the **pulmonary artery,** which branches, carrying blood to both lungs. Freshly oxygenated blood is returned to the left atrium through the **pulmonary veins.**

The oxygenated blood passes from the left atrium to the powerful left ventricle. When the left ventricle contracts, blood is forced into the aorta, the largest artery in the body, and from there through the vessels of the body.

Heart Valves. As the atria and ventricles contract, blood is prevented from flowing backward by four 1-way valves (Figure 40.4). The *tricuspid valve* lies between the right atrium and right ventricle, while its counterpart, the *bicuspid valve,* is between the left atrium and left ventricle. The "cusps" are extremely strong flaps of fibrous connective tissues that are pushed upward and come together as blood in the ventricle starts to back up. The cusps are anchored so that they cannot collapse completely into the atria when they snap shut. The anchors are tough tendinous cords, the *chordae tendineae,* that are attached to the cusps at one end and to cone-shaped *papillary muscles* arising from the chamber walls at the other. (Are the chordae tendineae the heartstrings of poem and song?)

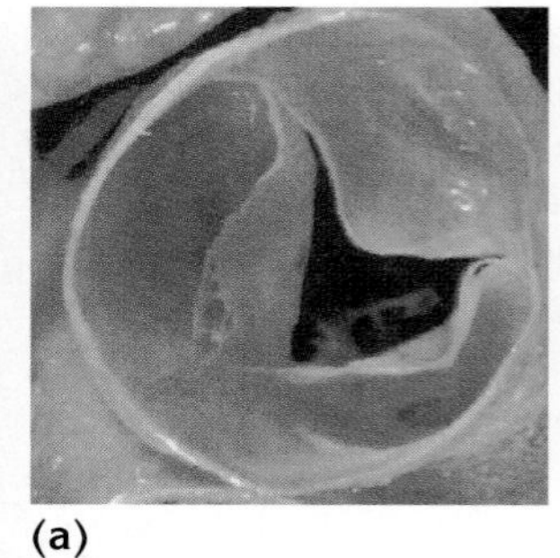

(a)

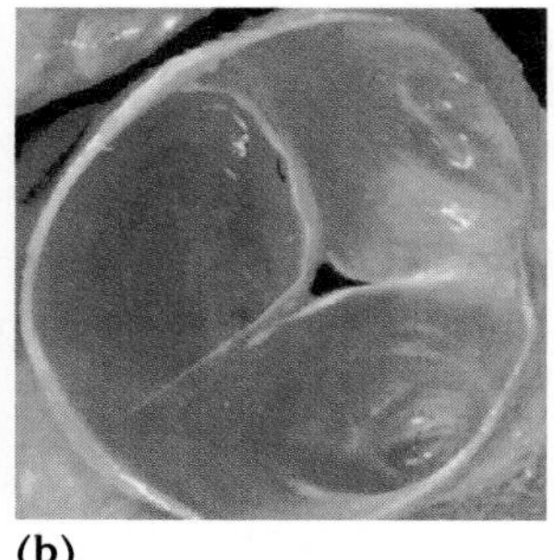

(b)

FIGURE 40.4 A VALVE IN ACTION.
Above are photos of the aortic semilunar valves **(a)** open, and **(b)** closed. The pressure doesn't force them backward because they are anchored by chordae tendineae, as we see at left.

The two other valves, the *pulmonary semilunar valve* and the *aortic semilunar valve,* lie at the base of the pulmonary artery and the aorta, respectively. These valves each consist of three small pockets attached to the inside artery wall. They fill with blood and balloon out (like parachutes), thereby preventing the backflow of blood from the arteries into the ventricles.

The valves have a great deal to do with the familiar "lub-dup" heart sounds. The first—the "lub"—is produced primarily by the closing of the tricuspid and bicuspid valves as they respond to the increase in pressure caused by the two contracting ventricles. The second, louder, sound—the "dup"—is the closing of the semilunar valves, which tells us when blood has filled the pulmonary artery and the aorta. (The sound is actually created by the turbulence of the blood in the heart chambers as the valves close.)

THE HEARTBEAT

Some might say that the subject of the human heart has been overworked. After all, not only have volumes been devoted to describing its relentless and tireless activity, but we have elevated and venerated it as the center of love and emotion. In reality, though, the heart needs no romanticizing, since no discussion can be cold and clinical enough to drain it of its wonder.

One of the surprising facts about heart muscle is that it is *intrinsically contractile;* that is, it can contract without external influences. An excised heart placed on a dish will go on beating—at least for a time. Because of its intrinsic contractility, a transplanted heart, cut off from the signals from the nervous system, will go on beating quite rhythmically. It is also extremely *conductive,* with many of the characteristics of nerve tissue. Recall (Chapter 33) the presence of intercalated disks—those extensive, interlocking regions between fibers that enhance cell-to-cell conduction.

The heartbeat originates in a group of specialized muscle cells in the wall of the right atrium. This control center is technically known as the *sinoatrial (SA) node,* but it is usually called the *pacemaker* (Figure 40.5). The pacemaker generates repeated action potentials similar to those of neural impulses. These are transmitted to both atria, which respond by contracting simultaneously. Then the action potential reaches a second node, the *atrioventricular (AV) node.* There is enough delay in the transmission of the impulse for the atria to complete their contraction before the ventricles begin theirs. Impulses from the AV node pass through a strand of specialized muscle in the interventricular septum known as the *Bundle of His.* The bundle branches into right and left halves that travel to the pointed apex of the heart, where they branch into *Purkinje fibers* that initiate contraction there. The ventricular muscle fibers are arranged in a spiraling fashion so that their contraction produces a twisting, wringing motion that squeezes the blood out, and into the aorta.

Should the pacemaker fail, the heart rhythm will go awry. In such cases, it may be necessary to install an artificial pacemaker, an electronic device about the size of a pocket watch. Such surgery is now fairly routine.

While the heart can pump blood without outside influence, it must respond to varying oxygen needs and changing blood pressure in the organs it serves. (Sometimes the heart can fail, as we see in Essay 40.1.) Input from the nervous system, specifically from the autonomic division, permits the necessary adjustments to be made.

You may recall (Chapter 34) that the heart is innervated by two groups of nerves from the autonomic nervous system—one sympathetic, the other parasympathetic. The parasympathetic nerves slow the heart rate by releasing the neurotransmitter acetylcholine into the pacemaker and the cardiac muscle. The sympathetic nerves accelerate the heart rate by releasing the neurotransmitter *norepinephrine*. The release of extrinsic neurotransmitters must be coordinated very precisely to control the heart's response to the body's changing demands. *Epinephrine* (or adrenalin) and norepinephrine (released by the adrenal medulla and carried by the bloodstream) can also elevate the heart rate (see Chapter 36).

Blood Vessels and Circulatory Routes

With the contraction of the ventricles, the blood begins its journey through the body. Let's now see where the blood goes, and what happens when it gets there.

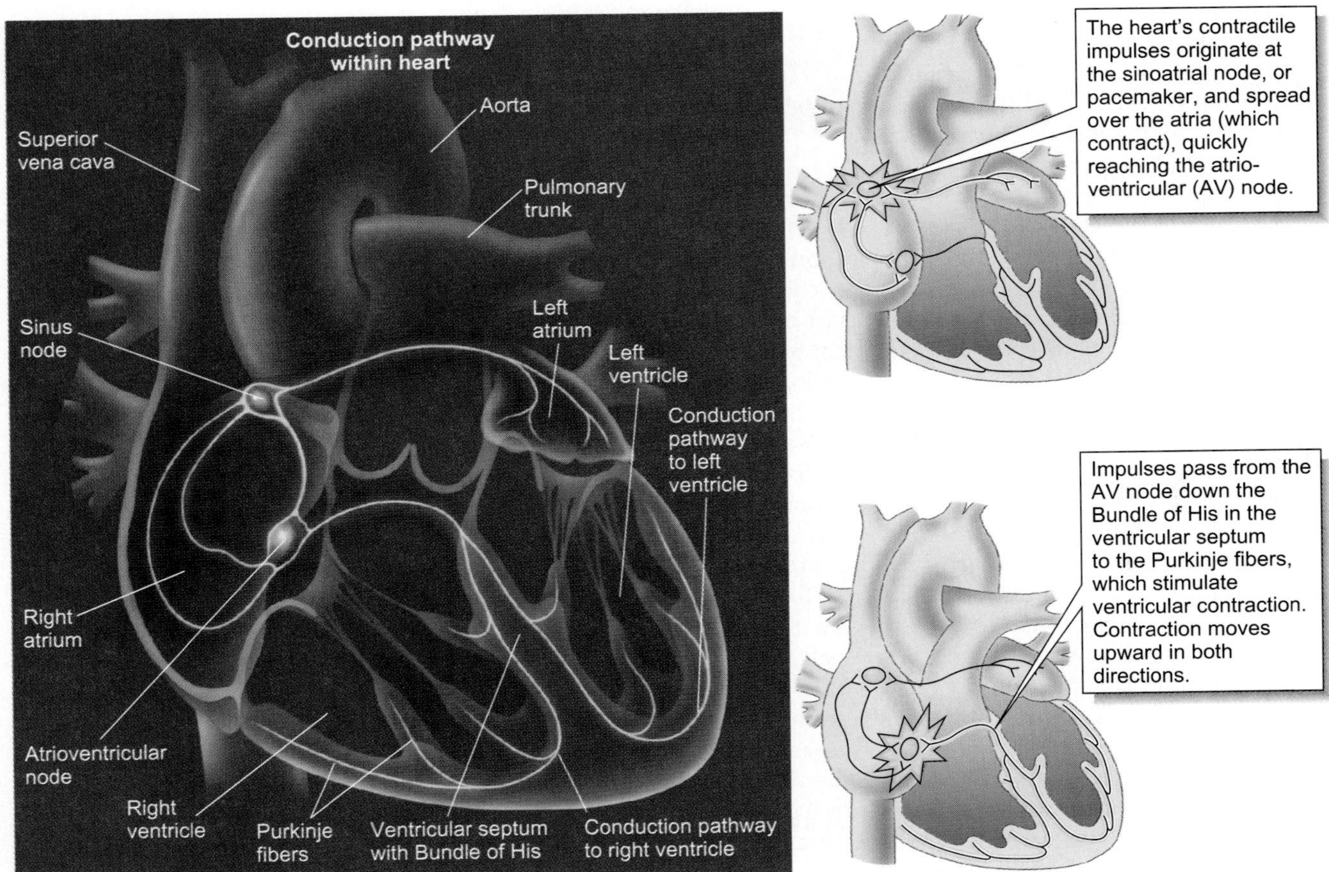

Figure 40.5 Control of Heart Contraction. The heart's contractile impulses originate in the sinoatrial node, or pacemaker, and spread over the atria (which contract), quickly reaching the atrioventricular (AV) node. From the AV node, impulses pass down the Bundle of His in the interventricular septum to the Purkinje fibers, whereupon ventricular contraction begins. Contraction proceeds upward in both ventricles producing a squeezing effect that directs the blood into the large arteries leaving the heart.

Essay 40.1 When the Heart Fails

At the time the Declaration of Independence was signed, the most common cause of death in this new country was, ostensibly, indigestion. Either the quality of our food has improved since then, or (more likely) the problem was misdiagnosis. What the founding fathers and mothers were really suffering from were heart attacks.

We're all familiar with the term *heart attack.* (You've heard about the faint-hearted guard dog who when told "Attack!" had one.) But what, exactly, *is* a heart attack? In most instances, it is heart failure caused by an insufficient blood supply to the heart muscle. The heart attack was once referred to as *coronary thrombosis* ("blood clot in a coronary vessel"). A coronary artery may indeed be blocked by a moving blood clot that has lodged there, but the blockage is more likely to be due to *coronary arteriosclerosis,* a hardening and narrowing of the vessel. A specific form of this condition, *atherosclerosis,* occurs when *plaque* coats the inside of arteries, reducing the inside diameter and restricting blood flow. Plaque is a flaky, crumbly substance that sometimes breaks away from artery walls to be carried through the bloodstream, increasing the risk of a blockage. It may also break away only partially, forming a flap that blocks blood flow. A blockage in a coronary artery near the aorta is particularly dangerous because these vessels feed large areas of the heart; indeed, most deaths from heart attacks result from such incidents. Blockages farther along in the coronary arteries, deeper in the heart muscle, affect less tissue.

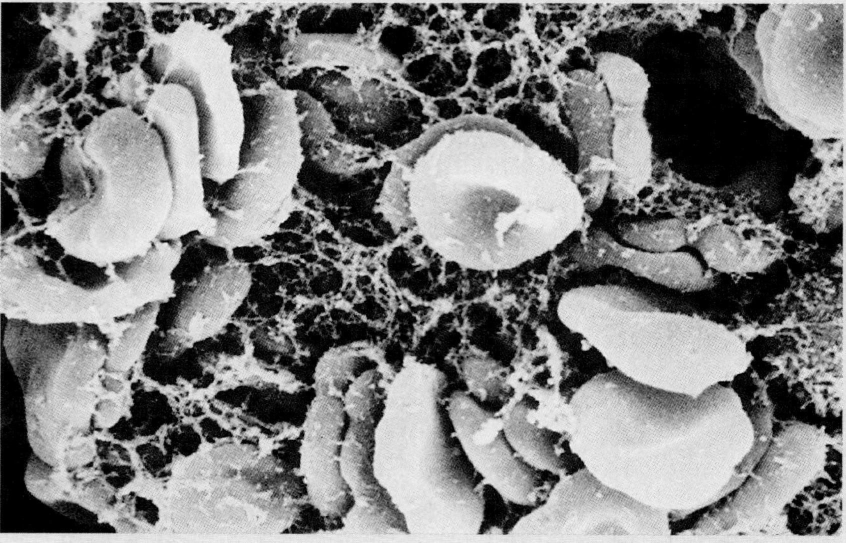

However a vessel is blocked, the heart muscle deprived of blood soon dies. The death of heart tissue is called *myocardial infarction*—and not only does this tissue cease to contribute to the heart's pumping action, but more seriously, it disrupts the rhythm of the heart. Impulse conduction and contraction of the heart muscle are precisely coordinated, so a loss of function in one area affects the coordinated contraction of the entire heart.

Today, blocked coronary vessels are routinely treated, either medically or surgically. For example, blocked vessels can be surgically repaired in what is known as *coronary bypass surgery.* In this procedure, the damaged arteries are replaced by corresponding lengths of vein, usually taken from the leg. Double, triple, and quadruple bypass procedures are now common, as surgeons seek to restore the blood supply to heart muscle that is not yet dead. Enzymes, such as *streptokinase,* can also be used to dissolve a clot, and are particularly effective if administered within an hour of a heart attack. A very physical approach to a blocked coronary vessel is *angioplasty,* in which a balloon catheter is inserted into the heart and then expanded, pressing the occluding material against the vessel wall and restoring the flow. Perhaps the most promising new technique in pinpointing problems is *intravascular ultrasonography.* Here a catheter carries an ultrasound generator into the heart, and the sound reflections are translated to visual images by computer. The physician then knows precisely where the clot is, and how it lies.

Researchers also now tell us that an aspirin every other day, and a little wine every day seems to stave off heart problems—problems that might not arise at all if people would do the obvious: cut down on fat, lose weight, and get plenty of exercise—and *then* celebrate their good health with a glass of wine.

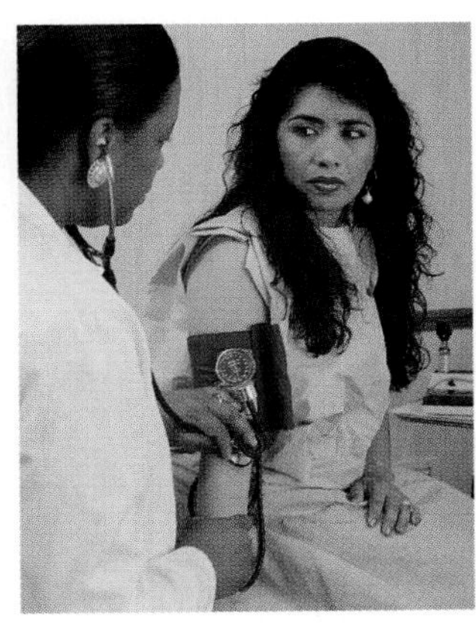

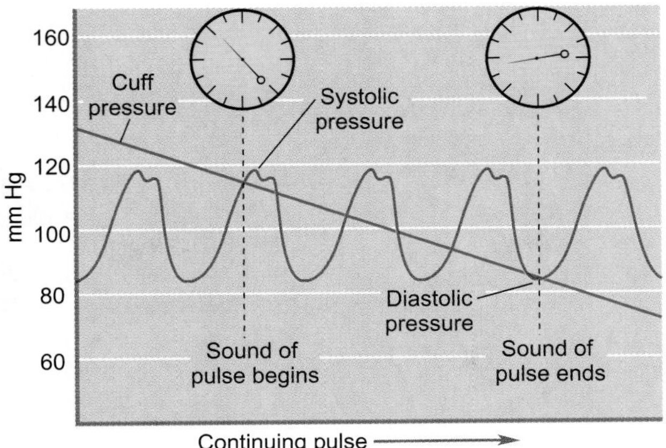

FIGURE 40.6 **BLOOD PRESSURE.**
The sphygnomanometer consists of an inflatable pressure cuff and a pressure gauge or mercury column. The cuff is wrapped around the upper arm and inflated until its pressure exceeds the pressure in the brachial artery. As the air in the cuff is gradually released, the first pulse sounds can be detected through a stethoscope. The gauge at this point will give the systolic pressure. With the continued release of air, the sounds become louder, but then disappear. At this instant the gauge will reveal the diastolic pressure.

Arteries. Arteries carry blood away from the heart, but they are not simple conduits; one of their major functions is in maintaining blood pressure by contracting and exerting pressure on the volume of blood inside. The great vessel that receives blood from the left ventricle, the aorta, receives the full impact of the heart's powerful surges. The sudden swell of blood during *ventricular systole* (contraction) expands the elastic walls of the aorta, causing it to automatically contract, helping to raise blood pressure and to push the blood along. What, though, *is* blood pressure?

Blood pressure is the force of blood against the vessel walls. The highest pressure occurs during ventricular *systole,* when the blood is pumped from the ventricle into arteries with great force. This pressure is called the **systolic pressure.** During *diastole* (relaxation, when the ventricle is filling), blood pressure remains relatively high because of the force of the recoiling aorta on the remaining blood in the vessel. However, there is some drop in pressure as the ventricles fill. This is called the **diastolic pressure.** A typical systolic pressure is 120 mm Hg; a typical diastolic value, 80 mm Hg. The blood pressure in this case would be "120 over 80" (Figure 40.6). (Millimeters of mercury—mm Hg—is the standard way of expressing blood pressure.)

As the aorta leaves the heart, it gives rise at once to its first branches, the *coronary arteries,* which go directly to the heart muscle, providing it with nutrients and oxygen. The aorta then curves to the left and forms the aortic arch from which additional arteries arise, their branches extending into the head and arms. From the arch, the aorta proceeds downward, sending branches into viscera, trunk muscles, and the vertebral column. It then divides, in the lower abdomen, to form the major arteries of the legs.

The arteries eventually form *arterioles* (small arteries), which branch once again to form capillaries. Many arterioles contain *precapillary sphincters,* rings of smooth muscle that regulate the flow of blood into the capillary beds (Figure 40.7). The sphincters respond to the autonomic nervous system and to certain hormones.

Capillaries. The capillaries are fascinating structures, some so small that blood cells must squeeze through them in single file. In spite of (or because of) their small size, this is where the blood does its work. Capillaries are so pervasive in active tissues that virtually no cell is far from these tiny rivers of life. This distance is critical, because the cells must draw sustenance from these vessels and deposit metabolic wastes into them.

The capillary wall is formed by a single layer of interlocking cells (Figure 40.7). It is so thin that some substances can cross by simple diffusion. The movement of other substances may be encouraged by the high hydrostatic pressure that forces smaller molecules across the thin walls. Such movement is facilitated by tiny pores at cell junctions. In addition, some movement is due to active transport and pinocytosis (see Chapter 5). In this process, fluids and suspended materials are drawn into a pocketlike indentation

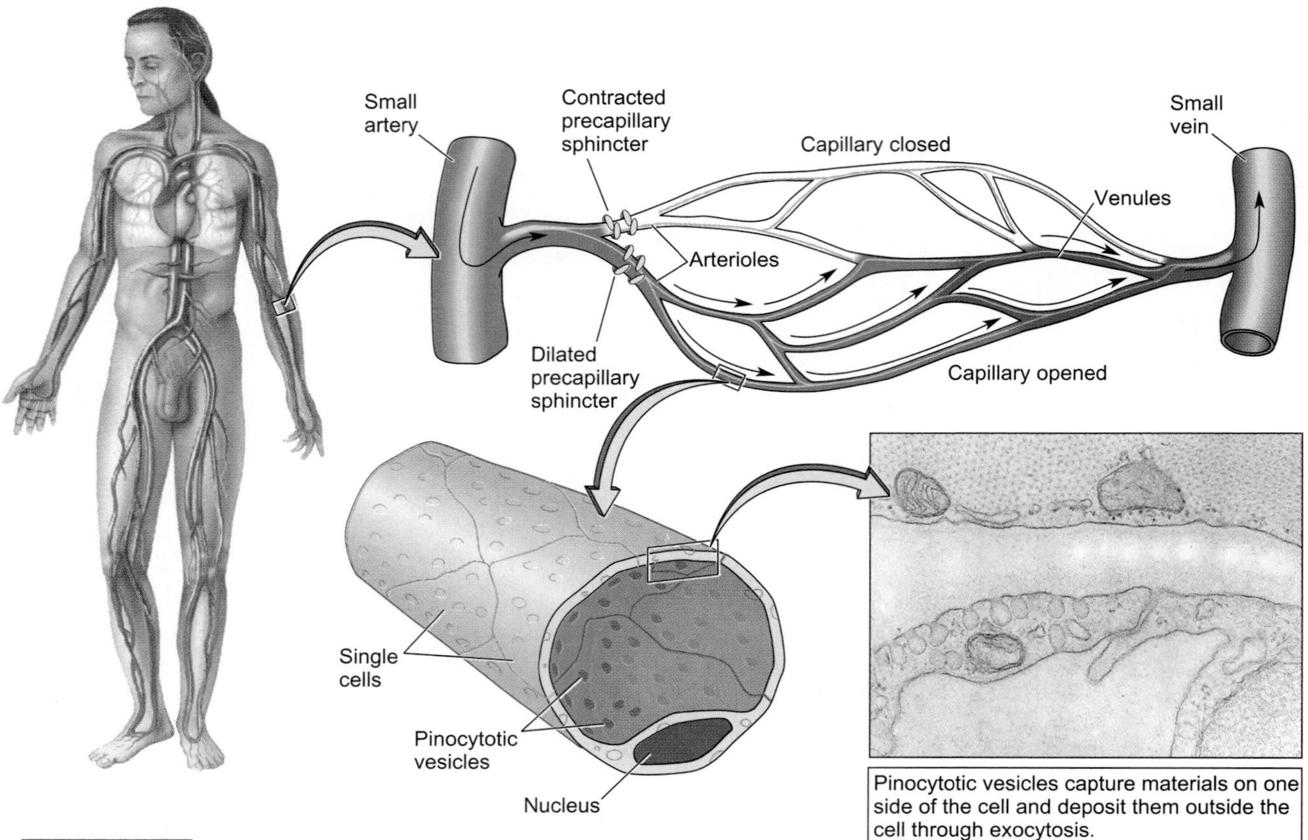

FIGURE 40.7 CAPILLARIES.
Capillaries are thin-walled, consisting of a single layer of cells. Some substances move readily through the capillary walls, while others must be actively transported. Capillaries usually occur in highly branched beds and arise from arterioles. They pass through tissue layers and among individual cells. Blood may be shunted from one bed to another by smooth-muscle sphincters in the arterioles. The mitochondria assist in providing the energy for the contractions. Many substances are taken in capillary cells through pinocytosis. Pinocytic vesicles capture materials on one side of the cell, pass to the other side in the form of vacuoles, and deposit the materials outside the cell through exocytosis.

formed by the membrane of a capillary cell. The pocket then breaks off, forming a vesicle that is transported across the cytoplasm, and the materials are squeezed out through exocytosis on the opposite side (Figure 40.7).

Veins. The capillaries—now depleted of food and oxygen, and laden with waste (especially carbon dioxide and nitrogenous waste)—make their way to small vessels called *venules,* which merge to form veins—the vessels that carry blood toward the heart. Veins have thinner walls than arteries, largely because they lack the heavy musculature of the arterial wall, but they are composed of the same tissues as we see in Figure 40.8.

Blood has reached the capillary beds under great pressure, forced along by the heart and squeezed by arterial muscles. But in its passage through the capillary beds, its pressure has been dissipated, and it enters the venules sluggishly. From here, though, it must reach the heart, and lacking the means to push blood along, veins must rely on a number of mechanisms to return blood to the heart. For example, many veins have one-way flap valves that allow the blood to move in only one direction—toward the heart. Also, veins in or near skeletal muscles are squeezed when the muscles contract, pushing the blood along. These two mechanisms work together in some veins. The muscles push blood along, and the valves ensure that the blood moves in the right direction. In addition, during breathing, changing pressure in the thoracic cavity helps blood move through the veins there.

BLOOD

The sheer volume of blood in the human body is remarkable; 5 to 6 liters in adult men and 4 to 5 liters in adult women; on the average, blood makes up about 8% of a person's total body weight. The blood volume includes a safety factor, since any healthy

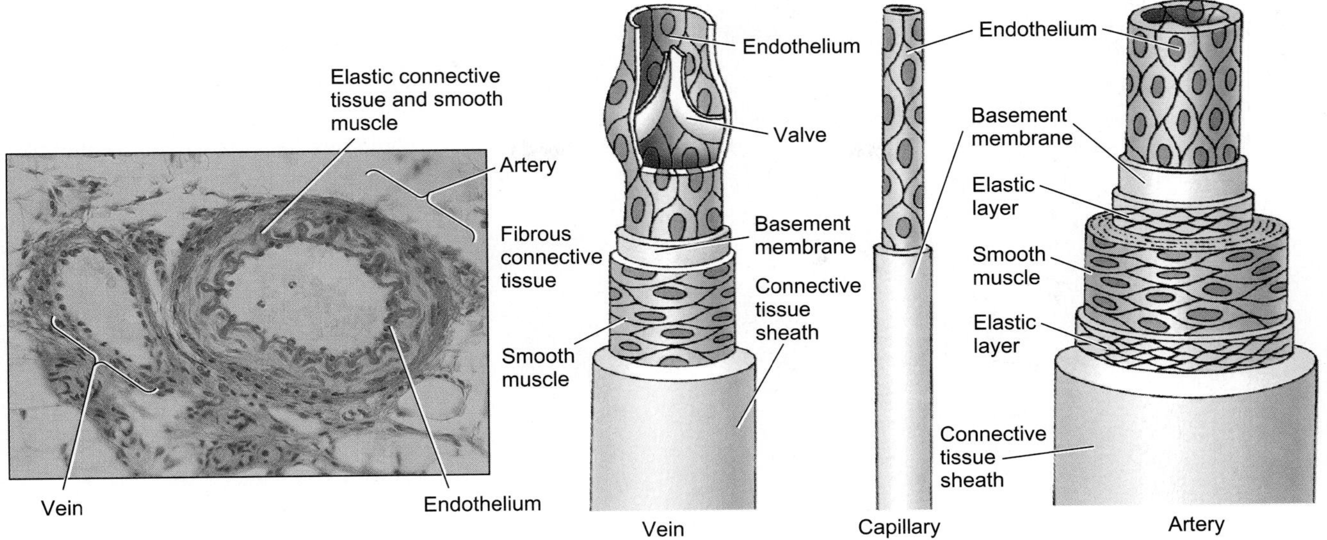

FIGURE 40.8 COMPARISON OF BLOOD VESSELS.
Both arteries and veins are composed of three layers of tissue, including a smooth inner endothelium, a middle layer of circular smooth muscle and elastic connective tissue, and an outer layer of fibrous connective tissue. The veins are thinner-walled and more irregular in shape.

adult can lose—or donate—about 1/10 of the total volume (roughly a pint) without ill effects.

COMPONENTS OF BLOOD

Structurally, the blood can be divided into two parts: the straw-colored liquid *plasma* (about 55% by volume) and the *formed elements* (about 45%). Formed elements include the *erythrocytes* (red blood cells), *leukocytes* (white blood cells), and *platelets* (also called thrombocytes—disklike cell fragments important in blood clotting). (See Figure 40.9.)

Plasma. Blood plasma is about 90% water; about 8% is made up of three groups of proteins: *albumin, globulin,* and *fibrinogen.* Albumins are important in maintaining osmotic conditions in the blood; globulins function as part of the immune defenses; and fibrinogen operates in blood clotting. Less than 2% of the plasma is formed from ions, hormones, vitamins, urea, and various nutrients, particularly sugars.

Erythrocytes. Erythrocytes or red blood cells (*erythro,* red; *cyte,* cell) are among the smallest and most specialized of human cells. They are only about 8 μm in diameter, and in most mammals take the form of a biconcave disk (Figure 40.9). Developing erythrocytes have the usual cell constituents, but mature cells lose them, including nuclei, mitochondria, and ribosomes. Erythrocytes function essentially as little bags of hemoglobin.

Normally, there are about 5 million red cells per microliter (μl) of human blood. Each cell lives about four months, after which it is destroyed in the liver or spleen. So the erythrocytes must be constantly replaced. In fact (are you ready for this?), we must replace about 2½ million erythrocytes every *second.* In adults, new red cells are produced by *stem cells* in the red bone marrow. Their rate of production (about 2 million per second) is controlled by a kidney hormone known as *erythropoietin,* which is released into the blood when cells in the kidney detect a drop in blood oxygen levels—since oxygen levels are a function of the number of red cells. As red cells increase, the stimulus is removed, and erythropoietin secretion diminishes (a typical negative feedback mechanism).

Leukocytes. Unlike erythrocytes, **leukocytes,** or white blood cells (*leuko,* clear or white), do not discard their nuclei as they mature (Figure 40.9). We will discuss the role of the various leukocytes in the immune responses in the next chapter; for now, just remember that *neutrophils,* whose numbers make up the majority of leukocytes, are important phagocytic cells. They aggregate at infection sites, engulfing invading microorganisms. *Basophils* are involved in the inflammatory and allergic responses. *Eosinophils* are also involved in inflammatory responses, but in addition, they help destroy larger parasites. *Lymphocytes* are the backbone of the immune system. In lymphatic tissue they carry out many tasks that aid in fighting off infectious organisms, destroying foreign molecules and providing long-term immunity from disease. The *monocytes,* once activated at infection sites, develop into *macrophages,* which, like

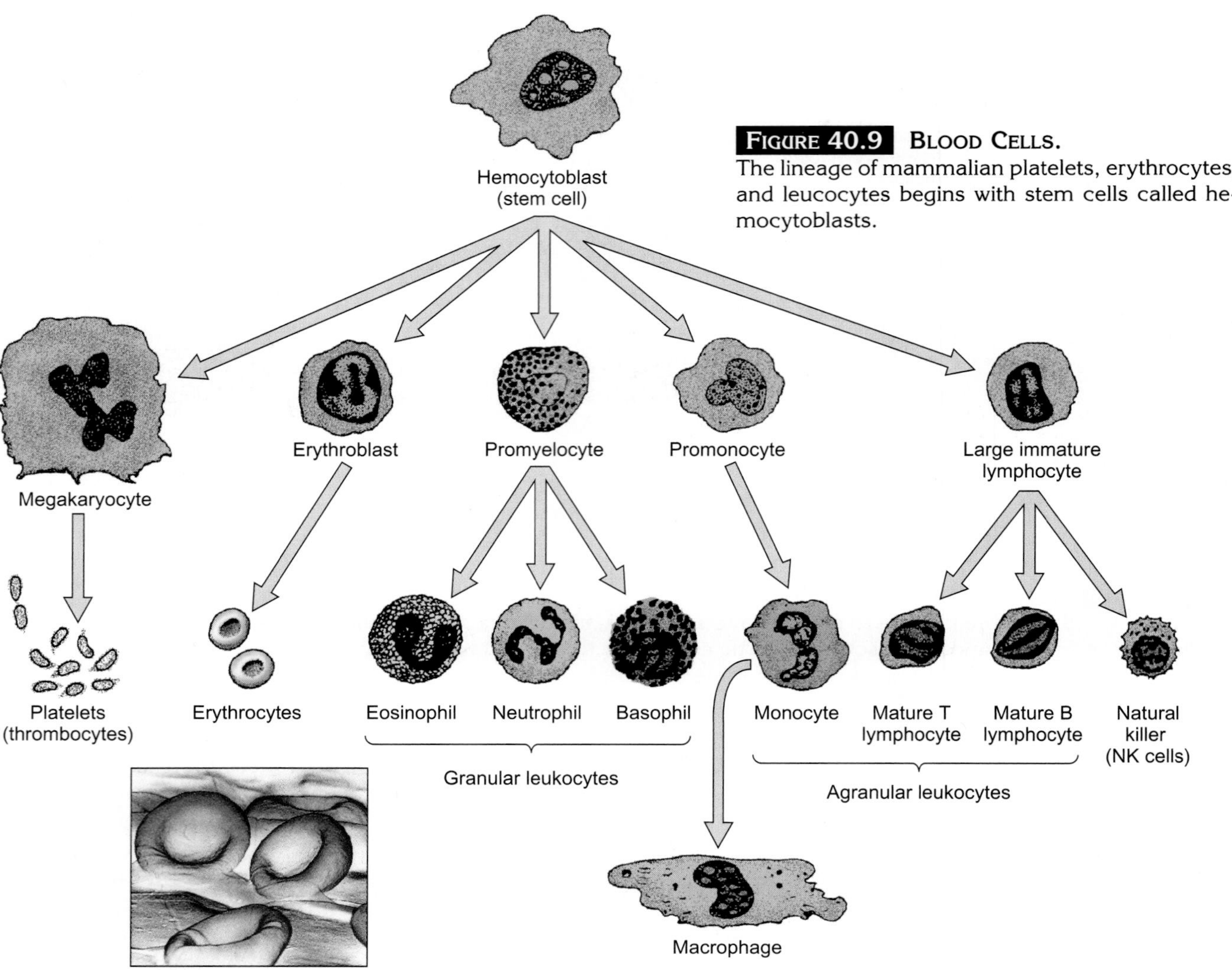

FIGURE 40.9 BLOOD CELLS. The lineage of mammalian platelets, erythrocytes, and leucocytes begins with stem cells called hemocytoblasts.

neutrophils, specialize in phagocytizing foreign cells and any cellular debris they encounter. They also play a special role in activating the immune system, as we will see in Chapter 41.

White blood cells vary in diameter from 9 μm—only slightly larger than an erythrocyte—to 15 μm or more, about the size of most other cells of the body. Under normal conditions there is about one white blood cell for every 700 red blood cells (an average of about 7,000 leukocytes per μl of blood). Low white cell counts (below 5,000 per μl) indicate damage to elements in the bone marrow that form the white cells. High counts (about 10,000 per μl) are a sign of infection or of more serious conditions such as leukemia (in which the count may increase to well over 100,000).

CLOTTING

It sometimes seems that we humans are a rowdy lot, considering how often we must rely on our clotting mechanisms. Indeed, we seem to put ourselves in harm's way, often resulting in pierced bodies, accompanied by a loss of blood that must somehow be stopped. Fortunately, our clotting mechanisms are usually quite efficient at stemming the flow.

As important as it is, blood clotting is only partially understood. We do know that the process is complex, and that at least 15 substances are involved, some of which are part of an intricate system that prevents accidental clotting. Two proteins are basic to the process: *prothrombin,* an inactive clotting protein present in the plasma, and *fibrinogen,* one of the major plasma proteins. **Platelets** are fragments of certain large cells called *megakaryocytes* and contain important enzymes and other agents that are critical to the clotting mechanism. Generally, the clotting process proceeds as follows:

1. A vessel is damaged.
2. Platelets attach at the wound site, form lengthy extensions, and adhere to collagen fibers, forming what is essentially a "plug."

3. The platelets release (a) vasoconstrictors that cause nearby vessels to constrict, reducing blood loss, and (b) enzymes called *thromboplastins.*
4. In the presence of thromboplastins and calcium ions, prothrombin in the plasma becomes *thrombin,* a kind of endopeptidase enzyme.
5. Thrombin breaks apart the large fibrinogen molecules of the plasma, the smaller pieces joining end to end to form a fibrous, sticky protein called *fibrin.*
6. Fibrin fibers, along with damaged platelets, red cells, and white cells, form a network that solidifies, becoming a clot that stops the bleeding.

Lymph

When we think of a circulatory system, we usually think of blood. However, there is a circulatory system of another kind, the one that circulates *lymph.* This one, logically enough, is called the *lymphatic system* (Figure 40.10). Lymph ultimately comes from the blood. Some of the liquid portion of the blood (plasma) filters out through the walls of the tiny capillaries. Here it bathes the cells, moves between them, and finally winds up in small channels called lymph capillaries, where it is now referred to as lymph. The lymph may then move into larger channels, which may eventually bring the fluid to one of the many **lymph nodes** scattered throughout the body. In the nodes it is filtered, thus removing any cellular debris or bacteria. Frogs and some other vertebrates have "lymph hearts" that move the lymph along, but mammals rely chiefly on muscular movements to squeeze the vessels and circulate the fluid. (Again, we see that humans are adapted to being on the move.)

Lymphocytes are found in great numbers in the lymph nodes. As we mentioned, the lymphocytes are important in defending the body against disease organisms. When such a battle is raging, the lymph

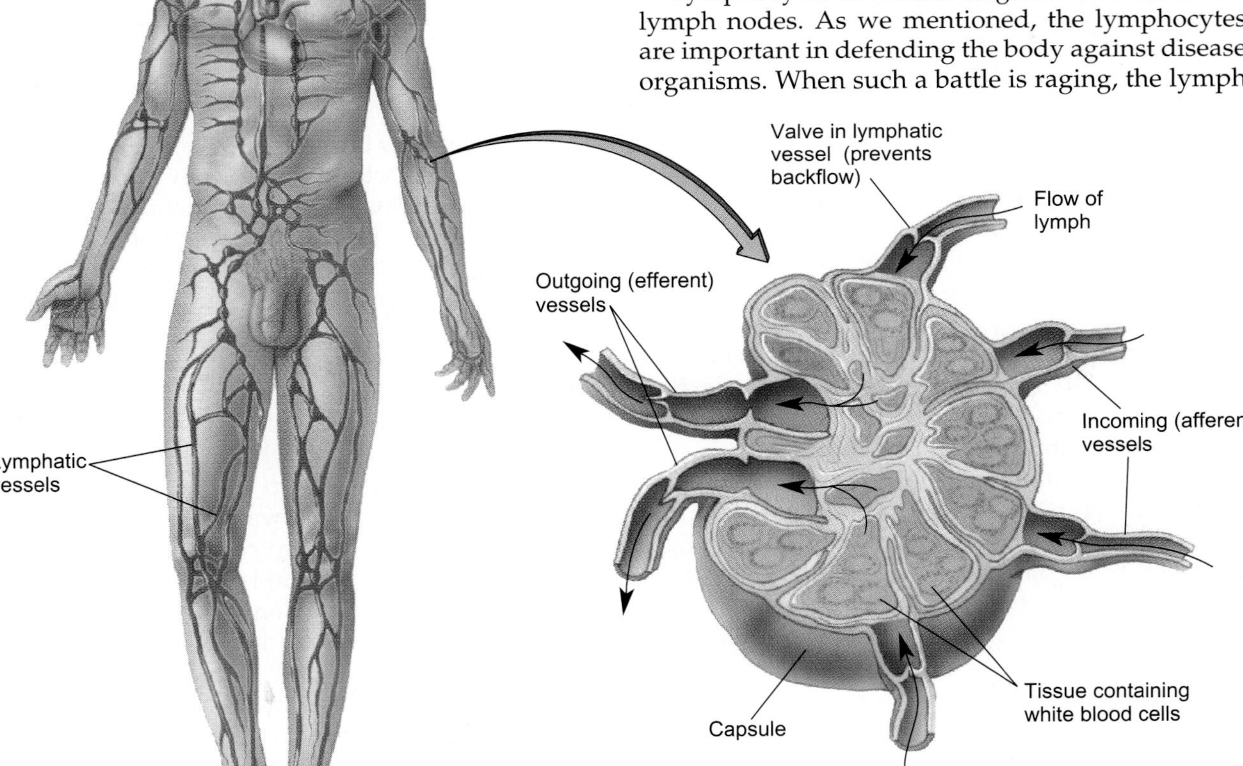

FIGURE 40.10 **THE LYMPHATIC SYSTEM.**
The lymphatic system is an extensive network of vessels, nodes, and ducts. Lymph, collected in various tissue spaces, reenters the circulatory system through special ducts. The lymph nodes cluster in several distinct regions (groin, abdomen, armpits, neck, and head). They are the sites of ongoing battles between the immune system and invading organisms.

nodes enlarge (a sign of infection). As the lymph moves along, it flows into increasingly larger channels, or **lymph ducts,** until, finally, it enters large veins near the heart, rejoining the blood.

The lymphatic system, then, has four essential roles: (1) by accepting fluids from the blood, it helps maintain proper fluid levels in that system; (2) it transports certain fatty acids from the intestinal villi to the blood (Chapter 38); (3) it assists with the work of the immune system; and (4) it provides a route by which interstitial (intercellular) fluids can be returned to the blood.

PERSPECTIVES

We have now seen how the products of digestion enter the bloodstream, and how the circulatory system allows oxygen and carbon dioxide to be exchanged with the environment. We have noted the development of the four-chambered heart, and how the resulting efficiency can promote homothermy.

Now we will move to one of the most dynamic areas in all of biology—the body's defense mechanisms. As we will see, these also largely depend on the circulatory system.

SUMMARY

THE ARRAY OF CIRCULATORY SYSTEMS

The circulatory system transports many substances, aids in thermoregulation, and supports the immune system.

Although simpler invertebrates rely on cell-to-cell transport, more complex types require circulatory systems. Annelids have a closed system with hemoglobin-rich blood. Exchanges occur through capillary walls. Arthropods have open systems in which blood leaves vessels to percolate through body spaces.

Vertebrates have essentially closed systems. The vertebrate heart consists of two chambers in fish, three in amphibians and most reptiles, and four in birds and mammals.

The four-chambered heart provides for separate pulmonary and systematic circuits. Deoxygenated blood is received and sent to the lungs by the right side, and oxygenated blood from the lungs is received and sent to the body by the left.

The superior and inferior venae cavae return deoxygenated blood to the right atrium. It then passes through the tricuspid valve to the right ventricle, which pumps it past the pulmonary semilunar valve into the pulmonary arteries. They direct it into the pulmonary circulation for gas exchange. Following this, pulmonary veins carry oxygenated blood to the left atrium, which sends it past the bicuspid valve into the thick-walled left ventricle. (Both tricuspid and bicuspid valves have thin flaps held in place by tough chordae tendineae.) From there blood is pumped past the aortic semilunar valve into the aorta for distribution. The two heart sounds are produced by the two sets of valves closing after the blood passes through.

Cardiac muscle is intrinsically contractile. The heartbeat originates in the sinoatrial (SA) node (pacemaker), which directs impulses across the atria, which contract. The impulses are then relayed to the atrioventricular (AV) node, which directs them to the ventricles, bringing on their contraction.

Heart rate is also regulated by the autonomic nervous system. Decreases in heart rate are brought about by acetylcholine secreted by parasympathetic nerves. Increases occur when norepinephrine is secreted by sympathetic nerves.

Arteries carry blood away from the heart. Blood pressure is the outward force on vessel walls. Maintenance of blood pressure is aided by the elasticity of arteries. Arterial pressure peaks during ventricular systole (contraction) and falls during diastole (relaxation).

The aorta's first branches go to the heart itself, after which the curving arch sends branches to the rest of the body. Arteries branch into arterioles, vessels with smooth-muscle precapillary sphincters that control flow into capillary beds.

All exchanges occur through the one-cell-thick capillary walls. Exchange occurs through hydrostatic pressure, diffusion, osmosis, and pinocytosis.

Capillaries form venules, which join to form veins. Venous blood pressure is low, and blood return to the heart is assisted by one-way valves, muscle movement, and pressure changes in the chest cavity.

Blood

Plasma (55% of blood volume) consists of water, blood proteins, ions, hormones, vitamins, urea, and nutrients.

Formed elements (about 45% of blood volume) include erythrocytes, leukocytes, and platelets. Erythrocytes (red blood cells) are biconcave, hemoglobin-filled cells that when mature lack organelles. Their formation from stem cells is regulated by an automated hormonal system involving erythropoietin.

The leukocytes (white blood cells) are larger cells that retain their organelles, and occur in far fewer numbers. They are involved in immune functions, and accordingly their numbers increase during infections.

Clotting

Platelets are essential to blood-clotting, releasing the enzyme thromboplastin that acts in fibrin (clot) formation.

Lymph

The lymphatic system includes simple ducts, vessels, and nodes. Its roles include maintaining fluid balances, transporting digested fats, aiding in immune responses, and returning interstitial (intercellular) fluids to the blood.

Key Terms

closed circulatory system • **692**
open circulatory system • **692**
ostia • **692**
atrium • **695**
ventricle • **695**
aorta • **695**
artery • **695**
vein • **695**
pulmonary artery • **695**
pulmonary vein • **695**
blood pressure • **699**
systolic pressure • **699**
diastolic pressure • **699**
erythrocyte • **701**
leukocyte • **701**
platelet • **702**
lymph nodes • **703**
lymph ducts • **704**

Review Questions

1. List six substances transported in the circulatory system.
2. List several examples of invertebrates that have no need for a circulatory system, and explain how they transport fluids.
3. What is a closed circulatory system? How can a closed system exchange materials with tissues?
4. Describe the anatomy of the closed circulatory system of the earthworm.
5. Describe the flow of blood through the insect open circulatory system.
6. Describe the fish heart and the complete circuit of blood in its circulatory system.
7. Explain how the four-chambered heart represents two pumps in one. What does it provide that is not possible with a three-chambered heart?
8. Starting at the venae cavae, trace the flow of blood through the human heart, finishing at the aorta. Name all chambers, valves, and vessels.
9. Compare the appearance and operation of the tricuspid and bicuspid valves with that of the semilunar valves.
10. Relate the heart sounds to activity in the heart valves.
11. Discuss how the structure and work of arteries affects blood pressure in the circulatory system.
12. A typical blood pressure is 120 over 80. What does each measurement actually reflect?
13. Relate the structure of a capillary to its principal functions.
14. Characterize blood pressure in the veins, and cite three factors that aid in returning blood to the heart.
15. Make a list of the blood plasma components, indicating whether each is generally permanent or transient.
16. How long do erythrocytes live? Explain the mechanism that assures an ongoing supply.
17. Compare the size, structure, and relative numbers of leukocytes with that of erythrocytes.
18. What would a white cell count of 10,000 or higher indicate? A count of over 100,000?
19. Discuss the general role of platelets. What triggers their activity?
20. List the main structures of the lymphatic system. What propels the lymph along?
21. What are the four principal functions of the lymphatic system?

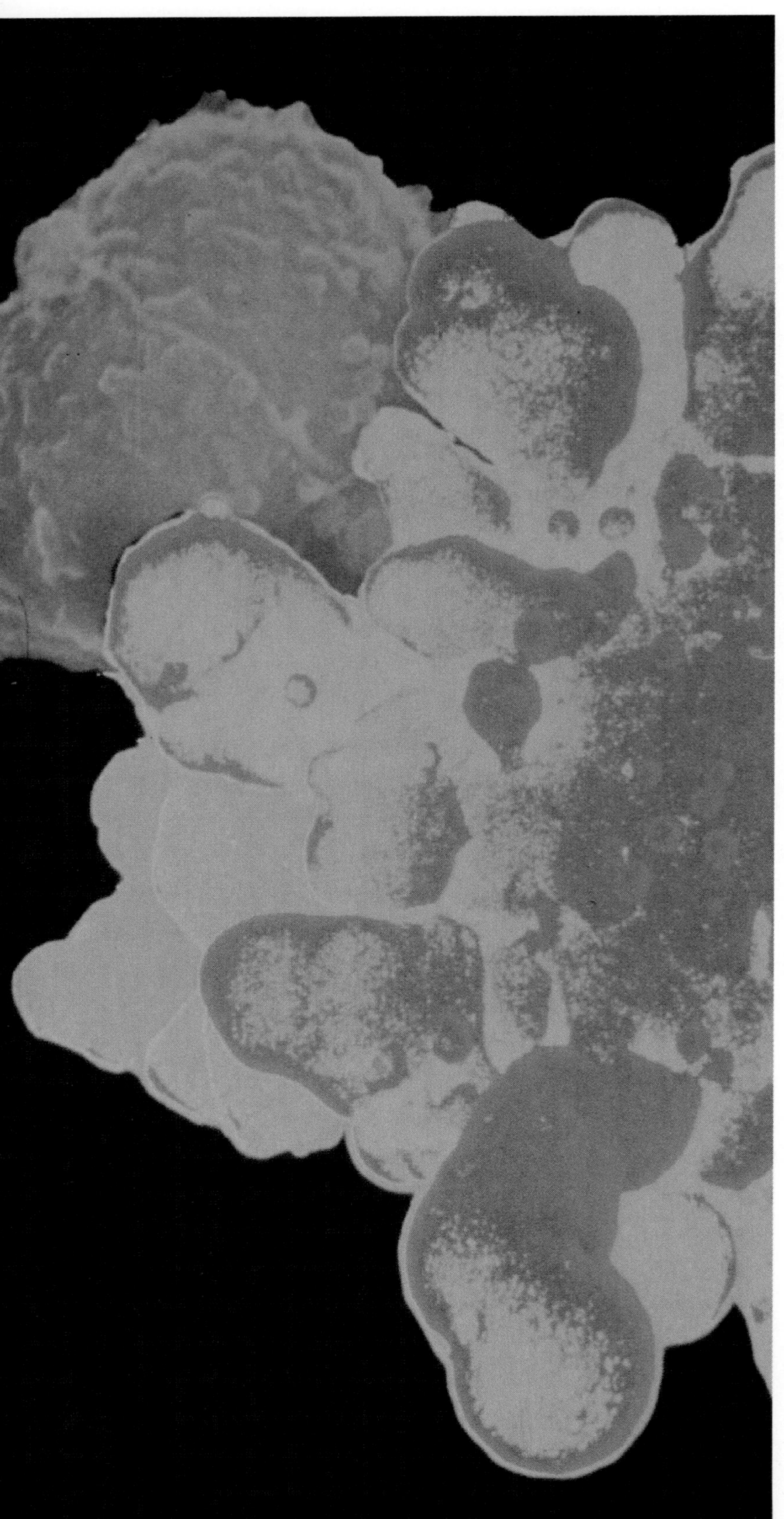

CHAPTER

41

Immunity

THE FACT THAT YOU ARE ALIVE VERY LIKELY MEANS YOUR IMMUNE SYSTEM *is functioning. After all, your molecules are valuable, not only to you but to a host of other living things that are prepared to invade your tissues. But you have, for the most part, been able to fend them off. Your devices have been as simple as an impermeable skin, and as complex as blood cells that remember previous invaders and quickly mobilize cellular armies if they should appear again. You, though, are living in an increasingly uncertain world since AIDS has crept upon the scene. We frequently hope our information regarding a cure or vaccine is dated by the time you read it. But never so much as with this topic.*

If an alien from another world landed on our planet, his biggest worry might not be a trigger-happy farmer or even our corrosive, oxygen-laden atmosphere. A greater problem might be the vast army of tiny living things that relentlessly attack other kinds of life. We, of course, evolved in a sea of these bacteria, viruses, and other parasitic invaders, and we usually have ways to repel their advances. The visitor probably would have developed no such protection and might quickly fall to their wheedling intrusions.

Even among our own species, some people have better defenses against a certain microbe than others do. Americans go to work with the sniffles without endangering their colleagues around the water cooler, but in the Ecuadorian Amazon, small, remote groups of the Waorani tribe have been devastated by outsiders who carried in common cold viruses.

How, then, do humans manage to survive surrounded by invisible opportunistic exploiters? Obviously, some of us don't; but many of us do, and in large part that's because we're the descendants of people who survived. As natural selection worked its way, each generation was left with stronger defenses and more resistance (even as the invaders were developing better ways to attack, sometimes with devastating success). So let's take a look at some of the ways our bodies defend themselves against their microscopic attackers.

SURFACE BARRIERS

The first line of defense in our own bodies and those of many other animals is the **integument,** or body covering (Figure 41.1). In humans, the integument is composed of the *skin* and *mucous membranes,* a covering that presents an effective barrier against most invaders. The skin itself has two remarkable defenses. It is tough, and it is poisonous. (You didn't know your skin was poisonous, did you?) It is tough because it contains a protein called *keratin* that resists the disruptive enzymes of would-be invaders such as bacteria. It is poisonous because sweat, oil, and the fatty acids of the skin are toxic to many bacteria. Furthermore, cells are constantly sloughed or worn off and replaced from below, the sloughed cells carrying bacteria away with them. Thus, as long as the skin is intact and unbroken, it is an effective first-line defense.

The body passages are defended by a lining of mucous membranes. Because the vertebrate body is essentially a "tube within a tube," technically, anything in the gut, respiratory, and reproductive passages is still *outside* the body, and in the case of dangerous organisms must be kept there. So we find that the mucous membranes are not only an effective physical barrier, but their secretions also flow along,

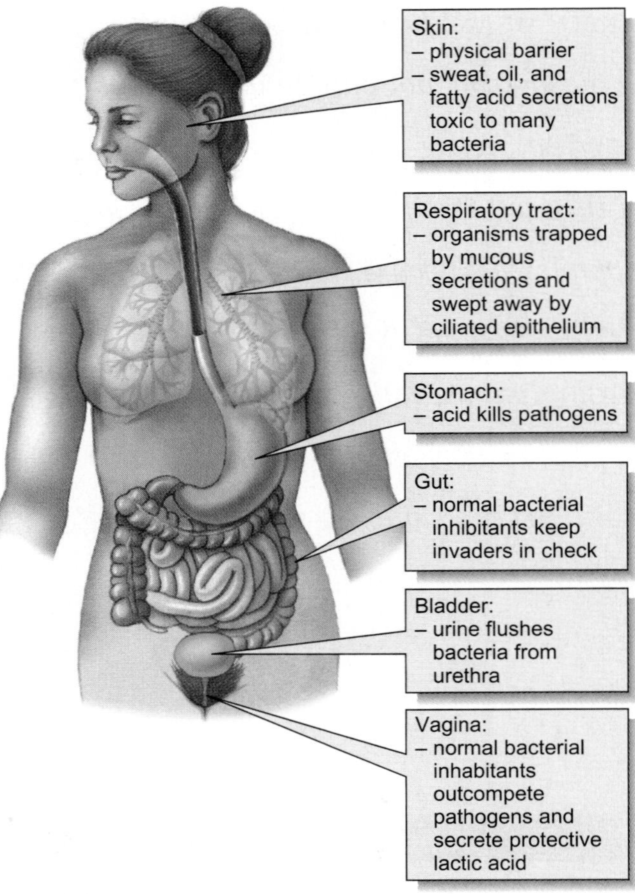
Skin:
– physical barrier
– sweat, oil, and fatty acid secretions toxic to many bacteria
Respiratory tract:
– organisms trapped by mucous secretions and swept away by ciliated epithelium
Stomach:
– acid kills pathogens
Gut:
– normal bacterial inhibitants keep invaders in check
Bladder:
– urine flushes bacteria from urethra
Vagina:
– normal bacterial inhabitants outcompete pathogens and secrete protective lactic acid

FIGURE 41.1 FIRST-LINE DEFENSES.
To gain entry to the body, invaders must first penetrate the external and internal body coverings, both of which provide physical and chemical barriers.

carrying microorganisms away. For example, mucus secretions of the respiratory system entrap countless microorganisms, holding them until they can be swept away by ciliary motion (such as in the trachea).

The body surfaces have yet other defenses. Most organisms that enter the stomach are killed by its strongly acidic environment. Antibacterial enzymes called *lysozymes,* found in tears and other body secretions, quickly dispatch some bacteria by cleaving chemical bonds in the cell wall. The urinary passages, another potential avenue of invasion, are naturally protected from bacteria by the flushing action of urine. We also get a little help from our friends. Over the long eons of our evolution, a variety of bacterial, fungal, and even animal species have adapted to living harmlessly on certain of our surfaces. Whereas they are no threat to us, they do present strong competition for parasites. In the large intestine, massive colonies of friendly colon bacteria, such as *E. coli,* simply use up most of the available nutrients, keeping invaders in check essentially by starving them out. In the vagina, populations of harmless bacteria feed on the glycogen secretions and secrete protective lactic acid that renders the vagina inhospitable to many other organisms. As an aside, such defenses are often lost after prolonged antibiotic therapy that kills the good bacteria along with the bad. With the competition down, invaders may then establish themselves. So an antibiotic may solve one problem, only to create others. We see, then, that our skin secretes toxic sweat, oil, and fatty acids, and constantly sloughs off cells as new cells move up to take their place—and that our gut, respiratory, and reproductive passages utilize moving mucus, lysozymes, urine, and competition from harmless bacteria as our first line of defense.

NONSPECIFIC SECONDARY DEFENSES

When the first lines of defense are penetrated, and the *pathogens* (disease-causers) enter the body, the secondary defenses come into play. Secondary defenses are of two types: *nonspecific* and *specific* (Figure 41.2). **Nonspecific defenses** are effective against any invader; they act immediately, and they involve both chemical agents and certain white blood cells. All of these are in a fully prepared state, needing only to be activated.

Specific defenses target particular pathogens; they require time for their preparation, and involve chemical agents and white blood cells, but both must be prepared. So, we see that both nonspecific and specific defenses have chemical and cellular aspects. Table 41.1 lists the white blood cells and their functions in nonspecific and specific defenses. Let's begin with the nonspecific defenses, first considering chemical, then cellular, defenses.

NONSPECIFIC CHEMICAL DEFENSES

The first responses to invaded or injured cells stem from the cells themselves, which release specific chemicals that call the body's defenses into action. Among these chemicals are the **histamines,** which bring on the redness and swelling associated with inflammation and infection by dilating the tiny arterioles at the site (thereby increasing the flow of blood into the area) and by increasing the permeability of blood vessels in the area, allowing fluids (such as plasma) to seep out. This increased flow allows defensive substances to enter the area. *Antihistamines,* then, reverse these reactions. This is why we

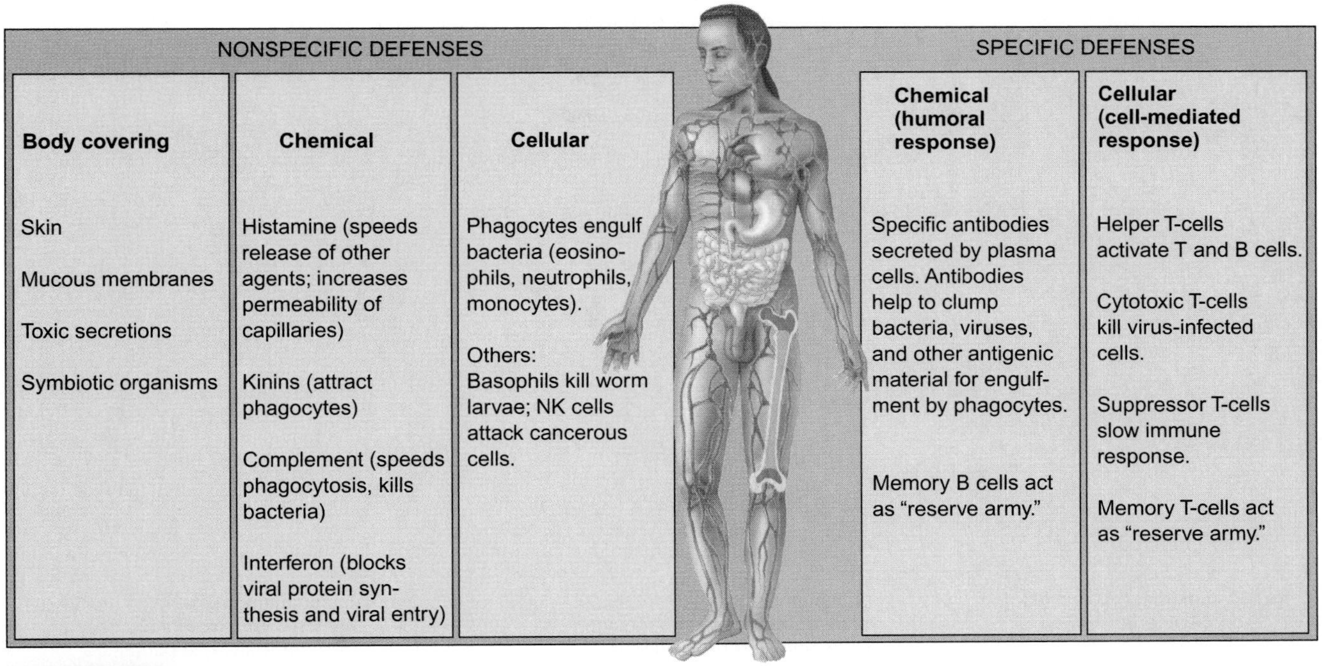

NONSPECIFIC DEFENSES			SPECIFIC DEFENSES	
Body covering	**Chemical**	**Cellular**	**Chemical (humoral response)**	**Cellular (cell-mediated response)**
Skin Mucous membranes Toxic secretions Symbiotic organisms	Histamine (speeds release of other agents; increases permeability of capillaries) Kinins (attract phagocytes) Complement (speeds phagocytosis, kills bacteria) Interferon (blocks viral protein synthesis and viral entry)	Phagocytes engulf bacteria (eosinophils, neutrophils, monocytes). Others: Basophils kill worm larvae; NK cells attack cancerous cells.	Specific antibodies secreted by plasma cells. Antibodies help to clump bacteria, viruses, and other antigenic material for engulfment by phagocytes. Memory B cells act as "reserve army."	Helper T-cells activate T and B cells. Cytotoxic T-cells kill virus-infected cells. Suppressor T-cells slow immune response. Memory T-cells act as "reserve army."

FIGURE 41.2 NONSPECIFIC AND SPECIFIC DEFENSES.
If organisms penetrate the body's protective barriers, the body mounts nonspecific and specific responses. The nonspecific defenses are in a constant ready state, whereas the specific ones require time for the immune reaction to be mounted.

race to the drugstore for medicines containing this class of medicines when we have hay fever or that charming condition known as the common cold.

Other active substances include polypeptides from a group called *kinins,* molecules that further increase local swelling and the permeability of vessels. Kinins also attract phagocytic cells that may engulf any invading organisms.

Infections also trigger activity in the **complement,** a chemical defense system composed of at least twenty plasma proteins. Activation begins as one kind of protein attaches to the surface of an invading bacterium, attracting phagocytes to the cell and making the bacterium far more susceptible to them because they fit receptor sites on the phagocyte, thereby helping the phagocyte to bind to the bacterium. More dramatically, the complement system destroys invaders directly, altering cell permeability so that the inrush of water and ions literally bursts the cell (Figure 41.3).

Another chemical defense involves the antiviral protein **interferon** that can slow the increase of viruses that have successfully invaded the body. As we saw in Chapter 22, viruses can produce many new viral particles by transcribing their own genes using the host's genetic machinery. Interferon—one of the proteins that can be produced through genetic engineering—blocks this action in some viruses.

TABLE 41.1 WHITE BLOOD CELLS AND THEIR FUNCTIONS

Cell Type	Function
Phagocytes	
Neutrophil	Participates in early stages of defense against microorganisms
Monocyte	Arrives at site after neutrophils, transforms into macrophages; engulfs foreign materials, presents antigens to lymphocytes, stimulates lymphocyte proliferation
Eosinophil	Responds to allergies and parasitic infections
Lymphocytes	
Cytotoxic T-cell	Destroys virus-infected and cancerous cells
Helper T-cell	Stimulates B-cell and killer T-cell proliferation
Suppressor T-cell	Slows down immune response
B-cell	When activated by foreign molecules, produces plasma and memory cells
Plasma cell	Secretes antibodies
Memory cell	Responds to antigens during secondary response
Natural killer cell	Directly destroys virus-infected cells and cancerous cells
Basophils	Release histamine in inflammatory response (as do damaged body cells)

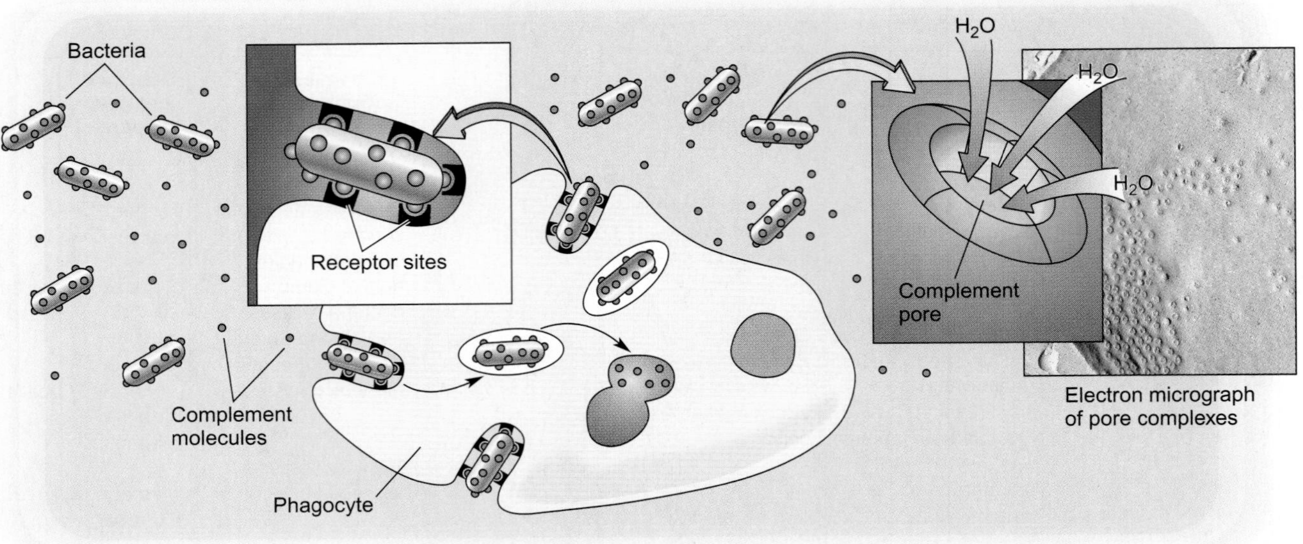

(a) Complement molecules coat bacteria.

Complement on bacteria binds to receptor sites on phagocyte. Phagocyte engulfs bacteria.

(b) Complement may also produce pore complexes in bacteria. Water rushing in bursts the bacterial cell.

FIGURE 41.3 COMPLEMENT DEFENSES.
When activated, the battery of proteins known as complement attacks invaders in two ways. **(a)** Complement binds to bacterial surfaces and then to receptor sites on phagocytes. This stimulates the phagocyte to engulf the invader. **(b)** Complement can kill some bacterial cells directly by allowing an inrush of water that bursts the cell open.

Further, interferon produced in one cell can protect other cells from viral invasion. Interferon is also believed to offer protection against cancer.

NONSPECIFIC CELLULAR DEFENSES

The body has a veritable army of cells roaming its tissues, defending against invading pathogens or the onset of cancer. They operate in a variety of ways, but in the nonspecific defenses, most simply eat the attackers. Phagocytosis—engulfing and destroying foreign debris or microbes—is one of the body's most important nonspecific cellular defenses (Figure 41.4). Key phagocytic cells are the white blood cells *neutrophils* and *macrophages,* the latter formed from other phagocytic white blood cells called *monocytes.*

The **neutrophil** have been referred to as expendable, frontline foot soldiers. Perhaps 100 billion neutrophils are produced each day. They survive only a few days and suffer great casualties with every invasion. They are the first to arrive at any invasion site, drawn toward chemicals emitted from bacteria, injured tissue, and clotting blood. They quickly swarm at an infection site, engulfing any intruder they can manage. (Sometimes, though, the invaders win the fight.) The monocytes arrive next, and once at the site they undergo remarkable changes, growing and swelling until they are transformed into huge macrophages, whose numbers further increase in the later stages of infection.

Macrophage means "big eater." Macrophages eat not only microorganisms but old, worn-out blood cells and dead tissue. Further, they do it with astonishing speed. They can engulf a foreign particle in 1/100 second. In some cases, even if they can't digest the offending particle, they engulf it anyway and hold it within their bodies where it can do no harm. Although they may live for years, as the phagocytes continue to engulf bacteria, they accumulate the bacterial poisons and eventually begin to die in great numbers, their tiny corpses forming pus.

Another cell that is important in nonspecific defenses is the **natural killer cell,** or *NK cell.* These cells routinely rove the body, touching and identifying other cells as they go. In ways we have yet to fully understand, the NK cells uncover abnormal cells—those infected by viruses or those in the early stages of cancer—and promptly kill them. As more becomes known about the NK cells, we may find that it is the failure of these first-line defenders that permits the early spread of cancerous cells. We already know that cancer patients have fewer NK

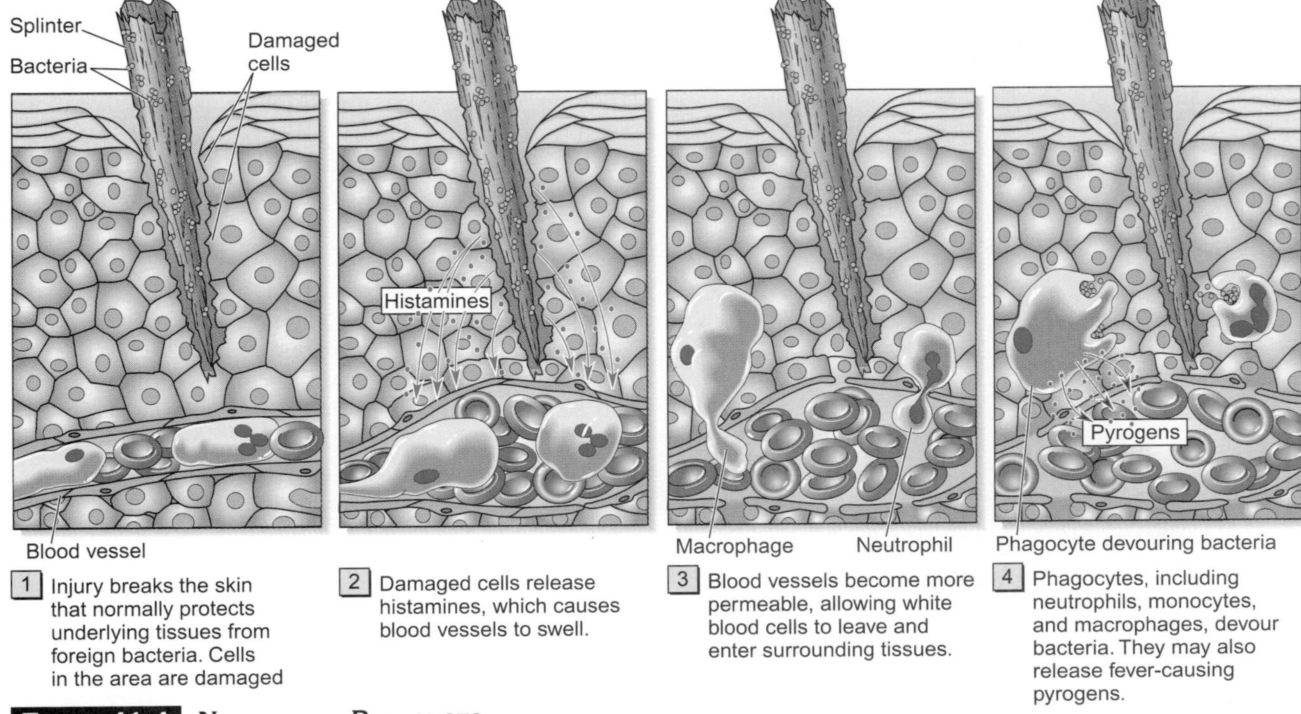

Splinter
Bacteria
Damaged cells
Blood vessel
1 Injury breaks the skin that normally protects underlying tissues from foreign bacteria. Cells in the area are damaged
Histamines
2 Damaged cells release histamines, which causes blood vessels to swell.
Macrophage
Neutrophil
3 Blood vessels become more permeable, allowing white blood cells to leave and enter surrounding tissues.
Pyrogens
Phagocyte devouring bacteria
4 Phagocytes, including neutrophils, monocytes, and macrophages, devour bacteria. They may also release fever-causing pyrogens.

FIGURE 41.4 NONSPECIFIC RESPONSES.
Here, defense reactions are triggered by the trauma and infection caused by a splinter.

cells, and that the severity of cancer is related to the extent of reduction in NK cells.

SPECIFIC SECONDARY DEFENSES

As the nonspecific secondary defenses destroy pathogens and prevent the spread of infection, the specific secondary defenses are being mobilized. There are two types of such responses. The first is the **humoral response** (*humor,* fluid), which produces **immunoglobins** (or antibodies) in response to the presence of an *antigen,* which can be a foreign substance or microbial invader. (An **antigen** is anything that triggers the production of immunoglobins.) Most antigens are proteins or large carbohydrates, molecules generally associated with microbic invaders. The second is the **cell-mediated response,** where certain kinds of cells, called lymphocytes, attack the invader. We will say more about each of these shortly, but first let's consider the two kinds of immune response.

Specific responses, those directed against particular microbes, form the *immune responses.* The *primary immune response* is triggered by the very first exposure to the invading agent. Since the body hasn't encountered this agent before, the response takes time, so primary immune responses can be slow to start. (When you received those painful immunizing shots just before starting school, or when going abroad to certain third-world countries, your body was being encouraged to produce just such defenses.) Once the body has been mobilized against a particular agent (once that agent has triggered a primary immune response), if it encounters that agent again, the *secondary immune response* is mobilized. The secondary immune response is rapid, because the body has been attacked by that invader previously, and is ready the next time. We will deal with both of these responses more specifically below.

Actually, people have been aware of the immune responses for a long time, both historically and individually. In the fifth century, the Greeks wrote that people who had recovered from the plague never suffered from it again. As school children we are aware that once we've had chicken pox or mumps, that's it; we don't catch them twice. Such resistance is due to very precise cellular processes that are triggered by the first exposure—processes that enable the body to "learn" the nature of the invader and to defend itself subsequently against those specific traits. This kind of immunity begins with the action of lymphocytes.

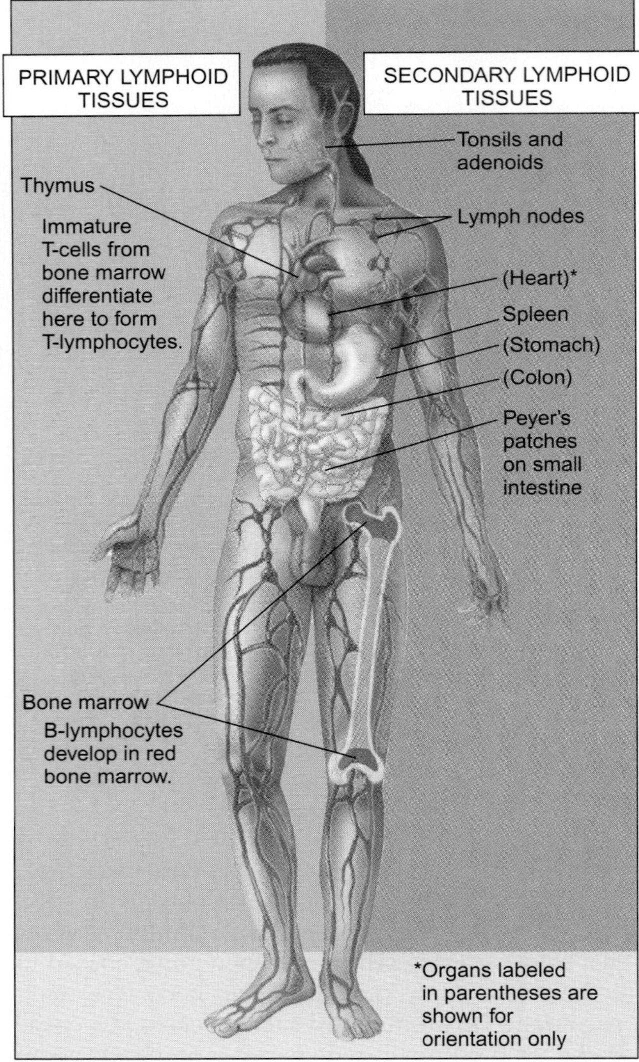

FIGURE 41.5 **LYMPHOID TISSUES OF HUMANS.** Lymphocytes arise in primary lymphoid tissues (thymus and red bone marrow), where they go through a maturation process. Later, as virgin B- and T-cells, they migrate to secondary lymphoid tissues, where they await selective activation. The activated cells then migrate out to the infection sites.

LYMPHOID TISSUES AND LYMPHOCYTES

Lymphocytes are white blood cells that arise in the red bone marrow early in fetal development. Regions where lymphocytes congregate are known collectively as *lymphoid tissues.* Functionally, there are two kinds of lymphoid tissues, primary and secondary (Figure 41.5). *Primary lymphoid tissue* exists mainly in the thymus and red bone marrow. (The thymus is a bilobed organ located underneath the breastbone). *Secondary lymphoid tissue* includes the lymph nodes, tonsils, adenoids, spleen, and Peyer's patches (clustered tissues on the wall of the small intestine). In the primary lymphoid tissue, lymphocytes develop and mature. In the secondary lymphoid tissue, the mature lymphocytes begin their critical roles in defense.

B- AND T-LYMPHOCYTES

Two kinds of lymphocytes are involved in the immune response: B-cells and T-cells. **B-cells** originate in the red bone marrow and in the fetal liver. (The "B" is from the *bursa of Fabricius,* a structure in the chicken, where they were first found.) In mammals, **T-cells** arise in bone marrow, then differentiate in the thymus, the organ for which they are named.

The B- and T-cells have quite different roles in the immune response. B-cells do two things: Most form plasma cells that make and secrete *antibodies* (discussed shortly), and a few act as "memory cells," which confer lasting immunity, thus helping the body respond quickly to future encounters with the same offender.

The T-cells also have multiple roles. Some T-cells specialize in the *cell-mediated response.* This includes a direct attack against diseased body cells, particularly those that have been invaded by a virus and those that are cancerous. Other T-cells activate B-cells, and still others moderate and suppress the immune response when the threat diminishes. In addition, some T-cells from each of these specialized groups become memory cells, which, like memory B-cells, confer lasting immunity.

THE PRIMARY IMMUNE RESPONSE

The initial immune response against an invading microbe, the **primary immune response,** is the slower, initial response of the body against invasion, during which inactive lymphocytes are activated into specialized B- and T-cell lymphocytes. Let's follow the events to see how the specific defense system mounts its attack against invaders never before encountered.

THE MACROPHAGE AS AN ANTIGEN-PRESENTING CELL

In addition to engulfing invaders and hydrolyzing them with their potent enzymes, the macrophages have another critical role. While most of what they consume is simply digested, some of the ingested molecules are saved and later brought to the cell surface where they are incorporated into the macrophage's plasma membrane. These molecules, called *antigens,* stimulate the immune responses. The macrophages then roam through the body wearing bits and pieces of what they ate. Like natural killer cells, they touch surfaces with the various cells they encounter. But, unlike the NK cells that search for diseased cells, these **antigen-presenting macrophages** are interested only in friendly cells. Specifically, their role is to find and activate **helper T-cell lymphocytes** whose cell surfaces carry molecules that match the antigen worn by the macrophage (Figure 41.6).

When an antigen-presenting macrophage finds the right helper T-cell, the immune system quickly responds by forming a veritable army of highly specific cells, a massive clone with just one task. Like angry bears aroused from hibernation, they track down and deal with the antigen that aroused them. The method by which the immune system prepares its leukocytes for the key step of antigen recognition is an interesting story that we will return to shortly, but for now let's look more closely at this amazing, single-minded army.

Arousing the Lymphocytes

When an antigen-presenting macrophage encounters a matching helper T-cell lymphocyte, the macrophage secretes a powerful activating substance called **interleukin-1,** a chemical from a group known as *lymphokines* ("lymphocyte activators"). The helper T-cell responds to the chemical agent by at once seeking out and contacting any inactive lymphocytes with similar surface recognition molecules. Upon contact, the excited helper T-cells secrete their own lymphokine, logically called **interleukin-2.** Interleukin-2 activates other inactive cells, which, once aroused, quickly mature and enter into a frenzy of mitotic activity, producing more just like them.

According to the **clonal selection theory,** there are enough kinds of inactive B- and T-cells to react against almost any potential antigen imaginable. Clonal selection involves the development of single lines (clones) selected from the vast lymphocyte army. The theory maintains that any time helper T-cells are activated and the alarm sounded, only a few virgin B- and T-cells are aroused, those with matching cell-surface recognition molecules. Upon arousal, they give rise to two lines of cells, B-cells and three kinds of T-cells (Figure 41.7).

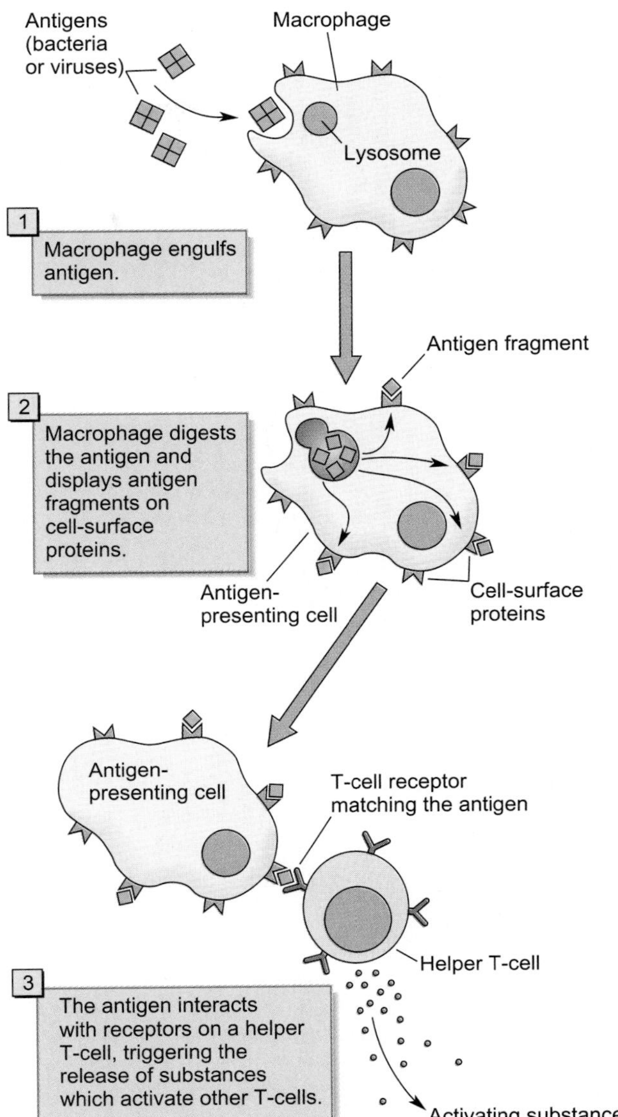

Antigens (bacteria or viruses)
Macrophage
Lysosome
1
Macrophage engulfs antigen.
Antigen fragment
2
Macrophage digests the antigen and displays antigen fragments on cell-surface proteins.
Antigen-presenting cell
Cell-surface proteins
Antigen-presenting cell
T-cell receptor matching the antigen
Helper T-cell
3
The antigen interacts with receptors on a helper T-cell, triggering the release of substances which activate other T-cells.
Activating substance

FIGURE 41.6 **ANTIGEN-PRESENTING MACROPHAGE.** Macrophages incorporate either free antigen or antigen from partially digested invaders into their cell surfaces. This produces a dual recognition site that will complement the dual receptor of some specific virgin T-cells.

Aroused B-cell Lymphocytes

The B-cells produce two kinds of cells, **plasma cells** and **memory cells.** The plasma cells are short-lived, lasting only a few days, but during that period they begin the humoral (chemical) response. Each plasma cell synthesizes and secretes copious amounts of antibodies (or more technically, immunoglobulins) constructed according to specifications in the original antigen captured by a macrophage.

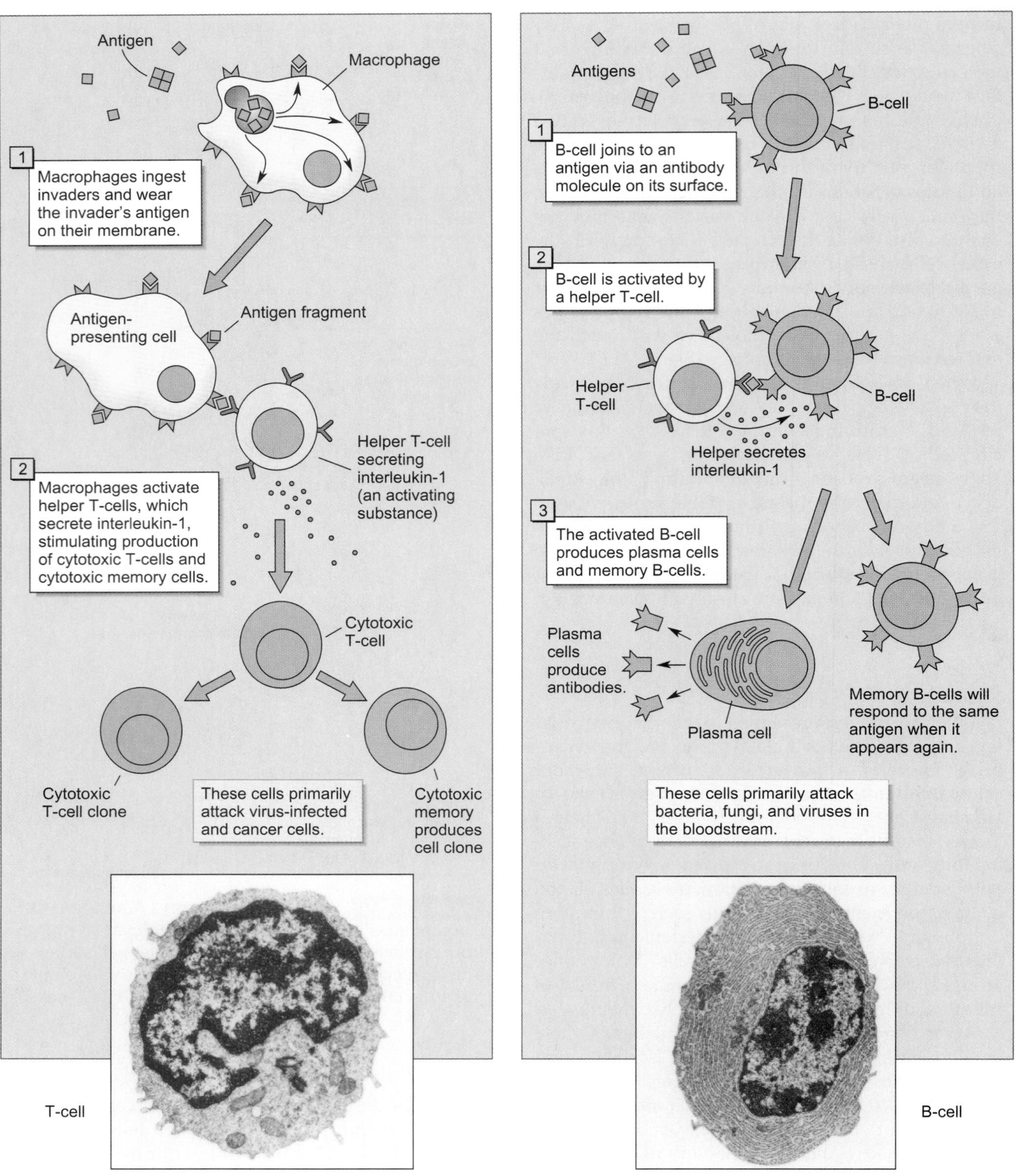
Antigen
Macrophage
1
Macrophages ingest invaders and wear the invader's antigen on their membrane.
Antigen-presenting cell
Antigen fragment
Helper T-cell secreting interleukin-1 (an activating substance)
2
Macrophages activate helper T-cells, which secrete interleukin-1, stimulating production of cytotoxic T-cells and cytotoxic memory cells.
Cytotoxic T-cell
Cytotoxic T-cell clone
These cells primarily attack virus-infected and cancer cells.
Cytotoxic memory produces cell clone
T-cell
Antigens
B-cell
1
B-cell joins to an antigen via an antibody molecule on its surface.
2
B-cell is activated by a helper T-cell.
Helper T-cell
B-cell
Helper secretes interleukin-1
3
The activated B-cell produces plasma cells and memory B-cells.
Plasma cells produce antibodies.
Plasma cell
Memory B-cells will respond to the same antigen when it appears again.
These cells primarily attack bacteria, fungi, and viruses in the bloodstream.
B-cell

FIGURE 41.7 **LYMPHOCYTE ACTIVATION.**
Different processes are involved in the attack of virus-infected and cancer cells, as opposed to those involved in the attack of bacteria, fungi, and viruses in the bloodstream.

Immunoglobulins circulate freely in the blood. When they encounter matching antigens, whether on an invader's cell surface or as loose molecules, they become firmly attached. As the reaction goes on, massive antigen-antibody complexes are formed. Such complexes, as we will see, appear as immobile clumps that are readily cleaned up by phagocytes.

The interaction of immunoglobin and antigen also activates the complement system as discussed earlier. Memory cells live much longer, forming a small but effective residual force that, if aroused, can quickly reproduce and mount new attacks.

Aroused T-cell Lymphocytes

Activated T-cell lymphocytes give rise to several specialized subpopulations. These include *cytotoxic T-cells* (sometimes called "killer T-cells"), *helper T-cells, suppressor T-cells,* and *memory T-cells.*

Their tasks, respectively, are: destroying infected cells, arousing virgin lymphocytes, slowing down and stopping the immune response, and retaining a "memory" of the invader in case of future attacks.

Cytotoxic T-cells. The role of identifying and killing invading cells and friendly but infected body cells goes to the **cytotoxic T-cell.** Upon identifying such cells, cytotoxic T-cells cluster about, releasing substances that lyse, or rupture, the cell (Figure 41.8). Recognition occurs when the antibody-like protein carried by the T-cell forms a match with the antigen site on the invader. If the infection is viral, the aroused cytotoxic T-cells will be carrying a surface antibody that matches an antigen that has formed on the cell surface. (Recall that when viruses invade cells, they make specific molecular changes in the host membrane, rendering it impenetrable by other viruses—a viral adaptation that eliminates competition.) Upon contact, the two sites bind together, and the cytotoxic T-cells then destroy the infected cell and its invader.

In addition, all T-cells have surface receptors that recognize "self." The term "self," of course, refers to one's own tissues. This is also important since it safeguards against inadvertent attacks on normal cells and tissues. In the case of infected or invading cells, this permits a dual recognition to go on (see Essay 41.1).

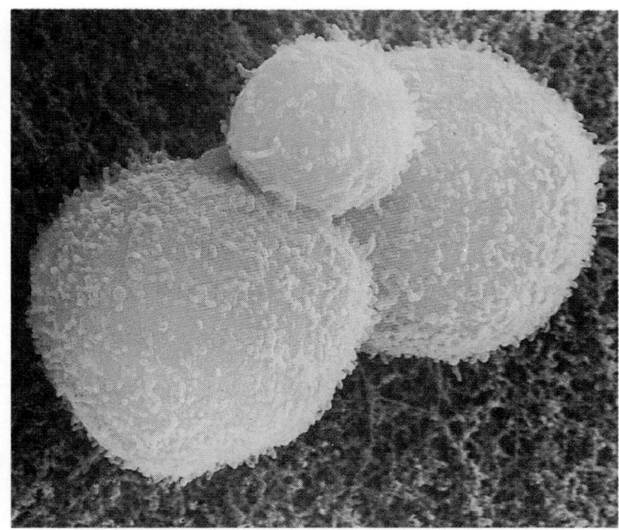

FIGURE 41.8 CELL-MEDIATED IMMUNITY.
Small T-lymphocyte killer cell shown attacking two large tumor cells.
During the killing process, granules from the T-cell fuse with the T-cell plasma membrane and release the protein perforin (SEM ×2500).

Helper T-cells. As we have seen, mature **helper T-cells** assist other lymphocytes in responding to antigen. These helper T-cells amplify the original activating process by continuing to secrete interleukin-2, prompting even more virgin B- and T-cells to divide and mature.

While the name "helper" doesn't sound very impressive, immunologists are now convinced that the helper T-cells' tasks are quite pivotal, indeed critical, to the overall operation of the immune system. The helper T-cells, for example, stimulate not only the formation of cytotoxic T-cells, but memory T-cells as well. Memory T-cells (like memory B-cells) are long-lived survivors that are readily activated if stimulated by a second invasion of the same antigen. They quickly produce more cytotoxic T-cells.

Chilling verification of the importance of helper T-cells comes to us from a relatively new saga in the story of infectious disease. Immunologists now believe that the deadly agent of AIDS (acquired immune deficiency syndrome) tends specifically to attack helper T-cells, thus devastating the immune

Essay 41.1 Recognizing "Self"

The very capability that makes the immune system so versatile has within it the potential for disaster. Thus arises yet another of those endless questions: What keeps the immune system from reacting against proteins and other molecules in the very body in which it resides? It turns out that the immune system must learn *not* to react against "self," a process called immunological tolerance. Sometimes, in fact, the immune system does react against self; but when all goes well, how and what has the immune system "learned"?

At some determined time during the embryo's development, the many kinds of virgin lymphocytes begin to rove the body, essentially becoming familiar with their parent organism. But when a wandering cell finds a chemical match with its specific antigen recognition protein, it is not activated (as happens in the primary immune response). Instead, it becomes permanently suppressed. The workings of the suppression process are unknown, but some evidence suggests that suppressor T-cells are involved. Once they have learned to identify self, they interact with B- and T-cells, preventing them from attacking the body.

Considering the enormous complexity of the immune system, it should be no surprise that it sometimes goes awry and reacts against self. Such a condition is called an autoimmune disease. Among the many known or suspected autoimmune diseases are arthritis, nephritis, rheumatic fever, systemic lupus erythematosus, various hormone disorders, certain forms of diabetes, and possibly even schizophrenia.

The source of at least some autoimmune diseases is the body's own response to infection. When antigens closely resemble the body's own chemistry, the immunoglobulins produced may affect the body's own tissues, bringing on a severe autoimmune reaction. The *Streptococcus* bacteria that cause "strep throat" are notorious for this sort of reaction. An infection in the throat creates antibodies that can attack tissue elsewhere, notably in the kidneys or heart valves, with serious and sometimes fatal results.

system. The defending army is still intact and could attack the AIDS virus, but by taking out the helper T-cells the invader has removed the body's ability to sound the alarm.

Suppressor T-cells. Once an invader has been successfully subdued, the task of shutting down the immune response goes to **suppressor T-cells.** The role is vital because, as we've seen, the immune response is self-amplifying (an example of positive feedback) and could easily soar out of control. Once the arousal subsides, the immense clones answering the call die off, and eventually the only B- and T-cells remaining will be the "quieter," long-lived memory cells.

The events described so far involved the primary immune response. While the primary response is being organized by the body, we may become quite ill as the invaders have their way. But given a little time and if all goes well, the offenders will be dealt with, and we will recover. We are left weaker but "wiser"—that is, our immune system is wiser. Such new wisdom is to be found in the memory B- and T-cells, which are mature, long-lived lymphocytes that retain information regarding the characteristics of previous antigenic invaders. (Essay 41.2 describes a common situation when the response goes awry.)

The Secondary Immune Response

The **secondary immune response** is a rapid response to a subsequent invasion of the body during which memory cells quickly produce large numbers of certain B- and T-cell lymphocytes.

Memory cells have the ability to recognize the specific antigen that earlier aroused their sister cells. While the other activated B- and T-cell clones live short, busy lives, memory cells live on and on, perhaps for decades. This is important, for should the same invader show up a second time, it is immediately recognized, and the body's response is much quicker than before. The selected memory B- and T-cells undergo round after round of cell division, and soon a massive new army of active lymphocytes arises to repel the second invasion. We may not even be aware of the renewed struggle, for symptoms of illness are often absent or quite mild. So, thanks to our memory cells, we suffer many diseases just once.

When you got chickenpox in grade school, your immune system developed a battery of memory cells that would not allow the disease to get a toehold again. You were ill for a while, but you don't have to worry about that disease again, because of your immunity. But you never had to worry about smallpox in the first place, because the previous generation was vaccinated against it. Let's look at the differences in how we became immune to each threat.

Active and Passive Immunization. When you got your first vaccinations, your body was provided with shortcuts to the secondary immune response. You skipped most of the primary immune response because you had received **active immunization** in which the vaccine contained weakened or killed disease agents that promoted the primary immune response with little of the misery and risk of the actual disease. During this shortened response, banks of memory cells were produced as usual, but the aroused lymphocyte army, having found itself with little real work to do, quickly retired. However, had you later been confronted with the real disease agent, your immune system would have gone right into the streamlined secondary immune response, and the invader would have been attacked immediately.

Where vaccines aren't available or where the disease has already begun, alternatives are possible. One is the injection of an *antiserum* containing the specific immunoglobulins against the agent. Such immunoglobulins are routinely obtained from animals exposed to the disease agent. When injected, the immunoglobulins go about the task of immobilizing the invader's antigens. Since the immune system is not activated, this process is known as **passive immunization.** Antibodies are short-lived, so the protective effects are temporary. Passive immunization, incidentally, is the only treatment available against the rabies virus.

We've left a few questions unanswered, as you may have noticed. For instance, what is an immunoglobulin, and how does it actually work? How can so many different types arise? Further, how does the body provide for the incredible diversity needed in virgin lymphocytes? After all, there is a seemingly endless number of different immunoglobins, or antigens, around. Where do they come from? Simple recognition is the key to many of the answers, and recognition involves the immunoglobulins.

The Immunoglobulins

Immunoglobulins (or antibodies) are globular proteins, each containing at least four polypeptide chains. Because of the many ways the amino acids can be arranged in the chains, there is a great range of variety possible (and that's an understatement). The basic four-part structure includes two identical *heavy* (long) *chains* of amino acids and two identical *light* (short) *chains,* arranged in a Y shape (Figure 41.9). The two double chains making up the Y contain a *constant region* (the stalk of the Y) and a *variable region* (the separate forks). Whereas the constant regions may be the same in thousands of different immunoglobulins, the variable regions are different in each kind of immunoglobulin. Variable regions, like the active sites of enzymes, have a highly specific shape, the *antigen recognition region,* which is capable of binding to its matching antigen only.

Because of their Y shape and two variable regions, immunoglobulins can attach to two antigens at once. When many immunoglobulins attach to many antigens, the foreign molecules may agglutinate into great clumps (Figure 41.10). (The clumping process often gets a boost by the presence of complement system plasma proteins, as mentioned earlier.)

The Source of Cell Surface and Immunoglobulin Diversity

Now let's look into the question of the seemingly unlimited diversity of virgin lymphocytes and immunoglobulins. Remember that antigen-presenting macrophages can be carrying virtually any kind of foreign molecule, and after touching many, many virgin lymphocytes will almost invariably find one that recognizes the foreign molecule. How can this be?

As we saw earlier, specific immunoglobulins and cell-surface recognition sites can be prepared that will react against virtually any large foreign molecule. For instance, when African crocodile hemoglobin is injected into Texas jackrabbits, the rabbit's B-cell lymphocytes respond by forming specific opposing immunoglobulins. Does this mean that, somehow, *the jackrabbit has lymphocytes that are preprogrammed to combat hemoglobin from African crocodiles?*

The basic question is, are there enough antibody-coding genes to go around to enable rabbits to resist everything, including crocodile molecules? Or, put another way, are there enough genes to code for antibodies against millions upon millions of other potential antigens?

ESSAY 41.2 Allergy: An Overreaction

The ragweed season brings on misery to countless sufferers each year. Other people must at all times avoid certain wines, cheeses, or oysters. Yet others cannot be around dust or cat dander. The reason is, these people are allergic to something associated with these items. (You won't be pleased to learn that household dust often contains tiny, allergy-producing mites that are kicked up into the air by vacuuming or sweeping, and then drawn into the delicate respiratory tract by breathing the dust.)

The body has ways of cleansing itself of intruders or incompatible substances, such as by violent rushes of air or profuse production of cleansing fluids. And so we sneeze and cough and our eyes water and our noses run when we encounter some irritant.

In some cases, though, the body overreacts to one of these irritants. This extreme sensitivity may occur after the body has been repeatedly exposed to the antigen in the irritant, so that the memory B-cells are always standing by. For example, certain kinds of pollen can bring on the reddening eyes, the sneezing and snuffling, that many of us associate with spring. The reaction is definitely extreme for the minor threat presented by the pollen. So what triggers such a violent allergic response? We don't know why the response is so extreme, but we do know, generally, what happens.

The pollen lands on stationary granulated cells called *mast cells,* which then explosively release their granules. These granules contain histamine, which causes capillary walls to become leaky. Fluid then leaks from the capillaries into the tissue spaces, causing the tissue to swell. The respiratory tract is often the first to encounter any airborne irritant, and the swelling tissues are accompanied by coughing, wheezing, sneezing, and a profusely runny nose—usually during a job interview. Although the irritant usually gets removed, the removal mechanism has been extreme, both physically and socially.

Allergic reactions can also produce a dangerous condition called anaphylactic shock. This happens when large areas of the body produce an allergic reaction, as when an individual allergic to bee stings or penicillin suddenly receives those antigens and they are rapidly transported throughout the body. In such a case, capillaries throughout the body become leaky, and blood pressure suddenly drops, reducing blood flow to the brain and heart. Death can result unless epinephrine (adrenalin) is quickly administered, thereby constricting blood vessels and stopping the leakage.

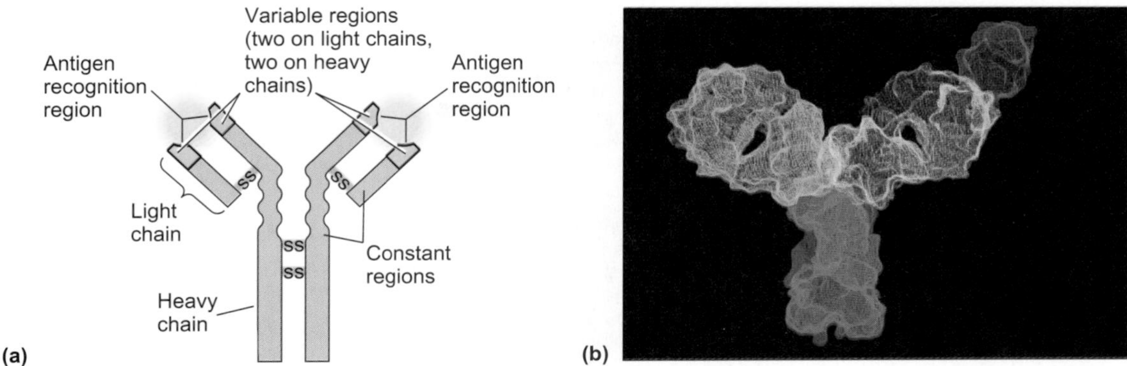

FIGURE 41.9 IMMUNOGLOBULINS.
(**a**) In the Y-shaped immunoglobulin, short and long amino acid chains are joined through disulfide bridges. Each has two specific antigen recognition regions at the tip of the Y. (**b**) A space-filling model of the molecule.

The answer is complex, but let's start by emphasizing that there are not nearly enough genes in any genome to provide for such diversity. To make all these antibodies, certain genes must be modified. The genes that code for the variable regions of antibodies occur in about 300 DNA segments located throughout the chromosomes. These segments are rearranged during the immune system's development—that is, the genes coding for the antibody variable regions are actually broken apart and recombined in a seemingly infinite number of ways (estimated at about 18 billion). You may recall that we once said that all the cells of the body have the same genetic information. Obviously, a correction is in order. You can now see that each clone of differentiated B-cells had its DNA rearranged so that it has certain tailor-made genes that other cells don't have.

We see, then, that during B-cell maturation, the DNA of the chromosome itself is permanently rearranged to create new variable region genes (Figure 41.11). Since there are many ways to break and rearrange the chromosome, any one of a vast array of possible antibody genes can be created in a given B-cell.

But gene rearrangement is not the only way to create immunoglobulin diversity. Another way is to rearrange the gene products—the proteins themselves. In addition, somatic mutations can produce single DNA base substitutions, causing slight variations in the antibody.

Now what about the T-cells? How do they produce surface receptors so varied that they can recognize virtually any antigen? Much less is known about the genetics of the T-cell's recognition system.

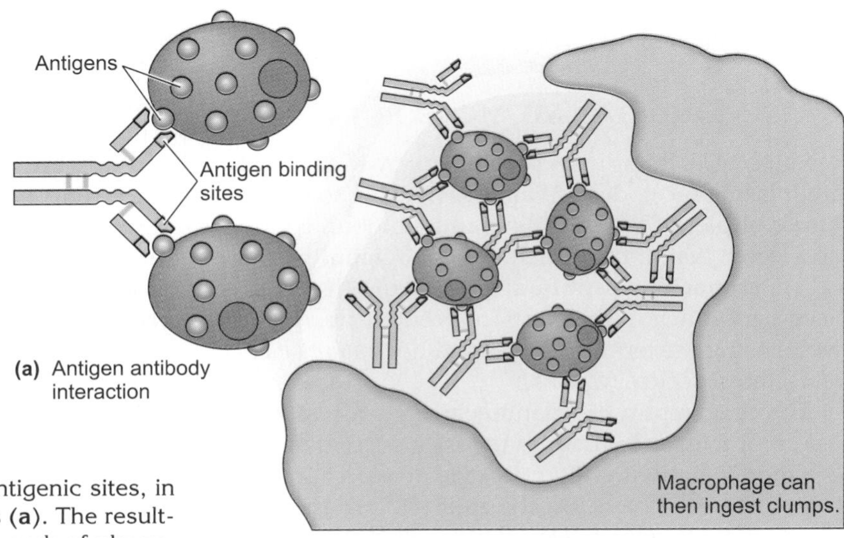

FIGURE 41.10 BINDING OF IMMUNOGLOBULINS TO ANTIGENS.
Each immunoglobin can bind to two antigenic sites, in this instance on the surface of two cells (**a**). The resulting clumps of invading cells make the work of phagocytic cells much more efficient (**b**).

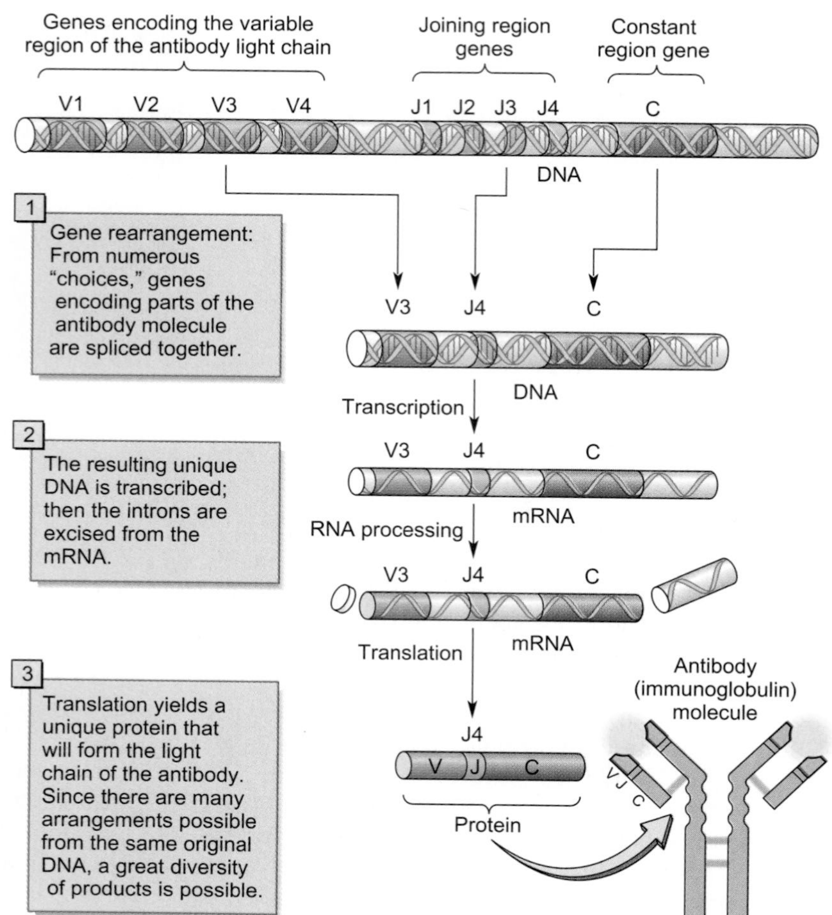
Genes encoding the variable region of the antibody light chain
Joining region genes
Constant region gene
V1 V2 V3 V4
J1 J2 J3 J4
C
DNA
1
Gene rearrangement: From numerous "choices," genes encoding parts of the antibody molecule are spliced together.
V3 J4 C
DNA
Transcription
2
The resulting unique DNA is transcribed; then the introns are excised from the mRNA.
V3 J4 C
mRNA
RNA processing
V3 J4 C
mRNA
Translation
3
Translation yields a unique protein that will form the light chain of the antibody. Since there are many arrangements possible from the same original DNA, a great diversity of products is possible.
J4
V J C
Protein
Antibody (immunoglobulin) molecule

FIGURE 41.11 IMMUNOGLOBULIN DIVERSITY.
During their development, lymphocytes undergo seemingly unlimited kinds of gene rearrangement, resulting in their great diversity. A variable region gene (here, V3) is excised and joined to one of several joining genes. The two are then fused to the end of a larger gene, designated "C," the one responsible for the constant region of the immunoglobulin. The new unique gene is then transcribed into mRNA, its introns removed, and translated into the polypeptides of an immunoglobulin.

However, immunologists strongly suspect that such diversity must be produced in a manner similar to that responsible for the diversity of immunoglobulins—that is, essentially through rearrangements of DNA segments within the genes responsible.

MONOCLONAL ANTIBODIES

Since 1976, it has been possible to grow clones of any plasma B-cell desired and, as a result, to obtain pure samples of the specific immunoglobulins produced by that B-cell. Such immunoglobulins are called **monoclonal antibodies,** proteins that have been very much in the news in recent years. (Here, we're using the term *antibodies* and not *immunoglobulins,* for historical reasons.)

The first step in the manufacturing process is to use recombinant techniques to fuse a desired line of B-lymphocytes with B-lymphocyte myeloma cells—tumor cells that have lost the ability to produce antibody. Tumor cells do especially well in tissue culture—so well that their cell lines are sometimes referred to as "immortal." The fused cells, called *hybridomas,* then go on through their cell cycles, producing clones. And, as is the case with plasma cells, each hybridoma produces only one specific antibody, a "monoclonal" antibody.

There are many exciting uses for monoclonal antibodies. They can be made radioactive or fluorescent and thus can be used to locate readily the antigen that brought about their formation in the first place. For example, the molecule of interest (the antigen) could be a hormone, or even its cell receptor site; either could be identified. Recent applications also include the clinical diagnosis of viral and bacterial diseases. For example, antigens associated with a sexually transmitted disease can now be detected in a matter of minutes, whereas the diagnosis once required days for culturing pathogens sampled from the infection site. Monoclonal antibodies can also be used as drug delivery systems—that is, as a means for getting medication to the right place in the body. In the treatment of cancer, for example, physicians are experimenting with the concept of linking potent but toxic cancer-treatment drugs to monoclonal antibodies that will bind to cancer cell-surface antigens. The hope is that the drug will go directly to its target and do its job. Thus, dosages of these harsh drugs could be greatly reduced, and the difficult side effects eliminated. It seems the potential applications for monoclonal antibodies are limited only by the imagination of researchers. (Essay 41.3 describes a new threat emanating from our efforts to combat a variety of infections.)

AIDS and the Crippled Immune System

Over the past few years the new and terrible disease called **acquired immune deficiency syndrome (AIDS)** has gained worldwide attention. The problem results from suppression of the immune system. The symptoms may begin as a simple but persistent cold, but as the disease progresses, people with AIDS grow increasingly susceptible to all sorts of diseases, even very rare ones. In fact, the appearance of some rare diseases, such as Kaposi's sarcoma and pneumocystic pneumonia, assists in identifying the condition in its later stages. By this time, unfortunately, such information is often more useful to the pathologist than to the patient.

The Growing AIDS Epidemic

The first cases of AIDS were recognized late in 1979. By mid-1986 the disease had appeared throughout most of the United States and in 100 other nations. Worldwide, by mid-1992 it was estimated that 10–12 million persons had been infected, and it is feared that probably most of them will develop AIDS. (People can carry the virus for 10 years without expressing symptoms other than the presence of specific viral antibodies.

It is now generally believed that the agent is a human retrovirus called HIV (human immunodeficiency virus).[1] Although HIV can affect several types of cells, it preferentially attacks helper T-cells. As the HIV infection progresses, the number of these pivotal lymphocytes dwindles. Initially there are enough helper T-cells to stimulate B-cell activity, and antibodies are formed against the viral agent. In fact, the presence of the antibody is how AIDS carriers and those exposed to AIDS are identified. Keep in mind that the antibody itself cannot destroy the virus. Once the virus invades a cell, the intervention of T-cells is needed. Let's see why the T-cells fail to act.

When the virus encounters its target cell, it invades the cell by penetrating the plasma membrane (see Figure 41.12). Once within the cell, the virus sheds its protective protein coat and releases two copies of single-stranded RNA (its genome), along with the enzyme reverse transcriptase. This versatile enzyme converts the RNA to double-stranded DNA, which is then inserted into one of the host's chromosomes. There the viral genome may remain inactive, simply replicating along with the host chromosome, perhaps for years. Or, the new viral DNA may immediately become active, transcribing new viral protein, replicating viral genes, and assembling many new viral particles. The cell soon ruptures. The new infective agents are released to infect still more cells or to be passed on to a new host should the occasion arise (Figure 41.12).

In its attack, the AIDS virus first binds to a cell-surface protein called CD4, located on the plasma membrane of helper T-cells. Certain glycoproteins on the viral coat form a perfect match with the CD4 molecule. The CD4 protein is also present on other immune cells, including macrophages, about 5% of the B-cells, and on certain cells outside the immune system. Included among the latter are selected duodenal and colon cells, certain skin cells, and the brain's macrophages and glial cells (the latter make up much of the brain's mass). In fact, HIV invasions of the brain have now been well documented. The point is, the susceptible cells of the body have one thing in common—they all bear the CD4 protein.

The "Achilles' heel" of HIV may well be its use of the CD4 protein as a binding site. Researchers now believe that one effective therapy might be to flood the patient with monoclonal antibodies that can bind with CD4 and thereby block the entry of the AIDS virus into body cells. Another possible therapy may be to introduce large amounts of CD4, which would then bind to the viruses themselves, thereby preventing them from binding to target cells. Such procedures are now being attempted, but so far with very limited success, and immunologists guess that we are still years away from an AIDS solution. Further, HIV is known to mutate frequently, introducing some knotty problems for medical research.

[1] *One prominent virologist insists that AIDS is not caused by HIV. He cites its remarkably slow development for a retrovirus, cases in which people with AIDS do not show the virus, and the widespread appearance of the virus in the general population apparently unrelated to the onset of AIDS. He suggests that AIDS is due to behavior that weakens the immune system, such as repeated incidence of sexually transmitted disease, drug use, and the use of HIV-fighting drugs such as AZT.*

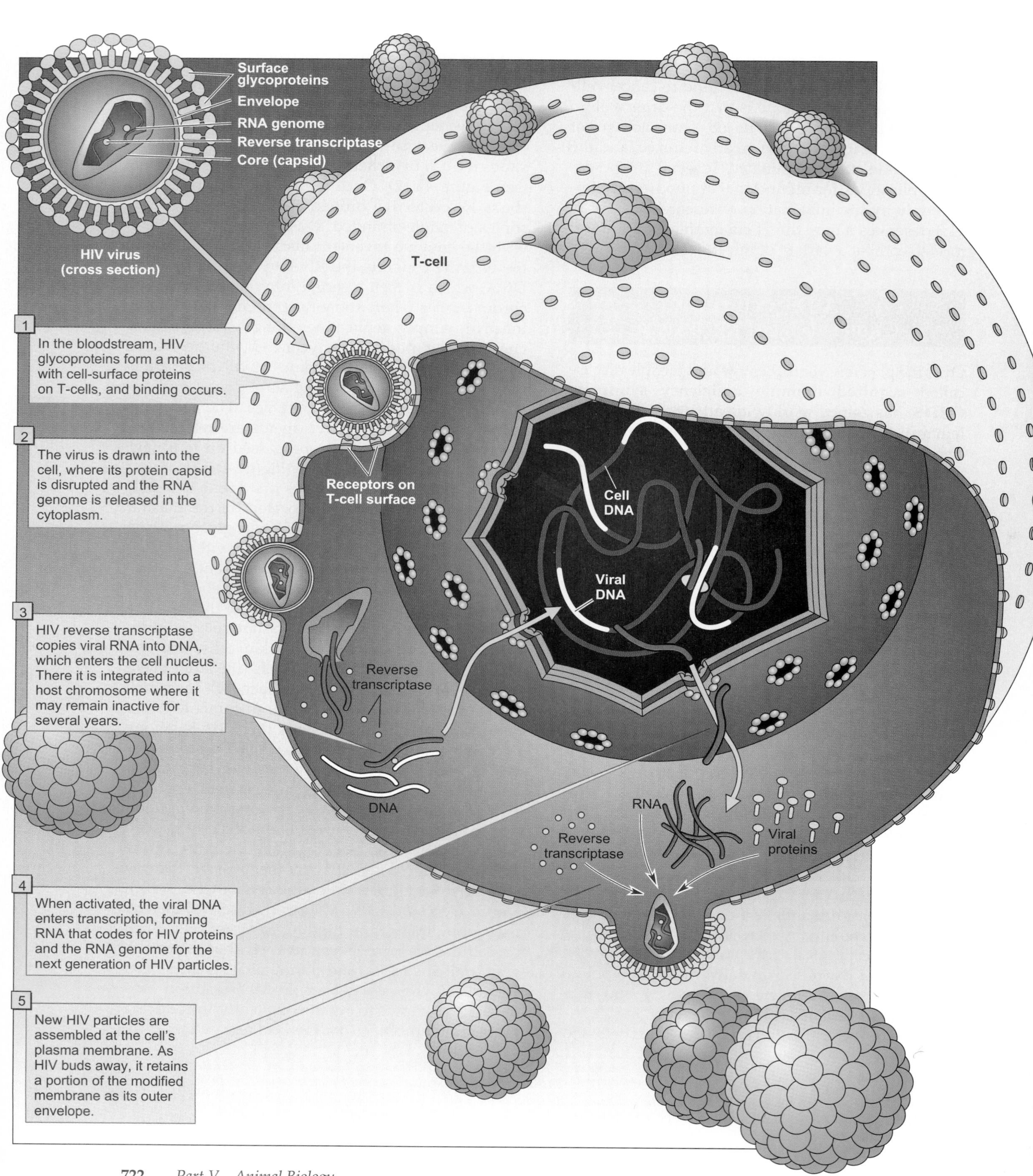
Surface glycoproteins
Envelope
RNA genome
Reverse transcriptase
Core (capsid)
HIV virus
(cross section)
T-cell
1
In the bloodstream, HIV glycoproteins form a match with cell-surface proteins on T-cells, and binding occurs.
2
The virus is drawn into the cell, where its protein capsid is disrupted and the RNA genome is released in the cytoplasm.
Receptors on T-cell surface
Cell DNA
Viral DNA
3
HIV reverse transcriptase copies viral RNA into DNA, which enters the cell nucleus. There it is integrated into a host chromosome where it may remain inactive for several years.
Reverse transcriptase
DNA
RNA
Reverse transcriptase
Viral proteins
4
When activated, the viral DNA enters transcription, forming RNA that codes for HIV proteins and the RNA genome for the next generation of HIV particles.
5
New HIV particles are assembled at the cell's plasma membrane. As HIV buds away, it retains a portion of the modified membrane as its outer envelope.

In the meantime, among the primary treatments for HIV infection in the United States is the antiviral drug AZT (azidothymidine). AZT does not cure HIV infection; however, it appears to prolong the lives of some people. AZT acts by blocking the enzyme reverse transcriptase, the one responsible for converting viral RNA to DNA. During reverse transcription, the enzyme incorporates AZT instead of the usual thymine nucleotide into the growing DNA chain. This immediately blocks the addition of the next DNA nucleotide, so reverse transcription stops and the viral life cycle cannot proceed. Unfortunately, AZT's effectiveness is temporary, and it has serious side effects in some people. Now, researchers fear that new mutant strains of HIV have emerged that are somehow resistant to the effects of AZT. Nevertheless, encouraged by such partial successes, researchers are working intensely at developing drugs with similar properties, and some, in fact, are already in use.

The Progression of AIDS

It is important to distinguish between the presence of HIV in a person and the onset of active AIDS. For each 100 people identified as having been infected (that is, their blood tests positive for antibodies against HIV), only one of them, on the average, has full-blown, active AIDS. These active cases are the people who are usually hospitalized for Kaposi's sarcoma, pneumocystic pneumonia, and other rare diseases. A second group, about 10% of the HIV-infected population, will have early symptoms, including weight loss, swollen lymph nodes, prolonged colds and other viral infections, fatigue, diarrhea, and persistent fever (all of which could be from other causes). The remainder, some 89%, will harbor the virus but have no outward symptoms (although they are infectious). The outcome for this latter group is uncertain, but we know that active AIDS can crop up in people infected as far back as 10 years, maybe longer. It is expected that they will all come down with the disease, although some people seem to have more resistance than others.

With more and more data available, immunologists are now finding a pattern in the response of the immune system to HIV. Following the progress of the virus over a 10-year period, researchers note that it multiplies rapidly during the first year, but then, as the immune system responds, the virus diminishes drastically (Figure 41.13). The immune system remains more or less in control of the infection for another 5 or so years, after which its effectiveness undergoes a decline, reaching a low in the ninth and tenth years. The decline period correlates closely with a rapid depletion of helper T-cells, those lymphocytes responsible for activating and coordinating the immune response. The decline in helper T's is accompanied by a new peak in the number of HIV viruses in the body. It is at this time that the HIV-infected person begins to suffer the fatal secondary infections.

FIGURE 41.12 The AIDS and the HIV Virus. On entering the bloodstream, the AIDS virus seeks out T-cell lymphocytes, the white blood cells that serve as master controls for the body's immune system. Using the various chemical markers on its surface, the AIDS virus binds easily to a receptor on the surface of the T-cell. The virus then enters the T-cell, in the process shedding its protective protein coat and exposing its core, which contains RNA and the enzyme reverse transcriptase. The enzyme in the virus helps translate the invading RNA into DNA. The DNA is treated by the host cell as its own, and inserted into its chromosomes. The AIDS virus DNA may remain dormant for weeks, months, even years inside the usurped T-cell before it begins to cause disease. Once activated, the viral DNA directs the T-cell to make copies of the virus. The new viruses bud from the T-cell's surface. Eventually the host cell dies.

The Transmission of AIDS

About 94% of HIV infections occur via semen or blood. HIV may also be spread in other ways, but the frequency of such spread is low and the means of transmission are not clear. HIV has been found in several other body fluids, including tears, saliva, sweat, mucus, and mother's milk.

Sexual transmission of HIV infection almost always involves viruses from infected semen finding their way into the blood. AIDS is common in people who engage in anal intercourse. The rectal wall is fragile, and intercourse invariably causes small abrasions through which the viruses enter the bloodstream. For women, vaginal intercourse with an infected man is less risky because the vagina is not normally abraded during the sexual act—although any minute abrasion, whatever its source, is all that is required for transmission, and there has been an alarming increase in AIDS among heterosexual women in the United States. Transmission from heterosexual women to heterosexual men is even less likely, since heterosexual intercourse does not commonly involve access to the man's bloodstream.

The risks accompanying sexual contact can be reduced through the use of condoms, since they capture the semen and prevent the escape of the virus. But it is important to know that condoms must be of

ESSAY 41.3 Germ Warfare

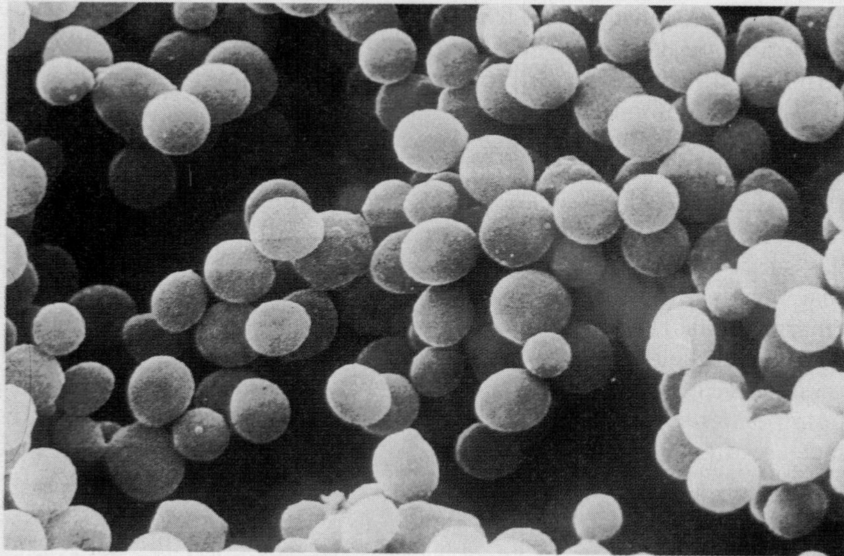

You get a sore throat. You go to the doctor. He prescribes an antibiotic. You take the pills a few days. Your throat gets better, and so you save some of the pills for next time. Suppose that next time, though, those pills don't work; so you go back to the doctor. He gives you some other pills. They don't work either. Your throat gets worse, your glands swell, the infection moves to other parts of your body. And then one bright spring day the doctor walks quietly into your room and tells you there is nothing more he can do. You are going to die.

Unfortunately, this morbid scene is being played out with increasing frequency, even in the United States, one of the most medically advanced countries on earth. In fact, the Centers for Disease Control in Atlanta reported that in 1992, 13,300 patients in hospitals died of bacterial infections that resisted every antibiotic administered. Put simply, nothing worked.

Recently, researchers found that at least one strain of *every* disease-causing bacterium is able to withstand at least one antibiotic. The problem is that some deadly bacteria respond to only one antibiotic, and some strains are showing resistance to it.

At this moment, there is a good chance that your skin harbors the bacterium *Staphylococcus aureus* in some relatively benign form. However, it has ample opportunity to mutate since a single bacterium can leave almost 17 million progeny in a single day. Some forms of staph can cause pneumonia and blood poisoning in surgical wounds, and have been known to sweep through hospitals, leaving death in their wake.

Unfortunately, about 40% of infectious staph are resistant to all antibiotics except one—vancomycin. Researchers say it is only a matter of time until a vancomycin-resistant strain shows up and then. . . . you can imagine the rest.

Another likely candidate for "Supergerm" is one called *enterococcus,* a deadly, blood-poisoning microbe. About 20% of enterococcal infections are resistant to vancomycin, and the number is increasing rapidly. Unfortunately, the resistant gene can be transferred from enterococcus to *Staphylococcus aureus*—a terrifying prospect.

The immune genes are generally located on a plasmid and travel from one bacterium to the next through a pilus, an extension that acts as a bridge for the genetic material of two bacteria (see Chapter 15). Most bacterial immunity arises by natural selection, as bacteria are exposed to nonlethal doses of antibiotics until they build a defense against them. However, because of the transfer of genes conferring immunity against antibiotics, bacteria can also develop immunity against an antibiotic they have never encountered.

Normally, antibiotics work by dissolving the bacterium's cell wall, by breaking down its DNA, or by interfering with its protein production. The bacterium defends itself by developing genes that strengthen the cell wall, protect its own protein-making genetic apparatus, or break down the drug. Obviously, the race is on—and at this point, the lowly bacterium is threatening to undo one of the most complex life forms the earth has ever known.

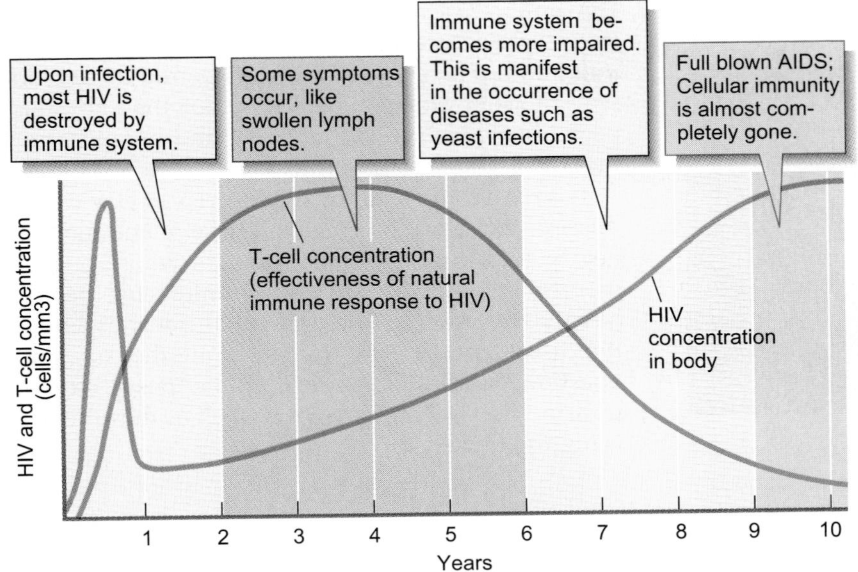
Upon infection, most HIV is destroyed by immune system.
Some symptoms occur, like swollen lymph nodes.
Immune system becomes more impaired. This is manifest in the occurrence of diseases such as yeast infections.
Full blown AIDS; Cellular immunity is almost completely gone.
T-cell concentration (effectiveness of natural immune response to HIV)
HIV concentration in body
HIV and T-cell concentration (cells/mm3)
1
2
3
4
5
6
7
8
9
10
Years

FIGURE 41.13 PROGRESSION OF AIDS.
Here we see the ebb and flow of T-cells and HIV concentrations in the blood. The telltale signs of infection will vary according to the length of time since the infection.

high quality, and used properly. There is evidence that condoms coated with nonoxynol-9 are particularly effective because the chemical coating kills the AIDS virus on contact (Figure 41.14). You should be aware, if you are not, that a startling number of condoms break during use. Animal membrane condoms are of little value, since they are readily penetrated by the AIDS virus.

Intravenous drug users and their sex partners have the fastest-growing rate of infection in the United States today. AIDS in this group is spread largely through the use of contaminated hypodermic needles used to inject "street drugs." Many drug users are among the least-informed people in our society, and so they often continue sharing needles, oblivious to the risk. Some use liquid laundry bleach to sterilize shared needles, and while use of undiluted bleach is certainly to be recommended, it is not entirely effective. Some, using their own personal needles, inject themselves with the AIDS virus picked up from contaminated drug batches shared with others as more than one needle is dipped into the same prepared solution (Figure 41.15).

The rate of transmission of the virus among non-IV-drug-using heterosexuals is also increasing. There are two main sources of infection connecting high-risk and low-risk groups. One is through "respectable, middle-class" IV-drug users who do not know or do not tell their sex partners—including their spouses—that they might be—or are—infected. The other is the male bisexual whose sex partners alternate between high-risk homosexual men and low-risk heterosexual women.

The clinical blood supply is another source of infection, albeit a much less frequent one. Whereas blood is always screened for HIV, the screening is not entirely effective. Some people, anticipating surgery, are now storing their own blood in advance; others bring their own donors, usually family members, to the hospital just in case.

A growing number of people with AIDS are infants and children. While no one is sure how infants become infected, it seems that HIV is able to cross the placenta. Alternatively, it may be passed to infants through breast-feeding. We also know that

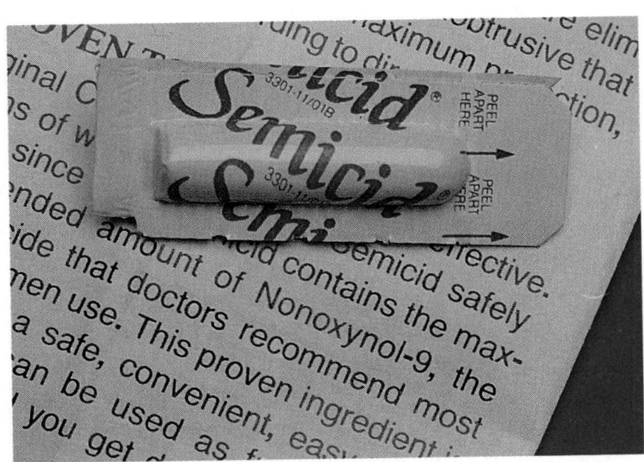
Semicid
Semicid
PEEL APART HERE
amount of Nonoxynol-9, the
ide that doctors recommend most
men use. This proven ingredient
a safe, convenient, easy
an be used as
Semicid contains the max-
Semicid safely
effective.

FIGURE 41.14 EFFORTS AT STEMMING INFECTION.
Here, a prophylactic display touting nonoxynol-9 is an example of increasing awareness that condom use can reduce the spread of AIDS.

FIGURE 41.15 **DRUG USE AS A THREAT.**
Drug users who use needles run a high risk of contracting HIV, the AIDS-causing virus, through contaminated needles or contaminated drugs.

HIV antibodies generated by an infected mother do cross the placental barrier and can be detected in the infant for up to a year. During this period there is no way of determining whether or not the baby has been infected, since the mother's and infant's antibodies are indistinguishable.

It is reassuring to know that the risk of HIV transmission in ways other than those mentioned is nearly zero. Studies of households where one or more infected people live with uninfected people have, so far, revealed no nonsexual transmission of the disease agent. And to answer another nagging question, studies indicate that the possibility of transmission of HIV by insects, such as mosquitoes, is highly unlikely.

PERSPECTIVES

Now that we have learned something about how the body defends itself against invaders in both general and specific terms, and how it has not yet been able to meet the challenge of AIDS, we move to a discussion of the reproductive systems.

Here, not only will we consider the structure and function of males and females and their gametes, but we will again address the subject of protecting one's self from the ravages of sexually transmitted diseases, including AIDS.

SUMMARY

THE FIRST LINE OF DEFENSE: SURFACE BARRIERS

Physical and chemical surface barriers provide a first line of defense against invading microbes. These include intact skin that is constantly replaced, its keratin layer and sweat and oil glands, mucus secretions, stomach acids, body fluids with lysozymes, urine flow, and populations of microorganisms.

NONSPECIFIC SECONDARY DEFENSES

When organisms penetrate the body's surface barriers, secondary defenses are mounted. Some of these are nonspecific; that is, they are effective against any invader. They are direct and immediate.

Histamine release from injured tissues produces an increase in blood flow resulting in inflammation. Kinins increase swelling, usually bringing on pain. Complement lyses bacterial cells, prompts inflammation, and attracts neutrophils. Interferon blocks protein synthesis in virus-infected cells.

Phagocytosis, engulfing and destroying foreign debris or microbes, is one of the body's most important nonspecific cellular defenses. Key phagocytic cells are the neutrophils and the macrophages. Neutrophils, the most numerous white cells, act early in infections, engulfing invading cells. Monocytes undergo transformation into phagocytic macrophages. In addition, natural killer cells identify and kill abnormal cells, those infected by viruses or in the early stages of cancer.

SPECIFIC SECONDARY DEFENSES

Specific cellular defenses involve lymphocytes. Specific chemical defenses involve immunoglobulin (antigen). Immune responses are specific secondary responses that target particular invaders. These require exposure to the invading agent and can be slow to start.

Lymphocytes arise and mature in primary lymphoid tissue (red bone marrow and thymus). They reside as virgin cells in secondary lymphoid tissues (lymph nodes, tonsils, adenoids, spleen, small intestine), where they are activated.

T-cells differentiate in the thymus and B-cells in the red bone marrow.

B-cells respond to invaders with the humoral response, the formation and secretion of immunoglobins. Some T-cells carry out the cell-mediated response, the activation of specialized cells that attack infected body cells. Other T-cells activate B-cells.

The Primary Immune Response

The primary immune response, the first one mounted against an invading agent, involves several types of cells and their products.

Macrophages incorporate invader molecules in their plasma membranes, thereby becoming antigen-presenting cells. They activate T-cell lymphocytes with matching surface molecules.

In the primary immune response, antigen-presenting cells making contact with helper T-cells secrete interleukin-1. The activated helper T-cells then seek out matching virgin lymphocytes, whereupon they secrete interleukin-2. In response, the virgin cells grow and differentiate into B-cells and T-cells, which then produce immense clones.

B-cells synthesize and secrete specific immunoglobulins that recognize and attach to invading cell-surface antigens and free antigens, forming masses that are easily engulfed by phagocytes. Memory cells are held in reserve, activated only if reinfection occurs.

Cytotoxic T-cells kill any invading and diseased cells whose surface antigens match their surface antibodies. Because of dual recognition complexes, friendly cells ("self") are avoided.

The immune response is amplified as helper T-cells arouse virgin lymphocytes, which give rise to additional helper T-cells. As invaders are subdued, suppressor T-cells inhibit the primary response.

The Secondary Immune Response

When a second invasion occurs, memory B- and T-cells quickly form immense clones whose members immediately subdue the invader.

Artificial active immunity is brought about by the use of vaccines. They contain killed or weakened disease organisms that initiate the primary immune response without most of the disease symptoms. Passive immunity is conferred by the injection of antiserum, which contains active immunoglobulins.

The Immunoglobulins

Immunoglobulins contain paired heavy and light chains of amino acids that have constant and variable regions. The latter is the highly specific antigen recognition region that binds to matching antigen.

The vast number of different cell-surface recognition proteins and potential immunoglobulins is attributed to the behavior of lymphocyte genes during development. Certain genes undergo rearrangement, during which each kind of lymphocyte has its variable gene segments rearranged in its own specific way. Because of the great numbers of rearrangements carried out, virtually any antigen can be matched by a lymphocyte.

Using recombinant DNA techniques, researchers have been able to produce clones of B-lymphocytes that produce any desired immunoglobulin.

Aids and the Crippled Immune System

AIDS (acquired immunodeficiency syndrome) is a condition in which the immune system fails to fight off disease. People with AIDS often die of rare diseases.

AIDS was recognized late in 1979 when the incidence of rare diseases rose suddenly. AIDS is a condition in which the immune system is suppressed.

The agent of AIDS, believed to be human immunodeficiency virus (HIV), a retrovirus that devastates primarily helper T-cells and others. It invades cells by binding to a cell surface protein, CD4. Most of the Americans stricken with AIDS are homosexuals, bisexuals, intravenous street drug users, blood transfusion recipients, and babies born to women with AIDS. Upon entry into host cells, the single-stranded RNA genome of HIV is converted to double-stranded DNA and inserted into the cell's genome. There, it may remain inactive or begin at once to reproduce, thereby crippling the immune system.

Only 1% of people testing positive for HIV antibodies has full-blown AIDS. About 10% of those testing positive will show early symptoms that can be attributed to other diseases. AIDS can appear in people infected 10 years previously. The immune responses appear to contain the infection after its appearance, but then fails.

About 94% of infections occur via semen or blood. HIV has been found in tears, saliva, sweat, mucus, and mother's milk. Anal intercourse transmits the virus more readily than does vaginal intercourse. Use of condoms with nonoxynol-9 are recommended to limit infection. Clinical screening of blood is not always effective. HIV can cross the placenta and infect fetuses.

Key Terms

integument • **707**
nonspecific defense • **708**
specific defense • **708**
histamine • **708**
complement • **709**
interferon • **709**
neutrophil • **710**
macrophage • **710**
natural killer cell • **710**
humoral response • **711**
immunoglobulin • **711**
antigen • **711**
cell-mediated response • **711**
B-cell • **712**
T-cell • **712**
primary immune response • **712**
antigen-presenting macrophage • **713**
helper T-cell lymphocyte • **713**
interleukin-1 • **713**
interleukin-2 • **713**
clonal selection theory • **713**
plasma cell • **713**
memory cell • **713**
cytotoxic T-cell • **715**
helper T-cell • **715**
suppressor T-cell • **716**
secondary immune response • **716**
active immunization • **717**
passive immunization • **717**
monoclonal antibody • **720**
acquired immune deficiency syndrome (AIDS) • **721**

Review Questions

1. Describe the physical and chemical barriers that serve as the first-line defense.
2. List four chemical agents that are nonspecific mechanisms to combat infection, and briefly summarize their role.
3. Name two key phagocytic cells involved in nonspecific defense. Contrast the defensive roles of NK cells and neutrophils.
4. Describe the change that occurs when a monocyte is activated at an infection site. What are two roles of these transformed cells?
5. List two primary and four secondary lymphoid tissues. In general, what happens in each?
6. Distinguish between humoral and cell-mediated specific immune responses. Which lymphocytes are responsible for each?
7. Explain how a macrophage becomes an antigen-presenting cell. What does such a cell seek?
8. In what two general ways do activated lymphocytes respond?
9. What happens to an antigen-presenting cell when it identifies a matching helper T-cell? What is the helper T-cell's response?
10. Describe the effects of interleukin-2 on virgin lymphocytes.
11. What kinds of cells make up an activated B-cell clone? What is the specialized role of each of its kind?
12. In what way do immunoglobulins participate in the immune response?
13. Discuss the work of the cytotoxic T-cell.
14. Describe the dual recognition mechanism of T-cells. Why is such a system vital?
15. Explain how helper T-cells amplify the primary immune response.
16. In what two different situations do suppressor T-cells carry out their functions?
17. Why is the secondary immune response so much faster than that of the primary response?
18. Briefly contrast active and passive immunization.
19. Using a simple diagram, describe the typical immunoglobulin. Include light and heavy chains, constant regions and variable regions. Label the antigen recognition regions.
20. What, if any, appear to be the limits in cell surface recognition and immunoglobin capabilities in lymphocytes?
21. Explain how gene rearrangement provides for the vast molecular diversity of lymphocytes. In what way does this seem to violate the central dogma?
22. What are potential ways in which monoclonal antibodies may be used?
23. Is AIDS itself a disease? What actually kills the person who has AIDS?
24. List, in descending order of risk, the groups in which AIDS is most prevalent.
25. Specifically, what is believed to be the AIDS agent? What specific cells does the agent favor?
26. How does this agent become part of the host genome? What is its effect on the host cell's functions?

CHAPTER

42

Reproduction

THROUGH THE EONS OF LIFE ON EARTH, SOME INDIVIDUALS HAVE REPRO-*duced, some have not. Each generation, then, is made up of the offspring of individuals that did reproduce. However, we, as descendants of reproducers and generally interested in reproducing ourselves, are often woefully ignorant of the cold facts. So here let's have a look at how we reproduce, how to avoid it, and what kinds of diseases we can get from the processes leading to reproduction. Here's a question for you: What is the most common bacterial disease in the United States? You may be surprised at the answer.*

The reproductive imperative of all living things is: "Reproduce or your genes will disappear from the population." This threat expresses nothing more than the cool, unprejudiced arithmetic of evolution, and its meaning is deceptively simple: Each generation is made up of the descendants of the reproducers of previous generations. Furthermore, most of the individuals of the next generation will be descended from the best reproducers of this generation. Stated another way, natural selection favors traits that result in greater reproductive success. Obviously, then, reproduction has been and will always be important to the continuance of life. So let's look into some of the ways animals reproduce. As you might expect, they do so in a great number of ways and with a variety of strategies and techniques (some quite startling). Nonetheless, as is usual among the diverse forms of life on earth, there are underlying themes.

ASEXUAL REPRODUCTION

We are aware by now that animals may reproduce asexually (*a,* without) or sexually. **Asexual reproduction,** reproduction without sex, involves just one parent. Animals that reproduce asexually have the great advantage of each of their offspring carrying 100% of their genes. From an evolutionary perspective, the basic task of reproduction is simply to get one's genes into the next generation, and a parent that can produce offspring bearing all of its genes, instead of half, would be more efficient. Also, such a parent doesn't have to spend time seeking out, courting, feeding, or defending a partner.

A variety of animals avail themselves of the advantages of asexual reproduction. Asexual reproduction occurs primarily among invertebrates, and the processes can involve some elaborate and complex patterns. We will consider three rather distinct cases here.

BINARY FISSION

A number of organisms reproduce by simply splitting in two, a process called **binary fission.** The planarian flatworm, for example (Figure 42.1), can reproduce itself in this way. (It can also reproduce sexually, but we'll ignore that small inconvenience for now.) In the process of fission, the planarian simply begins to constrict about midway along its

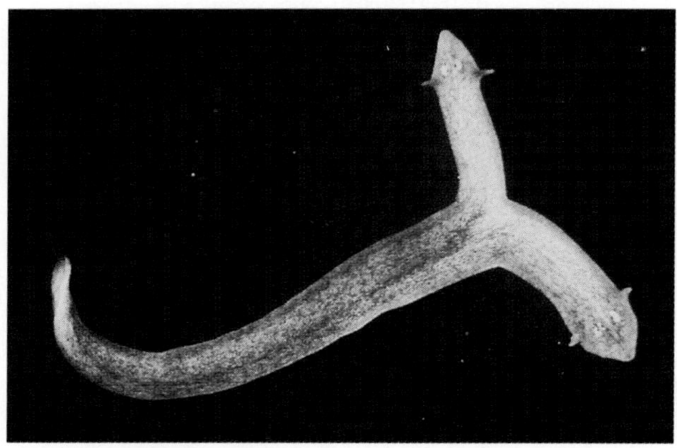

FIGURE 42.1 FISSION.
Planarian worms reproduce through binary fission. Basically, the body pinches in two roughly halfway along its length. Each end then regenerates the missing parts. Even when surgically split down the middle, the two halves can regenerate.

length until it pinches in two. Cells from each part then move into the severed area and regenerate the missing parts. The result is two new planarians.

It is apparent that the cells that rebuild the missing structures in such cases must undergo some sort of fundamental reorganization. After all, these cells have already differentiated and specialized and have been functioning in some specific role. Now they must somehow change and take on new characteristics and a new role. All the processes involved in such changes are not clear, but we know that in some cases, cells can reverse their development; that is, after having proceeded along one very specific developmental pathway, they reverse the process, revert to an earlier stage, and then follow a different developmental pathway.

Regeneration and Fragmentation

Regeneration, the replacement of missing parts, may be considered another form of asexual reproduction. For example, if a planarian is cut in half, each half can regenerate missing parts, presumably by the same reorganizational pathway that follows binary fission. In nature, other species may lose body parts through accident or injuries caused by predators. For instance, sea stars (you may know them as starfish—some biologists seem to change the rules from time to time, apparently just to see if the rest of us are paying attention) can replace a lost arm. In fact, as long as a part of the body (the "central disc") is present, an arm can regenerate into a whole new individual. Not long ago, before this fact was discovered, vengeful oystermen hauling up a sea star in a dredge would chop up the unfortunate creature and throw the pieces back in the water, quite sure that the sea star would eat no more oysters. However, they were wrong—the sea star population flourished as the oyster catches dwindled, causing the oystermen to chop away with even greater diligence until biologists were able to convince them to axe their behavior instead of the sea star.

Some species reproduce through **fragmentation,** the development of an entire individual from a body part. The body of the parent may break into several pieces; each piece then regenerates the missing parts and develops into a whole animal. Fragmentation is common among certain flatworms.

Budding

Another method of asexual production is **budding,** a process in which a new organism is produced as an outgrowth of the parent organism. The new appendage then pinches off and moves away on its own. A number of aquatic invertebrates reproduce this way. An example is found in the tiny *Hydra,* a fresh-water relative of the sea anemone (Figure 42.2).

Sexual Reproduction

In **sexual reproduction,** a parent's genes are combined with those of another individual. There are great arguments over just why sexual reproduction ever evolved to begin with, but the main advantage of such dilution seems to be in increasing the variation in one's offspring so that, as a group, some of them are likely to succeed under a wide range of environmental conditions. Sexually reproducing organisms have the means to produce great variation in their offspring (through processes like meiosis, crossing over, and random participation of gametes). In addition, at meiosis the chromosomes have an

FIGURE 42.2 ASEXUAL REPRODUCTION. Asexual reproduction in *Hydra* includes budding, wherein miniature individuals arise from the bodies of others.

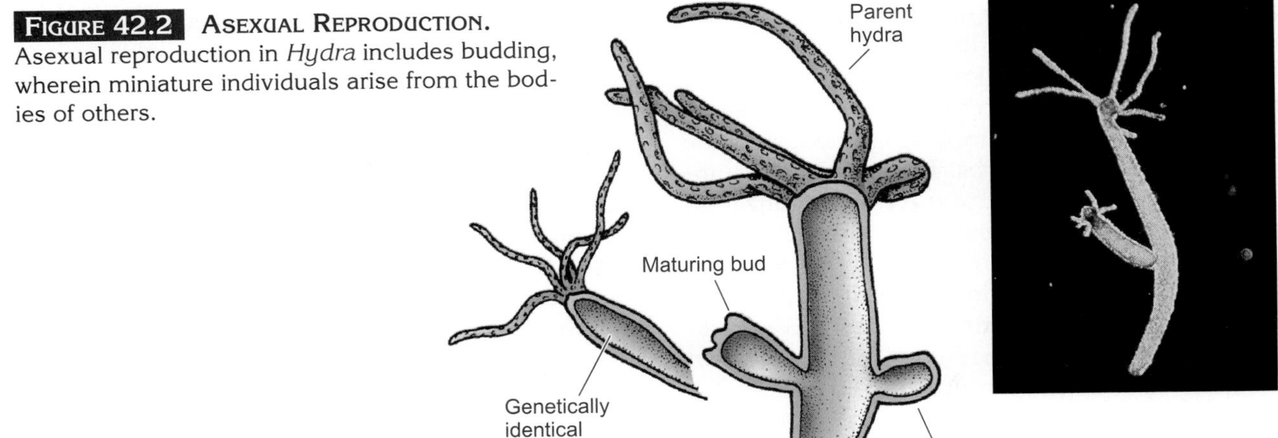

opportunity to line up side by side and to make changes in chromosomal material that might correct any defect or aberration in a single chromosome. (The advantages of sex are discussed in Essay 42.1.)

Most animals reproduce sexually (attesting to its advantages). So let's have a look at them and see how they have become increasingly efficient at (1) getting eggs and sperm together and then (2) protecting the developing embryo. We will find that sexual reproduction has seemingly endless variations. One fundamental variation involves where the gametes get together. It happens outside the bodies of some animals, inside others.

External Fertilization

Fertilization is the fusion of male and female gametes. (We will be more precise later.) **External fertilization**—fertilization that occurs outside the body—is, in its simplest form, probably a primitive trait and is certainly the least efficient form of sexual reproduction. Efficiency, here, refers to the numbers of gametes needed to ensure the simple replacement of the individual. External fertilization is not very efficient precisely because the gametes (eggs and sperm) are released into the environment outside the body. Its a big world out there and anything can happen, so many of the gametes die or are lost before they can join. The practice of external fertilization is almost exclusively aquatic because releasing gametes on dry land would present the certain risk of desiccation. Still, in spite of its inefficiency and high risk, external fertilization is common in both invertebrates and vertebrates.

Some species release their gametes into the water without taking further measures to ensure fertilization. Others, though, increase the odds of successful fertilization by coordinating the release of gametes. Sea stars, for example, often live in dense groups along the rocky sea bed or on wharf pilings. The release of eggs by one individual prompts others nearby to release great clouds of sperm, and the presence of sperm induces still others to release eggs. Thus, there is an amplifying effect, and the water is filled with gametes. This coordination, indeed, improves the chance of successful fertilization since some sea stars routinely shed as many as 2.5 million eggs at a time.

Other species also provide a measure of precision in gamete release, both in time and space. For example, the California grunion (Figure 42.3) is a fish that seasonally spawns three or four days after a new or full moon—the times of the highest tides. Shortly after the tide begin to recede, the females, hotly pursued by the males, ride the incoming waves, thrash their way up onto the wet beach, wiggle their tails into the sand, and deposit their eggs. The males curl around the females, releasing their sperm. Then, if all goes well, the fish catch the next wave out. (In southern California all does not always go so well; a favorite sport there is "grunion hunting," all very legal as long as only the hands are used.) The grunion zygotes are left to develop for a month, protected in the sand, until the next peak night tide stimulates hatching and washes the emerging young out to sea, completely unprotected by the parents.

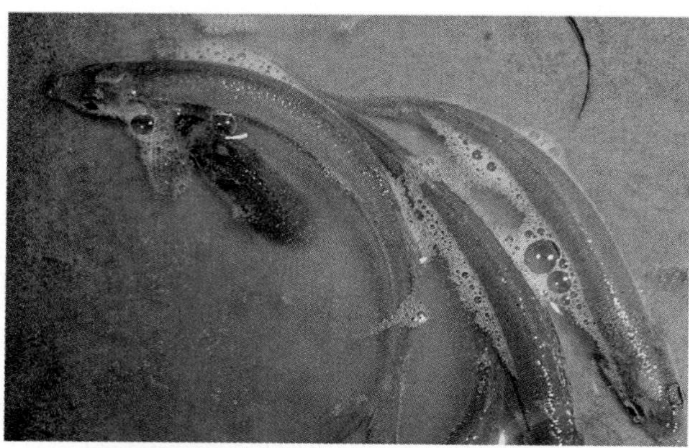

FIGURE 42.3 EXTERNAL FERTILIZATION. Sexual reproduction may involve external fertilization as seen in the Pacific grunion, a fish that comes ashore to reproduce.

Internal Fertilization

In the course of the evolution of sexual reproduction, we might assume that the first sexual animals released their gametes willy-nilly into the environment. Then others probably increased their efficiency by the sexes releasing the gametes at the same time, as do some echinoderms. The next step was to release them at the same time and place, as when grunion press their reproductive openings close together. Even these improvements on external fertilization, though, leave a high degree of risk for the gametes. Many are wasted, and natural selection is, by its very nature, a process that tends to minimize waste. (It minimizes waste as an end result—but natural selection itself is very wasteful. Do you see why this apparent paradox is true?)

The next step, then, would be to decrease the risk by developing ways to put the sperm directly into the female reproductive tract, where the eggs are. Getting the sperm into the female's reproductive tract has at least three advantages: (1) it reduces the

ESSAY 42.1 Why Have Sex?

Evolutionary biologists have long been puzzled by the prevalence of sex among the various species. (About 99% of vertebrates reproduce sexually.) Sex just doesn't make sense, they argue. (Evolutionary biologists are incurable romantics.) After all, species that reproduce asexually (essentially by forming clones), pass on 100% of their genes to their offspring, and they don't have to waste time and energy finding mates to bring it off. Sexual reproducers, on the other hand, not only have to find mates, but their offspring then contain only half of each parent's genes. This dilution is a high price to pay, the evolutionists argue, because after all, evolution rewards only reproduction. Each new generation will be composed of the best reproducers of previous generations, and reproduction success is measured by how many of its genes an individual can get into the next generation. So why would an animal ever reproduce by diluting its genes through sex? Strangely enough, the question cannot yet be answered in definite terms.

Evolutionary biologists have some tentative answers—and the traditional response is that sex results in a mixing of genes that produces variable offspring, and this variation is insurance in a variable and changing world. That is, in a variable group of offspring, at least some of them will be prepared to face new challenges as the environment changes.

Other biologists have extended the theoretical advantages of sex to include, for example, DNA repair. They say that life on earth was set off on the sexual path early in the evolution of life when early cells incorporated bits of DNA left over from dead and disintegrating cells. This DNA could then be used to repair any breaks or faults in the living cell's DNA. In time, the cells would come to accept DNA from living cells through tubelike extensions of the cell wall.

Others suggest that early sex was based on nutrition. After all, DNA is composed of bases, phosphates, and sugars. They say that when a cell, such as a bacterium, becomes low on energy, it tends to accept DNA from other cells for its nutritive value. Any leftover bits and pieces could eventually become incorporated into the host's DNA, helping to repair its own DNA and increasing variation in the host's offspring.

However sex started, its continuance attests to its benefits—and one major benefit may be in producing rapidly changing populations that cannot be easily tracked by infectious invaders such as pathogens and parasites. A stable, uniform population would be a sitting duck, enabling infectious organisms to draw a fine bead and adapt very specifically to the host cell's situation. But if sex keeps jumbling the host's genes, the pathogens can never adapt so finely. They have to chase a changing situation.

Whatever the benefits of sex, the cheerful consensus is that it is here to stay.

number of eggs necessary; (2) it vastly increases the probability that all the eggs will be fertilized; and (3), in species with internal development, it gives the developing eggs a measure of protection—as the mother looks after herself, she protects the embryos within her body.

There is an ecological advantage to internal fertilization as well. With the embryos safely protected in an egg, or in the mother's moist internal environs, reproduction on dry land is possible. This step, taken hundreds of millions of years ago, opened up a host of new evolutionary opportunities for animals. Those species that lay eggs are called *oviparous* (chickens and stingrays fall into this group). The young are protected by some form of external covering (the "shell"). Other species offer more protection. In *ovoviviparous* species, the eggs are incubated within the mother's body, as in the case of some sharks. The young are then born alive. In *viviparous* species, the young develop within a uterus, taking their primary nourishment not from yolk storages but from the mother's blood.

Although there are various means of mating and reproduction (see Essay 42.2), the principle remains the same in each case: perpetuating one's genes by passing them to a new generation. Now we will focus on a representative viviparous species and see, up close and personal, how it goes about reproducing.

Human Reproduction

Compared to some mammalian species, the reproductive system of *Homo sapiens* seems rather ordinary. It is, in fact, almost identical to those of most other primates. However, humans do have some unique traits, such as an extremely high level of sexuality, nonseasonal sexual behavior, and an inordinately large penis for a primate.

Males

In men, the external genitalia include the *penis* and the paired *testes* (singular, *testis*), or testicles, suspended in the saclike *scrotum* (Figure 42.4). The penis consists of a cylindrical shaft ending in the enlarged *glans,* with the urethral opening at its tip. It is an extremely sensitive organ with an abundance of touch receptors, especially around the glans.

During sexual excitement the penis becomes erect, lengthening and thickening and curving upward as

TABLE 42.1 MAJOR PARTS OF THE MALE REPRODUCTIVE SYSTEM

Structure	Function
Testes	
Seminiferous tubules	Sperm production
Interstitial cells	Production of male sex hormone (testosterone)
Epididymis	Storage and maturation of sperm
Vas deferens (sperm duct)	Tube that conducts sperm from epididymis to urethra
Urethra	Tube that conducts sperm (and urine) out of body
Penis	Transfer of sperm to vagina
Accessory glands	
Seminal vesicles	Secrete most of volume of semen
Prostate gland	Adds secretion to semen
Bulbourethral glands (Cowper's glands)	Secrete small amounts of fluid before ejaculation for lubrication and rinsing urine from urethra

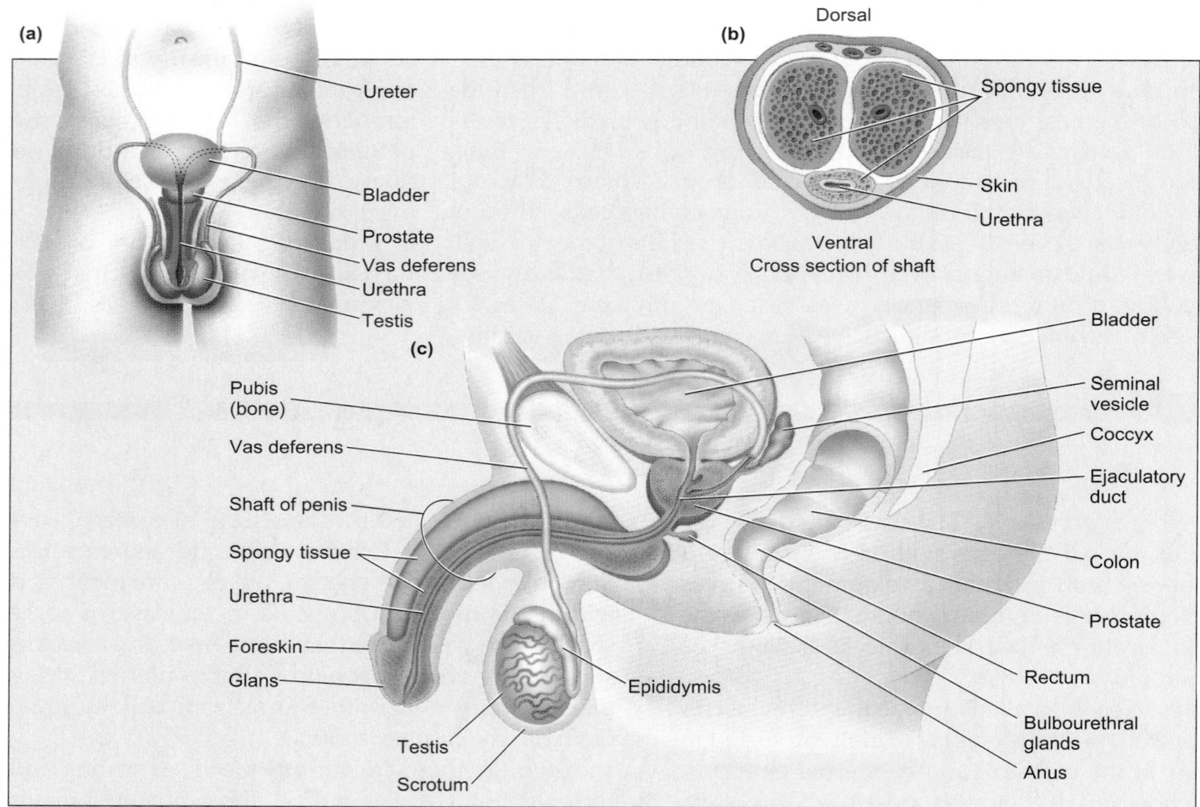

FIGURE 42.4 THE HUMAN MALE REPRODUCTIVE ANATOMY.
(**a**) Front view of the male genitalia, emphasizing the route followed by sperm. (**b**) Cross-section through the penis showing the major blood vessels and spongy tissue. (**c**) The male reproductive tract with associated structures in side view.

ESSAY 42.2 How They Do It

Many people seem to be under the impression that sex is sex—that basically, all the species must do it the same way. But the truth is that some animals mate in ways that humans would recognize, but most of them don't. For example, male bedbugs will often masquerade as females, enticing other males to copulate with them. However, neither male nor female bedbugs can survive many matings because the male pierces his mate's back with his sharp penis and ejaculates directly into the body cavity. Special cells then capture and ingest many of the sperm as they roam the recipient's tissues. Thus, the recipient is nutritionally rewarded, whether male or female.

In some mites, brothers and sisters copulate before they are born, so the little females are born pregnant and undoubtedly disillusioned. Certain snails have an enormous penis just over the right eye. They are hermaphroditic (each possessing the organs of both sexes), but they can't exchange sperm until they have pierced each other with a chalky dart that usually acts as a sexual stimulant but can also kill. Other snails begin life as wandering males, but eventually settle down and become sedentary females. If a wandering male mounts a female, he must copulate before his masculinity fades. The female praying mantis sometimes devours the male even while he's copulating with her. In fact, after she eats his head and brain, his sex becomes more intense (but we'll try not to extrapolate from that).

Some snails, fish, and amphibians change gender, so they may be a father at one time and a mother at another. In some fish, amphibians, and reptiles, there are no males at all. The egg is stimulated to develop when the female mates with a male of a different species. His sperm only activates the egg; it doesn't join with it.

The variation among the species goes on. We find geese that form ménages à trois (a threesome, usually with two males). And then forcible sex is common among lobsters, skunks, and orangutans. Speaking of force, the female of some species, such as rhinos, must fight before they will allow mating. Rhino females are apparently stimulated by a good fight.

Not only are reproductive behaviors sometimes startling, but so are the reproductive structures. Males of many insects have penises that look like instruments of torture with points, hooks, barbs, and impossible angles. Male opossums have split penises, with grooves instead of tubes, that match the divided vagina of the females. Pigs have a corkscrew-shaped penis that locks tightly into the female's vagina as they ejaculate about a pint of semen. Snakes and lizards can copulate with either of two penises that evert from the cloaca like turning the finger of a glove inside out. Each is covered with backwardly directed barbs, as are the penises of cats and skunks. And, finally, as the old jokes goes, how do porcupines do it? They, indeed, do it very, very carefully.

its spongy *erectile tissues* fill with blood. These changes occur because the flow of arterial blood entering the spongy tissue of the penis is increased, while venous outflow is decreased. Erection is brought about chiefly through a reflex involving the parasympathetic nervous system, especially in response to touch, but higher brain centers intercede significantly through erotic thought, odors, sounds, and visual images. When erection occurs, the glans emerges from a cuff of skin called the *foreskin* (or prepuce), unless it has been removed by circumcision. Physical stimulation may then trigger a reflex involving the sympathetic nervous system that brings on **ejaculation,** the release of sperm-bearing *semen.* The structures and functions of the male reproductive system are listed in Table 42.1.

The Testes and Sperm Production. Sperm and male sex hormones are produced in the testes. In many mammals, these dense oval bodies descend during the mating season and are withdrawn into the safer region of the body cavity when the season ends. But in humans the testes normally descend permanently shortly before birth. Their descent into the scrotum is essential, because developing sperm are quite heat-sensitive and the scrotum is cooler than the higher internal body temperatures. In response to heat, slender muscles within the scrotum relax, permitting the testes to descend away from the body, where it is cooler. Cold causes muscles in the scrotum to tighten and contract, drawing the testes closer to the warm body. (Threat or fear can also cause them to draw closer to the body where it is safer.)

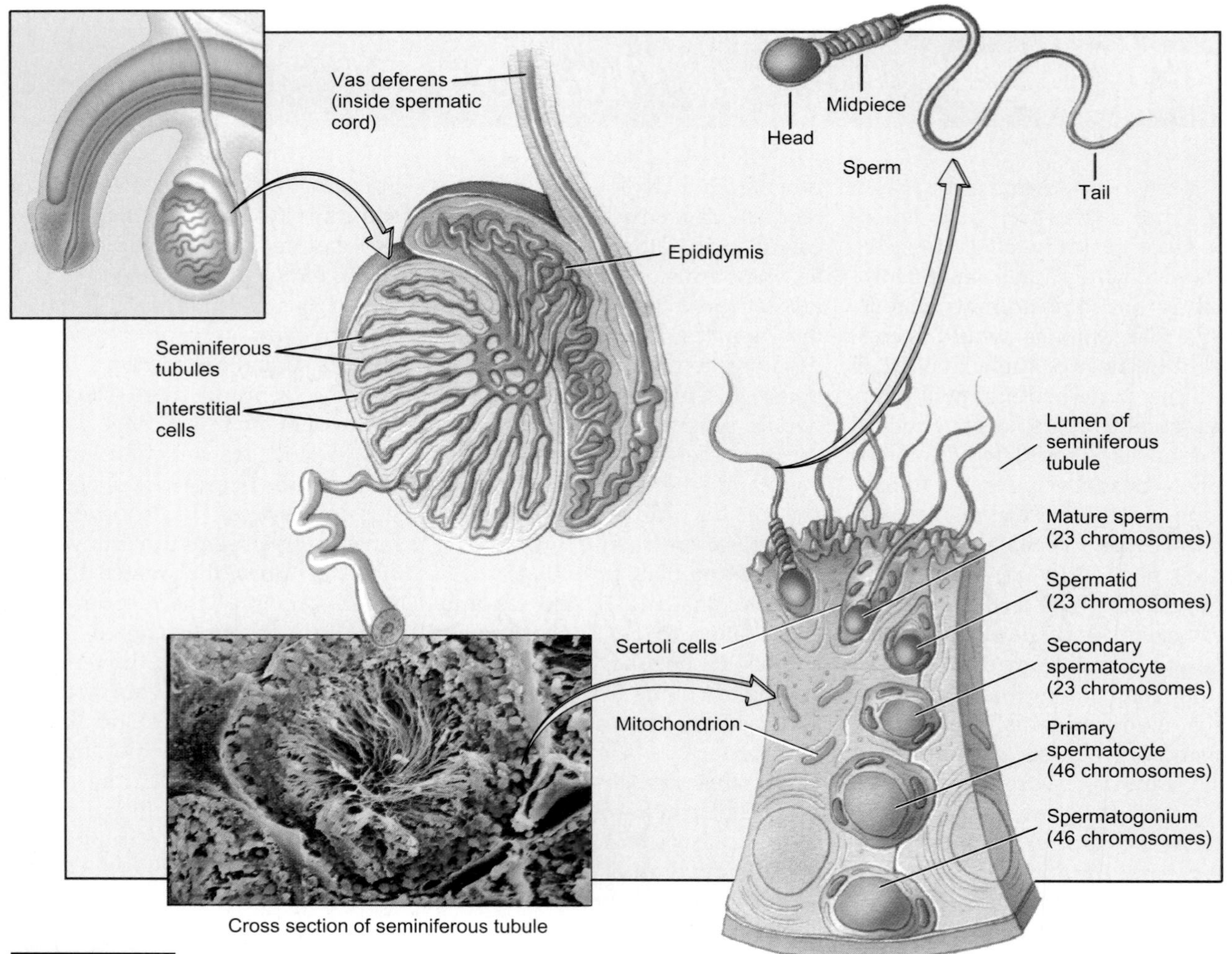

FIGURE 42.5 SPERM PRODUCTION.
The seminiferous tubules lead to the epididymis, whose duct continues as the vas deferens. Spermatogenesis occurs in the germinal epithelium where cells in all stages of meiosis can be seen. Spermatids come to reside within Sertoli cells as they complete their development. The scanning electron micrograph of a seminiferous tubule nicely illustrates the germinal cells and sperm (tails).

Much of the tissue of the testes consists of highly coiled *seminiferous tubules* (Figure 42.5). It is from very active tissue lining the seminiferous tubules—the *germinal epithelium*—that sperm are formed. Following meiosis, each haploid cell becomes a **spermatid,** an immature form of sperm cell. Spermatids lie embedded in supporting and nourishing *Sertoli cells* while they complete their development (Figure 42.5). Even then, sperm are not ready for ejaculation until they undergo a final period of maturation in the **epididymis,** a sperm storage structure perched above and along one side of each testis. The epididymis is formed by the union of many seminiferous tubules. Supporting tissue surrounding the seminiferous tubules contains *interstitial cells,* whose specialization is manufacturing the male sex hormone *testosterone.*

Sperm Pathway. Prior to ejaculation, the sperm are propelled to the urethra by the wavelike action of smooth muscles and the pressure of sperm and fluid that follow into the *vas deferens,* a slender duct that extends from the epididymis to the urethra (joined, of course, by its counterpart from the other testis). The vas deferens, along with nerves and blood vessels, and an outer sheath make up the *spermatic cord.*

Near the urethra, the sperm receive fluids from two glands, the *seminal vesicles* and the *prostate gland.* The seminal vesicles add fluid containing fructose (fruit sugar), which provides the sperm with an energy source. The prostate produces an alkaline secretion that gives semen its characteristic thickness and odor. (The alkalinity helps neutralize the vagina's acidic fluids.) The male urethra carries

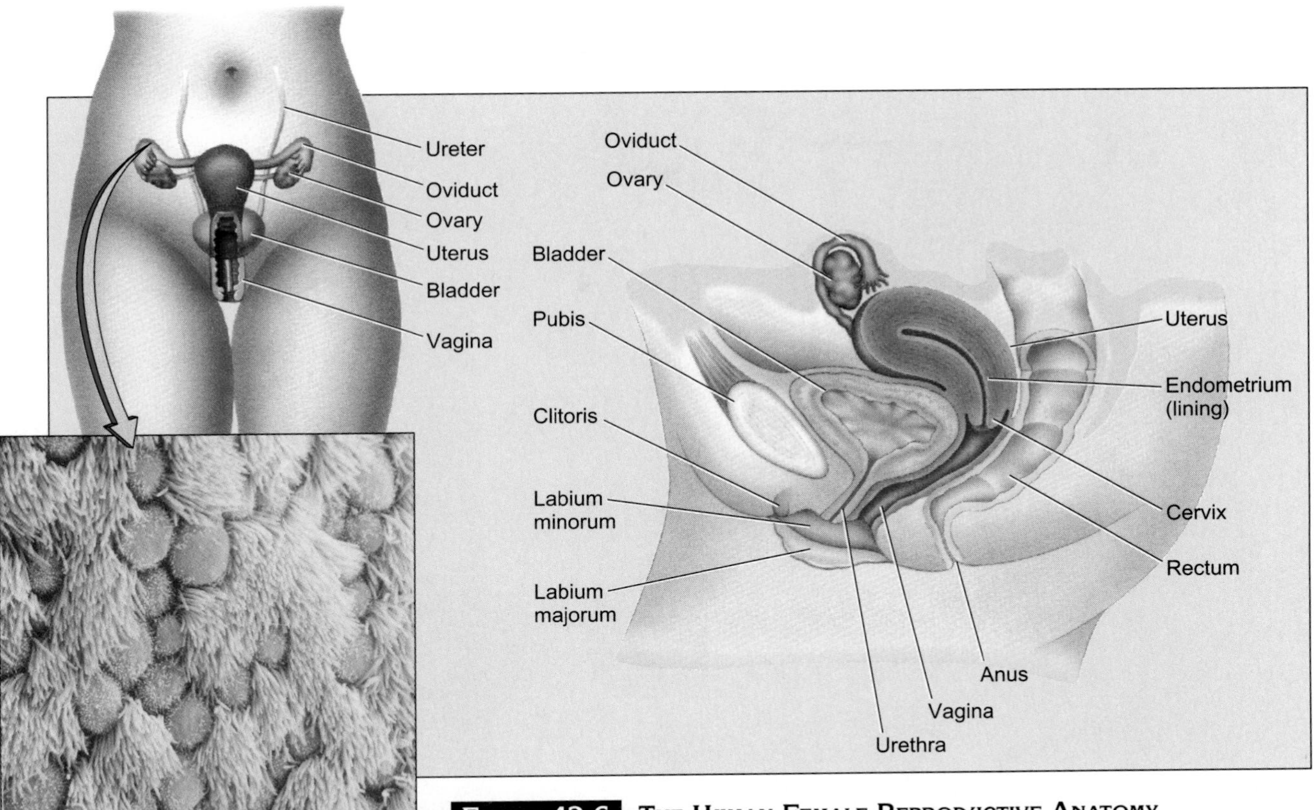

FIGURE 42.6 THE HUMAN FEMALE REPRODUCTIVE ANATOMY. The internal reproductive structures in the human female. The scanning electron microscope reveals the numerous cilia that line the oviduct and help move the egg or early zygote toward the uterus.

both urine and semen, but during sexual excitement the urine-releasing muscle sphincter below the urinary bladder is involuntarily contracted. During sexual excitement and prior to ejaculation, the *bulbourethral glands* secrete clear slippery mucus into the urethra, rinsing urine from the urethra, lubricating the glans, and adding lubricant to the semen. During ejaculation, powerful rhythmic contractions propel semen through the urethra.

FEMALES

In women, the external genitalia are collectively known as the *vulva* (Figure 42.6). The most prominent part is the hair-covered *mons veneris* (love mound), a fatty mound overlying the bony pubic arch. Below the mons veneris lie the outer folds of the vulva, the *labia majora* (major lips), which cover a number of sensitive structures. Just within the labia majora are the less prominent, thinner folds of the *labia minora* (minor lips). These join at their upper margins, forming a kind of hood over a small, sensitive prominence, the **clitoris.** The clitoris is derived from essentially the same embryonic tissue as is the glans of the penis. The two are both erectile and have a rich supply of sensory receptors. In an excited state, the clitoris becomes erect, firm, and highly sensitive.

Enclosed by the labia minora, and near their lower border, lies the vaginal opening, which may be partially blocked by a membrane known as the *hymen.* The strength of this membrane varies considerably, and its rupture, perhaps by the first intercourse, may produce discomfort and bleeding. The urethral opening in females lies just above the vaginal opening.

Internal Anatomy. The internal anatomy of the human female reproductive system includes the *vagina,* the *uterus,* the *ovaries,* and the *oviducts* (or Fallopian tubes) (Figure 42.6). The *vagina* is a distendable tube about 8 cm (3 in.) long when relaxed; it receives the penis during intercourse and is the passageway through which birth occurs. It is well adapted for both functions, with its highly folded, muscular walls and a lining that can stand up to friction.

The vagina leads to a soft, muscular, pear-shaped organ, the *uterus,* which will house the developing fetus. The lower tip of the uterus, called the *cervix,*

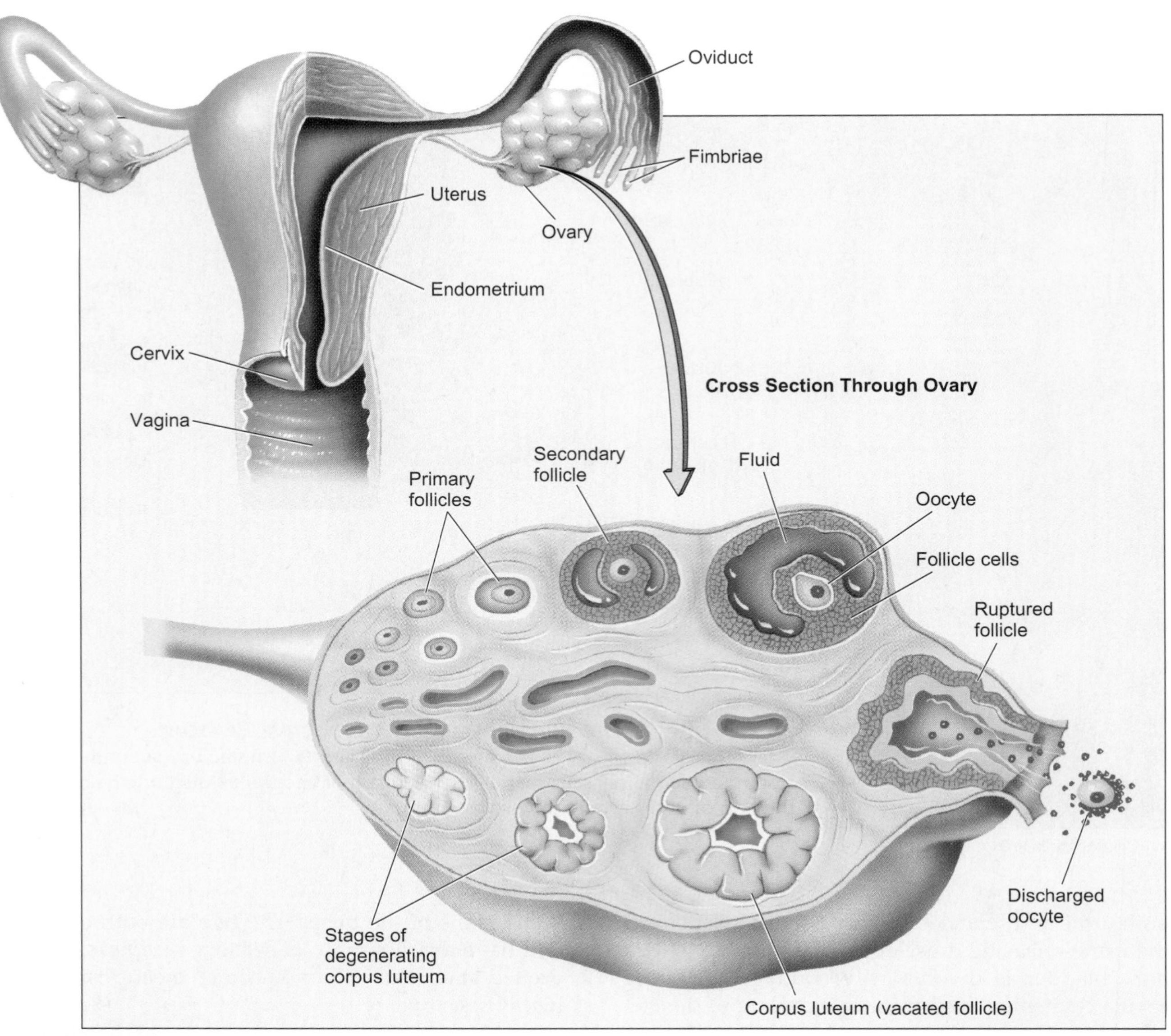

FIGURE 42.7 **THE OVARIES AND EGG PRODUCTION.**
The ovary's germinal epithelium contains all of the oocytes, each arrested in prophase of meiosis I. Each month a follicle develops around one or more of the oocytes and meiosis resumes. The enlarging, fluid-filled follicle supports the oocyte and secretes estrogen. On ovulation the ovum, now entering meiosis II, is released, and the empty follicle forms a corpus luteum.

extends slightly into the vagina. The uterus is lined by a soft, vascular *endometrium,* which will receive the embryo should fertilization occur.

The *oviducts* are tubes that emerge from each side of the upper end of the uterus, extend to the *ovaries,* the sites of egg production, and terminate in movable, fingerlike *fimbriae.* Although the ovaries produce the eggs, they do not directly connect with the oviducts. Instead, currents produced by the beating fimbriae draw eggs into the oviduct, where they are then swept along toward the uterus by cilia and by muscular contractions. The structures and functions of the female reproductive system are listed in Table 42.2.

The Ovary and Egg Production. Like the testes, the ovaries produce both gametes and hormones. Each oval-shaped ovary is about 2.5 cm (1 in.) long. (Figure 42.7). The egg cells, actually *oocytes,* are produced in the ovarian cortex. At puberty the ovaries contain a total of about 400,000 oocytes, a surprisingly large number considering how few can ever be fertilized. You may recall from our discussion of meiosis (see Chapter 10) that the oocytes are produced during embryonic development; they remain inactive until puberty, suspended in prophase I of meiosis. Beginning at puberty, a few are activated each month by pituitary hormones. Through some largely unknown process, the activated oocytes tend to suppress each other hormonally so that usually only one gains a developmental advantage, and it alone matures. The release of eggs from the ovary is referred to as *ovulation*. We will look into fertilization in the next chapter.

Hormones and Human Reproduction

Sex hormones undoubtedly play an extremely important role in our lives. They determine not only our sex itself but our sexual behavior, our sexual development, and our sexual physiology as well. An increasing body of evidence implies that they also play a role in aggressiveness. Their effects are particularly obvious during puberty, when sexual differences between girls and boys are emphasized. At full sexual maturity, these hormones cause the continued production of gametes as well as an increased sex drive. (The human sexual response is discussed in Essay 42.3.)

The primary sites of human reproductive hormonal activity are the pituitary gland, the gonads, and the placenta. These centers of hormone production are, in turn, under the influence of the hypothalamus. The role of the hypothalamus is described in detail in Chapter 36, so we will just say here that essentially the hypothalamus does two things: It initiates the release of certain hormones, and it reads the blood level of other hormones.

When prompted by gonadotropic-releasing hormones from the hypothalamus, the anterior pituitary secretes hormones known as *gonadotropins* (*trope*, change) (see Chapter 36). As their name implies, gonadotropins stimulate growth or activity in the gonads (both the ovaries and the testes). The gonadotropins stimulate the production of the sex hormones and initiate the development of sperm and egg cells.

Two specific gonadotropins of the pituitary are *FSH (follicle-stimulating hormone)* and *LH (luteinizing hormone)*, both of which are polypeptides. Their targets are the ovaries in women and the testes in men. (LH in men is sometimes called *ICSH* (interstitial cell-stimulating hormone), named for its targets, the interstitial cells of the testes. We will begin with hormonal action in males.

Male Hormonal Action

FSH is continually secreted throughout the life of the male. Its specific targets are the seminiferous tubules of the testes, where it stimulates sperm production. In addition to promoting sperm production, FSH teams up with LH to prompt the testes to produce *testosterone*, the primary male sex hormone. Recently, it became known that FSH secretion in both males and females is inhibited by another hormone, dubbed *inhibin*, which in males is produced in the seminiferous tubules. Inhibin is thought to stabilize testosterone levels through a delicate negative feedback loop to the anterior pituitary, where it slows FSH secretion.

TABLE 42.2 Major Parts of the Female Reproductive System

Structure	Function
Ovaries, follicles, corpus luteum	Source of eggs, estrogens, and progesterone
Oviducts (fallopian tubes)	Receive and transport egg to uterus
Uterus	
Endometrium	Attachment for developing embryo
Cervix	Narrow lower region of uterus opening into the vagina
Vagina	Receives penis during copulation; birth canal

Testosterone is produced throughout the testes in the interstitial cells that lie outside the seminiferous tubules. There are indications that the production of testosterone may rise and fall with increased and decreased sexual activity—or even with social interactions. Testosterone levels in many men commonly decline in all-male crews of ships at sea, as evidenced, for instance, by slower beard growth. (More recently, researchers found that the mere anticipation of sex can produce an elevation in body testosterone.)

Testosterone is vital to the development of the characteristics of "maleness." Although the appearance of the embryonic genitalia and gonads is genetic, their subsequent development requires testosterone. If it is absent during the eighth or ninth month of development, the testes fail to descend into the scrotum. Later on, testosterone is essential for those perturbing events associated with puberty, such as voice changes, the growth of body hair, bone and muscle development, and the enlargement of the testes and penis. Failure of these developmental events to transpire resulting from low testosterone levels in the developing male are readily corrected by administering the hormone. Finally, the sex drive itself appears to be greatly influenced by the presence of testosterone.

The testosterone level is apparently detected by the hypothalamus, which, if the level is too high, inhibits the release of LH by slowing its stimulation of

ESSAY 42.3 The Sexual Response

Because of the pioneering research efforts of William H. Masters and Virginia E. Johnson of the Reproductive Biology Research Foundation in St. Louis, our knowledge of human sexual behavior is far more extensive than it was before. Using volunteers, Masters and Johnson carefully monitored and recorded the physiological changes that occur during intercourse and orgasm.

The researchers described four phases of sexual response during intercourse. They labeled them *arousal, plateau, orgasm,* and *resolution.*

Arousal is characterized in both sexes by increased heart rate, blood pressure, and breathing rate. In women, the clitoris enlarges and becomes more sensitive, the labia majora elevate and part, and the labia minora redden and increase in size. Meanwhile, the vagina becomes lubricated. In men, excitement causes the penis to become erect. Secretions of the bulbourethral glands moisten the glans. The testes may elevate and become larger. Now the penis may be inserted into the vagina.

The plateau phase is reached during sexual intercourse. This phase may last for some minutes and is marked by an increased intensity in the level of pleasure. Either partner may begin involuntary thrusts of the pelvis. The uterus may elevate and tilt backward at this time. Interestingly, the clitoris may become smaller and recede into its hood, having become extremely sensitive to touch.

Orgasm is an intensely pleasurable sensation that accompanies complex contractions of several voluntary and involuntary muscles. In women, the contractions begin in the pelvic floor and surge through the vagina and uterus. In both men and women, involuntary orgasmic contractions occur at about 0.8 second intervals. In women, the upper part of the vagina may expand as the cervix moves downward (called "tenting"), a response that may help draw semen into the uterus.

In men, orgasm accompanies ejaculation, which involves rhythmic, involuntary contractions of the vas deferens, seminal vesicles, and prostate gland. The semen, containing its hordes of sperm, is forced into the urethra. The semen may be ejected from the penis in spurts, often propelled with considerable force by powerful muscles at the base of the penis.

Orgasm in men and women is compared in the accompanying graph. The most obvious difference between female (a) and male (b) orgasm is in the orgasmic peaks. Some women normally experience a single orgasm (3), but many can have multiple orgasms. These may occur as several minor episodes (2), or as fewer but more intense ones (1). In men, a single orgasm is the rule, followed by at least a partial state of resolution.

Resolution is apparently more pronounced in men than in women. At this time, erection may be lost rapidly, and most men cannot be aroused again for a time. The length of the resolution phase is highly variable and usually depends on a number of factors. Resolution is more gradual in women, and immediate rearousal is often possible.

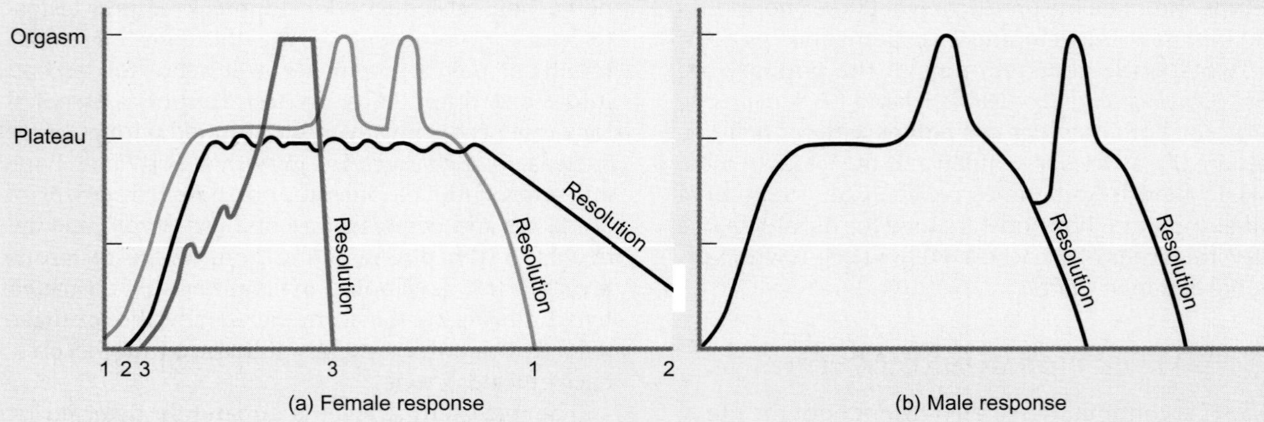

(a) Female response

(b) Male response

the pituitary. The pituitary, in response, slows its release of LH, thus lowering the production of testosterone by the interstitial cells. So, as you can see, a feedback mechanism regulates testosterone levels. Actually, though, things aren't quite that simple. For example, we mentioned earlier that sexual and social activity can influence hormonal levels. Apparently, the frequency of ejaculation and/or sexual arousal has some complex influence on the hypothalamus, or perhaps directly on the pituitary, or the testes—or all three.

Now let's see if we can simplify some of this. Under the influence of releasing hormone from the hypothalamus, the pituitary secretes the gonadotropins FSH and LH. These hormones stimulate sperm production and initiate the production and release of testosterone by the testes. Levels of testosterone are sensed by the hypothalamus, which responds by adjusting the release of gonadotropins. The result is a fairly steady level of hormone production. As we will see, all this is quite straightforward when compared to what happens in the human female.

Female Hormonal Action

The onset of puberty in girls most often occurs between 9 and 12 years of age. The beginning of fertility generally follows in 2 or 3 years. Both events are initiated by a rise in the pituitary gonadotropins FSH and LH, which act on the ovaries. The ovaries respond to FSH by producing *estrogens,* the primary female hormones. In turn, estrogens influence the growth of the breasts and nipples, broadening of the hips, and in general, the development of the adult female contours. Estrogen also influences the growth of the uterus, the vaginal lining, the labia, and the clitoris. In addition, the adrenals produce androgens (essentially, male hormones), which stimulate the growth of the coarser hair of the genitals and underarms.

Fertility in the female is marked by the start of the *ovarian* and *menstrual cycles,* in which the gonadotropins and ovarian hormones rise and fall with some regularity and oocytes mature. Other female vertebrates may experience somewhat analogous *estrous cycles* with a regular frequency or perhaps only once or twice each year; the remainder of the time they are sexually unresponsive and infertile. The human female is considered to be unusual among mammals in always being potentially able to copulate.

The Menstrual Cycle

On the average, in women, a mature ovum is released from one of the two ovaries about every 28 days. Coinciding with this event are intense preparatory activities in the uterus. Within a week after ovulation, its temporary lining, the endometrium, must be ready to support the young embryo. These events are closely correlated through the action of pituitary and ovarian hormones, which are themselves subject to timely negative feedback events in the hypothalamus.

In discussing the menstrual cycle, it is common to begin counting on the first day of menstrual flow. Ovulation generally occurs about 14 days before the beginning of the next cycle, so in a 28-day cycle it would occur on approximately day 14, about the midpoint of the cycle. Figure 42.8 will help in understanding how the events are organized in a 28-day cycle.

Days 1 to 14: Proliferation

The first half of a menstrual cycle is known as the *proliferative* or preovulatory phase. During the first 3 or 4 days (sometimes longer), blood and tissues of the endometrium are shed in what is called **menstruation** (the "period"). Since we are talking about a cycle, we'll come back to menstruation again, but first let's see what the hormones are doing.

During the proliferative phase, releasing hormones from the hypothalamus stimulate the pituitary to increase its output of FSH and LH. The targets of FSH are the ovary's primary follicles, the arrested oocytes, and the cells clustered about them. At first, the hormone stimulates several primary follicles to grow, although usually just one reaches maturity, migrating towards the ovary's surface as it does. It is at this time that meiosis resumes in the oocyte. As a follicle grows, it forms a fluid-filled cavity around the oocyte and becomes known as a *Graafian follicle.* By the midcycle, the Graafian follicle will protrude like a blister from the ovary's surface (see Figure 42.7).

During this time, the growing follicle secretes estrogens, which steadily increase in the bloodstream. Their principal target, the lining of the uterus, responds by producing the new endometrium (see Figure 42.8). Cells of the lining undergo rapid cell division, forming dense layers over the entire inner surface. The cell layers are soon penetrated by a profusion of blood vessels from the uterus. The result is a soft, highly vascularized lining, exactly what is required to support the early human embryo. The cervix also responds to rising estrogen levels, secreting a thin, alkaline mucus, thus changing the usually acidic vaginal environment to one that is more hospitable to sperm cells.

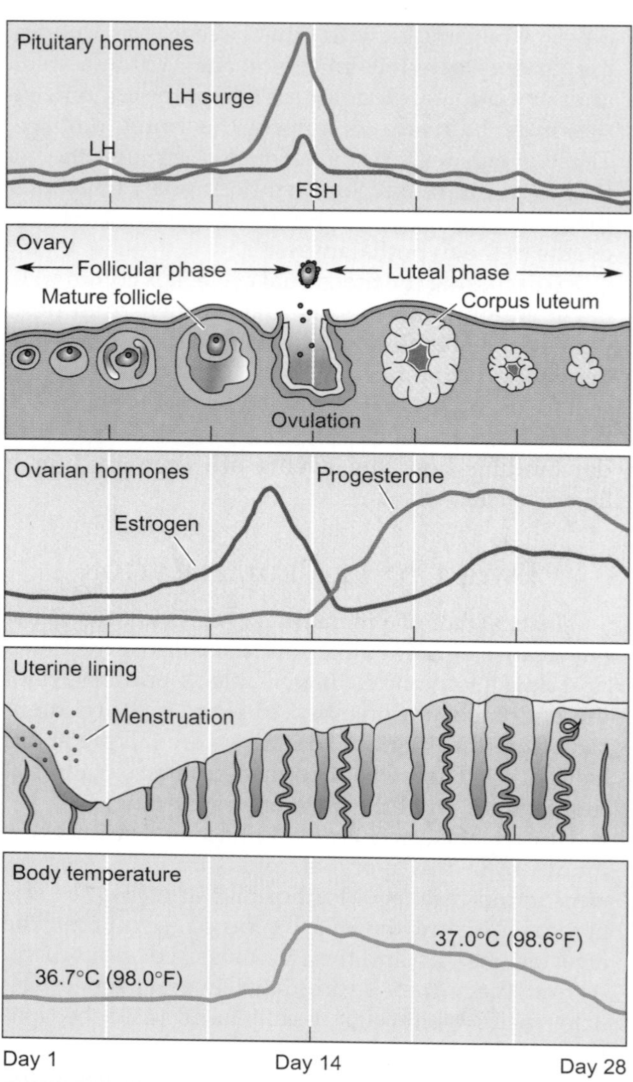

FIGURE 42.8 GRAPHIC REPRESENTATION OF A 28-DAY MENSTRUAL CYCLE.
The endometrium undergoes growth and repair during the first 14 days and reaches its fullest development a short time later. At midcycle (day 14) body temperature is slightly elevated, a fairly reliable indicator of ovulation.

As you can see, the reproductive hormones tightly coordinate the events in the ovary with those in the uterus. One result is that at the time of ovulation the uterus is nearing a fully receptive state. Should an egg be fertilized, the young embryo will reach the uterus in a few days and there find itself in a highly nurturing environment.

DAY 14: OVULATION

On the 14th day of the cycle, there is a rapid rise in LH, often referred to as the "LH surge." The mechanism behind this surge is not clear, but rising levels of estrogens are believed to form a *positive* feedback loop, one that leads the anterior pituitary to increase its output of LH. The target of LH is the enlarged, fluid-filled Graafian follicle, which responds by secreting still more estrogens—and as the positive loop increases, a final surge in LH secretion occurs, as mentioned above (see Figure 42.8).

The oocyte within the Graafian follicle, now having completed meiosis I, lies within a cluster of surrounding cells. Then, as the fluid content of the follicle increases still further, the follicle bursts, and ovulation occurs. The oocyte and its surrounding cells, the corona radiata ("radiating crown"), are released and begin the journey to the oviduct. The event is followed by a slight, temporary elevation in body temperature (about 1°F). This temperature elevation is an important indicator for women trying to establish the "fertile period," either in the interest of conception or contraception.

DAYS 15 THROUGH 28: SECRETION

Following ovulation, the secretory, or postovulatory phase of the cycle begins. The vacated ovarian follicle, under the influence of LH, forms a new structure called the **corpus luteum** (*corpus,* body; *luteum,* yellow). The corpus luteum continues its secretion of estrogens and, in addition, secretes the hormone *progesterone.* The target of both the estrogens and the progesterone is the endometrium, where progesterone brings on the secretory phase, prompting blood vessel growth and proliferation and glycogen accumulation.

Thus, progesterone instigates the final preparation of the uterus for the reception of a young embryo. And should fertilization occur, progesterone will help maintain the endometrium throughout the ensuing pregnancy. Progesterone also stimulates a temporary thickening and an increase in secretion in the vaginal lining and the growth of milk ducts in the breasts. Finally, the appearance of progesterone is believed to bring on the slight temperature rise following ovulation.

Progesterone levels rise in the blood for about 6 days following ovulation. During this time, the endometrium of the uterus will have reached its greatest development. Its now loosely packed cell layers become fluid-filled sinuses, penetrated with glandular tissue that produces other secretions. If fertilization has occurred, a tiny embryo, now just a ball of cells about 6 days old, will have formed. Through the action of its own enzymes, it will become implanted deep in this receptive tissue, literally digesting its way deep into the fluid-filled spaces, where it continues its development. The slight bleeding caused by the embryo settling into the uterus may cause the woman to believe her period has started and that she isn't pregnant.

The uterus remains receptive for about 10 to 12 days following ovulation, and unless an embryo is snugly implanted by that time and putting out its own signals to the corpus luteum, the corpus luteum will start to degenerate. Its production of estrogen and progesterone will decrease, and the endometrium will lose its receptive state. In a few days it will slough away, beginning menstruation and the start of the next 28-day cycle.

Why does the corpus luteum fail? While definitive answers aren't available, there is some consensus on the following. Unless pregnancy has occurred, the signals from the pituitary start to fall off during the fourth week of the cycle. The decrease in LH and FSH may occur because of negative feedback, primarily that of progesterone. Figure 42.8 shows that the levels of progesterone peak on about day 21. The high blood hormone levels may produce a negative effect on the hypothalamus. It slows its stimulation of the pituitary, which, in response, slows its LH and FSH secretion. The resulting drop causes its own chain reaction. Without the gonadotropins, the corpus luteum recedes, which causes a drop in the secretion of progesterone and estrogens. And finally, without their support, the endometrium simply breaks down. Alternatively, some evidence suggests that the corpus luteum may be programmed to degenerate after about two weeks.

However, if pregnancy has occurred, things are very different. The extraembryonic tissue of the fetus becomes an endocrine gland, secreting *human chorionic gonadotropin* (HCG). The latter constitutes the progesterone-secreting signal mentioned earlier—the signal directed to the corpus luteum. So the aging corpus luteum continues to secrete hormones that keep the uterine lining hospitable for the demanding little embryo. (The presence of HCG in the urine, incidentally, is the basis for many pregnancy tests.) Later—about the third month of pregnancy—the HCG will diminish as the placenta begins secreting its own progesterone and estrogen.

Let us now summarize the important events of the menstrual cycle. The cycle involves the production of a mature ovum along with the simultaneous preparation of the uterus to receive it. During the first half of the cycle, the anterior pituitary is producing FSH, which causes the development of follicles in the ovary. The growing follicles secrete estrogen, causing the uterine lining to thicken. Ovulation is brought about by a surge in LH at midcycle. In addition, LH converts the follicle cells remaining after the ovum was released to an endocrine structure called the corpus luteum. Besides secreting estrogen, the corpus luteum produces progesterone (Figure 42.9), summarizes the hormonal changes in the menstrual cycle. The effect is the continued preparation of the uterus for implantation. In the absence of pregnancy, decreasing levels of estrogen and progesterone end the cycle.

CONTRACEPTION

Contraception, the avoidance of pregnancy, at least in the more developed nations, has increasingly involved decision-making processes. Moreover, some forms of contraception can reduce risk of infection with sexually transmitted diseases (see Essay 42.4). The general acceptance of birth control has produced a great deal of technological research and development. As a result, there are many options in contraception today (see Table 42.3).

HISTORICAL AND FOLK METHODS

Historically, contraception has been attempted in a variety of ways, some effective, some worthless. (One wonders just how our ancestors reacted when they first discovered what causes pregnancies—it's still a bit hard to believe!) The ancient Egyptians blocked the cervix with leaves, cotton, or cloth. In the Middle Ages, condoms that fit, sheathlike, over the penis were fashioned from such materials as linen and fish skin. Douching has long been practiced as a means of flushing sperm out of the vagina. The finger has also been used to remove semen after it has developed a stringy quality through being exposed to the vaginal environment. The Old Testament refers to removing the penis just prior to ejaculation. The most effective method, of course, has always been total abstention, a method that has generally met with little applause.

There are also a number of useless "folk" devices that are employed even today. These range from regulating intercourse according to the phases of the moon, to stepping over graves, to using cellophane sandwich wrappings as condoms, to douching with soft drinks (the carbonic acid and sugar are believed to be spermicidal, and a shaken drink provides the propulsion for the agents to reach far into the vagina). Since the effectiveness of these folk methods is so low, we might consider a few other contraceptives that we know more about.

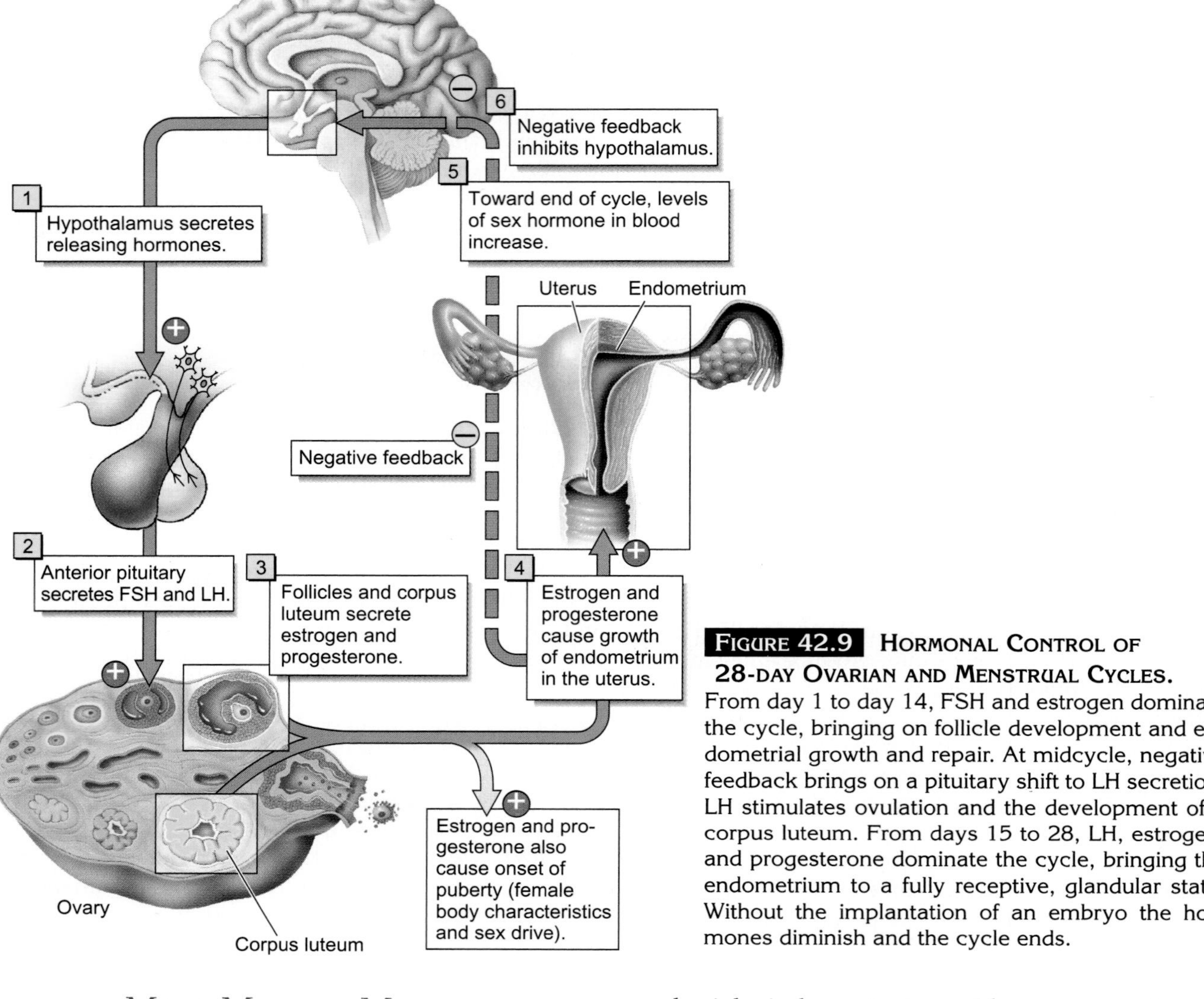
6
Negative feedback inhibits hypothalamus.
5
Toward end of cycle, levels of sex hormone in blood increase.
1
Hypothalamus secretes releasing hormones.
Uterus
Endometrium
Negative feedback
2
Anterior pituitary secretes FSH and LH.
3
Follicles and corpus luteum secrete estrogen and progesterone.
4
Estrogen and progesterone cause growth of endometrium in the uterus.
Estrogen and progesterone also cause onset of puberty (female body characteristics and sex drive).
Ovary
Corpus luteum

FIGURE 42.9 HORMONAL CONTROL OF 28-DAY OVARIAN AND MENSTRUAL CYCLES. From day 1 to day 14, FSH and estrogen dominate the cycle, bringing on follicle development and endometrial growth and repair. At midcycle, negative feedback brings on a pituitary shift to LH secretion. LH stimulates ovulation and the development of a corpus luteum. From days 15 to 28, LH, estrogen, and progesterone dominate the cycle, bringing the endometrium to a fully receptive, glandular state. Without the implantation of an embryo the hormones diminish and the cycle ends.

More Modern Methods

Douching, or rinsing the vagina, after intercourse was a fairly common method of contraception in past times. It usually involves flushing the vagina with plain water or some household or commercial solutions. Diluted vinegar was a common douche because its acid properties tend to kill sperm. We mention this crude method only because of its great failure rate (about 40%) and to discourage its use. (Failure rate refers to percentages of women who become pregnant per year of dependence on a given method.) Even the most swiftly administered douche can fail; sperm cells can pass through the cervix and into the safe confines of the uterus only moments after ejaculation.

An old joke goes, "What do you call couples who use the rhythm method?" The answer was "parents," although the joke was funnier to some than others. Still, the **rhythm method**—the timing of physiological events to avoid pregnancy—is perhaps the most biologically interesting way of practicing birth control. The principle itself is simple: There must be no intercourse during the period in which conception is likely. Ovulation usually occurs 14 (plus or minus 2) days before the next menstrual period begins. Sperm is thought to live for two days after it has been released into the female reproductive tract, and an egg remains fertile for about a day following ovulation. The couple must therefore avoid intercourse at least 2 days before and 1 day after ovulation, although, for safety, this interval is usually increased. There are changes in the consistency of the cervical mucus accompanying ovulation, and body temperature increases slightly on the days following ovulation. These changes can help pinpoint the day of ovulation. The difficulty, of course, is knowing when ovulation will occur 2 days *before* the event. Although the rhythm method requires the use of thermometers, calendars, paper,

TABLE 42.3 SUMMARY OF CONTRACEPTIVE METHODS

Method	Action	User	Failure rate (pregnancy per 100 women per year)*	Disadvantages
Douche	Washes away sperm in vagina	Female	40	None
Rhythm	Abstinence during fertile period	Male and female	24	Requires meticulous recordkeeping of women's body cycle (e.g., temperature)
Coitus interruptus	Removal of penis from vagina before ejaculation	Male	23	Frustration; some sperm may be released before withdrawal
Condom	Traps ejaculated sperm	Male	7 to 30	Some loss of sensation for male; condom material may tear if not used properly
Diaphragm (with spermicide)	Prevents sperm from entering cervix; kills sperm	Female	2 to 20	Must be applied before intercourse and inserted correctly
Cervical cap	Prevents sperm from entering cervix	Female	13	Infection; allergic reactions
Spermicide (foam, jelly, creams, suppositories)	Kills sperm	Female	10 to 25	May cause irritation; must be applied before intercourse and left in for at least six hours
IUD	Prevents implantation	Female	5	Infection; menstrual discomfort; expulsion of device possible
"Combination" pill (estrogen and progesterone)	Inhibits production of FSH and LH, preventing ovulation	Female	2	May produce unwanted side effects (e.g., risk of cardiovascular disease, water retention, nausea)
Minipill (progestin alone)	Thickens cervical mucus; inhibits sperm movement; prevents implantation	Female	3	None known
"Morning after" pill (50× normal dose of estrogen)	Arrests pregnancy	Female	**	Physical discomforts increase (e.g., nausea, water retention)
Norplant	Thickens cervical mucus; inhibits sperm movement; prevents implantation	Female	1	None known
Sponge	Mechanical barrier	Female	15	May cause irritation or allergic reactions
Sterilization	Surgical control of sperm and egg movement	Male and female	0	May produce irreversible sterility

**There is a wide range; these figures are taken to be an average.*
***Insufficient data*

and pencils, the claim is sometimes made that this is the only "natural" means of birth control. In any case, the method is not particularly reliable (about 76% under ideal conditions—so, for every 100 couples using this method for a year, at least 24 of the women are likely to become pregnant). It is unreliable partly because the menstrual periods of many women are highly irregular.

Coitus interruptus—withdrawing the penis just before ejaculation—is a surprisingly common method of contraception, particularly among novices, the poor, and the unprepared. But it is risky because the secretions of the bulbourethral glands, which are released before ejaculation, may contain sperm. Also, couples with sufficient stamina and time on their hands may engage in intercourse more than once, ignorant of the presence of residual sperm cells in the male urethra. But most significantly, coitus interruptus involves the psychological frustration of the man's having to do exactly what he is least inclined to do. On balance, then, coitus interruptus is considered to rank "poor to fair" in terms of dependability (Table 42.3).

The **condom,** or "rubber," is one of the most widely used contraceptive devices in the United States. Basically, it is a balloonlike sheath made of rubber, animal membrane, or other material, which fits tightly over the penis and traps the ejaculated sperm. There are many grades of condoms, ranging from the two-for-a-quarter specials sold in restroom vending machines to the more expensive ones made from animal membranes.

Condoms are only about 70% to 93% effective (partly because of differences in quality and the fact that they are often incorrectly used). The condom remains popular because not only is it an effective means of birth control if used properly, but it also provides the extra bonus of protection against sexually transmitted diseases. Some are covered with nonoxonyl-9, a spermicide that can also kill the AIDS virus. However, the various types of condoms have different degrees of effectiveness against conception and against disease (see Essay 42.4) and the two are not necessarily correlated. Animal membranes, for example, may allow a heightened sensitivity, and they may trap sperm effectively, but their relatively large pores allow the passage of viruses such as those that cause herpes and AIDS.

Female condoms have recently been marketed, and are meeting with mixed reactions. They are larger and looser than male condoms with a flanged opening that reduces the chance of contact in areas around the genitals.

The **diaphragm** (Figure 42.10) is a dome-shaped rubber device with a flexible steel spring enclosed in the rim. It is inserted into the vagina and fitted over the cervix to prevent sperm from entering the uterus. Unlike the condom, the diaphragm must be individually fitted. It is partially filled with a spermicidal cream or jelly and should be left in place for at least six hours after intercourse. The device is about 80% to 98% effective if well-fitted and used correctly. If the diaphragm is inserted in advance, sexual activities can proceed without interruption or delay.

Like the diaphragm, the **cervical cap** (Figure 42.10) covers the cervix and prevents sperm from entering the uterus. Essentially, it is a small rubber cap that fits tightly over the cervix and stays in place partly by suction. The cap should be partially filled with spermicidal cream or jelly. It should be left in place for at least 6 hours after intercourse, although one of its advantages is that it may be left in place for up to 2 days. It is about as effective as the diaphragm. One problem is that it can move out of position without being noticed.

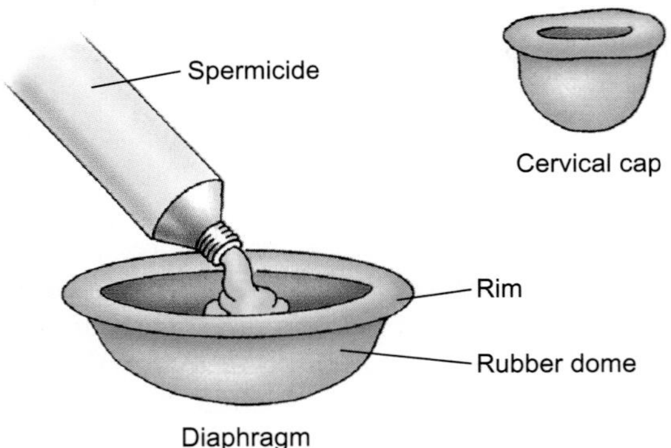

FIGURE 42.10 THE DIAPHRAGM AND CERVICAL CAP.
They both operate on the same principle—that is, blocking the cervix. However, the diaphragm is normally inserted each time before intercourse, while the cervical cap may be worn for up to two days. Both devices should be used with spermicidal cream or jelly.

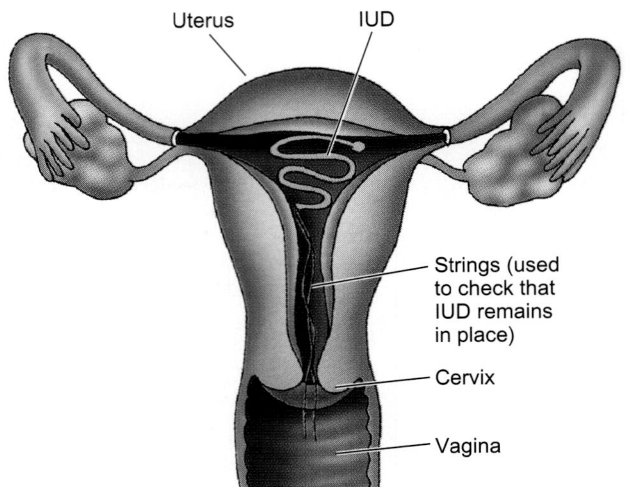

FIGURE 42.11 THE IUD.
IUDs are produced in a number of shapes and sizes. Although it is not clear exactly how the IUD works, somehow the presence of a foreign body in the uterus prevents pregnancy.

Spermicides (*-cide,* kill), are sperm-killing foams, jellies, suppositories, and creams. Creams and jellies may be used to increase the effectiveness of other contraceptives such as the diaphragm or cervical cap. The *spermicides* are usually quickly and easily applied inside the vagina and require neither fitting nor prescription by a doctor. Used alone, they are 75% to 90% effective, the variance depending on whether they are used properly.

The **intrauterine device,** or **IUD,** is a device that is placed in the uterus where its presence inhibits pregnancies. (Figure 42.11). It was recognized as an effective contraceptive for many years. The principle has apparently been utilized for a long time. It is said that Arab camel drivers on long, hard caravan marches prevented pregnancies in their camels by placing apricot seeds in the females' uteruses. (One wonders how the practice got started.) Modern

IUDs have been made from several materials and come in a variety of sizes and shapes. No one knows exactly how the IUD works beyond the fact that it seems that a foreign body in the uterus somehow prevents pregnancy. It has been suggested that the IUD prevents the implantation of the embryo, that it may cause the egg to move too rapidly through the oviduct, or that it may alter the condition of the uterine endometrium. In any case, however, it is technically not a contraceptive device since it functions not by preventing conception, but by preventing implantation.

The IUD has several major advantages. Once it has been installed in the uterus, it requires little attention for up to several years. Also, it is inexpensive and about 95% effective.

The IUD has, in recent years, fallen into disrepute because of the many problems it can cause. One problem is that the device may be expelled by the uterus in some women, perhaps without being noticed, leaving the woman unprotected against pregnancy. For some reason, IUDs are more likely to be rejected by the uterus in women who have never been pregnant. There may also be side effects just after insertion, including bleeding and pain. If these persist, the IUD must be removed. The IUD has also been associated with a number of ectopic pregnancies (those that occur in the body cavity, outside the uterus). Most important, however, the IUD has been implicated in a number of serious disorders of the pelvic region, perhaps brought on because the string hanging from the uterus into the vagina serves as a "wick" for vaginal bacteria, permitting them to ascend into the uterus and on into the oviduct. These bacteria may then cause *pelvic inflammatory disease (PID)*, an infection of the female reproductive structures that can result in reduced fertility or even sterility. PID may also increase the likelihood that the embryo will implant in the oviduct (an ectopic pregnancy) rather than in the uterus, a condition that is very dangerous for the mother.

The combination **birth control pill** contains either natural or synthetic forms of the hormones estrogen and progesterone that, together, inhibit the development of the egg in the ovary and ovulation. The great advantage of the birth control pill is its effectiveness. The combination pill, when used properly, may be more than 99% effective.

The so-called "minipill" contains only synthetic progesterone. It is thought to work by increasing the thickness of cervical mucus, interfering with the movement of sperm or making implantation more difficult. The minipill is slightly less effective, about 97%, but it has fewer side effects.

Birth control pills may cause undesirable side effects similar to those of pregnancy, especially during the first few months of use. They include weight gain, fluid retention, nausea, headaches, depression, irritability, increased facial pigment in certain areas, and enlargement of the breasts. Some or all of these symptoms may diminish after the first few months of use.

Although the pill has produced no serious medical problems for the overwhelming majority of its users, some argue that it is too early to assess its long-term use. There are some rather serious side effects in a small percentage of pill users. The most serious are those that affect the circulatory system. For example, some users develop high blood pressure, and this makes it more likely that they may have a heart attack or stroke. However, there are also some beneficial side effects. For instance, the pill is thought to decrease the risk of ovarian and endometrial cancer.

Massive doses of estrogen may be given when one suspects pregnancy may have occurred (the "morning after" pill). Finally, a great deal of research attention has been directed toward developing a contraceptive pill for men, and field tests are now underway.

Norplant implants consist of six thin, 1 to 2-inch rods containing a form of progesterone. These rods are inserted just under the skin through a tiny incision, usually in the upper arm. It is a hormonal means of contraception that eliminates the major reason for failure of the birth control pill—remembering to take one every day. The hormone continuously leaches out through the walls of the rod and enters the bloodstream, preventing pregnancy in much the same way as the minipill. The contraceptive effectiveness is thought to be significantly better than any other means of contraception, except for sterilization, and its effectiveness lasts for about five years (although new results suggest a more spotty effectiveness in some cases). Although Norplant was not approved until late 1990 for general use in the United States, new Norplant systems have already been designed that consist of fewer hormone-containing rods and that are effective for a shorter period of time. The rods can be removed easily if any complications develop or if the woman decides she would like to become pregnant.

The **sponge** is a small, absorbent polyurethane sponge that is saturated with a spermicide. It is inserted into the upper area of the vagina so that it

Essay 42.4 Sexually Transmitted Diseases

Not long ago, say a few decades past, Americans were warned of the two great hazards associated with promiscuity: gonorrhea and syphilis. Actually, there were other problems, even then, but those were the two greatest risks by far. They were called VD (venereal disease[1]). The picture has changed a bit now, and today, sexually active people face a virtual battery of sexually transmitted diseases (STDs), to the expressed delight of certain television evangelists. Let's briefly review a few of the threats. You may be surprised at what's going on out there. We'll begin with the old, familiar duo.

Gonorrhea, or clap, is caused by the bacterium *Neisseria gonorrhea.* It is transmitted by crossing mucosal surfaces and may affect the reproductive tract, the throat, or the rectum. Afflicted individuals, especially women, often show no symptoms. When a woman does show symptoms, they generally include bleeding and discharge between periods. When genitally infected, men show a discharge from the penis and severe burning on urination. (One diagnostic procedure suggests placing a nail between the man's teeth before he urinates. If he bites the nail in two, it's gonorrhea.) Symptoms usually develop within ten days of exposure. They generally subside on their own after several weeks. Drug-resistant strains are known.

Syphilis afflicted Benjamin Franklin, Florence Nightingale, Isak Dinesen, and Napoleon and can rightly be said to have influenced the course of history. It is caused by the bacterium *Treponema pallidum.* There are three stages. (1) Primary stage: a dry, dull red, and painless craterlike bump, called a chancre, that may appear anywhere on the skin, and then disappears. (2) Secondary stage: after a few weeks, a rash of discoloration over the body (a rash on the hands and feet is a strong sign of syphilis), a general malaise, hoarseness, fever, headache, muscle soreness, and possibly hepatitis and hair loss. These, after 2–10 weeks, disappear for 1–40 years (although there may be reoccurrences of these symptoms). (3) Tertiary stage: symptoms that range from benign lesions over the body, sharp pains along the legs and trunk as large nerves degenerate, blindness, weakened blood vessels, and a changing brain accompanied by insanity. It can be treated with antibiotics in the early stages—with less effectiveness in later stages.

Chlamydia (you probably didn't know this) is the most commonly transmitted bacterial disease in America. It is particularly prevalent on college campuses. It is caused by a bacterium, *Chlamydia trachomatis.* It is easy to cure but may cause no symptoms. If left alone, it can persist for years. In men, symptoms may mimic gonorrhea, but chlamydia also causes aching in the testicles and groin. Initially, women may have no symptoms; but as the disease progresses, there may be great discomfort, including pelvic pain, and pain in the lower left abdomen, often with vaginal discharge. The symptoms are especially intense during menstruation and in deep contact during sexual intercourse. Chlamydia can produce severe secondary complications. It is spread almost exclusively by vaginal intercourse. Chances of infection increase dramatically with multiple sex partners and, of course, if sex is unprotected.

Venereal warts are more common than most of us would like to believe. They appear on the genitals or rectal area and can reach the size of marbles. They are caused by a papilloma virus and are extremely contagious. Most appear within about a month after exposure. Young people are especially vulnerable. In women, the virus may later be linked to cervical cancer.

1. *venereal—from Venus, goddess of love*

covers the cervix up to 24 hours before intercourse, and is left there for about 8 to 10 hours afterward. Its success rate approaches 85%.

Sterilization is rendering an individual unable to become a parent. It is 100% effective. Furthermore, either the male or female can be sterilized, although it is a much simpler matter in the male.

A **vasectomy** (Figure 42.12) for a male normally takes about 15 minutes and can be performed in a doctor's office. A small incision is made in the side of the scrotum, a short section is cut from the vas deferens, the ends are tied off, and then the incision is closed. The operation can be reversed in 50% to 80% of the cases in those areas of the country where doctors have had the greatest experience with the operation. Another method, in which a removable plastic plug is inserted in the vas deferens, shows promise of being reversible in almost all cases.

Vaginitis is "simply" an infection of the vagina. It can be caused by a variety of organisms, such as fungi (yeast), trichomonas (a protistan), and bacteria. Yeast causes a discharge like cottage cheese; trichomonas causes a fishy smell and a greenish, watery discharge. Bacteria produce similar but more subtle symptoms. The woman may also be asymptomatic but contagious. Vaginitis is not always due to sexual intercourse, but can be caused by anything that increases vaginal moisture, such as sitting around in a wet swimming suit. Douching and taking antibiotics kills off the vagina's normal protective flora and can contribute to infection. Vaginitis can also be caused by soaps, powders, and deodorants, involving no organisms at all.

Honeymoon cystitis is caused by unusually frequent and vigorous intercourse. Usually the culprit is the intestinal bacterium *E. coli*, which is transferred from the anus to the urethra in women. The bacterium can then reach the bladder where it causes pain, and possibly even the kidneys where it can be dangerous. If not treated, it will reappear again and again, with increasing severity. Risk of infection can be reduced by drinking a quart of water after each sexual episode (probably a moderating effect in itself), and for some reason, drinking cranberry juice seems to help.

Herpes is caused by the *Herpes simplex* virus. Type I usually causes fever blisters (cold sores) on the lips. Type II causes the same sort of blisters on the genitals. It is generally not dangerous but can be aggravating to the afflicted. It usually appears as small blisters in the genital area, accompanied by a "tingling" sensation. When it erupts, the blister is extremely sore; it then disappears in about two weeks. Herpes is transmitted across mucosal surfaces, or openings in the skin. The virus never leaves the body but simply retreats along nerves where it resides in the spinal cord, to blossom forth on your vacation. Its appearance can be triggered by stress, either emotional or physical. The average American male has a 40% to 50% chance of becoming infected. (It is reported that 3% of nuns show antibodies.) In rare cases, the virus can inflame the brain and spinal cord. Herpes of the eye can cause blindness. People showing no symptoms *can* transmit the disease, but this is probably unusual.

Molluscum contagiosum sounds like a case of the snails, but it's actually known as "the hugging disease" since it is transmitted by a virus via skin-to-skin contact. It is not a serious condition, causing only a bump that reappears with annoying frequency if left untreated. Young people are particularly susceptible.

AIDS stands alone as a threat to humanity and is discussed in Chapter 41.

A final note: Most of these diseases can be largely avoided by a common sense approach to sex (if there is such a thing). Minimizing the number of partners is the safest bet, as well as developing an "index of suspicion," and avoiding those who do not appear to be responsible or forthcoming. Condoms, especially latex types rather than those made of animal membranes, reduce the transmission of STDs. Using condoms, and not engaging in any sexual activity that permits contact or exchange of body fluids, especially blood and semen, is the safest bet. Additional reduction of risk is obtained by minimizing the number of partners—each of whom brings his or her entire sexual history to bed with them—as well as avoiding any partner who gives any sign of being less than fully responsible and forthcoming.

Sperm continue to be produced after a vasectomy, but since they cannot be ejaculated, they are resorbed by the body. Semen is ejaculated as before, but it contains no sperm. Since the secretions of the accessory glands make up at least 80% of the normal ejaculate, and hormonal levels are not affected, there is no noticeable difference in the male's sexual performance.

Sterilization in women is usually accomplished by **tubal ligation** (Figure 42.13), where the oviduct is tied off or cut and both ends sealed. The oviduct can be sealed by being cauterized, or seared, causing the opening to seal shut as it heals, or it can be pinched together with a tiny clip or rubber band. This operation is much more complicated than a vasectomy since the abdominal wall must be opened. In some cases, the abdomen can be entered through a small incision at the navel or just above the pubic bone. It is also possible to enter through the vagina, but this

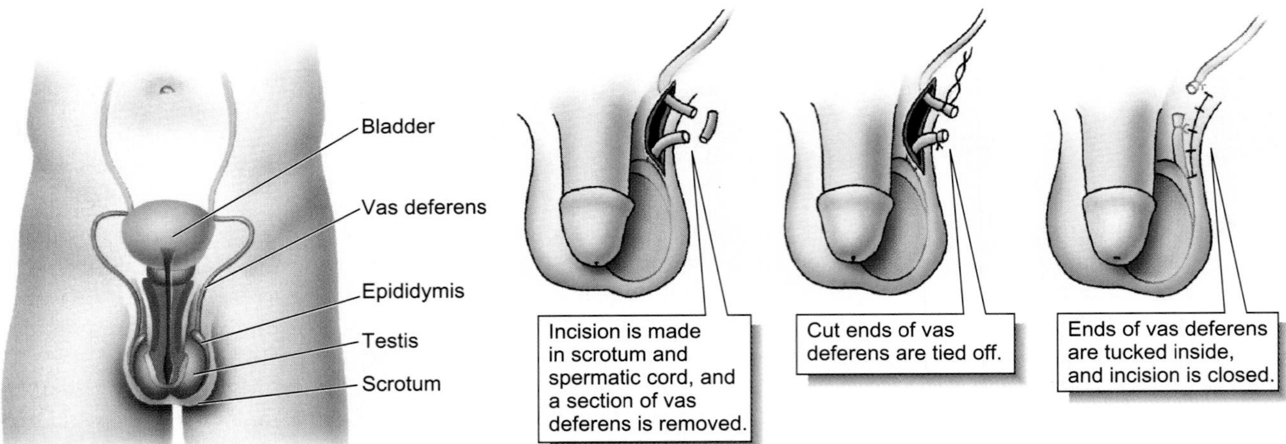

Bladder
Vas deferens
Epididymis
Testis
Scrotum
Incision is made in scrotum and spermatic cord, and a section of vas deferens is removed.
Cut ends of vas deferens are tied off.
Ends of vas deferens are tucked inside, and incision is closed.

FIGURE 42.12 THE SIMPLE PROCEDURE OF A VASECTOMY.
The operation has been described as similar to a visit to the dentist—with a few differences. Normally, under local anesthesia, the vas deferens is exposed by a small cut in the scrotum, and a section is removed from the tube. Both ends are tied in case they show a tendency to rejoin.

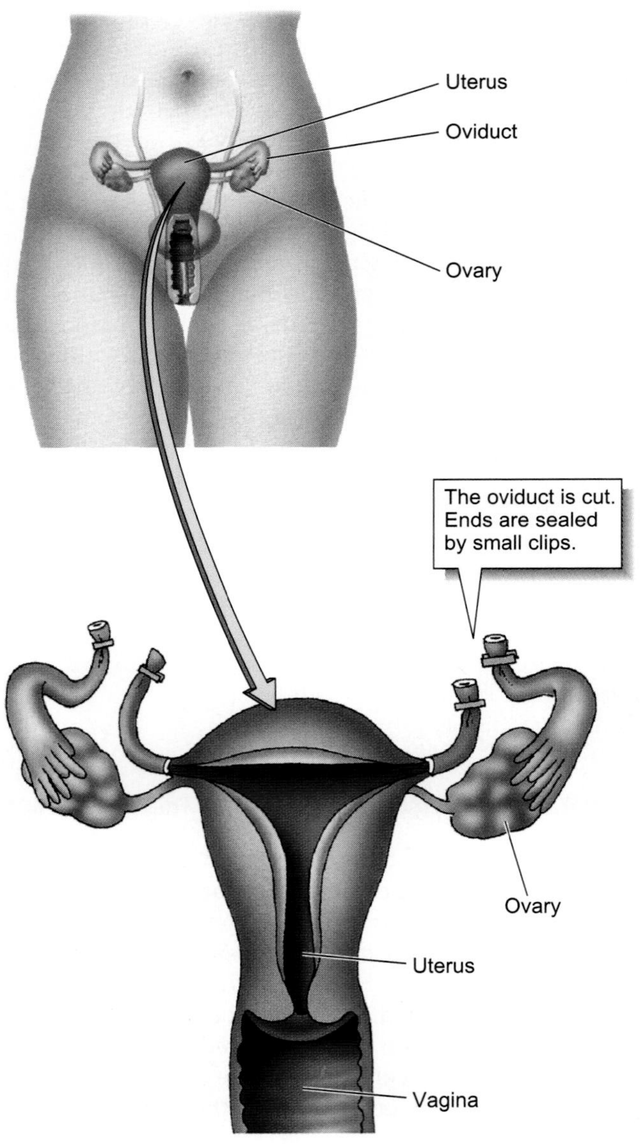

The oviduct is cut. Ends are sealed by small clips.

is riskier because of the increased possibility of infection. As with vasectomy, there is no evidence that, in humans, tubal ligation causes any change in hormone production, sex drive, or sexual performance. It has also been reported that none of these surgical techniques hurts at all if it's done on someone else.

ABORTION

Abortion—surgically removing an embryo or fetus, or inducing its expulsion—is by far the most controversial means of preventing an unwanted birth. However repugnant it is to many, the fact is that abortion is a widespread practice throughout the world.

In early pregnancies (less than three months), surgical removal of the embryo or fetus is commonly done with a suction or vacuum device, or through dilation and curettage ("D and C"). D and C is somewhat more severe since a curette (a spoon-shaped

FIGURE 42.13 TUBAL LIGATION.
Tubal ligation, or "tying the tubes," is a means of sterilization in women. It is a more serious operation than the male's vasectomy. It has traditionally been accomplished by opening the body wall, cutting the oviduct, and tying back the ends. Newer methods, however, include tying with a device that is pushed up the oviduct from the uterus. Ligation does not affect the ovaries, thus they continue to produce their hormones, so there is no change in the physiology of the woman other than totally negating the possibility of pregnancy. Some women have reported increased sexual satisfaction with the knowledge that they cannot become pregnant, but such reactions are strictly a personal matter and cannot be predicted.

surgical knife) is used to scrape the uterine wall, dislodging the embryo.

If the fetus has passed its third month, a saline treatment may be used to induce labor. A 20% saline (NaCl) solution is injected into the uterus. Labor and delivery of the fetus and its supporting placenta follow. Saline treatments are riskier than the other means. There is an occasional problem of salt poisoning and also a greater risk of infection. No one takes saline abortion lightly. The performance of an abortion after the second trimester can be extremely dangerous to the woman, and is illegal in most states.

PERSPECTIVES

We see, then, that the field of reproductive biology has a number of aspects that have attracted researchers, particularly in recent years. Theorists are interested in the biological advantage of sex. Physiologists are fascinated by the remarkable hormonal control of reproductive processes. Anatomists wonder at animals with two penises. And many of the rest of us are drawn to such discussions because we want to better understand ourselves and to gain more control over our own reproductive activity. The advances and technical information are increasingly hard-won—and each step seems to call forth yet another compelling question relating to this critical aspect of life.

If reproduction is successful, the processes we have just described are only the beginning. A long period of development must follow. This stage is fascinating and fraught with risk. So let's now take a look at the stage between the reproductive act and the appearance of a new individual.

SUMMARY

ASEXUAL REPRODUCTION

Asexual reproduction can occur through binary fission, regeneration, fragmentation, and budding. One parent is involved, so there is no genetic dilution.

SEXUAL REPRODUCTION

Sexual reproduction involves two parents, and with genetic mixing (dilution), increased variation in offspring results. In external fertilization, many gametes, each with little energy investment, are shed into the surroundings where fertilization occurs by chance.

Internal fertilization increases the chance of fertilization, so fewer gametes need to be produced, and the zygotes are protected within the mother's body for a time.

HUMAN REPRODUCTION

Male external genitalia includes the penis, testes, and scrotum. When spongy erectile tissue in the penile shaft and glans fill with blood, the penis becomes erect.

Sperm originate in germinal epithelium that lines the seminiferous tubules of the testes and develop within Sertoli cells. Storage and further maturation occur in the epididymis. Testosterone is produced in interstitial tissue surrounding the tubules. Sperm pass through the vas deferens, joined by secretions of the seminal vesicles and the prostate gland, forming semen. In ejaculation, muscular contraction propels semen through the urethra.

The female external genitalia (vulva) includes the mons veneris, labia majora, labia minora, clitoris, and vaginal opening. The highly sensitive clitoris contains erectile tissue.

Internal anatomy includes the distendable, tubelike vagina that extends to the cervix—the thickened opening of the uterus. The uterine endometrium is a temporary lining in which an embryo can implant. Oocytes, formed in the fetus, mature in the cortex of the ovary prior to ovulation. They are carried to the oviducts by waving fimbriae and cilia, where fertilization takes place.

HORMONES AND HUMAN REPRODUCTION

The hormonal control of reproduction involves an interaction between the hypothalamus, pituitary, and gonads in both sexes, with the addition of the placenta in females. Gonadotropin-releasing hormones from the hypothalamus prompt the anterior pituitary to release the gonadotropins, follicle-stimulating hormone (FSH), and luteinizing hormone (LH) (in males, interstitial cell-stimulating hormone—ICSH), which act on the ovaries or testes.

In males, FSH stimulates sperm production and works with LH to initiate testosterone production. Testosterone levels in the blood create a negative feedback loop back to the hypothalamus. Testosterone influences the development of maleness, the onset of puberty, and sex drive.

Puberty in females is associated with FSH secretion, which prompts estrogen production and release by the ovaries. Estrogens influence development in the breasts, reproductive organs, and uterus, and general body growth. Androgens from the adrenal cortex influence the growth of pubic and underarm hair. The onset of fertility begins with the monthly ovarian or menstrual cycle. In other female mammals, cycles of fertility (estrous cycles) occur less frequently, usually annually or semiannually.

The menstrual cycle includes the release of an ovum every 28 days, an event that corresponds to growth and thickening of the uterine endometrium, a preparation for the reception of an embryo. The cycle includes the following.

Days 1–14: The proliferative or preovulatory phase begins with the menses or menstruation. The pituitary secretes FSH, which prompts follicle development and estrogen secretion. Follicle cells surround the oocyte, supporting its growth into a mature Graafian follicle. Under estrogen's influence, the endometrium enlarges and becomes vascularized.

Day 14: Estrogens form a negative feedback loop to the hypothalamus, pituitary FSH secretions slow, and LH increases ("LH surge"). Upon reaching the follicle, LH brings about ovulation—the release of the ovum.

Days 15–28: In the secretory or postovulatory phase, LH and FSH maintain the corpus luteum, which secretes both estrogens and progesterone, further influencing endometrial growth. During the days following ovulation, the endometrium is ideally suited to receive the embryo. Terminating events in the cycle are still unclear, but is suspected that ovarian hormones produce a second negative feedback, suppressing LH and FSH secretion. The corpus luteum then fails, and without hormonal support, the endometrium breaks down.

The next cycle may be prompted by the lessening of FSH and LH suppression. But if fertilization and implantation of an embryo have occurred, cells associated with the embryo secrete human chorionic gonadotropin, which supports the corpus luteum for the first two months. Later progesterone and estrogens from the placenta maintain the uterus.

Natural methods of contraception include historical and folk methods, coitus interruptus (withdrawal prior to ejaculation), douching (attempting to flush semen from the vagina), and the rhythm method, in which intercourse is avoided during fertile periods. None is very reliable. The variables in the rhythm method include difficulties in determining the fertile period and the longevity of egg and sperm.

Mechanical methods include blocking sperm movement with sheathlike condoms worn over the penis; rubber diaphragms and cervical caps inserted with or without spermicide over the cervix; and intrauterine devices or IUDs. The latter is a device placed in the uterus, where it is thought to prevent implantation, but in ways not understood. Some IUDs contain spermicides.

Chemical methods of birth control include spermicides and the pill. The pill contains synthetic estrogens and progesterone, which are believed to suppress ovulation by overriding the natural rise and fall of fertility hormones. Side effects are noted in certain users, including high blood pressure.

Surgical intervention includes the vasectomy (cutting and tying the vas deferens) and tubal ligation (cutting and tying—or otherwise blocking—the oviducts).

The most controversial method of birth control is abortion—surgically removing an embryo or fetus, or inducing its expulsion. Surgical methods in the early months of pregnancy include suction or vacuuming of the uterus, and dilation and curettage (D and C), scraping the uterus. In later pregnancy, a saline (salt) solution can be injected into the uterine cavity to induce labor and expulsion of the fetus.

Key Terms

asexual reproduction • **730**
binary fission • **730**
regeneration • **731**
fragmentation • **731**
budding • **731**
sexual reproduction • **731**
external fertilization • **732**
ejaculation • **735**
spermatid • **736**
menstruation • **741**
corpus luteum • **742**
contraception • **743**
douching • **744**
rhythm method • **744**
coitus interruptus • **745**
condom • **745**
diaphragm • **746**
cervical cap • **746**
spermicide • **746**
intrauterine device (IUD) • **746**
birth control pill • **747**
Norplant • **747**
sponge • **747**
sterilization • **749**
vasectomy • **749**
tubal ligation • **749**
abortion • **750**

Review Questions

1. Describe how asexual reproduction occurs in *Hydra* and planaria. What, if any, are the advantages of asexual reproduction?
2. Describe three types of asexual reproduction. Compare the genetic makeup of the adult form with that of its offspring.
3. List three advantages of internal fertilization.
4. Describe the anatomical basis for the human male erection. What is the neural basis?
5. State the functions of the seminiferous tubules, the interstitial cells, and the epididymis.
6. List the structures through which sperm pass from their point of origin, and indicate which structures contribute to semen production. What is the significance of the semen's alkaline state?
7. Describe the manner in which an egg moves from the ovary into the oviduct and from the oviduct to the uterus.
8. In what specialized tissues do oocytes reside? When do they initially form?
9. Where does fertilization usually occur?
10. List two pituitary hormones in males and females, and describe their effects.
11. List several important roles of testosterone.
12. Describe several important pubertal effects of estrogen.
13. List three structures that play a part in the hormonal control of reproduction, starting with one located in the brain.
14. Explain how negative feedback works in controlling testosterone production in males.
15. What two events does the human menstrual cycle closely coordinate? Why is such coordination needed?
16. Describe events in the ovary, hypothalamus, uterus, and anterior pituitary through the first thirteen days of the human menstrual cycle. Include any negative feedback loops that may form.
17. Explain how shifting hormones bring on the process of ovulation.
18. Describe the output of hormones that maintains the endometrium through the first part of the postovulatory phase. What events end the cycle?
19. Once fertilization and implantation have occurred, what is the source of hormonal support during the pregnancy?
20. Describe the content of the combination type of birth control pill, and suggest how it prevents conception.
21. List several reasons why the rhythm system of birth control is subject to failure.
22. List the methods of birth control with the highest rates of failure.
23. List an advantage and a disadvantage of vasectomy and tubal ligation.

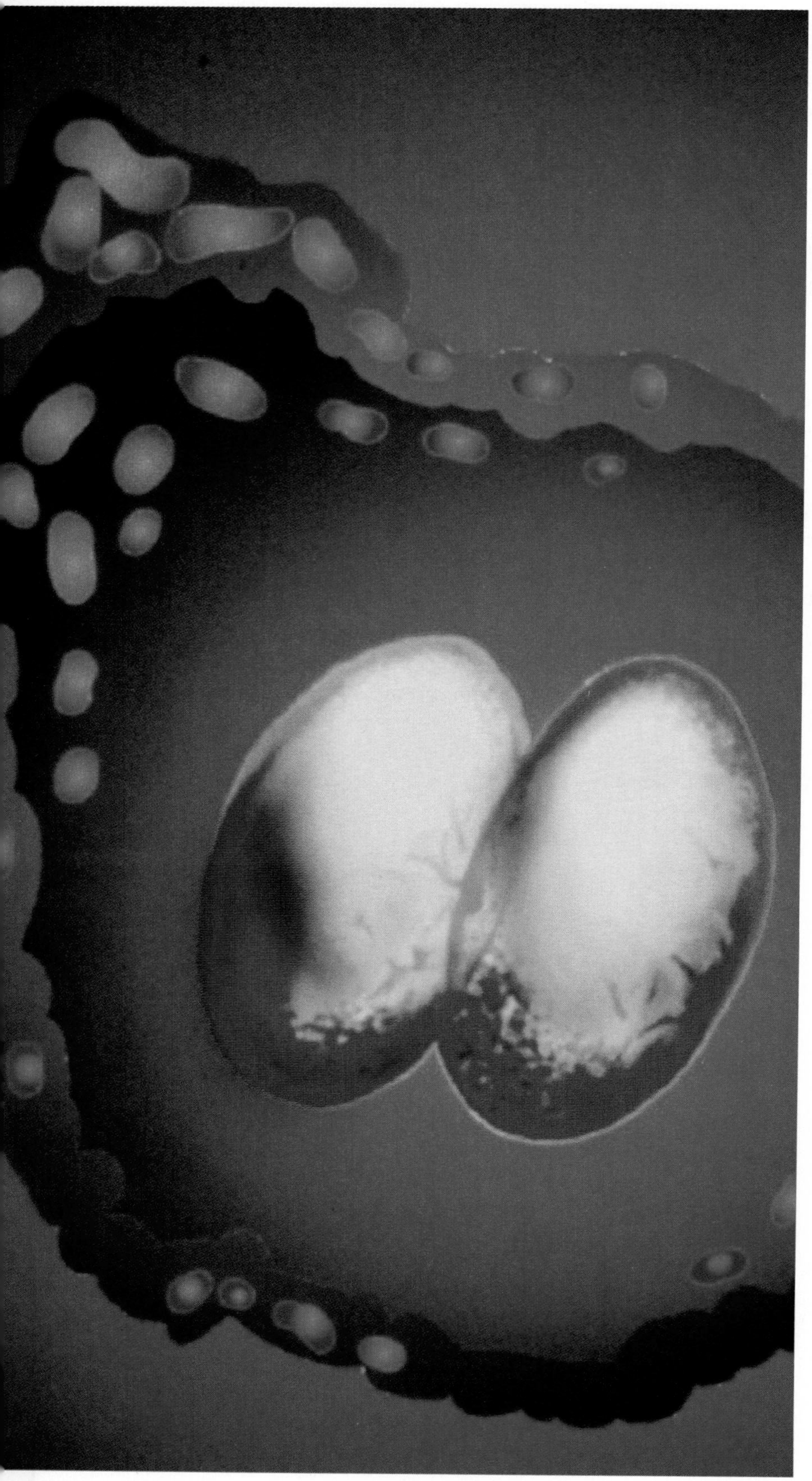

43

Development

THE PERIOD CALLED DEVELOPMENT BEGINS WITH FERTILIZATION AND *marks a time of remarkable transformations, the kinds of changes that produced you and that character sitting next to you. The period of development can be described in coldly clinical terms, but even then it can't be entirely drained of its fascination. How does a cell "know" when to die so that the body as a whole can continue its normal development? How can one kind of tissue direct the fate of another? How is it that virtually no chemical signals have been found? And what about humans? When does life begin? Why is it so hard to answer that question?*

One of the most fascinating biological moments is that brief instant when sperm and egg join in fertilization. Clearly, that instant triggers a remarkable series of events that makes us pause in wonder and realize how little we really understand about life.

Yet it is here, at the very beginning, that many believe we will come closest to understanding just how life occurs at all. This, then, is our chance to begin at the beginning, to see what happens at the union of delicate "half-cells" and during the momentous changes that come afterward. It is a precisely timed and finely choreographed ritual that is remarkably similar across the range of life.

The study of the developmental period historically has been called *embryology* (the study of the embryo), although now most people favor the term *developmental biology.* Developmental biologists may not have yet been able to answer some of the fundamental questions of life, but they have provided us with a broad view, scattered with some remarkable bits of very detailed information. They begin the broad view by telling us there are three processes in the development of an animal: *growth, cellular differentiation,* and *morphogenesis.* **Growth** is an increase in size generally brought about through repeated cycles of cell enlargement, mitosis, and cell division. **Cellular differentiation** is the process whereby cells become different—that is, the process by which various cells of an organism become increasingly specialized in shape, chemical makeup, and function. **Morphogenesis** is the emergence of an organism's overall recognizable form as individual structures become increasingly developed. Morphogenesis involves not only the movement, division, and change in shape of specialized cells, but also, in some cases, their programmed death.

GAMETES AND FERTILIZATION

SPERM

Except for a few differences in size and shape, the sperm cells of most animals are essentially similar. Perhaps this similarity stems from their common mission: to encounter the egg and penetrate it. The principal structures of a sperm are the *head, midpiece,* and *tail* (Figure 43.1). The head contains highly condensed *chromatin* (chromosomal DNA and its related proteins). At the tip of the head is an enzyme-laden *acrosome,* which helps the sperm penetrate the egg. The midpiece contains curiously shaped, spiraling mitochondria that generate the ATP needed to sustain the action of the flagellum, along with a centriole and the roots of the microtubules that make up the tail, or flagellum.

EGGS

Although eggs of different species vary greatly in size, they are generally larger than sperm. Even among vertebrates the range in size of eggs is enormous. A human egg is smaller than the period at the end of this sentence, but it would be hard to hide an ostrich egg with this entire book. The size of the vertebrate egg mainly depends on the quantity of yolk,

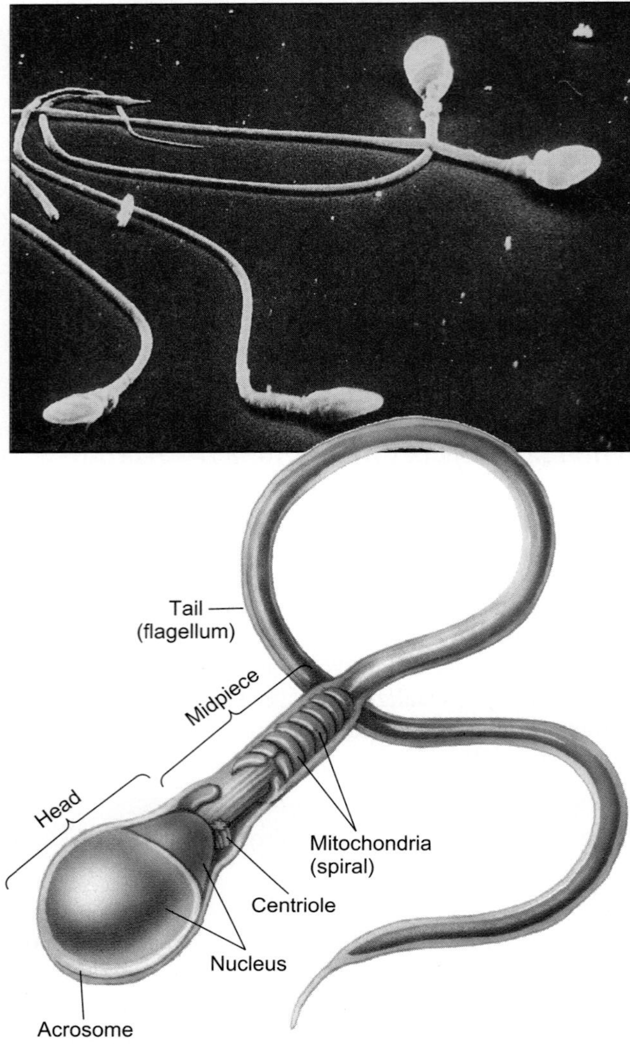

FIGURE 43.1 HUMAN SPERM.
The human sperm is similar to the sperm of other mammals.

the embryo's food supply, The food supply, in turn, relates to the mode of development. Mammals require only a small quantity of yolk, because the young embryo soon implants itself in its mother's nutrient-rich uterus, where its needs are provided by the placenta. Birds, though, develop outside the mother's body, and so they must have their yolky food supply with them until they hatch.

The egg cell has the usual cellular organelles but is quite unlike other cells in many respects. For example, in nearly all cases, eggs show polarity. This means that the cytoplasmic constituents are not equally distributed, but occur in a gradient. The unequal distribution results in essentially two hemispheric divisions. In most kinds of eggs, the metabolic "machinery" (mitochondria, Golgi bodies, endoplasmic reticulum, etc.) is concentrated in the more metabolically active *animal hemisphere,* while most of the yolky food reserve lies in a less active *vegetal hemisphere.*

Humans, like other mammals, produce eggs that, while far larger than the sperm, are still tiny—barely visible to the unaided eye. (At ovulation the human egg is only about 0.15 mm in diameter.) The human egg (Figure 43.2) emerges from the ovary surrounded by a dense covering of follicle cells known collectively as the *corona radiata* ("radiating crown"). Below the corona radiata lies a thick, glassy area, the *zona pellucida* ("clear zone"), which is secreted by corona cells. Finally, under the zona pellucida is the plasma membrane.

FERTILIZATION

Fertilization is the union of the haploid sperm nucleus and the haploid egg nucleus. They are each at this stage, before fertilization, called the *pronucleus.* Fortunately, the process is somewhat similar in all animals, so we can apply what we know from studies of species that happen to lend themselves to experimentation. Echinoderms, such as sea urchins, sand dollars, and sea stars, are good subjects because they can be readily induced to shed eggs and sperm, which survive and function well in the laboratory.

Echinoderm Fertilization. At fertilization the echinoderm sperm get a lot of help from the egg, as we see in Figure 43.3. To begin with, the egg has specialized membranal receptor sites to which a sperm may attach by its *acrosomal process,* an enzyme-producing extension of the acrosome that produces enzymes which help dissolve the jelly coat. Then the egg's plasma membrane suddenly extends its many microvilli, surrounding the sperm head, and forming what is called a **fertilization cone.** Once this cone is formed, the egg draws the sperm head, with the sperm nucleus, into its cytoplasm.

At this time a number of cortical granules, storage bodies lying just below the echinoderm egg membrane, rise toward the surface and rupture, releasing substances that bring on the formation of a **fertilization membrane.** As it forms, the membrane's role becomes apparent. It rises up and carries with it the unsuccessful sperm that had also attached to the egg, effectively preventing their penetration into the egg's interior. This barrier is critical because penetration by more than one sperm can cause the zygote to fail.

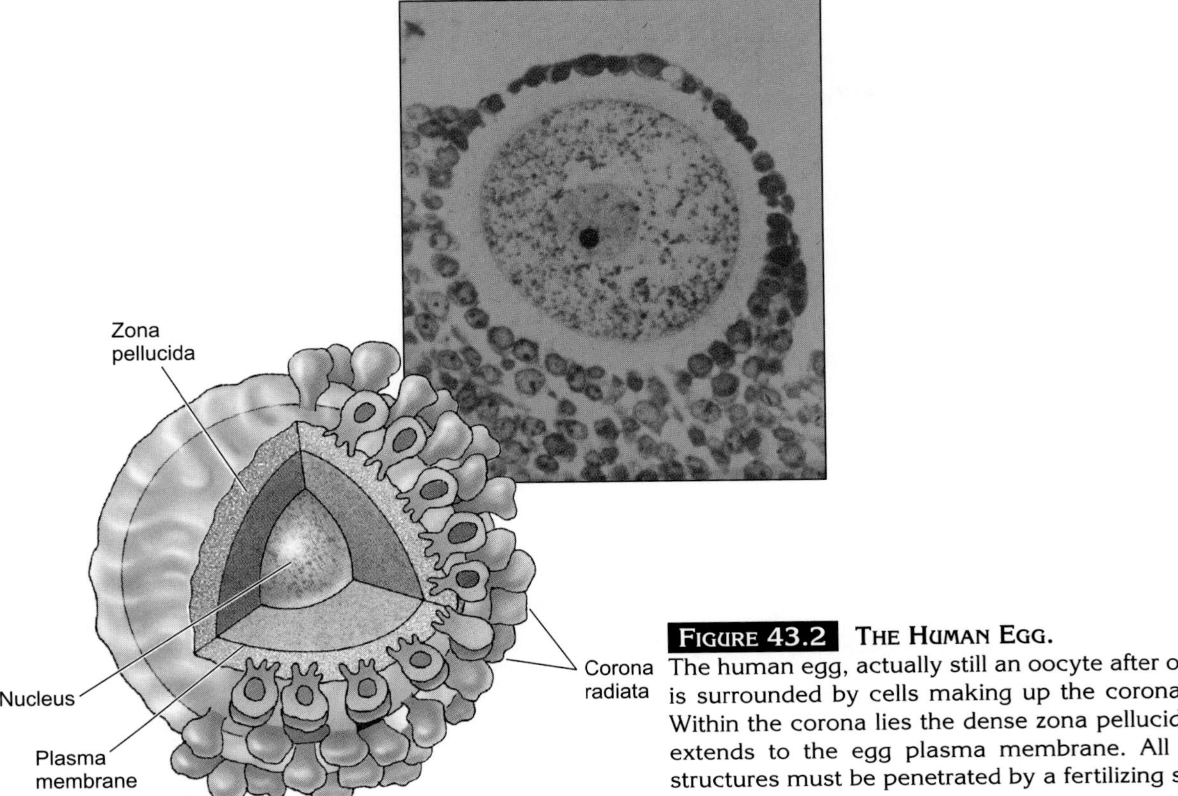

FIGURE 43.2 THE HUMAN EGG. The human egg, actually still an oocyte after ovulation, is surrounded by cells making up the corona radiata. Within the corona lies the dense zona pellucida, which extends to the egg plasma membrane. All of these structures must be penetrated by a fertilizing sperm.

Later, the single, successful sperm will release its pronucleus, which will fuse with the egg pronucleus, completing fertilization.

Human Fertilization. The human sperm usually reaches the egg somewhere in the upper oviduct. Even though only one sperm normally fuses with an egg, a great many are required to complete the process because the acrosomes contain enzymes needed to dissolve the corona radiata that surrounds the egg. Once penetration has occurred, changes in the dense zona pellucida prevent the entry of other sperm. It may seem odd, but it isn't until the egg and sperm pronuclei fuse in fertilization that the egg completes its second meiotic division (in fact, medical researchers use the appearance of another polar body as a signal that fertilization has occurred). The **zygote,** or fertilized egg, then begins a round of DNA replication as it prepares for mitosis, the first of countless mitotic events that will eventually produce a new voter. The early stages of development are summarized in Table 43.1.

EARLY DEVELOPMENT

The events that follow fertilization in animals differ from group to group, but still there are fundamental

TABLE 43.1 EARLY STAGES OF DEVELOPMENT

Stage	Description	Result
Fertilization	Fusion of nuclei of egg and sperm	Creates a cell (the zygote) with two copies of each chromosome; prompts reactions in the egg
Cleavage	Rapid series of mitotic division	Divides the zygote into smaller cells of varying size, shape, and activity
Gastrulation	Migration of cells	Forms three primary germ layers (ectoderm, mesoderm, endoderm), each of which will give rise to specific tissues

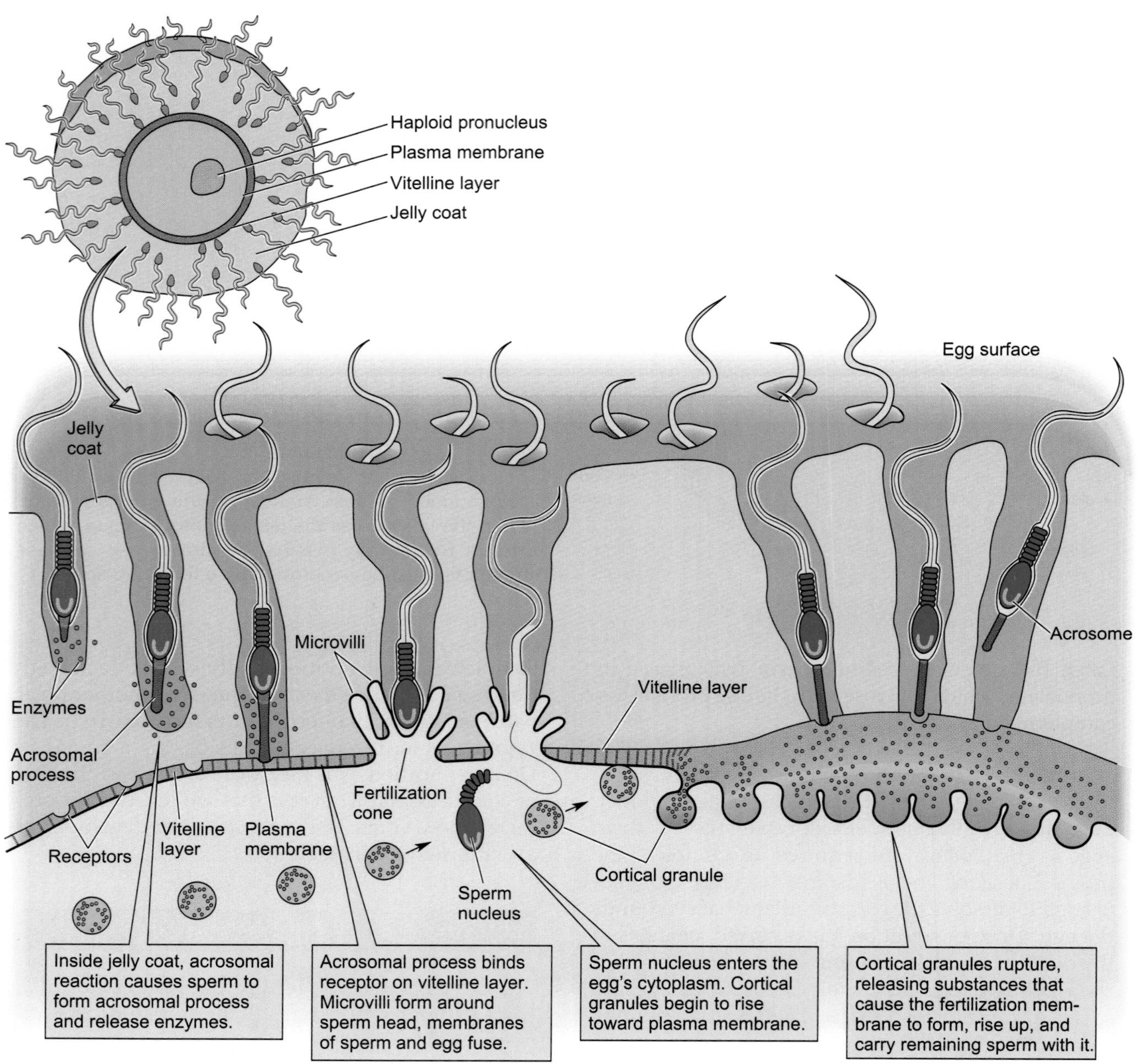

Haploid pronucleus
Plasma membrane
Vitelline layer
Jelly coat
Egg surface
Jelly coat
Acrosome
Microvilli
Vitelline layer
Enzymes
Acrosomal process
Fertilization cone
Receptors
Vitelline layer
Plasma membrane
Cortical granule
Sperm nucleus
Inside jelly coat, acrosomal reaction causes sperm to form acrosomal process and release enzymes.
Acrosomal process binds receptor on vitelline layer. Microvilli form around sperm head, membranes of sperm and egg fuse.
Sperm nucleus enters the egg's cytoplasm. Cortical granules begin to rise toward plasma membrane.
Cortical granules rupture, releasing substances that cause the fertilization membrane to form, rise up, and carry remaining sperm with it.

FIGURE 43.3 FERTILIZATION.
After passing through the jelly coat, the echinoderm sperm attaches to a receptor site on the egg membrane by forming an acrosomal process. Following attachment, the membrane forms numerous microvilli that ensnare the sperm head, drawing it into the cytoplasm. During this time, cortical granules below the membrane rupture, releasing a chemical into the membrane. As a result, a fertilization membrane forms, rising up from the egg surface and carrying unsuccessful sperm along with it.

similarities across the species. Let's review some of these differences and similarities, then get on to the human condition.

Cleavage

The zygote immediately begins to divide, a process called **cleavage.** The first division produces two cells, which then form four, which form eight, and so on, until at some point a solid ball of about 32 cells, called a **morula,** is produced.

Mitosis occurs before each of these early divisions, but there is no cell growth. This is because the cells become smaller with each division. Furthermore, since the contents of the egg cytoplasm were unequally distributed to begin with, cleavage results in cells that contain different components of the egg cytoplasm. Some scientists believe that this inequality is a source of variation in the population. They argue that if various cytoplasmic components interact in their own way with the genetic material, the developmental fate of the cell will be altered.

Patterns of cleavage vary greatly among different animal groups. As we saw earlier (Chapter 26, Figure 26.2), many protostomes undergo a *spiral, determinate cleavage.* Here, in the earliest stages of development, each new generation of cells arises at an angle from the parent cell. Furthermore, as each new cell arises in the protostome, its future is determined. There is no flexibility in its fate.

Most deuterostomes undergo cleavage that is *radial* and *indeterminate.* Each new cell lies in the same plane as its parent cell, and each will continue to have a great deal of developmental flexibility for some time. That is, it can take any of a number of developmental routes, depending on conditions.

Patterns of cleavage also vary according to the amount of yolk present in the fertilized egg. Yolk, an inert material, is inactive in the cleavage process. As you see in Figure 43.4, cleavage is complete (the entire embryo divides; the cleavage goes all the way across) in the amphibian embryo. But even in this group the distribution of yolk has a telling effect. A more rapid cleavage rate on the side with less yolk produces smaller, active cells that form the *animal hemisphere.* Meanwhile, at the opposite side—the *vegetal hemisphere,* where there is more yolk—the rate is much slower, and larger cells are produced.

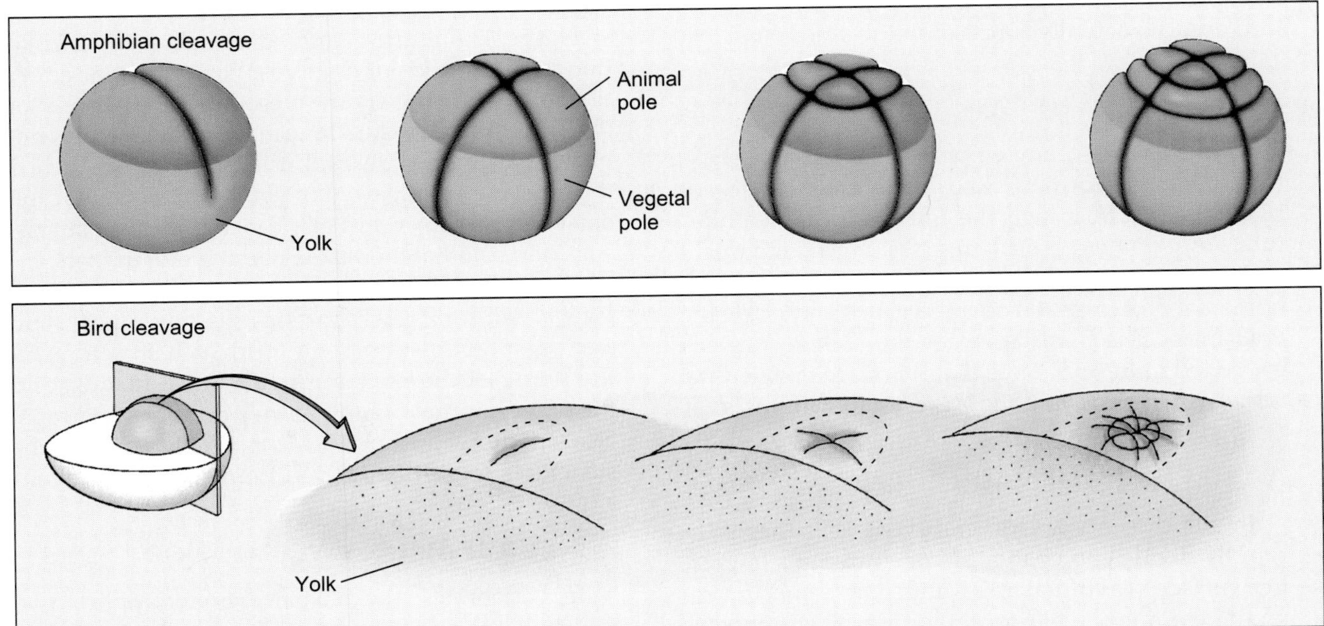

Figure 43.4 Early Amphibian and Bird Development.
Cleavages in the amphibian embryo begin simply enough, with the first two producing cells of equal size and content. The cleavages following, however, are highly unequal. This trend will continue, with the upper cells, or animal pole, dividing much more rapidly than the lower, vegetal pole cells. The immense yolk of the bird makes complete cell division impossible, so at first, a small group of cells lying at the surface actively divides. This disk of cells will form the embryo proper.

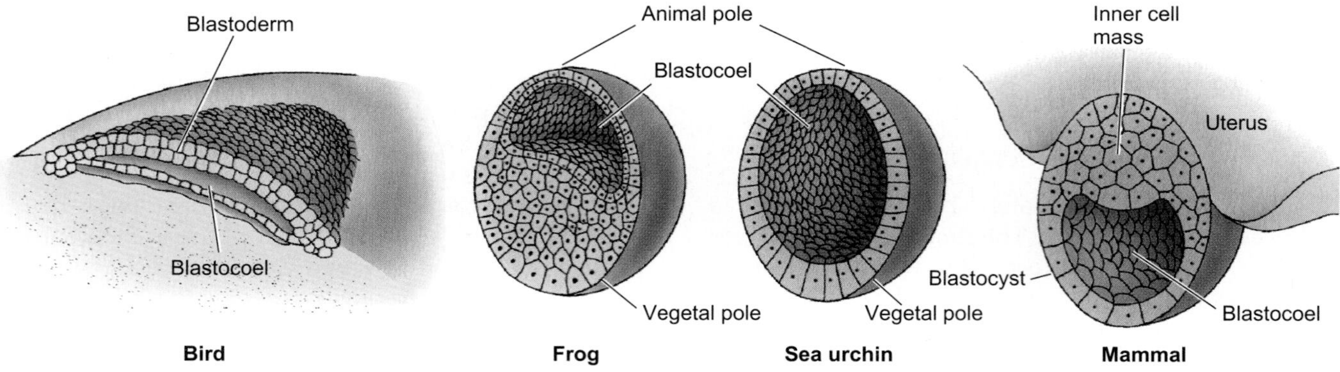

FIGURE 43.5 THE BLASTULA STAGE.
The blastulalike stage in the bird includes a two- to three-layered blastoderm, a slender cavity, and a thin layer of cells overlying the yolk. The frog blastula is a dense sphere of cells that occur in a definite size gradient. The blastocoel is offset toward the animal pole. The sea urchin blastula is fairly simple, with some vegetal-animal hemisphere orientation and a single layer of cells. The mammalian blastocyst represents the blastula stage. It has three regions: a thin trophoblast, a denser, inner cell mass, and the blastocyst cavity.

This pattern will continue and will affect events in the immediate future. In mammals, there is little yolk to impede cleavage, so the cells are all roughly the same size. In birds, unequal cleavage reaches the extreme. The immense bulk of the yolk present in the bird egg prevents complete cleavage. Thus, the bird embryo actually forms atop the yolk mass.

THE BLASTULA

As the cells of many embryos continue to divide, they begin to form a fluid-filled inner cavity or **blastocoel.** At this stage, the embryo is called a **blastula.** Development in bird, frog, sea urchin, and mammal embryos is similar in many ways, but the developmental process is still adapted to accommodate the yolk mass. The blastula stages are shown for each group in Figure 43.5. Note that this stage is called the *blastoderm* in birds, and the *blastocyst* in mammals.

THE GASTRULA

In all three types of embryos, cells begin to rapidly divide and change shape, forming a migrating surface wave that rolls under at a specific region, invading the blastocoel within. The process is called **gastrulation,** and it marks the formation of the **gastrula** stage. The gastrula stage is especially important because this is when the three embryonic *germ layers* first appear. You may recall from our discussions of animal evolution (see Chapter 26 that animals with three germ layers in their embryos are capable of developing complex organs and systems. The important point here is that gastrulation produces the first pronounced embryonic tissue organization. Furthermore, this new organization marks the beginning of the differentiation between internal and external parts of the animal.

Figure 43.6 illustrates the differences in gastrulation in three kinds of embryos. Pay particular attention to the formation of the mesoderm, the germ layer that appears for the first time in gastrulation. "It is not birth, marriage or death, but gastrulation, which is the most important event of your life," said embryologist Lewis Wolpert in 1903. The reason is, if your tissues don't shift at this stage, nothing else matters.

You may recall that we learned earlier that the *ectoderm* will form such parts as the nervous system and skin; the *endoderm* will form such structures as the lining of the gut, lungs, and most glands; and the other internal organs, as well as bones and muscles, are derived from *mesoderm.* (Table 43.2 lists the structures primarily derived from each germ layer.)

NEURULATION AND THE BODY AXIS

In vertebrates, the next events accomplish two things. First, the embryonic axis (the anterior-posterior organization) will become clear—that is, distinct head and tail ends are established. Second, the crude outlines of the nervous system begin to form. Both occur through a fascinating process called **neurulation,** the first appearance of the nervous system, marking the **neurula** stage of the embryo. Neurulation pro-

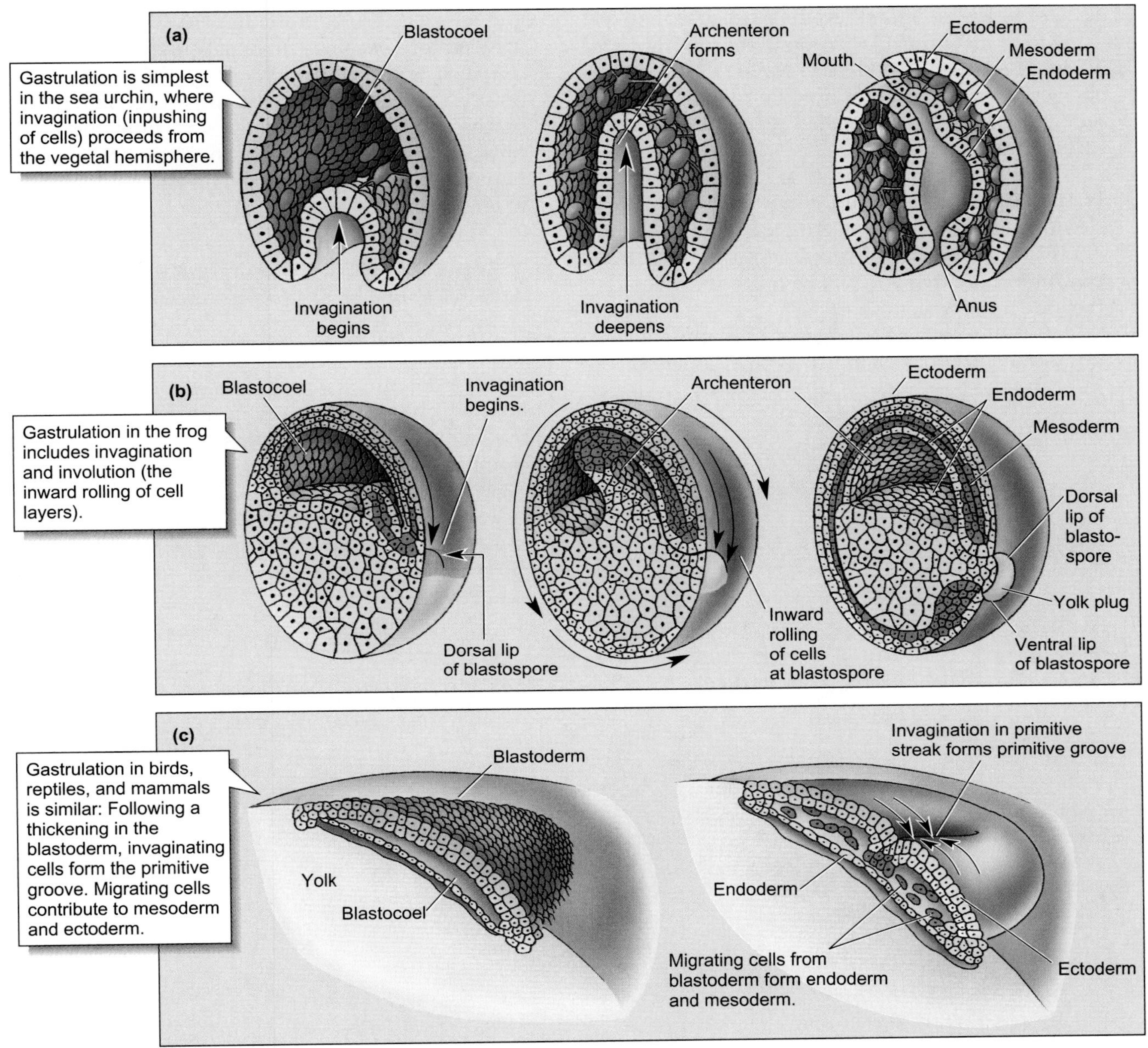
(a)
Gastrulation is simplest in the sea urchin, where invagination (inpushing of cells) proceeds from the vegetal hemisphere.
Blastocoel
Archenteron forms
Mouth
Ectoderm
Mesoderm
Endoderm
Invagination begins
Invagination deepens
Anus
(b)
Gastrulation in the frog includes invagination and involution (the inward rolling of cell layers).
Blastocoel
Invagination begins.
Archenteron
Ectoderm
Endoderm
Mesoderm
Dorsal lip of blasto-spore
Yolk plug
Dorsal lip of blastospore
Inward rolling of cells at blastospore
Ventral lip of blastospore
(c)
Gastrulation in birds, reptiles, and mammals is similar: Following a thickening in the blastoderm, invaginating cells form the primitive groove. Migrating cells contribute to mesoderm and ectoderm.
Blastoderm
Yolk
Blastocoel
Invagination in primitive streak forms primitive groove
Endoderm
Migrating cells from blastoderm form endoderm and mesoderm.
Ectoderm

FIGURE 43.6 **GASTRULATION IN THREE KINDS OF ANIMALS.**
Gastrulation in animals involves forces that bring about a vast relocation of tissues, resulting in the formation of an embryo with three germ layers. Here gastrulation is compared in the sea urchin **(a)**, the frog **(b)**, and the bird **(c)**.

ceeds in a similar fashion in most vertebrate groups, so we will consider frog neurulation as representative of the process (Figure 43.7).

Neurulation begins with the formation of a thickened strip of ectoderm called the *neural plate.* The strip extends part way around the gastrula, basically from the head to the tail region. The two edges of the neural plate thicken, forming *neural folds.* Between the folds a depression, the *neural groove,* also takes form. Next, the neural folds rise up along their length, curve inward above the neural grove, and close, thus forming the *neural tube.* In the closing, clusters of cells known as the *neural crest* become isolated between the neural tube and the overlying ectoderm. We mention these cells because they will later migrate from this location, contributing to a variety of structures. Finally, we should note that the neural tube is not uniform, but widens out considerably at one end, forming the crude outlines of an emerging brain region.

The neural plate is induced to begin its formation by a rod-shaped structure lying beneath the ectoderm. This is the *notochord.* You may recall (Chapter 27) that the notochord is present in all chordates at one time or another. We will have more to say about the induction process shortly.

Later, the neurula elongates and undergoes a number of internal changes. With elongation, the gut, which is formed from the early cavity called the archenteron, becomes more tubelike and later will form anal and mouth openings. On either side of the notochord, the mesoderm has organized into paired blocks called **somites,** which will form the trunk muscles and the axial skeleton. Their organization reminds us again of the segmental plan so common to all vertebrates.

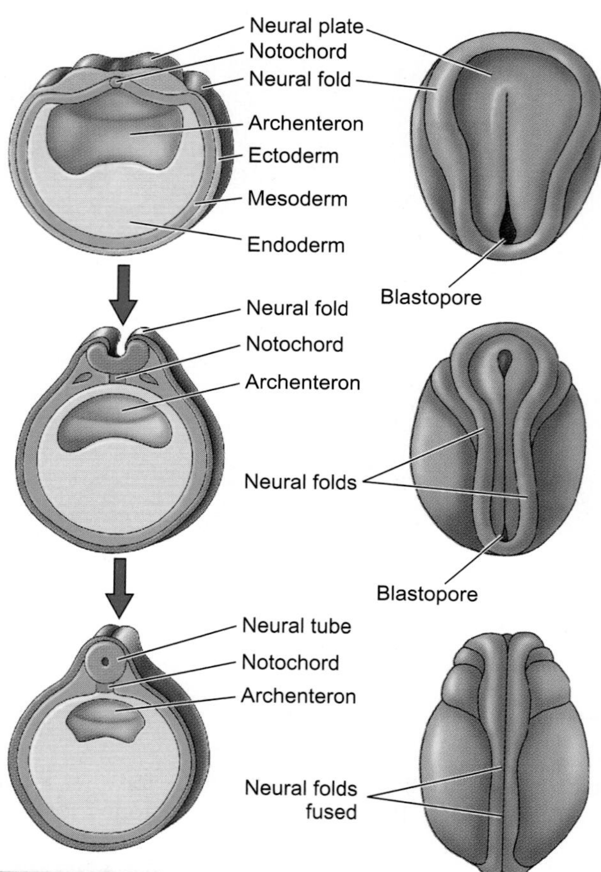

FIGURE 43.7 NEURULATION.
At the start of neurulation in the frog ectodermal cells form a thickened plate along the embryo. The edges then grow, rising up into folds. These will join above, grow together, and form the dorsal tubular nerve cord. This then sinks below the surface and becomes covered by new ectoderm. With our surface view, we see that the closure of the neural folds, NF, first begins at the center, proceeding in both directions, but lagging behind at the anterior end. The folds there become greatly enlarged as they form the crude outlines of the rudimentary brain.

HUMAN DEVELOPMENT

EARLY EVENTS

After the human sperm and egg join in fertilization, which usually occurs in the upper reaches of the oviduct, the zygote begins to develop even as it continues its journey down the tube (Figure 43.8). This development is slow at first, with the first cleavage occurring about a day and a half after fertilization. The next four cleavages, though, occur more rapidly, forming a morula, a rough ball of about 32 cells.

TABLE 43.2 DERIVATIVES OF THE THREE GERM LAYERS

Ectoderm	Mesoderm	Endoderm
Epidermis	Skeleton	Linings of:
Hair	Muscle	Gut
Milk glands	Skeletal	Pancreas
Oil glands	Smooth	Respiratory system
Sweat glands	Cardiac	Pharynx
Mouth lining	Dermis	Liver
Retina, cornea, lens of eye	Blood	Urinary bladder
Inner ear	Gonads	
Nervous system	Kidney	
Brain		
Spinal cord		
Spinal nerves		
Adrenal medulla		

FIGURE 43.8 HUMAN FERTILIZATION AND EARLY DEVELOPMENT.
After ovulation, the human oocyte is drawn into the oviduct, where fertilization can occur. From the time of fertilization, about six days are required for the transit to the receptive uterus and implantation. Following gastrulation the embryo takes on an elongated shape with a definite head and tail organization. The allantois contributes to the body stalk, where blood vessels will form. The primary chorionic villi also expand, increasing exchange surfaces.

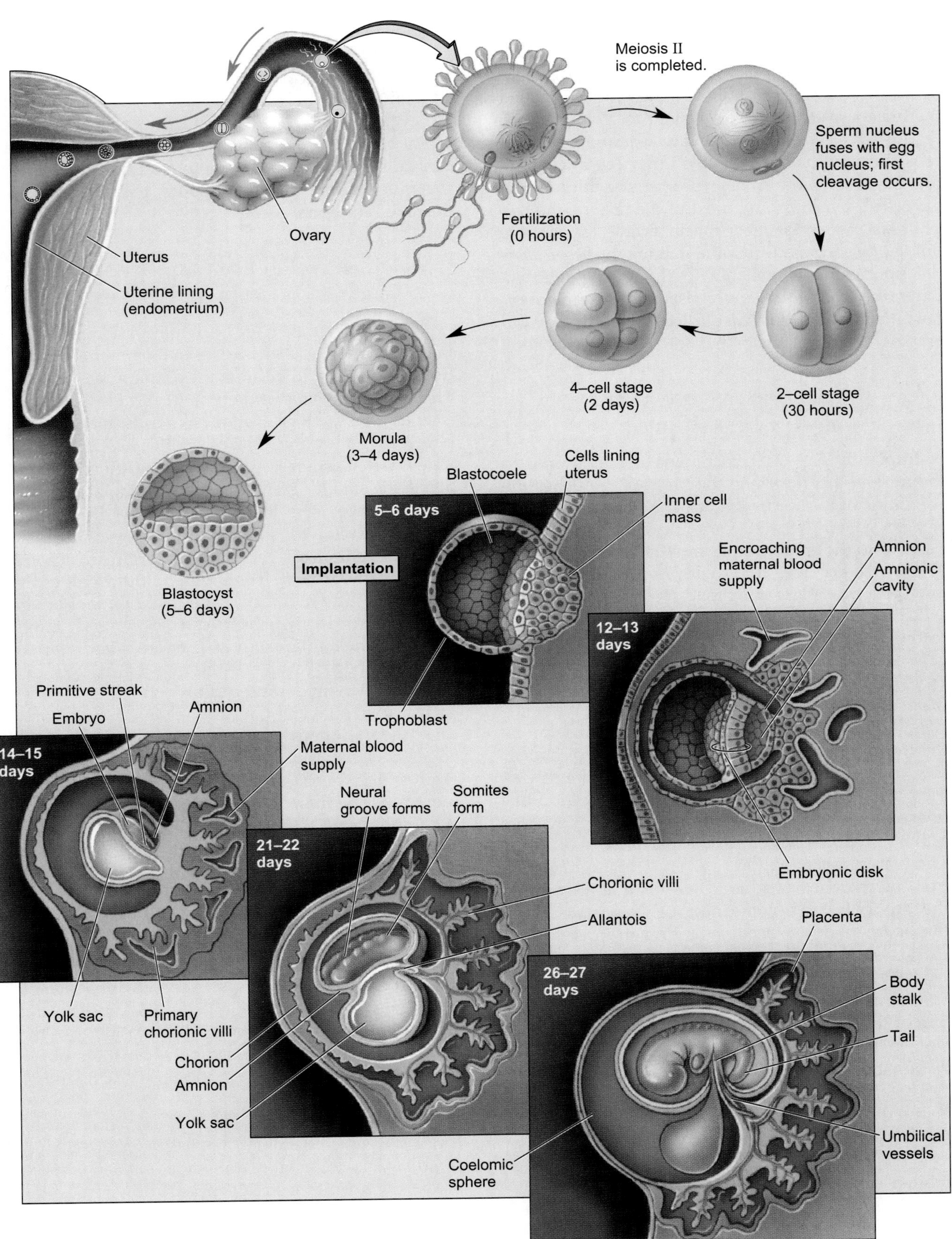

Meiosis II
is completed.
Sperm nucleus
fuses with egg
nucleus; first
cleavage occurs.
Fertilization
(0 hours)
Ovary
Uterus
Uterine lining
(endometrium)
4–cell stage
(2 days)
2–cell stage
(30 hours)
Morula
(3–4 days)
Blastocyst
(5–6 days)
Blastocoele
Cells lining
uterus
Inner cell
mass
5–6 days
Implantation
Trophoblast
Encroaching
maternal blood
supply
Amnion
Amnionic
cavity
12–13
days
Embryonic disk
Primitive streak
Embryo
Amnion
14–15
days
Maternal blood
supply
Yolk sac
Primary
chorionic villi
Neural
groove forms
Somites
form
21–22
days
Chorionic villi
Allantois
Chorion
Amnion
Yolk sac
Placenta
26–27
days
Body
stalk
Tail
Umbilical
vessels
Coelomic
sphere

As the embryo moves on toward the uterus, its cells continue to divide. About six days after fertilization, certain cell rearrangements of the morula usher in the next stage—the hollow sphere of cells called the **blastocyst** (roughly equivalent to a blastula; see Figure 43.8). The blastocyst has a denser group of cells clustered on one side, the **inner cell mass,** destined to become the embryo proper. The thinner, single-celled layer forming most of the sphere, the **trophoblast,** is specialized for invading the uterine lining (the endometrium) and making the first maternal nourishment available to the young embryo. Later, the trophoblast gives rise to the extraembryonic membranes. It is the blastocyst, the six-day old embryo, that implants in the mother's endometrium.

Implantation and Early Life Support

Implantation—the attachment of the embryo to the uterus—begins when the blastocyst touches the soft lining of the uterus (Figure 43.8). The trophoblast secretes enzymes that digest some of the endometrial tissue, and the embryo sinks into the resulting cavity. The cavity becomes filled with nutrient-rich blood, and for a brief time this blood will adequately provide the necessary food and oxygen and the removal of the embryo's meager wastes. But as the embryo grows, its food and oxygen demands increase, and more metabolic wastes are generated. Thus, it needs faster and more efficient means of exchanging materials with the mother's body. Rapid cell division in the trophoblast sends slender fingerlike projections, the *primary chorionic villi,* deeper into the endometrium. For a time this will be the connection between the embryo and the mother's tissues. But soon a more complex structure, the **placenta,** will form. The placenta is a matlike, spongy structure formed from membranes and blood vessels that arise from the trophoblast to become intimately associated with the soft, highly vascular endometrium of the mother's uterus. Throughout the pregnancy it provides for the nourishment of the embryo as food crosses it from the mother's blood, for the elimination of the embryo's waste, and for the exchange of oxygen and carbon dioxide between mother and developing fetus.

The continued maintenance of the endometrium is urgently required by the growing embryo. So the trophoblast begins to secrete *human chorionic gonadotropin (HCG),* which is picked up by the mother's blood and carried to its target, the corpus luteum within the mother's ovary (the same structure that once housed the egg; see Chapter 42). The corpus luteum responds by continuing to secrete progesterone and estrogens, hormones that support the endometrium. Because HCG overrides the ovarian cycle, the embryo is secure. Some HCG is normally present in the mother's urine at this time and can be detected through chemical tests (including the much publicized "pregnancy test kits," those designed for home use).

Early Development and Formation of the Placenta

Meanwhile, there have been many changes in the inner cell mass. A cavity arises in the inner cell mass (forming what will become the amnion; see below). The remaining cells of the inner cell mass form the *embryonic disk,* which is about two cells thick. At this point, the embryonic disk resembles a shelf and divides the blastocyst into two fluid-filled chambers—the smaller amniotic cavity and a larger blastocyst cavity (see Figure 43.8). The fluid-filled amniotic cavity will surround the embryo throughout its development, constantly enlarging to accommodate fetal growth. The amniotic fluid will cushion the delicate embryo and provide a watery environment in which it can grow and move around. (The amnion and its fluid are later referred to as the "bag of water"—the one that breaks, announcing the impending birth.)

The lower layer of the two-layered embryonic disk, the endoderm, soon spreads outward at the edges and then folds under to form a **yolk sac** (see Figure 43.8). In the bird egg these same cell movements enclose the yolk mass, but in the human embryo, they form a third fluid-filled chamber. Blood cells and blood vessels form in the yolk sac, which remains small and becomes increasingly insignificant as the embryo itself grows.

The yolk sac is one of four membranes produced by the embryo as it develops. The **amnion,** a fold of mesoderm and ectoderm, will enclose the embryo in protective amniotic fluid. The **chorion,** formed from the trophoblast, lies outside the amnion and produces fingerlike chorionic villi that increase the area of the exchange surface between the embryo and mother. The fourth membrane is the allantois, described below.

Some three weeks into development, the unstructured, two-layered embryonic disk begins to change rapidly. The disk takes on an elongated slipper shape, the **primitive streak** appears down its center, and the mammalian version of gastrulation occurs.

By the third week of development, neurulation begins. In the usual vertebrate manner, neural folds rise up on either side of the dorsal surface, their tips growing toward each other and finally closing to form the tubelike spinal cord. As this happens, the blocklike somites begin to appear along its length. Progress is rapid, and at 26 days, the brain is outlined, the crude S-shaped heart has formed, and the embryo has elongated, its postanal tail announcing its vertebrate status. At this time the embryo is about 3.6 mm (⅛ in.) long.

As these fast-moving events progress, the **allantois,** the fourth and final extraembryonic membrane, develops. The allantois will contribute blood vessels during the formation of the placenta. This is its only function in primates, although in other mammals and in birds and reptiles, it collects wastes and fuses with the chorion to form the chorioallantois. The connections between the embryo proper and the extraembryonic membranes narrow somewhat to become the **body stalk.** Later, the body stalk will become the **umbilical cord.** The blood vessels of the allantois become the arteries and veins of the umbilical cord (see Figure 43.8).

With development of the chorion, allantois, and the embryo's circulatory system, the placenta emerges. The chorionic villi enlarge, each giving rise to numerous microvilli, and blood vessels form extensive capillary beds (Figure 43.9). The maternal and embryonic blood are separated by only a thin layer of cells, which permits oxygen and growth-supporting substances to be readily exchanged for metabolic wastes.

Despite the great changes that have occurred so far, the embryo will not be clearly recognizable as human until seven or eight weeks of development (see Figure 43.10). At eight weeks the embryo technically becomes a fetus. It is now about 25 mm (1 in.) in length and growing rapidly. The eight-week-old fetus floats in a sea of amniotic fluid. As growth progresses, the placenta comes to contain so many blood vessels that it takes on a fibrous appearance. Its exchange surface by now consists of numerous mounds of tissue, each containing many villi that drastically increase its surface area; each villus contains its own capillaries from the placental vessels.

By the third month of pregnancy, the placenta assures its own survival by secreting estrogen and progesterone, replacing the corpus luteum as the primary source of these hormones.

Further Human Development

Clinically, the 280-day gestation period of human development is divided into three parts called **trimesters.** Most of the significant morphological events occur during the first trimester.

The First Trimester

The first trimester is marked by a number of sweeping developmental events. Upon the completion of neurulation, organogenesis has begun, and a more familiar form takes shape, marking the onset of the fetus stage.

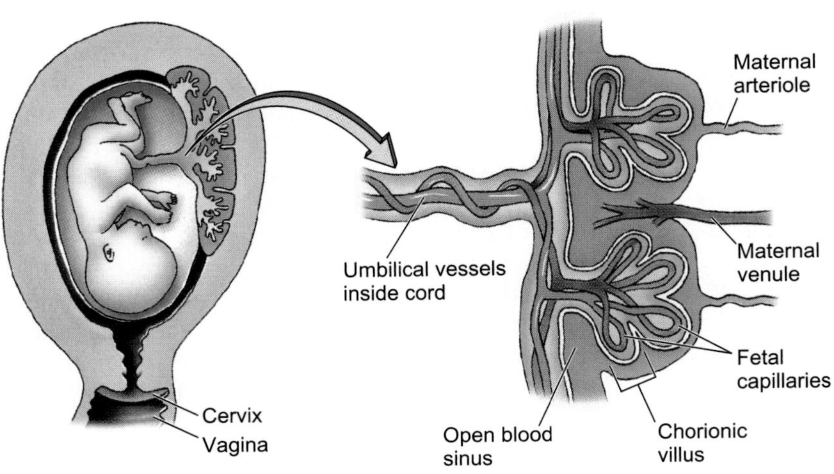

Figure 43.9 The Placenta. Eventually, the chorionic villi will have been well penetrated by blood vessels emerging from the umbilical cord. The villi lie within blood sinuses where the exchanges between mother and fetus go on.

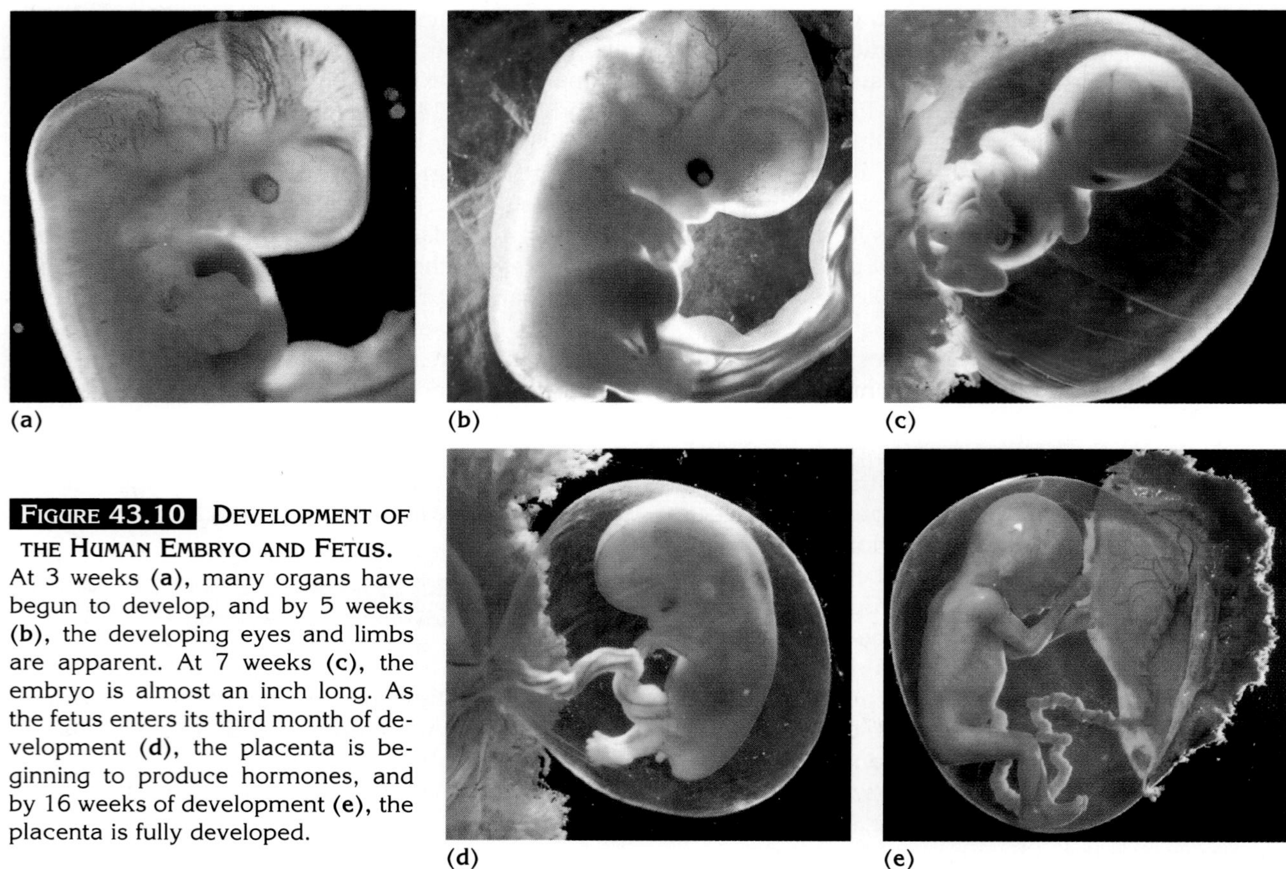

FIGURE 43.10 DEVELOPMENT OF THE HUMAN EMBRYO AND FETUS. At 3 weeks (**a**), many organs have begun to develop, and by 5 weeks (**b**), the developing eyes and limbs are apparent. At 7 weeks (**c**), the embryo is almost an inch long. As the fetus enters its third month of development (**d**), the placenta is beginning to produce hormones, and by 16 weeks of development (**e**), the placenta is fully developed.

In typical vertebrate fashion, the nervous system develops early. Neurulation begins at about day 18 or 19, closely following gastrulation. As with the amphibian embryo discussed earlier, in humans the neural folds arise on the embryonic disk, reaching upward and then folding together to form the dorsal, tubular nerve cord. Its anterior end enlarges to form the beginning of the brain.

The nervous system continues its rapid progression in the first trimester. While it is the earliest system to begin development, it will not be completed until long after the birth process—perhaps not until the moment of death—since learning is, in a sense, a development process. By the fourth week, the major regions of the brain and spinal cord are recognizable. When the first trimester ends, these are already well-defined. The still-smooth cerebrum now extends over much of the embryonic brain, and the cerebellum and medulla have become distinct.

As the neural ridges begin to break the contour of the human embryo, the heart and circulatory system make an early appearance. By day 22, the first timorous palpitations of the primitive heart begin. In vertebrates, this great organ is formed as cylinders of mesenchyme converge, producing a single tubelike structure. Within four to five days, the tube will have developed into a functional organ; it is crude, but it moves blood. As the blood is pushed along, it enters sinuses, forming channels that later become lined with endodermal cells, giving rise to blood vessels. Within another two weeks (or about 40 days from fertilization), the tubular heart will have looped back on itself, paving the way for its four-chambered pattern, which by now is nearly completed. An opening between the atrial chambers will remain until sometime after birth, since there is no reason to send blood to the lungs.

The respiratory and digestive systems develop fairly rapidly, so that by five to six weeks their basic patterns are clearly established. Once the basic tube has been outlined, a few blind pouches form, then more and more outpocketings, until finally the indistinct outlines of the gut, liver, and pancreas can be seen. The trachea, bronchi, and lungs begin as a small outpocketing in the pharynx. The outpocketing then branches to form the two lung buds, which will give rise to the first individual lung lobes.

The limbs of humans appear as rounded buds during the fourth week. The arms and legs are distinguishable at six weeks, but fingers and toes require an additional week. The rudimentary hands and feet actually begin as simple webbed paddles that take form through a kind of developmentally programmed cell death, as the tissue between the fingers and toes is broken down and absorbed.

The reproductive system begins to develop during the first few weeks, but until the eighth week even a trained observer can't determine the sex of the embryo (without a chromosome test). It is true that before this time the genitals have begun to develop, but the genitals of the two sexes start off in much the same way. The factors controlling sexual differentiation are complex. Certainly, the genetic sex of an individual is determined at fertilization by the chromosomal complement of the sperm cell. Fertilizing sperm cells that carry X chromosomes produce females, while those bearing Y chromosomes produce males. However, this is only a beginning. A number of genes are involved in sexual development, and its control may stem from different mechanisms, such as genetic or hormonal. For example, it appears that the Y chromosome of mammals initiates the development of the testes, and that the testes take over from there, producing hormones that determine the male's primary sexual characteristics.

The Second Trimester

In the second trimester, the fetus grows rapidly. By the end of the sixth month, it may be about a foot long, but it will weigh only about a pound and a half. Whereas the predominant growth of the fetus during the first trimester was in the head and brain areas, during the second trimester, the rapid growth of the body begins to catch up with the head.

By the fourth month, the fetus is moving vigorously, kicking and thrashing in its amniotic fluid—movements clearly felt by the mother. Interestingly, it must sleep now, and the thankful mother can also get some rest. (Unfortunately, they may not be on the same sleep schedule.) As time passes, the fetus becomes sensitive to more stimuli. For example, by the fifth month, the eyes are sensitive to light, although there is still no hearing. The synapses in the brain do not form until the 24th week, so there can be no integration of sensory input and no perception until this time. Other organs, such as the lungs, seem to be complete, but are still nonfunctional. The digestive organs are present, but cannot digest food. The skin is well formed, but it cannot adjust to any temperature changes. By the end of the fifth month, the skin is covered by a protective, cheesy paste consisting of wax and sweatlike secretions mixed with loosened skin cells. The fetus is still incapable, in nearly all instances, of surviving alone.

The skeleton has been developing rapidly during the second trimester, with some bones arising anew from undifferentiated embryonic cells, and others forming through the gradual replacement of cartilage cells by bone cells. Now, the mother must supply large amounts of calcium and other bone constituents as building materials for the fetal skeleton.

By the sixth month, the fetus is kicking and turning so constantly that the mother often must time her own sleep periods to coincide with its schedule. The distracting effect has been described as feeling somewhat similar to continually being tapped on the shoulder, but not exactly. The fetus now moves so vigorously that its movements can be seen clearly from the outside. To add to the mother's distraction, the fetus may even have periods of hiccups. It is now so large and demanding that it places a tremendous drain on the mother's reserves.

At the end of the second trimester, the fetus is clearly human, but it resembles a very old person because its skin is loose and wrinkled. In the event of a premature birth around the end of this trimester, the fetus may be able to survive.

The Third Trimester

During the third trimester, the fetus grows until it fills its available space and is no longer floating free in its amniotic pool; even in the greatly enlarged uterus, its movement is restricted. In these last three months, the mother's abdomen becomes greatly distended and heavy, and her posture and gait may be noticeably altered in response to the shift in her center of gravity. (However, occasionally a markedly overweight woman may go through much of this stage without realizing she is pregnant at all.) The mass of tissue and amniotic fluid that accompanies the fetus ordinarily weighs about twice as much as the fetus itself. Toward the end of this period, milk begins to form in the woman's mammary glands, which in the previous trimester may have undergone a sudden surge of growth.

At this time, the mother is at a great physical disadvantage in several ways. About 85% of the calcium she eats goes to the fetal skeleton, and about the same percentage of her iron intake goes to the fetal blood cells. Much of the protein she eats goes to the brain and other nerve tissues of the fetus.

During the third trimester, the fetus grows quite large. It requires more food each day, and it produces more wastes for the mother's body to carry away. Her heart must work harder to provide food and oxygen for two bodies. She must breathe now for two individuals. Her blood pressure and heart rate rise. The fetus and the tissues maintaining it form a large mass that crowds her internal organs. In fact, the fetus pressing against her diaphragm may make breathing difficult for her in these months. Several weeks before delivery, however, the fetus will change its position, dropping lower in the pelvis (a process called *lightening*), which relieves the pressure against the mother's lungs.

The "finishing" of the fetus proceeds rapidly in the last three months. Such changes are reflected in the survival rate of babies delivered by cesarean section (an incision through the mother's abdomen). In the seventh month, only 10% survive; in the eight month, 70%; and in the ninth month, 95%.

Interestingly, there is a change in the relationship of the fetus and mother during the last trimester. In the first trimester, measles and certain other infectious diseases would have affected the embryo. However, during the third trimester, the mother's antibodies confer an immunity to the fetus, a protection that may last through the first weeks of infancy.

About 255 to 265 days after conception, the life-sustaining placenta begins to break down. Parts of it begin to shrink and change, and the capillaries begin to disintegrate. The fetal environment becomes rather inhospitable, and premature births at this time are not unusual. At about this time, the fetus slows its growth and changes its position so that its head is directed toward the bottom of the uterus. Its internal organs undergo some final changes that will soon enable it to survive in an entirely different kind of world. So far, its home has been warm, sustaining, protected, and confining. It is not likely to encounter anything quite so secure again.

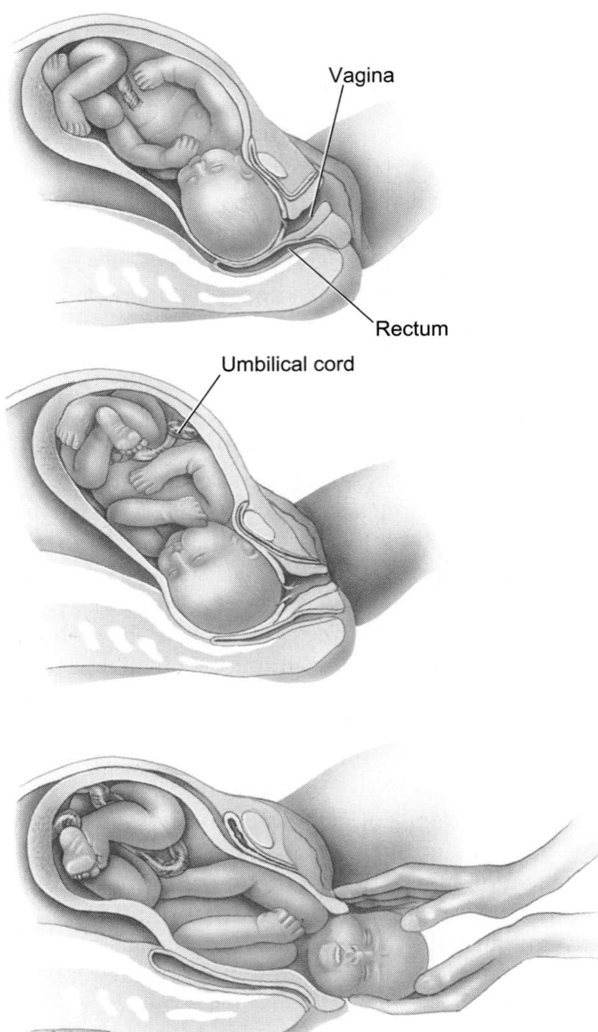

FIGURE 43.11 BEGINNING THE BIRTH PROCESS. The birth process begins as the cervix softens, followed by uterine contractions that push the baby's head into the vagina. Further contractions, that last from minutes to hours, move the baby through the birth canal into its new world.

BIRTH

It might be argued that the birth process is far too emotional and momentous an event to be analyzed clinically. Nonetheless, we're dealing with science here, so we'll begin by dividing the process into three stages (Figure 43.11). The first stage is *dilation*, marked by a softening and dilation of the cervical tissue. The period of dilation is highly variable, lasting from a few to many hours. It is accompanied by periodic contractions of the uterus called "labor pains." These contractions increase in frequency through the first stage.

Stage two is *expulsion*, during which the fetus is delivered. The expulsion process may last from a few minutes to hours. An anesthetic may be re-

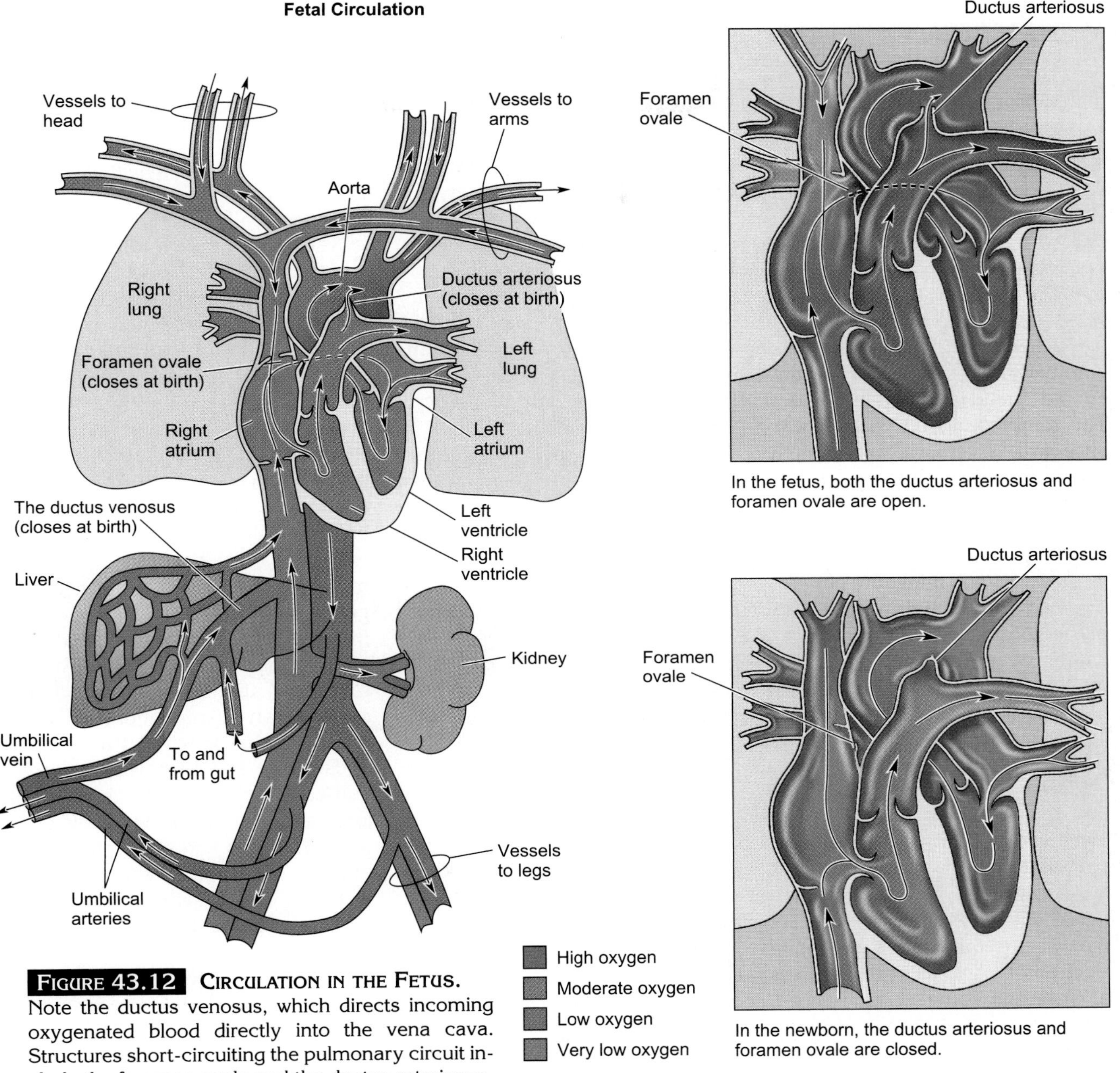

FIGURE 43.12 CIRCULATION IN THE FETUS. Note the ductus venosus, which directs incoming oxygenated blood directly into the vena cava. Structures short-circuiting the pulmonary circuit include the foramen ovale and the ductus arteriosus.

quired at this time. The pain is highly variable, depending on the mother's emotional state, pain threshold, cultural conditioning, and preparedness.

The final stage of birth is the *separation* of the placenta and its expulsion from the uterus as the afterbirth. In most sophisticated medical facilities, the afterbirth is examined for any abnormalities and for completeness of expulsion.

Physiological Changes in the Newborn

Because babies are propelled into a harsh new world with startling speed, their systems must be prepared for a new and more threatening existence. They have left the most secure world they will ever know. The vital exchange of gases, for example, for-

ESSAY 43.1 Aging

It somehow seems highly unlikely to the young, but under the best of circumstances, they will grow old—old and ugly, some say. The changes will be almost imperceptible, normally just a slow accumulation of small changes, a wrinkle here, a backache there. Perhaps the best barometer of the change is meeting an old friend after several years. The compliments on your appearance will be effusive and directly proportional to your state of deterioration. What brings about these changes? Why do our bodies let us down? What's happening anyway? Actually, several things are happening. Primarily, though, cells are simply becoming less efficient. They haltingly carry out their tasks of removing wastes, destroying poisons, repairing genes, and manufacturing proteins. As cells weaken, so do the bodies they serve.

There are a number of factors that may contribute to the problems of aging cells. There is an accumulation of free radicals (atoms with only a single electron in their outer shells). They may satisfy their shell requirements by taking electrons that were in the process of meeting some cellular need, such as generating ATP. Certain pigments may also accumulate and pose a threat. Lipofusion, a brown pigment produced by the metabolism of fat, may choke cells and impair their delicate functions. Even glucose can be a problem. In time, glucose can damage proteins, including collagen, one of the primary supportive materials of the body. Glucose can even clog the DNA helix and halt the orderly formation of proteins.

The causes of such changes remain a mystery, but there are two major theoretical approaches to the problem: (1) the time bomb theory and (2) the wear-and-tear theory. The time bomb theory is based on the idea that we are genetically programmed to fade out. The mechanisms include a genetic clock that causes healthy cell lines to die out after about 50 doublings (cancer cells do not). A hormone clock has also been suggested. It could operate, for example, if the pituitary gland began to secrete a "killer hormone" after puberty. (Aging has been retarded in some animals by removing the pituitary and dosing them with hormones artificially.) There may also be a group of clocks, all running at different rates, but interdependent, so that when one slows (perhaps due to some environmental change or signal), the others slow as well. (*Environmental*, in this sense, implies anything that is not internally based.)

The wear-and-tear theory suggests a genetic mechanism in which random mutations can be expected to eventually alter a few important genes that control an array of other genes or that manufacture certain vital proteins. The mutation rate, of course, can be increased by environmental abuses. The level of free radicals can also build up due to environmental abuses (and can also be reduced environmentally, some researchers say, by ingesting vitamins E and A). Finally, cells may give way to the continual assault from a number of sources, such as the constant (and increasing) barrage of radiation and chemicals that can cause an accumulation of small mutations that slowly destroy us.

The rate at which we age, no matter what the causes, seems to be largely genetic. Some people remain vigorous and alert into their eighties, while others show the effect of time much earlier and walk a degenerative and painful path to death. It has been suggested that the best way to live long and well is to choose an elderly set of healthy grandparents.

merly provided by the placenta, now must occur independently of the mother. The infant's respiratory system must function on its own for the first time, and it must do it right the first time.

The circulatory system must also change in order to accommodate the newly functioning lungs. After all, as we see in Figure 43.12, the fetus receives oxygenated blood from the placenta via the umbilical vein. From this vein it enters the inferior vena cava and then the right atrium. Since the mother provides oxygen across the placenta, there is no reason for the blood to travel to the collapsed, nonfunctional lungs. So instead, blood bypasses the lungs over two different routes.

First, an opening between the atria called the **foramen ovale** lets blood cross into the left atrium, and second, a vessel called the **ductus arteriosus** permits blood to pass from the pulmonary artery to the aorta. Of course, both must close at birth in order for the lungs to take over. The newly inflating lungs help solve the problem by suddenly permitting the free flow of blood into the pulmonary circuit. The return of this blood to the left atrium produces enough pressure to close the foramen ovale so that it can grow shut, and the ductus arteriosus soon constricts, eliminating this short circuit.

THE ANALYSIS OF DEVELOPMENT

It is indeed fascinating to watch the changes in a developing embryo. Tissues appear, grow, move, reorganize, and disappear in a genetically choreographed ballet that arouses wonder in us all. That wonder increases, though, when one moves from *when* and *what* happens in development to *why* and *how.* At this point, we start asking questions that many biologists view as the most challenging and exciting in developmental biology.

Such questions include, how does differentiation occur? Why do cells actually change? How are the developmental fates of cells determined? How do genes operate to produce the final phenotype? Why do we age (Essay 43.1)? We are often short on precise answers, either because we know too little, or, paradoxically, because we know too much. That is, we may have a great deal of specific information but haven't yet learned what to do with much of it.

TOTIPOTENCY AND SEALED FATES

Ultimately, the developmental program that unfolds in the embryo originates in the genome of the animal. Thus, it follows that the fertilized egg contains all of the hereditary material needed to carry out a total developmental program. We say, then, that the fertilized egg is *totipotent* ("all-powerful"): that is, it can give rise to any kind of cell. But as development continues, more and more cells commit to a specific fate—to becoming certain kinds of specialized cells. Some genes are deactivated, others activated, until only those that are required by the specific kind of cell are active.

But how permanent is this commitment? Can it be reversed? In other words, are the nuclear changes associated with development reversible? The question has been explored in great depth, beginning with certain fascinating experiments in the 1950s.

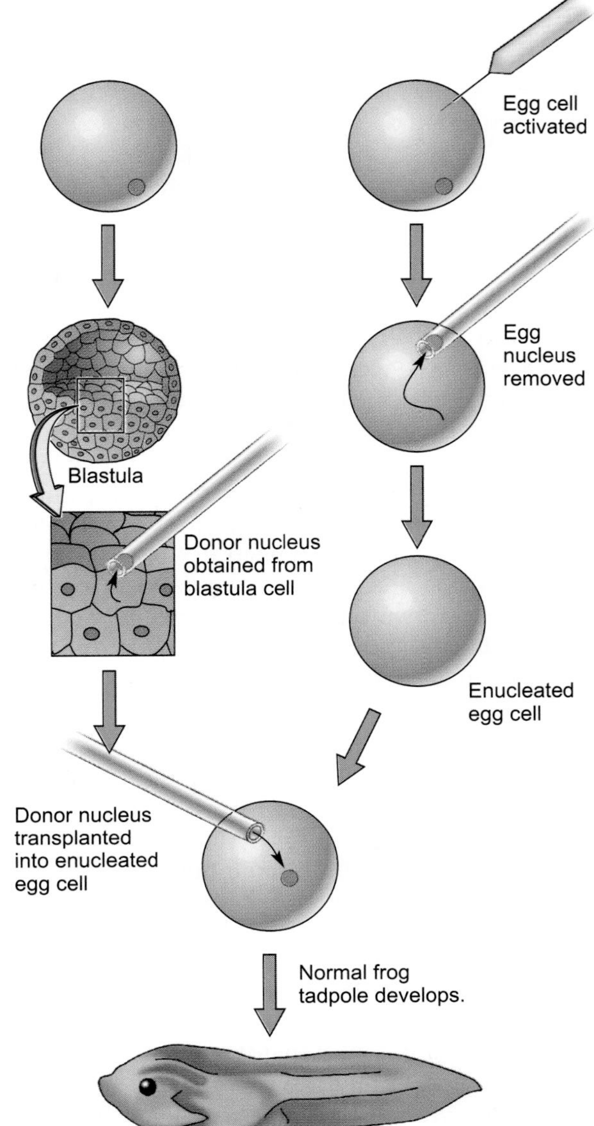

FIGURE 43.13 TOTIPOTENCY IN THE NUCLEUS. In their experiments, Briggs and King transplanted blastula nuclei into frogs' eggs whose nuclei had been removed. Most of the treated eggs developed normally.

Among the first of the great experiments of this era were those by R. W. Briggs and T. J. King, in which cells were enucleated (their nuclei removed) and then replaced by nuclei taken from other cells in various stages of development (Figure 43.13). First, an oocyte (egg cell) from a leopard frog (*Rana pipiens*) was pricked with a glass needle. This sort of stimulation will often induce some kinds of unfertilized eggs to undergo cleavage, in some cases producing a haploid individual. Once the oocyte began showing signs of mitotic activity, the membrane was punctured and the chromosomes removed. The result was an activated and enucleated egg.

Then, using a very fine glass pipette, Briggs and King carefully removed nuclei from other frog cells and inserted them into the enucleated eggs. They found that when the nuclei were obtained from cells of the frog blastula stage, most of the recipient eggs went on to develop into normal frogs. Obviously, blastula cell nuclei were still totipotent. Nuclei taken from older cells showed less potency, until nuclei taken from cells after the embryos had formed tail-buds would not prompt development in the egg cell at all. However, even at this stage, nuclei taken from germ cells (those that would have formed gametes) produced normal development in 40% of the oocytes. Briggs and King, then, told us two things. First, nuclei tend to lose their totipotency, their ability to form different kinds of cells, as time goes by. Second, nuclei of certain kinds of cells lose their totipotency before other kinds do.

The Role of Tissue Interaction in Determination

Much of what we know about how tissues interact during development has grown out of work done in the 1920s by Nobelist Hans Spemann and his protégée, Hilde Mangold. Their experimental subject was the amphibian embryo.

In experiments with the late blastula and early gastrula, Spemann found that the embryo could be divided through certain planes into two halves and still produce two normal embryos. The plane of division, he observed, had to pass through the dorsal lip of the blastopore (see Figure 43.6). If the cut was made along any other plane, only one normal embryo formed. Not surprisingly, interest suddenly focused on the dorsal lip.

In the next series of experiments, tissue from the dorsal lip of the blastopore from one blastula was transplanted to another embryo, specifically, to the recipient's ventral (belly) side (Figure 43.14). The transplanted tissue went on to form a perfectly good notochord, which in itself wasn't too surprising, since that is what it would have formed in the donor. But what was surprising was that a second neural plate formed over the notochord, and then, following the usual events of neurulation, a second rudimentary nervous system formed. The startling reorganization continued until the area of the transplant formed sort of a Siamese twin to the first (Figure

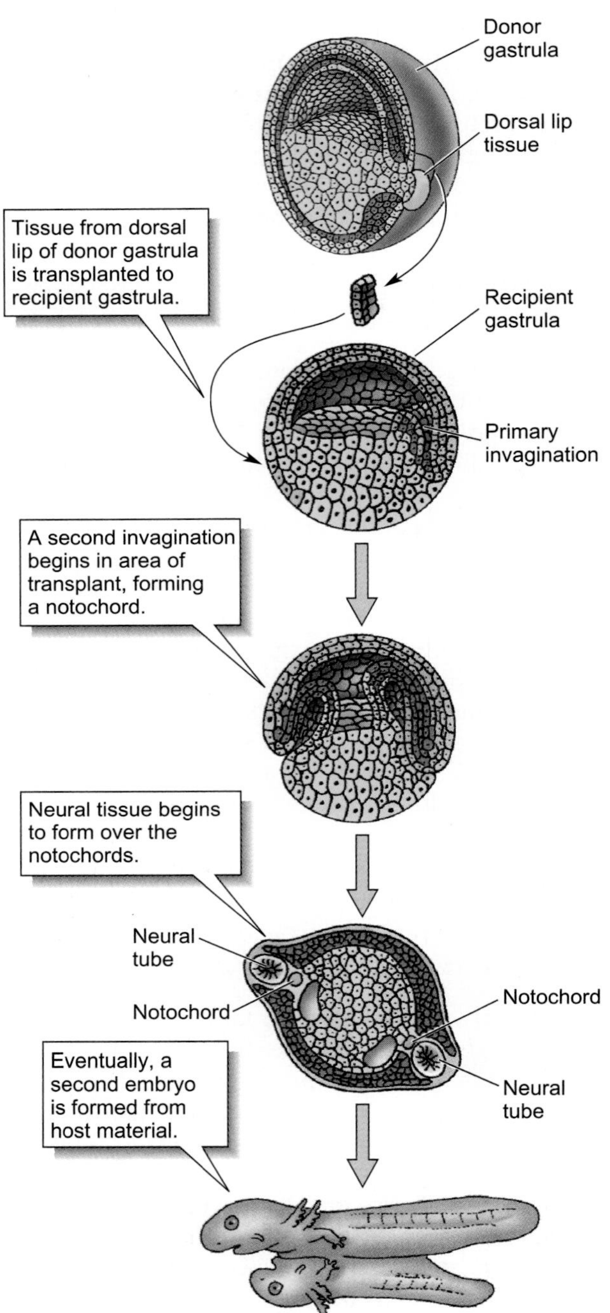

Figure 43.14 Induction in the Older Gastrula.
Spemann and Mangold found that the results of dorsal lip transplants varied with the age of the gastrula. When a source of the graft was an early gastrula, it produced the head region of the embryo, while grafts from the dorsal lip of substantially older gastrulas produced only tail regions.

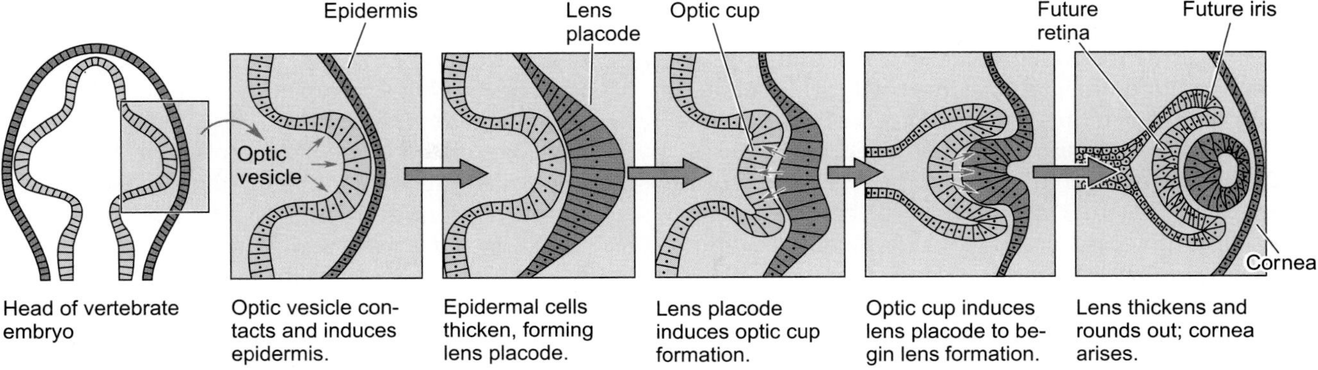
Epidermis
Lens placode
Optic cup
Future retina
Future iris
Optic vesicle
Cornea
Head of vertebrate embryo
Optic vesicle contacts and induces epidermis.
Epidermal cells thicken, forming lens placode.
Lens placode induces optic cup formation.
Optic cup induces lens placode to begin lens formation.
Lens thickens and rounds out; cornea arises.

FIGURE 43.15 TISSUE INDUCTION IN EYE DEVELOPMENT.
The lens of the eye is induced to develop from ectoderm by the influence of the brain underneath. In this case one kind of ectoderm influences the developmental pathway of another kind of ectoderm.

43.15). The question was, where did the material that formed the second embryo come from? Did it all come from grafted tissue?

Using dorsal lip transplant tissue from an amphibian that was highly pigmented, the researchers found the answer. The Siamese twin that developed was without the pigment of the donor. That meant it was formed from host tissues. The experiments were repeated again and again, but the results were always the same. Somehow the dorsal lip tissue was prompting the surrounding tissues to enter neurulation and go on to form the second embryo.

The researchers deduced that the dorsal lip tissue had an organizing role in normal development. Spemann therefore called the dorsal lip region the primary organizer. The ability of one tissue to *induce,* or determine, the developmental future of another became known as **embryonic induction.** Spemann and Mangold's work led to a veritable barrage of experiments, and it soon became clear that development involved many episodes of embryonic induction. One of the most complex involves the vertebrate eye, as we see in Figure 43.15.

The obvious question from the tissue interaction research is, what precisely is an organizer? Obviously, it is something that moves from the organizing tissue to the target tissue. Although we don't know what it is, we have given it a name. Such a substance is called a *morphogen.* Many substances have been suggested, from mRNAs to simple inorganic ions, all having had their movements tracked from one cell layer to the next as development progressed. Still, although we've got a name for it, no one is exactly sure what a morphogen is.

Recently, a chemical that has the ability to organize tissue along a new developmental pathway has been discovered. This putative morphogen, retinoic acid, is a derivative of vitamin A. When retinoic acid is dabbed on the limb bud of a chick embryo, it causes the formation of a second set of digits. This is the same result obtained by transplanting an extra "zone of polarizing activity" (ZPA) to the limb bud. The ZPA is a region near the attachment of the limb bud to the body that provides positional information to cells in the developing limb. Those cells nearest the ZPA appear to receive information telling them that they are located at the posterior end of the limb bud. Experiments have suggested that retinoic acid diffuses from the ZPA, and that its concentration gradient determines the anterior-posterior axis of the developing limb bud.

THE HOMEOBOX: MOLECULAR BIOLOGY OF DEVELOPMENT

Homeotic genes are genes that control the expression of blocks of other genes. Thus, their activity is critical in normal development. Molecular biologists have recently discovered that each of the homeotic genes contains a variable region of DNA and a constant, unvarying region of DNA, some 180 nucleotides long. The unvarying region of DNA that controls other genes they called a **homeobox.** Every organism has several homeotic genes. The amino acid sequence directed by the homeobox is very similar for the different homeotic gene loci. Because homeotic genes control so many other genes, a single homeotic change can have far-reaching effects. This greatly simplifies our perception of development, since it follows that perhaps a single environmental cue, something that alters a nucleotide in a homeobox, can greatly alter the developmental progress of the organism. We are confident that in the fruit fly, at least, such environmental influences can affect the homeobox.

Researchers working on homeotic mutations in *Drosophila* found that homeotic genes seem to control developmental events along the body length. In essence, they discovered that a homeotic mutation causes the body parts to appear in the wrong place. For example, a homeotic mutation in *Drosophila* can cause antennas to grow where the eyes should be, or extra pairs of wings to appear (Figure 43.16).

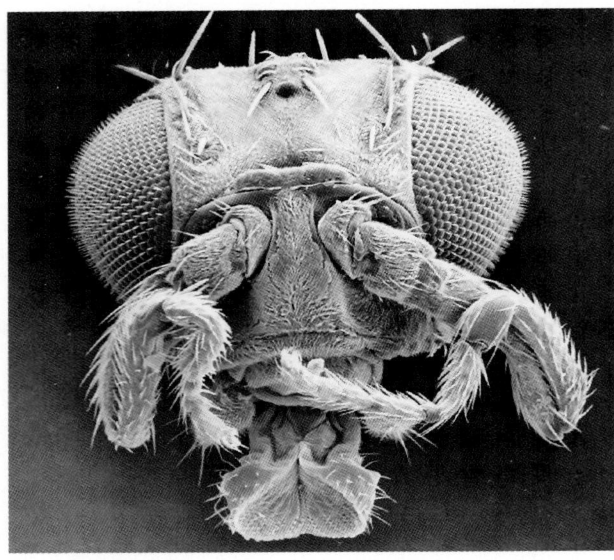

FIGURE 43.16 MUTATIONS CAN INFLUENCE DEVELOPMENTAL PROCESSES.
Here a mutation in a gene that governs the development of a specific organ has resulted in legs occurring where antennae should be.

Even greater implications arise from our new knowledge of the homeobox. Molecular biologists, using RNA probes containing the homeobox sequence, have found strikingly similar stretches of nucleotides in other insects. This discovery prompted the search for the sequence in other kinds of animals. The homeobox was then found in the genome of annelid worms. Not long after, to the great surprise of some, the homeobox was found in vertebrates, and a similar sequence has now been discovered in humans. Some theorists suggest that the homeobox occurs in all segmented animals.

How similar are the homeoboxes in different animal groups? The homeobox of the one gene complex of the fruit fly *Drosophila* is identical in 59 of its 60 amino acids with the homeobox of the frog *Xenopus*. Since the evolutionary lines of fruit flies and frogs must have diverged well over half a billion years ago, it looks as though natural selection has conserved the homeobox almost intact—and to biologists, this suggests that its function is precise and vital.

We've seen that cells in certain embryos pass through increasingly determined states, eventually to be committed to a specific fate in the embryo. We've also seen where determination has a strong genetic basis, involving the selective activation and inactivation of genes, and how specific DNA sequences called homeoboxes may play key roles in activating gene complexes that contribute to the formation of tissues and organs. In addition, we have found that homeoboxes probably react to external cues from their environment, and that the cellular environment in the embryo is a constantly shifting, changing entity.

PERSPECTIVES

The processes of fertilization, growth, and development are, indeed, remarkable. Each of us goes through these processes, which may partially explain our fascination with the newcomers to our small planet. We generally want them to get off to a good start, so a great deal of attention has been focused on ensuring normality through the developmental stages—but this area remains one of the most mysterious and compelling in biology.

We now move to another area loaded with mystery and intrigue—the area called animal behavior. This is a discipline with a long and contentious history, partly because some effort has been made to apply certain findings to the human condition.

Summary

Development includes growth (increase in the number of cells accompanied by an increase in body size), differentiation (cell and tissue specialization), and morphogenesis (emergence of form).

Gametes and Fertilization

Sperm structures include the head (chromatin and acrosome), midpiece (containing organelles including the centriole and mitochondria), and tail (flagellum).

Egg size depends on yolk quantity, which relates to mode of development. The egg shows polarity—that is, its cytoplasmic constituents are unequally distributed. At ovulation, the human egg is surrounded by cells that form the corona radiata, which merges with the zona pellucida.

Fertilization is the fusion of haploid pronuclei. Following attachment by the acrosomal process, echinoderm sperm are actively drawn into the egg by microvilli. Cortical granules rupture, and the fertilization membrane rises up, lifting the extra, unsuccessful sperm with it. Inside the egg, male and female pronuclei fuse.

Human fertilization requires thousands of sperm, but only one penetrates the egg. Upon completion of fertilization, a second polar body appears.

Early Development

A rapid series of mitotic cell divisions, called cleavages, follows fertilization and produces a solid ball of cells called a morula. Because of the egg polarity (differences in the hemispheres), cleavage produces cells that differ in cytoplasmic contents. This difference may influence the cell's developmental fate. Cleavage in protostomes is spiral and determinate, while in deuterostomes it is radial and indeterminate. Because of the large yolk reserves in the bird egg, cleavage occurs in one small region only. In frogs, more rapid cleavage at one end produces an active animal pole, whereas slower cleavage at the yolky opposite end results in a vegetal pole with larger cells.

A blastocoel cavity forms next inside the hollow blastula.

Gastrulation is an inward migration of cells resulting in an embryo with three germ layers: ectoderm, endoderm, and mesoderm. Each has its developmental role in producing body tissues.

The notochord, a mesodermally derived structure, lies along the embryonic axis. Neurulation adds to the axis by giving rise to the rudimentary central nervous system. Ectoderm thickens, forming the neural plate. Its edges rise up as folds and then close, forming a neural tube, which sinks below the surface. The wider end forms the brain. Somites, forerunners of the vertebral column and segmented muscles, form on each side of the notochord.

Human Development

From fertilization to implantation requires about 6 days, during which the human embryo progresses to the blastocyst stage. The inner cell mass forms the embryo while trophoblast cells provide for implantation and later form extraembryonic membranes.

Nutrient uptake is increased as fingerlike primary chorionic villi emerge from the trophoblast. Later, nutrients will enter the embryo through the placenta. The blastocyst releases a hormone that stimulates continued progesterone release, thus maintaining the assuring endometrium.

Cells of the inner cell mass form the amnion. Cells below form a fluid-filled yolk sac in which blood cells and vessels form.

Gastrulation and neurulation in the embryo follow, and by the fourth week the embryo takes on definite form. The emerging allantois contributes blood vessels to the placenta. Chorionic villi enlarge and are invaded by blood vessels, soon forming the matlike placenta.

At eight weeks the embryo is called a fetus. The placenta rapidly increases in size and soon secretes its own progesterone.

Further Human Development

All systems form during the first trimester. The heart, a simple tube at first, forms an S-shape, and the chambers emerge. With neurulation, the crude outlines of the brain arise, and by the fourth week the major regions and the spinal cord are recognizable. The heart begins as a simple tube, which by seven weeks has changed to a four-chambered pump. The digestive and respiratory systems emerge from outpocketings and refinements in the primitive gut.

Limb buds appear in the fourth week, with the limbs becoming discernible by six weeks. Fingers and toes go through a paddlelike stage. The reproductive system begins in an indifferent state, with both sexes having the genital tubercle. The second trimester is marked by rapid growth and the beginning of movement. During the third trimester, the fetus grows to fill the available space. Development is completed, and the fetus assumes a head-down position.

Birth

Stage one involves dilation of the cervix. In stage two, expulsion, the fetus passes through the birth canal, and in stage three, separation, the placenta (afterbirth) separates from the uterus and is expelled.

The first breaths are accompanied by rapid changes in circulatory structures that earlier shunted blood away from the fetal lungs. The foramen ovale, a flap-covered opening between atria, closes and later fuses. The ductus arteriosus, a vessel connecting the pulmonary artery and aorta, constricts and disappears.

The Analysis of Development

In the totipotent state, the cell nucleus can direct the cell or embryo along any developmental pathway. Nuclear transplant studies by Briggs and King revealed that the frog nucleus remains totipotent through the blastula stage but becomes increasingly less capable after that, and also that some cells keep their totipotency longer than others.

Spemann and Mangold transplanted tissue from the dorsal lip of the blastopore in the late blastula to the opposite side of a recipient blastula, and a second rudimentary nervous system arose. The transplanted tissue induced the host tissue nearby to undergo neurulation. The dorsal lip tissue became known as the primary organizer and its effect on the host as embryonic induction.

Homeotic genes have unvarying gene sequences called homeoboxes, through which homeotic gene control may be exercised. Thus, any change in a homeobox drastically affects development. Similar sequences are seen in many segmented animals, including humans, indicating the evolutionary conservancy of these gene sequences.

Key Terms

growth • **755**
cellular differentiation • **755**
morphogenesis • **755**
fertilization • **756**
fertilization cone • **756**
fertilization membrane • **756**
zygote • **757**
cleavage • **759**
morula • **759**
blastocoel • **760**
blastula • **760**
gastrulation • **760**
gastrula • **760**
neurulation • **760**
neurula • **760**
somite • **762**
blastocyst • **764**
inner cell mass • **764**
trophoblast • **764**
implantation • **764**
placenta • **764**
amnion • **764**
yolk sac • **764**
primitive streak • **764**
allantois • **765**
body stalk • **765**
umbilical cord • **765**
trimester • **765**
foramen ovale • **770**
ductus arteriosus • **771**
embryonic induction • **773**
homeotic gene • **773**
homeobox • **773**

Review Questions

1. Distinguish among growth, differentiation, and morphogenesis.
2. To what aspect of vertebrate reproduction and development does the amount of egg yolk relate?
3. Prepare a simple line drawing of the human egg at ovulation and label: nucleus, cytoplasm, corona radiata, zona pellucida, and plasma membrane.
4. Why are so many sperm required for human fertilization? What evidence is there that meiosis has been completed?
5. Summarize the events accompanying the formation of an amphibian morula, blastula, and gastrula.
6. Name the germ tissues that form during gastrulation. What is the significance of the germ tissues?
7. Starting with the neural plate, explain how the neural tube forms. To what two structures does it give rise?
8. Describe changes in the human embryo from time of conception to implantation. How many days does this involve?
9. Draw a simple diagram of the human blastocyst, labeling the inner cell mass, cavity, and trophoblast. What roles do the two cellular regions play?
10. How are the primary chorionic villi formed? What is their function?

11. List the three extraembryonic membranes, and describe their formation.
12. When do the following occur in the human: gastrulation, neurulation, heart formation?
13. Describe the structures of the fully formed placenta.
14. What is the significance of the eighth week of human development?
15. What functions do the foramen ovale and ductus arteriosus have in the fetus? What must happen to them directly after birth?
16. How did embryologists determine that many of the blastula's cells had already been committed to specific fates by the gastrula stage?
17. Describe Spemann's critical transplant experiment.
18. How did Spemann determine whether the newly formed nervous system originated from host or donor?
19. In what ways are the terms primary organizer and embryonic induction fitting?
20. What is the apparent role of the homeobox? Why might this be of such great importance to evolution?

BEHAVIOR AND ECOLOGY

Someone has said, "In order to understand how a cathedral is built, you must first know something about bricks and stones." *We have, indeed learned something about the bricks and stones of biology, and now it is time to bring the ideas together under the umbrella of ecology, first considering how the environment can shape behavior and ending with our own impact on the planet and its life.*

44

Animal Behavior

YOUR FOREBEARS PROBABLY KNEW THAT IF THEY CAME UPON A DEER *lying in the grass, they had a chance to catch it if they ran straight at the animal and, just before they reached it, veered off to the right or left. These animals almost always wait until a predator is upon them and then head off one way or another. The odds of guessing which way an animal will bolt are about 50 : 50, and early hunters probably reached the same conclusion from their own knowledge of animal behavior. Today, people may study animal behavior for other reasons, but underlying it all is a keen desire simply to know more about how the other animals behave. Such information has, from time to time, given us insights regarding our own behavior, as we shall see.*

An owl sits quietly on the beam of a barn, upright, eyes closed, until two mares pass by and one thumps the gate to the stall. The owl suddenly lunges forward, feathers abristle and eyes wide. The mares walk past and on out into the pasture. They stop, standing together head to tail, the tail of each swishing across the face of the other, driving flies away. Overhead a crow methodically beats its way toward the nearby woods, two small songbirds swooping and diving around it. As evening draws on, the songbirds fall silent as first one kind of insect sound, then another and another, fills the cool air. The mares are eating by now, but each immediately surrenders her mound of hay as the stallion resolutely approaches, his own hay unfinished.

The animals around us behave as animals will, each according to its tendencies, abilities, and perceptions. We often form easy explanations for their behavior, based on simple observation, but in so doing, we may miss the mark. For example, we may deduce that the owl lunged as a threat, that the horses stood so as to have insects brushed from their faces, and that the songbirds chased the egg-eating crow to protect their young. We can also imagine that the more aggressive stallion was making sure the mares were aware of his dominance, while the chorusing insects sang to attract mates in their own way. We may be absolutely correct, but there are two problems with such explanations. First, such reasoning suggests that each animal knew its behavior would achieve a desired result, and that may not be the case.

We must keep in mind that the animals may be behaving without intention and, instead, only responding in some programmed way to a specific stimulus. For example, when an owl is startled, it lunges. When songbirds see something of a certain shape and size, they attack. Something in the environment may trigger singing by insects, and the stallion just can't leave the mares alone. Also, we cannot know the adaptive response—the benefit—of any behavior without developing a hypothesis and testing it through careful experimentation. Does the behavior of the startled owl chase away intruders? Does the behavior of the stallion actually establish his rank? We can answer such questions only through careful testing. In this chapter, then, we'll see how some animals behave, how their specific behavioral patterns are important to their survival, and, finally, how we know. Some of our information, indeed, will be derived from rigid experimentation, while other conclusions will be more tentative, untested—based largely on observation and still open to analysis.

PROXIMATE AND ULTIMATE CAUSATION

In the Caribbean, each of several islands harbors its own species of woodpecker. The Guadeloupe woodpecker lives on the French resort island of Guadeloupe, the Hispaniolan woodpecker inhabits the island of Hispaniola (Haiti and the Dominican Republic), and the Puerto Rican woodpecker, you

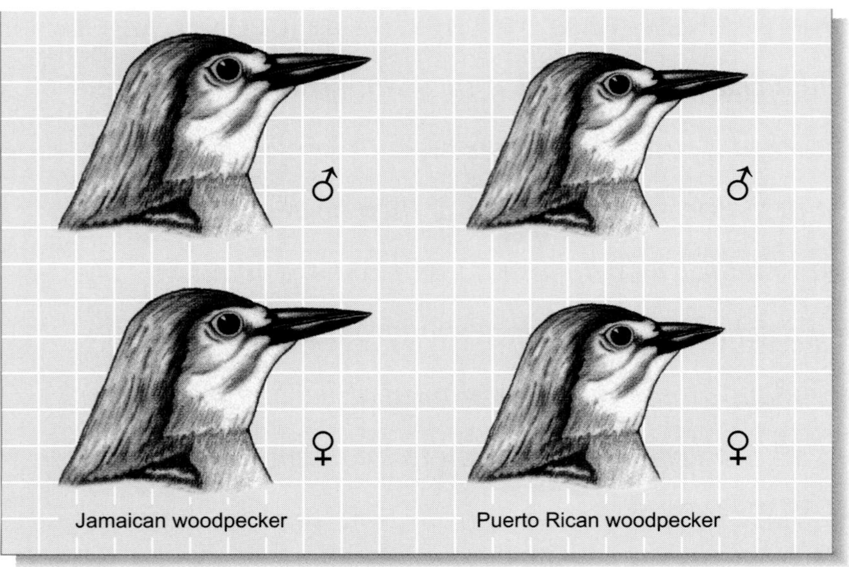

FIGURE 44.1 **JAMAICAN AND PUERTO RICAN WOODPECKERS.** The Jamaican birds are large, sexually monomorphic in bill size, and forage far apart. The Puerto Rican birds are smaller, sexually dimorphic in bill size, and forage closer together.

will not be shocked to learn, lives in Puerto Rico. In each of these species the male is larger than the female, and because of differences in bill size, the members of a pair take different foods and thus (since they don't compete with each other) are able to forage close together and to interact intensively.

Another species of woodpecker inhabits the island of Jamaica. The Jamaican woodpecker is slightly larger than its Caribbean cousins, and there is little sexual difference in body size or bill size (Figure 44.1). Both sexes take the same kind of food, and because of their intense competition for food, they must forage farther apart, interacting less. The Jamaican woodpecker vocalizes less than the other species, but when it calls, it calls more loudly.

The foraging behavior of the sexually monomorphic ("one form") Jamaican woodpecker and that of the sexually dimorphic ("two forms") species are clearly different. So we may ask about the *causes* of such differences. The answers may be formed in two ways.

Proximate causation is an explanation involving the more immediate bases for a behavior, particularly psychological or physiological mechanisms. Proximate questions usually involve *how* those mechanisms cause an animal to behave in a certain way. So we might answer the question of how sexually monomorphic species come to forage farther apart by considering whether the sexes have an innate (inborn) aversion for each other, or if they learned through experience that they could find more food if they competed less by looking for food in different places. We would be asking what sorts of internal processes are going on within the body of the animals that result in the performance of a certain behavior.

Ultimate causation is an explanation involving the evolutionary and adaptive bases for a behavior. Ultimate questions involve *why* an animal behaves in a certain way. In other words, ultimate questions ask why the proximate mechanisms evolved to begin with. Thus, if we try to answer—in ultimate terms—why sexually monomorphic and sexually dimorphic species came to forage differently we might consider whether Puerto Rican woodpecker food sizes were highly variable, but sparse, so that smaller-billed females could take the smaller food and larger-billed males the larger food as they foraged side by side. Or perhaps whether predatory pressures are different on the two types of animals. Maybe the sexually monomorphic Jamaican species evolved a weaker sexual attraction because if male and female foraged closer together, they would be more likely to attract the attention of hawks. The separate feeding behavior, then, could have been the result of predation or food sizes—both ultimate factors. (If you should propose that the sexually dimorphic species foraged close together as a means of repelling predators, would you be considering ultimate or proximate factors? Can you think of a way to test your hypothesis?)

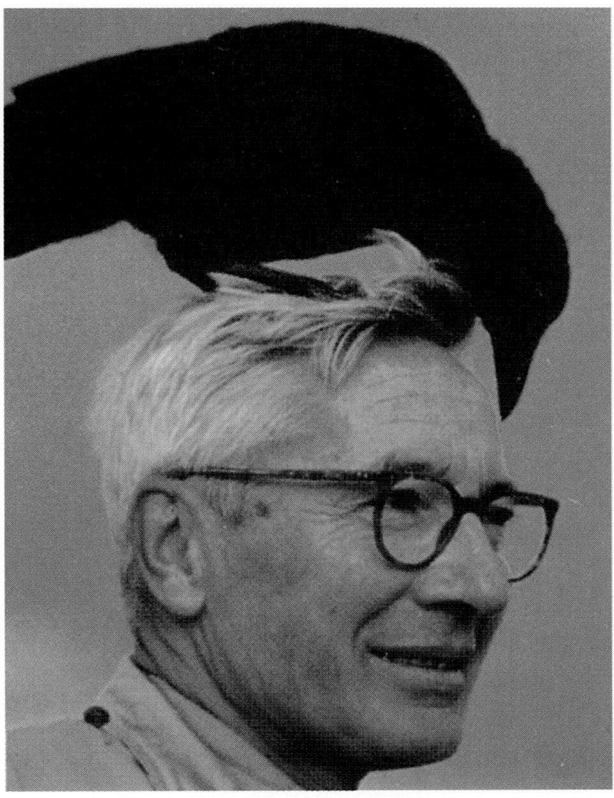

FIGURE 44.2 NIKO TINBERGEN. A founder of modern ethology.

ETHOLOGY AND COMPARATIVE PSYCHOLOGY

The modern study of animal behavior has taken two distinct routes in this century—one European, one American. Only in recent decades have the grand ideas begun to merge, blur, fuse, and finally crystallize into an encompassing discipline.

The European approach to the study of animal behavior was called **ethology.** Its goal was to understand behavior by studying its cause, development, evolution, and function through the observation of animals in the wild, or under somewhat natural conditions. It was the ethologists, particularly Niko Tinbergen (Figure 44.2) and Konrad Lorenz (Figure 44.11), who developed the idea upon which the modern concept of instinct is based. Their ideas were forcefully stated, for the first time, in 1951.

The American approach was called **comparative psychology.** Its goal was to understand behavior by studying animals in the laboratory under carefully controlled conditions. Its focus, at mid-century, was learning, and its primary animal was the Norway rat.

Unfortunately, the two groups were constantly at odds, especially in the early years. However, as each field matured and as researchers learned more about each other's work, ethologists came into the lab, and comparative psychologists learned where birds go in winter. Young students of animal behavior, particularly, are well-versed in either approach these days.

Let's now look at the classical ethological concept of instinct, and at the comparative psychology traditional view of learning. Then let's see how the two ideas have merged, forming the more modern explanation of animal behavior.

THE DEVELOPMENT OF THE INSTINCT IDEA

It was once believed that human behavior results entirely from learning, but that other species behave according to "instinct." But the term *instinct* was never precisely defined, and because of this imprecision, the entire concept fell into disrepute. Many behavioral scientists began to hesitate even to use the word. The problem was compounded by the fact that, even as ethologists struggled to clarify the concept, it found its way into the general population where it was twisted around, tossed about, and used to explain everything from a baby's grip to homemaking and swimming.

Today, **instinct** is defined as an inborn, unlearned behavior that involves *releasers, innate releasing mechanisms,* and *fixed action patterns.*

According to ethologists, instinctive behavior is triggered by the perception of certain very specific signals from the environment. Environmental signals that trigger instinctive behavior are called **releasers.** The releaser itself may be only a small part of any appropriate situation. For example, fighting behavior may be released in territorial male European robins not only by the sight of another male but even by the sight of a tuft of red feathers at a certain height within their territories (Figure 44.3). Of course, such a response is usually adaptive because tufts of red feathers at that height are normally on the breast of a competitor. The point is that the releaser may be only a small part of the environment that is perceived.

Following ethological theory, there are certain centers in the brain called **innate releasing mechanisms (IRMs),** which, when stimulated by impulses

FIGURE 44.3 RELEASER.
A male European robin in breeding condition will attack a tuft of red feathers placed in his territory.

The releaser is perceived by a receptor which activates the IRM.
Releaser
The IRM activates certain muscles, producing instinctive stereotyped movement

FIGURE 44.4 SIMPLIFIED DIAGRAM OF HOW A FIXED ACTION PATTERN CAN BE TRIGGERED.
The releaser is perceived by some sort of receptor, which triggers the IRM to activate certain muscles, thereby producing an instinctive movement that usually involves fixed action patterns.

set up by the perception of a releaser, trigger instinctive behavior. The instinctive behavior usually involves fixed action patterns. As Tinbergen described it, the IRM acts as a kind of filter that ignores all but the appropriate releaser, and when it is activated, it removes the inhibitions that lead to the performance of the stereotyped, fixed behavior called the fixed action pattern.

Fixed action patterns are movements that are innate (inborn, appearing without training), independent of the environment, stereotyped (always performed the same way), and characteristic of a given species (Figure 44.4). Animals are born with such patterns indelibly stamped into their behavioral repertoire. For example, birds build nests by using peculiar sideways swipes of their heads to jam twigs into the nest mass. And all dogs (and even birds) scratch their heads the same way, by moving their rear leg outside the foreleg (Figure 44.5). When a releaser triggers an IRM that allows signals to be sent to the muscles, thereby producing a stereotyped behavior, we say that an instinctive act has occurred.

LEARNING

The very word *learning* is revered among humans. (In fact, it's what you're supposed to be doing at this very moment.) We even like to see it in other species. ("I tell you, Chester, that dog's *smart*.") But what is learning? **Learning** is a change in behavior brought about by experience.

The problem is, we don't know very much about how learning occurs, in spite of a great deal of research effort in that direction. Not only are we vague on how it occurs in humans, but each species seems to have its own learning propensities, and each may learn in its own way; so establishing general principles is exceedingly difficult.

Although there may be a number of fundamentally distinct ways animals learn, we will consider three major types: *habituation, classical conditioning,* and *operant conditioning.* Then, after noting how instinct and learning can interact, in the modern view, we will look at the curious phenomenon called *imprinting.*

Habituation

Habituation is learning *not* to respond to a stimulus. The first time an animal encounters a stimulus, it may respond vigorously. But if the stimulus is presented over and over without consequence, the response to it gradually lessens and may finally disappear altogether (Figure 44.6). Habituation is not necessarily permanent, however. If an animal habituated to some stimulus does not encounter it for a period of time, the animal may respond if the stimulus later reappears.

Habituation may be quite important in the lives of many animals. For example, a bird must learn not to waste energy by taking flight at the sight of every skittering leaf. A coral-inhabiting fish may come to accept its neighbors, but will immediately attack a strange fish (Figure 44.7). Such behavior is adaptive because whereas neighbors are likely to be established on their own territories, a stranger might be seeking to displace the resident fish. Thus, the response is beneficial. Habituation may also help to explain why animals respond to the sight of a furtive predator they see only rarely, while ignoring the harmless species they see more often. Habituation is often not given the attention it deserves in studies of learning, perhaps because it seems so simple. But it may well be one of the more important learning phenomena in nature.

Classical Conditioning

Classical conditioning was first described by the famed Russian biologist Ivan Pavlov. In classical conditioning, the response to a normal stimulus comes to be elicited by a substitute stimulus. In his experiments, Pavlov found that dogs would salivate at the sight of food. He then began to switch on a light five seconds before food was dropped onto the feeding tray. After doing this a few times, he presented the light without the food and found the dogs would salivate in response to the light alone. Pavlov found that the number of drops of saliva elicited by the light alone was in direct proportion to the number of previous trials in which the light had been followed by food.

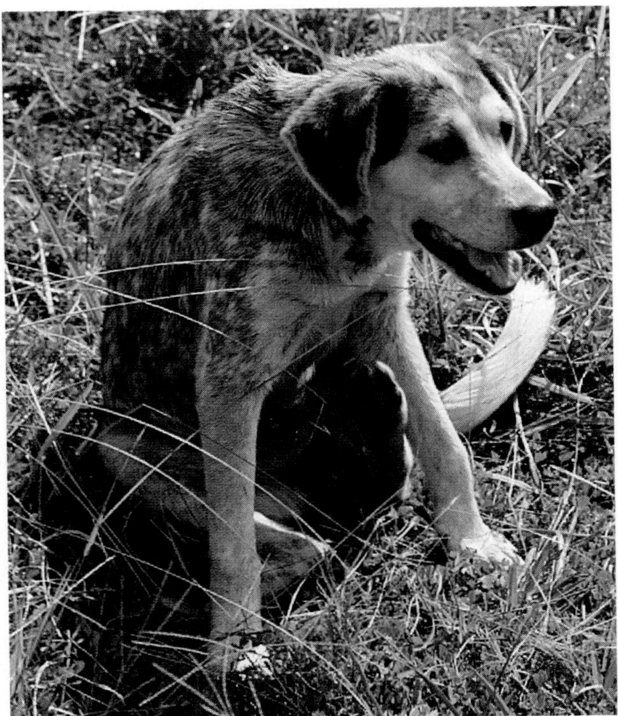

Figure 44.5 Fixed Action Pattern.
A fixed action pattern can be seen in the scratching movements of dogs, all done in the same way with the hindleg outside the foreleg. Interestingly, many birds show the same pattern.

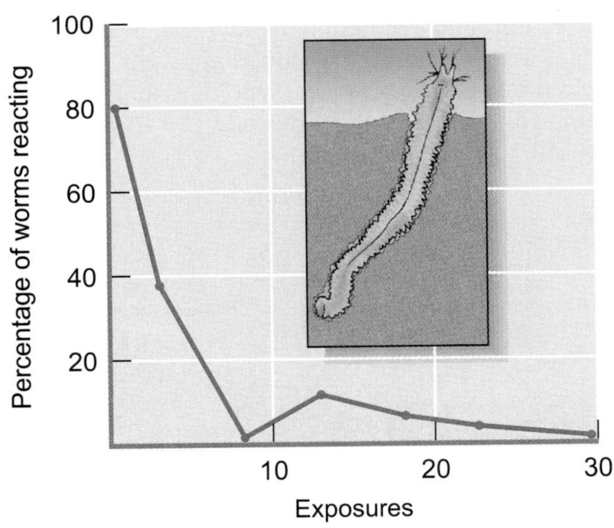

Figure 44.6 Habituation.
The marine worm *Nereis* withdraws into its protective tube if a shadow passes over. With repeated exposures to the stimulus, however, the response wanes.

FIGURE 44.7 **TERRITORIAL CORAL FISH.**
The fish become habituated to neighbors but will immediately attack a stranger.

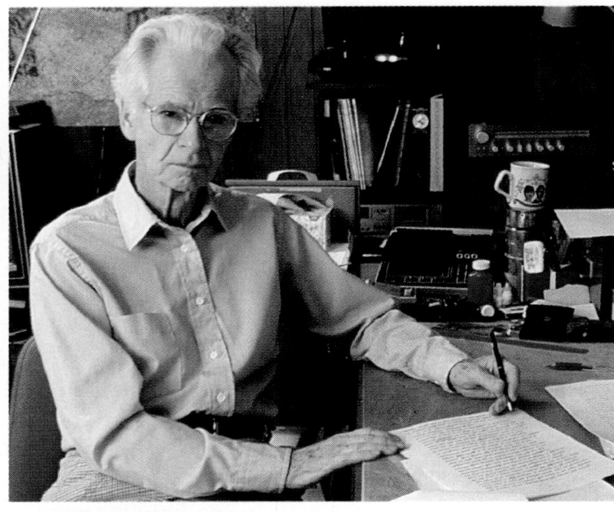

FIGURE 44.8 **B. F. SKINNER.**
Skinner, one of the most important twentieth-century psychologists, building a "Skinner box," which is used to demonstrate operant conditioning.

OPERANT CONDITIONING

Operant conditioning is learning to perform an act in order to receive a reward. It differs from classical conditioning in several important ways. Whereas in classical conditioning the reinforcement, such as food, follows the stimulus, in operant conditioning the reward follows a particular behavioral response. Also, in classical conditioning the experimental animal has no control over the situation. In Pavlov's experiment, all the dog could do was wait for lights to go on and food to appear. There was nothing it could do one way or the other to make it happen. In operant conditioning the animal's own behavior determines whether or not the reward appears.

In the 1930s, the noted psychologist B. F. Skinner demonstrated operant conditioning by employing a device now called a *Skinner box* (Figure 44.8). An animal placed inside a Skinner box must learn to press a small bar in order to receive a pellet of food from an automatic dispenser. Skinner found that when an experimental animal (usually a hungry rat) was first placed in the box, it ordinarily began a random investigation of its surroundings. When it accidentally pressed the bar, a food pellet was delivered. The animal did not immediately show any signs of associating the two events, bar pressing and food, but in time it began to hang around near the bar. As food pellets repeatedly appeared immediately after the bar was pressed, the animal's behavior became

CLASSICAL CONDITIONING

Food (unconditioned stimulus) → Salivation (unconditioned response)

Food & Light (conditioned stimulus) → Salivation

Light → Salivation (conditioned response)

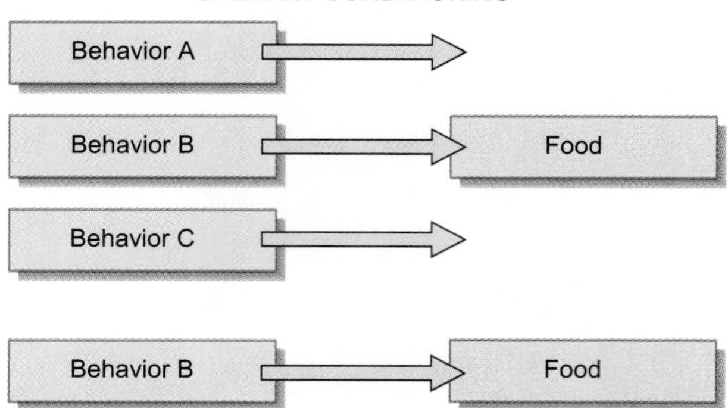

FIGURE 44.9 **CLASSICAL AND OPERANT CONDITIONING.**
In classical conditioning, a desirable commodity, such as food, comes to be associated with an irrelevant signal until the irrelevant signal alone can elicit an involuntary response normally associated with the commodity. Here, the animal learns passively. In operant conditioning, the animal can act when given a signal, but only one action is rewarded. This action then comes to predominate.

less random until finally it learned to press the bar to obtain food. Eventually, it spent most of its time just sitting and pressing the bar. Skinner called learning through such a sequence operant conditioning. (The essential differences between classical and operant conditioning are shown in Figure 44.9).

The relative importance of each type of learning to animals in the wild isn't known at this point. It is likely that most adaptive, or beneficial, behavior patterns arise in nature as interactions of several types of learning.

How Innate Behavior and Learning Can Interact

At one time it was acceptable to ask if a certain behavior was innate (inborn) or due to learning. Today, however, the question is regarded as a bit simplistic. In a sense, it's like asking whether the area of a triangle is due to its height or its base. A better question is: How do innate and learned patterns *interact* to produce an adaptive behavior?

Figure 44.10 Learning and Practice. Although young birds have the innate ability to fly, they can improve with practice. Here, an adult eagle has landed gracefully, followed by a younger bird that crashes headfirst into the ground.

Consider the development of flight in birds. Flight is usually considered a largely innate pattern. A young bird must be able to manage it pretty well on the first attempt or it will crash to the ground as surely as would a launched mouse. It was once believed that the little fluttering hops of nestling songbirds were incipient flight movements, and that the birds were, in effect, learning to fly by practicing before they left the nest. But in a set of experiments, some nestlings were allowed to flutter and hop, while others were reared in boxes which prevented any such movement. Then, at the time the young birds would have normally begun to fly, both groups were released. The restricted birds flew just as well as the ones that had practiced!

In at least some species, then, flight behavior is largely an innate, or unlearned, pattern. However, learning *is* involved in flight. Generally, young birds do not fly as well as adults (Figure 44.10). With flight, as with many other forms of behavior, an innate pattern can be improved upon by learning through practice.

We should keep in mind, then, that there may be both innate and learned components to any behavior. A rather remarkable kind of learning that involves both these components is called *imprinting*.

Imprinting

On your annual visit to the farm, you may have seen young ducklings waddling along after their mother, perhaps on the way to the pond. It's a quaint sight, but you're there to see where milk comes from and so you think no more about it. However, if you had visited a farm in southern Germany some years ago, you might have seen a more unusual sight—a column of young ducklings following a white-haired Austrian down to the pond. The man was Konrad Lorenz, a future Nobel prize winner, and the following behavior of the ducklings was the product of an experiment he devised.

In his experiment he let the ducklings see him moving around and making noises a few hours after they had hatched. If they had not seen and heard him until later, they would have treated him like any other human. But Lorenz was the figure they encountered during their "critical period," a window of time when the young are particularly sensitive to certain aspects of their environment. During this period, ducklings—and the young of many other species as well—learn the traits of whatever is around them. Normally, of course, this would be

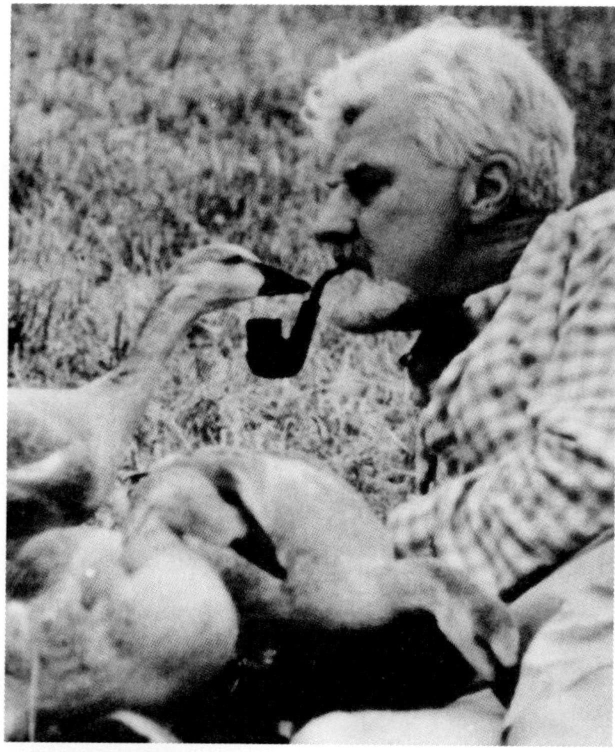

FIGURE 44.11 **KONRAD LORENZ.**
The Nobel Prize-winning ethologist hand-reared a jackdaw and became a foster parent to the bird.

FIGURE 44.12 **A CRANE IMPRINTED ON HUMANS.**
Tex, the only female whooping crane at the international Crane Foundation breeding area in 1982, had been hand-reared and therefore had imprinted on humans. She rejected the mate provided for her, but could be enticed to lay eggs (artificially fertilized) by "dancing" with humans. She preferred Caucasian men of average size with dark hair.

their mother; but these ducklings developed from eggs that were artificially incubated, and so the figure that they saw during their critical period was Konrad Lorenz, and forevermore they would regard him as one of their own.

Lorenz called this kind of learning **imprinting,** and he defined it as learning that occurs over a defined, relatively brief period of time in which an animal learns to make a specific response to certain aspects of its environment.

A great deal of research has been done on imprinting in the last few decades, and we have learned that it is difficult to draw hard-and-fast rules about its development. It is regarded, though, as a curious interaction of learned and innate patterns. In the case of the ducklings, they learned the general characteristics of the stimulus that would serve as a releaser of their instinctive following behavior.

Many animals also learn species identification during this critical period—that is, they learn the image of an appropriate mate. As they approach their first breeding season, they seek out an individual with traits generally similar to those of the individual they followed soon after hatching. If they are raised by parents of another species, when it is time to breed, they focus on individuals with traits similar to those of their foster parents. Lorenz once had a tame jackdaw (a European crow) that he had hand-reared, and it would try to "courtship feed" him during mating season (Figure 44.11). On occasion, when Lorenz turned his mouth away, he would receive an earful of worm pulp! The story of Tex, the dancing whooping crane, provides another example of this type of learning (Figure 44.12).

BEHAVIORAL ECOLOGY

Behavioral ecology is the study of how the environment affects behavior. Remember, the environment can be viewed in the historical sense (leading to ultimate causation) or the immediate sense (leading to proximate causation). Behavioral ecology can involve studies of the physical habitat, interactions with other individuals, or less conspicuous influences on

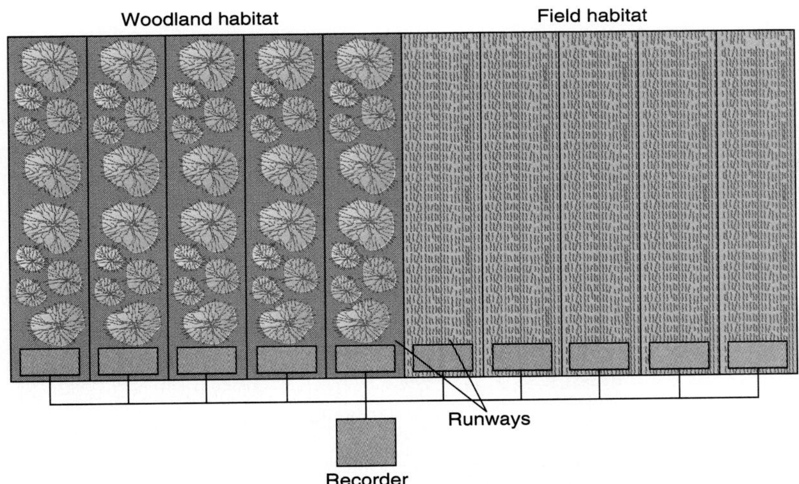

FIGURE 44.13 WECKER'S EXPERIMENTAL SETUP. Mice with different genetic makeups and different experiences were tested to see which kind of habitat they preferred. They were allowed to enter either simulated grassland or forest conditions from a runway. The results are described in the text.

behavior, such as weather, oxygen levels, and humidity—or any environmental variable that can influence behavior. As examples, we will consider the behavioral ecology of how animals choose a habitat and how they get food.

HABITAT SELECTION

One of the questions regarding habitat selection is: Does an animal choose its habitat based on genetic propensities, or through learning which kind of place is best to live in? Stanley Wecker conducted a series of fascinating experiments on prairie deer mice to see to what degree their habitat preferences were generally influenced. In his experiments, he used two populations of prairie deer mice—one that lives in woodland, the other in grassland. Wecker had found that, in the laboratory, the grassland mice *could* do quite well under simulated forest conditions. He wondered, then, why do they prefer grasslands in the wild? In one of his experiments he tested two groups of grassland mice, one reared in the laboratory, the other wild. The laboratory group was further subdivided into two groups—one of which had been reared under simulated grassland conditions, the other under simulated forest conditions.

Not surprisingly, Wecker found that the wild-caught field mice preferred the field-end of the enclosure (Figure 44.13). The "field"-reared laboratory mice made the same choice. However, the "forest"-reared laboratory mice were as likely to prefer one place as the other. Subjecting them to the forest had obliterated any tendency to move toward the field but had not caused them to prefer the forest. Wecker's experiment suggested, then, that habitat preference in prairie deer mice, at least, has a strong genetic component that can be influenced by experience (Table 44.1).

We can also imagine that the field mice would develop certain preferences while in the field, based primarily on experience. For example, it wouldn't take long to learn not to climb around in thorns, that certain kinds of grasses taste bad while others are quite palatable, and that sleeping with ants doesn't work. By experience, then, the mouse would fine-tune its behavior through proximate factors. The

TABLE 44.1 THE RESULTS OF WECKER'S EXPERIMENT.

Number of Mice Tested	Hereditary Background	Early Experience	Habitat Preference
12	Grassland	Grassland	Grassland
13	Laboratory	Grassland	Grassland
12	Grassland	Laboratory	Grassland
7	Grassland	Forest	Grassland
13	Laboratory	Laboratory	None
9	Laboratory	Forest	None

Source: *Data from S. C. Wecker. "The Role of Early Experience in Habitat Selection by the Prairie Deer Mouse, Perimaniculatus bairdi," Ecological Monographs 33:307–325.*

principle here might be phrased as: Ultimate factors help to set limits, and proximate factors influence the behavior of the animal within those limits.

Foraging Behavior

When asked why he robbed banks, the infamous Willie Sutton is reputed to have replied, "Because that's where the money is." For years, biologists seemed to have tacitly assumed that the same reasoning applies to where animals forage or search for food: They forage where the food is.

However, that answer has been found wanting as behavioral ecologists have asked not only why animals forage where they do, but how they make food choices once there.

Of course, different species forage in different ways. Basically, animals can be described as either generalists or specialists. **Generalists** are those species with a broad range of acceptable food items. They are often opportunists and will take advantage of whatever is available, with certain preferences, depending on the situation. Crows are an example of feeding generalists; they will eat anything from corn to carrion. **Specialists** are those with narrow ranges of acceptable food items. Some species are extremely specialized, such as the Everglade kite, or snail kite, which feeds almost exclusively on freshwater snails. There is a wide range of intermediate types between the two extremes, and in some species an animal will switch from being one type to being another depending on conditions, such as food availability or the demands of offspring.

Whatever the strategy, the question is, how does an animal forage most efficiently? More precisely, how does it maximize its gain relative to its expenditure? The gain is measured quite simply as the food value of the item, measured as the energy it contains. Expenditure is measured two ways—by the energy the animal spends in searching for food, and the energy it expends in handling the food once it has been found.

Obviously, the animal will maximize its foraging success by spending the most time where the most food is. But is this what they do? J.N.M. Smith and H.P.A. Sweatman examined the feeding behavior of the great tit (see Figure 44.14) by setting up a series of grids in an aviary. Each grid contained different densities of mealworms, a favorite food. The food density in each grid could be altered by experimenters. When a certain grid continued to hold the greatest food density for a time, the researchers found that the tits were soon spending almost all their time on that grid. However, the birds continued to hop over and sample from the other grids. The researchers wondered why the birds would waste time in these suboptimal sites, but the answers soon became clear. When the food density was suddenly reduced on the best grid, the birds immediately switched to the second-best grid. They spent little time in making a decision when their primary food source failed. Sampling all the grids, even after the best site had been determined, was adaptive after all.

FIGURE 44.14 THE GREAT TIT.
Parus major, the great tit, has been the object of a great deal of study in Great Britain.

Another question is, how do animals maximize their foraging efficiency when they are given a choice of food items with different food values? For example, all things being equal, larger food items are more profitable than smaller ones. Thus, the animal will be more successful if it takes larger items. Bluegill sunfish (Figure 44.15), for example, feed largely on *Daphnia,* the small crustaceans known as water fleas. Researchers found that sunfish will bypass very small water fleas, even if they are nearby, in favor of large water fleas farther away. In other words, they behave as if the extra travel were worth the gain.

Sunfish, of course, don't have to wrestle with the *Daphnia* once they find it, but striped bass may have to consider ease of handling. Bass feed on a variety of foods, including small fish called "shiners" and crayfish. Shiners are taken with a gulp, usually after a short chase; so after the bass catches the shiner, handling it is minimal. The crayfish, on the other hand,

FIGURE 44.15 **BLUEGILL SUNFISH.**
As with many other species, bluegill sunfish forage according to principles of efficiency.

are easier to catch but harder to handle. Not only do the little rascals fight back, requiring careful manipulation by the fish, but much of its body is indigestible chitin, which further reduces its value. So all things being equal, the bass is better off chasing the shiner.

Why, then, are crayfish taken at all? A predator's decision is often based on food availability. It will tend to take the items with the greatest net gain until their numbers are depleted to the degree that less desirable but more numerous food items become more attractive. Then it will generally begin to switch to the less desirable items—in which the food value may not be as high, but less time and energy are spent in searching.

Foraging, for most species, then, is a cost-benefit proposition. They tend to maximize the benefit while minimizing the cost to themselves. Since such studies have been undertaken, researchers have often been amazed at what appear to be analytical abilities of foragers. Of course, any such ability is largely programmed genetically, but learning may also play an important role. For example, young fish are often only fair optimal foragers, but with age and experience their efficiency improves.

We will now turn our attention to two phenomena that are critical to the adaptation of life on earth. They have to do with how animals adjust with respect to time and place.

ORIENTATION AND NAVIGATION

South Americans probably wonder where their robins go every spring. The robins, of course, are "our" birds; they simply vacation in the south each winter. Furthermore, they fly to very specific places in South America and will often come back to the same tree in your yard the following spring. So, how do robins and other birds find their way around? The question persisted for years, until, in the 1950s, a German scientist named Gustav Kramer provided some answers, and, in the process, raised new questions about orientation and navigation, two different phenomena. **Orientation** is simply facing in the right direction; **navigation** involves finding one's way from one place to another.

Early in his research, Kramer found that caged migratory birds became very restless at about the time they would normally have begun migration in the wild. Furthermore, he noticed that as they fluttered around in the cage, they often launched themselves in the direction of their normal migratory route. He then set up experiments with caged starlings and found that their orientation was, in fact, in the proper migratory direction—except when the sky was overcast. At these times, there was no clear direction to their restless movements. Kramer surmised, therefore, that they were orienting according to the position of the sun. To test this idea, he blocked their view of the sun and used mirrors to change its apparent position. He found that the birds then oriented with respect to the position of the new "sun." They seemed to be using the sun as a compass. Of course, this was preposterous. How could a bird navigate by the sun when we lose our way with road maps? Obviously, more testing was in order.

So, in another set of experiments, Kramer put identical food boxes around the cage, with food in only one of the boxes (Figure 44.16). The boxes were stationary, and the one containing food was always at the same point of the compass. However, its position with respect to the surroundings could be changed by revolving either the inner cage containing the birds or the outer walls, which served as the background. As long as the birds could see the sun, no matter how their surroundings were altered, they went directly to the correct food box. Whether the box appeared in front of the right wall or the left wall, they showed no signs of confusion. On overcast days, however, the birds were disoriented and had trouble locating their food box.

In experimenting with artificial suns, Kramer made another interesting discovery. If the artificial

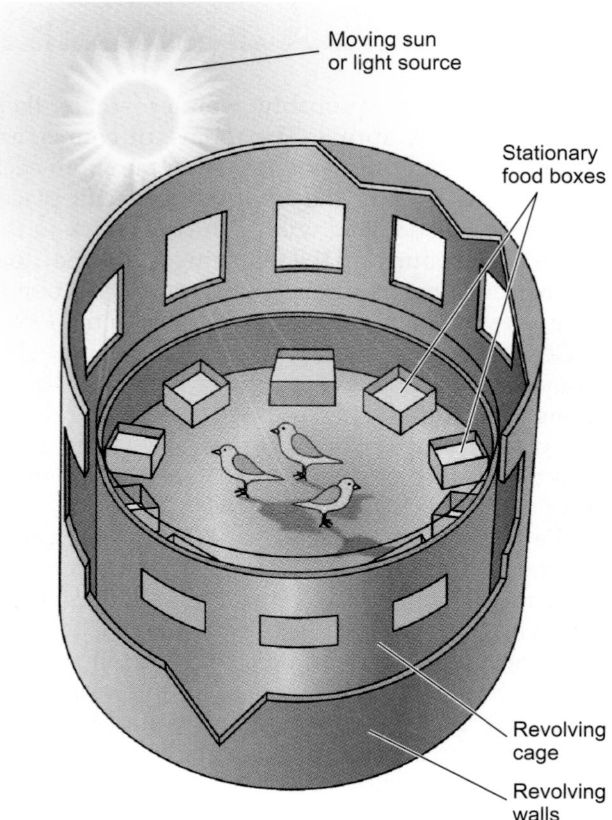

FIGURE 44.16 KRAMER'S ORIENTATION CAGE.
The birds can see only sky through the glass roof. The apparent direction of the sun can be shifted with mirrors.

sun remained stationary, the birds would shift their direction with respect to it at a rate of about 15 degrees per hour, the sun's rate of movement across the sky. (The ability to measure time is critical to many species, as we see in Essay 44.1.) Apparently, the birds were assuming that the artificial stationary "sun" they saw was moving at that rate. When the real sun was visible, however, the birds maintained a constant direction as it moved across the sky. In other words, they were able to compensate for the sun's movement. This meant that some sort of biological clock was operating—and a very precise clock at that.

Researchers began to wonder, what about birds that migrate at night? Could they navigate by the night sky? To answer the question, caged night-migrating birds were placed on the floor of a planetarium during their migratory period. (A planetarium is essentially a theater with a domelike ceiling onto which a night sky can be projected that duplicates the star pattern for any night of the year.) When the planetarium sky matched the sky outside, the birds fluttered in the direction of their normal migration. But when the dome was rotated, the birds changed their direction to match the artificial sky. The results clearly indicated that the birds were orienting according to the stars.

There is accumulating evidence indicating that birds navigate by using a wide variety of environmental cues. Other areas under investigation include magnetism, landmarks, coastlines, sonar, and even smells. The studies are complicated by the fact that the data are sometimes contradictory and the mechanisms apparently *change* from time to time. The questions associated with orientation and navigation have, indeed, presented an array of challenging questions for biologists.

SOCIAL BEHAVIOR

The famous student of animal behavior Jane Goodall (Figure 44.17), who has spent much of her life among the chimpanzees of East Africa's Gombe Stream Preserve, once said, "One chimpanzee is no chimpanzee at all." Her point was that researchers should not attempt to study chimpanzee behavior by observing a single chimpanzee in a cage, because an isolated chimpanzee will behave quite abnormally. Chimpanzees, she noted, are highly social creatures that interact intensively in complex ways. If you want to know what chimpanzees are like, according to Goodall, you must watch them when they are with other chimpanzees.

FIGURE 44.17 JANE GOODALL.
Jane Goodall's years of fieldwork in East Africa have taught us a great deal about the behavior and the needs of chimpanzees.

The same statement might be made of any of a number of other creatures, including us. What would a termite be like without other termites? Or a human without other humans? Many of the earth's animals are indeed highly social species, and they interact with each other in subtle and complex ways. Yet, there are some underlying themes. We will look at some of these principles here. In particular, we will consider how animals can show aggression, and how they can cooperate. We will also see why we should not attach values to either form of interaction, since both are adaptive devices that help the animal survive and reproduce in its own world.

Aggression

The old image of "Nature, red in tooth and claw" has somehow been replaced by the notion that only humans regularly kill members of their own species. Is this true? Let's see.

First, we should note that **aggression** is belligerent behavior that normally arises as a result of competition. An animal shows aggression mainly toward other individuals that tend to utilize the same resources. Thus, it is likely to be aggressive toward those most like itself. Those most like itself are, of course, of the same species and sex. And, in fact, most aggressive interactions occur within species, between members of the same sex. Predatory behavior, by the way, is not aggressive. A cheetah is about as aggressive toward an antelope as you are toward a hamburger.

Fighting

The most blatant form of aggression is fighting. We can discount the old films of leopards and pythons battling to the death; such fights simply aren't likely to happen. What does a python have that a leopard would risk its life for, or vice versa? Fighting is more likely between competitors. Therefore, interspecific (between species) fighting is much less likely than is intraspecific (within species) fighting. This is because the strongest competitors are likely to be of the same species.

Animals of different species do sometimes fight, of course. For example, golden-fronted and red-bellied woodpeckers that fail to find enough in common to interbreed, find enough similarities to fight over territories where their ranges overlap in central Texas (Figure 44.18). As another example, lions may attack and kill African Cape dogs at the site of a kill. The lions don't eat the dogs; they just exclude them. Under certain circumstances, such as at a kill, powerful

(a)

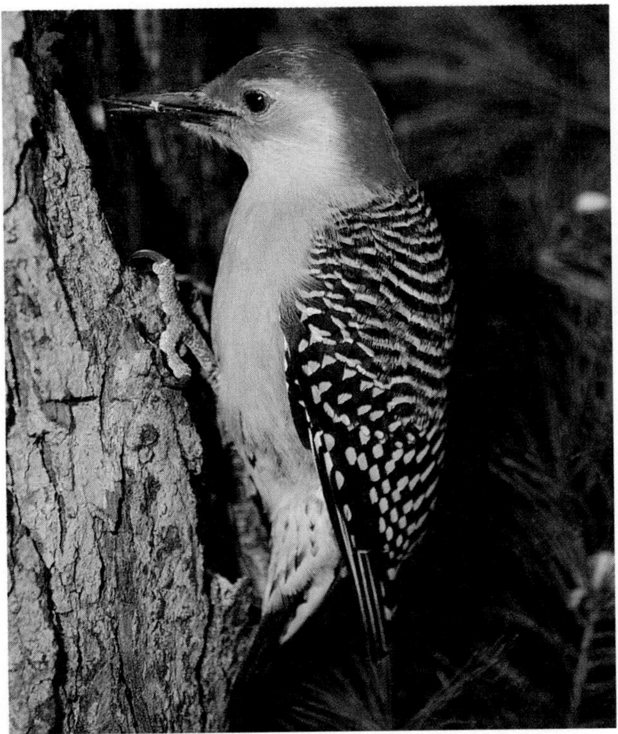

(b)

FIGURE 44.18 WOODPECKER TERRITORIALITY. Golden-fronted (**a**) and red-bellied (**b**) woodpeckers are similar enough to exclude each other from their territories; however, they do not interbreed.

Essay 44.1 Biological Clocks

It's the night before a big test. You are trying to study, but your pet hamster, active this night as usual, has a squeaky running wheel. The sound is so distracting that you decide to put your furry friend in the closet. A few nights later, the squeak coming from the closet reminds you that the hamster hasn't seen the sun, or anything else, for three days. Although it has had no environmental time cues, such as a light-dark or temperature cycle, the hamster is still active at night, the customary time.

How did the hamster do it? How did it know when it was night? Most researchers assume many living things can measure the passage of time by an internal, or biological, clock.

Hamsters are not the only organisms with a biological clock. Indeed, the behavior and physiology of most organisms from protists to humans are rhythmic. In fact, rhythms are so common that rhythmicity should be considered a fundamental property of life.

The prevalence of biological rhythms is not surprising when we remember that life evolved in a cyclic environment. Behavior fluctuates in a repeating pattern so that any pattern ideally occurs at the appropriate time of day, in keeping, say, with the state of the tides, phase of the moon, or the season of the year. The rhythmicity on earth reflects the rhythmic movements of certain heavenly bodies, such as the earth, the moon, and the sun. The relative movements of the earth, moon, and sun cause regular changes in such things as light, temperature, geomagnetism, barometric pressure, humidity, and cosmic radiation. Because these environmental changes have been so regular and so predictable, evolution has been able to adjust behavior and physiology to match these cycles.

It has been argued that living things do not possess an internal clock, but are merely responding to environmental stimuli such as day length or temperature. But the evidence does not support this argument. If you recorded the activity of your hamster while it was in the perpetual darkness of your closet, you would notice that, although bouts of activity regularly alternate with rest, the length of this activity cycle varies slightly from 24 hours. In other words, in constant conditions, daily rhythms are "about a day" in length, or circadian (*circa,* about, *diem,* day). If the hamster were responding to environmental cues, it would stay on a 24-hour cycle. Instead, environmental cues seem to keep its clock precisely "set."

Although biological rhythms continue without environmental cues, they are not completely independent of such cues. For example, light–dark cycles will set, or entrain, the rhythm so that its period length matches that of the en-

predators such as bears and cougars may fight. In each of these cases, one species is reducing its competition for food, and so the aggression is understandable. In some cases, however (Figure 44.19), the basis for the aggression is less clear.

The precise methods of fighting vary widely, but however animals fight, most species have means of avoiding injury to each other. Such avoidance has several benefits. First, no one is likely to get hurt. Although the opponent is permitted to continue its existence, the possibility of having to compete with him again is less risky than serious fighting.

Fighting between dangerous combatants is usually a stylized ritual and relatively harmless. For example, horned antelope may gore an attacking lion, but when antelopes fight each other, the horns are never directed toward the exposed flank of the opponent (Figure 44.20). Such stylized fighting enables the combatants to establish which is the stronger animal, and once dominance is established, the loser usually signals its submission and is permitted to retreat.

All-out fighting is most frequent between animals that do not normally injure each other seriously, such as hornless female antelope (Figure 44.21). It may also occur between animals that are so fast that the loser can escape before serious injury, as is the case with house cats.

Why don't antelopes gore each other? An incurable romantic might assume that it is because they don't want to hurt each other. In all likelihood, what the two antelope "want" has little to do with it. The fact is, they *can't* hurt each other. When the system works, an antelope could no more gore an opponent than fly! Perhaps the sight of an opponent's exposed

vironment. Internal clocks have another advantage over clocks set by external cues. For example, some animals must be able to *anticipate* critical changes in their surroundings so that they have adequate time to prepare. A fiddler crab scurrying along the beach must return to its burrow before the tide returns, or the waves will wash it away. Other animals use clocks to synchronize their behavior to an event that they cannot sense directly. This is the case for honeybees that travel to distant patches of flowers to gather nectar. Different types of flowers open their petals at different times of the day. The bees' clocks allow them to time their nectar-gathering forays so that they arrive when the flowers are open. They might visit morning glories early in the day but wait to visit four-o'clocks in the afternoon.

Clocks can be used not only to determine the time of day, but also how long it has been since some event. For animals such as the birds and the bees, the second function of a clock, measuring the passage of time, is particularly important because it is essential for sun compass orientation (see text). A homing pigeon flying south would be required to keep its path of flight at a 45-degree angle to the right of the sun at 9:00 A.M., but would have to change that angle by about 15 degrees an hour as the sun moved across the sky.

What makes the biological clock tick? We don't know. However, because rhythms exist in single cells and protists, we conclude that the clock must be intracellular. Cellular processes that may play a role in the timing process are protein synthesis on the cytoplasmic ribosomes, transport of ions across the plasma membrane, or perhaps proton transport in the mitochondrion.

If a single cell can tell time, does every cell in a multicellular organism have its own "wristwatch"? Apparently so. Also, isolated tissues often remain rhythmic. For example, it has been found that if the heart of a hamster is removed and kept alive in a tissue culture, it will continue to beat more rapidly at night than during the day. Even a single heart cell will display a daily rhythm.

If there are many clocks in an animal, then they must be set to the same time or there would be internal chaos. Indeed, animals may have master clocks that synchronize the timepieces in individual cells. In mammals such as the rat, the master clock seems to be in a region of the brain called the suprachiasmatic nucleus. When neurosurgeons destroy this tiny group of cells, the rat's running activity, its drinking patterns, and several of its normal hormone rhythms disappear.

Judith Goodenough

flank inhibits butting behavior, while the sight of an opponent head-on might release the stereotyped fighting behavior.

There are species, by the way, in which combatants do fight to the death. If a strange rat is placed in a cage with an established group of rats, the group may chase and sniff at the newcomer carefully for a period, but eventually they will attack and continue the attacks until it is dead. Guinea pigs and mice may also fight to the death. The males of a pride of lions are likely to kill any strange male they find within their hunting area, and a pack of hyenas may kill any member of another pack that they can catch. Even roving bands of male chimpanzees may attack and kill a male from another band. The chimpanzee findings were quite surprising since it had long been assumed that chimpanzees were essentially peaceful animals. In spite of such occurrences, fights to the death are rare in most species.

Cooperation

Cooperation may seem to be at the opposite end of the behavioral spectrum from aggression—aggression is not nice, cooperation is nice. But we will see that both aggression and cooperation might be termed "enabling devices." After all, they both function in enabling individual animals to survive and reproduce. Furthermore, just as humans are not unique in their aggressive behavior, neither are they alone in cooperating with each other.

Cooperative behavior occurs both within and between species. As an example of interspecific

FIGURE 44.19 **INTERSPECIFIC FIGHTING.**
In some cases, fighting may occur between species that rarely interact at any level. Here, a raccoon approaches a hare in the darkness and in the scuffle, the hare leaped over the raccoon and administered a severe drubbing with its powerful back legs to the raccoon's head.

FIGURE 44.21 **FEMALE ANTELOPE FIGHTING.**
Hornless females of the Nilgai antelope have no inhibitions against attacking the flank of a competitor, but their butts are quite harmless.

cooperation, consider the relationship of the rhinoceros and the tick bird that picks ticks and insects from the rhino's skin. The bird gets food while the rhinoceros rids itself of ticks and benefits from a wary little lookout. In spite of such instances of interspecific cooperation, though, the highest levels of cooperation are found among members of the same species.

As an example of intraspecific cooperation, consider the behavior of porpoises. Porpoises are air-breathing mammals, much vaunted in the popular press for their intelligence, and their behavior often seems to support the claim (Figure 44.22). Groups of porpoises will protectively circle a female in the process of giving birth, driving away any predatory sharks that might be attracted by the blood. They have also been known to carry a wounded comrade to the surface where it can breathe. Their behavior in such cases is highly flexible and is influenced by what is going on at the moment. Such flexibility indicates that their behavior is not simply a blind response to innate genetic influences.

Cooperation among mammals is probably most commonly found in their defensive and hunting behavior. For example, the adult musk oxen of the Arctic form a defensive circle around the young at the approach of danger, standing shoulder to shoulder with their massive horns directed outward (Figure 44.23). The defense is generally effective against all predators except humans, since the unfortunate animals try to maintain this stance while they are shot one by one. Wolves, African Cape dogs, jackals, and hyenas often hunt in packs and cooperate in

FIGURE 44.20 **MALE IMPALAS FIGHTING.**
Neither impala will attack the vulnerable flank of the other. Instead, a harmless pushing contest ensues as the tips of the ridged horns are engaged.

FIGURE 44.22 **YOUNG AND ADULT PORPOISES.**
These animals are intelligent and often cooperative creatures.

FIGURE 44.23 DEFENSIVE MUSK OXEN. Musk oxen live above the Arctic circle and are preyed on by wolves. If attacked, they immediately form a circle with the adults facing outward and the calves inside.

bringing down their prey. In addition, the hunting animals may bring food to those mates or young that were unable to participate in the hunt.

It might seem logical that cooperation would be most pronounced among such intelligent animals as mammals. It may be a bit surprising, therefore, to learn that cooperation is actually most highly developed among the insects. Perhaps the best-known example is found in honeybees.

In honeybees (Figure 44.24), the queen lays the eggs, and all the other duties are performed by the workers—sterile females. Each worker has a specific job at any given time. For example, newly emerged workers prepare cells in the hive to receive eggs and food. Then, in a day or so, their brood glands develop, and they begin to feed larvae. Later, they begin to accept nectar from field-workers and to pack pollen loads into cells. At about this time, their wax glands develop, and they begin to build combs. Some of these "house bees" may become guards, patrolling the area around the hive. Eventually, each bee becomes a field-worker, or forager. She flies afield and collects nectar, pollen, or water, according to the needs of the hive. These needs are indicated by the eagerness with which the field bees' different loads are accepted by the house bees.

If a large number of bees with a particular duty are removed from the hive, the normal sequence of duties of the remaining bees can be altered. Young bees may shorten or omit certain duties and begin to fill in where they are needed. Other bees may revert to a previous job that is now required again.

The watchword in a beehive is *efficiency*. In some bee species, the drones (males) exist only as sex objects, reproductive partners. Once the queen has been inseminated by a drone, the rest of the drones are quickly killed by the workers. They are of no further use. The sterile females live only to work. They tend the queen, rear the young, and maintain and defend the hive. When their wings are so torn and tattered that they can no longer fly, they either die or are killed by their sisters. But the hive goes on.

ALTRUISM

We don't wish to disillusion anyone, but, well . . . most of the Lassie stories aren't really true. Consider what would happen to the genes of any dog that was given to rushing in front of speeding trains to save baby chickens. The reproductive advantages would be considerable to chickens, but dogs with those tendencies might be selected out of the population by the action of fast trains. In contrast, a dog that spends its energy, not in chivalrous deeds, but in seeking out a mate, is more likely to be around to have puppies. So what kinds of dogs are likely to predominate in the next generation?

If an animal *is* going to engage in "unselfish" deeds, its best reproductive bet lies in those deeds that advance the genes of other members of its own species. And as we will see, it is here that we are most likely to see selfless behavior.

Altruism may be defined in a biological sense as an activity that benefits another organism at the individual's own expense. It seems to be common among animals, but is it? A more difficult question is, how is it maintained in populations (Figure 44.25)?

FIGURE 44.24 ALTRUISTIC BEES. Honeybee workers surround the queen.

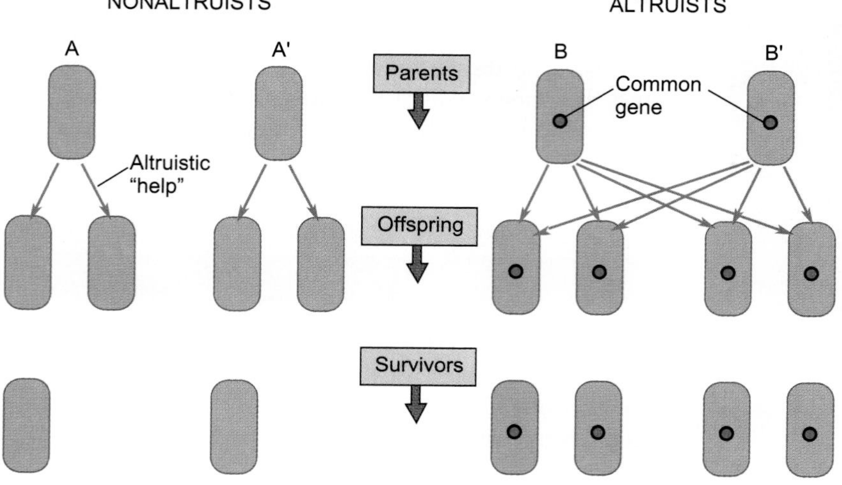

Hypothetical Mechanism for Maintenance of Altruism in Populations
NONALTRUISTS
ALTRUISTS
A
A'
B
B'
Parents
Common gene
Altruistic "help"
Offspring
Survivors
Populations A and A' do not have genes in common. Since they do not show altruistic behavior to each other's offspring, fewer offspring from each population survive.
Populations B and B' have a gene for altruism in common. Since they show altruistic behavior to each other's offspring, more offspring from each population survive.

FIGURE 44.25 A GENETIC BASIS FOR ALTRUISM.
In this population, A and A' are nonaltruists. They behave in such a way as to maximize their own reproductive success, but do nothing to benefit the offspring of other individuals. In another segment of the population (B and B'), a gene for altruism has appeared that results in individuals benefiting the offspring of others in some way. Assuming the altruistic behavior is only minimally disadvantageous to the altruist, it can be seen that generations springing from B and B' are likely to increase in the population over those from A and A'. It should be apparent that the altruistic behavior should be more common where B and B' are most strongly related, so that B shares a maximum number of genes in common with the offspring of B' and vice versa. The idea is that B, for example, can increase its own reproductive success by caring for the offspring of a relative with whom it has some genes in common. After all, reproduction is simply a way of continuing one's own kind of genes.

KIN SELECTION

Altruism, as we've considered it so far—that is, as a means of increasing one's reproductive output—doesn't explain why an animal would put itself at risk to help the offspring of *another* individual. Why would a bird feed the young of *another* pair, or why would an African hunting dog regurgitate food to almost *any* puppy in the group? Why, also, would a ground squirrel that has no offspring of its own give a warning cry to alert other ground squirrels of the approach of a predator, even at the risk of drawing the predator's attention to itself (Figure 44.25)?

To answer such questions, we must look past the answers that first come to mind. It may seem cynical, but we must start with the premise that ground squirrels don't give a hoot about each other. An animal that issues a warning call isn't thinking, "I must save the others." At least, there is a simpler explanation for its behavior.

Keep in mind that the biologically "successful" individual is the one that maximizes its reproductive output. One way of accomplishing this is for the organisms to behave in such a way as to leave as many individuals carrying *its own types of genes* as possible in the next generation. You can see right away that this would explain parental care. However—and this may not be so readily apparent—an individual can also leave its types of genes in the next generation by helping a *relative's* offspring to survive.

Kin selection is the process by which an individual increases its kinds of genes in the population by helping relatives. (Keep in mind that relatives have genes in common.) For example, an individual shares genes in common with a cousin, although fewer of course, than with a son or a daughter. Hence, there is, theoretically, a point at which an individual could leave more copies of its genes by saving its nieces and nephews (provided there were enough of them) rather than its own offspring. From the standpoint of reproductive output (the number of copies of one's genes that make it into the next generation), the organism would be better off leaving a hundred nephews than one son.

That's the basic idea behind kin selection. The mechanism by which it could operate, though, presents new questions. Suppose a gene for altruism appears in a population (notice that this sets up the mechanisms for the continuance of the behavior, but also consider whether a behavioral tendency might also be passed along culturally in a verbal and social species such as our own). As you can see from Figure 44.25, altruistic behavior would most likely be maintained only in groups in which the individuals are related (that is, in which they have some kinds of genes in common). The behavior might be expected, then, in populations in which there is little mixing with outside groups—in other words, where there is a high probability that proximity indicates kinship, or in populations in which individuals have some way of recognizing kin. In many species, by the way, individuals can recognize their kin (Essay 44.2).

Keep in mind that *no conscious decision* on the part of the altruist is necessary. It simply works out that when conditions are right, those individuals that behave altruistically increase their kinds of genes in the population, including the "altruism gene." Nonrelatives would be benefited by the behavior of altruists, of course, but individuals near enough to receive the benefits of an act are likely, to some degree, to be relatives.

It has been determined mathematically that the probability of altruism increasing in a population depends on how closely the altruist and the recipient are related (Figure 44.26). In other words, the advantage to the recipient must increase as the kinship becomes more remote, or the behavior will disappear from the population. For instance, altruism toward siblings (brothers and sisters) that results in the death of the altruist will be selected for if the net genetic gain is *more* than twice the loss; for half-siblings, four times the loss; and so on. To put it another way, an altruistic animal would gain reproductively if it sacrificed its life for more than two siblings, but not for fewer; or for more than four half-siblings, but not for fewer; and so on. Therefore, we can deduce that in highly related groups, such as a small troop of baboons, a male might fight a leopard to the death in defense of the troop. By the same token, a ground squirrel will give a warning cry when the chance of attracting a predator to itself is not too great and when the average neighbor is not too distantly related.

Kin selection helps to explain the extreme altruism shown by social insects. In some species, such as honeybees, the queen is inseminated only once, so all the workers in the hive are sisters. Because of the system of sex determination, honeybee sisters are likely to have more genes in common than you would with your sister. Furthermore, the workers share more genes with their sisters than they would with their own offspring! Why? Male honeybees (drones) are haploid (carrying one set of chromosomes), and females are diploid (carrying two sets—see Chapter 11). Therefore, sisters share three-fourths of their genes, not just half as they would with their offspring. (Can you explain the difference?) In such a system, then, the workers would leave more copies of their genes by helping to raise sisters than by reproducing themselves. So almost any sacrifice is worth any net gain to the hive and the queen.

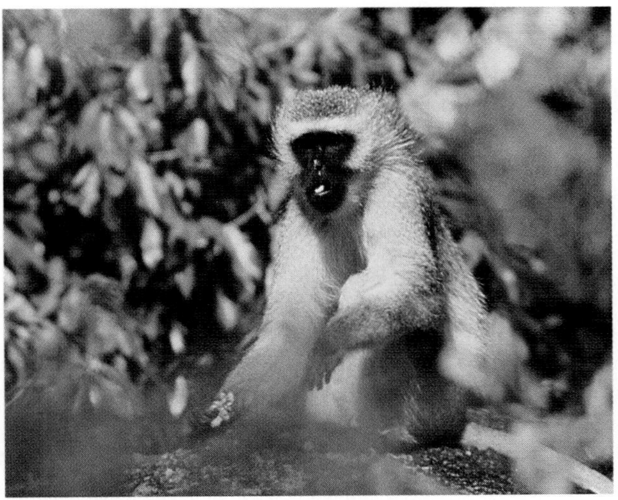

FIGURE 44.26 ALTRUISM IN A PRIMATE.
An African vervet monkey giving an alarm call to warn others of the presence of a dangerous predator even though it puts him directly in danger. Such risky behavior can be expected where the defender is likely to be strongly related to others in the group.

The development of altruistic behavior is an admittedly somewhat esoteric topic. However, it seems important to consider even our most cherished and most despised behaviors in the context of their evolutionary history.

RECIPROCAL ALTRUISM

In a brilliant essay, Robert Trivers expanded our understanding of the evolution of altruism by suggesting that in certain species, notably humans, altruism (outside of parental behavior) depends on the expectation of reciprocation: "I'll scratch your back if you'll scratch mine." **Reciprocal altruism** is selfless behavior that is performed when it is likely that the favor will be returned. It is an evolutionary strategy with some complex rules. Help (altruistic acts) is given to others—even offered to strangers—with the expectation that aid will be returned to the altruist, if needed, at some other time.

Reciprocal altruism would be expected in those groups that are highly social (so the altruist is likely to encounter the beneficiary again) and relatively intelligent (so that individuals are recognized and their behavior can be remembered and repaid).

ESSAY 44.2 How to Recognize Kin

A cartoon once depicted an old gentleman sauntering out the door after a grand Sunday dinner with the two startled hosts exclaiming, "*My* uncle! I thought he was *your* uncle!" The implication was that he would have been treated differently if they had known he was unrelated. And so it is in the animal world. In many species, animals behave differently toward kin.

Andrew Blaustein at Oregon State University and his coworkers have pointed out that any such recognition is likely to be based on familiarity. In other words, an animal comes to recognize those individuals it grew up with and to treat them as if they were related. If animals disperse soon after birth or hatching, though, there may be no opportunity for them to become familiar with each other and so other means of identification must be employed. One such means is through maternal "labeling." Individuals could acquire, for example, a specific scent while in the uterus or in the same clutch of eggs. Later, they would be programmed to treat individuals with that label as if they were relatives. Another way is perhaps the most remarkable of all. Theoretically, it is accomplished via "genetic markers." In this case, it is proposed that an individual carries a gene that enables it to recognize others with the same gene.

What does it mean to treat others as if they were relatives? For one thing, it means you don't eat them. Cannibalism is startlingly common among many kinds of animals, and so an individual needs to be able to recognize those bearing its same kinds of genes. A second advantage of recognizing kin is to avoid inbreeding (see Chapter 15). Thirdly, animals may warn or protect those to whom they are related when danger threatens, and they may help them in conflicts or in acquiring commodities.

Some evidence of intraspecific reciprocal altruism has been reported in troops of social mammals, such as hunting dogs and higher primates, but the evidence is not very strong for such behavior in any animals but humans (Figure 44.27). Trivers implies, in fact, that reciprocal altruism is the key to human evolution. The complexity of such behavior, entailing as it does memory of past actions, the calculation of risk, the foreseeing of the probable consequences of present actions, the possibility of advantageous cheating, and the need to be able to detect such cheating, all require a level of intelligence that is beyond most species.

SOCIOBIOLOGY AND SOCIETY

In the mid-1970s, an idea that had been around for decades was revived, reviewed, reanalyzed, refined, reconsidered, and brought to public attention by the noted biologist Edward O. Wilson of Harvard University. Because it was a generally familiar concept and was stated very carefully in a professional format, the scientific community was surprised by the turmoil that followed.

Sociobiology is the study of the effects of natural selection on social behavior. The idea is quite a simple one and has been around long enough that no one is really shocked by its revival; the whole idea might have generated little discussion except for the reaction of a group of people who believe that the concept is based on a politically dangerous premise. They immediately set about to attack it, and their attack was so vigorous and well publicized that it drew a great deal of attention to the very idea they didn't want to receive attention.

The defenders of the idea quickly rallied, and the arguments grew vitriolic as both sides firmed up their position and stiffened their resolution. It was an emotional argument—and some people apparently entered the fray with more opinions than information, and rather strange things were said. Furthermore, because of sloganeering and loose definitions and because there was so little dialogue between the two groups, semantics became a large part of the problem.

So what was the problem? The opponents of the idea feared that a sociobiological explanation of human behavior would lead to a revival of the notion of biological determination. This is the idea that we are primarily the product of our evolution and that we therefore respond to the primitive calling of our heritage. As a result, our social behavior cannot be changed in any fundamental way, and it is a waste

FIGURE 44.27 **RECIPROCAL ALTRUISM.** In a formalized reciprocal altruism, society at large pays a small cost in order to offer help. Here, food is distributed to hurricane victims at public expense. Implicit is the understanding that each member of society can expect such public assistance in a crisis.

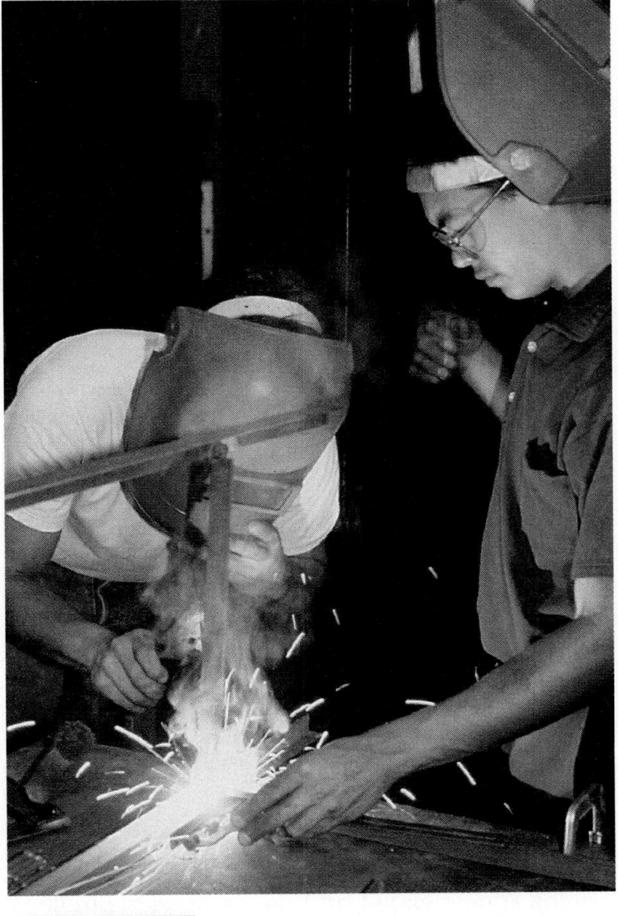

FIGURE 44.28 **ALTRUISM AS A SOCIAL PROGRAM.** Social programs such as this, where inmates at the California Youth Authority are being trained to be productive citizens, are offered in order to try to change the fabric of society.

of time to try. The opponents felt that the acceptance of sociobiology would mean resignation to the status quo. Furthermore, they said it supports such undesirable patterns as racism and sexism. They preferred to believe that our social patterns are molded by culture and learning, plain and simple, and that we can change any undesirable trait through education, incentive, and social programs (Figure 44.28).

Sociobiologists strongly deny supporting biological determination and certainly denounce racism and sexism, but they believe that their approach holds real promise for building a better society. Perhaps it does seem a bit irreverent to some to suggest that human behavior, to any great degree, is programmed through evolution and that the forces of natural selection mold our own species as well as others. But the sociobiologists argue that if our behavior is, to *any* degree, genetically controlled or influenced, then we should know it. They note that we can't hope to find solutions if we have ignored the role of our biological heritage. We must, they say, try to understand our social problems at every level and to use every tool we have to improve our lot. (We didn't develop the ability to fly by simply denying the existence of gravity.)

Sociobiology, in its refurbished form, is quickly maturing as data are now appearing from long-term studies and as new researchers approach the problem from different angles. The next few years should be interesting as sociobiologists present us with new approaches to a weathered idea.

PERSPECTIVES

Much of what we learned in previous chapters about the physiology of animals helps us better understand animal behavior. We know why animals will strive for enough oxygen, why they will face predators to get at water, and why some species are able to move so fast and how this ability affects the prey they choose.

Now we apply our information on animal behavior to understanding better how they fit into their environment in the larger sense. We begin with discussions of ecology and the biosphere, then move on to ecosystems and communities. We will be led to consider plants first in some cases, and then how animals live among these plants. We round out our discussion with a look at how populations behave, and then focus on the human condition. As always, we will look for the common themes of life within its great diversity.

SUMMARY

PROXIMATE AND ULTIMATE CAUSATION

Proximate causation involves psychological or physiological mechanisms. Ultimate causation is based on adaptation and evolution.

ETHOLOGY AND COMPARATIVE PSYCHOLOGY

The theory of instinct was formulated by Konrad Lorenz and amended by Niko Tinbergen. Ethology, a school of animal behavior founded in Europe during the 1930s, attempted to understand animal behavior by studying its cause, development, evolution, and function through observation of animals in the wild. At the same time, American comparative psychologists were in their labs studying learning processes in animals, typically the Norway rat.

THE DEVELOPMENT OF THE INSTINCT IDEA

An instinctive behavior is a species-specific, genetically based behavior often formed from fixed action patterns that is performed when the perception of something in the environment (a releaser) triggers a response in a center in the central nervous system (the innate releasing mechanism) that sends the signal to the muscles to perform the behavior (the fixed action pattern).

LEARNING

Animals are generally able to learn the kinds of things that are relevant to their lifestyles. There are often species differences in learning ability. Habituation is learning not to respond to a frequently encountered stimulus. It is not permanent. Habituation may save energy by eliminating responses to common harmless stimuli while allowing responses to rare stimuli. In classical conditioning, a response to a normal stimulus comes to be elicited by a substitute stimulus when the two occur together many times. The conditioned response can be lost if the normal stimulus, which reinforces the response, repeatedly fails to follow the conditioned stimulus. In operant conditioning, the animal is rewarded after performing some behavior. The frequency of the behavior increases when it is repeatedly rewarded.

HOW INNATE BEHAVIOR AND LEARNING CAN INTERACT

Innate and learned patterns can interact to produce an adaptive response. Many innate behavior patterns can be improved by learning. Some species learn certain behaviors only if they are exposed to the relevant stimulus during a brief critical period shortly after birth. This type of learning is called imprinting.

BEHAVIORAL ECOLOGY

Behavioral ecology is the study of how the environment affects behavior. Habitat selection has been shown to have a genetic component that can be influenced by experience. Foraging behavior involves maximizing strategies that yield the best results while minimizing the costs.

ORIENTATION AND NAVIGATION

Orientation involves facing in the right direction. Navigation is finding one's way from one place to another. Some daytime-migrating birds, such as starlings, navigate by using the sun as a compass. They use a biological clock to compensate for the sun's movement across the sky. Night-migrating birds may navigate by using the stars as a guide. Other cues, which may change over time, may also be involved.

Social Behavior

Aggression is belligerent behavior that normally arises from competition. It is most commonly shown toward members of the same species and sex. Members of many species avoid fights by threatening one another through displays. Fighting generally stops when the weaker individual shows submission. Fights between dangerous animals are usually stylized rituals that establish which combatant is dominant.

Cooperation

Cooperative behavior occurs both within (intraspecific) and between (interspecific) species. Cooperation among mammals of the same species often occurs in protective, defensive, and hunting behaviors. Cooperation is most highly developed among social insects such as the honeybee.

Altruism

Altruism is an activity that benefits another organism at the individual's own expense. It occurs most commonly among related individuals. Generally, altruistic behavior increases the frequency of the altruistic gene in the next generation.

Kin selection, the process by which an individual increases its type of genes in a population by helping relatives, may explain the continuance of certain altruistic behaviors. It would be most likely to occur in populations in which there is a high probability that proximity indicates kinship or in which individuals can recognize kin.

Reciprocal altruism is an evolutionary strategy in which help is given to another with the expectation that the favor will be returned. Reciprocal altruism would be expected in highly social groups of intelligent individuals, such as humans.

Sociobiology and Society

Sociobiology is the study of the effects of natural selection on social behavior. Opponents to the idea feared that it would revive the notion of biological determination. Proponents argue that our only hope of finding solutions to our social problems lies in understanding social behavior at every level.

Key Terms

proximate causation • **782**
ultimate causation • **782**
ethology • **783**
comparative psychology • **783**
instinct • **783**
releaser • **783**
innate releasing mechanism (IRM) • **783**
fixed action pattern • **784**
learning • **784**
habituation • **785**
classical conditioning • **785**
operant conditioning • **786**
imprinting • **788**
behavioral ecology • **788**
generalist • **790**
specialist • **790**
orientation • **791**
navigation • **791**
aggression • **793**
altruism • **797**
kin selection • **798**
reciprocal altruism • **799**
sociobiology • **800**

Review Questions

1. Certain squirrels will begin to disregard danger calls from other squirrels if they repeatedly prove to be false alarms. What behavioral mechanism is at work here? How can it be reversed?
2. Describe the role of reward in operant conditioning.
3. Describe the role of reward in classical conditioning.
4. Do you consider imprinting to be a kind of learning? Explain.
5. What are the presumed prerequisites for the development of reciprocal altruism?
6. Do processes within the body that produce behavioral patterns involve proximate or ultimate causation? Explain.
7. In what kinds of groups would you expect to see the highest levels of competition? Of aggression?
8. In what kinds of groups would you expect kin selection?
9. What are the essential differences in comparative psychology and ethology?
10. Describe an instinctive action using the terms *releaser, IRM,* and *fixed action pattern.*
11. How is biological determinism used to deter investigations in sociobiology?
12. Give three factors involved in choosing a prey item.
13. In experiments using caged migratory birds, why do the birds change the directions of their movements at the rate of 15 per hour?
14. Describe how learning and instinct can interact to produce an adaptive behavior.

45

Ecology and the Biosphere

SOMEONE HAS SAID THE ECONOMIC CONFLICT WITH THE OTHER GREAT *powers will be won by us only if our ecologists are better than their ecologists—interesting notion that underscores the role of those people who focus on the big picture, who make sense of things. Even big pictures, though, have to be divided into little pictures if we are to understand them. And so we will consider the earth as composed of areas called biomes, such as rain forest or desert. Then we will divide the earth's waters into freshwater and saltwater. In each case, though, we will be looking at the smaller picture to better understand the big picture—a daunting but fascinating task, indeed.*

If you glance at a photograph of the earth taken from a satellite, you might get the impression that the earth is a small place—and in the cosmic sense, perhaps it is. But in the biological sense, it is large, complex, and diverse. You can get an idea of its real dimensions by taking a hike along Alaska's North Slope, spending an afternoon in a Louisiana woods, or even going for a walk to the next town. Biologically speaking, the earth is decidedly a vast place. Furthermore, it is very different from one area to the next. The consideration of these differences and what they mean to life leads us to the broadest of all biological disciplines—ecology.

Ecology (*oikos,* the house; *logos,* knowledge of) is the study of the interaction between organisms and their environment. The key word here is *interaction.* Interaction implies reciprocity, and the two are indeed reciprocal—they shape each other. The ecologist's task is formidable, since such relationships are immensely complex, and understanding them requires expertise in a number of scientific areas. In a real sense, then, ecology is where the sciences come together. We will begin our consideration of ecology with a brief look at some rather grand concepts, then focus more narrowly as we approach the human condition on this enormous but fragile planet.

BIOSPHERE: THE REALM OF LIFE

The **biosphere** is where life exists. It is the thin veil over the earth in which the wondrous properties of air, light, water, and minerals interact to permit that life. The biosphere includes not only land areas and their subterranean realms, but the waters as well. It is marked, however, not only by its expanse, but by its shallowness as well—for life cannot exist in nature very far above or below the earth's surface. To be precise, the habitable regions of the earth lie within an amazingly thin layer of approximately 14 miles, from the tops of the highest mountains to the bottoms of the deepest ocean trenches. If the earth were the size of a basketball, the biosphere would be about the thickness of one coat of paint. Within these thin limits, the biosphere is a place with very special conditions.

THE ATMOSPHERE

The earth's atmosphere may indeed be wispy and ethereal, but virtually all life depends on this array of gases. Chemically, it is composed of 78% nitrogen, 21% oxygen, 0.04% carbon dioxide, and a number of other quite rare gases. The atmosphere also holds water in concentrations that vary greatly from place to place.

Most of the earth's atmosphere clings close to the planet, not extending more than 5 to 7 miles high; but it acts as a protective screen, shielding the earth from dangerous ultraviolet radiation that would be disruptive to life. In particular, the ozone (O_3) of the upper atmosphere protects us from the sun's harmful rays. (Unfortunately, the ozone is becoming seriously depleted in certain areas, and there may be other problems ahead; see Chapter 47.)

The atmosphere also helps hold heat close to the earth's surface, largely due to the presence of CO_2

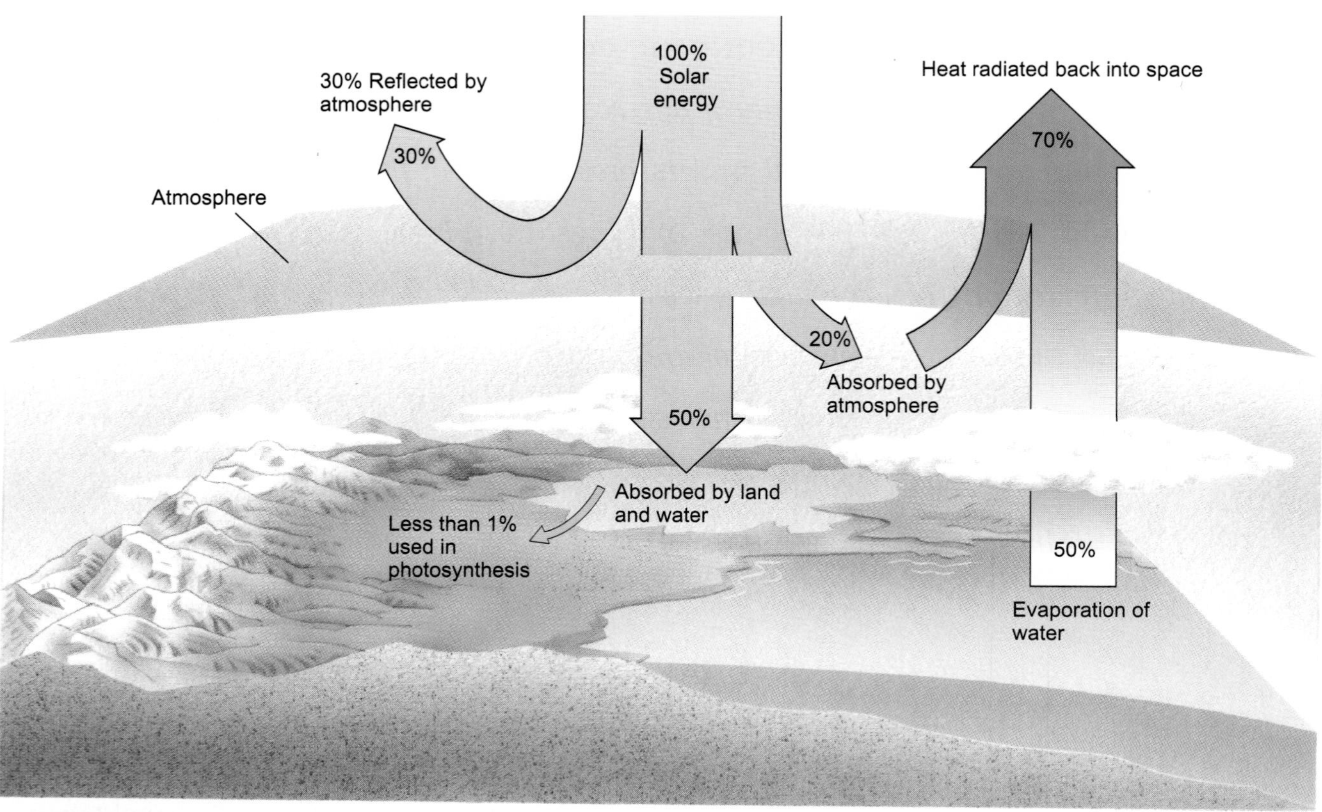

FIGURE 45.1 THE EARTH'S ENERGY BUDGET.
The relative constancy of conditions in the biosphere depends ultimately on an equilibrium between energy entering and energy leaving. Most of the absorbed energy reenters the atmosphere through evaporation from the earth's water. The thin arrow represents energy released by organisms.

(see Essay 45.1) and water vapor. Water absorbs heat slowly and releases it slowly. Its ability to change from liquid to gas permits it to move rather freely through the biosphere and greatly facilitates the wide distribution of heat. Its presence in the atmosphere slows the dissipation of radiant heat from the earth's surface, helping the temperature of the earth's surface to remain relatively constant.

SOLAR ENERGY

For the most part, the surface of the earth is heated by solar energy, energy from the sun. Only about half of the incoming solar energy ever actually reaches the earth's surface; about 30% is reflected back into space, while 20% is absorbed by the atmosphere. The 50% that does reach earth is absorbed by the land and waters, from which it radiates back into the atmosphere as heat (Figure 45.1). However, a great deal of work is accomplished during the time it interacts with the biosphere.

You are aware, of course (if all is going well), that photosynthesis drives much of the activity of life by enabling the energy of sunlight to be captured in the molecules of food manufactured by plants. However, you may be surprised to hear that on this green planet, less than 1% of incoming solar energy is used in photosynthesis. Actually, most of the solar energy reaching the earth is used to shuffle water around in the **hydrologic cycle,** which is the evaporation and condensation of the earth's waters (Figure 45.2). This cycle, of course, is responsible for the earth's rainfall pattern. In addition to distributing water more equitably over the earth's surface, the cycle redistributes heat.

CLIMATE AND THE TILTED EARTH

Although solar energy strikes the equatorial regions of the earth fairly evenly, in the northern and southern latitudes this energy varies markedly from one season to the next because of the earth's tilted rotational axis (about 23.5 degrees from vertical). As the earth makes its appointed rounds, circling the sun, first the Northern Hemisphere, then the Southern, is struck broadside by the sun's rays, creating

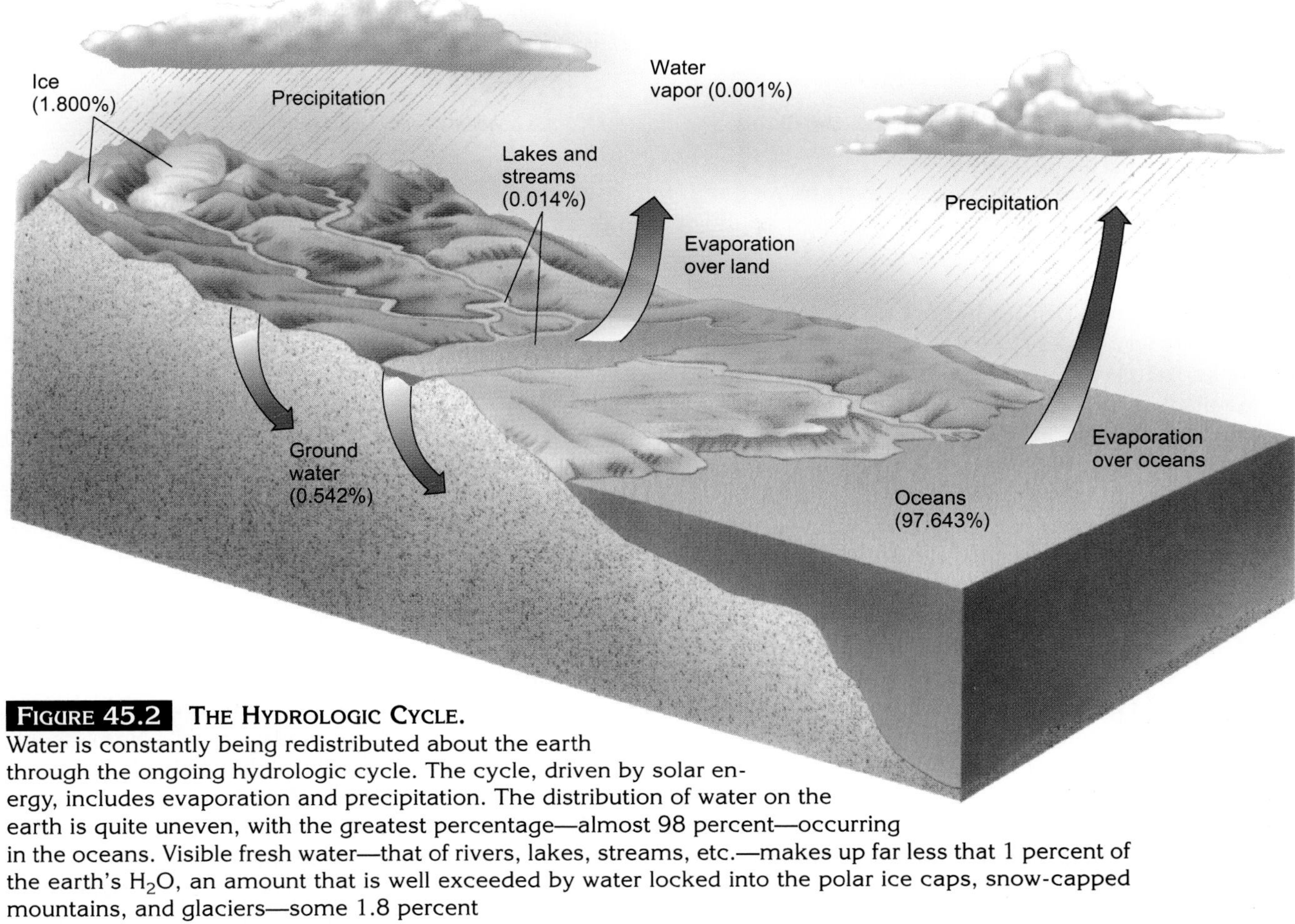

FIGURE 45.2 THE HYDROLOGIC CYCLE.
Water is constantly being redistributed about the earth through the ongoing hydrologic cycle. The cycle, driven by solar energy, includes evaporation and precipitation. The distribution of water on the earth is quite uneven, with the greatest percentage—almost 98 percent—occurring in the oceans. Visible fresh water—that of rivers, lakes, streams, etc.—makes up far less that 1 percent of the earth's H_2O, an amount that is well exceeded by water locked into the polar ice caps, snow-capped mountains, and glaciers—some 1.8 percent

the changing seasons. Without these fluctuations, the temperate zones, where much of the earth's human population now lives, would be perpetually frozen.

BIOMES AND THE WATERS OF THE EARTH

Because of the earth's varying climate and topography, the distribution of life—both in amount and kind—varies. Ecologists recognize a number of major plant associations in the terrestrial environment, which they refer to as biomes. Specifically, a **biome** is a group of ecological communities characterized by a particular array of plants, animals, and microorganisms. A **community,** in turn, is all of the living things in a particular area (including plants and animals) that live close enough to each other to potentially interact. The factors that influence the formation of biomes are chiefly climatic (precipitation, temperature, seasonal extremes, winds), although topography, soil composition, and light are also important.

There is some disagreement over how many biomes should be designated, but we will discuss eight (Figure 45.3). We should first note that biomes tend to be arranged along particular latitudes, especially in the Northern Hemisphere. Starting at the equator and moving northward, the order of major biomes is equatorial forests, grasslands and deserts, temperate forests, northern coniferous forests, and tundra. Interestingly, we find a similar distribution arranged vertically in high mountain ranges. There, we see that the effects of altitude can mimic those of latitude (Figure 45.4). Let's now consider the major biomes of the earth, beginning at the equator and moving through the temperate regions toward the poles.

THE TROPICAL RAIN FOREST BIOME

The first biome we encounter is the **tropical rain forest.** Typically, this lush and varied biome receives over 250 cm (100 in.) of annual rainfall. There are a great number of plant species packed into such areas. Rain is usually rather evenly distributed throughout the year, but there are "dry seasons" when it rains less frequently (say, three out of four

ESSAY 45.1 The Greenhouse Effect

In recent years, some atmospheric scientists have become greatly concerned about what they view as increasing levels of carbon dioxide in the air. Atmospheric CO_2 comes from a variety of sources. As we know, it is produced by the metabolism of most living things. In addition, the enormous amount of carbon dioxide released by the burning of fossil fuels has been a concern for many years; but today we face a new source of CO_2 production—the carbon dioxide that was once part of the living mass of the tropical rain forests. Such forests have always been a great CO_2 *sink* (region of concentration), but they are being cleared at the alarming rate of 1% each year (roughly the area of the state of Washington). After the trees are felled, they are burned, and the carbon they held in their tissue is released into the air as CO_2.

Why the alarm? It has to do with the storage of heat in the atmosphere. One of the characteristics of CO_2 in the atmosphere is that it allows the energy of sunlight to reach the earth but does not allow the heat produced to be radiated back. It "traps " the heat, just as a greenhouse does. Molecules other than CO_2 that may produce the effect include water, ozone, chlorofluorocarbons, nitrous oxide and methane, (all of which have increased in concentration since at least 1980).

Some scientists are also concerned about any increase in ocean temperatures. As the oceans warm even a little, their ability to hold CO_2 in solution decreases. The oceans contain much greater reserves of CO_2 than does the atmosphere, and a rise of one or two degrees would unload more of the gas into the air.

Others wonder whether global warming is occurring at all. Charles Rubin and Marc Landy, two political scientists, say that it may well be, but that we certainly don't have clear evidence of it. In some cases, they say, misinterpretations arise because *scenarios*—"what if" situations, played out in the scientists' imaginations—have been misinterpreted as taken for *predications.* They note that the World Resource Institute published a book called *The Greenhouse Trap,* wherein it summarized predictions by the Environmental Protection Agency, of global warming causing "greatly decreased forest yields, crop losses and 'moving' grain belts." They argue that these were not *predications* but *scenarios.* Others, of course, might respond that scenarios are based on assumptions and that if the Greenhouse Effect is, indeed, contributing to global warming, then the conditions are met for the scenario to have predictive powers.

Rubin and Landy also note that scientists do not agree on their predications of future climatic conditions. And they imply that those who want to promote *action now* often seek the refuge of some ambiguous consensus by saying things like "scientists believe that. . ." They argue that polls show that most scientists think that we don't know enough about global warming to make laws to prevent it.

How did all the hubbub get started? Rubin and Landy place part of the blame on "environmental administrators," bureaucrats who believe that their job is to "create policy," and so they bolster their case by seeking out scientists who agree with the policy they'd like to develop. They add, "To get their piece of the pie, greenhouse policy entrepreneurs must continually fan the flames of public hysteria."

The authors seem to weaken their case when they imply that even though we don't really know if we're causing global warming, even if we are, we're smart enough to adapt, to change our behavior, when the time comes and that may be the cheapest way to deal with it in the end. How smart is that?

days). In tropical rain forests there is little seasonal variation in temperature, often less than the change between day and night temperatures. The largest tropical rain forest is in the Amazon River Basin in South America; the second largest is in the wilds of the Indonesian Archipelago (see Figure 45.3).

The great amount of rainfall of the tropical rain forests can be traced to equatorial heat, where heated air masses rise rapidly and are replaced by cooler air from the north or south. This activity results in the formation of enormous rotating **air cells,** which are thrown into a diagonal direction by the earth's rotation (Figure 45.5). The rising equatorial air cells lift great amounts of moisture, but as they gain altitude, they cool, and the moisture condenses and begins to fall as rain. By the time the air cells reach their northern and southern limits (roughly 30 degrees north and south), most of the moisture has been released.

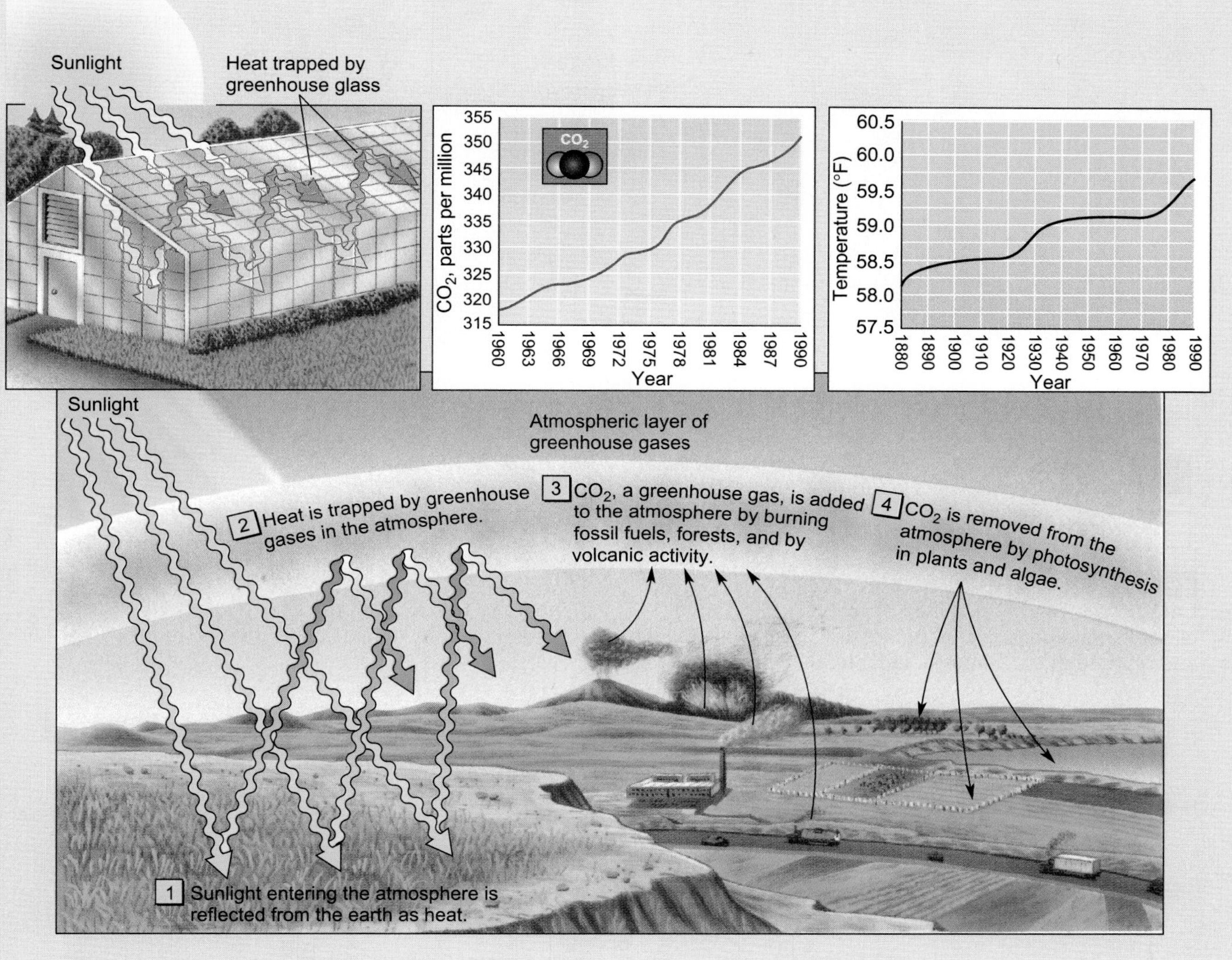

The behavior of equatorial air cells explains why many of the earth's great deserts lie just above and below the equatorial forests.

Plant and Animal Associations

A tropical rain forest is a verdant and beautiful place. The floor is dark and wet, the damp air ripe with smells. Unlike the situation in most other biomes, no single kind of plant is predominant. Any tree is likely to be a different species from its neighbors; the nearest tree of the same species may even be miles away.

The forest floor is usually shadowed by the nearly continuous leafy canopy of trees some 30 to 44 m tall. Below, smaller trees form rather distinct layers of subcanopies, interwoven by large numbers of vines. Both trees and vines may be festooned with

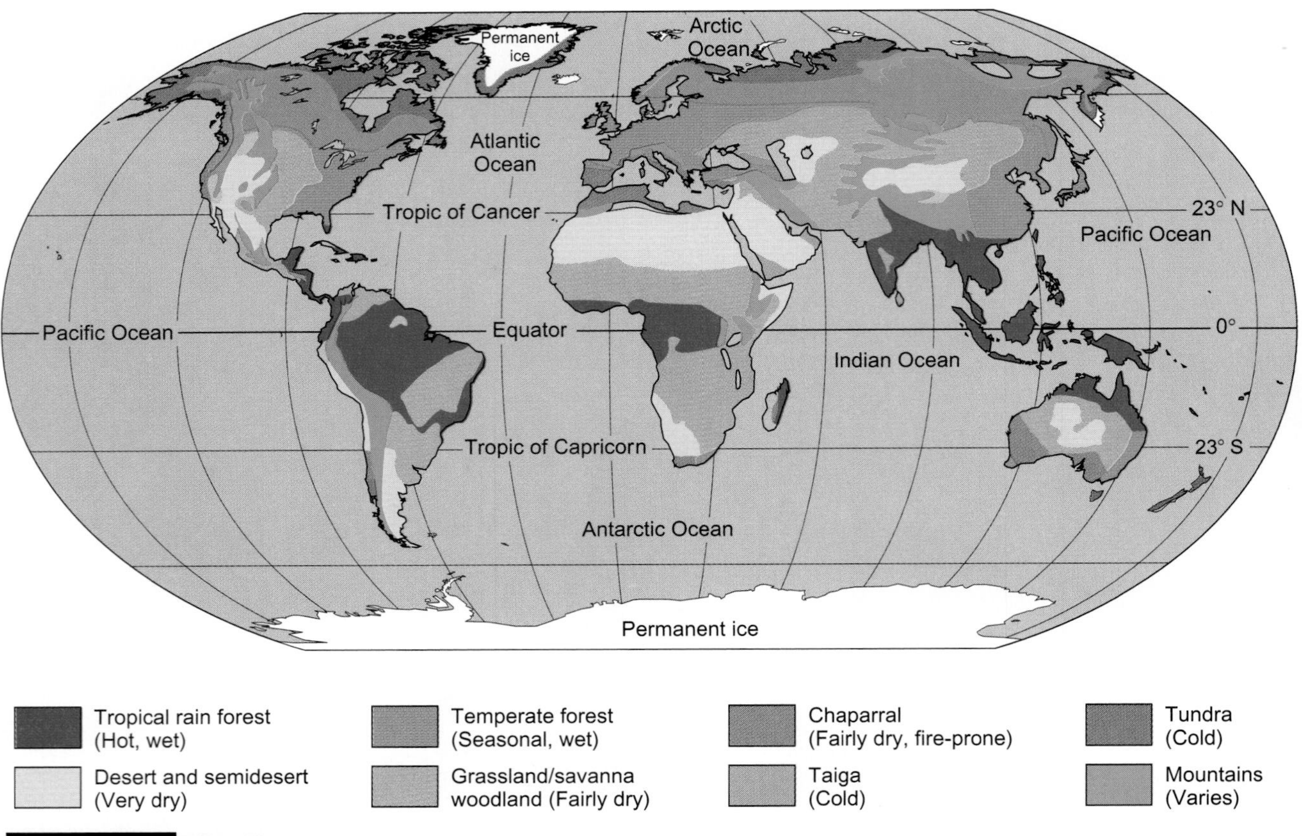

FIGURE 45.3 THE BIOMES.
Each biome can be identified primarily by its plant life. The plant life has adapted to the specific biome's climatic conditions, including precipitation, availability of light, and, of course, temperature. Both latitude and altitude affect all these variables.

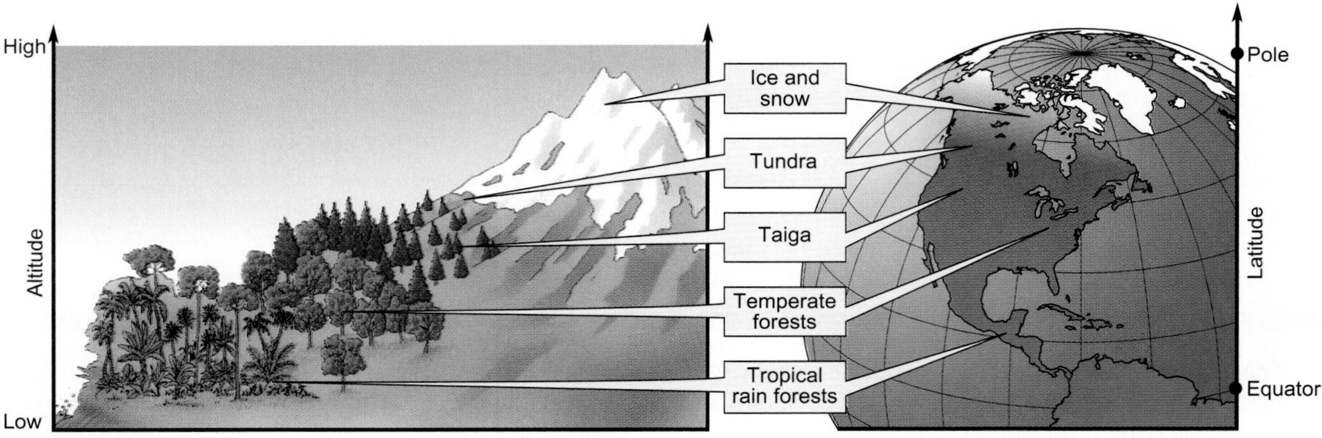

FIGURE 45.4 ALTITUDE VARIATION AND BIOME DISTRIBUTION.
The altitudinal variation provided by mountains often mimics latitudinal biome distributions. On the leeward side of the mountain, air masses holding little moisture rush down the slopes and across the barren terrain, leaving typical desert conditions

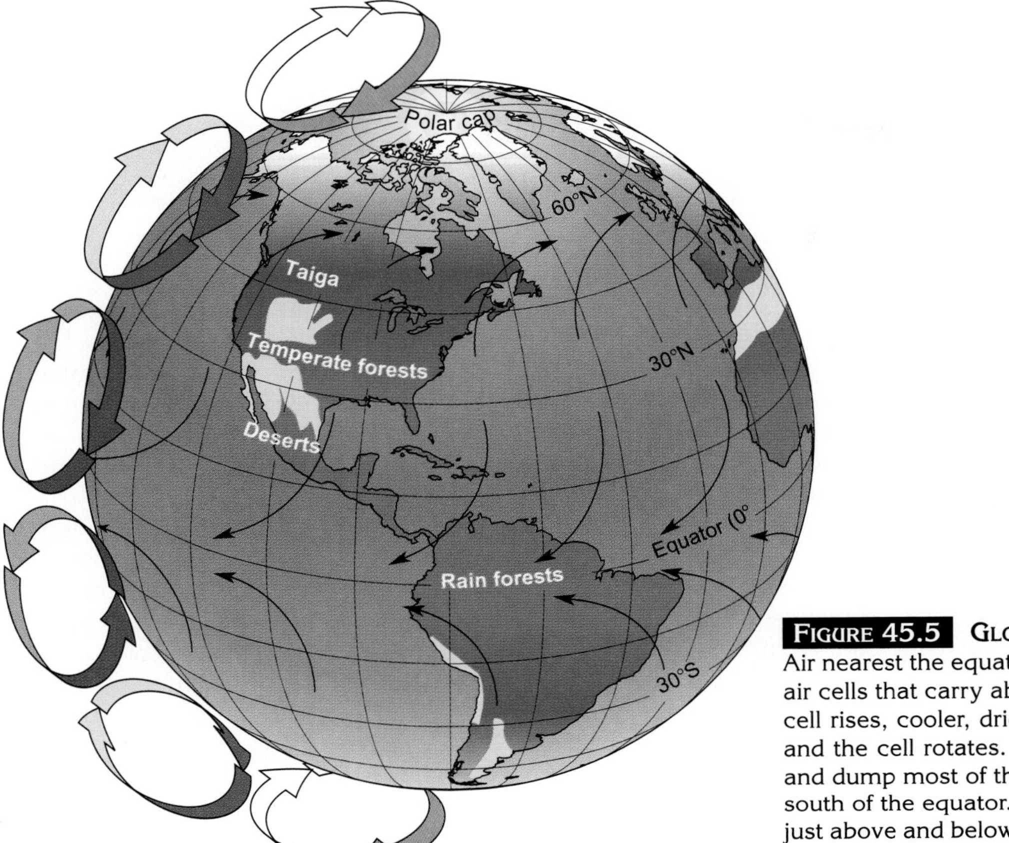

FIGURE 45.5 **GLOBAL WIND PATTERNS.** Air nearest the equator rises, forming equatorial air cells that carry abundant moisture. As an air cell rises, cooler, drier air rushes in underneath and the cell rotates. The rising air masses cool and dump most of their water in belts north and south of the equator. Rainfall is far more limited just above and below these belts; here are found some of the earth's great deserts.

epiphytes, plants that live on trees and that do not touch the soil below. They absorb water directly from the surrounding humid air.

The forest floor is often nearly devoid of foliage (Figure 45.6), but it is teeming with fungal and bacterial decomposers and insect scavengers. The darkness, warmth, and dampness are ideal for rapid decomposition. However, the products of decomposition do not accumulate as humus (partially decayed organic material) and enrich the soil as they do in other forests. Instead, they rapidly cycle back into the living plants, leaving the soil notoriously poor. The continual rain contributes to the poverty of the soil by percolating through it and leaching away valuable nutrients. Because the soil is so poor, clearing tropical rain forests for any major agricultural undertaking is foolhardy—the benefits are short-lived and marginal. Yet people continue to clear such areas for agriculture, causing yet other problems. For example, burning the trees cleared from such areas places an added CO_2 burden on the atmosphere (Essay 45.2).

The tropical rain forest harbors a great number of species. In fact, the species diversity here is greater than in any other biome. Insects and birds are par-

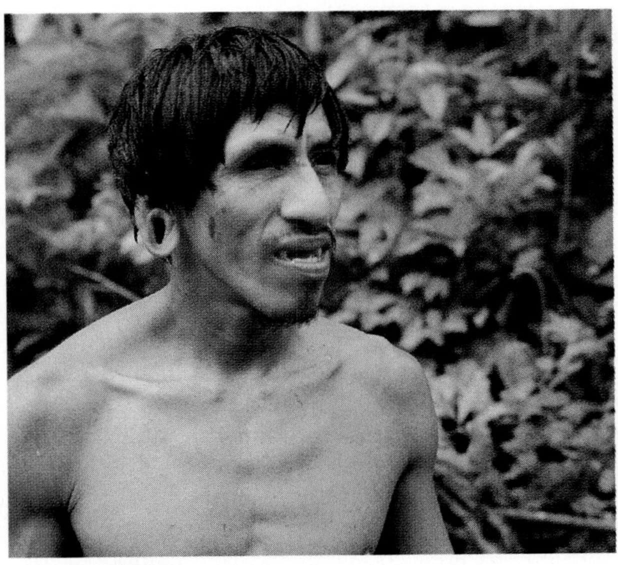

FIGURE 45.6 **THE FOREST FLOOR.** The indigenous people of the rain forest can move across the forest floor because the heavy canopy above allows little light, which would promote plant growth, to reach the floor. Here is a Waorani hunter of Ecuador's rain forests, who hunts the monkeys and parrots that inhabit the canopy above.

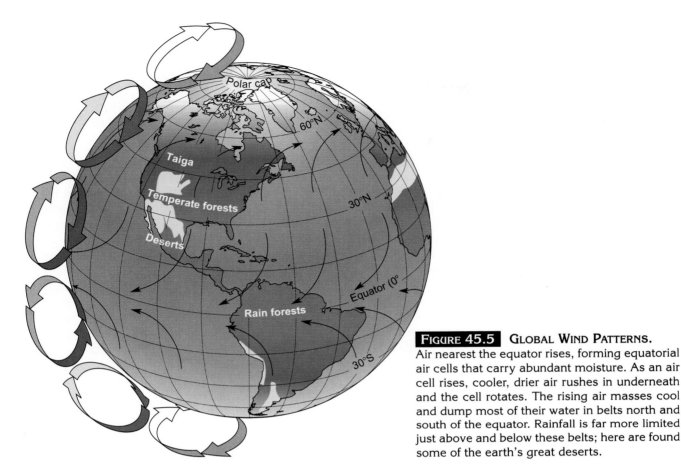

Essay 45.2

The Destruction of Tropical Rain Forests

More than half the world's people live in tropical and subtropical areas. About a third of these people are extremely poor. These two facts have set the stage for a potential worldwide disaster. The problem is, these are the people with direct access to the world's tropical rain forests. Those rain forests can meet some of these people's immediate needs, and so the forests are utilized on a pell-mell, devil-take-the-hindmost basis. Locally, the great forests are yielding to the unending search for firewood and new planting and grazing areas. In recent years, much of the Central and South American rain forest was cleared for grazing areas in order to supply North American fast-food places with beef. (This low-cost beef reduced the price of United States fast-food hamburgers by about a nickel.) Rain forests are also being cut to clear areas for gold mining and for the export of hardwood logs. Unfortunately, such logging operations are often highly inefficient. (In some Malaysian forests, over 50% of the trees in an area must be cut to obtain 3% that can be sold.) By 1980, almost half the world's tropical forests had been destroyed or severely disturbed. Between 1981 and 1990, about 9% of the world's rain forests was lost. In 1990, an additional 42 million acres (17 million hectares) were deforested, an area equal to that of Washington state. By 2000, it is expected that virtually no tropical rain forests will have escaped some level of devastation.

The greatest impact occurs through simply clearing the land. Worldwide, 100 acres of tropical rain forest are cleared or degraded (changed so that it cannot recover) each minute. Bulldozed areas, in particular, will never be recovered because the thin layer of topsoil, with its nutrients and germinating seeds is removed. In addition, with the clearing, watersheds are disrupted (see figure). Forests that are simply cut may be able to recover because seeds, seedlings, and a protective undergrowth are sometimes left. Land that has been cleared by burning recovers at an intermediate rate—that is, burned-over land will take about 80 years to recover.

The burning is done in order to release the nutrients the trees have stored back to the earth. The practice has long been carried out by the Indians of the forest through what is called "slash-and-burn agriculture." Here, the trees are cut over a small area and burned, thereby releasing their stored nutrients back to the soil. The problem is, the land is fertile for only a few years, and so the Indians must then clear new areas. Such small-scale destruction, though, has a negligible impact on the biome, unlike the enormous destruction brought by bulldozers and chain saws.

The extremely heavy rainfall that gave rise to this rain forest in the first place is almost entirely composed of moisture sent aloft from the steaming forest itself. Little additional moisture moves in from the ocean. Cutting down the trees greatly reduces the amount of moisture returned to the atmosphere within the Amazon basin, and this will ultimately reduce the rainfall there by a substantial amount. The rain that does fall into the cleared land will tend to run off uselessly into the river system and then to the ocean.

Finally, tropical soils are very poor and infertile. The nutrients formed by decomposition are immediately recycled back into plant growth, so no reservoir of humus remains. In addition, the rains continually leach precious nutrients from the porous soil.

One of the greatest problems resulting from the destruction of the rain forest is in destroying a form of promise we didn't even know we had. We have identified only about one-sixth of the 3 million species believed to exist there. Some of these unknown species may have medicinal properties that could remedy specific problems of humankind in ways we can now only imagine. (One of your authors has brought out a number of plants that have had medicinal value against human maladies, and the collecting and testing goes on.) There may be resistant genes in plants related to our vulnerable crops—genes that could lend their resistance to our crops through genetic recombination. There may also be new kinds of food, antibiotics, and therapies awaiting our discovery. But some scientists believe that, at current rates, 1 million tropical species may be gone by the year 2010. And with them, sadly, go promises untold.

FIGURE 45.7 **ANIMALS OF THE RAIN FOREST.**
Each stratum of the rain forest harbors specific kinds of animals.

ticularly abundant, as are reptiles, small mammals, and amphibians. Many of the animal species are stratified according to the layers established by the plants (Figure 45.7), having become specialists at occupying certain levels of the canopy and subcanopy. In one study of the Costa Rican rain forest, ecologists found 14 ground-foraging animals, 59 species occupying the subcanopy, and 69 in the upper canopy. They further found that about two-thirds of the mammals were arboreal (tree dwellers), as were a number of frogs, lizards, and snakes. To give you a better idea of the species density in such places, in *one tree* in the Peruvian rain forest, scientists discovered 43 species of ants, the same number of ant species as exists in all of Great Britain.

Jungles, by the way, are formed where the sunlight strikes the ground contributing to heavy growth. Such growth is found in areas where the great canopy of trees has been removed, perhaps by being cleared by humans, and the area reclaimed by nature. Jungles also appear along river banks. Jungle trees are often draped in great lianas, or vines. The myth of the dense, impenetrable rain forest arose from river travelers passing through such areas, people who had little interest in getting out of their boats and traipsing around with snakes to see what the forest was really like.

THE TROPICAL SAVANNA BIOME

The **tropical savanna** biome is a special kind of grassland that often borders tropical rain forests. Unlike other grasslands, the savanna contains scattered

FIGURE 45.8 SAVANNA.
A mixed herd walking past a clump of trees, characteristic of the savanna biome.

trees or clumps of trees (Figure 45.8). And unlike tropical rain forests, savannas have a prolonged dry season and an annual rainfall of only 100 to 150 cm (40 to 60 in.). The dry season is often marked by frequent and extensive fires, a phenomenon to which plants have had to adapt.

PLANT AND ANIMAL ASSOCIATIONS

The largest tropical savannas occur in Africa, but they are also found in South America and Australia. In Africa, while grasses are the dominant form of plant life, the strange, oddly shaped baobab trees, along with palms and colorful acacias, accent drab landscape. The number and variety of hooved animal species exceeds that of any other biome in Africa, and includes zebras, wildebeests, giraffes, and antelopes. This is also the domain of such predators as lions, cheetahs, hunting dogs, leopards, and hyenas.

THE DESERT BIOME

The desert is both a dreaded place and a setting for uncounted fables, but the image most people have of deserts probably reflects only a few parts of the actual desert biome. A **desert** receives less than 25 cm (10 in.) of rain per year. That usually falls in a very brief season and even that moisture quickly evaporates or runs off. Deserts are actually quite varied places. The world's largest desert is the Sahara, followed by the vast Australian desert. While all deserts are dry and most are hot, there are cool northern deserts where winter snows are common. All deserts have dramatic day–night extremes in temperature. Without the buffering effect of atmospheric moisture, the desert heats rapidly during the day and cools just as rapidly at night.

PLANT AND ANIMAL ASSOCIATIONS

Plants adapted to the extremely dry climates of deserts are known as **xerophytes** (*xeric,* dry). In the American desert, perennials such as the cactus, ocotillo, Joshua tree, creosote bush, sagebrush, and palo verde (Figure 45.9) are adapted to living for long periods with little water. The various cacti have greatly reduced water loss by replacing leaves with spines and producing a thick, waxy cuticle over their photosynthetic stems. Other perennials shed their small leathery leaves and remain dormant throughout the drier seasons.

Leafy desert perennials save water by having small leaves and relatively few stomata, which are mostly located on the protected underside of the leaf. There is a cost to such adaptation, however, since CO_2 uptake and photosynthesis are retarded.

(a)

(b)

FIGURE 45.9 DESERT.
(a) Both of the North American desert plants shown here (the saguaro cactus, *left,* and the Joshua tree, *right*) have adapted to limited water; such plants are called *xerophytes.* Their thorny epidermis discourages browsing and helps shield the green surfaces from the harsh direct sunlight. **(b)** The desert blooms in brilliant color when conditions are appropriate in its brief reproductive season.

Some desert perennials even close their stomata during the day and open them at night, storing CO_2 temporarily in intermediate compounds (see the CAM plants in Chapter 24).

The tiny but colorful desert annuals reveal another strategy. Their life cycle is short, an adaptation to the desert's brief period of rainfall. In many species, the tough, resistant seeds require a critical amount of water before germinating. If there is not enough water, the seed coats fail to split, and they just sit it out until it rains, perhaps for several seasons.

The animals of the desert are primarily arthropods, reptiles, birds, and mammals. Animals have adapted not only anatomically and physiologically, but behaviorally as well (see thermoregulation in Chapter 37). Many desert animals avoid the heat by simply staying out of the daytime sun and becoming nocturnal (active at night) (Figure 45.10).

Mammals are largely represented by rodents, for which nocturnality has very clear advantages. (Rodents tend to lose water quickly, because of rapid breathing and because small animals have a large surface area relative to their volume.) The desert kangaroo rat (*Dipodomys deserti*) of the southern California deserts is particularly interesting because it *never* drinks. It survives on the water contained in its food and the metabolic water it produces through cell respiration. Its remarkably efficient kidney produces small quantities of highly concentrated urine, but most of the rat's water loss occurs through simple breathing.

Of course, predators such as owls and rattlesnakes must be active when their prey is out and about, so they hunt mostly at night or in the cool of the evening. The relatively few species that hunt in the daytime—such as the swift, long-legged desert lizards—are preyed on by hawks and roadrunners, which are also adapted to daylight conditions. Even the daytime animals, however, restrict most of their activity to the morning and evening hours (Figure 45.10).

THE CHAPARRAL BIOME

The **chaparral** biome (also called Mediterranean scrub forest) is rather insignificant among the forests of the world. It does, however, have unique characteristics and its own peculiar plant associations. For example, it is exclusively coastal, found mainly along the Pacific coast of North America and the coastal hills of Chile, the Mediterranean, southern Africa, and southern Australia. This biome is unique in that it consists of broad-leaved evergreens in regions marked by a very limited winter rainfall—an average of 25 cm (10 in.)—followed by drought that extends through the rest of the year. The climate is moderated, however, by moist, cool air from the oceans (Figure 45.11).

During the dry season, the chaparral places the same demands on its inhabitants as does the desert; both plants and animals must adapt to long dry spells. In fact, some species in the two biomes are quite similar. But many of the chaparral plants are adapted to another factor—fire. Fire is a natural phenomenon encouraged by drought, resinous plants that burn easily, and a deep layer of dry, slowly decomposing litter on the forest floor. Brush fires

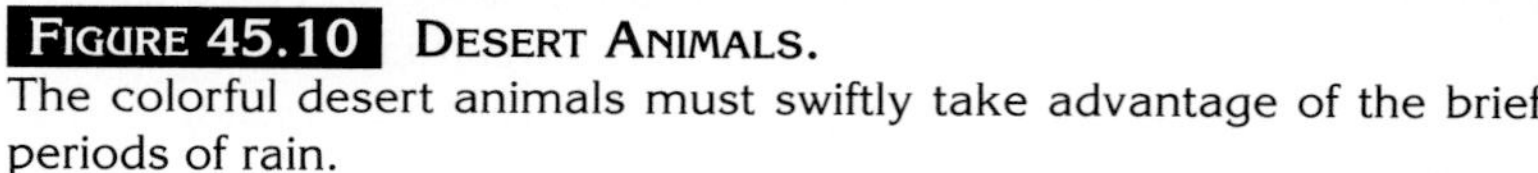

FIGURE 45.10 DESERT ANIMALS.
The colorful desert animals must swiftly take advantage of the brief periods of rain.

periodically sweep across the terrain, leaving behind the charred remains of plants and animals. Since the chaparral is made up chiefly of fire-adapted plant species, in most instances recovery is rapid, with shoots sprouting quickly from burned stumps and fire-resistant seeds. Some species require the heat and scarring action of fire to stimulate germination of seeds. Ecologists describe the chaparral as a *fire-disclimax* community, which means that because of fire it never reaches a state of maturity. Virtually every stand of trees is in some state of recovery. (Unfortunately, in southern California, many homes are built in the chaparral, and the fires that periodically sweep the area are very destructive.)

THE GRASSLAND BIOME

In the Northern Hemisphere, **grasslands** can form huge inland plains, such as the North American prairie and the vast Asian steppes. Extensive grasslands also occur in South America (pampas) and in Australia, where their area equals that of the desert (Figure 45.12). There are several similarities between grassland and desert; in fact, grassland often gradually fades into desert. The chief climatic difference between the two is precipitation; grasslands get more rain (25 to 100 cm per year—10 to 40 in.). Forests do not encroach on such areas because the rain is sea-

FIGURE 45.11 CHAPARRAL.
The chaparral in southern California may lack the glamour and luster of many other forests, but its plants are tenacious and hardy. These scrubby plants resist fire and annual drought that would discourage most other plants.

sonal, often not sustained enough to support forests. Another factor that prevents the formation of forests is fire, a natural and recurring event in grasslands.

PLANT AND ANIMAL ASSOCIATIONS

As you might expect, the dominant plants of grasslands are grasses. But since there are many kinds of grasses, grasslands may be very different from one place to the next. For example, on the American prairie (at least the part that is still identifiable), the grasses east of the Mississippi are generally tall, perhaps reaching 3 m (the tallgrass prairie), while those in the West rarely exceed 0.5 m (the shortgrass prairie). The difference is principally due to variation in rainfall.

Since rains are commonly seasonal in most grasslands, the plants have developed strategies for drought survival. In lowland regions, roots may penetrate as far as 2 m below the surface to the permanent water table. Some grasses, however, rely on a shallow, diffuse root system that can take advantage of light rains. In addition, during drought, grasses readily become dormant, reviving when water is once again available. Some grasses produce underground stems (rhizomes) that remain alive after all the foliage has died. Historically this matlike growth, or sod, prevented agricultural intrusion into the prairie until plowing implements were improved—hence the term "sodbuster." There is little virgin grassland in the world today, because most of the natural grasses have been replaced by grains and other crops. (The United States, by the way, has shown little interest in protecting natural grasslands. To see this biome growing wild in North America, one generally has to go to Canada.)

FIGURE 45.12 GRASSLANDS.
Grasslands, like savannas, support many herbivores. This South American grassland is made up mostly of bunchgrasses, the primary food of llamas, alpacas, and many smaller animals.

FIGURE 45.13 **TEMPERATE DECIDUOUS FOREST.**
The temperate deciduous forest changes its appearance with the seasons. The gently blushing spring turns to the lovely green hillside of summer that explodes in a riot of colors when autumn arrives, all in sharp contrast to the starkness of winter.

Grasses are highly efficient at rapidly converting solar energy into the chemical-bond energy of their living matter, so it is not surprising that the grasslands can support large animal populations. Chief among the mammals are the hooved and burrowing types. In North America the hooved animals include bison and pronghorn, animals whose teeth and digestive systems are well adapted to a diet of tough grassland plants. Both groups were drastically reduced in number by hunting and agriculture, and are now seen almost entirely in protected areas.

The Temperate Deciduous Forest Biome

If you live east of the Mississippi River, you are well acquainted with the **temperate deciduous forest** biome. (A deciduous forest is one in which trees lose their leaves seasonally.) The temperate deciduous biome extends over much of the eastern United States and northward into southeast Canada. It is also found in Europe and parts of Asia. Rainfall in the deciduous forest is rather evenly distributed throughout the year, often averaging more than 100 cm (40 in.)—enough to support a variety of plant life.

Temperate deciduous forests are usually characterized by marked seasonal changes, but unlike more northerly biomes, they have quite a long growing season. With the onset of winter, much of the water becomes frozen and unavailable to plants, so most perennial deciduous species become dormant at this time (Figure 45.13).

Plant and Animal Associations

The American deciduous forest is generally subdivided into a number of forest types in which certain species dominate. For example, beech and maple forests dominate in the north, while oak-hickory and oak-chestnut complexes are prevalent farther south (mostly oak today, since the chestnuts have been decimated by fungi).

White-tailed deer are plentiful (sometimes too plentiful) in temperate deciduous forests, their numbers no longer reflecting the pressure of the great predators. The temperate deciduous forests of North America were once the home of the gray wolf, the mountain lion, and black bear. Now these predators are mainly restricted to protected areas or, as in the case of the wolf, have been virtually eliminated from all but the more inaccessible areas of the north. Carnivores and omnivores of most North American deciduous forests today include the bobcat, raccoon, opossum, skunk, and an occasional red fox.

Animals of these forests adapt to the drastic seasonal changes in a number of ways. Some—primarily birds—migrate to southerly winter habitats. Other animals lapse into the long, chilled stupor called *hibernation,* their metabolism drastically lowered. Yet others—bears, for example—fall into a deep sleep, perhaps occasionally rousing themselves for brief foraging expeditions. Some species like the bobcat and fox must simply brave the cold and its food shortages.

The Taiga Biome

The **taiga,** or boreal forest, is almost exclusively confined to the Northern Hemisphere. There is nothing comparable in the Southern Hemisphere. It is made up primarily of great evergreen forests of pine, spruce, hemlock, and fir that extend across the North American and Asian continents. Smaller forests are found at higher elevations in many mountain ranges (Figure 45.14). The taiga is subject to long, frigid winters and short summer growing seasons.

Figure 45.14 Taiga.
The taiga is an extensive biome, found almost exclusively in the Northern Hemisphere. Since the conditions of the taiga are duplicated in high mountains, similar communities are found here. Shown are an elk, a grizzly bear, and a lynx.

Plant and Animal Associations

Although the taiga is characterized by conifers, communities of poplar, alder, willow, and birch may be found in disturbed places. The taiga is interrupted in places by extensive bogs, or muskegs, the remnants of very large ponds. The most common trees are spruce; low-lying shrubs, mosses, and grasses form the spongy ground cover.

Conifers have adapted to their cold dry environment through reduced needlelike or scalelike leaves that are covered by a waxy secretion. This specialization retards water loss, an important factor since arid conditions are common and ground water is frozen and unavailable throughout most of the year. (One is reminded of the spines and waxy cuticles of desert plants.)

The taiga harbors such large herbivores as moose, elk, and deer, and the omnivorous grizzly and black bears. Wolves still roam here, as do lynx and wolverines. Rabbits, porcupines, hares, and rodents abound, but insect populations aren't as large as they are in deciduous forests. However, there is a disconcerting abundance of mosquitoes and flies (including the voracious black fly) in some regions during the summer.

Coniferous forests are also found in high mountains such as the Sierra Nevada, where conditions mimic those of the taiga, and in regions of more moderate climate. Examples include the famous coastal redwood stands of Oregon and California, the sequoia forests of central California, and vast pine forests in our southeastern coastal plain.

The taiga is a continuing target of the lumber industry, as are the more southerly coniferous forests. The forests have been partially protected so far by their very size, but the lumber companies seem to be taking whatever they can reach, including ancient stands of redwoods. Their much-publicized replanting programs usually replace mixed forests of genetically diverse and disease-resistant plants with artificially developed and genetically homogeneous trees that are fast-growing and can quickly be reharvested. Biologists are still trying to assess the dangers of such genetic uniformity in a large system.

The Tundra Biome

Tundra, the northernmost biome, has no equivalent in the Southern Hemisphere (except for a few alpine meadows). The annual precipitation is meager, often less than 15 cm (6 in.), and much of this falls as snow. During the greater part of the year, much of the moisture exists as ice and is unavailable to most forms of life.

The tundra's growing season lasts about two months. During this time, the frozen surface waters thaw, and ponds begin to form everywhere. This is because the soil remains permanently frozen as *permafrost* a few feet below, so the surface water cannot percolate down. Although the growing season is short, the summer days are long, permitting an extended daily growing period.

Plant and Animal Associations

Tall trees and shrubs are entirely absent from the tundra, except around streams. While the tundra at higher elevations may be composed of only scattered and rather drab plants, in the lower areas growth may be luxurious, dense, and colorful. Pioneering lichens and mosses are common in the tundra, as are dwarfed versions of some familiar trees, among them willows and birches. Grasses, rushes, sedges, and other low-lying plants complete the summer ground cover (Figure 45.15)

Surprisingly, animal life isn't rare in this peculiar and rugged northern biome. In fact, the tundra supports some rather large herbivores. In North America we find caribou and musk oxen, and in Europe and Asia, reindeer. Other animals of the tundra include ptarmigans, snowshoe hares, arctic ground squirrels, arctic foxes, and the ever-present, legendary lemmings (see Figure 45.15).

Winter comes early in the tundra, and with the rapidly shortening days the migratory animals begin their southerly trek. The caribou, for example, leave for the forested taiga, where winter food is more plentiful. Those species that remain prepare for survival in a number of ways. For example, lemmings retreat to food-laden burrows; ptarmigans tunnel into snowbanks, to emerge only periodically on foraging expeditions. The larger resident herbivores rove the barren, windswept landscape to feed on subsistence food such as mosses and lichens.

The Marine Environment

Even a casual glance at a globe reveals that most (about 71%) of the earth's surface is oceanic. Furthermore, the greatest depths of these oceans are deeper than the peaks of the highest mountains are high. The **marine environment** is indeed extensive and complex, containing a vast array of communities. These are divided among two major provinces: the deeper, open sea or **oceanic province,** and the

FIGURE 45.15 **TUNDRA.**
In the summer, the treeless low tundra becomes a marsh as the snow melts. With little runoff, the landscape becomes dotted with small ponds. Plants include a number of dwarfed trees, grasses, and abundant lichens called reindeer moss. Shown are a snowy owl, a lemming, and caribou.

shallower seas along the coastlines, the **neritic province.** The marine environment can also be subdivided vertically into the light-penetrating **euphotic zone** (*eu*, true; *photo*, light) and the perpetually dark **aphotic zone** (*a*, without) (Figure 45.16).

The Oceanic Province

The seas are not uniformly filled with life; in fact, life may be quite sparse in the oceanic province. Nevertheless, because of the vastness of the open seas, the *total* amount of life there is enormous. Much of this life is found in the sunlit euphotic zone, which contains populations of minute **plankton,** floating and drifting organisms, and **nekton,** the swimmers that feed on the plankton.

Perhaps the greatest mysteries on earth lie in the deepest waters of the oceanic province, the depths called the **abyss.** The deepest part of the ocean is the Marianas Trench, which is 10,680 m (over 6 mi) deep. These mysterious depths are places of tremendous pressure and chilling cold. Nevertheless, the abyss supports a surprising number of peculiar, highly specialized scavengers and predators.

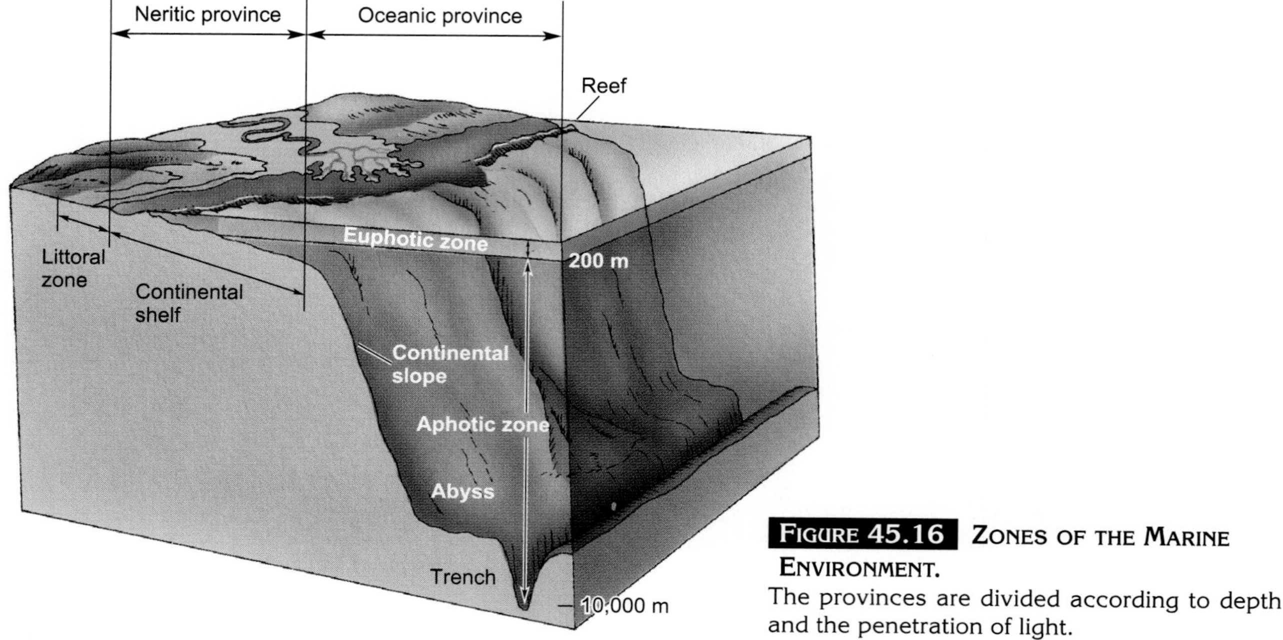

FIGURE 45.16 **ZONES OF THE MARINE ENVIRONMENT.**
The provinces are divided according to depth and the penetration of light.

Generally lacking a producer population (although we will encounter a startling exception in the next chapter), these **benthic** (bottom-dwelling) creatures rely on the continuous rain of the remains of life forms, such as plankton and nekton, from the euphotic zone above.

THE NERITIC PROVINCE

In the neritic province, the land masses extend outward below the sea, forming the highly variable **continental shelf.** The neritic province ends at the **continental slope,** where the shelf drops off, often abruptly. In the shallower areas of the shelf, light penetrates to the ocean bottom. Such regions are constantly stirred by waves, winds, and tides, which keep nutrients suspended and available to many forms of swimming and bottom-dwelling life. Just offshore, giant kelps and other seaweeds form extensive beds, offering hiding places for many fish.

One of the more productive regions in the marine environment is found farther offshore, particularly in the province's colder waters. Here we find regions of **upwellings,** where deep, nutrient-laden colder waters move to the surface. Little is known about such vertical movement of water in the open sea, but we have a better idea of what causes coastal upwellings. They are generally seasonal and occur when coastal winds blow either seaward or parallel to the coast, moving the surface layers, which are then replaced by deeper layers. This stirring brings up nutrients that would otherwise be forever locked in the bottom sediments. The nutrients can then help support photosynthetic organisms, which provide the base for marine food chains—the passage of energy and essential molecules from one trophic (food) level to the next. (The vast anchovy fishing grounds off the coast of Peru are dependent upon such upwellings.) In the oceans, as elsewhere, food chains begin with the producers.

Marine Producers. As light passes through the waters of the euphotic zone, its energy is utilized by a variety of microscopic photosynthesizing organisms—chiefly diatoms and dinoflagellates—that are collectively called **phytoplankton** (*phyto,* plant).

Phytoplankton, seaweeds, and other photosynthetic organisms are the *primary producers* of the sea. This means that the energy of sunlight enters the marine ecosystem via these minute creatures. It has been estimated that 80% to 90% of the earth's photosynthetic activity is carried on by marine organisms. Thus, the sea's food chain begins with tiny marine protists capturing the sun's energy within their fragile bodies. The photosynthesizers are fed upon by organisms only a little larger than themselves, the **zooplankton** (*zoo,* animal). These phytoplankton, then, initiate the food chain as organisms at increasingly higher trophic levels utilize the energy from their tiny bodies (Figure 45.17). Organisms that live on other organisms, dead or alive, are called *consumers.*

Coastal Communities

The varying physical makeup of the coast—sandy beaches, rocky shores, bays, estuaries, tidal flats, and reefs—provides for a number of coastal communities. Such communities fall within the *littoral zone,* the shallow waters along the shore. Life is quite diverse here because of the presence of shelter and hiding places, abundant sunlight, and nutrients swept in by water runoff from the land. In fact, coastal communities are among the most biologically productive of the marine environments. (Some coastal communities comprise wetlands—see Essay 45.3). However, the very nature of the shallow littoral zone imposes critical survival problems on its inhabitants. Included are violently surging waves and surf, and in the intertidal areas (that lie between high and low tide), cyclic flooding and drying as the tides come and go. The littoral zone may be marked by significant variations in salinity because of evaporation at low tide and the presence of freshwater runoff from the land.

Estuaries, where rivers run into oceans, can produce the problem of changing salinity, while low tides in mud flats require that their inhabitants be burrowers. Along rocky coasts (Figure 45.18), a number of plants and animals have adapted to the surging waves by developing means of holding fast to the rocks. In other cases, animals may seek refuge in burrows, or they may lodge themselves in crevices and on the underside of rocks.

Among the most fascinating of the shore communities are coral reefs. *Coral reefs* are common in tropical and subtropical waters where the temperatures average between 23°C and 25°C (Figure 45.19). Corals are vast colonies of cnidarians that secrete heavy walls of calcium carbonate around themselves (Chapter 25). Their irregular growth provides natural refuges for marine animals, including sponges, mollusks (such as the octopus), and many kinds of fishes. Sharks commonly patrol the deep waters alongside the reefs. Where these formations appear along coastlines, they are called *barrier reefs.* The largest is

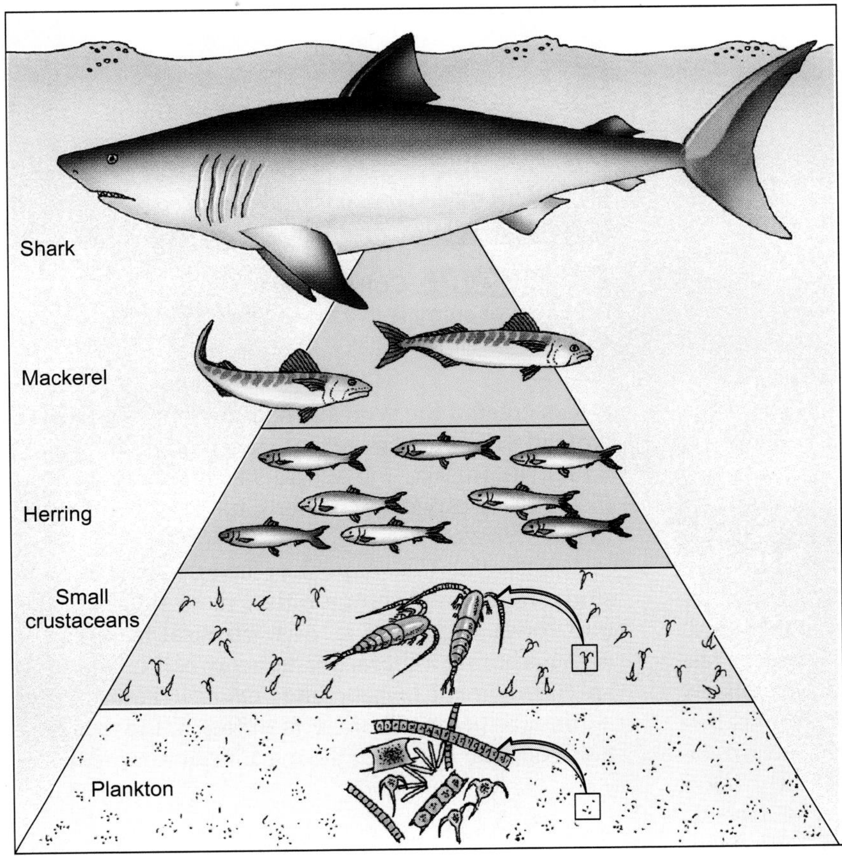

Figure 45.17 The Oceanic Food Chain.
Tiny phytoplankton capture the energy of the sun. They are eaten by animals larger than themselves, which are, in turn, eaten by later animals. At the top are the largest carnivores of the sea.

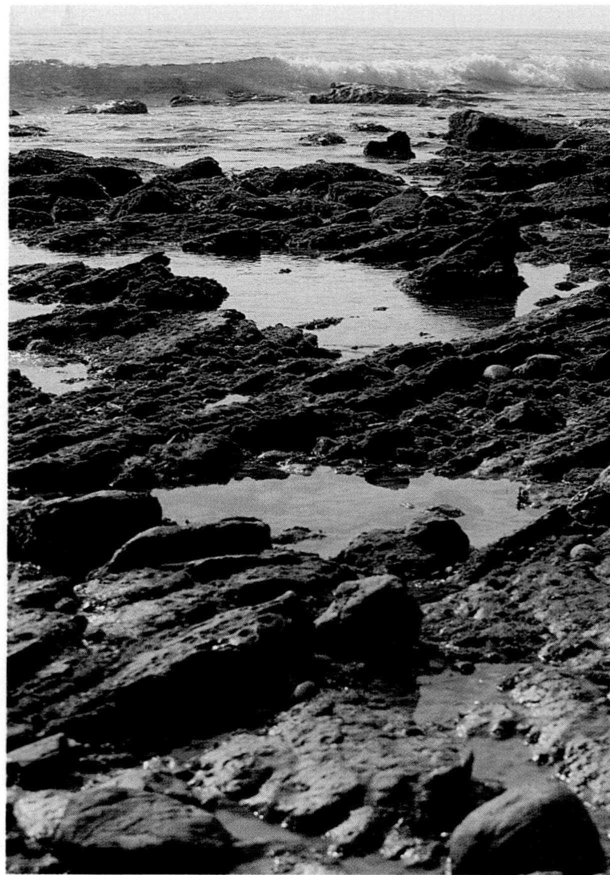

FIGURE 45.18 **ROCKY TIDE POOLS.**
The rocky coast is home for numerous marine animals. Each is adapted in some way to withstand both the surging and pounding of waves and intermittent periods of exposure to the air at low tide.

FIGURE 45.19 **CORAL REEFS.**
A variety of animals live among the heavy walls of calcium carbonate.

the Great Barrier Reef, which extends for 1,200 miles along the east coast of Queensland, Australia.

THE FRESHWATER ENVIRONMENT

The freshwater communities of rivers, streams, lakes, and ponds are among the most heavily studied places on earth, and what we have learned is fascinating, indeed. Some researchers are particularly captivated by those great, still bodies of water called lakes, even though most of them have had a relatively brief history. In fact, most northerly lakes were born at the time of the last glacial retreats (10,000 to 12,000 years ago) when the melting ice left behind great pools. Lakes are also produced through earthquake activity, as was Reelfoot Lake in Tennessee; through volcanic activity, as was Crater Lake in Oregon; and through uplifting of the land, which created the typically shallow lakes of Florida. Not all lakes are as recent, though. In fact, Lake Baikal in Russia, the world's deepest at 1,750 m (5,742 ft), is especially ancient, having been formed in the Mesozoic era.

Limnologists (*limne,* pool or lake), ecologists who study freshwater communities, have divided lakes into zones, each with its own physical features and each harboring a characteristic array of life. The zones are called *littoral, limnetic,* and *profundal* (Figure 45.20). Although the terminology is different, the organization is similar to that of the marine environment.

LIFE IN THE LAKE ZONES

The **littoral zone** of lakes is the shallow area along the shore where light penetrates to the bottom and rooted plants can grow. Producers in the littoral zone include a variety of free-floating and rooted

ESSAY 45.3 Wetlands

A recent administration in Washington promised before the election to protect America's wetlands. Then it enraged environmentalists by simply redefining wetlands so that much of these areas no longer fell under the definition, allowing developers to go ahead and destroy them. Even on purely economic grounds, the administration's decisions were ill-advised. Ecologist Eugene Odum, for example, estimated that an acre of Georgia coastal wetland is worth \$20,000 to \$50,000 an acre as a fish and shellfish producer, but drained and filled in as farmland, only \$4,000 an acre. But there are other reasons for protecting these delicate areas—and our record is not particularly enviable. Since the 17th century, the United States has lost over half its wetlands, with California leading the way, having destroyed over 90% of its wetlands (on a par with Australia and New Zealand).

If they can be so easily redefined for political reasons, the question arises, what *are* wetlands? In the broadest sense, they are simply transitional zones between water and land. They are characterized by the presence of water, but the abundance of that water can vary from standing water several feet deep to damp soil. And they can change from damp to submerged, for example, in tidal areas. The constancy of the water can deplete the underlying soil of its oxygen, and so wetlands vegetation has to be able to tolerate both the water and anaerobic conditions.

Wetlands can be categorized two ways: coastal and inland. Coastal wetlands are affected by tides, although they may be covered by either fresh or salt water. Examples include stands of mangroves, tidal salt marshes and tidal freshwater marshes. Inland wetlands are, of course, freshwater systems and include swamps, marshes, floodplains, and bogs.

Wetlands, along with rain forests and coral reefs, are among the most biologically diverse areas on earth and are ecologically important on a number of grounds. For example, they act in storing runoff from other areas (helping to prevent flooding), in allowing sediments to precipitate out from water passing through, in removing nitrates and phosphates from water, and in acting as "nurseries" for both fish and bird life.

Because they are often associated with boating and great views, they are coveted by both boaters and builders. Yet they clearly serve a greater purpose than as harbors or condo sites. And they clearly must be protected from the whims and dealings of shortsighted government.

plants that form a progression of types as the water deepens. Some rooted plants break the lake's surface; others are completely submerged. As in marine waters, freshwater producers include numerous species of photosynthetic bacteria, protists, and algae (the phytoplankton). Consumers in the littoral zone include protists, snails, mussels, aquatic insects, and insect larvae. Salamanders and frogs also prefer the littoral zone, as do both herbivorous and carnivorous fish and turtles (Figure 45.21). And here we find a number of wading birds and birds that step gingerly along the mats of broad-leaved plants.

The **limnetic zone** is in open water, but includes only the depths penetrated by light. Almost all the producers here are microscopic. They include the phytoplankton that extend from the littoral zone, along with flagellated algal forms such as *Euglena* and *Volvox.* In northern lakes, phytoplankton populations undergo seasonal *blooms* during which their productivity exceeds that of the plants of the littoral zone. These blooms are a response to available nutrients, sufficient light, and favorable temperatures, and often follow a bloom of nitrogen-fixing cyanobacteria. The primary consumers are the zooplankton, composed largely of tiny crustaceans. Their great numbers are made up of relatively few species, whose populations rise and fall in response to the numbers of producers. At the higher consumer levels, we find principally the lake fishes, which consist mainly of plankton-feeding species and the carnivorous species that feed upon them.

The **profundal zone** begins where light ceases to penetrate the lake waters and extends to the muddy, sediment-rich lake floor. There are, of course, no producers here. Life is represented mainly by decomposers, such as bacteria and fungi, and by a few detritus-feeding clams and wormlike insect larvae. All of the profundal species are adapted to periods

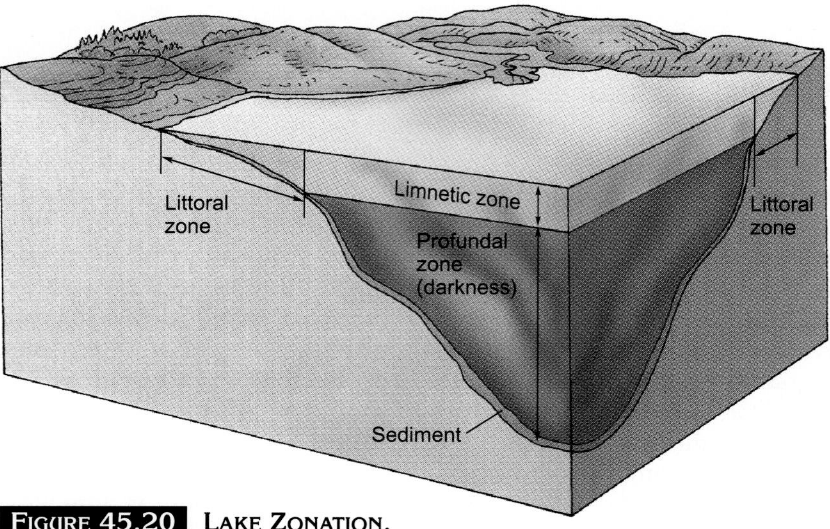

FIGURE 45.20 LAKE ZONATION.
Each zone differs in its physical conditions and inhabitants.

of very low oxygen concentrations. The sparse life here is critical to the organisms above, because decomposers convert deposits formed of corpses raining from above into mineral nutrients. As in marine environments, the distribution of these nutrients throughout the lake depends on the vertical movement of the waters.

TURNOVER AND LAKE PRODUCTIVITY

There is no aerobic (oxygen-using) life in the deeper waters of the tropical Lake Tanganyika (at 1,440 m, or almost a mile), but there is abundant life in the depths of temperate Lake Baikal, which is about 300 m deeper. This is because the amount of turnover in the two lakes is different. **Turnover** is the vertical movement of water masses brought on by seasonal temperature changes. Turnovers carry dissolved oxygen to the lake depths and bring nutrients to the surface. The seasons, of course, are more marked in temperate regions, and so life thrives in the deeper regions of the temperate Baikal.

The turnover in temperate lakes occurs in the fall and spring, when surface waters undergo drastic changes in temperature. It is a peculiar characteristic of water that, as its temperature decreases, its density increases—it gets heavier—until it reaches 4°C. Below 4°C water's density *decreases* sharply, and at 0°C (the freezing point) it is at its lightest.

These properties mean that cooling surface waters tend to sink, displacing and stirring the warmer layers below, which subsequently rise to the surface. But later in the season, the sinking and stirring stop as surface waters cool below 4°C (reaching their lightest, least dense state at freezing). In case you've ever wondered, that's why lakes start freezing at the surface, and why ice skating is more popular than it otherwise might be. The effects of changes in temperatures and density are seen in Figure 45.22.

The conditions that prevent overturn in Lake Tanganyika are similar to those that occur in temperate-zone lakes during the summer. A temperature gradient exists between top and bottom waters. The lighter, warmer upper waters may be well mixed by wind action, but they cannot move into the dense colder waters below. The result is that many temperate-zone lakes have three distinct summer temperature regions. There is an upper warm-water region, the *epilimnion* (upper lake), and a lower cold-water region, the *hypolimnion* (lower lake), each with relatively constant temperatures throughout. Between the two is a region called the *thermocline,* where temperatures drop as depth increases. Typically, oxygen depletion begins just below the thermocline.

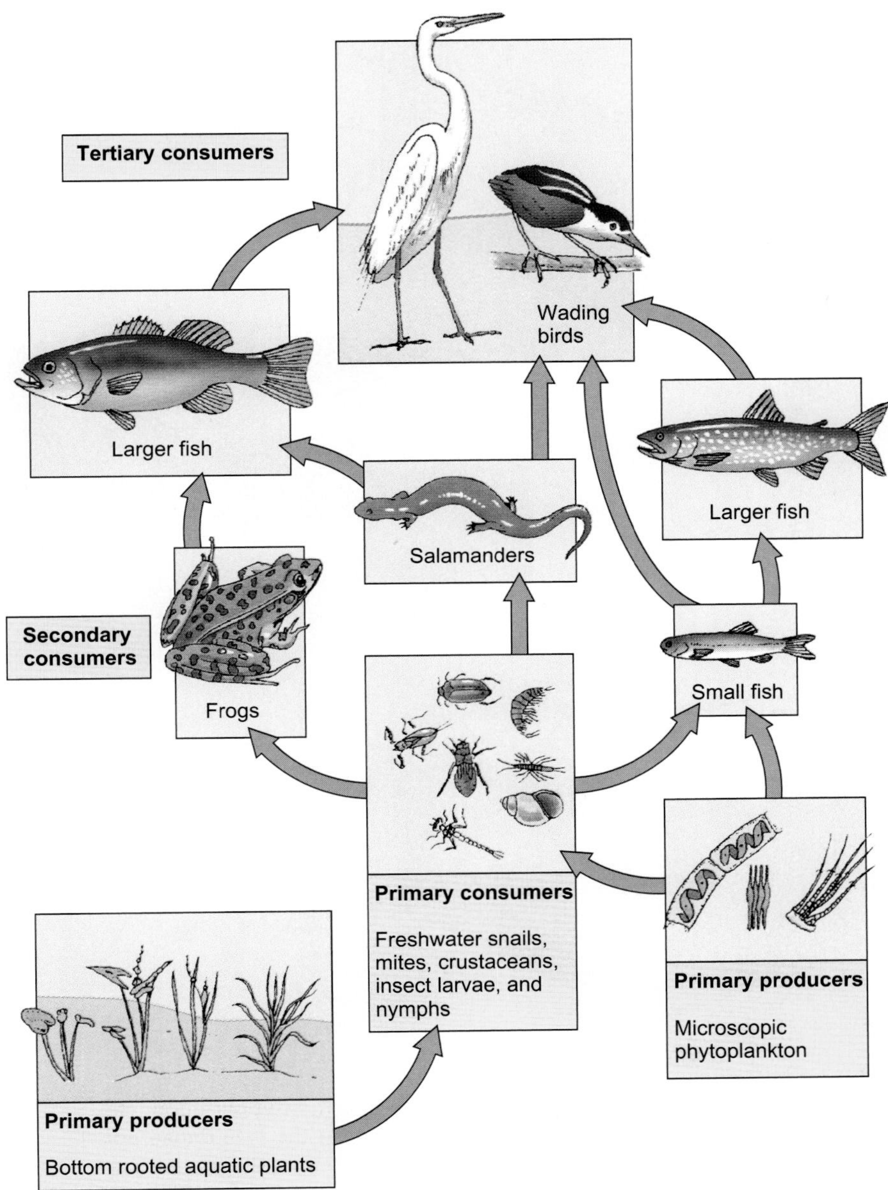

FIGURE 45.21 LIFE IN THE LAKES.
Producers of the littoral zone include bottom-rooted, aquatic plants such as cattails, bulrushes, and arrowheads; floating water lilies; and pond weeds. Microscopic producers (phytoplankton) include many filamentous and single-celled green algae, along with diatoms and cyanobacteria. Primary consumers include the familiar freshwater snails, bottom-dwelling water mites, crustaceans, insect larvae, and nymphs. Smaller aquatic predators of the littoral zone include the diving beetles, water scorpions, and nymphs of the damsel- and dragonflies. There are also larger carnivorous species such as fish, frogs, turtles, and wading birds.

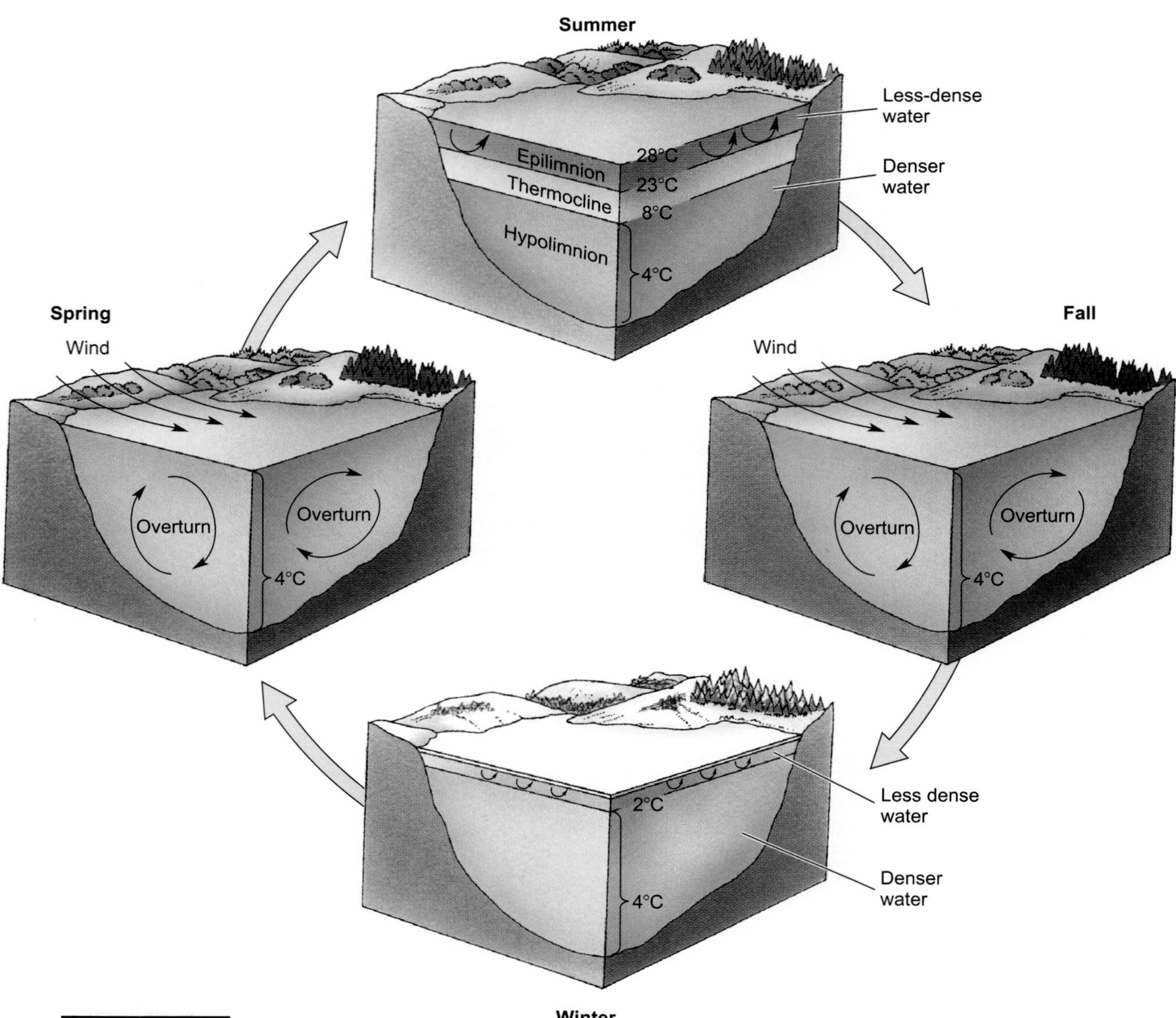

FIGURE 45.22 LAKE TURNOVER.
In summer the highest water temperature is in the epilimnion and the lowest is in the hypolimnion. Between the two, a steep temperature gradient, or thermocline, occurs. Movement in the less dense epilimnion cannot disturb the denser cooler layers below, so overturn cannot occur. In the fall, cooling surface waters approach 4°C (maximum density), sink to the bottom, and permit an overturn. In winter, the summer temperature gradient is reversed, with the coldest temperatures at the surface (no thermocline forms), yet the density gradient is roughly similar to that of summer, preventing overturn. In early spring, warming surface waters reach 4°C (maximum density) and sink, starting another overturn.

PERSPECTIVES

In this chapter, we have seen that the vastness of this small planet permits a wide range of environments and a great array of habitats. Keeping in mind the constant exceptions we encounter in biology, it is possible to categorize the earth's environment according to the arrangement of the life it supports. Such generalization may help us to understand the present status of life on the planet and may also yield clues regarding the historical progression that led to what we see today.

We will focus on relationships again in the next chapter as we discuss ecosystems and communities. And again, we will divide the big picture into smaller, more manageable units.

Summary

Biosphere: The Realm of Life

Ecology is the study of the interaction between organisms and their environment. The biosphere is the part of the earth that can support living things. It includes land areas, water habitats, and life-supporting air.

The atmospheric gases are primarily water vapor, nitrogen, oxygen, and carbon dioxide. The layer of gases is relatively thin, but it effectively screens out potentially harmful radiation. The presence of water confers a heat-retaining characteristic that provides a moderate climate.

Of the incoming solar energy, 30% is reflected back, 20% is retained in the atmosphere, and 50% reaches the surface. The resulting heat energy powers the hydrologic cycle, redistributing water and heat over the earth.

The earth, spinning on a tilted earth's axis as it orbits the sun, experiences continuously changing seasons. Without these seasonal fluctuations, the present temperate areas of the earth would be perpetually frozen.

Biomes and the Waters of the Earth

The uneven heating of the earth and the variable climate it produces result in the varied distribution of life across the planet. A biome is a group of ecological communities characterized by a particular array of plants, animals, and microorganisms.

The Tropical Rain Forest Biome

Tropical rain forests receive over 250 cm (100 in.) of rainfall per year. Equatorial heat produces rising air cells that rotate in northerly and southerly directions dumping most of their water in the tropics. The great variety of plant species precludes the predominance of any single species. The soil is poor because the products of decomposition are washed away or rapidly recycled into living plants. The tropical forest harbors a great number of animal species, many of which have become adapted to life in specific layers of the vegetation.

The Tropical Savanna Biome

The tropical savanna is a special kind of grassland that often borders on tropical rain forests. It contains scattered trees or clumps of trees and has a prolonged dry season, with an annual rainfall over 100 cm (40 in.). The largest tropical savannas occur in Africa, where grasses are the dominant form of plant life. There are many hooved animals, as well as predators such as lions and cheetahs.

The Desert Biome

Deserts receive less than 25 cm (10 in.) of seasonal precipitation per year. They also have dramatic day–night extremes in temperature. Xerophytes have adapted to the dry climate and temperature extremes with waxy cuticles, fewer stomata and leaves, and seeds that can remain dormant for several seasons. Animals adapt by conserving water, such as by staying out of the sun, producing concentrated urine, and being active mainly at night.

The Chaparral Biome

The chaparral is exclusively coastal and consists of broad-leaved evergreens in regions marked by winter rainfall and summer droughts. Plants and animals have adapted in much the same way as those in desert biomes, with additional plant adaptations for fire.

The Grassland Biome

Grasslands receive more moisture than deserts but resemble them in other respects. Plants must adapt to periods of drought and be resistant to fire. Many have vast diffuse root systems; some produce rhizomes that remain alive after foliage has died. Abundant plant life supports a variety of insects, birds, and hooved and burrowing mammals.

The Temperate Deciduous Forest Biome

Temperate deciduous forests cover large sections of the world's landmasses. Precipitation is evenly distributed throughout the year and supports a variety of plant life. The forests have a long growing season, but many deciduous species become dormant in winter. Beech and maple trees dominate northerly American forests, while oak, hickory, and chestnut trees are prevalent further south. Animals in deciduous forests adapt to seasonal changes by migrating, by hibernating, or by altering hunting patterns to adjust to food shortages.

The Taiga Biome

The taiga, made up of coniferous forests, is almost exclusively confined to the Northern Hemisphere. It is subject to long, severe winters and short growing seasons. Taiga is marked in places by muskegs, the remnants of large ponds. Conifers are prominent with needlelike leaves that help conserve water. The taiga supports large herbivores and a variety of other mammals.

The Tundra Biome

Tundra is the northernmost land biome. During most of the year much of its water is tied up as ice; a few feet below the surface, the soil is permanently frozen as permafrost. The growing season is short, but its days are long. The tundra can support large herbivores and predators as well as smaller animals. Animals must migrate or otherwise adapt to the long, frigid winters.

The Marine Environment

The marine environment can be divided into the neritic province along coastlines and the oceanic province in the deeper open sea. Euphotic zones are reached by light; aphotic zones are not.

The oceanic province's deep abyssal regions are characterized by tremendous pressure and dark, cold waters, yet support a surprising array of scavengers and predators.

The neritic province covers the continental shelf, ending at the continental slope. In shallower areas, suspended nutrients support many forms of swimming and bottom-dwelling life. The most productive areas are farther offshore, in regions of upwellings, where deep, nutrient-laden waters move to the surface. The nutrients help support photosynthesizing organisms (primarily phytoplankton) that are the base of the marine food chain. These primary producers are eaten by zooplankton, which are in turn eaten by many fish and marine mammals.

Coastal communities include sandy beaches, rocky shores, bays, estuaries (where rivers meet the ocean), tidal flats, and reefs. These areas can change dramatically with tides and weather, so the inhabitants have adapted in a number of ways to such forces as pounding waves, changing salinity, and periodic dry conditions.

The Freshwater Environment

The freshwater environment includes lakes, ponds, rivers, and streams. Most lakes are recent in origin, products of glaciation. Limnologists divide lakes into littoral, limnetic, and profundal zones.

Lakes, like terrestrial communities, have producers, consumers, and decomposers, with nutrients and gases cycling through the trophic layers. Littoral zones, where light penetrates to the bottom, include shorelines and shallow waters where sunlight and nutrients are abundant. Plants form a progression from the shore outward. Limnetic zones, in which there is still light, can support many types of producers and consumers. The profundal zone is the deepest level, where light ceases to penetrate the water. Although few producers are found there, it is rich in nutrients filtering down from above and supports a large population of bacteria and fungi.

Turnover, the vertical movement of water masses brought on by seasonal climatic changes, carries dissolved oxygen to lake depths and brings nutrients to the surface. In summer, many temperate-zone lakes have a warm epilimnion, a cold hypolimnion, and a thermocline between the two.

Key Terms

ecology • **805**
biosphere • **805**
hydrologic cycle • **806**
biome • **807**
community • **807**
tropical rain forest • **807**
air cell • **808**
epiphyte • **811**
tropical savanna • **813**
desert • **814**
xerophyte • **814**
chaparral • **815**
grassland • **816**
temperate deciduous forest • **818**
taiga • **819**
tundra • **820**
marine environment • **820**
oceanic province • **820**
neritic province • **821**
euphotic zone • **821**
aphotic zone • **821**
plankton • **821**
nekton • **821**
abyss • **821**
benthic • **822**
continental shelf • **822**
continental slope • **822**
upwelling • **822**
phytoplankton • **822**
zooplankton • **822**
estuary • **823**
littoral zone • **824**
limnetic zone • **825**
profundal zone • **825**
turnover • **826**

Review Questions

1. Define the term *biosphere.*
2. List three physical factors that make the presence of a tropical rain forest biome possible.
3. Suggest reasons why modern agriculture fails in the tropical rain forest biome. In what ways is clearing of the tropical rain forest affecting the biosphere?
4. Specifically, how does the tropical savanna differ from the grassland biome? List several familiar animal inhabitants of the tropical savanna.
5. List the physical characteristics of deserts.
6. Where do deserts occur? Summarize two major factors that characterize deserts.
7. Discuss three ways plants adapt to desert conditions.
8. Describe the kangaroo rat's physiological and behavioral adaptations to desert life.
9. List four specific locations of the chaparral forest biome. What physical characteristics do all have in common?
10. Explain what the term *fire-disclimax* means. To what biome does this primarily apply?
11. List four regions where major grasslands occur, and summarize the contributing climatic factors.
12. What is it about grasses that enables grasslands to support vast numbers of large herbivores?
13. Where is the tundra located? List several of its physical characteristics. How do these affect plant life?
14. Prepare a simple scheme showing the organization of the marine biome into provinces and zones.
15. List several animals that inhabit the oceanic abyss. Is there a producer population? What is the basic source of energy?
16. Where is most ocean life concentrated? Describe the phenomenon that assures the producers of a continuing supply of nutrients.
17. To what special conditions must marine animals adapt in the estuaries? Along the rocky coast?
18. What geological force produced most of the world's northern lakes? In which regions do lakes experience the least amount of turnover.
19. Prepare a diagram showing the zonation of a lake. In which of the zones are producers most numerous? In which are they absent? Why?
20. Characterize temperature conditions of tropical lakes or temperate zone lakes in summer. How do these conditions affect the distribution of oxygen?
21. In what season does thermal overturn occur? Explain what brings this on and how it affects life in the lake.

46

Ecosystems and Communities

You will probably find that the ecosystem is not an intuitively *obvious concept. Understanding it requires thinking in terms of energy and how it is channeled around, here and there over the earth. So here we will first look at the grand concept of energy and cycling—and then at communities of living things and why beef is more expensive than beans. We will also see life forms that do not depend on the sun for their energy. So this chapter will be a mixture of concept, fascinating detail and challenge.*

Over the eons in which life has existed, it has invaded, groped, held, thrived, yielded, and died. Its watchword, in a very real sense, has been exploitation, as it reached into every available nook and cranny of the earth, exploiting whatever was available and interacting with other life forms all the while. The result has been myriad life forms that have adapted not only to the habitats offered by the earth but also to each other. Thus we find *constellations* of organisms that have specific roles in the larger picture. To best understand living things and their places in nature, we must consider them not singly but rather as part of a system, a grand interacting group.

We have considered those grand interacting groups called biomes, but anything so large and encompassing presents problems in understanding interactions going on within the system. So within the biome we will look at smaller groups called *communities.* A **community** is a group of interacting plants and animals within a biome. A community, together with the physical, nonliving environment in which it exists, is called an **ecosystem.** It is often fruitful to look at ecosystems in terms of the energy flow through them, so that is where we will begin—with a consideration of **trophic levels**—that is, the positions of organisms in a food system.

Energy in Ecosystems

In tracing the flow of energy in ecosystems, we begin where the energy enters the system. In the last chapter, we saw that about half of the solar energy reaching the earth's atmosphere makes it to the planet's surface. Most of this energy is involved in warming the earth and shifting its waters about. Only a comparatively small amount (about one-tenth of one percent, as a worldwide average) is captured by photosynthesizers. But, meager as it seems, this energy is enough to produce 150 to 200 billion metric tons of organic matter each year. So we will see, now, what happens to that energy after it is captured by the photosynthesizers (and the chemotrophs).

Trophic Levels

Energy entering ecosystems is first captured by **producers.** The producers are autotrophs, the photosynthesizers and chemotrophs (such as the vent organisms at the bottom of the sea). The producers convert energy from light or chemicals into the energy of the chemical bond. From there, energy passes to various **consumers,** heterotrophic organisms that must rely on others for their energy requirements. The bodies of both the producers and the consumers must finally yield to the **decomposers,** who break down the complex bodies of other living things into simpler chemical components. Figure 46.1, a rather generalized view, traces the flow of energy through these trophic levels.

Producers. The phototrophic producers account for the preponderance of food molecules. They include plants, algal protists, and phototrophic bacteria. These producers use light energy captured by various pigments to produce organic molecules from carbon dioxide and water (see Chapter 7). Far less significant energetically, the chemotrophic producers are bacteria that obtain energy from inorganic substances in the earth's crust (Essay 46.1). The combined mass (called biomass) of the earth's producers is about 99% of the earth's total biomass. Read that sentence again.

Essay 46.1 The Oceanic Rift: An Unusual Community

Deep down on the ocean floor, well over a mile beneath the surface, a number of benthic (bottom) communities thrive in the cold blackness. There are no producers here, but there is life. The source of nutrients and energy is the constant deluge of organic debris raining down from the food webs above. Such sources of nutrients rarely go unexploited in nature.

Until 1977, this might well have summarized our knowledge of life on the ocean floor. Scientists believed that scavengers and decomposers were the only inhabitants of benthic communities. But in that year the research submarine *Alvin,* cruising at a depth of 2,500 m near the Galápagos Islands, came upon a startling sight.

The scientists aboard *Alvin* saw some sort of turbulence on the ocean floor. On closer examination, it was found to be a vent in the seabed spewing forth hot matter that, upon analysis, turned out to be hydrogen sulfide and carbon dioxide. This in itself was not entirely unexpected, but what they saw nearby was surprising, indeed. The waters around the vent were cloudy (a condition later attributed to dense aggregations of bacteria). Enormous, strange tubeworms—blood-red, nearly 3 m long, and as thick as a man's wrist—bristled from the ocean floor. These mouthless and gutless animals were named *Riftia pachyptila. Riftia*'s mode of feeding represents one of the most recently discovered and truly fascinating examples of mutualism. An extensive organ in the worm, called the *trophosome,* houses vast numbers of chemotrophic vent bacteria, a type capable of extracting energy from hydrogen sulfide, oxygen, and carbon dioxide, and of using that energy to reduce the carbon dioxide, forming carbohydrates much in the manner of plants and other phototrophs (their own version of the Calvin cycle). The raw materials—hydrogen sulfide, oxygen, and carbon dioxide—are absorbed by the worm's feathery plume and transported to the trophosome bacteria. In return, the bacteria provide a steady source of nutrients for use by the worm. Not surprisingly, a number of other rift animals have developed similar symbiotic relationships with the vent bacteria.

Smaller, previously unknown wormlike animals, arranged in spaghettilike masses, were found with more familiar filter-feeding crabs, mussels, barnacles, and a variety of other animals common to benthic communities. The seabed around the vents was also inhabited by large clusters of huge, smooth-shelled clams with blood-red flesh caused by high levels of hemoglobin, believed to be an adaptation to occasional periods of low oxygen.

The discovery of the vent communities intrigued the scientific world, and an immediate effort was made to collect and study specimens and to sort out the organization of this bizarre assemblage. Finally, the pieces of the puzzle began to fit together. It was then that scientists learned that primary consumers in this community do not feed upon detritus falling from above, but are totally independent of life in the upper realms. The source of energy around the vents was found to be the enormous populations of chemosynthetic bacteria that crowd the vent opening. These bacteria thrive on the chemicals boiling forth, producing carbohydrates on the spot.

The rifts have a seemingly endless supply of food. Oxygen is abundant in the cold abyssal water, so metabolic activity can be rapid. The large Galápagos rift clams, for instance, grow at a rate of 4 cm per year, about 500 times faster than their relatives in other waters.

At first, scientists aboard *Alvin* believed that the rift community they had discovered was unique to the Galápagos area, but other rift communities were discovered, and now it appears that they are quite extensive, cropping up here and there among the innumerable faults in the ocean floor. In fact, in time they may prove to be among the most widespread and richest of marine communities. Rich and active though they are, the rift communities can pass quickly. Like volcanoes, the hot springs eventually die down, leaving behind ghostly monuments of empty shells. But as some vents close, others appear in cracks torn in the sea bottom by the earth's restless, shifting crust.

Here, then, is a system that does not rely on energy emanating from the sun, but from primordial heat captured by our small planet when the sun was young.

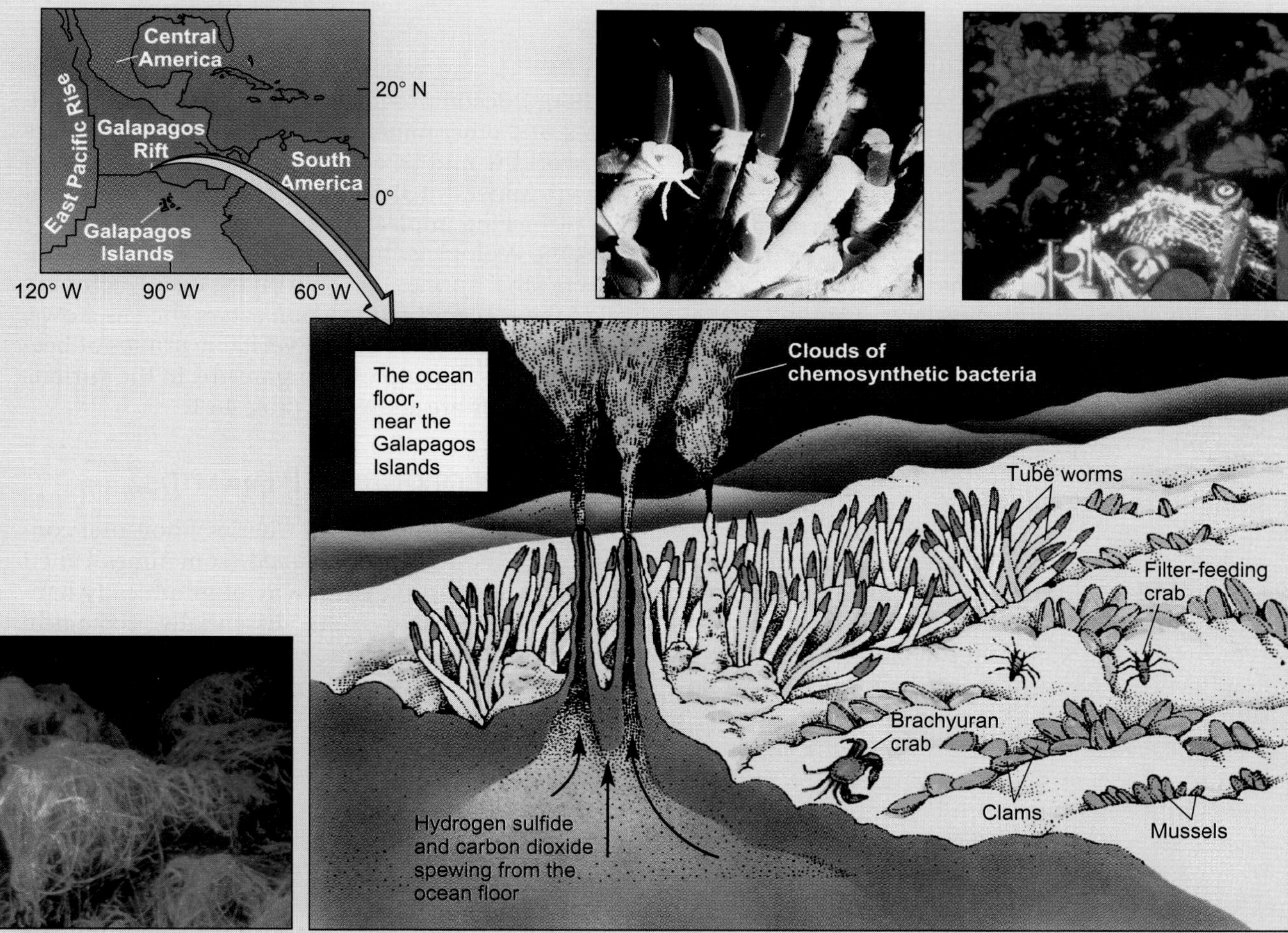

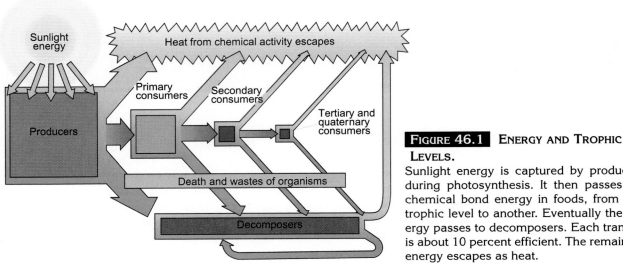

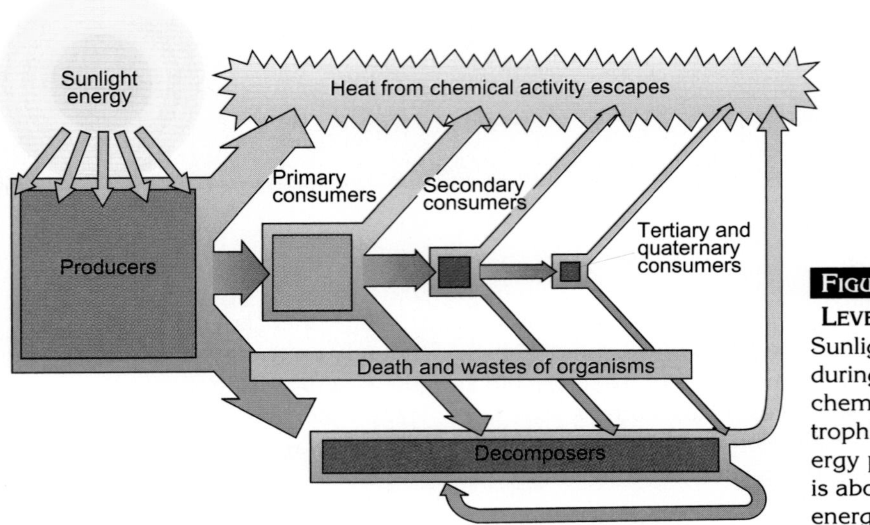

FIGURE 46.1 ENERGY AND TROPHIC LEVELS.
Sunlight energy is captured by producers during photosynthesis. It then passes, as chemical bond energy in foods, from one trophic level to another. Eventually the energy passes to decomposers. Each transfer is about 10 percent efficient. The remaining energy escapes as heat.

Consumers. Consumers—which comprise the remaining 1% of the earth's biomass—include animals, fungi, predatory protists, and bacteria. In other words, these are the earth's heterotrophs. (The fungi and bacteria make up a special category, which we will come to next.) Since some consumers eat producers, others eat other consumers, and some eat both, the flow of energy through the consumers involves several divisions. Thus we have *primary consumers,* the herbivores that feed directly on producers; *secondary consumers,* carnivores that feed on primary consumers; *tertiary consumers, quaternary,* and even higher consumer levels (with increasingly rare representatives).

Organizing the trophic levels in this manner can be done by simple food chains (A eats B which eats C, and so on). But in reality, nothing is ever that simple. Most consumers cross trophic levels. For example, there are few true carnivores (some sharks and flies are examples). Consider humans. At how many trophic levels do we feed? Do we have salad with our steak? The steak comes from a herbivore, of course—but what if we eat a tuna sandwich? Tuna eat other carnivores. Because of such complexities, feeding patterns in a community are better represented by *food webs,* such as we see in Figure 46.2.

Decomposers. Decomposers break down organic matter into simple products that can then be recycled through living systems. Decomposers are great in number but usually small in size—not surprising since they are primarily fungi and bacteria. Decomposers generally feed by secreting digestive enzymes into their food and then absorbing the breakdown products (Figure 46.3). In so doing, they produce carbon dioxide and water, as do all consumers; but, in addition, decomposers release sulfates, nitrites, nitrates, and other mineral ions. Without decomposers the world would be a far different place—a corpse-strewn, mineral-deficient wasteland that would render picnicking unpleasant at best.

Some ecologists, by the way, add to the decomposers other species that feed on the dead, including all of the scavengers or carrion eaters such as crows, jackals, crabs, vultures, and veritable armies of beetles and ants. Examples of organisms in the various trophic levels are shown in Table 46.1.

ECOLOGICAL PYRAMIDS

The pioneering ecologist Charles Elton first conceived of the *ecological pyramid* (sometimes called the Eltonian pyramid) as a way of graphically handling some complex data. Essentially, ecological pyramids show the relationships among the different elements in an ecosystem. Their simplicity is deceiving, however, because the data are often hard won. The most common ecological pyramids are *pyramids of numbers, pyramids of biomass,* and *pyramids of energy.*

Pyramids of Numbers. Pyramids of numbers involve counts of individuals at each trophic level. Consider, for example, a grassland community. The graph of the numbers of individuals in a summer grassland at each trophic level produces a typical "stepped" pyramidal shape (as seen in Figure 46.4a). In other systems, such as temperate forests, the pyramid can become more complex, even inverted, as we see in Figure 46.4b.

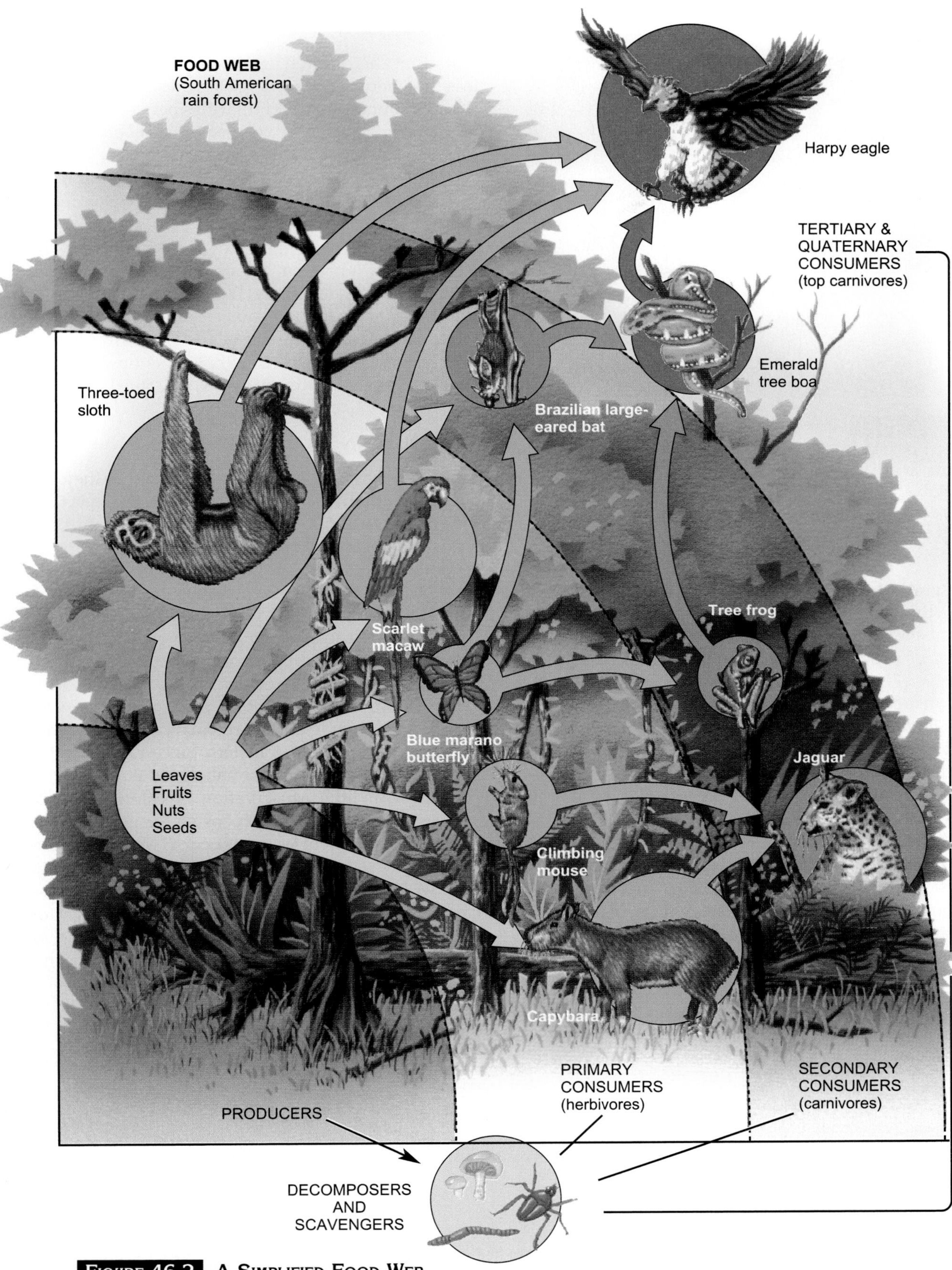

FIGURE 46.2 **A SIMPLIFIED FOOD WEB.**
Food webs reflect the complex feeding patterns in an ecosystem. The arrows indicate the direction of the flow of energy and matter. In nature, some animals feed from more than one trophic level, particularly during shortages when usual food sources become reduced.

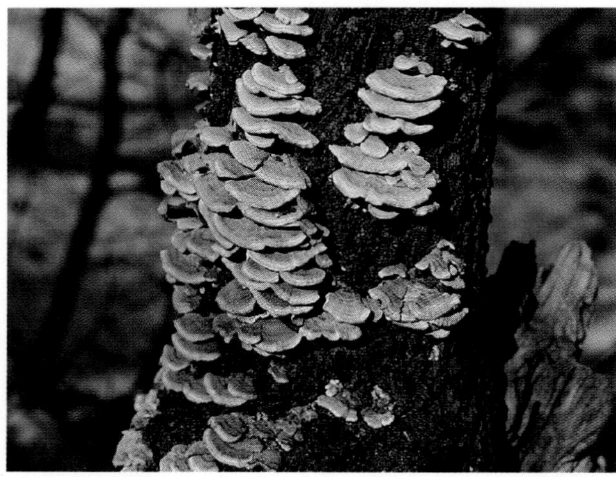

FIGURE 46.3 DECOMPOSERS.
Decomposers break down the wastes and remains of plants and animals, making simple mineral nutrients available again to the ecosystem. Although many are microscopic, the bracket fungi often seen on fallen trees are quite large.

Pyramids of Biomass. Pyramids of biomass depict the weight of the total living matter in an ecosystem at each trophic level. Typically, the biomass of producers is far greater than that of consumers (Figure 46.4c), and the biomass of any level of consumer is less than that of the level below. As was mentioned earlier, 99% of the earth's biomass is tied up in the producer level. But surprisingly little of the producer biomass is transferred to the primary consumer (herbivore) level. After all, of the plant biomass that is eaten by herbivores, some is not actually absorbed but is disposed of as waste. And then, of the part that is absorbed and utilized, a considerable amount is broken down during cell respiration. The result is that very little of the plant biomass becomes primary consumer biomass, and even less becomes secondary consumer biomass. Note that the differences between C_1 and C_2 is far less than between P and C_1. Do you see why that is? (Keep in mind that the similarities between the bodies of different animals are greater than between animals and plants.)

Pyramids of Energy. Pyramids of energy depict the transfer of energy (measured in calories) from one trophic level to the next. Again, we find the stepped configuration, simply because no energy transformation is 100% efficient. Energy is lost at each transfer. In fact, as you can see in Figure 46.4d, the transfers are extremely inefficient, bordering on 90% loss at each step. Nonetheless, enough energy is passed along to power the next trophic level.

HUMANS AND TROPHIC LEVELS

Energy pyramids may seem abstract, but they can suggest fundamental lessons in economics when they are applied to human populations. We know that humans are basically *omnivores,* capable of feeding at several trophic levels, and the availability of food determines which trophic level is most heavily utilized. As economic conditions worsen, people tend to increasingly shift to lower trophic levels. The poorest people tend to act largely as primary consumers; they eat more plant foods and less meat. The reason is clear—the primary level holds more energy, or calories. To feed people at the secondary level or higher requires costly energy transfers. Since raising cattle, sheep, hogs, and other animal stocks is a wasteful process with little conservation of calories or biomass, meat becomes too expensive for poor people, and protein deficiencies may occur, (Figure 46.5). Exceptions are seen where food is available from the sea and where livestock includes scavengers such as chickens and hogs, animals that tend to fend for themselves. People who are trying to impress other people rarely invite them over for a big plate of beans.

TABLE 46.1 EXAMPLES OF TROPHIC LEVELS

Type of Organisms	Energy Source	Example
Producers		
Photosynthetic autotrophs	Sunlight	Plants, algae
Chemosynthetic autotrophs	Inorganic materials	Vent bacteria
Primary Consumers		
Herbivores	Producers (e.g., plants)	Cows, antelopes, grasshoppers
Secondary Consumers		
Primary carnivores	Herbivores (e.g., antelopes, deer)	Lions, wolves, spiders
Tertiary Consumers		
Secondary carnivores	Primary carnivores (e.g., predatory fish)	Penguins
Decomposers	Dead organisms	Bacteria

Beans are low in the energy pyramid and hence not expensive. In fact, the most "impressive" foods are the most inefficiently produced ones, and hence the most expensive. (Perhaps they should invite their status-conscious friends over for a nice vulture.)

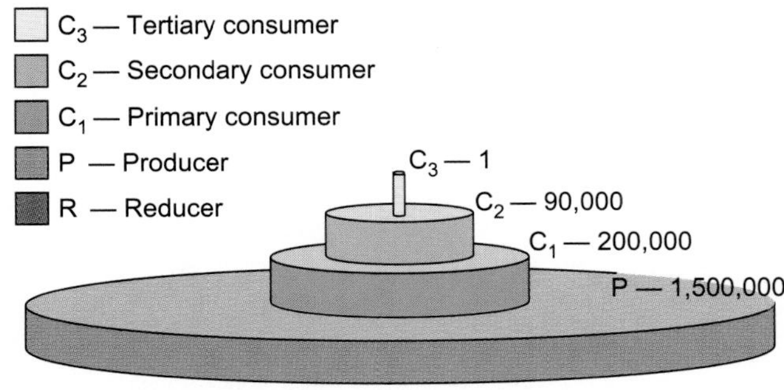

(a) Number of organisms: summer grassland

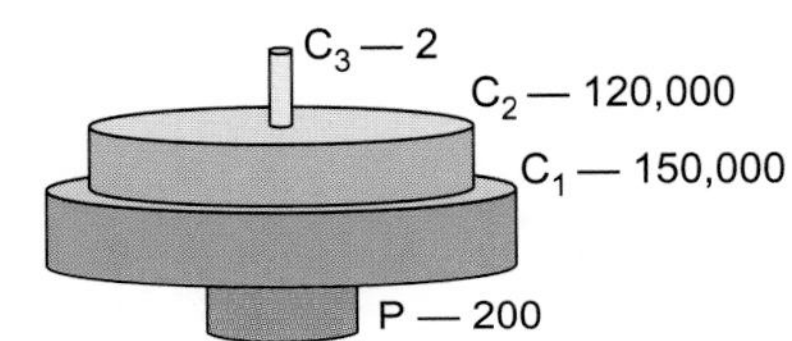

(b) Number of organisms: summer temperature forest

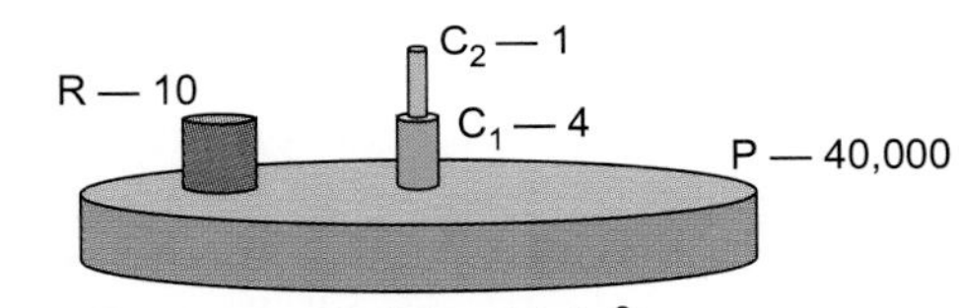

(c) Biomass: Panama tropical forest (g/m^2)

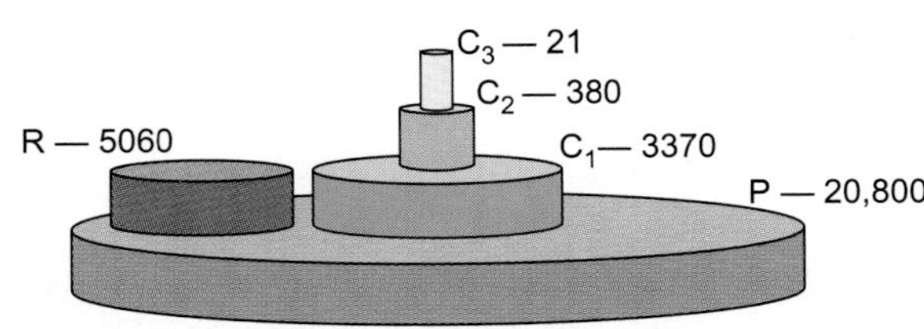

(d) Energy: freshwater organisms (kcal m^2/yr)

FIGURE 46.4 ECOLOGICAL PYRAMIDS.
(a) Pyramid of numbers. Partly inverted **(b)** pyramid of numbers where most of the producers (P) are large trees and shrubs, and their numbers are far exceeded by the numbers of consumers (C) they support. The numbers represents counts for each 0.1 hectare (quarter acre). **(c)** Pyramid of biomass. Here, 40,000 g/m^2 producer mass supports a total consumer mass of only 5 g/m^2. The large decomposer biomass should be expected in a community where nutrient cycling is very rapid. **(d)** An energy pyramid from a freshwater aquatic community at Silver Springs, Florida. Energy flow is expressed as kcal/m^2[pd]year.

PRODUCTIVITY IN ECOSYSTEMS

The rate at which producers in an ecosystem store energy is referred to as the system's **primary productivity,** a concept that has three main aspects: *gross productivity, net productivity,* and *net community productivity.*

Gross primary productivity is a measurement of the total amount of sunlight converted to chemical energy by producers. This energy is then available to be used by the producers themselves. Gross primary productivity doesn't really tell us how much energy the photosynthesizers are storing or how fast they are growing.

Net primary productivity is the energy the producer has left for storage after its own energy needs are met. In other words, the rate of energy stored minus the energy used in respiration equals net primary productivity. Net primary productivity is reflected in such things as new growth, seed production, and the storage of such energy-rich compounds as lipids and carbohydrates. It represents the biomass that is available to heterotrophs.

Net community productivity is the net primary productivity (the energy stored by producers) minus that used in respiration by heterotrophs. By measuring the net community productivity, ecologists can compare energy efficiency within systems and can answer such questions as: Is it growing? Has it reached a climax state (full maturity)? Is it declining? An increase in net community productivity, as you might expect, occurs only in communities that are growing. Eventually, though, the rates of energy productivity and energy use begin to equalize, until there is no net community productivity. This is the **climax state.** A net decrease in productivity signals a dying or declining community (or perhaps, in temperate zones, just the approach of winter). The results of productivity studies in several major regions of the earth are summarized in Table 46.2.

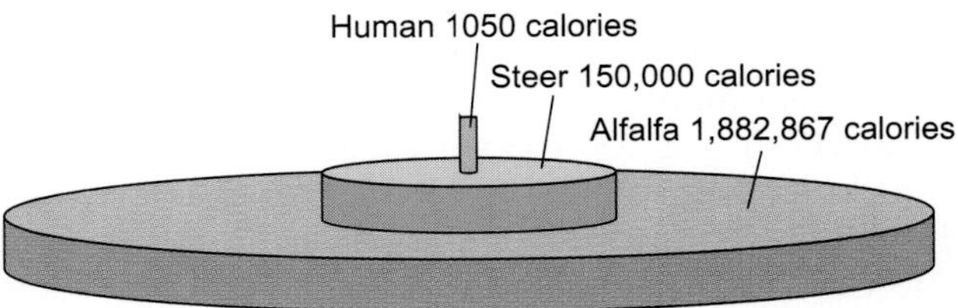

FIGURE 46.5 HUMANS AS SECONDARY CONSUMERS.
Humans feeding on beef are secondary consumers. The energy differences between trophic levels are rather typical. Note that for every 1,000 calories gained by humans, about 2 million would have been processed at the producer level.

TABLE 46.2 Estimated Gross Primary Production (Annual Basis) of the Biosphere and Its Distribution Among Major Ecosystems

Ecosystem	Area millions of km^2	Gross Primary Productivity kcal/m^2/yr	Total Gross Production 10^{16} kcal/yr
Marine			
Open ocean	326.0	1,000	32.6
Coastal zones	34.0	2,000	6.8
Upwelling zones	0.4	6,000	0.2
Estuaries and reefs	2.0	20,000	4.0
Subtotal	362.4	29,000	43.6
Terrestrial			
Deserts and tundras	40.0	200	0.8
Grasslands and pastures	42.0	2,500	10.5
Dry forests	9.4	2,500	2.4
Northern coniferous forests	10.0	3,000	3.0
Cultivated lands with little or no energy subsidy	10.0	3,000	3.0
Moist temperate forests	4.9	8,000	3.9
Fuel-subsidized (mechanized) agriculture	4.0	12,000	4.8
Wet tropical and subtropical (broad-leaved evergreen) forests	14.7	51,200	29.0
Subtotal	135.0	82,400	57.4
Total for biosphere (round figure; not including ice caps)	500.0	1,000	100.0

Source: *From Fundamentals of Ecology, 3rd Edition by Eugene P. Odum. Copyright © 1971 by W. B. Saunders Company. Reprinted by permission of Holt, Rinehart and Winston, CBS College Publishing.*

Productivity in a Forest Community. In 1969, ecologist George Woodwell and his associates completed a 10-year study of productivity in a scrub oak-pine forest community near the Brookhaven National Laboratory on Long Island. Oak-pine forests are common in this area, a product of the disturbance of the great deciduous forests that once graced the rural landscape. Woodwell's study was enormous in scope, but it is through such efforts that we are beginning to understand the basic principles of community ecology.

Woodwell and his group concluded that the annual gross primary productivity of the forest community was 2,650 g/m^2 (about 5.8 lbs), while the annual net community productivity was 1,200 g/m^2. Further tests revealed that the rate at which new organic material was appearing—the net community productivity (net primary productivity minus heterotroph consumption) was 550 g/m^2/year (about 1.2 lbs). The forest study is summarized in Figure 46.6.

Similar studies have since been done in other forests, and we now know that, compared to some communities, the net productivity of Woodwell's oak-pine forest is modest. Annual net community productivity in each square meter of some tropical rain forests, for example, can reach several thousand grams. Net community productivity is also high in estuaries, wetlands, coral reefs, and where intensive agriculture is carried out. For instance, in tropically grown sugarcane, an efficient C_4 plant (see Chapter 7), the annual net community productivity can exceed 9,000 g/m^2, while the productivity of grain fields ranges between 6,000 and 10,000 g/m^2.

One more observation—perhaps an obvious one—was made by Woodwell. Because the oak-pine community was growing, it had not reached its climax state. As we mentioned earlier, communities that have reached their climax state have no net community productivity. Their respiratory output equals their photosynthetic input.

Nutrient Cycling in Ecosystems

While energy *flows through* ecosystems, the chemical elements essential to life *cycle within* ecosystems. Some of these elements are incorporated into the molecular makeup of the organism, while others are shuffled through complex metabolic pathways and drained of their bond energy. Ultimately, though, all the molecules will become available for recycling either as the organism's waste or as its corpse. At that time, molecules are subjected to the decomposers that release the products of decomposition into the environment where they once again become available to the producers. Most of the elements cycle in

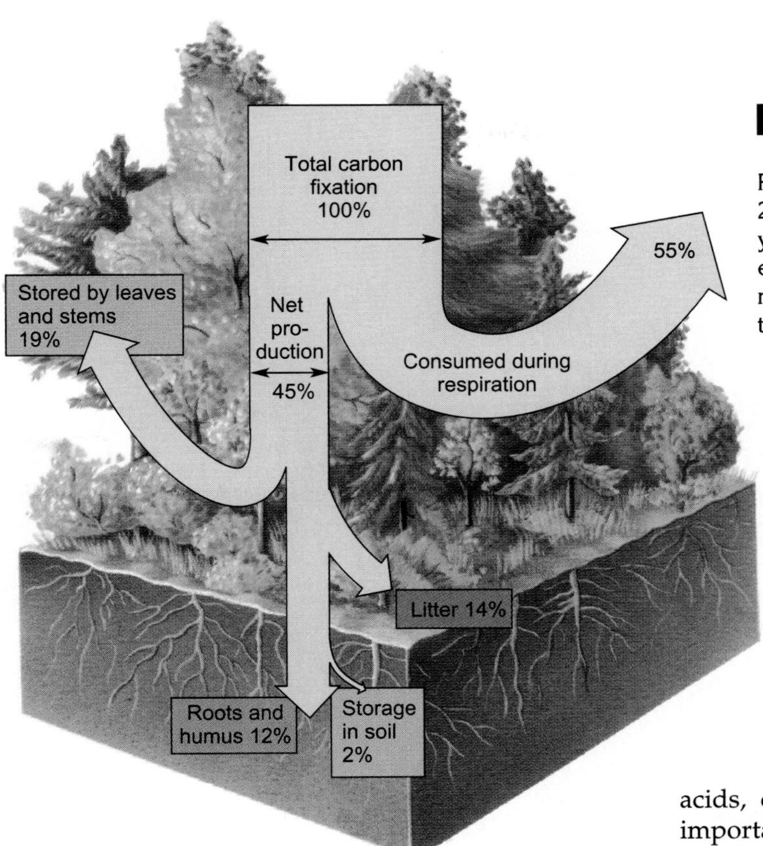

FIGURE 46.6 **THE RATE OF BIOMASS FORMED IN AN OAK-PINE FOREST.** Rate of biomass production is expressed as 2,650 g of organic matter per square meter per year. The pathways point out where the matter ended up. Only 21 percent occurs as new biomass. While a small amount was leached away, the rest was used in respiration.

the form of mineral nutrients (mineral ions) such as nitrogen compounds. Since the cycling of such chemicals involves both biological and geological activity in the ecosystem, the pathways are cleverly called *biogeochemical cycles.*

Some biogeochemical cycles can be rather simple, involving only a few steps. For example, some of the water a plant takes in is simply transpired, released through the leaves back to the atmosphere (see Chapter 7). However, water also may be used in photosynthesis. In this case, the water molecules are split, hydrogen is used to reduce carbon (forming foods), and oxygen is released into the atmosphere. Animals and other consumers may then breathe in the oxygen, using it in cell respiration where it is reunited with hydrogen, forming water, which is then returned to the environment. While we will focus on the nitrogen cycle, two other important cycles, the *phosphorus cycle* and the *carbon cycle,* are outlined and described in Figure 46.7.

THE NITROGEN CYCLE

One of the best-known biogeochemical cycles is the *nitrogen cycle.* We previously referred to the cycle in another context, but here let's consider it in more detail. Nitrogen, you recall, is essential to life, because it is a principal constituent of proteins, nucleic acids, chlorophyll, coenzymes, and several other important kinds of molecules.

Essential nutrients such as nitrogen are generally held in one of two places: *exchange pools,* where they are readily available to living systems, and *reservoirs,* where they are less available to living systems. The largest nitrogen reservoir is the atmosphere, about 79% of which is nitrogen gas (N_2). But atmospheric, or molecular, nitrogen as such is not available to the earth's organisms, except to certain nitrogen-fixing bacteria. Most life can be literally bathed in nitrogen without interacting with it, because molecular nitrogen is extremely stable, resisting any chemical change. Plants and most other producers must incorporate nitrogen primarily in the form of nitrate and ammonium ions ($NO^-{}_3$ and $NH^+{}_4$), which are produced by soil and water bacteria over two complex pathways: *decomposition* and *nitrogen fixation.* These ions in the soil and water constitute the major exchange pool of nitrogen. (Consumers, of course, must obtain their nitrogen from the producers.)

Role of Decomposers. The general principles of the nitrogen cycle are summarized in Figure 46.8, using a simple system involving plants, animals, and decomposers. As we see, plants take in nitrate ions and incorporate them into their own amino acids, where they are used to make plant protein. The molecules then pass through the consumer levels from herbivore to carnivore, where some of the amino acids are used in the production of the animal's proteins while many of the rest are metabolized. The

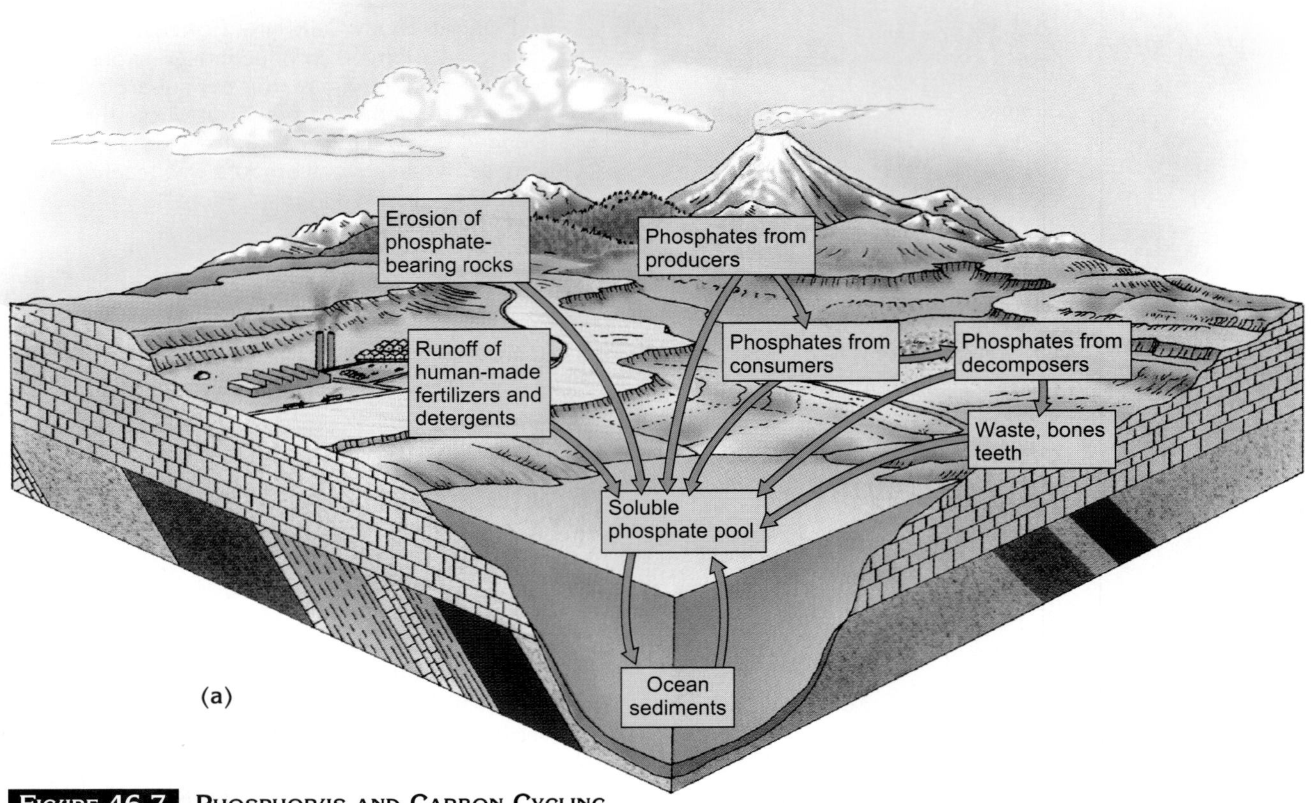

FIGURE 46.7 PHOSPHORUS AND CARBON CYCLING.
(a) The phosphorus cycle. Usable phosphorus in the form of soluble phosphates is found in soil water and in aquatic systems. Some phosphates pass to consumers through the trophic levels, while some are taken in through drinking water. During decomposition, reducers make some phosphates available again, but some—locked in animal remains such as bones, teeth, and shells—are unavailable for long periods. The loss of phosphates through leaching (from soil water) and through runoff (to the sea) is considerable. Some end up as insoluble phosphorus in deep sediments. A gain in available phosphates occurs through erosion and from pollutants introduced by humans. **(b)** The carbon cycle. Carbon dioxide must enter the trophic levels through producers that fix carbon into organic molecules during photosynthesis. CO^e is released during respiration by producers, consumers, and decomposers and some is released from the earth's crust through volcanic action.

nitrogen by-product is excreted in urine. Eventually, all organisms and their nitrogen wastes enter the realm of the decomposers, where they are broken down into their component parts.

During decomposition, several populations of microorganisms, each with a specific role in the process, carry out *ammonification* (production of ammonia) and *nitrification* (production of nitrites, which may become nitrates). The nitrates (the end product) can then join the exchange pool, from which they can return to the plant. This may all seem quite efficient, but in fact there are complications that lead to losses. For example, certain anaerobic soil bacteria—the *denitrifiers*—metabolize nitrates, converting them to nitrogen gas that diffuses from the soil, reentering the atmospheric reservoir.

Role of Nitrogen Fixers. The denitrifiers remove nitrogen from the soil, but the element is never depleted because of the action of **nitrogen fixers.** Nitrogen fixers are soil bacteria and cyanobacteria that are able to "fix" atmospheric nitrogen (N_2), by combining it with hydrogen, producing ammonium ions ($NH^+{}_4$). Nitrogen fixation is energetically costly because of the atmospheric nitrogen stability. Reducing nitrogen (by adding hydrogen) requires powerful enzymes and a considerable expenditure of ATP.

Cyanobacteria and other free-living nitrogen fixers release surplus ammonia into their watery surroundings where much of it is converted to nitrite and nitrate by soil bacteria. Bacteria living mutualistically (in a mutually beneficial association) in the roots of plants make ammonium ions available

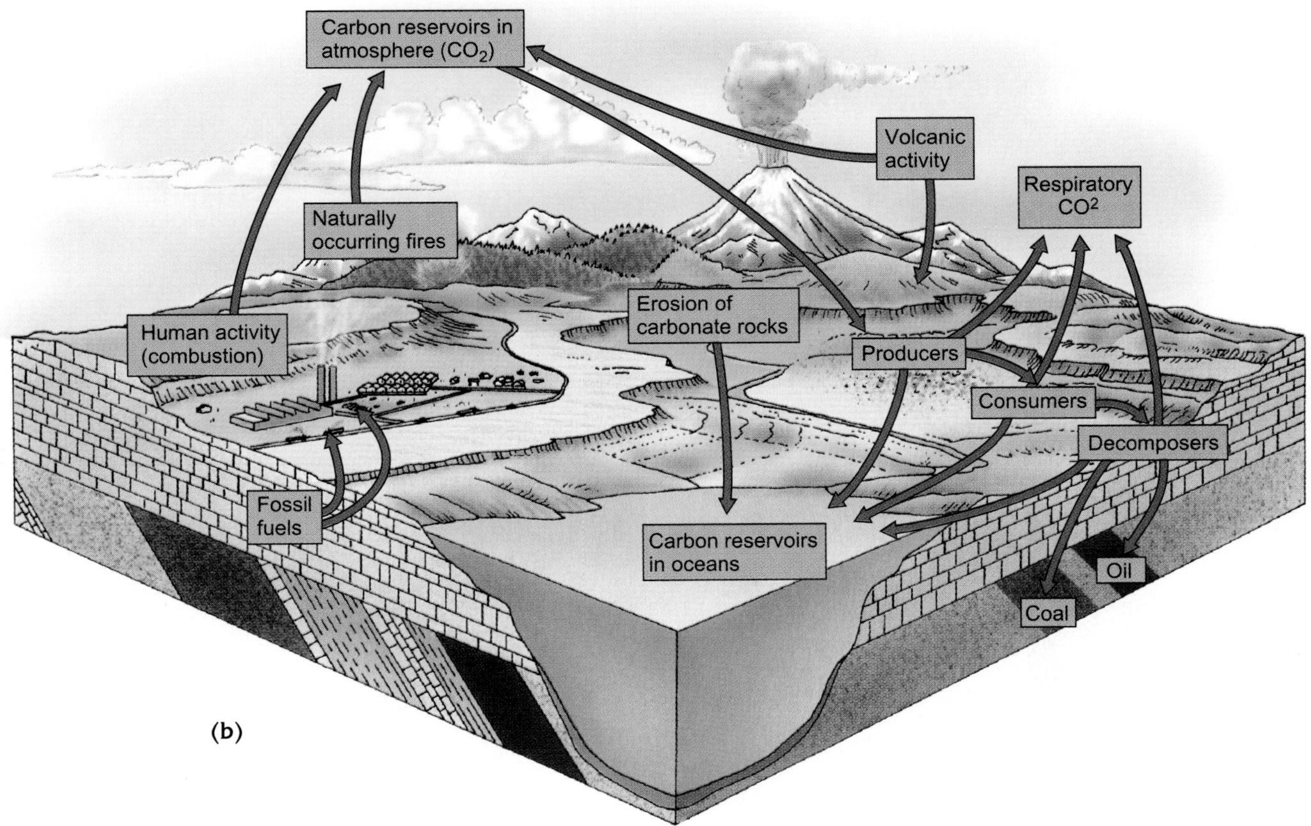

(b)

directly to their host's cells, where it is used in generating amino acids and other essentials.

While cyanobacteria generate their own ATP through photosynthesis, mutualistic bacteria use the host's supply. In fact, fully one-fifth of the ATP generated by the pea plant is used for nitrogen fixation by its guest bacteria.

Farmers have employed the nitrogen fixers for many years. For example, crop rotation (planting different crops in a rotating sequence) commonly includes such legumes as alfalfa. Nitrogen-fixing bacteria of the genus *Rhizobium* invade the alfalfa roots, which respond by forming cystlike nodules around the bacterial colony (Figure 46.9). The result is, nitrogen is returned to the soil from where it can enter the chain of life.

Communities over Time: Ecological Succession

As time passes, communities change. In some cases they grow as their biomass increases. In other cases more energy is released in respiration than is stored, indicating a state of decline. But even when biomass and energy are stabilized, the community may be changing. Sometimes the change produces a sequence of organisms within the community itself. Community change over time is known as **ecological succession.** Eventually, a climax state may be reached, whereupon no further change will occur as long as the environment remains stable.

As populations within a community alter the environment, they set the stage for invasions by different species. The result is that the community takes on new traits, and with new species altering the environment in their own manner, the change goes on. The entire sequence from start to climax is called a *sere,* while its parts are *seral stages.* Ecological succession is somewhat predictable and sequential, although ecologists are finding it less so than they once believed. Not only do invasions by unexpected, opportunistic species break up the orderly progression, but unanticipated physical changes—perhaps a drought—can alter the usual events of succession.

Ecological succession can involve both primary succession and secondary succession. **Primary succession** is the appearance of life where no community previously existed, such as on rocky outcroppings, newly formed deltas, sand dunes, emerging volcanic islands, and lava flows. **Secondary succession** is the

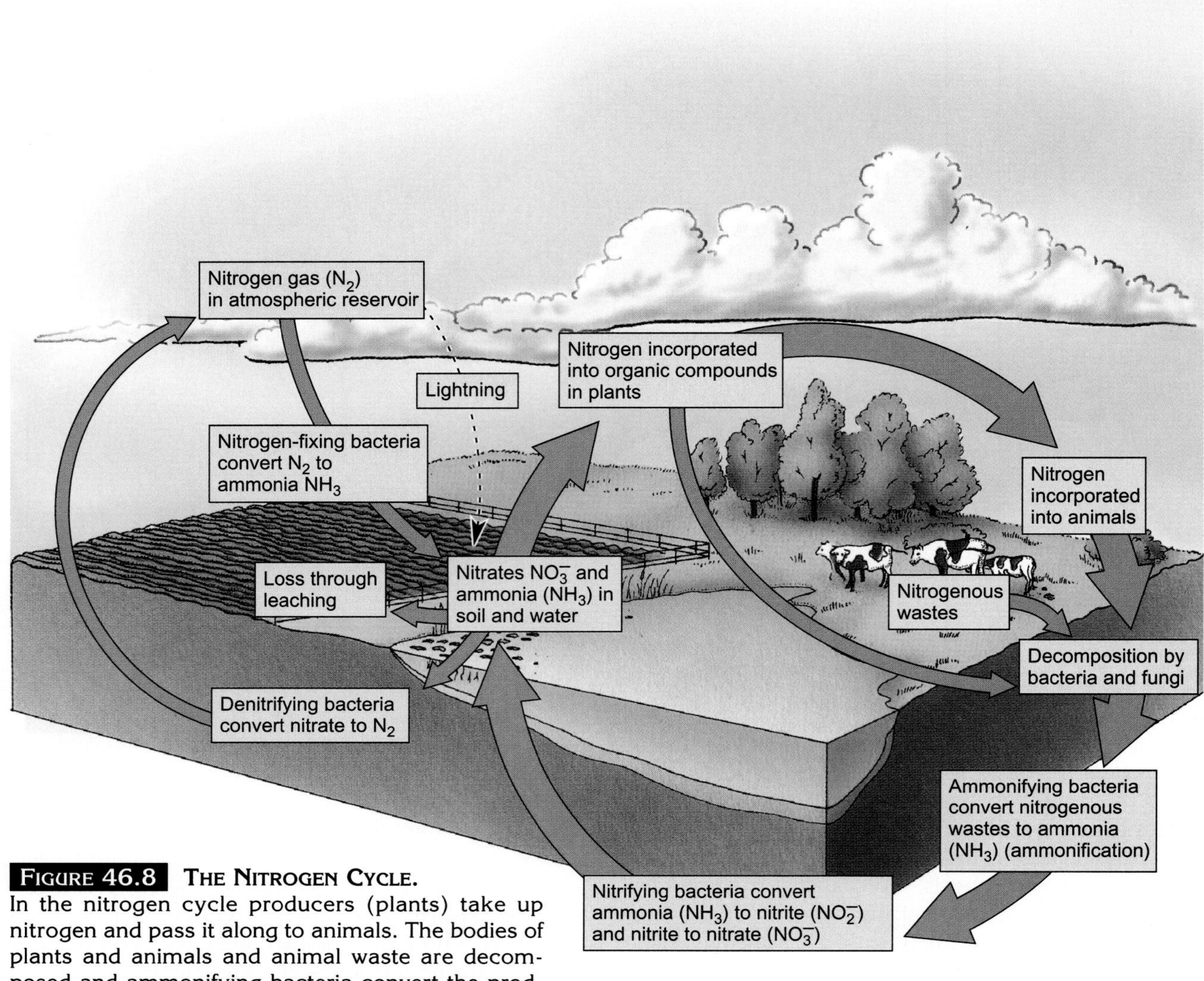

FIGURE 46.8 THE NITROGEN CYCLE.
In the nitrogen cycle producers (plants) take up nitrogen and pass it along to animals. The bodies of plants and animals and animal waste are decomposed and ammonifying bacteria convert the products to ammonia. The ammonia is converted to nitrite and then to nitrate which, with some ammonia, is returned to the soil and water. Denitrifying bacteria convert nitrate to N_2 (atmospheric nitrogen) which then reenters the biological system by the action of nitrogen-fixing bacteria. Some nitrates are also produced by lightning, even as nitrates and ammonia may be leached from the soil.

establishment of a new community where another community has been destroyed or disrupted. We find it, for example, where a neglected farm is reverting to the wild, or in a forest community that has been subjected to clear cutting, the controversial lumbering practice in which all trees are removed from a once-flourishing woodland.

PRIMARY SUCCESSION

In primary succession (Figure 46.10), the first organisms to invade are usually hardy, drought-resistant species, called *pioneer organisms.* For example, lichens are often the first to invade rocky outcroppings, held fast by their tenacious, water-seeking fungal component while the algal component provides food as its chloroplasts are exposed to the sun. Lichens gradually erode the rock surface as they probe into tiny crevices and help pry them open. Sand then accumulates in tiny fissures, and the bodies of dead lichens add to the humus. In this way, soil is born, and with it come opportunities for such plants as grasses and mosses to establish themselves.

In subsequent seral stages the plant roots penetrate the rocky crevices, exerting a remarkable pressure, prying at the rocks and gradually widening the fis-

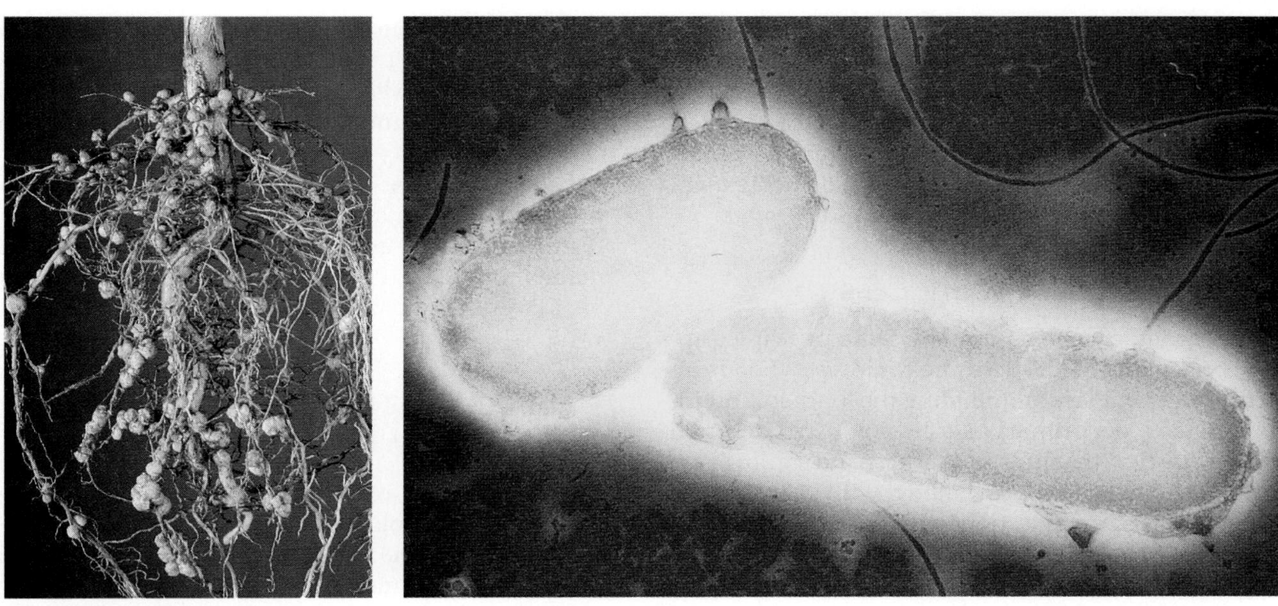

FIGURE 46.9 ROOT NODULES IN A LEGUMINOUS PLANT.
Such nodules contain large colonies of irregularly shaped nitrogen-fixing bacteria from the genus *Rhizobium*. Shown here are root nodules and a light micrograph of a nodule section.

sures. By now certain insect and decomposer populations will also have established themselves. As these early species gain a toehold, the lichens that made their penetration possible begin to give way; they cannot compete with the emerging plants for light, water, and minerals. Similarly, after the first plants—grasses and mosses—have contributed significantly to the soil-building efforts, they will be replaced by fast-growing shrubs, and new kinds of animals will continue to invade the ever-changing community.

FIGURE 46.10 PRIMARY SUCCESSION.
Succession begins here with a bare rock outcropping and ends with a fir-birch-spruce community. Pioneering lichens and mosses begin the soil-building process, followed by the invasion of increasingly larger plants until a more stable, long-lived, climax forest community emerges.

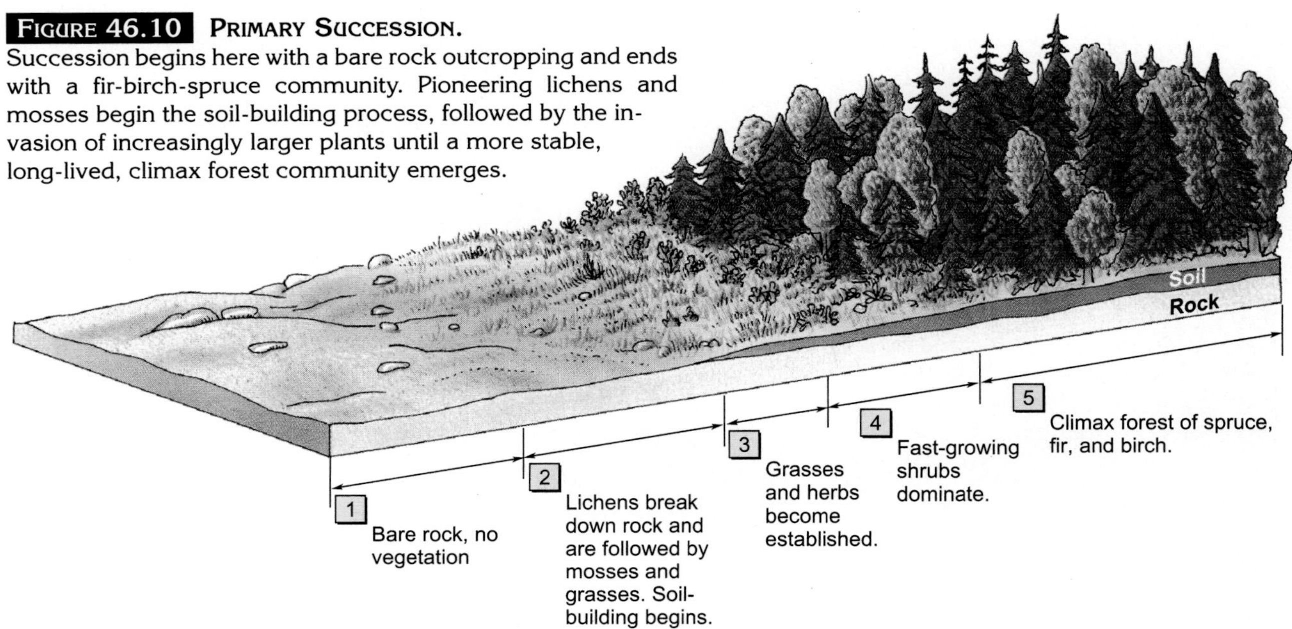

Soil
Rock
1 Bare rock, no vegetation
2 Lichens break down rock and are followed by mosses and grasses. Soil-building begins.
3 Grasses and herbs become established.
4 Fast-growing shrubs dominate.
5 Climax forest of spruce, fir, and birch.

Primary succession in bare rock outcroppings is an extremely slow process, often requiring hundreds of years, primarily because soil formation is so slow. But once the soil has formed, the process can accelerate. Studies of succession from sand dune to climax forest community on the shores of Lake Michigan indicate that this transition took about 1,000 years.

Secondary Succession

In secondary succession, the same principles apply, but events occur at a more rapid pace. Soil is often already in place, eliminating the long, soil-building stages of primary succession. On deserted farms, we find that grasses, shrubs, and saplings are often the first to appear, along with a variety of weeds—fast-growing, opportunistic plants. In balanced, stable communities that are undisturbed, weeds quickly invade disturbed communities.

As secondary succession progresses, the initial invaders are eventually replaced by plants from the surrounding community. Larger, faster-growing trees block the sunlight, and a new generation of shade-tolerant shrubs emerges below the canopy. Eventually the demarcation between the area in succession and the surrounding community begins to fade (Figure 46.11).

The time involved in secondary succession varies widely. In grassland, for example, succession to a climax community after a disturbance may take only 20 to 40 years. At the other extreme, fragile tundra may require hundreds of years to recover from a disturbance—if it ever recovers at all. Wagon tracks made more than 100 years ago are still clearly visible in this delicate biome.

Succession in the Freshwater Aquatic Community

Like terrestrial communities, freshwater communities undergo ecological succession. A major determining factor in the rate at which succession occurs is the availability of nutrients. Lakes and ponds that are rich in nutrients and high in productivity are called **eutrophic** ("true foods"), while those that have a more limited nutrient supply and little productivity are called **oligotrophic** ("few foods"). The general trend in freshwater bodies is toward increased nutrient enrichment, a process called **eutrophication**—

FIGURE 46.11 SECONDARY SUCCESSION. Four stages of secondary succession are revealed in a series of photos taken in the same locale.

Early in succession, aquatic plants begin to spread from the edges of the pond.

Eventually, these plants extend across the open water as bottom sediment begins to build up and floating plants appear.

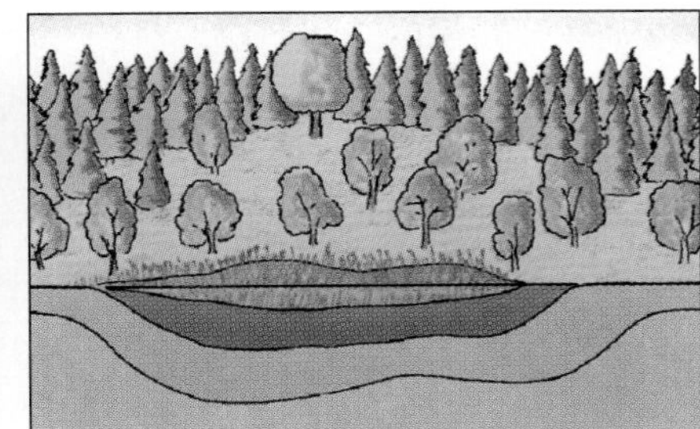

As the pond's waters disappear, invading marsh grasses, cattails, and sedges replace floating plants, converting the pond into a marsh.

FIGURE 46.12 AQUATIC SUCCESSION.
Early in succession, aquatic plants begin to spread from the edges of the pond. Eventually these plants extend across the open water. As the pond's waters disappear, invading marsh grasses, cattails, and sedges replace the floating plants, converting the ponds into marsh.

but if any essential nutrient becomes unavailable, the trend can be quickly reversed.

As lake succession progresses, new organic materials are produced at an increasing rate, sediments increase, and the lake depth decreases. Littoral zone plants crowd the shallows along the lakeshore, gradually extending across the water. They are followed by increasing numbers of weather-tolerant shore plants (Figure 46.12). Unless the trend is interrupted, the lake will eventually convert to a wetland, such as a marsh, and with the invasion of terrestrial plants from the surrounding community, the last traces of the aquatic environment will disappear. (Essay 46.2 describes another problem besetting lakes in industrial areas.)

EUTROPHICATION: HUMAN IMPACT

The ancient oligotrophic Lake Baikal (Figure 46.13) has shown alarming indications of eutrophication, but not through natural means. Wastes from the human community along its shores are artificially

FIGURE 46.13 LAKE BAIKAL.
One of the oldest and most fascinating lakes on earth, Lake Baikal is showing signs of eutrophication due to human activity.

Essay 46.2 What Have They Done to the Rain?

In the 1970s, it became clear that the rain was changing. In fact, in some areas the gentle raindrops were downright dangerous. The rain was becoming a dilute mixture of acids. It was first noticed in Scandinavia, then in the northeast United States and southeast Canada, then in Northern Europe and Japan.

Rainwater, of course, had always been slightly acidic because the water dissolved atmospheric carbon dioxide, forming carbonic acids. But now the rain was showing alarming concentrations of the more dangerous sulfuric acid and nitric acid. Where were they coming from? They were the result of accumulations of nitrous oxides and sulfuric oxides in the atmosphere. The nitrous oxides, it turned out, were from power plant and automobile emissions; the sulfuric oxides, mainly from power plants and smelters. Dissolved in the water of cloud formations, they formed nitric acid and sulfuric acid, then fell to earth to bathe our forests and cities and to fill our lakes with the corrosive mix.

The relative proportions of the two acids in our rain depends on where one lives. In the northeastern United States, the acidity is primarily due to sulfuric acid; in California, to nitric acid. So we do have a choice.

The rain has caused the reduction and even the elimination of fish in many of our lakes. The rain apparently doesn't kill the fish; it just keeps them from reproducing. So no young fish are found as the old ones gradually go the way of all flesh. In fact, about 700 lakes in southern Norway are now *entirely devoid* of fish, and our own northeastern lakes are following one by one. As our Adirondack lakes reach pH levels of 5 (not uncommon), 90% have no fish whatever. They are also curiously devoid of frogs and salamanders.

Entire patches of forests worldwide are sickening and dying as ecologists busily try to find out just what effects the rain is having.

In 1990, the results of a study were released, indicating that in the United States, the effect of acid rain in nature was not as bad as we thought, partly because relatively few lakes are affected, partly because the effect on such lakes is reversible.

Interestingly, the solution is clear to everyone. We simply need to reduce the levels of our effluent from power plants, smelters, and automobiles. Most of the technology exists, but its implementation would be too expensive for the polluters to willingly bear, or the public to pay for. So, in general, business continues as usual while our drinking water becomes increasingly unusual.

fertilizing the water and enabling the rapid production of organic material.

Eutrophication is ordinarily a slow process in lakes. Its progress is typically checked by a scarcity of two nutrients, phosphates and nitrates. In recent years, however, these nutrients have become readily available as humans dump sewage into water systems or permit the runoff of surface water from heavily fertilized farms or cattle feedlots. The addition of phosphates to laundry detergents has also added a heavy burden to natural water systems. The sudden increase of nitrates and phosphates in fresh waters produces unprecedented algal blooms, a first step in eutrophication through pollution.

PERSPECTIVES

The impact of humans on nature is profound, and our ability to alter such systems grows as our technological abilities increase. Even as we learn about the effects of our previous alterations, we continue to manipulate the intricate and myriad factors that contribute to the equilibrium of ecosystems. Of course, we are aware of our impact to some degree, and we often make well-intentioned, if unsuccessful, efforts to set things right. Our efforts often falter simply for lack of basic information. As we turn to the scientific community for answers, ecologists will play an increasingly important role in changing the ways in which we interact with the biosphere. In fact, ecologists may one day be recognized as the most important scientists on earth.

Next, we will see how populations change in nature, then focus on the dilemma of the human condition. We will continue to ask questions centering on ethics. As you will see, we won't always have the answers.

SUMMARY

An ecosystem is an interacting unit, or community, of plants and animals, together with the physical nonliving environment in which it exists. Each ecosystem contains characteristic trophic levels through which energy flows.

ENERGY IN ECOSYSTEMS

Energy reaching the biosphere accounts for 150 to 200 billion metric tons of organic matter per year.

Energy flows from producer through consumer levels. Producers, the autotrophs (phototrophs and chemotrophs) account for 99% of the earth's biomass.

Consumers, the heterotrophs, include primary consumers (herbivores), secondary and higher levels (carnivores). Diagrams of food webs portray feeding interactions.

Decomposers include bacteria and fungi, heterotrophs that secrete enzymes into food, breaking it down and releasing mineral ions.

Ecological pyramids include pyramids of numbers, biomass, and energy. Most are broad-based and stepped, indicating a loss at each level. The losses are represented by matter and energy not actually assimilated in the level above, or matter and energy expended in the ongoing metabolic processes. Energy transfer through the trophic levels is about 10% efficient.

Affluent humans feed partly at secondary or higher consumer levels, while the more impoverished feed mainly at the primary or herbivore level. At the latter level, protein deficiencies may occur.

Gross primary productivity measures the rate of production of energy in an ecosystem by producers. Net primary productivity is the energy the producer has left for storage after its own energy needs are met. Net community productivity is the rate of energy stored by producers minus that consumed and used in respiration by heterotrophs.

Productivity measurements are useful in determining growth status and making community comparisons. The highest productivity occurs in estuaries, wetlands, coral reefs, tropical forests, and in intensive agriculture.

NUTRIENT CYCLING IN ECOSYSTEMS

Mineral nutrients in ecosystems cycle between organisms and the environment in biogeochemical cycles. Examples include water, phosphorus, carbon, and nitrogen.

Nitrogen, an element required in most molecules of life, occurs as nitrogen gas in reservoirs and as nitrate and nitrite ions in exchange pools. Nitrogen gas is fixed into biologically useful compounds by nitrogen-fixing bacteria, and nitrogen compounds are cycled through decomposition.

A series of decomposers break down nitrogen compounds through ammonification and nitrification. Some loss to a reservoir occurs through bacterial denitrification, which yields nitrogen gas.

Nitrogen-fixing soil bacteria, living mutualistically in leguminous plant root nodules, and cyanobacteria living in water, convert nitrogen gas to ammonium ions. Excesses are made available to exchange pools.

Communities over Time: Ecological Succession

Community change occurs through ecological succession. Basically, as transitional communities develop, they produce changes that lead to their being succeeded by other communities until a climax state is reached. The sequence is a sere, and its parts, seral stages.

Primary succession (which occurs where no community previously existed) begins with soil-building, often by lichens and other pioneer organisms. Grasses and mosses may follow, with these eventually replaced by a succession of more complex communities until a climax state is reached.

Secondary succession (the establishment of a new community where another community has been destroyed or disrupted. It occurs much more rapidly than primary succession, as the disturbed area is invaded by surrounding species.

Change in eutrophic lakes is limited by the availability of mineral nutrients. Their increasing availability (eutrophication) speeds succession. As succession goes on, plant and other life rapidly increases, and the lake diminishes until it blends in with the surrounding terrestrial community.

Limited phosphates and nitrates generally restrict succession in lakes, but their presence in pollutants introduced by humans can promote rapid eutrophication.

Key Terms

community • **833**
ecosystem • **833**
trophic level • **833**
producer • **833**
consumer • **833**
decomposer • **833**
pyramid of numbers • **836**
pyramid of biomass • **838**
pyramid of energy • **838**
primary productivity • **839**
gross primary productivity • **839**
net primary productivity • **839**
net community productivity • **839**
climax state • **839**
nitrogen fixer • **842**
ecological succession • **843**
primary succession • **843**
secondary succession • **843**
eutrophic • **846**
oligotrophic • **846**
eutrophication • **846**

Review Questions

1. List the two fundamental components that make up any ecosystem.
2. Describe in general the fate of solar energy in the biosphere. How much do living organisms capture?
3. List the two fundamental trophic levels and describe them.
4. How do food webs differ from food chains?
5. List two important groups of decomposers. In what ways do their activities differ from those of other consumers?
6. What is the general shape taken by number pyramids? Cite an exception.
7. Explain the great difference between the amount of biomass in producers and biomass in consumers.
8. Relate the economic status of humans to their trophic levels, generally.
9. What did Woodwell conclude about the growth state of the scrub oak-pine community? How did he come to this conclusion?
10. What is a biogeochemical cycle? Describe one involving carbon.
11. Distinguish between exchange pools and reservoirs in biogeochemical cycles.
12. In general terms, explain how losses and gains occur in the nitrogen cycle.
13. In general, why does ecological succession occur? When, if ever, does it stop?
14. Where would one expect to look for signs of secondary succession? When does secondary succession end?
15. Essentially, what determines the rate of succession in freshwater bodies? Name a process that greatly speeds succession.
16. Lake Baikal has long been considered as oligotrophic. What does this mean, and why is this categorization changing?
17. List three human activities that increase eutrophication.

CHAPTER

47

Population Dynamics

THERE ARE A LOT OF REASONS TO CONSIDER POPULATION DYNAMICS—*how populations change. For one thing, such processes may be interesting as an intellectual topic. But even if you are not of such a bent, you may be curious about the fate of populations that place too much pressure on their environment. You may also wonder about what happens when population densities become too high, or how dominance and territoriality can save (and take) lives. Such questions take on an enormous importance, of course, when we apply them to our own species, as we will do in our last chapter.*

Some population changes are short-term or periodic and easily explainable. For example, grasshoppers, which are plentiful in the summer, reproduce and then disappear by late fall, leaving their fertilized eggs hidden in the soil. The disappearance of the adults is easy to understand—they die. By late fall, too, certain birds begin to disappear from North American forests. This disappearance is also easy to explain. As winter approaches, they simply migrate to more hospitable southern areas. These are familiar and rather predictable phenomena. We will expect the next generation of grasshoppers and the migratory birds to show up in spring.

Other sorts of changes are less predictable and more puzzling. As some species dwindle, we hear of others burgeoning in the form of "plagues" or "invasions." However, such dramatic descriptions may not accurately reflect what is going on. This is because most species seem to be under some sort of control or regulating influence. In order to consider any such control, we must first understand something about how populations generally behave. What determines the numbers of a population? What evolutionary and ecological factors are at work? What are the characteristics of population growth? What causes populations to fluctuate, and what sort of stabilizing influences might be at work? Finally, what influences the distribution of populations?

POPULATION DYNAMICS AND GROWTH PATTERNS

The human population can be divided into two groups—those who see the world mathematically, and those who don't. With an acknowledgment that both groups exist, we would like to look briefly at certain population changes through a rather idealized mathematical model. We can begin by considering the simple population equation

$$I = (b - d)N$$

which translates into: "I, the rate of change in the number of individuals in a population, is equal to b, the average birth rate, minus d, the average death rate, times N, the number of individuals in the population."

One can see from the equation that for populations to increase, that is, for I to be positive, the value of b must exceed the value of d—that is, births must exceed deaths. Should they be the same, I will be zero; or should the average death rate, d, exceed the average birth rate, I will be negative. So, the value of $b - d$ seems to determine what happens to population size (N). The simple difference between the two values we will call r, or **realized rate of increase.** The determination of this per-individual rate is simply $r = b - d$. We can now rewrite I to incorporate the new term r. Thus,

$$I = rN$$

or the population rate of change (I) is equal to the per-individual rate of change (r) times the population size (N).

This new equation reminds us of something important in population dynamics. In a growing population, the rate of change in population size is determined not only by r, but by N. For instance, when r is a positive number, N will grow with each new generation of offspring. Thus, I will increase in each generation. The story of population increase can be summarized in two simple curves, the *J-shaped curve* and the *S-shaped curve.* The letters *J* and *S* represent the direction taken by lines when population growth patterns are plotted. The J-shaped curve is the one environmentalists talk about when they try to interest people in the human population growth problem. We will consider this curve first.

Exponential Increase and Population Curves

If a few reproductively active organisms are placed in an idealized environment—one with unlimited resources and space and without danger from disease, predation, or other hazards—they may be expected to reproduce at their maximum physiological rate. This theoretical rate implies maximum gamete production, mating, fertilization, and survival of offspring. (If we allow a 30-year reproductive period for humans, then our species' maximum physiological birth rate would be a little over 30 offspring for each female.) This maximum birth rate, together with a physiologically minimum death rate, will produce a maximum value of *r*, which is symbolized r_m and is called the **intrinsic rate of natural increase.**

Even under r_m, numbers rise slowly at first, simply because we are dealing with a few individuals, say from 2 to 4, to 8, to 16, and so on. In plotting such numbers, at first we see a gently increasing slope. But soon, as the numbers continue to double, the curve arches sharply upward until it seems to approach (but never reaches) the vertical (Figure 47.1). Such increases are called *exponential increases.* They are unlike linear or *arithmetic increases,* those that progress from 1 to 2 to 3, and so on. We have described the exponential growth and J-shaped curve shown by populations growing at their maximum per-individual rate of increase, r_m. (We must add that any constant positive *r* will also produce a smooth J-shaped exponential growth curve, but with various rises in the curve.)

One of the clearest biological examples of exponential increase is seen when a small number of bacteria are introduced into a rich laboratory culture medium. Under such ideal conditions, the familiar *E. coli,* for example, will divide every 20 minutes. At this rate, after only 24 hours, just one bacterium would give rise to 40 septillion descendants (2^n, where the exponent equals the numbers of generations in 24 hours). But as simple as *E. coli's* growth requirements are, such rapid expansion could not be sustained. Very likely, sometime during the 24-hour period, the resources that were supporting such phenomenal growth would be reduced to a point where the r_m could no longer be reached. *E. coli* and the rapid rate of increase would begin to slow, producing the S-shaped curve. Those factors that impede population growth, keeping populations from reaching their intrinsic rate of natural increase, are collectively called **environmental resistance.**

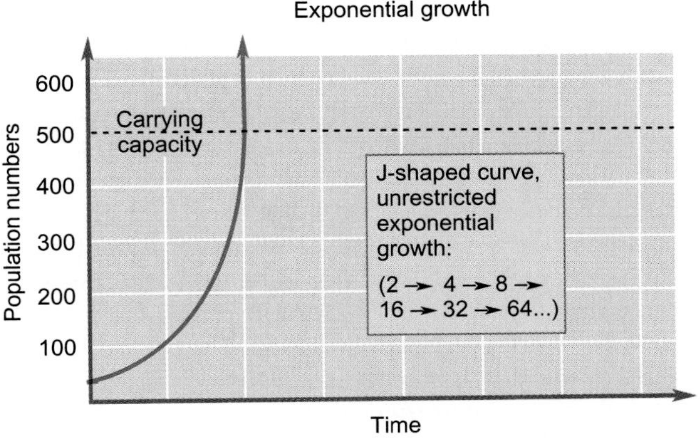

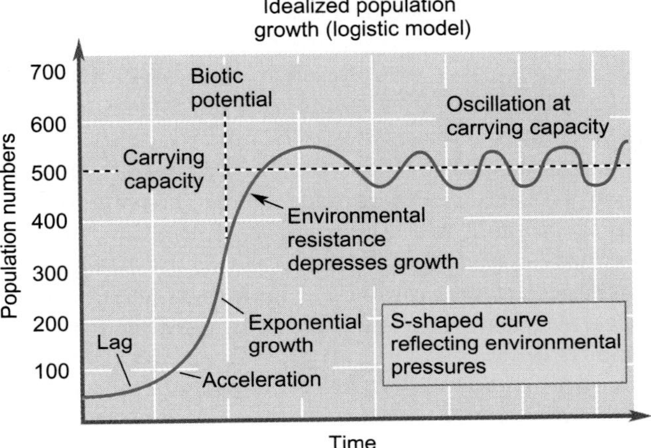

FIGURE 47.1 POPULATION GROWTH CURVES. Exponential growth produces the J-shaped growth curve. It is produced when organisms reproduce at or near their intrinsic rate of natural increase. When population growth is responsive to environmental resistance, it forms the S-shaped or logistic growth curve. Such populations hover at or below the carrying capacity. A population may overshoot its carrying capacity and cause such damage that it then undergoes a population crash.

Rapidly growing populations sometimes *overshoot* (quickly rise above) the environment's **carrying capacity** (the size of a population that a specific environment can maintain indefinitely) so drastically that they bring harsh pressures on their own numbers. The population may suddenly decrease, falling well below the original carrying capacity. Such a rapid decline is know as a **population crash.** A crash, induced by earlier rapid population growth, can be due to a number of factors, such as a depletion of resources or damage to the environment. Populations that overshoot the carrying capacity can drastically abuse the environment—to such a degree that they permanently decrease its carrying capacity. For example, the elephants crowded into reserves in East Africa have destroyed acres of the slow-growing trees (such as acacias and baobabs) on which they feed, and some ecologists believe that the area will never recover. (What is there to be said about a species that destroys its own environment? Can you think of any others?)

In some cases, population crashes are part of the normal cycle of life. Winter is a harsh season that seasonally decreases the environment's carrying capacity. Species that do not simply die out, as do many insects, have adapted in a number of ways—for example, by laying on fat, by migrating, by hibernating, by altering social patterns (huddling, for instance), or by simply making the best of it (as deer do). The seasonal return of warm weather restores the carrying capacity to higher levels, and populations normally respond by increasing their numbers quickly.

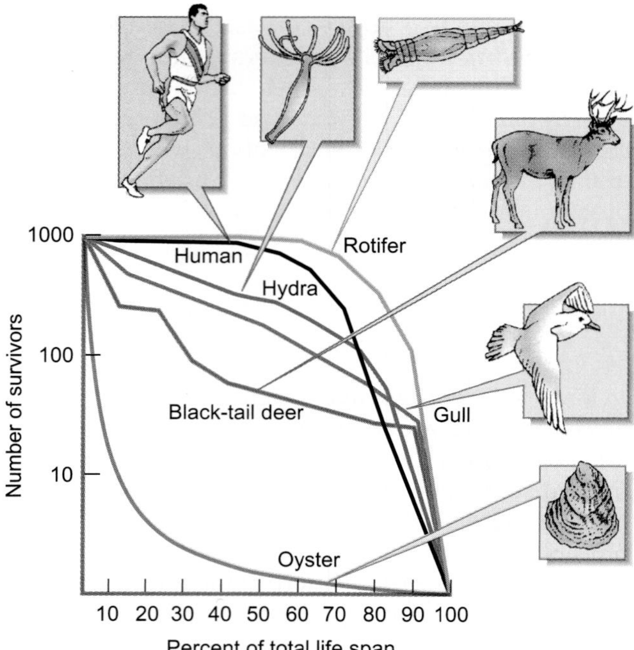

FIGURE 47.2 SURVIVORSHIP CURVES FOR SIX SPECIES.
Each species begins with a population of 1,000. Points along each line represent the percentage of the life span reached as the populations age. In blacktail deer, a species that experiences a high death rate among its young, the numbers drop off rapidly, with the average individual achieving only 6 percent of its potential life span.

LIFE SPAN AND THE POPULATION

Information about life span and age structure is important in characterizing a population and predicting its course. How many individuals in a population are below, within, or beyond the reproductive age? What is their life span? Do most individuals survive through their entire reproductive period? These questions are extremely important in determining probable future trends in our own population. For example, 1990 censuses revealed that 36% of the earth's humans were below 15 years of age. That is, many were not yet of reproductive age. And most of these people were statistically likely to live through their reproductive period and beyond.

Death now claims people later in life than ever before. But this information in itself is not particularly useful. In order to make meaningful predictions, we need to know which age groups are most vulnerable—that is, subjected to the greatest mortality. In many natural populations, death occurs quite frequently in the very young. But once an individual has survived the rigors of early life, the probability of living to old age increases. Obviously, mortality increases again in the aging segment of the population. **Survivorship curves,** which are graphs of the numbers of individuals surviving to different ages, indicate the probability of any individual living to a given age. Survivorship curves for six types of animals are shown in Figure 47.2.

INFLUENCES ON POPULATION DENSITY

Thus far, we have lumped the various factors that tend to affect population size under the catchall terms of intrinsic rates of natural increase and environmental resistance. Now let's take a closer look at some of these factors and how they operate. First, we should note that populations are influenced in two major ways, through density-independent and density- dependent controls.

FIGURE 47.3 NATURAL DISASTERS.
The effects of natural disasters such as fire, flood, landslide, and volcanic eruption are usually density-independent.

Density-Independent and Density-Dependent Population Controls

Density-independent controls are influences on populations that are not related to the density of the population. As an example, the effect of parching sun is not related to the density of drought-stricken corn plants in a field; no matter how many corn plants are present, the impact on individual plants will be the same. If that field should somehow recover, late spring rains could cause massive flooding and wipe them all out, whether few or many, densely or sparsely planted. Drought and floods, then, are density-independent population controls. Such controls are almost always *abiotic* (nonliving)—physical forces such as weather patterns or geological events (Figure 47.3).

On the other hand, if that field of corn does not meet with flood or drought, but is densely planted, the plants may compete for sunlight, nutrients, or water. In the competition, then, some will die as others thrive. Those that die, we can say, have succumbed to the effects of **density-dependent controls,** where the effect of environmental control on a population is related to the density of the population. Such controlling influences are usually *biotic,* (living).

More About Biotic Controls

Let's have a closer look at some ways in which populations can be controlled by biotic influences. Specifically, we will consider four such influences:

1. Competition, including that which produces territoriality and dominance hierarchies
2. Poisoning
3. Disease and parasitism
4. Predation

Control Through Competition. Competition exists where two or more individuals attempt to use some resources that may be in short supply, thereby

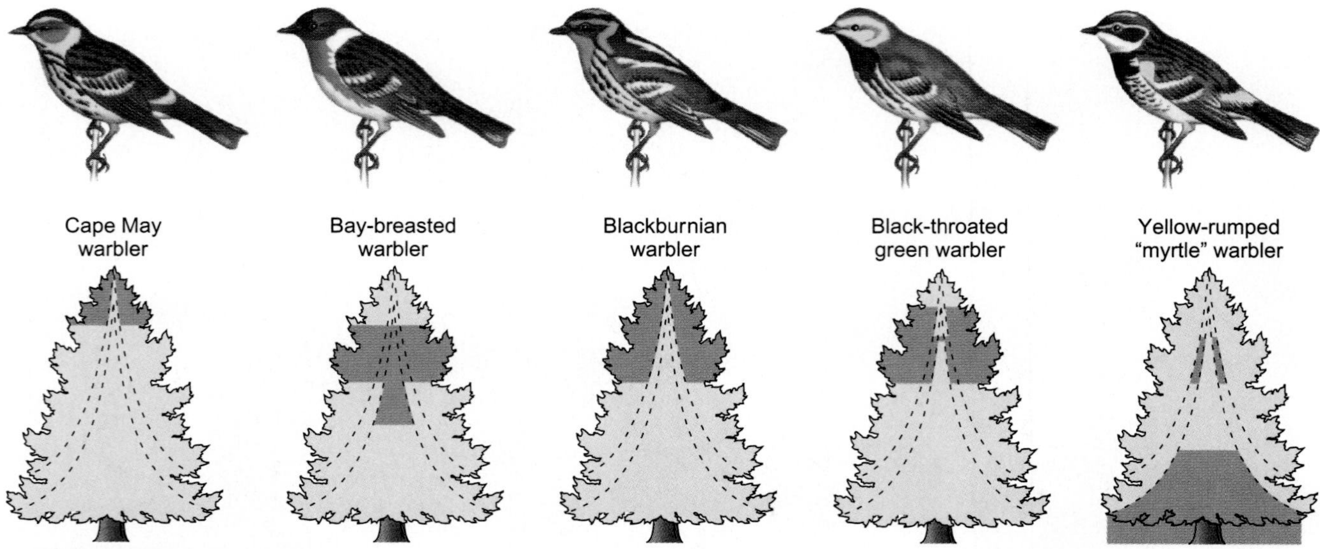

FIGURE 47.4 DIFFERENT SPECIES IN THE SAME TREE.
Five species of the North American warbler often use the same spruce trees for feeding and nesting, but each species tends to use a specific zone. The darkened areas indicate where each species spends at least half of its feeding time.

interfering with each other. Such competition can be *interspecific* (between species) or *intraspecific* (within a species). We will look at interspecific competition first, although intraspecific competition is far more common and generally more important in nature.

Often, two species will attempt to use the same limited resources in the same site. This, of course, leads to intense competition. If the level of competition is overly intense, one species will simply replace the other in the area of geographic overlap (due to the operation of the *principle of competitive exclusion,* where one species is replaced by a superior competitor). In the wild, the situation is usually not so desperate. In fact, very similar species often coexist as neighbors without undue effect. This is because, over long periods of adjustment, they have come to use different resources. The result is that similar species that would appear to be strongly competitive have instead become specialists, each species exploiting its own narrow part of the environment.

Probably the classic study of such specialization was performed by the American ecologist R. H. MacArthur. He found that five species of North American warblers coexisted in the same spruce tree groves by feeding in different parts of the trees, thereby reducing competition among themselves (Figure 47.4). But at the same time, by restricting themselves to certain feeding areas, the warblers effectively reduced their food supply. The reduced food supply could be expected to have an influence on their reproduction and thereby on their population sizes.

Intraspecific competition is most clearly apparent in two forms of social organization common among animals: *territoriality* and *dominance hierarchies.*

Territoriality is the defense of any area. The behavior was first discussed in the writings of the ancient Greek philosophers, who had observed the behavior in a variety of animals, including birds. Territorial birds, they saw, occupy an area and defend it, particularly against intruders of the same species and sex—animals that would offer the most competition. Territoriality has a number of advantages. A territory (defended area) may reserve a food supply and a nest site; it enables the territorial animal to become very familiar with a specific area; and it advertises that the territory holder is a healthy and able animal. In some species of birds, the males must have a good territory in order to attract females; even a splendid male cannot attract a mate if he holds an inferior territory. If good territories are in short supply, those birds unable to acquire one will not be able to reproduce. In such species, the availability of attractive territories would clearly affect population size (see Chapter 43).

Dominance hierarchies, sometimes called pecking orders (Figure 47.5), are rankings in which animals respond to each other differently, depending on their social position. In hierarchical groups, the dominant individuals have freer access to resources

FIGURE 47.5 **SOCIAL INTERACTIONS.** Complex social interaction characterizes baboons, providing a measure of both safety and order. Individuals and subgroups are organized in a hierarchy, with dominant members receiving first priority in feeding and reproduction.

than subordinate individuals do. The rankings may be established by combat, play, or by formal threats and confrontation without contact. While dominance hierarchies encourage stability by decreasing the likelihood of confrontation, this social structure also can influence population size. For instance, when food is scarce, the low-ranking individuals are often the first to starve, since higher-ranking individuals have priority for commodities. Lower-ranking individuals, though, do better by going hungry, seeking other resources, and risking starvation than by confronting a superior animal and provoking a conflict they would surely lose, resulting perhaps in injury or death.

Control Through Poisoning. Almost all species produce toxic (poisonous) wastes of some kind, and these wastes can drastically reduce population growth, because those that cannot escape their own wastes may well be poisoned by them. For example, yeast or bacteria growing in a petri dish will finally die of their own waste.

Toxic waste may also interfere with the populations of other species. For example, some ecologists have suggested that by rendering food rotten, bacteria make it unpalatable to most other organisms. Thus, the bacteria live for a time steeped in their own wastes, but with their decomposing food reserved for themselves.

FIGURE 47.6 PLANT TOXINS.
The sandy, grass-free zone bordering the cluster of purple sage is a product of volatile growth-inhibiting poisons released by the sage.

Some plants manufacture toxins that they secrete from their roots. The toxins do somewhat retard the growth of the plant that produced them, but they have an even stronger effect on other plants, including those of their own species. In this way, such plants reduce the level of competition for water and minerals (Figure 47.6).

Humans have, in recent years, managed to escape many of the biological checks on our population growth. Now we, like yeast and bacteria, are faced with the possibility of polluting ourselves into extinction. Never mind that the poisons are made by mines, factories, and automobiles, and not directly by our simple primate bodies; the principle is the same. It's difficult to say whether the effects on humans will be (or are) density-dependent or density-independent. If, in the future, finding an appropriate shelter with filtered air or bottled oxygen spells the difference between life and death during "smog alerts," then the control could well be density-dependent. If we are caught in a wave of some manufactured poison, then its effects will likely be density-independent. In 1985, thousands of people were killed or maimed by the accidental release of poisonous gases from a Union Carbide plant in Bhopal, India. The population density had no influence on the gas's effect. We are also now beginning to see the more subtle and widespread effects of a different kind of problem as we change the rain (see Essay 46.2).

Control Through Disease and Parasitism. Among the most important density-dependent regulators of many populations are disease and parasitism. One reason is that increased density means greater proximity, which in turn facilitates the transmission of diseases and parasites. Also, disease and parasitism interact in times of hunger and starvation, eliminating any weakened individuals. In some instances, disease reduces a population by increasing the incidence of predation. For example, a two-week-old caribou can already outsprint a full-grown wolf, and healthy adult caribou seldom fall prey to wolves. But caribou are susceptible to contagious and laming hoof disease, and it is these lamed animals that a wolf is likely to cull out of a herd. So both factors can operate together to influence population size.

Control Through Predation. There is little doubt that predators can influence the numbers of their prey. If a lion eats a zebra, that's one less zebra. But in the long run, the effect of a predator on its prey (and vice versa) may be quite complex. Predators normally do not eliminate their prey species altogether (barring human intervention, of course). This is because of their *reciprocal* effects—the influence of each species on the other.

The conventional explanation for the reciprocal influence is that, under natural conditions, a feedback principle governs the interaction of predator and prey. As a prey population increases, it provides more food for predators. As a result, the well-fed predators increase in number. Their numbers do not rise immediately, however, because it takes time for the energy derived from food to be converted into successful reproductive efforts. When the number of predators finally does rise, predators begin to exert increasing pressure on the prey. But as they begin to kill off the prey, they find themselves with less food, and so their own numbers fall, due perhaps to less successful reproductive efforts or perhaps to outright starvation. Again, the response lags a bit. Because of such lags, the prey population may be well on the road to recovery before the predator population begins to rise again (Figure 47.7).

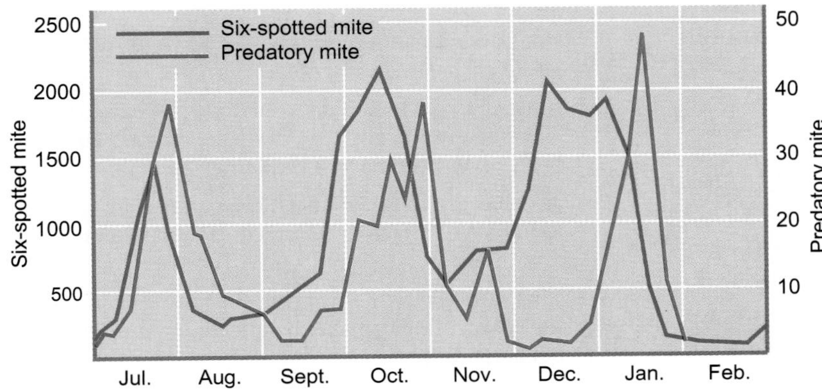

FIGURE 47.7 POPULATION OSCILLATIONS.
The oscillation of population numbers of the predatory mite and its prey the six-spotted mite is almost classic in its characteristics. Prey numbers increase first, followed at once by predator numbers. Then as the predator increase continues, the prey species diminishes in what is reminiscent of a population crash. The predator's fate is quite similar as its numbers fall rapidly.

PERSPECTIVES

We see, then, that the size and density of populations are influenced by a wide range of factors, from simple abiotic influences to the complex and interacting biotic factors. We are aware of some of the forces operating on populations, but we are still woefully short of understanding the underlying, fundamental principles of population change. The sooner we learn about such things the better, though, because our own numbers are rising at a rate unprecedented in human history. In the next chapter, we will consider that history and what we can expect of our future.

SUMMARY

POPULATION DYNAMICS AND GROWTH PATTERNS

Population growth can follow J-shaped or S-shaped curves. J-shaped population growth is called exponential growth and occurs when organisms survive and reproduce at their maximum rates. Because nature rarely provides such ideal conditions, J-shaped population growth is seldom seen.

In S-shaped growth, increases are rapid at first, but they begin tapering off, due to environmental resistance, effectively reaching zero as the environment's carrying capacity is approached. Overshoots may produce a crash in which most or all members die. Some crashes occur as a part of normal cycles, as seen in short-lived seasonal insect species.

Life Span and the Population

Survivorship curves are plots of numbers or proportions of individuals against age groupings. They reveal that portion of the life span in which most individuals die.

Influences on Population Density

Density-independent population controls are not influenced by the number of individuals per unit area. Most density-independent population controls are abiotic. Density-dependent population controls are influenced by the number of individuals per unit area. Most density-dependent controls are biotic.

Competition, which may be interspecific (between species) or intraspecific (within species), occurs when two or more individuals attempt to use the same resources that may be in short supply, thereby interfering with each other.

Intraspecific competition can involve territories and dominance hierarchies. In hard times, individuals without territories and farther down in hierarchies may die off or fail to breed, while superior individuals will survive and reproduce.

Population density can be reduced by self-poisoning—the production of toxic molecules. In the long run, humans are undoubtedly subject to the effects of their own toxic wastes.

Disease and parasitism, and their effects on reproduction, are partly influenced by population density, illustrating a negative feedback aspect.

Predators and prey have complex effects on each other's populations. There is a lag time between the rise and fall of predator numbers and the predator numbers that attack them.

Key Terms

realized rate of increase • **852**
intrinsic rate of natural increase • **853**
environmental resistance • **853**
carrying capacity • **853**
population crash • **854**
survivorship curve • **854**
density-independent control • **855**
density-dependent control • **855**
competition • **855**
territoriality • **856**
dominance hierarchy • **856**

Review Questions

1. What does the equation $I = (b - d)N$ refer to? What would cause I to be negative?
2. Construct graphs that illustrate J-shaped and S-shaped growth patterns. Does the upward curve of the "J" ever become a vertical line? Explain.
3. Describe the manner in which exponential growth occurs. What factors limit continued exponential growth?
4. What does the term "carrying capacity" mean? How can the carrying capacity be affected by populations that surpass it?
5. What is an overshoot? What commonly happens to populations that experience this? Is this result always considered abnormal? Explain.
6. Can two different species use the same limited resource in the same way? What are two common results of such an interaction?
7. Why would competition be greatest between members of the same species? The same species and sex?
8. Name four advantages of territoriality.
9. How might the establishment of dominance hierarchies affect survival?

CHAPTER

48

The Human Population

SOME TOPICS IN BIOLOGY ARE OBVIOUSLY OF MORE INTEREST TO YOU THAN *others and we believe this must be one of them. After all, here we will be discussing, in large part, just what you can expect in the coming years and how you may live out the rest of your life among your fellow humans. Here, we consider the human population, a group of which you are a member (if we are reaching our intended readership). We will first go over the history of human numbers. Then a few predictions that you may find startling. However, because you are human, you are a problem solver. So consider these findings as opportunities to solve problems. Don't expect it to be easy. You will immediately see that in many cases there are no simple solutions.*

Now we turn our attention to a set of numbers that has stimulated a great deal of concern and controversy in recent years: 5,420, 26, 9, 1.7, 41, and 36. You may not at first glance be particularly impressed. However, the numbers represent, respectively, the human population (in millions) our crude birth rate, crude death rate, rate of natural increase, population doubling time (in years), and percentage of people below the age of 15. (See Table 48.1 to see how these numbers are derived.)

It still may not be clear why we are bringing up these figures, but the reasons will soon be obvious. For example, we will see that the population of the world will double between the time most of today's college students were born and about the year 2012.

The gloom-and-doom statistics are abundant, and anyone who is at all interested has probably heard enough of them by now. But just in case, here is one more: There are 27 more humans living now than there were 10 seconds ago when you began reading this paragraph. By tomorrow at this time, 230,000 will have been added, and by next year, 84 million. That is slightly more than the population of Mexico—and most of them will live in a style similar to that of the average citizen of that struggling nation (Figure 48.1). Table 48.2 gives some idea of how other regions of the earth are faring.

We can generate such statements all day, but once we understand the problem, do we wallow in depression? Do we simply look away? Or do we join the ranks of hopeful and determined people who intend to learn as much as they can about the problem and then try to find ways to help? Obviously, we will assume you are in this last group, and so we'll begin by delving into the history of our numbers. The first thing we should know is that throughout most of our 3- to 4-million-year history, populations remained fairly stable, but in the past million years that stability has been interrupted by three significant growth surges (Figure 48.2).

HUMAN POPULATION HISTORY

Let's begin our overview with a look at the early population of *Homo erectus,* an ancestor that lived about 1 million years ago. In those days we might have seen occasional small bands roaming distant, grassy plains, perhaps a hunched figure stopping occasionally to dig at a root or pick at the soft parts of an insect. The hominid population, estimated to have been about 125,000 at that time, was not having much impact on the environment.

The lives of these unimpressive creatures must have been rigorous, indeed. Yet they survived. They adapted and they changed. In time, they gave rise to creatures of a different sort, the earliest individuals of the species *Homo sapiens.*

These ancient peoples were probably not much to look at either. They, too, roamed the earth as hunters and gatherers. And life for them was undoubtedly tough. Infants and children probably suffered high

FIGURE 48.1 **HUMAN POPULATION IMPACT.**
What does population growth mean to a nation? The two largest urban centers in the world are located in Japan and Mexico, nations that have far different social and economic prospects. What would happen to each of these vast urban areas if population doubling were to occur in 27 years? Japan has its population growth under control, so the question is academic. In the case of Mexico, a 27-year doubling time is precisely what demographers predict, so the question is frightening.

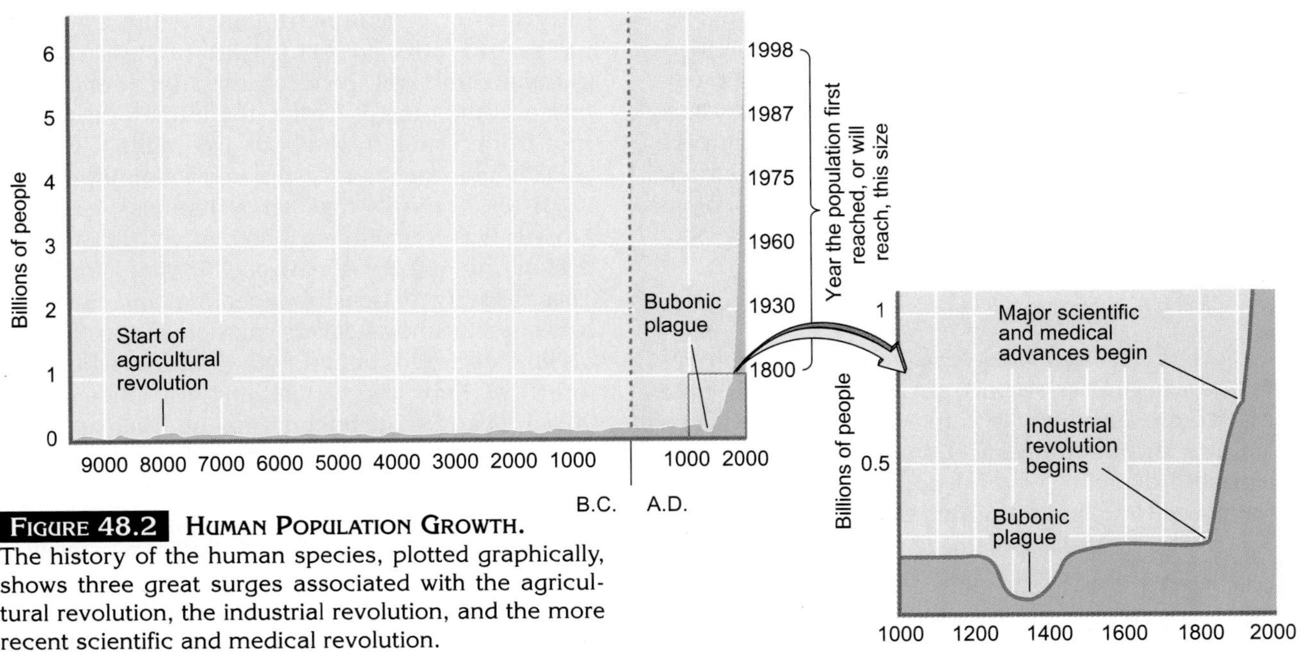

FIGURE 48.2 **HUMAN POPULATION GROWTH.**
The history of the human species, plotted graphically, shows three great surges associated with the agricultural revolution, the industrial revolution, and the more recent scientific and medical revolution.

TABLE 48.1 *BASIC POPULATION ARITHMETIC*

General fertility rate = number of births per number of women of reproductive age per year

Total fertility rate = average number of children a woman can be expected to have during her lifetime

Crude birth rate = number of births per year per 1,000 population

(determined by: $\frac{\text{total births}}{\text{midyear population}} \times 1{,}000$)

Crude death rate = number of deaths per year per 1,000 population

(determined by: $\frac{\text{total deaths}}{\text{midyear population}} \times 1{,}000$)

Rate of natural increase (or decrease) = crude BR – crude DR

% annual growth = $\frac{\text{rate of natural increase}}{10}$

Doubling time (approximately) = $\frac{70}{\text{\% annual growth}}$

Example: % annual growth in the world in 1986 was 1.7:

$$\frac{70}{1.7} = 41.18 \text{ years}$$

mortality rates, but such losses were quickly replaced in a species where fertility was not a seasonal event. The average life span is estimated to have been 30 years; but there were undoubtedly a number of old codgers in the group, their numbers balanced by the high infant mortality rate.

THE FIRST POPULATION SURGE

The first growth surge in the human population was probably due to the development of increasingly efficient tools. With them, early people could more effectively modify and exploit their environment. In addition, humans—inveterate wanderers—had by then penetrated and established themselves on all of the continents. By 10,000 years ago, the earth probably supported about 5 million people. (Estimates of these early human populations are based on limited data, partly from anthropological studies of surviving primitive cultures.)

THE SECOND POPULATION SURGE

About 8,000 to 10,000 years ago, the human population began its second growth surge, this time with more authority. With the development of agriculture and the domestication of animals came increasing densities of local populations. There was less need to roam the countryside in search of food; in fact, there was a great need to stay put and tend the fields and livestock. With the storage of surplus food, winter and drought no longer exacted such a great toll on human life.

With the increased quantity and dependability of the food supply, humans probably experienced a lower death rate, particularly among the young. Furthermore, there was quite possibly an increase in the birth rate because of better nutrition. Large families may have been encouraged because they meant more hands to till the fields. But life was by no means simple, since the crops were subject to the inconsistencies of weather and to infestation by insects and other herbivores whose own numbers responded to the novel food supply.

The unprecedented population growth of the early days of agriculture did not continue at its initial soaring rate, but settled into a steadier, more gradual climb. Yet, between the advent of agriculture and the time of Christ, the human population rose from 5 million to about 133 million. By 1650 A.D., it had reached an estimated 500 million.

History reveals that on a regional level this growth was interrupted many, many times by the decimating effects of disease, famine, and war. These are largely density-dependent and closely interrelated factors. A severe example of the effects of disease occurred in the 14th century, when one-fourth of Europe's population was killed by the "Black Death"—the bubonic plague. Such other diseases as typhus, influenza, and syphilis also took their toll on the crowded and incredibly filthy towns of the medieval period.

Interestingly, the loss of such numbers is insignificant in view of population growth today. At today's population growth rate, for example, the number of deaths from the 14th-century bubonic plague could

TABLE 48.2 **WORLD POPULATION DATA: 1970 AND 1992**

Region	Year	Total (millions)	Crude Birth Rate	Crude Death Rate	Natural Increase (annual %)	Doubling Time (years)	% Below 15 Years of Age
World	1970	3,632	34	14	2.0	35	37
	1992	5,420	26	9	1.7	41	36
Africa	1970	344	47	20	2.6	27	44
	1992	654	43	14	3.0	23	42
Asia	1970	2,045	38	15	2.0	31	40
	1992	2,042	30	10	2.0	34	36
North America	1970	228	18	9	1.0	63	20
	1992	283	16	8	0.8	89	21
Latin America	1970	283	38	9	3.0	24	42
	1992	453	28	7	2.1	34	40
Europe	1970	462	18	10	0.8	88	25
	1992	511	12	10	0.2	338	20
Nations of Special Interest							
United States	1970	205	17.5	9.6	1.0	70	30
	1992	255	16.0	9.0	0.8	89	22
People's Rep. of China (estimate)	1970	760	34.0	15.0	1.8	39	?
	1992	1,166	20.0	7.0	1.0	53	28
India	1970	554	42.0	17.0	2.6	27	41
	1992	883	30.0	10.0	2.3	34	36

Source of Data: Population Reference Bureau

be recouped in one year, while the number of people killed in all the wars of the last 500 years could be replaced in about six months.

THE THIRD POPULATION SURGE

The third population growth surge began in Europe in the mid-17th century, after the unexplained decline of the plague (perhaps only those who were naturally immune were left alive). A number of explanations have been advanced to account for this third surge. For one thing, the crowded populations of Europe expanded into the New World, with its array of opportunities and unexploited resources. More significantly, Pasteur, and others postulated the "germ theory" and people were becoming aware of some of the causes of disease. By the 19th century, public sanitation programs had begun, vaccines were developed, and rapid advances in food storage and transportation technologies led to a marked increase in food supplies. Between 1750 and 1850, the population of Europe doubled; that of the New World increased fivefold.

Populations were surging in other parts of the world as well, for reasons that are not completely understood. In China, the most heavily populated nation at that time, agriculture had made great gains, and a long period of comparative political stability followed the overthrow of the Ming dynasty in 1644. India, however, had known little rest from turmoil and periodic famines. In 1770, the worst famine of all reportedly killed three million people i n India. African populations are believed to have remained stable until about 1850, when the impact of imported European medical advances began to depress the death rate. However, recent experiences in Africa—namely the Ethiopian drought and famine—remind us of our often fragile relationship with the environment, and that we are not yet free of ancient hazards (Figure 48.3).

The third surge has continued into modern times, and the rate of the rise has constantly accelerated. This is due largely to a host of innovations in industry, agriculture, and public health. Many of these innovations arose in industrialized, developed countries and were exported to heavily populated developing regions. In the developed nations, famine was all but eradicated with the advent of pesticides, chemical fertilizers, and high-yield crops. Potential disease epidemics were routinely controlled by vaccines, antibiotics, and insecticides.

In conclusion, we can note that the human population grew from about 5 million at the dawn of agriculture to 500 million by 1650. In the next 200 years, the world population doubled, reaching 1 billion. (**Doubling time** is the number of years re

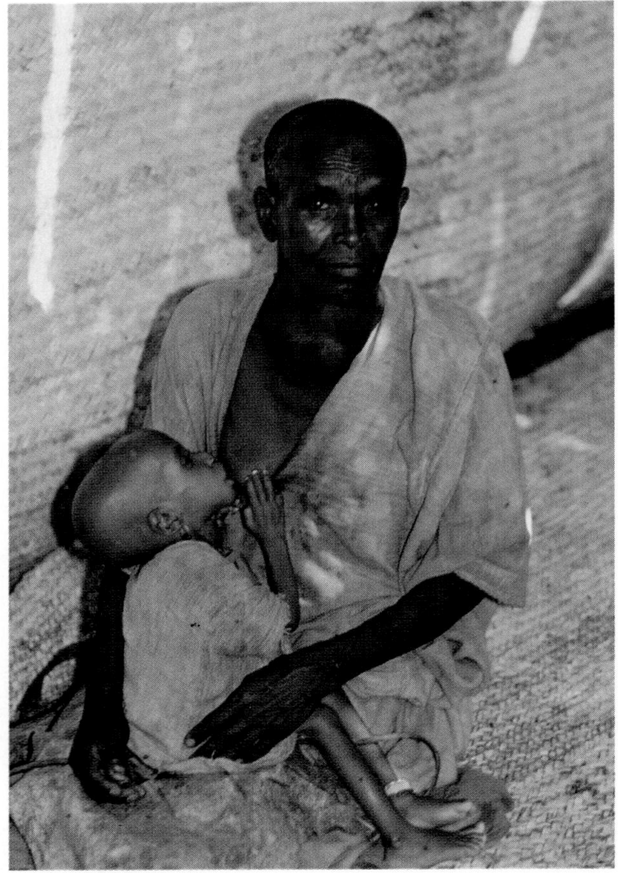

FIGURE 48.3 **FAMINE.**
In spite of modern achievements, undeveloped and developing nations are still subject to disruptive episodes of famine.

quired for a population to double in numbers.) In the 80 years between 1850 and 1930, the numbers doubled again, to 2 billion. The next doubling took only 45 years; and in 1975, our world population stood at 4 billion. By 1970, thoroughly alarmed population experts were predicting another doubling, to 8 billion, by the year 2000.

THE HUMAN POPULATION TODAY

To the surprise of nearly everyone, the *rate* of natural increase in the world population slowed toward the end of the 1970s, and that deceleration continues today (see Table 48.1). (Note the emphasis on the word *rate,* a measure of change over time. The occupants of a car approaching a cliff at 30 mph might take little comfort in knowing that the car's rate is slowing and that it will be traveling at only 15 mph by the time it goes over the edge.)

At the end of 1992, demographers estimated the world's population to be near 5½ billion. The annual growth rate had fallen slightly from that of 1970, and the doubling time had increased from 35 to 41 years. These data suggest some headway in our attempts to control our population, which can be at least partially attributed to changing attitudes of women toward their role in society, changes in preferences in family size, advances in methods of birth control, and the liberalization of abortion laws in developed nations.

GROWTH IN THE DEVELOPING REGIONS

The new data generated a measure of relief and even a growing optimism. But it turns out that the lower rates of natural increase occur primarily in the most developed nations, those that could best support increased populations, such as the United States, Japan, Western Europe, and the former Soviet Union. Most poorer, developing nations of the world show few signs of controlling their populations. The annual natural increase in those regions (excluding China) is an ominous 2.2%, and the doubling time is almost 32 years. While the crude birth and death rates in the United States in 1992 were 16 and 9, respectively, those rates in Africa were 43 and 14. Further, the doubling time in Africa is an alarmingly brief 23 years. How do you suppose these shifting population trends will affect political, social, and economic stability in the near future? What effects are they having now?

At first glance, the news from Latin America seemed hopeful: In the past few years, there has been a decrease in the crude birth rate. But again, the numbers are misleading. In this region, the birth rate is already four times the death rate. Furthermore, many Latin American nations already troubled by political unrest and dismal economic conditions face a doubling time of only 34 years.

Asia has traditionally troubled demographers because so little reliable statistical information is available about it, and of course it has enormous reproductive potential. (It is instructive to keep in mind that well over half of the people walking the earth today are Asians, and one in every five humans alive today is Chinese.) However, within the last decade the birth rate in China (reportedly) declined 53%, with an equivalent fall in the death rate. Partly because of this shift in China, the doubling time for Asian populations has increased from 31 to 34 years. Massive birth control programs in China and several other Asian nations have been instrumental in the declining birth rates. Similar efforts in

India have been somewhat successful, but India's annual increase is still over 2%, and its doubling time is only 34 years.

The successes Asia is currently enjoying, however, may be swamped by another problem, one of momentum. Specifically, 36% of Asian people are under the age of 15 and have yet to enter the breeding population. Asia could well be in for another population explosion.

It is difficult if not impossible to predict what the future holds, but there are ways of making educated guesses. Let's gaze into the c rystal ball of demography and see how population changes are forecast.

Demographic Transition

Developed nations are those that have a slow rate of population growth; a stable, industrialized economy; a low percentage of workers employed in the agricultural sector; a high per capita income; and a high degree of literacy. *Developing* nations have the opposite traits. Of course, many countries have intermediate conditions, and some have a combination—a privileged upper class and a poor lower class.

There is an interesting relationship between a country's developmental progress and its population structure: As nations undergo economic and technological development, their population growth tends to decrease. According to the theory of demographic transition, nations go through several developmental phases, the earliest of which is characterized by high birth and death rates and slow growth. As they begin to develop, the birth rate remains high, but the death rate falls. The result is that the population begins to grow rapidly. Then, as industrialization peaks, the birth rate falls and begins to approximate the death rate. The population enters a fluctuating equilibrium state (Figure 48.4).

One major prediction based on the theory of demographic transition is that population growth in developing parts of Asia, Africa, and Latin America will slow as they become further industrialized. But can the answer to the enigma of world population growth be that simple? Not quite. First, it is the developed nations that must pay the enormous costs of such a massive industrialization program, and this could place a great burden on their own economies. (Also, who needs more competition in the marketplace?) Second, there are severe risks in encouraging developing countries to go through such transitions. For example, if massive development programs stall in the second and most difficult phase, soaring populations will preclude the third stage and send the hopeful developing nation into a deadly population surge. Thus, any such developmental program must be approached with extreme caution.

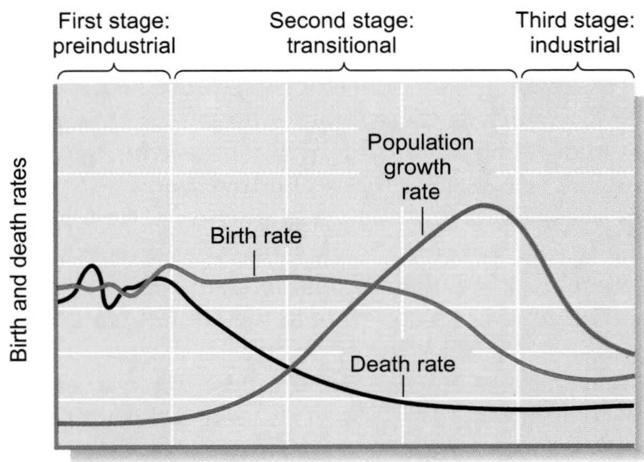

FIGURE 48.4 BIRTH RATES.
According to demographic transition theory, populations have three characteristics depending on their developmental state. Accordingly, the key to curbing world population growth lies in managing the birth rate during a nation's preindustrial period.

The Future of the Human Population

Our track record in predicting population growth is not very impressive; demographers in the United States were taken by surprise by the baby boom of the 1950s, and they are still trying to explain the latest downward trend. However, increasingly sophisticated and precise calculations show some promise of improving the accuracy of our projections.

One of the more useful statistics in discussing human populations is the **general fertility rate (GFR).** Unlike the birth rate, which reflects the number of babies born per total population in a given year, the GFR reflects the number of births per number of women of *reproductive age* per given year. (See Table 48.1 for this and other calculations.) Since the number of women of reproductive age varies from region to region and from nation to nation, comparisons based on the general fertility rate can be considerably more accurate. Reproductive age is somewhat arbitrarily defined as age 15–44, a figure that is most accurate for American women. The calculations for the GFR, together with other population statistics, can be seen in Table 48.1.

One of the highest general fertility rates ever recorded was in Iran—an incredible 200 live births per 1,000 women of reproductive age. At the same time, in the 1980s, the fertility rate in the Netherlands was 48.

A second useful indicator is the **total fertility rate (TFR),** which is the average number of children a woman will have during her lifetime. More simply, the TFR tells us how many children women are having these days. It is really a prediction of how many children any woman is likely to have if the known general fertility rate remains constant.

The general fertility rate in the United States and much of Western Europe dipped sharply in the Great Depression years, falling close to the actual replacement level by 1936. The total fertility rate in 1936 was about 2.2. (Replacement level is considered to be 2.1). This trend was short-lived, however, for by the end of the 1930s, the fertility rate had begun to climb, with the GFR peaking in the post-World War II years (the peak year: 1957) at 125 births per 1,000 women, with the TFR reaching 3.7 (Figure 48.5). In the United States, this era became known as the "baby-boom" years. This was a time when the "good life" meant a new car, a home in the suburbs, terrific appliances, and three or four kids playing in the yard.

Then came the awakening. Environmentally concerned scientists began to spread the alarm about the unprecedented population growth throughout the world. Soon after, reproductive rates in developed countries began to slow. The reasons may never be known, but perhaps people were taking the warnings seriously. By 1975, the total fertility rate in the United States, Japan, and most European nations had fallen to below 2.0. The fertility rates in most of these nations show no signs of beginning a new upward trend.

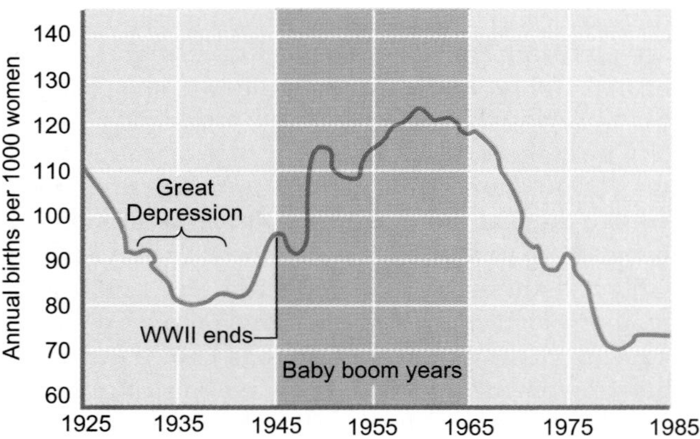

FIGURE 48.5 FERTILITY IN THE UNITED STATES.
The general fertility rate. The Depression years are seen as a valley, followed by a reproductive peak—the "baby boom" years—that followed World War II. The more recent valley is good news to those alarmed by population growth, but keep in mind that attitudes toward family size change, and women born in the baby-boom years are still in the reproductive period of life.

Demographers have tried to understand what caused such a drastic change in reproductive behavior in the developed world. We can begin to guess at what happened, at least in the United States. It was a matter of attitude. The attitude of American women in their reproductive period toward domestic life and family size changed radically in the 1960s and 1970s (as did that of women in many other developed nations). Women began to assume new roles in our industrialized society. They entered new areas of the work force, challenged male enclaves, demanded rights and privileges previously reserved for men, and placed less emphasis on raising children. Their new attitudes were duly noted by the demographers, who then predicted a reduced rate of population growth. But demographers are only too aware that such trends can change quickly, and if attitude is an important variable in forecasting population sizes, their data must be constantly updated and revised. In any case, measuring attitudes is a very risky business.

Whether or not it was the demographers' warnings that the developed nations heeded, their rate of population growth has slowed. But what about the developing countries, those with poorer, often illiterate populations? After all, these are where most of the earth's teeming billions reside. It turns out, somewhat unexpectedly, that the rates of increase in the developing countries are also declining, but the effect there is much more difficult to see. The problem is one of population momentum. With a declining death rate (particularly among the young), an increasing life span, and with great numbers of children who have yet to enter their reproductive years, downward trends in fertility will not have an appreciable effect for many years.

Using Mexico as an example once more, let's note that even at the peak of the baby boom in the United States, Mexican women experienced twice as many births per woman. Yet the subsequent dramatic decline seen in the United States also occurred in Mexico, and by the mid-1980s the total fertility rate in Mexico had gone from nearly 7 to below 5. Demographers estimate that it will continue to decline, reaching about 2.3 in 2025. Again, the problem is that, because of the large proportion of younger Mexicans in the population, the fertility *decline* will be accompanied by a whopping population *increase* as great numbers of youths enter the breeding population. The Mexican population, by some estimates, will go from 88.5 million in 1988 to 174 million in 2025.

Population Age Structure

Knowing the age structure of any population is critical to understanding growth patterns and making predictions. One way to portray such data is by *age structure histograms.* In Figure 48.6, you can see how such diagrams are formed. Note the marked differences in the shapes of such histograms between developed and developing nations. In developed areas, recent population increases have been comparatively slow, so the base is not very wide. Also, people tend to live longer and thus to occupy the upper levels in greater proportions. In developing nations, on the other hand, the rate of increase is still expanding, swelling the lower levels. Obviously, these lower levels are important to population forecasting, since they represent future reproducers. Finally, such shapes illustrate that people in developing regions have fewer health advantages and are more likely to die sooner, resulting in a significant decrease in the upper third of these regions' age brackets.

Growth Predictions and the Earth's Carrying Capacity

The fundamental question of how large the human population can become is irrevocably tied to what the earth can support—its human carrying capacity. If we have learned anything from population studies of other species, it is that the carrying capacity of any environment can be changed (see Essay 48.1). The worry is that a combination of such factors could set the stage for a devastating population crash.

What is the earth's carrying capacity for humans? The range of estimates varies enormously. The experts simply cannot agree (Figure 48.7). Some population biologists believe that we have already exceeded our limits and that our present population represents a drastic overshoot. Then there are those who believe the human population can increase from the current 5½ billion to 50 billion and still survive easily. (Biologists don't take this latter estimate very seriously.)

Recent estimates by more moderate population experts suggest that the human population could be sustained *temporarily* at 8 to 15 billion. From there, they suggest, our numbers could gradually decrease to new, more stable levels. In any event, if we continue on our present course, we can expect not a

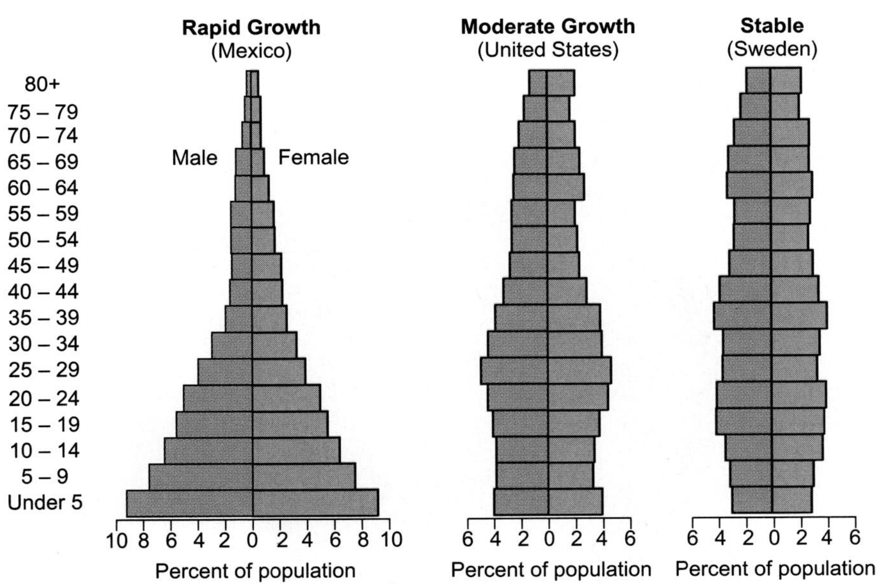

Figure 48.6 **Age Structure Histograms.**
Age structure histograms break the population down into 5-year age groups, revealing much about past history and permitting predictions of future trends to be made. *Left,* expanding nation, e.g., Mexico. *Middle,* moderately stable nation, e.g., the United States. *Right,* long-term stable nation, e.g., Sweden. *Source: Population Reference Bureau.*

Essay 48.1 Holes in the Sky

Now let's explore the relationship between underarm deodorants and the death of the oceans. Believe it or not, one may exist. The propellant in underarm sprays is, in many cases, a class of molecules called chlorofluorocarbons (CFCs). These are essentially carbon molecules to which are attached chorine and/or fluorine atoms. Chlorofluorocarbons are used in a variety of manufactured products, such as air-conditioning, refrigeration, insulating foams, plastics, and industrial solvents.

The problem is, these molecules are very stable. So after you spray under your arms, or after the refrigerant escapes, these long-lived little molecules are released into the air. Because they're light, they eventually, perhaps a few years later, end up in the upper atmosphere.

Paradoxically, these molecules would be safer for life if they stayed closer to earth mingling with living things. It turns out, though, that they threaten life precisely because they drift upward, away from it. The reason is, at an altitude of 15 miles, the CFCs break down the ozone layer. Ozone is O_3, formed, in nature, by the sun breaking down atmospheric O_2 molecules, allowing them to rejoin as ozone. The chlorine in the CFCs attacks the ozone, breaking it back down into its components. There isn't much ozone up there to begin with. At sea level, all of it together would form a layer over the earth about as deep as a pencil lead is thick.

The ozone, though, is critical to life on earth. Primarily, it functions by blocking destructive ultraviolet light from the sun. Those rays are destructive on three primary bases. First, they increase the risk of skin cancer, particularly among light-skinned people. Second, they depress the immune systems of humans, setting the stage for a host of illnesses. Third, they destroy the algae that form the first step in the ocean's food chains. Ultraviolet light may be more destructive, in fact, than we now imagine. Andrew Blaustein and his group at Oregon State University have correlated frog egg failure in ponds with the amount of UV light they receive, signaling what may be an impending crisis for other life forms.

Although ozone depletion was first discovered over the Antarctic, "holes" have now been found at midlatitude, in northern temperate regions. In fact, the ozone depletion in this area is now reaching 3–5% per decade. It has been calculated that a loss of 1% translates into 12,000 to 30,000 new skin cancers in the United States alone.

The manufacturers of CFCs have been reluctant to take action to reduce the levels of these chemicals over the earth. Some have tried to paint a brighter face on it.

which makes CFCs, took out ads in newspapers saying that the danger to the ozone layer has just recently improved. The company has now agreed to phase out the manufacture of CFCs, but the phasing out will not be complete until the year 2000. Since the United States makes only 30% of the world's CFCs, the effort will have to be global, demanding more cooperation that one often finds among industrial nations.

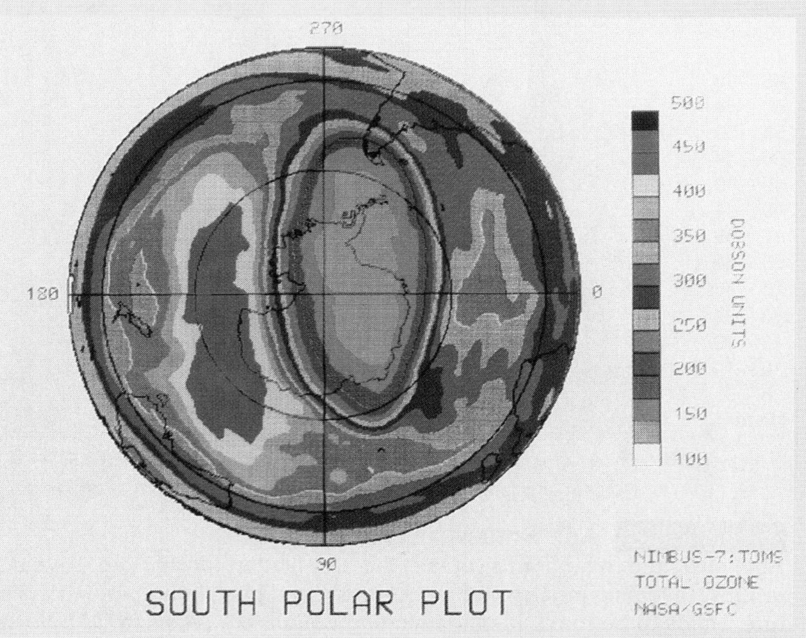

gradual decrease in numbers but a crash—a massive increase in our death rate. It has been calculated that such a crash might kill 50% to 80% of the human population, and at that point the human population will stabilize. Such a crash, it is believed, would likely be due to a combination of famine, war, disease, and ecological disruption, such as increasingly acid precipitation and a further reduction in our protective ozone.

To leave the subject on a note of cautious optimism, we reiterate that world population growth is slowing down somewhat and that some of the more heavily populated nations have recently joined this trend. Further, it is within the power of the world's family of nations to see that the decreasing trend is continued and reinforced. One goal that has been suggested is to reach *replacement level.* Essentially, it means that each couple simply replaces itself, by having only two children. (The popular belief that the United States has achieved this goal is inaccurate.) Because of the burgeoning numbers of young people, the future size of the world's population depends largely upon when that replacement level is reached.

Scenarios portraying a stable world population are admittedly utopian, but they must not be regarded as impossible. Some population experts maintain that many slow-growing developed nations are now reaching the desired level and that others can do so within a few decades. The trend toward family planning and use of birth control is increasing, and one hopes that efforts to stimulate such interest in developing nations will continue. Under the best of circumstances, the world could reach some state of population stability in about 50 years (Figure 48.7).

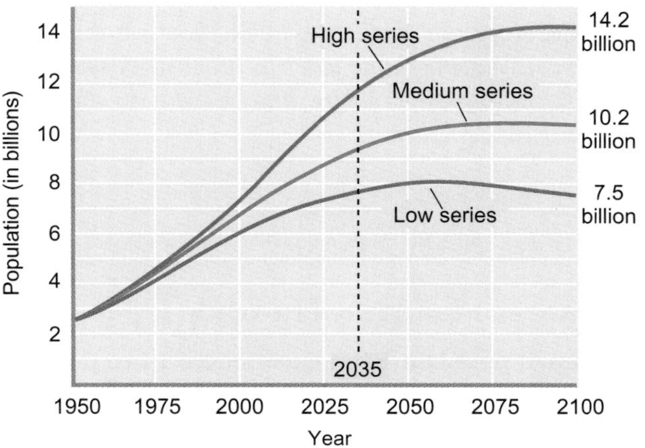

FIGURE 48.7 WORLD POPULATION GROWTH, 1950–2100: THREE SCENARIOS.
Three projections of world population based on different assumptions. The middle series assumes that by 2035 each woman will have about two children. Even then, the population will double today's level. If "two-child families" are achieved by 2010, the lower curve results; the higher figure results if the "two-child family" is delayed until 2065.
Source: Population Reference Bureau Data Sheet.

PERSPECTIVES

The bottom line is that we must assume our species is not exempt from the natural laws that govern population control in other species. Our only special feature is our mental capacity. We have the ability to analyze, predict, imagine, and finally to choose. Of course, we can choose by deciding not to choose, not to take a stand, not to be involved. But the time for that luxury is past. We must now learn as much as possible about the nature of overpopulation, apply ourselves to solving the problem, and be ready to stand accountable for our actions when we are judged by future generations.

SUMMARY

HUMAN POPULATION HISTORY

The human population has experienced three great surges—the first correlated with the advent of tools; the second with the development of agriculture and the domestication of animals; the third less well understood but probably related to advances in health practices, food storage, and transportation. The third surge continues today.

THE HUMAN POPULATION TODAY

The world's rate of natural increase slowed in the 1970s and is still slowing. The current human population is about 5½ billion with a doubling time of 41 years.

More developed nations usually show a lower rate of natural increase. Developing regions may have a higher rate of natural increase with an alarmingly brief doubling time. Many of the people of developing countries have yet to enter the reproductive group.

As nations become more economically, industrially, and technologically developed, their populations tend to decrease.

THE FUTURE OF THE HUMAN POPULATION

The general fertility rate (GFR) is the number of babies born per number of women of reproductive age in a given year. The total fertility rate (TFR) is the average number of children a woman will have during her lifetime. These rates in the United States dropped during the depression years, then rebounded after World War II, producing the "baby boom." They have now begun to slow again, partly due to the changing roles of women.

Knowing the age structure of any population is crucial to making predictions about it. One way to summarize such structures is through age structure histograms.

We do not know the earth's carrying capacity for humans. Some scientists say we can expect a crash soon while others say our population can go from 5½ billion to 50 billion.

KEY TERMS

doubling time • **865**
general fertility rate (GFR) • **867**
total fertility rate (TFR) • **868**

REVIEW QUESTIONS

1. How many people are *added* to the world's population each day (births minus deaths)? Each year?
2. To what can we ascribe the three human population surges?
3. If population growth is 1.5, what is its doubling time? (See table 48.1.)
4. What was the world's population by 1650? By the end of 1992?
5. High birth and death rates with slow growth is characteristic of what phase in a country's development?
6. What calculation best tells us how many children women are having these days? What is the calculation?
7. Define general fertility rate.
8. Is it true that by 1975 the total fertility rate in the United States was below 2.0, but above 2.0 in Japan?
9. How can age structure histograms predict future population growth?
10. How do CFCs affect anyone?

Appendix
Geologic Timetable

Eras (Years Since Start	Periods and Epochs	Extent in Millions of Years	Geological Events
Cenozoic	Quaternary		
	Holocene (present)	Last 10,000 years	
	Pleistocene	.01–2	Four ice ages, glaciation in N. Hemisphere; uplift of Sierras
	Tertiary		
	Pliocene	2–6	Continued uplift; drastic cooling
	Miocene	6–23	More uplift in Rockies; isthmus of Panama formed; climate drying
	Oligocene	23–36	Mountain uplift in Europe, Asia; volcanic action in Rockies
	Eocene	35–54	Inland seas diminished
65,000,000	Paleocene	54–65	Continents formed, separations continuing
(Cretaceous-Paleocene discontinuity. Asteroid colllides with earth; dust obscures sun [recent hypothesis])			
Mesozoic	Cretaceous	65–135	Two major land forms (Laurasia and Gondwana); continental separation occurs through era; Rockies forming, other continents low; seas over Europe
	Jurassic	135–197	Continued mountain building, most continents low, inland seas
225,000,000	Triassic	197–225	Mountain building in Pangaca (North America); most recent continental drift begins
Paleozoic	Permian	225–280	Very cool; mountain building, glaciation in south, seas diminish, Appalachians formed; vast extinction
	Carboniferous	280–345	Lowlands, shallow seas, coal swamps, mountain building in Pangaca (North America)
	Devonian	345–405	Landscape varies, Appalachians forming
	Silurian	405–500	Flattened landscape, some mountains; shallow seas, lowlands
	Ordovician	425–500	
570,000,000	Cambrian	500–570	Temperature increases
Precambrian		1000	Cooling; atmosphere has become oxidizing rather than reducing
		2000	Low-lying, vast, inland seas; tropical climate with little latitudinal variation
Origin of earth, 4.5–5.7 billion years		3000	Oxygen production
		4000	Oldest rock formation, crust hardening
		4500	Chemical evolution; solar system forms

SUGGESTED READINGS

For each part of the book, texts and other primary sources are listed first, then a selection of relevant *Scientific American* articles is given. The readings listed here are of three types. The selection includes classical books (such as *On the Origin of Species*) and articles (such as Watson and Crick's brief report from 1953); historical articles from the last thirty years or so that were important in leading to current thinking; and recent material that describes leading-edge findings, ideas, or positions.

PART I: MOLECULES AND CELLS

Alberts, B., et al. 1989. *Molecular Biology of the Cell,* 2d ed. New York: Garland Publishing Co.

Calvin, M., ed. 1973. *Organic Chemistry of Life: Readings from* Scientific American. San Francisco: W.H. Freeman Co.

The Chemical Basis of Life: An Introduction to Molecular and Cell Biology: Readings from Scientific American. Introduction by P.C. Hanawalt and R.H. Haynes. 1973. San Francisco: W.H. Freeman Co.

Darwin, C. 1859. *On the Origin of Species through Natural Selection.* A facsimile of the first edition. Cambridge, Mass.: Harvard University Press.

de Beer, G. 1965. *Charles Darwin: A Scientific Biography.* New York: Doubleday.

Moorehead, A. 1969. *Darwin and the Beagle.* New York: Harper & Row.

Porter, E. 1971. *Galapagos.* New York: Ballantine.

Skinner, B.F. 1928. *The Behavior of Organisms: An Experimental Analysis.* New York: Appleton-Century-Crofts.

Stryer, L. 1989. Biochemistry, 3d ed. San Francisco: W.H. Freeman Co.

Weissmann, G., and Clairborne, R., eds. 1975. *Cell Membranes: Biochemistry, Cell Biology and Pathology.* New York: HP Publishing Co

---

Scientific American articles. New York: W.H. Freeman Co.

Allen, R.D. 1987. "The Microtubule as an Intracellular Engine." February.

Dautry-Varsat, A., and Lodish, H. 1984 "How Receptors Bring Proteins and Particles into Cells." May.

Dustin, P. 1980. "Microtubules." August.

Gingerich, O. 1982. "The Galileo Affair." July.

Govindjee, W., and Coleman, W.J. 1990. "How Plants Make Oxygen." February.

Lederman, L. 1984. "The Value of Fundamental Science." November.

Lerner, L.S., and Gosselin, E.A. 1986. "Galileo and the Specter of Bruno." November.

Scientific American. 1985. "The Molecules of Life." Entire October issue.

Vollrath, F. 1992. "Spider Webs and Silk." March.

Youvan, D.C., and Marrs, B.L. 1987. "Molecular Mechanisms of Photosynthesis." June.

PART II: CONTINUITY OF LIFE

Avery, O.T., et al. 1944. "Studies of the Chemical Nature of the Substance Inducing Transformation of Pneumococcal Types." *Journal of Experimental Medicine* 79:137.

Dawkins, R. 1976. *The Selfish Gene.* New York: Oxford University Press.

Joravsky, D. 1970. *The Lysenko Affair.* Cambridge, Mass.: Harvard University Press.

Kleinsmith, L.J., and Kish, V.M. 1995. *Principles of Cell and Molecular Biology,* 2d ed. New York: HarperCollins.

Lack, D. 1947. *Darwin's Finches.* New York: Cambridge University Press.

Mendel, G. 1965. "Experiments in Plant Hybridization (1865)." Translated by Eva Sherwood. In *The Origin of Genetics,* edited by C. Stern and E. Sherwood. San Francisco: W.H. Freeman Co.

Meselson, M., and Stahl, F.W. 1958. "The Replication of DNA in *E. coli." Proceedings of the National Academy of Sciences* (U.S.). 44:671.

Shine, I., and Wrobel, S. 1976. *Thomas Hunt Morgan: Pioneer of Genetics.* Lexington, Ky.: University of Kentucky Press.

Watson, J.D. 1968. *The Double Helix.* New York: Atheneum.

Watson, J.D., and Crick, F.H.C. 1953. "Molecular Structure of Nucleic Acids: A structure of deoxyribose nucleic acid." Nature 171:737.

Scientific American articles. New York: W.H. Freeman Co.

Aharonowitz, Y., and Cohen, G. 1981. "The Microbiological Production of Pharmaceuticals." September.

Anderson, W.F., and Diacumakos, E.G. 1981. "Genetic Engineering in Mammalian Cells." July.

Beardsley, T. 1992. "Diagnosis by DNA." October.

Brill, W.J. 1981. "Agricultural Microbiology." September.

Chilton, M. 1983. "A Vector for Introducing New Genes into Plants." June.

Cohen, S.N., and Shapiro, J.A. 1980. "Transposable Genetic Elements." February.

Grant, P.R. 1991. "Natural Selection and Darwin's Finches." October.

Grivell, L.A. 1983. "Mitochondrial DNA." March.

Kornberg, R.D., and Klug, A. 1981. "The Nucleosome." February.

Lake, J.A. 1981. "The Ribosome." August.

Novick, R.P. 1980. "Plasmids." December.

Stahl, F.W. 1987. "Genetic Recombination." February.

PART III: EVOLUTION AND DIVERSITY

Ahmadkian, Vernon. 1982. "The Nature of Lichens." *Natural History*, March, pp. 30–37.

Alexopoulos, C.J. 1979. *Introduction to Mycology.* New York: Wiley.

Bold, Harold C., and Wynne, Michael J. 1985. *Introduction to the Algae: Structure and Reproduction,* 2d ed. Englewood Cliffs, N.J.: Prentice Hall.

Eldredge, N., and Gould, S.J. 1972. "Punctuated Equilibria: An Alternative to Phyletic Gradualism." In *Models in Paleobiology*, edited by T.J.M. Schopf. New York: W.H. Freeman Co.

Futuyuma, E.J. 1986. *Evolutionary Biology*, 2d ed. Sunderland, Mass.: Sinauer Associates.

Gilbert, L.E., and Raven, P.H. 1975. *Coevolution of Plants and Animals.* Austin, Texas: University of Texas Press.

Large, E.C. 1962. *The Advance of the Fungi.* New York: Dover.

Lee, J.J., Hutner, S.H., and Bovee, E.C, eds. 1985. *An Illustrated Guide to the Protozoa.* Society of Protozoologists, Lawrence, Kan.

Margulis, Lynn. 1981. *Symbiosis in Cell Evolution: Life and Its Environment on the Early Earth.* New York: W.H. Freeman Co.

Miller, S.L. 1935. "Production of Some Organic Compounds under Possible Primitive Earth Conditions." *Journal of the American Chemical Society* 77:2351.

Schopf, J.W., and Oehler, D.Z. 1971. "How Old Are the Eukaryotes?" *Science* 193:47.

Scientific American Editors. 1978. *Evolution: A* Scientific American *Book.* San Francisco: W.H. Freeman Co.

Stanier, R.Y., et al. 1986. *The Microbial World*, 5th ed. Englewood Cliffs, N.J.: Prentice Hall.

Tortora, Gerard J. 1989. *Microbiology: An Introduction,* 3d ed. Menlo Park, Calif.: Benjamin/Cummings.

Volpe, P. 1981. *Understanding Evolution*, 4th ed. Dubuque, Iowa: Wm. C. Brown.

Yates, G.T. 1986. "How Microorganisms Move through Water." *American Scientist* 74:358-365.

Zinsser, Hans. 1935. *Rats, Lice, and History.* Boston: Atlantic Monthly Press/Little, Brown.

Scientific American articles. New York: W.H. Freeman Co.

Beardsley, R. 1990. "Oravske Kuru." August.

Bishop, J.A., and Cook, Laurence M. 1975. "Moths, Mechanism and Clean Air." January.

Blakemore, R.P., and Frankel, R.B. 1981. "Magnetic Navigation in Bacteria." December.

Dickerson, R.E. 1978. "Chemical Evolution and the Origin of Life." September.

Eigen, M., et al. 1981 "The Origin of Genetic Information." April.

Fischetti, V.A. 1991. "Streptococcal Mprotein." June.

Gallo, Robert C. 1987. "The AIDS Virus." January

Gilbert, L.E. 1982."The Coevolution of a Butterfly and a Vine." August.

Groves, D. J., et al. 1981. "An Early Habitat of Life." October.

Kantor, F.S. 1994. "Disarming Lyme Disease." September.

Kosikowski, Frank V. 1965. "Cheese." May.

Litten, W. 1975. "The Most Poisonous Mushrooms." March.

Margulis, L. 1971. "Symbiosis and Evolution." August.

Prusiner, S.B. 1984. "Prions." October.

Shapiro, James A. 1988. "Bacteria as Multicellular Organisms." June.

Sibley, C.G., and Ahlquist, J.F. 1986. "Reconstructing Bird Phylogeny by Comparing DNAs." February.

Stanley, S.M. 1984. "Mass Extinction in the Ocean." June.

Vidal, G. 1984. "The Oldest Eukaryotic Cells." February.

Wilson, A. 1985. "The Molecular Basis of Evolution." October.

Woese, C.R. 1981. "Archaebacteria." June.

PART IV: PLANT BIOLOGY

Barth, Friedrich G. 1985. *Insects and Flowers: The Biology of a Partnership.* Princeton, N.J.: Princeton University Press.

Conard, Henry S., and Redfearn, Paul L., Jr. 1979. *How to Know the Mosses and Liverworts*, 2d ed. Dubuque, Iowa: Wm. C. Brown.

Esau, K. 1977. *Anatomy of Seed Plants*, 2d ed. New York: Wiley.

Heywood, Vernon H., ed. 1985. *Flowering Plants of the World.* Englewood Cliffs, N.J.: Prentice Hall.

Hitch, Charles J. 1982. "Dendrochronology and Serendipity." *American Scientist* 70:300-305.

Lewis, W., and Elvin-Lewis, M. 1977. *Medical Botany.* New York: Wiley.

Mulcahy, David L. 1981. "Rise of the Angiosperms." *Natural History*, September, pp. 30–35.

Norstog, Knut. 1987. "Cycads and the Origin of Insect Pollination." *American Scientist* 75:270–279.

Simpson, B.B., and Connor-Ogorzaly, M. 1977. "Ecology and Evolution of Flowering Plant Dominance." *Science* 196:622.

Sporne, K.R. 1971. *The Mysterious Origin of Flowering Plants.* Carolina Biological Supply Company, Burlington, N.C.

Temple, S. 1977. "The Dodo and the Tambalaoque Tree." *Science* 197:885.

Wardlaw, I.F. 1974. "Phloem Transport: Physical, Chemical or Impossible?" *Annual Review of Plant Physiology* 25:515.

Zimmermann, Martin H. 1983. *Xylem Structure and the Ascent of Sap.* New York: Springer-Verlag.

Zimmermann, Martin H., and Brown, Claude L. 1975. *Trees: Structure and Function.* New York: Springer-Verlag.

Scientific American articles. New York: W.H. Freeman Co.

Albersheim, P. 1975. "The Walls of Growing Plant Cells.: April.

Chilton, Mary-Dell. 1983. "A Vector for Introducing New Genes into Plants." June.

Evans, Michael L.; Moore, Randy, and Hasenstein, Karl-Heinz. 1986. "How Roots Respond to Gravity." December.

Kaplan, D.R. 1983. "The Development of Palm Leaves." July.

Niklas, Karl J. 1987. "Aerodynamics of Wind Pollilnation." July.

Rosenthal, Gerald A. 1986. "The Chemical Defenses of Higher Plants." January.

Shepart, J.F. 1982. "The Regeneration of Potato Plants from Leaf-Cell Protoplasts." May.

Shigo, Alex L. 1985. "Compartmentalization of Decay in Trees." April.

PART V: ANIMAL BIOLOGY

Alexander, T. 1975. "A Revolution Called Plate Tectonics Has Given Us a Whole New Earth." *Smithsonian* 5:30.

Barnes, Robert D. 1987. *Invertebrate Zoology*, 5th ed. Philadelphia: Saunders College/Holt, Rinehart and Winston.

Buchsbaum, Ralph, et al. 1987. *Animals without Backbones*, 3d ed. Chicago: University of Chicago Press.

Cornejo, D. 1982. "Night of the Spadefoot Toad." *Science 82*, September.

Eckert, Roger, and Randall, David. 1988. *Animal Physiology: Mechanisms and Adaptations*, 3d ed. New York: W.H. Freeman Co.

Hood, Leroy E., et al. 1984. *Immunology.* Menlo Park, Calif.: Benjamin/Cummings.

Johanson, D.C., and Edey, M.A. 1981. "Lucy: The Inside Story." *Science 81*, March.

Karp, Gerald, and Berrill, N.J. 1981. *Development*, 2d ed. New York: McGraw-Hill.

Lewin, Roger. 1988. *In the Age of Mankind.* Washington, D.C.: Smithsonian.

MacArthur, R.H. 1972. *Geographical Ecology: Patterns in the Distribution of Species.* New York: Harper & Row.

Mayr, E. 1963. *Animal Species and Evolution.* Cambridge, Mass.: Harvard University Press.

Miller, W.H. 1974. "Photoreception." In *The New Encyclopaedia Brittanica*, 15th ed. 14:353.

Money, John, and Ehrhardt, Anke A. 1972. *Man and Woman, Boy and Girl: The Differentiation and Dimorphism of Gender Identity from Conception to Maturity.* Baltimore: Johns Hopkins University Press.

Nilsson, Lennart, et al. 1986. *A Child Is Born: The Drama of Life before Birth.* New York: Dell.

Petit, C., and Ehrman, L. 1969. "Sexual Selection in *Drosophilia*." *Evolutionary Biology* 3:177.

Romer, Alfred, and Parsons, Thomas S. 1985. *The Vertebrate Body*, 6th ed. Philadelphia: Saunders College/Holt, Rinehart and Winston.

Schmidt-Nielsen, Knut. 1983. *Animal Physiology: Adaptation and Environment*, 3d ed. New York: Cambridge University Press.

Scientific American Editors. 1979. *The Brain.* New York: W.H. Freeman Co.

Scientific American Editors. 1985. Progress in Neuroscience. New York: W.H. Freeman Co.

Simmons, J.A.; Fenton, M.B., and O'Farrell, M.J. 1979. "Echolocation and Pursuit of Prey by Bats." *Science* 203:16.

Sperry, R. 1982. "Some Effects of Disconnecting the Cerebral Hemispheres." *Science* 217:1223.

Wallace, R.A. 1980. *How They Do It.* New York: Morrow.

Wilson, A.C.; Carlson, S.S., and White, T.J. 1977. "Biochemical Evolution." *Annual Review of Biochemistry* 46:573.

Woolacott, R.M., and Zimmer, R.L. 1979. *Biology of Bryozoans.* New York: McGraw-Hill.

Scientific American articles. New York: W.H. Freeman Co.

Ada, Gordon L, and Nossal, Gustav. 1987. "The Clonal Selection Theory." August.

Alvarez, W., and Asaro, F. 1990. "An Extraterrestrial Impact." October.

Baker, M.A. 1979. "A Brain-Cooling System in Mammals." April.

Beaconsfield, P., et al. 1980. "The Placenta." August.

Berridge, Michael J. 1985. "The Molecular Basis of Communication within the Cell." October.

Brown, B.E., and Ogden, J.C. 1993. "Coral Bleaching." January.

Buisseret, P.D. 1982. "Allergy." August.

Cantin, Marc, and Genest, Jacques. 1980. "The Heart as an Endocrine Gland." February.

Cohen, Irun R. 1988. "The Self, the World and Autoimmunity." April.

Dunant, Yves, and Israel, Maurice. 1985. "The Release of Acetylcholine." April.

Eastman, Joseph T., and DeVries, Arthur L. 1986. "Antarctic Fishes." November.

Edelson, Richard L., and Fink, Joseph M. 1985. "The Immunologic Function of Skin." June.

Feder, Martin E., and Burggren, Warren W. 1985. "Skin Breathing in Vertebrates." November.

Goldstein, Gary W., and Betz, A. Lorris. 1986. "The Blood-Brain Barrier." September.

Gurdon, J.B. 1968. "Transplanted Nuclei and Cell Differentiation." December.

Horner, J.R. 1984. "The Nesting Behavior of Dinosaurs." April.

Hudspeth, A.J. 1983. "The Hair Cells of the Inner Ear." January.

Koretz, Jane F., and Handelman, George H. 1988. "How the Human Eye Focuses." July.

Lawn, R.M. 1992. "Lipoprotein (a) in Heart Disease." June.

Levinton, J.S. 1992. "The Big Bang of Animal Evolution." November.

Llinas, R.R. 1982. "Calcium Synaptic Transmission." October.

McMenamin, Mark A.S. 1987. "The Emergence of Animals." April.

Milstein, Cesar. 1980. "Monoclonal Antibodies." October.

Mishkin, Mortimer, and Appenzeller, Tim. 1987. "The Anatomy of Memory." June.

Moog, F. 1981. "The Lining of the Small Intestine." November.

Morrell, P., and Norton, W.T. 1980. "Myelin." May.

Morrison, A. 1984. "A Window on the Sleeping Brain." April.

Mossman, D., and Sarjeant, W. 1983. "The Footprints of Extinct Animals."

Newman, E., and Harline, P. 1982. "The Infrared 'Vision' of Snakes." March.

Roper, C., and Boss, K. 1982. "The Giant Squid." April.

Rukang, W., and Shenglong, L. 1983. "Peking Man." June.

Russell, D. 1982. "The Mass Extinctions of the Late Mesozoic." January.

Schmidt-Nielsen, K. 1981. "Countercurrent Systems in Animals." May.

Schnapf, Julie L., and Baylor, Denis A. 1987. "How Photoreceptor Cells Respond to Light." April.

Scrimshaw, N.S. 1991. "Iron Deficiency." October.

Short, R.V. 1984. "Breastfeeding." April.

Snyder, Solomon H. 1985. "The Molecular Basis of Communication between Cells." October.

Stryer, Lubert. 1987. "The Molecules of Visual Excitation." July.

Tattersall, I. 1992. "Evolution Comes to Life." August.
Tonegawa, Susumu. 1985. "The Molecules of the Immune System." October.
Vollrath, F. 1992. "Spider Webs and Silks." March.
Webb, P.W. 1984. "Form and Function in Fish Swimming." July.
Winfree, A. 1983. "Sudden Cardiac Death: A Problem of Topology." May.
Zucker, M. 1980. "The Functioning of Blood Platelets." June.

PART VI: BEHAVIOR AND ECOLOGY

Behavior

Darwin, C. 1871. *The Descent of Man, and Selection in Relation to Sex.* New York: Appleton.
________. 1872. *The Expression of the Emotions in Man and Animals*. New York: Appleton.
Dawkins, R. 1976. *The Selfish Gene.* New York: Oxford University Press.
Ehrman, L., and Pasons, P.A. 1976. *The Genetics of Behavior.* Sunderland, Mass.: Sinauer Associates.
Jansen, D.H. 1966. "Coevolution of Mutualisms between Ants and Acacias in Central America." *Evolution* 20:249.
Johnsgard, P.A. 1967. "Dawn Rendezvous on the Lek." *Natural History* 76:16.
Jolly, A. 1985. "The Evolution of Primate Behavior." *American Scientist* 70:230-239.
Maynard Smith, J. 1964 "Group Selection and Kin Selection." *Nature* 201:1145.
_________. 1974. "The Theory of Games and the Evolution of Animal Conflicts." *Journal of Theoretical Biology* 47:209.
Mech, L.D. 1970. *The Wolf: The Ecology and Behavior of an Endangered Species.* Garden City, N.Y.: Natural History Press.
Skinner, B.F. 1938. *The Behavior of Organisms: An Experimental Analysis.* New York: Appleton-Century-Crofts.
Trivers, R.L. 1971. "The Evolution of Reciprocal Altruism." *Quarterly Review of Biology* 46:35.
Wallace, R.A. 1979. *The Genesis Factor*. New York: Morrow.
Wilson, E.O. 1971. *The Insect Societies.* Cambridge, Mass.: Harvard University Press.
_________. 1975. *Sociobiology: The New Synthesis.* Cambridge, Mass.: Harvard University Press.
_________. 1976. "Academic Vigilantism and the Political Significance of Sociobiology." *Bioscience* 26:183.
_________.1978. *On Human Nature.* Cambridge, Mass.: Harvard University Press.

Scientific American articles. New York: W.H. Freeman Co.

Alkon, D. 1983. "Learning in a Marine Snail." July.
Fitzgerald, G.J. 1993. "The Reproductive Behavior of the Stickleback." April.
Heinrich, B. 1981. "The Regulation of Temperature in the Honeybee Swarm." June.
Lloyd, J. 1981. "Mimicry in the Sexual Signals of Fireflies." July.
Maynard Smith, J. 1978. "The Evolution of Behavior." September.
Van Dyke, C., and Byck, R. 1982. "Cocaine." March.
Wilson, E.O. 1975. "Slavery in Ants." June.

Ecology

Ayensu, E., ed. 1980. *Jungles.* New York: Crown.
Begon, M.; Harper, J.L., and Townsend, C.R. 1992. *Ecology: Individuals and Communities*, 2d ed. Sunderland, Mass.: Sinauer Associates.
Borgstrom, G. 1976. "Never Before Has Humankind Had to Face the Problem of Feeding So Many People with So Little Food." *Smithsonian* 7:70.
Connell, J.H. 1978. "Diversity in Tropical Rain Forests and Coral Reefs." *Science* 199:1302.
Corliss, J.B., et al. 1979. "Submarine Thermal Springs on the Galapagos Rift." *Science* 203:1073.
Egbert, G., et al., eds. 1982. *The Ecology of a Tropical Forest.* Washington, D.C.: Smithsonian.
Hardin, G. 1993. *Living within the Limits: Ecology, Economics, and Population Taboos.* New York: Oxford University Press.
________. 1968. "The Tragedy of the Commons." *Science* 162:1243.
Hutchinson, G.E. 1959. "Homage to Santa Rosalia, or Why Are There So Many Kinds of Animals?" *American Naturalist* 93:145.
Moore, J.A. 1985. "Science as a Way of Knowing." *American Zoologist* 25:1-155.
Schaller, G.B. 1972. *The Serengeti Lion: A Study of Predator-Prey Relations.* Chicago: University of Chicago Press.
Terborgh, J. 1974. "Preservation of Natural Diversity, the Problem of Extinction-prone Species." *Bioscience* 24:715.
Wilson, E.O., ed. 1974. *Ecology, Evolution and Population Biology: Readings from* Scientific American. New York: W.H. Freeman Co.
Wilson, E.O., and Bossert, W.H. 1971. *A Primer of Population Biology.* Sunderland, Mass.: Sinauer Associates.
Wilson, E.O., ed. 1988. *Biodiversity.* Washington, D.C.: National Academy Press.

Scientific American articles. New York: W.H. Freeman Co.

Bergerud, A., 1984. "Prey Switching in a Simple Ecosystem." December.
Calhoun, J.R. 1962. "Population Density and Social Pathology." February.
Edmond, J., and von Damm, K. 1984. "Hot Springs on the Ocean Floor." April.
French, H.F. 1994. "Making Environmental Treaties Work." December.
Horn, H.H. 1975. "Forest Succession." May.
Ingersoll, A. 1983. "The Atmosphere." September.
Mohnen, Volker A. 1988. "The Challenge of Acid Rain." August.
Perry, D. 1984. "The Canopy of the Tropical Rain Forest." November.

GLOSSARY

A band a band in striated muscle corresponding to the region of myosin filaments.

abdomen 1. in mammals, the body cavity between the diaphragm and the pelvis. 2. in other vertebrates, the body cavity containing the stomach, intestines, liver, and reproductive organs. 3. in arthropods, the posterior section of the body.

abiotic *adj.,* characterized by the absence of life.

abortion the expulsion or removal of a fetus or embryo before it can survive on its own.

abscisic acid in plants, the hormone that causes leaves to separate from a plant. Also *ABA.*

abscission zone in plants, a layer of specialized, cutinized parenchyma cells at the base of a leaf petiole, fruit stalk, or branch in which normal separation occurs.

absorption spectrum a graph indicating the wave lengths of light absorbed by a molecule.

abyss the lowest depths of the ocean.

abyssal region the lowest depths of the ocean, especially the bottom water; also called *abysmal region.*

acellular not composed of cells; also called *noncellular.*

acellular slime mold a kind of organism characterized by a noncellular, multi nucleate, creeping somatic phase, and a reproductive phase in which fruiting bodies are produced bearing spores.

acetylcholine a chemical agent that when released by one neuron crosses the synapse to activate a second neuron as effector.

acetylcholinesterase a membrane-bound enzyme that hydrolyzes acetylcholine in the course of synaptic nerve impulse transmission.

acetyl-coA an intermediate compound in cell respiration, to which most fuels are converted before entering the citric acid cycle.

acid a proton donor that yields hydrogen ions in water; a compound capable of neutralizing bases and of lowering the pH of solutions.

acquired immune deficiency syndrome (AIDS) a condition that results in severely reduced immunity.

acrosomal process the part of a sperm cell that contains enzymes that aid in penetrating the egg.

acrosome an organelle in the tip of a sperm which ruptures in fertilization with the release of enzymes and (in echinoderms) of an acrosomal filament.

actin a cytoplasmic contractile protein and a constituent of muscle.

action potential in a neuron, a traveling depolarizing wave; a short-lived change in membrane potential that produces a neural impulse or, in a muscle, a contraction. Compare *resting potential.*

action-specific energy (ASE) in behavior, a hypothetical endogenous tension that increases in the nervous system until it is discharged, either on its own or due to the perception of an appropriate stimulus.

activation energy the energy input needed before an exergonic chemical reaction can proceed.

active immunization the conferring of immunity against a specific disease through the introduction of weakened or killed disease agents or their antigens.

active site the part of an enzyme that interacts directly with the substrate molecule.

active transport energy-requiring transport of a substance across the cell membrane, usually against the concentration gradient.

adaptive radiation evolutionary divergence of two or more species from an ancestral form with the establishment of new, more specialized niches.

ADH. See *antidiuretic hormone.*

adipose tissue fatty tissue.

adenine (A) a purine, one of the nitrogenous bases found in both DNA and RNA, as well as in ATP and several coenzymes.

adenylate cyclase an enzyme, usually incorporated into the cell membrane, that is capable of transforming ATP into cyclic ATP and pyrophosphate.

adhesion the attraction between two dissimilar substances, such as a solid and a liquid.

ADP (adenosine diphosphate) a compound of adenine, ribose, and two phosphate groups; a degraded form of ATP.

adrenal cortex the outer portion of the adrenal gland, producing steroid hormones.

adrenal gland a vertebrate endocrine gland; the outer area (*adrenal cortex*) produces steroid hormones, the inner area (*adrenal medulla*) produces *epinephrine.*

adrenaline See *epinephrine.*

adrenal cortex the outer portion of the adrenal gland, produces steroid hormones.

adrenal medulla the inner portion of the adrenal gland, produces *epinephrine* and *norepinephrine (adrenaline* and *noradrenaline)* as its principle hormonal products.

adrenocorticotropic hormone (ACTH) a pituitary hormone that stimulates the production of hormones of the adrenal cortex.

adventitious root a secondary root growing from stem tissue.

aerobic *adj.,* requiring oxygen.

afferent vessel any blood vessel carrying blood toward a specific structure such as a gill filament or kidney nephron.

age structure histogram also *age profile, population pyramid;* a graph of a population divided into age groups. Each age group is represented by a horizontal bar with that of the youngest group forming the base.

aggression in behavior, hostility (attack or threat), especially unprovoked, usually against a competitor or potential competitor.

aggregate a number of independent organisms grouped together either in a casual, temporary way or for more permanent mutual reasons.

aggregate fruit a fruit developing from multiple carpels of the same flower.

agnatha the class of vertebrates comprising the hagfishes, lampreys, and several extinct forms, having neither jaws or paired appendages.

AIDS (acquired immune deficiency syndrome) a viral disease that results in severely reduced immunity. No cure is known.

air cell a body of air that behaves largely as a single entity.

albumin 1. any of a class of clear, water-soluble plant or animal proteins. 2. *serum albumin,* a clear, water-soluble constituent of blood plasma, thought to serve detoxifying and osmotic functions. 3. *albumen,* egg white.

aldehyde any of a class of organic compounds containing the group—CHO, which yields acids when oxidized and alcohols when reduced.

aldosterone a steroid hormone produced by the adrenal cortex, involved in potassium reabsorption by the kidney.

algae plantlike, photosynthetic autotrophs that may be single-celled, colonial, or multicellular; protists.

algin a structural polysaccharide of the brown algae.

alkali any of various bases which neutralize acids to form salts and turn red litmus paper blue.

allantois one of the extraembryonic membranes; in birds and reptiles, serves as a repository for the embryo's nitrogenous wastes.

allele a particular form of a gene at a gene locus.

allopatric speciation formation of new species from populations that become geographically separated. Compare *sympatric speciation.*

allopolyploid *adj.,* having more than two haploid sets of chromosomes that are dissimilar and derived from different species. See also *allotetraploid, polyploid, tetraploid.*

allosteric site in certain enzymes, a secondary binding site for small metabolites involved in the regulation of enzymatic activity.

allotetraploid *adj.,* derived from the hybridization of two distinct species, carrying the full diploid chromosome complements of both. See also *polyploid, tetraploid,* chromosome complements of both. See also *polyploid, tetraploid, allopolyploid.*

alpha amylase a starch-digesting enzyme.

alpha cell an endocrine cell within the pancreas, which secretes insulin.

alpha helix the right-handed helical configuration spontaneously formed by certain polymers. See *secondary level.*

alternation of generations the existence in the life cycle of an individual of a haploid (IN) gametophyte stage that alternates with a diploid (2N) sporophyte stage.

altruism behavior that is directly beneficial to others at some cost or risk to the altruistic individual.

Alvarez hypothesis a hypothesis which proposes that much of the massive extinction of life that accompanied the end of the Mesozoic era was produced by the aftereffects of a gigantic asteroid's collision with the earth.

alveolus (*pl.,* alveoli) one of the air sacs that occur in grapelike clusters at the ends of bronchioles in the vertebrate lung. Alveolar walls contain dense capillaries where gas exchange occurs.

Alzheimer's disease presenile dementia usually occurring in a middle-aged person and associated with sclerosis and nerve degeneration.

amebocyte in many individuals, an ameboid cell that functions in reproduction, digestion, and so on.

amino acid 1. any organic molecule of the general formula R—$CH(NH_2)COOH$, having both acidic and basic properties. 2. any of the twenty molecular subunits that make up protein.

ammonification the formation of ammonia or its compounds, as in soil, by soil organisms.

amnion the innermost extraembryonic membrane of reptiles, birds, and animals.

amniotic cavity the smaller, upper division of the blastocyst, formed by the embryonic disk.

amniotic fluid the watery fluid in the amniotic sac, in which the embryo is suspended.

AMP (adenosine monophosphate) a molecule consisting of adenine, ribose and one phosphate group.

Amphibia the class of cold-blooded vertebrates (frogs, salamanders, etc.) the larv[ue] of which are typically aquatic, breathing by gills, and the adults of which are terrestrial, breathing by lungs and/or through moist, glandular skin.

amphibian any member of the class of cold-blooded vertebrates (frogs, salamanders, etc.), the larva of which are typically aquatic, breathing by gills, and the adult of which are terrestrial, breathing by lungs and through moist, glandular skin.

amygdala an almost-shaped part of the mid-brain.

anabolic *adj.,* referring to metabolic activity that results in the chemical building up of molecules. See also *catabolic.*

anaerobic *adj.,* in the absence of oxygen.

analogous in comparative morphology, similar in form or function but derived from different evolutionary or embryonic precursors (e.g., the wings of insects and birds).

anaphase the stage of mitosis or meiosis II in which the centromeres divide and separate and the two daughter chromosomes travel to opposite poles of the cell.

androecium a whorl of stamens; the stamens of a flower taken together.

angiosperm a plant in which the seeds are enclosed in an ovary; a flowering plant.

animal hemisphere the part of the ovum with the more metabolically active cytoplasm and where the greatest amount of cell division occurs. Compare *vegetal hemisphere.*

annelid any member of phylum Annelida, comprising segmented worms and leeches.

Annelida the phylum that includes the segmented worms, including earthworms, leeches, and various marine forms.

annual a plant that completes its life cycle in a single season.

annual plant a plant that completes its life cycle—germination, growth, reproduction, and death—in a year. Compare *perennial.*

annual rings the concentric rings seen in the cross-section of a woody stem, each of which corresponds to one season's growth.

antagonistic muscles a pair of skeletal muscles (or groups of muscles) whose actions cooperatively oppose one another.

anterior pituitary part of an endocrine gland communicating with the hypothalamus only by way of a small vascular portal system; secretes several hormones.

anther in a flower, the pollen-producing organ of the stamen.

antheridium (*pl.,* antheridia) in plants, a male reproductive organ that produces motile sperm.

anthophyte a plant of the division Anthophyta; a flowering plant.

Anthozoa the class of marine coelenterates that includes the corals, sea anemones, sea pens, and so on.

antibody a protein molecule of the immune system that can recognize and bind to a foreign substance or invader, such as a bacterium or virus. See *immunoglobulin.*

anticodon a region of a tRNA molecule consisting of three sequential nucleotides that will have a matching codon in mRNA.

anticodon loop the region of a tRNA molecule containing the anti-codon; one of the three or four loops of a tRNA molecule.

antidiuretic any substance that helps the body conserve water by increasing water reabsorption in the nephrons.

antidiuretic hormone (ADH) also called *vasopressin;* a polypeptide hormone secreted by the posterior pituitary that increases the resorption of fluid from the kidney filtrate.

antigen 1. any large molecule, such as a cell-surface protein or carbohydrate, that stimulates the production of specific antibodies or that binds specifically with such antibodies. 2. any antibody-specific site on such a molecule.

antigen-presenting macrophage a cell, carrying an antigen that can activate a matching helper T-cell.

antigen recognition region the separate part of a Y-shaped antibody that matches specific antigens.

antiserum blood serum containing antibodies specific to some particular antigen.

aorta in vertebrates, the principal or largest artery; carries oxygenated blood from the heart to the body.

aortic semilunar valve in the heart, a one-way valve at the base of the aorta that prevents back-flow into the left ventricle during systole.

aphotic zone "without light"; in a body of water, the depth to which light never penetrates.

apical dominance in the uppermost growing tip of a plant stem, the hormonal inhibition of the upward growth or formation of other branches.

apical meristem See *meristem.*

appendicular skeleton in vertebrates, bones of the pectoral and pelvic girdles and of the appendages. Compare *axial skeleton.*

appetitive stage a variable, nonstereotyped part of instinctive behavior that involves searching (for food, water, a mate) for the opportunity to perform. Compare *consummatory stage.*

arachnid any member of the class of arthropods which contains the spiders, scorpions, ticks, mites, etc.

Arachnida the class of arthropods that contains spiders, scorpions, mites, ticks, and others with eight legs.

arboreal *adj.,* pertaining to dwelling in trees.

Archaebacteria one of the two suggested prokaryote kingdoms (or phyla in some schemes); contains bacteria such as methanogens (methane producers), thermaphiles (heat lovers), and acidophiles (acid lovers). See also *Eubacteria.*

archegonium (*pl.,* archegonia) the female (egg-producing) reproductive structure in the gametophytes of ferns and bryophytes.

arteriole a small artery, usually giving rise directly to capillaries.

artery a vessel carrying blood away from the heart toward arterioles and capillary beds.

arthropod any member of the phylum containing segmented invertebrates with joined legs, primarily insects, arachnids, and crustaceans.

Arthropoda the phylum containing segmented invertebrates with jointed legs, including insects, arachnids, crustaceans, and myriapods.

artificial active immunity immunity developed within the body as the result of innoculation with a vaccine.

artificial selection the intentional selection (by humans) of domesticated animals or plants for breeding according to desired characteristics. See also *natural selection.*

ascending limb the part of the kidney's loop of Henle that leads to the distal convoluted tubule.

ascocarp a cuplike or saclike body in the Ascomycota in which ascospores are produced.

ascomycete a fungus of the division Ascomycota, which includes the yeasts, mildews, truffles, and so on; characterized by bearing the sexual spores in a sac, the *ascus.*

ascospore a spore, containing a single set of genes produced after sexual reproduction within the ascus of a sac fungus.

ascus (*pl.,* asci) in ascomycetes, the sac in which meiosis occurs and in which four or eight ascospores are subsequently formed.

asexual reproduction reproduction by one individual that does not involve the union of genetic material from two sources.

Asteroidea the class of echinoderms that contains sea stars.

asters "stars"; paired, radiating mitotic structures, each consisting of microtubules and microfilaments.

atom the smallest indivisible unit of an element still retaining the element's characteristics.

atomic mass the average weight of the atoms of an element; also called *atomic weight.*

atomic number the number of protons in an atomic nucleus, specific to each element.

ATP (adenosine triphosphate) a common molecule involved in many biological energy-exchange reactions; consists of the nitrogenous base adenine, the sugar ribose, and three phosphate groups joined by two energy-rich bonds.

atrial natriuretic factor (ANF) a peptide hormone produced by the atria of the heart, in response to stretching, that inhibits aldosterone production and therefore lowers blood pressure.

atrioventricular (AV) node a small mass of specialized muscle fibers at the base of the wall between the atria of the heart; conducts impulses to the bundle of His.

atrium (*pl.,* atria) a thinner-walled compartment of the heart that receives venous blood.

audition the act, sense, or power of hearing.

auditory canal the open, bony canal from the outer ear to the eardrum.

auditory nerve one of the cranial nerves, carrying sensory information to the brain from the cochlea (hearing) and the vestibule (sense of balance).

australopithecine *adj.,* pertaining to members of the extinct hominid genus *Australopithecus.*

autoimmune disease an abnormal condition in which the organism's immune system attacks and destroys one or more of the organism's own tissues.

autonomic nervous system (ANS) the system of motor nerves and their ganglia arising from the brain and spinal cord; controls involuntary functions, chiefly in the internal organs. Compare *somatic nervous system.*

autosome any chromosome other than a sex (**X** or **Y**) chromosome.

autotroph "self-feeder"; an organism capable of using simple compounds (such as carbon dioxide) as its only source of carbon and sunlight (phototrophs) or simple inorganic compounds (chemotrophs) as its only source of energy. Compare *heterotroph.*

auxin a class of natural or artificial substances that act as the principle growth hormone in plants.

Aves the class containing the birds.

axial skeleton in vertebrates, the skull, vertebral column, and bones of the chest. Compare *appendicular skeleton.*

axon the extension of a neuron that conducts nerve impulses away from the cell body.

axonal knob the bulbs at the terminal endings of an axon.

axonal tree the branching, terminal region of an axon. See *neuron.*

axopod in sarcodines, slender, microtubular spines upon which feeding pseudopods move. See also *pseudopod.*

bacillus a rod-shaped bacterium.

bacteriochlorophyll a light-sensitive pigment used by certain eubacteria for capturing light energy to be used in photosynthesis.

bacteriophage also called *phage;* a virus that infects and destroys bacteria.

bacteriorhodopsin a purple photosynthetic pigment used by photosynthetic archaebacteria.

ball-and-socket joint a joint allowing maximum rotation and flexion; consists of a ball-like termination on one part, held within a concave, spherical socket on the other (e.g., the hip joint).

bark the portion of a stem outside of the vascular cambium; consists of phloem, cortex, epidermis, cork cambium, and cork.

Barr body a dark-staining feature in the nuclei of the cells of female animals; represents the condensed X chromosome.

barrier reef a reef of coral running roughly parallel to the shore and separated from it by deep water.

basal body a structure found beneath each eukaryotic flagellum or cilium; consists of a circle of nine short triplets of microtubules.

basal metabolic rate (BMR) a measure of oxygen consumption of an individual at rest.

base 1. a compound that reacts with an acid to form a salt. 2. a substance that releases hydroxide ions when dissolved in water. 3. a nitrogenous base (purine or pyrimidine) of nucleic acids.

basement membrane in animals, a sheet of collagen that underlies and supports the cells of a tissue. Also called *basal membrane, basilar membrane.*

base pairing the specific manner in which the nitrogenous bases of nucleic acids pair up and form hydrogen bonds; adenine always opposes thymine or uracil (A-T or T-A, A-U or U-A) and cytosine always opposes guanine (C-G or G-C).

base substitution a mutation in DNA in which one nucleotide is replaced by another, or modified into another.

basidiocarp "club fruit"; the spore-producing organ in mushrooms and other Basidiomycota.

basidiomycete a fungus of the division Basidiomycota, which includes the smuts, ruts, mushrooms, puffballs, etc.; characterized by bearing the spores on a basidium.

basidium "little pedestal"; the clublike structure of basidiomycetes such as mushrooms; produces basidiospores by meiosis.

basilar membrane in the vertebrate ear, a membrane whose movement stimulates the sensitive hair cells that transmit neural impulses to the brain.

B-cell lymphocyte a lymphocyte that circulates in the blood and is involved in the immune response, especially in the production of free antibodies.

behavioral ecology the study of the interactions of behavior, environmental, and evolutionary forces.

benthic *adj.*, referring to the benthos, or bottom-dwelling, community of organisms.

berry a simple fruit, including a fleshy ovary and one or more carpels and seeds.

beta cell an endocrine cell in the pancreas; secretes glucogon.

bicuspid valve also *mitral valve;* the two-flapped valve between the left atrium and the left ventricle of the heart.

bilateral symmetry having left and right sides that are approximate mirror images; having a single plane of symmetry.

biennial a plant that completes its normal term of life in two years, flowering and fruiting the second year.

bile a pigmented, alkaline liquid secreted by the liver; contains bile salts and bile pigments and functions in fat digestion.

binary fission a form of asexual reproduction; fission (splitting) into two organisms of roughly the same size.

binomial nomenclature the tradition (introduced by Linnaeus) whereby each species is given two Latin taxonomic names, a generic term and a specific term (genus and species).

biodegradable capable of being rendered harmless upon exposure to the elements and organisms of the soil or water.

biogenetic law the fact that, under the current planetary conditions, life can arise only from preexisting life.

biogeochemical cycle the pathway of elements (e.g., carbon, nitrogen) or compounds (water) as they are taken up and released by organisms into the physical environment.

biogeography the study of the geographical distribution of living things.

biological determinism the influence of genetics, particularly on the social behavior of an organism.

biomass the total weight (dry) of all the organisms in a prescribed area.

biome a major plant association, e.g., a tundra.

biosphere the entire part of the earth's land, soil, water, and atmosphere in which the living organisms are found.

biotic *adj.*, pertaining to life.

bipedal gait walking on two feet.

birth control pill an oral steroid contraceptive that inhibits ovulation, fertilization, or implantation causing temporary infertility in women.

bivalve a mollusk having two shells hinged together, such as the oyster, clam, or mussel.

bivalvia the class of mollusks which includes clams and oysters.

bladder any membranous sac serving as a receptacle for fluid or gas, such as urinary bladder, gall bladder.

blade any broad, thin part of the thallus (body) of a red, green, or brown alga or of a leaf.

blastocoel in embryonic development, the cavity of a blastula, arising in the course of cleavage.

blastocyst the early preimplantation stage in the mammalian embryo.

blastula an early embryonic stage in many animals, consisting of a single layer of cells that forms a hollow ball enclosing a central cavity, the blastocoel.

blending inheritance a theory of inheritance which states that parental characteristics blend to produce an intermediate character in the offspring.

blood pressure pressure exerted by blood as it presses against blood vessels, especially arteries.

blue-green elga See *cyanobacteria.*

body stalk in embryology, the connection between the early embryo and the extraembryonic tissues.

Bohr effect the accelerated release of oxygen by hemoglobin when carbon dioxide is present.

book lungs the respiratory organ of a spider, scorpion, or other terrestrial arachnid; consists of thin, membranous structures arranged like the leaves of a book.

boreal forest a northern forest; see *taiga.*

bottleneck in population genetics, a period of time when the size of a population becomes small, resulting in a random change in allele frequencies. See also *genetic drift, founder effect.*

botulism a bacterial disease caused by the presence in foods (usually canned foods) of *Clostridium botulinum,* whose nerve toxins are among the most powerful poisons known.

Bowman's capsule a curved sac at the beginning of a nephron that surrounds the glomerulus.

bread mold a member of the fungal phylum Zygomycota.

bronchial tree. See *respiratory tree.*

bronchiole a minute airway in the lung that is a small branch of a bronchus and part of the respiratory tree.

bronchus (*pl.,* bronchi) either of the two main branches of the trachea.

brown algae any alg[ue]of the division Phaeophyta, usually brown due to fucoxanthin pigment, e.g., kelp.

brush border. See *microvillus.*

bryophyte a moss, liverwort, or hornwort of the division Bryophyta; nonvascular terrestrial plants.

budding asexual reproduction seen in yeasts and other organisms where smaller cells or entire offspring bud or grow from a parent cell.

buffer a solution of chemical compounds capable of neutralizing both acids and bases, thus resisting changes in pH.

bulbourethral glands a pair of small glands that secrete a mucous substance into the urethra in males, especially on sexual arousal. Also called *Cowper's glands.*

bulk flow the one-way movement of water or another liquid brought about by pressure, gravity, or solutes, generally from areas of greater water potential to areas of lesser water potential.

bundle sheath cell a specialized parenchyma cell, one of many forming compact layers to enclose the small veins in a leaf.

calcitonin one of the two major hormones of the thyroid gland whose major action is to inhibit the release of calcium from the bone.

calmodulin a protein responsible for activating many kinds of enzymes, itself activated by its union with calcium ions.

calorie 1. *small calorie* (calorie proper), the amount of heat (or equivalent chemical energy) needed to raise the temperature of one gram (roughly one milliliter) of water by 1°C. 2. *large calorie, kilocalorie* (1000 small calories), the heat needed to raise the temperature of a kilogram of water by 1°C.

calpain a chemical important in the formation of memory.

Calvin cycle the biochemical cycle in photosynthesis in which carbon dioxide, with the aid of NADPH and ATP, is fixed into carbohydrate. Also called *CO_2 fixation.* See also *light-independent reaction.*

calyx (*pl.,* calyces) the outermost whorl of floral parts (the sepals), usually green and leaflike.

canaliculus (*pl.,* canaliculi) a small canal or tubular passage, as in bone.

capillary a small blood vessel (with walls one cell thick) in which exchanges between the blood and tissues occur.

carbaminohemoglobin hemoglobin complexed with carbon dioxide.

carbohydrate a class of organic compounds with the empirical formula $(CH_2O)_n$, characterized by many—OH and—H side groups and an aldehyde or ketone group; sugars, starches, cellulose, and chitin.

carbon cycle the pathway of carbon as it is taken up and released by organisms into the physical environment.

carbon dioxide fixation. See *Calvin cycle.*

carbonic anhydrase an enzyme that catalyzes the reversible conversion of carbonic acid to carbon dioxide gas and water.

carboxyl group the proton-releasing acid group,—COOH, present in organic acids.

carcinogen any substance that tends to cause cancer.

cardiac circuit the blood vessels of the heart, including the coronary arteries and veins and the capillaries.

cardiac muscle specialized involuntary muscle of the heart whose fibers are striated and branching.

cardiac sphincter also *gastro-esophageal sphincter;* the ring of muscle that closes the passageway between the lower esophagus and the stomach.

carotenoid any of a group of red, yellow, and orange plant pigments chemically and functionally similar to carotene; light-gathering accessory pigments associated with chlorophyll.

carotid body a mass of cells and nerve endings on the carotid arteries that senses blood carbon dioxide and pH levels and responds by affecting the rate of breathing and the heartbeat.

carpel in flowers, a simple pistil, or a single member of a compound pistil; one sector or chamber of a compound fruit.

carrageenan a polysaccharide produced by Irish moss, a red alga; used as a thickening and smoothing agent.

carrying capacity in ecology, a property of the environment defined as the size of a population that can be maintained indefinitely.

cartilage a firm, elastic, flexible, translucent type of connective tissue; in development, a precursor of bone formation.

Casparian strip in plants, a waxy, waterproof strip on cell walls of root tip endodermis that permits some active control of water uptake by directing incoming water through the cytoplasm of the endodermal cells.

catabolic *adj.,* referring to metabolic activity that results in the chemical breakdown of molecules and the release of their energy. See also *anabolism.*

catalyst an agent that greatly accelerates a chemical reaction while not being permanently altered itself.

cecum in vertebrates, a blind pouch of the intestine at the juncture of the small and large intestine.

cell the structural unit of life; consists of metabolically active cytoplasm and genetic macromolecules (DNA and RNA) enclosed in a semipermeable membrane. See also *prokaryote, eukaryote.*

cell body the region of a neuron containing the cell nucleus and most of the cytoplasm and organelles.

cell cycle the typical cycling events in the life of a cell, including the G1, S, G2, and M phases.

cell division in a general sense, both nuclear (mitosis or meiosis) and cytoplasmic division; more specifically, the division of cytoplasm. See *cytokinesis.*

cell plate the plant cell division, the forming plasma membrane between newly forming daughter cells.

cell theory the universally accepted proposal that cells are the functional units of organization in living organisms and that all cells today come from preexisting cells.

cellular differentiation during development, the commitment of immature cells or tissues to a specific morphological and functional type.

cellular slime mold a mold of the phylum Acrasiomycota.

cellulose an inert, insoluble carbohydrate; a principal constituent of plant cell walls; consists of unbranched chains of beta glucose.

cell-mediated response the immune response carried out by cells that arouses other cells to seek out and kill specific antigen-bearing invading organisms and infected body cells.

cell respiration the energy-yielding metabolism of foods in which oxygen is used. See also *respiration.*

cell wall the semi-rigid enclosure of a plant, fungal, algal, or bacterial cell that gives it support and a definite shape.

central dogma the proposition that all biological information is encoded in DNA, transmitted by DNA replication, transcribed into RNA, and translated into protein; includes several exceptions for certain viruses.

central nervous system in vertebrates, the brain and spinal cord to which sensory impulses are transmitted and from which motor impulses are sent.

central sulcus the dividing depression between the frontal and parietal lobes of the brain.

centriole a paired, microtubular organelle in animals, protists, fungi, and lower plant cells; absent in the cells of seed plants.

centromere the specialized region of a chromosome to which spindle fibers are attached; also called *kinetochore.*

centromeric spindle fiber any of a group of microtubules attached to each centromere and proceeding to a spindle pole in mitosis or meiosis.

cephalic ganglia neural aggregations at the anterior ends of some invertebrates.

cephalochordata a chordate subphyllum in which the notochord persists throughout life and extends into the anterior regions.

cephalochordate "head cord animal"; a lancelet of the chordate subphylum Cephalochordata.

Cephalopoda the phylum of mollusks which includes octopuses, squid, and cattle fishes.

cerebellum the double, walnut-shaped portion of the hindbrain that coordinates voluntary movement, posture, and balance.

cerebral cortex the outermost region of the cerebrum, the "gray matter," consisting of several dense layers of neural cell bodies and including numerous conscious centers as well as regions specializing in voluntary movement and sensory reception.

cerebral hemisphere either the right or left half of the cerebrum.

cerebrospinal fluid a cushioning fluid that surrounds the central nervous system.

cerebrum the anterior portion of the vertebrate brain; the largest portion in humans, consisting of two *cerebral hemispheres* and controlling many localized functions (among others voluntary movement, perception, speech, memory and thought).

cervical cap a birth control device consisting of a small cap that covers the cervix.

cervix the opening to the uterus and the surrounding ring of firm, muscular tissue.

Cestoda the class composed of parasitic flatworms, such as the tapeworm.

CF1 particle an enzyme-containing body on the outer surface of the thylakoid in which chemiosmotic phosphorylation occurs.

C4 plant a plant in which carbon dioxide is fixed into four-carbon compounds in leaf mesophyll cells, transported to the bundle sheath cells, and released and concentrated for the Calvin cycle in chloroplasts there. Compare *C3 plant.*

chain-terminating mutation a base substitution mutation in which the new codon created is a chain termination codon.

chaparral a vegetation type common in coastal California, characterized by a dense growth of low, evergreen shrubs and trees.

character displacement where similar species share a niche, the tendency for natural selection to favor and accentuate physical differences.

Chargaff's rule in DNA, the amount of adenine present always equals the amount of thymine and the amount of cytosine always equals the amount of guanine.

chelicerae the first pair of usually pincerlike appendages of spiders and other arachnids.

Chelicerata arthropods with no jaws, no antennae, four pairs of legs, and book lungs, including the spiders.

chelicerate any member of the subphylum Chelicerata.

chemical bond any of several forms of attraction between atoms in a molecule.

chemical communication the transmission of information between individuals by the use of pheromones.

chemical reaction the reciprocal action of chemical agents on one another; chemical change.

chemiosmosis the process in mitochondria, chloroplasts, and aerobic bacteria in which an electron transport system uses the energy of photosynthesis or oxidation to pump hydrogen ions across a membrane, resulting in an electrochemical proton concentration gradient, or chemiosmotic differential, that can be used to produce ATP.

chemiosmotic gradient in ATP-generating systems, a steep proton gradient whose free energy is utilized in driving chemiosmotic phosphorylation.

chemiosmotic phosphorylation the phosphorylation of ADP to ATP using the free energy of the chemiosmotic differential in the CF1 and F1 particles.

chemoreceptor a neural receptor sensitive to a specific chemical or class of chemicals.

chemotroph an organism that obtains energy from inorganic chemical reactions. Compare *autotroph, heterotroph.*

chiasmata (*sing.,* chiasma) after repulsion during prophase I of meiosis, X-shaped points of residual connection between homologous chromatids; they are believed to represent prior crossover events.

chitin a structural carbohydrate that is the principle organic component of arthropod exoskeletons and fungal cell walls.

chlorophyll a green photosynthetic pigment found in chloroplasts and in some photosynthetic prokaryotes; occurs in several forms, *chlorophyll a, b,* and *c.*

chloroplast photosynthetic organelle (plastid) common in plants and algal protists; numerous internal membranous thylakoids containing photosystems with chlorophyll, electron/proton transport systems, and CF1 particles.

choanocyte See *collar cell.*

cholesterol a common steroid occurring in all animal fats; a vital component of cell membranes; an important constituent for bile for fat absorption; a precursor of vitamin D and a controversial dietary constituent.

Chondrichthyes the class of vertebrates containing the cartilaginous fishes, such as sharks.

chordae tendineae in the heart, tough, cordlike tendons that prevent backflow by holding the tricuspid and bicuspid valves in place where the ventricles contract.

Chordata the phylum which contains the true vertebrates and those animals having a notochord, such as the lancelets.

chordate belonging to the phylum Chordata, which contains the true vertebrates and those animals having a notochord, like the lancelet.

chorion the outermost extraembryonic membrane of birds, reptiles, and mammals; contributes to the formation of the placenta in placental mammals.

human chorionic gonadotropin (HCG) the hormone released by an implanted embryo that prompts the continued release of progesterone from the corpus luteum.

choroid coat one of the vascular coverings of the eyeball.

chromatid in the G2 chromosome, one of the two identical chromosome replicas held together by the centromere prior to mitosis or meiosis.

chromatin the substance of chromosomes; consists of DNA and chromosomal proteins.

chromatin net the netlike appearance of chromatin in the interphase nucleus.

chromosomal mutation a massive spontaneous change in DNA; generally, breakage involving a whole chromosome that is not repaired or has been repaired improperly. Also *chromosomal rearrangement.*

chromosome 1. in eukaryotes, a linear DNA molecule (or two DNA molecules when replication has occurred), one centromere, and associated proteins. 2. in prokaryotes, a naked, circular DNA molecule (lacking in protein). 3. the DNA or genetic RNA molecule of a virus.

chromosome puff in polytene chromosomes, an enlargement of one band associated with transcriptional activity (mRNA production).

chylomicron in digestion, a minute, protein-coated fat droplet formed during lipid transport in the intestinal villi.

cilia fine, hairlike, motile organelles found in groups on the surface of some cells.

ciliary muscles ring of muscles around the lens which bring about changes in lens shape.

ciliate any protozoan of the phylum Ciliophora, having cilia on part or all of the body.

Ciliophora the ciliated protozoans.

circulatory system the system consisting of blood-forming organs or tissue, vessels, the heart, and blood; also called *vascular system.*

cisterna slender, membranous channels making up much of the Golgi apparatus.

citric acid cycle in cell respiration, the biochemical pathway in the mitochondrion or bacterial cell where pyruvate, newly converted to acetyl CoA, is combined with oxaloacetate and sequentially oxidized and decarboxylated, thereby producing CO_2, reducing NAD^+ and FAD^+ to NADH and FADH, and producing a small amount of ATP. Also called *Krebs cycle.*

claspers a pair of specialized grooved pelvic fins used by male sharks and rays as a penis.

class a major taxonomic grouping intermediate between phylum and order.

classical conditioning in behavior, the process by which a conditioned response is learned and elicited. Compare *operant conditioning.* See also *conditioned response.*

cleavage 1. the breaking down of a molecule or compound into simpler substances. 2. the cytoplasmic division of a zygote and young embryo.

climax state in ecological succession, the stage of a plant or animal community that is stable and self-perpetuating.

cline 1. a simple geographic gradient, or regular change, in given character. 2. a regular change in allele frequency over geographic space.

clitoris a touch-sensitive, erogenous organ of the female genitalia considered to be homologous to the penis in the male.

cloaca the common cavity into which the intestinal, urinary, and reproductive canals open in many vertebrates.

clonal selection theory in immunology, the proposition that all potential antibody specificity is present in differentiated cells early in development and that the specific immune response consists of inducing appropriately differentiated cells to proliferate clonally.

cloning the production of genetically identical individuals from a single cell.

closed circulatory system a circulatory system in which the blood elements remain in blood vessels. Compare *open circulatory system.*

club fungi fungi of the phylum Basidiomycota.

cnidocyte in coelenterates, a stinging cell.

coacervates droplets of protein or other substances that form spontaneously in colloidal suspensions, surrounding themselves with a lipid shell; important in some theories of the origin of life.

coccus (*pl.,*. cocci) any spherical bacterium.

coccyx the tailbone or lowest portion of the vertebral column.

cochlea in mammals, a spiral cavity of the inner ear containing fluid, vibrating membranes, and sound-sensitive receptors.

codon 1. a series of three nucleotides in mRNA that specify a particular amino acid (or chain termination) in protein synthesis. 2. the colinear, complementary series of three nucleotides or nucleotide pairs in the DNA from which the mRNA is transcribed.

codominance the individual expression of both alleles in a heterozygote.

Coelacanth a fish thought to have been extinct since the Cretaceous period, but found in 1938 off the southern coast of Africa, near Madagascar.

coelenterate an animal of phylum coelenterata, characterized by a thin, sac-like body wall and only one digestive opening, such as the jelly fish.

coelom a principal body cavity, or one of several such cavities, between the body wall and gut entirely lined with mesodermal epithelium. Compare *pseudocoelom.*

coelomate *adj.,* having a true coelom.

coenzyme a small organic molecule required for an enzymatic reaction in which it is often reduced (i.e., NAD, NADP).

coevolution evolutionary change in one species that is influenced by the activities of another (i.e., the behavior of a predator and its prey).

cofactor any organic or inorganic substance, especially an ion, that is required for the function of an enzyme.

cohesion the attraction between molecules of a single substance.

coitus the act of sexual intercourse, especially between humans.

Coitus interruptus coitus that is interrupted by withdrawal of the penis before ejaculation.

coleoptile in the embryos and young shoots of grasses, a temporary tubular leaf structure that completely encloses and protects the plumule during emergence.

coleorhiza a temporary protective sheath over the radicle in the embryo of grasses and grains.

colinearity the principle that the linear arrangement of nucleotides in DNA corresponds to the linear arrangement of nucleotides in RNA, which in turn corresponds to the linear arrangement of amino acids in a polypeptide.

collagen a common, tough, fibrous animal protein occurring principally in connective tissue.

collar cell also called *choanocyte;* one of certain flagellated cells in sponges that create a current and ingest food particles from the water.

collecting duct the part of a nephron that collects fluids from the nephron and discharges it into the renal pelvis.

collenchyma in plants, a strengthening tissue; a modified parenchyma consisting of elongated cells with greatly thickened cells walls.

colon the large intestine.

colony 1. a group of animals or plants of the same kind living in a close semidependent association. 2. an aggregation of bacteria growing together as the descendents of a single individual, usually on a culture plate.

columnar epithelium epithelium consisting of one (simple columnar) or more (stratified columnar) layers of elongated, cylindrical cells.

commensalism a symbiotic relationship in which one partner benefits while the other is neither helped nor hurt. Compare *mutualism, parasitism.*

community in ecology, an assemblage of interacting populations forming an identifiable group within a biome (i.e., a sage desert community or a beech-maple deciduous forest community).

compact bone dense, hard bone with spaces of microscopic size.

companion cell in plants, a nucleated cell adjacent to a sieve tube member and believed to assist in its functions.

comparative anatomy the science of comparing the anatomy of animals, drawing conclusions about their evolution and relationships.

comparative psychology the comparison of behavior in animals, usually in a laboratory environment.

competition in ecology, the utilization of the same limiting resource by two or more individuals or species.

complement a group of blood proteins that interact with antibody-antigen complexes to destroy foreign cells.

complement system a group of proteins that function in the immune system. They combine with antigen-antibody complexes and help destroy the antigen-carrying systems.

complete digestive system a tubular digestive system generally with a mouth and an anus.

complete dominance the expression of a gene with the complete lack of expression of its allele.

complete gut a digestive tract which has both a mouth and an anus.

compound in chemistry, a pure substance of a single molecular type consisting of two or more elements in a fixed ratio.

compound eye an arthropod eye consisting of many simple eyes closely crowded together, each with an individual lens and a restricted field of vision, so that a mosaic image is formed.

concentration gradient a slow, consistent decrease in the concentration of a substance along a line in space. Also called *diffusion gradient.*

conditioned response an involuntary response (mouth-watering) that becomes associated with an arbitrary, previously unrelated stimulus (bell-ringing) through repeated presentation of the arbitrary stimulus (bell-ringing) simultaneously with a stimulus normally yielding the response (pizza).

conditioned stimulus in animal behavior, an environmental situation or condition that is normally irrelevant to a desired behavior but can come to stimulate the behavior. Compare *unconditioned stimulus.*

condom a thin sheath of rubber or animal membrane worn over the penis during sexual intercourse to prevent conception or venereal disease.

cone 1. a reproductive structure of conifers consisting of a cluster of scalelike modified leaves and either pollen or ovules, or the seed. 2. one of a class of conical photoreceptors in the retina that detect color.

conidium (pl. *conidia*) an asexual spore borne on the tip fungal hypha.

conifer an evergreen gymnospore of the division Coniferophyta, bearing ovules and pollen in cones; included are spruce, fir, pine, cedar, and juniper.

Coniferophyta the class of gymnosperms which includes the cone-bearing pines, firs, cedars, hemlocks, redwoods, etc.

conjugated protein protein containing one or more nonprotein substances; e.g., hemoglobin, which contains four heme groups.

conjugation sexual reproduction in which organisms (usually single-celled) fuse to exchange genetic materials.

connective tissue a primary animal tissue of mesodermal origin, with scattered cells and abstract extracellular substance or interlacing fibers, usually including collagen.

constant region the C-terminal portion of an immunoglobulin (antibody) light or heavy chain, not involved in antigen-specific binding.

consumer in ecology, an animal that feeds on plants (primary consumer) or other animals (secondary consumer).

consummatory stage a part of instinctive behavior that involves highly stereotyped (in variable) behavior and the performance of a fixed-action pattern, e.g. swallowing food. Compare *appetitive stage.*

continental drift the slow movement of the continents relative to one another on the earth's surface. See also *plate tectonics.*

continental shelf the part of a continent that is submerged in relatively shallow water.

continental slope the relatively shallow offshore waters of an ocean.

continuous spindle fiber one of two partially overlapping spindle fibers (not actually continuous), each originating near a pole and extending across the chromosomes without attaching. Also called *polar spindle fiber.*

continuous variation continuous gradation in the expression of a trait where two or more pairs of alleles contribute to such expression, i.e., height and skin color in humans.

contraception any process intended to prevent the sperm from reaching and fertilizing the egg, or preventing ovulation or implantation.

contractile vacuole an organelle of many freshwater protists that maintain the cell's osmotic equilibrium by expelling excess water through an active, ATP-powered process.

control 1. a standard of comparison in scientific experiment; a replicate of the experiment in which a possibly crucial factor being studied is omitted. 2. a function of the nucleus of a cell—the direction of virtually all cellular activity.

control group a part of an experiment in which a group is treated identically to the experimental group except for the one factor under examination.

controlled experiment a scientific study in which a number of subjects are randomly divided into two or more groups, and each group is treated exactly the same in all respects except one, which becomes the variable.

conus arteriosus (kōn'[ue]s är tir'ē ō' s[ue]s) in a two-chambered vertebrate heart, the heavy-walled blood vessel that leaves the ventricle.

convergent evolution the independent evolution of similar structures in distantly related organisms; often found in organisms that occur in similar ecological niches, such as marsupial moles and placental moles.

cooperation working or acting together for a common purpose or benefit; joint action by several individuals.

copulation sexual union or sexual intercourse.

coral reef a buildup of material by the activities of coral animals.

cork in plants, secondary tissue produced by the *cork cambium,* consisting of cells that become heavily suberized and die at maturity, resistant to the passage of moisture and gases; the outermost layer of bark.

cork cambium in plants, the outermost meristematic layer of the stem of woody plants, from which the outermost layer of bark is produced.

cornea the nonvascular, transparent fibrous coat of the eye through which the iris is visible.

corona radiata "radiating crown"; an aggregation of follicle cells surrounding the mammalian egg at ovulation.

coronary arteriosclerosis an inelasticity and thickening of the arterial walls of the heart, resulting in a decreased blood flow.

corpus callosum a broad, white neural tract that connects the cerebral hemispheres and correlates their activities.

corpus luteum "yellow body"; a temporary endocrine body that develops in the ovarian follicle after ovulation; secretes estrogen and progesterone.

cortex 1. in animals, the outer layer or rind of an organ, such as the adrenal cortex. 2. in plants, the portion of stem between the epidermis and the stele.

cortical granule a membranous vesicle beneath the surface of unfertilized echinoderm eggs; at fertilization these rupture to form the fertilization membrane during the cortical reaction.

Cotyledon a food-string structure in dicot seeds, sometimes emerging as first leaves; a food-digesting organ in most monocot seeds; first leaves in a gymnosperm embryo.

countercurrent exchange a heat or molecule concentrating mechanism in endothermic animals, e.g., the presence of parallel arteries and veins permit an efficient heat exchange as the outward flow of warmer blood opposes the inward flow of cooler blood from the extremities.

covalent bond a relatively strong chemical bond in which an electron pair is shared by two atoms.

cranial capacity the volume of the brain vault of the skull and thus the brain, expressed in cubic centimeters.

cranial nerve in humans, one of 12 pairs of major nerves that emerge directly from the brain.

cranium 1. the skull. 2. the part of the skull enclosing the brain.

creatine phosphate Also called *phosphocreatine.* In vertebrate muscles, a source of high-energy phosphate for restoring ATP expended during muscle contraction.

crista (*pl.,* cristae) a shelflike fold of the inner mitochondrial membrane, containing numerous electron transport systems, proton pumps, and F_1 particles.

cross current exchange in the bird lung, the flow of blood at right angles to the passage of air.

crossing over 1. the exchange of chromatid (DNA) segments by enzymatic breakage and reunion during meiotic prophase. 2. a specific instance of such an exchange; a crossover.

crossopterygian any fish of the group Crossopterygii, extinct except for the coelacanth, regarded as being ancestral to amphibians and other land vertebrates.

crowning in the human birth process, the appearance of the head of the fetus.

crude filtrate in the kidney, the fluid entering the nephron.

Crustacea the class composed of aquatic arthropods, typically having the body covered with a hard shell or crust, such as lobsters, shrimps, and crabs.

crustacean any member of the class Crustacea.

cryptogam any of the Cryptogamia, an old primary division of plants without true flowers and seeds, such as ferns and horsetails.

Ctenophorans members of the phylum Ctenophora, which contains the marine invertebrates having gelatinous bodies and eight rows of plates (combs) of fused cilia.

C_3 plant a plant in which carbon dioxide is fixed into carbohydrates directly in the Calvin cycle. Compare *C_4 plant.*

cuticle a tough, often waterproof, nonliving covering, usually secreted by epidermal cells.

cutin A waxy substance covering the epidermis of leaves and stems.

cyanobacteria in older terminology, *cyanophytes* and *blue-green algae,* a photosynthetic eubacterium that uses water as a source of protons and whose cells exhibit membranous chlorophyll-containing lamellae; many are capable of nitrogen fixation.

cycad a plant of the division Cycadophyta.

Cycadophyta the gymnosperm class composed of short palmlike plants.

cyclic adenosine monophosphate (cAMP) "second messenger"; synthesized in response to certain hormones arriving at the cell membrane; stimulates further activity in the target cell.

cyclic photophosphorylation light reactions employing only photosystem I, during which electrons from the P700 reaction center pass through the associated electron transport system and cycle back to P700, thus serving chemiosmosis only.

cytochrome iron-containing compounds including carrier proteins, found in electron/hydrogen transport chains in the thylakoid, mitochondrion, and bacterial membranes.

cytokinesis division of the cell cytoplasm following mitosis or meiosis.

cytokinin a plant cell hormone, mitogen, and plant tissue culture growth factor that interacts with other plant hormones in the control of cell differentiation.

cytological map a map locating genes on the physical chromosome by means other than recombination or genetic mapping.

cytoplasm in eukaryotes, that region of the cell between the cell membrane and nuclear membrane, including the cytoplasmic organelles and matrix.

cytophyge the fixed position on the surface of a protist from which wastes are discharged by exocytosis.

cytosine (C) a pyrimidine, one of the four nucleotide bases of DNA and RNA.

cytosome the mouth of a ciliate.

cytotoxic T-cell See *T-cell lymphocyte.*

daughter cell either of the two cells formed when one cell divides.

day-neutral plants plants that can flower at any time of the year regardless of the relative amounts of light and darkness.

deamination the removal of an amino group.

deciduous rain forest a lush forest biome, typically receiving 250 to 450 cm (100 to 180 in) of annual rainfull, with seasonal variations, in which trees drop their leaves in the "dry" season.

decomposer an organism that breaks down organic wastes and the remains of dead organisms into simpler compounds, such as carbon dioxide, ammonia, and water. Also called *reducer.*

decomposition the act or process of breaking down large molecules into simpler ones.

deductive reasoning the logical process in which specific conclusions are drawn from generalities. Compare *inductive reasoning.*

dehydration chemically, an enzymatic reaction during which water is lost and a covalent bond forms between the reactants.

dehydration synthesis the process by which a covalent bond is formed between two compounds by the removal of one oxygen atom and two hydrogens, which form water.

deletion 1. the removal of any segment of a chromosome or gene. 2. the site of such a removal after chromosome healing—a mutation. 3. the deleted chromosome.

delta cell an endocrine cell in the pancreas, which secretes somatostatin.

demographic transition also *theory of demographic transition;* the proposal that in the development of nations, population growth has distinct stages: (1) high birth and death rates and slow growth; (2) high birth rate, low death rate, and very rapid growth; (3) low birth and death rates, and very slow growth.

denatured *adj.*, of a protein, altered so as to destroy its properties, through heating or chemical treatment.

dendrite an extension of a neuron that receives impulses and conducts them toward the cell body.

dendrograph a device for recording fluctuations in the diameter of a tree trunk.

denitrifier any one of several common soil and manure bacteria that break down nitrates and nitrites into nitrogen gas.

density-dependent control a factor affecting population size, the severity of which depends on the density of the population in question, i.e., competition.

density-independent control a factor affecting population size that is independent of population density, i.e., temperature, salincty, fire, meteorites, etc.

deoxynucleotide any nucleotide of DNA.

deoxyribose a five-carbon sugar identical to ribose except that one hydroxyl group is replaced by a hydrogen atom.

deoxyribose nucleic acid. See DNA.

depolarized loss of a net-charged state.

dermal brachiae "skin gills" of echinoderms, consisting of outpocketings of the coelom and body wall.

dermis in animals, the inner mesodermally derived layer of skin beneath the ectodermally derived epidermis.

descending limb in the kidney nephron the part of the loop of Henle that goes down into the renal medulla.

desert a region characterized by scanty rainfall, especially less than 25 cm (10 in) annually.

desmotubule within the plasmodesmata of plant cells a continuation of the endoplasmic reticulum of one cell with that of the adjacent cell.

determinate cleavage in animal embryos the very early designation of cells to specific developmental pathways resulting in a loss of flexibility in individual cells and tissues. Also *determinate development.* See *indeterminate cleavage.*

determinate development development or growth that ceases at a certain age or stage.

deuterostome a bilateral animal (i.e., an echinoderm or chordate) whose anus arises early and whose mouth arises later as a second embryonic opening. Compare *protostome.*

diabetes mellitus a genetic disease of carbohydrate metabolism characterized by abnormally high levels of glucose in the blood and urine and the inadequate secretion of utilization or insulin.

diaphragm 1. in mammals, a dome-shaped, muscularized body partition that separates the chest and abdominal cavities and is involved in breathing movements. 2. a birth control device; a thin rubber cup with a springlike rim that, when inserted to cover the cervix, acts as a physical barrier to sperm.

diastole the period of expansion and dilation of the heart during which it fills with blood; the period between forceful contractions (systole) of the heart. See *systole.*

diastolic pressure the blood pressure between the heart's contractions; see also *blood pressure* and *systolic pressure.*

diatom an acellular or colonial yellow-green photosynthetic protist, having a silicon-impregnated cell wall in two parts.

dicot See *dicotyledon.*

dicotyledon a flowering plant of the angiosperm class Dicotyledonae, characterized by producing seeds with two cotyledons. Compare *monocot.*

Dicotyledonae the class of flowering plants (angiosperms) characterized by seeds with two cotyledons.

differentially permeable. See *selectively permeable.*

diffuse root system also *fibrous root system.* A root system of a plant in which there are many roots all about the same size.

diffusion the random movement of a gas or solute under thermal agitation, resulting in a net movement from regions of higher concentration to regions of lower concentration.

diffusion gradient. See *concentration gradient.*

digestion the hydrolytic cleavage (adding of water), through enzyme action to food molecules, permitting absorption.

dihybrid cross in Mendelian genetics, a cross involving two unlinked pairs of factors or alleles.

dikaryotic *adj.* pertaining to a condition in many fungi arising after conjugation, wherein parental plus and minus haploid nuclei remain separated for a time before fusion occurs.

dilation an enlargement or widening; in the human birth process, the initial stage in which the cervix softens and widens.

dinoflagellate a flagellated, photosynthetic, marine protist of the group Dinoflagellata.

dipeptide two amino acids united by a peptide linkage; a common intermediate product of digestion.

diploid (2n) *adj.*, doubled; having a double set of genes and chromosomes, one set originating from each parent. Compare *haploid, polyploid.*

disaccharide "two sugars"; a carbohydrate consisting of two monosaccharide simple sugar subunits, e.g. sucrose and lactose.

disruptive selection in population genetics, selection that through changing or cycling conditions shifts back and forth, favoring first one end of a phenotypic range and then another.

dissociating coming apart into discrete units; ionizing.

disulfide linkage in protein, a covalent link formed between sulfur groups in one polypeptide or adjacent polypeptides.

directional selection in population genetics, selection favoring one extreme of a continuous phenotypic distribution (e.g., the darkest of several colors present).

disease a condition of an organ, part, structure, or system of the body in which there is incorrect function resulting from the effect of heredity, infection, diet, or environment; illness, sickness.

distal convoluted tubule a portion of the nephron between the loop of Henle and the collecting duct.

diuretic a substance that increases the volume of urine excreted.

divergent evolution branching evolutionary change away from the ancestral type, with selection favoring differences in newly arising species.

divergent speciation. See *divergent evolution.*

division a major primary category or taxon of the plant kingdom, sometimes fungal moneran and protist kingdoms, equivalent to *phylem.*

DNA (deoxyribonucleic acid) a double-stranded nucleic acid polymer; the genetic material of all organisms (except RNA viruses).

DNA polymerase an enzyme that catalyzes the replication of DNA.

DNA probe a short, specific segment of single-stranded DNA nucleotides, some of which are radioactive, used in identifying and retrieving desired gene segments from DNA libraries.

DNA repair system a complex of enzymes that identifies and repairs spontaneous changes in the nucleotide bases of DNA.

DNA replication the semiconservative synthesis of DNA in which the double helix opens, the two strands separate, and each is used as a template for producing a new opposing strand.

dominance the phenotypic expression of only one of the two alleles in a heterozygote.

dominance hierarchy behavioral interactions established in a troop, flock, or other species group in which every individual is dominant to those lower on the order and submissive to those above. Also called *pecking order.*

dorsal hollow nerve cord a characteristic neural structure of chordates, which in vertebrates forms the *brain* and *spinal cord.* "Hollow" refers to the presence of fluid-filled foldings and cavities.

dorsal tubular nerve cord a characteristic of chordates, arising in development as a flat, dorsal plate of neurectoderm that sinks beneath the surface to form, in vertebrates, the brain and spinal cord.

double helix the configuration of the native DNA molecule which consists of two strands of nucleotides wound spirally around each other.

doubling time the number of years required for a population to double in size.

douching washing the vagina or cervix by flooding with a rinse.

Down syndrome. See *trisomy 21.*

drumstick a small projection (not unlike a chicken leg) of the nucleus of the polymorphonuclear leukocytes of human females, attributed to the condensed second X chromosome.

drupt in plants, a simple fleshy fruit, usually with a single seed, derived from one carpel; i.e., olive, peach, cherry.

ductus arteriosus in the mammalian fetus, a short, broad vessel conducting blood from the pulmonary artery to the aorta, thus bypassing the fetal lungs.

duodenum the first segment of the small intestine posterior to the stomach.

eardrum See *tympanic membrane.*

ecdysis also *molting;* 1. in arthropods, the shedding of the outer, noncellular cuticle. 2. this shedding together with associated changes in size, shape, or function.

ecdysone the molting hormone of insects.

echinoderm any organism of the marine, coelomate, deuterostome phylum Echinodermata, i.e., starfishes, sea urchins, sea cucumbers, etc.

Echinodermata the phylum containing marine, coelomate, deuterostamic organisms, such as the sea urchins and sea stars.

ecological niche the position or function of an organism in a community of plants and animals; the totality of adaptations, specializations, tolerance limits, functions, biological interactions, and behaviors of a species.

ecological pyramid See *pyramid of numbers, energy* and *biomass.*

ecological succession the gradual progression or change in the species composition of a community; the progression of one community into another.

ecology the branch of biology dealing with the relations between organisms and their environment.

ecosystem in ecology, a unit of interaction among organisms and between organisms and their physical environment.

ectoderm in animal development, the outermost of the three primary germ layers of the embryo; the source of all nerve tissue, sense organs and outer skin.

ectotherm an animal that lacks the ability to thermoregulate physiologically. Compare *endotherm.*

ectothermic *adj.* pertaining to a condition in which the body temperature cannot be metabolically regulated.

effector any structure that elicits a response to neural stimulation, e.g., a muscle gland.

efferent vessel any blood vessel carrying blood away from a structure under consideration, specifically in a gill filament or kidney nephron.

egg the ovum or female reproductive cell. See *egg cell.*

eidetiker a person with a "photographic" memory.

ejaculation the forceful expulsion of semen from the penis.

elastin an extracellular structural protein that forms long, elastic fibers in connective tissue.

electroencephalogram (EEG) a clinical record of electrical activity within the brain.

electroencephalograph a clinical device for detecting and recording electroencephalograms (EEGs) of electrical activity within the brain.

electromagnetic spectrum the range of electromagnetic radiation from low-energy, low-frequency radio waves to high-energy, high-frequency gamma rays.

electron one of the three common components of an atom; its mass is 1/1837 of a proton's, and its electrostatic charge is −1.

electron acceptor a molecule that accepts one or more electrons in an oxidation-reduction reaction (and thus becomes reduced), i.e., a cytochrome or coenzyme.

electron carrier a molecule that behaves cyclically as an electron acceptor and an electron donor; a common constituent of electron transport systems.

electron donor a molecule that loses one or more electrons to an electron acceptor in an oxidation-reduction reaction (and thereby becomes oxidized).

electron orbital 1. a path traveled by an electron. 2. the space in which an electron is found 90% of the time.

electron transport system (ETS) a series of cytochromes and other proteins, bound within a membrane of a thylakoid, mitochrondrion, or prokaryotic cell, that passes electrons or hydrogen atoms or both in a series of oxidation-reduction reactions that result in a net movement of hydrogen ions across the membrane.

element 1. a substance that cannot be separated into simpler substances by purely chemical means. 2. in plants, any one of the hollow xylem cells with thick, pitted walls.

elongation growth by lengthening; in plant development, the expansion in one direction of stem or root cells under turgor and the subsequent growth in length of the plant.

eltonian pyramid See *pyramid of numbers.*

embryology the scientific study of early development in plants and animals.

embryonic disk the part of the inner cell mass from which a mammalian embryo develops.

embryonic induction an influencing action, employing chemical or physical agents, during which one embryonic tissue influences the developmental fate of another.

embryo sac in flowering plants, the mature negagametophyte after division into six haploid cells and one binucleate cell, enclosed in a common cell wall.

endergonic *adj.* relating to a chemical reaction that requires an input of energy from some outside source and in which the products contain greater energy than the reactants (e.g., the synthesis of ATP and ADP and P_1). Compare *exergonic.*

endocrine system a system of cells, tissues, and organs whose chemical secretions help control body functioning.

endocytosis the process of taking food, solutes, or invading cells into the cell (into vacuoles) by engulfment (a form of active transport). See also *phagocytosis.*

endoderm in animal development, the innermost of the three primary germ layers of a metazoan embryo and the source of the gut epithelium and its embryonic outpocketings (in vertebrates, the liver, pancreas, and lung); also called *entoderm.*

endodermis a single layer of cells around the stele of vascular plant roots that forms a moisture barrier—its lateral cell walls are pressed tightly together and waterproofed, forming the *Casparian strip.*

endometrium in mammals, the tissue lining the cavity of the uterus; it responds cyclically to ovarian hormones by thickening in preparation for the implantation of an embryo, and is shed as the menstrual flow.

endoplasmic reticulum (ER) extensive and dynamic membranes of the cell cytoplasm, usually a site of synthesis, packaging and transport.
Rough ER with bound ribosomes, the site of protein synthesis. **Smooth ER** without bound ribosomes, the site of synthesis of nonprotein materials.

endorphin a brain neurotransmitter that reduces pain and produces euphoria.

endoskeleton a mesodermally derived supporting skeleton, inside the organism and surrounded by living tissue, as in vertebrates.

endosperm a nutritive tissue of angiosperm seeds, formed around the embryo.

endospore a resistant, thick-walled spore formed from a bacterial cell.

endotherm an organism with the ability to metabolically thermoregulate; also called "warm-blooded."

endothermic *adj.*, pertaining to the condition in which animals regulate their body temperature through the metabolism of cellular fuels, usually maintaining it at some constant optimal level.

energy the capacity or potential to accomplish work (with work defined as the movement of matter); it exists in several forms, including heat, chemical, electrical, magnetic, and radiant (electromagnetic).

energy pyramid the ecological pyramid comparing the energy content of producers, consumers, and decomposers.

energy shell one of several distinct energy levels at which the electrons of any element may be found; also called *electron shell.*

enkephalin a brain neurotransmitter that reduces pain and produces euphoria.

entrophy in thermodynamics, the energy in a closed system that is not available for doing work; also a measure of the randomness or disorder of such a system. See also *energy, potential energy,* and *second law of thermodynamics.*

environmental resistance the sum of environmental factors (e.g., limited resources, drought, disease) that restrict the growth of a population below its biotic potential (maximum possible population size).

enzyme a protein that catalyzes chemical reactions.

enzyme-substrate complex (ES) the unit formed by an enzyme bound by noncovalent bonds to its substrate.

epicotyl in the embryo of a seed plant, the part of the shoot above the attachment of the cotyledon.

epidermis 1. in plants, the outer protective cell layer in leaves and in the primary root and stem. 2. in animals, the outer epithelial layer of the skin.

epididymis in mammals, an elongated soft mass lying along side each testis, consisting of convoluted tubules; the site of sperm maturation.

epiglottis in the lower pharynx, a flexible cartilaginous flap that folds over the glottis during swallowing.

epilimnion in lakes, the layer of water above the termocline.

epinephrine a hormone with numerous effects produced by the adrenal medulla and by nerve synapses of the autonomic nervous system; also called *adrenaline.*

epiphyte a plant that grows nonparasitically on another plant, or sometimes, on an object, e.g., orchids.

epistasis the masking of a trait ordinarily determined by one gene locus by the action of a gene or genes at another locus.

epithelial tissue See *epithelium.*

epithelium a basic animal tissue type, which covers a surface or lines a canal or cavity; it serves to enclose and protect.

erectile tissue spongy tissue of the penis, which fills with blood during sexual arousal, causing the penis to become erect, lengthen, and thicken.

erythroblastosis fetalis during Rh incompatibility, the destruction of red blood cells in the Rh^+ fetus or newborn through the action of maternal Rh^- antibodies that have crossed the placenta.

erythrocyte a red blood cell, oxygen carrier.

erythropoietin a hormone that stimulates the production and differentiation of erythrocytes during erythropoiesis.

esophagus the muscular tube connecting the pharynx to the stomach.

estrogen a female hormone; one of the hormones involved in the production of secondary sex characteristics and the menstrual cycle.

estuary the part of the mouth of a river where the river meets the ocean.

ethology the scientific study of animal behavior, usually under natural conditions.

ethylene a colorless, sweet-smelling, unsaturated hydrocarbon gas emitted by ripening fruit and inducing further ripening.

ETS See *electron transport system.*

eubacteria "true bacteria"; the better-known prokaryote group containing many familiar human pathogens and important soil and water bacteria. See also *Archaebacteria.*

eukaryote a cell or organism with cells containing a membrane-bound nucleus, chromosomes complexed with histones, and other proteins and membrane-bounded cytoplasmic organelles. Compare *prokaryote.*

Eumetazoa a subkingdom of Animalia containing all animals except the sponges.

euphotic zone the area of a body of water that receives sufficient light for photosynthesis to occur.

eustachian tubes a canal extending from the middle ear to the pharynx.

eutrophic *adj.* pertaining to fresh waters, rich in nutrients essential to plant algal growth, thus rich in life and usually in rapid ecological succession. Compare *oligotrophic.*

eutrophication the aging process whereby a body of water supports increasing numbers of organisms, often causing lakes to become marshes and terrestrial communities.

evolution 1. any gradual process of formation, growth, or change. 2. descent with modification. 3. long-term change and speciation (division into discrete species) of biological entities. 4. the continuous genetic adaptation to populations through mutation, hybridization, random drift and natural selection.

exchange pool in biogeochemical cycles, the readily available reserves of a mineral nutrient, such as a soluble phosphate or nitrate pool in the soil water; compare *reservoir.*

excitatory synapse a synapse in which the secretion of neuro-transmitter stimulates neural impulses in the receiving neuron.

excretion the removal of cellular metabolic wastes from the body.

exergonic *adj.*, describing a chemical reaction that releases energy, with the products containing less free energy that the reactants; compare *endergonic.*

exocrine gland also called *ductless gland;* a gland that releases its secretions externally, into the gut or anywhere other than the bloodstream.

exocytosis the process of expelling material from vacuoles through the cell membrane; compare *endocytosis.*

exon See *expressed sequence.*

exoskeleton an external skeleton and supportive covering (as in arthropods, where it consists of chitin).

experiment a procedure conducted for the purpose of discovering something unknown or for testing a hypothesis, supposition, etc.

experimental group in a controlled experiment, the group subjected to the variable being tested or studied.

expiratory center part of the respiratory control center in the medulla oblongata, increases the expiratory rate and volume during strenuous activity.

exponential curve See *J-shaped curve.*

exponential growth also *geometric growth;* population growth in which the population size increases by a fixed proportion in each time period, successive values forming an exponential series.

expressed sequence also called *exon;* the nucleotide sequences that remain in a messenger RNA molecule after tailoring and thus are expressed in a polypeptide. See also *intron.*

expulsion in the human birth process, the phase of labor lasting from minutes to hours, during which the fetus is pushed out of the uterus.

external ear all parts of the ear external to the eardrum, comprising the external ear canal, the external auditory meatus, and the pinna.

external fertilization fertilization outside the body, as found in echinoderms and most bony fish.

extinction behaviorally, the loss of a conditioned response as a result of the absence of reinforcement.

extracellular digestion digestion outside of cells (usually in the gut).

extraembryonic membrane any of several external life-supporting structures produced by the vertebrate embryo (i.e., amnion, chorion, allantois, and yolk sac).

eyeball the ball of the eye.

facilitated diffusion diffusion of molecules across a cell membrane assisted by carrier molecules.

FAD (flavin adenine dinucleotide) a coenzyme that is a hydrogen carrier in mitochondrial metabolism.

fascia a heavy sheet of connective tissue covering or binding together muscles or other internal structures of the body, often connecting with ligaments or tendons.

fatty acid an organic acid consisting of a linear hydrocarbon "tail" and one terminal carboxyl group.

feedback the return of part of the output of a system to be reintroduced as input that regulates or affects the system's functioning.

negative feedback the output damps the activity of the system, reducing the flow of output.

positive feedback the output stimulates the activity of the system, further increasing the flow of output.

female gametophyte See *megagametophyte.*

fern any of the pteridophytes, having few leaves large in proportion to the stems and bearing sporangia on the undersurface or margin.

fertility rate See *general fertility rate.*

fertilization the sexual process by which gametes or gamete nuclei are united.

fertilization cone a cone-shaped mound of egg cytoplasm that, upon activation of the egg by a fertilizing sperm, rises to engulf the sperm.

fertilization membrane in echinoderms, a membrane formed upon egg activation by the rupture of cortical granules, preventing further sperm penetration.

fetus in vertebrates, an unborn ovum-hatched individual past the embryo stage; in humans, a developing unborn individual past the first eight weeks of pregnancy.

fibrin an insoluble fibrous protein forming blood clots and also contributing to the viscosity of blood. See *fibrinogen.*

fibrinogen a globular blood protein (globulin) that is converted into fibrin by the action of thrombin as part of the normal blood clotting process.

fibroblast a cell in the connective tissue group that produces fibers and matrix substances such as collagen.

fibroin the insoluble fibrous protein of silk and spider webs.

fibrous root system. See *diffuse root system.*

fiddlehead the coiled new leaves of a fern, having a form similar to the curved head of a fiddle or violin.

filament in flowers, the slender stalk of the stamen, on which the anther is situated. See also *myofilament.*

filamentous alga a group of green alg[ue] that tend to form lengthy filaments of cells.

filter feeder an animal that obtains its food by filtering minute organisms from a current of water.

fimbriae (*sing.*, fimbria) a fringe of ciliated tissue surrounding the opening of the oviduct into the peritoneal cavity.

fire-disclimax community an ecological community in continual transition because of recurring fires, such as chaparral.

first law of thermodynamics the physical law that states that energy cannot be created or destroyed and that the total amount of energy in a "closed" system remains constant.

first messenger a hormone, as distinguished from a *second messenger,* which causes changes within a cell.

fission the division of an organism into two (binary fission) or more organisms, as a process of asexual reproduction.

fixed-action pattern in behavior, a precise and identifiable set of movements, innate and characteristic of a given species. See also *instinct.*

flagellate any protozoan of the phylum Sarcomastigophora having one or more flagella.

flagellum (*pl.* flagella) 1. long, whiplike, motile eukaryote cell organelles projecting from the cell that propel the cell by undulations. 2. in prokaryotes, consisting of a solid, helical protein fiber that passes through the cell wall and propels the cell by rotating.

flame cells Also called *protonephridia;* in flatworms, rotifers, and lancelets, a cup-shaped cell with a tuft of cilia in its depression, the flamelike beating of which creates suction that draws waste materials into an excretory duct.

fluid mosaic model a description of the cell membrane as a phospholipid bilayer that has a fluidlike core and contains an assortment of specifically oriented proteins and other surface molecules.

fodrin one of the chemicals involved in memory formation.

follicle-stimulating hormone (FSH) a hormone produced by the anterior pituitary that stimulates the growth and maturation of eggs and sperm.

F_1 generation in genetics, the first generation of offspring in cross-breeding. Subsequent generations are designated F_2, F_3, etc.

F_1 particle an enzyme-containing body of the mitochondrial membrane in which chemiosmotic phosphyrylation takes place.

F plasmid a smaller bacterial chromosome, responsible for producing the sex pilus whose formation precedes conjugation.

food chain a sequence of organisms in an ecological community, each of which is a food for the next higher organism, from the primary producer to the top producer.

food vacuole in animals and protists, an intracellular vacuole that arises by the phagocytosis of solid food materials and in which digestive processes occur.

foodweb a group of interacting food chains; all of the feeding relations of a community taken together.

foramen (*pl.*, foramina) openings or perforations.

foramen magnum the large opening at the base of the skull through which the spinal cord passes.

foramen ovale in the fetal mammalian heart, an opening in the septum between the atria; it normally closes at the time of birth.

foraminiferana shelled marine sarcodine protist with one or more nuclei in a single plasma membrane, the carbonate shell having numerous minute openings (foramina) through which slender, branching pseudopods are extended.

forebrain the anterior of the three primary divisions of the vertebrate embryonic brain.

foregut the upper part of the embryonic alimentary canal from which the pharynx, esophagus, stomach, and part of the duodenum develop. A similar region in the gut of insects.

foreskin the fold of skin that covers the head of the uncircumcised penis.

founder effect in population genetics, the chance assortment of genes carried out of the original population by colonizers (or founders) who subsequently give rise to a large population.

fragmentation reproduction by the development of a part of the parent animal.

frame-shift mutation an insertion or deletion of a nucleotide which results in a frame's change and the misreading during translation of all mRNA codons "downstream."

free nerve ending one of the sensory elements of the skin.

frontal lobe an anterior portion of the cerebrum believed to be a site of higher cognition.

fruit the mature, seed-bearing ovary of a flowering plant; it may be swollen and sweet or starchy.

fucoxanthin a brown carotenoid pigment characteristic of brown algae.

fungi imperfecti yeast-like fungi of the phylum Deuteronmycota which cause skin and vaginal infections.

fungi (*sing.*, fungus) organism of the kingdom Fungi; included are yeasts, mushrooms, molds, mildew, puffballs, etc.

gall bladder a muscular saclike organ that is the temporary receptacle of bile.

gametes haploid cells that unite in sexual reproduction, producing a diploid zygote; typically the haploid sperm or egg cell.

gametic cycle in all animals (and some protists) a life cycle dominated by the diploid chromosome state, with the haploid state present only in the gametes.

gametophyte in plants with alternation of generations, the haploid form, in which gametes are produced; compare *sporophyte.*

ganglia (*sing.* ganglion) a mass of nerve tissue containing the cell bodies of neurons.

ganglion cell one of a group of nerve cells that exist as a clump.

gap junction dense structure that physically connects membranes of adjacent cells along with channels for cell-to-cell transport.

gastric juice the digestive fluid, containing pepsin and other enzymes, secreted by the glands of the stomach.

gastropod any mollusk of the class Gastropode, comprising the snails.

Gastropoda the class of mollusks including the snails.

gastrovascular cavity the cavity of coelenterates, ctenophores, and flatworms which opens to the outside only via the mouth and functions as a digestive cavity and crude circulatory system.

gastrula a young metazoan embryo consisting of three germ tissue layers, with an inner cavity (archenteron).

gastrulation the process of cellular migration by ectoderm whereby a blastula becomes a gastrula.

gated channels a cellular opening through which ions pass, that can be opened or closed.

gel electrophoresis an analytical procedure in biochemistry whereby a mixture of molecules enters a gel and separates by the differential migration of its constituents in an imposed electrical field.

gene (variously defined) 1. the unit of heredity that controls the development of a hereditary character. 2. a continuous length of DNA with a single genetic function.

gene cloning technique whereby pieces of DNA from any source are spliced into plasmid DNA, cultured in growing bacteria, purified and recovered in quantity. See also *recombinant DNA technology.*

gene flow the exchange of genes between populations through migration, pollen dispersal, chance encounters, and the like.

gene interaction occurring when the genes at one locus affect the phenotypic expression of genes at another locus.

gene locus specific place on a chromosome where a gene is located.

gene pool all the genetic information of a population considered collectively.

generalist an animal that can take advantage of a variety of opportunities.

generalization in behavior, the act or process of responding to a stimulus similar to but distinct from the conditioned stimulus.

general fertility rate the average number of live births per 1,000 females in their reproductive years (in the U.S., women 15 to 44).

generative cell a cell in the microgametophyte of a seed plant, capable of dividing to form two sperm cells.

generator potential in sensory receptors, depolarizations that differ from action potentials in that they may be graded in intensity.

generic name 1. the first part of a scientific name as in the binomial system. 2. a group of closely related and ecologically similar species, always italicized and capitalized.

gene amplification the presence in the genome of many copies of a specific gene, particularly where that gene's product is needed in large quantities.

gene sequencing determining the specific sequence of nucleotides in a gene.

gene splicing using recombinant DNA techniques to insert selected DNA fragments or genes into foreign strands of DNA or bacterial plasmids.

genetic code the specific groupings of nucleotides in DNA and RNA that specify the order of amino acids in a polypeptide or protein.

genetic continuity in generation after generation of cells and individuals of a species the preservation of the specific genetic code through preciseness in the processes of replication, mitosis, and meiosis.

genetic drift in population genetics, changes in gene frequencies through chance rather than through natural selection. See also *founder effect* and *bottleneck.*

genetic engineering modern techniques of gene management; including gene cloning, gene splicing, amino acid, DNA and RNA sequencing, and gene synthesis.

genetic equilibrium the genetic state of a population wherein the frequency of certain alleles remains constant generation after generation.

genetic recombination the exchange and subsequent recombination of genes as in crossing over in meiosis I.

genetics 1. the science of heredity, dealing with the resemblances and differences of related organisms resulting from the interaction of their genes and the environment. 2. the study of the structure, function, and transmission of genes.

genital tubercle in the sexually undifferentiated mammalian embryo, a structure that, in the male, will form the head and shaft of the penis, and in the female, the *glans clitoris* and the *labia majora.*

genome the full haploid complement of genetic information of a diploid organism, usually viewed as a property of a species.

genotype 1. the genetic constitution of an organism (specific genes present). 2. *total genotype,* the sum total of genetic information of an organism.

genus in taxonomy, a major subdivision of a family, consisting of one or more species.

geotaxis oriented movement of a motile organism toward or away from a gravitational force.

geotropism See *gravitropism.*

germinal epithelium the epithelium of the gonads, which through mitosis gives rise to gametocytes, i.e., sperm and eggs.

germ layer any of the three layers of cells formed at gastrulation: *ectoderm, endoderm,* and *mesoderm.*

germ theory the proposal by Pasteur and others that bacteria were the primary cause of disease.

gibberellin any of a family of plant growth hormones that control cell elongation, bud development, differentiation and other growth effects.

gills 1. in fungi, the thin, flattened structures on the underside of a mushroom (basidiocarp) that bear the spore-forming basida. 2. in animals, a thin-walled organ of great surface area, for obtaining oxygen from water.

gill arches also called *pharyngeal arches.* 1. in fish, a row of bony curved bars extending vertically between the gill slits on either side of the pharynx, supporting the gills. 2. in land vertebrate embryos, a row of similar rudimentary ridges that give rise to jaw, tongue, and ear bones.

gill chamber in fishes, one of the two chambers housing the gills, covered by a bony flap, the *operculum.*

gill slit a chordate characteristic, permanent openings into the pharyngeal cavity of primitive chordates; associated with gills in fishes and represented only as transitory structures in the embryos of terrestrial vertebrates; also called *pharyngeal gill slits.*

ginkgo a plant of the division Ginkgophyta.

Gingkophyta the gymnosperm class, once abundant, that now includes only the maidenhair tree.

glans the enlarged vascular, sensitive body forming the end of the penis.

glandular epithelium simple columnar epithelial tissue that lines glands and the intestine, specializing in synthesis and secretion.

glial cells numerous nonconducting supporting cells of the central nervous system that may play some role in information storage; also called *neuroglia.*

gliding joint a skeletal joint in which the articulating surfaces glide over one another without twisting.

globulin any of a class of globular proteins occurring in plant and animal tissues and in blood.

glomerulus (*pl.* glomeruli) the tuft or mass of capillaries within Bowman's capsule.

glucagon a polypeptide hormone secreted by the pancreatic islets of Langerhans, whose action increases the blood glucose level by stimulating the breakdown of glycogen in the liver.

glucocorticoid one of the hormones of the adrenal cortex.

gluconeogenesis in the liver, a biochemical pathway in which, at the expense of ATP lactate from active muscle is converted back to glucose or glycogen.

glucose a six-carbon sugar, occurring in an open chain form or either of two ring forms; the subunit of many carbohydrate polymers; also called *dextrose, blood sugar, corn sugar,* and *grape sugar.*

glue cell a glandular, thread-bearing cell, found only in ctenophores, used to capture prey by adhesion.

glycerol the three-carbon backbone with a hydroxyl group on each carbon; a component of neutral fats and of phospholipids.

glycogen animal starch, a highly branched polysaccharide consisting of alpha glucose subunits; a storage carbohydrate in the liver and muscles of animals.

glycolysis the enzymatic anaerobic breakdown of glucose in cells yielding ATP (from ADP), pyruvate, and NADH.

glycolytic pathway the sequence of enzymes and reactions responsible for the conversion of glucose to pyruvate and for the accompanying substrate level phosphorylation.

glycoprotein any of a group of complex proteins containing a carbohydrate combined with a simple protein.

Gnetophyta the gymnosperm class containing certain trees and climbing vines that may be related to gymnosperms and angiosperms.

gnetophyte a plant of the division Gnetophyta.

goblet cells a mucus-secreting epithelial cell; common in the lining of the respiratory system.

golden brown alga any of the single-celled and colonial algae of the phylum Chrysophyta, found in both marine and fresh waters, deriving its color from carotenoid pigments.

Golgi body a membraneous organelle found especially in the cytoplasm of secretory cells, involved in the packaging of cell products from the endoplasmic reticulum into secretion granules; also called *Golgi complex* or *Golgi apparatus.*

gonadotropin a hormone that stimulates growth or activity in the gonads; in vertebrates, a specific peptide hormone of the anterior pituitary.

G_1 (gap one) in the cell cycle, the first part of interphase, during which the cell doubles in size.

gonorrhea a sexually transmitted bacterial disease caused by the diplococcus *Neisseria gonorrhoeae.*

Graafian follicle a fluid-filled body in the ovary containing a mature oocyte within a cluster of follicular cells.

gradualism the Darwinian proposal that the pace of evolution is slow but steady with an ongoing accumulation of minor changes leading eventually to the formation of new species; compare *punctuated equilibrium.*

granum (*pl.,* grana) a stack of thylakoid disks in chloroplast.

grassland one of the natural biomes of the earth, characterized by perennial grasses, limited seasonal rainfall, and great numbers of herbivorous mammals, birds, and insects.

gravitropism in plants, a growth response to gravity.

gray matter a butterfly-shaped region within the spinal cord, composed of cell bodies of interneurons and motor neurons; the cerebral cortex.

green algae any algae of the phylum Chlorophyta, biochemically similar to plants.

greenhouse effect the warming principle of greenhouses, in which high-energy solar rays enter easily, while less energetic heat waves are not radiated outward; now especially applied to the analogous effect of increasing atmospheric concentrations of carbon dioxide through the burning of fossil fuels and forests.

gross productivity (GP) the amount of biochemical energy captured by photosynthesis in a particular area per unit of time.

ground meristem the primary plant tissue from which the ground tissues, collenchyma, parenchyma, and sclerenchyma, are developed.

growth increase in size through cell division and/or enlargement.

growth factor a hormone known to be necessary for the proliferation of cells in tissue culture.

growth hormone (GH) 1. a polypeptide hormone of the anterior pituitary that regulates growth invertebrates. 2. any hormone that regulates growth, e.g., auxin or gibberellin in plants.

growth ring a visible ring or line produced by changes in growth rate, usually age-related.

G_2 (gap two) in the cell cycle, a period of renewed protein synthesis during which tubulin is made.

guanine (G) one of the nitrogenous bases of DNA and RNA.

guard cell in a plant leaf or stem, one of a pair of crescent-shaped cells of a stoma whose changes in turgor regulate the size of stomatal opening.

gustation the sense of taste; the act of tasting.

gutation the normal exuding of moisture from the tip of a leaf or stem in certain plants, presumably due to root pressure.

gymnosperm a nonflowering seed plant of the group Gymnospermae, producing seeds that lack fruit; included are conifers, cycads, gnetophytes, and ginkgos.

gynoecium the pistil, or the pistils, collectively, of a flower.

habituation a simple form of learning in which response to a stimulus diminishes over time.

hair cell See *organ of Corti.*

half-life in a radioisotope, the time it takes for half of the atoms in a sample to undergo spontaneous decay.

halobacteria a photosynthetic bacteria that thrives in a salty environment, capturing light energy with bacteriorhodopsin.

haploid (n) *adj.* halved; having a single set of genes and chromosomes; compare *diploid, polyploid.*

Hardy-Weinberg Law in population genetics, a statement of the genotype frequencies expected given allele frequencies in a population of randomly mating diploid individuals; also called *binomial law.*

Haversian system a Haversian canal together with its surrounding concentrically arranged layers of bone, canaliculi, lacunae, and osteocytes.

HCE See *human chorionic gonadotropin.*

head the upper part of a human, cephalic to the neck; the anterior part of a structure.

heart a hollow, muscular organ which keeps the blood circulating throughout the body.

heavy chain the longer pair of polypeptides in an immunoglobulin.

helicase also called *unwinding enzyme;* one of the enzymes required to break the hydrogen bonds of the double helix of DNA during replication.

heliozoan a protist of a group in the class Sarcodina, consisting of free-living, freshwater forms looking rather like tiny suns, with multiple thin, stiff, radiating pseudopods.

helper T-cell a cell that activates elements of the immune system.

heme group an iron-containing, oxygen-binding, porphyrin ring present in all hemoglobin chains and in myoglobin.

Hemichordata the phylum containing deuterostome animals such as acorn worms.

hemichordate any member of the phylum Hemichordata.

hemizygous *adj.,* the condition of x-linked genes in males, which are neither homozygous nor heterozygous.

hemocyanin a copper-containing respiratory pigment occurring in solution in the blood plasma of certain invertebrates.

hemoglobin a protein, a respiratory pigment consisting of four polypeptide chains, each associated with a heme (iron-containing) group.

hemophilia also called *bleeder's disease;* uncontrolled bleeding in humans due to the lack of a necessary blood-clotting constituent; a genetic disease caused by recessive alleles at either of the two sex-linked loci.

hepatic portal circuit in the circulatory system, the assemblage of vessels that direct blood from the intestine to the liver.

hepatic portal vessel in vertebrates, a vein that collects blood from the capillaries and venules of the esophagus, stomach, and intestine, and carries it to the liver, where the vein again divides into capillaries.

hermaphroditic *adj.,* pertaining to an individual animal or plant with both male and female reproductive organs.

herpes simplex II a highly contagious, sexually transmitted virus, causing intermittent outbreaks of painful, infective blisters on the mouth and genitals, followed by a period of remission.

heterogamete male and female gametes differing in form, size, or structure. See *isogametes.*

heteromorphic *adj.,* dissimilar in shape, structure, or magnitude.

heterosporous in plants, producing morphologically distinct spores, microspores, and megaspores.

heterotherm animals that periodically switch from homeothermy to poikilothermy (insects).

heterotroph an organism that requires organic compounds as an energy and/or carbon source (i.e., all animals and fungi).

heterozygous *adj.,* having two different alleles, commonly a dominant and a recessive, for a specific trait.

Hfr strain (high frequency of recombination) strain of bacteria with a strong tendency to conjugate and transfer the main chromosome.

high-energy bond a molecular bond in ATP and ADP that releases a comparatively large amount of energy when it is cleaved.

hindbrain 1. the posterior of the three primary divisions of the embryonic vertebrate brain. 2. the parts of the adult brain derived from the embryonic hindbrain, including the *cerebellum, pons,* and *medulla oblongata.*

hindgut the lower part of the embryonic alimentary canal from which the colon and rectum develop.

hinge joint a joint that moves in one plane like a door hinge, e.g., the knee.

hippocampus a long, curved ridge of gray and white matter on the inner surface of each lateral ventricle of the brain, involved in memory consolidation, visual memory, and other functions.

Hirudinea the class composed of leeches.

histamine an amino compound released in allergic reactions that dilates blood vessels and reduces blood pressure.

histocompatibility protein a protein which is highly specific for each individual organism, which is the basis for organ or tissue transplant rejection.

histone any of a class of protein substances, as globin, having marked basic properties.

holdfast a rhizoidal base of a seaweed, serving to anchor it to the ocean floor.

homeobox a region within homeotic, or control, genes, whose base sequence is very similar in a variety of organisms.

homeostatic mechanism any mechanism involved in maintaining a stable internal equilibrium, especially one involving self-correcting negative feedback.

homeostasis the maintenance of a stable internal environment in the body of a higher animal through interacting physiological processes involving negative feedback control.

homeotherm an organism capable of maintaining a stable high internal body temperature; also called *warm-blooded.*

homeotic gene a gene that controls the expression of entire blocks of genes and is important in the development of certain major structures.

hominid the primate group composed of humans and related extinct forms.

hominoidea a primate group composed of humans, apes, and related extinct forms.

Homo erectus an extinct species of primitive hominid, followed *Homo habilis.*

Homo habilis an extinct species of hominid, probably stemmed from *Australopithecus afarensis.*

homologue either of the two members of each pair of chromosomes in a diploid cell.

homologous *adj.,* 1. similar because of a common evolutionary origin. 2. derived from a common embryological source, i.e., the *glans clitoris* and the *glans penis.*

Homo sapiens modern humans, the wise ape.

homosporus a state in plants in which all spores are the same.

homospory in plants, a primitive condition in which only one form of spore is produced.

homozygous *adj.,* having identical pairs of alleles for any given pair of hereditary characteristics.

hormone a chemical messenger transmitted from one part of the organism to another, producing a specific effect on target cells.

Human Chorionic Gonadotropin (HCG) a hormone secreted by the human chorion, causing the corpus luteum to continue producing progesterone.

humoral immune response the immune response carried out by immunoglobulins, which find and immobilize matching antigens in the body.

Huntington's disease a lethal hereditary disease of the nervous system developing in adult life and attributed to a dominant allele. Formerly, *Huntington's chorea.*

hybridization sexual reproduction between different species.

hydrogen bond a weak electrostatic attraction between the positively polar hydrogen of a side group and a negatively polar oxygen of another side group.

hydrologic cycle the cyclic evaporation and condensation of the earth's waters, driven by heat in the atmosphere including the biogeochemical cycling of water through photosynthesis and perspiration; also called *water cycle.*

hydrolytic cleavage the breaking of a covalent bond with hydrozide and hydrogen of water joining the subunits; also called *hydrolysis.*

hydrolysis. See *hydrolytic cleavage.*

hydrophilic *adj.,* "water-loving"; pertaining to the tendency of polar molecules or their polar side groups to mix readily with water.

hydrophobic *adj.,* "water-fearing"; pertaining to the tendency of nonpolar molecules or their nonpolar side groups to dissolve readily in organic solvents but not in water; resisting wetting.

hydrostatic pressure the pressure exerted in all directions within a liquid at rest.

hydrostatic skeleton a supporting and locomotory mechanism involving a fluid confined in a space within layers of muscle and connective tissue; movement occurring when muscle contractions increase or decrease the hydrostatic pressure of the fluid.

hydroxyl *adj.,* containing the univalent group, -OH.

Hydrozoa the class composed of the coelenterates.

hymen also *maidenhead;* a fold of thick mucous membrane partially closing the orifice of the vagina, especially in virgins.

hyperthyroidism an abnormal condition caused by an excess of circulating thyroid hormone.

hypertonic *adj.,* having a higher osmotic potential (e.g., higher solute concentration and lower water potential) than the cytoplasm of a living cell (or other reference solution).

hypha (*pl.,* hyphae) one of the individual filaments that make up a fungal mycelium.

hypocotyl the part of a plant embryo below the point of attachment of the cotyledon.

hypocotyl hook in some seedlings, the curved region of the hypocotyl that acts as a bumper as the seedling emerges from the soil.

hypolimnion in lakes, the layer of water below the thermocline.

hypothalamus the region that lies in front of the thalamus in the forebrain; it regulates the internal environment, pituitary secretions, and some of the basic drives.

hypothesis a conjecture set forth as a possible explanation for some observation or phenomenon and that serves as a basis for experimentation or argument; compare *theory.*

hypothyroidism a condition caused by abnormally low levels of circulating thyroid hormones.

hypotonic *adj.,* having a lower osmotic potential (e.g., lower solute concentration and greater water potential) than the cytoplasm of a living cell (or other reference solution).

H zone clearer, centermost zone of the relaxed contractile unit consisting of myosin alone.

I band one of the striations of striated muscle, of variable width, corresponding to the distance between the ends of the myosin filaments of adjacent units.

ileum the region of the small intestine that joins the large intestine.

immune system in vertebrates, widely dispersed tissues that respond to the presence of the antigens of invading microorganisms or foreign chemical substances.

immunoglobulin a protein antibody produced by B-cells in response to specific foreign substances; it consists of four subunits that are joined through disulfide linkages and have specific antigen-binding sites.

immunological tolerance the process by which the immune system becomes inhibited from reacting against "self."

implantation in mammalian reproduction, the invasion of the uterine endometrium and attachment therein of the mammalian embryo (blastocyst).

imprinting in behavior, learning occurring rapidly and very early in life, characterized chiefly by resistance to extinction or forgetfulness.

inclusive fitness increasing the frequency of one's genes in the next generation through reproduction by relatives.

incomplete digestive system a digestive system in which food enters and exits from the same opening.

incomplete gut a digestive system consisting of a blind sac, as in flatworms.

incomplete penetrance a variability in the expression of a dominant allele where it may not be expressed or may be only partially expressed (i.e., polydactyly).

incus the middle bone among three small bones in the middle ear of humans and other animals. Also called the *anvil.*

independent assortment the rule which indicates that the segregation of one pair of unlinked factors, or alleles, has no influence over the way any pair segregates.

indeterminate cleavage in newly forming cells in early animal embryos, the retention of developmental flexibility. Also *determinate development.* See *determinate cleavage.*

indeterminate development continued development at all ages and stages.

indeterminate growth continued growth at all ages and stages.

indeterminate life span the potentially endless life span in organisms such as perennial plants, where new growth is ongoing.

induced fit the proposition that enzymatic actions proceed because their active site forms an inexact fit with their substrate, producing physical stress.

inductive reasoning the logical process that moves from the specific to the general, reaching a conclusion based on a number of observations; compare *deductive reasoning.*

indusium in ferns, a covering over the sorus (spore forming organ).

inferior vena cava large vein that collects blood from ports of the body beneath the heart and returns it to the right atrium.

inhibiting hormone a chemical messenger produced by the hypothalamus that inhibits the release of the major pituitary hormones.

inhibitory neuron a nerve cell that functions by increasing the polarity of the next nerve cell, thus inhibiting it from firing.

inhibitory synapse the space between an inhibitory nerve cell and the next nerve cell.

initiation the first steps in the translation of the genetic code; the beginning of polypeptide synthesis.

initiation complex the physical association of a messenger RNA, two ribosomal subunits and methionine-charged transfer RNA required to begin synthesis of a polypeptide.

innate releasing mechanism (IRM) an ethological term referring to a neural mechanism that produces a specific behavioral event when triggered by a particular stimulus from the environment.

inner cell mass in a mammalian *blastocyst,* the portion that is destined to become the embryo proper.

inner ear the part of the ear enclosed within the temporal bone, consisting of a labyrinth and the associated auditory nerve; including the *semicircular canals,* the *cochlea,* and the *round* and *oval windows.*

insect any animal of the class Insecta.

Insecta the class composed of insects.

insertion 1. in genetics, the addition of extra genetic material into a chromosome, gene, or other DNA sequence. 2. in anatomy, the distal attachment of a tendon or muscle, i.e., the attachment on the part to be moved; compare *origin.*

inspiratory center part of the respiratory control center in the medulla oblongata, rhythmically stimulates inspiration by causing contraction in the diaphragm and rib muscles.

instinct any inherent, unlearned behavioral pattern that is functional the first time it happens that can occur in animals reared in total isolation.

insulin a polypeptide hormone secreted by the islet of Langerhans of the pancreas, whose principal action is to facilitate the transport of glucose into cells; it has many other functions as well.

integument a covering or envelope; in the flower, the part enclosing the nucleus of an ovule, and later forms part of the seed coat; in animals, the skin, exoskeleton, cuticle, or other coverings.

intercalated disk in heart muscle, the highly convoluted double plasma membrane between two adjacent cells in heart muscle.

interferon a cellular substance that interferes with viral replication.

interleukin-1 a powerful activating substance secreted by an antigen-presenting macrophage, which causes a helper T-cell to seek out any inactive lymphocytes and arouse them with interleukin-2.

interleukin-2 a lymphokine secreted by helper T-cells to arouse inactive lymphocytes.

internal ear the portion of the ear enclosed within the temporal bone; included are the three *semicircular canals,* the *cochlea,* and the *round* and *oval windows;* compare *external ear, middle ear.*

internal fertilization in animals, fertilization involving copulation, in which sperm release and fertilization occur within female.

interneuron a neuron of the central nervous system that makes connection between sensory and motor neurons.

interphase all of the cell cycle between successive mitoses; included are G_1, *S,* and G_2 *phases.*

interspecific *adj.,* pertaining to interaction between species.

interspecific cooperation cooperation between members of different species.

interstitial cells cells of the testis that have an endocrine function.

interstitial cell-stimulating hormone (ICSH) in males, an anterior pituitary hormone (identical to LH) that stimulates testosterone production in the interstitial cells of the testis.

intervening sequence also called *intron;* nonexpressed portions of messenger RNA, removed from the message prior to translation; compare *expressed sequence (exon).*

intervertebral disc one of the tough, elastic, fibrous disks situated between adjacent vertebrae.

intracellular digestion digestion within cells following phagocytosis.

intrauterine device (IUD) a plastic device, sometimes containing copper, that is inserted into the uterus as a means of preventing conception.

intrinsic heart rhythm rhythmic contraction of the heart originating through the inherent capacity of cardiac muscle to contract without outside influence.

intrinsic rate of natural increase the rate of population increase calculated by subtracting the average death rate from the average birth rate.

introitus the external opening of the vagina.

intron. See *interventing sequence.*

inversion the transposition of a portion of a chromosome following breakage and repair, altering the relative order of gene loci.

invertebrate any animal lacking a vertebral column.

involuntary muscle See *smooth muscle.*

ion any electrostatically charged atom or molecule.

ion channels in the neural membrane, sodium and potassium (and chloride) channels that, through opening and closing of gates, selectively admit or reject ions.

ion gate an area along a neuron that opens and closes, regulating the passage of ions.

ionic bond a chemical attraction between ions of opposite charge.

ionizing the dissociating of a molecule into oppositely charged ions in solution.

ionizing radiation radiation, including x-ray and gamma rays, energetic enough to create free radicals within the cell; may cause genetic damage.

iris the colored portion of the eyeball seen through the cornea.

islets of Langerhans clumps of alpha, beta, and delta endocrine cells in the *pancreas,* which secrete *insulin, glucagon,* and *somatostatin,* respectively.

isogametes gametes that are identical in size and appearance, such as the (+) and (–) isogametes of *Chlamydomonas.*

isotonic *adj.,* having the same osmotic potential (e.g., same solute concentration and water potential) as the cytoplasm of a living cell or other reference solution.

isotope a particular form of an element in terms of the number of neurons in the nucleus; see also *radioisotopes.*

jejunum the area of the small intestine between the duodenum and the ileum.

J-shaped curve a plot of population growth where growth approaches the biotic potential of the species and generally exceeds the environment's carrying capacity; compare *S-shaped curve.*

jungle dense parts of tropical rain forests penetrated by sunlight.

karyotype a mounted display of enlarged light photomicrographs of the chromosomes of an individual, arranged in order of decreasing size.

keel the enlarged breastbone to which the flight muscles of birds are attached.

kidney in vertebrates, paired excretory organs serving to excrete nitrogenous wastes and regulate water and ion balances.

kinesis in behavior, responsive movement with the vigor of the movement being related to the strength of the stimulus.

kinetic energy energy associated with motion (accomplishing work).

kingdom the largest taxonomic category.

kin selection the selection for traits that benefit individuals carrying some of the same kinds of genes, thereby helping those kinds of genes to be moved into the next generation.

Krause corpuscle a sensory receptor in the skin that responds to temperature changes; also called *Krause end bulb.*

K-selection reproductive strategy of a species adapted to a fairly constant environment, with a small number of offspring and considerable parental care; compare *r-selection.*

labia majora in human females, the outer, fatty, often hairy pair of folds bounding the vulva.

labia minora the smaller inner folds of skin that border the vagina.

lac operon (an inducible operon) in the chromosome of *E. coli,* composed of genes that code for three lactose-metabolizing enzymes and the region that controls their transcription. See also *opeon.*

lactose a digestive enzyme that hydrolyzes lactose to glucose and galactose.

lactation the secretion of milk.

lacteal lymphatic vessels of the intestinal villi.

lactose a disaccharide consisting of glucose and galactose subunits; also called *milk sugar.*

lacuna (*pl.,* lacunae) a minute cavity in bone or cartilage that holds an osteocyte or chondrocyte.

lagging strand also called *discontinuous strand* and *following region;* in DNA replication, the 5′ end of a DNA strand, where nucleotides are first assembled into Okazaki fragments and then added in using the enzyme ligase.

lamella (*pl.*, lamellae) 1. within chloroplasts, membranes that extend between grana (stacked thylakoids). 2. one of the bony concentric layers (ringlike in cross-section) that surround a Haversian canal in bone.

lamprey any eellike, marine or freshwater jawless fish, having a circular, suctioning mouth with horny teeth for boring into the flesh of other fish.

larva in animals, an early, active, feeding stage of development during which the offspring may be quite unlike the adult.

larvaceae a class of urochordates in which a larval-like state is retained although sexual maturity occurs. The process, called neotony, suggests how the chordate line may be evolved.

larynx in terrestrial vertebrates, the expanded part of the respiratory passage at the top of the trachea.

lateral bud primordia one of the patches of shoot meristem which give rise to branches.

lateral line organ in nearly all fish, a canal-like sensory organ containing a line of pitted openings and believed to be responsive to water currents and vibrations; distantly homologous to the mammalian inner ear.

lateral root in plants, a root which originates in the cells of the pericycle well above the root tip, growing perpendicular to the primary root.

law a statement of a relationship or a sequence of phenomena invariable under the same conditions.

leading strand also called *continuous strand* and *leading region;* in DNA replication, the 3′ end of a DNA strand, where nucleotides are added one at a time to the growing strand.

leaf hairs hairlike projections from leaf epidermal cells, generally those on the underside of the leaf.

leaf mesophyll cell photosynthetic parenchymal cells within the leaf.

learning in animal behavior, the process of acquiring a persistent change in a behavioral response as a result of experience.

left atrium the chamber of the heart which receives oxygenated blood from the lungs and pumps it to the left ventricle.

left ventricle the chamber of the heart which receives oxygenated blood from the left atrium and pumps it to the body via the arteries.

lens a transparent organ lying behind the pupil and iris of the eyeball, in front of the vitreous body.

lesion random spontaneous changes in DNA, subject to repair.

leukocyte a vertebrate white blood cell, it aids in resisting infection.

leukoplast a colorless body in plant cells that stores proteins, lipids, and starch.

lichen a combination of a fungus and an alga growing in a symbiotic relationship.

ligament a tough, flexible, but inelastic band of connective tissue that connects bones or supports an organ in place; compare *tendon.*

ligase an enzyme that joins broken DNA strands or Okazaki fragments; used in gene-splicing technology.

light chain the shorter two of the four polypeptides making up an immunoglobin.

light harvesting antenna part of a photosystem; clustered chlorophyll *a*, chlorophyll *b*, carotene molecules, and an associated reaction center; functions in absorbing light energy.

light-independent reaction the part of photosynthesis not immediately requiring light; specifically the fixation of CO_2 into carbohydrate using the NADPH and ATP produced by the light reactions; also called *Calvin cycle* and *dark reaction.*

light reaction the part of photosynthesis directly dependent on the capture of photons, specifically the photolysis of water, electron and proton transport, and the chemiosmotic synthesis of ATP and NADPH; also called *light dependent reaction.*

lignin an amorphous substance that helps give wood its rigidity.

limbic system a region of the brain concerned primarily with emotions.

limnetic zone the open waters of a lake beyond the littoral zone, but including only the depths through which light penetrates and in which photosynthesis can occur.

limnologist an ecologist who specializes in the study of interaction in the fresh waters.

linkage group also called *linked genes;* a group of gene loci located on the same chromosome; ultimately an entire chromosome.

linked *adj.*, of two gene loci, not segregating independently.

lipase a fat-digesting enzyme of the stomach or pancreas.

lipid an organic molecule that tends to be more soluble in nonpolar solvents (such as petroleum products) than in polar solvents (such as water.)

lipoprotein any of the class of proteins that contain a lipid combined with a simple protein.

littoral zone 1. a coastal region including both the land along the coast, the water along the shore, and the intertidal area. 2. a similar area in lakes.

liver an accessory digestive organ that serves many metabolic functions including glycogen storage, detoxification, production of blood proteins, food storage, and production of bile.

lobe-finned fishes any fish with fleshy pectoral and pelvic fins, including lung fishes and crossopterygians.

logistic growth curve. See S-*shaped growth curve.*

long-day plants plants that begin flowering at some specific time before the summer solstice when day length exceeds night; flowering is triggered by a critical established period of darkness; compare *short-day plants.*

long-term memory 1. learning that persists more than a few hours, the memory trace of which is physically located in a different part of the brain than short term memory. 2. the part of the brain and the general neural function with which such persistent memory traces are associated.

loop of Henle the prominent u-shaped loop in the nephron of the mammalian kidney.

lower epidermis in plants, an outer layer of leaf cells, commonly containing numerous stomata and projecting leaf hairs.

lumen the cavity or channel of a hollow tubular organ or organelle.

lung in land vertebrates, one of a pair of compound, saclike organs that function in the exchange of gases between the atmosphere and the bloodstream.

lung fish any fish of the group having a functional, lunglike air bladder as well as gills.

luteinizing hormone (LH) a pituitary hormone that causes ovulation and stimulates hormone production in the corpus luteum.

lymph watery intercellular fluid in the lymphatic system.

lymphatic system the system of lymphatic vessels and ducts and lymph nodes that serves to redistribute excess tissue fluids and to combat infections.

lymph duct the main trunk of the lymphatic system, passing along the spinal column in the thoracic cavity.

lymph node a cluster of cells and blind channels in the lymphatic system in which foreign cells and material are attacked by the immune system. Commonly called *lymph glands.*

lymphocytes any of the several varieties of similar-looking leukocytes involved in the production of antibodies and in other aspects of the immune responses; see also *B-cell, T-cell.*

lymphokine a lymphocyte activator channel.

Lyon effect the mosaiclike gene expression of X-linked genes in human females resulting from the random inactivation of one X chromosome in each cell of the human female embryo.

lysis the destruction or lysing of a cell by rupture of the plasma membrane.

lysogenic cycle the cycle during which a bacteriophage chromosome is incorporated into the bacterial host chromosome, together with mechanisms preventing further infection and lysis; in later cell generations, the incorporated DNA may excise, replicate, and eventually cause cell lysis. See also *lytic cycle.*

lysosome a small, membrane-bounded cytoplasmic organelle, generally containing strong digestive enzymes or other cytotoxic materials.

lytic cycle the short cycle following bacteriophage invasion during which viral replication, capsid synthesis, viral assembly and cell lysis occur, the latter releasing new infective phage particles. See also *lysogenic cycle.*

macromolecule any large biological polymer, such as a protein, nucleic acid, or complex polysaccharide (for example, cellulose, or glycogen).

macronucleus the larger of the two nuclei of *Paramecium* and certain other ciliate protozoans, carrying somatic line DNA; compare *micronucleus.*

macronutrient a plant nutrient required in relatively substantial quantities, e.g., nitrogen, phosphorus, potassium, sulfur, magnesium, and calcium. Compare *micronutrient.*

macrophage a large phagocyte, i.e., leukocyte, previously a monocyte.

madreporite See *water vascular system.*

male gametophyte in seed plant reproduction, the pollen grain.

malleus the outermost of a group of three small bones in the middle ear of humans and other mammals; also called the *hammer.*

Malpighian tubules numerous blind, hollow tubular structures that empty into the insect midget and function as a nitrogenous excretory system.

maltose a disaccharide consisting of two glucose subunits in an apha linkage; a digestion product of starch.

mammal any member of the vertebrate class Mammalia, most of which are characterized by the presence of hair, a muscular diaphragm, milk secretion, and placental development.

Mammalia the class of vertebrates that gives milk that has hair, and, with the exception of the monotremes, that bears living young rather than eggs.

Mandibulata the jawed arthropods.

mandibulate any member of the subphylum Mandibulata containing jawed arthropods, including crustaceans, insects, centipedes, and millipedes.

mantle in mollusks, a fleshy covering that secretes material to form the external shell.

marine environment the saltwater environment including oceans and seas.

marsupial a mammal of the subclass Metatheria; the female has a pouch (marsupium); included are the kangaroo, wombat, koala, Tasmanian devil, opossum, and wallaby.

mating types genetically different clones or strains of fungi, protists, and certain plant algae, where sexual reproduction can occur between clones (mating types) but not within a clone; males and females do not exist.

mechanism the idea that the processes of life are based on the same physical and chemical principles that apply to nonliving phenomena.

mechanist one who believes biological phenomena, and everything else in the universe, can be explained by physical laws; compare *vitalism.*

mechanoreceptor a sensory neuron specialized in detecting touch or pressure against the skin or hair.

medulla the inner portion of a gland or organ; compare *cortex.*

medulla oblongata a part of the brainstem developed from the posterior portion of the hindbrain and tapering into the spinal cord.

medusa the motile, free-swimming jelly fish form of coelenterate; compare *polyp.*

megagametophyte the female gametophyte produced from a megaspore in flowering plants. See also *embryo sac.*

megaspore in flowering plants, the "female" spore, a product of meiosis that gives rise to the megagametophyte.

megaspore mother cell in flowering plants, a large diploid cell in the ovule that will give rise to the megaspore by meiosis.

meiosis in all sexually reproducing eukaryotes, the process of chromosome reduction, in which a diploid cell or diploid cell nucleus is transformed into four haploid cells or four haploid nuclei (the usual manner of sperm and egg production); also called *reduction division.*

meiospore any haploid spore produced by meiosis.

Meissner's corpuscle in mammals, a small touch-responsive neural end organ.

memory cell a mature, long-lived, B- or T-cell lymphocyte, specialized in retaining specific antigen information.

Mendel's first law the law of paired factors and segregation; includes the concepts of paired factors, segregation, chance recombination, and dominance.

Mendel's second law the inheritance of one pair of factors (alleles) in an individual will occur independently of the simultaneous inheritance of a second pair of factors (alleles) (except where gene linkage groups occur).

meninges the tough, protective, connective tissues covering the brain and spinal cord.

mensenteries folds of the membrane which hold the organs of the human digestive system in place, carrying blood vessels, nerves, and lymphatic vessels.

menstrual cycle the cycle of hormonal and physiological events and changes involving growth of the uterine endometrium, ovulation, and the subsequent breakdown and discharge of the endometrium in menstruation (menses); the cycle averages 28 days; also *ovarian cycle.*

menstruation in women, the periodic discharge of blood, secretions, and tissue resulting from the temporary breakdown of the uterine mucosa in the absence of implantation following ovulation.

meristem also called *apical meristem;* the undifferentiated tissue at the stem and root tip that contributes cells for primary growth.

Merkel's disk a sensory structure of the skin.

mesentery the membranous folds of peritoneum to which the intestines are attached.

mesoderm in animal development, the middle layer of the three primary germ layers of the gastrula, giving rise in development to the skeletal, muscular, vascular, renal and connective tissues and to the inner layer of skin and the epithelium of the coelom (peritoneum).

mesoglea the loose, gelatinous middle layer of the bodies of coelenterates between the outer ectoderm and the inner endoderm.

mesosome in many bacteria, an inward extension of the plasma membrane that forms a spherical membranous network.

messenger RNA (mRNA) in eukaryotes, RNA directly transcribed from a gene modified and transported to the cytoplasm; specifying a polypeptide sequence.

metabolic pathways a sequence of enzymatic reactions through which a metabolite passes before the formation of the final product or products.

metamorphosis a change or successive changes of form during the postembryonic or embryonic growth of an animal; in insects, the transition from egg to larva, to pupa, and adult.

metaphase the stage of mitosis or meiosis in which the chromosomes are brought to a well-defined plane in the middle of the mitotic spindle prior to separation in anaphase.

metaphase plate the equatorial plane of the mitotic spindle on which the centromeres are oriented in mitosis.

metastasis the spread of malignant cells to other parts of the body by way of the blood vessels, lymphatics, or membranous surfaces.

metazoa all animals other than sponges; that is, all animals whose bodies are composed of cells differentiated into tissues and organs and who usually have a digestive cavity lined with specialized cells.

methanogen a methane-generating archaebacterium.

microfilament a submicroscopic filament in the cytoskeleton, involved in cell movement and shape.

microgametophyte a sperm-producing body developed from a microspore.

micronucleus in ciliate protists, the smaller of the two nuclei and the one carrying germ-like DNA; compare *macronucleus.*

micronutrient an element necessary for plant growth but needed only in extremely small quantities; compare *macronutrient.*

micropyle in seed plants, a minute opening in the integument of an ovule through which sperm enter.

microspore in seed plants, one of the four haploid cells formed from meiosis of the microspore mother cell; it undergoes mitosis and differentiation to form a pollen grain.

microspore mother cell in the anther of a flowering plant, the diploid cells that will undergo meiosis to form the haploid microspores.

microtrabecular lattice in the cell cytoplasm, a weblike system of microtubules and microfilaments that form a cytoskeletal framework upon which many organelles are suspended.

microtubular organizing center (MOC) in certain mitotic cells an amorphous region surrounding the centrioles, believed to be responsible for assembly and disassembly of microtubular proteins into spindle fibers or microtubules.

microtubule a cytoplasmic hollow tubule composed of spherical molecules of tubulin, found in the cytoskeleton, the spindle, centrioles, basal bodies, cilia, and flagella.

microvilla 1. tiny, fingerlike outpocketings of the cell membrane of various epithelial secretory or absorbing cells, such as those of the kidney tubule epithelium and the intestinal epithelium. 2. cellular projections form in the placenta and containing capillaries.

midbrain 1. the middle of the three primary divisions of the vertebrate embryonic brain. 2. the parts of the adult brain derived from the embryonic midbrain.

middle ear the middle portion of the ear consisting of the tympanic membrane and an air-filled chamber lined with mucus, which contains the malleus, incus, and stapes.

middle ear bones the *malleus, incus,* and *stapes.*

middle lamella a layer of cementing material between adjacent plant cell walls.

midgut 1. the middle part of the embryonic alimentary canal from which the intestines develop. 2. the central portion of the three-part insect gut.

midpiece the central region of a spermatozoan, containing the mitochondria and centriole.

midrib in plants, the large central vein of a dicot leaf, containing vascular and supporting fibrous tissue.

migration the act or process of moving periodically from one region to another, as in certain animals.

millipede a nonpoisonous, herbivorous, terrestrial anthropod of the class Diplopoda, with a long cylindrical, segmented body, and two pairs of legs per segment.

mineral nutrient an inorganic compound, element or ion needed for normal growth of all organisms.

mineralocorticoid any mineral-regulating steroid hormone of the adrenal cortex.

minimal medium the least complex medium capable of sustaining the growth of a specific microorganism.

minimum mutation tree in systematics, a hypothetical phylogenic tree selected because it represents the smallest number of evolutionary changes needed to account for known relationships.

mitochondrian (*pl.,* mitochondria) self-replicating, membrane-bounded eukaryotic organelle, functions in oxidative respiration and chemiosmotic phosphorylation.

mitosis nuclear division in eukaryotes, involving chromosome condensation, spindle formation, precise alignment of centromeres, and the regular segregation of daughter chromosomes to produce identical daughter nuclei; followed by cytoplasmic division.

model a contrived biological mechanism that, when applied, is expected to yield data consistent with past observations; a biological hypothesis with mathematical predictions.

mole in chemistry, the quantity of a substance whose weight in grams is equal to the substance's molecular weight, or Avogardro's number: 6.023×10^{23} molecules of a substance; the combined atomic weights of all the atoms of a substance, expressed in grams.

molecular orbital the shell formed by the shared pair of electrons in a covalent bond.

molecule a unit of chemical substance consisting of atoms bound by covalent bonding.

molecular clock an analogy based on the proposition that the rates at which specific mutations occur is regular and clocklike and can thus be used to establish an evolutionary time framework.

mollusca the phylum of invertebrates typically having a hard shell to some degree enclosing the body, such as the chitons, snails, and squids.

mollusk any invertebrate of the phylum Molluska, typically having a hard shell that wholly or partly encloses the soft, unsegmented body, such as the chitons, snails, bivalves, squids, etc.

molting See *ecdysis.*

Monocotyledonae the class of flowering plants characterized by seeds with only one cotyledon (for example, grasses, palms, and lilies).

monosaccharide "single sugar"; the molecular carbohydrate subunit; a simple sugar (e.g., glucose, fructose).

monotreme a platypus, echidna, or extinct egg-laying mammal of the order Monotremata.

mons veneris in women, a rounded usually hairy bulge of fatty tissue over the pubic symphysis and above the vulva.

morph one of the particular forms of an organism that exists in two or more distinct forms in a single population.

morphogenesis the emergence of final form and structure in the embryo.

morula an early embryonic state consisting of a ball of cells.

motor neuron a neuron that innervates muscle fibers; its impulses stimulating muscle contraction.

Monera the kingdom of prokaryotes, comprised of the bacteria.

monoamine a brain neurotransmitter, composed of a single amino acid group, such as norepinephrine, dopamine, histamine, and serotonin.

monoclonal antibody a specific immunoglobin produced by hybrid cells artificially cloned in the laboratory.

monocot a flowering plant of the angiosperm class Monocotyledonae, characterized by producing seeds with one cotyledon; included are palms, grasses, orchids, lilies, irises, and others.

monocotyledon See *monocot.*

monocyte a large, phagocytic leukocyte, formed in bone marrow and in the spleen.

monophyletic *adj.,* of a taxonomic group, deriving entirely from a single ancestral species.

Morula an early embryonic stage in which the cells are a solid ball.

motor neuron a neuron that innervates effectors such as muscles or glands.

motor unit a motor neuron together with the muscle fibers it innervates, which contract as a unit.

M phase See *cell cycle.*

mucin any of a class of muco-proteins that bind water and form thick, slimy, viscid fluids in various secretions.

multiple alleles the alleles of a single gene locus when there are more than two alternatives in a population.

multiplicative law of probability in mathematical probability theory, the statement that the probability of two independent outcomes both occurring is equal to the product of their individual probabilities.

muscle a contractile tissue, in vertebrates including skeletal, smooth, and cardiac muscle.

muscle fiber in skeletal muscle, one of the multinucleate cells, which takes the form of a long contractile cylinder.

mutagen a chemical or physical agent that causes mutations.

mutation any abnormal, heritable change in genetic material.

mutualism a mutually beneficial association between different kinds of organisms; a form of symbiosis beneficial to both partners; compare *commensalism, parasitism.*

mutualistic symbiosis. See *mutualism.*

mycelium (*pl., mycelia*) the mass of interwoven hyphae that forms the vegetative body of a fungus.

mycoplasma a tiny, nonmotile, wall-less bacterium of irregular shape, occurring as an intracellular parasite of animals and plants; smallest cellular organism.

mycorrhiza a mutualistic fungus-root association with the fungal mycelium either surrounding or penetrating the roots of a plant; also *mycorrhizal association.*

myelin a soft, white, somewhat fatty, material derived from plasma membrane, that forms a myelin sheath around certain nerve axons.

myelin sheath a fatty sheath surrounding the axons of some vertebrate neurons.

myocardial infarction the necrosis of the muscular substance of the heart caused by blood deprivation.

myofibril a tubular subunit of muscle fiber structure, consisting of many myofilaments organized into sacromeres, the contractile units.

myofilaments the highly organized fibrous proteins of striated muscle, including the thin, movable *actin myofilaments* and the thicker, stationary *myosin myofilaments.*

myosin a protein involved in cell movement and structure, especially in muscle cells; see also *myofilament.*

myosin cross bridge an ATP-activated, movable connection between myosin and actin which is responsible for muscle contractions.

myosin head a globular projection of the protein myosin, which forms the myosin cross bridge.

myotome 1. the portion of a vertebrate embryonic somite from which skeletal musculature is developed. 2. one of the muscular segments of the body wall of a fish or lancelet.

NAD (nicotinamide adenine dinucleotide) "nad"; a coenzyme that is an electron and hydrogen carrier in glycolysis and cell respiration.

NADP (nicotinamide adenine dinucleotide phosphate) "nad-phosphate"; a co-enzyme which is an electron and hydrogen carrier in photosynthesis.

nasal epithelium. See *olfactory epithelium.*

natural killer cell (NK cell) a nonspecific phagocyte that destroys cancerous and otherwise diseased cells.

natural selection the differential survival and reproduction in nature of organisms having different heritable characteristics, resulting in the perpetuation of those characteristics and/or organisms that are best adapted to a specific environment. Also called *survival of the fittest.*

navigation the act or process of finding one's way.

Neanderthal an extinct, highly variable type or race of *Homo sapiens* of the middle Paleolithic.

nectary a flower gland whose nectar secretions attract animal pollinators.

negative feedback. See *feedback.*

negative feedback loop a mechanism in which an action directly or indirectly reduces the stimulus.

nekton aquatic animals that can swim.

nematocyst one of the minute stinging cells of the coelenterates, consisting of a hollow thread coiled within a capsule and an external hair trigger.

nematode an unsegmented worm of the phylum or class Nematoda, having an elongated, cylindrical body; roundworm.

nephridium (*pl.,* nephridia) an annelid excretory organ, occurring paired in each body segment and typically consisting of a ciliated funnel and a duct to the exterior.

nephron a single excretory unit of a kidney, consisting of a glomerulus, Bowman's capsule, proximal convoluted tubule, loop of Henle, and distal convoluted tubule, loop of Henle, and distal convoluted tubule.

neritic province the coastal sea from the low-tide line to a depth of 100 fathoms; generally, waters of the continental shelf.

nerve a number of neurons following a common pathway, covered by a protective sheath and supporting tissue.

nerve cell See *neuron.*

nervous system the brain, spinal cord, nerves, ganglia, and the neural parts of receptor organs, considered as an integrated whole.

net community productive (NCP) in ecology, the rate at which a biotic community gains in biomass or stored energy; determined by subtracting the total energy released in respiration from the total energy incorporated.

net productivity (NP) in ecology, the rate at which producers store energy or biomass; determined by subtracting the rate of energy release or biomass during respiration from the rate of energy or biomass incorporation.

neural crest a ridge of embryonic cells that will contribute to the central nervous system.

neural fold In vertebrate embryos, one of a pair of longitudinal ridges that arise from the neural plate on either side of the neural groove later folding over to form the neural tube.

neural ganglia clumps of neurons.

neural folds in early vertebrate embryology, a pair of longitudinal ridges that arise from the neural plate on either side of the neural groove and that fold over and give rise to the neural tube, which eventually becomes the spinal cord.

neural groove an ingrowth of the neural plate of a vertebrate embryo; eventually forms the spinal cord in mammals.

neural impulse a transient membrane depolarization, followed by immediate repolarization, traveling in a wavelike manner along a neuron.

neural plate in embryology, a thick plate formed by rapid cell division over the notochord on the dorsal side of the embryo; eventually forms the nervous system in mammals.

neural tree 1. in early vertebrate embryology, the hollow dorsal tube formed by the infusion of the neural folds over the neural groove. 2. the spinal cord.

neural tube 1. in early vertebrate embryology, the tubular dorsal structure formed by the fusion of the neural folds over the neural groove.

neuroglia See *glial cell.*

neuromuscular junction the synapse between a neural motor end plate and a muscle fiber.

neuron a cell specialized for the transmission of nerve impulses.

neuropeptide a brain neurotransmitter composed of short chains of amino acids varying in number from 2 to 40.

neurotransmitter a short-lived, hormone-like chemical such as acetylcholine that stimulates a second neuron to transmit a nerve impulse. See also *synapse, synaptic cleft.*

neurula an early vertebrate embryo slightly older than a gastrula, in which the neural tube has formed.

neurulation the development of a neurula from a gastrula including formation of the neural plate, neural folds, and neural tube.

neutralist one who argues that a substantial proportion of the protein and DNA changes in evolution have been due to selectively neutral mutations fixed by random drift.

neutral mutation mutational change in DNA that has no measurable effect on the fitness of an organism, and which may sometimes be incorporated into the genome of a species. Also called *selectively neutral mutation.*

neutron one of the two common components of an atomic nucleus; has no charge and no effect on chemical reactions.

neutrophil the most common mammalian phagocytic leukocyte.

niche See *ecological niche.*

nitrification the chemical conversion of ammonia to nitrites and nitrates by the action of soil bacteria.

nitrogenase an enzyme that catalyzes nitrogen fixation.

nitrogen base a purine or pyrimidine used in the synthesis of nucleic acids and other molecules; specifically adenine, cytosine, guanine, thymine or uracil.

nitrogen cycle the cyclic transfer of nitrogen compounds in a biogeochemical cycle, including the gain of nitrogen compounds through nitrogen fixation and lightning and their loss through denitrification.

nitrogen fixation the conversion of atmospheric nitrogen (N_2) to more readily utilized forms; usually by the action of cyanobacteria and other nitrogen-fixing bacteria, but sometimes by lightning, automobile engines, or the industrial preparation of synthetic fertilizers.

nitrogen fixer any of the bacteria or cyanobacteria which are able to use atmospheric nitrogen, combining it with hydrogen to produce ammonium ions.

node 1. in plants, region of leaf attachment in the stem; see *internode.* 2. in neurons, a constriction in myelin sheath, corresponding to a gap between successive Schwann cells; see *saltatory propagation.*

noncyclic photophosphorylation light reactions employing both photosystem II and photosystem I during which electrons from water pass through both photosystems reducing NADP to NADPH and H^+, thus serving both chemiosmosis and carbohydrate synthesis. Also *noncyclic events.*

nondisjunction the failure of homologous chromosomes to segregate properly in meiosis, resulting in daughter cells with extra or missing chromosomes.

nonspecific defenses any of several cellular and chemical responses by the body against foreign substances, cancerous cells, or invading organisms, but not involving B-cell and T-cell lymphocytes.

norepinephrine a compound that serves as a synaptic neurotransmitter and as an adrenal hormone. Also called *noradrenaline.*

normal distribution the idealized, symmetrical distribution taken by a population of values centering on a mean, when departures from the mean are due to the chance occurrences of a large number of individually small independent effects; often approached in real populations. Also called *bell-shaped curve.*

Norplant a contraceptive measure in which slow release hormones are implanted beneath the skin.

notochord a turgid, flexible rod running along the back beneath the nerve cord and serving as a body axis; exists in all chordates at some point in development, but is replaced in most vertebrates by the vertebral column.

nuclear envelope the double cellular membrane surrounding the eukaryote nucleus, the outermost of which is continuous with the endoplasmic reticulum. Also called *nuclear membrane.*

nucleic acid either DNA or RNA; DNA is a double polymer of deoxynucleotides and RNA is a single polymer of nucleotides.

nucleic acid hybridization through base-pairing the union of single strands of DNA from different sources, as occurs when DNA probes are used. See *DNA probe.*

nucleolar organizing region within the nucleus, multiple copies of a DNA loop that transcribes ribosomal RNA.

nucleolus (*pl.,* nucleoli) a conspicuous, dark-staining region of DNA, RNA, and protein within the cell nucleus.

nucleotide a compound consisting of a nitrogenous base and a phosphate group linked to the l′ and s′ carbons of ribose or deoxyribose, respectively; the repeating subunit of DNA or RNA.

nucleus 1. in all eukaryote cells, a prominent, usually spherical or elipsoidal double membrane-bounded organelle containing the chromosomes and providing physical separation of the DNA and the cytoplasm. 2. a clump of neural cell bodies that develop in the same way and that innervates the same region.

occipital lobe one of the four major lobes of the brain; involved in the reception and processing of visual information.

oceanic province the open sea, as distinguished from the neritic province.

Okazaki fragment during DNA replication in the lagging strand, an assembly of some 200–300 bases, incorporated through use of the enzyme ligase.

olfaction 1. the sense of smell. 2. the process of smelling.

olfactory bulb an extension of the brain that receives neurons from the olfactory receptors in the nasal passage.

olfactory epithelium the chemically sensitive neural tissue lining of the nasal cavity.

oligochaeta a member of the class of annelids that have locomotory setae sunk directly into the body wall, such as earthworms.

oligodendrocytes a neuroglial cell that forms the myelin sheath over axons of the central nervous system.

oligotrophic *adj.,* a lake rich in dissolved oxygen and poor in plant and algal requirements; producer growth is thus held in check, and the lake ages very slowly. Compare *eutrophic.*

ommatidium (*pl.,* ommatidia) one of the elements of compound eye, consisting of a corneal lens, crystalline cone; rhabdome, light-sensitive retinula, and sheathing pigment cells.

omnivore "eating everything"; an organism that feeds on both animal and plant material.

oocyte an egg cell before maturation: *primary oocyte,* a diploid cell precursor of an egg before meiosis; *secondary oocyte,* an egg cell after formation of the first polar body.

oogenesis the meiotic process that results in the production of an egg cell or ovum.

oogonium a cell that gives rise to oocytes, the large, spherical, unicellular female sex organ of some protists in which egg cells are produced.

open circulatory system a circulatory system in which blood passes from vessels into intercellular spaces before returning to the heart. Compare *closed circulatory system.*

open life span See *indeterminate life span.*

operant conditioning in behavior, learning through conditioning in which the reward or reinforcement follows a particular response that is desired over other possible responses.

operator See *operon.*

operculum a lid or cover, e.g., the skin-covered bony plates that cover the gill chambers of a bony fish.

operon in prokaryotes, a gene-controlling mechanism that includes structural genes and the genes controlling them; transcription may be *inducible,* remaining shut down until activated by an inducer substance, or *repressible,* remaining active until shut down by a repressor substance. See *lac operon.*

opposable thumb a condition in which the thumb can touch the other fingers.

opsin one of the two chemicals that rhodopsin breaks down into when exposed to light.

optic nerve the nerve carrying impulses from the retina to visual centers in the brain.

orbital the approximate and hypothetical path on which an electron moves around the nucleus of an atom.

organ a distinct structure that consists of a number of tissues and carries out a specific function.

organelle a specialized part of a cell.

organismis respiration the physical processes by which oxygen and carbon dioxide are exchanged by the body.

organ of Corti on the basilar membrane in the cochlea, an organ containing the neural receptors for hearing.

organ system a number of organs participating jointly in carrying out a basic function of life (e.g., respiration, excretion, reproduction, digestion).

orgasm in humans, the climax of sexual excitement and pleasure, usually accompanied in men by ejaculation and in women by rhythmic contractions of the cervix.

orientation 1. the ability to determine direction in one's environment; may be part of an instinctive action.

orienting movement in behavior, the directing of bodily position according to the location of a particular stimulus; may be part of an instinctive action.

origin 1. evolutionary ancestry. 2. the nonmoving, skeletal base to which a muscle or tendon attaches. Compare *insertion.*

osculum an opening in a sponge through which water exits.

osmoconformer an aquatic organism that does not regulate the solute content of its body tissues, but allows it to fluctuate with that of the environment.

osmoregulation in aquatic organisms, the retention of a specific solute content in the body in spite of fluctuations in the salinity of the environment.

osmoregulator an aquatic organism that osmoregulates. See *osmoregulation.*

osmosis the movement of water across a membrane from an area of high water potential to an area of lower water potential when the difference was brought about by the presence of solutes.

osmotic pressure the actual hydrostatic pressure that builds up in a confined fluid because of osmosis; the amount of force necessary to equal the water potential.

Osteichthyes the class composed of bony fishes.

osteoblast a bone-forming cell.

osteoclast a bone-destroying ameboid cell that dissolves calcium phosphate.

osteocyte a bone cell isolated in a lacuna of bone tissue.

ostia in an arthropod's open circulatory system, the tiny openings through which blood reenters the heart.

ostracoderm any of a group of extinct jawless fish.

outer ear the normally visible part of the ear.

oval window in the cochlea, a membrane articulating with the stapes that moves in response to its vibrations.

ovarian cycle See *menstrual cycle.*

ovarian follicle the cluster of cells in which a mammalian oocyte matures, and which later gives rise to a corpus luteum.

ovary 1. in animals, the (usually paired) organ in which oogenesis occurs and eggs mature. 2. in flowering plants, the enlarged, rounded base of a pistil, consisting of a carpel or several united carpels, in which ovules mature and megasporogenesis occurs.

overshoot the condition in which a rapidly growing population rises above the environment's carrying capacity so drastically that it places harsh pressure on its own numbers, frequently leading to a population crash.

oviduct a tube, usually paired for the passage of eggs from the ovary toward the exterior or to a uterus often modified for the secretion of a shell or a protective membrane; in humans it is also known as a *fallopian tube.*

oviparity *n.,* the condition of producing eggs that mature and hatch outside the body, common to many vertebrates other than mammals.

oviparous See *oviparity.*

ovipositor in invertebrates, an organ specialized for the depositing of eggs and often for boring holes in which eggs may be deposited.

ovoviviparity *n.,* the condition of producing eggs which are hatched within the body so that the young are born alive but without placental attachment, as in certain reptiles, fishes, etc.

ovoviviparous. See *ovoviviparity.*

ovulation the release of one or more eggs from an ovary.

ovule in seed plants, an oval body in the ovary that contains the megagametophyte mother cell and later the female gametophyte.

ovum the female reproductive cell or gamete of animals which is capable of developing, usually only after fertilization, into a new individual.

oxidation 1. the loss of electrons from an element or compound. 2. the addition of oxygen to an element or compound. 3. a multistep process in which oxygen is added or hydrogen or electrons are removed (e.g., the oxidation of glucose).

oxidative *adj.,* requiring oxygen.

oxygen debt a state of oxygen depletion after extreme physical exertion; measured by the amount of oxygen required to restore the system to its original state.

oxytocin a polypeptide hormone of the posterior pituitary that stimulates the contraction of the uterus and the release of milk.

pacemaker See *sinoatrial (SA) node.*

pacinian corpuscle an oral pressure receptor of the skin, containing the ends of sensory neurons, especially in the hands and feet.

palisade parenchyma in plants, a tissue of a lengthy, vertically arranged photosynthetic cell in the leaf, forming a closely packed layer (or layers) just below the upper epidermis.

pancreas a large digestive and endocrine gland of vertebrates, which secretes various digestive enzymes and hormones.

pancreatic islet cells See *islets of Langerhans.*

papillary muscle one of the small bundles of muscles attached to the ventricle walls and to the chordae tendineae that tighten those tendons during ventricular contractions.

paradigm a broad, major concept in science, usually representing a new way of viewing natural phenomena (e.g., the theory of relativity in physics, continental drift in geology, the central dogma in biology).

parasitism a symbiotic relationship in which an organism of one kind (the parasite) lives in or on an organism of another kind (the host), generally to the host's detriment. Compare *commensalism, mutualism.*

parasympathetic division that part of the autonomic nervous system whose nerves emerge from the brain and lower spinal cord and whose action generally shows activity in the viscera. Compare *sympathetic division.*

parathyroid glands four small endocrine glands embedded in or adjacent to the thyroid gland and involved in the regulation of body calcium.

parathyroid hormone (PTH) the internal secretion of the parathyroid glands, involved in maintaining calcium balance.

Parazoa phylum Porifera, the sponges. See also *Metazoa.*

parenchyma the fundamental tissue of plants, composed of thin-walled cells and commonly specializing in photosynthesis and storage.

parietal cells large cells of the stomach lining that secrete hydrochloric acid.

parietal lobe one of the four major lobes of the cortex; the detection of body division and sensory input are the major functions.

partial dominance where the combined expression of two alleles in the heterozygote produces an intermediate trait (i.e., in the blossoms of four o'clocks, red + white = pink).

passive immunization short-lived immunity resulting from the injection of an antiserum containing specific immunoglobins against the invading agent.

passive transport in organisms, the movement of fluids, solutes, or other materials without the expenditure of ATP energy (e.g., by diffusion), especially across a membrane.

pathogen any disease-producing organism.

pathogenic *adj.,* disease-producing.

pectoral girdle the bones and cartilage supporting and articulating with the vertebrate forelimb; in humans, consisting of the clavicle and scapula.

pedipalps in arachnids, the pair of usually longer appendages immediately behind the chelicerae.

pelvic girdle the bones and cartilage supporting and articulating with the vertebrate hindlimbs; in humans, consisting of the fused bones of the pelvis.

penis the male intromittent sex organ in species where internal fertilization is carried out through copulation.

pentaradial symmetry a modified radial symmetry in which there are five repeated radial parts; seen in echinoderm adults.

pentose a five-carbon monosaccharide (i.e., ribose, deoxyribose).

pepsin an enzyme secreted as *pepsinogen* by glands of the stomach lining that is active only at low pH and that acts primarily to reduce complex proteins to simple polypeptides.

pepsinogen the initial, inactive form of pepsin as occurring in and secreted by gastric glands that readily converts to pepsin in an acid medium.

peptide bond the dehydration linkage formed between the carboxyl group of one amino acid and the amino group of another; also *peptide linkage,* a covalent bond.

peptidoglycan chemical material of the cell wall in Eubacteria; consists of sugars and short peptide strands cross-linked by strands of the amino acid glycine.

peptidyl transferase during translation, the enzyme involved in the formation of peptide bond between adjacent amino acids in the growing polypeptide.

perennial in plants, a species that lives for an indefinite number of years. Compare *annual, biennial.*

pericycle a layer of tissue that sheaths the stele of the root and is associated with the formation of lateral roots.

periosteum tough connective covering of bone.

peripheral nervous system in vertebrates, all neurons outside the central nervous system.

peristalsis successive waves of involuntary contractions passing along the walls of the esophagus, intestine, or other hollow muscularized tubes, forcing the contents onward.

peritoneum the smooth, transparent membrane lining the abdominal cavity of a mammal.

peritubular capillaries in the kidney, a capillary bed surrounding the nephron and involved in tubular reabsorption and secretion, related to osmoregulation and excretion.

permafrost in arctic and high-altitude tundra, the permanently frozen layer of soil or subsoil or both.

permeability the degree to which materials are able to pass through a substance, membrane, or barrier.

permease a membranal carrier that functions in facilitated transport of a specific substance across a plasma membrane.

peroxisome a cytoplasmic organelle involved in the detoxification of peroxides.

petal one of the usually white or brightly colored leaflike elements of the corolla of a flower.

petiole in dicot plants, the small stalk that emerges from the stem and supports the leaf.

phagocytosis engulfment of solid materials into the cell and the subsequent pinching off of the cell membrane to form a digestive vacuole.

phanerogam any of the Phanerogamia, a former primary division of plants comprising those having readily visible reproductive organs such as cones or flowers; a seed plant.

pharyngeal gill arches See *gill arches.*

pharynx 1. in most vertebrates, the cavity between the mouth and the esophagus (contains the gills in fishes); 2. an analogous region in the alimentary canals of various invertebrates, including some in which it is reversible and toothed.

phenotype the final effect of the total interaction of genes with each other and with the environment, expressed in the individual.

pheromone a substance that is released into the environment by one individual and affects the behavior of another (i.e., the sex attraction in moths).

phloem a complex vascular tissue of higher plants that consists of sieve tubes, companion cells, and phloem fibers, and functions in transport of sugars and other solutes, e.g., phloem sap.

phloem parenchyma part of the phloem tissue complex; parenchymal cells that lie adjacent to phloem sieve tubes.

phosphate ion an ion of phosphoric acid.

phospholipid any of a class of phosphate-esterified lipids, including *lecithin, cephalin,* and *sphingomyelin;* a major component of cell membranes.

phosphorus cycle the pathway of phosphorus as it is taken up and released by organisms into the physical environment.

phosphorylation reaction the enzymatic addition of a phosphate group to a compound, e.g., the addition of phosphate to ADP yielding ATP and water.

photon a "packet" (quantum) of electromagnetic radiant energy; a unit of light energy.

photoperiodism the response of an organism to the length of daylight or dark periods (photoperiods), involving sensitivity to the onset of light or darkness and a capacity to measure time.

photophosphorylation phosphorylation that utilizes light as an initial source of energy.

photorespiration in C3 plants, the addition of oxygen rather than carbon dioxide to ribulose diphosphate and the subsequent loss of Calvin cycle components.

photosynthesis in phototrophic organisms, the organized capture of light energy in photosystems and its transformation into chemical bond energy in carbohydrates (glucose and other compounds).

photosystem in the thylakoid, light-harvesting antennas and electron/hydrogen transport systems, both of which function in the light reactions of photosynthesis; also *photosystem I (P700)* and *photosystem II (P680).*

photosystem I (P700) the second in the two photosystems in the electron pathway of photosynthesis in cyanobacteria and chloroplasts, and in the cyclic reactions, the one involving the reduction of NADP to NADPH2.

photosystem II (P680) the first of two photosystems in the electron pathway of photosynthesis in cyanobacteria and all photosynthetic eukaryotes, and the one involving the photolysis of water.

phototaxis a tendency to move toward light.

phototroph an autotrophic organism that derives its energy initially from light; a photosynthetic organism.

phototropic response See *phototropism.*

phototropism the growth response of a plant to light.

pH scale a common measure of the acidity or alkalinity of a liquid, based on hydrogen ion concentration.

phyletic speciation the linear change over time of one species into another.

phylum 1. a major taxonomic unit of related, similar classes of animals, e.g., phylum Annelida; 2. a division of the plant kingdom.

phytochrome a red-light and far-red-light-sensitive protein complex of certain plant cell membranes; involved in many light-induced phenomena, including flowering, leaf formation, and seed germination.

phytoplankton in the aquatic environment (marine and fresh waters), minute photosynthesizing organisms, such as diatoms; the base of the marine food chain.

piezoelectric crystal crystalline structures, including the keratin of skin and hair, that when touched or otherwise deformed generate weak electrical currents.

piloerection the lifting up of hair by tiny involuntary muscles in response to cold or fright; bristling.

pineal gland also called *pineal organ, pineal body;* a small body in the brain directly sensitive to light in reptiles and birds.

pinna in common terms, the "ear."

pinocytic vesicle inpocketing of the plasma membrane of a cell formed as pinocytosis (cell "drinking") occurs.

pinocytosis a form of endocytosis in which dissolved molecular food materials are taken into the cell by an inpocketing of the plasma membrane. See also *pinocytic vesicle.*

pioneer organism during ecological succession, a type of organism specialized for the initial invasion of an uninhabited or seriously disturbed area (e.g., a rocky outcropping, landslide, or burned-out region).

pistil in flowering plants, the female reproductive structure, composed of one or more carpels and ovaries, and a style and a stigma. See also *carpel* and *gynoecium.*

pith thin-walled parenchymous tissue in the central strand of a stem's primary growth; the dead remains of such tissue at the center of a woody stem.

pit pairs also called *pits;* the thin, porelike regions in adjacent plant cell walls where only the primary cell wall remains.

pituitary also called *hypophysis;* a small, double endocrine gland lying just below the brain and intimately associated with the hypothalamus in all vertebrates, consisting of an anterior and a posterior lobe. See also *anterior pituitary* and *posterior pituitary.*

pituitary dwarf an abnormally small individual resulting from the failure of the anterior pituitary to secrete growth hormone.

pituitary giant an abnormally large individual whose excessive growth is due to excessive secretions of growth hormone.

placenta in mammals other than monotremas and marsupials, the organ formed by the union of the endometrium and the extraembryonic membranes of the fetus; it provides for nourishment of the fetus, elimination of waste products, and the exchange of dissolved gases.

placental mammal mammals that provide support to the embryo via the placenta.

Placodermata an extinct class of vertebrates containing the armored fishes; the first jawed fishes.

plankton tiny, drifting plant, algal, and animal life in both marine and fresh waters.

Plantae the plant kingdom. Plants are nonmotile (stationary), multicellular, photosynthetic autotrophs with cell walls of cellulose, having highly specialized tissues and organs.

planula larva the early, ciliated, free-swimming form of a coelenterate.

plaque 1. pathological deposit of lipid, fibrous material and often calcium salts in the inner wall of a blood vessel; 2. a film of bacterial polysaccharide, mucus, and detritus harboring dental bacteria.

plasma the fluid matrix of blood tissues; it is 90% water and 10% various other substances, including plasma proteins, ions, and foods.

plasma cell a mature, short-lived B-cell lymphocyte specialized in secreting antibodies.

plasma membrane the external, semipermeable, limiting layer of the cytoplasm.

plasmid in bacteria, a small ring of DNA that occurs in addition to the main bacterial chromosome and is transferred from host to host; also used in recombinant DNA techniques.

plasmodesma (*pl.,* plasmodesmata) in plants, minute cytoplasmic junctions between cells, occurring at pores in the cell wall through which the plasma membrane of one cell becomes continuous with that of the next; highly significant to transport.

platelets minute, fragile noncellular discs present in the vertebrate blood; upon injury they are ruptured, releasing factors that initiate blood clotting and wound healing.

plastid any of several forms of self-replicating, semiautonomous plant cell organelles, including *chloroplast* (specialized for photosynthesis), *chromoplast* (specialized for photosynthesis), *chromoplast* (specialized for pigmentation), and *leukoplast* (specialized for starch storage).

platelet any one of the tiny fragile, noncellular discs present in the vertebrate blood. Upon injury, they are ruptured, releasing substances that initiate blood clotting.

plate tectonics the movement of great land and ocean floor masses (plates) on the surface of the earth relative to one another. See also *continental drift.*

Platyhelminthes the phylum including worms with bilateral symmetry and a soft, usually flattened body, such as the planarians, tapeworms, and trematodes.

pleura *pl.,* pleurae in mammals, the tough, clear, serous connective tissue membrane covering a lung and lining the cavity in which the lung lies.

plumule the apex of certain plant embryos, consisting of immature leaves and an epicotyl.

pogonophoran one of the large, red tubeworms found near gas vents on the ocean bottom.

poikilotherms organisms with body temperatures that fluctuate with the temperature of the environment.

point mutation a mutation involving a minor change in a DNA sequence, such as a base substitution, addition, or deletion.

polar referring to the charged portion of a chemical or part of a molecule capable of forming hydrogen bonds with water and other polar molecules.

polar body a small, functionless daughter cell produced by the highly unequal cleavage during meiosis I and II in oogenesis.

polarity in chemistry, the presence or manifestation of two opposite or contrasting charges.

polarization the process in which the outside of a resting neuron becomes positively charged relative to the inside.

polarized in a neuron, the period when the region outside a neuron is positive and the region inside is negative, producing a membrane voltage potential of –60 mV. See also *resting state.*

pollen the male gametophyte of a seed plant; contains a generative nucleus and a tube nucleus and is enclosed in a hardened, resistant case.

pollen grain See *pollen.*

pollen sac one of two or four chambers in an anther in which pollen develops and is held.

pollen tube a tube that extends from a germinating pollen grain and grows down through the style to the embryo sac, into which it releases sperm nuclei.

pollen tube cell in plant reproduction, the cell directly behind the pollen tube as it grows through the style of a flower before actual fertilization by one of the nuclei that follow it.

pollination the transfer of pollen from a stamen to a stigma, preceding fertilization in a flowering plant.

Polychaeta the class including annelids with unsegmented swimming appendages with many chaetae or bristles.

polychaete a member of the class of annelids having unsegmented appendages with many chaetae or bristles.

polydactyly the genetically derived state of having extra fingers or toes.

polygenic inheritance inheritance involving many interacting variable genes, each having a small effect on a specific trait.

polymer a molecule made up of a string of more or less identical subunits (e.g., starch, nucleic acid, polypeptide).

polymerization the combination of many like or unlike molecules to form a more complex product of higher molecular weight.

polyp the typical attached, nonswimming form of coelenterate. Compare *medusa.*

polypeptide a strand of amino acids linked by peptide bonds, longer than a peptide but not usually a complete, functional protein. See also *protein.*

polyphyletic of a taxonomic group, having two or more ancestral lines of origin.

polyploidy the condition of having a chromosome number that is more than double the basic or haploid number.

polyribosome several ribosomes simultaneously transcribing the same messenger RNA strand; also called *polysome.*

polysaccharide "many sugar"; a polymer of sugar subunits.

polysome See *polyribosome.*

polytene chromosome in certain insects, enlarged chromosomes created by chromatin replication without chromosome division.

pome a simple fleshy fruit, formed in its outer portions by the floral parts themselves, i.e., apple, pear.

P_1 generation in genetics, the first generation of parents in cross-breeding; subsequent generations are designated P_2, P_3, etc.

pons a broad mass of nerve fibers running across the ventral surface of the mammalian brain.

population 1. the total number of persons inhabiting a given geographical or political area (demography); 2. an aggregate of individuals of one species, interbreeding or closely related through interbreeding and recent common descent, and evolving as a unit (genetics); 3. the assemblage of plants or animals or both living in a given area; or all of the individuals of one species in a given area (ecology).

population crash the usually rapid depletion of a population that has exceeded the carrying capacity of its environment.

population genetics the scientific study of genetic variation within populations, of the genetic correlation between related individuals in a population, and of the genetic basis of evolutionary change.

population growth rate in population dynamics, the change in the percentage of growth over time. See *exponential growth.*

Porifera the phylum composed of the sponges.

porocyte a pore-bearing cell in the body wall of a sponge, specialized for admitting water.

positive feedback a system in which the output of a system further increases the flow of output.

positive feedback mechanism occurring when the result of an activity causes that activity increase, thereby further increasing the action; relatively uncommon in nature.

postanal tail a tail that extends from the anus posteriorally.

posterior pituitary also called *neurohypophysis;* not actually a gland but the enlarged, glandlike termini of axons of cell bodies in the hypothalamus, the hormones being synthesized in the cell bodies and translocated to the posterior pituitary within the axons, to be stored pending release; among these are *oxytocin* and *antidiuretic hormone.*

postsynaptic membrane the receptive surface of a dendrite or receiving cell body adjacent to a synapse. See also *synapse.*

postsynaptic neuron the neuron that receives an impulse from a synapse.

posttranscriptional processing following RNA transcription, modification of raw transcripts through the removal of introns and other changes.

postzygotic the stages after the zygote stage.

postzygotic reproductive isolating mechanisms biological mechanisms that function after fertilization to create reproductive incompatibility between species; i.e., embryo death, sterile adults.

potassium gates See *ion channel.*

potential energy energy stored in chemical bonds, nonrandom organization, elastic bodies, elevated weight, or any other static form in which it can theoretically be transformed into another form or into work. Also called *free energy.*

precapillary sphincter in arterioles, rings of smooth muscle capable of regulating blood flow into capillaries; controlled by the autonomic nervous system.

prefrontal area in the brain, the anterior portion of the frontal lobe which functions in sorting sensory input.

prehensile adapted for seizing, grasping, or wrapping around, as the tail of a New World monkey or the upper lip of a rhinoceros.

pressure flow hypothesis an explanation of sap movement in the phloem based on the active transport of sugars into the phloem stream from the source, followed by an inward movement of water and the formation of hydrostatic pressure which pushes the stream to the sink.

presynaptic membrane the membrane of an axon or of the synaptic knob of an axon in the region of a synaptic cleft, into which it secretes neurotransmitters in the course of the transmission of a nerve impulse. See *synapse, synaptic cleft.*

presynaptic neuron a neuron that stimulates another neuron at a synapse.

prezygotic the stage before the zygote forms.

prezygotic reproductive isolating mechanisms biological mechanisms that operate through altered behavior, anatomy, and physiology to prevent fertilization, thereby producing reproductive incompatibility between species.

primary bronchus the left or right division of the trachea. See also *respiratory tree.*

primary chorionic villi slender, fingerlike projections sent into the endometrium by the trophoblast, providing connection between the embryo and the mother's tissues prior to placental formation.

primary consumer in ecology, a herbivore; an organism that feeds directly on producers.

primary growth the initial growth or elongation of a plant stem or root, resulting mainly in an increase in length and the addition of leaves, buds, and branches. Compare *secondary growth.*

primary immune response the relatively slow response of the immune system upon its first contact with an invading organism or foreign protein. Compare *secondary immune response.*

primary lesion a damaged or mismatched segment of DNA, subject to repair. See *DNA repair system.*

primary level in proteins, the first level of structure, the arrangement of amino acids into a simple, linear polypeptide.

primary oocyte a diploid cell of egg-producing potential prior to its first meiotic division. See also *oocyte.*

primary organizer in development, any tissue that induces or influences the future development of other subservient tissues.

primary phloem phloem developed from procambium, the phloem of primary growth.

primary producers in ecology, a plant, alga, or other photosynthetic (or chemosynthetic) organism that forms the base of a food chain.

primary productivity See *gross productivity.*

primary root the root of a plant that grows through the production of primary tissues.

primary succession the succession of vegetational states that occurs as an area changes from bare earth to a climax community.

primary xylem in primary growth, xylem produced by procambium rather than vascular cambium.

primate any member of the order Primates including humans, apes, monkeys, lemurs, tarsiers, and marmosets.

primitive streak in bird, reptile, and mammalian embryos, a thickening in the blastoderm formed by convergence of cells in preparation for gastrulation.

principle of competitive exclusion the ecological principle that no two species can coexist while occupying the same ecological niche.

procambium in plants, the primary tissue that gives rise to primary xylem and primary phloem.

producer See *primary producer.*

profundal zone the depths of a lake below the penetration of light.

progeny testing determination of an organism's genotype by crossing it with one of a known genotype and observing the resulting offspring.

progesterone an ovarian hormone, produced by the corpus luteum; assists in the preparation of the uterus for implantation of a fertilized egg and maintains the placenta during gestation.

proglottid one of many segments of a mature tapeworm, containing male and female organs and being shed when full of mature fertilized eggs.

prokaryote any organism of the kingdom Monera, including *Archaebacteria* or *Eubacteria;* its cells lack a membrane-bounded nucleus and membrane-bounded organelles. Compare *eukaryote.*

prolactin a hormone of the anterior pituitary that in mammals induces milk production.

promoter a DNA sequence to which RNA polymerase must bind in order for transcription to begin. See *operon.*

pronucleus the haploid nucleus of either gamete after the entry of a sperm into an egg and before nuclear fusion.

prophase the first stage of mitosis or meiosis, characterized by the condensation of chromosomes.

prop root in plants, an aerial adventitious root which functions to help keep the main stem erect.

prostacyclin a prostaglandin.

prostaglandin any of a group of hormone-like substances derived from long-chain fatty acids and produced in most animal tissues.

prostate gland a pale, firm, partly muscular and partly glandular organ that surrounds and connects with the base of the urethra in male animals; its viscid, opalescent secretion is a major component of semen.

protein one of the molecules of life, a functional macromolecule consisting of one or more polypeptides, often joined by disulfide linkages and frequently including one or more prosthetic groups.

proteinoid microspheres minute globes of protein that form spontaneously when proteins are concentrated in solution; possible precursors to the first cellular life.

prothrombin a blood plasma protein that is converted to the protein throbin during clotting.

Protista the polyphylitic kingdom of protozoa, algae, slime molds, and water molds.

protocell in the origin of life, a hypothetical cell containing the simplest organization and chemical substances needed to carry on life.

protoderm in plants, primary meristematic tissue; it gives rise to the epidermis of roots, leaves, and stems.

proton 1. one of the two particles composing the atomic nucleus in ordinary matter; it has an electrostatic charge of +1 and a mass 1837 times that of an electron; 2. a hydrogen ion.

proton pump an active transport system that uses energy to move hydrogen ions (protons) from one side of a membrane to the other against a concentration gradient, as in chemiosmosis.

protostome an animal in which the mouth derives from the region of the first embryonic opening (the blastopore). Compare *deuterostome.*

Protozoa the phylum containing single-celled, animal-like heterotrophs of the kingdom Protista.

proximal convoluted tubule See *nephron.*

proximate causation a causative factor that has a relatively immediate basis, such as phynotogical.

pseudocoelom in nematodes and rotifers, the body cavity between the body wall and the intestine that is not entirely lined with mesodermal epithelium.

pseudocoelomate animals that form pseudocoeloms.

pseudopod "false foot"; any temporary protrusion of the protoplasm of a cell serving as a structure of locomotion or engulfment.

P700. See *photosystem I.*

psilophyte any whisk fern; among the earliest known vascular plants, lacking roots, cambium, leaves, and leaf traces.

P680. See *photosystem II.*

pterophyta "winged plants"; ferns.

pubic joint. See *pubic symphysis.*

pubic symphysis in mammals, the usually semirigid fibrous articulation of the two pubic bones.

pubis also called *pubic bone;* one of the constituent bones of the coxa or innominate bone.

pulmonary artery the artery that carries deoxygenated blood from the right ventricle to the lungs.

pulmonary circuit blood vessels carrying blood to and from the lungs.

pulmonary semilunar valve a one-way valve at the base of the pulmonary artery that prevents back flow of blood into the right ventricle during systole.

pulmonary vein in birds and mammals, the vein that carries oxygenated blood, returning it from the lungs to the left atrium.

punctuated equilibrium a theory stating that evolution does not proceed in a gradual manner but rather in sudden bursts of activity, followed by very long time-intervals during which little evolutionary activity is seen.

Punnett square in genetics, a grid used to predict the outcome of cross-breeding.

pupa in insects, the period of development between the larval and adult stages, during which time extensive body transformations occur prior to the emergence of the adult; the pupa generally is enclosed in a hardened pupal case or cocoon.

pupil the contractile aperture in the iris of the eye.

purine a nitrogenous, double-ringed base of DNA or RNA consisting of five-membered and six-membered rings (e.g., adenine, guanine).

pyloric sphincters also *pyloric valve;* a ring of muscle capable of closing off the opening between the stomach and the small intestine.

pyramid of biomass ecological pyramid based on the dry weight of the total living matter.

pyramid of energy See *energy pyramid.*

pyramid of numbers an ecological pyramid based on relative numbers of producers and consumers.

pyrimidine a nitrogenous, single-ringed base of DNA or RNA consisting of a six-membered ring (e.g., cytosine, uracil, thymine).

pyruvate the ionized form of pyruvic acid, the final organic product of glycolysis in animal cells and the initial substrate of the citric acid cycle; also *pyruvic acid* (nonionized).

pyruvic acid See *pyruvate.*

quadrupedal *adj.,* four-footed.

quaternary structure in protein, the interaction of two or more polypeptides through disulfide linkages; the fourth level of structure.

radial cleavage in deuterostome embryos, a radial pattern in the early cleavages in which the cleavage planes come to lie one atop the other. See also *spiral cleavage.*

radial symmetry circular or spherical body symmetry where a radius cut through any part will intersect the same body parts (e.g., jelly fish, comb jelly).

radiates those organisms which are disk-shaped, spherical, or cylindrical; a line across any radius divides the animal into similar halves.

radicle the lower portion of the axis of a plant embryo, including the part that will become the root.

radioactive relating to the phenomenon exhibited by certain elements of spontaneously emitting radiations resulting from changes in the nuclei of atoms of the elements.

radioisotope an unstable isotope that spontaneously breaks down with the release of ionizing radiation; also called *radioactive isotope.*

radiolarian any minute marine protozoan of the phylum Sarcomastigophora, having an amebalike body with radiating pseudopods and an elaborate skeleton

radius one of the two bones of the forearm of land vertebrates, rotating about the ulna and articulating with the wrist.

radula in all mollusks except bivalves, a toothed, chitinous band that slides backward and forward, scraping and tearing food and bringing it into the mouth.

random drift See *genetic drift.*

range the geographical area occupied by a species.

rapid eye movement (REM) sleep that part of sleep characterized by a high degree of relaxation in the voluntary muscles, but rapid movement of the closed eyes and considerable alpha activity registering on an electroencephalogram.

rate of natural increase in population dynamics, the difference between the crude birth rate and the crude death rate.

ray-finned fish fish in which the fins have supportive structures in the shape of rays running roughly parallel.

reabsorption in the kidney, the return of water, ions, amino acids, sugars, and other valuable substances from the crude kidney filtrate back into the blood following force filtration.

reaction the reciprocal action of chemical agents on one another; chemical change.

reaction center the part of a light-harvesting antenna in which light-activated chlorophylla transfers an electron to the electron transport system. See also *photosystem.*

reafference in behavior, the presumed reward associated with the performance of a particular movement when an animal is performing an adaptive or instinctive pattern; behavior as its own reward.

receptor potential a graded neural signal, generated by a sensory receptor.

receptor site a specific site on a cell membrane, usually a protein, that is capable of recognizing and binding with a specific hormone or other informational molecule.

recessive *adj.,* of an allele, not expressed in a heterozygote.

recessivity the lack of expression by one allele of a gene when a different, dominant allele for that gene is present.

reciprocal altruism in behavior, an action by one individual at some cost that benefits another individual in the expectation and probability that the favor will be returned.

recombinant DNA technology general term for laboratory manipulation of DNA; includes gene splicing, sequencing, and gene cloning.

recombinant virus technique in genetic engineering, the splicing of donor genes into a viral genome and using the virus to transfer such genes into a host cell.

red alga any alga of the division Rhodophyta.

red marrow regions within the ribs, sternum, vertebrae, and hip bones where red blood cells are produced.

red tide sea water discolored by a dinoflagellate bloom in a density fatal to many forms of life.

reducer See *decomposer.*

reduction the addition of electrons or hydrogen atoms to a substance.

reductionism in science, the theory which attempts to divide phenomena and mechanisms into their most elemental parts, isolating as far as possible the effects of individual factors and testing these effects in separate controlled experiments.

reductionist in science, one who attempts to divide mechanisms into their most elemental parts.

reflex arc the simplest form of a complete neural reaction, involving a *sensory neuron, interneuron,* and *motor neuron,* where the integration of information involves only the spinal cord.

refractory period a brief period of depolarization when a neuron cannot generate a second impulse.

regeneration the regrowth of a lost body part through differentiation and repetition of the original developmental events.

regulator. See *operon.*

releaser Also *sign stimulus;* a stimulus that acts as a cue, releasing a certain behavior in an animal.

releasing hormone Also *releasing factor;* a chemical messenger produced by the hypothalamus that stimulates hormonal release by the pituitary gland.

REM sleep. See *rapid eye movement (REM) sleep.*

renal circuit in mammals, the renal arteries, the glomerulus, the capillary beds of the nephron, and the renal veins.

renal cortex the outer region of the kidney, containing the upper portions of the nephrons.

renal medulla the region of the kidney below the cortex, consisting mainly of loops of Henle, capillaries, and collecting ducts.

renal pelvis the cavity of the kidney into which the collecting ducts empty.

rennin an enzyme produced by the stomach linings of young mammals, the action of which is to coagulate milk.

replication DNA synthesis; the process whereby a DNA helix is unwound, the hydrogen bonds between adjacent nitrogen bases broken, and a new strand assembled through base pairing along each old strand.

replication complex during replication, a grouping of the essential enzymes of that process including the unwinding enzymes, helicase, and DNA polymerase.

replication fork the point at which unwinding proteins separate the two DNA strands in the course of DNA replication.

repolarization in the neuron, reestablishment of the resting potential or polarized state following an action potential.

repressible operon the point at which unwinding proteins separate the two DNA strands in the course of DNA replication.

repressor protein in bacterial operons, a protein that binds the operator and prevents transcription.

reproduction the natural process among organisms by which new individuals are generated and the species perpetuated. See *asexual reproduction* and *sexual reproduction.*

reproductive fitness a measure of one's success at leaving one's genes in the next generation.

reproductive isolating mechanisms biological factors which prevent interbreeding between species.

reproductive potential the maximum physiological rate at which any population can increase.

reptile any member of the class Reptilia, composed of cold-blooded vertebrates such as turtles, lizards, snakes, crocodilians, and the tuatara.

Reptilia the class composed of scaled, cold-blooded vertebrates, such as the turtles, lizards, snakes and crocodilians.

repulsion during prophase I, the process during which homologous chromosomes move apart after crossing over.

reservoir in biogeochemical cycles, the less readily available reserves of mineral nutrients, such as atmospheric nitrogen, which is only useful to nitrogen-fixing organisms. Compare *exchange pool.*

residual volume. Also *residual air;* the volume of air remaining in the lungs following maximum, forceful exhalation.

resolution the final phase of sexual intercourse, marked by the loss of erection in males.

resolving power the ability of the eye to distinguish objects near each other as distinct and separate; also the chief factor limiting useful magnification in light microscopes.

respiration See *cell respiration* and *organismic respiration.*

respiratory interface a gas exchange surface; generally a moist, thin-walled membrane, such as the skin of an earthworm, or the inner lining of a lung.

respiratory tree in air breathing vertebrates, the passageways of the respiratory system, including the trachea, paired bronchi, highly branched tracheoles, and alveoli.

response threshold in behavior, the minimum stimulus required to elicit a response.

resting potential the charge difference across the membrane of a polarized neuron (while it is not transmitting an impulse). Compare *action potential.*

restriction enzyme in bacteria, a defensive enzyme that recognizes and cuts out specific, short, viral DNA sequences, thus protecting the cell against most viruses; useful in genetic engineering.

reticular activating system (RAS) the portion of the reticular system involved in activating the appropriate parts of the brain upon receiving a stimulus.

reticular system. Also called *reticular formation;* a major neural tract in the brainstem containing neural pathways to other parts of the brain and to the *reticular activating system (RAS),* and arousal center.

retina the layer of light sensitive cells in the vertebrate eye.

retinol one of the two chemicals formed from *rhodopsin* when it is broken down by light.

retrovirus an infectious single-stranded RNA that copies itself into double-stranded DNA which is then inserted into a host chromosome; represents a rare instance of RNA producing DNA.

reverse transcriptase an enzyme of retroviruses that copies RNA sequences into single-stranded and double-stranded DNA sequences in a minor reversal of the central dogma. Also called *RNA-dependent DNA polymermase.* See *retrovirus.*

reward in behavior, reinforcement; something positive given to strengthen the probability of a specific response to a given stimulus.

R-group a chemistry shorthand where R stands for the variable part of an amino acid.

rhizoid 1. a rootlike structure that serves to anchor the gametophyte of a fern or bryophyte to the soil. 2. a portion of a fungal mycelium that penetrates its food medium.

rhizome an underground, horizontal plant stem that produces shoots above and roots below, and is distinguished from a true root in possessing buds and nodes.

rhodopsin the light-sensitive protein pigment of retinal rods that bleaches in the presence of light, somehow starting an action potential, and is restored in darkness. Also called *visual purple.*

rhythm method a method of contraception whereby copulation is avoided during periods when conception is likely. Also called *natural birth control.*

riboflavin a derivative of vitamin B_2 and an active group in the coenzyme FAD (flavin adenine dinucleotide).

ribonucleic acid See *RNA.*

ribose a five-carbon aldose sugar, a constituent of many nucleosides and nucleotides.

ribosome a two-part cytoplasmic organelle upon which translation occurs, consisting of ribosomal RNA and proteins.

ribosomal RNA (rRNA) the RNA that forms the matrix of ribosome structure.

right atrium the chamber of the heart which receives the deoxygenated blood from the body via the veins, and then pumps it to the right ventricle.

right ventricle the chamber of the heart which receives deoxygenated blood from the right atrium and then pumps it to the lungs.

ring chromosomes a genetic mutation in which a broken chromosome is repaired improperly with its two ends fused, forming a circle; may be associated with hereditary mental retardation.

RNA (ribonucleic acid) a single-stranded nucleic acid macromolecule consisting of adenine, guanine, cytosine, and uracil; divided functionally into rRNA (ribosomal RNA), mRNA (messenger RNA) and tRNA (transfer RNA).

RNA-dependent DNA polymerase See *reverse transcriptase.*

RNA polymerase the enzyme or enzyme complex catalyzing transcription.

rod one of the numerous, long, rod-shaped sensory bodies in the vertebrate retina; responsive to faint light but not to variations in color. Compare *cone.*

root the portion of a vascular plant that functions as an organ of absorption, anchorage, and sometimes food storage, and differs from the stem in lacking nodes, buds, and leaves.

root apical meristem in plants, region of undifferentiated tissue just above the root cap; gives rise to primary growth in the root.

root cap a protective mass of parenchymal cells that covers the root apical meristem.

root hair one of the many tiny tubular outgrowths of root epidermal cells, especially just behind the root apex, that function in absorption.

root pressure modest pressure in the root prompted by osmosis, causing water to rise in the stem.

root tip in plants, the actively growing end of a primary or secondary root, where root apical meristem is found.

rotifer any microscopic animal of the phylum Rotifera, found in fresh and salt waters, having a ciliary apparatus on the anterior end.

round window at the termination of the cochlea, a membranous window that moves in response to perilymph movement started by the oval window. See also *oval window.*

roundworm any nematode that infests the intestine of humans and other animals.

r-selection the reproductive strategy seen in disruptive environments, where the organism produces large numbers of offspring but offers little if any parental care. Compare *K-selection.*

Ruffini corpuscle sensory receptors of the skin that respond to temperature changes.

rugae (*sing.,* ruga) wrinkles, folds, or ridges, especially of the lining of the small intestine.

ruminants hooved grazing mammals that digest cellulose through the action of microorganisms in a four-part stomach.

saccule 1. a small sac. 2. the smaller of two chambers of the membranous labyrinth of the ear. Compare *utricle.*

sac fungi fungi of the phylum Ascomycota.

sacrum the part of the vertebral column that connects with the pelvis; in humans, consists of five fused vertebrae with transverse processes fused into a solid bony mass on either side.

saliva a viscous, colorless, mucoid fluid secreted into the mouth by ducted salivary glands.

salivary amylase an enzyme in the mouth that begins starch digestion.

salivary glands in land vertebrates, any of several glands secreting saliva into the mouth.

salp any free-swimming, oceanic tunicate of the genus *Salpa,* having a transparent, more or less fusiform body.

saltatory propagation in myelinated neurons, the skipping movement of an impulse from one node to another.

sap the watery fluid transported by the phloem that transports dissolved sugars, other organic compounds, and mineral nutrients from one part of the plant to another.

sarcolemma the membranous sheath enclosing a muscle fiber.

satellite DNA unexplained, frequently occurring short segments of chromosomal DNA containing mainly the base pairs adenine and thymine.

sarcomers the contractile unit of striated muscle.

sarcoplasmic reticulum a membranous, hollow tubule in the cytoplasm of a muscle fiber; similar to the *endoplasmic reticulum* of other cells; calcium ions are sequestered here when the muscle is at rest.

saturated *adj.,* in lipid chemistry, having accepted as many hydrogens as possible.

Schwann cells one of the many cells that constitute the myelin sheath, wrapped around the axon of a myelinated neuron.

scientific method a research system in which a problem is identified, relevant data are gathered, hypotheses are formulated, and predictions are made and tested through experimentation or additional observation.

scientific name in biological taxonomy, the two Latin names (generic and specific) used to identify a species.

sclera the heavy, white connective tissue enclosing most of the eyeball; the white of the eye.

sclerenchyma in plants, a protective or supporting tissue composed of cells with greatly thickened, lignified, and often mineralized cell walls.

scolex the hook-bearing head of a larval or adult tapeworm, from which the proglottids are produced by strobilation.

scrotum the external pouch of skin that contains the testis in most adult male mammals.

scutellum in monocots, the cotyledon in its specialized form as a digestive and absorptive organ.

Scyphozoa the class consisting of marine jellyfishes.

secondary consumer a carnivorous animal that feeds upon herbivores (primary consumers).

secondary growth growth in dicot plants that results from the activity of secondary meristem, producing chiefly an increase in the diameter of stem or root. Compare *primary growth.*

secondary immune response the more rapid arousal of lymphocytes and conquest of an invader during a second or subsequent infection, due to memory cells. Compare *primary immune response.*

secondary level the second level of organization of a protein, the formation of helices or sheets by polypeptides.

secondary oocyte. See also *oocyte;* a developing egg cell after the first polar body has been produced and before the second polar body appears after meiosis I.

secondary phloem phloem produced by the vascular cambium during secondary growth.

secondary succession ecological succession occurring in a disturbed community.

secondary xylem xylem produced by the vascular cambium during secondary growth.

second law of thermodynamics the statement that all systems proceed toward entropy and that all chemical transformations are imperfect, that is, energy is lost as it goes from one form to another; the free energy in any closed system constantly decreases.

second messenger an intracellular chemical compound activated by a hormone that becomes active in the cytoplasm or nucleus.

secretion vesicle a small, bladderlike cavity which periodically breaks away from the maturing face of the Golgi body and moves to the plasma membrane where the substances it contains are released outside the cell.

seed the fertilized and ripened ovule of a seed plant, comprising an embryo, including one or two cotyledons, and usually a supply of food in a protective seed coat.

seedless plants any species that does not produce seeds in its reproduction. See *cryptogam.*

seedling any recently germinated seed plant.

seed plants any plant capable of seed production; includes gymnosperms and angiosperms.

segmented body plan a plan in which the body is divided into segments, originally repetitions of nearly identical parts (as still seen in some annelids, chilopods, diplopods, and *Peripatus*); frequently followed in evolution by the specialization of different segments as seen in most arthropods, some annelids, and vertebrates.

segregation the random separation of alleles in different gametes during meiosis.

selectionists biologists who attribute most, if not all, evolutionary change to natural selection.

selectively permeable in cellular membranes, the characteristic of permitting selected substances to pass through while rejecting others. Also called *differentially permeable.*

SEM (scanning electron microscope) a device for visualizing microscopic objects in three dimensions by scanning them with a moving beam of electrons, recording impulses from scattered electrons, and displaying the image by means of the synchronized scan of an electron beam in a cathode ray (television) tube.

semen in mammals, a viscous white sperm-bearing fluid produced in the male reproductive tract and released by ejaculation.

semicircular canals any of the three curved, tubular, fluid-filled canals in the labyrinth of the ear, associated with the sense of equilibrium.

semiconservative replication replication of DNA molecule in which the original molecule divides into two complimentary parts, both halves being preserved while each half promotes the synthesis of a new complement to itself.

seminal receptacle a storage organ for sperm in certain invertebrate females.

seminal vesicle 1. in various invertebrates, a pouch in the male reproductive tract serving as the temporary storage of sperm. 2. in male mammals, paired outpocketings of the vas deferens producing much of the fluid substance of semen.

seminiferous tubule any of the coiled, thread-like tubules that make up the bulk of a testis and are lined with germinal epithelium from which sperm are produced.

sensory neuron also *afferent neuron;* a neuron that conducts impulses carrying sensory information from a receptor to the brain or spinal cord.

sensory receptor a cell or tissue, specialized in responding to specific kinds of stimuli.

sepal the green, leaflike floral parts that surround the flower bud before it opens and later forms a whorl (the calyx) beneath and outside the petals.

seral stage one stage in a *sere.*

sere a recognizable sequence of changes, sometimes predictable, occurring during community development or succession.

Sertoli cells also *nurse cell;* elongated cells of the tubules of the testis that support spermatid development.

seta (*pl.,* setae) the bristlelike, chitinous structure in the body wall of annelids, arthropods, and certain other invertebrates.

sex-influenced trait a genetic trait that can occur in either sex but is more common in one (i.e., breast cancer in women, baldness in men).

sex-limited trait a variable trait that affects members of one sex only.

sex linkage inheritance of traits based on genes present exclusively in either the **X** or the **Y** chromosome.

sex pilus in some prokaryotes, an enlarged pilus (tube) through which a DNA replica can presumably pass from one cell to another.

sexual dimorphism differences in size, color, anatomy, etc. between the sexes.

sexual reproduction reproduction involving the union of genetic material from two individuals.

shaft of bone the long, comparatively straight part of a bone.

shell 1. the space occupied by the orbits of a group of electrons of approximately equal energy. 2. a hard, rigid, usually calcareous covering of an animal. 3. the covering of an egg. 4. the hard, rigid outer covering of a fruit or seed; e.g., walnut shell.

shoot the plant stem and foliage.

shoot apical meristem in plants, undifferentiated tissue at the stem tip that produces primary growth in the stem.

shoot tip in plants, the actively growing end of a primary or secondary stem, where shoot apical meristem is found.

short-day plant a plant that begins flowering after the summer solstice and in which flowering is triggered by periods of dark longer than some innately determined minimum. Compare *long-day plant.*

short-term memory transient, newly acquired memory that has not been consolidated through long-term memory processing.

sickle cell anemia a severe recessive condition attributable to homozygosity for an allele producing an abnormal, crescent-shaped erythrocyte.

sieve element a thin-walled phloem cell having no nucleus at maturity and forming a continuous cytoplasm with other such cells to form *sieve cells* and *sieve tubes,* functioning in the transport of organic solutes, hormones, and mineral elements.

sigmoid curve See *S-shaped curve.*

sign stimulus See *releaser.*

silent mutation a change in a DNA codon in which a synonymous codon forms (third letter change), having no effect on the amino acid sequence of the polypeptide.

simple eye in invertebrates, light receptors that gauge intensity.

simple goiter a visible enlargement of the thyroid gland resulting from iodine deficiency.

sinoatrial (SA) node conducting tissue embedded in the musculature of the right atrium, serving as an intrinsic source of regular contractile impulses to the heart.

sinus arteriosus in the heart of some vertebrates, the widened blood vessel that leads into the atrium.

skeletal muscle muscle under direct and conscious control; striated with multinucleate unbranched fibers. Also called *voluntary muscle, striated muscle.* Compare *cardiac muscle, smooth muscle.*

skin breather any organism in which a significant proportion of the exchange of respiratory gases occurs through a vascularized moist skin (for example, earthworms, most amphibians).

skin exchange. See *skin breather.*

Skinner box in behavior, a device for investigating operant conditioning, named after B. F. Skinner, who invented it.

sliding filament theory the widely accepted explanation of skeletal muscle contraction in which actin myofilaments in the sarcomere are actively drawn past myosin myofilaments, thus shortening the contractile unit.

slime mold a funguslike protist with an ameboid feeding stage and a funguslike spore-forming stage.

small intestine the region of the digestive tract between the stomach and the cecum; the region in which most food digestion and absorption occurs.

smooth muscle the muscle tissue of the glands, viscera, iris, pilo-erectors, and other involuntary structures; consists of masses of uninucleate, unstriated, spindle-shaped cells, usually occurring in thin sheets. Also called *involuntary muscle.*

sociobiology the area of biological science concerned with the genetic basis of human individual and group behavior.

sodium activation gate See *sodium gate.*

sodium gate either of the two gates controlling sodium ion passage through an ion channel, including a *sodium activation gate* and *sodium inactivation gate.* See *ion channel.*

sodium potassium ion exchange pump a membrane active transport mechanism that utilizes ATP energy to move sodium ions out of the cell and potassium into it.

sodium-potassium pump a molecular mechanism in the plasma membrane, capable of activity transporting sodium out of the cell and potassium in, using the energy of ATP.

solar energy energy obtained from the sun.

somatic system the voluntary or conscious part of the peripheral nervous system. Compare *autonomic nervous system.*

somatostatin a hormonal secretion of the *hypothalamus* and of *delta cells* in the *islets of Langerhans.*

somite in the early vertebrate embryo, one of a longitudinal series of paired blocks of tissue that are forerunners of body muscles and the axial skeleton.

sorus (*pl.,* sori) one of the clusters of sporangia on the underside of a fern frond.

specialist an organism that requires very specific conditions.

specialization the process of narrowing abilities so that fewer roles are performed with greater efficiency; the process of becoming increasingly differentiated.

speciation an evolutionary process by which new species are formed, often by the division of one species into two.

species 1. the major subdivision of a genus, regarded as the basic category of biological classification; 2. related individuals that resemble one another through recent common ancestry and that share a single ecological niche; 3. in sexual organisms, a group whose members are potentially able to breed with one another but are unable to breed with members of any other group.

specific defenses highly specific cellular and chemical responses by the B-cell and T-cell lymphocytes against foreign substances, cancerous cells, or invading organisms.

specific heat the heat, expressed in calories, required to raise the temperature of 1 gram of some substance 1°C; a way of considering the quantity of heat that various substances are capable of holding (for example, water has great specific heat).

specific name the second part of a formal scientific name, indicating a species with a genus, always italicized, but usually not capitalized.

sperm 1. a male gamete. 2. a spermatozoan. 3. a spermatozoid. 4. *adj.,* pertaining to the male gamete or male gamete function.

spermatic cord in the male reproductive system, a cordlike structure consisting of a tough, fibrous coat containing the vas deferens, the nerves and blood vessels supplying the testis, and a retractory muscle, the last being vestigial in humans.

spermatid one of the cells that result from the meiotic division of a spermatocyte and mature into spermatozoa.

spermatogenesis the meiotic process resulting in the production of haploid sperm.

spermicide an agent that kills sperm.

S phase in the cell cycle, the stage during which DNA replication occurs.

sphincter a ring of muscle surrounding a body opening or channel that is able to close off the opening; e.g., oral sphincter, anal sphincter.

spicule a tiny calcareous or siliceous pointed body embedded in and serving to stiffen and support the tissues of sponges, sea cucumbers, and other invertebrates.

spinal cord the coplex band of neurons that runs through the spinal column of vertebrates to the brain.

spinal nerves any of the many nerves that enter and leave the spinal cord, including both somatic and autonomic. Compare *cranial nerve.*

spindle a system of microtubules present in the cell during mitosis and meiosis, resembling the spindle of a primitive loom; it serves in the separation of chromosomes during the division process. Also called *spindle apparatus.*

spindle fiber one of the microtubule filaments constituting the mitotic spindle.

spindle pole one of the two points of origin of the mitotic apparatus; represents the microtubular organizing center and contains the centriole in cells where these bodies appear.

spinneret an arthropod organ for producing threads of silk from the secretion of a silk gland.

spiracles the external openings to the respiratory system of terrestrial arthropods.

spiral cleavage in protostome embryos, a spiral pattern in the early cleavages in which the cleavage planes become offset from those below. See also *radial cleavage.*

spiral valve a helical fold of the intestinal wall in the short intestine of sharks and certain bony fishes; slows the passage of food and provides additional absorptive surfaces.

spirillum spiral-shaped bacteria, primarily of the genus *Spirillum.*

spirochaete any of an order of slender, corkscrew-shaped bacteria.

sponge any animal of the phylum Porifera.

spongin a tough, insoluble protein that makes up the skeleton of a class of sponges including the once commercially important bath sponge.

spongy bone bone with a network of thin, hard walls and numerous spaces; spongelike in appearance.

spongy parenchyma in a leaf, the loosely arranged photosynthetic tissue, containing many air spaces, below the palisade parenchyma.

spontaneous generation the production of living organisms from inanimate matter.

sporangium (*pl.,* sporangia) a structure in which spores are produced; found in algae, fungi, bryophytes, and ferns.

spore a minute unicellular reproductive or resistant body, specialized for dispersal, for surviving unfavorable environmental conditions, and for germinating to produce a new vegetative individual when conditions improve. See also *endospore.*

sporic cycle in some algae and all plants, a life cycle in which meiosis leads to a haploid generation of cells that later produces gametes through simple mitosis. The cycle is often described as an alternation of generations, wherein a diploid sporophyte generation alternates with a haploid gametophyte generation.

sporophyte in plants having an alternation of generations, a diploid individual capable of producing haploid spores by meiosis; the prominent form of ferns and seed plants. Compare *gametophyte.*

sporozoan any parasitic protozoan of the phylum Sporozoa, certain species of which cause malaria.

S-shaped curve a plot of population growth where growth is rapid at first but then slows when *environmental resistance* is met, and levels off at some point near or below the *carrying capacity.* Compare *J-shaped curve.* Also *logistic growth curve.*

stabilizing selection in population genetics, selection against both extremes of a continuous phenotype, favoring the intermediate.

stamen the male reproductive structure of a flower, consisting of a pollen-bearing anther and the filament on which it is borne.

stapes the outer, stirrup-shaped bone of a chain of three small bones in the middle ear of mammals. Also called *stirrup.*

staphylococcus spherical bacteria arranged in grape-like clusters.

stele in roots and stems of vascular plants, a central cylinder containing vascular tissue.

stem 1. the ascending axis of a plant, whether above- or belowground; 2. the stalk which supports a leaf, flower, or fruit; 3. the main body of that portion of a plant which is aboveground. Also *shoot.*

stem cell also *hemocytoblast;* generalized cell of red bone marrow from which all blood cells form.

sterilization the destruction of the ability to reproduce.

steroid any of a class of lipid-soluble compounds, some of which are hormones, consisting of four interlocking saturated hydrocarbon rings and their side groups; included are *cholesterol, estrogen, testosterone, cortisol,* and others.

steroid sex hormone part of a class of hormones consisting of the steroid molecule with various side group substitutions; e.g., estrogen and testosterone.

stigma in the floral pistil, the top, slightly enlarged and often hairy or sticky end of the style, on which pollen grains adhere and germinate.

stipe the stemlike structure in red or brown algae that supports the blades or blade.

stoma (*pl.,* stomata) minute pores in the *epidermis* of leaves, stems and other plant organs, including *guard cells;* allows the diffusion of gases into and out of intercellular spaces.

stomach a saclike enlargement of the digestive tract functioning in storage, churning, and digestion.

storage vesicle See *lysosome.*

stratified squamous epithelium multilayered, flattened epithelial cells as seen in the epidermis of the skin, esophagus, and vagina.

streptococcus spherical bacteria arranged in a chain.

stretch receptor a sensory receptor that is stimulated by stretching, as in a tendon, muscle, or bladder wall.

striated muscle. See *skeletal muscle.*

stroma the enzyme-containing fluid region that surrounds the thylakoids of a chloroplast.

stromatolite a macroscopic living or fossil geological structure of layered domes of deposited material, attributed to the presence of shallow-water photosynthetic prokaryotes.

style the stalk of a pistil in a flower, connecting the stigma with the ovary.

suberin a complex fatty substance of cork cell walls and other waterproofed cell walls.

subspecies a more or less clearly defined, morphologically distinct, named, geographic variety of a species; a third part of the scientific name (*genus, species, subspecies*).

substrate a substance acted upon by an enzyme.

substrate level phosphorylation the production in one or more steps of ATP from ADP, P_1, and an appropriate organic substrate; the capture of high-energy phosphate bonds directly from metabolic transformations. Compare *chemiosmotic phosphorylation.*

subunit disassembly a hypothetical explanation of spindle shortening, an anaphase process during which spindle fiber microtubules are broken down into tubulin molecules.

sucrase an enzyme that digests sucrose, forming fructose and glucose.

sucrose a sweet, twelve-carbon disaccharide consisting of glucose and fructose subunits. Also called *table sugar.*

sun-compass orientation a behavioral mechanism utilizing the angle of the sun and the time of day to compute direction for navigation.

superior vena cava the large vein by which blood is returned to the right atrium from the areas above the heart.

super-normal releaser in behavior, an environmental stimulus with exaggerated features that produces an instinctive response; not normally encountered in nature.

suppressor T-cell a type of white blood cell that shuts down the immune response after the risk of infection has passed.

surface volume hypothesis the proposal that cells are restricted to a size that assures a surface-volume ratio that provides a sufficient membrane area to support the transport needed to maintain metabolic activity.

survival of the fittest See *natural selection;* the principle of the survival of the forms of animal and vegetable life best fitted for existing conditions, while related but less fit forms become extinct.

survivorship curve a graph with numbers of individuals plotted on the X axis and percentage of total life span completed plotted on the Y axis; useful in comparing periods of greatest mortality in several different species.

suture in anatomy, immovable joints formed by the articulation of skull bones.

swim bladder also called *air bladder;* a gas-filled sac giving controlled buoyancy to most bony fish; homologous with the lungs of land vertebrates and lungfish.

syllogism in logic, an argument the conclusion of which is supported by two premises, of which one contains the term that is the predicate of the conclusion, and the other contains the term that is the subject of the conclusion; common to both premises is a term that is excluded from the conclusion; e.g., all A is B; all B is C; therefore all A is C.

symbiont a symbiotic organism, one of two intimately associated species.

symbiosis the living together in intimate association of two species; includes three categories: *mutualism* (both organisms gain), *commensalism* (one gains at little or no expense to the other), and *parasitism* (one gains at the expense of the other).

symbiosis hypothesis the hypothesis that the eukaryotic cell evolved from the mutualistic union of various prokaryotic organisms, one of which gave rise basically to the cytoplasm, nucleus, and motile membranes; a second to mitochondria; a third to chloroplasts and other plastids; and a fourth to cilia, eukaryotic flagella, basal bodies, centrioles, and spindle, and all other microtubule structures.

sympathetic division that portion of the autonomic nervous system whose nerves emerge only from the spinal cord and function to speed up the usual pace of the visceral organs served. Compare *parasympathetic division.*

sympatric speciation speciation in populations without geographic separation. Compare *allopatric speciation.*

synapse the junction between the axon of one neuron and the dendrite or cell body of another; crossed by neural impulses.

synapsis the pairing up and fusing of homologous chromosomes during the first meiotic prophase, whereby preparations are made for crossing over.

synaptic cleft the minute space between communicating neurons or between motor neurons and effector organs; the space crossed by neurotransmitters.

synaptic knobs one of the multiple bulbous swellings at the axonal endings, containing neurotransmitters in secretion granules and forming one side of the synaptic cleft.

synaptonemal complex in crossing over, a complex, zipperlike structure composed of protein and RNA formed between homologous chromatids in meiotic prophase.

synonymous codons in the genetic code, codons that represent the same amino acid.

synovial fluid a transparent, viscid lubricating fluid secreted by the synovial membranes of joints, bursae, and tendon sheaths.

synovial joint a freely movable joint surrounded by a fibrous capsule lined by a synovial membrane that secretes lubricating synovial fluid.

synthesist in science, one who attempts to draw broad general principles from widely disparate observations, or one who reinterprets established relationships in support of a new general idea or synthesis.

syphilis a sexually transmitted disease caused by the spirochete *Treponema pallidum.*

system an assemblage of organs cooperatively performing a body process (i.e., the *nervous system* or *digestive system*).

systole the period of heart contraction, particularly of the ventricles. Compare *diastole.*

systolic pressure the highest arterial blood pressure of the cardiac cycle, a product of ventricular contraction.

tactile receptors See *mechanoreceptor.*

taiga a subarctic forest biome dominated by spruce and fir trees; found in Europe and North American and at high altitudes elsewhere.

tap root system a root system consisting of a large primary root and its secondary and lateral branches.

target cell a cell acted upon by a specific chemical messenger, generally containing or bearing specific receptor proteins not found in other cells.

taste bud a sensory receptor sensitive to taste, found chiefly in the epithelium of the tongue.

taxis the movement of an organism toward or away from a stimulus.

taxonomy the science of identifying, naming, and classifying organisms.

T-cell lymphocyte a lymphocyte specializing in cell-mediated responses; interacts in complex ways with other types of cells in the immune system; eliminates foreign or infected cells it encounters.

TDF gene the "testis determining factor" which stimulates the development of the testes in male embryos.

tectorial membrane a membrane of the cochlea, overlying and contacting the hair cells of the organ of Corti.

telphase the stage of mitosis or meiosis in which new nuclear membranes around each group of daughter chromosomes, the nucleoli appear, and the chromosomes decondense; at this time the cell membrane and cytoplasm usually divide to form two daughter cells.

TEM (transmission electron microscope) a device that uses an electron beam, rather than light, to form magnified images.

temperate deciduous forest a forest biome of the temperate zone, in which the dominant tree species and most other trees are deciduous (shedding) and are bare in winter months.

temporal lobe a large lobe on the lateral portion of each cerebral hemisphere.

tendon a dense tough cord of fibrous connective tissue that is attached at one end to a muscle and at the other to that part of the skeleton that moves when the muscle contracts.

termination the end of polypeptide synthesis, when "stop" codons on the messenger RNA are encountered by the ribosome, which releases the polypeptide and comes apart at its subunits.

territory the space defended by a territorial animal.

territoriality the behavior of an animal defending its territory.

tertiary consumer a carnivorous animal that feeds on secondary consumers, which in turn feed on herbivores.

tertiary level the third level of organization of a protein, the pattern of folding of a polypeptide upon itself, which is generally quite specific for each protein type.

test cross the cross of a dominant individual with a homozygous recessive individual to determine whether recessive alleles exist.

testes (*sing.,* testis) the male gonads, in which spermatozoa are produced by meiosis.

testicles See *testes.*

testosterone a male hormone, produced in the testes, important in the sex drive and producing secondary sex characteristics.

tetrahedron in geometry, a solid contained by four plane faces; a triangular pyramid; the shape taken by methane (CH_4).

tetraploid *adj.,* having four complete sets of chromosomes in each cell. Compare *allotetraploid, allopolyploid, polyploid.*

tetrapod. Also *land vertebrate;* any vertebrate of the classes Amphidia, Reptilia, Mammalia, and Aves, including some that don't have four feet (e.g., birds, people).

thalamus a subdivision of the forebrain, just above the brainstem, concerned with relaying information.

theory 1. a coherent group of general propositions used as principles of explanation for a class of phenomena (i.e., Darwin's theory of the origin of species). 2. a more or less verified explanation accounting for a body of known facts or phenomena. Compare *hypothesis.*

thermal overturn the seasonal overturn of lake waters brought about by decreasing surface temperatures that increase the density of surface waters, allowing them to sink, thereby disrupting the thermocline and permitting wind-blown revolving of water.

thermal proteinoids polymers spontaneously generated using dry heat and amino acids, which when placed in water cluster to form proteinoid microspheres.

thermocline in a body of water, a temperature gradient where the temperature changes rapidly as a function of depth; thermoclines act as barriers to the vertical movement of water.

thermodynamics 1. the branch of physics that deals with the interconversions of energy as heat, potential energy, kinetic energy, radiant energy, entropy, and work. 2. the processes and phenomena of energy interconversions. See *first and second law of thermodynamics.*

thermonastic movement in plants, a sudden reaction to touch, associated with rapid ion transport and sudden changes in turgor.

thermoreceptor a sensory receptor that responds to changes in temperature.

thermoregulation an animal's control over its internal temperature: the behavioral or physiological mechanisms that maintain a body at a particular temperature.

thoracic region the portion of the vertebral column that articulates with ribs, between the cervical and lumbar regions.

thorax 1. in animals, the part of the body anterior to the diaphragm and posterior to the neck, containing the lungs and the heart; 2. the middle of the three parts of an insect body, bearing the legs and wings.

thrombin a proteolyfic enzyme that catalyzes the conversion of fibrinogen to fibrin, and in turn is produced from prothrombin by the action of thromboplastin.

thromboplastin in the blood-clotting reactions, a protein released from damaged platelets that catalyzes the conversion of the prothrombin into the active enzyme thrombin.

thromboxane a prostaglandin.

thylakoid the membranous structure of chloroplasts consisting of a *thylakoid membrane,* containing light-harvesting antennas and the photosynthetic electron transport chain; an inner lumen that collects protons during active photosynthesis; and CF_1 particles, the sites of chemiosmotic phosphorylation.

thymine a pyrimidine, one of four nitrogenous bases of DNA.

thymus a glandular body above the lungs, believed to stimulate T-cell lymphocyte development through the secretion of thymosin.

thyroid gland a large endocrine gland in the lower neck region of all vertebrates, the thyroxin secretions of which regulate the rate of metabolism.

thyroid-stimulating hormone (TSH). Also *thyrotropin;* a peptide hormone of the anterior pituitary, the action of which is to stimulate the release of thyroxin from the thyroid.

thyrotropin. See *thyroid-stimulating hormone* (TSH).

thyroxin a thyroid hormone that functions in regulating metabolism.

tissue a grouping of cells of similar origin, structure, and function. Compare *organ.*

total fertility rate the predicted general fertility rate of a population, based on the average number of children women aged 14–44 will have.

totipotent *adj.,* a cell or tissue retaining developmental flexibility; undifferentiated; having the ability to produce the entire original organism.

trace element. See *micronutrient.*

tracer a radioactive element or compound that is put into a biological system to be later located by detection of the radioactivity.

trachea (*pl.,* tracheae) 1. in land vertebrates, air passages between the lungs and the larynx; 2. one of the air conveying tubules in the respiratory system of an insect, millipede, or centipede.

tracheal system the respiratory system of insects, composed of thin-walled air-conducting tubules opening to *spiracles* and extending to finer, branched *tracheoles,* some terminating in *air sacs.*

tracheid a long, tubular xylem element that functions in support and water conduction; distinguished from *xylem vessels* by having tapered, pitted end walls.

tracheoles. See *tracheal system.*

tracheophyte a vascular terrestrial plant.

transcription RNA sequence in which the RNA nucleotide sequence is determined by specific base pairing with the nucleotide sequence of DNA.

transcription complex enzymes essential to the process through which DNA is transcribed into RNA, including helicase and RNA polymerase.

transducer a device that receives energy in one form and retransmits it in a different form, e.g., the conversion of touch (pressure) to action potentials.

transfer RNA (tRNA) in the synthesis of polypeptides, a class of RNA molecules with the task of identifying and bonding with specific amino acids and then, on the ribosome, identifying and bonding with corresponding sites on the messenger RNA.

transformation in a bacterium, the direct incorporation of a DNA fragment from its medium into its own chromosome.

translation polypeptide synthesis as it is directed by mRNA and assisted by ribosomes and tRNA; the transfer of linear information from a nucleotide sequence to an amino acid sequence according to the genetic code.

translocation 1. during protein synthesis in which a transfer RNA molecule is moved (translocated) from one ribosomal tRNA attachment site (pocket) to the other; 2. in chromosomes, the breakage and improper (nonhomologous) rejoining of chromosome segments; 3. the movement of solutes through the phloem from one part of a plant to another.

translocation Down syndrome. See *trisomy 21.*

transpiration 1. in plants, the evaporation of water vapor from leaves, especially through the stomata; 2. the physical effects of such evaporation taken together.

transpiration pull the pulling of water up through the xylem of a plant using the energy of evaporation, the water potential gradient in the leaf, and the tensile strength of water.

transposon a segment of DNA capable of being moved from one chromosomal location to another, containing insertion sequences and forming a complete transposable unit. Also *transposable gene.*

transposable gene See *transposon.*

transverse tubule also called *T-tubule;* one of a specialized system of tubules in a muscle fiber, transmitting the contractile impulse from the sarcolemma to the sarcomeres.

Trematoda the class composed of parasitic flatworms having one or more suckers.

tricuspid valve the valve between the right atrium and the right ventricle consisting of three triangular membranous flaps.

triglyceride a nonpolar, hydrophobic lipid consisting of three fatty acids, covalently bonded to one molecule of glycerol.

trimester one of the three-month periods of the nine months of human gestation.

trinucleate cell in plant reproduction, the triploid cell formed when a sperm penetrates the binucleate central cell in the embryo sac.

triphosphate nucleoside one of the components occurring in a nuclear pool, having two high-energy bonds available for use in DNA replication.

trisomy 21 a severe human congenital pathology attributable to the presence of three rather than two homologues of chromosome 21. Also *Down syndrome* and *translocation Down syndrome.*

trophic level relating to nutrition; a level in a food pyramid.

trophoblast in mammals, the thin-wall side of a blastocyst that forms the chorion when implantation occurs. See also *blastocyst.*

trophosome an extensive organ in a tubeworm which houses vast numbers of chemotrophic vent bacteria capable of extracting energy from simple chemicals and forming carbohydrates.

tropical rain forest a tropical woodland biome that has an annual rainfall of at least 250 cm (100 in.) and often much more; typically restricted to lowland areas and characterized by a mix of many species of tall, broad-leaved evergreen trees that form a continuous canopy, with vines and woody epiphytes, and by a dark, nearly bare forest floor.

tropical savanna a biome that is primarily grassy but frequently interrupted by groves of drought-resistant trees; commonly occurs between tropical rain forest and desert.

tropic hormone a hormone whose target is another endocrine gland.

tropism growth toward or away from an external stimulus, in plants usually accomplished by differential cell elongation in the stem or root. See also *gravitotropism* and *phototropism.*

tropomyosin a low-molecular weight filamentous protein that accompanies the globular protein actin in making up actin microfilaments.

troponin a protein of low molecular weight that binds calcium in muscle contraction, a specific variant of the ubiquitous calcium-binding molecule *calmodulin.*

true-breeding homozygous; containing one form of an allele only so the offspring always resemble the parents and each other for a particular trait.

trypanosome a parasitic, flagellated protozoan of the genus *Trypanosoma,* infecting mammals and transmitted by insect vectors, responsible for *chagas disease* and *sleeping sickness.*

trypsin a powerful proteolytic enzyme secreted by the pancreas in the inactive form *trypsinogen* and activated in the intestine.

tryptophan operon (a repressible operon) a region of DNA in *E. coli* that includes structural genes coding for enzymes that synthesize tryptophan, and a region that controls their transcription. See also *operon.*

tubal ligation female sterilization by cutting and tying off the Fallopian tubes.

tubular secretion in the nephron, the active transport of certain substances from the crude filtrate back into the blood.

tubulin a protein consisting of two dissimilar spherical polypeptides making up the subunits of microtubules.

tundra a biome characterized by level or gently undulating treeless plains of the arctic and subarctic that support dense growth of mosses and lichens as well as dwarf herbs and shrubs; underlain by permafrost and seasonally covered by snow.

tunica in plants, the protective layer covering the short apical meristem.

tunicate any marine chordate having a saclike body enclosed in a thick membrane (or tunic) from which protrude two openings or siphons for the entry and exit of water.

Turbellaria a class of platyhelminthes mostly aquatic, having cilia on the body surface.

turgor the normal state of turgidity and tension in living plant cells by the uptake of water through osmosis.

turgor pressure the actual hydrostatic pressure developed by the fluid of a swollen plant cell.

tympanal organs See *tympanum.*

tympanic membrane also *tympanum, eardrum;* a thin, clear, tense double membrane dividing the middle ear from the external auditory canal; on the surface in frogs and toads, and intermediate in birds and reptiles.

tympanum See *tympanic membrane.*

typhlosole in the earthworm, an infolding in the roof of the intestine that greatly increases the digestive and absorptive surface.

ultimate causation a causation factor that developed due to long term influences, such as evolutionary.

ultrastructure the submicroscopic structure of organelles of the cell.

umbilical cord in placental mammals, a vascular cord connecting the fetus with the placenta.

umbilicus the navel; point on the abdomen at which the umbilical cord was attached to the embryo. Also "belly-button."

unconditioned response in behavior, an involuntary reaction to a common stimulus.

unconditioned stimulus the normal or usual stimulus that produces a certain predictable behavior. Compare *conditioned stimulus.*

unlinked *adj.,* gene loci of different chromosomes, independently assorted.

unsaturated *adj.,* in lipid chemistry, capable of accepting hydrogens.

upper epidermis in plants, the upper (light-exposed) layer of cutinized epidermal cells in a leaf.

upwelling in oceans, the wind-driven rise of deep water layers to the surface; associated with the circulation of mineral nutrients and a consequent increase in productivity and biomass.

uracil one of the nitrogenous bases of RNA.

urea a highly soluble compound that is the principle nitrogenous waste of the urine of animals.

ureter the tube that conducts urine from the kidney to the bladder.

urethra the tube that conducts urine from the urinary bladder to the outside of the body.

uric acid a relatively insoluble purine; a principal nitrogenous excretion product of reptiles, birds, and insects.

urinary bladder a membranous sac in which urine is stored until it is discharged from the body.

urine the liquid nitrogenous matter excreted by the kidneys.

Urochordata A subphylum of the chordates that includes the tunicates, salps, and larvacea.

urochordate any member of the subphylum *Urochordata:* tunicates, salps, larvaceae.

uterus 1. in female mammals, a muscular, vascularized, mucous membrane-lined organ for containing and nourishing the developing young; 2. an enlarged section of the oviduct of various vertebrates and invertebrates.

utricle the chamber of the membranous labyrinth of the middle ear into which the semicircular canals open.

vaccine killed or weakened disease agents, injected to stimulate the primary immune reaction.

vacuole a general term for any fluid-filled, membrane-bounded body within the cytoplasm of a cell.

vagina the female copulatory organ and birth canal in mammals and other animals.

variability the general qualitative term for the presence of genetic differences between individuals in a population.

variable experimental variable; the focus of an experiment, tube tested and compared with a control.

variable region the portion of an immunoglobulin polypeptide concerned with the binding of an antigen, which varies with the specific immunoglobin.

vascular cambium the cylinder of meristematic tissue that in secondary growth produces xylem on its inner side and phloem on its outer side, thus contributing to growth in circumference.

vascular plant a plant with xylem and phloem; a tracheophyte.

vascular rays in woody stems, radiating, spokelike lines of parenchyma and collenchyma tissue that conduct materials laterally and help relieve pressure caused by expansion during circumferential growth.

vascular system 1. the xylem and phloem of a vascular plant; 2. the circulatory system of an animal.

vascular tissue any tissue that contains vessels through which fluids are passed.

vas deferens the duct of the testis which transports the sperm from the epididymis to the penis.

vasectomy male sterilization by cutting and tying off the seminal ducts.

vasoconstrictor a nerve or drug that causes the constriction of the blood vessels.

vasopressin. See *antidiuretic hormone.*

vegetal hemisphere in eggs with poles, the less metabolically active, yolky region.

vegetative stage. Also *vegetative state;* in fungi, a spreading mycelium specialized in feeding and growth.

vein 1. a vessel returning blood toward the heart. 2. a vascular bundle in a leaf or petiole.

ventral nerve cord a common feature of many invertebrate phyla, the main longitudinal nerve cord of the body; solid and paired, with a series of ganglionic masses.

ventricle one of the large muscular chambers of the four-chambered heart.

venule a small vein.

vermiform appendix a hollow, finger-like, blind extension of the cecum, having no known function in humans.

vertebral canal the pathway of the spinal cord through the vertebral column.

vertebral column the articulated series of vertebrae connected by ligaments and separated by intervertebral discs that in vertebrates form the supporting axis of the body and of the tail in most forms.

vertebrata the subphylum of the chordates with a vertebral column, or backbone.

vertebrate an animal in the subphylum Vertebrata, phylum Chordata; an animal with a vertebral column of bone or cartilage.

vessel a conducting tube in a dicot formed in the xylem by the end-to-end fusion of a series of cells (vessel elements) followed by the loss of adjacent end walls and of cell protoplasm. Compare *tracheid.*

vestibular apparatus that portion of the inner ear involved in sensing position and movement of the head.

vestigial *adj.,* pertaining to a degenerate or imperfectly developed organ or structure no longer functional.

villi (*sing.,* villus) minute cellular processes that cover the inner surface of the small intestine, containing blood vessels and lacteals and serving in the absorption of nutrients.

virus a noncellular, parasitic organism transmitted as DNA or RNA enclosed in a membrane or protein coat, often together with one or several enzymes; replicates only within a host cell, using host ribosomes and enzymes of synthesis.

viscera the internal organs within the body cavity, e.g., heart, lungs, intestines, liver.

visible light electromagnetic wave lengths longer than about 400 nm and shorter than about 750 nm, which can serve as visual stimuli to most photoreceptive organisms.

vital capacity the maximum amount of air that can be exhaled after a fully forced inhalation.

vital force a discredited notion of mysterious life-giving force that animates and perpetuates living beings.

vitalism an untestable doctrine that attributes the functions of a living organism to a *vital principle* or *vital force* distinct from chemical and physical forces; no longer taken seriously.

vitamin an organic substance taken in with food that is essential to the metabolic activity of an organism, usually because it supplies part of a coenzyme not made by the organism.

vitelline membrane the membrane surrounding the egg yolk in some animals.

viviparity *n.*, the condition of bringing forth living young rather than eggs, as most mammals and some reptiles and fishes.

viviparous *adj.*, producing young from within the uterus; during development, nourishment is supplied by the mother's tissues, usually via a placenta. Compare *oviparous, ovoviviparous.*

voluntary muscle See *skeletal muscle.*

vulva the external female genitalia.

water mold an aquatic protist of the oomycetes.

water potential the potential energy of water to move, as a result of concentration, gravity, pressure, or solute content.

water potential gradient the difference in potential energy of water between a region of greater water potential and a region of lesser water potential. Because of gravity, pressure, or an osmotic pressure, water moves down its water potential gradient.

water vacuole See *contractile vacuole.*

water vascular system in echinoderms, a system of vessels that contains sea water and is used as a hydraulic system in the movement of tube feet.

white matter the portion of the spinal cord consisting of myelinated axons and dendrites. See *gray matter.*

wilting the loss of turgor, especially because of an inadequate supply of water to a plant or plant cell.

X chromosome one of the two heteromorphic sex chromosomes of mammals, flies, and certain other insects. In most cases **XX** = female and **XY** = male.

xerophyte a plant adapted to dry areas.

x-ray crystallography a procedure of directing x-rays at a crystal of a protein or other large molecule, recording the pattern produced as the rays are bent by the regular, repeating molecular structures within the crystal, and using this pattern to reconstruct the three-dimensional structure of the molecule.

xylem one of the two complex tissues in the vascular system of plants; consists of the dead cell walls of vessels, tracheids, or both, often together with sclerenchyma and parenchyma cells; functions chiefly in water conduction and strengthening the plant. See also *tracheid.* Compare *phloem.*

Y chromosome one of the two heteromorphic sex chromosomes of mammals, flies, and certain other insects. In most cases, **XY** = male and **XX** = female; in general, nearly devoid of genes not concerned with maleness.

yellow marrow yellow, fatty material within the central cavity of long bones.

yolk one of the extraembryonic membranes of a bird, reptile or mammal.

yolk sac an extraembryonic membrane of a bird, reptile, or mammal, that in birds and reptiles encloses the yolk mass, and in placental mammals encloses a fluid-filled space.

zeatin a plant cytokinin hormone and growth factor extracted from corn.

Z-line in striated muscle, the partition between adjacent contractile units to which actin filaments are anchored.

zona pellucida "clear zone"; a thick, transparent, elastic membrane or envelope secreted around an ovum by follicle cells in some vertebrates.

zooplankton the nonphotosynthetic, minute, animal life drifting at or near the surface of the open sea.

Z-scheme a graphic presentation of the energy levels in the oxidation-reduction reactions occurring in the light reaction of photosynthesis.

zygospore a diploid fungal or algal spore formed by the union of two similar sexual cells; has a thickened wall and serves as a resistant resting spore.

zygote a cell formed by the union of two gametes; a fertilized egg.

zygotic *adj.*, pertaining to the cell formed by the union of two gametes or the fertilized egg.

zygotic cycle in some protists and all fungi, a life cycle in which the haploid state dominates. The diploid state, which follows fertilization, is usually a very brief interlude followed by meiosis and resumption of the haploid state.

ILLUSTRATION ACKNOWLEDGMENTS

CHAPTER 1

1.1/© Joseph-Nicolas Robert-Fleury/Photo RMN 1.2/© NASA 1.5(top left)/© Manfred Kage/Peter Arnold, Inc. 1.5(left center)/© 1980 Dwight R. Kuhn 1.5(bot left)/© Dr. E. R. Degginger 1.5(top right)/© 1983 Dwight R. Kuhn 1.5(center right)/© Dwight R. Kuhn 1.5F-I/© Stephen J. Krasemann/DRK Photo **Page 4**/© Stephen J. Krasemann/Peter Arnold, Inc.

CHAPTER 2

2.1B/© Archive Photos 2.2/Courtesy of Eric M. Reiman, M.D. 2.7(top)/© Cabisco/ Visuals Unlimited 2.7(bottom)/© Ken Wagner/Phototake NYC 2.11(left)/© Richard C. Walters/Visuals Unlimited 2.11(left center)/ © Bruce Iverson 2.11(right center)/© Les Christman/Visuals Unlimited 2.11(right)/ © Bruce Iverson 2.12B/© Larry Lefever/ Grant Heilman Photography, Inc. **Page 18**/ © Science VU/IBMRL/Visuals Unlimited **Page 24**/© The Bettmann Archive

CHAPTER 3

3.1B/© C. J. Olivares, Jr./Allsport USA 3.3(left)/© Dr. Don W. Fawcett/Photo Researchers, Inc. 3.3(center)/© S. E. Frederick/E. H. Newcomb/Biological Photo Service 3.3(right)/© Biophoto Associates/ Photo Researchers, Inc. 3.8(left)/© Cabisco/ Visuals Unlimited 3.8(right)/© Cabisco/Visuals Unlimited **Page 32**/© John Gerlach/ Animals Animals

CHAPTER 4

4.1A/© Kevin Collins/Visuals Unlimited 4.1B/© Science VU/Visuals Unlimited 4.3(top right)/Robert Knauft/Photo Researchers, Inc. 4.3(top center)/© Andrew Syred/Science Photo Library/Photo Researchers, Inc. 4.3(bottom center)/© Chuck Brown/Photo Researchers, Inc. 4.3(top left)/ © Biophoto Associates/Science Source 4.3 (bottom right)/© Biophoto Associates/Science Source 4.3(bottom left)/© Philip Harris/Biological Ltd./Biophoto Associates 4.4 (top right)/© John Taylor/University of California, Berkeley/Scott Foresman 4.4(right center)/© Dr. B. R. Brinkley and Donna Turner/Department of Cell Biology, Baylor College of Medicine 4.4(botttom right)/© K. R. Porter/Photo Researchers, Inc. 4.4(top left)/© Dr. Don W. Fawcett/Roy Jones/ Photo Researchers, Inc. 4.4(left center)/© Dwight R. Kuhn 4.5(bottom right)/© 1993 Dwight R. Kuhn 4.5(bottom left)/© Dr. Don W. Fawcett/Micrograph by G. Raviola/ Photo Researchers, Inc. 4.7C/© Dr. Don W. Fawcett/Photo Researchers, Inc. 4.8B/© John Taylor/University of California, Berkeley/Scott Foresman 4.9(left)/© Birgit H. Satir/Department of Anatomy, Albert Einstein College of Medicine 4.9(right)/© Dr. Don W. Fawcett/Roy Jones/Photo Researchers, Inc. 4.10(top left)/© Runk/ Schoenberger/Grant Heilman Photography, Inc. 4.1O(top right)/© Biophoto Associates/Science Source/Photo Researchers, Inc. 4.11(bottom)/© K. R. Porter/Photo Researchers, Inc. 4.12(top)/© M. Schliwa/Visuals Unlimited 4.12(left center)/© John J. Wolosewick/Department of Anatomy, University of Illinois at Chicago/Biological Photo Service 4.12(center)/© Paul R. Burton/Biological Photo Service 4.12(right)/© L. Evans Roth/Biological Photo Service 4.13(left)/© Fred Hossler/Visuals Unlimited 4.13(right)/© K. G. Murti/Visuals Unlimited 4.13(center)/© 1993 Dwight R. Kuhn 4.13(bottom)/© William L. Dentler/Biological Photo Service 4.14A/© Dr. B. R. Brinkley and Donna Turner/Department of Cell Biology, Baylor College of Medicine 4.15B/© Stanley C. Holt/Biological Photo Service **Page 48**/© 1993 Science Photo Library/Custom Medical Stock Photo **Page 61**/4A2(top left center) © John Gerlach/Visuals Unlimited (top left) © Triarch/Visuals Unlimited (bottom right center) © Biophoto Associates/Photo Researchers, Inc. (top right) From *Living Images* by Gene Shih and Richard Kessel/Reprinted courtesy of Jones & Bartlett Publishers, Inc., Boston/Scott Foresman (right center) © Dr. Jeremy Burgess/ Science Photo Library/Photo Researchers, Inc. (bottom right) © KP/Beck

CHAPTER 5

5.1/Wallace et al., *Biology: Science of Life*, 3d ed., New York: HarperCollins **Page 74**/© Dwight R. Kuhn

CHAPTER 6

6.1/© Rod Allin/Tom Stack & Associates 6.2/© David R. Frazier 6.3A/© Don and Pat Valenti 6.3B/© Lawrence Gilbert/Biological Photo Service 6.4A/© Archive Photos 6.6B/ © Science VU/IBMUKOU/Visuals Unlimited **Page 86**/© Roy Morsch/The Stock Market

CHAPTER 7

7.1/© Dave Millert/Tom Stack & Associates 7.2(top)/© Runk/Schoenberger/Grant Heilman Photography, Inc. 7.2(bottom)/© Dr. Kenneth R. Miller/Science Photo Library/ Photo Researchers, Inc. **Page 102**/© Wendell Metzen/Bruce Coleman Inc. **Page 114**/(top left) © E. F. Anderson/Visuals Unlimited (bottom right) © Biophoto Associates/Photo Researchers, Inc.

CHAPTER 8

8.8/© M.H. Sharp/Photo Researchers, Inc. **Page 119**/© Pam Taylor/Bruce Coleman Inc.

CHAPTER 9

9.2/© NASA (Ames Research Center)/Scott Foresman 9.3(left page: left)/© Howard Sochurek/Medical Images Inc. 9.3(right page: left)/© Dr. Benjamin Lewin/Cell Press 9.3(left page: right)/© Barbara Hamkalo 9.3(right page: right)/© 1994 Jim Cummins/FPG International 9.6(all)/Carolina Biological Supply Company 9.7(center)/

J Pickett-Heaps/Science Source/Photo Researchers, Inc. 9.7(top)/© Conly Rieder/Biological Photo Service 9.8/From *Scanning Electron Microscopy In Biology: A Student's Atlas On Biological Organization* by Richard G. Kessel and Gene Y. Shih) © 1974 Springer-Verlag/Scott Foresman 9.9(all)/Gimenez-Martin, G. C. de la Torre and J. F. Lopez-Saez. In *Mechanisms and Control of Cell Division*, ed. T. L. Rost and E. M. Gifford, Jr. 267-283 **Page 136**/© Gene Shih/Richard Kessel/Visuals Unlimited **Page 142**/© Scott Foresman

CHAPTER 10
Page 153/© David M. Phillips/Visuals Unlimited **Page 162**/© Richard Hutchings/Photo Researchers, Inc.

CHAPTER 11
11.1/© Historical Pictures/Stock Montage, Inc. **Page 167**/© Gerard Lacz/Animals Animals

CHAPTER 12
12.8B/Courtesy Macmillan Science Co., Inc. 12.9A-B/Courtesy of Dr. Murray L. Barr **Page 182**/© Dr. Gopal Murti/Science Photo Library/Photo Researchers, Inc. **Page 192**/© Hulton Deutsch Collection Limited, Unique House **Page 189**/(bottom left) Dr. Henry L. Nadler, Genetics Dept., Children's Memorial Hospital (bottom right)© Biophoto Associates/Science Source/Photo Researchers, Inc.

CHAPTER 13
13.1/© Carl Purcell/Photo Researchers, Inc. 13.2B/© Dr. E. R. Degginger 13.3/© John D. Cunningham/Visuals Unlimited 13.4B/© Runk/Schoenberger/Grant Heilman Photography, Inc. 13.7/© Walter Chandoha 13.8A/© Yoav Levy/Phototake NYC 13.8B/© 1994 Lester Bergman/L.V. Bergman & Associates, Inc. **Page 197**/© Mark J. Stouffer/Animals Animals

CHAPTER 14
14.4/© From "The Double Helix," by James D. Watson, p. 168/Cold Spring Harbor Laboratory Archives 14.5B/© 1994 Dan Richardson 14.7A/National Institutes of Health **Page 209**/© K. G. Murti/Visuals Unlimited

CHAPTER 15
15.3/© O. L. Miller, B. R. Beatty, D. W. Fawcett/Visuals Unlimited 15.6B/© Science Source/Photo Researchers, Inc. 15.7C/© 1994 Goodsell 15.13/Miller, O.L., Jr., B.A. Hamkalo, and C.A. Thomas, Jr. 1970. Science 169:392-395. 15.19A/© 1962 J. Edstrom & W. Beermann, Journal of Cell Biology, 14.374, Reprinted by permission of The Rockefeller University Press/Scott Foresman **Page 225**/© Zig Leszczynski/Animals Animals **Page 241**/(top right) © Dr. Jeremy Burgess/Science Photo Library/Photo Researchers, Inc. (top left) © F. R. Turner/Dept. of Biology, Indiana University/ Biological Photo Service

CHAPTER 16
16.2(left)/© CNRI/Science Photo Library/Photo Researchers, Inc. 16.2(right)/© Tom Broker/Rainbow 16.5/© Dan McCoy/ Rainbow 16.6A/© Genentech 16.10/© Jon Gordon/Phototake NYC 16.11/© Dr. R. L. Brinster/Laboratory of Reproductive Physiology, School of Veterinary Medicine, University of Pennsylvania **Page 246**/© Dan McCoy/Rainbow **Page 250**/© Chuen-mo To and C. C. Brinton, Jr. **Page 256**/© Douglas Kirkland/Sygma **Page 255**/Monsanto

CHAPTER 17
17.1/© Scott Foresman 17.3/© 1988 L. V. Bergman & Associates, Inc. 17.9(top)/© UPI/The Bettmann Archive 17.9(bottom)/© G. R. Roberts **Page 261**/© William J. Weber/Visuals Unlimited **Page 266**/Dr. Gopal Murti/Science Photo Library/Photo Researchers, Inc.

CHAPTER 18
18.1/Painting by George Richmond/© The Granger Collection 18.3/© Stock Montage, Inc. 18.4/© Gardiner Waterbury/Visuals Unlimited 18.6/© Publications Department, National Portrait Gallery 18.7/Painting by Sir Luke Fildes/© The Granger Collection 18.8/© Science Source/Photo Researchers, Inc. 18.9/© Office of News & Public Affairs, Harvard University 18.10B/© Charlie Heidecker/Visuals Unlimited 18.12B/© Office of News & Public Affairs, Harvard University 18.14A-C/© Tom McHugh/Photo Researchers, Inc. 18.15B/© Don and Pat Valenti 18.16(top)/© Frederic/Jacana/Photo Researchers, Inc. 18.16(bottom)/© 1994 Don Mason/The Stock Market **Page 276**/© Francois Gohier/Photo Researchers, Inc. **Page 282**/(top right)© George H. Harrison (left center) © Elliott Varner Smith (right center) © Elliott Varner Smith (bottom left) © David M. Stone (bottom right) © David M. Stone **Page 283**/(top left) © Elliott Varner Smith (top right) © George H. Harrison (left center) © Dr. E. R. Degginger (bottom left) © Dr. E. R. Degginger (right center) © Dr. E. R. Degginger (bottom right) © Dr. E. R. Degginger

CHAPTER 19
19.1(top)/© Charles Sykes/Visuals Unlimited 19.1(bottom)/© 1989 Guy Gillette/Photo Researchers, Inc. 19.2/© Alan Oddie/PhotoEdit 19.4B-C/© Walter Chandoha 19.5A-B/© Michael Tweedie/Photo Researchers, Inc. 19.9/© 1973 Tom McHugh/Photo Researchers, Inc. 19.11/Milwaukee Journal Photo **Page 299**/© John Shaw/Tom Stack & Associates

CHAPTER 20
20.2/© C. Haagner /Bruce Coleman Inc. 20.3(left)/© 1987 Gregory K. Scott / Photo Researchers, Inc. 20.3(right)/© Stephen J. Lang/Visuals Unlimited 20.4/Jungle Larry's Safari Land, Inc., © Daniel L. Feicht 20.6(top) /© William H. Beatty / Visuals Unlimited 20.6(bottom)/© Dale Jackson / Visuals Unlimited 20.11(top right)/© 1972 Tom McHugh / Photo Researchers, Inc. 20.11 (top right center)/© Jany Sauvanet / Photo Researchers, Inc. 20.11(top left center)/© Melinda Berge 20.11(top left)/© Tom McHugh/National Audubon Society/ Photo Researchers, Inc. 20.11(bottom left center)/© 1987 John Cancalosi / Tom Stack & Associates 20.11(bottom left)/© Loren A. McIntyre 20.11(bottom right center)/© G. R. Roberts 20.11(bottom right)/© Tom McHugh / Photo Researchers 20.12(left)/© Lawrence Gilbert/ Biological Photo Service 20.12(right)/© Lawrence Gilbert / Biological Photo Service **Page 316**/© Michael Dick/ Animals Animals **Page 330** (top)/© Robert Noonan/Photo Reesearchers, Inc. **Page 330** (bottom)/© Thomas Dimock/The Stock Market

CHAPTER 21
21.2A/© Kjell B. Sandved/Visuals Unlimited 21.3/Dr. M.P. Kahl 21.4A/© Jen and Des Bartlett/Bruce Coleman Inc. 21.7/© Institute of Human Origins 21.9/FPG International **Page 333**/© John Reader/Science Photo Library/Photo Researchers, Inc.

CHAPTER 22
22.1/© Herb Orth/Life Magazine/Time Warner Inc. 22.2/© Science VU/Sidney Fox/Visuals Unlimited 22.3/© 1988 Roger Ressmeyer/Starlight 22.4/© Dr. Sidney W. Fox. In Fox and Dose, *Molecular Evolution and the Origins of Life*. 22.7/© David M. Phillips/Photo Researchers, Inc. 22.9B/© Dr. R. Wyckoff/National Institutes of Health 22.10B/© David M. Phillips/Visuals Unlimited 22.11(top left)/© Dr. Tony Brain/Photo Researchers, Inc. 22.11(left center)/© The Upjohn Company 22.11(top right)/© John D. Cunningham/Visuals Unlimited 22.11(bottom right)/© Armed Forces Institute of Pathology 22.11(bottom left)/© David M. Phillips/Photo Researchers, Inc. 22.12/© Dr. Maria Costa and Dr. George B. Chapman/Biology Department, Georgetown University 22.13/© David R. Frazier Photolibrary 22.14(left)/© C. C. Brinton, University of Pittsburgh 22.14(right)/© Centers for Disease Control 22.15(top left)/© Sinclair Stammers/Science Photo Library/Photo Re-

searchers, Inc. 22.15(top right)/© J. Robert Waaland/Biological Photo Service 22.15(bottom left)/© Science VU/Sherman Thomson/Visuals Unlimited 22.15(bottom right)/© Courtesy of D. L. Findley, P. L. Walne and R. W. Holton, University of Tennessee, Knoxville/From *J. Psychology* 6:182-188, 1970 **Page 345**/© Alfred Pasieka/Bruce Coleman Inc. **Page 352**/(top right)© Cabisco/Visuals Unlimited (top left)© NASA (bottom left)© Francois Gohier/Photo Researchers (bottom right)© Robert Calentine/Visuals Unlimited

CHAPTER 23

23.2(left)/© Fred Marsik/Visuals Unlimited 23.2(right)/© Martin Dohrn/Science Photo Library/Photo Researchers, Inc. 23.3(top left)/© John D. Cunningham/Visuals Unlimited 23.3(top right)/© Biophoto Associates/Science Source/Photo Researchers, Inc. 23.3(bottom right)/© T.E. Adams/Visuals Unlimited 23.3(bottom left)/© Jan Hinsch/Science Photo Library/Photo Researchers, Inc. 23.5(top right)/© William Patterson/Tom Stack & Associates 23.5(top left)/© Eric Grave/Phototake NYC 23.5(bottom right)/© Manfred Kage/Peter Arnold, Inc. 23.5(bottom left)/© Eric Grave/Phototake NYC 23.7(top)/© David M. Phillips/Visuals Unlimited 23.7(bottom)/© Eric Grave/Phototake NYC 23.8(center)/© A. M. Siegelman/Visuals Unlimited 23.8(bottom left)/© Stanley Flegler/Visuals Unlimited 23.8(upper center left)/© R. Howard Berg/Visuals Unlimited 23.8(top left)/© Biophoto Associates/Photo Researchers, Inc. 23.8(upper center right)/© Cabisco/Visuals Unlimited 23.8(top right)/© Science VU/Visuals Unlimited 23.8(lower center left)/© Biophoto Associates/Photo Researchers, Inc. 23.8(bottom right)/© Biophoto Associates/Photo Researchers, Inc. 23.9B/© T.E. Adams/Visuals Unlimited 23.10/© Gary R. Robinson/Visuals Unlimited 23.11/© 1993 Tammy Peluso/Tom Stack & Associates 23.13/© Cabisco/Visuals Unlimited 23.14(top)/© Macmillan Science Co., Inc. 23.14(bottom)/© Andrew J. Martinez / Photo Researchers, Inc. 23.16(top)/© Runk/Schoenberger/Grant Heilman Photography, Inc. 23.16B-D/ William F. Loomis, *Dictyoslelium discoideum: A Developmental System.* © 1975 by Academic Press, Inc. 23.17A/Carolina Biological Supply Company 23.18B/© L. West/Valenti Photo 23.19(top left)/© Loren A. McIntyre 23.19(bottom left)/© Runk/Schoenberger/Grant Heilman Photography, Inc. 23.19(center)/© Stanley Flegler/Visuals Unlimited 23.19(right)/© John D. Cunningham/Visuals Unlimited 23.20B/© Glenn Oliver/Visuals Unlimited 23.21(left)/© Eric Grave/Phototake NYC 23.21(right)/© 1964 D. Pramer, *Science* 144:382–388 23.22/© Dwight R. Kuhn **Page 368**/© Cabisco/Visuals Unlimited

CHAPTER 24

24.2(left)/© John Trager/Visuals Unlimited 24.2(right)/© James W. Richardson/Visuals Unlimited 24.5(top)/© Richard Thom/Visuals Unlimited 24.5(bottom)/© 1986 Dwight R. Kuhn 24.5(right)/© G. R. Roberts 24.6(right)/© Breck P. Kent/Earth Scenes 24.6(left)/© George Loun/Visuals Unlimited 24.7B/© Kjell B. Sandved/Visuals Unlimited 24.9A/© Dr. Jeremy Burgess/John Innes Institute/Science Photo Library/Photo Researchers, Inc. 24.10(left)/© Bernd Wittich/Visuals Unlimited 24.10(center)/© John D. Cunningham/Visuals Unlimited 24.10 (right)/© Tim Davis/Photo Researchers, Inc. 24.11/© Harald Sund 24.13(top center)/© Don and Pat Valenti 24.13(top right)/© Terry Donnelly/Tom Stack & Associates 24.13(bottom right)/© Alex Bartel 1993/FPG International 24.13(bottom center)/© Rod Planck/Tom Stack & Associates 24.13(bottom left)/© Jeff Foott/Tom Stack & Associates **Page 394**/© Larry Ditto/Bruce Coleman Inc.

CHAPTER 25

25.7B/© Kim Taylor/Bruce Coleman Inc. 25.8(top right)/© Oxford Scientific Films/Animals Animals 25.8(bottom left)/© Douglas Faulkner 25.8(bottom right)/© Daniel Gotshall/Visuals Unlimited 25.9/© Jeffrey L. Rotman 25.10B/© 1985 Ed Reschke 25.11B/© Stanley Flegler/Visuals Unlimited 25.13/© Larry Jensen/Visuals Unlimited 25.14B/© Roland Birke/Peter Arnold, Inc. **Page 441**/© Randy Morse/Tom Stack & Associates

CHAPTER 26

26.3(top right)/© Kjell B. Sandved/Photo Researchers, Inc. 26.3(right center)/© James H. Carmichael, Jr./Photo Researchers, Inc. 26.3(bottom right)/© Harold W. Pratt/Biological Photo Service 26.4(left)/© Brian Parker/Tom Stack & Associates 26.4(center)/© David M. Dennis/Tom Stack & Associates 26.4(right)/© C. P. Hickman/Visuals Unlimited 26.7(top left)/© Hans Pfletschinger/Peter Arnold, Inc. 26.7(bottom left)/© Phillip Brownell/Department of Zoology, Oregon State University 26.8(top left)/© A. Kerstitch 26.8(center left)/© Darlyne Murawski/Tony Stone Images 26.8(bottom left)/© Stephen Dalton/Photo Researchers, Inc. 26.8(top center)/© M.P. Kahl/DRK Photo 26.8(top right)/© Maria Zorn/ Animals Animals 26.8(bottom right)/© John Ebeling 26.9(left, right)/© Robert P. Carr 26.10(left)/© Robert Calentine/Visuals Unlimited 26.10(right)/© Manfred Kage/Peter Arnold, Inc. 26.11(top)/© Gianni Tortoli/Photo Researchers, Inc. 26.11(center)/© Richard Endris/Visuals Unlimited 26.11 (bottom)/© Ken Greer/Visuals Unlimited 26.14(center)/© Robert Dunne/Photo Researchers, Inc. 26.14(bottom)/© 1979 Gilbert Grant/Photo Researchers, Inc. 26.15B/© Dave B. Fleetham/Tom Stack & Associates 26.17E/© 1984 Mike Neumann/Photo Researchers, Inc. **Page 429**/© Randy Morse/Tom Stack & Associates

CHAPTER 27

27.1(top left)/© Ken Lucas/Biological Photo Service 27.1(bottom left)/© Science VU/Visuals Unlimited 27.1(top right)/© 1985 Patrice Ceisel/Visuals Unlimited 27.4(top right)/© 1977 Tom McHugh/Photo Researchers, Inc. 27.4(center right)/© Dave B. Fleetham/Tom Stack & Associates 27.4(bottom right)/© Dave B. Fleetham/Tom Stack & Associates 27.5B/© Jeffrey L. Rotman/Peter Arnold, Inc. 27.6(left)/© Dave B. Fleetham/ Tom Stack & Associates 27.6(top right)/© Tom McHugh/Photo Researchers, Inc. 27.6 (center)/© Kjell B. Sandved/Visuals Unlimited 27.6(bottom right)/© 1992 Andrew J. Martinez/Photo Researchers, Inc. 27.8B/© Peter Scoones/Seaphoto 27.12 (top)/© S. & J. Collins/Photo Researchers, Inc. 27.12(center)/© Gregory G. Dimijian/Photo Researchers, Inc. 27.12 (bottom)/Steinhart Aquarium/© 1978 Tom McHugh/Photo Researchers, Inc. 27.13B/© James P. Rowan/ Tony Stone Images 27.15(top left)/ © John Cancalosi/Tom Stack and Associates 27.15 (top center)/© 1995 Lynn M. Stone 27.15(top right)/© Dwight R. Kuhn 27.15 (bottom left)/© Lynn M. Stone 27.15(bottom right)/© 1995 Lynn M. Stone 27.21(top left)/ © Fritz Prenzel/Animals Animals 27.21(top right)/ © Dave Watts/Tom Stack & Associates 27.21(bottom left)/© 1988 VU/SIU/Visuals Unlimited 27.22(right)/ © 1989 Michele Burgess/The Stock Market 27.22 (bottom left)/© 1994 Tom Brakefield/The Stock Market 27.22(top left)/© 1994 Tom Brakefield/ The Stock Market **Page 451**/© Roger and Donna Aitkenhead/Animals Animals

CHAPTER 28

28.5(left)/© 1975 George Holton / Photo Researchers, Inc. 28.5(right)/© Marie Francois Michel-Ange SEETAVE **Page 476**/© 1989 Dwight R. Kuhn **Page 478**/(left)Jim Steinberg/Photo Researchers, Inc. **Page 478**/(center) © Robert P. Carr/Bruce Coleman Inc. (right) © 1979 Ken Brate/Photo Researchers, Inc. **Page 479**/(left)© E. S. Ross (center) © Oxford Scientific Films/Animals Animals (right)

© Merlin D. Tuttle **Page 484**/(left) © G. R. Roberts (right) © G. R. Roberts **Page 485**/(top left)© Robert P. Carr/Bruce Coleman Inc. (left center) © G. R. Roberts (center) © 1974 John Ebeling (right center) © G. R. Roberts (bottom left)© John Shaw (bottom right) © John N. A. Lott/Biological Photo Service

CHAPTER 29

29.1/© Pat & Tom Leeson/Photo Researchers, Inc. 29.2/© Gerald and Buff Corsi/Tom Stack & Associates 29.6B/© Bruce Iverson 29.7/© Cabisco/Visuals Unlimited 29.9/© J. Robert Waaland/Biological Photo Service 29.10(top right)/© 1989 Dwight R. Kuhn 29.10(bottom right)/© Dr. E. R. Degginger/Photo Researchers, Inc. 29.11B/© J. Robert Waaland/Biological Photo Service 29.12(top right)/© Bruce Iverson 29.12(bottom right)/© Bruce Iverson 29.14/© John D. Cunningham/Visuals Unlimited 29.15/© 1988 John Gerlach/Tom Stack & Associates 29.16(left)/© Kjell B. Sandved/Visuals Unlimited 29.16(right)/© John D. Cunningham/Visuals Unlimited 29.17B/© Ed Reschke **Page 491**/© Gregory G. Dimijian/Photo Researchers, Inc.

CHAPTER 30

30.1/© Tom Stack/Tom Stack & Associates 30.2/© Grant Heilman/Grant Heilman Photography, Inc. 30.6/© P.W. Grace/Photo Researchers, Inc. 30.8B/© Dr. Jeremy Burgess/Science Photo Library/Photo Researchers, Inc. 30.9/© Walt Anderson/Visuals Unlimited 30.10/© Peter Arnold, Inc. 30.11(top)/© D.H. Marx/Visuals Unlimited 30.11(center)/© Stanley Flegler/Visuals Unlimited 30.11(bottom)/© Dana Richter/Visuals Unlimited **Page 510**/© John Eastcott/YVA Momatiuk/Earth Scenes

CHAPTER 31

31.2B/© Biophoto Associates/Photo Researchers, Inc. 31.3/© John D. Cunningham/Visuals Unlimited 31.4/© E. Webber/Visuals Unlimited 31.9(left)/© Ken Wagner/Phototake NYC 31.9(right)/© Zig Leszczynski/Earth Scenes 31.10(top)/© Derek Fell 31.10(bottom)/© 1982 Richard Parker/Photo Researchers, Inc. **Page 523**/© Porterfield-Chickering/Photo Researchers.

CHAPTER 32

32.2(left)/© Ed Reschke 32.2(center)/© 1985 Ed Reschke 32.2(right)/© Ed Reschke 32.3(top)/© 1988 Ed Reschke 32.3(bottom)/© Ed Reschke 32.4(left)/© Prof. P. Motta/Department of Anatomy, University "La Sapienza," Rome/Science Photo Library/Photo Researchers, Inc. 32.4(right)/© 1985 Ed Reschke 32.5/© Biophoto Associates/Science Source/Photo Researchers, Inc. 32.6(top)/© Cosmos Blank/National Audubon Society/Photo Researchers, Inc. 32.6 (bottom)/© Tom Stack/Tom Stack & Associates 32.7(top right)/© Science VU/Dr. Don W. Fawcett/Visuals Unlimited 32.7(bottom right)/© 1990 Science Photo Library/Custom Medical Stock Photo 32.9(top left)/© Brian Parker/Tom Stack & Associates 32.9(top center)/© Oxford Scientific Films/Animals Animals 32.9(top right)/© Bruce Watkins/Animals Animals 32.9(bottom left)/© 1989 R. Calentine/Visuals Unlimited 32.12(left, center, right)/Carolina Biological Supply 32.14(top right)/Biophoto Associates/Photo Researchers, Inc. 32.14(center right)/From *Tissues and Organs: A Text-Atlas of Scanning Electron Microscopy* by Richard G. Kessel and Randy H. Kardon. Copyright © 1979, W. H. Freeman and Company 32.14 (bottom left)/© Franzini Armstrong/Photo Researchers, Inc. 32.15(top right)/© Franzini Armstrong/Photo Researchers, Inc. 32.15 (bottom right)/© Dr. Don W. Fawcett/Visuals Unlimited **Page 538**/© Dwight R. Kuhn

CHAPTER 33

33.1/© Marty Stouffer/Animals Animals 33.2B/© Science Photo Library/Custom Medical Stock Photo 33.3B/© Prof. Dr. S. Cedric Rouine 33.5/From *Tissues and Organs: A Text-Atlas of Scanning Electron Microscopy* by Richard G. Kessel and Randy N. Kardon. © 1979 W. H. Freeman & Company/Scott Foresman 33.9(left)/Science VU/J. Heuser/D.W. Fawcett/Visuals Unlimited 33.9 (right)/Science VU/E.R. Lewis, T.E. Everhart, and Y. Y. Zeevi/Visuals Unlimited 33.11/© 1991 Inga Spence/Tom Stack & Associates **Page 559**/© Marty Snyderman/Visuals Unlimited

CHAPTER 34

34.3/© A. Glauberman/Photo Researchers, Inc. 34.6A/© I. Shoulson/Visuals Unlimited 34.11/© SIU/Visuals Unlimited **Page 574**/© Joe McDonald/Bruce Coleman Inc. **Page 589**/© Cecil Fox/Science Source/Photo Researchers, Inc.

CHAPTER 35

35.2B/© SIU/Visuals Unlimited 35.4A/© 1994 Richard L. Carlton/Visuals Unlimited 35.6B/© 1992 Science Photo Library/Custom Medical Stock Photo 35.7(left)/© Dave B. Fleetham/Tom Stack & Associates 35.7 (right)/© Rod Planck/Tom Stack & Associates 35.8B/© 1992 Science Photo Library/Custom Medical Stock Photo **Page 598**/© Zig Leszczynski/Animals Animals.

CHAPTER 36

36.7/© UPI/Bettmann Newsphotos 36.10 (top)/© Ken Greer/Visuals Unlimited 36.10(bottom)/© 1988 Ed Reschke 36.11B/© AFIP/Science Source/Photo Researchers, Inc. **Page 614**/© Eric Dragesco/Bruce Coleman Inc.

CHAPTER 37

37.2(top left)/© Kevin Schafer/Martha Hill/Tom Stack & Associates 37.2(bottom left)/© Reinhard Siegel/Tony Stone Images 37.2(right)/© Keith Gunnar/Bruce Coleman Inc. 37.6/© 1988 SIU/Photo Researchers, Inc. 37.9B/© CNRI/Science Photo Library/Photo Researchers, Inc. **Page 632**/© John Gerlach/Animals Animals

CHAPTER 38

38.8B/From *Tissues and Organs: A Text-Atlas of Scanning Electron Microscopy* by Richard G. Kessel and Randy N. Kardon. © 1979 W. H. Freeman & Company/Scott Foresman 38.15/© Jeff Greenberg/Peter Arnold, Inc. **Page 652**/© Joe McDonald/Bruce Coleman Inc.

CHAPTER 39

39.1(left)/Carolina Biological Supply 39.1 (top center)/© Dale Jackson / Visuals Unlimited 39.1(right)/© Douglas Faulkner 39.1 (cont. left)/© D. Wilder/West Virginia Hardy Co. 39.1(cont. right)/© Thomas Kitchin/Tom Stack & Associates 39.2(lower left)/© Fred Hossler/Visuals Unlimited 39.2(upper right)/© William H. Beatty/Visuals Unlimited 39.3(left)/© Bruce Iverson 39.3(right)/© Bruce Iverson 39.4D/© Science Photo Library/Photo Researchers, Inc. **Page 673**/© Jane Burton/Bruce Coleman Inc. 39B/© David M. Phillips/Visuals Unlimited

CHAPTER 40

40.4C/© Lennart Nilsson 40.6B/© Blair Seitz/Photo Researchers, Inc. 40.7B/© Dr. Don W. Fawcett/Photo Researchers, Inc. 40.8B/© 1985 Ed Reschke 40.9B/© Lennart Nilsson **Page 690**/© Kim Taylor/Bruce Coleman Inc. **Page 698**/© Tony Brain/Science Photo Library/Photo Researchers, Inc.

CHAPTER 41

41.1A/© 1990 Kent Wood/Photo Researchers, Inc. 41.3B/© Dr. Don W. Fawcett/Hektor Chemes/Photo Researchers, Inc. 41.4/© David M. Phillips/Visuals Unlimited 41.7(left, right)/D. Zucker-Franklin, M. F. Greaves, C. E. Grossi, and A. M. Marmont. *1981. Atlas of Blood Cell: Function and Pathology*, vol. 2. Philadelphia. Lea & Febiger. © 1981 by Edi. Ermes s.r.l.-Milan, Italy 41.8/© Dr. A. Liepins/Photo Researchers, Inc. 41.9B/© Jean-Claude Revy/Phototake NYC 41.14/© 1994 Joel Gordon 41.15/© Oscar Burriel/Latin Stock/Science Photo Library/

Photo Researchers, Inc. **Page 706**/© 1990 Science Photo Library/Custom Medical Stock Photo **Page 718**/© William H. Beatty/Visuals Unlimited **Page 724**/© David M. Phillips/Visuals Unlimited

CHAPTER 42
42.1/© T.E. Adams/Visuals Unlimited 42.2B/© Biophoto Associates/Photo Researchers, Inc. 42.3/© Walter E. Harvey/National Audubon Society/Photo Researchers, Inc. 42.5B/© Gene Shih/Richard Kessel/Visuals Unlimited 42.6B/From *Tissues and Organs: A Text-Atlas of Scanning Electron Microscopy* by Richard G. Kessel and Randy N. Kardon. © 1979 W. H. Freeman & Company/Scott Foresman **Page 729**/© Karl Ammann/Bruce Coleman Inc.

CHAPTER 43
43.1B/© David M. Phillips/Visuals Unlimited 43.2B/© L. Zamboni/Dr. Don W. Fawcett/Visuals Unlimited 43.10A-B/John D. Cunningham/Visuals Unlimited **Page 754**/© 1994 BSIP/Custom Medical Stock Photo

CHAPTER 44
44.5/© Sydney Thomson/Animals Animals 44.7/© 1989 Al Grotell 44.8/AP/Wide World 44.10(top)/© Stephen J. Krasemann/DRK Photo 44.10(bottom)/© Stephen J. Krasemann/DRK Photo 44.14/© 1973 A. Rider/Photo Researchers, Inc. 44.15/© Breck P. Kent/Animals Animals 44.17/© Penelope Breese/Gamma Liaison 44.18(top)/© 1987 Anthony Mercieca/Photo Researchers, Inc. 44.18(bottom)/© 1994 Gregory K. Scott/Photo Researchers, Inc. 44.19/© Lynwood Chace/National Audubon Society/Photo Researchers, Inc. 44.20/© Peter Davey/Bruce Coleman Inc. 44.22/© Francois Gohier/Photo Researchers, Inc. 44.23/© Erwin & Peggy Bauer/Bruce Coleman Inc. 44.24/© L. Connor/Visuals Unlimited 44.26/© Tom Tietz/Natural Selection 44.27/© Allan Tannenbaum/Sygma 44.28/© Bob Daemmrich/Stock Boston **Page 780**/© Brian Parker/Tom Stack & Associates

CHAPTER 45
45.6/© James D. Nations/DDB Stock Photo 45.7(left)/© Loren McIntyre/Woodfin Camp & Associates, Inc. 45.7(top right)/© Jany Sauvanet/Photo Researchers, Inc. 45.7(center right) /© Michael DeMacker/Visuals Unlimited 45.7(bottom right)/© Ken Schafer and Martha Hill/Tom Stack and Associates 45.8/© Dr. E. R. Degginger/Photo Researchers, Inc. 45.9(top left)/© John Shaw 45.9(top right)/© James Tallon/Outdoor Exposures 45.9(bottom)/© Doug Sokell/Tom Stack & Associates 45.10(bottom left)/© Jeff Foott/Tom Stack & Associates 45.10(top right)/© John D. Cunningham/Visuals Unlimited 45.10(top left)/© 1990 John Cancalosi/Tom Stack & Associates 45.10(bottom right)/© Craig Lorenz/Photo Researchers, Inc. 45.11/© Bruce Iverson 45.12/© Luis Villota/The Stock Market 45.13(top left, top right, bottom right)/© Bill Binzen/The Stock Market 45.13(bottom left)/© Glenn M. Oliver/Visuals Unlimited 45.14(left)/© Wayne Lankinen/DRK Photo 45.14(top right)/© Dominique Braud/Tom Stack & Associates 45.14(center right)/© George J. Sanker/DRK Photo 45.14(bottom right)/© Gary Milburn/Tom Stack & Associates 45.15(top left)/© Stephen J. Krasemann/DRK Photo 45.15(bottom left)/© Thomas Kitchin/Tom Stack & Associates 45.15(top right)/© Tom McHugh/Photo Researchers, Inc. 45.15(bottom right)/© Patrick J. Endres/Visuals Unlimited 45.18/© D. Long/Visuals Unlimited 45.19(bottom right)/© Chris Huss/The Wildlife Collection 45.19 (top right)/© Manfred Gottschalk/Tom Stack & Associates **Page 804**/© Breck P. Kent/Animals Animals

CHAPTER 46
46.3/© William E. Ferguson 46.9(left)/© 1992 David M. Dennis/Tom Stack & Associates 46.9(right)/© George Musil/Visuals Unlimited 46.11(top left, top right, bottom left, bottom right)/© Bitterroot National Forest/U.S. Forest Service 46.13/© Jeanette Thomas/Visuals Unlimited **Page 832**/© Henry Ausloos/Animals Animals **Page 835**/(top left)© Science VU/WHOI/D. Foster/Visuals Unlimited (top right)Alvin External Camera/Woods Hole Oceanographic Institution (bottom)James Childress/Woods Hole Oceanographic Institution

CHAPTER 47
47.3/© Milton Rand/Tom Stack & Associates 47.5(bottom)/© Zig Leszczynski/Animals Animals 47.5(top)/© Tim Davis/Photo Researchers, Inc. 47.6/© Frank T. Awbrey/Visuals Unlimited **Page 851**/© Y. Arthus-Bertrand/Peter Arnold, Inc.

CHAPTER 48
48.1(left)/© 1991 David Madison/Bruce Coleman Inc. 48.1(right)/© Monkmeyer/LeDuc/Monkmeyer Press 48.3/© 1985 Thomas S. England/Photo Researchers, Inc. **Page 861**/© NASA Page 870/© NASA/Science Photo Library/Photo Researchers, Inc.

INDEX

A Biological Lexicon

A list of Greek and Latin prefixes, suffixes, and word roots commonly used in biological terms.

a-, an- [Gk. *an-*, not, without, lacking]: anaerobic, atom, abiotic, anonymous, anaphrodisiac, anemia

ad- [L. *ad-*, toward, to]: adhesion, admixture, adopt, adrenaline

amphi- [Gk. *amphi-*, two, both, both sides of]: amphibian, Amphineura

ana- [Gk. *ana-*, up, up against]: anaphase, analogy, anabolic

andro- [Gk. *andros,* an old man]: androecium, androgen

anti- [Gk. *anti-*, against, opposite, opposed to]: antibiotic, antibody, antigen, antidiuretic hormone

archeo- [Gk. *archaios,* beginning]: archegonium, archenteron, archaic, menarche

arthro- [Gk. *arthron,* a joint]: arthropod, arthritis, arthrodire, condylarthra

auto- [Gk. *auto-*, self, same]: autoimmune, autotroph, autosome, autonomic

bi-, bin- [L. *bis,* twice; *bini,* two-by-two]: binary fission, binocular vision, biennial, bicarbonate, bilateral, binomial, bipolar

bio- [Gk. *bios,* life]: biology, biomass, biome, biosphere, biosynthesis, biotic

blasto-, -blast [Gr. *blastos,* sprout; now "pertaining to the embryo"]: blastoderm, blastodisc, blastopore, blastula, trophoblast, osteoblast

brachi- [Gk. *brachion,* arm]: brachiation, brachiopod

broncho- [Gk. *bronchos,* windpipe]: bronchus, bronchi, bronchiole, bronchitis

carcino- [Gk. *karkin,* a crab, cancer]: carcinogen, carcinoma

cardio- [Gk. *kardia,* heart]: cardiac, myocardium, electrocardiogram

cephalo- [Gk. *kephale,* head]: cephalization, cephalochordate, cephalopod, cephalothorax

chloro- [Gk. *chloros,* green]: chlorophyll, chloroplast, chlorine

chromo- [Gk. *chroma,* color]: chromosome, chromoplast, chromatin

coelo-, -coel [Gk. *koilos,* hollow, cavity]: coelacanth, coelenteron, coelenterate, coelom

com-, con-, col-, cor-, co- [L. *cum,* with, together]: coenzyme, commensal, conjugation, convergence, covalent

cranio- [Gk. *kranios,* L. *cranium,* skull]: cranial, cranium

cuti- [L. *cutis,* skin]: cutaneous, cuticle, cutin

cyclo-, -cycle [Gk. *kyklos,* circle, ring, cycle]: cyclostome, pericycle

cyto-, -cyte [Gk. *kytos,* vessel or container; now, "cell"]: cytoplasm, cytology, cytochrome, cytokinesis, erythrocyte, leucocyte

de- [L. *de-*, "away, off"; deprivation, removal, separation, negation]: dedicuous, decomposer, decarboxylase, dehydration

derm-, dermato- [Gk. *derma,* skin]: dermis, epidermis, ectoderm, endoderm, mesoderm

di- [Gk. *dis,* twice]: dicaryon, dicotyledon, dioxide, dipole

dia- [Gk. through, passing through, thorough, thoroughly]: diabetes, dialysis, diaphragm

diplo- [Gk. *diploos,* two-fold]: diploid, diploblastic

eco- [Gk. *oikos,* house, home]: ecology, androecium, ecosphere, ecosystem, economy

ecto- [Gk. *ektos,* outside]: ectoderm, ectoplasm, ectoparasite

endo- [Gk. *endon,* within]: endocrine, endoderm, endodermis, endometrium, endoskeleton

epi- [Gk. *epi,* on, upon, over]: epicotyl, epidermis, epididymis, epiglottis, epiphyte, epithelium

eu- [Gk. *eus,* good; *eu,* well; now "true"]: eubacterium, eucaryote

ex-, exo-, ec, e- [Gk., L. out, out of, from, beyond]: emission, ejaculation, excretion, exergonic, exhale, exocytosis, exon, exoskeleton

extra- [L. outside of, beyond]: extracellular, extraembryonic

-fer [L. *ferre,* to bear]: fertile, fertilization, conifer, rotifer

galacto- [Gk. *galakt-*, milk]: galactose, galactic [Milky Way]

gam-, gameto- [Gk. gamos, marriage; now usually in reference to gametes (sex cells)]: gamete, polygamy, isogamete, cryptogam

gastro- [Gk. *gaster,* stomach]: gastric, gastrula, gastrin, gastrovascular cavity, gastropod

gen- [Gk. *gen,* born, produced by; Gk. *genos,* race, kind; L. *genus, generare,* to beget]: polygenic, genotype, geneology, glycogen, florigen, pyrogen, estrogen, heterogenous

gluco, glyco- [Gk. *glykys,* sweet; now pertaining to sugar]: glucose, glycogen, glycolysis, glycoprotein

gyn-, gyno-, gyneco- [Gk. *gyne,* woman]: gynecology, gynoecium, polygyny, misogyny

hemo-, hemato-, -hemia, -emia [Gk. *haima,* blood]: hematology, hemodialysis, hemoglobin, hemophilia, anemia, leukemia

hepato- [Gk. *hepar, hepat-*, liver]: hepatitis, hepatic portal system

hetero- [Gk. *heteros,* other, different]: heterogeneous, heterogamous, heterozygote

histo- [Gk. *histos,* web of a loom, tissue; now pertaining to biological tissues]: histology, histone, histamine, antihistamine

homo-, homeo- [Gk. *homos,* same; Gk. *homios,* similar]: homeostasis, homeothermy, homogeneous, homologous, homozygote